Miller & Levine Biology

Kenneth R. Miller, Ph.D.
Professor of Biology, Brown University
Providence, Rhode Island

Joseph S. Levine, Ph.D.
Science Writer and Producer
Concord, Massachusetts

Pearson's Biology program has been developed to meet teachers' goals and help every student achieve success. Print and digital program components help students connect science to the world around them.

Blooming Sonoran Desert at Picacho Peak State Park, Arizona

Pearson

GLENVIEW, ILLINOIS • BOSTON, MASSACHUSETTS
CHANDLER, ARIZONA • NEW YORK, NEW YORK

Miller & Levine

Biology

Print Components

- Student Edition
- Teacher's Edition
- Study Workbook A
- Study Workbook A, Teacher's Edition
- Study Workbook B: Reading Foundations
- Study Workbook B: Reading Foundations, Teacher's Edition
- Laboratory Manual A
- Laboratory Manual A, Teacher's Edition
- Laboratory Manual B: Skill Foundations
- Laboratory Manual B: Skill Foundations, Teacher's Edition
- Probeware Lab Manual
- Assessment Resources
- Transparencies

Technology Components

- Biology.com
- Untamed Science® Video Series: BioAdventures DVD
- ExamView® CD-ROM
- Teacher Edition eText DVD
- Virtual BioLab DVD-ROM with Lab Manual
- Miller & Levine Biology iBook

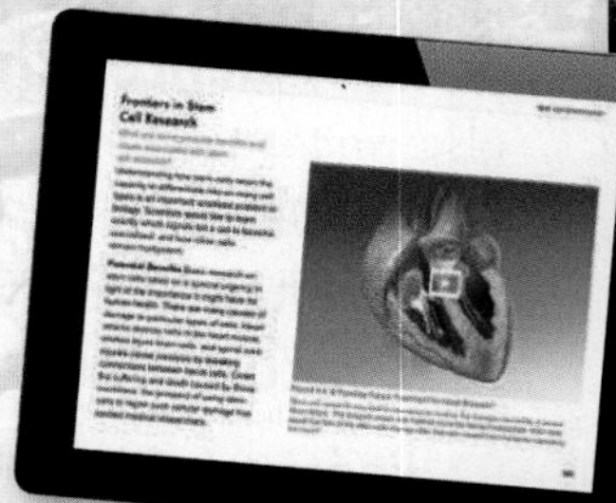

English Language Learners

- Teacher's ELL Handbook
- Multilingual Glossary

Spanish Components

- Spanish Student Edition (with online Spanish audio)
- Spanish Teacher's Guide
- Spanish Study Workbook

Photographs Every effort has been made to secure permission and provide appropriate credit for photographic material. The publisher deeply regrets any omission and pledges to correct errors called to its attention in subsequent editions.

Unless otherwise acknowledged, all photographs are the property of Pearson Education, Inc.

Photo locators denoted as follows: Top (T), Center (C), Bottom (B), Left (L), Right (R), Background (Bkgd).

Front Matter Acknowledgments i: Anton Foltin/Shutterstock. **ii–iii:** topseller/Shutterstock.

All other credits appear on pages C–0 to C–2, which constitute an extension of this copyright page.

ISBN-13: 978-1-32-320585-3
ISBN-10: 1-32-320585-3

9 18

About the Authors

Kenneth R. Miller grew up in Rahway, New Jersey, attended the local public schools, and graduated from Rahway High School in 1966. Miller attended Brown University on a scholarship and graduated with honors. He was awarded a National Defense Education Act fellowship for graduate study, and earned his Ph.D. in Biology at the University of Colorado. Miller is Professor of Biology at Brown University in Providence, Rhode Island, where he teaches courses in general biology and cell biology.

Miller's research specialty is the structure of biological membranes. He has published more than 70 research papers in journals such as *Cell*, *Nature*, and *Scientific American*. He has also written the popular trade books *Finding Darwin's God* and *Only a Theory*. He is a fellow of the American Association for the Advancement of Science (AAAS). His honors include the Public Service Award from the American Society for Cell Biology, the Distinguished Service Award from the National Association of Biology Teachers, the AAAS Award for Public Engagement with Science, the Gregor Mendel Medal from Villanova University, and most recently, the Stephen Jay Gould Prize from the Society for the Study of Evolution.

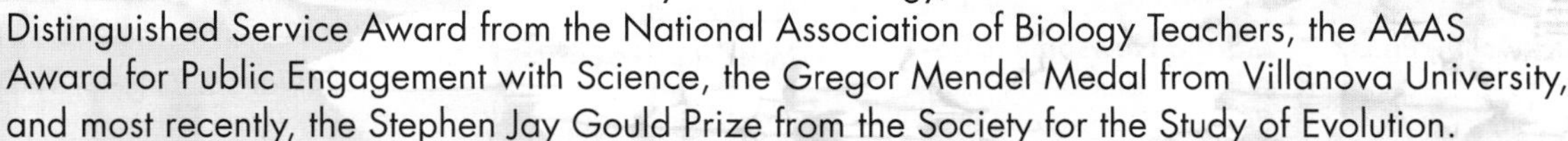

Miller lives with his wife, Jody, on a small farm in Rehoboth, Massachusetts. He is the father of two daughters, one a wildlife biologist and the other a high-school history teacher. He swims competitively in the masters' swimming program and umpires high school and collegiate softball.

Joseph S. Levine was born in Mount Vernon, New York, where he attended public schools, and graduated from Mount Vernon High School in 1969. He studied at Tufts University, and earned his Ph.D. working between Harvard University and the Marine Biological Laboratory in Woods Hole. His writing has appeared in periodicals ranging from *Science* to *Scientific American* and *Smithsonian*, and he has taught lecture and field courses at Boston College and Boston University.

Following a fellowship in Science Broadcast Journalism at WGBH-TV, Levine served as science correspondent for National Public Radio's *Morning Edition* and *All Things Considered*, and helped launch Discovery Channel's *Discover Magazine*. He served as scientific advisor to NOVA for programs including *Judgment Day*, and as Science Editor for the OMNI-MAX films *Cocos: Island of Sharks* and *Coral Reef Adventure*, and for several PBS series, including *The Secret of Life* and *The Evolution Project*.

Levine has led seminars and professional development workshops for teachers across the United States, Mexico, Puerto Rico, the U.S. Virgin Islands, Indonesia, and Malaysia. He currently serves on the Board of Overseers at the Marine Biological Laboratory, and on the Board of Visitors of the Organization for Tropical Studies. Since 2004, he has been teaching *Inquiry in Rain Forests*, a graduate-level field course for middle- and high-school teachers through the Organization for Tropical Studies.

Consultants/Reviewers

Content Reviewers

Lily Chen
Associate Professor
Department of Biology
San Francisco State University
San Francisco, CA

Elizabeth Coolidge-Stolz, MD
Medical/Life Science Writer/Editor
North Reading, MA

Elizabeth A. De Stasio, Ph.D.
Raymond H. Herzog Professor of Science
Associate Professor of Biology
Lawrence University
Appleton, WI

Jennifer C. Drew, Ph.D.
Lecturer/Scientist
University of Florida
Kennedy Space Center, FL

Donna H. Duckworth, Ph.D.
Professor Emeritus
College of Medicine
University of Florida
Gainesville, FL

Alan Gishlick, Ph.D.
Assistant Professor
Gustavus Adolphus College
St. Peter, MN

Deborah L. Gumucio, Ph.D.
Professor
Department of Cell and Developmental Biology
University of Michigan
Ann Arbor, MI

Janet Lanza, Ph.D.
Professor of Biology
University of Arkansas at Little Rock
Little Rock, AR

Charles F. Lytle, Ph.D.
Professor of Zoology
North Carolina State University
Raleigh, NC

Martha Newsome, DDS
Adjunct Instructor of Biology
Cy-Fair College, Fairbanks Center
Houston, TX

Jan A. Pechenik, Ph.D.
Professor of Biology
Tufts University
Medford, MA

Imara Y. Perera, Ph.D.
Research Assistant, Professor
Department of Plant Biology
North Carolina State University
Raleigh, NC

Daniel M. Raben, Ph.D.
Professor
Department of Biological Chemistry
Johns Hopkins University
Baltimore, MD

Megan Rokop, Ph.D.
Educational Outreach Program Director
Broad Institute of MIT and Harvard
Cambridge, MA

Gerald P. Sanders
Former Biology Instructor
Grossmont College
Julian, CA

Ronald Sass, Ph.D.
Professor Emeritus
Rice University
Houston, TX

Linda Silveira, Ph.D.
Professor
University of Redlands
Redlands, CA

Richard K. Stucky, Ph.D.
Curator of Paleontology and Evolution
Denver Museum of Nature and Science
Denver, CO

Robert Thornton, Ph.D.
Senior Lecturer Emeritus
Department of Plant Biology
College of Biological Sciences
University of California at Davis
Davis, CA

Edward J. Zalisko, Ph.D.
Professor of Biology
Blackburn College
Carlinville, IL

ESL Lecturer

Nancy Vincent Montgomery, Ed.D.
Southern Methodist University
Dallas, TX

High-School Reviewers

Christine Bill
Sayreville War Memorial High School
Parlin, NJ

Jean T. (Caye) Boone
Central Gwinnett High School
Lawrenceville, GA

Samuel J. Clifford, Ph.D.
Biology Teacher
Round Rock High School
Round Rock, TX

Jennifer Collins, M.A.
South County Secondary School
Lorton, VA

Roy Connor, M.S.
Science Department Head
Muncie Central High School
Muncie, IN

Norm Dahm, Jr.
Belleville East High School
Belleville, IL

Cora Nadine Dickson
Science Department Chair
Jersey Village High School
Cypress Fairbanks ISD
Houston, TX

Dennis M. Dudley
Science Department Chair/Teacher
Shaler Area High School
Pittsburgh, PA

Mary K. Dulko
Sharon High School
Sharon, MA

Erica Everett, M.A.T., M.Ed.
Science Department Chair
Manchester-Essex Regional High School
Manchester, MA

Heather M. Gannon
Elisabeth Ann Johnson High School
Mt. Morris, MI

Virginia Glasscock
Science Teacher
California High School
Whittier, CA

Ruth Gleicher
Biology Teacher
Niles West High School
Skokie, IL

Lance Goodlock
Biology Teacher/Science Department Chairperson
Sturgis High School
Sturgis, MI

W. Tony Heiting, Ph.D.
State Science Supervisor (retired)
Iowa Department of Education
Panora, IA

Patricia Anne Johnson, M.S.
Biology Teacher
Ridgewood High School
Ridgewood, NJ

Judith Decherd Jones, M.A.T.
NBCT AYA Science
East Chapel Hill High School
Chapel Hill, NC

Shellie Jones
Science Teacher
California High School
Whittier, CA

Michelle Lauria, M.A.T.
Biology Teacher
Hopkinton High School
Hopkinton, MA

Kimberly Lewis
Science Department Chair
Wellston High School
Wellston, OH

Lenora Lewis
Teacher
Creekview High School
Canton, GA

JoAnn Lindell-Overton, M.Ed.
Supervisor of Secondary Science
Chesapeake Public Schools
Chesapeake, VA

Lender Luse
H.W. Byers High School
Holly Springs, MS

Molly J. Markey, Ph.D.
Science Teacher
Newton Country Day School of the Sacred Heart
Newton, MA

Rebecca McLelland-Crawley
Biological Sciences Teacher
Piscataway, NJ

Mark L. Mettert, M.S. Ed.
Science Department Chair
New Haven High School
New Haven, IN

Jane Parker
Lewisville High School North
Lewisville, TX

Ian Pearce
Educator
Austin, TX

Jim Peters
Science Resource Teacher
Carroll County Public Schools
Westminster, MD

Michelle Phillips, M.A.T.
Secondary Science: Education Science Teacher
Jordan High School
Durham, NC

Randy E. Phillips
Science Teacher/Department Chair
Green Bay East High School
Green Bay, WI

Nancy Richey
Educator
Longmont, CO

Linda Roberson
Department Chairman
Jenks Freshman Academy
Jenks, OK

Sharon D. Spencer
Assistant Principal
Bronx Center for Science and Math
Bronx, NY

Stephen David Wright, M.S.
Biology Teacher
Montgomery County Public Schools
Columbia, MD

Alan W. Zimroth, M.S.
Science Teacher/Department Chairperson
Hialeah-Miami Lakes High School
Hialeah, FL

Miller & Levine Biology

Along with the trusted content that has made **Miller & Levine Biology** the most respected high school biology program on the market, this brand-new edition has added several key features that will help support classrooms as they transition to implementing the Next Generation Science Standards and Common Core Standards. This new update also offers an increased focus on STEM education as more and more classrooms attempt to address the engineering and design process.

NGSS Problem-based Learning

Spotlight on NGSS
- **Core Idea LS1.A** Structure and Function
- **Practice** Developing and Using Models
- **Crosscutting Concept** Systems and System Models

UNIT PROJECT 8

Body Mechanics

Read the following information about medical technologies. As you learn about human body systems, research how these technologies work and think about how they might be improved or adapted.

The human body is a complex machine made up of many interconnected parts that work together. But what happens if one of these parts is missing or becomes damaged over time? How can you keep the "machine" running?

The mechanics charged with solving this problem include physicians, biologists, chemists, physicists, engineers, inventors, and designers. They collaborate to design solutions that prolong or improve the lives of people with disabilities and chronic illnesses.

Various technologies are available to replace a *part* of the human body or a *function* that a diseased body is no longer able to do on its own. Some examples are shown here.

For people who live with a disability such as limb loss, a prosthetic limb can greatly improve quality of life.

859a The Human Body • Problem-based Learning Project

Keep an eye out for this logo that will identify opportunities to interweave NGSS principles.

New Standards for More Successful Schools—and Students

The National Research Council's (NRC) Framework calls for a full integration of the practices of science with its ideas and concepts. That is, students should learn the ideas of science through actually doing science. The Framework describes a vision of what it means to be proficient in science; it rests on a view of science as both a body of knowledge and an evidence-based, model- and theory-building enterprise that continually extends, refines, and revises knowledge.

Pearson is prepared to offer complete and cohesive support to implement the new Next Generation Science Standards (NGSS) Initiative. We'll be with you every step of the way to provide the easiest possible transition to the NGSS with a coherent, phased approach to implementation.

Pearson's NGSS Commitment

- Help you understand what the new science standards mean, diagnose where you are, and help you get where you need to be.
- Deliver the NGSS vision of personalized learning for every student by using technology to link instructional materials with formative and summative assessment.

In the Miller & Levine Biology Program

- Each unit provides Next Generation Science Standards support and activities for classrooms that are transitioning to a Project-Based Learning model.
- In-person and fully digital professional development opportunities are available to provide teachers with the tools they need to successfully implement the Next Generation Science Standards.

Keep an eye out for this logo that will identify opportunities to implement STEM activities.

STEM

Through confidence-building professional development and the project-based instructional approach, STEM activities and support empower teachers to raise expectations and improve student performance. In an area where teachers often lack training and confidence, STEM guides and supports teachers in developing the tools and confidence they need to be successful.

Students' natural interests in science and engineering blossom through real-world engineering design problems and hands-on inquiry. These types of activities and assessments promote higher-order critical thinking skills that will result in improved student performance.

- Print and digital STEM activities support the implementation of the engineering and design process in an engaging and hands-on fashion.
- Teachers are provided with point-of-use STEM activities and teaching strategies in the Teacher's Edition.

Your Task: Refine the Design

STEM

valuate a technology modeled on human ody systems, and propose a way to improve e design. A useful technology is one that solves problem. Your task is to research a medical roblem and the technology available to solve it. he solution may be mechanical, or it may be a iochemical treatment for a functional problem, uch as a hormone deficiency. After studying the ature of the problem and the range of possible olutions, you will suggest an optimized solution. t may be a refinement of an existing solution, or t may be an alternative solution that you design. Use the rubric at the end of the unit to evaluate your work for this task.

Define the Problem Discuss with your teacher the technology you wish to focus on. What is the problem or need that the technology was designed to solve? List the strengths and weaknesses of the technology. Make sure to consider the people who use it. What else might they add to your list? Explain why the current solution to the problem is incomplete.

Identify the requirements that must be met in order to solve the problem or need. These requirements are called design criteria. Then list the factors that might limit the success of your design. These factors are called design constraints.

Design a Solution Generate ideas for solving the problem. How would you design an improvement or alternative to the solution(s) currently in use? Evaluate your ideas for a solution based on the design criteria and constraints that you have specified. Select one or more of these ideas to develop further.

Develop Models Use sketches, physical models, or simulations to help you visualize and test your design. What does the model tell you about the strengths and limitations of your design? Ask your classmates for feedback on your proposed solution, and revise the design as needed. What feedback might you expect from the people in need of your solution? What flaws might they identify based on your model?

Obtain, Evaluate, and Communicate Information Create a Web page or presentation that summarizes your work. With your teacher's help, identify an expert on the problem you are solving, and ask that person to provide feedback on your design. Use the feedback to further improve your design.

Medical implants are devices that are placed surgically inside the body. These include artificial joints (such as knees or hips), artificial heart valves, pacemakers, cochlear implants, and dental implants.

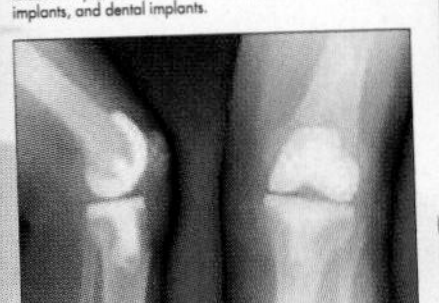

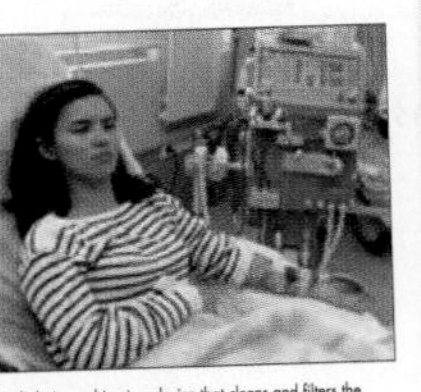

A dialysis machine is a device that cleans and filters the blood. It is used to treat people suffering from kidney failure.

Find resources on medical technologies research to help inform your solution design.

859b

Problem-based Learning • Problem-based Learning • Problem-based Learning • Problem-based Learning

Common Core

Keep an eye out for this logo that will identify where Common Core Standards are addressed.

The Common Core State Standards are the result of a state-led effort to establish a set of common educational standards. Developed by the National Governor's Association (NGA) and the Council of Chief State School Officers (CCSSO), these standards define the knowledge and skills students need for college and career success.

The Common Core State Standards for English Language Arts & Literacy in History/Social Studies, Science, and Technical Subjects focus on the knowledge and skills students need to achieve Literacy in Science. Students have many opportunities in the Miller & Levine Biology program to practice and develop scientific literacy skills that are embedded in the standards.

CHECKING YOUR *Scientific Literacy*

Refer to the lesson content and digital assets below as you prepare for your chapter assessment. Then, evaluate your understanding of the evolution of populations by answering these questions.

1. Scientific and Engineering Practice

3. Text Types and Purposes Write an essay for the science section of a newspaper explaining to the general public what science is and what scientific methodology involves.

4. STEM You are a physician being interviewed about how technology affects your work. Describe three examples of how tools or technological advances might affect work or research in your field of expertise.

BUILDING *Scientific Literacy*

Deepen your understanding of cell growth and division by engaging in key practices that allow you to make connections across concepts.

NGSS You will construct and use models to demonstrate how cells grow and divide during the cell cycle.

STEM How can cancer be treated? Research new technologies and frontiers in treating cancer.

Common Core You will use quantitative reasoning and cite textual evidence to support your analysis of the processes of cell growth and division.

- Common Core call-outs are provided at point-of-use to help support the implementation of these standards. in a science classroom
- Each chapter opens and closes with a statement and/or question that addresses Common Core State Standards.

Increase Student Achievement

Every student will succeed with the **Miller & Levine Biology** program. Results from an independent study indicate that students of all ethnicities and language proficiency levels experienced significant gains in science achievement when using Miller & Levine. **Miller & Levine Biology** is contributing to closing the achievement gap for students in the area of science achievement.

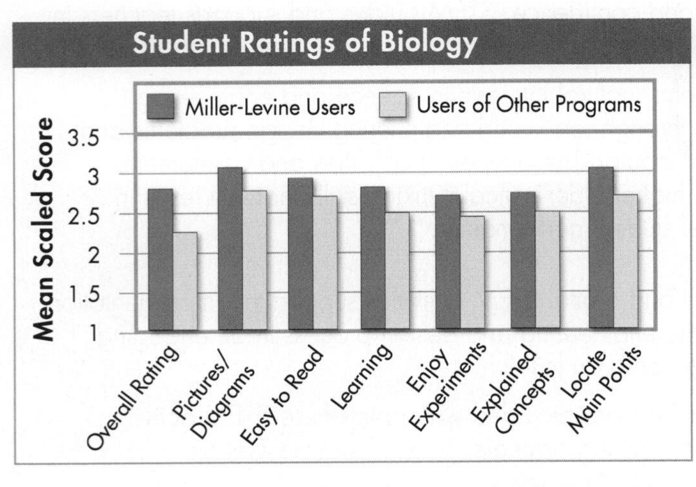

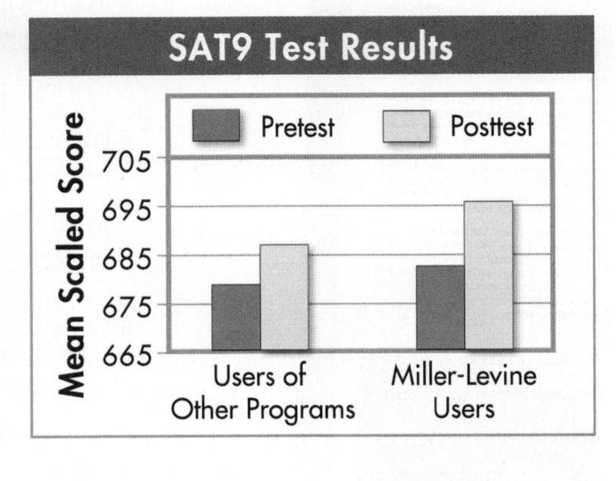

The Effects of Miller & Levine Biology (2010) on Student Performance. Cobblestone Applied Research & Evaluation, Inc.

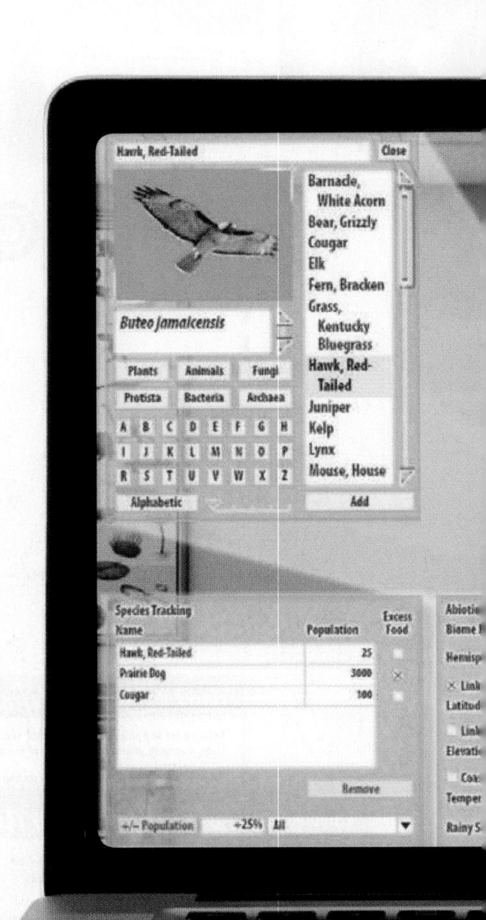

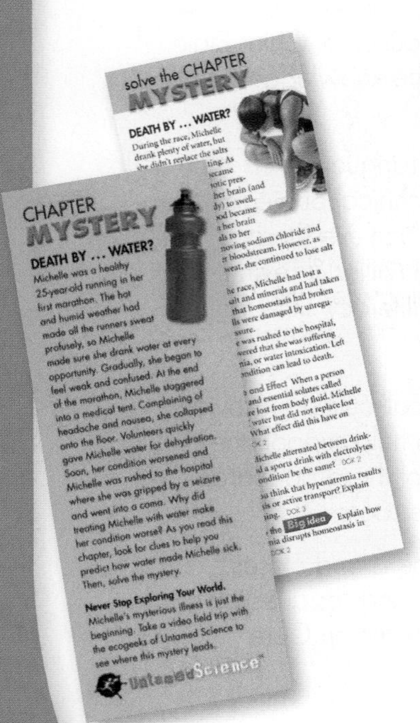

Deepen Understanding Through Integrated Inquiry

Inquiry-based activities are woven throughout the program to encourage students to think like scientists. Each chapter in **Miller & Levine Biology** opens with a real-world mystery that can be solved using planted clues and basic biological concepts. Engaging real-world examples such as the marathon runner who passes out during a race (Chapter 7) and the football player caught taking steroids (Chapter 30) help students make connections.

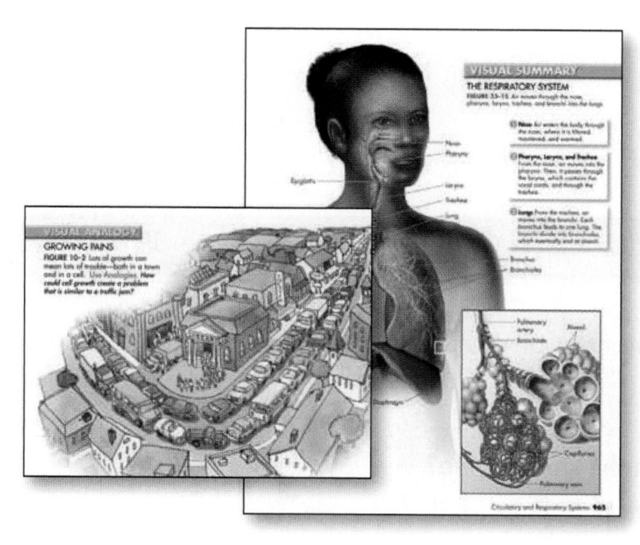

Enhance Comprehension through Visual Learning

Miller & Levine Biology offers a visual learning strand that helps make difficult topics more accessible. Real-world analogies provide a familiar context to support student understanding while visual summaries break down complex concepts into key components in a visually appealing way.

Discover Biology through Video

The Untamed Science video crew brings biology to life as these young scientists travel the globe in search of answers to the big questions of science. Swimming with sharks, jumping out of planes, and visiting an erupting volcano are just a few examples of how they engage students in the process.

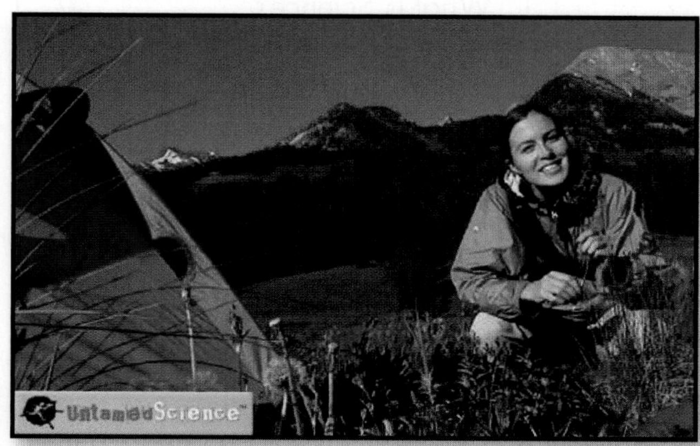

Virtual BioLabs put students into a virtual world of inquiry.

Explore Your Complete Online Course

Biology.com, the digital companion to **Miller & Levine Biology,** is the next generation of online science learning. A wealth of resources enables both teachers and students to personalize their online learning experience. **Biology.com** contains digital lessons, videos, animations, simulations, ebooks, editable worksheets, and a comprehensive set of online classroom management tools.

Biology.com Explore Biology.com for the next-generation of digital learning

Contents

UNIT 1 The Nature of Life 1–60

Problem-based Learning Harnessing the Fear of Water 1a

1 The Science of Biology 2

Big idea **What role does science play in the study of life?**

Chapter Mystery: *Height by Prescription*

1.1 What Is Science? 4
1.2 Science in Context. 10
1.3 Studying Life. 17

2 The Chemistry of Life 32

Big idea **What are the basic chemical principles that affect living things?**

Chapter Mystery: *The Ghostly Fish*

2.1 The Nature of Matter 34
2.2 Properties of Water 40
2.3 Carbon Compounds 45
2.4 Chemical Reactions and Enzymes 50

UNIT 2 Ecology 61–186

Problem-based Learning Disappearing Mussels! 61a

3 The Biosphere

Big idea **How do living and nonliving parts of the Earth interact and affect the survival of organisms?**

Chapter Mystery: *Changes in the Bay*

3.1 What Is Ecology? 64
3.2 Energy, Producers, and Consumers 69
3.3 Energy Flow in Ecosystems 73
3.4 Cycles of Matter 79

4 Ecosystems and Communities. 94

Big idea **How do abiotic and biotic factors shape ecosystems?**

Chapter Mystery: *The Wolf Effect*

4.1 Climate 96
4.2 Niches and Community Interactions. 99
4.3 Succession 106
4.4 Biomes. 110
4.5 Aquatic Ecosystems 117

5 Populations 128
Big idea What factors contribute to changes in populations?
Chapter Mystery: *A Plague of Rabbits*
5.1 How Populations Grow 130
5.2 Limits to Growth 137
5.3 Human Population Growth 142

6 Humans in the Biosphere 152
Big idea How have human activities shaped local and global ecology?
Chapter Mystery: *Moving the Moai*
6.1 A Changing Landscape 154
6.2 Using Resources Wisely 158
6.3 Biodiversity 166
6.4 Meeting Ecological Challenges 173

UNIT 3 Cells 187–304

Problem-based Learning Maxed Out Muscles! 187a

7 Cell Structure and Function
Big idea How are cell structures adapted to their functions?
Chapter Mystery: *Death by...Water?*
7.1 Life Is Cellular 190
7.2 Cell Structure 196
7.3 Cell Transport 208
7.4 Homeostasis and Cells 214

8 Photosynthesis 224
Big idea How do plants and other organisms capture energy from the sun?
Chapter Mystery: *Out of Thin Air?*
8.1 Energy and Life 226
8.2 Photosynthesis: An Overview 230
8.3 The Process of Photosynthesis 235

9 Cellular Respiration and Fermentation 248
Big idea How do organisms obtain energy?
Chapter Mystery: *Diving Without a Breath*
9.1 Cellular Respiration: An Overview 250
9.2 The Process of Cellular Respiration 254
9.3 Fermentation 262

10 Cell Growth and Division 272
Big idea How does a cell produce a new cell?
Chapter Mystery: Pet Shop Accident
10.1 Cell Growth, Division, and Reproduction 274
10.2 The Process of Cell Division 279
10.3 Regulating the Cell Cycle 286
10.4 Cell Differentiation 292

UNIT 4 Genetics 305–446

Problem-based Learning Food Fight! 305a

11 Introduction to Genetics 306
Big idea How does cellular information pass from one generation to another?
Chapter Mystery: Green Parakeets
11.1 The Work of Gregor Mendel 308
11.2 Applying Mendel's Principles 313
11.3 Other Patterns of Inheritance 319
11.4 Meiosis 323

12 DNA 336
Big idea What is the structure of DNA, and how does it function in genetic inheritance?
Chapter Mystery: UV Light
12.1 Identifying the Substance of Genes 338
12.2 The Structure of DNA 344
12.3 DNA Replication 350

13 RNA and Protein Synthesis 360
Big idea How does information flow from the cell nucleus to direct the synthesis of proteins in the cytoplasm?
Chapter Mystery: Mouse-Eyed Fly
13.1 RNA 362
13.2 Ribosomes and Protein Synthesis 366
13.3 Mutations 372
13.4 Gene Regulation and Expression 377

14 Human Heredity 390
Big idea How can we use genetics to study human inheritance?
Chapter Mystery: The Crooked Cell
14.1 Human Chromosomes 392
14.2 Human Genetic Disorders 398
14.3 Studying the Human Genome 403

15 Genetic Engineering 416
Big idea How and why do scientists manipulate DNA in living cells?
Chapter Mystery: A Case of Mistaken Identity
15.1 Selective Breeding 418
15.2 Recombinant DNA 421
15.3 Applications of Genetic Engineering 428
15.4 Ethics and Impacts of Biotechnology 436

UNIT 5 Evolution 447–570

Problem-based Learning The Alpine Chipmunk's Genetic Decline 447a

16 Darwin's Theory of Evolution 448
Big idea What is natural selection?
Chapter Mystery: Such Varied Honeycreepers
16.1 Darwin's Voyage of Discovery 450
16.2 Ideas That Shaped Darwin's Thinking 454
16.3 Darwin Presents His Case 460
16.4 Evidence of Evolution 465

17 Evolution of Populations 480
Big idea How can populations evolve to form new species?
Chapter Mystery: Epidemic
17.1 Genes and Variation 482
17.2 Evolution as Genetic Change in Populations 487
17.3 The Process of Speciation 494
17.4 Molecular Evolution 498

18 Classification 508
Big idea What is the goal of biologists who classify living things?
Chapter Mystery: Grin and Bear It
18.1 Finding Order in Diversity 510
18.2 Modern Evolutionary Classification 516
18.3 Building the Tree of Life 523

19 History of Life 536
Big idea How do fossils help biologists understand the history of life on Earth?
Chapter Mystery: Murder in the Permian
19.1 The Fossil Record 538
19.2 Patterns and Processes of Evolution 546
19.3 Earth's Early History 553

Biology.com Explore Biology.com for the next-generation of digital learning

UNIT 6 From Microorganisms to Plants 571–726

Problem-based Learning A Living Roof 571a

20 Viruses and Prokaryotes 572

Big idea Are all microbes that make us sick made of living cells?

Chapter Mystery: *The Mad Cows*

20.1 Viruses 574
20.2 Prokaryotes 580
20.3 Diseases Caused by Bacteria and Viruses 586

21 Protists and Fungi 600

Big idea How do protists and fungi affect the homeostasis of other organisms and ecosystems?

Chapter Mystery: *"A Blight of Unusual Character"*

21.1 Protist Classification—The Saga Continues 602
21.2 Protist Structure and Function 606
21.3 The Ecology of Protists 610
21.4 Fungi 618

22 Introduction to Plants 632

Big idea What are the five main groups of plants, and how have four of these groups adapted to life on land?

Chapter Mystery: *Stone Age Storytellers*

22.1 What Is a Plant? 634
22.2 Seedless Plants 639
22.3 Seed Plants 646
22.4 Flowering Plants 650

23 Plant Structure and Function 662

Big idea How are cells, tissues, and organs organized into systems that carry out the basic functions of a seed plant?

Chapter Mystery: *The Hollow Tree*

23.1 Specialized Tissues in Plants 664
23.2 Roots 669
23.3 Stems 674
23.4 Leaves 680
23.5 Transport in Plants 685

24 Plant Reproduction and Response 694
Big idea **How do changes in the environment affect the reproduction, development, and growth of plants?**
Chapter Mystery: *The Green Lemons*
24.1 Reproduction in Flowering Plants 696
24.2 Fruits and Seeds 704
24.3 Plant Hormones 708
24.4 Plants and Humans 715

UNIT 7 Animals 727–858

Problem-based Learning Biomimicry **727a**

25 Introduction to Animals 728
Big idea **What characteristics and traits define animals?**
Chapter Mystery: *Slime Day at the Beach*
25.1 What Is an Animal? 730
25.2 Animal Body Plans and Evolution 737

26 Animal Evolution and Diversity 750
Big idea **How have animals descended from earlier forms through the process of evolution?**
Chapter Mystery: *Fossil Quest*
26.1 Invertebrate Evolution and Diversity 752
26.2 Chordate Evolution and Diversity 757
26.3 Primate Evolution 765

27 Animal Systems I 780
Big idea **How do the structures of animals allow them to obtain essential materials and eliminate wastes?**
Chapter Mystery: *(Near) Death by Salt Water*
27.1 Feeding and Digestion 782
27.2 Respiration 787
27.3 Circulation 791
27.4 Excretion 794

Biology.com Explore Biology.com for the next-generation of digital learning

28 Animal Systems II 806

Big idea How do the body systems of animals allow them to collect information about their environments and respond appropriately?

Chapter Mystery: She's Just Like Her Mother!

28.1 Response 808
28.2 Movement and Support 814
28.3 Reproduction 819
28.4 Homeostasis 827

29 Animal Behavior 838

Big idea How do animals interact with one another and their environments?

Chapter Mystery: Elephant Caller ID?

29.1 Elements of Behavior 840
29.2 Animals in Their Environments 847

UNIT 8 The Human Body 859–1034

Problem-based Learning Body Mechanics **859a**

30 Digestive and Excretory Systems 860

Big idea How are the materials that go into your body and the materials that come from your body related to homeostasis?

Chapter Mystery: The Telltale Sample

30.1 Organization of the Human Body 862
30.2 Food and Nutrition 868
30.3 The Digestive System 875
30.4 The Excretory System 882

31 Nervous System 894

Big idea How does the structure of the nervous system allow it to control functions in every part of the body?

Chapter Mystery: Poisoning on the High Seas

31.1 The Neuron 896
31.2 The Central Nervous System 901
31.3 The Peripheral Nervous System 906
31.4 The Senses 909

32 Skeletal, Muscular, and Integumentary Systems 920

Big idea **What systems form the structure of the human body?**

Chapter Mystery: *The Demise of a Disease*

32.1 The Skeletal System 922
32.2 The Muscular System 928
32.3 Skin—The Integumentary System 935

33 Circulatory and Respiratory Systems 946

Big idea **How do the structures of the circulatory and respiratory systems allow for their close functional relationship?**

Chapter Mystery: *In the Blood*

33.1 The Circulatory System 948
33.2 Blood and the Lymphatic System 954
33.3 The Respiratory System 963

34 Endocrine and Reproductive Systems 976

Big idea **How does the body use chemical signals to maintain homeostasis?**

Chapter Mystery: *Out of Stride*

34.1 The Endocrine System 978
34.2 Glands of the Endocrine System 982
34.3 The Reproductive System 988
34.4 Fertilization and Development 995

35 Immune System and Disease 1008

Big idea **How does the body fight against invading organisms that may disrupt homeostasis?**

Chapter Mystery: *The Search for a Cause*

35.1 Infectious Disease 1010
35.2 Defenses Against Infection 1014
35.3 Fighting Infectious Disease 1020
35.4 Immune System Disorders 1024

STEM Activities 1035

Biology.com Explore Biology.com for the next-generation of digital learning

A Visual Guide to the Diversity of Life DOL•1–DOL•64

Bacteria . . . . DOL•6
Archaea . . . . DOL•8
Protists . . . . DOL•10
Fungi . . . . DOL•16
Plants . . . . DOL•20
Animals . . . . DOL•30

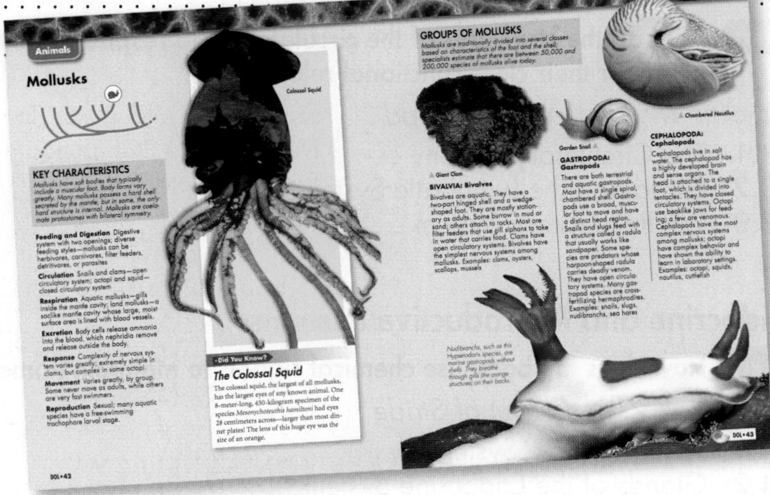

Animals

Mollusks

KEY CHARACTERISTICS
Mollusks have soft bodies that typically include a muscular foot. Body forms vary greatly. Many mollusks possess a hard shell secreted by the mantle, but in some, the only hard structure is internal. Mollusks are coelomate protostomes with bilateral symmetry.

Feeding and Digestion Digestive system with two openings; diverse feeding styles—mollusks can be herbivores, carnivores, filter feeders, detritivores, or parasites

Circulation Snails and clams—open circulatory system; octopi and squid—closed circulatory system

Respiration Aquatic mollusks—gills inside the mantle cavity; land mollusks—a saclike mantle cavity whose large, moist surface area is lined with blood vessels.

Excretion Body cells release ammonia into the blood, which nephridia remove and release outside the body.

Response Complexity of nervous system varies greatly; extremely simple in clams, but complex in some octopi.

Movement Varies greatly, by group. Some never move as adults, while others are very fast swimmers.

Reproduction Sexual; many aquatic species have a free-swimming trochophore larval stage.

Colossal Squid

Did You Know?

The Colossal Squid

The colossal squid, the largest of all mollusks, has the largest eyes of any known animal. One 8-meter-long, 450-kilogram specimen of the species *Mesonychoteuthis hamiltoni* had eyes 28 centimeters across—larger than most dinner plates! The lens of this huge eye was the size of an orange.

DOL•42

GROUPS OF MOLLUSKS
Mollusks are traditionally divided into several classes based on characteristics of the foot and the shell; specialists estimate that there are between 50,000 and 200,000 species of mollusks alive today.

Giant Clam

BIVALVIA: Bivalves
Bivalves are aquatic. They have a two-part hinged shell and a wedge-shaped foot. They are mostly stationary as adults. Some burrow in mud or sand; others attach to rocks. Most are filter feeders that use gill siphons to take in water that carries food. Clams have open circulatory systems. Bivalves have the simplest nervous systems among mollusks. Examples: clams, oysters, scallops, mussels

Garden Snail

GASTROPODA: Gastropods
There are both terrestrial and aquatic gastropods. Most have a single spiral, chambered shell. Gastropods use a broad, muscular foot to move and have a distinct head region. Snails and slugs feed with a structure called a radula that usually works like sandpaper. Some species are predators whose harpoon-shaped radula carries deadly venom. They have open circulatory systems. Many gastropod species are cross-fertilizing hermaphrodites. Examples: snails, slugs, nudibranchs, sea hares

Chambered Nautilus

CEPHALOPODA: Cephalopods
Cephalopods live in salt water. The cephalopod has a highly developed brain and sense organs. The head is attached to a single foot, which is divided into tentacles. They have closed circulatory systems. Octopi use beaklike jaws for feeding; a few are venomous. Cephalopods have the most complex nervous systems among mollusks; octopi have complex behavior and have shown the ability to learn in laboratory settings. Examples: octopi, squids, nautilus, cuttlefish

Nudibranchs, such as this Hypselodoris species, are marine gastropods without shells. They breathe through gills (the orange structures) on their backs.

DOL•43

Appendices

Appendix A: Science Skills
Data Tables and Graphs A–1
Reading Diagrams A–3
Basic Process Skills A–4
Organizing Information A–6

Appendix B: Lab Skills
Conducting an Experiment A–8
The Metric System A–10
Safety Symbols A–11
Science Safety Rules A–12
Use of the Microscope A–14

Appendix C: Technology & Design . . A–16
Appendix D: Math Skills A–18
Appendix E: Periodic Table A–24
English/Spanish Glossary G–1
Index I–1
Credits C–0

Labs and Activities

Quick Lab

Replicating Procedures 13
Model an Ionic Compound 36
Acidic and Basic Foods 43
How Do Abiotic Factors Affect Different Plant Species? 67
How Do Different Types of Consumers Interact? . . 72
Successful Succession? 108
How Does Competition Affect Growth? 138
Reduce, Reuse, Recycle 155
What Is a Cell? 193
Making a Model of a Cell 203
What Waste Material Is Produced During Photosynthesis? 234
How Does Exercise Affect Disposal of Wastes From Cellular Respiration? 264
Modeling the Relationship Between Surface Area and Volume 275
Mitosis in Action 283
Classroom Variation 311
How Are Dimples Inherited? 315

Modeling DNA Replication 352
How Does a Cell Interpret Codons? 367
Modeling Mutations 374
How Is Colorblindness Transmitted? 395
Modeling Restriction Enzymes 405
Inserting Genetic Markers 425
Survey Biotechnology Opinions 438
Darwin's Voyage 451
Variation in Peppers 457
Classifying Fruits 513
Constructing a Cladogram 520
Modeling Half-Life 541
How Do Viruses Differ in Structure? 575
What Are Protists? 603
How Does a Paramecium Eat? 612
What Is the Structure of Bread Mold? 620
Are All Plants the Same? 635
What Forms Do Fruits Take? 651
What Parts of Plants Do We Eat? 665
Examining Stomata 683
What Is the Role of Leaves in Transpiration? 686
What Is the Structure of a Flower? 698
How Hydra Feed 732
Binocular Vision 766
Breathing in Clams and Crayfishes 788
Water and Nitrogen Excretion 797
Does a Planarian Have a Head? 810
What Are Some Adaptations of Vertebrae? 816
What Kind of Learning Is Practice? 844
Maintaining Temperature 866
Modeling Bile Action 878
How Do You Respond to an External Stimulus? . . 908
Observe Calcium Loss 924
What Do Tendons Do? 932
What Factors Affect Heart Rate? 951
What's in the Air? 964
Tracing Human Gamete Formation 990
Embryonic Development 1000
How Do Diseases Spread? 1021

Analyzing Data

What's in a Diet? 20
Comparing Fatty Acids 48
The 10 Percent Rule 77
Predator-Prey Dynamics 102
Which Biome? 115
Multiplying Rabbits 135
American Air Pollution Trends 164
Saving the Golden Lion Tamarin 172
Mitochondria Distribution in the Mouse 216
Rates of Photosynthesis 240
You Are What You Eat 251
The Rise and Fall of Cyclins 288
Cellular Differentiation of *C. elegans* 294
Human Blood Types 320
Calculating Haploid and Diploid Numbers 327
Base Percentages 345
The Discovery of RNA Interference 381
The Geography of Malaria 400
Genetically Modified Crops in the United States 429
Molecular Homology in *Hoxc8* 470
Allele Frequency 491
Fishes in Two Lakes 500
Comparing the Domains 524
Extinctions Through Time 548
Comparing Atmospheres 556
MRSA on the Rise 591
Mycorrhizae and Tree Height 624
Keeping Ferns in Check 644
Reading a Tree's History 678
Temperature and Seed Germination 706
Auxins and Plant Growth 710
Differences in Differentiation 740
Feather Evolution 763
Protein Digestion 784
Comparing Ectotherms and Endotherms 828
Caring for Young 850
The Composition of Urine 883
Sound Intensity 910
The Rising Rate of Melanoma 938
Blood Transfusions 956
Immune System "Memory" 1017
Food Allergies 1025

Features

Visual Analogies

Unlocking Enzymes 53
Earth's Recycling Center 74
The Matter Mill 79
Interlocking Nutrients 86
The Greenhouse Effect 97
Ecological Footprints 173
The Cell as a Living Factory 196
ATP as a Charged Battery 227
Carrying Electrons 232
Growing Pains 276
The Main Functions of DNA 342
Master Plans and Blueprints 363
Finch Beak Tools 472
Geologic Time as a Clock 543
How a Lytic Virus Is Like an Outlaw 576
How Cells Move Like Boats 607
Transpirational Pull 685
Specialized Teeth 785
Excretion in Aquatic Animals 796
A Chain Reaction 899
The Skeleton 923
A City's Transportation System 948

Technology & Biology

A Nature-Inspired Adhesive 39
Global Ecology From Space 87
Fluorescence Microscopy 291
Artificial Life? 435
Bar-Coding Life 529
Low-Tech Weapons Against a High-Tech Parasite 617
Bioartificial Kidneys 799
Studying the Brain and Addiction 905
Testing for Heart Disease 962

Biology & History

Understanding Photosynthesis 229
Discovering the Role of DNA 349
Origins of Evolutionary Thought 459
The Evolution of Agriculture 719
Human-Fossil Seekers 773
Emerging Diseases 1023

Careers & Biology

Marine Biologist, Park Ranger, Wildlife Photographer 105
Laboratory Technician, Microscopist, Pathologist 195
Forensic Scientist, Plant Breeder, Population Geneticist 322
Fossil Preparator, Museum Guide, Paleontologist 559
Farmer, Plant Pathologist, Botanical Illustrator 655
Zoo Curator, Beekeeper, Invertebrate Biologist 736

Biology & Society

Who Should Fund Product Safety Studies? 16
What Can Be Done About Invasive Mussels? 136
Should Creatine Supplements Be Regulated? 261
Are Laws Protecting Genetic Privacy Necessary? 402
Should Antibiotic Use Be Restricted? 493
Should More Vaccinations Be Required? 593
Head for the Hills? 831
Should Marine Mammals Be Kept in Captivity? 846
Who Should Solve America's Obesity Problem? 874
Should Student Athletes Be Tested for Steroids? 934

Dear Student:

Welcome to our world—the endlessly fascinating world of biology.

I can guess what some of you are thinking right now. "Fascinating? Yeah, right. Totally." Well, give us—and biology—a chance to show you that the study of the natural world really is more exciting, more fascinating, and more important to you personally than you've ever realized. In fact, biology is more important to our daily lives today than it has ever been.

Why? Three words: "We are one." This isn't meant in a "touchy-feely" or "New Age" way. "We" includes all forms of life on Earth. And "are one" means that all of us are tied together more tightly, in more different ways, than anyone imagined until recently.

Both our "hardware" (body structures) and our "software" (genetic instructions and biochemical processes that program body functions) are incredibly similar to those of all other living things. Genetic instructions in our bodies are written in the same universal code as instructions in bacteria and palm trees. As biologists "read" and study that code, they find astonishingly similar processes in all of us. That's why medical researchers can learn about human diseases that may strike you or your family by studying not only apes and pigs and mice, but even yeasts. We are one on the molecular level.

All organisms interact with one another and with the environment to weave our planet's web of life. Organisms make rain forests and coral reefs, prairies and swamps—and farms and cities. We interact, too, with the winds and ocean currents that tie our planet together. Human activity is changing local and global environments in ways that we still don't understand... and that affect our ability to produce food and protect ourselves from diseases. We are one ecologically with the rest of life on Earth.

All organisms evolve over time, adapting to their surroundings. If humans alter the environment, other organisms respond to that change. When we use antibiotics against bacteria, they develop resistance to our drugs. If we use pesticides against insects, they become immune to our poisons. We are one in our ability to evolve over time.

Those are the kinds of connections you will find in this book. Microscopic. Enormous. Amusing. Threatening. But always fascinating. That's why—no matter where you start off in your attitude about biology—we think you are in for some surprises!

Sincerely,

Joe Levine

Dear Student:

Biology is one of the subjects you're going to study this year, but I hope you'll realize from the very first pages of this book that biology is a lot more than just a "subject." Biology is what makes an eagle fly, a flower bloom, or a caterpillar turn into a butterfly. It's the study of ourselves—of how our bodies grow and change and respond to the outside world, and it's the study of our planet, a world transformed by the actions of living things. Of course, you might have known some of this already. But there's something more—you might call it a "secret" that makes biology unique.

That secret is that you've come along at just the right time. In all of human history, there has never been a moment like the present, a time when we stood so close to the threshold of answering the most fundamental questions about the nature of life. You belong to the first generation of students who can read the human genome almost as your parents might have read a book or a newspaper. You are the first students who will grow up in a world that has a chance to use that information for the benefit of humanity, and you are the very first to bear the burden of using that knowledge wisely.

If all of this seems like heavy stuff, it is. But there is another reason we wrote this book, and we hope that is not a secret at all. Science is fun! Biologists aren't a bunch of serious, grim-faced, middle-aged folks in lab coats who think of nothing but work. In fact, most of the people we know in science would tell you honestly, with broad grins on their faces, that they have the best jobs in the world. They would say there's nothing that compares to the excitement of doing scientific work, and that the beauty and variety of life make every day a new adventure.

We agree, and we hope that you'll keep something in mind as you begin the study of biology. You don't need a lab coat or a degree or a laboratory to be a scientist. What you do need is an inquiring mind, the patience to look at nature carefully, and the willingness to figure things out. We've filled this book with some of the latest and most important discoveries about living things, but we hope we've also filled it with something else: our wonder, our amazement, and our sheer delight in the variety of life itself. Come on in, and enjoy the journey!

Sincerely,

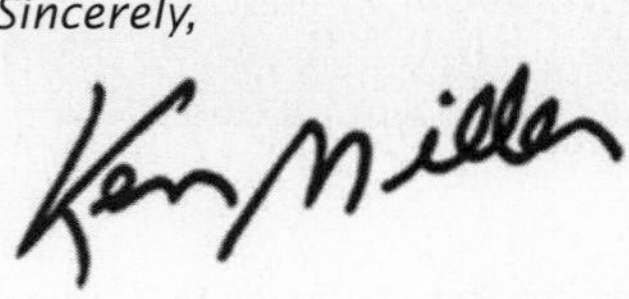

NGSS Standards for Biology

The Next Generation Science Standards for Biology identify the key scientific ideas and practices that all students should learn by the time they graduate from high school. Each standard is written as a performance expectation that integrates disciplinary core ideas with science and engineering practices, such as *asking questions* or *constructing explanations*, and crosscutting concepts, such as *cause and effect* and *stability and change*. This table lists the Life Sciences, the Earth Sciences, and the engineering design expectations for Biology.

NGSS Performance Expectations

HS-LS1 From Molecules to Organisms: Structures and Processes

HS-LS1-1. Construct an explanation based on evidence for how the structure of DNA determines the structure of proteins which carry out the essential functions of life through systems of specialized cells.

HS-LS1-2. Develop and use a model to illustrate the hierarchical organization of interacting systems that provide specific functions within multicellular organisms.

HS-LS1-3. Plan and conduct an investigation to provide evidence that feedback mechanisms maintain homeostasis.

HS-LS1-4. Use a model to illustrate the role of cellular division (mitosis) and differentiation in producing and maintaining complex organisms.

HS-LS1-5. Use a model to illustrate how photosynthesis transforms light energy into stored chemical energy.

HS-LS1-6. Construct and revise an explanation based on evidence for how carbon, hydrogen, and oxygen from sugar molecules may combine with other elements to form amino acids and/or other large carbon-based molecules.

HS-LS1-7. Use a model to illustrate that cellular respiration is a chemical process whereby the bonds of food molecules and oxygen molecules are broken and the bonds in new compounds are formed resulting in a net transfer of energy.

HS-LS2 Ecosystems: Interactions, Energy, and Dynamics

HS-LS2-1. Use mathematical and/or computational representations to support explanations of factors that affect carrying capacity of ecosystems at different scales.

HS-LS2-2. Use mathematical representations to support and revise explanations based on evidence about factors affecting biodiversity and populations in ecosystems of different scales.

HS-LS2-3. Construct and revise an explanation based on evidence for the cycling of matter and flow of energy in aerobic and anaerobic conditions.

HS-LS2-4. Use mathematical representations to support claims for the cycling of matter and flow of energy among organisms in an ecosystem.

HS-LS2-5. Develop a model to illustrate the role of photosynthesis and cellular respiration in the cycling of carbon among the biosphere, atmosphere, hydrosphere, and geosphere.

HS-LS2-6. Evaluate the claims, evidence, and reasoning that the complex interactions in ecosystems maintain relatively consistent numbers and types of organisms in stable conditions, but changing conditions may result in a new ecosystem.

HS-LS2-7. Design, evaluate, and refine a solution for reducing the impacts of human activities on the environment and biodiversity.*

HS-LS2-8. Evaluate the evidence for the role of group behavior on individual and species' chances to survive and reproduce.

HS-LS3 Heredity: Inheritance and Variation of Traits

HS-LS3-1. Ask questions to clarify relationships about the role of DNA and chromosomes in coding the instructions for characteristic traits passed from parents to offspring.

HS-LS3-2. Make and defend a claim based on evidence that inheritable genetic variations may result from: (1) new genetic combinations through meiosis, (2) viable errors occurring during replication, and/or (3) mutations caused by environmental factors.

HS-LS3-3. Apply concepts of statistics and probability to explain the variation and distribution of expressed traits in a population.

HS-LS4 Biological Evolution: Unity and Diversity

HS-LS4-1. Communicate scientific information that common ancestry and biological evolution are supported by multiple lines of empirical evidence.

HS-LS4-2. Construct an explanation based on evidence that the process of evolution primarily results from four factors: (1) the potential for a species to increase in number, (2) the heritable genetic variation of individuals in a species due to mutation and sexual reproduction, (3) competition for limited resources, and (4) the proliferation of those organisms that are better able to survive and reproduce in the environment.

HS-LS4-3. Apply concepts of statistics and probability to support explanations that organisms with an advantageous heritable trait tend to increase in proportion to organisms lacking this trait.

HS-LS4-4. Construct an explanation based on evidence for how natural selection leads to adaptation of populations.

HS-LS4-5. Evaluate the evidence supporting claims that changes in environmental conditions may result in: (1) increases in the number of individuals of some species, (2) the emergence of new species over time, and (3) the extinction of other species.

HS-LS4-6. Create or revise a simulation to test a solution to mitigate adverse impacts of human activity on biodiversity.*

HS-ESS2 Earth's Systems

HS-ESS2-2. Analyze geoscience data to make the claim that one change to Earth's surface can create feedbacks that cause changes to other Earth's systems.

HS-ESS2-4. Use a model to describe how variations in the flow of energy into and out of Earth's systems result in changes in climate.

HS-ESS2-6. Develop a quantitative model to describe the cycling of carbon among the hydrosphere, atmosphere, geosphere, and biosphere.

HS-ESS2-7. Construct an argument based on evidence about the simultaneous coevolution of Earth's systems and life on Earth.

HS-ESS3 Earth and Human Activity

HS-ESS3-1. Construct an explanation based on evidence for how the availability of natural resources, occurrence of natural hazards, and changes in climate have influenced human activity.

HS-ESS3-3. Create a computational simulation to illustrate the relationships among management of natural resources, the sustainability of human populations, and biodiversity.

HS-ESS3-4. Evaluate or refine a technological solution that reduces impacts of human activities on natural systems.*

HS-ESS3-5. Analyze geoscience data and the results from global climate models to make an evidence-based forecast of the current rate of global or regional climate change and associated future impacts to Earth systems.

HS-ESS3-6. Use a computational representation to illustrate the relationships among Earth systems and how those relationships are being modified due to human activity.

HS-ETS1 Engineering Design

HS-ETS1-1. Analyze a major global challenge to specify qualitative and quantitative criteria and constraints for solutions that account for societal needs and wants.

HS-ETS1-2. Design a solution to a complex real-world problem by breaking it down into smaller, more manageable problems that can be solved through engineering.

HS-ETS1-3. Evaluate a solution to a complex real-world problem based on prioritized criteria and trade-offs that account for a range of constraints, including cost, safety, reliability, and aesthetics, as well as possible social, cultural, and environmental impacts.

**The performance expectations marked with an asterisk integrate traditional science content with engineering through a Practice or Disciplinary Core Idea.*

UNIT 1

Chapters

1 The Science of Biology

2 The Chemistry of Life

INTRODUCE the **Big ideas**

- **Science as a Way of Knowing**
- **Matter and Energy**

"Science is 'a way of knowing'—a way of explaining the natural world through observations, questions, and experiments. But science isn't just dry old data, pressed between pages of this book like prom flowers in a school yearbook. Science is a living adventure story, aimed at understanding humans and the world around us. That story begins with the relationship between the matter that forms our bodies and the energy that powers life's processes."

NGSS Problem-based Learning

UNIT 1 PROJECT

Spotlight on NGSS
- **Core Idea LS1.A** Structure and Function
- **Practice** Obtaining, Evaluating, and Communicating Information
- **Crosscutting Concept** Structure and Function

Harnessing the Fear of Water

If you've ever been caught in a rainstorm, you know how quickly your clothes can get drenched with water. Most of your everyday clothes are made of materials that absorb water. But some of your clothes—think of your raincoat or your windbreaker—are made of materials that are waterproof.

The leaves of plants are naturally waterproof. The top and bottom surfaces of leaves are covered with a thin, waxy layer that acts as a protective barrier. Some leaves are more waterproof than others. In fact, some are so waterproof that scientists call them hydrophobic, or "water fearing."

The leaves of the lotus flower are a well-known example. Water on a lotus leaf immediately beads up and rolls off the leaf surface, almost as if the leaf itself were pushing the water away. Try doing a video search for "lotus effect" and "hydrophobic effect," and you can see for yourself.

Lotus plants, sometimes called water lilies, grow in calm, freshwater habitats in temperate and tropical regions of the world.

Your Task: Take a Deep Dive Into Hydrophobicity

STEM

What makes a hydrophobic substance "fear" water? As you learn about the properties of water and the chemistry of living things, investigate the phenomenon of hydrophobicity on your own or with a partner. Find out what causes the lotus effect, and discuss the technological innovations that have been inspired by it. Use the rubric at the end of the unit to evaluate your work for this task.

Construct an Explanation Search for and watch an online video clip that shows the lotus effect. What do you think causes this phenomenon? Discuss possible answers with your classmates. Then do your own research to deepen your understanding of what you observed.

Write a scientific explanation of the lotus effect. Your explanation should include descriptions of the microscopic structure of the leaf surface and the physical properties of the materials involved. You should also identify the function that the lotus effect provides to the plant. What problem or need does it solve?

Develop Models Draw a diagram or build a model that illustrates the main points of your explanation. Ask your classmates to identify any gaps or weaknesses in your diagram or model. Then refine your work based on the feedback.

Obtain, Evaluate, and Communicate Information Hydrophobic phenomena in the natural world have inspired inventions in the fields of materials science and nanotechnology. Identify a product that uses hydrophobic technology, and specify the problem or need that it solves. What claims have been made about the product's effectiveness? Explain how you could scientifically test the validity of such claims and how the data generated from your procedure would be used.

Design a Solution Think of a problem or need in your own world that could be solved by hydrophobic technology. What are the criteria for a successful solution to this problem? Discuss ideas for possible solutions, and then write a proposal outlining how you would design a solution to the problem.

Water on a lotus leaf will bead up and roll off, carrying away dirt particles and keeping the leaf's surface clean.

Fabric treated with hydrophobic technology repels water and dirt for as long as the coating on the fabric lasts.

Biology.com

Go online to learn more about hydrophobic phenomena and their applications.

1 The Science of Biology

Big idea Science as a Way of Knowing

Q: What role does science play in the study of life?

Paleontologists, such as these students at the Academy of Natural Sciences in Philadelphia, study fossils to learn about ancient life. Scientists use the scientific skills of observation and inference to generate questions and hypotheses about ancient organisms. They then design experiments to test their hypotheses.

INSIDE:

- 1.1 What Is Science?
- 1.2 Science in Context
- 1.3 Studying Life

BUILDING *Scientific Literacy*

Deepen your understanding of science and the study of biology by engaging in key practices that allow you to make connections across concepts.

NGSS You will learn how to use scientific methodology to plan and carry out scientific investigations.

STEM Microscopes are powerful tools in biology, used to gather data and test hypotheses. How do they work? Design, build, test, and redesign your own microscope.

Common Core You will write explanatory texts to communicate your understanding of the way science works.

CHAPTER MYSTERY

HEIGHT BY PRESCRIPTION

A doctor injects a chemical into the body of an eight-year-old boy named David. This healthy boy shows no signs of disease. The "condition" for which he is being treated is quite common—David is short for his age. The medication he is taking is human growth hormone, or HGH.

HGH, together with genes and diet, controls growth during childhood. People who produce little or no HGH are abnormally short and may have other related health problems. But David has normal HGH levels. He is short simply because his parents are both healthy, short people.

But if David isn't sick, why does his doctor prescribe HGH? Where does medicinal HGH come from? Is it safe? What does this case say about science and society? As you read this chapter, look for clues about the nature of science, the role of technology in our modern world, and the relationship between science and society. Then, solve the mystery.

Biology.com

Finding the solution to the growth hormone mystery is only the beginning. Take a video field trip with the ecogeeks of Untamed Science to see where this mystery leads.

Go online to access additional resources including:
• eText • Flash Cards • Lesson Overviews • Chapter Mystery

1.1 What Is Science?

Key Questions

What are the goals of science?

What procedures are at the core of scientific methodology?

Vocabulary

science • observation • inference • hypothesis • controlled experiment • independent variable • dependent variable • control group • data

Taking Notes

Flowchart As you read, create a flowchart showing the steps scientists use to answer questions about the natural world.

THINK ABOUT IT One day long ago, someone looked around and wondered: Where did plants and animals come from? How did I come to be? Since then, humans have tried to answer those questions in different ways. Some ways of explaining the world have stayed the same over time. Science, however, is always changing.

What Science Is and Is Not

What are the goals of science?

This book contains lots of facts and ideas about living things. Many of those facts are important, and you will be tested on them! But you shouldn't think that biology, or any science, is just a collection of never-changing facts. For one thing, you can be sure that some "facts" presented in this book will change soon—if they haven't changed already. What's more, science is not a collection of unchanging beliefs about the world. Scientific ideas are open to testing, discussion, and revision. So, some ideas presented in this book will also change.

These statements may puzzle you. If "facts" and ideas in science change, why should you bother learning them? And if science is neither a list of facts nor a collection of unchanging beliefs, what is it?

FIGURE 1–1 Studying the Natural World How do endangered gelada baboons communicate? How far do they travel? How are they affected by environmental changes? Researchers can use science to answer these questions.

Science as a Way of Knowing **Science** is an organized way of gathering and analyzing evidence about the natural world. It is a way of observing, a way of thinking, and "a way of knowing" about the world. In other words, science is a *process,* not a "thing." The word *science* also refers to the body of knowledge that scientific studies have gathered over the years.

Several features make science different from other human endeavors. First, science deals only with the natural world. Scientific endeavors never concern, in any way, supernatural phenomena of any kind. Second, scientists collect and organize information in an orderly way, looking for patterns and connections among events. Third, scientists propose explanations that are based on evidence, not belief. Then they test those explanations with more evidence.

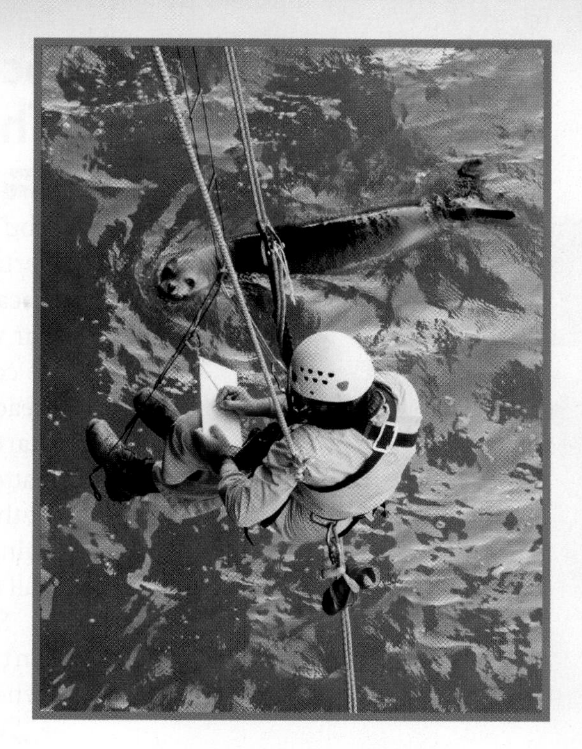

FIGURE 1–2 Science in Action This biologist is recording information about Mediterranean monk seals.

The Goals of Science The scientific way of knowing includes the view that the physical universe is a system composed of parts and processes that interact. From a scientific perspective, all objects in the universe, and all interactions among those objects, are governed by universal natural laws. The same natural laws apply whether the objects or events are large or small.

Aristotle and other Greek philosophers were among the first to try to view the universe in this way. They aimed to explain the world around them in terms of events and processes they could observe. Modern scientists continue that tradition. **One goal of science is to provide natural explanations for events in the natural world. Science also aims to use those explanations to understand patterns in nature and to make useful predictions about natural events.**

BUILD Vocabulary

WORD ORIGINS The word **science** derives from the Latin word *scientia,* which means "knowledge." Science represents knowledge that has been gathered over time.

Science, Change, and Uncertainty Over the centuries, scientists have gathered an enormous amount of information about the natural world. Scientific knowledge helps us cure diseases, place satellites in orbit, and send instantaneous electronic communications. Yet, despite all we know, much of nature remains a mystery. It is a mystery because science never stands still; almost every major scientific discovery raises more questions than it answers. Often, research yields surprises that point future studies in new and unexpected directions. This constant change doesn't mean science has failed. On the contrary, it shows that science continues to advance.

That's why learning about science means more than just understanding what we know. It also means understanding what we don't know. You may be surprised to hear this, but science rarely "proves" anything in absolute terms. Scientists aim for the best understanding of the natural world that current methods can reveal. Uncertainty is part of the scientific process and part of what makes science exciting! Happily, as you'll learn in later chapters, science has allowed us to build enough understanding to make useful predictions about the natural world.

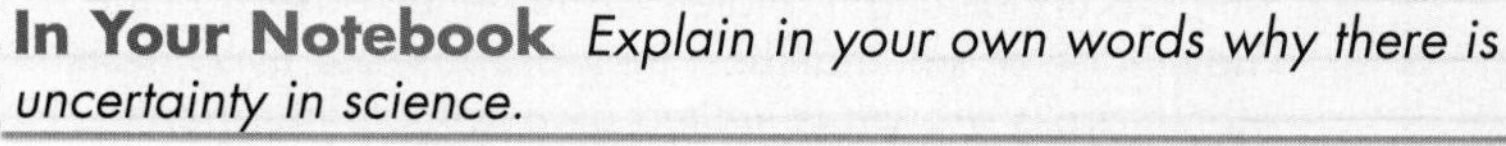

In Your Notebook *Explain in your own words why there is uncertainty in science.*

Scientific Methodology: The Heart of Science

What procedures are at the core of scientific methodology?

You might think that science is a mysterious process, used only by certain people under special circumstances. But that's not true, because you use scientific thinking all the time. Suppose your family's car won't start. What do you do? You use what you know about cars to come up with ideas to test. At first, you might think the battery is dead. So you test that idea by turning the key in the ignition. If the starter motor works but the engine doesn't start, you reject the dead-battery idea. You might guess next that the car is out of gas. A glance at the fuel gauge tests that idea. Again and again, you apply scientific thinking until the problem is solved—or until you run out of ideas and call a mechanic!

Scientists approach research in pretty much the same way. There isn't any single, cut-and-dried "scientific method." There is, however, a general style of investigation that we can call scientific methodology. **Scientific methodology involves observing and asking questions, making inferences and forming hypotheses, conducting controlled experiments, collecting and analyzing data, and drawing conclusions. Figure 1–3** shows how one research team used scientific methodology in its study of New England salt marshes.

Observing and Asking Questions Scientific investigations begin with **observation,** the act of noticing and describing events or processes in a careful, orderly way. Of course, scientific observation involves more than just looking at things. A good scientist can, as the philosopher Arthur Schopenhauer put it, "Think something that nobody has thought yet, while looking at something that everybody sees." That kind of observation leads to questions that no one has asked before.

FIGURE 1–3 Salt Marsh Experiment Salt marshes are coastal environments often found where rivers meet the sea. Researchers made an interesting observation on the way marsh grasses grow. Then, they applied scientific methodology to answer questions that arose from their observation.

OBSERVING AND ASKING QUESTIONS

Location A

Location B

Researchers observed that marsh grass grows taller in some places than others. This observation led to a question: *Why do marsh grasses grow to different heights in different places?*

INFERRING AND HYPOTHESIZING

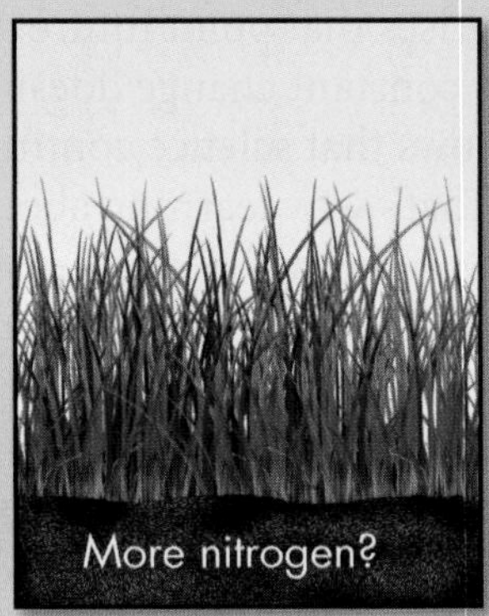

The researchers inferred that something limits grass growth in some places. It could be any environmental factor—temperature, sunlight, water, or nutrients. Based on their knowledge of salt marshes, they proposed a hypothesis: *Marsh grass growth is limited by available nitrogen.*

Inferring and Forming a Hypothesis After posing questions, scientists use further observations to make inferences. An **inference** is a logical interpretation based on what scientists already know. Inference, combined with a creative imagination, can lead to a hypothesis. A **hypothesis** is a scientific explanation for a set of observations that can be tested in ways that support or reject it.

Designing Controlled Experiments Testing a scientific hypothesis often involves designing an experiment that keeps track of various factors that can change, or variables. Examples of variables include temperature, light, time, and availability of nutrients. Whenever possible, a hypothesis should be tested by an experiment in which only one variable is changed. All other variables should be kept unchanged, or controlled. This type of experiment is called a **controlled experiment.**

▶ ***Controlling Variables*** Why is it important to control variables? The reason is that if several variables are changed in the experiment, researchers can't easily tell which variable is responsible for any results they observe. The variable that is deliberately changed is called the **independent variable** (also called the manipulated variable). The variable that is observed and that changes in response to the independent variable is called the **dependent variable** (also called the responding variable).

▶ ***Control and Experimental Groups*** Typically, an experiment is divided into control and experimental groups. A **control group** is exposed to the same conditions as the experimental group except for one independent variable. Scientists always try to reproduce or replicate their observations. Therefore, they set up several sets of control and experimental groups, rather than just a single pair.

In Your Notebook *What is the difference between an observation and an inference? List three examples of each.*

DESIGNING CONTROLLED EXPERIMENTS

The researchers selected similar plots of marsh grass. All plots had similar plant density, soil type, input of freshwater, and height above average tide level. The plots were divided into control and experimental groups.

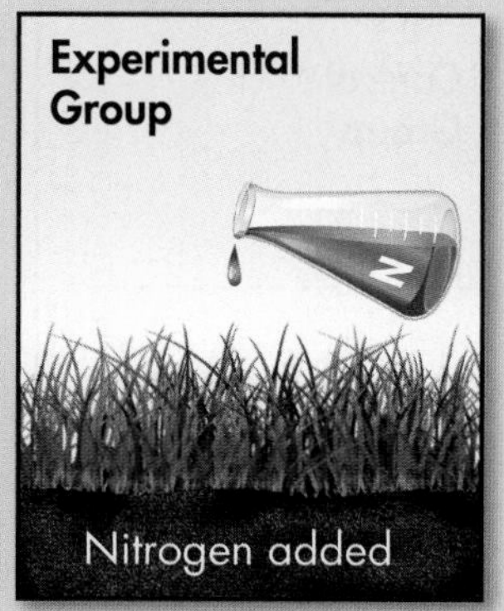

The researchers added nitrogen fertilizer (the independent variable) to the experimental plots. They then observed the growth of marsh grass (the dependent variable) in both experimental and control plots.

Collecting and Analyzing Data Scientists make detailed records of experimental observations, gathering information called **data.** There are two main types of data. Quantitative data are numbers obtained by counting or measuring. In the marsh grass experiment, quantitative data could include the number of plants per plot, the length, width, and weight of each blade of grass, and so on. Qualitative data are descriptive and involve characteristics that cannot usually be counted. Qualitative data in the marsh grass experiment might include notes about foreign objects in the sample plots or information on whether the grass was growing upright or sideways.

▶ ***Research Tools*** Scientists choose appropriate tools for collecting and analyzing data. The tools may range from simple devices such as metersticks and calculators to sophisticated equipment such as machines that measure nitrogen content in plants and soil. Charts and graphs are also tools that help scientists organize their data. In the past, data were recorded by hand, often in notebooks or personal journals. Today, researchers typically enter data into computers, which make organizing and analyzing data easier. Many kinds of data are now gathered directly by computer-controlled equipment.

▶ ***Sources of Error*** Researchers must be careful to avoid errors in data collection and analysis. Tools used to measure the size and weight of marsh grasses, for example, have limited accuracy. Data analysis and sample size must be chosen carefully. In medical studies, for example, both experimental and control groups should be quite large. Why? Because there is always variation among individuals in control and experimental groups. The larger the sample size, the more reliably researchers can analyze that variation and evaluate the differences between experimental and control groups.

FIGURE 1–3 Continued

COLLECTING AND ANALYZING DATA

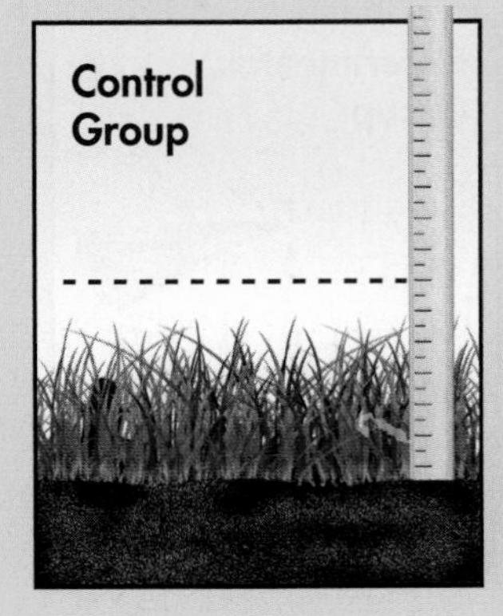

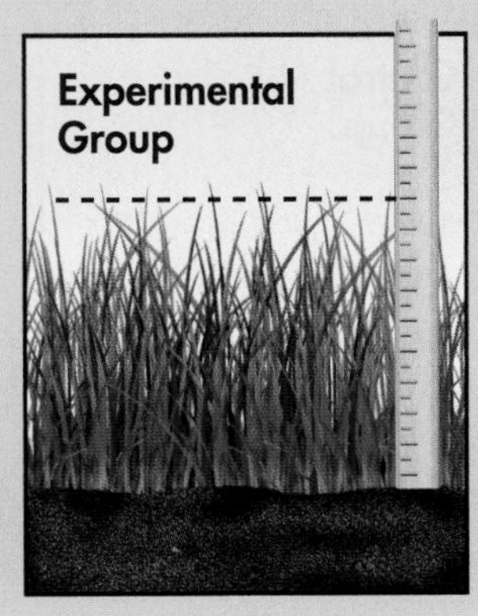

The researchers sampled all the plots throughout the growing season. They measured growth rates and plant sizes, and analyzed the chemical composition of living leaves.

DRAWING CONCLUSIONS

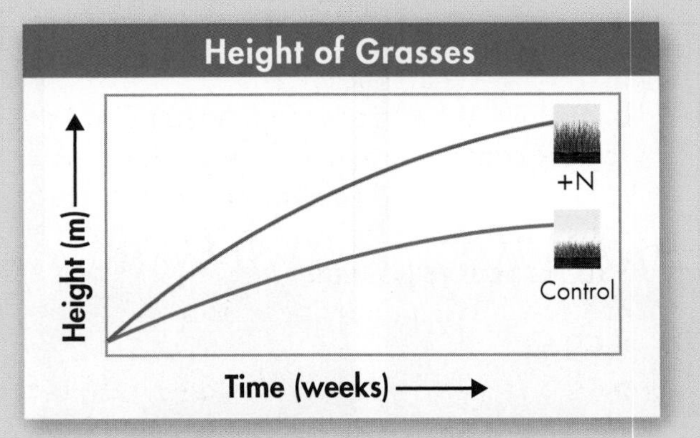

Data from all plots were compared and evaluated by statistical tests. Data analysis confirmed that marsh grasses in experimental plots with additional nitrogen did, in fact, grow taller and larger than controls. The hypothesis and its predictions were supported.

Drawing Conclusions Scientists use experimental data as evidence to support, refute, or revise the hypothesis being tested, and to draw a valid conclusion. Hypotheses are often not fully supported or refuted by one set of experiments. Rather, new data may indicate that the researchers have the right general idea but are wrong about a few particulars. In that case, the original hypothesis is reevaluated and revised; new predictions are made, and new experiments are designed. Those new experiments might suggest changes in the experimental treatment or better control of more variables. As shown in **Figure 1–4,** many circuits around this loop are often necessary before a final hypothesis is supported and conclusions can be drawn.

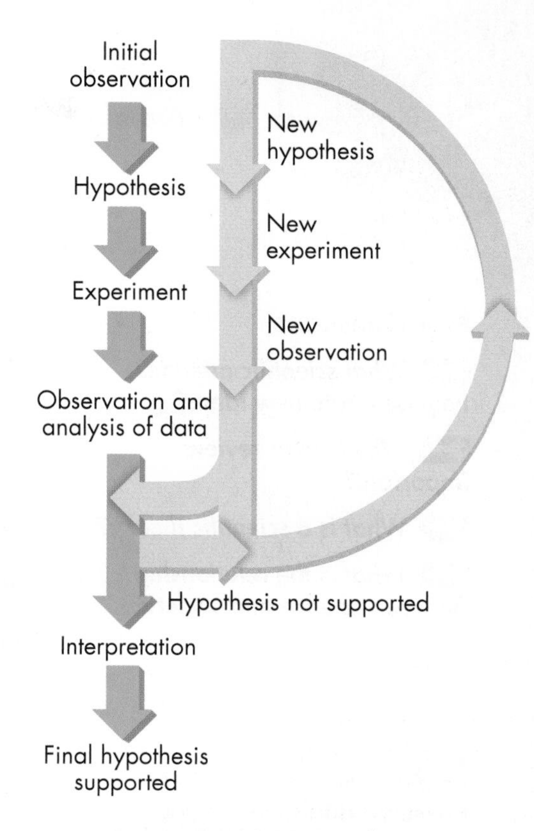

FIGURE 1–4 Revising Hypotheses During the course of an investigation, hypotheses may have to be revised and experiments redone several times.

When Experiments Are Not Possible It is not always possible to test a hypothesis with an experiment. In some of these cases, researchers devise hypotheses that can be tested by observations. Animal behavior researchers, for example, might want to learn how animal groups interact in the wild. Investigating this kind of natural behavior requires field observations that disturb the animals as little as possible. When researchers analyze data from these observations, they may devise hypotheses that can be tested in different ways.

Sometimes, ethics prevents certain types of experiments—especially on human subjects. Medical researchers who suspect that a chemical causes cancer, for example, would not intentionally expose people to it! Instead, they search for volunteers who have already been exposed to the chemical. For controls, they study people who have not been exposed to the chemical. The researchers still try to control as many variables as possible. For example, they might exclude volunteers who have serious health problems or known genetic conditions. Medical researchers always try to study large groups of subjects so that individual genetic differences do not produce misleading results.

1.1 Assessment

Review Key Concepts

1. **a. Review** What is science?
   b. Explain What kinds of understandings does science contribute about the natural world?
   c. Form an Opinion Do you think that scientists will ever run out of things to study? Explain your reasoning.
2. **a. Review** What does scientific methodology involve?
   b. Craft and Structure Why are hypotheses so important to controlled experiments? Why do you think the authors chose the marsh grass experiment as an example of the importance of hypotheses?

WRITE ABOUT SCIENCE

Creative Writing

3. **Production and Distribution of Writing** A few hundred years ago, observations seemed to indicate that some living things could just suddenly appear: maggots showed up on meat; mice were found on grain; and beetles turned up on cow dung. Those observations led to the incorrect idea of spontaneous generation—the notion that life could arise from nonliving matter. Write a paragraph for a history magazine evaluating the spontaneous generation hypothesis. Why did it seem logical at the time? What evidence was overlooked or ignored?

Science in Context

Key Questions

What scientific attitudes help generate new ideas?

Why is peer review important?

What is a scientific theory?

What is the relationship between science and society?

Vocabulary

theory • bias

Taking Notes

Preview Visuals Before you read, study **Figure 1–10.** As you read, use the figure to describe the role science plays in society.

THINK ABOUT IT Scientific methodology is the heart of science. But that vital "heart" is only part of the full "body" of science. Science and scientists operate in the context of the scientific community and society at large.

Exploration and Discovery: Where Ideas Come From

What scientific attitudes help generate new ideas?

Scientific methodology is closely linked to exploration and discovery, as shown in **Figure 1–5.** Recall that scientific methodology starts with observations and questions. But where do those observations and questions come from in the first place? They may be inspired by scientific attitudes, practical problems, and new technology.

Scientific Attitudes

Good scientists share scientific attitudes, or habits of mind, that lead them to exploration and discovery. **Curiosity, skepticism, open-mindedness, and creativity help scientists generate new ideas.**

▶ ***Curiosity*** A curious researcher, for example, may look at a salt marsh and immediately ask, "What's that plant? Why is it growing here?" Often, results from previous studies also spark curiosity and lead to new questions.

▶ ***Skepticism*** Good scientists are skeptics, which means that they question existing ideas and hypotheses, and they refuse to accept explanations without evidence. Scientists who disagree with hypotheses design experiments to test them. Supporters of hypotheses also undertake rigorous testing of their ideas to confirm them and to address any valid questions raised.

▶ ***Open-Mindedness*** Scientists must remain open-minded, meaning that they are willing to accept different ideas that may not agree with their hypothesis.

▶ ***Creativity*** Researchers also need to think creatively to design experiments that yield accurate data.

FIGURE 1–5 The Process of Science As the arrows indicate, the different aspects of science are interconnected—making the process of science dynamic, flexible, and unpredictable.

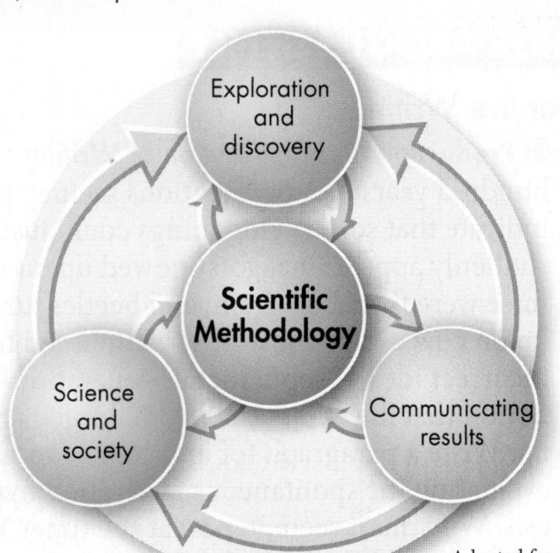

Adapted from *Understanding Science*, UC Berkeley, Museum of Paleontology

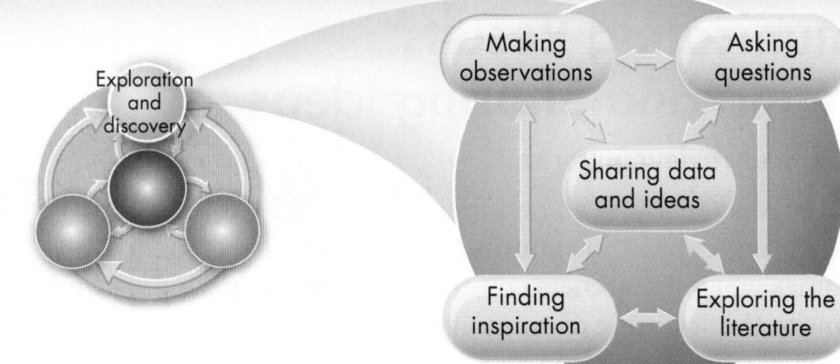

Adapted from *Understanding Science*, UC Berkeley, Museum of Paleontology

FIGURE 1–6 Exploration and Discovery Ideas in science can arise in many ways—from simple curiosity or from the need to solve a particular problem. Scientists often begin investigations by making observations, asking questions, talking with colleagues, and reading about previous experiments.

Practical Problems Sometimes, ideas for scientific investigations arise from practical problems. Salt marshes, for example, play vital roles in the lives of many ecologically and commercially important organisms, as you will learn in the next unit. Yet they are under intense pressure from industrial and housing development. Should marshes be protected from development? If new houses or farms are located near salt marshes, can they be designed to protect the marshes? These practical questions and issues inspire scientific questions, hypotheses, and experiments.

The Role of Technology Technology, science, and society are closely linked. Discoveries in one field of science may lead to new technologies. Those technologies, in turn, enable scientists in other fields to ask new questions or to gather data in new ways. For example, the development of new portable, remote data-collecting equipment enables field researchers to monitor environmental conditions around the clock, in several locations at once. This capability allows researchers to pose and test new hypotheses. Technological advances can also have big impacts on daily life. In the field of genetics and biotechnology, for instance, it is now possible to mass-produce complex substances—such as vitamins, antibiotics, and hormones—that before were only available naturally.

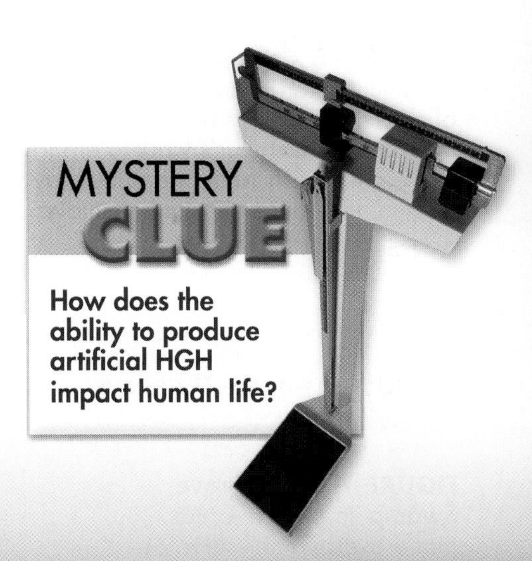

In Your Notebook *Describe a situation where you were skeptical of a "fact" you had seen or heard.*

FIGURE 1–7 Ideas From Practical Problems People living on a strip of land like this one in Murrells Inlet, South Carolina, may face flooding and other problems.
Pose Questions ***What are some scientific questions that can arise from a situation like this one?***

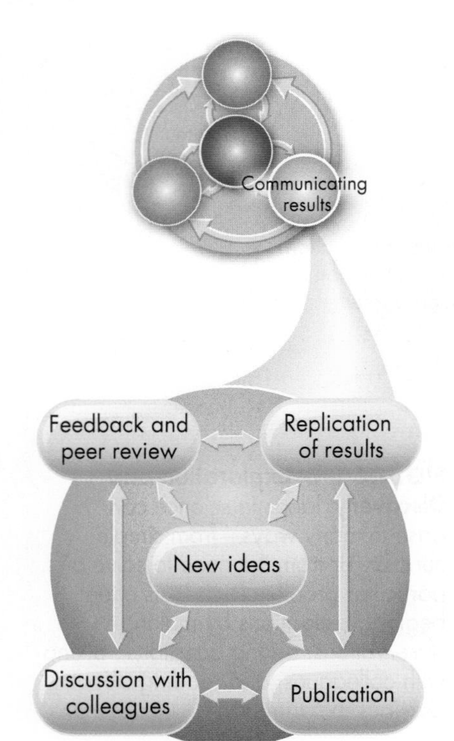

Adapted from *Understanding Science*, UC Berkeley, Museum of Paleontology

FIGURE 1–8 Communicating Results Communication is an important part of science. Scientists review and evaluate one another's work to ensure accuracy. Results from one study may lead to new ideas and further studies.

Communicating Results: Reviewing and Sharing Ideas

Why is peer review important?

Data collection and analysis can be a long process. Scientists may focus intensely on a single study for months or even years. Then, the exciting time comes when researchers communicate their experiments and observations to the scientific community. Communication and sharing of ideas are vital to modern science.

Peer Review Scientists share their findings with the scientific community by publishing articles that have undergone peer review. In peer review, scientific papers are reviewed by anonymous, independent experts. **Publishing peer-reviewed articles in scientific journals allows researchers to share ideas and to test and evaluate each other's work.** Scientific articles are like high-powered versions of your high school lab reports. They contain details about experimental conditions, controls, data, analysis, and conclusions. Reviewers read them looking for oversights, unfair influences, fraud, or mistakes in techniques or reasoning. They provide expert assessment of the work to ensure that the highest standards of quality are met. Peer review does not guarantee that a piece of work is correct, but it does certify that the work meets standards set by the scientific community.

Sharing Knowledge and New Ideas Once research has been published, it enters the dynamic marketplace of scientific ideas, as shown in **Figure 1–8.** How do new findings fit into existing scientific understanding? Perhaps they spark new questions. For example, the finding that growth of salt marsh grasses is limited by available nitrogen suggests other hypotheses: Is the growth of other plants in the same habitat also limited by nitrogen? What about the growth of different plants in similar environments, such as the mangrove swamp shown in **Figure 1–9?** Each of these logical and important questions leads to new hypotheses that must be independently confirmed by controlled experiments.

In Your Notebook *Predict what might happen if an article is published without undergoing peer review.*

FIGURE 1–9 Mangrove Swamp In tropical areas, mangrove swamps serve as the ecological equivalents of temperate salt marshes. The results of the salt marsh experiment suggest that nitrogen might be a limiting nutrient for mangroves and other plants in these similar habitats. **Design an Experiment** ***How would you test this hypothesis?***

Quick Lab
GUIDED INQUIRY

RST.9-10.3 Key Ideas and Details

Replicating Procedures

1. Working with a partner behind a screen, assemble ten blocks into an unusual structure. Write directions that others can use to replicate that structure without seeing it.
2. Exchange directions with another team. Replicate the team's structure by following its directions.
3. Compare each replicated structure to the original. Identify which parts of the directions were clear and accurate, and which were unclear or misleading.

Analyze and Conclude

1. Evaluate How could you have written better directions?

2. Infer Why is it important that scientists write procedures that can be replicated?

Scientific Theories

What is a scientific theory?

Evidence from many scientific studies may support several related hypotheses in a way that inspires researchers to propose a scientific **theory** that ties those hypotheses together. As you read this book, you will often come across terms that will be new to you because they are used only in science. But the word *theory* is used both in science and in everyday life. It is important to understand that the meaning you give the word *theory* in daily life is very different from its meaning in science. When you say, "I have a theory," you may mean, "I have a hunch." When a friend says, "That's just a theory" she may mean, "People aren't too certain about that idea." In those same situations, a scientist would probably use the word *hypothesis*. But when scientists talk about gravitational theory or evolutionary theory, they mean something very different from *hunch* or *hypothesis*.

In science, the word *theory* applies to a well-tested explanation that unifies a broad range of observations and hypotheses and that enables scientists to make accurate predictions about new situations. Charles Darwin's early observations and hypotheses about change over time in nature, for example, grew and expanded for years before he collected them into a theory of evolution by natural selection. Today, evolutionary theory is the central organizing principle of all biological and biomedical science. It makes such a wide range of predictions about organisms—from bacteria to whales to humans—that it is mentioned throughout this book.

A useful theory that has been thoroughly tested and supported by many lines of evidence may become the dominant view among the majority of scientists, but no theory is considered absolute truth. Science is always changing; as new evidence is uncovered, a theory may be revised or replaced by a more useful explanation.

BUILD Vocabulary

ACADEMIC WORDS A scientific **theory** describes a well-tested explanation for a range of phenomena. Scientific theories are different from scientific laws and it is important to understand that theories do not *become* laws. Laws, such as ideal gas laws in chemistry or Newton's laws of motion, are concise, specific descriptions of how some aspect of the natural world is expected to behave in a certain situation. In contrast, scientific theories, such as cell theory or the theory of evolution, are more dynamic and complex. Scientific theories encompass a greater number of ideas and hypotheses than laws, and are constantly fine-tuned through the process of science.

Science and Society

What is the relationship between science and society?

Make a list of health-related things that you need to understand to protect your life and the lives of others close to you. Your list may include drugs and alcohol, smoking and lung disease, AIDS, cancer, and heart disease. Other topics focus on social issues and the environment. How much of the information in your genes should be kept private? Should communities produce electricity using fossil fuels, nuclear power, solar power, wind power, or hydroelectric dams? How should chemical wastes be disposed of?

All these questions require scientific information to answer, and many have inspired important research. But none of these questions can be answered by science alone. These questions involve the society in which we live, our economy, and our laws and moral principles. **Using science involves understanding its context in society and its limitations. Figure 1–10** shows the role science plays in society.

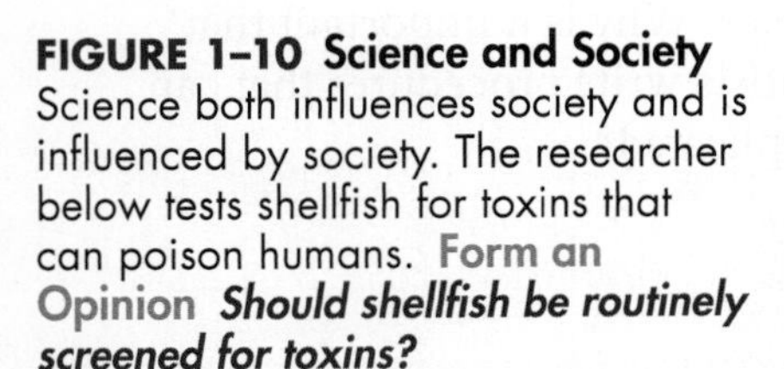

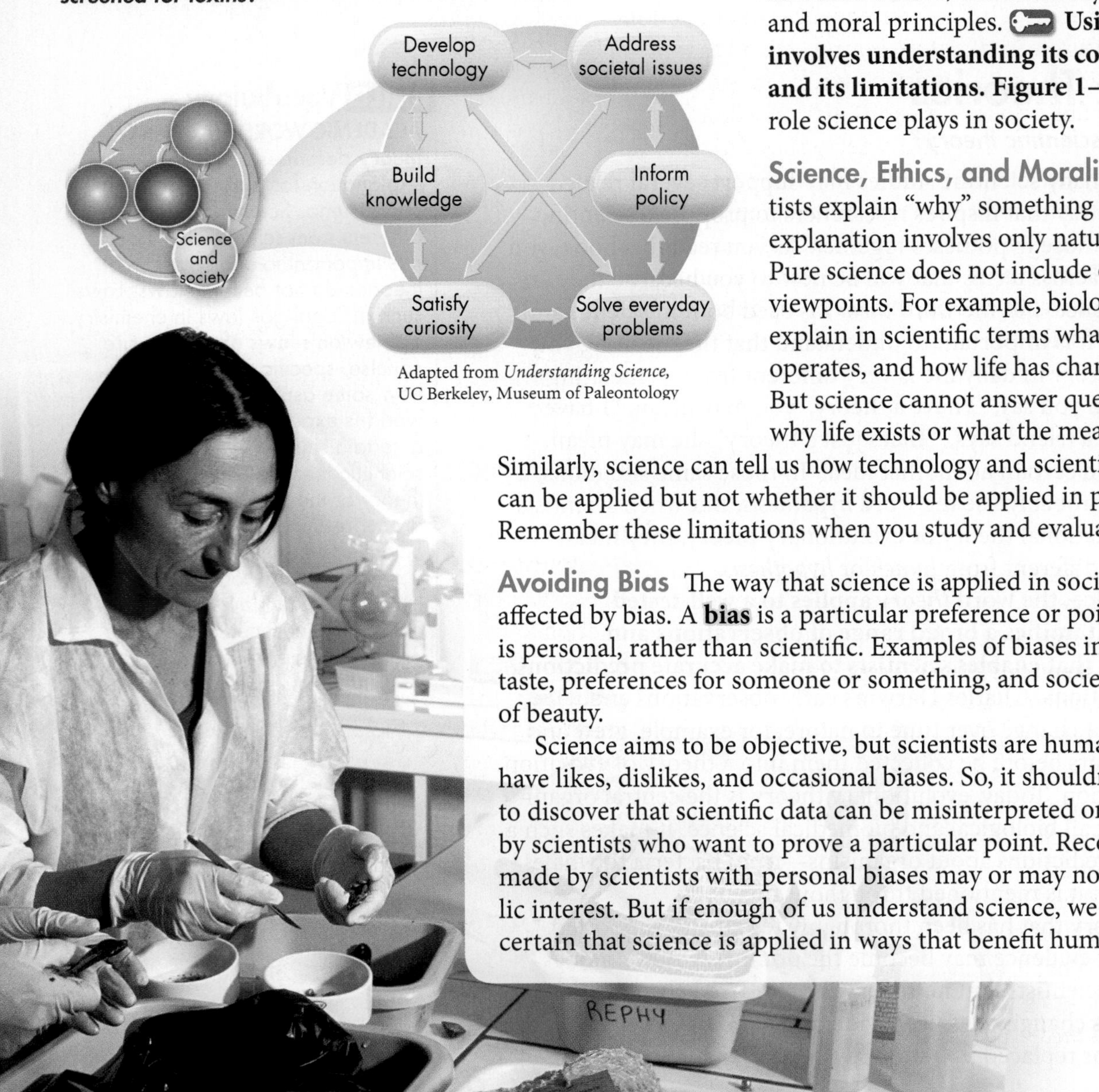

Adapted from *Understanding Science,* UC Berkeley, Museum of Paleontology

FIGURE 1–10 Science and Society Science both influences society and is influenced by society. The researcher below tests shellfish for toxins that can poison humans. **Form an Opinion** ***Should shellfish be routinely screened for toxins?***

Science, Ethics, and Morality When scientists explain "why" something happens, their explanation involves only natural phenomena. Pure science does not include ethical or moral viewpoints. For example, biologists try to explain in scientific terms what life is, how life operates, and how life has changed over time. But science cannot answer questions about why life exists or what the meaning of life is. Similarly, science can tell us how technology and scientific knowledge can be applied but not whether it should be applied in particular ways. Remember these limitations when you study and evaluate science.

Avoiding Bias The way that science is applied in society can be affected by bias. A **bias** is a particular preference or point of view that is personal, rather than scientific. Examples of biases include personal taste, preferences for someone or something, and societal standards of beauty.

Science aims to be objective, but scientists are human, too. They have likes, dislikes, and occasional biases. So, it shouldn't surprise you to discover that scientific data can be misinterpreted or misapplied by scientists who want to prove a particular point. Recommendations made by scientists with personal biases may or may not be in the public interest. But if enough of us understand science, we can help make certain that science is applied in ways that benefit humanity.

Understanding and Using Science Science will keep changing as long as humans keep wondering about nature. We invite you to join us in that wonder and exploration as you read this book. Think of this text, not as an encyclopedia, but as a "user's guide" to the study of life. Don't just memorize today's scientific facts and ideas. And please don't *believe* them! Instead, try to *understand* how scientists developed those ideas. Try to see the thinking behind experiments we describe. Try to pose the kinds of questions scientists ask.

If you learn to think as scientists think, you will understand the process of science and be comfortable in a world that will keep changing throughout your life. Understanding science will help you make complex decisions that also involve cultural customs, values, and ethical standards.

Furthermore, understanding biology will help you realize that we humans can predict the consequences of our actions and take an active role in directing our future and that of our planet. In our society, scientists make recommendations about big public policy decisions, but they don't make the decisions. Who makes the decisions? Citizens of our democracy do. In a few years, you will be able to exercise the rights of a voting citizen, influencing public policy by the ballots you cast and the messages you send public officials. That's why it is important that you understand how science works and appreciate both the power and the limitations of science.

FIGURE 1–11 Using Science in Everyday Life These student volunteers are planting mangrove saplings as part of a mangrove restoration project.

1.2 Assessment

Review Key Concepts

1. a. **Review** List the attitudes that lead scientists to explore and discover.
   b. **Explain** What does it mean to describe a scientist as skeptical? Why is skepticism an important quality in a scientist?
2. a. **Review** What is peer review?
   b. **Apply Concepts** An advertisement claims that studies of a new sports drink show it boosts energy. You discover that none of the study results have been peer-reviewed. What would you tell consumers who are considering buying this product?
3. a. **Review** What is a scientific theory?
   b. **Compare and Contrast** How does use of the word *theory* differ in science and in daily life?
4. a. **Review** How is the use of science related to its context in society?
   b. **Explain** Describe some of the limitations of science.
   c. **Apply Concepts** A study shows that a new pesticide is safe for use on food crops. The researcher who conducted the study works for the pesticide company. What potential biases may have affected the study?

Apply the Big idea

Science as a Way of Knowing

5. Explain in your own words why science is considered a "way of knowing."

WHST.9-10.1d Text Types and Purposes, **RST.9-10.10** Level of Text Complexity, **WHST.9-10.10** Range of Writing

Biology & Society

Who Should Fund Product Safety Studies?

Biology plays a major role in the research, development, and production of food, medicine, and other consumer items. Companies that make these items profit by selling reliable and useful products in the marketplace. For example, the plastics industry provides countless products for everyday use.

But sometimes questions arise concerning product safety. Bisphenol-A (BPA), for instance, is a chemical found in hard plastics. Those plastics are used to make baby bottles, reusable water bottles, and the linings of many food and soft drink cans. Is BPA safe? This type of question can be posed as a scientific hypothesis to be tested. But who does the testing? Who funds the studies and analyzes the results?

Ideally, independent scientists test products for safety and usefulness. That way, the people who gather and analyze data can remain objective—they have nothing to gain by exaggerating the positive effects of products and nothing to lose by stating any risks. However, scientists are often hired by private companies to develop or test their products.

Often, test results are clear: A product is safe or it isn't. Based on these results, the Food and Drug Administration (FDA) or another government agency makes recommendations to protect and promote public health. Sometimes, though, results are tough to interpret.

More than 100 studies have been done on BPA—some funded by the government, some funded by the plastics industry. Most of the independent studies found that low doses of BPA could have negative health effects on laboratory animals. A few studies, mostly funded by the plastics industry, concluded that BPA is safe. In this case, the FDA ultimately declared BPA to be safe. When the issue of BPA safety hit the mass media, government investigations began. So, who should sponsor product safety studies?

The Viewpoints

Independent Organizations Should Fund Safety Studies Scientists performing safety studies should have no affiliation with private industries, because conflict of interest seems unavoidable. A company, such as a BPA manufacturer, would naturally benefit if its product is declared to be safe. Rather, safety tests should be funded by independent organizations such as universities and government agencies, which should be as independent as possible. This way, recommendations for public health can remain free of biases.

Private Industries Should Fund Safety Studies There are an awful lot of products out there! Who would pay scientists to test all those products? There are simply too many potentially useful and valuable products being developed by private industry for the government to keep track of and test adequately with public funds. It is in a company's best interest to produce safe products, so it would be inclined to maintain high standards and perform rigorous tests.

Research and Decide

1. Analyze the Viewpoints To make an informed decision, research the current status of the controversy over BPA by using the Internet and other resources. Compare this situation with the history of safety studies on cigarette smoke and the chemical Teflon.

2. Form an Opinion Should private industries pay scientists to perform product safety studies? Write an argument to support your claim, with attention to the issue of potential bias in interpreting results. To help maintain an objective tone, avoid the pronoun "I."

1.3 Studying Life

THINK ABOUT IT Think about important and exciting news stories you've seen or heard. Bird flu spreads around the world, killing thousands of birds and threatening a human epidemic. Users of certain illegal drugs experience permanent damage to their brains and other parts of their nervous systems. Reports surface about efforts to clone human cells to grow new organs to replace those lost to disease or injury. These and many other stories involve biology—the science that employs scientific methodology to study living things. (The Greek word *bios* means "life," and *-logy* means "study of.")

Key Questions

What characteristics do all living things share?

What are the central themes of biology?

How do different fields of biology differ in their approach to studying life?

How is the metric system important in science?

Vocabulary

biology • DNA • stimulus • sexual reproduction • asexual reproduction • homeostasis • metabolism • biosphere

Taking Notes

Concept Map As you read, draw a concept map showing the big ideas in biology.

Characteristics of Living Things

What characteristics do all living things share?

Biology is the study of life. But what is life? What distinguishes living things from nonliving matter? Surprisingly, it isn't as simple as you might think to describe what makes something alive. No single characteristic is enough to describe a living thing. Also, some nonliving things share one or more traits with organisms. For example, a firefly and fire both give off light, and each moves in its own way. Mechanical toys, automobiles, and clouds (which are not alive) move around, while mushrooms and trees (which are alive) stay in one spot. To make matters more complicated, some things, such as viruses, exist at the border between organisms and nonliving things.

Despite these difficulties, we can list characteristics that most living things have in common. **Living things are made up of basic units called cells, are based on a universal genetic code, obtain and use materials and energy, grow and develop, reproduce, respond to their environment, maintain a stable internal environment, and change over time.**

FIGURE 1–12 Is It Alive? The fish are clearly alive, but what about the colorful structure above them? Is it alive? As a matter of fact, it is. The antlerlike structure is actually a marine animal called elkhorn coral. Corals show all the characteristics common to living things.

VISUAL SUMMARY

THE CHARACTERISTICS OF LIVING THINGS

FIGURE 1–13 Apple trees share certain characteristics with other living things. **Compare and Contrast** ***How are the apple tree and the grass growing below similar? How are they different?***

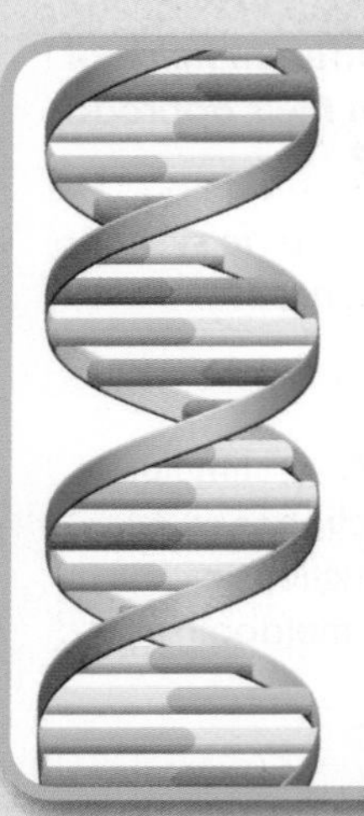

Living things are based on a universal genetic code. All organisms store the complex information they need to live, grow, and reproduce in a genetic code written in a molecule called **DNA**. That information is copied and passed from parent to offspring. With a few minor variations, life's genetic code is almost identical in every organism on Earth.

◀ *The growth, form, and structure of an apple tree are determined by information in its DNA.*

Living things grow and develop. Every organism has a particular pattern of growth and development. During development, a single fertilized egg divides again and again. As these cells divide, they differentiate, which means they begin to look different from one another and to perform different functions.

◀ *An apple tree develops from a tiny seed.*

Living things respond to their environment. Organisms detect and respond to stimuli from their environment. A **stimulus** is a signal to which an organism responds.

▼ *Some plants can produce unsavory chemicals to ward off caterpillars that feed on their leaves.*

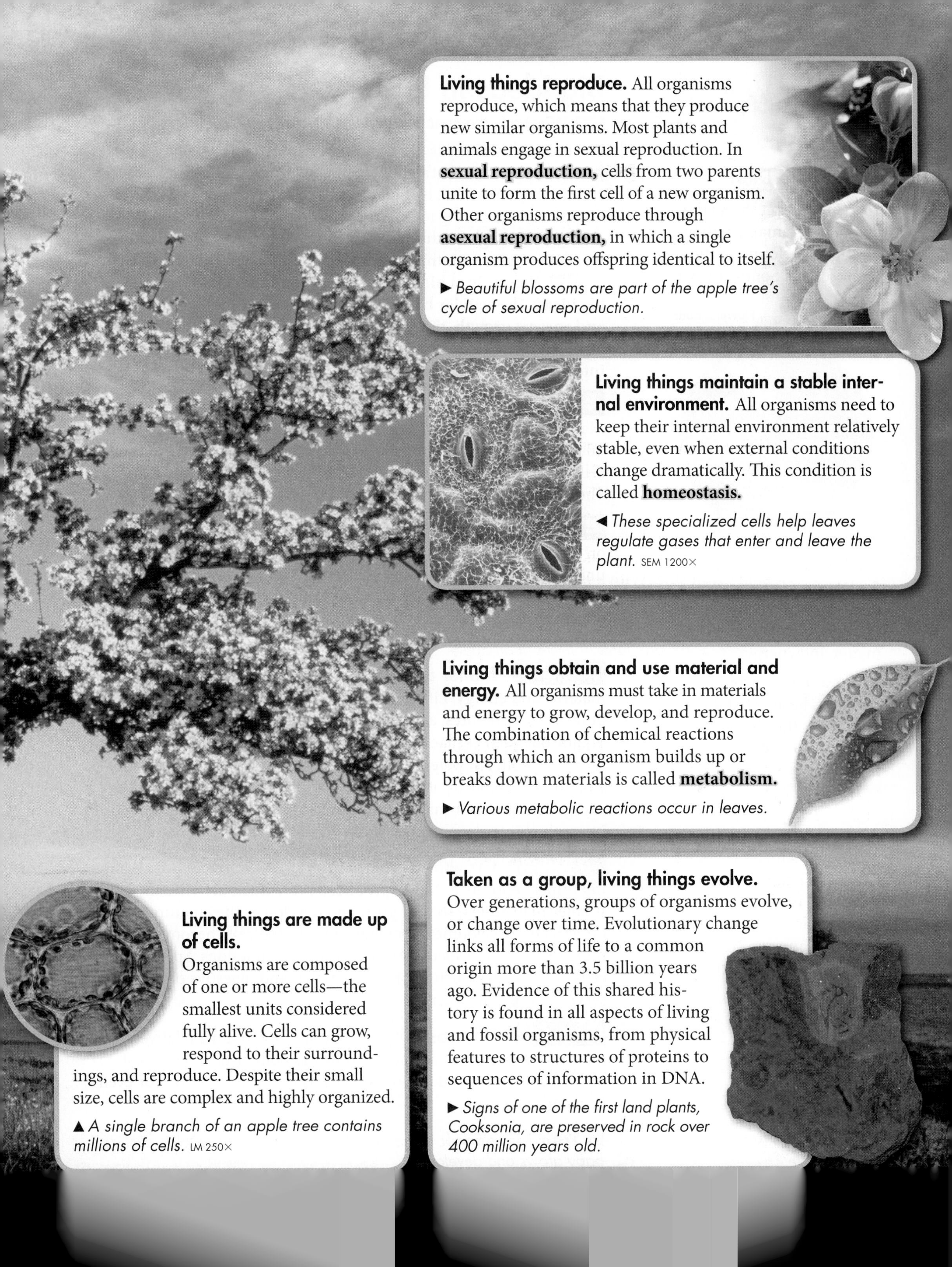

Living things reproduce. All organisms reproduce, which means that they produce new similar organisms. Most plants and animals engage in sexual reproduction. In **sexual reproduction,** cells from two parents unite to form the first cell of a new organism. Other organisms reproduce through **asexual reproduction,** in which a single organism produces offspring identical to itself.

► *Beautiful blossoms are part of the apple tree's cycle of sexual reproduction.*

Living things maintain a stable internal environment. All organisms need to keep their internal environment relatively stable, even when external conditions change dramatically. This condition is called **homeostasis.**

◄ *These specialized cells help leaves regulate gases that enter and leave the plant.* SEM 1200×

Living things obtain and use material and energy. All organisms must take in materials and energy to grow, develop, and reproduce. The combination of chemical reactions through which an organism builds up or breaks down materials is called **metabolism.**

► *Various metabolic reactions occur in leaves.*

Living things are made up of cells. Organisms are composed of one or more cells—the smallest units considered fully alive. Cells can grow, respond to their surroundings, and reproduce. Despite their small size, cells are complex and highly organized.

▲ *A single branch of an apple tree contains millions of cells.* LM 250×

Taken as a group, living things evolve. Over generations, groups of organisms evolve, or change over time. Evolutionary change links all forms of life to a common origin more than 3.5 billion years ago. Evidence of this shared history is found in all aspects of living and fossil organisms, from physical features to structures of proteins to sequences of information in DNA.

► *Signs of one of the first land plants, Cooksonia, are preserved in rock over 400 million years old.*

What's in a Diet?

The circle graph shows the diet of the siamang gibbon, a type of ape found in the rainforests of Southeast Asia.

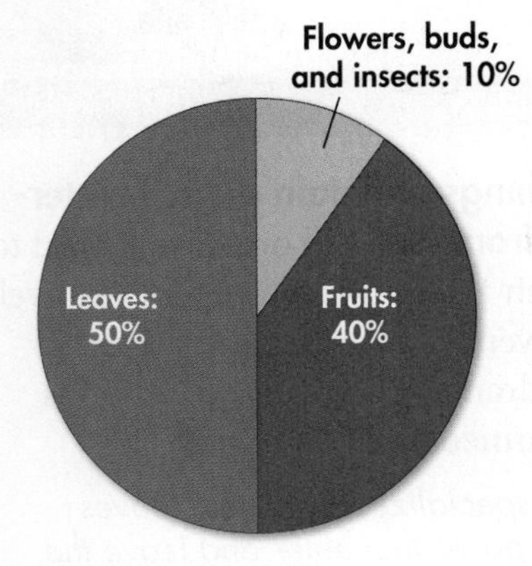

Analyze and Conclude

1. Interpret Graphs Which plant parts do siamangs rely on most as a source of their matter and energy?

2. Predict How would siamangs be affected if the rainforests they live in were cut down?

Big Ideas in Biology

What are the central themes of biology?

The units of this book seem to cover different subjects. But we'll let you in on a secret: That's not how biology works. All biological sciences are tied together by themes and methods of study that cut across disciplines. These "big ideas" overlap and interlock, and crop up again and again throughout the book. You'll also notice that several of these big ideas overlap with the characteristics of life or the nature of science.

The study of biology revolves around several interlocking big ideas: The cellular basis of life; information and heredity; matter and energy; growth, development, and reproduction; homeostasis; evolution; structure and function; unity and diversity of life; interdependence in nature; and science as a way of knowing.

Big idea **Cellular Basis of Life** Living things are made of cells. Many living things consist of only a single cell; they are called unicellular organisms. Plants and animals are multicellular. Cells in multicellular organisms display many different sizes, shapes, and functions. The human body contains 200 or more different cell types.

Big idea **Information and Heredity** Living things are based on a universal genetic code. The information coded in DNA forms an unbroken chain that stretches back roughly 3.5 billion years. Yet, the DNA inside your cells right now can influence your future—your risk of getting cancer, the amount of cholesterol in your blood, and the color of your children's hair.

Big idea **Matter and Energy** Living things obtain and use material and energy. Life requires matter that serves as nutrients to build body structures, and energy that fuels life's processes. Some organisms, such as plants, obtain energy from sunlight and take up nutrients from air, water, and soil. Other organisms, including most animals, eat plants or other animals to obtain both nutrients and energy. The need for matter and energy link all living things on Earth in a web of interdependent relationships.

Big idea **Growth, Development, and Reproduction** All living things reproduce. Newly produced individuals are virtually always smaller than adults, so they grow and develop as they mature. During growth and development, generalized cells typically become more and more different and specialized for particular functions. Specialized cells build tissues, such as brains, muscles, and digestive organs, that serve various functions.

Big idea **Homeostasis** Living things maintain a relatively stable internal environment, a process known as homeostasis. For most organisms, any breakdown of homeostasis may have serious or even fatal consequences.

In Your Notebook *Describe what happens at the cellular level as a baby grows and develops.*

Big idea **Evolution** Taken as a group, living things evolve. Evolutionary change links all forms of life to a common origin more than 3.5 billion years ago. Evidence of this shared history is found in all aspects of living and fossil organisms, from physical features to structures of proteins to sequences of information in DNA. Evolutionary theory is the central organizing principle of all biological and biomedical sciences.

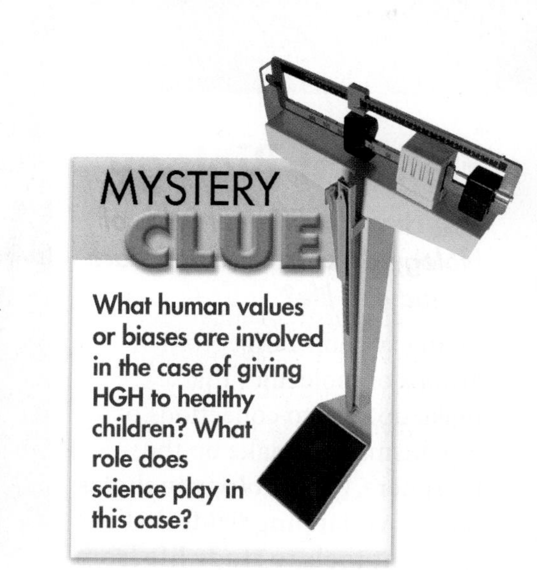

MYSTERY CLUE

What human values or biases are involved in the case of giving HGH to healthy children? What role does science play in this case?

Big idea **Structure and Function** Each major group of organisms has evolved its own particular body part "tool kit,"—a collection of structures that have evolved in ways that make particular functions possible. From capturing food to digesting it, and from reproducing to breathing, organisms use structures that have evolved into different forms as species have adapted to life in different environments. The structures of wings, for example, enable birds and insects to fly. The structures of legs enable horses to gallop and kangaroos to hop.

Big idea **Unity and Diversity of Life** Although life takes an almost unbelievable variety of forms, all living things are fundamentally similar at the molecular level. All organisms are composed of a common set of carbon-based molecules, store information in a common genetic code, and use proteins to build their structures and carry out their functions. One great contribution of evolutionary theory is that it explains both this unity of life and its diversity.

Big idea **Interdependence in Nature** All forms of life on Earth are connected into a **biosphere,** which literally means "living planet." Within the biosphere, organisms are linked to one another and to the land, water, and air around them. Relationships between organisms and their environments depend on the cycling of matter and the flow of energy. Human life and the economies of human societies also require matter and energy, so human life depends directly on nature.

Big idea **Science as a Way of Knowing**
Science is not a list of facts, but "a way of knowing." The job of science is to use observations, questions, and experiments to explain the natural world in terms of natural forces and events. Successful scientific research reveals rules and patterns that can explain and predict at least some events in nature. Science enables us to take actions that affect events in the world around us. To make certain that scientific knowledge is used for the benefit of society, all of us must understand the nature of science—its strengths, its limitations, and its interactions with our culture.

FIGURE 1–14 Different But Similar
The colorful keel-billed toucan is clearly different from the plant on which it perches. Yet, the two organisms are fundamentally similar at the molecular level. Unity and diversity of life is an important theme in biology.

Fields of Biology

How do different fields of biology differ in their approach to studying life?

Living systems range from groups of molecules that make up cells to collections of organisms that make up the biosphere. **Biology includes many overlapping fields that use different tools to study life from the level of molecules to the entire planet.** Here's a peek into a few of the smallest and largest branches of biology.

Global Ecology Life on Earth is shaped by weather patterns and processes in the atmosphere so large that we are just beginning to understand them. We are also learning that activities of living organisms—including humans—profoundly affect both the atmosphere and climate. Humans now move more matter and use more energy than any other multicellular species on Earth. Global ecological studies, aided by satellite technology and supercomputers, are enabling us to learn about our global impact, which affects all life on Earth.

▶ *Scientists in Brazil record data to understand how environmental factors affect tree growth.*

Biotechnology This field, created by the molecular revolution, is based on our ability to "edit" and rewrite the genetic code—in a sense, redesigning the living world to order. We may soon learn to correct or replace damaged genes that cause inherited diseases. Other research seeks to genetically engineer bacteria to clean up toxic wastes. Biotechnology also raises enormous ethical, legal, and social questions. Dare we tamper with the fundamental biological information that makes us human?

▶ *A plant biologist analyzes genetically modified rice plants.*

Building the Tree of Life Biologists have discovered and identified roughly 1.8 million different kinds of living organisms. That may seem like an incredible number, but researchers estimate that somewhere between 2 and 100 million more forms of life are waiting to be discovered around the globe—from caves deep beneath the surface, to tropical rainforests, to coral reefs and the depths of the sea. Identifying and cataloguing all these life forms is enough work by itself, but biologists aim to do much more. They want to combine the latest genetic information with computer technology to organize all living things into a single universal "Tree of All Life"—and put the results on the Web in a form that anyone can access.

▶ *Paleontologists study the fossilized bones of dinosaurs.*

Ecology and Evolution of Infectious Diseases
HIV, bird flu, and drug-resistant bacteria seem to have appeared out of nowhere, but the science behind their stories shows that relationships between hosts and pathogens are dynamic and constantly changing. Organisms that cause human disease have their own ecology, which involves our bodies, medicines we take, and our interactions with each other and the environment. Over time, disease-causing organisms engage in an "evolutionary arms race" with humans that creates constant challenges to public health around the world. Understanding these interactions is crucial to safeguarding our future.

▶ *An entomologist (center) and other researchers inspect mosquito traps lining an area between a neighborhood and a mosquito breeding area in Florida.*

Genomics and Molecular Biology These fields focus on studies of DNA and other molecules inside cells. The "molecular revolution" of the 1980s created the field of genomics, which is now looking at the entire sets of DNA code contained in a wide range of organisms. Ever-more-powerful computer analyses enable researchers to compare vast databases of genetic information in a fascinating search for keys to the mysteries of growth, development, aging, cancer, and the history of life on Earth.

▶ *A molecular biologist analyzes a DNA sequence.*

Performing Biological Investigations

How is the metric system important in science?

During your study of biology, you will have the opportunity to perform scientific investigations. Biologists, like other scientists, rely on a common system of measurement and practice safety procedures when conducting studies. As you study and experiment, you will become familiar with scientific measurement and safety procedures.

Scientific Measurement Because researchers need to replicate one another's experiments, and because many experiments involve gathering quantitative data, scientists need a common system of measurement. **Most scientists use the metric system when collecting data and performing experiments.** The metric system is a decimal system of measurement whose units are based on certain physical standards and are scaled on multiples of 10. A revised version of the original metric system is called the International System of Units, or SI. The abbreviation *SI* comes from the French *Le Système International d'Unités.*

Because the metric system is based on multiples of 10, it is easy to use. Notice in **Figure 1–15** how the basic unit of length, the meter, can be multiplied or divided to measure objects and distances much larger or smaller than a meter. The same process can be used when measuring volume and mass. You can learn more about the metric system in Appendix B.

BUILD Vocabulary

PREFIXES The SI prefix *milli-* means "thousandth." Therefore, 1 millimeter is one-thousandth of a meter, and 1 milligram is one-thousandth of a gram.

Common Metric Units

Length	Mass
1 meter (m) = 100 centimeters (cm) 1 meter = 1000 millimeters (mm) 1000 meters = 1 kilometer (km)	1 kilogram (kg) = 1000 grams (g) 1 gram = 1000 milligrams (mg) 1000 kilograms = 1 metric ton (t)
Volume	**Temperature**
1 liter (L) = 1000 milliliters (mL) 1 liter = 1000 cubic centimeters (cm^3)	0°C = freezing point of water 100°C = boiling point of water

FIGURE 1–15 The Metric System Scientists usually use the metric system in their work. This system is easy to use because it is based on multiples of 10. This penguin in China has been trained to hop onto the scale to be weighed. **Predict** ***What unit of measurement would you use to express the penguin's mass?***

Safety Scientists working in a laboratory or in the field are trained to use safe procedures when carrying out investigations. Laboratory work may involve flames or heating elements, electricity, chemicals, hot liquids, sharp instruments, and breakable glassware. Laboratory work and fieldwork may involve contact with living or dead organisms—not just potentially poisonous plants and venomous animals but also disease-carrying mosquitoes and water contaminated with dangerous microorganisms.

Whenever you work in your biology laboratory, you must follow safe practices as well. Careful preparation is the key to staying safe during scientific activities. Before performing any activity in this course, study the safety rules in Appendix B. Before you start each activity, read all the steps and make sure that you understand the entire procedure, including any safety precautions.

The single most important safety rule is to always follow your teacher's instructions and directions in this textbook. Any time you are in doubt about any part of an activity, ask your teacher for an explanation. And because you may come in contact with organisms you cannot see, it is essential that you wash your hands thoroughly after every scientific activity. Remember that you are responsible for your own safety and that of your teacher and classmates. If you are handling live animals, you are responsible for their safety too.

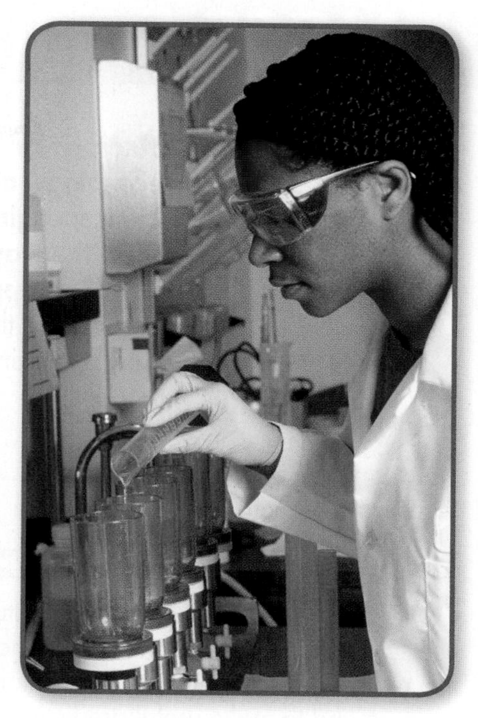

FIGURE 1–16 Science Safety Wearing appropriate protective gear is important while working in a laboratory.

1.3 Assessment

Review Key Concepts

1. a. **Review** List the characteristics that define life.

   b. **Applying Concepts** Suppose you feel hungry, so you reach for a plum you see in a fruit bowl. Explain how both external and internal stimuli are involved in your action.

2. a. **Review** What are the themes in biology that come up again and again?

   b. **Predict** Suppose you discover a new organism. What would you expect to see if you studied it under a microscope?

3. a. **Review** At what levels do biologists study life?

   b. **Classify** A researcher studies why frogs are disappearing in the wild. What field of biology does the research fall into?

4. a. **Review** Why do scientists use a common system of measurement?

   b. **Relate Cause and Effect** Suppose two scientists are trying to perform an experiment that involves dangerous chemicals. How might their safety be affected by not using a common measurement?

PRACTICE PROBLEM

5. In an experiment, you need 250 grams of potting soil for each of 10 plant samples. How many kilograms of soil in total do you need? MATH

NGSS Smart Guide The Science of Biology

Disciplinary Core Idea ETS2.B Influence of engineering, technology, and science on society and the natural world: How do science, engineering, and the technologies that result from them affect the ways in which people live? How do they affect the natural world? In applying scientific methodology, biologists can find answers to questions that arise in the study of life. These answers can then be used by engineers to design and develop technological advances.

1.1 What is Science?

- One goal of science is to provide natural explanations for events in the natural world. Science also aims to use those explanations to understand patterns in nature and to make useful predictions about natural events.
- Scientific methodology involves observing and asking questions, making inferences and forming hypotheses, conducting controlled experiments, collecting and analyzing data, and drawing conclusions.

science 5 • observation 6 • inference 7 • hypothesis 7 • controlled experiment 7 • independent variable 7 • dependent variable 7 • control group 7 • data 8

Biology.com

Untamed Science Video Be prepared for some surprise answers as the Untamed Science crew hits the streets to ask people basic questions about science and biology.

Art in Motion Learn about the steps scientists use to solve problems. Change the variables, and watch what happens!

Art Review Review your understanding of the various steps of experimental processes.

ART IN MOTION Experimental Design
The Experiment
The researchers add nitrogen fertilizer (the independent variable) to the experimental plots. Then they observe the growth of marsh plants (the dependent variable) in both experimental and control plots.
No N added
Control
Experimental

1.2 Science in Context

- Curiosity, skepticism, open-mindedness, and creativity help scientists generate new ideas.
- Publishing peer-reviewed articles in scientific journals allows researchers to share ideas and to test and evaluate each other's work.
- In science, the word *theory* applies to a well-tested explanation that unifies a broad range of observations and hypotheses and that enables scientists to make accurate predictions about new situations.
- Using science involves understanding its context in society and its limitations.

theory 13 • bias 14

Biology.com

Lesson Notes Describe the role science plays in society.

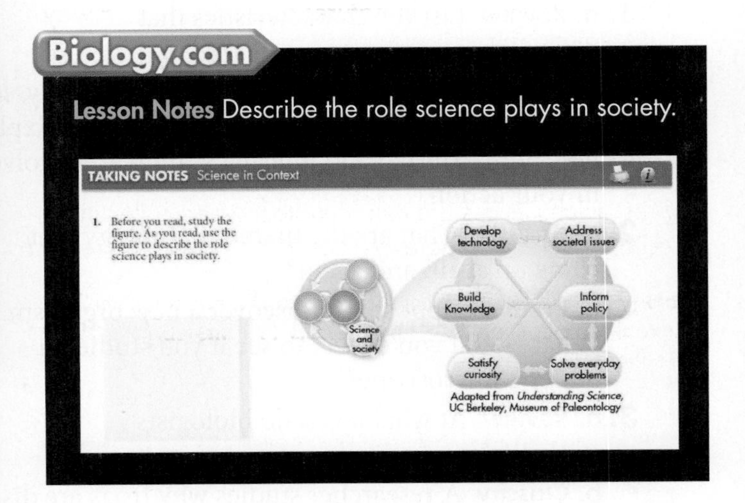

CHECKING YOUR *Scientific Literacy*

Refer to the lesson content and digital assets below as you prepare for your chapter assessment. Then, evaluate your understanding of scientific methodology by answering these questions.

1. **Scientific and Engineering Practice Planning and Carrying Out Investigations** Suppose you are trying to determine how temperature affects the rate at which sliced apples turn brown. Design a controlled experiment and carry out the investigation.

2. Crosscutting Concept **Systems and System Models** Using the model for the process of science illustrated in Lesson 2, explain how science and society influence each other.

3. Text Types and Purposes Write an essay for the science section of a newspaper explaining to the general public what science is and what scientific methodology involves.

4. STEM You are a physician being interviewed about how technology affects your work. Describe three examples of how tools or technological advances might affect work or research in your field of expertise.

1.3 Studying Life

- Living things are made up of units called cells, are based on a universal genetic code, obtain and use materials and energy, grow and develop, reproduce, respond to their environment, maintain a stable internal environment, and change over time.
- The study of biology revolves around several interlocking big ideas: the cellular basis of life; information and heredity; matter and energy; growth, development, and reproduction; homeostasis; evolution; structure and function; unity and diversity of life; interdependence in nature; and science as a way of knowing.
- Biology includes many overlapping fields that use different tools to study life from the level of molecules to the entire planet.
- Most scientists use the metric system when collecting data and performing experiments.

biology 17 • DNA 18 • stimulus 18 • sexual reproduction 19 • asexual reproduction 19 • homeostasis 19 • metabolism 19 • biosphere 21

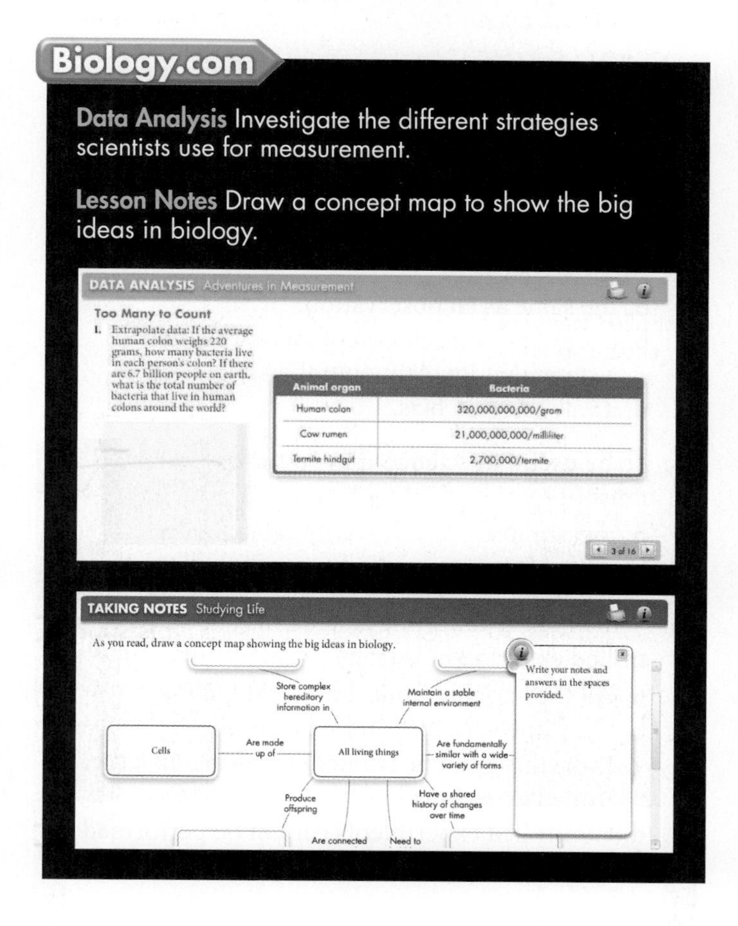

Skills Lab

Using a Microscope to Estimate Size This chapter lab is available in your lab manual and at Biology.com

1 Assessment

1.1 What Is Science?

Understand Key Concepts

1. Which of the following statements about the image shown below is NOT an observation?
   a. The insect has three legs on the left side.
   b. The insect has a pattern on its back.
   c. The insect's pattern shows that it is poisonous.
   d. The insect is green, white, and black.

2. The statement "The worm is 2 centimeters long" is a(n)
   a. observation.
   b. theory.
   c. inference.
   d. hypothesis.

3. An inference is
   a. the same as an observation.
   b. a logical interpretation of an observation.
   c. a statement involving numbers.
   d. a way to avoid bias.

4. To be useful in science, a hypothesis must be
   a. measurable.
   b. observable.
   c. testable.
   d. correct.

5. Which of the following statements about a controlled experiment is true?
   a. All the variables must be kept the same.
   b. Only one variable is tested at a time.
   c. Everything can be studied by setting up a controlled experiment.
   d. Controlled experiments cannot be performed on living things.

6. What are the goals of science?

7. How does an observation about an object differ from an inference about that object?

8. How does a hypothesis help scientists understand the natural world?

9. Why does it make sense for scientists to test just one variable at a time in an experiment?

10. Distinguish between an experimental group and a control group.

11. What steps are involved in drawing a conclusion?

12. How can a graph of data be more informative than a table of the same data?

Think Critically

13. **Design an Experiment** Suggest an experiment that would show whether one food is better than another at speeding an animal's growth.

14. **Control Variables** Explain why you cannot draw a conclusion about the effect of one variable in an investigation when the other key variables are not controlled.

1.2 Science in Context

Understand Key Concepts

15. A skeptical attitude in science
    a. prevents scientists from accepting new ideas.
    b. encourages scientists to readily accept new ideas.
    c. means a new idea will only be accepted if it is backed by evidence.
    d. is unimportant.

16. The purpose of peer review in science is to ensure that
    a. all scientific research is funded.
    b. the results of experiments are correct.
    c. all scientific results are published.
    d. published results meet standards set by the scientific community.

17. A scientific theory is
    a. the same as a hypothesis.
    b. a well-tested explanation that unifies a broad range of observations.
    c. the same as the conclusion of an experiment.
    d. the first step in a controlled experiment.

18. Why are scientific theories useful?

19. Why aren't theories considered absolute truths?

Think Critically

20. **Evaluate** Why is it misleading to describe science as a collection of facts?
21. **Propose a Solution** How would having a scientific attitude help you in everyday activities, for example, in trying to learn a new skill?
22. **Conduct Peer Review** If you were one of the anonymous reviewers of a paper submitted for publication, what criteria would you use to determine whether or not the paper should be published?

1.3 Studying Life

Understand Key Concepts

23. The process in which two cells from different parents unite to produce the first cell of a new organism is called
    a. homeostasis.
    b. development.
    c. asexual reproduction.
    d. sexual reproduction.
24. The process by which organisms keep their internal conditions relatively stable is called
    a. metabolism.
    b. a genome.
    c. evolution.
    d. homeostasis.
25. How are unicellular and multicellular organisms alike? How are they different?
26. Give an example of changes that take place as cells in a multicellular organism differentiate.
27. List three examples of stimuli that a bird responds to.

Think Critically

28. **Measure** Use a ruler to find the precise length and width of this book in millimeters.
29. **Craft and Structure** Each of the following safety symbols might appear in a laboratory activity in this book. Describe what each symbol stands for. (*Hint:* Refer to Appendix B.)

solve the CHAPTER MYSTERY

HEIGHT BY PRESCRIPTION

Although scientific studies have not proved that HGH treatment significantly increases adult height, they do suggest that extra HGH may help some short kids grow taller sooner. Parents who learn about this possibility may want treatment for their children. David's doctor prescribed HGH to avoid criticism for not presenting it as an option.

This situation is new. Many years ago, HGH was available only from cadavers, and it was prescribed only for people with severe medical problems. Then, genetic engineering made it possible to mass-produce safe, artificial HGH for medical use—safe medicine for sick people.

However, many people who are shorter than average often face prejudice in our society. This led drug companies to begin marketing HGH to parents of healthy, short kids. The message: "Help your child grow taller!"

As David's case illustrates, science has the powerful potential to change lives, but new scientific knowledge and advances may raise more questions than they answer. Just because science makes something *possible*, does that mean it's *right* to do it? This question is difficult to answer. When considering how science should be applied, we must consider both its limitations and its context in society.

1. **Research to Build and Present Knowledge** Search the Internet for the latest data on HGH treatment of healthy children. What effect does early HGH treatment have on adult height?
2. **Predict** HGH was among the first products of the biotechnology revolution. Many more are in the pipeline. As products become available that could change other inherited traits, what challenges await society?
3. **Connect to the Big idea** Why would it be important for scientists to communicate clearly the results of HGH studies? How might parents benefit by understanding the science behind the results?

Connecting Concepts

Use Science Graphics

The following graphs show the size of four different populations over a period of time. Use the graphs to answer questions 30–32.

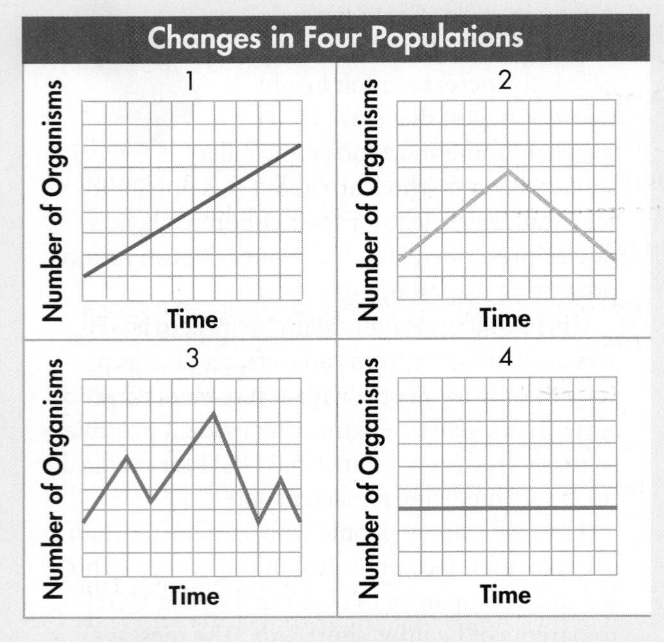

30. **Integration of Knowledge and Ideas** Write a sentence summarizing what each graph shows.

31. **Interpret Graphs** Before any of the graphs could be used to make direct comparisons among the populations, what additional information would be necessary?

32. **Compare and Contrast** Graphs of completely different events can have the same appearance. Select one of the graphs and explain how the shape of the graph could apply to a different set of events.

Write About Science

33. **Text Types and Purposes** Suppose you have a pet cat and want to determine which type of cat food it prefers. Write an explanation of how you could use scientific methodology to determine the answer. (*Hint*: Before you start writing, list the steps you might take. Then, arrange them in order, beginning with the first step, to link the ideas together cohesively.)

34. **Assess the Big idea** Many people add fertilizer to their house and garden plants. Make a hypothesis about whether you think fertilizers really help plants grow. Next, design an experiment to test your hypothesis. Include in your plan what variable you will test and what variables you will control.

Integration of Knowledge and Ideas

A researcher studied two groups of fruit flies: Population A was kept in a 0.5 L container; Population B was kept in a 1 L container.

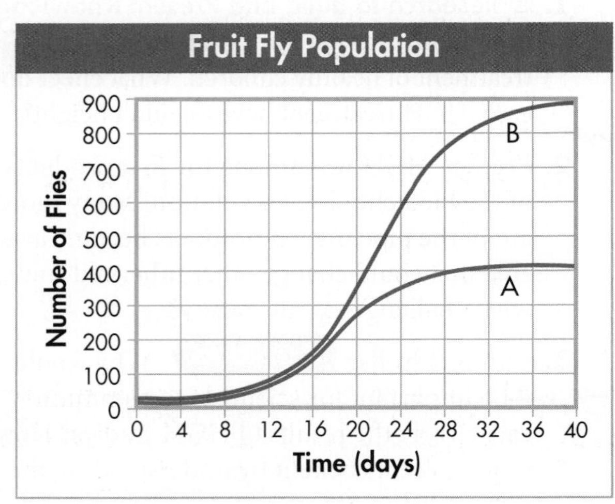

35. **Interpret Graphs** The independent variable in the controlled experiment was the

a. number of flies.
b. number of groups studied.
c. number of days.
d. size of the containers.

36. **Infer** Which of the following is a logical inference based on the content of the graph?

a. The flies in Group B were healthier than those in Group A.
b. A fly population with more available space will grow larger than a population with less space.
c. If Group B was observed for 40 more days, the size of the population would double.
d. In 40 more days, the size of both populations would decrease at the same rate.

Standardized Test Prep

Multiple Choice

1. To ensure that a scientific work is free of bias and meets standards set by the scientific community, a research group's work is peer reviewed by
A anonymous scientific experts.
B the general public.
C the researchers' friends.
D lawmakers.

2. Which of the following characteristics is NOT shared by both a horse and the grass it eats?
A uses energy
B response to stimulus
C movement from place to place
D stable internal environment

3. Which of the following statements about a scientific theory is NOT true?
A It has the same meaning in science as it does in daily life.
B It enables scientists to make accurate predictions about new situations.
C Scientific theories tie many hypotheses together.
D It is based on a large body of evidence.

4. A bird-watcher sees an unusual bird at a feeder. He takes careful notes on the bird's color, shape, and other physical features and then goes to a reference book to see if he can identify the species. What aspect of scientific thinking is most apparent in this situation?
A observation
B inference
C hypothesis formation
D controlled experimentation

5. Unlike sexual reproduction, asexual reproduction involves
A two cells.
B two parents.
C one parent.
D one nonliving thing.

6. One meter is equal to
A 1000 millimeters.
B 1 millimeter.
C 10 kilometers.
D 1 milliliter.

Questions 7–8

Once a month, a pet owner recorded the mass of her puppy in a table. When the puppy was 3 months old, she started to feed it a "special puppy food" she saw advertised on TV.

Change in a Puppy's Mass Over Time

Age (months)	Mass at Start of Month (kg)	Change in Mass per Month (kg)
2	5	—
3	8	+3
4	13	+5

7. According to the table, which statement is true?
A The puppy's mass increased at the same rate for each month shown.
B The puppy's mass was less than 5 kg at the start of the new diet.
C The puppy gained 5 kg between age 3 and 4 months.
D The puppy had gained 13 kg as a result of the new diet.

8. All of the following statements about the pet owner's study are true EXCEPT
A The owner used the metric system.
B The owner recorded data.
C The owner could graph the data.
D The owner conducted a controlled experiment.

Open-Ended Response

9. Explain how a controlled experiment works.

If You Have Trouble With . . .

Question	1	2	3	4	5	6	7	8	9
See Lesson	1.2	1.3	1.2	1.1	1.3	1.3	1.1	1.1	1.1

2 The Chemistry of Life

Matter and Energy

Q: What are the basic chemical principles that affect living things?

Water is locked in ice in the Svalbard islands of Norway—home to the polar bear. Even in such an extreme environment, organisms are able to obtain the matter and energy they need to survive.

INSIDE:

- 2.1 The Nature of Matter
- 2.2 Properties of Water
- 2.3 Carbon Compounds
- 2.4 Chemical Reactions and Enzymes

BUILDING *Scientific Literacy*

Deepen your understanding of the chemistry of life by engaging in key practices that allow you to make connections across concepts.

NGSS You will use models to visualize how the structure of life's molecules connects to their function.

STEM How important is temperature regulation in the manufacture of a commercial product where enzymes are required? Analyze data on enzyme activity at various temperatures and communicate your findings.

Common Core You will translate technical information to support your analysis of the molecules involved in life's processes.

CHAPTER MYSTERY

THE GHOSTLY FISH

Most fish, just like you and other vertebrates, have red blood. Red blood cells carry oxygen, a gas essential for life. The cells' red color comes from an oxygen-binding protein called hemoglobin.

But a very small number of fish don't have such cells. Their blood is clear—almost transparent. Because they live in cold antarctic waters and have a ghostly appearance, they are nicknamed "ice fish." How do these animals manage to survive without red blood cells?

As you read this chapter, look for clues to help you explain the ice fish's unusual feature. Think about the chemistry that might be involved. Then, solve the mystery.

Finding the solution to the fishy mystery is only the beginning. Take a video field trip with the ecogeeks of Untamed Science to continue exploring your world.

Go online to access additional resources including:
- eText • Flash Cards • Lesson Overviews • Chapter Mystery

2.1 The Nature of Matter

Key Questions

What three subatomic particles make up atoms?

How are all of the isotopes of an element similar?

In what ways do compounds differ from their component elements?

What are the main types of chemical bonds?

Vocabulary

atom • nucleus • electron • element • isotope • compound • ionic bond • ion • covalent bond • molecule • van der Waals forces

Taking Notes

Outline Before you read, make an outline of the major headings in the lesson. As you read, fill in main ideas and supporting details under each head.

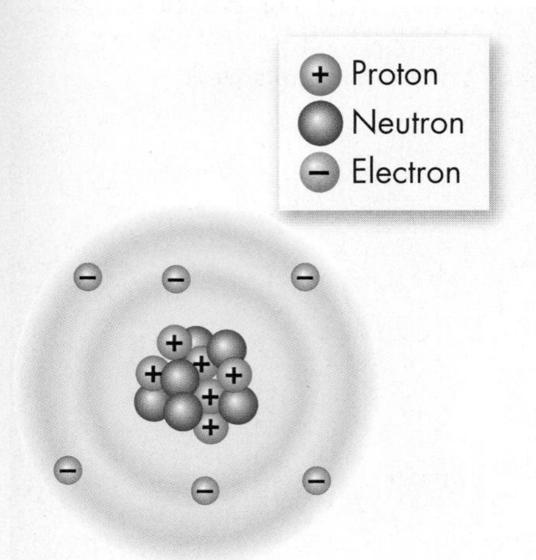

FIGURE 2–1 A Carbon Atom

THINK ABOUT IT What are you made of? Just as buildings are made from bricks, steel, glass, and wood, living things are made from chemical compounds. But it doesn't stop there. When you breathe, eat, or drink, your body uses the substances in air, food, and water to carry out chemical reactions that keep you alive. If the first task of an architect is to understand building materials, then what would be the first job of a biologist? Clearly, it is to understand the chemistry of life.

Atoms

What three subatomic particles make up atoms?

The study of chemistry begins with the basic unit of matter, the **atom.** The concept of the atom came first from the Greek philosopher Democritus, nearly 2500 years ago. Democritus asked a simple question: If you take an object like a stick of chalk and break it in half, are both halves still chalk? The answer, of course, is yes. But what happens if you break it in half again and again and again? Can you continue to divide without limit, or does there come a point at which you cannot divide the fragment of chalk without changing it into something else? Democritus thought that there had to be a limit. He called the smallest fragment the atom, from the Greek word *atomos,* which means "unable to be cut."

Atoms are incredibly small. Placed side by side, 100 million atoms would make a row only about 1 centimeter long—about the width of your little finger! Despite its extremely small size, an atom contains subatomic particles that are even smaller. **Figure 2–1** shows the subatomic particles in a carbon atom. **The subatomic particles that make up atoms are protons, neutrons, and electrons.**

Protons and Neutrons Protons and neutrons have about the same mass. However, protons are positively charged particles (+) and neutrons carry no charge at all. Strong forces bind protons and neutrons together to form the **nucleus,** at the center of the atom.

Electrons The **electron** is a negatively charged particle (–) with only 1/1840 the mass of a proton. Electrons are in constant motion in the space surrounding the nucleus. They are attracted to the positively charged nucleus but remain outside the nucleus because of the energy of their motion. Because atoms have equal numbers of electrons and protons, their positive and negative charges balance out, and atoms themselves are electrically neutral.

Elements and Isotopes

How are all of the isotopes of an element similar?

A chemical **element** is a pure substance that consists entirely of one type of atom. More than 100 elements are known, but only about two dozen are commonly found in living organisms. Elements are represented by one- or two-letter symbols. C, for example, stands for carbon, H for hydrogen, Na for sodium, and Hg for mercury. The number of protons in the nucleus of an element is called its atomic number. Carbon's atomic number is 6, meaning that each atom of carbon has six protons and, consequently, six electrons. See Appendix E, The Periodic Table, which shows the elements.

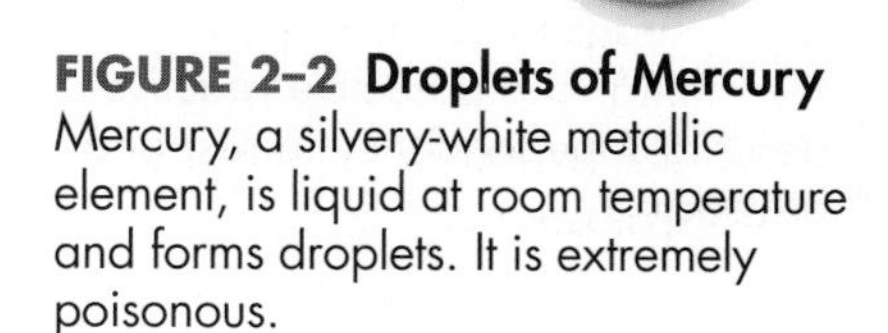

FIGURE 2–2 Droplets of Mercury Mercury, a silvery-white metallic element, is liquid at room temperature and forms droplets. It is extremely poisonous.

Isotopes Atoms of an element may have different numbers of neutrons. For example, although all atoms of carbon have six protons, some have six neutrons, some seven, and a few have eight. Atoms of the same element that differ in the number of neutrons they contain are known as **isotopes.** The total number of protons and neutrons in the nucleus of an atom is called its mass number. Isotopes are identified by their mass numbers. **Figure 2–3** shows the subatomic composition of carbon-12, carbon-13, and carbon-14 atoms. The weighted average of the masses of an element's isotopes is called its atomic mass. "Weighted" means that the abundance of each isotope in nature is considered when the average is calculated. **Because they have the same number of electrons, all isotopes of an element have the same chemical properties.**

Isotopes of Carbon

Isotope	Number of Protons	Number of Electrons	Number of Neutrons
Carbon–12 (nonradioactive)	6	6	6
Carbon–13 (nonradioactive)	6	6	7
Carbon–14 (radioactive)	6	6	8

FIGURE 2–3 Carbon Isotopes Isotopes of carbon all have 6 protons but different numbers of neutrons—6, 7, or 8. They are identified by the total number of protons and neutrons in the nucleus: carbon–12, carbon–13, and carbon–14. **Classify** ***Which isotope of carbon is radioactive?***

Radioactive Isotopes Some isotopes are radioactive, meaning that their nuclei are unstable and break down at a constant rate over time. The radiation these isotopes give off can be dangerous, but radioactive isotopes have a number of important scientific and practical uses.

Geologists can determine the ages of rocks and fossils by analyzing the isotopes found in them. Radiation from certain isotopes can be used to detect and treat cancer and to kill bacteria that cause food to spoil. Radioactive isotopes can also be used as labels or "tracers" to follow the movements of substances within organisms.

In Your Notebook *Draw a diagram of a helium atom, which has an atomic number of 2.*

Chemical Compounds

In what ways do compounds differ from their component elements?

In nature, most elements are found combined with other elements in compounds. A chemical **compound** is a substance formed by the chemical combination of two or more elements in definite proportions. Scientists show the composition of compounds by a kind of shorthand known as a chemical formula. Water, which contains two atoms of hydrogen for each atom of oxygen, has the chemical formula H_2O. The formula for table salt, NaCl, indicates that the elements that make up table salt—sodium and chlorine—combine in a 1 : 1 ratio.

The physical and chemical properties of a compound are usually very different from those of the elements from which it is formed. For example, hydrogen and oxygen, which are gases at room temperature, can combine explosively and form liquid water. Sodium is a silver-colored metal that is soft enough to cut with a knife. It reacts explosively with water. Chlorine is very reactive, too. It is a poisonous, yellow-greenish gas that was used in battles during World War I. Sodium chloride, table salt, is a white solid that dissolves easily in water. As you know, sodium chloride is not poisonous. In fact, it is essential for the survival of most living things.

BUILD Vocabulary

RELATED WORD FORMS The verb *react* means to act in response to something. The adjective *reactive* describes the tendency to respond or react.

Chemical Bonds

What are the main types of chemical bonds?

The atoms in compounds are held together by various types of chemical bonds. Much of chemistry is devoted to understanding how and when chemical bonds form. Bond formation involves the electrons that surround each atomic nucleus. The electrons that are available to form bonds are called valence electrons. **The main types of chemical bonds are ionic bonds and covalent bonds.**

Quick Lab
GUIDED INQUIRY

Model an Ionic Compound

1. You will be assigned to represent either a sodium atom or a chlorine atom.
2. Obtain the appropriate number of popcorn kernels to represent your electrons.
3. Find a partner with whom you can form the ionic compound sodium chloride—table salt.
4. In table salt, the closely packed sodium and chloride ions form an orderly structure called a crystal. With all your classmates, work as a class to model a sodium chloride crystal.

Analyze and Conclude

1. **Relate Cause and Effect** Describe the exchange of popcorn kernels (electrons) that took place as you formed the ionic bond. What electrical charges resulted from the exchange?
2. **Use Models** How were the "ions" arranged in the model of the crystal? Why did you and your classmates choose this arrangement?

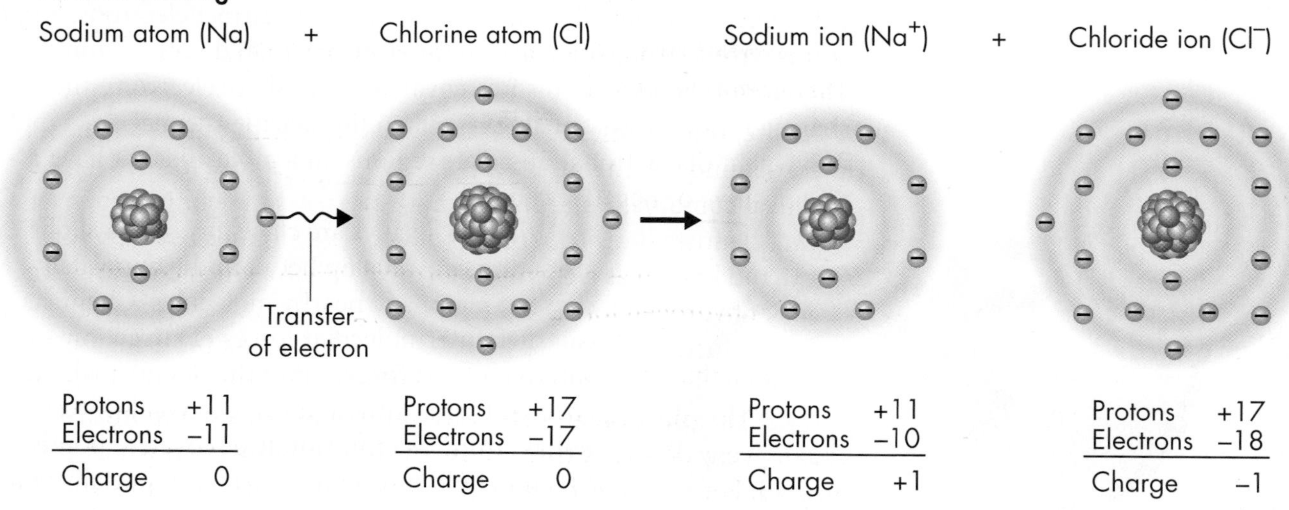

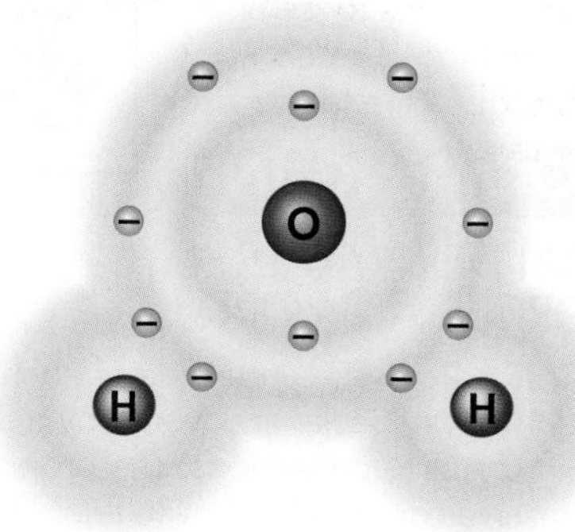

Water molecule (H_2O)

FIGURE 2–4 Ionic Bonding and Covalent Bonding A. The compound sodium chloride forms when sodium loses its valence electron to chlorine. **B.** In a water molecule, each hydrogen atom shares two electrons with the oxygen atom.

Ionic Bonds An **ionic bond** is formed when one or more electrons are transferred from one atom to another. Recall that atoms are electrically neutral because they have equal numbers of protons and electrons. An atom that loses electrons becomes positively charged. An atom that gains electrons has a negative charge. These positively and negatively charged atoms are known as **ions.**

Figure 2–4A shows how ionic bonds form between sodium and chlorine in table salt. A sodium atom easily loses its one valence electron and becomes a sodium ion (Na^+). A chlorine atom easily gains an electron and becomes a chloride ion (Cl^-). In a salt crystal, there are trillions of sodium and chloride ions. These oppositely charged ions have a strong attraction, forming an ionic bond.

Covalent Bonds Sometimes electrons are shared by atoms instead of being transferred. What does it mean to share electrons? It means that the moving electrons actually travel about the nuclei of both atoms, forming a **covalent bond.** When the atoms share two electrons, the bond is called a single covalent bond. Sometimes the atoms share four electrons and form a double bond. In a few cases, atoms can share six electrons, forming a triple bond. The structure that results when atoms are joined together by covalent bonds is called a molecule. The **molecule** is the smallest unit of most compounds. The diagram of a water molecule in **Figure 2–4B** shows that each hydrogen atom is joined to water's lone oxygen atom by a single covalent bond. When atoms of the same element join together, they also form a molecule. Oxygen molecules in the air you breathe consist of two oxygen atoms joined by covalent bonds.

In Your Notebook *In your own words, describe the differences between ionic and covalent bonds.*

Fish do not break water molecules into their component atoms to obtain oxygen. Rather, they use oxygen gas dissolved in the water. How are the atoms in an oxygen molecule (O_2) joined together?

Van der Waals Forces Because of their structures, atoms of different elements do not all have the same ability to attract electrons. Some atoms have a stronger attraction for electrons than do other atoms. Therefore, when the atoms in a covalent bond share electrons, the sharing is not always equal. Even when the sharing is equal, the rapid movement of electrons can create regions on a molecule that have a tiny positive or negative charge.

When molecules are close together, a slight attraction can develop between the oppositely charged regions of nearby molecules. Chemists call such intermolecular forces of attraction **van der Waals forces,** after the scientist who discovered them. Although van der Waals forces are not as strong as ionic bonds or covalent bonds, they can hold molecules together, especially when the molecules are large.

ZOOMING IN

VAN DER WAALS FORCES AT WORK

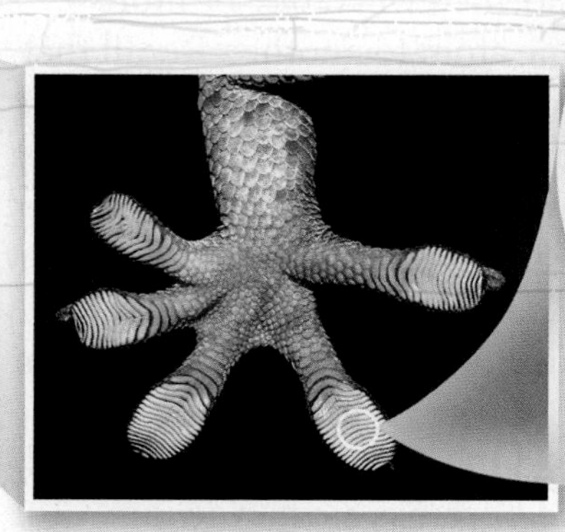

FIGURE 2–5 The underside of each foot on this Tokay gecko is covered by millions of tiny hairlike projections. The projections themselves are made of even finer fibers, creating more surface area for "sticking" to surfaces at the molecular level. This allows geckos to scurry up walls and across ceilings.

2.1 Assessment

Review Key Concepts

1. a. Review Describe the structure of an atom.

b. Infer An atom of calcium contains 20 protons. How many electrons does it have?

2. a. Review Why do all isotopes of an element have the same chemical properties?

b. Compare and Contrast Compare the structure of carbon–12 and carbon–14.

3. a. Review What is a compound?

b. Apply Concepts Water (H_2O) and hydrogen peroxide (H_2O_2) both consists of hydrogen and oxygen atoms. Explain why they have different chemical and physical properties.

4. a. Review What are two types of bonds that hold the atoms within a compound together?

b. Classify A potassium atom easily loses its one valence electron. What type of bond will it form with a chlorine atom?

Apply the Big idea

Matter and Energy

5. Why do you think it is important that biologists have a good understanding of chemistry?

WHST.9-10.9 Research to Build and Present Knowledge, RST.9-10.10 Level of Text Complexity. Also WHST.9-10.10

A Nature-Inspired Adhesive

People who keep geckos as pets have always marveled at the way these little lizards can climb up vertical surfaces, even smooth glass walls, and then hang on by a single toe despite the pull of gravity. How do they do it? No, they do not have some sort of glue on their feet and they don't have suction cups. Incredibly, they use van der Waals forces.

A gecko foot is covered by as many as half a million tiny hairlike projections. Each projection is further divided into hundreds of tiny, flat-surfaced fibers. This design allows the gecko's foot to come in contact with an extremely large area of the wall at the molecular level. Van der Waals forces form between molecules on the surface of the gecko's foot and molecules on the surface of the wall. This allows the gecko to actually balance the pull of gravity.

If it works for the gecko, why not for us? That's the thinking of researchers at the Massachusetts Institute of Technology, who have now used the same principle to produce a bandage. This new bandage is held to tissue by van der Waals forces alone. Special materials make it possible for the new bandage to work even on moist surfaces, which means that it may be used to reseal internal tissues after surgery. By learning a trick or two from the gecko, scientists may have found a way to help heal wounds, and even save lives in the process.

WRITING **Suppose you are a doctor reviewing this new bandage for its potential applications. In what ways might you use such a bandage? Point to evidence in the text supporting the use of this new bandage.**

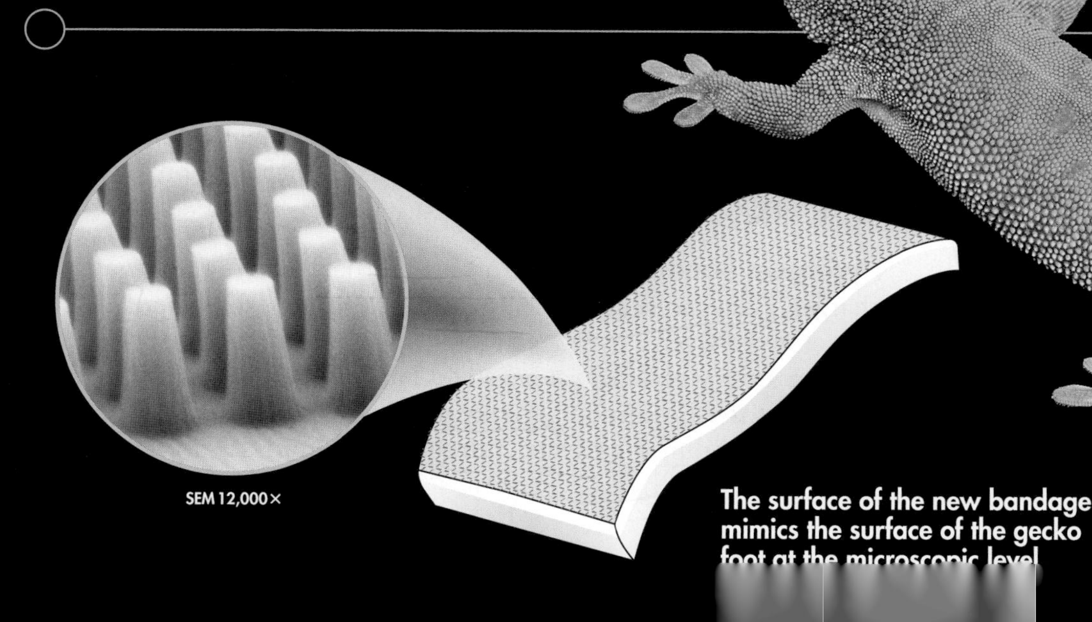

The surface of the new bandage mimics the surface of the gecko foot at the microscopic level.

2.2 Properties of Water

Key Questions

How does the structure of water contribute to its unique properties?

How does water's polarity influence its properties as a solvent?

Why is it important for cells to buffer solutions against rapid changes in pH?

Vocabulary

hydrogen bond • cohesion • adhesion • mixture • solution • solute • solvent • suspension • pH scale • acid • base • buffer

Taking Notes

Venn Diagram As you read, draw a Venn diagram showing the differences between solutions and suspensions and the properties that they share.

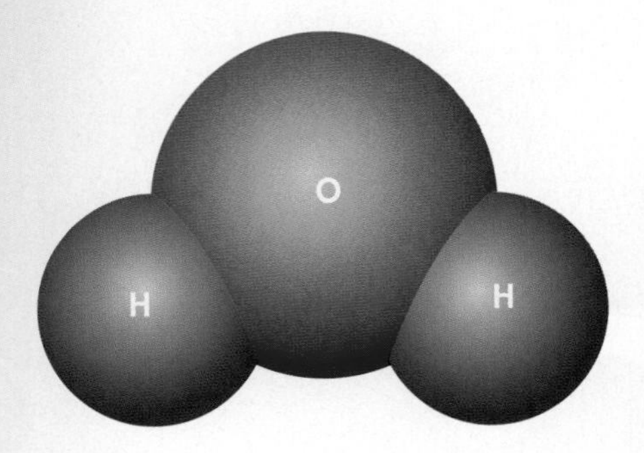

FIGURE 2–6 A Water Molecule A water molecule is polar because there is an uneven distribution of electrons between the oxygen and hydrogen atoms. The negative pole is near the oxygen atom and the positive pole is between the hydrogen atoms.

THINK ABOUT IT Looking back at our beautiful planet, an astronaut in space said that if other beings have seen the Earth, they must surely call it "the blue planet." He referred, of course, to the oceans of water that cover nearly three fourths of Earth's surface. The very presence of liquid water tells a scientist that life may also be present on such a planet. Why should this be so? Why should life itself be connected so strongly to something so ordinary that we often take it for granted? The answers to those questions suggest that there is something very special about water and the role it plays in living things.

The Water Molecule

How does the structure of water contribute to its unique properties?

Water is one of the few compounds found in a liquid state over most of the Earth's surface. Like other molecules, water (H_2O) is neutral. The positive charges on its 10 protons balance out the negative charges on its 10 electrons. However, there is more to the story.

Polarity With 8 protons, water's oxygen nucleus attracts electrons more strongly than the single protons of water's two hydrogen nuclei. As a result, water's shared electrons are more likely to be found near the oxygen nucleus. Because the oxygen nucleus is at one end of the molecule, as shown in **Figure 2–6**, water has a partial negative charge on one end, and a partial positive charge on the other.

A molecule in which the charges are unevenly distributed is said to be "polar," because the molecule is a bit like a magnet with two poles. The partial charges on a polar molecule are written in parentheses, (−) or (+), to show that they are weaker than the charges on ions such as Na^+ and Cl^-.

Hydrogen Bonding Because of their partial positive and negative charges, polar molecules such as water can attract each other. The attraction between a hydrogen atom with a partial positive charge and another atom with a partial negative charge is known as a **hydrogen bond.** The most common partially negative atoms involved in hydrogen bonding are oxygen, nitrogen, and fluorine.

Hydrogen bonds are not as strong as covalent or ionic bonds, but they give one of life's most important molecules many of its unique characteristics. **Because water is a polar molecule, it is able to form multiple hydrogen bonds, which account for many of water's special properties.** These include the fact that water expands slightly upon freezing, making ice less dense than liquid water. Hydrogen bonding also explains water's ability to dissolve so many other substances, a property essential in living cells.

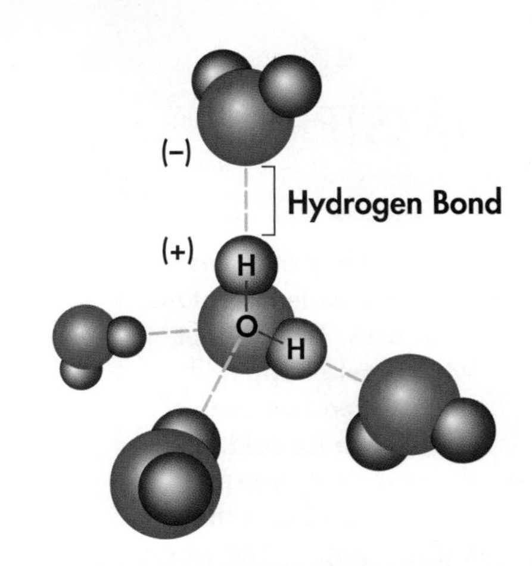

Cohesion **Cohesion** is an attraction between molecules of the same substance. Because a single water molecule may be involved in as many as four hydrogen bonds at the same time, water is extremely cohesive. Cohesion causes water molecules to be drawn together, which is why drops of water form beads on a smooth surface. Cohesion also produces surface tension, explaining why some insects and spiders can walk on a pond's surface, as shown in **Figure 2–7.**

FIGURE 2–7 Hydrogen Bonding and Cohesion Each molecule of water can form multiple hydrogen bonds with other water molecules. The strong attraction between water molecules produces a force sometimes called "surface tension," which can support very lightweight objects, such as this raft spider. Apply Concepts ***Why are water molecules attracted to one another?***

Adhesion On the other hand, **adhesion** is an attraction between molecules of different substances. Have you ever been told to read the volume in a graduated cylinder at eye level? As shown in **Figure 2–8,** the surface of the water in the graduated cylinder dips slightly in the center because the adhesion between water molecules and glass molecules is stronger than the cohesion between water molecules. Adhesion between water and glass also causes water to rise in a narrow tube against the force of gravity. This effect is called capillary action. Capillary action is one of the forces that draws water out of the roots of a plant and up into its stems and leaves. Cohesion holds the column of water together as it rises.

Heat Capacity Another result of the multiple hydrogen bonds between water molecules is that it takes a large amount of heat energy to cause those molecules to move faster, which raises the temperature of the water. Therefore, water's heat capacity, the amount of heat energy required to increase its temperature, is relatively high. This allows large bodies of water, such as oceans and lakes, to absorb large amounts of heat with only small changes in temperature. The organisms living within are thus protected from drastic changes in temperature. At the cellular level, water absorbs the heat produced by cell processes, regulating the temperature of the cell.

In Your Notebook *Draw a diagram of a meniscus. Label where cohesion and adhesion occur.*

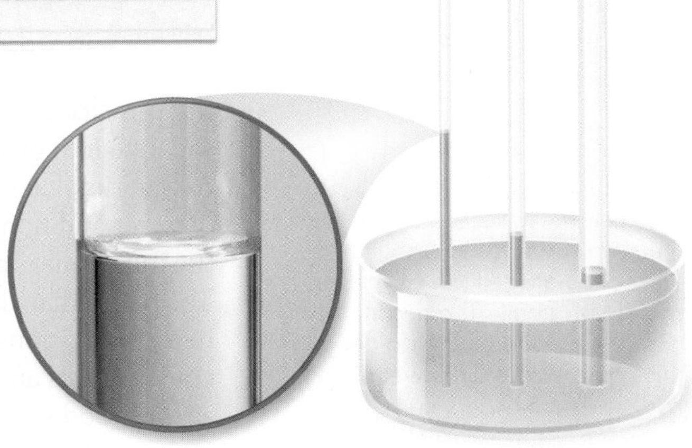

FIGURE 2–8 Adhesion Adhesion between water and glass molecules is responsible for causing the water in these columns to rise. The surface of the water in the glass column dips slightly in the center, forming a curve called a meniscus.

MYSTERY
CLUE

The solubility of gases increases as temperature decreases. Think about what would happen if you opened two cans of soda—one ice cold, and the other very warm. Which would give the bigger "pop"? The warm can, of course! That's because the gas in the soda is much less soluble at warm temperatures. How might the temperature of Antarctic waters affect the amount of dissolved oxygen available for ice fish?

Solutions and Suspensions

How does water's polarity influence its properties as a solvent?

Water is not always pure; it is often found as part of a mixture. A **mixture** is a material composed of two or more elements or compounds that are physically mixed together but not chemically combined. Salt and pepper stirred together constitute a mixture. So do sugar and sand. Earth's atmosphere is a mixture of nitrogen, oxygen, carbon dioxide, and other gases. Living things are in part composed of mixtures involving water. Two types of mixtures that can be made with water are solutions and suspensions.

Solutions If a crystal of table salt is placed in a glass of warm water, sodium and chloride ions on the surface of the crystal are attracted to the polar water molecules. Ions break away from the crystal and are surrounded by water molecules, as illustrated in **Figure 2–9.** The ions gradually become dispersed in the water, forming a type of mixture called a solution. All the components of a **solution** are evenly distributed throughout the solution. In a saltwater solution, table salt is the **solute**—the substance that is dissolved. Water is the **solvent**—the substance in which the solute dissolves. **Water's polarity gives it the ability to dissolve both ionic compounds and other polar molecules.**

Water easily dissolves salts, sugars, minerals, gases, and even other solvents such as alcohol. Without exaggeration, water is the greatest solvent on Earth. But even water has limits. When a given amount of water has dissolved all of the solute it can, the solution is said to be saturated.

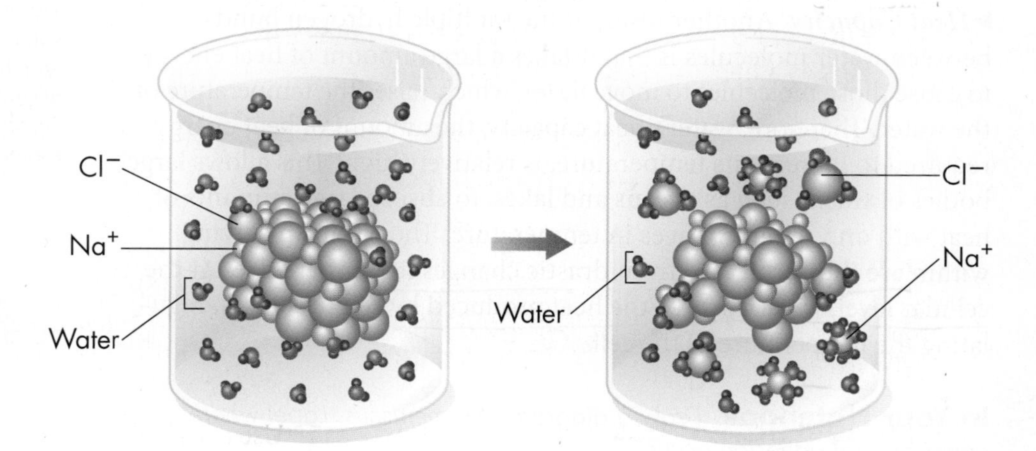

FIGURE 2–9 A Salt Solution When an ionic compound such as sodium chloride is placed in water, water molecules surround and separate the positive and negative ions. Interpret Visuals *What happens to the sodium ions and chloride ions in the solution?*

Suspensions Some materials do not dissolve when placed in water, but separate into pieces so small that they do not settle out. The movement of water molecules keeps the small particles suspended. Such mixtures of water and nondissolved material are known as **suspensions.** Some of the most important biological fluids are both solutions and suspensions. The blood that circulates through your body is mostly water. The water in the blood contains many dissolved compounds. However, blood also contains cells and other undissolved particles that remain in suspension as the blood moves through the body.

Acids, Bases, and pH

Why is it important for cells to buffer solutions against rapid changes in pH?

Water molecules sometimes split apart to form ions. This reaction can be summarized by a chemical equation in which double arrows are used to show that the reaction can occur in either direction.

$$H_2O \rightleftharpoons H^+ + OH^-$$

water $\rightleftharpoons$ hydrogen ion + hydroxide ion

How often does this happen? In pure water, about 1 water molecule in 550 million splits to form ions in this way. Because the number of positive hydrogen ions produced is equal to the number of negative hydroxide ions produced, pure water is neutral.

The pH Scale Chemists devised a measurement system called the **pH scale** to indicate the concentration of H^+ ions in solution. As **Figure 2–10** shows, the pH scale ranges from 0 to 14. At a pH of 7, the concentration of H^+ ions and OH^- ions is equal. Pure water has a pH of 7. Solutions with a pH below 7 are called acidic because they have more H^+ ions than OH^- ions. The lower the pH, the greater the acidity. Solutions with a pH above 7 are called basic because they have more OH^- ions than H^+ ions. The higher the pH, the more basic the solution. Each step on the pH scale represents a factor of 10. For example, a liter of a solution with a pH of 4 has 10 times as many H^+ ions as a liter of a solution with a pH of 5.

In Your Notebook *Order these items in order of increasing acidity: soap, lemon juice, milk, acid rain.*

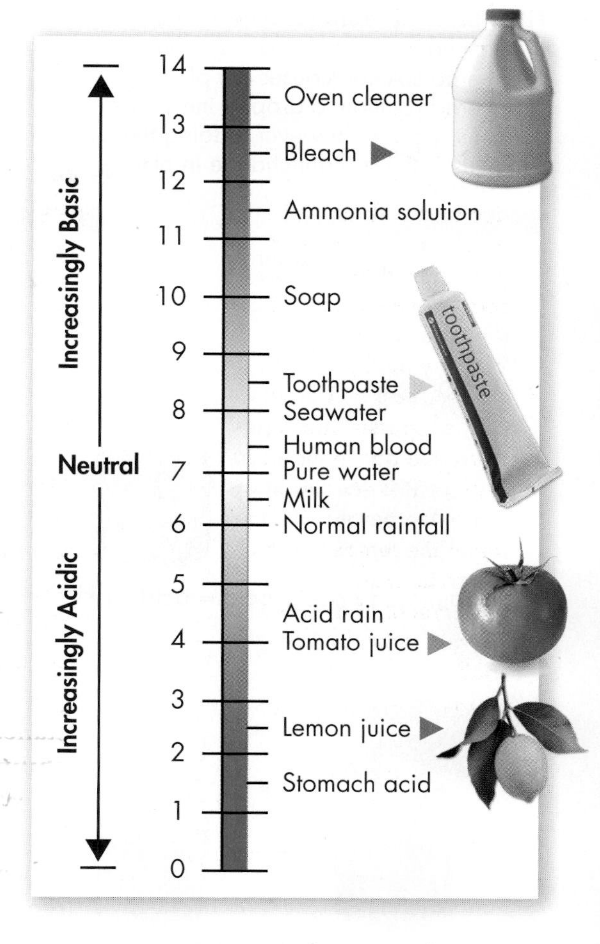

FIGURE 2–10 The pH Scale The concentration of H^+ ions determines whether solutions are acidic or basic. The most acidic material on this pH scale is stomach acid. The most basic material on this scale is oven cleaner.

Quick Lab
GUIDED INQUIRY

Acidic and Basic Foods

❶ Predict whether the food samples provided are acidic or basic.

❷ Tear off a 2-inch piece of pH paper for each sample you will test. Place these pieces on a paper towel.

❸ Construct a data table in which you will record the name and pH of each food sample.

❹ Use a scalpel to cut a piece off each solid. **CAUTION:** *Be careful not to cut yourself. Do not eat the food.* Touch the cut surface of each sample to a square of pH paper. Use a dropper pipette to place a drop of any liquid sample on a square of pH paper. Record the pH of each sample in your data table.

Analyze and Conclude

1. Analyze Data Were most of the samples acidic or basic?

2. Evaluate Was your prediction correct?

FIGURE 2–11 Buffers Buffers help prevent drastic changes in pH. Adding acid to an unbuffered solution causes the pH of the unbuffered solution to drop. If the solution contains a buffer, however, adding the acid will cause only a slight change in pH.

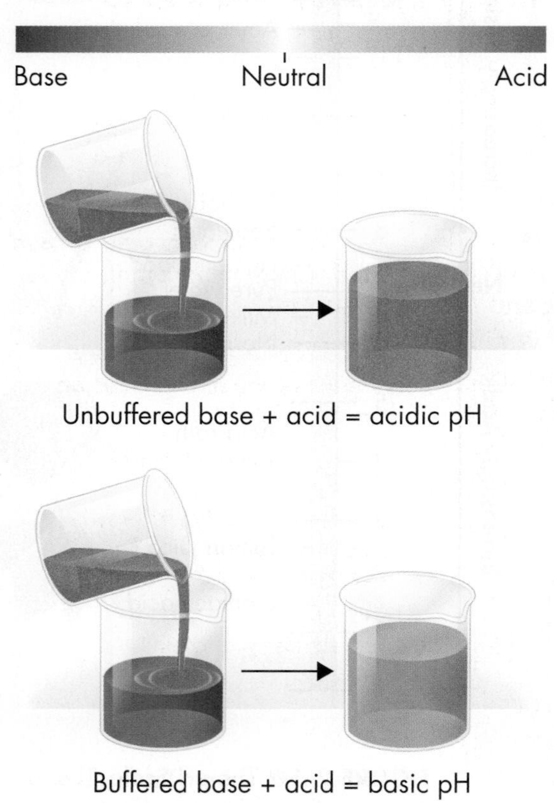

Acids Where do all those extra H^+ ions in a low-pH solution come from? They come from acids. An **acid** is any compound that forms H^+ ions in solution. Acidic solutions contain higher concentrations of H^+ ions than pure water and have pH values below 7. Strong acids tend to have pH values that range from 1 to 3. The hydrochloric acid (HCl) produced by the stomach to help digest food is a strong acid.

Bases A **base** is a compound that produces hydroxide (OH^-) ions in solution. Basic, or alkaline, solutions contain lower concentrations of H^+ ions than pure water and have pH values above 7. Strong bases, such as the lye (commonly NaOH) used in soapmaking, tend to have pH values ranging from 11 to 14.

Buffers The pH of the fluids within most cells in the human body must generally be kept between 6.5 and 7.5. If the pH is lower or higher, it will affect the chemical reactions that take place within the cells. Thus, controlling pH is important for maintaining homeostasis. One of the ways that organisms control pH is through dissolved compounds called buffers. **Buffers** are weak acids or bases that can react with strong acids or bases to prevent sharp, sudden changes in pH. Blood, for example, has a normal pH of 7.4. Sudden changes in blood pH are usually prevented by a number of chemical buffers, such as bicarbonate and phosphate ions. **Buffers dissolved in life's fluids play an important role in maintaining homeostasis in organisms.**

2.2 Assessment

Review Key Concepts

1. **a. Review** What does it mean when a molecule is said to be "polar"?
   b. Explain How do hydrogen bonds between water molecules occur?
   c. Use Models Use the structure of a water molecule to explain why it is polar.
2. **a. Review** Why is water such a good solvent?
   b. Compare and Contrast What is the difference between a solution and a suspension?
3. **a. Review** What is an acid? What is a base?
   b. Explain The acid hydrogen fluoride (HF) can be dissolved in pure water. Will the pH of the solution be greater or less than 7?
   c. Infer During exercise, many chemical changes occur in the body, including a drop in blood pH, which can be very serious. How is the body able to cope with such changes?

WRITE ABOUT SCIENCE

Creative Writing

4. Suppose you are a writer for a natural history magazine for children. This month's issue will feature insects. Write a paragraph explaining why some bugs, such as the water strider, can walk on water.

2.3 Carbon Compounds

THINK ABOUT IT In the early 1800s, many chemists called the compounds created by organisms "organic," believing they were fundamentally different from compounds in nonliving things. Today we understand that the principles governing the chemistry of living and nonliving things are the same, but the term "organic chemistry" is still around. Today, organic chemistry means the study of compounds that contain bonds between carbon atoms, while inorganic chemistry is the study of all other compounds.

Key Questions

What elements does carbon bond with to make up life's molecules?

What are the functions of each of the four groups of macromolecules?

Vocabulary

monomer • polymer • carbohydrate • monosaccharide • lipid • nucleic acid • nucleotide • protein • amino acid

Taking Notes

Compare/Contrast Table As you read, make a table that compares and contrasts the four groups of organic compounds.

The Chemistry of Carbon

What elements does carbon bond with to make up life's molecules?

Why is carbon so interesting that a whole branch of chemistry should be set aside just to study carbon compounds? There are two reasons for this. First, carbon atoms have four valence electrons, allowing them to form strong covalent bonds with many other elements. **Carbon can bond with many elements, including hydrogen, oxygen, phosphorus, sulfur, and nitrogen to form the molecules of life.** Living organisms are made up of molecules that consist of carbon and these other elements.

Even more important, one carbon atom can bond to another, which gives carbon the ability to form chains that are almost unlimited in length. These carbon-carbon bonds can be single, double, or triple covalent bonds. Chains of carbon atoms can even close up on themselves to form rings, as shown in **Figure 2–12.** Carbon has the ability to form millions of different large and complex structures. No other element even comes close to matching carbon's versatility.

FIGURE 2–12 Carbon Structures Carbon can form single, double, or triple bonds with other carbon atoms. Each line between atoms in a molecular drawing represents one covalent bond. **Observing** ***How many covalent bonds are there between the two carbon atoms in acetylene?***

Methance **Acetylene** **Butadiene** **Benzene** **Isooctane**

Macromolecules

What are the functions of each of the four groups of macromolecules?

Many of the organic compounds in living cells are so large that they are known as macromolecules, which means "giant molecules." Macromolecules are made from thousands or even hundreds of thousands of smaller molecules.

Most macromolecules are formed by a process known as polymerization (pah lih mur ih ZAY shun), in which large compounds are built by joining smaller ones together. The smaller units, or **monomers**, join together to form **polymers.** The monomers in a polymer may be identical, like the links on a metal watch band; or the monomers may be different, like the beads in a multicolored necklace. **Figure 2–13** illustrates the process of polymerization.

Biochemists sort the macromolecules found in living things into groups based on their chemical composition. The four major groups of macromolecules found in living things are carbohydrates, lipids, nucleic acids, and proteins. As you read about these molecules, compare their structures and functions.

BUILD Vocabulary

WORD ORIGINS **Monomer** comes from the Greek words *monos*, meaning "single," and *meros*, meaning "part." *Monomer* means "single part." The prefix *poly-* comes from the Greek word *polus*, meaning "many," so **polymer** means "many parts."

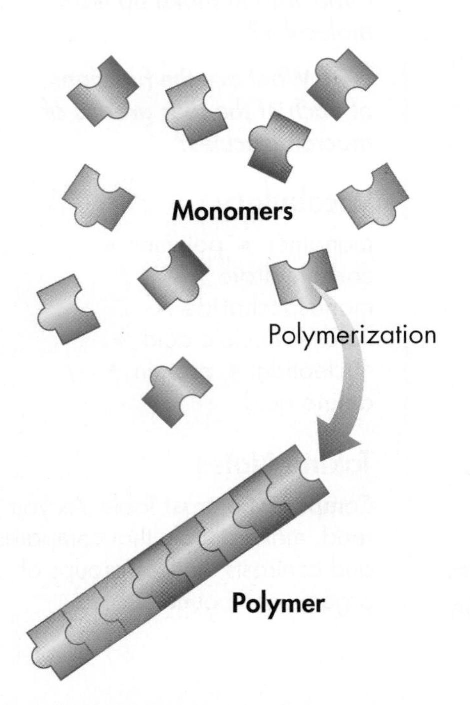

FIGURE 2–13 Polymerization When monomers join together, they form polymers. **Using Analogies** ***How are monomers similar to links in a chain?***

Carbohydrates **Carbohydrates** are compounds made up of carbon, hydrogen, and oxygen atoms, usually in a ratio of 1 : 2 : 1. **Living things use carbohydrates as their main source of energy. Plants, some animals, and other organisms also use carbohydrates for structural purposes.** The breakdown of sugars, such as glucose, supplies immediate energy for cell activities. Many organisms store extra sugar as complex carbohydrates known as starches. As shown in **Figure 2–14,** the monomers in starch polymers are sugar molecules.

▶ ***Simple Sugars*** Single sugar molecules are also known as **monosaccharides** (mahn oh SAK uh rydz). Besides glucose, monosaccharides include galactose, which is a component of milk, and fructose, which is found in many fruits. Ordinary table sugar, sucrose, consists of glucose and fructose. Sucrose is a disaccharide, a compound made by joining two simple sugars together.

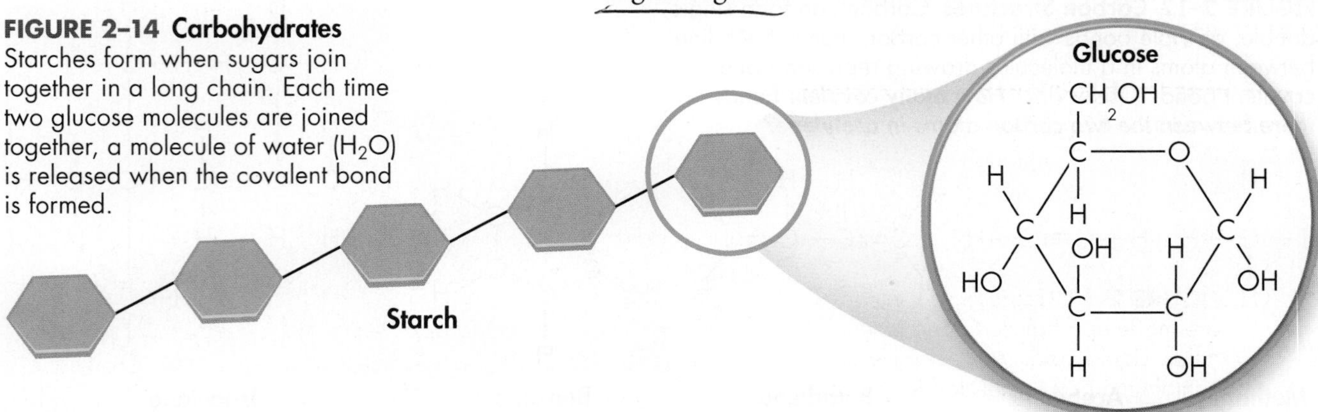

FIGURE 2–14 Carbohydrates Starches form when sugars join together in a long chain. Each time two glucose molecules are joined together, a molecule of water (H_2O) is released when the covalent bond is formed.

▶ ***Complex Carbohydrates*** The large macromolecules formed from monosaccharides are known as polysaccharides. Many animals store excess sugar in a polysaccharide called glycogen, which is sometimes called "animal starch." When the level of glucose in your blood runs low, glycogen is broken down into glucose, which is then released into the blood. The glycogen stored in your muscles supplies the energy for muscle contraction and, thus, for movement.

Plants use a slightly different polysaccharide, called starch, to store excess sugar. Plants also make another important polysaccharide called cellulose. Tough, flexible cellulose fibers give plants much of their strength and rigidity. Cellulose is the major component of both wood and paper, so you are actually looking at cellulose as you read these words!

Lipids Lipids are a large and varied group of biological molecules that are generally not soluble in water. **Lipids** are made mostly from carbon and hydrogen atoms. The common categories of lipids are fats, oils, and waxes. **Lipids can be used to store energy. Some lipids are important parts of biological membranes and waterproof coverings.** Steroids synthesized by the body are lipids as well. Many steroids, such as hormones, serve as chemical messengers.

Many lipids are formed when a glycerol molecule combines with compounds called fatty acids, as shown in **Figure 2–15.** If each carbon atom in a lipid's fatty acid chains is joined to another carbon atom by a single bond, the lipid is said to be saturated. The term *saturated* is used because the fatty acids contain the maximum possible number of hydrogen atoms.

If there is at least one carbon-carbon double bond in a fatty acid, the fatty acid is said to be unsaturated. Lipids whose fatty acids contain more than one double bond are said to be polyunsaturated. If the terms *saturated* and *polyunsaturated* seem familiar, you have probably seen them on food package labels. Lipids that contain unsaturated fatty acids, such as olive oil, tend to be liquid at room temperature. Other cooking oils, such as corn oil, sesame oil, canola oil, and peanut oil, contain polyunsaturated lipids.

In Your Notebook *Compare and contrast saturated and unsaturated fats.*

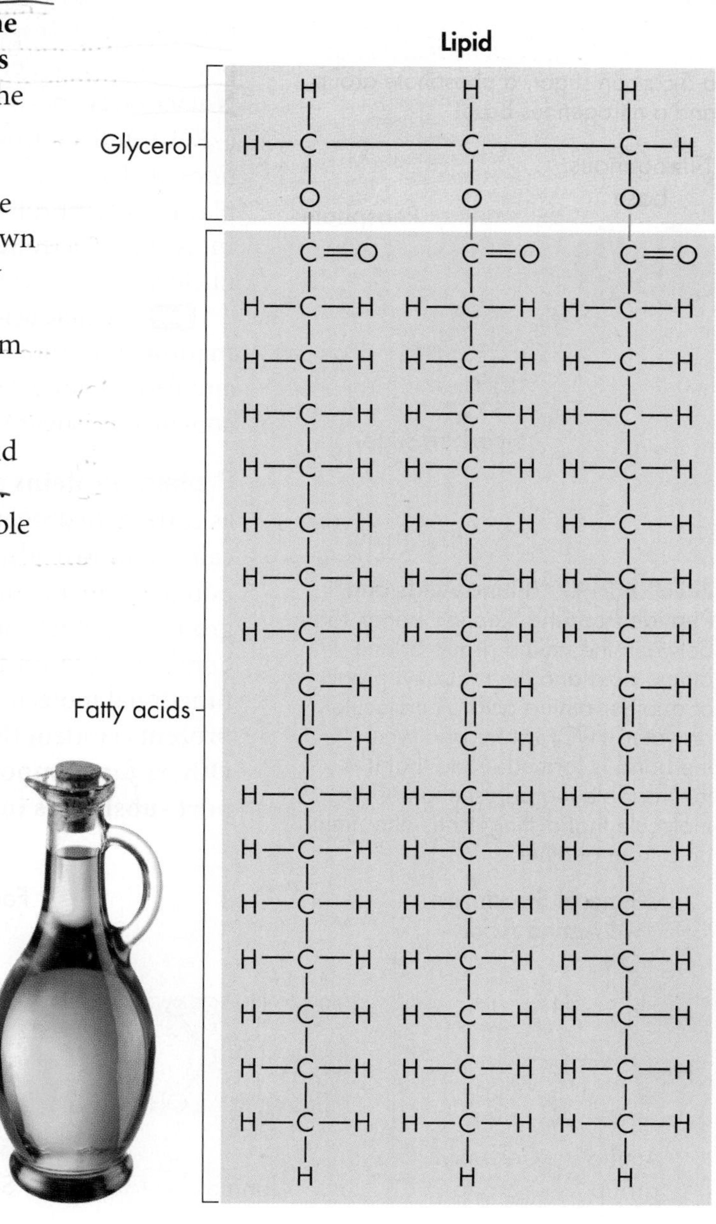

FIGURE 2–15 Lipids Lipid molecules are made up of glycerol and fatty acids. Liquid lipids, such as olive oil, contain mainly unsaturated fatty acids.

Comparing Fatty Acids

The table compares four different fatty acids. Although they all have the same number of carbon atoms, their properties vary.

1. Interpret Data Which of the four fatty acids is saturated? Which are unsaturated?

2. Observe How does melting point change as the number of carbon-carbon double bonds increases?

Effect of Carbon Bonds on Melting Point

Fatty Acid	Number of Carbons	Number of Double Bonds	Melting Point (°C)
Stearic acid	18	0	69.6
Oleic acid	18	1	14
Linoleic acid	18	2	–5
Linolenic acid	18	3	–11

3. Infer If room temperature is 25°C, which fatty acid is a solid at room temperature? Which is liquid at room temperature?

FIGURE 2–16 Nucleic Acids The monomers that make up a nucleic acid are nucleotides. Each nucleotide has a 5-carbon sugar, a phosphate group, and a nitrogenous base.

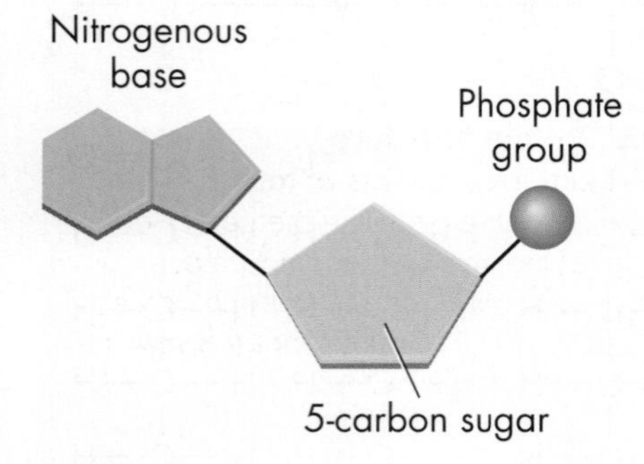

Nucleic Acids **Nucleic acids** are macromolecules containing hydrogen, oxygen, nitrogen, carbon, and phosphorus. Nucleic acids are polymers assembled from individual monomers known as nucleotides. **Nucleotides** consist of three parts: a 5-carbon sugar, a phosphate group ($-PO_4$), and a nitrogenous base, as shown in **Figure 2–16.** Some nucleotides, including the compound known as adenosine triphosphate (ATP), play important roles in capturing and transferring chemical energy. Individual nucleotides can be joined by covalent bonds to form a polynucleotide, or nucleic acid.

Nucleic acids store and transmit hereditary, or genetic, information. There are two kinds of nucleic acids: ribonucleic acid (RNA) and deoxyribonucleic acid (DNA). As their names indicate, RNA contains the sugar ribose and DNA contains the sugar deoxyribose.

Protein **Proteins** are macromolecules that contain nitrogen as well as carbon, hydrogen, and oxygen. Proteins are polymers of molecules called amino acids, shown in **Figure 2–17. Amino acids** are compounds with an amino group ($-NH_2$) on one end and a carboxyl group ($-COOH$) on the other end. Covalent bonds called peptide bonds link amino acids together to form a polypeptide. A protein is a functional molecule built from one or more polypeptides. **Some proteins control the rate of reactions and regulate cell processes. Others form important cellular structures, while still others transport substances into or out of cells or help to fight disease.**

FIGURE 2–17 Amino Acids and Peptide Bonding Peptide bonds form between the amino group of one amino acid and the carboxyl group of another amino acid. A molecule of water (H_2O) is released when the bond is formed. Note that it is the variable R-group section of the molecule that distinguishes one amino acid from another.

General Structure of Amino Acids

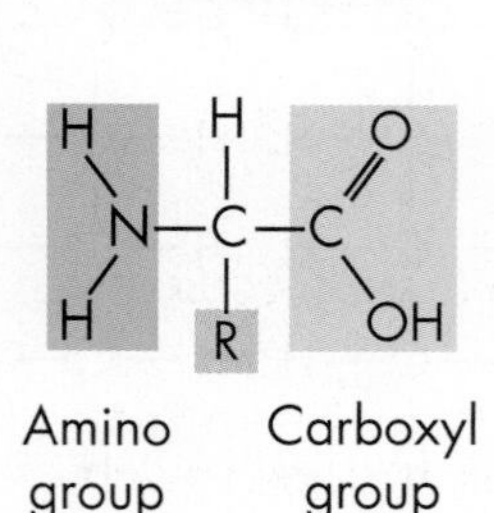

Formation of Peptide Bond

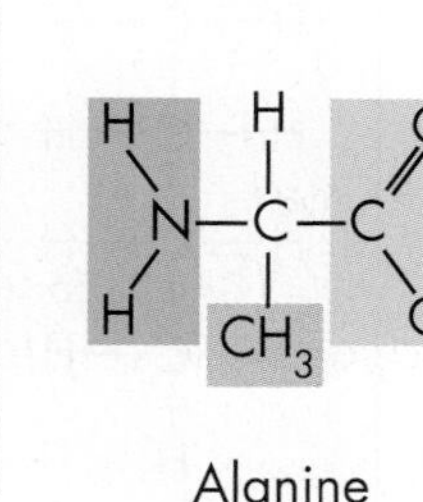

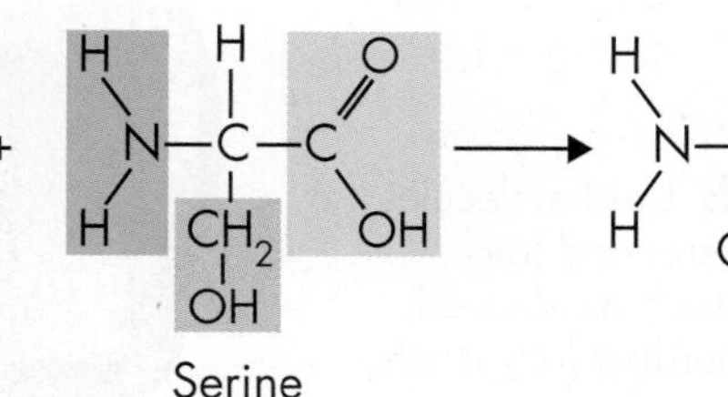

▶ ***Structure and Function*** More than 20 different amino acids are found in nature. All amino acids are identical in the regions where they may be joined together by covalent bonds. This uniformity allows any amino acid to be joined to any other amino acid—by bonding an amino group to a carboxyl group. Proteins are among the most diverse macromolecules. The reason is that amino acids differ from each other in a side chain called the R-group, which have a range of different properties. Some R-groups are acidic and some are basic. Some are polar, some are nonpolar, and some even contain large ring structures.

▶ ***Levels of Organization*** Amino acids are assembled into polypeptide chains according to instructions coded in DNA. To help understand these large molecules, scientists describe proteins as having four levels of structure. A protein's primary structure is the sequence of its amino acids. Secondary structure is the folding or coiling of the polypeptide chain. Tertiary structure is the complete, three-dimensional arrangement of a polypeptide chain. Proteins with more than one chain are said to have a fourth level of structure, describing the way in which the different polypeptides are arranged with respect to each other. **Figure 2–18** shows these four levels of structure in hemoglobin, a protein found in red blood cells that helps to transport oxygen in the bloodstream. The shape of a protein is maintained by a variety of forces, including ionic and covalent bonds, as well as van der Waals forces and hydrogen bonds. In the next lesson, you will learn why a protein's shape is so important.

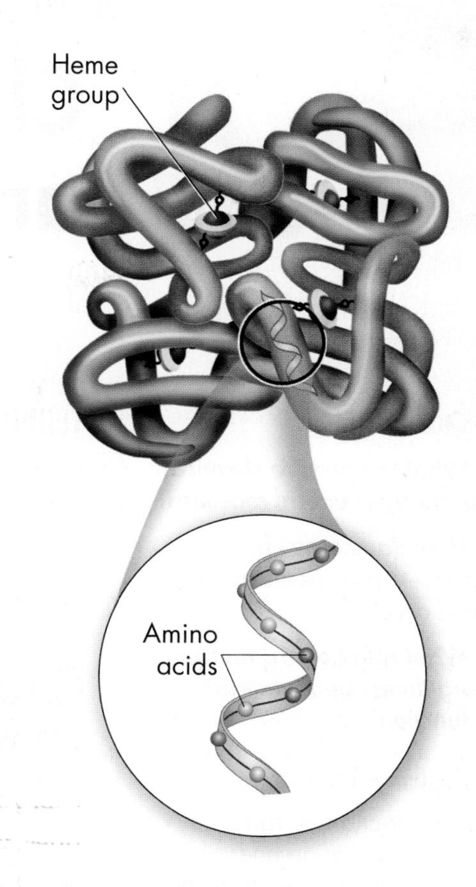

FIGURE 2–18 Protein Structure
The protein hemoglobin consists of four subunits. The iron-containing heme group in the center of each subunit gives hemoglobin its red color. An oxygen molecule binds tightly to each heme molecule. **Interpret Visuals** ***How many levels of organization does hemoglobin have?***

2.3 Assessment

Review Key Concepts

1. a. Review What are the major elements of life?

b. Relate Cause and Effect What properties of carbon explain carbon's ability to form different large and complex structures?

2. a. Review Name four groups of organic compounds found in living things.

b. Explain Describe at least one function of each group of organic compound.

c. Infer Why are proteins considered polymers but lipids not?

VISUAL THINKING

3. A structural formula shows how the atoms in a compound are arranged.

a. Observe What atoms constitute the compound above?

b. Classify What class of macromolecule does the compound belong to?

2.4 Chemical Reactions and Enzymes

Key Questions

What happens to chemical bonds during chemical reactions?

How do energy changes affect whether a chemical reaction will occur?

What role do enzymes play in living things and what affects their function?

Vocabulary

chemical reaction • reactant • product • activation energy • catalyst • enzyme • substrate

Taking Notes

Concept Map As you read, make a concept map that shows the relationship among the vocabulary terms in this lesson.

THINK ABOUT IT Living things, as you have seen, are made up of chemical compounds—some simple and some complex. But chemistry isn't just what life is made of—chemistry is also what life does. Everything that happens in an organism—its growth, its interaction with the environment, its reproduction, and even its movement—is based on chemical reactions.

Chemical Reactions

What happens to chemical bonds during chemical reactions?

A **chemical reaction** is a process that changes, or transforms, one set of chemicals into another. An important scientific principle is that mass and energy are conserved during chemical transformations. This is also true for chemical reactions that occur in living organisms. Some chemical reactions occur slowly, such as the combination of iron and oxygen to form an iron oxide called rust. Other reactions occur quickly. The elements or compounds that enter into a chemical reaction are known as **reactants.** The elements or compounds produced by a chemical reaction are known as **products.** **Chemical reactions involve changes in the chemical bonds that join atoms in compounds.** An important chemical reaction in your bloodstream that enables carbon dioxide to be removed from the body is shown in **Figure 2–19.**

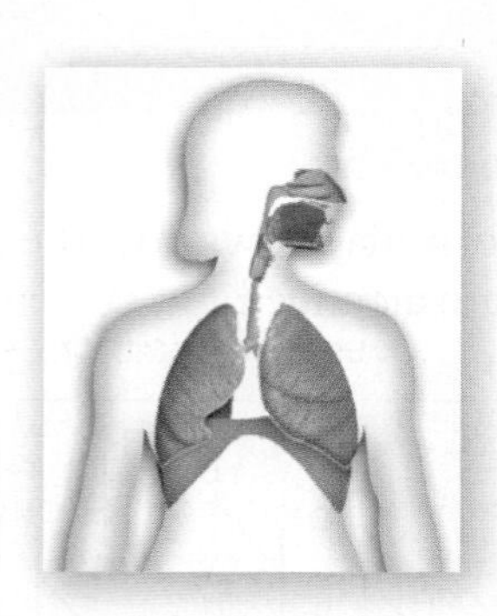

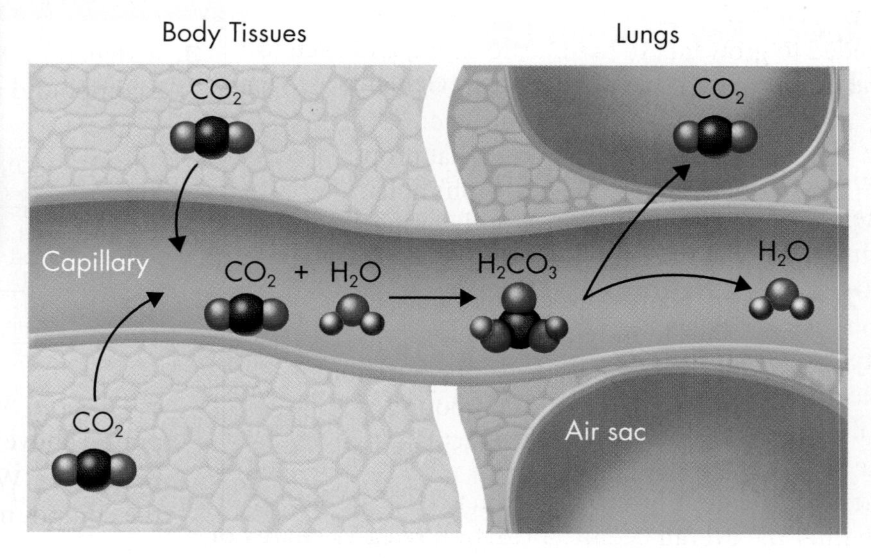

FIGURE 2–19 Carbon Dioxide in the Bloodstream As it enters the blood, carbon dioxide reacts with water to produce carbonic acid (H_2CO_3), which is highly soluble. This reaction enables the blood to carry carbon dioxide to the lungs. In the lungs, the reaction is reversed and produces carbon dioxide gas, which you exhale.

Energy in Reactions

How do energy changes affect whether a chemical reaction will occur?

Energy is released or absorbed whenever chemical bonds are formed or broken. This means that chemical reactions also involve changes in energy.

Energy Changes Some chemical reactions release energy, and other reactions absorb it. Energy changes are one of the most important factors in determining whether a chemical reaction will occur. **Chemical reactions that release energy often occur on their own, or spontaneously. Chemical reactions that absorb energy will not occur without a source of energy.** An example of an energy-releasing reaction is the burning of hydrogen gas, in which hydrogen reacts with oxygen to produce water vapor.

$$2H_2 + O_2 \longrightarrow 2H_2O$$

The energy is released in the form of heat, and sometimes—when hydrogen gas explodes—light and sound.

The reverse reaction, in which water is changed into hydrogen and oxygen gas, absorbs so much energy that it generally doesn't occur by itself. In fact, the only practical way to reverse the reaction is to pass an electrical current through water to decompose water into hydrogen gas and oxygen gas. Thus, in one direction the reaction produces energy, and in the other direction the reaction requires energy.

Energy Sources In order to stay alive, organisms need to carry out reactions that require energy. Because matter and energy are conserved in chemical reactions, every organism must have a source of energy to carry out chemical reactions. Plants get that energy by trapping and storing the energy from sunlight in energy-rich compounds. Animals get their energy when they consume plants or other animals. Humans release the energy needed to grow tall, to breathe, to think, and even to dream through the chemical reactions that occur when we metabolize, or break down, digested food.

Activation Energy Chemical reactions that release energy do not always occur spontaneously. That's a good thing because if they did, the pages of this book might burst into flames. The cellulose in paper burns in the presence of oxygen and releases heat and light. However, paper burns only if you light it with a match, which supplies enough energy to get the reaction started. Chemists call the energy that is needed to get a reaction started the **activation energy.** As **Figure 2–20** shows, activation energy is involved in chemical reactions regardless of whether the overall chemical reaction releases energy or absorbs energy.

FIGURE 2–20 Activation Energy The peak of each graph represents the energy needed for the reaction to go forward. The difference between this required energy and the energy of the reactants is the activation energy. **Interpret Graphs** ***How do the energy of the reactants and products differ between an energy-absorbing reaction and an energy-releasing reaction?***

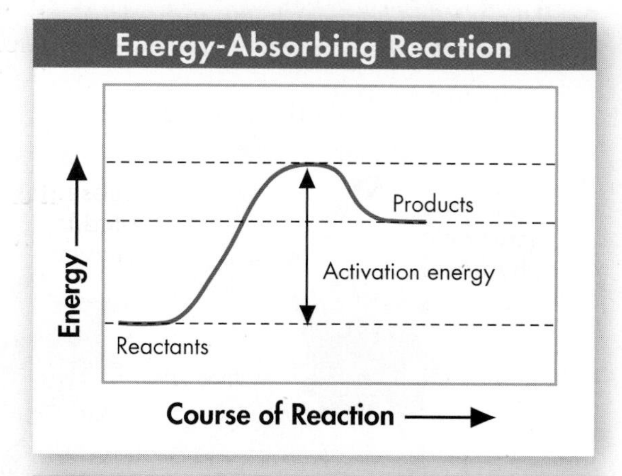

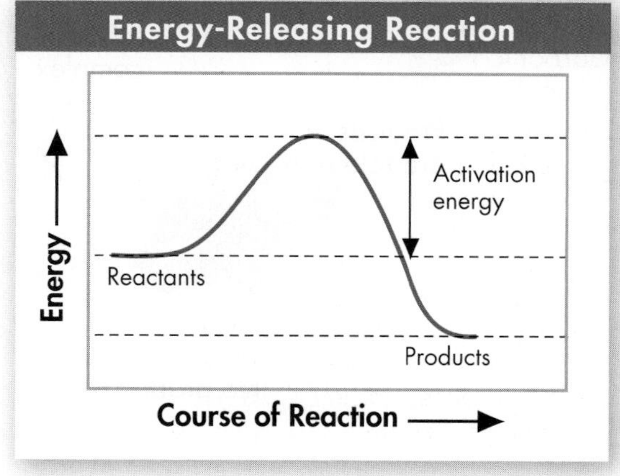

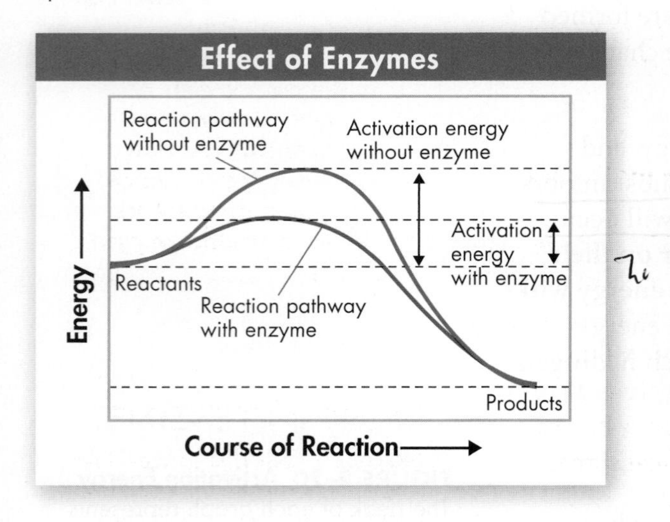

FIGURE 2–21 Effect of Enzymes Notice how the addition of an enzyme lowers the activation energy in this reaction. The enzyme speeds up the reaction.

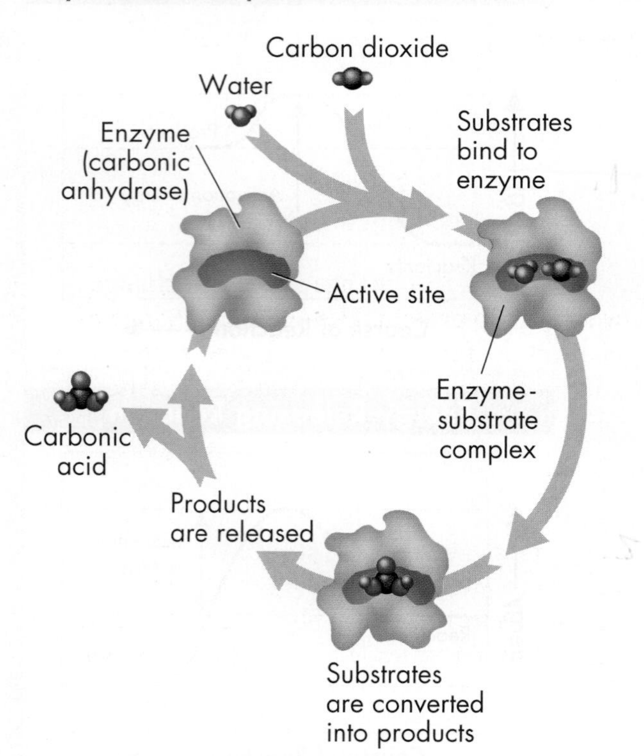

FIGURE 2–22 An Enzyme-Catalyzed Reaction The enzyme carbonic anhydrase converts the substrates carbon dioxide and water into carbonic acid (H_2CO_3). **Predicting** ***What happens to the carbonic anhydrase after the products are released?***

Enzymes

What role do enzymes play in living things and what affects their function?

Some chemical reactions that make life possible are too slow or have activation energies that are too high to make them practical for living tissue. These chemical reactions are made possible by a process that would make any chemist proud—cells make catalysts. A **catalyst** is a substance that speeds up the rate of a chemical reaction. Catalysts work by lowering a reaction's activation energy.

Nature's Catalysts **Enzymes** are proteins that act as biological catalysts. **Enzymes speed up chemical reactions that take place in cells.** Like other catalysts, enzymes act by lowering the activation energies, as illustrated by the graph in **Figure 2–21.** Lowering the activation energy has a dramatic effect on how quickly the reaction is completed. How big an effect does it have? Consider the reaction in which carbon dioxide combines with water to produce carbonic acid.

$$CO_2 + H_2O \longrightarrow H_2CO_3$$

Left to itself, this reaction is so slow that carbon dioxide might build up in the body faster than the bloodstream could remove it. Your bloodstream contains an enzyme called carbonic anhydrase that speeds up the reaction by a factor of 10 million. With carbonic anhydrase on the job, the reaction takes place immediately and carbon dioxide is removed from the blood quickly.

Enzymes are very specific, generally catalyzing only one chemical reaction. For this reason, part of an enzyme's name is usually derived from the reaction it catalyzes. Carbonic anhydrase gets its name because it also catalyzes the reverse reaction that removes water from carbonic acid.

The Enzyme-Substrate Complex How do enzymes do their jobs? For a chemical reaction to take place, the reactants must collide with enough energy so that existing bonds will be broken and new bonds will be formed. If the reactants do not have enough energy, they will be unchanged after the collision.

Enzymes provide a site where reactants can be brought together to react. Such a site reduces the energy needed for reaction. The reactants of enzyme-catalyzed reactions are known as **substrates. Figure 2–22** provides an example of an enzyme-catalyzed reaction.

The substrates bind to a site on the enzyme called the active site. The active site and the substrates have complementary shapes. The fit is so precise that the active site and substrates are often compared to a lock and key, as shown in **Figure 2–23.**

Regulation of Enzyme Activity Enzymes play essential roles in controlling chemical pathways, making materials that cells need, releasing energy, and transferring information. Because they are catalysts for reactions, enzymes can be affected by any variable that influences a chemical reaction. **Temperature, pH, and regulatory molecules can affect the activity of enzymes.**

Many enzymes are affected by changes in temperature. Not surprisingly, those enzymes produced by human cells generally work best at temperatures close to 37°C, the normal temperature of the human body. Enzymes work best at certain ionic conditions and pH values. For example, the stomach enzyme pepsin, which begins protein digestion, works best under acidic conditions. In addition, the activities of most enzymes are regulated by molecules that carry chemical signals within cells, switching enzymes "on" or "off" as needed.

MYSTERY CLUE

The chemical reactions of living things, including those that require oxygen, occur more slowly at low temperatures. How would frigid antarctic waters affect the ice fish's need for oxygen?

VISUAL ANALOGY

UNLOCKING ENZYMES

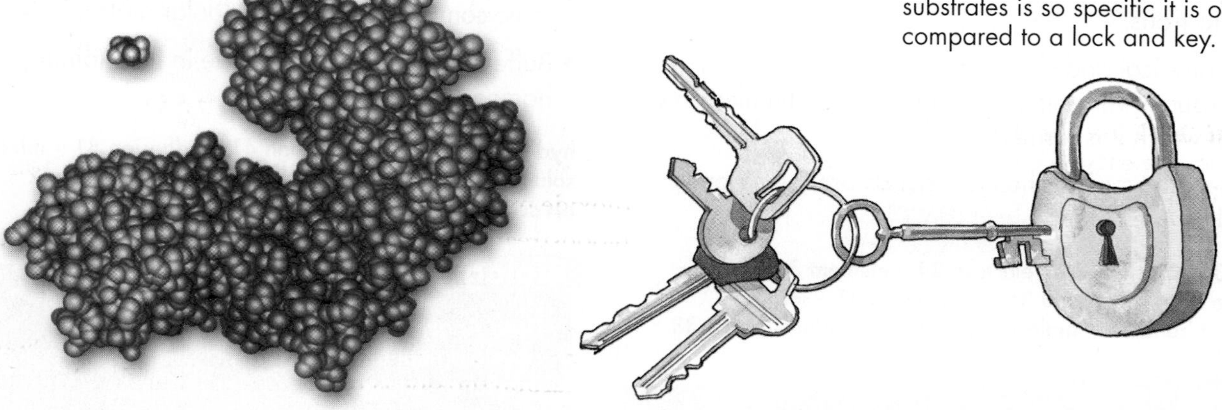

FIGURE 2–23 This space-filling model shows how a substrate binds to an active site on an enzyme. The fit between an enzyme and its substrates is so specific it is often compared to a lock and key.

2.4 Assessment

Review Key Concepts

1. a. **Review** What happens to chemical bonds during chemical reactions?
   b. **Apply Concepts** Why is the melting of ice not a chemical reaction?
2. a. **Review** What is activation energy?
   b. **Compare and Contrast** Describe the difference between a reaction that occurs spontaneously and one that does not.
3. a. **Review** What are enzymes?
   b. **Explain** Explain how enzymes work, including the role of the enzyme-substrate complex.
   c. **Use Analogies** A change in pH can change the shape of a protein. How might a change in pH affect the function of an enzyme such as carbonic anhydrase? (*Hint:* Think about the analogy of the lock and key.)

VISUAL THINKING

4. Make a model that demonstrates the fit between an enzyme and its substrate. Show your model to a friend or family member and explain how enzymes work using your model.

NGSS Smart Guide The Chemistry of Life

Disciplinary Core Idea LS1.A Structure and Function: How do the structures of organisms enable life's functions? Chemical bonds join together the molecules and compounds of life. Water and carbon compounds play essential roles in organisms, which carry out chemical reactions in their daily life processes.

2.1 The Nature of Matter

- The subatomic particles that make up atoms are protons, neutrons, and electrons.
- All isotopes of an element have the same chemical properties, because they have the same number of electrons.
- The physical and chemical properties of a compound are usually very different from those of the elements from which it is formed.
- The main types of chemical bonds are ionic bonds and covalent bonds.

atom 34 • nucleus 34 • electron 34 • element 35 • isotope 35 • compound 36 • ionic bond 37 • ion 37 • covalent bond 37 • molecule 37 • van der Waals forces 38

Biology.com

Untamed Science Video Watch the Untamed Science crew find answers to the mystery of why water is such a special compound.

Art Review Learn about ionic and covalent bonding.

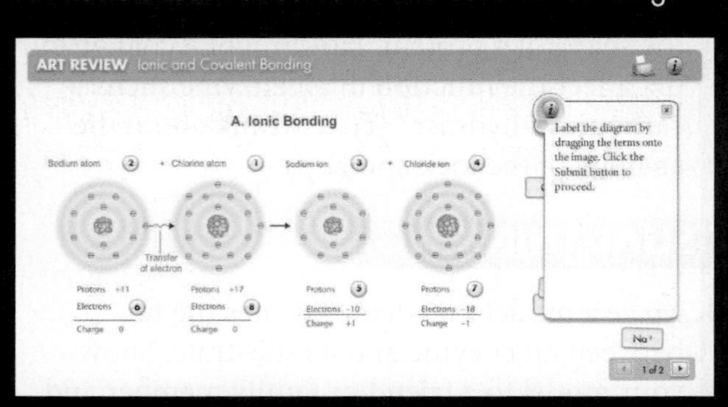

2.2 Properties of Water 

- Water is a polar molecule. Therefore, it is able to form multiple hydrogen bonds, which account for many of its special properties.
- Water's polarity gives it the ability to dissolve both ionic compounds and other polar molecules.
- Buffers play an important role in maintaining homeostasis in organisms.

hydrogen bond 41 • cohesion 41 • adhesion 41 • mixture 42 • solution 42 • solute 42 • solvent 42 • suspension 42 • pH scale 43 • acid 44 • base 44 • buffer 44

Biology.com

Data Analysis Analyze data that explain the physiological effects of low pH and the ecological impact of acid rain.

Art in Motion View an animation that shows the process of a salt crystal dissolving in water.

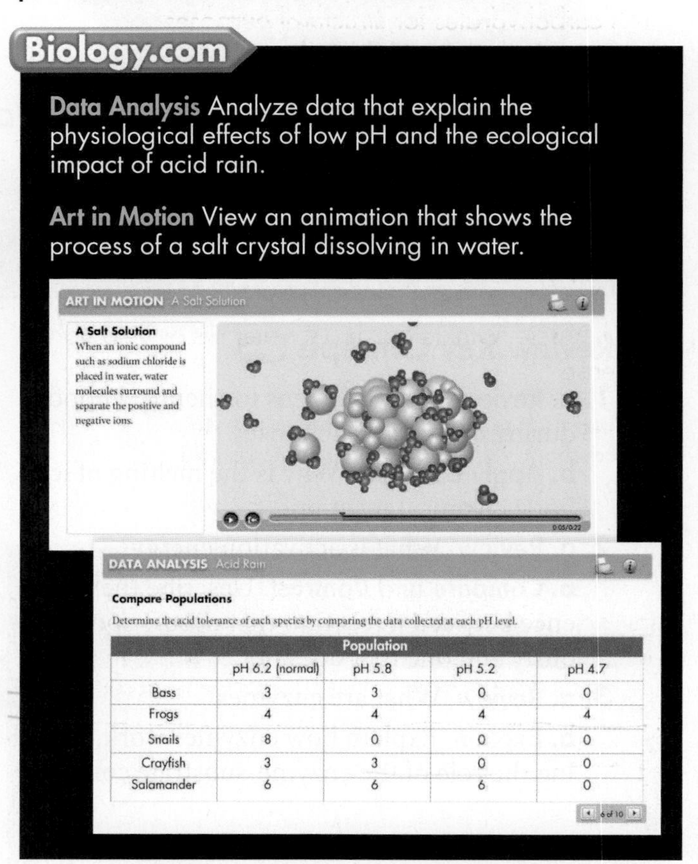

	Population			
	pH 6.2 (normal)	pH 5.8	pH 5.2	pH 4.7
Bass	3	3	0	0
Frogs	4	4	4	4
Snails	8	0	0	0
Crayfish	3	3	0	0
Salamander	6	6	6	0

CHECKING YOUR *Scientific Literacy*

Refer to the lesson content and digital assets below as you prepare for your chapter assessment. Then, evaluate your understanding of the chemistry of life by answering these questions.

1. Scientific and Engineering Practice **Developing and Using Models** Using common materials, build a model of a water molecule. Then build additional water molecules and show the interaction between them.

2. Crosscutting Concept **Structure and Function** Use your model to explain how water's structure contributes to its unique properties and the role it plays in organisms.

3. Integration of Knowledge and Ideas Translate technical information in the text to describe the four groups of macromolecules and the role they play in life's processes.

4. STEM Using common materials, build a model of an enzyme and its substrate. Use your model to show how the active site of an enzyme is specific to its substrate, and how the enzyme acts as a catalyst for the chemical reaction.

2.3 Carbon Compounds

- Carbon can bond with many elements, including hydrogen, oxygen, phosphorus, sulfur, and nitrogen, to form the molecules of life.
- Living things use carbohydrates as their main source of energy. Plants, some animals, and other organisms also use carbohydrates for structural purposes.
- Lipids can be used to store energy. Some lipids are important parts of biological membranes and waterproof coverings.
- Nucleic acids store and transmit hereditary, or genetic, information.
- Some proteins control the rate of reactions and regulate cell processes. Some proteins build tissues such as bone and muscle. Others transport materials or help fight disease.

monomer 46 • polymer 46 • carbohydrate 46 • monosaccharide 46 • lipid 47 • nucleic acid 48 • nucleotide 48 • protein 48 • amino acid 48

Biology.com

Lesson Notes Compare and contrast the four groups of organic compounds.

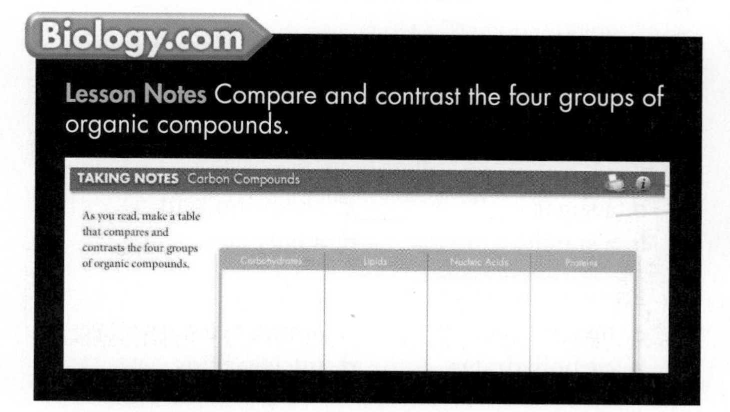

2.4 Chemical Reactions and Enzymes

- Chemical reactions always involve changes in the chemical bonds that join atoms in compounds.
- Chemical reactions that release energy often occur spontaneously. Chemical reactions that absorb energy will not occur without a source of energy.
- Enzymes speed up chemical reactions that take place in cells.
- Temperature, pH, and regulatory molecules can affect the activity of enzymes.

chemical reaction 50 • reactant 50 • product 50 • activation energy 51 • catalyst 52 • enzyme 52 • substrate 52

Biology.com

Visual Analogy Compare enzymes and substrates to a lock and key.

TAKING NOTES Chemical Reactions and Enzymes

As you read, make a concept map that shows the relationship among the vocabulary terms in this lesson.

Reactants — Chemical Reactions — Require — Can be

Design Your Own Lab

Temperature and Enzymes This chapter lab is available in your lab manual and at **Biology.com**

2 Assessment

2.1 The Nature of Matter

Understand Key Concepts

1. The positively charged particle in an atom is called the
   a. neutron.
   b. ion.
   c. proton.
   d. electron.

2. Two or more different atoms are combined in definite proportions in any
   a. symbol.
   b. isotope.
   c. element.
   d. compound.

3. A covalent bond is formed by the
   a. transfer of electrons.
   b. sharing of electrons.
   c. gaining of electrons.
   d. losing of electrons.

4. **Craft and Structure** Explain the relationship among atoms, elements, and compounds.

5. What is a radioactive isotope? Describe two scientific uses of radioactive isotopes.

6. Describe how the atoms in a compound are held together.

7. Distinguish among single, double, and triple covalent bonds.

Think Critically

8. **Use Models** Make a diagram like the one in **Figure 2–4** to show how chlorine and hydrogen form from the compound hydrogen chloride, HCl.

9. **Calculate** A nanometer (nm) is one billionth of a meter (1 nm = 10^{-9} m). If 100 million atoms make a row 1 cm in length, what is the diameter of one atom in nanometers? MATH

2.2 Properties of Water

Understand Key Concepts

10. When you shake sugar and sand together in a test tube, you cause them to form a
    a. compound.
    b. mixture.
    c. solution.
    d. suspension.

11. A compound that produces hydrogen ions in solution is a(n)
    a. salt.
    b. acid.
    c. base.
    d. polymer.

12. Compared to most other substances, a great deal of heat is needed to raise the temperature of water by a given amount. This is because water
    a. is an acid.
    b. readily forms solutions.
    c. has a high heat capacity.
    b. acts as a buffer.

13. Explain the properties of cohesion and adhesion. Give an example of each property.

14. **Craft and Structure** What is the relationship among solutions, solutes, and solvents?

15. How are acids and bases different? How do their pH values differ?

Think Critically

16. **Propose a Solution** Silica is a hard, glassy material that does not dissolve in water. Suppose sodium chloride is accidentally mixed with silica. Describe a way to remove the sodium chloride.

17. **Predict** As part of the digestive process, the human stomach produces hydrochloric acid, HCl. Sometimes excess acid causes discomfort. In such a case, a person might take an antacid such as magnesium hydroxide, Mg(OH)2. Explain how this substance can reduce the amount of acid in the stomach.

2.3 Carbon Compounds

Understand Key Concepts

18. What does the following formula represent?

```
    H   H   O
    |   |   ||
H — N — C — C — OH
        |
    H — C — H
        |
        H
```

    a. a sugar
    b. a starch
    c. an amino acid
    d. a fatty acid

19. Proteins are polymers formed from
    a. lipids.
    b. carbohydrates.
    c. amino acids.
    d. nucleic acids.

20. Explain the relationship between monomers and polymers, using polysaccharides as an example.

21. Identify three major roles of proteins.

22. Describe the parts of a nucleotide.

Think Critically

23. **Design an Experiment** Suggest one or two simple experiments to determine whether a solid white substance is a lipid or a carbohydrate. What evidence would you need to support each hypothesis?

24. **Infer** Explain what the name "carbohydrate" might indicate about the chemical composition of sugars.

2.4 Chemical Reactions and Enzymes

Understand Key Concepts

25. An enzyme speeds up a reaction by
    a. lowering the activation energy.
    b. raising the activation energy.
    c. releasing energy.
    d. absorbing energy.

26. In a chemical reaction, a reactant binds to an enzyme at a region known as the
    a. catalyst.
    b. product.
    c. substrate.
    d. active site.

27. Describe the two types of energy changes that can occur in a chemical reaction.

28. **Craft and Structure** What relationship exists between an enzyme and a catalyst?

29. Describe some factors that may influence enzyme activity.

Think Critically

30. **Infer** Why is it important that energy-releasing reactions take place in living organisms?

31. **Text Types and Purposes** Changing the temperature or pH can change an enzyme's shape. Explain how changing the temperature or pH might affect the function of an enzyme.

32. **Use Analogies** Explain why a lock and key are used to describe the way an enzyme works. Describe any ways in which the analogy is not perfect.

solve the CHAPTER MYSTERY

THE GHOSTLY FISH

The oxygen-binding abilities of hemoglobin enable the blood of most fish to carry nearly 50 times the oxygen it would without the protein. The ghostly white appearance of the antarctic ice fish results from its clear blood—blood without hemoglobin. Ice fish, however, are able to survive without hemoglobin because of the properties of water at low temperatures.

Oxygen from the air dissolves in seawater, providing the oxygen that fish need to survive. Fish absorb dissolved oxygen directly through their gills, where it passes into their bloodstream. The solubility of oxygen is much greater at low temperatures. Therefore, the icy cold antarctic waters are particularly rich in oxygen.

The large, well-developed gills and scaleless skin of ice fishes allow them to absorb oxygen efficiently from the water. Compared to red-blooded fishes, ice fishes have a higher blood volume, thinner blood, and larger hearts. So, their blood can carry more dissolved oxygen and the large hearts can pump the thinner blood through the body faster. These and other physical features, combined with the chemistry of oxygen in water at low temperatures, enable ice fish to survive where many other organisms cannot.

1. **Relate Cause and Effect** Ice fish produce antifreeze proteins to keep their blood from freezing; their body temperature stays below 0°C. How does low body temperature affect the blood's ability to carry dissolved oxygen?

2. **Infer** People living at high altitudes generally have more hemoglobin in their blood than people living at sea level. Why do you think this is so?

3. **Predict** If the antarctic oceans were to warm up, how might this affect ice fish?

4. **Connect to the Big idea** The chemical reactions in all living things slow down at low temperatures. Since some of the most important reactions in our body require oxygen, how would low temperatures affect the ice fish's need for oxygen?

Connecting Concepts

Use Science Graphics

The following graph shows the total amount of product from a chemical reaction performed at three different temperatures. The same enzyme was involved in each case. Use the graph to answer questions 33–35.

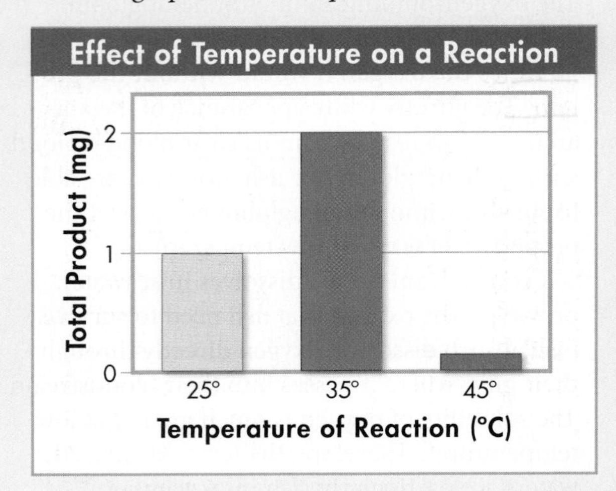

33. **Integration of Knowledge and Ideas** At which temperature was the greatest amount of product formed?

34. **Integration of Knowledge and Ideas** Describe the results of each reaction. How can you explain these results?

35. **Predict** A student performs the same chemical reaction at 30°C. Approximately how much product can she expect to obtain?

Write About Science

36. **Key Ideas and Details** Write a summary that includes the following: **(a)** a description of the four major classes of organic compounds found in living things, and **(b)** a description of how these organic compounds are used by the human body.

37. **Assess the Big idea** Based on the text, what can you conclude about the properties of carbon that allow it to play such a major role in the chemistry of living things? Explain.

Integration of Knowledge and Ideas

A student measured the pH of water from a small pond at several intervals throughout the day. Use the graph to answer questions 38 and 39.

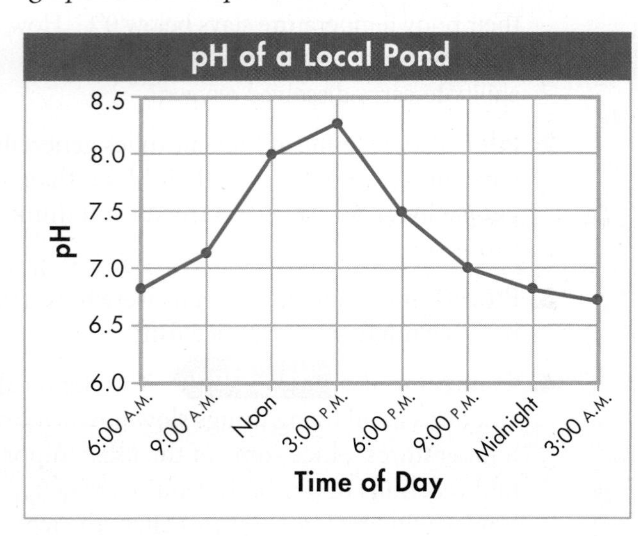

38. **Interpret Graphs** At what time of day is the pond most acidic?

a. between noon and 6:00 P.M.
b. at noon
c. between midnight and 6:00 A.M.
d. at 6:00 P.M.

39. **Form a Hypothesis** Which of the following is the most reasonable hypothesis based on the results obtained?

a. Pond water maintains constant pH throughout the day.
b. pH rises with increasing daylight and falls with decreasing daylight.
c. Living things cannot survive in this pond because enzymes will be destroyed.
d. pH is higher at night than during the day.

Standardized Test Prep

Multiple Choice

1. The elements or compounds that enter into a chemical reaction are called
A products.
B catalysts.
C active sites.
D reactants.

2. Chemical bonds that involve the total transfer of electrons from one atom or group of atoms to another are called
A covalent bonds.
B ionic bonds.
C hydrogen bonds.
D van der Waals bonds.

3. Which of the following is NOT an organic molecule found in living organisms?
A protein
B nucleic acid
C sodium chloride
D lipid

4. Which combination of particle and charge is correct?
A proton: positively charged
B electron: positively charged
C neutron: negatively charged
D electron: no charge

5. In which of the following ways do isotopes of the same element differ?
A in number of neutrons only
B in number of protons only
C in numbers of neutrons and protons
D in number of neutrons and in mass

6. Which of the following molecules is made up of glycerol and fatty acids?
A sugars
B starches
C lipids
D nucleic acids

7. Nucleotides consist of a phosphate group, a nitrogenous base, and a
A fatty acid.
B lipid.
C 5-carbon sugar.
D 6-carbon sugar.

Questions 8–9

The enzyme catalase speeds up the chemical reaction that changes hydrogen peroxide into oxygen and water. The amount of oxygen given off is an indication of the rate of the reaction.

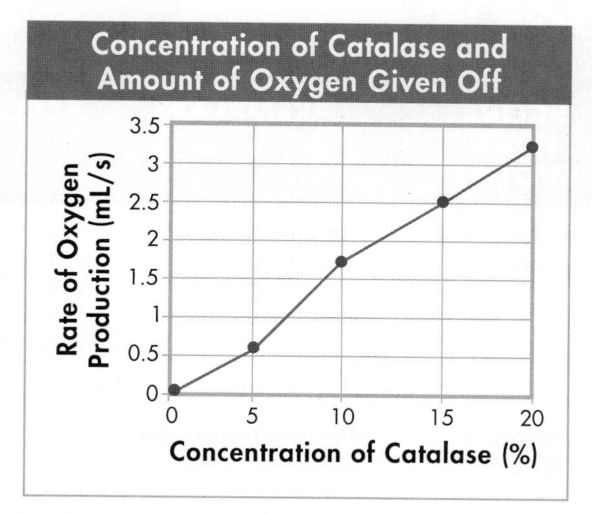

8. Based on the graph, what can you conclude about the relationship between enzyme concentration and reaction rate?
A Reaction rate decreases with increasing enzyme concentration.
B Reaction rate increases with decreasing enzyme concentration.
C Reaction rate increases with increasing enzyme concentration.
D The variables are indirectly proportional.

9. Which concentration of catalase will produce the fastest reaction rate?
A 5%
B 10%
C 15%
D 20%

Open-Ended Response

10. List some of the properties of water that make it such a unique substance.

If You Have Trouble With . . .

Question	1	2	3	4	5	6	7	8	9	10
See Lesson	2.4	2.1	2.3	2.1	2.1	2.3	2.3	2.4	2.4	2.2

NGSS Problem-based Learning

UNIT 1 PROJECT

Your Self-Assessment

The rubric below will help you evaluate your understanding of hydrophobic phenomena and their technological applications.

SCORE YOUR WORK!	EXEMPLARY Score your work a 4 if:	ACCOMPLISHED Score your work a 3 if:	DEVELOPING Score your work a 2 if:	BEGINNING Score your work a 1 if:
Construct an Explanation ▶	You have shared your revised explanation with classmates, and have refined it based on the feedback you received.	You have revised and edited your explanation of the phenomenon. Your explanation produces a sense of understanding.	You understand the observed phenomenon, and you have written the first draft of a plausible explanation.	You have begun researching the phenomenon, but you do not yet understand it well enough to explain it.
Develop Models ▶	You have refined your diagram or model through iterative rounds of questioning, analysis, and revision. You have identified and discussed the strengths and limitations of your diagram or model.	You have revised your diagram or model, analyzed it, and identified its strengths and limitations.	You have produced a draft diagram or model that conveys the main points of your explanation.	You have identified a way to realize your explanation as a diagram or model.
	TIP The model development process is iterative, which means you won't get it right on the first try. Keep asking questions and evaluating your ideas, and try to steadily improve them.			
Obtain, Evaluate, and Communicate Information ▶	You have explained how the data from your procedure can be used to evaluate the product and inform consumers about its use, as well as why such information is important.	You have outlined a procedure for obtaining and analyzing data on the product's effectiveness.	You have identified the problem or need that the product was designed to solve, as well as claims about the product's effectiveness.	You have identified a product that uses hydrophobic technology.
	TIP Think about the criteria that the product was designed to satisfy. Ask, "How can these criteria be measured? What data do you need in order to evaluate whether or not these criteria have been fulfilled?"			
Design a Solution ▶	You have discussed, revised, and refined your design proposal based on feedback from teachers and/or classmates. The final draft makes a convincing argument for your design solution.	You have written a proposal for designing a solution to the defined problem. Your proposal outlines both the strengths and limitations of your design.	You have defined the problem and specified the criteria for a solution. You have also discussed possible solutions.	You have identified a problem that is in need of a solution.

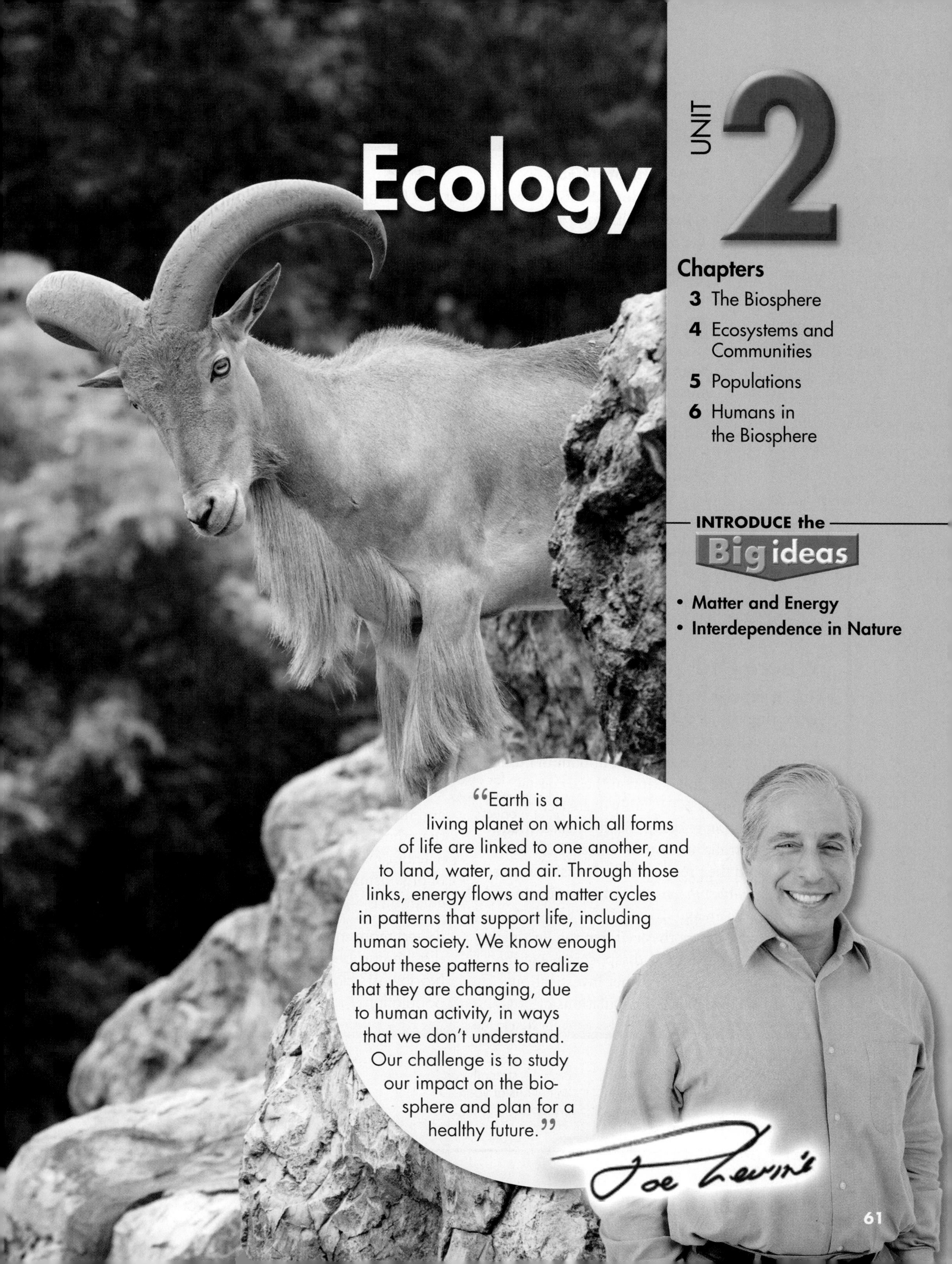

UNIT 2 Ecology

Chapters

3 The Biosphere

4 Ecosystems and Communities

5 Populations

6 Humans in the Biosphere

INTRODUCE the **Big ideas**

- **Matter and Energy**
- **Interdependence in Nature**

“Earth is a living planet on which all forms of life are linked to one another, and to land, water, and air. Through those links, energy flows and matter cycles in patterns that support life, including human society. We know enough about these patterns to realize that they are changing, due to human activity, in ways that we don’t understand. Our challenge is to study our impact on the biosphere and plan for a healthy future.”

[Signature: Joe Levine]

NGSS Problem-based Learning

Spotlight on NGSS
- **Core Idea LS2.B** Cycles of Matter and Energy Transfer in Ecosystems
- **Practice** Engage in Argument from Evidence
- **Crosscutting Concept** Systems and Systems Models

UNIT 2 PROJECT

Disappearing Mussels!

Your friend is writing a magazine article about the disappearance of mussels living in the coastal tide pools off the Pacific Coast. However, the research, and therefore the article, isn't complete. Read the draft of the article below, then use various science practices to solve this mystery and help finish the article. Develop an argument as to the cause as well as the prevention plan for other disappearing marine life.

Coastal Magazine AND Outdoor Guide

Published since 1967

Where Did You Come From?

Once a popular Pacific Coast activity, mussel harvesting has all but stopped because people are having a hard time finding the mussels. In the past, mussels have been abundant in number, but recently the population of these coastal tide pool animals has steadily decreased. A similar situation exists in the northeast Atlantic Ocean, Australia, and South Africa. Research has revealed the presence of an invasive animal called *Carcinus maenas,* or European green crab, in all of these areas.

Experts have hypothesized that the European green crab is causing the disappearance of the Pacific Coast mussels. Further investigation exposed well-established populations of *Carcinus maenas* in the Pacfic Coast's Bodega Harbor and Willapa Bay. These creatures are described as a voracious predator that is generally quicker and more dexterous than the native crabs of the Pacific Northwest. The European green crab feeds on many types of organisms and is capable of learning and adapting to improve its ability to search for food.

European green crab

(Need more information for this article! How did the European green crab get to the Pacific Northwest? How can people help the mussel population?)

Done

Your Task: Solving a Crabby Case!

You are going to help your friend finish the magazine article. You must investigate and present an argument to your friend as to how the European green crabs got to the Pacific Northwest. You must cite evidence from your research to support your argument. You must also develop a workable solution for increasing the mussel population and containing or decreasing the European green crab population. Use the rubric at the end of the unit to evaluate your work for this task.

Ask Questions Begin by asking questions. What do I already know? What do I need to ask?

Plan and Carry Out an Investigation Write a plan to help you identify the information you need and where to find it. What details should your plan include to investigate and solve this mystery? Conduct research and then record your data. What important information can you learn about the coastal tide pool's ecosystems and invasive species? Are there other factors, besides the European green crab, that could be causing the decline of the mussel population?

Analyze and Interpret Data Organize your research and data by using charts, graphs, spreadsheets, or other tools. What information should you include in your chart? Interpret the data by making connections between pieces of information. Do you need any more data to bring you closer to finding a possible explanation or developing a solution?

Construct an Explanation Use the evidence you collected to write an explanation for the declining mussel population. What evidence is strongest in supporting your argument? What do you recommend that the community do to prevent other marine creatures from the same fate?

Engage in Argument from Evidence Develop an argument to defend your explanation using reasoning and evidence. Create a report, video, podcast, or blog to present your explanation and argument to your friend. Ask for feedback from your teacher, a classmate, or someone in the community. What else could you have thought about? Do you think your friend will be pleased?

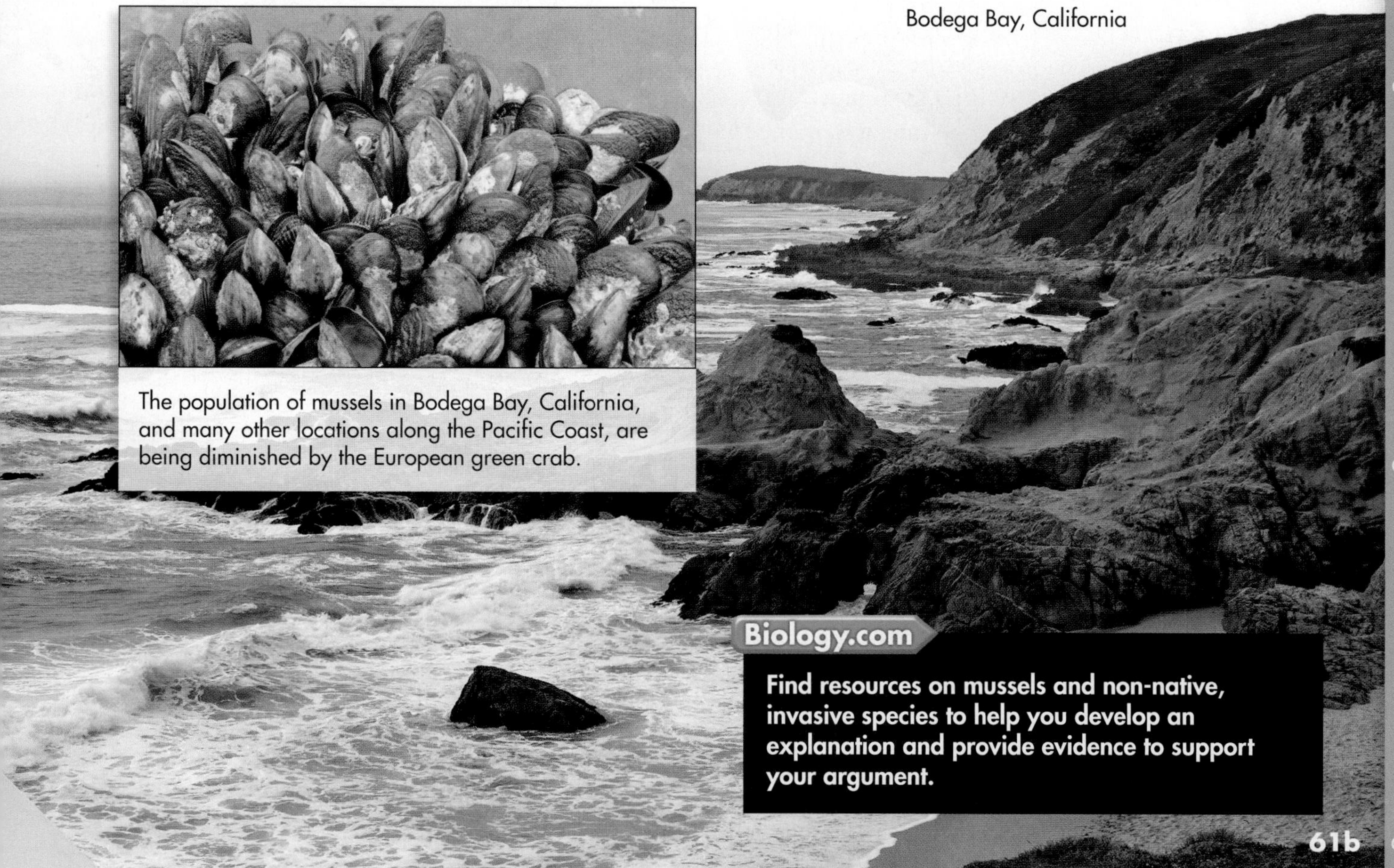

Bodega Bay, California

The population of mussels in Bodega Bay, California, and many other locations along the Pacific Coast, are being diminished by the European green crab.

Biology.com

Find resources on mussels and non-native, invasive species to help you develop an explanation and provide evidence to support your argument.

3 The Biosphere

Matter and Energy, Interdependence in Nature

Q: How do Earth's living and nonliving parts interact and affect the survival of organisms?

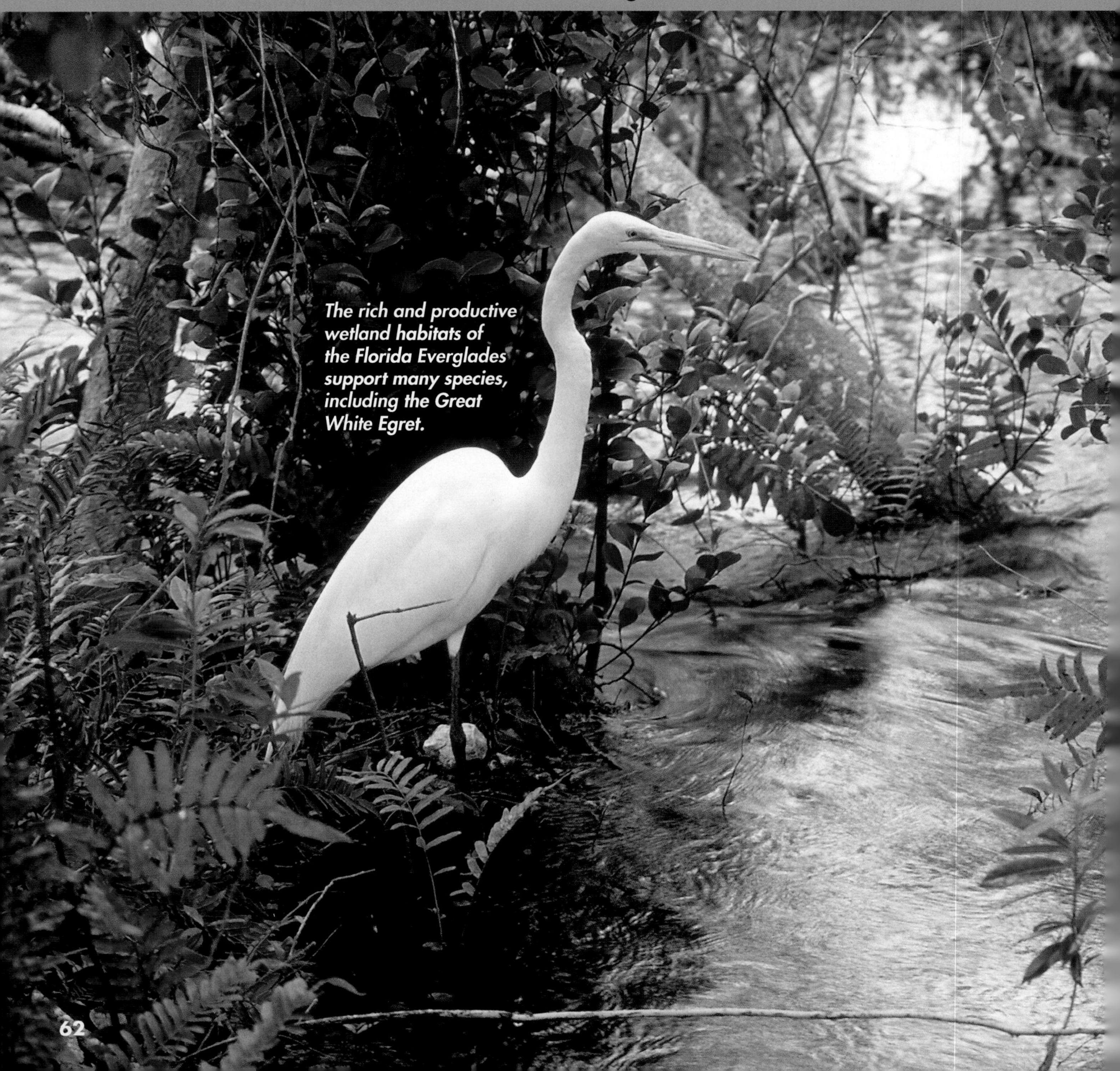

The rich and productive wetland habitats of the Florida Everglades support many species, including the Great White Egret.

INSIDE:

- 3.1 What Is Ecology?
- 3.2 Energy, Producers, and Consumers
- 3.3 Energy Flow in Ecosystems
- 3.4 Cycles of Matter

BUILDING *Scientific Literacy*

Deepen your understanding of the biosphere by engaging in key practices that allow you to make connections across concepts.

NGSS Tabulate processes in natural cycles of matter and evaluate the relationship of those processes in an ecosystem.

STEM What do living systems need for survival? Find out what biotic and abiotic factors are necessary for an ecosystem to be self-sustaining.

Common Core Cite evidence from the text to explain how disrupting a cycle of matter can affect an ecosystem.

CHAPTER MYSTERY

CHANGES IN THE BAY

Marine life in Rhode Island's Narragansett Bay is changing. One clue to those changes comes from fishing boat captains who boast about catching bluefish in November—a month after those fish used to head south for winter. Catches of winter flounder, however, are not as plentiful as they once were. These changes in fish populations coincide with the disappearance of the annual spring increase in plant and animal growth. Researchers working in the bay, meanwhile, report puzzling changes in the activities of bacteria living in mud on the bay floor. What's going on?

Farms, towns, and cities surround the bay, but direct human influence on the bay has not changed much lately. So why are there so many changes to the bay's plant and animal populations? Could these changes be related to mud-dwelling bacteria?

As you read the chapter, look for clues to help you understand the interactions of plants, animals, and bacteria in Narragansett Bay. Then, solve the mystery.

Biology.com

Finding out about Narragansett Bay is only the beginning. Take a video field trip with the ecogeeks of Untamed Science to see where this mystery leads.

Go online to access additional resources including:
- **eText** • **Flash Cards** • **Lesson Overviews** • **Chapter Mystery**

3.1 What Is Ecology?

Key Questions

- *What is ecology?*
- *What are biotic and abiotic factors?*
- *What methods are used in ecological studies?*

Vocabulary

biosphere • species • population • community • ecology • ecosystem • biome • biotic factor • abiotic factor

Taking Notes

Venn Diagram Make a Venn diagram that shows how the environment consists of biotic factors, abiotic factors, and some components that are truly a mixture of both. Use examples from the lesson.

THINK ABOUT IT Lewis Thomas, a twentieth-century science writer, was sufficiently inspired by astronauts' photographs of Earth to write: "Viewed from the distance of the moon, the astonishing thing about the earth ... is that it is alive." Sounds good. But what does it mean? Was Thomas reacting to how green Earth is? Was he talking about how you can see moving clouds from space? How is Earth, in a scientific sense, a "living planet"? And how do we study it?

Studying Our Living Planet

What is ecology?

When biologists want to talk about life on a global scale, they use the term *biosphere*. The **biosphere** consists of all life on Earth and all parts of the Earth in which life exists, including land, water, and the atmosphere. The biosphere contains every organism, from bacteria living underground to giant trees in rain forests, whales in polar seas, mold spores drifting through the air—and, of course, humans. The biosphere extends from about 8 kilometers above Earth's surface to as far as 11 kilometers below the surface of the ocean.

Individual Organism
A **species** is a group of similar organisms that can breed and produce fertile offspring.

A **population** is a group of individuals that belong to the same species and live in the same area.

An assemblage of different populations that live together in a defined area is called a **community.**

The Science of Ecology Organisms in the biosphere interact with each other and with their surroundings, or environment. The study of these interactions is called **ecology.** **Ecology is the scientific study of interactions among organisms and between organisms and their physical environment.** The root of the word *ecology* is the Greek word *oikos*, which means "house." So, ecology is the study of nature's "houses" and the organisms that live in those houses.

Interactions within the biosphere produce a web of interdependence between organisms and the environments in which they live. Organisms respond to their environments and can also change their environments, producing an ever-changing, or dynamic, biosphere.

BUILD Vocabulary

PREFIXES The prefix *inter-* means "between or among." *Interdependence* is a noun that means "dependence between or among individuals or things." The physical environment and organisms are considered interdependent because changes in one cause changes in the other.

Ecology and Economics The Greek word *oikos* is also the root of the word *economics*. Economics is concerned with human "houses" and human interactions based on money or trade. Interactions among nature's "houses" are based on energy and nutrients. As their common root implies, human economics and ecology are linked. Humans live within the biosphere and depend on ecological processes to provide such essentials as food and drinkable water that can be bought and sold or traded.

Levels of Organization Ecologists ask many questions about organisms and their environments. Some ecologists focus on the ecology of individual organisms. Others try to understand how interactions among organisms (including humans) influence our global environment. Ecological studies may focus on levels of organization that include those shown in **Figure 3–1.**

In Your Notebook *Draw a circle and label it "Me." Then, draw five concentric circles and label each of them with the appropriate level of organization. Describe your population, community, etc.*

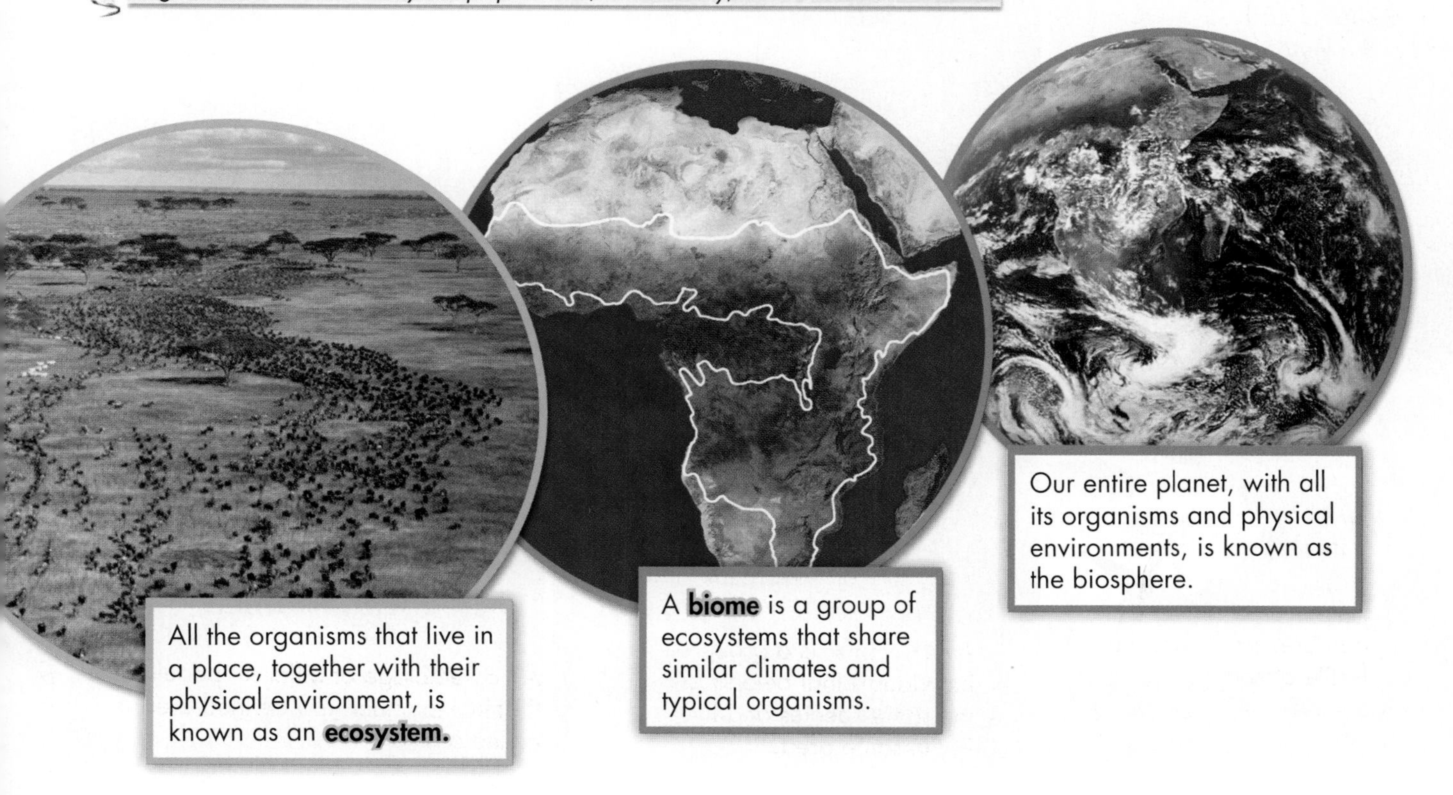

FIGURE 3–1 Levels of Organization The kinds of questions that ecologists may ask about the living environment can vary, depending on the level at which the ecologist works. **Interpret Visuals** ***What is the difference between a population and a community?***

Biotic and Abiotic Factors

What are biotic and abiotic factors?

Ecologists use the word *environment* to refer to all conditions, or factors, surrounding an organism. Environmental conditions include biotic factors and abiotic factors, as shown in **Figure 3–2.**

Biotic Factors **The biological influences on organisms are called biotic factors.** A **biotic factor** is any living part of the environment with which an organism might interact, including animals, plants, mushrooms, and bacteria. Biotic factors relating to a bullfrog, for example, might include algae it eats as a tadpole, insects it eats as an adult, herons that eat bullfrogs, and other species that compete with bullfrogs for food or space.

Abiotic Factors **Physical components of an ecosystem are called abiotic factors.** An **abiotic factor** is any nonliving part of the environment, such as sunlight, heat, precipitation, humidity, wind or water currents, soil type, and so on. For example, a bullfrog could be affected by abiotic factors such as water availability, temperature, and humidity.

FIGURE 3–2 Biotic and Abiotic Factors Like all ecosystems, this pond is affected by a combination of biotic and abiotic factors. Some environmental factors, such as the "muck" around the edges of the pond, are a mix of biotic and abiotic components. Biotic and abiotic factors are dynamic, meaning that they constantly affect each other. **Classify** ***What biotic factors are visible in this ecosystem?***

Biotic and Abiotic Factors Together The difference between biotic and abiotic factors may seem to be clear and simple. But if you think carefully, you will realize that many physical factors can be strongly influenced by the activities of organisms. Bullfrogs hang out, for example, in soft "muck" along the shores of ponds. You might think that this muck is strictly part of the physical environment, because it contains nonliving particles of sand and mud. But typical pond muck also contains leaf mold and other decomposing plant material produced by trees and other plants around the pond. That material is decomposing because it serves as "food" for bacteria and fungi that live in the muck.

Taking a slightly wider view, the "abiotic" conditions around that mucky shoreline are strongly influenced by living organisms. A leafy canopy of trees and shrubs often shade the pond's shoreline from direct sun and protect it from strong winds. In this way, organisms living around the pond strongly affect the amount of sunlight the shoreline receives and the range of temperatures it experiences. A forest around a pond also affects the humidity of air close to the ground. The roots of trees and other plants determine how much soil is held in place and how much washes into the pond. Even certain chemical conditions in the soil around the pond are affected by living organisms. If most trees nearby are pines, their decomposing needles make the soil acidic. If the trees nearby are oaks, the soil will be more alkaline. This kind of dynamic mix of biotic and abiotic factors shapes every environment.

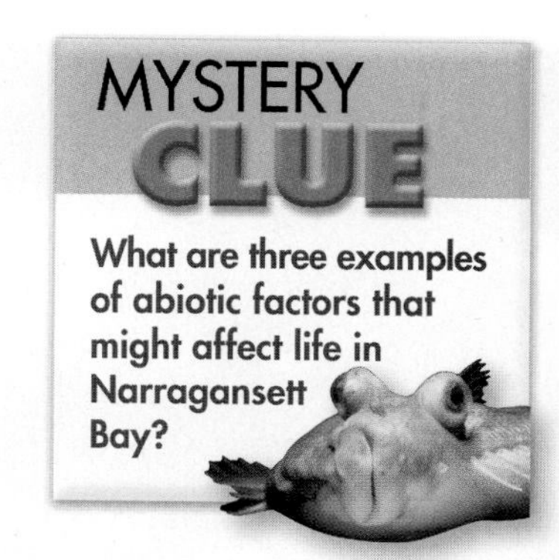

What are three examples of abiotic factors that might affect life in Narragansett Bay?

In Your Notebook *In your own words, explain the difference between biotic and abiotic factors. Give three examples of each.*

Quick Lab
GUIDED INQUIRY

How Do Abiotic Factors Affect Different Plant Species?

1. Gather four paper cups. Use a pencil to punch three holes in the bottom of each cup. Fill two cups with equal amounts of sand and two cups with the same amount of potting soil. **CAUTION:** *Wash your hands well with soap and warm water after handling soil or plants.*
2. Plant five rice seeds in one sand-filled cup and five rice seeds in one soil-filled cup. Plant five rye seeds in each of the other two cups. Label each cup with the type of seeds and soil it contains.
3. Place all the cups in a warm, sunny location. Each day for two weeks, water the cups equally and record your observations of any plant growth.

Analyze and Conclude

1. **Analyze Data** In which medium did the rice grow better—sand or soil? Which was the better medium for the growth of rye?
2. **Infer** Soil retains more water than sand does, providing a moister environment. What can you infer from your observations about the kind of environment that favors the growth of rice? What kind of environment favors the growth of rye?
3. **Draw Conclusions** Which would compete more successfully in a dry environment—rice or rye? Which would be more successful in a moist environment?

FIGURE 3–3 Ecology Field Work The three fundamental approaches to ecological research involve observing, experimenting, and modeling. This ecologist is measuring a Mediterranean tortoise.

Ecological Methods

What methods are used in ecological studies?

Some ecologists, like the one in **Figure 3–3,** use measuring tools to assess changes in plant and wildlife communities. Others use DNA studies to identify bacteria in marsh mud. Still others use data gathered by satellites to track ocean surface temperatures. **Regardless of their tools, modern ecologists use three methods in their work: observation, experimentation, and modeling. Each of these approaches relies on scientific methodology to guide inquiry.**

Observation Observation is often the first step in asking ecological questions. Some observations are simple: Which species live here? How many individuals of each species are there? Other observations are more complex: How does an animal protect its young from predators? These types of questions may form the first step in designing experiments and models.

Experimentation Experiments can be used to test hypotheses. An ecologist may, for example, set up an artificial environment in a laboratory or greenhouse to see how growing plants react to different conditions of temperature, lighting, or carbon dioxide concentration. Other experiments carefully alter conditions in selected parts of natural ecosystems.

Modeling Many ecological events, such as effects of global warming on ecosystems, occur over such long periods of time or over such large distances that they are difficult to study directly. Ecologists make models to help them understand these phenomena. Many ecological models consist of mathematical formulas based on data collected through observation and experimentation. Further observations by ecologists can be used to test predictions based on those models.

3.1 Assessment

Review Key Concepts

1. a. Review What are the six different major levels of organization, from smallest to largest, that ecologists commonly study?
   b. Apply Concepts Give an example of two objects or activities in your life that are interdependent. Explain your choice.
2. a. Review Is weather a biotic or abiotic factor?
   b. Compare and Contrast How are biotic and abiotic factors related? What is the difference between them?
3. a. Review Describe the three basic methods of ecological research.
   b. Apply Concepts Give an example of an ecological phenomenon that could be studied by modeling. Explain why modeling would be useful.

PRACTICE PROBLEM

4. Suppose you want to know if the water in a certain stream is safe to drink. Which ecological method(s) would you use in your investigation? Explain your reasoning and outline your procedure.

3.2 Energy, Producers, and Consumers

THINK ABOUT IT At the core of every organism's interaction with the environment is its need for energy to power life's processes. Ants use energy to carry objects many times their size. Birds use energy to migrate thousands of miles. You need energy to get out of bed in the morning! Where does energy in living systems come from? How is it transferred from one organism to another?

Key Questions

What are primary producers?

How do consumers obtain energy and nutrients?

Vocabulary

autotroph • primary producer • photosynthesis • chemosynthesis • heterotroph • consumer • carnivore • herbivore • scavenger • omnivore • decomposer • detritivore

Taking Notes

Concept Map As you read, use the highlighted vocabulary words to create a concept map that organizes the information in this lesson.

Primary Producers

What are primary producers?

Living systems operate by expending energy. Organisms need energy for growth, reproduction, and their own metabolic processes. In short, if there is no energy, there are no life functions! Yet, no organism can create energy—organisms can only use energy from other sources. You probably know that you get your energy from the plants and animals you eat. But where does the energy in your food come from? For most life on Earth, sunlight is the ultimate energy source. Over the last few decades, however, researchers have discovered that there are other energy sources for life. For some organisms, chemical energy stored in inorganic chemical compounds serves as the ultimate energy source for life processes.

Only algae, certain bacteria, and plants like the one in **Figure 3–4** can capture energy from sunlight or chemicals and convert it into forms that living cells can use. These organisms are called **autotrophs.** Autotrophs use solar or chemical energy to produce "food" by assembling inorganic compounds into complex organic molecules. But autotrophs do more than feed themselves. Autotrophs store energy in forms that make it available to other organisms that eat them. That's why autotrophs are also called **primary producers.** **Primary producers are the first producers of energy-rich compounds that are later used by other organisms.** Primary producers are, therefore, essential to the flow of energy through the biosphere.

BUILD Vocabulary

PREFIXES The prefix *auto-* means "by itself." The Greek word *trophikos* means "to feed." An **autotroph** can, therefore, be described as a "self feeder," meaning that it does not need to eat other organisms for food.

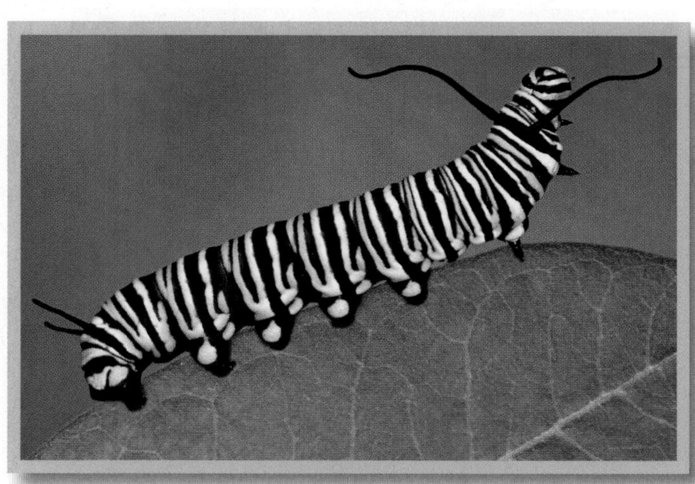

FIGURE 3–4 Primary Producers Plants obtain energy from sunlight and turn it into nutrients that can, in turn, be eaten and used for energy by animals such as this caterpillar.

Energy From the Sun The best-known and most common primary producers harness solar energy through the process of photosynthesis. **Photosynthesis** captures light energy and uses it to power chemical reactions that convert carbon dioxide and water into oxygen and energy-rich carbohydrates such as sugars and starches. This process, shown in **Figure 3–5** (below left), adds oxygen to the atmosphere and removes carbon dioxide. Without photosynthetic producers, the air would not contain enough oxygen for you to breathe! Plants are the main photosynthetic producers on land. Algae fill that role in freshwater ecosystems and in the sunlit upper layers of the ocean. Photosynthetic bacteria, most commonly cyanobacteria, are important primary producers in ecosystems such as tidal flats and salt marshes.

Life Without Light About 30 years ago, biologists discovered thriving ecosystems around volcanic vents in total darkness on the deep ocean floor. There was no light for photosynthesis, so who or what were the primary producers? Research revealed that these deep-sea ecosystems depended on primary producers that harness chemical energy from inorganic molecules such as hydrogen sulfide. These organisms carry out a process called **chemosynthesis** (kee moh SIN thuh sis) in which chemical energy is used to produce carbohydrates as shown in **Figure 3–5** (below right). Chemosynthetic organisms are not only found in the deepest, darkest ocean, however. Several types of chemosynthetic producers have since been discovered in more parts of the biosphere than anyone expected. Some chemosynthetic bacteria live in harsh environments, such as deep-sea volcanic vents or hot springs. Others live in tidal marshes along the coast.

In Your Notebook *In your own words, explain the differences and similarities between photosynthetic and chemosynthetic producers.*

FIGURE 3–5 Photosynthesis and Chemosynthesis Plants use the energy from sunlight to carry out the process of photosynthesis. Other autotrophs, such as sulfur bacteria, use the energy stored in chemical bonds in a process called chemosynthesis. In both cases, energy-rich carbohydrates are produced. **Compare and Contrast** ***How are photosynthesis and chemosynthesis similar?***

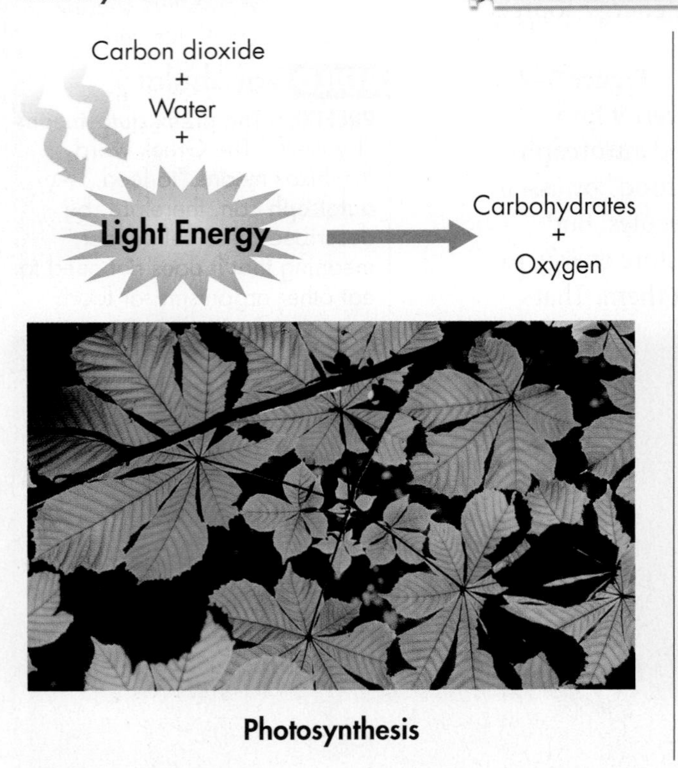

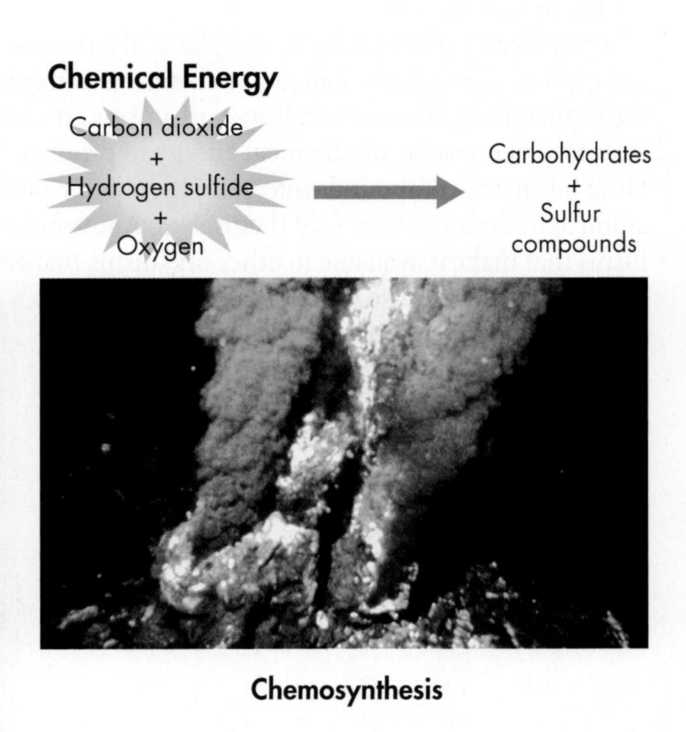

Consumers

How do consumers obtain energy and nutrients?

Animals, fungi, and many bacteria cannot directly harness energy from the environment as primary producers do. These organisms, known as **heterotrophs** (HET uh roh trohfs) must acquire energy from other organisms—by ingesting them in one way or another. Heterotrophs are also called **consumers. Organisms that rely on other organisms for energy and nutrients are called consumers.**

Types of Consumers Consumers are classified by the ways in which they acquire energy and nutrients, as shown in **Figure 3–6.** As you will see, the definition of *food* can vary quite a lot among consumers.

FIGURE 3–6 Consumers Consumers rely on other organisms for energy and nutrients. The Amazon rain forest shelters examples of each type of consumer as shown here.

Quick Lab
GUIDED INQUIRY

How Do Different Types of Consumers Interact?

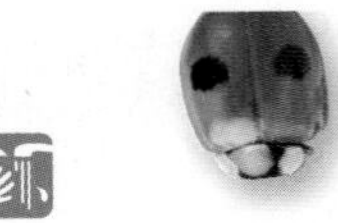

1. Place a potted bean seedling in each of two jars.
2. Add 20 aphids to one jar and cover the jar with screening to prevent the aphids from escaping. Use a rubber band to attach the screening to the jar.
3. Add 20 aphids and 4 ladybird beetles to the second jar. Cover the second jar as you did the first one.
4. Place both jars in a sunny location. Observe the jars each day for one week and record your observations each day. Water the seedlings as needed.

Analyze and Conclude

1. Observe What happened to the aphids and the seedling in the jar without the ladybird beetles? What happened in the jar with the ladybird beetles? How can you explain this difference?

2. Classify Identify each organism in the jars as a producer or a consumer. If the organism is a consumer, what kind of consumer is it?

MYSTERY CLUE

Bacteria are important members of the living community in Narragansett Bay. How do you think the bacterial communities on the floor of the bay might be linked to its producers and consumers?

Beyond Consumer Categories Categorizing consumers is important, but these simple categories often don't express the real complexity of nature. Take herbivores, for instance. Seeds and fruits are usually rich in energy and nutrients, and they are often easy to digest. Leaves are generally poor in nutrients and are usually very difficult to digest. For that reason, herbivores that eat different plant parts often differ greatly in the ways they obtain and digest their food. In fact, only a handful of birds eat leaves, because the kind of digestive system needed to handle leaves efficiently is heavy and difficult to fly around with!

Moreover, organisms in nature often do not stay inside the tidy categories ecologists place them in. For example, some animals often described as carnivores, such as hyenas, will scavenge if they get a chance. Many aquatic animals eat a mixture of algae, bits of animal carcasses, and detritus particles—including the feces of other animals! So, these categories make a nice place to start talking about ecosystems, but it is important to expand on this topic by discussing the way that energy and nutrients move through ecosystems.

3.2 Assessment

Review Key Concepts

1. a. Review What are the two primary sources of energy that power living systems?

b. Pose Questions Propose a question that a scientist might ask about the variety of organisms found around deep-sea vents.

2. a. Review Explain how consumers obtain energy.

b. Compare and Contrast How are detritivores different from decomposers? Provide an example of each.

BUILD VOCABULARY

3. The word *autotroph* comes from the Greek words *autos,* meaning "self," and *trophe,* meaning "food or nourishment." Knowing this, what do you think the Greek word *heteros,* as in *heterotroph,* means?

3.3 Energy Flow in Ecosystems

THINK ABOUT IT What happens to energy stored in body tissues when one organism eats another? That energy moves from the "eaten" to the "eater." You've learned that the flow of energy through an ecosystem always begins with either photosynthetic or chemosynthetic primary producers. Where it goes from there depends literally on who eats whom!

Key Questions

How does energy flow through ecosystems?

What do the three types of ecological pyramids illustrate?

Vocabulary

food chain • phytoplankton • food web • zooplankton • trophic level • ecological pyramid • biomass

Taking Notes

Preview Visuals Before you read, look at **Figure 3–7** and **Figure 3–9.** Note how they are similar and how they are different. Based on the figures, write definitions for *food chain* and *food web.*

Food Chains and Food Webs

How does energy flow through ecosystems?

In every ecosystem, primary producers and consumers are linked through feeding relationships. Despite the great variety of feeding relationships in different ecosystems, energy always flows in similar ways. **Energy flows through an ecosystem in a one-way stream, from primary producers to various consumers.**

Food Chains You can think of energy as passing through an ecosystem along a food chain. A **food chain** is a series of steps in which organisms transfer energy by eating and being eaten. Food chains can vary in length. For example, in a prairie ecosystem, a primary producer, such as grass, is eaten by an herbivore, such as a grazing antelope. A carnivore, such as a coyote, in turn feeds upon the antelope. In this two-step chain, the carnivore is just two steps removed from the primary producer.

In some aquatic food chains, primary producers are a mixture of floating algae called **phytoplankton** and attached algae. As shown in **Figure 3–7,** these primary producers may be eaten by small fishes, such as flagfish. Larger fishes, like the largemouth bass, eat the small fishes. The bass are preyed upon by large wading birds, such as the anhinga, which may ultimately be eaten by an alligator. There are four steps in this food chain. The top carnivore is therefore four steps removed from the primary producer.

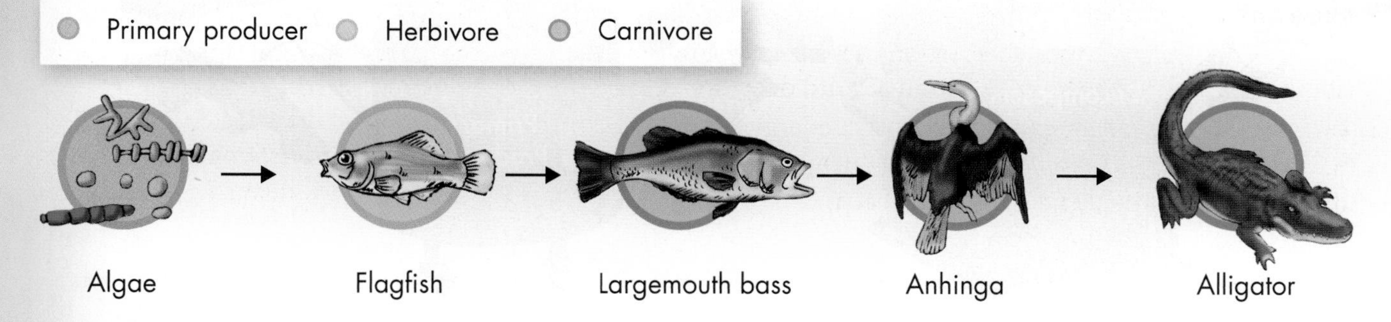

FIGURE 3–7 Food Chains Food chains show the one-way flow of energy in an ecosystem. **Apply Concepts** ***What is the ultimate source of energy for this food chain?***

Food Webs In most ecosystems, feeding relationships are much more complicated than the relationships described in a single, simple chain. One reason for this is that many animals eat more than one kind of food. For example, on Africa's Serengeti Plain, herbivores, such as zebras, gazelles, and buffaloes, often graze upon several different species of grasses. Several predators such as lions, hyenas, and leopards, in turn, often prey upon those herbivores! Ecologists call this network of feeding interactions a **food web.**

▶ ***Food Chains Within Food Webs*** The Everglades are a complex marshland ecosystem in southern Florida. Here, aquatic and terrestrial organisms interact in many overlapping feeding relationships that have been simplified and represented in **Figure 3–9.** Starting with a primary producer (algae or plants), see how many different routes you can take to reach the alligator, vulture, or anhinga. One path, from the algae to the alligator, is the same food chain you saw in **Figure 3–7.** In fact, each path you trace through the food web is a food chain. You can think of a food web, therefore, as linking together all of the food chains in an ecosystem. Realize, however, that this is a highly simplified representation of this food web, in which many species have been left out. Now, you can begin to appreciate how complicated food webs are!

▶ ***Decomposers and Detritivores in Food Webs*** Decomposers and detritivores are as important in most food webs as other consumers are. Look again at the Everglades web. Although white-tailed deer, moorhens, raccoons, grass shrimp, crayfish, and flagfish feed at least partly on primary producers, most producers die without being eaten. In the detritus pathway, decomposers convert that dead material to detritus, which is eaten by detritivores, such as crayfish, grass shrimp, and worms. At the same time, the decomposition process releases nutrients that can be used by primary producers. Thus, decomposers recycle nutrients in food webs as seen in **Figure 3–8.** Without decomposers, nutrients would remain locked within dead organisms.

BUILD Vocabulary

ACADEMIC WORDS The verb **convert** means "to change from one form to another." Decomposers convert, or change, dead plant matter into a form called detritus that is eaten by detritivores.

In Your Notebook *Explain how food chains and food webs are related.*

VISUAL ANALOGY

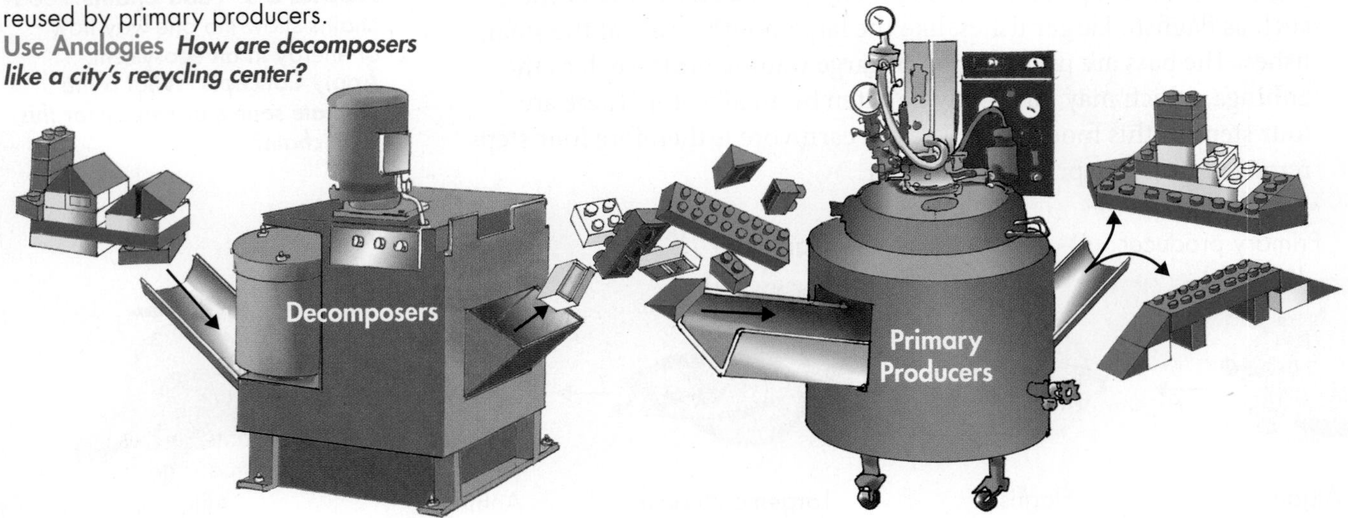

FIGURE 3–8 Earth's Recycling Center Decomposers break down dead and decaying matter and release nutrients that can be reused by primary producers. **Use Analogies** ***How are decomposers like a city's recycling center?***

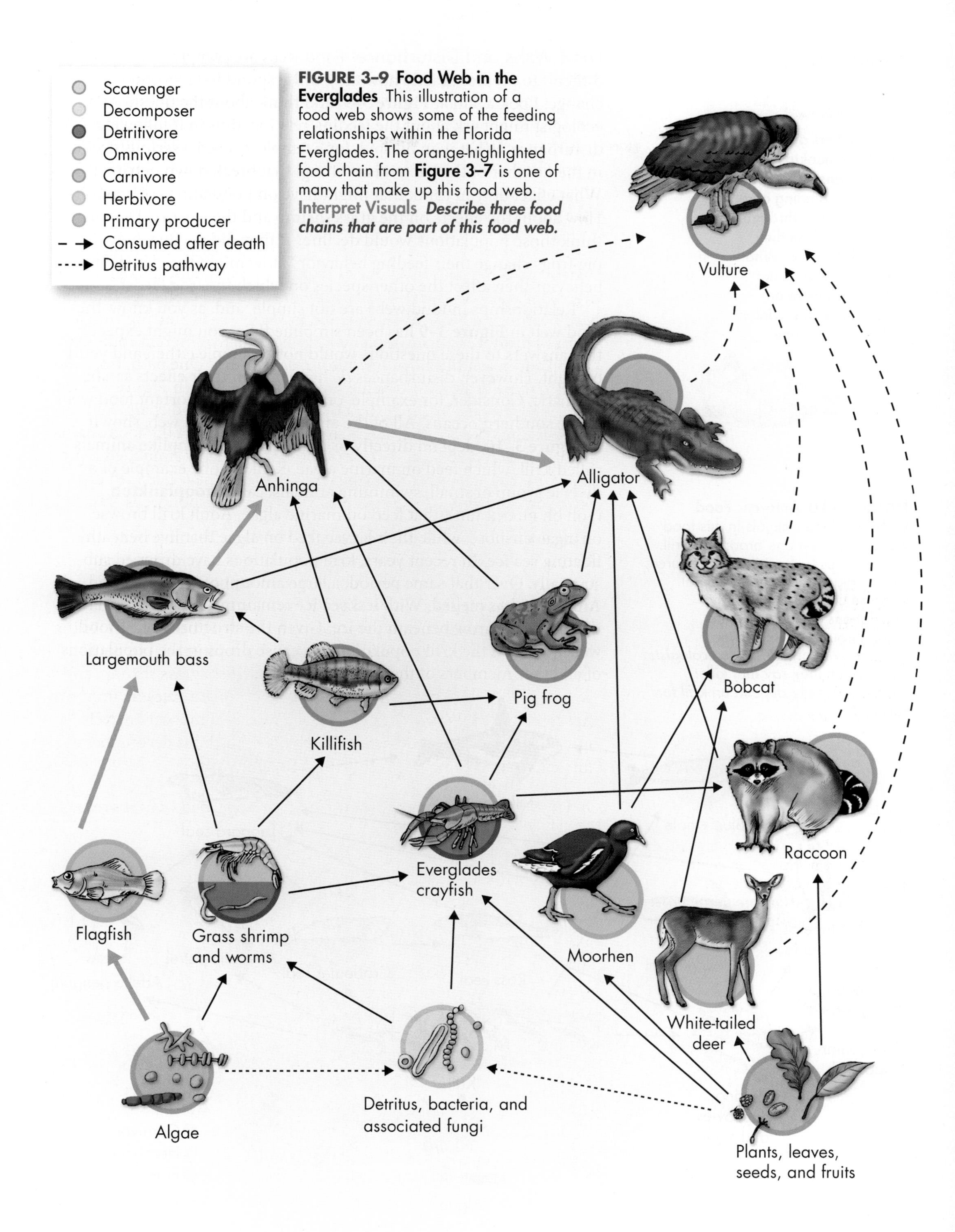

FIGURE 3–9 Food Web in the Everglades This illustration of a food web shows some of the feeding relationships within the Florida Everglades. The orange-highlighted food chain from **Figure 3–7** is one of many that make up this food web. **Interpret Visuals** ***Describe three food chains that are part of this food web.***

Food Webs and Disturbance Food webs are complex, so it is often difficult to predict exactly how they will respond to environmental change. Look again at **Figure 3–9,** and think about the questions an ecologist might ask about the feeding relationships in it following a disturbance. What if an oil spill, for example, caused a serious decline in the number of the bacteria and fungi that break down detritus? What effect do you think that might have on populations of crayfish? How about the effects on the grass shrimp and the worms? Do you think those populations would decline? If they did decline, how might pig frogs change their feeding behavior? How might the change in frog behavior then affect the other species on which the frog feeds?

Relationships in food webs are not simple, and, as you know, the food web in **Figure 3–9** has been simplified! So, you might expect that answers to these questions would not be simple either, and you'd be right. However, disturbances *do* happen, and their effects can be dramatic. Consider, for example, one of the most important food webs in the southern oceans. All of the animals in this food web, shown in **Figure 3–10,** depend directly or indirectly on shrimplike animals called krill, which feed on marine algae. Krill are one example of a diverse group of small, swimming animals, called **zooplankton** (zoh oh PLANK tun), that feed on marine algae. Adult krill browse on algae offshore, while their larvae feed on algae that live beneath floating sea ice. In recent years, krill populations have dropped substantially. Over that same period, a large amount of sea ice around Antarctica has melted. With less sea ice remaining, there are fewer of the algae that grow beneath the ice. Given the structure of this food web, a drop in the krill population can cause drops in the populations of all other members of the food web shown.

FIGURE 3–10 Antarctic Food Web All of the animals in this food web depend on one organism: krill. Disturbances to the krill's food source, marine algae, have the potential to cause changes in all of the populations connected to the algae through this food web. Interpret Visuals *What do ecologists mean when they say that killer whales indirectly depend on krill for survival?*

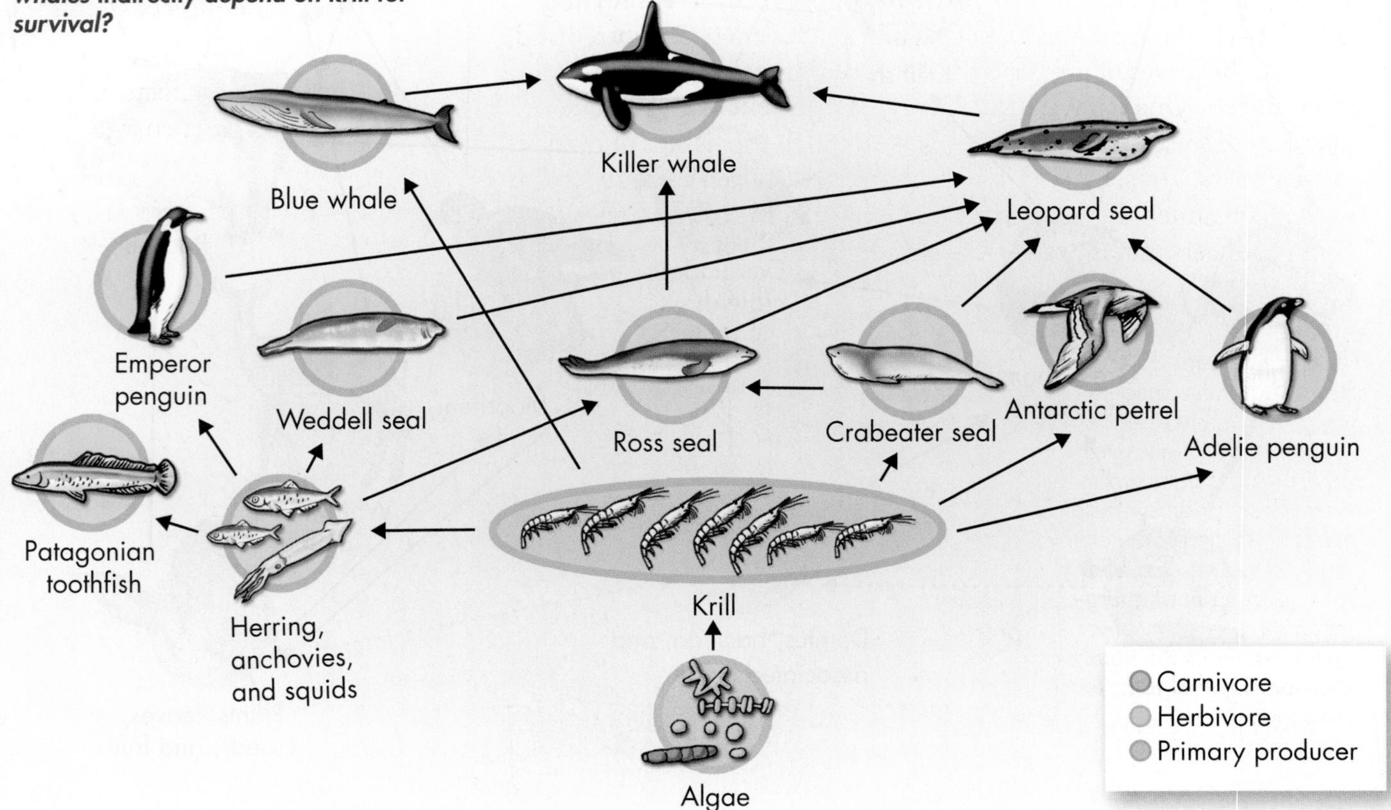

Trophic Levels and Ecological Pyramids

What do the three types of ecological pyramids illustrate?

Each step in a food chain or food web is called a **trophic level.** Primary producers always make up the first trophic level. Various consumers occupy every other level. One way to illustrate the trophic levels in an ecosystem is with an ecological pyramid. **Ecological pyramids** show the relative amount of energy or matter contained within each trophic level in a given food chain or food web. There are three different types of ecological pyramids: pyramids of energy, pyramids of biomass, and pyramids of numbers.

In Your Notebook *Make a two-column chart to compare the three types of ecological pyramids.*

Pyramids of Energy Theoretically, there is no limit to the number of trophic levels in a food web or the number of organisms that live on each level. But there is one catch. Only a small portion of the energy that passes through any given trophic level is ultimately stored in the bodies of organisms at the next level. This is because organisms expend much of the energy they acquire on life processes, such as respiration, movement, growth, and reproduction. Most of the remaining energy is released into the environment as heat—a byproduct of these activities. **Pyramids of energy show the relative amount of energy available at each trophic level of a food chain or food web.**

The efficiency of energy transfer from one trophic level to another varies. On average, about 10 percent of the energy available within one trophic level is transferred to the next trophic level, as shown in **Figure 3–11.** For instance, one tenth of the solar energy captured and stored in the leaves of grasses ends up stored in the tissues of cows and other grazers. One tenth of *that* energy—10 percent of 10 percent, or 1 percent of the original amount—gets stored in the tissues of humans who eat cows. Thus, the more levels that exist between a producer and a given consumer, the smaller the percentage of the original energy from producers that is available to that consumer.

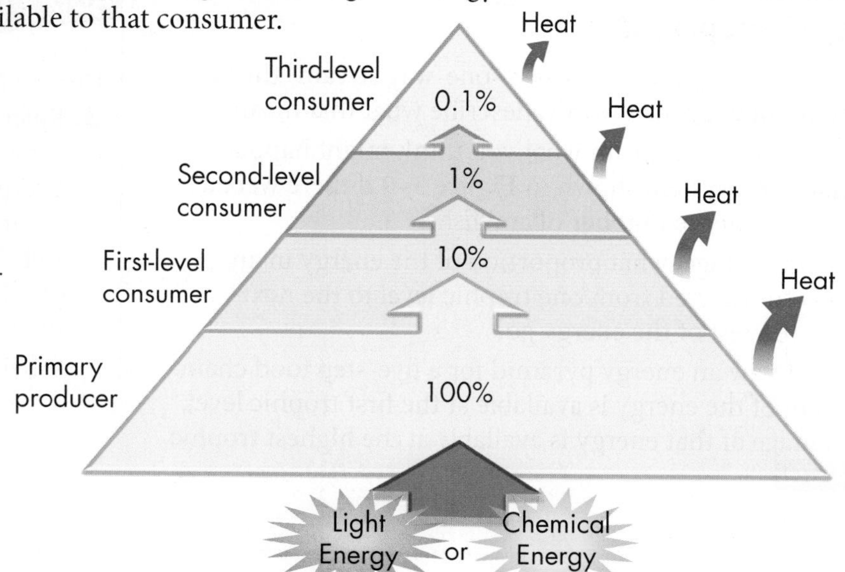

FIGURE 3–11 Pyramid of Energy Pyramids of energy show the relative amount of energy available at each trophic level. An ecosystem requires a constant supply of energy from photosynthetic or chemosynthetic producers. **Apply Concepts** ***Explain how the amount of energy available at each trophic level often limits the number of organisms that each level can support.***

Analyzing Data

The 10 Percent Rule

As shown in **Figure 3–11,** an energy pyramid is a diagram that illustrates the transfer of energy through a food chain or food web. In general, only 10 percent of the energy available in one level is stored in the level above. Look at **Figure 3–11** and answer the questions below.

1. Calculate If there are 1000 units of energy available at the producer level of the energy pyramid, approximately how many units of energy are available to the third-level consumer? MATH

2. Interpret Diagrams What is the original source of the energy that flows through most ecosystems? Why must there be a continuous supply of energy into the ecosystem?

3. Infer Why are there usually fewer organisms in the top levels of an energy pyramid?

FIGURE 3–12 Pyramids of Biomass and Numbers In most cases, pyramids of biomass and numbers follow the same general pattern. In the field modeled here, there are more individual primary producers than first-level consumers. Likewise, the primary producers collectively have more mass. The same patterns hold for the second and third-level consumers. With each step to a higher trophic level, biomass and numbers decrease.

Pyramids of Biomass and Numbers The total amount of living tissue within a given trophic level is called its **biomass.** Biomass is usually measured in grams of organic matter per unit area. The amount of biomass a given trophic level can support is determined, in part, by the amount of energy available. **A pyramid of biomass illustrates the relative amount of living organic matter available at each trophic level in an ecosystem.**

Ecologists interested in the number of organisms at each trophic level uses a pyramid of numbers. **A pyramid of numbers shows the relative number of individual organisms at each trophic level in an ecosystem.** In most ecosystems, the shape of the pyramid of numbers is similar to the shape of the pyramid of biomass for the same ecosystem. In this shape, the numbers of individuals on each level decrease from the level below it. To understand this point more clearly, imagine that an ecologist marked off several square meters in a field, and then weighed and counted every organism in that area. The result might look something like the pyramid in **Figure 3–12.**

In some cases, however, consumers are much less massive than organisms they feed upon. Thousands of insects may graze on a single tree, for example, and countless mosquitos can feed off a few deer. Both the tree and deer have a lot of biomass, but they each represent only one organism. In such cases, the pyramid of numbers may be turned upside down, but the pyramid of biomass usually has the normal orientation.

3.3 Assessment

Review Key Concepts

1. a. Review Energy is said to flow in a "one-way stream" through an ecosystem. In your own words, describe what that means.

b. Form a Hypothesis Explain what you think might happen to the Everglades ecosystem shown in **Figure 3–9** if there were a sudden decrease in the number of crayfish.

2. a. Review On average, what proportion of the energy in an ecosystem is transferred from one trophic level to the next? Where does the rest of the energy go?

b. Calculate Draw an energy pyramid for a five-step food chain. If 100 percent of the energy is available at the first trophic level, what percentage of that energy is available at the highest trophic level? MATH

Apply the Big idea

Interdependence In Nature

3. Refer to **Figure 3–9,** which shows a food web in the Everglades. Choose one of the food chains within the web. Then, write a paragraph describing the feeding relationships among the organisms in the food chain.

3.4 Cycles of Matter

THINK ABOUT IT Living organisms are composed mostly of four elements: oxygen, carbon, hydrogen, and nitrogen. These four elements (and a few others, such as sulfur and phosphorus) are the basis of life's most important compounds: water, carbohydrates, lipids, nucleic acids, and proteins. In short, a handful of elements combine to form the building blocks of all known organisms. And yet, organisms cannot manufacture these elements and do not "use them up." So, where do essential elements come from? How does their availability affect ecosystems?

Recycling in the Biosphere

How does matter move through the biosphere?

Matter moves through the biosphere differently than the way in which energy moves. Solar and chemical energy are captured by primary producers and then pass in a one-way fashion from one trophic level to the next—dissipating in the environment as heat along the way. But while energy in the form of sunlight is constantly entering the biosphere, Earth doesn't receive a significant, steady supply of new matter from space. **Unlike the one-way flow of energy, matter is recycled within and between ecosystems.** Elements pass from one organism to another and among parts of the biosphere through closed loops called **biogeochemical cycles,** which are powered by the flow of energy as shown in **Figure 3–13.** As that word suggests, cycles of matter involve *bio*logical processes, *geo*logical processes, and *chem*ical processes. Human activity can also play an important role. As matter moves through these cycles, it is transformed. It is never created or destroyed—just changed.

Key Questions

How does matter move through the biosphere?

How does water cycle through the biosphere?

What is the importance of the main nutrient cycles?

How does nutrient availability relate to the primary productivity of an ecosystem?

Vocabulary

biogeochemical cycle • nutrient • nitrogen fixation • denitrification • limiting nutrient

Taking Notes

Outline Make an outline using the green and blue headings in this lesson. Fill in details as you read to help you organize the information.

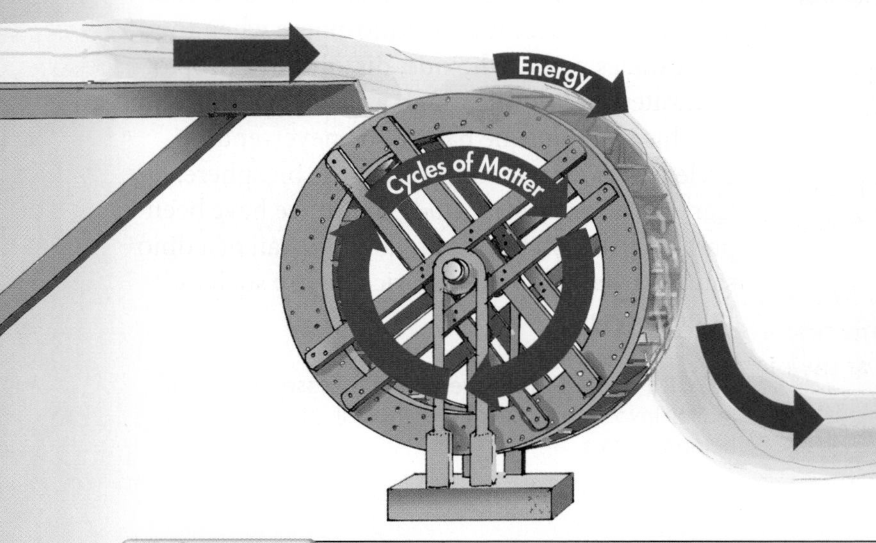

VISUAL ANALOGY

THE MATTER MILL

FIGURE 3–13 Nutrients are recycled through biogeochemical cycles. These cycles are powered by the one-way flow of energy through the biosphere.
Use Analogies ***How is the water flowing over the water wheel similar to the flow of energy in the biosphere?***

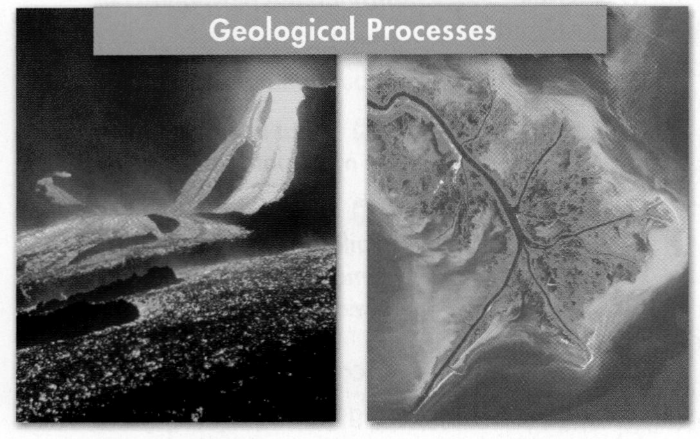

There are many ways in which the processes involved in biogeochemical cycles can be classified. Here, we will use the following guidelines:

▶ ***Biological Processes*** Biological processes consist of any and all activities performed by living organisms. These processes include eating, breathing, "burning" food, and eliminating waste products.

▶ ***Geological Processes*** Geological processes include volcanic eruptions, the formation and breakdown of rock, and major movements of matter within and below the surface of the earth.

▶ ***Chemical and Physical Processes*** Chemical and physical processes include the formation of clouds and precipitation, the flow of running water, and the action of lightning.

▶ ***Human Activity*** Human activities that affect cycles of matter on a global scale include the mining and burning of fossil fuels, the clearing of land for building and farming, the burning of forests, and the manufacture and use of fertilizers.

These processes, shown in **Figure 3–14,** pass the same atoms and molecules around again and again. Imagine, for a moment, that you are a carbon atom in a molecule of carbon dioxide that has just been shot out of a volcano. The leaf of a blueberry bush in a nearby mountain range absorbs you during photosynthesis. You become part of a carbohydrate molecule in a blueberry. A caribou eats the fruit, and within a few hours, you pass out of the animal's body. You are soon swallowed by a dung beetle, which gets eaten by a hungry shrew. You are combined into the body tissues of the shrew, which is then eaten by an owl. You are released back into the atmosphere when the owl exhales carbon dioxide, dissolve in a drop of rainwater, and flow through a river into the ocean.

This could just be part of the never-ending cycle of a carbon atom through the biosphere. Carbon atoms in your body may once have been part of a rock on the ocean floor, the tail of a dinosaur, or even part of a historical figure such as Julius Caesar!

FIGURE 3–14 Biogeochemical Processes Cycles of matter involve biological, geological, chemical, and human factors.

The Water Cycle

How does water cycle through the biosphere?

Every time you see rain or snow, or watch a river flow, you are witnessing part of the water cycle. **Water continuously moves between the oceans, the atmosphere, and land—sometimes outside living organisms and sometimes inside them.** As **Figure 3–15** shows, water molecules typically enter the atmosphere as water vapor, a gas, when they evaporate from the ocean or other bodies of water. Water can also enter the atmosphere by evaporating from the leaves of plants in the process of transpiration (tran spuh RAY shun).

Water vapor may be transported by winds over great distances. If the air carrying it cools, water vapor condenses into tiny droplets that form clouds. When the droplets become large enough, they fall to Earth's surface as precipitation in the form of rain, snow, sleet, or hail. On land, some precipitation flows along the surface in what scientists call runoff, until it enters a river or stream that carries it to an ocean or lake. Precipitation can also be absorbed into the soil and is then called groundwater. Groundwater can enter plants through their roots, or flow into rivers, streams, lakes, or oceans. Some groundwater penetrates deeply enough into the ground to become part of underground reservoirs. Water that re-enters the atmosphere through transpiration or evaporation begins the cycle anew.

In Your Notebook *Define each of the following terms and describe how they relate to the water cycle:* evaporation, transpiration, precipitation, *and* runoff.

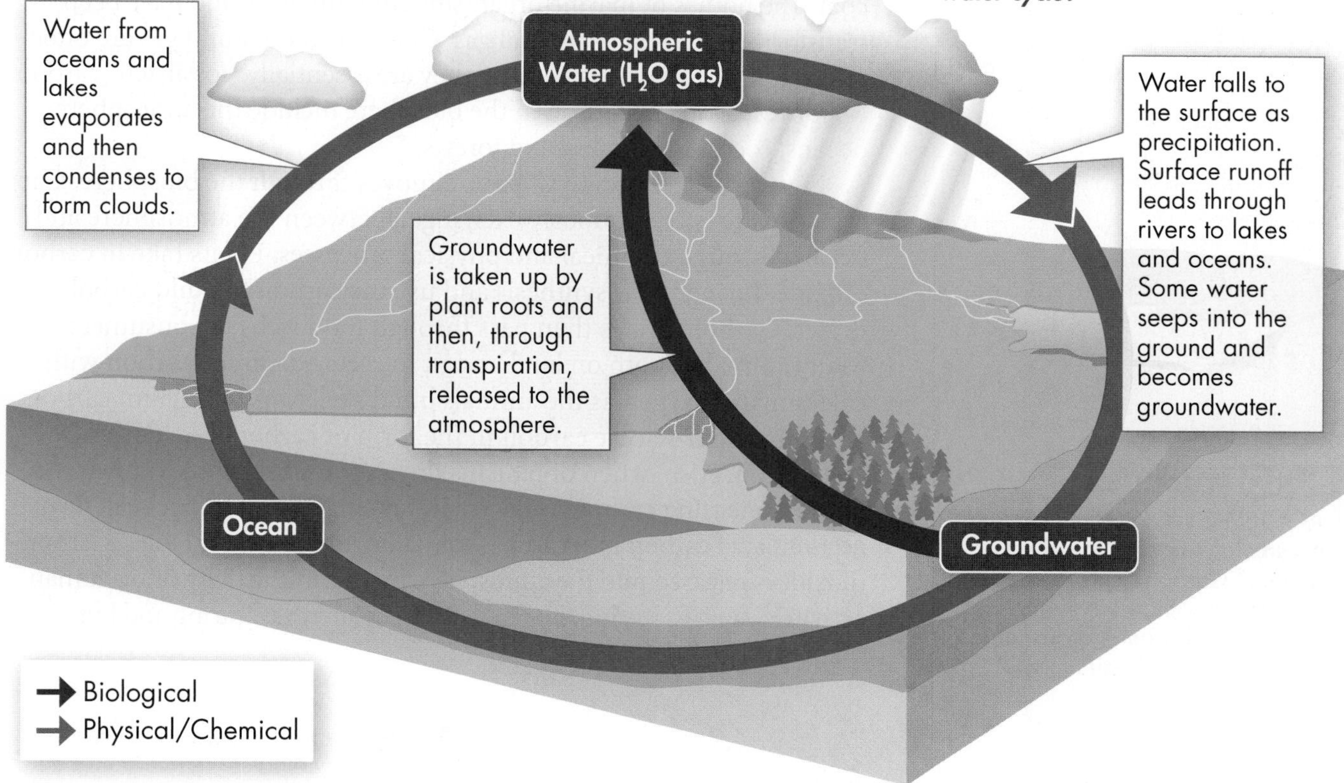

FIGURE 3–15 The Water Cycle This diagram shows the main processes involved in the water cycle. Scientists estimate that it can take a single water molecule as long as 4000 years to complete one cycle. **Interpret Visuals** ***What are the two primary ways in which water that falls to Earth as precipitation passes through the water cycle?***

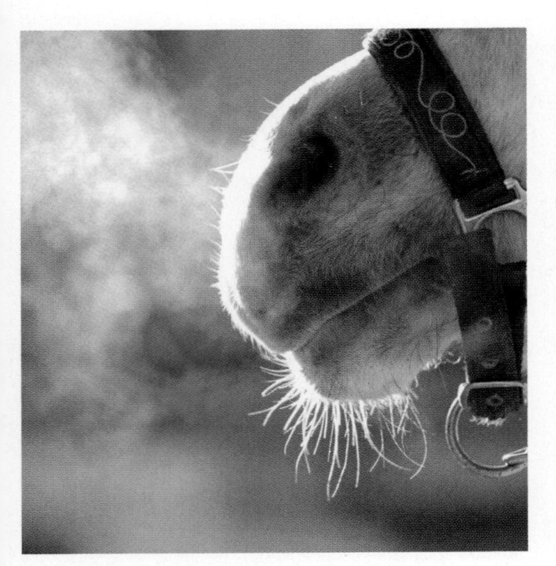

FIGURE 3–16 Oxygen in the Biosphere The oxygen contained in the carbon dioxide exhaled by this horse may be taken up by producers and re-released as oxygen gas. Together, respiration and photosynthesis contribute to oxygen's cycling through the biosphere.

Nutrient Cycles

What is the importance of the main nutrient cycles?

The chemical substances that an organism needs to sustain life are called **nutrients.** **Every organism needs nutrients to build tissues and carry out life functions. Like water, nutrients pass through organisms and the environment through biogeochemical cycles. The three pathways, or cycles that move carbon, nitrogen, and phosphorus through the biosphere are especially critical for life.**

Another element, oxygen, participates in parts of the carbon, nitrogen, and phosphorus cycles by combining with these elements and cycling with them through parts of their journeys. Oxygen gas in the atmosphere is released by one of the most important of all biological activities: photosynthesis. Oxygen is used in respiration by all multicellular forms of life, and many single-celled organisms as well.

The Carbon Cycle Carbon is a major component of all organic compounds, including carbohydrates, lipids, proteins, and nucleic acids. In fact, carbon is such a key ingredient of living tissue and ecosystems that life on Earth is often described as "carbon-based life." Carbon in the form of calcium carbonate ($CaCO_3$) is an important component of many different kinds of animal skeletons and is also found in several kinds of rocks. Carbon and oxygen form carbon dioxide gas (CO_2), which is an important component of the atmosphere and is dissolved in oceans.

Some carbon-containing compounds that were once part of ancient forests have been buried and transformed by geological processes into coal. The bodies of marine organisms containing carbon have been transformed into oil or natural gas. Coal, oil, and natural gas are often referred to as fossil fuels because they are essentially "fossilized" carbon. Major reservoirs of carbon in the biosphere include the atmosphere, oceans, rocks, fossil fuels, and forests.

Figure 3–17 shows how carbon moves through the biosphere. Carbon dioxide is continuously exchanged between the atmosphere and oceans through chemical and physical processes. Plants take in carbon dioxide during photosynthesis and use the carbon to build carbohydrates. Carbohydrates then pass through food webs to consumers. Many animals—both on land and in the sea—combine carbon with calcium and oxygen as the animals build skeletons of calcium carbonate. Organisms release carbon in the form of carbon dioxide gas by respiration. Also, when organisms die, decomposers break down the bodies, releasing carbon to the environment. Geologic forces can turn accumulated carbon into carbon-containing rocks or fossil fuels. Carbon dioxide is released into the atmosphere by volcanic activity or by human activities, such as the burning of fossil fuels and the clearing and burning of forests.

BUILD Vocabulary

ACADEMIC WORDS The verb **accumulate** means "to collect or gather." Carbon accumulates, or collects, in soil and in the oceans where it cycles among organisms or is turned into fossil fuels.

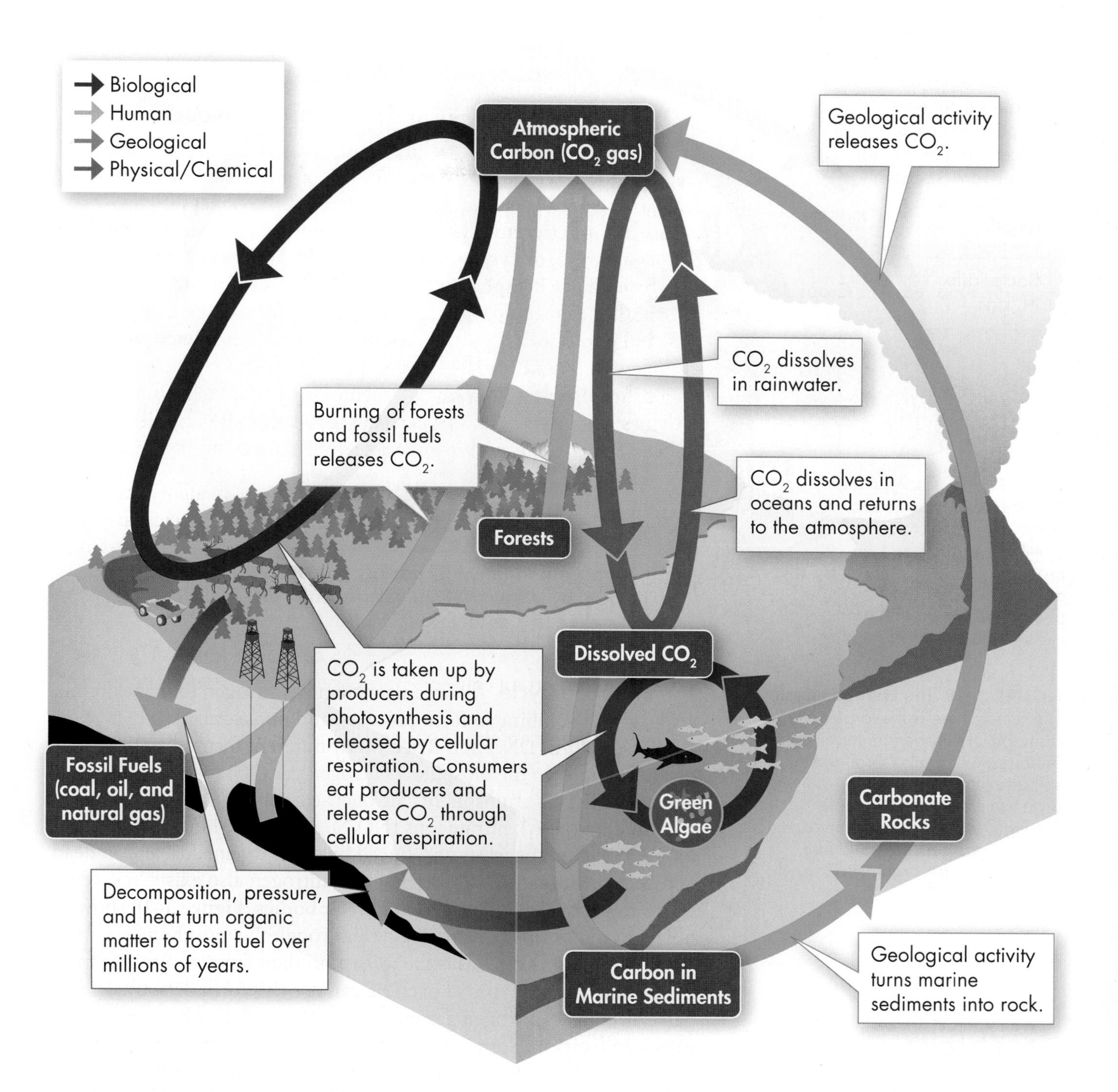

FIGURE 3–17 The Carbon Cycle Carbon is found in several large reservoirs in the biosphere. In the atmosphere, it is found as carbon dioxide gas (CO_2); in the oceans, as dissolved carbon dioxide; on land, in organisms, rocks, and soil; and underground, as coal, petroleum, and calcium carbonate. **Interpret Visuals** ***What is one of the processes that takes carbon dioxide out of the atmosphere?***

Scientists know a great deal about the biological, geological, chemical, and human processes that are involved in the carbon cycle, but important questions remain. How much carbon moves through each pathway? How do ecosystems respond to changes in atmospheric carbon dioxide concentration? How much carbon dioxide can the ocean absorb? Later in this unit, you will learn why answers to these questions are so important.

In Your Notebook *Describe one biological, one geological, one chemical, and one human activity that is involved in the carbon cycle.*

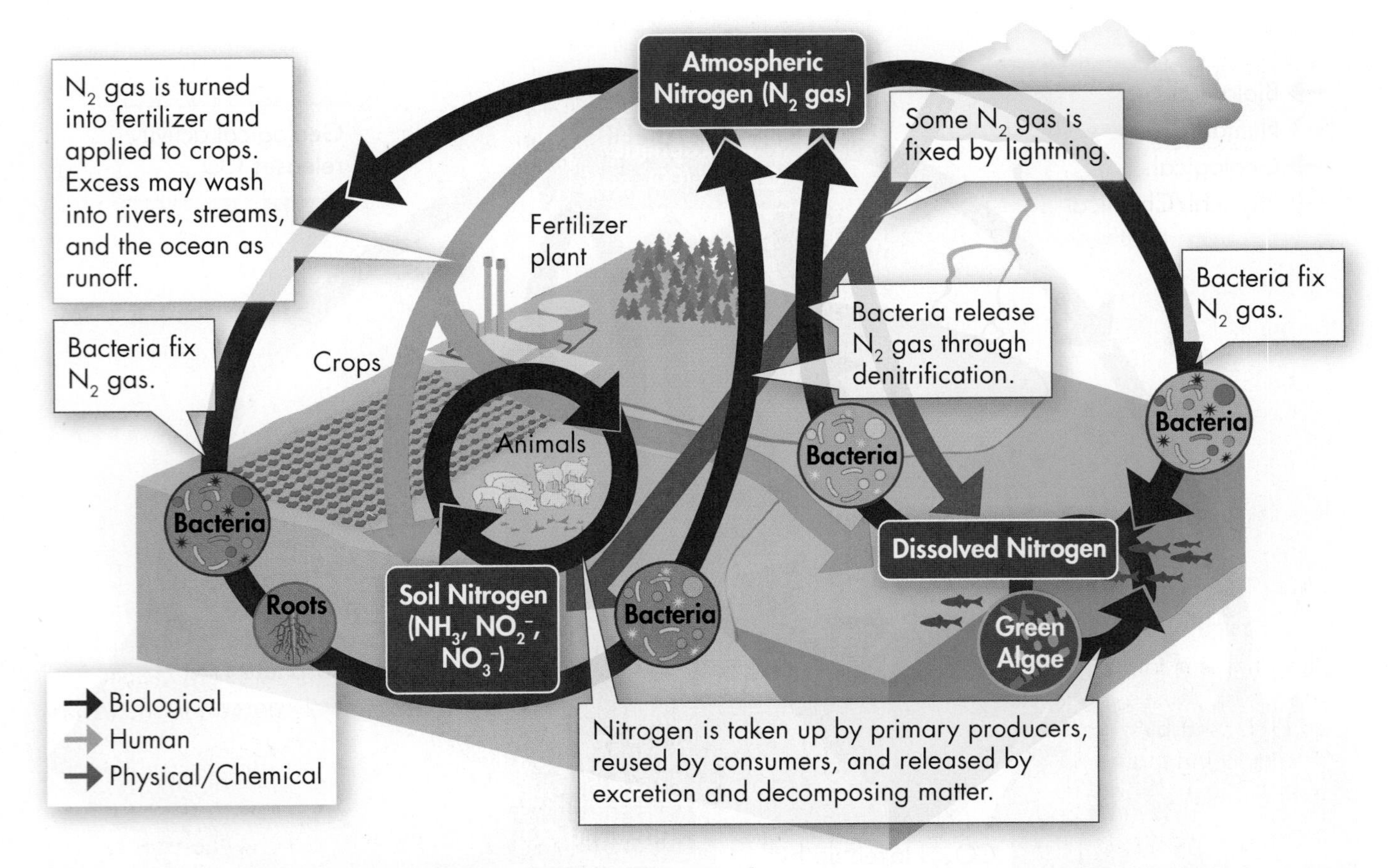

FIGURE 3–18 The Nitrogen Cycle The atmosphere is the largest reservoir of nitrogen in the biosphere. Nitrogen also cycles through the soil and through the tissues of living organisms. **Interpret Visuals** ***Through which two processes does nitrogen gas get converted into usable forms for organisms?***

The Nitrogen Cycle All organisms require nitrogen to make amino acids, which combine to form proteins, and nucleic acids, which combine to form DNA and RNA. Many different forms of nitrogen occur naturally in the biosphere. Nitrogen gas (N_2) makes up 78 percent of Earth's atmosphere. Nitrogen-containing substances such as ammonia (NH_3), nitrate ions (NO_3^-), and nitrite ions (NO_2^-) are found in soil, in the wastes produced by many organisms, and in dead and decaying organic matter. Dissolved nitrogen also exists in several forms in the ocean and other large water bodies. **Figure 3–18** shows how different forms of nitrogen cycle through the biosphere.

Although nitrogen gas is the most abundant form of nitrogen on Earth, only certain types of bacteria can use this form directly. These bacteria convert nitrogen gas into ammonia, a process known as **nitrogen fixation.** Some of these nitrogen-fixing bacteria live in the soil and on the roots of certain plants, such as peanuts and peas, called legumes. Other bacteria convert that fixed nitrogen into nitrates and nitrites. Once these forms of nitrogen are available, primary producers can use them to make proteins and nucleic acids. Consumers eat the producers and reuse nitrogen to make their own nitrogen-containing compounds. Decomposers release nitrogen from waste and dead organisms as ammonia, nitrates, and nitrites that producers may take up again. Other bacteria obtain energy by converting nitrates into nitrogen gas, which is released into the atmosphere in a process called **denitrification.** A relatively small amount of nitrogen gas is converted to usable forms by lightning in a process called atmospheric nitrogen fixation. Humans add nitrogen to the biosphere through the manufacture and use of fertilizers. Excess fertilizer is often carried into surface water or groundwater by precipitation.

MYSTERY CLUE

Recently, researchers discovered that levels of dissolved nitrogen in the bay have increased. Given that human activity hasn't changed much, which organisms in the bay do you think might be responsible?

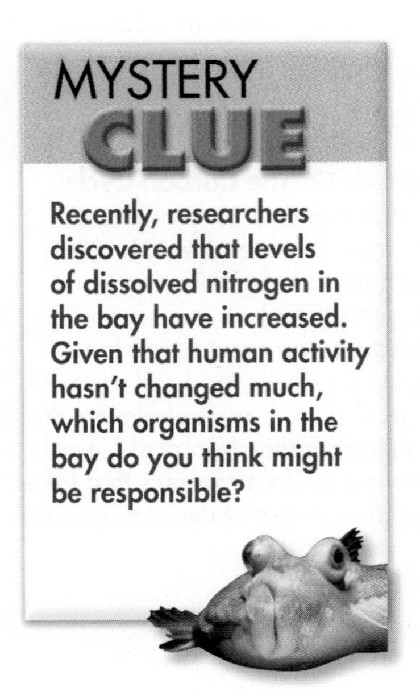

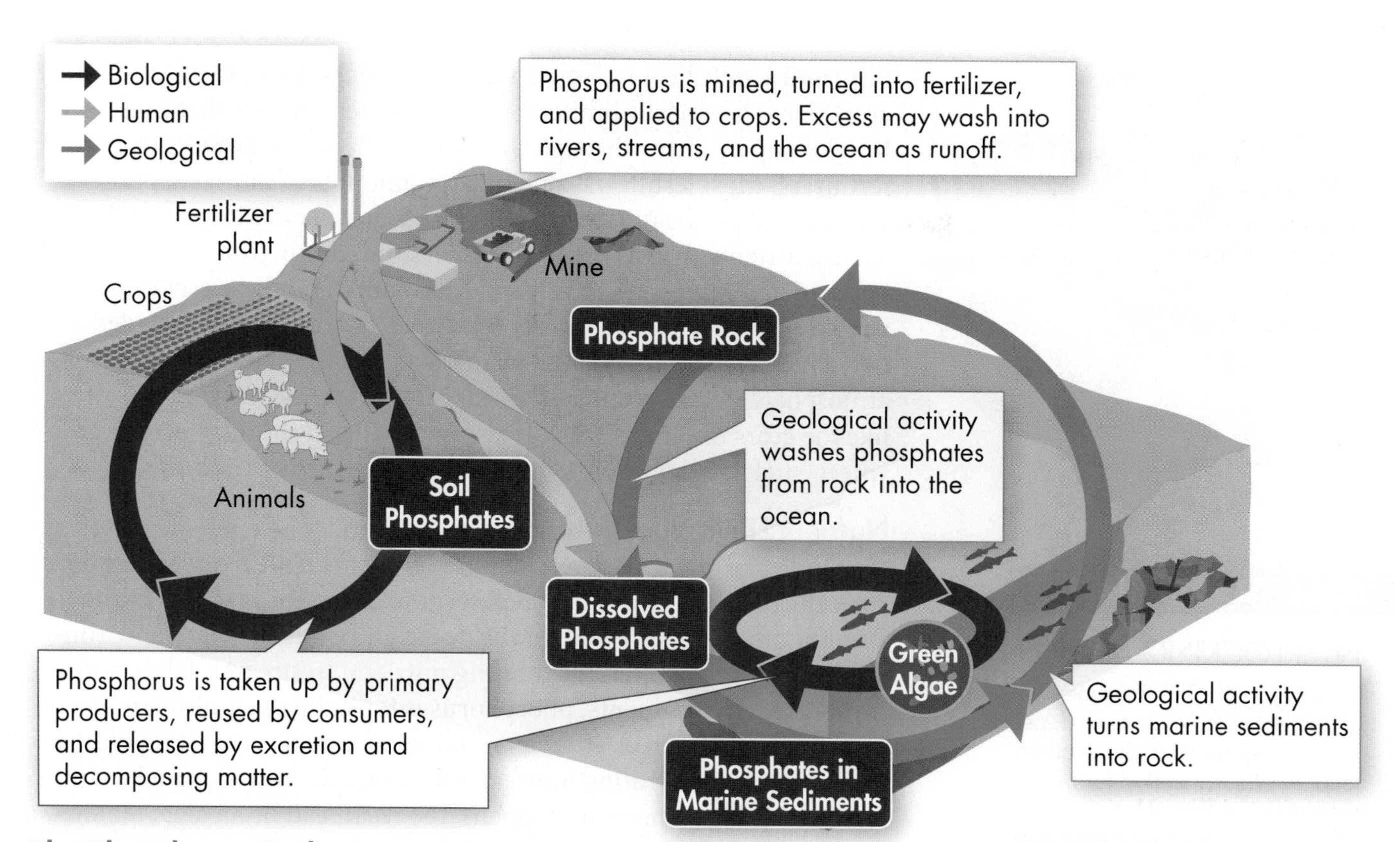

FIGURE 3–19 The Phosphorus Cycle Phosphorus in the biosphere cycles among the land, ocean sediments, and living organisms. Unlike other nutrients, phosphorus is not found in significant quantities in the atmosphere.

The Phosphorus Cycle Phosphorus is essential to living organisms because it forms a part of vital molecules such as DNA and RNA. Although phosphorus is of great biological importance, it is not abundant in the biosphere. Unlike carbon, oxygen, and nitrogen, phosphorus does not enter the atmosphere in significant amounts. Instead, phosphorus in the form of inorganic phosphate remains mostly on land, in the form of phosphate rock and soil minerals, and in the ocean, as dissolved phosphate and phosphate sediments, as seen in **Figure 3–19.**

As rocks and sediments gradually wear down, phosphate is released. Some phosphate stays on land and cycles between organisms and soil. Plants bind phosphate into organic compounds when they absorb it from soil or water. Organic phosphate moves through the food web, from producers to consumers, and to the rest of the ecosystem. Other phosphate washes into rivers and streams, where it dissolves. This phosphate may eventually makes its way to the ocean, where marine organisms process and incorporate it into biological compounds.

Nutrient Limitation

How does nutrient availability relate to the primary productivity of an ecosystem?

Ecologists are often interested in an ecosystem's primary productivity—the rate at which primary producers create organic material. **If ample sunlight and water are available, the primary productivity of an ecosystem may be limited by the availability of nutrients.** If even a single essential nutrient is in short supply, primary productivity will be limited. The nutrient whose supply limits productivity is called the **limiting nutrient.**

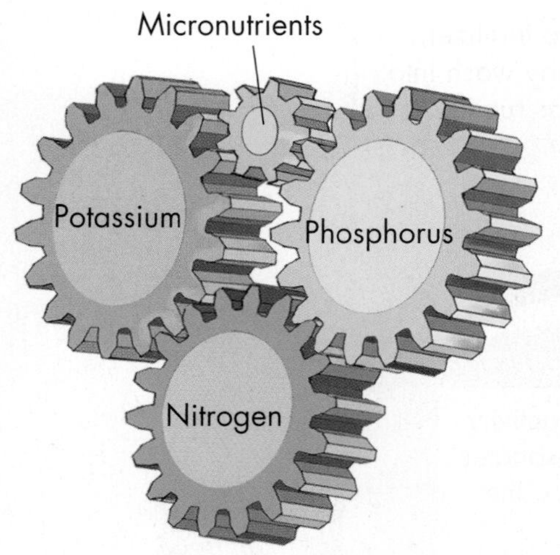

VISUAL ANALOGY

INTERLOCKING NUTRIENTS

FIGURE 3–20 The movement of each nutrient through ecosystems depends on the movements of all the others, because all are needed for living systems to function. **Use Analogies** ***If these gears were modeling nutrient cycling in the ocean, which gear would typically determine how quickly—or slowly—all the other gears turn?***

Nutrient Limitation in Soil In all but the richest soil, the growth of crop plants is typically limited by one or more nutrients that must be taken up by plants through their roots. That's why farmers use fertilizers! Most fertilizers contain large amounts of nitrogen, phosphorus, and potassium, which help plants grow better in poor soil. Micronutrients such as calcium, magnesium, sulfur, iron, and manganese are necessary in relatively small amounts, and these elements are sometimes included in specialty fertilizers. (Carbon is not included in chemical fertilizers because plants acquire carbon dioxide from the atmosphere during photosynthesis.) All nutrient cycles work together like the gears in **Figure 3–20.** If any nutrient is in short supply—if any wheel "sticks"—the whole system slows down or stops altogether.

Nutrient Limitation in Aquatic Ecosystems The open oceans of the world are nutrient-poor compared to many land areas. Seawater typically contains only 0.00005 percent nitrogen, or 1/10,000 of the amount often found in soil. In the ocean and other saltwater environments, nitrogen is often the limiting nutrient. In streams, lakes, and freshwater environments, phosphorus is typically the limiting nutrient.

Sometimes, such as after heavy rains, an aquatic ecosystem receives a large input of a limiting nutrient—for example, runoff from heavily fertilized fields. When this happens, the result can be an algal bloom—a dramatic increase in the amount of algae and other primary producers. Why can runoff from fertilized fields produce algal blooms? More nutrients are available, so producers can grow and reproduce more quickly. If there are not enough consumers to eat the algae, an algal bloom can occur, in which case algae can cover the water's surface and disrupt the functioning of an ecosystem.

3.4 Assessment

Review Key Concepts

1. a. Review How does the way that matter flows through an ecosystem differ from the way that energy flows?

b. Apply Concepts What are the four types of processes that cycle matter through the biosphere? Give an example of each.

2. a. Review By what two processes is water cycled from land to the atmosphere?

b. Sequence Describe one way in which water from the ocean may make one complete cycle through the atmosphere and back to the ocean. Include the names of each process involved in your cycle.

3. a. Review Why do living organisms need nutrients?

b. Predict Based on your knowledge of the carbon cycle, what do you think might happen if humans were to continue to clear and burn vast areas of forests for building?

4. a. Review Explain how a nutrient can be a limiting factor in an ecosystem.

b. Apply Concepts Look back at the nitrogen and phosphorus cycles (**Figures 3–18** and **3–19**). How is fertilizer runoff related to algal blooms?

WRITE ABOUT SCIENCE

Explanation

5. Describe how oxygen, although it does not have an independent cycle, moves through the biosphere as part of the carbon cycle. Include a description of the various forms that oxygen takes.

WHST.9-10.7 Production and Distribution of Writing, WHST.9-10.10 Range of Writing, RST.9-10.10 Level of Text Complexity

Technology & BIOLOGY

Global Ecology From Space

Can ecologists track plant growth around the world? Can they follow temperature change in oceans from day to day, or the amount of polar ice from year to year? Yes! Satellites can provide these data, essential for understanding global ecology. Satellite sensors can be programmed to scan particular bands of the electromagnetic spectrum to reveal global patterns of temperature, rainfall, or the presence of plants on land or algae in the oceans. The resulting false-color images are both beautiful and filled with vital information.

Changes in Polar Ice Cover Sea ice around the North Pole has been melting more each summer since satellites began gathering data in 1979. The image below shows in white the amount of ice remaining at the end of the summer in 2007. The amount of ice at the same time of year for an average year between 1979 and 2007 is shown in green.

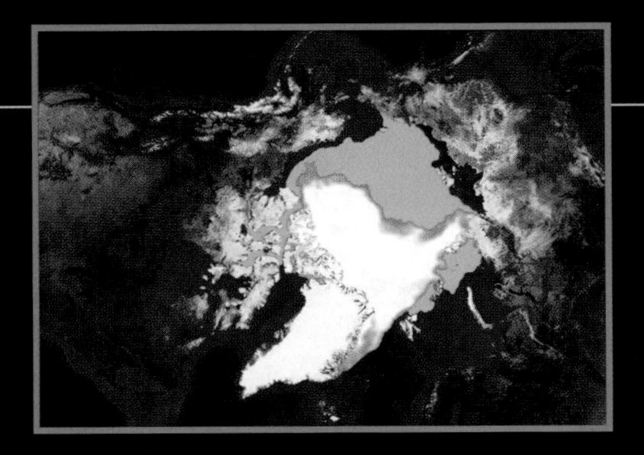

2007 White areas show the average minimum amount of arctic ice cover at the end of the summer, 2007.
1979–2007 Green areas show the average minimum amount of ice cover between 1979 and 2007.

Plant and Algal Growth These data were gathered by NASA's Sea-viewing Wide Field-of-view Sensor (SeaWiFS), which is programmed to monitor the color of reflected light. In the image below, you can see how actively plants on land and algae in the oceans were harnessing solar energy for photosynthesis when these data were taken. A measurement of photosynthesis gives a measure of growth rates and the input of energy and nutrients into the ecosystem.

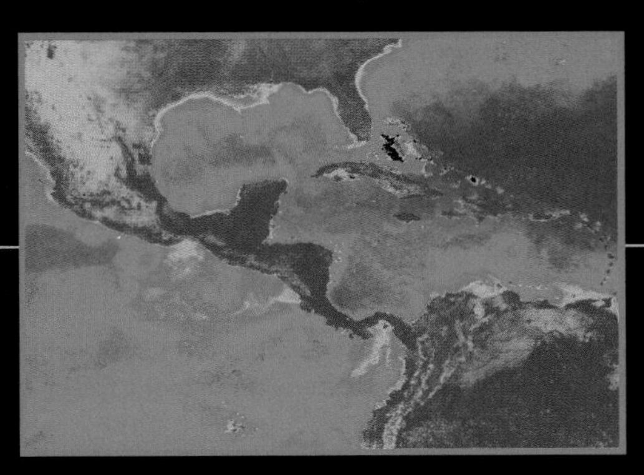

On Land Dark green indicates active plant growth; yellow areas indicate barren deserts or mountains.
In the Sea Dark blue indicates very low active growth of algae. Red indicates the highest active growth.

WRITING **Visit the Web site for the Goddard Space Flight Center Scientific Visualization service and select a set of satellite data to examine. Write a brief paragraph explaining what you learned from looking at those data.**

NGSS Smart Guide The Biosphere

Disciplinary Core Idea LS2.B Cycles of Matter and Energy Transfer in Ecosystems: How do matter and energy move through an ecosystem? Energy flows in one direction through ecosystems from producers to consumers. In contrast, matter is recycled within and between ecosystems.

3.1 What Is Ecology?

- Ecology is the scientific study of interactions among organisms and between organisms and their physical environment.
- The biological influences on organisms are called biotic factors.
- Physical components of an ecosystem are called abiotic factors.
- Modern ecologists use three methods in their work: observation, experimentation, and modeling. Each of these approaches relies on scientific methodology to guide inquiry.

biosphere 64 • species 64 • population 64 • community 64 • ecology 65 • ecosystem 65 • biome 65 • biotic factor 66 • abiotic factor 66

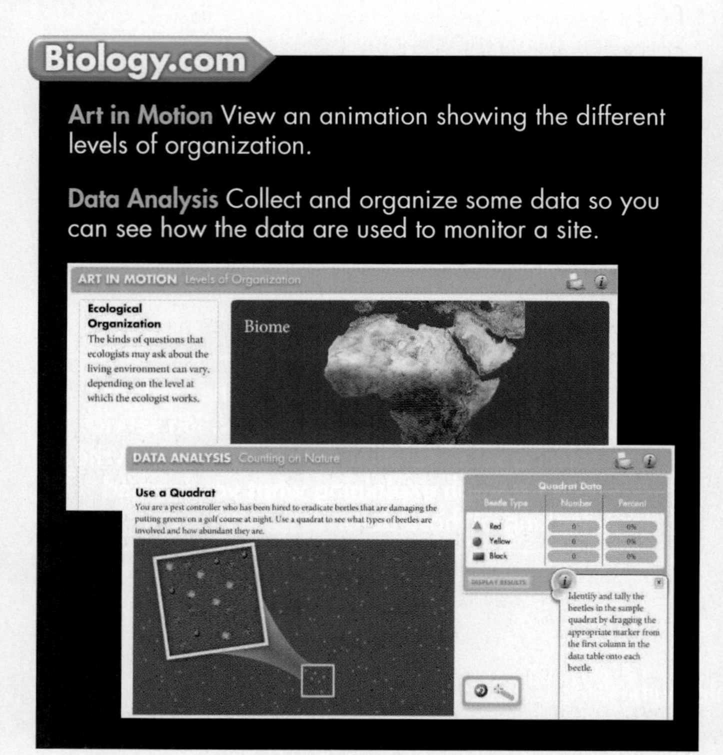

Biology.com

Art in Motion View an animation showing the different levels of organization.

Data Analysis Collect and organize some data so you can see how the data are used to monitor a site.

3.2 Energy, Producers, and Consumers

- Primary producers are the first producers of energy-rich compounds that are later used by other organisms.
- Organisms that rely on other organisms for energy and nutrients are called consumers.

autotroph 69 • primary producer 69 • photosynthesis 70 • chemosynthesis 70 • heterotroph 71 • consumer 71 • carnivore 71 • herbivore 71 • scavenger 71 • omnivore 71 • decomposer 71 • detritivore 71

Art Review Review your understanding of which organisms are producers and which are consumers with this drag-and-drop activity.

Tutor Tube Get some clarification on producers and consumers and learn how the flow of matter and energy is not what you may think!

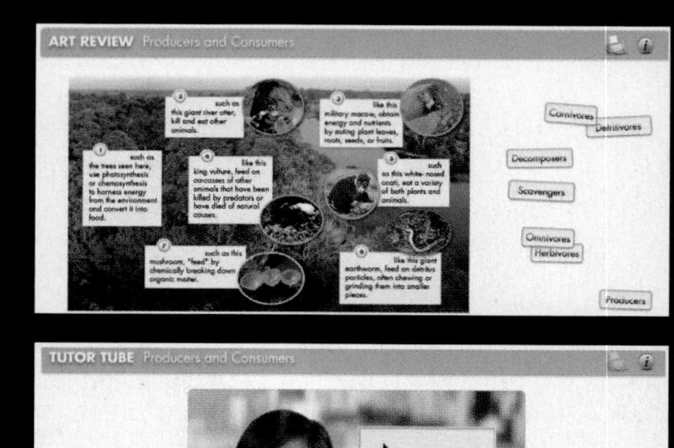

CHECKING YOUR *Scientific Literacy*

Refer to the lesson content and digital assets below as you prepare for your chapter assessment. Then, evaluate your understanding of the biosphere by answering these questions.

1. **Scientific and Engineering Practice Analyzing Data** Construct a table with a row for each cycle of matter (water, carbon, nitrogen, and phosphorus) and a column for each process (physical/chemical, biological, geological, and human). Fill in the table with examples of each process, using the text and figures in Lesson 3.4.

2. **Crosscutting Concept Energy and Matter** White-tailed deer eat plants. Anhingas eat animals. What percentage of the energy originally captured by primary producers is available to each of these animals? Explain your answer. Hint: See the food web in Lesson 3.3.

3. **Craft and Structure** Why are normally unseen members of the food web, such as soil microorganisms, essential to the nitrogen cycle? Hint: Review the nitrogen cycle in Figure 3-18.

4. **STEM** The owners of a game park in Africa would like their fenced-in natural reserve to house impala, warthogs, and other grazers, along with carnivores such as cheetahs and lions—which would prey on the grazers. Research and present the biotic and abiotic factors that need to be considered when determining how much room and resources are necessary for those animals.

3.3 Energy Flow in Ecosystems

- Energy flows through an ecosystem in a one-way stream, from primary producers to various consumers.
- Pyramids of energy show the relative amount of energy available at each trophic level of a food chain or food web. A pyramid of biomass illustrates the relative amount of living organic matter available at each trophic level of an ecosystem. A pyramid of numbers shows the relative number of individual organisms at each trophic level in an ecosystem.

food chain 73 • phytoplankton 73 • food web 74 • zooplankton 76 • trophic level 77 • ecological pyramid 77 • biomass 78

Biology.com

Untamed Science Video Help the Untamed Science crew explore food relationships as they turn the ecological pyramid upside down.

Visual Analogy Compare a recycling center to decomposers in this activity.

3.4 Cycles of Matter

- Unlike the one-way flow of energy, matter is recycled within and between ecosystems.
- Water continuously moves between the oceans, the atmosphere, and land—sometimes outside living organisms and sometimes inside them.
- Every organism needs nutrients to build tissues and carry out life functions. Like water, nutrients pass through organisms and the environment through biogeochemical cycles. The carbon, nitrogen, and phosphorus cycles are especially critical for life.
- If ample sunlight and water are available, the primary productivity of an ecosystem may be limited by the availability of nutrients.

biogeochemical cycle 79 • nutrient 82 • nitrogen fixation 84 • denitrification 84 • limiting nutrient 85

Biology.com

Visual Analogy Compare nutrient limitation to a series of cogs in this activity.

Real-World Lab

The Effect of Fertilizer on Algae This chapter lab is available in your lab manual and at Biology.com

3 Assessment

3.1 What Is Ecology?

Understand Key Concepts

1. All of life on Earth exists in
   a. an ecosystem.
   b. a biome.
   c. the biosphere.
   d. ecology.

2. Which term describes a group of different species that live together in a defined area?
   a. a population
   b. a community
   c. an ecosystem
   d. a biosphere

3. Name the different levels of organization within the biosphere, from smallest to largest.

4. How do ecologists use modeling?

5. Give an example of how a biotic factor might influence the organisms in an ecosystem.

Think Critically

6. **Text Types and Purposes** Ecologists have discovered that the seeds of many plants that grow in forests cannot germinate unless they have been exposed to fire. Design an experiment to test whether a particular plant has seeds with this requirement. Include your hypothesis statement, a description of control and experimental groups, and an outline of your procedure.

7. **Pose Questions** You live near a pond that you have observed for years. One year you notice the water is choked with a massive overgrowth of green algae. What are some of the questions you might have about this unusual growth?

3.2 Energy, Producers, and Consumers

Understand Key Concepts

8. Primary producers are organisms that
   a. rely on other organisms for their energy and food supply.
   b. consume plant and animal remains and other dead matter.
   c. use energy they take in from the environment to convert inorganic molecules into complex organic molecules.
   d. obtain energy by eating only plants.

9. Which of the following organisms is a decomposer?

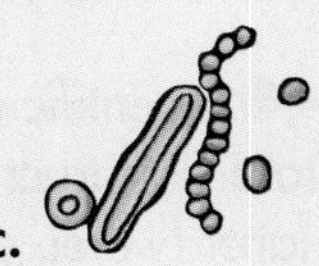

a. c.
b. d.

10. Which of the following describes how ALL consumers get their energy?
    a. directly from the sun
    b. from eating primary producers
    c. from inorganic chemicals like hydrogen sulfide
    d. from eating organisms that are living or were once living

11. What is chemosynthesis?

Think Critically

12. **Craft and Structure** Classify each of the following as an herbivore, a carnivore, an omnivore, or a detritivore: earthworm, bear, cow, snail, owl, human.

13. **Form a Hypothesis** People who explore caves where there is running water but no sunlight often find them populated with unique types of fishes and insects. What hypothesis can you make to explain the ultimate source of energy for these organisms?

3.3 Energy Flow in Ecosystems

Understand Key Concepts

14. The series of steps in which a large fish eats a small fish that has eaten algae is a
    a. food web.
    b. food chain.
    c. pyramid of numbers.
    d. pyramid of biomass.

15. The total amount of living tissue at each trophic level in an ecosystem can be shown in a(n)
    a. energy pyramid.
    b. pyramid of numbers.
    c. biomass pyramid.
    d. biogeochemical cycle.

Think Critically

16. Which group of organisms is always found at the base of a food chain or food web?

17. Apply Concepts Why is the transfer of energy in a food chain usually only about 10 percent efficient?

18. Use Models Describe a food chain of which you are a member. You may draw or use words to describe the chain.

19. Use Models Create flowcharts that show four different food chains in the food web shown below.

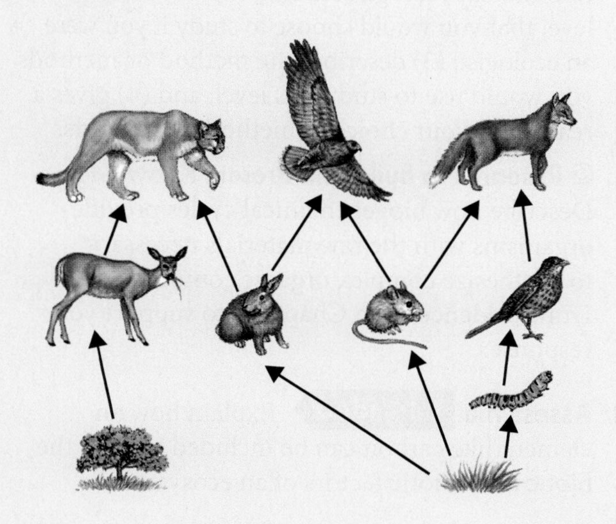

3.4 Cycles of Matter

Understand Key Concepts

20. Nutrients move through an ecosystem in
- **a.** biogeochemical cycles.
- **b.** water cycles.
- **c.** energy pyramids.
- **d.** ecological pyramids.

21. Which biogeochemical cycle does NOT include a major path in which the substance cycles through the atmosphere?
- **a.** water cycle
- **b.** carbon cycle
- **c.** nitrogen cycle
- **d.** phosphorus cycle

22. List two ways in which water enters the atmosphere in the water cycle.

23. Explain the process of nitrogen fixation.

24. What is meant by "nutrient limitation"?

solve the CHAPTER MYSTERY

CHANGES IN THE BAY

According to one hypothesis, rising water temperatures have caused most of the changes reported in Narragansett Bay. The bay's temperature has risen more than 1.5°C (3°F) since 1960. This warmth encourages bluefish to stay in the bay later in the fall. It also allows predatory warm-water shrimp to remain in the bay all winter, feeding on baby flounder. Warmer water also enables zooplankton to graze heavily on marine algae. This eliminates the late-winter algal bloom whose primary production used to provide organic carbon to the entire food web.

Those food web changes, in turn, seem to be driving unexpected shifts in the activities of bacteria that transform nitrogen. When the spring bloom provided organic carbon, bacteria denitrified the water, releasing nitrogen into the atmosphere. Now, the bacterial community has changed and actually fixes nitrogen, bringing more of it into the water. It is still not clear what this change means for the long-term health of the bay and adjacent coastal waters.

1. Key Ideas and Details Compare the original situation in the bay with the current situation, taking note of changes in both the food web and the nitrogen cycle. Cite textual evidence to support your response.

2. Infer Narragansett Bay harbors sea jellies that prefer warm water and have previously been present only in summer and early fall. These sea jellies eat fish eggs, fish larvae, and zooplankton. If the bay continues to warm, what do you think might happen to the population of sea jellies in the bay? What might that mean for the organisms the jellies feed on?

3. Connect to the Big idea Explain how the Narragansett Bay example demonstrates interconnections among members of a food web and abiotic environmental factors. Can you find similar studies in other aquatic habitats, such as Chesapeake Bay, the Everglades, or the Mississippi River delta? Explain.

Think Critically

25. Form a Hypothesis Ecologists discovered that trout were dying in a stream that ran through some farmland where nitrogen fertilizer was used on the crops. How might you explain what happened?

26. Apply Concepts Using a flowchart, trace the flow of energy in a simple marine food chain. Then, show where nitrogen is cycled through the chain when the top-level carnivore dies and is decomposed.

Connecting Concepts

Use Science Graphics

The graph below shows the effect of annual rainfall on the rate of primary productivity in an ecosystem. Use the graph to answer questions 27–29.

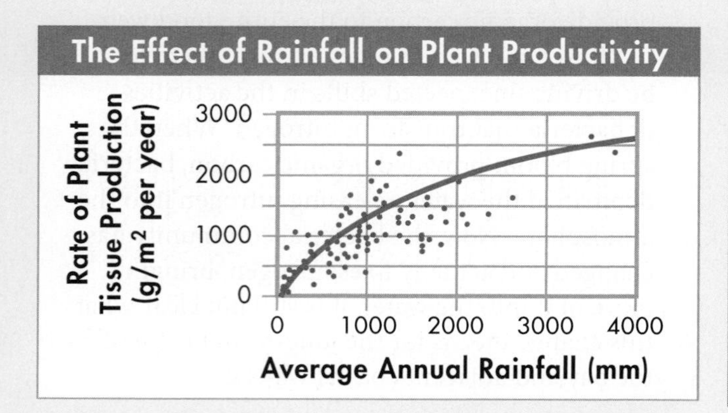

27. Integration of Knowledge and Ideas What happens to productivity as rainfall increases?

28. Integration of Knowledge and Ideas What do you think the graph would look like if the x-axis were extended out to 6000 mm? Represent your prediction in a graph and explain your answer.

29. Apply Concepts What factors other than water might affect primary productivity?

Write About Science

30. Text Types and Purposes Write a paragraph that (1) names and defines the levels of organization that an ecologist studies; (2) identifies the level that you would choose to study if you were an ecologist; (3) describes the method or methods you would use to study this level; and (4) gives a reason for your choice of method or methods.

31. Research to Build and Present Knowledge Describe how biogeochemical cycles provide organisms with the raw materials necessary to synthesize complex organic compounds. Draw evidence from Chapter 2 to support your response.

32. Assess the 

Explain how an element like carbon can be included in both the biotic and abiotic factors of an ecosystem.

Integration of Knowledge and Ideas

Samples of ocean water are taken at different depths, and the amount of oxygen in the water at each depth is measured. The results are shown in the table.

Concentration of Oxygen

Depth of Sample (m)	Oxygen Concentration (ppm)
0	7.5
50	7.4
100	7.4
150	4.5
200	3.2
250	3.1
300	2.9

33. Interpret Tables Which of the following is the best description of what happens to the amount of available oxygen as you get deeper in the ocean?

a. Available oxygen decreases at a constant rate.
b. Available oxygen increases at a constant rate.
c. Available oxygen remains steady until about 100 m, then drops rapidly.
d. Oxygen is available at all ocean depths.

34. Draw Conclusions Light can penetrate to only a depth of between 50 and 100 m in most ocean water. What effect does this have on the water's oxygen concentration? Explain.

Standardized Test Prep

Multiple Choice

1. A group of individuals that belong to a single species and that live together in a defined area is termed a(n)
   A population.
   B ecosystem.
   C community.
   D biome.

2. Which of the following is NOT true about matter in the biosphere?
   A Matter is recycled in the biosphere.
   B Biogeochemical cycles transform and reuse molecules.
   C The total amount of matter decreases over time.
   D Water and nutrients pass between organisms and the environment.

3. Which is a source of energy for Earth's living things?
   A wind energy only
   B sunlight only
   C wind energy and sunlight
   D sunlight and chemical energy

4. Which of the following is a primary producer?
   A a producer, like algae
   B a carnivore, like a lion
   C an omnivore, like a human
   D a detritivore, like an earthworm

5. Human activities, such as the burning of fossil fuels, move carbon through the carbon cycle. Which other processes also participate in the carbon cycle?
   A biological processes only
   B geochemical processes only
   C chemical processes only
   D a combination of biological, geological, and chemical processes

6. What are the physical, or nonliving components of an ecosystem called?
   A abiotic factors
   B temperate conditions
   C biotic factors
   D antibiotic factors

Questions 7–8

The diagrams below represent the amount of biomass and the numbers of organisms in an ecosystem.

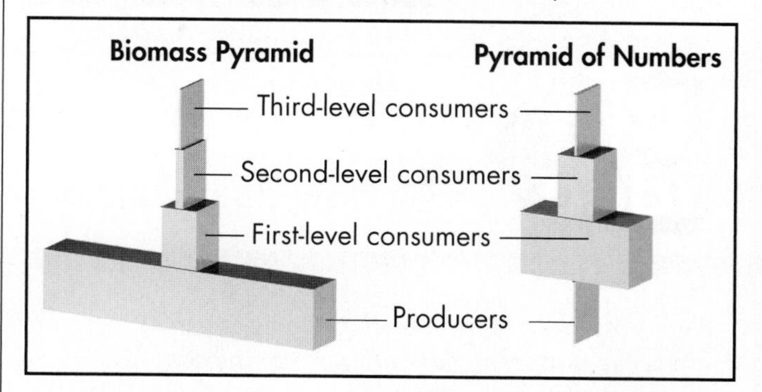

7. What can you conclude about the ecosystem from the pyramid of numbers shown?
   A There are more first-level consumers than producers.
   B There are more third-level consumers than second-level consumers.
   C There are more producers than first-level consumers.
   D There are more second-level consumers than first-level consumers.

8. What can you conclude about the producers in the ecosystem based on the two pyramids shown?
   A The producers in the ecosystem are probably very small organisms.
   B There are no producers in the ecosystem.
   C The producers in the ecosystem are probably large organisms.
   D Decomposers in the ecosystem outnumber the producers in the ecosystem.

Open-Ended Response

9. What ultimately happens to the bulk of matter in any trophic level of a biomass pyramid—that is, the matter that does not get passed to the trophic level above?

If You Have Trouble With . . .

Question	1	2	3	4	5	6	7	8	9
See Lesson	3.1	3.4	3.2	3.2	3.4	3.1	3.3	3.3	3.3

Ecosystems and Communities

Interdependence in Nature

Q: How do abiotic and biotic factors shape ecosystems?

Predators often choose resting places that allow them a good view of their surroundings. This cheetah at the Masai Mara National Reserve in Kenya has chosen a termite mound.

INSIDE:

- 4.1 Climate
- 4.2 Niches and Community Interactions
- 4.3 Succession
- 4.4 Biomes
- 4.5 Aquatic Ecosystems

BUILDING *Scientific Literacy*

Deepen your understanding of ecosystems and communities by engaging in key practices that allow you to make connections across concepts.

NGSS You will use the concept of competition to explain how related species can survive in the same habitat.

STEM Could placing two species in the same aquarium or zoo exhibit benefit both species? Use your understanding of symbiosis to complete an activity that explores this question.

Common Core You will translate visual evidence to explain the geographic location of certain biomes.

CHAPTER MYSTERY

THE WOLF EFFECT

During the 1920s, hunting and trapping eliminated wolves from Yellowstone National Park. For decades, ecologists hypothesized that the loss of wolves—important predators of elk and other large grazing animals—had changed the park ecosystem. But because there were no before-and-after data, it was impossible to test that hypothesis directly.

Then, in the mid-1990s, wolves were reintroduced to Yellowstone. Researchers watched park ecosystems carefully and sure enough, the number of elk in parts of the park began to fall just as predicted. But, unpredictably, forest and stream communities have changed, too. Could a "wolf effect" be affecting organisms in the park's woods and streams?

As you read this chapter, look for connections among Yellowstone's organisms and their environment. Then, solve the mystery.

Biology.com

Solving the mystery of the wolves is just the beginning. Take a video field trip with the ecogeeks of Untamed Science to continue exploring your world.

Go online to access additional resources including:
- **eText**
- **Flash Cards**
- **Lesson Overviews**
- **Chapter Mystery**

4.1 Climate

Key Questions

What is climate?

What factors determine global climate?

Vocabulary

weather
climate
microclimate
greenhouse effect

Taking Notes

Preview Visuals Before you read, look at **Figure 4–2.** What questions do you have about this diagram? Write a prediction that relates this figure to climate.

BUILD Vocabulary

PREFIXES The prefix *hemi-* in *hemisphere* means "half." The Northern Hemisphere encompasses the northern half of Earth.

THINK ABOUT IT When you think about climate, you might think of dramatic headlines: "Hurricane Katrina floods New Orleans!" or "Drought parches the Southeast!" But big storms and seasonal droughts are better described as *weather* rather than *climate.* So, what *is* climate, and how does it differ from weather? How do climate and weather affect organisms and ecosystems?

Weather and Climate

What is climate?

Weather and climate both involve variations in temperature, precipitation, and other environmental factors. **Weather** is the day-to-day condition of Earth's atmosphere. Weather where you live may be clear and sunny one day but rainy and cold the next. **Climate,** on the other hand, refers to average conditions over long periods. **A region's climate is defined by year-after-year patterns of temperature and precipitation.**

It is important to note that climate is rarely uniform even within a region. Environmental conditions can vary over small distances, creating **microclimates.** For example, in the Northern Hemisphere, south-facing sides of trees and buildings receive more sunlight, and are often warmer and drier, than north-facing sides. We may not notice these differences, but they can be very important to many organisms.

Factors That Affect Climate

What factors determine global climate?

A person living in Orlando, Florida, may wear shorts and a T-shirt in December, while someone in Minneapolis, Minnesota, is still wearing a heavy coat in April. It rarely rains in Phoenix, Arizona, but it rains often in Mobile, Alabama. Clearly, these places all have different climates—but why? What causes differences in climate? **Global climate is shaped by many factors, including solar energy trapped in the biosphere, latitude, and the transport of heat by winds and ocean currents.**

In Your Notebook *Describe the climate where you live. What factors influence it?*

Solar Energy and the Greenhouse Effect

Solar Energy and the Greenhouse Effect The main force that shapes our climate is solar energy that arrives as sunlight and strikes Earth's surface. Some of that energy is reflected back into space, and some is absorbed and converted into heat. Some of that heat, in turn, radiates back into space, and some is trapped in the biosphere. The balance between heat that stays in the biosphere and heat lost to space determines Earth's average temperature. This balance is largely controlled by concentrations of three gases found in the atmosphere—carbon dioxide, methane, and water vapor.

As shown in **Figure 4–1,** these gases, called greenhouse gases, function like glass in a greenhouse, allowing visible light to enter but trapping heat. This phenomenon is called the **greenhouse effect.** If greenhouse gas concentrations rise, they trap more heat, so Earth warms. If their concentrations fall, more heat escapes, and Earth cools. Without the greenhouse effect, Earth would be about 30° Celsius cooler than it is today. Note that all three of these gases pass in and out of the atmosphere as part of nutrient cycles.

VISUAL ANALOGY

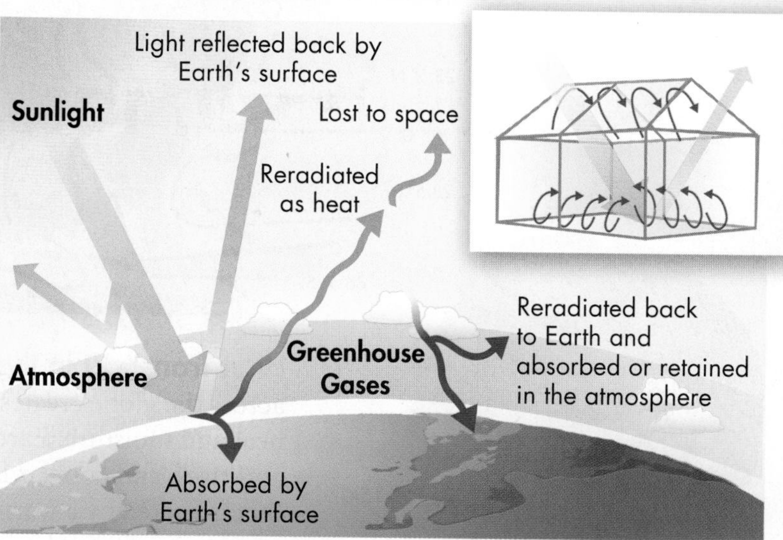

THE GREENHOUSE EFFECT

FIGURE 4–1 Greenhouse gases in the atmosphere allow solar radiation to enter the biosphere but slow down the loss of reradiated heat to space. **Use Analogies** ***What part of a greenhouse is analogous to the greenhouse gases in Earth's atmosphere?***

Latitude and Solar Energy Near the equator, solar energy is intense as the sun is almost directly overhead at noon all year. That's why equatorial regions are generally so warm. As **Figure 4–2** shows, the curvature of Earth causes the same amount of solar energy to spread out over a much larger area near the poles than near the equator. Thus, Earth's polar areas annually receive less intense solar energy, and therefore heat, from the sun. This difference in heat distribution creates three different climate zones: tropical, temperate, and polar.

The tropical zone, or tropics, which includes the equator, is located between 23.5° north and 23.5° south latitudes. This zone receives nearly direct sunlight all year. On either side of the tropical zone are the two temperate zones, between 23.5° and 66.5° north and south latitudes. Beyond the temperate zones are the polar zones, between 66.5° and 90° north and south latitudes. Temperate and polar zones receive very different amounts of solar energy at different times of the year because Earth's axis is tilted. As Earth revolves around the sun, solar radiation strikes different regions at angles that vary from summer to winter. During winter in the temperate and polar zones, the sun is much lower in the sky, days are shorter, and solar energy is less intense.

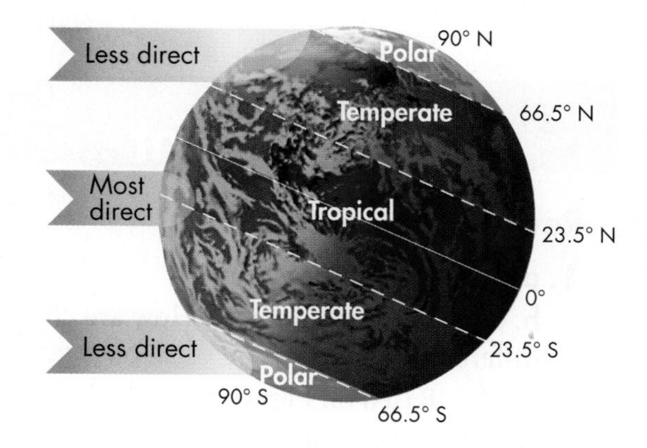

FIGURE 4–2 Climate Zones Earth's climate zones are produced by unequal distribution of the sun's heat on Earth's surface. Polar regions receive less solar energy per unit area, and so less heat, than tropical regions do. The tilt of Earth's axis causes the distribution of sunlight to change over the course of the year.

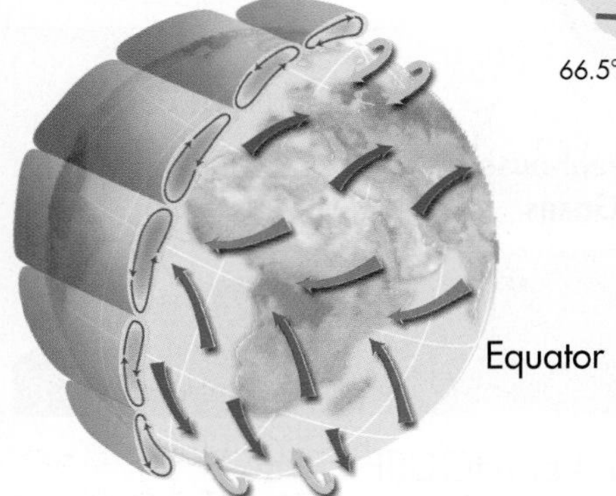

FIGURE 4–3 Winds and Currents Earth's winds (above left) and ocean currents (above right) interact to help produce climate patterns. The paths of winds and currents are the result of heating and cooling, Earth's rotation, and geographic features. **Interpret Visuals** ***In what direction do cold currents in the Northern Hemisphere generally move?***

Heat Transport in the Biosphere The unequal distribution of heat across the globe creates wind and ocean currents, which transport heat and moisture. Earth has winds because warm air is less dense and rises, and cool air is more dense and sinks. For this reason, air that is heated by a warm area of Earth's surface—such as air near the equator, for example—rises. As this warm air rises, it expands and spreads north and south, losing heat along the way. As it cools, the air sinks. At the same time, in cooler regions, near the poles, chilled air sinks toward Earth's surface, pushing air at the surface outward. This air warms as it travels over the surface. And as the air warms, it rises. These upward and downward movements of air create winds, as shown in **Figure 4–3** (above left). Winds transport heat from regions of rising warmer air to regions of sinking cooler air. Earth's rotation causes winds to blow generally from west to east over the temperate zones and from east to west over the tropics and the poles.

Similar patterns of heating and cooling occur in the oceans. Surface water is pushed by winds. These ocean currents transport enormous amounts of heat. Warm surface currents add moisture and heat to air that passes over them. Cool surface currents cool air that passes over them. In this way, surface currents affect the weather and climate of nearby landmasses. Deep ocean currents are caused by cold water near the poles sinking and flowing along the ocean floor. This water rises in warmer regions through a process called upwelling.

4.1 Assessment

Review Key Concepts

1. a. **Review** What is climate?

   b. **Compare and Contrast** How are climate and weather different?

   c. **Infer** Based on **Figure 4–3**, which do you think has a cooler climate: the east or west coast of southern Africa? Why?

2. a. **Review** What are the main factors that determine climate?

   b. **Relate Cause and Effect** Explain what would likely happen to global climate if there was a dramatic decrease in greenhouse gases trapped in the atmosphere.

ANALYZING DATA

3. Research average monthly precipitation (in mm) and temperature (in °C) for Quito, Ecuador, a city on the equator. Create a bar graph for the precipitation data. Plot the temperature data in a line graph.

4.2 Niches and Community Interactions

THINK ABOUT IT If you ask someone where an organism lives, that person might answer "on a coral reef" or "in the desert." These answers are like saying that a person lives "in Miami" or "in Arizona." The answer gives the environment or location. But ecologists need more information to understand fully why an organism lives where it does and how it fits into its surroundings. What else do they need to know?

Key Questions

What is a niche?

How does competition shape communities?

How do predation and herbivory shape communities?

What are the three primary ways that organisms depend on each other?

Vocabulary

tolerance • habitat • niche • resource • competitive exclusion principle • predation • herbivory • keystone species • symbiosis • mutualism • parasitism • commensalism

Taking Notes

Concept Map Use the highlighted vocabulary words to create a concept map that organizes the information in this lesson.

The Niche

What is a niche?

Organisms occupy different places in part because each species has a range of conditions under which it can grow and reproduce. These conditions help define where and how an organism lives.

Tolerance Every species has its own range of **tolerance,** the ability to survive and reproduce under a range of environmental circumstances, as shown in **Figure 4–4.** When an environmental condition, such as temperature, extends in either direction beyond an organism's optimum range, the organism experiences stress. Why? Because it must expend more energy to maintain homeostasis, and so has less energy left for growth and reproduction. Organisms have an upper and lower limit of tolerance for every environmental factor. Beyond those limits, the organism cannot survive. A species' tolerance for environmental conditions, then, helps determine its "address" or **habitat**—the general place where an organism lives.

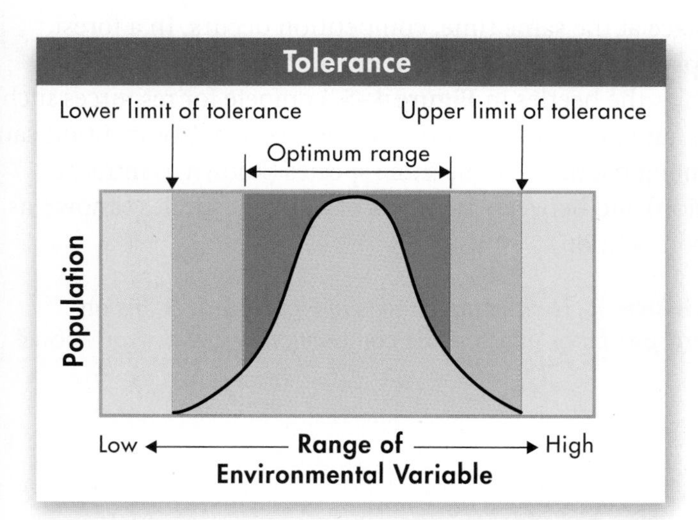

FIGURE 4–4 Tolerance This graph shows the response of a hypothetical organism to different values of a single environmental variable such as sunlight or temperature. At the center of the optimum range, organisms are likely to be most abundant. They become more rare in zones of physiological stress (medium blue), and are absent from zones of intolerance (light blue).

Defining the Niche Describing a species' "address" tells only part of its story. Ecologists also study a species' ecological "occupation"—where and how it "makes a living." This idea of occupation is encompassed in the idea of an organism's niche (nich). A **niche** describes not only what an organism does, but also how it interacts with biotic and abiotic factors in the environment. **A niche is the range of physical and biological conditions in which a species lives and the way the species obtains what it needs to survive and reproduce.** Understanding niches is important to understanding how organisms interact to form a community.

▶ ***Resources and the Niche*** The term **resource** can refer to any necessity of life, such as water, nutrients, light, food, or space. For plants, resources can include sunlight, water, and soil nutrients—all of which are essential to survival. For animals, resources can include nesting space, shelter, types of food, and places to feed.

BUILD Vocabulary

ACADEMIC WORDS The noun **aspect** means "part." There are two aspects—or parts—of an organism's niche: physical aspects and biological aspects.

▶ ***Physical Aspects of the Niche*** Part of an organism's niche involves the abiotic factors it requires for survival. Most amphibians, for example, lose and absorb water through their skin, so they must live in moist places. If an area is too hot and dry, or too cold for too long, most amphibians cannot survive.

▶ ***Biological Aspects of the Niche*** Biological aspects of an organism's niche involve the biotic factors it requires for survival. When and how it reproduces, the food it eats, and the way in which it obtains that food are all examples of biological aspects of an organism's niche. Birds on Christmas Island, a small island in the Indian Ocean, for example, all live in the same habitat but they prey on fish of different sizes and feed in different places. Thus, each species occupies a distinct niche.

FIGURE 4–5 Competition Animals such as these two male stag beetles compete for limited resources. **Infer** ***What resource do you think these two males are fighting over?***

Competition

How does competition shape communities?

If you look at any community, you will probably find more than one kind of organism attempting to use various essential resources. When organisms attempt to use the same limited ecological resource in the same place at the same time, competition occurs. In a forest, for example, plant roots compete for water and nutrients in the soil. Animals, such as the beetles in **Figure 4–5,** compete for resources such as food, mates, and places to live and raise their young. Competition can occur both among members of the same species (known as intraspecific competition) and between members of different species (known as interspecific competition).

In Your Notebook *Look at the beetles in* ***Figure 4–5.*** *Is this an example of intraspecific or interspecific competition? How do you know?*

The Competitive Exclusion Principle Direct competition between different species almost always produces a winner and a loser—and the losing species dies out. One series of experiments demonstrated this using two species of single-celled organisms. When the species were grown in separate cultures under the same conditions, each survived, as shown in **Figure 4–6.** But when both species were grown together in the same culture, one species outcompeted the other. The less competitive species did not survive.

Experiments like this one, along with observations in nature, led to the discovery of an important ecological rule. The **competitive exclusion principle** states that no two species can occupy exactly the same niche in exactly the same habitat at exactly the same time. If two species attempt to occupy the same niche, one species will be better at competing for limited resources and will eventually exclude the other species. As a result, if we look at natural communities, we rarely find species whose niches overlap significantly.

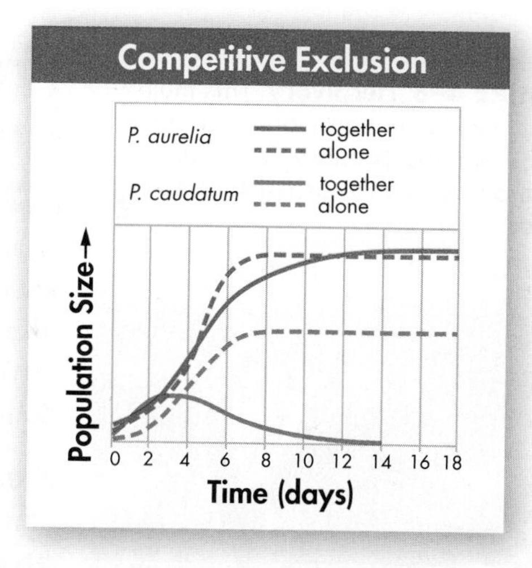

FIGURE 4–6 Competitive Exclusion
The two species of paramecia *P. aurelia* and *P. caudatum* have similar requirements. When grown in cultures separately (dashed lines), both populations grow quickly and then level off. When grown together under certain conditions (solid lines), however, *P. aurelia* outcompetes *P. caudatum* and drives it to extinction.

Dividing Resources Instead of competing for similar resources, species usually divide them. For instance, the three species of North American warblers shown in **Figure 4–7** all live in the same trees and feed on insects. But one species feeds on high branches, another feeds on low branches, and another feeds in the middle. The resources utilized by these species are similar yet different. Therefore, each species has its own niche. This division of resources was likely brought about by past competition among the birds. **By causing species to divide resources, competition helps determine the number and kinds of species in a community and the niche each species occupies.**

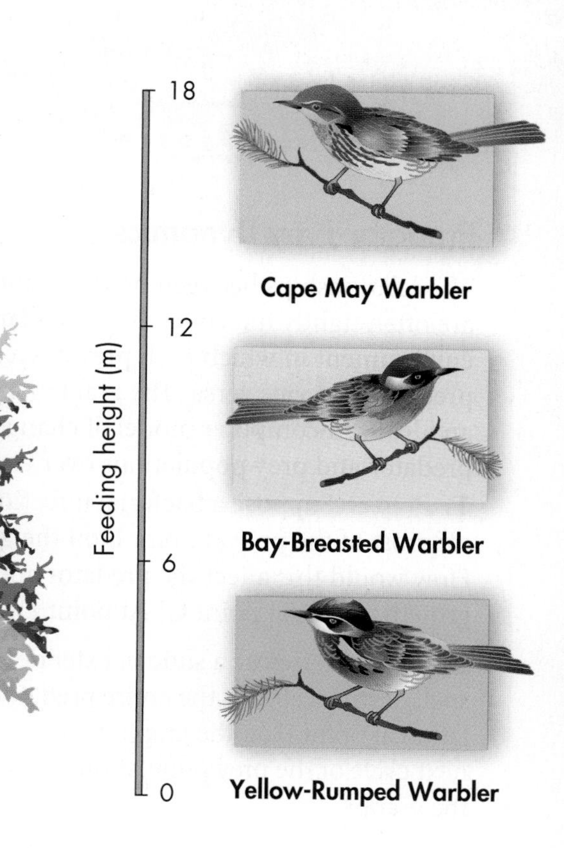

FIGURE 4–7 Resource Sharing
Each of these warbler species has a different niche in its spruce tree habitat. By feeding in different areas of the tree, the birds avoid competing directly with one another for food.
Infer ***What would happen if two of the warbler species tried to occupy the same niche in the same tree at the same time?***

FIGURE 4–8 Herbivory This mule deer buck is an herbivore—meaning that it obtains its energy and nutrients from plants like the grasses it's eating here.

Predation, Herbivory, and Keystone Species

How do predation and herbivory shape communities?

Virtually all animals, because they are not primary producers, must eat other organisms to obtain energy and nutrients. Yet if a group of animals devours all available food in the area, they will no longer have anything to eat! That's why predator-prey and herbivore-plant interactions are very important in shaping communities.

Predator-Prey Relationships An interaction in which one animal (the predator) captures and feeds on another animal (the prey) is called **predation** (pree DAY shun). **Predators can affect the size of prey populations in a community and determine the places prey can live and feed.** Birds of prey, for example, can play an important role in regulating the population sizes of mice, voles, and other small mammals.

Herbivore-Plant Relationships Interactions between herbivores and plants, like the one shown in **Figure 4–8,** are as important as interactions between predators and prey. An interaction in which one animal (the herbivore) feeds on producers (such as plants) is called **herbivory. Herbivores can affect both the size and distribution of plant populations in a community and determine the places that certain plants can survive and grow.** Herbivores ranging from caterpillars to elk can have major effects on plant survival. For example, very dense populations of white-tailed deer are eliminating their favorite food plants from many places across the United States.

Analyzing Data

Predator-Prey Dynamics

The relationships between predator and prey are often tightly intertwined, particularly in an environment in which each prey has a single predator and vice versa. The graph here shows an idealized computer model of changes in predator and prey populations over time.

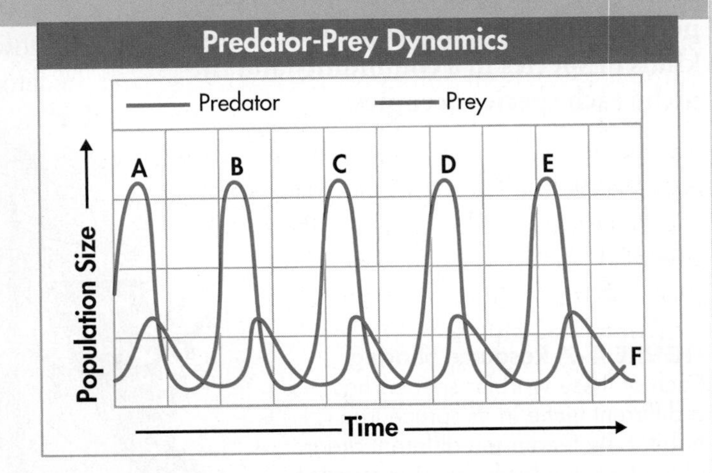

1. Predict Suppose a bacterial infection kills off most of the prey at point B on the graph. How would this affect the predator and prey growth curves at point C? At point D?

2. Predict Suppose a sudden extended cold spell destroys almost the entire predator population at point F on the graph. How would the next cycle of the prey population appear on the graph?

3. Relate Cause and Effect Suppose a viral infection kills all the prey at point D on the graph. What effect would this have on the predator and prey growth curves at point E? What will happen in future years to the predator population? How could ecologists ensure the continued survival of the predators in this ecosystem?

Keystone Species Sometimes changes in the population of a single species, often called a **keystone species,** can cause dramatic changes in the structure of a community. In the cold waters off the Pacific coast of North America, for example, sea otters devour large quantities of sea urchins. Urchins, in turn, are herbivores. Their favorite food is kelp, giant algae that grow in undersea "forests."

A century ago, sea otters were nearly eliminated by hunting. Unexpectedly, the kelp forest nearly vanished. What happened? Without otters as predators, the sea urchin population skyrocketed. Armies of urchins devoured kelp down to bare rock. Without kelp to provide habitat, many other animals, including seabirds, disappeared. Clearly, otters were a keystone species in this community. After otters were protected as an endangered species, their population began to recover. As otters returned, the urchin populations dropped, and kelp forests began to thrive again. Recently, however, the otter population has been falling again, and no one knows why.

In Your Notebook *Not all keystone-species effects are due to predation. Describe the dramatic effects that the dam-building activities of beavers, a keystone species, might have on other types of organisms.*

MYSTERY CLUE

One of the favorite prey species of the wolves in Yellowstone is elk. How do you think this relationship could affect the ability of certain *plants* to grow in Yellowstone?

Symbioses

What are the three primary ways that organisms depend on each other?

Any relationship in which two species live closely together is called **symbiosis** (sim by OH sis), which means "living together." **Biologists recognize three main classes of symbiotic relationships in nature: mutualism, parasitism, and commensalism.**

Mutualism The sea anemone's sting has two functions: to capture prey and to protect the anemone from predators. Even so, certain fish manage to snack on anemone tentacles. The clownfish, however, is immune to anemone stings. When threatened by a predator, clownfish seek shelter by snuggling deep into tentacles that would be deadly to most other fish, as seen in **Figure 4–9.** But if an anemone-eating species tries to attack their living home, the spunky clownfish dart out and fiercely chase away fish many times their size. This kind of relationship between species in which both benefit is known as **mutualism.**

FIGURE 4–9 Mutualism Clownfish live among the sea anemone's tentacles and protect the sea anemone by chasing away would-be attackers. The sea anemone, in turn, protects the clownfish from their predators. **Infer** ***What could happen to the sea anemone if the clownfish died?***

FIGURE 4–10 Parasitism This brown leech is feeding on the blood of its host, a human. In a parasitic relationship, the parasite benefits while the host is harmed.

Parasitism Tapeworms live in the intestines of mammals, where they absorb large amounts of their hosts' food. Fleas, ticks, lice, and leeches live on the bodies of mammals, feeding on their blood and skin, as seen in **Figure 4–10.** These are examples of **parasitism** (PAR uh sit iz um), relationships in which one organism lives inside or on another organism and harms it. The parasite obtains all or part of its nutritional needs from the host organism. Generally, parasites weaken but do not kill their host, which is usually larger than the parasite.

Commensalism Small marine animals called barnacles often attach themselves to a whale's skin, as seen in **Figure 4–11.** The barnacles perform no known service to the whale, nor do they harm it. Yet the barnacles benefit from the constant movement of water—that is full of food particles—past the swimming whale. This is an example of **commensalism** (kuh MEN sul iz um), a relationship in which one organism benefits and the other is neither helped nor harmed.

FIGURE 4–11 Commensalism The barnacles attached to the skin of this grey whale are feeding on food in the water that passes over them as the whale swims. Although the barnacles clearly benefit from their relationship with the whale, they do not appear to affect the whale positively or negatively.

4.2 Assessment

Review Key Concepts

1. a. Review What is the difference between a habitat and a niche?

b. Use Analogies How is a niche like a profession? In ecological terms, describe your niche.

2. a. Review What is competition? Why can't two organisms compete if they live in different habitats?

b. Interpret Visuals Look at **Figure 4–7** and describe how the three species of warblers have divided their resources. Does each warbler have its own niche?

3. a. Review What is a keystone species?

b. Infer How might a dramatic decrease in vegetation lead to a decrease in a prey species? (*Hint:* Think of how the vegetation, prey, and predator could be connected in a food chain.)

4. a. Review What is symbiosis? What are the three major types of symbiosis?

b. Explain Bacteria living in a cow's stomach help the cow break down the cellulose in grass, gaining nutrients in the process. Is this an example of commensalism or mutualism? Explain your answer.

c. Apply Concepts What is the difference between a predator and a parasite? Explain your answer.

BUILD VOCABULARY

5. The suffix *-ism* means "the act, practice, or result of." Look up the meaning of *mutual,* and write a definition for *mutualism.*

WHST.9-10.4 Production and Distribution of Writing, WHST.9-10.10 Range of Writing, RST.9-10.10 Level of Text Complexity

Careers & BIOLOGY

Do you enjoy being outdoors? If you do, you might want to consider one of these careers.

MARINE BIOLOGIST

Ocean ecosystems cover over 70 percent of Earth's surface. Marine biologists study the incredible diversity of ocean life. Some marine biologists study organisms found in deep ocean trenches to understand how they survive in extreme conditions. Others work in aquariums, where they might conduct research, educate the public, or rehabilitate rescued marine wildlife.

PARK RANGER

For some people, camping and hiking aren't just recreational activities—they're work. Park rangers work in national, state, and local parks caring for the land and ensuring the safety of visitors. Park rangers perform a variety of tasks, including maintaining campsites and helping with search and rescue. Rangers are also responsible for looking after park wildlife.

WILDLIFE PHOTOGRAPHER

Wildlife photographers capture nature "in action." Their photographs can be used in books, magazines, and on the Internet to educate and entertain the public. Successful wildlife photographers need to be observant and adventurous. They also need to be patient enough to wait for the perfect shot.

CAREER CLOSE-UP

Dudley Edmondson, Wildlife Photographer

Dudley Edmondson began bird-watching at a young age. After high school, he began traveling and photographing the birds he observed. Mr. Edmondson has since been all over the United States taking pictures of everything from the landscapes and grizzly bears of Yellowstone Park to the butterflies that inhabit his own backyard. Through his work, he hopes to inspire people to travel and experience nature for themselves. This, he believes, will encourage a sense of responsibility to protect and preserve the environment.

"What I like most about my work is the unique perspective it gives me on the world. Birds, insects, and plants are totally unaware of things like clocks, deadlines, and technology. When you work with living things, you work on their terms."

WRITING Where have you seen nature photography used or displayed? How do those photos, or Mr. Edmondson's, help the public learn about the natural world?

4.3 Succession

Key Questions

How do communities change over time?

Do ecosystems return to "normal" following a disturbance?

Vocabulary

ecological succession
primary succession
pioneer species
secondary succession

Taking Notes

Compare/Contrast Table As you read, create a table comparing primary and secondary succession.

THINK ABOUT IT In 1883, the volcanic island of Krakatau in the Indian Ocean was blown to pieces by an eruption. The tiny island that remained was completely barren. Within two years, grasses were growing. Fourteen years later, there were 49 plant species, along with lizards, birds, bats, and insects. By 1929, a forest containing 300 plant species had grown. Today, the island is blanketed by mature rain forest. How did the island ecosystem recover so quickly?

Primary and Secondary Succession

How do communities change over time?

The story of Krakatau after the eruption is an example of **ecological succession**—a series of more-or-less predictable changes that occur in a community over time. **Ecosystems change over time, especially after disturbances, as some species die out and new species move in.** Over the course of succession, the number of different species present typically increases.

Primary Succession Volcanic explosions like the ones that destroyed Krakatau in 1883 and blew the top off Mount Saint Helens in Washington State in 1980 can create new land or sterilize existing areas. Retreating glaciers can have the same effect, leaving only exposed bare rock behind them. Succession that begins in an area with no remnants of an older community is called **primary succession.** An example of primary succession is shown in **Figure 4–12.**

FIGURE 4–12 Primary Succession Primary succession occurs on newly exposed surfaces. In Glacier Bay, Alaska, a retreating glacier exposed barren rock. Over the course of more than 100 years, a series of changes has led to the hemlock and spruce forest currently found in the area. Changes in this community will continue for centuries.

The first species to colonize barren areas are called **pioneer species**—named after rugged human pioneers who first settled the wilderness. After pioneers created settlements, different kinds of people with varied skills and living requirements moved into the area. Pioneer species function in similar ways. One ecological pioneer that grows on bare rock is lichen—a mutualistic symbiosis between a fungus and an alga. Over time, lichens convert, or fix, atmospheric nitrogen into useful forms for other organisms, break down rock, and add organic material to form soil. Certain grasses, like those that colonized Krakatau early on, are also pioneer species.

BUILD Vocabulary

WORD ORIGINS The origin of the word *succession* is the Latin word *succedere,* meaning "to come after." **Ecological succession** involves changes that occur one after the other as species move into and out of a community.

Secondary Succession Sometimes, existing communities are not completely destroyed by disturbances. In these situations, where a disturbance affects the community without completely destroying it, **secondary succession** occurs. Secondary succession proceeds faster than primary succession, in part because soil survives the disturbance. As a result, new and surviving vegetation can regrow rapidly. Secondary succession often follows a wildfire, hurricane, or other natural disturbance. We think of these events as disasters, but many species are adapted to them. Although forest fires kill some trees, for example, other trees are spared, and fire can stimulate their seeds to germinate. Secondary succession can also follow human activities like logging and farming. An example of secondary succession is shown in **Figure 4–13.**

Why Succession Occurs Every organism changes the environment it lives in. One model of succession suggests that as one species alters its environment, other species find it easier to compete for resources and survive. As lichens add organic matter and form soil, for example, mosses and other plants can colonize and grow. As organic matter continues to accumulate, other species move in and change the environment further. For example, as trees grow, their branches and leaves produce shade and cooler temperatures nearer the ground. Over time, more and more species can find suitable niches and survive.

In Your Notebook *Summarize what happens in primary and secondary succession.*

FIGURE 4–13 Secondary Succession Secondary succession occurs in disturbed areas where remnants of previous ecosystems—soil and even plants—remain. This series shows changes taking place in abandoned fields of the Carolinas' Piedmont. Over the last century, these fields have passed through several stages and matured into oak forests. Changes will continue for years to come.

Quick Lab
GUIDED INQUIRY

Successful Succession? 

1. Place a handful of dried plant material into a clean jar.
2. Fill the jar with boiled pond water or sterile spring water. Determine the initial pH of the water with pH paper.
3. Cover the jar and place it in an area that receives indirect light.
4. Examine the jar every day for the next few days.
5. When the water in the jar appears cloudy, prepare microscope slides of water from various levels of the jar. Use a pipette to collect the samples.
6. Look at the slides under the low-power objective lens of a microscope and record your observations.

Analyze and Conclude

1. **Infer** Why did you use boiled or sterile water?
2. **Infer** Where did the organisms you saw come from?
3. **Draw Conclusions** Was ecological succession occurring? Give evidence to support your answer.
4. **Evaluate and Revise** Check your results against those of your classmates. Do they agree? How do you explain any differences?

Climax Communities

Do ecosystems return to "normal" following a disturbance?

Ecologists used to think that succession in a given area always proceeds through the same stages to produce a specific and stable climax community like the mature spruce and hemlock forest that is developing in Glacier Bay. Recent studies, however, have shown that succession doesn't always follow the same path, and that climax communities are not always uniform and stable.

FIGURE 4–14 Recovery From a Natural Disaster These photos show El Yunque Rain Forest in Puerto Rico, immediately following Tropical Storm Jeanne in September 2004, and then again in May, 2007. **Apply Concepts** *What kind of succession occurred in this rain forest? How do you know?*

Succession After Natural Disturbances Natural disturbances are common in many communities. Healthy coral reefs and tropical rain forests recover from storms, as shown in **Figure 4–14.** Healthy temperate forests and grasslands recover from wildfires. **Secondary succession in healthy ecosystems following natural disturbances often reproduces the original climax community.** But detailed studies show that some climax communities are not uniform. Often, they look more like patchwork quilts with areas in varying stages of secondary succession following multiple disturbances that took place at different times. Some climax communities are disturbed so often that they can't really be called stable.

In Your Notebook *Describe what causes instability in some climax communities.*

Succession After Human-Caused Disturbances In North America, land cleared for farming and then abandoned often passes through succession that restores the original climax community. But this is not always the case. **Ecosystems may or may not recover from extensive human-caused disturbances.** Clearing and farming of tropical rain forests, for example, can change the microclimate and soil enough to prevent regrowth of the original community.

Studying Patterns of Succession Ecologists, like the ones seen in **Figure 4–15,** study succession by comparing different cases and looking for similarities and differences. Researchers who swarmed over Mount Saint Helens as soon as it was safe might also have studied Krakatau, for example. In both places, primary succession proceeded through predictable stages. The first plants and animals that arrived had seeds, spores, or adult stages that traveled over long distances. Hardy pioneer species helped stabilize loose volcanic debris, enabling later species to take hold. Historical studies in Krakatau and ongoing studies on Mount Saint Helens confirm that early stages of primary succession are slow, and that chance can play a large role in determining which species colonize at different times.

FIGURE 4–15 Studying Succession
These Forest Service rangers are surveying some of the plants and animals that have returned to the area around Mount Saint Helens. The volcano erupted in 1980, leaving only barren land for miles.

4.3 Assessment

Review Key Concepts

1. a. Review What effects do pioneer species have on an environment undergoing primary succession?
   b. Explain Why do communities change over time?
   c. Apply Concepts When a whale or other large marine mammal dies and falls to the ocean floor, different waves of decomposers and scavengers feed off the carcass until nothing remains. Do you think this is an example of succession? Explain your reasoning.
2. a. Review What is a climax community?
   b. Relate Cause and Effect What kinds of conditions might prevent a community from returning to its predisturbance state?

VISUAL THINKING

3. Look at the photo below. If you walked from this dune in a straight line away from the beach, what kinds of changes in vegetation would you expect to see? What sort of succession is this?

4.4 Biomes

Key Questions

What abiotic and biotic factors characterize biomes?

What areas are not easily classified into a major biome?

Vocabulary

canopy • understory • deciduous • coniferous • humus • taiga • permafrost

Taking Notes

Preview Visuals Before you read, preview **Figure 4–18.** Write down the names of the different biomes. As you read, examine the photographs and list the main characteristics of each biome.

THINK ABOUT IT Why does the character of biological communities vary from one place to another? Why, for example, do temperate rain forests grow in the Pacific Northwest while areas to the east of the Rocky Mountains are much drier? How do similar conditions shape ecosystems elsewhere?

The Major Biomes

What abiotic and biotic factors characterize biomes?

In Lesson 1, you learned that latitude and the heat transported by winds are two factors that affect global climate. But Oregon, Montana, and Vermont have different climates and biological communities, even though those states are at similar latitudes and are all affected by prevailing winds that blow from west to east. Why? The reason is because other factors, among them an area's proximity to an ocean or mountain range, can influence climate.

Regional Climates Oregon, for example, borders the Pacific Ocean. Cold ocean currents that flow from north to south have the effect of making summers in the region cool relative to other places at the same latitude. Similarly, moist air carried by winds traveling west to east is pushed upward when it hits the Rocky Mountains. This air expands and cools, causing the moisture in the air to condense and form clouds. The clouds drop rain or snow, mainly on the upwind side of the mountains—the side that faces the winds, as seen in **Figure 4–16.** West and east Oregon, then, have very different regional climates, and different climates mean different plant and animal communities.

FIGURE 4–16 The Effect of Coastal Mountains As moist ocean air rises over the upwind side of coastal mountains, it condenses, cools, and drops precipitation. As the air sinks on the downwind side of the mountain, it expands, warms, and absorbs moisture.

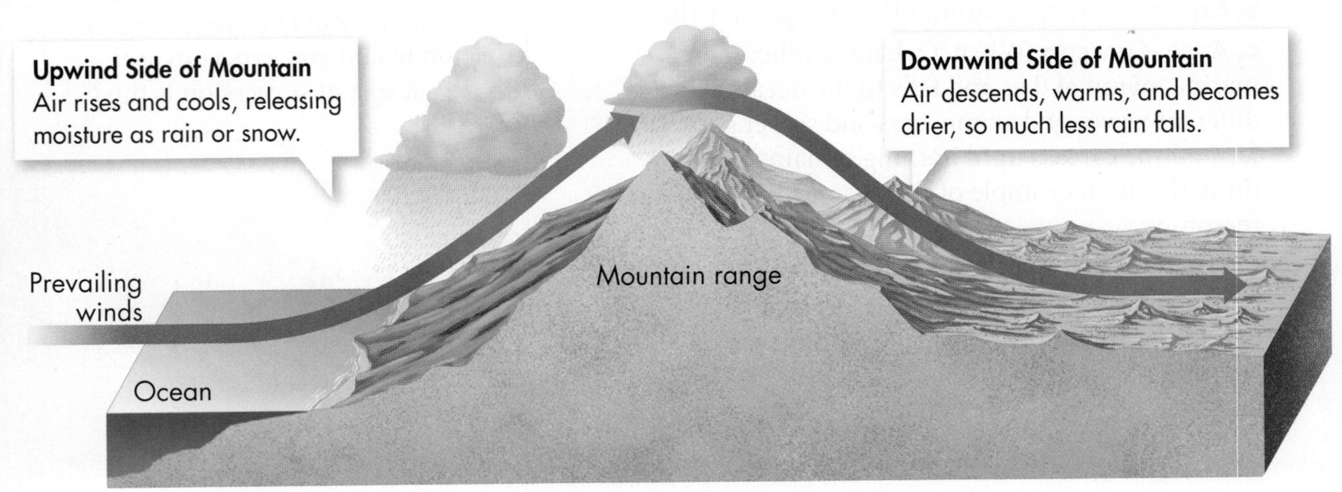

Defining Biomes Ecologists classify Earth's terrestrial ecosystems into at least ten different groups of regional climate communities called biomes. **Biomes are described in terms of abiotic factors like climate and soil type, and biotic factors like plant and animal life.** Major biomes include tropical rain forest, tropical dry forest, tropical grassland/savanna/shrubland, desert, temperate grassland, temperate woodland and shrubland, temperate forest, northwestern coniferous forest, boreal forest/taiga, and tundra. Each biome is associated with seasonal patterns of temperature and precipitation that can be summarized in a graph called a climate diagram, like the one in **Figure 4–17.** Organisms within each biome can be characterized by adaptations that enable them to live and reproduce successfully in the environment. The pages that follow discuss these adaptations and describe each biome's climate.

The distribution of major biomes is shown in **Figure 4–18.** Note that even within a defined biome, there is often considerable variation among plant and animal communities. These variations can be caused by differences in exposure, elevation, or local soil conditions. Local conditions also can change over time because of human activity or because of the community interactions described in this chapter and the next.

In Your Notebook *On the biome map in* ***Figure 4–18,*** *locate the place where you live. Which biome do you live in? Do your climate and environment seem to match the description of the biome on the following pages?*

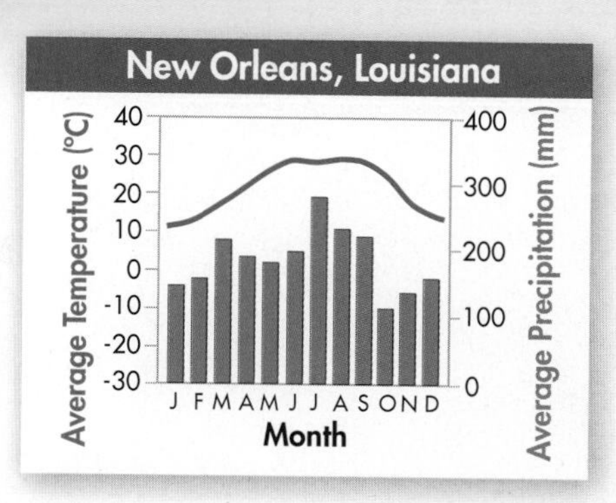

FIGURE 4–17 Climate Diagram A climate diagram shows the average temperature and precipitation at a given location during each month of the year. In this graph, and those to follow, temperature is plotted as a red line, and precipitation is shown as vertical blue bars.

VISUAL SUMMARY

BIOMES

FIGURE 4–18 This map shows the locations of the world's major biomes. Each biome has a characteristic climate and community of organisms.

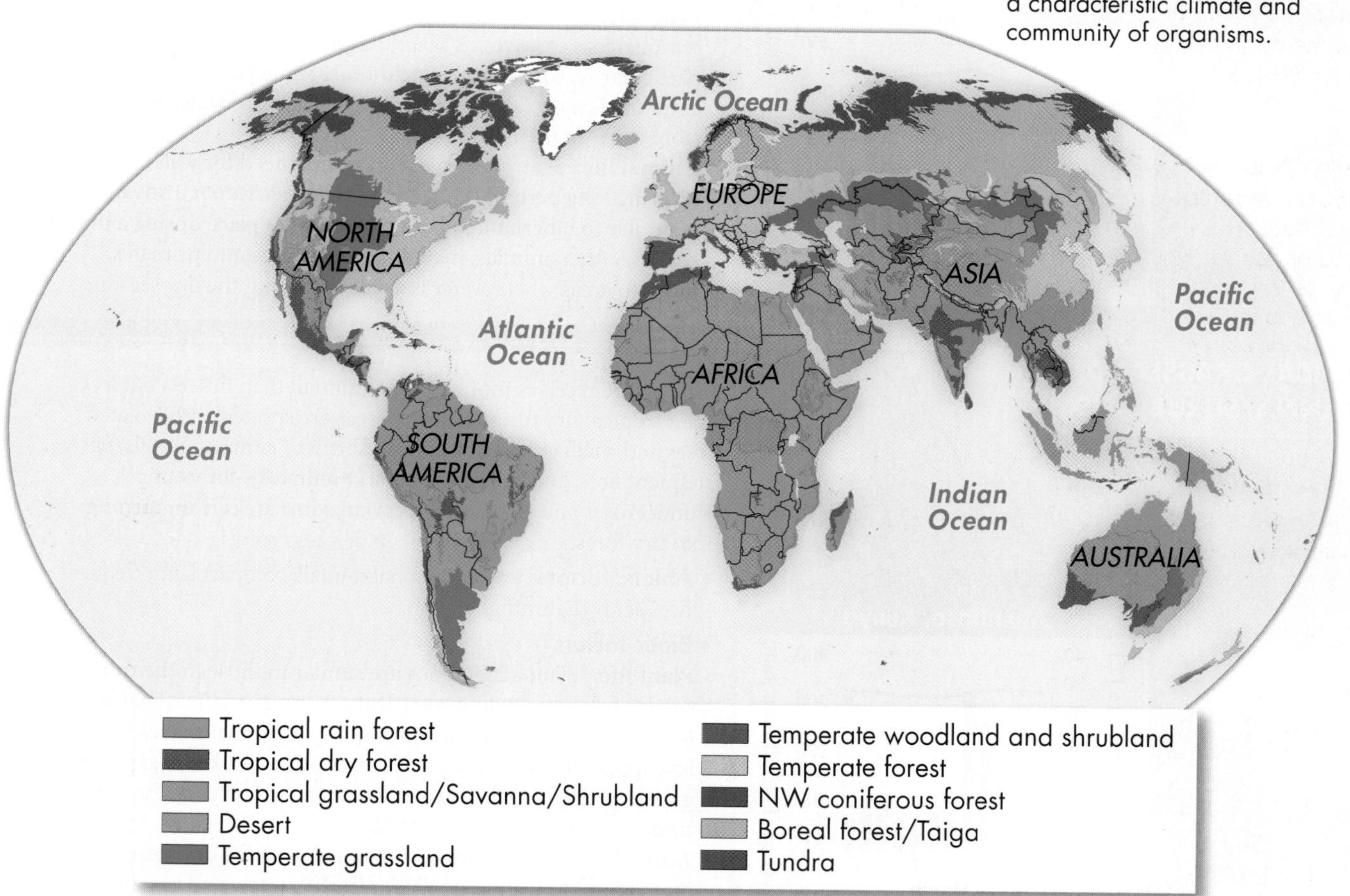

TROPICAL RAIN FOREST

Tropical rain forests are home to more species than all other biomes combined. As the name suggests, rain forests get a lot of rain—at least 2 meters of it a year! Tall trees form a dense, leafy covering called a **canopy** from 50 to 80 meters above the forest floor. In the shade below the canopy, shorter trees and vines form a layer called the **understory.** Organic matter on the forest floor is recycled and reused so quickly that the soil in most tropical rain forests is not very rich in nutrients.

- **Abiotic factors** hot and wet year-round; thin, nutrient-poor soils subject to erosion
- **Biotic factors**
  Plant life: Understory plants compete for sunlight, so most have large leaves that maximize capture of limited light. Tall trees growing in poor shallow soil often have buttress roots for support. Epiphytic plants grow on the branches of tall plants as opposed to soil. This allows epiphytes to take advantage of available sunlight while obtaining nutrients through their host.
  Animal life: Animals are active all year. Many animals use camouflage to hide from predators; some can change color to match their surroundings. Animals that live in the canopy have adaptations for climbing, jumping, and/or flight.

TROPICAL DRY FOREST

Tropical dry forests grow in areas where rainy seasons alternate with dry seasons. In most places, a period of rain is followed by a prolonged period of drought.

- **Abiotic factors** warm year-round; alternating wet and dry seasons; rich soils subject to erosion
- **Biotic factors**
  Plant life: Adaptations to survive the dry season include seasonal loss of leaves. A plant that sheds its leaves during a particular season is called **deciduous**. Some plants also have an extra thick waxy layer on their leaves to reduce water loss, or store water in their tissues.
  Animal life: Many animals reduce their need for water by entering long periods of inactivity called *estivation.* Estivation is similar to hibernation, but typically takes place during a dry season. Other animals, including many birds and primates, move to areas where water is available during the dry season.

TROPICAL GRASSLAND/ SAVANNA/SHRUBLAND

Mombasa, Kenya

Average Temperature (°C): 40, 30, 20, 10, 0, −10, −20, −30
Average Precipitation (mm): 400, 300, 200, 100, 0
J F M A M J J A S O N D
Month

This biome receives more seasonal rainfall than deserts but less than tropical dry forests. Grassy areas are spotted with isolated trees and small groves of trees and shrubs. Compacted soils, fairly frequent fires, and the action of large animals—for example, rhinoceroses and elephants—prevent some areas from turning into dry forest.

- **Abiotic factors** warm; seasonal rainfall; compact soils; frequent fires set by lightning
- **Biotic factors**
  Plant life: Plant adaptations are similar to those in the tropical dry forest, including waxy leaf coverings and seasonal leaf loss. Some grasses have a high silica content that makes them less appetizing to grazing herbivores. Also, unlike most plants, grasses grow from their bases, not their tips, so they can continue to grow after being grazed.
  Animal life: Many animals migrate during the dry season in search of water. Some smaller animals burrow and remain dormant during the dry season.

DESERT

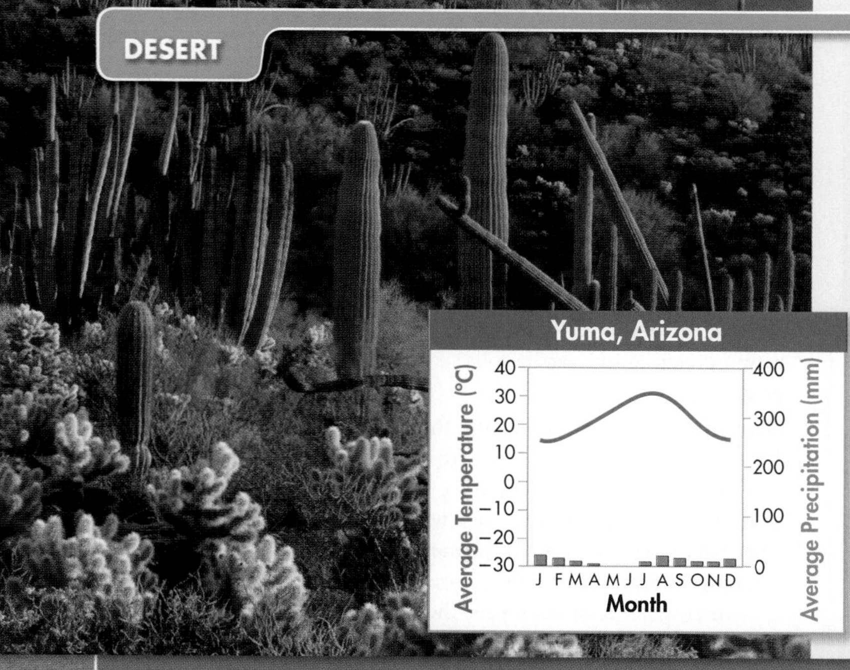

Deserts have less than 25 centimeters of precipitation annually, but otherwise vary greatly, depending on elevation and latitude. Many deserts undergo extreme daily temperature changes, alternating between hot and cold.

- **Abiotic factors** low precipitation; variable temperatures; soils rich in minerals but poor in organic material
- **Biotic factors**

Plant life: Many plants, including cacti, store water in their tissues, and minimize leaf surface area to cut down on water loss. Cactus spines are actually modified leaves. Many desert plants employ special forms of photosynthesis that enable them to open their leaf pores only at night, allowing them to conserve moisture on hot, dry days.

Animal life: Many desert animals get the water they need from the food they eat. To avoid the hottest parts of the day, many are nocturnal—active only at night. Large or elongated ears and other extremities are often supplied with many blood vessels close to the surface. These help the animal lose body heat and regulate body temperature.

TEMPERATE GRASSLAND

Plains and prairies, underlain by fertile soils, once covered vast areas of the midwestern and central United States. Periodic fires and heavy grazing by herbivores maintained plant communities dominated by grasses. Today, most have been converted for agriculture because their soil is so rich in nutrients and is ideal for growing crops.

- **Abiotic factors** warm to hot summers; cold winters; moderate seasonal precipitation; fertile soils; occasional fires
- **Biotic factors**

Plant life: Grassland plants—especially grasses, which grow from their base—are resistant to grazing and fire. Dispersal of seeds by wind is common in this open environment. The root structure and growth habit of native grassland plants helps establish and retain deep, rich, fertile topsoil.

Animal life: Because temperate grasslands are such open, exposed environments, predation is a constant threat for smaller animals. Camouflage and burrowing are two common protective adaptations.

TEMPERATE WOODLAND AND SHRUBLAND

In open woodlands, large areas of grasses and wildflowers such as poppies are interspersed with oak and other trees. Communities that are more shrubland than forest are known as chaparral. Dense low plants that contain flammable oils make fire a constant threat.

- **Abiotic factors** hot dry summers; cool moist winters; thin, nutrient-poor soils; periodic fires
- **Biotic factors**

Plant life: Plants in this biome have adaptated to drought. Woody chaparral plants have tough waxy leaves that resist water loss. Fire resistance is also important, although the seeds of some plants need fire to germinate.

Animal life: Animals tend to be browsers—meaning they eat varied diets of grasses, leaves, shrubs, and other vegetation. In exposed shrubland, camouflage is common.

TEMPERATE FOREST

Temperate forests are mostly made up of deciduous and evergreen coniferous (koh NIF ur us) trees. **Coniferous** trees, or conifers, produce seed-bearing cones, and most have leaves shaped like needles, which are coated in a waxy substance that helps reduce water loss. These forests have cold winters. In autumn, deciduous trees shed their leaves. In the spring, small plants burst from the ground and flower. Fertile soils are often rich in **humus,** a material formed from decaying leaves and other organic matter.

- **Abiotic factors** cold to moderate winters; warm summers; year-round precipitation; fertile soils
- **Biotic factors**
  Plant life: Deciduous trees drop their leaves and go into a state of dormancy in winter. Conifers have needlelike leaves that minimize water loss in dry winter air.
  Animal life: Animals must cope with changing weather. Some hibernate; others migrate to warmer climates. Animals that do not hibernate or migrate may be camouflaged to escape predation in the winter when bare trees leave them more exposed.

NORTHWESTERN CONIFEROUS FOREST

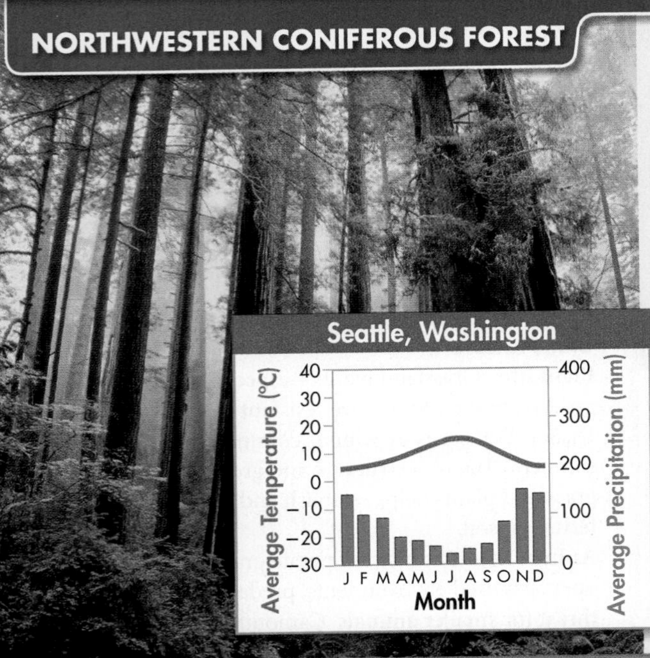

Mild moist air from the Pacific Ocean influenced by the Rocky Mountains provides abundant rainfall to this biome. The forest includes a variety of conifers, from giant redwoods to spruce, fir, and hemlock, along with flowering trees and shrubs such as dogwood and rhododendron. Moss often covers tree trunks and the forest floor. Because of its lush vegetation, the northwestern coniferous forest is sometimes called a "temperate rain forest."

- **Abiotic factors** mild temperatures; abundant precipitation in fall, winter, and spring; cool dry summers; rocky acidic soils
- **Biotic factors**
  Plant life: Because of seasonal temperature variation, there is less diversity in this biome than in tropical rain forests. However, ample water and nutrients support lush, dense plant growth. Adaptations that enable plants to obtain sunlight are common. Trees here are among the world's tallest.
  Animal life: Camouflage helps insects and ground-dwelling mammals avoid predation. Many animals are browsers—they eat a varied diet—an advantage in an environment where vegetation changes seasonally.

BOREAL FOREST

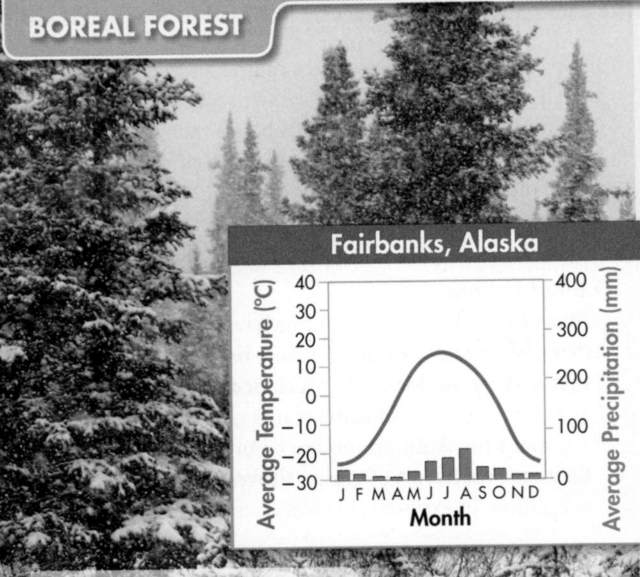

Dense forests of coniferous evergreens along the northern edge of the temperate zone are called boreal forests, or **taiga** (TY guh). Winters are bitterly cold, but summers are mild and long enough to allow the ground to thaw. The word *boreal* comes from the Greek word for "north," reflecting the fact that boreal forests occur mostly in the northern part of the Northern Hemisphere.

- **Abiotic factors** long cold winters; short mild summers; moderate precipitation; high humidity; acidic, nutrient-poor soils
- **Biotic factors**
  Plant life: Conifers are well suited to the boreal-forest environment. Their conical shape sheds snow, and their wax-covered needlelike leaves prevent excess water loss. In addition, the dark green color of most conifers absorbs heat energy.
  Animal life: Staying warm is the major challenge for animals. Most have small extremities and extra insulation in the form of fat or downy feathers. Some migrate to warmer areas in winter.

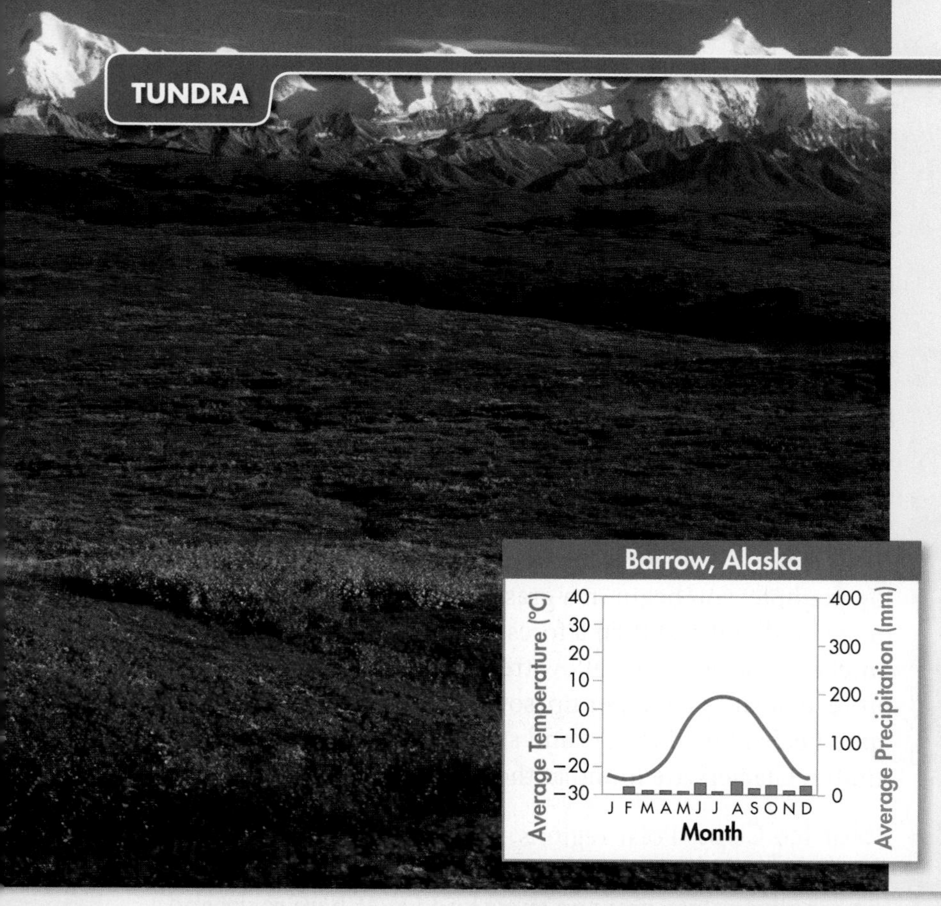

The tundra is characterized by **permafrost,** a layer of permanently frozen subsoil. During the short cool summer, the ground thaws to a depth of a few centimeters and becomes soggy. In winter, the top layer of soil freezes again. This cycle of thawing and freezing, which rips and crushes plant roots, is one reason that tundra plants are small and stunted. Cold temperatures, high winds, a short growing season, and humus-poor soils also limit plant height.

- **Abiotic factors** strong winds; low precipitation; short and soggy summers; long, cold, dark winters; poorly developed soils; permafrost
- **Biotic factors**
  Plant life: By hugging the ground, mosses and other low-growing plants avoid damage from frequent strong winds. Seed dispersal by wind is common. Many plants have adaptated to growth in poor soil. Legumes, for example, have nitrogen-fixing bacteria on their roots.
  Animal life: Many animals migrate to avoid long harsh winters. Animals that live in the tundra year-round display adaptations, among them natural antifreeze, small extremities that limit heat loss, and a varied diet.

Analyzing Data

Which Biome?

An ecologist collected climate data from two locations. The graph shows the monthly average temperatures in the two locations. The total yearly precipitation in Location A is 273 cm. In Location B, the total yearly precipitation is 11 cm.

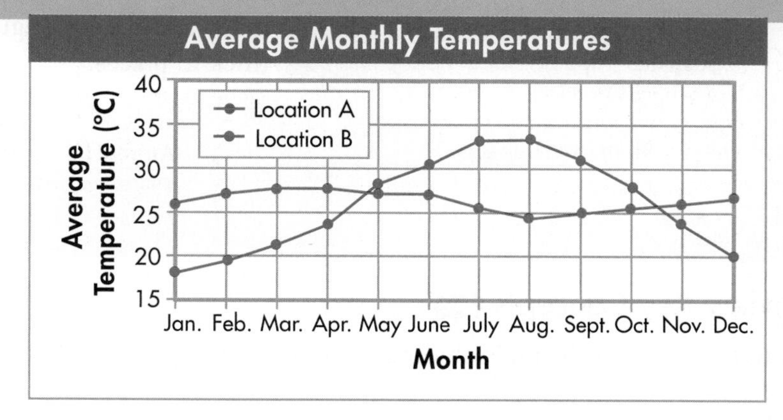

1. Interpret Graphs What variable is plotted on the horizontal axis? On the vertical axis?

2. Interpret Graphs How would you describe the temperature over the course of the year in Location A? In Location B?

3. Draw Conclusions In which biome would you expect to find each location, given the precipitation and temperature data? Explain your answer.

4. Analyze Data Look up the average monthly temperature last year in the city you live in. Plot the data. Then look up the monthly rainfall for your city, and plot those data. Based on your results, which biome do you live in? Did the data predict the biome correctly?

MYSTERY CLUE

Yellowstone has high mountain slopes and valleys with streams. Can you think of any reason why moose and elk might prefer to graze in one of those places rather than the other? How do you think their preference might affect Yellowstone's plant communities?

Other Land Areas

What areas are not easily classified into a major biome?

Some land areas do not fall neatly into one of the major biomes. **Because they are not easily defined in terms of a typical community of plants and animals, mountain ranges and polar ice caps are not usually classified into biomes.**

Mountain Ranges Mountain ranges exist on all continents and in many biomes. On mountains, conditions vary with elevation. From river valley to summit, temperature, precipitation, exposure to wind, and soil types all change, and so do organisms. If you climb the Rocky Mountains in Colorado, for example, you begin in a grassland. You then pass through pine woodland and then a forest of spruce and other conifers. Thickets of aspen and willow trees grow along streambeds in protected valleys. Higher up, soils are thin. Strong winds buffet open fields of wildflowers and stunted vegetation resembling tundra. Glaciers are found at the peaks of many ranges.

Polar Ice Caps Polar regions, like the one in **Figure 4–19,** border the tundra and are cold year-round. Plants are few, though some algae grow on snow and ice. Where rocks and ground are exposed seasonally, mosses and lichens may grow. Marine mammals, insects, and mites are the typical animals. In the north, where polar bears live, the Arctic Ocean is covered with sea ice, although more and more ice is melting each summer. In the south, the continent of Antarctica, inhabited by many species of penguins, is covered by ice nearly 5 kilometers thick in places.

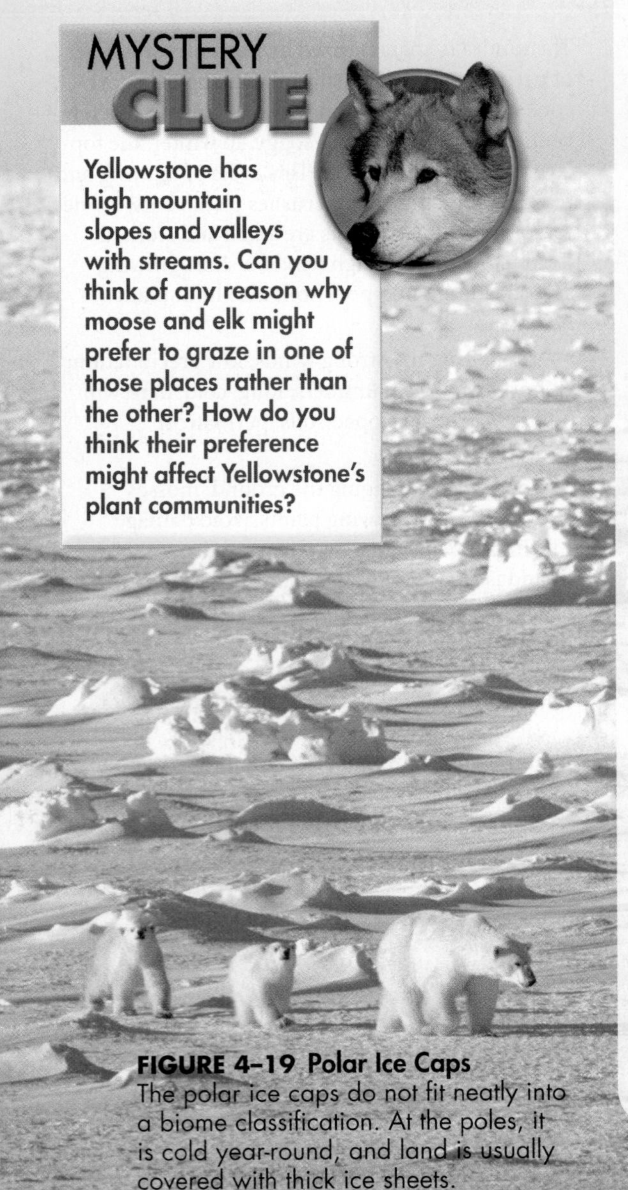

FIGURE 4–19 Polar Ice Caps
The polar ice caps do not fit neatly into a biome classification. At the poles, it is cold year-round, and land is usually covered with thick ice sheets.

4.4 Assessment

Review Key Concepts

1. **a. Review** List the major biomes, and describe one characteristic of each.

   b. Explain How are biomes classified?

   c. Compare and Contrast Choose two very different biomes. For each biome, select a common plant and animal. Compare how the plants and animals have adaptated to their biomes.

2. **a. Review** Why aren't mountain ranges or polar ice caps classified as biomes?

   b. Sequence Imagine that you are hiking up a mountain in the temperate forest biome. Describe how the plant life might change as you climb toward the summit.

Apply the Big idea

Interdependence in Nature

3. Choose one of the biomes discussed in this lesson. Then, sketch the biome. Include the biome's characteristic plant and animal life in your sketch. Add labels to identify the organisms, and write a caption describing the content of the sketch.

4.5 Aquatic Ecosystems

THINK ABOUT IT We call our planet "Earth," yet nearly three-fourths of Earth's surface is covered with water. Despite the vital roles aquatic ecosystems play in the biosphere, many of these ecosystems are only partly understood. What's life like underwater?

Key Questions

What factors affect life in aquatic ecosystems?

What are the major categories of freshwater ecosystems?

Why are estuaries so important?

How do ecologists usually classify marine ecosystems?

Vocabulary

photic zone • aphotic zone • benthos • plankton • wetland • estuary

Taking Notes

Compare/Contrast Table As you read, note the similarities and differences between the major freshwater and marine ecosystems in a compare/contrast table.

Conditions Underwater

What factors affect life in aquatic ecosystems?

Like organisms living on land, underwater organisms are affected by a variety of environmental factors. **Aquatic organisms are affected primarily by the water's depth, temperature, flow, and amount of dissolved nutrients.** Because runoff from land can affect some of these factors, distance from shore also shapes marine communities.

Water Depth Water depth strongly influences aquatic life because sunlight penetrates only a relatively short distance through water, as shown in **Figure 4–20.** The sunlit region near the surface in which photosynthesis can occur is known as the **photic zone.** The photic zone may be as deep as 200 meters in tropical seas, but just a few meters deep or less in rivers and swamps. Photosynthetic algae, called phytoplankton, live in the photic zone. Zooplankton—tiny free-floating animals—eat phytoplankton. This is the first step in many aquatic food webs. Below the photic zone is the dark **aphotic zone,** where photosynthesis cannot occur.

Many aquatic organisms live on, or in, rocks and sediments on the bottoms of lakes, streams, and oceans. These organisms are called the **benthos,** and their habitat is the benthic zone. Where water is shallow enough for the benthos to be within the photic zone, algae and rooted aquatic plants can grow. When the benthic zone is below the photic zone, chemosynthetic autotrophs are the only primary producers.

FIGURE 4–20 The Photic Zone Sunlight penetrates only a limited distance into aquatic ecosystems. Whatever the depth of this photic zone, it is the only area in which photosynthesis can occur. **Infer** ***Why do you think some photic zones are only a few meters deep and others are as much as 200 meters deep?***

Temperature and Currents Aquatic habitats, like terrestrial habitats, are warmer near the equator and colder near the poles. Temperature in aquatic habitats also often varies with depth. The deepest parts of lakes and oceans are often colder than surface waters. Currents in lakes and oceans can dramatically affect water temperature because they can carry water that is significantly warmer or cooler than would be typical for any given latitude, depth, or distance from shore.

Nutrient Availability As you learned in Chapter 3, organisms need certain substances to live. These include oxygen, nitrogen, potassium, and phosphorus. The type and availability of these dissolved substances vary within and between bodies of water, greatly affecting the types of organisms that can survive there.

What is one way in which life in Yellowstone's streams might be affected by the presence or absence of plants along stream banks?

Freshwater Ecosystems

What are the major categories of freshwater ecosystems?

Only 3 percent of Earth's surface water is fresh water, but that small percentage provides terrestrial organisms with drinking water, food, and transportation. Often, a chain of streams, lakes, and rivers begins in the interior of a continent and flows through several biomes to the sea. **Freshwater ecosystems can be divided into three main categories: rivers and streams, lakes and ponds, and freshwater wetlands.** Examples of these ecosystems are shown in **Figure 4–21.**

Rivers and Streams Rivers, streams, creeks, and brooks often originate from underground water sources in mountains or hills. Near a source, water has plenty of dissolved oxygen but little plant life. Downstream, sediments build up and plants establish themselves. Still farther downstream, water may meander slowly through flat areas. Animals in many rivers and streams depend on terrestrial plants and animals that live along their banks for food.

In Your Notebook *What kinds of adaptations would you expect in organisms living in a fast-flowing river or stream?*

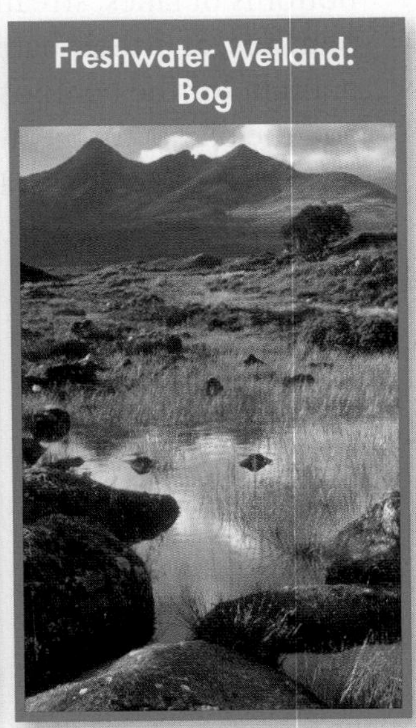

FIGURE 4–21 Freshwater Ecosystems and Estuaries Freshwater ecosystems include streams, lakes, and freshwater wetlands (bogs, swamps, and marshes). Salt marshes and mangrove swamps are estuaries—areas where fresh water from rivers meets salt water. **Interpret Visuals** ***Based on these photos, what are two differences between streams and bogs?***

Lakes and Ponds The food webs in lakes and ponds often are based on a combination of plankton and attached algae and plants. **Plankton** is a general term that includes both phytoplankton and zooplankton. Water typically flows in and out of lakes and ponds and circulates between the surface and the benthos during at least some seasons. This circulation distributes heat, oxygen, and nutrients.

Freshwater Wetlands A **wetland** is an ecosystem in which water either covers the soil or is present at or near the surface for at least part of the year. Water may flow through freshwater wetlands or stay in place. Wetlands are often nutrient-rich and highly productive, and they serve as breeding grounds for many organisms. Freshwater wetlands have important environmental functions: They purify water by filtering pollutants and help to prevent flooding by absorbing large amounts of water and slowly releasing it. Three main types of freshwater wetlands are freshwater bogs, freshwater marshes, and freshwater swamps. Saltwater wetlands are called estuaries.

Estuaries

Why are estuaries so important?

An **estuary** (es tyoo er ee) is a special kind of wetland, formed where a river meets the sea. Estuaries contain a mixture of fresh water and salt water, and are affected by the rise and fall of ocean tides. Many are shallow, which means that enough sunlight reaches the benthos to power photosynthesis. Estuaries support an astonishing amount of biomass—although they usually contain fewer species than freshwater or marine ecosystems—which makes them commercially valuable. **Estuaries serve as spawning and nursery grounds for many ecologically and commercially important fish and shellfish species including bluefish, striped bass, shrimp, and crabs.**

Salt marshes are temperate estuaries characterized by salt-tolerant grasses above the low-tide line and seagrasses below water. One of the largest salt marshes in America surrounds the Chesapeake Bay in Maryland (shown below). Mangrove swamps are tropical estuaries characterized by several species of salt-tolerant trees, collectively called mangroves. The largest mangrove area in America is in Florida's Everglades National Park (shown below).

Marine Ecosystems

How do ecologists usually classify marine ecosystems?

Just as biomes typically occupy certain latitudes and longitudes, marine ecosystems may typically occupy specific areas within the ocean. **Ecologists typically divide the ocean into zones based on depth and distance from shore.** Starting with the shallowest and closest to land, marine ecosystems include the intertidal zone, the coastal ocean, and the open ocean, as shown in **Figure 4–22.** Within these zones live a number of different communities.

In Your Notebook *How would you expect communities of organisms in the open ocean to differ from those along the coast?*

BUILD Vocabulary

MULTIPLE MEANINGS The noun *subject* has many meanings, including "the main theme of a piece of work such as a novel" or "a course of study." The verb *subject,* however, means "to expose" or "to tend toward." Organisms are subjected, or exposed, to extreme conditions in the rocky intertidal zone.

Intertidal Zone Organisms in the intertidal zone are submerged in seawater at high tide and exposed to air and sunlight at low tide. These organisms, then, are subjected to regular and extreme changes in temperature. They also are often battered by waves and currents. There are many different types of intertidal communities. A typical rocky intertidal community exists in temperate regions where exposed rocks line the shore. There, barnacles and seaweed permanently attach themselves to the rocks.

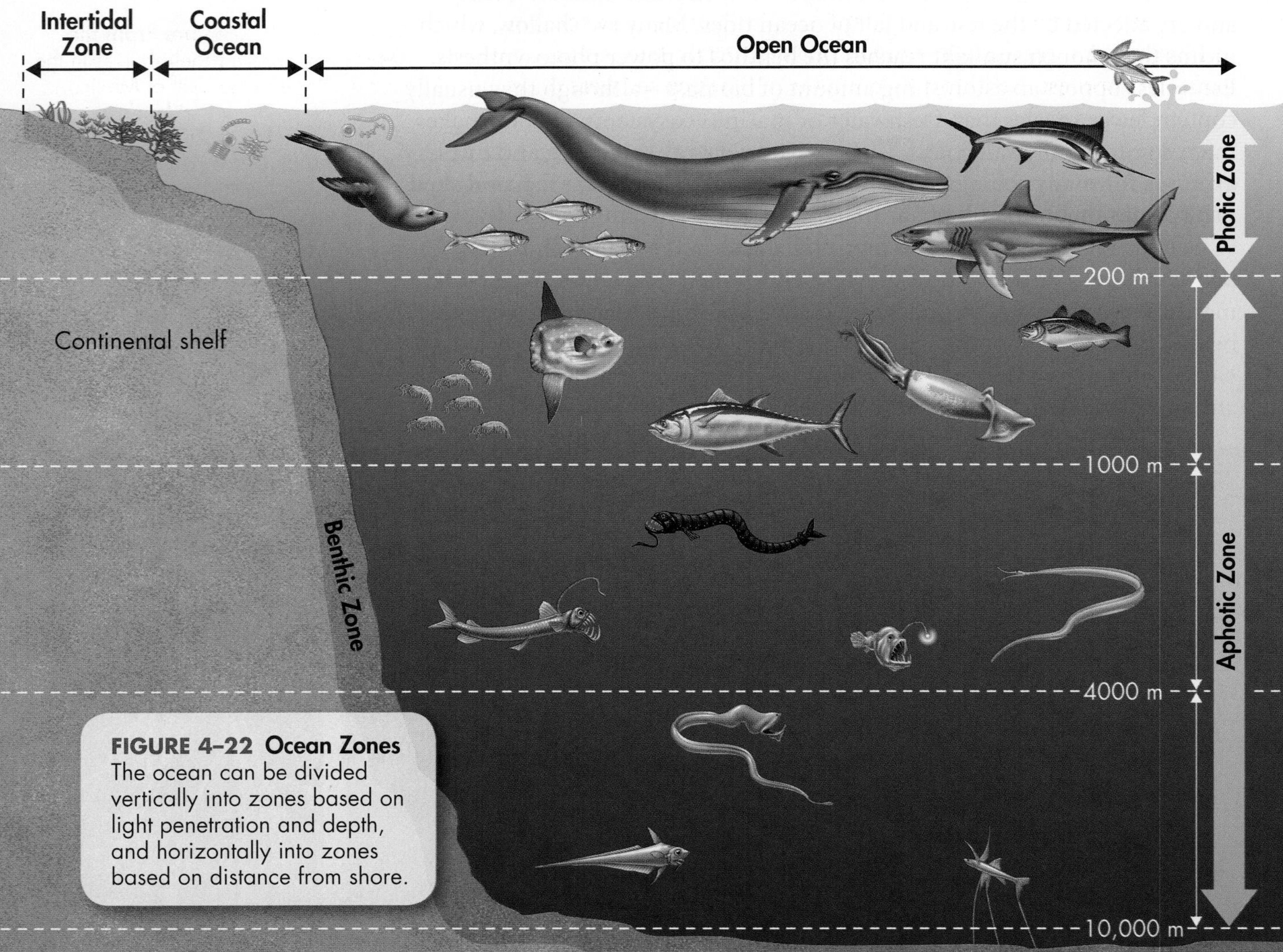

FIGURE 4–22 Ocean Zones The ocean can be divided vertically into zones based on light penetration and depth, and horizontally into zones based on distance from shore.

Coastal Ocean The coastal ocean extends from the low-tide mark to the outer edge of the continental shelf—the relatively shallow border that surrounds the continents. Water here is brightly lit, and is often supplied with nutrients by freshwater runoff from land. As a result, coastal oceans tend to be highly productive. Kelp forests and coral reefs are two exceptionally important coastal communities.

Open Ocean The open ocean begins at the edge of the continental shelf and extends outward. More than 90 percent of the world's ocean area is considered open ocean. Depth ranges from about 500 meters along continental slopes to more than 10,000 meters in deep ocean trenches. The open ocean can be divided into two main zones according to light penetration: the photic zone and the aphotic zone.

▶ ***The Open Ocean Photic Zone*** The open ocean typically has low nutrient levels and supports only the smallest species of phytoplankton. Still, because of its enormous area, most photosynthesis on Earth occurs in the sunlit top 100 meters of the open ocean.

▶ ***The Open Ocean Aphotic Zone*** The permanently dark aphotic zone includes the deepest parts of the ocean. Food webs here are based either on organisms that fall from the photic zone above, or on chemosynthetic organisms. Deep ocean organisms, like the fish in **Figure 4–23,** are exposed to high pressure, frigid temperatures, and total darkness. Benthic environments in the deep sea were once thought to be nearly devoid of life but are now known to have islands of high productivity. Deep-sea vents, where superheated water boils out of cracks on the ocean floor, support chemosynthetic primary producers.

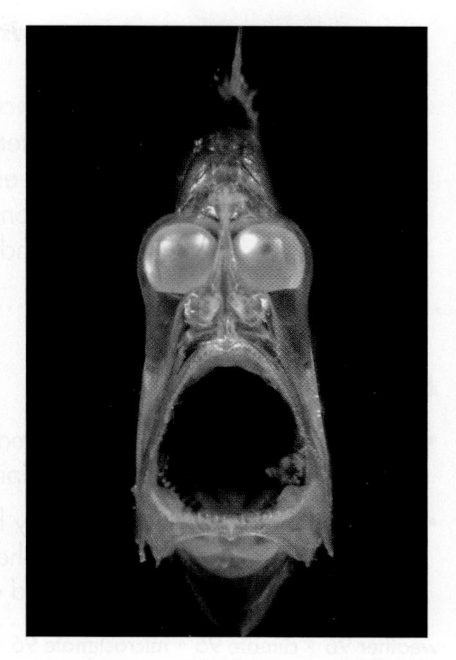

FIGURE 4–23 Creature From the Deep This silver hatchetfish lives in the aphotic zone of the Gulf of Mexico. **Apply Concepts** ***What kinds of adaptations do you think this fish has that enable it to live in the harsh deep-ocean environment?***

4.5 Assessment

Review Key Concepts

1. a. Review What are the primary abiotic factors that affect life underwater?

b. Compare and Contrast What are some ways in which life in an aphotic zone might differ from life in a photic zone?

2. a. Review What are the major categories of freshwater ecosystems?

b. Apply Concepts What is a wetland? Why are wetlands important?

3. a. Review Where are estuaries found? Why is it important to protect estuaries?

b. Predict How might a dam upriver affect an estuary at the river's mouth?

4. a. Review List the three major marine ecological zones. Give two abiotic factors for each zone.

b. Apply Concepts Using **Figure 4–22** as a guide, draw a cross section of the ocean starting with a beach and ending with an ocean trench. Label the intertidal zone, coastal ocean, and open ocean. Subdivide the open ocean into photic and aphotic zones.

WRITE ABOUT SCIENCE

Explanation

5. Choose three different aquatic ecosystems. For each of these ecosystems, select a plant and an animal, and explain how the organisms have adapted to their environment.

NGSS Smart Guide Ecosystems and Communities

Disciplinary Core Idea LS2.A Interdependent Relationships in Ecosystems: How do organisms interact with the living and nonliving environments to obtain matter and energy? Biotic factors such as competition and predation help shape communities, as do natural and human-caused disturbances.

4.1 Climate

- A region's climate is defined by year-after-year patterns of temperature and precipitation.
- Global climate is shaped by many factors, including solar energy trapped in the biosphere, latitude, and the transport of heat by winds and ocean currents.

weather 96 • climate 96 • microclimate 96 • greenhouse effect 97

Biology.com

Visual Analogy Compare Earth's atmosphere to a greenhouse.

4.2 Niches and Community Interactions

- A niche is the range of physical and biological conditions in which a species lives and the way the species obtains what it needs to survive and reproduce.
- By causing species to divide resources, competition helps determine the number and kinds of species in a community and the niche each species occupies.
- Predators can affect the size of prey populations in a community and determine the places prey can live and feed.
- Herbivores can affect both the size and distribution of plant populations in a community and can determine the places that certain plants can survive and grow.
- Biologists recognize three main classes of symbiotic relationships in nature: mutualism, parasitism, and commensalism.

tolerance 99 • habitat 99 • niche 100 • resource 100 • competitive exclusion principle 101 • predation 102 • herbivory 102 • keystone species 103 • symbiosis 103 • mutualism 103 • parasitism 104 • commensalism 104

Biology.com

Data Analysis Use data to explain the zonation patterns of intertidal species.

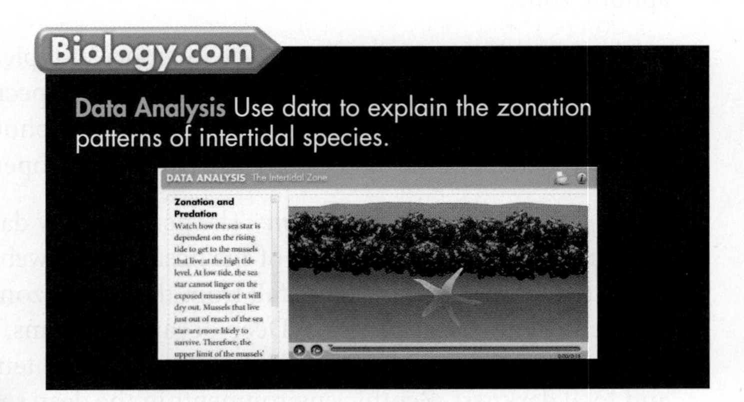

4.3 Succession

- Ecosystems change over time, especially after disturbances, as some species die out and new species move in.
- Secondary succession in healthy ecosystems following natural disturbances often reproduces the original climax community. Ecosystems may or may not recover from human-caused disturbances.

ecological succession 106 • primary succession 106 • pioneer species 107 • secondary succession 107

Biology.com

Untamed Science Video Join the Untamed Science crew as they explore succession after a volcanic eruption.

Art in Motion View a short animation that brings succession to life.

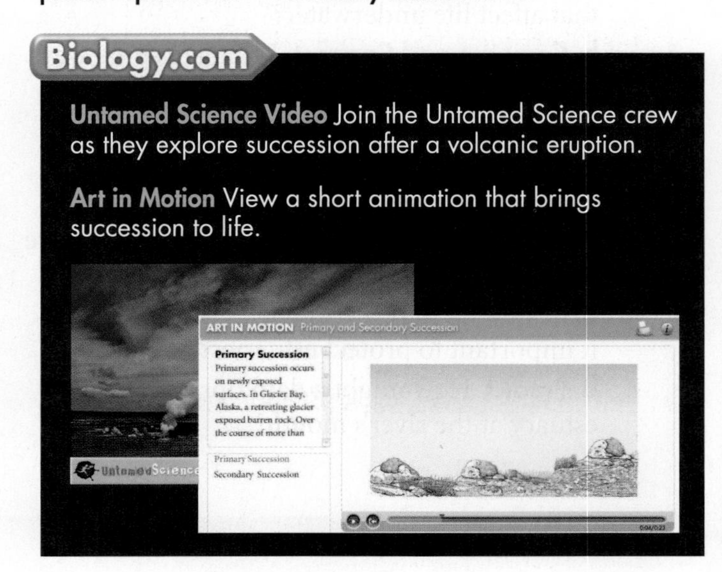

CHECKING YOUR *Scientific Literacy*

Refer to the lesson content and digital assets below as you prepare for your chapter assessment. Then, evaluate your understanding of ecosystems and communities by answering these questions.

1. **Scientific and Engineering Practice Constructing Explanations** Joshua Tree National Park has a variety of insect-eating lizards. Some lizards rise at sunrise; others are active at night. Some spend a lot of time underground, some bask on rocks, and some live in the branches of desert plants. Explain how these differences in activity times and locations allow the species to coexist while depending on similar resources.

2. **Crosscutting Concept Stability and Change** Primary succession and secondary succession are processes that occur after an environment is disturbed. What are the main differences between the two types of succession?

3. **Integration of Knowledge and Ideas** Use the biome map in Figure 4-18 to determine which biome exists to the west of the coastal mountain range in Washington State. Explain why this biome makes sense given the effect of winds, ocean currents, and mountains on climate.

4. **STEM** Wildlife biologists can use tracking technology to gather data about the movement of animals. Research the types of tags that are used. Focus on the advantages and disadvantages of each type of tag.

4.4 Biomes

- Biomes are described in terms of abiotic factors like climate and soil type, and biotic factors like plant and animal life.
- Mountain ranges and polar ice caps are not easy to define in terms of a typical community of plants and animals. So they are not usually classified into biomes.

canopy 112 • **understory 112** • **deciduous 112** • **coniferous 114** • **humus 114** • **taiga 114** • **permafrost 115**

Biology.com

In Your Notebook Use a biome map to determine the biome for where you live.

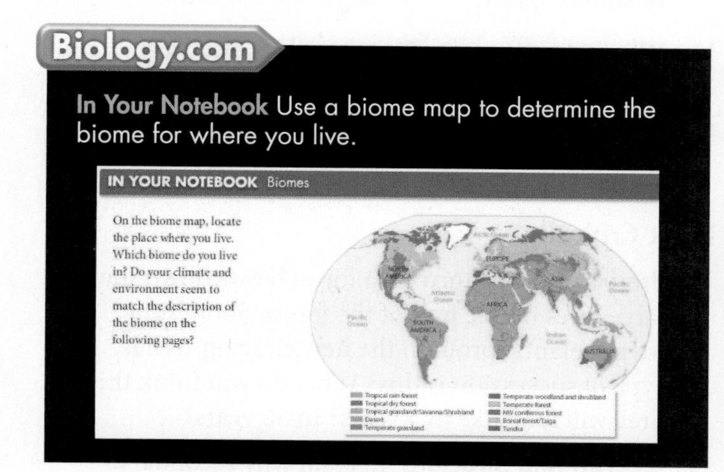

Real World Lab

Abiotic Factors and Plant Selection This chapter lab is available in your lab manual and at Biology.com

4.5 Aquatic Ecosystems

- Aquatic organisms are affected primarily by the water's depth, temperature, flow, and amount of dissolved nutrients.
- Freshwater ecosystems can be divided into three main categories: rivers and streams, lakes and ponds, and freshwater wetlands.
- Estuaries serve as spawning and nursery grounds for many ecologically and commercially important fish and shellfish species.
- Ecologists typically divide the ocean into zones based on depth and distance from shore.

photic zone 117 • **aphotic zone 117** • **benthos 117** • **plankton 119** • **wetland 119** • **estuary 119**

Biology.com

Art Review Review your understanding of ocean zones with this drag-and-drop activity.

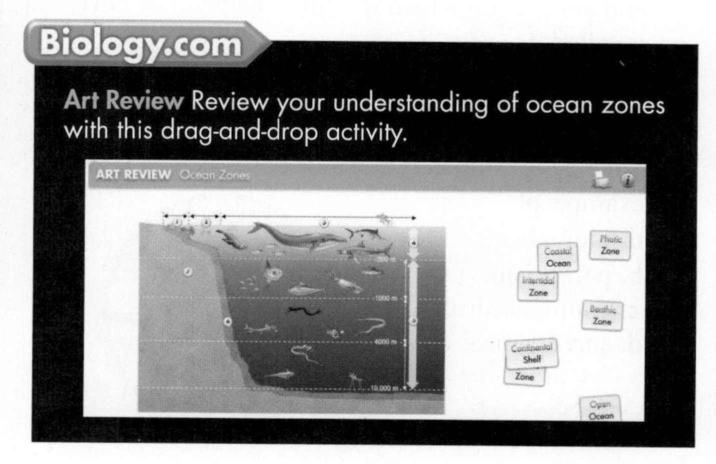

4 Assessment

4.1 Climate

Understand Key Concepts

1. An increase in the greenhouse effect causes an increase in
   a. carbon dioxide.
   b. temperature.
   c. oxygen.
   d. water.
2. A small valley where the average temperature is usually higher than that of the surrounding countryside has its own
   a. weather.
   b. climate.
   c. rainfall.
   d. microclimate.
3. Distinguish between weather and climate.
4. Describe the three primary abiotic factors that produce Earth's major climate zones.

Think Critically

5. **Apply Concepts** Based on the relative positions of the sun and Earth, explain why Earth has climate zones and seasons.
6. **Infer** A plant grower has a greenhouse where she grows plants in the winter. The greenhouse is exposed to direct sunlight and often gets too hot for the plants. She paints the inside of the glass with a chalky white paint, and the temperature drops to comfortable levels. Explain why this procedure works.

4.2 Niches and Community Interactions

Understand Key Concepts

7. A relationship in which one organism is helped and another organism is neither helped nor hurt is called
   a. parasitism.
   b. mutualism.
   c. competition.
   d. commensalism.
8. The relationship between a tick and its host is an example of
   a. mutualism.
   b. parasitism.
   c. commensalism.
   d. succession.
9. What is the difference between an organism's habitat and its niche?
10. What is the competitive exclusion principle?

Think Critically

11. **Craft and Structure** How are predation and parasitism similar? How are they different?
12. **Infer** Competition for resources in an area is usually more intense within a single species than between two different species. How would you explain this observation?
13. **Production and Distribution of Writing** Write a description of your niche in the environment. Include details about your ecosystem, and the biotic and abiotic factors around you. Describe your feeding habits as well as any interactions you have with members of other species.

4.3 Succession

Understand Key Concepts

14. Fires, hurricanes, and other natural disturbances can result in
    a. commensalism.
    b. competition.
    c. parasitism.
    d. succession.
15. The first organisms to repopulate an area affected by a volcanic eruption are called
    a. keystone species.
    b. climax species.
    c. primary producers.
    d. pioneer species.
16. What type of succession takes place after lava from a volcanic eruption covers an area?
17. Describe two major causes of ecological succession.

Think Critically

18. **Predict** A windstorm in a forest blows down the large trees in one part of the forest. Soon, sun-loving plants sprout in the new clearing. What type of succession is this? What do you think this area will look like in 5 years? In 50 years?
19. **Craft and Structure** Explain why secondary succession usually proceeds faster than primary succession.

4.4 Biomes

Understand Key Concepts

20. In a tropical rain forest, the dense covering formed by the leafy tops of tall trees is called the
a. canopy. **c.** niche.
b. taiga. **d.** understory.

21. Permafrost characterizes the biome called
a. taiga. **c.** savanna.
b. boreal forest. **d.** tundra.

22. What is a biome?

23. Why are plants generally few and far between in a desert?

Think Critically

24. Apply Concepts Although the amount of precipitation is low, most parts of the tundra are very wet during the summer. How would you explain this apparent contradiction?

25. Craft and Structure Deciduous trees in tropical dry forests lose water through their leaves every day. During summers with adequate rain, the leaves remain on the trees. During the cold dry season, the trees drop their leaves. In an especially dry summer, how might the adaptation of dropping leaves enable a tree to tolerate the drought?

26. Infer Consider these two biomes: (1) the temperate grassland and (2) the temperate woodland and shrubland. Coyotes live in both biomes. Describe two adaptations that might enable coyotes to tolerate conditions in both biomes.

4.5 Aquatic Ecosystems

Understand Key Concepts

27. Organisms that live near or on the ocean floor are called
a. parasites. **c.** plankton.
b. benthos. **d.** mangroves.

28. What is the meaning of the term *plankton*? Name the two types of plankton.

29. What are three types of freshwater wetlands?

30. How are salt marshes and mangrove swamps alike? How are they different?

solve the CHAPTER MYSTERY

THE WOLF EFFECT

Eliminating wolves from Yellowstone National Park contributed to an increase in the number of elk. These elk grazed so heavily, especially along streams, that the seedlings and shoots of aspens and willows, and other trees, could not grow. Fewer trees led to fewer dams being built by beavers and to an increase in runoff and erosion. Aquatic food webs broke down, affecting birds, fish, and other animals. The recent reintroduction of wolves has caused a decrease in the overall elk population and seems to have reduced elk grazing along certain streams. That may be in part because wolves are killing more elk and in part because elk have learned to stay away from places like stream banks and valleys, where wolves can attack them most easily.

In recent years, researchers have shown that streamside vegetation is exhibiting secondary succession and that aspen and willow trees are starting to grow back. There have been other changes as well. Fewer elk mean more food for smaller animals. The increase in small prey has brought diverse predators into the community. Carcasses abandoned by the wolves provide food for scavengers. In short, organisms from every trophic level have been affected by the Yellowstone wolves.

1. Predict The Yellowstone wolf and elk are linked through a predator-prey relationship. If a disease were to strike the elk population, how would the wolves be affected?

2. Text Types and Purposes Yellowstone is owned by the federal government. The reintroduction of wolves there angered nearby farmers because they feared their animals would be hunted. What level of responsibility do you think national parks should have toward their neighbors? Write an argument to support your claim.

3. Connect to the Big idea Draw a food chain that connects Yellowstone's wolves, aspen and willow trees, and elk. Then write a paragraph that explains why the Yellowstone wolves are a keystone species.

Think Critically

31. Form a Hypothesis The deep ocean lies within the aphotic zone and is very cold. Suggest some of the unique characteristics that enable animals to live in the deep ocean.

32. Form an Opinion A developer has proposed filling in a salt marsh to create a coastal resort. What positive and negative effects do you think this proposal would have on wildlife and local residents? Would you support the proposal?

Connecting Concepts

Use Science Graphics

The following table presents primary productivity (measured in grams of organic matter produced per year per square meter) for several ecosystems. Use the table below to answer questions 33–36.

Productivity of Aquatic and Land Ecosystems

Ecosystem	Average Primary Productivity
Aquatic Ecosystems	
Coral reef	2500
Estuary	1800
Open ocean	125
Land Ecosystems	
Tropical rain forest	2200
Tropical savanna	900
Tundra	90

33. Integration of Knowledge and Ideas According to the table, which ecosystem is most productive? Use what you know to explain that fact.

34. Infer The open ocean is among the least productive ecosystems, yet it contributes greatly to the overall productivity of the biosphere. How do you explain this paradox?

35. Apply Concepts For each set of ecosystems, aquatic and land, explain how abiotic factors may account for the differences in primary productivity seen. Give two examples.

36. Infer Review the description of the Northwest coniferous forest on page 114. Do you think its average primary productivity is greater or less than that of the tropical savanna? Explain your answer.

Write About Science

37. Research to Build and Present Knowledge Choose one of the ten major biomes, and write an overview of its characteristics. Explain how abiotic factors and common plants and wildlife are interrelated. Support your analysis with specific examples.

38. Assess the Big idea How do abiotic factors influence what kinds of organisms are involved in the primary succession in an area following a volcanic eruption?

Integration of Knowledge and Ideas

The graph here summarizes the changes in the total volume of ice in all the world's glaciers since 1960. Note that the volume changes on the y-axis are negative, meaning an overall loss of volume.

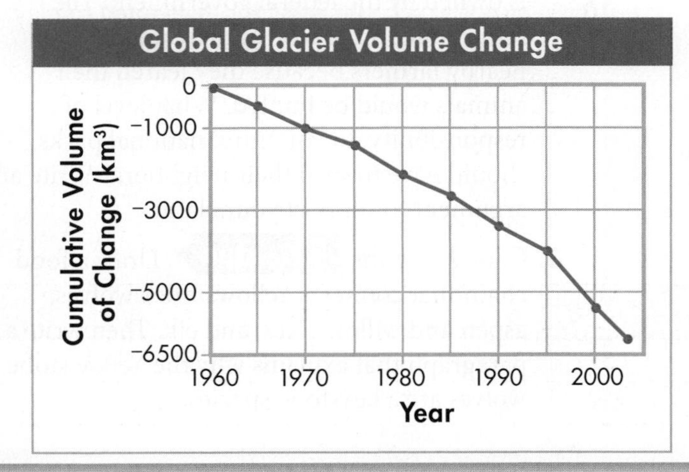

39. Interpret Graphs The greatest volume of glacial ice was lost

a. between 1960 and 1970.
b. between 1980 and 1990.
c. between 1995 and 2000.
d. before 1960.

40. Relate Cause and Effect The most reasonable explanation for the loss of glacier mass since 1960 is

a. an increase in the total productivity of the world's oceans.
b. a gradual rise in Earth's average temperature.
c. an increase in the total amount of ice at Earth's poles.
d. an increase in the sun's output of radiant energy.

Standardized Test Prep

Multiple Choice

1. The factor that generally has the greatest effect on determining a region's climate is its
A longitude.
B abundant plant species.
C distance from the equator.
D closeness to a river.

2. All of the following are abiotic factors that affect global climate EXCEPT
A latitude. **C** solar energy.
B longitude. **D** ocean currents.

3. The way an organism makes its living, including its interactions with biotic and abiotic factors of its environment, is called the organism's
A habitat. **C** lifestyle.
B niche. **D** biome.

4. If a newly introduced species fills a niche that is normally occupied by a native species, the two species compete. One of the species may die out as a result of
A competitive exclusion.
B predation.
C commensalism.
D mutualism.

5. Photosynthetic algae are MOST likely to be found in
A the open-ocean benthic zone.
B the aphotic zone.
C the photic zone.
D ocean trenches.

6. The water in an estuary is
A salt water only.
B poor in nutrients.
C fresh water only.
D a mixture of fresh water and salt water.

7. In which biome do organisms have the greatest tolerance to dry conditions?
A tundra **C** tropical savanna
B desert **D** boreal forest

Questions 8–9

Month-by-month climate data for the city of Lillehammer, Norway, is shown in the table below.

Climate Data for Lillehammer, Norway

Month	Average Temperature (°C)	Average Precipitation (mm)
Jan.	−8.1	38.1
Feb.	−6.2	27.9
Mar.	−3.9	30.5
Apr.	3.3	35.6
May	8.9	45.7
June	13.9	63.5
July	16.4	81.3
Aug.	14.2	88.9
Sept.	9.5	58.4
Oct.	3.9	63.5
Nov.	−3.8	50.8
Dec.	−6.1	48.3

8. Which type of graph would be BEST suited to showing the precipitation data from the table?
A bar graph **C** pie chart
B pictograph **D** scatter plot

9. For a given set of data, the range is the difference between highest and lowest points. The average annual temperature range, in °C, for Lillehammer is approximately
A −8.
B 8.5.
C 16.5.
D 24.5.

Open-Ended Response

10. Why are lichens especially well adapted to play the role of pioneer organisms in an ecological succession?

If You Have Trouble With . . .

Question	1	2	3	4	5	6	7	8	9	10
See Lesson	4.1	4.1	4.2	4.2	4.5	4.5	4.4	4.1	4.1	4.3

5 Populations

Big idea: Interdependence in Nature

Q: What factors contribute to changes in populations?

Millions of red crabs live on Christmas Island in the Indian Ocean. Each year the entire adult crab population migrates from forest to sea to breed, causing roads to be closed to traffic.

INSIDE:

- 5.1 How Populations Grow
- 5.2 Limits to Growth
- 5.3 Human Population Growth

BUILDING *Scientific Literacy*

Deepen your understanding of population growth by engaging in key practices that allow you to make connections across concepts.

NGSS Use quantitative thinking and construct models to show how growth patterns for populations can vary.

STEM What could cause a dramatic and widespread drop in the size of bee populations? View a video about bees to learn more about this problem.

Common Core You will gather information about an invasive species from multiple authoritative sources.

CHAPTER MYSTERY

A PLAGUE OF RABBITS

In 1859, an Australian farmer released 24 wild European rabbits from England on his ranch. "A few rabbits," he said, "could do little harm and might provide a touch of home, in addition to a spot of hunting."

Seven years later, he and his friends shot 14,253 rabbits. In ten years, more than 2 million rabbits were hunted on that farm alone! But hunters' glee turned into nationwide despair. That "touch of home" was soon covering the countryside like a great gray blanket. The millions of rabbits devoured native plants and pushed native animals to near extinction. They made life miserable for sheep and cattle ranchers.

These cute, fuzzy creatures weren't a problem in England. Why did they turn into a plague in Australia? Could they be stopped? How? As you read this chapter, look for clues on factors that affect population growth. Then, solve the mystery.

Biology.com

Finding the solution to the rabbit population mystery is only the beginning. Take a video field trip with the ecogeeks of Untamed Science to see where this mystery leads.

Go online to access additional resources including:
• eText • Flash Cards • Lesson Overviews • Chapter Mystery

5.1 How Populations Grow

Key Questions

- *How do ecologists study populations?*
- *What factors affect population growth?*
- *What happens during exponential growth?*
- *What is logistic growth?*

Vocabulary

population density • age structure • immigration • emigration • exponential growth • logistic growth • carrying capacity

Taking Notes

Concept Map As you read, use the highlighted vocabulary words to create a concept map that organizes the information in this lesson.

THINK ABOUT IT In the 1950s, a fish farmer in Florida tossed a few plants called hydrilla into a canal. Hydrilla was imported from Asia for use in home aquariums because it is hardy and adaptable. The fish farmer assumed that hydrilla was harmless. But the few plants he tossed away reproduced quickly . . . and kept on reproducing. Today, their offspring strangle waterways across Florida and many other states. Tangled stems snag boats in rivers and overtake habitats; native water plants and animals are disappearing. Why did these plants get so out of control? Is there any way to get rid of them?

Meanwhile, people in New England who fish for a living face a different problem. Despite hard work and new equipment, their catch has dropped dramatically. The cod catch in one recent year was 3048 metric tons. Back in 1982, it was 57,200 metric tons—almost 19 times higher! Where did all the fish go? Can anything be done to increase their numbers?

Describing Populations

How do ecologists study populations?

At first glance, the stories of hydrilla and cod may seem unrelated. One is about plants growing out of control, and the other is about fish disappearing. Yet both involve dramatic changes in the size of a population. Recall that a population is a group of organisms of a single species that lives in a given area. **Researchers study populations' geographic range, density and distribution, growth rate, and age structure.**

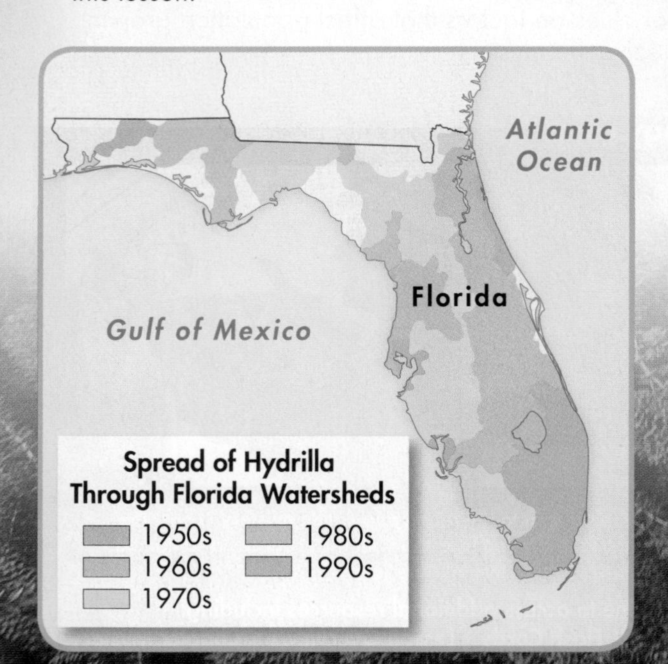

FIGURE 5–1 Invasive Hydrilla Hydrilla has spread through most of Florida in just a few decades. Efforts to control the waterweed cost millions of dollars a year.

Geographic Range The area inhabited by a population is called its geographic range. A population's range can vary enormously in size, depending on the species. A bacterial population in a rotting pumpkin, for example, may have a range smaller than a cubic meter. The population of cod in the western Atlantic, on the other hand, covers a range that stretches from Greenland down to North Carolina. The natural range of one hydrilla population includes parts of southern India and Sri Lanka. The native range of another hydrilla population was in Korea. But humans have carried hydrilla to so many places that its range now includes every continent except Antarctica, and it is found in many places in the United States.

Density and Distribution **Population density** refers to the number of individuals per unit area. Populations of different species often have very different densities, even in the same environment. For example, a population of ducks in a pond may have a low density, while fish in the same pond community may have a higher density. *Distribution* refers to how individuals in a population are spaced out across the range of the population—randomly, uniformly, or mostly concentrated in clumps, as shown in **Figure 5–2.**

Growth Rate A population's growth rate determines whether the size of the population increases, decreases, or stays the same. Hydrilla populations in their native habitats tend to stay more or less the same size over time. These populations have a growth rate of around zero. In other words, they neither increase nor decrease in size. The hydrilla population in Florida, by contrast, has a high growth rate—which means that it increases in size. Populations can also decrease in size, as cod populations have been doing. The cod population has a negative growth rate.

Age Structure To fully understand a plant or animal population, researchers need to know more than just the number of individuals it contains. They also need to know the population's **age structure**—the number of males and females of each age a population contains. Why? Because most plants and animals cannot reproduce until they reach a certain age. Also, among animals, only females can produce offspring.

A. Random

B. Uniform

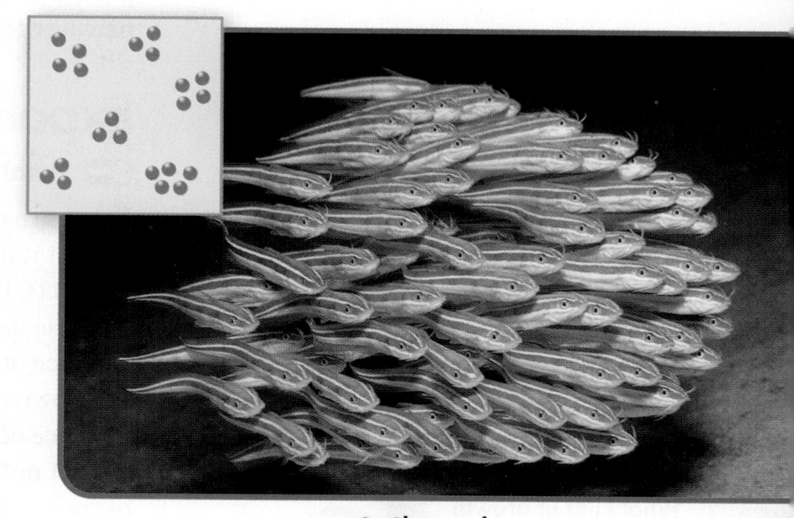

C. Clumped

FIGURE 5–2 Patterns of Distribution The dots in the inset illustrations represent individual members of a population. **A.** Purple lupines grow randomly in a field of wildflowers. **B.** King penguin populations show uniform spacing between individuals. **C.** Striped catfish form tight clumps.

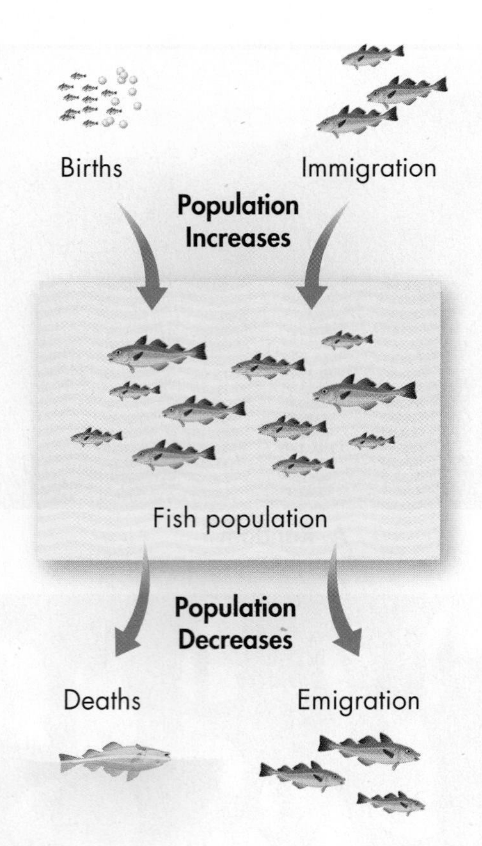

FIGURE 5–3 Natural Factors That Affect the Growth of a Fish Population The numbers of fish that hatch, die, enter, or leave the population affect the growth of the population. **Use Models** ***How would you expand this model to include the effects of fishing?***

Population Growth

What factors affect population growth?

What determines whether a population grows, shrinks, or stays the same size? A population will increase or decrease in size depending on how many individuals are added to it or removed from it, as shown in **Figure 5–3.** **The factors that can affect population size are the birthrate, death rate, and the rate at which individuals enter or leave the population.**

Birthrate and Death Rate Populations can grow if more individuals are born than die in any period of time. In other words, a population can grow when its birthrate is higher than its death rate. If the birthrate equals the death rate, the population may stay the same size. If the death rate is greater than the birthrate, the population is likely to shrink. Note that *birth* means different things in different species. Lions are born much like humans are born. Codfish, however, release eggs that hatch into new individuals.

Immigration and Emigration A population may grow if individuals move into its range from elsewhere, a process called **immigration** (im uh GRAY shun). Suppose, for example, that an oak grove in a forest produces a bumper crop of acorns one year. The squirrel population in that grove may increase as squirrels immigrate in search of food. On the other hand, a population may decrease in size if individuals move out of the population's range, a process called **emigration** (em uh GRAY shun). For example, a local food shortage or overcrowding can cause emigration. Young animals approaching maturity may emigrate from the area where they were born to find mates or establish new territories.

Exponential Growth

What happens during exponential growth?

If you provide a population with all the food and space it needs, protect it from predators and disease, and remove its waste products, the population will grow. Why? The population will increase because members of the population will be able to produce offspring. After a time, those offspring will produce their own offspring. Then, the offspring of *those* offspring will produce offspring. So, over time, the population will grow.

But notice that something interesting will happen: The size of each generation of offspring will be larger than the generation before it. This situation is called exponential (eks poh NEN shul) growth. In **exponential growth,** the larger a population gets, the faster it grows. **Under ideal conditions with unlimited resources, a population will grow exponentially.** Let's examine why this happens under different situations.

Organisms That Reproduce Rapidly We begin a hypothetical experiment with a single bacterium that divides to produce two cells every 20 minutes. We supply it with ideal conditions—and watch. After 20 minutes, the bacterium divides to produce two bacteria. After another 20 minutes, those two bacteria divide to produce four cells. At the end of the first hour, those four bacteria divide to produce eight cells.

Do you see what is happening here? After three 20-minute periods, we have $2 \times 2 \times 2$, or 8 cells. Another way to say this is to use an exponent: 2^3 cells. In another hour (six 20-minute periods), there will be 2^6, or 64 bacteria. In just one more hour, there will be 2^9, or 512. In one day, this bacterial population will grow to an astounding 4,720,000,000,000,000,000,000 individuals. What would happen if this growth continued without slowing down? In a few days, this bacterial population would cover the planet!

If you plot the size of this population on a graph over time, you get a J-shaped curve that rises slowly at first, and then rises faster and faster, as shown in **Figure 5–4.** If nothing interfered with this kind of growth, the population would become larger and larger, faster and faster, until it approached an infinitely large size.

Organisms That Reproduce Slowly Of course, many organisms grow and reproduce much more slowly than bacteria. For example, a female elephant can produce a single offspring only every 2 to 4 years. Newborn elephants take about 10 years to mature. But as you can see in **Figure 5–4,** if exponential growth continued, the result would be impossible. In the unlikely event that all descendants of a single elephant pair survived and reproduced, after 750 years there would be nearly 20 million elephants!

Organisms in New Environments Sometimes, when an organism is moved to a new environment, its population grows exponentially for a time. That's happening with hydrilla in the United States. It also happened when a few European gypsy moths were accidentally released from a laboratory near Boston. Within a few years, these plant-eating pests had spread across the northeastern United States. In peak years, they devoured the leaves of thousands of acres of forest. In some places, they formed a living blanket that covered the ground, sidewalks, and cars.

In Your Notebook *Draw a growth curve for a population of waterweed growing exponentially.*

BUILD Vocabulary

RELATED WORD FORMS An exponent indicates the number of times a number is multiplied by itself. The adjective *exponential* describes something that is expressed using exponents—such as the rate of growth.

Models of Exponential Growth

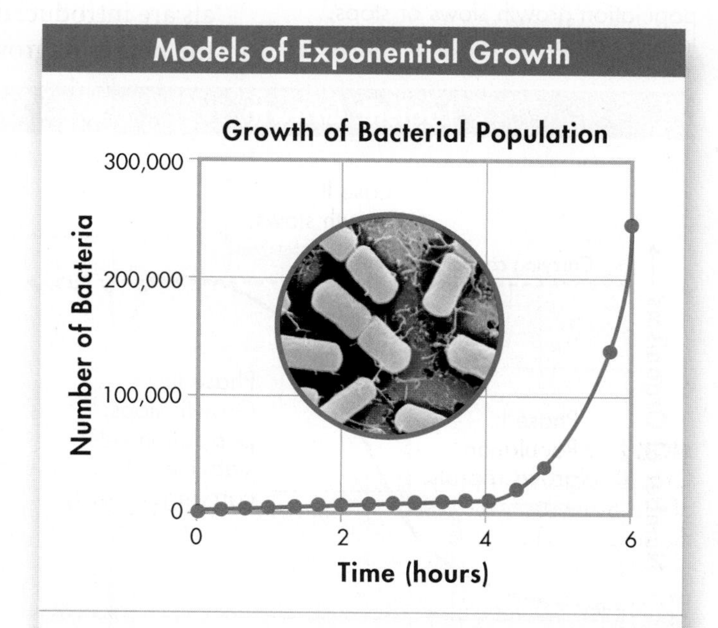

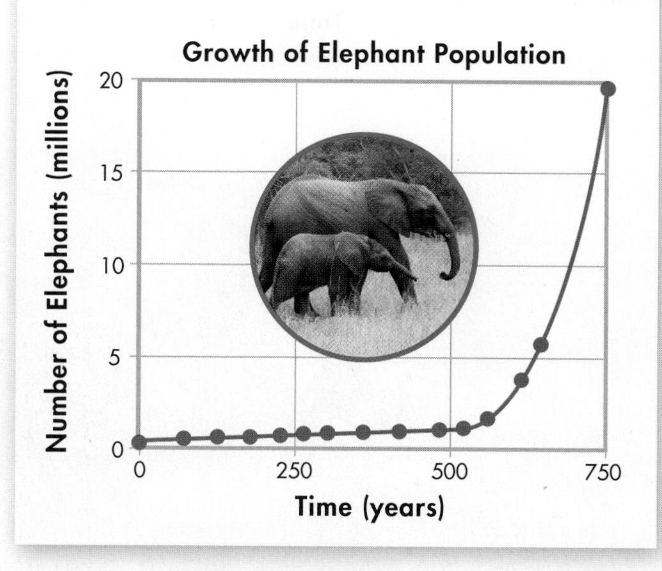

FIGURE 5–4 Exponential Growth In the presence of unlimited resources and in the absence of predation and disease, populations will grow exponentially. Bacteria, which reproduce rapidly, can produce huge populations in a matter of days. It would take elephants, which reproduce slowly, a few hundred years. Both hypothetical graphs show the characteristic J-shape of exponential growth.

Logistic Growth

What is logistic growth?

This ability of populations to grow exponentially presents a puzzle. Obviously, bacteria, elephants, hydrilla, and gypsy moths don't cover the Earth. This means that natural populations don't grow exponentially for long. Sooner or later, something—or several "somethings"—stops exponential growth. What happens?

Phases of Growth One way to begin answering this question is to watch how populations behave in nature. Suppose that a few individuals are introduced into a real-world environment. **Figure 5–5** traces the phases of growth that the population goes through.

VISUAL SUMMARY

LOGISTIC GROWTH

FIGURE 5–5 Real-world populations, such as those of the rhinoceros, show the characteristic S-shaped curve of logistic growth. As resources become limited, population growth slows or stops, leveling off at the carrying capacity.

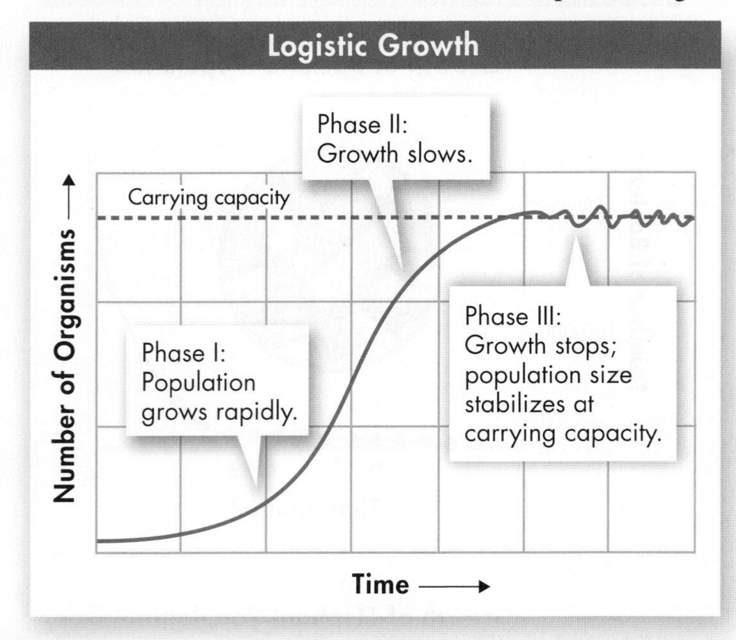

▶ ***Phase 1: Exponential Growth*** After a short time, the population begins to grow exponentially. During this phase, resources are unlimited, so individuals grow and reproduce rapidly. Few individuals die, and many offspring are produced, so both the population size and the rate of growth increase more and more rapidly.

▶ ***Phase 2: Growth Slows Down.*** In real-world populations, exponential growth does not continue for long. At some point, the rate of population growth begins to slow down. This does not mean that the population size decreases. The population still grows, but the rate of growth slows down, so the population size increases more slowly.

▶ ***Phase 3: Growth Stops.*** At some point, the rate of population growth drops to zero. This means that the size of the population levels off. Under some conditions, the population will remain at or near this size indefinitely.

The Logistic Growth Curve The curve in **Figure 5–5** has an S-shape that represents what is called **logistic growth.** **Logistic growth occurs when a population's growth slows and then stops, following a period of exponential growth.** Many familiar plant and animal populations follow a logistic growth curve.

What kinds of changes in a population's characteristics can produce logistic growth? Remember that a population grows when more organisms are born (or added to it) than die (or leave it). Thus, population growth may slow for several reasons. Growth may slow because the population's birthrate decreases. Growth may also slow if the death rate increases—or if births fall and deaths rise together. Similarly, population growth may slow if the rate of immigration decreases, the rate of emigration increases, or both. There are several reasons why these rates might change in a population, as you will see in the next lesson.

Carrying Capacity When the birthrate and the death rate are the same, and when immigration equals emigration, population growth stops. The population may still rise and fall somewhat, but the ups and downs average out around a certain population size. If you look again at **Figure 5–5,** you will see a broken, horizontal line through the region of the graph where population growth levels off. The point at which that line intersects the y-axis represents what ecologists call the carrying capacity. **Carrying capacity** is the maximum number of individuals of a particular species that a particular environment can support. Once a population reaches the carrying capacity of its environment, a variety of factors act to stabilize it at that size.

Analyzing Data

Multiplying Rabbits

Suppose that a pair of rabbits produces six offspring: three males and three females. Assume that no offspring die.

1. Calculate If each pair of rabbits breeds only once, how many offspring would be produced each year for five generations? MATH

2. Interpret Graphs Construct a graph of your data. Plot time on the x-axis and population on the y-axis. What type of growth is the rabbit population going through after 5 years?

5.1 Assessment

Review Key Concepts

1. a. **Review** List four characteristics that are used to describe a population.
   b. **Infer** On your travels through eastern Canada and the United States, you notice gray squirrels everywhere. What can you infer about the squirrels' geographic range?
2. a. **Review** What natural factors can change a population's size?
   b. **Relate Cause and Effect** More dandelion seedlings develop in a lawn than dandelion plants are removed. What is likely to happen to the lawn's dandelion population?
3. a. **Review** When do populations grow exponentially?
   b. **Apply Concepts** Why does exponential growth show a characteristic J-shaped curve?
4. a. **Review** What is the characteristic shape of a logistic growth curve?
   b. **Explain** Describe when logistic growth occurs.
   c. **Form a Hypothesis** What factors might cause the carrying capacity of a population to change?

PRACTICE PROBLEM

5. Suppose you are studying a population of sunflowers growing in a small field. How would you determine the population density of sunflowers in a square meter of the field and in the entire field? Describe your procedure.

WHST.9-10.7 Research to Build and Present Knowledge. Also RST.9-10.10, WHST.9-10.10

Biology & Society

What Can Be Done About Invasive Mussels?

It's hard to imagine that shellfish could cause millions of dollars worth of trouble every year. Meet the zebra mussel and the quagga mussel. Both species were carried to the Great Lakes in the mid-1980s from Eastern Europe in ships' ballast waters (water carried inside boats for balance). As adults, these mussels attach to almost any hard surface, including water pipes and boat hulls. After just a few years, both species colonized the entire Great Lakes region. Since then, they have been spread by recreational boaters who unknowingly carry mussels attached to their boats. By 2008, zebra mussels had been reported in 24 states; quagga mussels are already known in 14 states.

Why have these mussels become such pests? In American waterways, they escape whatever environmental factors keep their numbers in check in their native European habitats. As a result, these introduced species have become invasive species whose exponential growth produces huge populations at high densities—over 10,000 mussels per square meter of water in some places! These mussels grow in layers up to 20 centimeters thick, clogging water pipes that supply power plants and water treatment facilities. They also upset aquatic food webs, filtering so much plankton from the water that some native fishes and shellfish starve. What can be done to control such invasive species?

The Viewpoints

Invasive Species Should Be Destroyed A number of groups contend that zebra mussels should be removed completely. Some engineers are developing robotic submarines that can remove mussels from pipes. Some chemists are testing chemicals for the potential to destroy or disrupt the life cycle of zebra mussels. Other scientists are adding chemicals to paints and plastics to prevent mussels from attaching to new surfaces.

Zebra mussels clog water intake pipes.

Invasive Species Management Should Focus on Control and Prevention Others argue that efforts to physically remove or chemically poison invasive mussels offer only temporary control. The population bounces right back. These removal efforts are also incredibly expensive. In the Great Lakes alone, more than $200 million is spent each year in efforts to get rid of zebra and quagga mussels.

Therefore, many scientists believe that there is no way to remove these mussels and other established invasive species. Instead, these scientists attempt to control the growth of populations and prevent transfer of invasive species to new areas. One regulation, for example, could require boaters to filter and chemically clean all ballast water. Meanwhile, the search continues for some kind of control that naturally limits mussel numbers when they rise.

Research and Decide

1. Analyze the Viewpoints Research the current status of invasive mussel populations and the approaches being used to prevent the spread of these and other invasive aquatic species. What trends are zebra mussel populations showing?

2. Form an Opinion What kinds of natural population controls do you think would manage these invasive mussels most effectively? Why?

5.2 Limits to Growth

THINK ABOUT IT Now that you've seen *how* populations typically grow in nature, we can explore *why* they grow as they do. If populations tend to grow exponentially, why do they often follow logistic growth? In other words, what determines the carrying capacity of an environment for a particular species? Think again about hydrilla. In its native Asia, populations of hydrilla increase in size until they reach carrying capacity, and then population growth stops. But here in the United States, hydrilla grows out of control. The same is true of gypsy moths and many other introduced plant and animal species. Why does a species that is "well-behaved" in one environment grow out of control in another?

Key Questions

What factors determine carrying capacity?

What limiting factors depend on population density?

What limiting factors do not typically depend on population density?

Vocabulary

limiting factor
density-dependent limiting factor
density-independent limiting factor

Taking Notes

Outline Make an outline using the green and blue headings in this lesson. Fill in details as you read to help you organize the information.

Limiting Factors

What factors determine carrying capacity?

Recall that the productivity of an ecosystem can be controlled by a limiting nutrient. A limiting nutrient is an example of a general ecological concept: a limiting factor. In the context of populations, a **limiting factor** is a factor that controls the growth of a population.

As shown in **Figure 5–6,** there are several kinds of limiting factors. Some—such as competition, predation, parasitism, and disease—depend on population density. Others—including natural disasters and unusual weather—do not depend on population density. **Acting separately or together, limiting factors determine the carrying capacity of an environment for a species.** Limiting factors keep most natural populations somewhere between extinction and overrunning the planet.

Charles Darwin recognized the importance of limiting factors in shaping the history of life on Earth. As you will learn in Unit 5, the limiting factors we describe here produce the pressures of natural selection that stand at the heart of evolutionary theory.

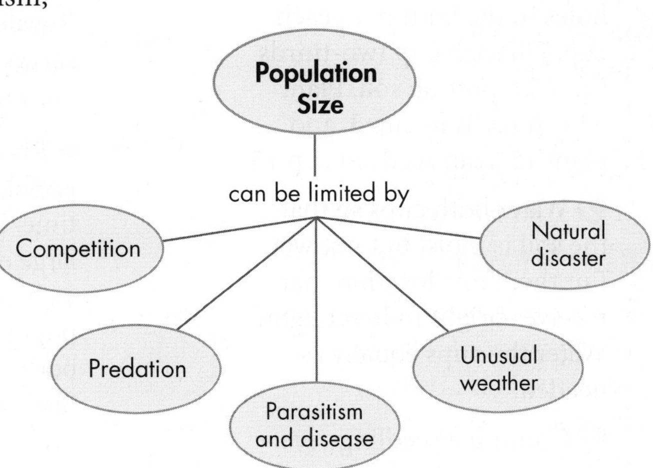

FIGURE 5–6 Limiting Factors Many different factors can limit population growth. Some of these factors depend on population density, while others do not. **Infer** ***How might each of these factors increase the death rate in a population?***

FIGURE 5–7 Competition Male wolves may fight one another for territory or access to mates.

Density-Dependent Limiting Factors

🔑 ***What limiting factors depend on population density?***

Density-dependent limiting factors operate strongly only when population density—the number of organisms per unit area—reaches a certain level. These factors do not affect small, scattered populations as much. 🔑 **Density-dependent limiting factors include competition, predation, herbivory, parasitism, disease, and stress from overcrowding.**

Competition When populations become crowded, individuals compete for food, water, space, sunlight, and other essentials. Some individuals obtain enough to survive and reproduce. Others may obtain just enough to live but not enough to enable them to raise offspring. Still others may starve to death or die from lack of shelter. Thus, competition can lower birthrates, increase death rates, or both.

Competition is a density-dependent limiting factor, because the more individuals living in an area, the sooner they use up the available resources. Often, space and food are related to one another. Many grazing animals compete for territories in which to breed and raise offspring. Individuals that do not succeed in establishing a territory find no mates and cannot breed.

Competition can also occur among members of different species that are attempting to use similar or overlapping resources. This type of competition is a major force behind evolutionary change.

Predation and Herbivory The effects of predators on prey and the effects of herbivores on plants are two very important density-dependent population controls. One classic study focuses on the relationship between wolves, moose, and plants on Isle Royale, an island in Lake Superior. The graph in **Figure 5–8** shows that populations of wolves and moose have **fluctuated** over the years. What drives these changes in population size?

▶ ***Predator-Prey Relationships*** In a predator-prey relationship, populations of predators and prey may cycle up and down over time. Sometimes, the moose population on Isle Royale grows large enough that moose become easy prey for wolves. When wolves have plenty to eat, their population grows. As the wolf population grows, the wolves begin to kill more moose than are born. This causes the moose death rate to rise higher than its birthrate, so the moose population falls. As the moose population drops, wolves begin to starve. Starvation raises the wolves' death rate and lowers their birthrate, so the wolf population also falls. When only a few predators are left, the moose death rate drops, and the cycle repeats.

In Your Notebook *Describe conditions that lead to competition in a population.*

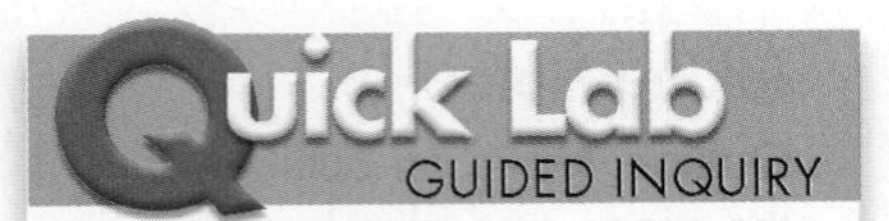

How Does Competition Affect Growth?

1. Label two paper cups 3 and 15. Make several small holes in the bottom of each cup. Fill each cup two-thirds full with potting soil. Plant 3 bean seeds in cup 3, and plant 15 bean seeds in cup 15.
2. Water both cups so that the soil is moist but not wet. Put them in a location that receives bright indirect light. Water the cups equally as needed.
3. Count the seedlings every other day for two weeks.

Analyze and Conclude

1. Observe What differences did you observe between the two cups?

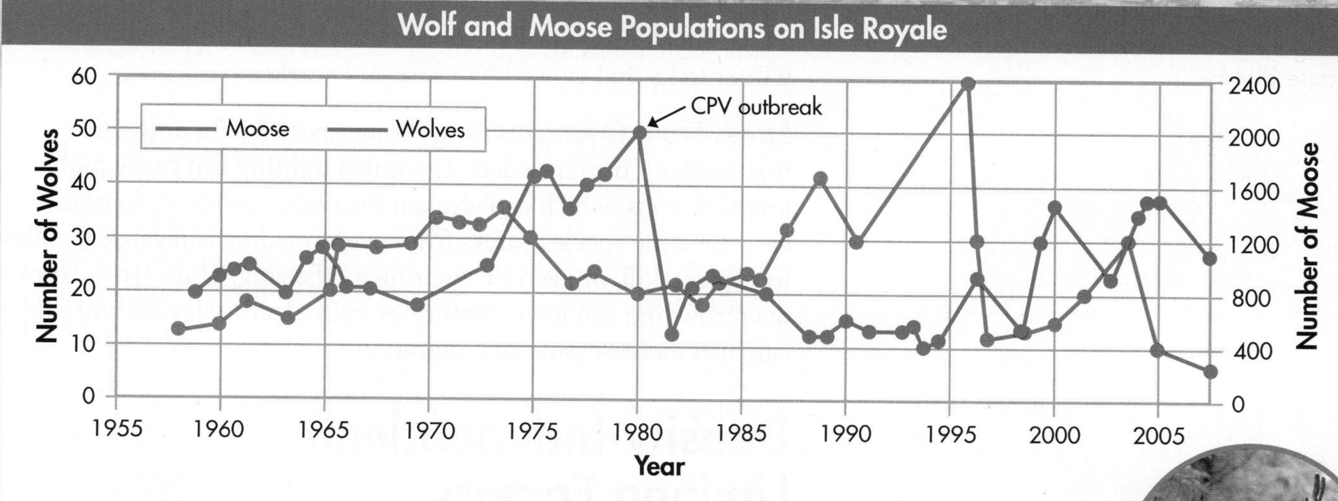

FIGURE 5–8 Moose-Wolf Populations on Isle Royale The relationship between moose and wolves on Isle Royale illustrates how predation can affect population growth. In this case, the moose population was also affected by changes in food supply, and the wolf population was also impacted by a canine parvovirus (CPV) outbreak.

▶ ***Herbivore Effects*** Herbivory can also contribute to changes in population numbers. From a plant's perspective, herbivores are predators. So it isn't surprising that populations of herbivores and plants cycle up and down, just like populations of predators and prey. On parts of Isle Royale, large, dense moose populations can eat so much balsam fir that the population of these favorite food plants drops. When this happens, the moose may suffer from lack of food.

▶ ***Humans as Predators*** In some situations, human activity limits populations. For example, humans are major predators of codfish in New England. Fishing fleets, by catching more and more fish every year, have raised cod death rates so high that birthrates cannot keep up. As a result, the cod population has been dropping. Is there any way to solve the problem? Think of predator-prey interactions. The cod population can recover if we scale back fishing to lower the death rate sufficiently. Biologists are studying birthrates and the age structure of the cod population to determine how many fish can be taken without threatening the survival of the population.

BUILD Vocabulary

ACADEMIC WORDS The verb **fluctuate** means to "rise and fall as if in waves." A population that fluctuates is unstable: Its numbers go up and down irregularly.

FIGURE 5–9 Parasitism The ticks feeding on the blood of this hedgehog can transmit bacteria that cause disease.

Parasitism and Disease Parasites and disease-causing organisms feed at the expense of their hosts, weakening them and often causing disease or death. The ticks on the hedgehog in **Figure 5–9,** for example, can carry diseases. Parasitism and disease are density-dependent effects because the denser the host population, the more easily parasites can spread from one host to another.

If you look back at the graph in **Figure 5–8,** you can see a sudden and dramatic drop in the wolf population around 1980. At that time, a viral disease of wolves was accidentally introduced to the island. This virus killed all but 13 wolves on the island—and only three of the survivors were females. The removal of wolves caused moose populations to skyrocket to 2400. The densely packed moose then became infested with winter ticks that caused hair loss and weakness.

Stress From Overcrowding Some species fight amongst themselves if overcrowded. Too much fighting can cause high levels of stress, which can weaken the body's ability to resist disease. In some species, stress from overcrowding can cause females to neglect, kill, or even eat their own offspring. Thus, stress from overcrowding can lower birthrates, raise death rates, or both. It can also increase rates of emigration.

Density-Independent Limiting Factors

What limiting factors do not typically depend on population density?

Density-independent limiting factors affect all populations in similar ways, regardless of population size and density. **Unusual weather such as hurricanes, droughts, or floods, and natural disasters such as wildfires, can act as density-independent limiting factors.** In response to such factors, a population may "crash." After the crash, the population may build up again quickly, or it may stay low for some time.

For some species, storms can nearly extinguish local populations. For example, thrips, aphids, and other insects that feed on leaves can be washed out by a heavy rainstorm. Waves whipped up by hurricanes can devastate shallow coral reefs. Extremes of cold or hot weather also can take their toll, regardless of population density. A severe drought, for example, can kill off great numbers of fish in a river, as shown in **Figure 5–10.**

True Density Independence? Sometimes, however, the effects of so-called density-independent factors can actually vary with population density. On Isle Royale, for example, the moose population grew exponentially for a time after the wolf population crashed. Then, a bitterly cold winter with very heavy snowfall covered the plants that moose feed on, making it difficult for the moose to move around to find food.

What factors do you think could limit the size of a rabbit population?

Because this was an island population, emigration was not possible; the moose weakened and many died. So, in this case, the effects of bad weather on the large, dense population were greater than they would have been on a small population. (In a smaller population, the moose would have had more food available because there would have been less competition.) This situation shows that it is sometimes difficult to say that a limiting factor acts *only* in a density-independent way.

Human activities can also place ecological communities under stress in ways that can hamper a population's ability to recover from natural disturbance. You will learn more about that situation in the next chapter.

Controlling Introduced Species In hydrilla's natural environment, density-dependent population limiting factors keep it under control. Perhaps plant-eating insects or fishes devour it. Or perhaps pests or diseases weaken it. Whatever the case, those limiting factors are not found in the United States. The result is runaway population growth!

Efforts at artificial density-independent control measures—such as herbicides and mechanical removal—offer only temporary solutions and are expensive. Researchers have spent decades looking for natural predators and pests of hydrilla. The best means of control so far seems to be an imported fish called grass carp, which view hydrilla as an especially tasty treat. These grass carp are not native to the United States. Only sterilized grass carp can be used to control hydrilla. Can you understand why?

FIGURE 5–10 Effects of a Severe Drought on a Population Dead fish lie rotting on the banks of the once-flowing Paraná de Manaquiri River in Brazil.

5.2 Assessment

Review Key Concepts

1. a. Review What is a limiting factor?

b. Apply Concepts How do limiting factors affect the growth of populations?

2. a. Review List three density-dependent limiting factors.

b. Relate Cause and Effect What is the relationship between competition and population size?

3. a. Review What is a density-independent limiting factor?

b. Apply Concepts Give three examples of density-independent factors that could severely limit the growth of a population of bats living in a cave.

Apply the Big idea

Interdependence in Nature

4. Study the factors that limit population growth shown in **Figure 5–6.** Classify each factor as biotic or abiotic. (*Hint:* Refer to Lesson 3.1 for information on biotic and abiotic factors.)

5.3 Human Population Growth

Key Questions

How has human population size changed over time?

Why do population growth rates differ among countries?

Vocabulary

demography
demographic transition

Taking Notes

Preview Visuals Before you read, preview the graphs in **Figures 5–11, 5–12,** and **5–13.** Make a list of questions about the graphs. Then, as you read, write down the answers to your questions.

THINK ABOUT IT How quickly is the global human population growing? In the United States and other developed countries, the population growth rate is low. But in some developing countries, the population is growing very rapidly. Worldwide, there are more than four human births every second. At this birthrate, the human population is well on its way to reaching 9 billion in your lifetime. What do the present and future of human population growth mean for our species and its interactions with the rest of the biosphere?

Historical Overview

How has human population size changed over time?

The human population, like populations of other organisms, tends to increase. The rate of that increase has changed dramatically over time. For most of human existence, the population grew slowly because life was harsh. Food was hard to find. Predators and diseases were common and life-threatening. These limiting factors kept human death rates very high. Until fairly recently, only half the children in the world survived to adulthood. Because death rates were so high, families had many children, just to make sure that some would survive.

Exponential Human Population Growth As civilization advanced, life became easier, and the human population began to grow more rapidly. That trend continued through the Industrial Revolution in the 1800s. Food supplies became more reliable, and essential goods could be shipped around the globe. Several factors, including improved nutrition, sanitation, medicine, and healthcare, dramatically reduced death rates. Yet, birthrates in most parts of the world remained high. The combination of lower death rates and high birthrates led to exponential growth, as shown in **Figure 5–11.**

The Predictions of Malthus As you've learned, this kind of exponential growth cannot continue forever. Two centuries ago, this problem troubled English economist Thomas Malthus. Malthus suggested that only war, famine, and disease could limit human population growth. Can you see what Malthus was suggesting? He thought that human populations would be regulated by competition (war), limited resources (famine), parasitism (disease), and other density-dependent factors. Malthus's work was vitally important to the thinking of Charles Darwin.

BUILD Vocabulary

ACADEMIC WORDS The adverb **dramatically** means "forcefully" or "significantly." When something is described as having changed dramatically, it means it has changed in a striking way.

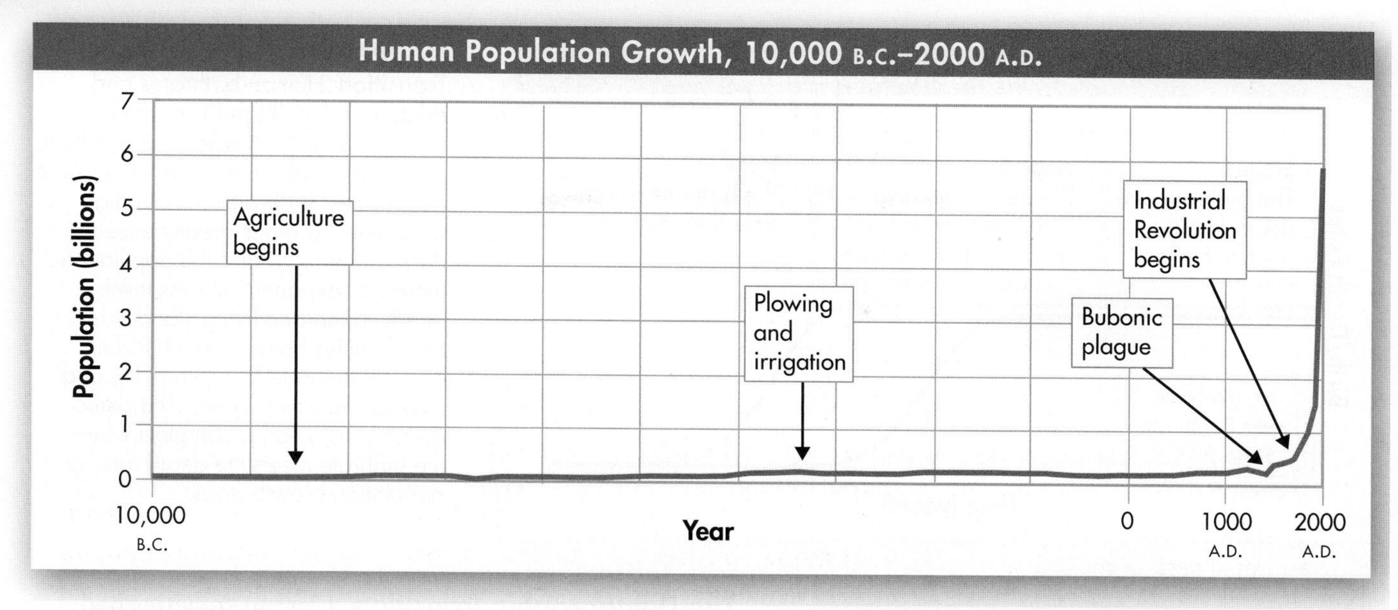

FIGURE 5–11 Human Population Growth Over Time After a slow start, the human population grew exponentially following advances in civilization. Change can be dramatic; these photos of Katmandu, Nepal, were taken from the same position in 1969 and 1999—just 30 years apart!

World Population Growth Slows So what is happening to human population growth today? Exponential growth continued up to the second half of the twentieth century. The human population growth rate reached a peak around 1962–1963, and then it began to drop. The size of the global human population is still growing rapidly, but the rate of growth is slowing down.

It took 123 years for the human population to double from 1 billion in 1804 to 2 billion in 1927. Then it took just 33 years for it to grow by another billion people. The time it took for the population to increase each additional billion continued to fall until 1999, when it began, very slowly, to rise. It now takes longer for the global human population to grow by 1 billion than it did 20 years ago. What has been going on?

Patterns of Human Population Growth

Why do population growth rates differ among countries?

Scientists have identified several social and economic factors that affect human population growth. The scientific study of human populations is called **demography.** Demography examines characteristics of human populations and attempts to explain how those populations will change over time. **Birthrates, death rates, and the age structure of a population help predict why some countries have high growth rates while other countries grow more slowly.**

In Your Notebook *Explain how the size of the global human population can increase while the rate of growth decreases.*

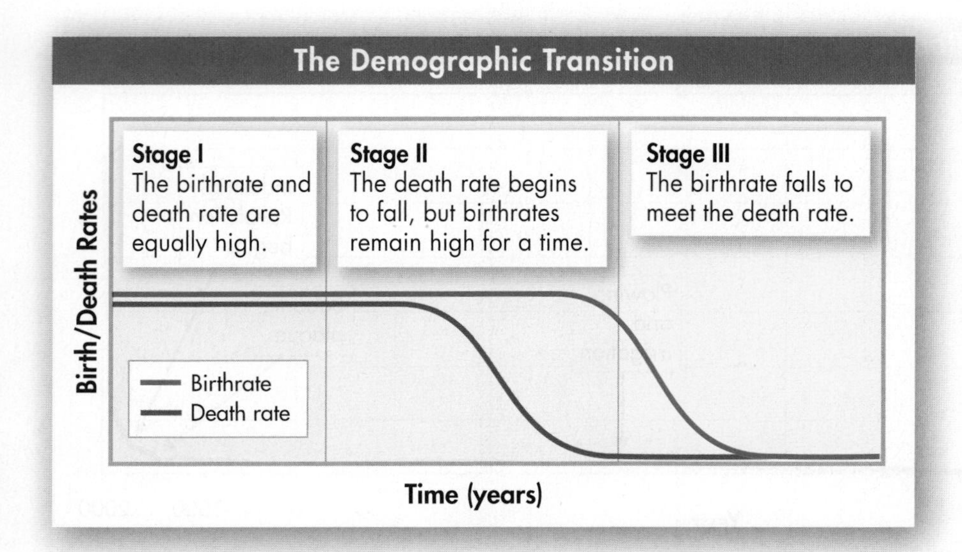

FIGURE 5–12 The Demographic Transition Human birthrates and death rates are high for most of history (Stage I). Advances in nutrition, sanitation, and medicine lead to lower death rates. Birthrates remain high for a time, so births greatly exceed deaths (Stage II), and the population increases exponentially. As levels of education and living standards rise, families have fewer children and the birthrate falls (Stage III), and population growth slows. The demographic transition is complete when the birthrate meets the death rate, and population growth stops.

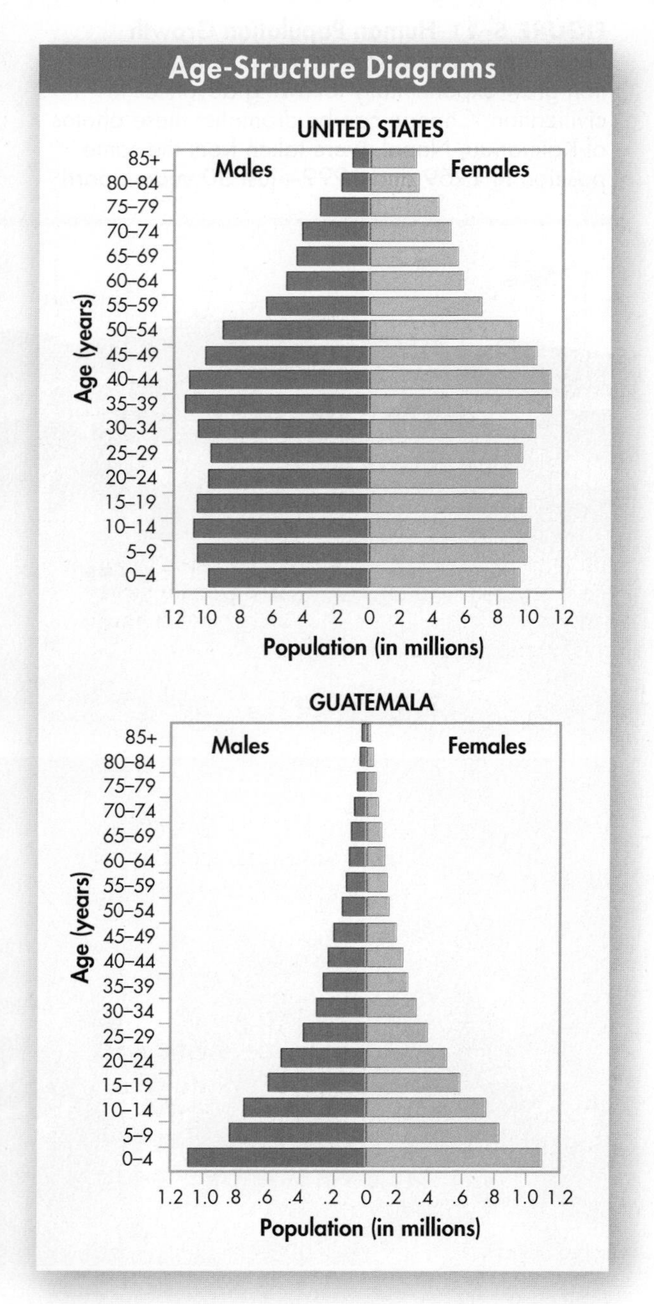

The Demographic Transition Human societies had equally high birthrates and death rates during most of history. But over the past century, population growth in the United States, Japan, and much of Europe slowed dramatically. Demographers developed a hypothesis to explain this shift. According to this hypothesis, these countries have completed the **demographic transition,** a dramatic change from high birthrates and death rates to low birthrates and death rates. The demographic transition is divided into three stages, as shown in **Figure 5–12.**

To date, the United States, Japan, and Europe have completed the demographic transition. Parts of South America, Africa, and Asia are passing through Stage II. (The United States passed through Stage II between 1790 and 1910.) A large part of ongoing human population growth is happening in only ten countries, with India and China in the lead. Globally, human population is still growing rapidly, but the rate of growth is slowing down. Our J-shaped growth curve may be changing into a logistic growth curve.

Age Structure and Population Growth To understand population growth in different countries, we turn to age-structure diagrams. **Figure 5–13** compares the age structure of the U.S. population with that of Guatemala, a country in Central America. In the United States, there are nearly equal numbers of people in each age group. This age structure predicts a slow but steady growth rate for the near future. In Guatemala, on the other hand, there are many more young children than teenagers, and many more teenagers than adults. This age structure predicts a population that will double in about 30 years.

FIGURE 5–13 Comparison of Age Structures These diagrams compare the populations of the United States and Guatemala. Notice the difference in their *x*-axis scales. **Analyze Data** ***How do the two countries differ in the percentages of 10–14-year-olds in their populations?***

Future Population Growth To predict how the world's human population will grow, demographers consider many factors, including the age structure of each country and the effects of diseases on death rates—especially AIDS in Africa and parts of Asia. Current projections suggest that by 2050 the world population will reach 9 billion people. Will the human population level out to a logistic growth curve and become stable? This may happen if countries that are currently growing rapidly complete the demographic transition.

Current data suggest that global human population will grow more slowly over the next 50 years than it grew over the last 50 years. But because the growth rate will still be higher than zero in 2050, our population will continue to grow. In the next chapter, we will examine the effect of human population growth on the biosphere.

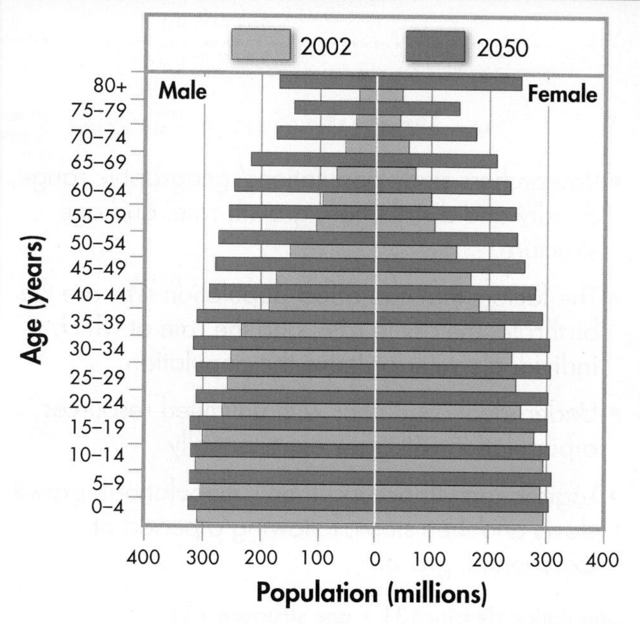

FIGURE 5–14 A Growing Population This graph (from the U.S. Census Bureau, International Database) shows the projected age structure of the world population in 2050. As population numbers climb, cities face various challenges, such as housing. The photo above shows a housing complex in Hong Kong; each apartment building is home to thousands of residents.

5.3 Assessment

Review Key Concepts

1. a. Review Describe the general trend of human population growth over time.

b. Relate Cause and Effect What factors contributed to the pattern of growth shown in **Figure 5–11**?

c. Form an Opinion Are age-structure diagrams useful in predicting future population trends?

2. a. Review Why do populations in different countries grow at different rates?

b. Explain Describe the demographic transition and explain how it could affect a country's population growth rate.

VISUAL THINKING

3. Describe the changes in human population predicted by **Figure 5–14.** How do you think those changes will affect society?

NGSS Smart Guide Populations

Disciplinary Core Idea LS2.A Interdependent Relationships in Ecosystems: How do organisms interact with the living and nonliving environments to obtain matter and energy? The way a population changes depends on many things, including its age structure, the rates at which individuals are added or removed from the population, and factors in the environment that limit its growth.

5.1 How Populations Grow

- Researchers study populations' geographic range, density and distribution, growth rate, and age structure.
- The factors that can affect population size are the birthrate, the death rate, and the rate at which individuals enter or leave the population.
- Under ideal conditions with unlimited resources, a population will grow exponentially.
- Logistic growth occurs when a population's growth slows and then stops, following a period of exponential growth.

population density 131 • age structure 131 • immigration 132 • emigration 132 • exponential growth 132 • logistic growth 135 • carrying capacity 135

Biology.com

Untamed Science Video Join the Untamed Science crew as they learn the latest techniques for counting populations.

Data Analysis Analyze logistic growth curves in order to make predictions about zebra mussel growth.

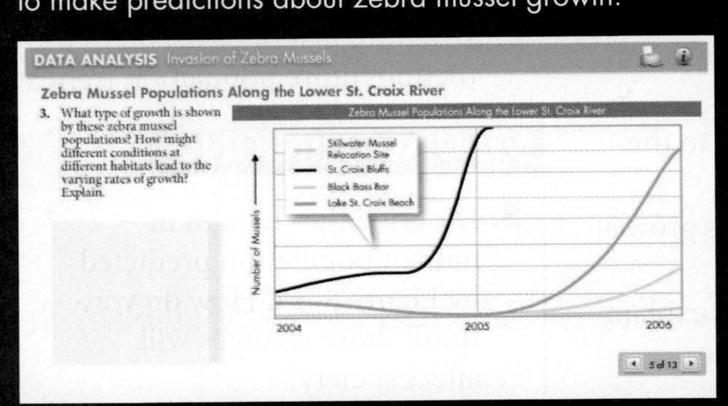

5.2 Limits to Growth

- Acting separately or together, limiting factors determine the carrying capacity of an environment for a species.
- Density-dependent limiting factors include competition, predation, herbivory, parasitism, disease, and stress from overcrowding.
- Unusual weather, such as hurricanes, droughts, or flood and natural disasters such as wildfires, can act as density-independent limiting factors.

limiting factor 137 • density-dependent limiting factor 138 • density-independent limiting factor 140

Biology.com

Art Review Review your understanding of limiting factors with this drag-and-drop activity.

InterActive Art Manipulate factors such as birthrate and death rate to see how they would impact moose and wolf populations over time.

STEM Activity View a video about the decline of bee populations and suggest causes for the decline.

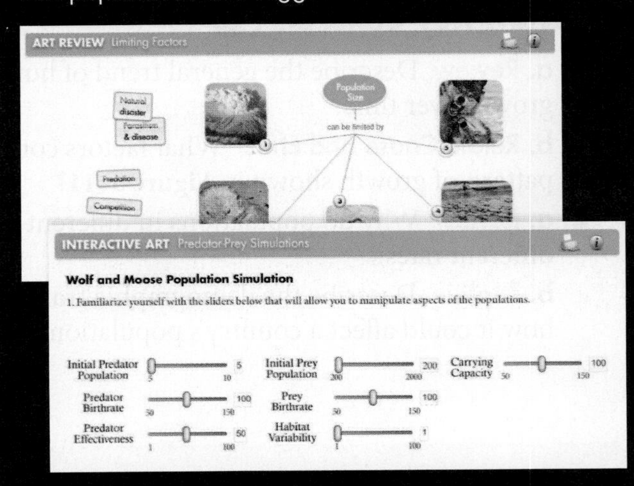

CHECKING YOUR *Scientific Literacy*

Refer to the lesson content and digital assets below as you prepare for your chapter assessment. Then, evaluate your understanding of populations by answering these questions.

1. **Scientific and Engineering Practice Using Mathematics and Computational Thinking** How can the size of a population continue to increase even as its rate of growth is decreasing? Assume that the increase in population is not due to immigration.

2. **Crosscutting Concept Systems and System Models** Suppose all the wolves were removed from Isle Royale, leaving the moose with no natural predators. What would a graph of the moose population versus time look like, and why?

3. **Research to Build and Present Knowledge** Research an invasive animal species other than zebra mussels that is a problem in your state or region. When and how did this species invade the United States? What is its geographic range? What effect has it had on native species? What methods have scientists used to try to control or eliminate the invasive populations?

4. **STEM** Wildlife biologists usually do not get to set up a controlled experiment to test a hypothesis. What they can do is choose a system where there are some natural limits on the number of variables. Explain why a biologist studying population growth might choose a small island, such as Isle Royale. Hint: Think about the different factors that affect population size and limit population growth.

5.3 Human Population Growth

- The human population, like populations of other organisms, tends to increase. The rate of that increase has changed dramatically over time.
- Birthrates, death rates, and the age structure of a population help predict why some countries have high growth rates while other countries grow more slowly.

demography 143 • demographic transition 144

Biology.com

Art in Motion View a short animation that brings age-structure diagrams to life.

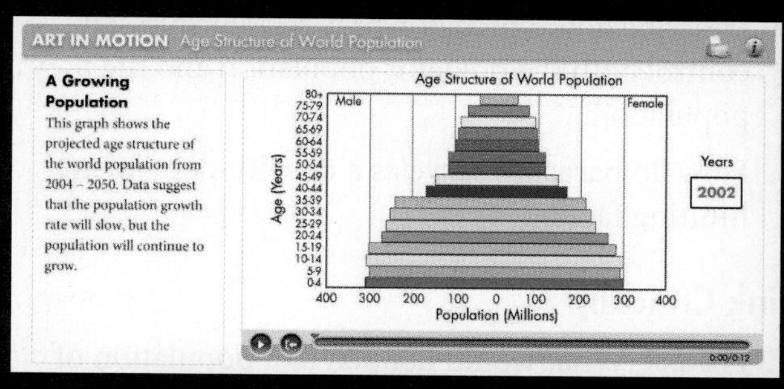

Flash Cards Review the meaning of a vocabulary term.

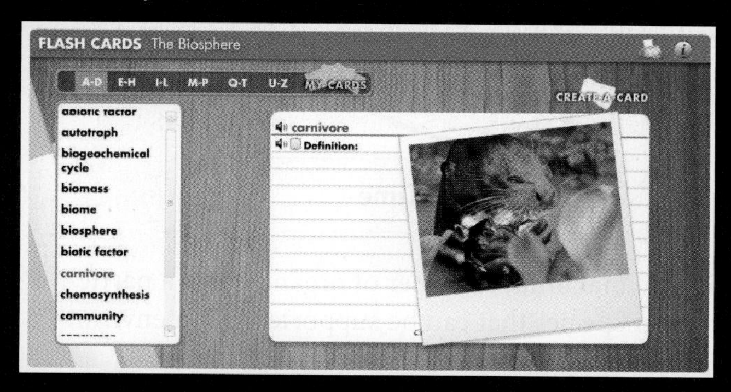

Skills Lab

The Growth Cycle of Yeast This chapter lab is available in your lab manual and at Biology.com

5 Assessment

5.1 How Populations Grow

Understand Key Concepts

1. The number of individuals of a single species per unit area is known as
   a. carrying capacity.
   b. logistic growth.
   c. population density.
   d. population growth rate.

2. The movement of individuals into an area is called
   a. demography.
   b. carrying capacity.
   c. immigration.
   d. emigration.

3. The area inhabited by a population is known as its
   a. growth rate.
   b. geographic range.
   c. age structure.
   d. population density.

4. The graph below represents
   a. carrying capacity.
   b. exponential growth.
   c. logistic growth.
   d. age structure.

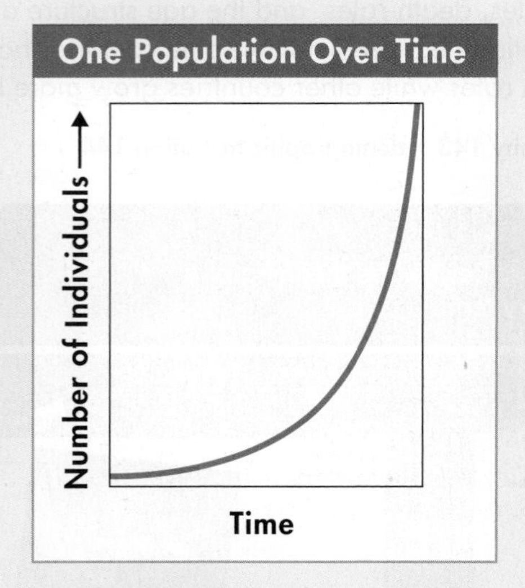

5. The maximum number of organisms of a particular species that can be supported by an environment is called
   a. logistic growth.
   b. carrying capacity.
   c. exponential growth.
   d. population density.

6. What is the difference between immigration and emigration?

7. Sketch the exponential growth curve of a hypothetical population.

8. Describe the conditions under which logistic growth occurs.

9. What is carrying capacity? Give an example.

Think Critically

10. **Use Analogies** How is the carrying capacity of a city's roads similar to the carrying capacity of an ecosystem?

5.2 Limits to Growth

Understand Key Concepts

11. A limiting factor that depends on population size is called a
    a. density-dependent limiting factor.
    b. density-independent limiting factor.
    c. predator-prey relationship.
    d. parasitic relationship.

12. One example of a density-independent limiting factor is
    a. predation.
    b. hurricanes.
    c. competition.
    d. parasitism.

13. How might increasing the amount of a limiting nutrient in a pond affect the carrying capacity of the pond?

14. Describe the long-term effects of competition on populations of two different species competing for the same resources.

15. Describe how a predator-prey relationship can control both the predator population and the prey population.

16. How do parasites serve as a density-dependent limiting factor?

Think Critically

17. **Predict** What would happen to a population of predators if there was a sudden increase in food for the prey? Explain your answer.

18. **Apply Concepts** Why would a contagious virus that causes a fatal disease be considered a density-dependent limiting factor?

19. Infer Would a density-independent limiting factor have more of an effect on population size in a large ecosystem or in a small ecosystem?

20. Craft and Structure How is the relationship between parasites and their hosts similar to a predator-prey relationship?

21. Apply Concepts How would a drop in the water level of a river affect a fish population living in that river?

5.3 Human Population Growth

Understand Key Concepts

22. The scientific study of human populations is called
- **a.** immigration.
- **b.** emigration.
- **c.** demographic transition.
- **d.** demography.

23. The demographic transition is considered complete when
- **a.** population growth stops.
- **b.** the birthrate is greater than the death rate.
- **c.** the death rate begins to fall.
- **d.** the death rate is greater than the birthrate.

24. How can you account for the fact that the human population has grown more rapidly during the past 500 years than it has at any other time in history?

25. What is the significance of the demographic transition in studies of human population around the world?

26. How does the age structure of a population affect its growth rate?

27. What factors did Thomas Malthus think would eventually limit the human population?

Think Critically

28. Compare and Contrast What shape population growth curve would you expect to see in a small town made up mainly of senior citizens? Compare this growth curve to that of a small town made up of newly married couples in their twenties.

29. Pose Questions What questions would a demographer need to answer to determine whether a country is approaching the demographic transition?

solve the CHAPTER MYSTERY

A PLAGUE OF RABBITS

Australia had no native rabbit population when the European rabbits arrived, so there were no density-dependent controls to keep their numbers in check. The rabbits' new environment provided many favorable conditions for survival, including fewer predators, parasites, and diseases. The initial small number of rabbits—which can reproduce rapidly—soon multiplied into millions.

High rabbit numbers caused serious environmental and agricultural damage. In an effort to manage the problem, many methods have been tried, including fencing, poisoning, the destruction of burrows, and the use of parasites and disease. In the 1950s, a rabbit virus that causes the fatal rabbit disease myxomatosis was deliberately introduced as a form of biological control. It killed countless rabbits. But the virus and rabbits soon reached an equilibrium that allowed host and parasite to coexist, and the rabbit population rose. Later, a new virus that causes rabbit hemorrhagic disease (RHD) was introduced, and the rabbit population dropped again. In several places, environmental recovery was dramatic: Native animals recovered, and native trees and shrubs thought to be locally extinct began to grow again. But the RHD virus and rabbits appear to have reached a new balance, and the rabbit population is rising again!

1. Predict Populations of wildcats and foxes (both also introduced to Australia) have come to depend on rabbits as prey. How do you think wildcats and foxes would be affected by a crash in the rabbit population?

2. Craft and Structure Analyze the author's purpose in choosing the Australian rabbit population for the Chapter Mystery.

3. Connect to the Big idea Why should people be cautious about introducing organisms into new environments?

Connecting Concepts

Use Science Graphics

The following actual and projected data, from the United Nations Department of Economic and Social Affairs, Population Division, show when the global population reached or will reach an additional billion. Use the data table to answer questions 30 and 31.

World Population Milestones

Population (billion)	Year	Time Interval (years)
1	1804	—
2	1927	123
3	1960	33
4	1974	14
5	1987	13
6	1999	12
7	2012	13
8	2027	15
8.9	2050	23

30. Integration of Knowledge and Ideas When did the world population reach 1 billion people? When did it reach 6 billion?

31. Integration of Knowledge and Ideas Describe the trend in population growth since the 1-billion-people mark.

Write About Science

32. Text Types and Purposes Write a paragraph on the human population. Include the characteristics of a population, factors that affect its size, and changes in the size of the population from about 500 years ago to the present. Give a projection of how large the world population might be in the year 2050 and of how the growth rate in 2050 might compare to that in 2000. (*Hint*: As you outline your ideas, identify phrases such as *for example* and *as a result* that you can use to transition from one idea to another.)

33. Assess the Big idea Choose a specific organism and explain how the population of that organism depends on a number of factors that may cause it to increase, decrease, or remain stable in size.

Analyzing Data

Integration of Knowledge and Ideas

The following graph shows the "boom-and-bust" pattern of regular rises and falls in the rabbit population in South Australia. The points at which various population control measures were introduced are indicated. Use the graph to answer questions 34 and 35.

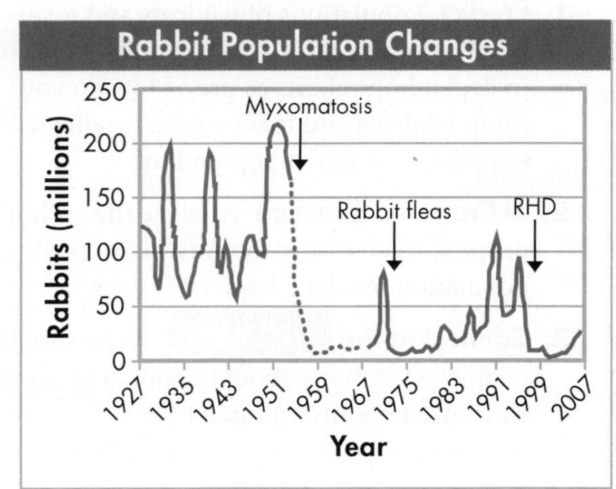

34. Interpret Graphs In which of the following years was the rabbit population density in South Australia most dense?

a. 1936
b. 1952
c. 1975
d. 2000

35. Infer European rabbit fleas were introduced in the late 1960s to help spread the effects of the rabbit disease myxomatosis. Based on the graph, what can you infer about the rabbit population after the fleas were introduced?

a. The rabbit birthrate increased.
b. The rabbit death rate increased.
c. The rabbit death rate decreased.
d. The fleas had no effect on the rabbit population.

Standardized Test Prep

Multiple Choice

1. The movement of individuals into an area is called
A immigration.
B emigration.
C population growth rate.
D population density.

2. All other things being equal, the size of a population will decrease if
A birthrate exceeds the death rate.
B immigration rate exceeds emigration rate.
C death rate exceeds birthrate.
D birthrate equals death rate.

3. Which of the following is NOT an example of a density-dependent limiting factor?
A natural disaster
B predator
C competition
D disease

4. A population like that of the United States with an age structure of roughly equal numbers in each of the age groups can be predicted to
A grow rapidly over a 30-year-period and then stabilize.
B grow little for a generation and then grow rapidly.
C fall slowly and steadily over many decades.
D show slow and steady growth for some time into the future.

5. In the presence of unlimited resources and in the absence of disease and predation, what would probably happen to a bacterial population?
A logistic growth
B exponential growth
C endangerment
D extinction

6. Which of the following statements best describes human population growth?
A The growth rate has remained constant over time.
B Growth continues to increase at the same rate.
C Growth has been exponential in the last few hundred years.
D Birthrate equals death rate.

7. Which of the following refers to when a population's birthrate equals its death rate?
A limiting factor
B carrying capacity
C exponential growth
D population density

Questions 8–9

Use the graph below to answer the following questions.

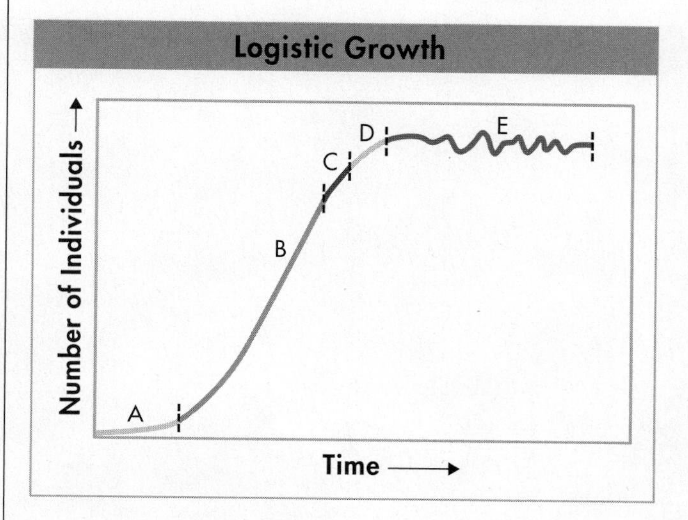

8. Which time interval(s) in the graph shows exponential growth?
A D and E
B A and B
C C and D
D E only

9. Which time interval(s) in the graph depicts the effects of limiting factors on the population?
A A only
B A and B
C C, D, and E
D C and D

Open-Ended Response

10. When a nonnative species is imported into a new ecosystem, the population sometimes runs wild. Explain why this might be the case.

If You Have Trouble With . . .

Question	1	2	3	4	5	6	7	8	9	10
See Lesson	5.1	5.1	5.2	5.3	5.1	5.3	5.1	5.1	5.2	5.2

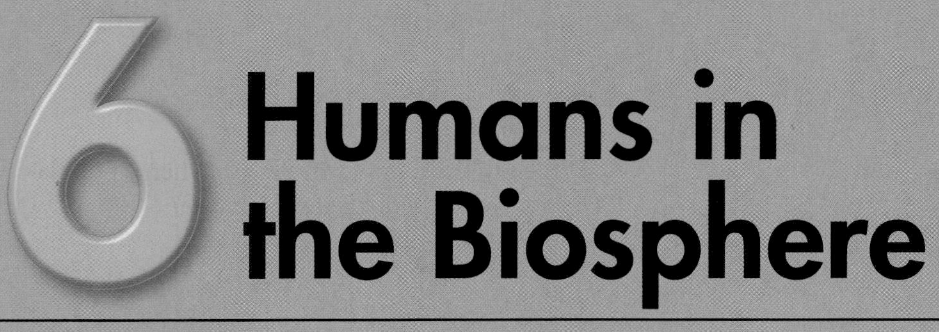

6 Humans in the Biosphere

Big idea: Interdependence in Nature

Q: How have human activities shaped local and global ecology?

The lights of human settlements, which are visible from space, remind us that humans far exceed any other species in shaping both local and global ecology. As the world population grows from about 7 billion in 2012 to an estimated 9 billion by 2050, think about the impact on the biosphere.

INSIDE:

- 6.1 A Changing Landscape
- 6.2 Using Resources Wisely
- 6.3 Biodiversity
- 6.4 Meeting Ecological Challenges

BUILDING *Scientific Literacy*

Deepen your understanding of humans in the biosphere by engaging in key practices that allow you to make connections across concepts.

NGSS Ask questions about, and identify, the effects of different agricultural practices on humans and on ecosystems.

STEM How can the amount of solid waste in landfills be reduced? Work with a team to find out if new designs for product packaging are part of the solution.

Common Core You will use data and other evidence to fairly argue both positions on a local land-use issue.

CHAPTER MYSTERY

MOVING THE *MOAI*

Easter Island is a tiny speck of land in the vast Pacific Ocean off the coast of Chile with a harsh tropical climate. The original islanders, who called themselves Rapa Nui, came from Polynesia. They carved hundreds of huge stone statues called *moai* (moh eye). Starting around 1200 A.D., the Rapa Nui somehow moved these mysterious statues, each of which weighed between 10 and 14 tons, from quarries to locations around the island. Nearly all theories about this process suggest that strong, large logs were necessary to move the *moai*. Yet by the time Europeans landed on the island in 1722, there was no sign of any trees large enough to provide such logs.

What had happened? As you read this chapter, look for clues about the interactions of the Rapa Nui with their island environment. Then, solve the mystery.

Biology.com

The mystery of how the Rapa Nui moved the *moai* is only the beginning. Take a video field trip with the ecogeeks of Untamed Science to continue exploring your world.

Go online to access additional resources including:
- eText • Flash Cards • Lesson Overviews • Chapter Mystery

6.1 A Changing Landscape

Key Questions

How do our daily activities affect the environment?

What is the relationship between resource use and sustainable development?

Vocabulary

monoculture
renewable resource
nonrenewable resource
sustainable development

Taking Notes

Outline As you read, create an outline using the green and blue heads in this lesson. As you read, fill in key words, phrases, and ideas about each heading.

MYSTERY CLUE

Easter Island's first colonists brought with them banana trees, taro root, and chickens—and possibly some small mammalian "stowaways." What impact might these new organisms have had on the island's ecosystems?

THINK ABOUT IT The first humans to settle Hawaii came from Polynesia about 1600 years ago. These island people had customs that protected the natural resources of their new home. For example, they were prohibited from catching certain fish during spawning season and, for every coconut palm tree cut down, they had to plant two palms in its place. But Hawaiians did not treat their islands entirely like nature reserves. They cut trees to plant farms, and they introduced nonnative plants, pigs, chickens, dogs, and rats. This combination drove many native plant and animal species to extinction. Yet for centuries Hawaii's ecosystems provided enough fresh water, fertile soil, fish, and other resources to keep the society self-sufficient. What happened next is a lesson on managing limited resources—a lesson that is as important today as it was over 1000 years ago.

The Effect of Human Activity

How do our daily activities affect the environment?

Beginning in the late 1700s, new waves of settlers arrived in Hawaii. These people did not seem to understand the limits of island ecosystems. They imported dozens more plants and animals that became invasive pests. They cleared vast tracts of forest to grow sugar cane, pineapples, and other crops that required lots of water. And as the island's human population grew, they converted untouched land for other uses, including housing and tourism, as shown in **Figure 6–1.** The effect of these activities on Hawaii's ecosystems and its human inhabitants offers a window onto a globally important question: What happens when a growing human population does not adequately manage natural resources that are both vital and limited?

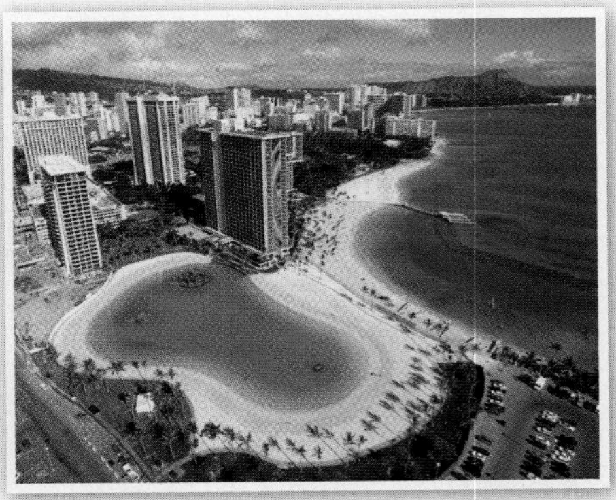

FIGURE 6–1 The Lesson of Hawaii Kalalau Valley along the Na Pali coast of Kauai looks almost untouched by humans. In contrast, Waikiki Beach on the island of Oahu is surrounded by built-up areas that support tourism.

Living on Island Earth Humans, like all forms of life, rely on Earth's life-support systems. And like all other organisms, we affect our environment when we obtain food, eliminate waste products, and build places to live. The effects of these activities can be most obvious on islands such as Hawaii because of their small size. Living on an island also can make people aware of limited resources and of an area's carrying capacity for humans because anything not available locally must be brought in from far away.

Most of us who live on large continents, however, probably don't think of land, food, and water as limited resources. In the past, environmental problems were local. There was always new land to settle and new sources of food and water. But today human activity has used or altered roughly half of all the land that's not covered with ice and snow. Some people suggest that as the global population reaches 7 billion people, we may be approaching the carrying capacity of the biosphere for humans. **Humans affect regional and global environments through agriculture, development, and industry in ways that have an impact on the quality of Earth's natural resources, including soil, water, and the atmosphere.**

In Your Notebook *Explain how Earth is like an island.*

Agriculture Agriculture is one of the most important inventions in human history. A dependable supply of food that can be stored for later use enabled humans to gather in settlements that grew into towns and cities. Settlements, in turn, encouraged the growth of modern civilization—government, laws, writing, and science. Modern agricultural practices have enabled farmers to double world food production over the last 50 years. **Monoculture,** for example, is the practice of clearing large areas of land to plant a single highly productive crop year after year, like the soybeans in **Figure 6–2.** Monoculture enables efficient sowing, tending, and harvesting of crops using machines. However, providing food for nearly 7 billion people impacts natural resources, including fresh water and fertile soil. Fertilizer production and farm machinery also consume large amounts of fossil fuels.

Reduce, Reuse, Recycle

1. Collect one day's worth of dry trash.
2. Sort the trash into items that can be reused, recycled, or discarded because they can't be reused or recycled.

Analyze and Conclude

1. **Analyze Data** Look at the trash you've sorted. Roughly what percentage of the total does each type represent?
2. **Predict** What do you think happens to the trash you produce? Think of at least three ways trash can impact living things.
3. **Evaluate** List three ways you can reduce the amount of trash you produce.

BUILD Vocabulary

PREFIXES The prefix *mono-* in **monoculture** means "one, alone, single." Monoculture is the practice of planting a single productive crop, year after year.

FIGURE 6–2 Monoculture Soybean fields dominate this landscape. **Apply Concepts** ***How has agriculture helped shape civilization?***

Development As modern society developed, many people chose to live in cities. In the United States, as urban centers became crowded, people moved to, and built up, suburbs. The growth of cities and suburbs is tied to the high standard of living that Americans enjoy. Yet this development has environmental effects. Dense human communities produce lots of wastes. If these wastes are not disposed of properly, they affect air, water, and soil resources. In addition, development consumes farmland and divides natural habitats into fragments.

Industrial Growth Human society was transformed by the Industrial Revolution of the 1800s. Today, industry and scientific know-how provide us with the conveniences of modern life—from comfortable homes and clothes to electronic devices for work and play. Of course these conveniences require a lot of energy to produce and power. We obtain most of this energy by burning fossil fuels—coal, oil, and natural gas—and that affects the environment. In addition, industries have traditionally discarded wastes from manufacturing and energy production directly into the air, water, and soil.

FIGURE 6–3 Ecosystem Services The Hennepin and Hopper Lakes wetland is managed by The Wetlands Initiative—an organization dedicated to protecting and restoring Illinois's wetlands. The area, originally drained and leveed for farming in 1900, is shown in the inset before its 2003 restoration. **Apply Concepts** ***What ecological services do wetlands provide?***

Sustainable Development

What is the relationship between resource use and sustainable development?

In the language of economics, *goods* are things that can be bought and sold, that have value in terms of dollars and cents. *Services* are processes or actions that produce goods. Ecosystem goods and services are the goods and services produced by ecosystems that benefit the human economy.

Ecosystem Goods and Services Some ecosystem goods and services—like breathable air and drinkable water—are so basic that we often take them for granted. Healthy ecosystems provide many goods and services naturally and largely free of charge. But, if the environment can't provide these goods and services, society must spend money to produce them. In many places, for example, drinkable water is provided naturally by streams, rivers, and lakes, and filtered by wetlands like the one in **Figure 6–3.** But if water sources or wetlands are polluted or damaged, water quality may fall. In such cases, cities and towns must pay for mechanical or chemical treatment to provide safe drinking water.

In Your Notebook *Describe three ecosystem goods and services you've used today.*

Renewable and Nonrenewable Resources Ecosystem goods and services are classified as either renewable or nonrenewable, as shown in **Figure 6–4.** A **renewable resource** can be produced or replaced by a healthy ecosystem. A single southern white pine is an example of a renewable resource because a new tree can grow in place of an old tree that dies or is cut down. But some resources are **nonrenewable resources** because natural processes cannot replenish them within a reasonable amount of time. Fossil fuels like coal, oil, and natural gas are nonrenewable resources formed from buried organic materials over millions of years. When existing deposits are depleted, they are essentially gone forever.

Sustainable Resource Use Ecological science can teach us how to use natural resources to meet our needs without causing long-term environmental harm. Using resources in such an environmentally conscious way is called **sustainable development.** **Sustainable development provides for human needs while preserving the ecosystems that produce natural resources.**

What should sustainable development look like? It should cause no long-term harm to the soil, water, and climate on which it depends. It should consume as little energy and material as possible. Sustainable development must be flexible enough to survive environmental stresses like droughts, floods, and heat waves or cold snaps. Finally, sustainable development must take into account human economic systems as well as ecosystem goods and services. It must do more than just enable people to survive. It must help them improve their situation.

FIGURE 6–4 Natural Resources Natural resources are classified as renewable or nonrenewable. Wind and coal are both natural resources that can provide energy. But wind is renewable, while coal—like other fossil fuels—is not.

6.1 Assessment

Review Key Concepts

1. a. **Review** List the three primary types of human activities that have affected regional and global environments. For each, give one benefit and one environmental cost.

   b. **Relate Cause and Effect** How might more productive agricultural practices affect a developing nation's population? Its environmental health?

2. a. **Review** What is sustainable development? How can it help minimize the negative impacts of human activities?

   b. **Explain** Explain why energy from the sun is a renewable resource but energy from oil is a nonrenewable resource.

   c. **Apply Concepts** In addition to filtering water, wetlands provide flood control by absorbing excess water. Explain how society would provide these services (for a cost) if the ecosystem could not.

WRITE ABOUT SCIENCE

Description

3. What signs of growth do you see in your community? Write a paragraph telling how this growth might affect local ecosystems.

6.2 Using Resources Wisely

Key Questions

Why is soil important, and how do we protect it?

What are the primary sources of water pollution?

What are the major forms of air pollution?

Vocabulary

desertification
deforestation
pollutant
biological magnification
smog
acid rain

Taking Notes

Concept Map As you read, create a concept map to organize the information in this lesson.

THINK ABOUT IT Our economy is built on the use of natural resources, so leaving those resources untouched is not an option. Humans need to eat, for example, so we can't just stop cultivating land for farming. But the goods and services provided by healthy ecosystems are essential to life. We can't grow anything in soil that has lost its nutrients due to overfarming. If we don't properly manage agriculture, then, we may one day lose the natural resource on which it depends. So how do we find a balance? How do we obtain what we need from local and global environments without destroying those environments?

Soil Resources

Why is soil important, and how do we protect it?

When you think of natural resources, soil may not be something that comes to mind. But many objects you come into contact with daily rely on soil—from the grain in your breakfast cereal, to the wood in your home, to the pages of this textbook. **Healthy soil supports both agriculture and forestry.** The mineral- and nutrient-rich portion of soil is called topsoil. Good topsoil absorbs and retains moisture yet allows water to drain. It is rich in organic matter and nutrients, but low in salts. Good topsoil is produced by long-term interactions between soil and the plants growing in it.

Topsoil can be a renewable resource if it is managed properly, but it can be damaged or lost if it is mismanaged. Healthy soil can take centuries to form but can be lost very quickly. And the loss of fertile soil can have dire consequences. Years of poorly managed farming in addition to severe drought in the 1930s badly eroded the once-fertile soil of the Great Plains. Thousands upon thousands of people lost their jobs and homes. The area essentially turned to desert, or, as it came to be known, a "dust bowl," as seen in **Figure 6–5.** What causes soil erosion, and how can we prevent it?

FIGURE 6–5 The Dust Bowl A ranch in Boise City, Idaho, is about to be hit by a cloud of dry soil on April 15, 1935.

Soil Erosion The dust bowl of the 1930s was caused, in part, by conversion of prairie land to cropland in ways that left soil vulnerable to erosion. Soil erosion is the removal of soil by water or wind. Soil erosion is often worse when land is plowed and left barren between plantings. When no roots are left to hold soil in place, it is easily washed away. And when soil is badly eroded, organic matter and minerals that make it fertile are often carried away with the soil. In parts of the world with dry climates, a combination of farming, overgrazing, seasonal drought, and climate change can turn farmland into desert. This process is called **desertification,** and it is what happened to the Great Plains in the 1930s. Roughly 40 percent of Earth's land is considered at risk for desertification. **Figure 6–6** shows vulnerable areas in North and South America.

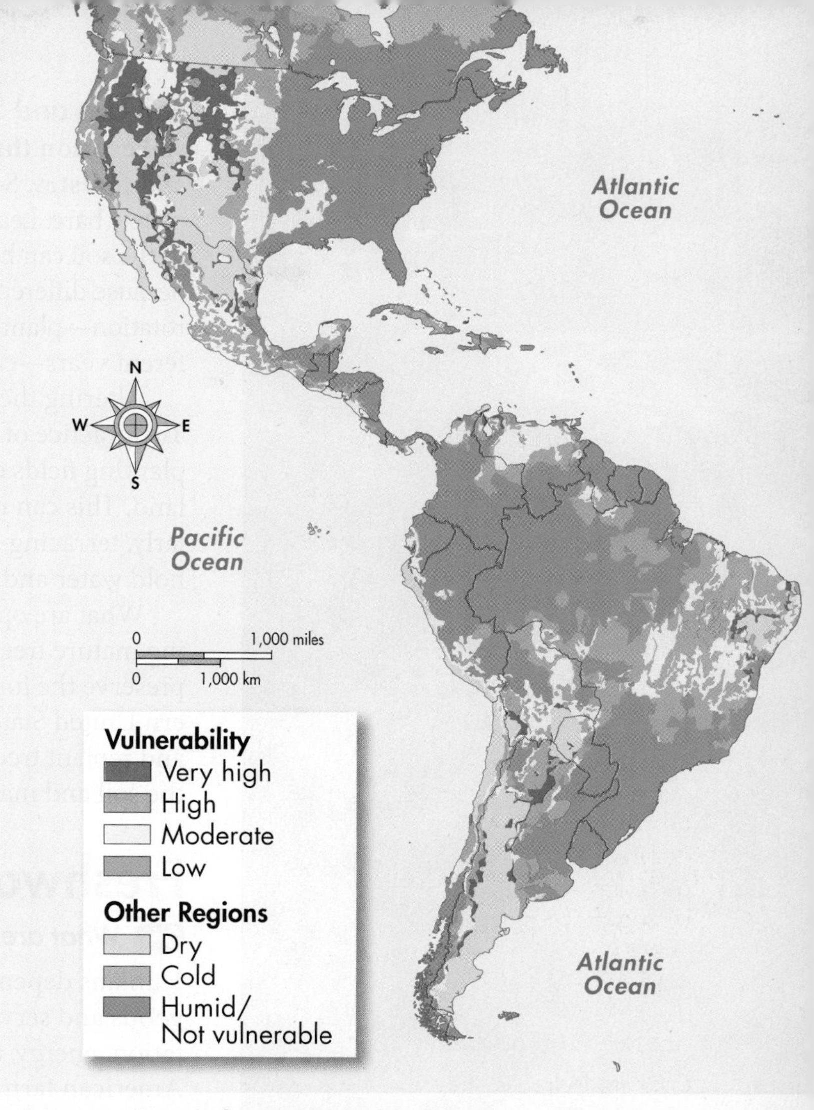

FIGURE 6–6 Desertification Risk The U.S. Department of Agriculture assigns desertification risk categories based on soil type and climate. **Interpret Visuals** ***Find your approximate location on the map. What category of desertification risk is your area in?***

Deforestation, or loss of forests, can also have a negative effect on soil quality. Healthy forests not only provide wood, but also hold soil in place, protect the quality of fresh water supplies, absorb carbon dioxide, and help moderate local climate. Unfortunately, more than half of the world's old-growth forests (forests that had never been cut) have already been lost to deforestation. In some temperate areas, such as the Eastern United States, forests can regrow after cutting. But it takes centuries for succession to produce mature, old-growth forests. In some places, such as in parts of the tropics, forests don't grow back at all after logging. This is why old-growth forests are usually considered nonrenewable resources.

Deforestation can lead to severe erosion, especially on mountainsides. Grazing or plowing after deforestation can permanently change local soils and microclimates in ways that prevent the regrowth of trees. Tropical rain forests, for example, look lush and rich, so you might assume they would grow back after logging. Unfortunately, topsoil in these forests is generally thin, and organic matter decomposes rapidly under high heat and humidity. When tropical rain forests are cleared for timber or for agriculture, their soil is typically useful for just a few years. After that the areas become wastelands, the harsh conditions there preventing regrowth.

In Your Notebook *Describe the relationship between agriculture and soil quality.*

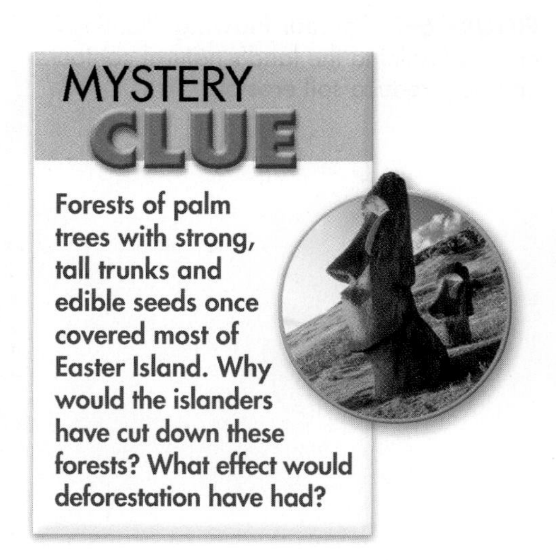

Forests of palm trees with strong, tall trunks and edible seeds once covered most of Easter Island. Why would the islanders have cut down these forests? What effect would deforestation have had?

FIGURE 6–7 Contour Plowing Planting crops parallel to the land's natural contours can help reduce soil erosion.

Soil Use and Sustainability **It is possible to minimize soil erosion through careful management of both agriculture and forestry.** Soil is most vulnerable to erosion when it is completely bare. Leaving stems and roots of the previous year's crop in the soil can help hold soil in place between plantings. And because different plants take different nutrients from the soil, crop rotation—planting different crops at different seasons or in different years—can help prevent both erosion and nutrient loss.

Altering the shape of the land is another way to limit erosion. The practice of contour plowing, shown in **Figure 6–7,** involves planting fields of crops across, instead of down, the slope of the land. This can reduce water runoff and therefore erosion. Similarly, terracing—shaping the land to create level "steps"—helps hold water and soil.

What are options for sustainable forestry? Selectively harvesting mature trees can promote the growth of younger trees and preserve the forest ecosystem, including its soil. In the southeastern United States, conditions enable foresters to plant, harvest, and replant tree farms. A well-managed tree farm both protects the soil and makes the trees themselves a renewable resource.

Freshwater Resources

What are the primary sources of water pollution?

Humans depend on fresh water and freshwater ecosystems for goods and services, including drinking water, industry, transportation, energy, and waste disposal. Some of the most productive American farmland relies heavily on irrigation, in which fresh water is brought in from other sources.

While fresh water is usually considered a renewable resource, some sources of fresh water are not renewable. The Ogallala aquifer, for example, spans eight states from South Dakota to Texas. The aquifer took more than a million years to collect and is not replenished by rainfall today. So much water is being pumped out of the Ogallala that it is expected to run dry in 20 to 40 years. In many places, freshwater supplies are limited. Only 3 percent of Earth's water is fresh water—and most of that is locked in ice at the poles. Since we can't infinitely expand our use of a finite resource, we must protect the ecosystems that collect and purify fresh water.

Water Pollution Freshwater sources can be affected by different kinds of pollution. A **pollutant** is a harmful material that can enter the biosphere. Sometimes pollutants enter water supplies from a single source—a factory or an oil spill, for example. This is called point source pollution. Often, however, pollutants enter water supplies from many smaller sources—the grease and oil washed off streets by rain or the chemicals released into the air by factories and automobiles. These pollutants are called nonpoint sources.

Pollutants may enter both surface water and underground water supplies that we access with wells. Once contaminants are present, they can be extremely difficult to get rid of. **The primary sources of water pollution are industrial and agricultural chemicals, residential sewage, and nonpoint sources.**

▶ ***Industrial and Agricultural Chemicals*** One industrial pollutant is a class of organic chemicals called PCBs that were widely used in industry until the 1970s. After several large-scale contamination events, PCBs were banned. However, because PCBs often enter mud and sand beneath bodies of water, they can be difficult, if not impossible, to eliminate. Parts of the Great Lakes and some coastal areas, for example, are still polluted with PCBs. Other harmful industrial pollutants are heavy metals like cadmium, lead, mercury, and zinc.

Large-scale monoculture has increased the use of pesticides and insecticides. These chemicals can enter the water supply in the form of runoff after heavy rains, or they can seep directly into groundwater. Pesticides can be very dangerous pollutants. DDT, which is both cheap and long lasting, effectively controls agricultural pests and disease-carrying mosquitoes. But when DDT gets into a water supply, it has disastrous effects on the organisms that directly and indirectly rely on that water—a function of a phenomenon called biological magnification.

Biological magnification occurs if a pollutant, such as DDT, mercury, or a PCB, is picked up by an organism and is not broken down or eliminated from its body. Instead, the pollutant collects in body tissues. Primary producers pick up a pollutant from the environment. Herbivores that eat those producers concentrate and store the compound. Pollutant concentrations in herbivores may be more than ten times the levels in producers. When carnivores eat the herbivores, the compound is still further concentrated. Thus, pollutant concentration increases at higher trophic levels. In the highest trophic levels, pollutant concentrations may reach 10 million times their original concentration in the environment, as shown in **Figure 6–8.**

These high concentrations can cause serious problems for wildlife and humans. Widespread DDT use in the 1950s threatened fish-eating birds like pelicans, osprey, falcons, and bald eagles. It caused females to lay eggs with thin, fragile shells, reducing hatching rates and causing a drop in bird populations. Since DDT was banned in the 1970s, bird populations have recovered. Still a concern is mercury, which accumulates in the bodies of certain marine fish such as tuna and swordfish.

In Your Notebook *In your own words, explain the process of biological magnification.*

FIGURE 6–8 Biological Magnification In the process of biological magnification, the concentration of a pollutant like DDT—represented by the orange dots—is multiplied as it passes up the food chain from producers to consumers. Calculate ***By what number is the concentration of DDT multiplied at each successive trophic level?*** MATH

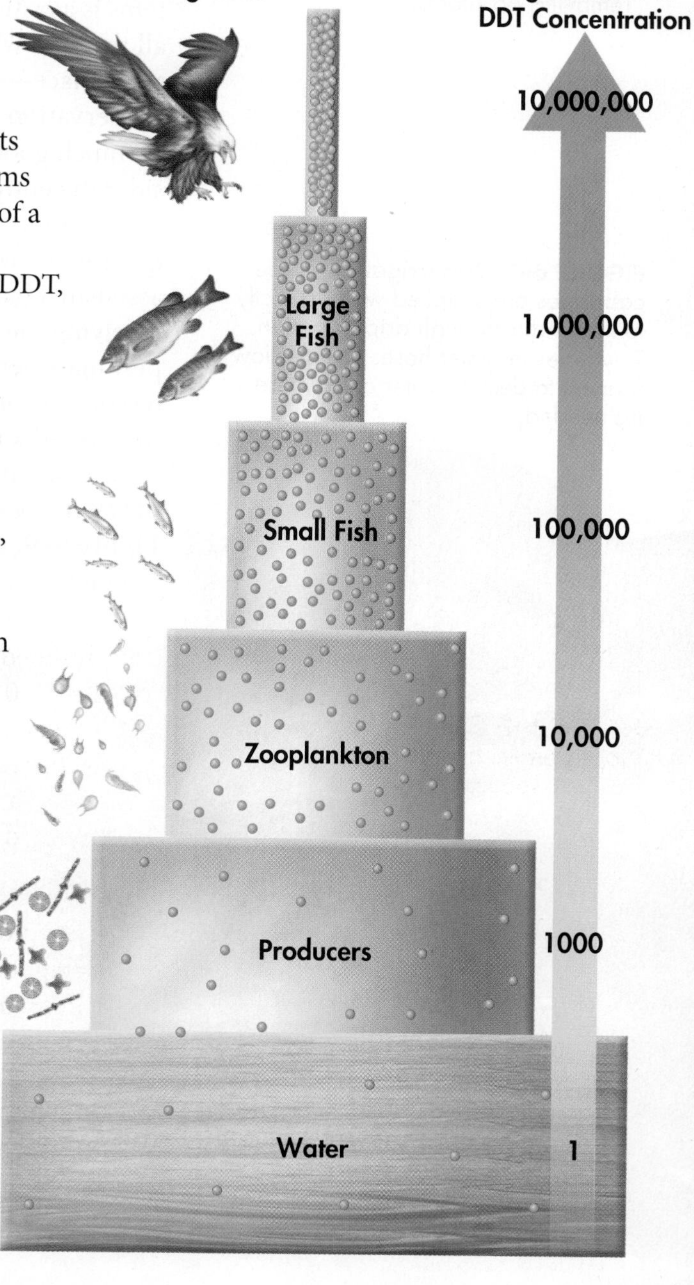

▶ ***Residential Sewage*** Have you ever stopped to think what happens after you flush your toilet? Those wastes don't disappear! They become residential sewage. Sewage isn't poisonous, but it does contain lots of nitrogen and phosphorus. Reasonable amounts of these nutrients can be processed by and absorbed into healthy ecosystems. But large amounts of sewage can stimulate blooms of bacteria and algae that rob water of oxygen. Oxygen-poor areas called "dead zones" can appear in both fresh and salt water. Raw sewage also contains microorganisms that can spread disease.

BUILD Vocabulary

RELATED WORD FORMS The verb *purify* is related to the noun *pure*. *To purify* means "to make pure or clean." Wetlands purify water by removing pollutants.

Water Quality and Sustainability One key to sustainable water use is to protect the natural systems involved in the water cycle. For example, as water flows slowly through a wetland, densely growing plants absorb some excess nutrients and filter out certain pollutants. Similarly, forests and other vegetation help purify water that seeps into the ground or runs off into rivers and lakes. Protecting these ecosystems is a critical part of watershed conservation. A watershed includes all the land whose groundwater, streams, and rivers drain into the same place—such as a large lake or river. The idea behind watershed conservation is simple: Cleaning up the pollution in a local area can't do much good if the water running into it is polluted. You must consider the entire watershed to achieve long-lasting results.

Pollution control can have direct and positive effects on the water quality in a watershed. Sewage treatment can lower levels of sewage-associated bacteria and help prevent dead zones in bodies of water receiving the runoff. In some situations, agriculture can use integrated pest management (IPM) instead of pesticides. IPM techniques include biological control—using predators and parasites to regulate for pest insects—the use of less-poisonous sprays, and crop rotation.

Conserving water is, of course, also important. One example of water conservation in agriculture is drip irrigation, shown in **Figure 6–9**, which delivers water drop by drop directly to the roots of plants that need it.

FIGURE 6–9 Drip Irrigation These cabbages are supplied water directly to their roots through drip irrigation. Tiny holes in water hoses (inset) allow farmers to deliver water only where it's needed.

Atmospheric Resources

What are the major forms of air pollution?

The atmosphere is a common resource whose quality has direct effects on health. After all, the atmosphere provides the oxygen we breathe! In addition, ozone, a form of oxygen that is found naturally in the upper atmosphere, absorbs harmful ultraviolet radiation from sunlight before it reaches Earth's surface. It is the ozone layer that protects our skin from damage that can cause cancer.

The atmosphere provides many other services. For example, the atmosphere's greenhouse gases, including carbon dioxide, methane, and water vapor, regulate global temperature. As you've learned, without the greenhouse effect, Earth's average temperature would be about 30° Celsius cooler than it is today.

The atmosphere is never "used up." So, classifying it as a renewable or nonrenewable resource is not as important as understanding how human activities affect the quality of the atmosphere. For most of Earth's history, the quality of the atmosphere has been naturally maintained by biogeochemical cycles. However, if we disrupt those cycles, or if we overload the atmosphere with pollutants, the effects on its quality can last a very long time.

FIGURE 6–10 Smog Despite closing factories and restricting vehicle access to the city, Beijing remained under a blanket of dense smog just days before the 2008 Summer Olympics. Apply Concepts ***What component of smog is beneficial when part of the atmosphere, but harmful when at ground level?***

Air Pollution What happens when the quality of Earth's atmosphere is reduced? For one thing, respiratory illnesses such as asthma are made worse and skin diseases tend to increase. Globally, climate patterns may be affected. What causes poor air quality? Industrial processes and the burning of fossil fuels can release pollutants of several kinds. **Common forms of air pollution include smog, acid rain, greenhouse gases, and particulates.**

▶ ***Smog*** If you live in a large city, you've probably seen **smog,** a gray-brown haze formed by chemical reactions among pollutants released into the air by industrial processes and automobile exhaust. Ozone is one product of these reactions. While ozone high up in the atmosphere helps protect life on Earth from ultraviolet radiation, at ground level, ozone and other pollutants threaten the health of people, especially those with respiratory conditions. Many athletes participating in the 2008 Summer Olympics in Beijing, China, expressed concern over how the intense smog, seen in **Figure 6–10,** would affect their performance and health.

In Your Notebook *Compare and contrast the atmosphere as a resource with fresh water as a resource.*

American Air Pollution Trends

Each year, the U.S. Environmental Protection Agency (EPA) estimates emissions from a variety of sources. Look at the graph in **Figure 6–12.** The combined emissions of six common pollutants are plotted along with trends in energy consumption and automobile travel between 1980 and 2007. The values shown are the total percentage change. For example, in 1995, aggregate emissions had dropped about 30 percent from their level in 1980.

1. Interpret Data Describe the overall trend in emissions since 1980. Is this what you would expect given the trends in energy consumption and automobile travel? Explain your answer.

2. Interpret Data How does this graph differ from one that shows *absolute* values for emissions? Would that graph start at zero as this one does?

3. Infer What do you think has contributed to the trends you see in this graph? Why would the EPA be particularly interested in these data?

▶ ***Acid Rain*** When we burn fossil fuels in our factories and homes, we release nitrogen and sulfur compounds. When those compounds combine with water vapor in the air, they form nitric and sulfuric acids. These airborne acids can drift for many kilometers before they fall as **acid rain.** Acidic water vapor can also affect ecosystems as fog or snow. In some areas, acid rain kills plants by damaging their leaves and changing the chemistry of soils and surface water. Examples of its effects are shown in **Figure 6–11.** Acid precipitation also can dissolve and release mercury and other toxic elements from soil, freeing those elements to enter other parts of the biosphere.

In Your Notebook *Create a flowchart that shows the steps in acid rain formation.*

FIGURE 6–11 Acid Rain Acid rain results from the chemical transformation of nitrogen and sulfur products that come from human activities. These reactions can cause damage to stone statues and plant life.

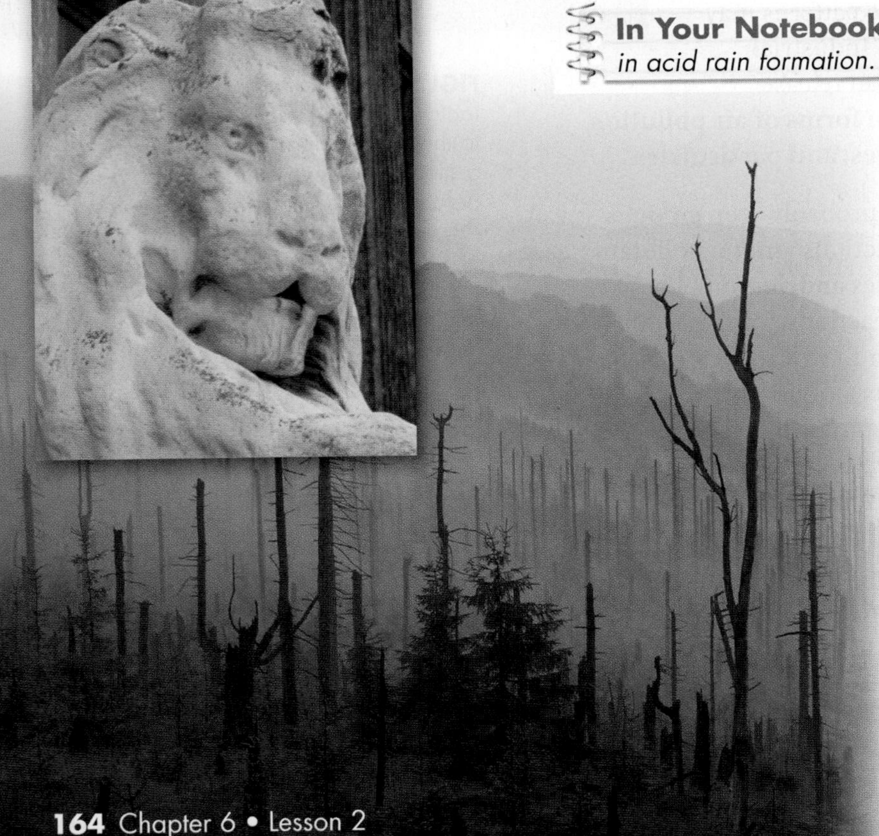

▶ ***Greenhouse Gases*** Burning fossil fuels and forests releases stored carbon into the atmosphere as carbon dioxide, a greenhouse gas. Agricultural practices from raising cattle to farming rice release methane, another greenhouse gas. Although some greenhouse gases are necessary, when excess greenhouse gases accumulate in the atmosphere, they contribute to global warming and climate change.

▶ ***Particulates*** Particulates are microscopic particles of ash and dust released by certain industrial processes and certain kinds of diesel engines. Very small particulates can pass through the nose and mouth and enter the lungs, where they can cause serious health problems.

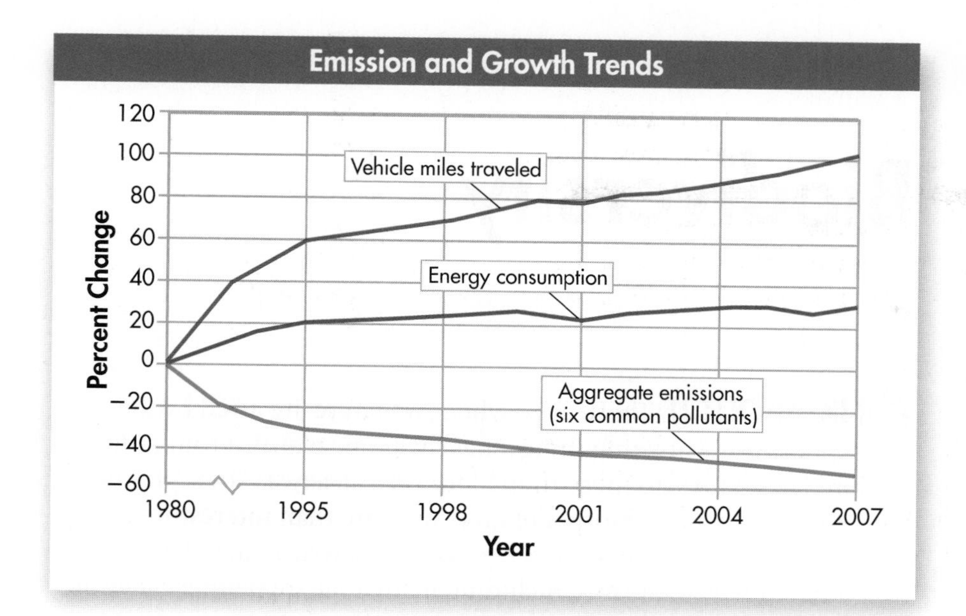

FIGURE 6–12 Air Pollution Trends This graph summarizes EPA findings of the total percentage change from 1980 to 2007 in vehicle miles traveled, energy consumption, and the combined emissions of six common pollutants—carbon monoxide, lead, nitrogen oxides, organic compounds, particulates, and sulfur dioxide. **Calculate** ***In 1980, motorists in the Puget Sound region of Washington State traveled 36.4 million miles. Assuming that these motorists increased their miles traveled at the national rate, approximately how many miles did they travel in 2007?*** MATH

Air Quality and Sustainability Improving air quality is difficult. Air doesn't stay in one place and doesn't "belong" to anyone. Automobile emission standards and clean-air regulations have improved air quality in some regions, however, and seem to be having a net positive effect, as shown in **Figure 6–12.** Efforts like these also have improved the atmosphere globally. At one time, for example, all gasoline was enriched with lead. But as leaded gasoline burned, lead was released in exhaust fumes and ultimately washed onto land and into rivers and streams. U.S. efforts to phase out leaded gasoline started in 1973 and were completed in 1996 when the sale of leaded gasoline was banned. Now that unleaded gasoline is used widely across the United States, lead levels in soils, rivers, and streams around the country have dropped significantly from earlier, higher levels.

6.2 Assessment

Review Key Concepts

1. a. Review What causes soil erosion? Why is soil erosion a problem?

b. Apply Concepts What are three ways in which the agriculture and forestry industries can improve the sustainability of soil?

2. a. Review How is fresh water both a renewable and a limited resource?

b. Explain Why are some pollutants more harmful to organisms at higher trophic levels?

c. Propose a Solution Pick one source of water pollution and describe a way in which we can reduce its effect.

3. a. Review What ecological goods and services does the atmosphere provide?

b. Relate Cause and Effect How does the use of fossil fuels negatively impact Earth's atmosphere?

ANALYZING DATA

4. Look at **Figure 6–8.** If the concentration of DDT in zooplankton measures 0.04 parts per million, what is the approximate concentration of DDT at each other trophic level shown? MATH

6.3 Biodiversity

Key Questions

Why is biodiversity important?

What are the most significant threats to biodiversity?

How do we preserve biodiversity?

Vocabulary

biodiversity
ecosystem diversity
species diversity
genetic diversity
habitat fragmentation
ecological hot spot

Taking Notes

Preview Visuals Before you read, look at **Figure 6–20.** Record three questions you have about the map. When you've finished reading, answer the questions.

THINK ABOUT IT Those of us who love nature are awed by the incredible variety of living things that share our planet. From multicolored coral reefs to moss-draped forests, *variety,* is "the spice of life." But variety in the biosphere gives us more than interesting things to look at. Our well-being is closely tied to the well-being of a great number of other organisms, including many that are neither majestic nor beautiful to our eyes.

The Value of Biodiversity

Why is biodiversity important?

Biological diversity, or **biodiversity,** is the total of all the genetically based variation in all organisms in the biosphere. To biologists, biodiversity is precious, worth preserving for its own sake. But what kinds of biodiversity exist, and what value do they offer society?

Types of Biodiversity Biodiversity exists on three levels: ecosystem diversity, species diversity, and genetic diversity. **Ecosystem diversity** refers to the variety of habitats, communities, and ecological processes in the biosphere. The number of different species in the biosphere, or in a particular area, is called **species diversity.** To date, biologists have identified and named more than 1.8 million species, and they estimate that at least 30 million more are yet be discovered. Much of this diversity exists among single-celled organisms. But new species of vertebrates, like the snake in **Figure 6–13,** are still being found.

Genetic diversity can refer to the sum total of all different forms of genetic information carried by a particular species, or by all organisms on Earth. Within each species, genetic diversity refers to the total of all different forms of genes present in that species. In many ways, genetic diversity is the most basic kind of biodiversity. It is also the hardest kind to see and appreciate. Yet, genetic diversity is vitally important to the survival and evolution of species in a changing world.

FIGURE 6–13 A New Species This tiny snake, native to the island of Barbados, is one of many recently discovered species. Photos of the snake were released in 2008. **Infer** ***Why are you more likely to discover a new vertebrate species in a tropical area than in a desert?***

Valuing Biodiversity You can't touch, smell, or eat biodiversity, so many people don't think of it as a natural resource. But biodiversity is one of Earth's greatest natural resources. **Biodiversity's benefits to society include contributions to medicine and agriculture, and the provision of ecosystem goods and services.** When biodiversity is lost, significant value to the biosphere and to humanity may be lost along with it.

▶ ***Biodiversity and Medicine*** Wild species are the original source of many medicines, including painkillers like aspirin and antibiotics like penicillin. The chemicals in wild species are used to treat diseases like depression and cancer. For example, the foxglove, shown in **Figure 6–14,** contains compounds called digitalins that are used to treat heart disease. These plant compounds are assembled according to instructions coded in genes. So the genetic information carried by diverse species is like a "natural library" from which we have a great deal to learn.

▶ ***Biodiversity and Agriculture*** Genetic diversity is also important in agriculture. Most crop plants have wild relatives, like the potatoes in **Figure 6–15.** These wild plants may carry genes we can use—through plant breeding or genetic engineering—to transfer disease or pest resistance, or other useful traits, to crop plants.

▶ ***Biodiversity and Ecosystem Services*** The number and variety of species in an ecosystem can influence that ecosystem's stability, productivity, and value to humans. Sometimes the presence or absence of a single keystone species, like the sea otter in **Figure 6–16,** can completely change the nature of life in an ecosystem. Also, healthy and diverse ecosystems play a vital role in maintaining soil, water, and air quality.

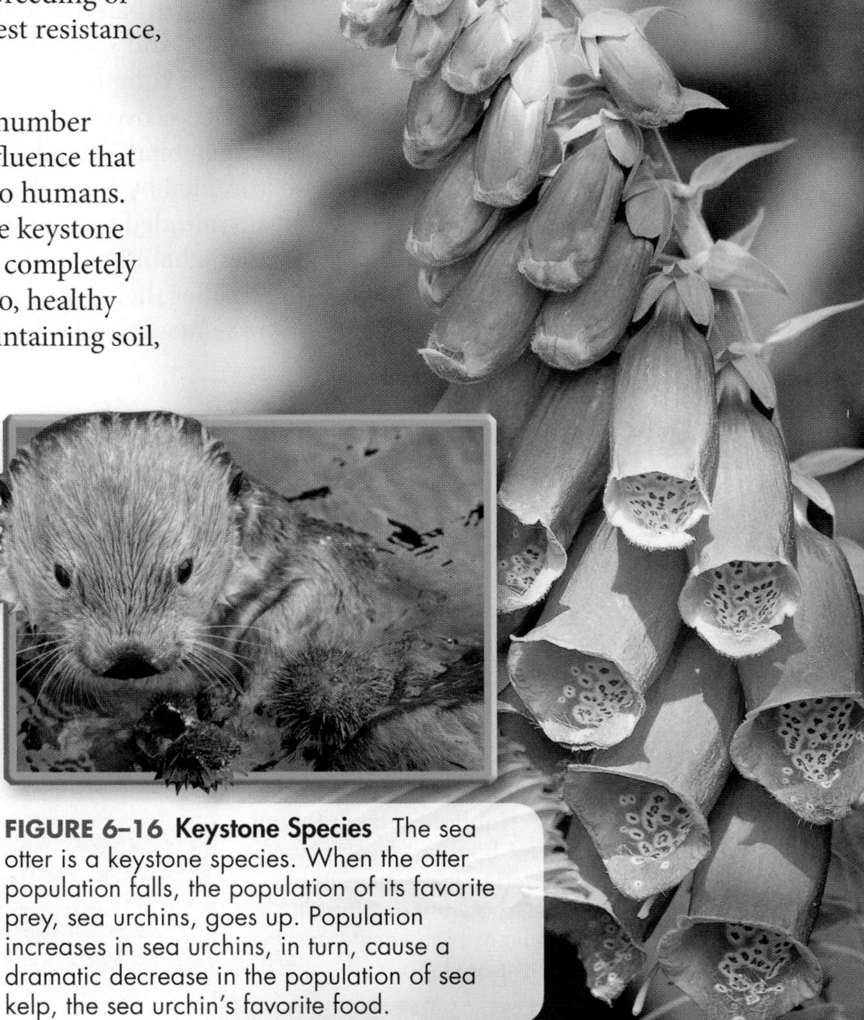

FIGURE 6–14 Medicinal Plants Digoxin, a drug derived from digitalin compounds in the foxglove plant, is used to treat heart disease.

FIGURE 6–15 Potato Diversity The genetic diversity of wild potatoes in South America can be seen in the colorful varieties shown here. The International Potato Center, based in Peru, houses a "library" of more than 4500 tuber varieties.

FIGURE 6–16 Keystone Species The sea otter is a keystone species. When the otter population falls, the population of its favorite prey, sea urchins, goes up. Population increases in sea urchins, in turn, cause a dramatic decrease in the population of sea kelp, the sea urchin's favorite food.

Threats to Biodiversity

What are the most significant threats to biodiversity?

Species have been evolving, changing, and dying out since life began. In fact scientists estimate that over 99 percent of the species that have ever lived are now extinct. So extinction is not new. But human activity today is causing the greatest wave of extinctions since dinosaurs disappeared. The current rate of species loss is approaching 1000 times the "typical" rate. And as species disappear, the potential contribution to human knowledge that is carried in their genes is lost.

Species diversity is related to genetic diversity. The more genetically diverse a species is, the greater its chances of surviving disturbances. So, as human activity reduces genetic diversity, species are put at a greater risk for extinction. Species diversity, in turn, is linked to ecosystem diversity. Therefore, as ecosystems are damaged, the organisms that inhabit them become more vulnerable to extinction.

How are humans influencing biodiversity? **Humans reduce biodiversity by altering habitats, hunting, introducing invasive species, releasing pollution into food webs, and contributing to climate change.** Biologists compare loss of biodiversity to destroying a library before its books are ever read.

MYSTERY CLUE

Easter Island probably never had the biological diversity of Hawaii and many other tropical islands. How would this affect its ability to rebound after a disturbance?

Altered Habitats When natural habitats are eliminated for agriculture or for urban development, the number of species in those habitats drops, and some species may become extinct. But, habitats don't need to be completely destroyed to put species at risk. Development often splits ecosystems into pieces, a process called **habitat fragmentation,** leaving habitat "islands." You probably think of islands as bits of land surrounded by water, but a biological island can be any patch of habitat surrounded by a different habitat, as shown in **Figure 6–17.** The smaller a habitat island, the fewer the species that can live there and the smaller their populations. Both changes make habitats and species more vulnerable to other disturbance.

BUILD Vocabulary

ACADEMIC WORDS The adjective **vulnerable** means "open to attack or damage." Fragmented habitats are more vulnerable, or more apt to be damaged, than larger undisturbed habitats because they contain fewer species and smaller populations of organisms.

FIGURE 6–17 Habitat Fragmentation Deforestation for housing developments in Florida has led to the pattern of forest "islands" shown here. Habitat fragmentation limits biodiversity and the potential size of populations.

FIGURE 6–18 Hunted and Sold as Pets *These caged green parrots were captured in the Amazon rainforest and brought to a market in Peru.* **Infer** ***What impact do you think hunting has on the animals left behind?***

Hunting and the Demand for Wildlife Products Humans can push species to extinction by hunting. In the 1800s, hunting wiped out the Carolina parakeet and the passenger pigeon. Today endangered species in the United States are protected from hunting, but hunting still threatens rare animals in Africa, South America, and Southeast Asia. Some animals, like many birds, are hunted for meat. Others are hunted for their commercially valuable hides or skins or because people believe their body parts have medicinal properties. Still others, like the parrots in **Figure 6–18,** are hunted to be sold as pets. Hunted species are affected even more than other species by habitat fragmentation because fragmentation increases access for hunters and limits available hiding spaces for prey. The Convention on International Trade in Endangered Species (CITES) bans international trade in products from a list of endangered species. Unfortunately, it's difficult to enforce laws in remote wilderness areas.

Introduced Species Recall that organisms introduced to new habitats can become invasive and threaten biodiversity. For example, more than 130 introduced species live in the Great Lakes, where they have been changing aquatic ecosystems and driving native species close to extinction. One European weed, leafy spurge, infests millions of hectares across the Northern Great Plains. On rangelands, leafy spurge displaces grasses and other food plants, and its milky latex can sicken or kill cattle and horses. Each year, ranchers and farmers suffer losses of more than $120 million because of this single pest.

MYSTERY CLUE

Almost all the coconut shells found by researchers on Easter Island show signs of having been gnawed on by nonnative rats. Coconuts contain the seeds of the coconut palm. What effect do you think the rats had on the coconut palm population?

Pollution Many of the pollutants described in the last lesson also threaten biodiversity. DDT, for example, prevents birds from laying healthy eggs. In the United States, brown pelican, peregrine falcon, and other bird populations plummeted with widespread use of the chemical. Acid rain places stress on land and water organisms. Increased carbon dioxide in the atmosphere is dissolving in oceans, making them more acidic, which threatens biodiversity on coral reefs and in other marine ecosystems.

In Your Notebook *Why is acidic water harmful to coral?*

Climate Change Climate change (a topic in the next lesson) is a major threat to biodiversity. Remember that organisms are adapted to their environments and have specific tolerance ranges to temperature and other abiotic conditions. If conditions change beyond an organism's tolerance, the organism must move to a more suitable location or face extinction. Species in fragmented habitats are particularly vulnerable to climate change because if conditions change they may not be able to move easily to a suitable habitat. Estimates vary regarding the effects of climate change on biodiversity. If global temperatures increase 1.5°C–2.5°C over late twentieth-century temperatures, 30 percent of species studied are likely to face increased risk of extinction. If the global temperature increase goes beyond 3.5°C, it is likely that 40–70 percent of species studied will face extinction.

Conserving Biodiversity

How do we preserve biodiversity?

What can we do to protect biodiversity? Should we focus on a particular organism like the scarlet macaw? Or should we try to save an entire ecosystem like the Amazon rain forest? We must do both. At the same time, conservation efforts must take human interests into account. **To conserve biodiversity, we must protect individual species, preserve habitats and ecosystems, and make certain that human neighbors of protected areas benefit from participating in conservation efforts.**

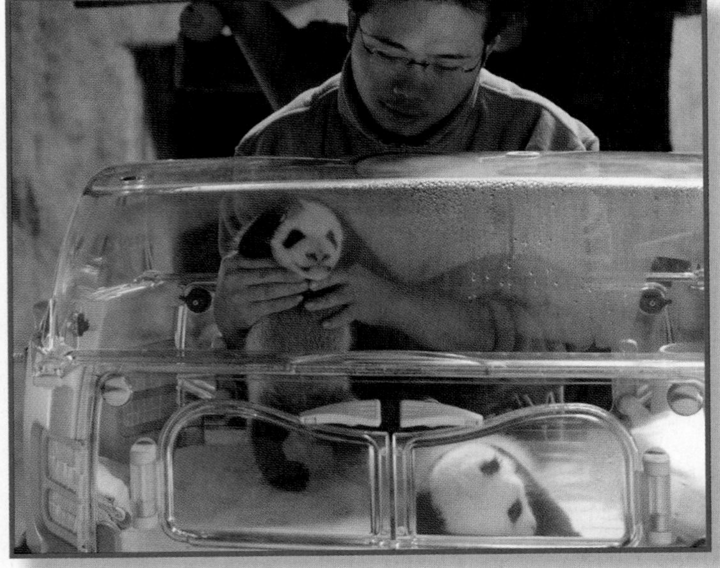

FIGURE 6–19 Saving an Individual Species Efforts to save the giant panda include a comprehensive captive breeding and reintroduction program. Here, a specialist from China holds one of twin pandas born at the zoo in Madrid, Spain. **Apply Concepts** ***How does captive breeding affect a population's genetic diversity?***

Protecting Individual Species In the past, most conservation efforts focused on individual species, and some of this work continues today. The Association of Zoos and Aquariums (AZA), for example, oversees species survival plans (SSPs) designed to protect threatened and endangered species. A key part of those plans is a captive breeding program. Members of the AZA carefully select and manage mating pairs of animals to ensure maximum genetic diversity. The ultimate goal of an SSP is to reintroduce individuals to the wild. Research, public education, and breeding programs all contribute to that goal. More than 180 species, including the giant panda shown in **Figure 6–19,** are currently covered by SSPs.

Preserving Habitats and Ecosystems The main thrust of global conservation efforts today is to protect not just individual species but entire ecosystems. The goal is to preserve the natural interactions of many species at once. To that end, governments and conservation groups work to set aside land as parks and reserves. The United States has national parks, forests, and other protected areas. Marine sanctuaries are being created to protect coral reefs and marine mammals.

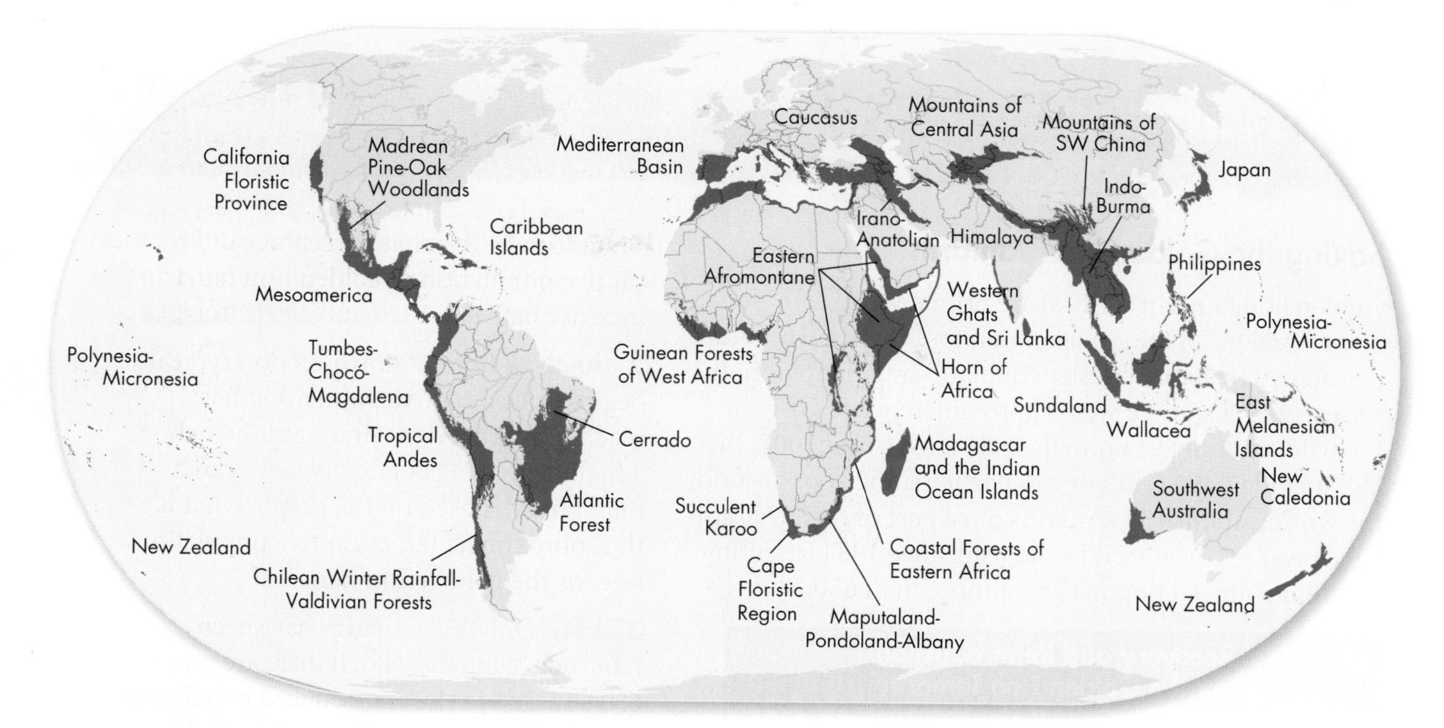

The challenge is protecting areas that are large enough and that contain the right resources to protect biodiversity. To make sure that conservation efforts are concentrated in the most important places, conservation biologists have identified ecological "hot spots," shown in red in **Figure 6–20.** An **ecological hot spot** is a place where significant numbers of species and habitats are in immediate danger of extinction. By identifying these areas, ecologists hope that scientists and governments can better target their efforts to save as many species as possible.

FIGURE 6–20 Ecological Hot Spots Conservation International identifies biodiversity hot spots using two criteria. The area (1) must contain at least 1500 species of native vascular plants, and (2) it must have lost at least 70 percent of its original habitat. The 34 hot spots seen here cover just 2.3 percent of Earth's land surface, but they contain over 50 percent of the world's plant species and 42 percent of its terrestrial vertebrates.

Considering Local Interests Protecting biodiversity often demands that individuals change their habits or the way they earn their living. In these cases it is helpful to offer some reward or incentive to the people or communities involved. The United States government, for example, has offered tax credits to people who've installed solar panels or bought hybrid cars. Similarly, many communities in Africa, Central America, and Southeast Asia have set aside land for national parks and nature reserves, like the park shown in **Figure 6–21,** to attract tourist dollars. In some Australian communities, farmers were paid to plant trees along rivers and streams as part of wildlife corridors connecting forest fragments. Not only did the trees help improve local water quality; they also improved the health of the farmers' cows, which were able to enjoy shade on hot days!

The use of carbon credits is one strategy aimed at encouraging industries to cut fossil fuels use. Companies are allowed to release a certain amount of carbon into the environment. Any unused carbon may be sold back at a set market value or traded to other companies. This strategy encourages industries to pay for lower-emission machinery and to adopt carbon-saving practices. In this way, pollution is capped or cut without adding a financial burden to the industry involved. This helps protect the economy while reducing biodiversity loss due to pollution. These examples show that conservation efforts work best when they are both informed by solid scientific information and benefit the communities affected by them.

FIGURE 6–21 Ecotourism A tourist gets an elephant-size kiss from one of the over 30 rescued elephants at Thailand's Elephant Nature Park.

Saving the Golden Lion Tamarin

Golden lion tamarins (GLTs) are primates native to the coastal regions of the Amazon rain forest. They have been threatened by habitat destruction and fragmentation. In the early 1970s, there were approximately 200 GLTs in the wild and only 91 animals in 26 zoos. As of 2007, the SSP included 496 GLTs in 145 participating zoos around the world. About 153 tamarins once part of the program have been reintroduced to the wild since 1984, resulting in a reintroduced population of more than 650.

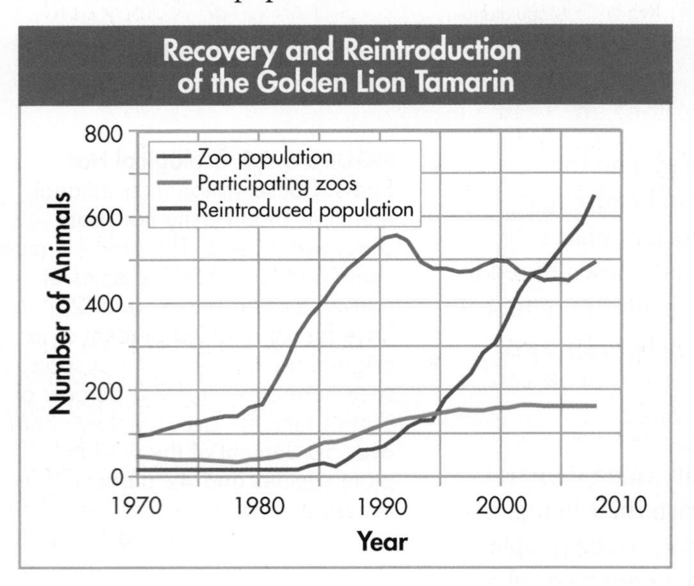

1. **Calculate** By what percentage did the captive population of golden lion tamarins increase between 1970 and 2007? MATH

2. **Analyze Data** Reintroduction typically begins once a captive population has reached a target size—the size at which a high degree of genetic diversity can be maintained. Based on the graph, what is the approximate target captive population size for the golden lion tamarin?

3. **Infer** Only 153 tamarins have been reintroduced to the wild. If there are now 650 tamarins in the reintroduced population, where did the other 497 come from?

4. **Form an Opinion** When populations of wild animals get very small, do you think they should be removed from the wild and brought into captivity? Why or why not?

Adapted from J. D. Ballou and J. Mickelberg, *International Studbook for Golden Lion Tamarins* (Washington, D.C.: National Zoological Park, Smithsonian Institution, 2007). B. Holst et al., *Lion Tamarin Population and Habitat Viability Assessment Workshop 2005, Final Report* (Apple Valley, MN: IUCN/SSC Conservation Breeding Specialist Group, 2006.)

6.3 Assessment

Review Key Concepts

1. a. **Review** Describe the different components of global biodiversity.
   b. **Apply Concepts** What benefits does society get from biodiversity?
2. a. **Review** What are the major threats to biodiversity?
   b. **Relate Cause and Effect** Explain the relationship between habitat size and species diversity.
3. a. **Review** What is the goal of a species survival plan?
   b. **Form an Opinion** Do you think that the hot spot strategy is a good one? Explain your answer.

VISUAL THINKING

4. Look back at the biome map on page 111. Compare it to the map in **Figure 6–20.** Are there any similarities among the biomes the hot spots belong to? Using what you know about biomes, are you surprised by what you've found? Explain your answer.

6.4 Meeting Ecological Challenges

THINK ABOUT IT Every year, the EPA awards up to ten President's Environmental Youth Awards. Past winners have included an Eagle Scout from Massachusetts who encouraged people who fish to stop using lead weights that contaminate water and poison organisms, students from Washington State who reduced waste at their school and saved more than half a million dollars in the process, and a student from Florida who developed an outreach program to protect local sea turtles. What do these award winners have in common? They came up with ideas that protect the environment while satisfying both present and future needs. This kind of leadership is what will help us chart a new course for the future.

Key Questions

How does the average ecological footprint in America compare to the world's average?

How can ecology guide us toward a sustainable future?

Vocabulary

ecological footprint
ozone layer
aquaculture
global warming

Taking Notes

Compare/Contrast Table As you read, create a table comparing the challenges associated with the ozone layer, fisheries, and global climate. Note the problem observed, the causes identified, and the solutions implemented.

Ecological Footprints

How does the average ecological footprint in America compare to the world's average?

What is our impact on the biosphere today? To answer that question, think about the kind and amount of resources each of us uses. Ecologists refer to the human impact on the biosphere using a concept called the ecological footprint. The **ecological footprint** describes the total area of functioning land and water ecosystems needed both to provide the resources an individual or population uses and to absorb and make harmless the wastes that individual or population generates. Ecological footprints take into account the need to provide resources such as energy, food, water, and shelter, and to absorb such wastes as sewage and greenhouse gases. Ecologists use footprint calculations to estimate the biosphere's carrying capacity for humans. An artist's rendition of an ecological footprint is shown in **Figure 6–22.**

Footprint Limitations Ecologists talk about the ecological footprints of individuals, of countries, and of the world's population. Calculating actual numbers for ecological footprints, however, is complicated. The concept is so new that there is no universally accepted way to calculate footprint size. What's more, footprints give only a "snapshot" of the situation at a particular point in time.

VISUAL ANALOGY

ECOLOGICAL FOOTPRINTS

FIGURE 6–22 The food you eat, the miles you travel, and the electricity you use all contribute to your—and the population's—ecological footprint.

Comparing Footprints Although calculating *absolute* footprints is difficult, ecological footprints can be useful for making *comparisons* among different populations, as shown in **Figure 6–23.** **According to one data set, the average American has an ecological footprint over four times larger than the global average.** The per person use of resources in America is almost twice that in England, more than twice that in Japan, and almost six times that in China. To determine the ecological footprint of an entire country, researchers calculate the footprint for a typical citizen and then multiply that by the size of the population.

In Your Notebook *How have you contributed to your ecological footprint today? Give at least ten examples.*

FIGURE 6–23 Relative Footprints This world map shows each country in proportion to its ecological footprint. The United States has an ecological footprint about twice the world's average. By contrast, the African nation of Zambia has a footprint a little over one-fourth the global average. Compare each country's "footprint" size to its actual size on the smaller map below.

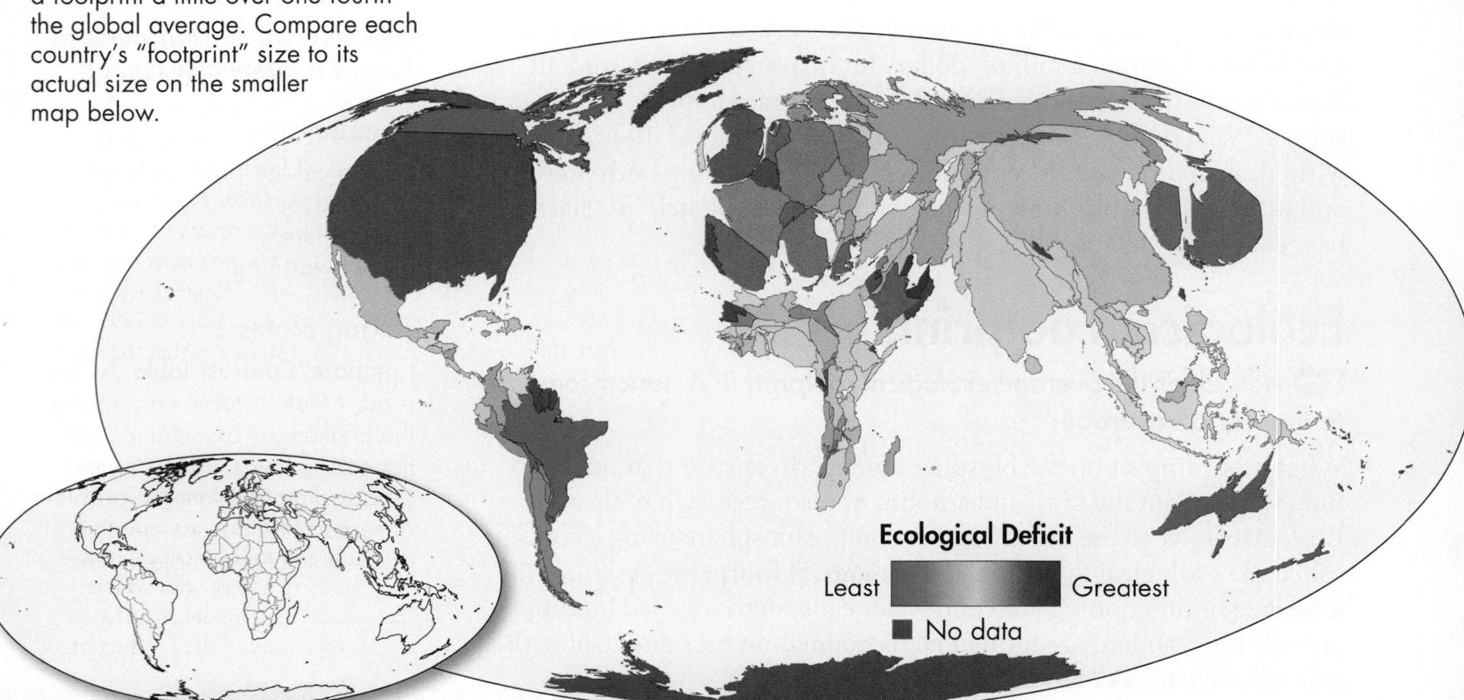

Ecology in Action

How can ecology guide us toward a sustainable future?

The future of the biosphere depends on our ecological footprints, global population growth, and technological development. Right now it's more common to hear stories of ecological challenges than successes. Given the size of those challenges, you might be tempted to give up, to feel that things are getting worse, and that there is nothing we can do about it. But ecological research, properly collected, analyzed, and applied, can help us make decisions that will produce profoundly positive effects on the human condition. The basic principles of ecology can guide us toward a sustainable future. **By (1) recognizing a problem in the environment, (2) researching that problem to determine its cause, and then (3) using scientific understanding to change our behavior, we can have a positive impact on the global environment.** The following case studies illustrate the importance of the steps.

Case Study #1: Atmospheric Ozone

Between 20 and 50 kilometers above Earth's surface, the atmosphere contains a relatively high concentration of ozone called the **ozone layer.** Ozone at ground level is a pollutant, but the natural ozone layer absorbs harmful ultraviolet (UV) radiation from sunlight. Overexposure to UV radiation is the main cause of sunburn. It also can cause cancer, damage eyes, and lower resistance to disease. And intense UV radiation can damage plants and algae. By absorbing UV light, the ozone layer serves as a global sunscreen.

The following is an ecological success story. Over four decades, society has recognized a problem, identified its cause, and cooperated internationally to address a global issue.

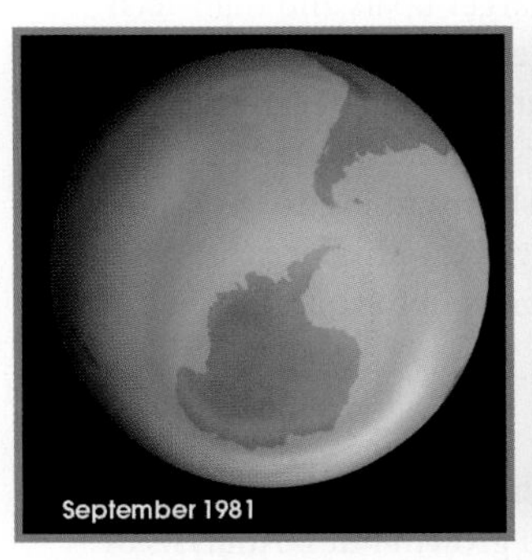

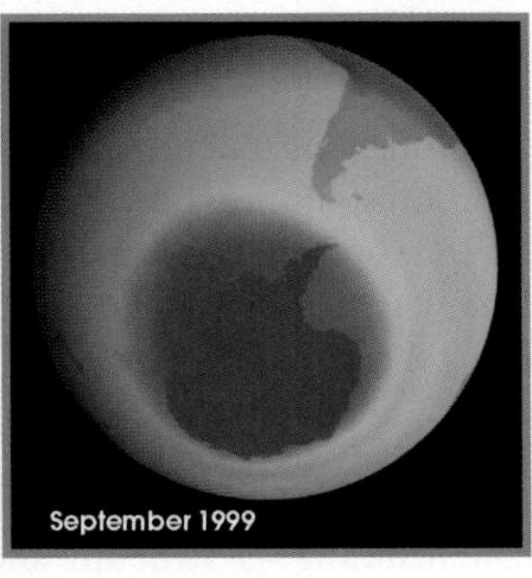

FIGURE 6–24 The Disappearing Ozone

1 Recognizing a Problem: "Hole" in the Ozone Layer Beginning in the 1970s, satellite data revealed that the ozone concentration over Antarctica was dropping during the southern winter. An area of lower ozone concentration is commonly called an ozone hole. It isn't really a "hole" in the atmosphere, of course, but an area where little ozone is present. For several years after the ozone hole was first discovered, it grew larger and lasted longer each year. **Figure 6–24** shows the progression from 1981 to 1999. The darker blue color in the later image indicates that the ozone layer had thinned since 1981.

FIGURE 6–25 CFC-Containing Refrigerators

2 Researching the Cause: CFCs In 1974 a research team led by Mario Molina, F. Sherwood Rowland, and Paul J. Crutzen demonstrated that gases called chlorofluorocarbons (CFCs) could damage the ozone layer. This research earned the team a Nobel Prize in 1995. CFCs were once widely used as propellants in aerosol cans; as coolant in refrigerators, freezers, and air conditioners; and in the production of plastic foams.

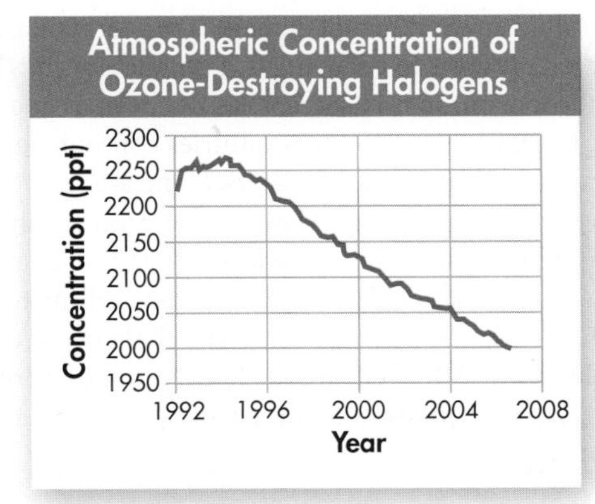

FIGURE 6–26 The Decline of CFCs

3 Changing Behavior: Regulation of CFCs Once the research on CFCs was published and accepted by the scientific community, the rest was up to policymakers—and in this case, their response was tremendous. Following the recommendations of ozone researchers, 191 countries signed a major agreement, the Montreal Protocol, which banned most uses of CFCs. Because CFCs can remain in the atmosphere for a century, their effects on the ozone layer are still visible. But ozone-destroying halogens from CFCs have been steadily decreasing since about 1994, as shown in **Figure 6–26,** evidence that the CFC ban has had positive long-term effects. In fact, current data predict that although the ozone hole will continue to fluctuate in size from year to year, it should disappear for good around the middle of this century.

Case Study #2: North Atlantic Fisheries

From 1950 to 1997, the annual world seafood catch grew from 19 million tons to more than 90 million tons. This growth led many to believe that the fish supply was an endless, renewable resource. However, recent dramatic declines in commercial fish populations have proved otherwise. This problem is one society is still working on.

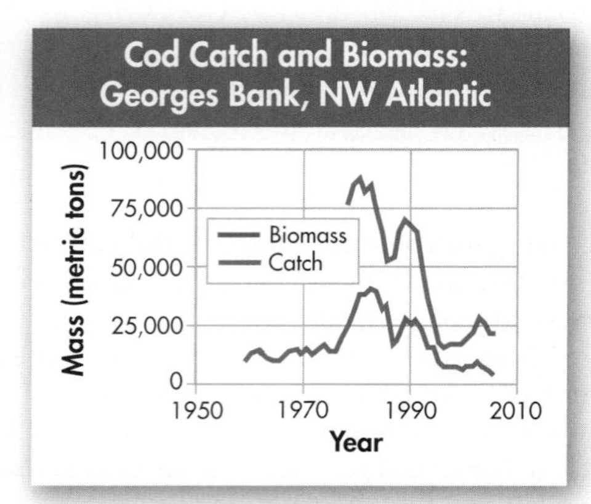

FIGURE 6–27 The Decline of Cod

FIGURE 6–28 Overfishing

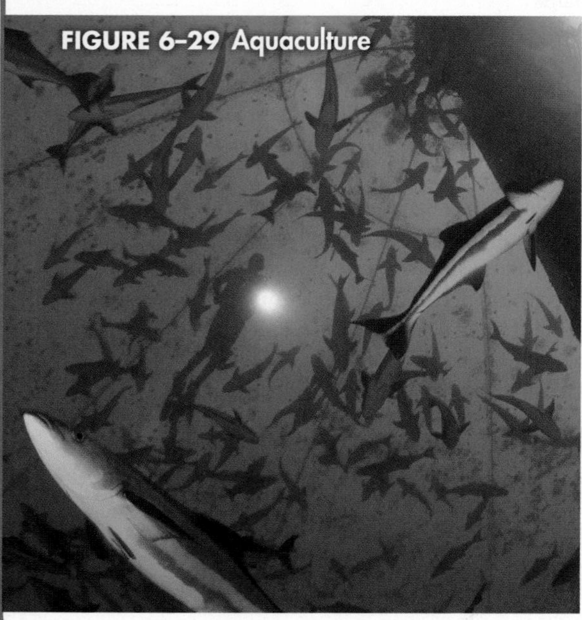

FIGURE 6–29 Aquaculture

1 Recognizing a Problem: More Work, Fewer Fish The cod catch has been rising and falling over the last century. Some of that fluctuation has been due to natural variations in ocean ecosystems. But often, low fish catches resulted when boats started taking too many fish. From the 1950s through the 1970s, larger boats and high-tech fish-finding equipment made the fishing effort both more intense and more efficient. Catches rose for a time but then began falling. The difference this time, was that fish catches continued to fall despite the most intense fishing effort in history. As shown in **Figure 6–27,** the total mass of cod caught has decreased significantly since the 1980s because of the sharp decrease of cod biomass in the ocean. You can't catch what isn't there.

2 Researching the Cause: Overfishing Fishery ecologists gathered data including age structure and growth rates. Analysis of these data showed that fish populations were shrinking. By the 1990s, cod and haddock populations had dropped so low that researchers feared these fish might disappear for good. It has become clear that recent declines in fish catches were the result of overfishing, as seen in **Figure 6–28.** Fish were being caught faster than they could be replaced by reproduction. In other words, the death rates of commercial fish populations were exceeding birth rates.

3 Changing Behavior: Regulation of Fisheries The U.S. National Marine Fisheries Service used its best data to create guidelines for commercial fishing. The guidelines specified how many fish of what size could be caught in U.S. waters. In 1996, the Sustainable Fisheries Act closed certain areas to fishing until stocks recover. Other areas are closed seasonally to allow fish to breed and spawn. These regulations are helping some fish populations recover, but not all. **Aquaculture**—the farming of aquatic animals—offers a good alternative to commercial fishing with limited environmental damage if properly managed.

Overall, however, progress in restoring fish populations has been slow. International cooperation on fisheries has not been as good as it was with ozone. Huge fleets from other countries continue to fish the ocean waters outside U.S. territorial waters. Some are reluctant to accept conservation efforts because regulations that protect fish populations for the future cause job and income losses today. Of course, if fish stocks disappear, the result will be even more devastating to the fishing industry than temporary fishing bans. The challenge is to come up with sustainable practices that ensure the long-term health of fisheries with minimal short-term impact on the fishing industry. Exactly how to meet that challenge is still up for debate.

Case Study #3: Climate Change

Global climate involves cycles of matter across the biosphere and everything modern humans do—from cutting and burning forests to manufacturing, driving cars, and generating electricity. The most reliable current information available on this subject comes from the 2007 report of the Intergovernmental Panel On Climate Change (IPCC). The IPCC is an international organization established in 1988 to provide the best possible scientific information on climate change. IPCC reports contain data and analyses that have been agreed upon and accepted by 2500 climate scientists from around the world and the governments participating in the study.

1 Recognizing a Problem: Global Warming The IPCC report confirms earlier observations that global temperatures are rising. This increase in average temperature is called **global warming.** Remember that winds and ocean currents, which are driven by differences in temperature across the biosphere, shape climate. Given this link between temperature and climate, it isn't surprising that the IPCC report discusses more than warming. The report also discusses climate change—changes in patterns of temperature, rainfall, and other physical environmental factors that can result from global warming. There are many lines of evidence, both physical and biological, that have contributed to our current understanding of the climate change issue.

- **Physical Evidence** Physical evidence of global warming comes from several sources. The graphs in **Figure 6–30,** taken from data in the 2007 IPCC report, show that Earth's temperatures are getting warmer, its sea ice is melting, and its sea levels are rising. Eleven of the twelve years between 1995 and 2006 were among the warmest years since temperature recording began in 1850. Between 1906 and 2005, Earth's average global temperature rose 0.74°C. The largest changes are occurring in and near the Arctic Circle. Average temperatures in Alaska, for example, increased 2.4°C over the last 50 years. Sea level has risen since 1961 at a rate of 1.8 mm each year. This increase is caused by warmer water expanding and by melting glaciers, ice caps, and polar ice sheets. Satellite data confirm that arctic sea ice, glaciers, and snow cover are decreasing.

FIGURE 6–30 A Warming Earth

A.

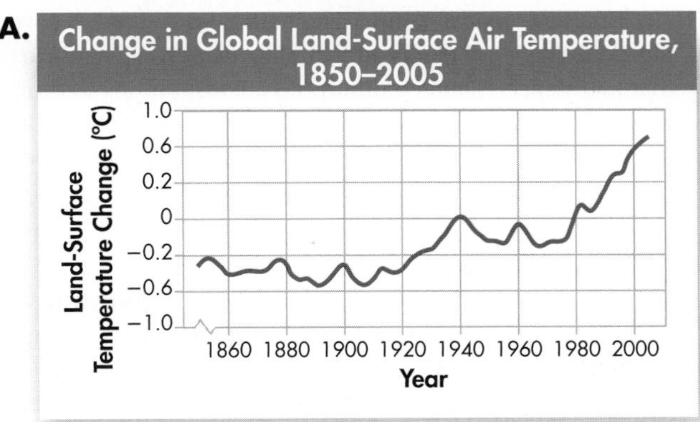

change from average 1961–1990 temperature

B.

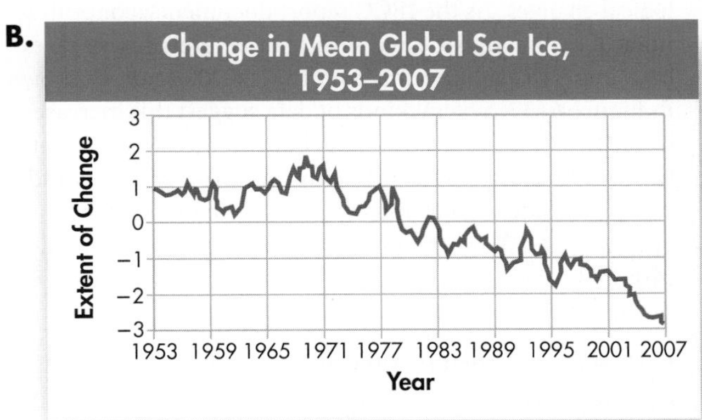

change from average 1953–2007 sea ice extent

C.

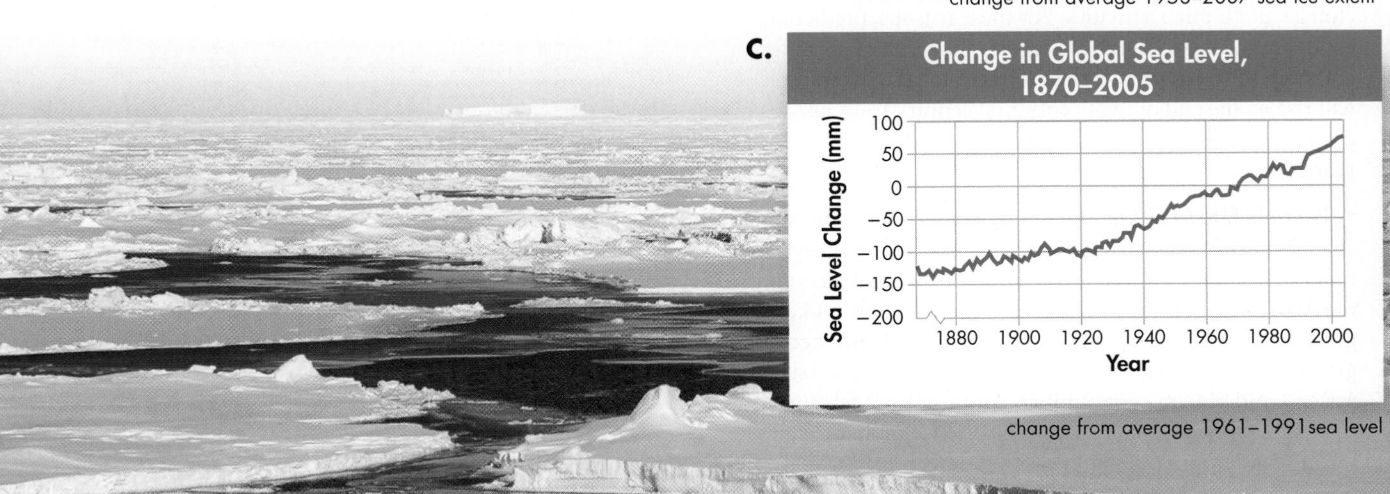

change from average 1961–1991 sea level

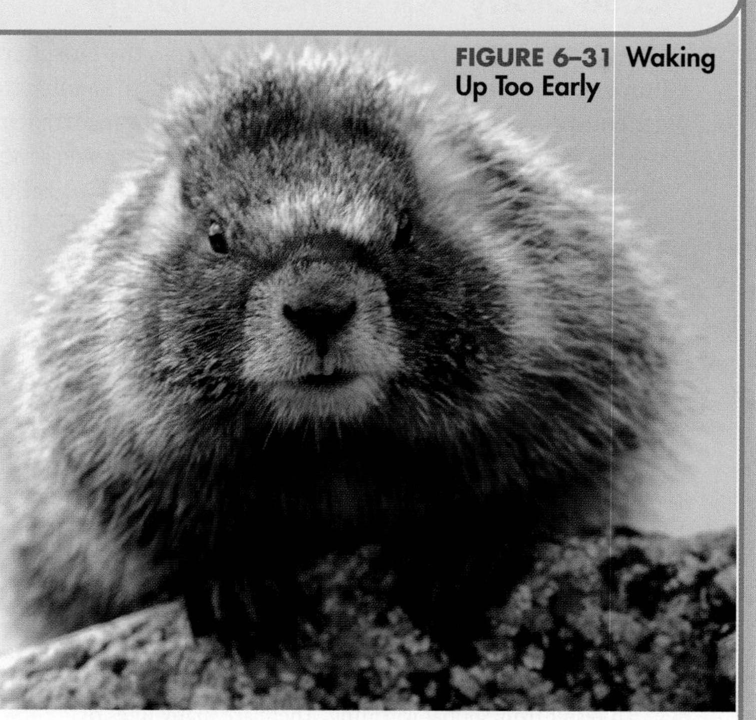

FIGURE 6–31 Waking Up Too Early

• **Biological Evidence** Small changes in climate that humans scarcely notice can be important to other organisms. Remember that each organism's range is determined by factors like temperature, humidity, and rainfall. If those conditions change, the organisms can be affected. If temperature rises, for example, organisms would usually move toward cooler places away from the equator and from warm lowlands to cooler, higher altitudes. In addition, plant flowering and animal breeding are often cued by seasonal changes. If warming is occurring, these organisms should respond as though spring begins earlier.

The IPCC report summarizes data from 75 studies covering 1700 species of plants and animals. These data confirm that many species and communities are responding as though they are experiencing rising temperatures. The yellow-bellied marmot in **Figure 6–31,** for example, is coming out of hibernation over a month earlier than it used to.

❷ Researching the Cause: Models and Questions

What is causing global warming? Earth's climate has changed often during its history. So researchers had to determine whether current warming is part of a natural cycle or whether it is caused by human activity or by astronomical and geological changes. As the IPCC report documents, concentrations of carbon dioxide and several other greenhouse gases have increased significantly over the last 200 years, as shown in **Figure 6–32.** Several kinds of data suggest this increase is due to the burning of fossil fuels, combined with the cutting and burning of forests worldwide. These activities add carbon dioxide to the atmosphere faster than the carbon cycle removes it. Most climate scientists agree that this added carbon dioxide is strengthening the natural greenhouse effect, causing the biosphere to retain more heat.

• **How Much Change?** How much warming is expected? For answers, researchers turn to computer models based on data. The models are complex and involve assumptions about climate and human activities. For these reasons, predictions are open to debate. The IPCC reports the result of six different models, which predict that average global temperatures will rise by the end of the twenty-first century from just under 2°C to as much as 6.4°C higher than they were in the year 2000.

• **Possible Effects of Climate Change** What does climate change mean? Some changes are likely to threaten ecosystems ranging from tundra and northern forests to coral reefs and the Amazon rain forest. The western United States is likely to get drier. The Sahara Desert, on the other hand, may become greener. Sea level may rise enough to flood some coastal ecosystems and human communities. And some models suggest that parts of North America may experience more droughts during the summer growing season.

FIGURE 6–32 Greenhouse Gases

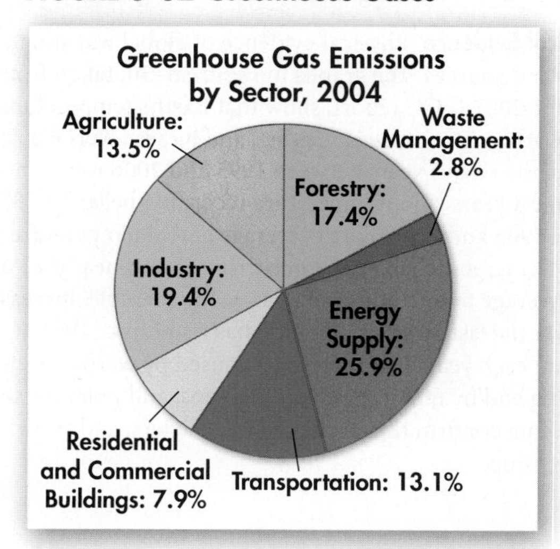

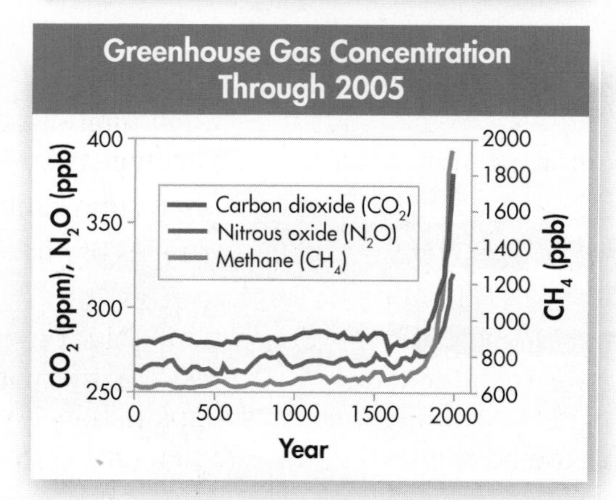

3 Changing Behavior: The Challenges Ahead You have seen how research has led to actions that are preserving the ozone layer and attempting to restore fisheries. In terms of global climate, great challenges lie ahead of us. Scientists have been saying for more than two decades that the world needs to recognize the importance of climate change and take steps to minimize further warming. The changes in behavior needed to cut back on greenhouse gas emissions will be major and will require input from economics and many other fields beyond biology. Some changes will rely on new technology for renewable energy and more efficient energy use. Because changing our use of fossil fuels and other behaviors will be difficult, researchers continue to gather data as they try to make more accurate models. In the meantime, we have begun to see the emergence of electric cars, recycled products, and green buildings.

Nations of the world have begun holding international climate summits, at which they attempt to work out agreements to protect the atmosphere and climate—both of which are truly global issues. As the world, and our own government, tries to work through these challenges, remember that the purpose of ecology is not to predict disaster or to prevent people from enjoying modern life. The world is our island of life. Hopefully, humanity can work toward a day when scientific information and human ingenuity help us reach the common goal of preserving the quality of life on Earth.

FIGURE 6–33 Little Changes, Big Results

6.4 Assessment

Review Key Concepts

1. a. **Review** What are ecological footprints?

   b. **Apply Concepts** What are the limitations of the ecological footprint model, and how can ecologists best use it?

2. a. **Review** Why is the ozone layer important to living things?

   b. **Explain** What are the major types of physical and biological evidence for climate change?

   c. **Propose a Solution** Suggest one solution for the fisheries problem. Your solution can be at the international, national, regional, or individual level. Explain how it would help, and what challenges you see in implementing it.

Apply the Big idea

Interdependence in Nature

3. Refer to the carbon cycle on page 83. Describe how extensive burning of fossil fuels is affecting other reservoirs of carbon in the biosphere.

NGSS Smart Guide Humans in the Biosphere

Disciplinary Core Idea LS2.C Ecosystem Dynamics, Functioning, and Resilience: What happens to ecosystems when the environment changes? Humans affect natural ecological processes through agriculture, urban development, and industry. But ecological science provides strategies for sustainable development and ways to protect the environment.

6.1 A Changing Landscape

- Humans affect regional and global environments through agriculture, development, and industry in ways that have an impact on the quality of Earth's natural resources, including soil, water, and the atmosphere.
- Sustainable development provides for human needs while preserving the ecosystems that produce natural resources.

monoculture 155 • renewable resource 157 • nonrenewable resource 157 • sustainable development 157

Biology.com

Untamed Science Video The Untamed Science crew visits a zoo to learn about the important work that goes on behind the scenes.

Flash Cards Review the meaning of a vocabulary term.

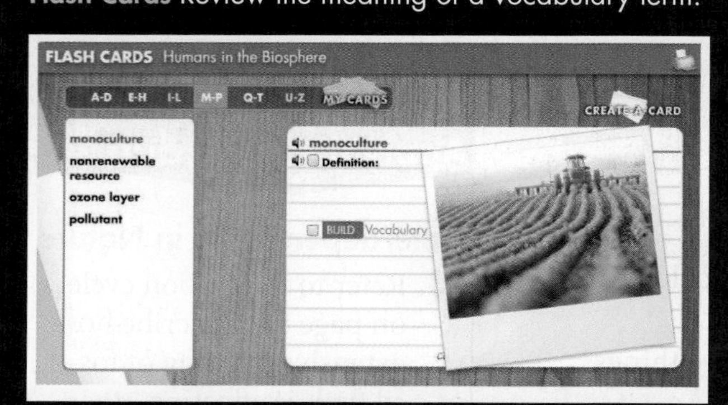

6.2 Using Resources Wisely

- Healthy soil supports both agriculture and forestry.
- It is possible to minimize soil erosion through careful management of both agriculture and forestry.
- The primary sources of water pollution are industrial and agricultural chemicals, residential sewage, and nonpoint sources.
- Common forms of air pollution include smog, acid rain, greenhouse gases, and particulates.

desertification 159 • deforestation 159 • pollutant 160 • biological magnification 161 • smog 163 • acid rain 164

Biology.com

Art in Motion View a short animation of biological magnification.

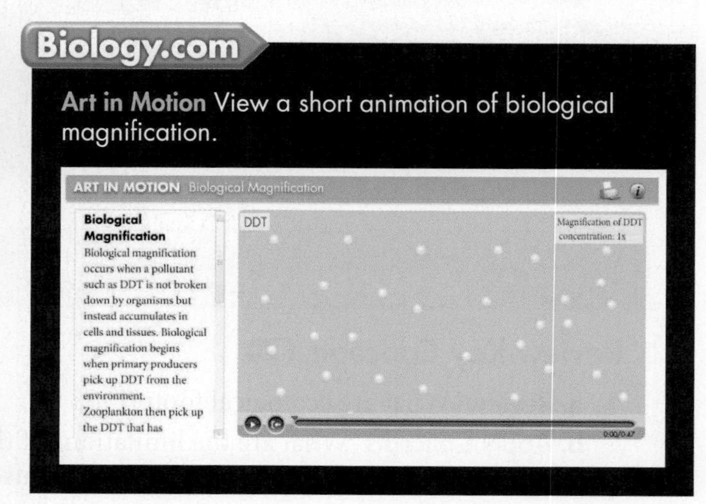

6.3 Biodiversity

- Biodiversity's benefits to society include contributions to medicine and agriculture, and the provision of ecosystem goods and services.
- When humans alter habitats, hunt, introduce invasive species, release pollution into food webs, and contribute to climate change, they reduce biodiversity.

CHECKING YOUR *Scientific Literacy*

Refer to the lesson content and digital assets below as you prepare for your chapter assessment. Then, evaluate your understanding of humans in the biosphere by answering these questions.

1. Scientific and Engineering Practice **Asking Questions** List three questions you would ask to determine whether a farmer is actively involved in sustainable agriculture.

2. Crosscutting Concept **Cause and Effect** What is the advantage of using fertilizers and pesticides to grow crops? What are the risks?

3. Text Types and Purposes Identify a land-use issue in your community on which people disagree. Present the reasons and evidence for each position. Be sure to point out the strengths and limitations of each argument.

4. STEM As a class, make a list of all the ways that water is used at your school. Then, determine what technologies your school currently uses or could use in the future to help conserve water.

- To conserve biodiversity, we must protect individual species, preserve habitats and ecosystems, and make certain that human neighbors of protected areas benefit from participating in conservation efforts.

biodiversity 166 • ecosystem diversity 166 • species diversity 166 • genetic diversity 166 • habitat fragmentation 168 • ecological hot spot 171

Biology.com

Art Review Review your understanding of the various threats to biodiversity with this activity.

Data Analysis Simulate data collection in order to compare two sites, and learn how to calculate a biodiversity index to quantify biodiversity.

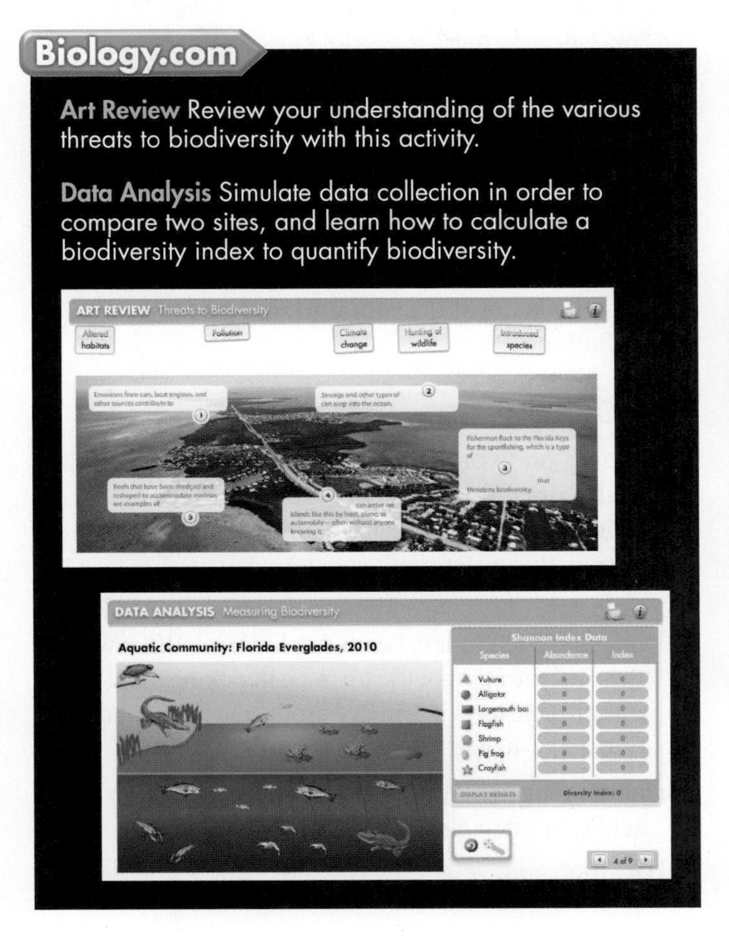

6.4 Meeting Ecological Challenges

- According to one data set, the average American has an ecological footprint over four times larger than the global average.
- By (1) recognizing a problem in the environment, (2) researching that problem to determine its cause, and then (3) using scientific understanding to change our behavior, we can have a positive impact on the global environment.

ecological footprint 173 • ozone layer 175 • aquaculture 176 • global warming 177

Biology.com

Visual Analogy Compare human impact on the biosphere to a footprint in this activity.

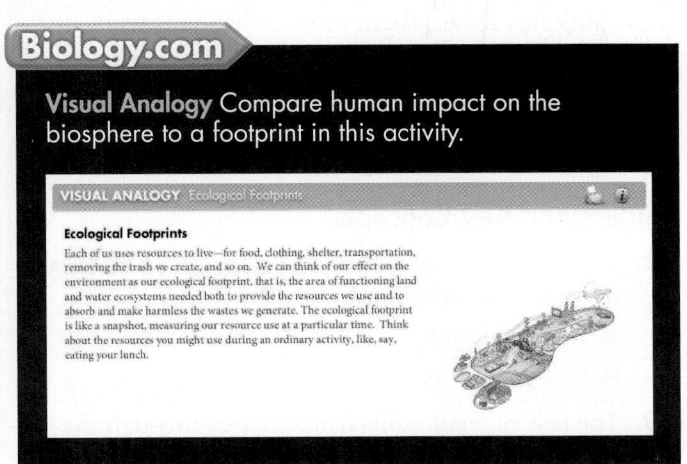

Design Your Own Lab

Acid Rain and Seeds This chapter lab is available in your lab manual and at Biology.com

6 Assessment

6.1 A Changing Landscape

Understand Key Concepts

1. Which of the following human activities has NOT had an important role in transforming the biosphere to date?

a. agriculture **c.** development
b. industry **d.** aquaculture

2. A resource that cannot easily be replenished by natural processes is called

a. common. **c.** nonrenewable.
b. renewable. **d.** conserved.

3. Describe how Hawaiian settlers negatively affected the islands after the 1700s.

4. Name four services that ecosystems provide for the biosphere.

Think Critically

5. **Production and Distribution of Writing** Devise clear and coherent guidelines your biology class can use to dispose of its nonlab trash in a safe, "environmentally friendly" way.

6. **Craft and Structure** How are renewable and nonrenewable resources alike? How are they different?

7. **Form a Hypothesis** Monoculture fields are usually very large and homogeneous. Do you think this makes them more or less vulnerable to disease and pests? Explain.

6.2 Using Resources Wisely

Understand Key Concepts

8. The conversion of a once soil-rich area to an area of little to no vegetation is called

a. fragmentation. **c.** desertification.
b. deforestation. **d.** acid rain.

9. The loss of fertile soils from an area through the action of water or wind is called

a. acid rain. **c.** desertification.
b. erosion. **d.** monoculture.

10. The concept of using natural resources at a rate that does not deplete them is called

a. conservation.
b. sustainable development.
c. reforestation.
d. successful use.

11. Examine the food web below. Which of the following organisms would accumulate the highest levels of a pesticide?

a. hawk **c.** frog
b. rabbit **d.** grasses

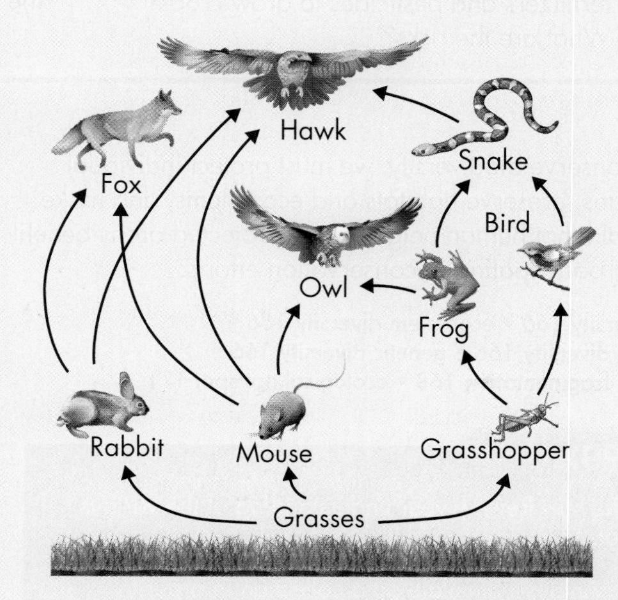

12. **Craft and Structure** What is the difference between sustainable forestry and deforestation?

13. Identify some of the common sources of water pollution.

Think Critically

14. **Design an Experiment** Can covering soil with mulch or compost near the bases of plants help reduce soil erosion? Design an experiment to answer this question.

15. **Calculate** The concentration of a toxic chemical is magnified ten times at each trophic level. What will the concentration of the toxin be in organisms at the fifth trophic level if primary producers have concentrations of 40 parts per million? MATH

16. **Infer** Why are lakes that have been affected by acid rain often clear and blue?

6.3 Biodiversity

Understand Key Concepts

17. A species that is introduced to an environment where it has not lived before is described as

a. native. **c.** threatened.
b. nonnative. **d.** predatory.

18. What is a habitat fragment?

19. List three different kinds of biodiversity that might be described in a given biome.

Think Critically

20. **Predict** How do you think the loss of biodiversity would adversely affect humans?

21. **Craft and Structure** Explain the difference between species diversity and ecosystem diversity.

6.4 Meeting Ecological Challenges

Understand Key Concepts

22. The burning of fossil fuels is a direct cause of each the following EXCEPT

a. acid rain. **c.** smog.
b. global warming. **d.** the ozone hole.

23. The total impact a person has on the biosphere can be represented by his or her

a. contribution to climate change.
b. ecological footprint.
c. consumption of fossil fuel.
d. production of carbon dioxide.

24. Cite three examples of physical evidence for global warming.

25. What are some of the biological effects of climate change?

Think Critically

26. **Relate Cause and Effect** Why hasn't the ozone layer repaired itself fully since the widespread ban of CFCs in 1987?

27. **Key Ideas and Details** Describe some of the steps taken to counter the effects of overfishing cod in the North Atlantic. Why is overfishing such a complex environmental issue?

solve the CHAPTER MYSTERY

MOVING THE *MOAI*

Easter Island's environment was not as biologically diverse, and not as resistant to ecological damage, as the Hawaiian Islands. The Rapa Nui cut palm trees for agriculture, for logs to move *moai*, and for wood to make fishing canoes. They mismanaged cleared fields, so fertile topsoil washed away.

Meanwhile, rats they brought to the island became invasive. Hordes of the rodents destroyed palm seedlings, ate coconuts, and digested palm seeds before they could germinate. Hawaiians also brought rats to their islands, and rats did serious damage to native Hawaiian plants. But in Hawaii's more diverse forests, some plant species were not as hard hit by rats and survived.

The combination of human activity and the effects of an invasive species led to the destruction of virtually all of Easter Island's forests. This combination, along with the effects of a harsh climate, limited the island's carrying capacity for humans from then on.

1. **Relate Cause and Effect** How did the small size of the island (about half the size of Long Island, New York) affect the outcome of deforestation and pest invasion?

2. **Research to Build and Present Knowledge** Gather information on differences in geography, climate, and biological diversity between Hawaii and Easter Island. How do you think those differences made the islands respond differently to human settlement? Write a summary of your analysis.

3. **Connect to the Big idea** All human cultures throughout history have interacted with their environments. Do you think that global human society has any lessons to learn from the experiences of the Rapa Nui, the Hawaiians, and other historic cultures?

Connecting Concepts

Use Science Graphics

The graph shows the amount of bluefin tuna caught by the United States in the Atlantic Ocean between 2002 and 2006. Use the graph to answer questions 28 and 29.

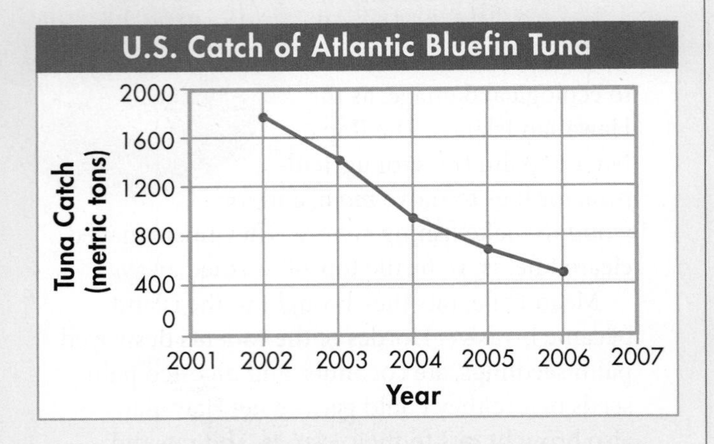

28. Predict What trend would you expect to see in the annual catch from 2006 to 2007?

29. Propose a Solution What recommendations would you make to help the bluefin tuna population recover in the next decade or two?

Write About Science

30. Text Types and Purposes Write a paragraph explaining the value of wetlands to human societies. In your paragraph, include the concept of biodiversity as well as the role of wetlands in maintaining water resources for human use.

31. Assess the  Why is it important to maintain species diversity in areas where humans live?

32. Assess the Big idea What environmental factors make high levels of biodiversity possible in most coastal waters? Refer to the discussion of abiotic and biotic factors in Chapter 4 if you need help answering this question.

Integration of Knowledge and Ideas

The following graph shows the number of species introduced to new habitats in the United States in the last century. Some of the species were relocated to new habitats within the United States while others were imported from other countries.

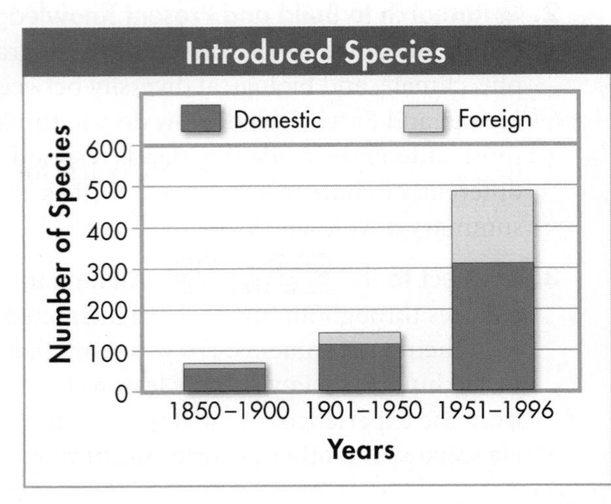

33. Interpret Graphs Of domestic species and foreign species, which showed the greatest percentage increase between the 1901–1950 period and the 1951–1996 period?

a. domestic species
b. foreign species
c. Both increased the same amount.
d. There is not enough information to tell.

34. Draw Conclusions Which of the following statements about introduced species is most likely true based on the data shown?

a. Species introduced from foreign countries are always more harmful than species relocated within the country.
b. All introduced species are brought into this country by accident.
c. It is likely that the increase in the number of introduced species is due to increased global travel, trade, and communication.
d. The number of introduced species is likely to fall in the next half-century.

Standardized Test Prep

Multiple Choice

1. Which of the following statements about renewable resources is TRUE?
A They are found only in tropical climates.
B They can never be depleted.
C They are replaceable by natural means.
D They can never regenerate.

2. Which of the following is a nonrenewable resource?
A wind
B fresh water
C coal
D topsoil

3. Which of the following is NOT a direct effect of deforestation?
A decreased productivity of the ecosystem
B soil erosion
C biological magnification
D habitat destruction

4. The total variety of organisms in the biosphere is called
A biodiversity.
B species diversity.
C ecosystem diversity.
D genetic diversity.

5. Ozone is made up of
A hydrogen.
B oxygen.
C nitrogen.
D chlorine.

6. Ozone depletion in the atmosphere has been caused by
A monoculture.
B CFCs.
C suburban sprawl.
D soil erosion.

7. In a food chain, concentrations of harmful substances increase in higher trophic levels in a process is known as
A biological magnification.
B genetic drift.
C biological succession.
D pesticide resistance.

Questions 8 and 9

Fire ants first arrived in the United States in 1918, probably on a ship traveling from South America to Alabama. The maps below show the geographic location of the U.S. fire ant population in 1953 and 2001.

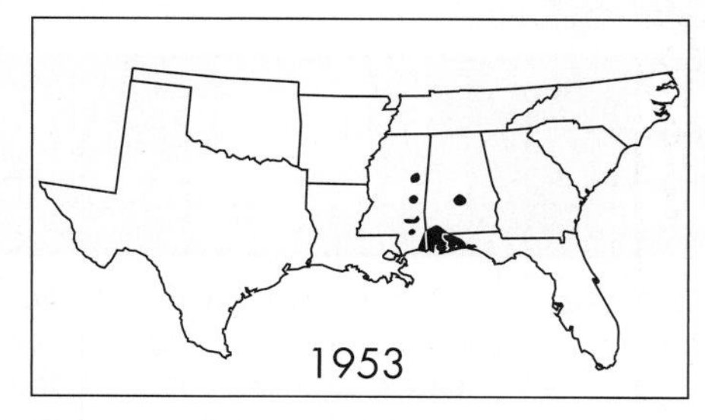

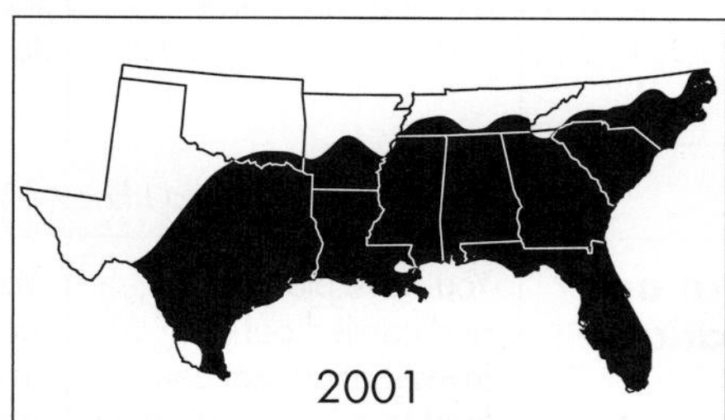

8. Which of the following statements about fire ants in the United States is TRUE?
A They reproduce slowly.
B They are a native species of the United States.
C They are an invasive species.
D They do not compete with other ant species.

9. By 2010, fire ants are MOST likely to
A have spread to a larger area.
B have reached their carrying capacity.
C die out.
D return to South America.

Open-Response

10. Describe how ecologists use the ecological footprint concept.

If You Have Trouble With . . .

Question	1	2	3	4	5	6	7	8	9	10
See Lesson	6.1	6.2	6.2	6.3	6.2	6.4	6.2	6.3	6.3	6.4

NGSS Problem-based Learning

UNIT PROJECT 2

Your Self-Assessment

The rubric below will help you evaluate your work as you construct an explanation and communicate information.

SCORE YOUR WORK!	EXEMPLARY Score your work a 4 if:	ACCOMPLISHED Score your work a 3 if:	DEVELOPING Score your work a 2 if:	BEGINNING Score your work a 1 if:
Ask Questions ▶	You have asked and refined questions that lead to explanations and can be tested.	You have asked and revised most questions that can lead to explanations and be tested.	You have asked multiple questions, but they do not lead to an explanation or cannot be tested. Questions need to be revised.	You've asked only one question. Your question cannot be tested in an investigation and should be revised.
	TIP Ask, *"What do I know? What do I need to find out?"*			
Carry Out an Investigation ▶	You have planned and carried out an investigation that can lead to a full explanation of phenomena. You selected appropriate tools to collect, record, analyze and evaluate data.	You have planned and carried out an investigation that provides a partial explanation of phenomena. You selected appropriate tools to collect, record, analyze and evaluate data.	You have planned an investigation, but have not carried it out. You have selected some tools that are not appropriate to collect, record, analyze, and evaluate data.	You have planned an investigation. You need to carry out the investigation. You have not selected appropriate tools to collect, record, analyze, and evaluate data.
Analyze and Interpret Data ▶	You used tools, technologies, and/or models to generate and analyze data in order to make a valid and reliable scientific claim or explanation.	You used tools, technologies, and/or models to generate and analyze data, however your claim or explanation is not completely valid.	You used some tools, technologies, and/or models to generate minimal data. Your claim or explanation is not valid.	You used some tools, technologies, and/or models to generate minimal data. You have not developed a claim or scientific explanation.
Construct an Explanation ▶	You have applied scientific reasoning and theory to link evidence to claims and show why the data are adequate for the explanation.	You have applied scientific reasoning and theory to link evidence to claims and show that some of your data is adequate for the explanation.	You have applied some scientific reasoning and theory to link evidence to claims. You haven't shown why the data are adequate for the explanation.	You have applied minimal scientific reasoning and theory to link evidence to claims. You have not shown why the data are adequate for the explanation.
Engage in Argument from Evidence ▶	You have constructed an argument supported by empirical evidence and reasoning. You have evaluated the claims, evidence, and reasoning of currently accepted explanations as a basis for the merits of the argument.	You have constructed an argument supported by empirical evidence and reasoning. You have evaluated some of the claims, evidence, and reasoning of currently accepted explanations as a basis for the merits of the argument.	You have constructed an argument, but it is supported by minimal empirical evidence and reasoning. You have evaluated some of the claims, evidence, and reasoning for currently accepted explanations as a basis for the merits of argument.	You have constructed an argument, but it is supported by minimal empirical evidence and reasoning. You have not evaluated the claims, evidence, and reasoning for currently accepted explanations as a basis for merits of argument.
	TIP Empirical evidence and reasoning come from your research and data analysis.			

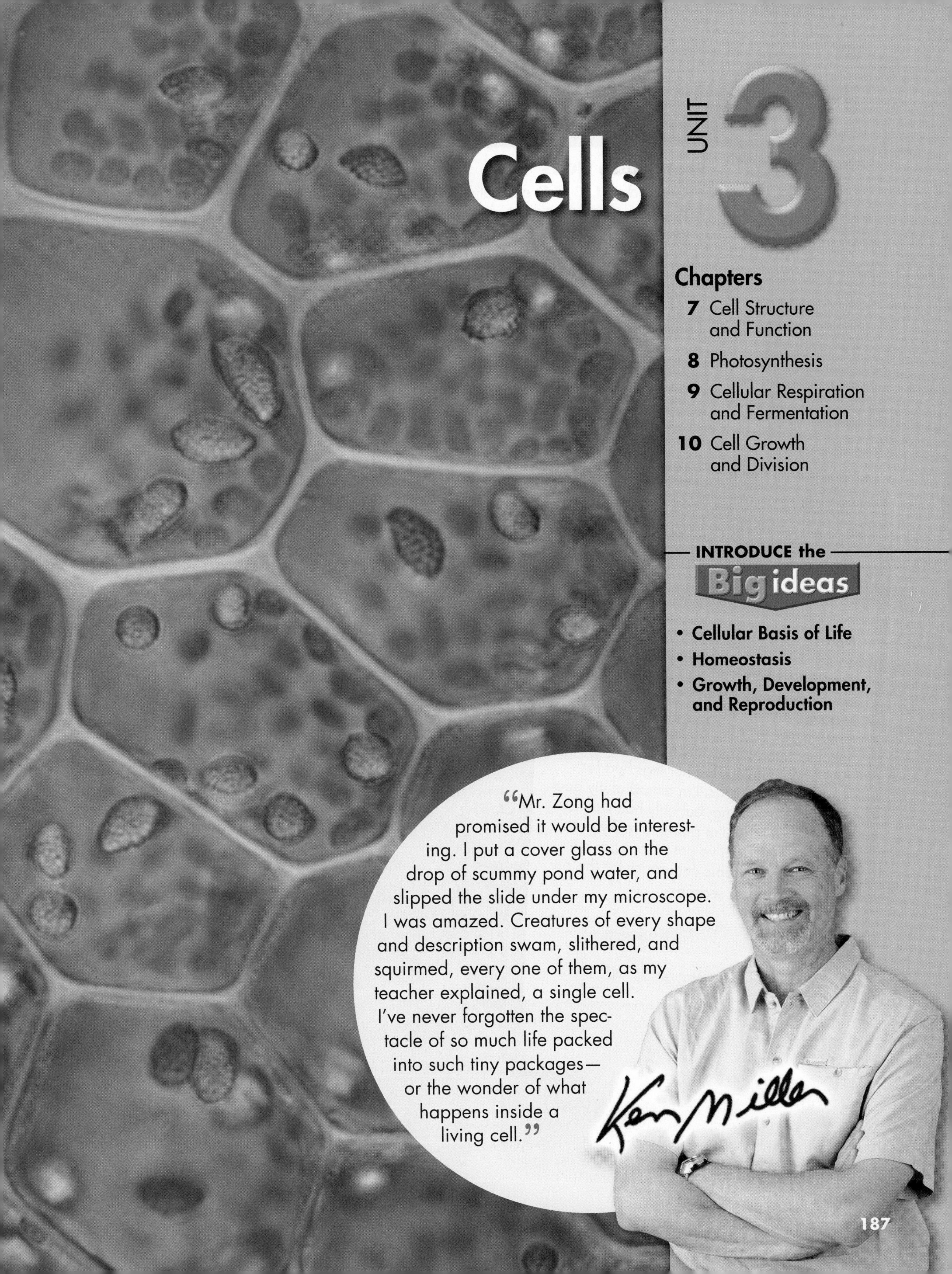

Cells

UNIT 3

Chapters

7 Cell Structure and Function

8 Photosynthesis

9 Cellular Respiration and Fermentation

10 Cell Growth and Division

INTRODUCE the Big ideas

- **Cellular Basis of Life**
- **Homeostasis**
- **Growth, Development, and Reproduction**

"Mr. Zong had promised it would be interesting. I put a cover glass on the drop of scummy pond water, and slipped the slide under my microscope. I was amazed. Creatures of every shape and description swam, slithered, and squirmed, every one of them, as my teacher explained, a single cell. I've never forgotten the spectacle of so much life packed into such tiny packages—or the wonder of what happens inside a living cell."

Ken Miller

NGSS Problem-based Learning

UNIT PROJECT 3

Spotlight on NGSS

- **Core Idea LS1.C** Organization for Matter and Energy Flow in Organisms
- **Practice** Analyzing and Interpreting Data
- **Crosscutting Concept** Energy and Matter

Maxed Out Muscles!

Hi there. I'm Mimi. Soccer season is almost here, and my friend Tricia and I have been training like crazy! My coach has all of the players record their training and progress using a mobile fitness app. Check it out...

Running is one form of aerobic exercise.

Your Task: Stop the Soreness!

You are a student research assistant at a university's sports medicine department. You've been contacted by a local student named Mimi to find out why her muscles become extremely tired and sore during aerobic exercise. She also wants to know why her friend Tricia continues to feel sore for two days after working out. Use various resources, research articles, and data, to develop an explanation of what may be the cause of Mimi's and Tricia's muscle fatigue and soreness. To help each student athlete with their fitness goals, communicate the information through a written fitness plan.

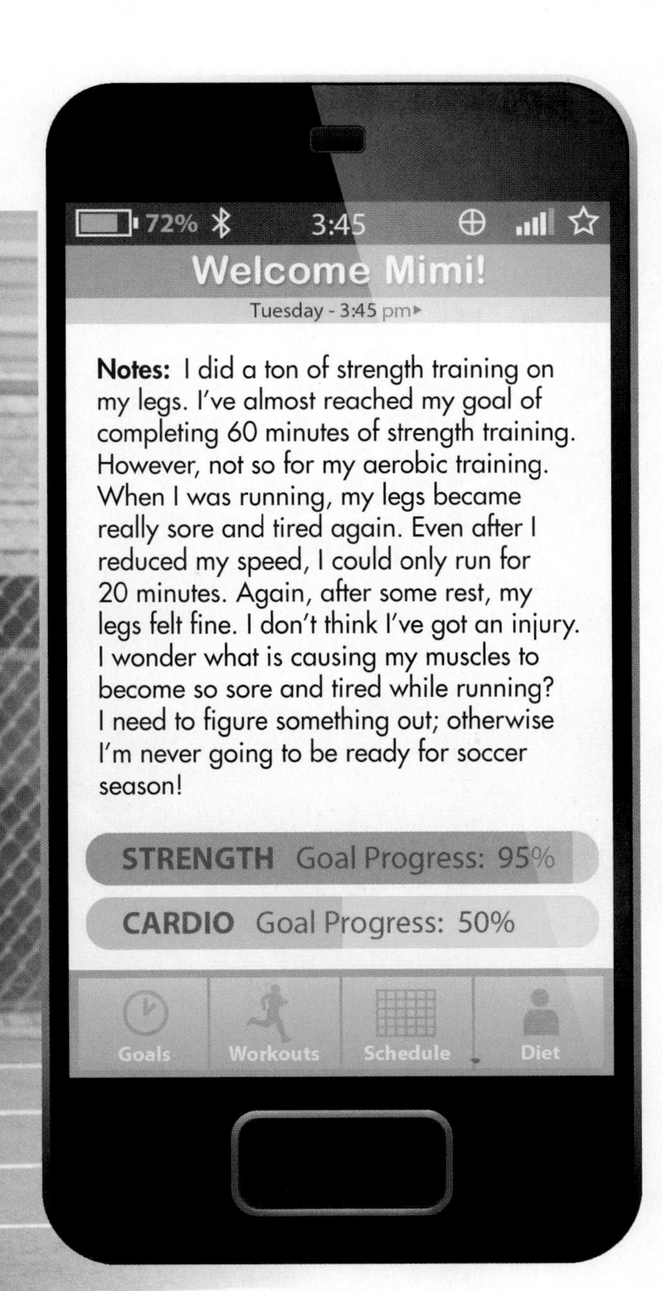

Ask Questions What do I already know? What do I need to find out? Read the resources and analyze data to help you.

Plan an Investigation Develop a plan to identify the kind of resources and data you need to answer your questions. Determine what information you need to construct an explanation on the relationship between muscle fatigue and aerobic exercise. What resources and research articles are valuable? How will you evaluate and analyze the types, amount, and accuracy of data?

Analyze and Interpret Data Develop a table to compare and contrast various types of data. Analyze the data for consistency. Interpret the data. What connections or explanations about muscle fatigue and pain during or after aerobic exercise can be derived from the data?

Construct an Explanation Use the resources and data you collected to construct two to three possible explanations. Decide which explanation best addresses Mimi's problem. Does this explanation also address her friend Tricia's problem? Explain why or why not. Identify the information that supports each explanation.

Communicate Information Create fitness plans for Mimi and Tricia that explain the causes of muscle fatigue due to aerobic exercise. Each plan should offer a solution to help Mimi and Tricia reach their fitness goals.

Biology.com

Find resources on muscle fatigue and exercise to help you construct explanations and create solutions to address Mimi and Tricia's fitness needs.

7 Cell Structure and Function

Big ideas: Cellular Basis of Life, Homeostasis

Q: How are cell structures adapted to their functions?

INSIDE:

- 7.1 Life Is Cellular
- 7.2 Cell Structure
- 7.3 Cell Transport
- 7.4 Homeostasis and Cells

BUILDING *Scientific Literacy*

Deepen your understanding of cell structure and function by engaging in key practices that allow you to make connections across concepts.

NGSS You will use models to learn about the structure and function of cells.

STEM What makes a good sports drink? Determine the best solute concentration for a new sports drink in this activity.

Common Core You will analyze the relationship among structures and processes as you learn about cells.

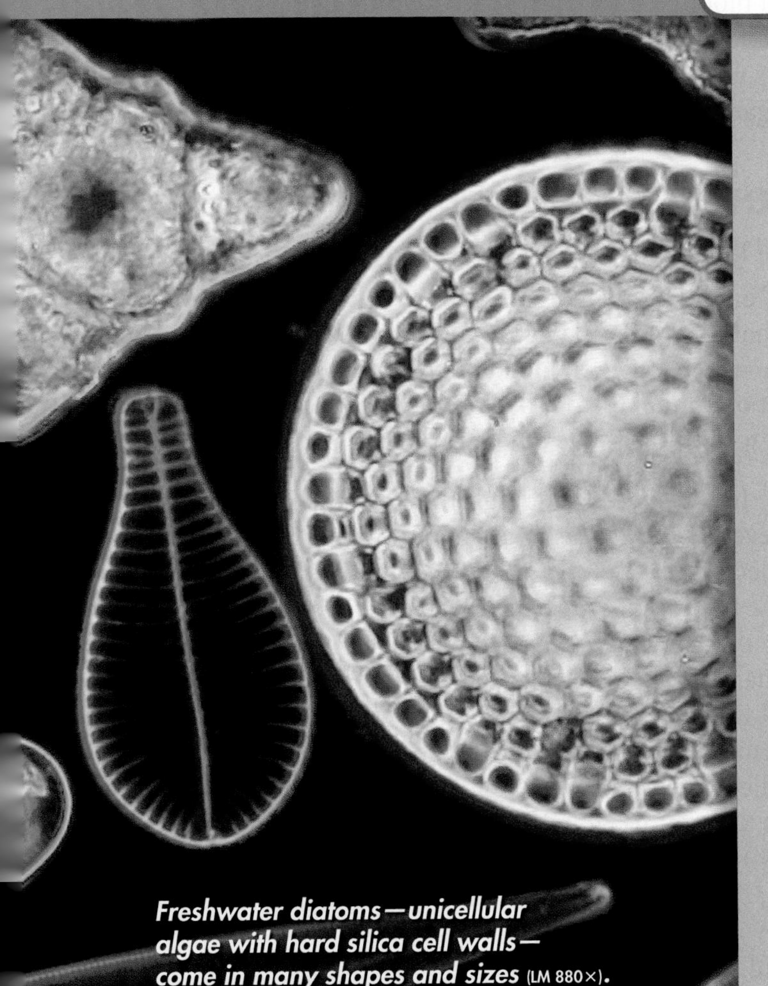

Freshwater diatoms—unicellular algae with hard silica cell walls—come in many shapes and sizes (LM 880×).

CHAPTER MYSTERY

DEATH BY ... WATER?

Michelle was a healthy 25-year-old running in her first marathon. The hot and humid weather had made all the runners sweat profusely, so Michelle made sure she drank water at every opportunity. Gradually, she began to feel weak and confused. At the end of the marathon, Michelle staggered into a medical tent. Complaining of headache and nausea, she collapsed onto the floor. Volunteers quickly gave Michelle water for dehydration. Soon, her condition worsened and Michelle was rushed to the hospital, where she was gripped by a seizure and went into a coma. Why did treating Michelle with water make her condition worse? As you read this chapter, look for clues to help you predict how water made Michelle sick. Then, solve the mystery.

Michelle's mysterious illness is just the beginning. Take a video field trip with the ecogeeks of Untamed Science to see where this mystery leads.

Go online to access additional resources including:
- **eText** • **Flash Cards** • **Lesson Overviews** • **Chapter Mystery**

7.1 Life Is Cellular

Key Questions

What is the cell theory?

How do microscopes work?

How are prokaryotic and eukaryotic cells different?

Vocabulary

cell • cell theory • cell membrane • nucleus • eukaryote • prokaryote

Taking Notes

Outline Before you read, make an outline using the green and blue headings in the text. As you read, fill in notes under each heading.

THINK ABOUT IT What's the smallest part of any living thing that still counts as being "alive"? Is a leaf alive? How about your big toe? How about a drop of blood? Can we just keep dividing living things into smaller and smaller parts, or is there a point at which what's left is no longer alive? As you will see, there is such a limit, the smallest living unit of any organism—the cell.

The Discovery of the Cell

What is the cell theory?

"Seeing is believing," an old saying goes. It would be hard to find a better example of this than the discovery of the cell. Without the instruments to make them visible, cells remained out of sight and, therefore, out of mind for most of human history. All of this changed with a dramatic advance in technology—the invention of the microscope.

Early Microscopes In the late 1500s, eyeglass makers in Europe discovered that using several glass lenses in combination could magnify even the smallest objects to make them easy to see. Before long, they had built the first true microscopes from these lenses, opening the door to the study of biology as we know it today.

In 1665, Englishman Robert Hooke used an early compound microscope to look at a nonliving thin slice of cork, a plant material. Under the microscope, cork seemed to be made of thousands of tiny empty chambers. Hooke called these chambers "cells" because they reminded him of a monastery's tiny rooms, which were called cells. The term *cell* is used in biology to this day. Today we know that living cells are not empty chambers, that in fact they contain a huge array of working parts, each with its own function.

In Holland around the same time, Anton van Leeuwenhoek used a single-lens microscope to observe pond water and other things. To his amazement, the microscope revealed a fantastic world of tiny living organisms that seemed to be everywhere, in the water he and his neighbors drank, and even in his own mouth. Leeuwenhoek's illustrations of the organisms he found in the human mouth—which today we call bacteria—are shown in **Figure 7–1.**

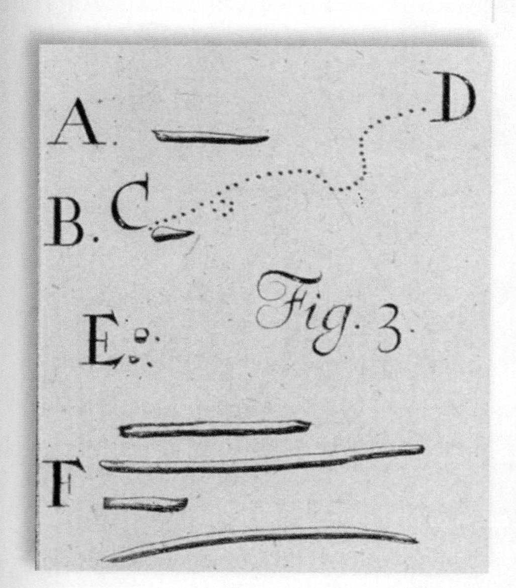

FIGURE 7–1 Early Microscope Images Using a simple microscope, Anton van Leeuwenhoek was the first to observe living microorganisms. These drawings, taken from one of his letters, show bacteria in the human mouth.

The Cell Theory Soon after van Leeuwenhoek, observations by scientists made it clear that **cells** are the basic units of life. In 1838, German botanist Matthias Schleiden concluded that all plants are made of cells. The next year, German biologist Theodor Schwann stated that all animals are made of cells. In 1855, German physician Rudolf Virchow concluded that new cells can be produced only from the division of existing cells, confirming a suggestion made by German Lorenz Oken 50 years earlier. These discoveries, confirmed by many biologists, are summarized in the **cell theory,** a fundamental concept of biology. **The cell theory states:**

- **All living things are made up of cells.**
- **Cells are the basic units of structure and function in living things.**
- **New cells are produced from existing cells.**

Exploring the Cell

How do microscopes work?

A microscope, as you know, produces an enlarged image of something very small. **Most microscopes use lenses to magnify the image of an object by focusing light or electrons.** Following in the footsteps of Hooke, Virchow, and others, modern biologists still use microscopes to explore the cell. But today's researchers use technology more powerful than the pioneers of biology could ever have imagined.

Light Microscopes and Cell Stains The type of microscope you are probably most familiar with is the compound light microscope. A typical light microscope allows light to pass through a specimen and uses two lenses to form an image. The first lens, called the objective lens, is located just above the specimen. This lens enlarges the image of the specimen. Most light microscopes have several objective lenses so that the power of magnification can be varied. The second lens, called the ocular lens, magnifies this image still further. Unfortunately, light itself limits the detail, or resolution, of images in a microscope. Like all forms of radiation, lightwaves are diffracted, or scattered, as they pass through matter. Because of this, light microscopes can produce clear images of objects only to a magnification of about 1000 times.

Another problem with light microscopy is that most living cells are nearly transparent. Using chemical stains or dyes, as in **Figure 7–2,** can usually solve this problem. Some of these stains are so specific that they reveal only certain compounds or structures within the cell. Many of the slides you'll examine in your biology class laboratory will be stained this way.

A powerful variation on these staining techniques uses dyes that give off light of a particular color when viewed under specific wavelengths of light, a property called fluorescence. Fluorescent dyes can be attached to specific molecules and can then be made visible using a special fluorescence microscope. New techniques, in fact, enable scientists to engineer cells that attach fluorescent labels of different colors to specific molecules as they are produced. Fluorescence microscopy makes it possible to see and identify the locations of these molecules and even allows scientists to watch them move around in a living cell.

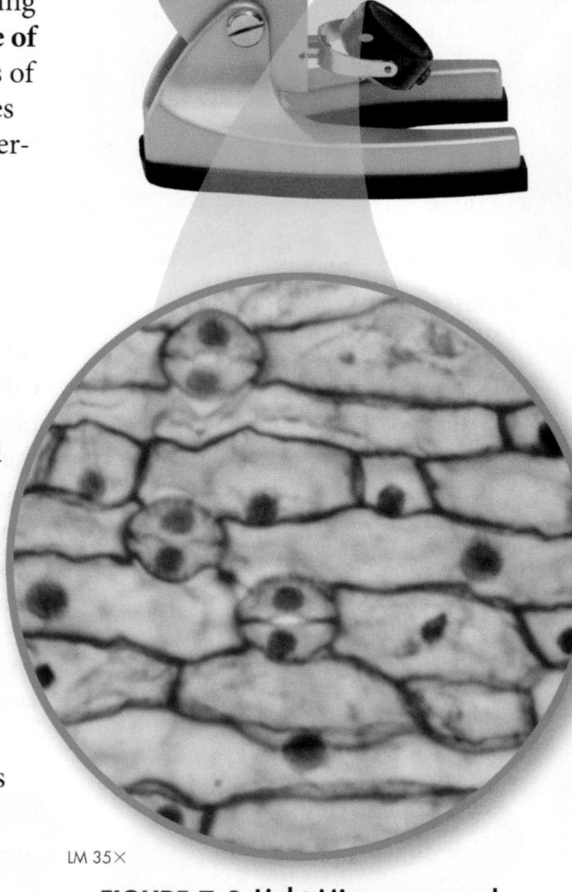

FIGURE 7–2 Light Microscope and Cell Stains This specimen of onion leaf skin has been stained with a compound called toluidine blue. The dye makes the cell boundaries and nuclei clearly visible.

Electron Microscopes Light microscopes can be used to see cells and cell structures as small as 1 millionth of a meter—certainly pretty small! But what if scientists want to study something smaller than that, such as a virus or a DNA molecule? For that, they need electron microscopes. Instead of using light, electron microscopes use beams of electrons that are focused by magnetic fields. Electron microscopes offer much higher resolution than light microscopes. Some types of electron microscopes can be used to study cellular structures that are 1 billionth of a meter in size.

A Researcher Working With a Transmission Electron Microscope

There are two major types of electron microscopes: transmission and scanning. Transmission electron microscopes make it possible to explore cell structures and large protein molecules. But because beams of electrons can only pass through thin samples, cells and tissues must be cut into ultrathin slices before they can be examined. This is the reason that such images often appear flat and two dimensional.

In scanning electron microscopes, a pencil-like beam of electrons is scanned over the surface of a specimen. Because the image is formed at the specimen's surface, samples do not have to be cut into thin slices to be seen. The scanning electron microscope produces stunning three-dimensional images of the specimen's surface.

Electrons are easily scattered by molecules in the air, which means samples must be placed in a vacuum to be studied with an electron microscope. As a result, researchers must chemically preserve their samples. Electron microscopy, then, can only be used to examine nonliving cells and tissues.

Look at **Figure 7–3,** which shows yeast cells as they might look under a light microscope, transmission electron microscope, and scanning electron microscope. You may wonder why the cells appear to be different colors in each micrograph. (A micrograph is a photo of an object seen through a microscope.) The colors in light micrographs come from the cells themselves, or from the stains and dyes used to highlight them. Electron micrographs, however, are actually black and white. Electrons, unlike light, don't come in colors. So scientists often use computer techniques to add "false color" to make certain structures stand out.

In Your Notebook *You are presented with a specimen to examine. What are two questions you would ask to determine the best microscope to use?*

FIGURE 7–3 Micrographs Different types of microscopes can be used to examine cells. Here, yeast cells are shown in a light micrograph (LM 500×), transmission electron micrograph (TEM 4375×), and a scanning electron micrograph (SEM 3750×).
Infer ***If scientists were studying a structure found on the surface of yeast, which kind of microscope would they likely use?***

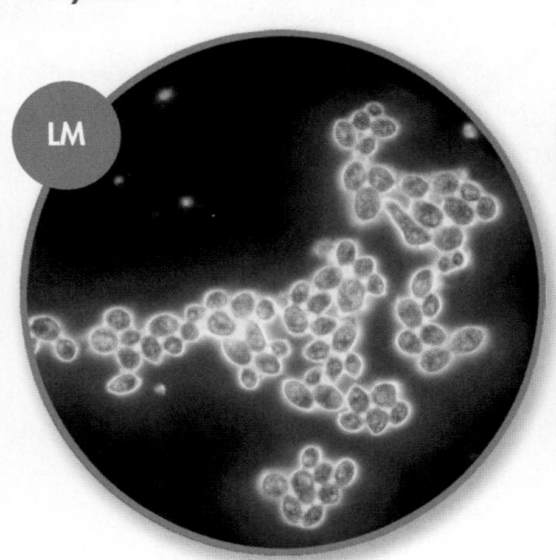

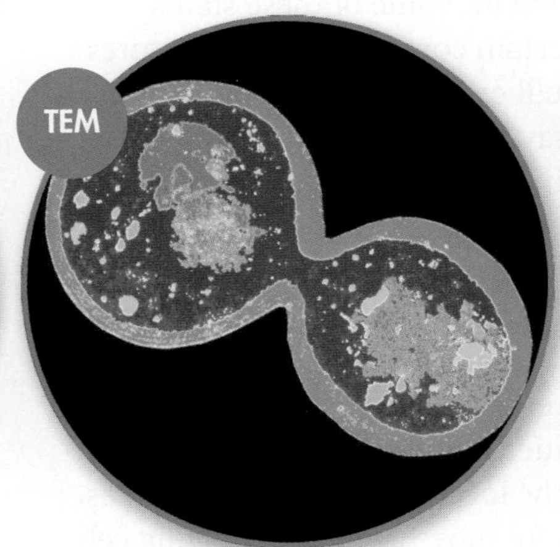

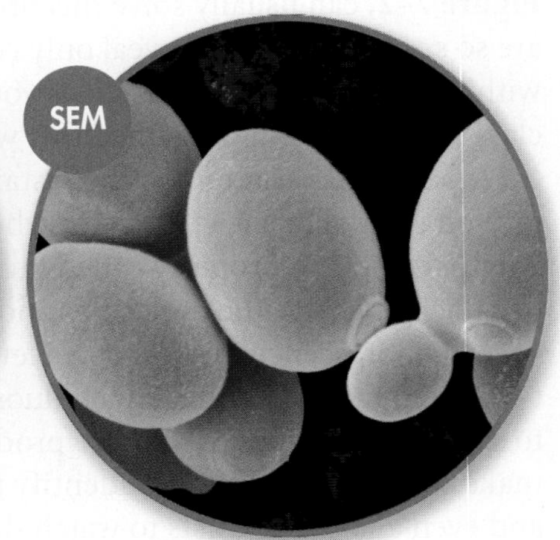

Quick Lab
GUIDED INQUIRY

What Is a Cell?

❶ Look through a microscope at a slide of a plant leaf or stem cross section. Sketch one or more cells. Record a description of their shape and internal parts.

❷ Repeat step 1 with slides of nerve cells, bacteria, and paramecia.

❸ Compare the cells by listing the characteristics they have in common and some of the differences among them.

Analyze and Conclude

1. Classify Classify the cells you observed into two or more groups. Explain what characteristics you used to put each cell in a particular group.

Prokaryotes and Eukaryotes

How are prokaryotic and eukaryotic cells different?

Cells come in an amazing variety of shapes and sizes, some of which are shown in **Figure 7–4.** Although typical cells range from 5 to 50 micrometers in diameter, the smallest *Mycoplasma* bacteria are only 0.2 micrometer across, so small that they are difficult to see under even the best light microscopes. In contrast, the giant amoeba *Chaos chaos* can be 1000 micrometers (1 millimeter) in diameter, large enough to be seen with the unaided eye as a tiny speck in pond water. Despite their differences, all cells, at some point in their lives, contain DNA, the molecule that carries biological information. In addition, all cells are surrounded by a thin flexible barrier called a **cell membrane.** (The cell membrane is sometimes called the *plasma membrane* because many cells in the body are in direct contact with the fluid portion of the blood—the plasma.) There are other similarities as well, as you will learn in the next lesson.

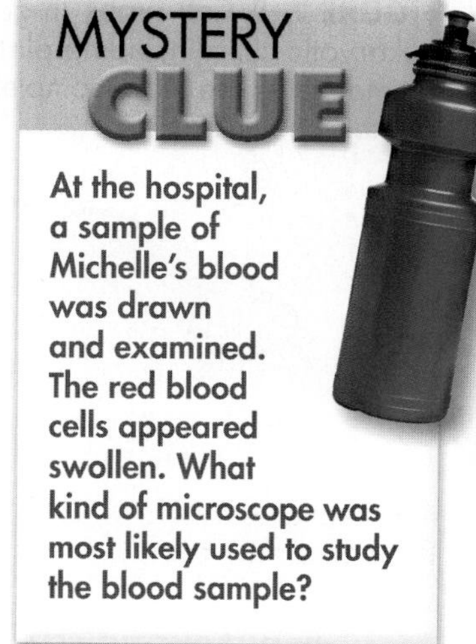

Cells fall into two broad categories, depending on whether they contain a nucleus. The **nucleus** (plural: nuclei) is a large membrane-enclosed structure that contains genetic material in the form of DNA and controls many of the cell's activities. **Eukaryotes** (yoo KAR ee ohts) are cells that enclose their DNA in nuclei. **Prokaryotes** (pro KAR ee ohts) are cells that do not enclose DNA in nuclei.

FIGURE 7–4 Cell Size Is Relative The human eye can see objects larger than about 0.5 mm. Most of what interests cell biologists, however, is much smaller than that. Microscopes make seeing the cellular and subcellular world possible.

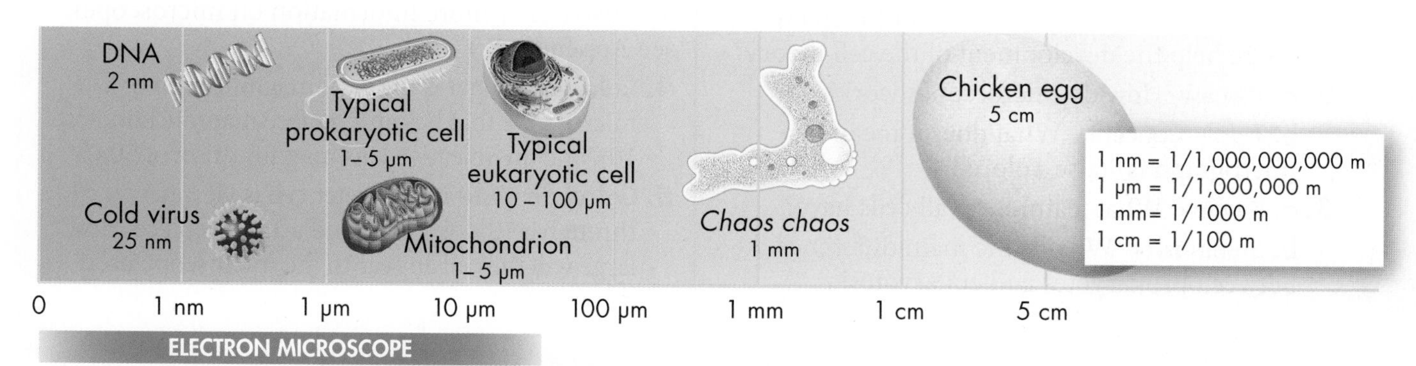

BUILD Vocabulary

WORD ORIGINS The noun prokaryote comes from the Greek word *karyon*, meaning "kernel," or nucleus. The prefix *pro-* means "before." Prokaryotic cells first evolved before nuclei developed.

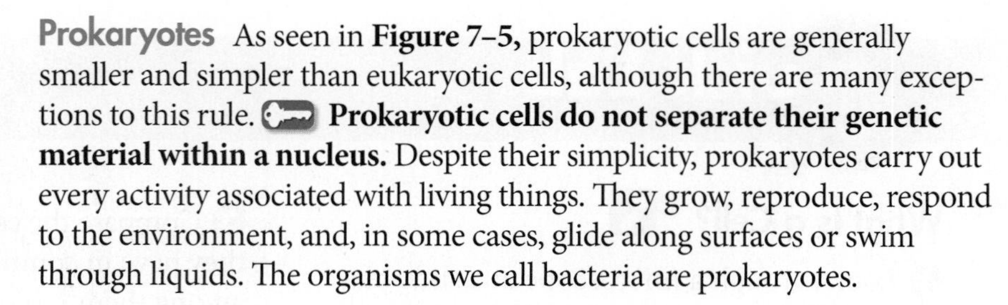

Prokaryotes As seen in **Figure 7–5,** prokaryotic cells are generally smaller and simpler than eukaryotic cells, although there are many exceptions to this rule. **Prokaryotic cells do not separate their genetic material within a nucleus.** Despite their simplicity, prokaryotes carry out every activity associated with living things. They grow, reproduce, respond to the environment, and, in some cases, glide along surfaces or swim through liquids. The organisms we call bacteria are prokaryotes.

Eukaryotes Eukaryotic cells are generally larger and more complex than prokaryotic cells. Most eukaryotic cells contain dozens of structures and internal membranes, and many are highly specialized. **In eukaryotic cells, the nucleus separates the genetic material from the rest of the cell.** Eukaryotes display great variety: some, like the ones commonly called "protists," live solitary lives as unicellular organisms; others form large, multicellular organisms—plants, animals, and fungi.

FIGURE 7–5 Cell Types In general, eukaryotic cells (including plant and animal cells) are more complex than prokaryotic cells.

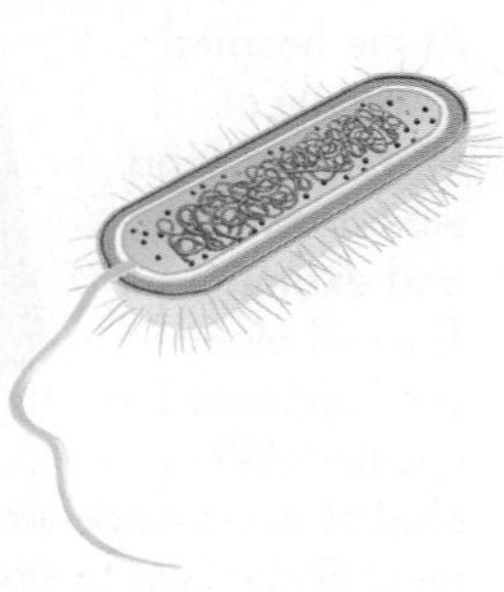

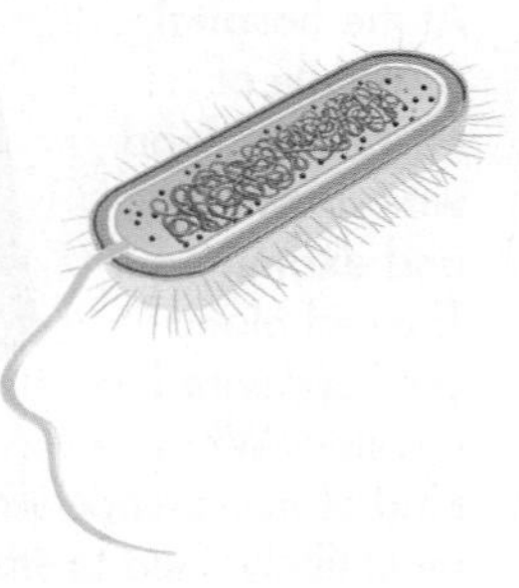

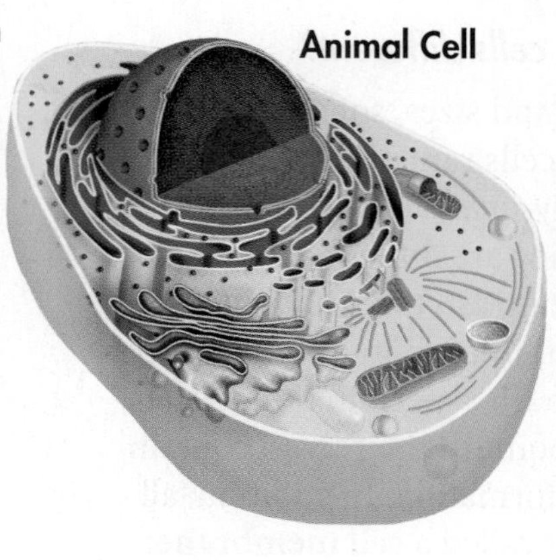

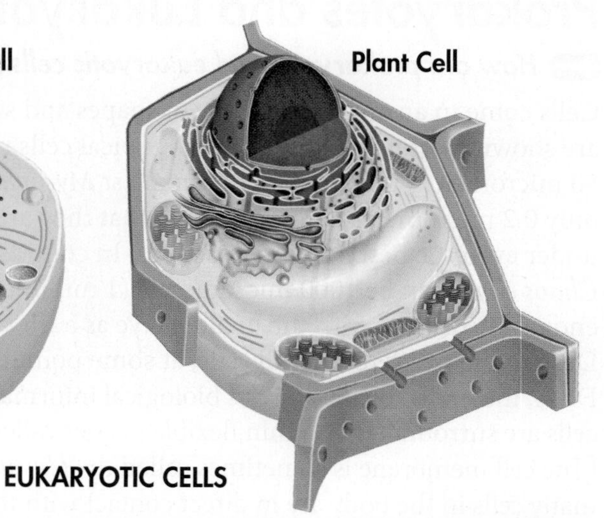

7.1 Assessment

Review Key Concepts

1. a. **Review** What is a cell?
   b. **Explain** What three statements make up the cell theory?
   c. **Infer** How did the invention of the microscope help the development of the cell theory?
2. a. **Review** How do microscopes work?
   b. **Apply Concepts** What does it mean if a micrograph is "false-colored?"
3. a. **Review** What features do all cells have?
   b. **Summarize** What is the main difference between prokaryotes and eukaryotes?

PRACTICE PROBLEMS MATH

A light microscope can magnify images up to 1000 times. To calculate the total magnification of a specimen, multiply the magnification of the eyepiece lens by the magnification of the objective lens used. (For more information on microscopes, see Appendix B.)

4. **Calculate** What is the total magnification of a microscope that has an eyepiece magnification of 10× and an objective lens magnification of 50×.
5. **Calculate** A 10 micrometer cell is viewed through a 10× objective and a 10× eyepiece. How large will the cell appear to the microscope user?

Careers & BIOLOGY

Cells are the basic unit of all known life. If cells interest you, you might want to consider one of the following careers.

LABORATORY TECHNICIAN

Ever wonder what happens to the blood your doctor collects during your annual physical? It goes to a laboratory technician. Laboratory technicians perform routine procedures using microscopes, computers, and other equipment. Many laboratory technicians work in the medical field, evaluating and analyzing test results.

MICROSCOPIST

The images in **Figure 7–3** were captured by a microscopist. Microscopists make it possible to study structures too small to be seen without magnification. There are a variety of microscopy techniques, including staining and fluorescence, that microscopists can use to make images clear and informative for researchers. Some of these images are so striking that they have become a form of scientific art.

PATHOLOGIST

Pathologists are like detectives: They collect cellular information and tissue evidence to diagnose illness. Using a broad knowledge of disease characteristics and the best-available technology, pathologists analyze cells and tissues under a microscope and discuss their diagnoses with other doctors.

CAREER CLOSE-UP

Dr. Tanasa Osborne, Veterinary Pathologist

Dr. Tanasa Osborne studies osteosarcoma, the most common malignant bone tumor in children and adolescents. Her research with the National Institutes of Health and the National Cancer Institute is focused on improving outcomes for patients whose cancer has spread from one organ or system to another. Dr. Osborne is not a medical doctor, however—she is a veterinarian. Animals are often used as models to study human disease. Dr. Osborne's research, therefore, contributes to both animal and human health. Veterinary pathologists investigate many important issues in addition to cancer, including West Nile virus, avian flu, and other emerging infectious diseases that affect humans as well as animals.

"My distinctive background allows me to approach science from a global (or cross-species) and systemic perspective."

WRITING Based on this career close-up, explain how Dr. Osborne's research is an example of the effect science can have on society.

7.2 Cell Structure

Key Questions

What is the role of the cell nucleus?

What are the functions of vacuoles, lysosomes, and the cytoskeleton?

What organelles help make and transport proteins?

What are the functions of chloroplasts and mitochondria?

What is the function of the cell membrane?

Vocabulary

cytoplasm • organelle • vacuole • lysosome • cytoskeleton • centriole • ribosome • endoplasmic reticulum • Golgi apparatus • chloroplast • mitochondrion • cell wall • lipid bilayer • selectively permeable

Taking Notes

Venn Diagram Create a Venn diagram that illustrates the similarities and differences between prokaryotes and eukaryotes.

THINK ABOUT IT At first glance, a factory is a puzzling place. Machines buzz and clatter; people move quickly in different directions. So much activity can be confusing. However, if you take the time to watch carefully, what might at first seem like chaos begins to make sense. The same is true for the living cell.

Cell Organization

What is the role of the cell nucleus?

The eukaryotic cell is a complex and busy place. But if you look closely at eukaryotic cells, patterns begin to emerge. For example, it's easy to divide each cell into two major parts: the nucleus and the cytoplasm. The **cytoplasm** is the portion of the cell outside the nucleus. As you will see, the nucleus and cytoplasm work together in the business of life. Prokaryotic cells have cytoplasm too, even though they do not have a nucleus.

In our discussion of cell structure, we consider each major component of plant and animal eukaryotic cells—some of which are also found in prokaryotic cells—one by one. Because many of these structures act like specialized organs, they are known as **organelles,** literally "little organs." Understanding what each organelle does helps us understand the cell as a whole. A summary of cell structure can be found on pages 206–207.

VISUAL ANALOGY

THE CELL AS A LIVING FACTORY

FIGURE 7–6 The specialization and organization of work and workers contribute to the productivity of a factory. In much the same way, the specialized parts in a cell contribute to the cell's overall stability and survival.

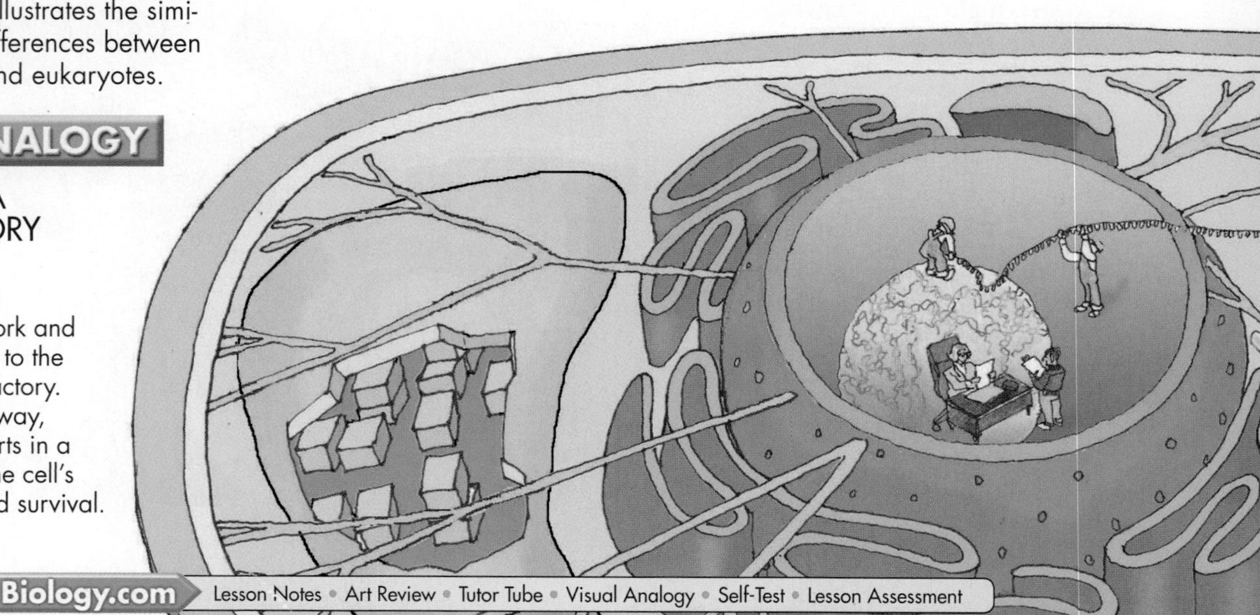

Comparing the Cell to a Factory In some respects, the eukaryotic cell is much like a living version of a modern factory **(Figure 7–6).** The different organelles of the cell can be compared to the specialized machines and assembly lines of the factory. In addition, cells, like factories, follow instructions and produce products. As we look through the organization of the cell, we'll find plenty of places in which the comparison works so well that it will help us understand how cells work.

The Nucleus In the same way that the main office controls a large factory, the nucleus is the control center of the cell. **The nucleus contains nearly all the cell's DNA and, with it, the coded instructions for making proteins and other important molecules.** Prokaryotic cells lack a nucleus, but they do have DNA that contains the same kinds of instructions.

The nucleus, shown in **Figure 7–7,** is surrounded by a nuclear envelope composed of two membranes. The nuclear envelope is dotted with thousands of nuclear pores, which allow material to move into and out of the nucleus. Like messages, instructions, and blueprints moving in and out of a factory's main office, a steady stream of proteins, RNA, and other molecules move through the nuclear pores to and from the rest of the cell.

Chromosomes, which carry the cell's genetic information, are also found in the nucleus. Most of the time, the threadlike chromosomes are spread throughout the nucleus in the form of chromatin—a complex of DNA bound to proteins. When a cell divides, its chromosomes condense and can be seen under a microscope. You will learn more about chromosomes in later chapters.

Most nuclei also contain a small dense region known as the nucleolus (noo KLEE uh lus). The nucleolus is where the assembly of ribosomes begins.

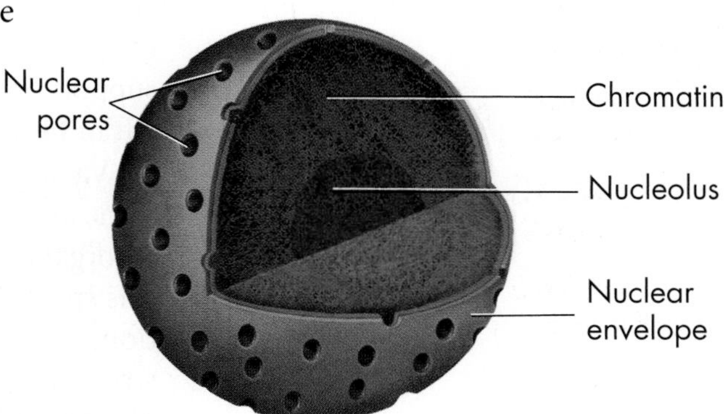

FIGURE 7–7 The Nucleus The nucleus controls most cell processes and contains DNA. The small, dense region in the nucleus is known as the nucleolus.

In Your Notebook *Describe the structure of the nucleus. Include the words* nuclear envelope, nuclear pore, chromatin, chromosomes, *and* nucleolus *in your description.*

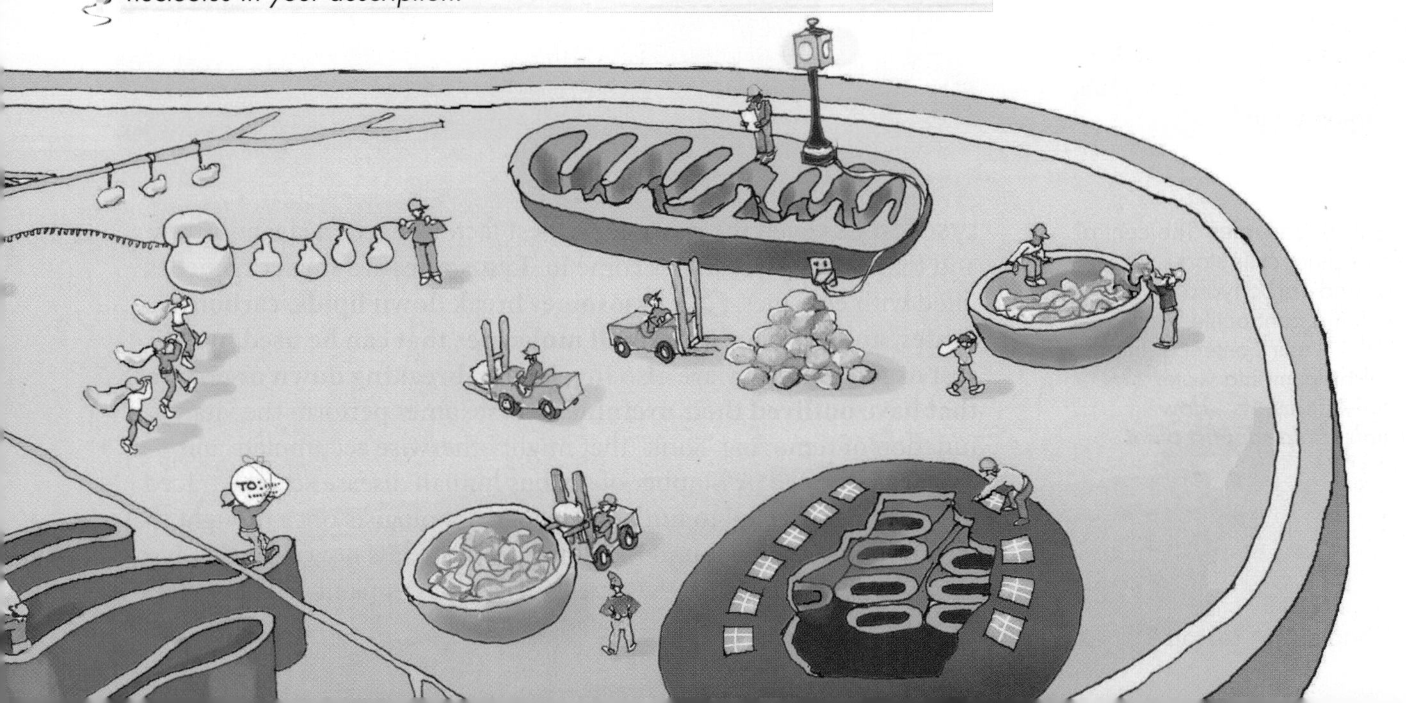

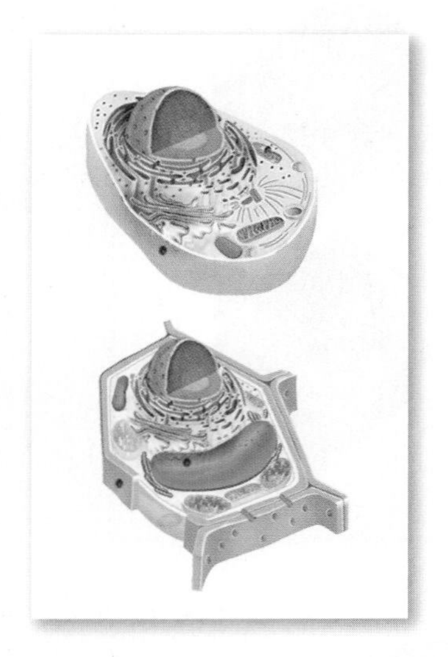

Organelles That Store, Clean Up, and Support

What are the functions of vacuoles, lysosomes, and the cytoskeleton?

Many of the organelles outside the nucleus of a eukaryotic cell have specific functions, or roles. Among them are structures called vacuoles, lysosomes, and cytoskeleton. These organelles represent the cellular factory's storage space, cleanup crew, and support structures.

Vacuoles and Vesicles Every factory needs a place to store things, and so does every cell. Many cells contain large, saclike, membrane-enclosed structures called **vacuoles. Vacuoles store materials like water, salts, proteins, and carbohydrates.** In many plant cells, there is a single, large central vacuole filled with liquid. The pressure of the central vacuole in these cells increases their rigidity, making it possible for plants to support heavy structures, such as leaves and flowers. The image on the left in **Figure 7–8** shows a typical plant cell's large central vacuole.

Vacuoles are also found in some unicellular organisms and in some animals. The paramecium on the right in **Figure 7–8** contains an organelle called a contractile vacuole. By contracting rhythmically, this specialized vacuole pumps excess water out of the cell. In addition, nearly all eukaryotic cells contain smaller membrane-enclosed structures called vesicles. Vesicles store and move materials between cell organelles, as well as to and from the cell surface.

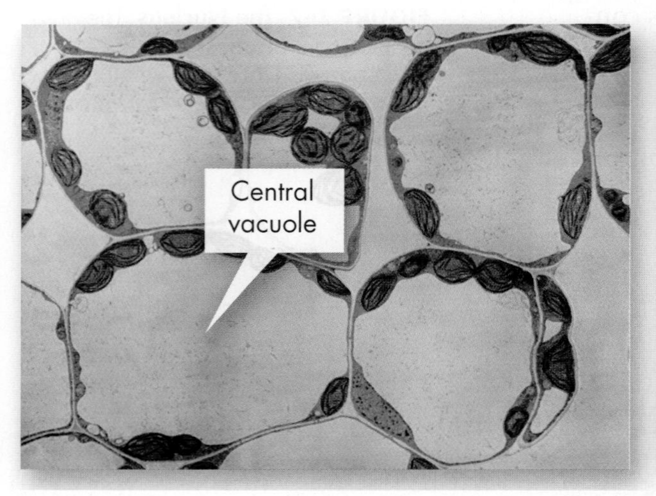

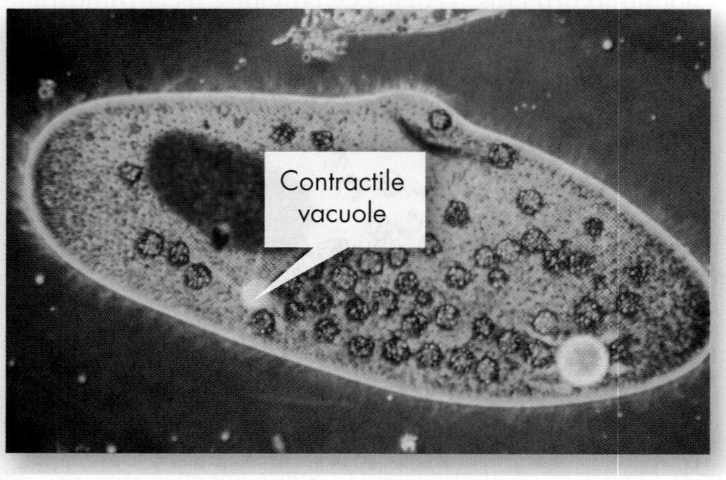

TEM 7000×

LM 500×

FIGURE 7–8 Vacuoles The central vacuole of plant cells stores salts, proteins, and carbohydrates. A paramecium's contractile vacuole controls the water content of the organism by pumping water out. **Apply Concepts** ***How do vacuoles help support plant structures?***

Lysosomes Even the neatest, cleanest factory needs a cleanup crew, and that's where lysosomes come in. **Lysosomes** are small organelles filled with enzymes. **Lysosomes break down lipids, carbohydrates, and proteins into small molecules that can be used by the rest of the cell. They are also involved in breaking down organelles that have outlived their usefulness.** Lysosomes perform the vital function of removing "junk" that might otherwise accumulate and clutter up the cell. A number of serious human diseases can be traced to lysosomes that fail to function properly. Biologists once thought that lysosomes were only found in animal cells, but it is now clear that lysosomes are also found in a few specialized types of plant cells as well.

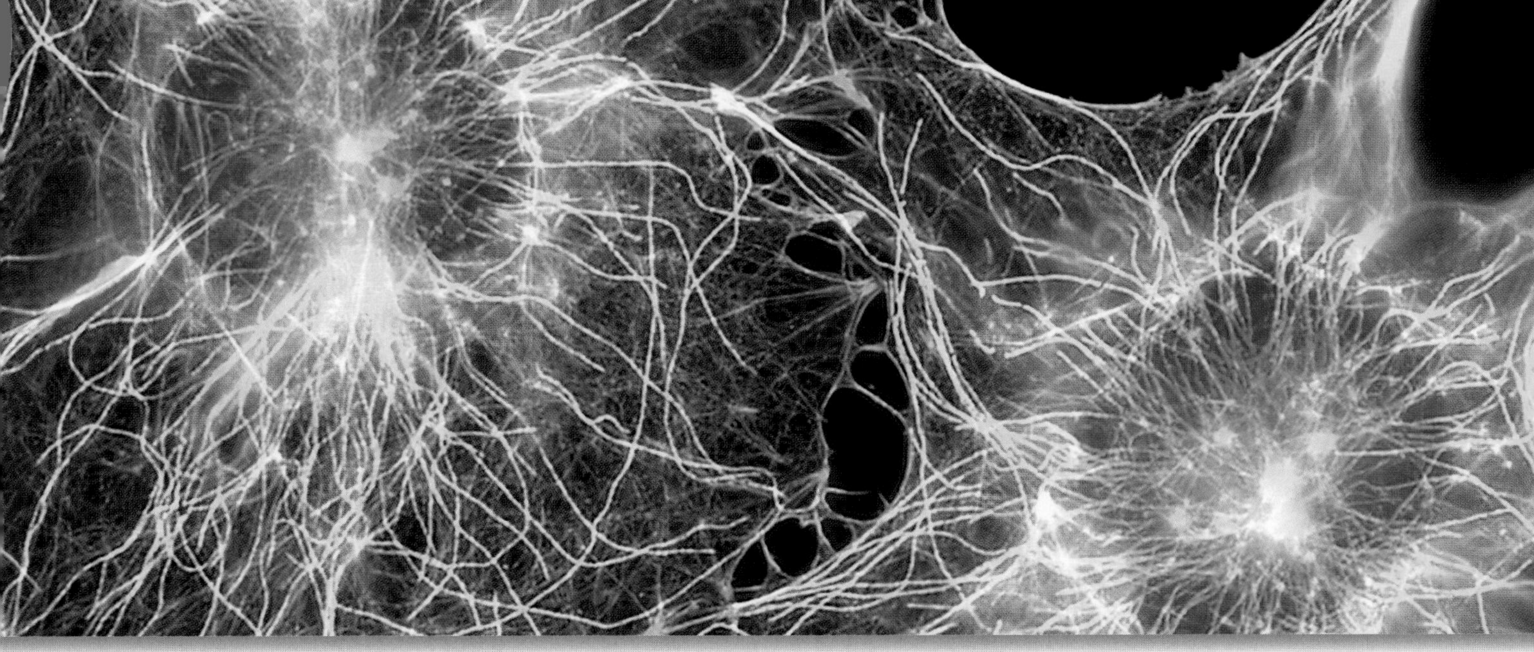

LM 1175×

FIGURE 7–9 Cytoskeleton The cytoskeleton supports and gives shape to the cell, and is involved in many forms of cell movement. These connective tissue fibroblast cells have been treated with fluorescent tags that bind to certain elements. Microfilaments are pale purple, microtubules are yellow, and the nuclei are green.

The Cytoskeleton As you know, a factory building is supported by steel or cement beams and by columns that hold up its walls and roof. Eukaryotic cells are given their shape and internal organization by a network of protein filaments known as the **cytoskeleton.** Certain parts of the cytoskeleton also help transport materials between different parts of the cell, much like the conveyor belts that carry materials from one part of a factory to another. Cytoskeletal components may also be involved in moving the entire cell as in cell flagella and cilia. **The cytoskeleton helps the cell maintain its shape and is also involved in movement.** Fluorescence imaging, as seen in **Figure 7–9,** clearly shows the complexity of a cell's cytoskeletal network. Microfilaments (pale purple) and microtubules (yellow) are two of the principal protein filaments that make up the cytoskeleton.

▶ ***Microfilaments*** Microfilaments are threadlike structures made up of a protein called actin. They form extensive networks in some cells and produce a tough flexible framework that supports the cell. Microfilaments also help cells move. Microfilament assembly and disassembly are responsible for the cytoplasmic movements that allow amoebas and other cells to crawl along surfaces.

▶ ***Microtubules*** Microtubules are hollow structures made up of proteins known as tubulins. In many cells, they play critical roles in maintaining cell shape. Microtubules are also important in cell division, where they form a structure known as the mitotic spindle, which helps to separate chromosomes. In animal cells, organelles called centrioles are also formed from tubulins. **Centrioles** are located near the nucleus and help organize cell division. Centrioles are not found in plant cells.

Microtubules also help build projections from the cell surface—known as cilia (singular: cilium) and flagella (singular: flagellum)—that enable cells to swim rapidly through liquid. The microtubules in cilia and flagella are arranged in a "9 + 2" pattern, as shown in **Figure 7–10.** Small cross-bridges between the microtubules in these organelles use chemical energy to pull on, or slide along, the microtubules, producing controlled movements.

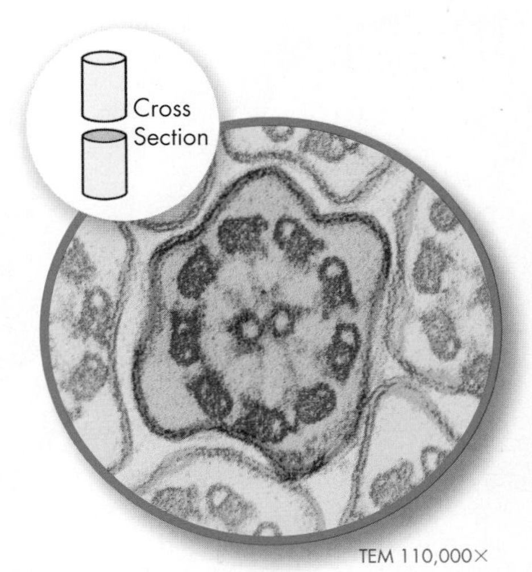

TEM 110,000×

FIGURE 7–10 The "9 + 2" Pattern of Microtubules In this micrograph showing the cross section of a cilium, you can clearly see the 9 + 2 arrangement of the red microtubules.
Apply Concepts *What is the function of cilia?*

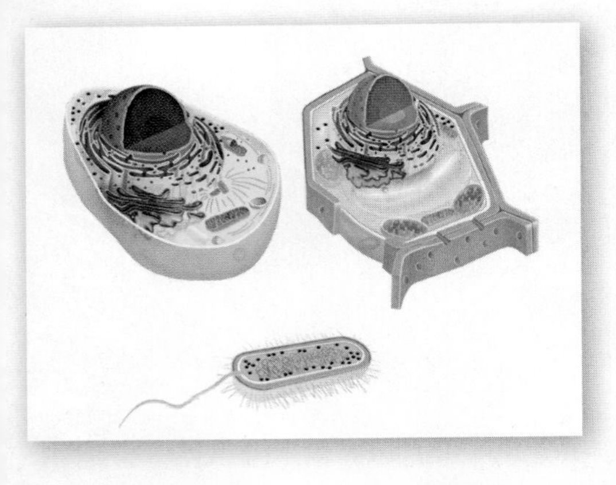

Organelles That Build Proteins

What organelles help make and transport proteins?

Life is a dynamic process, and living things are always working, building new molecules all the time, especially proteins, which catalyze chemical reactions and make up important structures in the cell. Because proteins carry out so many of the essential functions of living things, a big part of the cell is devoted to their production and distribution. Proteins are synthesized on ribosomes, sometimes in association with the rough endoplasmic reticulum in eukaryotes. The process of making proteins is summarized in **Figure 7–11.**

Ribosomes One of the most important jobs carried out in the cellular "factory" is making proteins. **Proteins are assembled on ribosomes. Ribosomes** are small particles of RNA and protein found throughout the cytoplasm in all cells. Ribosomes produce proteins by following coded instructions that come from DNA. Each ribosome, in its own way, is like a small machine in a factory, turning out proteins on orders that come from its DNA "boss." Cells that are especially active in protein synthesis often contain large numbers of ribosomes.

Endoplasmic Reticulum Eukaryotic cells contain an internal membrane system known as the **endoplasmic reticulum** (en doh PLAZ mik rih TIK yuh lum), or ER. The endoplasmic reticulum is where lipid components of the cell membrane are assembled, along with proteins and other materials that are exported from the cell.

The portion of the ER involved in the synthesis of proteins is called rough endoplasmic reticulum, or rough ER. It is given this name because of the ribosomes found on its surface. Newly made proteins leave these ribosomes and are inserted into the rough ER, where they may be chemically modified.

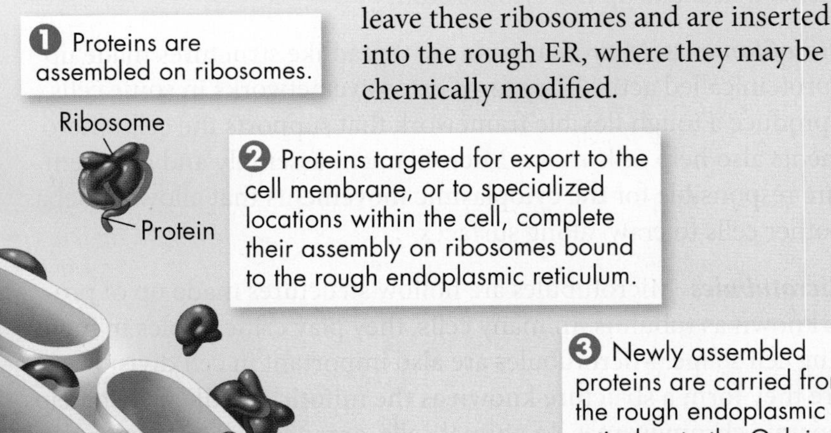

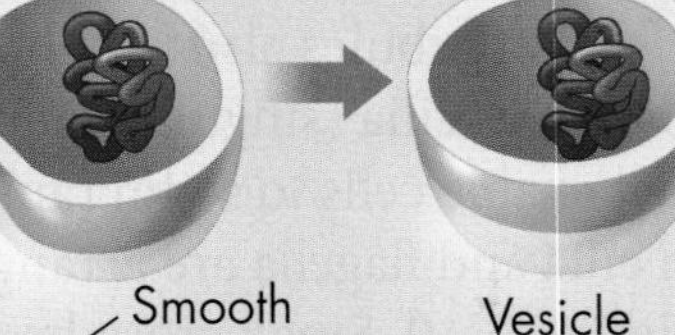

Proteins made on the rough ER include those that will be released, or secreted, from the cell as well as many membrane proteins and proteins destined for lysosomes and other specialized locations within the cell. Rough ER is abundant in cells that produce large amounts of protein for export. Other cellular proteins are made on "free" ribosomes, which are not attached to membranes.

The other portion of the ER is known as smooth endoplasmic reticulum (smooth ER) because ribosomes are not found on its surface. In many cells, the smooth ER contains collections of enzymes that perform specialized tasks, including the synthesis of membrane lipids and the detoxification of drugs. Liver cells, which play a key role in detoxifying drugs, often contain large amounts of smooth ER.

Golgi Apparatus In eukaryotic cells, proteins produced in the rough ER move next into an organelle called the **Golgi apparatus,** which appears as a stack of flattened membranes. As proteins leave the rough ER, molecular "address tags" get them to the right destinations. As these tags are "read" by the cell, the proteins are bundled into tiny vesicles that bud from the ER and carry them to the Golgi apparatus. **The Golgi apparatus modifies, sorts, and packages proteins and other materials from the endoplasmic reticulum for storage in the cell or release outside the cell.** The Golgi apparatus is somewhat like a customization shop, where the finishing touches are put on proteins before they are ready to leave the "factory." From the Golgi apparatus, proteins are "shipped" to their final destination inside or outside the cell.

In Your Notebook *Make a flowchart that shows how proteins are assembled in a cell.*

VISUAL SUMMARY

MAKING PROTEINS

FIGURE 7–11 Together, ribosomes, the endoplasmic reticulum, and the Golgi apparatus synthesize, modify, package, and ship proteins. **Infer** ***What can you infer about a cell that is packed with more than the typical number of ribosomes?***

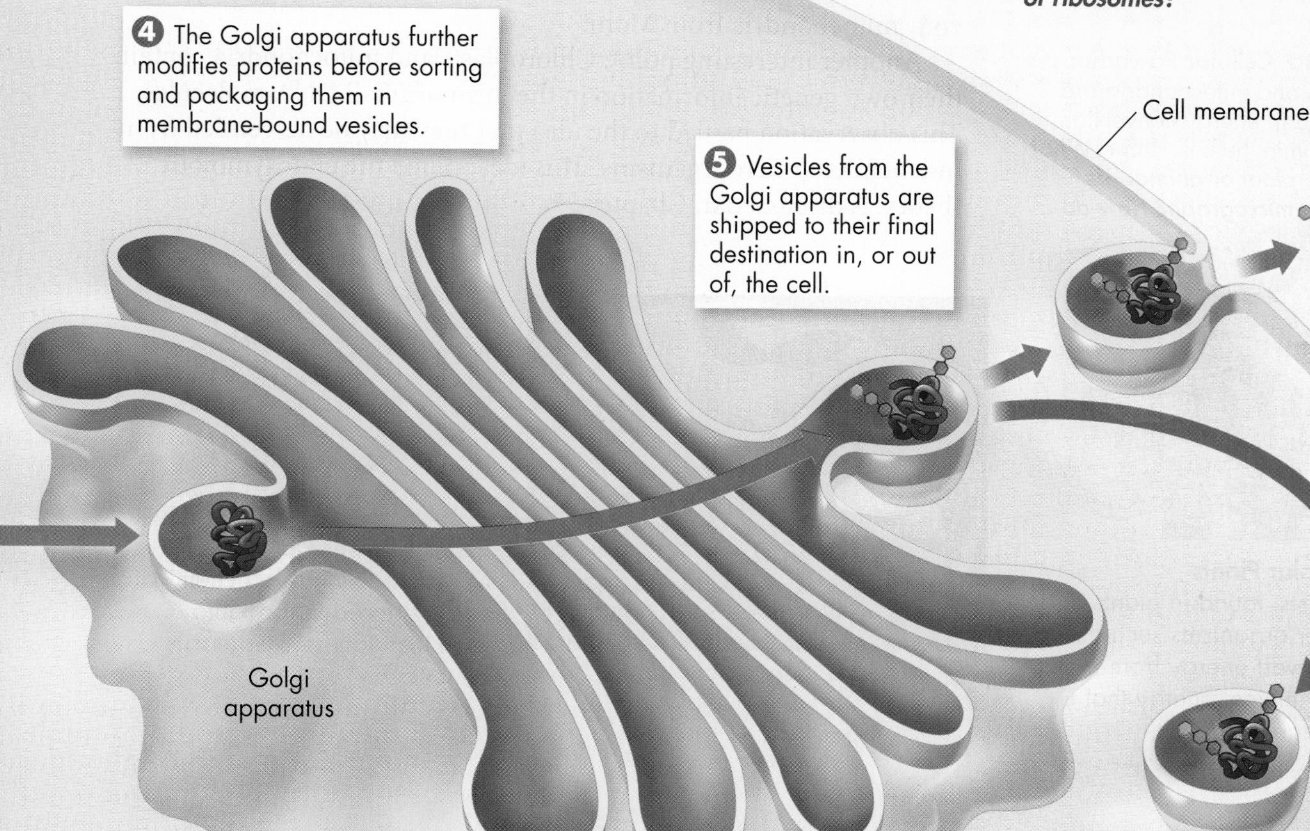

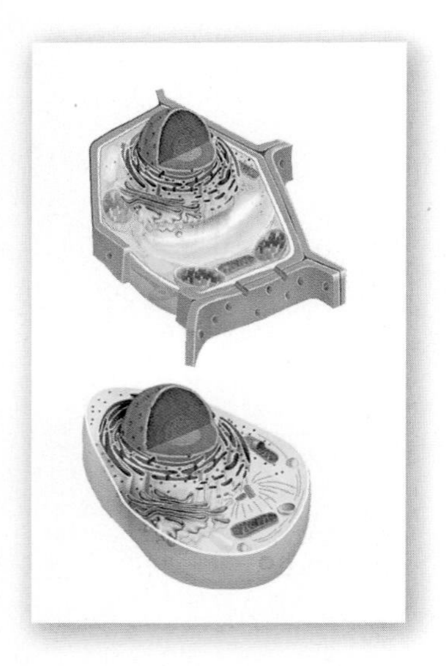

Organelles That Capture and Release Energy

What are the functions of chloroplasts and mitochondria?

All living things require a source of energy. Factories are hooked up to the local power company, but how do cells get energy? Most cells are powered by food molecules that are built using energy from the sun.

Chloroplasts Plants and some other organisms contain chloroplasts (KLAWR uh plasts). **Chloroplasts** are the biological equivalents of solar power plants. **Chloroplasts capture the energy from sunlight and convert it into food that contains chemical energy in a process called photosynthesis.** Two membranes surround chloroplasts. Inside the organelle are large stacks of other membranes, which contain the green pigment chlorophyll.

Mitochondria Nearly all eukaryotic cells, including plants, contain mitochondria (myt oh KAHN dree uh; singular: mitochondrion). **Mitochondria** are the power plants of the cell. **Mitochondria convert the chemical energy stored in food into compounds that are more convenient for the cell to use.** Like chloroplasts, two membranes—an outer membrane and an inner membrane—enclose mitochondria. The inner membrane is folded up inside the organelle, as shown in **Figure 7–12.**

One of the most interesting aspects of mitochondria is the way in which they are inherited. In humans, all or nearly all of our mitochondria come from the cytoplasm of the ovum, or egg cell. This means that when your relatives are discussing which side of the family should take credit for your best characteristics, you can tell them that you got your mitochondria from Mom!

Another interesting point: Chloroplasts and mitochondria contain their own genetic information in the form of small DNA molecules. This observation has led to the idea that they may be descended from independent microorganisms. This idea, called the endosymbiotic theory, is discussed in Chapter 19.

FIGURE 7–12 Cellular Powerhouses Chloroplasts and mitochondria are both involved in energy conversion processes within the cell. **Infer** ***What kind of cell—plant or animal—is shown in the micrograph? How do you know?***

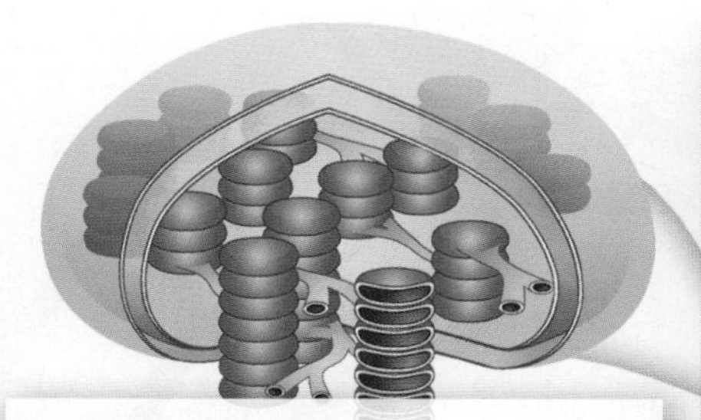

Cellular Solar Plants
Chloroplasts, found in plants and some other organisms such as algae, convert energy from the sun into chemical energy that is stored as food.

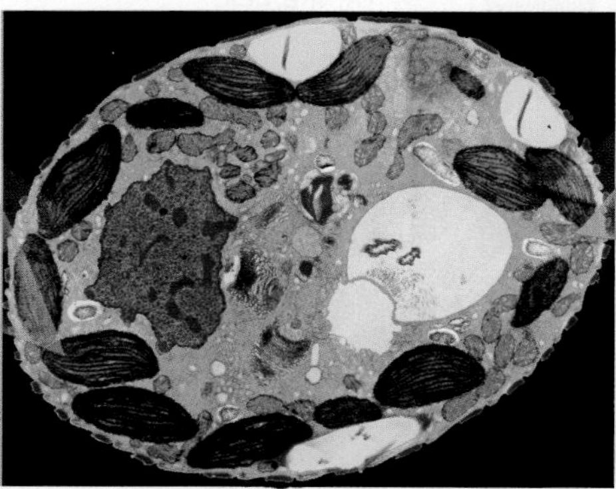

TEM 4500×

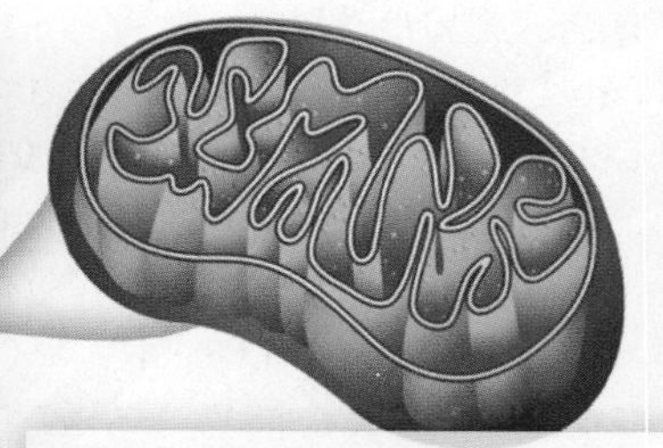

Cellular Power Plants
Mitochondria convert chemical energy stored in food into a form that can be used easily by the cell.

Quick Lab
OPEN-ENDED INQUIRY

Making a Model of a Cell

❶ Your class is going to make a model of a plant cell using the whole classroom. Work with a partner or in a small group to decide what cell part or organelle you would like to model. (Use **Figure 7–14** on pages 206–207 as a starting point. It gives you an idea of the relative sizes of various cell parts and their possible positions.)

❷ Using materials of your choice, make a three-dimensional model of the cell part or organelle you chose. Make the model as complete and as accurate as you can.

❸ Label an index card with the name of your cell part or organelle, and list its main features and functions. Attach the card to your model.

❹ Attach your model to an appropriate place in the room. If possible, attach your model to another related cell part or organelle.

Analyze and Conclude

1. **Calculate** Assume that a typical plant cell is 50 micrometers wide (50×10^{-6} m). Calculate the scale of your classroom cell model. (*Hint:* Divide the width of the classroom by the width of a cell, making sure to use the same units.) MATH
2. **Compare and Contrast** How is your model cell part or organelle similar to the real cell part or organelle? How is it different?
3. **Evaluate** Based on your work with this model, describe how you could make a better model. What new information would your improved model demonstrate?

Cellular Boundaries

What is the function of the cell membrane?

A working factory needs walls and a roof to protect it from the environment outside, and also to serve as a barrier that keeps its products safe and secure until they are ready to be shipped out. Cells have similar needs, and they meet them in a similar way. As you have learned, all cells are surrounded by a barrier known as the cell membrane. Many cells, including most prokaryotes, also produce a strong supporting layer around the membrane known as a **cell wall.**

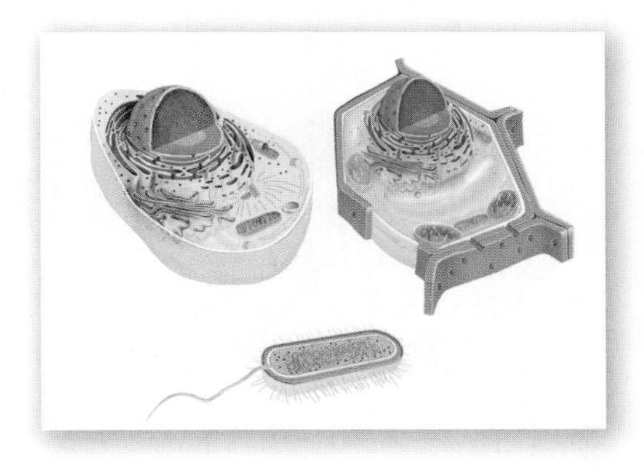

Cell Walls Many organisms have cell walls in addition to cell membranes. The main function of the cell wall is to support, shape, and protect the cell. Most prokaryotes and many eukaryotes have cell walls. Animal cells do not have cell walls. Cell walls lie outside the cell membrane. Most cell walls are porous enough to allow water, oxygen, carbon dioxide, and certain other substances to pass through easily.

Cell walls provide much of the strength needed for plants to stand against the force of gravity. In trees and other large plants, nearly all of the tissue we call wood is made up of cell walls. The cellulose fiber used for paper as well as the lumber used for building comes from these walls. So if you are reading these words off a sheet of paper from a book resting on a wooden desk, you've got cell walls all around you.

BUILD Vocabulary

ACADEMIC WORDS The adjective **porous** means "allowing materials to pass through." A porous cell wall allows substances like water and oxygen to pass through it.

Cell Membranes All cells contain cell membranes, which almost always are made up of a double-layered sheet called a lipid bilayer, as shown in **Figure 7–13.** The **lipid bilayer** gives cell membranes a flexible structure that forms a strong barrier between the cell and its surroundings. **The cell membrane regulates what enters and leaves the cell and also protects and supports the cell.**

▶ ***The Properties of Lipids*** The layered structure of cell membranes reflects the chemical properties of the lipids that make them up. You may recall that many lipids have oily fatty acid chains attached to chemical groups that interact strongly with water. In the language of a chemist, the fatty acid portions of this kind of lipid are hydrophobic (hy druh FOH bik), or "water-hating," while the opposite end of the molecule is hydrophilic (hy druh FIL ik), or "water-loving." When these lipids, including the phospholipids that are common in animal cell membranes, are mixed with water, their hydrophobic fatty acid "tails" cluster together while their hydrophilic "heads" are attracted to water. A lipid bilayer is the result. As you can see in **Figure 7–13,** the head groups of lipids in a bilayer are exposed to the outside of the cell, while the fatty acid tails form an oily layer inside the membrane that keeps water out.

ZOOMING IN

THE CELL MEMBRANE

FIGURE 7–13 Every cell has a membrane that regulates the movement of materials. Nearly all cell membranes are made up of a lipid bilayer in which proteins and carbohydrates are embedded. **Apply Concepts** ***Explain why lipids "self-assemble" into a bilayer when exposed to water.***

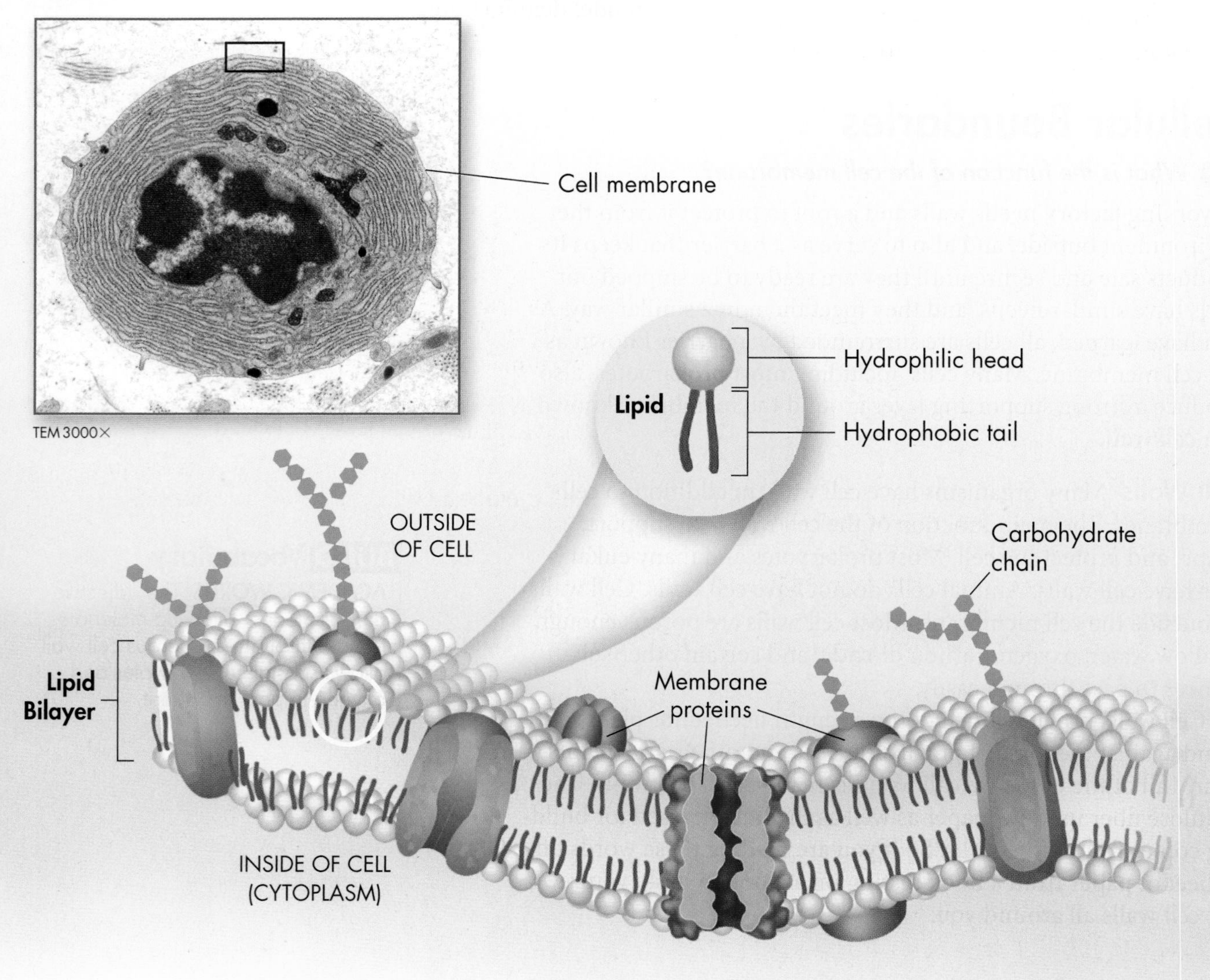

▶ ***The Fluid Mosaic Model*** Embedded in the lipid bilayer of most cell membranes are protein molecules. Carbohydrate molecules are attached to many of these proteins. Because the proteins embedded in the lipid bilayer can move around and "float" among the lipids, and because so many different kinds of molecules make up the cell membrane, scientists describe the cell membrane as a "fluid mosaic." A mosaic is a kind of art that involves bits and pieces of different colors or materials. What are all these different molecules doing? As you will see, some of the proteins form channels and pumps that help to move material across the cell membrane. Many of the carbohydrate molecules act like chemical identification cards, allowing individual cells to identify one another. Some proteins attach directly to the cytoskeleton, enabling cells to respond to their environment by using their membranes to help move or change shape.

As you know, some things are allowed to enter and leave a factory, and some are not. The same is true for living cells. Although many substances can cross biological membranes, some are too large or too strongly charged to cross the lipid bilayer. If a substance is able to cross a membrane, the membrane is said to be permeable to it. A membrane is impermeable to substances that cannot pass across it. Most biological membranes are **selectively permeable,** meaning that some substances can pass across them and others cannot. Selectively permeable membranes are also called semipermeable membranes.

7.2 Assessment

Review Key Concepts

1. a. Review What are the two major parts of the cell?

b. Use Analogies How is the role of the nucleus in a cell similar to the role of the captain on a sports team?

2. a. Review What is the function of lysosomes?

b. Apply Concepts How do contractile vacuoles help maintain water balance?

3. a. Review What is the difference between rough and smooth ER?

b. Sequence Describe the steps involved in the synthesis, packaging, and export of a protein from a cell.

4. a. Review What is the function of mitochondria?

b. Infer You examine an unknown cell under a microscope and discover that the cell contains chloroplasts. From what type of organism does the cell likely come?

5. a. Review Why is the cell membrane sometimes referred to as a fluid mosaic? What part of the cell membrane acts like a fluid? And what makes it like a mosaic?

b. Explain How do the properties of lipids help explain the structure of a cell membrane?

c. Infer Why do you think it's important that cell membranes are *selectively* permeable?

VISUAL THINKING

6. Using the cells on the next page as a guide, draw your own models of a prokaryotic cell, a plant cell, and an animal cell. Then use each of the vocabulary words from this lesson to label your cells.

VISUAL SUMMARY

TYPICAL CELLS

FIGURE 7–14 Eukaryotic cells contain a variety of organelles, a few of which they have in common with prokaryotic cells. Note in the table on the facing page that while prokaryotic cells lack cytoskeleton and chloroplasts, they accomplish their functions in other ways as described. **Interpret Visuals** ***What structures do prokaryotic cells have in common with animal cells? With plant cells?***

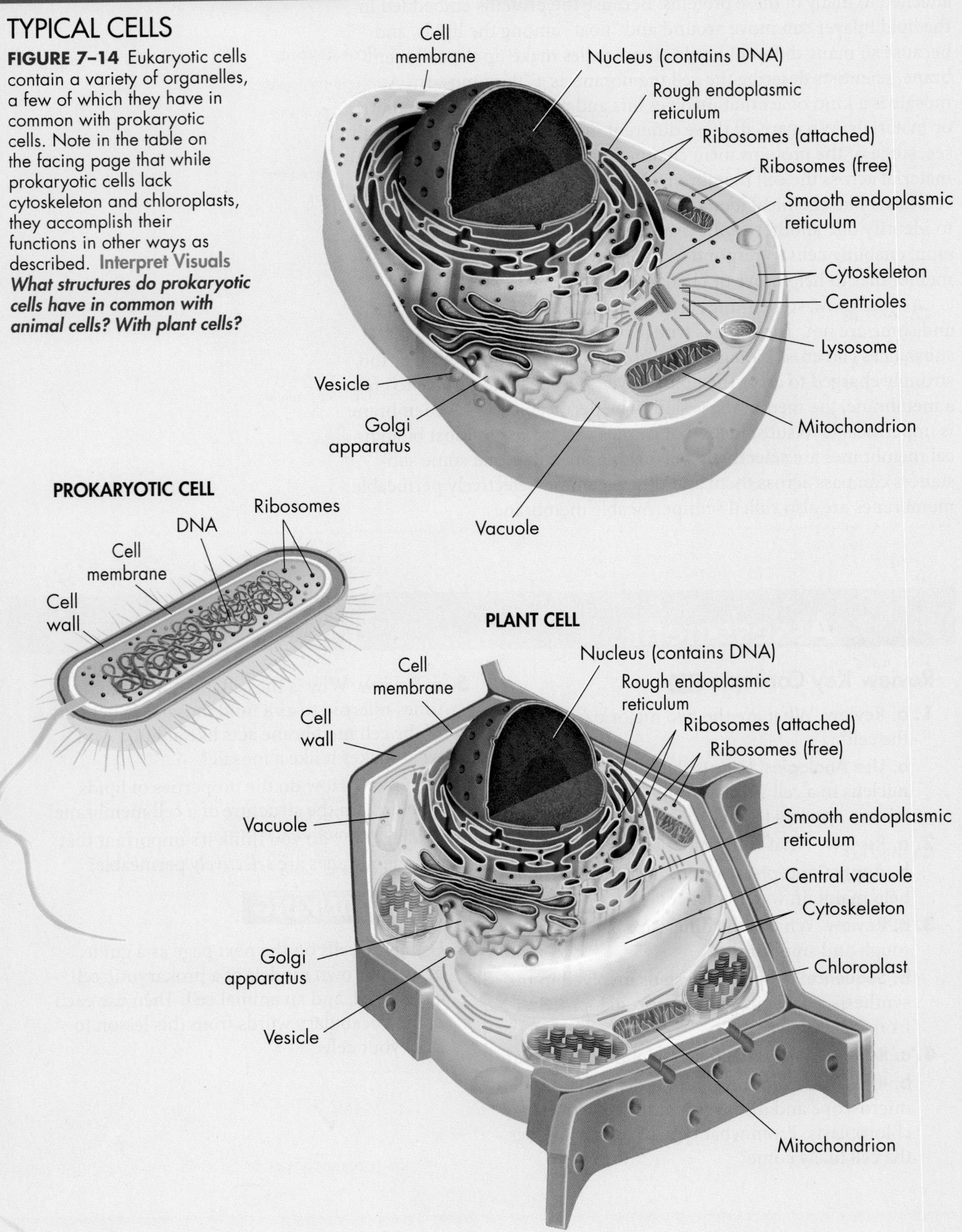

	Structure	Function	Prokaryote	Eukaryote: Animal	Eukaryote: Plant
Cellular Control Center	Nucleus	Contains DNA	*Prokaryote DNA is found in cytoplasm.*	✓	✓
Organelles That Store, Clean-Up, and Support	Vacuoles and vesicles	Store materials		✓	✓
	Lysosomes	Break down and recycle macromolecules		✓	✓ (rare)
	Cytoskeleton	Maintains cell shape; moves cell parts; helps cells move	*Prokaryotic cells have protein filaments similar to actin and tubulin.*	✓	✓
	Centrioles	Organize cell division		✓	
Organelles That Build Proteins	Ribosomes	Synthesize proteins	✓	✓	✓
	Endoplasmic reticulum	Assembles proteins and lipids		✓	✓
	Golgi apparatus	Modifies, sorts, and packages proteins and lipids for storage or transport out of the cell		✓	✓
Organelles That Capture and Release Energy	Chloroplasts	Convert solar energy to chemical energy stored in food	*In some prokaryotic cells, photosynthesis occurs in association with internal photosynthetic membranes.*		✓
	Mitochondria	Convert chemical energy in food to usable compounds	*Prokaryotes carry out these reactions in the cytoplasm rather than in specialized organelles.*	✓	✓
Cellular Boundaries	Cell wall	Shapes, supports, and protects the cell	✓		✓
	Cell membrane	Regulates materials entering and leaving cell; protects and supports cell	✓	✓	✓

7.3 Cell Transport

Key Questions

 *What is passive transport?*

 *What is active transport?*

Vocabulary

diffusion • facilitated diffusion • aquaporin • osmosis • isotonic • hypertonic • hypotonic • osmotic pressure

Taking Notes

Compare/Contrast Table As you read, create a compare/contrast table for passive and active transport.

MYSTERY CLUE

As Michelle ran, she perspired, losing salts from her bloodstream. And as she drank more and more water during the race, the concentration of dissolved salts and minerals in her bloodstream decreased. How do you think these phenomena contributed to Michelle's condition?

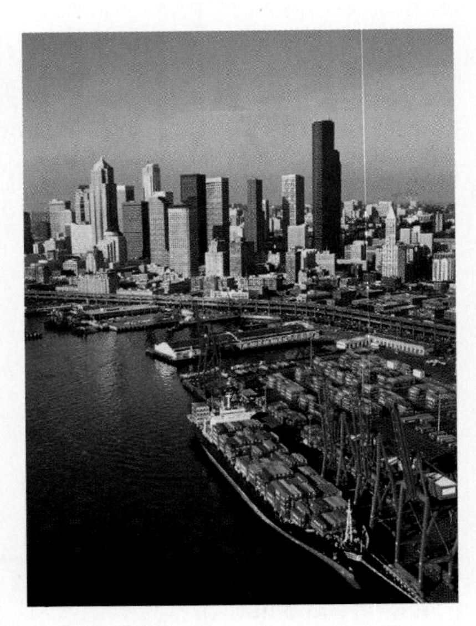

THINK ABOUT IT In the previous lesson, cell walls and cell membranes were compared to the roof and walls of a factory. When you think about how cells move materials in and out, it can be helpful to think of a cell as a nation. Before you can learn anything about a nation, it's important to understand where it begins and where it ends. The boundaries of a nation are its borders, and nearly every country tries to regulate and control the goods that move across those borders, like the shipping containers seen here entering and leaving the port of Seattle. Each cell has its own border, which separates the cell from its surroundings and also determines what comes in and what goes out. How can a cell separate itself from its environment and still allow material to enter and leave? That's where transport across its border, the cell membrane, comes in.

Passive Transport

What is passive transport?

Every living cell exists in a liquid environment. One of the most important functions of the cell membrane is to keep the cell's internal conditions relatively constant. It does this by regulating the movement of molecules from one side of the membrane to the other.

Diffusion Cellular cytoplasm consists of many different substances dissolved in water. In any solution, solute particles move constantly. They collide with one another and tend to spread out randomly. As a result, the particles tend to move from an area where they are more concentrated to an area where they are less concentrated. When you add sugar to coffee or tea, for example, the sugar molecules move away from their original positions in the sugar crystals and disperse throughout the hot liquid. The process by which particles move from an area of high concentration to an area of lower concentration is known as **diffusion** (dih FYOO zhun). Diffusion is the driving force behind the movement of many substances across the cell membrane.

What does diffusion have to do with the cell membrane? Suppose a substance is present in unequal concentrations on either side of a cell membrane, as shown in **Figure 7–15.** If the substance can cross the cell membrane, its particles will tend to move toward the area where it is less concentrated until it is evenly distributed. Once the concentration of the substance on both sides of the cell membrane is the same, equilibrium is reached.

Even when equilibrium is reached, particles of a solution continue to move across the membrane in both directions. However, because almost equal numbers of particles move in each direction, there is no further net change in the concentration on either side.

Diffusion depends on random particle movements. Therefore, substances diffuse across membranes without requiring the cell to use additional energy. **The movement of materials across the cell membrane without using cellular energy is called passive transport.**

FIGURE 7–15 Diffusion Diffusion is the process by which molecules of a substance move from an area of higher concentration to an area of lower concentration. It does not require the cell to use energy. **Predict** ***How would the movement of solute particles seen here be different if the initial area of high concentration had been on the inside of the cell instead of the outside?***

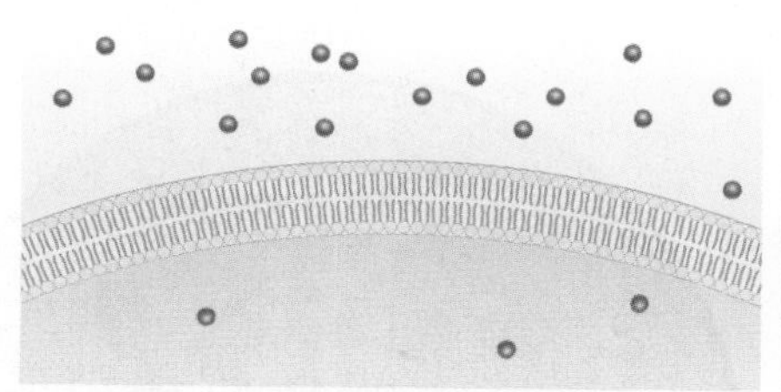

There is a higher concentration of solute on one side of the membrane than on the other.

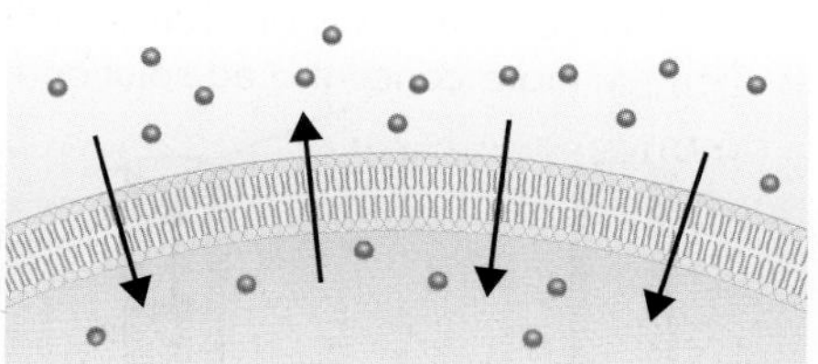

Diffusion causes a net movement of solute particles from the side of the membrane with the higher solute concentration to the side with the lower solute concentration.

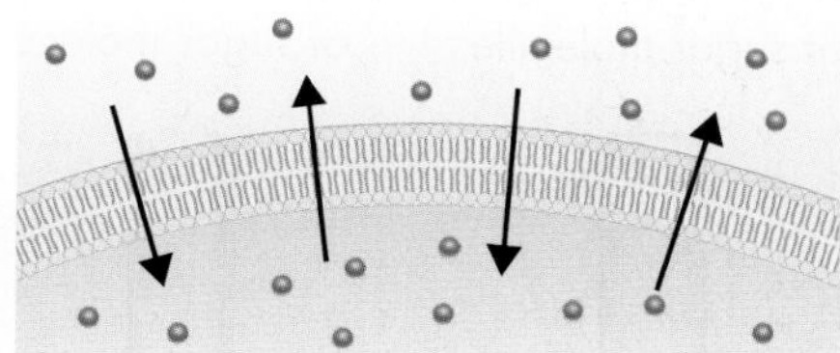

Once equilibrium is reached, solute particles continue to diffuse across the membrane in both directions but at approximately equal rates, so there is no net change in solute concentration.

Facilitated Diffusion Since cell membranes are built around lipid bilayers, the molecules that pass through them most easily are small and uncharged. These properties allow them to dissolve in the membrane's lipid environment. But many ions, like Cl^-, and large molecules, like the sugar glucose, seem to pass through cell membranes much more quickly than they should. It's almost as if they have a shortcut across the membrane.

How does this happen? Proteins in the cell membrane act as carriers, or channels, making it easy for certain molecules to cross. Red blood cells, for example, have protein carriers that allow glucose to pass through them in either direction. Only glucose can pass through these protein carriers. These cell membrane channels facilitate, or help, the diffusion of glucose across the membrane. This process, in which molecules that cannot directly diffuse across the membrane pass through special protein channels, is known as **facilitated diffusion.** Hundreds of different proteins have been found that allow particular substances to cross cell membranes. Although facilitated diffusion is fast and specific, it is still diffusion, so it does not require any additional use of the cell's energy.

In Your Notebook *Explain how you can demonstrate diffusion by spraying air freshener in a large room.*

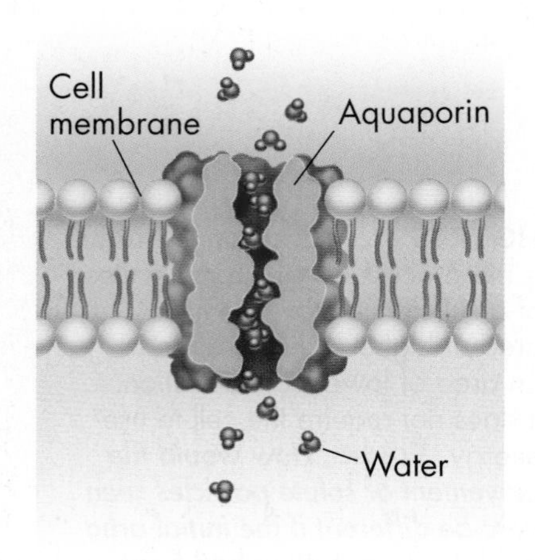

FIGURE 7–16 An Aquaporin

Osmosis: An Example of Facilitated Diffusion Surprising new research has added water to the list of molecules that enter cells by facilitated diffusion. Recall that the inside of a cell's lipid bilayer is hydrophobic, or "water-hating." Because of this, water molecules have a tough time passing through the cell membrane. However, many cells contain water channel proteins, known as **aquaporins** (ak wuh PAWR inz), that allow water to pass right through them, as shown in **Figure 7–16.** The movement of water through cell membranes by facilitated diffusion is an extremely important biological process—the process of osmosis.

Osmosis is the diffusion of water through a selectively permeable membrane. In osmosis, as in other forms of diffusion, molecules move from an area of higher concentration to an area of lower concentration. The only difference is that the molecules that move in the case of osmosis are water molecules, not solute molecules. The process of osmosis is shown in **Figure 7–17.**

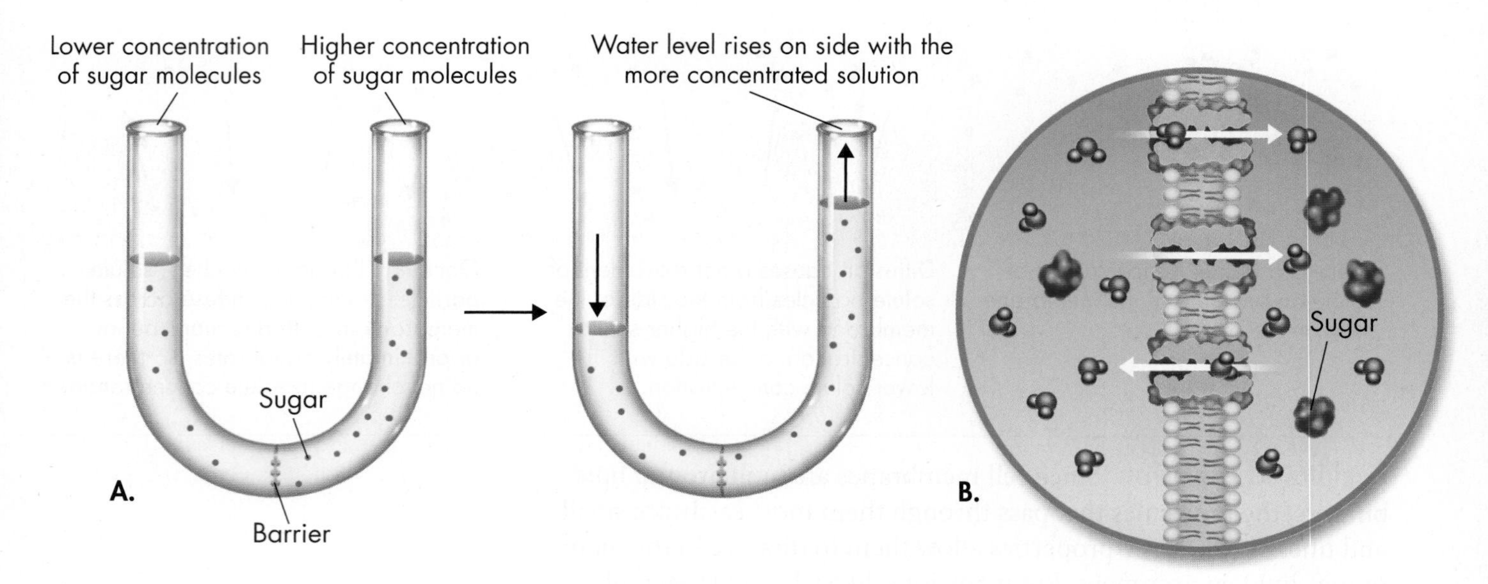

FIGURE 7–17 Osmosis Osmosis is a form of facilitated diffusion. **A.** In a laboratory experiment, water moves through a selectively permeable barrier from an area of lower to higher solute concentration. **B.** In the cell, water passes in through aquaporins embedded in the cell membrane. Although water moves in both directions through aquaporins, there is a net movement of water from an area of lower to higher sugar concentration. Apply Concepts ***Does osmosis require the cell to use energy?***

▶ ***How Osmosis Works*** Look at the experimental setup in **Figure 7–17A.** The barrier is permeable to water but not to sugar. This means that water can cross the barrier in both directions, but sugar cannot. To start, there are more sugar molecules on the right side of the barrier than on the left side. Therefore, the concentration of water is lower on the right, where more of the solution is made of sugar. Although water molecules move in both directions across the membrane, there is a net movement of water toward the concentrated sugar solution.

Water will tend to move across the membrane until equilibrium is reached. At that point, the concentrations of water and sugar will be the same on both sides of the membrane. When this happens, the two solutions will be **isotonic,** which means "same strength." Note that "strength" refers to the amount of solute, not water. When the experiment began, the more concentrated sugar solution (right side of the tube) was **hypertonic,** or "above strength," compared to the left side. So the dilute sugar solution (left side of the tube) was **hypotonic,** or "below strength," compared to the right side. **Figure 7–17B** shows how osmosis works across a cell membrane.

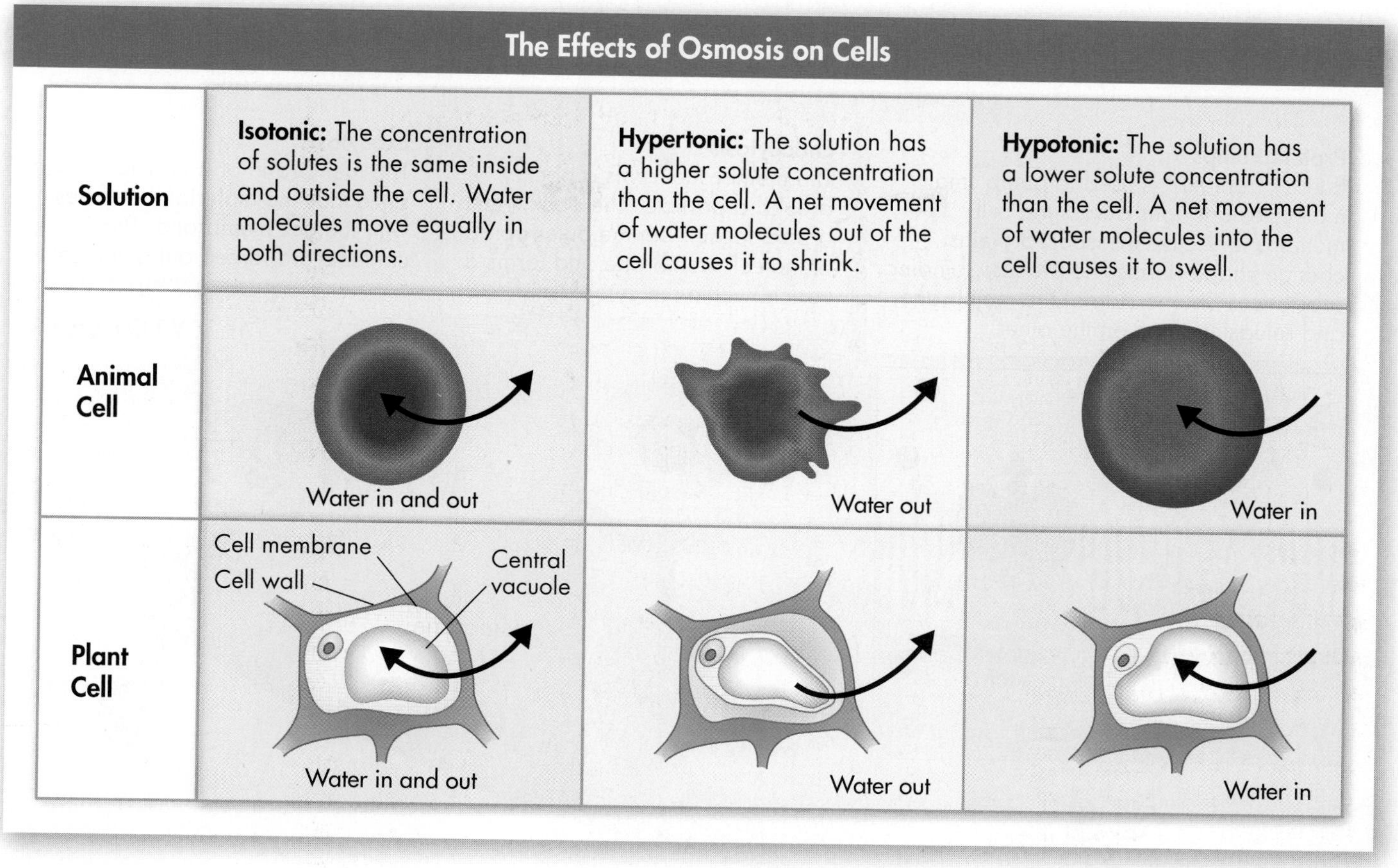

The Effects of Osmosis on Cells

Solution	**Isotonic:** The concentration of solutes is the same inside and outside the cell. Water molecules move equally in both directions.	**Hypertonic:** The solution has a higher solute concentration than the cell. A net movement of water molecules out of the cell causes it to shrink.	**Hypotonic:** The solution has a lower solute concentration than the cell. A net movement of water molecules into the cell causes it to swell.
Animal Cell	Water in and out	Water out	Water in
Plant Cell	Cell membrane Cell wall Central vacuole Water in and out	Water out	Water in

▶ ***Osmotic Pressure*** Driven by differences in solute concentration, the net movement of water out of or into a cell produces a force known as **osmotic pressure.** As shown in **Figure 7–18,** osmotic pressure can cause an animal cell in a hypertonic solution to shrink, and one in a hypotonic solution to swell. Because cells contain salts, sugars, proteins, and other dissolved molecules, they are almost always hypertonic to fresh water. As a result, water tends to move quickly into a cell surrounded by fresh water, causing it to swell. Eventually, the cell may burst like an overinflated balloon. In plant cells, osmotic pressure can cause changes in the size of the central vacuole, which shrinks or swells as water moves into or out of the cell.

Fortunately cells in large organisms are not in danger of bursting because most of them do not come in contact with fresh water. Instead, the cells are bathed in blood or other isotonic fluids. The concentrations of dissolved materials in these isotonic fluids are roughly equal to those in the cells themselves.

What happens when cells do come in contact with fresh water? Some, like the eggs laid in fresh water by fish and frogs, lack water channels. As a result, water moves into them so slowly that osmotic pressure is not a problem. Others, including bacteria and plant cells, are surrounded by tough walls. The cell walls prevent the cells from expanding, even under tremendous osmotic pressure. Notice how the plant cell in **Figure 7–18** holds its shape in both hypertonic and hypotonic solutions while the animal red blood cell does not. However, increased osmotic pressure makes plant cells extremely vulnerable to cell wall injuries.

FIGURE 7–18 Osmotic Pressure Water molecules move equally into and out of cells placed in an isotonic solution. In a hypertonic solution, animal cells, like the red blood cell shown, shrink, and plant cell central vacuoles collapse. In a hypotonic solution, animal cells swell and burst. The central vacuoles of plant cells also swell, pushing the cell contents out against the cell wall. **Predict** ***What would happen to the cells of a saltwater plant if the plant were placed in fresh water?***

In Your Notebook *In your own words, explain why osmosis is really just a special case of facilitated diffusion.*

Protein Pumps
Energy from ATP is used to pump small molecules and ions across the cell membrane. Active transport proteins change shape during the process, binding substances on one side of the membrane, and releasing them on the other.

Endocytosis
The membrane forms a pocket around a particle. The pocket then breaks loose from the outer portion of the cell membrane and forms a vesicle within the cytoplasm.

Exocytosis
The membrane of a vesicle surrounds the material then fuses with the cell membrane. The contents are forced out of the cell.

CYTOPLASM

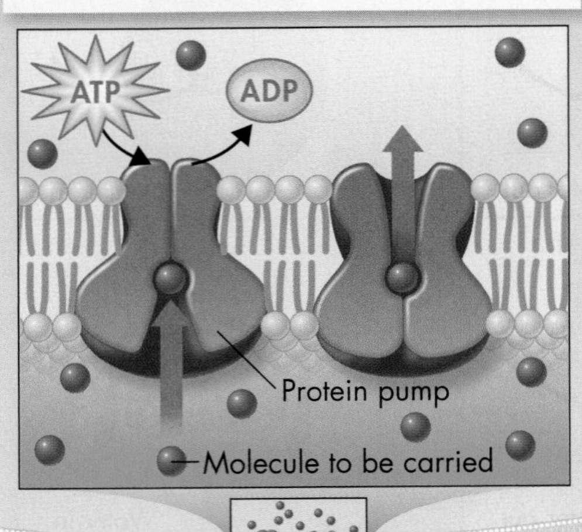

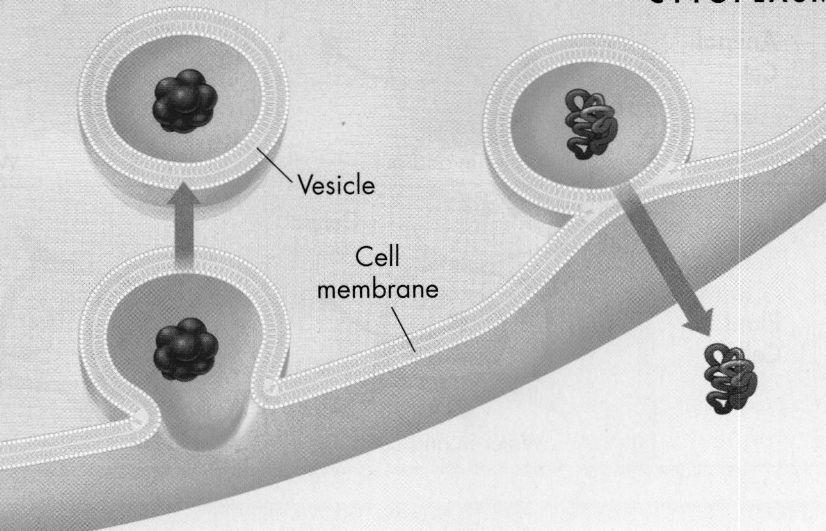

VISUAL SUMMARY

ACTIVE TRANSPORT

FIGURE 7–19 Energy from the cell is required to move particles against a concentration gradient. Compare and Contrast *What are the similarities and differences between facilitated diffusion and active transport by protein pump?*

Active Transport

What is active transport?

As powerful as diffusion is, cells sometimes must move materials against a concentration difference. **The movement of materials against a concentration difference is known as active transport. Active transport requires energy.** The active transport of small molecules or ions across a cell membrane is generally carried out by transport proteins—protein pumps—that are found in the membrane itself. Larger molecules and clumps of material can also be actively transported across the cell membrane by processes known as endocytosis and exocytosis. The transport of these larger materials sometimes involves changes in the shape of the cell membrane. The major types of active transport are shown in **Figure 7–19.**

Molecular Transport Small molecules and ions are carried across membranes by proteins in the membrane that act like pumps. Many cells use protein pumps to move calcium, potassium, and sodium ions across cell membranes. Changes in protein shape seem to play an important role in the pumping process. A considerable portion of the energy used by cells in their daily activities is spent providing the energy to keep this form of active transport working. The use of energy in these systems enables cells to concentrate substances in a particular location, even when the forces of diffusion might tend to move these substances in the opposite direction.

Bulk Transport Larger molecules and even solid clumps of material can be transported by movements of the cell membrane known as bulk transport. Bulk transport can take several forms, depending on the size and shape of the material moved into or out of the cell.

BUILD Vocabulary

PREFIXES The prefix *endo-* in *endocytosis* comes from a Greek word meaning "inside" or "within." The prefix *exo-* in *exocytosis* means "outside."

▶ ***Endocytosis*** Endocytosis (en doh sy TOH sis) is the process of taking material into the cell by means of infoldings, or pockets, of the cell membrane. The pocket that results breaks loose from the outer portion of the cell membrane and forms a vesicle or vacuole within the cytoplasm. Large molecules, clumps of food, even whole cells can be taken up in this way.

Phagocytosis (fag oh sy TOH sis) is a type of endocytosis, in which extensions of cytoplasm surround a particle and package it within a food vacuole. The cell then engulfs it. Amoebas use this method for taking in food, and white blood cells use phagocytosis to "eat" damaged cells, as shown in **Figure 7–20.** Engulfing material in this way requires a considerable amount of energy and is considered a form of active transport.

In a process similar to phagocytosis, many cells take up liquid from the surrounding environment. Tiny pockets form along the cell membrane, fill with liquid, and pinch off to form vacuoles within the cell. This type of endocytosis is known as pinocytosis (py nuh sy TOH sis).

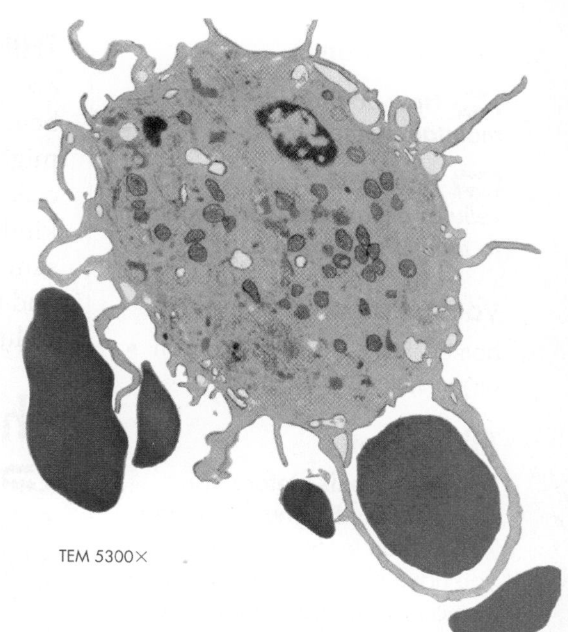

FIGURE 7–20 Endocytosis The white blood cell seen here is engulfing a damaged red blood cell by phagocytosis—a form of endocytosis. Extensions, or "arms," of the white blood cell's cell membrane have completely surrounded the red blood cell.

▶ ***Exocytosis*** Many cells also release large amounts of material, a process known as exocytosis (ek soh sy TOH sis). During exocytosis, the membrane of the vacuole surrounding the material fuses with the cell membrane, forcing the contents out of the cell. The removal of water by means of a contractile vacuole is one example of this kind of active transport.

7.3 Assessment

Review Key Concepts

1. a. **Review** What happens during diffusion?
   b. **Explain** Describe the process of osmosis.
   c. **Compare and Contrast** What is the difference between diffusion and facilitated diffusion?
2. a. **Review** How is active transport different from passive transport?
   b. **Explain** Describe the two major types of active transport.
   c. **Compare and Contrast** How is endocytosis different from exocytosis?

BUILD VOCABULARY

3. Based on the meanings of *isotonic, hypertonic,* and *hypotonic,* write definitions for the prefixes *iso-, hyper-,* and *hypo-*. Then come up with another set of words that uses these prefixes (the words do not need to have the same suffixes).
4. The prefix *phago-* means "to eat." The prefix *pino-* means "to drink." Look up the definition of *-cytosis,* and write definitions for *phagocytosis* and *pinocytosis.*

7.4 Homeostasis and Cells

Key Questions

How do individual cells maintain homeostasis?

How do the cells of multicellular organisms work together to maintain homeostasis?

Vocabulary

homeostasis • tissue • organ • organ system • receptor

Taking Notes

Preview Visuals Before you read, look at **Figures 7–22** and **7–23.** Then write two questions you have about the micrographs. As you read, write answers to your questions.

THINK ABOUT IT From its simple beginnings, life has spread to every corner of our planet, penetrating deep into the earth and far beneath the surface of the seas. The diversity of life is so great that you might have to remind yourself that all living things are composed of cells, have the same basic chemical makeup, and even contain the same kinds of organelles. This does not mean that all living things are the same: Differences arise from the ways in which cells are specialized and the ways in which cells associate with one another to form multicellular organisms.

The Cell as an Organism

How do individual cells maintain homeostasis?

Cells are the basic living units of all organisms, but sometimes a single cell is the organism. In fact, in terms of their numbers, unicellular organisms dominate life on Earth. A single-celled organism does everything you would expect a living thing to do. Just like other living things, unicellular organisms must maintain **homeostasis,** relatively constant internal physical and chemical conditions. **To maintain homeostasis, unicellular organisms grow, respond to the environment, transform energy, and reproduce.**

Unicellular organisms include both prokaryotes and eukaryotes. Prokaryotes, especially bacteria, are remarkably adaptable. Bacteria live almost everywhere—in the soil, on leaves, in the ocean, in the air, even within the human body.

Many eukaryotes, like the protozoan in **Figure 7–21,** also spend their lives as single cells. Some types of algae, which contain chloroplasts and are found in oceans, lakes, and streams around the world, are single celled. Yeasts, or unicellular fungi, are also widespread. Yeasts play an important role in breaking down complex nutrients, making them available for other organisms. People use yeasts to make bread and other foods.

Don't make the mistake of thinking that single-celled organisms are always simple. Prokaryote or eukaryote, homeostasis is still an issue for each unicellular organism. That tiny cell in a pond or on the surface of your pencil still needs to find sources of energy or food, to keep concentrations of water and minerals within certain levels, and to respond quickly to changes in its environment. The microscopic world around us is filled with unicellular organisms that are successfully maintaining that homeostatic balance.

FIGURE 7–21 Unicellular Life Single-celled organisms, like this freshwater protozoan, must be able to carry out all of the functions necessary for life (SEM 600×).

Multicellular Life

How do the cells of multicellular organisms work together to maintain homeostasis?

Unlike most unicellular organisms, the cells of human beings and other multicellular organisms do not live on their own. They are interdependent; and like the members of a winning baseball team, they work together. In baseball, each player plays a particular position: pitcher, catcher, infielder, outfielder. And to play the game effectively, players and coaches communicate with one another, sending and receiving signals. Cells in a multicellular organism work the same way. **The cells of multicellular organisms become specialized for particular tasks and communicate with one another to maintain homeostasis.**

BUILD Vocabulary

PREFIXES The prefix *homeo-* in **homeostasis** means "the same." Organisms are constantly trying to maintain homeostasis, to keep their internal physical and chemical conditions relatively constant despite changes in their internal and external environments.

Cell Specialization The cells of a multicellular organism are specialized, with different cell types playing different roles. Some cells are specialized to move; others, to react to the environment; still others, to produce substances that the organism needs. No matter what its role, each specialized cell, like the ones in **Figures 7–22** and **7–23,** contributes to homeostasis in the organism.

In Your Notebook *Where in the human body do you think you would find cells that are specialized to produce digestive enzymes? Why?*

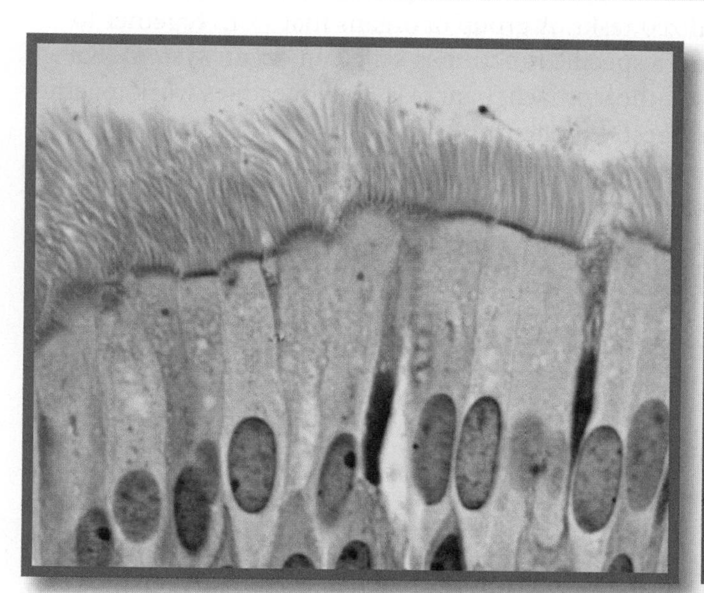

FIGURE 7–22 Specialized Animal Cells: Human Trachea Epithelium (LM 1000×)

▶ ***Specialized Animal Cells*** Even the cleanest, freshest air is dirty, containing particles of dust, smoke, and bacteria. What keeps this bad stuff from getting into your lungs? That's the job of millions of cells that work like street sweepers. These cells line the upper air passages. As you breathe, they work night and day sweeping mucus, debris, and bacteria out of your lungs. These cells are filled with mitochondria, which produce a steady supply of the ATP that powers the cilia on their upper surfaces to keep your lungs clean.

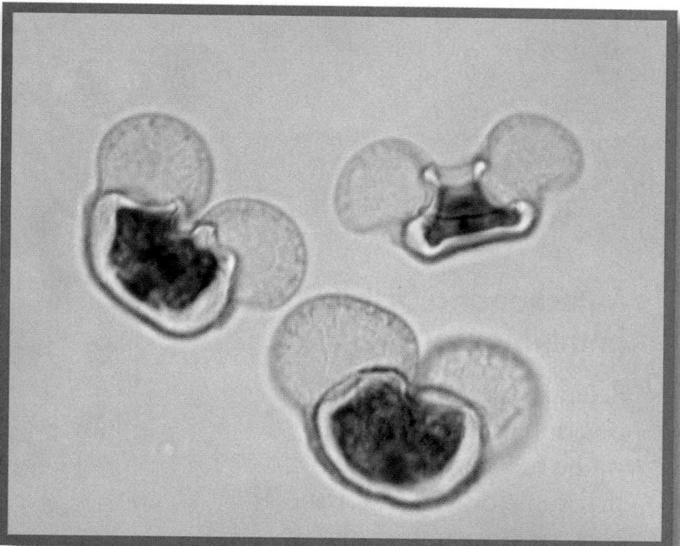

FIGURE 7–23 Specialized Plant Cells: Pine Pollen (LM 430×)

▶ ***Specialized Plant Cells*** How can a pine tree, literally rooted in place, produce offspring with another tree hundreds of meters away? It releases pollen grains, some of the world's most specialized cells. Pollen grains are tiny and light, despite tough walls to protect the cells inside. In addition, pine pollen grains have two tiny wings that enable them to float in the slightest breeze. Pine trees release millions of pollen grains like these to scatter in the wind, land on seed cones, and begin the essential work of starting a new generation.

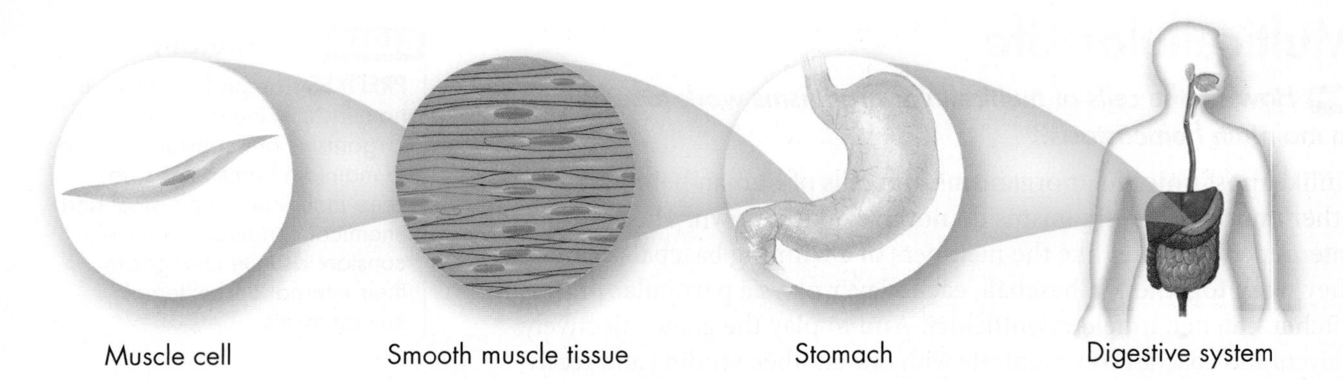

FIGURE 7–24 Levels of Organization From least complex to most complex, the levels of organization in a multicellular organism include cells, tissues, organs, and organ systems.

Levels of Organization The specialized cells of multicellular organisms are organized into tissues, then into organs, and finally into organ systems, as shown in **Figure 7–24.** A **tissue** is a group of similar cells that performs a particular function. Many tasks in the body are too complicated to be carried out by just one type of tissue. In these cases, many groups of tissues work together as an **organ.** For example, each muscle in your body is an individual organ. Within a muscle, however, there is much more than muscle tissue. There are nervous tissues and connective tissues too. Each type of tissue performs an essential task to help the organ function. In most cases, an organ completes a series of specialized tasks. A group of organs that work together to perform a specific function is called an **organ system.** For example, the stomach, pancreas, and intestines work together as the digestive system.

Analyzing Data

Mitochondria Distribution in the Mouse

Scientists studied the composition of several organs in the mouse. They found that some organs and tissues contain more mitochondria than others. They described the amount of mitochondria present as a percentage of total cell volume. The higher the percentage volume made up of mitochondria, the more mitochondria present in the cells of the organ. The data are shown in the graph.

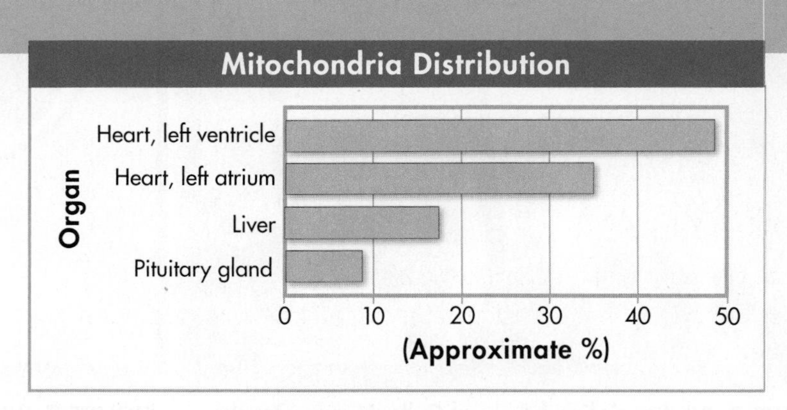

1. **Interpret Graphs** What approximate percentage of cell volume in the mouse liver is composed of mitochondria?

2. **Calculate** Approximately how much more cellular volume is composed of mitochondria in the left ventricle than in the pituitary gland? MATH

3. **Infer** There are four chambers in the mouse heart, the right and left ventricles, and the right and left atria. Based on the data given, which chamber, the left ventricle or left atrium, do you think pumps blood from the heart to the rest of the body? Explain your answer.

The organization of the body's cells into tissues, organs, and organ systems creates a division of labor among those cells that allows the organism to maintain homeostasis. Specialization and interdependence are two of the remarkable attributes of living things. Appreciating these characteristics is an important step in understanding the nature of living things.

Cellular Communication Cells in a large organism communicate by means of chemical signals that are passed from one cell to another. These cellular signals can speed up or slow down the activities of the cells that receive them and can even cause a cell to change what it is doing in a most dramatic way.

Certain cells, including those in the heart and liver, form connections, or cellular junctions, to neighboring cells. Some of these junctions, like those in **Figure 7–25,** hold cells together firmly. Others allow small molecules carrying chemical messages or signals to pass directly from one cell to the next. To respond to one of these chemical signals, a cell must have a **receptor** to which the signaling molecule can bind. Some receptors are on the cell membrane; receptors for other types of signals are inside the cytoplasm. The chemical signals sent by various types of cells can cause important changes in cellular activity. For example, the electrical signal that causes heart muscle cells to contract begins in a region of the muscle known as the pacemaker. Ions carry that electrical signal from cell to cell through a special connection known as a gap junction, enabling millions of heart muscle cells to contract as one in a single heartbeat. Other junctions hold the cells together, so the force of contraction does not tear the muscle tissue. Both types of junctions are essential for the heart to pump blood effectively.

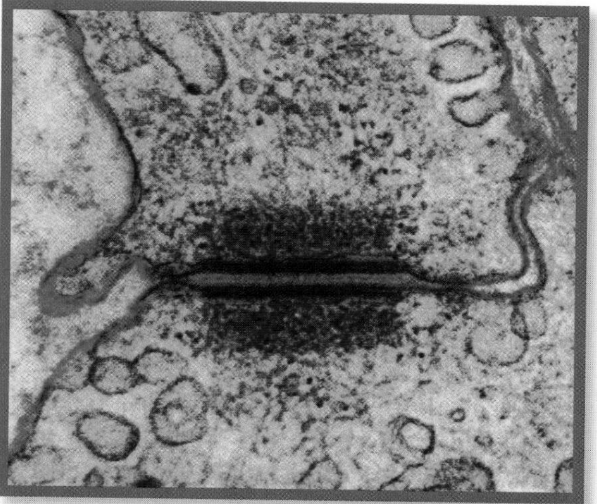

FIGURE 7–25 Cellular Junctions Some junctions, like the one seen in brown in this micrograph of capillary cells in the gas bladder of a toadfish, hold cells together in tight formations (TEM 21,600×).

7.4 Assessment

Review Key Concepts

1. a. Review What is homeostasis?
   b. Explain What do unicellular organisms do to maintain homeostasis?
   c. Apply Concepts The contractile vacuole is an organelle found in paramecia, a group of unicellular organisms. Contractile vacuoles pump out fresh water that accumulates in the organisms by osmosis. Explain how this is an example of the way paramecia maintain homeostasis.
2. a. Review What is cellular specialization?
   b. Explain How do cellular junctions and receptors help an organism maintain homeostasis?
   c. Predict Using what you know about the ways muscles move, predict which organelles would be most common in muscle cells.

WRITE ABOUT SCIENCE

Description

3. Use an area in your life—such as school, sports, or extracurricular activities—to construct an analogy that explains why specialization and communication are necessary for you to function well.

NGSS Smart Guide Cell Structure and Function

Disciplinary Core Idea LS1.A Structure and Function: How do the structures of organisms enable life's functions? Cells are the basic units of life. Their structures are specifically adapted to their function and the overall goal of maintaining homeostasis. In multicellular organisms, cells may become specialized to carry out a particular function.

7.1 Life Is Cellular

- The cell theory states that (1) all living things are made up of cells, (2) cells are the basic units of structure and function in living things, and (3) new cells are produced from existing cells.
- Most microscopes use lenses to magnify the image of an object by focusing light or electrons.
- Prokaryotic cells do not separate their genetic material within a nucleus. In eukaryotic cells, the nucleus separates the genetic material from the rest of the cell.

cell 191 • cell theory 191 • cell membrane 193 • nucleus 193 • eukaryote 193 • prokaryote 193

Biology.com

Untamed Science Video Travel to the ocean's depths with the Untamed Science crew to explore how fish maintain water homeostasis.

7.2 Cell Structure

- The nucleus contains nearly all the cell's DNA and, with it, the coded instructions for making proteins and other important molecules.
- Vacuoles store materials like water, salts, proteins, and carbohydrates. Lysosomes break down large molecules into smaller ones that can be used by the cell. They are also involved in breaking down organelles that have outlived their usefulness. The cytoskeleton helps the cell maintain its shape and is also involved in movement.
- Proteins are assembled on ribosomes.
- Proteins made on the rough ER include those that will be released from the cell as well as many membrane proteins and proteins destined for specialized locations within the cell. The Golgi apparatus then modifies, sorts, and packages proteins and other materials for storage in the cell or release outside the cell.
- Chloroplasts capture the energy from sunlight and convert it into food that contains chemical energy in a process called photosynthesis. Mitochondria convert the chemical energy stored in food into compounds that are more convenient for the cell to use.
- The cell membrane regulates what enters and leaves the cell and also protects and supports the cell.

cytoplasm 196 • organelle 196 • vacuole 198 • lysosome 198 • cytoskeleton 199 • centriole 199 • ribosome 200 • endoplasmic reticulum 200 • Golgi apparatus 201 • chloroplast 202 • mitochondrion 202 • cell wall 203 • lipid bilayer 204 • selectively permeable 205

Biology.com

Art Review Review your understanding of plant and animal cell structures with this activity.

Tutor Tube Hear suggestions from the tutor for help in remembering cell structures.

Visual Analogy Compare the structures of the cell to the parts of a factory.

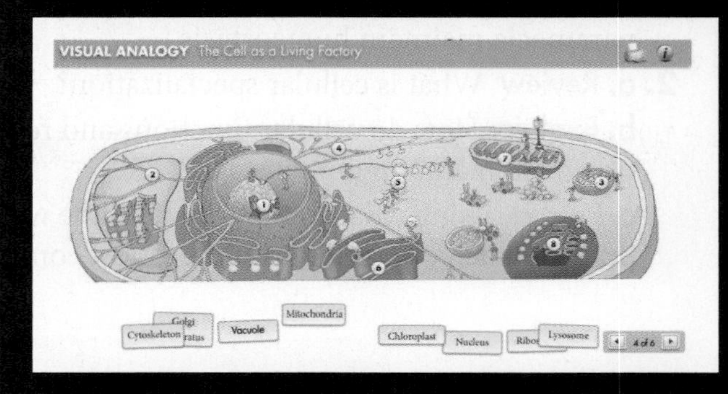

CHECKING YOUR *Scientific Literacy*

Refer to the lesson content and digital assets below as you prepare for your chapter assessment. Then, evaluate your understanding of cell structure and function by answering these questions.

1. **Scientific and Engineering Practice Developing and Using Models** Draw a diagram of the cell membrane based on the fluid mosaic model. Label the key structures of the membrane.

2. **Crosscutting Concept Structure and Function** Use your model of the cell membrane to explain how the cell regulates what enters or leaves the cell.

3. **Text Types and Purposes** Explain how water molecules move across the cell membrane in relation to solute concentration.

4. **STEM** Research how to dissolve the shells on raw chicken eggs to produce shell-less eggs. Describe how these eggs, held together only by the inner membrane, can be used to demonstrate the concepts of osmosis and homeostasis.

7.3 Cell Transport

- Passive transport (including diffusion and osmosis) is the movement of materials across the cell membrane without cellular energy.
- The movement of materials against a concentration difference is known as active transport. Active transport requires energy.

diffusion 208 • facilitated diffusion 209 • aquaporin 210 • osmosis 210 • isotonic 210 • hypertonic 210 • hypotonic 210 • osmotic pressure 211

Biology.com

Art in Motion View a short animation that explains the different types of active transport.

InterActive Art Build your understanding of osmosis and diffusion with these animations.

7.4 Homeostasis and Cells

- To maintain homeostasis, unicellular organisms grow, respond to the environment, transform energy, and reproduce.
- The cells of multicellular organisms become specialized for particular tasks and communicate with one another to maintain homeostasis.

homeostasis 214 • tissue 216 • organ 216 • organ system 216 • receptor 217

Biology.com

Data Analysis Analyze data that explain why some cell types have more mitochondria than others.

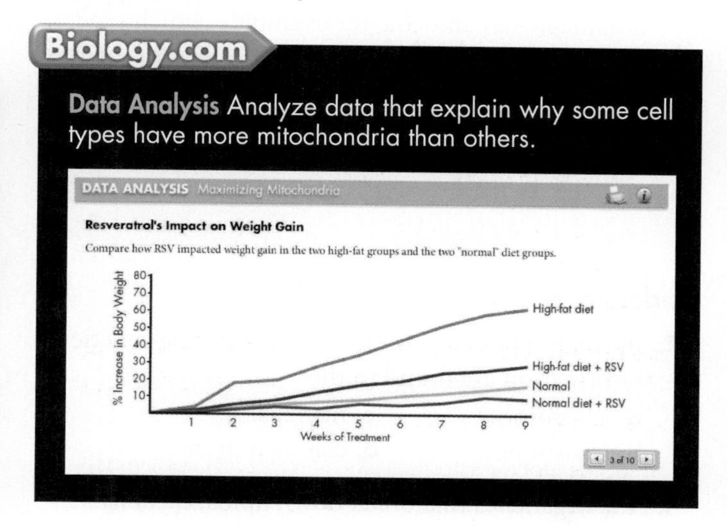

Skills Lab

Detecting Diffusion This chapter lab is available in your lab manual and at Biology.com

7 Assessment

7.1 Life Is Cellular

Understand Key Concepts

1. In many cells, the structure that controls the cell's activities is the
   a. cell membrane. c. nucleolus.
   b. organelle. d. nucleus.

2. Despite differences in size and shape, at some point all cells have DNA and a
   a. cell wall. c. mitochondrion.
   b. cell membrane. d. nucleus.

3. What distinguishes a eukaryotic cell from a prokaryotic cell is the presence of
   a. a cell wall. c. DNA.
   b. a nucleus. d. ribosomes.

4. **Integration of Knowledge and Ideas** Create a table that summarizes the contributions made to the cell theory by Robert Hooke, Matthias Schleiden, Theodor Schwann, and Rudolf Virchow.

Think Critically

5. **Apply Concepts** If you wanted to observe a living organism—an amoeba, for example—which type of microscope would you use?

6. **Craft and Structure** How are prokaryotic and eukaryotic cells alike? How do they differ?

7.2 Cell Structure

Understand Key Concepts

7. In eukaryotic cells, chromosomes carrying genetic information are found in the
   a. ribosomes. c. nucleus.
   b. lysosomes. d. cell membrane.

8. The organelles that break down lipids, carbohydrates, and proteins into small molecules that can be used by the cell are called
   a. vacuoles.
   b. lysosomes.
   c. ribosomes.
   d. microfilaments.

9. Cell membranes contain a central bilayer formed by
   a. lipids. c. carbohydrates.
   b. protein pumps. d. proteins.

10. Draw a cell nucleus. Label and give the function of the following structures: chromatin, nucleolus, and nuclear envelope.

11. What is the function of a ribosome?

12. Describe the role of the Golgi apparatus.

Think Critically

13. **Infer** The pancreas, an organ present in certain animals, produces enzymes used elsewhere in the animals' digestive systems. Which type of cell structure(s) might produce those enzymes? Explain your answer.

14. **Craft and Structure** For each of the following, indicate if the structure is found only in eukaryotes, or if it is found in eukaryotes and prokaryotes: cell membrane, mitochondria, ribosome, Golgi apparatus, nucleus, cytoplasm, and DNA.

7.3 Cell Transport

Understand Key Concepts

15. The movement of water molecules across a selectively permeable membrane is known as
    a. exocytosis. c. endocytosis.
    b. phagocytosis. d. osmosis.

16. A substance that moves by passive transport tends to move
    a. away from the area of equilibrium.
    b. away from the area where it is less concentrated.
    c. away from the area where it is more concentrated.
    d. toward the area where it is more concentrated.

17. **Text Types and Purposes** Describe the process of diffusion, including a detailed explanation of equilibrium.

18. **Craft and Structure** What is the relationship between diffusion and osmosis? By definition, what's the only substance that undergoes osmosis?

19. **Craft and Structure** What is the difference between passive transport and active transport?

Think Critically

20. **Predict** The beaker in the diagram below has a selectively permeable membrane separating two solutions. Assume that both water and salt can pass freely through the membrane. When equilibrium is reached, will the fluid levels be the same as they are now? Explain.

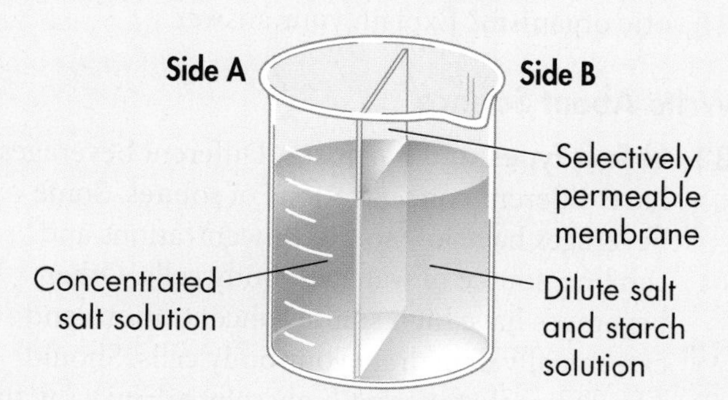

21. **Craft and Structure** What would happen to a sample of your red blood cells if they were placed in a hypotonic solution? Explain.

22. **Design Experiments** You are given food coloring and three beakers. The first beaker contains water at room temperature, the second beaker contains ice water, and the third beaker contains hot water. Design a controlled experiment to determine the effects of temperature on the rate of diffusion. Be sure to state your hypothesis.

7.4 Homeostasis and Cells

Understand Key Concepts

23. Which of the following is true of ALL single-celled organisms?
    a. They are all prokaryotes.
    b. They are all bacteria.
    c. They all reproduce.
    d. They all have a nucleus.

24. A tissue is composed of a group of
    a. similar cells.
    b. related organelles.
    c. organ systems.
    d. related organs.

25. **Craft and Structure** Explain the relationship among cell specialization, multicellular organisms, and homeostasis.

26. **Craft and Structure** Describe the relationship among cells, tissues, organs, and organ systems.

solve the CHAPTER MYSTERY

DEATH BY ... WATER?

During the race, Michelle drank plenty of water, but she didn't replace the salts she lost due to sweating. As a result, her blood became hypotonic, and osmotic pressure led the cells in her brain (and throughout her body) to swell.

As Michelle's blood became more dilute, cells in her brain sent chemical signals to her kidneys to stop removing sodium chloride and other salts from her bloodstream. However, as she continued to sweat, she continued to lose salt through her skin.

By the end of the race, Michelle had lost a large quantity of salt and minerals and had taken in so much water that homeostasis had broken down, and her cells were damaged by unregulated osmotic pressure.

When Michelle was rushed to the hospital, the doctors discovered that she was suffering from hyponatremia, or water intoxication. Left untreated, this condition can lead to death.

1. **Relate Cause and Effect** When a person sweats, water and essential solutes called electrolytes are lost from body fluid. Michelle drank lots of water but did not replace lost electrolytes. What effect did this have on her cells?
2. **Infer** Had Michelle alternated between drinking water and a sports drink with electrolytes would her condition be the same?
3. **Infer** Do you think that hyponatremia results from osmosis or active transport? Explain your reasoning.
4. **Connect to the Big idea** Explain how hyponatremia disrupts homeostasis in the body.

Think Critically

27. Infer Would you expect skin cells to contain more or fewer mitochondria than muscle cells? Explain your answer.

28. Infer Pacemakers are devices that help keep heart muscle cells contracting at a steady rate. If a person needs a pacemaker, what does that suggest about his or her heart cells' ability to send and receive chemical messages?

Connecting Concepts

Use Science Graphics

Use the data table to answer questions 29–31.

Cell Sizes

Cell	Approximate Diameter
Escherichia coli (bacterium)	0.5–0.8 μm
Human erythrocyte (red blood cell)	6–8 μm
Human ovum (egg cell)	100 μm
Saccharomyces cerevisiae (yeast)	5–10 μm
Streptococcus pneumoniae (bacterium)	0.5–1.3 μm

29. Classify Classify each of the cells listed as prokaryotic or eukaryotic.

30. Compare and Contrast Compare the sizes of the prokaryotic cells and eukaryotic cells.

31. Infer *Chlamydomonas reinhardtii* is a single-celled organism with an approximate diameter of 10 μm. Is it more likely a prokaryotic or eukaryotic organism? Explain your answer.

Write About Science

32. Text Types and Purposes Different beverages have different concentrations of solutes. Some beverages have low solute concentrations and can be a source of water for body cells. Other beverages have high solute concentrations and can actually dehydrate your body cells. Should companies that market high-solute drinks say that the drinks quench thirst? Write an argument to support your claim.

33. Assess the Big idea What is the relationship between active transport and homeostasis? Give one example of active transport in an organism, and explain how the organism uses energy to maintain homeostasis.

Integration of Knowledge and Ideas

Most materials entering the cell pass across the cell membrane by diffusion. In general, the larger the molecule, the slower the molecule diffuses across the membrane. The graph shows the sizes of several molecules that can diffuse across a lipid bilayer.

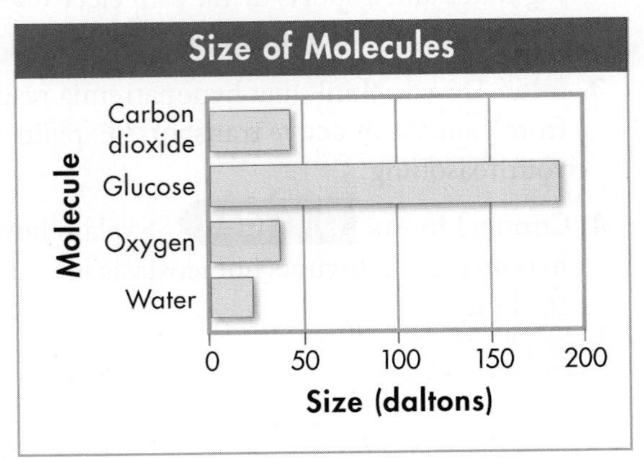

34. Calculate By approximately what percentage is a molecule of carbon dioxide smaller than a molecule of glucose? MATH

a. 25% **b.** 50% **c.** 75% **d.** 100%

35. Formulate Hypotheses Which of the following is a logical hypothesis based on the graph shown?

a. Cells contain more glucose than oxygen.

b. Oxygen molecules diffuse across the cell membrane faster than water molecules.

c. Glucose molecules must cross the cell membrane by active transport.

d. Carbon dioxide crosses the cell membrane faster than glucose.

Standardized Test Prep

Multiple Choice

1. Animal cells have all of the following EXCEPT
A mitochondria.
B chloroplasts.
C a nucleus.
D a cell membrane.

2. The nucleus includes all of the following structures EXCEPT
A cytoplasm. **C** DNA.
B a nuclear envelope. **D** a nucleolus.

3. The human brain is an example of a(n)
A cell.
B tissue.
C organ.
D organ system.

4. Which cell structures are sometimes found attached to the endoplasmic reticulum?
A chloroplasts
B nuclei
C mitochondria
D ribosomes

5. Which process always involves the movement of materials from inside the cell to outside the cell?
A phagocytosis
B exocytosis
C endocytosis
D osmosis

6. Which of the following is an example of active transport?
A facilitated diffusion
B osmosis
C diffusion
D endocytosis

7. The difference between prokaryotic and eukaryotic cells involves the presence of
A a nucleus.
B genetic material in the form of DNA.
C chloroplasts.
D a cell membrane.

Questions 8–10

In an experiment, plant cells were placed in sucrose solutions of varying concentrations, and the rate at which they absorbed sucrose from the solution was measured. The results are shown in the graph below.

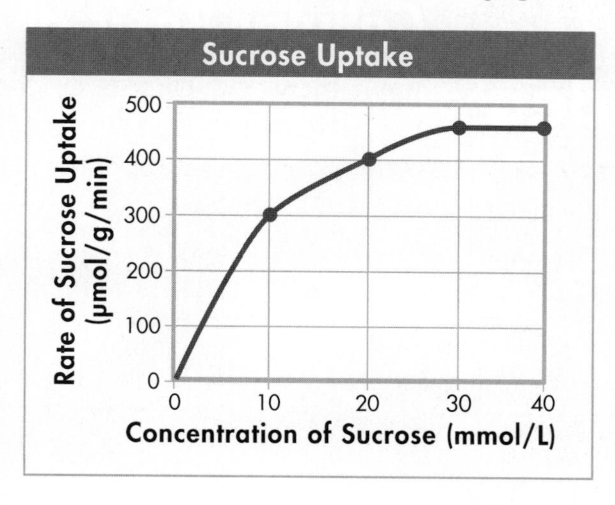

8. In this experiment, sucrose probably entered the cells by means of
A endocytosis. **C** osmosis.
B phagocytosis. **D** active transport.

9. The graph shows that as the concentration of sucrose increased from 10 to 30 mmol/L, the plant cells
A took in sucrose more slowly.
B took in sucrose more quickly.
C failed to take in more sucrose.
D secreted sucrose more slowly.

10. Based on the graph, the rate of sucrose uptake
A increased at a constant rate from 0 to 30 mmol/L.
B decreased at varying rates from 0 to 30 mmol/L.
C was less at 25 mmol/L than at 5 mmol/L.
D was constant between 30 and 40 mmol/L.

Open-Ended Response

11. What would you expect to happen if you placed a typical cell in fresh water?

If You Have Trouble With . . .

Question	1	2	3	4	5	6	7	8	9	10	11
See Lesson	7.2	7.2	7.4	7.2	7.3	7.3	7.1	7.3	7.3	7.3	7.3

Photosynthesis

Cellular Basis of Life

Q: How do plants and other organisms capture energy from the sun?

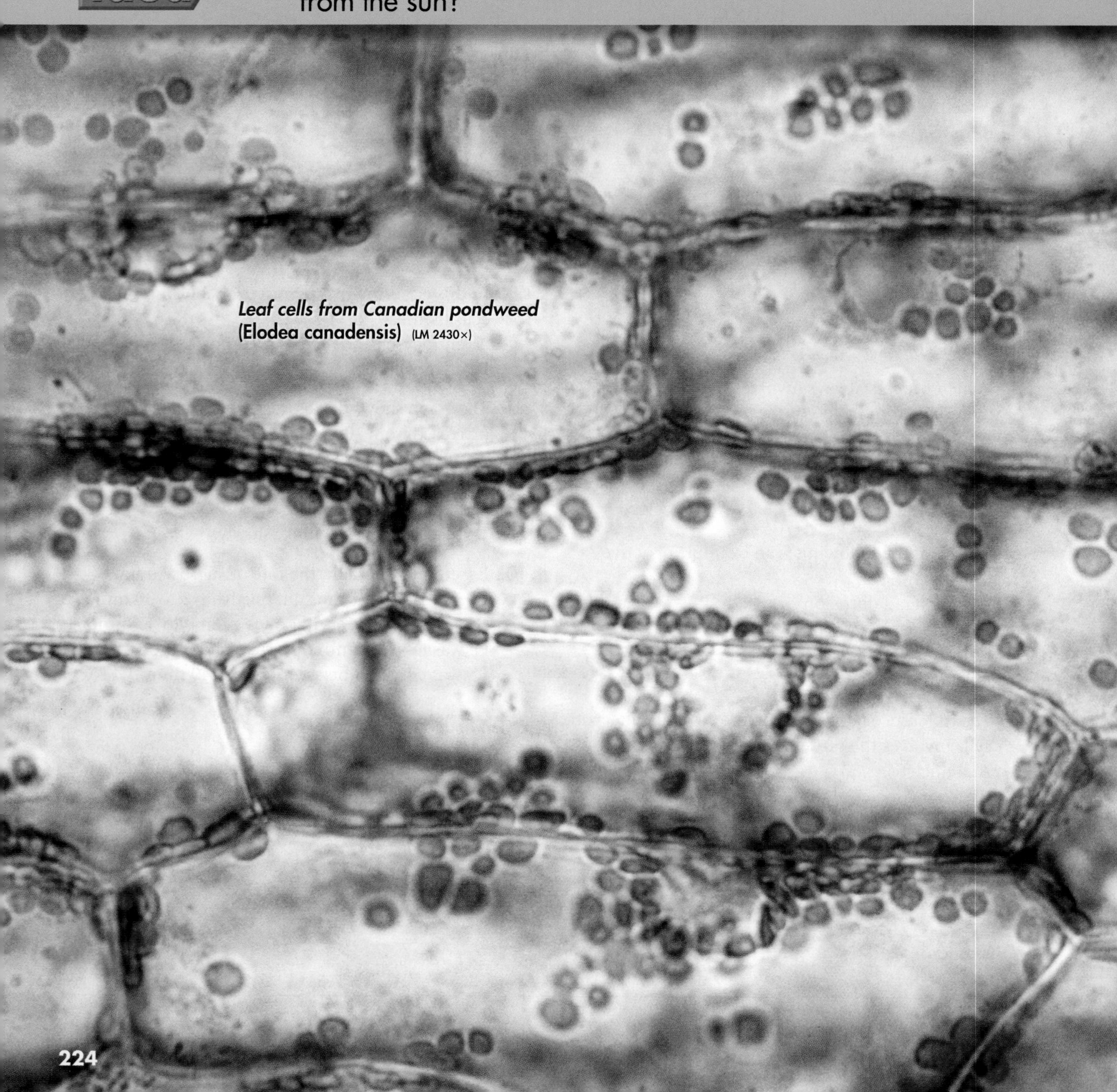

Leaf cells from Canadian pondweed **(Elodea canadensis)** (LM 2430×)

INSIDE:

- 8.1 Energy and Life
- 8.2 Photosynthesis: An Overview
- 8.3 The Process of Photosynthesis

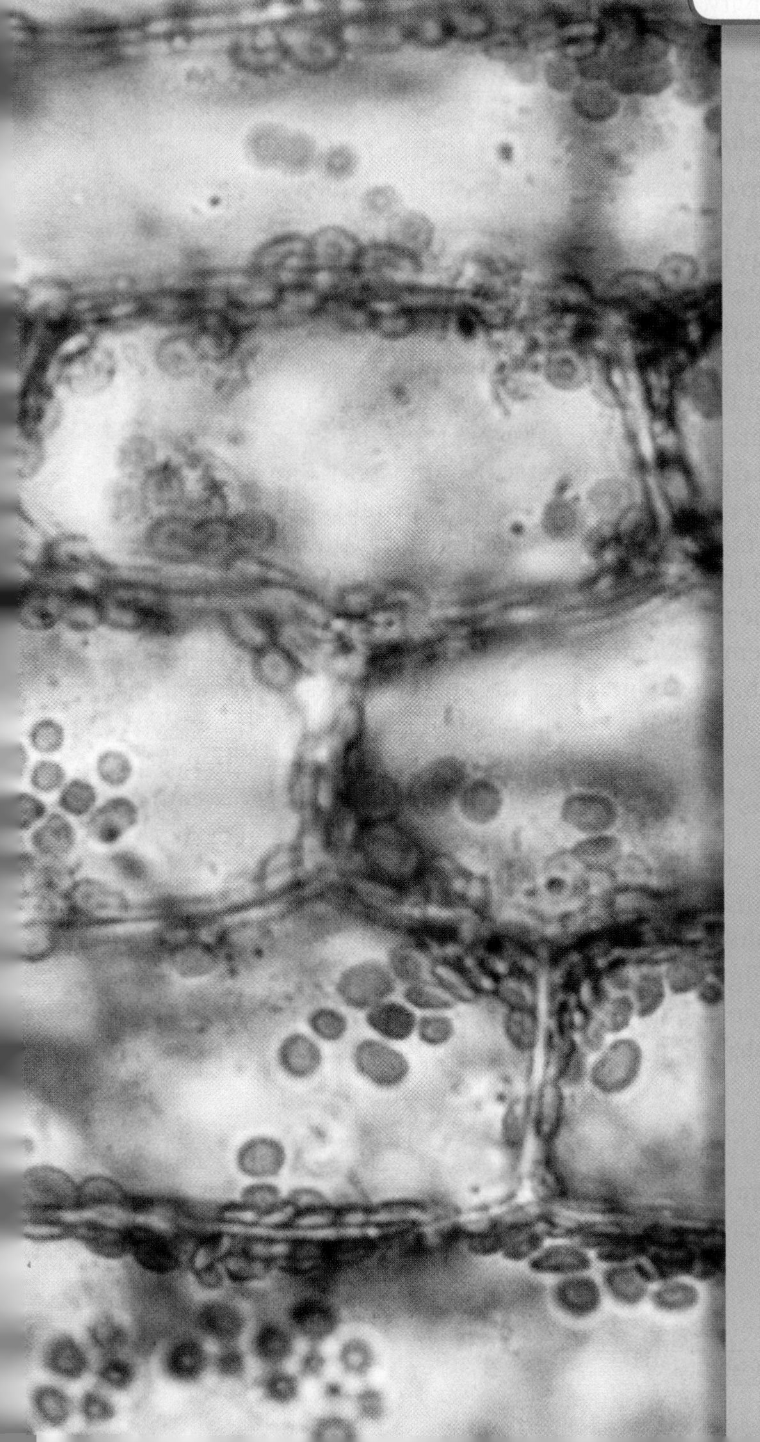

BUILDING *Scientific Literacy*

Deepen your understanding of photosynthesis by engaging in key practices that allow you to make connections across concepts.

NGSS You will use models to demonstrate how energy is converted and transferred during the process of photosynthesis.

STEM How can the conditions for plants growing in growth chambers be optimized? Analyze data on factors affecting photosynthesis and communicate your findings.

Common Core You will integrate knowledge to explain the reactions that occur during photosynthesis.

CHAPTER MYSTERY

OUT OF THIN AIR?

One of the earliest clues as to how photosynthesis works came from a simple study of plant growth. When a tiny seed grows into a massive tree, where does all its extra mass come from? More than 300 years ago, a Flemish physician named Jan van Helmont decided to find out. He planted a young willow tree, with a mass of just 2 kilograms, in a pot with 90 kilograms of dry soil. He watered the plant as needed and allowed it to grow in bright sunlight. Five years later, he carefully removed the tree from the pot and weighed it. It had a mass of about 77 kilograms. Where did the extra 75 kilograms come from? The soil, the water—or, maybe, right out of thin air? As you read this chapter, look for clues to help you discover where the willow tree's extra mass came from. Then, solve the mystery.

Biology.com

Understanding Jan van Helmont's experiments is just the beginning. Take a video field trip with the ecogeeks of Untamed Science to see where this mystery leads.

Go online to access additional resources including:
• eText • Flash Cards • Lesson Overviews • Chapter Mystery

8.1 Energy and Life

Key Questions

Why is ATP useful to cells?

What happens during the process of photosynthesis?

Vocabulary

adenosine triphosphate (ATP) • heterotroph • autotroph • photosynthesis

Taking Notes

Compare/Contrast Table As you read, create a table that compares autotrophs and heterotrophs. Think about how they obtain energy, and include a few examples of each.

BUILD Vocabulary

ACADEMIC WORDS The verb **obtain** means "to get" or "to gain." Organisms must obtain energy in order to carry out life functions.

THINK ABOUT IT Homeostasis is hard work. Just to stay alive, organisms and the cells within them have to grow and develop, move materials around, build new molecules, and respond to environmental changes. Plenty of energy is needed to accomplish all this work. What powers so much activity, and where does that power come from?

Chemical Energy and ATP

Why is ATP useful to cells?

Energy is the ability to do work. Nearly every activity in modern society depends upon energy. When a car runs out of fuel—more precisely, out of the chemical energy in gasoline—it comes to a sputtering halt. Without electrical energy, lights, appliances, and computers stop working. Living things depend on energy, too. Sometimes the need for energy is easy to see. It takes plenty of energy to play soccer or other sports. However, there are times when that need is less obvious. Even when you are sleeping, your cells are quietly busy using energy to build new molecules, contract muscles, and carry out active transport. Simply put, without the ability to **obtain** and use energy, life would cease to exist.

Energy comes in many forms, including light, heat, and electricity. Energy can be stored in chemical compounds, too. For example, when you light a candle, the wax melts, soaks into the wick, and is burned. As the candle burns, chemical bonds between carbon and hydrogen atoms in the wax are broken. New bonds then form between these atoms and oxygen, producing CO_2 and H_2O (carbon dioxide and water). These new bonds are at a lower energy state than the original chemical bonds in the wax. The energy lost is released as heat and light in the glow of the candle's flame.

Living things use chemical fuels as well. One of the most important compounds that cells use to store and release energy is **adenosine triphosphate** (uh DEN uh seen try FAHS fayt), abbreviated **ATP.** As shown in **Figure 8–1,** ATP consists of adenine, a 5-carbon sugar called ribose, and three phosphate groups. As you'll see, those phosphate groups are the key to ATP's ability to store and release energy.

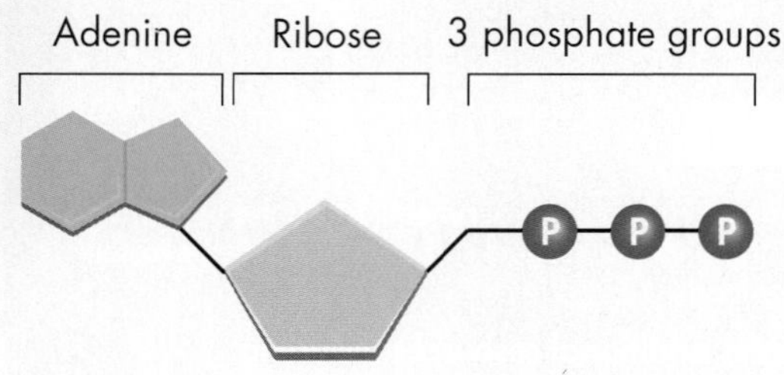

FIGURE 8–1 ATP ATP is the basic energy source used by all types of cells.

Storing Energy Adenosine diphosphate (ADP) is a compound that looks almost like ATP, except that it has two phosphate groups instead of three. This difference is the key to the way in which living things store energy. When a cell has energy available, it can store small amounts of it by adding phosphate groups to ADP molecules, producing ATP. As seen in **Figure 8–2,** ADP is like a rechargeable battery that powers the machinery of the cell.

Releasing Energy Cells can release the energy stored in ATP by the controlled breaking of the chemical bonds between the second and third phosphate groups. Because a cell can add or subtract these phosphate groups, it has an efficient way of storing and releasing energy as needed. **ATP can easily release and store energy by breaking and re-forming the bonds between its phosphate groups. This characteristic of ATP makes it exceptionally useful as a basic energy source for all cells.**

Using Biochemical Energy One way cells use the energy provided by ATP is to carry out active transport. Many cell membranes contain sodium-potassium pumps, membrane proteins that pump sodium ions (Na^+) out of the cell and potassium ions (K^+) into it. ATP provides the energy that keeps this pump working, maintaining a carefully regulated balance of ions on both sides of the cell membrane. In addition, ATP powers movement, providing the energy for motor proteins that contract muscle and power the wavelike movement of cilia and flagella.

Energy from ATP powers other important events in the cell, including the synthesis of proteins and responses to chemical signals at the cell surface. The energy from ATP can even be used to produce light. In fact, the blink of a firefly on a summer night comes from an enzyme that is powered by ATP!

ATP is such a useful source of energy that you might think cells would be packed with ATP to get them through the day—but this is not the case. In fact, most cells have only a small amount of ATP—enough to last for a few seconds of activity. Why? Even though ATP is a great molecule for transferring energy, it is not a good one for storing large amounts of energy over the long term. A single molecule of the sugar glucose, for example, stores more than 90 times the energy required to add a phosphate group to ADP to produce ATP. Therefore, it is more efficient for cells to keep only a small supply of ATP on hand. Instead, cells can regenerate ATP from ADP as needed by using the energy in foods like glucose. As you will see, that's exactly what they do.

In Your Notebook *With respect to energy, how are ATP and glucose similar? How are they different?*

VISUAL ANALOGY

ATP AS A CHARGED BATTERY

FIGURE 8–2 When a phosphate group is added to an ADP molecule, ATP is produced. ADP contains some energy, but not as much as ATP. In this way, ADP is like a partially charged battery that can be fully charged by the addition of a phosphate group.
Use Analogies ***Explain the difference between the beams of light produced by the flashlight "powered" by ADP and the flashlight "powered" by ATP.***

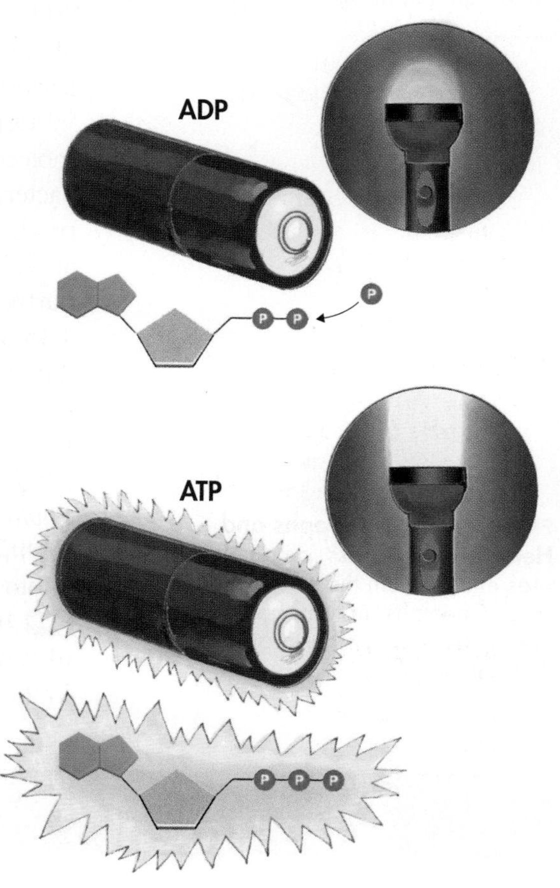

MYSTERY CLUE

Like all plants, the willow tree van Helmont planted was an autotroph. What might its ability to harness the sun's energy and store it in food have to do with the tree's gain in mass?

Heterotrophs and Autotrophs

What happens during the process of photosynthesis?

Cells are not "born" with a supply of ATP—they must somehow produce it. So, where do living things get the energy they use to produce ATP? The simple answer is that it comes from the chemical compounds that we call food. Organisms that obtain food by consuming other living things are known as **heterotrophs.** Some heterotrophs get their food by eating plants such as grasses. Other heterotrophs, such as the cheetah in **Figure 8–3,** obtain food from plants indirectly by feeding on plant-eating animals. Still other heterotrophs—mushrooms, for example—obtain food by absorbing nutrients from decomposing organisms in the environment.

Originally, however, the energy in nearly all food molecules comes from the sun. Plants, algae, and some bacteria are able to use light energy from the sun to produce food. Organisms that make their own food are called **autotrophs.** Ultimately, nearly all life on Earth, including ourselves, depends on the ability of autotrophs to capture the energy of sunlight and store it in the molecules that make up food. The process by which autotrophs use the energy of sunlight to produce high-energy carbohydrates—sugars and starches—that can be used as food is known as **photosynthesis.** *Photosynthesis* comes from the Greek words *photo*, meaning "light," and *synthesis*, meaning "putting together." Therefore, photosynthesis means "using light to put something together." **In the process of photosynthesis, plants convert the energy of sunlight into chemical energy stored in the bonds of carbohydrates.** In the rest of this chapter, you will learn how this process works.

FIGURE 8–3 Autotrophs and Heterotrophs Grass, an autotroph, uses energy from the sun to produce food. Cheetahs, in turn, get their energy by eating other organisms that eat the grass.

8.1 Assessment

Review Key Concepts

1. a. **Review** What is ATP and what is its role in the cell?
   b. **Explain** How does the structure of ATP make it an ideal source of energy for the cell?
   c. **Use Analogies** Explain how ADP and ATP are each like a battery. Which one is "partially charged" and which one is "fully charged?" Why?
2. a. **Review** What is the ultimate source of energy for plants?
   b. **Explain** How do heterotrophs obtain energy? How is this different from how autotrophs obtain energy?
   c. **Infer** Why are decomposers, such as mushrooms, considered heterotrophs and not autotrophs?

Apply the Big idea

Interdependence in Nature

3. Recall that energy flows—and that nutrients cycle—through the biosphere. How does the process of photosynthesis impact both the flow of energy and the cycling of nutrients? You may wish to refer to Chapter 3 to help you answer this question.

Biology & HISTORY

WHST.9-10.8 Research to Build and Present Knowledge, WHST.9-10.2b Text Types and Purposes. Also WHST.9-10.10

Understanding Photosynthesis **Many scientists have contributed to understanding how plants carry out photosynthesis. Early research focused on the overall process. Later, researchers investigated the detailed chemical pathways.**

1650 1700 1750 1800 1850 1900 1950 2000

1643

▲ After analyzing his measurements of a willow tree's water intake and mass increase, Jan van Helmont concludes that trees gain most of their mass from water.

1771

Joseph Priestley experiments with a bell jar, a candle, and a plant and concludes that the plant releases oxygen. ▼

1779

Jan Ingenhousz finds that aquatic plants produce oxygen bubbles in the light but not in the dark. He concludes that plants need sunlight to produce oxygen. ▼

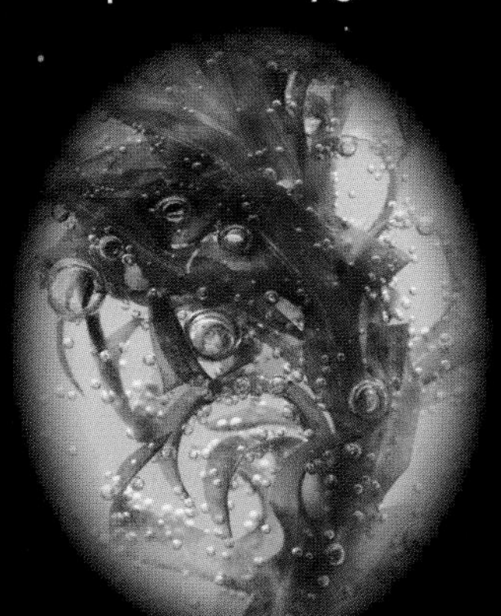

1845

Julius Robert Mayer proposes that plants convert light energy into chemical energy.

1948

Melvin Calvin traces the chemical path that carbon follows to form glucose. These reactions are also known as the Calvin cycle.

1992

Rudolph Marcus wins the Nobel Prize in chemistry for describing the process by which electrons are transferred from one molecule to another in the electron transport chain.

2004

▲ So Iwata and Jim Barber identify the precise mechanism by which water molecules are split in the process of photosynthesis. Their research may one day be applied to artificial photosynthesis technologies in order to produce a cheap supply of hydrogen gas that can be used as fuel.

WRITING **Use the Internet or library resources to research the historical experiments conducted by one of these scientists. Then, write a summary describing how the scientist contributed to the modern understanding of photosynthesis.**

8.2 Photosynthesis: An Overview

Key Questions

What role do pigments play in the process of photosynthesis?

What are electron carrier molecules?

What are the reactants and products of photosynthesis?

Vocabulary

pigment • chlorophyll • thylakoid • stroma • NADP$^+$ • light-dependent reactions • light-independent reactions

Taking Notes

Outline Make an outline using the green and blue headings in this lesson. Fill in details as you read to help you organize the information.

THINK ABOUT IT How would you design a system to capture the energy of sunlight and convert it into a useful form? First, you'd have to collect that energy. Maybe you'd spread out lots of flat panels to catch the light. You might then coat the panels with light-absorbing compounds, but what then? How could you take the energy, trapped ever so briefly in these chemical compounds, and get it into a stable, useful, chemical form? Solving such problems may well be the key to making solar power a practical energy alternative. But plants have already solved all these issues on their own terms—and maybe we can learn a trick or two from them.

Chlorophyll and Chloroplasts

What role do pigments play in the process of photosynthesis?

Our lives, and the lives of nearly every living thing on the surface of Earth, are made possible by the sun and the process of photosynthesis. In order for photosynthesis to occur, light energy from the sun must somehow be captured.

Light Energy from the sun travels to Earth in the form of light. Sunlight, which our eyes perceive as "white" light, is actually a mixture of different wavelengths. Many of these wavelengths are visible to our eyes and make up what is known as the visible spectrum. Our eyes see the different wavelengths of the visible spectrum as different colors: shades of red, orange, yellow, green, blue, indigo, and violet.

FIGURE 8–4 Light Absorption

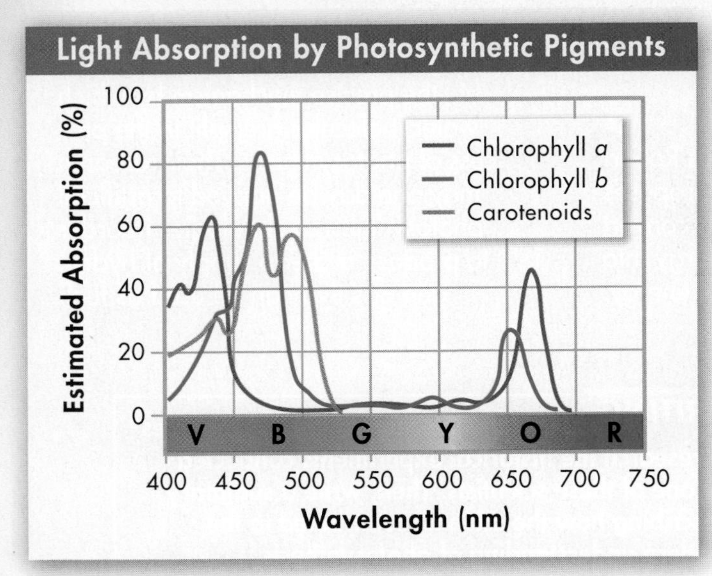

Pigments Plants gather the sun's energy with light-absorbing molecules called **pigments.** **Photosynthetic organisms capture energy from sunlight with pigments.** The plants' principal pigment is **chlorophyll** (KLAWR uh fil). The two types of chlorophyll found in plants, chlorophyll *a* and chlorophyll *b*, absorb light very well in the blue-violet and red regions of the visible spectrum. However, chlorophyll does not absorb light well in the green region of the spectrum, as shown in **Figure 8–4.**

Leaves reflect green light, which is why plants look green. Plants also contain red and orange pigments such as carotene that absorb light in other regions of the spectrum. Most of the time, the intense green color of chlorophyll overwhelms the accessory pigments, so we don't notice them. As temperatures drop late in the year, however, chlorophyll molecules break down first, leaving the reds and oranges of the accessory pigments for all to see. The beautiful colors of fall in some parts of the country are the result of this process.

Chloroplasts Recall from Chapter 7 that in plants and other photosynthetic eukaryotes, photosynthesis takes place inside organelles called chloroplasts. Chloroplasts contain an abundance of saclike photosynthetic membranes called **thylakoids** (THY luh koydz). Thylakoids are interconnected and arranged in stacks known as grana (singular: granum). Pigments such as chlorophyll are located in the thylakoid membranes. The fluid portion of the chloroplast, outside of the thylakoids, is known as the **stroma.** The structure of a typical chloroplast is shown in **Figure 8–5.**

Energy Collection What's so special about chlorophyll that makes it important for photosynthesis? Because light is a form of energy, any compound that absorbs light absorbs energy. Chlorophyll absorbs visible light especially well. In addition, when chlorophyll absorbs light, a large fraction of that light energy is transferred directly to electrons in the chlorophyll molecule itself. By raising the energy levels of these electrons, light energy can produce a steady supply of high-energy electrons, which is what makes photosynthesis work.

In Your Notebook *In your own words, explain why most plants will not grow well if kept under green light.*

ZOOMING IN

THE CHLOROPLAST

FIGURE 8–5 In plants, photosynthesis takes place inside chloroplasts. **Observe** ***How are thylakoids arranged in the chloroplast?***

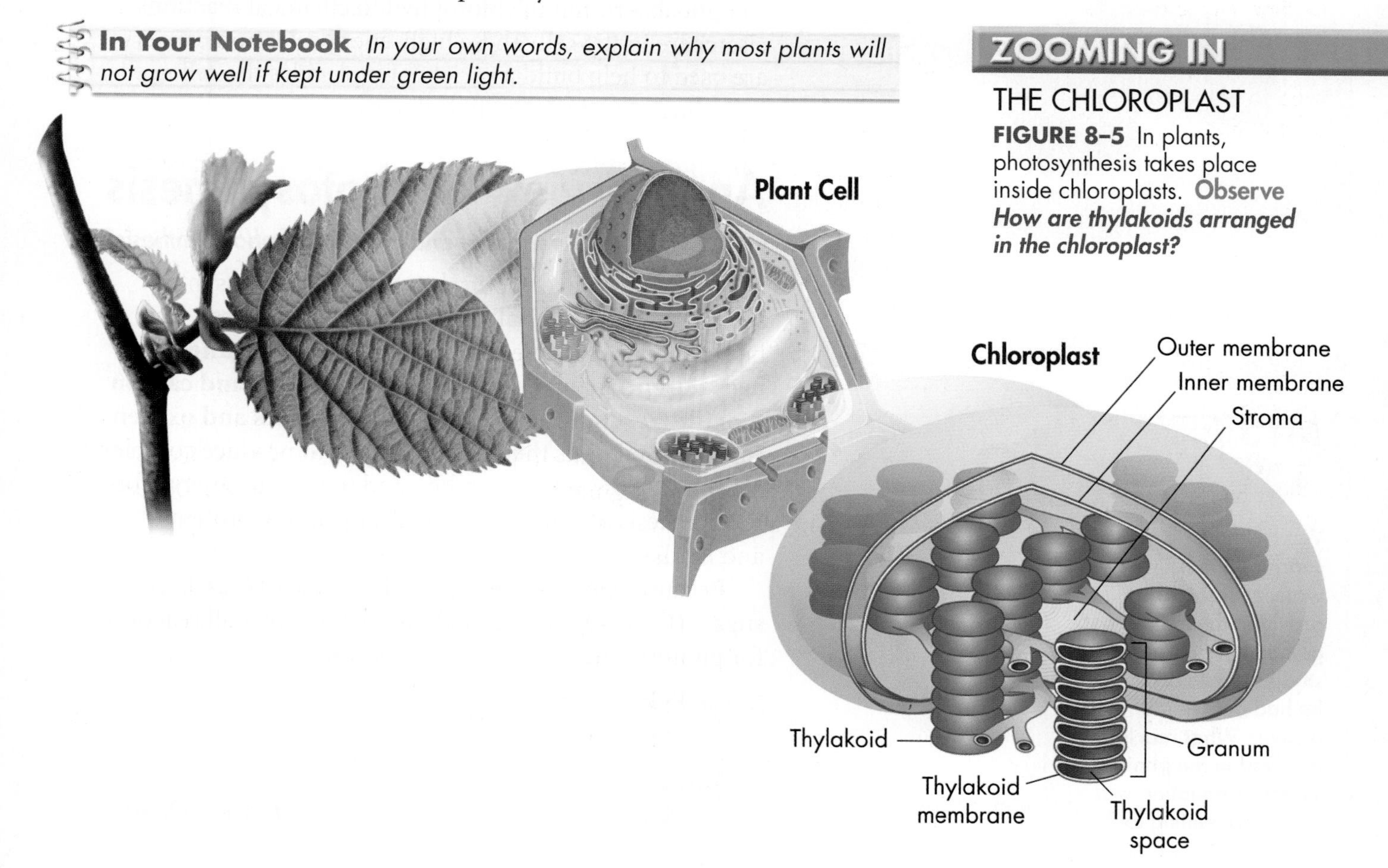

VISUAL ANALOGY

CARRYING ELECTRONS

FIGURE 8–6 $NADP^+$ is a carrier molecule that transports pairs of electrons (and an H^+ ion) in photosynthetic organisms, similar to how an oven mitt is used to transport a hot object such as a baked potato.

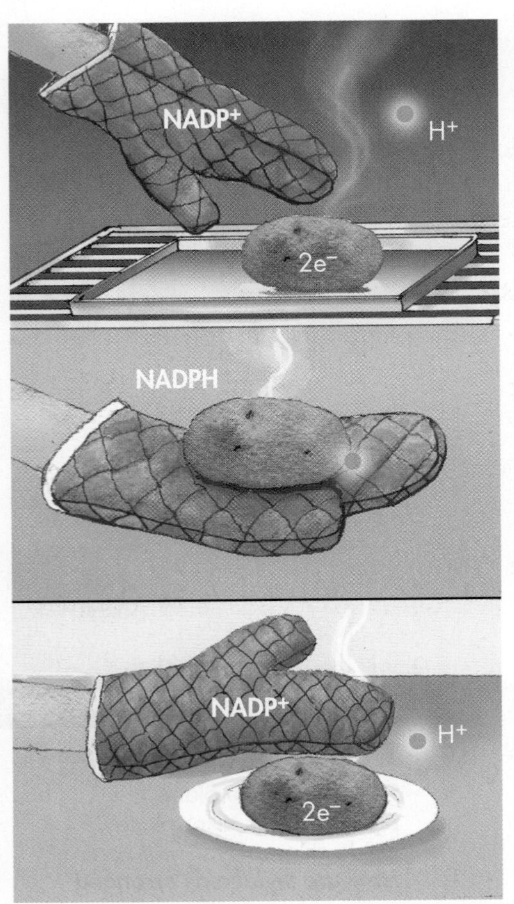

MYSTERY CLUE

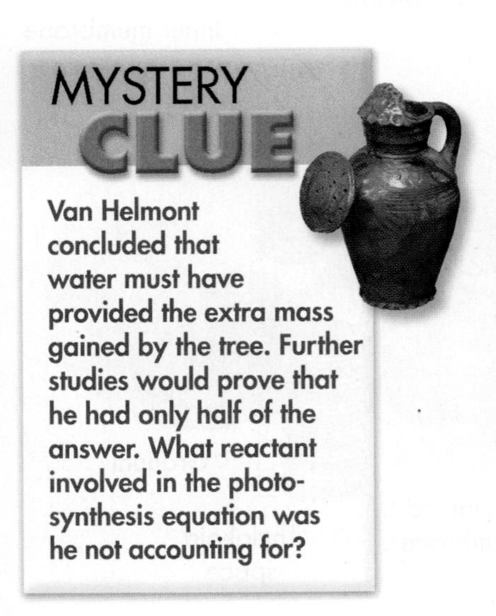

Van Helmont concluded that water must have provided the extra mass gained by the tree. Further studies would prove that he had only half of the answer. What reactant involved in the photosynthesis equation was he not accounting for?

High-Energy Electrons

What are electron carrier molecules?

In a chemical sense, the high-energy electrons produced by chlorophyll are highly reactive and require a special "carrier." Think of a high-energy electron as being similar to a hot potato straight from the oven. If you wanted to move the potato from one place to another, you wouldn't pick it up in your hands. You would use an oven mitt—a carrier—to transport it, as shown in **Figure 8–6.** Plant cells treat high-energy electrons in the same way. Instead of an oven mitt, however, they use electron carriers to transport high-energy electrons from chlorophyll to other molecules. **An electron carrier is a compound that can accept a pair of high-energy electrons and transfer them, along with most of their energy, to another molecule.**

One of these carrier molecules is a compound known as **$NADP^+$** (nicotinamide adenine dinucleotide phosphate). The name is complicated, but the job that $NADP^+$ has is simple. $NADP^+$ accepts and holds 2 high-energy electrons, along with a hydrogen ion (H^+). This converts the $NADP^+$ into NADPH. The conversion of $NADP^+$ into NADPH is one way in which some of the energy of sunlight can be trapped in chemical form. The NADPH can then carry the high-energy electrons that were produced by light absorption in chlorophyll to chemical reactions elsewhere in the cell. These high-energy electron carriers are used to help build a variety of molecules the cell needs, including carbohydrates like glucose.

An Overview of Photosynthesis

What are the reactants and products of photosynthesis?

Many steps are involved in the process of photosynthesis. However, the overall process of photosynthesis can be summarized in one sentence. **Photosynthesis uses the energy of sunlight to convert water and carbon dioxide (reactants) into high-energy sugars and oxygen (products).** Plants then use the sugars to produce complex carbohydrates such as starches, and to provide energy for the synthesis of other compounds, including proteins and lipids.

Because photosynthesis usually produces 6-carbon sugars ($C_6H_{12}O_6$) as the final product, the overall reaction for photosynthesis can be shown as follows:

In Symbols:

$$6CO_2 + 6H_2O \xrightarrow{\text{light}} C_6H_{12}O_6 + 6O_2$$

In Words:

$$\text{Carbon dioxide + Water} \xrightarrow{\text{light}} \text{Sugars + Oxygen}$$

Light-Dependent Reactions Although the equation for photosynthesis looks simple, there are many steps to get from the reactants to the final products. In fact, photosynthesis actually involves two sets of reactions. The first set of reactions is known as the **light-dependent reactions** because they require the direct involvement of light and light-absorbing pigments. The light-dependent reactions use energy from sunlight to produce energy-rich compounds such as ATP. These reactions take place within the thylakoids—specifically, in the thylakoid membranes—of the chloroplast. Water is required in these reactions as a source of electrons and hydrogen ions. Oxygen is released as a byproduct.

Light-Independent Reactions Plants absorb carbon dioxide from the atmosphere and complete the process of photosynthesis by producing carbon-containing sugars and other carbohydrates. During the **light-independent reactions,** ATP and NADPH molecules produced in the light-dependent reactions are used to produce high-energy sugars from carbon dioxide. As the name implies, no light is required to power the light-independent reactions. The light-independent reactions take place outside the thylakoids, in the stroma.

The interdependent relationship between the light-dependent and light-independent reactions is shown in **Figure 8–7.** As you can see, the two sets of reactions work together to capture the energy of sunlight and transform it into energy-rich compounds such as carbohydrates.

In Your Notebook *Create a two-column compare/contrast table that shows the similarities and differences between the light-dependent and light-independent reactions of photosynthesis.*

BUILD Vocabulary

ACADEMIC WORDS The noun **byproduct** means "anything produced in the course of making another thing." Oxygen is considered a byproduct of the light-dependent reactions of photosynthesis because it is produced as a result of extracting electrons from water. Also, unlike ATP and NADPH, oxygen is not used in the second stage of the process, the light-independent reactions.

FIGURE 8–7 The Stages of Photosynthesis There are two stages of photosynthesis: light-dependent reactions and light-independent reactions. **Interpret Diagrams** ***What happens to the ATP and NADPH produced in the light-dependent reactions?***

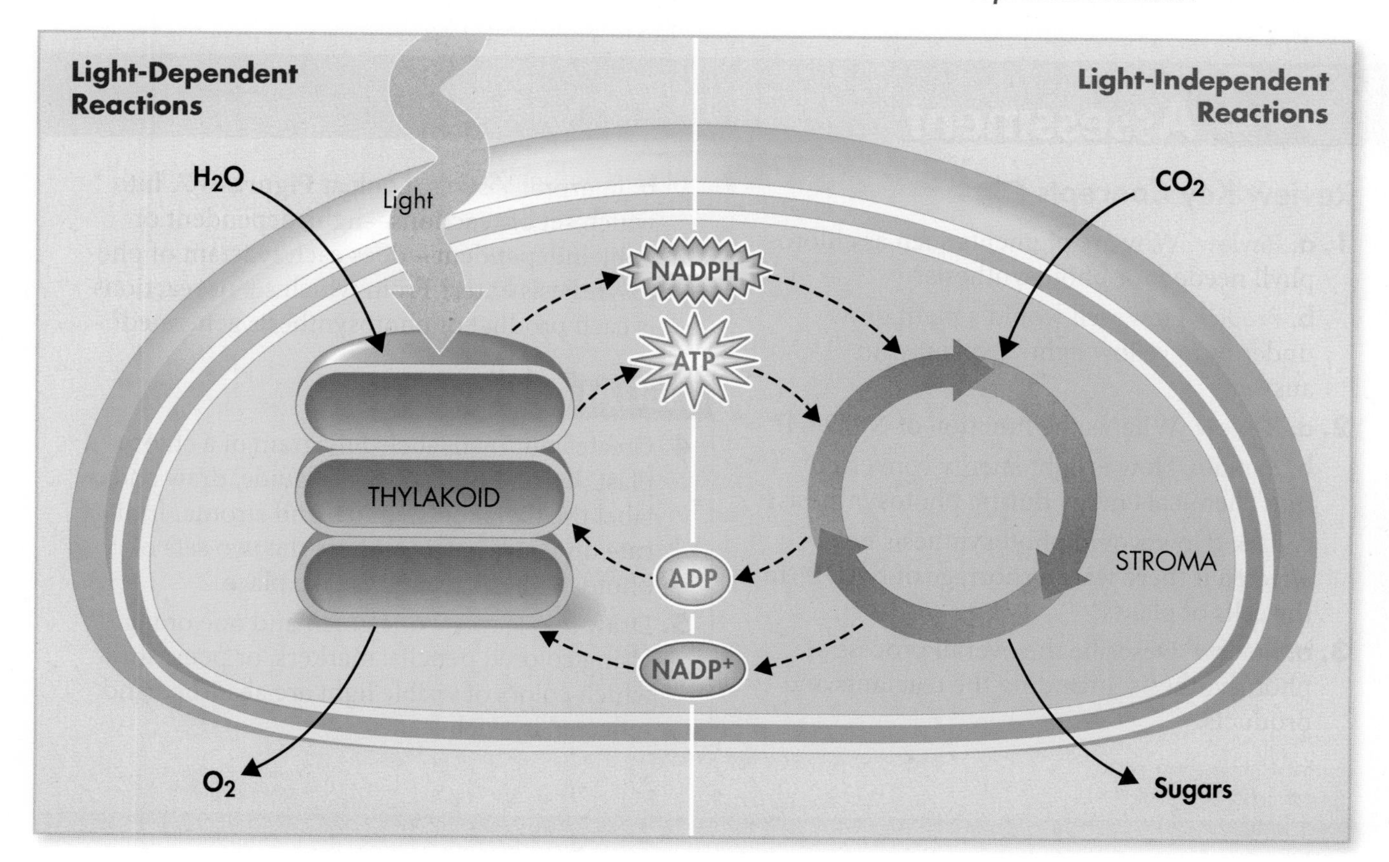

Quick Lab
GUIDED INQUIRY

What Waste Material Is Produced During Photosynthesis?

1. Fill a large, clear, plastic cup about halfway full with sodium bicarbonate solution. The sodium bicarbonate solution is a source of carbon dioxide.

2. Place a freshly cut *Elodea* plant (with the cut stem at the bottom) in a large test tube. Fill the tube with sodium bicarbonate solution. **CAUTION:** *Handle the test tube carefully.*

3. Hold your finger over the mouth of the test tube. Turn the test tube over, and lower it to the bottom of the cup. Make sure no air is trapped in the test tube.

4. Place the cup in bright light.

5. After no fewer than 20 minutes, look closely at the elodea leaves. Record your observations.

Analyze and Conclude

1. Observe What did you observe on the *Elodea* leaves?

2. Infer What substance accumulated on the leaves? Should that substance be considered a waste product? Explain.

3. Apply Concepts Which plant organelle carries out photosynthesis and produces the gas?

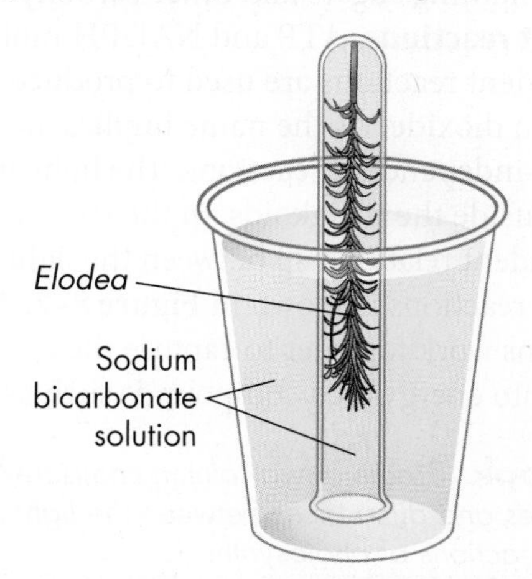

8.2 Assessment

Review Key Concepts

1. a. Review Why are pigments such as chlorophyll needed for photosynthesis?

b. Predict How well would a plant grow under pure yellow light? Explain your answer.

2. a. Review What is the function of NADPH?

b. Explain How is light energy converted into chemical energy during photosynthesis?

c. Infer How would photosynthesis be affected if there were a shortage of $NADP^+$ in the cells of plants?

3. a. Review Describe the overall process of photosynthesis, including the reactants and products.

b. Interpret Visuals Look at **Figure 8–7.** Into which set of reactions—light-dependent or light-independent—does each reactant of photosynthesis enter? From which set of reactions is each product of photosynthesis generated?

VISUAL THINKING

4. Create your own labeled diagram of a chloroplast. Using **Figure 8–5** as a guide, draw and label the thylakoids, grana, and stroma. Indicate on your drawing where the two sets of photosynthesis reactions take place.

5. Draw two leaves—one green and one orange. Using colored pencils, markers, or pens, show which colors of visible light are absorbed and reflected by each leaf.

8.3 The Process of Photosynthesis

THINK ABOUT IT Why membranes? Why do chloroplasts contain so many membranes? Is there something about biological membranes that makes them absolutely essential for the process of photosynthesis? As you'll see, there is. When most pigments absorb light, they eventually lose most of that energy as heat. In a sense, the "trade secret" of the chloroplast is how it avoids such losses, capturing light energy in the form of high-energy electrons—and membranes are the key. Without them, photosynthesis simply wouldn't work.

Key Questions

What happens during the light-dependent reactions?

What happens during the light-independent reactions?

What factors affect photosynthesis?

Vocabulary

photosystem • electron transport chain • ATP synthase • Calvin cycle

Taking Notes

Flowchart As you read, create a flowchart that clearly shows the steps involved in the light-dependent reactions.

The Light-Dependent Reactions: Generating ATP and NADPH

What happens during the light-dependent reactions?

Recall that the process of photosynthesis involves two primary sets of reactions: the light-dependent and the light-independent reactions. The light-dependent reactions encompass the steps of photosynthesis that directly involve sunlight. These reactions explain why plants need light to grow. **The light-dependent reactions use energy from sunlight to produce oxygen and convert ADP and $NADP^+$ into the energy carriers ATP and NADPH.**

The light-dependent reactions occur in the thylakoids of chloroplasts. Thylakoids are saclike membranes containing most of the machinery needed to carry out these reactions. Thylakoids contain clusters of chlorophyll and proteins known as **photosystems.** The photosystems, which are surrounded by accessory pigments, are essential to the light-dependent reactions. Photosystems absorb sunlight and generate high-energy electrons that are then passed to a series of electron carriers embedded in the thylakoid membrane. Light absorption by the photosystems is just the beginning of this important process.

FIGURE 8–8 The Importance of Light Like most plants, this rice plant needs light to grow.
Apply Concepts ***Which stage of photosynthesis requires light?***

FIGURE 8–9 **Why Green?** The green color of most plants is caused by the reflection of green light by the pigment chlorophyll. Pigments capture light energy during the light-dependent reactions of photosynthesis.

Photosystem II The light-dependent reactions, shown in **Figure 8–10,** begin when pigments in photosystem II absorb light. (This first photosystem is called photosystem II simply because it was discovered after photosystem I.) Light energy is absorbed by electrons in the pigments found within photosystem II, increasing the electrons' energy level. These high-energy electrons (e^-) are passed to the electron transport chain. An **electron transport chain** is a series of electron carrier proteins that shuttle high-energy electrons during ATP-generating reactions.

As light continues to shine, more and more high-energy electrons are passed to the electron transport chain. Does this mean that chlorophyll eventually runs out of electrons? No, the thylakoid membrane contains a system that provides new electrons to chlorophyll to replace the ones it has lost. These new electrons come from water molecules (H_2O). Enzymes on the inner surface of the thylakoid break up each water molecule into 2 electrons, 2 H^+ ions, and 1 oxygen atom. The 2 electrons replace the high-energy electrons that have been lost to the electron transport chain. As plants remove electrons from water, oxygen is left behind and is released into the air. This reaction is the source of nearly all of the oxygen in Earth's atmosphere, and it is another way in which photosynthesis makes our lives possible. The hydrogen ions left behind when water is broken apart are released inside the thylakoid.

In Your Notebook *Explain in your own words why photosynthetic organisms need water and sunlight.*

Electron Transport Chain What happens to the electrons as they move down the electron transport chain? Energy from the electrons is used by the proteins in the chain to pump H^+ ions from the stroma into the thylakoid space. At the end of the electron transport chain, the electrons themselves pass to a second photosystem called photosystem I.

Photosystem I Because some energy has been used to pump H^+ ions across the thylakoid membrane, electrons do not contain as much energy as they used to when they reach photosystem I. Pigments in photosystem I use energy from light to reenergize the electrons. At the end of a short second electron transport chain, $NADP^+$ molecules in the stroma pick up the high-energy electrons, along with H^+ ions, at the outer surface of the thylakoid membrane, to become NADPH. This NADPH becomes very important, as you will see, in the light-independent reactions of photosynthesis.

Hydrogen Ion Movement and ATP Formation Recall that in photosystem II, hydrogen ions began to accumulate within the thylakoid space. Some were left behind from the splitting of water at the end of the electron transport chain. Other hydrogen ions were "pumped" in from the stroma. The buildup of hydrogen ions makes the stroma negatively charged relative to the space within the thylakoids. This gradient, the difference in both charge and H^+ ion concentration across the membrane, provides the energy to make ATP.

BUILD Vocabulary

ACADEMIC WORDS The noun **gradient** refers to "an area over which something changes." There is a charge gradient across the thylakoid membrane because there is a positive charge on one side and a negative charge on the other.

H^+ ions cannot cross the membrane directly. However, the thylakoid membrane contains a protein called **ATP synthase** that spans the membrane and allows H^+ ions to pass through it. Powered by the gradient, H^+ ions pass through ATP synthase and force it to rotate, almost like a turbine being spun by water in a hydroelectric power plant. As it rotates, ATP synthase binds ADP and a phosphate group together to produce ATP. This process, which is known as chemiosmosis (kem ee ahz MOH sis), enables light-dependent electron transport to produce not only NADPH (at the end of the electron transport chain), but ATP as well.

Summary of Light-Dependent Reactions The light-dependent reactions produce oxygen gas and convert ADP and $NADP^+$ into the energy carriers ATP and NADPH. What good are these compounds? As we will see, they have an important role to play in the cell: They provide the energy needed to build high-energy sugars from low-energy carbon dioxide.

ZOOMING IN

LIGHT-DEPENDENT REACTIONS

FIGURE 8–10 The light-dependent reactions of photosynthesis take place in the thylakoids of the chloroplast. They use energy from sunlight to produce ATP, NADPH, and oxygen. **Interpret Visuals** ***How many molecules of NADPH are produced per water molecule used in photosynthetic electron transport?***

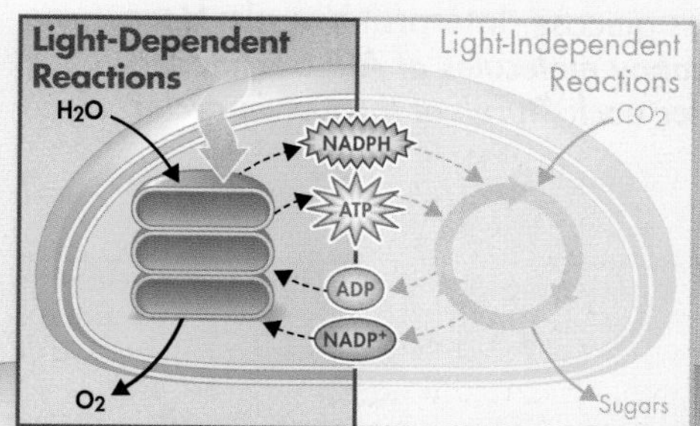

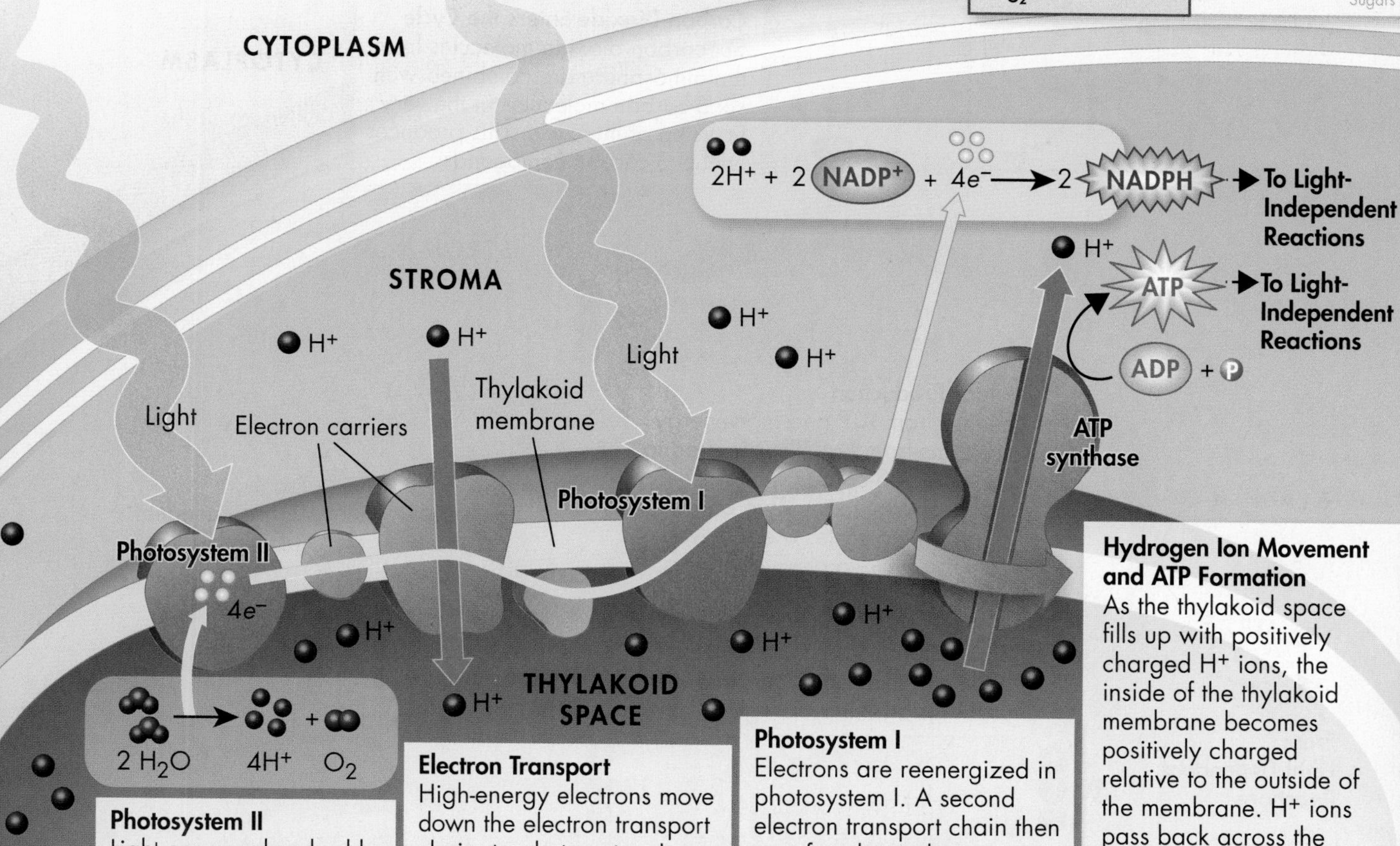

The Light-Independent Reactions: Producing Sugars

What happens during the light-independent reactions?

The ATP and NADPH formed by the light-dependent reactions contain an abundance of chemical energy, but they are not stable enough to store that energy for more than a few minutes. During the light-independent reactions, commonly referred to as the **Calvin cycle,** plants use the energy that ATP and NADPH contain to build stable high-energy carbohydrate compounds that can be stored for a long time. **During the light-independent reactions, ATP and NADPH from the light-dependent reactions are used to produce high-energy sugars.** The Calvin cycle is named after the American scientist Melvin Calvin, who worked out the details of this remarkable cycle. Follow **Figure 8–11** to see each step in this set of reactions.

ZOOMING IN

LIGHT-INDEPENDENT REACTIONS

FIGURE 8–11 The light-independent reactions of photosynthesis take place in the stroma of the chloroplast. The reactions use ATP and NADPH from the light-dependent reactions to produce high-energy sugars such as glucose. **Interpret Visuals** ***How many molecules of ATP are needed for each "turn" of the Calvin cycle?***

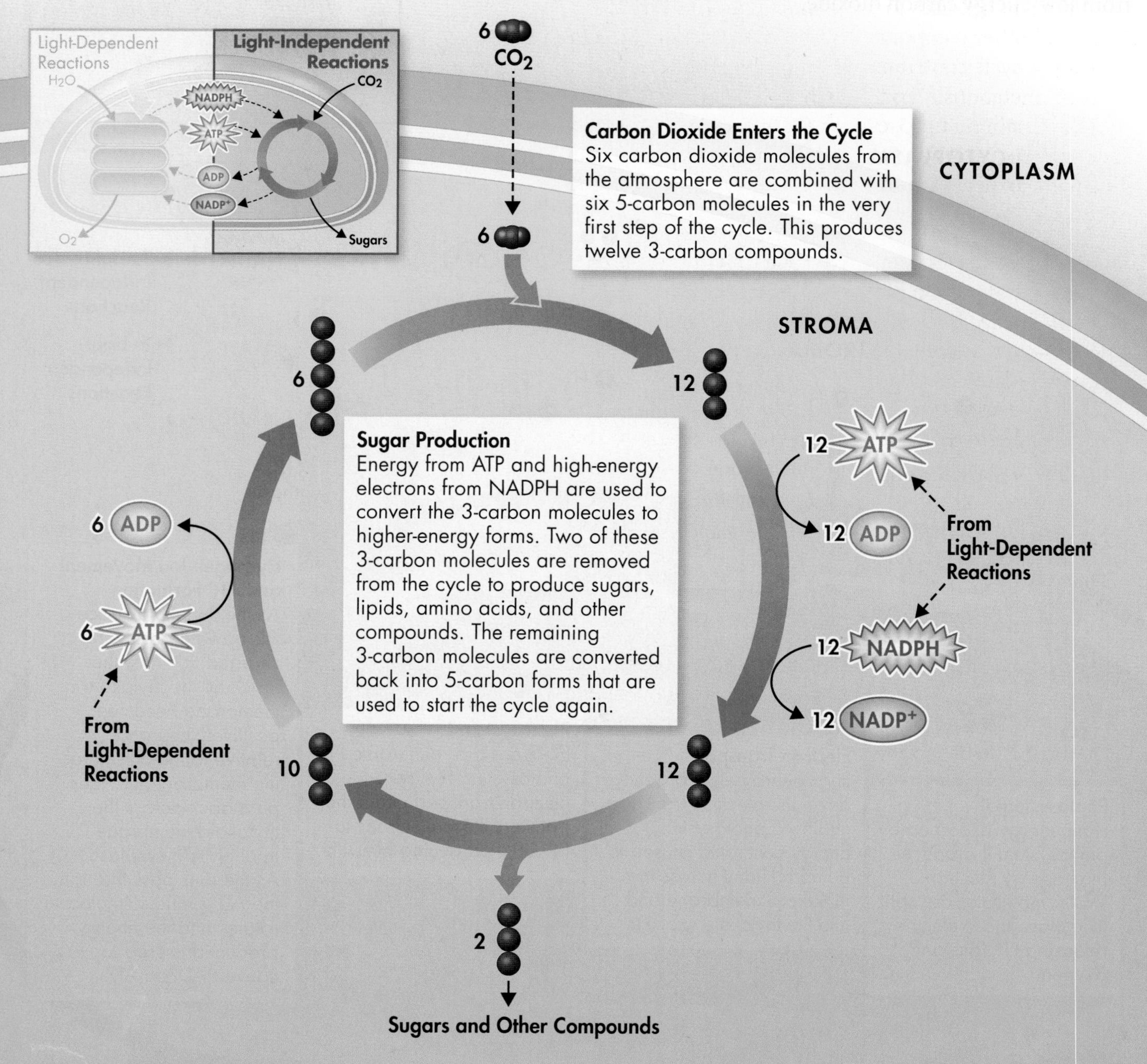

Carbon Dioxide Enters the Cycle Carbon dioxide molecules enter the Calvin cycle from the atmosphere. An enzyme in the stroma of the chloroplast combines these carbon dioxide molecules with 5-carbon compounds that are already present in the organelle, producing 3-carbon compounds that continue into the cycle. For every 6 carbon dioxide molecules that enter the cycle, a total of twelve 3-carbon compounds are produced. Other enzymes in the chloroplast then convert these compounds into higher-energy forms in the rest of the cycle. The energy for these conversions comes from ATP and high-energy electrons from NADPH.

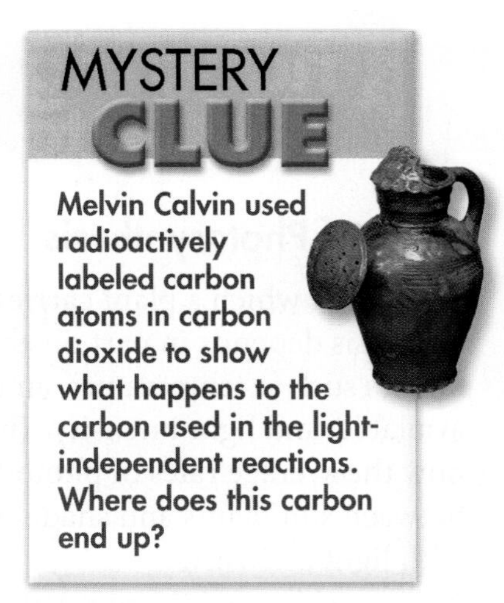

Melvin Calvin used radioactively labeled carbon atoms in carbon dioxide to show what happens to the carbon used in the light-independent reactions. Where does this carbon end up?

Sugar Production At midcycle, two of the twelve 3-carbon molecules are removed from the cycle. This is a very special step because these molecules become the building blocks that the plant cell uses to produce sugars, lipids, amino acids, and other compounds. In other words, this step in the Calvin cycle contributes to all of the products needed for plant metabolism and growth.

The remaining ten 3-carbon molecules are converted back into six 5-carbon molecules. These molecules combine with six new carbon dioxide molecules to begin the next cycle.

Summary of the Calvin Cycle The Calvin cycle uses 6 molecules of carbon dioxide to produce a single 6-carbon sugar molecule. The energy for the reactions that make this possible is supplied by compounds produced in the light-dependent reactions. As photosynthesis proceeds, the Calvin cycle works steadily, removing carbon dioxide from the atmosphere and turning out energy-rich sugars. The plant uses the sugars to meet its energy needs and to build macromolecules needed for growth and development, including lipids, proteins, and complex carbohydrates such as cellulose. When other organisms eat plants, they, too, can use the energy and raw materials stored in these compounds.

The End Results The two sets of photosynthetic reactions work together—the light-dependent reactions trap the energy of sunlight in chemical form, and the light-independent reactions use that chemical energy to produce stable, high-energy sugars from carbon dioxide and water. And, in the process, animals, including ourselves, get plenty of food and an atmosphere filled with oxygen. Not a bad deal at all!

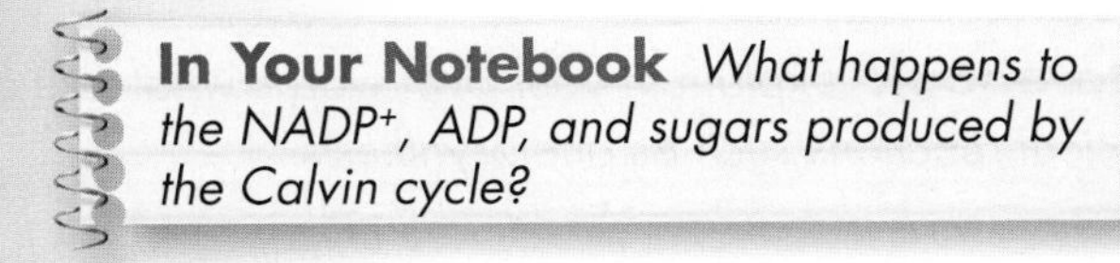

In Your Notebook *What happens to the NADP+, ADP, and sugars produced by the Calvin cycle?*

Analyzing Data

Rates of Photosynthesis

The rate at which a plant carries out photosynthesis depends in part on environmental factors such as temperature, amount of water available, and light intensity. The graph shows how the average rates of photosynthesis between sun plants and shade plants changes with light intensity.

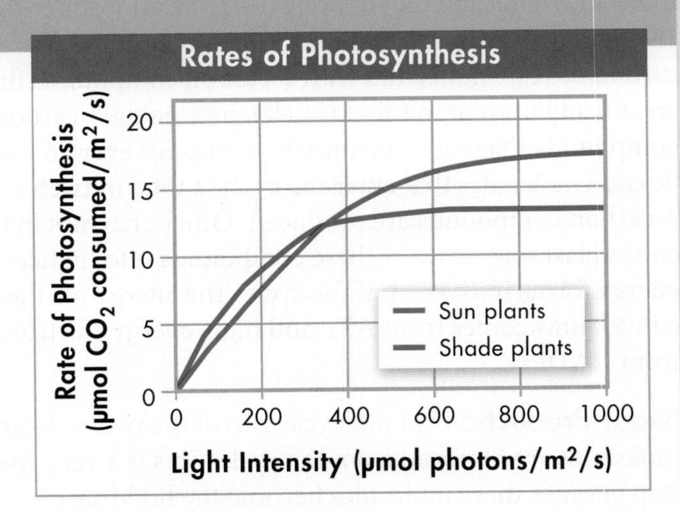

1. **Use Tables and Graphs** When light intensity is below 200 μmol photons/m^2/s, do sun plants or shade plants have a higher rate of photosynthesis?

2. **Infer** Light intensity in the Sonoran Desert averages about 400 μmol photons/m^2/s. According to the graph, what would be the approximate rate of photosynthesis for sun plants that grow in this environment?

3. **Form a Hypothesis** Suppose you transplant a sun plant to a shaded forest floor that receives about 100 μmol photons/m^2/s. Do you think this plant will grow and thrive? Why or why not? How does the graph help you answer this question?

Factors Affecting Photosynthesis

What factors affect photosynthesis?

Temperature, Light, and Water Many factors influence the rate of photosynthesis. **Among the most important factors that affect photosynthesis are temperature, light intensity, and the availability of water.** The reactions of photosynthesis are made possible by enzymes that function best between 0°C and 35°C. Temperatures above or below this range may affect those enzymes, slowing down the rate of photosynthesis. At very low temperatures, photosynthesis may stop entirely.

The intensity of light also affects the rate at which photosynthesis occurs. As you might expect, high light intensity increases the rate of photosynthesis. After the light intensity reaches a certain level, however, the plant reaches its maximum rate of photosynthesis.

Because water is one of the raw materials of photosynthesis, a shortage of water can slow or even stop photosynthesis. Water loss can also damage plant tissues. To deal with these dangers, plants (such as desert plants and conifers) that live in dry conditions often have waxy coatings on their leaves that reduce water loss. They may also have biochemical adaptations that make photosynthesis more efficient under dry conditions.

BUILD Vocabulary

MULTIPLE MEANINGS The noun *intensity* is commonly used to refer to something or someone who is very emotional, focused, or active. In science, however, *intensity* refers to energy. Thus, light intensity is a measure of the amount of energy available in light. More intense light has more energy.

In Your Notebook *Explain in your own words what role enzymes play in chemical reactions such as photosynthesis.*

Photosynthesis Under Extreme Conditions In order to conserve water, most plants under bright, hot conditions (of the sorts often found in the tropics) close the small openings in their leaves that normally admit carbon dioxide. While this keeps the plants from drying out, it causes carbon dioxide within the leaves to fall to very low levels. When this happens to most plants, photosynthesis slows down or even stops. However, some plants have adapted to extremely bright, hot conditions. There are two major groups of these specialized plants: C4 plants and CAM plants. C4 and CAM plants have biochemical adaptations that minimize water loss while still allowing photosynthesis to take place in intense sunlight.

► ***C4 Photosynthesis*** C4 plants have a specialized chemical pathway that allows them to capture even very low levels of carbon dioxide and pass it to the Calvin cycle. The name "C4 plant" comes from the fact that the first compound formed in this pathway contains 4 carbon atoms. The C4 pathway enables photosynthesis to keep working under intense light and high temperatures, but it requires extra energy in the form of ATP to function. C4 organisms include important crop plants like corn, sugar cane, and sorghum.

► ***CAM Plants*** Other plants adapted to dry climates use a different strategy to obtain carbon dioxide while minimizing water loss. These include members of the family Crassulaceae. Because carbon dioxide becomes incorporated into organic acids during photosynthesis, the process is called Crassulacean Acid Metabolism (CAM). CAM plants admit air into their leaves only at night. In the cool darkness, carbon dioxide is combined with existing molecules to produce organic acids, "trapping" the carbon within the leaves. During the daytime, when leaves are tightly sealed to prevent the loss of water, these compounds release carbon dioxide, enabling carbohydrate production. CAM plants include pineapple trees, many desert cacti, and also the fleshy "ice plants" shown in **Figure 8–12,** which are frequently planted near freeways along the west coast to retard brush fires and prevent erosion.

FIGURE 8–12 CAM Plants Plants like this ice plant can survive in dry conditions due to their modified light-independent reactions. Air is allowed into the leaves only at night, minimizing water loss.

8.3 Assessment

Review Key Concepts

1. a. Review Summarize what happens during the light-dependent reactions of photosynthesis.

b. Sequence Put the events of the light-dependent reactions in the order in which they occur and describe how each step is dependent on the step that comes before it.

2. a. Review What is the Calvin cycle?

b. Compare and Contrast List at least three differences between the light-dependent and light-independent reactions of photosynthesis.

3. a. Review What are the three primary factors that affect the rate of photosynthesis?

b. Interpret Graphs Look at the graph on page 240. What are the independent and dependent variables being tested?

BUILD VOCABULARY

4. The word *carbohydrate* comes from the prefix *carbo-*, meaning "carbon," and the word *hydrate*. Based on the reactants of the photosynthesis equation, what does *hydrate* mean?

NGSS Smart Guide Photosynthesis

Disciplinary Core Idea LS1.C Organization for Matter and Energy Flow in Organisms: How do organisms obtain and use the matter and energy they need to live and grow? Photosynthesis is the process by which organisms convert light energy into chemical energy that all organisms can use directly, or indirectly, to carry out life functions.

8.1 Energy and Life

- ATP can easily release and store energy by breaking and re-forming the bonds between its phosphate groups. This characteristic of ATP makes it exceptionally useful as a basic energy source for all cells.
- In the process of photosynthesis, plants convert the energy of sunlight into chemical energy stored in the bonds of carbohydrates.

adenosine triphosphate (ATP) 226 • heterotroph 228 • autotroph 228 • photosynthesis 228

Biology.com

Untamed Science Video Journey to Panama with the Untamed Science crew to discover how CO_2 affects plant growth.

Visual Analogy See how the electron transport chain is like passing a hot potato.

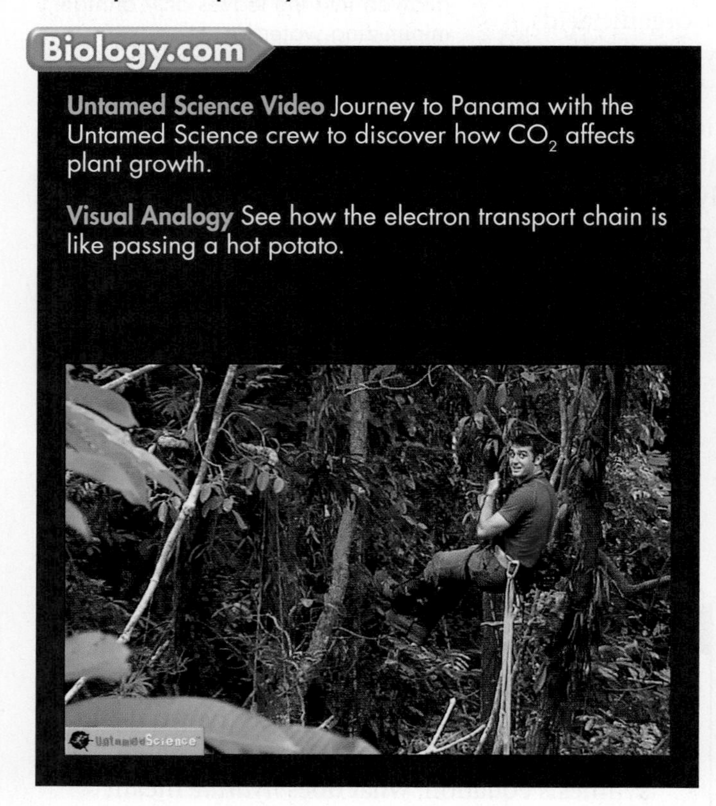

8.2 Photosynthesis: An Overview

- Photosynthetic organisms capture energy from sunlight with pigments.
- An electron carrier is a compound that can accept a pair of high-energy electrons and transfer them, along with most of their energy, to another molecule.
- Photosynthesis uses the energy of sunlight to convert water and carbon dioxide (reactants) into high-energy sugars and oxygen (products).

pigment 230 • chlorophyll 230 • thylakoid 231 • stroma 231 • $NADP^+$ 232 • light-dependent reactions 233 • light-independent reactions 233

Biology.com

InterActive Art Bring the components of photosynthesis together to run an animation.

Tutor Tube Learn how to sort out the products and reactants in both the light-dependent and light-independent reactions.

Data Analysis Look at pigment color data in the ocean to find out how marine algae photosynthesize in the blue light available underwater.

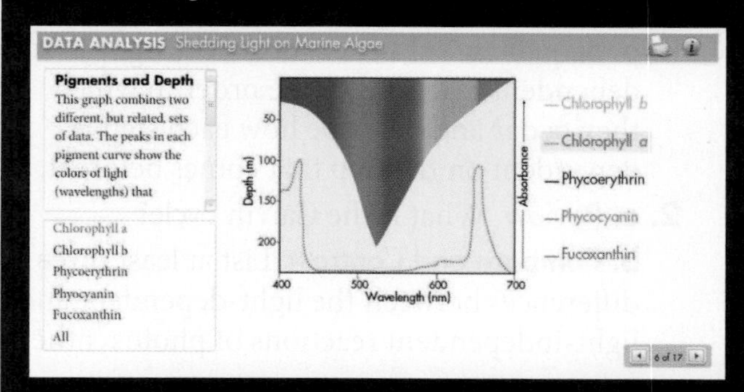

CHECKING YOUR *Scientific Literacy*

Refer to the lesson content and digital assets below as you prepare for your chapter assessment. Then, evaluate your understanding of photosynthesis by answering these questions.

1. **Scientific and Engineering Practice Developing and Using Models** Draw a model of a chloroplast. In your model, label the structures and indicate the key events that occur in the conversion of sunlight energy into chemical energy.

2. **Crosscutting Concept Structure and Function** Expand your model to show the location of photosystems. Explain how the location of the electron transport chain and structure of thylakoid membranes contribute to energy production.

3. **Integration of Knowledge and Ideas** Create a table that summarizes the process of photosynthesis. Include mathematical and quantitative information related to the reactions involved.

4. **STEM** You are an engineer assigned to construct an interactive museum exhibit on how plants harness the sun's energy. Define the problems the exhibit will tackle and list ideas on how to communicate the process of photosynthesis to the general public.

8.3 The Process of Photosynthesis

- The light-dependent reactions use energy from sunlight to produce oxygen and convert ADP and $NADP^+$ into the energy carriers ATP and NADPH.
- During the light-independent reactions, ATP and NADPH from the light-dependent reactions are used to produce high-energy sugars.
- Among the most important factors that affect photosynthesis are temperature, light intensity, and the availability of water.

photosystem 235 • electron transport chain 236 • ATP synthase 237 • Calvin cycle 238

Biology.com

Art in Motion Watch the steps of the light-dependent reactions in motion at the molecular level.

Art Review Focus on the thylakoid membrane to review your knowledge of the light-dependent reactions.

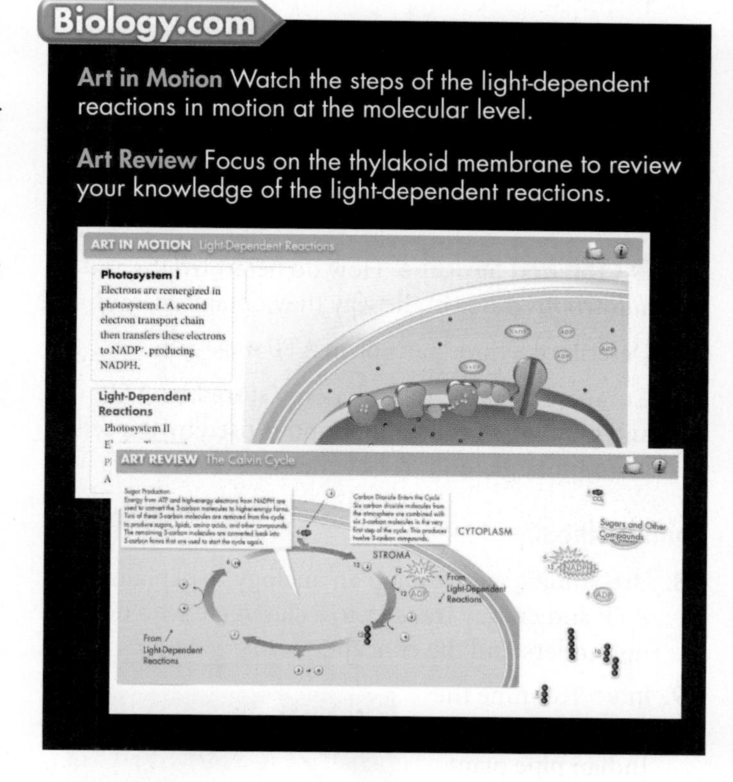

Skills Lab

Plant Pigments and Photosynthesis This chapter lab is available in your lab manual and at Biology.com

8 Assessment

8.1 Energy and Life

Understand Key Concepts

1. Which of the following are autotrophs?
   a. deer
   b. plants
   c. leopards
   d. mushrooms
2. Which of the following is used by cells to store and release the energy needed to power cellular processes?
   a. DNA.
   b. ATP.
   c. H_2O.
   d. CO_2.
3. The amount of energy stored in a molecule of ATP compared to the amount stored in a molecule of glucose is
   a. greater.
   b. less.
   c. the same.
   d. variable.
4. When a candle burns, energy is released in the form of
   a. carbon dioxide and water.
   b. the chemical substance ATP.
   c. light and heat.
   d. electricity and motion.
5. **Craft and Structure** How do heterotrophs and autotrophs differ in the way they obtain energy?
6. Describe the three parts of an ATP molecule.
7. Compare the amounts of energy stored by ATP and glucose. Which compound is used by the cell as an immediate source of energy?

Think Critically

8. **Use Analogies** Develop an analogy to explain ATP and energy transfer to a classmate who does not understand the concept.
9. **Infer** Examine the photograph of the Indian pipe plant shown here. What can you conclude about the ability of the Indian pipe plant to make its own food? Explain your answer.

8.2 Photosynthesis: An Overview

Understand Key Concepts

10. In addition to light and chlorophyll, photosynthesis requires
    a. water and oxygen.
    b. water and sugars.
    c. oxygen and carbon dioxide.
    d. water and carbon dioxide.
11. The leaves of a plant appear green because chlorophyll
    a. reflects blue light.
    b. absorbs blue light.
    c. reflects green light.
    d. absorbs green light.
12. Write the basic equation for photosynthesis using the names of the starting and final substances of the process.
13. What role do plant pigments play in the process of photosynthesis?
14. Identify the chloroplast structures labeled A, B, and C. In which structure(s) do the light-dependent reactions occur? In which structure(s) do the light-independent reactions take place?

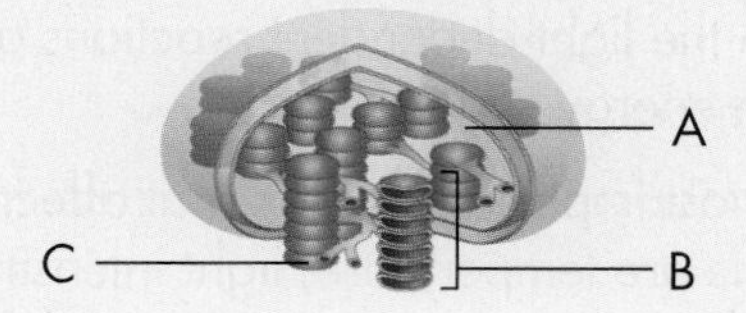

Think Critically

15. **Form a Hypothesis** Although they appear green, some plant leaves contain yellow and red pigments as well as chlorophyll. In the fall, those leaves may become red or yellow. Suggest an explanation for these color changes.
16. **Design an Experiment** Design an experiment that uses pond water and algae to demonstrate the importance of light energy to pond life. Be sure to identify the variables you will control and the variable you will change.
17. **Text Types and Purposes** Suppose you water a potted plant and place it by a window in a transparent, airtight jar. Predict how the rate of photosynthesis might be affected over the next few days. What might happen if the plant were left there for several weeks? Explain.

8.3 The Process of Photosynthesis

Understand Key Concepts

18. The first process in the light-dependent reactions of photosynthesis is
a. light absorption. **c.** oxygen production.
b. electron transport. **d.** ATP formation.

19. Which substance from the light-dependent reactions of photosynthesis is a source of energy for the Calvin cycle?
a. ADP **c.** H_2O
b. NADPH **d.** pyruvic acid

20. The light-independent reactions of photosynthesis are also known as the
a. Calvin cycle. **c.** carbon cycle.
b. sugar cycle. **d.** ATP cycle.

21. ATP synthase in the chloroplast membrane makes ATP, utilizing the energy of highly concentrated
a. chlorophyll. **c.** hydrogen ions.
b. electrons. **d.** NADPH.

22. CAM plants are specialized to survive under what conditions that would harm most other kinds of plants?
a. low temperatures **c.** hot, dry conditions
b. excess water **d.** long day lengths

23. Explain the role of $NADP^+$ as an energy carrier in photosynthesis.

24. Describe the role of ATP synthase and explain how it works.

25. **Key Ideas and Details** Summarize the events of the Calvin cycle.

26. Discuss three factors that affect the rate at which photosynthesis occurs.

Think Critically

27. **Key Ideas and Details** Study **Figure 8–11** on page 238 and give evidence from the text to support the idea that the Calvin cycle does not depend on light.

28. **Apply Concepts** How do the events in the Calvin cycle depend on the light-dependent reactions of photosynthesis?

29. **Form a Hypothesis** Many of the sun's rays may be blocked by dust or clouds formed by volcanic eruptions or pollution. What are some possible short-term and long-term effects of this on photosynthesis? On other forms of life?

solve the CHAPTER MYSTERY

OUT OF THIN AIR?

Most plants grow out of the soil, of course, and you might hypothesize, as Jan van Helmont did, that soil contributes to plant mass. At the conclusion of his experiment with the willow tree, however, van Helmont discovered that the mass of the soil was essentially unchanged, but that the tree had increased in mass by nearly 75 kilograms. Van Helmont concluded that the mass must have come from water, because water was the only thing he had added throughout the experiment. What he didn't know, however, was that the increased bulk of the tree was built from carbon, as well as from the oxygen and hydrogen in water. We now know that most of that carbon comes from carbon dioxide in the air. Thus, mass accumulates from two sources: carbon dioxide and water. What form does the added mass take? Think about the origin of the word *carbohydrate*, from *carbo-*, meaning "carbon," and *hydrate*, meaning "to combine with water," and you have your answer.

1. Infer Although soil does not significantly contribute to plant mass, how might it help plants grow?

2. Infer If a scientist were able to measure the exact mass of carbon dioxide and water that entered a plant, and the exact mass of the sugars produced, would the masses be identical? Why or why not?

3. Apply Concepts What do plants do with all of the carbohydrates they produce by photosynthesis? (*Hint*: Plant cells have mitochondria in addition to chloroplasts. What do mitochondria do?)

4. Connect to the Big idea Explain how the experiments carried out by van Helmont and Calvin contributed to our understanding of how nutrients cycle in the biosphere.

Connecting Concepts

Use Science Graphics

A water plant placed under bright light gives off bubbles of oxygen. The table below contains the results of an experiment in which the distance from the light to the plant was varied. Use the data table to answer questions 30–33.

Oxygen Production

Distance From Light (cm)	Bubbles Produced per Minute
10	39
20	22
30	8
40	5

30. Graph Use the data in the table to make a line graph. MATH

31. Integration of Knowledge and Ideas Describe the observed trend. How many bubbles would you predict if the light was moved to 50 cm away? Explain.

32. Draw Conclusions What relationship exists between the plant's distance from the light and the number of bubbles produced? What process is occuring? Explain your answer.

33. Apply Concepts Based on the results of this experiment, explain why most aquatic primary producers live in the uppermost regions of deep oceans, lakes, and ponds.

Write About Science

34. Production and Distribution of Writing Imagine that you are an oxygen atom and two of your friends are hydrogen atoms. Together, you make up a water molecule. Describe the events and changes that happen to you and your friends as you journey through the light-dependent reactions and the Calvin cycle of photosynthesis. Include illustrations with your description.

35. Assess the Big idea In eukaryotic plants, chlorophyll is found only in chloroplasts. Explain how the function of chlorophyll is related to its very specific location in the cell.

Integration of Knowledge and Ideas

An experimenter subjected corn plants and bean plants to different concentrations of carbon dioxide and measured the amount of CO_2 taken up by the plants and used in photosynthesis. Data for the two plants are shown in the following graph.

Rates of Photosynthesis

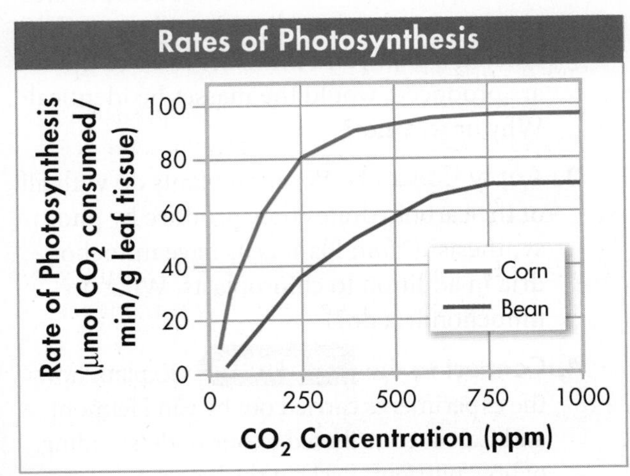

36. Interpret Graphs Bean plants reach their maximum rate of photosynthesis at what concentration of carbon dioxide?

a. about 50 ppm
b. about 200 ppm
c. about 750 ppm
d. 1000 ppm

37. Draw Conclusions From the data it is possible to conclude that

a. beans contain more chlorophyll than corn contains.
b. corn reaches its maximum photosynthetic rate at lower concentrations than beans do.
c. beans reach their maximum photosynthetic rate at lower concentrations than corn does.
d. beans use carbon dioxide more efficiently than corn does.

Standardized Test Prep

Multiple Choice

1. Autotrophs differ from heterotrophs because they
   A utilize oxygen to burn food.
   B do not require oxygen to live.
   C make carbon dioxide as a product of using food.
   D make their own food from carbon dioxide and water.

2. The principal pigment in plants is
   A chlorophyll.
   B oxygen.
   C ATP.
   D NADPH.

3. Which of the following is NOT produced in the light-dependent reactions of photosynthesis?
   A NADPH
   B sugars
   C hydrogen ions
   D ATP

4. Which of the following correctly summarizes the process of photosynthesis?
   A $H_2O + CO_2 \xrightarrow{\text{light}} \text{sugars} + O_2$
   B $\text{sugars} + O_2 \xrightarrow{\text{light}} H_2O + CO_2$
   C $H_2O + O_2 \xrightarrow{\text{light}} \text{sugars} + CO_2$
   D $\text{sugars} + CO_2 \xrightarrow{\text{light}} H_2O + O_2$

5. The color of light that is LEAST useful to a plant during photosynthesis is
   A red.
   B blue.
   C green.
   D violet.

6. The first step in photosynthesis is the
   A synthesis of water.
   B production of oxygen.
   C breakdown of carbon dioxide.
   D absorption of light energy.

7. In a typical plant, all of the following factors are necessary for photosynthesis EXCEPT
   A chlorophyll.
   B light.
   C oxygen.
   D water.

Questions 8–10

Several drops of concentrated pigment were extracted from spinach leaves. These drops were placed at the bottom of a strip of highly absorbent paper. After the extract dried, the paper was suspended in a test tube containing alcohol so that only the tip of the paper was in the alcohol. As the alcohol was absorbed and moved up the paper, the various pigments contained in the extract separated as shown in the diagram.

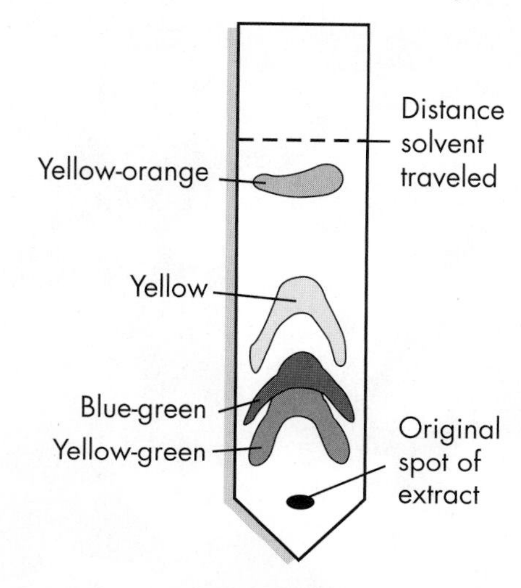

8. Which pigment traveled the shortest distance?
   A yellow-orange
   B yellow
   C blue-green
   D yellow-green

9. A valid conclusion that can be drawn from this information is that spinach leaves
   A use only chlorophyll during photosynthesis.
   B contain several pigments.
   C contain more orange pigment than yellow pigment.
   D are yellow-orange rather than green.

10. In which organelle would MOST of these pigments be found?
    A vacuoles
    B centrioles
    C mitochondria
    D chloroplasts

Open-Ended Response

11. Describe how high-energy electrons are ultimately responsible for driving the photosynthetic reactions.

If You Have Trouble With . . .

Question	1	2	3	4	5	6	7	8	9	10	11
See Lesson	8.1	8.2	8.2	8.2	8.2	8.3	8.3	8.2	8.2	8.2	8.3

9 Cellular Respiration and Fermentation

Cellular Basis of Life

Q: How do organisms obtain energy?

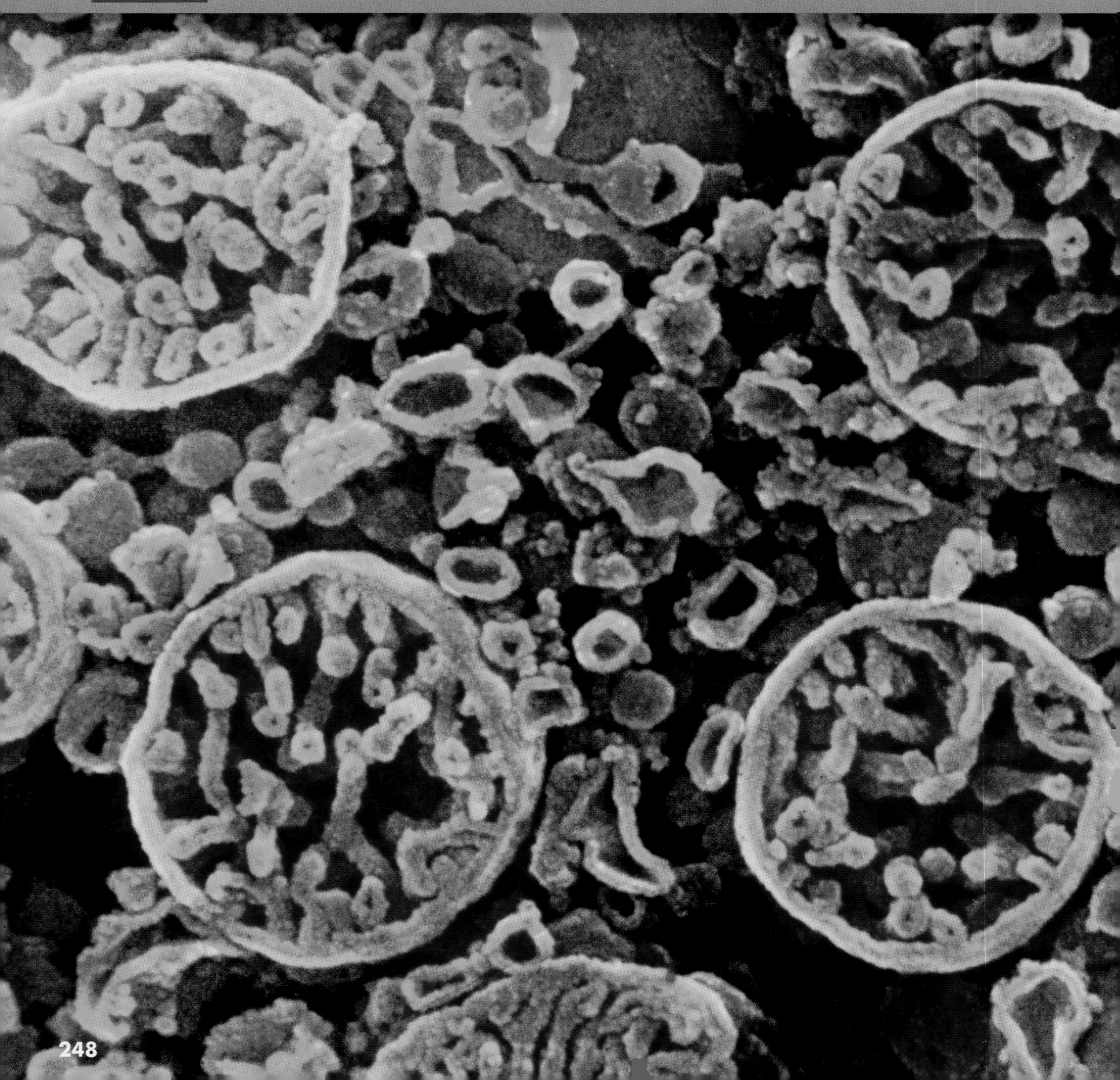

INSIDE:

- 9.1 Cellular Respiration: An Overview
- 9.2 The Process of Cellular Respiration
- 9.3 Fermentation

BUILDING *Scientific Literacy*

Deepen your understanding of cellular respiration and fermentation by engaging in key practices that allow you to make connections across concepts.

NGSS You will use mathematics to account for the energy produced during the breakdown of glucose.

STEM Can cellular respiration provide energy for the future? Learn how scientists are applying cellular respiration in bacteria to drive energy production.

Common Core You will demonstrate your understanding of cellular respiration and fermentation by translating technical information expressed visually and mathematically.

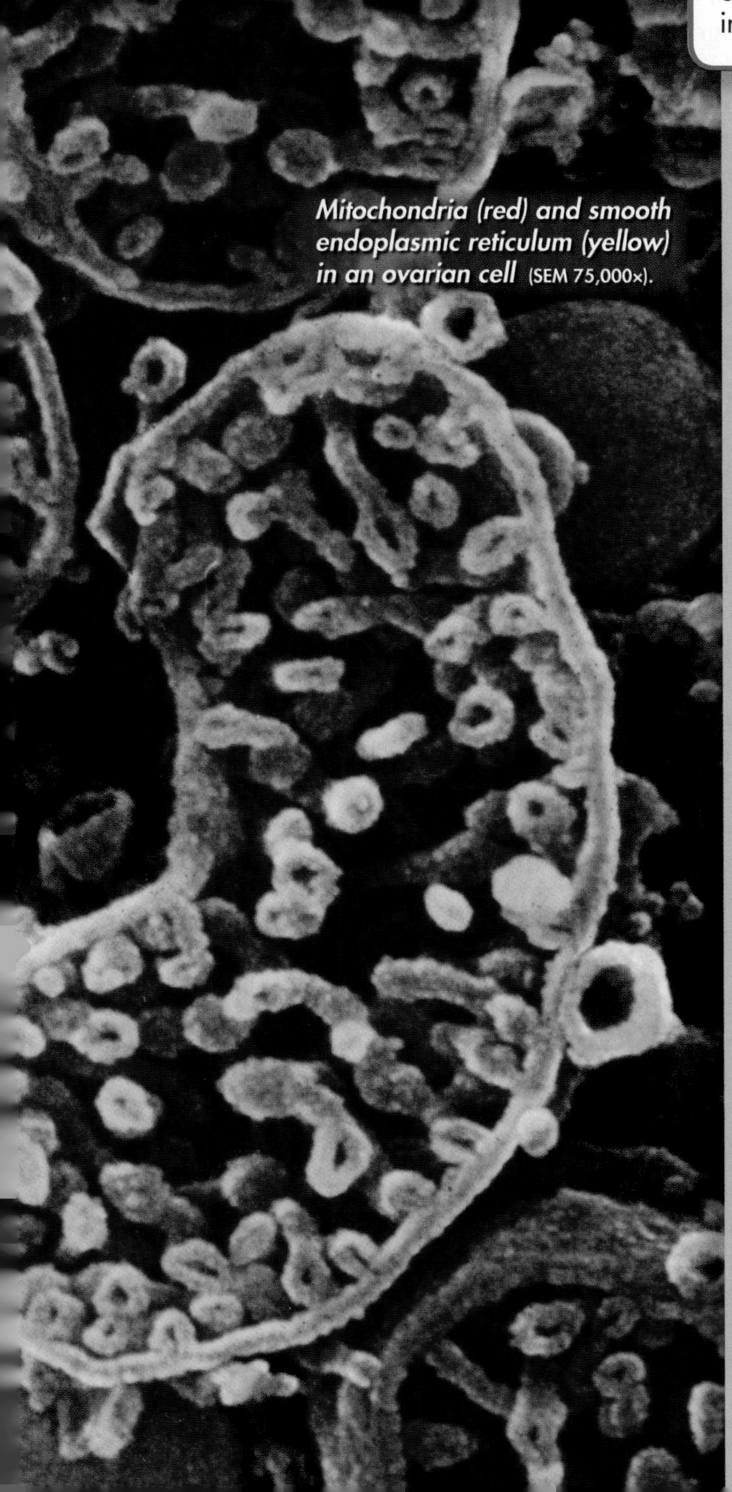

Mitochondria (red) and smooth endoplasmic reticulum (yellow) in an ovarian cell (SEM 75,000×).

CHAPTER MYSTERY

DIVING WITHOUT A BREATH

Everyone is familiar with the sensation of being "out of breath." Just a few minutes of vigorous exercise can have humans huffing and puffing for air. But what if you couldn't get air? What if you were asked to hold your breath and exercise? Before too long, you'd pass out due to a lack of oxygen. This may seem like a silly thought experiment, but there are animals that exercise without breathing and without passing out all the time—whales. Unlike most animals that live their entire lives in water, whales still rely on oxygen obtained from air when they surface. Amazingly, sperm whales routinely stay underwater for 45 minutes or more when diving. Some scientists suspect that they can stay underwater for 90 minutes! How is that possible? Diving takes a lot of energy. How do whales stay active for so long on only one breath? As you read this chapter, look for clues. Then, solve the mystery.

Biology.com

Learning about whales and their extraordinary ability to hold their breath is just the beginning. Take a video field trip with the ecogeeks of Untamed Science to see where this mystery leads.

Go online to access additional resources including:
• eText • Flash Cards • Lesson Overviews • Chapter Mystery

9.1 Cellular Respiration: An Overview

Key Questions

Where do organisms get energy?

What is cellular respiration?

What is the relationship between photosynthesis and cellular respiration?

Vocabulary

calorie • cellular respiration • aerobic • anaerobic

Taking Notes

Preview Visuals Before you read, study **Figure 9–2** on page 252. Make a list of questions that you have about the diagram. As you read, write down the answers to the questions.

THINK ABOUT IT When you are hungry, how do you feel? If you are like most people, you might feel sluggish, a little dizzy, and—above all—weak. Weakness is a feeling triggered by a lack of energy. You feel weak when you are hungry because food serves as a source of energy. Weakness is your body's way of telling you that your energy supplies are low. But how does food get converted into a usable form of energy? Car engines have to burn gasoline in order to release its energy. Do our bodies burn food the way a car burns gasoline, or is there something more to it?

Chemical Energy and Food

Where do organisms get energy?

Food provides living things with the chemical building blocks they need to grow and reproduce. Recall that some organisms, such as plants, are autotrophs, meaning that they make their own food through photosynthesis. Other organisms are heterotrophs, meaning that they rely on other organisms for food. For all organisms, food molecules contain chemical energy that is released when their chemical bonds are broken. **Organisms get the energy they need from food.**

How much energy is actually present in food? Quite a lot, although it varies with the type of food. Energy stored in food is expressed in units of calories. A **calorie** is the amount of energy needed to raise the temperature of 1 gram of water 1 degree Celsius. The Calorie (capital C) that is used on food labels is a kilocalorie, or 1000 calories. Cells can use all sorts of molecules for food, including fats, proteins, and carbohydrates. The energy stored in each of these macromolecules varies because their chemical structures, and therefore their energy-storing bonds, differ. For example, 1 gram of the sugar glucose releases 3811 calories of heat energy when it is burned. By contrast, 1 gram of the triglyceride fats found in beef releases 8893 calories of heat energy when its bonds are broken. In general, carbohydrates and proteins contain approximately 4000 calories (4 Calories) of energy per gram, while fats contain approximately 9000 calories (9 Calories) per gram.

Cells, of course, don't simply burn food and release energy as heat. Instead, they break down food molecules gradually, capturing a little bit of chemical energy at key steps. This enables cells to use the energy stored in the chemical bonds of foods like glucose to produce compounds such as ATP that directly power the activities of the cell.

BUILD Vocabulary

PREFIXES The prefix *macro-* means "large" or "elongated." Macromolecules are made up of many smaller molecular subunits. Carbohydrates, proteins, and lipids are important macromolecules found in living things.

Analyzing Data MATH

You Are What You Eat

Organisms get energy from the food they eat, but the energy contained in foods varies greatly. Most foods contain a combination of proteins, carbohydrates, and fats. One gram of protein or a carbohydrate such as glucose contains roughly 4 Calories. One gram of fat, however, contains about 9 Calories. The accompanying table shows the approximate composition of one serving of some common foods.

1. **Interpret Data** Per serving, which of the foods included in the table has the most protein? Which has the most carbohydrates? Which has the most fat?

Composition of Some Common Foods

Food	Protein (g)	Carbohydrate (g)	Fat (g)
Apple, 1 medium	0	22	0
Bacon, 2 slices	5	0	6
Chocolate, 1 bar	3	23	13
Eggs, 2 whole	12	0	9
2% milk, 1 cup	8	12	5
Potato chips, 15 chips	2	14	10
Skinless roasted turkey, 3 slices	11	3	1

2. **Calculate** Approximately how many more Calories are there in 2 slices of bacon than there are in 3 slices of roasted turkey? Why is there a difference?
3. **Calculate** Walking at a moderate pace consumes around 300 Calories per hour. At that rate, how many minutes would you have to walk to burn the Calories in one chocolate bar? (*Hint:* Start by calculating the number of Calories consumed per minute by walking.)

Overview of Cellular Respiration

What is cellular respiration?

If oxygen is available, organisms can obtain energy from food by a process called **cellular respiration.** **Cellular respiration is the process that releases energy from food in the presence of oxygen.** Although cellular respiration involves dozens of separate reactions, an overall chemical summary of the process is remarkably simple:

In Symbols:

$$6O_2 + C_6H_{12}O_6 \longrightarrow 6CO_2 + 6H_2O + \text{Energy}$$

In Words:

Oxygen + Glucose ⟶ Carbon dioxide + Water + Energy

As you can see, cellular respiration requires oxygen and a food molecule such as glucose, and it gives off carbon dioxide, water, and energy. Do not be misled, however, by the simplicity of this equation. If cellular respiration took place in just one step, all of the energy from glucose would be released at once, and most of it would be lost in the form of light and heat. Clearly, a living cell has to control that energy. It can't simply start a fire—the cell has to release the explosive chemical energy in food molecules a little bit at a time. The cell needs to find a way to trap those little bits of energy by using them to make ATP.

FIGURE 9–1 A Controlled Release Cellular respiration involves a series of controlled reactions that slowly release the energy stored in food. If the energy were to be released too suddenly, most of it would be lost in the forms of light and heat—just as it is when a marshmallow catches fire.

In Your Notebook *Do plants undergo cellular respiration? What organelle(s) do they have that helps you determine the answer?*

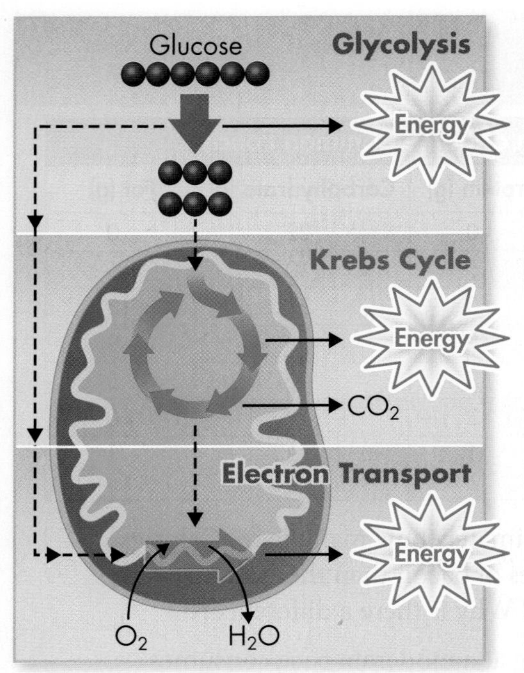

FIGURE 9–2 The Stages of Cellular Respiration There are three stages to cellular respiration: glycolysis, the Krebs cycle, and the electron transport chain. **Interpret Visuals** ***Which stage(s) of cellular respiration occur in the mitochondrion?***

Stages of Cellular Respiration Cellular respiration captures the energy from food in three main stages—glycolysis, the Krebs cycle, and the electron transport chain. Although cells can use just about any food molecule for energy, we will concentrate on just one as an example—the simple sugar glucose. Glucose first enters a chemical pathway known as glycolysis (gly KAHL ih sis). Only a small amount of energy is captured to produce ATP during this stage. In fact, at the end of glycolysis, about 90 percent of the chemical energy that was available in glucose is still unused, locked in chemical bonds of a molecule called pyruvic (py ROO vik) acid.

How does the cell extract the rest of that energy? First, pyruvic acid enters the second stage of cellular respiration, the Krebs cycle, where a little more energy is generated. The bulk of the energy, however, comes from the final stage of cellular respiration, the electron transport chain. This stage requires reactants from the other two stages of the process, as shown by dashed lines in **Figure 9–2.** How does the electron transport chain extract so much energy from these reactants? It uses one of the world's most powerful electron acceptors—oxygen.

Oxygen and Energy Oxygen is required at the very end of the electron transport chain. Any time a cell's demand for energy increases, its use of oxygen increases, too. As you know, the word *respiration* is often used as a synonym for *breathing*. This is why we have used the term *cellular respiration* to refer to energy-releasing pathways within the cell. The double meaning of respiration points out a crucial connection between cells and organisms: Most of the energy-releasing pathways within cells require oxygen, and that is the reason we need to breathe, to respire.

Pathways of cellular respiration that require oxygen are said to be **aerobic** ("in air"). The Krebs cycle and electron transport chain are both aerobic processes. Even though the Krebs cycle does not *directly* require oxygen, it is classified as an aerobic process because it cannot run without the oxygen-requiring electron transport chain. Glycolysis, however, does not directly require oxygen, nor does it rely on an oxygen-requiring process to run. Glycolysis is therefore said to be **anaerobic** ("without air"). Even though glycolysis is anaerobic, it is considered part of cellular respiration because its final products are key reactants for the aerobic stages.

Recall that mitochondria are structures in the cell that convert chemical energy stored in food to usable energy for the cell. Glycolysis actually occurs in the cytoplasm of a cell, but the Krebs cycle and electron transport chain, which generate the majority of ATP during cellular respiration, take place inside the mitochondria. If oxygen is not present, another anaerobic pathway, known as fermentation, makes it possible for the cell to keep glycolysis running, generating ATP to power cellular activity. You will learn more about fermentation later in this chapter.

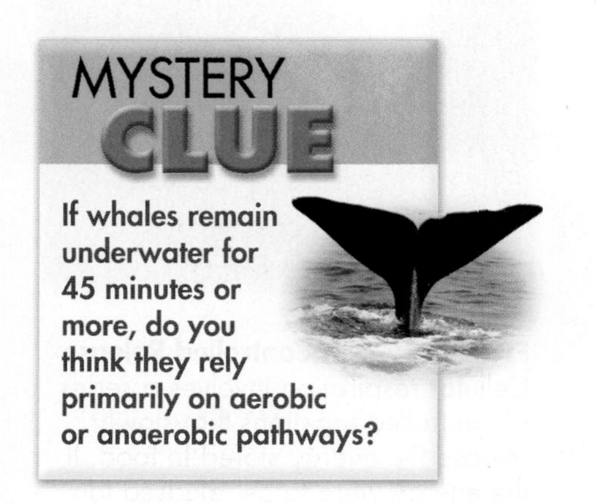

In Your Notebook *Make a flowchart that shows the different steps of cellular respiration.*

Comparing Photosynthesis and Cellular Respiration

What is the relationship between photosynthesis and cellular respiration?

If nearly all organisms break down food by the process of cellular respiration, why doesn't Earth run out of oxygen? Where does all of the carbon dioxide waste product go? How does the chemical energy stored in food get replaced? As it happens, cellular respiration is balanced by another process: photosynthesis. The energy in photosynthesis and cellular respiration flows in opposite directions. Look at **Figure 9–3** and think of the chemical energy in carbohydrates as money in the Earth's savings account. Photosynthesis is the process that "deposits" energy. Cellular respiration is the process that "withdraws" energy. As you might expect, the equations for photosynthesis and cellular respiration are the reverse of each other.

On a global level, photosynthesis and cellular respiration are also opposites. **Photosynthesis removes carbon dioxide from the atmosphere, and cellular respiration puts it back. Photosynthesis releases oxygen into the atmosphere, and cellular respiration uses that oxygen to release energy from food.** The release of energy by cellular respiration takes place in nearly all life: plants, animals, fungi, protists, and most bacteria. Energy capture by photosynthesis, however, occurs only in plants, algae, and some bacteria.

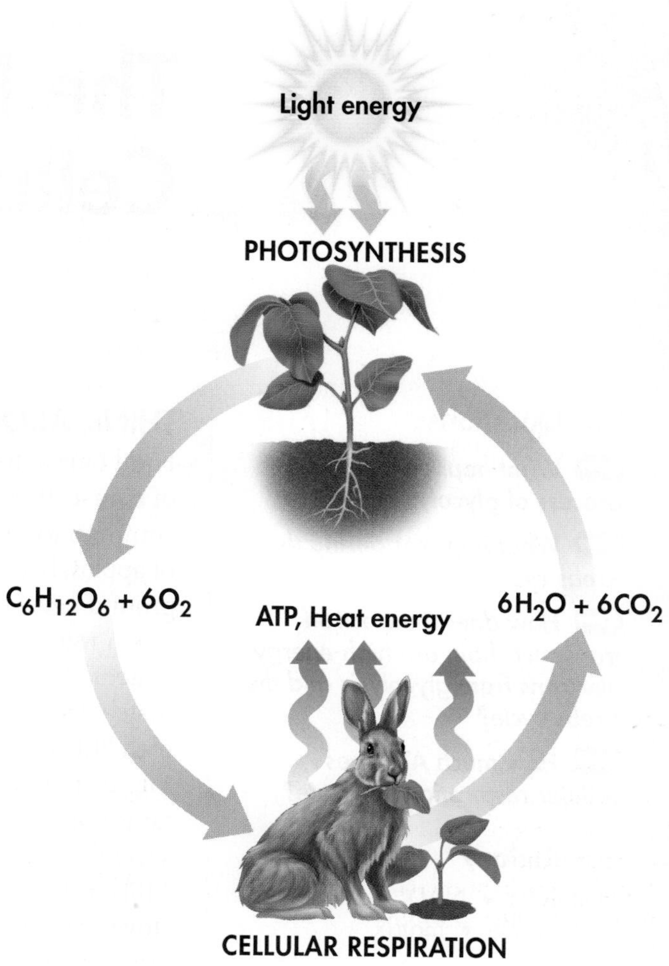

FIGURE 9–3 Opposite Processes Photosynthesis and cellular respiration can be thought of as opposite processes. **Compare and Contrast** ***Exactly how is the equation for photosynthesis different from the equation for cellular respiration?***

9.1 Assessment

Review Key Concepts

1. **a. Review** Why do all organisms need food?
   b. Relate Cause and Effect Why do macromolecules differ in the amount of energy they contain?
2. **a. Review** Write the overall reaction for cellular respiration.
   b. Apply Concepts How does the process of cellular respiration maintain homeostasis at the cellular level?
3. **a. Review** In what ways are cellular respiration and photosynthesis considered opposite processes?
   b. Use Analogies How is the chemical energy in glucose similar to money in a savings account?

BUILD VOCABULARY

4. The Greek word *glukus* means "sweet," and the Latin word *lysis* refers to a process of loosening or decomposing. Based on this information, write a definition for the word *glycolysis*.

9.2 The Process of Cellular Respiration

Key Questions

What happens during the process of glycolysis?

What happens during the Krebs cycle?

How does the electron transport chain use high-energy electrons from glycolysis and the Krebs cycle?

How much ATP does cellular respiration generate?

Vocabulary

glycolysis • NAD^+ • Krebs cycle • matrix

Taking Notes

Compare/Contrast Table As you read, make a compare/contrast table showing the location, starting reactants, and end products of glycolysis, the Krebs cycle, and the electron transport chain. Also include how many molecules of ATP are produced in each step of the process.

THINK ABOUT IT Food burns! It's true, of course, that many common foods (think of apples, bananas, and ground beef) have too much water in them to actually light with a match. However, foods with little water, including sugar and cooking oil, will indeed burn. In fact, flour, which contains both carbohydrates and protein, is so flammable that it has caused several explosions, including the one seen here at London's City Flour Mills in 1872 (which is why you're not supposed to store flour above a stove). So, plenty of energy is available in food, but how does a living cell extract that energy without setting a fire or blowing things up?

Glycolysis

What happens during the process of glycolysis?

The first set of reactions in cellular respiration is known as **glycolysis,** a word that literally means "sugar-breaking." Glycolysis involves many chemical steps that transform glucose. The end result is 2 molecules of a 3-carbon molecule called pyruvic acid. **During glycolysis, 1 molecule of glucose, a 6-carbon compound, is transformed into 2 molecules of pyruvic acid, a 3-carbon compound.** As the bonds in glucose are broken and rearranged, energy is released. The process of glycolysis can be seen in **Figure 9–4.**

ATP Production Even though glycolysis is an energy-releasing process, the cell needs to put in a little energy to get things going. At the pathway's beginning, 2 ATP molecules are used up. Earlier in this chapter, photosynthesis and respiration were compared, respectively, to a deposit to and a withdrawal from a savings account. Similarly, the 2 ATP molecules used at the onset of glycolysis are like an investment that pays back interest. In order to earn interest from a bank, first you have to put money into an account. Although the cell puts 2 ATP molecules into its "account" to get glycolysis going, glycolysis produces 4 ATP molecules. This gives the cell a net gain of 2 ATP molecules for each molecule of glucose that enters glycolysis.

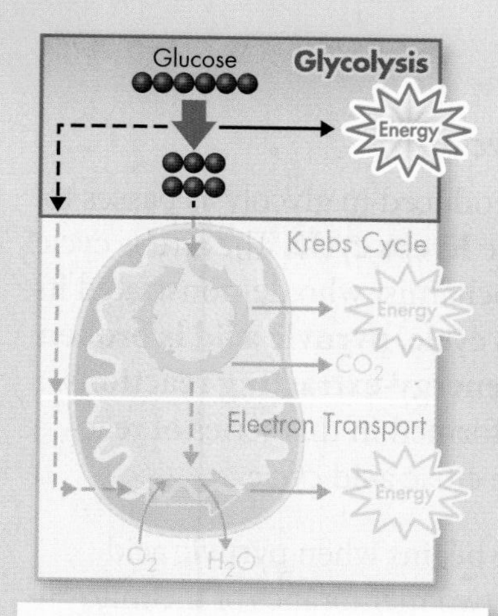

NADH Production
Four high-energy electrons are passed to the carrier NAD+ to produce NADH. NADH carries these electrons to the electron transport chain.

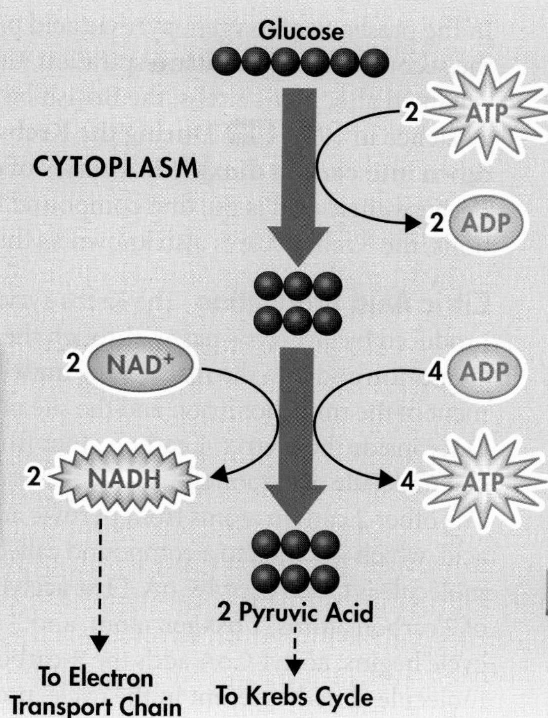

ATP Production
Two ATP molecules are "invested" to get the process of glycolysis going. Overall, 4 ATP molecules are produced, for a net gain of 2 ATP per molecule of glucose.

ZOOMING IN

GLYCOLYSIS

FIGURE 9–4 Glycolysis is the first stage of cellular respiration. During glycolysis, glucose is broken down into 2 molecules of pyruvic acid. ATP and NADH are produced as part of the process. **Interpret Visuals** ***How many carbon atoms are there in glucose? How many carbon atoms are in each molecule of pyruvic acid?***

NADH Production One of the reactions of glycolysis removes 4 electrons, now in a high-energy state, and passes them to an electron carrier called **NAD⁺,** or nicotinamide adenine dinucleotide. Like NADP⁺ in photosynthesis, each NAD⁺ molecule accepts a pair of high-energy electrons. This molecule, now known as NADH, holds the electrons until they can be transferred to other molecules. As you will see, in the presence of oxygen, these high-energy electrons can be used to produce even more ATP molecules.

The Advantages of Glycolysis In the process of glycolysis, 4 ATP molecules are synthesized from 4 ADP molecules. Given that 2 ATP molecules are used to start the process, there is a net gain of just 2 ATP molecules. Although the energy yield from glycolysis is small, the process is so fast that cells can produce thousands of ATP molecules in just a few milliseconds. The speed of glycolysis can be a big advantage when the energy demands of a cell suddenly increase.

Besides speed, another advantage of glycolysis is that the process itself does not require oxygen. This means that glycolysis can quickly supply chemical energy to cells when oxygen is not available. When oxygen is available, however, the pyruvic acid and NADH "outputs" generated during glycolysis become the "inputs" for the other processes of cellular respiration.

BUILD Vocabulary

ACADEMIC WORDS The verb **synthesize** means "to bring together as a whole." Therefore, a molecule of ATP is synthesized when a phosphate group combines with the molecule ADP, forming a high-energy bond.

In Your Notebook *In your own words, describe the advantages of glycolysis to the cell in terms of energy production.*

The Krebs Cycle

MYSTERY CLUE

The urge to surface and gasp for breath when underwater is a response to CO_2 buildup in the blood. The average human can hold his or her breath for only about a minute. Whales stay underwater for much longer. What does this suggest about a whale's tolerance of CO_2?

What happens during the Krebs cycle?

In the presence of oxygen, pyruvic acid produced in glycolysis passes to the second stage of cellular respiration, the **Krebs cycle.** The Krebs cycle is named after Hans Krebs, the British biochemist who demonstrated its existence in 1937. **During the Krebs cycle, pyruvic acid is broken down into carbon dioxide in a series of energy-extracting reactions.** Because citric acid is the first compound formed in this series of reactions, the Krebs cycle is also known as the citric acid cycle.

Citric Acid Production The Krebs cycle begins when pyruvic acid produced by glycolysis passes through the two membranes of the mitochondrion and into the matrix. The **matrix** is the innermost compartment of the mitochondrion and the site of the Krebs cycle reactions. Once inside the matrix, 1 carbon atom from pyruvic acid becomes part of a molecule of carbon dioxide, which is eventually released into the air. The other 2 carbon atoms from pyruvic acid rearrange and form acetic acid, which is joined to a compound called coenzyme A. The resulting molecule is called acetyl-CoA. (The acetyl part of acetyl-CoA is made up of 2 carbon atoms, 1 oxygen atom, and 3 hydrogen atoms.) As the Krebs cycle begins, acetyl-CoA adds the 2-carbon acetyl group to a 4-carbon molecule already present in the cycle, producing a 6-carbon molecule called citric acid.

Energy Extraction As the cycle continues, citric acid is broken down into a 4-carbon molecule, more carbon dioxide is released, and electrons are transferred to energy carriers. Follow the reactions in **Figure 9–5** and you will see how this happens. First, look at the 6 carbon atoms in citric acid. One is removed, and then another, releasing 2 molecules of carbon dioxide and leaving a 4-carbon molecule. Why is the Krebs cycle a "cycle"? Because the 4-carbon molecule produced in the last step is the same molecule that accepts the acetyl-CoA in the first step. The molecule needed to start the reactions of the cycle is remade with every "turn."

Next, look for ATP. For each turn of the cycle, a molecule of ADP is converted to a molecule of ATP. Recall that glycolysis produces 2 molecules of pyruvic acid from 1 molecule of glucose. So, each starting molecule of glucose results in two complete turns of the Krebs cycle and, therefore, 2 ATP molecules. Finally, look at the electron carriers, NAD^+ and FAD (flavine adenine dinucleotide). At five places, electron carriers accept a pair of high-energy electrons, changing NAD^+ to NADH and FAD to $FADH_2$. FAD and $FADH_2$ are molecules similar to NAD^+ and NADH, respectively.

What happens to each of these Krebs cycle products—carbon dioxide, ATP, and electron carriers? Carbon dioxide is not useful to the cell and is expelled every time you exhale. The ATP molecules are *very* useful and become immediately available to power cellular activities. As for the carrier molecules like NADH, in the presence of oxygen, the electrons they hold are used to generate huge amounts of ATP.

In Your Notebook *List the electron carriers involved in the Krebs cycle. Include their names before and after they accept the electrons.*

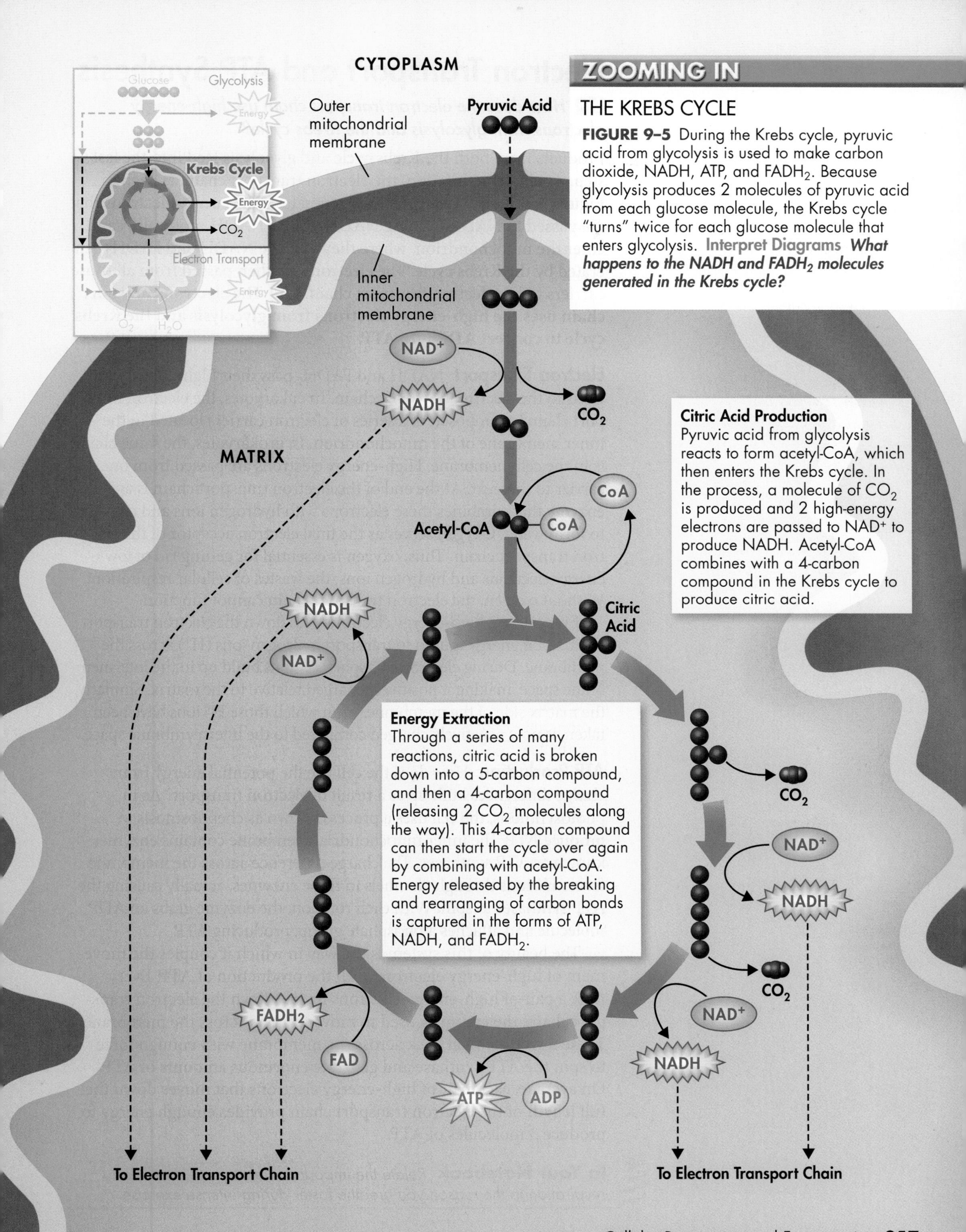

ZOOMING IN

THE KREBS CYCLE

FIGURE 9–5 During the Krebs cycle, pyruvic acid from glycolysis is used to make carbon dioxide, NADH, ATP, and $FADH_2$. Because glycolysis produces 2 molecules of pyruvic acid from each glucose molecule, the Krebs cycle "turns" twice for each glucose molecule that enters glycolysis. **Interpret Diagrams** ***What happens to the NADH and $FADH_2$ molecules generated in the Krebs cycle?***

Electron Transport and ATP Synthesis

How does the electron transport chain use high-energy electrons from glycolysis and the Krebs cycle?

Products from both the Krebs cycle and glycolysis feed into the last step of cellular respiration, the electron transport chain, as seen in **Figure 9–6.** Recall that glycolysis generates high-energy electrons that are passed to NAD^+, forming NADH. Those NADH molecules can enter the mitochondrion, where they join the NADH and $FADH_2$ generated by the Krebs cycle. The electrons are then passed from all those carriers to the electron transport chain. **The electron transport chain uses the high-energy electrons from glycolysis and the Krebs cycle to convert ADP into ATP.**

Electron Transport NADH and $FADH_2$ pass their high-energy electrons to the electron transport chain. In eukaryotes, the electron transport chain is composed of a series of electron carriers located in the inner membrane of the mitochondrion. In prokaryotes, the same chain is in the cell membrane. High-energy electrons are passed from one carrier to the next. At the end of the electron transport chain is an enzyme that combines these electrons with hydrogen ions and oxygen to form water. Oxygen serves as the final electron acceptor of the electron transport chain. Thus, oxygen is essential for getting rid of low-energy electrons and hydrogen ions, the wastes of cellular respiration. Without oxygen, the electron transport chain cannot function.

Every time 2 high-energy electrons pass down the electron transport chain, their energy is used to transport hydrogen ions (H^+) across the membrane. During electron transport, H^+ ions build up in the intermembrane space, making it positively charged relative to the matrix. Similarly, the matrix side of the membrane, from which those H^+ ions have been taken, is now negatively charged compared to the intermembrane space.

ATP Production How does the cell use the potential energy from charge differences built up as a result of electron transport? As in photosynthesis, the cell uses a process known as chemiosmosis to produce ATP. The inner mitochondrial membrane contains enzymes known as ATP synthases. The charge difference across the membrane forces H^+ ions through channels in these enzymes, actually causing the ATP synthases to spin. With each rotation, the enzyme grabs an ADP molecule and attaches a phosphate group, producing ATP.

The beauty of this system is the way in which it couples the movement of high-energy electrons with the production of ATP. Every time a pair of high-energy electrons moves down the electron transport chain, the energy is used to move H^+ ions across the membrane. These ions then rush back across the membrane with enough force to spin the ATP synthase and generate enormous amounts of ATP. On average, each pair of high-energy electrons that moves down the full length of the electron transport chain provides enough energy to produce 3 molecules of ATP.

In Your Notebook *Relate the importance of oxygen in cellular respiration to the reason you breathe faster during intense exercise.*

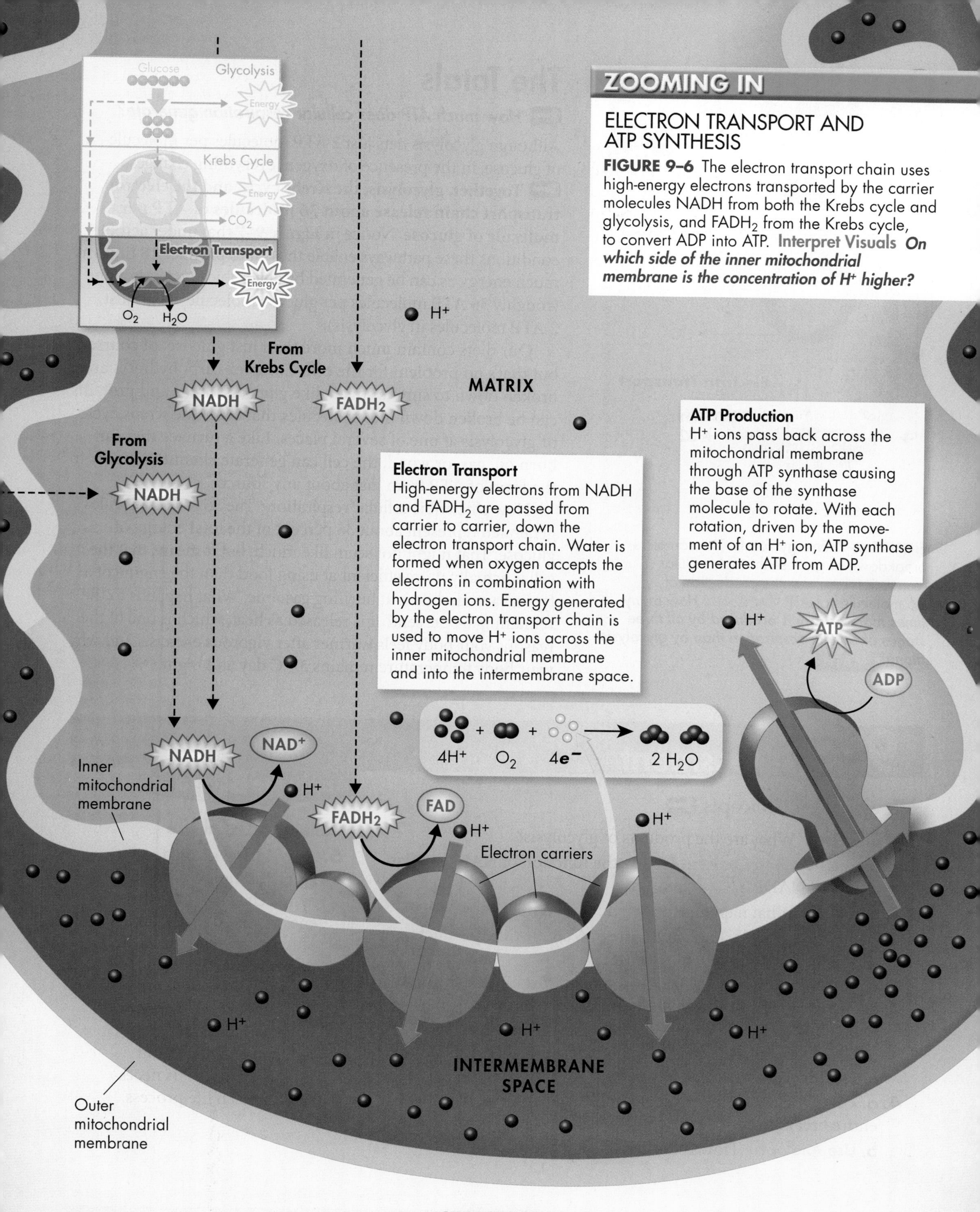

ZOOMING IN

ELECTRON TRANSPORT AND ATP SYNTHESIS

FIGURE 9–6 The electron transport chain uses high-energy electrons transported by the carrier molecules NADH from both the Krebs cycle and glycolysis, and $FADH_2$ from the Krebs cycle, to convert ADP into ATP. **Interpret Visuals** ***On which side of the inner mitochondrial membrane is the concentration of H^+ higher?***

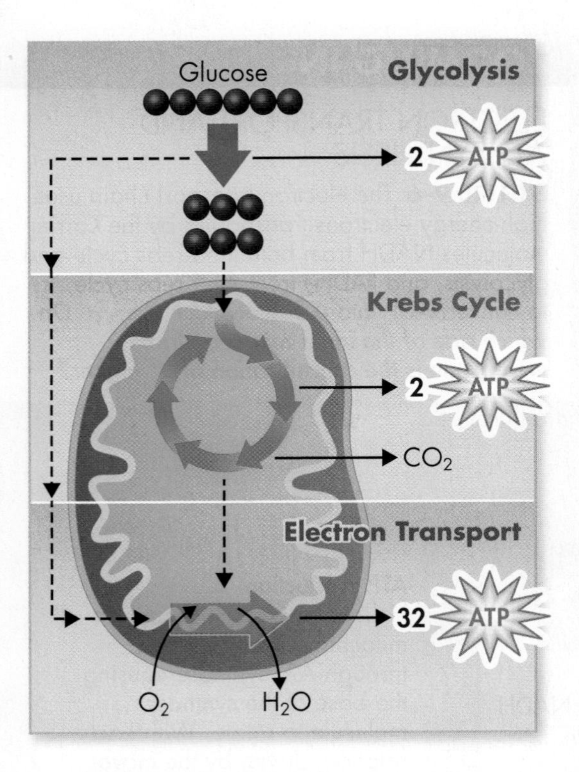

FIGURE 9–7 Energy Totals The complete breakdown of glucose through cellular respiration results in the production of 36 molecules of ATP. Calculate ***How many times more energy is produced by all three stages of cellular respiration than by glycolysis alone?*** MATH

The Totals

How much ATP does cellular respiration generate?

Although glycolysis nets just 2 ATP molecules per molecule of glucose, in the presence of oxygen, everything changes. **Together, glycolysis, the Krebs cycle, and the electron transport chain release about 36 molecules of ATP per molecule of glucose.** Notice in **Figure 9–7** that under aerobic conditions these pathways enable the cell to produce 18 times as much energy as can be generated by anaerobic glycolysis alone (roughly 36 ATP molecules per glucose molecule versus just 2 ATP molecules in glycolysis).

Our diets contain much more than just glucose, of course, but that's no problem for the cell. Complex carbohydrates are broken down to simple sugars like glucose. Lipids and proteins can be broken down into molecules that enter the Krebs cycle or glycolysis at one of several places. Like a furnace that can burn oil, gas, or wood, the cell can generate chemical energy in the form of ATP from just about any source.

How efficient is cellular respiration? The 36 ATP molecules generated represent about 36 percent of the total energy of glucose. That might not seem like much, but it means that the cell is actually more efficient at using food than the engine of a typical automobile is at burning gasoline. What happens to the remaining 64 percent? It is released as heat, which is one of the reasons your body feels warmer after vigorous exercise, and why your body temperature remains 37°C day and night.

9.2 Assessment

Review Key Concepts

1. a. Review What are the products of glycolysis?
   b. Compare and Contrast How is the function of NAD^+ similar to that of $NADP^+$?
2. a. Review What happens to pyruvic acid in the Krebs cycle?
   b. Interpret Visuals Look at **Figure 9–5** and list the products of the Krebs cycle. What happens to each of these products?
3. a. Review How does the electron transport chain use the high-energy electrons from glycolysis and the Krebs cycle?
   b. Relate Cause and Effect How does the cell use the charge differences that build up across the inner mitochondrial membrane during cellular respiration?
4. a. Review How many molecules of ATP are produced in the entire breakdown of glucose?
   b. Use Analogies How is the cell like a furnace?

Apply the Big idea

Cellular Basis of Life

5. As you have learned, cellular respiration is a process by which cells transform energy stored in the bonds of food molecules into the bonds of ATP. What does the body do with all of the ATP this process generates? Review the characteristics of life in Chapter 1 and explain why ATP is necessary for each life process.

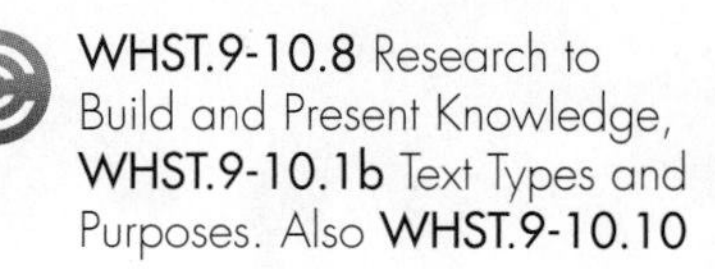

Biology & Society

Should Creatine Supplements Be Regulated?

ATP is the chemical compound that gives muscles the energy to contract, but the amount of ATP in most muscle cells is only enough for a few seconds of activity. Muscle cells have a chemical trick, however, that enables them to sustain maximum effort for several more seconds. They attach phosphate groups to a compound called creatine. As they contract, the cells quickly transfer phosphate from creatine to ADP, producing enough ATP to keep working. The creatine phosphate in skeletal muscles effectively doubles or triples the amount of ATP available for intense exercise.

If a little creatine is good, then more creatine would be even better, right? That's what many athletes think and that's why they take creatine supplements. Some studies do suggest that creatine may increase the body's capacity for strong, short-term muscle contractions. As a reason to regulate the use of creatine, however, critics point to potentially serious side effects—such as liver and kidney damage—when creatine is overused.

Because creatine occurs naturally in the body and in foods, testing for creatine use is nearly impossible; so, creatine is *not* banned in major sports leagues. However, due to a lack of long-term studies, the NCAA prohibits coaches from giving creatine to college athletes. Some schools argue that creatine should be banned altogether.

The Viewpoints

Creatine Supplements Should Not Be Regulated Taken in recommended doses, creatine helps build muscle strength and performance. Creatine supplements may help athletes train longer and build strength. No serious side effects have been reported in people who follow the instructions on container labels. Of course, anything can be harmful when abused, but creatine should not be treated any differently from other substances such as caffeine or sugar.

Creatine Supplements Should Be Regulated Scientists know that creatine can cause severe health problems when abused. But even when used properly, creatine is known to cause some problems, such as dehydration and stomach upset. There have been no adequate studies on creatine use by people younger than 18, and there are no good studies of its long-term effects. For these reasons, creatine supplements should be regulated like cigarettes and alcohol—no one under the age of 18 should be allowed to buy them, and schools should have the right to regulate or prohibit their use by athletes.

Research and Decide

1. **Analyze the Viewpoints** Learn more about this issue by consulting library or Internet resources. Then, list the key arguments of the proponents and critics of creatine use.
2. **Form an Opinion** Should creatine be regulated? Research examples of high schools or colleges that have banned creatine use by athletes. What were the reasons for these decisions? Do you agree with them?

9.3 Fermentation

Key Questions

How do organisms generate energy when oxygen is not available?

How does the body produce ATP during different stages of exercise?

Vocabulary

fermentation

Taking Notes

Outline Before you read, make an outline using the green and blue headings in the text. As you read, fill in notes under each heading.

THINK ABOUT IT We are air-breathing organisms, and we use oxygen to release chemical energy from the food we eat. But what if oxygen is not around? What happens when you hold your breath and dive under water, or use up oxygen so quickly that you cannot replace it fast enough? Do your cells simply stop working? And, what about microorganisms that live in places where oxygen is not available? Is there a pathway that allows cells to extract energy from food in the absence of oxygen?

Fermentation

How do organisms generate energy when oxygen is not available?

Recall from earlier in this chapter that two benefits of glycolysis are that it can produce ATP quickly and that it does not require oxygen. However, when a cell generates large amounts of ATP from glycolysis, it runs into a problem. In just a few seconds, all of the cell's available NAD^+ molecules are filled up with electrons. Without oxygen, the electron transport chain does not run, so there is nowhere for the NADH molecules to deposit their electrons. Thus, NADH does not get converted back to NAD^+. Without NAD^+, the cell cannot keep glycolysis going, and ATP production stops. That's where a process called fermentation comes in.

When oxygen is not present, glycolysis is followed by a pathway that makes it possible to continue to produce ATP without oxygen. The combined process of this pathway and glycolysis is called **fermentation.** **In the absence of oxygen, fermentation releases energy from food molecules by producing ATP.**

During fermentation, cells convert NADH to NAD^+ by passing high-energy electrons back to pyruvic acid. This action converts NADH back into the electron carrier NAD^+, allowing glycolysis to produce a steady supply of ATP. Fermentation is an anaerobic process that occurs in the cytoplasm of cells. Sometimes, glycolysis and fermentation are together referred to as anaerobic respiration. There are two slightly different forms of the process—alcoholic fermentation and lactic acid fermentation, as seen in **Figure 9–8.**

BUILD Vocabulary

RELATED WORD FORMS The noun **fermentation** and the verb *ferment* are related word forms. Dough that is beginning to ferment is just starting to undergo the process of fermentation.

In Your Notebook *Make a compare/contrast table in which you compare alcoholic fermentation to lactic acid fermentation.*

Alcoholic Fermentation Yeasts and a few other microorganisms use alcoholic fermentation, which produces ethyl alcohol and carbon dioxide. A summary of alcoholic fermentation after glycolysis is

$$\text{Pyruvic acid} + \text{NADH} \longrightarrow \text{Alcohol} + CO_2 + NAD^+$$

Alcoholic fermentation is used to produce alcoholic beverages. It is also the process that causes bread dough to rise. When yeast cells in the dough run out of oxygen, the dough begins to ferment, giving off tiny bubbles of carbon dioxide. These bubbles form the air spaces you see in a slice of bread. The small amount of alcohol produced in the dough evaporates when the bread is baked.

Lactic Acid Fermentation Most organisms carry out fermentation using a chemical reaction that converts pyruvic acid to lactic acid. Unlike alcoholic fermentation, lactic acid fermentation does not give off carbon dioxide. However, like alcoholic fermentation, lactic acid fermentation also regenerates NAD^+ so that glycolysis can continue. Lactic acid fermentation after glycolysis can be summarized as

$$\text{Pyruvic acid} + \text{NADH} \longrightarrow \text{Lactic acid} + NAD^+$$

Certain bacteria that produce lactic acid as a waste product during fermentation are important to industry. For example, prokaryotes are used in the production of a wide variety of foods and beverages—such as cheese, yogurt, buttermilk, and sour cream—to which the acid contributes the familiar sour taste. Pickles, sauerkraut, and kimchi are also produced using lactic acid fermentation.

Humans are lactic acid fermenters. During brief periods without oxygen, many of the cells in our bodies are capable of producing ATP by lactic acid fermentation. The cells best adapted to doing that, however, are muscle cells, which often need very large supplies of ATP for rapid bursts of activity.

FIGURE 9–8 Fermentation In alcoholic fermentation, pyruvic acid produced by glycolysis is converted into alcohol and carbon dioxide. Lactic acid fermentation converts the pyruvic acid to lactic acid. **Compare and Contrast** ***What reactants and products do the two types of fermentation have in common?***

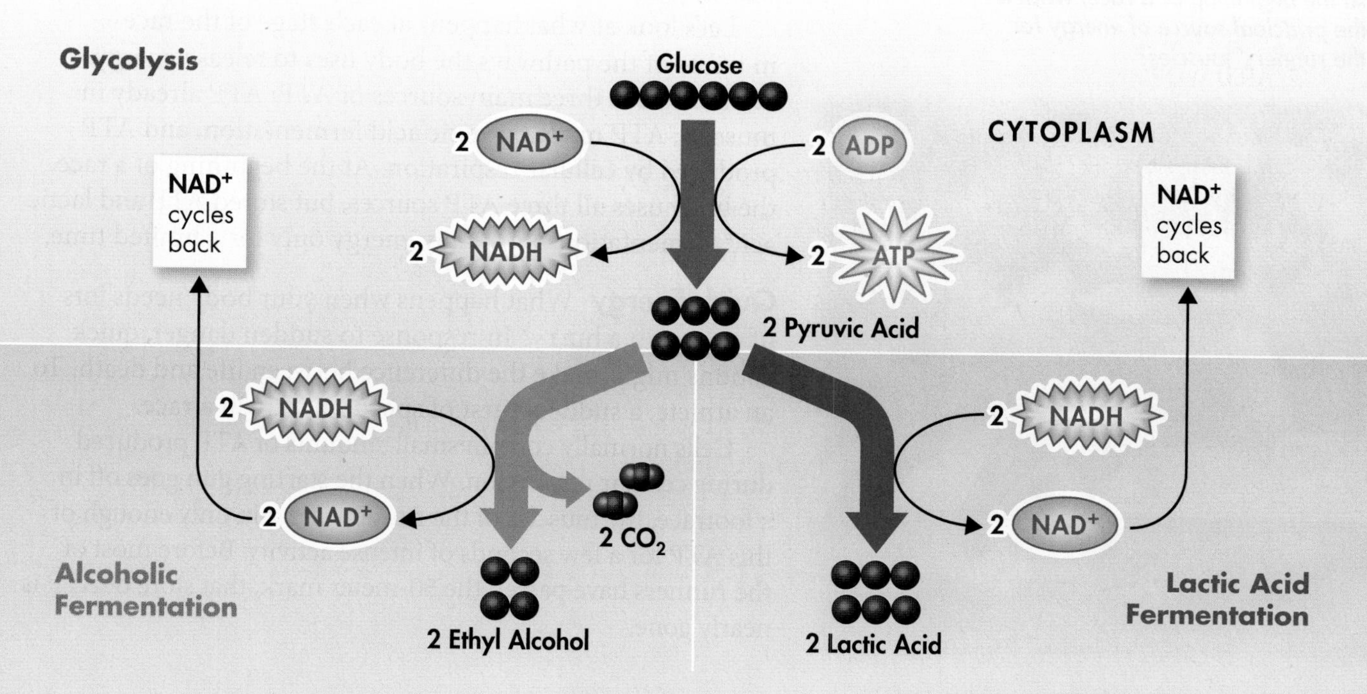

Quick Lab
GUIDED INQUIRY

How Does Exercise Affect Disposal of Wastes From Cellular Respiration?

1. Label two test tubes A and B. Put 10 mL of water and a few drops of bromthymol blue solution in each test tube. Carbon dioxide causes bromthymol blue to turn yellow or green.
2. Your partner will time you during this step. When your partner says "go," slowly blow air through a straw into the bottom of test tube A. **CAUTION:** *Do not inhale through the straw.*
3. When the solution changes color, your partner should say "stop" and then record how long the color change took.
4. Jog in place for 2 minutes. **CAUTION:** *Do not do this if you have a medical condition that interferes with exercise. If you feel faint or dizzy, stop immediately and sit down.*
5. Repeat steps 2–4 using test tube B.
6. Trade roles with your partner. Repeat steps 1 through 5.

Analyze and Conclude

1. Analyze Data How did exercise affect the time it took the solution to change color?

2. Infer What process in your body produces carbon dioxide? How does exercise affect this process?

Energy and Exercise

How does the body produce ATP during different stages of exercise?

Bang! The starter's pistol goes off, and the runners push off their starting blocks and sprint down the track, as seen in **Figure 9–9.** The initial burst of energy soon fades, and the runners settle down to a steady pace. After the runners hit the finish line, they walk around slowly and breathe deeply to catch their breath.

Let's look at what happens at each stage of the race in terms of the pathways the body uses to release energy. Humans have three main sources of ATP: ATP already in muscles, ATP made by lactic acid fermentation, and ATP produced by cellular respiration. At the beginning of a race, the body uses all three ATP sources, but stored ATP and lactic acid fermentation can supply energy only for a limited time.

FIGURE 9–9 Exercise and Energy During a race, runners rely on the energy supplied by ATP to make it to the finish line. **Apply Concepts** ***At the beginning of a race, what is the principal source of energy for the runners' muscles?***

Quick Energy What happens when your body needs lots of energy in a hurry? In response to sudden danger, quick actions might make the difference between life and death. To an athlete, a sudden burst of speed might win a race.

Cells normally contain small amounts of ATP produced during cellular respiration. When the starting gun goes off in a footrace, the muscles of the runners contain only enough of this ATP for a few seconds of intense activity. Before most of the runners have passed the 50-meter mark, that store of ATP is nearly gone.

At this point, the runners' muscle cells are producing most of their ATP by lactic acid fermentation, which can usually supply enough ATP to last about 90 seconds. In a 200- or 300-meter sprint, this may be just enough to reach the finish line.

Fermentation produces lactic acid as a byproduct. When the race is over, the only way to get rid of lactic acid is in a chemical pathway that requires extra oxygen. For that reason, you can think of a quick sprint as building up an oxygen debt that a runner has to repay with plenty of heavy breathing after the race. An intense effort that lasts just 10 or 20 seconds may produce an oxygen debt that requires several minutes of huffing and puffing to clear. **For short, quick bursts of energy, the body uses ATP already in muscles as well as ATP made by lactic acid fermentation.**

MYSTERY CLUE

Whales rely on lactic acid fermentation for much of their energy requirements during a deep dive. If they can't inhale to repay their oxygen debt, what are they doing with all of the lactic acid produced by fermentation?

Long-Term Energy What happens if a race is longer? How does your body generate the ATP it needs to run 2 kilometers or more, or to play in a soccer game that lasts more than an hour? **For exercise longer than about 90 seconds, cellular respiration is the only way to continue generating a supply of ATP.** Cellular respiration releases energy more slowly than fermentation does, which is why even well-conditioned athletes have to pace themselves during a long race or over the course of a game. Your body stores energy in muscle and other tissues in the form of the carbohydrate glycogen. These stores of glycogen are usually enough to last for 15 or 20 minutes of activity. After that, your body begins to break down other stored molecules, including fats, for energy. This is one reason why aerobic forms of exercise such as running, dancing, and swimming are so beneficial for weight control. Some organisms, like the bear in **Figure 9–10,** count on energy stored in fat to get them through long periods without food.

FIGURE 9–10 Energy Storage Animals that hibernate, such as this brown bear in Germany, rely on stored fat for energy when they sleep through the winter. **Predict** ***How will this bear look different when it wakes up from hibernation?***

9.3 Assessment

Review Key Concepts

1. a. **Review** Name the two main types of fermentation.

   b. **Compare and Contrast** How are alcoholic fermentation and lactic acid fermentation similar? How are they different?

2. a. **Review** Why do runners breathe heavily after a sprint race?

   b. **Sequence** List the body's sources of energy in the order in which they are used during a long-distance race.

PRACTICE PROBLEM

3. You have opened a bakery, selling bread made according to your family's secret recipe. Unfortunately, most customers find the bread too heavy. Review what you have learned about chemical reactions in Chapter 2 and make a list of factors such as temperature that might affect the enzyme-catalyzed fermentation reaction involved in baking bread. Predict how each factor will affect the rate of fermentation and propose a solution for making the bread lighter by adding more bubbles to your family bread recipe.

NGSS Smart Guide Cellular Respiration and Fermentation

Disciplinary Core Idea LS1.C Organization for Matter and Energy Flow in Organisms: How do organisms obtain and use the matter and energy they need to live and grow? Organisms obtain the energy they need from the breakdown of food molecules by cellular respiration and fermentation.

9.1 Cellular Respiration: An Overview

- Organisms get the energy they need from food.
- Cellular respiration is the process that releases energy from food in the presence of oxygen.
- Photosynthesis removes carbon dioxide from the atmosphere, and cellular respiration puts it back. Photosynthesis releases oxygen into the atmosphere, and cellular respiration uses that oxygen to release energy from food.

calorie 250 • cellular respiration 251 • aerobic 252 • anaerobic 252

Biology.com

Untamed Science Video Go underwater with the Untamed Science crew to discover why marine mammals can stay submerged for such a long time.

9.2 The Process of Cellular Respiration

- During glycolysis, 1 molecule of glucose, a 6-carbon compound, is transformed into 2 molecules of pyruvic acid, a 3-carbon compound.
- During the Krebs cycle, pyruvic acid is broken down into carbon dioxide in a series of energy-extracting reactions.
- The electron transport chain uses the high-energy electrons from glycolysis and the Krebs cycle to convert ADP into ATP.
- Together, glycolysis, the Krebs cycle, and the electron transport chain release about 36 molecules of ATP per molecule of glucose.

glycolysis 254 • NAD$^+$ 255 • Krebs cycle 256 • matrix 256

Biology.com

Art Review Review the components of electron transport and ATP synthesis.

Tutor Tube Improve your understanding of respiration by working "backward" from a breath of oxygen.

Interactive Art See glycolysis and the Krebs cycle in action.

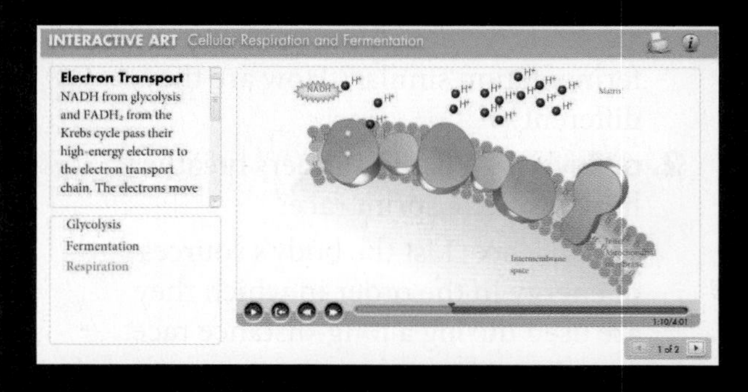

CHECKING YOUR *Scientific Literacy*

Refer to the lesson content and digital assets below as you prepare for your chapter assessment. Then, evaluate your understanding of cellular respiration and fermentation by answering these questions.

1. **Scientific and Engineering Practice Using Mathematics and Computational Thinking** Create a table that summarizes the reactants, products, and number of energy molecules as indicated in Figures 9-4 *Glycolysis*, 9-5 *The Krebs Cycle*, and 9-6 *Electron Transport and ATP Synthesis.*
2. **Crosscutting Concept Energy and Matter: Flows, Cycles, and Conservation** Expand your table to include the chemical equations for the reactions that occur during cellular respiration.
3. **Integration of Knowledge and Ideas** Using Figure 9-7 *Energy Totals* for reference, draw a diagram showing the locations where the three stages of cellular respiration take place. Include in your diagram chemical equations and labels indicating the various cell structures.
4. **STEM** You are a biochemist at a dairy company and have been asked to create a poster to be displayed at a yogurt festival. Your poster will explain to consumers the process of lactic acid fermentation and its role in yogurt production. Outline the contents of your poster.

9.3 Fermentation

- In the absence of oxygen, fermentation releases energy from food molecules by producing ATP.
- For short, quick bursts of energy, the body uses ATP already in muscles as well as ATP made by lactic acid fermentation.
- For exercise longer than about 90 seconds, cellular respiration is the only way to continue generating a supply of ATP.

fermentation 262

Biology.com

Data Analysis Analyze the role of lactic acid in exercise and learn about its effects on athletic performance.

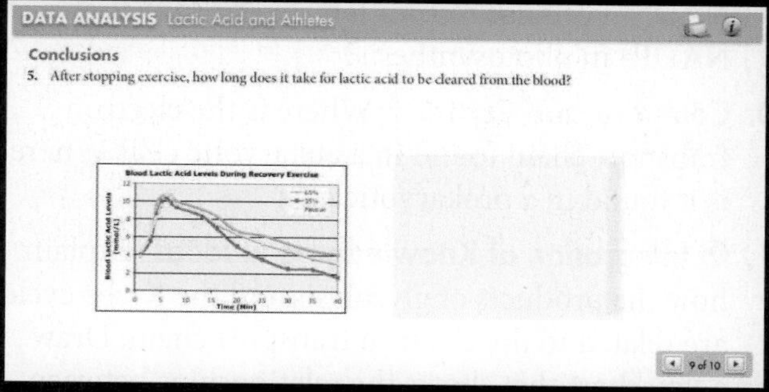

DATA ANALYSIS Lactic Acid and Athletes

Conclusions

5. After stopping exercise, how long does it take for lactic acid to be cleared from the blood?

Blood Lactic Acid Levels During Recovery Exercise

Blood Lactic Acid Levels (mmol/L)

Time (Min)

9 of 10

Lesson Notes Compare the location, reactants, products, and number of ATP molecules produced during the different stages of cellular respiration.

TAKING NOTES The Process of Cellular Respiration

As you read, make a compare/contrast table showing the location, starting reactants, and end products of glycolysis, the Krebs cycle, and the electron transport chain. Also include how many molecules of ATP are produced in each step of the process.

Cellular Respiration			
	Glycolysis	The Krebs Cycle	Electron Transport Chain
Location			
Starting Reactants			
End Products			
# of ATP Molecules			

Write your notes and answers in the spaces provided.

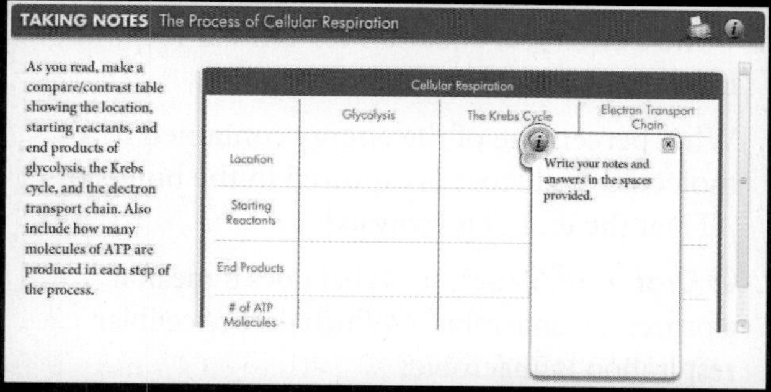

Real-World Lab

Comparing Fermentation Rates of Sugars This chapter lab is available in your lab manual and at Biology.com

9 Assessment

9.1 Cellular Respiration: An Overview

Understand Key Concepts

1. Cells use the energy available in food to make a final energy-rich compound called
   a. water. c. ATP.
   b. glucose. d. ADP.

2. Each gram of glucose contains approximately how much energy?
   a. 1 calorie c. 4 calories
   b. 1 Calorie d. 4 Calories

3. The process that releases energy from food in the presence of oxygen is
   a. synthesis. c. ATP synthase.
   b. cellular respiration. d. photosynthesis.

4. The first step in releasing the energy of glucose in the cell is known as
   a. fermentation. c. the Krebs cycle.
   b. glycolysis. d. electron transport.

5. Which of the following organisms perform cellular respiration?

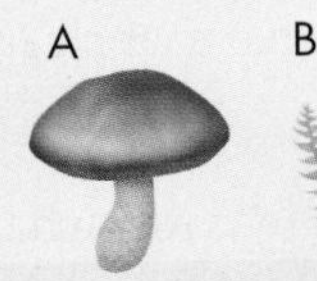

   a. only C c. only B and D
   b. only A and C d. all of the above

6. What is a calorie? Briefly explain how cells use a high-calorie molecule such as glucose.
7. Write a chemical equation for cellular respiration. Label the molecules involved.
8. What percentage of the energy contained in a molecule of glucose is captured in the bonds of ATP at the end of glycolysis?
9. **Craft and Structure** What does it mean if a process is "anaerobic"? Which part of cellular respiration is anaerobic?

Think Critically

10. **Use Analogies** Why is comparing cellular respiration to a burning fire a poor analogy?
11. **Craft and Structure** Why are cellular respiration and photosynthesis considered opposite reactions?

9.2 The Process of Cellular Respiration

Understand Key Concepts

12. The net gain of energy in glycolysis from one molecule of glucose is
    a. 4 ATP molecules. c. 8 ADP molecules.
    b. 2 ATP molecules. d. 3 pyruvic acid molecules.

13. In eukaryotes, the Krebs cycle takes place within the
    a. chloroplast. c. mitochondrion.
    b. nucleus. d. cytoplasm.

14. The electron transport chain uses the high-energy electrons from the Krebs cycle to
    a. produce glucose.
    b. move H^+ ions across the inner mitochondrial membrane.
    c. convert acetyl-CoA to citric acid.
    d. convert glucose to pyruvic acid.

15. How is glucose changed during glycolysis?
16. What is NAD^+? Why is it important?
17. **Key Ideas and Details** Summarize what happens during the Krebs cycle. What happens to high-energy electrons generated during the Krebs cycle?
18. How is ATP synthase involved in making energy available to the cell?

Think Critically

19. **Compare and Contrast** How is the function of NAD^+ in cellular respiration similar to that of $NADP^+$ in photosynthesis?
20. **Compare and Contrast** Where is the electron transport chain found in a eukaryotic cell? Where is it found in a prokaryotic cell?
21. **Integration of Knowledge and Ideas** Explain how the products of glycolysis and the Krebs cycle are related to the electron transport chain. Draw a flowchart that shows the relationships between these products and the electron transport chain.
22. **Use Models** Draw and label a mitochondrion surrounded by cytoplasm. Indicate where glycolysis, the Krebs cycle, and the electron transport chain occur in a eukaryotic cell.

9.3 Fermentation

Understand Key Concepts

23. Because fermentation takes place in the absence of oxygen, it is said to be

a. aerobic.
b. anaerobic.
c. cyclic.
d. oxygen-rich.

24. The process carried out by yeast that causes bread dough to rise is

a. alcoholic fermentation.
b. lactic acid fermentation.
c. cellular respiration.
d. yeast mitosis.

25. During heavy exercise, the buildup of lactic acid in muscle cells results in

a. cellular respiration.
b. oxygen debt.
c. fermentation.
d. the Krebs cycle.

26. **Craft and Structure** How are fermentation and cellular respiration similar?

27. Write equations to show how lactic acid fermentation compares with alcoholic fermentation. Which reactant(s) do they have in common?

Think Critically

28. **Infer** Certain types of bacteria thrive in conditions that lack oxygen. What does that fact indicate about the way they obtain energy?

29. **Infer** To function properly, heart muscle cells require a steady supply of oxygen. After a heart attack, small amounts of lactic acid are present. What does this evidence suggest about the nature of a heart attack?

30. **Predict** In certain cases, regular exercise causes an increase in the number of mitochondria in muscle cells. How might that situation improve an individual's ability to perform energy-requiring activities?

31. **Formulate Hypotheses** Yeast cells can carry out both fermentation and cellular respiration, depending on whether oxygen is present. In which case would you expect yeast cells to grow more rapidly? Explain.

32. **Key Ideas and Details** Carbon monoxide (CO) molecules bring the electron transport chain in a mitochondrion to a stop by binding to an electron carrier. Use this information to explain why carbon monoxide gas kills organisms.

solve the CHAPTER MYSTERY

DIVING WITHOUT A BREATH

To be able to sustain regular 45-minute intervals underwater, whales employ a number of special mechanisms. For example, whale blood is very tolerant of CO_2 buildup that results from the Krebs cycle. This allows whales to stay underwater for an extended period without triggering the reflex to surface. The Krebs cycle and electron transport rely on oxygen, of course. And once the oxygen is used—and it's used quickly!—whale muscles must rely on lactic acid fermentation to generate energy. In humans, lactic acid causes the pH of the blood to drop. If the blood gets too acidic, a dangerous condition called acidosis can occur. Whale muscles are extremely tolerant of lactic acid. The lactic acid remains in the muscles without causing acidosis. When whales resurface after a long dive, they inhale oxygen that clears away the lactic acid buildup.

1. **Relate Cause and Effect** Why must whales have blood that is tolerant of CO_2?

2. **Predict** Myoglobin, a molecule very similar to hemoglobin, stores oxygen in muscles. Would you expect to find more or less myoglobin than average in the muscle tissue of whales if you were to examine it under the microscope?

3. **Infer** How might being able to dive into very deep water be an advantage for whales such as the sperm whale?

4. **Connect to the Big idea** When swimming near the surface, whales breathe every time their heads break out of the water. How do you think the energy pathways used during this type of swimming differ from the ones used during long dives?

Connecting Concepts

Use Science Graphics

Use the nutritional information below to answer questions 33–35.

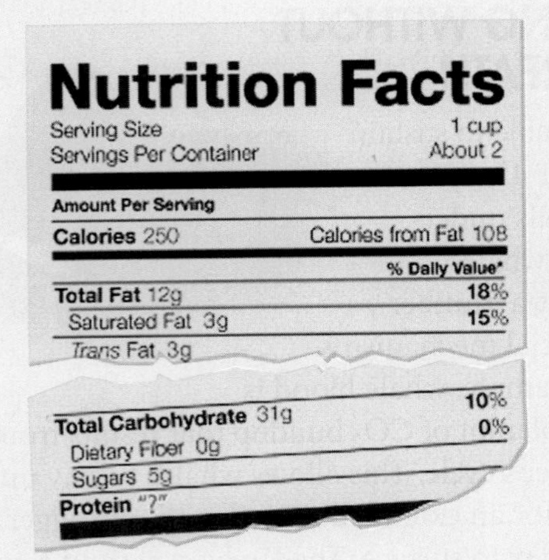

33. Apply Concepts On average, how many Calories are there in 1 gram of a lipid, carbohydrate, and protein? Why the differences?

34. Calculate How many grams of protein must there be in order to account for the number of Calories per serving indicated? MATH

35. Calculate Look at the percent daily value column on the food label. The percent daily value represents the proportion of a typical day's Calories that, on average, should be contributed from the category listed. For example, 31 g of carbohydrates is approximately 10 percent of a daily value. So, a typical person's daily diet should contain about 310 g of carbohydrates. How many Calories does this represent? What percentage of a typical 2000-Calories-per-day diet should therefore come from carbohydrates? MATH

Write About Science

36. Text Types and Purposes Expand the analogy of deposits and withdrawals of money that was used in the chapter to write a short paragraph that explains cellular respiration. (*Hint:* Think about what "inputs" or deposits are required and what "outputs" or returns are produced at each step.)

37. Assess the Big idea Draw a sketch that shows respiration (breathing) at the organismal, or whole animal, level. Draw another sketch that shows the overall process of cellular respiration. How do your sketches show breathing and cellular respiration as related processes?

Integration of Knowledge and Ideas

The volume of oxygen uptake was measured in liters per minute (L/min). The scientist collecting the data was interested in how the volume of oxygen breathed in was affected as the difficulty level of the exercise (measured in watts) increased. The data are summarized in the accompanying graph.

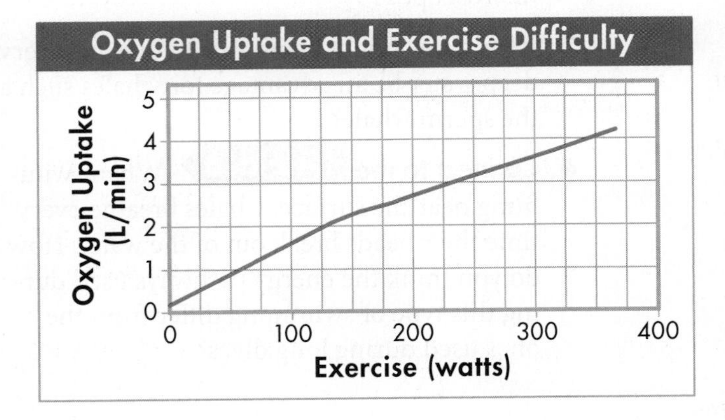

38. Interpret Graphs Based on the graph, at what level of exercise difficulty did oxygen uptake reach 3 L/min?

a. approximately 100 watts
b. approximately 200 watts
c. between 200 and 300 watts
d. between 300 and 400 watts

39. Formulate Hypotheses Which of the following is a valid hypothesis that explains the trend shown on the graph?

a. As exercise becomes more difficult, the body relies more and more on lactic acid fermentation.
b. Exercise below a level of 100 watts does not require increased oxygen uptake.
c. Difficult exercise requires additional oxygen intake in order to generate extra ATP for muscle cells.
d. The human body cannot maintain exercise levels above 500 watts.

Standardized Test Prep

Multiple Choice

1. Which of the following raw materials would be sufficient to allow cellular respiration to take place?
   A glucose and carbon dioxide
   B glucose and oxygen
   C carbon dioxide and oxygen
   D oxygen and lactic acid

2. During the Krebs cycle
   A hydrogen ions and oxygen form water.
   B the cell releases a small amount of energy through fermentation.
   C each glucose molecule is broken down into 2 molecules of pyruvic acid.
   D pyruvic acid is broken down into carbon dioxide in a series of reactions.

3. Which substance is needed to begin the process of glycolysis?
   A ATP C pyruvic acid
   B NADP D carbon dioxide

4. In eukaryotic cells, MOST of cellular respiration takes place in the
   A nuclei. C mitochondria.
   B cytoplasm. D cell walls.

5. Which substance is broken down during the process of glycolysis?
   A carbon C glucose
   B NAD^+ D pyruvic acid

6. The human body can use all of the following as energy sources EXCEPT
   A ATP in muscles.
   B glycolysis.
   C lactic acid fermentation.
   D alcoholic fermentation.

7. During cellular respiration, which of the following are released as byproducts?
   A CO_2 and O_2
   B H_2O and O_2
   C O_2 and H_2O
   D CO_2 and H_2O

8. Which of the following is an aerobic process?
   A the Krebs cycle C alcoholic fermentation
   B glycolysis D lactic acid fermentation

Questions 9 and 10

The graph below shows the rate of alcoholic fermentation for yeast at different temperatures.

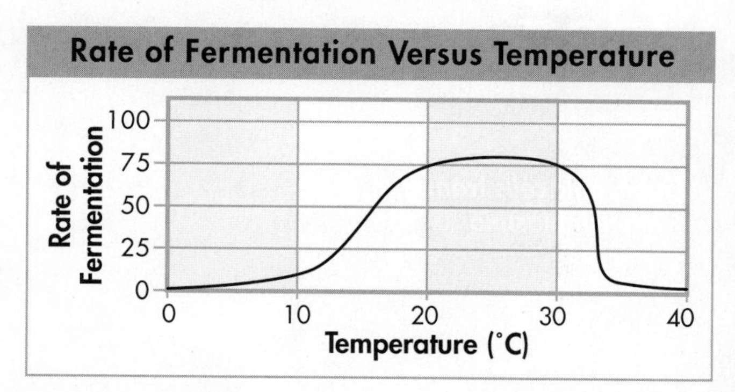

9. According to the graph, what is the relationship between the rate of fermentation and temperature?
   A The rate of fermentation continually increases as temperature increases.
   B The rate of fermentation continually decreases as temperature increases.
   C The rate of fermentation increases with temperature at first, and then it rapidly decreases.
   D The rate of fermentation decreases with temperature at first, and then it rapidly increases.

10. Which statement could explain the data shown in the graph?
   A The molecules that regulate fermentation perform optimally at temperatures above 30°C.
   B The yeast begins releasing carbon dioxide at 30°C.
   C The yeast cannot survive above 30°C.
   D The molecules that regulate fermentation perform optimally at temperatures below 10°C.

Open-Ended Response

11. Explain how a sprinter gets energy during a 30-second race. Is the process aerobic or anaerobic? How does it compare to a long-distance runner getting energy during a 5-kilometer race?

If You Have Trouble With . . .

Question	1	2	3	4	5	6	7	8	9	10	11
See Lesson	9.1	9.2	9.2	9.1	9.2	9.3	9.1	9.1	9.3	9.3	9.3

10 Cell Growth and Division

Growth, Development, and Reproduction

Q: How does a cell produce a new cell?

Embryonic cells from a whitefish blastula (LM 1250)

INSIDE:

- 10.1 Cell Growth, Division, and Reproduction
- 10.2 The Process of Cell Division
- 10.3 Regulating the Cell Cycle
- 10.4 Cell Differentiation

BUILDING *Scientific Literacy*

Deepen your understanding of cell growth and division by engaging in key practices that allow you to make connections across concepts.

NGSS You will construct and use models to demonstrate how cells grow and divide during the cell cycle.

STEM How can cancer be treated? Research new technologies and frontiers in treating cancer.

Common Core You will use quantitative reasoning and cite textual evidence to support your analysis of the processes of cell growth and division.

CHAPTER MYSTERY

PET SHOP ACCIDENT

Julia stared into the salamander tank in horror. As an assistant in a pet shop, Julia had mistakenly put a small salamander in the same tank as a large one. Just as she realized her error, the large salamander attacked and bit off one of the small salamander's limbs.

Acting quickly, Julia scooped up the injured salamander and put it in its own tank. She was sure it would die before her shift ended. But she was wrong! Days passed . . . then weeks. Every time Julia checked on the salamander, she was more amazed at what she saw. How did the salamander's body react to losing a limb? As you read this chapter, look for clues to help you predict the salamander's fate. Think about the cell processes that would be involved. Then, solve the mystery.

Biology.com

Finding the solution to the pet shop mystery is only the beginning. Take a video field trip with the ecogeeks of Untamed Science to see where the mystery leads.

Go online to access additional resources including:
• eText • Flash Cards • Lesson Overviews • Chapter Mystery

10.1 Cell Growth, Division, and Reproduction

Key Questions

What are some of the difficulties a cell faces as it increases in size?

How do asexual and sexual reproduction compare?

Vocabulary

cell division
asexual reproduction
sexual reproduction

Taking Notes

Outline As you read, create an outline about cell growth, division, and reproduction. As you read, fill in key phrases or sentences about each heading.

THINK ABOUT IT When a living thing grows, what happens to its cells? Does an organism get larger because each cell increases in size or because it produces more of them? In most cases, living things grow by producing more cells. What is there about growth that requires cells to divide and produce more of themselves?

Limits to Cell Size

What are some of the difficulties a cell faces as it increases in size?

Nearly all cells can grow by increasing in size, but eventually, most cells divide after growing to a certain point. There are two main reasons why cells divide rather than continuing to grow. **The larger a cell becomes, the more demands the cell places on its DNA. In addition, a larger cell is less efficient in moving nutrients and waste materials across the cell membrane.**

Information "Overload" Living cells store critical information in a molecule known as DNA. As a cell grows, that information is used to build the molecules needed for cell growth. But as a cell increases in size, its DNA does not. If a cell were to grow too large, an "information crisis" would occur.

To get a better sense of information overload, compare a cell to a growing town. Suppose a small town has a library with a few thousand books. As more people move in, more people will borrow books. Sometimes, people may have to wait to borrow popular books. Similarly, a larger cell would make greater demands on its genetic "library." After a while, the DNA would no longer be able to serve the needs of the growing cell—it might be time to build a new library.

Exchanging Materials There is another critical reason why cell size is limited. Food, oxygen, and water enter a cell through its cell membrane. Waste products leave a cell in the same way. The rate at which this exchange takes place depends on the surface area of the cell, which is the total area of its cell membrane. The rate at which food and oxygen are used up and waste products are produced depends on the cell's volume. Understanding the relationship between a cell's surface area and its volume is the key to understanding why cells must divide rather than continue to grow.

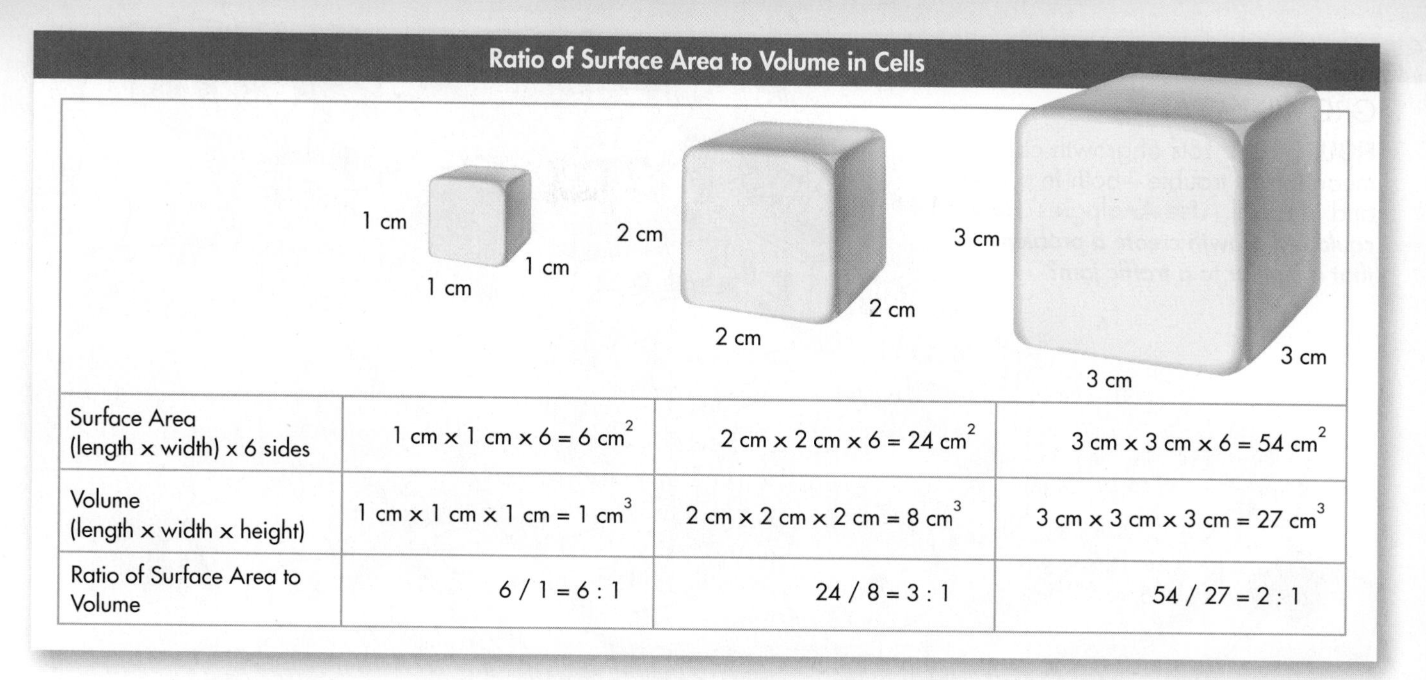

Surface Area (length x width) x 6 sides	1 cm x 1 cm x 6 = 6 cm^2	2 cm x 2 cm x 6 = 24 cm^2	3 cm x 3 cm x 6 = 54 cm^2
Volume (length x width x height)	1 cm x 1 cm x 1 cm = 1 cm^3	2 cm x 2 cm x 2 cm = 8 cm^3	3 cm x 3 cm x 3 cm = 27 cm^3
Ratio of Surface Area to Volume	6 / 1 = 6 : 1	24 / 8 = 3 : 1	54 / 27 = 2 : 1

FIGURE 10–1 Ratio of Surface Area to Volume As the length of the sides increases, the volume increases more than the surface area. **Interpret Tables** ***What are the ratios comparing?***

▶ ***Ratio of Surface Area to Volume*** Imagine a cell that is shaped like a cube, like those shown in **Figure 10–1**. The formula for area ($l \times w$) is used to calculate the surface area. The formula for volume ($l \times w \times h$) is used to calculate the amount of space inside. By using a ratio of surface area to volume, you can see how the size of the cell's surface area grows compared to its volume.

Notice that for a cell with sides that measure 1 cm in length, the ratio of surface area to volume is 6/1 or 6 : 1. Increase the length of the cell's sides to 2 cm, and the ratio becomes 24/8 or 3 : 1. What if the length triples? The ratio of surface area to volume becomes 54/27 or 2 : 1. Notice that the surface area is not increasing as fast as the volume increases. For a growing cell, a decrease in the relative amount of cell membrane available creates serious problems.

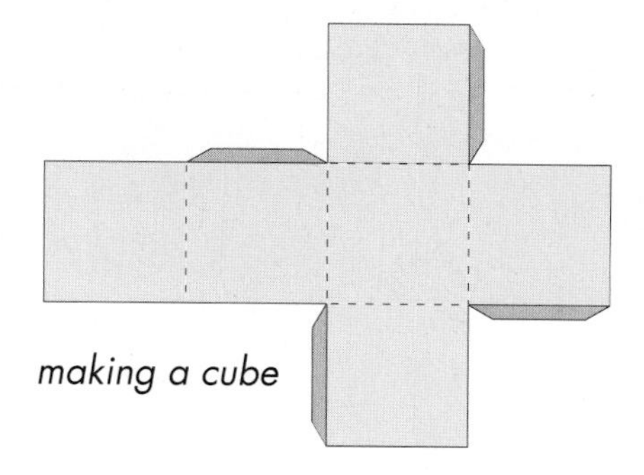

making a cube

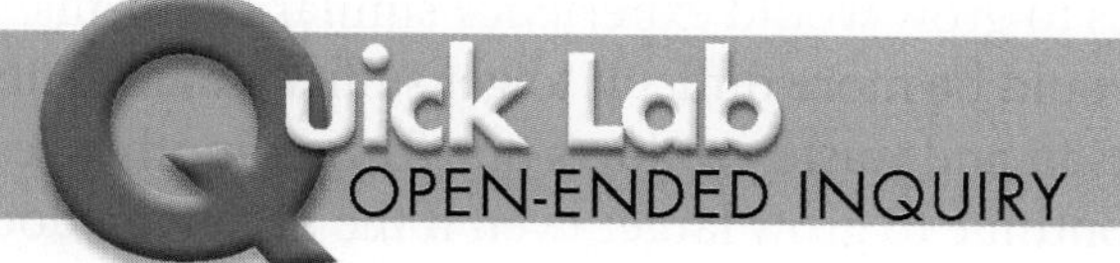

Modeling the Relationship Between Surface Area and Volume

❶ Use the drawing and grid paper to make patterns for a 6-cm cube, a 5-cm cube, a 4-cm cube, and a 3-cm cube.

❷ Cut out your patterns and fold them. Then use the tabs to tape or glue the sides together. Don't tape down the top side.

❸ Construct a data table to compare the volume, the surface area, and the ratio of surface area to volume of each cube.

❹ Use your data to calculate the number of 3-cm cubes that would fit in the same volume as the 6-cm cube. Also calculate the total surface area for the smaller cubes. MATH

Analyze and Conclude

1. Review Describe the function of a cell membrane and its relationship to what happens inside a cell.

2. Apply Concepts How does the surface area change when a large cell divides into smaller cells that have the same total volume?

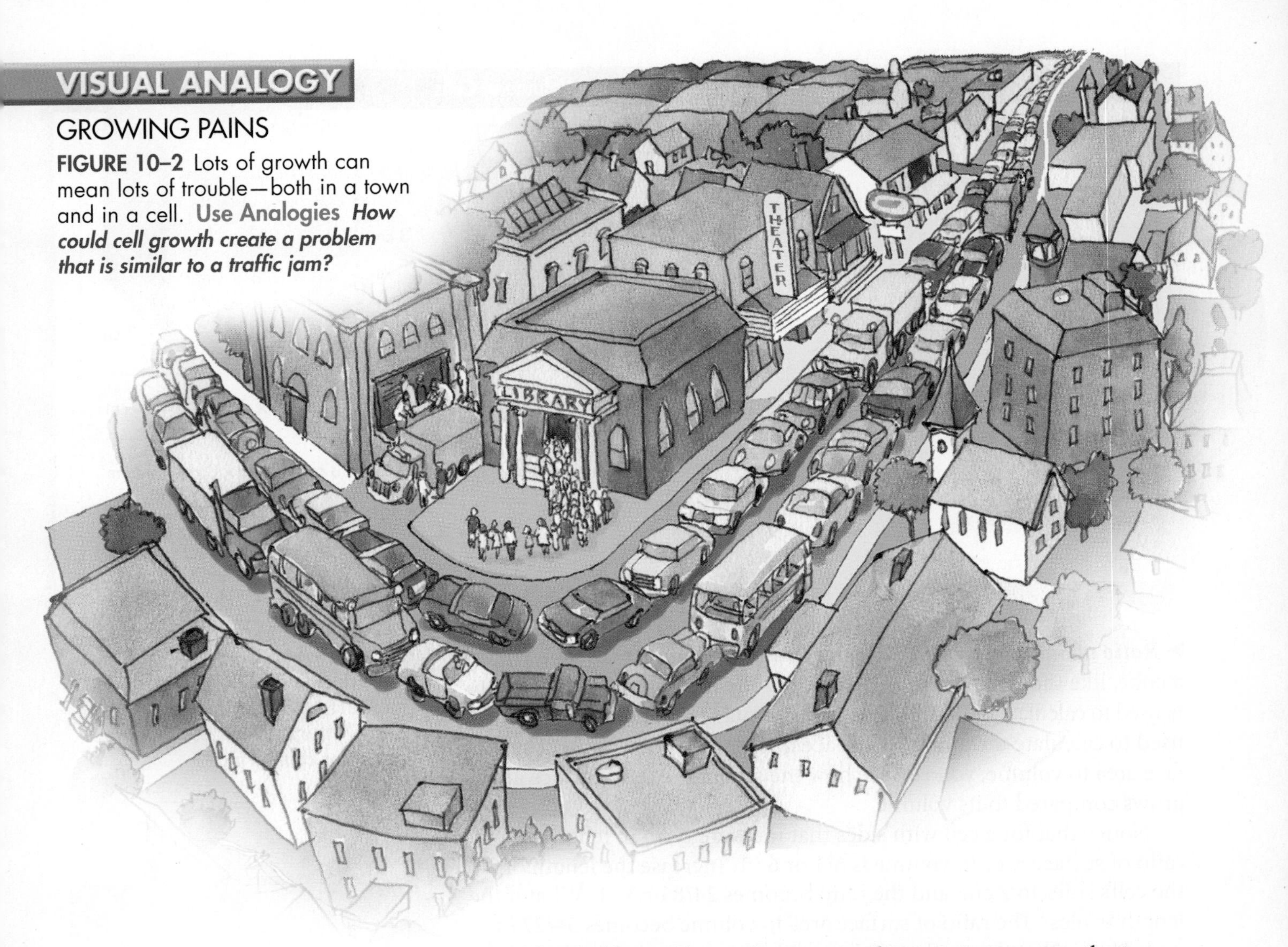

VISUAL ANALOGY

GROWING PAINS

FIGURE 10–2 Lots of growth can mean lots of trouble—both in a town and in a cell. **Use Analogies** ***How could cell growth create a problem that is similar to a traffic jam?***

▶ ***Traffic Problems*** To use the town analogy again, suppose the town has just a two-lane main street leading to the center of town. As the town grows, more and more traffic clogs the main street. It becomes increasingly difficult to move goods in and out.

A cell that continues to grow would experience similar problems. If a cell got too large, it would be more difficult to get sufficient amounts of oxygen and nutrients in and waste products out. This is another reason why cells do not continue to grow larger even if the organism does.

Division of the Cell Before it becomes too large, a growing cell divides, forming two "daughter" cells. The process by which a cell divides into two new daughter cells is called **cell division.**

Before cell division occurs, the cell replicates, or copies all of its DNA. This replication of DNA solves the problem of information overload because each daughter cell gets one complete copy of genetic information. Cell division also solves the problem of increasing size by reducing cell volume. Cell division results in an increase in the ratio of surface area to volume for each daughter cell. This allows for the efficient exchange of materials within a cell.

Cell Division and Reproduction

How do asexual and sexual reproduction compare?

Reproduction, the formation of new individuals, is one of the most important characteristics of living things. For an organism composed of just one cell, cell division can serve as a perfectly good form of reproduction. You don't have to meet someone else, conduct a courtship, or deal with rivals. All you have to do is to divide, and *presto*—there are two of you!

Bacterium
(TEM 32,800×)

Hydra
(LM 25×)

Kalanchoe

FIGURE 10–3 Asexual Reproduction Cell division leads to reproduction in single-celled organisms and some multicellular organisms. **Apply Concepts** ***What do the offspring of each of these organisms have in common?***

Asexual Reproduction For many single-celled organisms, such as the bacterium in **Figure 10–3,** cell division is the only form of reproduction. The process can be relatively simple, efficient, and effective, enabling populations to increase in number very quickly. In most cases, the two cells produced by cell division are genetically identical to the cell that produced them. This kind of reproduction is called **asexual reproduction. The production of genetically identical offspring from a single parent is known as asexual reproduction.**

Asexual reproduction also occurs in many multicellular organisms. The small bud growing off the hydra will eventually break off and become an independent organism, an example of asexual reproduction in an animal. Each of the small shoots or plantlets on the tip of the kalanchoe leaf may also grow into a new plant.

BUILD Vocabulary

PREFIXES The prefix *a-* in *asexual* means "without." **Asexual reproduction** is reproduction without the fusion of reproductive cells.

Sexual Reproduction Unlike asexual reproduction, where cells separate to form a new individual, **sexual reproduction** involves the fusion of two separate parent cells. In sexual reproduction, offspring are produced by the fusion of special reproductive cells formed by each of two parents. **Offspring produced by sexual reproduction inherit some of their genetic information from each parent.** Most animals and plants reproduce sexually, and so do some single-celled organisms. You will learn more about the form of cell division that produces reproductive cells in Chapter 11.

In Your Notebook *Use a Venn diagram to compare asexual and sexual reproduction.*

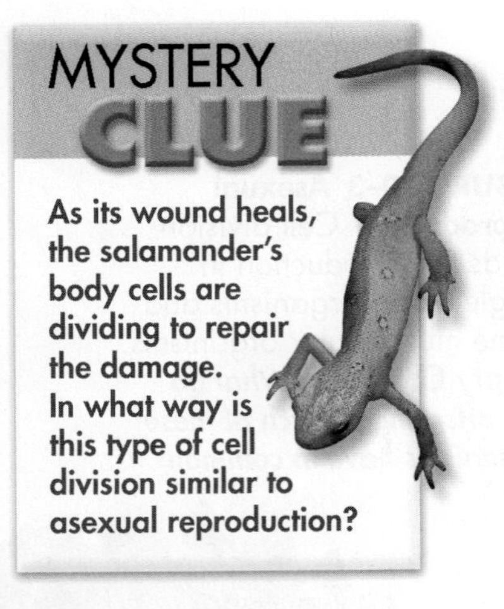

As its wound heals, the salamander's body cells are dividing to repair the damage. In what way is this type of cell division similar to asexual reproduction?

Comparing Asexual and Sexual Reproduction You can see that each type of reproduction has its advantages and disadvantages when you look at each one as a strategy for survival. Species survive by reproducing. The better suited a species is to its environment, the greater its chance of survival.

For single-celled organisms, asexual reproduction is a survival strategy. When conditions are right, the faster they reproduce, the better their chance of survival over other organisms using the same resources. Having offspring that are genetically identical is also an advantage as long as conditions remain favorable. However, a lack of genetic diversity becomes a disadvantage when conditions change in ways that do not fit the characteristics of an organism.

Sexual reproduction is a different type of survival strategy. The process of finding a mate and the growth and development of offspring require more time. However, this can be an advantage for species that live in environments where seasonal changes affect weather conditions and food availability. Sexual reproduction also provides genetic diversity. If an environment changes, some offspring may have the right combination of characteristics needed to survive.

Some organisms reproduce both sexually and asexually. Yeasts, for example, are single-celled eukaryotes that use both strategies. They reproduce asexually most of the time. However, under certain conditions, they enter a sexual phase. The different advantages of each type of reproduction may help to explain why the living world includes organisms that reproduce sexually, those that reproduce asexually, and many organisms that do both.

10.1 Assessment

Review Key Concepts

1. a. Review Identify two reasons why a cell's growth is limited.

b. Explain As a cell's size increases, what happens to the ratio of its surface area to its volume?

c. Applying Concepts Why is a cell's surface area-to-volume ratio important?

2. a. Review What is asexual reproduction? What is sexual reproduction?

b. Explain What types of organisms reproduce sexually?

c. Summarize What are the advantages and disadvantages of both asexual and sexual reproduction?

VISUAL THINKING MATH

3. The formula for finding the surface area of a sphere, such as a baseball or a basketball, is $A = 4\pi r^2$, where r is the radius. The formula for finding the volume of a sphere is $V = 4/3\pi r^3$.

a. Calculate Calculate the surface area and the volume of the baseball and the basketball. Then, write the ratio of surface area to volume for each sphere.

b. Infer If the baseball and basketball were cells, which would possess a larger ratio of area of cell membrane to cell volume?

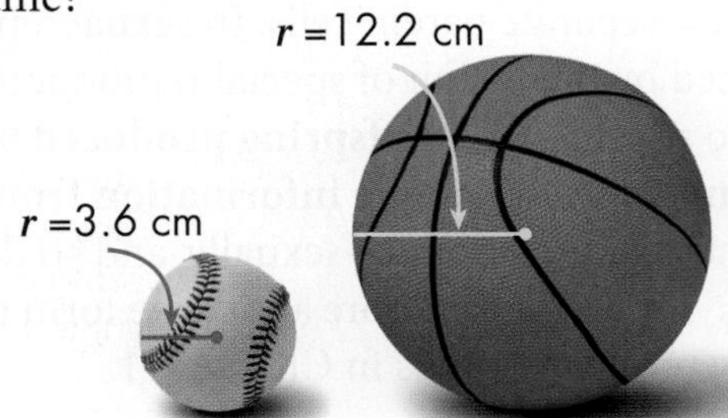

10.2 The Process of Cell Division

THINK ABOUT IT What role does cell division play in your life? You know from your own experience that living things grow, or increase in size, during particular stages of life or even throughout their lifetime. This growth clearly depends on the production of new cells through cell division. But what happens when you are finished growing? Does cell division simply stop? Think about what must happen when your body heals a cut or a broken bone. And finally, think about the everyday wear and tear on the cells of your skin, digestive system, and blood. Cell division has a role to play there, too.

Key Questions

What is the role of chromosomes in cell division?

What are the main events of the cell cycle?

What events occur during each of the four phases of mitosis?

How do daughter cells split apart after mitosis?

Vocabulary

chromosome • chromatin • cell cycle • interphase • mitosis • cytokinesis • prophase • centromere • chromatid • centriole • metaphase • anaphase • telophase

Taking Notes

Two-Column Chart As you read, create a two-column chart. In the left column, make notes about what is happening in each stage of the cell cycle. In the right column, describe what the process looks like or draw pictures.

Chromosomes

What is the role of chromosomes in cell division?

What do you think would happen if a cell were simply to split in two, without any advance preparation? The results might be disastrous, especially if some of the cell's essential genetic information wound up in one of the daughter cells, and not in the other. In order to make sure this doesn't happen, cells first make a complete copy of their genetic information before cell division begins.

Even a small cell like the bacterium *E. coli* has a tremendous amount of genetic information in the form of DNA. In fact, the total length of this bacterium's DNA molecule is 1.6 mm, roughly 1000 times longer than the cell itself. In terms of scale, imagine a 300-meter rope stuffed into a school backpack. Cells can handle such large molecules only by careful packaging. Genetic information is bundled into packages of DNA known as **chromosomes.**

Prokaryotic Chromosomes Prokaryotes lack nuclei and many of the organelles found in eukaryotes. Their DNA molecules are found in the cytoplasm along with most of the other contents of the cell. Most prokaryotes contain a single, circular DNA chromosome that contains all, or nearly all, of the cell's genetic information.

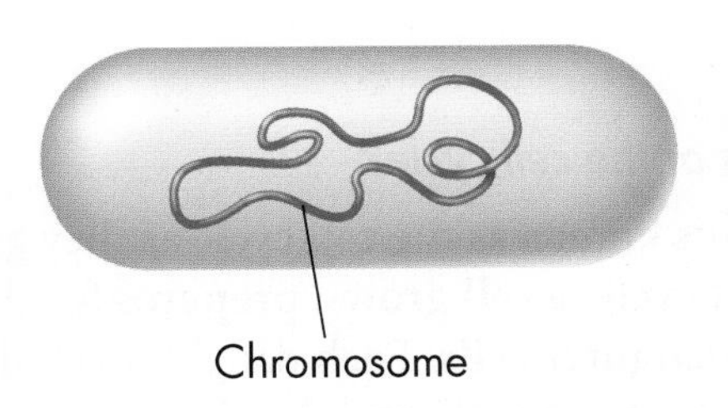

FIGURE 10–4 Prokaryotic Chromosome In most prokaryotes, a single chromosome holds most of the organism's DNA.

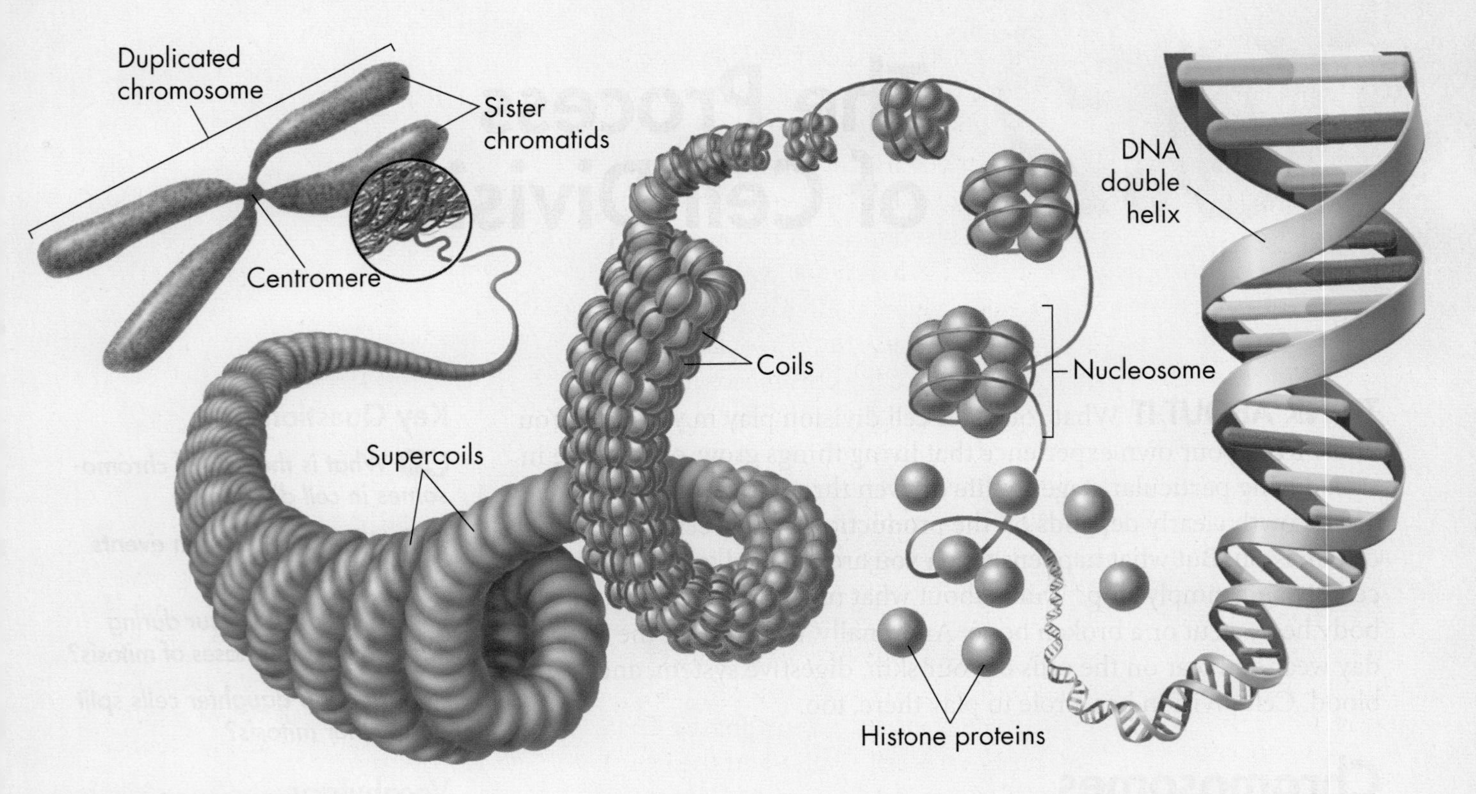

FIGURE 10–5 Eukaryotic Chromosome As a eukaryotic cell prepares for division, each chromosome coils more and more tightly to form a compact structure. **Interpret Visuals** ***Which side of the diagram, left or right, shows the smallest structures, and which shows the largest?***

Eukaryotic Chromosomes Eukaryotic cells generally have much more DNA than prokaryotes have and, therefore, contain multiple chromosomes. Fruit flies, for example, have 8 chromosomes per cell, human cells have 46, and carrot cells have 18. The chromosomes in eukaryotic cells form a close association with histones, a type of protein. This complex of chromosome and protein is referred to as **chromatin.** DNA tightly coils around the histones, and together, the DNA and histone molecules form beadlike structures called nucleosomes. Nucleosomes pack together to form thick fibers, which condense even further during cell division. Usually the chromosome shape you see drawn is a duplicated chromosome with supercoiled chromatin, as shown in **Figure 10–5.**

Why do cells go to such lengths to package their DNA into chromosomes? One of the principal reasons is to ensure equal division of DNA when a cell divides.  **Chromosomes make it possible to separate DNA precisely during cell division.**

In Your Notebook *Write instructions to build a eukaryotic chromosome.*

The Cell Cycle

What are the main events of the cell cycle?

Cells go through a series of events known as the **cell cycle** as they grow and divide. **During the cell cycle, a cell grows, prepares for division, and divides to form two daughter cells.** Each daughter cell then moves into a new cell cycle of activity, growth, and division.

The Prokaryotic Cell Cycle The prokaryotic cell cycle is a regular pattern of growth, DNA replication, and cell division that can take place very rapidly under ideal conditions. Researchers are only just beginning to understand how the cycle works in prokaryotes, and relatively little is known about its details. It is known that most prokaryotic cells begin to replicate, or copy, their DNA chromosomes once they have grown to a certain size. When DNA replication is complete, or nearly complete, the cell begins to divide.

The process of cell division in prokaryotes is a form of asexual reproduction known as binary fission. Once the chromosome has been replicated, the two DNA molecules attach to different regions of the cell membrane. A network of fibers forms between them, stretching from one side of the cell to the other. The fibers constrict and the cell is pinched inward, dividing the cytoplasm and chromosomes between two newly formed cells. Binary fission results in the production of two genetically identical cells.

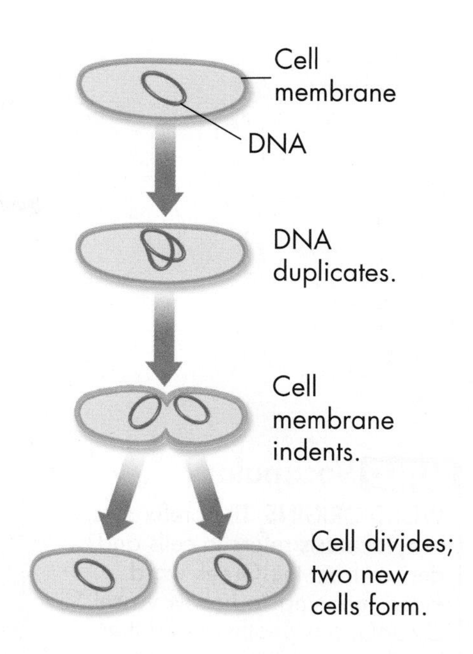

FIGURE 10–6 Binary Fission Cell division in a single-celled organism produces two genetically identical organisms.

The Eukaryotic Cell Cycle In contrast to prokaryotes, much more is known about the eukaryotic cell cycle. As you can see in **Figure 10–7,** the eukaryotic cell cycle consists of four phases: G_1, S, G_2, and M. The length of each part of the cell cycle—and the length of the entire cell cycle—varies depending on the type of cell.

At one time, biologists described the life of a cell as one cell division after another separated by an "in-between" period of growth called **interphase.** We now appreciate that a great deal happens in the time between cell divisions. Interphase is divided into three parts: G_1, S, and G_2.

▶ ***G_1 Phase: Cell Growth*** Cells do most of their growing during the G_1 phase. In this phase, cells increase in size and synthesize new proteins and organelles. The *G* in G_1 and G_2 stands for "gap," but the G_1 and G_2 phases are actually periods of intense growth and activity.

▶ ***S Phase: DNA Replication*** The G_1 phase is followed by the S phase. The *S* stands for "synthesis." During the S phase, new DNA is synthesized when the chromosomes are replicated. The cell at the end of the S phase contains twice as much DNA as it did at the beginning.

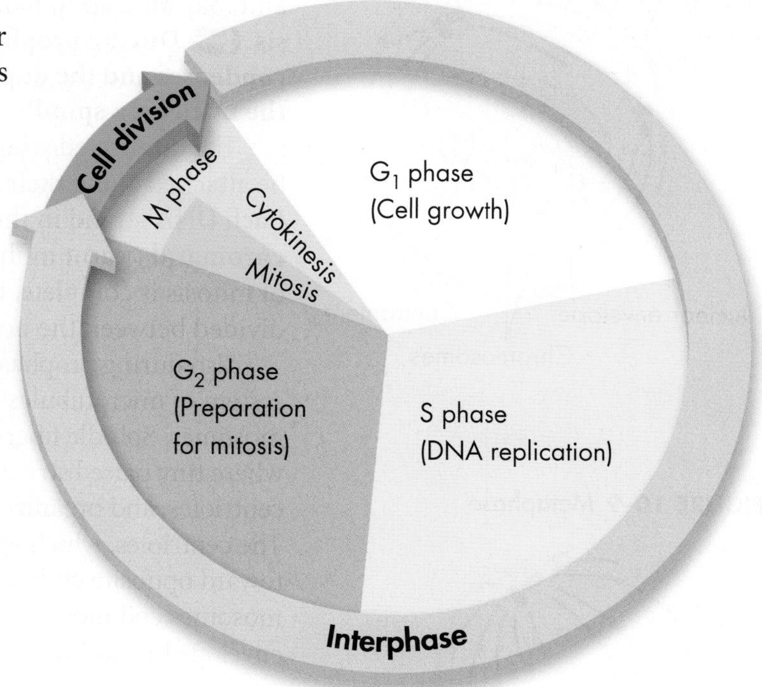

FIGURE 10–7 The Cell Cycle During the cell cycle, a cell grows, prepares for division, and divides to form two daughter cells. The cell cycle includes four phases—G_1, S, G_2, and M. **Infer** ***During which phase or phases would you expect the amount of DNA in the cell to change?***

▶ ***G_2 Phase: Preparing for Cell Division*** When DNA replication is completed, the cell enters the G_2 phase. G_2 is usually the shortest of the three phases of interphase. During the G_2 phase, many of the organelles and molecules required for cell division are produced. When the events of the G_2 phase are completed, the cell is ready to enter the M phase and begin the process of cell division.

▶ ***M Phase: Cell Division*** The M phase of the cell cycle, which follows interphase, produces two daughter cells. The M phase takes its name from the process of mitosis. During the normal cell cycle, interphase can be quite long. In contrast, the process of cell division usually takes place quickly.

In eukaryotes, cell division occurs in two main stages. The first stage of the process, division of the cell nucleus, is called **mitosis** (my TOH sis). The second stage, the division of the cytoplasm, is called **cytokinesis** (sy toh kih NEE sis). In many cells, the two stages may overlap, so that cytokinesis begins while mitosis is still taking place.

BUILD Vocabulary

WORD ORIGINS The prefix *cyto-* in **cytokinesis** refers to cells and derives from the Greek word *kytos*, meaning "a hollow vessel." *Cytoplasm* is another word that has the same root.

Mitosis

What events occur during each of the four phases of mitosis?

Biologists divide the events of mitosis into four phases: prophase, metaphase, anaphase, and telophase. Depending on the type of cell, mitosis may last anywhere from a few minutes to several days. **Figure 10–8** through **Figure 10–11** show mitosis in an animal cell.

FIGURE 10–8 Prophase

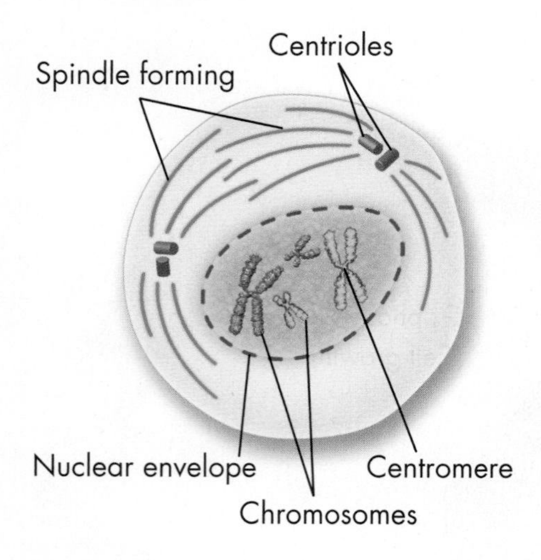

Prophase The first phase of mitosis, **prophase,** is usually the longest and may take up to half of the total time required to complete mitosis. **During prophase, the genetic material inside the nucleus condenses and the duplicated chromosomes become visible. Outside the nucleus, a spindle starts to form.**

The duplicated strands of the DNA molecule can be seen to be attached along their length at an area called the **centromere.** Each DNA strand in the duplicated chromosome is referred to as a **chromatid** (KROH muh tid), or sister chromatid. When the process of mitosis is complete, the chromatids will have separated and been divided between the new daughter cells.

Also during prophase, the cell starts to build a spindle, a fanlike system of microtubules that will help to separate the duplicated chromosomes. Spindle fibers extend from a region called the centrosome, where tiny paired structures called **centrioles** are located. Plant cells lack centrioles, and organize spindles directly from their centrosome regions. The centrioles, which were duplicated during interphase, start to move toward opposite ends, or poles, of the cell. As prophase ends, the chromosomes coil more tightly, the nucleolus disappears, and the nuclear envelope breaks down.

FIGURE 10–9 Metaphase

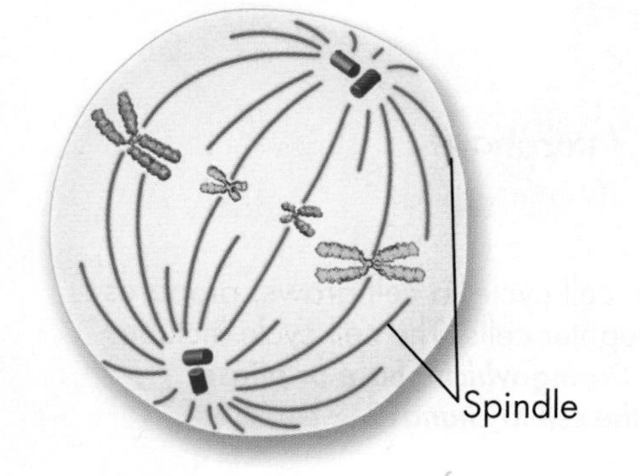

Metaphase The second phase of mitosis, **metaphase,** is generally the shortest. **During metaphase, the centromeres of the duplicated chromosomes line up across the center of the cell. Spindle fibers connect the centromere of each chromosome to the two poles of the spindle.**

Anaphase The third phase of mitosis, **anaphase,** begins when sister chromatids suddenly separate and begin to move apart. Once anaphase begins, each sister chromatid is now considered an individual chromosome. **During anaphase, the chromosomes separate and move along spindle fibers to opposite ends of the cell.** Anaphase comes to an end when this movement stops and the chromosomes are completely separated into two groups.

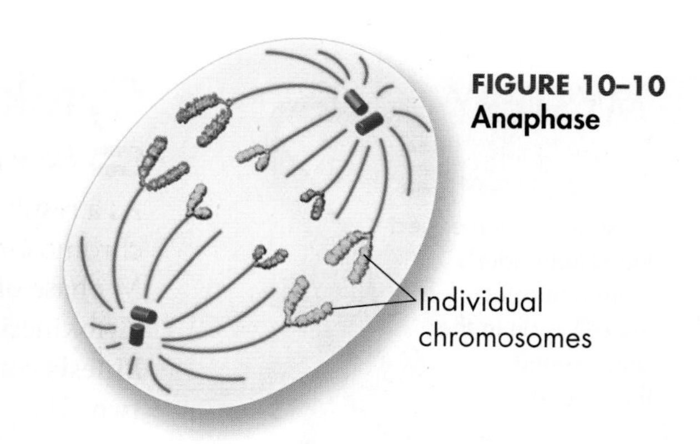

FIGURE 10–10 Anaphase

Telophase Following anaphase is **telophase,** the fourth and final phase of mitosis. **During telophase, the chromosomes, which were distinct and condensed, begin to spread out into a tangle of chromatin.** A nuclear envelope re-forms around each cluster of chromosomes. The spindle begins to break apart, and a nucleolus becomes visible in each daughter nucleus. Mitosis is complete. However, the process of cell division has one more step to go.

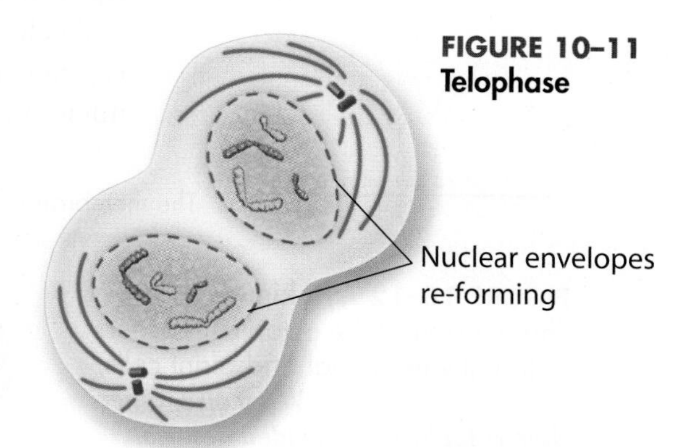

FIGURE 10–11 Telophase

In Your Notebook *Create a chart that lists the important information about each phase of mitosis.*

Quick Lab
GUIDED INQUIRY

Mitosis in Action 

1. Examine a slide of a stained onion root tip under a microscope. Viewing the slide under low power, adjust the stage until you find the boxlike cells just above the root tip.

2. Switch the microscope to high power and locate cells that are in the process of dividing.

3. Find and sketch cells that are in each phase of mitosis. Label each sketch with the name of the appropriate phase.

Analyze and Conclude

1. Observe In which phase of the cell cycle were most of the cells you observed? Why do you think this is?

2. Draw Conclusions What evidence did you observe that shows mitosis is a continuous process, not a series of separate events?

3. Apply Concepts Cells in the root divide many times as the root grows longer and thicker. With each cell division, the chromosomes are divided between two daughter cells, yet the number of chromosomes in each cell does not change. What processes ensure that the normal number of chromosomes is restored after each cell division?

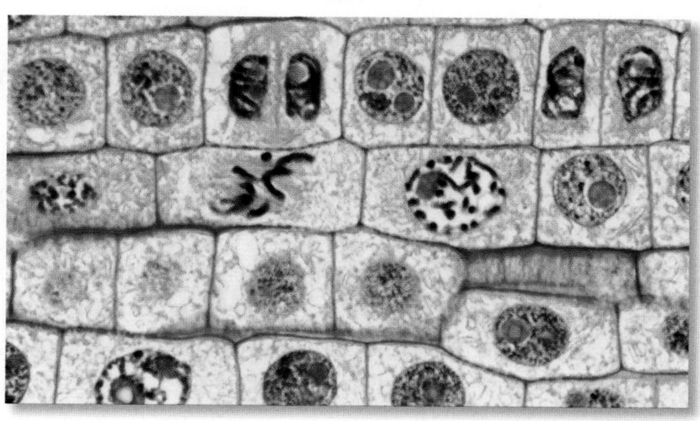

(LM 820×)

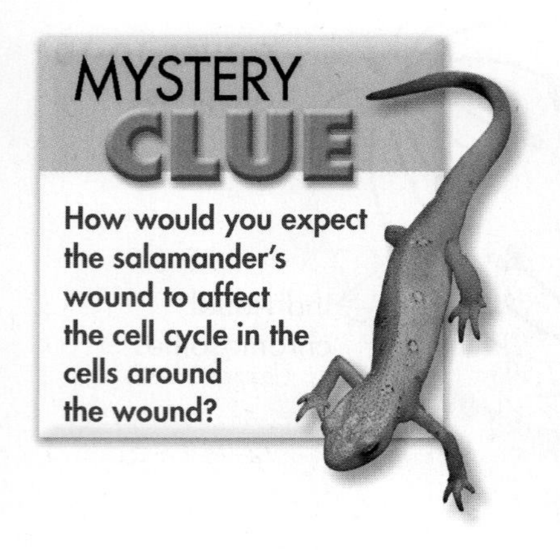

Cytokinesis

How do daughter cells split apart after mitosis?

As a result of mitosis, two nuclei—each with a duplicate set of chromosomes—are formed. All that remains to complete the M phase of the cycle is cytokinesis, the division of the cytoplasm itself. Cytokinesis usually occurs at the same time as telophase. **Cytokinesis completes the process of cell division—it splits one cell into two.** The process of cytokinesis differs in animal and plant cells.

Cytokinesis in Animal Cells During cytokinesis in most animal cells, the cell membrane is drawn inward until the cytoplasm is pinched into two nearly equal parts. Each part contains its own nucleus and cytoplasmic organelles.

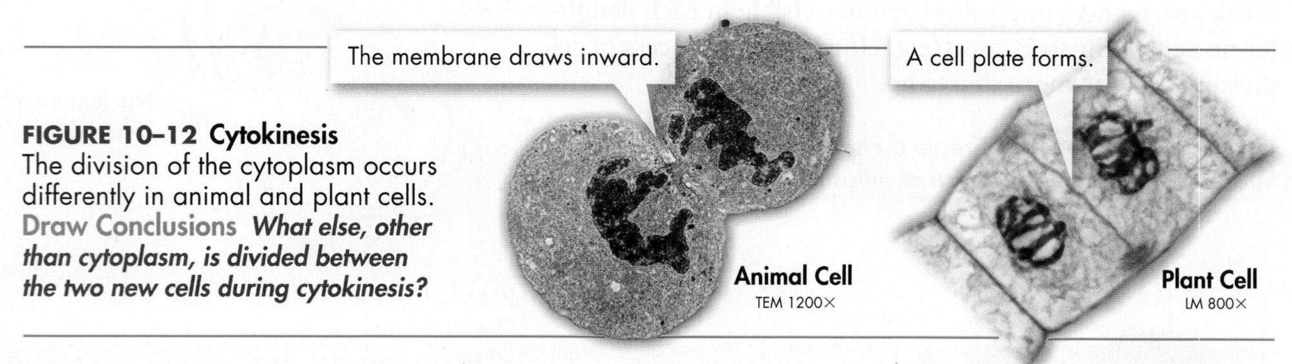

FIGURE 10–12 Cytokinesis The division of the cytoplasm occurs differently in animal and plant cells. Draw Conclusions ***What else, other than cytoplasm, is divided between the two new cells during cytokinesis?***

Cytokinesis in Plant Cells Cytokinesis in plant cells proceeds differently. The cell membrane is not flexible enough to draw inward because of the rigid cell wall that surrounds it. Instead, a structure known as the cell plate forms halfway between the divided nuclei. The cell plate gradually develops into cell membranes that separate the two daughter cells. A cell wall then forms in between the two new membranes, completing the process.

10.2 Assessment

Review Key Concepts

1. a. Review What are chromosomes?
b. Compare and Contrast How does the structure of chromosomes differ in prokaryotes and eukaryotes?

2. a. Review What is the cell cycle?
b. Sequence During which phase of the cell cycle are chromosomes replicated?

3. a. Review What happens during each of the four phases of mitosis? Write one or two sentences for each phase.
b. Predict What do you predict would happen if the spindle fibers were disrupted during metaphase?

4. a. Review What is cytokinesis and when does it occur?
b. Compare and Contrast How does cytokinesis differ in animal and plant cells?

WRITE ABOUT SCIENCE

Summary

5. Summarize what happens during interphase. Be sure to include all three parts of interphase. *Hint:* Include all of the main details in your summary.

VISUAL SUMMARY

MITOSIS

FIGURE 10–13 The phases of mitosis shown here are typical of eukaryotic cells. These light micrographs are from a developing whitefish embryo (LM 415×). **Infer** ***Why is the timing between what happens to the nuclear envelope and the activity of the mitotic spindle so critical?***

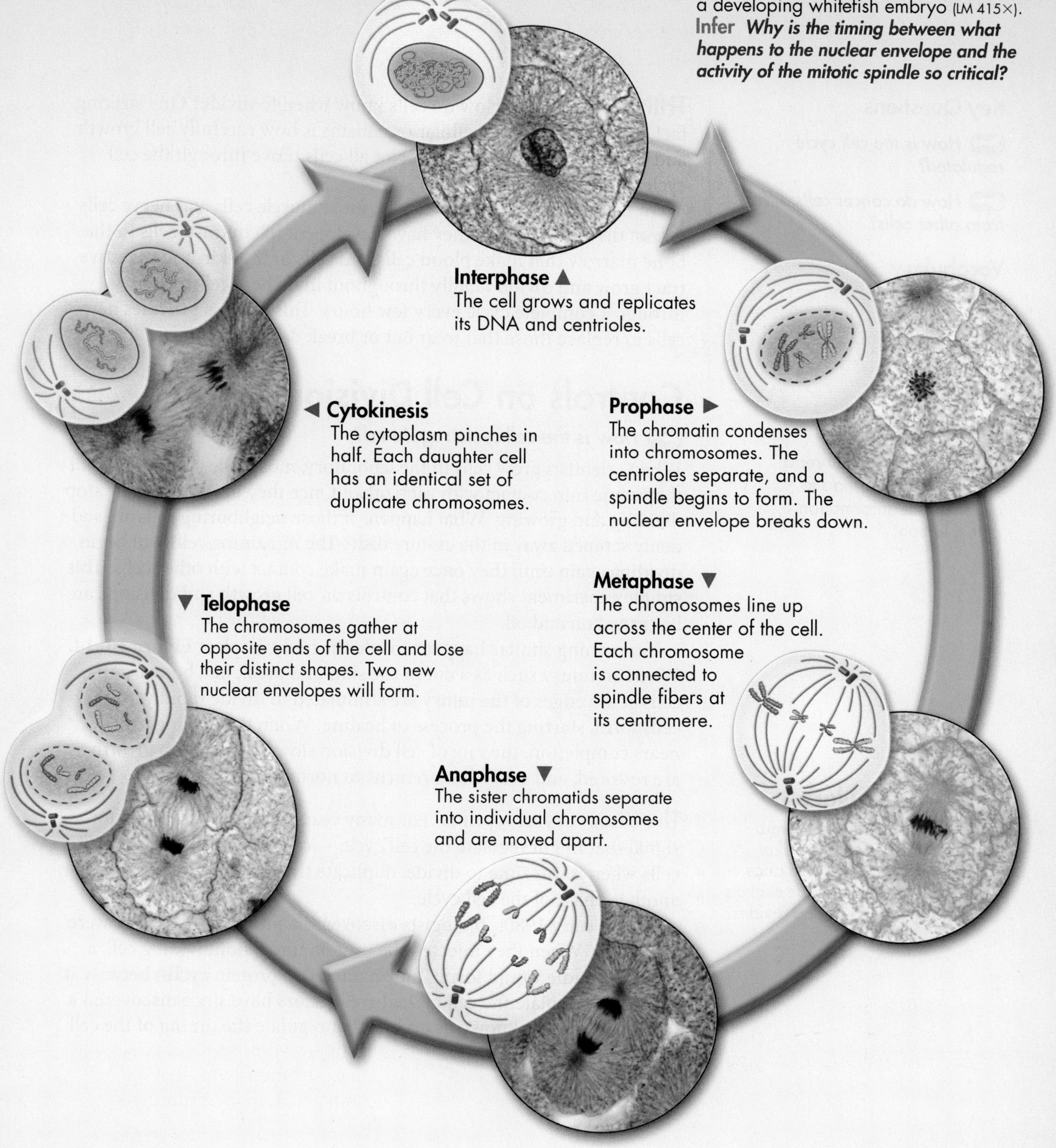

Interphase ▲
The cell grows and replicates its DNA and centrioles.

Prophase ▶
The chromatin condenses into chromosomes. The centrioles separate, and a spindle begins to form. The nuclear envelope breaks down.

Metaphase ▼
The chromosomes line up across the center of the cell. Each chromosome is connected to spindle fibers at its centromere.

Anaphase ▼
The sister chromatids separate into individual chromosomes and are moved apart.

▼ Telophase
The chromosomes gather at opposite ends of the cell and lose their distinct shapes. Two new nuclear envelopes will form.

◀ Cytokinesis
The cytoplasm pinches in half. Each daughter cell has an identical set of duplicate chromosomes.

10.3 Regulating the Cell Cycle

Key Questions

How is the cell cycle regulated?

How do cancer cells differ from other cells?

Vocabulary

cyclin
growth factor
apoptosis
cancer
tumor

Taking Notes

Concept Map As you read, create a concept map to organize the information in this lesson.

BUILD Vocabulary

ACADEMIC WORDS The verb **regulate** means "to control or direct." Therefore, a substance that regulates the cell cycle controls when the cell grows and divides.

THINK ABOUT IT How do cells know when to divide? One striking fact about cells in multicellular organisms is how carefully cell growth and cell division are controlled. Not all cells move through the cell cycle at the same rate.

In the human body, for example, most muscle cells and nerve cells do not divide at all once they have developed. In contrast, cells in the bone marrow that make blood cells and cells of the skin and digestive tract grow and divide rapidly throughout life. These cells may pass through a complete cycle every few hours. This process provides new cells to replace those that wear out or break down.

Controls on Cell Division

How is the cell cycle regulated?

When scientists grow cells in the laboratory, most cells will divide until they come into contact with each other. Once they do, they usually stop dividing and growing. What happens if those neighboring cells are suddenly scraped away in the culture dish? The remaining cells will begin dividing again until they once again make contact with other cells. This simple experiment shows that controls on cell growth and division can be turned on and off.

Something similar happens inside the body. Look at **Figure 10–14.** When an injury such as a cut in the skin or a break in a bone occurs, cells at the edges of the injury are stimulated to divide rapidly. New cells form, starting the process of healing. When the healing process nears completion, the rate of cell division slows, controls on growth are restored, and everything returns to normal.

The Discovery of Cyclins For many years, biologists searched for a signal that might regulate the cell cycle—something that would "tell" cells when it was time to divide, duplicate their chromosomes, or enter another phase of the cell cycle.

In the early 1980s, biologists discovered a protein in cells that were in mitosis. When they injected the protein into a nondividing cell, a mitotic spindle would form. They named this protein **cyclin** because it seemed to regulate the cell cycle. Investigators have since discovered a family of proteins known as cyclins that regulate the timing of the cell cycle in eukaryotic cells.

Regulatory Proteins The discovery of cyclins was just the start. Scientists have since identified dozens of other proteins that also help to regulate the cell cycle. **The cell cycle is controlled by regulatory proteins both inside and outside the cell.**

▶ ***Internal Regulators*** One group of proteins, internal regulatory proteins, respond to events occurring inside a cell. Internal regulatory proteins allow the cell cycle to proceed only when certain events have occurred in the cell itself. For example, several regulatory proteins make sure a cell does not enter mitosis until its chromosomes have replicated. Another regulatory protein prevents a cell from entering anaphase until the spindle fibers have attached to the chromosomes.

▶ ***External Regulators*** Proteins that respond to events outside the cell are called external regulatory proteins. External regulatory proteins direct cells to speed up or slow down the cell cycle.

One important group of external regulatory proteins is the group made up of the growth factors. **Growth factors** stimulate the growth and division of cells. These proteins are especially important during embryonic development and wound healing. Other external regulatory proteins on the surface of neighboring cells often have an opposite effect. They cause cells to slow down or stop their cell cycles. This prevents excessive cell growth and keeps body tissues from disrupting one another.

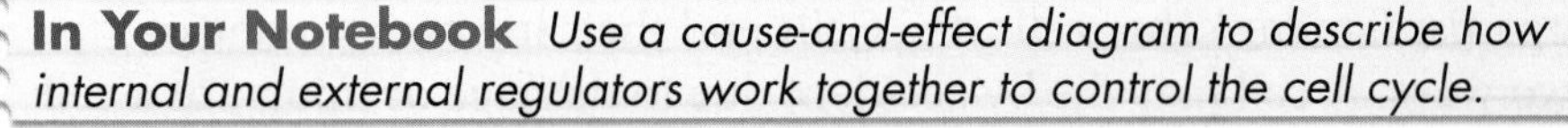
In Your Notebook *Use a cause-and-effect diagram to describe how internal and external regulators work together to control the cell cycle.*

ZOOMING IN

CELL GROWTH AND HEALING

FIGURE 10–14 When a person breaks a bone, cells at the edges of the injury are stimulated to divide rapidly. The new cells that form begin to heal the break. As the bone heals, the cells stop dividing and growing.

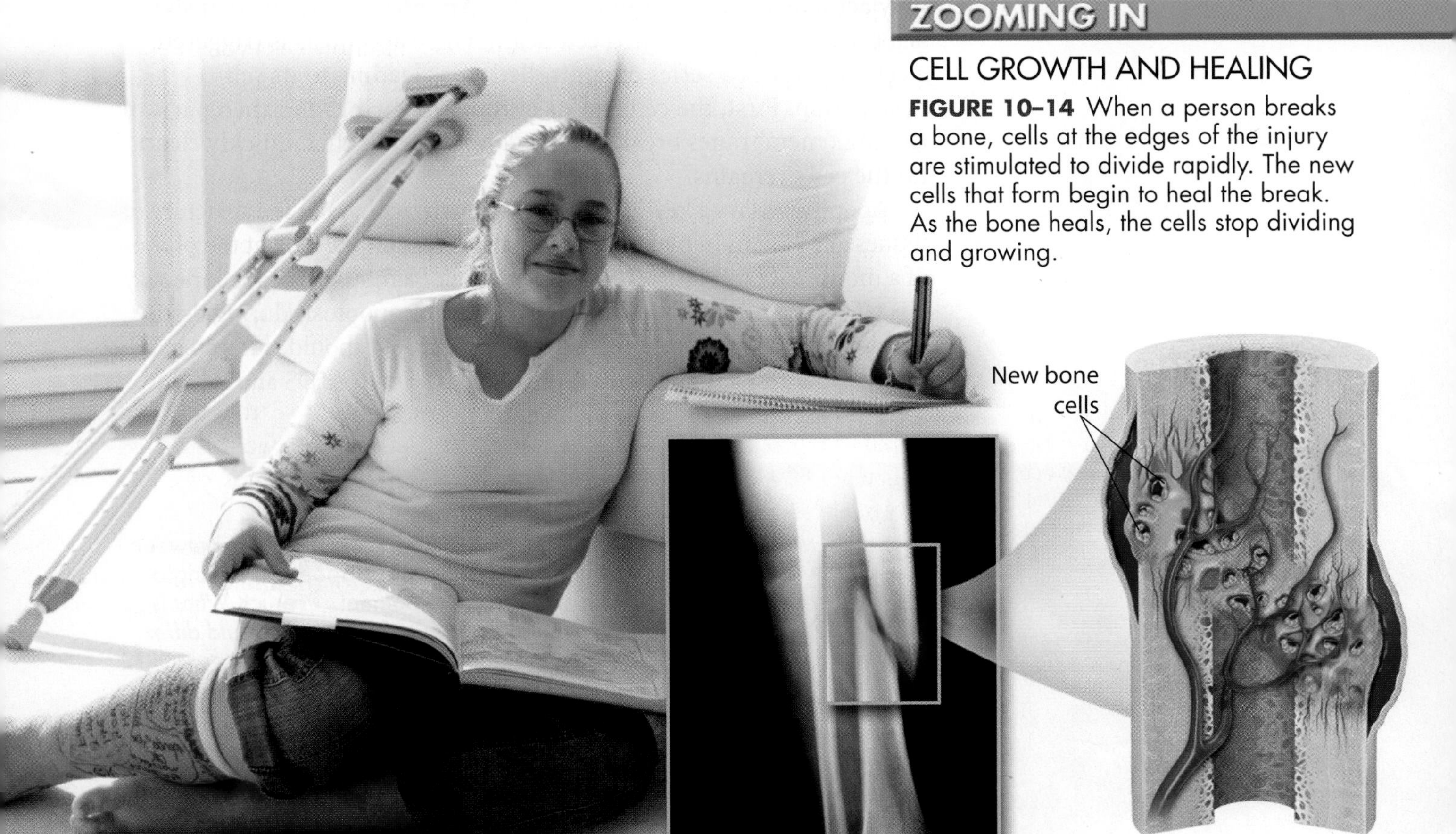

Analyzing Data

The Rise and Fall of Cyclins

Scientists measured cyclin levels in clam egg cells as the cells went through their first mitotic divisions after fertilization. The data are shown in the graph.

Cyclins are continually produced and destroyed within cells. Cyclin production signals cells to enter mitosis, while cyclin destruction signals cells to stop dividing and enter interphase.

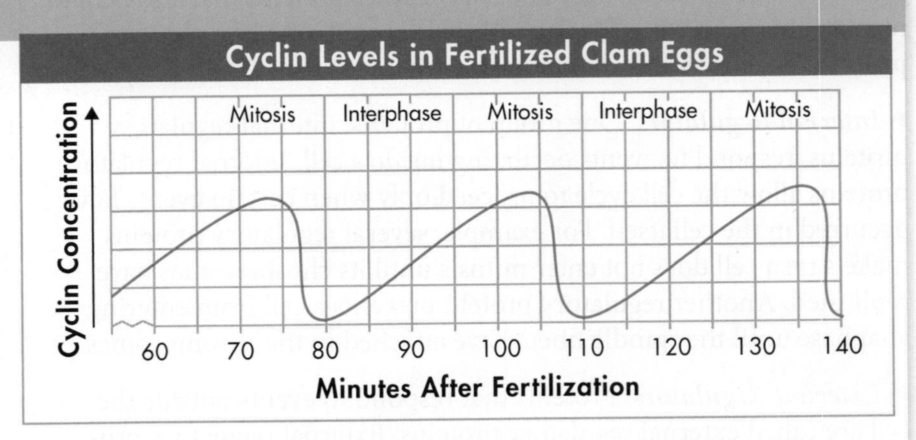

1. Interpret Graphs How long does cyclin production last during a typical cell cycle in fertilized clam eggs?

2. Infer During which part of the cell cycle does cyclin production begin? How quickly is cyclin destroyed?

3. Predict Suppose that the regulators that control cyclin production are no longer produced. What are two possible outcomes?

Apoptosis Just as new cells are produced every day in a multicellular organism, many other cells die. Cells end their life cycle in one of two ways. A cell may die by accident due to damage or injury, or a cell may actually be "programmed" to die. **Apoptosis** (AYP up TOH sis) is a process of programmed cell death. Once apoptosis is triggered, a cell undergoes a series of controlled steps leading to its self-destruction. First, the cell and its chromatin shrink, and then parts of the cell's membranes break off. Neighboring cells then quickly clean up the cell's remains.

Apoptosis plays a key role in development by shaping the structure of tissues and organs in plants and animals. For example, look at the photos of a mouse foot in **Figure 10–15.** Each foot of a mouse is shaped the way it is partly because cells between the toes die by apoptosis during tissue development. When apoptosis does not occur as it should, a number of diseases can result. For example, the cell loss seen in AIDS and Parkinson's disease can result if too much apoptosis occurs.

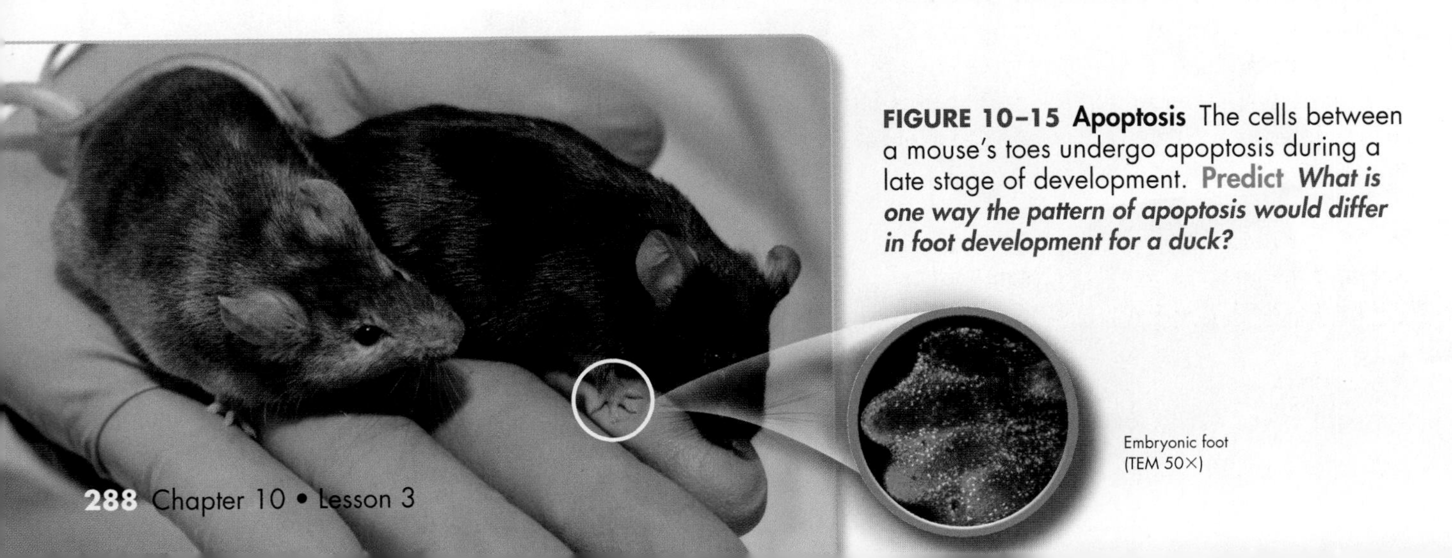

FIGURE 10–15 Apoptosis The cells between a mouse's toes undergo apoptosis during a late stage of development. **Predict** ***What is one way the pattern of apoptosis would differ in foot development for a duck?***

Cancer: Uncontrolled Cell Growth

How do cancer cells differ from other cells?

Why is cell growth regulated so carefully? The principal reason may be that the consequences of uncontrolled cell growth in a multicellular organism are very severe. **Cancer,** a disorder in which body cells lose the ability to control growth, is one such example.

Cancer cells do not respond to the signals that regulate the growth of most cells. As a result, the cells divide uncontrollably. Cancer cells form a mass of cells called a **tumor.** However, not all tumors are cancerous. Some tumors are benign, or noncancerous. A benign tumor does not spread to surrounding healthy tissue or to other parts of the body. Cancerous tumors, such as the one in **Figure 10–16,** are malignant. Malignant tumors invade and destroy surrounding healthy tissue.

As the cancer cells spread, they absorb the nutrients needed by other cells, block nerve connections, and prevent the organs they invade from functioning properly. Soon, the delicate balances that exist in the body are disrupted, and life-threatening illness results.

What Causes Cancer? Cancers are caused by defects in the genes that regulate cell growth and division. There are several sources of such defects, including: smoking or chewing tobacco, radiation exposure, other defective genes, and even viral infection. All cancers, however, have one thing in common: The control over the cell cycle has broken down. Some cancer cells will no longer respond to external growth regulators, while others fail to produce the internal regulators that ensure orderly growth.

An astonishing number of cancer cells have a defect in a gene called p53, which normally halts the cell cycle until all chromosomes have been properly replicated. Damaged or defective p53 genes cause cells to lose the information needed to respond to signals that normally control their growth.

In Your Notebook *Use a two-column chart to compare the controls that regulate normal cell growth to the lack of control seen in cancer cells.*

FIGURE 10–16 Growth of Cancer Cells Normal cells grow and divide in a carefully controlled fashion. Cells that are cancerous lose this control and continue to grow and divide, producing tumors.

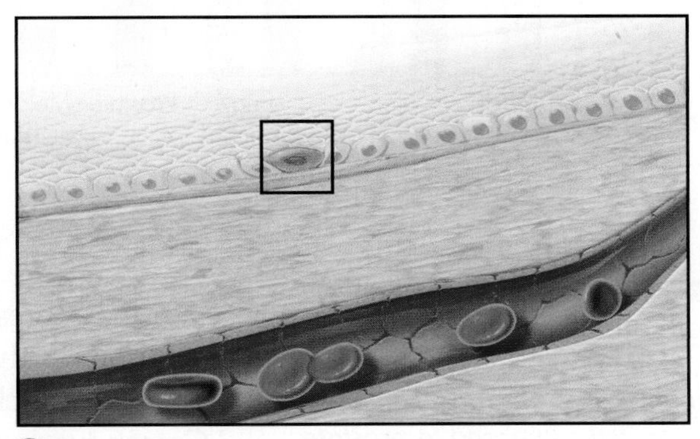

1 A cell begins to divide abnormally.

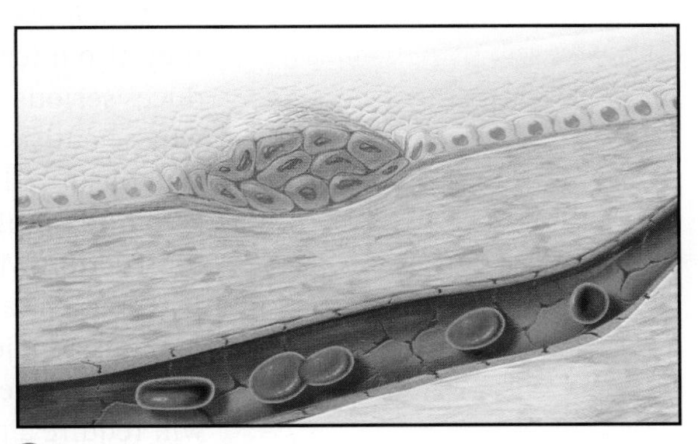

2 The cancer cells produce a tumor, which begins to displace normal cells and tissues.

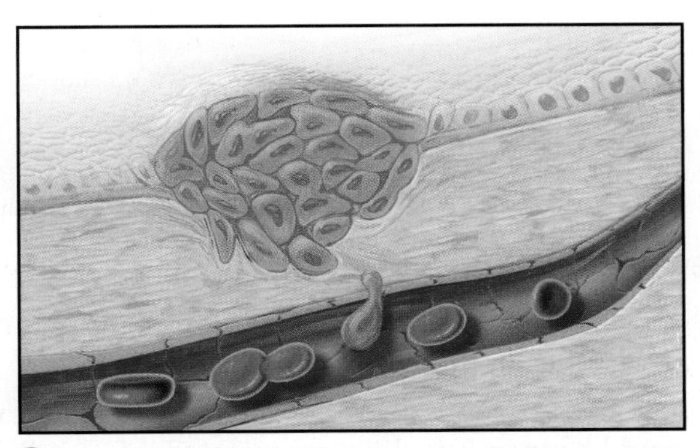

3 Cancer cells are particularly dangerous because of their tendency to spread once they enter the bloodstream or lymph vessels. The cancer then moves into other parts of the body and forms secondary tumors, a process called metastasis.

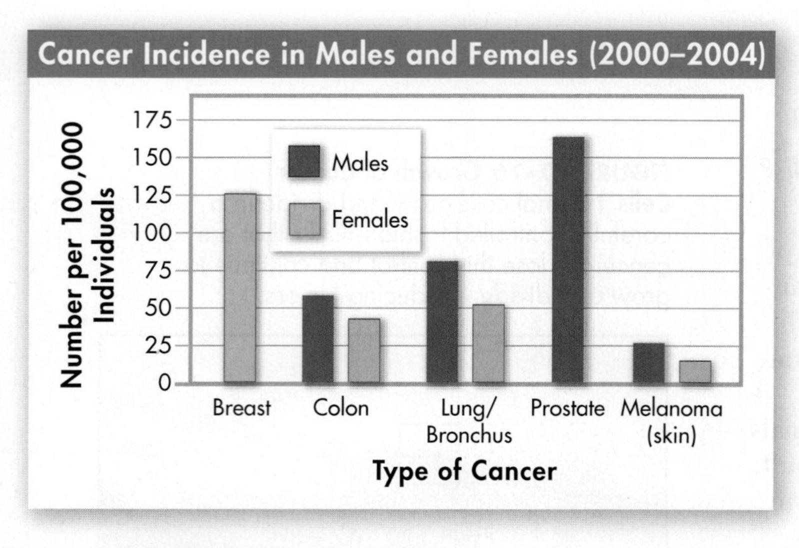

FIGURE 10–17 Cancer Incidence Cancer can affect almost every organ in the body. **Interpret Graphs** ***How many cases of breast cancer were reported compared to prostate cancer for the time period shown?***

Treatments for Cancer When a cancerous tumor is localized, it can often be removed by surgery. Skin cancer, the most common form of the disease, can usually be treated this way. Melanomas, the most serious form of skin cancer, can be removed surgically, but only if spotted very early.

Other forms of treatment make use of the fact that cancer cells grow rapidly and, therefore, need to copy their DNA more quickly than do most normal cells. This makes them especially vulnerable to damage from radiation. As a result, many tumors can be effectively treated with carefully targeted beams of radiation.

Medical researchers have worked for years to develop chemical compounds that would kill cancer cells, or at least slow their growth. The use of such compounds against cancer is known as chemotherapy. Great advances in chemotherapy have taken place in recent years and have even made it possible to cure some forms of cancer. However, because most chemotherapy compounds target rapidly dividing cells, they also interfere with cell division in normal, healthy cells. This produces serious side effects in many patients, and it is one of the reasons why scientists are so interested in gaining a better understanding of the role of cell cycle proteins in cancer. The goal of many researchers is to find highly specific ways in which cancer cells can be targeted for destruction while leaving healthy cells unaffected.

Cancer is a serious disease. Understanding and combating cancer remains a major scientific challenge, but scientists at least know where to start. Cancer is a disease of the cell cycle, and conquering cancer will require a much deeper understanding of the processes that control cell division.

10.3 Assessment

Review Key Concepts

1. a. Review Name the two types of proteins that regulate the cell cycle. How do these proteins work?

b. Form a Hypothesis Write a hypothesis about what you think would happen if cyclin were injected into a cell during mitosis. How could you test your hypothesis?

2. a. Review Why is cancer considered a disease of the cell cycle?

b. Compare and Contrast How are the growth of a tumor and the repair of a scrape on your knee similar? How are they different?

Apply the Big idea

Growth, Development, and Reproduction

3. Why do you think it is important that cells have a "control system" to regulate the timing of cell division?

WHST.9-10.4 Production and Distribution of Writing, WHST.9-10.10 Range of Writing

Technology & BIOLOGY

Fluorescence Microscopy

Imagine being able to "see" proteins at work inside a cell, or to track proteins from where they are made to where they go. Scientists can now do all of these things, thanks to advances in fluorescence microscopy. One advance came from the discovery that crystal jellyfish, properly known as *Aequorea victoria*, produce a protein that glows. By fusing the gene for this protein to other genes, scientists can label different parts of the cell with fluorescence. Other advances include the development of additional highly specific fluorescent labels and the invention of powerful laser microscopes. As the images on this page show, the view is clearly amazing.

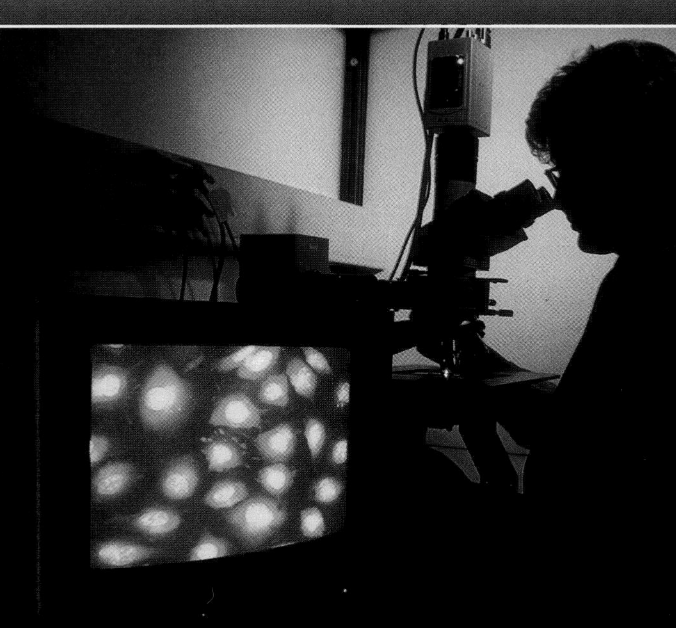

▲ Viewing Labeled Specimens
In fluorescence microscopy, a specimen is labeled with a molecule that glows under a specific wavelength of light. Different fluorescent labels give off different colors. This way, biologists can easily see exactly where a protein is located within a cell or tissue.

WRITING **Suppose you are a cell biologist studying cell division and cancer. What might you use a fluorescence microscope to study? Describe your ideas in a paragraph.**

▼ Normal Spindle
Different fluorescent labels enable biologists to track how spindle fibers (green) form and how proteins help distribute chromosomes (red) evenly during mitosis.

▼ Abnormal Spindle
Cell cycle control has gone awry in this cell, causing an abnormal mitotic spindle to form.

10.4 Cell Differentiation

Key Questions

How do cells become specialized for different functions?

What are stem cells?

What are some possible benefits and issues associated with stem cell research?

Vocabulary

embryo • differentiation • totipotent • blastocyst • pluripotent • stem cell • multipotent

Taking Notes

Compare/Contrast Table As you read, create a table comparing the ability of different cell types to differentiate.

THINK ABOUT IT The human body contains an estimated 100,000,000,000,000 (one hundred trillion) cells. That's a staggering number, but in one respect it's not quite as large as you might think. Why? Try to estimate how many times a single cell would have to divide through mitosis to produce that many cells. It may surprise you to learn that as few as 47 rounds of cell division can produce that many cells.

The results of those 47 cell cycles are truly amazing. The human body contains hundreds of distinctly different cell types, and every one of them develops from the single cell that starts the process. How do the cells get to be so different from each other?

From One Cell to Many

How do cells become specialized for different functions?

Each of us started life as just one cell. So, for that matter, did your pet dog, an earthworm, and the petunia on the windowsill. These living things pass through a developmental stage called an **embryo,** from which the adult organism is gradually produced. During the development process, an organism's cells become more and more differentiated and specialized for particular functions. **Figure 10–18** shows some of the specialized cells found in the roots, stems, and leaves of a plant.

FIGURE 10–18 Specialized Plant Cells

Cells that transport materials

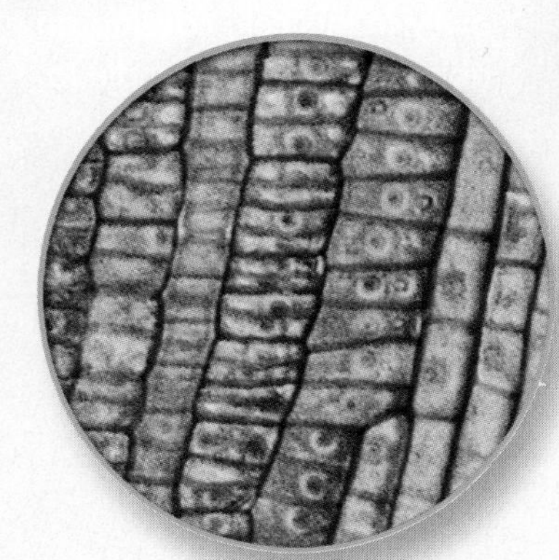
Cells that store sugar

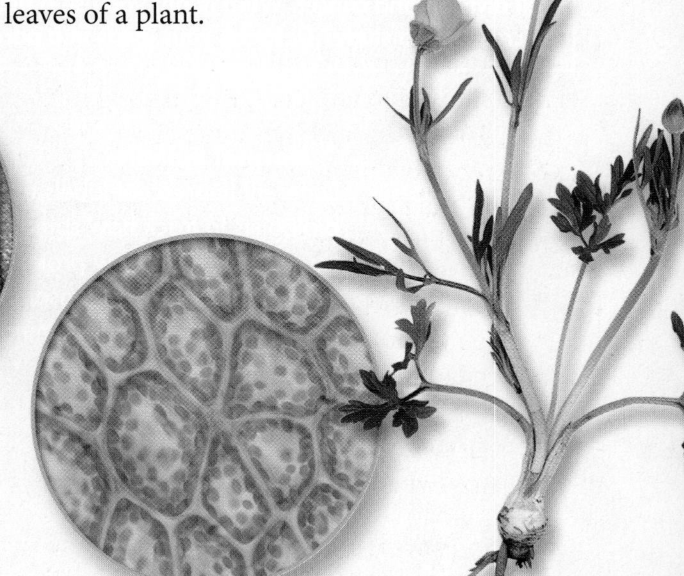
Cells that carry out photosynthesis

Defining Differentiation The process by which cells become specialized is known as **differentiation** (dif ur en shee AY shun). **During the development of an organism, cells differentiate into many types of cells.** A differentiated cell has become, quite literally, different from the embryonic cell that produced it, and specialized to perform certain tasks, such as contraction, photosynthesis, or protection. Our bodies, and the bodies of all multicellular organisms, contain highly differentiated cells that carry out the jobs we need to perform to stay alive.

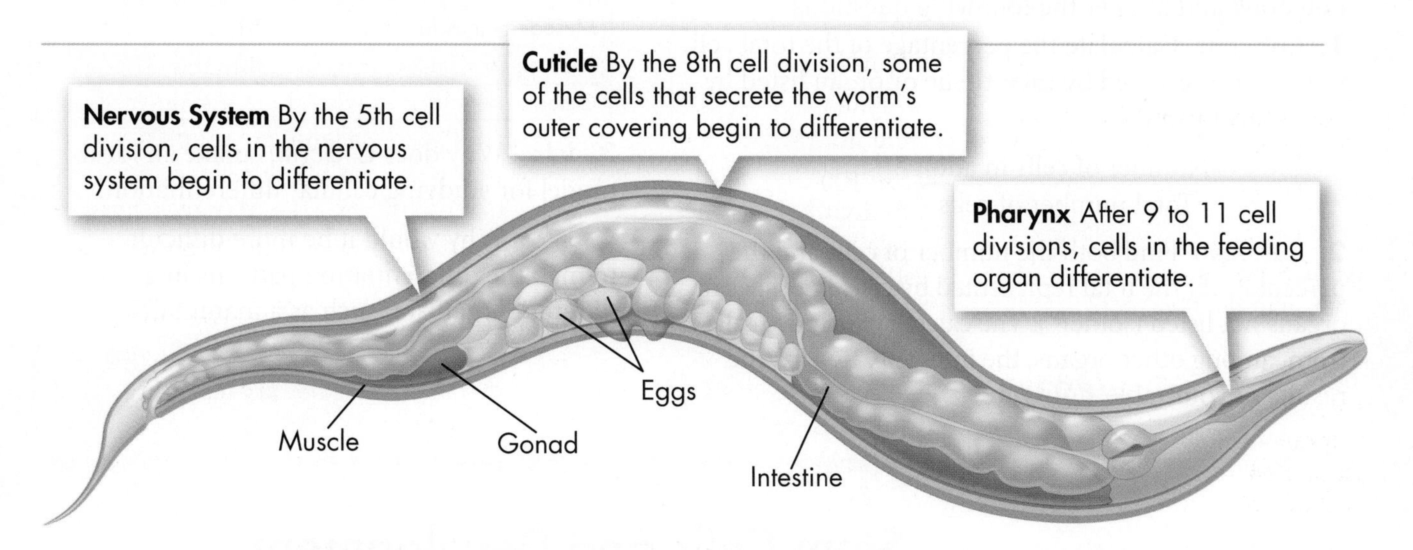

FIGURE 10–19 Differentiation in *C. elegans* A fertilized egg develops into an adult worm after many cell divisions. Daughter cells from each cell division follow a specific path toward a role as a particular kind of cell.

Mapping Differentiation The process of differentiation determines a cell's ultimate identity, such as whether it will spend its life as a nerve cell or a muscle cell. In some organisms, a cell's role is rigidly determined at a specific point in the course of development. In the microscopic worm *Caenorhabditis elegans*, for example, biologists have mapped the outcome of each and every cell division from fertilized egg to adult.

The process of cell differentiation in *C. elegans* begins with the very first division and continues throughout embryonic development. **Figure 10–19** shows when some of the cells found in the adult begin to differentiate during development. Each and every time a new worm develops, the process is the same, resulting in 959 cells with precisely determined functions.

Differentiation in Mammals Other organisms, including mammals like us, go through a more flexible process in which cell differentiation is controlled by a number of interacting factors in the embryo, many of which are still not well understood. What is known, however, is that adult cells generally do reach a point at which their differentiation is complete—when they can no longer become other types of cells.

In Your Notebook *Starting with a single cell, calculate how many cells might result after 4, 8, and 10 cell divisions.*

Cellular Differentiation of *C. elegans*

The adult microscopic worm *C. elegans* contains 959 cells. The data table shows some of the different cell types in this worm. Copy the data table into your notebook and answer the following questions.

1. Calculate Calculate the percentage of the total cell number represented by each tissue or organ listed by using this formula:

$$\frac{\text{Number of cells in adult}}{\text{Total number of cells}} \times 100$$

2. Calculate Find both the number of cells and the percentage of the total represented by cells in tissues or organs not listed ("other"). The category includes cells from, among other organs, the intestine. Record the results in your table. MATH

Cell Type	Number of Cells in Adult	Percent of Total
Cuticle	213	22%
Gonad (excluding germ line cells)	143	
Mesoderm muscle	81	
Pharynx	80	
Other		

3. Infer Why does *C. elegans* make an ideal model for studying cellular differentiation?

4. Infer Why would it be more difficult to map the differentiation patterns in a different organism, such as a mammal?

Stem Cells and Development

What are stem cells?

One of the most important questions in biology is how all of the specialized, differentiated cell types in the body are formed from just a single cell, the fertilized zygote. Biologists say that the zygote is **totipotent** (toh TIP uh tunt), literally able to do everything, to develop into any type of cell in the body (including the cells that make up the extraembryonic membranes and placenta). Only the fertilized egg and the cells produced by the first few cell divisions of embryonic development are truly totipotent. If there is a "secret" by which cells start the process of differentiation, these are the cells that know that secret.

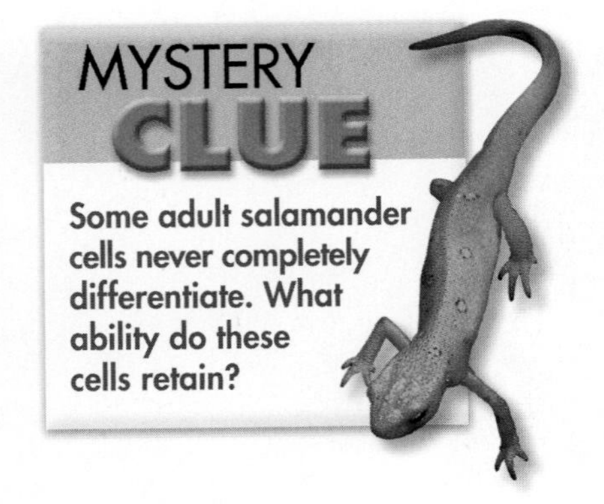

MYSTERY CLUE

Some adult salamander cells never completely differentiate. What ability do these cells retain?

Human Development After about four days of development, a human embryo forms into a **blastocyst,** a hollow ball of cells with a cluster of cells inside known as the inner cell mass. Even at this early stage, the cells of the blastocyst have begun to specialize. The outer cells form tissues that attach the embryo to its mother, while the inner cell mass becomes the embryo itself. The cells of the inner cell mass are said to be pluripotent (plu RIP uh tunt). **Pluripotent** cells can develop into any of the body's cell types, although they generally cannot form the tissues surrounding the embryo.

In Your Notebook *Look up the roots that form the words* totipotent, pluripotent, *and* multipotent. *How do the roots relate to each cell's ability to differentiate?*

Stem Cells **The unspecialized cells from which differentiated cells develop are known as stem cells.** As the name implies, **stem cells** sit at the base of a branching "stem" of development from which different cell types form. As you might expect, stem cells are found in the early embryo, but they are also found in many places in the adult body.

► ***Adult Stem Cells*** Cells in some tissues, like blood and skin, have a limited life span and must be constantly replaced. Pools of adult stem cells, found in various locations throughout the body, produce the new cells needed for these tissues. New blood cells differentiate from stem cells found in the bone marrow, while many skin stem cells are found in hair follicles. Small clusters of adult stem cells are even found in the brain, in the heart, and in skeletal muscle. These adult cells are referred to as **multipotent** (muhl TIP uh tunt) stem cells, because the types of differentiated cells they can form are usually limited to replacing cells in the tissues where they are found.

► ***Embryonic Stem Cells*** Pluripotent embryonic stem cells are more versatile than adult stem cells, since they are capable of producing every cell type in the body. In laboratory experiments, scientists have managed to coax embryonic stem cells to differentiate into nerve cells, muscle cells, and even sperm and egg cells. In 1998, when researchers at the University of Wisconsin found a way to grow human embryonic stem cells in culture, it became possible to explore the potential of these remarkable cells.

FIGURE 10–20 Embryonic Stem Cells After fertilization, the human embryo develops into a hollow ball of cells known as a blastocyst. The actual body of the embryo develops from the inner cell mass, a cluster of cells inside the blastocyst. Because of their ability to differentiate into each of the body's many cell types, these cells are known as embryonic stem cells.

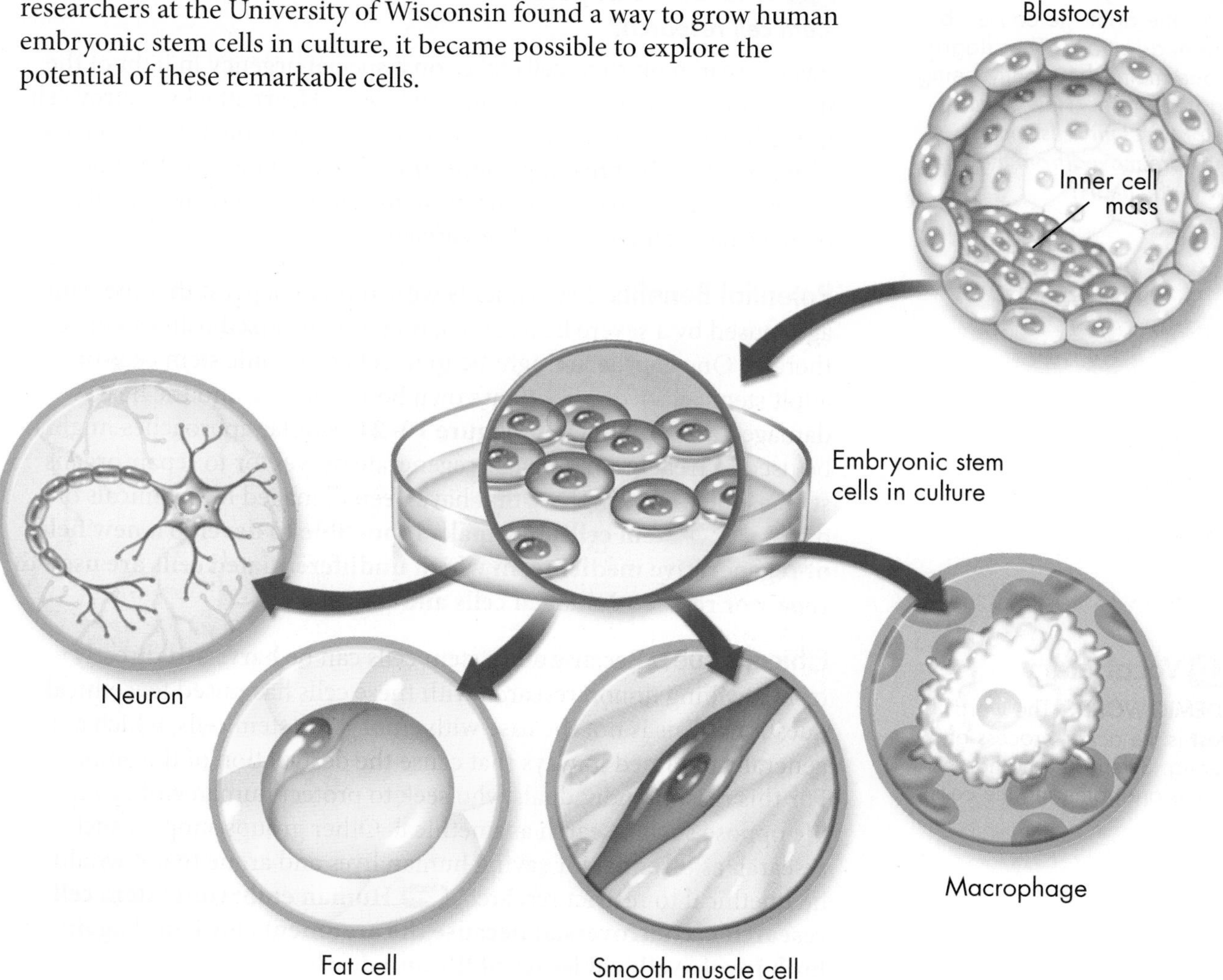

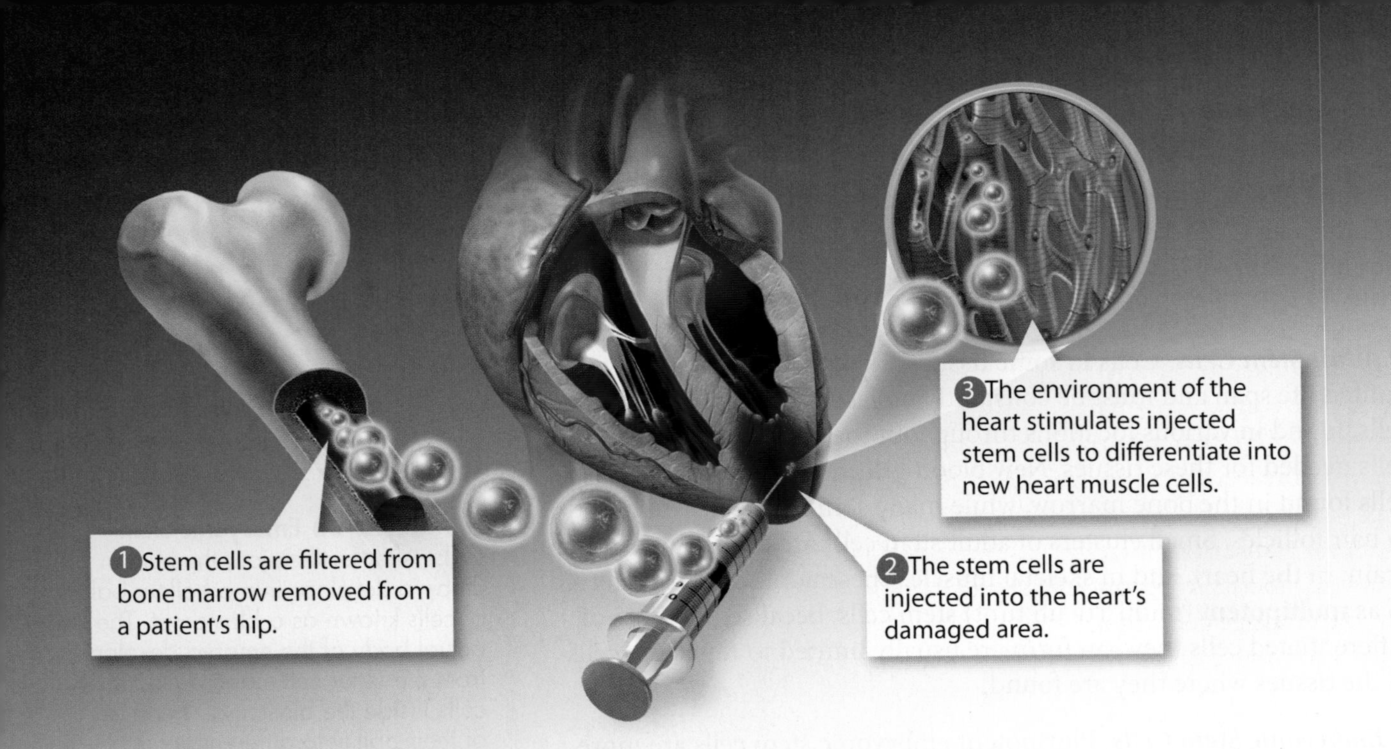

FIGURE 10–21 A Possible Future Treatment for Heart Disease? Stem cell research may lead to new ways to reverse the damage caused by a severe heart attack. The diagram shows one method currently being investigated. **Infer** ***How would the fate of the stem cells change after they are moved from the bone marrow to the heart?***

Frontiers in Stem Cell Research

What are some possible benefits and issues associated with stem cell research?

Basic research on stem cells takes on a special urgency in light of the importance it might have for human health. Heart attacks destroy cells in the heart muscle, strokes injure brain cells, and spinal cord injuries cause paralysis by breaking connections between nerve cells. Not surprisingly, the prospect of using stem cells to repair such cellular damage has excited medical researchers.

Potential Benefits Experiments with animals suggest that the damage caused by a severe heart attack might be reversed using stem cell therapy. One approach might be to inject embryonic stem cells or adult stem cells from a patient's own bone marrow into the heart's damaged area, as shown in **Figure 10–21.** Similar approaches might be used to treat brain damage, regenerate nerves, or to repair organs such as the liver or kidney that have been damaged by chemicals or disease. **Stem cells may make it possible to develop a new field of regenerative medicine, in which undifferentiated cells are used to repair or replace damaged cells and tissues.**

BUILD Vocabulary

ACADEMIC WORDS The word **harvest** is the act or process of gathering. Scientists who harvest stem cells are gathering the cells.

Ethical Issues Because adult stem cells can be **harvested** directly from a willing donor, research with these cells has raised few ethical questions. This is not the case with embryonic stem cells, which are generally obtained in ways that cause the destruction of the embryo. For this reason, individuals who seek to protect human embryonic life oppose such research as unethical. Other groups support such research as essential for saving human lives and argue that it would be unethical to restrict research. **Human embryonic stem cell research is controversial because the arguments for it and against it both involve ethical issues of life and death.**

Cellular Reprogramming Intensive research in the last few years has begun to address some of the most basic biological questions about the nature of stem cells. In particular, scientists have begun to figure out the genetic mechanisms that enable stem cells to produce so many types of differentiated cells.

By studying the genes that are expressed when stem cells differentiate, researchers have now developed laboratory "recipes" that can remake cells into certain other cell types. In most cases, these experiments have involved the use of transcription factors, proteins that bind to DNA and activate other genes. In 2012, for example, several groups of researchers inserted genes for certain transcription factors into fibroblasts, a common cell found in connective tissue. By using different transcription factors, these researchers were able to reprogram fibroblasts to become stem cells capable of producing nerve cells or heart muscle cells. In one study, they were actually able to repair a damaged mouse heart by using these reprogrammed cells.

Induced Pluripotent Stem Cells An even more fundamental breakthrough took place in 2007 when Shinya Yamanaka (**Figure 10–22**) of Kyoto University in Japan was able to convert human fibroblast cells into cells that closely resembled embryonic stem cells. These *induced pluripotent stem cells (iPS cells),* as they are known, may make it possible to tailor specific therapies to the needs of each individual patient by using their own cells. Approaches like these, if successful, might allow potentially lifesaving research to go forward while avoiding the destruction of embryonic life. Therefore, it is possible that new research may solve the very ethical problems that have made stem cell research controversial.

FIGURE 10–22 Shinya Yamanaka Dr. Shinya Yamanaka's breakthrough research on induced pluripotent stem cells has made a huge impact on the field of regenerative medicine. In 2012, Dr. Yamanaka was awarded the Nobel Prize for his discoveries.

In Your Notebook *Make a two-column chart that lists the benefits and issues related to stem cell research.*

10.4 Assessment

Review Key Concepts

1. a. **Review** What happens during differentiation?

   b. **Apply Concepts** What does "mapping" refer to in the process of cell differentiation?

2. a. **Review** What are stem cells?

   b. **Compare and Contrast** How are embryonic stem cells and adult stem cells alike? How are they different?

3. a. **Review** Summarize the potential benefits and issues of stem cell research.

   b. **Form an Opinion** How might new developments help address the ethical concerns surrounding stem cell research?

Apply the Big idea

Cellular Basis of Life

4. Use what you learned in this lesson to discuss how cells become specialized for different functions. Include an explanation of how the potential for specialization varies with cell type and how it varies over the life span of an organism.

NGSS Smart Guide Cell Growth and Division

Disciplinary Core Idea LS1.B Growth and Development of Organisms: How do organisms grow and develop? Cells undergo cell division to produce new cells. In eukaryotic cells, cell division is part of a highly regulated cycle known as the cell cycle.

10.1 Cell Growth, Division, and Reproduction

- The larger a cell becomes, the more demands the cell places on its DNA. In addition, a larger cell is less efficient in moving nutrients and waste materials across the cell membrane.
- Asexual reproduction is the production of genetically identical offspring from a single parent.
- Offspring produced by sexual reproduction inherit some of their genetic information from each parent.

cell division 276 • asexual reproduction 277 • sexual reproduction 277

Biology.com

Untamed Science Video Journey with the Untamed Science crew to a research facility in Sweden to learn why scientists are studying regeneration in brittle stars.

Visual Analogy Compare a growing cell to a growing city to understand limits on cell size.

10.2 The Process of Cell Division

- Chromosomes make it possible to separate DNA precisely during cell division.
- During the cell cycle, a cell grows, prepares for division, and divides to form two daughter cells.
- During prophase, the genetic material inside the nucleus condenses. During metaphase, the chromosomes line up across the center of the cell. During anaphase, the chromosomes separate and move along spindle fibers to opposite ends of the cell. During telophase, the chromosomes, which were distinct and condensed, begin to spread out into a tangle of chromatin.
- Cytokinesis completes the process of cell division—it splits one cell into two.

chromosome 279 • chromatin 280 • cell cycle 280 • interphase 281 • mitosis 282 • cytokinesis 282 • prophase 282 • centromere 282 • chromatid 282 • centriole 282 • metaphase 282 • anaphase 283 • telophase 283

Biology.com

InterActive Art See the phases of mitosis in action.

Data Analysis Learn how to time the cell cycle by counting cells in mitosis.

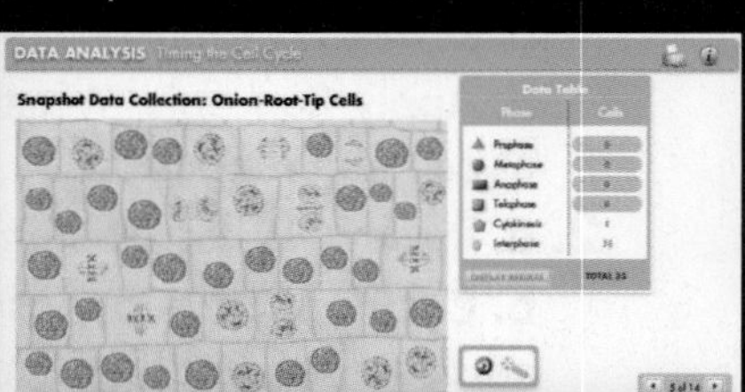

Art Review Test your knowledge of the structure of a eukaryotic chromosome.

Tutor Tube Sort out chromosome structure vocabulary with this simple tutorial video.

CHECKING YOUR *Scientific Literacy*

Refer to the lesson content and digital assets below as you prepare for your chapter assessment. Then, evaluate your understanding of cell growth and development by answering these questions.

1. Scientific and Engineering Practice **Developing and Using Models** Draw a model of the cell cycle for a eukaryotic cell with four chromosomes. In your model, label the key events that result in growth and division of a cell.
2. Crosscutting Concept **Patterns** Expand your model of the eukaryotic cell cycle to include two more rounds of cell division. Explain the patterns you observe in the genetic material of the parent cell and all eight daughter cells.
3. Key Ideas and Details Cite strong and thorough textual evidence to explain the role of cell division in producing a complex organism.
4. STEM You are a biomedical engineer that has been tasked with developing an artificial cell. Explain the constraints you need to consider when designing the cell.

10.3 Regulating the Cell Cycle

- The cell cycle is controlled by regulatory proteins both inside and outside the cell.
- Cancer cells do not respond to the signals that regulate the growth of most cells. As a result, the cells divide uncontrollably.

cyclin 286 • **growth factor 287** • **apoptosis 288** • **cancer 289** • **tumor 289**

Biology.com

Art in Motion See what happens when cancerous cells invade normal tissue.

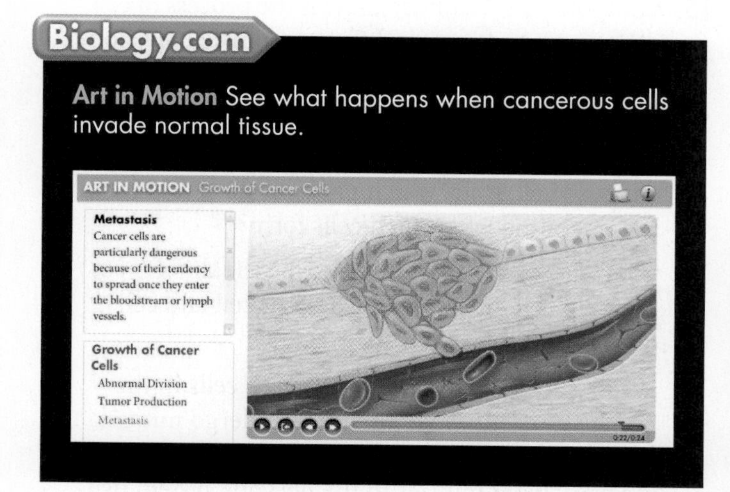

10.4 Cell Differentiation

- During the development of an organism, cells differentiate into many types of cells.
- The unspecialized cells from which differentiated cells develop are known as stem cells.
- Stem cells may make it possible to develop a new field of regenerative medicine, in which undifferentiated cells are used to repair damaged cells and tissues.
- Human embryonic stem cell research is controversial because the arguments for it and against it both involve ethical issues of life and death.

embryo 292 • **differentiation 293** • **totipotent 294** • **blastocyst 294** • **pluripotent 294** • **stem cell 295** • **multipotent 295**

Biology.com

Lesson Notes Compare the ability of different cells to differentiate.

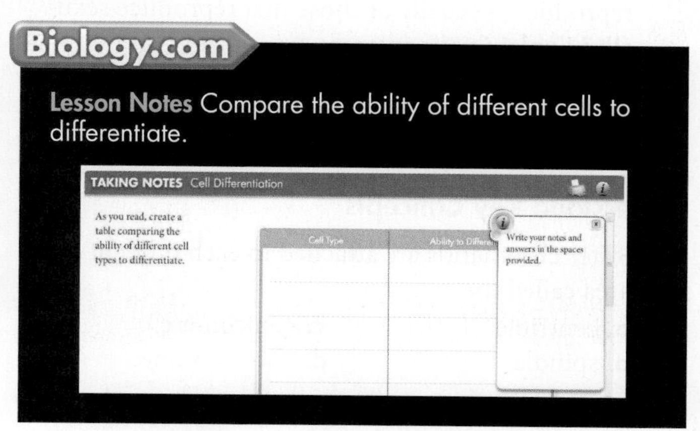

Design Your Own Lab

Regeneration in Planaria This chapter lab is available in your lab manual and at Biology.com

10 Assessment

10.1 Cell Growth, Division, and Reproduction

Understand Key Concepts

1. The rate at which materials enter and leave the cell depends on the cell's
   a. volume.
   b. weight.
   c. speciation.
   d. surface area.

2. In order for a cell to divide successfully, the cell must first
   a. duplicate its genetic information.
   b. decrease its volume.
   c. increase its number of chromosomes.
   d. decrease its number of organelles.

3. The process that increases genetic diversity within a population is
   a. asexual reproduction.
   b. sexual reproduction.
   c. cell division.
   d. binary fission.

4. **Craft and Structure** Describe what is meant by each of the following terms: *cell volume, cell surface area, ratio of surface area to volume.*

5. Describe asexual and sexual reproduction as survival strategies.

Think Critically

6. **Calculate** Calculate the ratio of surface area to volume of an imaginary cubic cell measuring 4 mm long on each side. MATH

7. **Form a Hypothesis** In a changing environment, which organisms have an advantage—those that reproduce asexually or those that reproduce sexually? Explain your answer.

10.2 The Process of Cell Division

Understand Key Concepts

8. Sister chromatids are attached to each other at an area called the
   a. centriole.
   b. spindle.
   c. centromere.
   d. chromosome.

9. If a cell has 12 chromosomes, how many chromosomes will each of its daughter cells have after mitosis and cytokinesis?
   a. 4 b. 6 c. 12 d. 24

10. Which of the illustrations below best represents metaphase of mitosis?

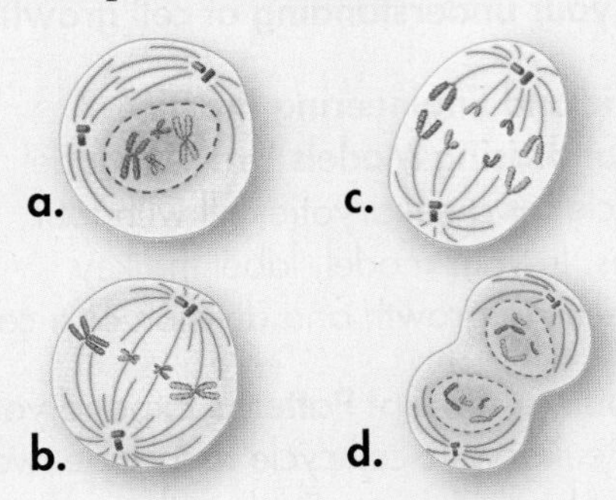

11. In plant cells, what forms midway between the divided nuclei during cytokinesis?
    a. nuclear membrane
    b. centromere
    c. cell membrane
    d. cell plate

12. Describe how a eukaryotic cell's chromosomes change as a cell prepares to divide.

13. **Craft and Structure** What is the relationship between interphase and cell division?

14. **Key Ideas and Details** List the following stages of mitosis in the correct sequence, and describe what happens during each stage: anaphase, metaphase, prophase, and telophase.

Think Critically

15. **Compare and Contrast** How is the process of cell division in prokaryotes different from cell division in eukaryotes?

16. **Form a Hypothesis** Some cells have several nuclei within their cytoplasm. Considering the events in a typical cell cycle, which phase of the cell cycle is not operating when such cells form?

17. **Compare and Contrast** Describe the differences between cell division in an animal cell and cell division in a plant cell.

18. **Relate Cause and Effect** The nerve cells in the human nervous system seldom undergo mitosis. Based on this information, explain why complete recovery from injuries to the nervous system usually does not occur.

19. **Apply Concepts** A scientist treats cells with a chemical that prevents DNA synthesis. In which stage of the cell cycle will these cells remain?

20. **Integration of Knowledge and Ideas** The diagram shows a phase of mitosis. Use the diagram to answer the following questions.

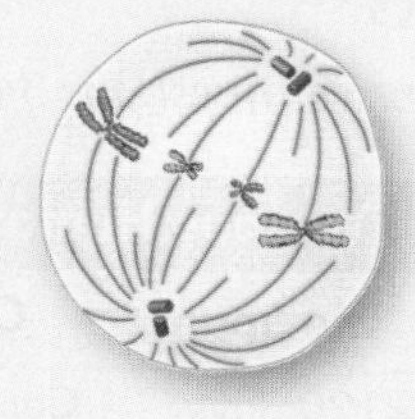

a. Identify the phase of mitosis shown in the diagram.
b. Is this a plant or animal cell? How do you know?
c. The four chromosomes shown in the center of this cell each have two connected strands. Explain how the two strands on the same chromosome compare with regard to the genetic information they carry. In your answer, be sure to explain why this is important to the cell.

10.3 Regulating the Cell Cycle

Understand Key Concepts

21. The timing in the cell cycle in eukaryotic cells is believed to be controlled by a group of closely related proteins known as
a. chromatids.
b. cyclins.
c. centromeres.
d. centrioles.

22. Which statement does NOT describe external regulatory proteins?
a. They respond to events occurring inside a cell.
b. Growth factors are one group of external regulatory proteins.
c. They can speed up or slow down the cell cycle.
d. Some can cause cells to slow down or stop their cell cycles.

23. When some cells are removed from the center of a tissue culture, will new cells replace the cells that were removed? Explain.

24. **Text Types and Purposes** Describe the role of cyclins.

Think Critically

25. **Compare and Contrast** How do cancer cells differ from noncancerous cells? How are they similar?

26. **Research to Build and Present Knowledge** A cell will usually undergo apoptosis if the cell experiences DNA damage that could lead to a tumor. Predict what may happen if a gene that controls apoptosis is damaged.

solve the CHAPTER MYSTERY

PET SHOP ACCIDENT

Julia kept a close eye on the injured salamander. About a month after the accident, Julia realized that a new limb was growing to replace the lost one! Salamanders are one of only a few vertebrates that can regenerate a complete limb. Examine the illustrations that show how a new limb develops. Then answer the questions.

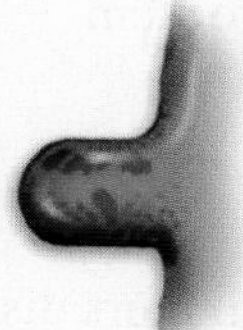

Week 1: Dedifferentiation
At first, cells in the injured limb undergo dedifferentiation. During this process, cells such as muscle cells and nerve cells lose the characteristics that make them specialized.

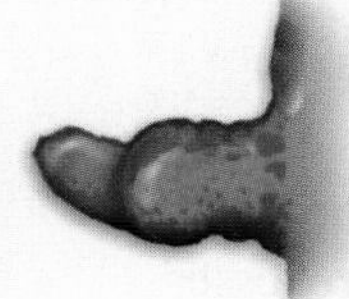

Week 3: Blastema Formation
The dedifferentiated cells migrate to the wounded area and form a blastema—a growing mass of undifferentiated cells.

Week 5: Redifferentiation
Cells in the blastema then redifferentiate and form the tissues needed for a mature limb. The limb will continue to grow until it is full size.

1. **Relate Cause and Effect** Why is dedifferentiation of the salamander's limb cells necessary before regeneration can occur?

2. **Classify** What type of cells do you think are contained in the blastema? Explain.

3. **Connect to the Big idea** Unlike salamanders, planarians contain undifferentiated cells throughout their adult bodies. How might the regeneration process in salamanders and planarians differ?

10.4 Cell Differentiation

Understand Key Concepts

27. Bone marrow cells that produce blood cells are best categorized as

a. embryonic stem cells. **c.** pluripotent.
b. adult stem cells. **d.** totipotent cells.

28. Which type of cell has the potential to develop into any type of cell?

a. totipotent **c.** multipotent
b. pluripotent **d.** differentiated

29. **Craft and Structure** What is a blastocyst?

30. **Craft and Structure** What is cell differentiation and how is it important to an organism's development?

31. Describe two ways that technology may address the ethical concerns related to stem cell research.

Think Critically

32. **Relate Cause and Effect** When researchers discovered how to make skin stem cells pluripotent, how did they apply their discovery to the treatment for heart attack patients?

33. **Compare and Contrast** How does embryonic development and cell differentiation in *C. elegans* differ from how these processes work in mammals?

Connecting Concepts

Use Science Graphics

Use the data table to answer questions 34 and 35.

Life Spans of Various Human Cells

Cell Type	Life Span	Cell Division
Red blood cells	<120 days	Cannot divide
Cardiac (heart) muscle	Long-lived	Cannot divide
Smooth muscle	Long-lived	Can divide
Neuron (nerve cell)	Long-lived	Most do not divide

34. **Compare and Contrast** Based on the data, in what ways might injuries to the heart and spinal cord be similar? How might they differ from injuries to smooth muscles?

35. **Predict** If cancer cells were added to the table, predict what would be written in the Life Span and Cell Division columns. Explain.

Write About Science

36. **Research to Build and Present Knowledge** Recall what you learned about the characteristics of life in Chapter 1. Explain how cell division is related to two or more of those characteristics.

37. **Assess the Big idea** How is cancer an example of how changes to a single cell can affect the health of an entire organism?

Integration of Knowledge and Ideas

A scientist performed an experiment to determine the effect of temperature on the length of the cell cycle in onion cells. His data are summarized in the table below.

Effect of Temperature on Length of Onion Cell Cycle

Temperature (°C)	Length of Cell Cycle (hours)
10	54.6
15	29.8
20	18.8
25	13.3

38. **Interpret Tables** On the basis of the data in the table, how long would you expect the cell cycle to be at 5°C?

a. less than 13.3 hours
b. more than 54.6 hours
c. between 29.8 and 54.6 hours
d. about 20 hours

39. **Draw Conclusions** Given this set of data, what is one valid conclusion the scientist could state?

Standardized Test Prep

Multiple Choice

1. Which statement is true regarding a cell's surface area-to-volume ratio?
- **A** As the size of a cell increases, its volume decreases.
- **B** As the size of a cell decreases, its volume increases.
- **C** Larger cells will have a greater surface area-to-volume ratio.
- **D** Smaller cells will have a greater surface area-to-volume ratio.

2. Which of the following is NOT an advantage of asexual reproduction?
- **A** simple and efficient
- **B** produces large number of offspring quickly
- **C** increases genetic diversity
- **D** requires one parent

3. At the beginning of cell division, a chromosome consists of two
- **A** centromeres.
- **B** centrioles.
- **C** chromatids.
- **D** spindles.

4. What regulates the timing of the cell cycle in eukaryotes?
- **A** chromosomes
- **B** cyclins
- **C** nutrients
- **D** DNA and RNA

5. The period between cell divisions is called
- **A** interphase.
- **B** prophase.
- **C** G_3 phase.
- **D** cytokinesis.

6. Which of the following is TRUE about totipotent cells?
- **A** Embryonic stem cells are totipotent cells.
- **B** Totipotent cells are differentiated cells.
- **C** Totipotent cells can differentiate into any type of cell and tissue.
- **D** Adult stem cells are totipotent cells.

7. A cell enters anaphase before all of its chromosomes have attached to the spindle. This may indicate that the cell is not responding to
- **A** internal regulators.
- **B** mitosis.
- **C** growth factors.
- **D** apoptosis.

Questions 8–10

The spindle fibers of a dividing cell were labeled with a fluorescent dye. At the beginning of anaphase, a laser beam was used to mark a region of the spindle fibers about halfway between the centrioles and the chromosomes. The laser beam stopped the dye from glowing in this region, as shown in the second diagram. The laser did not inhibit the normal function of the fibers.

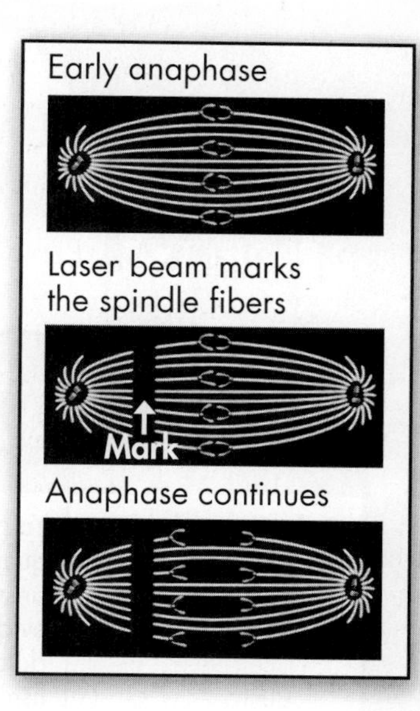

8. This experiment tests a hypothesis about
- **A** how chromosomes migrate during cell division.
- **B** how fluorescent dyes work in the cell.
- **C** the effect of lasers on cells.
- **D** why cells divide.

9. The diagrams show that chromosomes move to the poles of the cell as the spindle fibers
- **A** shorten on the chromosome side of the mark.
- **B** lengthen on the chromosome side of the mark.
- **C** shorten on the centriole side of the mark.
- **D** lengthen on the centriole side of the mark.

10. A valid conclusion that can be drawn from this experiment is that the spindle fibers break down
- **A** at the centrioles.
- **B** in the presence of dye.
- **C** when marked by lasers.
- **D** where they are attached to chromosomes.

Open-Ended Response

11. Explain why careful regulation of the cell cycle is important to multicellular organisms.

If You Have Trouble With . . .

Question	1	2	3	4	5	6	7	8	9	10	11
See Lesson	10.1	10.1	10.2	10.3	10.2	10.4	10.3	10.2	10.2	10.2	10.3

NGSS Problem-based Learning

UNIT PROJECT 3

Your Self-Assessment

The rubric below will help you evaluate your work as you construct an explanation and communicate information.

SCORE YOUR WORK!	EXEMPLARY Score your work a 4 if:	ACCOMPLISHED Score your work a 3 if:	DEVELOPING Score your work a 2 if:	BEGINNING Score your work a 1 if:
Ask Questions ▶	You have formulated questions that can be answered using evidence and observed data. Questions establish what is already known, and determined what questions have yet to be answered.	You have formulated questions that can be answered using evidence and observed data. Questions establish what is known, but you need to determine what questions have yet to be answered.	You have written some questions that can be answered using evidence, but you still need to establish what is known and unknown.	You have written some questions, but are unsure if they can be answered using evidence. You need help identifying what is known and unknown.
Plan an Investigation ▶	You have planned an investigation that identifies the important resources and data needed to construct or revise explanations.	You have planned an investigation and have identified resources needed, but you need to identify the important data needed to construct an explanation.	You have planned an investigation, but you need to identify the important observations and data needed to construct an explanation.	You have planned an investigation, but are unsure what important observations and data are needed to construct an explanation.
Analyze and Interpret Data ▶	You have collected, analyzed, and interpreted data using a range of tools to identify the significant features and patterns in the data.	You have collected, analyzed, and interpreted data using a limited number of tools.	You have collected data, and analyzed data, but you are unsure how to interpret the data.	You have started to collect data, but you are unsure of what tools to use to analyze and interpret the data.
	TIP Use a range of tools such as tabulation, graphical interpretation, and statistical analysis.			
Construct an Explanation ▶	You constructed and revised several clear and logical explanations that incorporate a current understanding of science, and are consistent with available evidence.	You constructed and revised a clear and logical explanation that incorporates a current understanding of science and is consistent with available evidence.	You have constructed an explanation that incorporate a current understanding of science, but need further evidence to revise or support it.	You have constructed an explanation, but it doesn't incorporate an understanding of science and needs further evidence to support it.
	TIP Revise your explanation based on evidence from a variety of sources including scientific principles, theories, and peer review.			
Communicate Information ▶	You have communicated clearly the ideas and the results of your inquiry. You have reviewed multiple scientific texts, evaluated the scientific validity of the information, and have integrated that information.	You have communicated clearly the ideas and the results of your inquiry. You have reviewed a limited number of scientific texts, evaluated them for scientific validity, and integrated that information.	You have communicated your ideas and results of your inquiry, but you need to review scientific texts, evaluate the scientific validity of the information, and integrate the information.	You have your ideas and results, but you are unsure how to communicate them.

Genetics

Chapters

11 Introduction to Genetics

12 DNA

13 RNA and Protein Synthesis

14 Human Heredity

15 Genetic Engineering

INTRODUCE the **Big ideas**

- **Information and Heredity**
- **Cellular Basis of Life**
- **Science as a Way of Knowing**

"Do you look more like mom or dad? I once ran my daughter's DNA on a fingerprinting gel. It didn't settle whether she had her mother's eyes or mine, but half of the bands on that gel were identical to mine, and half, of course, to her mom's. It made me think just how remarkable human genetics really is. Our genes may come from our parents, but each of us gets a fresh shuffle and a brand-new deal of those genetic cards as we start our lives."

Ken Miller

NGSS Problem-based Learning

UNIT PROJECT 4

Spotlight on NGSS

- **Core Idea LS3.B** Variation of Traits
- **Practice** Obtaining, Evaluating, and Communicating Information
- **Crosscutting Concept** Cause and Effect

Food Fight!

To kick off the summer, Olivia's family invited their neighbors over for the first barbeque of the season. Olivia's dad Tim fired up the grill as her mom brought out the chicken and corn. Olivia noticed her friend Brandon eyeing the corn warily.

"What's wrong Brandon. Don't you like grilled corn on the cob?"

"It looks good. But a lot of corn is genetically modified. Do you know if this corn is?"

"I don't know. How can your tell? Does it taste any different?" Olivia asked.

"You can't tell by looking at it—or tasting it actually," Brandon responded.

"I've heard that GM foods are good because they reduce the need for pesticides," Tim chimed in as he put the corn on the grill.

"That may be true," Brandon replied. "But some people say we don't really know the long-term effects of GM foods on humans or on the environment."

"Well ..." said Olivia. "That corn looks too good not to eat. But tomorrow I'm going to research this issue and see what I can find out."

"Me too!" Brandon declared as he shook Olivia's hand.

Those who support GM foods focus on how these foods can help feed hungry people. Those who oppose GM foods focus on the risks to human health of food with altered genes. ▶

Your Task: To Eat or Not to Eat

Some people strongly oppose the development and distribution of GM foods, some people strongly support the use of GM foods, and some people do not know enough about GM foods to have an opinion. Before you begin this task, decide which category you belong in. Then, follow the steps below. Use the rubric at the end of the unit to evaluate your work for this task.

Ask Questions Begin by asking questions that will help you understand the controversy. For example, how are plants modified to produce GM foods? Is there a benefit to using GM foods? Is there evidence that GM foods can harm humans? As you do the task, add to your list of questions.

Plan and Carry Out an Investigation Make a plan to find the information you will need to answer your questions. Decide which questions you should answer first. Then use the sources provided to investigate both sides of the issue. As you do your research, you may notice that the type of information varies with the source. Use the following questions to decide whether an article you read is biased. Make a chart to record the data you collect about the sources.

1. What person or group developed the article?
2. What do you think the goal of the article is?
3. What is the main idea of the article?
4. Are the details used to support the main idea facts or opinions?

Analyze and Interpret Data After you finish collecting your data, analyze the process you used to gather and evaluate the articles. Which sources did you think were unbiased, and why? In what way were the sources in support of GM foods similar? In what way were the sources opposed to GM foods similar? Did your opinion of any source change after you read other articles? If so, provide a specific example.

Engage in Argument from Evidence Recall the category you chose when you began—strongly oppose, strongly support, or do not know enough. Based on the data you evaluated, would you choose the same category now? Write a paragraph to explain your position on the GM controversy.

Communicate Information Construct a persuasive visual summary of your position. A cartoon, a poster, or the home page for a Web site are possible formats. Choose a target audience for your visual. Gather feedback on your visual from your target audience. Based on the feedback, you may want to revise your summary.

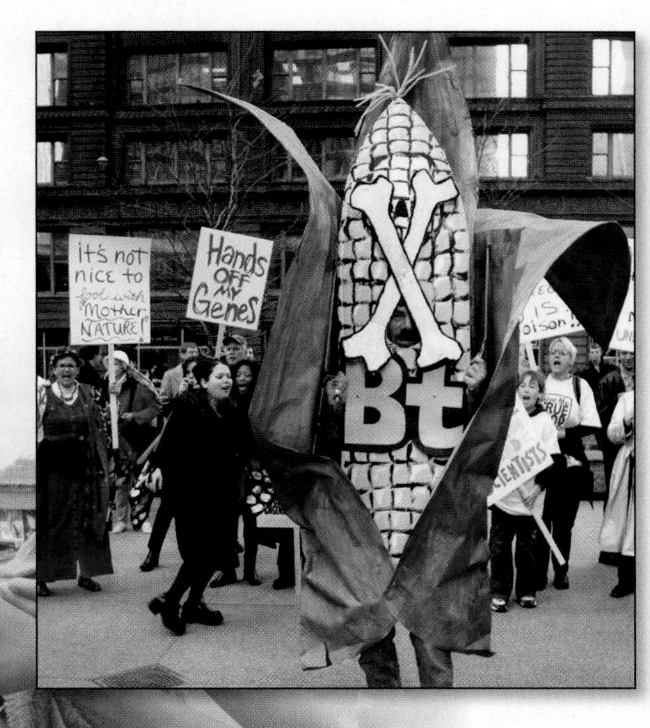

Biology.com

Use this website for information on genetically modified foods. Find sources to help you investigate the GM foods controversy.

11 Introduction to Genetics

Big idea: Information and Heredity

Q: How does biological information pass from one generation to another?

Genetics is the study of biological inheritance. The different coat colors of these Labrador retrievers are an example of the inherited characteristics that geneticists try to understand.

INSIDE:

- 11.1 The Work of Gregor Mendel
- 11.2 Applying Mendel's Principles
- 11.3 Other Patterns of Inheritance
- 11.4 Meiosis

BUILDING Scientific Literacy

Deepen your understanding of genetic principles by engaging in key practices that allow you to make connections across concepts.

NGSS You will use mathematics and computational thinking to analyze patterns of inheritance.

STEM How could you grow flowers that match a school's colors for a fundraiser? Do a simulation based on genetic principles to find out.

Common Core You will analyze the relationship between meiosis and the principle of independent assortment.

CHAPTER MYSTERY

GREEN PARAKEETS

Susan's birthday was coming up. Parakeets make great pets, so Susan's parents decided to give two birds to her as a birthday present. At the pet store, they selected two healthy green parakeets—one male and one female. They knew that green was Susan's favorite color.

Susan was delighted about her birthday present. She fed the birds and kept their cage clean. A few weeks later, Susan found three small eggs in the birds' nest. She couldn't wait to welcome three new green parakeets. When the eggs finally hatched, however, Susan was amazed. None of the chicks was green—one chick was white, one was blue, and one was yellow. Why weren't any of them green? What had happened to the green color of the birds' parents? As you read this chapter, look for clues to help you identify why the parakeet chicks were differently colored than their parents.

Biology.com

Finding the solution to the Green Parakeets mystery is only the beginning. Take a video field trip with the ecogeeks of Untamed Science to see where the mystery leads.

Go online to access additional resources including:
- eText • Flash Cards • Lesson Overviews • Chapter Mystery

11.1 The Work of Gregor Mendel

Key Questions

Where does an organism get its unique characteristics?

How are different forms of a gene distributed to offspring?

Vocabulary

genetics • fertilization • trait • hybrid • gene • allele • principle of dominance • segregation • gamete

Taking Notes

Two-Column Chart Before you read, draw a line down the center of a sheet of paper. On the left side, write the main ideas in this lesson. On the right side, note the details and examples that support each of those ideas.

THINK ABOUT IT What is an inheritance? To many people, it is money or property left to them by relatives who have passed away. That kind of inheritance matters, of course, but there is another kind that matters even more. It is something we each receive from our parents—a contribution that determines our blood type, the color of our hair, and so much more. Most people leave their money and property behind by writing a will. But what kind of inheritance makes a person's face round or their hair curly?

The Experiments of Gregor Mendel

Where does an organism get its unique characteristics?

Every living thing—plant or animal, microbe or human being—has a set of characteristics inherited from its parent or parents. Since the beginning of recorded history, people have wanted to understand how that inheritance is passed from generation to generation. The delivery of characteristics from parent to offspring is called heredity. The scientific study of heredity, known as **genetics,** is the key to understanding what makes each organism unique.

The modern science of genetics was founded by an Austrian monk named Gregor Mendel. Mendel, shown in **Figure 11–1,** was born in 1822 in what is now the Czech Republic. After becoming a priest, Mendel spent several years studying science and mathematics at the University of Vienna. He spent the next 14 years working in a monastery and teaching high school. In addition to his teaching duties, Mendel was in charge of the monastery garden. In this simple garden, he was to do the work that changed biology forever.

Mendel carried out his work with ordinary garden peas, partly because peas are small and easy to grow. A single pea plant can produce hundreds of offspring. Today we call peas a "model system." Scientists use model systems because they are convenient to study and may tell us how other organisms, including humans, actually function. By using peas, Mendel was able to carry out, in just one or two growing seasons, experiments that would have been impossible to do with humans and that would have taken decades—if not centuries—to do with pigs, horses, or other large animals.

FIGURE 11–1 Gregor Mendel

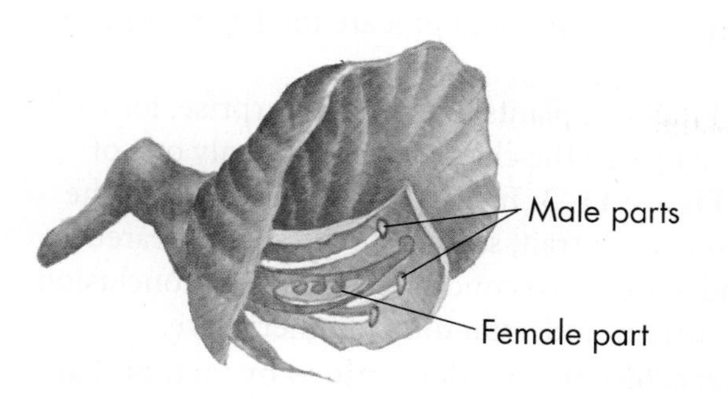

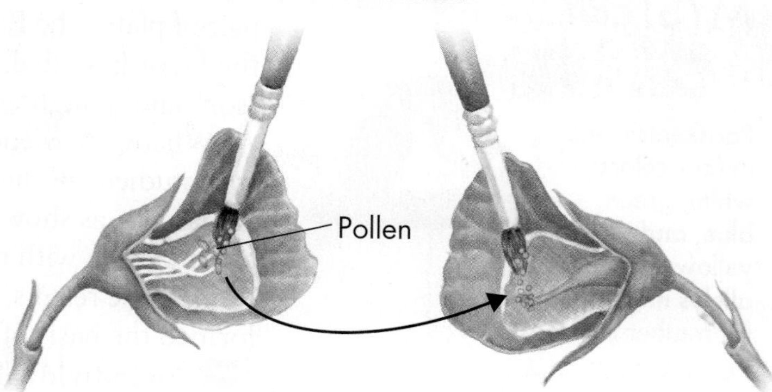

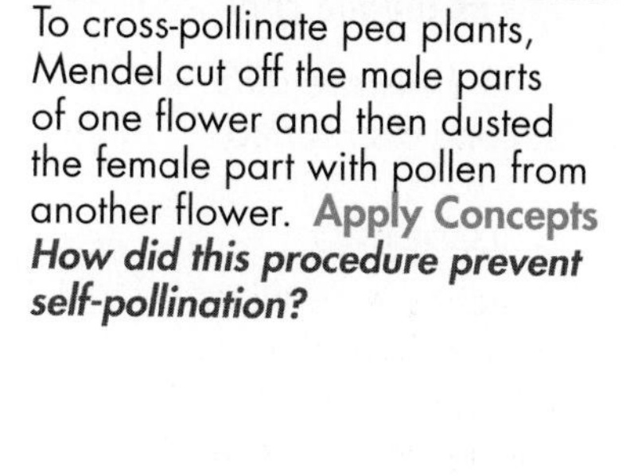

FIGURE 11–2 Cross-Pollination To cross-pollinate pea plants, Mendel cut off the male parts of one flower and then dusted the female part with pollen from another flower. **Apply Concepts** ***How did this procedure prevent self-pollination?***

The Role of Fertilization When Mendel began his experiments, he knew that the male part of each flower makes pollen, which contains the plant's male reproductive cells, called sperm. Similarly, Mendel knew that the female portion of each flower produces reproductive cells called eggs. During sexual reproduction, male and female reproductive cells join in a process known as **fertilization** to produce a new cell. In peas, this new cell develops into a tiny embryo encased within a seed.

Pea flowers are normally self-pollinating, which means that sperm cells fertilize egg cells from within the same flower. A plant grown from a seed produced by self-pollination inherits all of its characteristics from the single plant that bore it; it has a single parent.

Mendel's monastery garden had several stocks of pea plants. These plants were "true-breeding," meaning that they were self-pollinating, and would produce offspring identical to themselves. In other words, the traits of each successive generation would be the same. A **trait** is a specific characteristic, such as seed color or plant height, of an individual. Many traits vary from one individual to another. For instance, one stock of Mendel's seeds produced only tall plants, while another produced only short ones. One line produced only green seeds, another produced only yellow seeds.

To learn how these traits were determined, Mendel decided to "cross" his stocks of true-breeding plants—that is, he caused one plant to reproduce with another plant. To do this, he had to prevent self-pollination. He did so by cutting away the pollen-bearing male parts of a flower. He then dusted the pollen from a different plant onto the female part of that flower, as shown in **Figure 11–2.** This process, known as cross-pollination, produces a plant that has two different parents. Cross-pollination allowed Mendel to breed plants with traits different from those of their parents and then study the results.

Mendel studied seven different traits of pea plants. Each of these seven traits had two contrasting characteristics, such as green seed color or yellow seed color. Mendel crossed plants with each of the seven contrasting characteristics and then studied their offspring. The offspring of crosses between parents with different traits are called **hybrids.**

In Your Notebook *Explain, in your own words, what fertilization is.*

Parakeets come in four colors: white, green, blue, and yellow. How many alleles might there be for feather color?

Genes and Alleles When doing genetic crosses, we call each original pair of plants the P, or parental, generation. Their offspring are called the F_1, or first filial, generation. (*Filius* and *filia* are the Latin words for "son" and "daughter.")

What were Mendel's F_1 hybrid plants like? To his surprise, for each trait studied, all the offspring had the characteristics of only one of its parents, as shown in **Figure 11–3.** In each cross, the nature of the other parent, with regard to each trait, seemed to have disappeared. From these results, Mendel drew two conclusions. His first conclusion formed the basis of our current understanding of inheritance. **An individual's characteristics are determined by factors that are passed from one parental generation to the next.** Today, scientists call the factors that are passed from parent to offspring **genes.**

Each of the traits Mendel studied was controlled by a single gene that occurred in two contrasting varieties. These variations produced different expressions, or forms, of each trait. For example, the gene for plant height occurred in one form that produced tall plants and in another form that produced short plants. The different forms of a gene are called **alleles** (uh LEELZ).

Dominant and Recessive Alleles Mendel's second conclusion is called the **principle of dominance.** This principle states that some alleles are dominant and others are recessive. An organism with at least one dominant allele for a particular form of a trait will exhibit that form of the trait. An organism with a recessive allele for a particular form of a trait will exhibit that form only when the dominant allele for the trait is not present. In Mendel's experiments, the allele for tall plants was dominant and the allele for short plants was recessive. Likewise, the allele for yellow seeds was dominant over the recessive allele for green seeds.

FIGURE 11–3 Mendel's F_1 Crosses When Mendel crossed plants with contrasting traits, the resulting hybrids had the traits of only one of the parents.

Mendel's Seven F_1 Crosses on Pea Plants

	Seed Shape	Seed Color	Seed Coat	Pod Shape	Pod Color	Flower Position	Plant Height
P	Round X Wrinkled	Yellow X Green	Gray X White	Smooth X Constricted	Green X Yellow	Axial X Terminal	Tall X Short
F_1	Round	Yellow	Gray	Smooth	Green	Axial	Tall

Quick Lab
GUIDED INQUIRY

Classroom Variation

1. Copy the data table into your notebook.
2. Write a prediction of whether the traits listed in the table will be evenly distributed or if there will be more dominant than recessive traits.
3. Examine your features, using a mirror if necessary. Determine which traits you have for features A–E.
4. Interview at least 14 other students to find out which traits they have. Tally the numbers. Record the totals in each column.

Trait Survey				
Feature	**Dominant Trait**	**Number**	**Recessive Trait**	**Number**
A	Free ear lobes		Attached ear lobes	
B	Hair on fingers		No hair on fingers	
C	Widow's peak		No widow's peak	
D	Curly hair		Straight hair	
E	Cleft chin		Smooth chin	

Analyze and Conclude

1. **Calculate** Calculate the percentages of each trait in your total sample. How do these numbers compare to your prediction? MATH
2. **Form a Hypothesis** Why do you think recessive traits are more common in some cases?

In Your Notebook *Make a diagram that explains Mendel's principle of dominance.*

Segregation

How are different forms of a gene distributed to offspring?

Mendel didn't just stop after crossing the parent plants, because he had another question: Had the recessive alleles simply disappeared, or were they still present in the new plants? To find out, he allowed all seven kinds of F_1 hybrids to self-pollinate. The offspring of an F_1 cross are called the F_2 (second filial) generation. In effect, Mendel crossed the F_1 generation with itself to produce the F_2 offspring, as shown in **Figure 11–4.**

The F_1 Cross When Mendel compared the F_2 plants, he made a remarkable discovery: The traits controlled by the recessive alleles reappeared in the second generation. Roughly one fourth of the F_2 plants showed the trait controlled by the recessive allele. Why, then, did the recessive alleles seem to disappear in the F_1 generation, only to reappear in the F_2 generation?

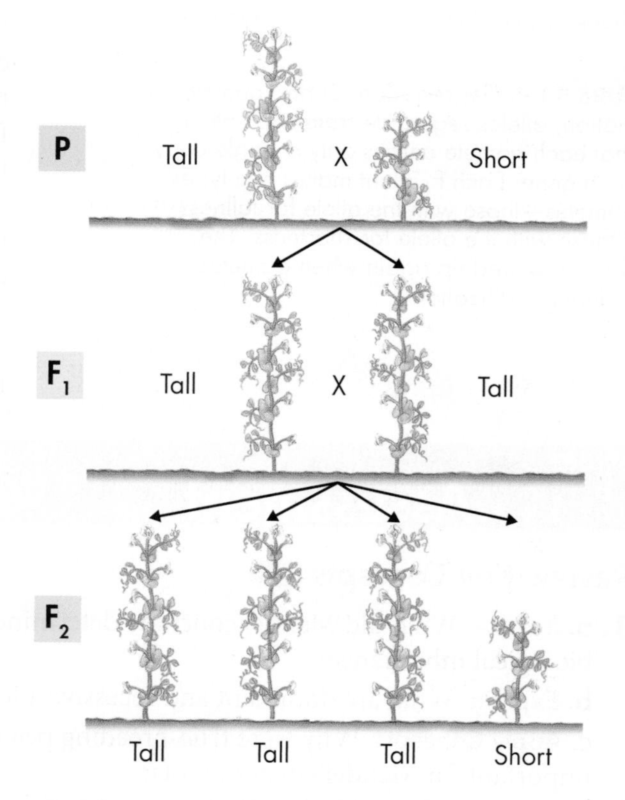

FIGURE 11–4 Results of the F_1 Cross When Mendel allowed the F_1 plants to reproduce by self-pollination, the traits controlled by recessive alleles reappeared in about one fourth of the F_2 plants in each cross. **Calculate** ***What proportion of the F_2 plants had a trait controlled by a dominant allele?*** MATH

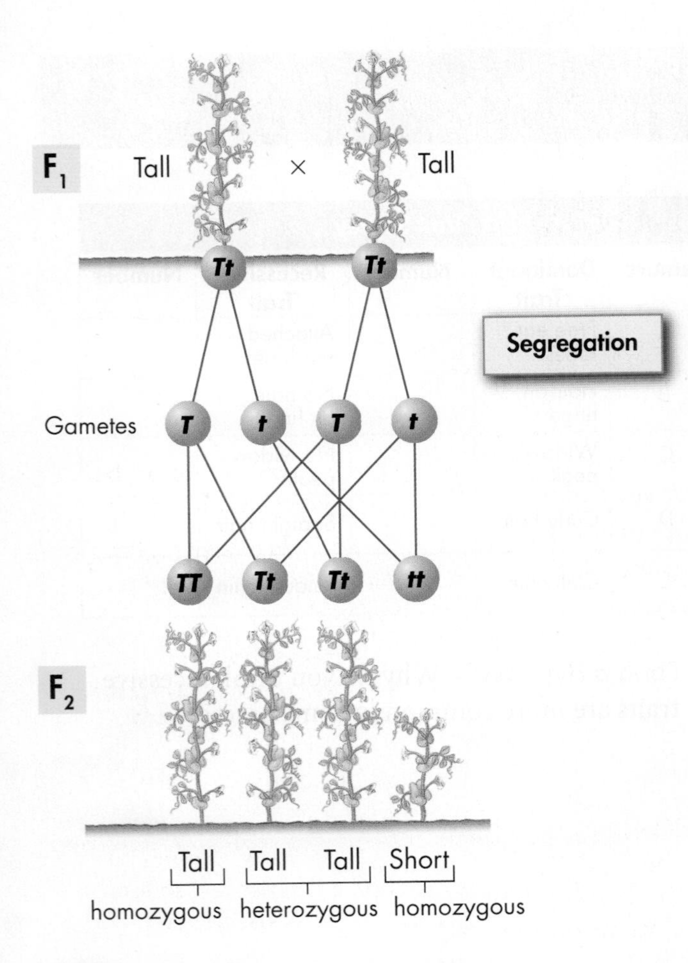

FIGURE 11–5 Segregation During gamete formation, alleles segregate from each other so that each gamete carries only a single copy of each gene. Each F_1 plant makes two types of gametes—those with the allele for tallness and those with the allele for shortness. The alleles are paired up again when gametes fuse during fertilization.

Explaining the F_1 Cross To begin with, Mendel assumed that a dominant allele had masked the corresponding recessive allele in the F_1 generation. However, the trait controlled by the recessive allele did show up in some of the F_2 plants. This reappearance indicated that, at some point, the allele for shortness had separated from the allele for tallness. How did this separation, or **segregation,** of alleles occur? Mendel suggested that the alleles for tallness and shortness in the F_1 plants must have segregated from each other during the formation of the sex cells, or **gametes** (GAM eetz). Did that suggestion make sense?

The Formation of Gametes Let's assume, as Mendel might have, that all the F_1 plants inherited an allele for tallness from the tall parent and one for shortness from the short parent. Because the allele for tallness is dominant, all the F_1 plants are tall. **During gamete formation, the alleles for each gene segregate from each other, so that each gamete carries only one allele for each gene.** Thus, each F_1 plant produces two kinds of gametes—those with the tall allele and those with the short allele.

Look at **Figure 11–5** to see how alleles separate during gamete formation and then pair up again in the F_2 generation. A capital letter represents a dominant allele. A lowercase letter represents a recessive allele. Now we can see why the recessive trait for height, *t*, reappeared in Mendel's F_2 generation. Each F_1 plant in Mendel's cross produced two kinds of gametes—those with the allele for tallness and those with the allele for shortness. Whenever a gamete that carried the *t* allele paired with the other gamete that carried the *t* allele to produce an F_2 plant, that plant was short. Every time one or both gametes of the pairing carried the *T* allele, a tall plant was produced. In other words, the F_2 generation had new combinations of alleles.

11.1 Assessment

Review Key Concepts

1. a. Review What did Mendel conclude determines biological inheritance?

b. Explain What are dominant and recessive alleles?

c. Apply Concepts Why were true-breeding pea plants important for Mendel's experiments?

2. a. Review What is segregation?

b. Explain What happens to alleles between the P generation and the F_2 generation?

c. Infer What evidence did Mendel use to explain how segregation occurs?

VISUAL THINKING

3. Use a diagram to explain Mendel's principles of dominance and segregation. Your diagram should show how alleles segregate during gamete formation.

11.2 Applying Mendel's Principles

THINK ABOUT IT *Nothing in life is certain.* There's a great deal of wisdom in that old saying, and genetics is a fine example. If a parent carries two different alleles for a certain gene, we can't be sure which of those alleles will be inherited by any one of the parent's offspring. However, think carefully about the nature of inheritance and you'll see that even if we can't predict the exact future, we can do something almost as useful—we can figure out the odds.

Probability and Punnett Squares

How can we use probability to predict traits?

Whenever Mendel performed a cross with pea plants, he carefully categorized and counted the offspring. Consequently, he had plenty of data to analyze. For example, whenever he crossed two plants that were hybrids for stem height (*Tt*), about three fourths of the resulting plants were tall and about one fourth were short.

Upon analyzing his data, Mendel realized that the principles of probability could be used to explain the results of his genetic crosses. **Probability** is a concept you may have learned about in math class. It is the likelihood that a particular event will occur. As an example, consider an ordinary event, such as flipping a coin. There are two possible outcomes of this event: The coin may land either heads up or tails up. The chance, or probability, of either outcome is equal. Therefore, the probability that a single coin flip will land heads up is 1 chance in 2. This amounts to 1/2, or 50 percent.

If you flip a coin three times in a row, what is the probability that it will land heads up every time? Each coin flip is an independent event with a 1/2 probability of landing heads up. Therefore, the probability of flipping three heads in a row is:

$$1/2 \times 1/2 \times 1/2 = 1/8$$

As you can see, you have 1 chance in 8 of flipping heads three times in a row. The multiplication of individual probabilities illustrates an important point: Past outcomes do not affect future ones. Just because you've flipped three heads in a row does not mean that you're more likely to have a coin land tails up on the next flip. The probability for that flip is still 1/2.

Key Questions

How can we use probability to predict traits?

How do alleles segregate when more than one gene is involved?

What did Mendel contribute to our understanding of genetics?

Vocabulary

probability • homozygous • heterozygous • phenotype • genotype • Punnett square • independent assortment

Taking Notes

Preview Visuals Before you read, preview **Figure 11–7.** Try to infer the purpose of this diagram. As you read, compare your inference to the text. After you read, revise your statement if needed or write a new one about the diagram's purpose.

FIGURE 11–6 Probability Probability allows you to calculate the likelihood that a particular event will occur. The probability that the coin will land heads up is ½, or 50 percent.

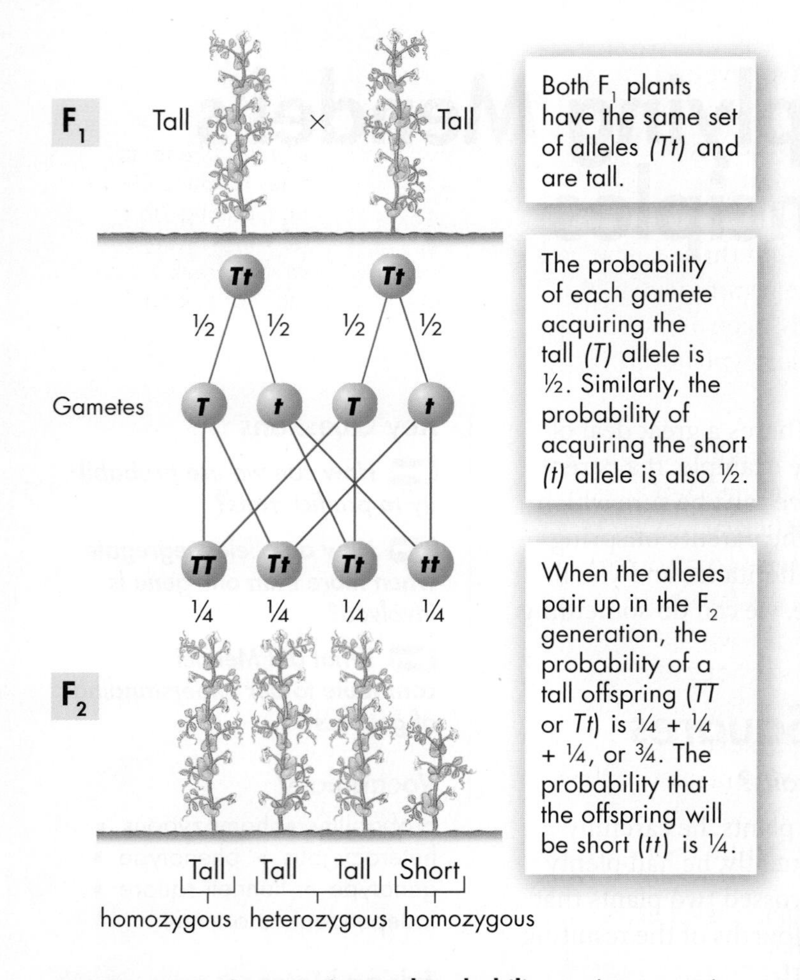

FIGURE 11–7 Segregation and Probability In this cross, the *TT* and *Tt* allele combinations produced three tall pea plants, while the *tt* allele combination produced one short plant. These quantities follow the laws of probability. **Predict** ***If you crossed a*** **TT** ***plant with a*** **Tt** ***plant, would the offspring be tall or short?***

Using Segregation to Predict Outcomes

The way in which alleles segregate during gamete formation is every bit as random as a coin flip. Therefore, the principles of probability can be used to predict the outcomes of genetic crosses.

Look again at Mendel's F_1 cross, shown in **Figure 11–7.** This cross produced a mixture of tall and short plants. Why were just 1/4 of the offspring short? Well, the F_1 plants were both tall. If each plant had one tall allele and one short allele (*Tt*), and if the alleles segregated as Mendel thought, then 1/2 of the gametes produced by the plants would carry the short allele (*t*). Yet, the *t* allele is recessive. The only way to produce a short (*tt*) plant is for two gametes, each carrying the *t* allele, to combine.

Like the coin toss, each F_2 gamete has a one in two, or 1/2, chance of carrying the *t* allele. There are two gametes, so the probability of both gametes carrying the *t* allele is $1/2 \times 1/2 = 1/4$. In other words, roughly one fourth of the F_2 offspring should be short, and the remaining three fourths should be tall. This predicted ratio—3 offspring exhibiting the dominant trait to 1 offspring exhibiting the recessive trait—showed up consistently in Mendel's experiments. For each of his seven crosses, about 3/4 of the plants showed the trait controlled by the dominant allele. About 1/4 showed the trait controlled by the recessive allele. Segregation did occur according to Mendel's model.

As you can see in the F_2 generation, not all organisms with the same characteristics have the same combinations of alleles. Both the *TT* and *Tt* allele combinations resulted in tall pea plants, but only one of these combinations contains identical alleles. Organisms that have two identical alleles for a particular gene—*TT* or *tt* in this example—are said to be **homozygous** (hoh moh ZY gus). Organisms that have two different alleles for the same gene—such as *Tt*—are **heterozygous** (het ur oh ZY gus).

Probabilities Predict Averages Probabilities predict the average outcome of a large number of events. If you flip a coin twice, you are likely to get one heads and one tails. However, you might also get two heads or two tails. To get the expected 50 : 50 ratio, you might have to flip the coin many times. The same is true of genetics.

The larger the number of offspring, the closer the results will be to the predicted values. If an F_2 generation contains just three or four offspring, it may not match Mendel's ratios. When an F_2 generation contains hundreds or thousands of individuals, the ratios usually come very close to matching predictions.

Genotype and Phenotype One of Mendel's most revolutionary insights followed directly from his observations of F_1 crosses: Every organism has a genetic makeup as well as a set of observable characteristics. All of the tall pea plants had the same **phenotype,** or physical traits. They did not, however, have the same **genotype,** or genetic makeup. Look again at **Figure 11–7** and you will find three different genotypes among the F_2 plants: *TT, Tt,* and *tt.* The genotype of an organism is inherited, and the phenotype is largely determined by the genotype. Two organisms may share the same phenotype but have different genotypes.

BUILD Vocabulary

PREFIXES The prefix *pheno-* in **phenotype** comes from the Greek word *phainein,* meaning "to show." *Geno-,* the prefix in **genotype,** is derived from the Greek word *genus,* meaning "race, kind."

Using Punnett Squares One of the best ways to predict the outcome of a genetic cross is by drawing a simple diagram known as a **Punnett square.** **Punnett squares use mathematical probability to help predict the genotype and phenotype combinations in genetic crosses.** Constructing a Punnett square is fairly easy. You begin with a square. Then, following the principle of segregation, all possible combinations of alleles in the gametes produced by one parent are written along the top edge of the square. The other parent's alleles are then segregated along the left edge. Next, every possible genotype is written into the boxes within the square, just as they might appear in the F_2 generation. **Figure 11–8** on the next page shows step-by-step instructions for constructing Punnett squares.

In Your Notebook *In your own words, write definitions for the terms* homozygous, heterozygous, phenotype, *and* genotype.

Quick Lab
GUIDED INQUIRY

How Are Dimples Inherited?

1. Write the last four digits of any telephone number. These four random digits represent the alleles of a gene that determines whether a person will have dimples. Odd digits represent the allele for the dominant trait of dimples. Even digits represent the allele for the recessive trait of no dimples.

2. Use the first two digits to represent a father's genotype. Use the symbols *D* and *d* to write his genotype as shown in the example.

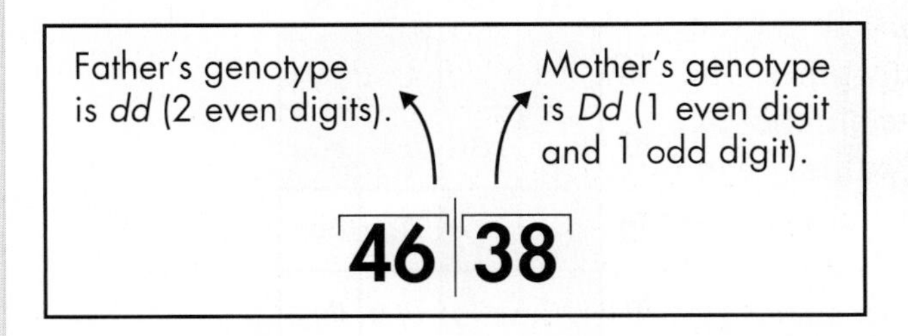

3. Use the last two digits the same way to find the mother's genotype. Write her genotype.

4. Use **Figure 11–8** on the next page to construct a Punnett square for the cross of these parents. Then, using the Punnett square, determine the probability that their child will have dimples.

5. Determine the class average of the percent of children with dimples.

Analyze and Conclude

1. **Apply Concepts** How does the class average compare with the result of a cross of two heterozygous parents?

2. **Draw Conclusions** What percentage of the children will be expected to have dimples if one parent is homozygous for dimples (*DD*) and the other is heterozygous (*Dd*)?

VISUAL SUMMARY

HOW TO MAKE A PUNNETT SQUARE

FIGURE 11–8 By drawing a Punnett square, you can determine the allele combinations that might result from a genetic cross.

One-Factor Cross | Two-Factor Cross

① Start With the Parents

One-Factor Cross: Write the genotypes of the two organisms that will serve as parents in a cross. In this example we will cross a male and female osprey, or fish hawk, that are heterozygous for large beaks. They each have genotypes of *Bb*.

Bb* and *Bb

Two-Factor Cross: In this example we will cross two pea plants that are heterozygous for size (tall and short alleles) and pod color (green and yellow alleles). The genotypes of the two parents are *TtGg* and *TtGg*.

TtGg* and *TtGg

② Figure Out the Gametes

One-Factor Cross: Determine what alleles would be found in all of the possible gametes that each parent could produce.

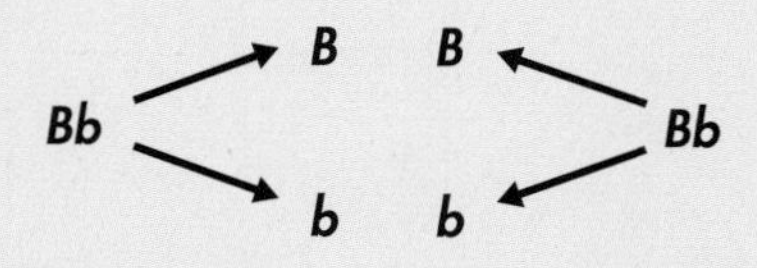

Two-Factor Cross: Determine what alleles would be found in all of the possible gametes that each parent could produce.

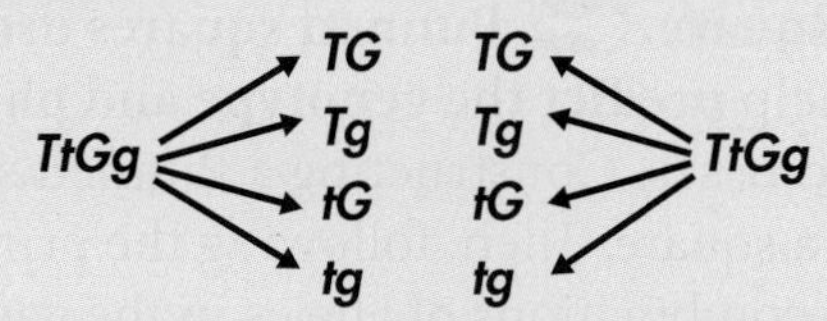

One-Factor Cross: Draw a table with enough squares for each pair of gametes from each parent. In this case, each parent can make two different types of gametes, *B* and *b*. Enter the genotypes of the gametes produced by both parents on the top and left sides of the table.

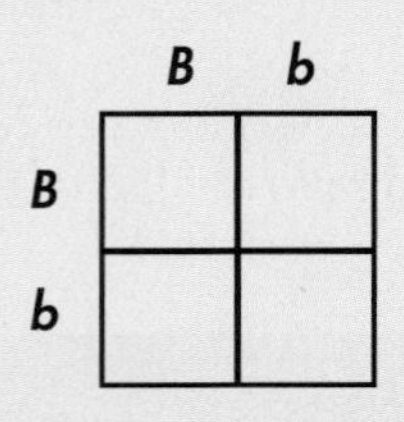

Two-Factor Cross: In this case, each parent can make 4 different types of gametes, so the table needs to be 4 rows by 4 columns, or 16 squares.

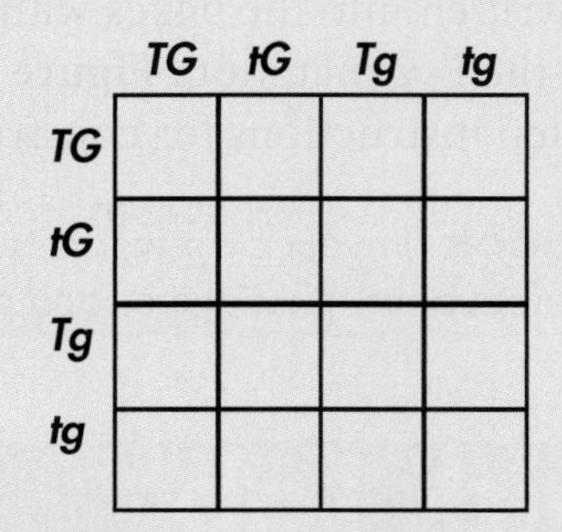

④ Write Out the New Genotypes

One-Factor Cross: Fill in the table by combining the gametes' genotypes.

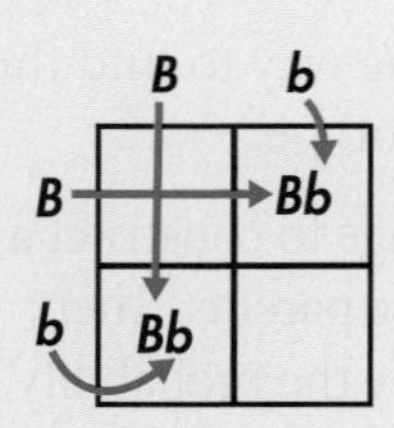

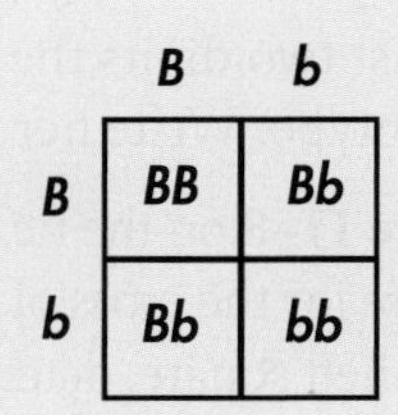

Two-Factor Cross: Fill in the table by combining the gametes' genotypes.

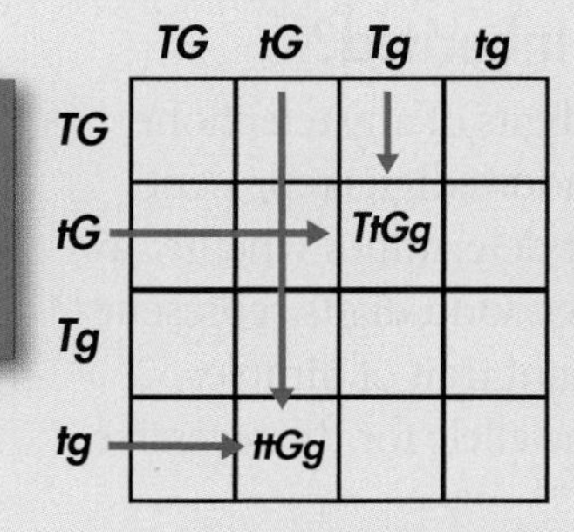

	TG	tG	Tg	tg
TG	TTGG	TtGG	TTGg	TtGg
tG	TtGG	ttGG	TtGg	ttGg
Tg	TTGg	TtGg	TTgg	Ttgg
tg	TtGg	ttGg	Ttgg	ttgg

⑤ Figure Out the Results

One-Factor Cross: Determine the genotype and phenotype of each offspring. Calculate the percentage of each. In this example, 3/4 of the chicks will have large beaks, but only 1/2 will be heterozygous for this trait *(Bb)*.

	B	b
B	BB	Bb
b	Bb	bb

Two-Factor Cross: In this example, the color of the squares represents pod color. Alleles written in black indicate short plants, while alleles written in red indicate tall plants.

	TG	tG	Tg	tg
TG	TTGG	TtGG	TTGg	TtGg
tG	TtGG	ttGG	TtGg	ttGg
Tg	TTGg	TtGg	TTgg	Ttgg
tg	TtGg	ttGg	Ttgg	ttgg

Independent Assortment

How do alleles segregate when more than one gene is involved?

After showing that alleles segregate during the formation of gametes, Mendel wondered if the segregation of one pair of alleles affects another pair. For example, does the gene that determines the shape of a seed affect the gene for seed color? To find out, Mendel followed two different genes as they passed from one generation to the next. Because it involves two different genes, Mendel's experiment is known as a two-factor, or "dihybrid," cross. (Single-gene crosses are "monohybrid" crosses.)

The Two-Factor Cross: F_1 First, Mendel crossed true-breeding plants that produced only round yellow peas with plants that produced wrinkled green peas. The round yellow peas had the genotype *RRYY*, and the wrinkled green peas had the genotype *rryy*. All of the F_1 offspring produced round yellow peas. These results showed that the alleles for yellow and round peas are dominant. As the Punnett square in **Figure 11–9** shows, the genotype in each of these F_1 plants is *RrYy*. In other words, the F_1 plants were all heterozygous for both seed shape and seed color. This cross did not indicate whether genes assort, or segregate independently. However, it provided the hybrid plants needed to breed the F_2 generation.

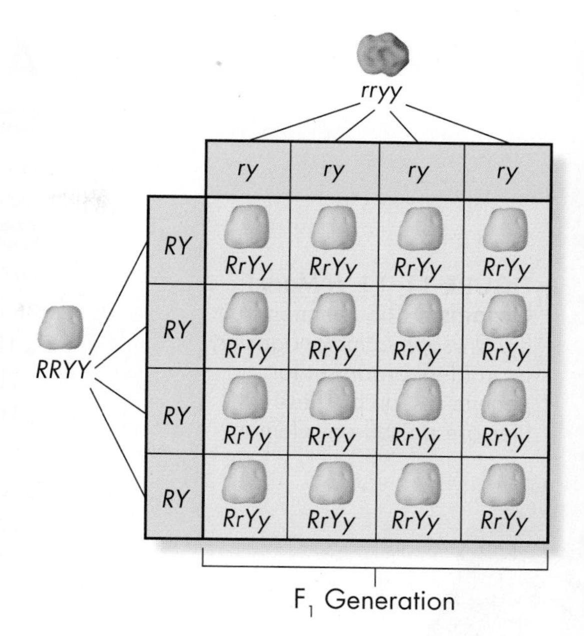

FIGURE 11–9 Two-Factor Cross: F_1 Mendel crossed plants that were homozygous dominant for round yellow peas with plants that were homozygous recessive for wrinkled green peas. All of the F_1 offspring were heterozygous dominant for round yellow peas. **Interpret Graphics** ***How is the genotype of the offspring different from that of the homozygous dominant parent?***

The Two-Factor Cross: F_2 In the second part of this experiment, Mendel crossed the F_1 plants to produce F_2 offspring. Remember, each F_1 plant was formed by the fusion of a gamete carrying the dominant *RY* alleles with another gamete carrying the recessive *ry* alleles. Did this mean that the two dominant alleles would always stay together, or would they segregate independently, so that any combination of alleles was possible?

In Mendel's experiment, the F_2 plants produced 556 seeds. Mendel compared their variation. He observed that 315 of the seeds were round and yellow, while another 32 seeds were wrinkled and green—the two parental phenotypes. However, 209 seeds had combinations of phenotypes, and therefore combinations of alleles, that were not found in either parent. This clearly meant that the alleles for seed shape segregated independently of those for seed color. Put another way, genes that segregate independently (such as the genes for seed shape and seed color in pea plants) do not influence each other's inheritance.

Mendel's experimental results were very close to the 9 : 3 : 3 : 1 ratio that the Punnett square shown in **Figure 11–10** predicts. Mendel had discovered the principle of **independent assortment.** **The principle of independent assortment states that genes for different traits can segregate independently during the formation of gametes.** Independent assortment helps account for the many genetic variations observed in plants, animals, and other organisms—even when they have the same parents.

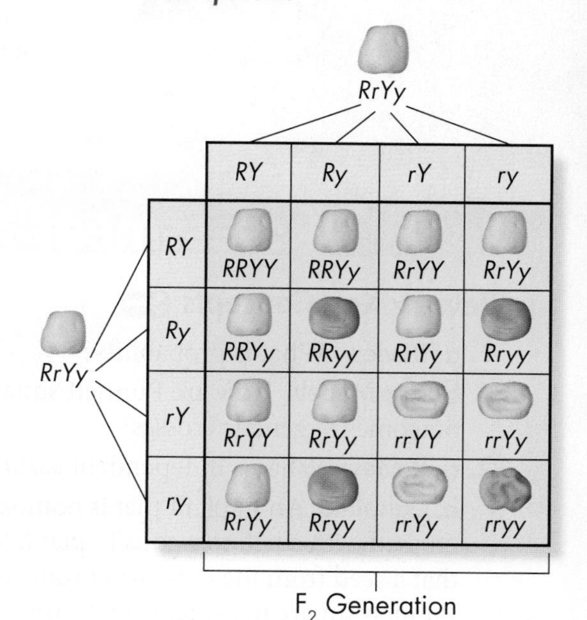

FIGURE 11–10 Two-Factor Cross: F_2 When Mendel crossed F_1 plants that were heterozygous dominant for round yellow peas, he found that the alleles segregated independently to produce the F_2 generation.

A Summary of Mendel's Principles

What did Mendel contribute to our understanding of genetics?

As you have seen, Mendel's principles of segregation and independent assortment can be observed through one- and two-factor crosses. **Mendel's principles of heredity, observed through patterns of inheritance, form the basis of modern genetics.** These principles are as follows:

- The inheritance of biological characteristics is determined by individual units called genes, which are passed from parents to offspring.
- Where two or more forms (alleles) of the gene for a single trait exist, some alleles may be dominant and others may be recessive.
- In most sexually reproducing organisms, each adult has two copies of each gene—one from each parent. These genes segregate from each other when gametes are formed.
- Alleles for different genes usually segregate independently of each other.

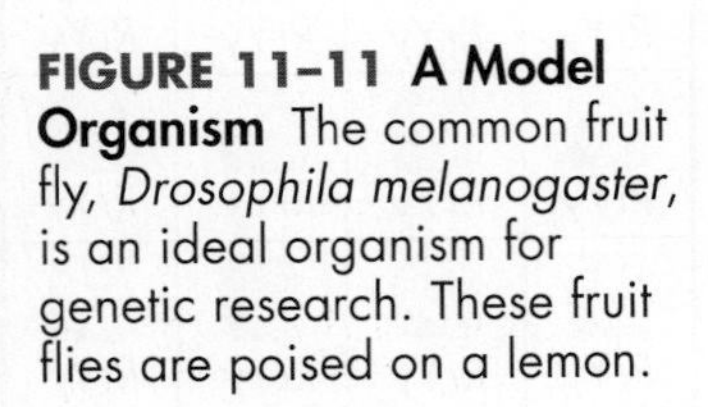

FIGURE 11–11 A Model Organism The common fruit fly, *Drosophila melanogaster*, is an ideal organism for genetic research. These fruit flies are poised on a lemon.

Mendel's principles don't apply only to plants. At the beginning of the 1900s, the American geneticist Thomas Hunt Morgan wanted to use a model organism of another kind to advance the study of genetics. He decided to work on a tiny insect that kept showing up, uninvited, in his laboratory. The insect was the common fruit fly, *Drosophila melanogaster,* shown in **Figure 11–11.** *Drosophila* can produce plenty of offspring—a single pair can produce hundreds of young. Before long, Morgan and other biologists had tested all of Mendel's principles and learned that they applied to flies and other organisms as well. In fact, Mendel's basic principles can be used to study the inheritance of human traits and to calculate the probability of certain traits appearing in the next generation. You will learn more about human genetics in Chapter 14.

11.2 Assessment

Review Key Concepts

1. a. **Review** What is probability?
   b. **Use Models** How are Punnett squares used to predict the outcomes of genetic crosses?
2. a. **Review** What is independent assortment?
   b. **Calculate** An F_1 plant that is homozygous for shortness is crossed with a heterozygous F_1 plant. What is the probability that a seed from the cross will produce a tall plant? Use a Punnett square to explain your answer and to compare the probable genetic variations in the F_2 plants. MATH
3. a. **Review** How did Gregor Mendel contribute to our understanding of inherited traits?
   b. **Apply Concepts** Why is the fruit fly an ideal organism for genetic research?

Apply the Big idea

Information and Heredity

4. Suppose you are an avid gardener. One day, you come across a plant with beautiful lavender flowers. Knowing that the plant is self-pollinating, you harvest its seeds and plant them. Of the 106 plants that grow from these seeds, 31 have white flowers. Using a Punnett square, draw conclusions about the nature of the allele for lavender flowers.

11.3 Other Patterns of Inheritance

THINK ABOUT IT Mendel's principles offer a tidy set of rules with which to predict various patterns of inheritance. Unfortunately, biology is not a tidy science. There are exceptions to every rule, and exceptions to the exceptions. What happens if one allele is not completely dominant over another? What if a gene has several alleles?

Key Questions

What are some exceptions to Mendel's principles?

Does the environment have a role in how genes determine traits?

Vocabulary

incomplete dominance • codominance • multiple allele • polygenic trait

Taking Notes

Outline Make an outline using the green and blue headings. As you read, write bulleted notes below each heading to summarize its topic.

Beyond Dominant and Recessive Alleles

What are some exceptions to Mendel's principles?

Despite the importance of Mendel's work, there are important exceptions to most of his principles. For example, not all genes show simple patterns of inheritance. In most organisms, genetics is more complicated, because the majority of genes have more than two alleles. Also, many important traits are controlled by more than one gene. Understanding these exceptions allows geneticists to predict the ways in which more complex traits are inherited.

Incomplete Dominance A cross between two four o'clock (*Mirabilis jalapa*) plants shows a common exception to Mendel's principles. **Some alleles are neither dominant nor recessive.** As shown in **Figure 11–12,** the F_1 generation produced by a cross between red-flowered (*RR*) and white-flowered (*WW*) *Mirabilis* plants consists of pink-colored flowers (*RW*). Which allele is dominant in this case? Neither one. Cases in which one allele is not completely dominant over another are called **incomplete dominance.** In incomplete dominance, the heterozygous phenotype lies somewhere between the two homozygous phenotypes.

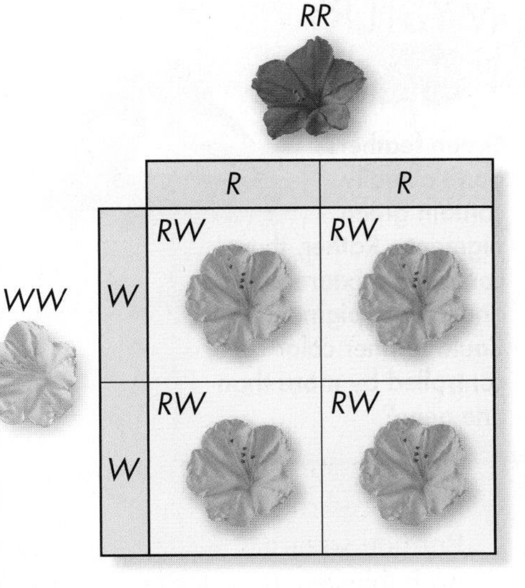

FIGURE 11–12 Incomplete Dominance In four o'clock plants, the alleles for red and white flowers show incomplete dominance. Heterozygous (*RW*) plants have pink flowers—a mix of red and white coloring.

Codominance A similar situation arises from **codominance,** in which the phenotypes produced by both alleles are clearly expressed. For example, in certain varieties of chicken, the allele for black feathers is codominant with the allele for white feathers. Heterozygous chickens have a color described as "erminette," speckled with black and white feathers. Unlike the blending of red and white colors in heterozygous four o'clocks, black and white colors appear separately in chickens. Many human genes, including one for a protein that controls cholesterol levels in the blood, show codominance, too. People with the heterozygous form of this gene produce two different forms of the protein, each with a different effect on cholesterol levels.

Analyzing Data

Human Blood Types

Red blood cells carry antigens, molecules that can trigger an immune reaction, on their surfaces. Human blood type A carries an A antigen, type B has a B antigen, type AB has both antigens, and type O carries neither antigen. The gene for these antigens has three alleles; A, B, and O.

For a transfusion to succeed, it must not introduce a new antigen into the body of the recipient. So, a person with type A blood may receive type O, but not vice versa.

Another gene controls a second type of antigen, known as Rh factor. Rh^+ individuals carry this antigen, while Rh^- ones don't. This chart of the U.S. population shows the percentage of each blood type.

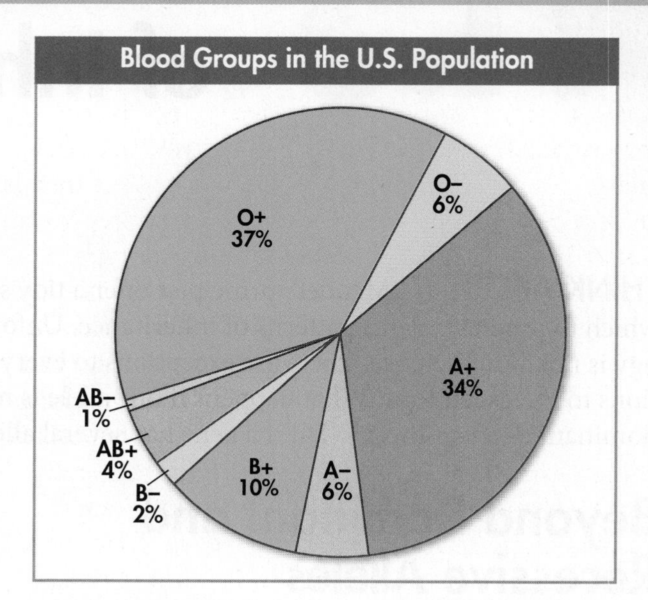

1. **Interpret Graphs** Which blood type makes up the greatest percentage of the U.S. population?
2. **Calculate** What percentage of the total U.S. population has a positive Rh factor? What percentage has a negative Rh factor?
3. **Infer** Which blood type can be used for transfusion into the largest percentage of individuals? Which type has the smallest percentage of possible donors available?
4. **Predict** Could a person with O^+ blood have two parents with O^- blood? Could that person have a daughter with AB^+ blood? Explain your answers.

MYSTERY CLUE

Green feathers don't actually contain green pigments. Rather, they contain a mixture of blue and yellow pigments. Could feather color be controlled by more than one gene?

Multiple Alleles So far, our examples have described genes for which there are only two alleles, such as *a* and *A*. In nature, such genes are the exception rather than the rule. **Many genes exist in several different forms and are therefore said to have multiple alleles.** A gene with more than two alleles is said to have **multiple alleles.** An individual, of course, usually has only two copies of each gene, but many different alleles are often found within a population. One of the best-known examples is coat color in rabbits. A rabbit's coat color is determined by a single gene that has at least four different alleles. The four known alleles display a pattern of simple dominance that can produce four coat colors. Many other genes have multiple alleles, including the human genes for blood type.

Polygenic Traits **Many traits are produced by the interaction of several genes.** Traits controlled by two or more genes are said to be **polygenic traits.** *Polygenic* means "many genes." For example, at least three genes are involved in making the reddish-brown pigment in the eyes of fruit flies. Polygenic traits often show a wide range of phenotypes. The variety of skin color in humans comes about partly because more than four different genes probably control this trait.

In Your Notebook *In your own words, describe multiple alleles and polygenic traits. How are they similar? How are they different?*

Genes and the Environment

Does the environment have a role in how genes determine traits?

The characteristics of any organism—whether plant, fruit fly, or human being—are not determined solely by the genes that organism inherits. Genes provide a plan for development, but how that plan unfolds also depends on the environment. In other words, the phenotype of an organism is only partly determined by its genotype.

Consider the buckeye butterfly, *Precis coenia.* It is found throughout North America. Butterfly enthusiasts had noted for years that buckeyes hatching in the summer had different color patterns on their wings than those hatching in the fall. Scientific studies suggested a reason—butterflies hatching in the shorter days of autumn had greater levels of pigment in their wings, making their markings appear darker than those hatching in the longer days of summer. In other words, the environment in which the butterflies develop influences the expression of their genes for wing coloration. **Environmental conditions can affect gene expression and influence genetically determined traits.** An individual's actual phenotype is determined by its environment as well as its genes.

Studies on another species, the western white butterfly, have shown the importance of changes in wing pigmentation. In order to fly effectively, the body temperature of the butterfly must be 28°C–40°C (about 84°F–104°F) as shown in **Figure 11–13.** Since the spring months are cooler in the west, greater pigmentation helps them reach the body temperature needed for flight. Similarly, in the hot summer months, less pigmentation enables the butterflies to avoid overheating.

Environmental Temperature and Butterfly Needs		
Temp. Needed for Flight	Average Spring Temp.	Average Summer Temp.
28–40°C	26.5°C	34.8°C

FIGURE 11–13 Temperature and Wing Color Western white butterflies that hatch in the spring have darker wing patterns than those that hatch in summer. The dark wing color helps increase their body heat. This trait is important because the butterflies need to reach a certain temperature in order to fly. The buckeye butterflies shown above are darker in the autumn than they are in the summer. **Calculate** ***What is the difference between the minimum temperature western white butterflies need to fly and the average spring temperature? Would the same calculation apply to butterflies developing in the summer? How about the buckeye butterflies developing in the autumn?*** MATH

11.3 Assessment

Review Key Concepts

1. **a. Review** What does *incomplete dominance* mean? Give an example.

   b. Design an Experiment Design an experiment to determine whether the pink flowers of petunia plants result from incomplete dominance.

2. **a. Review** What is the relationship between the environment and phenotype?

   b. Infer What might be the result of an exceptionally hot spring on wing pigmentation in the western white butterfly?

PRACTICE PROBLEM

3. Construct a genetics problem to be given as an assignment to a classmate. The problem must test incomplete dominance, codominance, multiple alleles, or polygenic traits. Your problem must have an answer key that includes all of your work.

Careers & BIOLOGY

If you enjoy learning about genetics, you may want to pursue one of the careers listed below.

FORENSIC SCIENTIST

Do you enjoy solving puzzles? That's what forensic scientists do when they solve crimes. Local, state, and federal agencies employ forensic scientists to use scientific approaches that support criminal investigations. Criminalists are forensic scientists who specialize in the analysis of physical evidence, such as hair, fiber, DNA, fingerprints, and weapons. They are often called to testify in trials as expert witnesses.

PLANT BREEDER

Did you ever wonder how seedless watermelons become seedless? They are the product of a plant breeder. Plant breeders use genetic techniques to manipulate crops. Often, the goal is to make a crop more useful by increasing yield or nutritional value. Some breeders introduce new traits, such as pesticide resistance, to the plant's genetic makeup.

POPULATION GENETICIST

Why are certain populations more susceptible to particular diseases? This is the kind of question that population geneticists answer. Their goal is to figure out why specific traits of distinct groups of organisms occur in varying frequencies. The patterns they uncover can lead to an understanding of how gene expression changes as a population evolves.

CAREER CLOSE-UP:

Sophia Cleland, Population Geneticist and Immunologist

Sophia Cleland, a Ph.D. student in immunology at George Washington University, studies the molecular, cellular, and genetic mechanisms that contribute to autoimmune diseases. One of only a few Native Americans with an advanced degree in genetics, Ms. Cleland became interested in autoimmune diseases when she noticed that the frequencies of these illnesses, such as rheumatoid arthritis and lupus, were several times higher among her tribal communities (Lakota-Sioux and California Mission Indian) than among Caucasians. Furthermore, she observed that such diseases progressed more rapidly among these communities than in any other human group in the world. Because of the frequency and severity of these diseases among indigenous tribal groups, Ms. Cleland is spreading the word about the need for focused research in this area.

"A compromise is needed between the world views of indigenous tribal groups and modern scientific approaches to gathering knowledge. We will encounter difficulties, but by working together with an open mind to learn, balanced and just results are possible."

WRITING How do you think a high frequency of genetic illness can affect a population? Explain.

11.4 Meiosis

THINK ABOUT IT As geneticists in the early 1900s applied Mendel's principles, they wondered where genes might be located. They expected genes to be carried on structures inside the cell, but *which* structures? What cellular processes could account for segregation and independent assortment, as Mendel had described?

Key Questions

How many sets of genes are found in most adult organisms?

What events occur during each phase of meiosis?

How is meiosis different from mitosis?

How can two alleles from different genes be inherited together?

Vocabulary

homologous • diploid • haploid • meiosis • tetrad • crossing-over • zygote

Taking Notes

Compare/Contrast Table Before you read, make a compare/contrast table to show the differences between mitosis and meiosis. As you read, complete the table.

Chromosome Number

How many sets of genes are found in most adult organisms?

To hold true, Mendel's principles require at least two events to occur. First, an organism with two parents must inherit a single copy of every gene from each parent. Second, when that organism produces gametes, those two sets of genes must be separated so that each gamete contains just one set of genes. As it turns out, chromosomes—those strands of DNA and protein inside the cell nucleus—are the carriers of genes. The genes are located in specific positions on chromosomes.

Diploid Cells Consider the fruit fly that Morgan used, *Drosophila.* A body cell in an adult fruit fly has eight chromosomes, as shown in **Figure 11–14.** Four of the chromosomes come from its male parent, and four come from its female parent. These two sets of chromosomes are **homologous** (hoh MAHL uh gus), meaning that each of the four chromosomes from the male parent has a corresponding chromosome from the female parent. A cell that contains both sets of homologous chromosomes is said to be **diploid,** meaning "two sets." **The diploid cells of most adult organisms contain two complete sets of inherited chromosomes and two complete sets of genes.** The diploid number of chromosomes is sometimes represented by the symbol 2N. Thus, for *Drosophila,* the diploid number is 8, which can be written as 2N = 8, where N represents the single set of chromosomes found in a sperm or egg cell.

Haploid Cells Some cells contain only a single set of chromosomes, and therefore a single set of genes. Such cells are **haploid,** meaning "one set." The gametes of sexually reproducing organisms, including fruit flies and peas, are haploid. For *Drosophila* gametes, the haploid number is 4, which can be written as N = 4.

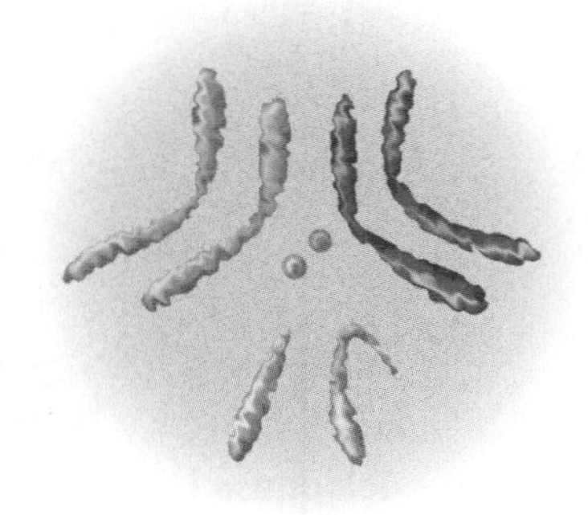

FIGURE 11–14 Fruit Fly Chromosomes These chromosomes are from a fruit fly. Each of the fruit fly's body cells is diploid, containing eight chromosomes.

FIGURE 11–15 Meiosis I During meiosis I, a diploid cell undergoes a series of events that results in the production of two daughter cells. Neither daughter cell has the same sets of chromosomes that the original diploid cell had. **Interpret Graphics** ***How does crossing-over affect the alleles on a chromosome?***

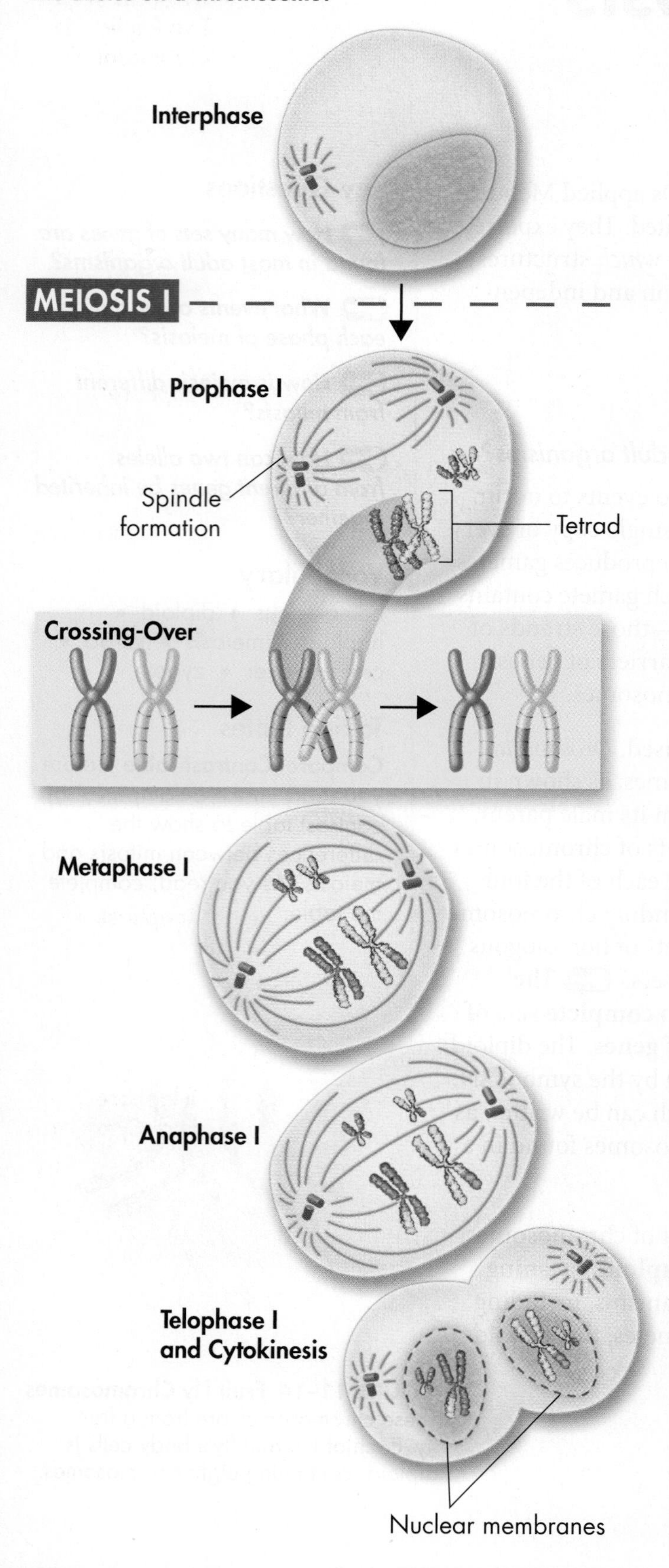

Phases of Meiosis

What events occur during each phases of meiosis?

How are haploid (N) gamete cells produced from diploid (2N) cells? That's where meiosis (my OH sis) comes in. **Meiosis** is a process in which the number of chromosomes per cell is cut in half through the separation of homologous chromosomes in a diploid cell. Meiosis usually involves two distinct divisions, called meiosis I and meiosis II. By the end of meiosis II, the diploid cell becomes four haploid cells. Let's see how meiosis takes place in a cell that has a diploid number of 4 (2N = 4).

Meiosis I Just prior to meiosis I, the cell undergoes a round of chromosome replication during interphase. As in mitosis, which was discussed in Chapter 10, each replicated chromosome consists of two identical chromatids joined at the center. Follow the sequence in **Figure 11–15** as you read about meiosis I.

▶ ***Prophase I*** After interphase I, the cell begins to divide, and the chromosomes pair up. **In prophase I of meiosis, each replicated chromosome pairs with its corresponding homologous chromosome.** This pairing forms a structure called a **tetrad,** which contains four chromatids. As the homologous chromosomes form tetrads, they undergo a process called **crossing-over.** First, the chromatids of the homologous chromosomes cross over one another. Then, the crossed sections of the chromatids—which contain alleles—are exchanged. Crossing-over therefore produces new combinations of alleles in the cell.

▶ ***Metaphase I and Anaphase I*** As prophase I ends, a spindle forms and attaches to each tetrad. **During metaphase I of meiosis, paired homologous chromosomes line up across the center of the cell.** As the cell moves into anaphase I, the homologous pairs of chromosomes separate. **During anaphase I, spindle fibers pull each homologous chromosome pair toward opposite ends of the cell.**

▶ ***Telophase I and Cytokinesis*** When anaphase I is complete, the separated chromosomes cluster at opposite ends of the cell. **The next phase is telophase I, in which a nuclear membrane forms around each cluster of chromosomes. Cytokinesis follows telophase I, forming two new cells.**

Meiosis I results in two cells, called daughter cells. However, because each pair of homologous chromosomes was separated, neither daughter cell has the two complete sets of chromosomes that it would have in a diploid cell. Those two sets have been shuffled and sorted almost like a deck of cards. The two cells produced by meiosis I have sets of chromosomes and alleles that are different from each other and from the diploid cell that entered meiosis I.

Meiosis II The two cells now enter a second meiotic division. Unlike the first division, neither cell goes through a round of chromosome replication before entering meiosis II.

▶ ***Prophase II*** **As the cells enter prophase II, their chromosomes—each consisting of two chromatids—become visible.** The chromosomes do not pair to form tetrads, because the homologous pairs were already separated during meiosis I.

▶ ***Metaphase II, Anaphase II, Telophase II, and Cytokinesis*** During metaphase of meiosis II, chromosomes line up in the center of each cell. As the cell enters anaphase, the paired chromatids separate. **The final four phases of meiosis II are similar to those in meiosis I. However, the result is four haploid daughter cells.** In the example shown here, each of the four daughter cells produced in meiosis II receive two chromosomes. These four daughter cells now contain the haploid number (N)—just two chromosomes each.

Gametes to Zygotes The haploid cells produced by meiosis II are the gametes that are so important to heredity. In male animals, these gametes are called sperm. In some plants, pollen grains contain haploid sperm cells. In female animals, generally only one of the cells produced by meiosis is involved in reproduction. The female gamete is called an egg in animals and an egg cell in some plants. After it is fertilized, the egg is called a **zygote** (ZY goht). The zygote undergoes cell division by mitosis and eventually forms a new organism.

In Your Notebook *Describe the difference between meiosis I and meiosis II. How are the end results different?*

FIGURE 11–16 Meiosis II The second meiotic division, called meiosis II, produces four haploid daughter cells.

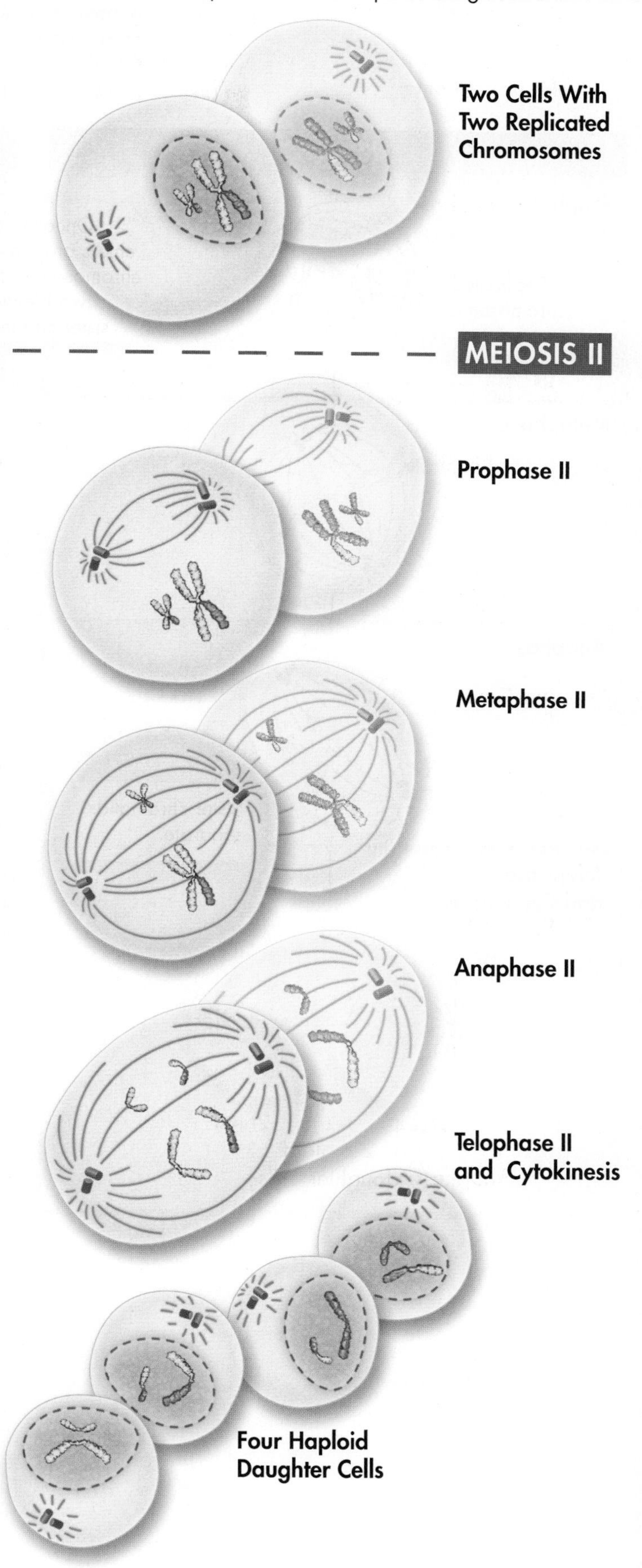

VISUAL SUMMARY

COMPARING MITOSIS AND MEIOSIS

FIGURE 11–17 Mitosis and meiosis both ensure that cells inherit genetic information. Both processes begin after interphase, when chromosome replication occurs. However, the two processes differ in the separation of chromosomes, the number of cells produced, and the number of chromosomes each cell contains.

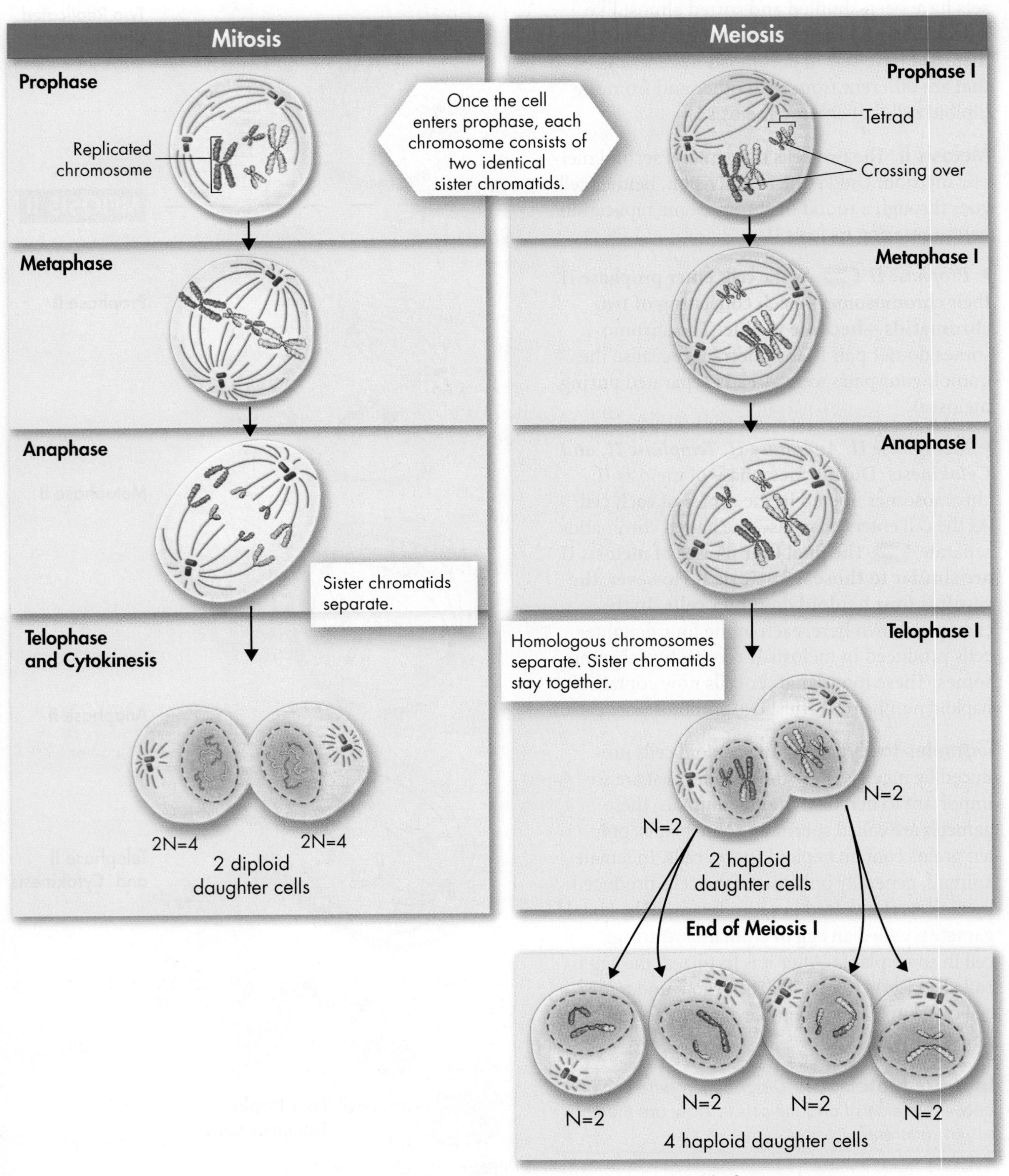

Comparing Meiosis and Mitosis

How is meiosis different from mitosis?

The words *mitosis* and *meiosis* may sound similar, but the two processes are very different, as you can see in **Figure 11–17.** Mitosis can be a form of asexual reproduction, whereas meiosis is an early step in sexual reproduction. There are three other ways in which these two processes differ.

Replication and Separation of Genetic Material Mitosis and meiosis are both preceded by a complete copying, or replication, of the genetic material of chromosomes. However, the next steps differ dramatically. **In mitosis, when the two sets of genetic material separate, each daughter cell receives one complete set of chromosomes. In meiosis, homologous chromosomes line up and then move to separate daughter cells.** As a result, the two alleles for each gene are segregated, and end up in different cells. The sorting and recombination of genes in meiosis result in a greater variety of possible gene combinations than could result from mitosis.

Changes in Chromosome Number **Mitosis does not normally change the chromosome number of the original cell. This is not the case for meiosis, which reduces the chromosome number by half.** A diploid cell that enters mitosis with eight chromosomes will divide to produce two diploid daughter cells, each of which also has eight chromosomes. On the other hand, a diploid cell that enters meiosis with eight chromosomes will pass through two meiotic divisions to produce four haploid gamete cells, each with only four chromosomes.

Calculating Haploid and Diploid Numbers

Haploid and diploid numbers are designated by the algebraic notations N and 2N, respectively. Either number can be calculated when the other is known. For example, if the haploid number (N) is 3, the diploid number (2N) is 2×3, or 6. If the diploid number (2N) is 12, the haploid number (N) is 12/2, or 6.

The table shows haploid or diploid numbers of a variety of organisms. Copy the table into your notebook and complete it. Then, use the table to answer the questions that follow.

Trait Survey

Organism	Haploid Number	Diploid Number
Amoeba	N=25	
Chimpanzee	N=24	
Earthworm	N=18	
Fern		2N=1010
Hamster	N=22	
Human		2N=46
Onion		2N=16

1. **Calculate** What are the haploid numbers for the fern and onion plants? MATH
2. **Interpret Data** In the table, which organisms' diploid numbers are closest to that of a human?
3. **Apply Concepts** Why is a diploid number always even?
4. **Evaluate** Which organism's haploid and diploid numbers do you find the most surprising? Why?

Number of Cell Divisions Mitosis is a single cell division, resulting in the production of two identical daughter cells. On the other hand, meiosis requires two rounds of cell division, and, in most organisms, produces a total of four daughter cells. **Mitosis results in the production of two genetically identical diploid cells, whereas meiosis produces four genetically different haploid cells.**

Gene Linkage and Gene Maps

How can two alleles from different genes be inherited together?

If you think carefully about Mendel's principle of independent assortment in relation to meiosis, one question might bother you. Genes that are located on different chromosomes assort independently, but what about genes that are located on the same chromosome? Wouldn't they generally be inherited together?

Gene Linkage The answer to this question, as Thomas Hunt Morgan first realized in 1910, is yes. Morgan's research on fruit flies led him to the principle of gene linkage. After identifying more than 50 *Drosophila* genes, Morgan discovered that many of them appeared to be "linked" together in ways that, at first glance, seemed to violate the principle of independent assortment. For example, Morgan used a fly with reddish-orange eyes and miniature wings in a series of test crosses. His results showed that the genes for those two traits were almost always inherited together. Only rarely did the genes separate from each other. Morgan and his associates observed so many genes that were inherited together that, before long, they could group all of the fly's genes into four linkage groups. The linkage groups assorted independently, but all of the genes in one group were inherited together. As it turns out, *Drosophila* has four linkage groups and four pairs of chromosomes.

FIGURE 11–18 Gene Map This gene map shows the location of a variety of genes on chromosome 2 of the fruit fly. The genes are named after the problems that abnormal alleles cause, *not* after the normal structures. **Interpret Graphics** ***Where on the chromosome is the "purple eye" gene located?***

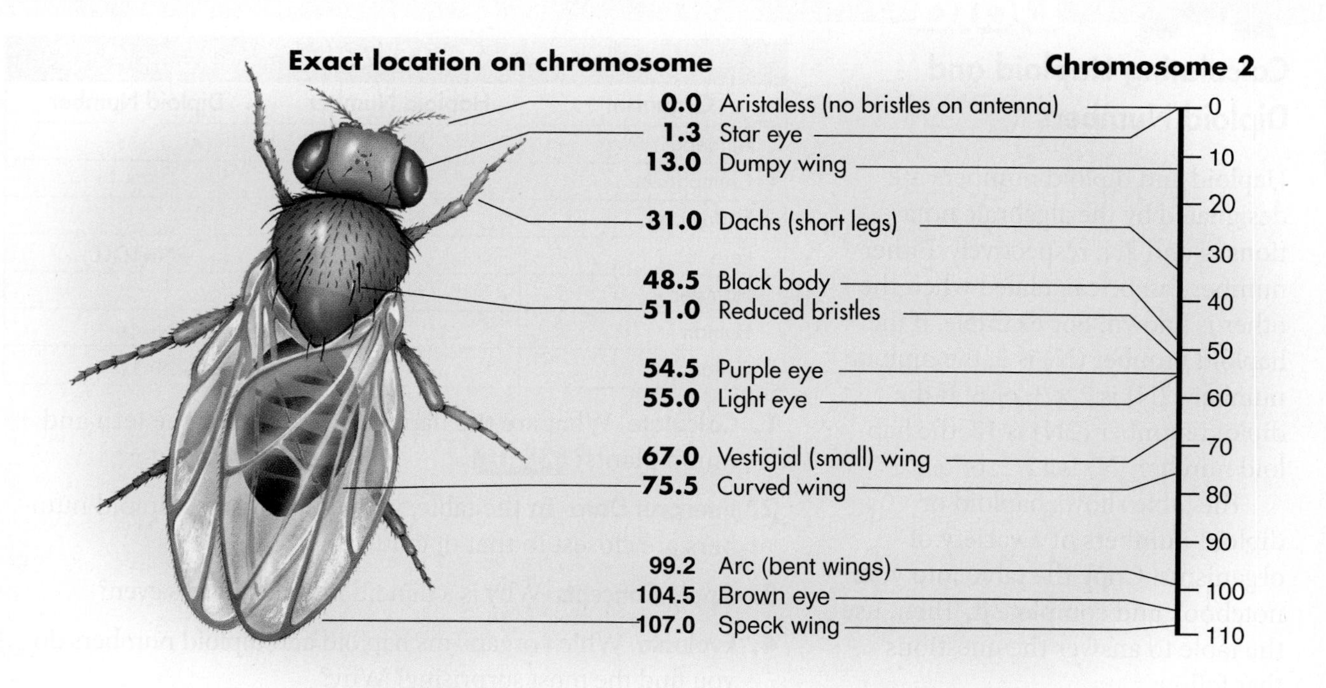

Morgan's findings led to two remarkable conclusions. First, each chromosome is actually a group of linked genes. Second, Mendel's principle of independent assortment still holds true. It is the chromosomes, however, that assort independently, not individual genes. **Alleles of different genes tend to be inherited together from one generation to the next when those genes are located on the same chromosome.**

How did Mendel manage to miss gene linkage? By luck, or design, several of the genes he studied are on different chromosomes. Others are so far apart that they also assort independently.

Gene Mapping In 1911, a Columbia University student was working part time in Morgan's lab. This student, Alfred Sturtevant, wondered if the frequency of crossing-over between genes during meiosis might be a clue to the genes' locations. Sturtevant reasoned that the farther apart two genes were on a chromosome, the more likely it would be that crossing-over would occur between them. If two genes are close together, then crossovers between them should be rare. If two genes are far apart, then crossovers between them should be more common. By this reasoning, he could use the frequency of crossing-over between genes to determine their distances from each other.

Sturtevant gathered up several notebooks of lab data and took them back to his room. The next morning, he presented Morgan with a gene map showing the relative locations of each known gene on one of the *Drosophila* chromosomes. Sturtevant's method has been used to construct gene maps, like the one in **Figure 11–18,** ever since this discovery.

MYSTERY CLUE

White is the least common color found in parakeets. What does this fact suggest about the genotypes of both green parents?

11.4 Assessment

Review Key Concepts

1. a. **Review** Describe the main results of meiosis.
   b. **Calculate** In human cells, 2N = 46. How many chromosomes would you expect to find in a sperm cell? How many would you expect to find in an egg cell? MATH
2. a. **Review** Write a summary of each phase of meiosis.
   b. **Use Analogies** Compare the chromosomes of a diploid cell to a collection of shoes in a closet. How are they similar? What would make the shoe collection comparable to the chromosomes of a haploid cell?
3. a. **Review** What are the principal differences between mitosis and meiosis?
   b. **Apply Concepts** Is there any difference between sister chromatids and homologous pairs of chromosomes? Explain.
4. a. **Review** How does the principle of independent assortment apply to chromosomes?
   b. **Infer** If two genes are on the same chromosome but usually assort independently, what does that tell you about how close together they are?

Apply the Big idea

Information and Heredity

5. In asexual reproduction, mitosis occurs but meiosis does not occur. Which type of reproduction—sexual or asexual—results in offspring with greater genetic variation? Explain your answer.

NGSS Smart Guide Introduction to Genetics

Disciplinary Core Idea LS3.B Variation of Traits: Why do individuals of the same species vary in how they look, function, and behave? Genetic information passes from parent to offspring during meiosis when gametes, each containing one representative from each chromosome pair, unite.

11.1 The Work of Gregor Mendel

- An individual's characteristics are determined by factors that are passed from one parental generation to the next.
- During gamete formation, the alleles for each gene segregate from each other so that each gamete carries only one allele for each gene.

genetics 308 • fertilization 309 • trait 309 • hybrid 309 • gene 310 • allele 310 • principle of dominance 310 • segregation 312 • gamete 312

Biology.com

Untamed Science Video Travel back in time with the Untamed Science explorers as they prove Mendel was no pea brain!

11.2 Applying Mendel's Principles

- Punnett squares use mathematical probability to help predict the genotype and phenotype combinations in genetic crosses.
- The principle of independent assortment states that genes for different traits can segregate independently during the formation of gametes.
- Mendel's principles of heredity, observed through patterns of inheritance, form the basis of modern genetics.

probability 313 • homozygous 314 • heterozygous 314 • phenotype 315 • genotype 315 • Punnett square 315 • independent assortment 317

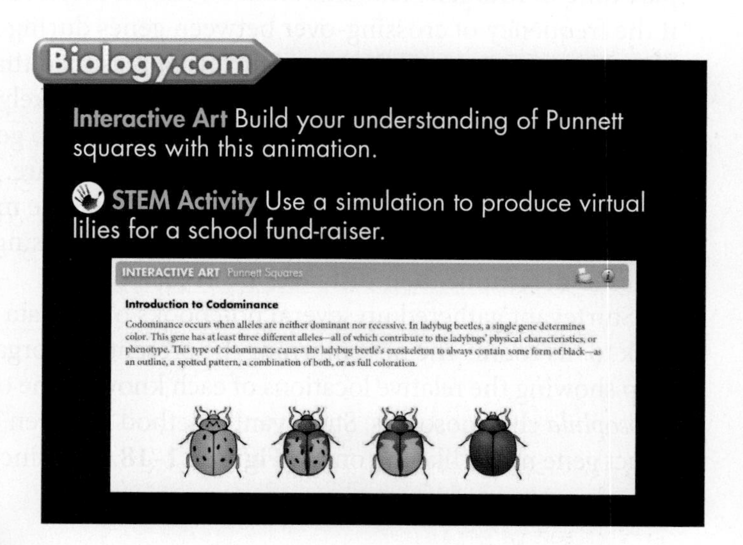

Biology.com

Interactive Art Build your understanding of Punnett squares with this animation.

STEM Activity Use a simulation to produce virtual lilies for a school fund-raiser.

11.3 Other Patterns of Inheritance

- Many genes exist in several different forms and are therefore said to have multiple alleles. Many traits are produced by the interaction of several genes.
- Environmental conditions can affect gene expression and influence genetically determined traits.

incomplete dominance 319 • codominance 319 • multiple alleles 320 • polygenic trait 320

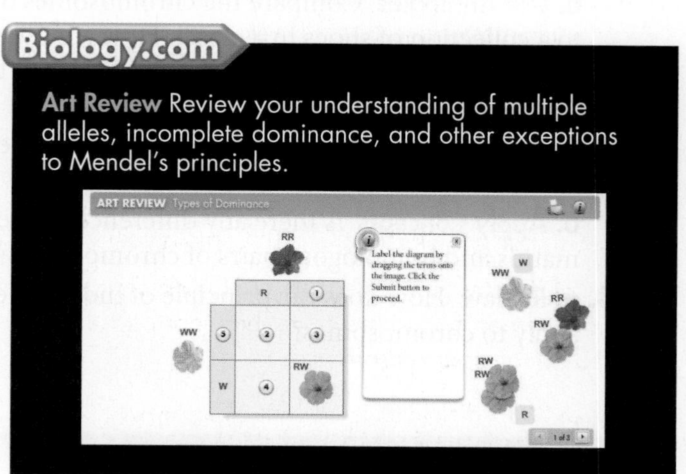

Biology.com

Art Review Review your understanding of multiple alleles, incomplete dominance, and other exceptions to Mendel's principles.

CHECKING YOUR *Scientific Literacy*

Refer to the lesson content and digital assets below as you prepare for your chapter assessment. Then, evaluate your understanding of the principles of genetics by answering these questions.

1. **Scientific and Engineering Practice Using Mathematics and Computational Thinking** In sweet pea plants, purple flowers (P) are dominant over red flowers (p). Long grains of pollen (L) are dominant over round grains (l). Use a Punnett square to show the expected results from a two-factor cross between plants that are heterozygous for both traits (PpLl).

2. **Crosscutting Concept Cause and Effect** In an actual cross between the pea plants described in Question 1, the number of plants with both purple flowers and long grains was much higher than expected. The same was true for red flowers and round grains. Use a concept you studied in Chapter 11 to explain this result.

3. **Craft and Structure** Identify the steps in meiosis that support the independent assortment of inherited genes. Write a brief paragraph summarizing your results.

4. **STEM** Research the technology of blood typing. What problem was this technology designed to solve? Look for any recent improvements in the design of testing kits.

11.4 Meiosis

- The diploid cells of most adult organisms contain two complete sets of inherited chromosomes and two complete sets of genes.
- In prophase I, replicated chromosomes pair with corresponding homologous chromosomes. At metaphase I, paired chromosomes line up across the center of the cell. In anaphase I, chromosome pairs move toward opposite ends of the cell. In telophase I, a nuclear membrane forms around each cluster of chromosomes. Cytokinesis then forms two new cells. As the cells enter prophase II, their chromosomes become visible. The final four phases of meiosis II result in four haploid daughter cells.
- In mitosis, each daughter cell receives one complete set of chromosomes. In meiosis, homologous chromosomes line up and move to separate daughter cells. Mitosis does not normally change the chromosome number of the original cell. Meiosis reduces the chromosome number by half. Mitosis results in the production of two genetically identical diploid cells, whereas meiosis produces four genetically different haploid cells.
- Alleles of different genes tend to be inherited together from one generation to the next when those genes are located on the same chromosome.

homologous 323 • diploid 323 • haploid 323 • meiosis 324 • tetrad 324 • crossing-over 324 • zygote 325

Biology.com

Art in Motion View a short animation that brings the process of meiosis to life.

Tutor Tube Tune into the tutor to review the connection between setting up Punnett squares and meiosis.

Data Analysis Determine gene linkage and construct a gene map by examining the phenotypic frequencies of offspring.

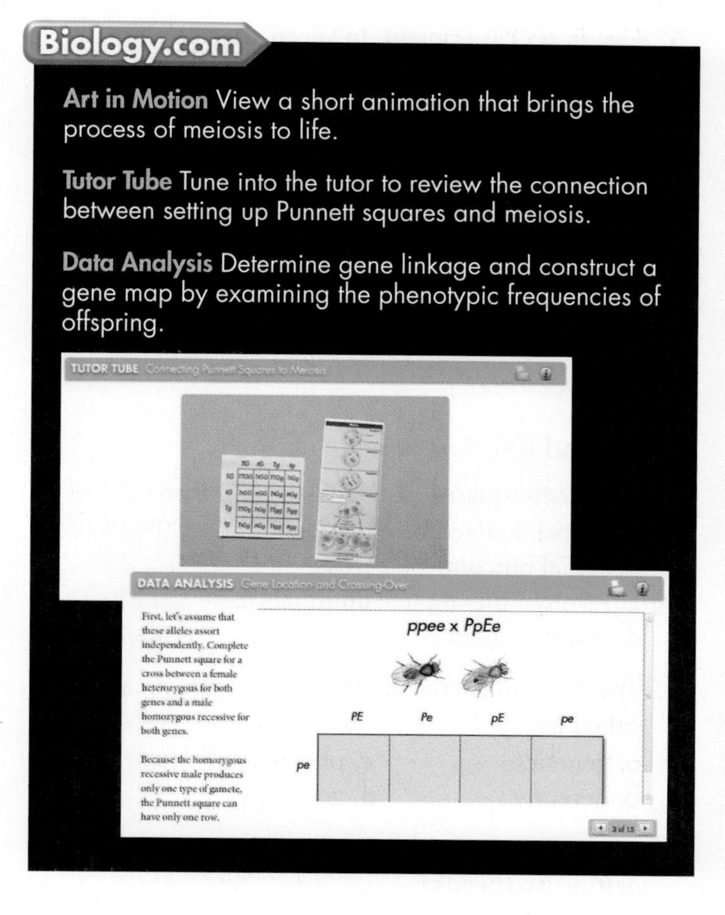

Skills Lab

Modeling Meiosis This chapter lab is available in your lab manual and at Biology.com

11 Assessment

11.1 The Work of Gregor Mendel

Understand Key Concepts

1. Different forms of a gene are called
a. hybrids.
b. dominant factors.
c. alleles.
d. recessive factors.

2. Organisms that have two identical alleles for a particular trait are said to be
a. hybrid.
b. heterozygous.
c. homozygous.
d. dominant.

3. **Craft and Structure** Mendel had many stocks of pea plants that were true-breeding. What is meant by this term?

4. Explain how Mendel kept his pea plants from self-pollinating.

Think Critically

5. **Design an Experiment** In sheep, the allele for white wool (A) is dominant over the allele for black wool (a). A ram is a male sheep. How would you determine the genotype of a white ram?

6. **Infer** Suppose Mendel crossed two pea plants and got both tall and short offspring. What could have been the genotypes of the two original plants? What genotype could *not* have been present?

11.2 Applying Mendel's Principles

Understand Key Concepts

7. A Punnett square is used to determine the
a. probable outcome of a cross.
b. actual outcome of a cross.
c. result of incomplete dominance.
d. result of meiosis.

8. The physical characteristics of an organism are called its
a. genetics.
b. heredity.
c. phenotype.
d. genotype.

9. The probability of flipping a coin twice and getting two heads is
a. 1.
b. 1/2.
c. 1/4.
d. 3/4.

10. **Key Ideas and Details** Summarize the four basic principles of genetics that Mendel discovered in his experiments.

11. In pea plants, the allele for yellow seeds is dominant over the allele for green seeds. Predict the genotypic ratio of offspring produced by crossing two parents that are heterozygous for this trait. Draw a Punnett square to illustrate your prediction.

Think Critically

12. **Apply Concepts** In guinea pigs, the allele for a rough coat (R) is dominant over the allele for a smooth coat (r). A heterozygous guinea pig (Rr) and a homozygous recessive guinea pig (rr) have a total of nine offspring. The Punnett square for this cross shows a 50 percent chance that any particular offspring will have a smooth coat. Explain how all nine offspring can have smooth coats.

	R	*r*
r	*Rr*	*rr*
r	*Rr*	*rr*

11.3 Other Patterns of Inheritance

Understand Key Concepts

13. A situation in which a gene has more than two alleles is known as
a. complete dominance.
b. codominance.
c. polygenic dominance.
d. multiple alleles.

14. A pink-flowered *Mirabilis* plant (RW) is crossed with a white-flowered *Mirabilis* (WW). What is the chance that a seed from this cross will produce a red-flowered plant?
a. 0
b. 1/4
c. 1/2
d. 1

15. What is the difference between multiple alleles and polygenic traits?

16. Why can multiple alleles result in many different phenotypes for a trait?

17. Are an organism's characteristics determined only by its genes? Explain.

Think Critically

18. Interpret Visuals Genes that control hair or feather color in some animals are expressed differently in the winter than in the summer. How might such a difference be beneficial to the ptarmigan shown here?

11.4 Meiosis

Understand Key Concepts

19. The illustration below represents what stage of meiosis?

a. prophase I
b. anaphase II
c. telophase I
d. metaphase I

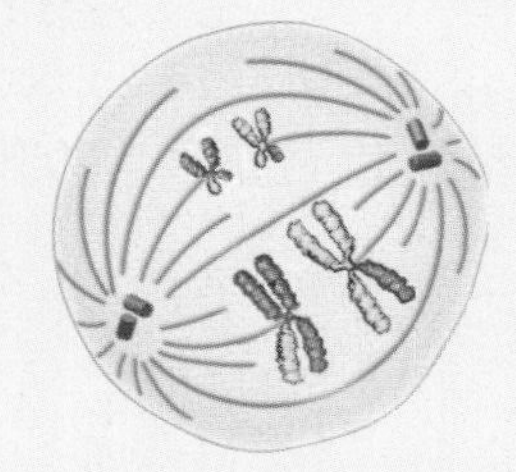

20. Unlike mitosis, meiosis in male mammals results in the formation of

a. one haploid gamete.
b. three diploid gametes.
c. four diploid gametes.
d. four haploid gametes.

21. A gene map shows

a. the number of possible alleles for a gene.
b. the relative locations of genes on a chromosome.
c. where chromosomes are in a cell.
d. how crossing-over occurs.

22. Suppose that an organism has the diploid number 2N = 8. How many chromosomes do this organism's gametes contain?

23. Describe the process of meiosis.

24. Explain why chromosomes, not individual genes, assort independently.

Think Critically

25. Craft and Structure Compare the phases of meiosis I and meiosis II in terms of number and arrangement of the chromosomes.

solve the CHAPTER MYSTERY

GREEN PARAKEETS

After consulting with the owner of the pet store, Susan realized she had a rare gift. White parakeets are very uncommon. The pet shop owner told Susan that two genes control feather color. A dominant Y allele results in the production of a yellow pigment. The dominant B allele controls melanin production. If the genotype contains a capital Y (either YY or Yy) and a capital B, the offspring will be green. If the genotype contains two lowercase y alleles, and a capital B, the offspring will be blue. If the genotype contains two lowercase y's and two lowercase b's, the offspring will be white.

1. Use Models Draw a Punnett square that accounts for the inheritance of blue pigment.

2. Use Models Construct a Punnett square that explains the inheritance of a white pigment.

3. Apply Concepts Solve the mystery by determining the genotypes and phenotypes of the parents and offspring.

4. Connect to the Big idea What ratio of colored offspring would you expect if Susan breeds her original pair of parakeets in the years ahead? Would any offspring be green?

Connecting Concepts

Use Science Graphics

Seed coat was one trait that Mendel studied in pea plants. The coat, or covering, of the seed is either smooth or wrinkled. Suppose a researcher has two plants—one that makes smooth seeds and another that makes wrinkled seeds. The researcher crosses the wrinkled-seed plants and the smooth-seed plants, obtaining the following data. Use the data to answer questions 26–28.

Results of Seed Experiment

Phenotype	Number of Plants in the F_1 Generation	
	Expected	Observed
Smooth seeds		60
Wrinkled seeds		72

26. Predict Mendel knew that the allele for smooth (*R*) seeds was dominant over the allele for wrinkled (*r*) seeds. If this cross was $Rr \times rr$, what numbers would fill the middle column?

27. Analyze Data Are the observed numbers consistent with the hypothesis that the cross is $Rr \times rr$? Explain your answer.

28. Integration of Knowledge and Ideas Are the data from this experiment alone sufficient to make a claim that the allele for smooth seeds is dominant over the allele for wrinkled seeds?

Write About Science

29. Production and Distribution of Writing Write an explanation of dominant and recessive alleles that would be appropriate for an eighth-grade science class. Assume that the students know the meanings of *gene* and *allele*. (*Hint*: Use examples to make your explanation clear.)

30. Text Types and Purposes Explain why the alleles for reddish-orange eyes and miniature wings in *Drosophila* are usually inherited together. Describe the pattern of inheritance these alleles follow, and include the idea of gene linkage. (*Hint*: To organize your ideas, draw a cause-effect diagram that shows what happens to the two alleles during meiosis.)

31. Assess the Big idea Explain why the gene pairs described by Mendel behave in a way that is consistent with the behavior of chromosomes during gamete formation, fertilization, and reproduction.

Integration of Knowledge and Ideas

A researcher studying fruit flies finds a mutant fly with brown-colored eyes. Almost all fruit flies in nature have bright red eyes. When the researcher crosses the mutant fly with a normal red-eyed fly, all of the F_1 offspring have red eyes. The researcher then crosses two of the F_1 red-eyed flies and obtains the following results in the F_2 generation.

Eye Color in the F_2 Generation	
Red eyes	37
Brown eyes	14

32. Calculate What is the ratio of red-eyed flies to brown-eyed flies? MATH

a. 1 : 1
b. 1 : 3
c. 3 : 1
d. 4 : 1

33. Draw Conclusions The allele for red eyes in fruit flies is

a. dominant over brown eyes.
b. recessive to brown eyes.
c. codominant with the brown-eyed gene.
d. a multiple allele with the brown-eyed gene and others.

Standardized Test Prep

Multiple Choice

1. What happens to the chromosome number during meiosis?
   A It doubles.
   B It stays the same.
   C It halves.
   D It becomes diploid.

2. Which ratio did Mendel find in his F_2 generation?
   A 3 : 1
   B 1 : 3 : 1
   C 1 : 2
   D 3 : 4

3. During which phase of meiosis is the chromosome number reduced?
   A anaphase I
   B metaphase I
   C telophase I
   D telophase II

4. Two pink-flowering plants are crossed. The offspring flower as follows: 25% red, 25% white, and 50% pink. What pattern of inheritance does flower color in these flowers follow?
   A dominance
   B multiple alleles
   C incomplete dominance
   D polygenic traits

5. Which of the following is used to construct a gene map?
   A chromosome number
   B mutation rate
   C rate of meiosis
   D recombination rate

6. Alleles for the same trait are separated from each other during the process of
   A cytokinesis.
   B meiosis I.
   C meiosis II.
   D metaphase II.

7. Which of the following is NOT one of Gregor Mendel's principles?
   A The alleles for different genes usually segregate independently.
   B Some forms of a gene may be dominant.
   C The inheritance of characteristics is determined by factors (genes).
   D Crossing-over occurs during meiosis.

Questions 8–9

Genes A, B, C, and D are located on the same chromosome. After calculating recombination frequencies, a student determines that these genes are separated by the following map units: C–D, 25 map units; A–B, 12 map units; B–D, 20 map units; A–C, 17 map units.

8. How many map units apart are genes A and D?
   A 5
   B 8
   C 10
   D 12.5

9. Which gene map best reflects the student's data?
   A A 5 B 12 C 8 D
   B A 8 B 20 C 5 D
   C A 8 B 17 C 12 D
   D C 5 B 12 A 8 D

Open-Ended Response

10. Explain why meiosis allows organisms to maintain their chromosome numbers from one generation to the next.

If You Have Trouble With . . .

Question	1	2	3	4	5	6	7	8	9	10
See Lesson	11.4	11.1	11.4	11.3	11.4	11.4	11.2	11.4	11.4	11.4

12 DNA

Big idea: Information and Heredity, Cellular Basis of Life

Q: What is the structure of DNA, and how does it function in genetic inheritance?

This sculpture, outside the Lawrence Hall of Science at the University of California at Berkeley, models the structure of DNA—the substance that genes are made of.

INSIDE:

- 12.1 Identifying the Substance of Genes
- 12.2 The Structure of DNA
- 12.3 DNA Replication

BUILDING Scientific Literacy

Deepen your understanding of DNA by engaging in key practices that allow you to make connections across concepts.

NGSS You will explain how the functions of the DNA molecule are made possible by its unique structure.

STEM What feature of DNA replication makes it possible for cells to make near-perfect copies of DNA molecules? Do a hands-on activity to find out.

Common Core You will develop and strengthen your writing as you relate DNA's structure to its functions.

CHAPTER MYSTERY

UV LIGHT

"Put on your sunscreen!" This familiar phrase can be heard at most beaches on a sunny day. It's an important directive because sunlight—for all its beneficial effects—can readily damage the skin. The most dangerous wavelengths of sunlight are the ones we can't see: the ultraviolet (UV) region of the electromagnetic spectrum. Not only can excess exposure to UV light damage skin cells, it can cause a deadly form of skin cancer that kills nearly 10,000 Americans each year.

Why is UV light so dangerous? How can these particular wavelengths of light damage our cells to the point of causing cell death and cancer? As you read this chapter, look for clues to help you solve the question of why UV light is so damaging to skin cells. Then, solve the mystery.

Biology.com

Finding the connection between UV light and DNA is only the beginning. Take a video field trip with the ecogeeks of Untamed Science to continue exploring your world.

Go online to access additional resources including:
• eText • Flash Cards • Lesson Overviews • Chapter Mystery

Identifying the Substance of Genes

Key Questions

What clues did bacterial transformation yield about the gene?

What role did bacterial viruses play in identifying genetic material?

What is the role of DNA in heredity?

Vocabulary

transformation
bacteriophage

Taking Notes

Flowchart As you read this section, make a flowchart that shows how scientists came to understand the molecule known as DNA.

THINK ABOUT IT How do genes work? To answer that question, the first thing you need to know is what genes are made of. After all, you couldn't understand how an automobile engine works without understanding what the engine is made of and how it's put together. So, how would you go about figuring out what molecule or molecules go into making a gene?

Bacterial Transformation

What clues did bacterial transformation yield about the gene?

In the first half of the twentieth century, biologists developed the field of genetics to the point where they began to wonder about the nature of the gene itself. To truly understand genetics, scientists realized they first had to discover the chemical nature of the gene. If the molecule that carries genetic information could be identified, it might be possible to understand how genes actually control the inherited characteristics of living things.

Like many stories in science, the discovery of the chemical nature of the gene began with an investigator who was actually looking for something else. In 1928, the British scientist Frederick Griffith was trying to figure out how bacteria make people sick. More specifically, Griffith wanted to learn how certain types of bacteria produce the serious lung disease known as pneumonia.

Griffith had isolated two very similar types of bacteria from mice. These were actually two different varieties, or strains, of the same bacterial species. Both strains grew very well in culture plates in Griffith's lab, but only one of them caused pneumonia. The disease-causing bacteria (S strain) grew into smooth colonies on culture plates, whereas the harmless bacteria (R strain) produced colonies with rough edges. The difference in appearance made the two strains easy to tell apart.

Griffith's Experiments When Griffith injected mice with disease-causing bacteria, the mice developed pneumonia and died. When he injected mice with harmless bacteria, the mice stayed healthy. Griffith wondered what made the first group of mice get pneumonia. Perhaps the S-strain bacteria produced a toxin that made the mice sick? To find out, he ran the series of experiments shown in **Figure 12–1.** First, Griffith took a culture of the S strain, heated the cells to kill them, then injected the heat-killed bacteria into laboratory mice. The mice survived, suggesting that the cause of pneumonia was not a toxin from these disease-causing bacteria.

In Griffith's next experiment, he mixed the heat-killed, S-strain bacteria with live, harmless bacteria from the R strain. This mixture he injected into laboratory mice. By themselves, neither type of bacteria should have made the mice sick. To Griffith's surprise, however, the injected mice developed pneumonia, and many died. When he examined the lungs of these mice, he found them to be filled not with the harmless bacteria, but with the disease-causing bacteria. How could that happen if the S-strain cells were dead?

Transformation Somehow, the heat-killed bacteria passed their disease-causing ability to the harmless bacteria. Griffith reasoned that, when he mixed the two types of bacteria together, some chemical factor transferred from the heat-killed cells of the S strain into the live cells of the R strain. This chemical compound, he hypothesized, must contain information that could change harmless bacteria into disease-causing ones. He called this process **transformation,** because one type of bacteria (the harmless form) had been changed permanently into another (the disease-causing form). Because the ability to cause disease was inherited by the offspring of the transformed bacteria, Griffith concluded that the transforming factor had to be a gene.

In Your Notebook *Write a summary of Griffith's experiments.*

FIGURE 12–1 Griffith's Experiments Griffith injected mice with four different samples of bacteria. When injected separately, neither heat-killed, disease-causing bacteria nor live, harmless bacteria killed the mice. The two strains injected together, however, caused fatal pneumonia. From this experiment, Griffith inferred that genetic information could be transferred from one bacterial strain to another. **Infer** ***Why did Griffith test to see whether the bacteria recovered from the sick mice in his last experiment would produce smooth or rough colonies in a petri dish?***

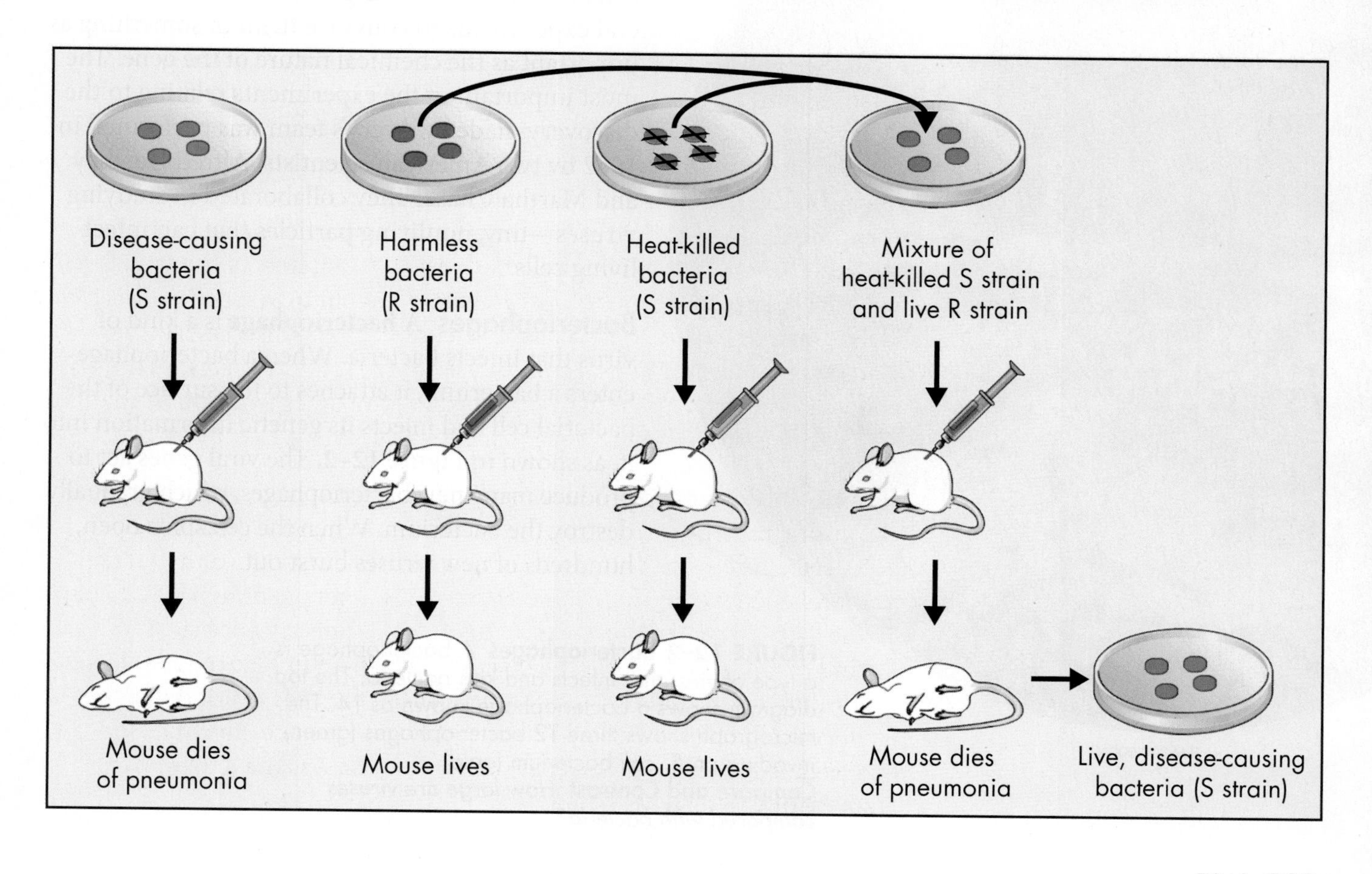

The Molecular Cause of Transformation In 1944, a group of scientists at the Rockefeller Institute in New York decided to repeat Griffith's work. Led by the Canadian biologist Oswald Avery, the scientists wanted to determine which molecule in the heat-killed bacteria was most important for transformation. They reasoned that if they could find this particular molecule, it might reveal the chemical nature of the gene.

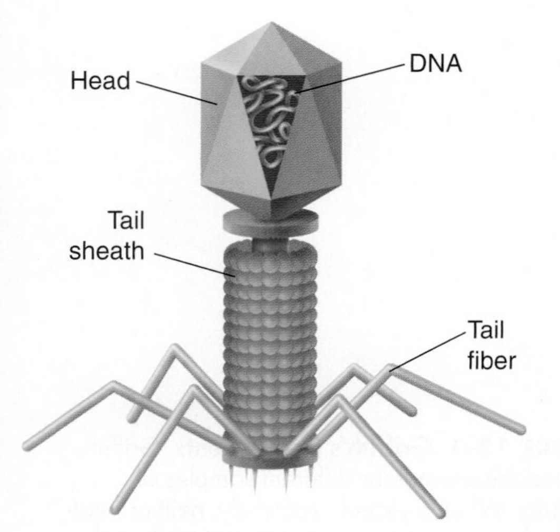

Avery and his team extracted a mixture of various molecules from the heat-killed bacteria. They carefully treated this mixture with enzymes that destroyed proteins, lipids, carbohydrates, and some other molecules, including the nucleic acid RNA. Transformation still occurred. Clearly, since those molecules had been destroyed, none of them could have been responsible for transformation.

Avery's team repeated the experiment one more time. This time, they used enzymes that would break down a different nucleic acid—DNA. When they destroyed the DNA in the mixture, transformation did not occur. There was just one possible explanation for these results: *DNA was the transforming factor.* **By observing bacterial transformation, Avery and other scientists discovered that the nucleic acid DNA stores and transmits genetic information from one generation of bacteria to the next.**

Bacterial Viruses

What role did bacterial viruses play in identifying genetic material?

Scientists are a skeptical group. It usually takes several experiments to convince them of something as important as the chemical nature of the gene. The most important of the experiments relating to the discovery made by Avery's team was performed in 1952 by two American scientists, Alfred Hershey and Martha Chase. They collaborated in studying viruses—tiny, nonliving particles that can infect living cells.

Bacteriophages A **bacteriophage** is a kind of virus that infects bacteria. When a bacteriophage enters a bacterium, it attaches to the surface of the bacterial cell and injects its genetic information into it, as shown in **Figure 12–2.** The viral genes act to produce many new bacteriophages, which gradually destroy the bacterium. When the cell splits open, hundreds of new viruses burst out.

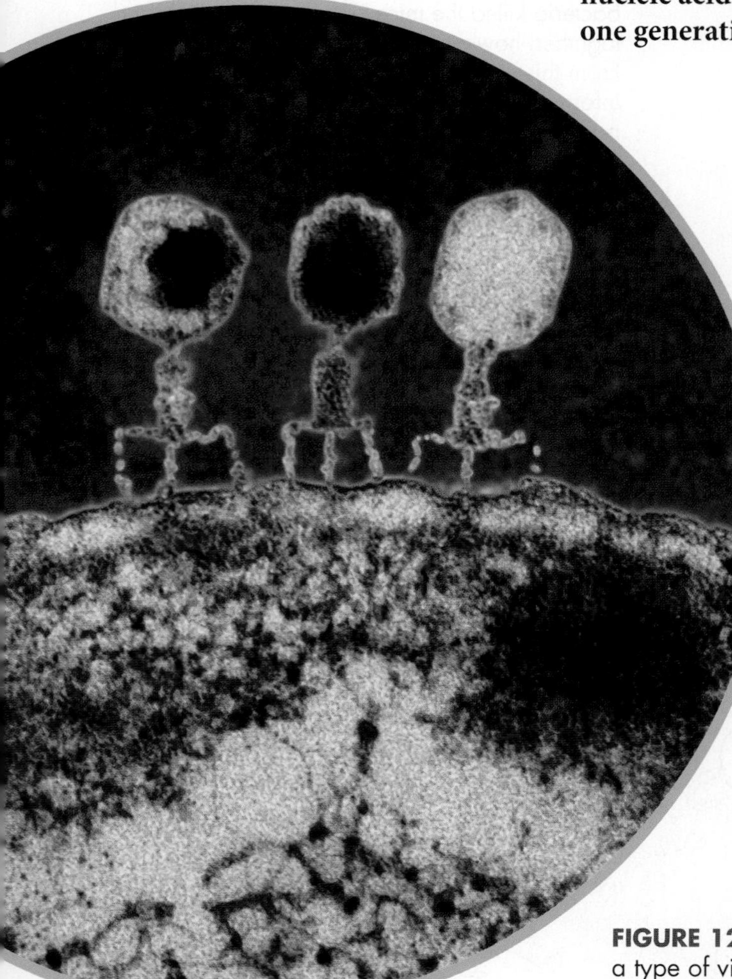

FIGURE 12–2 Bacteriophages A bacteriophage is a type of virus that infects and kills bacteria. The top diagram shows a bacteriophage known as T4. The micrograph shows three T2 bacteriophages (green) invading an *E. coli* bacterium (gold).
Compare and Contrast ***How large are viruses compared with bacteria?***

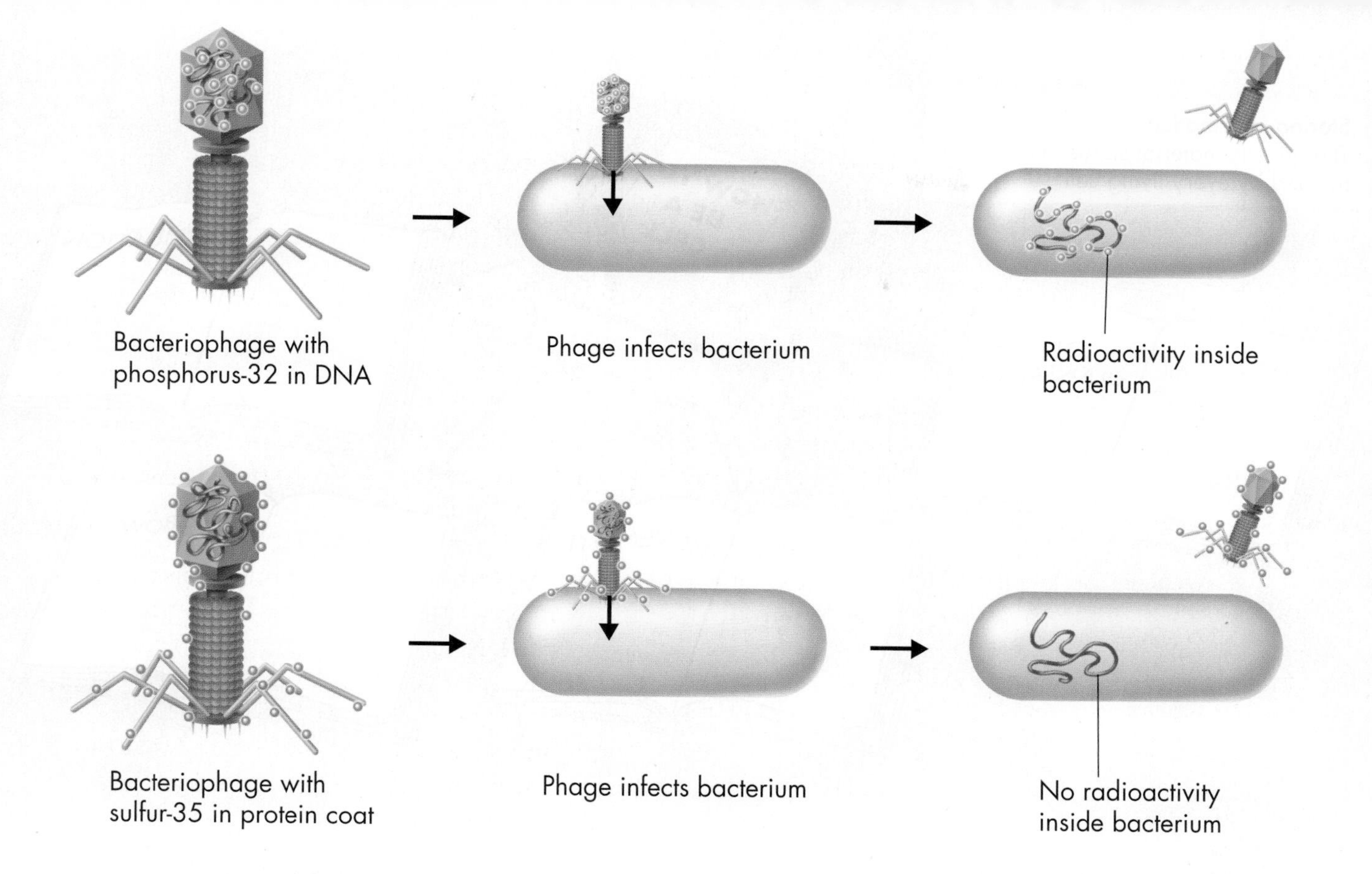

FIGURE 12–3 Hershey-Chase Experiment Alfred Hershey and Martha Chase used different radioactive markers to label the DNA and proteins of bacteriophages. The bacteriophages injected only DNA, not proteins, into bacterial cells.

The Hershey-Chase Experiment Hershey and Chase studied a bacteriophage that was composed of a DNA core and a protein coat. They wanted to determine which part of the virus—the protein coat or the DNA core—entered the bacterial cell. Their results would either support or disprove Avery's finding that genes were made of DNA.

The pair grew viruses in cultures containing radioactive isotopes of phosphorus-32 (^{32}P) and sulfur-35 (^{35}S). This was a clever strategy, because proteins contain almost no phosphorus, and DNA contains no sulfur. Therefore, these radioactive substances could be used as markers, enabling the scientists to tell which molecules actually entered the bacteria, carrying the genetic information of the virus. If they found radioactivity from ^{35}S in the bacteria, it would mean that the virus's protein coat had been injected into the bacteria. If they found ^{32}P, then the DNA core had been injected.

The two scientists mixed the marked viruses with bacterial cells. They waited a few minutes for the viruses to inject their genetic material. Next, they separated the viruses from the bacteria and tested the bacteria for radioactivity. **Figure 12–3** shows the steps in this experiment. What were the results? Nearly all the radioactivity in the bacteria was from phosphorus (^{32}P), the marker found in DNA. Hershey and Chase concluded that the genetic material of the bacteriophage was indeed DNA, not protein. **Hershey and Chase's experiment with bacteriophages confirmed Avery's results, convincing many scientists that DNA was the genetic material found in genes—not just in viruses and bacteria, but in all living cells.**

In Your Notebook *Identify the independent and dependent variables in the Hershey-Chase experiment, and list some possible control variables.*

VISUAL ANALOGY

THE MAIN FUNCTIONS OF DNA

FIGURE 12–4 Like DNA, the book in this diagram contains coded instructions for a cell to carry out important biological processes, such as how to move or transport ions. The book, like DNA, can also be copied and passed along to the next generation. These three tasks—storing, copying, and transmitting information—are also the three main functions of DNA.

The Role of DNA

What is the role of DNA in heredity?

You might think that scientists would have been satisfied knowing that genes were made of DNA, but that was not the case at all. Instead, they wondered how DNA, or any molecule for that matter, could do the critical things that genes were known to do. The next era of study began with one crucial assumption. **The DNA that makes up genes must be capable of storing, copying, and transmitting the genetic information in a cell.** These three functions are analogous to the way in which you might share a treasured book, as pictured in **Figure 12–4.**

Storing Information The foremost job of DNA, as the molecule of heredity, is to store information. The genes that make a flower purple must somehow carry the information needed to produce purple pigment. Genes for blood type and eye color must have the information needed for their jobs as well, and other genes have to do even more. Genes control patterns of development, which means that the instructions that cause a single cell to develop into an oak tree, a sea urchin, or a dog must somehow be written into the DNA of each of these organisms.

Copying Information Before a cell divides, it must make a complete copy of every one of its genes. To many scientists, the most puzzling aspect of DNA was how it could be copied. The solution to this and other puzzles had to wait until the structure of the DNA molecule became known. Within a few weeks of this discovery, a copying mechanism for the genetic material was put forward. You will learn about this mechanism later in the chapter.

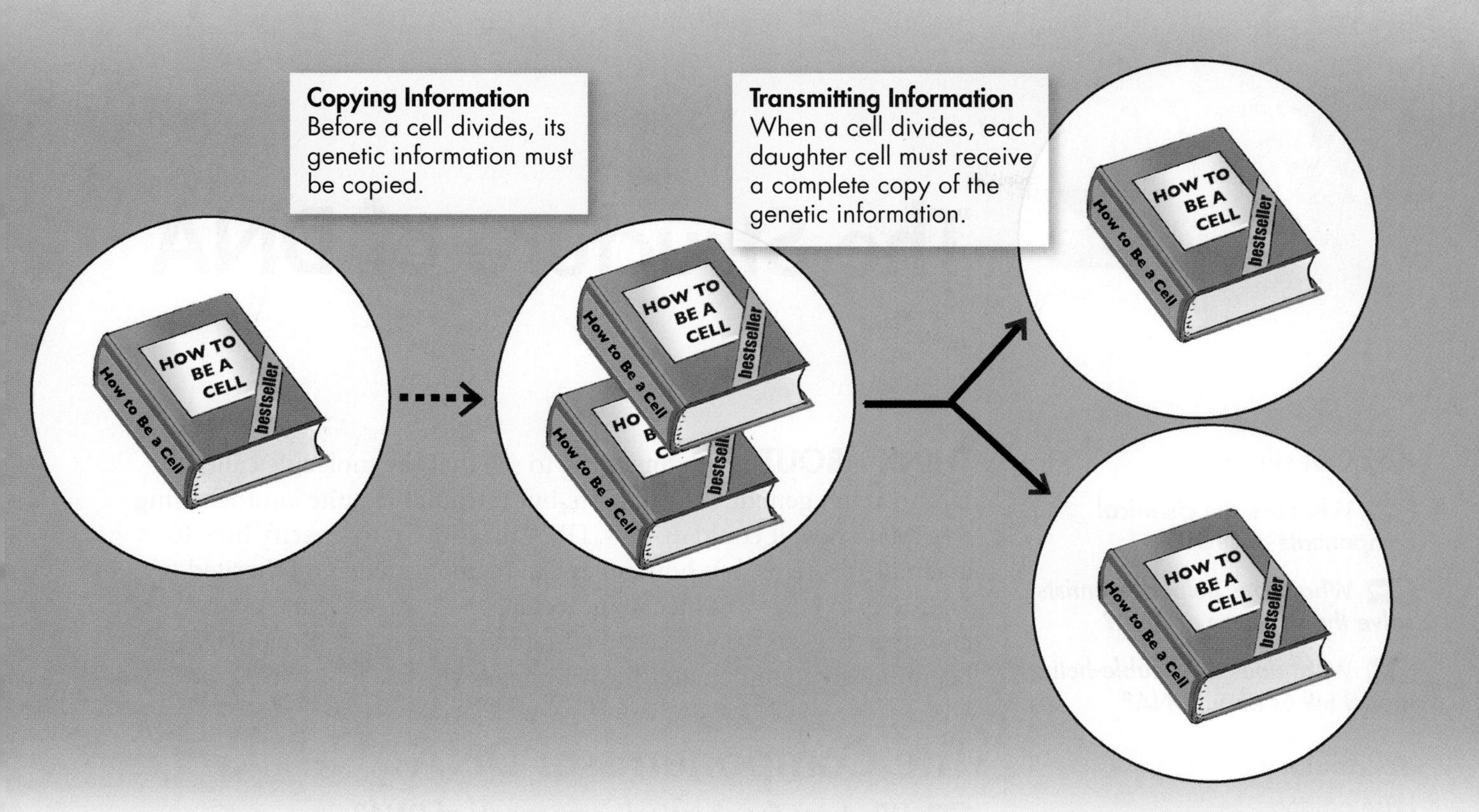

Transmitting Information As Mendel's work had shown, genes are transmitted from one generation to the next. Therefore, DNA molecules must be carefully sorted and passed along during cell division. Such careful sorting is especially important during the formation of reproductive cells in meiosis. Remember, the chromosomes of eukaryotic cells contain genes made of DNA. The loss of any DNA during meiosis might mean a loss of valuable genetic information from one generation to the next.

12.1 Assessment

Review Key Concepts

1. a. Review List the conclusions that Griffith and Avery drew from their experiments.

b. Identify Variables What was the experimental variable that Avery used when he repeated Griffith's work?

2. a. Review What conclusion did Hershey and Chase draw from their experiments?

b. Infer Why did Hershey and Chase grow viruses in cultures that contained both radioactive phosphorus and radioactive sulfur? What might have happened if they had used only one radioactive substance?

3. a. Review What are the three key roles of DNA?

b. Apply Concepts Why would the storage of genetic information in genes help explain why chromosomes are separated so carefully during mitosis?

Apply the Big idea

Science as a Way of Knowing

4. Research to Build and Present Knowledge Choose Griffith, Avery, or Hershey and Chase. Select evidence from the text to develop a flowchart that shows how that scientist or team of scientists used scientific methods. Be sure to identify each method. Use your flowchart from Taking Notes and content from Chapter 1 as a guide.

12.2 The Structure of DNA

Key Questions

What are the chemical components of DNA?

What clues helped scientists solve the structure of DNA?

What does the double-helix model tell us about DNA?

Vocabulary

base pairing

Taking Notes

Outline As you read, find the key ideas for the text under each green heading. Write down a few key words from each main idea. Then, use these key words to summarize the information about DNA.

MYSTERY CLUE

The energy from UV light can excite electrons in the absorbing substance to the point where the electrons cause chemical changes. What chemical changes might occur in the nitrogenous bases of DNA?

THINK ABOUT IT It's one thing to say that the molecule called DNA carries genetic information, but it would be quite another thing to explain how it could do this. DNA must not only specify how to assemble proteins, but how genes can be replicated and inherited. DNA has to be a very special molecule, and it's got to have a very special structure. As we will see, understanding the structure of DNA has been the key to understanding how genes work.

The Components of DNA

What are the chemical components of DNA?

Deoxyribonucleic acid, or DNA, is a unique molecule indeed. **DNA is a nucleic acid made up of nucleotides joined into long strands or chains by covalent bonds.** Let's examine each of these components more closely.

Nucleic Acids and Nucleotides As you may recall, nucleic acids are long, slightly acidic molecules originally identified in cell nuclei. Like many other macromolecules, nucleic acids are made up of smaller subunits, linked together to form long chains. Nucleotides are the building blocks of nucleic acids. **Figure 12–5** shows the nucleotides in DNA. These nucleotides are made up of three basic components: a 5-carbon sugar called deoxyribose, a phosphate group, and a nitrogenous base.

Nitrogenous Bases and Covalent Bonds Nitrogenous bases, simply put, are bases that contain nitrogen. DNA has four kinds of nitrogenous bases: adenine (AD uh neen), guanine (GWAH neen), cytosine (SY tuh zeen), and thymine (THY meen). Biologists often refer to the nucleotides in DNA by the first letters of their base names: A, G, C, and T. The nucleotides in a strand of DNA are joined by covalent bonds formed between the sugar of one nucleotide and the phosphate group of the next. The nitrogenous bases stick out sideways from the nucleotide chain. The nucleotides can be joined together in any order, meaning that any sequence of bases is possible. These bases, by the way, have a chemical structure that makes them especially good at absorbing ultraviolet (UV) light. In fact, we can determine the amount of DNA in a solution by measuring the amount of light it absorbs at a wavelength of 260 nanometers (nm), which is in the UV region of the electromagnetic spectrum.

If you don't see much in **Figure 12–5** that could explain the remarkable properties of DNA, don't be surprised. In the 1940s and early 1950s, the leading biologists in the world thought of DNA as little more than a string of nucleotides. They were baffled, too. The four different nucleotides, like the 26 letters of the alphabet, could be strung together in many different sequences, so it was possible they could carry coded genetic information. However, so could many other molecules, at least in principle. Biologists wondered if there were something more to the structure of DNA.

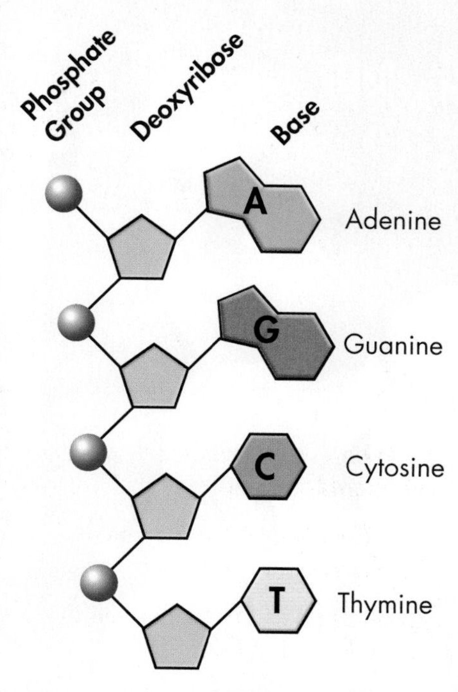

FIGURE 12–5 DNA Nucleotides DNA is made up of nucleotides, each with a deoxyribose molecule, a phosphate group, and a nitrogen-containing base. The four bases are adenine (A), guanine (G), cytosine (C), and thymine (T). Interpret Visuals ***How are these four nucleotides joined together to form part of a DNA chain?***

Solving the Structure of DNA

What clues helped scientists solve the structure of DNA?

Knowing that DNA is made from long chains of nucleotides was only the beginning of understanding the structure of this molecule. The next step required an understanding of the way in which those chains are arranged in three dimensions.

Chargaff's Rule One of the puzzling facts about DNA was a curious relationship between its nucleotides. Years earlier, Erwin Chargaff, an Austrian-American biochemist, had discovered that the percentages of adenine [A] and thymine [T] bases are almost equal in any sample of DNA. The same thing is true for the other two nucleotides, guanine [G] and cytosine [C]. The observation that [A] = [T] and [G] = [C] became known as "Chargaff's rule." Despite the fact that DNA samples from organisms as different as bacteria and humans obeyed this rule, neither Chargaff nor anyone else had the faintest idea why.

Analyzing Data

Base Percentages

In 1949, Erwin Chargaff discovered that the relative amounts of A and T, and of G and C, are almost always equal. The table shows a portion of the data that Chargaff collected.

Percentages of Bases in Five Organisms

Source of DNA	A	T	G	C
Streptococcus	29.8	31.6	20.5	18.0
Yeast	31.3	32.9	18.7	17.1
Herring	27.8	27.5	22.2	22.6
Human	30.9	29.4	19.9	19.8
E.coli	24.7	23.6	26.0	25.7

1. Interpret Tables Which organism has the highest percentage of adenine?

2. Calculate If a species has 35 percent adenine in its DNA, what is the percentage of the other three bases? MATH

3. Draw Conclusions What did the fact that A and T, and G and C, occurred in equal amounts suggest about the relationship among these bases?

VISUAL SUMMARY

CLUES TO THE STRUCTURE OF DNA

FIGURE 12–6 Erwin Chargaff, Rosalind Franklin, James Watson, and Francis Crick were among the many scientists who helped solve the puzzle of DNA's molecular structure. Franklin's X-ray diffraction photograph shows the pattern that indicated the structure of DNA is helical.

Erwin Chargaff

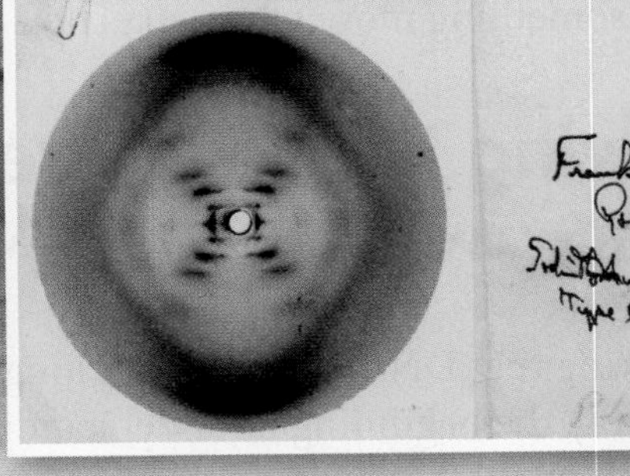

Franklin's X-ray diffraction photograph, May 1952

Rosalind Franklin

Franklin's X-Rays In the early 1950s, the British scientist Rosalind Franklin began to study DNA. Franklin used a technique called X-ray diffraction to get information about the structure of the DNA molecule. First, she purified a large amount of DNA, then stretched the DNA fibers in a thin glass tube so that most of the strands were parallel. Next, she aimed a powerful X-ray beam at the concentrated DNA samples and recorded the scattering pattern of the X-rays on film. Franklin worked hard to obtain better and better patterns from DNA until the patterns became clear. The result of her work is the X-ray photograph shown in **Figure 12–6,** taken in the summer of 1952.

By itself, Franklin's X-ray pattern does not reveal the structure of DNA, but it does carry some very important clues. The X-shaped pattern shows that the strands in DNA are twisted around each other like the coils of a spring, a shape known as a helix. The angle of the X suggests that there are two strands in the structure. Other clues suggest that the nitrogenous bases are near the center of the DNA molecule.

BUILD Vocabulary

ACADEMIC WORDS In biochemistry, the noun **helix** refers to an extended spiral chain of units in a protein, nucleic acid, or other large molecule. The plural term is *helices*.

The Work of Watson and Crick While Franklin was continuing her research, James Watson, an American biologist, and Francis Crick, a British physicist, were also trying to understand the structure of DNA. They built three-dimensional models of the molecule that were made of cardboard and wire. They twisted and stretched the models in various ways, but their best efforts did nothing to explain DNA's properties.

Then, early in 1953, Watson was shown a copy of Franklin's remarkable X-ray pattern. The effect was immediate. In his book *The Double Helix*, Watson wrote: "The instant I saw the picture my mouth fell open and my pulse began to race."

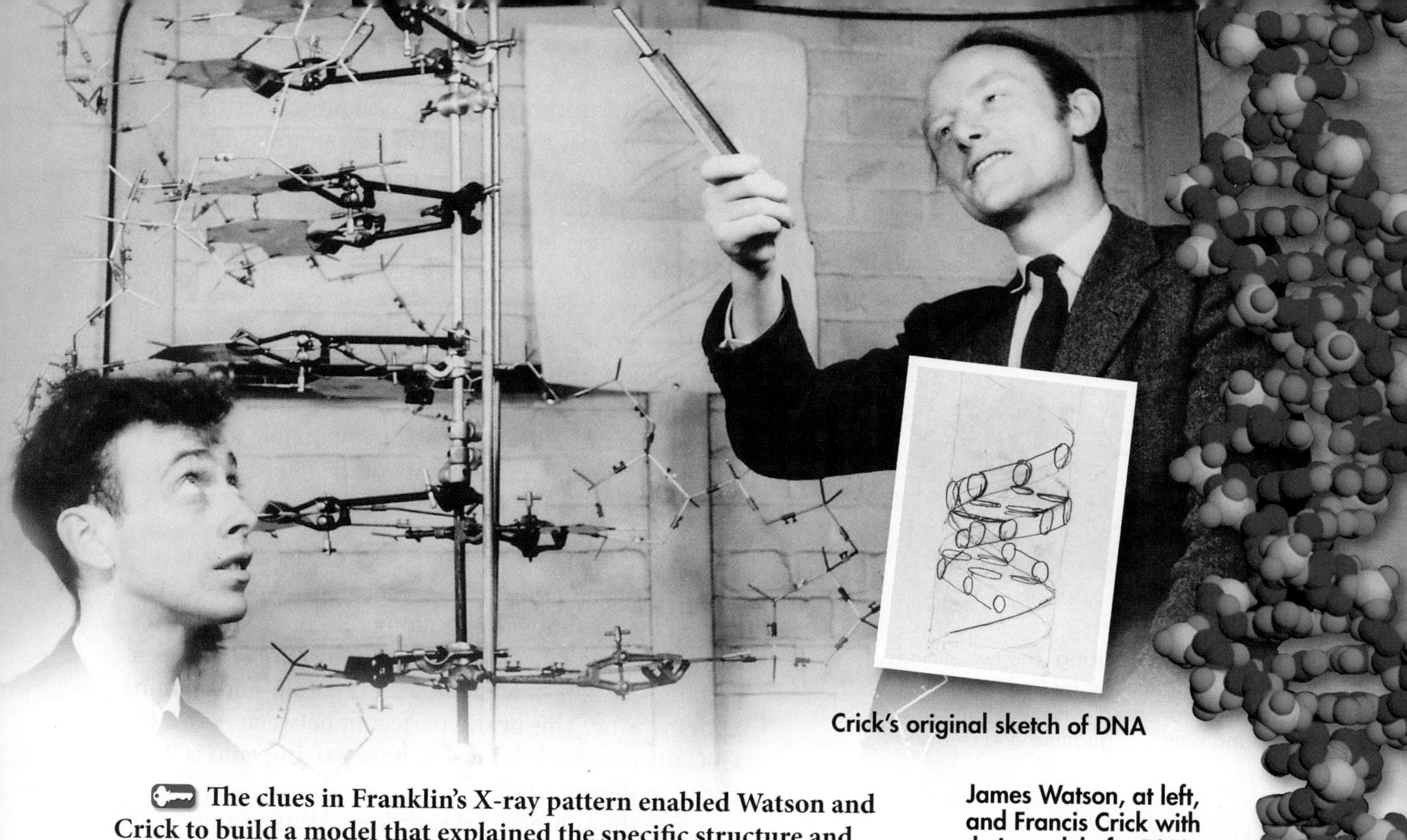

Crick's original sketch of DNA

James Watson, at left, and Francis Crick with their model of a DNA molecule in 1953

A computer model of DNA

The clues in Franklin's X-ray pattern enabled Watson and Crick to build a model that explained the specific structure and properties of DNA. The pair published their results in a historic one-page paper in April of 1953, when Franklin's paper describing her X-ray work was also published. Watson and Crick's breakthrough model of DNA was a double helix, in which two strands of nucleotide sequences were wound around each other.

The Double-Helix Model

What does the double-helix model tell us about DNA?

A double helix looks like a twisted ladder. In the double-helix model of DNA, the two strands twist around each other like spiral staircases. Watson and Crick realized that the double helix accounted for Franklin's X-ray pattern. Further still, it explained many of the most important properties of DNA. **The double-helix model explains Chargaff's rule of base pairing and how the two strands of DNA are held together.** This model can even tell us how DNA can function as a carrier of genetic information.

Antiparallel Strands One of the surprising aspects of the double-helix model is that the two strands of DNA run in opposite directions. In the language of biochemistry, these strands are "antiparallel." This arrangement enables the nitrogenous bases on both strands to come into contact at the center of the molecule. It also allows each strand of the double helix to carry a sequence of nucleotides, arranged almost like letters in a four-letter alphabet.

In Your Notebook *Draw and label your own model of the DNA double-helix structure.*

MYSTERY CLUE

Our skin cells are exposed to UV light whenever they are in direct sunlight. How might this exposure affect base pairing in the DNA of our skin cells?

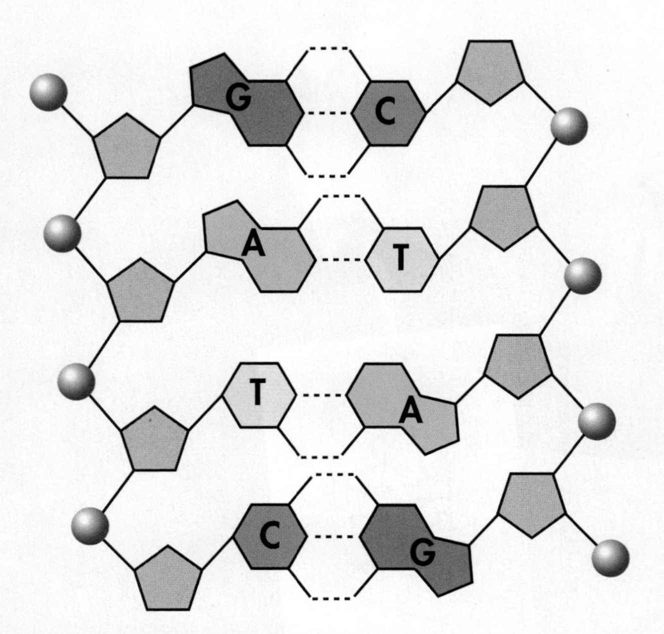

FIGURE 12–7 Base Pairing The two strands of DNA are held together by hydrogen bonds between the nitrogenous bases adenine and thymine, and between guanine and cytosine.

Hydrogen Bonding At first, Watson and Crick could not explain what forces held the two strands of DNA's double helix together. They then discovered that hydrogen bonds could form between certain nitrogenous bases, providing just enough force to hold the two strands together. As you may recall, hydrogen bonds are relatively weak chemical forces.

Does it make sense that a molecule as important as DNA should be held together by weak bonds? Indeed, it does. If the two strands of the helix were held together by strong bonds, it might well be impossible to separate them. As we will see, the ability of the two strands to separate is critical to DNA's functions.

Base Pairing Watson and Crick's model showed that hydrogen bonds could create a nearly perfect fit between nitrogenous bases along the center of the molecule. However, these bonds would form only between certain base pairs—adenine with thymine, and guanine with cytosine. This nearly perfect fit between A–T and G–C nucleotides is known as **base pairing,** and is illustrated in **Figure 12–7.**

Once they observed this process, Watson and Crick realized that base pairing explained Chargaff's rule. It gave a reason why [A] = [T] and [G] = [C]. For every adenine in a double-stranded DNA molecule, there has to be exactly one thymine. For each cytosine, there is one guanine. The ability of their model to explain Chargaff's observations increased Watson and Crick's confidence that they had come to the right conclusion, with the help of Rosalind Franklin.

12.2 Assessment

Review Key Concepts

1. a. Review List the chemical components of DNA.

b. Relate Cause and Effect Why are hydrogen bonds so essential to the structure of DNA?

2. a. Review Describe the discoveries that led to the modeling of DNA.

b. Infer Why did scientists have to use tools other than microscopes to solve the structure of DNA?

3. a. Review Describe Watson and Crick's model of the DNA molecule.

b. Apply Concepts Did Watson and Crick's model account for the equal amounts of thymine and adenine in DNA? Explain.

VISUAL THINKING

4. Make a three-dimensional model showing the structure of a DNA molecule. Your model should include the four base pairs that help form the double helix.

RST.9-10.10 Level of Text Complexity, WHST.9-10.6 Production and Distribution of Writing. Also WHST.9-10.7

Biology & HISTORY

Discovering the Role of DNA **Genes and the principles of genetics were discovered before scientists identified the molecules that genes are made of. With the discovery of DNA, scientists have been able to explain how genes are replicated and how they function.**

1860 1880 1900 1920 1940 1960 1980 2000

1865

Gregor Mendel shows that the characteristics of pea plants are passed along in a predictable way. His discovery begins the science of genetics.

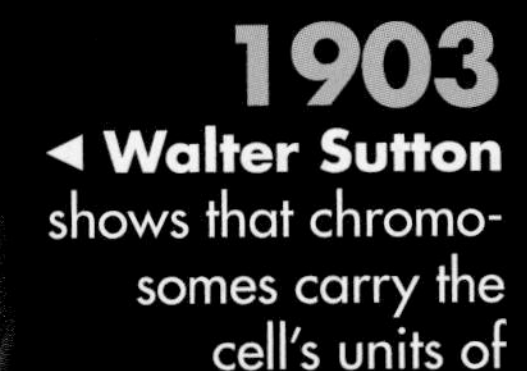

1903

◀ **Walter Sutton** shows that chromosomes carry the cell's units of inheritance.

1911

Thomas Hunt Morgan ▲ demonstrates that genes are arranged in linear fashion on the chromosomes of the fruit fly.

1928

▼ **Frederick Griffith** discovers that bacteria contain a molecule that can transfer genetic information from cell to cell.

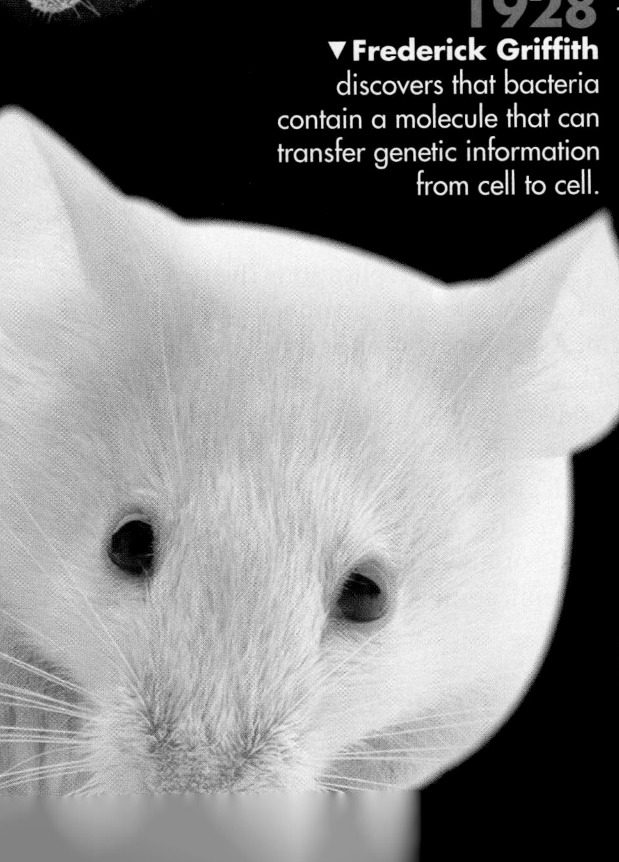

1944

Oswald Avery, Colin Macleod, and Maclyn McCarty show the substance that Griffith discovered is DNA.

1950

Erwin Chargaff analyzes the base composition of DNA in cells. He discovers that the amounts of adenine and thymine are almost always equal, as are the amounts of guanine and cytosine.

1952

Alfred Hershey and Martha Chase confirm that the genetic material of viruses is DNA, not protein. **Rosalind Franklin** records a critical X-ray diffraction pattern, demonstrating that DNA is in the form of a helix.

1953

James Watson and Francis Crick publish their model of the DNA double helix. The model was made possible by Franklin's work.

2000

▼ **Craig Venter and Francis Collins** announce the draft DNA sequence of the human genome at a White House ceremony in Washington, D.C. The final version is published in 2003.

WRITING **Use library or Internet resources to find out what James Watson or Francis Crick worked on after discovering the structure of DNA. Organize your findings about the scientist's work and make a multimedia presentation for the class.**

12.3 DNA Replication

Key Questions

What role does DNA polymerase play in copying DNA?

How does DNA replication differ in prokaryotic cells and eukaryotic cells?

Vocabulary

replication
DNA polymerase
telomere

Taking Notes

Preview Visuals Before you read, study the diagram in **Figure 12–8.** Make a list of questions about the diagram. As you read, write down the answers to your questions.

BUILD Vocabulary

WORD ORIGINS The prefix *re-* means "back" or "again." *Plicare* is a Latin verb meaning "to fold." To replicate something is, in a sense, to repeat it, or to fold back again.

THINK ABOUT IT Before a cell divides, its DNA must first be copied. How might the double-helix structure of DNA make that possible? What might happen if one of the nucleotides were damaged or chemically altered just before the copying process? How might this affect the DNA inherited by each daughter cell after cell division?

Copying the Code

What role does DNA polymerase play in copying DNA?

When Watson and Crick discovered the structure of DNA, they immediately recognized one genuinely surprising aspect of the structure. Base pairing in the double helix explains how DNA can be copied, or replicated, because each base on one strand pairs with one—and only one—base on the opposite strand. Each strand of the double helix therefore has all the information needed to reconstruct the other half by the mechanism of base pairing. Because each strand can be used to make the other strand, the strands are said to be complementary.

The Replication Process Before a cell divides, it duplicates its DNA in a copying process called **replication.** This process, which occurs during late interphase of the cell cycle, ensures that each resulting cell has the same complete set of DNA molecules. During replication, the DNA molecule separates into two strands and then produces two new complementary strands following the rules of base pairing. Each strand of the double helix of DNA serves as a template, or model, for the new strand.

Figure 12–8 shows the process of DNA replication. The two strands of the double helix have separated, or "unzipped," allowing two replication forks to form. As each new strand forms, new bases are added following the rules of base pairing. If the base on the old strand is adenine, then thymine is added to the newly forming strand. Likewise, guanine is always paired to cytosine. For example, a strand that has the base sequence TACGTT produces a strand with the complementary base sequence ATGCAA. The result is two DNA molecules identical to each other and to the original molecule. Note that each DNA molecule resulting from replication has one original strand and one new strand.

In Your Notebook *In your own words, describe the process of DNA replication.*

The Role of Enzymes DNA replication is carried out by a series of enzymes. These enzymes first "unzip" a molecule of DNA by breaking the hydrogen bonds between base pairs and unwinding the two strands of the molecule. Each strand then serves as a template for the attachment of complementary bases. You may recall that enzymes are proteins with highly specific functions. For this reason, they are often named for the reactions they catalyze. The principal enzyme involved in DNA replication is called **DNA polymerase** (PAHL ih mur ayz). **DNA polymerase is an enzyme that joins individual nucleotides to produce a new strand of DNA.** Besides producing the sugar-phosphate bonds that join nucleotides together, DNA polymerase also "proofreads" each new DNA strand, so that each molecule is a near-perfect copy of the original.

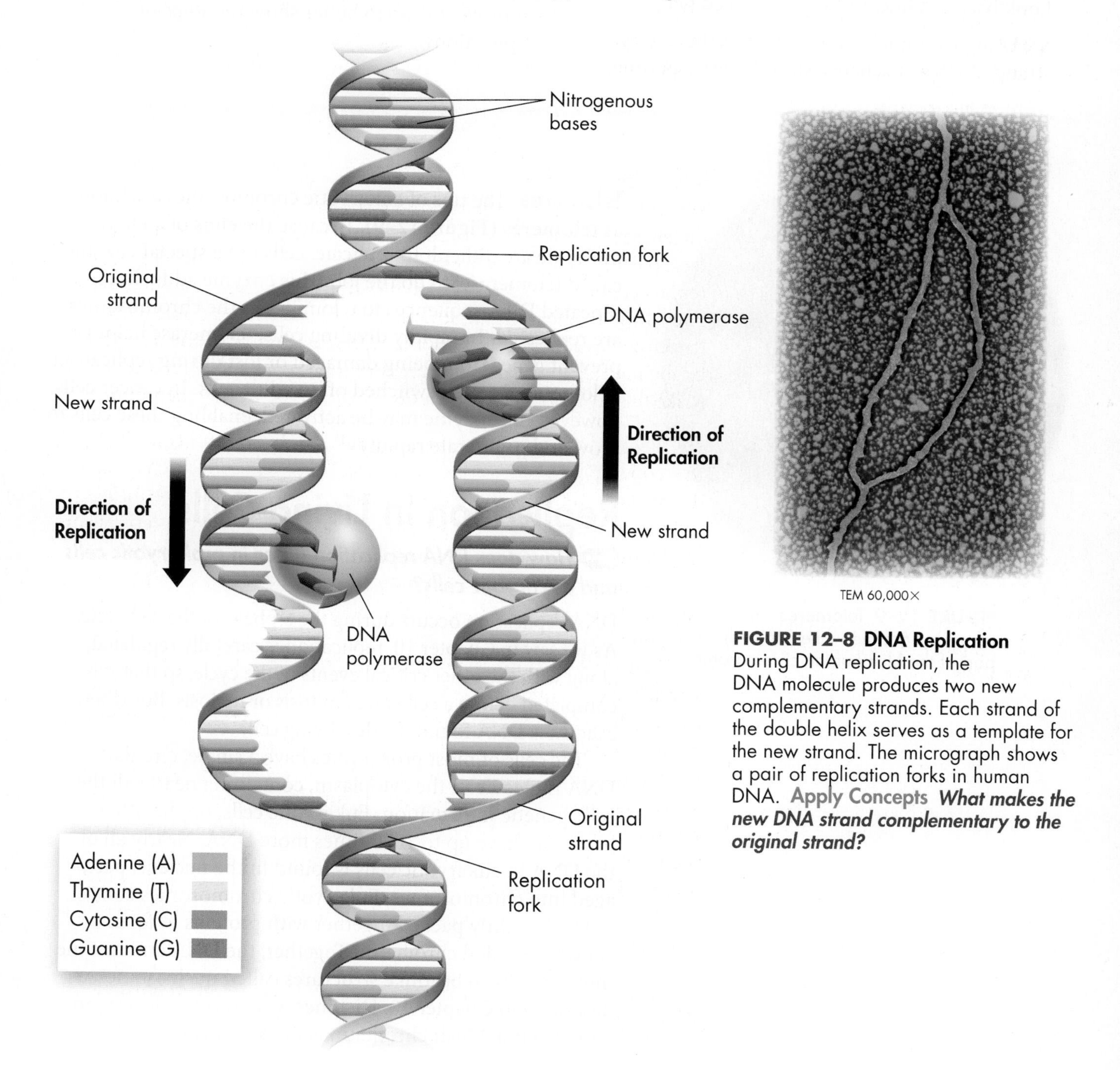

FIGURE 12–8 DNA Replication During DNA replication, the DNA molecule produces two new complementary strands. Each strand of the double helix serves as a template for the new strand. The micrograph shows a pair of replication forks in human DNA. **Apply Concepts** ***What makes the new DNA strand complementary to the original strand?***

Quick Lab
OPEN-ENDED INQUIRY

Modeling DNA Replication

❶ Cut out small squares of white and yellow paper to represent phosphate and sugar molecules. Then, cut out small strips of blue, green, red, and orange paper to represent the four nitrogenous bases. Build a set of five nucleotides using your paper strips and tape. Look back at **Figure 12–5** if you need help.

❷ Using your nucleotides, tape together a single strand of DNA. Exchange strands with a partner.

❸ Model DNA replication by creating a strand that is complementary to your partner's original strand.

Analyze and Conclude

1. **Use Models** Taping together the nucleotides models the action of what enzyme?
2. **Evaluate** In what ways does this lab accurately represent DNA replication? How could you improve the lab to better show the steps of replication?

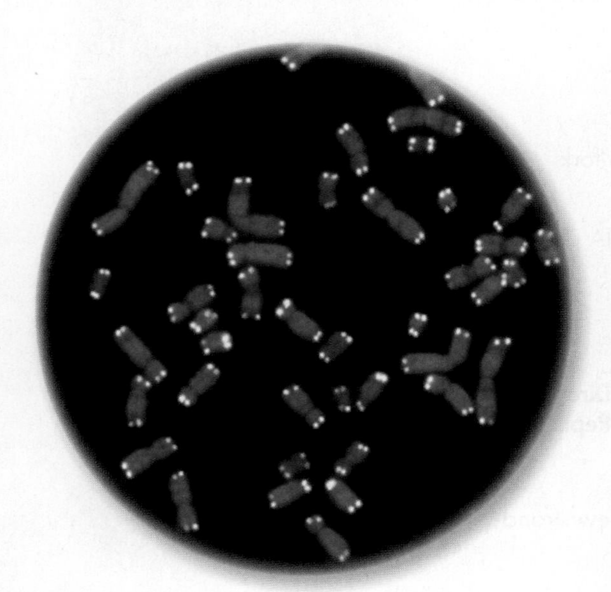

FIGURE 12–9 Telomeres The telomeres are the white (stained) part of the blue human chromosomes.

Telomeres The tips of eukaryotic chromosomes are known as **telomeres (Figure 12–9)**. Because the ends of a DNA molecule are difficult to replicate, cells use a special enzyme, called telomerase, to do the job. This enzyme adds short, repeated DNA sequences to telomeres as the chromosomes are replicated. In rapidly dividing cells, telomerase helps to prevent genes from being damaged or lost during replication. Telomerase is often switched off in adult cells. In cancer cells, however, the enzyme may be activated, enabling these cells to grow and proliferate rapidly.

Replication in Living Cells

How does DNA replication differ in prokaryotic cells and eukaryotic cells?

DNA replication occurs during the S phase of the cell cycle. As we saw in Chapter 10, replication is carefully regulated, along with the other critical events of the cycle, so that it is completed before a cell enters mitosis or meiosis. But where, exactly, is DNA found inside a living cell?

The cells of most prokaryotes have a single, circular DNA molecule in the cytoplasm, containing nearly all the cell's genetic information. Eukaryotic cells, on the other hand, can have up to 1000 times more DNA. Nearly all of the DNA of eukaryotic cells is found in the nucleus, packaged into chromosomes. Eukaryotic chromosomes consist of DNA, tightly packed together with proteins to form a substance called chromatin. Together, the DNA and histone molecules form beadlike structures called nucleosomes, as described in Chapter 10. Histones, you may recall, are proteins around which chromatin is tightly coiled.

Prokaryotic DNA Replication In most prokaryotes, DNA replication does not start until regulatory proteins bind to a single starting point on the chromosome. These proteins then trigger the beginning of the S phase, and DNA replication begins. **Replication in most prokaryotic cells starts from a single point and proceeds in two directions until the entire chromosome is copied.** This process is shown in **Figure 12–10.** Often, the two chromosomes produced by replication are attached to different points inside the cell membrane and are separated when the cell splits to form two new cells.

Eukaryotic DNA Replication Eukaryotic chromosomes are generally much bigger than those of prokaryotes. **In eukaryotic cells, replication may begin at dozens or even hundreds of places on the DNA molecule, proceeding in both directions until each chromosome is completely copied.** Although a number of proteins check DNA for chemical damage or base pair mismatches prior to replication, the system is not foolproof. Damaged regions of DNA are sometimes replicated, resulting in changes to DNA base sequences that may alter certain genes and produce serious consequences.

The two copies of DNA produced by replication in each chromosome remain closely associated until the cell enters prophase of mitosis. At that point, the chromosomes condense, and the two chromatids in each chromosome become clearly visible. They separate from each other in anaphase of mitosis, as described in Chapter 10, producing two cells, each with a complete set of genes coded in DNA.

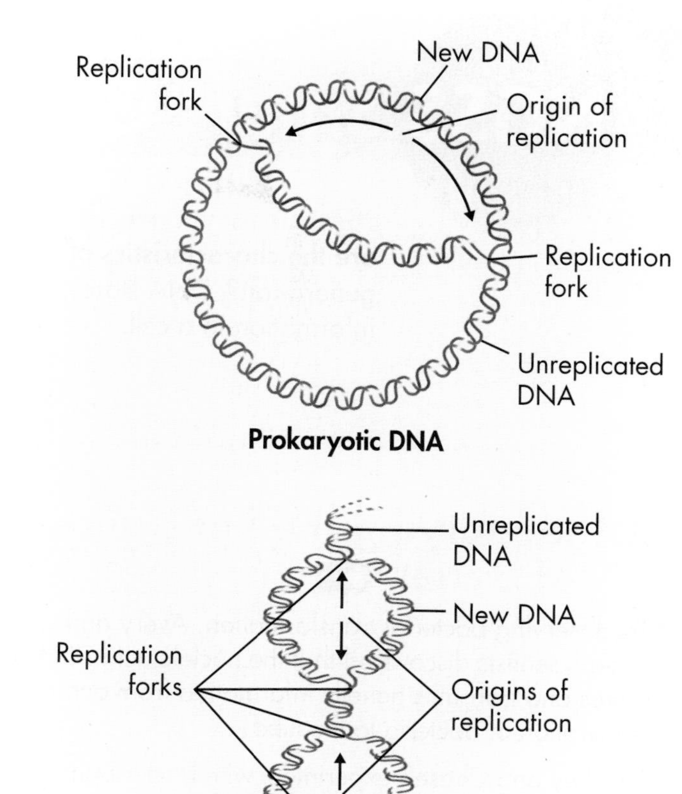

FIGURE 12–10 Differences in DNA Replication Replication in most prokaryotic cells (top) begins at a single starting point and proceeds in two directions until the entire chromosome is copied. In eukaryotic cells (bottom), replication proceeds from multiple starting points on individual chromosomes and ends when all the chromosomes are copied.

12.3 Assessment

Review Key Concepts

1. a. Review How is DNA replicated?

b. Apply Concepts What is the role of DNA polymerase in DNA replication?

2. a. Review Where and in what form is prokaryotic DNA found? Where is eukaryotic DNA found?

b. Infer What could be the result of damaged DNA being replicated?

VISUAL THINKING

3. Make a Venn diagram that compares the process of DNA replication in prokaryotes and eukaryotes. Compare the location, steps, and end products of the process in each kind of cell.

NGSS Smart Guide DNA

Disciplinary Core Idea LS3.A Inheritance of Traits: How are the characteristics of one generation related to the previous generation? DNA stores, copies, and transmits the genetic information in a cell.

12.1 Identifying the Substance of Genes

- By observing bacterial transformation, Avery and other scientists discovered that the nucleic acid DNA stores and transmits genetic information from one generation of bacteria to the next.
- Hershey and Chase's experiment with bacteriophages confirmed Avery's results, convincing many scientists that DNA was the genetic material found in genes—not just in viruses and bacteria, but in all living cells.
- The DNA that makes up genes must be capable of storing, copying, and transmitting the genetic information in a cell.

transformation 339 • bacteriophage 340

Biology.com

Untamed Science Video The Untamed Science CSI crew unravels the secrets of DNA left at the scene of a crime.

Art in Motion View an animation that re-creates the Hershey-Chase experiments.

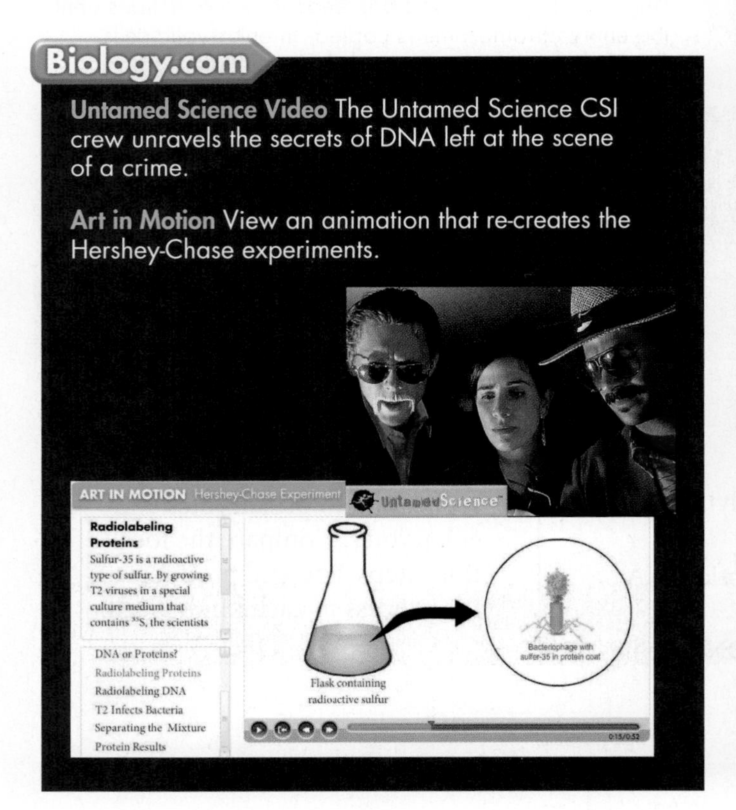

12.2 The Structure of DNA 

- DNA is a nucleic acid made up of nucleotides joined into long strands or chains by covalent bonds.
- The clues in Franklin's X-ray pattern enabled Watson and Crick to build a model that explained the specific structure and properties of DNA.
- The double-helix model explains Chargaff's rule of base pairing and how the two strands of DNA are held together.

base pairing 348

Biology.com

Data Analysis Learn how analysis of DNA base sequences can be used to track animal poaching.

Tutor Tube Tune into the tutor to find hints for remembering which bases pair together.

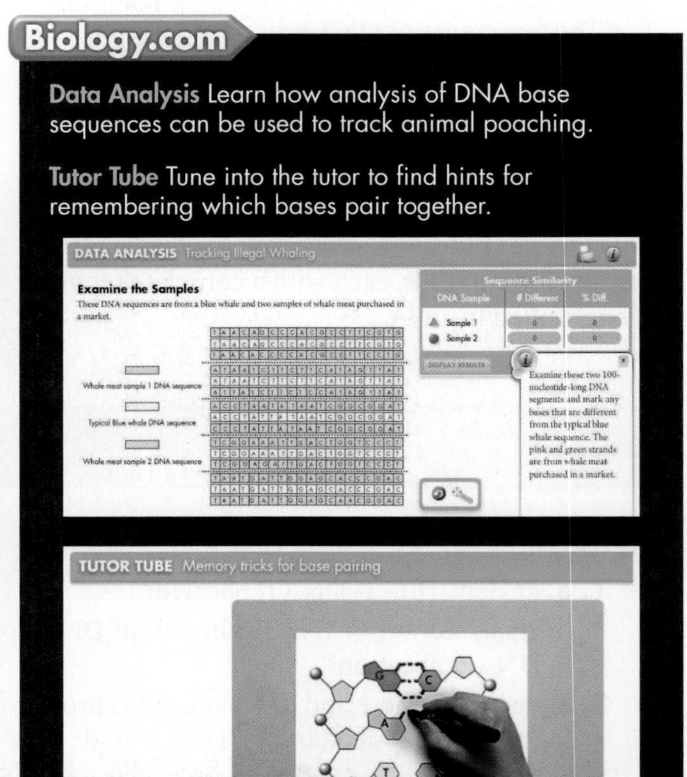

CHECKING YOUR *Scientific Literacy*

Refer to the lesson content and digital assets below as you prepare for your chapter assessment. Then, evaluate your understanding of DNA by answering these questions.

1. **Scientific and Engineering Practice Asking Questions** Review the experiments described in Lesson 12.1. What question was Griffith trying to answer? Based on his results, what was the answer to his question? Do the same thing for Avery's experiment and for the Hershey-Chase experiment.

2. **Crosscutting Concept Structure and Function** Two of DNA's functions are storing information and copying information. Explain how the structure of DNA supports each of these functions.

3. **Production and Distribution of Writing** Evaluate your answer to Question 2. Use the following questions as part of your evaluation. Is the order of your sentences logical? Does your explanation include specific, concrete details about DNA's structure? Are there any misspelled words or grammatical errors? Based on this evaluation, revise your answer to Question 2.

4. **STEM** The technology that scientists use in their experiments isn't always as complex as X-ray diffraction. Identify at least three technologies that Griffith used in his experiments, in addition to a microscope.

12.3 DNA Replication

- DNA polymerase is an enzyme that joins individual nucleotides to produce a new strand of DNA.
- Replication in most prokaryotic cells starts from a single point and proceeds in two directions until the entire chromosome is copied.
- In eukaryotic cells, replication may begin at dozens or even hundreds of places on the DNA molecule, proceeding in both directions until each chromosome is completely copied.

replication 350 • DNA polymerase 351 • telomere 352

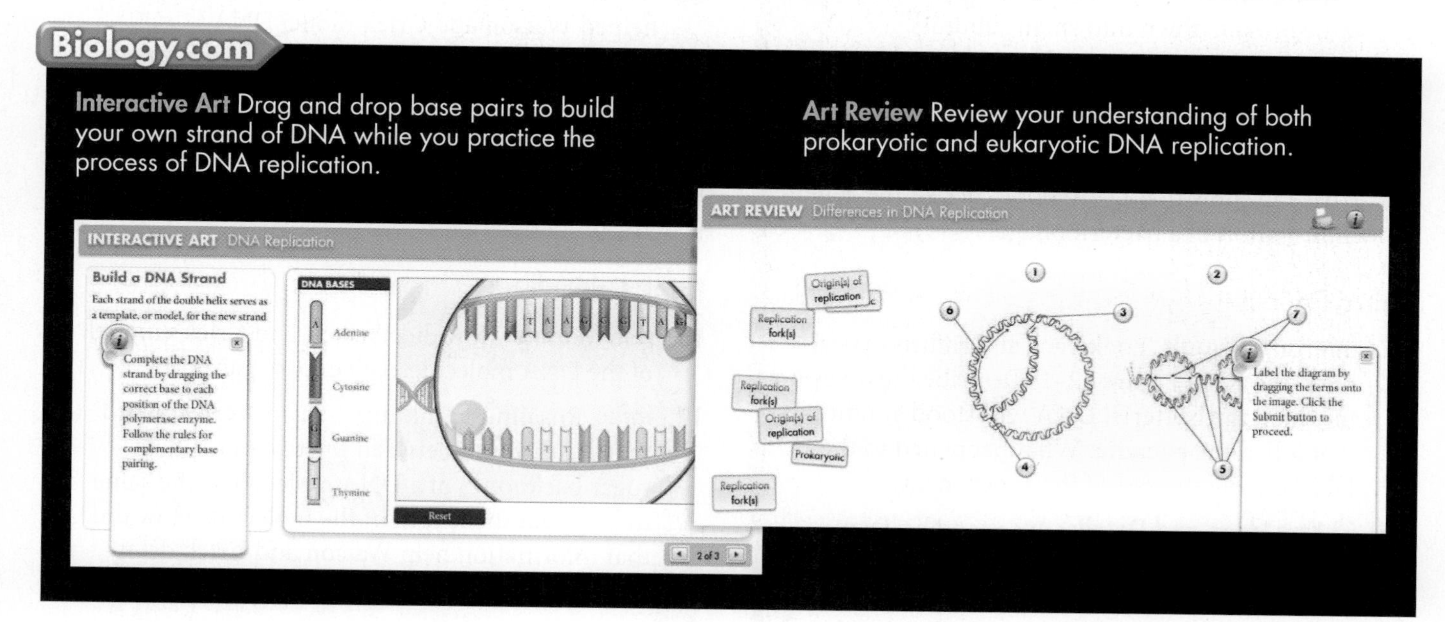

Skills Lab

Extracting DNA This chapter lab is available in your lab manual and at Biology.com

12 Assessment

12.1 Identifying the Substance of Genes

Understand Key Concepts

1. The process by which one strain of bacterium is changed into another strain is called
   a. transcription.
   b. transformation.
   c. duplication.
   d. replication.

2. Bacteriophages are
   a. a form of bacteria.
   b. enzymes.
   c. coils of DNA.
   d. viruses.

3. Which of the following researchers used radioactive markers in experiments to show that DNA was the genetic material in cells?
   a. Frederick Griffith
   b. Oswald Avery
   c. Alfred Hershey and Martha Chase
   d. James Watson and Francis Crick

4. Before DNA could definitively be shown to be the genetic material in cells, scientists had to show that it could
   a. tolerate high temperatures.
   b. carry and make copies of information.
   c. be modified in response to environmental conditions.
   d. be broken down into small subunits.

5. Briefly describe the conclusion that could be drawn from the experiments of Frederick Griffith.

6. What was the key factor that allowed Hershey and Chase to show that DNA alone carried the genetic information of a bacteriophage?

Think Critically

7. **Interpret Visuals** Look back at Griffith's experiment shown in **Figure 12–1.** Describe the occasion in which the bacterial DNA withstood conditions that killed the bacteria. What happened to the DNA during the rest of the experiment?

8. **Key Ideas and Details** Avery and his team identified DNA as the molecule responsible for the transformation seen in Griffith's experiment. Cite textual evidence of how they controlled variables to make sure that only DNA caused the effect.

12.2 The Structure of DNA

Understand Key Concepts

9. A nucleotide does NOT contain
   a. a 5-carbon sugar.
   b. an amino acid.
   c. a nitrogen base.
   d. a phosphate group.

10. According to Chargaff's rule of base pairing, which of the following is true about DNA?
    a. A = T, and C = G
    b. A = C, and T = G
    c. A = G, and T = C
    d. A = T = C = G

11. The bonds that hold the two strands of DNA together come from
    a. the attraction of phosphate groups for each other.
    b. strong bonds between nitrogenous bases and the sugar-phosphate backbone.
    c. weak hydrogen bonds between nitrogenous bases.
    d. carbon-to-carbon bonds in the sugar portion of the nucleotides.

12. Describe the components and structure of a DNA nucleotide.

13. Explain how Chargaff's rule of base pairing helped Watson and Crick model DNA.

14. What important clue from Rosalind Franklin's work helped Watson and Crick develop their model of DNA?

15. Why is it significant that the two strands of DNA are antiparallel?

Think Critically

16. **Use Models** How did Watson and Crick's model of the DNA molecule explain base pairing?

17. **Infer** Rosalind Franklin's X-ray pattern showed that the distance between the two phosphate-sugar backbones of a DNA molecule is the same throughout the length of the molecule. How did that information help Watson and Crick determine how bases are paired?

12.3 DNA Replication

Understand Key Concepts

18. In prokaryotes, DNA molecules are located in the
a. nucleus. **c.** cytoplasm.
b. ribosomes. **d.** histones.

19. In eukaryotes, nearly all the DNA is found in the
a. nucleus. **c.** cytoplasm.
b. ribosomes. **d.** histones.

20. The diagram below shows the process of DNA
a. replication. **c.** transformation.
b. digestion. **d.** transpiration.

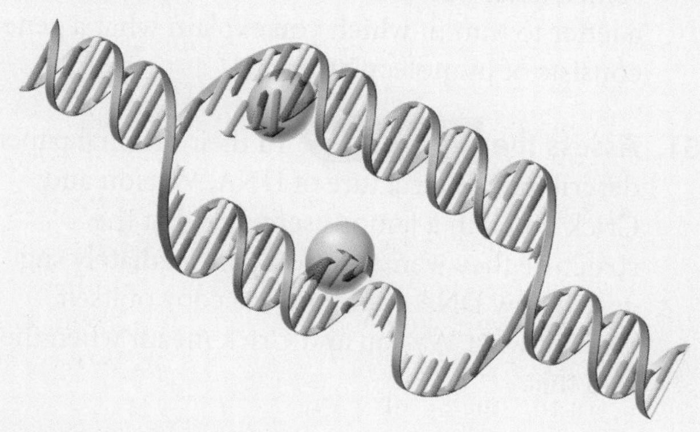

21. The main enzyme involved in linking individual nucleotides into DNA molecules is
a. DNA protease. **c.** carbohydrase.
b. ribose. **d.** DNA polymerase.

22. What is meant by the term *base pairing*? How is base pairing involved in DNA replication?

23. Describe the appearance of DNA in a typical prokaryotic cell.

24. Explain the process of replication. When a DNA molecule is replicated, how do the new molecules compare to the original molecule?

Think Critically

25. Use Analogies Is photocopying a document similar to DNA replication? Think of the original materials, the copying process, and the final products. Explain how the two processes are alike. Identify major differences.

26. Craft and Structure Describe the similarities and differences between DNA replication in prokaryotic cells and in eukaryotic cells.

solve the CHAPTER MYSTERY

UV LIGHT

The nucleotides in DNA include the nitrogenous bases adenine, cytosine, guanine, and thymine (A, C, G, and T). The energy from UV light can produce chemical changes in these bases, damaging the DNA molecule and producing errors when DNA is replicated.

1. Predict Use your understanding of the structure of DNA to predict what sorts of problems excessive UV light might produce in the DNA molecule. How might these changes affect the functions of DNA?

2. Infer All cells have systems of enzymes that repair UV-induced damage to their DNA. Some cellular systems block DNA replication if there are base pairing problems in the double helix. Why are these systems important? How might they work?

3. Relate Cause and Effect Analyze the effects that UV light might have on skin cells. Why is UV light so dangerous? Why is the skin particularly vulnerable to it?

4. Connect to the Big idea Among humans who inherit genetic defects in their DNA-repair systems, the incidence of skin cancer is as much as 1000 times greater than average. Based on this information, what can you infer about the effect of UV light on DNA?

Connecting Concepts

Use Science Graphics

A scientist studied the effect of exposing DNA to various wavelengths of ultraviolet light. The scientist determined the number of copying errors made after exposure to ultraviolet rays. The graph shows the results. Use the graph to answer questions 27 and 28.

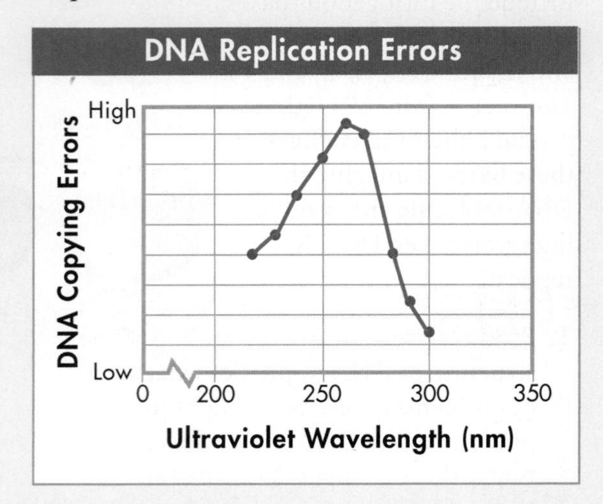

27. Interpret Graphs The most damaging effects of ultraviolet light on DNA replication occur at which wavelength?

28. Infer What conclusion would you draw from the graph about the effect of ultraviolet light on living organisms?

29. Pose Questions Ozone is a molecule that is very effective at absorbing ultraviolet light from the sun. Evidence indicates that human activities have contributed to the destruction of ozone in the atmosphere. What question would you ask about the effect of removing ozone from the atmosphere?

Write About Science

30. Production and Distribution of Writing Gregor Mendel concluded that factors, which we now call genes, determine the traits that pass from one generation to the next. Imagine that you could send a letter backward in time to Mendel. Write a letter to him in which you explain what a gene consists of in molecular terms.

31. Assess the Big idea In their original paper describing the structure of DNA, Watson and Crick noted in a famous sentence that the structure they were proposing immediately suggested how DNA could make a copy of itself. Explain what Watson and Crick meant when they said this.

Integration of Knowledge and Ideas

The following table shows the results of measuring the percentages of the four bases in the DNA of several different organisms. Some of the values are missing from the table.

Nitrogenous Bases (%)				
Organism	A	G	T	C
Human		19.9	29.4	
Chicken	28.8			21.5
Bacterium (*S. lutea*)	13.4			

32. Predict Based on Chargaff's rule, the percentage of adenine bases in human DNA should be around

a. 30.9%.
b. 19.9%.
c. 21.5%.
d. 13.4%.

33. Calculate The value for the percent of guanine bases in the bacterium would be expected to be about MATH

a. 13.4%.
b. 28.8%.
c. 36.6%.
d. There is not enough information given.

34. Predict If the two DNA strands of the bacterium were separated and the base composition of just one of the strands was determined, you could expect

a. the amount of A to equal the amount of T.
b. the amount of C to equal the amount of G.
c. the amount of A to equal the amounts of T, C, and G.
d. the four nitrogenous bases to have any value.

Standardized Test Prep

Multiple Choice

1. During replication, which sequence of nucleotides would bond with the DNA sequence TATGA?
   A TATGA
   B ATACT
   C CACTA
   D AGTAT

2. The scientist(s) responsible for the discovery of bacterial transformation is (are)
   A Watson and Crick.
   B Avery.
   C Griffith.
   D Franklin.

3. Which of the following does NOT describe the structure of DNA?
   A double helix
   B nucleotide polymer
   C contains adenine-guanine pairs
   D sugar-phosphate backbone

4. What did Hershey and Chase's work show?
   A Genes are probably made of DNA.
   B Genes are probably made of protein.
   C Viruses contain DNA but not protein.
   D Bacteria contain DNA but not protein.

5. The two "backbones" of the DNA molecule consist of
   A adenines and sugars.
   B phosphates and sugars.
   C adenines and thymines.
   D thymines and sugars.

6. In eukaryotic chromosomes, DNA is tightly coiled around proteins called
   A DNA polymerase.
   B chromatin.
   C histones.
   D nucleotides.

7. When prokaryotic cells copy their DNA, replication begins at
   A one point on the DNA molecule.
   B two points on opposite ends of the DNA molecule.
   C dozens to hundreds of points along the molecule.
   D opposite ends of the molecule.

8. Compared to eukaryotic cells, prokaryotic cells contain
   A about 1000 times more DNA.
   B about one thousandth as much DNA.
   C twice as much DNA.
   D the same amount of DNA.

Questions 9–10

Under ideal conditions, a single bacterial cell can reproduce every 20 minutes. The graph shows how the total number of cells under ideal conditions can change over time.

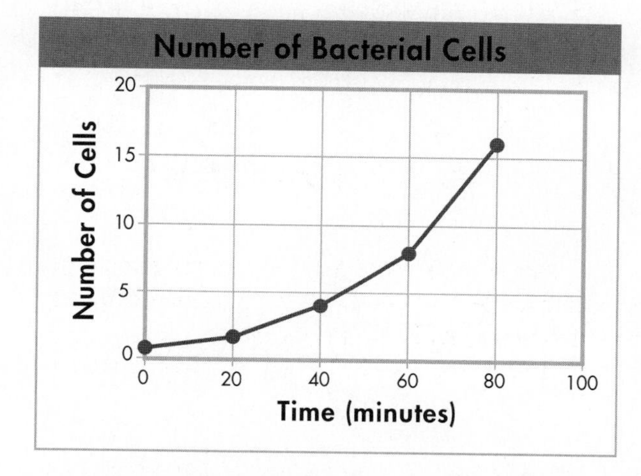

9. How many cells are present after 80 minutes?
   A 1
   B 2
   C 16
   D 32

10. If the DNA of this bacterium is 4 million base pairs in length, how many total molecules of A, T, C, and G are required for replication to be successful?
    A 2 million
    B 4 million
    C 8 million
    D 32 million

Open-Ended Response

11. Describe how eukaryotic cells are able to keep such large amounts of DNA in the small volume of the cell nucleus.

If You Have Trouble With . . .

Question	1	2	3	4	5	6	7	8	9	10	11
See Lesson	12.3	12.1	12.2	12.1	12.2	12.3	12.3	12.3	12.3	12.3	12.3

13 RNA and Protein Synthesis

Information and Heredity

Q: How does information flow from DNA to RNA to direct the synthesis of proteins?

The striking difference in appearance between these Bengal tigers is the result of mutations affecting fur color.

INSIDE:

- 13.1 RNA
- 13.2 Ribosomes and Protein Synthesis
- 13.3 Mutations
- 13.4 Gene Regulation and Expression

BUILDING *Scientific Literacy*

Deepen your understanding of RNA and protein synthesis by engaging in key practices that allow you to make connections across concepts.

NGSS You will explain the effects of transcription and translation on gene expression.

STEM What makes an antibiotic a valuable weapon against harmful bacteria? Do an activity to find the connection between antibiotics and protein synthesis in bacteria.

Common Core You will cite textual evidence to support your analysis of the role of mutations in gene expression.

CHAPTER MYSTERY

MOUSE-EYED FLY

It was definitely not a science fiction movie. The animal in the laboratory was real. Besides having two forward-looking eyes, it also had eyes on its knees and eyes on its hind legs. It even had eyes in the back of its head! Yet as strange as it looked, this animal was not a monster. It was simply a fruit fly with eyes in very strange places. These eyes looked like the fly's normal compound eyes, but a mouse gene transplanted into the fly's DNA had produced them. How could a mouse gene produce extra eyes in a fly?

As you read this chapter, look for clues to explain how a gene that normally controls the growth of eyes in mice could possibly cause a fly to grow extra eyes in unusual places. Then, solve the mystery.

Biology.com

Finding the solution to the Mouse-eyed Fly mystery is only the beginning. Take a video field trip with the ecogeeks of Untamed Science to see where the mystery leads.

Go online to access additional resources including:
• eText • Flash Cards • Lesson Overviews • Chapter Mystery

13.1 RNA

Key Questions

How does RNA differ from DNA?

How does the cell make RNA?

Vocabulary

RNA
messenger RNA
ribosomal RNA
transfer RNA
transcription
RNA polymerase
promoter
intron
exon

Taking Notes

Preview Visuals Before you read, look at **Figure 13–3.** Write a prediction of how you think a cell makes RNA based on the figure. Then as you read, take notes on how a cell makes RNA. After you read, compare your notes and your prediction.

THINK ABOUT IT We know that DNA is the genetic material, and we know that the sequence of nucleotide bases in its strands must carry some sort of code. For that code to work, the cell must be able to understand it. What exactly do those bases code for? Where is the cell's decoding system?

The Role of RNA

How does RNA differ from DNA?

When Watson and Crick solved the double-helix structure of DNA, they understood right away how DNA could be copied. All a cell had to do was to separate the two strands and then use base pairing to make a new complementary strand for each. But the structure of DNA by itself did not explain how a gene actually works. That question required a great deal more research. The answer came from the discovery that another nucleic acid—ribonucleic acid, or RNA—was involved in putting the genetic code into action. **RNA,** like DNA, is a nucleic acid that consists of a long chain of nucleotides.

In a general way, genes contain coded DNA instructions that tell cells how to build proteins. The first step in decoding these genetic instructions is to copy part of the base sequence from DNA into RNA. RNA then uses these instructions to direct the production of proteins, which help to determine an organisms's characteristics.

Comparing RNA and DNA Remember that each nucleotide in DNA is made up of a 5-carbon sugar, a phosphate group, and a nitrogenous base. This is true for RNA as well. **But there are three important differences between RNA and DNA: (1) the sugar in RNA is ribose instead of deoxyribose, (2) RNA is generally single-stranded and not double-stranded, and (3) RNA contains uracil in place of thymine.** These chemical differences make it easy for enzymes in the cell to tell DNA and RNA apart.

You can compare the different roles played by DNA and RNA molecules in directing the production of proteins to the two type of plans builders use. A master plan has all the information needed to construct a building. But builders never bring a valuable master plan to the job site, where it might be damaged or lost. Instead, as **Figure 13–1** shows, they work from blueprints, inexpensive, disposable copies of the master plan.

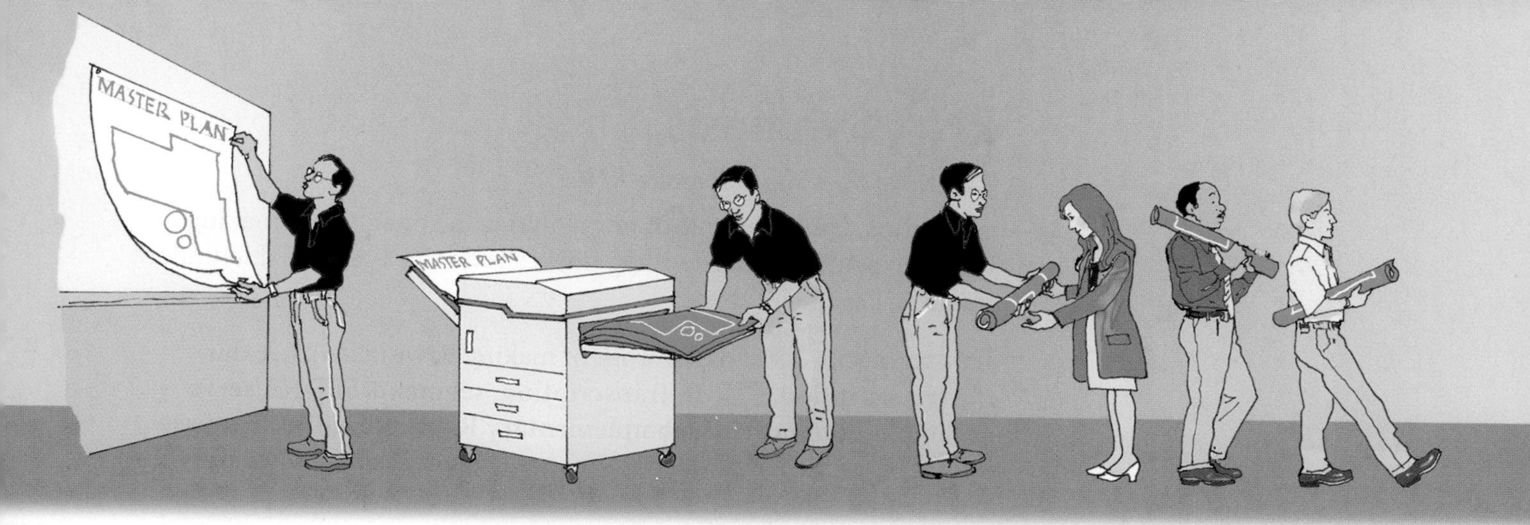

Similarly, the cell uses the vital DNA "master plan" to prepare RNA "blueprints." The DNA molecule stays safely in the cell's nucleus, while RNA molecules go to the protein-building sites in the cytoplasm—the ribosomes.

VISUAL ANALOGY

MASTER PLANS AND BLUEPRINTS

FIGURE 13–1 The different roles of DNA and RNA molecules in directing protein synthesis can be compared to the two types of plans used by builders: master plans and blueprints.

Functions of RNA You can think of an RNA molecule as a disposable copy of a segment of DNA, a working facsimile of a single gene. RNA has many functions, but most RNA molecules are involved in just one job—protein synthesis. RNA controls the assembly of amino acids into proteins. Like workers in a factory, each type of RNA molecule specializes in a different aspect of this job. **Figure 13–2** shows the three main types of RNA: messenger RNA, ribosomal RNA, and transfer RNA.

▶ ***Messenger RNA*** Most genes contain instructions for assembling amino acids into proteins. The RNA molecules that carry copies of these instructions are known as **messenger RNA** (mRNA). They carry information from DNA to other parts of the cell.

▶ ***Ribosomal RNA*** Proteins are assembled on ribosomes, small organelles composed of two subunits. These subunits are made up of several **ribosomal RNA** (rRNA) molecules and as many as 80 different proteins.

▶ ***Transfer RNA*** When a protein is built, a third type of RNA molecule transfers each amino acid to the ribosome as it is specified by the coded messages in mRNA. These molecules are known as **transfer RNA** (tRNA).

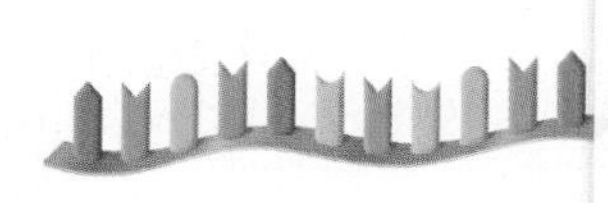

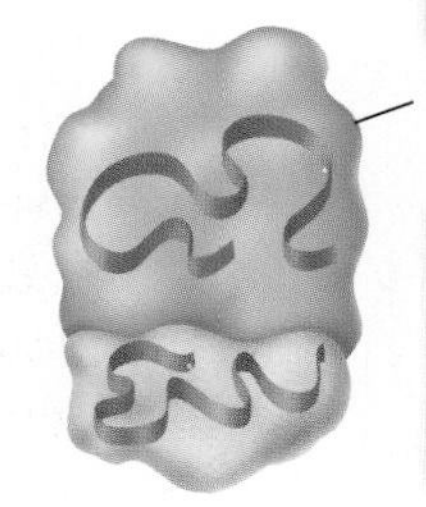

FIGURE 13–2 Types of RNA The three main types of RNA are messenger RNA, ribosomal RNA, and transfer RNA.

RNA Synthesis

How does the cell make RNA?

Cells invest large amounts of raw material and energy into making RNA molecules. Understanding how cells do this is essential to understanding how genes work.

Transcription Most of the work of making RNA takes place during **transcription.** **In transcription, segments of DNA serve as templates to produce complementary RNA molecules.** The base sequences of the transcribed RNA complement the base sequences of the template DNA.

In prokaryotes, RNA synthesis and protein synthesis take place in the cytoplasm. In eukaryotes, RNA is produced in the cell's nucleus and then moves to the cytoplasm to play a role in the production of protein. Our focus here is on transcription in eukaryotic cells.

Transcription requires an enzyme, known as **RNA polymerase,** that is similar to DNA polymerase. RNA polymerase binds to DNA during transcription and separates the DNA strands. It then uses one strand of DNA as a template from which to assemble nucleotides into a complementary strand of RNA, as shown in **Figure 13–3.** The ability to copy a single DNA sequence into RNA makes it possible for a single gene to produce hundreds or even thousands of RNA molecules.

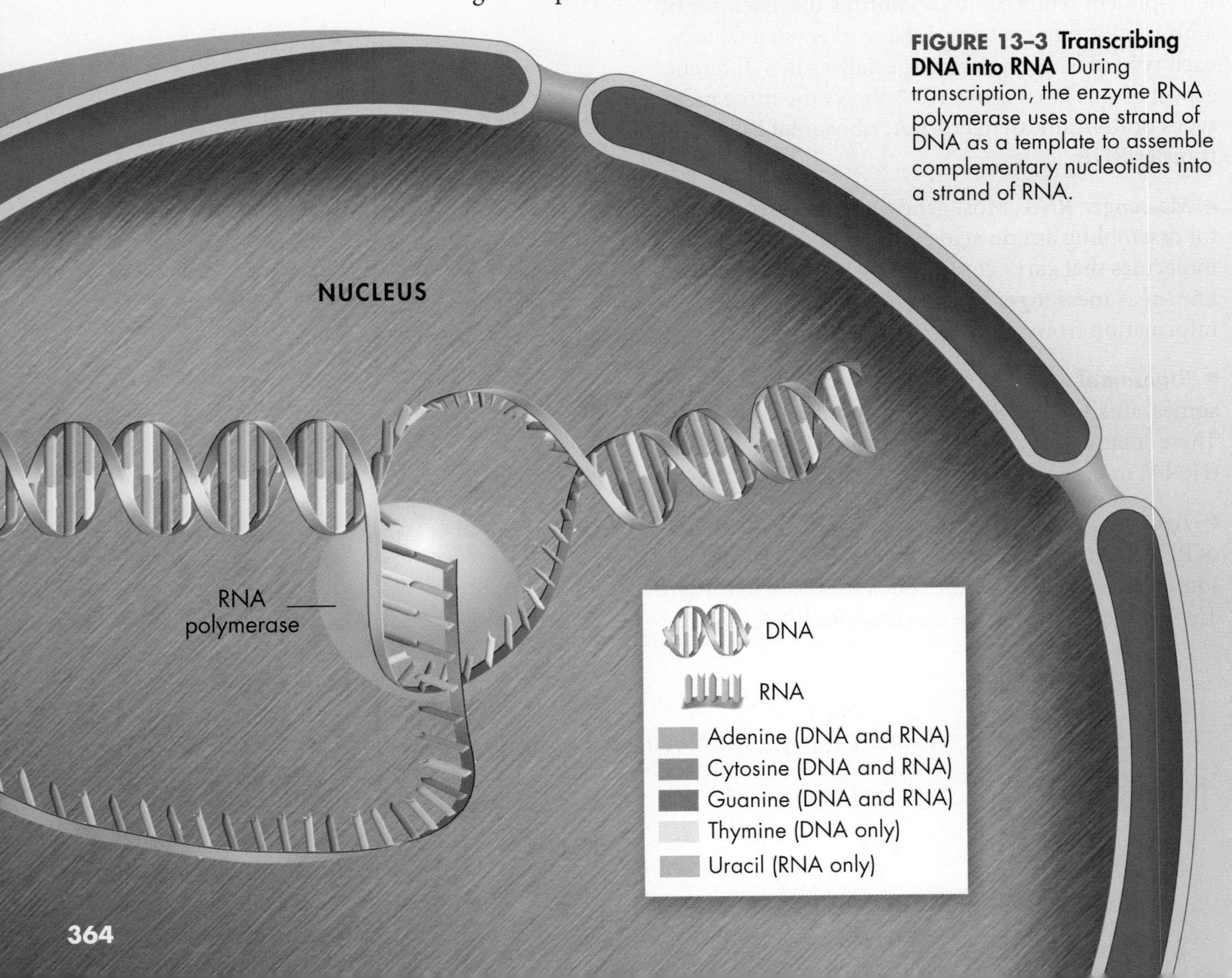

FIGURE 13–3 Transcribing DNA into RNA During transcription, the enzyme RNA polymerase uses one strand of DNA as a template to assemble complementary nucleotides into a strand of RNA.

Promoters How does RNA polymerase know where to start and stop making a strand of RNA? The answer is that RNA polymerase doesn't bind to DNA just anywhere. The enzyme binds only to **promoters,** regions of DNA that have specific base sequences. Promoters are signals in the DNA molecule that show RNA polymerase exactly where to begin making RNA. Similar signals in DNA cause transcription to stop when a new RNA molecule is completed.

RNA Editing Like a writer's first draft, RNA molecules sometimes require a bit of editing before they are ready to be read. These pre-mRNA molecules have bits and pieces cut out of them before they can go into action. The portions that are cut out and discarded are called **introns.** In eukaryotes, introns are taken out of pre-mRNA molecules while they are still in the nucleus. The remaining pieces, known as **exons,** are then spliced back together to form the final mRNA, as shown in **Figure 13–4.**

Why do cells use energy to make a large RNA molecule and then throw parts of that molecule away? That's a good question, and biologists still don't have a complete answer. Some pre-mRNA molecules may be cut and spliced in different ways in different tissues, making it possible for a single gene to produce several different forms of RNA. Introns and exons may also play a role in evolution, making it possible for very small changes in DNA sequences to have dramatic effects on how genes affect cellular function.

FIGURE 13–4 Introns and Exons Before many mRNA molecules can be read, sections called introns are "edited out." The remaining pieces, called exons, are spliced together. Then, an RNA cap and tail are added to form the final mRNA molecule.

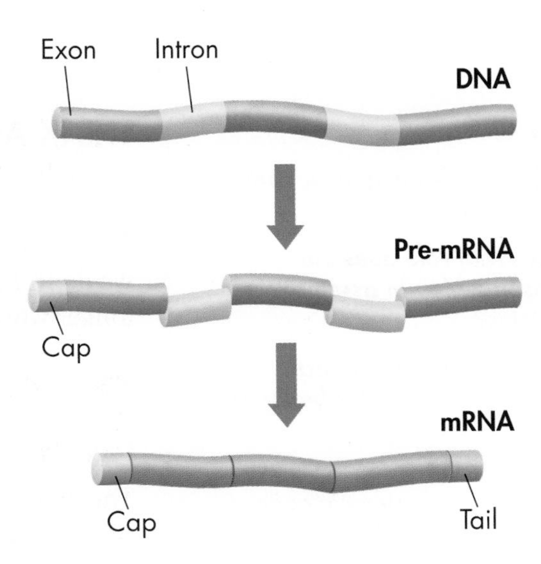

13.1 Assessment

Review Key Concepts

1. a. **Review** Describe three main differences between RNA and DNA.

   b. **Explain** List the three main types of RNA, and explain what they do.

   c. **Infer** Why is it important for a single gene to be able to produce hundreds or thousands of the same RNA molecules?

2. a. **Review** Describe what happens during transcription.

   b. **Predict** What do you think would happen if introns were not removed from pre-mRNA?

WRITE ABOUT SCIENCE

Creative Writing

3. An RNA molecule is looking for a job in a protein synthesis factory. It asks you to write its résumé. This RNA molecule is not yet specialized and could, with some structural changes, function as mRNA, rRNA, or tRNA. Write a résumé for this molecule that reflects the capabilities of each type of RNA.

13.2 Ribosomes and Protein Synthesis

Key Questions

What is the genetic code, and how is it read?

What role does the ribosome play in assembling proteins?

What is the "central dogma" of molecular biology?

Vocabulary

polypeptide • genetic code • codon • translation • anticodon • gene expression

Taking Notes

Outline Before you read, write down the green headings in this lesson. As you read, keep a list of the main points, and then write a summary for each heading.

THINK ABOUT IT How would you build a system to read the messages that are coded in genes and transcribed into RNA? Would you read the bases one at a time, as if the code were a language with just four words—one word per base? Perhaps you would read them, as we do in English, as individual letters that can be combined to spell longer words.

The Genetic Code

What is the genetic code, and how is it read?

The first step in decoding genetic messages is to transcribe a nucleotide base sequence from DNA to RNA. This transcribed information contains a code for making proteins. You learned in Chapter 2 that proteins are made by joining amino acids together into long chains, called **polypeptides.** As many as 20 different amino acids are commonly found in polypeptides.

The specific amino acids in a polypeptide, and the order in which they are joined, determine the properties of different proteins. The sequence of amino acids influences the shape of the protein, which in turn determines its function. How is the order of bases in DNA and RNA molecules translated into a particular order of amino acids in a polypeptide?

As you know from Lesson 13.1, RNA contains four different bases: adenine, cytosine, guanine, and uracil. In effect, these bases form a "language" with just four "letters": A, C, G, and U. We call this language the **genetic code.** How can a code with just four letters carry instructions for 20 different amino acids? **The genetic code is read three "letters" at a time, so that each "word" is three bases long and corresponds to a single amino acid.** Each three-letter "word" in mRNA is known as a **codon.** As shown in **Figure 13–5,** a codon consists of three consecutive bases that specify a single amino acid to be added to the polypeptide chain.

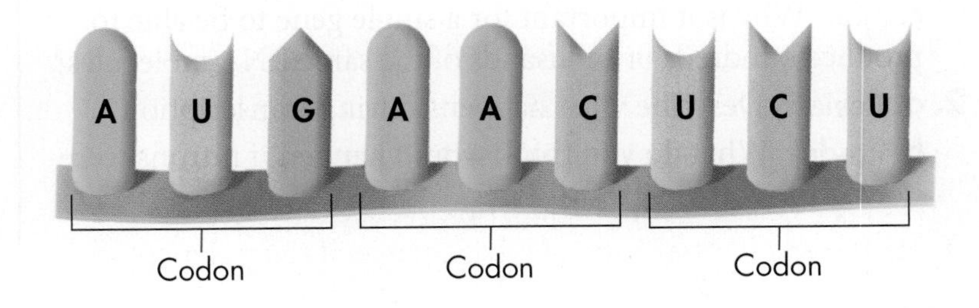

FIGURE 13–5 Codons A codon is a group of three nucleotide bases in messenger RNA that specifies a particular amino acid. **Observe** *What are the three-letter groups of the codons shown here?*

How to Read Codons Because there are four different bases in RNA, there are 64 possible three-base codons ($4 \times 4 \times 4 = 64$) in the genetic code. **Figure 13–6** shows these possible combinations. Most amino acids can be specified by more than one codon. For example, six different codons—UUA, UUG, CUU, CUC, CUA, and CUG—specify leucine. But only one codon—UGG—specifies the amino acid tryptophan.

Decoding codons is a task made simple by use of a genetic code table. Just start at the middle of the circle with the first letter of the codon, and move outward. Next, move out to the second ring to find the second letter of the codon. Find the third and final letter among the smallest set of letters in the third ring. Then read the amino acid in that sector.

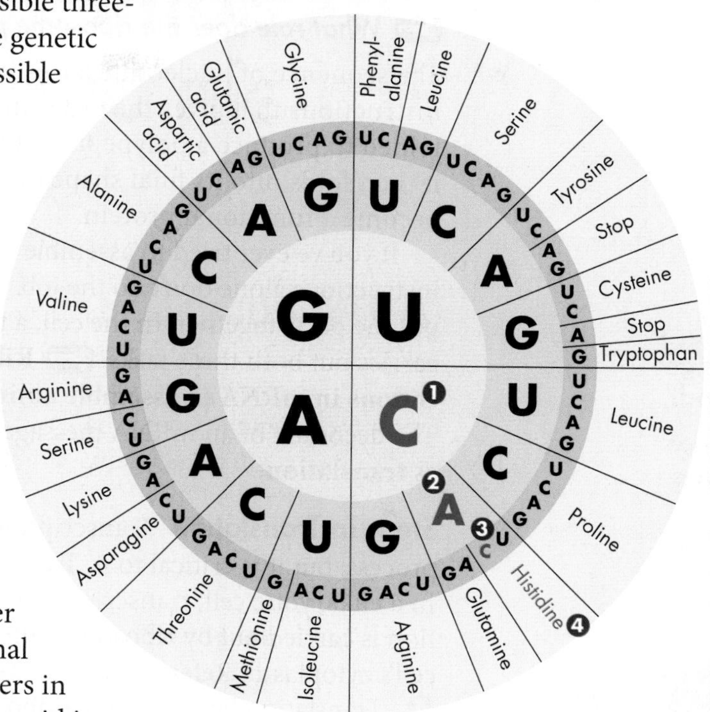

❶ To decode the codon CAC, find the first letter in the set of bases at the center of the circle.

❷ Find the second letter of the codon A, in the "C" quarter of the next ring.

❸ Find the third letter, C, in the next ring, in the "C-A" grouping.

❹ Read the name of the amino acid in that sector—in this case histidine.

FIGURE 13–6 Reading Codons
This circular table shows the amino acid to which each of the 64 codons corresponds. To read a codon, start at the middle of the circle and move outward.

Start and Stop Codons Any message, whether in a written language or the genetic code, needs punctuation marks. In English, punctuation tells us where to pause, when to sound excited, and where to start and stop a sentence. The genetic code has punctuation marks, too. The methionine codon AUG, for example, also serves as the initiation, or "start," codon for protein synthesis. Following the start codon, mRNA is read, three bases at a time, until it reaches one of three different "stop" codons, which end translation. At that point, the polypeptide is complete.

Quick Lab
GUIDED INQUIRY

How Does a Cell Interpret Codons?

❶ A certain gene has the following base sequence:

GACAAGTCCACAATC

Write this sequence on a separate sheet of paper.

❷ From left to right, write the sequence of the mRNA molecule transcribed from this gene.

❸ Using **Figure 13–6,** read the mRNA codons from left to right. Then write the amino acid sequence of the polypeptide.

❹ Repeat step 2, reading the sequence of the mRNA molecule from right to left.

Analyze and Conclude

1. Apply Concepts Why did steps 3 and 4 produce different polypeptides?

2. Infer Do cells usually decode nucleotides in one direction only or in either direction?

Translation

What role does the ribosome play in assembling proteins?

The sequence of nucleotide bases in an mRNA molecule is a set of instructions that gives the order in which amino acids should be joined to produce a polypeptide. Once the polypeptide is complete, it then folds into its final shape or joins with other polypeptides to become a functional protein.

If you've ever tried to assemble a complex toy, you know that instructions alone don't do the job. You need to read them and then put the parts together. In the cell, a tiny factory—the ribosome—carries out both these tasks. **Ribosomes use the sequence of codons in mRNA to assemble amino acids into polypeptide chains.** The decoding of an mRNA message into a protein is a process known as **translation.**

Steps in Translation Transcription isn't part of the translation process, but it is critical to it. Transcribed mRNA directs that process. In a eukaryotic cell, transcription goes on in the cell's nucleus; translation is carried out by ribosomes after the transcribed mRNA enters the cell's cytoplasm. Refer to **Figure 13–7** as you read about translation.

Ⓐ Translation begins when a ribosome attaches to an mRNA molecule in the cytoplasm. As each codon passes through the ribosome, tRNAs bring the proper amino acids into the ribosome. One at a time, the ribosome then attaches these amino acids to the growing chain.

VISUAL SUMMARY

TRANSLATION

FIGURE 13–7 During translation, or protein synthesis, the cell uses information from messenger RNA to produce proteins.

Messenger RNA
Messenger RNA is transcribed in the nucleus and then enters the cytoplasm.

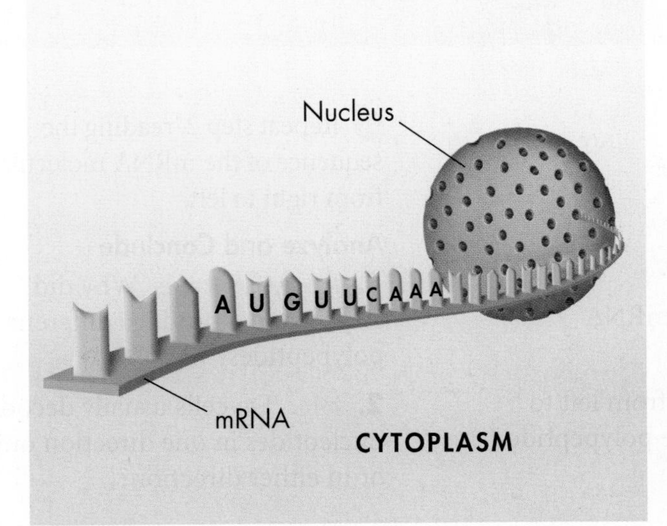

Ⓐ **Transfer RNA**
Translation begins at AUG, the start codon. Each transfer RNA has an anticodon whose bases are complementary to the bases of a codon on the mRNA strand. The ribosome positions the start codon to attract its anticodon, which is part of the tRNA that binds methionine. The ribosome also binds the next codon and its anticodon.

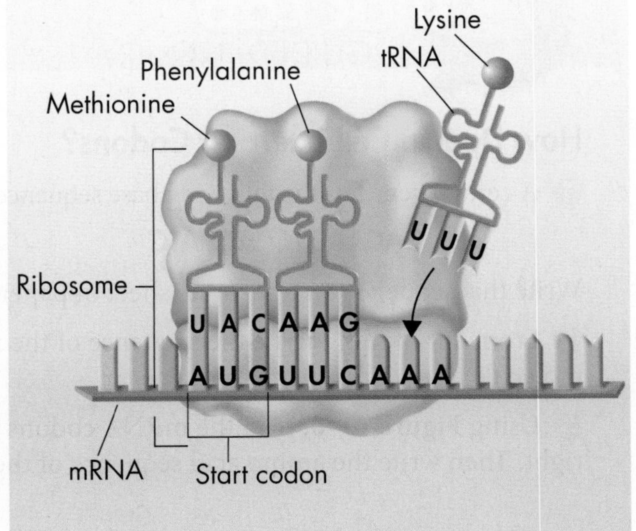

Each tRNA molecule carries just one kind of amino acid. In addition, each tRNA molecule has three unpaired bases, collectively called the **anticodon.** Each tRNA anticodon is complementary to one mRNA codon.

In the case of the tRNA molecule for methionine, the anticodon is UAC, which pairs with the methionine codon, AUG. The ribosome has a second binding site for a tRNA molecule for the next codon. If that next codon is UUC, a tRNA molecule with an AAG anticodon fits against the mRNA molecule held in the ribosome. That second tRNA molecule brings the amino acid phenylalanine into the ribosome.

B Like an assembly-line worker who attaches one part to another, the ribosome helps form a peptide bond between the first and second amino acids—methionine and phenylalanine. At the same time, the bond holding the first tRNA molecule to its amino acid is broken. That tRNA then moves into a third binding site, from which it exits the ribosome. The ribosome then moves to the third codon, where tRNA brings it the amino acid specified by the third codon.

C The polypeptide chain continues to grow until the ribosome reaches a "stop" codon on the mRNA molecule. When the ribosome reaches a stop codon, it releases both the newly formed polypeptide and the mRNA molecule, completing the process of translation.

In Your Notebook *Briefly summarize the three steps in translation.*

FIGURE 13–8 Molecular Model of a Ribosome This model shows ribosomal RNA and associated proteins as colored ribbons. The large subunit is blue, green, and purple. The small subunit is shown in yellow and orange. The three solid elements in the center are tRNA molecules.

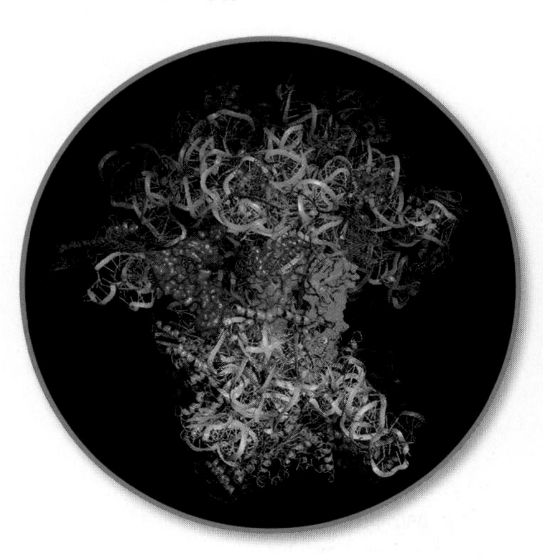

B The Polypeptide "Assembly Line"

The ribosome joins the two amino acids—methionine and phenylalanine—and breaks the bond between methionine and its tRNA. The tRNA floats away from the ribosome, allowing the ribosome to bind another tRNA. The ribosome moves along the mRNA, from right to left, binding new tRNA molecules and amino acids.

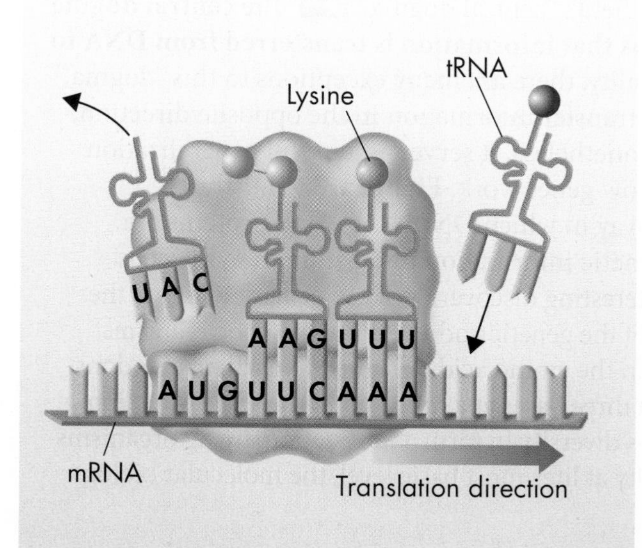

C Completing the Polypeptide

The process continues until the ribosome reaches one of the three stop codons. Once the polypeptide is complete, it and the mRNA are released from the ribosome.

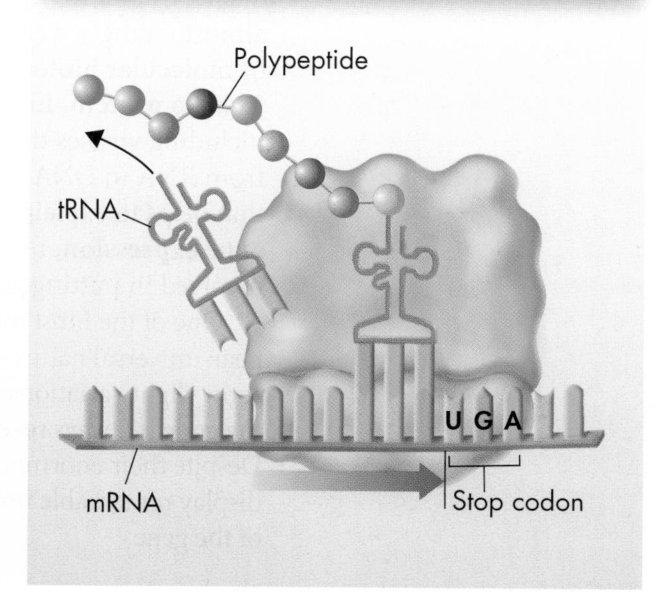

The Roles of tRNA and rRNA in Translation All three major forms of RNA—mRNA, tRNA, and rRNA—come together in the ribosome during translation. The mRNA molecule, of course, carries the coded message that directs the process. The tRNA molecules deliver exactly the right amino acid called for by each codon on the mRNA. The tRNA molecules are, in effect, adaptors that enable the ribosome to "read" the mRNA's message accurately and to get the translation just right.

Ribosomes themselves are composed of roughly 80 proteins and three or four different rRNA molecules. These rRNA molecules help hold ribosomal proteins in place and help locate the beginning of the mRNA message. They may even carry out the chemical reaction that joins amino acids together.

The Molecular Basis of Heredity

What is the "central dogma" of molecular biology?

Gregor Mendel might have been surprised to learn that most genes contain nothing more than instructions for assembling proteins. He might have asked what proteins could possibly have to do with the color of a flower, the shape of a leaf, or the sex of a newborn baby. The answer is that proteins have everything to do with these traits. Remember that many proteins are enzymes, which catalyze and regulate chemical reactions. A gene that codes for an enzyme to produce pigment can control the color of a flower. Another gene produces proteins that regulate patterns of tissue growth in a leaf. Yet another may trigger the female or male pattern of development in an embryo. In short, proteins are microscopic tools, each specifically designed to build or operate a component of a living cell.

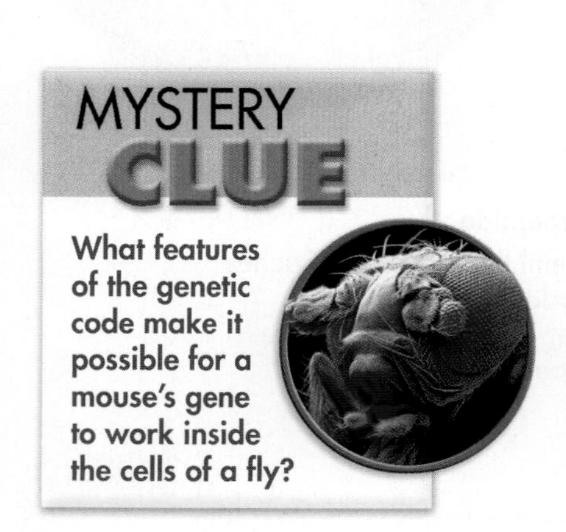

As you've seen, once scientists learned that genes were made of DNA, a series of other discoveries soon followed. Before long, with the genetic code in hand, a new scientific field called molecular biology had been established. Molecular biology seeks to explain living organisms by studying them at the molecular level, using molecules like DNA and RNA. One of the earliest findings came to be known, almost jokingly, as the field's "central dogma." **The central dogma of molecular biology is that information is transferred from DNA to RNA to protein.** In reality, there are many exceptions to this "dogma," including viruses that transfer information in the opposite direction, from RNA to DNA. Nonetheless, it serves as a useful generalization that helps to explain how genes work. **Figure 13–9** illustrates **gene expression,** the way in which DNA, RNA, and proteins are involved in putting genetic information into action in living cells.

One of the most interesting discoveries of molecular biology is the near-universal nature of the genetic code. Although some organisms show slight variations in the amino acids assigned to particular codons, the code is always read three bases at a time and in the same direction. Despite their enormous diversity in form and function, living organisms display remarkable unity at life's most basic level, the molecular biology of the gene.

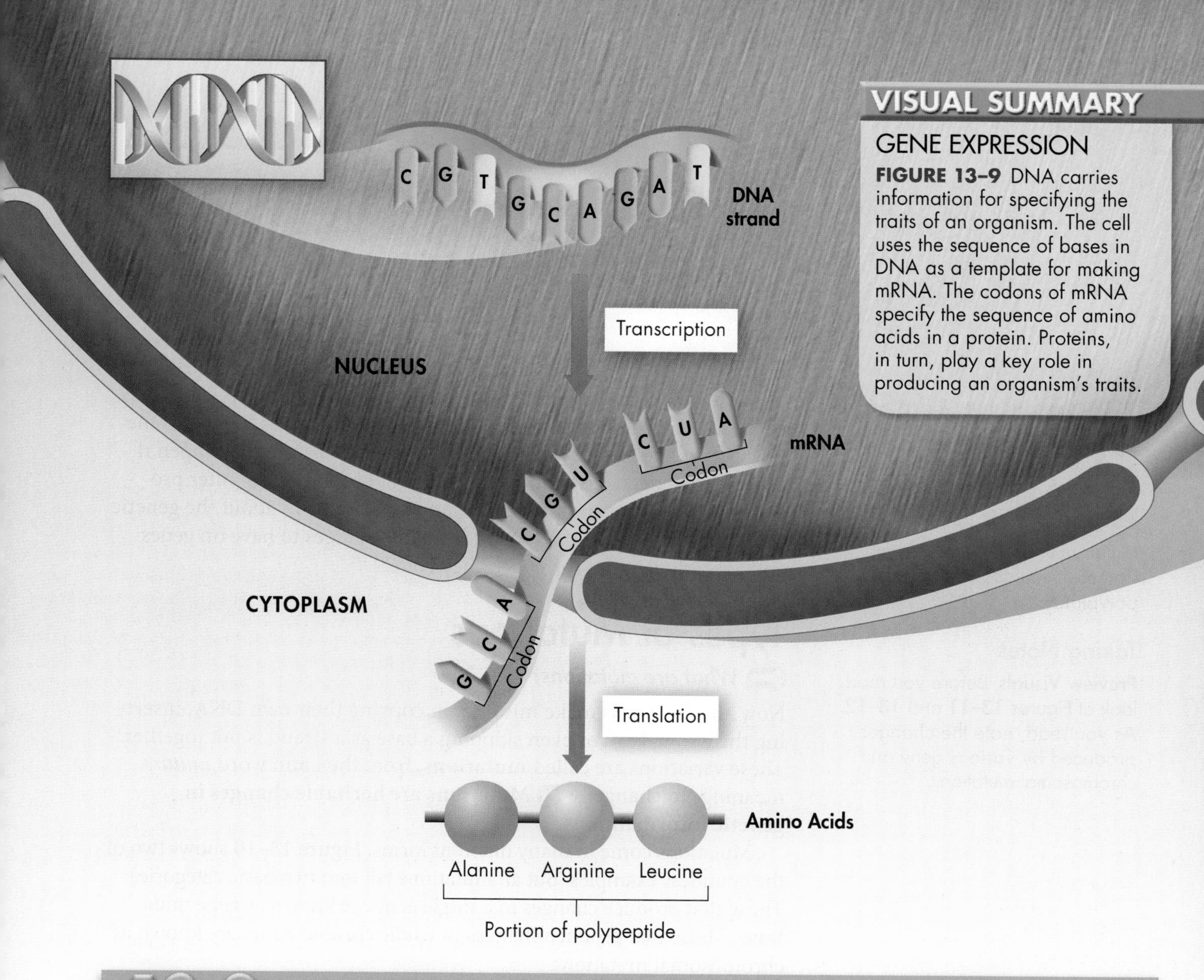

VISUAL SUMMARY

GENE EXPRESSION

FIGURE 13–9 DNA carries information for specifying the traits of an organism. The cell uses the sequence of bases in DNA as a template for making mRNA. The codons of mRNA specify the sequence of amino acids in a protein. Proteins, in turn, play a key role in producing an organism's traits.

13.2 Assessment

Review Key Concepts

1. a. Review How does a cell interpret the genetic code?

b. Explain What are codons and anticodons?

c. Apply Concepts Using the table in **Figure 13–6,** identify the amino acids specified by codons: UGG, AAG, and UGC.

2. a. Review What happens during translation?

b. Compare and Contrast How is protein synthesis different from DNA replication? (*Hint*: Revisit Lesson 12.3.)

3. a. Review Why is the genetic code considered universal?

b. Explain What does the term *gene expression* mean?

c. Infer In what way does controlling the proteins in an organism control the organism's characteristics?

Apply the Big idea

Information and Heredity

4. Choose one component of translation to consider in depth. For instance, you might choose to consider one form of RNA or one step in the process. Then write a question or a series of questions about that component. Select one question, and use it to form a hypothesis that could be tested in an experiment.

13.3 Mutations

Key Questions

🔑 ***What are mutations?***

🔑 ***How do mutations affect genes?***

Vocabulary

mutation • point mutation • frameshift mutation • mutagen • polyploidy

Taking Notes

Preview Visuals Before you read, look at **Figures 13–11** and **13–12.** As you read, note the changes produced by various gene and chromosomal mutations.

THINK ABOUT IT The sequence of bases in DNA are like the letters of a coded message, as we've just seen. But what would happen if a few of those letters changed accidentally, altering the message? Could the cell still understand its meaning? Think about what might happen if someone changed at random a few lines of code in a computer program that you rely on. Knowing what you already do about the genetic code, what effects would you predict such changes to have on genes and the polypeptides for which they code?

Types of Mutations

🔑 ***What are mutations?***

Now and then cells make mistakes in copying their own DNA, inserting the wrong base or even skipping a base as a strand is put together. These variations are called **mutations,** from the Latin word *mutare,* meaning "to change." 🔑 **Mutations are heritable changes in genetic information.**

Mutations come in many different forms. **Figure 13–10** shows two of the countless examples. But all mutations fall into two basic categories: Those that produce changes in a single gene are known as gene mutations. Those that produce changes in whole chromosomes are known as chromosomal mutations.

FIGURE 13–10 Plant and Animal Mutations

The elongated shape of this flower is caused by a mutation that affects the growing regions of the flower tissue.

A genetic condition called leucism leaves this lion without pigments in its hair, skin, and eyes.

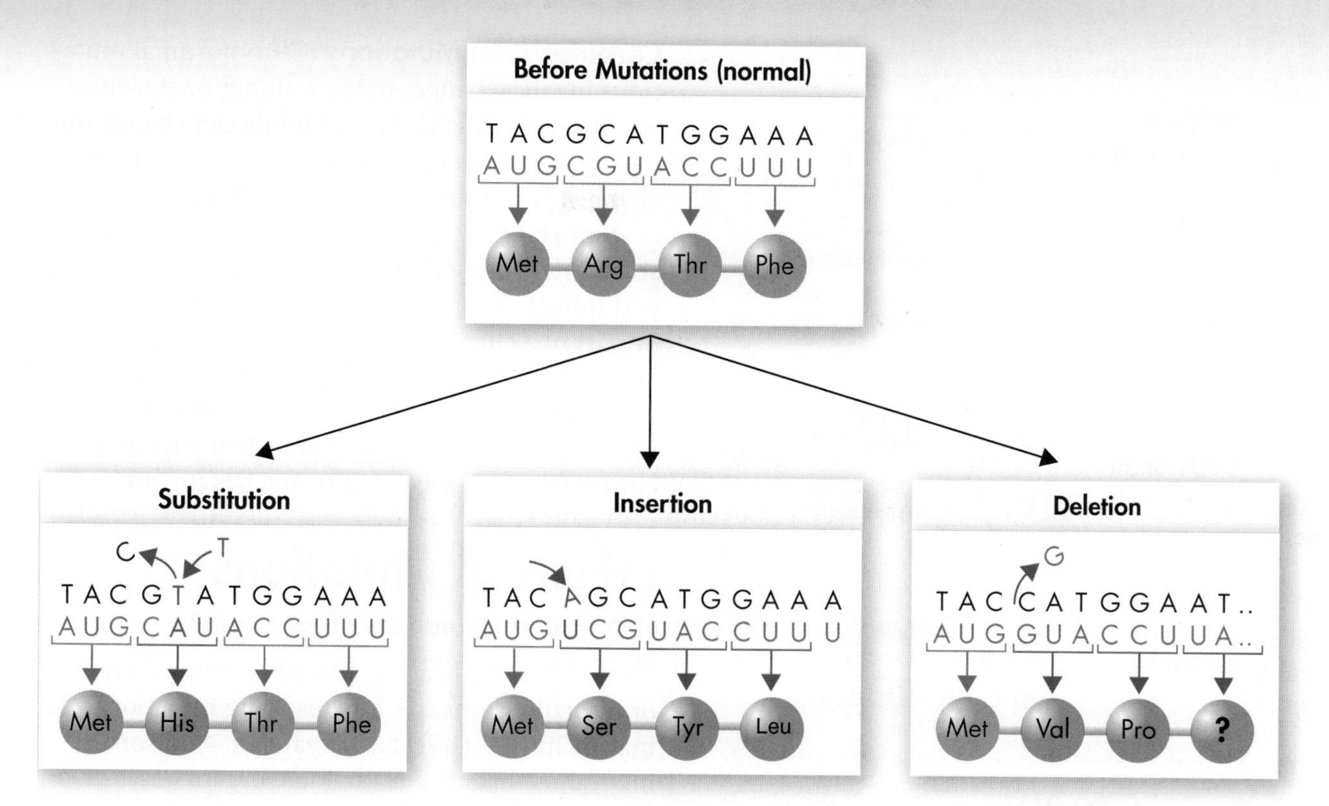

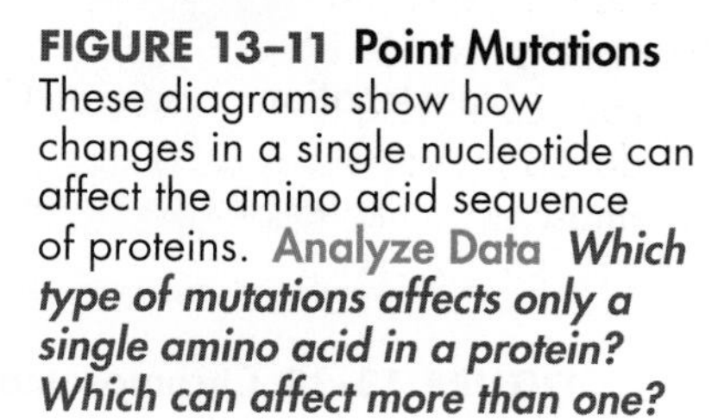

FIGURE 13–11 Point Mutations These diagrams show how changes in a single nucleotide can affect the amino acid sequence of proteins. **Analyze Data** ***Which type of mutations affects only a single amino acid in a protein? Which can affect more than one?***

Gene Mutations Gene mutations that involve changes in one or a few nucleotides are known as **point mutations** because they occur at a single point in the DNA sequence. Point mutations include substitutions, insertions, and deletions. They generally occur during replication. If a gene in one cell is altered, the alteration can be passed on to every cell that develops from the original one. Refer to **Figure 13–11** as you read about the different forms of point mutations.

▶ ***Substitutions*** In a substitution, one base is changed to a different base. Substitutions usually affect no more than a single amino acid, and sometimes they have no effect at all. For example, if a mutation changed one codon of mRNA from CCC to CCA, the codon would still specify the amino acid proline. But a change in the first base of the codon—changing CCC to ACC—would replace proline with the amino acid threonine.

▶ ***Insertions and Deletions*** Insertions and deletions are point mutations in which one base is inserted or removed from the DNA sequence. The effects of these changes can be dramatic. Remember that the genetic code is read three bases at a time. If a nucleotide is added or deleted, the bases are still read in groups of three, but now those groupings shift in every codon that follows the mutation.

Insertions and deletions are also called **frameshift mutations** because they shift the "reading frame" of the genetic message. By shifting the reading frame, frameshift mutations can change every amino acid that follows the point of the mutation. They can alter a protein so much that it is unable to perform its normal functions.

In Your Notebook *Use a cause/effect diagram to describe the different types of gene mutations.*

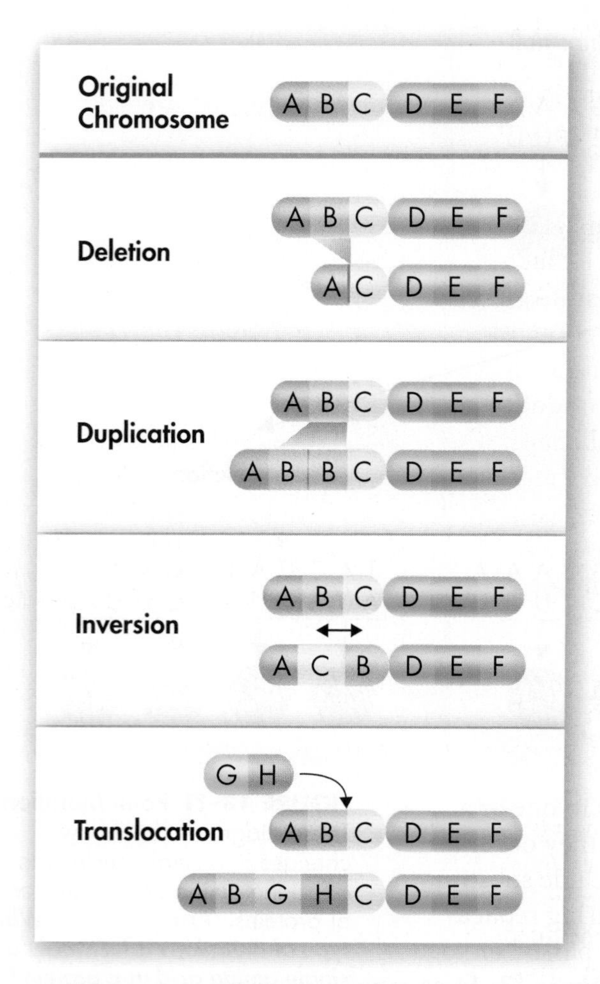

FIGURE 13–12 Chromosomal Mutations Four types of mutations cause changes in whole chromosomes. **Use Diagrams** ***What is the difference between inversion and translocation?***

Chromosomal Mutations Chromosomal mutations involve changes in the number or structure of chromosomes. These mutations can change the location of genes on chromosomes and can even change the number of copies of some genes.

Figure 13–12 shows four types of chromosomal mutations: deletion, duplication, inversion, and translocation. Deletion involves the loss of all or part of a chromosome; duplication produces an extra copy of all or part of a chromosome; and inversion reverses the direction of parts of a chromosome. Translocation occurs when part of one chromosome breaks off and attaches to another.

Effects of Mutations

How do mutations affect genes?

Genetic material can be altered by natural events or by artificial means. The resulting mutations may or may not affect an organism. And some mutations that affect individual organisms can also affect a species or even an entire ecosystem.

Many mutations are produced by errors in genetic processes. For example, some point mutations are caused by errors during DNA replication. The cellular machinery that replicates DNA inserts an incorrect base roughly once in every 10 million bases. But small changes in genes can gradually accumulate over time.

Quick Lab
GUIDED INQUIRY

Modeling Mutations

Small mutations in DNA can cause huge changes in the proteins that are synthesized. Similarly, small changes in a word can dramatically alter its meaning. Look at the following sequence of words:

milk mile wile wise wisp wasp

Notice that each word differs from the previous word by just one letter and that none of the words is meaningless. Think of these changes as "point mutations" that affect word meaning.

Analyze and Conclude

1. Apply Concepts Start with the word *gene*, and change it letter by letter to make new words. Make sure each new word is an actual word but not a proper noun. Write at least four "point mutations" of the word *gene*.

2. Apply Concepts Show how you could use words to model a frameshift mutation. (*Hint*: You can use a sentence.)

3. Use Models Use the words in this sentence to model a substitution mutation.

Stressful environmental conditions may cause some bacteria to increase mutation rates. This can actually be helpful to the organism, since mutations may sometimes give such bacteria new traits, such as the ability to consume a new food source or to resist a poison in the environment.

Mutagens Some mutations arise from **mutagens,** chemical or physical agents in the environment. Chemical mutagens include certain pesticides, a few natural plant alkaloids, tobacco smoke, and environmental pollutants. Physical mutagens include some forms of electromagnetic radiation, such as X-rays and ultraviolet light. If these agents interact with DNA, they can produce mutations at high rates. Cells can sometimes repair the damage; but when they cannot, the DNA base sequence changes permanently. Some compounds interfere with base-pairing, increasing the error rate of DNA replication. Others weaken the DNA strand, causing breaks and inversions that produce chromosomal mutations.

BUILD Vocabulary

WORD ORIGINS The word **mutagen** is a Latin word that means "origin of change." Mutagens change an organism's genetic information.

In Your Notebook *Make a table to keep track of both the helpful and harmful results of mutations. As you read, fill it in.*

Harmful and Helpful Mutations As you've already seen, some mutations don't even change the amino acid specified by a codon, while others may alter a complete protein or even an entire chromosome. **The effects of mutations on genes vary widely. Some have little or no effect; and some produce beneficial variations. Some negatively disrupt gene function.** Many if not most mutations are neutral; they have little or no effect on the expression of genes or the function of the proteins for which they code. Whether a mutation is negative or beneficial depends on how its DNA changes relative to the organism's situation. Mutations are often thought of as negative, since they can disrupt the normal function of genes. However, without mutations, organisms could not evolve, because mutations are the source of genetic variability in a species.

▶***Harmful Effects*** Some of the most harmful mutations are those that dramatically change protein structure or gene activity. The defective proteins produced by these mutations can disrupt normal biological activities, and result in genetic disorders. Some cancers, for example, are the product of mutations that cause the uncontrolled growth of cells. Sickle cell disease is a disorder associated with changes in the shape of red blood cells. You can see its effects in **Figure 13–13.** It is caused by a point mutation in one of the polypeptides found in hemoglobin, the blood's principal oxygen-carrying protein. Among the symptoms of the disease are anemia, severe pain, frequent infections, and stunted growth.

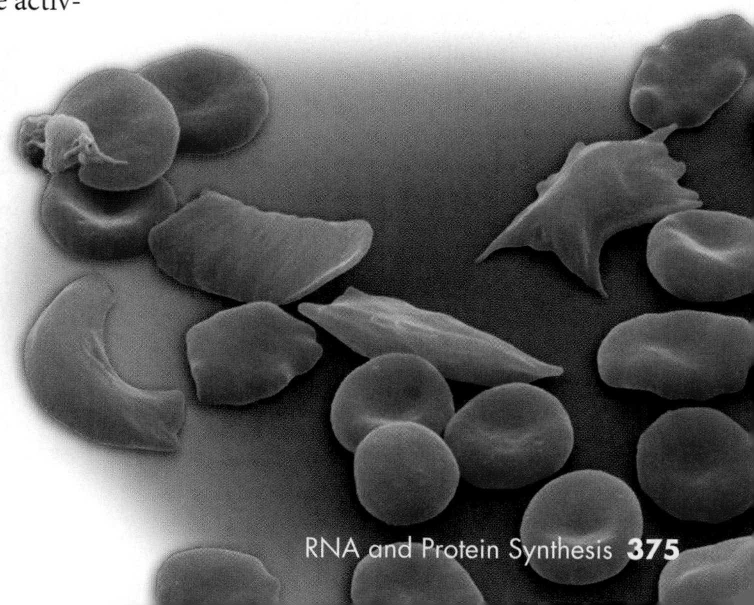

FIGURE 13–13 Effects of a Point Mutation Sickle cell disease affects the shape of red blood cells. The round cells in this false-colored SEM are normal red blood cells. The crescent and star-shaped cells are sickled cells. (SEM 1700×)

FIGURE 13–14 Polyploid Plants The fruit of the Tahiti lime is seedless, a result of polyploidy. Changes to the ploidy number of citrus plants can affect the size and strength of the trees as well as the quality and seediness of their fruit.

▶ ***Beneficial Effects*** Some of the variation produced by mutations can be highly advantageous to an organism or species. **Mutations often produce proteins with new or altered functions that can be useful to organisms in different or changing environments.** For example, mutations have helped many insects resist chemical pesticides. And some have enabled microorganisms to adapt to new chemicals in the environment.

Over the past 20 years, mutations in the mosquito genome have made many African mosquitoes resistant to the chemical pesticides once used to control them. This may be bad news for humans, but it is highly beneficial to the insects themselves. Beneficial mutations occur in humans, too, including ones that increase bone strength and density, making fractures less likely, and mutations that increase resistance to HIV, the virus that causes AIDS.

Plant and animal breeders often make use of "good" mutations. For example, when a complete set of chromosomes fails to separate during meiosis, the gametes that result may produce triploid (3N) or tetraploid (4N) organisms. The condition in which an organism has extra sets of chromosomes is called **polyploidy.** Polyploid plants are often larger and stronger than diploid plants. Important crop plants—including bananas and the limes shown in **Figure 13–14**—have been produced this way. Polyploidy also occurs naturally in citrus plants, often through spontaneous mutations.

In Your Notebook *List five examples of mutations. Classify each as neutral, harmful, or helpful, and explain your reasoning.*

13.3 Assessment

Review Key Concepts

1. a. Review Describe the two main types of mutations.

b. Explain What is a frameshift mutation? Give an example.

c. Infer The effects of a mutation are not always visible. Choose a species, and explain how a biologist might determine whether a mutation has occurred and, if so, what type of mutation it is.

2. a. Review List three effects mutations can have on genes.

b. Apply Concepts What is the significance of mutations to living things?

VISUAL THINKING

3. Make a compare/contrast table to organize your ideas about gene mutations and chromosomal mutations. Then use your table to write a paragraph comparing and contrasting these two kinds of mutations.

13.4 Gene Regulation and Expression

THINK ABOUT IT Think of a library filled with how-to books. Would you ever need to use all of those books at the same time? Of course not. If you wanted to know how to fix a leaky faucet, you'd open a book about plumbing but would ignore the one on carpentry. Now picture a tiny bacterium like *E. coli*, which contains more than 4000 genes. Most of its genes code for proteins that do everything from building cell walls to breaking down food. Do you think *E. coli* uses all 4000-plus volumes in its genetic library at the same time?

Key Questions

How are prokaryotic genes regulated?

How are genes regulated in eukaryotic cells?

What controls the development of cells and tissues in multicellular organisms?

Vocabulary

operon
operator
RNA interference
differentiation
homeotic gene
homeobox gene
Hox gene

Taking Notes

Outline Before you read, use the headings in this lesson to make an outline. As you read, fill in the subtopics and smaller topics. Then add phrases or a sentence after each subtopic that provides key information.

Prokaryotic Gene Regulation

How are prokaryotic genes regulated?

As it turns out, bacteria and other prokaryotes do not need to transcribe all of their genes at the same time. To conserve energy and resources, prokaryotes regulate their activities, using only those genes necessary for the cell to function. For example, it would be wasteful for a bacterium to produce enzymes that are needed to make a molecule that is readily available from its environment. By regulating gene expression, bacteria can respond to changes in their environment—the presence or absence of nutrients, for example. How? **DNA-binding proteins in prokaryotes regulate genes by controlling transcription.** Some of these regulatory proteins help switch genes on, while others turn genes off.

How does an organism know when to turn a gene on or off? One of the keys to gene transcription in bacteria is the organization of genes into operons. An **operon** is a group of genes that are regulated together. The genes in an operon usually have related functions. *E. coli*, shown in **Figure 13–15,** provides us with a clear example. The 4288 genes that code for proteins in *E. coli* include a cluster of 3 genes that must be turned on together before the bacterium can use the sugar lactose as a food. These three lactose genes in *E. coli* are called the *lac* operon.

FIGURE 13–15
Small Cell, Many Genes
This *E. coli* bacterium has been treated with an enzyme enabling its DNA, which contains more than 4000 genes, to spill out.

TEM 27,000×

The *Lac* Operon Why must *E. coli* be able to switch the *lac* genes on and off? Lactose is a compound made up of two simple sugars, galactose and glucose. To use lactose for food, the bacterium must transport lactose across its cell membrane and then break the bond between glucose and galactose. These tasks are performed by proteins coded for by the genes of the *lac* operon. This means, of course, that if the bacterium grows in a medium where lactose is the only food source, it must transcribe these genes and produce these proteins. If grown on another food source, such as glucose, it would have no need for these proteins.

Remarkably, the bacterium almost seems to "know" when the products of these genes are needed. When lactose is not present, the *lac* genes are turned off by proteins that bind to DNA and block transcription.

Promoters and Operators On one side of the operon's three genes are two regulatory regions. The first is a promoter (P), which is a site where RNA-polymerase can bind to begin transcription. The other region is called the **operator** (O). The O site is where a DNA-binding protein known as the *lac* repressor can bind to DNA.

▶***The* Lac *Repressor Blocks Transcription*** As **Figure 13–16** shows, when the *lac* repressor binds to the O region, RNA polymerase cannot reach the *lac* genes to begin transcription. In effect, the binding of the repressor protein switches the operon "off" by preventing the transcription of its genes.

▶***Lactose Turns the Operon "On"*** If the repressor protein is always present, how can the *lac* genes ever be switched on? Besides its DNA binding site, the *lac* repressor protein has a binding site for lactose itself. When lactose is added to the medium, it diffuses into the cell and attaches to the *lac* repressor. This changes the shape of the repressor protein in a way that causes it to fall off the operator. Now, with the repressor no longer bound to the O site, RNA polymerase can bind to the promoter and transcribe the genes of the operon. As a result, in the presence of lactose, the operon is automatically switched on.

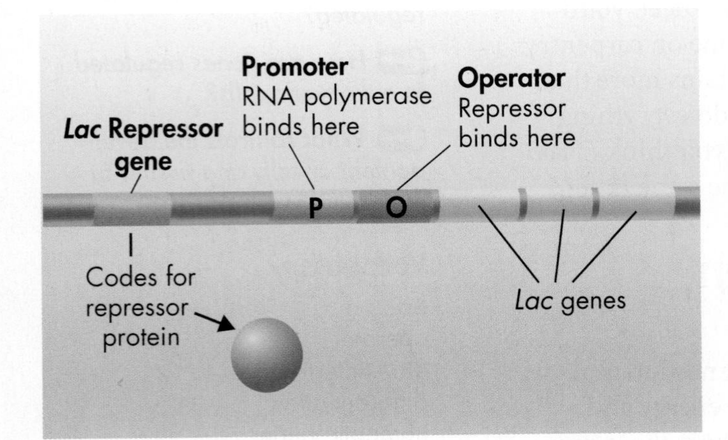

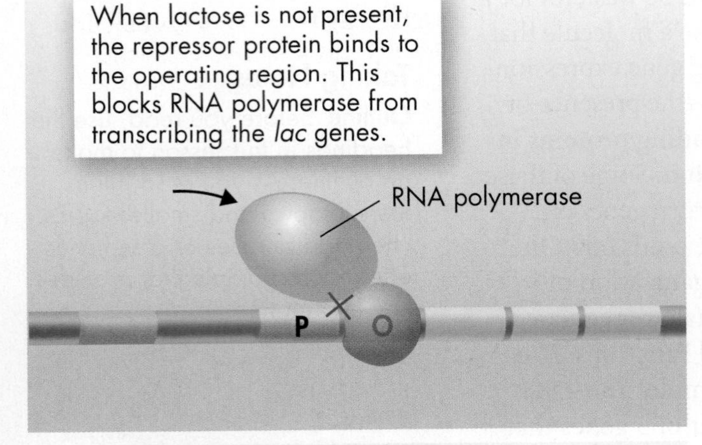

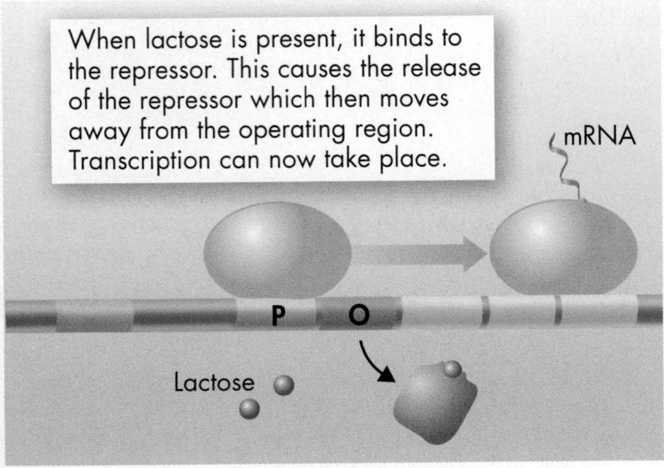

FIGURE 13–16 Gene Expression in Prokaryotes The *lac* genes in *E. coli* are turned off by *lac* repressors and turned on in the presence of lactose. **Use Analogies** ***How is the way lactose turns genes on and off similar to the way cold air signals a furnace to turn on or off?***

Eukaryotic Gene Regulation

How are genes regulated in eukaryotic cells?

The general principles of gene regulation in prokaryotes also apply to eukaryotes, although there are differences. Most eukaryotic genes are controlled individually and have more complex regulatory sequences than those of the *lac* repressor system.

Figure 13–17 shows several features of a typical eukaryotic gene. One of the most interesting is the TATA box, a short region of DNA, about 25 or 30 base pairs before the start of a gene, containing the sequence TATATA or TATAAA. The TATA box binds a protein that helps position RNA polymerase by marking a point just before the beginning of a gene.

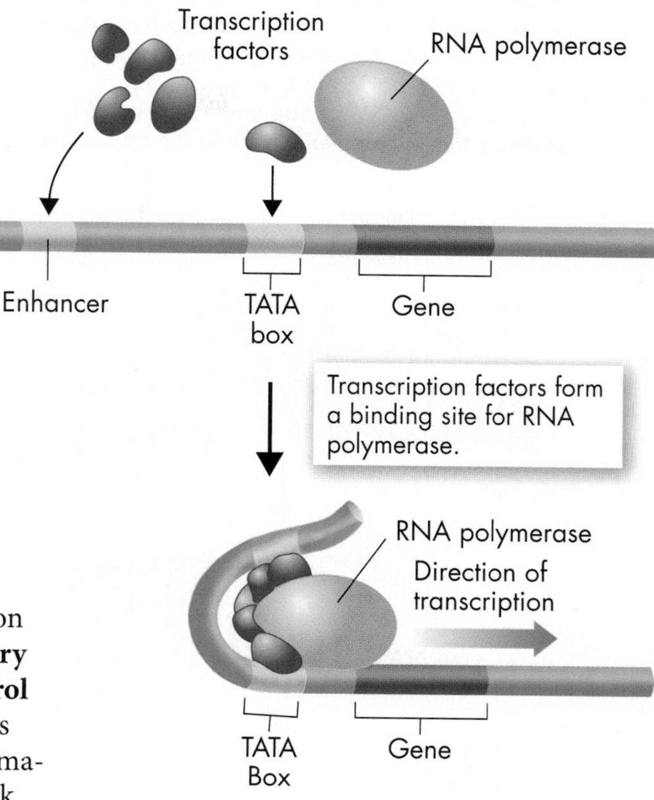

FIGURE 13–17 The TATA Box and Transcription Many eukaryotic genes include a region called the TATA box that helps position RNA polymerase.

Transcription Factors Gene expression in eukaryotic cells can be regulated at a number of levels. One of the most critical is the level of transcription, by means of DNA-binding proteins known as transcription factors. **By binding DNA sequences in the regulatory regions of eukaryotic genes, transcription factors control the expression of those genes.** Some transcription factors enhance transcription by opening up tightly packed chromatin. Others help attract RNA polymerase. Still others block access to certain genes, much like prokaryotic repressor proteins. In most cases, multiple transcription factors must bind before RNA polymerase is able to attach to the promoter region and start transcription.

Promoters have multiple binding sites for transcription factors, each of which can influence transcription. Certain factors activate scores of genes at once, dramatically changing patterns of gene expression in the cell. Other factors form only in response to chemical signals. Steroid hormones, for example, are chemical messengers that enter cells and bind to receptor proteins. These "receptor complexes" then act as transcription factors that bind to DNA, allowing a single chemical signal to activate multiple genes. Eukaryotic gene expression can also be regulated by many other factors, including the exit of mRNA molecules from the nucleus, the stability of mRNA, and even the breakdown of a gene's protein products.

In Your Notebook *Compare gene regulation in single-cell organisms and multicellular organisms.*

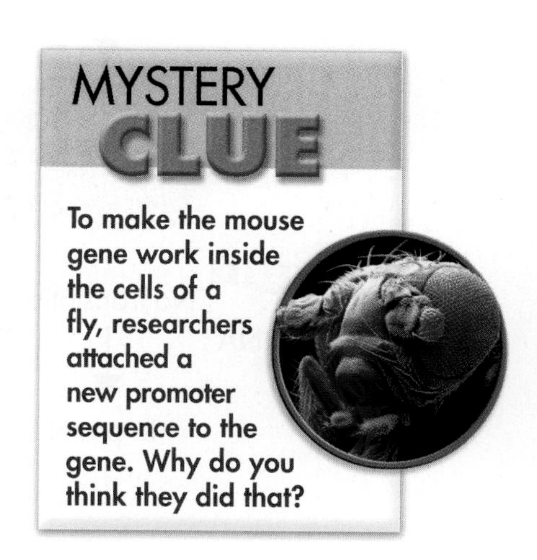

To make the mouse gene work inside the cells of a fly, researchers attached a new promoter sequence to the gene. Why do you think they did that?

FIGURE 13–18 Blocking Gene Expression Like tiny pieces of sticky tape, microRNAs attach to certain mRNA molecules and stop them from passing on their protein-making instructions.
Interpret Visuals ***What happens to the mRNA sequence that is complementary to the bound miRNA?***

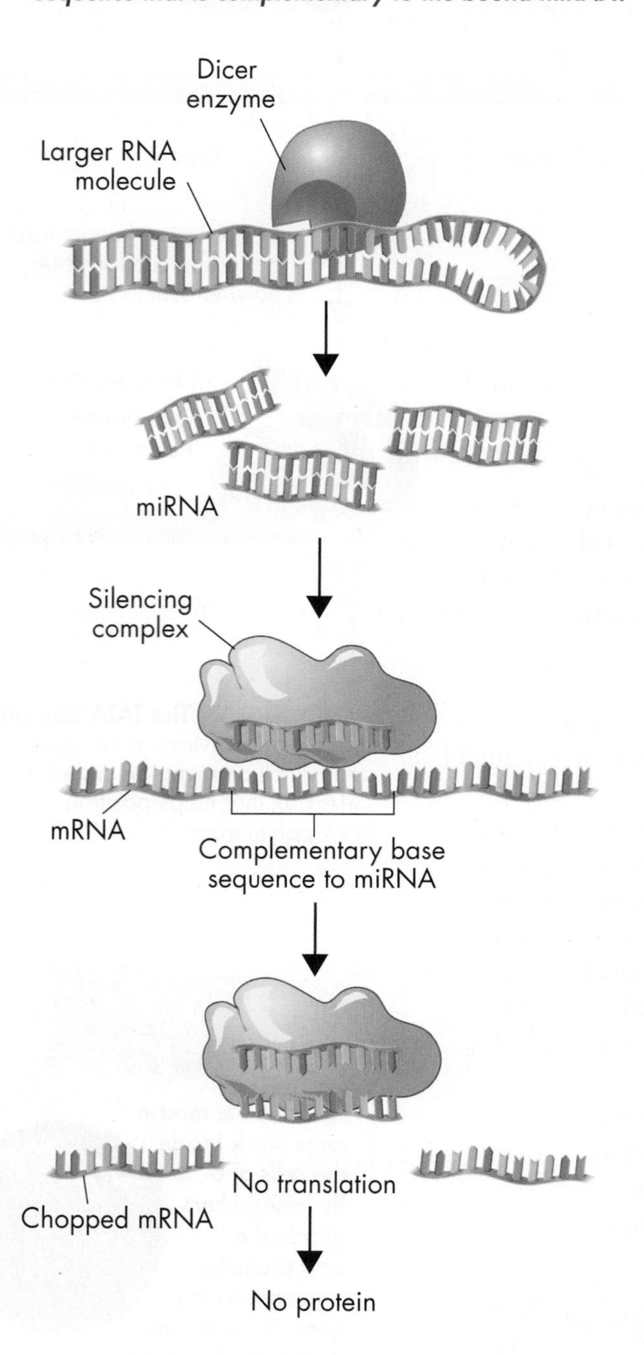

Cell Specialization Why is gene regulation in eukaryotes more complex than in prokaryotes? Think for a moment about the way in which genes are expressed in a multicellular organism. The genes that code for liver enzymes, for example, are not expressed in nerve cells. Keratin, an important protein in skin cells, is not produced in blood cells. Cell specialization requires genetic specialization, yet all of the cells in a multicellular organism carry the same genetic code in their nucleus. Complex gene regulation in eukaryotes is what makes specialization possible.

RNA Interference For years biologists wondered why cells contain lots of small RNA molecules, only a few dozen bases long, that don't belong to any of the major groups of RNA (mRNA, tRNA, or rRNA). In the last decade, a series of important discoveries has shown that these small RNA molecules play a powerful role in regulating gene expression. And they do so by interfering with mRNA.

As **Figure 13–18** shows, after they are produced by transcription, the small interfering RNA molecules fold into double-stranded hairpin loops. An enzyme called the "Dicer" enzyme cuts, or dices, these double-stranded loops into microRNA (miRNA), each about 20 base pairs in length. The two strands of the loops then separate. Next, one of the miRNA pieces attaches to a cluster of proteins to form what is known as a silencing complex. The silencing complex binds to and destroys any mRNA containing a sequence that is complementary to the miRNA. In effect, miRNA sticks to certain mRNA molecules and stops them from passing on their protein-making instructions.

The silencing complex effectively shuts down the expression of the gene whose mRNA it destroys. Blocking gene expression by means of an miRNA silencing complex is known as **RNA interference.** At first, RNA interference (RNAi) seemed to be a rare event, found only in a few plants and other species. It's now clear that RNA interference is found throughout the living world and that it even plays a role in human growth and development.

The Discovery of RNA Interference

In 1998, Andrew Fire and Craig Mello carried out an experiment that helped explain the mechanism of RNA interference. They used RNA from a large gene called unc-22, which codes for a protein found in muscle cells. They prepared short mRNA fragments corresponding to two exon regions of the gene and injected them into egg cells of the worm *C. elegans*. Some of their results are shown in the table.

Injections of mRNA into *C. elegans* Eggs

Portion of Gene Used to Produce mRNA	Strand Injected	Result in Adult Worm
Unc-22 (exon 21–22)	Sense	Normal
	Antisense	Normal
	Sense + Antisense	Twitching
Unc-22 (exon 27)	Sense	Normal
	Antisense	Normal
	Sense + Antisense	Twitching

1. Draw Conclusions How did the adult worms' responses differ to injections of single-stranded mRNA (the "sense" strand), its complementary strand ("antisense"), and double-stranded RNA ("sense + antisense")?

2. Form a Hypothesis Twitching results from the failure of muscle cells to control their contractions. What does this suggest about the unc-22 protein in some of the worms? How would you test your hypothesis?

3. Infer The injected fragments came from two different places in the gene and were only a few hundred bases long. The unc-22 mRNA is thousands of bases long. What does this suggest about the mechanism of RNA interference?

The Promise of RNAi Technology The discovery of RNAi has made it possible for researchers to switch genes on and off at will, simply by inserting double-stranded RNA into cells. The Dicer enzyme then cuts this RNA into miRNA, which activates silencing complexes. These complexes block the expression of genes producing mRNA complementary to the miRNA. Naturally this technology is a powerful way to study gene expression in the laboratory. However, RNAi technology also holds the promise of allowing medical scientists to turn off the expression of genes from viruses and cancer cells, and it may provide new ways to treat and perhaps even cure diseases.

Genetic Control of Development

What controls the development of cells and tissues in multicellular organisms?

Regulating gene expression is especially important in shaping the way a multicellular organism, like the mouse embryo in **Figure 13–19,** develops. Each of the specialized cell types found in the adult originates from the same fertilized egg cell. Cells don't just grow and divide during embryonic development. As the embryo develops, different sets of genes are regulated by transcription factors and repressors. Gene regulation helps cells undergo **differentiation,** becoming specialized in structure and function. The study of genes that control development and differentiation is one of the most exciting areas in biology today.

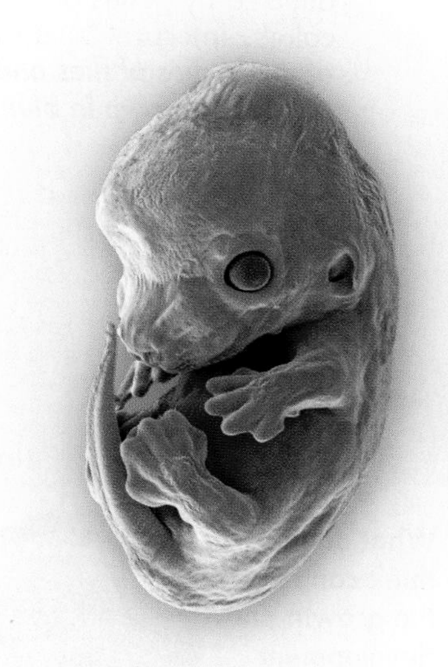

FIGURE 13–19 Differentiation This scanning electron micrograph shows a mouse embryo undergoing cell differentiation 23 days after conception.

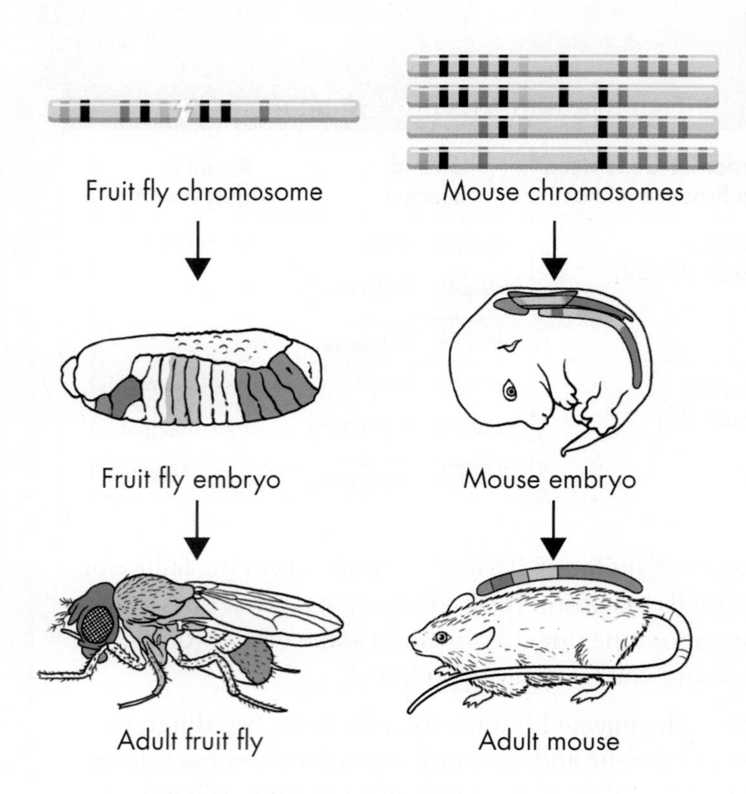

FIGURE 13–20 Hox Genes and Body Development In fruit flies, a series of Hox genes along a chromosome determines the basic body structure. Mice have similar genes on four different chromosomes. The colored areas on the fly and mouse show the approximate body areas affected by genes of the corresponding colors. **Interpret Visuals** ***What section of the bodies of flies and mice is coded by the genes shown in blue?***

What do you think controls the growth and development of eyes in flies and mice?

Homeotic Genes The American biologist Edward B. Lewis was the first to show that a specific group of genes controls the identities of body parts in the embryo of the common fruit fly. Lewis found that a mutation in one of these genes actually resulted in a fly with a leg growing out of its head in place of an antenna! From Lewis's work it became clear that a set of master control genes, known as **homeotic genes,** regulates organs that develop in specific parts of the body.

Homeobox and Hox Genes Molecular studies of homeotic genes show that they share a very similar 180-base DNA sequence, which was given the name homeobox. **Homeobox genes** code for transcription factors that activate other genes that are important in cell development and differentiation. Homeobox genes are expressed in certain regions of the body, and they determine factors like the presence of wings or legs.

In flies, a group of homeobox genes known as **Hox genes** are located side by side in a single cluster, as shown in **Figure 13–20.** Hox genes determine the identities of each segment of a fly's body. They are arranged in the exact order in which they are expressed, from anterior to posterior. A mutation in one of these genes can completely change the organs that develop in specific parts of the body.

Remarkably, clusters of Hox genes exist in the DNA of other animals, including humans. These genes are arranged in the same way—from head to tail. The function of Hox genes in humans seems to be almost the same as it is in fruit flies: They tell the cells of the body how to differentiate as the body grows. What this means, of course, is that nearly all animals, from flies to mammals, share the same basic tools for building the different parts of the body.

The striking similarity of master control genes—genes that control development—has a simple scientific explanation. Common patterns of genetic control exist because all these genes have descended from the genes of common ancestors. **Master control genes are like switches that trigger particular patterns of development and differentiation in cells and tissues.** The details can vary from one organism to another, but the switches are nearly identical. Recent studies have shown that the very same Hox gene that triggers the development of hands and feet is also active in the fins of certain fish.

Environmental Influences You've seen how cell differentiation is controlled at least in part by the regulation of gene expression. Conditions in an organism's environment play a role too. In prokaryotes and eukaryotes, environmental factors like temperature, salinity, and nutrient availability can influence gene expression. One example: The *lac* operon in *E. coli* is switched on only when lactose is the only food source in the bacteria's environment.

Metamorphosis is another well-studied example of how organisms can modify gene expression in response to change in their environment. Metamorphosis involves a series of transformations from one life stage to another. It is typically regulated by a number of external (environmental) and internal (hormonal) factors. As organisms move from larval to adult stages, their body cells differentiate to form new organs. At the same time, old organs are lost through cell death.

Consider the metamorphosis of a tadpole into a bullfrog, as shown in **Figure 13–21.** Under less than ideal conditions—a drying pond, a high density of predators, low amounts of food—tadpoles may speed up their metamorphosis. In other words, the speed of metamorphosis is determined by various environmental changes that are translated into hormonal changes, with the hormones functioning at the molecular level. Other environmental influences include temperature and population size.

FIGURE 13–21 Metamorphosis Environmental factors can affect gene regulation. If the bullfrog's environment changes for the worse, its genes will direct the production of hormones to speed the transformation of the tadpole (top photo) to the adult bullfrog (bottom photo).

13.4 Assessment

Review Key Concepts

1. a. **Review** How is the *lac* operon regulated?
   b. **Explain** What is a promoter?
   c. **Use Analogies** Write an analogy that demonstrates how the *lac* repressor functions.
2. a. **Review** Describe how most eukaryotic genes are controlled.
   b. **Compare and Contrast** How is gene regulation in prokaryotes and eukaryotes similar? How is it different?
3. a. **Review** What genes control cell differentiation during development?
   b. **Compare and Contrast** How is the way Hox genes are expressed in mice similar to the way they are expressed in fruit flies? How is it different?

PRACTICE PROBLEM

4. A hormone is a chemical that is produced in one part of the body, travels through the blood, and affects cells in other parts of the body. Many hormones are proteins. How might the production of a hormone affect the expression of genes in a eukaryotic cell? Write a hypothesis that could be tested to answer this question. (*Hint*: Include promoters in your hypothesis.)

NGSS Smart Guide RNA and Protein Synthesis

Disciplinary Core Idea LS3.B Variation of Traits: Why do individuals of the same species vary in how they look, function, and behave? Mutations introduce variation into DNA. These changes are passed along when RNA translates DNA's genetic code into functional proteins, which direct the expression of genes.

13.1 RNA

- The main differences between RNA and DNA are that (1) the sugar in RNA is ribose instead of deoxyribose; (2) RNA is generally single-stranded, not double-stranded; and (3) RNA contains uracil in place of thymine.
- In transcription, segments of DNA serve as templates to produce complementary RNA molecules.

RNA 362 • messenger RNA 363 • ribosomal RNA 363 • transfer RNA 363 • transcription 364 • RNA polymerase 364 • promoter 365 • intron 365 • exon 365

Biology.com

Visual Analogy Compare DNA and RNA to the master plans and blueprints of a builder.

Art in Motion Watch how RNA is processed to make mRNA.

VISUAL ANALOGY Master Plans and Blueprints

Dissect the Analogy

1. Classify The copier and the blueprints it produces are analogous to

a. DNA and RNA.

b. DNA and proteins.

c. RNA polymerase and RNA.

d. RNA polymerase and DNA.

ART IN MOTION RNA Processing

mRNA Processing
Before many mRNA molecules can be read, sections called introns must be "edited out." Before editing occurs, a cap and a tail are added. Then, the introns are removed. The remaining pieces, called exons, are spliced together.

13.2 Ribosomes and Protein Synthesis

- The genetic code is read three "letters" at a time, so that each "word" is three bases long and corresponds to a single amino acid.
- Ribosomes use the sequence of codons in mRNA to assemble amino acids into polypeptide chains.
- The central dogma of molecular biology is that information is transferred from DNA to RNA to protein.

polypeptide 366 • genetic code 366 • codon 366 • translation 368 • anticodon 369 • gene expression 370

Biology.com

Interactive Art Build your understanding of transcription and translation with these animations.

Tutor Tube Tune into the tutor to find out why proteins are so important!

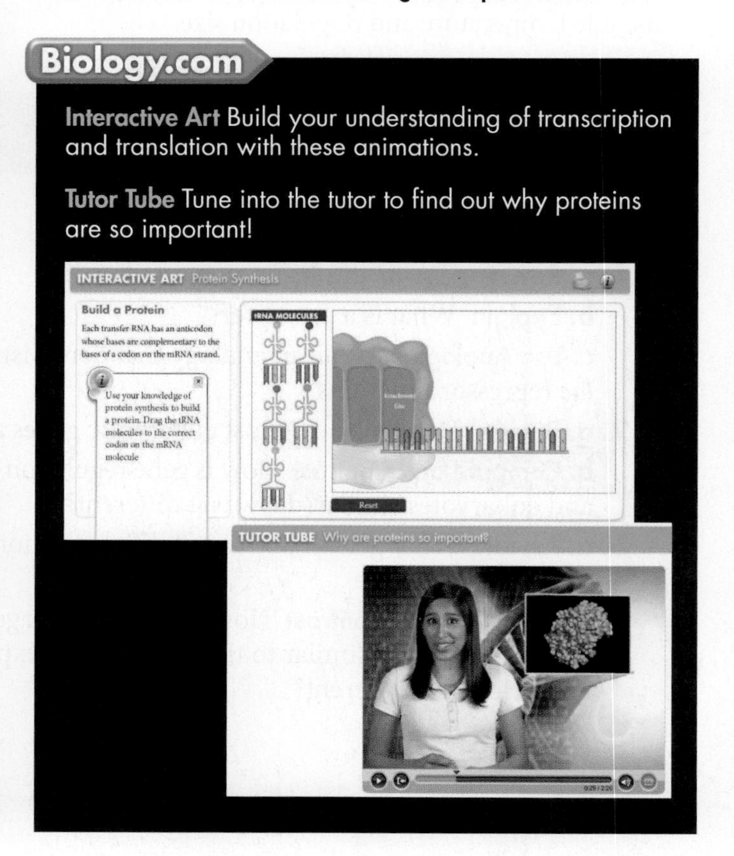

CHECKING YOUR *Scientific Literacy*

Refer to the lesson content and digital assets below as you prepare for your chapter assessment. Then, evaluate your understanding of RNA and protein synthesis by answering these questions.

1. **Scientific and Engineering Practice Communicating Information** Make a flowchart to summarize the transcription process. Remember that a summary should be accurate and brief. Then, make a second flowchart to summarize the translation process.

2. **Crosscutting Concept Cause and Effect** Sometimes the RNA produced during transcription has a sequence of nucleotides that does not match the sequence in the DNA template. Identify the step in RNA processing where this change is most likely to occur. What effect could this change have on protein synthesis?

3. **Key Ideas and Details** Cite specific evidence from Chapter 13 to support the concept that mutations can be helpful.

4. **STEM** Traditionally, companies have built and stored large quantities of their products so they could respond quickly to a purchase order. The trick with this approach is being able to accurately predict sales. An alternative approach is to wait until a customer places an order and then build the product to match exactly what the customer wants. In what ways are the processes used in cells to regulate gene expression similar to a "build to order" manufacturing model?

13.3 Mutations

- Mutations are heritable changes in genetic information.
- The effects of mutations on genes vary widely. Some have little or no effect; some produce beneficial variations. Some negatively disrupt gene function.
- Mutations often produce proteins with new or altered functions that can be useful to organisms in different or changing environments.

mutation 372 • point mutation 373 • frameshift mutation 373 • mutagen 375 • polyploidy 376

Biology.com

Untamed Science Video Watch the Untamed Science explorers as they search for examples of how mutations have benefited a species.

Art Review Review your understanding of different types of mutations with this drag-and-drop activity.

13.4 Gene Regulation and Expression

- DNA-binding proteins in prokaryotes regulate genes by controlling transcription.
- By binding DNA sequences in the regulatory regions of eukaryotic genes, transcription factors control the expression of those genes.
- Master control genes are like switches that trigger particular patterns of development and differentiation in cells and tissues.

operon 377 • operator 378 • RNA interference 380 • differentiation 381 • homeotic gene 382 • homeobox gene 382 • Hox gene 382

Biology.com

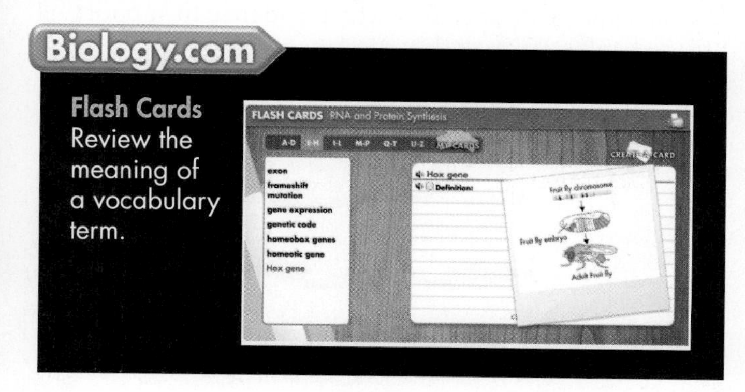

Flash Cards Review the meaning of a vocabulary term.

Skills Lab

From DNA to Protein Synthesis This chapter lab is available in your lab manual and at Biology.com

13 Assessment

13.1 RNA

Understand Key Concepts

1. The process by which the genetic code of DNA is copied into a strand of RNA is called

a. translation.
b. transcription.
c. transformation.
d. replication.

2. Which of the following describes RNA?

a. RNA is usually double-stranded and contains the base thymine.
b. RNA is usually single-stranded and contains the base uracil.
c. RNA is longer than DNA and uses five bases to encode information.
d. RNA is made in the nucleus of eukaryotic cells and stays there to carry out its functions.

3. Describe the function of each of the three types of RNA.

4. How does the enzyme that makes RNA know where to start transcribing the DNA?

5. Compare introns and exons.

Think Critically

6. **Apply Concepts** Suppose you start with the DNA strand ACCGTCAC. Use the rules of base pairing to list the bases on a messenger RNA strand transcribed from this DNA strand.

7. **Key Ideas and Details** Look at the first intron in the diagram below. What would happen to the protein produced by the mRNA molecule if the intron were not removed but functioned instead as an exon? Cite specific textual evidence to support your answer.

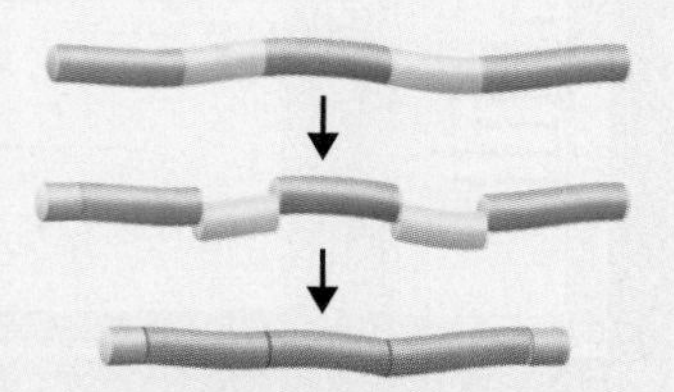

13.2 Ribosomes and Protein Synthesis

Understand Key Concepts

8. In messenger RNA, each codon specifies a particular

a. nucleotide.
b. enzyme.
c. amino acid.
d. promoter.

9. The number of codons in the genetic code is

a. 3. **b.** 4. **c.** 20. **d.** 64.

10. Which of the following statements about the genetic code is true?

a. A codon can specify more than one amino acid.
b. Every codon specifies a different amino acid.
c. Some codons specify the same amino acid.
d. Some codons have no function at all.

11. The process of making proteins on the ribosome based on instructions from messenger RNA is called

a. transcription.
b. transformation.
c. translation.
d. molecular biology.

12. What is a codon?

13. How do anticodons function?

14. If a code on a DNA molecule for a specific amino acid is CTA, what would the messenger RNA codon be? The transfer RNA codon?

15. Explain why controlling the proteins in an organism controls the organism's characteristics.

Think Critically

16. **Craft and Structure** The word *transcribe* means "to write out." The word *translate* means "to express in another language." Review the meanings of *transcription* and *translation* in genetics. How do the technical meanings of these words relate to the everyday meanings of the words?

17. **Text Types and Purposes** A researcher identifies the nucleotide sequence AAC in a long strand of RNA inside a nucleus. In the genetic code, AAC codes for the amino acid asparagine. When that RNA becomes involved in protein synthesis, will asparagine necessarily appear in the protein? Your concluding statement should follow from the facts you use to construct your explanation.

13.3 Mutations

Understand Key Concepts

18. Changes in DNA sequences that affect genetic information are known as
a. replications.
b. mutations.
c. transformations.
d. translations.

19. A single-base mutation in a region of DNA coding for mRNA could transcribe the DNA sequence CAGTAT into
a. GTCATA.
b. GUCAUA.
c. GTCUTU.
d. GUAAUA.

20. A substance that can cause a change in the DNA code of an organism is called a
a. toxin.
b. mutagen.
c. nitrogenous base.
d. nucleotide.

21. Name and give examples of two major types of mutations. What do they have in common? How are they different?

22. How does a deletion mutation differ from a substitution mutation?

23. Can mutations have a positive effect?

Think Critically

24. **Craft and Structure** How does the possible impact of a chromosomal mutation that occurs during meiosis differ from that of a similar event that occurs during mitosis of a body cell that is not involved in reproduction?

25. **Apply Concepts** A mutation in the DNA of an organism changes one base sequence in a protein-coding region from CAC to CAT. What is the effect of the mutation on the final protein? Explain your answer.

13.4 Gene Regulation and Expression

Understand Key Concepts

26. An expressed gene
a. functions as a promoter.
b. is transcribed into RNA.
c. codes for just one amino acid.
d. is made of mRNA.

solve the CHAPTER MYSTERY

MOUSE-EYED FLY

Years ago geneticists discovered a fly gene they called eyeless. Mutations that inactivate this gene cause flies to develop without eyes. Geneticists later discovered a mouse gene, called *Pax6,* that was homologous to eyeless. Transplanting an activated Pax6 gene into a fruit fly can cause the fly to grow eyes in odd places. This happens despite the fact that mouse eyes and fly eyes are very different. In fact the only reason we describe them as "eyes" is because they make vision possible.

How can the Pax6 gene perform the same role in such diverse animals? It probably began very early in the history of life, when eyes were just patches of light-sensitive cells on the skin of the common ancestors of all animals. As those organisms evolved and diversified, master control genes like Pax6 kept working, but with altered functions. Many genes like Pax6 are shared, not only by insects, but by all animals, including worms, sea urchins, and humans.

1. **Craft and Structure** How are fly eyes and mouse eyes different? Similar?

2. **Infer** The Pax6 and eyeless genes code for transcription factors, not for parts of the actual eye. Why does this make sense in light of the effect of Pax6 when it is inserted into a fly?

3. **Connect to the Big idea** What feature of the genetic code makes it possible for a mouse gene to work inside the cell of a fly?

27. A group of genes that are regulated together is called a(n)
   a. promoter.   c. intron.
   b. operon.   d. allele.

28. To turn on the lactose-digesting enzymes of *E. coli*, the lactose must first
   a. bind to the repressor.
   b. bind to the DNA of the bacterium.
   c. separate from the repressor.
   d. initiate the synthesis of messenger RNA.

29. Blocking gene expression in eukaryotes with microRNA strands is called RNA
   a. transcription.   c. interference.
   b. translation.   d. digestion.

30. How is gene expression controlled in prokaryotes?

31. What is meant by the term *cell specialization*? How is cell specialization controlled?

32. Describe how a TATA box helps position RNA polymerase in a eukaryotic cell.

33. What is a homeobox gene?

Think Critically

34. **Apply Concepts** The number of promoter sequences, enhancer sites, and the TATA box in eukaryotes makes gene regulation in these organisms far more complex than regulation in prokaryotes. Why is regulation in eukaryotes so much more sophisticated?

Connecting Concepts

Use Science Graphics

Use the data table to answer questions 35 and 36.

Codon Translation

Amino Acid	mRNA Codons
Alanine (Ala)	GCA, GCG, GCU, GCC
Valine (Val)	GUA, GUG, GUU, GUC
Leucine (Leu)	CUA, CUG, CUU, CUC, UUA, UUG

35. **Relate Cause and Effect** The table shows RNA codons for three amino acids. How would a substitution mutation in the third nucleotide position of the codons for alanine and valine affect the resulting protein?

36. **Infer** The three amino acids shown in the table have very similar—though not identical—properties. What substitution mutations could result in switching one of these amino acids for another? What might be the result?

Write About Science

37. **Explanation** Write a paragraph explaining why the effect of mutations can vary widely—from neutral to harmful to beneficial.

38. **Assess the**  Explain the roles of the three types of RNA in taking the information in DNA and using it to make proteins.

Integration of Knowledge and Ideas

RNA is the genetic material of many viruses. Scientists analyzed RNA from four different types of viruses. The content of the four nitrogenous bases is shown below.

Base Percentages in Four Viruses

Virus	A	U	C	G
A	26.3	29.3	20.6	23.8
B	x	x	17.6	17.5
C	21.9	12.8	34.3	31.1
D	29.8	26.3	18.5	25.3

39. **Interpret Graphics** Which of the four types of viruses is most likely to use double-stranded RNA as its genetic material?
   a. Virus A   c. Virus C
   b. Virus B   d. Virus D

40. **Infer** The values in the two boxes labeled with an *x* would most likely be about
   a. 32.5 % A and 32.5% U.
   b. 17.5% A and 17.5% U.
   c. 26.3% A and 29.3% U.
   d. 32.5% A and 17.5% U.

Standardized Test Prep

Multiple Choice

1. How does RNA differ from DNA?
   A RNA contains uracil and deoxyribose.
   B RNA contains ribose and thymine.
   C RNA contains uracil and ribose.
   D RNA contains adenine and ribose.

2. How would the DNA sequence GCTATA be transcribed to mRNA?
   A GCUAUA
   B CGATAT
   C CGAUAU
   D GCUTUT

Questions 3–4

Use the chart below to answer the questions.

First Base in Code Word	Second Base in Code Word: A	G	U	C	Third Base in Code Word
A	Lys	Arg	Ile	Thr	A
A	Lys	Arg	Met	Thr	G
A	Asn	Ser	Ile	Thr	U
A	Asn	Ser	Ile	Thr	C
G	Glu	Gly	Val	Ala	A
G	Glu	Gly	Val	Ala	G
G	Asp	Gly	Val	Ala	U
G	Asp	Gly	Val	Ala	C
U	"Stop"	"Stop"	Leu	Ser	A
U	"Stop"	Trp	Leu	Ser	G
U	Tyr	Cys	Phe	Ser	U
U	Tyr	Cys	Phe	Ser	C
C	Gln	Arg	Leu	Pro	A
C	Gln	Arg	Leu	Pro	G
C	His	Arg	Leu	Pro	U
C	His	Arg	Leu	Pro	C

3. Which of the following codons signifies the end of translation?
   A CAA
   B UGA
   C AUC
   D CCA

4. Which of the chains of amino acids corresponds to the nucleotide sequence UCAAGCGUA?
   A glu-cys-pro
   B glu-asp-"stop"
   C thr-arg-met
   D ser-ser-val

5. In the *lac* operon, the promoter is
   A exons spliced together after introns are removed.
   B introns spliced together after exons are removed.
   C a region of DNA that has specific base sequences.
   D long pieces of RNA shortened by the Dicer enzyme.

6. Promoters are
   A genes that code for individual proteins.
   B proteins that bind with DNA and prevent transcription.
   C DNA sequences near operons that regulate transcription.
   D small molecules that bind with repressor proteins.

Questions 7–8

Use the diagrams below to answer the questions.

Normal Chromosome M N O P Q R S

Mutant 1 M P O N Q R S

Mutant 2 M N N O P Q R S

7. Mutant 1 is a(n)
   A deletion.
   B translocation.
   C inversion.
   D duplication.

8. Mutant 2 is a(n)
   A deletion.
   B translocation.
   C inversion.
   D duplication.

Open-Ended Response

9. What is the function of the *lac* repressor system in *E. coli*?

If You Have Trouble With . . .

Question	1	2	3	4	5	6	7	8	9
See Lesson	13.1	13.1	13.2	13.2	13.1	13.1	13.3	13.3	13.4

14 Human Heredity

Big idea **Information and Heredity**

Q: How can we use genetics to study human inheritance?

One thing to notice about this group of students is that none of them looks alike. The diversity of traits among the human race stems from one microscopic molecule—DNA.

INSIDE:

- 14.1 Human Chromosomes
- 14.2 Human Genetic Disorders
- 14.3 Studying the Human Genome

BUILDING *Scientific Literacy*

Deepen your understanding of human heredity by engaging in key practices that allow you to make connections across concepts.

NGSS You will construct a model to demonstrate the pattern of inheritance for a recessive human gene.

STEM How is it possible for medical treatments to be tailored to individual patients? Analyze data to explain the growing reliance on personal genomes in medicine.

Common Core You will cite evidence from the text as you evaluate the role computer technology played in the Human Genome Project.

CHAPTER MYSTERY

THE CROOKED CELL

When Ava visited her uncle, Eli, in the hospital, he appeared tired and pale. He complained of sharp pains in his bones. "I've got sickle cell disease," Eli explained, short of breath. "I just hope it doesn't run in your side of the family."

That evening, Ava searched the Internet for information about her uncle's disease. She saw photos of red blood cells shaped like the letter C—a far cry from normal, round blood cells. Ava learned that these sickle-shaped cells are rigid and sticky. In blood vessels, they form clumps that can block blood flow and even cause organ damage. "Am I at risk?" Ava wondered. To find out, she would need to investigate her family history—and her own cells. As you read this chapter, look for clues that would help Ava discover whether she might carry the sickle cell trait. Then, solve the mystery.

Finding out about Ava's risk of sickle cell disease is only the beginning. Take a video field trip with the ecogeeks of Untamed Science to see where the mystery leads.

Go online to access additional resources including:
• eText • Flash Cards • Lesson Overviews • Chapter Mystery

14.1 Human Chromosomes

Key Questions

🔑 ***What is a karyotype?***

🔑 ***What patterns of inheritance do human traits follow?***

🔑 ***How can pedigrees be used to analyze human inheritance?***

Vocabulary

genome • karyotype • sex chromosome • autosome • sex-linked gene • pedigree

Taking Notes

Outline Before you read, make an outline of the major headings in the lesson. As you read, fill in main ideas and supporting details for each heading.

THINK ABOUT IT If you had to pick an ideal organism for the study of genetics, would you choose one that produced lots of offspring? How about one that was easy to grow in the lab? Would you select one with a short life span in order to do several crosses per month? How about all of the above? You certainly would not choose an organism that produced very few offspring, had a long life span, and could not be grown in a lab. Yet, when we study human genetics, this is exactly the sort of organism we deal with. Given all of these difficulties, it may seem a wonder that we know as much about human genetics as we do.

Karyotypes

🔑 ***What is a karyotype?***

What makes us human? We might try to answer that question by looking under the microscope to see what is inside a human cell. Not surprisingly, human cells look much like the cells of other animals. To find what makes us uniquely human, we have to look deeper, into the genetic instructions that build each new individual. To begin this undertaking, we have to explore the human genome. A **genome** is the full set of genetic information that an organism carries in its DNA.

The study of any genome starts with chromosomes—those bundles of DNA and protein found in the nuclei of eukaryotic cells. To see human chromosomes clearly, cell biologists photograph cells in mitosis, when the chromosomes are fully condensed and easy to view. Scientists then cut out the chromosomes from the photographs and arrange them in a picture known as a **karyotype** (KAR ee uh typ). 🔑 **A karyotype shows the complete diploid set of chromosomes grouped together in pairs, arranged in order of decreasing size.**

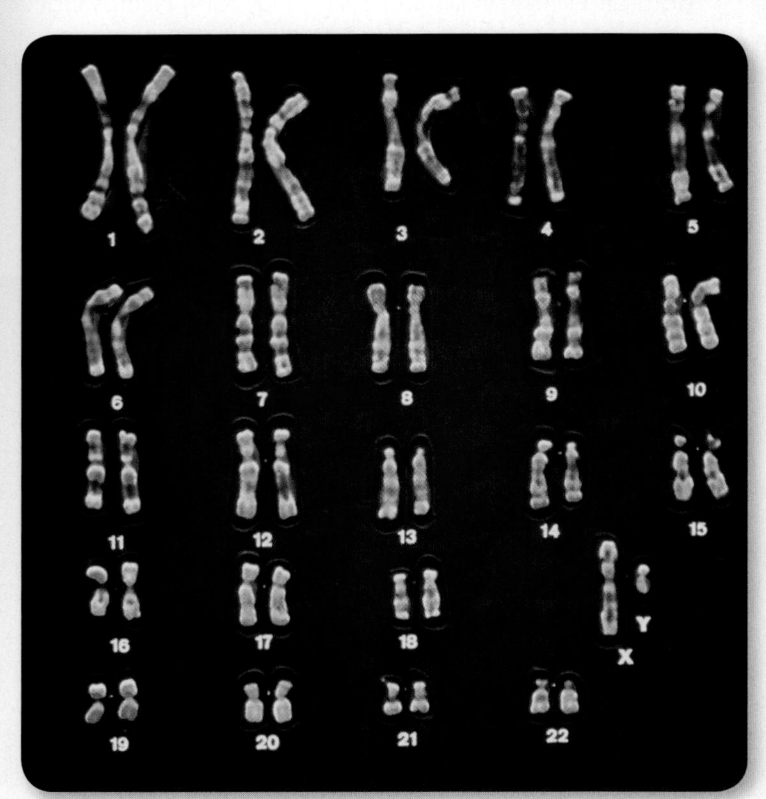

FIGURE 14–1 A Human Karyotype A typical human cell has 23 pairs of chromosomes. These chromosomes have been cut out of a photograph and arranged to form a karyotype.

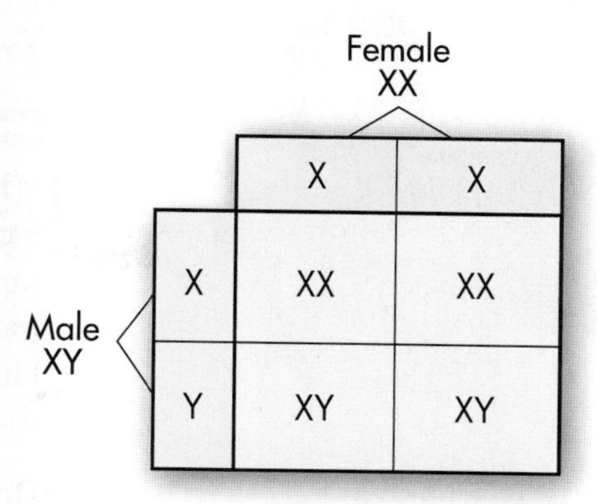

FIGURE 14–2 Sex Ratios Human egg cells contain a single X chromosome. Sperm cells contain either one X chromosome or one Y chromosome. Interpret Tables ***What does this Punnett square suggest about the sex ratio of the human population?***

The karyotype in **Figure 14–1** is from a typical human cell, which contains 46 chromosomes, arranged in 23 pairs. Why do our chromosomes come in pairs? Remember that we begin life when a haploid sperm, carrying just 23 chromosomes, fertilizes a haploid egg, also with 23 chromosomes. The resulting diploid cell develops into a new individual and carries the full complement of 46 chromosomes—two sets of 23.

Sex Chromosomes Two of the 46 chromosomes in the human genome are known as **sex chromosomes,** because they determine an individual's sex. Females have two copies of the X chromosome. Males have one X chromosome and one Y chromosome. As you can see in **Figure 14–2,** this is the reason why males and females are born in a roughly 50 : 50 ratio. All human egg cells carry a single X chromosome (23,X). However, half of all sperm cells carry an X chromosome (23,X) and half carry a Y chromosome (23,Y). This ensures that just about half the zygotes will be males and half will be females.

More than 1200 genes are found on the X chromosome, some of which are shown in **Figure 14–3.** Note that the human Y chromosome is much smaller than the X chromosome and contains only about 140 genes, most of which are associated with male sex determination and sperm development.

Autosomal Chromosomes To distinguish them from the sex chromosomes, the remaining 44 human chromosomes are known as autosomal chromosomes, or **autosomes.** The complete human genome consists of 46 chromosomes, including 44 autosomes and 2 sex chromosomes. To quickly summarize the total number of chromosomes present in a human cell—both autosomes and sex chromosomes—biologists write 46,XX for females and 46,XY for males.

In Your Notebook *Describe what makes up a human karyotype.*

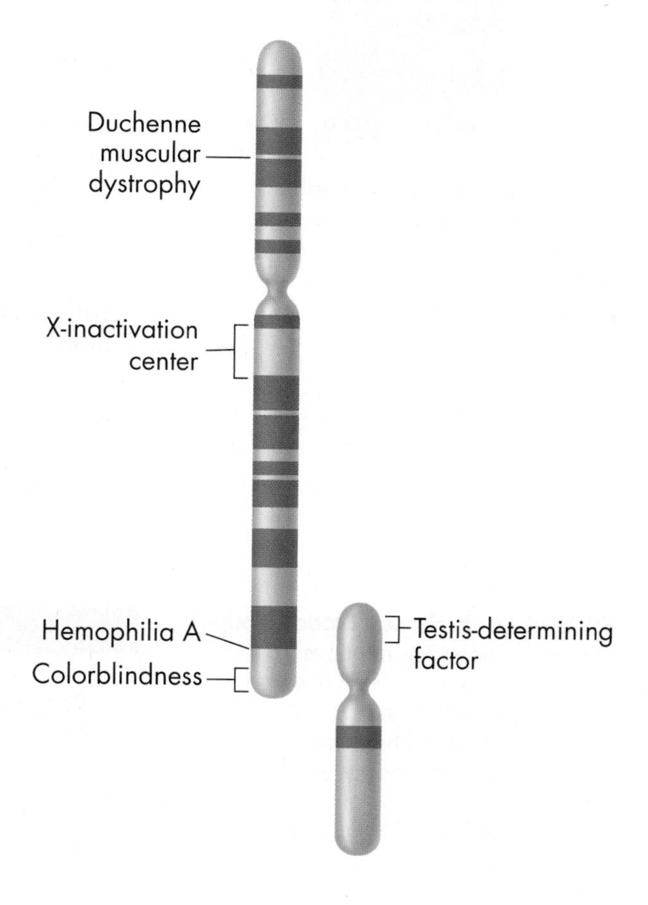

FIGURE 14–3 X and Y Chromosomes The human Y chromosome is smaller and carries fewer genes than the human X chromosome.

FIGURE 14–4 Recessive Alleles Some of the recessive alleles of the *MC1R* gene cause red hair. An individual with red hair usually has two of these recessive alleles.

Transmission of Human Traits

What patterns of inheritance do human traits follow?

It has not been easy studying our species using traditional genetic techniques. Despite the difficulties, human genetics has progressed rapidly, especially in recent years, with the use of molecular techniques to study human DNA. What have these studies shown? Human genes follow the same Mendelian patterns of inheritance as the genes of other organisms.

Dominant and Recessive Alleles **Many human traits follow a pattern of simple dominance.** For instance, a gene known as *MC1R* helps determine skin and hair color. Some of *MC1R*'s recessive alleles produce red hair. An individual with red hair usually has two of these recessive alleles, inheriting a copy from each parent. Dominant alleles for the *MC1R* gene help produce darker hair colors.

Another trait that displays simple dominance is the Rhesus, or Rh blood group. The allele for Rh factor comes in two forms: Rh^+ and Rh^-. Rh^+ is dominant, so an individual with both alleles (Rh^+/Rh^-) is said to have Rh positive blood. Rh negative blood is found in individuals with two recessive alleles (Rh^-/Rh^-).

Codominant and Multiple Alleles **The alleles for many human genes display codominant inheritance.** One example is the ABO blood group, determined by a gene with three alleles: I^A, I^B, and *i*. Alleles I^A and I^B are codominant. They produce molecules known as antigens on the surface of red blood cells. As **Figure 14–5** shows, individuals with alleles I^A and I^B produce both A and B antigens, making them blood type AB. The *i* allele is recessive. Individuals with alleles I^AI^A or I^Ai produce only the A antigen, making them blood type A. Those with I^BI^B or I^Bi alleles are type B. Those homozygous for the *i* allele (*ii*) produce no antigen and are said to have blood type O. If a patient has AB-negative blood, it means the individual has I^A and I^B alleles from the ABO gene and two Rh^- alleles from the Rh gene.

FIGURE 14–5 Human Blood Groups This table shows the relationship between genotype and phenotype for the ABO blood group. It also shows which blood types can safely be transfused into people with other blood types. **Apply Concepts** ***How can there be four different phenotypes even though there are six different genotypes?***

Blood Groups

Phenotype (Blood Type)	Genotype	Antigen on Red Blood Cell	Safe Transfusions	
			To	From
A	I^AI^A or I^Ai	A	A, AB	A, O
B	I^BI^B or I^Bi	B	B, AB	B, O
AB	I^AI^B	A and B	AB	A, B, AB, O
O	*ii*	None	A, B, AB, O	O

Sex-Linked Inheritance **Because the X and Y chromosomes determine sex, the genes located on them show a pattern of inheritance called sex-linkage.** A **sex-linked gene** is a gene located on a sex chromosome. As you might expect, genes on the Y chromosome are found only in males and are passed directly from father to son. Genes located on the X chromosome are found in both sexes, but the fact that men have just one X chromosome leads to some interesting consequences.

For example, humans have three genes responsible for color vision, all located on the X chromosome. In males, a defective allele for any of these genes results in colorblindness, an inability to distinguish certain colors. The most common form, red-green colorblindness, occurs in about 1 in 12 males. Among females, however, colorblindness affects only about 1 in 200. Why is there such a difference? In order for a recessive allele, like colorblindness, to be expressed in females, it must be present in two copies—one on each of the X chromosomes. This means that the recessive phenotype of a sex-linked genetic disorder tends to be much more common among males than among females.

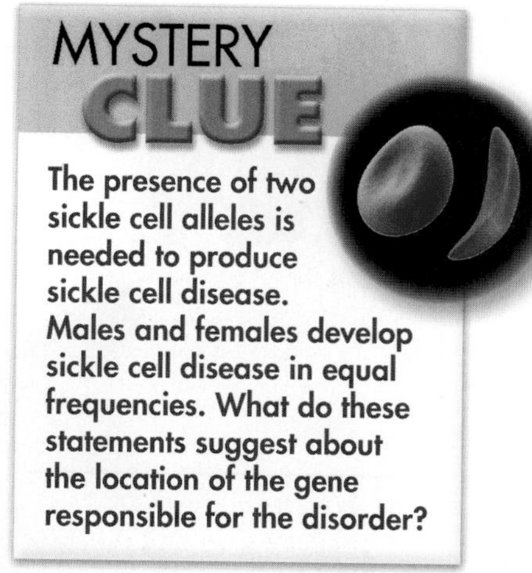

The presence of two sickle cell alleles is needed to produce sickle cell disease. Males and females develop sickle cell disease in equal frequencies. What do these statements suggest about the location of the gene responsible for the disorder?

Quick Lab
GUIDED INQUIRY

How Is Colorblindness Transmitted?

1. Make a data table with the column headings Trial, Colors, Sex of Individual, and Number of X-Linked Alleles. Draw ten rows under the headings and fill in the numbers 1 through 10 in the Trial column. Label one plastic cup Mother and a second plastic cup Father.
2. The white beans represent X chromosomes. Use a black marker to make a dot on 1 white bean to represent the X-linked allele for colorblindness. Place this bean, plus 1 unmarked white bean, into the cup labeled Mother.
3. Mark a black dot on 1 more white bean. Place this bean, plus 1 red bean, into the cup labeled Father. The red bean represents a Y chromosome.

4. Close your eyes and pick one bean from each cup to represent how each parent contributes to a sex chromosome and a fertilized egg.
5. In your data table, record the color of each bean and the sex of an individual who would carry this pair of sex chromosomes. Also record how many X-linked alleles the individual has. Put the beans back in the cups they came from.
6. Determine whether the individual would have colorblindness.
7. Repeat steps 4 to 6 for a total of 10 pairs of beans.

Analyze and Conclude

1. **Draw Conclusions** How do human sex chromosomes keep the numbers of males and females roughly equal?
2. **Calculate** Calculate the class totals for each data column. How many females were colorblind? How many males? Explain these results. MATH
3. **Use Models** Evaluate your model. How accurately does it represent the transmission of colorblindness in a population? Why?

FIGURE 14–6 X-Chromosome Inactivation Female calico cats are tri-colored. The color of spots on their fur is controlled by a gene on the X chromosome. Spots are either orange or black, depending on which X chromosome is inactivated in different patches of their skin.

X-Chromosome Inactivation If just one X chromosome is enough for cells in males, how does the cell "adjust" to the extra X chromosome in female cells? The answer was discovered by the British geneticist Mary Lyon. In female cells, most of the genes in one of the X chromosomes are randomly switched off, forming a dense region in the nucleus known as a Barr body. Barr bodies are generally not found in males because their single X chromosome is still active.

The same process happens in other mammals. In cats, for example, a gene that controls the color of coat spots is located on the X chromosome. One X chromosome may have an allele for orange spots and the other X chromosome may have an allele for black spots. In cells in some parts of the body, one X chromosome is switched off. In other parts of the body, the other X chromosome is switched off. As a result, the cat's fur has a mixture of orange and black spots, like those in **Figure 14–6.** Male cats, which have just one X chromosome, can have spots of only one color. Therefore, if the cat's fur has three colors—white with orange and black spots, for example—you can almost be certain that the cat is female.

In Your Notebook *Write three quiz questions about the transmission of human traits and answer them.*

Human Pedigrees

How can pedigrees be used to analyze human inheritance?

Given the complexities of genetics, how would you go about determining whether a trait is caused by a dominant or recessive allele and whether the gene for that trait is autosomal or sex-linked? The answers, not surprisingly, can be found by applying Mendel's basic principles of genetics.

To analyze the pattern of inheritance followed by a particular trait, you can use a chart that shows the relationships within a family. Such a chart is called a **pedigree.** A pedigree shows the presence or absence of a trait according to the relationships between parents, siblings, and offspring. It can be used for any species, not just humans.

The pedigree in **Figure 14–7** shows how one human trait—a white lock of hair just above the forehead—passes through three generations of a family. The allele for the white forelock trait is dominant. At the top of the chart is a grandfather who had the white forelock trait. Two of his three children inherited the trait. Three grandchildren have the trait, but two do not.

BUILD Vocabulary

WORD ORIGINS The word **pedigree** combines the Latin words *pedem,* meaning "foot," and *gruem,* meaning "crane." A crane is a long-legged waterbird. On old manuscripts, a forked sign resembling a crane's footprint indicated a line of ancestral descent.

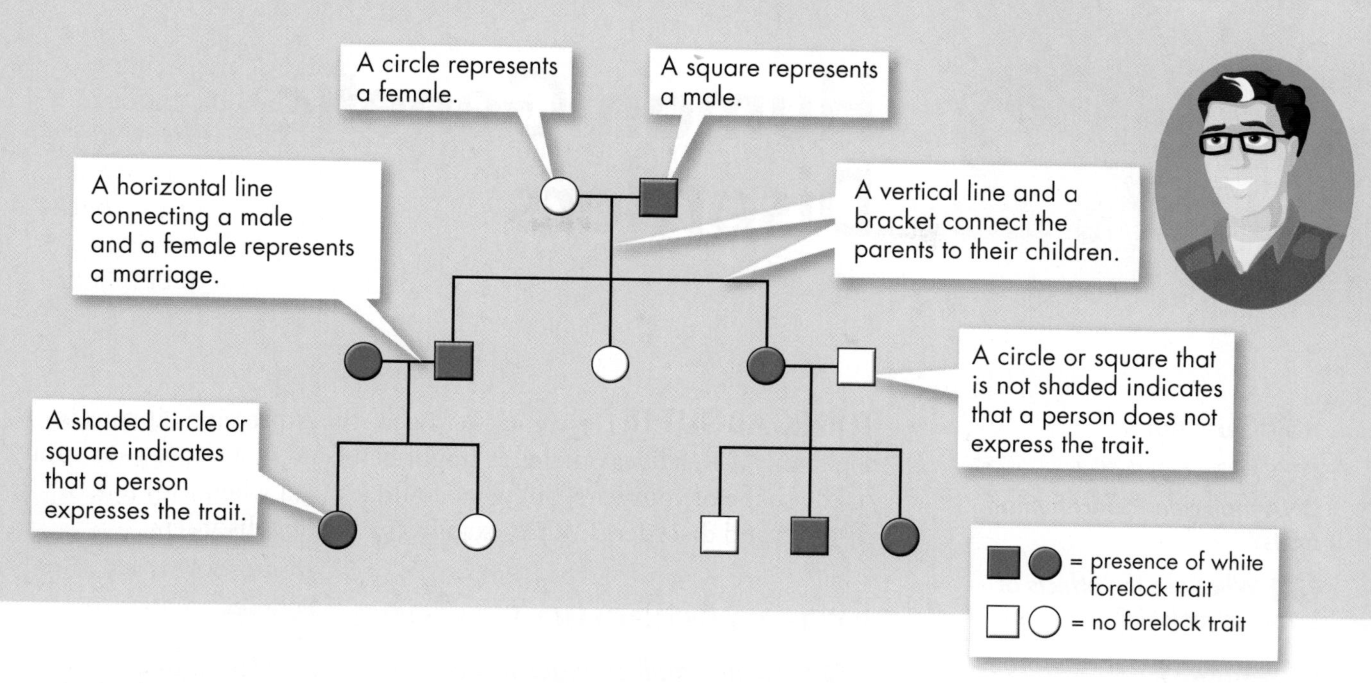

FIGURE 14–7 Pedigree Example This diagram shows what the symbols in a pedigree represent. Interpret Visuals ***What are the genotypes of both parents on the left in the second row? How do you know?***

By analyzing a pedigree, we can often infer the genotypes of family members. For example, because the white forelock trait is dominant, all the family members in **Figure 14–7** lacking this trait must have homozygous recessive alleles. One of the grandfather's children lacks the white forelock trait, so the grandfather must be heterozygous for this trait.

With pedigree analysis, it is possible to apply the principles of Mendelian genetics to humans. **The information gained from pedigree analysis makes it possible to determine the nature of genes and alleles associated with inherited human traits.** Based on a pedigree, you can often determine if an allele for a trait is dominant or recessive, autosomal or sex-linked.

14.1 Assessment

Review Key Concepts

1. a. Review What are autosomes?

b. Explain What determines whether a person is male or female?

c. Propose a Solution How can you use karyotypes to identify a species?

2. a. Review Explain how sex-linked traits are inherited.

b. Predict If a woman with type O blood and a man with type AB blood have children, what are the children's possible genotypes?

3. a. Review What does a pedigree show?

b. Infer Why would the Y chromosome be unlikely to contain any of the genes that are absolutely necessary for survival?

VISUAL THINKING

4. Choose a family and a trait, such as facial dimples, that you can trace through three generations. Find out who in the family has had the trait and who has not. Then, draw a pedigree to represent the family history of the trait.

14.2 Human Genetic Disorders

Key Questions

How do small changes in DNA molecules affect human traits?

What are the effects of errors in meiosis?

Vocabulary

nondisjunction

Taking Notes

Two-Column Chart Before you read, make a two-column chart. In the first column, write three questions you have about genetic disorders. As you read, fill in answers to your questions in the second column. When you have finished, research the answers to your remaining questions.

THINK ABOUT IT Have you ever heard the expression "It runs in the family"? Relatives or friends might have said that about your smile or the shape of your ears, but what could it mean when they talk of diseases and disorders? What, exactly, is a genetic disorder?

From Molecule to Phenotype

How do small changes in DNA molecules affect human traits?

We know that genes are made of DNA and that they interact with the environment to produce an individual organism's characteristics, or phenotype. However, when a gene fails to work or works improperly, serious problems can result.

Molecular research techniques have shown us a direct link between genotype and phenotype. For example, the wax that sometimes builds up in our ear canals can be one of two forms: wet or dry. People of African and European ancestry are more likely to have wet earwax—the dominant form. Those of Asian or Native American ancestry most often have the dry form, which is recessive. A single DNA base in the gene for a membrane-transport protein is the culprit. A simple base change from guanine (G) to adenine (A) causes this protein to produce dry earwax instead of wet earwax.

The connection between molecule and trait, and between genotype and phenotype, is often that simple, and just as direct. **Changes in a gene's DNA sequence can change proteins by altering their amino acid sequences, which may directly affect one's phenotype.** In other words, there is a molecular basis for genetic disorders.

Disorders Caused by Individual Genes Thousands of genetic disorders are caused by changes in individual genes. These changes often affect specific proteins associated with important cellular functions.

▶ ***Sickle Cell Disease*** This disorder is caused by a defective allele for beta-globin, one of two polypeptides in hemoglobin, the oxygen-carrying protein in red blood cells. The defective polypeptide makes hemoglobin a bit less soluble, causing hemoglobin molecules to stick together when the blood's oxygen level decreases. The molecules clump into long fibers, forcing cells into a distinctive sickle shape, which gives the disorder its name.

Sickle-shaped cells are more rigid than normal red blood cells, and, therefore, they tend to get stuck in the capillaries—the narrowest blood vessels in the body. If the blood stops moving through the capillaries, damage to cells, tissues, and even organs can result.

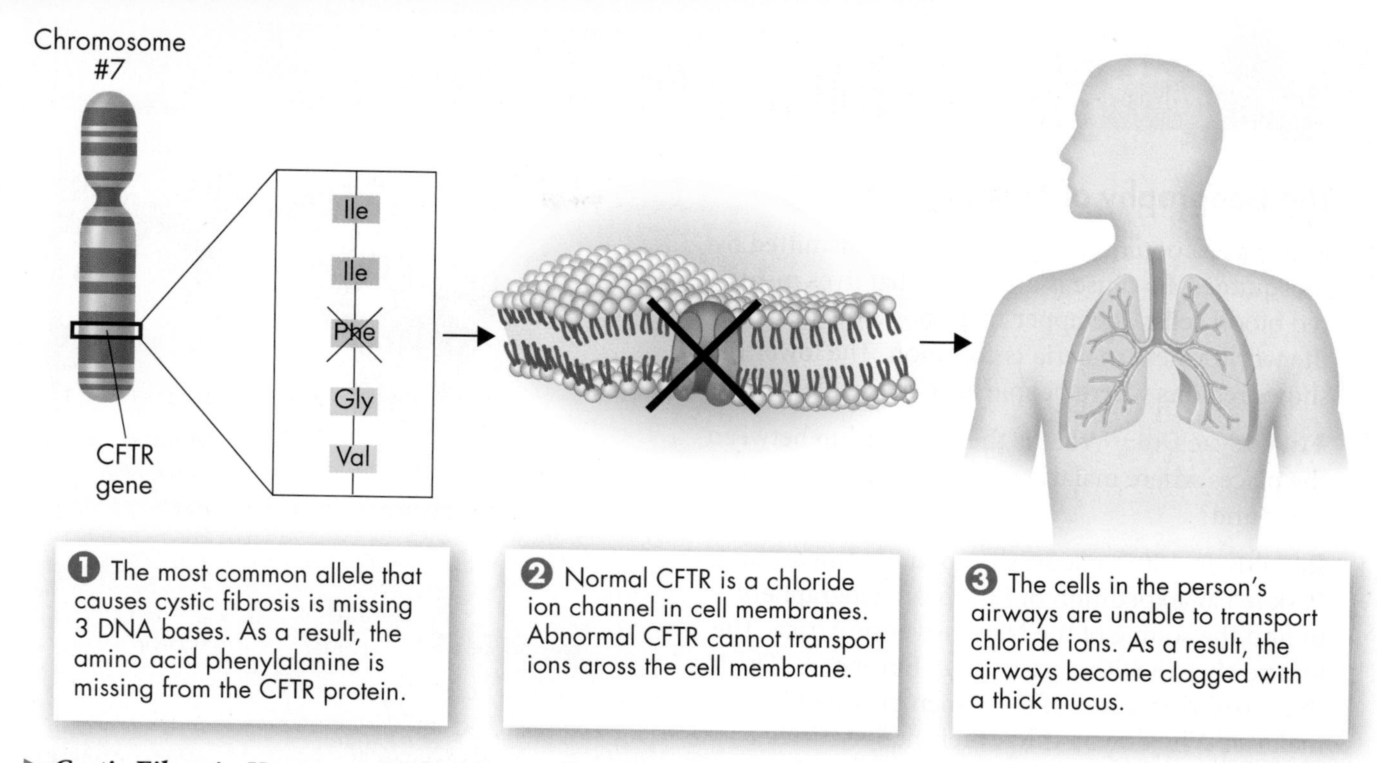

FIGURE 14–8 Mutations Cause Cystic Fibrosis CF is usually caused by the deletion of three bases in the DNA of a single gene. As a result, the body does not produce normal CFTR, a protein needed to transport chloride ions. **Infer** ***Why isn't the cause of CF considered a frameshift mutation?***

▶ ***Cystic Fibrosis*** Known as CF for short, cystic fibrosis is most common among people of European ancestry. CF is caused by a genetic change almost as small as the earwax allele. Most cases result from the deletion of just three bases in the gene for a protein called cystic fibrosis transmembrane conductance regulator (CFTR). CFTR normally allows chloride ions (Cl^-) to pass across cell membranes. The loss of these bases removes a single amino acid—phenylalanine—from CFTR, causing the protein to fold improperly. The misfolded protein is then destroyed. With cell membranes unable to transport chloride ions, tissues throughout the body malfunction.

People with one normal copy of the CF allele are unaffected by CF, because they can produce enough CFTR to allow their cells to work properly. Two copies of the defective allele are needed to produce the disorder, which means the CF allele is recessive. Children with CF have serious digestive problems and produce thick, heavy mucus that clogs their lungs and breathing passageways.

▶ ***Huntington's Disease*** Huntington's disease is caused by a dominant allele for a protein found in brain cells. The allele for this disease contains a long string of bases in which the codon CAG—coding for the amino acid glutamine—repeats over and over again, more than 40 times. Despite intensive study, the reason why these long strings of glutamine cause disease is still not clear. The symptoms of Huntington's disease, namely mental deterioration and uncontrollable movements, usually do not appear until middle age. The greater the number of codon repeats, the earlier the disease appears, and the more severe are its symptoms.

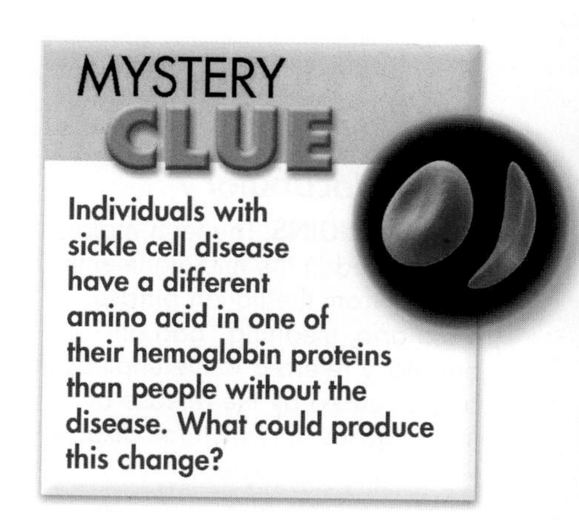

Individuals with sickle cell disease have a different amino acid in one of their hemoglobin proteins than people without the disease. What could produce this change?

The Geography of Malaria

Malaria is a potentially fatal disease transmitted by mosquitoes. Its cause is a parasite that lives inside red blood cells. The upper map shows the parts of the world where malaria is common. The lower map shows regions where people have the sickle cell allele.

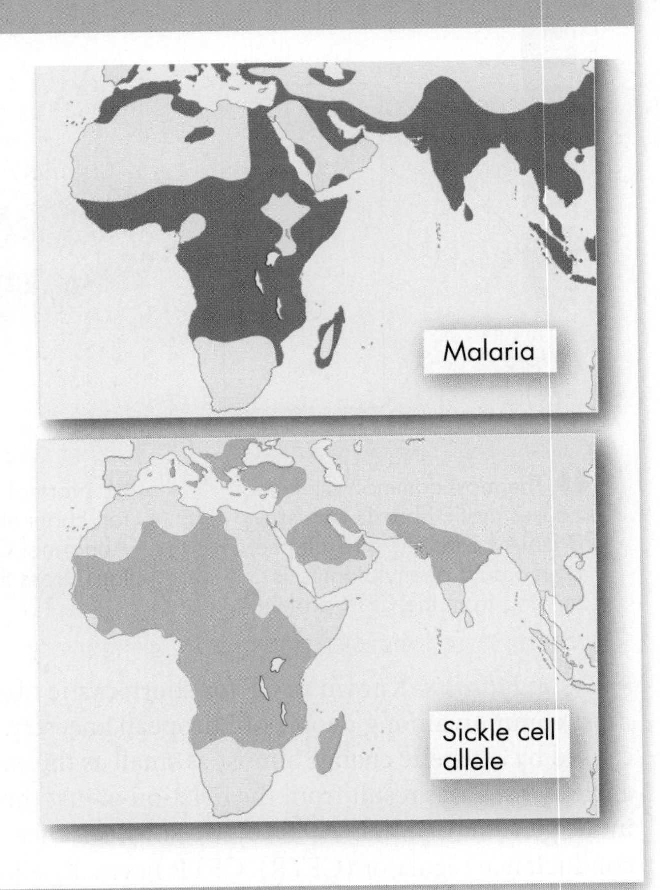

1. Analyze Data What is the relationship between the places where malaria and the sickle cell allele are found?

2. Infer In 1805, a Scottish explorer named Mungo Park led an expedition of European geographers to find the source of the Niger River in Africa. The journey began with a party of 45 Europeans. During the expedition, most of these men perished from malaria. Why do you think their native African guides survived?

3. Form a Hypothesis As the map shows, the sickle cell allele is not found in African populations that are native to southern Africa. Propose an explanation for this discrepancy.

Genetic Advantages Disorders such as sickle cell disease and CF are still common in human populations. In the United States, the sickle cell allele is carried by approximately 1 person in 12 of African ancestry, and the CF allele is carried by roughly 1 person in 25 of European ancestry. Why are these alleles still around if they can be fatal for those who carry them? The answers may surprise you.

Most African Americans today are descended from populations that originally lived in west central Africa, where malaria is common. Malaria is a mosquito-borne infection caused by a parasite that lives inside red blood cells. Individuals with just one copy of the sickle cell allele are generally healthy and are also highly resistant to the parasite. This resistance gives them a great advantage against malaria, which even today claims more than a million lives every year.

More than 1000 years ago, the cities of medieval Europe were ravaged by epidemics of typhoid fever. Typhoid is caused by a bacterium that enters the body through cells in the digestive system. The protein produced by the CF allele helps block the entry of this bacterium. Individuals heterozygous for CF would have had an advantage when living in cities with poor sanitation and polluted water, and—because they also carried a normal allele—these individuals would not have suffered from cystic fibrosis.

BUILD Vocabulary

WORD ORIGINS The term *malaria* was coined in the mid-eighteenth century from the Italian phrase, *mala aria*, meaning "bad air." It originally referred to the unpleasant odors caused by the release of marsh gases, to which the disease was initially attributed.

Chromosomal Disorders

What are the effects of errors in meiosis?

Most of the time, the process of meiosis works perfectly and each human gamete gets exactly 23 chromosomes. Every now and then, however, something goes wrong. The most common error in meiosis occurs when homologous chromosomes fail to separate. This mistake is known as **nondisjunction,** which means "not coming apart." **Figure 14–9** illustrates the process.

If nondisjunction occurs during meiosis, gametes with an abnormal number of chromosomes may result, leading to a disorder of chromosome numbers. For example, if two copies of an autosomal chromosome fail to separate during meiosis, an individual may be born with three copies of that chromosome. This condition is known as a trisomy, meaning "three bodies." The most common form of trisomy, involving three copies of chromosome 21, is Down syndrome, which is associated with a range of cognitive disabilities and a high frequency of certain birth defects.

Nondisjunction of the X chromosomes can lead to a disorder known as Turner's syndrome. A female with Turner's syndrome usually inherits only one X chromosome. Women with Turner's syndrome are sterile, which means that they are unable to reproduce. Their sex organs do not develop properly at puberty.

In males, nondisjunction may cause Klinefelter's syndrome, resulting from the inheritance of an extra X chromosome, which interferes with meiosis and usually prevents these individuals from reproducing. There have been no reported instances of babies being born without an X chromosome, indicating that this chromosome contains genes that are vital for the survival and development of the embryo.

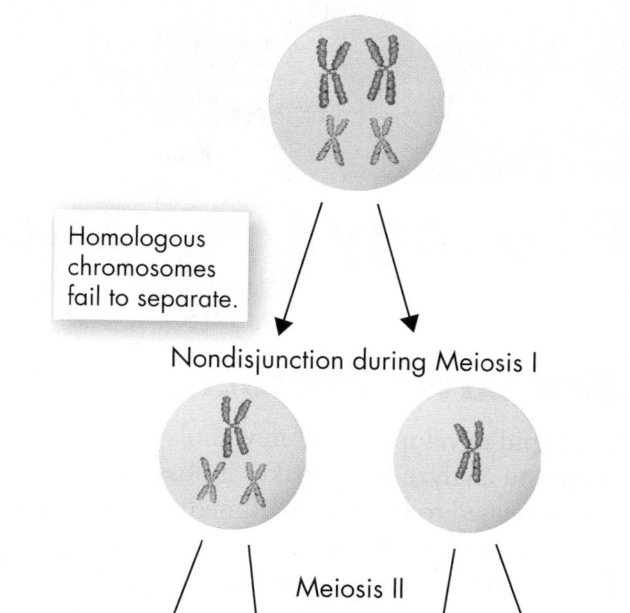

FIGURE 14–9 Nondisjunction This failure of meiosis causes gametes to have an abnormal number of chromosomes. **Apply Concepts** ***Which phase of meiosis is shown in the first cell?***

14.2 Assessment

Review Key Concepts

1. **a. Review** How can a small change in a person's DNA cause a genetic disorder?

   b. Infer How do genetic disorders such as CF support the theory of evolution?

2. **a. Review** Describe two sex chromosome disorders.

   b. Apply Concepts How does nondisjunction cause chromosomal disorders?

WRITE ABOUT SCIENCE

Description

3. Write a paragraph explaining the process of nondisjunction. (*Hint:* To organize your writing, create a flowchart that shows the steps in the process.)

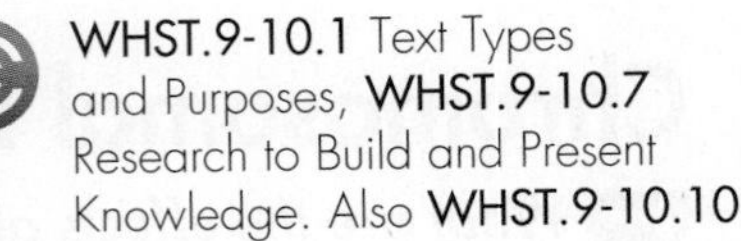

WHST.9-10.1 Text Types and Purposes, WHST.9-10.7 Research to Build and Present Knowledge. Also WHST.9-10.10

Biology & Society

Are Laws Protecting Genetic Privacy Necessary?

The rapid development of new tools and techniques to analyze DNA makes it possible to test for alleles related to thousands of medical conditions. In theory, the results of genetic testing should benefit everyone. Accurate genetic data helps physicians select the proper treatments for patients. It may allow people with genes that place them at risk of certain conditions to minimize those risks.

At issue, however, is individual privacy. Once a test is done, who has access to the data, and how can they use it? Could employers refuse to hire people who might drive up their medical costs? Might insurance companies refuse to renew the policies of individuals with genes for certain disorders? These are not hypothetical questions. In 2005, managers of a professional basketball team asked one of its players to be tested for a gene linked to heart ailments. When he refused, they traded the player to another team. Dr. Francis Collins, director of the National Human Genome Research Institute, worries that "the public is afraid of taking advantage of genetic testing." Is he correct? Should genetic data be protected by law, or should it be open to public view?

Many commercial laboratories test human DNA for genetic disorders.

The Viewpoints

Genetic Privacy Does Not Need Legal Protection

Other laws already protect individuals from discrimination on the basis of medical disability. Employers and insurance companies are nonetheless allowed to ask individuals if they smoke, use alcohol, or have a history of medical problems. Having this information allows employers to make intelligent choices about whom to hire. It also helps insurance companies maintain lower rates for their healthiest clients. Free access to genetic data should be a public right.

Genetic Privacy Should Be Protected by Law

The Genetic Information Nondiscrimination Act (GINA) went into effect in 2009, and it provides important protections to personal privacy. Individuals may not take advantage of today's advances in genetic medicine if they fear their personal information might be used to deny them employment or insurance. We need such laws to realize the full benefits of modern medicine and to protect otherwise healthy individuals from genetic discrimination.

Research and Decide

1. Analyze the Viewpoints To make an informed decision, learn more about genetic testing by consulting library or Internet resources. Then, list the key arguments expressed by the proponents and critics of both points of view. Find out if laws preventing genetic discrimination have been proposed or passed in your state.

2. Form an Opinion Should access and use of genetic data be regulated? Weigh both sides of the issue. Who will benefit from the sharing of genetic data? Will anyone suffer? Do some arguments outweigh others? If so, which ones? Explain your answers.

14.3 Studying the Human Genome

THINK ABOUT IT Just a few decades ago, computers were gigantic machines found only in laboratories and universities. Today, many of us carry small, powerful computers to school and work every day. Decades ago, the human genome was unknown. Today, we can see our entire genome on the Internet. How long will it be before having a copy of your own genome is as ordinary as carrying a cellphone in your pocket?

Manipulating DNA

What techniques are used to study human DNA?

Since discovering the genetic code, biologists have dreamed of a time when they could read the DNA sequences in the human genome. For a long time, it seemed impossible. DNA is a huge molecule—even the smallest human chromosome contains nearly 50 million base pairs. Manipulating such large molecules is extremely difficult. In the late 1960s, however, scientists found they could use natural enzymes in DNA analysis. From this discovery came many useful tools. **By using tools that cut, separate, and then replicate DNA base by base, scientists can now read the base sequences in DNA from any cell.** Such techniques have revolutionized genetic studies of living organisms, including humans.

Cutting DNA Nucleic acids are chemically different from other macromolecules such as proteins and carbohydrates. This difference makes DNA relatively easy to extract from cells and tissues. However, DNA molecules from most organisms are much too large to be analyzed, so they must first be cut into smaller pieces. Many bacteria produce enzymes that do exactly that. Known as **restriction enzymes,** these highly specific substances cut even the largest DNA molecule into precise pieces, called restriction fragments, that are several hundred bases in length. Of the hundreds of known restriction enzymes, each cuts DNA at a different sequence of nucleotides.

In Your Notebook *Make a flowchart that shows the processes scientists use to analyze DNA.*

Key Questions

What techniques are used to study human DNA?

What were the goals of the Human Genome Project, and what have we learned so far?

Vocabulary

restriction enzyme
gel electrophoresis
bioinformatics
genomics

Taking Notes

Preview Visuals Before you read, look at **Figure 14–10,** and write down three questions you have about the figure. As you read, find answers to your questions.

Cutting DNA
A restriction enzyme is like a key that fits only one lock. The *Eco*RI restriction enzyme can only recognize the base sequence GAATTC. It cuts each strand of DNA between the G and A bases, leaving single-stranded overhangs with the sequence AATT. The overhangs are called "sticky ends" because they can bond, or "stick," to a DNA fragment with the complementary base sequence.

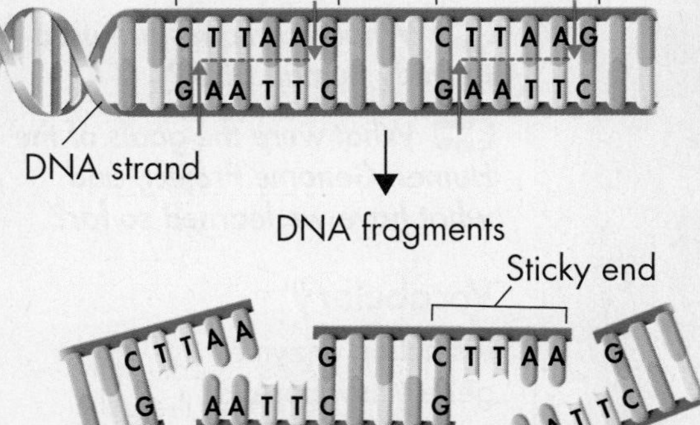

Separating DNA
Gel electrophoresis is used to separate DNA fragments. After being cut by restriction enzymes, the fragments are put into wells on a gel that is similar to a slice of gelatin. An electric voltage moves them across the gel. Shorter fragments move faster than longer fragments. Within an hour or two, the fragments all separate, each appearing as a band on the gel.

DNA plus restriction enzyme
Power source
Longer fragments
Shorter fragments
Mixture of DNA fragments
Gel

VISUAL SUMMARY

HOW SCIENTISTS MANIPULATE DNA

FIGURE 14–10 By using tools that cut, separate, and replicate DNA, scientists can read the base sequences in DNA from any cell. Knowing the sequence of an organism's DNA allows us to study specific genes.

Separating DNA Once DNA has been cut by restriction enzymes, scientists can use a technique known as **gel electrophoresis** to separate and analyze the differently sized fragments. **Figure 14–10** illustrates this simple, yet effective, method. A mixture of DNA fragments is placed at one end of a porous gel. When an electric voltage is applied to the gel, DNA molecules—which are negatively charged—move toward the positive end of the gel. The smaller the DNA fragment, the faster and farther it moves. The result is a pattern of bands based on fragment size. Specific stains that bind to DNA make these bands visible. Researchers can then remove individual restriction fragments from the gel and study them further.

Reading DNA After the DNA fragments have been separated, researchers use a clever chemical "trick" to read, or sequence, them. The single-stranded DNA fragments are placed in a test tube containing DNA polymerase—the enzyme that copies DNA—along with the four nucleotide bases, A, T, G, and C. As the enzyme goes to work, it uses the unknown strand as a template to make one new DNA strand after another. The tricky part is that researchers also add a small number of bases that have a chemical dye attached. Each time a dye-labeled base is added to a new DNA strand, the synthesis of that strand stops. When DNA synthesis is completed, the result is a series of color-coded DNA fragments of different lengths. Researchers can then separate these fragments, often by gel electrophoresis. The order of colored bands on the gel tells the exact sequence of bases in the DNA. The entire process can be automated and controlled by computers, so that DNA sequencing machines can read thousands of bases in a matter of seconds.

Reading DNA

A small proportion of dye-labeled nucleotides are used to make a complementary DNA strand. Each time a labeled nucleotide is added to the strand, DNA replication stops. Because each base was labeled with a different color, the result is color-coded DNA fragments of different lengths. When gel electrophoresis is used to separate the fragments, scientists can "read" the DNA sequence directly from the gel.

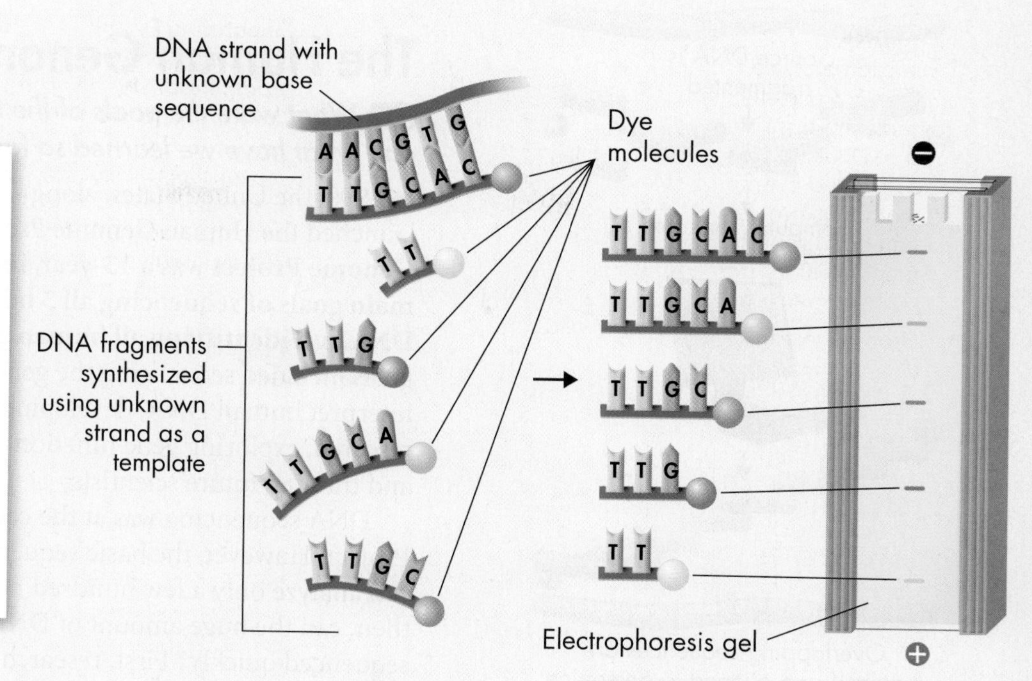

Base sequence as "read" from the order of the bands on the gel from bottom to top: **T G C A C**

Quick Lab
GUIDED INQUIRY

Modeling Restriction Enzymes

1. Write a 50-base, double-stranded DNA sequence using the bases A, C, G, and T in random order. Include each sequence shown below at least once in the sequence you write.
2. Make three copies of your double-stranded sequence on three different-colored strips of paper.
3. Use the drawings below to see how the restriction enzyme *Eco*RI would cut your DNA sequence. Use scissors to cut one copy of the sequence as *Eco*RI would.

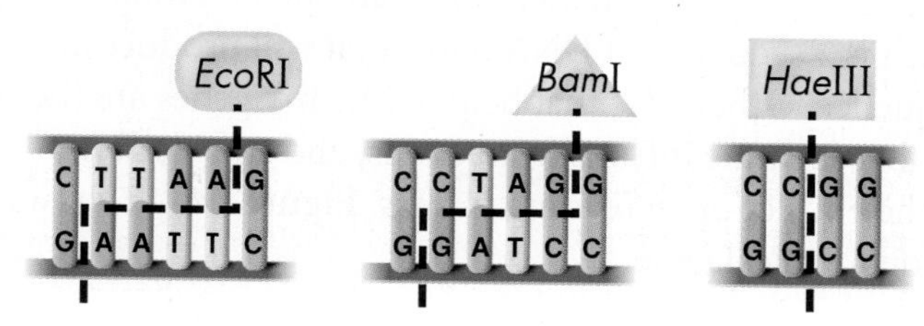

4. Use the procedure in Step 3 to cut apart another copy of your sequence as the restriction enzyme *Bam*I would. Then, cut the third copy as the restriction enzyme *Hae*III would.
5. Tape the single-stranded end of one of your DNA fragments to a complementary, single-stranded end of a classmate's fragment. This will form a single, double-stranded DNA molecule.

Analyze and Conclude

1. **Observe** Which restriction enzyme produced the most pieces? The fewest pieces?
2. **Integration of Knowledge and Ideas** How well did your model represent the actual process of using restriction enzymes to cut DNA? (*Hint*: Contrast the length of your model DNA sequence to the actual length of a DNA molecule.)

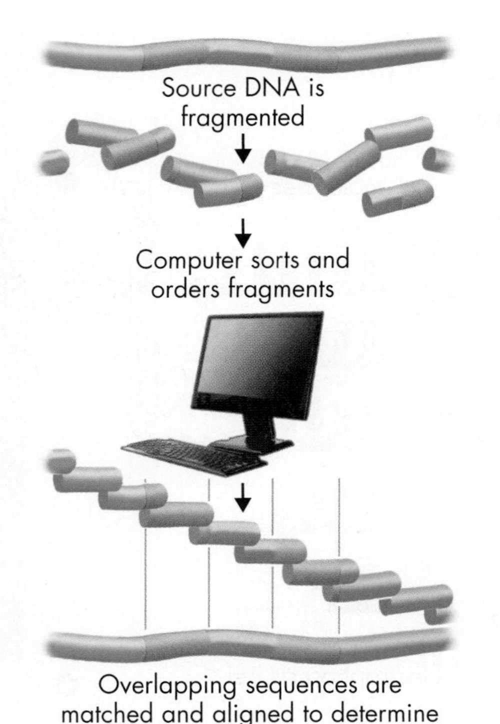

FIGURE 14–11 Shotgun Sequencing This method rapidly sorts DNA fragments by overlapping base sequences.

Insulin gene
Promoter
Intron
Start codon
Intron
Stop codon

FIGURE 14–12 Locating a Gene A typical gene, such as that for insulin, has several DNA sequences that can serve as locators. These include the promoter, sequences between introns and exons, and start and stop codons.

The Human Genome Project

What were the goals of the Human Genome Project, and what have we learned so far?

In 1990, the United States, along with several other countries, launched the Human Genome Project. **The Human Genome Project was a 13-year, international effort with the main goals of sequencing all 3 billion base pairs of human DNA and identifying all human genes.** Other important goals included sequencing the genomes of model organisms to interpret human DNA, developing technology to support the research, exploring gene functions, studying human variation, and training future scientists.

DNA sequencing was at the center of the Human Genome Project. However, the basic sequencing method you saw earlier can analyze only a few hundred nucleotides at a time. How, then, can the huge amount of DNA in the human genome be sequenced quickly? First, researchers must break up the entire genome into manageable pieces. By determining the base sequences in widely separated regions of a DNA strand, they can use the regions as markers, not unlike the mile markers along a road that is thousands of miles long. The markers make it possible for researchers to locate and return to specific locations in the DNA.

Sequencing and Identifying Genes Once researchers have marked the DNA strands, they can use the technique of "shotgun sequencing." This rapid sequencing method involves cutting DNA into random fragments, then determining the base sequence in each fragment. Computer programs take the sequencing data, find areas of overlap between fragments, and put the fragments together by linking the overlapping areas. The computers then align these fragments relative to the known markers on each chromosome, as shown in **Figure 14–11.** The entire process is like putting a jigsaw puzzle together, but instead of matching shapes, the computer matches DNA base sequences.

Reading the DNA sequence of a genome is not the same as understanding it. Much of today's research explores the vast amount of data from the Human Genome Project to look for genes and the DNA sequences that control them. By locating sequences known to be promoters—binding sites for RNA polymerase—scientists can identify many genes. Shortly after a promoter, there is usually an area called an open reading frame, which is a sequence of DNA bases that will produce an mRNA sequence. Other sites that help to identify genes are the sequences that separate introns from exons, and stop codons located at the ends of open reading frames. **Figure 14–12** shows these sites on a typical gene.

Comparing Sequences If you were to compare the genomes of two unrelated individuals, you would find that most—but not all—of their DNA matches base-for-base with each other. On average, one base in 1200 will not match between two individuals. Biologists call these single base differences SNPs (pronounced "snips"), which stands for single nucleotide polymorphisms. Researchers have discovered that certain sets of closely linked SNPs occur together time and time again. These collections of linked SNPs are called haplotypes—short for haploid genotypes. To locate and identify as many haplotypes in the human population as possible, the International HapMap Project began in 2002. The aim of the project is to give scientists a rapid way to identify haplotypes associated with various diseases and conditions and to pave the way to more effective life-saving medical care in the future.

MYSTERY CLUE

Scientists can detect the sickle cell allele with a test for SNPs in the genes for the polypeptides that make up hemoglobin. What does this tell you about the sickle cell mutation?

Sharing Data The Human Genome Project was completed in 2003. Copies of the human genome DNA sequence, and those of many other organisms, are now freely available on the Internet. Online computer access enables researchers and students to browse through databases of human DNA and study its sequence. More data from the human genome, and the genomes of other organisms, are added to these databases every day.

One of the key research areas of the Human Genome Project was a new field of study called **bioinformatics.** The root word, *informatics,* refers to the creation, development, and operation of databases and other computing tools to collect, organize, and interpret data. The prefix *bio-* refers to life sciences—specifically, molecular biology. Assembling the bits and pieces of the human genome would have been impossible without sophisticated computer programs that could recognize overlapping sequences and place them in the proper order, or immense databases where such information could be stored and retrieved. Without the tools of bioinformatics shown in **Figure 14–13,** the wealth of information gleaned from the Human Genome Project would hardly be useful. Bioinformatics also launched a more specialized field of study known as **genomics**—the study of whole genomes, including genes and their functions.

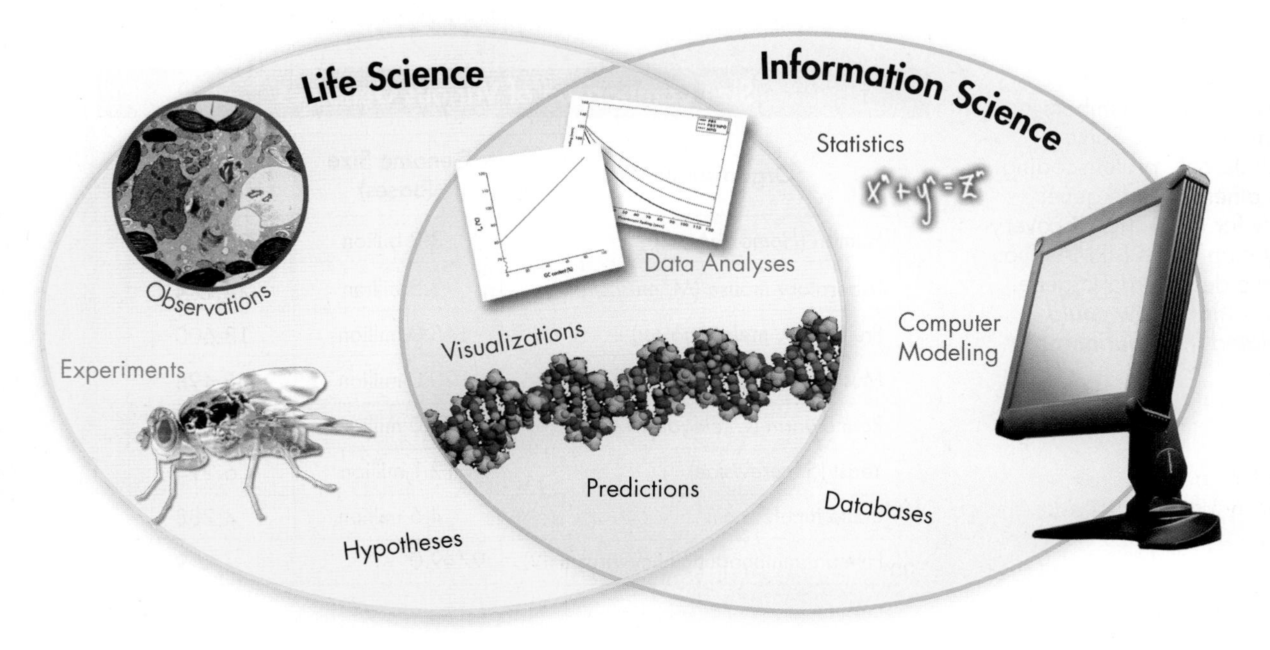

FIGURE 14–13 Bioinformatics Bioinformatics is a new field that combines molecular biology with information science. It is critical to studying and understanding the human genome.

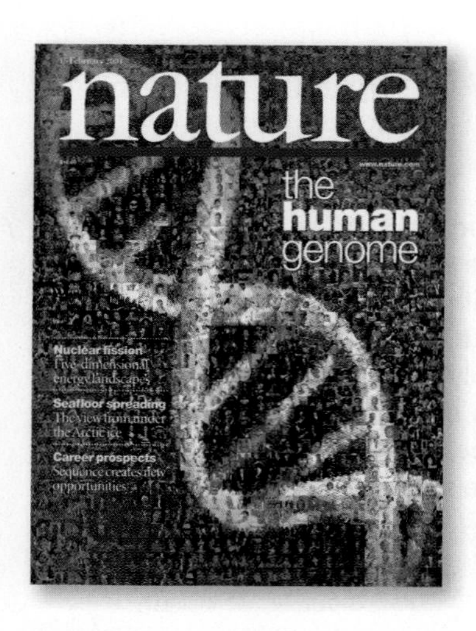

FIGURE 14–14 Announcement In February 2001, *Nature* was one of the first scientific journals to publish details about the completion of the Human Genome Project.

What We Have Learned The Human Genome Project has uncovered some surprising information. For instance, only about 2 percent of the human genome encodes instructions for the synthesis of proteins and many chromosomes contain large areas with very few genes. As much as half of the genome is made up of DNA sequences from viruses and other genetic elements. In addition, more than 40 percent of the proteins coded for by the human genome have strong similarity to proteins in other organisms, like those listed in **Figure 14–15.**

By any standard, the Human Genome Project has been a great scientific success. **The Human Genome Project pinpointed genes and associated particular sequences in those genes with numerous diseases and disorders. It also identified about three million locations where single-base DNA differences occur in humans.** This information may help us find DNA sequences associated with diseases such as diabetes and cancer. The Human Genome Project also provided important new technologies for the private sector, including agriculture and medicine. By doing so, the project catalyzed the U.S. biotechnology industry and fostered the development of new medical applications.

New Challenges Data from the Human Genome Project have raised a number of ethical and legal questions. For example, who owns and controls genetic information? Who should have access to personal genetic information? In response to some of these issues, in 2008, the U.S. Congress passed the Genetic Information Nondiscrimination Act. This act makes it illegal for insurance companies and employers to discriminate based on information from genetic tests. As the science advances, other protective laws may soon follow.

What's Next? Many more sequencing projects are underway, helped along by powerful new technologies. In medicine, treatments for a host of diseases are already being tailored to the individual genomes of patients. This approach allows physicians to quickly find the best drugs to fight diseases such as cancer, asthma, and diabetes.

FIGURE 14–15 Genome Size Comparisons The gene numbers in this table are not final. Some estimates include only protein-coding genes, while others include genes that code only for RNA. The discovery of small interfering RNAs (siRNAs) has complicated the definition of a gene. **Propose a Solution *How could you find updated information on genome sizes?***

Size Comparison of Various Genomes

Organism or Virus	Genome Size (bases)	Estimated Genes
Human (*Homo sapiens*)	3.2 billion	25,000
Laboratory mouse (*M. musculus*)	2.5 billion	24,174
Fruit fly (*D. melanogaster*)	165.0 million	13,600
Mustard weed (*A. thaliana*)	120.0 million	25,498
Roundworm (*C. elegans*)	97.0 million	19,000
Yeast (*S. cerevisiae*)	12.1 million	6,294
Bacterium (*E. coli*)	4.6 million	4,288
Human immunodeficiency virus (HIV)	9749.0	9

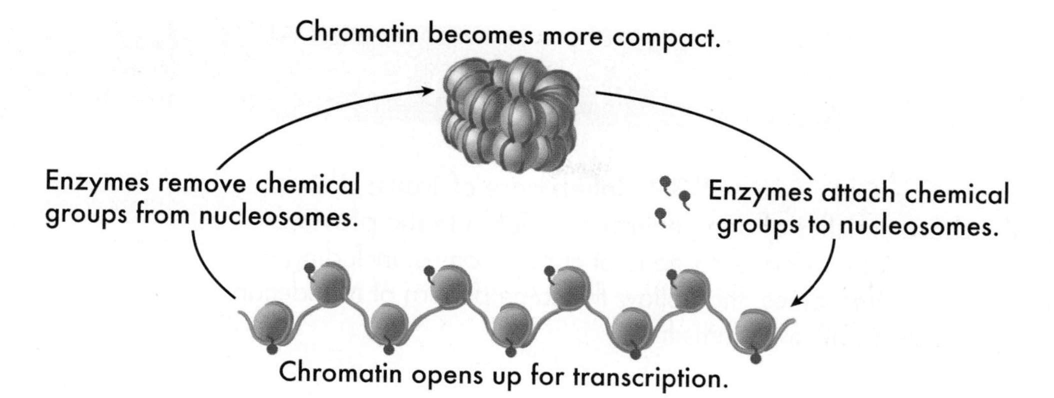

FIGURE 14–16 Effect of Chemical Marks The presence or absence of chemical marks determines how compact chromatin is in a given region of DNA. For transcription to occur, the chromatin must be less compact.

Beyond DNA You might think that DNA sequences are destiny, determining every detail of your development, appearance, and health. But things are not that simple. Recall that nuclear DNA is coiled around protein clusters called nucleosomes to form chromatin. In regions of DNA where chromatin is compact, expression of genes is switched off. In regions where chromatin opens up, gene expression is enhanced. **Figure 14–16** summarizes what happens. Cellular enzymes attach chemical marks to the DNA and proteins in chromatin that help to regulate gene expression. Surprisingly, these enzymes can be triggered by factors like stress, diet, and disease.

The chemical marks on chromatin are said to be *epigenetic,* that is, they are above the level of the genome. (The prefix *epi-* means "above.") The changes don't affect DNA base sequences, but they do affect gene expression. Some chemical marks are inherited by offspring. For example, studies show that people who have lived through famines pass epigenetic changes to their children that affect their growth and development.

During the formation of eggs and sperm, some genes are imprinted with tags that will "silence" those genes in the next generation. So the effect a particular gene has on you can depend, in part, on whether the copy you inherited came from your mother or your father.

14.3 Assessment

Review Key Concepts

1. a. Review How do molecular biologists identify genes in sequences of DNA?

b. Use Analogies How is shotgun sequencing similar to doing a jigsaw puzzle?

2. a. Review What is the Human Genome Project?

b. Form an Opinion Judge the potential impact of the Human Genome Project on both scientific thought and society. How might the project be used to benefit humankind? What potential problems might it create?

WRITE ABOUT SCIENCE

Persuasion

3. Scientists may one day be able to use genomics and molecular biology to alter a child's inherited traits. Under what circumstances, if any, should this ability be used? When should it not be used? Write a persuasive paragraph expressing your opinion. (*Hint:* Use specific examples of traits to support your ideas.)

NGSS Smart Guide Human Heredity

Disciplinary Core Idea LS3.A Inheritance of Traits: How are the characteristics of one generation related to the previous generation? Humans have 23 pairs of chromosomes, including one pair of sex chromosomes, that follow the same pattern of Mendelian inheritance as do other organisms.

14.1 Human Chromosomes

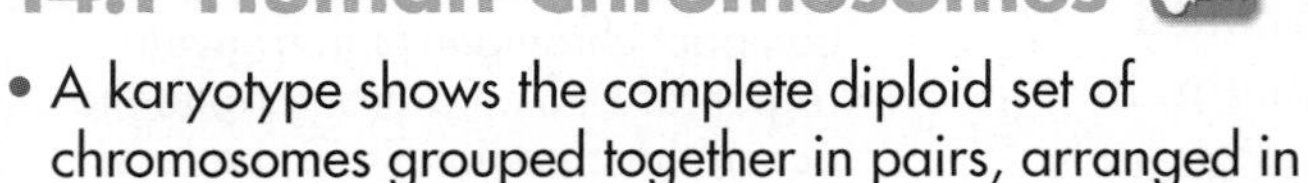

- A karyotype shows the complete diploid set of chromosomes grouped together in pairs, arranged in order of decreasing size.
- Many human traits follow a pattern of simple dominance. The alleles for other human genes display codominant inheritance. Because the X and Y chromosomes determine sex, the genes located on them show a pattern of inheritance called sex-linkage.
- The information gained from pedigree analysis makes it possible to determine the nature of genes and alleles associated with inherited human traits.

genome 392 • karyotype 392 • sex chromosome 393 • autosome 393 • sex-linked gene 395 • pedigree 396

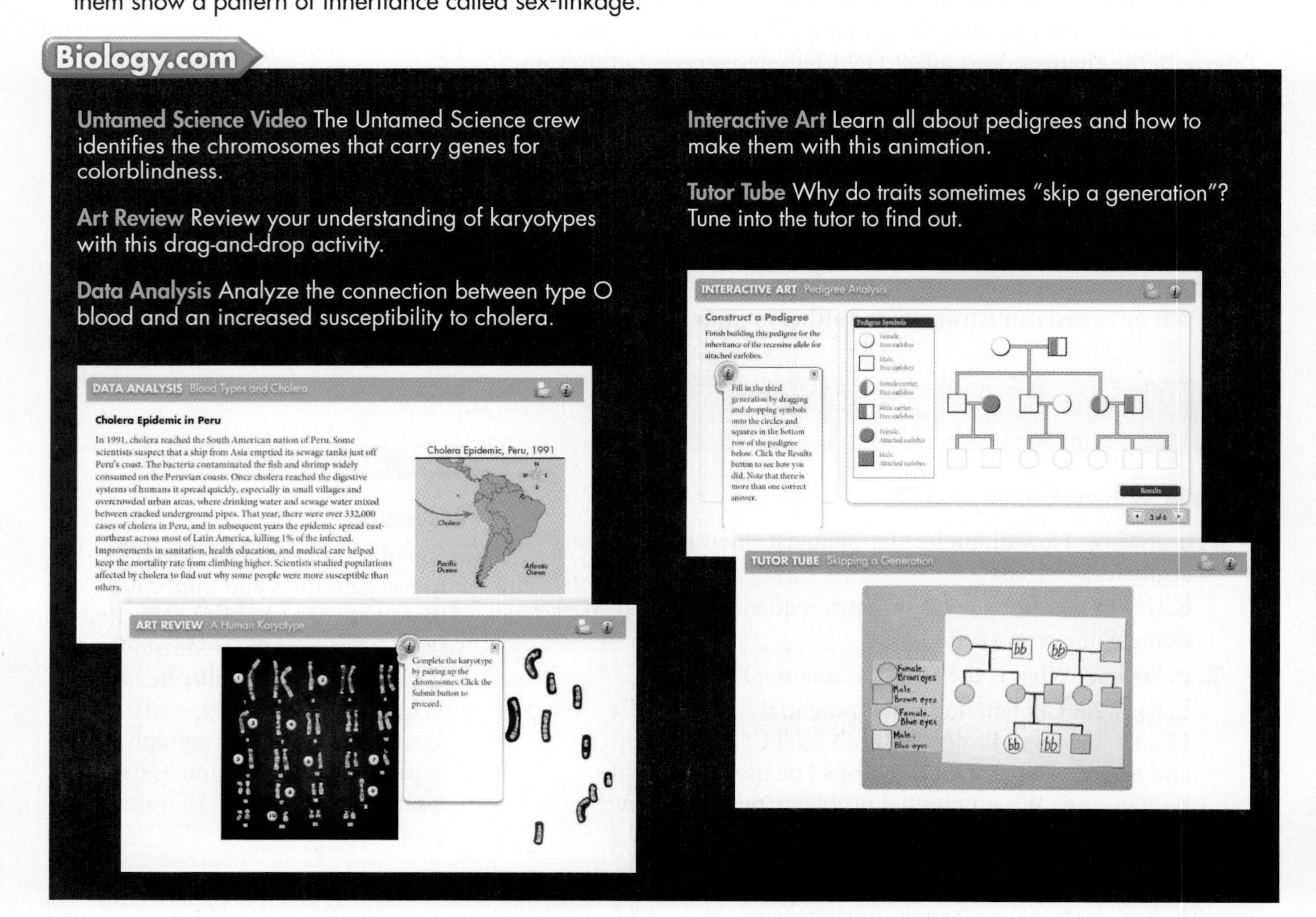

CHECKING YOUR *Scientific Literacy*

Refer to the lesson content and digital assets below as you prepare for your chapter assessment. Then, evaluate your understanding of human heredity by answering these questions.

1. **Scientific and Engineering Practice Developing and Using Models** The allele for cystic fibrosis is recessive. Construct a two-generation pedigree to show how this trait might be expressed when a mother with the genotype Cc and a father with the genotype cc have four children.

2. Crosscutting Concept **Cause and Effect** Use the fact that cystic fibrosis occurs equally often in men and women to explain why cystic fibrosis is not a sex-linked trait.

3. Text Types and Purposes Consider the following claim: Without computer technology, scientists could not have completed the Human Genome Project or organized its data. Cite specific evidence from Chapter 14 to support this claim.

4. STEM People can order DNA test kits. Do research to answer these questions. How does the kit work? What does it cost? List three reasons why people might buy a kit. What concerns might a consumer have about this technology?

14.2 Human Genetic Disorders

- Changes in a gene's DNA sequence can change proteins by altering their amino acid sequences, which may directly affect one's phenotype.
- If nondisjunction occurs during meiosis, gametes with an abnormal number of chromosomes may result, leading to a disorder of chromosome numbers.

nondisjunction 401

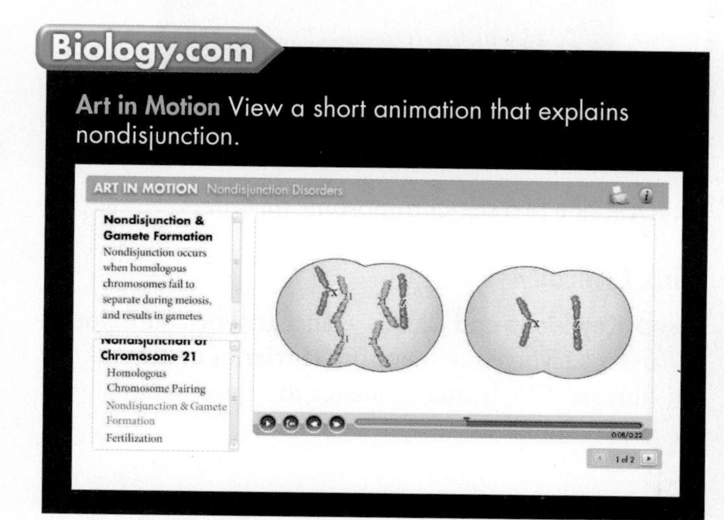

Forensics Lab

Using DNA to Identify Human Remains This chapter lab is available in your lab manual and at Biology.com

14.3 Studying the Human Genome

- By using tools that cut, separate, and then replicate DNA base by base, scientists can now read the base sequences in DNA from any cell.
- The Human Genome Project was a 13-year, international effort with the main goals of sequencing all 3 billion base pairs of human DNA and identifying all human genes.
- The Human Genome Project pinpointed genes and associated particular sequences in those genes with numerous diseases and disorders. It also identified about three million locations where single-base DNA differences occur in humans.

restriction enzyme 403 • gel electrophoresis 404 • bioinformatics 407 • genomics 407

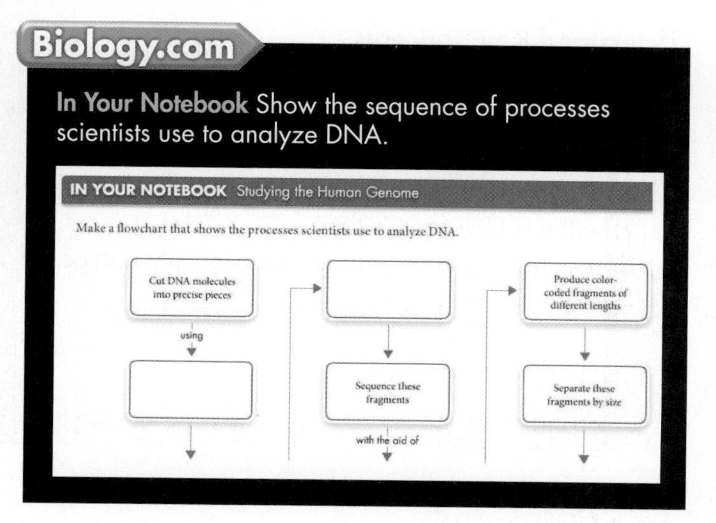

14 Assessment

14.1 Human Chromosomes

Understand Key Concepts

1. A normal human diploid zygote contains
   a. 23 chromosomes. c. 44 chromosomes.
   b. 46 chromosomes. d. XXY chromosomes.
2. A chart that traces the inheritance of a trait in a family is called a(n)
   a. pedigree. c. genome.
   b. karyotype. d. autosome.
3. An example of a trait that is determined by multiple alleles is
   a. cystic fibrosis. c. Down syndrome.
   b. ABO blood groups. d. colorblindness.
4. What is the difference between autosomes and sex chromosomes?
5. Is it possible for a person with blood type alleles I^A and I^B to have blood type A? Explain your answer. (Refer to Figure 14–5).

Think Critically

6. **Predict** What are the possible genotypes of the parents of a male child who is colorblind?
7. **Design an Experiment** Fruit fly sex is determined by X and Y chromosomes, just as it is in humans. Researchers suspect that a certain disease is caused by a recessive allele in a gene located on the X chromosome in fruit flies. Design an experiment to test this hypothesis.

14.2 Human Genetic Disorders

Understand Key Concepts

8. A mutation involving a change in a single DNA base pair
   a. will definitely result in a genetic disease.
   b. will have no effect on the organism's phenotype.
   c. will produce a positive change.
   d. may have an effect on the organism's phenotype.
9. Cystic fibrosis can be caused by
   a. nondisjunction of an autosome.
   b. a change of three base pairs in DNA.
   c. nondisjunction of a sex chromosome.
   d. deletion of an entire gene from a chromosome.
10. Malaria is a disease caused by a
    a. gene mutation.
    b. defect in red blood cells.
    c. bacterium found in water.
    d. parasite carried by mosquitoes.
11. Analyze the human karyotype below. Identify the chromosomal disorder that it shows.

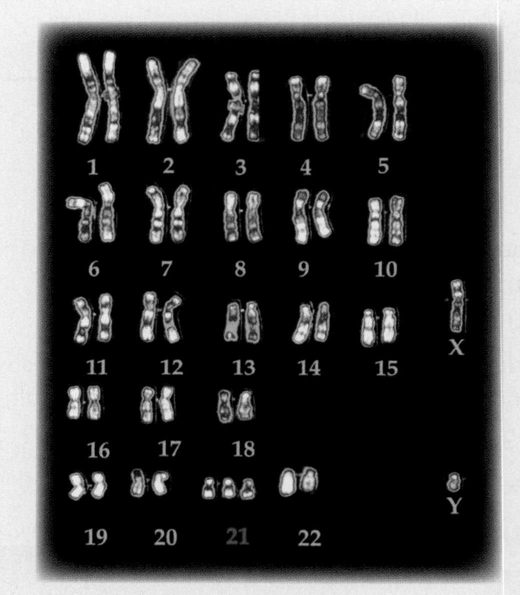

12. **Craft and Structure** What is a chromosomal disorder?
13. Describe two sex-chromosome disorders.

Think Critically

14. **Key Ideas and Details** Can a genetic counselor use a karyotype to identify a carrier of cystic fibrosis? Cite textual evidence to support your explanation.

15. **Interpret Graphs** What can you infer about the relationship between the age of the mother and the incidence of Down syndrome?

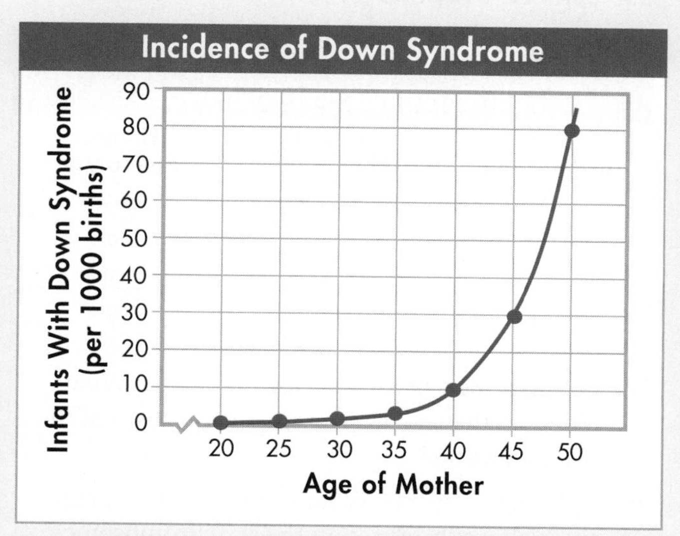

14.3 Studying the Human Genome

Understand Key Concepts

16. The human genome consists of approximately how many DNA base pairs?
    a. 30,000
    b. 3,000,000
    c. 300,000,000
    d. 3,000,000,000

17. The fraction of the human genome that actually codes for proteins is about
    a. 2%.
    b. 20%.
    c. 98%.
    d. 100%.

18. Cutting DNA into small pieces that can be sequenced is accomplished by
    a. restriction enzymes.
    b. DNA polymerase.
    c. gel electrophoresis.
    d. RNA polymerase.

19. If you sequence short pieces of DNA and then use a computer to find overlapping sequences that map to a much longer DNA fragment, you are using
    a. genomics.
    b. hapmaps.
    c. shotgun sequencing.
    d. open reading frame analysis.

20. Describe the tools and processes that scientists use to manipulate human DNA.

21. Explain why restriction enzymes are useful tools in sequencing DNA.

solve the CHAPTER MYSTERY

THE CROOKED CELL

When Ava inquired about her family's medical history, she found out that Uncle Eli's mother (Ava's grandmother) also had sickle cell disease, but Uncle Eli's father did not. One of her uncle's four children also had the disease. However, Ava's father, who is Eli's only sibling, did not have sickle cell disease, nor did Ava's mother. Ava's two siblings showed no signs of the disease, either.

1. **Key Ideas and Details** In general, what pattern of heredity does the sickle cell trait follow? Cite evidence from the chapter and its clues to support your conclusion.

2. **Draw Conclusions** Based on your answer to question 1, what can you conclude about the inheritance of sickle cell disease in Ava's family? What might be Ava's chances of being a carrier of the sickle cell trait?

3. **Classify** What kind of medical test could Ava request that would help determine whether or not she has the sickle cell trait? Explain your answer.

4. **Infer** The restriction enzyme *Mst*II, which cuts normal DNA at a particular site, will not recognize (and, therefore, will not cut) DNA that contains the sickle cell mutation. If Uncle Eli's DNA is cut with *Mst*II, will the restriction fragments be identical to those from his brother, Ava's father? Explain.

5. **Focus on the Big idea** Which technique(s) that you have read about in this chapter could be used to perform the kind of test described in question 4? Which technique could be used to analyze the results?

22. What is an SNP (single nucleotide polymorphism)?

23. What is bioinformatics?

Think Critically

24. **Draw Conclusions** Scientists have searched the human genome database to find possible promoter sequences. What is likely to be found near a promoter sequence?

25. **Infer** Why does DNA move toward the positive end of the gel during gel electrophoresis?

26. **Observe** The table below shows the DNA sequences that are recognized by five different restriction enzymes and the locations where those enzymes cut. Which enzymes produce DNA fragments with "sticky ends"? What is the common feature of the sequences cut by these enzymes?

DNA Sequences Cut by Enzymes

Enzyme	Recognition Sequence
*Alu*I	A G↓C T T C↑G A
*Hae*III	G G↓C C C C↑G G
*Bam*HI	G↓G A T C C C C T A G↑G
*Hind*III	A↓A G C T T T T C G A↑A
*Eco*RI	G↓A A T T C C T T A A↑G

Connecting Concepts

Use Science Graphics

Use the data table to answer questions 27 and 28.

Chromosomes and Phenotypes

Sex Chromosomes	Fruit Fly Phenotype	Human Phenotype
XX	Female	Female
XY	Male	Male
X	Male	Female
XXY	Female	Male

27. **Integration of Knowledge and Ideas** What differs in the sex-determining mechanism of the two organisms?

28. **Draw Conclusions** What can you logically conclude about the genes on the sex chromosomes of fruit flies and humans?

Write About Science

29. **Text Types and Purposes** Write a paragraph that tells how colorblindness is inherited. Describe the condition and explain why it is much more common in males. (*Hint*: Begin your paragraph with a topic sentence that expresses the paragraph's main idea.)

30. **Assess the Big idea** Explain the relationship between meiosis and Down syndrome, Turner's syndrome, and Klinefelter's syndrome. Make a plan, write a first draft, and then revise your answer.

Integration of Knowledge and Ideas

Hemophilia is an example of a sex-linked disorder. Two genes carried on the X chromosome help control blood clotting. A recessive allele in either of these two genes may produce hemophilia. The pedigree shows the transmission of hemophilia through three generations of a family.

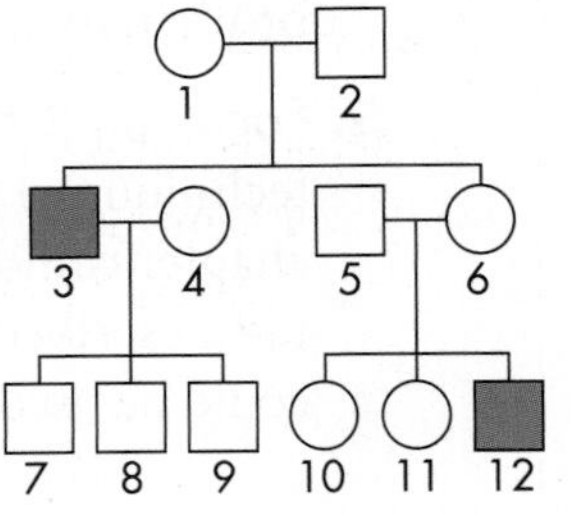

31. **Interpret Diagrams** Which mothers are definite carriers of the gene?

32. **Apply Concepts** Why did the sons of Person 3 not inherit the trait?

33. **Apply Concepts** How could Person 12 have hemophilia if neither of his parents had hemophilia?

Standardized Test Prep

Multiple Choice

1. Which of the following disorders can be observed in a human karyotype?
A colorblindness
B trisomy 21
C cystic fibrosis
D sickle cell disease

2. Which of the following disorders is a direct result of nondisjunction?
A sickle cell disease
B Turner's syndrome
C Huntington's disease
D cystic fibrosis

3. A woman is homozygous for A^- blood type. A man has AB^- blood type. What is the probability that the couple's child will have type B^- blood?
A 0%
B 50%
C 75%
D 100%

4. Cystic fibrosis is a genetic disorder caused by a
A single base substitution in the gene for hemoglobin.
B deletion of an amino acid from a chloride channel protein.
C defective gene found on the X chromosome.
D trisomy of chromosome 21.

5. The technique used to separate DNA strands of different lengths is
A gel electrophoresis.
B shotgun sequencing.
C restriction enzyme digestion.
D bioinformatics.

6. The study of whole genomes, including genes and their functions, is called
A bioinformatics.
B information science.
C life science.
D genomics.

7. DNA can be cut into shorter sequences by proteins known as
A haplotypes.
B polymerases.
C restriction enzymes.
D restriction fragments.

Questions 8–9

A student traced the recurrence of a widow's peak hairline in her family. Based on her interviews and observations, she drew the pedigree shown below.

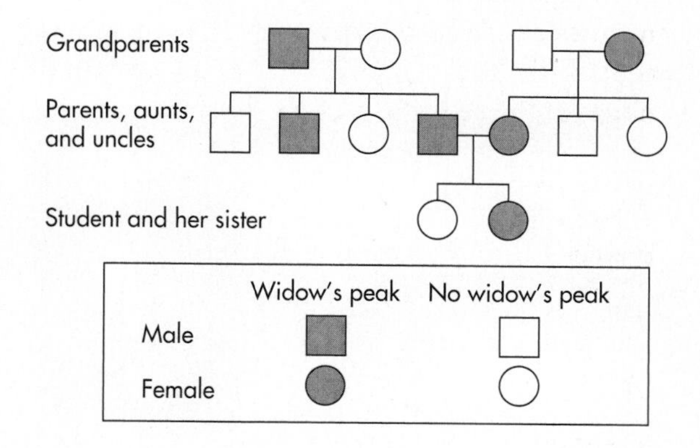

8. Which pattern of inheritance is consistent with the pedigree?
A sex-linked inheritance
B complete dominance
C codominance
D multiple alleles

9. What are the probable genotypes of the student's parents?
A Mother—*Ww*; Father—*ww*
B Mother—*ww*; Father—*ww*
C Mother—*WW*; Father—*Ww*
D Mother—*Ww*; Father—*Ww*

Open-Ended Response

10. Explain how the allele for sickle cell disease, which is a harmful allele when a person is homozygous, can be beneficial when a person is heterozygous.

If You Have Trouble With . . .

Question	1	2	3	4	5	6	7	8	9	10
See Lesson	14.1	14.2	14.1	14.2	14.3	14.3	14.3	14.1	14.1	14.2

15 Genetic Engineering

Big idea: Science as a Way of Knowing

Q: How and why do scientists manipulate DNA in living cells?

By cloning cells and modifying genes, scientists in Korea have developed cats that glow bright red in the dark. The cloned Turkish Angola on the left has a fluorescent protein in its skin cells. The protein gives off a red glow when exposed to ultraviolet light. The ordinary Turkish Angola on the right lacks the red fluorescent protein, so it appears green under ultraviolet light.

INSIDE:

- 15.1 Selective Breeding
- 15.2 Recombinant DNA
- 15.3 Applications of Genetic Engineering
- 15.4 Ethics and Impacts of Biotechnology

BUILDING *Scientific Literacy*

Deepen your understanding of genetic engineering by engaging in key practices that allow you to make connections across concepts.

NGSS You will describe and evaluate different methods used to engineer traits expressed in an organism.

STEM What are some benefits of recombinant DNA technology for humans? Use examples of genetically modified organisms and your imagination to answer this question.

Common Core You will research the rules that forensic scientists use to ensure that evidence is not contaminated before and during testing.

CHAPTER MYSTERY

A CASE OF MISTAKEN IDENTITY

In the summer of 1998, an elderly Indiana woman was brutally assaulted. In the predawn darkness, she didn't get a look at her assailant's face.

At first light, police found a man only a few blocks from the victim's house. He was unconscious, his clothing was stained with blood, and there were scratches on his forearms. The man claimed that he had passed out following a drunken brawl. He couldn't remember what had happened afterward. The blood type of the stains on his clothing matched the victim's blood type. The police thought they had their man.

Hours later, the police knew they had the wrong suspect. They resumed their search for the real attacker, who was subsequently caught, tried, and convicted. As you read this chapter, look for clues to help you determine how the police knew they had the wrong suspect. Then, solve the mystery.

Biology.com

Finding the solution to the mistaken identity mystery is only the beginning. Take a video field trip with the ecogeeks of Untamed Science to continue exploring your world.

Go online to access additional resources including:
• eText • Flash Cards • Lesson Overviews • Chapter Mystery

15.1 Selective Breeding

Key Questions

What is selective breeding used for?

How do people increase genetic variation?

Vocabulary

selective breeding
hybridization
inbreeding
biotechnology

Taking Notes

Outline Before you read this lesson, start an outline. Use the green headings in the lesson as first-level entries. Use the blue headings as second-level entries, leaving space after each entry. As you read, summarize the key ideas below your entries.

THINK ABOUT IT You've enjoyed popcorn at the movies, you've probably made it at home, and you've certainly seen it in stores. Where does it come from? Would you be surprised to learn that popcorn is one of the earliest examples of human efforts to select and improve living organisms for our benefit? Corn as we know it was domesticated at least 6000 years ago by Native Americans living in Mexico. A tiny kernel of popped corn found in a cave in New Mexico is more than 5000 years old!

Selective Breeding

What is selective breeding used for?

Visit a dog show, and what do you see? Striking contrasts are everywhere—from a tiny Chihuahua to a massive Great Dane, from the short coat of a Labrador retriever to the curly fur of a poodle, from the long muzzle of a wolfhound to the pug nose of a bulldog. The differences among breeds of dogs, like the ones in **Figure 15–1,** are so great that someone might think they are different species. They're not, of course, but where did these obvious differences come from?

The answer is that we did it. Humans have kept and bred dogs for thousands of years, always looking to produce animals that are better hunters, better retrievers, or better companions. We've done so by **selective breeding,** allowing only those animals with wanted characteristics to produce the next generation. **Humans use selective breeding, which takes advantage of naturally occurring genetic variation, to pass wanted traits on to the next generation of organisms.**

FIGURE 15–1 Dog Breeds There are more than 150 dog breeds, and many new breeds are still being developed.

For thousands of years, we've produced new varieties of cultivated plants and nearly all domestic animals—including horses, cats, and cows—by selectively breeding for particular traits. Long before Europeans came to the New World, Native Americans had selectively bred teosinte (tee oh SIN tee), a wild grass native to central Mexico, to produce corn, a far more productive and nutritious plant. **Figure 15–2** shows both plants. Corn is now one of the world's most important crops. There are two common methods of selective breeding—hybridization and inbreeding.

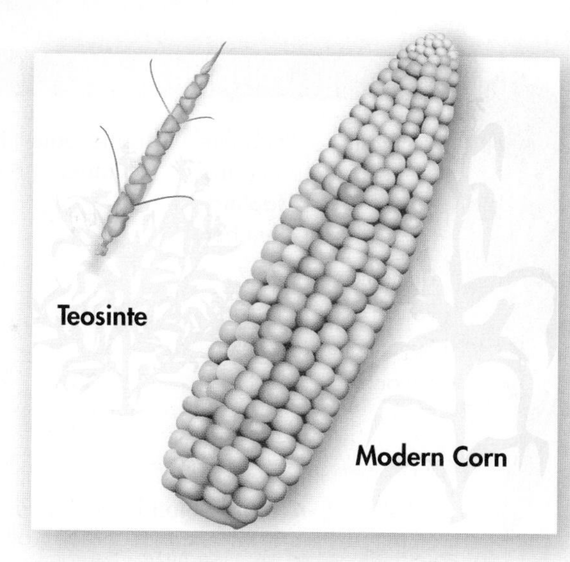

FIGURE 15–2 Corn From Teosinte Modern corn was selectively bred from teosinte at least 6000 years ago. During its domestication, corn lost the ability to survive in the wild but gained valuable agricultural traits. For example, the hard case around the kernel disappeared over time, leaving the rows of soft corn kernels we enjoy today. **Observe** ***What other differences can you see between the two plants?***

Hybridization American botanist Luther Burbank may have been the greatest selective breeder of all time. During his lifetime (1849–1926), he developed more than 800 varieties of plants. As one of his tools, Burbank used **hybridization,** crossing dissimilar individuals to bring together the best of both organisms. Hybrids—the individuals produced by such crosses—are often hardier than either of the parents. Many of Burbank's hybrid crosses combined the disease resistance of one plant with the food-producing capacity of another. The result was a new line of plants that had the traits farmers needed to increase food production. **Figure 15–3** shows a type of peach developed using Burbank's methods.

Inbreeding To maintain desirable characteristics in a line of organisms, breeders often use a technique known as inbreeding. **Inbreeding** is the continued breeding of individuals with similar characteristics. The many breeds of dogs—from beagles to poodles—are maintained using this practice. Inbreeding helps ensure that the characteristics that make each breed unique are preserved. Although inbreeding is useful in preserving certain traits, it can be risky. Most of the members of a breed are genetically similar, which increases the chance that a cross between two individuals will bring together two recessive alleles for a genetic defect.

In Your Notebook *Compare and contrast hybridization and inbreeding.*

FIGURE 15–3 Selectively Bred Fruit Luther Burbank used hybridization—a form of selective breeding—to develop a variety of plants. These July Elberta peaches, *Prunus persica*, are among his most successful varieties.

Increasing Variation

How do people increase genetic variation?

Selective breeding would be nearly impossible without the wide variation found in natural populations of plants and animals. But sometimes breeders want more variation than exists in nature. **Breeders can increase the genetic variation in a population by introducing mutations, which are the ultimate source of biological diversity.**

When scientists manipulate the genetic makeup of an organism, they are using biotechnology. **Biotechnology** is the application of a technological process, invention, or method to living organisms. Selective breeding is one form of biotechnology important in agriculture and medicine, but there are many others.

Plant	Probable Ancestral Haploid Number	Chromosome Number	Ploidy Level
Domestic oat	7	42	6N
Peanut	10	40	4N
Sugar cane	10	80	8N
Banana	11	22, 33	2N, 3N
Cotton	13	52	4N

Polyploid Crops

FIGURE 15–4 Ploidy Numbers Because polyploid plants are often larger than other plants, many farmers deliberately grow polyploid varieties of crops like those listed above. **Interpret Tables** ***Which plant has undergone the most dramatic changes in chromosome number?***

Bacterial Mutations Mutations—heritable changes in DNA—occur spontaneously, but breeders can increase the mutation rate of an organism by using radiation or chemicals. Many mutations are harmful to the organism. With luck and perseverance, however, breeders can often produce a few mutants—individuals with mutations—with useful characteristics that are not found in the original population. This technique has been particularly useful with bacteria. Because they are small, millions of bacteria can be treated with radiation or chemicals at the same time, which increases the chances of producing a useful mutant. This technique has allowed scientists to develop hundreds of useful bacterial strains. For instance, we have known for decades that certain strains of oil-digesting bacteria are effective for cleaning up oil spills. Today scientists are working to produce bacteria that can clean up radioactive substances and metal pollution in the environment.

Polyploid Plants Drugs that prevent the separation of chromosomes during meiosis are very useful in plant breeding. These drugs can produce cells that have many times the normal number of chromosomes. Plants grown from these cells are called polyploid because they have many sets of chromosomes. Polyploidy is usually fatal in animals. But, for reasons that are not clear, plants are much better at tolerating extra sets of chromosomes. Polyploidy can quickly produce new species of plants that are larger and stronger than their diploid relatives. A number of important crop plants, including bananas and many varieties of citrus fruits, have been produced in this way. **Figure 15 – 4** lists several examples of polyploid plants.

15.1 Assessment

Review Key Concepts

1. **a. Review** Give an example of selective breeding.
   b. Compare and Contrast Suppose you are a geneticist trying to develop a sunflower with red petals and a short stem. As you compare the sunflowers you have on hand, what genetic variations would you look for? What kinds of plants would you select for crossing?
   c. Draw Conclusions How is selective breeding a form of biotechnology?
2. **a. Review** What is the relationship between genetic variations and mutations?
   b. Explain How can breeders introduce mutations?

WRITE ABOUT SCIENCE

Explanation

3. Write a paragraph in which you suggest ways that plants could be genetically altered to improve the world's food supply. (*Hint*: The first sentence in your paragraph should express the paragraph's main idea.)

15.2 Recombinant DNA

THINK ABOUT IT Suppose you have an electronic game you want to change. Knowing that the game depends on a coded program in a computer microchip, how would you set about rewriting the program? First you'd need a way to get the existing program out of the microchip. Then you'd have to read the program, make the changes you want, and put the modified code back into the microchip. What does this scenario have to do with genetic engineering? Just about everything.

Key Questions

How do scientists copy the DNA of living organisms?

How is recombinant DNA used?

How can genes from one organism be inserted into another organism?

Vocabulary

polymerase chain reaction
recombinant DNA
plasmid
genetic marker
transgenic
clone

Taking Notes

Preview Visuals Before you read, preview **Figure 15–7** and write down any questions you may have about the figure. As you read, find answers to your questions.

Copying DNA

How do scientists copy the DNA of living organisms?

Until recently plant and animal breeders could only work with variations that already exist in nature. Even when breeders tried to add variation by introducing mutations, the changes they produced were unpredictable. Today genetic engineers can transfer certain genes at will from one organism to another, designing new living things to meet specific needs.

Recall from Chapter 14 that it is relatively easy to extract DNA from cells and tissues. The extracted DNA can be cut into fragments of manageable size using restriction enzymes. These restriction fragments can then be separated according to size using gel electrophoresis or another similar technique. That's the easy part. The tough part comes next: How do you find a specific gene?

The problem is huge. If we were to cut DNA from a bacterium like *E. coli* into restriction fragments averaging 1000 base pairs in length, we would have 4000 restriction fragments. In the human genome, we would have 3 million restriction fragments. How do we find the DNA of a single gene among millions of fragments? In some respects, it's the classic problem of finding a needle in a haystack—we have an enormous pile of hay and just one needle.

Actually, there is a way to find a needle in a haystack. We can toss the hay in front of a powerful magnet until something sticks. The hay won't stick, but a needle made of iron or steel will. Believe it or not, similar techniques can help scientists identify specific genes.

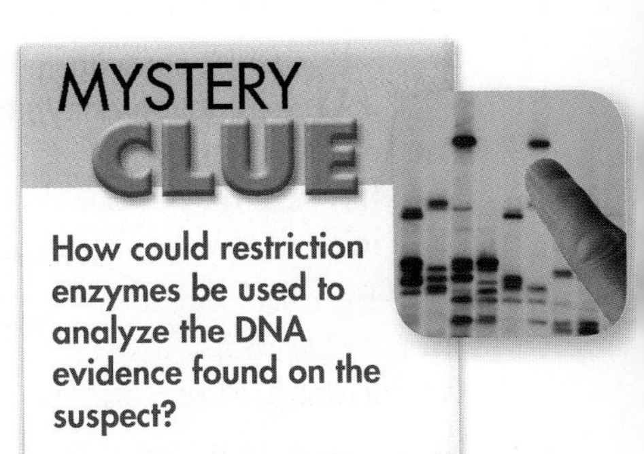

MYSTERY CLUE

How could restriction enzymes be used to analyze the DNA evidence found on the suspect?

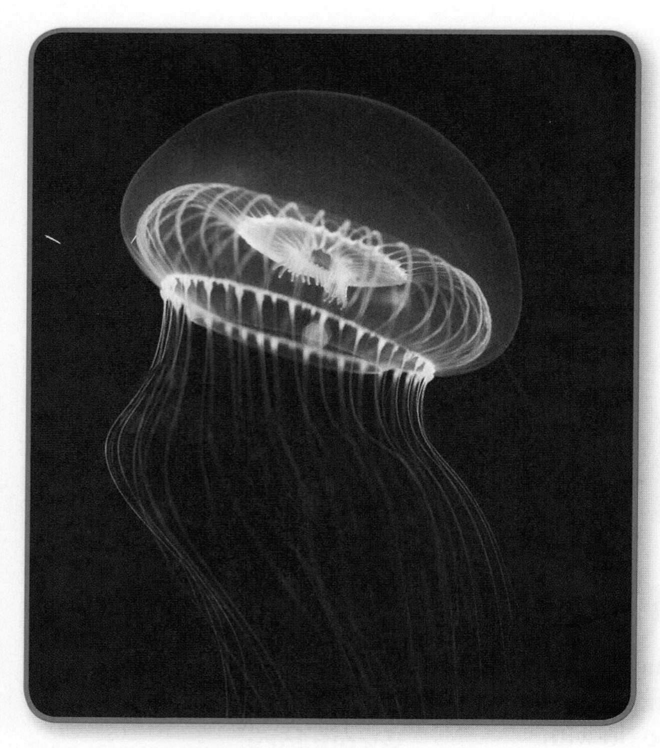

FIGURE 15–5 A Fluorescent Gene The Pacific Ocean jellyfish, *Aequoria victoria,* emits a bluish glow. A protein in the jellyfish absorbs the blue light and produces green fluorescence. This protein, called GFP, is now widely used in genetic engineering.

Finding Genes In 1987, Douglas Prasher, a biologist at Woods Hole Oceanographic Institute in Massachusetts, wanted to find a specific gene in a jellyfish. The gene he hoped to identify is the one that codes for a molecule called green fluorescent protein, or GFP. This natural protein, found in the jellyfish shown in **Figure 15–5,** absorbs energy from light and makes parts of the jellyfish glow. Prasher thought that GFP from the jellyfish could be used to report when a protein was being made in a cell. If he could somehow link GFP to a specific protein, it would be a bit like attaching a light bulb to that molecule.

To find the GFP gene, Prasher studied the amino acid sequence of part of the GFP protein. By comparing this sequence to a genetic code table, he was able to predict a probable mRNA base sequence that would have coded for this sequence of amino acids. Next, Prasher used a complementary base sequence to "attract" an mRNA that matched his prediction and would bind to that sequence by base pairing. After screening a genetic "library" with thousands of different mRNA sequences from the jellyfish, he found one that bound perfectly.

After Prasher located the mRNA that produced GFP, he set out to find the actual gene. Taking a gel in which restriction fragments from the jellyfish genome had been separated, he found that one of the fragments bound tightly to the mRNA. That fragment contained the actual gene for GFP, which is now widely used to label proteins in living cells. The method he used, shown in **Figure 15–6,** is called Southern blotting. Today it is often quicker and less expensive for scientists to search for genes in computer databases where the complete genomes of many organisms are available.

FIGURE 15–6 Southern Blotting Southern blot analysis, named after its inventor Edwin Southern, is a research technique for finding specific DNA sequences, among dozens. A labeled piece of nucleic acid serves as a probe among the DNA fragments.

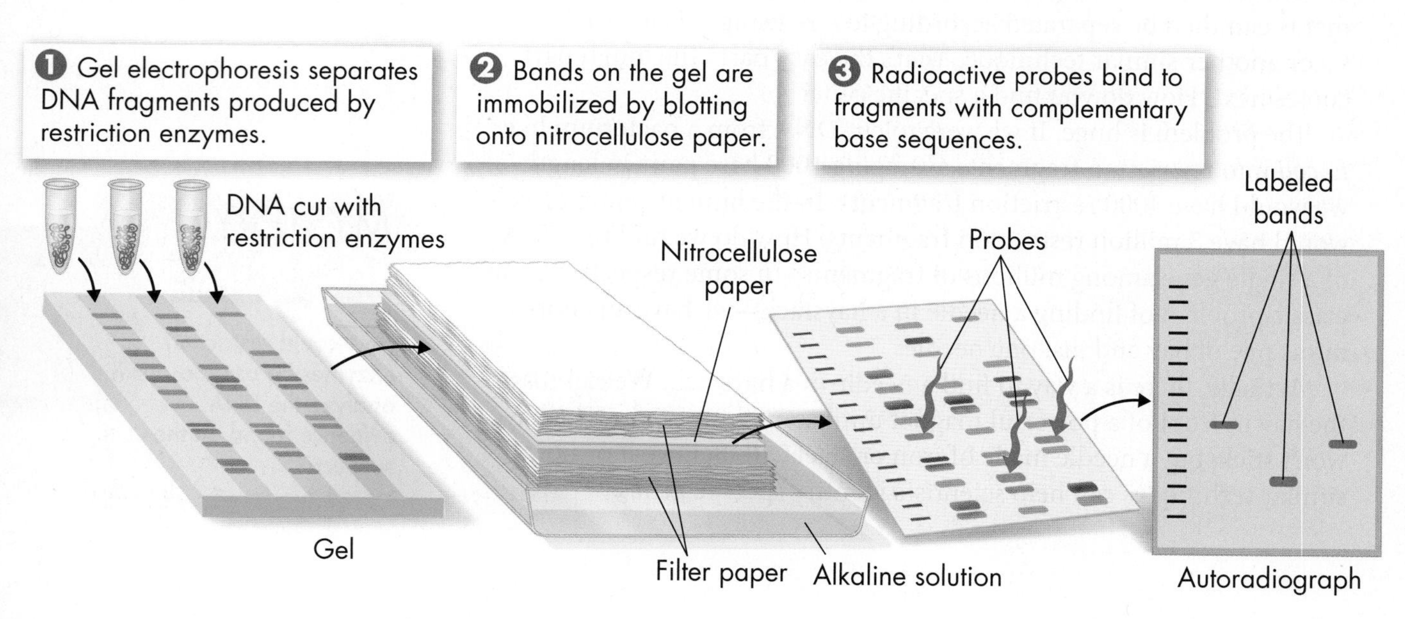

Polymerase Chain Reaction Once they find a gene, biologists often need to make many copies of it. A technique known as **polymerase chain reaction** (PCR) allows them to do exactly that. At one end of the original piece of DNA, a biologist adds a short piece of DNA that complements a portion of the sequence. At the other end, the biologist adds another short piece of complementary DNA. These short pieces are known as primers because they prepare, or prime, a place for DNA polymerase to start working.

As **Figure 15–7** suggests, the idea behind the use of PCR primers is surprisingly simple. **The first step in using the polymerase chain reaction method to copy a gene is to heat a piece of DNA, which separates its two strands. Then, as the DNA cools, primers bind to the single strands. Next, DNA polymerase starts copying the region between the primers. These copies can serve as templates to make still more copies.** In this way, just a few dozen cycles of replication can produce billions of copies of the DNA between the primers.

Where did Kary Mullis, the American scientist who invented PCR, find a DNA polymerase enzyme that could stand repeated cycles of heating and cooling? Mullis found it in bacteria from the hot springs of Yellowstone National Park in the northwestern United States—a powerful example of the importance of biodiversity to biotechnology!

 **In Your Notebook** *List the steps in the PCR method.*

Changing DNA

How is recombinant DNA used?

Just as they were beginning to learn how to read and analyze DNA sequences, scientists began wondering if it might be possible to change the DNA of a living cell. As many of them realized, this feat had already been accomplished decades earlier. Do you remember Griffith's experiments on bacterial transformation? During transformation, a cell takes in DNA from outside the cell, and that added DNA becomes a component of the cell's own genome. Today biologists understand that Griffith's extract of heat-killed bacteria contained DNA fragments. When he mixed those fragments with live bacteria, a few of them took up the DNA molecules, transforming them and changing their characteristics. Griffith, of course, could only do this with DNA extracted from other bacteria.

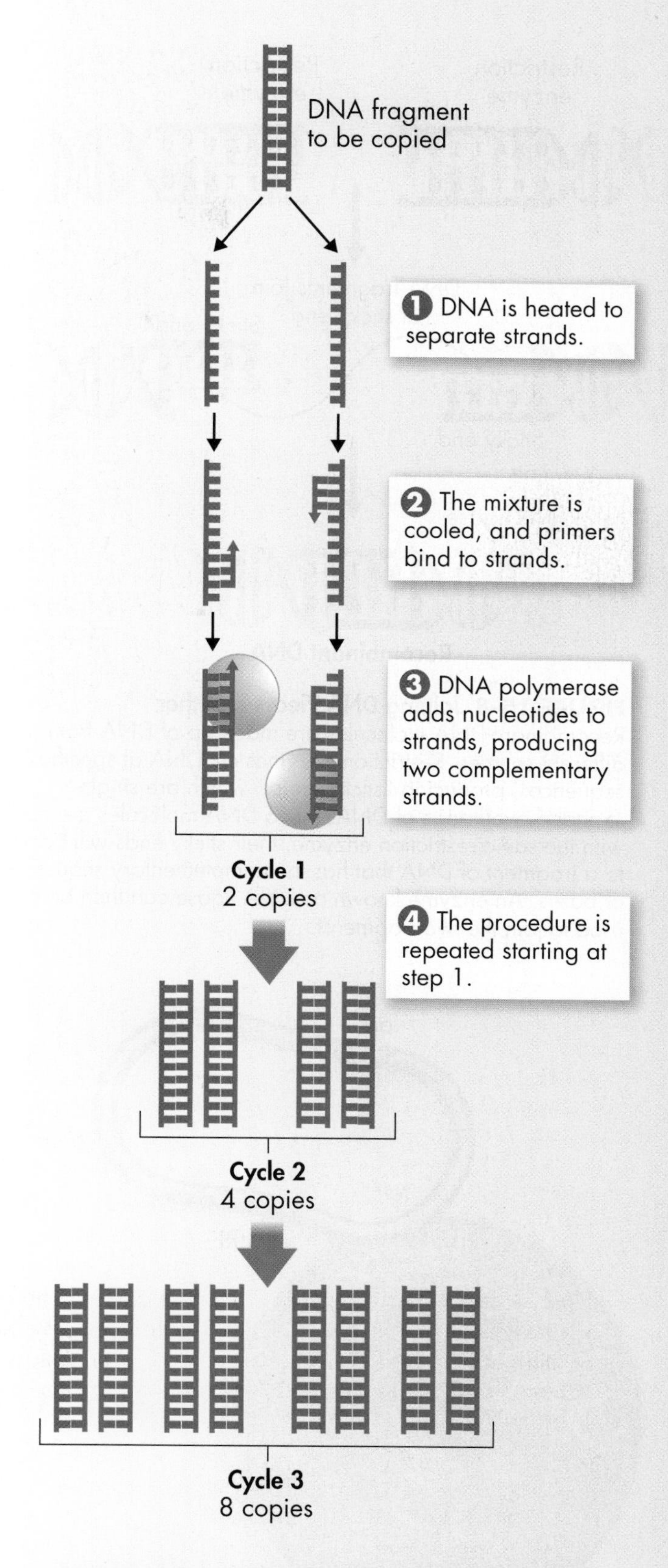

FIGURE 15–7 The PCR Method Polymerase chain reaction is used to make multiple copies of a gene. This method is particularly useful when only tiny amounts of DNA are available. **Calculate** ***How many copies of the DNA fragment will there be after six PCR cycles?*** MATH

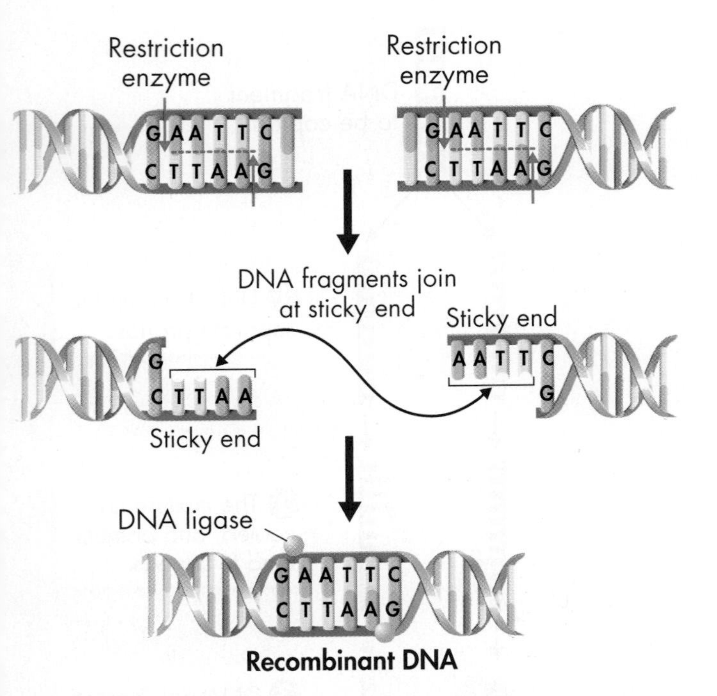

FIGURE 15–8 Joining DNA Pieces Together Recombinant DNA molecules are made up of DNA from different sources. Restriction enzymes cut DNA at specific sequences, producing "sticky ends," which are single-stranded overhangs of DNA. If two DNA molecules are cut with the same restriction enzyme, their sticky ends will bond to a fragment of DNA that has the complementary sequence of bases. An enzyme known as DNA ligase can then be used to join the two fragments.

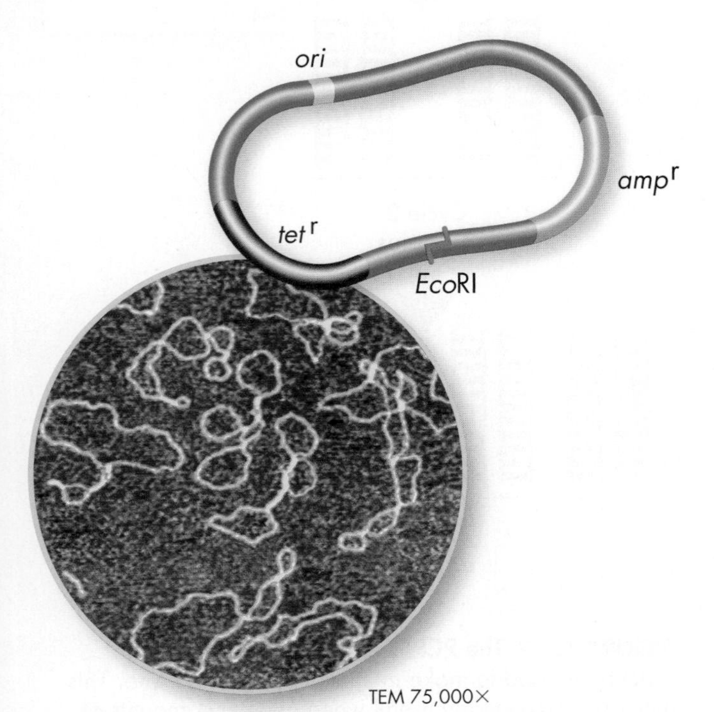

FIGURE 15–9 A Plasmid Map Plasmids used for genetic engineering typically contain a replication start signal, called the origin of replication (*ori*), and a restriction enzyme cutting site, such as *Eco*RI. They also contain genetic markers, like the antibiotic resistance genes tet^r and amp^r shown here.

Combining DNA Fragments With today's technologies, scientists can produce custom-built DNA molecules in the lab and then insert those molecules—along with the genes they carry—into living cells. The first step in this sort of genetic engineering is to build a DNA sequence with the gene or genes you'd like to insert into a cell. Machines known as DNA synthesizers can produce short pieces of DNA, up to several hundred bases in length. These synthetic sequences can then be joined to natural sequences using DNA ligase or other enzymes that splice DNA together. These same enzymes make it possible to take a gene from one organism and attach it to the DNA of another organism, as shown in **Figure 15–8.** The resulting molecules are called **recombinant DNA.** This technology relies on the fact that any pair of complementary sequences tends to bond, even if each sequence comes from a different organism. **Recombinant-DNA technology—joining together DNA from two or more sources—makes it possible to change the genetic composition of living organisms.** By manipulating DNA in this way, scientists can investigate the structure and functions of genes.

Plasmids and Genetic Markers Scientists working with recombinant DNA soon discovered that many of the DNA molecules they tried to insert into host cells simply vanished because the cells often did not copy, or replicate, the added DNA. Today scientists join recombinant DNA to another piece of DNA containing a replication "start" signal. This way, whenever the cell copies its own DNA, it copies the recombinant DNA too.

In addition to their own chromosomes, some bacteria contain small circular DNA molecules known as **plasmids.** Plasmids, like those shown in **Figure 15–9,** are widely used in recombinant DNA studies. Joining DNA to a plasmid, and then using the recombinant plasmid to transform bacteria, results in the replication of the newly added DNA along with the rest of the cell's genome.

Plasmids are also found in yeasts, which are single-celled eukaryotes that can be transformed with recombinant DNA as well. Biologists working with yeasts can construct artificial chromosomes containing centromeres, telomeres, and replication start sites. These artificial chromosomes greatly simplify the process of introducing recombinant DNA into the yeast genome.

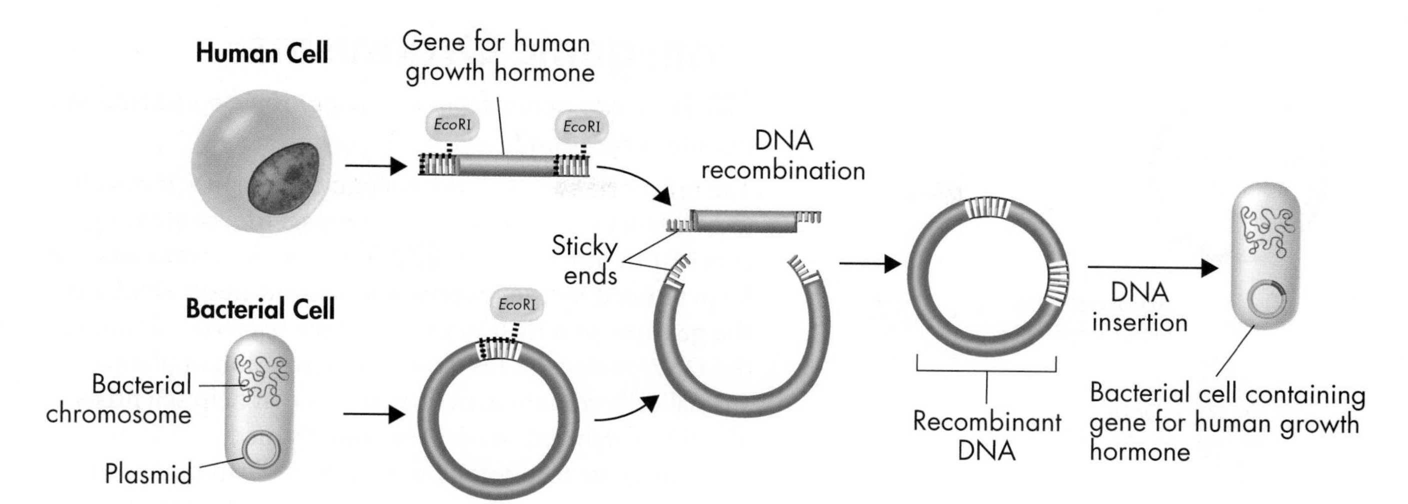

FIGURE 15–10 Plasmid DNA Transformation
Scientists can insert a piece of DNA into a plasmid if both the plasmid and the target DNA have been cut by the same restriction enzymes to create sticky ends. With this method, bacteria can be used to produce human growth hormone. First, a human gene is inserted into bacterial DNA. Then, the new combination of genes is returned to a bacterial cell, which replicates the recombinant DNA over and over again. Infer ***Why might scientists want to copy the gene for human growth hormone?***

Figure 15–10 shows how bacteria can be transformed using recombinant plasmids. First, the DNA being used for transformation is joined to a plasmid. The plasmid DNA contains a signal for replication, helping to ensure that if the DNA does get inside a bacterial cell, it will be replicated. In addition, the plasmid also has a genetic marker, such as a gene for antibiotic resistance. A **genetic marker** is a gene that makes it possible to distinguish bacteria that carry the plasmid from those that don't. Using genetic markers, researchers can mix recombinant plasmids with a culture of bacteria, add enough DNA to transform just one cell in a million, and still locate that one cell. After transformation, the culture is treated with an antibiotic. Only those rare cells that have been transformed survive, because only they carry the resistance gene.

In Your Notebook *Write a summary of the process of plasmid DNA transformation.*

Quick Lab
GUIDED INQUIRY

Inserting Genetic Markers

❶ Write a random DNA sequence on a long strip of paper to represent an organism's genome.

❷ Have your partner write a short DNA sequence on a short strip of paper to represent a marker gene.

❸ Using the chart your teacher gives you, work with your partner to figure out how to insert the marker gene into the genome.

Analyze and Conclude

1. Apply Concepts Which restriction enzyme did you use? Why?

2. Use Models What kind of molecule did you and your partner develop?

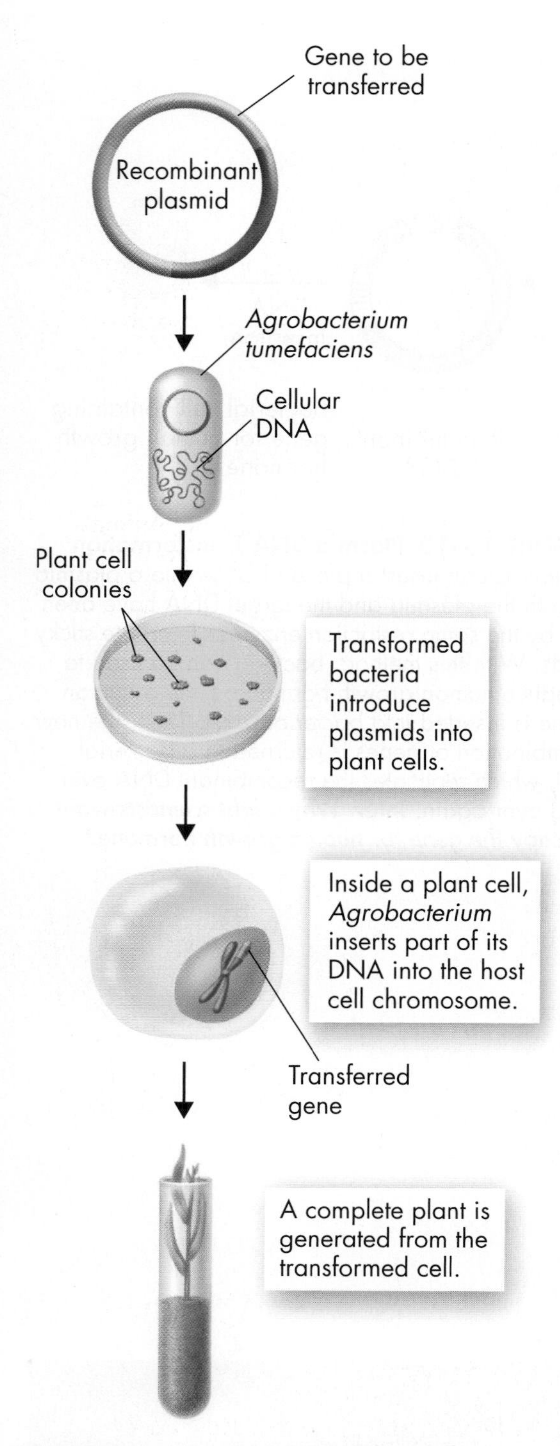

FIGURE 15–11 Transforming a Plant Cell *Agrobacterium* can be used to introduce bacterial DNA into a plant cell. The transformed cells can be cultured to produce adult plants.

Transgenic Organisms

How can genes from one organism be inserted into another organism?

The universal nature of the genetic code makes it possible to construct organisms that are **transgenic,** containing genes from other species. **Transgenic organisms can be produced by the insertion of recombinant DNA into the genome of a host organism.** Like bacterial plasmids, the DNA molecules used for transformation of plant and animal cells contain genetic markers that help scientists identify which cells have been transformed.

Transgenic technology was perfected using mice in the 1980s. Genetic engineers can now produce transgenic plants, animals, and microorganisms. By examining the traits of a genetically modified organism, it is possible to learn about the function of the transferred gene. This ability has contributed greatly to our understanding of gene regulation and expression.

Transgenic Plants Many plant cells can be transformed using *Agrobacterium.* In nature this bacterium inserts a small DNA plasmid that produces tumors in a plant's cells. Scientists can deactivate the plasmid's tumor-producing gene and replace it with a piece of recombinant DNA. The recombinant plasmid can then be used to infect and transform plant cells, as shown in **Figure 15–11.**

There are other ways to produce transgenic plants as well. When their cell walls are removed, plant cells in culture will sometimes take up DNA on their own. DNA can also be injected directly into some cells. If transformation is successful, the recombinant DNA is integrated into one of the plant cell's chromosomes.

Transgenic Animals Scientists can transform animal cells using some of the same techniques used for plant cells. The egg cells of many animals are large enough that DNA can be injected directly into the nucleus. Once the DNA is in the nucleus, enzymes that are normally responsible for DNA repair and recombination may help insert the foreign DNA into the chromosomes of the injected cell.

Recently it has become possible to eliminate particular genes by carefully engineering the DNA molecules that are used for transformation. The DNA molecules can be constructed with two ends that will sometimes recombine with specific sequences in the host chromosome. Once they do, the host gene normally found between those two sequences may be lost or specifically replaced with a new gene. This kind of gene replacement has made it possible to pinpoint the specific functions of genes in many organisms, including mice.

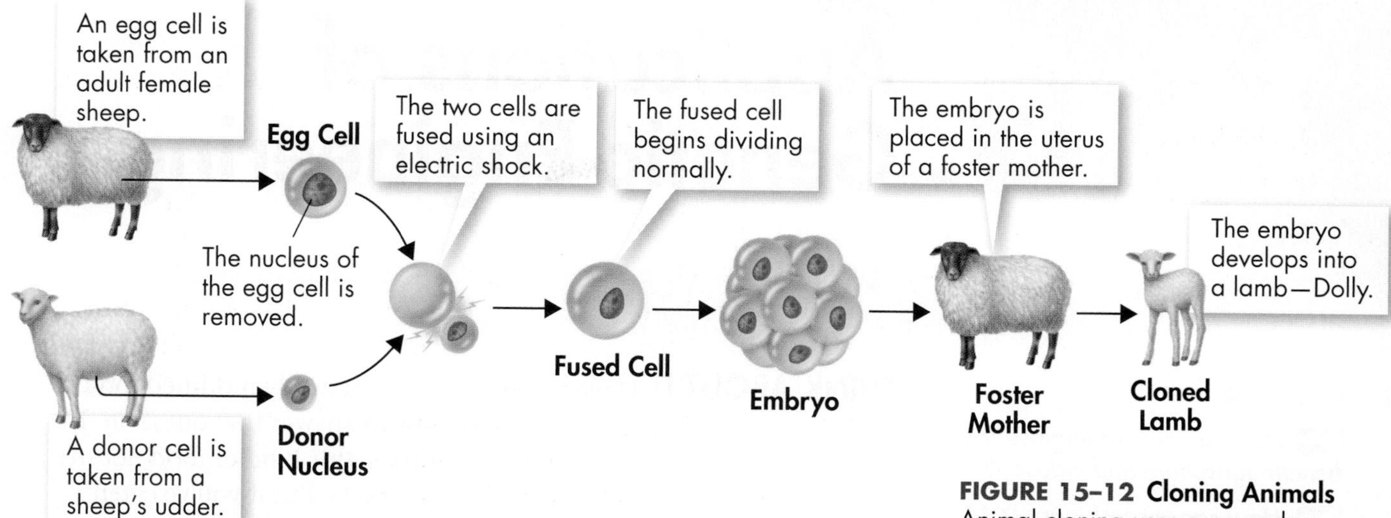

FIGURE 15–12 Cloning Animals Animal cloning uses a procedure called nuclear transplantation. The process combines an egg cell with a donor nucleus to produce an embryo. **Apply Concepts** ***Why won't the cloned lamb resemble its foster mother?***

Cloning A **clone** is a member of a population of genetically identical cells produced from a single cell. The technique of cloning uses a single cell from an adult organism to grow an entirely new individual that is genetically identical to the organism from which the cell was taken.

Cloned colonies of bacteria and other microorganisms are easy to grow, but this is not always true of multicellular organisms, especially animals. Clones of animals were first produced in 1952 using amphibian tadpoles. In 1997, Scottish scientist Ian Wilmut stunned biologists by announcing that he had produced a sheep, called Dolly, by cloning.

Figure 15–12 shows the basic steps by which an animal can be cloned. First, the nucleus of an unfertilized egg cell is removed. Next, the egg cell is fused with a donor cell that contains a nucleus, taken from an adult. The resulting diploid egg develops into an embryo, which is then implanted in the uterine wall of a foster mother, where it develops until birth. Cloned cows, pigs, mice, and even cats have since been produced using similar techniques.

15.2 Assessment

Review Key Concepts

1. a. **Review** Describe the process scientists use to copy DNA.
   b. **Infer** Why would a scientist want to know the sequence of a DNA molecule?
2. a. **Review** How do scientists use recombinant DNA?
   b. **Use Analogies** How is genetic engineering like computer programming?
3. a. **Review** What is a transgenic organism?
   b. **Compare and Contrast** Compare the transformation of a plant cell with the transformation of an animal cell.

PRACTICE PROBLEM

4. Design an experiment to find a way to treat disorders caused by a single gene. State your hypothesis and list the steps you would follow. (*Hint*: Think about the uses of recombinant DNA.)

15.3 Applications of Genetic Engineering

Key Questions

How can genetic engineering benefit agriculture and industry?

How can recombinant-DNA technology improve human health?

How is DNA used to identify individuals?

Vocabulary

gene therapy
DNA microarray
DNA fingerprinting
forensics

Taking Notes

Outline Make an outline of this lesson by using the green and blue headings. As you read, take notes on the different applications of genetic engineering.

THINK ABOUT IT Have you eaten any genetically modified food lately? Don't worry if you're not sure how to answer that question. In the United States and many other countries, this kind of food doesn't have to be labeled in grocery stores or markets. But if you've eaten corn, potatoes, or soy products in any of your meals this week, chances are close to 100 percent that you've eaten foods modified in some way by genetic engineering.

Agriculture and Industry

How can genetic engineering benefit agriculture and industry?

Everything we eat and much of what we wear come from living organisms. Not surprisingly, then, researchers have used genetic engineering to try to improve the products we get from plants and animals. **Ideally, genetic modification could lead to better, less expensive, and more nutritious food as well as less-harmful manufacturing processes.**

GM Crops Since their introduction in 1996, genetically modified (GM) plants, like the soybeans in **Figure 15–13,** have become an important component of our food supply. In 2007, GM crops made up 92 percent of soybeans, 86 percent of cotton, and 80 percent of corn grown in the United States. One type of modification, which has already proved particularly useful to agriculture, uses bacterial genes that produce a protein known as Bt toxin. While this toxin is harmless to humans and most other animals, enzymes in the digestive systems of insects convert Bt to a form that kills the insects. Plants with the Bt gene, then, do not have to be sprayed with pesticides. In addition, they produce higher yields of crops.

Resistance to insects is just one useful characteristic being engineered into crops. Others include resistance to herbicides, which are chemicals that destroy weeds, and resistance to viral infections. Some transgenic plants may soon produce foods that are resistant to rot and spoilage. And engineers are currently developing GM plants that may produce plastics for the manufacturing industry.

FIGURE 15–13 GM Soybeans Genetically modified soybeans are a popular crop in the United States.

Analyzing Data

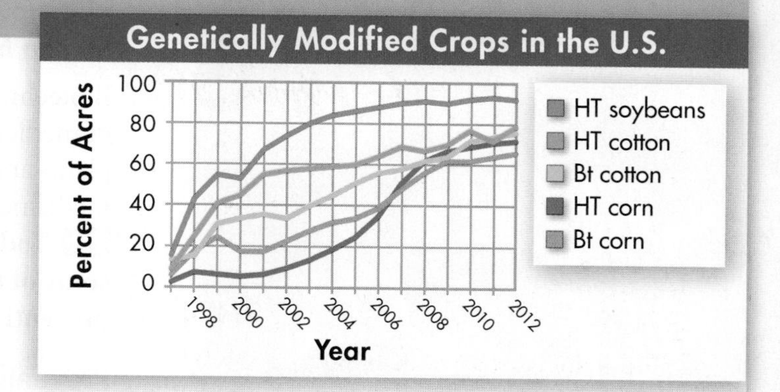

Source: U.S. Department of Agriculture Economic Research Service Data Sets

Genetically Modified Crops in the United States

U.S. farmers have adopted GM crops widely since their introduction in 1996. Soybeans, cotton, and corn have been modified to tolerate herbicides and resist insect damage. The graph at the right summarizes the extent to which these crops were adopted between 1997 and 2012. The modified traits shown here include herbicide tolerance (HT) and insect resistance (Bt).

1. Analyze Data Which two crops were most widely and rapidly adopted?

2. Draw Conclusions Why do you think the levels of adoption fell at certain points over the period?

3. Predict What do you think will happen to HT soybeans and HT corn over the next few years? Why? Use the graph to support your prediction.

4. Infer Why do you think an increasing number of farmers have chosen to grow crops with herbicide tolerance?

GM Animals Transgenic animals are also becoming more important to our food supply. For example, about 30 percent of the milk in U.S. markets comes from cows that have been injected with hormones made by recombinant-DNA techniques to increase milk production. Pigs can be genetically modified to produce more lean meat or high levels of healthy omega-3 acids. Using growth-hormone genes, scientists have developed transgenic salmon that grow much more quickly than wild salmon. This effort makes it practical to grow these nutritious fish in captive aquaculture facilities that do not threaten wild populations.

When scientists in Canada combined spider genes into the cells of lactating goats, the goats began to manufacture silk along with their milk. By extracting polymer strands from the milk and weaving them into thread, we can create a light, tough, and flexible material that could be used in such applications as military uniforms, medical sutures, and tennis racket strings. Scientists are now using human genes to develop antibacterial goat milk.

Researchers hope that cloning will enable them to make copies of transgenic animals, which would increase the food supply and could even help save endangered species. In 2008, the U.S. government approved the sale of meat and milk from cloned animals. Many farmers and ranchers hope that cloning technology will allow them to duplicate the best qualities of prize animals without the time and complications of traditional breeding.

FIGURE 15–14 Antibacterial Goat Milk Scientists are working to combine a gene for lysozyme — an antibacterial protein found in human tears and breast milk — into the DNA of goats. Milk from these goats may help prevent infections in young children who drink it.
Apply Concepts ***What action do scientists hope the lysozyme gene will take in genetically modified goats?***

In Your Notebook *Describe the ways in which GM organisms can benefit agriculture and industry.*

FIGURE 15–15 Vitamin-Rich Rice
Golden rice is a GM plant that contains increased amounts of provitamin A, or beta-carotene. Two genes engineered into the rice genome help the grains produce and accumulate beta-carotene. The intensity of the golden color indicates the concentration of beta-carotene in the edible part of the rice seed.

Health and Medicine

How can recombinant-DNA technology improve human health?

Biotechnology, in its broadest sense, has always been part of medicine. Early physicians extracted substances from plants and animals to cure their patients. Twentieth-century medicine saw the use of vaccination to save countless lives. **Today, recombinant-DNA technology is the source of some of the most important and exciting advances in the prevention and treatment of disease.**

Preventing Disease One interesting development in transgenic technology is golden rice, shown in **Figure 15–15.** This rice contains increased amounts of provitamin A, also known as beta-carotene—a nutrient that is essential for human health. Provitamin A deficiencies produce serious medical problems, including infant blindness. There is hope that provitamin A-rich golden rice will help prevent these problems. Other scientists are developing transgenic plants and animals that produce human antibodies to fight disease.

In the future, transgenic animals may provide us with an ample supply of our own proteins. Several laboratories have engineered transgenic sheep and pigs that produce human proteins in their milk, making it easy to collect and refine the proteins. Many of these proteins can be used in disease prevention.

Medical Research Transgenic animals are often used as test subjects in medical research. In particular they can simulate human diseases in which defective genes play a role. Scientists use models based on these simulations to follow the onset and progression of diseases and to construct tests of new drugs that may be useful for treatment. This approach has been used to develop models for disorders like Alzheimer's disease and arthritis.

Treating Disease When recombinant-DNA techniques were developed for bacteria, biologists realized almost immediately that the technology held the promise to do something that had never been done before—to make important proteins that could prolong and even save human lives. For example, human growth hormone, which is used to treat patients suffering from pituitary dwarfism, was once scarce. Human growth hormone is now widely available because it is mass-produced by recombinant bacteria. Other products now made in genetically engineered bacteria include insulin to treat diabetes, blood-clotting factors for hemophiliacs, and potential cancer-fighting molecules such as interleukin-2 and interferon.

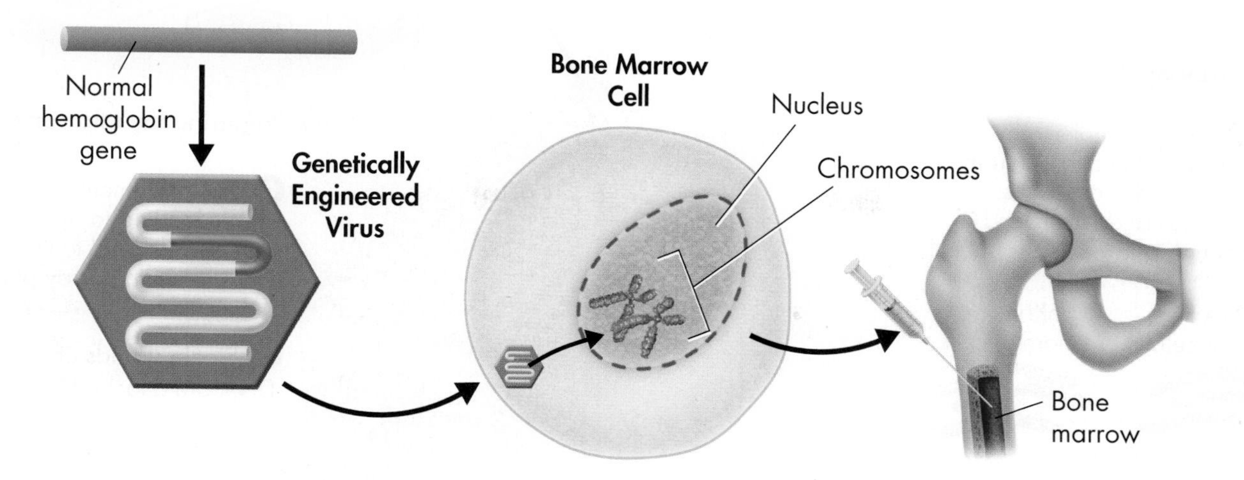

If an individual is suffering from a missing or defective gene, can we replace that gene with a healthy one and fix the problem? The experimental field of gene therapy is attempting to answer that question. **Gene therapy** is the process of changing a gene to treat a medical disease or disorder. In gene therapy, an absent or faulty gene is replaced by a normal, working gene. This process allows the body to make the protein or enzyme it needs, which eliminates the cause of the disorder.

The idea of using gene therapy to cure disease arose from the major advances in molecular biology made in the past 20 years, including the Human Genome Project. **Figure 15–16** shows one of the ways in which researchers have attempted to carry out gene therapy. To deliver the correct, or therapeutic, gene to the affected, or target, cells, researchers first engineer a virus that cannot reproduce or cause harmful effects. They place DNA containing the therapeutic gene into the modified virus, and then they infect the patient's cells with it. In theory the virus will insert the healthy gene into the target cell and correct the defect. The challenge, however, is to deliver a gene that works correctly over the long term. For all the promise it holds, in most cases gene therapy remains a high-risk experimental procedure. For gene therapy to become an accepted treatment, we need more reliable ways to insert working genes and to ensure that the DNA used in the therapy does no harm.

FIGURE 15–16 How Gene Therapy Can Be Used Gene therapy uses normal genes to add to or replace defective genes or to boost a normal function like immunity.
Interpret Visuals *How is the virus in this diagram being used?*

Genetic Testing If two prospective parents suspect they are carrying the alleles for a genetic disorder such as cystic fibrosis (CF), how could they find out for sure? Because the CF allele has slightly different DNA sequences from its normal counterpart, genetic tests using labeled DNA probes can distinguish it. Like many genetic tests, the CF test uses specific DNA sequences that detect the complementary base sequences found in the disease-causing alleles. Other genetic tests search for changes in cutting sites of restriction enzymes. Some use PCR to detect differences between the lengths of normal and abnormal alleles. Genetic tests are now available for diagnosing hundreds of disorders.

FIGURE 15–17 A Brave Volunteer Gene therapy can be risky. In 1999, 18-year-old Jesse Gelsinger volunteered for a gene therapy experiment designed to treat a genetic disorder of his liver. He suffered a massive reaction from the viruses used to carry genes into his liver cells, and he died a few days later. Jesse's case makes clear that experiments with gene therapy must be done with great caution.

1 Preparing the cDNA Probe

a mRNA samples are isolated from two different types of cells or tissues, such as cancer cells and normal cells.

mRNA from cancer cells

mRNA from normal cells

b Enzymes are used to prepare complementary DNA molecules (cDNA) from both groups of mRNA. Contrasting fluorescent labels are attached to both groups of cDNA (red to one, green to the other).

cDNA from cancer cells

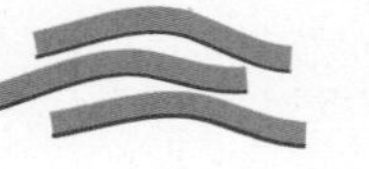

cDNA from normal cells

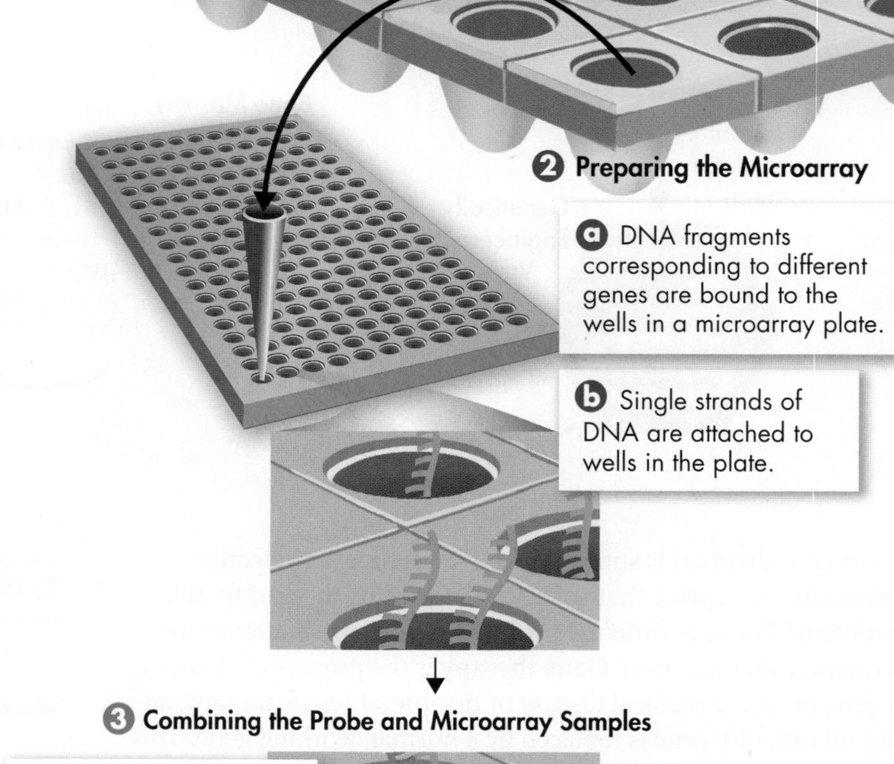

2 Preparing the Microarray

a DNA fragments corresponding to different genes are bound to the wells in a microarray plate.

b Single strands of DNA are attached to wells in the plate.

3 Combining the Probe and Microarray Samples

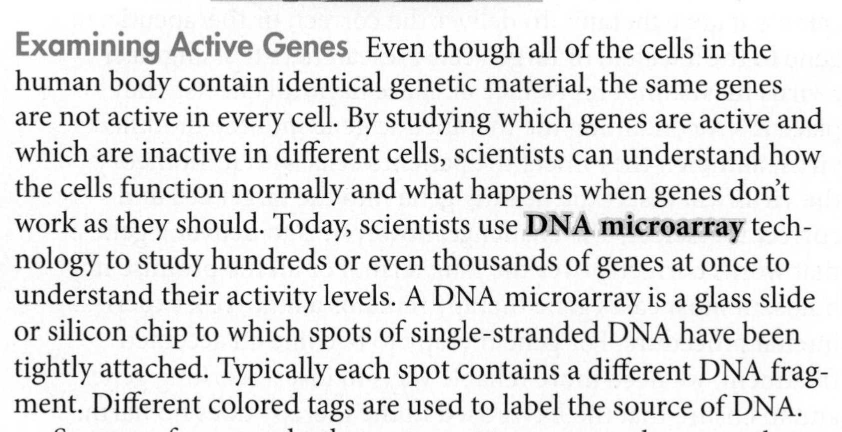

Labeled cDNA molecules bind to complementary sequences on the plate.

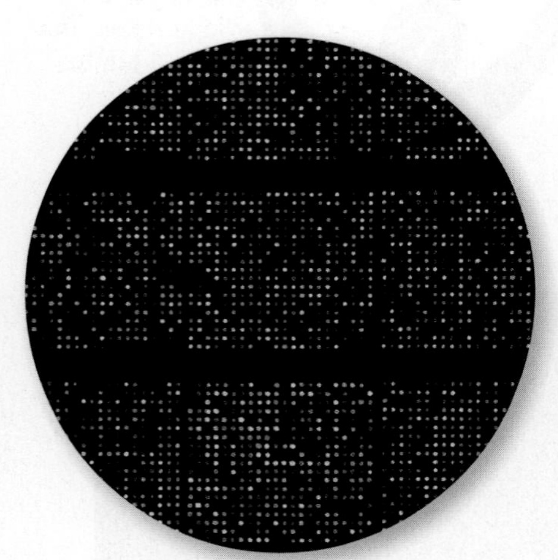

FIGURE 15–18 Analyzing Gene Activity DNA microarrays help researchers explore the underlying genetic causes of many human diseases.

Examining Active Genes Even though all of the cells in the human body contain identical genetic material, the same genes are not active in every cell. By studying which genes are active and which are inactive in different cells, scientists can understand how the cells function normally and what happens when genes don't work as they should. Today, scientists use **DNA microarray** technology to study hundreds or even thousands of genes at once to understand their activity levels. A DNA microarray is a glass slide or silicon chip to which spots of single-stranded DNA have been tightly attached. Typically each spot contains a different DNA fragment. Different colored tags are used to label the source of DNA.

Suppose, for example, that you want to compare the genes abnormally expressed in cancer cells with genes in normal cells from the same tissue. After isolating mRNA from both types of cells, you would use an enzyme to copy the mRNA base sequence into single-stranded DNA labeled with fluorescent colors—red for the cancer cell and green for the normal cell. Next you would mix both samples of labeled DNA together and let them compete for binding to the complementary DNA sequences already in the microarray. If the cancer cell produces more of a particular form of mRNA, then more red-labeled molecules will bind at the spot for that gene, turning it red. Where the normal cell produces more mRNA for another gene, that spot will be green. Where there is no difference between the two cell types, the spot will be yellow because it contains both colors. **Figure 15–18** shows how a DNA microarray is constructed and used.

❶ Chromosomes contain many regions with repeated DNA sequences that do not code for proteins. These vary from person to person. Here, one sample has 12 repeats between genes A and B, while the second sample has 9 repeats between the same genes.

❷ Restriction enzymes are used to cut the DNA into fragments containing genes and repeats. Note that the repeat fragments from these two samples are of different lengths.

❸ The restriction fragments are separated according to size using gel electrophoresis. The DNA fragments containing repeats are then labeled using radioactive probes. This labeling produces a series of bands–the DNA fingerprint.

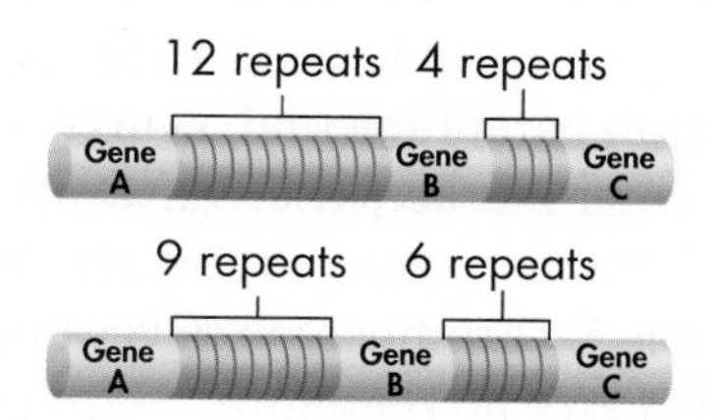

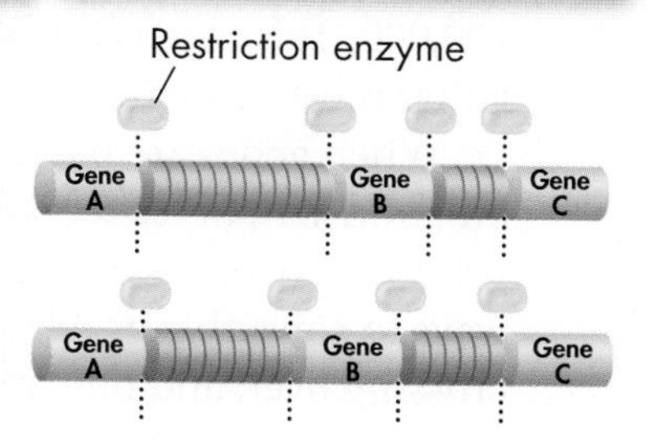

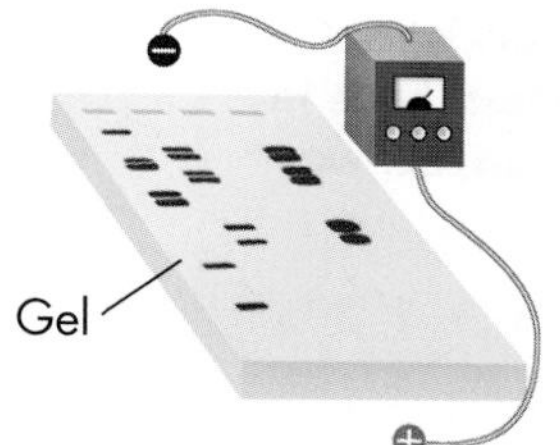

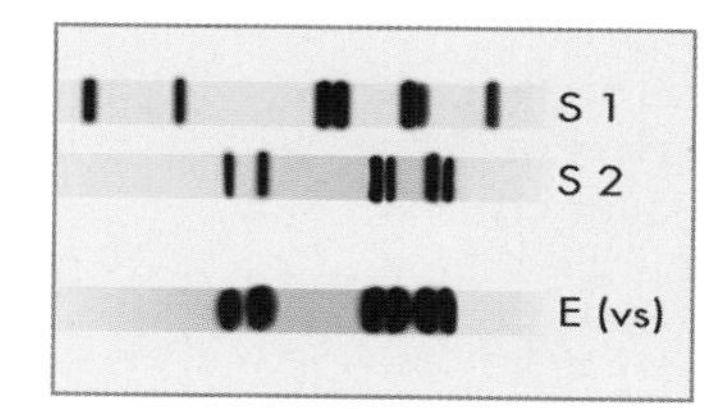

DNA fingerprint

FIGURE 15–19 Identifying Individuals DNA fingerprinting can be used to determine a person's identity. It is especially useful in solving crimes. The diagram above shows how scientists match DNA evidence from a crime scene with two possible suspects. **Interpret Graphics** ***Does the DNA fingerprint above match suspect 1 (S1) or suspect 2 (S2)? How can you tell?***

Personal Identification

How is DNA used to identify individuals?

The complexity of the human genome ensures that no individual is exactly like any other genetically—except for identical twins, who share the same genome. Molecular biology has used this fact to develop a powerful tool called **DNA fingerprinting** for use in identifying individuals. **DNA fingerprinting analyzes sections of DNA that may have little or no function but that vary widely from one individual to another.** This method is shown in **Figure 15–19.** First, restriction enzymes cut a small sample of human DNA. Next, gel electrophoresis separates the restriction fragments by size. Then, a DNA probe detects the fragments that have highly variable regions, revealing a series of variously sized DNA bands. If enough combinations of enzymes and probes are used, the resulting pattern of bands can be distinguished statistically from that of any other individual in the world. DNA samples can be obtained from blood, sperm, or tissue—even from a hair strand if it has tissue at the root.

Forensic Science DNA fingerprinting has been used in the United States since the late 1980s. Its precision and reliability have revolutionized **forensics**—the scientific study of crime scene evidence. DNA fingerprinting has helped solve crimes, convict criminals, and even overturn wrongful convictions. To date, DNA evidence has saved more than 110 wrongfully convicted prisoners from death sentences.

DNA forensics is used in wildlife conservation as well. African elephants are a highly vulnerable species. Poachers, who slaughter the animals mainly for their precious tusks, have reduced their population dramatically. To stop the ivory trade, African officials now use DNA fingerprinting to identify the herds from which black-market ivory has been taken.

In Your Notebook *Describe the process of DNA fingerprinting.*

MYSTERY CLUE

What kind of evidence do you think investigators collected at the crime scene? What kinds of tests would they have run on this evidence? What would the tests have to show before the suspect was released?

Establishing Relationships In cases of disputed paternity, how does our justice system determine the rightful father of a child? DNA fingerprinting makes it easy to find alleles carried by the child that do not match those of the mother. Any such alleles must come from the child's biological father, and they will show up in his DNA fingerprint. The probability that those alleles will show up in a randomly picked male is less than 1 in 100,000. This means the likelihood that a given male is the child's father must be higher than 99.99 percent to confirm his paternity.

When genes are passed from parent to child, genetic recombination scrambles the molecular markers used for DNA fingerprinting, so ancestry can be difficult to trace. There are two ways to solve this problem. The Y chromosome never undergoes crossing over, and only males carry it. Therefore, Y chromosomes pass directly from father to son with few changes. The same is true of the small DNA molecules found in mitochondria. These are passed, with very few changes, from mother to child in the cytoplasm of the egg cell.

Because mitochondrial DNA (mtDNA) is passed directly from mother to child, your mtDNA is the same as your mother's mtDNA, which is the same as her mother's mtDNA. This means that if two people have an exact match in their mtDNA, then there is a very good chance that they share a common maternal ancestor. Y-chromosome analysis has been used in the same way and has helped researchers settle longstanding historical questions. One such question—did President Thomas Jefferson father the child of a slave?—may have been answered in 1998. DNA testing showed that descendants of the son of Sally Hemings, a slave on Jefferson's Virginia estate, carried his Y chromosome. This result suggests Jefferson was the child's father, although the Thomas Jefferson Foundation continues to challenge that conclusion.

15.3 Assessment

Review Key Concepts

1. a. Review Give two practical applications for transgenic plants and two for transgenic animals.
b. Infer What might happen if genetically modified fish were introduced into an aquaculture facility?

2. a. Review Name three uses for recombinant-DNA technology.
b. Apply Concepts Medicines in the body interact with the body's proteins. How might normal variations in your genes affect your response to different medicines?

3. a. Review List the steps in DNA fingerprinting.
b. Infer Why is DNA fingerprinting more accurate if the samples are cut with more than one restriction enzyme?

PRACTICE PROBLEM

4. Using restriction enzymes and gel electrophoresis, write the steps of a protocol in which you test for the allele of a gene that causes a genetic disorder.

RST.9-10.10 Level of Text Complexity, WHST.9-10.2 Text Types and Purposes. Also WHST.9-10.10

Technology & BIOLOGY

Artificial Life?

In 2008, scientists at the J. Craig Venter Institute in Rockville, Maryland, produced a synthetic genome with more than half a million DNA base pairs. It may not be long before artificial cells containing similar genomes can be grown in the laboratory. How? First a complete DNA molecule, containing the minimum set of the genetic information needed to keep a cell alive, is produced in the laboratory. Then, that molecule is inserted into a living cell to replace the cell's DNA. The result is a cell whose genome is entirely synthetic. Scientists hope this technique can help them design cells for specific purposes, like capturing solar energy or manufacturing biofuels.

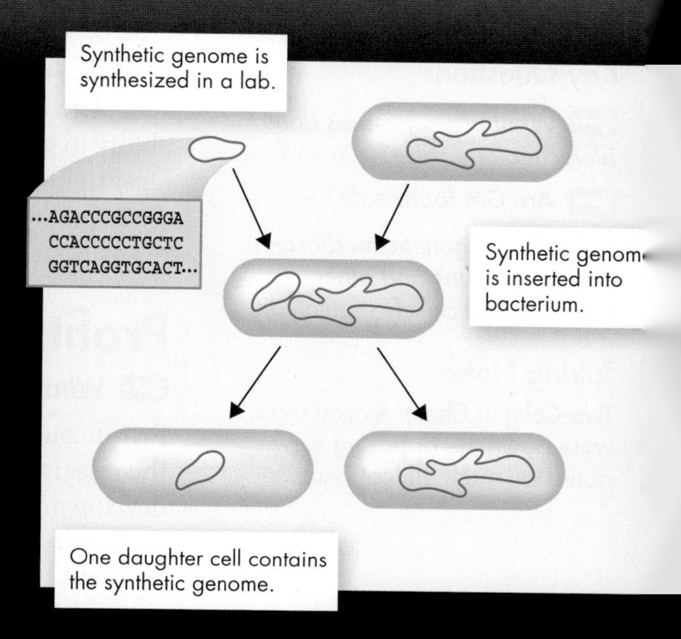

▲ **Synthesizing a Genome**
One way to synthesize life is to replace a cell's genome with an artificial DNA molecule. As a result, cell division may produce a daughter cell containing only the human-made genome.

WRITING **How could ethical and societal factors affect the production of synthetic organisms? If you were a scientist working on the latest breakthroughs, how would you address those issues? Describe your ideas in an essay.**

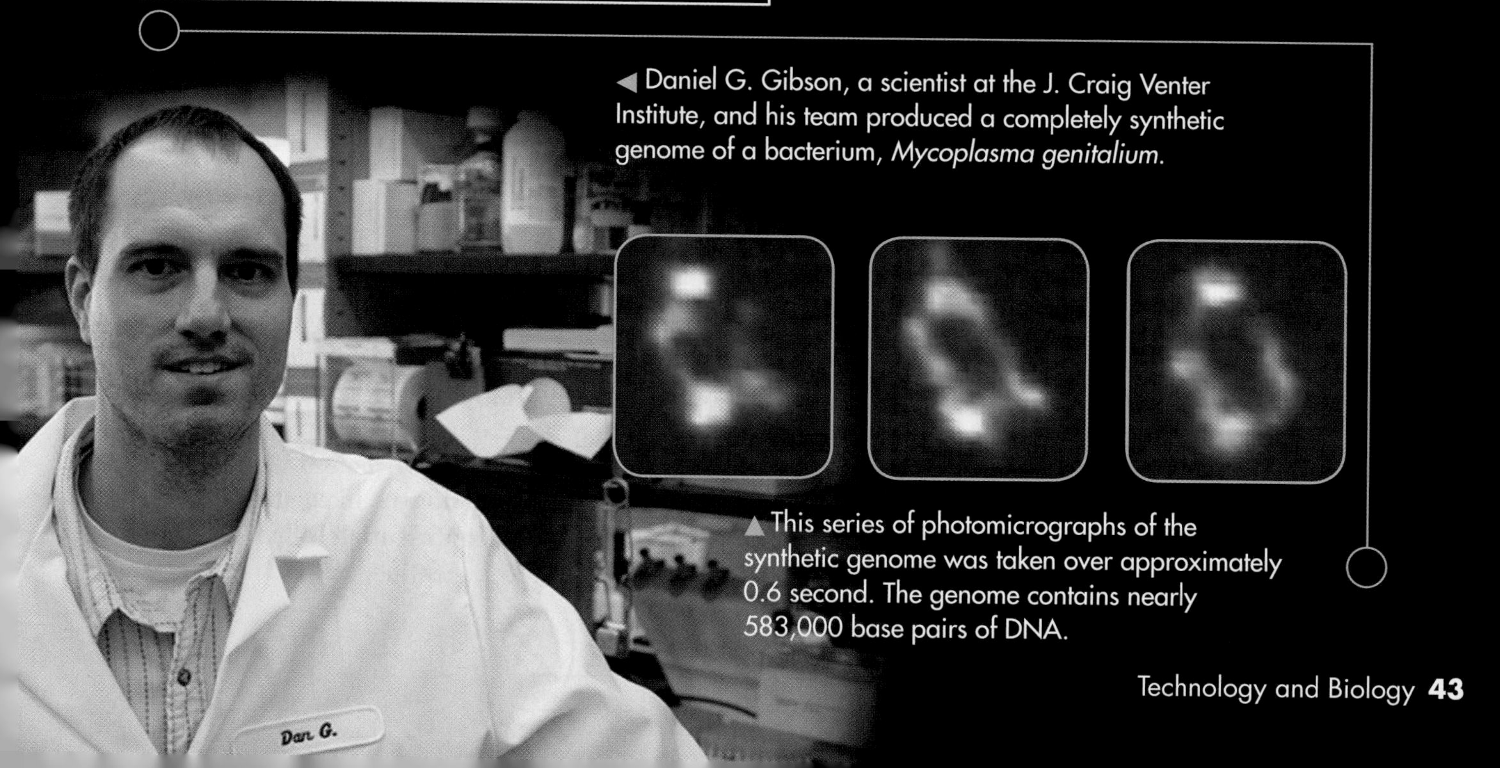

◀ Daniel G. Gibson, a scientist at the J. Craig Venter Institute, and his team produced a completely synthetic genome of a bacterium, *Mycoplasma genitalium*.

▲ This series of photomicrographs of the synthetic genome was taken over approximately 0.6 second. The genome contains nearly 583,000 base pairs of DNA.

15.4 Ethics and Impacts of Biotechnology

Key Questions

- What privacy issues does biotechnology raise?
- Are GM foods safe?
- Should genetic modifications to humans and other organisms be closely regulated?

Taking Notes

Two-Column Chart As you read, write down the opposing viewpoints on each ethical issue.

THINK ABOUT IT Years ago a science fiction movie titled *Gattaca* speculated about a future world in which genetics determines people's ability to get ahead in life. In the movie, schooling, job prospects, and legal rights are rigidly determined by an analysis of the individual's DNA on the day he or she is born. Are we moving closer to this kind of society?

Profits and Privacy

What privacy issues does biotechnology raise?

Private biotechnology and pharmaceutical companies do much of the research involving GM plants and animals. Their goal is largely to develop profitable new crops, drugs, tests, or other products. Like most inventors, they protect their discoveries and innovations with patents. A patent is a legal tool that gives an individual or company the exclusive right to profit from its innovations for a number of years.

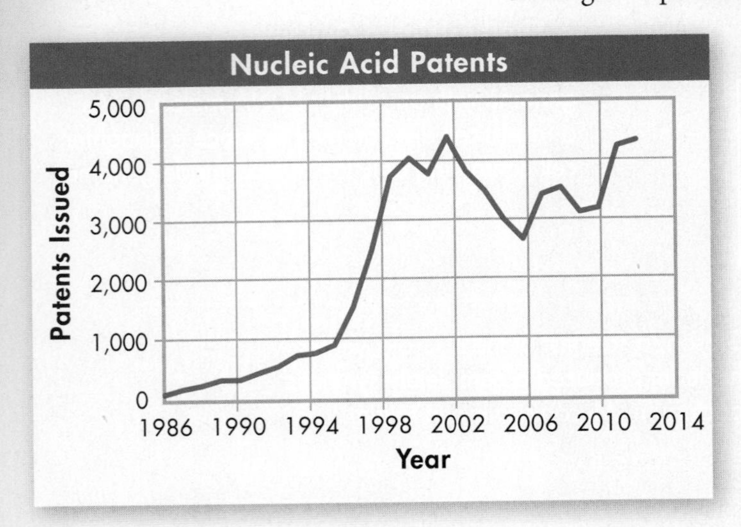

FIGURE 15–20 Patenting Nucleic Acids This graph shows the rise in the number of nucleic-acid patents between 1986 and 2012.

Patenting Life When you think about patents, you probably think about an inventor protecting a new machine or device. But molecules and DNA sequences can be patented, too. In fact, roughly one fifth of the known genes in the human genome are now patented commercially. Even laboratory techniques like PCR have been patented. When a scientist wants to run a PCR test, he or she must pay a fee for the license to use this process.

The ability to patent is meant to spur discovery and advancements in medicine and industry. After all, patent holders stand a good chance of reaping large financial rewards. Sometimes, though, patent holders demand high fees that block other scientists from exploring certain lines of research. That was the case with provitamin A-enriched golden rice, a GM plant described in Lesson 15.3. After the rice was developed, patent disputes kept it out of the hands of farmers for years. Based on **Figure 15–20,** patent disputes may be an ongoing issue.

Now consider the information held in your own genome. **Do you have exclusive rights to your DNA? Should you, like patent holders, be able to keep your genetic information confidential?** When it comes to your own DNA, how much privacy are you entitled to?

Genetic Ownership One of the most hallowed sites in the United States is the one shown in **Figure 15–21.** It is the Tomb of the Unknowns in Arlington National Cemetery, near Washington, D.C. Buried here are the remains of unidentified American soldiers who fought our nation's wars. The tomb also serves as a focal point for the honor and remembrance of those service members lost in combat whose bodies have never been recovered.

Biotechnology offers hope that there will never be another unknown soldier. The U.S. military now requires all personnel to give a DNA sample when they begin their service. Those DNA samples are kept on file and used, if needed, to identify the remains of individuals who perish in the line of duty. In many ways, this practice is a comfort to military families, who can be assured that the remains of a loved one can be properly identified for burial.

But what if the government wants to use an individual's DNA sample for another purpose, in a criminal investigation or a paternity suit? What if health-insurance providers manage their healthcare policies based on a genetic predisposition to disease? For example, suppose that, years after giving a DNA sample, an individual is barred from employment or rejected for health insurance because of a genetic defect detected in the sample. Would this be a fair and reasonable use of genetic information?

After considering this issue for years, United States Congress passed the Genetic Information Nondiscrimination Act, which became law in 2008. This act protects Americans against discrimination based on their genetic information. Physicians and ethicists hope this will lead to more effective use of personal genetic information, without fear of prejudice in obtaining health insurance or employment.

FIGURE 15–21 Unknown Identities The Tomb of the Unknowns in Arlington National Cemetery holds the remains of unknown American soldiers from World Wars I and II, the Korean War, and, until 1998, the Vietnam War. **Form an Opinion** ***Should DNA testing be used to identify the remaining soldiers buried here? Why or why not?***

Safety of Transgenics

Are GM foods safe?

Much controversy exists concerning foods that have had their DNA altered through genetic engineering. The majority of GM crops today are grown in the United States, although farmers around the world have begun to follow suit. Are the foods from GM crops the same as those prepared from traditionally bred crops?

Pros of GM Foods The companies producing seeds for GM crops would say that GM plants are actually better and safer than other crops. Farmers choose them because they produce higher yields, reducing the amount of land and energy that must be devoted to agriculture and lowering the cost of food for everyone.

Insect-resistant GM plants need little, if any, insecticide to grow successfully, reducing the chance that chemical residues will enter the food supply and lessening damage to the environment. In addition, GM foods have been widely available for more than a decade. **Careful studies of such foods have provided no scientific support for concerns about their safety, and it does seem that foods made from GM plants are safe to eat.**

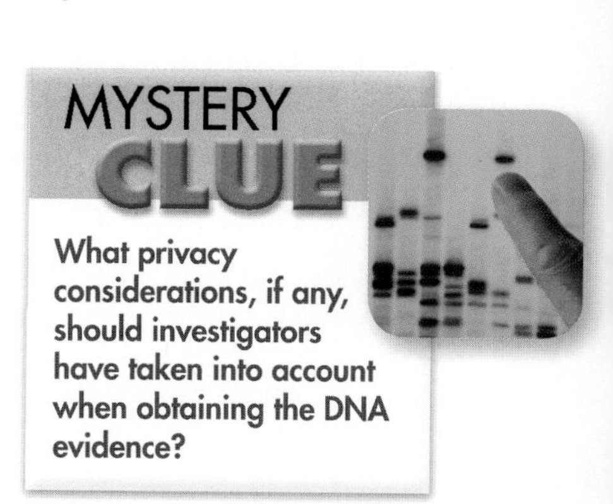

What privacy considerations, if any, should investigators have taken into account when obtaining the DNA evidence?

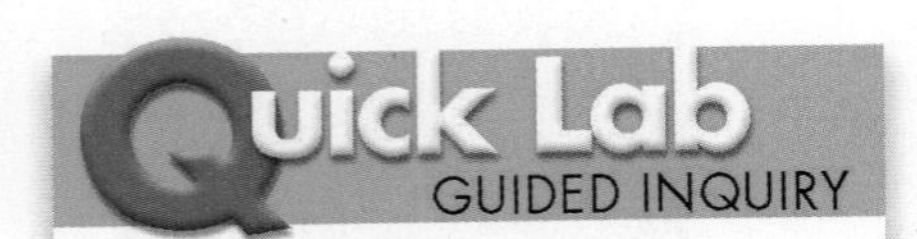

Survey Biotechnology Opinions

1. Select three safety, legal, or ethical issues related to genetic engineering.
2. Design a survey to ask people their opinions on these issues.
3. Find 15 people to answer your survey.
4. Collect the surveys and tabulate the answers.

Analyze and Conclude

1. Analyze Data Did all respondents agree on any issue? If so, which one(s)?

2. Draw Conclusions If you had surveyed more people, do you think you would have found more or less agreement in the responses? Why or why not?

3. Evaluate How informed about biotechnology issues were the people you surveyed? If you were a politician or government official, how would you act on the results of your survey?

Cons of GM Foods Critics acknowledge some benefits of genetically modified foods, but they also point out that no long-term studies have been made of the hazards these foods might present. **Even if GM food itself presents no hazards, there are many serious concerns about the unintended consequences that a shift to GM farming and ranching may have on agriculture.** Some worry that the insect resistance engineered into GM plants may threaten beneficial insects, killing them as well as crop pests. Others express concerns that use of plants resistant to chemical herbicides may lead to overuse of these weed-killing compounds.

Another concern is that the patents held on GM seeds by the companies that produce them may prove costly enough to force small farmers out of business, especially in the developing world. It is not clear whether any of these concerns should block the wider use of these new biotechnologies, but it is certain that they will continue to prove controversial in the years ahead.

In the United States, current federal regulations treat GM foods and non-GM foods equally. As a result, GM foods are not required to undergo special safety testing before entering the market. No additional labeling is required to identify a product as genetically modified unless its ingredients are significantly different from its conventional counterpart. The possibility that meat from GM animals may soon enter the food supply has heightened concerns about labeling. As a result, some states have begun to consider legislation to require the labeling of GM foods, thereby providing consumers with an informed choice.

In Your Notebook *List the pros and cons of GM foods.*

Ethics of the New Biology

Should genetic modifications to humans and other organisms be closely regulated?

"Know yourself." The ancient Greeks carved this good advice in stone, and it has been guiding human behavior ever since. Biotechnology has given us the ability to know ourselves more and more. With this knowledge, however, comes responsibility.

You've seen how easy it is to move genes from one organism to another. For example, the GFP gene can be extracted from a jellyfish and spliced onto genes coding for important cellular proteins. This ability has led to significant new discoveries about how cells function.

The same GFP technology was used to create the fluorescent zebra fish shown in **Figure 15–22.** These fish—along with fluorescent mice, tadpoles, rabbits, and even cats—have all contributed to our understanding of cells and proteins. But the ability to alter life forms for any purpose, scientific or nonscientific, raises important questions. **Just because we have the technology to modify an organism's characteristics, are we justified in doing so?**

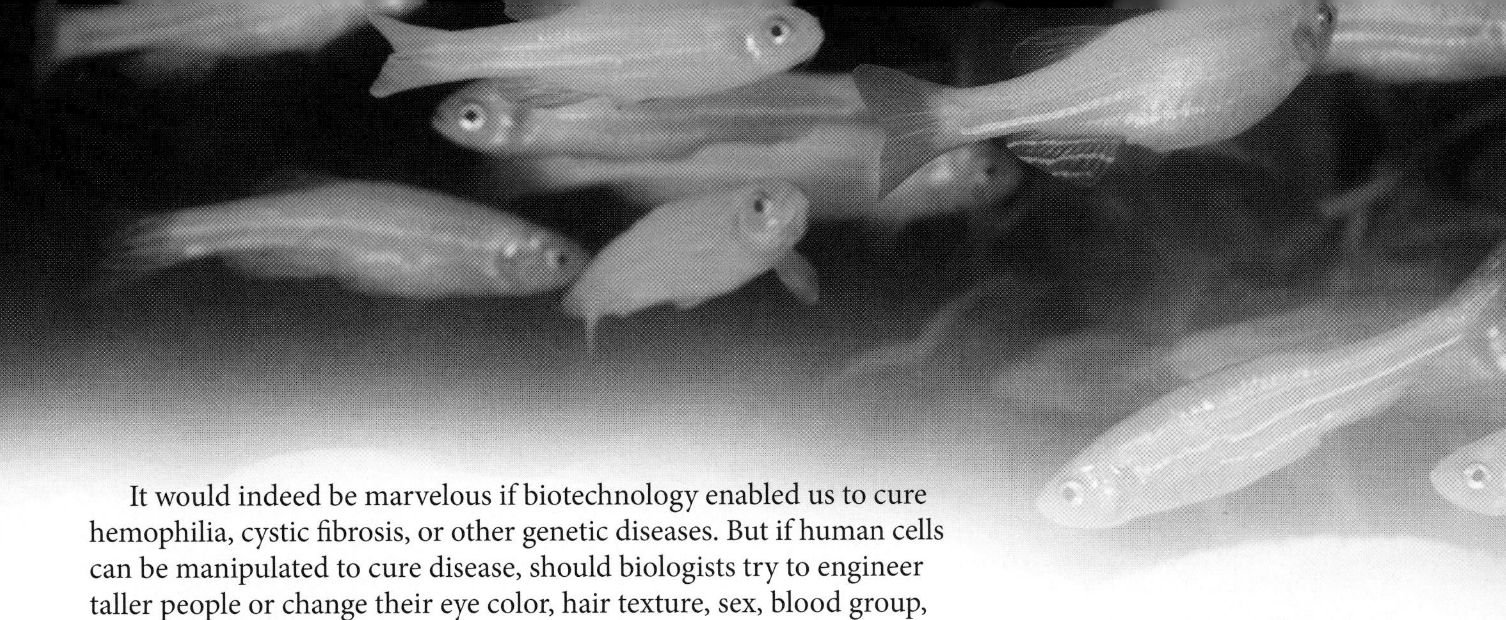

It would indeed be marvelous if biotechnology enabled us to cure hemophilia, cystic fibrosis, or other genetic diseases. But if human cells can be manipulated to cure disease, should biologists try to engineer taller people or change their eye color, hair texture, sex, blood group, or appearance? What will happen to the human species when we gain the opportunity to design our bodies or those of our children? What will be the consequences if biologists develop the ability to clone human beings by making identical copies of their cells? These are questions with which society must come to grips.

The goal of biology is to gain a better understanding of the nature of life. As our knowledge increases, however, so does our ability to manipulate the genetics of living things, including ourselves. In a democratic nation, all citizens—not just scientists—are responsible for ensuring that the tools science has given us are used wisely. This means that you should be prepared to help develop a thoughtful and ethical consensus of what should and should not be done with the human genome. To do anything less would be to lose control of two of our most precious gifts: our intellect and our humanity.

FIGURE 15–22 Gaining More Understanding These fluorescent zebra fish were originally bred to help scientists detect environmental pollutants. Today, studying fluorescent fish is helping us understand cancer and other diseases. The fish are also sold to the public at a profit.

15.4 Assessment

Review Key Concepts

1. a. Review What is a patent?

b. Apply Concepts How could biotechnology affect your privacy?

2. a. Review What are genetically modified foods?

b. Form an Opinion Should a vegetarian be concerned about eating a GM plant that contains DNA from a pig gene? Support your answer with details from the text.

3. a. Review What are the main concerns about genetic engineering discussed in this lesson or elsewhere in the chapter?

b. Pose Questions Write three specific questions about the ethical, social, or legal implications of genetic engineering that do not appear in this lesson. For example, how does personal genetic information affect self-identity?

WRITE ABOUT SCIENCE

Persuasion

4. Biologists may one day be able to use genetic engineering to alter a child's inherited traits. Under what circumstances, if any, should this ability be used? Write a persuasive paragraph expressing your opinion.

NGSS Smart Guide Genetic Engineering

Disciplinary Core Idea LS3.B Variation of Traits: Why do individuals of the same species vary in how they look, function, and behave? Genetic engineering allows scientists to manipulate the genomes of living things. But there are ethical, legal, safety, and societal issues surrounding the use of genetic engineering.

15.1 Selective Breeding

- Humans use selective breeding, which takes advantage of naturally occurring genetic variation, to pass wanted traits on to the next generation of organisms.
- Breeders can increase the genetic variation in a population by introducing mutations, which are the ultimate source of biological diversity.

selective breeding 418 • hybridization 419 • inbreeding 419 • biotechnology 419

Biology.com

Untamed Science Video Pigeon breeding helps the Untamed Science crew unravel the mysteries of genetic engineering.

15.2 Recombinant DNA

- The first step in using the polymerase chain reaction method to copy a gene is to heat a piece of DNA, which separates its two strands. Then, as the DNA cools, primers bind to the single strands. Next, DNA polymerase starts copying the region between the primers. These copies can serve as templates to make still more copies.
- Recombinant-DNA technology—joining together DNA from two or more sources—makes it possible to change the genetic composition of living organisms.
- Transgenic organisms can be produced by the insertion of recombinant DNA into the genome of a host organism.

polymerase chain reaction 423 • recombinant DNA 424 • plasmid 424 • genetic marker 425 • transgenic 426 • clone 427

Biology.com

Art in Motion View a short animation that brings bacterial transformation to life.

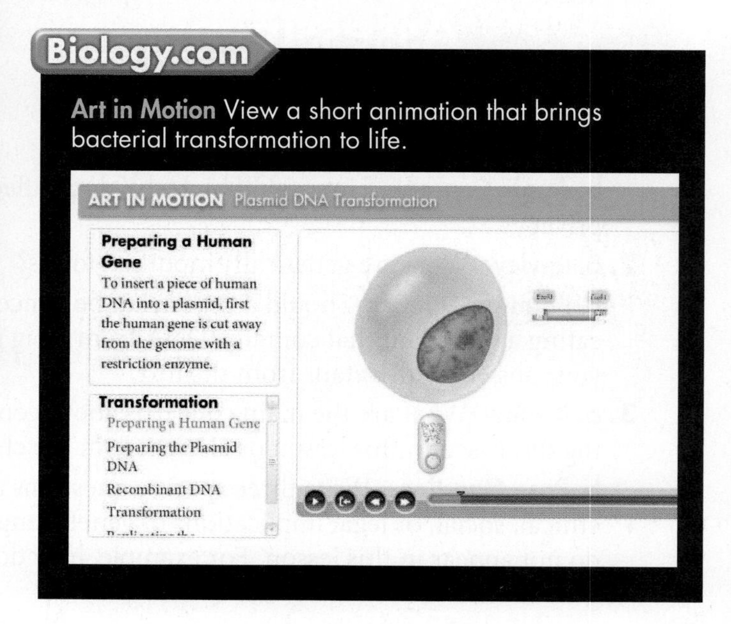

CHECKING YOUR *Scientific Literacy*

Refer to the lesson content and digital assets below as you prepare for your chapter assessment. Then, evaluate your understanding of genetic engineering by answering these questions.

1. Scientific and Engineering Practice **Evaluating and Communicating Information** Review the traditional breeding techniques of hybridization and inbreeding. Which of these techniques has more in common with transgenic technology, and why? Hint: What are the goals of each technique?

2. Crosscutting Concept **Structure and Function** How do the DNA sequences included in a plasmid help ensure that recombinant DNA molecules are reproduced when they are inserted into a host cell?

3. Research to Build and Present Knowledge The results of tests at a crime lab may be used as evidence in court. So, strict rules are in place to ensure that samples are not contaminated or tampered with before or during testing. Research what rules are used by people who gather evidence and people who test evidence. Write a few paragraphs to summarize what you learn.

4. STEM Both Southern Blotting and microarray technology were developed as solutions to similar problems. Briefly describe the problem that each technology was designed to address.

15.3 Applications of Genetic Engineering

- Ideally, genetic modification could lead to better, less expensive, and more nutritious food as well as less-harmful manufacturing processes.
- Recombinant-DNA technology is advancing the prevention and treatment of disease.
- DNA fingerprinting analyzes sections of DNA that vary widely from one individual to another.

gene therapy 431 • **DNA microarray 432** • **DNA fingerprinting 433** • **forensics 433**

Biology.com

Art Review Review your understanding of DNA fingerprinting with this drag-and-drop activity.

15.4 Ethics and Impacts of Biotechnology

- Should you, like patent holders, be able to keep your genetic information confidential?
- Careful studies of GM foods have provided no scientific support for concerns about their safety.
- There are many concerns about unintended consequences that a shift to GM farming and ranching may have on agriculture.
- Just because we have the technology to modify an organism's characteristics, are we justified in doing so?

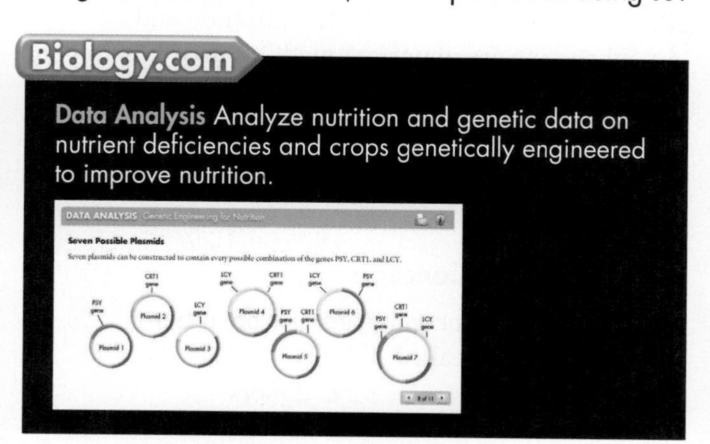

Biology.com

Data Analysis Analyze nutrition and genetic data on nutrient deficiencies and crops genetically engineered to improve nutrition.

Forensics Lab

Using DNA to Solve Crimes This chapter lab is available in your lab manual and at Biology.com

15 Assessment

15.1 Selective Breeding

Understand Key Concepts

1. Crossing dissimilar individuals to bring together their best characteristics is called
   a. domestication. c. hybridization.
   b. inbreeding. d. polyploidy.

2. Crossing individuals with similar characteristics so that those characteristics will appear in their offspring is called
   a. inbreeding. c. recombination.
   b. hybridization. d. polyploidy.

3. Taking advantage of naturally occurring variations in organisms to pass wanted traits on to future generations is called
   a. selective breeding. c. hybridization.
   b. inbreeding. d. mutation.

4. How do breeders produce genetic variations that are not found in nature?

5. What is polyploidy? When is this condition useful?

Think Critically

6. **Craft and Structure** Suppose a plant breeder has a thornless rose bush with scentless pink flowers, a thorny rose bush with sweet-smelling yellow flowers, and a thorny rose bush with scentless purple flowers. How might this breeder develop a pure variety of thornless, sweet-smelling purple roses?

7. **Compare and Contrast** Hybridization and inbreeding are important methods used in selective breeding. How are the methods similar? How are they different?

15.2 Recombinant DNA

Understand Key Concepts

8. Organisms that contain genes from other organisms are called
   a. transgenic. c. donors.
   b. mutagenic. d. clones.

9. What process is shown below?
   a. fingerprinting
   b. transformation
   c. hybridization
   d. polymerase chain reaction

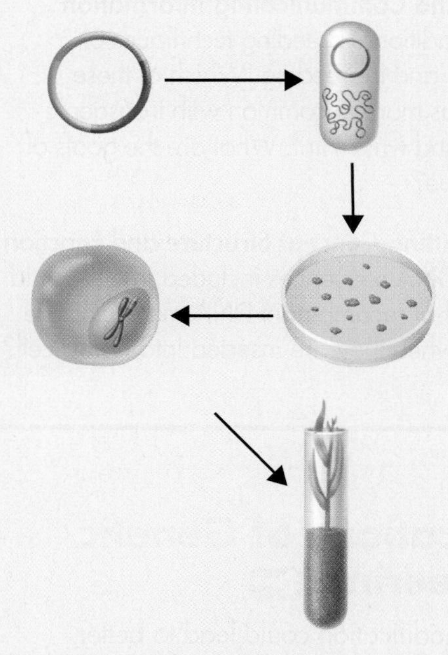

10. When cell transformation is successful, the recombinant DNA
    a. undergoes mutation.
    b. is treated with antibiotics.
    c. becomes part of the transformed cell's genome.
    d. becomes a nucleus.

11. Bacteria often contain small circular molecules of DNA known as
    a. clones. c. plasmids.
    b. chromosomes. d. hybrids.

12. A member of a population of genetically identical cells produced from a single cell is a
    a. clone. c. mutant.
    b. plasmid. d. sequence.

13. Describe what happens during a polymerase chain reaction.

14. **Craft and Structure** Explain what genetic markers are and describe how scientists use them.

15. How does a transgenic plant differ from a hybrid plant?

Think Critically

16. **Apply Concepts** Describe one or more advantages of producing insulin and other proteins through genetic engineering.

17. **Text Types and Purposes** Bacteria and humans are very different organisms. Why is it sometimes possible to combine their DNA and use a bacterium to make a human protein?

15.3 Applications of Genetic Engineering

Understand Key Concepts

18. Which of the following characteristics is often genetically engineered into crop plants?
    a. improved flavor
    b. resistance to herbicides
    c. shorter ripening times
    d. thicker stems

19. A substance that has been genetically engineered into transgenic rice has the potential to treat
    a. cancer.
    b. high blood pressure.
    c. vitamin A deficiency.
    d. malaria.

20. Which of the following techniques would scientists most likely use to understand the activity levels of hundreds of genes at once?
    a. a DNA microarray.
    b. PCR.
    c. restriction enzyme analysis.
    d. DNA sequencing.

21. Describe how a DNA microarray might be used to distinguish normal cells from cancer cells.

22. Describe two important uses for DNA fingerprinting.

Think Critically

23. **Infer** If a human patient's bone marrow was removed, altered genetically, and reimplanted, would the change be passed on to the patient's children? Explain your answer.

solve the CHAPTER MYSTERY

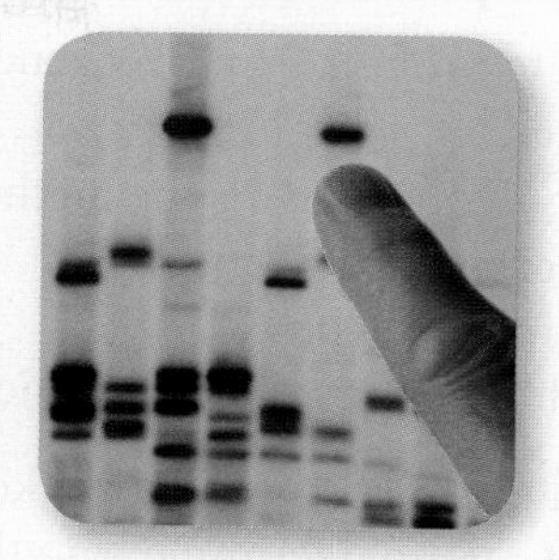

A CASE OF MISTAKEN IDENTITY

The first suspect was lucky: Twenty years earlier, it would have been an open-and-shut case. But by 1998, DNA fingerprinting was widely available. After the police took the suspect into custody, forensic scientists tested the DNA in the bloodstains on his shirt. Within a few hours, they knew they had the wrong suspect. Before long, the police caught the real attacker, who was subsequently tried and convicted of the crime.

1. **Infer** How did the investigators determine that the person they took into custody was not guilty of this crime?
2. **Apply Concepts** Did the DNA evidence from the bloodstains come from the red blood cells, the white blood cells, or both? Explain your answer.
3. **Predict** What if the initial suspect was related to the victim? Would that have changed the result? Why or why not?
4. **Connect to the Big idea** What might have happened if this crime were committed before DNA fingerprinting was discovered? Describe the series of events that might have taken place after police took in the first suspect.

15.4 Ethics and Impacts of Biotechnology

Understand Key Concepts

24. The right to profit from a new genetic technology is protected by
- **a.** getting a copyright for the method.
- **b.** discovering a new gene.
- **c.** obtaining a patent.
- **d.** publishing its description in a journal.

25. Which of the following is most likely to be used in a court case to determine who the father of a particular child is?
- **a.** microarray analysis
- **b.** DNA fingerprinting
- **c.** gene therapy
- **d.** genetic engineering

26. Give an example of a disadvantage associated with patenting genes.

27. What is one argument used by critics of genetically modified foods?

Think Critically

28. Predict List three ways in which genetically engineered organisms might be used in the future.

29. Evaluate Your friend suggests that genetic engineering makes it possible for biologists to produce an organism with any combination of characteristics—an animal with the body of a frog and the wings of a bat, for example. Do you think this is a reasonable statement? Explain your answer.

Connecting Concepts

Use Science Graphics

Use the table below to answer question 30.

DNA Restriction Enzymes

Enzyme	Recognition Sequence
*Bgl*III	A↓G A T C T T C T A G↑A
*Eco*RI	G↓A A T T C C T T A A↑G
*Hind*III	A↓A G C T T T T C G A↑A

30. Apply Concepts Copy the following DNA sequence and write its complementary strand.

ATGAGATCTACGGAATTCTCAAGCTTGAATCG

Where will each restriction enzyme in the table cut the DNA strand?

Write About Science

31. Text Types and Purposes A friend blogs about genetic engineering. She asserts that GM is too new, and traditional selective breeding can accomplish the same things as GM. Write a thoughtful response that supports or opposes her position.

32. Assess the Big idea Briefly describe the major steps involved in inserting a human gene into a bacterium.

Analyzing Data

Integration of Knowledge and Ideas

Questions 33–35 refer to the diagram, which shows the results of a criminal laboratory test.

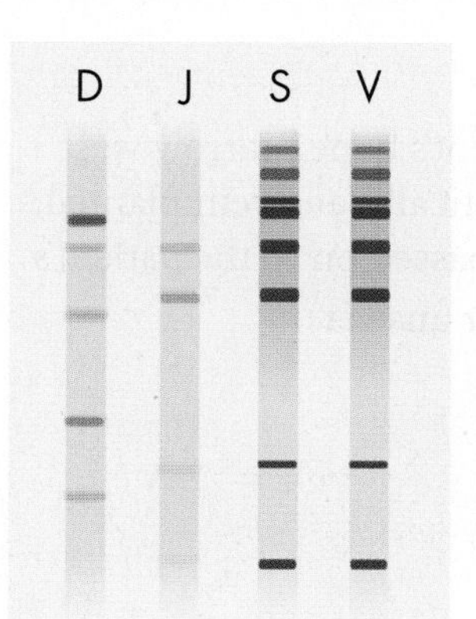

D = Defendant's blood
J = Blood from defendant's jeans
S = Blood from defendant's shirt
V = Victim's blood

33. Infer Briefly describe the biotechnological methods that would have been used to produce the results shown at the right.

34. Compare and Contrast How are the bands from the jeans and the shirt similar? How are they different?

35. Draw Conclusions Based on the results shown, what conclusions might a prosecutor present to a jury during a criminal trial?

Standardized Test Prep

Multiple Choice

1. Polyploidy may instantly produce new types of organisms that are larger and stronger than their diploid relatives in
- **A** animals.
- **B** plants.
- **C** bacteria.
- **D** fungi.

2. Which of the following characteristics does NOT apply to a plasmid?
- **A** made of DNA
- **B** found in bacterial cells
- **C** has circular loops
- **D** found in animal cells

3. To separate DNA fragments from one another, scientists use
- **A** polymerase chain reaction.
- **B** DNA microarrays.
- **C** gel electrophoresis.
- **D** restriction enzymes.

4. Restriction enzymes cut DNA molecules
- **A** into individual nucleotides.
- **B** at random locations.
- **C** at short sequences specific to each type of enzyme.
- **D** into equal-sized pieces.

5. The expression of thousands of genes at one time can be followed using
- **A** polymerase chain reaction.
- **B** plasmid transformation.
- **C** restriction enzymes.
- **D** DNA microarrays.

6. Genetically engineered crop plants can benefit farmers by
- **A** reducing the amount of land that is required to grow them.
- **B** introducing chemicals into the environment.
- **C** increasing an animal's resistance to antibiotics.
- **D** changing the genomes of other crop plants.

7. Genetic markers allow scientists to
- **A** clone animals.
- **B** separate strands of DNA.
- **C** synthesize antibiotics.
- **D** identify transformed cells.

Questions 8–9

The graph below shows the number of accurate copies of DNA produced by polymerase chain reaction.

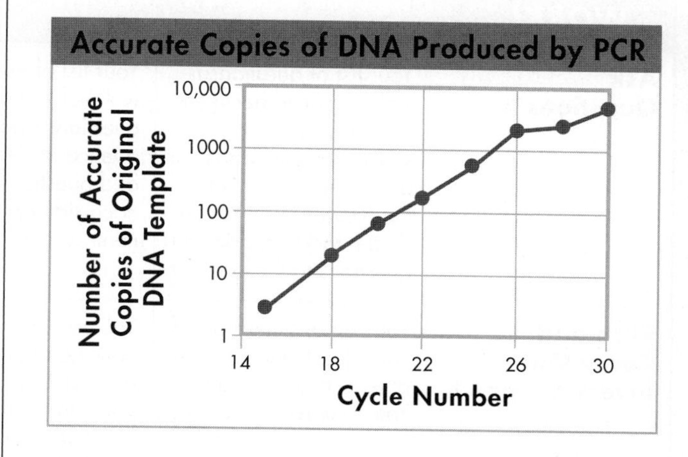

8. What can you conclude about cycles 18 through 26?
- **A** PCR produced accurate copies of template DNA at an exponential rate.
- **B** The amount of DNA produced by PCR doubled with each cycle.
- **C** The DNA copies produced by PCR were not accurate copies of the original DNA template.
- **D** The rate at which PCR produced accurate copies of template DNA fell in later cycles.

9. Based on the graph, which of the following might have happened between cycles 26 and 28?
- **A** PCR stopped producing accurate copies of the template.
- **B** The rate of reaction increased.
- **C** All of the template DNA was used up.
- **D** A mutation occurred.

Open-Ended Response

10. Why are bacteria able to make human proteins when a human gene is inserted in them with a plasmid?

If You Have Trouble With . . .

Question	1	2	3	4	5	6	7	8	9	10
See Lesson	15.1	15.2	15.2	15.2	15.3	15.4	15.2	15.2	15.2	15.3

NGSS Problem-based Learning

UNIT PROJECT 4

Your Self-Assessment

The rubric below will help you evaluate your work as you construct an explanation and communicate information.

SCORE YOUR WORK!	EXEMPLARY Score your work a 4 if:	ACCOMPLISHED Score your work a 3 if:	DEVELOPING Score your work a 2 if:	BEGINNING Score your work a 1 if:
Ask Questions ▶	Your list of questions is well balanced between questions about the science of GM foods and questions about the controversy. As you do the task you add questions to the list.	Your list of questions is well balanced between questions about the science of GM foods and questions about each side of the controversy.	You have questions about the science of GM foods and about both sides of the controversy. But the number of questions is tilted in favor of one side of the controversy.	Your list of questions is incomplete. You are missing either questions about the science of GM foods or your questions address only one side of the controversy.
Plan and Carry Out Investigations ▶	You can answer most of your questions. You use at least three sources. You complete the chart for every source you use. The data you collect is useful in determining bias.	You can answer most of your questions. You use at least three sources. You complete the chart for every source you use, but the data you collect is not useful in determining bias.	You can answer most of your questions. You use at least three sources. Either you do not complete the chart for every source you use or the data you collect is not useful in determining bias.	You cannot answer most of your questions. You use only one or two sources. Either you do not complete the chart for every source you use or the data you collect is not useful in determining bias.
	TIP Review the meaning of the term *bias* before you begin your investigation.			
Analyze and Interpret Data ▶	You cite evidence to support your classification. You identify similarities among sources. You explain how your opinion of a source changed.	You cite evidence to support your classification of sources as biased or unbiased. You identify similarities among sources in a category.	You cite evidence to support your classification of sources as biased or unbiased. You do not identify similarities among sources in a category.	You label sources as biased or unbiased but do not cite evidence to support your classification or identify similarities among sources in a category.
Engage in Argument from Evidence ▶	The factual evidence you cite is relevant and sufficient to support your argument.	The factual evidence you cite is relevant but not sufficient to support your argument.	The factual evidence you cite does not support your argument.	You use opinions, not facts, to support your argument.
	TIP Use specific evidence from your sources to support your opinion.			
Communicate Information ▶	Your target audience is defined. Your summary is persuasive and clearly communicates your position. You revise your summary based on feedback.	Your target audience is defined. Your visual summary is persuasive and clearly communicates your position on GM foods.	Your target audience is clearly defined, but your summary is not persuasive.	Your target audience is not clearly defined and your summary is not persuasive.
	TIP You may need to revise your work more than once.			

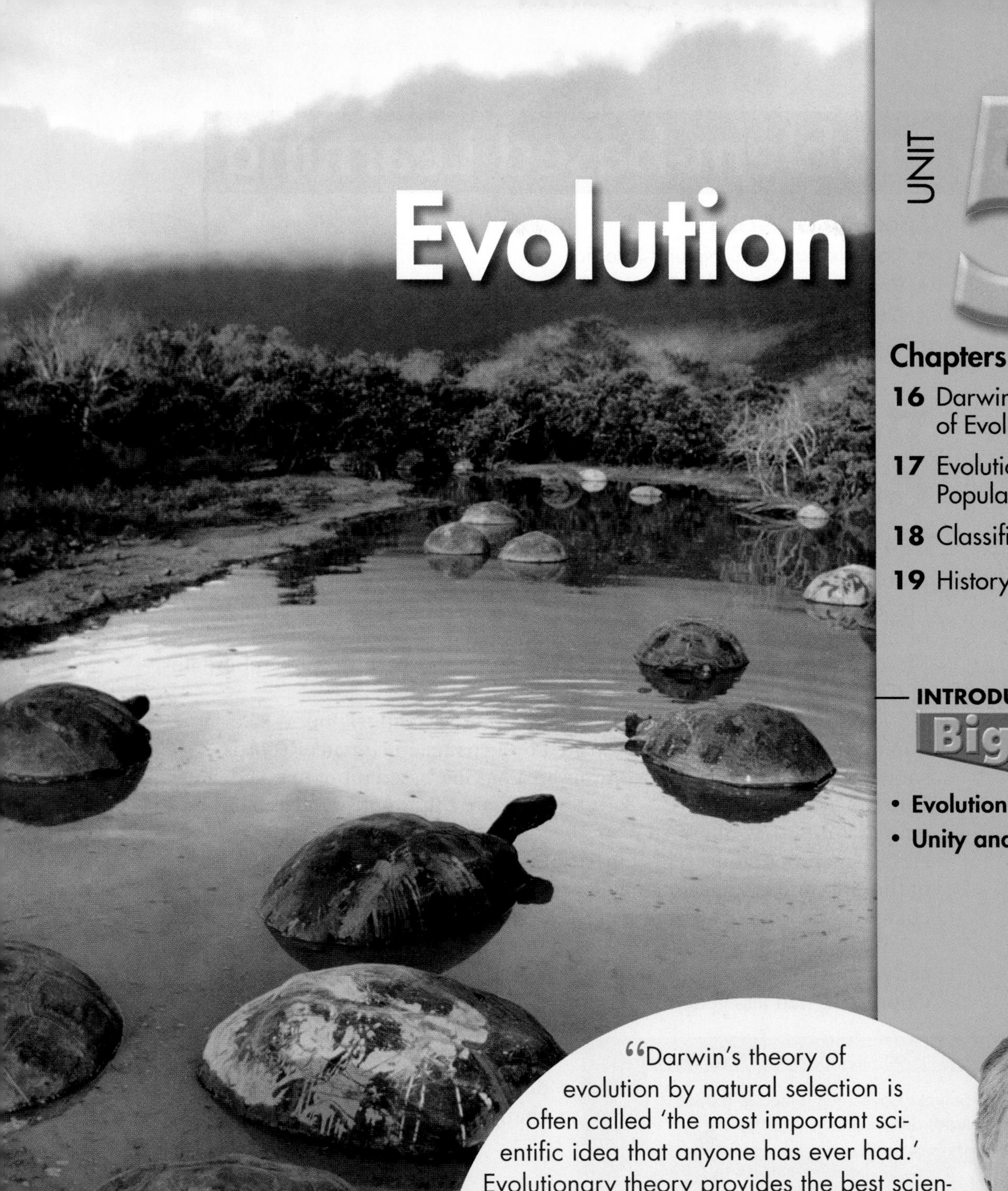

Evolution

UNIT 5

Chapters

16 Darwin's Theory of Evolution

17 Evolution of Populations

18 Classification

19 History of Life

INTRODUCE the **Big ideas**

- **Evolution**
- **Unity and Diversity of Life**

"Darwin's theory of evolution by natural selection is often called 'the most important scientific idea that anyone has ever had.' Evolutionary theory provides the best scientific explanation for the unity and diversity of life. It unites all living things in a single tree of life and reminds us that humans are part of nature. As researchers explore evolutionary mysteries, they continue to marvel at Darwin's genius and his grand vision of the natural world."

Joe Levine

NGSS Problem-based Learning

UNIT PROJECT 5

Spotlight on NGSS
- **Core Idea LS4.B** Natural Selection
- **Practice** Planning and Carrying Out Investigations
- **Crosscutting Concept** Patterns

The Alpine Chipmunk's Genetic Decline

Climate change is identified as one of the leading environmental problems facing our world. Current research shows climate change can impact genetic diversity and the natural selection of organisms. One example of climate change affecting genetic diversity can be seen in Tuolumne Meadows in Yosemite National Park, located in the Sierra Nevada mountains of California.

Research shows the alpine chipmunk (*Tamias alpinus*), found in Tuolumne Meadows, has been affected by a change in climate. Records show the average temperature of this area has increased more than 5 degrees Fahrenheit over the last 100 years. Since alpine chipmunks prefer cooler temperatures, they migrated from the lower land meadows to the upper elevation of mountains. This move caused both the range and population of the chipmunks to decrease. Researchers also found that there was a change in the genetic variation of the alpine chipmunk.

When scientists compared DNA from chipmunks collected in the early 1900s to those collected in the last 10 years, they found that modern chipmunks had less genetic variation than the chipmunks from 100 years ago. Scientists believe that as the population of alpine chipmunks decreased in number due to climate change, it also lost genetic variation.

Alpine chipmunks have been forced to move to higher altitudes due to climate change.

Your Task: Raise Local Awareness

The phrase *Think Globally, Act Locally* has been used in the context of environmental awareness since the late 1960s. How has climate change impacted the biomes and ecosystems in which you live? A biome is a group of ecosystems that share similar climates and typical organisms. Desert, Temperate Grassland, Temperate Forest, and Northwestern Coniferous Forest are examples of biomes found in the United States. Think about the part of the country where you live and identify the biome. Research how climate change has impacted the ecosystems in your biome. What can be done to raise local awareness?

Solving the Problem

Ask Questions Determine the part of the country, biome, and ecosystem in which you live. What impacts of climate change on biological diversity have been identified for this local region? In what ways can climate change affect genetic diversity and natural selection of plant and animal species? What, if any, action or awareness has been taken to help people in the region understand the effects of climate change on local ecosystems?

Develop and Use Models Construct drawings or diagrams that represent the impact of climate change on your local region. The model should be the foundation of an explanation or prediction about what is happening in your local ecosystem or biome.

Plan and Carry Out an Investigation Formulate a question and develop a hypothesis that can be investigated based on your model. Decide what and how much data are needed to produce reliable measurements. Determine the tools needed and how data will be recorded.

Use Mathematical and Computational Thinking Use mathematical and computational tools to help you statistically analyze data. Identify the significant features and patterns in the data. Is there a relationship between climate change and its impact on populations of local species?

Engage in Argument from Evidence Construct a scientific argument showing how data supports your hypothesis. Raise awareness! Write an article for a local newspaper or a blog that explains the role climate change may play in the evolution of species and its impact on local ecosystems. Include information that supports your claim and is easy for the reader to understand.

Biology.com

Go to biology.com for resources on climate change and its impact on genetic diversity.

16 Darwin's Theory of Evolution

Big idea **Evolution**
Q: What is natural selection?

INSIDE:

- 16.1 Darwin's Voyage of Discovery
- 16.2 Ideas That Shaped Darwin's Thinking
- 16.3 Darwin Presents His Case
- 16.4 Evidence of Evolution

The diversity of colors and banding patterns on these Cuban tree snails demonstrates the genetic variation that exists within most species. This kind of heritable variation provides the raw material for evolutionary change.

BUILDING Scientific Literacy

Deepen your understanding of evolution by engaging in key practices that allow you to make connections across concepts.

NGSS You will communicate information to explain several mechanisms that drive evolutionary change.

STEM Can you tell what a bird eats by looking at its beak? Determine how beak shape is adapted to a bird's diet.

Common Core You will write explanatory texts in your analysis of the structure of relationships among key terms and concepts of evolution.

CHAPTER MYSTERY

SUCH VARIED HONEYCREEPERS

The misty rain forests on the Hawaiian island of Kauai are home to birds found nowhere else on Earth. Hiking at dawn, you hear them before you see them. Their songs fill the air with beautiful music. Then you spot a brilliant red bird with black wings called an 'i'iwi. As you watch, it uses its long, curved beak to probe for nectar deep in the flowers of 'ohi'a trees.

The 'i'iwi is just one of a number of species of Hawaiian honeycreepers, all of which are related to finches. Various honeycreeper species feed on nectar, insects, seeds, or fruits. Many Hawaiian honeycreepers, however, feed only on the seeds or nectar of unique Hawaiian plants.

How did all these birds get to Hawaii? How did some of them come to have such specialized diets? As you read the chapter, look for clues that help explain the number and diversity of Hawaiian honeycreepers. Then, solve the mystery.

Biology.com

Finding the solution to the honeycreepers mystery is only the beginning. Take a video field trip to Hawaii with the ecogeeks of Untamed Science to see where the mystery leads.

Go online to access additional resources including:
• eText • Flash Cards • Lesson Overviews • Chapter Mystery

16.1 Darwin's Voyage of Discovery

Key Questions

What was Charles Darwin's contribution to science?

What three patterns of biodiversity did Darwin note?

Vocabulary

evolution
fossil

Taking Notes

Preview Visuals Before you read, look at **Figure 16–1.** Briefly summarize the route the *Beagle* took.

BUILD Vocabulary

RELATED WORD FORMS In biology, the noun **evolution** means "the process by which organisms have changed over time." The verb *evolve* means "to change over time."

THINK ABOUT IT If you'd met young Charles Darwin, you probably wouldn't have guessed that his ideas would change the way we look at the world. As a boy, Darwin wasn't a star student. He preferred bird-watching and reading for pleasure to studying. His father once complained, "You will be a disgrace to yourself and all your family." Yet Charles would one day come up with one of the most important scientific theories of all time—becoming far from the disgrace his father feared he would be.

Darwin's Epic Journey

What was Charles Darwin's contribution to science?

Charles Darwin was born in England on February 12, 1809—the same day as Abraham Lincoln. He grew up at a time when the scientific view of the natural world was shifting dramatically. Geologists were suggesting that Earth was ancient and had changed over time. Biologists were suggesting that life on Earth had also changed. The process of change over time is called **evolution.** **Darwin developed a scientific theory of biological evolution that explains how modern organisms evolved over long periods of time through descent from common ancestors.**

Darwin's journey began in 1831, when he was invited to sail on the HMS *Beagle*'s five-year voyage along the route shown in **Figure 16–1.** The captain and his crew would be mapping the coastline of South America. Darwin planned to collect specimens of plants and animals. No one knew it, but this would be one of the most important scientific voyages in history. Why? Because the *Beagle* trip led Darwin to develop what has been called the single best idea anyone has ever had.

If you think evolution is just about explaining life's ancient history, you might wonder why it's so important. But Darwin's work offers vital insights into today's world by showing how the living world is constantly changing. That perspective helps us understand modern phenomena like drug-resistant bacteria and newly emerging diseases like avian flu.

In Your Notebook *Using what you know about ecology, explain how the ideas of a changing Earth and evolving life forms might be related.*

Observations Aboard the *Beagle*

What three patterns of biodiversity did Darwin note?

A collector of bugs and shells in his youth, Darwin had always been fascinated by biological diversity. On his voyage, the variety and number of different organisms he encountered dazzled him. In a single day's trip into the Brazilian forest, he collected 68 species of beetles, and he wasn't particularly looking for beetles!

Darwin filled his notebooks with observations about the characteristics and habitats of the different species he saw. But Darwin wasn't content just to describe biological diversity. He wanted to explain it in a scientific way. He kept his eyes and mind open to larger patterns into which his observations might fit. As he traveled, Darwin noticed three distinctive patterns of biological diversity: (1) Species vary globally, (2) species vary locally, and (3) species vary over time.

Species Vary Globally Darwin visited a wide range of habitats on the continents of South America, Australia, and Africa and recorded his observations. For example, Darwin found flightless, ground-dwelling birds called rheas living in the grasslands of South America. Rheas look and act a lot like ostriches. Yet rheas live only in South America, and ostriches live only in Africa. When Darwin visited Australia's grasslands, he found another large flightless bird, the emu. **Darwin noticed that different, yet ecologically similar, animal species inhabited separated, but ecologically similar, habitats around the globe.**

Darwin also noticed that rabbits and other species living in European grasslands were missing from the grasslands of South America and Australia. What's more, Australia's grasslands were home to kangaroos and other animals that were found nowhere else. What did these patterns of geographic distribution mean? Why did different flightless birds live in similar grasslands across South America, Australia, and Africa, but not in the Northern Hemisphere? Why weren't there rabbits in Australian habitats that seemed ideal for them? And why didn't kangaroos live in England?

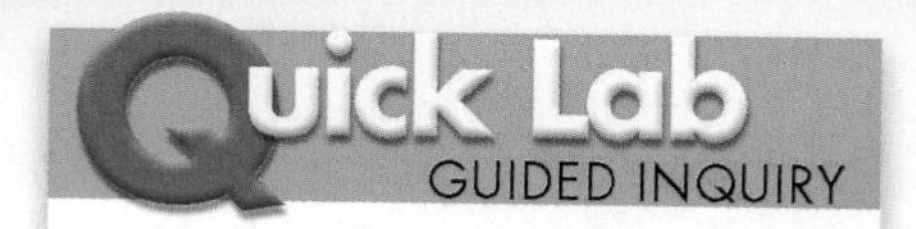

Darwin's Voyage

1. Using a world map and **Figure 16–1,** count the number of lines of 10° latitude the *Beagle* crossed.
2. Using the biome map from Chapter 4 as a reference, identify three different biomes Darwin visited on his voyage.

Analyze and Conclude

1. Infer How did the geography of Darwin's voyage give him far greater exposure to species variability than his fellow scientists back home had?

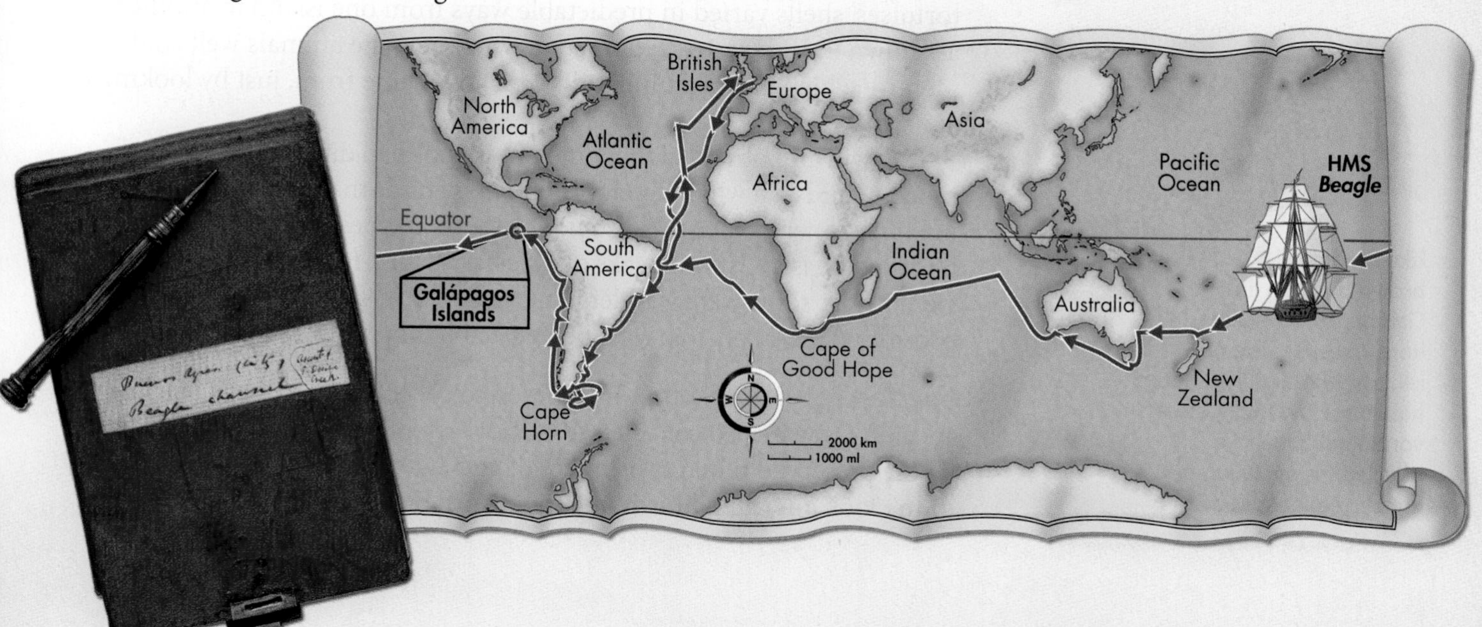

FIGURE 16–1 Darwin's Voyage On a five-year voyage aboard the *Beagle,* Charles Darwin visited several continents and many remote islands. **Draw Conclusions** ***Why is it significant that many of the stops the* Beagle *made were in tropical regions?***

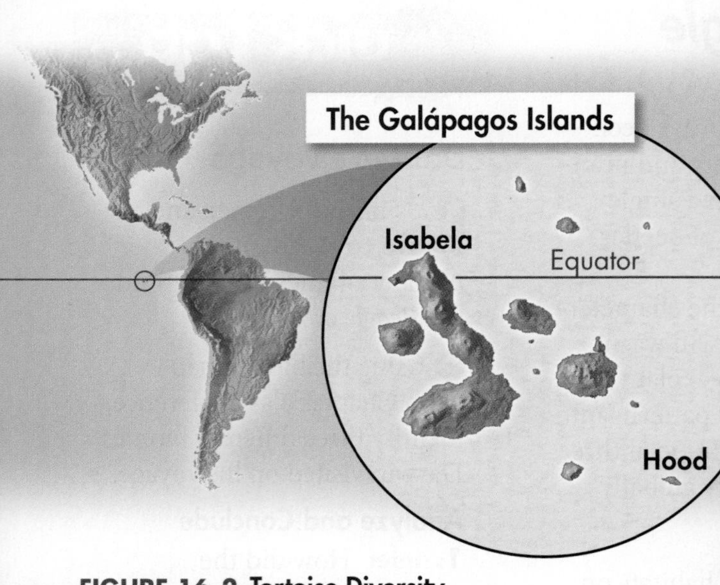

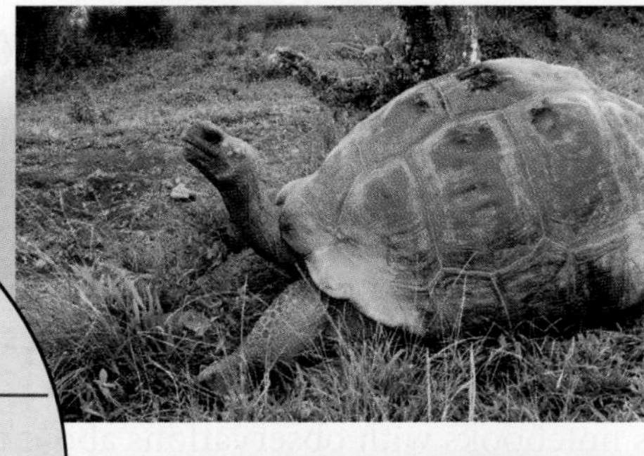

Isabela Island Tortoise Tortoises from Isabela Island have dome-shaped shells and short necks. Vegetation on this island is abundant and close to the ground.

Hood Island Tortoise The shells of Hood Island tortoises are curved and open around their long necks and legs. This enables them to reach the island's sparse, high vegetation.

FIGURE 16–2 Tortoise Diversity Among tortoises in the Galápagos Islands, shell shape corresponds to different habitats. Isabela Island has high peaks, is rainy, and has abundant vegetation. Hood Island, in contrast, is flat, dry, and has sparse vegetation.

Species Vary Locally There were other puzzles, too. For example, Darwin found two species of rheas living in South America. One lived in Argentina's grasslands and the other in the colder, harsher grass and scrubland to the south. **Darwin noticed that different, yet related, animal species often occupied different habitats within a local area.**

Other examples of local variation came from the Galápagos Islands, about 1000 km off the Pacific coast of South America. These islands are close to one another, yet they have different ecological conditions. Several islands were home to distinct forms of giant land tortoises. Darwin saw differences among the tortoises but didn't think much about them. In fact, like other travelers, Darwin ate several tortoises and tossed their remains overboard without studying them closely! Then Darwin learned from the islands' governor that the tortoises' shells varied in predictable ways from one island to another, as shown in **Figure 16–2.** Someone who knew the animals well could identify which island an individual tortoise came from, just by looking at its shell.

Darwin also observed that different islands had different varieties of mockingbirds, all which resembled mockingbirds that Darwin had seen in South America. Darwin also noticed several types of small brown birds on the islands with beaks of different shapes. He thought that some were wrens, some were warblers, and some were blackbirds. He didn't consider these smaller birds to be unusual or important—at first.

Species Vary Over Time In addition to collecting specimens of living species, Darwin also collected **fossils,** which scientists already knew to be the preserved remains or traces of ancient organisms. Some fossils didn't look anything like living organisms, but others did.

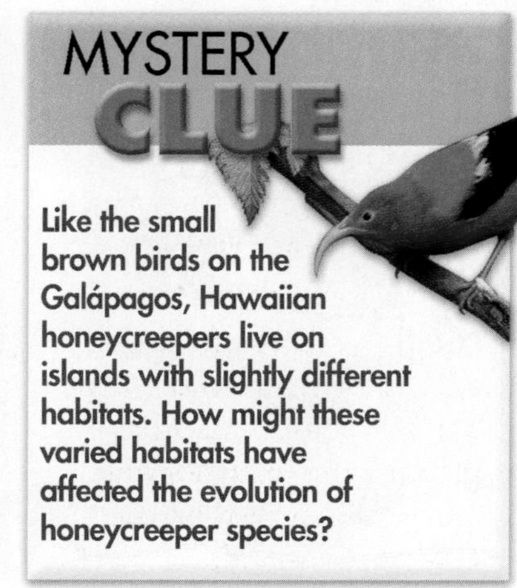

Like the small brown birds on the Galápagos, Hawaiian honeycreepers live on islands with slightly different habitats. How might these varied habitats have affected the evolution of honeycreeper species?

Darwin noticed that some fossils of extinct animals were similar to living species. One set of fossils unearthed by Darwin belonged to the long-extinct glyptodont, a giant armored animal. Currently living in the same area was a similar animal, the armadillo. You can see in **Figure 16–3** that the armadillo appears to be a smaller version of the glyptodont. Darwin said of the organisms: "This wonderful relationship in the same continent between the dead and the living, will, I do not doubt, hereafter throw more light on the appearance of organic beings on our earth, and their disappearance from it, than any other class of facts." So, why had glyptodonts disappeared? And why did they resemble armadillos?

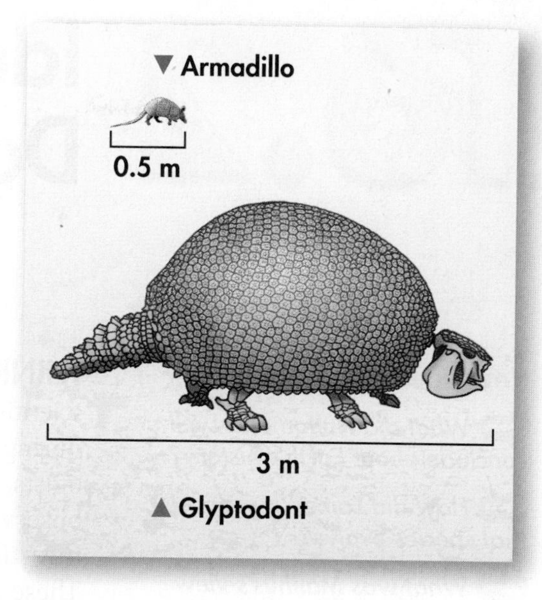

FIGURE 16–3 Related Organisms? Despite their obvious differences, Darwin wondered if the armadillo might be related to the ancient glyptodont. Compare and Contrast *What similarities and differences do you see between these two animals?*

Putting the Pieces of the Puzzle Together On the voyage home, Darwin thought about the patterns he'd seen. The plant and animal specimens he sent to experts for identification set the scientific community buzzing. The Galápagos mocking-birds turned out to belong to three separate species found nowhere else! And the little brown birds that Darwin thought were wrens, warblers, and blackbirds were actually all species of finches! They, too, were found nowhere else, though they resembled a South American finch species. The same was true of Galápagos tortoises, marine iguanas, and many plants that Darwin collected on the islands.

Darwin was stunned by these discoveries. He began to wonder whether different Galápagos species might have evolved from South American ancestors. He spent years actively researching and filling notebooks with ideas about species and evolution. The evidence suggested that species are not fixed and that they could change by some natural process.

16.1 Assessment

Review Key Concepts

1. a. Review What is evolution?

   b. Apply Concepts What ideas were changing in the scientific community at the time of Darwin's travels? How might those new ideas have influenced Darwin?

2. a. Review What three kinds of variations among organisms did Darwin observe during the voyage of the *Beagle?*

   b. Infer Darwin found fossils of many organisms that did not resemble any living species. How might this finding have affected his understanding of life's diversity?

Apply the Big idea

Interdependence in Nature

3. You have learned that both biotic and abiotic factors affect ecosystems. Give some examples of each, and explain how biotic and abiotic factors could have affected the tortoises that Darwin observed on the Galápagos Islands.

16.2 Ideas That Shaped Darwin's Thinking

Key Questions

What did Hutton and Lyell conclude about Earth's history?

How did Lamarck propose that species evolve?

What was Malthus's view of population growth?

How is inherited variation used in artificial selection?

Vocabulary

artificial selection

Taking Notes

Outline Make an outline of this lesson using the green headings as main topics and the blue headings as subtopics. As you read, fill in details under each heading.

THINK ABOUT IT All scientists are influenced by the work of other scientists, and Darwin was no exception. The *Beagle*'s voyage came during one of the most exciting periods in the history of science. Geologists, studying the structure and history of Earth, were making new observations about the forces that shape our planet. Naturalists were investigating connections between organisms and their environments. These and other new ways of thinking about the natural world provided the foundation on which Darwin built his ideas.

An Ancient, Changing Earth

What did Hutton and Lyell conclude about Earth's history?

Many Europeans in Darwin's day believed Earth was only a few thousand years old, and that it hadn't changed much. By Darwin's time, however, the relatively new science of geology was providing evidence to support different ideas about Earth's history. Most famously, geologists James Hutton and Charles Lyell formed important hypotheses based on the work of other researchers and on evidence they uncovered themselves. **Hutton and Lyell concluded that Earth is extremely old and that the processes that changed Earth in the past are the same processes that operate in the present.** In 1785, Hutton presented his hypotheses about how geological processes have shaped the Earth. Lyell, who built on the work of Hutton and others, published the first volume of his great work, *Principles of Geology,* in 1830.

FIGURE 16–4 Ancient Rocks These rock layers in the Grand Canyon were laid down over millions of years and were then slowly washed away by the river, forming a channel.

Hutton and Geological Change Hutton recognized the connections between a number of geological processes and geological features, like mountains, valleys, and layers of rock that seemed to be bent or folded. Hutton realized, for example, that certain kinds of rocks are formed from molten lava. He also realized that some other kinds of rocks, like those shown in **Figure 16–4,** form very slowly, as sediments build up and are squeezed into layers.

Hutton also proposed that forces beneath Earth's surface can push rock layers upward, tilting or twisting them in the process. Over long periods, those forces can build mountain ranges. Mountains, in turn, can be worn down by rain, wind, heat, and cold. Most of these processes operate very slowly. For these processes to have produced Earth as we know it, Hutton concluded that our planet must be much older than a few thousand years. He introduced a concept called *deep time*—the idea that our planet's history stretches back over a period of time so long that it is difficult for the human mind to imagine—to explain his reasoning.

BUILD Vocabulary

ACADEMIC WORDS The noun **process** means "a series of actions or changes that take place in a definite manner." The processes that shape Earth are the series of geological actions that do things such as build mountains and carve valleys.

Lyell's *Principles of Geology* Lyell argued that laws of nature are constant over time and that scientists must explain past events in terms of processes they can observe in the present. This way of thinking, called *uniformitarianism,* holds that the geological processes we see in action today must be the same ones that shaped Earth millions of years ago. Ancient volcanoes released lava and gases, just as volcanoes do now. Ancient rivers slowly dug channels, like the one in **Figure 16–5,** and carved canyons in the past, just as they do today. Lyell's theories, like those of Hutton before him, relied on there being enough time in Earth's history for these changes to take place. Like Hutton, Lyell argued that Earth was much, much older than a few thousand years. Otherwise, how would a river have enough time to carve out a valley?

Darwin had begun to read Lyell's books during the voyage of the *Beagle*, which was lucky. Lyell's work helped Darwin appreciate the significance of an earthquake he witnessed in South America. The quake was so strong that it threw Darwin onto the ground. It also lifted a stretch of rocky shoreline more than 3 meters out of the sea—with mussels and other sea animals clinging to it. Sometime later, Darwin observed fossils of marine animals in mountains thousands of feet above sea level.

Those experiences amazed Darwin and his companions. But only Darwin turned them into a startling scientific insight. He realized that he had seen evidence that Lyell was correct! Geological events like the earthquake, repeated many times over many years, could build South America's Andes Mountains—a few feet at a time. Rocks that had once been beneath the sea could be pushed up into mountains. Darwin asked himself, If Earth can change over time, could life change too?

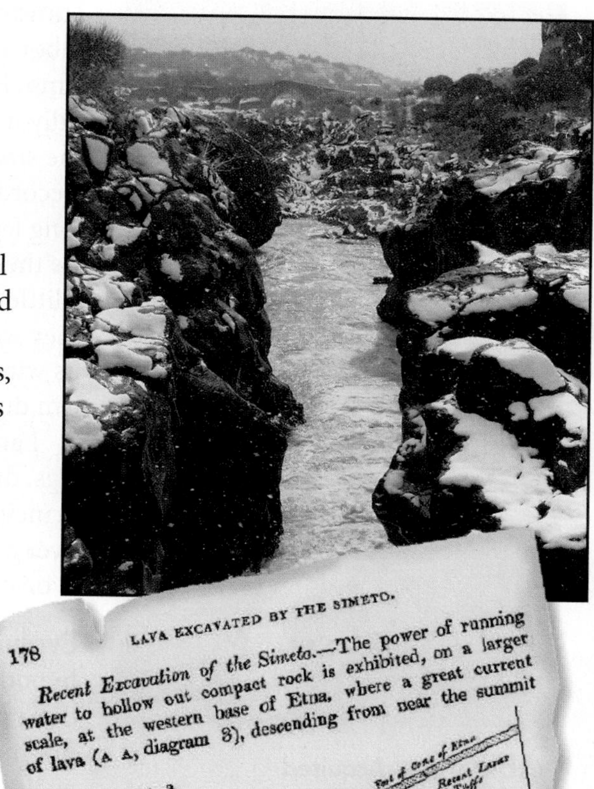

178 LAVA EXCAVATED BY THE SIMETO.

Recent Excavation of the Simeto.—The power of running water to hollow out compact rock is exhibited, on a larger scale, at the western base of Etna, where a great current of lava (A A, diagram 3), descending from near the summit

FIGURE 16–5 A woodcut from Lyell's *Principles of Geology* shows geological features near Italy's Mount Etna. Among them is a deep channel, labeled "B," carved into a bed of lava. The channel, shown in the photo, was formed gradually by the movement of water in the Simeto River.

Lamarck's Evolutionary Hypotheses

How did Lamarck propose that species evolve?

Darwin wasn't the first scientist to suggest that characteristics of species could change over time. Throughout the eighteenth century, a growing fossil record supported the idea that life somehow evolved. Ideas differed, however, about just *how* life evolved. The French naturalist Jean-Baptiste Lamarck proposed two of the first hypotheses. **Lamarck suggested that organisms could change during their lifetimes by selectively using or not using various parts of their bodies. He also suggested that individuals could pass these acquired traits on to their offspring, enabling species to change over time.** Lamarck published his ideas in 1809, the year Darwin was born.

FIGURE 16–6 Acquired Characteristics? According to Lamarck, this black-winged stilt's long legs were the result of the bird's innate tendency toward perfection. He claimed that if a water bird needs long legs to wade in deep water, it can acquire them by making an effort to stretch and use its legs in new ways. He also claimed that the bird can then pass the trait on to its offspring.

Lamarck's Ideas Lamarck proposed that all organisms have an inborn urge to become more complex and perfect. As a result, organisms change and acquire features that help them live more successfully in their environments. He thought that organisms could change the size or shape of their organs by using their bodies in new ways. According to Lamarck, for example, a water bird could have acquired long legs because it began to wade in deeper water looking for food. As the bird tried to stay above the water's surface, its legs would grow a little longer. Structures of individual organisms could also change if they were not used. If a bird stopped using its wings to fly, for example, its wings would become smaller. Traits altered by an individual organism during its life are called *acquired characteristics.*

Lamarck also suggested that a bird that acquired a trait, like longer legs, during its lifetime could pass that trait on to its offspring, a principle referred to as *inheritance of acquired characteristics.* Thus, over a few generations, birds such as the one in **Figure 16–6** could evolve longer and longer legs.

Evaluating Lamarck's Hypotheses Today, we know that Lamarck's hypotheses were incorrect in several ways. For one thing, organisms don't have an inborn drive to become more perfect. Evolution does not mean that over time a species becomes "better" somehow, and evolution does not progress in a predetermined direction. We now also know that traits acquired by individuals during their lifetime cannot be passed on to offspring. However, Lamarck was one of the first naturalists to suggest that species are not fixed. He was among the first to try to explain evolution scientifically using natural processes. He also recognized that there is a link between an organism's environment and its body structures. So, although Lamarck's explanation of evolutionary change was wrong, his work paved the way for later biologists, including Darwin.

In Your Notebook *Why are Lamarck's ideas called scientific hypotheses and not scientific theories?*

Population Growth

What was Malthus's view of population growth?

In 1798, English economist Thomas Malthus noted that humans were being born faster than people were dying, causing overcrowding, as shown in **Figure 16–7.** **Malthus reasoned that if the human population grew unchecked, there wouldn't be enough living space and food for everyone.** The forces that work against population growth, Malthus suggested, include war, famine, and disease.

Darwin realized that Malthus's reasoning applied even more to other organisms than it did to humans. A maple tree can produce thousands of seeds each summer. One oyster can produce millions of eggs each year. If all the descendants of almost any species survived for several generations, they would overrun the world. Obviously, this doesn't happen. Most offspring die before reaching maturity, and only a few of those that survive manage to reproduce.

Why was this realization so important? Darwin had become convinced that species evolved. But he needed a mechanism—a scientific explanation based on a natural process—to explain how and why evolution occurred. When Darwin realized that most organisms don't survive and reproduce, he wondered which individuals survive ... and why.

FIGURE 16–7 Overcrowding in London A nineteenth-century engraving shows the crowded conditions in London during Darwin's time. **Relate Cause and Effect** ***According to Malthus, what would happen if the population of London continued to grow?***

Artificial Selection

How is inherited variation used in artificial selection?

To find an explanation for change in nature, Darwin studied change produced by plant and animal breeders. Those breeders knew that individual organisms vary—that some plants bear larger or smaller fruit than average for their species, that some cows give more or less milk than others in their herd. They told Darwin that some of this variation could be passed from parents to offspring and used to improve crops and livestock.

Quick Lab
GUIDED INQUIRY

Variation in Peppers 

1. Obtain a green, yellow, red, or purple bell pepper.
2. Slice open the pepper and count the number of seeds it contains.
3. Compare your data with the data of other students who have peppers of a different color.

Analyze and Conclude

1. Calculate Find the average (mean) number of seeds in your class's peppers. Then determine by how much the number of seeds in each pepper differs from the mean number. MATH

2. Pose Questions Think of the kinds of variations among organisms that Darwin observed. If Darwin had seen your data, what questions might he have asked?

FIGURE 16–8 Artificial Selection Darwin used artificial selection in breeding fancy pigeons at his home outside London.

Farmers would select for breeding only trees that produced the largest fruit or cows that produced the most milk. Over time, this selective breeding would produce more trees with even bigger fruit and cows that gave even more milk. Darwin called this process **artificial selection.** **In artificial selection, nature provides the variations, and humans select those they find useful.** Darwin put artificial selection to the test by raising and breeding plants and fancy pigeon varieties, like those in **Figure 16–8.**

Darwin had no idea how heredity worked or what caused heritable variation. But he did know that variation occurs in wild species as well as in domesticated plants and animals. Before Darwin, scientists thought variations among individuals in nature were simply minor defects. Darwin's breakthrough was in recognizing that natural variation was very important because it provided the raw material for evolution. Darwin had all the information he needed. His scientific explanation for evolution was now formed—and when it was published, it would change the way people understood the living world.

16.2 Assessment

Review Key Concepts

1. a. **Review** What were Hutton's and Lyell's ideas about the age of Earth and the processes that shape the planet?

   b. **Apply Concepts** How would Hutton and Lyell explain the formation of the Grand Canyon?

2. a. **Review** What is an acquired characteristic? What role did Lamarck think acquired characteristics played in evolution?

   b. **Evaluate** What parts of Lamarck's hypotheses have been proved wrong? What did Lamarck get right?

3. a. **Review** According to Malthus, what factors limit human population growth?

   b. **Draw Conclusions** How did Malthus influence Darwin?

4. a. **Review** What is artificial selection?

   b. **Infer** Could artificial selection occur without inherited variation? Explain your answer.

WRITE ABOUT SCIENCE

Creative Writing

5. Imagine you are Thomas Malthus and the year is 1798. Write a newspaper article that explains your ideas about the impact of a growing population on society and the environment.

Biology & HISTORY

Origins of Evolutionary Thought The groundwork for the modern theory of evolution was laid during the 1700s and 1800s. Charles Darwin developed the central idea of evolution by natural selection, but others before and during his lifetime influenced his thinking.

1780 1790 1800 1810 1820 1830 1840 1850 1860

1785
▼ James Hutton
Hutton proposes that slow-acting geological forces shape the planet. He estimates Earth to be millions—not thousands—of years old.

1809
Jean-Baptiste Lamarck
Lamarck publishes his hypotheses of the inheritance of acquired traits. The ideas are flawed, but he is one of the first to propose a mechanism explaining how organisms change over time. ▼

1830–1833
Charles Lyell ▶
In his *Principles of Geology*, Lyell explains that over long periods, the same processes affecting Earth today have shaped Earth's ancient geological features.

1858
Alfred Russel Wallace
Wallace writes to Darwin, speculating on evolution by natural selection, based on his studies of the distribution of plants and animals.

1798
Thomas Malthus
In his *Essay on the Principle of Population*, Malthus predicts that left unchecked, the human population will grow beyond the space and food needed to sustain it.

1831
Charles Darwin
Darwin sets sail on the HMS *Beagle*, a voyage that will provide him with vast amounts of evidence to support his explanation of how evolution works. ▶

185
Darwin publish
*On the Orig
of Speci*

WRITING Use the library or the Internet to find out more about Darwin and Wallace. Then write a dialogue between these two men, in which the conversation shows the similarities in their careers and theories.

16.3 Darwin Presents His Case

Key Questions

Under what conditions does natural selection occur?

What does Darwin's mechanism for evolution suggest about living and extinct species?

Vocabulary

adaptation
fitness
natural selection

Taking Notes

Preview Visuals Before you read this lesson, look at **Figure 16–10.** Read the information in the figure, and then write three questions you have about it. As you read, answer your questions.

THINK ABOUT IT Soon after reading Malthus and thinking about artificial selection, Darwin worked out the main points of his theory about natural selection. Most of his scientific friends considered Darwin's arguments to be brilliant, and they urged him to publish them. But although he wrote up a complete draft of his ideas, he put the work aside and didn't publish it for another 20 years. Why? Darwin knew that many scientists, including some of Darwin's own teachers, had ridiculed Lamarck's ideas. Darwin also knew that his own theory was just as radical, so he wanted to gather as much evidence as he could to support his ideas before he made them public.

Then, in 1858, Darwin reviewed an essay by Alfred Russel Wallace, an English naturalist working in Malaysia. Wallace's thoughts about evolution were almost identical to Darwin's! Not wanting to get "scooped," Darwin decided to move forward with his own work. Wallace's essay was presented together with some of Darwin's observations at a scientific meeting in 1858. The next year, Darwin published his first complete work on evolution: *On the Origin of Species.*

Evolution by Natural Selection

Under what conditions does natural selection occur?

Darwin's great contribution was to describe a process in nature—a scientific mechanism—that could operate like artificial selection. In *On the Origin of Species,* he combined his own thoughts with ideas from Malthus and Lamarck.

The Struggle for Existence After reading Malthus, Darwin realized that if more individuals are produced than can survive, members of a population must compete to obtain food, living space, and other limited necessities of life. Darwin described this as *the struggle for existence.* But which individuals come out on top in this struggle?

Variation and Adaptation Here's where individual variation plays a vital role. Darwin knew that individuals have natural variations among their heritable traits. He hypothesized that some of those variants are better suited to life in their environment than others. Members of a predatory species that are faster or have longer claws or sharper teeth can catch more prey. And members of a prey species that are faster or better camouflaged can avoid being caught.

Any heritable characteristic that increases an organism's ability to survive and reproduce in its environment is called an **adaptation.** Adaptations can involve body parts or structures, like a tiger's claws; colors, like those that make camouflage or mimicry possible; or physiological functions, like the way a plant carries out photosynthesis. Many adaptations also involve behaviors, such as the complex avoidance strategies prey species use. Examples of adaptations are shown in **Figure 16–9.**

Survival of the Fittest Darwin, like Lamarck, recognized that there must be a connection between the way an organism "makes a living" and the environment in which it lives. According to Darwin, differences in adaptations affect an individual's fitness. **Fitness** describes how well an organism can survive and reproduce in its environment.

Individuals with adaptations that are well suited to their environment can survive and reproduce and are said to have high fitness. Individuals with characteristics that are not well suited to their environment either die without reproducing or leave few offspring and are said to have low fitness. This difference in rates of survival and reproduction is called *survival of the fittest.* Note that *survival* here means more than just staying alive. In evolutionary terms, *survival* means reproducing and passing adaptations on to the next generation.

In Your Notebook *If an organism produces many offspring, but none of them reach maturity, do you think the organism has high or low fitness? Explain your answer.*

BUILD Vocabulary

RELATED WORD FORMS The verb *inherited* and the adjective *heritable* are related word forms. Inherited traits are passed on to offspring from their parents. They are described as *heritable* (or sometimes *inheritable*) characteristics.

VISUAL SUMMARY

ADAPTATIONS

FIGURE 16–9 Adaptations take many forms.

▼ **A.** The scarlet king snake (bottom) is exhibiting mimicry—an adaptation in which an organism copies, or mimics, a more dangerous organism. Although the scarlet king snake is harmless, it looks like the poisonous eastern coral snake (top), so predators avoid it, too.

B. A scorpionfish's coloring is an example of camouflage—an adaptation that allows an organism to blend into its background and avoid predation. ▶

▼ **C.** Adaptations often involve many systems and even behavior. Here, a crane is displaying defensive behavior in an effort to scare off the nearby fox.

VISUAL SUMMARY

NATURAL SELECTION

FIGURE 16–10 This hypothetical population of grasshoppers changes over time as a result of natural selection. **Interpret Visuals** ***In the situation shown here, what characteristic is affecting the grasshoppers' fitness?***

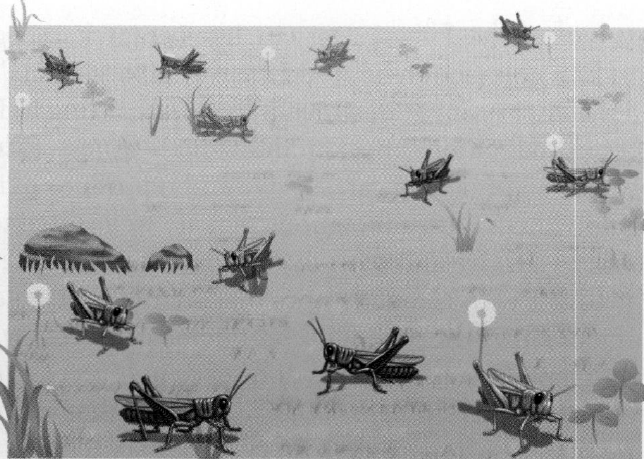

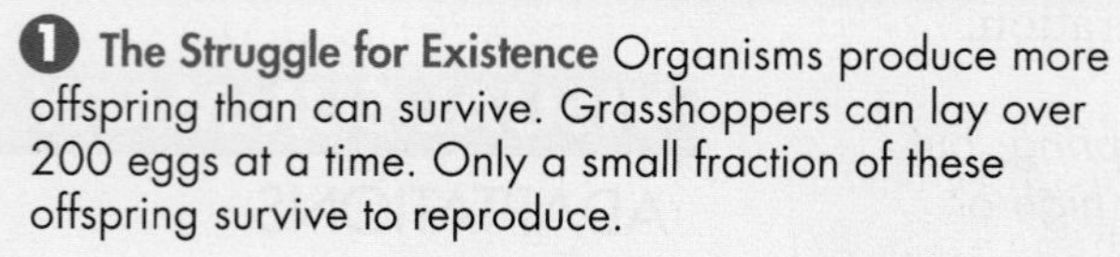

1 The Struggle for Existence Organisms produce more offspring than can survive. Grasshoppers can lay over 200 eggs at a time. Only a small fraction of these offspring survive to reproduce.

2 Variation and Adaptation There is variation in nature, and certain heritable variations—called adaptations—increase an individual's chance of surviving and reproducing. In this population of grasshoppers, heritable variation includes yellow and green body color. Green coloration is an adaptation: Green grasshoppers blend into their environment and so are less visible to predators.

3 Survival of the Fittest Because their green color serves to camouflage them from predators, green grasshoppers have a higher fitness than yellow grasshoppers. This means that green grasshoppers survive and reproduce more often than do yellow grasshoppers in this environment.

4 Natural Selection Green grasshoppers become more common than yellow grasshoppers in this population over time because: (1) more grasshoppers are born than can survive, (2) individuals vary in color and color is a heritable trait, and (3) green individuals have a higher fitness in their current environment.

Natural Selection Darwin named his mechanism for evolution *natural selection* because of its similarities to artificial selection. **Natural selection** is the process by which organisms with variations most suited to their local environment survive and leave more offspring. In both artificial and natural selection, only certain individuals in a population produce new individuals. But in natural selection, the environment—not a farmer or animal breeder—influences fitness.

When does natural selection occur? **Natural selection occurs in any situation in which more individuals are born than can survive (the struggle for existence), there is natural heritable variation (variation and adaptation), and there is variable fitness among individuals (survival of the fittest).** Well-adapted individuals survive and reproduce. From generation to generation, populations continue to change as they become better adapted, or as their environment changes. **Figure 16–10** uses a hypothetical example to show the process of natural selection. Notice that natural selection acts only on inherited traits because those are the only characteristics that parents can pass on to their offspring.

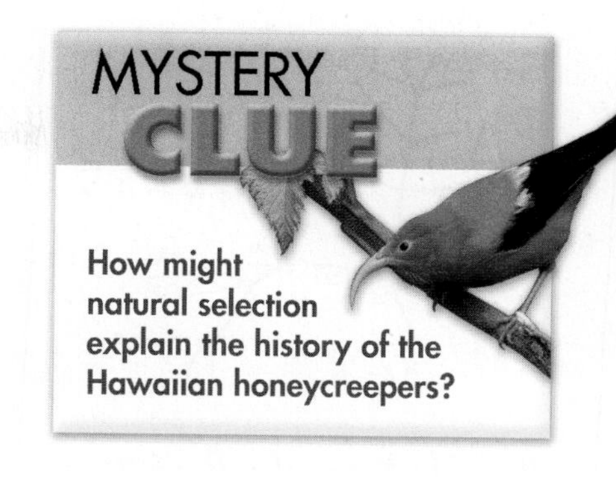

How might natural selection explain the history of the Hawaiian honeycreepers?

Natural selection does not make organisms "better." Adaptations don't have to be perfect—just good enough to enable an organism to pass its genes to the next generation. Natural selection also doesn't move in a fixed direction. There is no one, perfect way of doing something, as demonstrated by **Figure 16–11.** Natural selection is simply a process that enables species to survive and reproduce in a local environment. If local environmental conditions change, some traits that were once adaptive may no longer be useful, and different traits may become adaptive. And if environmental conditions change faster than a species can adapt to those changes, the species may become extinct. Of course, natural selection is not the only mechanism driving evolution. You will learn about other evolutionary mechanisms in the next chapter.

In Your Notebook *Give at least two reasons why the following statement is NOT true: "The goal of natural selection is to produce perfect organisms."*

FIGURE 16–11 No Such Thing as Perfect Many different styles of pollination have evolved among flowering plants. Oak tree flowers (right) are pollinated by wind. Apple tree flowers (left) are pollinated by insects. Neither method is "better" than the other. Both kinds of pollination work well enough for these plants to survive and reproduce in their environments.

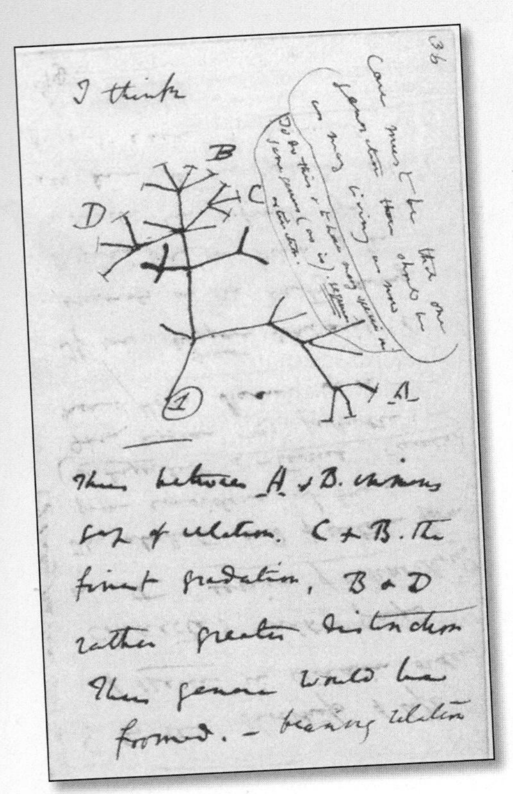

FIGURE 16–12 Descent With Modification This page from one of Darwin's notebooks shows the first evolutionary tree ever drawn. This sketch shows Darwin's explanation for how descent with modification could produce the diversity of life. Note that, just above the tree, Darwin wrote, "I think."

Common Descent

What does Darwin's mechanism for evolution suggest about living and extinct species?

Natural selection depends on the ability of organisms to reproduce, which means to leave descendants. Every organism alive today is descended from parents who survived and reproduced. Those parents descended from their parents, and so forth back through time.

Just as well-adapted individuals in a species survive and reproduce, well-adapted species survive over time. Darwin proposed that, over many generations, adaptation could cause successful species to evolve into new species. He also proposed that living species are descended, with modification, from common ancestors—an idea called *descent with modification.* Notice that this aspect of Darwin's theory implies that life has been on Earth for a very long time—enough time for all this descent with modification to occur! This is Hutton and Lyell's contribution to Darwin's theory: Deep time gave enough time for natural selection to act. For evidence of descent with modification over long periods of time, Darwin pointed to the fossil record.

Darwin based his explanation for the diversity of life on the idea that species change over time. To illustrate this idea, he drew the very first evolutionary tree, shown in **Figure 16–12.** This "tree-thinking" implies that all organisms are related. Look back in time, and you will find common ancestors shared by tigers, panthers, and cheetahs. Look farther back, and you will find ancestors that these felines share with dogs, then horses, and then bats. Farther back still is the common ancestor that all mammals share with birds, alligators, and fish. Far enough back are the common ancestors of all living things. **According to the principle of common descent, all species—living and extinct—are descended from ancient common ancestors.** A single "tree of life" links all living things.

16.3 Assessment

Review Key Concepts

1. a. **Review** What happens in the process of natural selection?
   b. **Explain** Why do organisms with greater fitness generally leave more offspring than organisms that are less fit?
   c. **Compare and Contrast** How are natural selection and artificial selection similar? How are they different?
2. a. **Review** Why were Hutton's and Lyell's ideas important to Darwin?
   b. **Apply Concepts** What do evolutionary trees show? What does a tree of life imply about all species living and extinct?

VISUAL THINKING

3. Look at the teeth in the lion's mouth. How is the structure of the lion's teeth an adaptation?

16.4 Evidence of Evolution

THINK ABOUT IT Darwin's theory depended on assumptions that involved many scientific fields. Scientists in some fields, including geology, physics, paleontology, chemistry, and embryology, did not have the technology or understanding to test Darwin's assumptions during his lifetime. And other fields, like genetics and molecular biology, didn't exist yet! In the 150 years since Darwin published *On the Origin of Species*, discoveries in all these fields have served as independent tests that could have supported or refuted Darwin's work. Astonishingly, every scientific test has supported Darwin's basic ideas about evolution.

Key Questions

How does the geographic distribution of species today relate to their evolutionary history?

How do fossils help to document the descent of modern species from ancient ancestors?

What do homologous structures and similarities in embryonic development suggest about the process of evolutionary change?

How can molecular biology be used to trace the process of evolution?

What does recent research on the Galápagos finches show about natural selection?

Vocabulary

biogeography
homologous structure
analogous structure
vestigial structure

Taking Notes

Concept Map Construct a concept map that shows the kinds of evidence that support the theory of evolution.

Biogeography

How does the geographic distribution of species today relate to their evolutionary history?

Darwin recognized the importance of patterns in the distribution of life—the subject of the field called biogeography. **Biogeography** is the study of where organisms live now and where they and their ancestors lived in the past. **Patterns in the distribution of living and fossil species tell us how modern organisms evolved from their ancestors.** Two biogeographical patterns are significant to Darwin's theory. The first is a pattern in which closely related species differentiate in slightly different climates. The second is a pattern in which very distantly related species develop similarities in similar environments.

Closely Related but Different To Darwin, the biogeography of Galápagos species suggested that populations on the island had evolved from mainland species. Over time, natural selection on the islands produced variations among populations that resulted in different, but closely related, island species.

Distantly Related but Similar On the other hand, similar habitats around the world are often home to animals and plants that are only distantly related. Darwin noted that similar ground-dwelling birds inhabit similar grasslands in Europe, Australia, and Africa. Differences in body structures among those animals provide evidence that they evolved from different ancestors. Similarities among those animals, however, provide evidence that similar selection pressures had caused distantly related species to develop similar adaptations.

The Age of Earth and Fossils

How do fossils help to document the descent of modern species from ancient ancestors?

Two potential difficulties for Darwin's theory involved the age of Earth and gaps in the fossil record. Data collected since Darwin's time have answered those concerns and have provided dramatic support for an evolutionary view of life.

The Age of Earth Evolution takes a long time. If life has evolved, then Earth must be very old. Hutton and Lyell argued that Earth was indeed very old, but technology in their day couldn't determine just how old. Half a century after Darwin published his theory, however, physicists discovered radioactivity. Geologists now use radioactivity to establish the age of certain rocks and fossils. This kind of data could have shown that Earth is young. If that had happened, Darwin's ideas would have been refuted and abandoned. Instead, radioactive dating indicates that Earth is about 4.5 billion years old—plenty of time for evolution by natural selection to take place.

VISUAL SUMMARY

EVIDENCE FROM FOSSILS

FIGURE 16–13 Recently, researchers have found more than 20 related fossils that document the evolution of modern whales from ancestors that walked on land. Several reconstructions based on fossil evidence are shown below in addition to the modern mysticete and odontocete. **Infer** ***Which of the animals shown was probably the most recent to live primarily on land?***

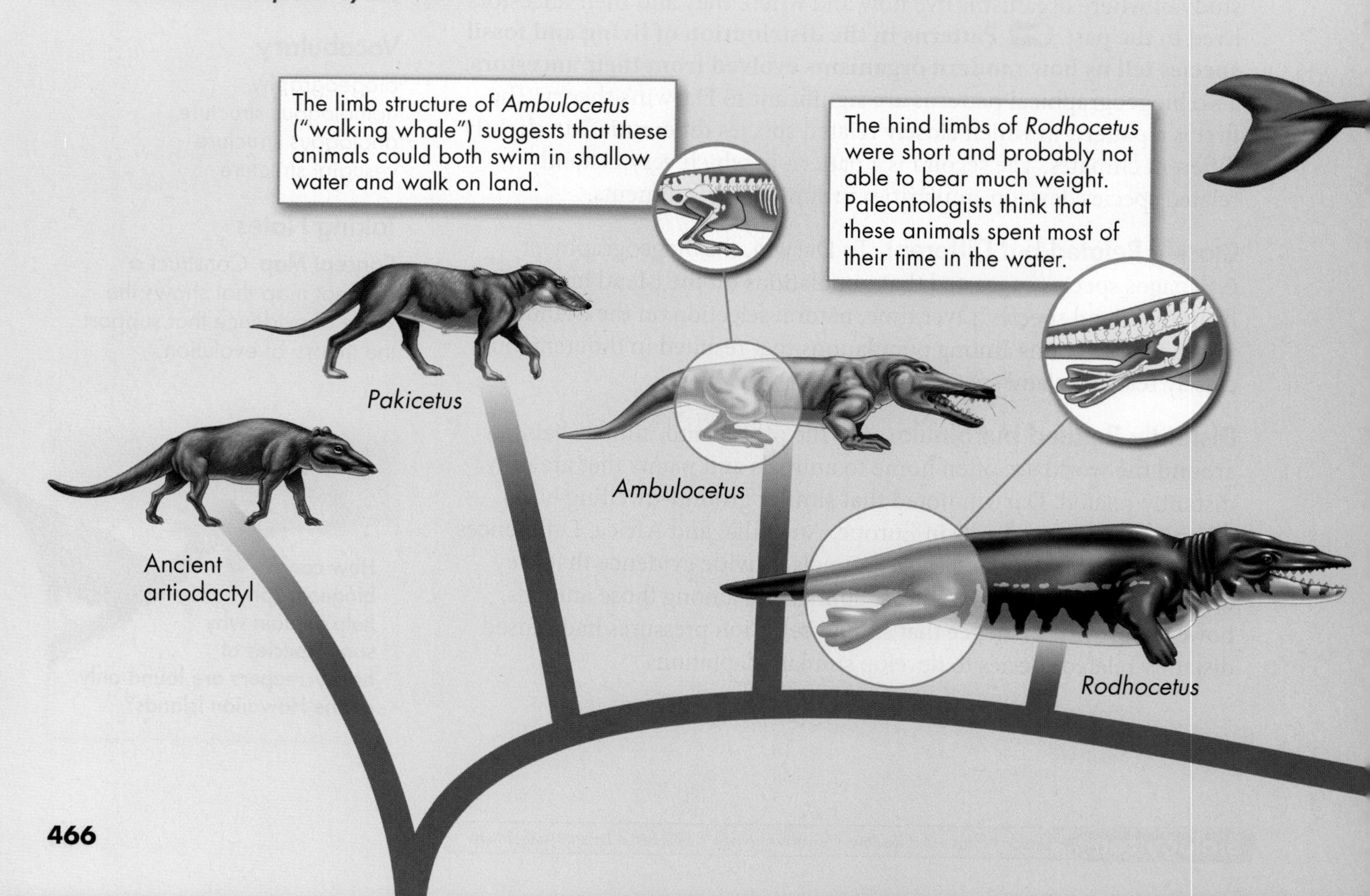

Recent Fossil Finds Darwin also struggled with what he called the "imperfection of the geological record." Darwin's study of fossils had convinced him and other scientists that life evolved. But paleontologists in 1859 hadn't found enough fossils of intermediate forms of life to document the evolution of modern species from their ancestors. **Many recently discovered fossils form series that trace the evolution of modern species from extinct ancestors.**

Since Darwin, paleontologists have discovered hundreds of fossils that document intermediate stages in the evolution of many different groups of modern species. One recently discovered fossil series documents the evolution of whales from ancient land mammals, as shown in **Figure 16–13.** Other recent fossil finds connect the dots between dinosaurs and birds, and between fish and four-legged land animals. In fact, so many intermediate forms have been found that it is often hard to tell where one group begins and another ends. All historical records are incomplete, and the history of life is no exception. The evidence we do have, however, tells an unmistakable story of evolutionary change.

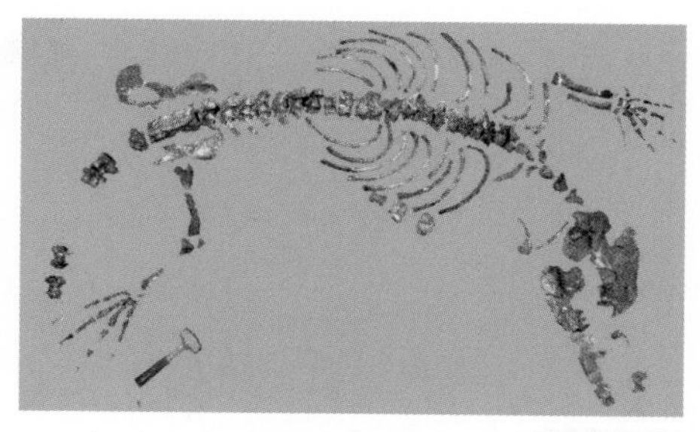

Fossil of the Eocene whale *Ambulocetus natans* (about 49 million years old)

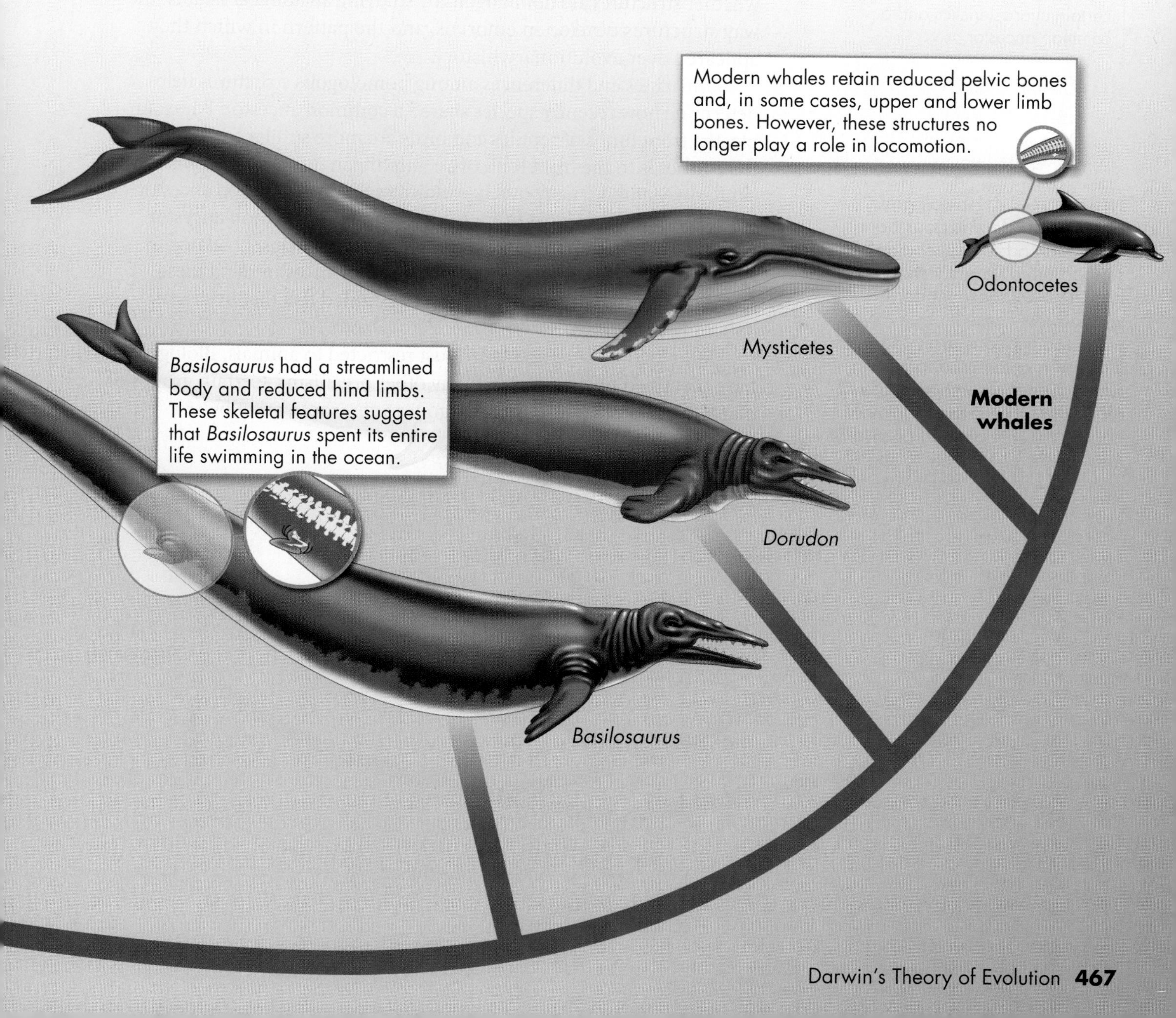

Comparing Anatomy and Embryology

What do homologous structures and similarities in embryonic development suggest about the process of evolutionary change?

By Darwin's time, scientists had noted that all vertebrate limbs had the same basic bone structure, as shown in **Figure 16–14.** Yet, some were used for crawling, some for climbing, some for running, and others for flying. Why should the same basic structures be used over and over again for such different purposes?

BUILD Vocabulary

WORD ORIGINS The word **homologous** comes from the Greek word *homos*, meaning "same." Homologous structures may not look exactly the same, but they share certain characteristics and a common ancestor.

Homologous Structures Darwin proposed that animals with similar structures evolved from a common ancestor with a basic version of that structure. Structures that are shared by related species and that have been inherited from a common ancestor are called **homologous structures. Evolutionary theory explains the existence of homologous structures adapted to different purposes as the result of descent with modification from a common ancestor.** Biologists test whether structures are homologous by studying anatomical details, the way structures develop in embryos, and the pattern in which they appeared over evolutionary history.

Similarities and differences among homologous structures help determine how recently species shared a common ancestor. For example, the front limbs of reptiles and birds are more similar to each other than either is to the front limb of an amphibian or mammal. This similarity—among many others—indicates that the common ancestor of reptiles and birds lived more recently than the common ancestor of reptiles, birds, and mammals. So birds are more closely related to crocodiles than they are to bats! The common ancestor of all these four-limbed animals was an ancient lobe-finned fish that lived over 380 million years ago.

Homologous structures aren't just restricted to animals. Biologists have identified homologies in many other organisms. Certain groups of plants, for example, share homologous stems, roots, and flowers.

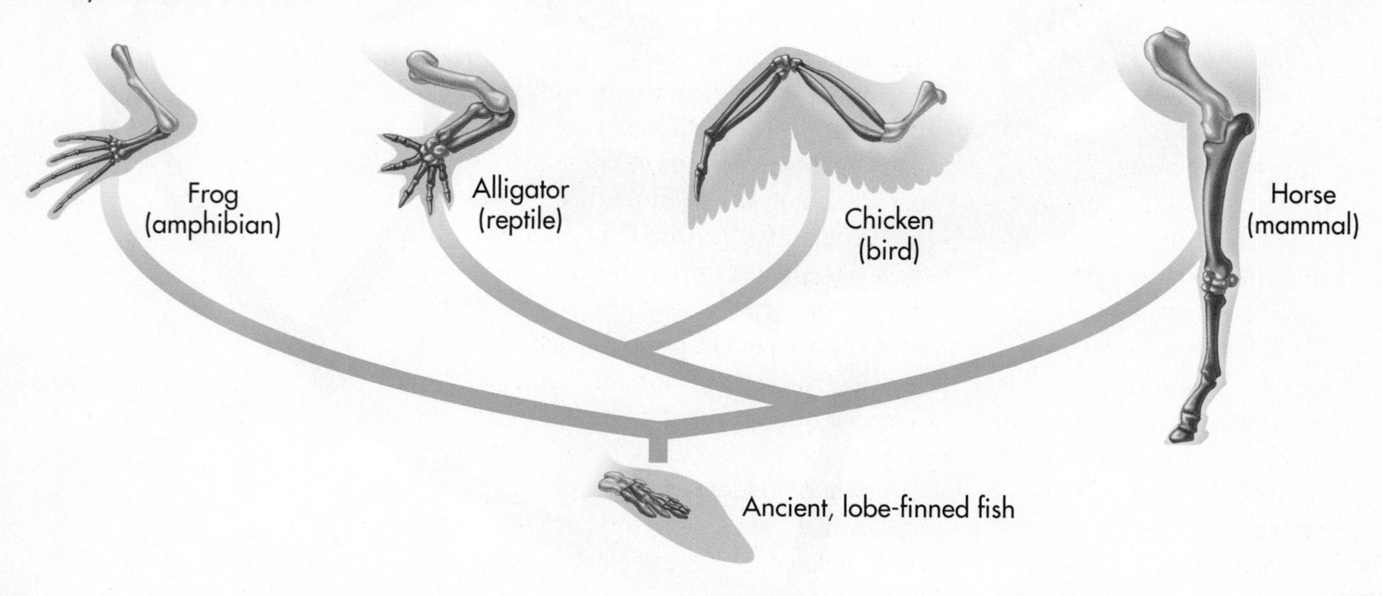

FIGURE 16–14 Homologous Limb Bones Homologous bones, as indicated by color-coding, support the differently shaped front limbs of these modern vertebrates. These limbs evolved, with modifications, from the front limbs of a common ancestor whose bones resembled those of an ancient fish. If these animals had no recent common ancestor, they would be unlikely to share so many common structures.

A.

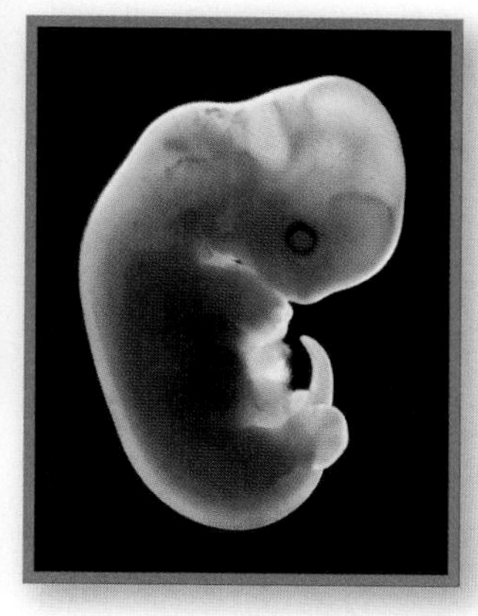

B.

FIGURE 16–15 Vestigial Organs and Embryology
A. The wings of the flightless cormorant and the legs of the Italian three-toed skink are vestigial structures. **B.** Because the early stages of development among vertebrates are so similar, it would take an expert to identify this as a cat embryo. **Infer** ***Looking at the legs of the skink, do you think its ancestors had functioning legs? Explain your answer.***

▶ ***Analogous Structures*** Note that the clue to common descent is common *structure*, not common *function*. A bird's wing and a horse's front limb have different functions but similar structures. Body parts that share common function, but not structure, are called **analogous structures.** The wing of a bee and the wing of a bird are analogous structures.

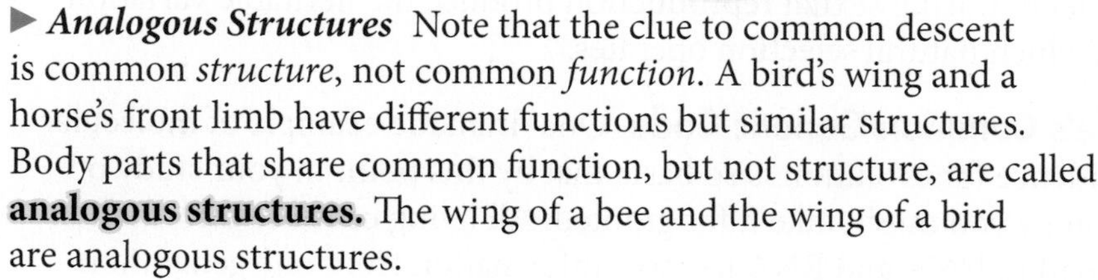

In Your Notebook *Do you think the shell of a clam and the shell of a lobster are homologous or analogous structures? Explain.*

▶ ***Vestigial Structures*** Not all homologous structures have important functions. **Vestigial structures** are inherited from ancestors but have lost much or all of their original function due to different selection pressures acting on the descendant. For example, the hipbones of the bottlenose dolphin, shown on page 467, are vestigial structures. In their ancestors, hipbones played a role in terrestrial locomotion. However, as the dolphin lineage adapted to life at sea, this function was lost. Why do dolphins and the organisms in **Figure 16–15** retain structures with little or no function? One possibility is that the presence of the structure does not affect an organism's fitness, and, therefore, natural selection does not act to eliminate it.

Embryology Researchers noticed a long time ago that the early developmental stages of many animals with backbones (called vertebrates) look very similar. Recent observations make clear that the same groups of embryonic cells develop in the same order and in similar patterns to produce many homologous tissues and organs in vertebrates. For example, despite the very different adult shapes and functions of the limb bones in **Figure 16–14,** all those bones develop from the same clumps of embryonic cells. Evolutionary theory offers the most logical explanation for these similarities in patterns of development. **Similar patterns of embryological development provide further evidence that organisms have descended from a common ancestor.**

Darwin realized that similar patterns of development offer important clues to the ancestory of living organisms. He could not have anticipated, however, the incredible amount of evidence for his theory that would come from studying the genes that control development—evidence from the fields of genetics and molecular biology.

Genetics and Molecular Biology

How can molecular biology be used to trace the process of evolution?

The most troublesome "missing information" for Darwin had to do with heredity. Darwin had no idea how heredity worked, and he was deeply worried that this lack of knowledge might prove fatal to his theory. As it happens, some of the strongest evidence supporting evolutionary theory comes from genetics. A long series of discoveries, from Mendel to Watson and Crick to genomics, helps explain how evolution works. **At the molecular level, the universal genetic code and homologous molecules provide evidence of common descent.** Also, we now understand how mutation and the reshuffling of genes during sexual reproduction produce the heritable variation on which natural selection operates.

Life's Common Genetic Code One dramatic example of molecular evidence for evolution is so basic that by this point in your study of biology you might take it for granted. All living cells use information coded in DNA and RNA to carry information from one generation to the next and to direct protein synthesis. This genetic code is nearly identical in almost all organisms, including bacteria, yeasts, plants, fungi, and animals. This is powerful evidence that all organisms evolved from common ancestors that shared this code.

Molecular Homology in *Hoxc8*

Molecular homologies can be used to infer relationships among organisms. The diagram below shows a small portion of the DNA for the same gene, *Hoxc8,* in three animals—a mouse, a baleen whale, and a chicken.

1. Calculate What percentage of the nucleotides in the baleen whale's DNA are different from those of the mouse? (*Hint*: First count the number of DNA nucleotides in one entire sequence. Then count the nucleotides in the whale DNA that differ from those in the mouse DNA. Finally, divide the number of nucleotides that are different by the total number of nucleotides, and multiply the result by 100.) MATH

2. Calculate What percentage of the nucleotides in the chicken are different from those of the mouse? MATH

3. Draw Conclusions Do you think a mouse is more closely related to a baleen whale or to a chicken? Explain your answer.

4. Evaluate Do you think that scientists can use small sections of DNA, like the ones shown here, to infer evolutionary relationships? Why or why not?

Animal	Sequence of Bases in Section of *Hoxc8*																																							
Mouse	C	A	G	A	A	A	T	G	C	C	A	C	T	T	T	T	A	T	G	G	C	C	C	T	G	T	T	T	G	T	C	T	C	C	C	T	G	C	T	C
Baleen whale	C	**C**	G	A	A	A	T	G	C	C	**T**	C	T	T	T	T	A	T	G	G	C	**G**	C	T	G	T	T	T	G	T	C	T	C	C	C	T	G	C	**G**	C
Chicken	**A**	A	**A**	A	A	A	T	G	C	C	**G**	C	T	T	T	T	A	**C**	**A**	G	C	**T**	C	T	G	T	T	T	G	T	C	T	C	**T**	C	T	G	C	T	**A**

Homologous Molecules In Darwin's day, biologists could only study similarities and differences in structures they could see. But physical body structures can't be used to compare mice with yeasts or bacteria. Today, we know that homology is not limited to physical structures. As shown in **Figure 16–16,** homologous proteins have been found in some surprising places. Homologous proteins share extensive structural and chemical similarities. One homologous protein is cytochrome c, which functions in cellular respiration. Remarkably similar versions of cytochrome c are found in almost all living cells, from cells in baker's yeast to cells in humans.

There are many other kinds of homologies at the molecular level. Genes can be homologous, too, which makes sense given the genetic code that all plants and animals share. One spectacular example is a set of ancient genes that determine the identities of body parts. Known as the Hox genes, they help to determine the head-to-tail axis in embryonic development. In vertebrates, sets of homologous Hox genes direct the growth of front and hind limbs. Small changes in these genes can produce dramatic changes in the structures they control. So, relatively minor changes in an organism's genome can produce major changes in an organism's structure and the structure of its descendants. At least some homologous Hox genes are found in almost all multicellular animals, from fruit flies to humans. Such profound biochemical similarities are best explained by Darwin's conclusion: Living organisms evolved through descent with modification from a common ancestor.

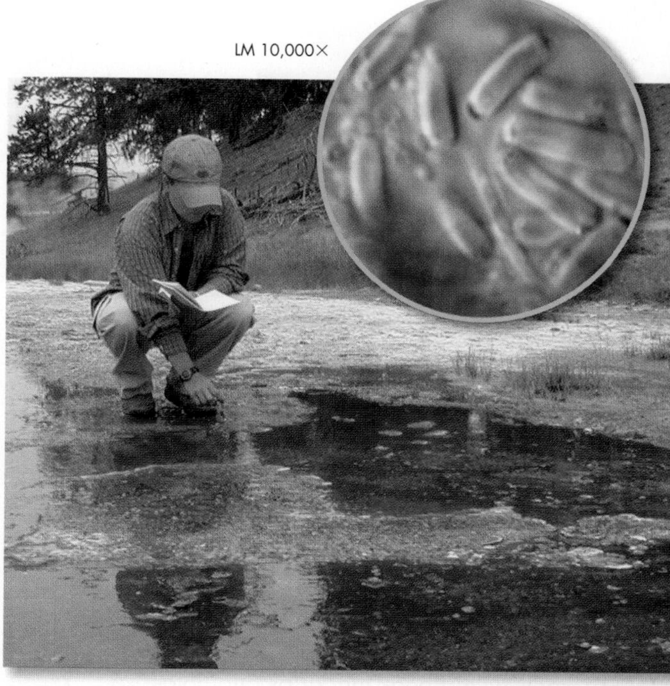

FIGURE 16–16 Similar Genes Bacteria in this hot spring live in near-boiling water—an inhospitable environment to animals. Their cells even look different from animal cells. Yet many of their genes, and therefore the proteins coded by those genes, are similar to those of animals. This is more evidence that all organisms share an ancient common ancestor.

Testing Natural Selection

What does recent research on the Galápagos finches show about natural selection?

One way to gather evidence for evolutionary change is to observe natural selection in action. But most kinds of evolutionary change we've discussed so far took place over millions of years—which makes it tough to see change actually happening. Some kinds of evolutionary change, however, have been observed and studied repeatedly in labs and in controlled outdoor environments. Scientists have designed experiments involving organisms from bacteria to guppies to test Darwin's theories. Each time, the results have supported Darwin's basic ideas. But one of the best examples of natural selection in action comes from observations on animals living in their natural environment. Fittingly, those observations focused on Galápagos finches.

A Testable Hypothesis Remember that when Darwin first saw the Galápagos finches, he thought they were wrens, warblers, and blackbirds because they looked so different from one another. Once Darwin learned that the birds were all finches, he hypothesized that they had descended from a common ancestor.

VISUAL ANALOGY

FINCH BEAK TOOLS

FIGURE 16–17 Finches use their beaks as tools to pick up and handle food. Different types of foods are most easily handled with beaks of different sizes and shapes.

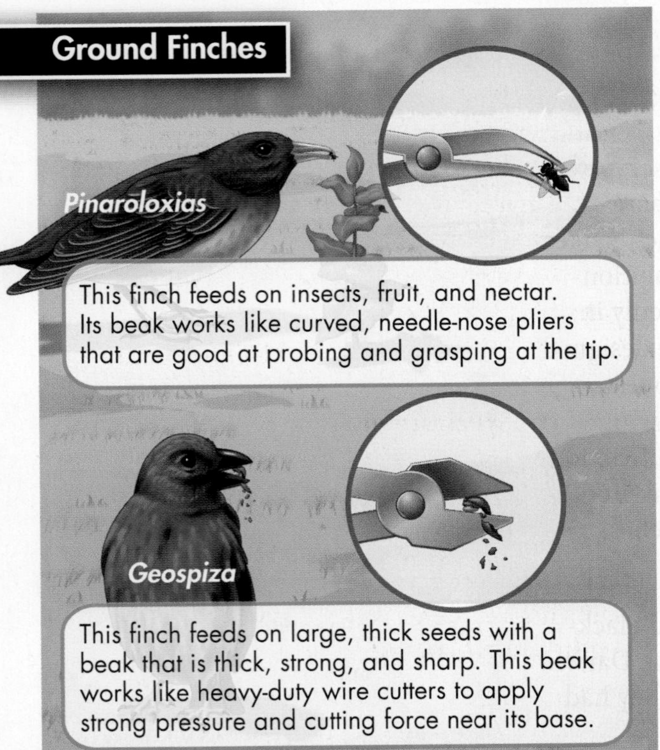

Darwin noted that several finch species have beaks of very different sizes and shapes. Each species uses its beak like a specialized tool to pick up and handle its food, as shown in **Figure 16–17.** Darwin proposed that natural selection had shaped the beaks of different bird populations as they became adapted to eat different foods. That was a reasonable hypothesis. But was there any way to test it? No one thought there was a way until Peter and Rosemary Grant of Princeton University came along.

The Grants have spent more than 35 years studying Galápagos finches. They realized that Darwin's hypothesis rested on two testable assumptions. First, for beak size and shape to evolve, there must be enough heritable variation in those traits to provide raw material for natural selection. Second, differences in beak size and shape must produce differences in fitness.

The Grants have tested these hypotheses on the medium ground finch (*Geospiza*) on the island of Daphne Major. This island is large enough to support good-sized finch populations, yet small enough to allow the Grants to catch, tag, and identify nearly every bird of the species.

During their study, the Grants periodically recapture the birds. They record which individuals are alive and which have died, which have reproduced and which have not. For each individual, the Grants record anatomical characteristics like wing length, leg length, beak length, beak depth, beak color, feather colors, and total mass. The data the Grants have recorded show that there is indeed great variation of heritable traits among Galápagos finches.

Natural Selection The Grants' data have shown that individual finches with different-size beaks have better or worse chances of surviving both seasonal droughts and longer dry spells. When food becomes scarce during dry periods, birds with the largest beaks are more likely to survive, as shown in **Figure 16–18.** As a result, average beak size in this finch population has increased dramatically. **The Grants have documented that natural selection takes place in wild finch populations frequently, and sometimes rapidly.** Changes in food supply created selection pressure that caused finch populations to evolve within decades. This evolutionary change occurred much faster than many researchers thought possible.

Not only have the Grants documented natural selection in nature, their data also confirm that competition and environmental change drive natural selection. Traits that don't matter much under one set of environmental conditions became adaptive as the environment changes during a drought. **The Grants' work shows that variation within a species increases the likelihood of the species' adapting to and surviving environmental change.** Without heritable variation in beak sizes, the medium ground finch would not be able to adapt to feeding on larger, tougher seeds during a drought.

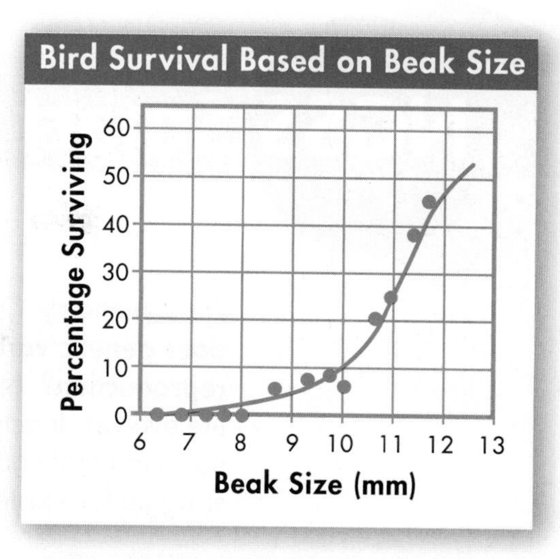

FIGURE 16–18 Survival and Beak Size This graph shows the survival rate of one species of ground finch, the medium ground finch, *Geospiza fortis*, during a drought period. Interpret Graphs ***What trend does the graph show?***

Evaluating Evolutionary Theory Advances in many fields of biology, along with other sciences, have confirmed and expanded most of Darwin's hypotheses. Today, evolutionary theory—which includes natural selection—offers insights that are vital to all branches of biology, from research on infectious disease to ecology. That's why evolution is often called the grand unifying theory of the life sciences.

Like any scientific theory, evolutionary theory is constantly reviewed as new data are gathered. Researchers still debate important questions such as precisely how new species arise and why species become extinct. And there is also significant uncertainty about exactly how life began. However, any questions that remain are about *how* evolution works—not *whether* evolution occurs. To scientists, evolution is the key to understanding the natural world.

16.4 Assessment

Review Key Concepts

1. a. Review What is biogeography?

b. Relate Cause and Effect Why do distantly related species in very different places sometimes share similar traits?

2. a. Review Why are fossils important evidence for evolution?

b. Interpret Visuals Use **Figure 16–13** to describe how a modern mysticete whale is different from *Ambulocetus*.

3. a. Review How do vestigial structures provide evidence for evolution?

b. Compare and Contrast Explain the difference between homologous and analogous structures. Which are more important to evolutionary biologists? Why?

4. a. Explain What is the relationship between Hox genes and embryological development?

b. Draw Conclusions Organisms A and B have very similar Hox genes, and their embryos, in the earliest stages of development, are also very similar. What do these similarities indicate about the ancestry of organisms A and B?

5. a. Explain What hypothesis have the Grants been testing?

b. Draw Conclusions How do the Grants' data show that genetic variation is important in the survival of a species?

WRITE ABOUT SCIENCE

Explanation

6. In your own words, write a paragraph that explains how evidence since Darwin's time has strengthened his theories.

NGSS Smart Guide Darwin's Theory of Evolution

Disciplinary Core Idea LS4.B Natural Selection: How does genetic variation among organisms affect survival and reproduction? Natural selection is a natural process through which life evolves. It acts on populations whose individuals exhibit heritable variation that causes differential survival and reproduction in the struggle for existence.

16.1 Darwin's Voyage of Discovery

- Darwin developed a scientific theory of biological evolution that explains how modern organisms evolved over long periods of time through descent from common ancestors.
- Darwin noticed that (1) different, yet ecologically similar, animal species inhabited separated, but ecologically similar, habitats around the globe; (2) different, yet related, animal species often occupied different habitats within a local area; and (3) some fossils of extinct animals were similar to living species.

evolution 450 • fossil 452

Biology.com

Untamed Science Video Islands are rich environments for evolution, as you will find out with the Untamed Science crew.

UntamedScience

16.2 Ideas That Shaped Darwin's Thinking

- Hutton and Lyell concluded that Earth is extremely old and that the processes that changed Earth in the past are the same processes that operate in the present.
- Lamarck suggested that organisms could change during their lifetimes by selectively using or not using various parts of their bodies. He also suggested that individuals could pass these acquired traits on to their offspring, enabling species to change over time.
- Malthus reasoned that if the human population grew unchecked, there wouldn't be enough living space and food for everyone.
- In artificial selection, nature provides the variations, and humans select those they find useful.

artificial selection 458

Biology.com

Art in Motion This animation shows how fossil layers accumulate and are later exposed.

ART IN MOTION Darwin and Geology

Uniformitarianism
Lyell argued that processes that changed Earth in the past are the same processes that operate in the present. This idea, known as

An Ancient Changing Earth
Lyell's Geology
Uniformitarianism
Lyell and Darwin

CHECKING YOUR *Scientific Literacy*

Refer to the lesson content and digital assets below as you prepare for your chapter assessment. Then, evaluate your understanding of natural selection by answering these questions.

1. **Scientific and Engineering Practice Developing and Using Models** Suppose a newly discovered species of bird feeds exclusively on walnuts and sunflower seeds. Build a model of a bird beak best adapted to picking up both kinds of seeds.

2. Crosscutting Concept **Cause and Effect: Mechanism and Explanation** Use your model to explain how the elimination of walnut trees from an area might affect the local population of this species of bird.

3. Text Types and Purposes Write an essay describing the process of natural selection in a fictional population of beetles facing pressure from a newly introduced predator. Include key terms and illustrations in your narration of the technical process.

4. STEM Focusing on the common tools illustrated in Figure 16-17 *Finch Beak Tools*, design a simple experiment to demonstrate how effective the various shaped "beaks" would be at handling various foods. Explain how heritable variation in beak shape plays a role in a bird species' evolution.

16.3 Darwin Presents His Case

- Natural selection occurs in any situation in which more individuals are born than can survive, there is natural heritable variation, and there is variable fitness among individuals.
- According to the principle of common descent, all species—living and extinct—are descended from ancient common ancestors.

adaptation 461 • fitness 461 • natural selection 463

Biology.com

Data Analysis Collect population data for several generations of grasshoppers and then analyze how the population changed due to natural selection.

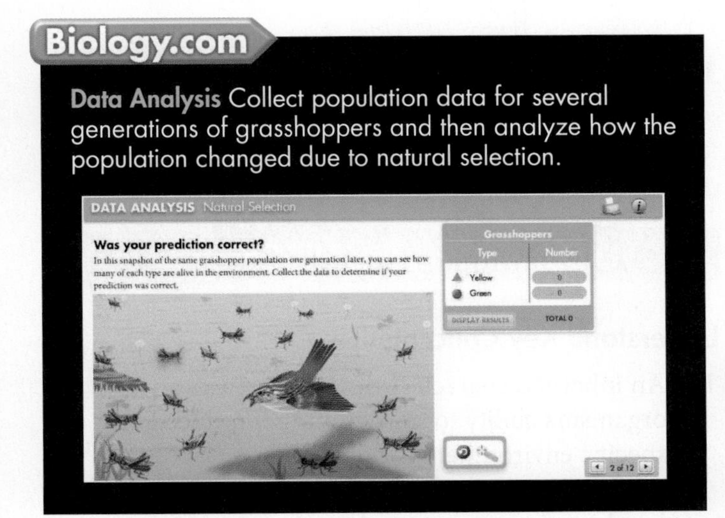

Skills Lab

Amino Acid Sequences: Indicators of Evolution This chapter lab is available in your lab manual and at Biology.com

16.4 Evidence of Evolution

- Patterns in the distribution of living and fossil species tell us how modern organisms evolved from their ancestors.
- Many recently discovered fossils form series that trace the evolution of modern species from extinct ancestors.
- Evolutionary theory explains the existence of homologous structures adapted to different purposes as the result of descent with modification from a common ancestor.
- The universal genetic code and homologous molecules provide evidence of common descent.
- The Grants have documented that natural selection takes place in wild Galápagos finch populations frequently, and sometimes rapidly, and that variation within a species increases the likelihood of the species adapting to and surviving environmental change.

biogeography 465 • homologous structure 468 • analogous structure 469 • vestigial structure 469

Biology.com

Visual Analogy See how different types of finch beaks function like tools.

Art Review Review homologous and analogous structures in vertebrates.

16 Assessment

16.1 Darwin's Voyage of Discovery

Understand Key Concepts

1. Who observed variations in the characteristics of plants and animals on different islands of the Galápagos?
   a. James Hutton
   b. Charles Lyell
   c. Charles Darwin
   d. Thomas Malthus

2. In addition to observing living organisms, Darwin studied the preserved remains of ancient organisms called
   a. fossils.
   b. adaptations.
   c. homologies.
   d. vestigial structures.

3. What pattern of variation did Darwin observe among rheas, ostriches, and emus?

4. What connection did Darwin make between the Galápagos tortoises and their environments?

Think Critically

5. **Craft and Structure** Explain what the term *evolution* means, and give an example.

6. **Relate Cause and Effect** Why was Darwin's trip aboard the *Beagle* so important to his development of the theory of natural selection?

7. **Infer** Why was Darwin puzzled by the fact that there were no rabbits in Australia?

16.2 Ideas That Shaped Darwin's Thinking

Understand Key Concepts

8. Which of the following ideas proposed by Lamarck was later found to be incorrect?
   a. Acquired characteristics can be inherited.
   b. All species are descended from other species.
   c. Living things change over time.
   d. There is a relationship between an organism and its environment.

9. Which of the following would an animal breeder use to increase the number of cows that give the most milk?
   a. overproduction
   b. genetic isolation
   c. acquired characteristics
   d. artificial selection

10. What accounts for the presence of marine fossils on mountaintops?

11. **Text Types and Purposes** How did Lyell's *Principles of Geology* influence Darwin?

12. According to Malthus, what factors limit population growth? Why did Malthus's ideas apply to other organisms better than they did to humans?

13. What is artificial selection? How did this concept influence Darwin's thinking?

Think Critically

14. **Relate Cause and Effect** A sunflower produces many seeds. Will all the seeds grow into mature plants? Explain your answer.

15. **Text Types and Purposes** Explain why Lamarck made a significant contribution to science even though his explanation of evolution was wrong.

16.3 Darwin Presents His Case

Understand Key Concepts

16. An inherited characteristic that increases an organism's ability to survive and reproduce in its specific environment is called a(n)
    a. vestigial structure.
    b. adaptation.
    c. speciation.
    d. analogous structure.

17. How well an organism survives and reproduces in its environment can be described as its
    a. fitness.
    b. homologies.
    c. common descent.
    d. analogies.

18. Craft and Structure How does natural variation affect evolution?

19. Production and Distribution of Writing Write a paragraph explaining in your own words the following statement: "Descent with modification explains the diversity of life we see today." Refine your explanation by rewriting with your audience in mind.

20. Describe the conditions necessary for natural selection to occur.

Think Critically

21. **Apply Concepts** How would Darwin explain the long legs of the water bird in **Figure 16–6**? How would Darwin's explanation differ from Lamarck's explanation?

22. Craft and Structure Distinguish between fitness and adaptation. How are the two concepts related?

23. **Infer** How does the process of natural selection account for the diversity of organisms that Darwin observed on the Galápagos Islands?

24. **Infer** Many species of birds build nests in which they lay eggs and raise the newly hatched birds. How might nest-building behavior be an adaptation that ensures reproductive fitness?

16.4 Evidence of Evolution

Understand Key Concepts

25. Structures that have different mature forms but develop from the same embryonic tissue are called
    a. analogous.
    b. adaptations.
    c. homologous.
    d. fossils.

26. Intermediate fossil forms are important evidence of evolution because they show
    a. how organisms changed over time.
    b. how animals behaved in their environments.
    c. how the embryos of organisms develop.
    d. molecular homologies.

27. How does the geographic distribution of organisms support the theory of evolution?

28. How do vestigial structures indicate that present-day organisms are different from their ancient ancestors?

29. How do DNA and RNA provide evidence for common descent?

solve the CHAPTER MYSTERY

SUCH VARIED HONEYCREEPERS

The 'i'iwi and other Hawaiian honeycreepers resemble Galápagos finches in a number of ways. They are species of small birds found nowhere else on Earth. They live on islands that are separated from one another by stretches of open sea and that are hundreds of miles from the nearest continent. They are also related to finches!

There are more than 20 known species of Hawaiian honeycreeper. Like the species of Galápagos finches, the honeycreeper species are closely related to one another. This is an indication that they are all descended, with modification, from a relatively recent common ancestor. Experts think the ancestor colonized the islands between 3 million and 4 million years ago. Many honeycreepers have specialized diets, evolutionary adaptations to life on the particular islands they call home. Today, habitat loss is endangering most of the honeycreepers. In fact, many species of honeycreeper are thought to have become extinct since humans settled on the islands.

1. **Infer** Suppose a small group of birds, not unlike the modern honeycreepers, landed on one of Hawaii's islands millions of years ago and then reproduced. Do you think all the descendants would have stayed on that one island? Explain your answer.

2. **Infer** Do you think that the climate and other environmental conditions are exactly the same everywhere on the Hawaiian Islands? How might environmental conditions have affected the evolution of honeycreeper species?

3. Text Types and Purposes Explain how the different species of honeycreepers in Hawaii today might have evolved from one ancestral species.

4. **Connect to the Big idea** Why are islands often home to species that exist nowhere else on Earth?

Think Critically

30. Infer Which animal—a cricket or a cat—would you expect to have cytochrome c more similar to that of a dog? Explain your answer.

31. Infer In all animals with backbones, oxygen is carried in blood by a molecule called hemoglobin. What could this physiological similarity indicate about the evolutionary history of vertebrates (animals with backbones)?

32. Apply Concepts Do you think some species of snake might have vestigial hip and leg bones? Explain your answer.

Connecting Concepts

Use Science Graphics

Use the illustration below to answer questions 33 and 34.

33. Integration of Knowledge and Ideas Based on what you can see, which mice—white or brown—are better adapted to their environment? Explain your answer.

34. Apply Concepts In what way is the coloring of the brown mice an adaptation? What other adaptations besides coloring might affect the mice's ability to survive and reproduce?

Write About Science

35. Text Types and Purposes Write a paragraph that explains how the age of Earth supports the theory of evolution.

36. Key Ideas and Details Summarize the conditions under which natural selection occurs. Then, describe three lines of evidence that support the theory of evolution by natural selection.

37. Assess the Big idea Write a newspaper article about the historical meeting at which Darwin's and Wallace's hypotheses of evolution were first presented. Explain the theory of evolution by natural selection for an audience that knows nothing about the subject.

38. Assess the Big idea Look back at **Figure 16–10** on page 462. Explain how conditions could change so that yellow coloring becomes adaptive. What would happen to the relative numbers of green and yellow grasshoppers in the population?

Analyzing Data

Integration of Knowledge and Ideas

Cytochrome c is a small protein involved in cellular respiration. The table compares the cytochrome c of various organisms to that of chimpanzees. The left column indicates the organism, and the right column indicates the number of amino acids that are different from those in chimpanzee cytochrome c.

Organism	Number of Amino Acids That Are Different From Chimpanzee Cytochrome c
Dog	10
Moth	24
Penguin	11
Yeast	38

39. Interpret Data Which of these organisms probably shares the most recent common ancestor with chimpanzees?

a. dog
b. moth
c. penguin
d. yeast

40. Calculate The primary structure of cytochrome c contains 104 amino acids. Approximately how many of these amino acids are the same in the chimpanzee and moth? MATH

a. 10
b. 24
c. 80
d. 128

Standardized Test Prep

Multiple Choice

1. Which scientist formulated the theory of evolution through natural selection?
A Charles Darwin
B Thomas Malthus
C James Hutton
D Jean-Baptiste Lamarck

2. Lamarck's ideas about evolution were wrong because he proposed that
A species change over time.
B species descended from other species.
C acquired characteristics can be inherited.
D species are adapted to their environments.

3. Lyell's *Principles of Geology* influenced Darwin because it explained how
A organisms change over time.
B adaptations occur.
C the surface of Earth changes over time.
D the Galápagos Islands formed.

4. A farmer's use of the best livestock for breeding is an example of
A natural selection.
B artificial selection.
C extinction.
D adaptation.

5. The ability of an individual organism to survive and reproduce in its natural environment is called
A natural selection.
B evolution.
C descent with modification.
D fitness.

6. Which of the following is an important concept in Darwin's theory of evolution by natural selection?
A descent with modification
B homologous molecules
C processes that change the surface of Earth
D the tendency toward perfection

7. Which of the following does NOT provide evidence for evolution?
A fossil record
B the function of structures in an individual organism
C geographical distribution of living things
D homologous structures of living organisms

8. DNA and RNA provide evidence of evolution because
A all organisms have nearly identical DNA and RNA.
B no two organisms have exactly the same DNA.
C each RNA codon specifies just one amino acid.
D in most organisms, the same codons specify the same amino acids.

9. A bird's wings are homologous to a(n)
A fish's tailfin.
B alligator's claws.
C dog's front legs.
D mosquito's wings.

Questions 10 and 11

The birds shown below are 2 of the species of finches Darwin found on the Galápagos Islands.

Woodpecker Finch

Large Ground Finch

10. What process produced the two different types of beaks shown?
A artificial selection
B natural selection
C geographical distribution
D disuse of the beak

11. The large ground finch obtains food by cracking seeds. Its short, strong beak is an example of
A the struggle for existence.
B the tendency toward perfection.
C an adaptation.
D a vestigial organ.

Open-Ended Response

12. Compare and contrast the processes of artificial selection and natural selection.

If You Have Trouble With . . .

Question	1	2	3	4	5	6	7	8	9	10	11	12
See Lesson	16.1	16.2	16.2	16.2	16.3	16.3	16.4	16.4	16.4	16.3	16.3	16.3

17 Evolution of Populations

Evolution

Q: How can populations evolve to form new species?

Poised on a flower, the two common blue butterflies (Polyommatus icarus) appear identical. However, if you look closely, you can see that the patterns on their wings are slightly different. Variations among individual members of a population provide the raw material for evolution and sometimes for the formation of new species.

INSIDE:

- 17.1 Genes and Variation
- 17.2 Evolution as Genetic Change in Populations
- 17.3 The Process of Speciation
- 17.4 Molecular Evolution

BUILDING *Scientific Literacy*

Deepen your understanding of the evolution of populations by engaging in key practices that allow you to make connections across concepts.

NGSS You will communicate information to demonstrate your understanding of how populations change.

STEM How can unintended cross-pollination affect the genetics of corn populations? Investigate the issue and how it affects farmers.

Common Core You will translate technical information expressed visually in the text into words to describe how selection acts on polygenic traits.

CHAPTER MYSTERY

EPIDEMIC

In 1918, an epidemic began that would go on to kill more than 40 million people. A doctor wrote: "Dead bodies are stacked about the morgue like cordwood."

What was this terrible disease? It was a variety of the same influenza virus that causes "the flu" you catch again and again. How did this strain of a common virus become so deadly? And could that kind of deadly flu epidemic happen again?

The answers to those questions explain why we can't make a permanent vaccine against the flu, as we can against measles or smallpox. They also explain why public health officials worry so much about something you may have heard referred to as "bird flu." As you read this chapter, look for evolutionary processes that might help explain how new strains of influenza virus appear all the time. Then, solve the mystery.

Biology.com

Finding the solution to the flu epidemic mystery is only the beginning. Take a video field trip with the ecogeeks of Untamed Science to see where the mystery leads.

Go online to access additional resources including:
- eText • Flash Cards • Lesson Overviews • Chapter Mystery

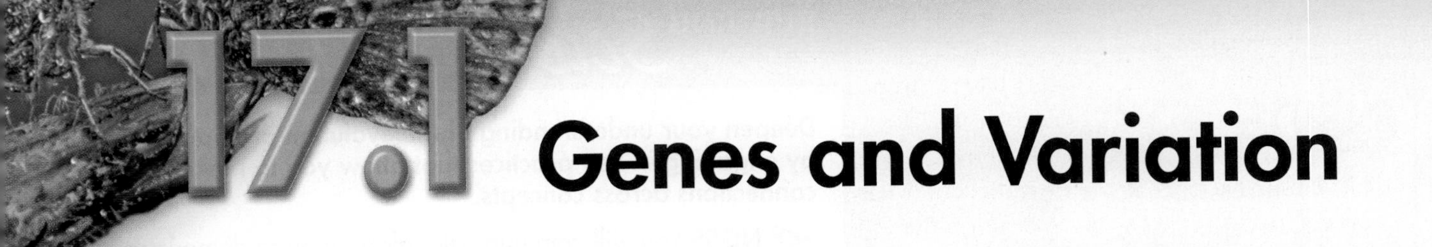

17.1 Genes and Variation

Key Questions

How is evolution defined in genetic terms?

What are the sources of genetic variation?

What determines the number of phenotypes for a given trait?

Vocabulary

gene pool
allele frequency
single-gene trait
polygenic trait

Taking Notes

Concept Map As you read about sources of genetic variation, construct a concept map to describe the sources.

THINK ABOUT IT Darwin developed his theory of natural selection without knowing how heredity worked. Mendel's studies on inheritance in peas were published during Darwin's lifetime, but no one (including Darwin) realized how important that work was. So Darwin had no idea how heritable traits pass from one generation to the next. What's more, although Darwin based his theory on heritable variation, he had no idea where that variation came from. What would happen when genetics answered those questions?

Genetics Joins Evolutionary Theory

How is evolution defined in genetic terms?

After Mendel's work was rediscovered around 1900, genetics took off like a rocket. Researchers discovered that heritable traits are controlled by genes that are carried on chromosomes. They learned how changes in genes and chromosomes generate variation.

All these discoveries in genetics fit perfectly into evolutionary theory. Variation is the raw material for natural selection, and finally scientists could study how and why variation occurs. Today, techniques of molecular genetics are used to form and test many hypotheses about heritable variation and natural selection. Modern genetics enables us to understand, better than Darwin ever could, how evolution works.

Genotype and Phenotype in Evolution Typical plants and animals contain two sets of genes, one contributed by each parent. Specific forms of a gene, called alleles, may vary from individual to individual. An organism's genotype is the particular combination of alleles it carries. An individual's genotype, together with environmental conditions, produces its phenotype. Phenotype includes all physical, physiological, and behavioral characteristics of an organism, such as eye color or height. Natural selection acts directly on phenotype, not genotype. In other words, natural selection acts on an organism's characteristics, not directly on its alleles.

FIGURE 17–1 Genes and Variation Why do biological family members resemble each other, yet also look so different? Similarities come from shared genes. Most differences come from gene shuffling during reproduction and environmental influences. A few differences may be caused by random mutations.

How does that work? In any population, some individuals have phenotypes that are better suited to their environment than are the phenotypes of other individuals. The better-suited individuals produce more offspring than the less fit individuals do. Therefore, organisms with higher fitness pass more copies of their genes to the next generation.

Natural selection never acts directly on genes. Why? Because it is an entire organism—not a single gene—that either survives and reproduces or dies without reproducing.

In Your Notebook *Describe how natural selection affects genotypes by acting on phenotypes.*

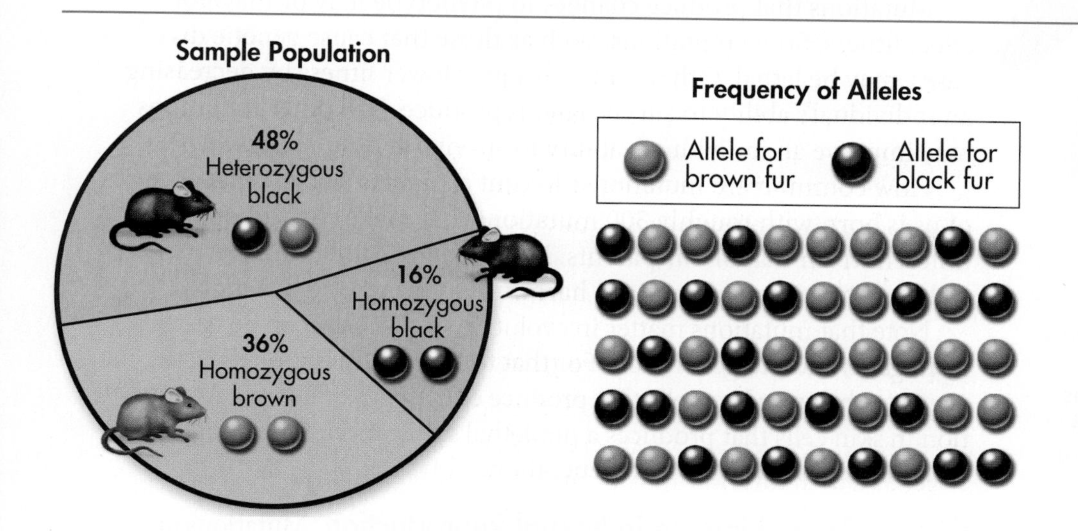

FIGURE 17–2 Alleles in a Population When scientists try to determine whether a population is evolving, they study its allele frequencies. This diagram shows allele frequencies for fur color in a mouse population. **Calculate** ***Here, in a total of 50 alleles, 20 alleles are*** **B** ***(black) and 30 are*** **b** ***(brown). How many of each allele would be present in a total of 100 alleles?*** **MATH**

Populations and Gene Pools Genetic variation and evolution are both studied in populations. A population is a group of individuals of the same species that mate and produce offspring. Because members of a population interbreed, they share a common group of genes called a gene pool. A **gene pool** consists of all the genes, including all the different alleles for each gene, that are present in a population.

Researchers study gene pools by examining the numbers of different alleles they contain. **Allele frequency** is the number of times an allele occurs in a gene pool, compared to the total number of alleles in that pool for the same gene. For example, in the mouse population in **Figure 17–2,** the allele frequency of the domiant *B* allele (black fur) is 40 percent, and the allele frequency of the recessive *b* allele (brown fur) is 60 percent. The allele frequency of an allele has nothing to do with whether the allele is dominant or recessive. In this mouse population, the recessive allele occurs more frequently than the dominant allele.

Evolution, in genetic terms, involves a change in the frequency of alleles in a population over time. For example, if the frequency of the *B* allele in **Figure 17–2** drops to 30 percent, the population is evolving. It's important to note that populations, not individuals, evolve. Natural selection operates on individual organisms, but the changes it causes in allele frequency show up in the population as a whole.

BUILD Vocabulary

MULTIPLE MEANINGS Perhaps the most common definition of the noun *pool* is a large man-made body of water in which you can swim. However, a *pool* can also refer to an available supply of a resource. In the case of a **gene pool,** the resource is genetic information.

FIGURE 17–3 Genetic Variation Genetic variation may produce visible variations in phenotype, such as the different-colored kernels in these ears of maize. Other kinds of genetic variation, such as resistance to disease, may not be visible, even though they are more important to evolutionary fitness.

Sources of Genetic Variation

What are the sources of genetic variation?

Genetics enables us to understand how heritable variation is produced. **Three sources of genetic variation are mutation, genetic recombination during sexual reproduction, and lateral gene transfer.**

Mutations A mutation is any change in the genetic material of a cell. Some mutations involve changes within individual genes. Other mutations involve changes in larger pieces of chromosomes. Some mutations—called neutral mutations—do not change an organism's phenotype.

Mutations that produce changes in phenotype may or may not affect fitness. Some mutations, such as those that cause genetic diseases, may be lethal. Other mutations may lower fitness by decreasing an individual's ability to survive and reproduce. Still other mutations may improve an individual's ability to survive and reproduce.

How common are mutations? Recent estimates suggest that each of us is born with roughly 300 mutations that make parts of our DNA different from that of our parents. Most of those mutations are neutral. One or two are potentially harmful. A few may be beneficial.

Note that mutations matter in evolution only if they can be passed from generation to generation. For that to happen, mutations must occur in the germ line cells that produce either eggs or sperm. A mutation in skin cells that produces a nonlethal skin cancer, for example, will not be passed to the next generation.

Genetic Recombination in Sexual Reproduction Mutations are not the only source of heritable variation. You do not look exactly like your biological parents, even though they gave you all your genes. You probably look even less like any brothers or sisters you may have. Yet no matter how you feel about your relatives, mutant genes are not primarily what makes them look so different from you. Most heritable differences are due not to mutations, but to genetic recombination during sexual reproduction. Remember that each chromosome in a pair moves independently during meiosis. In humans, who have 23 pairs of chromosomes, this process can produce 8.4 million gene combinations!

Crossing-over is another way in which genes are recombined. Recall that crossing-over occurs during meiosis. In this process, paired chromosomes often swap lengths of DNA at random. Crossing-over further increases the number of new genotypes created in each generation. You can now understand why, in species that reproduce sexually, no two siblings (except identical twins) ever look exactly alike. With all that independent assortment and crossing-over, you can easily end up with your mother's eyes, your father's nose, and hair that combines qualities from both your parents. You can also now understand why, as Darwin noted, individual members of a species differ from one another.

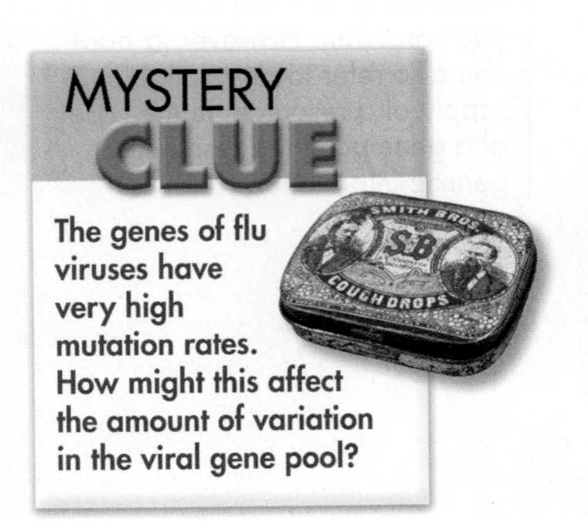

In Your Notebook *Which source of variation brings more diversity into a gene pool—mutation or sexual reproduction? Explain.*

Lateral Gene Transfer Most of the time, in most eukaryotic organisms, genes are passed only from parents to offspring (during sexual or asexual reproduction). Some organisms, however, pass genes from one individual to another, or even from individuals of one species to another. Recall, for example, that many bacteria swap genes on plasmids as though the genes were trading cards. This passing of genes from one organism to another organism that is not its offspring is called lateral gene transfer. Lateral gene transfer can occur between organisms of the same species or organisms of different species.

Lateral gene transfer can increase genetic variation in any species that picks up the "new" genes. This process is important in the evolution of antibiotic resistance in bacteria. Lateral gene transfer has been common, and important, in single-celled organisms during the history of life.

Single-Gene and Polygenic Traits

What determines the number of phenotypes for a given trait?

Genes control phenotype in different ways. In some cases, a single gene controls a trait. Other times, several genes interact to control a trait. **The number of phenotypes produced for a trait depends on how many genes control the trait.**

Single-Gene Traits In the species of snail shown below, some snails have dark bands on their shells, and other snails don't. The presence or absence of dark bands is a **single-gene trait**—a trait controlled by only one gene. The gene that controls shell banding has two alleles. The allele for a shell without bands is dominant over the allele for a shell with dark bands. All genotypes for this trait have one of two phenotypes—shells with bands or shells without bands. Single-gene traits may have just two or three distinct phenotypes.

The bar graph in **Figure 17–4** shows the relative frequency of phenotypes for this single gene in one population of snails. This graph shows that the presence of dark bands on the shells may be more common in a population than the absence of bands. This is true even though the allele for shells without bands is the dominant form. In populations, phenotypic ratios are determined by the frequency of alleles in the population as well as by whether the alleles are dominant or recessive.

FIGURE 17–4 Two Phenotypes In this species of snail, a single gene with two alleles controls whether or not a snail's shell has bands. The graph shows the percentages, in one population, of snails with bands and snails without bands.

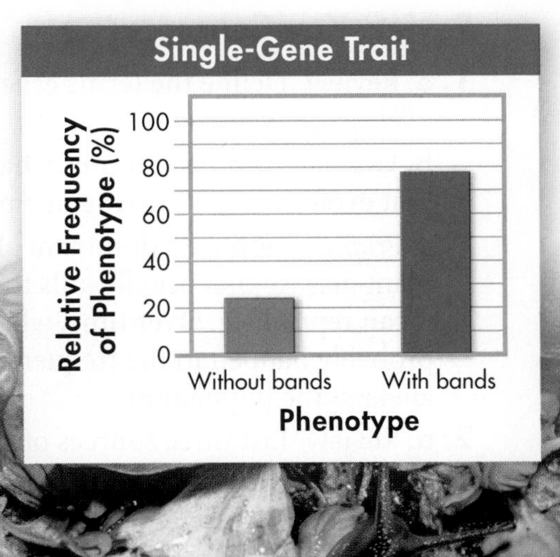

With bands ▶

Without bands ▶

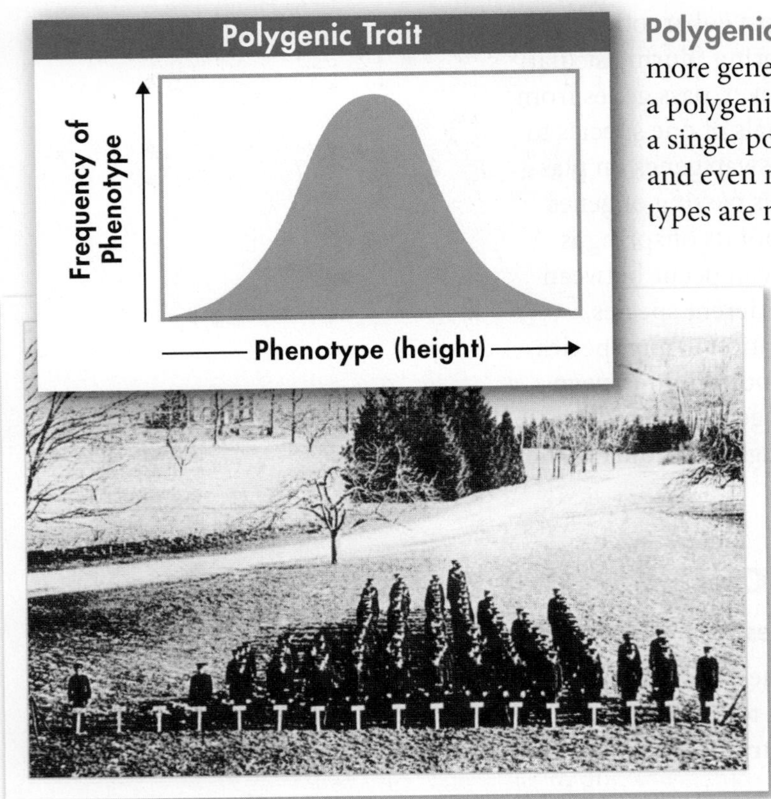

Polygenic Traits Many traits are controlled by two or more genes and are called **polygenic traits.** Each gene of a polygenic trait often has two or more alleles. As a result, a single polygenic trait often has many possible genotypes and even more different phenotypes. Often those phenotypes are not clearly distinct from one another.

Height in humans is one example of a polygenic trait. Height varies from very short to very tall and everywhere in between. You can sample phenotypic variation in this trait by measuring the height of all the students in your class. You can then calculate average height for this group. Many students will be just a little taller or shorter than average. Some, however, will be very tall or very short. If you graph the number of individuals of each height, you may get a graph similar to the one in **Figure 17–5.** The symmetrical bell-like shape of this curve is typical of polygenic traits. A bell-shaped curve is also called a normal distribution.

FIGURE 17–5 A Range of Phenotypes The graph above shows the distribution of phenotypes that would be expected for a trait if many genes contributed to the trait. The photograph shows the actual distribution of heights in a group of young men. **Interpret Graphs** ***What does the shape of the graph indicate about height in humans?***

17.1 Assessment

Review Key Concepts

1. a. Review Define the terms *gene pool* and *allele frequency*.

b. Explain In genetic terms, what indicates that evolution is occurring in a population?

c. Predict Suppose a dominant allele causes a plant disease that usually kills the plant before it can reproduce. Over time, what would probably happen to the frequency of that allele in the population?

2. a. Review List three sources of genetic variation.

b. Explain How does genetic recombination result in genetic variation?

c. Relate Cause and Effect Why does sexual reproduction provide more opportunities for genetic variation than asexual reproduction?

3. a. Review What is a single-gene trait? What is a polygenic trait?

b. Explain How does the range of phenotypes for single-gene traits differ from the range for polygenic traits?

c. Infer A black guinea pig and a white guinea pig mate and have offspring. All the offspring are black. Is the trait of coat color probably a single-gene trait or a polygenic trait? Explain.

WRITE ABOUT SCIENCE

Explanation

4. Explain how mutations are important in the process of biological evolution. (*Hint:* How does mutation affect genetic variation?)

17.2 Evolution as Genetic Change in Populations

THINK ABOUT IT Ever since humans began farming, they have battled insects that eat crops. Many farmers now use chemicals called pesticides to kill crop-destroying insects. When farmers first used modern pesticides such as DDT, the chemicals killed most insects. But after a few years, many pesticides stopped working. Today, farmers fight an ongoing "arms race" with insects. Scientists constantly search for new chemicals to control pests that old chemicals no longer control. How do insects fight back? By evolving.

At first, individual pesticides kill almost all insects exposed to them. But a few individual insects usually survive. Why? Because insect populations often contain enough genetic variation that a few individuals, just by chance, are resistant to a particular pesticide. By killing most of the susceptible individuals, farmers increase the relative fitness of the few individuals that can resist the poison. Those insects survive and reproduce, passing their resistance on to their offspring. After a few generations, the descendants of the original, resistant individuals dominate the population.

To understand completely how pesticide resistance develops, you need to know the relationship between natural selection and genetics.

Key Questions

How does natural selection affect single-gene and polygenic traits?

What is genetic drift?

What conditions are required to maintain genetic equilibrium?

Vocabulary

directional selection
stabilizing selection
disruptive selection
genetic drift
bottleneck effect
founder effect
genetic equilibrium
Hardy-Weinberg principle
sexual selection

Taking Notes

Preview Visuals Before you read, look at **Figure 17–6.** What evolutionary trend does it seem to show?

How Natural Selection Works

How does natural selection affect single-gene and polygenic traits?

Pesticide-resistant insects have a kind of fitness that protects them from a harmful chemical. In genetic terms, what does *fitness* mean? Each time an organism reproduces, it passes copies of its genes on to its offspring. We can, therefore, view evolutionary fitness as success in passing genes to the next generation. In the same way, we can view an evolutionary adaptation as any genetically controlled trait that increases an individual's ability to pass along its alleles.

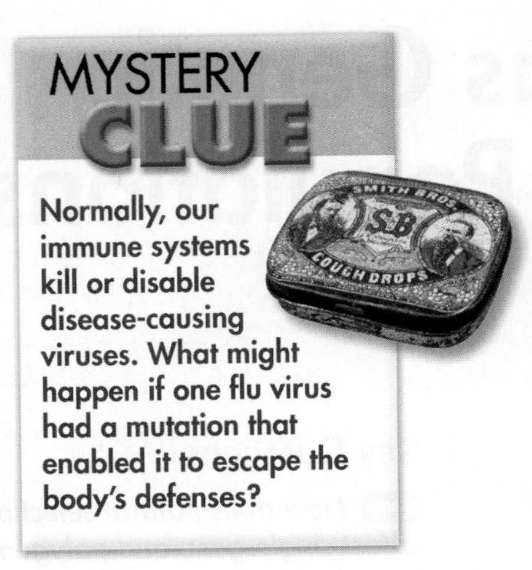

MYSTERY CLUE

Normally, our immune systems kill or disable disease-causing viruses. What might happen if one flu virus had a mutation that enabled it to escape the body's defenses?

Natural Selection on Single-Gene Traits Recall that evolution is any change over time in the allele frequency in a population. This process works somewhat differently for single-gene traits than for polygenic traits. **Natural selection on single-gene traits can lead to changes in allele frequencies and, thus, to changes in phenotype frequencies.** For example, imagine that a population of lizards experiences mutations in one gene that determines body color. The normal color of the lizards is brown. The mutations produce red and black forms, as shown in **Figure 17–6.** What happens to the new alleles? If red lizards are more visible to predators, they might be less likely to survive and reproduce. Therefore, the allele for red coloring might not become common.

Black lizards, on the other hand, might absorb more sunlight and warm up faster on cold days. If high body temperature allows the lizards to move faster to feed and avoid predators, they might produce more offspring than brown forms produce. The allele for black color might increase in frequency. The black phenotype would then increase in frequency. If color change has no effect on fitness, the allele that produces it will not be under pressure from natural selection.

Effect of Color Mutations on Lizard Survival

Initial Population	Generation 10	Generation 20	Generation 30
80%	80%	70%	40%
10%	0%	0%	0%
10%	20%	30%	60%

FIGURE 17–6 Selection on a Single-Gene Trait Natural selection on a single-gene trait can lead to changes in allele frequencies and, thus, to evolution. **Interpret Visuals** ***What has happened to produce the population shown in Generation 30?***

Natural Selection on Polygenic Traits When traits are controlled by more than one gene, the effects of natural selection are more complex. As you saw earlier, polygenic traits such as height often display a range of phenotypes that form a bell curve. The fitness of individuals may vary from one end of such a curve to the other. Where fitness varies, natural selection can act. **Natural selection on polygenic traits can affect the relative fitness of phenotypes and thereby produce one of three types of selection: directional selection, stabilizing selection, or disruptive selection.** These types of selection are shown in **Figure 17–7.**

In Your Notebook *As you read the text on the following page, summarize each of the three types of selection.*

▶ ***Directional Selection*** When individuals at one end of the curve have higher fitness than individuals in the middle or at the other end, **directional selection** occurs. The range of phenotypes shifts because some individuals are more successful at surviving and reproducing than are others.

Consider how limited resources, such as food, can affect individuals' fitness. Among seed-eating birds such as Darwin's finches, birds with bigger, thicker beaks can feed more easily on larger, harder, thicker-shelled seeds. Suppose the supply of small and medium-size seeds runs low, leaving only larger seeds. Birds with larger beaks would have an easier time feeding than would small-beaked birds. Big-beaked birds would therefore be more successful in surviving and passing genes to the next generation. Over time, the average beak size of the population would probably increase.

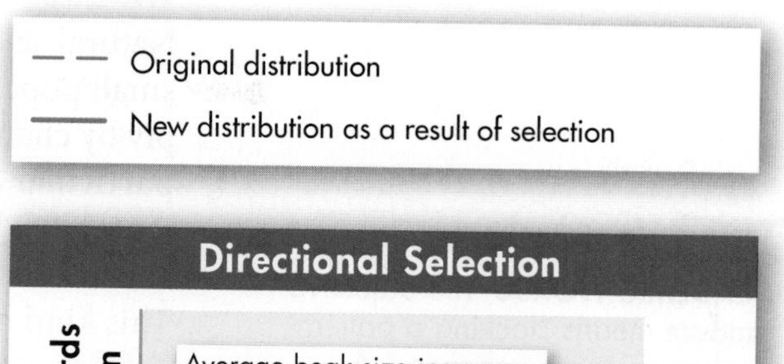

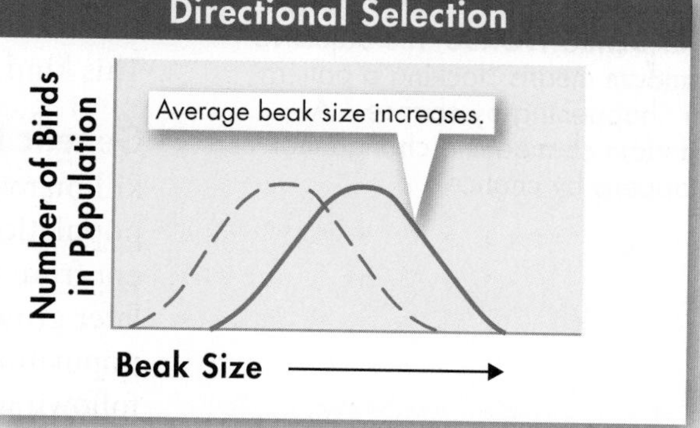

▶ ***Stabilizing Selection*** When individuals near the center of the curve have higher fitness than individuals at either end, **stabilizing selection** takes place. This situation keeps the center of the curve at its current position, but it narrows the curve overall.

For example, the mass of human infants at birth is under the influence of stabilizing selection. Very small babies are likely to be less healthy and, thus, less likely to survive. Babies who are much larger than average are likely to have difficulty being born. The fitness of these smaller or larger babies is, therefore, lower than that of more average-size individuals.

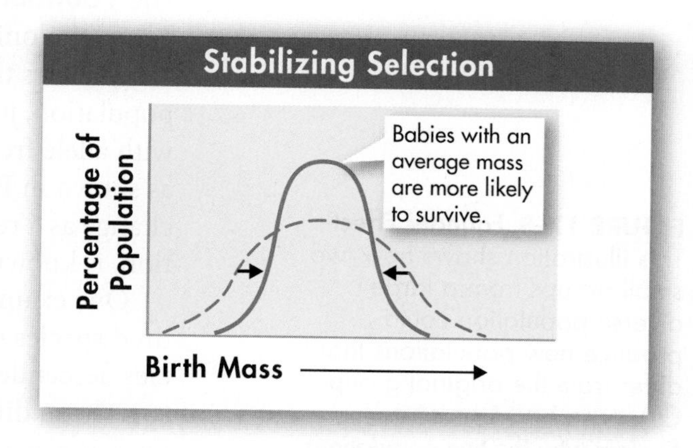

▶ ***Disruptive Selection*** When individuals at the outer ends of the curve have higher fitness than individuals near the middle of the curve, **disruptive selection** occurs. Disruptive selection acts against individuals of an intermediate type. If the pressure of natural selection is strong and lasts long enough, this situation can cause the single curve to split into two. In other words, disruptive selection creates two distinct phenotypes.

Suppose a bird population lives in an area where medium-size seeds become less common and large and small seeds become more common. Birds with unusually small or large beaks would have higher fitness. As shown in the graph, the population might split into two groups: one with smaller beaks and one with larger beaks.

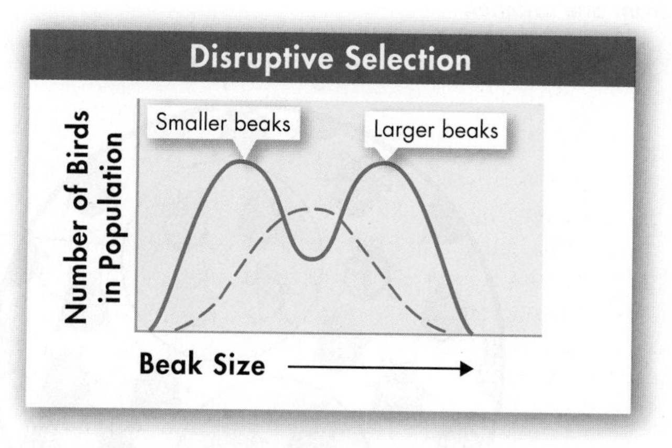

FIGURE 17–7 Selection on Polygenic Traits Natural selection on polygenic traits has one of three patterns—directional selection, stabilizing selection, or disruptive selection.

Genetic Drift

What is genetic drift?

Natural selection is not the only source of evolutionary change. In small populations, an allele can become more or less common simply by chance. **In small populations, individuals that carry a particular allele may leave more descendants than other individuals leave, just by chance. Over time, a series of chance occurrences can cause an allele to become more or less common in a population.** This kind of random change in allele frequency is called **genetic drift.**

BUILD Vocabulary

ACADEMIC WORDS The adjective **random** means "lacking a pattern" or "happening by chance." A random change is a change that happens by chance.

Genetic Bottlenecks Sometimes, a disaster, such as disease, can kill many individuals in a population. Just by chance, the smaller population's gene pool may have allele frequencies that are different from those of the original gene pool. If the reduced population later grows, its alleles will be different in frequency from the original population's. The **bottleneck effect** is a change in allele frequency following a dramatic reduction in the size of a population. A severe bottleneck effect can sharply reduce a population's genetic diversity.

The Founder Effect Genetic drift may also occur when a few individuals colonize a new habitat. These founding individuals may carry alleles that differ in relative frequencies from those of the main population, just by chance. The new gene pool may therefore start out with allele frequencies different from those of the parent gene pool, as shown in **Figure 17–8.** This situation, in which allele frequencies change as a result of the migration of a small subgroup of a population, is known as the **founder effect.**

One example of the founder effect is the evolution of several hundred species of fruit flies on different Hawaiian islands. All those species descended from the same mainland fruit fly population. However, species on different islands have allele frequencies that are different from those of the original species.

FIGURE 17–8 Founder Effect This illustration shows how two small groups from a large, diverse population could produce new populations that differ from the original group. **Compare and Contrast** ***Explain why the two populations of descendants are so different from one another.***

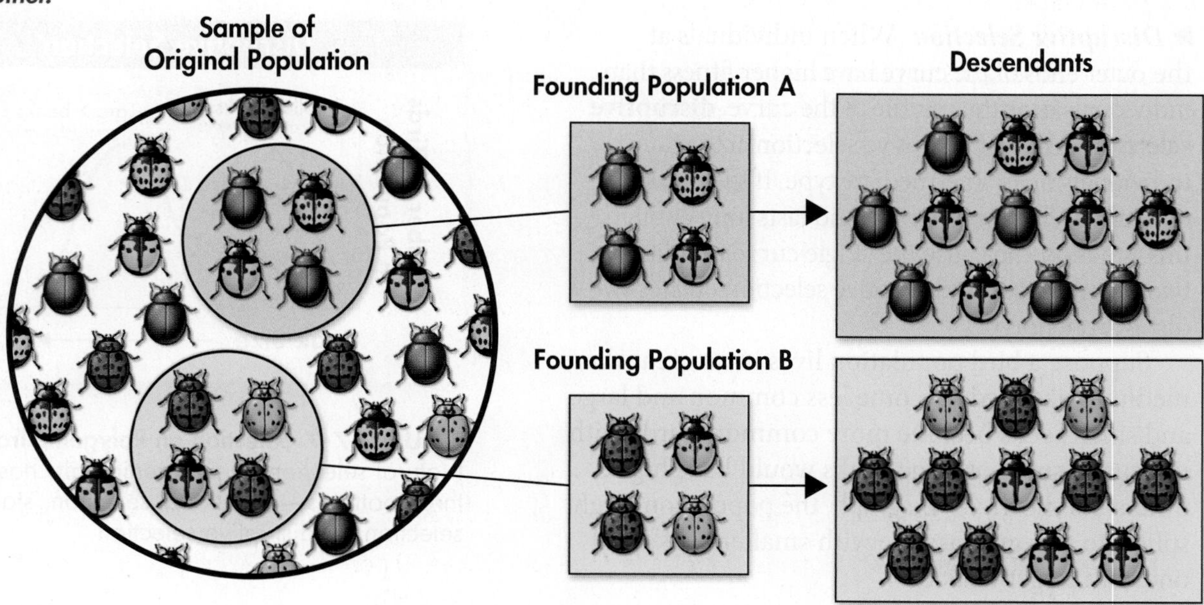

Allele Frequency

The Hardy-Weinberg principle can be used to predict the frequencies of certain genotypes if you know the frequency of other genotypes.

Imagine, for example, that you know of a genetic condition, controlled by two alleles *S* and *s*, which follow the rule of simple dominance at a single locus. The condition affects only homozygous recessive individuals. (The heterozygous phenotype shows no symptoms.) The population you are studying has a population size of 10,000 and there are 36 individuals affected by the condition ($q^2 = 0.0036$). Based on this information, use the Hardy-Weinberg equations to answer the following questions.

1. **Calculate** What are the frequencies of the *S* and *s* alleles?
2. **Calculate** What are the frequencies of the *SS*, *Ss*, and *ss* genotypes?
3. **Calculate** What percentage of people, in total, is likely to be carrying the *s* allele, whether or not they know it?

Evolution Versus Genetic Equilibrium

What conditions are required to maintain genetic equilibrium?

One way to understand how and why populations evolve is to imagine a model of a hypothetical population that does not evolve. If a population is not evolving, allele frequencies in its gene pool do not change, which means that the population is in **genetic equilibrium.**

Sexual Reproduction and Allele Frequency Gene shuffling during sexual reproduction produces many gene combinations. But a century ago, researchers realized that meiosis and fertilization, by themselves, do not change allele frequencies. So hypothetically, a population of sexually reproducing organisms could remain in genetic equilibrium.

The Hardy-Weinberg Principle The **Hardy-Weinberg principle** states that allele frequencies in a population should remain constant unless one or more factors cause those frequencies to change. The Hardy-Weinberg principle makes predictions like Punnett squares—but for populations, not individuals. Here's how it works. Suppose that there are two alleles for a gene: *A* (dominant) and *a* (recessive). A cross of these alleles can produce three possible genotypes: *AA*, *Aa*, and *aa*. The frequencies of genotypes in the population can be predicted by these equations, where *p* and *q* are the frequencies of the dominant and recessive alleles:

In symbols:

$$p^2 + 2pq + q^2 = 1 \text{ and } p + q = 1$$

In words:

(frequency of *AA*) + (frequency of *Aa*) + (frequency of *aa*) = 100% and (frequency of *A*) + (frequency of *a*) = 100%

Suppose that, in one generation, the frequency of the *A* allele is 40 percent ($p = 0.40$) and the frequency of the *a* allele is 60 percent ($q = 0.60$).

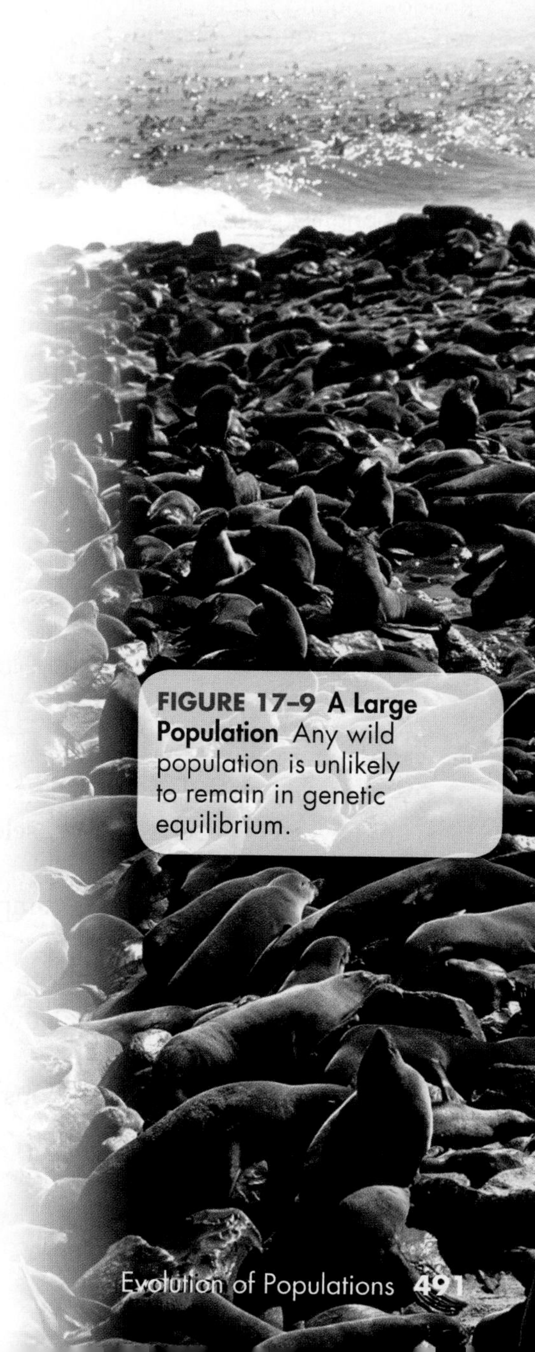

FIGURE 17–9 A Large Population Any wild population is unlikely to remain in genetic equilibrium.

FIGURE 17–10 Choosing a Mate Random mating is one condition required to maintain genetic equilibrium in a population. However, in many species, mating is not random. Female peacocks, for example, choose mates on the basis of physical characteristics such as brightly patterned tail feathers. This is a classic example of sexual selection.

If this population is in genetic equilibrium, chances of an individual in the next generation having genotype *AA* would be 16% ($p^2 = 0.40^2 = 0.16$ or 16%). The probability of genotype *aa* would be 36% ($q^2 = 0.60^2 = 0.36$). The probability of genotype *Aa* would be 48% ($2pq = 2\ (0.40)\ (0.60) = 0.48$). If a population doesn't show these predicted phenotype frequencies, evolution is taking place. **The Hardy-Weinberg principle predicts that five conditions can disturb genetic equilibrium and cause evolution to occur: (1) nonrandom mating; (2) small population size; and (3) immigration or emigration; (4) mutations; or (5) natural selection.**

▶ ***Nonrandom Mating*** In genetic equilibrium, individuals must mate with other individuals at random. But in many species, individuals select mates based on heritable traits, such as size, strength, or coloration, a practice known as **sexual selection.** When sexual selection is at work, genes for the traits selected for or against are not in equilibrium.

▶ ***Small Population Size*** Genetic drift does not usually have major effects in large populations, but can affect small populations strongly. Evolutionary change due to genetic drift thus happens more easily in small populations.

▶ ***Immigration or Emigration*** Individuals who join a population may introduce new alleles into the gene pool, and individuals who leave may remove alleles. Thus, any movement of individuals into (immigration) or out of (emigration) a population can disrupt genetic equilibrium, a process called *gene flow*.

▶ ***Mutations*** Mutations can introduce new alleles into a gene pool, thereby changing allele frequencies and causing evolution to occur.

▶ ***Natural Selection*** If different genotypes have different fitness, genetic equilibrium will be disrupted, and evolution will occur.

One or more of these conditions usually holds for real populations. So, most of the time, in most species, evolution happens.

17.2 Assessment

Review Key Concepts

1. a. **Review** How does natural selection affect a single-gene trait?
   b. **Compare and Contrast** Compare directional selection and disruptive selection.
2. a. **Review** Define genetic drift.
   b. **Relate Cause and Effect** How can the founder effect lead to changes in a gene pool?
3. a. **Review** What five conditions are necessary to maintain genetic equilibrium?
   b. **Infer** Why is genetic equilibrium uncommon in actual populations?

Apply the Big idea

Evolution

4. Do you think populations stay in genetic equilibrium after the environment has changed significantly? Explain your answer.

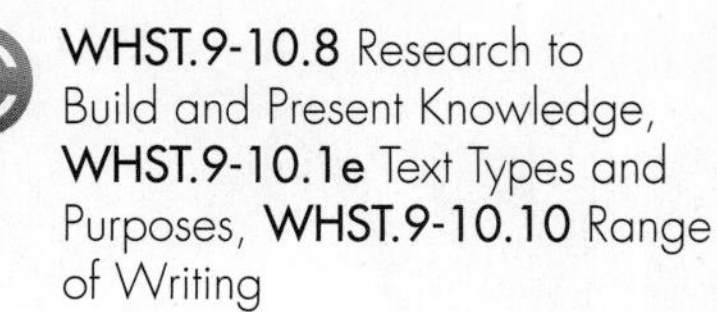

WHST.9-10.8 Research to Build and Present Knowledge, WHST.9-10.1e Text Types and Purposes, WHST.9-10.10 Range of Writing

Biology & Society

Should Antibiotic Use Be Restricted?

Natural selection and evolution aren't just about fossils and finches. Many disease-causing bacteria are evolving resistance to antibiotics—drugs intended to kill them or interfere with their growth.

During your lifetime, antibiotics have always been available and effective. So it is probably hard for you to imagine what life was like before antibiotics were discovered. It wasn't pleasant. During the 1930s, it was not unusual for half of all children in a family to die from bacterial infections that are considered trivial today.

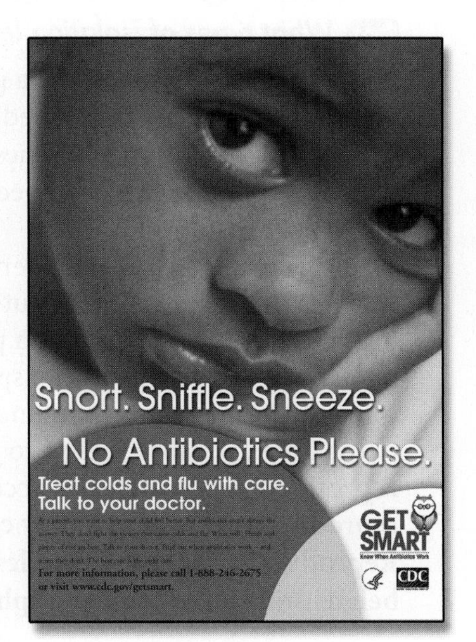

When antibiotics were developed, they rapidly became one of medicine's greatest weapons. Antibiotics saved thousands of lives during World War II by controlling bacterial infections among wounded soldiers. Soon, many bacterial diseases, such as pneumonia, posed much less of a threat. That's why antibiotics were called "magic bullets" and "wonder drugs." But the magic is fading as bacteria evolve.

Bacterial populations have always contained a few individuals with mutations that enabled them to destroy, inactivate, or eliminate antibiotics. But those individuals didn't have higher fitness, so those mutant alleles didn't become common.

Then, doctors began prescribing antibiotics widely, and farmers started feeding antibiotics to farm animals to prevent infections. As a result, antibiotics have become a regular part of the environment for bacteria.

In this new environment, individuals with resistance alleles have higher fitness, so the resistance alleles increase in frequency. Also, resistance alleles can be transferred from one bacterial species to another on plasmids. Thus, disease-causing bacteria can pick up resistance from harmless strains. Many bacteria, including those that cause tuberculosis and certain forms of staph infections, are evolving resistance to not just one antibiotic, but to almost all medicines known. Many doctors are terrified. They fear the loss of one of the vital weapons against bacterial disease. Given this problem, should government agencies restrict antibiotic use?

The Viewpoints

Restrict Antibiotic Use Some people think that the danger of an incurable bacterial epidemic is so high that the government must take action. Doctors overuse antibiotics because patients demand them. The livestock industry likes using antibiotics and will not change their practice unless forced to do so.

Don't Restrict Use Other people think that the doctors and the livestock industry need the freedom to find solutions that work best for them. Researchers are constantly developing new drugs. Some of these drugs can be reserved for human use only.

Research and Decide

1. Analyze the Viewpoints Learn more about this issue by consulting library and Internet resources. Then, list the advantages and disadvantages of restricting antibiotic use.

2. Form Your Opinion Should antibiotics be restricted? Would regulations be more appropriate in some situations than in others? Write an argument to support your claim.

17.3 The Process of Speciation

Key Questions

What types of isolation lead to the formation of new species?

What is a current hypothesis about Galápagos finch speciation?

Vocabulary

species
speciation
reproductive isolation
behavioral isolation
geographic isolation
temporal isolation

Taking Notes

Compare/Contrast Table
In a compare/contrast table, describe the three mechanisms of reproductive isolation.

THINK ABOUT IT How does one species become two? Natural selection and genetic drift can change allele frequencies, causing a population to evolve. But a change in allele frequency by itself does not lead to the development of a new species.

Isolating Mechanisms

What types of isolation lead to the formation of new species?

Biologists define a **species** as a population or group of populations whose members can interbreed and produce fertile offspring. Given this genetic definition of species, what must happen for one species to divide or give rise to a new species? The formation of a new species is called **speciation.**

Interbreeding links members of a species genetically. Any genetic changes can spread throughout the population over time. But what happens if some members of a population stop breeding with other members? The gene pool can split. Once a population has thus split into two groups, changes in one of those gene pools cannot spread to the other. Because these two populations no longer interbreed, **reproductive isolation** has occurred. **When populations become reproductively isolated, they can evolve into two separate species. Reproductive isolation can develop in a variety of ways, including behavioral isolation, geographic isolation, and temporal isolation.**

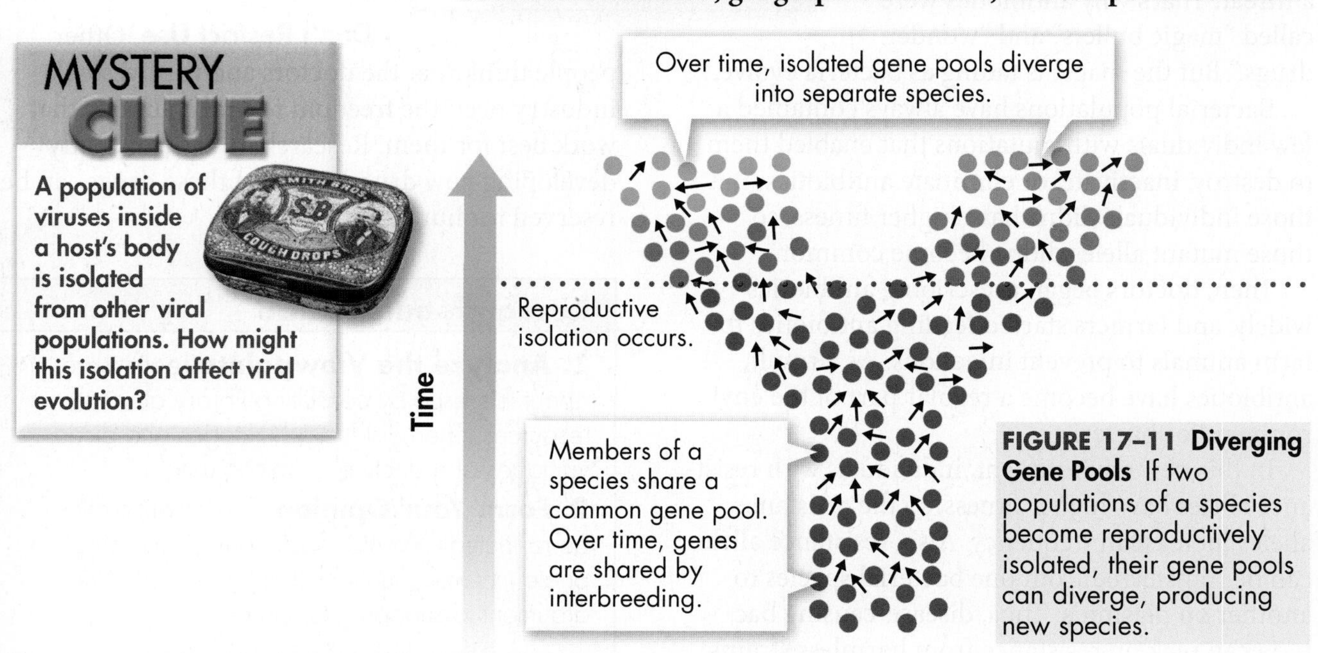

FIGURE 17–11 Diverging Gene Pools If two populations of a species become reproductively isolated, their gene pools can diverge, producing new species.

Behavioral Isolation Suppose two populations that are capable of interbreeding develop differences in courtship rituals or other behaviors. **Behavioral isolation** can then occur. For example, eastern and western meadowlarks are similar birds whose habitats overlap. But, members of the two species will not mate with each other, partly because they use different songs to attract mates. Eastern meadowlarks don't respond to western meadowlark songs, and vice versa.

Geographic Isolation When two populations are separated by geographic barriers such as rivers, mountains, or bodies of water, **geographic isolation** occurs. The Abert's squirrel in **Figure 17–12,** for example, lives in the Southwest. About 10,000 years ago, a small population became isolated on the north rim of the Grand Canyon. Separate gene pools formed. Genetic changes that appeared in one group were not passed to the other. Natural selection and genetic drift worked separately on each group and led to the formation of a distinct subspecies, the Kaibab squirrel. The Abert's and Kaibab squirrels are very similar, indicating that they are closely related. However, the Kaibab squirrel differs from the Abert's squirrel in significant ways, such as fur coloring.

Geographic barriers do not always guarantee isolation. Floods, for example, may link separate lakes, enabling their fish populations to mix. If those populations still interbreed, they remain a single species. Also, a geographic barrier may separate certain organisms but not others. A large river may keep squirrels and other small rodents apart but probably won't isolate bird populations.

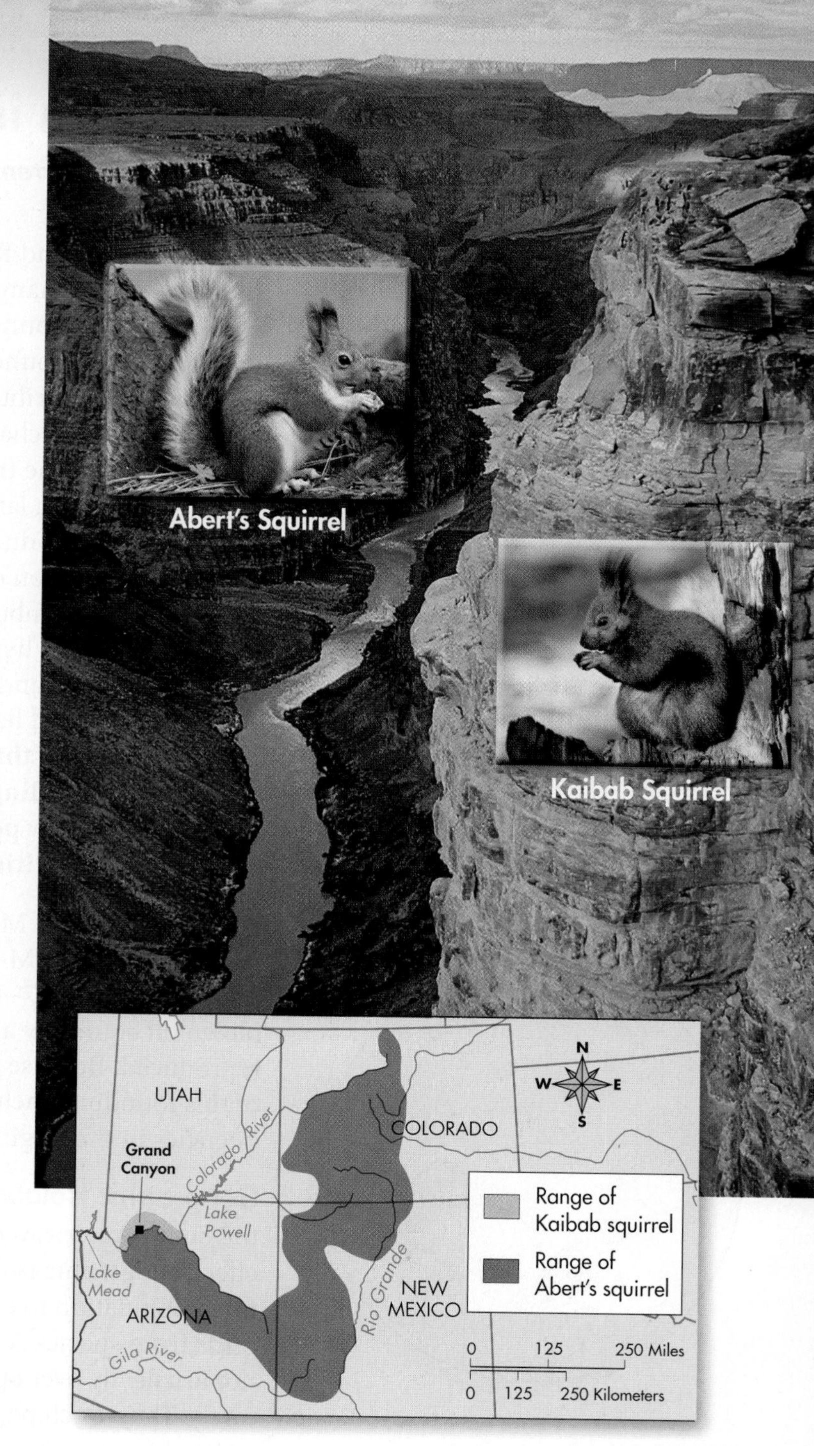

FIGURE 17–12 Geographic Isolation Abert's squirrel and the Kaibab squirrel are distinct subspecies within the same species. Their gene pools are separate. **Interpret Visuals** ***What geographic barrier separates the two populations of squirrels?***

Temporal Isolation A third isolating mechanism, known as **temporal isolation,** happens when two or more species reproduce at different times. For example, suppose three similar species of orchids live in the same rain forest. Each species has flowers that last only one day and must be pollinated on that day to produce seeds. Because the species bloom on different days, they cannot pollinate one another.

In Your Notebook *Explain how temporal isolation can lead to speciation.*

Speciation in Darwin's Finches

What is a current hypothesis about Galápagos finch speciation?

Recall that Peter and Rosemary Grant spent years on the Galápagos islands studying changes in finch populations. The Grants measured and recorded anatomical characteristics such as beak length of individual medium ground finches. Many of the characteristics appeared in bell-shaped distributions typical of polygenic traits. As environmental conditions changed, the Grants documented directional selection among the traits. When drought struck the island of Daphne Major, finches with larger beaks capable of cracking the thickest seeds survived and reproduced more often than others. Over many generations, the proportion of large-beaked finches increased.

We can now combine these studies by the Grants with evolutionary concepts to form a hypothesis that answers a question: How might the founder effect and natural selection have produced reproductive isolation that could have led to speciation among Galápagos finches? **According to this hypothesis, speciation in Galápagos finches occurred by founding of a new population, geographic isolation, changes in the new population's gene pool, behavioral isolation, and ecological competition.**

FIGURE 17-13

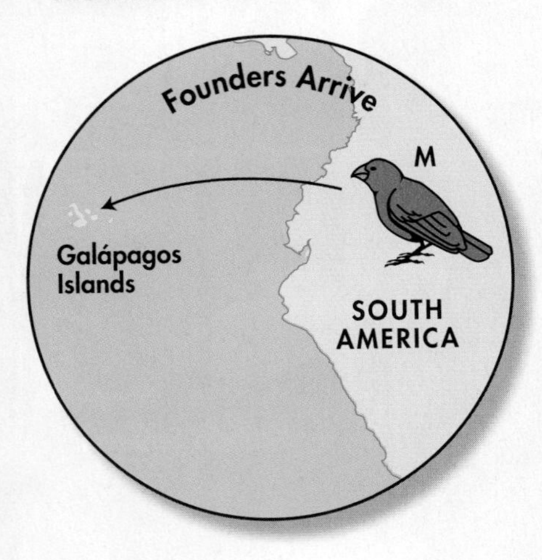

Founders Arrive Many years ago, a few finches from South America—species M—arrived on one of the Galápagos islands, as shown in **Figure 17-13.** These birds may have gotten lost or been blown off course by a storm. Once on the island, they survived and reproduced. Because of the founder effect, allele frequencies of this founding finch population could have differed from allele frequencies in the original South American population.

FIGURE 17-14

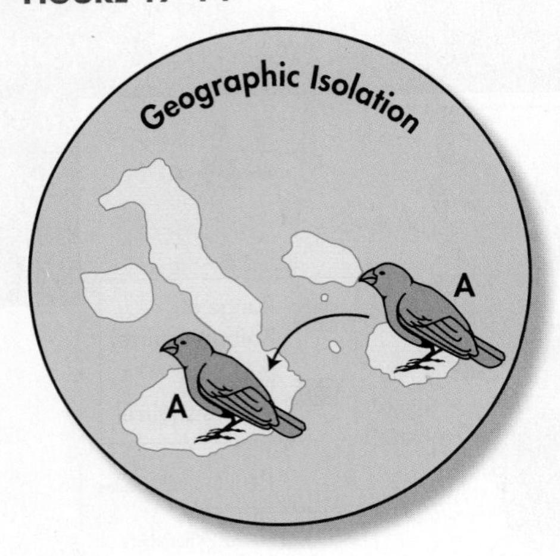

Geographic Isolation The island's environment was different from the South American environment. Some combination of the founder effect, geographic isolation, and natural selection enabled the island finch population to evolve into a new species—species A. Later, a few birds from species A crossed to another island. Because these birds do not usually fly over open water, they move from island to island very rarely. Thus, finch populations on the two islands were geographically isolated from each other and no longer shared a common gene pool.

FIGURE 17-15

Changes in Gene Pools Over time, populations on each island adapted to local environments. Plants on the first island may have produced small, thin-shelled seeds, whereas plants on the second island may have produced larger, thick-shelled seeds. On the second island, directional selection would have favored individuals with larger, heavier beaks. These birds could crack open and eat the large seeds more easily. Thus, birds with large beaks would be better able to survive on the second island. Over time, natural selection would have caused that population to evolve larger beaks, forming a distinct population, B, characterized by a new phenotype.

Behavioral Isolation Now, imagine that a few birds from the second island cross back to the first island. Will population-A birds breed with population-B birds? Probably not. These finches choose mates carefully. During courtship, they closely inspect a potential partner's beak. Finches prefer to mate with birds that have the same-size beak as they do. Big-beaked birds prefer to mate with other big-beaked birds, and smaller-beaked birds prefer to mate with other smaller-beaked birds. Because the populations on the two islands have evolved differently sized beaks, they would probably not mate with each other.

Thus, differences in beak size, combined with mating behavior, could lead to reproductive isolation. The gene pools of the two bird populations remain isolated—even when individuals live in the same place. The populations have now become two distinct species.

FIGURE 17–16

Competition and Continued Evolution As these two new species live together on the first island, they compete for seeds. During the dry season, birds that are most different from each other have the highest fitness. That is because the more specialized birds have less competition for certain kinds of seeds and other foods. Over time, species evolve in a way that increases the differences between them. The species-B birds on the first island may evolve into a new species, C.

The combined processes of geographic isolation on different islands, genetic change, and behavioral isolation could have repeated itself again and again across the Galápagos chain. Over many generations, the process could have produced the 13 different finch species found there today.

FIGURE 17–17

In Your Notebook *Explain how natural selection and behavioral isolation may have lead to reproductive isolation in Darwin's finches.*

17.3 Assessment

Review Key Concepts

1. a. Review What is geographic isolation?

b. Predict A newly formed lake divides a population of a beetle species into two groups. What other factors besides isolation might lead to the two groups becoming separate species?

2. a. Review What types of reproductive isolation may have been important in Galápagos finch speciation? Explain.

b. Apply Concepts Explain how the vegetarian tree finch, which feeds on fruit, might have evolved.

BUILD VOCABULARY

3. *Temporal* comes from the Latin word *tempus*, meaning "time." How is time a factor in temporal isolation?

4. *Isolation* is related to the Latin word *insula*, meaning "island." After reading about isolating mechanisms in this lesson, does the common origin of these two words make sense? Explain your answer.

Molecular Evolution

Key Questions

What are molecular clocks?

Where do new genes come from?

How may Hox genes be involved in evolutionary change?

Vocabulary

molecular clock

Taking Notes

Outline As you read, make an outline of this lesson. Use the green headings as the main topics and the blue headings as the subtopics.

BUILD Vocabulary

ACADEMIC WORDS The word **sequence** means "the order in which parts are put together." The sequence of DNA is the order in which its molecules are arranged.

THINK ABOUT IT Recall that an organism's genome is its complete set of genetic information. Thousands of ongoing projects are analyzing the genomes of organisms ranging from viruses to humans. The analysis of genomes enables us to study evolution at the molecular level. By comparing DNA sequences from all of these organisms, we can often solve important evolutionary puzzles. For example, DNA evidence may indicate how two species are related to one another, even if their body structures don't offer enough clues.

Timing Lineage Splits: Molecular Clocks

What are molecular clocks?

When researchers use a **molecular clock,** they compare stretches of DNA to mark the passage of evolutionary time. **A molecular clock uses mutation rates in DNA to estimate the time that two species have been evolving independently.**

Neutral Mutations as "Ticks" To understand molecular clocks, think about old-fashioned pendulum clocks. They mark time with a swinging pendulum. A molecular clock also relies on a repeating process to mark time—mutation. As you've learned, simple mutations occur all the time, causing slight changes in the sequence of DNA. Some mutations have a major positive or negative effect on an organism's phenotype. These types of mutations are under powerful pressure from natural selection.

Many mutations, however, have no effect on phenotype. These neutral mutations tend to accumulate in the DNA of different species at about the same rate. Researchers can compare such DNA sequences in two species. The comparison can reveal how many mutations have occurred independently in each group, as shown in **Figure 17–18.** The more differences there are between the DNA sequences of the two species, the more time has elapsed since the two species shared a common ancestor.

In Your Notebook *Which kind of mutation—neutral or negative—will most likely persist in a population over time? Explain.*

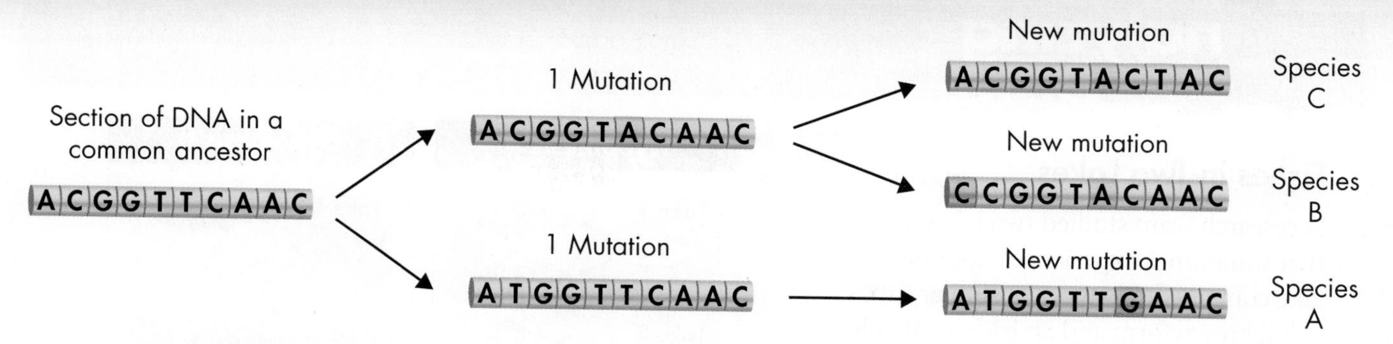

FIGURE 17–18 Molecular Clock
By comparing the DNA sequences of two or more species, biologists estimate how long the species have been separated. **Analyze Data** ***What evidence indicates that species C is more closely related to species B than to species A?***

Calibrating the Clock The use of molecular clocks is not simple, because there is not just one molecular clock in a genome. There are many different clocks, each of which "ticks" at a different rate. This is because some genes accumulate mutations faster than others. These different clocks allow researchers to time different evolutionary events. Think of a conventional clock. If you want to time a brief event, you use the second hand. To time an event that lasts longer, you use the minute hand or the hour hand. In the same way, researchers choose a different molecular clock to compare great apes than to estimate when mammals and fishes shared a common ancestor.

Researchers check the accuracy of molecular clocks by trying to estimate how often mutations occur. In other words, they estimate how often the clock they have chosen "ticks." To do this, they compare the number of mutations in a particular gene in species whose age has been determined by other methods.

Gene Duplication

Where do new genes come from?

Where did the roughly 25,000 working genes in the human genome come from? Modern genes probably descended from a much smaller number of genes in the earliest life forms. But how could that have happened? **One way in which new genes evolve is through the duplication, and then modification, of existing genes.**

Copying Genes Most organisms carry several copies of various genes. Sometimes organisms carry two copies of the same gene. Other times there may be thousands of copies. Where do those extra copies come from, and what happens to them?

Remember that homologous chromosomes exchange DNA during meiosis in a process called crossing-over. Sometimes crossing-over involves an unequal swapping of DNA. In other words, one chromosome in the pair gets extra DNA. That extra DNA can carry part of a gene, a full gene, or a longer length of chromosome. Sometimes, in different ways, an entire genome can be duplicated.

Fishes in Two Lakes

A research team studied two lakes in an area that sometimes experiences flooding. Each lake contained two types of similar fishes: a dull brown form and an iridescent gold form. The team wondered how all the fishes were related, and they considered the two hypotheses diagrammed on the right.

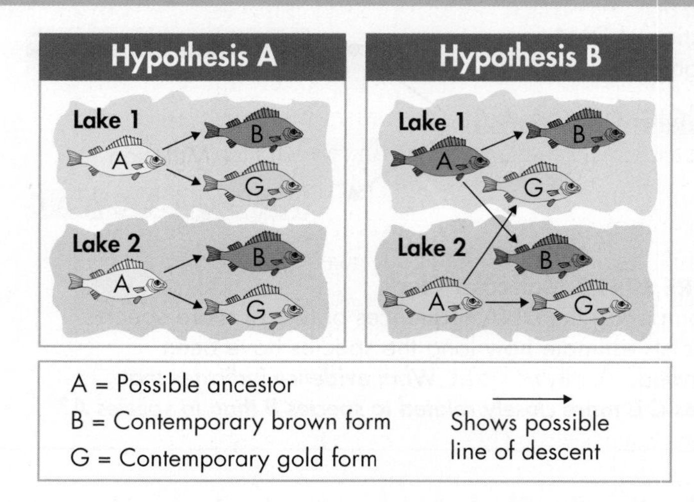

1. Interpret Visuals Study the two diagrams. What does hypothesis A indicate about the ancestry of the fishes in Lake 1 and Lake 2? What does hypothesis B indicate?

2. Compare and Contrast According to the two hypotheses, what is the key difference in the way the brown and gold fish populations might have formed?

3. Draw Conclusions A DNA analysis showed that the brown and gold fishes from Lake 1 are the most closely related. Which hypothesis does this evidence support?

Duplicate Genes Evolve What's so important about gene duplication? Think about using a computer to write an essay for English class. You then want to submit a new version of the essay to your school newspaper. So, you make an extra copy of the original file and edit it for the newspaper.

Duplicate genes can work in similar ways. Sometimes, extra copies of a gene undergo mutations that change their function. The original gene is still around, just like the original copy of your English essay. So, the new genes can evolve without affecting the original gene function or product. **Figure 17–19** shows how this happens.

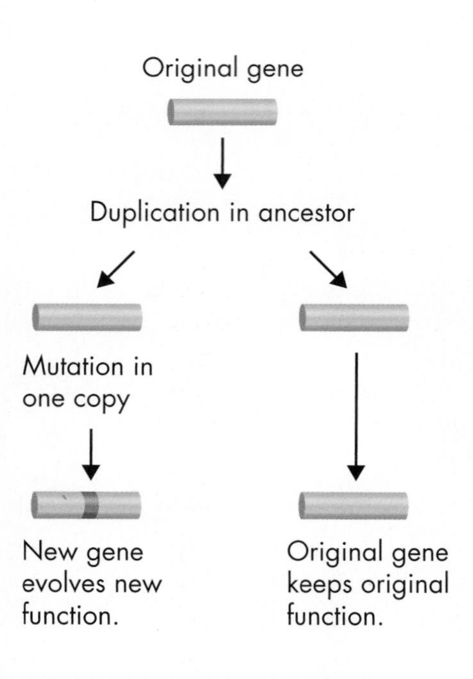

FIGURE 17–19 Gene Duplication In this diagram, a gene is first duplicated, and then one of the two resulting genes undergoes mutation.

Gene Families Multiple copies of a duplicated gene can turn into a group of related genes called a gene family. Members of a gene family typically produce similar, yet slightly different, proteins. Your body, for example, produces a number of molecules that carry oxygen. Several of these compounds—called globins—are hemoglobins. The globin gene family that produces them evolved, after gene duplication, from a single ancestral globin gene. Some of the most important evolution research focuses on another gene family—Hox genes.

Developmental Genes and Body Plans

How may Hox genes be involved in evolutionary change?

One exciting new research area is nicknamed "evo-devo" because it studies the relationship between evolution and embryological development. Darwin himself had a hunch that changes in the growth of embryos could transform adult body shape and size. Researchers now study how small changes in Hox gene activity could produce the kinds of evolutionary changes we see in the fossil record.

Hox Genes and Evolution As you read in Chapter 13, Hox genes determine which parts of an embryo develop arms, legs, or wings. Groups of Hox genes also control the size and shape of those structures. In fact, homologous Hox genes shape the bodies of animals as different as insects and humans—even though those animals last shared a common ancestor no fewer than 500 million years ago!

Small changes in Hox gene activity during embryological development can produce large changes in adult animals. For example, insects and crustaceans are related to ancient common ancestors that possessed dozens of legs. Today's crustaceans, including shrimp and lobsters, still have large numbers of paired legs, but insects have just 3 pairs of legs. What happened to those extra legs? Recent studies have shown that mutations in a single Hox gene, known as *Ubx*, turns off the growth of legs in the abdominal regions of insects. Thus, a change in one Hox gene accounts for a major evolutionary difference between two important animal groups.

Timing Is Everything Each part of an embryo starts to grow at a certain time, grows for a specific time, and stops growing at a specific time. Small changes in starting and stopping times can make a big difference in organisms. For example, small timing changes can make the difference between long, slender fingers and short, stubby toes. No wonder "evo-devo" is one of the hottest areas in evolutionary biology!

FIGURE 17–20 Change in a Hox Gene Insects such as fruit flies and crustaceans such as brine shrimp are descended from a common ancestor that had many legs. Due to mutations in the activity of a single Hox gene that happened millions of years ago, modern insects have fewer legs than do modern crustaceans. In the illustration, the legs of the fruit fly and the legs of the brine shrimp are the same color (red) because a variant of the same Hox gene, *Ubx*, directs the development of the legs of both animals.

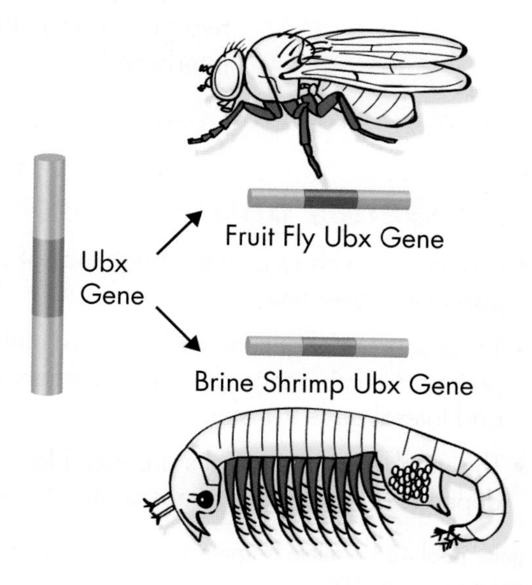

17.4 Assessment

Review Key Concepts

1. **a. Review** What is a molecular clock?
   b. Explain Why do molecular clocks use mutations that have no effect on phenotype?
2. **a. Review** How can crossing-over result in gene duplication?
   b. Explain Describe how duplicate genes form.
   c. Relate Cause and Effect Why is gene duplication important in evolution?
3. **a. Review** Use the evolution of the insect body plan to explain the significance of Hox genes in evolution.
   b. Infer In evolution, why have small changes in Hox genes had a great impact?

VISUAL THINKING

4. The colored bands in the diagrams below represent mutations in a segment of DNA in species A, B, and C. Which two of the three species probably share the most recent common ancestor?

Species A	Species B	Species C
C	C	C
T	G	T
A	A	A
A	G	G
C	C	C
G	G	G
T	T	T
T	T	C
G	A	G
C	C	C

NGSS Smart Guide The Evolution of Populations

Disciplinary Core Idea LS4.C Adaptation: How does the environment influence populations of organisms over multiple generations? A new species can form when a population splits into two groups that are isolated from each other. The gene pools of the two groups may become so different that the groups can no longer interbreed.

17.1 Genes and Variation

- Evolution is a change in the frequency of alleles in a population over time.
- Three sources of genetic variation are mutation, genetic recombination during sexual reproduction, and lateral gene transfer.
- The number of phenotypes produced for a trait depends on how many genes control the trait.

gene pool 483 • allele frequency 483 • single-gene trait 485 • polygenic trait 486

Biology.com

Untamed Science Video Climb the cliffs of Hawaii with the Untamed Science crew to discover how geographic isolation can aid in the development of a new species.

Art Review Review your understanding of alleles and allele frequencies in a population.

17.2 Evolution as Genetic Change in Populations

- Natural selection on single-gene traits can lead to changes in allele frequencies and, thus, to changes in phenotype frequencies.
- Natural selection on polygenic traits can affect the relative fitness of phenotypes and thereby produce one of three types of selection: directional selection, stabilizing selection, or disruptive selection.
- In small populations, individuals that carry a particular allele may leave more descendants than other individuals leave, just by chance. Over time, a series of chance occurrences can cause an allele to become more or less common in a population.
- The Hardy-Weinberg principle predicts that five conditions can disturb genetic equilibrium and cause evolution to occur: (1) nonrandom mating; (2) small population size; and (3) immigration or emigration; (4) mutations; or (5) natural selection.

directional selection 489 • stabilizing selection 489 • disruptive selection 489 • genetic drift 490 • bottleneck effect 490 • founder effect 490 • genetic equilibrium 491 • Hardy-Weinberg principle 491 • sexual selection 492

Biology.com

Art in Motion Watch how different types of selection change the types of individuals in a population.

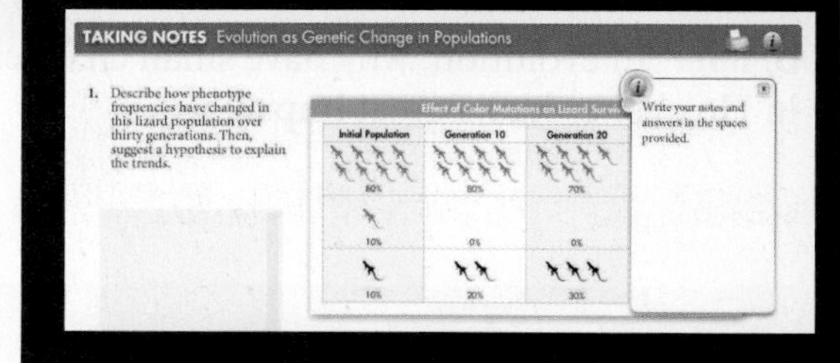

CHECKING YOUR *Scientific Literacy*

Refer to the lesson content and digital assets below as you prepare for your chapter assessment. Then, evaluate your understanding of the evolution of populations by answering these questions.

1. Scientific and Engineering Practice **Obtaining, Evaluating, and Communicating Information** Explain how population genetics and molecular genetics relate to modern evolutionary theory.
2. Crosscutting Concept **Stability and Change** How can differences in genotype affect change in a population?
3. Integration of Knowledge and Ideas Use Figures 17-13 to 17-17 to summarize the current hypothesis on how speciation of the Galápagos finches may have occurred.
4. STEM Design and build a physical model to demonstrate the concept of genetic bottlenecks to a science museum audience.

17.3 The Process of Speciation

- When populations become reproductively isolated, they can evolve into two separate species. Reproductive isolation can develop in a variety of ways, including behavioral isolation, geographic isolation, and temporal isolation.
- Speciation in Galápagos finches most likely occurred by founding of a new population, geographic isolation, changes in the new population's gene pool, behavioral isolation, and ecological competition.

species 494 • speciation 494 • reproductive isolation 494 • behavioral isolation 495 • geographic isolation 495 • temporal isolation 495

Biology.com

Data Analysis Find out what happened to Galápagos finches during a drought by comparing data on finches and their food sources.

Tutor Tube Learn more about the mechanisms of speciation from the tutor.

17.4 Molecular Evolution

- A molecular clock uses mutation rates in DNA to estimate the time that two species have been evolving independently.
- One way in which new genes evolve is through the duplication, and then modification, of existing genes.
- Small changes in Hox gene activity during embryological development can produce large changes in adult animals.

molecular clock 498

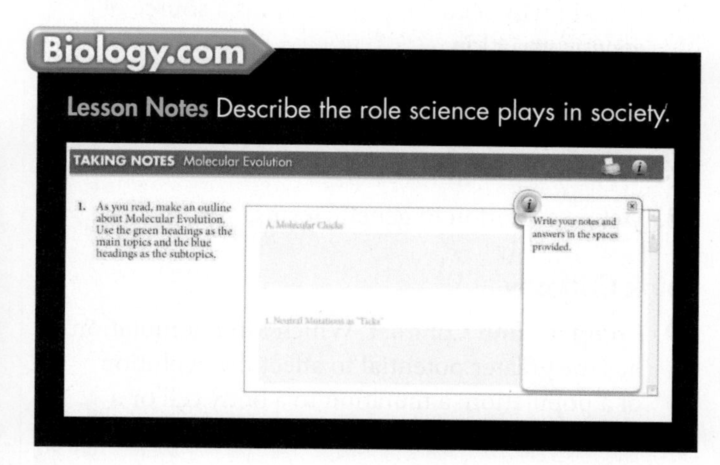

Lesson Notes Describe the role science plays in society.

Skills Lab

Competing for Resources This chapter lab is available in your lab manual and at Biology.com

17 Assessment

17.1 Genes and Variation

Understand Key Concepts

1. The combined genetic information of all members of a particular population forms a
   a. gene pool.
   b. niche.
   c. phenotype.
   d. population.

2. Mutations that improve an individual's ability to survive and reproduce are
   a. harmful.
   b. neutral.
   c. beneficial.
   d. chromosomal.

3. Traits, such as human height, that are controlled by more than one gene are known as
   a. single-gene traits.
   b. polygenic traits.
   c. recessive traits.
   d. dominant traits.

4. **Craft and Structure** Explain what the term *allele frequency* means. Include an example illustrating your answer.

5. Explain why sexual reproduction is a source of genetic variation.

6. Explain what determines the number of phenotypes for a given trait.

7. What is *lateral gene transfer*?

8. Define evolution in genetic terms.

Think Critically

9. **Compare and Contrast** Which kind of mutation has the greater potential to affect the evolution of a population: a mutation to a body cell or a mutation in an egg cell? Explain.

10. **Craft and Structure** Explain how natural selection is related to phenotypes and genotypes.

11. **Craft and Structure** Explain how natural selection is related to individuals and populations.

12. **Relate Cause and Effect** How does genetic recombination affect genetic variation?

17.2 Evolution as Genetic Change in Populations

Understand Key Concepts

13. The type of selection in which individuals of average size have greater fitness than small or large individuals have is called
    a. disruptive selection. c. directional selection.
    b. stabilizing selection. d. neutral selection.

14. If coat color in a rabbit population is a polygenic trait, which process might have produced the graph below?

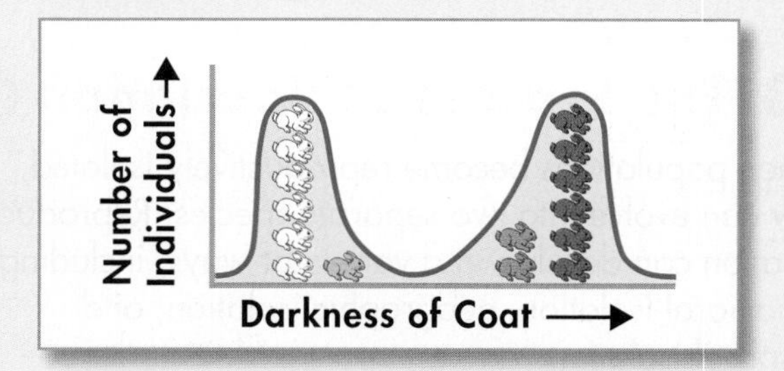

    a. disruptive selection c. directional selection
    b. stabilizing selection d. genetic equilibrium

15. A random change in a small population's allele frequency is known as
    a. a gene pool c. variation
    b. genetic drift d. fitness

16. **Craft and Structure** What is *fitness* in genetic terms?

17. How do stabilizing selection and disruptive selection differ?

18. What is genetic equilibrium? In what kinds of situations is it likely to occur?

Think Critically

19. **Compare and Contrast** Distinguish between the ways in which natural selection affects single-gene traits and the ways in which it affects polygenic traits. How are phenotype frequencies altered in each case?

20. **Infer** In a certain population of plants, flower size is a polygenic trait. What kind of selection is likely to occur if environmental conditions favor small flowers?

21. **Infer** A road built through a forest splits a population of frogs into two large groups. The allele frequencies of the two groups are identical. Has genetic drift occurred? Why or why not?

22. **Text Types and Purposes** DDT is an insecticide that was first used in the 1940s to kill mosquitoes and stop the spread of malaria. As time passed, people began to notice that DDT became less effective. Explain, in genetic terms, how the insects became resistant to the pesticide.

17.3 The Process of Speciation

Understand Key Concepts

23. Temporal isolation occurs when two different populations
    a. develop different mating behaviors.
    b. become geographically separated.
    c. reproduce at different times.
    d. interbreed.

24. When two populations no longer interbreed, what is the result?
    a. genetic equilibrium
    b. reproductive isolation
    c. stabilizing selection
    d. artificial selection

25. **Text Types and Purposes** Explain how the different species of Galápagos finches may have evolved.

Think Critically

26. **Text Types and Purposes** Explain why reproductive isolation usually must occur before a population splits into two distinct species.

27. **Form a Hypothesis** A botanist identifies two distinct species of violets growing in a field, as shown in the left of the illustration below. Also in the field are several other types of violets that, although somewhat similar to the two known species, appear to be new species. Develop a hypothesis explaining how the new species may have originated.

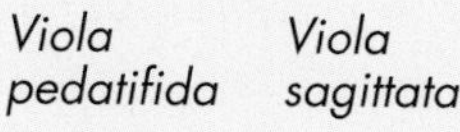

Viola pedatifida *Viola sagittata* Other violets

solve the CHAPTER MYSTERY

EPIDEMIC

The genes of flu viruses mutate often, and different strains can swap genes if they infect the same host at the same time. These characteristics produce genetic diversity that enables the virus to evolve.

Flu viruses also undergo natural selection. Think of our bodies as the environment for viruses. Our immune system attacks viruses by "recognizing" proteins on the surface of the viruses. Viruses whose proteins our bodies can recognize and destroy have low fitness. Viruses our bodies can't recognize have higher fitness.

Influenza Virus

Viral evolution regularly produces slightly different surface proteins that our immune systems can't recognize right away. These strains evade the immune system long enough to make people sick. That's why you can catch the flu every winter, and why new flu vaccines must be made every year.

But now and then, influenza evolution produces radically new molecular "disguises" that our immune systems can't recognize *at all.* These can be deadly, like the 1918 strain. If a strain like that were to appear today, it could kill many people. That's why researchers are worried about "bird flu"—a strain of flu that can pass from birds, such as chickens, to humans.

1. **Connect to the Big idea** Explain why mutation and natural selection make developing new flu vaccines necessary every year.
2. **Infer** People do not need to receive a new measles vaccination every year. What does this suggest about a difference between flu viruses and the measles virus?
3. **Apply Concepts** Can you think of any other issues in public health that relate directly to evolutionary change?

17.4 Molecular Evolution

Understand Key Concepts

28. A group of related genes that resulted from the duplication and modification of a single gene is called a

a. gene pool. **c.** lateral gene transfer.
b. molecular clock. **d.** gene family.

29. Each "tick" of a molecular clock is an occurrence of

a. genetic drift. **c.** DNA mutation.
b. crossing-over. **d.** mitosis.

30. How do chromosomes gain an extra copy of a gene during meiosis?

31. What are neutral mutations?

32. **Craft and Structure** What is the study of "evo-devo," and how is it related to evolution?

Think Critically

33. **Pose Questions** What kinds of questions would scientists who are studying the evolution of Hox genes most likely be asking?

34. **Craft and Structure** Describe the relationship between evolutionary time and the similarity of genes in two species.

Connecting Concepts

Use Science Graphics

Use the data table to answer questions 35 and 36.

Frequency of Alleles

Year	Frequency of Allele *B*	Frequency of Allele *b*
1910	0.81	0.19
1930	0.49	0.51
1950	0.25	0.75
1970	0.10	0.90

35. **Integration of Knowledge and Ideas** Describe the trend shown by the data in the table.

36. **Form a Hypothesis** What might account for the trend shown by the data?

Write About Science

37. **Text Types and Purposes** Explain the process that may have caused fruit flies to have fewer legs than their ancestors had.

38. **Assess the**  Sometimes, biologists say, "Evolution is ecology over time." Explain that statement.

Analyzing Data

Integration of Knowledge and Ideas

The graph shows data regarding the lengths of the beaks of three finch species. The percentage of individuals in each category of beak length is given.

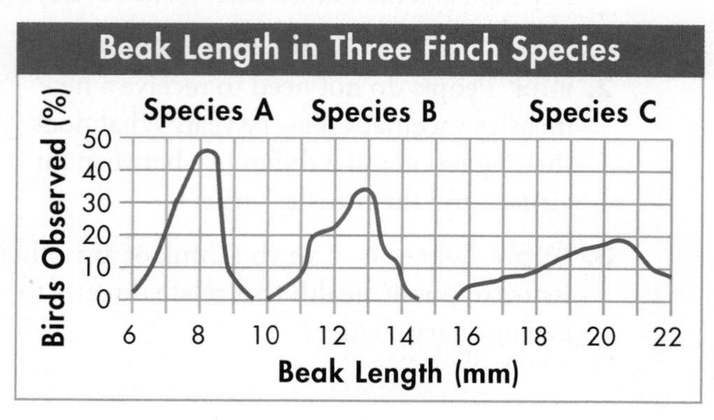

39. **Interpret Graphs** What is the shortest beak length observed in species A?

a. 3 mm **c.** 9 mm
b. 6 mm **d.** 12 mm

40. **Analyze Data** Which of the following is a logical interpretation of the data?

a. Species B eats the smallest seeds.
b. About 50 percent of species C eats seeds that are 20 mm long.
c. Species C eats the largest seeds.
d. All three species eat seeds of the same size.

Standardized Test Prep

Multiple Choice

1. Which of the following conditions is MOST likely to result in changes in allele frequencies in a population?
A random mating
B small population size
C no migrations into or out of a population
D absence of natural selection

2. Mutations and the genetic recombination that occurs during sexual reproduction are both sources of
A genetic variation.
B stabilizing selection.
C genetic equilibrium.
D genetic drift.

3. In a population of lizards, the smallest and largest lizards are more easily preyed upon than medium-size lizards. What kind of natural selection is MOST likely to occur in this situation?
A genetic drift **C** stabilizing selection
B sexual selection **D** directional selection

4. Populations of antibiotic-resistant bacteria are the result of the process of
A natural selection. **C** genetic drift.
B temporal isolation. **D** artificial selection.

5. If species A and B have very similar genes and proteins, what is probably true?
A Species A and B share a relatively recent common ancestor.
B Species A evolved independently of species B for a long period.
C Species A is younger than species B.
D Species A is older than species B.

6. When two species reproduce at different times, the situation is called
A genetic drift.
B temporal selection.
C temporal isolation.
D lateral gene transfer.

7. The length of time that two taxa have been evolving separately can be estimated using
A genetic drift. **C** a molecular clock.
B gene duplication. **D** Hox genes.

Questions 8–9

The graphs below show the changes in crab color at one beach.

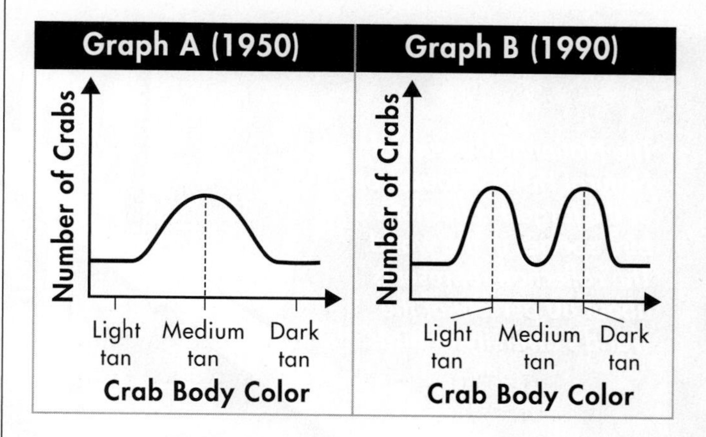

8. What process occurred over the 40-year period?
A artificial selection **C** stabilizing selection
B directional selection **D** disruptive selection

9. Which of the following is MOST likely to have caused the change in the distribution?
A A new predator arrived that preferred dark-tan crabs.
B A new predator arrived that preferred light-tan crabs.
C A change in beach color made medium-tan crabs the least visible to predators.
D A change in beach color made medium-tan crabs the most visible to predators.

Open-Ended Response

10. How does evolution change the relative frequency of alleles in a gene pool? Why does this happen?

If You Have Trouble With . . .

Question	1	2	3	4	5	6	7	8	9	10
See Lesson	17.3	17.1	17.2	17.2	17.4	17.3	17.4	17.2	17.2	17.1

18 Classification

Big idea Unity and Diversity of Life

Q: What is the goal of biologists who classify living things?

The National Museum of Natural History houses one of the largest collections of bird species in the world. The collection represents about 80 percent of the world's diversity of birds.

INSIDE:

- 18.1 Finding Order in Diversity
- 18.2 Modern Evolutionary Classification
- 18.3 Building the Tree of Life

Deepen your understanding of classification by engaging in key practices that allow you to make connections across concepts.

NGSS You will analyze and interpret evolutionary data represented by cladograms.

STEM How can newly discovered species be classified? Investigate the tools and skills used by scientists to classify organisms.

Common Core You will translate information expressed visually in cladograms to demonstrate your understanding of modern evolutionary classification.

CHAPTER MYSTERY

GRIN AND BEAR IT

If you simply looked at a polar bear and brown bear, you would probably never doubt that they are members of different species. Polar bears grow much larger than brown bears, and their paws have adapted to swimming long distances and to walking on snow and ice. Their white fur camouflages them, but the coats on brown bears are, well, brown—and their paws aren't adapted to water.

Clearly polar bears and brown bears are very different physically. But do physical characteristics tell the whole story? Remember the definition of *species:* "a group of similar organisms that can breed and produce fertile offspring." Well, polar bears and brown bears can mate and produce fertile offspring. They must be members of the same species, then. But are they? As you read this chapter, look for clues to whether polar bears are a separate species. Then, solve the mystery.

Solving the mystery of bear classification is only the beginning. Take a video field trip with the ecogeeks of Untamed Science to see where the mystery leads.

Go online to access additional resources including:
• eText • Flash Cards • Lesson Overviews • Chapter Mystery

18.1 Finding Order in Diversity

Key Questions

What are the goals of binomial nomenclature and systematics?

How did Linnaeus group species into larger taxa?

Vocabulary

binomial nomenclature • genus • systematics • taxon • family • order • class • phylum • kingdom

Taking Notes

Preview Visuals Before you read, look at **Figure 18–5.** Notice all the levels of classification. As you read, refer to the figure again.

THINK ABOUT IT Scientists have been trying to identify, name, and find order in the diversity of life for a long time. The first scientific system for naming and grouping organisms was set up long before Darwin. In recent decades, biologists have been completing a changeover from that older system of names and classification to a newer strategy that is based on evolutionary theory.

Assigning Scientific Names

What are the goals of binomial nomenclature and systematics?

The first step in understanding and studying diversity is to describe and name each species. To be useful, each scientific name must refer to one and only one species, and everyone must use the same name for that species. But what kind of name should be used? Common names can be confusing, because they vary among languages and from place to place. The animal in **Figure 18–1,** for example, can be called a cougar, a puma, a panther, or a mountain lion. Furthermore, different species may share a common name. In the United Kingdom, the word *buzzard* refers to a hawk, whereas in the United States, *buzzard* refers to a vulture.

Back in the eighteenth century, European scientists recognized that these kinds of common names were confusing, so they agreed to assign Latin or Greek names to each species. Unhappily, that didn't do much to clear up the confusion. Early scientific names often described species in great detail, so the names could be long. For example, the English translation of the scientific name of a tree might be "Oak with deeply divided leaves that have no hairs on their undersides and no teeth around their edges." It was also difficult to standardize these names, because different scientists focused on different characteristics. Many of these same characteristics can still be used to identify organisms when using dichotomous keys, as you can see in **Figure 18–2.**

FIGURE 18–1 Common Names You might recognize this as a cougar, a puma, a panther, or a mountain lion—all common names for the same animal. The scientific name for this animal is *Felis concolor.*

USING A DICHOTOMOUS KEY

FIGURE 18–2 A dichotomous key is used to identify organisms. It consists of a series of paired statements or questions that describe alternative possible characteristics of an organism. The paired statements usually describe the presence or absence of certain visible characteristics or structures. Each set of choices is arranged so that each step produces a smaller subset.

Suppose you found a leaf that you wanted to identify. The leaf looks like the one shown here. Use the key to identify this leaf.

Step	Leaf Characteristics	Tree
1a	Compound leaf (leaves divided into leaflets) . . . go to Step 2	
1b	Simple leaf (leaf not divided into leaflets) . . . go to Step 4	
2a	Leaflets all attached at a central point	Buckeye 
2b	Leaflets attached at several points . . . go to Step 3	
3a	Leaflets tapered with pointed tips	 Pecan
3b	Leaflets oval with rounded tips	Locust 
4a	Veins branched out from one central point . . . go to Step 5	
4b	Veins branched off main vein in middle of the leaf . . . go to Step 6	
5a	Heart-shaped leaf	Redbud 
5b	Star-shaped leaf 	Sweet gum
6a	Leaf with jagged edges	Birch
6b	Leaf with smooth edges	Magnolia 

Because your leaf is a simple leaf, you skip ahead to Step 4.

Continue reading the statements until you determine the identity of your leaf.

Because your leaf has jagged edges, you determine that it's from a birch tree.

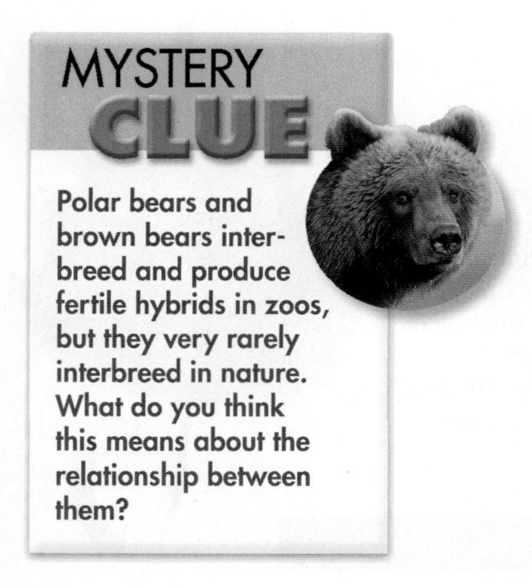

MYSTERY CLUE

Polar bears and brown bears interbreed and produce fertile hybrids in zoos, but they very rarely interbreed in nature. What do you think this means about the relationship between them?

Binomial Nomenclature In the 1730s, Swedish botanist Carolus Linnaeus, developed a two-word naming system called **binomial nomenclature.** **In binomial nomenclature, each species is assigned a two-part scientific name.** Scientific names are written in italic. The first word begins with a capital letter, and the second word is lowercased.

The polar bear in **Figure 18–3** is called *Ursus maritimus*. The first part of the name—*Ursus*—is the genus to which the organism belongs. A **genus** (plural: genera, JEN ur uh) is a group of similar species. The genus *Ursus* contains five other species of bears, including *Ursus arctos*, the brown bear or "grizzly."

The second part of a scientific name—in these examples, *maritimus* or *arctos*—is unique to each species. A species, remember, is generally defined as a group of individuals capable of interbreeding and producing fertile offspring. The species name is often a description of an important trait or the organism's habitat. The Latin word *maritimus*, refers to the sea, because polar bears often live on pack ice that floats in the sea.

In Your Notebook *The word* binomial *means "having two names." How does this meaning apply to binomial nomenclature?*

Classifying Species Into Larger Groups In addition to naming organisms, biologists also try to organize, or classify, living and fossil species into larger groups that have biological meaning. In a useful classification system, organisms in a particular group are more similar to one another than they are to organisms in other groups. The science of naming and grouping organisms is called **systematics** (sis tuh MAT iks). **The goal of systematics is to organize living things into groups that have biological meaning.** Biologists often refer to these groups as **taxa** (singular: taxon).

Whether you realize it or not, you use classification systems all the time. You may, for example, talk about "teachers" or "mechanics." Sometimes you refer to a smaller, more specific group, such as "biology teachers" or "auto mechanics." When you do this, you refer to these groups using widely accepted names and characteristics that many people understand. In the same way, when you hear the word *bird*, you immediately think of an animal with wings and feathers.

FIGURE 18–3 Binomial Nomenclature The scientific name of the polar bear is *Ursus maritimus*, which means "marine bear." The scientific name of the red maple is *Acer rubrum*. The genus *Acer* consists of all maple trees. The species *rubrum* describes the red maple's color.

Quick Lab
GUIDED INQUIRY

Classifying Fruits

1. Obtain five different fruits.
2. Use a knife to cut each fruit open and examine its structure. **CAUTION:** *Be careful with sharp instruments. Do not eat any of the cut fruits.*

3. Construct a table with five rows and four columns. Label each row with the name of a different fruit.
4. Examine the fruits, and choose four characteristics that help you tell the fruits apart. Label the columns in your table with the names of these characteristics.
5. Record a description of each fruit in the table.

Analyze and Conclude

1. **Classify** Based on your table, which fruits most closely resemble one another?

The Linnaean Classification System

How did Linnaeus group species into larger taxa?

In addition to creating the system of binomial nomenclature, Linnaeus also developed a classification system that organized species into taxa that formed a hierarchy or set of ordered ranks. Linnaeus's original system had just four levels. **Over time, Linnaeus's original classification system expanded to include seven hierarchical taxa: species, genus, family, order, class, phylum, and kingdom.**

We've already discussed the two smallest categories, species and genus. Now let's work our way up to the rank of kingdom by examining how camels are classified. The scientific name of a camel with two humps is *Camelus bactrianus.* (Bactria was an ancient country in Asia.) As you can see in **Figure 18–5,** the genus *Camelus* also includes another species, *Camelus dromedarius,* the dromedary, which has only one hump. In deciding how to place organisms into these larger taxa, Linnaeus grouped species according to anatomical similarities and differences.

FIGURE 18–4 Carolus Linnaeus

▶ ***Family*** The South American llama bears some resemblance to Bactrian camels and dromedaries. But the llama is more similar to other South American species than it is to European and Asian camels. Therefore, llamas are placed in a different genus, *Lama;* their species name is *Lama glama.* Several genera that share many similarities, like *Camelus* and *Lama,* are grouped into a larger category, the **family**—in this case, Camelidae.

▶ ***Order*** Closely related families are grouped into the next larger rank—an **order.** Camels and llamas (family Camelidae) are grouped with several other animal families, including deer (family Cervidae) and cattle (family Bovidae), into the order Artiodactyla, hoofed animals with an even number of toes.

BUILD Vocabulary

MULTIPLE MEANINGS The words **family, order, class,** and **kingdom** all have different meanings in biology than they do in common usage. For example, in systematics, a *family* is a group of genera. In everyday usage, a *family* is a group of people who are related to one another. Use a dictionary to find the common meanings of *order, class,* and *kingdom.*

▶ ***Class*** Similar orders, in turn, are grouped into the next larger rank, a **class.** The order Artiodactyla is placed in the class Mammalia, which includes all animals that are warmblooded, have body hair, and produce milk for their young.

▶ ***Phylum*** Classes are grouped into a **phylum.** A phylum includes organisms that are different but share important characteristics. The class Mammalia is grouped with birds (class Aves), reptiles (class Reptilia), amphibians (class Amphibia), and all classes of fish into the phylum Chordata. These organisms share important body-plan features, among them a nerve cord along the back.

▶ ***Kingdom*** The largest and most inclusive of Linnaeus's taxonomic categories is the **kingdom.** All multicellular animals are placed in the kingdom Animalia.

FIGURE 18–5 From Species to Kingdom This illustration shows how a Bactrian camel, *Camelus bactrianus*, is grouped within each taxonomic category. Only some representative organisms are illustrated for each taxon above the genus level. **Interpret Visuals** ***What phylum does*** **Camelus bactrianus** ***belong to?***

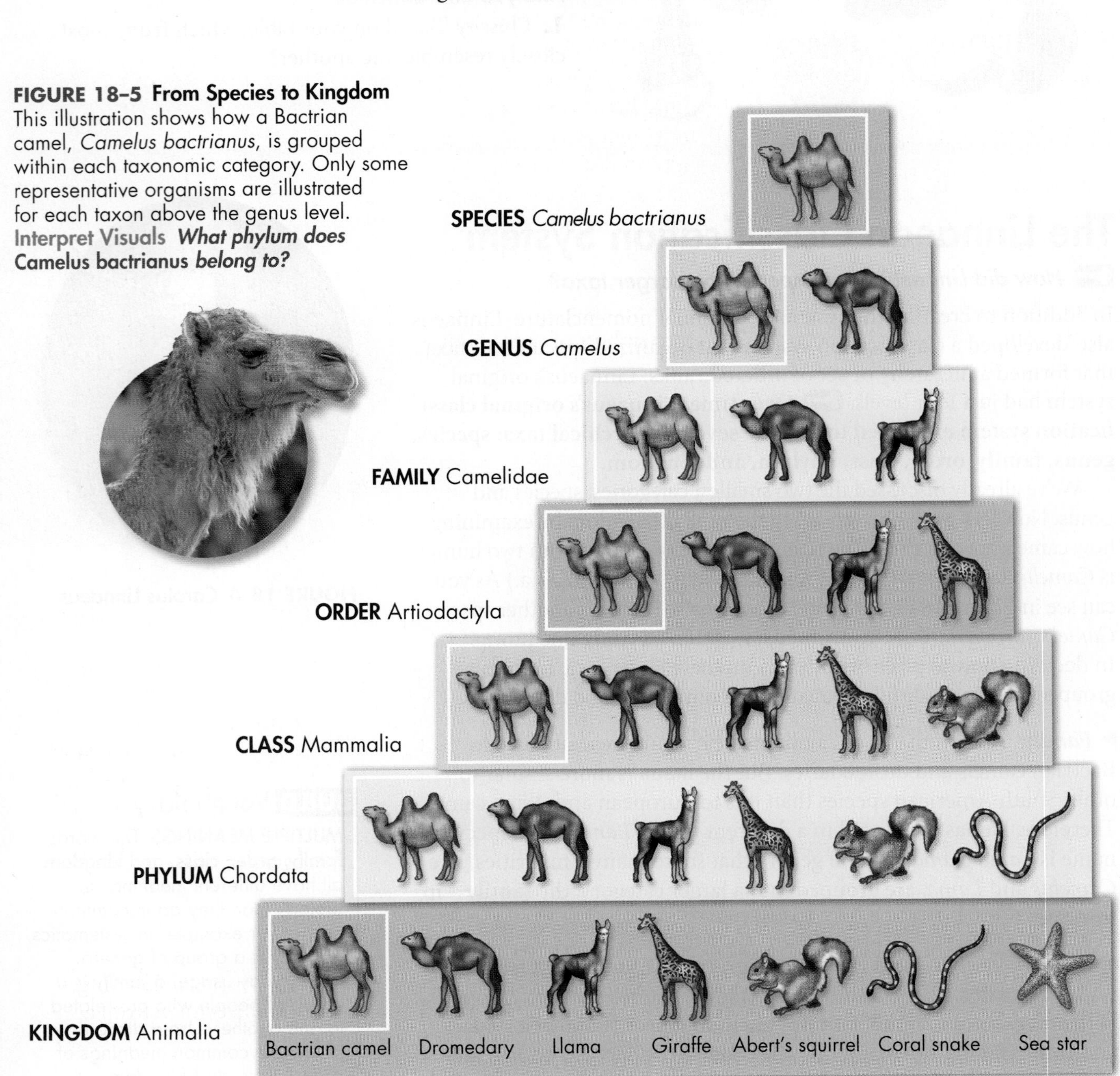

Problems With Traditional Classification In a sense, members of a species determine which organisms belong to that species by deciding with whom they mate and produce fertile offspring. There is thus a "natural" definition of species. Researchers, on the other hand, define Linnaean ranks above the level of species. Because, over time, systematists have emphasized a variety of characteristics, some of these groups have been defined in different ways at different times.

For example, Linnaeus's strategy of classifying organisms according to visible similarities and differences seems simple at first. But how should scientists decide which similarities and differences are most important? If you lived in Linnaeus's time, for example, how would you have classified the animals shown in **Figure 18–6**? Adult barnacles and limpets live attached to rocks and have similar-looking shells. Adult crabs look quite unlike both barnacles and limpets. Based on these features, would you place limpets and barnacles together, and crabs in a different group? As biologists attempted to classify more organisms over time, these kinds of questions arose frequently.

Linnaeus was a good scientist, and he chose his characteristics carefully. Many of his groups are still valid under modern classification schemes. But Linnaeus worked more than a century before Darwin published his ideas about descent with modification. Modern systematists apply Darwin's ideas to classification and try to look beyond simple similarities and differences to ask questions about evolutionary relationships. Linnaeus grouped organisms strictly according to similarities and differences. Scientists today try to assign species to a larger group in ways that reflect how closely members of those groups are related to each other.

FIGURE 18–6 Barnacles, Limpets, and Crabs Problems can arise when species are classified based on easily observed traits. Look closely at the barnacles (top), the limpets (bottom), and the crab (left). Notice their similarities and differences. **Compare and Contrast** ***Which animals seem most alike? Why?***

18.1 Assessment

Review Key Concepts

1. **a. Review** Identify two goals of systematics.
   b. Explain Why do the common names of organisms—like *daisy* or *mountain lion*—often cause problems for scientists?
   c. Classify The scientific name of the sugar maple is *Acer saccharum.* What does each part of the name designate?
2. **a. Review** List the ranks in the Linnaean system of classification, beginning with the smallest.
   b. Explain In which group of organisms are the members more closely associated—all of the organisms in the same kingdom or all of the organisms in the same order? Explain your answer.
   c. Apply Concepts What do scientists mean when they say that species is the only "natural" rank in classification?

Apply the Big idea

Unity and Diversity of Life

3. Which category has more biological meaning—all brown birds or all birds descended from a hawklike ancestor? Why?

18.2 Modern Evolutionary Classification

Key Questions

What is the goal of evolutionary classification?

What is a cladogram?

How are DNA sequences used in classification?

Vocabulary

phylogeny
clade
monophyletic group
cladogram
derived character

Taking Notes

Outline Make an outline of this lesson using the green headings as the main topics and the blue headings as subtopics. As you read, fill in details under each heading.

BUILD Vocabulary

WORD ORIGINS The word **cladogram** comes from two Greek words: *klados*, meaning "branch," and *gramma*, meaning "something that is written or drawn." A cladogram is an evolutionary diagram with a branching pattern.

THINK ABOUT IT Darwin's ideas about a "tree of life" suggests a new way to classify organisms—not just based on similarities and differences, but instead based on evolutionary relationships. Under this system, taxa are arranged according to how closely related they are. When organisms are rearranged in this way, some of the old Linnaean ranks fall apart. For example, the Linnaean class reptilia isn't valid unless birds are included—which means birds are reptiles! And not only are birds reptiles, they're also dinosaurs! Wondering why? To understand, we need to look at the way evolutionary classification works.

Evolutionary Classification

What is the goal of evolutionary classification?

The concept of descent with modification led to the study of **phylogeny** (fy LAHJ uh nee)—the evolutionary history of lineages. Advances in phylogeny, in turn, led to phylogenetic systematics. **The goal of phylogenetic systematics, or evolutionary classification, is to group species into larger categories that reflect lines of evolutionary descent, rather than overall similarities and differences.**

Common Ancestors

Phylogenetic systematics places organisms into higher taxa whose members are more closely related to one another than they are to members of any other group. The larger a taxon is, the farther back in time all of its members shared a common ancestor. This is true all the way up to the largest taxa.

Clades

Classifying organisms according to these rules places them into groups called clades. A **clade** is a group of species that includes a single common ancestor and all descendants of that ancestor—living and extinct. How are clades different from Linnaean taxa? A clade must be a monophyletic (mahn oh fy LET ik) group. A **monophyletic group** includes a single common ancestor and *all* of its descendants.

Some groups of organisms defined before the advent of evolutionary classification are monophyletic. Some, however, are paraphyletic, meaning that the group includes a common ancestor but excludes one or more groups of descendants. These groups are invalid under evolutionary classification.

In Your Notebook *In your own words, explain what makes a clade monophyletic or paraphyletic.*

Cladograms

What is a cladogram?

Modern evolutionary classification uses a method called cladistic analysis. Cladistic analysis compares carefully selected traits to determine the order in which groups of organisms branched off from their common ancestors. This information is then used to link clades together into a diagram called a **cladogram.** **A cladogram links groups of organisms by showing how evolutionary lines, or lineages, branched off from common ancestors.**

Building Cladograms To understand how cladograms are constructed, think back to the process of speciation. A speciation event, in which one ancestral species splits into two new ones, is the basis of each branch point, or node, in a cladogram. That node represents the last point at which the two new lineages shared a common ancestor. As shown in part 1 of **Figure 18–7,** a node splits a lineage into two separate lines of evolutionary ancestry.

Each node represents the last point at which species in lineages above the node shared a common ancestor. The bottom, or "root" of a cladogram, represents the common ancestor shared by all of the organisms in the cladogram. A cladogram's branching patterns indicate degrees of relatedness among organisms. Look at part 2 of **Figure 18–7.** Because lineages 3 and 4 share a common ancestor more recently with each other than they do with lineage 2, you know that lineages 3 and 4 are more closely related to each other than either is to lineage 2. The same is true for lineages 2, 3, and 4. In terms of ancestry, they are more closely related to each other than any of them is to lineage 1. Look at the cladogram shown in part 3 of **Figure 18–7.** Does it surprise you that amphibians are more closely related to mammals than they are to ray-finned fish? In terms of ancestry, it's true!

FIGURE 18–7 Building a Cladogram A cladogram shows relative degrees of relatedness among lineages.

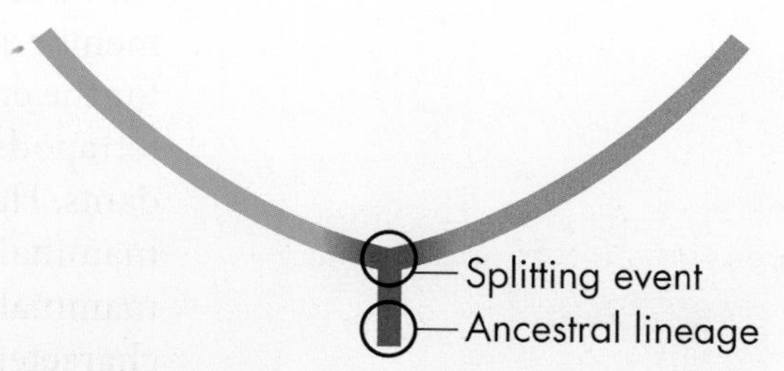

❶ Cladograms are diagrams showing how evolutionary lines, or lineages, split from each other over time. This diagram shows a single ancestral lineage splitting into two. The point of splitting is called a "node" in the cladogram.

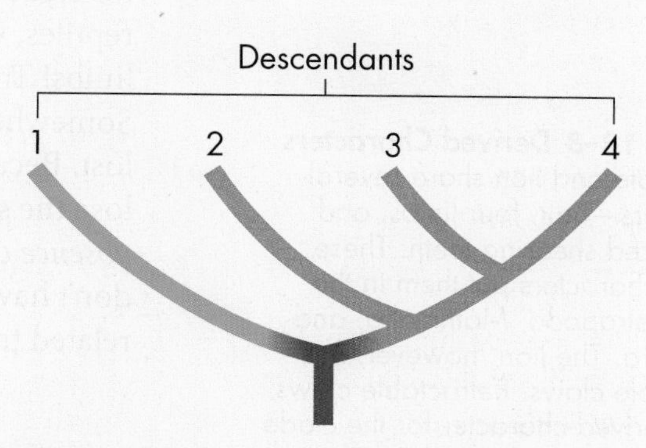

❷ How recently lineages share a common ancestor reflect how closely the lineages are related to one another. Here, lineages 3 and 4 are each more closely related to each other than any of them is to any other lineage.

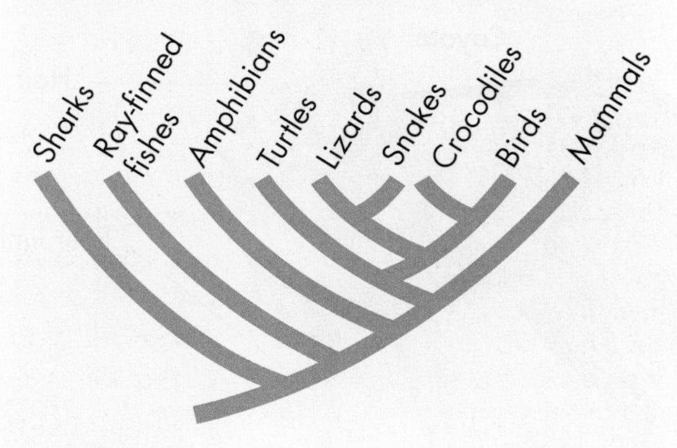

❸ This cladogram shows the evolutionary relationships among vertebrates, animals with backbones.

Derived Characters In contrast to Linnaean taxonomy, cladistic analysis focuses on certain kinds of characters, called derived characters, when assigning organisms into clades. A **derived character** is a trait that arose in the most recent common ancestor of a particular lineage and was passed along to its descendants.

Whether or not a character is derived depends on the level at which you're grouping organisms. Here's what we mean. **Figure 18–8** shows several traits that are shared by coyotes and lions, both members of the clade Carnivora. Four limbs is a derived character for the entire clade Tetrapoda because the common ancestor of all tetrapods had four limbs, and this trait was passed to its descendants. Hair is a derived character for the clade Mammalia. But for mammals, four limbs is *not* a derived character—if it were, only mammals would have that trait. Nor is four limbs or hair a derived character for clade Carnivora. Specialized shearing teeth, however, is. What about retractable claws? This trait is found in lions but not in coyotes. Thus, retractable claws is a derived character for the clade Felidae—also known as cats.

Losing Traits Notice above that four limbs is a derived character for clade Tetrapoda. But what about snakes? Snakes are definitely reptiles, which are tetrapods. But snakes certainly don't have four limbs! The *ancestors* of snakes, however, did have four limbs. Somewhere in the lineage leading to modern snakes, that trait was lost. Because distantly related groups of organisms can sometimes lose the same character, systematists are cautious about using the *absence* of a trait as a character in their analyses. After all, whales don't have four limbs either, but snakes are certainly more closely related to other reptiles than they are to whales.

FIGURE 18–8 Derived Characters The coyote and lion share several characters—hair, four limbs, and specialized shearing teeth. These shared characters put them in the clades Tetrapoda, Mammalia, and Carnivora. The lion, however, has retractable claws. Retractable claws is the derived character for the clade Felidae.

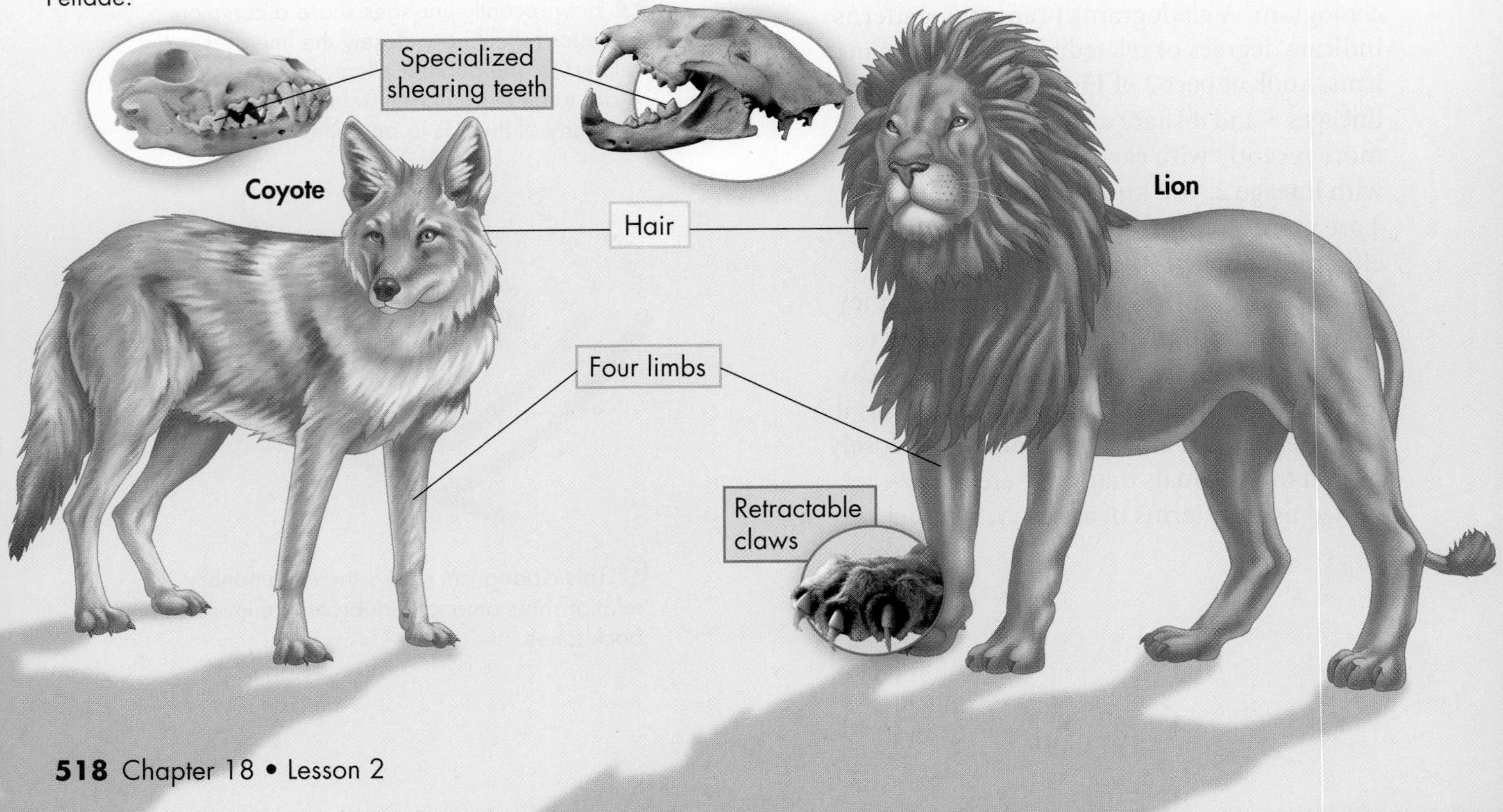

VISUAL SUMMARY

INTERPRETING A CLADOGRAM

FIGURE 18–9 This cladogram shows the evolutionary history of cats. In a cladogram, all organisms in a clade share a set of derived characters. Notice that smaller clades are nested within larger clades. **Interpret Visuals** ***For which clade in this cladogram is an amniotic egg a derived character?***

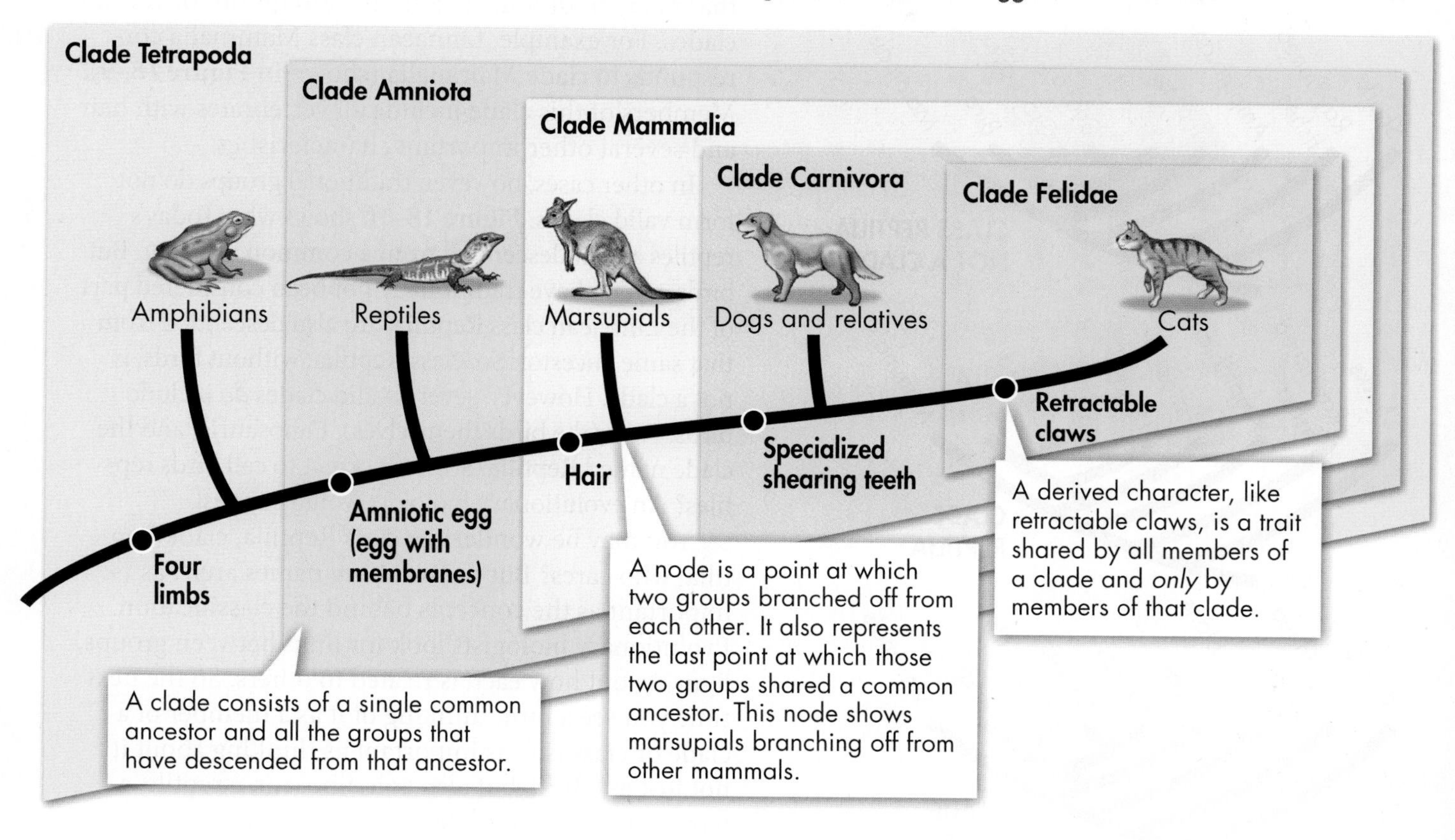

Interpreting Cladograms We can now put this information together to "read" a cladogram. **Figure 18–9** shows a simplified phylogeny of the cat family. The lowest node represents the last common ancestor of all four-limbed animals—members of the clade Tetrapoda. The forks in this cladogram show the order in which various groups branched off from the tetrapod lineage over the course of evolution. The positions of various characters in the cladogram reflect the order in which those characteristics arose in this lineage. In the lineage leading to cats, for example, specialized shearing teeth evolved before retractable claws. Furthermore, each derived character listed along the main trunk of the cladogram defines a clade. Hair, for example, is a defining character for the clade Mammalia. Retractable claws is a derived character shared only by the clade Felidae. Derived characters that occur "lower" on the cladogram than the branch point for a clade are not derived for that particular clade. Hair, for example, is not a derived character for the clade Carnivora.

In Your Notebook *List the derived characters in* ***Figure 18–9*** *and explain which groups in the cladogram have those characters.*

FIGURE 18–10 Clade or Not? A clade includes an ancestral species and all its descendants. Linnaean class Reptilia is not a clade because it does not include modern birds. Because it leaves this descendant group out, the class is paraphyletic. Clades Reptilia and Aves, however, are monophyletic and, therefore, valid clades. **Apply Concepts** ***Would a group that included all of clade Reptilia plus amphibians be monophyletic or paraphyletic? Explain.***

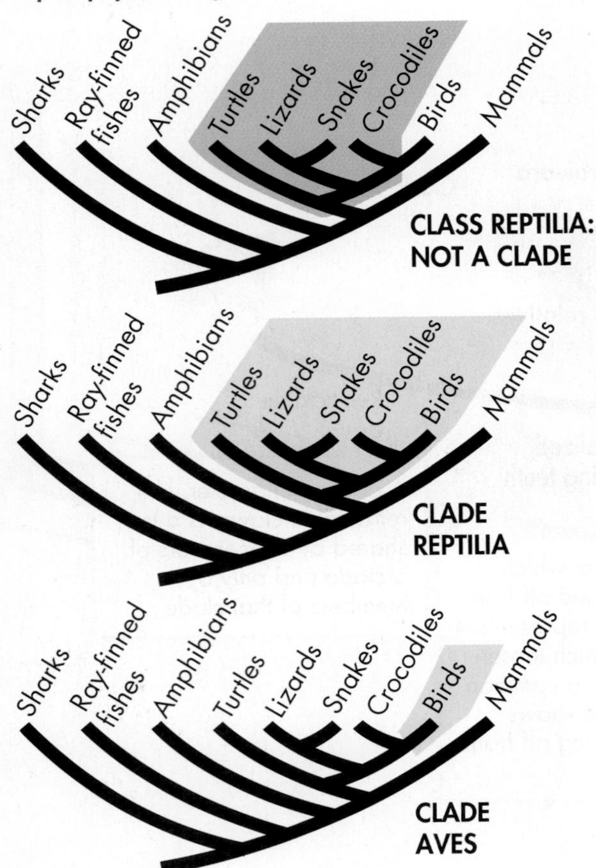

Clades and Traditional Taxonomic Groups Which of the Linnaean groupings form clades, and which do not? Remember that a true clade must be monophyletic, which means that it contains an ancestral species and *all* of its descendants—it can't leave any out. It also cannot include any species which are not descendants of that original ancestor. Cladistic analysis shows that many traditional taxonomic groups do form valid clades. For example, Linnaean class Mammalia corresponds to clade Mammalia (shown in **Figure 18–9**). Members of this clade include all vertebrates with hair and several other important characteristics.

In other cases, however, traditional groups do not form valid clades. **Figure 18–10** shows why. Today's reptiles are all descended from a common ancestor. But birds, which have traditionally not been considered part of the Linnaean class Reptilia, are also descended from that same ancestor. So, class Reptilia, without birds, is not a clade. However, several valid clades *do* include birds: Aves (the birds themselves), Dinosauria, and the clade named Reptilia. So, is it correct to call birds reptiles? An evolutionary biologist would say yes!

You may be wondering: class Reptilia, clade Reptilia, who cares? But the resulting names aren't as important as the concepts behind the classification. Evolutionary biologists look for links between groups, figuring out how each is related to others. So the next time you see a bird, thinking of it as a member of a clade or class isn't as important as thinking about it not just as a bird, but also as a dinosaur, a reptile, a tetrapod, and a chordate.

Quick Lab
GUIDED INQUIRY

Constructing a Cladogram

1. Identify the organism in the table that is least closely related to the others.

2. Use the information in the table to construct a cladogram of these animals.

Derived Characters in Organisms			
Organism	**Derived Character**		
	Backbone	**Legs**	**Hair**
Earthworm	Absent	Absent	Absent
Trout	Present	Absent	Absent
Lizard	Present	Present	Absent
Human	Present	Present	Present

Analyze and Conclude

1. **Interpret Tables** What trait separates the least closely related animal from the other animals?

2. **Apply Concepts** Do you have enough information to determine where a frog should be placed on the cladogram? Explain your answer.

3. **Draw Conclusions** Does your cladogram indicate that lizards and humans share a more recent common ancestor than either does with an earthworm? Explain your answer.

DNA in Classification

How are DNA sequences used in classification?

The examples of cladistic analysis we've discussed so far are based largely on physical characteristics like skeletons and teeth. But the goal of modern systematics is to understand the evolutionary relationships of all life on Earth—from bacteria to plants, snails, and apes. How can we devise hypotheses about the common ancestors of organisms that appear to have no physical similarities?

Genes as Derived Characters Remember that all organisms carry genetic information in their DNA passed on from earlier generations. A wide range of organisms share a number of genes and show important homologies that can be used to determine evolutionary relationships. For example, all eukaryotic cells have mitochondria, and all mitochondria have their own genes. Because all genes mutate over time, shared genes contain differences that can be treated as derived characters in cladistic analysis. For that reason, similarities and differences in DNA can be used to develop hypotheses about evolutionary relationships. **In general, the more derived genetic characters two species share, the more recently they shared a common ancestor and the more closely they are related in evolutionary terms.**

New Techniques Suggest New Trees The use of DNA characters in cladistic analysis has helped to make evolutionary trees more accurate. Consider, for example, the birds in **Figure 18–11.** The hooded vulture from Africa in the top photograph looks a lot like the American vulture in the middle photograph. Both were traditionally classified in the falcon clade. But American vultures have a peculiar behavior: When they get overheated, they urinate on their legs, relying on evaporation to cool them down. Storks share this behavior, while hooded vultures and other vultures from Africa do not. Could the behavior be a clue to the real relationships between these birds?

Biologists solved the puzzle by analyzing DNA from all three species. Molecular analysis showed that the DNA from American vultures is more similar to the DNA of storks than to the DNA of African vultures. DNA evidence therefore suggests that American vultures and storks share a more recent common ancestor than the American and African vultures do. Molecular analysis is a powerful tool that is now routinely used by taxonomists to supplement data from anatomy and answer questions like these.

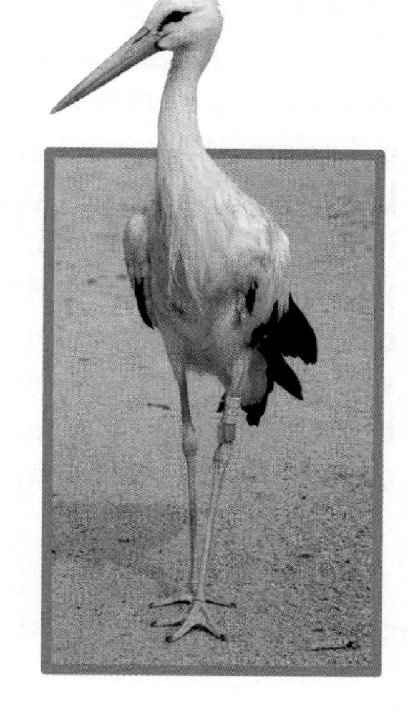

FIGURE 18–11 DNA and Classification Scientists use similarities in the genetic makeup of organisms to help determine classification. Traditionally African vultures and American vultures were classified together in the falcon family. But DNA analysis suggests that American vultures are actually more closely related to storks.

DNA comparisons show that some populations of brown bears are more closely related to polar bears than they are to other brown bears. What do you think this means for the classification of polar bears?

Often, scientists use DNA evidence when anatomical traits alone can't provide clear answers. Giant pandas and red pandas, for example, have given taxonomists a lot of trouble. These two species share anatomical similarities with both bears and raccoons, and both of them have peculiar wrist bones that work like a human thumb. DNA analysis revealed that the giant panda shares a more recent common ancestor with bears than with raccoons. DNA places red pandas, however, outside the bear clade. So pandas have been reclassified, placed with other bears in the clade Ursidae, as shown in **Figure 18–12.** What happened to the red panda? It is now placed in a different clade that also includes raccoons and other organisms such as seals and weasels.

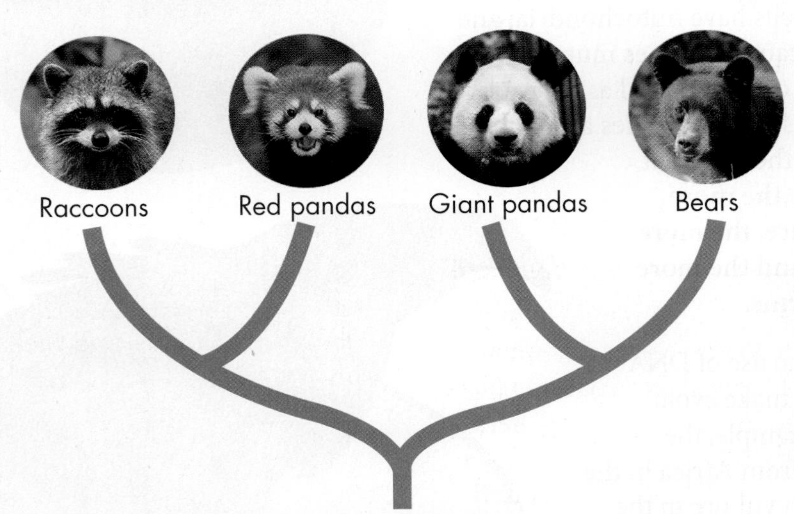

FIGURE 18–12 Classification of Pandas Biologists used to classify the red panda and the giant panda together. However, cladistic analysis using DNA suggests that the giant panda shares a more recent common ancestor with bears than with either red pandas or raccoons.

18.2 Assessment

Review Key Concepts

1. a. Explain How does evolutionary classification differ from traditional classification?
   b. Apply Concepts To an evolutionary taxonomist, what determines whether two species are in the same genus?
2. a. Explain What is a derived character?
   b. Interpret Diagrams Along any one lineage, what do the locations of derived characters on a cladogram show? In your answer, use examples from **Figure 18–9.**
3. a. Review How do taxonomists use the DNA sequences of species to determine how closely two species are related?
   b. Relate Cause and Effect Explain why the classification of American vultures has changed.

VISUAL THINKING

4. Examine the cladogram.
   a. Interpret Diagrams Which groups—X and Y, or X , Y, and Z—have the most recent common ancestor?
   b. Infer Which species—X and Y, or X and Z—share more derived characters?

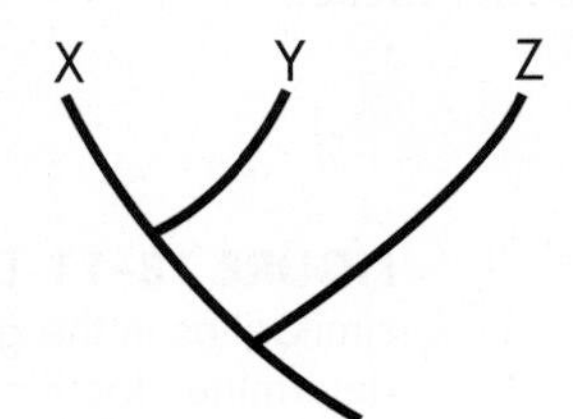

18.3 Building the Tree of Life

THINK ABOUT IT The process of identifying and naming all known organisms, living and extinct, is a huge first step toward the goal of systematics. Yet naming organisms is only part of the work. The real challenge is to group everything, from bacteria to dinosaurs to blue whales, in a way that reflects their evolutionary relationships. Over the years, new information and ways of studying organisms have produced major changes in Linnaeus's original scheme for organizing living things.

Key Questions

What are the six kingdoms of life as they are now identified?

What does the tree of life show?

Vocabulary

domain • Bacteria • Archaea • Eukarya

Taking Notes

Concept Map As you read, construct a concept map describing the characteristics of the three domains.

Changing Ideas About Kingdoms

What are the six kingdoms of life as they are now identified?

During Linnaeus's time, the only known differences among living things were the fundamental characteristics that separated animals from plants. Animals were organisms that moved from place to place and used food for energy. Plants were green organisms that generally did not move and got their energy from the sun.

As biologists learned more about the natural world, they realized that Linnaeus's two kingdoms—Animalia and Plantae—did not reflect the full diversity of life. Classification systems have changed dramatically since Linnaeus's time, as shown in **Figure 18–13.** And hypotheses about relationships among organisms are still changing today as new data are gathered.

Kingdoms of Life, 1700s–1990s

<table>
<tr><th>First Introduced</th><th colspan="6">Names of Kingdoms</th></tr>
<tr><td>1700s</td><td colspan="5">Plantae</td><td>Animalia</td></tr>
<tr><td>Late 1800s</td><td colspan="3">Protista</td><td colspan="2">Plantae</td><td>Animalia</td></tr>
<tr><td>1950s</td><td colspan="2">Monera</td><td>Protista</td><td>Fungi</td><td>Plantae</td><td>Animalia</td></tr>
<tr><td>1990s</td><td>Eubacteria</td><td>Archaebacteria</td><td>Protista</td><td>Fungi</td><td>Plantae</td><td>Animalia</td></tr>
</table>

FIGURE 18–13 From Two to Six Kingdoms This diagram shows some of the ways in which organisms have been classified into kingdoms since the 1700s.

Comparing the Domains

The table in **Figure 18–14** compares the three domains and six kingdoms. Use the information in the table to answer the following questions.

1. Interpret Tables Which kingdom has cells that lack cell walls?

2. Interpret Tables Which domain contains multicellular organisms?

3. Compare and Contrast On the basis of information in the table, how are the members of domain Archaea similar to those of domain Bacteria? How are organisms in domain Archaea similar to those in domain Eukarya?

Five Kingdoms As researchers began to study microorganisms, they discovered that single-celled organisms were significantly different from plants and animals. At first all microorganisms were placed in their own kingdom, named Protista. Then yeasts and molds, along with mushrooms, were placed in their own kingdom, Fungi.

Later still, scientists realized that bacteria lack the nuclei, mitochondria, and chloroplasts found in other forms of life. All prokaryotes (bacteria) were placed in yet another new kingdom, Monera. Single-celled eukaryotic organisms remained in the kingdom Protista. This process produced five kingdoms: Monera, Protista, Fungi, Plantae, and Animalia.

Six Kingdoms By the 1990s, researchers had learned a great deal about the genetics and biochemistry of bacteria. That knowledge made clear that the organisms in kingdom Monera were actually two genetically and biochemically different groups. As a result, the monerans were separated into two kingdoms, Eubacteria and Archaebacteria, bringing the total number of kingdoms to six. **The six-kingdom system of classification includes the kingdoms Eubacteria, Archaebacteria, Protista, Fungi, Plantae, and Animalia.** This system of classification is shown in the bottom row of **Figure 18–13** on the previous page.

FIGURE 18–14 Three Domains Today organisms are grouped into three domains and six kingdoms. This table summarizes the key characteristics used to classify organisms into these major taxonomic groups.

Classification of Living Things

DOMAIN	Bacteria	Archaea	Eukarya			
KINGDOM	Eubacteria	Archaebacteria	"Protista"	Fungi	Plantae	Animalia
CELL TYPE	Prokaryote	Prokaryote	Eukaryote	Eukaryote	Eukaryote	Eukaryote
CELL STRUCTURES	Cell walls with peptidoglycan	Cell walls without peptidoglycan	Cell walls of cellulose in some; some have chloroplasts	Cell walls of chitin	Cell walls of cellulose; chloroplasts	No cell walls or chloroplasts
NUMBER OF CELLS	Unicellular	Unicellular	Most unicellular; some colonial; some multicellular	Most multicellular; some unicellular	Most multicellular: some green algae unicellular	Multicellular
MODE OF NUTRITION	Autotroph or heterotroph	Autotroph or heterotroph	Autotroph or heterotroph	Heterotroph	Autotroph	Heterotroph
EXAMPLES	*Streptococcus, Escherichia coli*	Methanogens, halophiles	*Amoeba, Paramecium,* slime molds, giant kelp	Mushrooms, yeasts	Mosses, ferns, flowering plants	Sponges, worms, insects, fishes, mammals

Three Domains Genomic analysis has revealed that the two main prokaryotic groups are even more different from each other, and from eukaryotes, than previously thought. So biologists established a new taxonomic category—the domain. A **domain** is a larger, more inclusive category than a kingdom. Under this system, there are three domains—domain Bacteria (corresponding to the kingdom Eubacteria); domain Archaea (which corresponds to the kingdom Archaebacteria); and domain Eukarya (kingdoms Fungi, Plantae, and Animalia, and the "Protista").

Why do we put quotations around about the old kingdom Protista? Well, scientists now recognize that this is a paraphyletic group. This means that there is no way to put all unicellular eukaryotes into a clade that contains a single common ancestor, all of its descendants, and only those descendants. Since only monophyletic groups are valid under evolutionary classification, we use quotations to show that this is not a true clade. A summary of the three-domain system is shown in **Figure 18–14.**

The Tree of All Life

What does the tree of life show?

Remember that modern evolutionary classification is a rapidly changing science with a difficult goal—to present all life on a single evolutionary tree. As evolutionary biologists study relationships among taxa, they regularly change not only the way organisms are grouped, but also sometimes the names of groups. Remember that cladograms are visual presentations of hypotheses about relationships, and not hard and fast facts. **The tree of life shows current hypotheses regarding evolutionary relationships among the taxa within the three domains of life.**

Domain Bacteria Members of the domain **Bacteria** are unicellular and prokaryotic. Their cells have thick, rigid walls that surround a cell membrane. The cell walls contain a substance known as peptidoglycan (PEP tih doh gly kun). These bacteria are ecologically diverse, ranging from free-living soil organisms to deadly parasites. Some photosynthesize, while others do not. Some need oxygen to survive, while others are killed by oxygen. This domain corresponds to the kingdom Eubacteria.

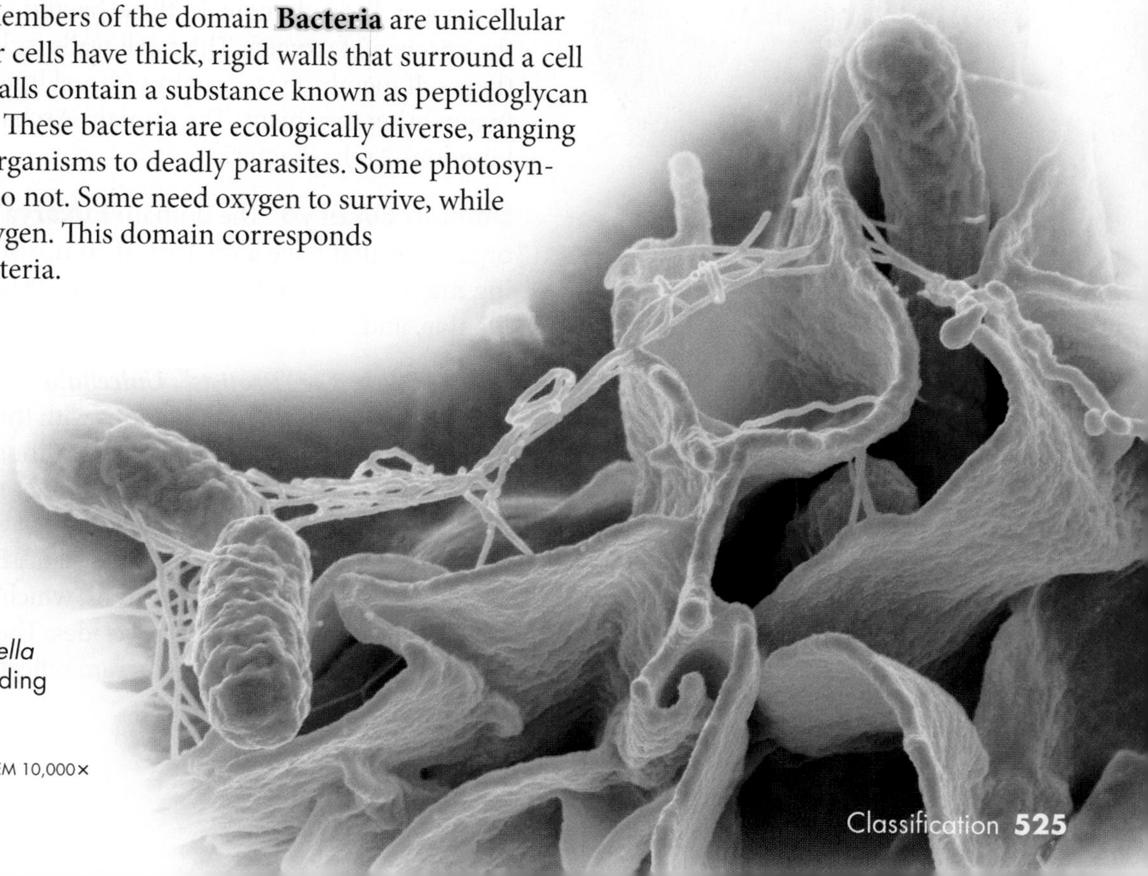

FIGURE 18–15 *Salmonella typhimurium* (green) invading human epithelial cells

SEM 10,000×

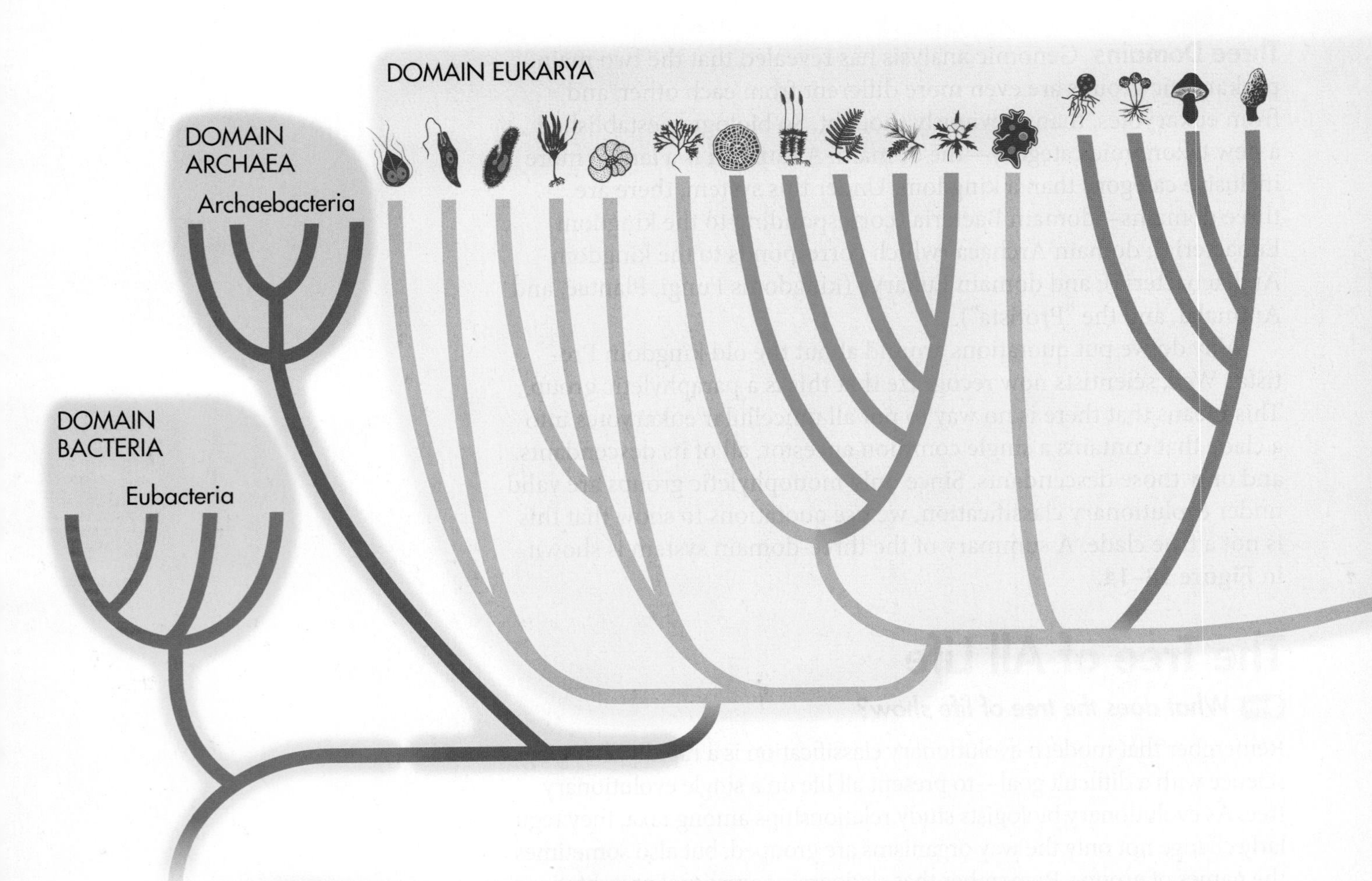

SEM 13,000×

FIGURE 18–16 ***Sulfolobus*** This member of the domain Archaea is found in hot springs and thrives in acidic and sulfur-rich environments.

Domain Archaea Also unicellular and prokaryotic, members of the domain **Archaea** (ahr KEE uh) live in some of the most extreme environments you can imagine—in volcanic hot springs, brine pools, and black organic mud totally devoid of oxygen. Indeed, many of these bacteria can survive only in the absence of oxygen. Their cell walls lack peptidoglycan, and their cell membranes contain unusual lipids that are not found in any other organism. The domain Archaea corresponds to the kingdom Archaebacteria.

Domain Eukarya The domain **Eukarya** consists of all organisms that have a nucleus. It comprises the four remaining major groups of the six-kingdom system: "Protista," Fungi, Plantae, and Animalia.

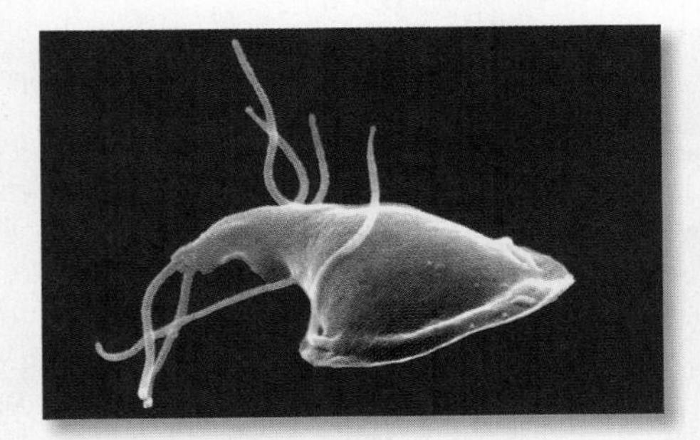

LM 900×

FIGURE 18–17 "Protists" "Protists" can live just about anywhere. *Giardia* is a parasitic freshwater ciliate.

► ***The "Protists": Unicellular Eukaryotes*** Recall that we are using quotations with this group to indicate that it is a paraphyletic group. Although some people still use the name "protists" to refer to these organisms, scientists who work with them have known for years that they do not form a valid clade. **Figure 18–18** reflects current cladistic analysis, which divides these organisms into at least five clades. The positions of these groups on the cladogram reflect its paraphyletic nature.

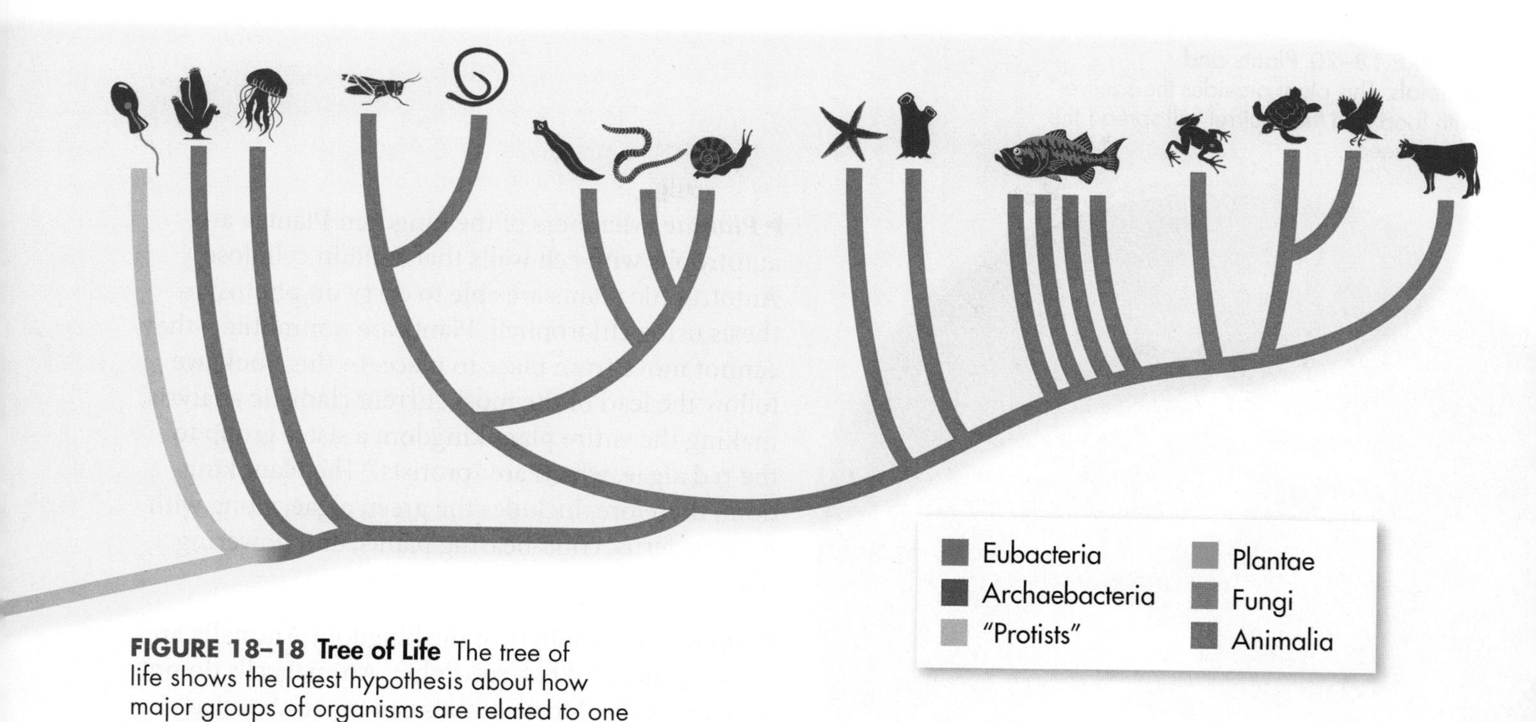

FIGURE 18–18 Tree of Life The tree of life shows the latest hypothesis about how major groups of organisms are related to one another. Note that both domain and kingdom designations are shown. **Classify** ***Which of the six kingdoms contains organisms that are not all in the same clade?***

Each group of "the eukaryotes formerly known as protists" is separate, and each shares closest common ancestors with other groups, rather than with each other. Most are unicellular, but one group, the brown algae, is multicellular. Some are photosynthetic, while others are heterotrophic. Some display characters that most closely resemble those of plants, fungi, or animals.

▶ ***Fungi*** Members of the kingdom Fungi are heterotrophs with cell walls containing chitin. Most feed on dead or decaying organic matter. Unlike other heterotrophs, fungi secrete digestive enzymes into their food source. After the digestive enzymes have broken down the food into smaller molecules, the fungi absorb the small molecules into their bodies. Mushrooms and other recognizable fungi are multicellular. Some fungi—yeasts, for example—are unicellular.

In Your Notebook *Explain why kingdom Protista is not valid under evolutionary classification.*

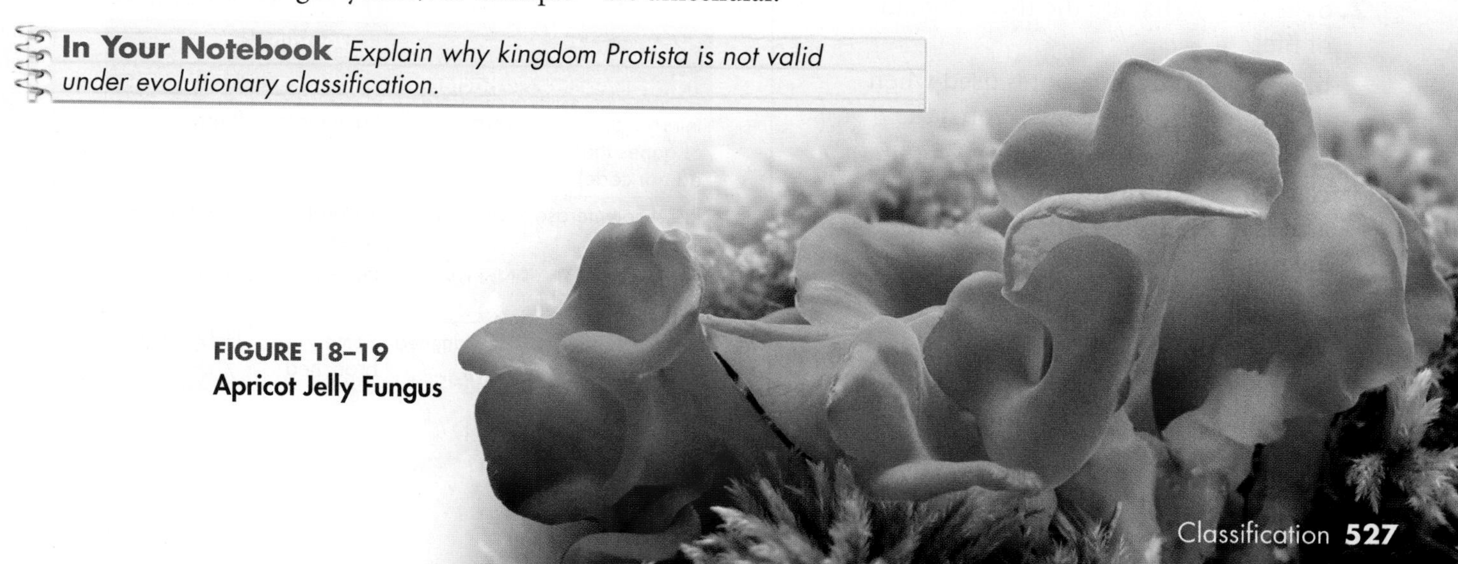

FIGURE 18–19
Apricot Jelly Fungus

FIGURE 18–20 Plants and Animals This plant provides the squirrel with food and the squirrel will spread the plant's seeds.

▶***Plantae*** Members of the kingdom Plantae are autotrophs with cell walls that contain cellulose. Autotrophic plants are able to carry on photosynthesis using chlorophyll. Plants are nonmotile—they cannot move from place to place. In this book, we follow the lead of the most current cladistic analysis, making the entire plant kingdom a sister group to the red algae, which are "protists." The plant kingdom, therefore, includes the green algae, along with mosses, ferns, cone-bearing plants, and flowering plants.

▶***Animalia*** Members of the kingdom Animalia are multicellular and heterotrophic. Animal cells do not have cell walls. Most animals can move about, at least for some part of their life cycle. As you will see in later chapters, there is incredible diversity within the animal kingdom, and many species of animals exist in nearly every part of the planet.

18.3 Assessment

Review Key Concepts

1. **a. Review** What are the six kingdoms of life as they are now identified?
   b. Explain Why did systematists establish the domain?
   c. Classify What were the monerans? Why did systematists split them into two kingdoms?
2. **a. Review** What are the three domains of life?
   b. Explain Why are quotes used when describing the kingdom "Protista"?
   c. Predict Do you think the tree of life cladogram will always stay the same as it is in **Figure 18–18?** Explain your answer.

ANALYZING DATA

3. The table compares some molecular characteristics of organisms in the three domains.
   a. Interpret Tables Which domains have unbranched lipids in their cell membranes?
   b. Interpret Tables Which domain has just one type of RNA polymerase?
   c. Analyze Data On the basis of this table, how are archaea different from bacteria?

Molecular Characteristic	Domain		
	Bacteria	Archaea	Eukarya
Introns (parts of genes that do not code)	Rare	Sometimes present	Present
RNA polymerase	One type	Several types	Several types
Histones found with DNA	Not present	Present	Present
Lipids in cell membrane	Unbranched	Some branched	Unbranched

WHST.9-10.8 Research to Build and Present Knowledge, WHST.9-10.10 Range of Writing

Technology & BIOLOGY

Bar-Coding Life

Until recently, classification has been a time-consuming process. A new project hopes to make identifying species as simple as scanning a supermarket bar code. It combines DNA sequencing with miniature computers, data processing, and the Internet.

To make this work, researchers picked a segment of DNA that all animals carry, the mitochondrial cytochrome oxidase (CO1) gene. (A chloroplast gene will probably be used for plants). Each base in the DNA sequence of CO1 is shown as a color-coded stripe, making it easy to spot differences between the barcodes of two specimens. The results are stored in a database.

WRITING **Learn more about DNA bar-coding on the Internet. Then write a paragraph describing another possible use for the DNA bar-coding technology.**

Closely related species have similar bar codes. Species that are not closely related have bar codes that are very different from one another.

In the future, a researcher will be able to take a tiny sample of tissue or hair, analyze it using a portable device, and get a report on closest matches. Recent versions of this software even use maps to show where similar specimens have been collected before.

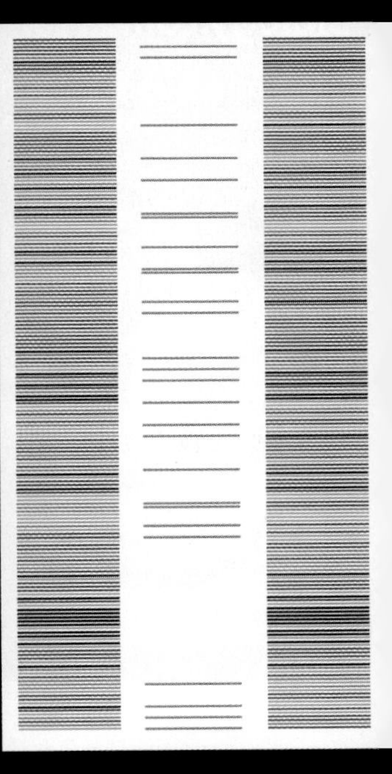

▶ The bar code on the left belongs to the hermit thrush and the bar code on the right belongs to the American robin. Differences between the two bar codes, shown as lines in the middle column, show the genetic distance between the two species.

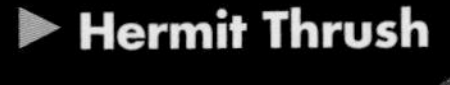

▶ **Hermit Thrush**

▶ **American Robin**

NGSS Smart Guide The Evolution of Populations

Disciplinary Core Idea LS4.A Evidence of Common Ancestry and Diversity: What evidence shows that different species are related? The goal of biologists who classify organisms is to construct a tree of life that shows how all organisms are related to one another.

18.1 Finding Order in Diversity

- In binomial nomenclature, each species is assigned a two-part scientific name.
- The goal of systematics is to organize living things into groups that have biological meaning.
- Over time, Linnaeus's original classification system expanded to include seven hierarchical taxa: species, genus, family, order, class, phylum, and kingdom.

binomial nomenclature 512 • genus 512 • systematics 512 • taxon 512 • family 513 • order 513 • class 514 • phylum 514 • kingdom 514

Biology.com

Untamed Science Video Hop on board with the Untamed Science crew to find out how organisms are classified.

Art in Motion View a short animation that explains how to use a dichotomous key.

Interactive Art Build your understanding of cladograms with this animation.

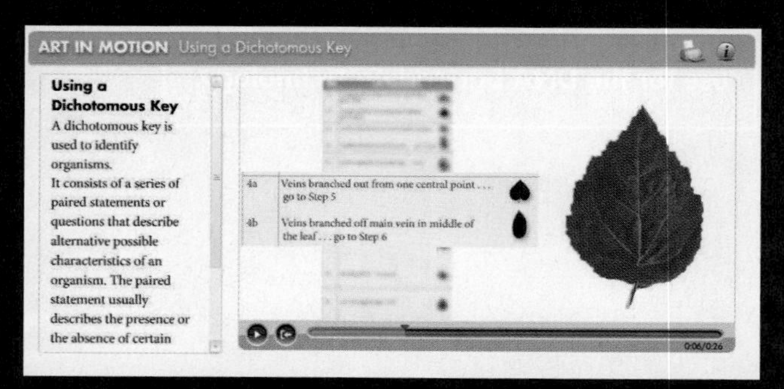

CHECKING YOUR *Scientific Literacy*

Refer to the lesson content and digital assets below as you prepare for your chapter assessment. Then, evaluate your understanding of classification by answering these questions.

1. **Scientific and Engineering Practice** **Analyzing and Interpreting Data** Use Figure18-9 *Interpreting a Cladogram* to point out some key similarities and differences in dogs and cats.

2. **Crosscutting Concept** **Patterns** Two mammals known as the ocelot and caracal, are classified as members of clade Felidae along with house cats. What similarity can you infer that ocelots and caracals share?

3. **Integration of Knowledge and Ideas** Summarize the information shown visually in Figure 18-9 about derived characters for mammals, carnivores, and felines.

4. **STEM** You are an engineer tasked with designing and building a tree of life display for a natural history museum. The display's goal is to show the evolutionary relationships among mammals in a clear, informative, and engaging way. Describe the materials, resources, and constraints you need to consider in your design and construction.

18.2 Modern Evolutionary Classification

- The goal of phylogenetic systematics, or evolutionary classification, is to group species into larger categories that reflect lines of evolutionary descent, rather than overall similarities and differences.
- A cladogram links groups of organisms by showing how evolutionary lines, or lineages, branched off from common ancestors.
- In general, the more derived genetic characters two species share, the more recently they shared a common ancestor and the more closely they are related in evolutionary terms.

phylogeny 516 • clade 516 • monophyletic group 516 • cladogram 517 • derived character 518

Biology.com

Interactive Art Build your understanding of cladograms with this animation.

18.3 Building the Tree of Life 

- The six-kingdom system of classification includes the kingdoms Eubacteria, Archaebacteria, Protista, Fungi, Plantae, and Animalia.
- The tree of life shows current hypotheses regarding evolutionary relationships among the taxa within the three domains of life.

domain 525 • Bacteria 525 • Archaea 526 • Eukarya 526

Biology.com

Art Review How well do you know the characteristics of the three domains? Test yourself in this activity.

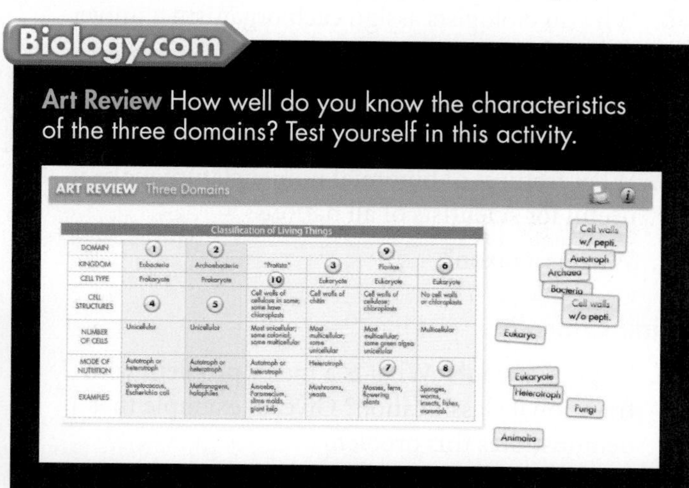

Design Your Own Lab

Dichotomous Keys This chapter lab is available in your lab manual and at **Biology.com**

18 Assessment

18.1 Finding Order in Diversity

Understand Key Concepts

1. The science of naming and grouping organisms is called
   a. anatomy.
   b. systematics.
   c. botany.
   d. paleontology.

2. Solely from its name, you know that *Rhizopus nigricans* must be
   a. a plant.
   b. an animal.
   c. in the genus *Nigricans.*
   d. in the genus *Rhizopus.*

3. A useful classification system does NOT
   a. show relationships.
   b. reveal evolutionary trends.
   c. use different scientific names for the same organism.
   d. change the taxon of an organism based on new data.

4. In Linnaeus's system of classifying organisms, orders are grouped together into
   a. classes.
   b. species.
   c. families.
   d. genera.

5. The largest and most inclusive of the Linnaean taxonomic ranks is the
   a. kingdom.
   b. order.
   c. phylum.
   d. domain.

6. Why do biologists assign each organism a universally accepted name?

7. Why is species the only Linnaean rank defined "naturally"?

8. What features of binomial nomenclature make it useful for scientists of all nations?

9. What is a taxon?

Think Critically

10. **Apply Concepts** What is a major problem with traditional classification? Give an example that demonstrates this problem.

11. **Use Analogies** Why is it important for a supermarket to have a classification scheme for displaying the foods that it sells?

12. **Craft and Structure** Venn diagrams can be used to make models of hierarchical classification schemes. A Venn diagram is shown below. Four groups are represented by circular regions—A, B, C, and D. Each region represents a collection of organisms or members of a taxonomic level. Regions that overlap, or intersect, share common members. Regions that do not overlap do not have members in common. Use the following terms to label the regions shown in the diagram: *kingdom Animalia, phylum Chordata, class Insecta,* and *class Mammalia.*

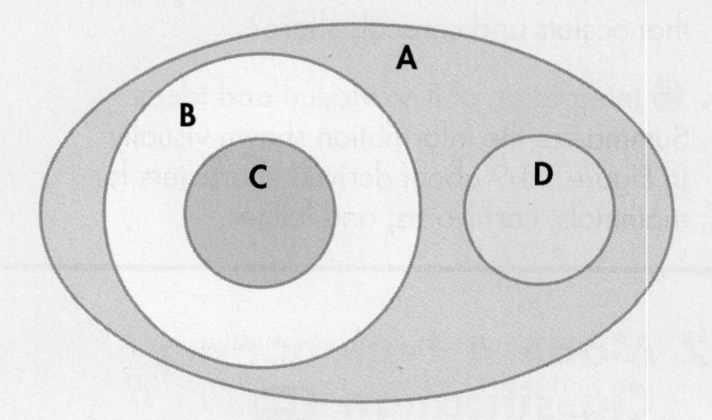

18.2 Modern Evolutionary Classification

Understand Key Concepts

13. A group that is limited to a common ancestor and all of its descendants is called a
    a. taxon.
    b. phylogeny.
    c. tree of life.
    d. monophyletic group.

14. A specific trait that is used to construct a cladogram is called a
    a. taxon.
    b. structural feature.
    c. clade.
    d. derived character.

15. A branch of a cladogram that consists of a single common ancestor and all the descendants of that ancestor is called
    a. cladistics.
    b. a kingdom.
    c. a clade.
    d. a class.

16. What does each individual node in a cladogram represent?

17. Why can differences in mitochondrial DNA be used as derived characters?

18. What is phylogeny?

Think Critically

19. **Apply Concepts** Both snakes and worms are tubular, with no legs. How could you determine whether their similarity in shape means that they share a recent common ancestor?

20. **Pose Questions** What questions would Linnaeus ask to determine a classification? What questions would a modern systematist ask?

21. **Key Ideas and Details** You are a reviewer of a paper on the discovery of new species in the Amazon jungle. Two new species of beetles have been found, beetle A and beetle B, that resemble each other closely but have somewhat different markings on their wings. In addition, both beetle A and beetle B resemble beetle C, a species that has already been identified. How could DNA similarities be used to help determine whether beetle A and beetle B are more closely related to each other or to beetle C? Cite evidence from the text to support your analysis.

22. **Craft and Structure** What is the relationship between natural selection and phylogeny?

23. **Apply Concepts** Explain why hair is a derived character for clade Mammalia but having four limbs is not. For which clade is four limbs a derived character?

18.3 Building the Tree of Life

Understand Key Concepts

24. The three domains are
    a. Animalia, Plantae, and Archaebacteria.
    b. Plantae, Fungi, and Eubacteria.
    c. Bacteria, Archaea, and Eukarya.
    d. Protista, Bacteria, and Animalia.

25. Which of the following kingdoms includes only heterotrophs?
    a. Protista
    b. Fungi
    c. Plantae
    d. Eubacteria

26. **Craft and Structure** How do domains and kingdoms differ?

27. What characteristics are used to place an organism in the domain Bacteria?

solve the CHAPTER MYSTERY

GRIN AND BEAR IT

Most biologists classify the polar bear, *Ursus maritimus,* as a separate species from the brown bear, *Ursus arctos*. The teeth, body shape, metabolism, and behavior of polar bears are very different from those of brown bears. But some systematists are now questioning that classification.

Are polar bears and brown bears two distinct species? The answer depends on what a species is. The usual definition of *species* is "a group of similar organisms that can breed and produce fertile offspring." Polar bears and brown bears can, in fact, mate and produce offspring that are fertile. However, in the natural environment, polar bears and brown bears almost never mate.

The question is complicated by DNA analysis. There are different populations of brown bears, and these different populations have somewhat different genetic makeups. DNA analysis has shown that some populations of brown bears are more closely related to polar bears than they are to other populations of brown bears. According to DNA analysis, if polar bears are indeed a separate species, brown bears by themselves do not form a single clade.

1. **Research to Build and Present Knowledge** List the evidence that supports classifying polar bears and brown bears into two different species. Then list the evidence that indicates that polar bears and brown bears belong to the same species.

2. **Infer** What evidence indicates that different populations of brown bears belong to different clades?

3. **Connect to the Big idea** Do you think that the classic definition of *species*—"a group of similar organisms that can breed and produce fertile offspring"—is still adequate? Why or why not?

28. Which domain consists of prokaryotes whose cell walls lack peptidoglycan?

29. Describe the four kingdoms that make up the domain Eukarya.

30. What characteristic(s) differentiate the kingdom Fungi from the kingdom Eubacteria?

31. What do the branches of the tree of life try to show?

Think Critically

32. **Production and Distribution of Writing** In terms of cladistic analysis, what is the problem with placing all members of kingdom Protista into the same clade? Write a paragraph explaining the issue with the general public as your audience.

33. **Classify** Study the descriptions of the following organisms, and place them in the correct kingdom.
Organism A: Multicellular eukaryote without cell walls
Organism B: Its cell walls lack peptidoglycan, and its cell membranes contain certain lipids that are not found in other organisms. It lives in an extreme environment and can survive only in the absence of oxygen.
Organism C: Unicellular eukaryote with cell walls of chitin

Connecting Concepts

Use Science Graphics

The cladogram below shows the relationships among three imaginary groups of organisms—groups A, B, and C. Use the cladogram to answer questions 34–36.

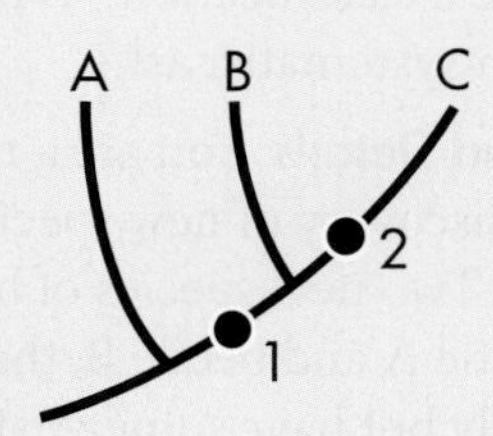

34. **Interpret Visuals** Which groups share derived character 1?

35. **Integration of Knowledge and Ideas** What does the node, or fork, between groups B and C represent?

36. **Apply Concepts** Which group split off from the other groups first?

Write About Science

37. **Text Types and Purposes** Write a short explanation of the way in which taxonomists use similarities and differences in DNA to help classify organisms and infer evolutionary relationships. (*Hint:* Use a specific example to help clarify your explanation.)

38. **Assess the**  Explain what the tree of life is and what its various parts represent. Also explain why the tree of life probably will change. (*Hint*: When you explain what the various parts represent, use the terms *base* and *branches*.)

Integration of Knowledge and Ideas

Use the table to answer questions 39–41.

	Turtle	Lamprey	Frog	Fish	Cat
Hair	No	No	No	No	Yes
Amniotic egg	Yes	No	No	No	Yes
Four legs	Yes	No	Yes	No	Yes
Jaw	Yes	No	Yes	Yes	Yes
Vertebrae	Yes	Yes	Yes	Yes	Yes

39. **Interpret Tables** The first column lists derived characters that can be used to make a cladogram of vertebrates. Which characteristic is shared by the most organisms? Which by the fewest?

40. **Sequence** From the information given, place the animals in sequence from the most recently evolved to the most ancient.

41. **Draw Conclusions** Of the following pairs—lamprey-turtle, fish-cat, and frog-turtle—which are probably most closely related?

Standardized Test Prep

Multiple Choice

1. Which of the following is NOT a characteristic of Linnaeus's system for naming organisms?
   A two-part name
   B multipart name describing several traits
   C name that identifies the organism's genus
   D name that includes the organism's species identifier

2. In which of the following are the Linnaean ranks in correct order?
   A phylum, kingdom, species
   B genus, order, family
   C kingdom, phylum, class
   D order, class, family

3. In the six-kingdom system of classifying living things, which kingdoms contain unicellular organisms?
   A Eubacteria only
   B Eubacteria and "Protista" only
   C Archaebacteria only
   D Eubacteria, Archaebacteria, Plantae, and "Protista"

4. If species A and B have very similar genes, which of the following statements is probably true?
   A Species A and B shared a relatively recent common ancestor.
   B Species A evolved independently of species B for a long period.
   C Species A and species B are the same species.
   D Species A is older than species B.

5. The taxon called Eukarya is a(n)
   A order.
   B phylum.
   C kingdom.
   D domain.

6. Members of the kingdom "Protista" are classified into
   A two domains.
   B three domains.
   C three species.
   D three kingdoms.

Questions 7–9

The cladogram below shows the evolutionary relationships among four groups of plants.

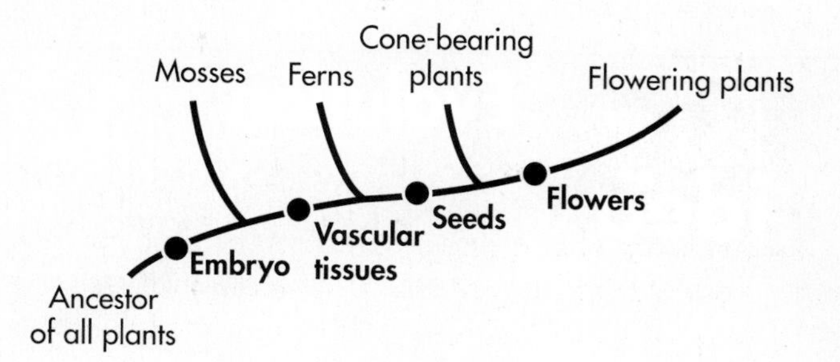

7. Which of the following groups, taken by themselves, do NOT form a clade?
   A cone-bearing plants and flowering plants
   B ferns, cone-bearing plants, and flowering plants
   C mosses and ferns
   D mosses, ferns, cone-bearing plants, and flowering plants

8. Which of the following groups share the most recent common ancestor?
   A cone-bearing plants and flowering plants
   B mosses and ferns
   C mosses and cone-bearing plants
   D ferns and flowering plants

9. Which derived character appeared first during the course of the plants' evolution?
   A seeds
   B flowers
   C embryo
   D vascular tissues

Open-Ended Response

10. Why have biologists changed many of Linnaeus's original classifications of organisms?

If You Have Trouble With . . .

Question	1	2	3	4	5	6	7	8	9	10
See Lesson	18.1	18.1	18.3	18.2	18.3	18.3	18.2	18.2	18.2	18.2

19 History of Life

Evolution

Q: How do fossils help biologists understand the history of life on Earth?

Ichthyosaurs were dolphinlike marine reptiles that prowled the seas in the Mesozoic. This ichthyosaur died around the time of giving birth.

INSIDE:

- 19.1 The Fossil Record
- 19.2 Patterns and Processes of Evolution
- 19.3 Earth's Early History

BUILDING *Scientific Literacy*

Deepen your understanding of the history of life by engaging in key practices that allow you to make connections across concepts.

NGSS You will analyze and interpret data in your study of ancient life.

STEM What has Earth looked like since the Cambrian Period? Investigate how current mapping technologies help scientists document Earth's early history.

Common Core You will write informative texts to explain the patterns and processes of evolution.

CHAPTER MYSTERY

MURDER IN THE PERMIAN

Just over 250 million years ago, during the Permian Period, life on Earth came as close as it has ever come to being wiped out. The Permian extinction may be the greatest murder mystery in the history of the world. Whatever happened back then killed off over 55 percent of all families on Earth, including about 96 percent of marine species and 70 percent of terrestrial vertebrate species. Ancient ecosystems were so completely disrupted that it took millions of years for them to be restored.

Researchers once thought that this "great dying" took place over a long time. But new fossil data suggest that it took no more than 200,000 years and possibly less. In geological terms, that's a short time. As you read this chapter, look for clues as to what could have killed so many different forms of life. Then, solve the mystery.

Biology.com

Finding the solution to this extinction mystery is only the beginning. Take a video field trip with the ecogeeks of Untamed Science to see where the mystery leads.

Go online to access additional resources including:
- **eText** • **Flash Cards** • **Lesson Overviews** • **Chapter Mystery**

19.1 The Fossil Record

Key Questions

What do fossils reveal about ancient life?

How do we date events in Earth's history?

How was the geologic time scale established, and what are its major divisions?

How have our planet's environment and living things affected each other to shape the history of life on Earth?

Vocabulary

extinct • paleontologist • relative dating • index fossil • radiometric dating • half-life • geologic time scale • era • period • plate tectonics

Taking Notes

Outline Make an outline using the green and blue headings in this lesson. Fill in details as you read to help you organize the information in the lesson.

THINK ABOUT IT Fossils, the preserved remains or traces of ancient life, are priceless treasures. They tell of life-and-death struggles and of mysterious worlds lost in the mists of time. Taken together, the fossils of ancient organisms make up the history of life on Earth called the fossil record. How can fossils help us understand life's history?

Fossils and Ancient Life

What do fossils reveal about ancient life?

Fossils are the most important source of information about extinct species. An **extinct** species is one that has died out. Fossils vary enormously in size, type, and degree of preservation, and they form only under certain conditions. For every organism preserved as a fossil, many died without leaving a trace, so the fossil record is not complete.

Types of Fossils Fossils can be as large and perfectly preserved as an entire animal, complete with skin, hair, scales, or feathers. They can also be as tiny as bacteria, developing embryos, or pollen grains. Many fossils are just fragments of an organism—teeth, pieces of a jawbone, or bits of leaf. Sometimes an organism leaves behind trace fossils—casts of footprints, burrows, tracks, or even droppings. Although most fossils are preserved in sedimentary rocks, some are preserved in other ways, like the insect shown in **Figure 19–1.**

FIGURE 19–1 Diversity of Fossils There are all different types of fossils. A fossil can be a single bone, some footprints, or entire organisms.

▲ Dimetrodon footprints

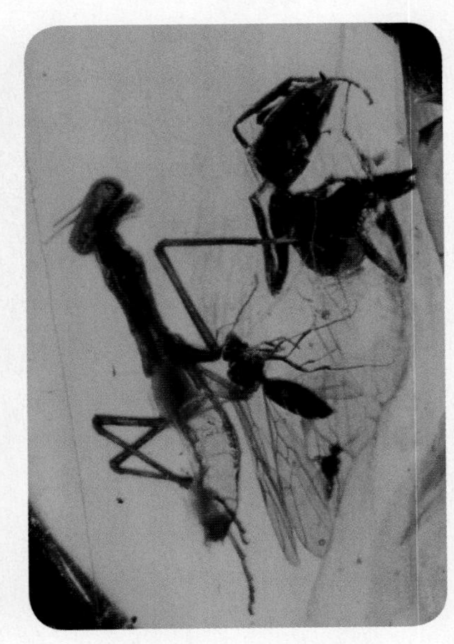

▲ Insect preserved in amber

Fossils in Sedimentary Rock Most fossils are preserved in sedimentary rock. **Figure 19–2** shows how. ❶ Sedimentary rock usually forms when small particles of sand, silt, clay, or lime muds settle to the bottom of a river, lake, ocean, or other body of water. Sedimentary rock can also form from compacted desert sands. ❷ As sediments build up, they bury dead organisms that have sunk to the bottom. If the remains of these organisms are buried relatively quickly, they may not be scattered by scavengers. Usually, soft body structures decay quickly after death, so only wood, shells, bones, or teeth remain. These hard structures can be preserved if they are saturated or replaced with mineral compounds. Sometimes, however, organisms are buried so quickly that soft tissues are protected from aerobic decay. When this happens, fossils may preserve incredibly detailed imprints of soft-bodied animals and structures like skin or feathers.

❸ As layers of sediment continue to build up over time, the remains are buried deeper and deeper. Over many years, water pressure gradually compresses the lower layers. This pressure, along with chemical activity, can turn the sediments into rock.

BUILD Vocabulary

WORD ORIGINS The words paleontology and **paleontologist** come from the Greek word *palaios*, meaning "ancient." A paleontologist studies the remains of ancient life.

What Fossils Can Reveal Although the fossil record is incomplete, it contains an enormous amount of information for **paleontologists** (pay lee un TAHL uh jists), researchers who study fossils to learn about ancient life. **From the fossil record, paleontologists learn about the structure of ancient organisms, their environment, and the ways in which they lived.** By comparing body structures in fossils—a backbone, for example—to body structures in living organisms, researchers can infer evolutionary relationships and form hypotheses about how body structures and species have evolved. Bone structure and footprints can indicate how animals moved. Fossilized plant leaves and pollen suggest whether an area was a swamp, a lake, a forest, or a desert. Also, when different kinds of fossils are found together, researchers can sometimes reconstruct entire ancient ecosystems.

In Your Notebook *Construct a flowchart to explain how the remains of a snail might become fossilized in sedimentary rock.*

Fossil fish *Diplomystus dentatus* (about 50 million years old)

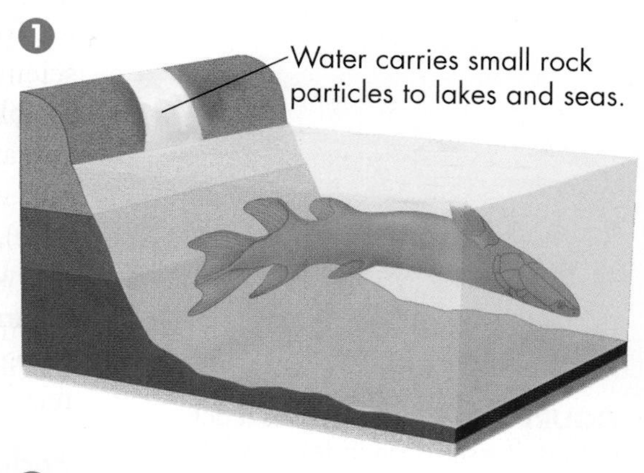

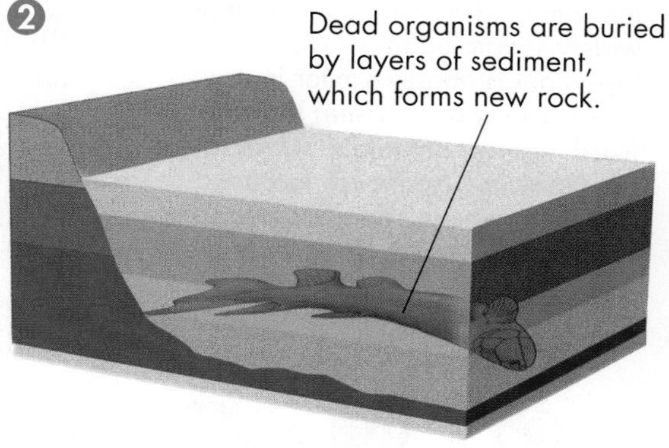

FIGURE 19–2 Fossil Formation Most fossils, like the fish shown here, form in sedimentary rock. **Interpret Photos** ***What part of the fish has been preserved as a fossil?***

Dating Earth's History

How do we date events in Earth's history?

The fossil record wouldn't be as useful without a time scale to tell us what happened when. Researchers use several techniques to date rocks and fossils.

Relative Dating Since sedimentary rock is formed as layers of sediment are laid on top of existing sediments, lower layers of sedimentary rock, and fossils they contain, are generally older than upper layers. **Relative dating** places rock layers and their fossils in a temporal sequence, as shown in **Figure 19–3.** **Relative dating allows paleontologists to determine whether a fossil is older or younger than other fossils.**

To help establish the relative ages of rock layers and their fossils, scientists use index fossils. **Index fossils** are distinctive fossils used to establish and compare the relative ages of rock layers and the fossils they contain. A useful index fossil must be easily recognized and will occur only in a few rock layers (meaning the organism lived only for a short time), but these layers will be found in many places (meaning the organism was widely distributed). Trilobites, a large group of distinctive marine organisms, are often used as index fossils. There are more than 15,000 recognized species of trilobite. Together, they can be used to establish the relative dates of rock layers spanning nearly 300 million years.

Radiometric Dating Relative dating is important, but provides no information about a fossil's absolute age in years. One way to date rocks and fossils is radiometric dating. **Radiometric dating** relies on radioactive isotopes, which decay, or break down, into stable isotopes at a steady rate. A **half-life** is the time required for half of the radioactive atoms in a sample to decay. After one half-life, half of the original radioactive atoms have decayed, as shown in **Figure 19–4.**

FIGURE 19–3 Index Fossils Each of these fossils is an index fossil. If the same index fossil is found in two widely separated rock layers, the rock layers are probably similar in age. **Draw Conclusions** ***Using the index fossils shown, determine which layers are "missing" from each location. Layers may be missing because they were never formed, or because they were eroded.***

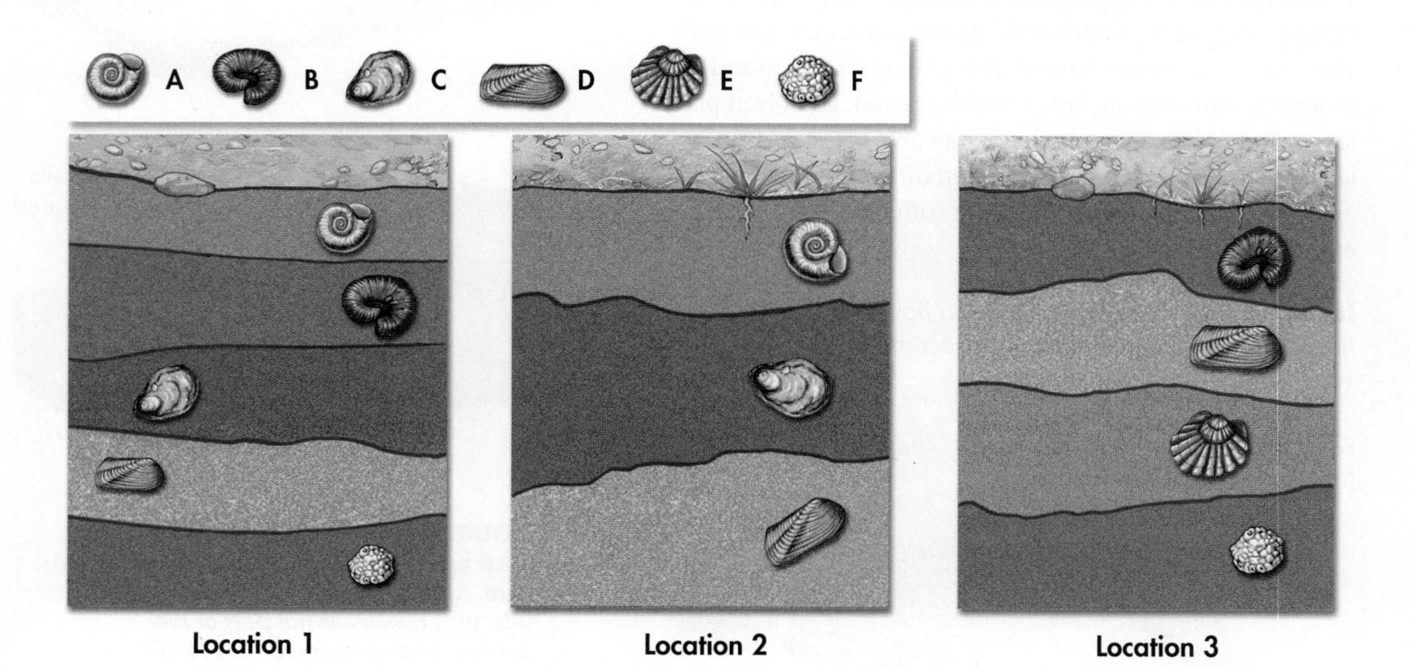

After another half-life, another half of the remaining radioactive atoms will have decayed. **Radiometric dating uses the proportion of radioactive to stable isotopes to calculate the age of a sample.**

Different radioactive isotopes decay at different rates, so they have different half-lives. Elements with short half-lives are used to date recent fossils. Elements with long half-lives are used for dating older fossils. To understand this, think of timing sports events. For a 50-yard dash, a coach depends on the fast-moving second hand of a stopwatch. To time a marathon, slower-moving hour and minute hands are also important.

A number of radioactive isotopes are used to determine the ages of rocks and fossils. An isotope known as carbon-14 is particularly useful for directly dating organisms that lived in the recent past. Carbon-14 is produced at a steady rate in the upper atmosphere, so air generally contains a tiny amount of it, in addition to the much more common stable, nonradioactive form, carbon-12. Plants take carbon-14 in when they absorb carbon dioxide during photosynthesis, and animals acquire it when they eat plants or other animals. Once an organism dies, it no longer takes in this isotope, so its age can be determined by the amount of carbon-14 still remaining in tissues such as bone, hair, or wood. Carbon-14 has a half-life of roughly 5730 years, so its use is limited to organisms that lived in the last 60,000 years.

Older fossils can be dated indirectly by dating the rock layers in which they are found. Isotopes with much longer half-lives are used for this purpose, including potassium-40 (half-life: 1.26 billion years, shown in **Figure 19–4**), uranium-238 (4.5 billion years), and rubidium-87 (48.8 billion years). Over many years, geologists have combined the use of these and other isotope methods to make increasingly accurate estimates of the ages of geological formations. These studies have provided direct physical evidence for the ages of the index fossils used to identify periods of Earth history.

In Your Notebook *Explain why carbon-14 can't be used to estimate the age of very old fossils.*

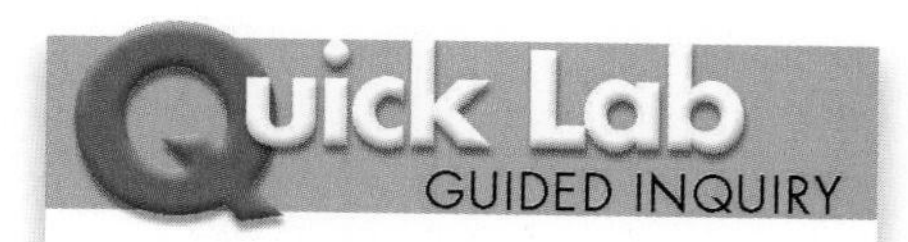

Modeling Half-Life

1. Construct a data table or spreadsheet with two columns and five rows. Label the columns Spill Number and Number of Squares Returned. Take a sheet of paper, and cut out 100 1-cm squares. Place an *X* on each square, and put all the squares in a cup.

2. Mix the squares in the cup, and spill them out.

3. Remove all the squares that have an *X* showing. Count the squares left, record the number, and return the remaining squares to the cup.

4. Repeat steps 2 and 3 until there are five or fewer squares left. Make a graph of your results with the number of spills on the *x*-axis and the number of squares remaining after each spill on the *y*-axis.

Analyze and Conclude

1. Analyze Data How many spills did you need to remove half the squares? To remove three fourths?

2. Calculate If each spill represents one year, what is the half-life of the squares? MATH

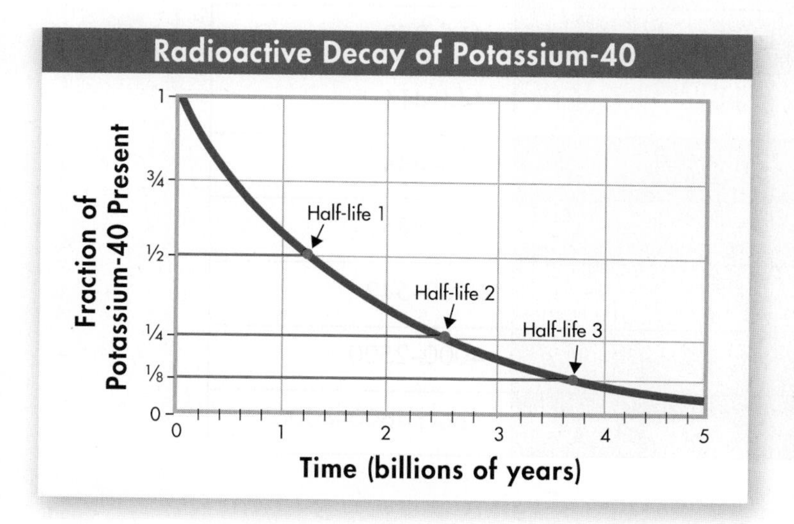

FIGURE 19–4 Radioactive Decay A half-life is the time it takes half the radioactive atoms in a sample to decay. The half-life of potassium-40 is 1.26 billion years.

Geologic Time Scale

How was the geologic time scale established, and what are its major divisions?

Geologists and paleontologists have built a time line of Earth's history called the **geologic time scale.** The most recent version is shown in **Figure 19–5. The geologic time scale is based on both relative and absolute dating. The major divisions of the geologic time scale are eons, eras, and periods.**

Establishing the Time Scale By studying rock layers and index fossils, early paleontologists placed Earth's rocks and fossils in order according to their relative age. As they worked, they noticed major changes in the fossil record at boundaries between certain rock layers. Geologists used these boundaries to determine where one division of geologic time ended and the next began. Years later, radiometric dating techniques were used to assign specific ages to the various rock layers. This time scale is constantly being tested, verified, and adjusted.

FIGURE 19–5 Geologic Time Scale The basic divisions of the geologic time scale are eons, eras, and periods. Precambrian time was the name originally given to all of Earth's history before the Phanerozoic Eon. Note that the Paleogene and Neogene are sometimes called the Tertiary period. However, this term is generally considered outdated.

Geologic Time Scale

	Eon	Era	Period	Time (millions of years ago)
	Phanerozoic	Cenozoic	Quaternary	1.8–present
			Neogene	23–1.8
			Paleogene	65.5–23
		Mesozoic	Cretaceous	146–65.5
			Jurassic	200–146
			Triassic	251–200
		Paleozoic	Permian	299–251
			Carboniferous	359–299
			Devonian	416–359
			Silurian	444–416
			Ordovician	488–444
			Cambrian	542–488
Precambrian Time	Proterozoic			2500–542
	Archean			4000–2500
	Hadean			About 4600–4000

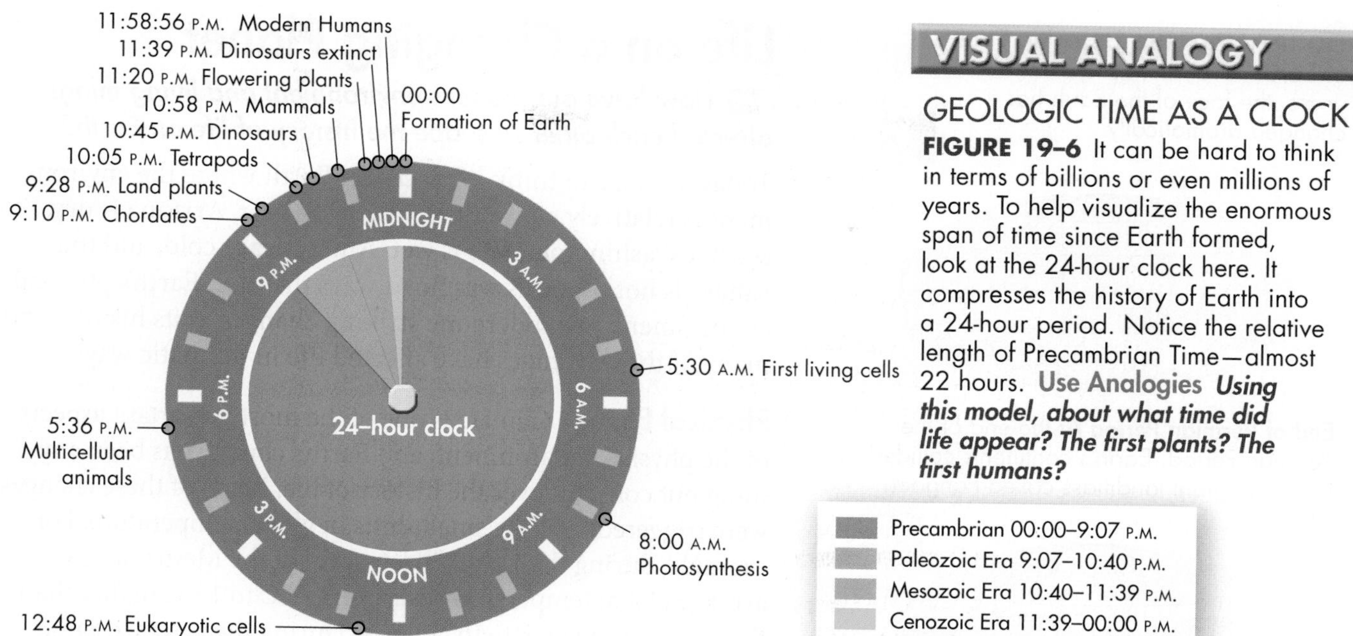

VISUAL ANALOGY

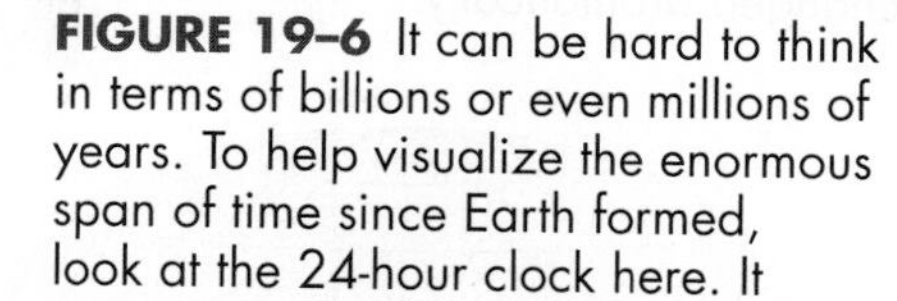

GEOLOGIC TIME AS A CLOCK

FIGURE 19–6 It can be hard to think in terms of billions or even millions of years. To help visualize the enormous span of time since Earth formed, look at the 24-hour clock here. It compresses the history of Earth into a 24-hour period. Notice the relative length of Precambrian Time—almost 22 hours. **Use Analogies** ***Using this model, about what time did life appear? The first plants? The first humans?***

Divisions of the Geologic Time Scale Divisions of geologic time have different lengths. The Cambrian Period, for example, began 542 million years ago and continued until 488 million years ago, which makes it 54 million years long. The Cretaceous Period was 80 million years long.

Geologists now recognize four eons. The Hadean Eon, during which the first rocks formed, spans the time from Earth's formation to about 4 billion years ago. The Archean Eon, during which life first appeared, followed the Hadean. The Proterozoic Eon began 2.5 billion years ago and lasted until 542 million years ago. The Phanerozoic (fan ur uh ZOH ic) Eon began at the end of the Proterozoic and continues to the present.

Eons are divided into **eras.** The Phanerozoic Eon, for example, is divided into the Paleozoic, Mesozoic, and Cenozoic Eras. And eras are subdivided into **periods,** which range in length from nearly 100 million years to just under 2 million years. The Paleozoic Era, for example, is divided into six periods, including the Permian Period.

Naming the Divisions Divisions of the geologic time scale were named in different ways. The Cambrian Period, for example, was named after Cambria—an old name for Wales, where rocks from that time were first identified. The Carboniferous ("carbon-bearing") Period is named for large coal deposits that formed during that time.

Geologists started to name divisions of the time scale before any rocks older than the Cambrian Period had been identified. For this reason, all of geologic time before the Cambrian was simply called Precambrian Time. Precambrian Time, however, actually covers about 90 percent of Earth's history, as shown in **Figure 19–6.**

MYSTERY CLUE

Paleontologists discovered dramatic changes in the fossil record at the end of the Permian Period. What methods do you think they used to date that change at 251 million years ago?

FIGURE 19–7 The Changing Face of Earth Over the last 225 million years, the face of the Earth has changed dramatically.

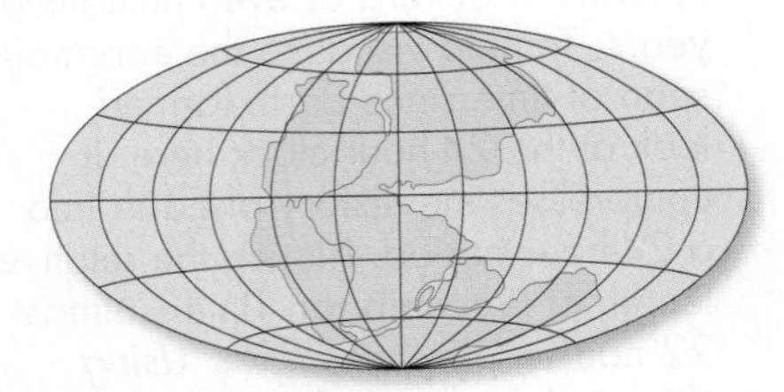

End of Permian Period At the end of the Permian Period, Earth's continents collided to form one giant landmass called Pangaea.

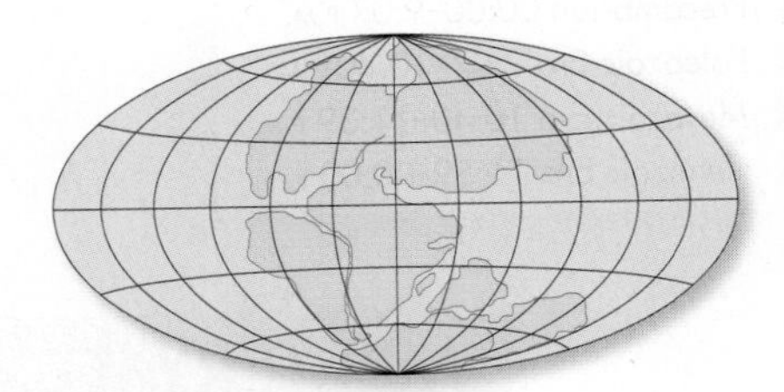

Triassic Period During the Triassic Period, Pangaea started to break apart and form separate land masses.

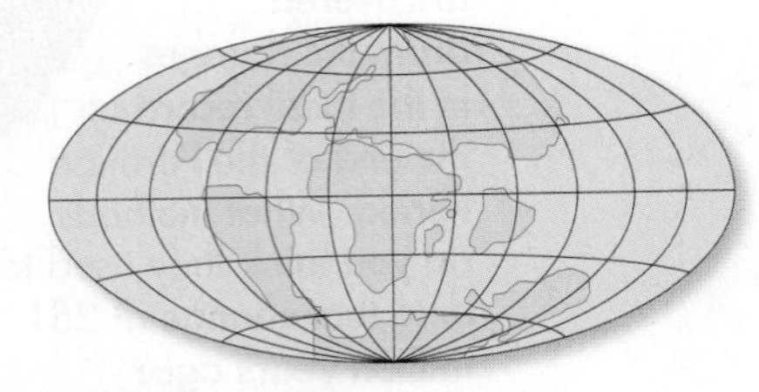

End of Cretaceous Period By the end of the Cretaceous Period, the continents as we know them began to drift apart.

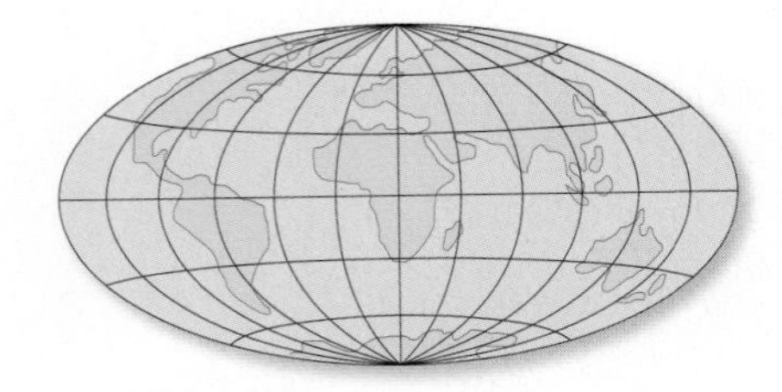

Present Day

Life on a Changing Planet

How have our planet's environment and living things affected each other to shape the history of life on Earth?

Today, it's easy to think of places on Earth where the environment is relatively constant from year to year. Arizona is dry, coastal Washington State is wet, Antarctica is cold, and the Sahara is hot. But this was not always the case. Earth's physical environment has undergone striking changes in its history, and many of these changes have affected life in dramatic ways.

Physical Forces Climate is one of the most important aspects of the physical environment, and Earth's climate has been anything but constant over the history of life. Many of these changes were triggered by fairly small shifts in global temperature. For example, during the global "heat wave" of the Mesozoic era, average global temperatures were only 6°C to 12°C higher than they were in the twentieth century. During the great ice ages, which swept across the globe as recently as 10,000 years ago, temperatures were only about 5°C cooler than they are now. Yet, these temperature shifts had far-reaching effects on living things.

Geological forces have also transformed life on Earth, building mountains and even moving whole continents. Remember that local climates are influenced by the interactions of wind and ocean currents with geological features like mountains and plains. Volcanic forces have altered landscapes over much of Earth, even producing entire islands that provide new habitats. The Hawaiian Islands, home to scores of unique plant and animal species, are a perfect example of how volcanic islands can alter the course of evolution. **Building mountains, opening coastlines, changing climates, and geological forces have altered habitats of living organisms repeatedly throughout Earth history.**

Over the long term, the process of continental drift has produced even more dramatic changes in Earth's biological landscape. As shown in **Figure 19–7,** continents have collided to form "super continents," and then drifted apart again, profoundly changing the flow of ocean currents. Continental drift has also affected the distribution of fossils and living organisms worldwide. For example, the continents of Africa and South America are now separated by the Atlantic Ocean. But fossils of *Mesosaurus*, an aquatic reptile, have been found in Africa and South America. The presence of these fossils on both continents reflects the fact that both were joined at one time. The theory of **plate tectonics** explains these movements as the result of solid "plates" moving slowly, as little as 3 cm a year, over Earth's mantle.

Forces from space have even altered Earth's physical environment. There is strong evidence that comets and large meteors have crashed into Earth many times in the past. Some of these impacts may have been so violent that they kicked enough dust and debris in the atmosphere to cause, or contribute to, worldwide extinctions of organisms on land and in the water.

Biological Forces Although we think of life as reacting to Earth's physical environment, in many cases life actually plays a major role in shaping that environment. Iron deposits in ancient sedimentary rock indicate that Earth's early oceans contained large amounts of soluble iron and little oxygen. The first photosynthetic organisms began absorbing carbon dioxide and releasing large amounts of oxygen. Our planet has never been the same since then. Earth cooled as carbon dioxide levels dropped. The iron content of the oceans fell, as iron ions reacted with oxygen to form insoluble compounds that settled to the ocean floor. These changes affected climate and ocean chemistry in many ways. **The actions of living organisms over time have changed conditions in the land, water, and atmosphere of planet Earth.**

Even today, organisms shape the landscape as they build soil from rock and sand. Plants, animals, and microorganisms are active players in global cycles of key elements, including carbon, nitrogen, and oxygen. Earth is a living planet, and its physical environment reflects that fact.

19.1 Assessment

Review Key Concepts

1. a. **Explain** What can a paleontologist learn from fossils?
   b. **Relate Cause and Effect** Why have so few organisms become fossilized?
2. a. **Review** What are the two ways in which geologists determine the age of fossils?
   b. **Draw Conclusions** Many more fossils have been found since Darwin's day, giving us a more complete record of life's history. How would this information make relative dating more accurate?
3. a. **Explain** How are eras and periods related?
   b. **Interpret Visuals** Use **Figure 19–5** to determine when the Silurian Period began and how long it lasted.
4. a. **Review** Describe three processes that have affected the history of life on Earth.
   b. **Relate Cause and Effect** Describe two ways in which continental drift has affected organisms.

VISUAL THINKING

5. Look at the fossil bat in the photograph below. Describe the fossil. What can you infer about how the organism moved? Explain your answer.

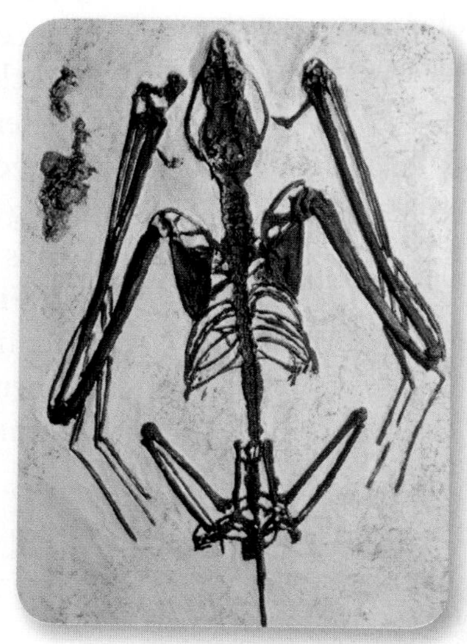

19.2 Patterns and Processes of Evolution

Key Questions

What processes influence whether species and clades survive or become extinct?

How fast does evolution take place?

What are two patterns of macroevolution?

What evolutionary characteristics are typical of coevolving species?

Vocabulary

macroevolutionary patterns
background extinction
mass extinction
gradualism
punctuated equilibrium
adaptive radiation
convergent evolution
coevolution

Taking Notes

Concept Map Construct a concept map that includes the patterns of macroevolution shown in this lesson.

FIGURE 19–8 Paleontologists at Work The white covering protects the fossils until they can reach a museum.

THINK ABOUT IT The fossil record shows a parade of organisms that evolved, survived for a time, and then disappeared. More than 99 percent of all species that have lived on Earth are extinct. How have so many different groups evolved? Why are so many now extinct?

Speciation and Extinction

What processes influence whether species and clades survive or become extinct?

The study of life's history leaves no doubt that life has changed over time. Many of those changes occurred within species, but others occurred in larger clades and over longer periods of time. These grand transformations in anatomy, phylogeny, ecology, and behavior, which usually take place in clades larger than a single species, are known as **macroevolutionary patterns.** The ways new species emerge through speciation, and the ways species disappear through extinction, are among the simplest macroevolutionary patterns. The emergence, growth, and extinction of larger clades, such as dinosaurs, mammals, or flowering plants are examples of larger macroevolutionary patterns.

Macroevolution and Cladistics Paleontologists study fossils to learn about patterns of macroevolution and the history of life. Part of this process involves classifying fossils. Fossils are classified using the same cladistic techniques, based on shared derived characters, that are used to classify living species. In some cases, fossils are placed in clades that contain only extinct organisms. In other cases, fossils are classified into clades that include living organisms.

Remember that cladograms illustrate hypotheses about how closely related organisms are. Hypothesizing that a fossil species is *related* to a living species is not the same thing as claiming that the extinct organism is a direct *ancestor* of that (or any other) living species. For example, **Figure 19–9** does not suggest that any of the extinct species shown are direct ancestors of modern birds. Instead, those extinct species are shown as a series of species that descended, over time, from a line of common ancestors.

In Your Notebook *Explain what macroevolution is and how fossils can show macroevolutionary trends.*

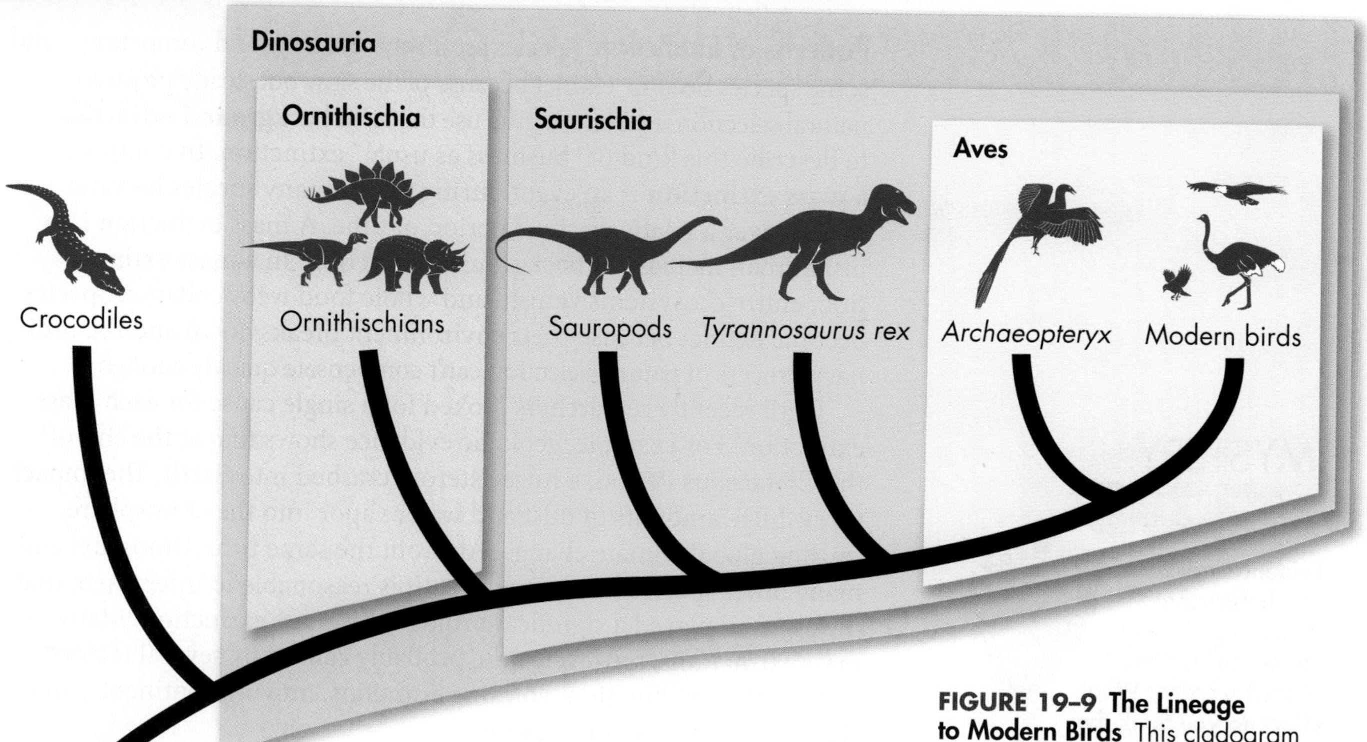

FIGURE 19–9 The Lineage to Modern Birds This cladogram shows some of the clades within the large clade Reptilia. Notice that clade Dinosauria is represented today by modern birds. **Classify** ***What are the two major clades of dinosaurs?***

Adaptation and Extinction Throughout the history of life, organisms have faced changing environments. When environmental conditions change, processes of evolutionary change enable some species to adapt to new conditions and thrive. Species that fail to adapt eventually become extinct. Interestingly, the rates at which species appear, adapt, and become extinct vary among clades, and from one period of geologic time to another.

Why have some clades produced many successful species that survived over long periods of time, while other clades gave rise to only a few species that vanished due to extinction? Paleontologists have tried to answer this question by studying macroevolutionary patterns of speciation and extinction in different clades over time.

One way to think about this process is in terms of species diversity. The emergence of new species with different characteristics can serve as the "raw material" for macroevolutionary change within a clade over long periods. In some cases, the more varied the species in a particular clade are, the more likely the clade is to survive environmental change. This is similar to the way in which genetic variation serves as raw material for evolutionary change for populations within a species. **If the rate of speciation in a clade is equal to or greater than the rate of extinction, the clade will continue to exist. If the rate of extinction in a clade is greater than the rate of speciation, the clade will eventually become extinct.**

The clade Reptilia (part of which is shown in **Figure 19–9**) is one example of a highly successful clade. It not only includes living organisms like snakes, lizards, turtles and crocodiles, but also dinosaurs that thrived for tens of millions of years. As you know, most species in the clade Dinosauria are now extinct. But the clade itself survived, because it produced groups of new species that successfully adapted to changing conditions. One of those groups survives and thrives today—we call them birds.

Patterns of Extinction Species are always evolving and competing—and some species become extinct because of the slow but steady process of natural selection. Paleontologists use the term **background extinction** to describe this kind of "business as usual" extinction. In contrast, a **mass extinction** is an event during which many species become extinct over a relatively short period of time. A mass extinction isn't just a small increase in background extinction. In a mass extinction, entire ecosystems vanish, and whole food webs collapse. Species become extinct because their environment breaks down and the ordinary process of natural selection can't compensate quickly enough.

Until recently researchers looked for a single cause for each mass extinction. For example, geologic evidence shows that at the end of the Cretaceous Period, a huge asteroid crashed into Earth. The impact threw huge amounts of dust and water vapor into the atmosphere, causing global climate change. At about the same time, dinosaurs and many other species became extinct. It is reasonable to infer, then, that the asteroid played a significant role in this mass extinction. Many mass extinctions, however, were probably caused by several factors, working in combination: volcanic eruptions, moving continents, *and* changing sea levels, for example.

After a mass extinction, biodiversity is dramatically reduced. But this is not bad for all organisms. Extinction offers new opportunities to survivors. And as speciation and adaptation produce new species to fill empty niches, biodiversity recovers. But this recovery takes a long time—typically between 5 and 10 million years. Some groups of organisms survive a mass extinction, while other groups do not.

MYSTERY CLUE

Evidence indicates that before the Permian extinction, the oceans lost most of their oxygen. What effect do you think the loss of oxygen had on most organisms?

Extinctions Through Time

The graph shows how the rate of extinction has changed over time. Study the graph, and then answer the questions.

1. **Interpret Graphs** What is plotted on the *y*-axis?
2. **Analyze Data** Which mass extinction killed off the highest percentage of genera?
3. **Draw Conclusions** Describe the overall pattern of extinction shown on the graph.
4. **Infer** What evidence is this graph probably based on?

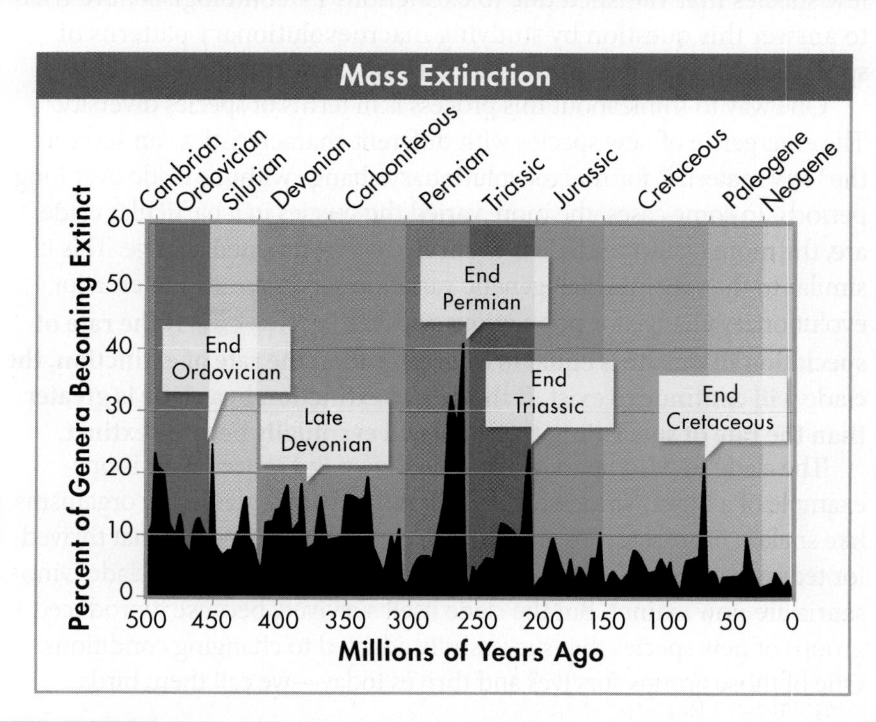

Rate of Evolution

How fast does evolution take place?

How quickly does evolution operate? Does it always take place at the same speed? **Evidence shows that evolution has often proceeded at different rates for different organisms at different times over the long history of life on Earth.** Two models of evolution—gradualism and punctuated equilibrium—are shown in **Figure 19–10.**

Gradualism Darwin was impressed by the slow, steady pace of geologic change. He suggested that evolution also needed to be slow and steady, an idea known as **gradualism.** The fossil record shows that many organisms have indeed changed gradually over time.

Punctuated Equilibrium However, numerous examples in the fossil record indicate that the pattern of slow, steady change does not always hold. Horseshoe crabs, for example, have changed little in structure from the time they first appeared in the fossil record. Much of the time, these species are said to be in a state of equilibrium. This means that their structures do not change much even though they continue to evolve genetically.

Every now and then something happens to upset this equilibrium for some species. **Punctuated equilibrium** is the term used to describe equilibrium that is interrupted by brief periods of more rapid change. (Remember that we use *rapid* here relative to the geologic time scale. For geologists, rapid change can take thousands of years!) The fossil record does reveal periods of relatively rapid change in particular groups of organisms. In fact, some biologists suggest that most new species are produced during periods of rapid change.

Rapid Evolution After Equilibrium There are several reasons why evolution may proceed at different rates for different organisms at different times. Rapid evolution may occur after a small population becomes isolated from the main population. This small population can evolve faster than the larger one because genetic changes spread more quickly among fewer individuals. Rapid evolution may also occur when a small group of organisms migrates to a new environment. That's what happened with the Galápagos finches. In addition, mass extinctions open many ecological niches, creating new opportunities for those organisms that survive. It's not surprising, then, that groups of organisms that survive mass extinctions evolve rapidly in the several million years after the extinction.

In Your Notebook *In your own words, describe gradualism and punctuated equilibrium.*

FIGURE 19–10 Models of Evolution Biologists have considered two different patterns for the rate of evolution, gradualism and punctuated equilibrium. These illustrations are simplified to show the general trend of each model. **Interpret Visuals** ***How do the diagrams illustrate these two models?***

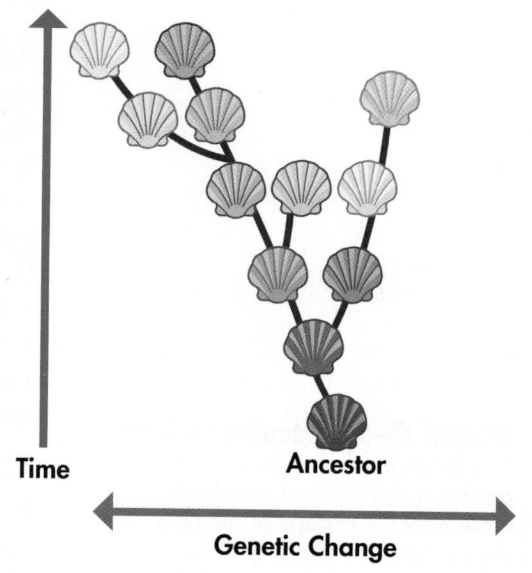

Gradualism involves a slow, steady change in a particular line of descent.

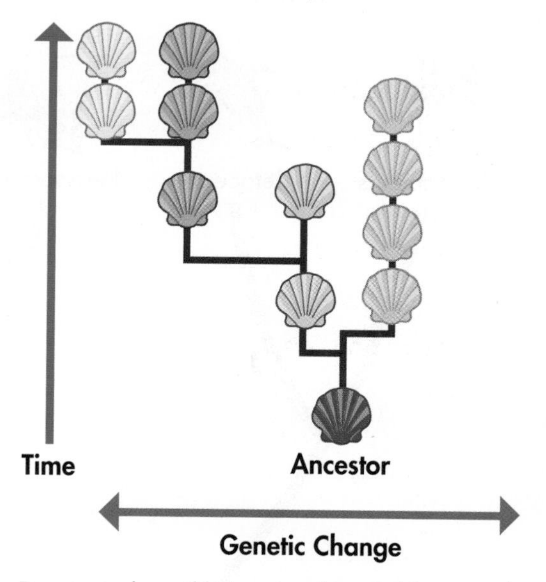

Punctuated equilibrium involves stable periods interrupted by rapid changes.

Adaptive Radiation and Convergent Evolution

What are two patterns of macroevolution?

As paleontologists study the fossil record, they look for patterns. **Two important patterns of macroevolution are adaptive radiation and convergent evolution.** As you'll see, Darwin noted both patterns while aboard the *Beagle.*

Adaptive Radiation Studies of fossils and living organisms often show that a single species or small group of species has diversified over time into a clade containing many species. These species display variations on the group's ancestral body plan, and often occupy different ecological niches. These differences are the product of an evolutionary process called adaptive radiation. **Adaptive radiation** is the process by which a single species or a small group of species evolves over a relatively short time into several different forms that live in different ways. An adaptive radiation may occur when species migrate to a new environment or when extinction clears an environment of a large number of inhabitants. In addition, a species may evolve a new feature that enables it to take advantage of a previously unused environment.

▶ ***Adaptive Radiations in the Fossil Record*** Dinosaurs—one of several spectacular adaptive radiations of reptiles—flourished for about 150 million years during the Mesozoic. The fossil record documents that in the dinosaurs' heyday, mammals diversified but remained small. After most dinosaurs became extinct, however, an adaptive radiation of mammals began. That radiation, part of which is shown in **Figure 19–11,** produced the great diversity of mammals of the Cenozoic Era.

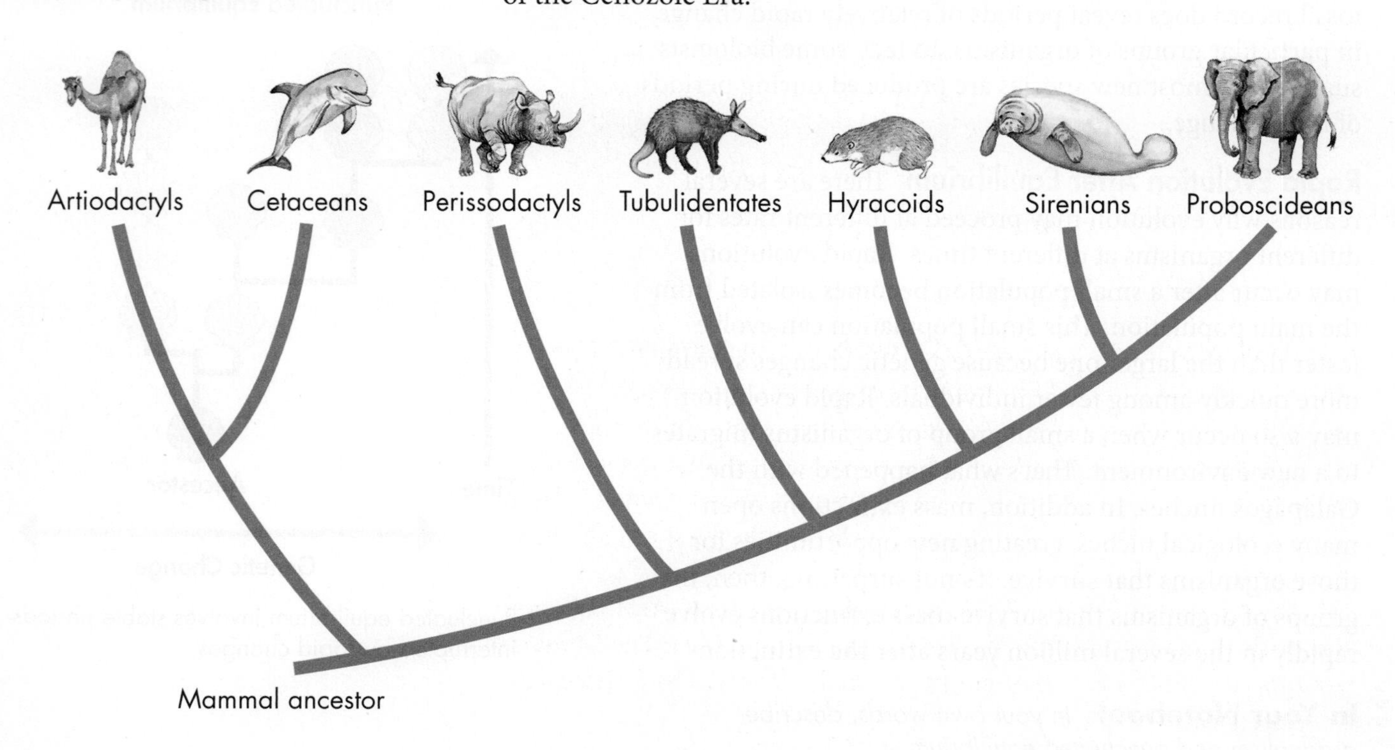

FIGURE 19–11 Adaptive Radiation This diagram shows part of the adaptive radiation of mammals. Note how the groups of animals shown have adapted to many different ways of life—including two groups which have become aquatic. **Interpret Visuals** ***According to this diagram, which mammal group is most closely related to elephants? Does this surprise you? Explain.***

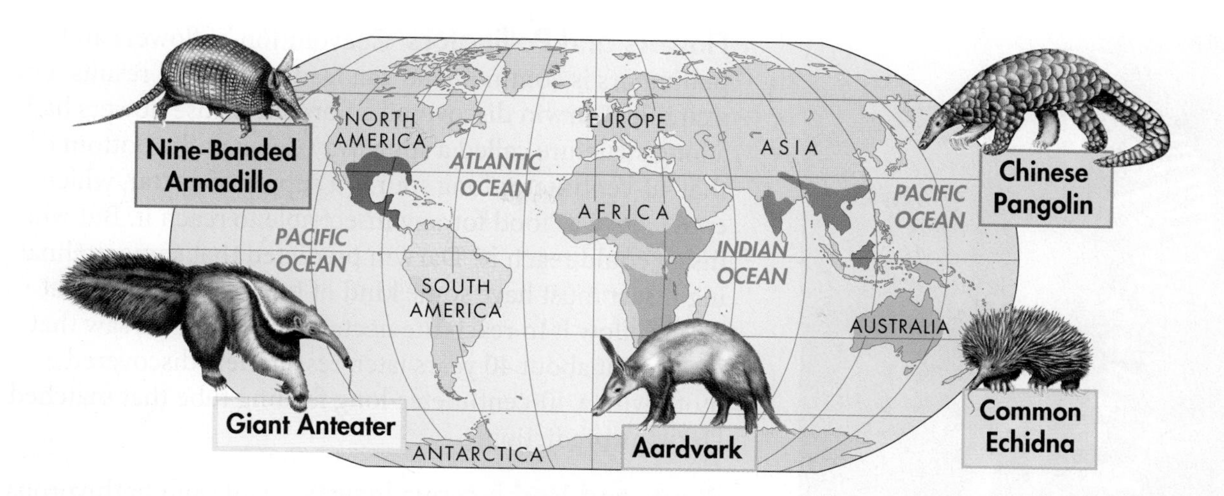

FIGURE 19–12 Convergent Evolution Mammals that feed on ants and termites evolved independently five times. Although each species is unique, each has evolved powerful front claws, a long hairless snout, and a tongue covered with sticky saliva. These adaptations are useful for hunting and eating insects.

▶ ***Modern Adaptive Radiations*** Galápagos finches and Hawaiian honeycreepers are two examples of adaptive radiations in modern organisms. In each of these cases, numerous species evolved from a single founding species. Both finches and honeycreepers evolved different beaks and behaviors that enable each of them to eat different kinds of food.

Convergent Evolution Sometimes, groups of organisms evolve in different places or at different times, but in similar environments. These organisms start out with different structures on which natural selection can operate. But they face similar selection pressures. In these situations, natural selection may mold different body structures in ways that perform similar functions. Because they perform similar functions, these body structures may look similar. Evolution produces similar structures and characteristics in distantly related organisms through the process of **convergent evolution.** Convergent evolution has occurred often in both plants and animals. For example, mammals that feed on ants and termites evolved not once, but five times, in different regions as shown in **Figure 19–12.** Remember how Darwin noted striking similarities among large, distantly related grassland birds? Emus, rheas, and ostriches are another example of convergent evolution.

Coevolution

What evolutionary characteristics are typical of coevolving species?

Sometimes the life histories of two or more species are so closely connected that they evolve together. Many flowering plants, for example, can reproduce only if their flowers attract a specific pollinator species. Pollinators, in turn, may depend on the flowers of certain plants for food in the form of pollen or nectar. The process by which two species evolve in response to changes in each other over time is called **coevolution.** **The relationship between two coevolving organisms often becomes so specific that neither organism can survive without the other. Thus, an evolutionary change in one organism is usually followed by a change in the other organism.**

FIGURE 19–13 Plants and Herbivorous Insects Milkweed plants produce toxic chemicals. But monarch caterpillars not only can tolerate this toxin, they also can store it in their body tissues to use as a defense against *their* predators.

Flowers and Pollinators Coevolution of flowers and pollinators is common and can lead to unusual results. For example, Darwin discovered an orchid whose flowers had a long structure called a spur. Way down at the bottom of that 40-centimeter-long spur is a supply of nectar, which could serve as food for any insect able to reach it. But what insect could reach it? Darwin predicted that some pollinating insect must have some kind of feeding structure that would allow it to reach the nectar. Darwin never saw that insect. But about 40 years later, researchers discovered a moth with a 40-centimeter-long feeding tube that matched Darwin's prediction!

Plants and Herbivorous Insects Plants and herbivorous insects also demonstrate close, albeit less "friendly," coevolutionary relationships. Insects have been feeding on flowering plants since both groups emerged. Over time, many plants evolved bad-tasting or poisonous compounds that discourage insects from eating them. Some of the most powerful natural poisons are compounds developed by plants in response to insect attacks. But once plants began to produce poisons, natural selection on herbivorous insects favored any variants that could alter, inactivate, or eliminate those poisons. Time and again, a group of insects, like the caterpillar in **Figure 19–13,** evolved a way to deal with the particular poisons produced by a certain group of plants.

19.2 Assessment

Review Key Concepts

1. a. Review How does variation within a clade affect the clade's chance of surviving environmental change?

b. Compare and Contrast How is mass extinction different from background extinction?

2. a. Review Explain how punctuated equilibrium is different from gradualism.

b. Relate Cause and Effect Why would evolution speed up when a small group of organisms migrates to a new environment?

3. a. Review What is adaptive radiation?

b. Relate Cause and Effect When might adaptive radiation result in convergent evolution?

4. a. Review What is coevolution?

b. Apply Concepts Describe an example of coevolution.

Apply the Big idea

Evolution

5. What role does the environment play in convergent evolution?

19.3 Earth's Early History

THINK ABOUT IT How did life on Earth begin? What were the earliest forms of life? How did life and the biosphere interact? Origin-of-life research is a dynamic field. But even though some current hypotheses likely will change, our understanding of other aspects of the story is growing.

Key Questions

What do scientists hypothesize about early Earth and the origin of life?

What theory explains the origin of eukaryotic cells?

What is the evolutionary significance of sexual reproduction?

Vocabulary

endosymbiotic theory

Taking Notes

Flowchart Construct a flowchart that shows what scientists hypothesize are the major steps from the origin of Earth to the appearance of eukaryotic cells.

The Mysteries of Life's Origins

What do scientists hypothesize about early Earth and the origin of life?

Geological and astronomical evidence suggests that Earth formed as pieces of cosmic debris collided with one another. While the planet was young, it was struck by one or more huge objects, and the entire globe melted. For millions of years, violent volcanic activity shook Earth's crust. Comets and asteroids bombarded its surface. About 4.2 billion years ago, Earth cooled enough to allow solid rocks to form and water to condense and fall as rain. Earth's surface became stable enough for permanent oceans to form.

This infant planet was very different from Earth today. **Earth's early atmosphere contained little or no oxygen. It was principally composed of carbon dioxide, water vapor, and nitrogen, with lesser amounts of carbon monoxide, hydrogen sulfide, and hydrogen cyanide.** If you had been there, a few deep breaths would have killed you! Because of the gases in the atmosphere, the sky was probably pinkish-orange. And because the oceans contained lots of dissolved iron, they were probably brown. This was the Earth on which life began.

FIGURE 19–14 Early Earth Violent volcanic eruptions helped shape Earth's early history.

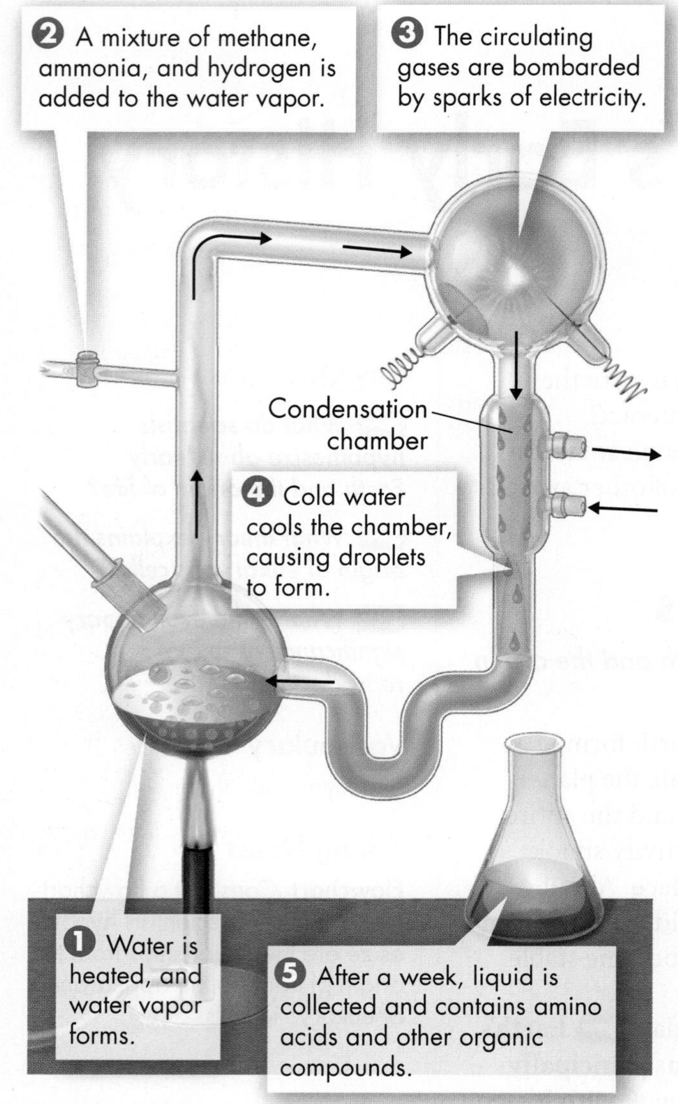

FIGURE 19–15 Miller-Urey Experiment Miller and Urey produced amino acids, which are needed to make proteins, by passing sparks through a mixture of hydrogen, methane, ammonia, and water vapor. Evidence now suggests that the composition of Earth's early atmosphere was different from their 1953 experiment. However, more recent experiments with different mixtures of gases have produced similar results.

The First Organic Molecules Could organic molecules assemble under conditions on early Earth? In 1953, chemists Stanley Miller and Harold Urey tried to answer that question. They filled a sterile flask with water, to simulate the oceans, and boiled it. To the water vapor, they added methane, ammonia, and hydrogen, to simulate what they thought had been the composition of Earth's early atmosphere. Then, as shown in **Figure 19–15,** they passed the gases through electrodes, to simulate lightning. Next, they passed the gases through a condensation chamber, where cold water cooled them, causing drops to form. The liquid circulated through the experimental apparatus for a week. The results were spectacular: They produced 21 amino acids—building blocks of proteins. **Miller and Urey's experiment suggested how mixtures of the organic compounds necessary for life could have arisen from simpler compounds on a primitive Earth.**

We now know that Miller and Urey's ideas on the composition of the early atmosphere were incorrect. But new experiments based on current ideas of the early atmosphere have also produced organic compounds. In fact, in 1995, one of Miller's more accurate mixtures produced cytosine and uracil, two bases found in RNA.

Formation of Microspheres A stew of organic molecules is a long way from a living cell, and the leap from nonlife to life is the greatest gap in scientific hypotheses of life's early history. Geological evidence suggests that during the Archean Eon, 200 to 300 million years after Earth cooled enough to carry liquid water, cells similar to bacteria were common. How might these cells have originated?

Large organic molecules form tiny bubbles called proteinoid microspheres under certain conditions. Microspheres are not cells, but they have some characteristics of living systems. Like cells, they have selectively permeable membranes through which water molecules can pass. Microspheres also have a simple means of storing and releasing energy. Several hypotheses suggest that structures similar to proteinoid microspheres acquired the characteristics of living cells as early as 3.8 billion years ago.

Evolution of RNA and DNA Another unanswered question is the origin of RNA and DNA. Remember that cells are controlled by information stored in DNA, which is transcribed into RNA and then translated into proteins. How could this complex biochemical machinery have evolved?

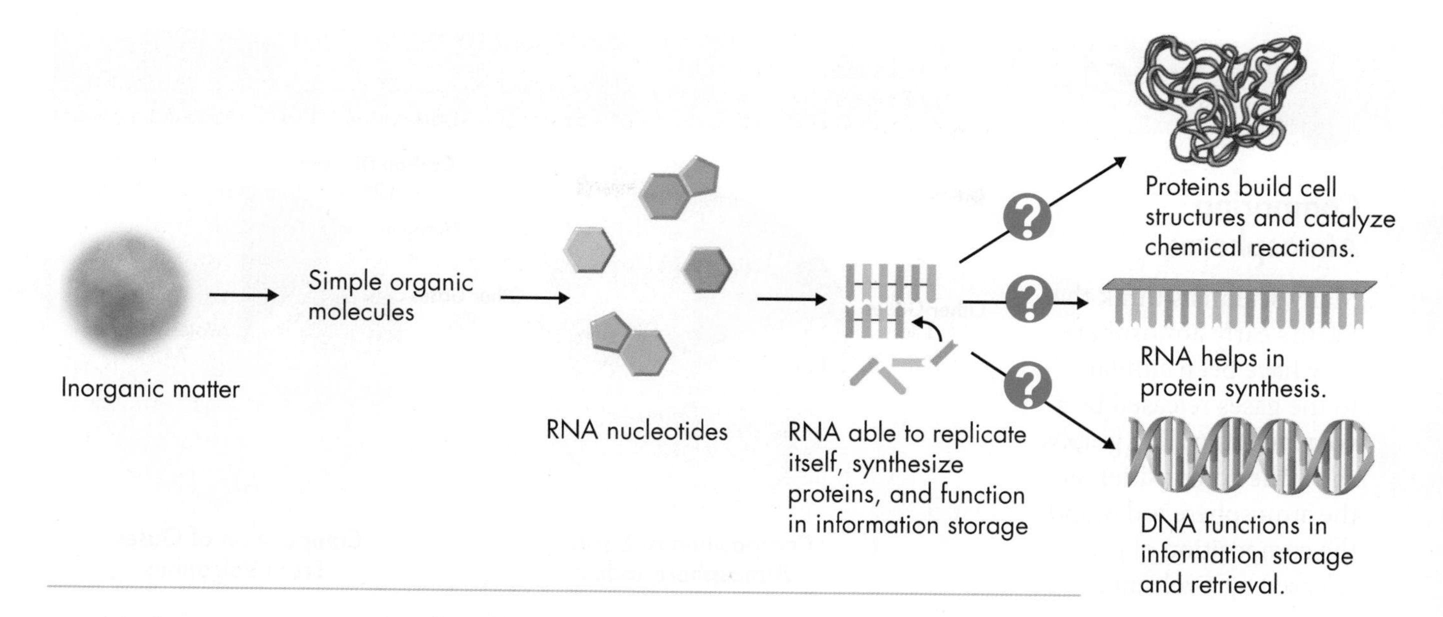

FIGURE 19–16 Origin of RNA and DNA The "RNA world" hypothesis about the origin of life suggests that RNA evolved before DNA. Scientists have not yet demonstrated the later stages of this process in a laboratory setting. **Interpret Visuals** ***How would RNA have stored genetic information?***

Scientists haven't solved this puzzle, but molecular biologists have generated intriguing hypotheses. A number of experiments that simulated conditions on early Earth suggest that small sequences of RNA could have formed from simpler molecules. Why is that interesting? It is interesting because we now know that, under the right conditions, some RNA sequences help DNA replicate. Other RNA sequences process messenger RNA after transcription. Still other RNA sequences catalyze chemical reactions, and some RNA molecules even grow and replicate on their own. **The "RNA world" hypothesis proposes that RNA existed by itself before DNA. From this simple RNA-based system, several steps could have led to DNA-directed protein synthesis.** This hypothesis, shown in **Figure 19–16,** is still being tested.

Production of Free Oxygen Microscopic fossils, or microfossils, of prokaryotes that resemble bacteria have been found in Archean rocks more than 3.5 billion years old. Those first life forms evolved in the absence of oxygen because at that time Earth's atmosphere contained very little of that highly reactive gas.

During the early Proterozoic Eon, photosynthetic bacteria became common. By 2.2 billion years ago, these organisms were churning out oxygen. At first, the oxygen combined with iron in the oceans, producing iron oxide, or rust. Iron oxide, which is not soluble in water, sank to the ocean floor, forming great bands of iron that are the source of most iron ore mined today. Without iron, the oceans changed color from brown to blue-green.

Next, oxygen gas began to accumulate in the atmosphere. The ozone layer began to form, and the skies turned their present shade of blue. Over several hundred million years, oxygen concentrations rose until they reached today's levels. In a sense, this increase in oxygen created the first global "pollution" crisis. To the first cells, which evolved in the absence of oxygen, this reactive gas was a deadly poison! The rise of oxygen in the atmosphere drove some early life forms to extinction. Some organisms, however, evolved new metabolic pathways that used oxygen for respiration. These organisms also evolved ways to protect themselves from oxygen's powerful reactive abilities.

FIGURE 19–17 Fossilized Bacteria Fossilized bacteria are the earliest evidence of life on Earth. These rod-shaped bacterial cells (red) are seen calcified on the shell of a single-celled protozoan.

Analyzing Data

Comparing Atmospheres

Many scientists think that Earth's early atmosphere may have been similar to the gases released by a volcano today. The graphs show the composition of the atmosphere today and the composition of gases released by a volcano.

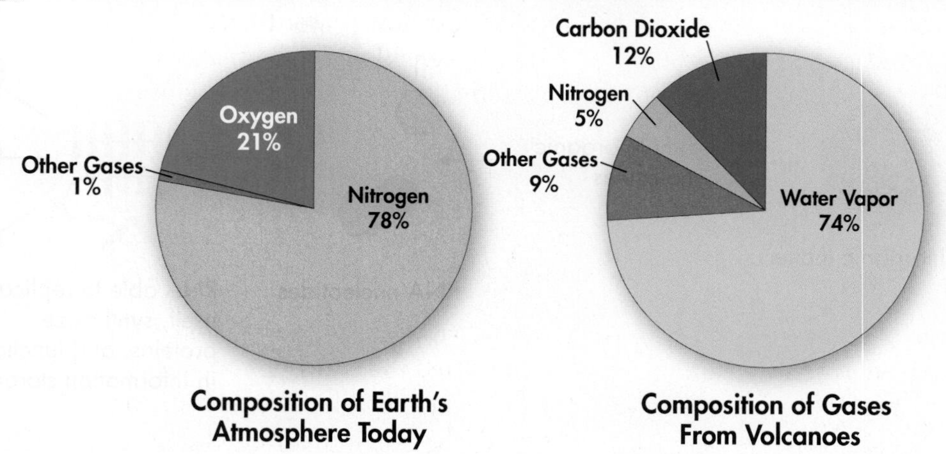

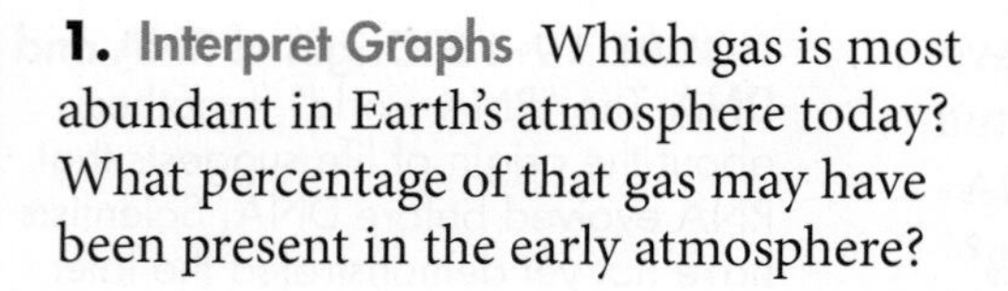

1. **Interpret Graphs** Which gas is most abundant in Earth's atmosphere today? What percentage of that gas may have been present in the early atmosphere?

2. **Interpret Graphs** Which gas was probably most abundant in the early atmosphere?

3. **Infer** Where did the water in today's oceans probably come from?

Origin of Eukaryotic Cells

What theory explains the origin of eukaryotic cells?

One of the most important events in the history of life was the evolution of eukaryotic cells from prokaryotic cells. Remember that eukaryotic cells have nuclei, but prokaryotic cells do not. Eukaryotic cells also have complex organelles. Virtually all eukaryotes have mitochondria, and both plants and algae also have chloroplasts. How did these complex cells evolve?

BUILD Vocabulary

PREFIXES The prefix *endo-* in **endosymbiotic theory** means "within" or "inner." The endosymbiotic theory involves a symbiotic relationship between eukaryotic cells and the prokaryotes within them.

Endosymbiotic Theory Researchers hypothesize that about 2 billion years ago, some ancient prokaryotes began evolving internal cell membranes. These prokaryotes were the ancestors of eukaryotic organisms. Then, according to **endosymbiotic** (en doh sim by AHT ik) **theory,** prokaryotic cells entered those ancestral eukaryotes. These intruders didn't infect their hosts, as parasites would have done, and the host cells didn't digest them, as they would have digested prey. Instead, the small prokaryotes began living inside the larger cells, as shown in **Figure 19–18.**

The endosymbiotic theory proposes that a symbiotic relationship evolved over time, between primitive eukaryotic cells and the prokaryotic cells within them. This idea was proposed more than a century ago. At that time, microscopists saw that the membranes of mitochondria and chloroplasts resembled the cell membranes of free-living prokaryotes. This observation led to two related hypotheses.

One hypothesis proposes that mitochondria evolved from endosymbiotic prokaryotes that were able to use oxygen to generate energy-rich ATP. Inside primitive eukaryotic cells, these energy-generating prokaryotes evolved into mitochondria that now power the cells of all multicellular organisms. Mitochondria enabled cells to metabolize oxygen. Without this ability, cells would have been killed by the free oxygen in the atmosphere.

Another hypothesis proposes that chloroplasts evolved from endosymbiotic prokaryotes that had the ability to photosynthesize. Over time, these photosynthetic prokaryotes evolved within eukaryotic cells into the chloroplasts of plants and algae.

Modern Evidence During the 1960s, Lynn Margulis of Boston University gathered evidence that supported the endosymbiotic theory. Margulis noted first that mitochondria and chloroplasts contain DNA similar to bacterial DNA. Second, she noted that mitochondria and chloroplasts have ribosomes whose size and structure closely resemble those of bacteria. Third, she found that mitochondria and chloroplasts, like bacteria, reproduce by binary fission when cells containing them divide by mitosis. Mitochondria and chloroplasts, then, share many features of free-living bacteria. These similarities provide strong evidence of a common ancestry between free-living bacteria and the organelles of living eukaryotic cells.

In Your Notebook *Describe two hypotheses relating to the endosymbiotic theory.*

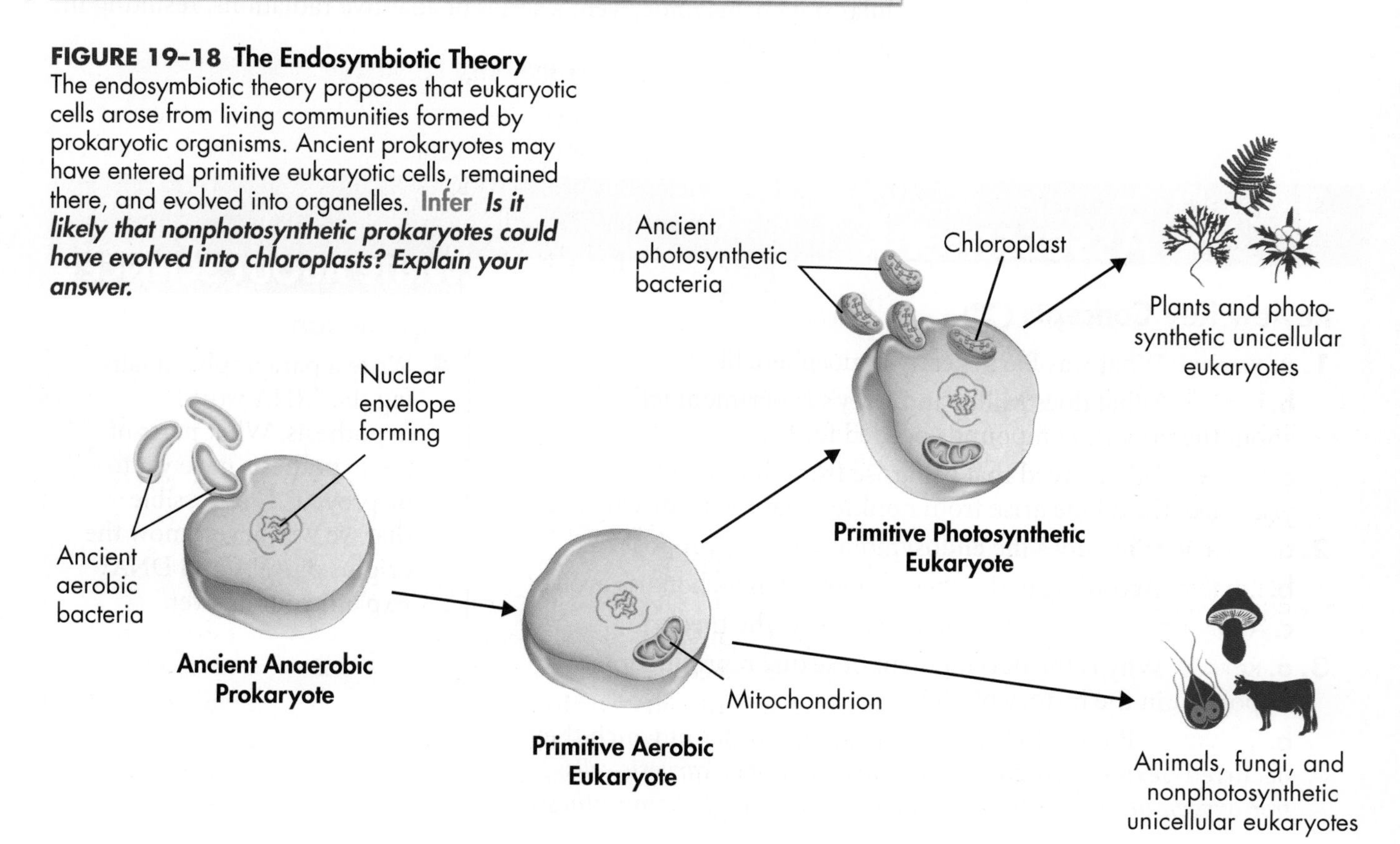

FIGURE 19–18 The Endosymbiotic Theory The endosymbiotic theory proposes that eukaryotic cells arose from living communities formed by prokaryotic organisms. Ancient prokaryotes may have entered primitive eukaryotic cells, remained there, and evolved into organelles. **Infer** ***Is it likely that nonphotosynthetic prokaryotes could have evolved into chloroplasts? Explain your answer.***

Sexual Reproduction and Multicellularity

What is the evolutionary significance of sexual reproduction?
Sometime after eukaryotic cells arose, they began to reproduce sexually. **The development of sexual reproduction sped up evolutionary change because sexual reproduction increases genetic variation.**

Significance of Sexual Reproduction When prokaryotes reproduce asexually, they duplicate their genetic material and pass it on to daughter cells. This process is efficient, but it yields daughter cells whose genomes duplicate their parent's genome. Genetic variation is basically restricted to mutations in DNA.

In contrast, when eukaryotes reproduce sexually, offspring receive genetic material from two parents. Meiosis and fertilization shuffle and reshuffle genes, generating lots of genetic diversity. That's why the offspring of sexually reproducing organisms are never identical to either their parents or their siblings (except for identical twins). The more heritable variation, the more "raw material" natural selection has to work on. Genetic variation increases the likelihood of a population's adapting to new or changing environmental conditions.

Multicellularity Multicellular organisms evolved a few hundred million years after the evolution of sexual reproduction. Early multicellular organisms underwent a series of adaptive radiations, resulting in great diversity.

19.3 Assessment

Review Key Concepts

1. a. **Review** What was Earth's early atmosphere like?
   b. **Explain** What does Miller and Urey's experiment tell us about the organic compounds needed for life?
   c. **Predict** You just read that life arose from nonlife billions of years ago. Could life arise from nonlife today? Why or why not?
2. a. **Review** What does the endosymbiotic theory propose?
   b. **Explain** According to this theory, how did mitochondria evolve?
   c. **Apply Concepts** What evidence supports the theory?
3. a. **Review** Why is the development of sexual reproduction so important in the history of life?
   b. **Sequence** Put the following events in the order in which they occurred: *sexual reproduction, development of eukaryotic cells, free oxygen in the atmosphere,* and *development of photosynthesis.*

WRITE ABOUT SCIENCE

Explanation

4. Write a paragraph explaining the "RNA world" hypothesis. What parts of the hypothesis have yet to be proved? Is it possible that we will never know the origins of RNA and DNA? Explain your answer.

WHST.9-10.2a Text Types and Purposes,
WHST.9-10.10 Range of Writing

Careers & BIOLOGY

More than 99 percent of the species that ever lived are now extinct. If studying past life interests you, you might consider one of the following careers.

FOSSIL PREPARATOR

If you believe what you see in the movies, fossils are usually found perfectly preserved and intact. But the truth is that fossils are almost always found jumbled and encased in rock. Using microscopes and delicate hand tools, fossil preparators remove fossils from the surrounding rock. Preparators carefully reconstruct damaged pieces and record information about fossil position and rock composition.

MUSEUM GUIDE

Museum guides are educators. But instead of using books to teach, they use museum exhibits. A museum guide at a natural history museum, for example, might have fossils that visitors can touch and manipulate. Museum guides also perform demonstrations and give informal talks.

PALEONTOLOGIST

Paleontologists study extinct and ancient life. It is not all about fossils, however. Today paleontolgists use everything from biochemistry to computer modeling to understand the evolutionary relationships among organisms. Living animals are also sometimes used to study movement, behavior, or development.

CAREER CLOSE-UP:

Dr. Kristi Curry Rogers, Curator of Paleontology, Science Museum of Minnesota

Dr. Curry Rogers' work is big—very big. Dr. Curry Rogers is a paleontologist who studies how the giant long-necked sauropod dinosaurs grew. How can you study how an extinct animal grew over 65 million years ago? By studying microscopic bone structure, Dr. Curry Rogers can estimate how long it took the animal to reach full size. This kind of research can help scientists understand how dinosaurs regulated their body temperature. In addition to questions about sauropod growth, Dr. Curry Rogers is also investigating how different sauropods are related.

"Unlike many kids who go through a 'fossil phase,' I never grew out of it!"

WRITING Choose one of the careers described here. Explain why this career is important to understanding the history of life.

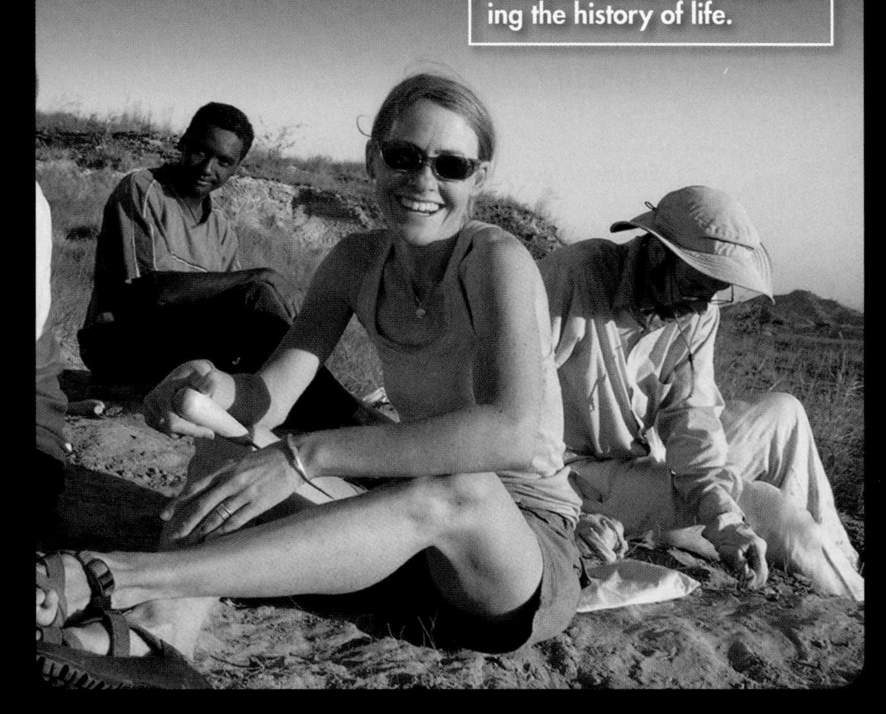

PHANEROZOIC EON

PALEOZOIC ERA

Cambrian Period | Ordovician Period | Silurian Period

Cambrian Period

▲ *Elrathia*

During the Cambrian Period, multicellular life experienced its greatest adaptive radiation in what is called the Cambrian Explosion. Many species were fossilized during this period because many organisms evolved hard body parts, including shells and outer skeletons. Landmasses moved in ways that created vast shallow marine habitats. Jawless fishes first appeared. The Cambrian ended with a large mass extinction in which nearly 30 percent of all animal groups died.

Silurian Period

During the Silurian Period, land areas rose, draining shallow seas and creating moist tropical habitats. Jawless fishes underwent an extensive radiation, and the first fish with true jaws appeared. The first multicellular land plants evolved from aquatic ancestors. Arthropods became the first animals to live on land.

▲ Sea Lily Fossil

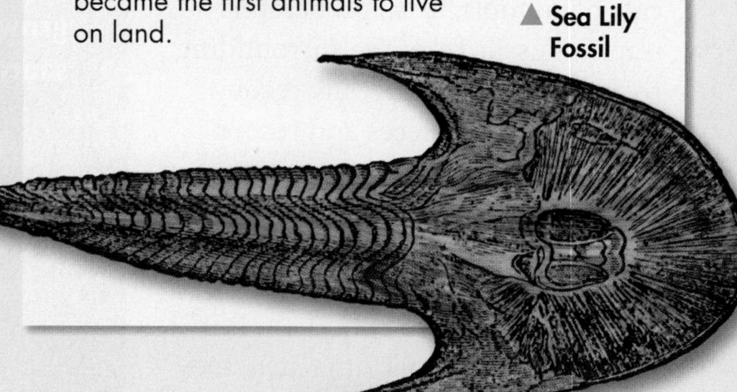

▲ *Cephalaspis* (raylike jawless fish)

▼ *Stenaster* (early sea star)

Pleurocysities (early echinoderms) ▼

Ordovician Period

Oceans flooded large land areas, creating more shallow marine habitats. Animal groups that survived the Cambrian extinction experienced dramatic adaptive radiations. These radiations generated great diversity in major animal phyla. Invertebrates dominated the seas. Early vertebrates evolved bony coverings.

Devonian Period | Carboniferous Period | Permian Period

Devonian Period

During the Devonian Period, invertebrates and vertebrates thrived in the seas. Fishes evolved jaws, bony skeletons, and scales. Sharks began their adaptive radiation. Certain groups of fishes evolved leglike fins, and some of these evolved into the first amphibians. Some land plants, such as ferns, adapted to drier areas. Insects began to radiate on land.

◀ **Fossil Fern From Carboniferous Period**

Permian Period

▲ **Crinoid**

During the Permian Period, invertebrates, vertebrates, and land plants continued to expand over Earth's continents. Reptiles experienced the first of several major adaptive radiations, which produced the ancestors of modern reptiles, dinosaurs, and mammals. The Permian Period ended with the biggest mass extinction of all time. More than 50 percent of terrestrial animal families and more than 95 percent of marine species became extinct.

Early Amphibian ▼

Carboniferous Period

During the Carboniferous Period, mountain building created a wide range of habitats, from swampy lowlands to drier upland areas. Giant ferns, club mosses, and horsetails formed vast swampy forests. Amphibians, insects, and land plants experienced major adaptive radiations. Winged insects evolved into many forms, including huge dragonflies and cockroaches. For early vertebrates, insects were food; for plants, insects were predators. The first reptiles evolved from ancient amphibians.

PHANEROZOIC EON

MESOZOIC ERA

Triassic Period | Jurassic Period | Cretaceous Period

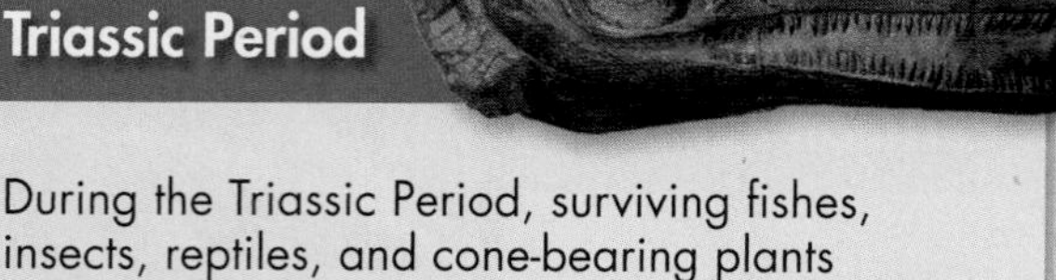

Triassic Period

During the Triassic Period, surviving fishes, insects, reptiles, and cone-bearing plants evolved rapidly. About 225 million years ago, the first dinosaurs evolved. The earliest mammals evolved during the late Triassic. Triassic mammals were very small, about the size of a mouse or shrew.

▲ Living Horsetail

▲ Horsetail Fossil

Jurassic Period

During the Jurassic Period, dinosaurs became the most diverse land animals. They "ruled" for about 150 million years, but different types lived at different times. One lineage of dinosaurs evolved feathers and ultimately led to modern birds. *Archaeopteryx*, the first feathered fossil to be discovered, evolved during this time.

◀ Pterodactyl Fossil

Cretaceous Period

During the Cretaceous Period, *Tyrannosaurus rex* roamed the land, while flying reptiles and birds soared in the sky. Turtles, crocodiles, and other, now-extinct reptiles like plesiosaurs swam among fishes and invertebrates in the seas. Leafy trees, shrubs, and flowering plants emerged and experienced adaptive radiations. The Cretaceous ended with another mass extinction. More than half of all plant and animal groups were wiped out, including all dinosaurs except the ancestors of modern birds.

▲ *T. rex*

▼ *Maiasaura* Nest

CENOZOIC ERA

Paleogene Period

Neogene Period

Quaternary Period

Paleogene Period

During the Paleogene Period, climates changed from warm and moist to cool and dry. Flowering plants, grasses, and insects flourished. After the dinosaurs and giant marine reptiles went extinct, mammals underwent a major adaptive radiation. As climates changed, forests were replaced by open woods and grasslands. Large mammals—ancestors of cattle, deer, and sheep and other grazers—evolved and spread across the grasslands. In the sea, the first whales evolved.

▲ Early Mammal

Neogene Period

During the Neogene Period, colliding continents pushed up modern mountain ranges, including the Alps in Europe and the Rockies, Cascades, and Sierra Nevadas in North America. As mountains rose, ice and snow built up at high elevations and in the Arctic. Falling sea levels and colliding continents created connections between North and South America, and between Africa, Europe, and Asia. Those connections led to great movements of land animals between continents. Climates continued a cooling and drying trend, and grasslands continued to expand. Modern grazing animals continued to coevolve with grasses, evolving specialized digestive tracts to deal with tough, low-nutrient grass tissue.

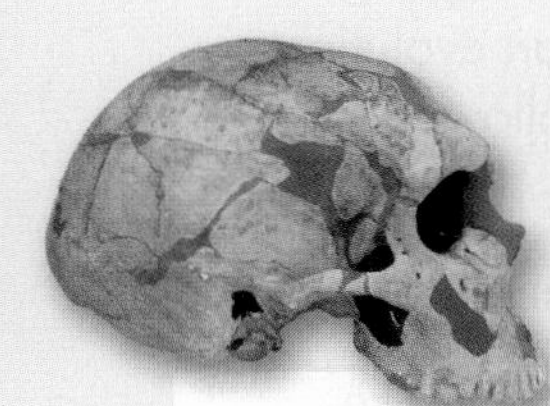

◀ Neanderthal Skull

Quaternary Period

During the Quaternary Period, Earth cooled. A series of ice ages saw thick glaciers advance and retreat over parts of Europe and North America. So much water was frozen in glaciers that sea levels fell by more than 100 meters. Then, about 20,000 years ago, Earth's climate began to warm. Over thousands of years, glaciers melted, and sea levels rose. In the oceans, algae, coral, mollusks, fishes, and mammals thrived. Insects and birds shared the skies. Land mammals—among them bats, cats, dogs, cattle, and mammoths—became common. Between 6 and 7 million years ago, one group of mammals began an adaptive radiation that led to the ancestors and close relatives of modern humans.

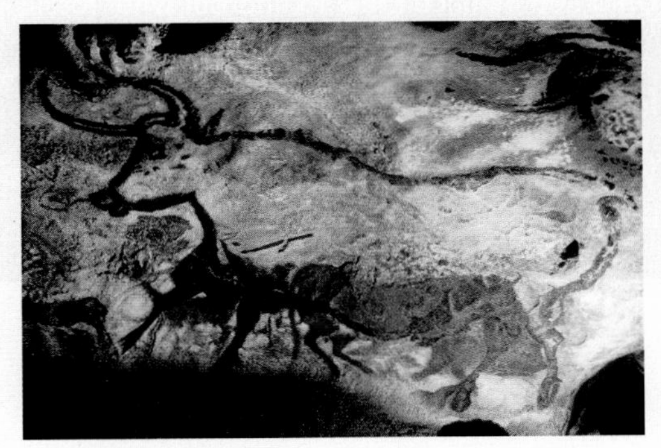

▲ Cave painting

NGSS Smart Guide History of Life

Disciplinary Core Idea LS4.C Adaptation: How does the environment influence populations of organisms over multiple generations? Paleontologists use fossils to learn about the structure and environments of ancient organisms. Fossils also give clues to events that happened during Earth's history.

19.1 The Fossil Record

- From the fossil record, paleontologists learn about the structure of ancient organisms, their environment, and the ways in which they lived.
- Relative dating allows paleontologists to determine whether a fossil is older or younger than other fossils. Radiometric dating uses the proportion of radioactive to stable isotopes to calculate the age of a sample.
- The geologic time scale is based on both relative and absolute dating. The major divisions of the geologic time scale are eons, eras, and periods.
- Building mountains, opening coastlines, changing climates, and geological forces have altered habitats of living organisms repeatedly throughout Earth's history. In turn, the actions of living organisms over time have changed conditions in the land, water, and atmosphere of planet Earth.

extinct 538 • paleontologist 539 • relative dating 540 • index fossil 540 • radiometric dating 540 • half-life 540 • geologic time scale 542 • era 543 • period 543 • plate tectonics 544

Biology.com

Untamed Science Video Go back in time with the Untamed Science crew to find out what fossils reveal.

Art in Motion View a short animation that shows how fossils form.

Visual Analogy Compare geologic time to a 24-hour clock.

19.2 Patterns and Processes of Evolution

- If the rate of speciation in a clade is equal to or greater than the rate of extinction, the clade will continue to exist. If the rate of extinction in a clade is greater than the rate of speciation, the clade will eventually become extinct.
- Evidence shows that evolution has often proceeded at different rates for different organisms at different times over the long history of life on Earth.
- Two important patterns of macroevolution are adaptive radiation and convergent evolution. Adaptive radiation occurs when a single species or a small group of species evolves over a relatively short time into several different forms that live in different ways. Convergent evolution occurs when unrelated organisms evolve into similar forms.
- The relationship between two coevolving organisms often becomes so specific that neither organism can survive without the other. Thus, an evolutionary change in one organism is usually followed by a change in the other organism.

macroevolutionary patterns 546 • background extinction 548 • mass extinction 548 • gradualism 549 • punctuated equilibrium 549 • adaptive radiation 550 • convergent evolution 551 • coevolution 551

Biology.com

Data Analysis Correlate data on extinction events with other types of data to identify likely causes of extinction.

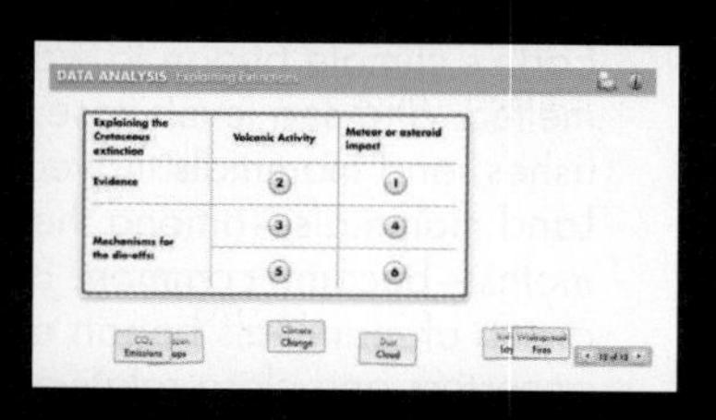

CHECKING YOUR *Scientific Literacy*

Refer to the lesson content and digital assets below as you prepare for your chapter assessment. Then, evaluate your understanding of the earth's early history by answering these questions.

1. Scientific and Engineering Practice **Developing and Using Models** Use Figure 19-10 *Models of Evolution* to summarize how species can change at different rates over time.

2. Crosscutting Concept **Patterns** Expand your summary to include how the gradualism and punctuated equilibrium models of evolution can be applied to interpreting patterns seen in the fossil record.

3. Text Types and Purposes Suppose an almost complete fossil of an ancient bird has just been discovered. Write a paragraph describing how the fossil may have formed.

4. STEM Using common materials such as clay, sand, and shells, develop and build a physical model to demonstrate how fossils are formed. Use your model to communicate the role index fossils play in helping paleontologists date rocks and fossils.

19.3 Earth's Early History

- Earth's early atmosphere contained little or no oxygen. It was principally composed of carbon dioxide, water vapor, and nitrogen, with lesser amounts of carbon monoxide, hydrogen sulfide, and hydrogen cyanide.
- Miller and Urey's experiment suggested how mixtures of the organic compounds necessary for life could have arisen from simpler compounds on a primitive Earth.
- The "RNA world" hypothesis proposes that RNA existed by itself before DNA. From this simple RNA-based system, several steps could have led to DNA-directed protein synthesis.
- The endosymbiotic theory proposes that a symbiotic relationship evolved over time between primitive eukaryotic cells and the prokaryotic cells within them.
- The development of sexual reproduction sped up evolutionary change because sexual reproduction increases genetic variation.

endosymbiotic theory 556

Biology.com

Art Review Review your understanding of the composition of Earth's early atmosphere as compared with the composition of Earth's current atmosphere.

Lesson Notes Sequence the hypothesized events from the origin of Earth to the appearance of eukaryotic cells.

Forensics Lab

Using Index Fossils This chapter lab is available in your lab manual and at Biology.com

19 Assessment

19.1 The Fossil Record

Understand Key Concepts

1. Scientists who specialize in the study of fossils are called
a. biologists.
b. paleontologists.
c. zoologists.
d. geologists.

2. Sedimentary rocks usually form when layers of small particles are compressed
a. in the atmosphere.
b. in a snow field.
c. in mountains.
d. under water.

3. Using C–14 to analyze rock layers
a. is a method of estimating absolute age.
b. is a method of estimating relative age.
c. can only be used on extremely ancient rock layers.
d. is impossible because rock layers do not contain carbon.

4. Half-life is the time required for half the atoms in a radioactive sample to
a. decay.
b. double.
c. expand.
d. be created.

5. According to the theory of plate tectonics,
a. Earth's climate has changed many times.
b. Earth's continents move very slowly.
c. evolution occurs at different rates.
d. giant asteroids crashed into Earth in the past.

6. How does relative dating enable paleontologists to estimate a fossil's age?

7. Explain how radioactivity is used to date rocks.

8. What is the geologic time scale, and how was it developed?

9. How have the activities of organisms affected Earth's environment?

Think Critically

10. **Calculate** The half-life of carbon-14 is 5730 years. What is the age of a fossil containing 1/16 the amount of carbon-14 of living organisms? Explain your calculation. MATH

11. **Apply Concepts** Evolutionary biologists say that there is a good reason for gaps in the fossil record. Can you explain why some extinct animals and plants were never fossilized?

19.2 Patterns and Processes of Evolution

Understand Key Concepts

12. The process that produces similar-looking structures in unrelated groups of organisms is
a. adaptive radiation.
b. coevolution.
c. convergent evolution.
d. mass extinction.

13. The general term for large-scale evolutionary changes that take place over long periods of time is called
a. macroevolution.
b. coevolution.
c. convergent evolution.
d. geologic time.

14. Cladograms that are based on the fossil record always show
a. which organisms are direct ancestors of the others.
b. relationships based on shared derived characteristics.
c. that clades are made up only of extinct species.
d. relative ages of organisms in the clade.

15. **Text Types and Purposes** Explain and give an example of the process of adaptive radiation.

16. **Text Types and Purposes** Explain the model of evolution known as punctuated equilibrium.

17. Use an example to explain the concept of coevolution.

Think Critically

18. **Infer** Major geologic changes often go hand in hand with mass extinctions. Why do you think this is true?

19. **Apply Concepts** Why is rapid evolution especially likely to occur in a small population that has been separated from the main population?

20. **Craft and Structure** What is the role of natural selection in adaptive radiation? How do these processes lead to diversity?

19.3 Earth's Early History

Understand Key Concepts

21. Earth's early atmosphere contained little or no

a. water vapor. **c.** nitrogen.
b. carbon dioxide. **d.** oxygen.

22. In their experiment that modeled conditions on ancient Earth, Miller and Urey used electric sparks to simulate

a. temperature.
b. sunlight.
c. atmospheric gases.
d. lightning.

23. Outlines of ancient cells that are preserved well enough to identify them as prokaryotes are

a. microfossils. **c.** autotrophs.
b. heterotrophs. **d.** phototrophic.

24. What hypotheses have scientists proposed to explain Earth's early atmosphere and the way the oceans formed?

25. The diagram below shows the apparatus that Miller and Urey used in their experiment. Explain both what water and gases were meant to represent and what Miller and Urey were hoping to accomplish.

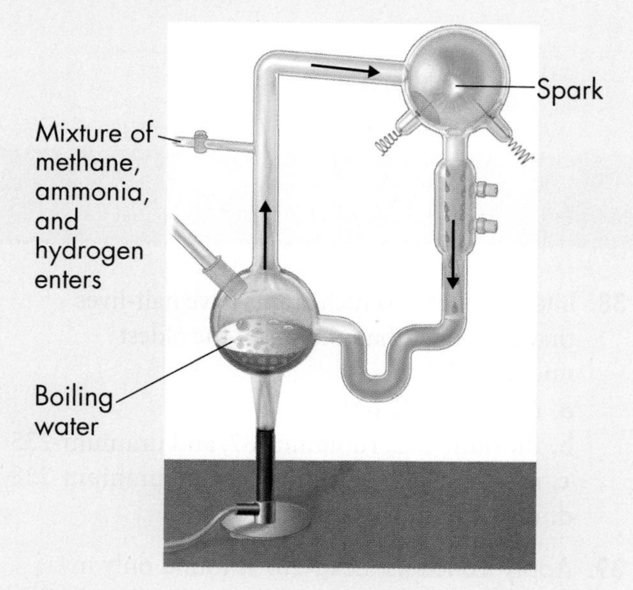

26. How are proteinoid microspheres similar to living cells?

27. How did the addition of oxygen to Earth's atmosphere affect the evolution of life?

28. According to the endosymbiotic theory, how did mitochondria originate?

solve the CHAPTER MYSTERY

MURDER IN THE PERMIAN

Solving a 250-million-year-old murder mystery isn't easy! In recent years, scientists have studied the chemistry of Permian rocks and changes in the fossil record. Some researchers determined that enormous and long-lasting volcanic eruptions in Siberia vented carbon dioxide into the atmosphere, causing a massive change in global climate. This put species and ecosystems under great environmental stress.

Other researchers used geochemical analyses to show that atmospheric oxygen levels dropped to roughly half of what they are today. Huge parts of the oceans lost all oxygen. Because of the reduction in available oxygen, land animals near sea level might have been gasping for breath as you would on top of Mount Everest.

Finally, there is evidence that an asteroid hit Earth! To this day, paleontologists are testing competing hypotheses that try to explain which of the events that occurred at this time caused the mass extinction. However, these hypotheses are constantly changing and have probably changed since this book was written.

1. Compare and Contrast How do current hypotheses about the Permian extinction compare with the predominant theory about the Cretaceous extinction?

2. Research to Build and Present Knowledge Drawing evidence from the information in this book, suggest an explanation for the Permian mass extinction.

3. Pose Questions What questions could you ask to find out whether your hypothesis is correct? What evidence would answer those questions?

4. Connect to the Big idea What role have mass extinctions played in the history of life?

Think Critically

29. **Use Models** What part of Miller and Urey's apparatus represents rain? What important role would rain play in chemical evolution?
30. **Relate Cause and Effect** How do you think the cells that took in the ancestors of mitochondria and chloroplasts benefited from the relationship?

Connecting Concepts

Use Science Graphics

The diagram shows rock layers in two different places. Use the diagram to answer questions 31 and 32.

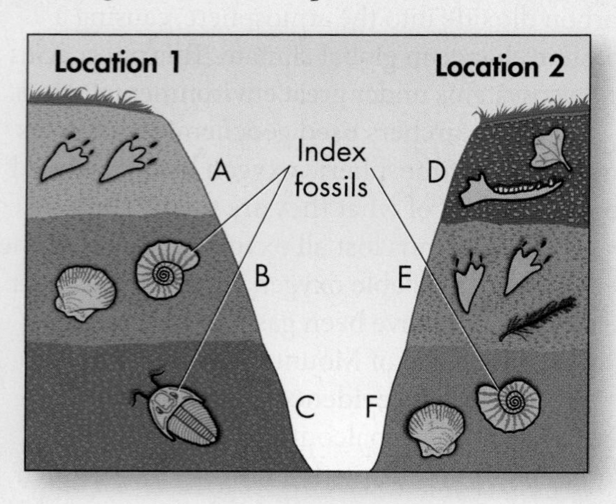

31. Which fossils are probably older—those in layer A or those in layer C? How do you know?
32. **Integration of Knowledge and Ideas** Which rock layer in location 2 is probably about the same age as layer C in location 1? How do you know?
33. What are the characteristics of a useful index fossil?

Write About Science

34. **Text Types and Purposes** Write a paragraph comparing conditions on early Earth with those on modern Earth.
35. **Text Types and Purposes** Use the example of body shape in sharks, dolphins, and penguins to explain convergent evolution.
36. **Assess the Big idea** Explain how the formation of sedimentary rock gives paleontologists information about the sequence in which life forms appeared on Earth.
37. **Assess the Big idea** When describing their theory of punctuated equilibrium, Stephen Jay Gould and Niles Eldredge often used the motto "stasis is data." Stasis is another word for equilibrium. Explain what Gould and Eldredge meant.

Analyzing Data

The table below compares the half-life of several radioactive atoms. Use the table to answer questions 38 and 39.

Isotope and Decay Product	Half-Life (years)
Rubidium-87 ⟶ Strontium-87	48.8 billion
Thorium-232 ⟶ Lead-208	14.0 billion
Uranium-235 ⟶ Lead-207	704.0 million
Uranium-238 ⟶ Lead-206	4.5 billion

38. **Interpret Data** Which atoms have half-lives that are longer than the age of the oldest microfossils?
    a. uranium-235 only
    b. thorium-232, rubidium-87, and uranium-235
    c. rubidium-87, thorium-232, and uranium-238
    d. uranium-235 and rubidium-87
39. **Apply Concepts** Lead-207 is found only in rocks that also contain uranium-235. Analysis of a sample shows that it has three times as many atoms of lead-207 as there are atoms of uranium-235. How many half-lives have passed since this rock formed?
    a. one   b. two   c. three   d. four

Standardized Test Prep

Multiple Choice

1. Useful index fossils are found
- **A** in a small area for a short time.
- **B** in a small area for a long time.
- **C** over a large area for a short time.
- **D** over a large area for a long time.

2. What happens if the rate of extinction in a clade is greater than the rate of speciation?
- **A** The clade will eventually become extinct.
- **B** The clade will continue to exist.
- **C** The species in the clade will become more varied.
- **D** The number of species in the clade will stay the same.

3. Which of the following is evidence for the endosymbiotic theory?
- **A** Mitochondria and chloroplasts contain DNA similar to bacterial DNA.
- **B** Mitochondria and chloroplasts have similar functions in the cell.
- **C** Mitochondria and chloroplasts have no DNA of their own.
- **D** Mitochondria and chloroplasts can live independently when removed from the eukaryotic cell.

4. Carbon–14 is NOT useful for dating most fossils because
- **A** it has a very long half-life.
- **B** it has a very short half-life.
- **C** most organisms contain more potassium than carbon.
- **D** it is found only in certain rock layers.

5. The movement of continents has played a significant role in evolution because
- **A** continents move rapidly and some organisms cannot adjust.
- **B** without the movement of continents, there would be no water on Earth.
- **C** the movement of continents has caused environments to change.
- **D** all mass extinctions are the result of continental drift.

Questions 6 and 7

The graph shows the decay of radioactive isotopes. Use the information in the graph to answer the questions that follow.

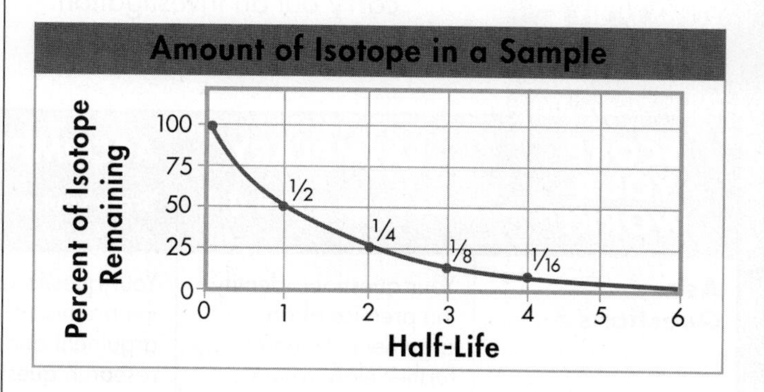

6. The half-life of thorium-230 is 75,000 years. How long will it take for $\frac{7}{8}$ of the original amount of thorium-230 in a sample to decay?
- **A** 75,000 years
- **B** 225,000 years
- **C** 25,000 years
- **D** 150,000 years

7. The half-life of potassium-40 is about 1.3 billion years. After four half-lives have passed, how much of the original sample will be left?
- **A** $\frac{1}{16}$
- **B** $\frac{1}{16} \times$ 1300 million grams
- **C** $\frac{1}{4}$
- **D** $\frac{1}{4} \times$ 1300 million grams

Open-Ended Response

8. How does the process by which sedimentary rock forms allow scientists to determine the relative ages of fossils?

If You Have Trouble With . . .

Question	1	2	3	4	5	6	7	8
See Lesson	19.1	19.2	19.3	19.1	19.1	19.1	19.1	19.1

NGSS Problem-based Learning

UNIT PROJECT 5

Your Self-Assessment

The rubric below will help you evaluate your work as you plan and carry out an investigation.

SCORE YOUR WORK!	EXEMPLARY Score your work a 4 if:	ACCOMPLISHED Score your work a 3 if:	DEVELOPING Score your work a 2 if:	BEGINNING Score your work a 1 if:
Ask Questions ▶	Your questions identify the premise of an argument, request further elaboration, and refine a research question.	Your questions identify the premise of an argument and refine a research question.	Your questions partially challenge the premise of an argument. Questions don't help refine a research question.	Your questions do not challenge the premise of an argument.
Develop and Use Models ▶	You constructed models that can be used as the basis of an explanation or prediction about what is happening in your region.	Your model can be used as the basis of an explanation or prediction but does not specifically apply to your region.	Your model is developing, but you need to simplify it to better visualize the scenario.	Your model is developing, but you need help in simplifying it so you can use it as the basis of an explanation.
Plan and Carry Out an Investigation ▶	You have formulated a question that can be investigated based on your model. You have identified what kind of data is needed and how the data should be gathered.	You have formulated a question that can be investigated and is based on your model. You have identified what data should be gathered.	You have formulated a question that can be investigated and is based on your model, but you need to identify what and how data should be gathered.	You have formulated a question that can be investigated, but it isn't based on your model.
	TIP A hypothesis should be based on a well-developed model.			
Use Mathematical and Computational Thinking ▶	You have used appropriate mathematical and computational tools in the statistical analysis of data.	You have used some appropriate computational tools in analyzing data.	You have identified which tools to use for analyzing data.	You need help identifying the tools needed for analyzing data.
	TIP Express relationship and quantities in appropriate mathematical forms for scientific modeling and investigation.			
Engage in Argument from Evidence ▶	You created an article or blog that includes a scientific argument showing how data support a claim.	The article or blog is based on one of your arguments and is persuasive. But it does not communicate enough information to help the audience understand the issue.	The article or blog is based on one of your arguments. But it is not persuasive and does not communicate enough information to help the audience understand the issue.	The article or blog does not communicate enough information. It is not based on one of your arguments and is not persuasive.
	TIP Use language needed to talk about the argument, with terms such as *claim, reason,* and *data.*			

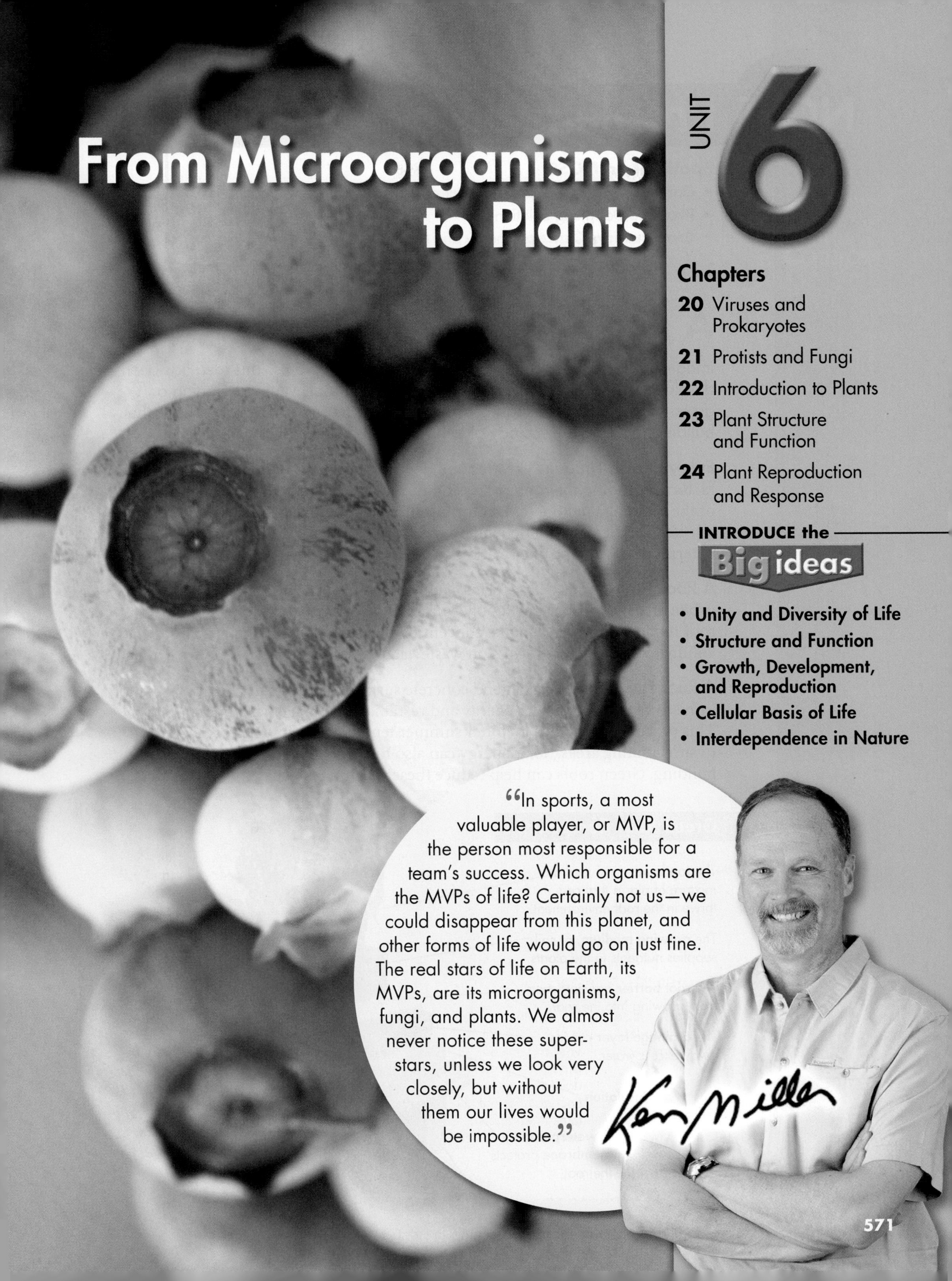

From Microorganisms to Plants
UNIT 6
Chapters
20 Viruses and Prokaryotes
21 Protists and Fungi
22 Introduction to Plants
23 Plant Structure and Function
24 Plant Reproduction and Response
INTRODUCE the Big ideas
• Unity and Diversity of Life
• Structure and Function
• Growth, Development, and Reproduction
• Cellular Basis of Life
• Interdependence in Nature
“In sports, a most valuable player, or MVP, is the person most responsible for a team’s success. Which organisms are the MVPs of life? Certainly not us—we could disappear from this planet, and other forms of life would go on just fine. The real stars of life on Earth, its MVPs, are its microorganisms, fungi, and plants. We almost never notice these superstars, unless we look very closely, but without them our lives would be impossible.”
Ken Miller

NGSS Problem-based Learning

Spotlight on NGSS
- **Core Idea LS2.B** Cycles of Matter and Energy Transfer in Ecosystems
- **Practice** Designing Solutions
- **Crosscutting Concept** Energy and Matter

UNIT PROJECT 6

A Living Roof

Imagine covering the rooftop of a building with live plants. This idea has been developed into the technology known as a green roof. When designing a green roof, an architect must consider not only the structural properties of the building, but also the ecology of the site and its surroundings.

Every Roof Is Unique The area and depth of a green roof depend on many factors, including the roof's accessibility, the plants selected, the load capacity of the building, and the budget for installation and upkeep. The types of plants selected for a green roof depend on the climate of the area. The climate also determines what type of irrigation system is needed to keep the plants healthy.

A Roof That Cleans Rainstorms can overwhelm wastewater systems in urban areas because water simply runs off the surfaces of buildings. A green roof retains more water than a conventional roof. This results in less runoff volume, which puts less stress on wastewater systems.

A Roof That Cools Asphalt and concrete surfaces of buildings and infrastructure absorb solar energy and re-radiate it to the environment. This process contributes to hotter summer temperatures in cities compared to surrounding areas. Solar energy can also raise the temperature inside a building. Green roofs can help reduce these warming effects.

Green Roof Components

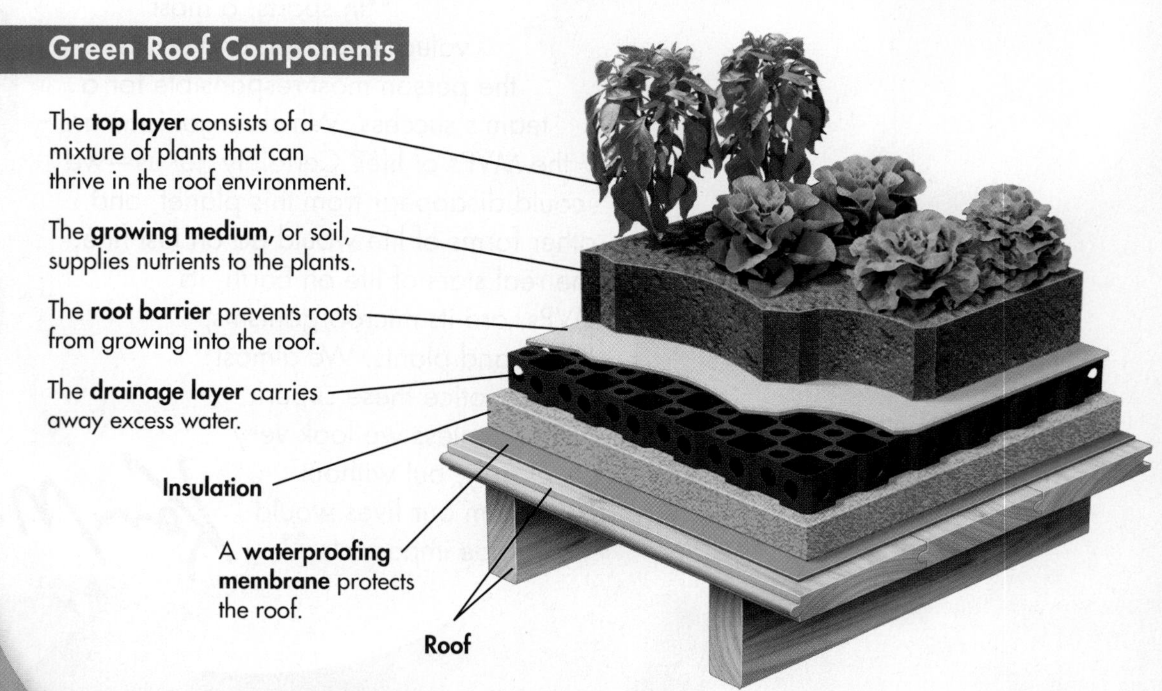

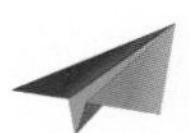

Your Task: Green That Roof!

STEM

Design a green roof system for a building of your choosing. The building may be one that you are already familiar with, such as your home or your school, or it may be a place that you have never visited before. Your design plan should explain the problem you aim to solve, the factors that guide your decision making, and details about the installation, materials, and cost of your green roof. Use the rubric at the end of the unit to evaluate your work for this task.

Define the Problem Research the site of your proposed green roof installation. What problem are you trying to solve? In what ways is the problem ecological? In what ways is the problem economic? Explain.

Develop Models Research the different types of green roof systems available and the materials used in their construction. Compare the advantages and disadvantages of each system. With the help of your class, brainstorm ways to "green" your roof. Create sketches or scale models of the systems you think will work for your green roof.

Design a Solution Your green roof must meet certain performance requirements in order to successfully solve the problem. These requirements are your design criteria. You also must specify the constraints that limit the scope of your design solution. Modify your model to meet your design criteria and constraints. Use your classmates' feedback to help you further modify and refine your roof design. Explain how your revised design optimizes the achievement of your design criteria.

Communicate Information Write a proposal to the owner of the site of your green roof design. Your proposal should outline the costs and benefits of installing the green roof and make a convincing case for adopting your design.

Biology.com

Go online to learn more about green roofs from around the world.

A green roof, such as this one atop Chicago's City Hall, does more than protect the human inhabitants underneath. It can also reduce the need for air conditioning, absorb rainwater, and provide beauty.

20 Viruses and Prokaryotes

Big idea **Cellular Basis of Life**

Q: Are all microbes that make us sick made of living cells?

Colonies of E. coli *bacteria*

Deepen your understanding of viruses and prokaryotes by engaging in key practices that allow you to make connections across concepts.

NGSS You will develop and use models to demonstrate how the structures of viruses and prokaryotes affect their functions.

STEM Not all bacteria are bad! Learn about engineering a specialized bacterium that can reduce the toxic effects of radioactive waste sites.

Common Core You will summarize, represent, and interpret data, and compare and contrast findings related to viruses and prokaryotes.

INSIDE:

- **20.1 Viruses**
- **20.2 Prokaryotes**
- **20.3 Diseases Caused by Bacteria and Viruses**

THE MAD COWS

In 1986, something strange began to happen to cattle in Great Britain. Without warning, the animals began acting strangely, losing control of their movements, staggering and stumbling, and eventually dying. Farmers watched helplessly as the disease they called "mad cow" spread through their cattle. The disease affected more than 30,000 cattle in 1991.

Studies of the brains of cattle killed by mad cow disease showed that large areas of the animals' brains had been destroyed. Under the microscope, the holes in the tissue made the brain resemble a sponge. Because of this, the disease was given the name bovine spongiform encephalopathy, or BSE. But the cause of the disease was a mystery. As you read this chapter, look for clues that explain the culprit behind this disease. Then, solve the mystery.

Biology.com

Finding the solution to The Mad Cows mystery is only the beginning. Take a video field trip with the ecogeeks of Untamed Science to see where the mystery leads.

Go online to access additional resources including:
• eText • Flash Cards • Lesson Overviews • Chapter Mystery

20.1 Viruses

Key Questions

How do viruses reproduce?

What happens after a virus infects a cell?

Vocabulary

virus
capsid
bacteriophage
lytic infection
lysogenic infection
prophage
retrovirus

Taking Notes

Venn Diagram Make a Venn diagram in which to record the similarities and differences between viruses and cells. Fill it in as you read the lesson.

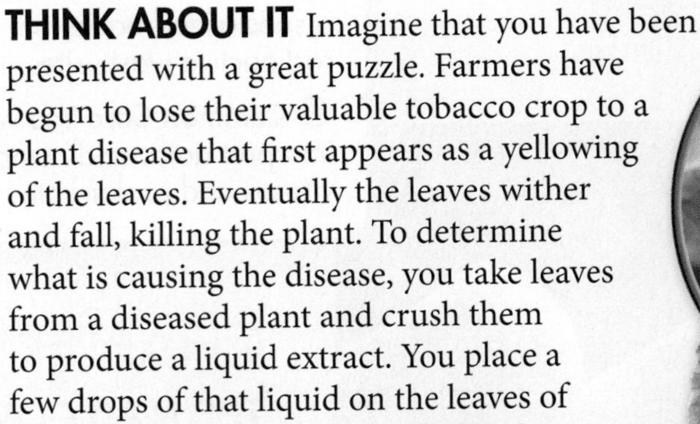

THINK ABOUT IT Imagine that you have been presented with a great puzzle. Farmers have begun to lose their valuable tobacco crop to a plant disease that first appears as a yellowing of the leaves. Eventually the leaves wither and fall, killing the plant. To determine what is causing the disease, you take leaves from a diseased plant and crush them to produce a liquid extract. You place a few drops of that liquid on the leaves of healthy plants. A few days later, the leaves turn yellow where you put the drops.

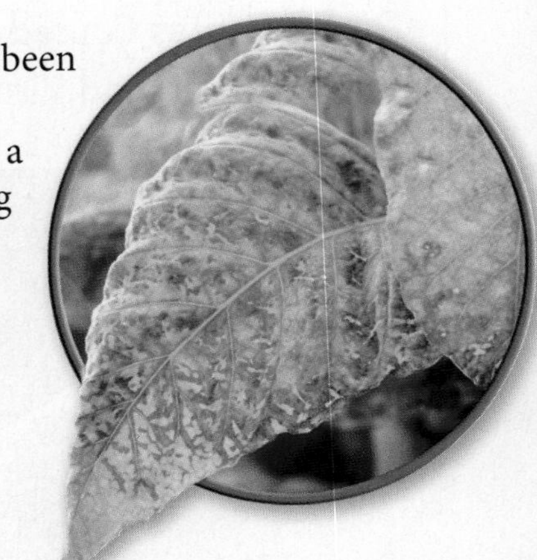

You use a light microscope to look for a germ that might cause the disease, but none can be seen. In fact, when even the tiniest of cells are filtered out of the liquid, the liquid still causes the disease. You hypothesize that the liquid must contain disease-causing agents so small that they are not visible under the microscope and can pass right through the filter. Although you cannot see the tiny disease-causing particles, you're sure they are there. What would you do next? How would you deal with the invisible?

The Discovery of Viruses

How do viruses reproduce?

If you think you could have carried out the investigation described above, congratulations! You're walking in the footsteps of a 28-year-old Russian biologist, Dmitri Ivanovski. In 1892, Ivanovski demonstrated that the cause of this particular plant disease—called tobacco mosaic disease—was found in the liquid extracted from infected plants. But he could not pin down the culprit.

Discovery of Viruses In 1897, Dutch scientist Martinus Beijerinck suggested that tiny particles in the juice caused the disease, and he named these particles *viruses,* after the Latin word for "poison." Then, in 1935, the American biochemist Wendell Stanley isolated crystals of tobacco mosaic virus. Living organisms do not crystallize, so Stanley inferred that viruses were not truly alive. This is a conclusion that biologists still recognize as being valid today. A **virus** is a nonliving particle made of proteins, nucleic acids, and sometimes lipids. **Viruses can reproduce only by infecting living cells.**

MYSTERY CLUE

British scientists carefully investigated the veterinary histories of 169 cattle with BSE. All 169 had been given feed enriched with meat and bone meal protein from slaughtered cattle. How could this practice spread a disease?

Structure and Composition Viruses differ widely in terms of size and structure, as you can see in **Figure 20–1.** Most viruses are so small they can be seen only with the aid of a powerful electron microscope. The protein coat surrounding a virus is called a **capsid.** In addition, some viruses, such as the influenza virus, have an additional membrane that surrounds the capsid. The simplest viruses contain only a few genes, whereas the most complex may have hundreds of genes.

To enter a host cell, most viruses have proteins on their surface membrane or capsid that bind to receptor proteins on the cell. In either case, the proteins "trick" the cell to take in the virus, or in some cases just its genetic material. Once inside the cell, the viral genes are eventually expressed and may destroy the cell.

Because viruses must bind precisely to proteins on the host cell surface and then use the host's genetic system, most viruses infect only a very specific kind of cell. Plant viruses infect plant cells; most animal viruses infect only certain related species of animals; bacterial viruses infect only certain types of bacteria. Viruses that infect bacteria are called **bacteriophages,** which literally means "bacteria eaters."

T4 Bacteriophage

Head
DNA
Tail fiber
Tail sheath
TEM 60,000×

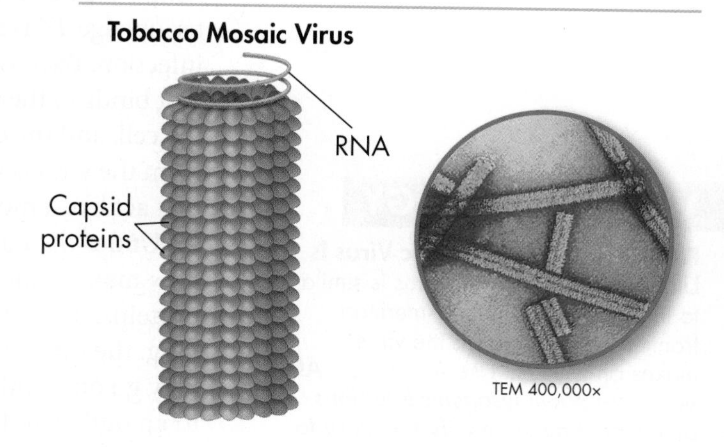

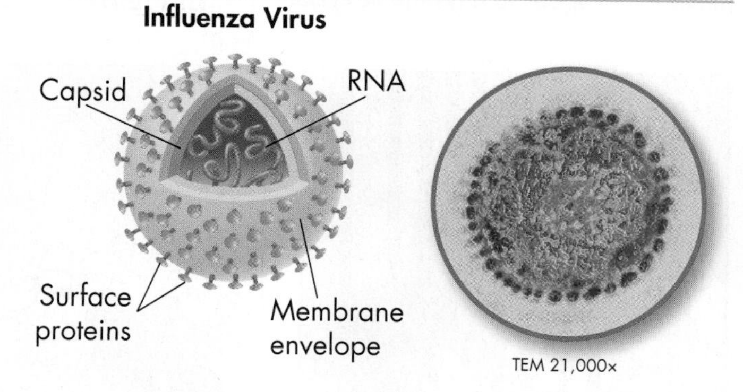

FIGURE 20–1 Diversity of Viral Forms Viruses come in a wide variety of sizes and shapes. Three types of viruses are shown here. **Interpret Diagrams** ***What kind of nucleic acid does each virus type have?***

Quick Lab
GUIDED INQUIRY

How Do Viruses Differ in Structure?

1. Make models of two of the viruses shown in **Figure 20–1.**
2. Label the parts of each of your virus models.
3. Measure and record the length of each of your virus models in centimeters. Convert the length of each model into nanometers by using the following formula: 1 cm = 10 million nm. MATH
4. Measure the length of each virus you modeled. Divide the length of each model by the length of the actual virus to determine how much larger each model is than the virus it represents. MATH

Analyze and Conclude

1. **Use Models** Which parts of your models are found in all viruses?
2. **Draw Conclusions** Which parts of one or both of your models are found in only some viruses?
3. **Calculate** How many times larger are your models than the viruses they represent? MATH

Viral Infections

What happens after a virus infects a cell?

After a virus has entered a host cell, what happens? **Inside living cells, viruses use their genetic information to make multiple copies of themselves. Some viruses replicate immediately, while others initially persist in an inactive state within the host.** These two patterns of infection are called lytic infection and lysogenic infection.

Lytic Infections In a **lytic infection,** a virus enters a bacterial cell, makes copies of itself, and causes the cell to burst, or lyse (LYS). Bacteriophage *T4* is an example of a bacteriophage that causes such an infection. Bacteriophage *T4* has a DNA core inside a protein capsid that binds to the surface of a host cell. The virus injects its DNA into the cell, and the cell then begins to make messenger RNA (mRNA) from the viral genes. The viral mRNA is translated into viral proteins that act like a molecular wrecking crew, chopping up the cell's DNA.

Under the control of viral genes, the host cell's metabolic system now makes thousands of copies of viral nucleic acid and capsid proteins. The viral DNA is assembled into new virus particles. Before long, the infected cell lyses, releasing hundreds of virus particles that may go on to infect other cells. In its own way, a lytic virus is similar to an outlaw in the Wild West of the American frontier, as illustrated in **Figure 20–2.**

VISUAL ANALOGY

FIGURE 20–2 How a Lytic Virus Is Like an Outlaw A lytic virus is similar to the Wild West of the American frontier in the demands the virus makes on its host. **Use Analogies** ***After you learn about lysogenic infections on the next page, modify this story to make it analogous to a lysogenic cycle.***

First, the outlaw eliminates the town's existing authority.
Lytic Infection The host cell's DNA is chopped up.

Next, the outlaw demands to be outfitted with new equipment from the local townspeople.
Lytic Infection Viruses use the host cell to make viral DNA and viral proteins.

Finally, the outlaw forms a gang that leaves the town to attack new communities.
Lytic Infection The host cell bursts, releasing hundreds of virus particles.

Lysogenic Infection Some bacterial viruses, including the bacteriophage *lambda,* cause a **lysogenic infection,** in which a host cell is not immediately taken over. Instead, the viral nucleic acid is inserted into the host cell's DNA, where it is copied along with the host DNA without damaging the host. Viral DNA multiplies as the host cells multiply. In this way, each generation of daughter cells derived from the original host cell is infected.

Bacteriophage DNA that becomes embedded in the bacterial host's DNA is called a **prophage.** The prophage may remain part of the DNA of the host cell for many generations. Influences from the environment—including radiation, heat, and certain chemicals—trigger the prophage to become active. It then removes itself from the host cell DNA and directs the synthesis of new virus particles. The lysogenic infection now becomes an active lytic infection, as shown in **Figure 20–3.**

The details of viral infection in eukaryotic cells differ in many ways from viral infection of bacteria by bacteriophages. But for the most part, the basic patterns of infection in animals and other eukaryotes are similar to the lytic and lysogenic infections of bacteria.

In Your Notebook *Describe how a lysogenic infection can change into a lytic infection.*

BUILD Vocabulary

WORD ORIGINS The adjective *lytic,* the verb *lyse,* and the prefix *lyso-* all come from the Greek word *lyein,* meaning "to loosen or break up."

FIGURE 20–3 Comparing Two Types of Bacteriophage Infection Viruses that infect bacteria, called bacteriophages, may infect cells in one of two ways: lytic infection or lysogenic infection.

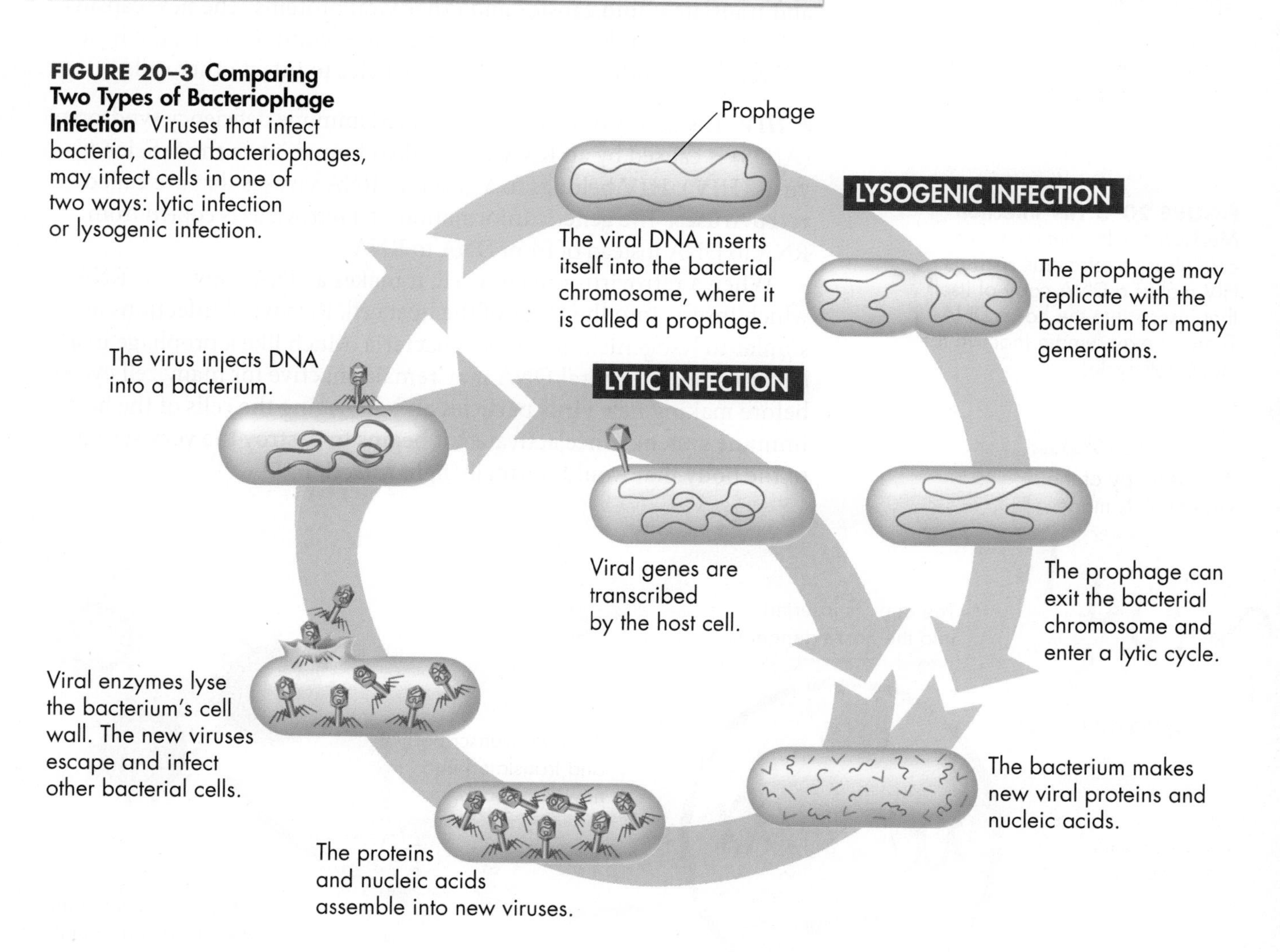

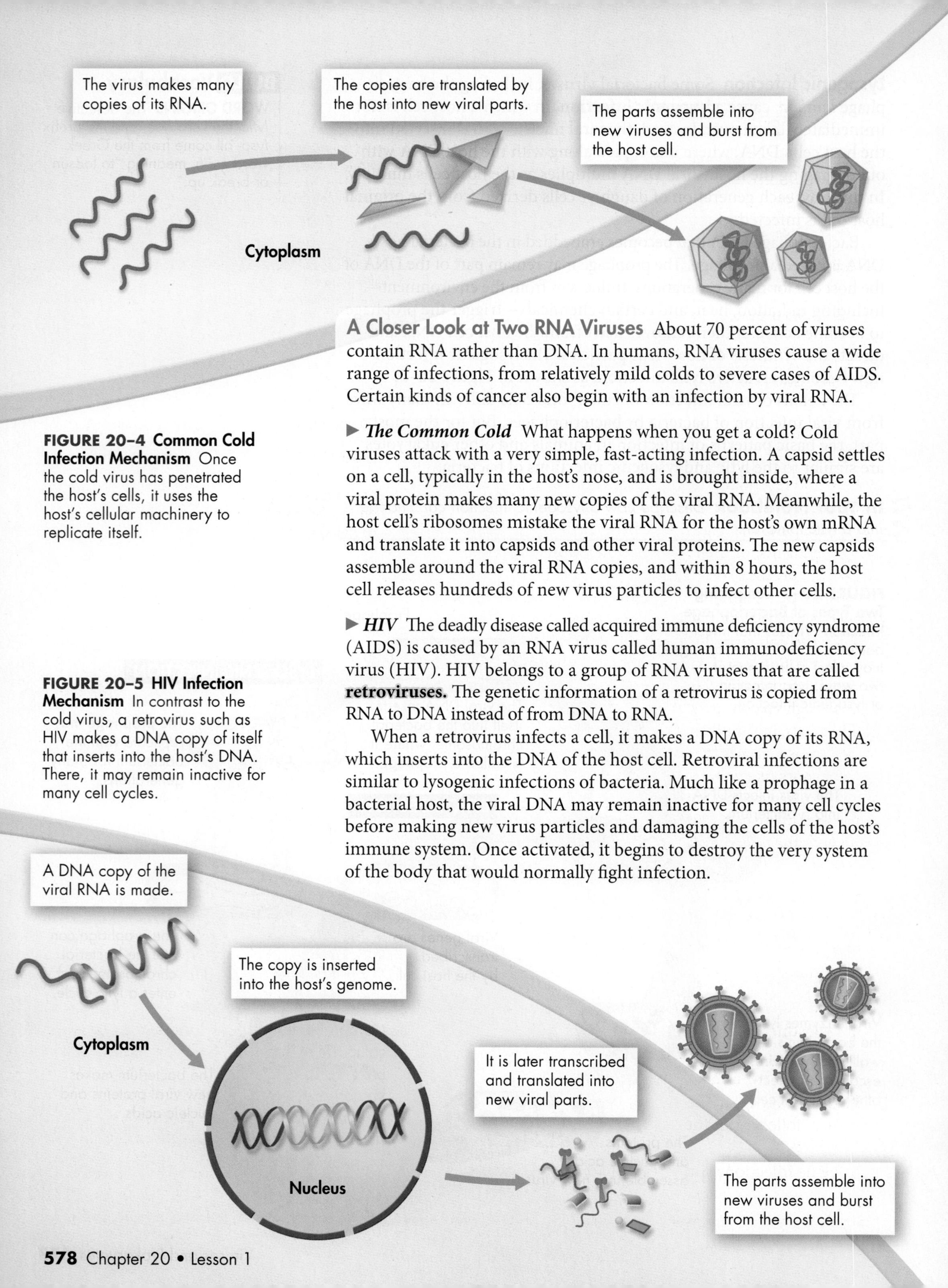

FIGURE 20–4 Common Cold Infection Mechanism Once the cold virus has penetrated the host's cells, it uses the host's cellular machinery to replicate itself.

A Closer Look at Two RNA Viruses About 70 percent of viruses contain RNA rather than DNA. In humans, RNA viruses cause a wide range of infections, from relatively mild colds to severe cases of AIDS. Certain kinds of cancer also begin with an infection by viral RNA.

▶ ***The Common Cold*** What happens when you get a cold? Cold viruses attack with a very simple, fast-acting infection. A capsid settles on a cell, typically in the host's nose, and is brought inside, where a viral protein makes many new copies of the viral RNA. Meanwhile, the host cell's ribosomes mistake the viral RNA for the host's own mRNA and translate it into capsids and other viral proteins. The new capsids assemble around the viral RNA copies, and within 8 hours, the host cell releases hundreds of new virus particles to infect other cells.

▶ ***HIV*** The deadly disease called acquired immune deficiency syndrome (AIDS) is caused by an RNA virus called human immunodeficiency virus (HIV). HIV belongs to a group of RNA viruses that are called **retroviruses.** The genetic information of a retrovirus is copied from RNA to DNA instead of from DNA to RNA.

When a retrovirus infects a cell, it makes a DNA copy of its RNA, which inserts into the DNA of the host cell. Retroviral infections are similar to lysogenic infections of bacteria. Much like a prophage in a bacterial host, the viral DNA may remain inactive for many cell cycles before making new virus particles and damaging the cells of the host's immune system. Once activated, it begins to destroy the very system of the body that would normally fight infection.

FIGURE 20–5 HIV Infection Mechanism In contrast to the cold virus, a retrovirus such as HIV makes a DNA copy of itself that inserts into the host's DNA. There, it may remain inactive for many cell cycles.

Viruses and Cells

Characteristic	Virus	Cell
Structure	DNA or RNA in capsid, some with envelope	Cell membrane, cytoplasm; eukaryotes also contain nucleus and many organelles
Reproduction	Only within a host cell	Independent cell division, either asexually or sexually
Genetic Code	DNA or RNA	DNA
Growth and Development	No	Yes; in multicellular organisms, cells increase in number and differentiate
Obtain and Use Energy	No	Yes
Response to Environment	No	Yes
Change Over Time	Yes	Yes

FIGURE 20–6 Comparing Viruses and Cells The differences between viruses and cells are listed in this chart. Form an Opinion ***Based on this information, would you classify viruses as living or nonliving? Explain.***

Viruses and Cells Viruses must infect living cells in order to grow and reproduce, taking advantage of the nutrients and cellular machinery of their hosts. This means that all viruses are parasites. Parasites depend entirely upon other living organisms for their existence, harming these organisms in the process.

Despite the fact that they are not alive, viruses have many of the characteristics of living things. After infecting living cells, viruses can reproduce, regulate gene expression, and even evolve. Some of the main differences between cells and viruses are summarized in **Figure 20–6.**

Although viruses are smaller and simpler than the smallest cells, it is unlikely that they were the first organisms. Because viruses are dependent upon living organisms, it seems more likely that viruses developed after living cells. In fact, the first viruses may have evolved from the genetic material of living cells. Viruses have continued to evolve, along with the cells they infect, for billions of years.

20.1 Assessment

Review Key Concepts

1. a. Review What do viruses depend on for their reproduction?

b. Compare and Contrast How is viral reproduction different from that of cell-based organisms?

2. a. Review Describe each of the two paths viruses may follow once they have entered a cell.

b. Compare and Contrast How are lytic and lysogenic infections similar? How are they different?

Apply the Big idea

Structure and Function

3. Compare the structure of a virus to the structure of both a prokaryotic cell and a eukaryotic cell. Use a graphic organizer of your choice to organize the information. You may wish to refer to Chapter 7, which discusses the structures of cells in detail.

20.2 Prokaryotes

Key Questions

How are prokaryotes classified?

How do prokaryotes vary in their structure and function?

What roles do prokaryotes play in the living world?

Vocabulary

prokaryote
bacillus
coccus
spirillum
binary fission
endospore
conjugation

Taking Notes

Preview Visuals Look at **Figure 20–9.** Describe in your own words the three shapes of prokaryotes shown.

For more on the diversity of Bacteria and Archaea, go to the Visual Guide. **DOL•6–DOL•9.**

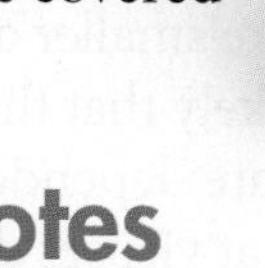

THINK ABOUT IT Imagine living all your life as a member of what you believe is the only family on your street. Then, one morning, you open the front door and discover houses all around you. You see neighbors tending their gardens and children walking to school. Where did all the people come from? What if the answer turned out to be that they had always been there—you just hadn't seen them? How would your view of the world change? The sudden appearance of the new neighbors would be quite a shock.

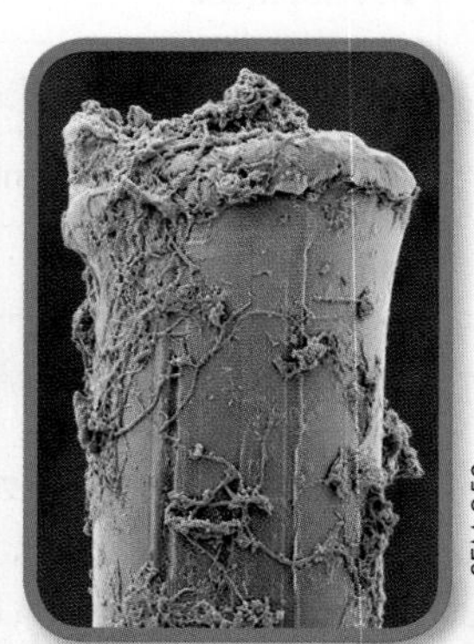

When the microscope was first invented, we humans had just such a shock. Suddenly, the street was very crowded! Far from being alone, we share every corner of our world with microorganisms. Even something that seems clean, like a toothbrush, may be covered with a film of bacteria.

Classifying Prokaryotes

How are prokaryotes classified?

Microscopic life covers nearly every square centimeter of Earth. The smallest and most abundant of these microorganisms are **prokaryotes** (pro KAR ee ohts)—unicellular organisms that lack a nucleus. Prokaryotes have DNA, like all other cells, but their DNA is not found in a membrane-bound nuclear envelope as it is in eukaryotes. Prokaryote DNA is located in the cytoplasm. For many years, most prokaryotes were simply called "bacteria." We now know, however, that the classification of prokaryotes is more complex.

Until recently, all prokaryotes were placed in a single kingdom. More recently, however, biologists have divided prokaryotes into two very distinct groups: Bacteria and Archaea. These groups are as different from each other as both are from eukaryotes. Therefore, biologists now consider each group of prokaryotes as a separate domain. **Prokaryotes are classified as Bacteria or Archaea—two of the three domains of life.** Eukarya is the third domain. The domain Bacteria corresponds to the kingdom Eubacteria. The domain Archaea corresponds to the kingdom Archaebacteria.

FIGURE 20–7 Typical Bacterial Structure A bacterium such as *E. coli* has the basic structure typical of most prokaryotes. *E. coli* also has an outer membrane composed of lipids. This outer membrane is not present in all bacteria. The micrograph shows *E. coli* undergoing binary fission, with pili visible.

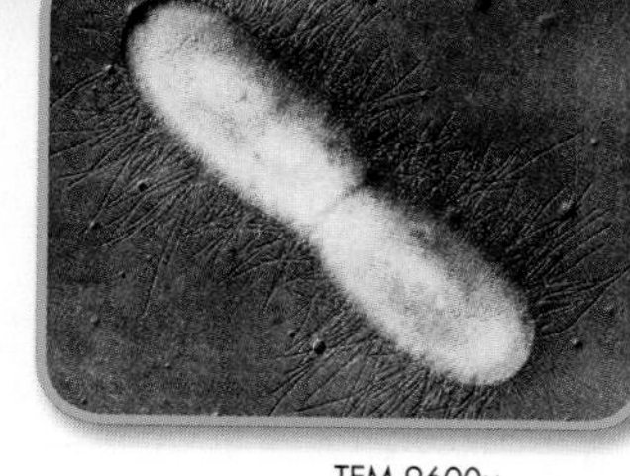

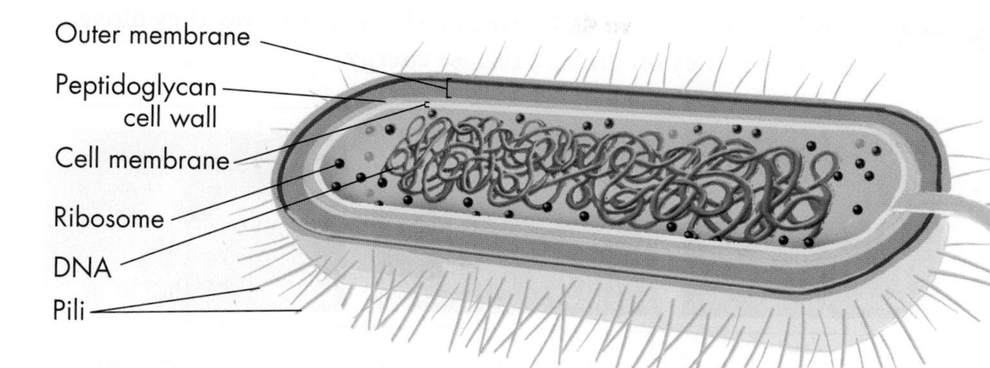

Bacteria The larger of the two domains of prokaryotes is the Bacteria. Bacteria include a wide range of organisms with lifestyles so different that biologists do not agree exactly how many phyla are needed to classify this group. Bacteria live almost everywhere. They live in fresh water, in salt water, on land, and on and within the bodies of humans and other eukaryotes. **Figure 20–7** shows a diagram of *Escherichia coli,* a typical bacterium that lives in human intestines.

Bacteria are usually surrounded by a cell wall that protects the cell from injury and determines its shape. The cell walls of bacteria contain peptidoglycan—a polymer of sugars and amino acids that surrounds the cell membrane. Some bacteria, such as *E. coli,* have a second membrane outside the peptidoglycan wall that makes the cell especially resistant to damage. In addition, some prokaryotes have flagella that they use for movement, or pili (PY ly; singular: pilus), which in *E. coli* serve mainly to anchor the bacterium to a surface or to other bacteria.

Archaea Under a microscope, archaea look very similar to bacteria. Both are equally small, lack nuclei, and have cell walls, but there are important differences. For instance, the walls of archaea lack peptidoglycan, and their membranes contain different lipids. Also, the DNA sequences of key archaea genes are more like those of eukaryotes than those of bacteria. Based on these and other observations, scientists have concluded that archaea and eukaryotes are related more closely to each other than to bacteria.

Many archaea live in extremely harsh environments. One group of archaea produce methane gas and live in environments with little or no oxygen, such as thick mud and the digestive tracts of animals. Other archaea live in extremely salty environments, such as Utah's Great Salt Lake, or in hot springs where temperatures approach the boiling point of water.

In Your Notebook *Create a Venn diagram in which you compare and contrast the characteristics of bacteria and archaea.*

FIGURE 20–8 Habitat for Archaea This hot spring in New Zealand is teeming with archaea that thrive at extremely hot temperatures.

Structure and Function

How do prokaryotes vary in their structure and function?

Because prokaryotes are so small, it may seem hard to tell them apart. **Prokaryotes vary in their size and shape, in the way they move, and in the way they obtain and release energy.**

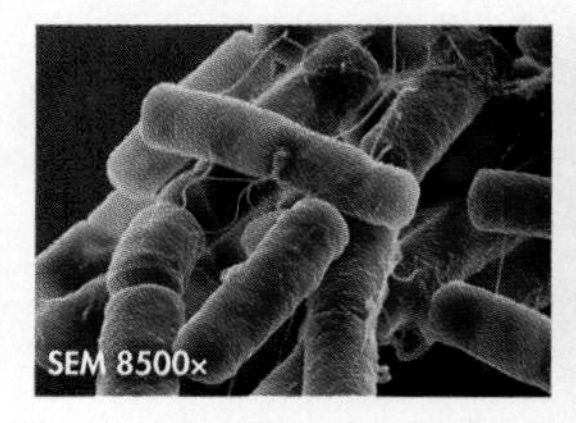

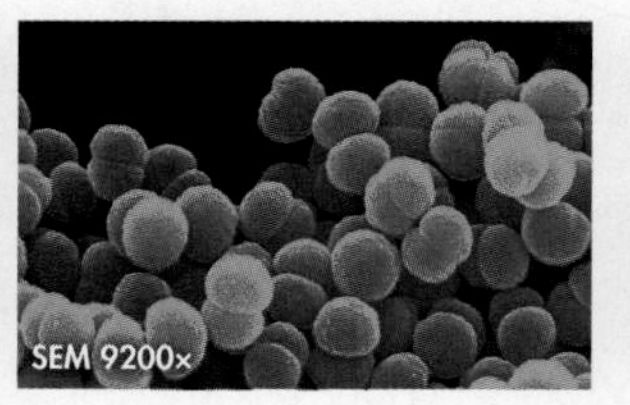

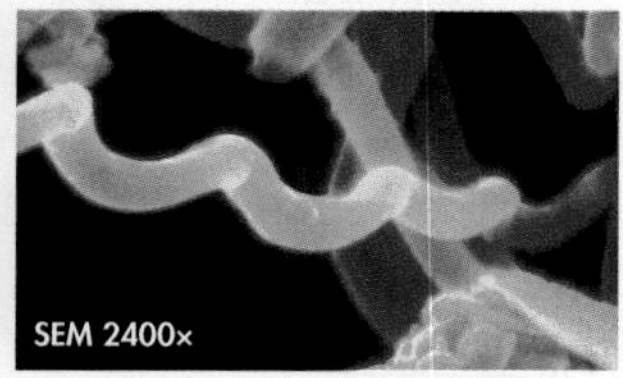

FIGURE 20–9 Prokaryotic Shapes Prokaryotes usually have one of three basic shapes: bacilli (left), cocci (middle), or spirilla (right).

Size, Shape, and Movement Prokaryotes range in size from 1 to 5 micrometers, making them much smaller than most eukaryotic cells. Prokaryotes come in a variety of shapes, as shown in **Figure 20–9.** Rod-shaped prokaryotes are called **bacilli** (buh SIL eye; singular: bacillus). Spherical prokaryotes are called **cocci** (KAHK sy; singular: coccus). Spiral and corkscrew-shaped prokaryotes are called **spirilla** (spy RIL uh; singular: spirillum). You can also distinguish prokaryotes by whether they move and how they move. Some prokaryotes do not move at all. Others are propelled by flagella. Some glide slowly along a layer of slimelike material they secrete.

Nutrition and Metabolism Like all organisms, prokaryotes need a supply of chemical energy, which they store in the form of fuel molecules such as sugars. Energy is released from these fuel molecules during cellular respiration, fermentation, or both. The diverse ways prokaryotes obtain and release energy are summarized in **Figure 20–10.** Notice that some species are able to change their method of energy capture or release depending on the conditions of their environment.

FIGURE 20–10 Energy Capture and Release by Prokaryotes Prokaryotes vary in the ways they obtain energy and the ways they release it. **Interpret Tables** ***What is the term for a prokaryote that uses only light as its energy source?***

Energy Capture by Prokaryotes

Mode of Nutrition	How Energy Is Captured	Habitat	Example
Heterotroph "other feeder"	Take in organic molecules from environment or other organisms to use as both energy and carbon supply	Wide range of environments	*Clostridium*
Photoheterotroph "light and other feeder"	Like basic heterotrophs, but also use light energy	Where light is plentiful	*Rhodobacter, Chloroflexus*
Photoautotroph "light self-feeder"	Use light energy to convert CO_2 into carbon compounds	Where light is plentiful	*Anabaena*
Chemoautotroph "chemical self-feeder"	Use energy released by chemical reactions involving ammonia, hydrogen sulfide, etc.	In chemically harsh and/or dark environments: deep in the ocean, in thick mud, in digestive tracts of animals, in boiling hot springs	*Nitrobacter* ▼ TEM 3000×

Growth, Reproduction, and Recombination When a prokaryote has grown so that it has nearly doubled in size, it replicates its DNA and divides in half, producing two identical cells. This type of reproduction is known as **binary fission.** Because binary fission does not involve the exchange or recombination of genetic information, it is a form of asexual reproduction. When conditions are favorable, prokaryotes can grow and divide at astonishing rates. Some divide as often as once every 20 minutes!

FIGURE 20–11
Endospore Formation

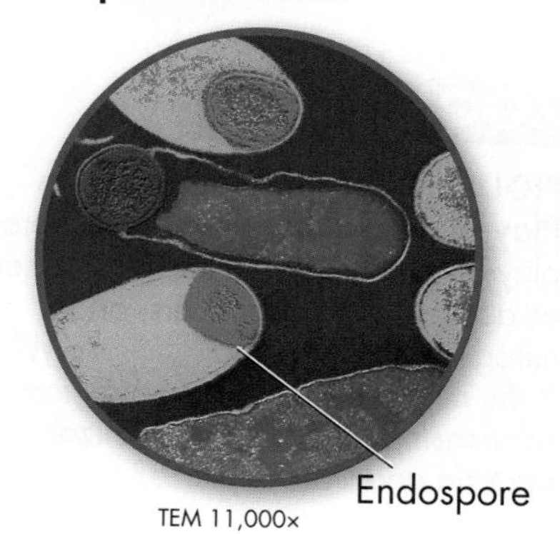

When growth conditions become unfavorable, many prokaryotic cells form an **endospore**—a thick internal wall that encloses the DNA and a portion of the cytoplasm. Endospores can remain dormant for months or even years. The ability to form endospores makes it possible for some prokaryotes to survive very harsh conditions. The bacterium *Bacillus anthracis*, which causes the disease anthrax, is one such bacterium.

As in any organism, adaptations that increase the survival and reproduction of a particular prokaryote are favored. Recall that in organisms that reproduce sexually, genes are shuffled and recombined during meiosis. But prokaryotes reproduce asexually. So, how do their populations evolve?

FIGURE 20–12
Conjugation

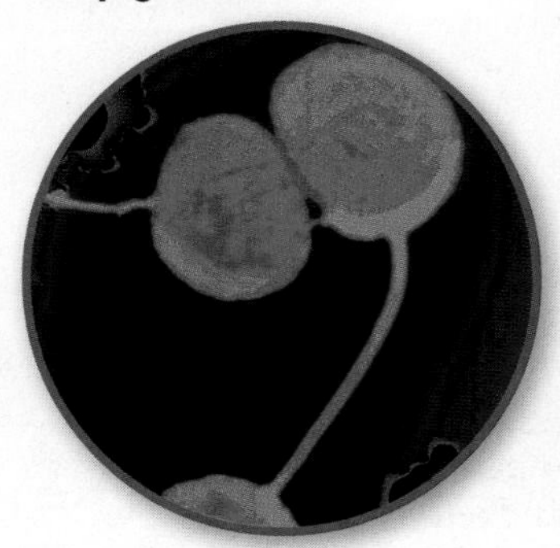

► ***Mutation*** Mutations are one of the main ways prokaryotes evolve. Recall from Chapter 13 that mutations are random changes in DNA that occur in all organisms. In prokaryotes, mutations are inherited by daughter cells produced by binary fission.

► ***Conjugation*** Many prokaryotes exchange genetic information by a process called conjugation. During **conjugation,** a hollow bridge forms between two bacterial cells, and genetic material, usually in the form of a plasmid, moves from one cell to the other. Many plasmids carry genes that enable bacteria to survive in new environments or to resist antibiotics that might otherwise prove fatal. This transfer of genetic information increases genetic diversity in populations of prokaryotes.

Energy Release by Prokaryotes

Mode of Metabolism	How Energy Is Released	Habitat	Example
Obligate aerobe "requiring oxygen"	Cellular respiration; must have ready supply of O_2 to release fuel energy	Oxygen-rich environments, such as near water surface or in animal lungs	*Mycobacterium tuberculosis*: Sometimes found in human lungs
Obligate anaerobe "requiring a lack of oxygen"	Fermentation; die in presence of oxygen	◄ Environments lacking O_2, such as deep soil, animal intestines, or airtight containers	*Clostridium botulinum*: Sometimes found in improperly sterilized canned food, causing food poisoning
Facultative anaerobe "surviving without oxygen when necessary"	Can use either cellular respiration or fermentation as necessary	Oxygen-rich or oxygen-poor environments	*E. coli*: Lives aerobically in sewage and anaerobically in human large intestine

The Importance of Prokaryotes

What roles do prokaryotes play in the living world?

You may remember the star actors in the last movie you saw, but have you ever stopped to consider whether there would be any film at all without the hundreds of workers who never appear on screen? Prokaryotes are just like those unseen workers. **Prokaryotes are essential in maintaining every aspect of the ecological balance of the living world. In addition, some species have specific uses in human industry.** Three roles played by prokaryotes in the environment are shown in **Figure 20–13.**

Decomposers Every living thing depends on a supply of raw materials for its survival. If these materials were not recovered when organisms died, life could not continue. By assisting in breaking down, or decomposing, dead organisms, prokaryotes supply raw materials and thus help to maintain equilibrium in the environment. Bacterial decomposers are also essential to industrial sewage treatment, helping to produce purified water and chemicals that can be used as fertilizers.

Producers Photosynthetic prokaryotes are among the most important producers on the planet. The tiny cyanobacterium *Prochlorococcus* is probably the most abundant photosynthetic organism in the world. This species alone may account for more than half of the primary production in the open ocean. Food chains everywhere are dependent upon prokaryotes as producers of food and biomass.

ZOOMING IN

FIGURE 20–13 Ecological Roles Played by Prokaryotes Prokaryotes play important roles in the environment as decomposers, producers, and nitrogen fixers. **Apply Concepts** ***What is the name of the type of ecological relationship found between Rhizobium bacteria and legume plants?***

Bacteria of the genus *Rhizobium* often live symbiotically within nodules attached to roots of legumes such as clover, where they convert atmospheric nitrogen into a form that is useable by plants.

Cyanobacteria in the genus *Anabaena* form filamentous chains in ponds and other aquatic environments, where they perform photosynthesis.

Bacteria called actinomycetes are present in soil and in rotting plant material such as fallen logs, where they decompose complex organic molecules into simpler molecules.

Nitrogen Fixers All organisms need nitrogen to make proteins and other molecules. But while nitrogen gas (N_2) makes up 80 percent of Earth's atmosphere, only a few kinds of organisms—all of them prokaryotes—can convert N_2 into useful forms. The process of nitrogen fixation converts nitrogen gas into ammonia (NH_3). Ammonia can then be converted to nitrates that plants use, or attached to amino acids that all organisms use. Nitrogen-fixing bacteria and archaea provide 90 percent of the nitrogen used by other organisms. The rest is provided by nitrogen-containing compounds from weathering rocks. Some even comes when lightning combines oxygen and nitrogen in the atmosphere.

Some plants even have symbiotic relationships with nitrogen-fixing prokaryotes. The bacterium *Rhizobium* grows in nodules, or knobs, on the roots of legume plants such as clover and soybean. The *Rhizobium* bacteria within these nodules convert nitrogen in the air into the nitrogen compounds essential for plant growth. In effect, these plants have fertilizer factories in their roots!

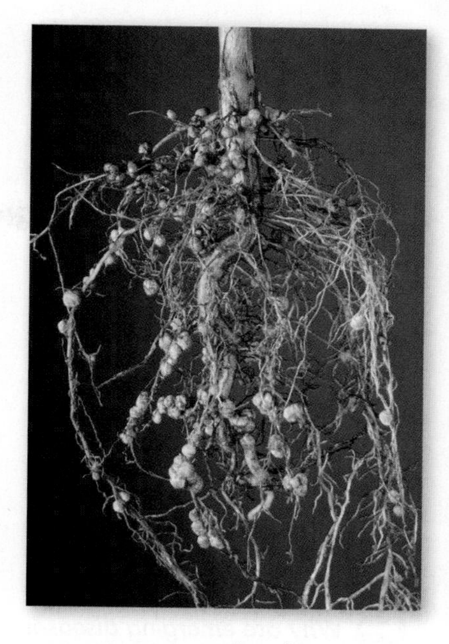

FIGURE 20–14 *Rhizobium* These soybean root nodules contain *Rhizobium* bacteria.

In Your Notebook *Outline three important functions of prokaryotes in the environment. Fill in supporting details for each of these functions.*

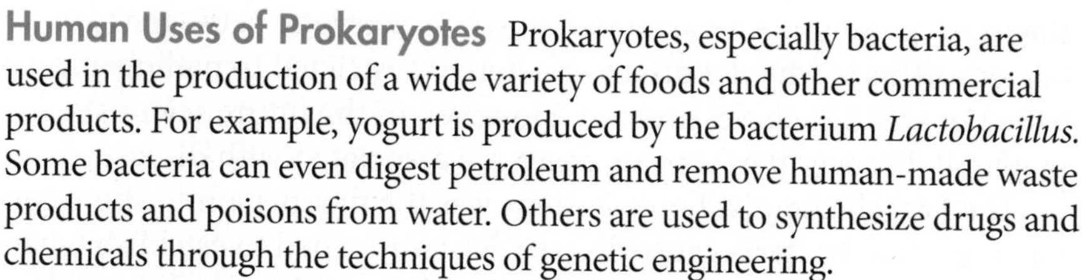

Human Uses of Prokaryotes Prokaryotes, especially bacteria, are used in the production of a wide variety of foods and other commercial products. For example, yogurt is produced by the bacterium *Lactobacillus.* Some bacteria can even digest petroleum and remove human-made waste products and poisons from water. Others are used to synthesize drugs and chemicals through the techniques of genetic engineering.

Biologists continue to discover new uses for prokaryotes. For example, bacteria and archaea adapted to extreme environments may be a rich source of heat-stable enzymes that can be used in medicine, food production, and industrial chemistry.

20.2 Assessment

Review Key Concepts

1. a. Review Which two domains of life contain only prokaryotes?

b. Interpret Diagrams Review **Figure 20–7.** Which feature of the cell wall is characteristic of bacteria but not of archaea?

2. a. Review In what ways do prokaryotes differ from one another?

b. Evaluate Review **Figure 20–10.** Which category of prokaryote is the most flexible in the energy sources it can use? Explain.

3. a. Review List three ecological roles played by prokaryotes.

b. Explain Why are nitrogen-fixing bacteria so important?

c. Apply Concepts Many farmers practice "crop rotation" by planting a field with corn one year and soybeans the next. Why might they do this?

WRITE ABOUT SCIENCE

Description

4. Suppose a new bacterial species is discovered that is spherical in shape, survives drought by forming a thick outer wall, and cannot survive without oxygen. Using vocabulary terms from this lesson, describe this species to a scientific audience.

20.3 Diseases Caused by Bacteria and Viruses

Key Questions

How do bacteria cause disease?

How do viruses cause disease?

Why are emerging diseases particularly threatening to human health?

Vocabulary

pathogen • vaccine • antibiotic • emerging disease • prion

Taking Notes

Outline Use the green and blue heads in this lesson to create an outline. Fill in details as you read the lesson.

MYSTERY CLUE

The meat and bone meal substances added to cattle feed in Britain were sterilized at high temperatures (in excess of 100°C) during processing. What does this suggest about the possibility that bacteria are the cause of BSE?

THINK ABOUT IT We share this planet with prokaryotes and viruses, and most of the time we are never aware of our relationships with them. Often, these relationships are highly beneficial, but in a few cases, sharing simply doesn't work—and disease is the result.

Bacterial Diseases

How do bacteria cause disease?

Disease-causing agents are called **pathogens.** Although pathogens can come from any taxonomic group, bacteria and viruses are among the most common. All currently known prokaryotic pathogens are bacteria. This is why the discussion here is restricted to pathogenic bacteria, and it excludes archaea. However, in the future scientists may well discover that some archaea are associated with disease.

The French chemist Louis Pasteur was the first person to show convincingly that bacteria cause disease. Pasteur helped to establish what has become known as the *germ theory of disease* when he showed that bacteria were responsible for a number of human and animal diseases.

Disease Mechanisms Bacteria produce disease in one of two general ways. **Bacteria cause disease by destroying living cells or by releasing chemicals that upset homeostasis.** Some bacteria destroy living cells and tissues of the infected organism directly, while some cause tissue damage when they provoke a response from the immune system. Other bacteria release toxins (poisons) that interfere with the normal activity of the host. **Figure 20–15** lists some common human diseases caused by bacteria.

▶ ***Damaging Host Tissue*** One example of a pathogen that damages host tissue is the bacterium that causes tuberculosis. This pathogen is inhaled into the lungs, where its growth triggers an immune response that can destroy large areas of tissue. The bacterium also may travel through blood vessels to other sites in the body, causing similar damage.

▶ ***Releasing Toxins*** Bacteria that produce toxins include the species that causes diphtheria, and the species responsible for a deadly form of food poisoning known as botulism. Diphtheria has largely been eliminated in developed countries by vaccination, but outbreaks of botulism still claim many lives.

Some Human Bacterial Diseases

Disease	Effect on Body	Transmission
Lyme disease	"Bull's-eye" rash at site of tick bite, fever, fatigue, headache	Ticks transmit the bacterium *Borrelia burgdorferi.* ▶
Tetanus	Lockjaw, stiffness in neck and abdomen, difficulty swallowing, fever, elevated blood pressure, severe muscle spasms	Bacteria enter the body through a break in the skin.
Tuberculosis	Fatigue, weight loss, fever, night sweats, chills, appetite loss, bloody sputum from lungs	Bacteria particles are inhaled.
Bacterial meningitis	High fever, headache, stiff neck, nausea, fatigue	Bacteria are spread in respiratory droplets caused by coughing and sneezing; close or prolonged contact with someone infected with meningitis
Strep throat	Fever, sore throat, headache, fatigue, nausea	Direct contact with mucus from an infected person or direct contact with infected wounds or breaks in the skin

SEM 7300×

FIGURE 20–15 Common Human Bacterial Diseases Some common bacterial diseases are shown in the table above. **Infer** ***Why do bacterial meningitis outbreaks sometimes occur in college dormitories?***

BUILD Vocabulary

WORD ORIGINS Pathogen comes from the Greek words *pathos*, meaning "suffering," and *genes*, meaning "produced."

Controlling Bacteria Although most bacteria are harmless, and many are beneficial, the everyday risks of any person acquiring a bacterial infection are great enough to warrant efforts to control bacterial growth. Various control methods are used.

▶ ***Physical Removal*** Washing hands or other surfaces with soap under running water doesn't kill pathogens, but it helps dislodge both bacteria and viruses.

▶ ***Disinfectants*** Chemical solutions that kill bacteria can be used to clean bathrooms, kitchens, hospital rooms, and other places where bacteria may flourish.

▶ ***Food Storage*** Low temperatures, like those inside a refrigerator, will slow the growth of bacteria and keep most foods fresher for a longer period of time than possible at room temperature.

▶ ***Food Processing*** Boiling, frying, or steaming can sterilize many kinds of food by raising the temperature of the food to a point where bacteria are killed.

▶ ***Sterilization by Heat*** Sterilization of objects such as medical instruments at temperatures well above 100° Celsius can prevent the growth of potentially dangerous bacteria. Most bacteria cannot survive such temperatures.

In Your Notebook *Relate the methods for controlling bacteria listed above to your everyday life. Which methods have you used this week? Give specific details.*

FIGURE 20–16 Pathogen Defense Hand washing is one of the most simple, inexpensive, and effective ways to prevent disease.

Preventing Bacterial Diseases Many bacterial diseases can be prevented by stimulating the body's immune system with vaccines. A **vaccine** is a preparation of weakened or killed pathogens or inactivated toxins. When injected into the body, a vaccine prompts the body to produce immunity to a specific disease. Immunity is the body's ability to destroy pathogens or inactivated toxins.

Treating Bacterial Diseases A number of drugs can be used to attack a bacterial infection. These drugs include **antibiotics,** such as penicillin and tetracycline, that block the growth and reproduction of bacteria. Antibiotics disrupt proteins or cell processes that are specific to bacterial cells. In this way, they do not harm the host's cells.

Viral Diseases

How do viruses cause disease?

Like bacteria, viruses produce disease by disrupting the body's normal homeostasis. **Figure 20–17** lists some common human diseases caused by viruses. Viruses produce serious animal and plant diseases as well.

Disease Mechanisms In many viral infections, viruses attack and destroy certain cells in the body, causing the symptoms of the associated disease. Poliovirus, for example, destroys cells in the nervous system, producing paralysis. Other viruses cause infected cells to change their patterns of growth and development, sometimes leading to cancer. **Viruses cause disease by directly destroying living cells or by affecting cellular processes in ways that upset homeostasis.**

FIGURE 20–17 Common Human Viral Diseases Some common viral diseases are shown in the table below. **Interpret Tables** ***Which virus can cause cancer?***

Some Human Viral Diseases

Disease	Effect on Body	Transmission
Common cold	Sneezing, sore throat, fever, headache, muscle aches	Contact with contaminated objects; droplet inhalation
Influenza	Body aches, fever, sore throat, headache, dry cough, fatigue, nasal congestion	Flu viruses spread in respiratory droplets caused by coughing and sneezing.
AIDS (HIV)	Helper T cells, which are needed for normal immune-system function, are destroyed.	Sexual contact; contact with contaminated blood or body fluids; can be passed to babies during delivery or during breastfeeding.
Chicken pox	Skin rash of blisterlike lesions	Virus particles are spread in respiratory droplets caused by coughing and sneezing; highly contagious
Hepatitis B	Jaundice, fatigue, abdominal pain, nausea, vomiting, joint pain	Contact with contaminated blood or bodily fluids
West Nile Virus	Fever, headache, body ache	Bite from an infected mosquito ▶
Human papillomavirus (HPV)	Genital or anal warts, also cancer of the cervix, penis, and anus	Sexual contact

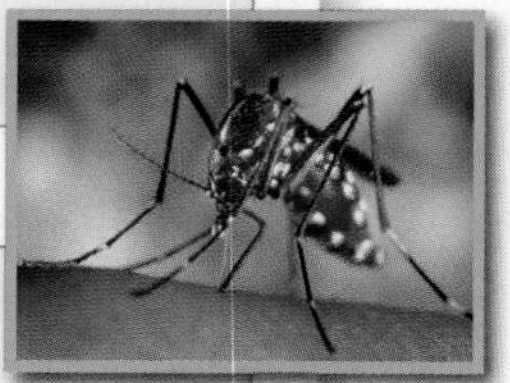

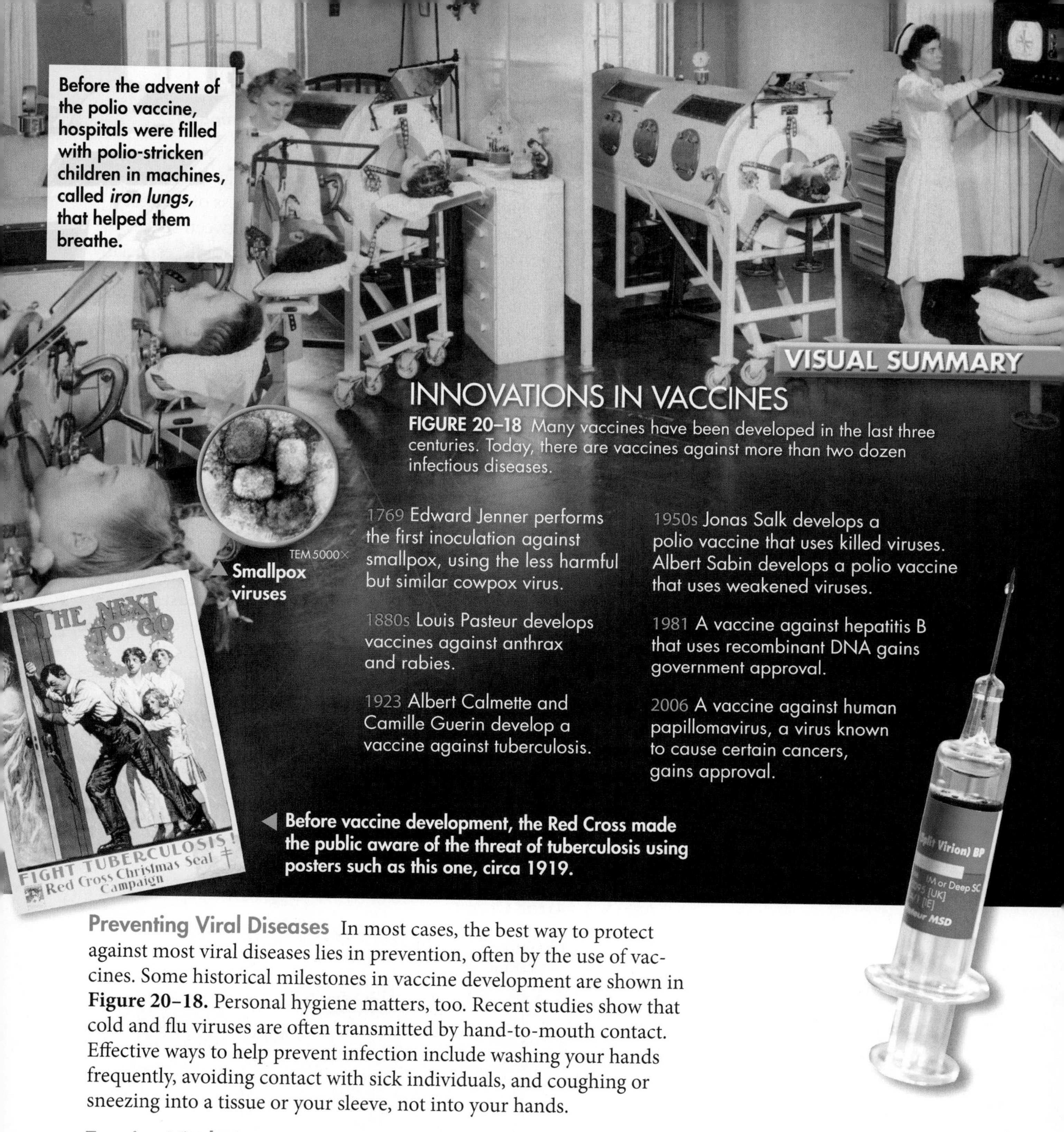

FIGURE 20–18 Many vaccines have been developed in the last three centuries. Today, there are vaccines against more than two dozen infectious diseases.

Preventing Viral Diseases In most cases, the best way to protect against most viral diseases lies in prevention, often by the use of vaccines. Some historical milestones in vaccine development are shown in **Figure 20–18.** Personal hygiene matters, too. Recent studies show that cold and flu viruses are often transmitted by hand-to-mouth contact. Effective ways to help prevent infection include washing your hands frequently, avoiding contact with sick individuals, and coughing or sneezing into a tissue or your sleeve, not into your hands.

Treating Viral Diseases Unlike bacterial diseases, viral diseases cannot be treated with antibiotics. In recent years, however, limited progress has been made in developing a handful of antiviral drugs that attack specific viral enzymes that host cells do not have. These treatments include an antiviral medication that can help speed recovery from the flu virus, and others that have helped prolong the lives of people infected with HIV.

Emerging Diseases

Why are emerging diseases particularly threatening to human health?

If pathogenic viruses and bacteria were unable to change over time—that is, if they could not evolve—they would pose far less of a threat than they actually do. Unfortunately, the short time between successive generations of these pathogens allows them to evolve rapidly, especially in response to human efforts to control them. An unknown disease that appears in a population for the first time or a well-known disease that suddenly becomes harder to control is called an **emerging disease.**

Figure 20–19 shows locations worldwide where specific emerging diseases have broken out in recent years. Changes in lifestyle and commerce have made emerging diseases even more of a threat. High-speed travel means that a person can move halfway around the world in a day. Huge quantities of food and consumer goods are now shipped between regions of the world that previously had little contact with each other. This brings human populations that were once isolated by oceans and mountain ranges into close contact with more developed parts of the world. The possibility of the rapid spread of new diseases is a risk of every trip a person takes and every shipment of food or goods.

The pathogens that cause emerging diseases are particularly threatening to human health because human populations have little or no resistance to them, and because methods of control have yet to be developed. Because of their sudden appearance and resistance to existing control methods, emerging diseases are of particular concern. Deeper understanding of the functions of the molecular structures and genetics of bacteria and viruses will be one key to defending against them.

FIGURE 20–19 Emerging Diseases In recent years, new diseases, such as severe acute respiratory syndrome (SARS) in Asia, have appeared. At the same time, some diseases thought to be under control have come back. Both examples are classified as emerging diseases. **Interpret Graphics** ***Which emerging diseases are found in Africa?***

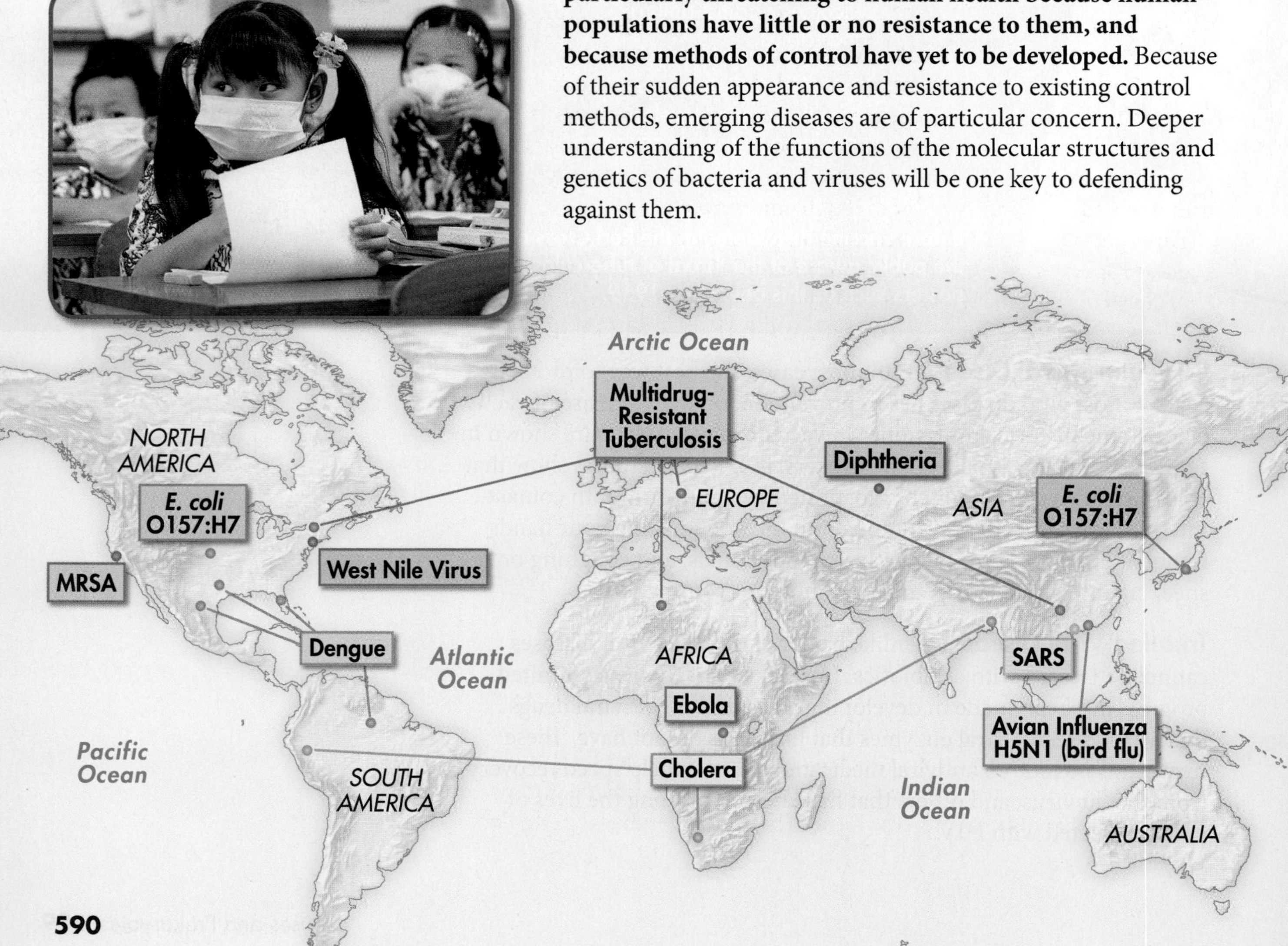

MRSA on the Rise

Infection by methicillin-resistant *Staphylococcus aureus* (MRSA) can spread very quickly in hospitals and nursing homes. The table at right shows the incidence of MRSA infections in U.S. hospitals during a 13-year period.

1. Graph Prepare a line graph showing the number of MRSA infections in U.S. hospitals over time. Describe the trend shown.

2. Calculate By what percentage did MRSA infections in U.S. hospitals increase between 1995 and 2005? MATH

3. Draw Conclusions A 2007 study reported that the average hospital stay in the United States lasted 4.6 days, while that of the average MRSA-infected patient was 10.0 days. If the trend shown by the data above continues, what effect will MRSA infections have on future hospital costs?

Incidence of MRSA

Year	Hospital Cases Reported
1993	1900
1995	38,100
1997	69,800
1999	108,600
2001	175,000
2003	248,300
2005	368,600

"Superbugs" When first introduced in the 1940s, penicillin, an antibiotic derived from fungi, was a miracle drug. Patients suffering from life-threatening infections were cured almost immediately by this powerful new drug. Conquest of bacterial diseases seemed to be in sight. Within a few decades, however, penicillin lost much of its effectiveness, as have other, more current antibiotics. The culprit is evolution.

The widespread use of antibiotics has led to a process of natural selection that favors the emergence of resistance to these powerful drugs. Physicians now must fight "superbugs" that are resistant to whole groups of antibiotics and that transfer drug-resistant genes from one bacterium to another through conjugation.

An especially dangerous form of multiple drug resistance has recently appeared in a common bacterium. Methicillin-resistant *Staphylococcus aureus*, known as MRSA (pronounced MURS uh), can cause infections that are especially difficult to control. MRSA skin infections can be spread by close contact, including the sharing of personal items such as towels and athletic gear, and can often spread in hospitals, where MRSA bacteria can infect surgical wounds and spread from patient to patient.

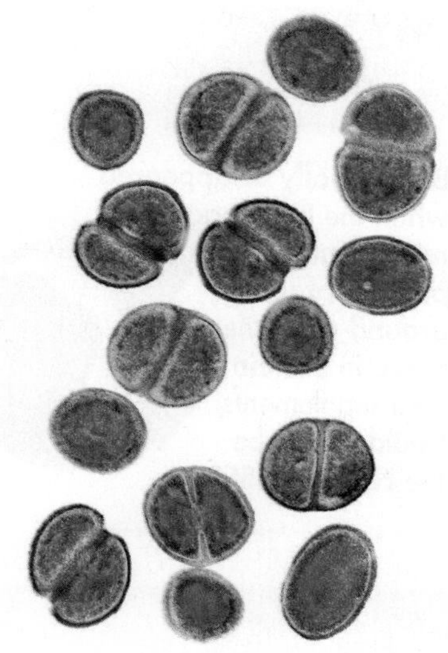

FIGURE 20–20 MRSA Methicillin-resistant *S. aureus* is a bacterium that is resistant to methicillin and other common antibiotics.

New Viruses Because viruses replicate so quickly, their genetic makeup can change rapidly, sometimes allowing a virus to jump from one host species to another. Researchers have evidence that this is how the virus that causes AIDS originated, moving from nonhuman primates into humans.

Public health officials are especially worried about the flu virus. Gene shuffling among different flu viruses infecting wild and domesticated bird populations has led to the emergence of a "bird flu" that is similar in many ways to the most deadly human versions of flu. In a few isolated cases, bird flu has indeed infected humans, and health officials warn that a major "jump" into the human population remains possible in the future.

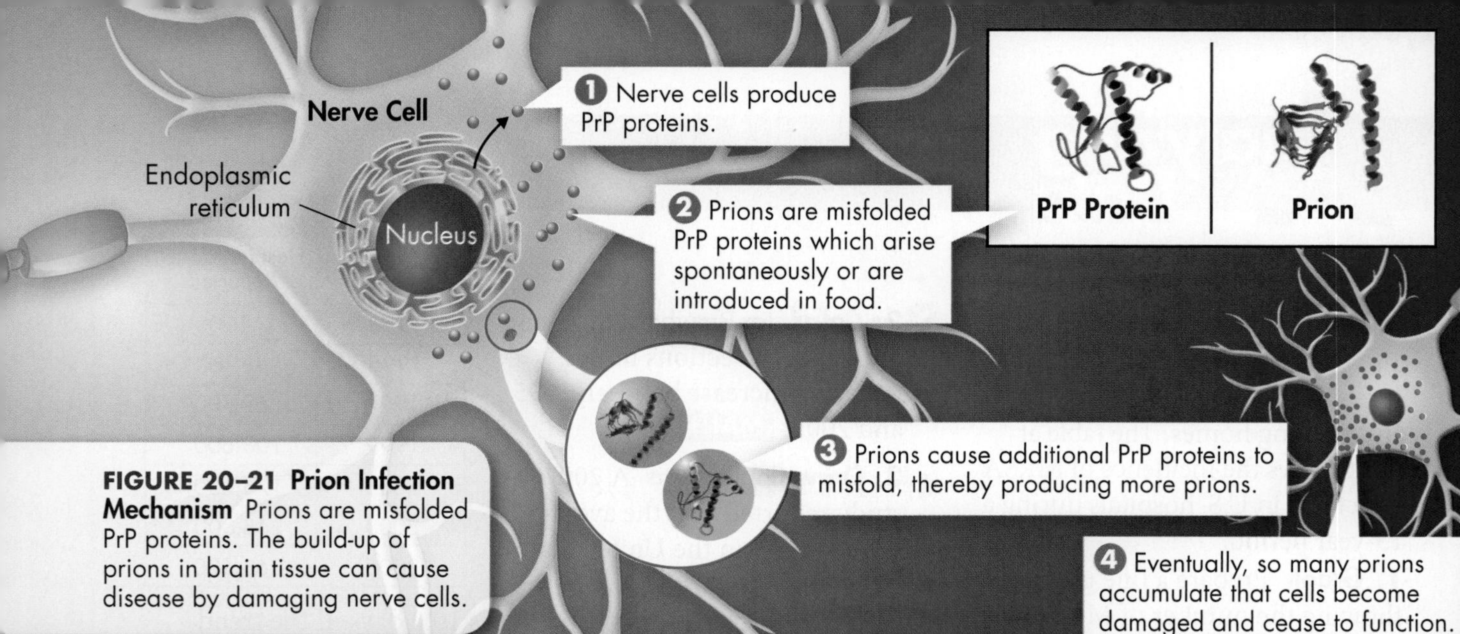

FIGURE 20–21 Prion Infection Mechanism Prions are misfolded PrP proteins. The build-up of prions in brain tissue can cause disease by damaging nerve cells.

MYSTERY CLUE

BSE virtually disappeared when the British government banned the practice of using ground-up cattle tissue in protein feed supplements. Could prions be the cause of BSE?

Prions In 1972, American scientist Stanley Prusiner became interested in scrapie, an infectious disease in sheep, the exact cause of which was unknown. At first, he suspected a viral cause, but experiments revealed clumps of tiny protein particles in the brains of infected sheep. Prusiner called these particles **prions,** short for "protein infectious particles." Although prions were first discovered in sheep, many animals, including humans, can become infected with prions. Prions are formed when a protein known as PrP is improperly folded. Prions themselves can cause PrP proteins to misfold, producing even more prions. An accumulation of prions can damage nerve cells, as shown in **Figure 20–21.**

20.3 Assessment

Review Key Concepts

1. a. Review Describe how bacteria cause disease.

b. Relate Cause and Effect Are vaccines effective before or after infection? Explain.

2. a. Review How do viruses cause disease?

b. Compare and Contrast How does the treatment of viral diseases contrast with the treatment of bacterial diseases?

3. a. Review Why are emerging diseases of particular concern?

b. Explain Why are "superbugs" difficult to control?

c. Propose a Solution What actions could your school take to help combat the evolution of "superbugs"? Explain how these actions could make an impact.

BUILD VOCABULARY

4. Research the word origins for the term *vaccine*. Which word in which language does it come from and why?

RST.9-10.10 Level of Text Complexity, RST.9-10.9 Integration of Knowledge and Ideas. Also WHST.9-10.1, WHST.9-10.7

Biology & Society

Should More Vaccinations Be Required?

In the 1800s, diphtheria was the scourge of American children. Each winter, tens of thousands of children fell ill with fever and sore throats caused by this airborne bacterium, and thousands died from it. But you may not have heard of it, and you certainly haven't had to worry about it. Only five cases of diphtheria have been recorded in the United States since 2000. The reason is vaccination. In 1920, a vaccine for the disease was introduced and is now mandatory for American schoolchildren. As a result, diphtheria is one of several diseases, including polio, that have all but vanished from our society.

Medical guidelines call for vaccinations against at least 14 childhood diseases. New vaccines have been introduced against diseases that are not usually fatal, such as chicken pox. Some authorities have even suggested that everyone should also be vaccinated against bacteria and viruses that might be used for germ warfare, including anthrax and smallpox. Should all of these vaccinations be required for everyone entering school?

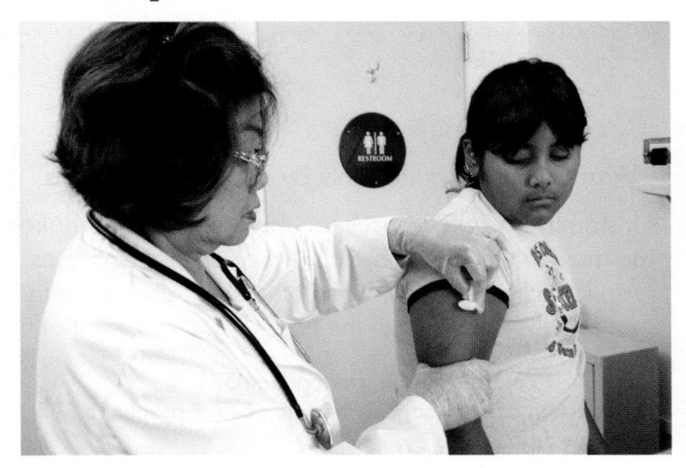

This girl is receiving a vaccination that is required before she starts school.

The Viewpoints

Expand Mandatory Vaccinations Infectious diseases are spread from person to person. When they work properly, vaccinations stimulate the immune system in a way that not only protects the vaccinated individual, but also indirectly protects others who may not have been immunized. This helps to improve public health, whether the specific disease in question is life-threatening or not. In addition, vaccination against germ warfare agents renders these potential terrorist weapons useless.

Limit Mandatory Vaccinations There is no question that a limited number of vaccinations against deadly and crippling diseases such as diphtheria and polio makes good sense. However, every vaccination carries with it the risk that the child being vaccinated will experience adverse reactions. Some reactions to vaccinations can be severe.

More young people now suffer from complications of polio vaccination than develop the disease itself, and smallpox vaccination was discontinued in the early 1970s due to deaths from the vaccine. Students should not be forced to be vaccinated for diseases that are not life-threatening nor for hypothetical threats like germ warfare.

Research and Decide

1. Analyze the Viewpoints Investigate the diseases for which vaccination is now required to enter school in your state. What would be the risks and benefits of expanding the number of required vaccines? Similarly, what would be the effects of limiting the number of required vaccines to just a handful?

2. Form an Opinion Compare the results of your research with statements published by U.S. government agencies. Prepare a list of the vaccines that you would make mandatory for all students in your school, and write an argument to support your claim.

NGSS Smart Guide Viruses and Prokaryotes

Disciplinary Core Idea LS1.A Structure and Function: How do the structures of organisms enable life's functions? Viruses are nonliving particles that reproduce by infecting cells. Bacteria and archaea are prokaryotes that play many important ecosystem roles. Bacterial infections are treated with medicines that disrupt prokaryotic cellular structure or function.

20.1 Viruses

- Viruses can reproduce only by infecting living cells.
- Inside living cells, viruses use their genetic information to make multiple copies of themselves. Some viruses replicate immediately, while others initially persist in an inactive state within the host.

virus 574 • capsid 575 • bacteriophage 575 • lytic infection 576 • lysogenic infection 577 • prophage 577 • retrovirus 578

Biology.com

Untamed Science Video Join the Untamed Science crew as they fire up the microscopes for a look at bacteria and all the ways they are good for us.

Visual Analogy Compare an Old West outlaw taking over a town to a lytic infection.

InterActive Art Build your understanding of lytic and lysogenic cycles.

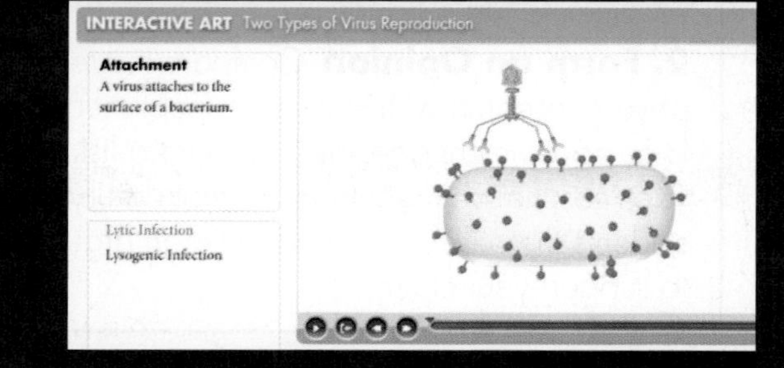

20.2 Prokaryotes

- Prokaryotes are classified as Bacteria or Archaea—two of the three domains of life.
- Prokaryotes vary in their size and shape, in the way they move, and in the way they obtain and release energy.
- Prokaryotes are essential in maintaining every aspect of the ecological balance of the living world. In addition, some species have specific uses in human industry.

prokaryote 580 • bacillus 582 • coccus 582 • spirillum 582 • binary fission 583 • endospore 583 • conjugation 583

Biology.com

Art Review Review your understanding of the structure and classification of prokaryotes.

CHECKING YOUR *Scientific Literacy*

Refer to the lesson content and digital assets below as you prepare for your chapter assessment. Then, evaluate your understanding of viruses and prokaryotes by answering these questions.

1. **Scientific and Engineering Practice Developing and Using Models** Construct a model to show how a bird infected with West Nile virus can lead to an outbreak of the virus in humans.

2. **Crosscutting Concept Structure and Function** Compare and contrast the ways that viruses and bacteria cause disease. How are these mechanisms related to the structures of viruses and bacteria?

3. **Integration of Knowledge and Ideas** Use the Internet and other sources to research ways to prevent or treat West Nile virus. Compare and contrast your findings to the methods of preventing and treating diseases discussed in this text.

4. **STEM** You are a member of a panel assigned to evaluate the safety and efficacy of a new vaccine. What questions would you ask the manufacturer of the vaccine? What experimental studies would you insist that the manufacturer do before asking for approval? Under what conditions would you approve the vaccine for use in humans? Under what conditions would you suggest that the vaccine not be approved? Explain your reasoning.

20.3 Diseases Caused by Bacteria and Viruses

- Bacteria cause disease by destroying living cells or by releasing chemicals that upset homeostasis.
- Viruses cause disease by destroying living cells or by affecting cellular processes in ways that upset homeostasis.
- The pathogens that cause emerging diseases are particularly threatening to human health because human populations have little or no resistance to them, and because methods of control have yet to be developed.

pathogen 586 • **vaccine 588** • **antibiotic 588** • **emerging disease 590** • **prion 592**

Biology.com

Data Analysis Analyze data on MRSA and identify whether the increase in prevalence is due to increased spread, virulence, or simply increased accuracy of diagnosis.

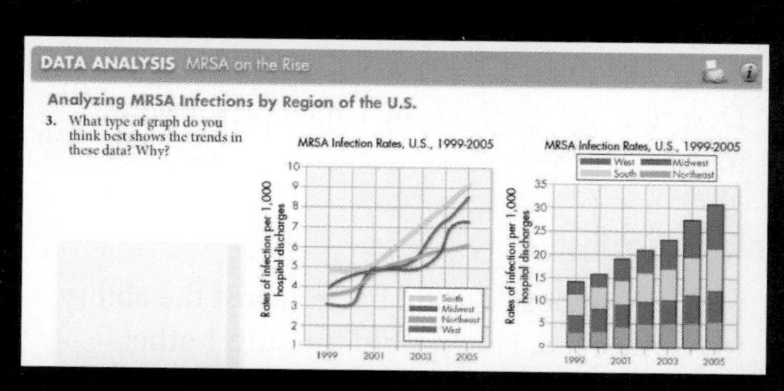

Art in Motion View a short animation of prion infection and see how misfolded proteins interact with normal proteins.

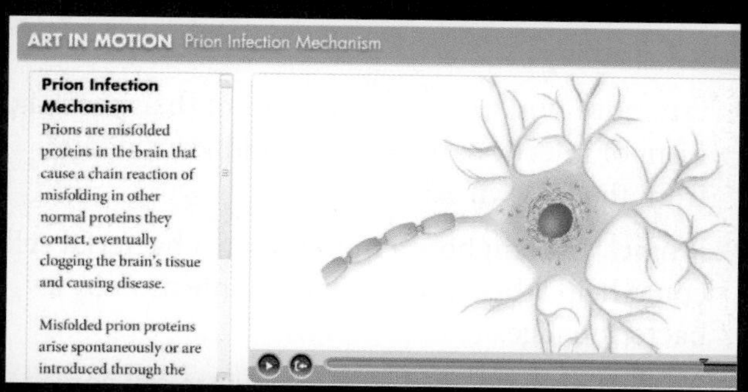

Real-World Lab

Controlling Bacterial Growth This chapter lab is available in your lab manual and at Biology.com

20 Assessment

20.1 Viruses

Understand Key Concepts

1. Particles made up of proteins, nucleic acids, and sometimes lipids that can reproduce only by infecting living cells are called
   a. bacteria.
   b. capsids.
   c. prophages.
   d. viruses.
2. The structure labeled "A" in the diagram of the virus below is called the
   a. viral genome.
   b. RNA envelope.
   c. capsid.
   d. nuclear membrane.

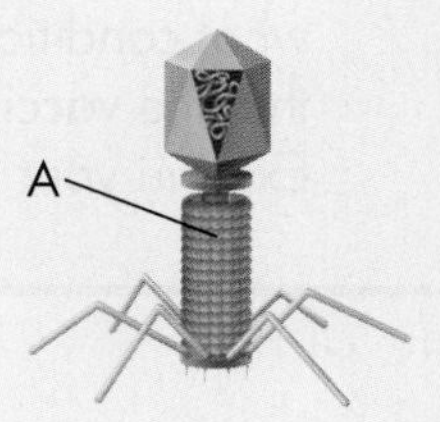

3. One group of viruses that contain RNA as their genetic information is the
   a. bacteriophages.
   b. retroviruses.
   c. capsids.
   d. prophages.
4. What characteristics do all viruses have in common?
5. How are capsid proteins important to the way a virus functions?
6. Describe the sequence of events that occurs during a lytic infection.
7. Explain what a prophage is.

Think Critically

8. **Compare and Contrast** In terms of their mechanism of infection, how does a cold virus differ from the HIV virus?
9. **Predict** Explain how a mutation in a bacterial cell could help it become resistant to infection by a bacteriophage.
10. **Apply Concepts** Explain how a virus can spread in a bacterial population during the lysogenic phase of infection.

20.2 Prokaryotes

Understand Key Concepts

11. Prokaryotes are unlike all other organisms in that their cells
    a. lack nuclei.
    b. have organelles.
    c. have cell walls.
    d. lack nucleic acids.
12. Prokaryotes that thrive in oxygen-free environments are called
    a. aerobes.
    b. retroviruses.
    c. anaerobes.
    d. heterotrophs.
13. Which micrograph shows bacillus bacteria?

a. 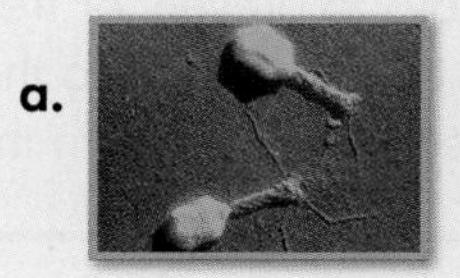

c. 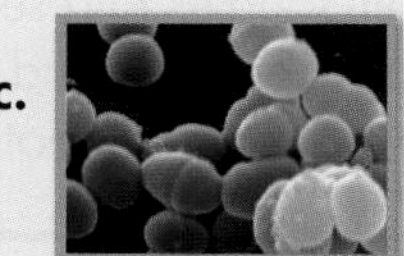

b. 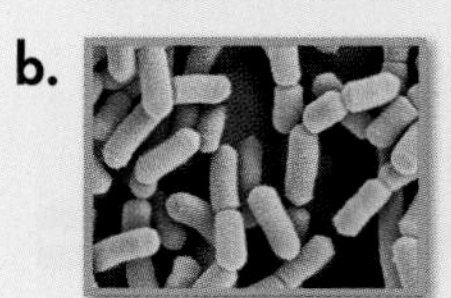

d. 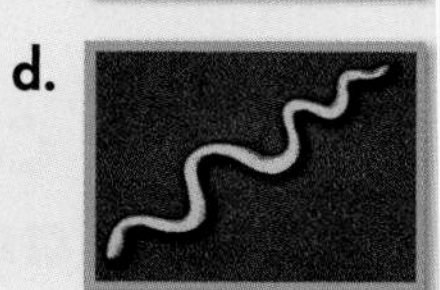

14. Prokaryotes reproduce asexually by
    a. binary fission.
    b. endospores.
    c. conjugation.
    d. mutation.
15. The process of converting nitrogen into a form plants can use is known as nitrogen
    a. formation.
    b. ammonification.
    c. decomposition.
    d. fixation.
16. What are the two distinguishing characteristics of prokaryotes?
17. Describe the three main cell shapes of prokaryotes.
18. Describe two methods by which prokaryotes move.

Think Critically

19. **Predict** Suppose certain bacteria lost the ability to fix nitrogen. How would this affect other organisms in their ecosystem?
20. **Apply Concepts** Why don't foods such as uncooked rice and raisins spoil?

21. **Form a Hypothesis** Bacteria that live on teeth produce an acid that causes decay. Why do people who do not brush their teeth regularly have more cavities than those who do?

22. **Classify** A newly discovered organism is unicellular, has a cell wall containing peptidoglycan, has a circular DNA molecule, and lacks a nucleus. Based on those characteristics, to which domain does it belong?

23. **Craft and Structure** Explain how the outcome of binary fission differs from that of both endospore formation and conjugation.

20.3 Diseases Caused by Bacteria and Viruses

Understand Key Concepts

24. Disease-causing organisms are known as
    a. cocci.
    b. bacteria.
    c. pathogens.
    d. archaea.

25. Which of the following scientists is responsible for developing the germ theory of disease?
    a. Ivanovski
    b. Beijerinck
    c. Pasteur
    d. Darwin

26. Viruses typically cause disease by
    a. releasing toxins.
    b. infecting and then destroying cells.
    c. causing mutations in the host cell DNA.
    d. destroying red blood cells.

27. Which of the following can be helpful in treating bacterial diseases but NOT viral diseases?
    a. vaccines.
    b. antibiotics.
    c. antiviral drugs.
    d. aspirin.

28. What is the best way for people to protect themselves against most viral diseases?

29. **Key Ideas and Details** List three different ways bacterial growth can be controlled.

30. **Craft and Structure** What is meant by the term *emerging disease*? Give three examples of emerging diseases in North America.

31. How do misfolded prions cause disease?

solve the CHAPTER MYSTERY

THE MAD COWS

The "mad cow" disease that appeared in 1986 in Britain spread quickly among cattle herds. Humans were afflicted by a similar disease, known as nvCJD (new variant Creutzfeld-Jacob Disease), and scores of people died. The disease virtually disappeared when the government banned the practice of using ground-up cattle tissue in protein feed supplements.

It now seems clear that "mad cow" and nvCJD were caused by prions in the meat and brain tissue of infected cattle. When these prions entered the food supply, the infection was able to spread to other cattle, and to humans eating meat from infected animals. Officials in Europe and the United States have instituted new controls on meat production to try to prevent further outbreaks of this prion-based disease.

1. **Infer** The rapid rise of BSE between 1986 and 1991 ended when British authorities banned the use of meat and bone meal in feed supplements for cattle. How does this support the hypothesis that BSE is caused by prions?

2. **Apply Concepts** Why did most scientists conclude that BSE was not caused by either viruses or bacteria?

3. **Connect to the Big idea** Prions are proteins, not organisms unto themselves. What characteristics do prions share with cell-based life? Cite evidence you learned from the text and explain.

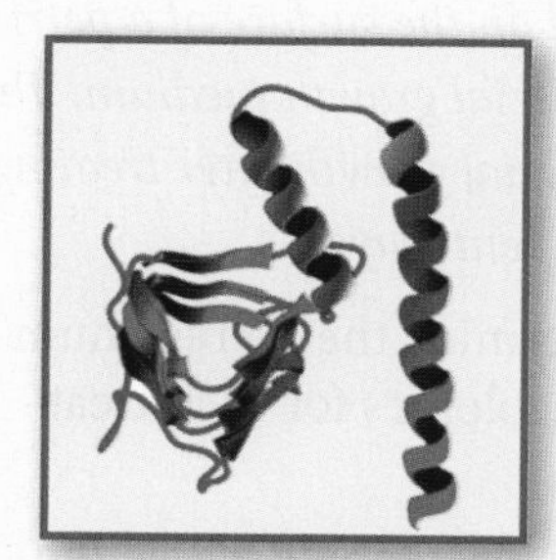

◀ Misfolded prion protein

Think Critically

32. Apply Concepts What advantages does the physical removal of infectious microbes by hand washing have over the use of disinfectants? Explain.

33. Predict Would antibiotics be effective in treating an outbreak of bird flu? Explain.

34. Biologists conducted an experiment to determine the effectiveness of several antibiotics against a certain strain of bacteria. Four disks, each soaked in a different antibiotic, were placed in a petri dish where the bacteria were growing. The results are summarized below.

Effects of Antibiotics

Antibiotic	Observation After One Week
A	Growth retarded for 6 mm diameter
B	Growth not retarded
C	Growth not retarded
D	Growth retarded for 2 mm diameter

a. Analyze Data Which antibiotics were the least effective at retarding the growth of the bacteria? Explain your answer using data from the experiment.

b. Infer Which antibiotic might be the most effective treatment for an infection caused by this strain of bacteria? Explain your answer using data from the experiment.

Connecting Concepts

Use Science Graphics

E. coli *bacteria can be grown on agar in a petri dish, clouding over the entire surface and forming a bacterial "lawn." The photograph shows a lawn over which a solution containing bacteriophage particles has been poured.*

35. Interpret Visuals What is the most reasonable explanation for the small, circular, clear areas on the bacterial lawn?

36. Form a Hypothesis Suppose you touched the tip of a glass rod to one of the clear areas and then touched it again to the surface of a petri dish with a fresh lawn of *E. coli.* What would happen to the new lawn of bacteria after several days?

Write About Science

37. Text Types and Purposes Write an article for a newspaper about the roles bacteria play in the biosphere. As you develop your explanation, think about the expertise of likely readers.

38. Assess the Big idea Compare and contrast the reproduction of viruses with that of prokaryotes.

Analyzing Data

Integration of Knowledge and Ideas

The table shows the number of bacteria colonies that grow after a person's treated hand is swabbed with a sterile cotton ball and the cotton ball is rubbed on the surface of a petri dish containing bacterial growth medium. The table compares bacterial growth after treating the hand in five different ways.

Hand Treatment	Trial 1: Number of Colonies	Trial 2: Number of Colonies
Unwashed	247	210
Rinsed in warm water	190	220
Washed with soap and warm water	21	15
Rinsed in alcohol and air-dried	3	0

39. Calculate Determine the average number of bacteria colonies for each treatment. MATH

40. Graph Make a bar graph to show the results of the experiment. Graph the averages you just calculated.

41. Analyze Data According to the data, what is the most effective method of preventing transfer of bacteria by hand contact?

42. Draw Conclusions What is the most logical explanation of how alcohol works?

Standardized Test Prep

Multiple Choice

1. A type of virus that infects bacterial cells is called a
- **A** capsid.
- **B** prion.
- **C** bacteriophage.
- **D** retrovirus.

2. Prokaryotic cells that have a spherical shape are called
- **A** cocci.
- **B** methanogens.
- **C** spirilli.
- **D** bacilli.

3. What is a capsid?
- **A** viral DNA that inserts into a host's DNA
- **B** a protein coat surrounding a virus
- **C** a type of plant virus
- **D** a rod-shaped bacterium

4. Which of the following is NOT used to identify specific prokaryotes?
- **A** type of nucleic acid
- **B** shape
- **C** movement
- **D** energy source

5. Which method is NOT used to protect food against microorganisms?
- **A** heating
- **B** freezing
- **C** sterilization
- **D** vaccination

6. Which illness is caused by a bacterium?
- **A** AIDS
- **B** polio
- **C** diphtheria
- **D** common cold

7. Which process is used for the exchange of genetic information between two bacterial cells?
- **A** endospore formation
- **B** lysogenic cycle
- **C** conjugation
- **D** binary fission

8. All bacteria are classified as
- **A** eukaryotes.
- **B** protists.
- **C** archaea.
- **D** prokaryotes.

Questions 9–10

Use the graph below to answer the questions.

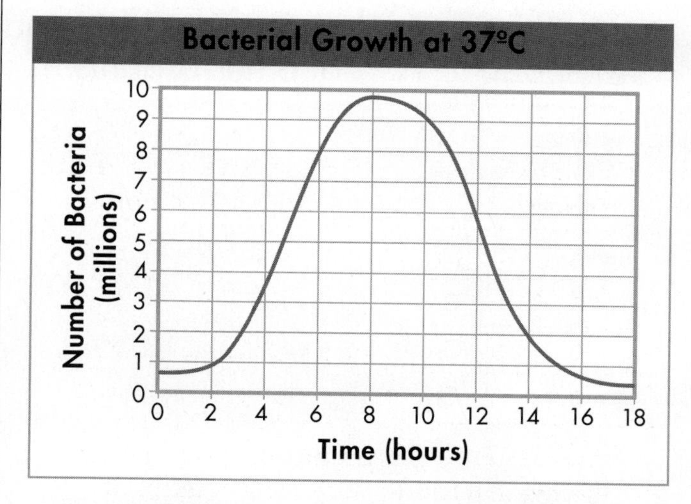

9. At which interval in the graph does the number of living bacteria increase at the greatest rate?
- **A** between hours 2 and 4
- **B** between hours 4 and 6
- **C** between hours 6 and 8
- **D** between hours 10 and 12

10. Which is the most likely reason for the decrease in bacteria shown?
- **A** The temperature of the bacterial culture was too high after 8 hours.
- **B** The bacteria stopped reproducing after 8 hours.
- **C** More nutrients were added to the culture at regular intervals.
- **D** Waste products from the bacteria accumulated in the nutrient solution.

Open-Ended Response

11. Explain why antibiotics can be useful in treating bacterial diseases but not in treating viral diseases.

If You Have Trouble With . . .

Question	1	2	3	4	5	6	7	8	9	10	11
See Lesson	20.1	20.2	20.1	20.2	20.3	20.3	20.2	20.2	20.2	20.2	20.3

21 Protists and Fungi

Interdependence in Nature

Q: How do protists and fungi affect the homeostasis of other organisms and ecosystems?

A goldcrest perched on branches covered with lichens

INSIDE:

- 21.1 Protist Classification—The Saga Continues
- 21.2 Protist Structure and Function
- 21.3 The Ecology of Protists
- 21.4 Fungi

BUILDING *Scientific Literacy*

Deepen your understanding of protists and fungi by engaging in key practices that allow you to make connections across concepts.

NGSS You will construct explanations on how protists and fungi interact with their environments.

STEM Design a mosquito net that works with a fungus to help combat malaria outbreaks.

Common Core You will make sense of problems and persevere in solving them, and provide an accurate summary of the text.

CHAPTER MYSTERY

"A BLIGHT OF UNUSUAL CHARACTER"

Within the first few decades of the nineteenth century, Ireland became heavily dependent on potato farming. Potatoes are nutritious and easy to grow, and they thrived in the damp soil and wet climate of the Emerald Isle. Tenant farmers began to grow potatoes as the primary source of food for themselves and their families.

Then, during the summer of 1845, something strange began to happen. A magazine called *The Gardener's Chronicle* reported that "a blight of unusual character" was attacking potatoes. Everywhere in Ireland, potatoes began to rot and turn black. By the beginning of the twentieth century, starvation and emigration would cut the population of Ireland in half, while the island's principal food crop rotted in the fields. As you read this chapter, look for clues to help you identify what caused the potato blight.

Biology.com

Finding the solution to the "A Blight of Unusual Character" mystery is only the beginning. Take a video field trip with the ecogeeks of Untamed Science to see where the mystery leads.

Go online to access additional resources including:
• eText • Flash Cards • Lesson Overviews • Chapter Mystery

21.1 Protist Classification—The Saga Continues

Key Questions

What are protists?

How are protists related to other eukaryotes?

Taking Notes

Preview Visuals Look at the names of the organisms in **Figure 21–2.** Are any of the organisms familiar to you? Formulate two questions you have about this diagram.

THINK ABOUT IT Some of the organisms we call "protists" live quietly on the bottom of shallow ponds, soaking up the energy of sunlight. Others swim vigorously in search of tiny prey. Some sparkle like diamonds in coastal waters, and others drift in the human bloodstream, destroying blood cells and killing nearly a million people a year, most of them children. What kind of life is this, capable of such beauty and such destruction?

The First Eukaryotes

What are protists?

More than a billion years ago, a new form of organism appeared on Earth. Subtle clues in the microscopic fossils of these single cells mark them as the very first eukaryotes. Single-celled eukaryotes are still with us today and are often called "protists"—a name that means "first." Traditionally, protists are classified as members of the kingdom Protista. **Protists are eukaryotes that are not members of the plant, animal, or fungi kingdoms.**

Although most protists are unicellular, quite a few are not. The largest protists—brown algae called kelp—contain millions of cells arranged in differentiated tissues. They are considered protists because they are related more closely to certain unicellular protists than to members of any other kingdom. Kelp and several other protists are shown in **Figure 21–1.**

In Your Notebook *Think about the things that define a group. What do you think defines protists as a group?*

FIGURE 21–1 Extreme Diversity of Protists Protists vary greatly in size, form, and function. Here are several examples.

◀ Otters wrap themselves in giant kelp, a multicellular protist species, to keep from drifting out to sea while they sleep.

Quick Lab
GUIDED INQUIRY

What Are Protists? 

1. Place a drop of water containing a variety of microorganisms on a microscope slide. Add a drop of methyl cellulose and a coverslip. Observe the slide under the microscope at low and high magnifications.
2. Record your observations by drawing and labeling each type of organism.
3. Make a chart listing each type of organism you observed and its characteristics.

Analyze and Conclude

1. Observe For any of the organisms that move, describe their motion and any structures involved in producing the motion.

2. Draw Conclusions Do you observe any structures that you think relate to food-gathering or reproduction? Explain why you think so.

3. Classify Are any of these organisms bacteria, plants, or animals? Explain your answer.

The "Protist" Dilemma In recent years, biologists have studied these eukaryotes closely, eager to learn what the organisms reveal about the history of life. They have discovered that the "protists" display a far greater degree of diversity than any other eukaryotic kingdom. Furthermore, they found that many of these organisms are far more closely related to members of other eukaryotic kingdoms than they are to other "protists."

This finding has created a dilemma. By definition, the members of a living kingdom, such as plants or animals, should be more like one another than like members of other kingdoms. This is not true of protists, which means that reclassification is necessary. Biologists continue to debate the best way to do this.

In the past, scientists sorted protists into three groups: plantlike protists, animal-like protists, and funguslike protists. This simple solution began to fail as biologists learned that many protists do not fit into any of these groups. To make matters worse, they discovered that many of the animal-like and funguslike protists are so similar that they belong in a single group, not split into two. Clearly, a new way of thinking about the "protists" is now needed.

At a certain stage of their life cycle, protists called slime molds aggregate into colonies like this one (SEM 15×). ▼

Photosynthetic, motile *Euglena* are common freshwater protists (LM 250×). ▶

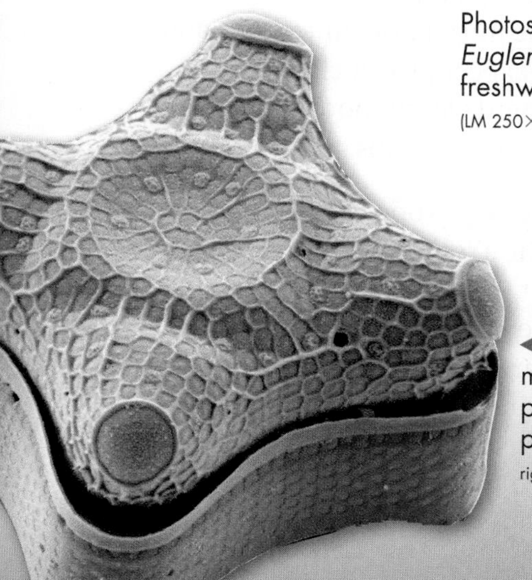

◀ The shells of diatoms, microscopic marine protists, are intricately patterned (left pair: SEM 960×; right: SEM 270×).

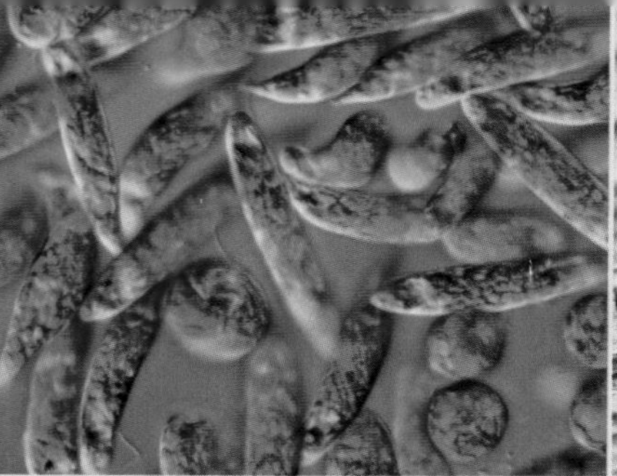

Euglena is classified as an excavate.

Brown algae and diatoms are examples of chromalveolates.

Slime molds are classified with the Amoebozoa.

Six Major Groups

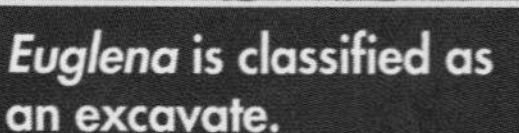

Excavata

Chromalveolata

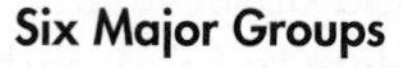

Cercozoa, Foraminifera, and Radiolaria

Rhodophyta (red algae)

Amoebozoa

Choanozoa

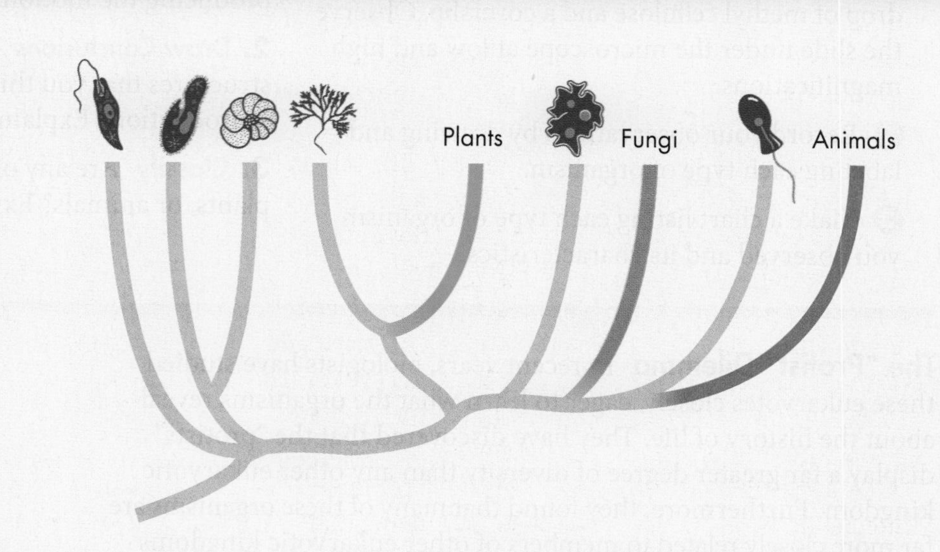

FIGURE 21–2 Protist Classification—A Work in Progress This cladogram represents an understanding of protist relationships supported by current research.
Interpret Diagrams ***Which group of "protists" is most closely related to plants? Which group is most closely related to animals? To fungi?***

For more on the diversity of protists, go to the Visual Guide. **DOL•10–DOL•15**

Multiple Kingdoms? The most recent studies of protists divide them into six major clades, shown in **Figure 21–2,** each of which could be considered a kingdom in its own right. Where would that leave the plant, animal, and fungi kingdoms? Surprisingly, they fit right into these six clades, and two of them, animals and fungi, actually emerge from the same protist ancestors.

In a way, this should have been expected. Protists were the first eukaryotes, and evolution has had far more time to develop differences among protists than among more recently evolved eukaryotes like plants and animals. In other words, by finding the fundamental divisions among protists, we also identify the most basic differences among all eukaryotes. You should expect protist classification to change yet again as biologists learn more about the genomes of these remarkable organisms.

What "Protist" Means Today Today biologists assembling what is often called the Tree of Life favor the classification shown above. But the word "protist" remains in such common usage, even among scientists, that we continue to use it here. Bear in mind, however, that the "protists" are not a single kingdom but a collection of organisms that includes several distinct clades. This is why the term is sometimes surrounded by quotation marks.

Protists—Ancestors and Descendants

How are protists related to other eukaryotes?

Protists were the first eukaryotes. How are they related to other eukaryotic organisms today? As tempting as it might be to look among living protists to find the ancestors to the first plants or the earliest fungi, it would be a scientific mistake to do so. The reason, of course, is that protists living today have been through a process of evolution just as extensive as the one that produced every other living organism.

Microscopic fossils of eukaryotic cells, like the one shown in **Figure 21–3,** have been found in rocks as old as 1.5 billion years. Genetic and fossil evidence indicates that eukaryotes evolved from prokaryotes and are more closely related to present-day Archaea than to Bacteria. The actual split between Archaea and Eukarya may have come as early as 2.5 billion years ago. Since that time, protists have diversified into as many as 300,000 species found in every corner of the planet.

Most of the major protist groups have remained unicellular, but two have produced organisms that developed true multicellularity. It is from the ancestors of these groups that plants, animals, and fungi arose.

Today's protists include groups whose ancestors were among the very last to split from the organisms that gave rise to plants, animals, and fungi. The roots of all eukaryotic diversity, from plants to animals to fungi, are found among the ancestors of the organisms that we call protists.

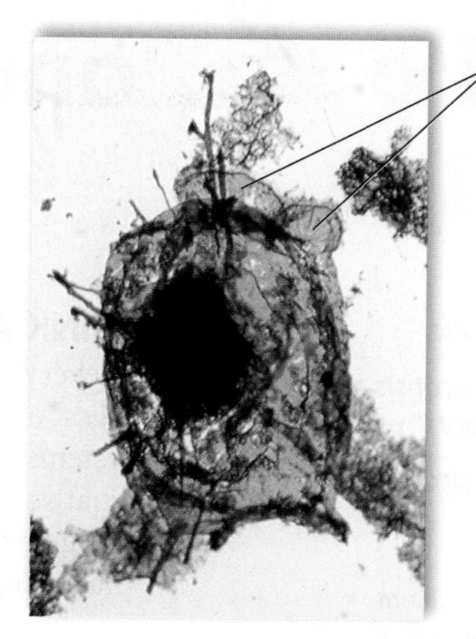

FIGURE 21–3 Fossil of an Early Eukaryote This 1.5-billion-year-old fossil of *Tappania plana* indicated to scientists that ancient eukaryotes already had the cytoskeletal structures characteristic of protists today. The bulbous projections on the cell are hypothesized to have functioned in asexual reproduction. (LM 285×)

21.1 Assessment

Review Key Concepts

1. a. Review What is a protist?
   b. Compare and Contrast Compare the updated classification of protists with the older one.
2. a. Review Which kingdoms arose from protist ancestors?
   b. Apply Concepts Why is it misguided to try to find our earliest eukaryotic ancestor among modern-day protists?

VISUAL THINKING

3. Compare **Figure 21–2** with the Tree of Life presented in Chapter 18. What simplification does **Figure 21–2** make? How could this simplification be misinterpreted? Explain your answer using your knowledge from Chapter 18 of how cladograms are constructed.

21.2 Protist Structure and Function

Key Questions

How do protists move in the environment?

How do protists reproduce?

Vocabulary

pseudopod • cilium • flagellum • spore • conjugation • alternation of generations • sporangium

Taking Notes

Compare/Contrast Table As you read, make a table that compares and contrasts the different ways protists move.

THINK ABOUT IT Our bodies are packed with specialized systems of every sort. Organ systems help us move, sense the environment, digest our food, and even reproduce. But protists have no such systems—they do it all within the confines of a single cell. Imagine what such cells would have to be like to succeed in the never-ending struggle for life on Earth. The protists we see today are winners in that struggle.

How Protists Move

How do protists move in the environment?

Before they gave rise to multicellular eukaryotes, protists evolved just about every form of cellular movement known to exist. **Some protists move by changing their cell shape, and some move by means of specialized organelles. Other protists do not move actively but are carried by wind, water, or other organisms.**

Amoeboid Movement Many unicellular protists move by changing their shape, a process that makes use of cytoplasmic projections known as **pseudopods** (soo doh pahdz). The best-known protists with this form of movement are the amoebas. In **Figure 21–4,** you can see how the cytoplasm of the amoeba streams into the pseudopod and the rest of the cell follows. This type of locomotion is called amoeboid movement and is found in many protists. It is powered by a cytoskeletal protein called actin. Actin is also found in the muscle cells of animals, where it plays an important role in muscle contraction.

BUILD Vocabulary

GREEK ROOTS The word **pseudopod** comes from the Greek roots *pseudo,* meaning "false," and *pod,* meaning "foot."

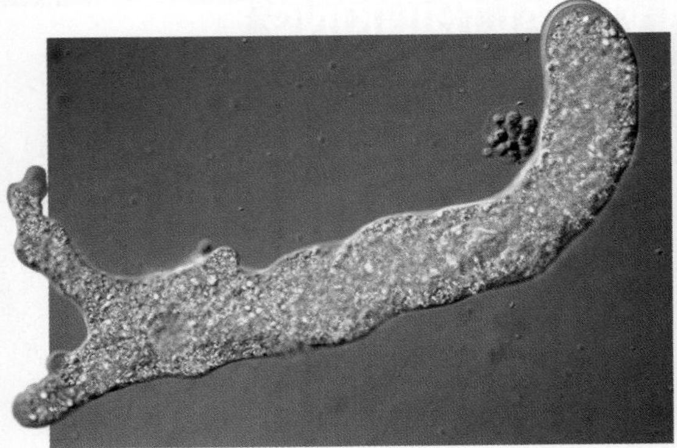
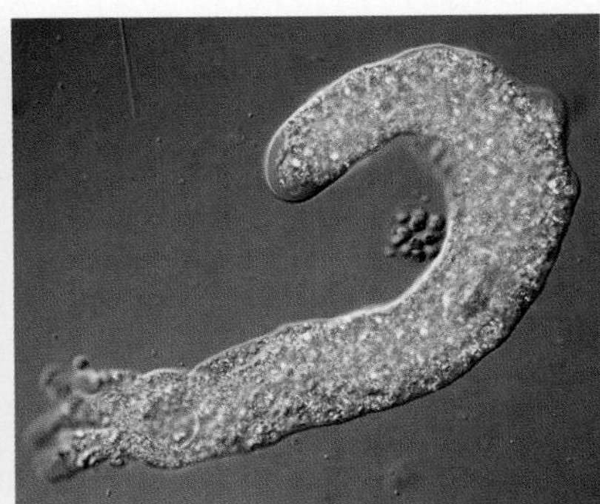
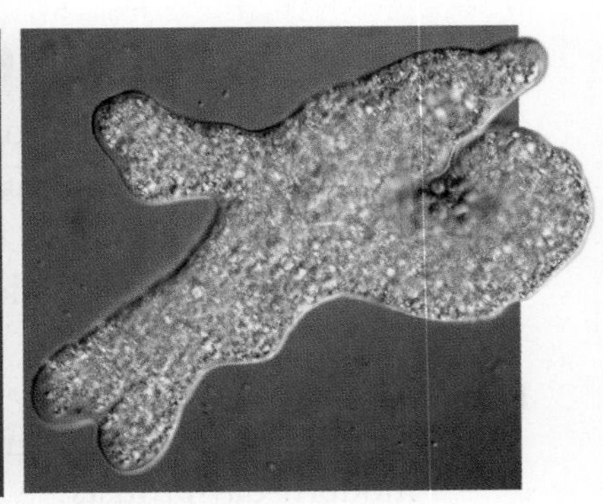

FIGURE 21–4 Amoeboid Movement An amoeba moves by first extending a pseudopod away from its body. The organism's cytoplasm then streams into the pseudopod. Amoebas also use pseudopods to surround and ingest prey. Here, the prey is a cluster of green algal cells. (LM 220×)

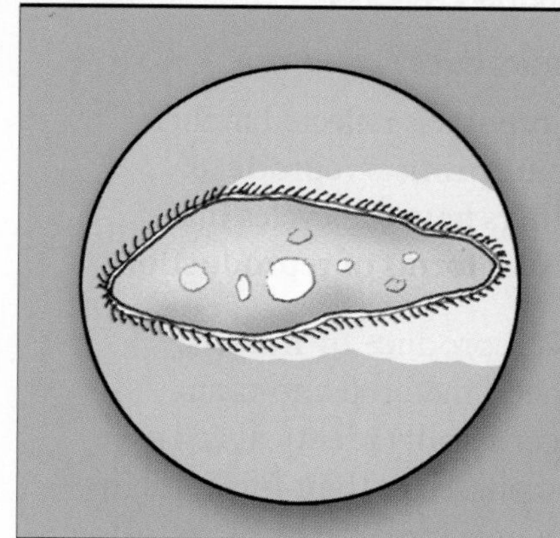

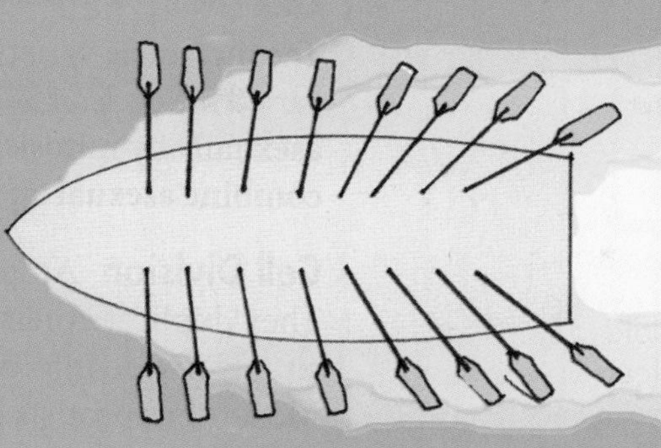

Motion by cilia is analogous to oars propelling a large rowboat forward through the water.

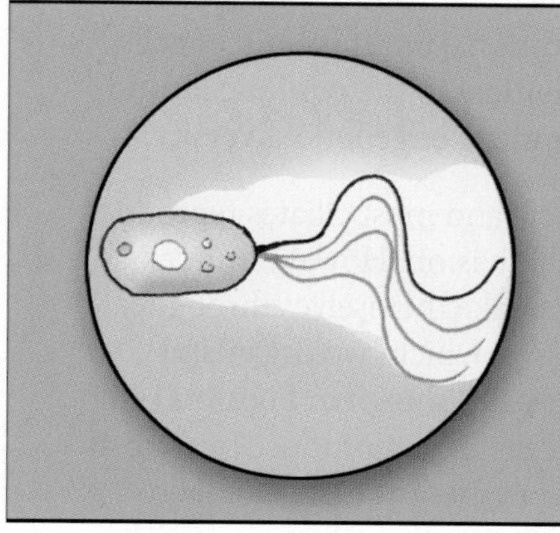

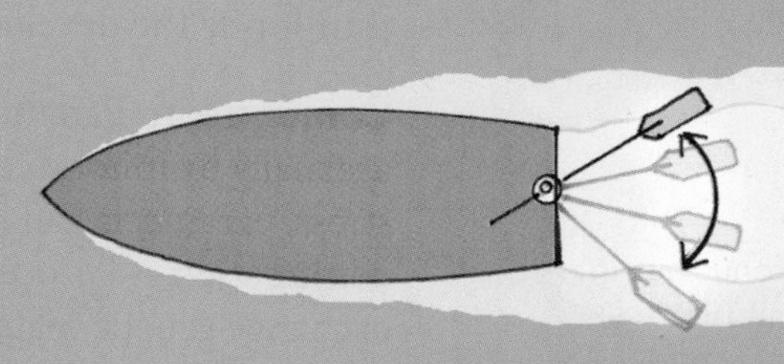

Motion by some flagella is analogous to the back-and-forth movement of a single long oar at the back of a boat, propelling it forward.

VISUAL ANALOGY

HOW CELLS MOVE LIKE BOATS

FIGURE 21–5 The forward motion provided by cilia or some flagella is similar to two ways by which oars propel a boat.

Cilia and Flagella Many protists move by means of cilia (sil ee uh) and flagella (fluh jel uh), structures supported by microtubules. Cilia and flagella have nearly identical internal structures, but they produce cellular motion differently. **Cilia** (singular: cilium) are short and numerous, and they move somewhat like oars on a boat. **Flagella** (singular: flagellum) are relatively long and usually number only one or two per cell. Some flagella spin like tiny propellers, but most produce a wavelike motion from base to tip. Compare these two types of motion in **Figure 21–5.** Protists that move using cilia are known as *ciliates,* and those that move with flagella are called *flagellates.*

Passive Movement It may surprise you to learn that some of the most important protists are nonmotile—they depend on air or water currents and other organisms to carry them around. These protists form reproductive cells called **spores** that can enter the cells of other organisms and live as parasites. Spore-forming protists include *Plasmodium,* which is carried by mosquitoes and causes malaria, and *Cryptosporidium,* which spreads through contaminated drinking water and causes severe intestinal disease.

In Your Notebook *Look up the word roots for* cilia *and* flagella, *and write an explanation of how each term relates to its root.*

MYSTERY CLUE

The Irish potato crop was propagated by cutting out the small buds on the potatoes—the eyes—and saving them for the next year's crop. This resulted in whole fields of genetically identical potatoes. How do you think this practice might have contributed to the spread of the blight?

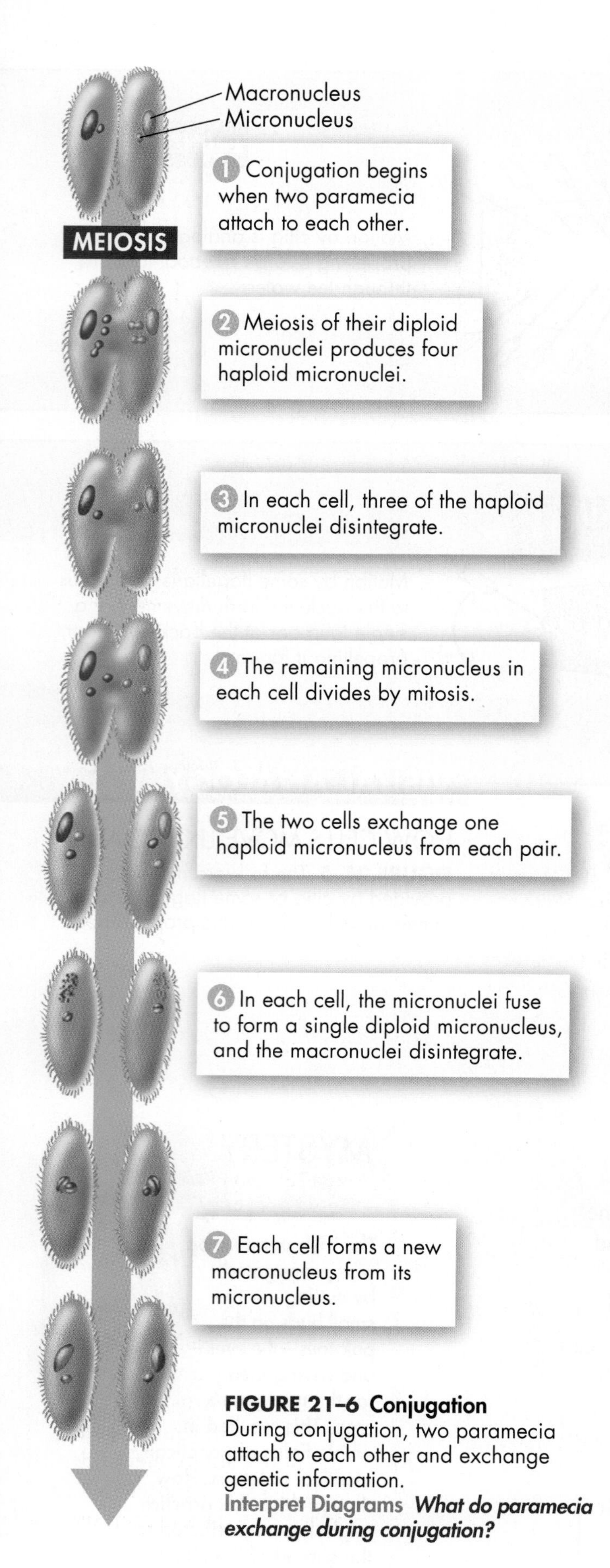

FIGURE 21–6 Conjugation
During conjugation, two paramecia attach to each other and exchange genetic information.
Interpret Diagrams ***What do paramecia exchange during conjugation?***

Protist Reproduction

How do protists reproduce?

The incredible variety of protists is reflected in their varied life cycles. **Some protists reproduce asexually by mitosis. Others have life cycles that combine asexual and sexual forms of reproduction.**

Cell Division Amoebas reproduce by mitosis: They duplicate their genetic material and then simply divide into two genetically identical cells. Most other protists have phases in their life cycle in which they also produce new individuals by mitosis. Mitosis enables protists to reproduce rapidly, especially under ideal conditions, but it produces cells that are genetically identical to the parent cell, and thus limits the development of genetic diversity.

Conjugation Paramecia and most ciliates reproduce asexually by mitotic cell division. However, under stress, paramecia can remake themselves through **conjugation**—a process in which two organisms exchange genetic material, as shown in **Figure 21–6.** After conjugating, the cells then reproduce by mitosis.

Paramecium has two types of nuclei: a macronucleus and one or more smaller micronuclei. The micronucleus is a bit like a reference library where books don't circulate—it holds a "reserve copy" of every gene in the cell. The macronucleus is more like a lending library—it has multiple copies of the genes the cell uses in its day-to-day activities.

Conjugation is not a type of reproduction because no new individuals are formed. It is, however, a sexual process, using meiosis to produce new combinations of genetic information. In a large population, conjugation helps produce and maintain genetic diversity, the raw material for evolution.

Sexual Reproduction Many protists have complex sexual life cycles in which they alternate between a diploid and a haploid phase, a process known as **alternation of generations.** An example is the life cycle of a type of protist known as a water mold. Water molds, or oomycetes (oh oh MY seets), thrive on dead and decaying organic matter in water or as parasites of plants on land.

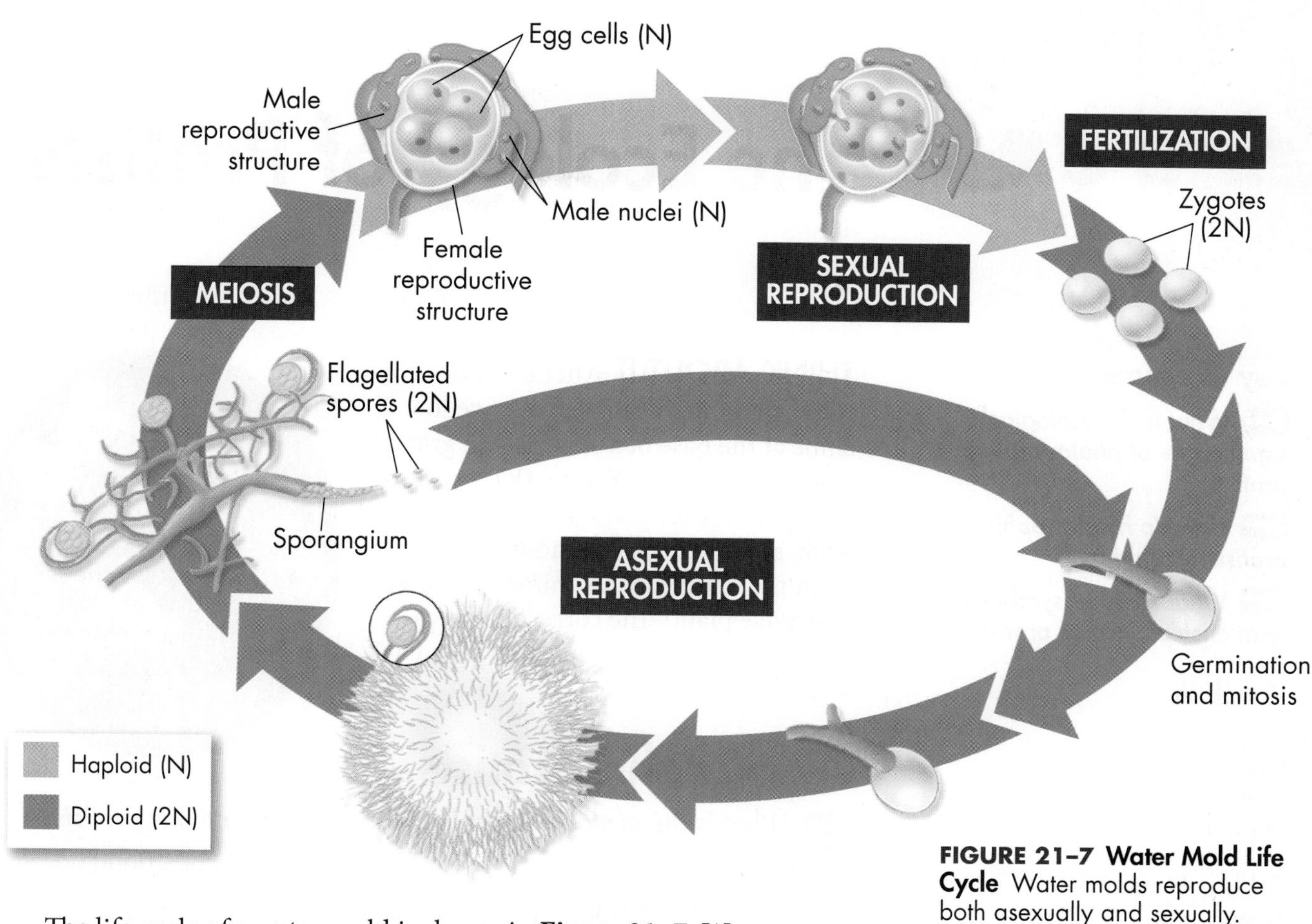

FIGURE 21–7 Water Mold Life Cycle Water molds reproduce both asexually and sexually.

The life cycle of a water mold is shown in **Figure 21–7.** Water molds grow into long branching filaments consisting of many cells formed by mitotic cell division. Water molds—and many other protists—reproduce asexually by producing spores in a structure called a **sporangium** (spoh RAN jee um). In water molds the spores are flagellated. Water molds also reproduce sexually by undergoing meiosis and forming male and female structures. These structures produce haploid nuclei that fuse during fertilization, forming a zygote that begins a new life cycle.

21.2 Assessment

Review Key Concepts

1. a. Review Summarize three ways in which protists move.
   b. Compare and Contrast How is movement by means of flagella different from movement by means of cilia?
2. a. Review Describe how protists reproduce.
   b. Explain How does conjugation produce genetic diversity in a population of *Paramecium*?
   c. Compare and Contrast How does a macronucleus differ in function from a micronucleus?

Apply the Big idea

Information and Heredity

3. Compare asexual and sexual processes in paramecia. Include the terms *mitosis* and *meiosis* in your answer. You may want to refer back to Chapters 10 and 11 to review mitosis and meiosis.

21.3 The Ecology of Protists

Key Questions

What is the ecological significance of photosynthetic protists?

How do heterotrophic protists obtain food?

What types of symbiotic relationships involve protists?

Vocabulary

algal bloom
food vacuole
gullet
plasmodium

Taking Notes

Outline Preview the heads of this lesson to construct an outline of different types of ways that protists obtain energy. Fill in your outline with specific examples as you read.

THINK ABOUT IT After a few days of rain, you notice a small spot of yellow slime at the base of a stand of tall grass. Is it some sort of rot? You mark its position with paint. A few days later, you come back, and it has grown and moved away from the mark. Is it an animal? A fungus? A strange plant? The correct answer is none of the above. It's a protist called a slime mold.

Autotrophic Protists

What is the ecological significance of photosynthetic protists?

If you've seen greenish scum growing along the banks of a pond or maybe even at the edges of a poorly maintained swimming pool, you might have called it "algae" without thinking. What you may not have realized at the time is that many of the organisms in that scum were, in fact, protists.

Diversity Biologists long ago realized that the organisms commonly called "algae" actually belong to many different groups. Some (the cyanobacteria) are prokaryotes, some (the green algae) belong to the plant kingdom, and some are protists. Photosynthetic protists include many phytoplankton species and the red and brown algae, as well as euglenas and dinoflagellates. These organisms share an autotrophic lifestyle, marked by the ability to use the energy from light to make a carbohydrate food source.

You might think that all photosynthetic protists are closely related to plants, but this is not the case. In fact, it is the red algae that are most closely related to plants. Many other photosynthetic protists are more closely related to nonphotosynthetic protists. In some cases certain species within a group have lost chloroplasts. In other cases endosymbiosis added a chloroplast to some species but not to their relatives.

In Your Notebook *Preview the next page. Then, make a four-column table using the headings listed. Use the table to record your notes as you read.*

Ecological Roles Photosynthetic protists play major ecological roles on Earth. **The position of photosynthetic protists at the base of the food chain makes much of the diversity of aquatic life possible.** Some examples of ecological roles played by photosynthetic protists are shown in **Figure 21–8.**

▶ ***Feeding Fish and Whales*** Photosynthetic protists make up a large portion of phytoplankton, the small, free-floating photosynthetic organisms found near the surface of oceans and lakes. About half of the photosynthesis that takes place on Earth is carried out by phytoplankton, which provide a direct source of nourishment for organisms as diverse as shrimp and baleen whales. And they are an indirect source of nourishment for humans. When you eat tuna fish, you are eating fish that fed on smaller fish that fed on still smaller animals that fed on photosynthetic protists.

▶ ***Supporting Coral Reefs*** Coral reefs, which are found in warm ocean waters throughout the world, provide food and shelter to large numbers of fish and other organisms. Protist algae known as zooxanthellae provide most of the coral's energy needs by photosynthesis. By nourishing coral animals, these algae help maintain the equilibrium of the coral ecosystem. Coralline red algae also help to provide calcium carbonate to stabilize growing coral reefs.

▶ ***Providing Shelter*** The largest known protist is giant kelp, a brown alga that can grow to more than 60 meters in length. Kelp forests provide shelter for many marine species, and the kelp itself is a source of food for sea urchins. Another brown alga, called *Sargassum,* forms huge floating mats many kilometers long in an area of the Atlantic Ocean near Bermuda known as the Sargasso Sea.

▶ ***Recycling Wastes*** Many protists grow rapidly in regions where sewage is discharged, where they play a vital role in recycling waste materials. When the amount of waste is excessive, however, populations of protists like *Euglena* can grow to enormous numbers and create an **algal bloom.** Algal blooms can disrupt ecosystem homeostasis. For example, an algal bloom in a pond or lake depletes nutrients from the water, and the decomposition of the dead protists can rob water of its oxygen, causing fish and invertebrates to die. In another example, blooms of marine protists called dinoflagellates create what is known as a red tide. The buildup of toxins produced by these protists can poison fish and shellfish.

FIGURE 21–8 Ecological Roles of Protists
Photosynthetic protists play many roles in the environment. **Apply Concepts** ***Which example is disruptive to ecosystem homeostasis?***

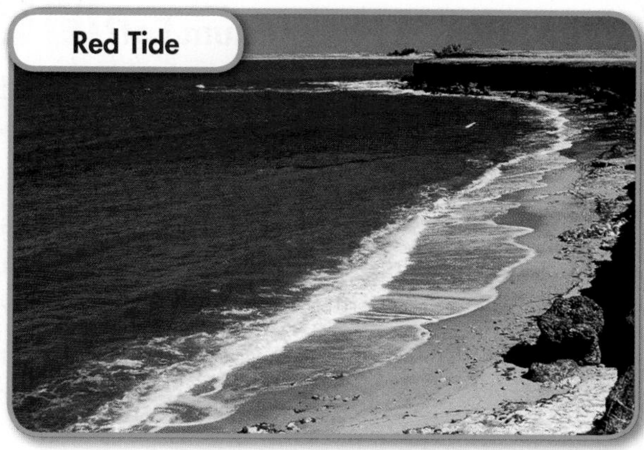

Heterotrophic Protists

How do heterotrophic protists obtain food?

Many protists are heterotrophs: They obtain food from other living organisms. **Some heterotrophic protists engulf and digest their food, while others live by absorbing molecules from the environment.**

BUILD Vocabulary

MULTIPLE MEANINGS The name *Amoeba* refers to a particular genus of protists, but it is also used generically (in lowercase and without italicization) to mean "any protists that move using pseudopods," including the unicellular form of slime molds.

Amoebas Amoebas can capture and digest their food, surrounding a cell or particle and then taking it inside themselves to form a food vacuole. A **food vacuole** is a small cavity in the cytoplasm that temporarily stores food. Once inside the cell, the material is digested rapidly, and the nutrients are passed along to the rest of the cell. Indigestible waste materials remain inside the vacuole until the vacuole releases them outside the cell.

LM 230×

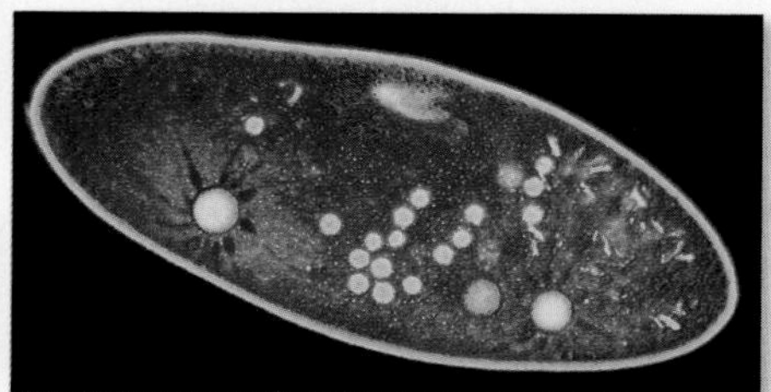

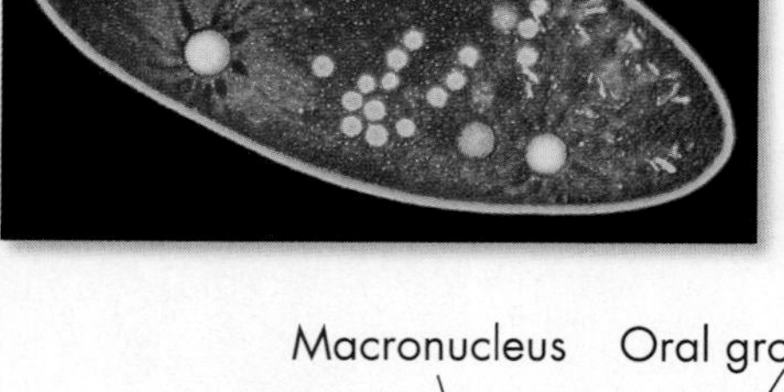

Ciliates *Paramecium* and other ciliates use their cilia to sweep food particles into the **gullet,** an indentation in one side of the organism, as shown in **Figure 21–9.** The particles are trapped in the gullet and forced into food vacuoles that form at its base. The food vacuoles pinch off into the cytoplasm and eventually fuse with lysosomes, which contain digestive enzymes. Waste materials are emptied into the environment when the food vacuole fuses with a region of the cell membrane called the anal pore.

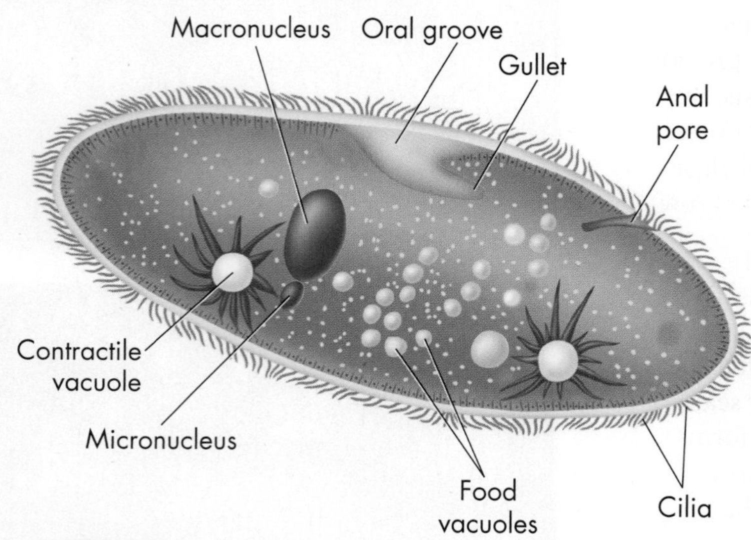

FIGURE 21–9 Feeding Structures of *Paramecium* Cilia lining the gullet move food to the organism's interior. There, the food particles are engulfed, forming food vacuoles.

Quick Lab
GUIDED INQUIRY

How Does a Paramecium Eat?

1. Use separate dropper pipettes to place a drop of paramecium culture and a drop of *Chlorella* culture next to each other on a microscope slide.
2. Use a toothpick to transfer a few granules of carmine dye to the drops on the slide. Add a coverslip so that the two drops mix.
3. Place the slide on the microscope stage and use the low-power objective to locate several paramecia.
4. Use the high-power objective to observe the contents and behavior of the paramecia.

Analyze and Conclude

1. **Observe** Where did the *Chlorella* cells and carmine dye granules accumulate?
2. **Infer** How do you think this accumulation of cells and dye granules occurs?
3. **Form a Hypothesis** What process in the paramecia do you think resulted in this change?

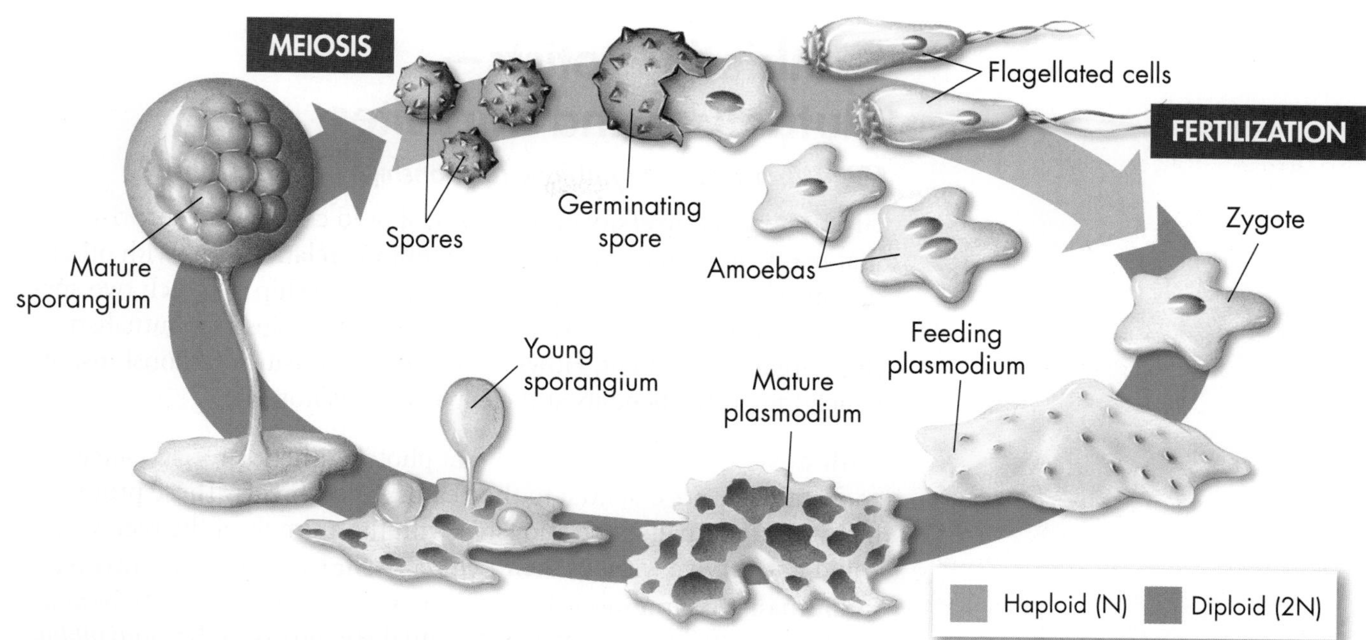

FIGURE 21–10 Slime Mold Life Cycle The feeding-plasmodium stage in the slime mold life cycle is a collection of many amoebalike organisms—in some species, these organisms are contained within a single cell membrane. The plasmodium eventually produces sporangia, which in turn undergo meiosis and produce haploid spores. The spores grow into amoebalike or flagellated cells. The flagellated cells then fuse to produce diploid zygotes that repeat the cycle.

MYSTERY CLUE

Nutrients from decomposing potatoes infected by potato blight enter the blight organism by diffusing through its cell walls. Which organism described on these pages feeds in a similar way?

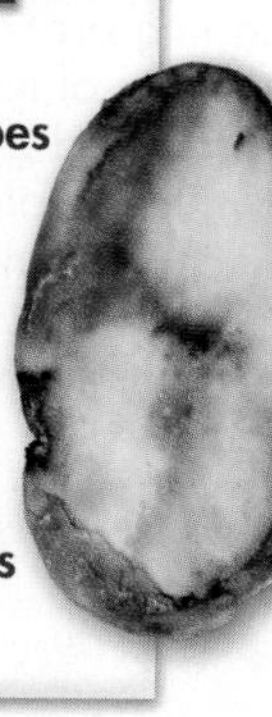

Slime Molds Another type of heterotrophic protist is a slime mold, which thrives on decaying organic matter. Slime molds are found in places that are damp and rich in organic matter—on the floor of a forest or a backyard compost pile, for example. Slime molds play key roles in recycling nutrients in an ecosystem.

At one stage in their life cycle, shown in **Figure 21–10,** slime molds exist as a collection of individual amoebalike cells. Eventually these aggregate to form a large structure known as a **plasmodium,** which may continue to move. The plasmodium eventually develops sporangia, in which meiosis produces haploid spores to continue the cycle.

Protists That Absorb Some protists survive by absorbing molecules that other organisms have released to the environment. Water molds like the one shown in **Figure 21–11,** for example, grow on dead or decaying plants and animals, absorbing food molecules through their cellulose cell walls and cell membranes. If you've seen white fuzz growing on the surface of a dead fish in the water, you've seen water molds in action.

FIGURE 21–11 Water Mold This dead goldfish is covered with the common water mold *Saprolegnia.* Compare and Contrast ***Contrast how a water mold and a paramecium obtain food.***

Symbiotic Protists—Mutualists and Parasites

What types of symbiotic relationships involve protists?

Given the great diversity of protists, it should come as no surprise that many of them are involved in symbiotic relationships with other organisms. As you know, symbiosis is a relationship in which two species live closely together. Many of these relationships are mutualistic: Both organisms benefit. However, some are parasitic relationships, in which the protist benefits at the expense of its host.

Mutualists Earlier you learned that photosynthetic protists called zooxanthellae are essential to the health of coral reefs. These protists maintain a mutualistic relationship with the animals of the reef, which could not survive without their help. **Many protists are involved in mutualistic symbioses, in which they and their hosts both benefit.**

Another striking example of a mutualistic protist is *Trichonympha*, shown in **Figure 21–12.** *Trichonympha* is a flagellated protist that lives within the digestive system of various species of termites and makes it possible for the insects to digest wood. Termites themselves do not have enzymes to break down the cellulose in wood. How, then, does a termite digest cellulose? In a sense, it doesn't. *Trichonympha* does.

Trichonympha and other organisms in the termite's gut manufacture an enzyme called cellulase that breaks the chemical bonds in cellulose, making it possible for termites to digest wood. With the help of their protist partners, then, termites can munch away, digesting all the wood they can eat.

ZOOMING IN

A PROTIST MUTUALIST

FIGURE 21–12 *Trichonympha* is a wood-digesting protist that lives in the digestive system of termites. Digestive enzymes produced by the protist break down the particles of wood, which you can see inside the protist's body. **Predict** ***What would happen to a termite if its*** Trichonympha ***colony died?***

▼ ***Trichonympha*** **inside a termite gut** (LM 250×)

◀ **Termite colony in a rotting log**

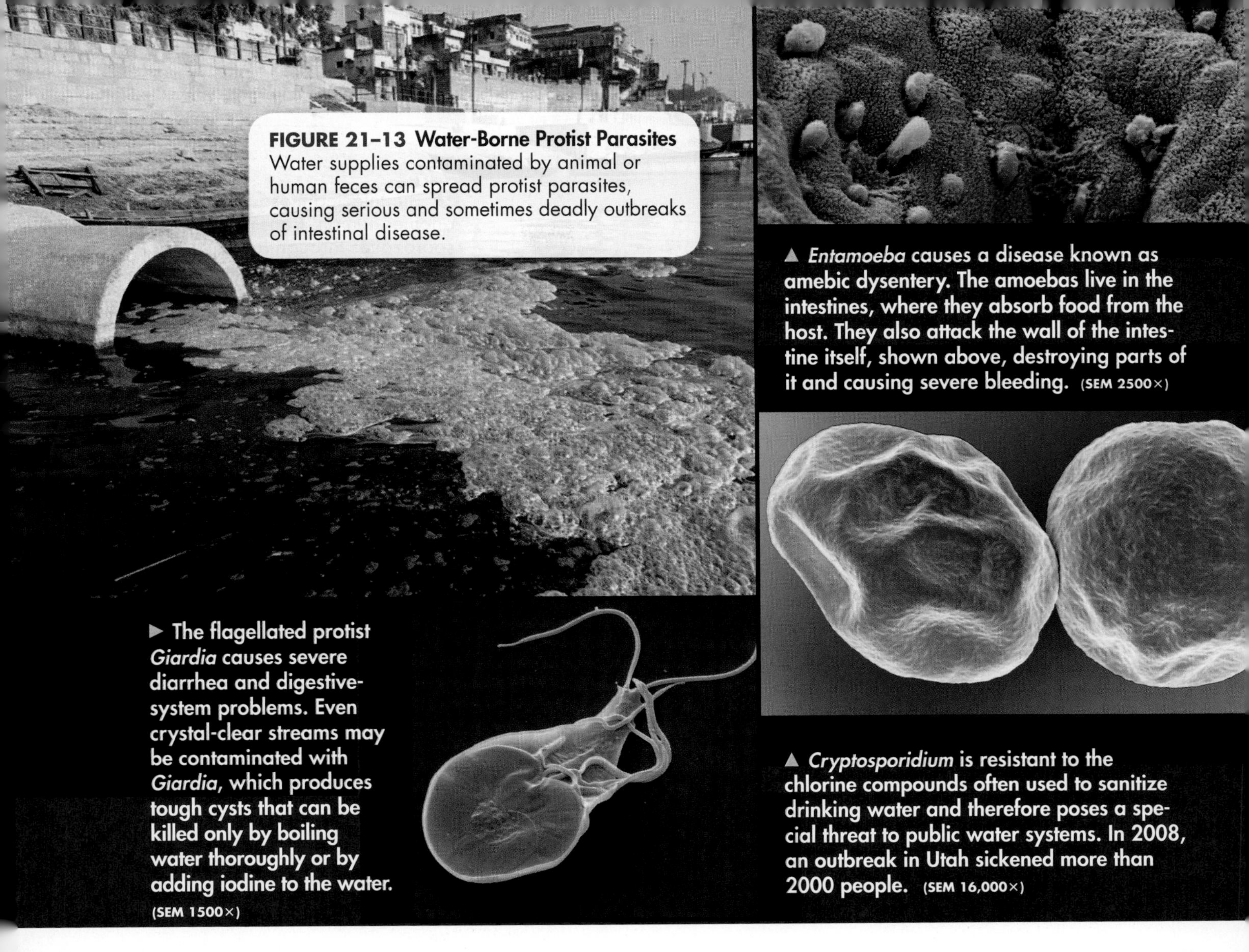

FIGURE 21–13 Water-Borne Protist Parasites Water supplies contaminated by animal or human feces can spread protist parasites, causing serious and sometimes deadly outbreaks of intestinal disease.

▲ *Entamoeba* causes a disease known as amebic dysentery. The amoebas live in the intestines, where they absorb food from the host. They also attack the wall of the intestine itself, shown above, destroying parts of it and causing severe bleeding. (SEM 2500×)

▶ The flagellated protist *Giardia* causes severe diarrhea and digestive-system problems. Even crystal-clear streams may be contaminated with *Giardia*, which produces tough cysts that can be killed only by boiling water thoroughly or by adding iodine to the water. (SEM 1500×)

▲ *Cryptosporidium* is resistant to the chlorine compounds often used to sanitize drinking water and therefore poses a special threat to public water systems. In 2008, an outbreak in Utah sickened more than 2000 people. (SEM 16,000×)

Parasites and Disease Unfortunately for humans and some other organisms, protists cause a number of very serious diseases. **Parasitic protists are responsible for some of the world's most deadly diseases, including several kinds of debilitating intestinal diseases, African sleeping sickness, and malaria.**

▶ ***Intestinal Diseases*** Water-borne protists are found in streams, lakes, and oceans. Most cause little harm to humans, but some of these organisms—like those shown in **Figure 21–13**—are parasites that cause serious problems.

▶ ***African Sleeping Sickness*** Flagellated protists of the genus *Trypanosoma* cause African sleeping sickness. Trypanosomes are spread from person to person by the bite of the tsetse fly. They destroy blood cells and infect other tissues in the body, including nerve cells. Severe damage to the nervous system causes some individuals to lose consciousness and lapse into a deep and sometimes fatal sleep, from which the disease gets its name. Control of the tsetse fly and the protist pathogens that it spreads is a major goal of health workers in Africa.

In Your Notebook *How do parasitic protists use passive movement to spread disease? Give examples.*

FIGURE 21–14
***Plasmodium* Life Cycle**
The parasite that causes malaria, *Plasmodium,* requires two hosts to complete its life cycle: an *Anopheles* mosquito and a human.

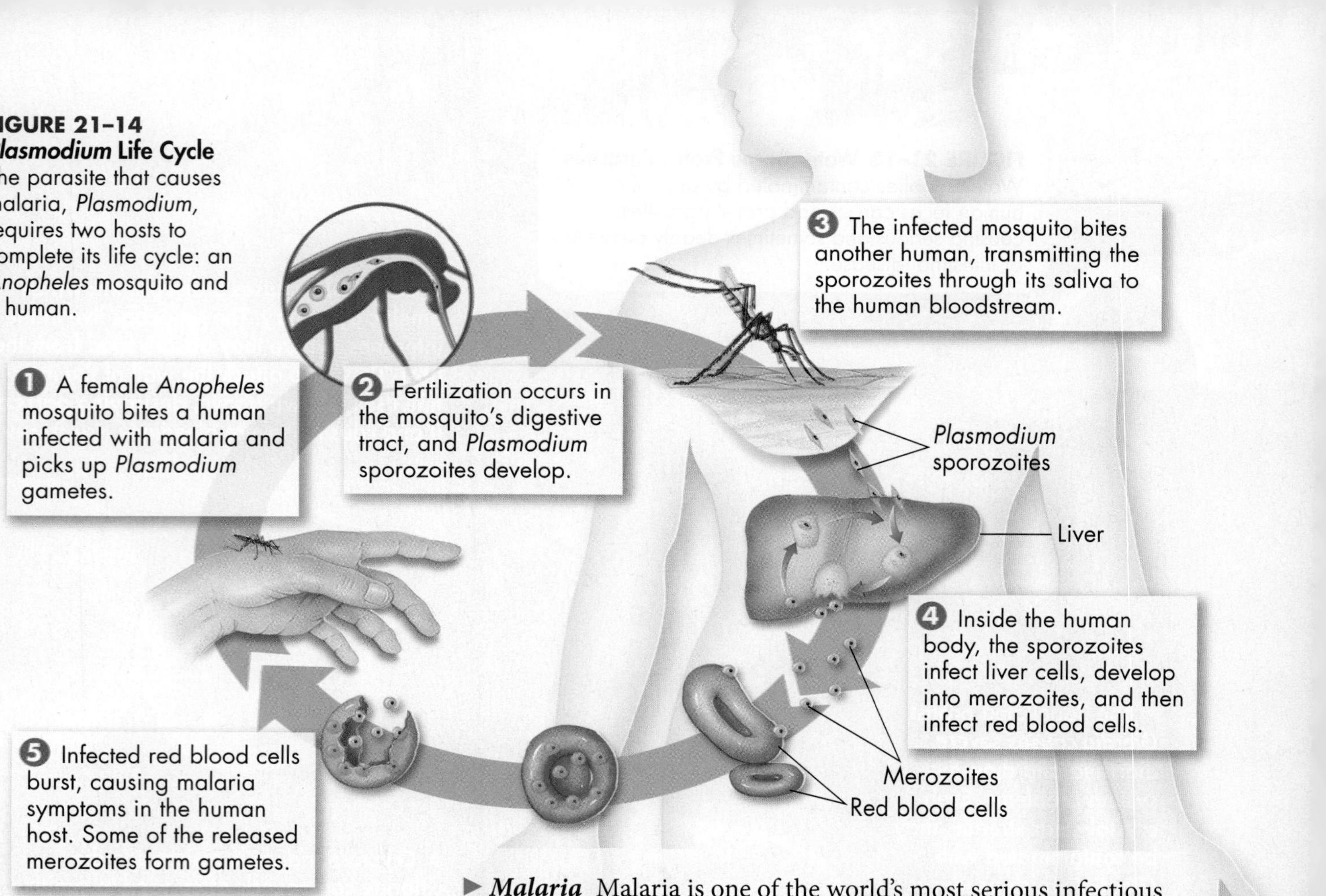

1. A female *Anopheles* mosquito bites a human infected with malaria and picks up *Plasmodium* gametes.
2. Fertilization occurs in the mosquito's digestive tract, and *Plasmodium* sporozoites develop.
3. The infected mosquito bites another human, transmitting the sporozoites through its saliva to the human bloodstream.
4. Inside the human body, the sporozoites infect liver cells, develop into merozoites, and then infect red blood cells.
5. Infected red blood cells burst, causing malaria symptoms in the human host. Some of the released merozoites form gametes.

Plasmodium sporozoites
Liver
Merozoites
Red blood cells

▶ ***Malaria*** Malaria is one of the world's most serious infectious diseases. More than 1 million people die from malaria every year, many of them children. Malaria is caused by *Plasmodium,* a spore-forming protist carried by the female *Anopheles* mosquito. The life cycle of *Plasmodium* is shown in **Figure 21–14.**

21.3 Assessment

Review Key Concepts

1. a. Review Give four examples of photosynthetic protists.
b. Predict How would ocean food chains change in the absence of photosynthetic protists?

2. a. Review Describe the ways amoebas, ciliates, and water molds obtain food.
b. Compare and Contrast In protist terms, how is engulfing food different from absorbing it?

3. a. Review Give two examples each of mutualism and parasitism involving protists.
b. Explain By what two methods do parasitic protists primarily spread disease?
c. Form an Opinion Why are protist-caused diseases more widespread in tropical areas of the world?

WRITE ABOUT SCIENCE

Creative Writing

4. Use the library or the Internet to investigate the number of algal blooms off the California coast over the last ten years. Be sure to note the causes, the types of protists identified, and the effects on wildlife and people. Present your findings to the class as an unbiased investigative report.

RST.9-10.10 Level of Text Complexity, WHST.9-10.7 Research to Build and Present Knowledge

Technology & BIOLOGY

Low-Tech Weapons Against a High-Tech Parasite

Malaria is one of the most serious infectious diseases in the world, claiming the life of a child every 30 seconds, day and night. The disease is caused by a spore-forming protist, *Plasmodium falciparum*, carried by the *Anopheles* mosquito. Efforts to produce vaccines against the disease have shown some promise, but even the best vaccines to date provide only marginal protection. To make matters worse, drugs that once kept the disease in check, including chloroquine, are almost useless today. In most regions of the world, the parasite has evolved resistance to these drugs.

Research on the disease continues, and scientists have now worked out the complete genome of both the parasite and the mosquito that carries it. The hope is that a better understanding of the genetics of the disease will allow scientists to fashion high-tech weapons against this killer. Surprisingly, however, one of the most promising approaches involves a very low-tech weapon—mosquito netting.

Most malaria infections are the result of mosquito bites that occur when people are sleeping. Studies have shown that when most of a village's residents sleep under insecticide-treated nets, mosquitoes begin to die off, and the likelihood of malaria infection goes down. The cost of the nets, about ten dollars, is remarkably small when measured against the human cost in lives and lost productivity. But it is estimated that as many as 250 million of the nets will be needed to protect populations at risk of this disease. Many charities are now featuring the nets in their fundraising appeals.

WRITING **Visit several Web sites run by organizations with an antimalaria mission. Then, describe how mosquito netting, a low-tech approach, might be combined effectively with antimalarial vaccines and drugs and other high-tech approaches.**

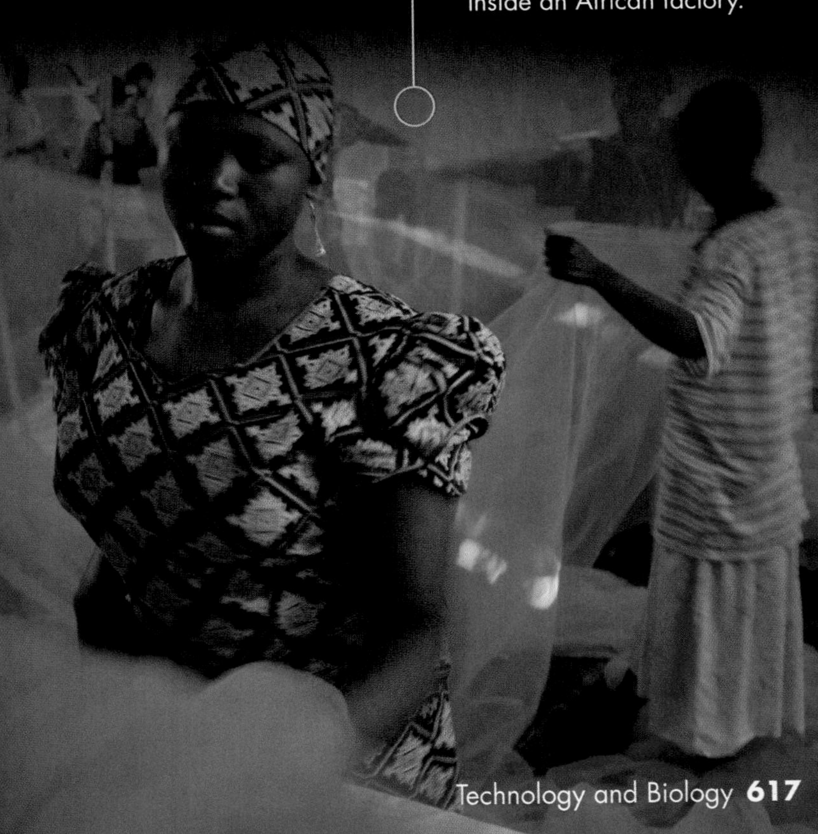

▼ A young woman folds insecticide-treated nets inside an African factory.

21.4 Fungi

Key Questions

What are the basic characteristics of fungi?

How do fungi affect homeostasis in other organisms and the environment?

Vocabulary

chitin • hypha • fruiting body • mycelium • lichen • mycorrhiza

Taking Notes

Concept Map As you read, develop a concept map showing the relationships of fungi to other organisms in their environment.

THINK ABOUT IT What is the largest organism in this photo? At first glance you might pick the tree, but in fact the largest organism is a fungus. The only trace of it is the ring of mushrooms that has popped up in the grass after a brief rainstorm.

The mushrooms are just the reproductive structures of a much larger organism. Most of the mass of the fungus is underground, spanning at least the width of the ring of mushrooms, and extending more than 2 meters into the ground! Hundreds of years ago, some cultures believed these rings of mushrooms marked spots where fairies danced in circles on warm summer nights. Today people still call them fairy rings.

What Are Fungi?

What are the basic characteristics of fungi?

Like the ring of mushrooms above, many fungi grow from the ground. This once led scientists to classify them as nonphotosynthetic plants. But they aren't plants at all. Instead of carrying out photosynthesis, fungi produce powerful enzymes that digest food outside their bodies. Then they absorb the small molecules released by the enzymes. Many fungi feed by absorbing nutrients from decaying matter in the soil. Others live as parasites, absorbing nutrients from the bodies of their hosts.

Another defining characteristic of fungi is the composition of their cell walls, which contain chitin (KY tun). **Chitin** is a polymer made of modified sugars that is also found in the external skeletons of insects. The presence of chitin is one of several features that show fungi are more closely related to animals than to plants. **Fungi are heterotrophic eukaryotes with cell walls that contain chitin.**

FIGURE 21–15
Scarlet Cup Fungus

Structure and Function There are two general growth patterns among fungi. Yeasts are tiny fungi that live most of their lives as single cells. Mushrooms and other fungi grow much larger, their bodies made up of cells that form long, slender branching filaments called **hyphae** (HY fee; singular: hypha), as shown in **Figure 21–16.** In most fungi, cross walls divide the hyphae into compartments resembling cells, each containing one or two nuclei. In the cross walls, there are openings through which cytoplasm and organelles such as mitochondria can move.

What you recognize as a mushroom is actually the **fruiting body,** the reproductive structure of the fungus. The fruiting body grows from the **mycelium** (my SEE lee um; plural: mycelia), the mass of branching hyphae below the soil. Clusters of mushrooms are often part of the same mycelium, which means that they are actually part of the same organism.

Some mycelia live for many years and grow very large. The mycelium of the soil fungus in a fairy ring has grown so large that it has used up all of the nutrients near its center. It grows and produces fruiting bodies—the mushrooms—only at its edges, where it comes in contact with fresh soil and abundant nutrients.

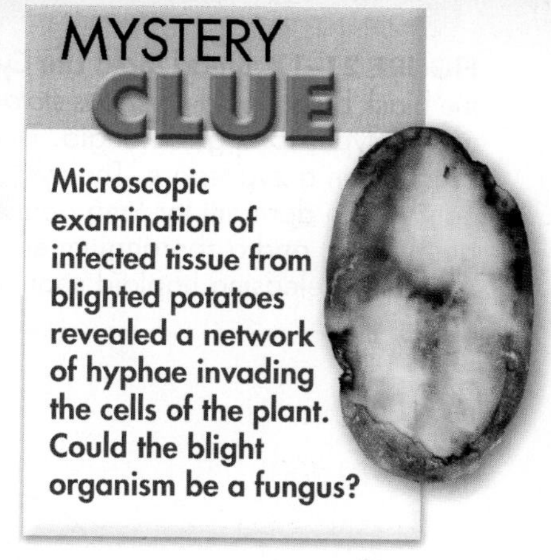

Microscopic examination of infected tissue from blighted potatoes revealed a network of hyphae invading the cells of the plant. Could the blight organism be a fungus?

In Your Notebook *How do fungi differ from other multicellular organisms?*

FIGURE 21–16 Structure of a Mushroom The body of a mushroom is actually its reproductive structure, also called a fruiting body. The major portion of the organism is the mycelium, which grows underground.

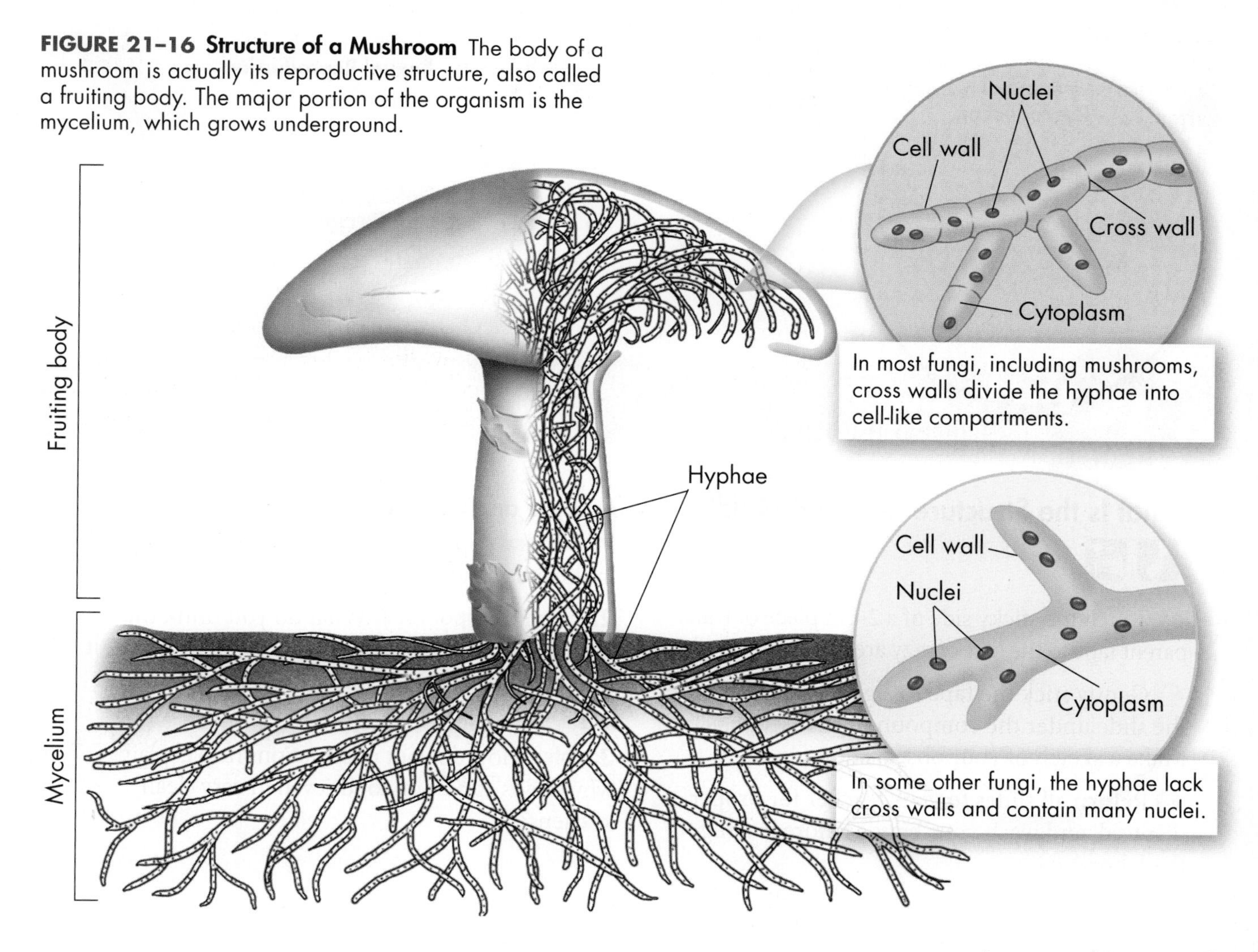

FIGURE 21–17 Bread Mold Life Cycle During sexual reproduction in the black bread mold *Rhizopus stolonifer,* hyphae from two different mating types form gametangia. The gametangia fuse, and zygotes form within a zygospore. The zygospore develops a thick wall and can remain dormant for long periods. The zygospore eventually germinates, and a sporangium emerges. The sporangium reproduces asexually, releasing haploid spores produced by meiosis.

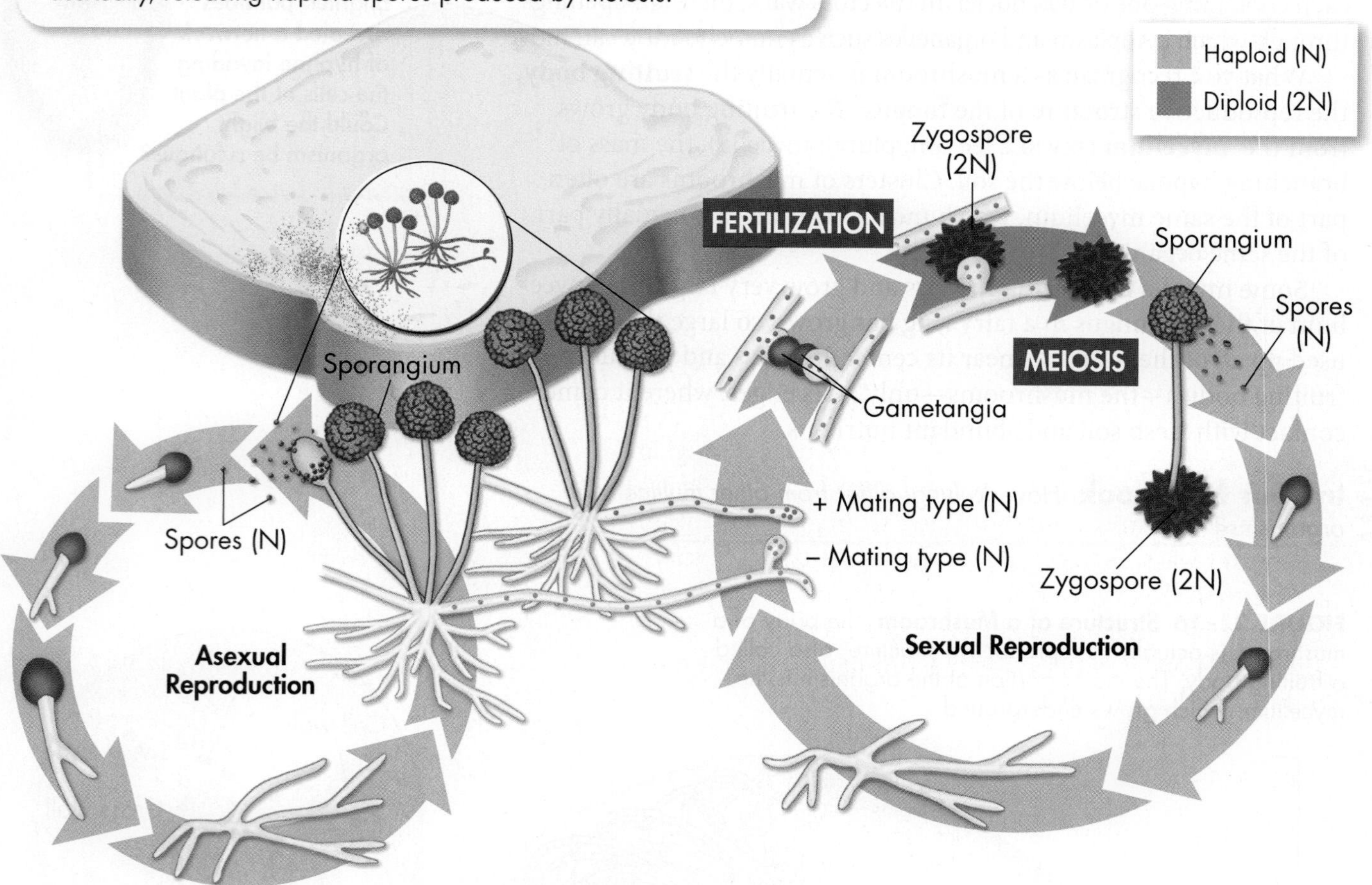

Quick Lab
GUIDED INQUIRY

What Is the Structure of Bread Mold?

❶ Touch the sticky side of a 2-cm piece of transparent tape to the black fuzzy area of a bread mold.

❷ Gently stick the tape to a glass slide. Observe the slide under the compound microscope. Make a sketch of your observations.

❸ Return all slides to your teacher for proper disposal, and wash your hands before leaving the laboratory.

Analyze and Conclude

1. Observe Describe the structures you observed in the bread mold.

2. Form a Hypothesis What do you think the function of the round structures is? Why might it be advantageous for a single mass of bread mold to produce so many of the round structures?

3. Infer How do your observations help explain why molds appear on foods even in very clean kitchens?

Reproduction Fungi can reproduce asexually, primarily by releasing spores that are adapted to travel through air and water. Simply breaking off a hypha or budding off a cell can also serve as asexual reproduction.

Most fungi also can reproduce sexually. **Figure 21–17** shows the life cycle of a type of bread mold, a fungus called *Rhizopus stolonifer.* Sexual reproduction in fungi often involves two different mating types. In *Rhizopus,* as in most fungi, gametes of both mating types are about the same size and are not usually called male and female. Instead, one mating type is called "+" (plus) and the other "–" (minus). When hyphae of opposite mating types meet, they start the process of sexual reproduction by fusing, bringing + and – nuclei together in the same cell. The + and – nuclei form pairs that divide in unison as the mycelium grows. Many of the paired nuclei fuse to form diploid zygote nuclei, which go through meiosis to make haploid spores. Each spore has a different combination of parental genes, and each can make a new mycelium.

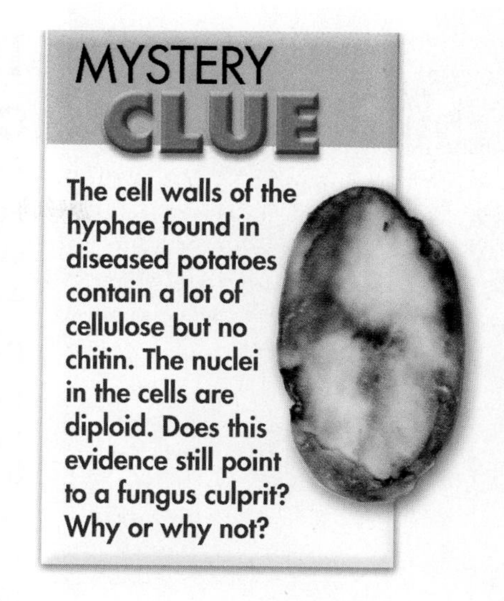

Diversity of Fungi More than 100,000 species of fungi are known. Of course, they all share the characteristics that define them as fungi, but they differ from one another in important ways. Biologists have used these similarities and differences, along with DNA comparisons, to place the fungi into several distinct groups. The major groups of fungi differ from one another in their reproductive structures, as summarized in **Figure 21–18.**

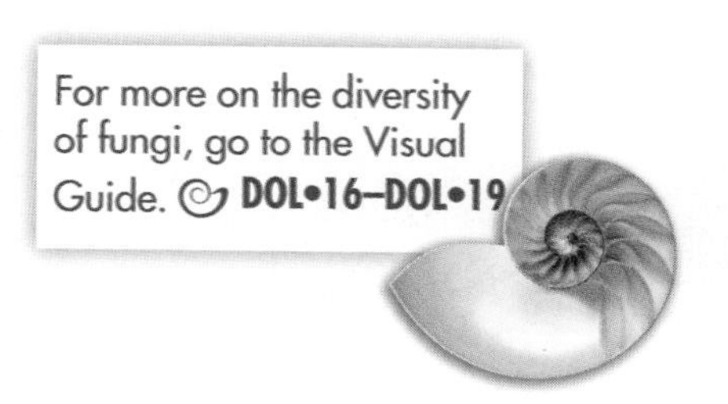

FIGURE 21–18 The Major Phyla of Fungi The table summarizes the main differences among the four major phyla of fungi. **Infer** ***Would you expect to find chytrids in aquatic or terrestrial habitats? Explain your answer.***

The Major Phyla of Fungi

Phylum	Distinguishing Features	Examples
Basidiomycota (club fungi)	Sexual spores found in club-shaped cell called a basidium	Mushrooms, puffballs, earthstars, shelf fungi, jelly fungi, rusts
Ascomycota (sac fungi)	Sexual spores found in saclike structure called an ascus	Morels, truffles, *Penicillium* species, yeasts
Zygomycota (common molds)	Tough zygospore produced during sexual reproduction that can stay dormant for long periods	*Rhizopus stolonifer* (black bread mold), molds found on rotting strawberries and other soft fruits, mycorrhizae associated with plant roots
Chytridomycota (chytrids)	Only fungi with flagellated spores	Many species are decomposers found in lakes and moist soil.

The Ecology of Fungi

How do fungi affect homeostasis in other organisms and the environment?

Fungi play an essential role in maintaining equilibrium in nearly every ecosystem. But there are some species that cause disease in plants and animals.

FIGURE 21–19 Champions of Decomposition The mycelia of these mushrooms have released enzymes that are breaking down the wood tissues of the decaying tree trunk.

Decomposition Many fungi feed by releasing digestive enzymes that break down leaves, fruit, and other organic material into simple molecules. These molecules then diffuse into the fungus. The mycelia of fungi produce digestive enzymes that speed the breakdown of wastes and dead organisms. Many organisms, especially plants, remove important trace elements and nutrients from the soil. If these materials were not returned, the soil would quickly be depleted, and Earth would become barren and lifeless. **Fungi are champions of decomposition. Many species help ecosystems maintain homeostasis by breaking down dead organisms and recycling essential elements and nutrients.**

Parasitism As useful as many fungi are, others can infect plants and animals. **Parasitic fungi can cause serious diseases in plants and animals by disrupting homeostasis.**

▶ ***Plant Diseases*** A number of parasitic fungi cause diseases that threaten food crops. Corn smut, for example, destroys corn kernels and wheat rust affects one of the most important crops grown in North America. Some mildews, which infect a wide variety of plants, are also fungi. Fungal diseases are responsible for the loss of approximately 15 percent of the crops grown in temperate regions of the world and even more of the crops grown in tropical areas.

FIGURE 21–20 Parasitic Fungi Corn smut infests the kernels of a corn plant, reducing the farmer's crop yield (left). A moth falls victim to the *Cordyceps* fungus (right).

▶ ***Animal Diseases*** Fungal diseases also affect insects, frogs, and mammals. One deadly example is caused by a fungus in the genus *Cordyceps*. This fungus infects grasshoppers in rain forests in Costa Rica. Microscopic spores become lodged in the grasshopper, where they germinate and produce enzymes that slowly penetrate the insect's tough external skeleton. The spores multiply in the insect's body, digesting all its cells and tissues until the insect dies. To complete the process of digestion, hyphae develop, cloaking the decaying exoskeleton in a web of fungal material. Reproductive structures, which will produce more spores and spread the infection, then emerge from the grasshopper's remains.

Parasitic fungi can also infect humans. The fungus that causes athlete's foot forms a mycelium in the outer layers of the skin, which produces a red, inflamed sore from which the spores can easily spread from person to person. The yeast *Candida albicans* can also disrupt the equilibrium in the human body. It is often responsible for vaginal yeast infections and for infections of the mouth called thrush. *Candida* is usually kept in check by competition from bacteria and by the body's immune system. This balance can be upset by the use of antibiotics, which kill bacteria, or by damage to the immune system.

Lichens The close relationships fungi form with members of other species are not always parasitic in nature. **Some fungi form mutualistic associations with photosynthetic organisms in which both partners benefit.** For example, a **lichen** (LY-kun) is a symbiotic association between a fungus and a photosynthetic organism. The photosynthetic organism is either a green alga or a cyanobacterium, or both. **Figure 21–21** shows the structure of a lichen.

Lichens are extremely resistant to drought and cold. Therefore, they can grow in places where few other organisms can survive—on dry bare rock in deserts and on the tops of mountains. Lichens are able to survive in these harsh environments because of the relationship between the two partner organisms. The green algae or cyanobacteria carry out photosynthesis, providing the fungus with a source of energy. The fungus, in turn, provides the green algae or cyanobacteria with water and minerals. Furthermore, the densely packed hyphae protect the delicate green cells from intense sunlight.

Lichens are often the first organisms to enter barren environments, gradually breaking down the rocks on which they grow. In this way, lichens help in the early stages of soil formation. Lichens are also remarkably sensitive to air pollution: They are among the first organisms to be affected when air quality deteriorates.

In Your Notebook *Summarize three roles of fungi in the environment. Compare these roles to those of protists.*

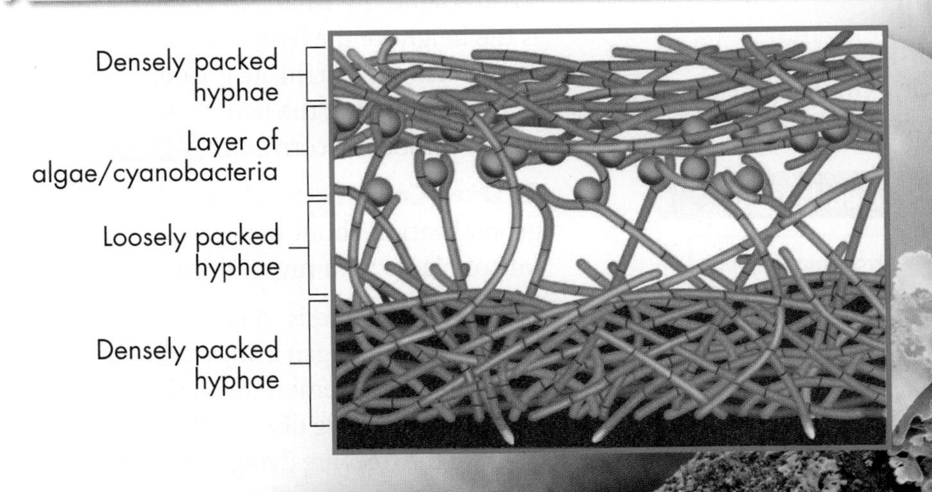

FIGURE 21–21 Inside a Lichen The protective upper surface of a lichen is made up of densely packed fungal hyphae. Below this are layers of green algae or cyanobacteria and loosely woven hyphae. The bottom layer contains small projections that attach the lichen to a rock or tree. **Infer** ***How do lichens assist in soil formation?***

Mycorrhizae Fungi also form mutualistic relationships with plant roots. Almost half of the tissues of trees are hidden beneath the ground in masses of tangled roots. These roots are woven into a partnership with an even larger web of fungal mycelia. These symbiotic associations of plant roots and fungi are called **mycorrhizae** (my koh RY zee; singular: mycorrhiza).

Scientists have known about this partnership for years, but recent research shows that it is more common and more important than was previously thought. Researchers now estimate that 80 to 90 percent of all plant species form mycorrhizae with fungi. The hyphae of the fungi form a network associated with the roots of the plants and extending into the soil. The hyphae collect water and minerals and bring them to the roots, greatly increasing the effective surface area of the root system. In addition, the fungi release enzymes that free nutrients in the soil. The plants, in turn, provide the fungi with the products of photosynthesis.

The presence of mycorrhizae is essential for the growth of many plants. The seeds of orchids, for example, cannot germinate in the absence of mycorrhizal fungi. Many trees are unable to survive without fungal symbionts. Interestingly, the partnership between plant and fungus does not end with a single plant. The roots of each plant are plugged into mycorrhizal networks that connect many plants. What's more astounding is that these networks appear to connect plants of different species.

BUILD Vocabulary

RELATED WORD FORMS *Symbiosis* is a noun for the condition of two unlike organisms living together in very close association. *Symbiotic* is an adjective referring to a symbiosis. *Symbiont* is a noun naming an organism in a symbiotic relationship, especially the smaller member of the pair.

In Your Notebook *Make a graphic organizer that illustrates the flow of materials between a fungus and a plant in a mycorrhizal symbiosis.*

Analyzing Data

Mycorrhizae and Tree Height

The graph below illustrates the growth rates of three species of trees—two individuals of each species. One tree of each species grew with mycorrhizae, and one grew without mycorrhizae.

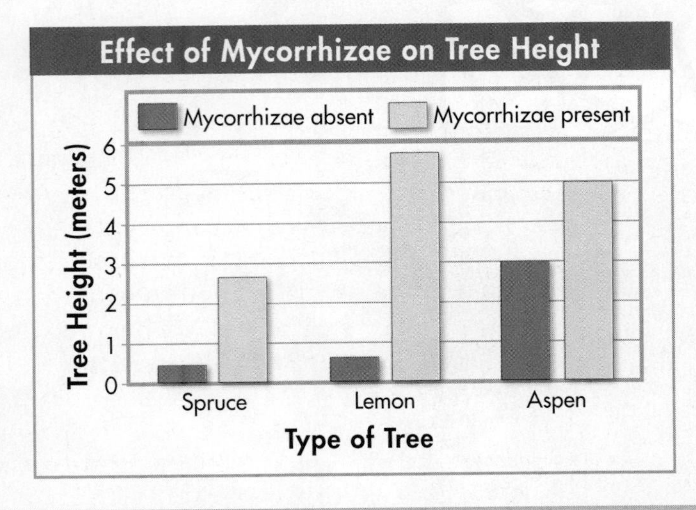

1. **Calculate** By what percentage is the height of the lemon tree grown with mycorrhizae greater than the height of the lemon tree grown without mycorrhizae? MATH

2. **Draw Conclusions** Make a generalization about the growth rate of plants with mycorrhizae.

3. **Form a Hypothesis** A citrus grower recently began using sterilized soil for repotting lemon trees with the goal of reducing disease. But many of the trees are dying in the new soil. Form a hypothesis to explain this observation.

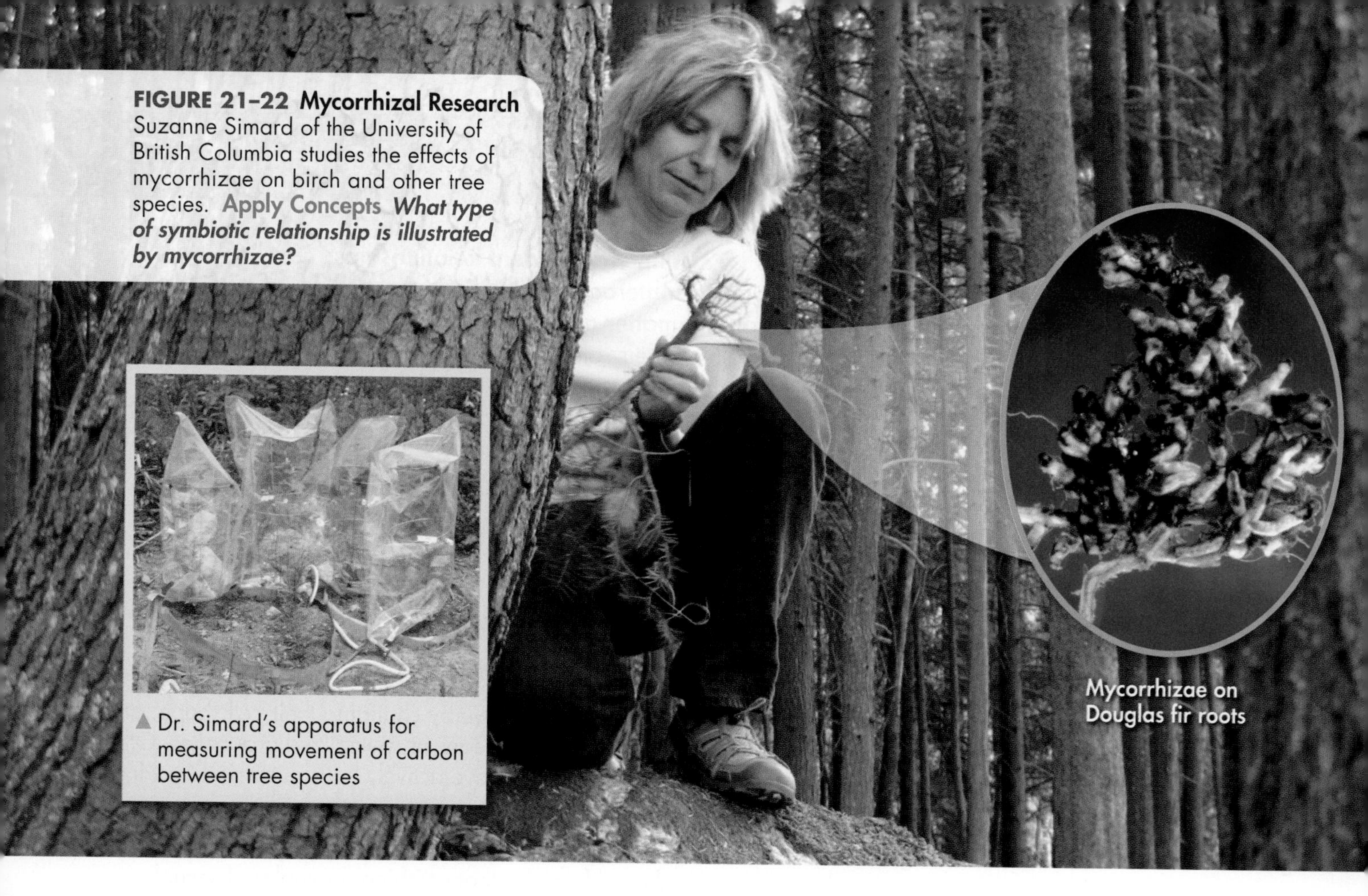

FIGURE 21–22 Mycorrhizal Research Suzanne Simard of the University of British Columbia studies the effects of mycorrhizae on birch and other tree species. Apply Concepts ***What type of symbiotic relationship is illustrated by mycorrhizae?***

▲ Dr. Simard's apparatus for measuring movement of carbon between tree species

A recent experiment showed that carbon atoms from one tree often end up in another tree nearby. In an experiment using isotopes to trace the movement of carbon, ecologist Suzanne Simard, shown in **Figure 21–22,** found that mycorrhizal fungi transferred carbon from paper birch trees growing in the sun to Douglas fir trees growing in the shade. As a result, the sun-starved fir trees thrived, basically by being "fed" carbon from the birches. Simard's findings suggest that plants—and their associated fungi—may be evolving as part of an ecological partnership.

21.4 Assessment

Review Key Concepts

1. a. Review Identify the characteristics all fungi have in common.

b. Explain What is the structure of the body of a typical fungus?

c. Apply Concepts Tissues from several mushrooms gathered near the base of a tree were tested and found to be genetically identical. How could you explain this?

2. a. Review Describe four ways fungi affect homeostasis in other organisms and the environment.

b. Apply Concepts Summarize the role of fungi in maintaining homeostasis in a forest ecosystem.

Apply the Big idea

Structure and Function

3. Both bacteria and fungi are decomposers. What characteristics do these two groups share that allow them to function in this ecological role? Use information from Lesson 20.2 to help you answer this question.

NGSS Smart Guide Protists and Fungi

Disciplinary Core Idea LS2.A Interdependent Relationships in Ecosystems: How do organisms interact with the living and nonliving environments to obtain matter and energy? Some species of protists and fungi—especially in their role as photosynthesizers and decomposers—are critical to maintaining equilibrium in ecosystems. However, certain species disrupt homeostasis in organisms by causing disease in various plants and animals, including humans.

21.1 Protist Classification—The Saga Continues

- Protists are eukaryotes that are not members of the plant, animal, or fungi kingdoms.
- Today's protists include groups whose ancestors were among the very last to split from the organisms that gave rise to plants, animals, and fungi.

Biology.com

Untamed Science Video Have a look at the fascinating world of mushrooms through the lenses of the Untamed Science crew.

UntamedScience

21.2 Protist Structure and Function

- Some protists move by changing their cell shape, and some move by means of specialized organelles. Other protists do not move actively but are carried by wind, water, or other organisms.
- Some protists reproduce asexually by mitosis. Others have life cycles that combine asexual and sexual forms of reproduction.

pseudopod 606 • cilium 607 • flagellum 607 • spore 607 • conjugation 608 • alternation of generations 608 • sporangium 609

Biology.com

InterActive Art Investigate the structures of an amoeba and paramecium.

Visual Analogy Compare the way boats move to the motion of flagella and cilia in a cell.

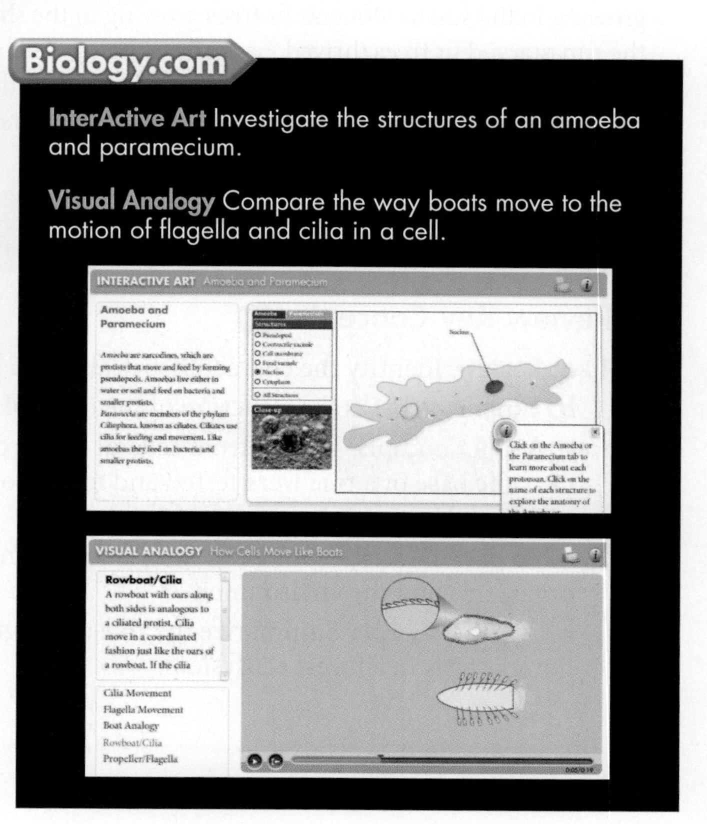

CHECKING YOUR *Scientific Literacy*

Refer to the lesson content and digital assets below as you prepare for your chapter assessment. Then, evaluate your understanding of protists and fungi by answering these questions.

1. Scientific and Engineering Practice **Constructing Explanations** A scientist used isotopes to trace the movement of carbon in a forest. She found that labeled carbon atoms from one tree were transferred to a different tree. What is a possible explanation for this observation?

2. Crosscutting Concept **Cause and Effect** Describe another example of a mutualistic relationship between a protist or fungus and its host. What happens when one of the species dies or disappears?

3. Key Ideas and Details Summarize the ways in which protists and fungi interact with other organisms in their environments.

4. STEM You are a commercial vegetable farmer and want to broaden your market by growing mushrooms. Write a plan describing how you will grow and harvest your mushrooms. Be sure to include in your plan the materials and equipment you need to purchase to ensure a profitable yield.

21.3 The Ecology of Protists

- The position of photosynthetic protists at the base of the food chain makes much of the diversity of aquatic life possible.
- Some heterotrophic protists engulf and digest their food, while others live by absorbing molecules from the environment.
- Many protists are involved in mutualistic symbioses, in which they and their hosts both benefit.
- Parasitic protists are responsible for some of the world's most deadly diseases, including several kinds of debilitating intestinal diseases, African sleeping sickness, and malaria.

algal bloom 611 • **food vacuole 612** • **gullet 612** • **plasmodium 613**

Biology.com

STEM Activity Design a better mosquito net to decrease the infection rate of malaria.

Data Analysis Learn how microfossils can be used to learn about conditions and changes in ancient oceans.

Art in Motion View a short animation that shows the plasmodium life cycle and how malaria is transmitted by the *Anopheles* mosquito.

21.4 Fungi

- Fungi are heterotrophic eukaryotes with cell walls that contain chitin.
- Fungi are champions of decomposition. Many species help ecosystems maintain homeostasis by breaking down dead organisms and recycling essential elements and nutrients.
- Parasitic fungi can cause serious diseases in plants and animals by disrupting homeostasis.
- Some fungi form mutualistic associations with photosynthetic organisms in which both partners benefit.

chitin 618 • **hypha 619** • **fruiting body 619** • **mycelium 619** • **lichen 623** • **mycorrhiza 624**

Biology.com

Art Review Review your understanding of the different structures of a mushroom.

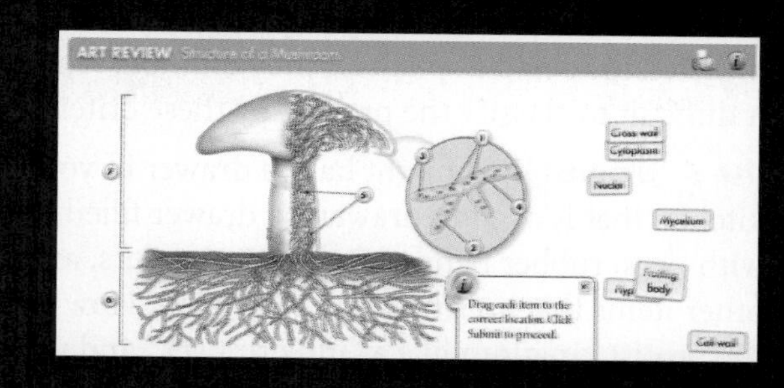

Design Your Own Lab

Mushroom Farming This chapter lab is available in your lab manual and at Biology.com

21 Assessment

21.1 Protist Classification

Understand Key Concepts

1. Which of the following descriptions applies to most protists?
   a. unicellular prokaryotes
   b. multicellular prokaryotes
   c. unicellular eukaryotes
   d. multicellular eukaryotes
2. The fossil record shows that the first eukaryotes may have appeared on Earth
   a. more than 4 billion years ago.
   b. more than 1 billion years ago.
   c. about 500 million years ago.
   d. about 100 million years ago.
3. Which of the following statements is most accurate?
   a. Protists are more closely related to one another than to other organisms in other kingdoms.
   b. Protists are the direct descendants of bacteria.
   c. The classification of protists is a work in progress.
   d. Scientists are debating between two classification schemes for protists.
4. What is the problem with the traditional classification of protists into plantlike, animal-like, and funguslike groups?
5. Why do scientists think that all modern plants, animals, and fungi can be traced to protist ancestors?

Think Critically

6. **Key Ideas and Details** Once, living things were classified as animals if they moved or ingested food and as plants if they did not. Why is it difficult to classify the protists by these criteria?
7. **Use Analogies** You might have a drawer in your kitchen that is a "junk drawer": a drawer filled with keys, rubber bands, pens, string, rulers, and other items that aren't easy to categorize. How is the protist kingdom like a "junk drawer," and why do you think scientists would like to change that situation?

21.2 Protist Structure and Function

Understand Key Concepts

8. Which is NOT true of amoebas?
   a. They reproduce by mitosis.
   b. They move using pseudopods.
   c. They have a rigid cell membrane.
   d. The protein actin powers their movement.
9. Which of the following protists moves by means of cilia?

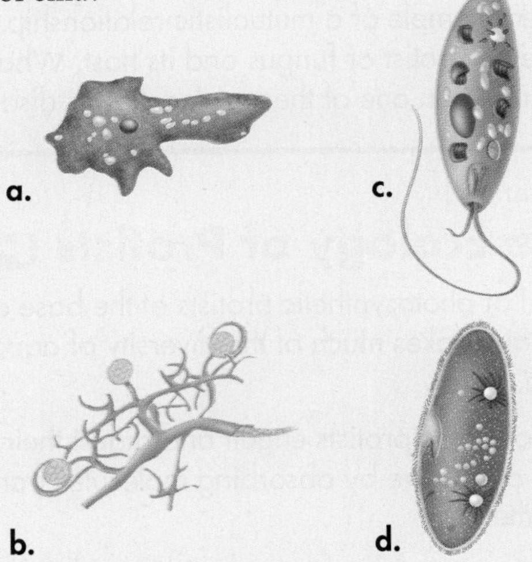

10. Alternation of generations is the process of alternating between
    a. mitosis and meiosis.
    b. asexual and sexual reproduction.
    c. male and female reproductive structures.
    d. diploid and haploid phases.
11. What function do the cilia and flagella in protists carry out? How do they differ in structure?
12. Summarize the process of conjugation. Is conjugation a form of reproduction? Explain.

Think Critically

13. **Apply Concepts** Some protists cannot move on their own. What generalization can you make about how these organisms survive?
14. **Infer** How do you think the ability to switch between asexual and sexual reproduction has aided the evolution of water molds and many other protists?

21.3 The Ecology of Protists

Understand Key Concepts

15. Which of the following statements about photosynthetic protists is most accurate?
- **a.** Most photosynthetic protists are heterotrophs.
- **b.** All photosynthetic protists are closely related to plants.
- **c.** Small photosynthetic organisms near the ocean's surface are called phytoplankton.
- **d.** Giant kelp play an important role in the formation of coral reefs.

16. Slime molds are found primarily in
- **a.** oceans.
- **b.** decaying organic matter.
- **c.** fast-moving streams.
- **d.** deserts.

17. African sleeping sickness is caused by
- **a.** *Trypanosoma.*
- **b.** *Plasmodium.*
- **c.** *Trichonympha.*
- **d.** *Amoeba.*

18. **Key Ideas and Details** Describe the nature of the relationship between a termite and the protists that live in its gut.

19. How do insects transmit malaria?

Think Critically

20. **Form a Hypothesis** A scientist observes that termites that are fed a certain antibiotic die of starvation after a few days. She also notices that the antibiotic affects certain protists that live inside the termite's gut: Although the protists continue to thrive, a certain kind of structure disappears from their cytoplasm. Develop a hypothesis to explain these observations.

21. **Infer** Slime molds produce sporangia and spores only when food is scarce. Why is this so? What advantages do slime molds gain from this?

22. **Predict** Holes in Earth's ozone layer may increase the amount of radiation that reaches the surface of the ocean. If this radiation affects the growth of phytoplankton, what do you think the long-term consequences would be for Earth's atmosphere? Explain your answer.

23. **Infer** Examine the life cycle of *Plasmodium* shown in **Figure 21–14.** Based on the illustration, do you think malaria could be transmitted through a blood transfusion? Why or why not?

solve the CHAPTER MYSTERY

"A BLIGHT OF UNUSUAL CHARACTER"

In Ireland in the 1840s, conditions were just right for the rapid and devastating spread of potato blight. The potato had become a staple crop—the main source of food for the Irish population. The disease persisted from year to year because the new crop was grown from potato eyes saved from the previous year's crop. Genetically identical crops offered no new resistance to the blight.

What clues throughout the chapter point to the culprit's identity? Some of its features—notably, the presence of hyphae and the fact that it obtains food by absorption through its cell walls—are traits shared by both water molds and fungi. The final clue, however, is telling—the culprit's cells lack chitin and the hyphae are diploid. These traits eliminate the fungus option. The organism that causes the blight is a water mold—the protist *Phytophthora*— which literally means "plant eater."

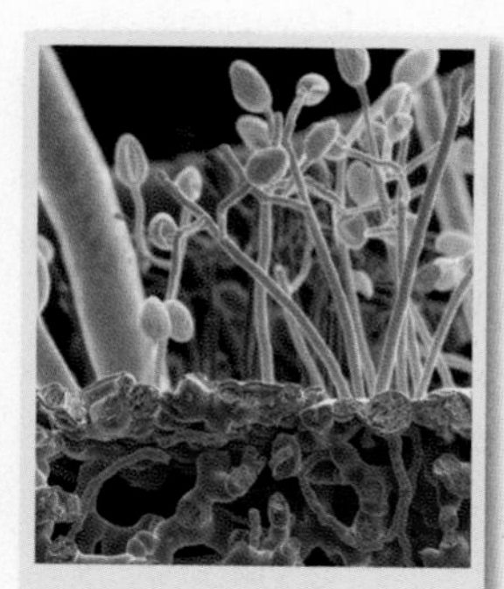

Phytophthora hyphae invading a potato (SEM 100×)

1. **Relate Cause and Effect** Observers in Ireland noted that the weather leading up to the summer of 1845 was unusually wet and cool. How might the weather conditions in 1845 have favored this organism's life cycle?

2. **Infer** Scientists now believe that *Phytophthora* came with some batches of potatoes from South America. Why hadn't the same organism caused such widespread destruction there?

3. **Connect to the Big idea** In the 1840s, one observer remarked that potatoes were grown so widely in Ireland that they formed "a continuous canopy of leaves." In the United States, the same disease appeared several years earlier, but did not cause widespread famine. Why not?

21.4 Fungi

Understand Key Concepts

24. Which of the following statements about fungi is false?
a. All fungi are unicellular.
b. All fungi have cell walls.
c. All fungi are eukaryotic.
d. All fungi are heterotrophs.

25. A symbiotic relationship between a fungus and a green alga or a cyanobacterium is a
a. mycorrhiza. **c.** lichen.
b. fruiting body. **d.** mushroom.

26. How are the cell walls of fungi similar to the exoskeleton of insects?

27. **Craft and Structure** Distinguish between the terms *hypha* and *mycelium*.

28. What is the evolutionary significance of mycorrhizae?

Think Critically

29. **Compare and Contrast** Both fungi and humans are heterotrophs. Compare the way fungi obtain food with the way humans do.

30. **Form a Hypothesis** The antibiotic penicillin is a natural secretion of a certain kind of fungus—a green mold called *Penicillium*. Penicillin kills bacteria. Why do you think a mold species has evolved a way to kill bacteria?

Connecting Concepts

Use Science Graphics

This photograph shows a comparison of corn plants grown without mycorrhizae (left) to plants of the same age that were grown with mycorrhizae. Use the photograph to answer questions 31–32.

31. **Compare and Contrast** Which corn plant shows more robust growth?

32. **Relate Cause and Effect** What is the most likely explanation for the treatment results shown?

Write About Science

33. **Text Types and Purposes** Write a paragraph explaining how fungi either maintain or disrupt the equilibrium of an ecosystem.

34. **Assess the Big idea** Choose one of the protists discussed in this chapter, and explain how the presence and the activity of that organism influence other species.

Analyzing Data

Integration of Knowledge and Ideas

Use the graph below to answer the questions.

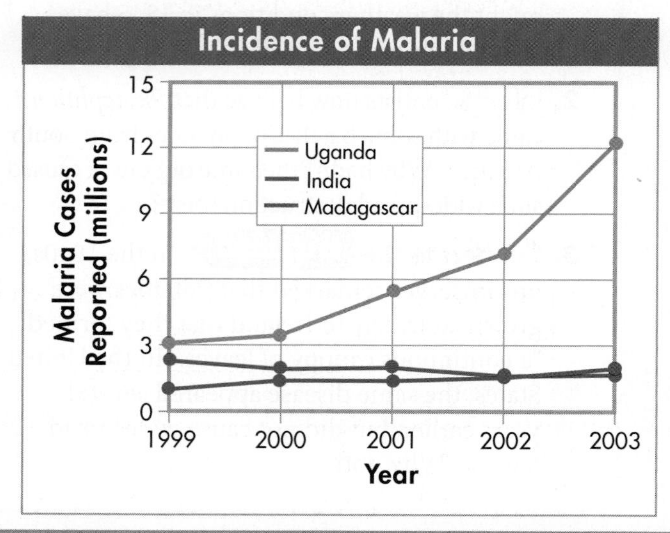

35. **Interpret Graphs** Which statement best describes the trend shown in the graph?
a. Globally, malaria cases are on the rise.
b. Malaria cases declined in India during the five-year period shown.
c. In 2003, Uganda had four times as many malaria cases as did Madagascar.
d. Annual changes in rainfall explain the patterns seen in the data.

36. **Calculate** By about what percentage did the number of malaria cases increase in Uganda between 1999 and 2003? MATH
a. 50 percent **c.** 300 percent
b. 100 percent **d.** 1000 percent

Standardized Test Prep

Multiple Choice

1. All of the following are characteristics of some protists EXCEPT
   A peptidoglycan in the cell walls.
   B a membrane-bound nucleus.
   C flagella.
   D cilia.

2. The structures in *Amoeba* that help the organism move and feed are the
   A flagella.
   B cilia.
   C food vacuoles.
   D pseudopods.

3. In protists, the process of conjugation
   A is linked to photosynthesis.
   B results in an exchange of some genetic material.
   C produces offspring that are genetically identical to the parent.
   D decreases the genetic diversity of a population.

4. Which of the following statements about slime molds is false?
   A Slime molds are eukaryotes.
   B Slime molds play an important part in recycling nutrients.
   C Slime molds are multicellular at some time during their life cycle.
   D Slime molds are photosynthetic protists.

5. Algal blooms can be caused by
   A paramecia.
   B lichens.
   C dinoflagellates.
   D *Trichonympha.*

6. Alternation of generations BEST describes sexual reproduction in
   A *Paramecium.*
   B water molds.
   C *Amoeba.*
   D yeast.

7. The primary carbohydrate found in the cell walls of fungi is
   A chitin.
   B actin.
   C cellulose.
   D starch.

Questions 8–10

Ripe grapes are covered with a grayish film, or bloom, that contains yeasts and sometimes other microorganisms. A group of students prepared three test tubes of fresh mashed grapes. They heated two of the test tubes to boiling and then cooled them. They inoculated one of those test tubes with live yeast, incubated all three test tubes at 30°C for 48 hours, and then examined the test tubes for signs of fermentation—an alcohol odor and bubbles. Their data are summarized in the table below.

Evidence of Fermentation

Test-Tube Contents	Alcohol Odor (yes or no)	Bubbles (yes or no)
Unheated grape mash	yes	yes
Boiled grape mash	no	no
Boiled grape mash inoculated with yeast	yes	yes

8. What is the independent variable in the students' investigation?
   A the presence of live yeast
   B an odor of alcohol
   C bubbles
   D time

9. What is the dependent variable in the students' investigation?
   A time
   B boiling
   C an odor of alcohol and the presence of bubbles
   D the presence of live yeast

10. What can you conclude based on the students' results?
    A Uninoculated, boiled grape mash does not seem to ferment over a 48-hour period.
    B Boiled grape mash that contains live yeast undergoes fermentation.
    C Grape mash does not ferment unless live yeast is added.
    D Both A and B are correct.

Open-Ended Response

11. How does each of the partners in the lichen symbiosis benefit from the relationship?

If You Have Trouble With . . .

Question	1	2	3	4	5	6	7	8	9	10	11
See Lesson	21.2	21.2	21.2	21.3	21.3	21.2	21.4	21.4	21.4	21.4	21.4

22 Introduction to Plants

Big idea: Unity and Diversity of Life

Q: What are the five main groups of plants, and how have four of these groups adapted to life on land?

An alpine meadow provides forage for deer and other herbivores.

INSIDE:

- 22.1 What Is a Plant?
- 22.2 Seedless Plants
- 22.3 Seed Plants
- 22.4 Flowering Plants

BUILDING *Scientific Literacy*

Deepen your understanding of plants by engaging in key practices that allow you to make connections across concepts.

NGSS You will obtain, evaluate, and communicate information about how plants have adapted to their environments.

STEM Plants are pretty good at keeping their seeds safe. Use what you learn about fruits to design, build, and test an effective storage container.

Common Core You will summarize, represent, and interpret data and determine the meanings of words related to plants.

CHAPTER MYSTERY

STONE AGE STORYTELLERS

Some 5300 years ago, a man died on a remote mountain pass in the Alps. In 1991, the glacier that had kept him frozen for so long melted, revealing his well-preserved corpse. The body of this "Iceman," along with his clothes, tools, and other artifacts, has provided an amazing amount of information about his society, daily life, and the circumstances of his death.

Some of the most interesting evidence found with Iceman comes from plant material. He had a storage container made from birch bark and a wooden bow and quiver of arrows. Scientists also found several species of tree pollen inside Iceman's digestive tract, which he most likely ingested by accident after the airborne pollen settled on his food. What might these and other botanical clues reveal about Iceman? As you read the chapter, look for clues to help you solve the mystery of Iceman.

Biology.com

Finding the solution to the Stone Age Storytellers mystery is only the beginning. Take a video field trip with the ecogeeks of Untamed Science to see where the mystery leads.

Go online to access additional resources including:
- eText • Flash Cards • Lesson Overviews • Chapter Mystery

22.1 What Is a Plant?

Key Questions

What do plants need to survive?

How did plants adapt to life on land?

What feature defines most plant life cycles?

Vocabulary

alternation of generations • sporophyte • gametophyte

Taking Notes

Preview Visuals Preview **Figure 22–4** and list five main groups of plants. For each group, list any characteristics or specific examples that you already know.

THINK ABOUT IT What color is life? Living things can be just about any color, of course. But imagine yourself in a place so abundant with life that living things actually blot out the sun. Now what color do you see? If you've imagined a thick forest or a teeming jungle, then just one color will fill the landscape of your mind: green—the color of plants. Plants have adapted so well to so many environments that they dominate much of the surface of our planet.

Characteristics of Plants

What do plants need to survive?

What are plants? You are already familiar with many examples, such as trees, shrubs, and grasses. But did you know that mosses and ferns are also types of plants? In the last several years, biologists have reclassified green algae as plants, too. (Green algae used to be considered protists.) What characteristics do all these organisms share?

The Plant Kingdom Traditionally, plants are classified as members of the kingdom Plantae. Plants are eukaryotes that have cell walls containing cellulose and carry out photosynthesis using chlorophyll *a* and *b*. While most plants are autotrophs, a few are parasites or saprobes.

FIGURE 22–1 Diagram of a Plant Cell Plant leaves appear green due to the photosynthetic pigments chlorophyll *a* and *b*, which are located in chloroplasts.

Quick Lab
GUIDED INQUIRY

Are All Plants the Same? 

1. From your teacher, obtain three plants, a metric ruler, and a hand lens.
2. Identify the major parts of each plant. Measure the heights of the plants and the sizes of their parts.
3. Use the hand lens to examine the plants. Record your observations.

Analyze and Conclude

1. **Compare and Contrast** What patterns and symmetries do you observe among the three plants? How are the three plants alike? How do they differ?
2. **Infer** What do the shapes of plant structures suggest about their functions?
3. **Classify** Use your observations to classify the three plants into two groups by outward characteristics alone. Explain your reasons for classifying them in these groups.

What Plants Need Surviving as stationary organisms on land is a difficult task that most plants face. Plants have developed a number of adaptations that enable them to succeed. **The lives of plants center on the need for sunlight, gas exchange, water, and minerals.** These basic needs are illustrated in **Figure 22–2.**

▶ ***Sunlight*** Plants use the energy from sunlight to carry out photosynthesis. As a result, every plant displays adaptations shaped by the need to gather sunlight. Photosynthetic organs such as leaves are typically broad and flat and are arranged on the stem so as to maximize light absorption.

▶ ***Gas Exchange*** Plants require oxygen to support cellular respiration as well as carbon dioxide to carry out photosynthesis. They also need to release excess oxygen made during photosynthesis. Plants must exchange these gases with the atmosphere and the soil without losing excessive amounts of water through evaporation.

▶ ***Water and Minerals*** On a hot sunny day, plants can lose a great deal of water to the air, just as we sweat when we are hot. Also, water is one of the raw materials of photosynthesis, so it is consumed when the sun is shining. Thus, land plants have evolved structures that limit water loss and speed the uptake of water from the ground.

As they absorb water, plants also absorb minerals. Minerals are nutrients in the soil needed for plant growth. Many plants have specialized tissues that carry water and nutrients upward from the soil and distribute the products of photosynthesis throughout the plant body. Simpler types of plants carry out these functions by diffusion.

In Your Notebook *Outline the basic needs of plants. Under each head, add supporting details.*

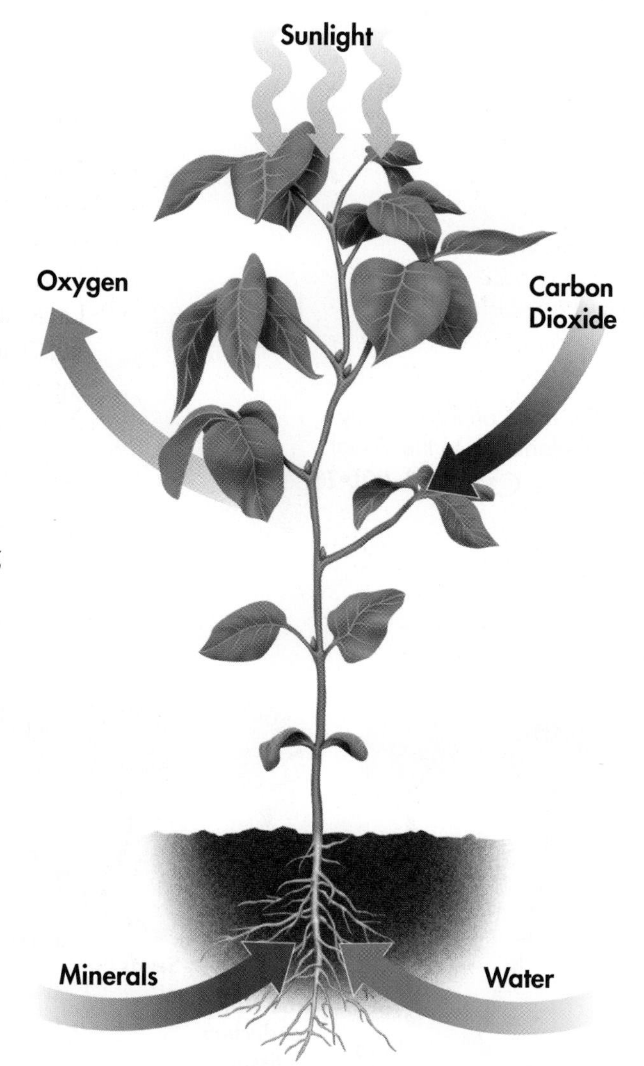

FIGURE 22–2 Basic Needs of a Plant All plants have the same basic needs: sunlight, a way to exchange gases with the surrounding air, water, and minerals. **Observe** ***Where do water and minerals enter the plant?***

FIGURE 22–3 A Fossilized Plant One of the earliest fossil vascular plants was *Cooksonia*. This fossil shows the branched stalks that bore reproductive structures at their tips.

The History and Evolution of Plants

How did plants adapt to life on land?

For most of Earth's history, land plants simply did not exist. Life was concentrated in oceans, lakes, and streams. Although photosynthetic prokaryotes added oxygen to our planet's atmosphere and provided food for animals and microorganisms, true plants had not yet appeared on the planet.

Origins in the Water The fossil record indicates that the ancestors of today's land plants were water-dwelling organisms similar to today's green algae. Most of these photosynthetic eukaryotes were unicellular, although a few were composed of multiple cells. At first, many biologists wondered if green algae should actually be included in a kingdom that includes organisms as large and complex as oak trees and orchids. But now it is clear that green algae have cell walls and photosynthetic pigments that are identical to those of plants. They also have reproductive cycles similar to those of plants. Finally, studies of their genomes suggest that they are so closely related to other plants that they should be considered part of the plant kingdom.

The First Land Plants Fossil spores of land plants occur in rocks 475 million years old, but the plants themselves from this time period left no fossils. The oldest fossils of land plants themselves are found roughly 50 million years later in the fossil record. Lacking leaves and roots, these plants were only a few centimeters tall. The greatest challenge that early land plants faced was obtaining water, which they achieved by growing close to the ground in damp locations. Fossils also suggest that the first true land plants were still dependent on water to complete their life cycles.
Over time, the demands of life on land favored the evolution of plants more resistant to the drying rays of the sun, more capable of conserving water, and more capable of reproducing without water.

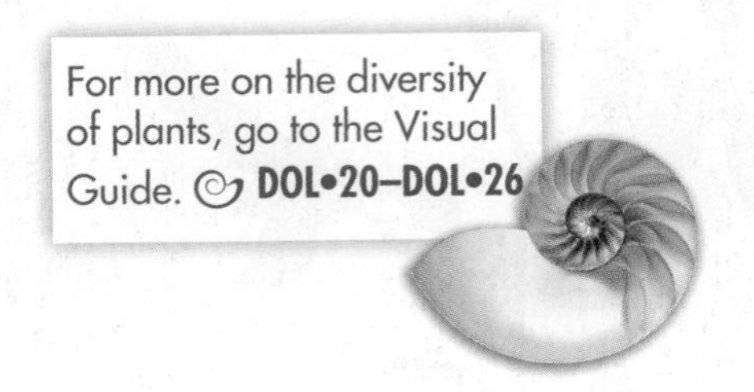

For more on the diversity of plants, go to the Visual Guide. DOL•20–DOL•26

Flowers; seeds enclosed in fruit

Seeds

True water-conducting tissue

Embryo formation

Plant ancestor

FIGURE 22–4 Major Groups of Plants There are five main groups of plants in existence today. **Interpret Diagrams** ***In the ancestor to which plant groups did seeds first evolve?***

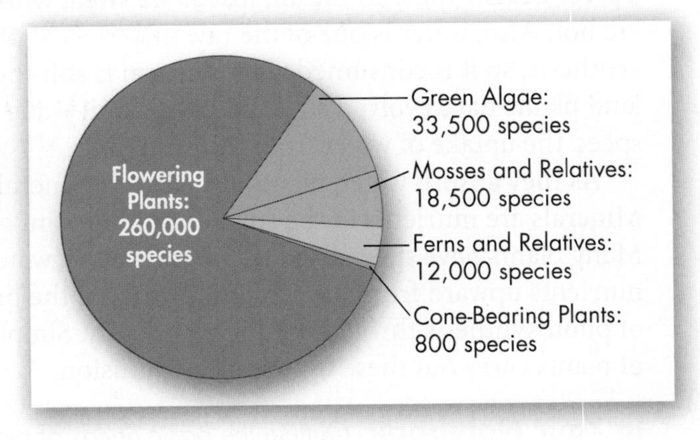

The appearance of plants on land changed the rest of life on Earth. As these new organisms colonized the land, they changed the environment in ways that enabled new species to evolve. New ecosystems emerged, and organic matter began to form soil.

Several groups of plants evolved from the first pioneering land plants. One group developed into mosses. Another lineage gave rise to ferns, cone-bearing plants, and flowering plants. All of these groups of plants are now successful in living on land, but they have evolved very different adaptations for a wide range of terrestrial environments.

An Overview of the Plant Kingdom Botanists divide the plant kingdom into five major groups based on four important features: embryo formation, specialized water-conducting tissues, seeds, and flowers. The relationship of these groups to one another is shown in **Figure 22–4.** Plants that form embryos are often referred to as "land plants," even though some of them now live in watery environments. Plant scientists classify plants into finer groups within these major branches by comparing the DNA sequences of various species.

MYSTERY CLUE

Scientists extracted high levels of chlorophyll from the maple leaves found with Iceman. How could this evidence be used to determine the season of his death?

In Your Notebook *Determine the key feature that a fern has in common with a flowering plant using information in **Figure 22–4.***

The Plant Life Cycle

What feature defines most plant life cycles?

Land plants have a distinctive sexual life cycle that sets them apart from most other living organisms. **The life cycle of land plants has two alternating phases, a diploid (2N) phase and a haploid (N) phase.** This shift between haploid and diploid is known as the **alternation of generations.**

The multicellular diploid (2N) phase is known as the **sporophyte** (SPOH ruh fyt), or spore-producing plant. The multicellular haploid (N) phase is known as the **gametophyte** (guh MEET uh fyt), or gamete-producing plant. Recall from Chapter 11 that haploid (N) organisms carry a single set of chromosomes in their cell nuclei, while diploid (2N) organisms have two sets of chromosomes.

You can follow the basic steps of the life cycle in **Figure 22–5,** starting from the top. A sporophyte produces haploid spores through meiosis. These spores grow into multicellular structures called gametophytes. Each gametophyte produces reproductive cells called gametes—sperm and egg cells. During fertilization, a sperm and egg fuse with each other, producing a diploid zygote. The zygote develops into a new sporophyte, and the cycle begins again.

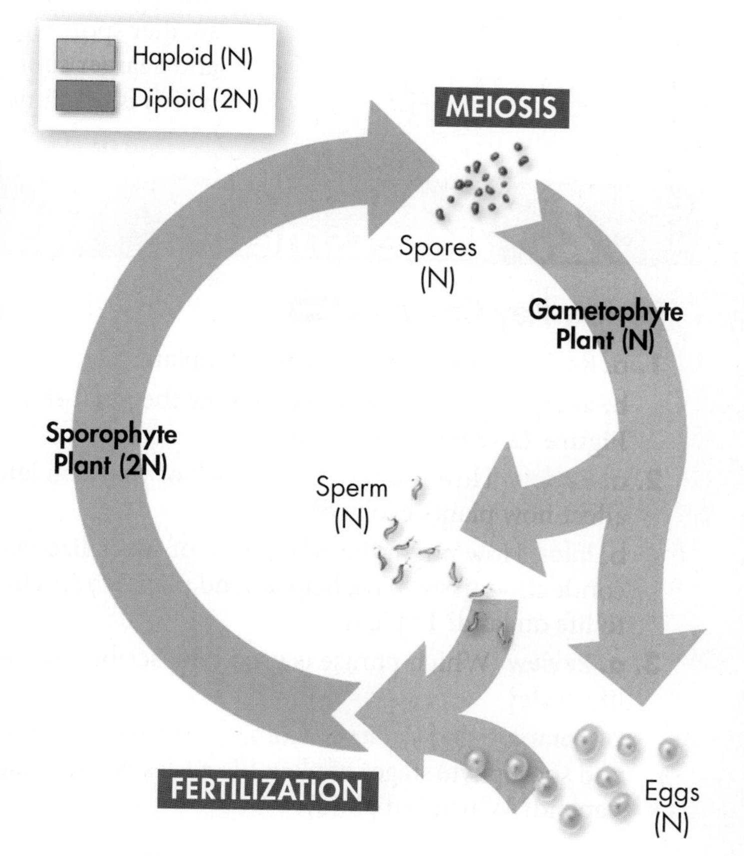

FIGURE 22–5 The Plant Life Cycle Most plants have a life cycle with alternation of generations, in which the haploid gametophyte phase alternates with the diploid sporophyte phase.

FIGURE 22–6 Trends in Plant Evolution An important trend in plant evolution is the reduction in size of the gametophyte and the increasing size of the sporophyte. **Interpret Visuals** ***How does the relative size of the haploid and diploid stages differ between mosses and seed plants?***

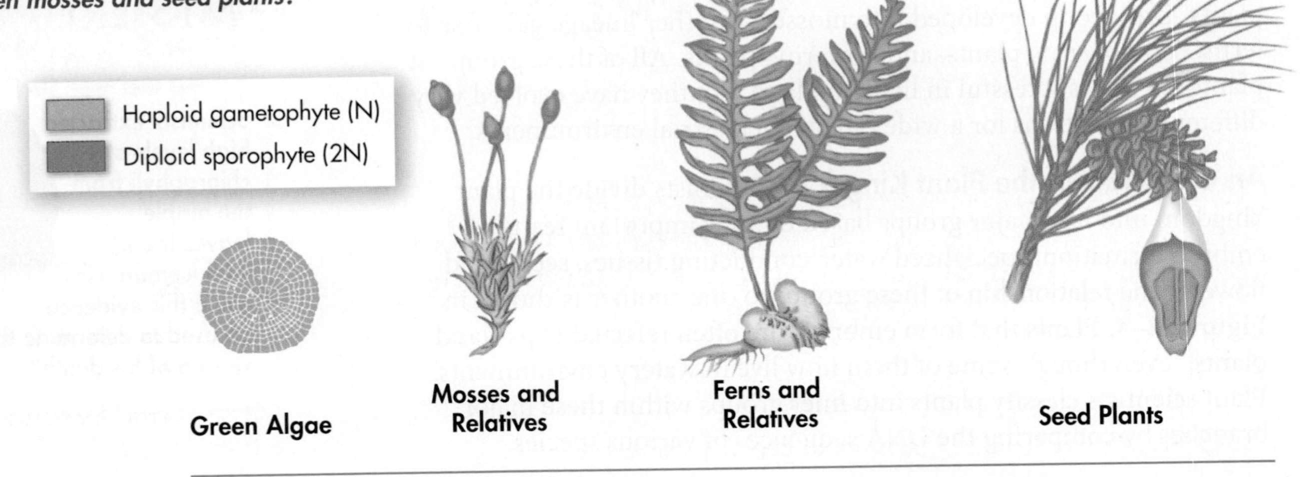

Figure 22–6 shows an important trend in plant evolution—the reduction in size of the gametophyte and the increasing size of the sporophyte. Although many green algae do have a diploid sporophyte phase, some do not; their only multicellular bodies are gametophytes. Mosses and their relatives consist of a relatively large gametophyte and smaller sporophytes. Ferns and their relatives have a small gametophyte and a larger sporophyte. Seed plants have an even smaller gametophyte, which is contained within sporophyte tissues.

22.1 Assessment

Review Key Concepts

1. a. **Review** List the basic needs of a plant.
   b. **Interpret Visuals** Summarize how the plant shown in **Figure 22–2** meets basic needs.
2. a. **Review** How did the relative lack of water on land affect how plants evolved?
   b. **Infer** How might the adaptation of specialized water-conducting tissue have helped land plants meet challenges to life on land? Explain.
3. a. **Review** Which phrase is used to describe a plant's life cycle?
   b. **Compare and Contrast** Compare the gametophyte and sporophyte stages of plant life cycles. Which stage is haploid? Which is diploid?

ANALYZING DATA

Use the circle graph in **Figure 22–4** to answer the questions below.

4. **Calculate** What percentage of all plants are flowering plants? MATH
5. **Interpret Graphs** What is the second largest group of plants?
6. **Pose Questions** What question about plant evolution would you ask based on the data in this graph?

22.2 Seedless Plants

THINK ABOUT IT We generally think of plants as growing from seeds. But there are plenty of plants, such as mosses, ferns, and green algae, that don't produce seeds at all. How do they manage to reproduce and grow without them?

Key Questions

What are the characteristics of green algae?

What factor limits the size of bryophytes?

How is vascular tissue important?

Vocabulary

bryophyte • vascular tissue • archegonium • antheridium • sporangium • tracheophyte • tracheid • xylem • phloem

Taking Notes

Venn Diagram Construct a Venn diagram in which to record similarities and differences among the three major groups of seedless plants. Fill it in as you read the lesson.

Green Algae

What are the characteristics of green algae?

What do you think of when you hear the word *algae*? You might think of seaweed, given that algae make up the most common forms of seaweed, and the singular form of the word, *alga*, actually means "seaweed" in Latin. As we use the word today, the algae are not a single group of organisms. Biologists apply the name to any photosynthetic eukaryote other than a land plant. As a result, some algae are classified as protists and others are classified as plants. Those algae that are grouped with plants are called *green algae.* DOL•21

The First Plants Fossil evidence suggests that the green algae were the first plants, appearing on Earth before plants first emerged on land. In fact, fossil formations from the Cambrian Period, more than 550 million years ago, show evidence of large mats of green algae.

The green algae share many characteristics—including their photosynthetic pigments and cell wall composition—with larger, more complex plants. **Green algae are mostly aquatic. They are found in fresh and salt water, and in some moist areas on land.** Because most green algae are single cells or branching filaments, they make direct contact with the water in which they grow. They are able to absorb moisture and nutrients directly from their surroundings. Therefore, most green algae do not contain the specialized tissues found in other plants.

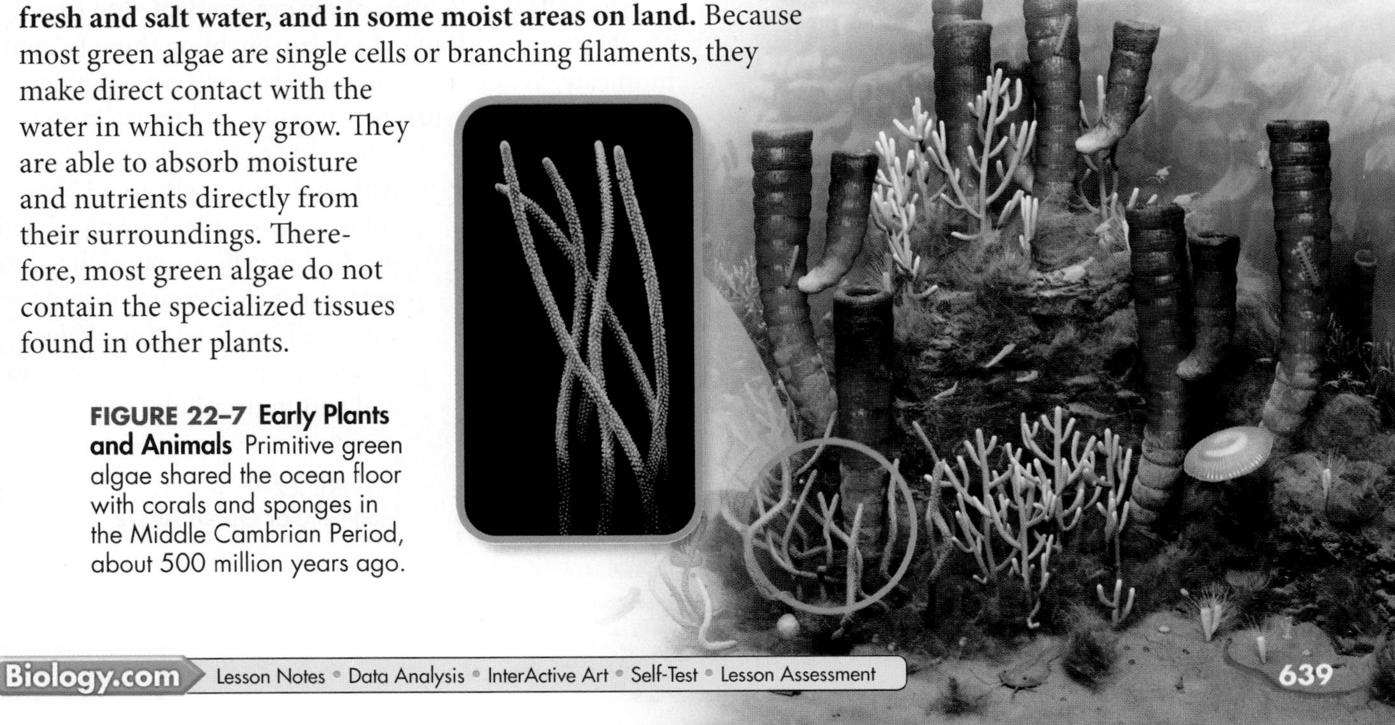

FIGURE 22–7 Early Plants and Animals Primitive green algae shared the ocean floor with corals and sponges in the Middle Cambrian Period, about 500 million years ago.

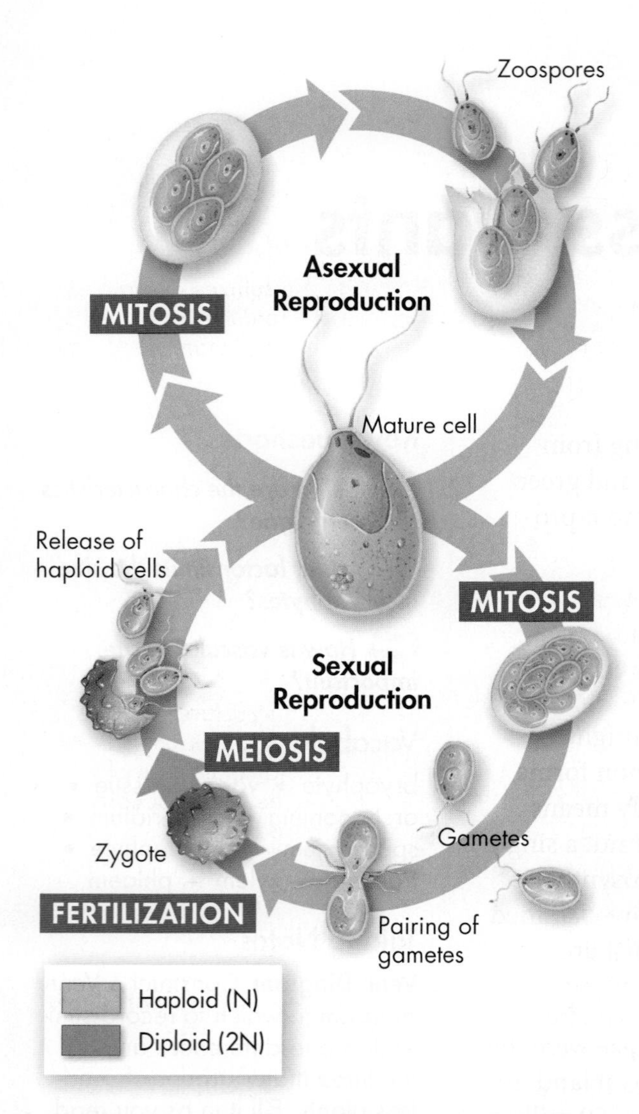

FIGURE 22–8 Life Cycle of *Chlamydomonas* The green alga *Chlamydomonas* can switch from asexual reproduction to sexual reproduction as environmental conditions change.
Interpret Visuals ***Which form of reproduction includes a diploid organism that can survive harsh conditions?***

Life Cycle Like land plants, many green algae have life cycles that switch back and forth between haploid and diploid phases. Some green algae may not alternate between haploid and diploid with each and every generation, however. Consider the single-celled green alga *Chlamydomonas*. It can stay in the haploid stage for multiple generations. As long as living conditions are suitable, the haploid cell reproduces asexually by mitosis, as shown in the top half of **Figure 22–8.**

If environmental conditions become unfavorable, *Chlamydomonas* can switch to a stage that reproduces sexually. The cell releases gametes that fuse into a diploid zygote—a sporophyte. The zygote has a thick protective wall, permitting survival in freezing or drying conditions that would ordinarily kill it. When conditions once again become favorable, the zygote begins to grow. It divides by meiosis to produce four flagellated haploid cells. These haploid cells then swim away, mature, and reproduce asexually.

In Your Notebook *Describe a more general advantage to an organism that can change its mode of reproduction under different environmental conditions.*

Multicellularity Many of the green algae form colonies, providing a hint as to how the first multicellular plants may have evolved. Two examples of colonial algae are shown in **Figure 22–9.** On the left is the freshwater alga *Spirogyra*, which forms long, threadlike colonies called filaments. The cells of a colony are stacked almost like soda cans placed end to end. The *Volvox* colonies shown on the right are more complex than those of *Spirogyra*, consisting of as few as 500 to as many as 50,000 cells arranged to form hollow spheres.

The cells in a *Volvox* colony are connected to one another by strands of cytoplasm, enabling them to communicate. When the colony moves, cells on one side of the colony "pull" with their flagella, and the cells on the other side of the colony "push." Although most cells in a *Volvox* colony are identical, a few gamete-producing cells are specialized for reproduction. Because it shows some cell specialization, *Volvox* straddles the fence between colonial and multicellular life.

Spirogyra (LM 140×)

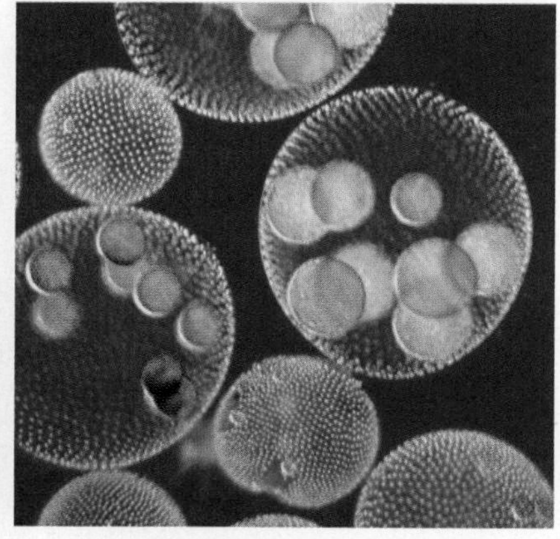

Volvox (LM 50×)

FIGURE 22–9
Multicellular Green Algae

Mosses and Other Bryophytes

What factor limits the size of bryophytes?

In the cool forests of the northern United States, the moist ground is covered with green. When you walk, the ground almost feels like a soft, spongy carpet. Look closely, however, and you will see that this forest carpet is made of short, soft plants known as mosses. Mosses have a thin waxy coating that makes it possible for them to resist drying, and thin filaments known as rhizoids (RY zoydz) that anchor them to the soil. Rhizoids also absorb water and minerals from the surrounding soil. **Figure 22–10** shows the structure of a typical moss.

Mosses belong to a group of plants that is known as **bryophytes** (BRY oh fyts). Unlike algae, the bryophytes have specialized reproductive organs enclosed by other, nonreproductive cells. The bryophytes show a higher degree of cell specialization than do the green algae and were among the very first plants to become established on land. In addition to mosses, the bryophytes include two other groups, known as hornworts and liverworts. Each of the three groups is generally considered to be a separate phylum. DOL•22

BUILD Vocabulary

SUFFIXES The suffixes *-phyta* and *-phyte* come from the Greek word *phyton*, meaning "plant."

Why Bryophytes Are Small Bryophytes are generally found in damp places where there is plenty of available water, and there are good reasons for this. Most other land plants carry water in a specialized tissue called **vascular tissue,** which contains tubes hardened with a substance called lignin. Bryophytes, however, do not make lignin and do not contain true vascular tissue. **Bryophytes are small because they lack vascular tissue.** They can draw up water no higher than a meter above the ground. Without strong cell walls hardened by lignin, bryophytes also cannot support a tall plant body against the pull of gravity. These factors limit the height of bryophytes and confine them to damp environments.

Life Cycle Like all land plants, bryophytes display alternation of generations. In bryophytes, the gametophyte is the dominant, recognizable stage of the life cycle. The gametophyte is also the stage that carries out most of the plant's photosynthesis. The sporophyte is dependent on the gametophyte for its supply of water and nutrients.

Bryophytes produce sperm cells that swim using flagella. For fertilization to occur successfully, these sperm must be released where there is enough water for them to swim to an egg cell. Because of this, bryophytes live only in damp habitats where there is standing water for at least part of the year.

FIGURE 22–10 Structure of Moss In bryophytes, the gametophyte is the dominant, more familiar stage of the life cycle and is the form that carries out photosynthesis.

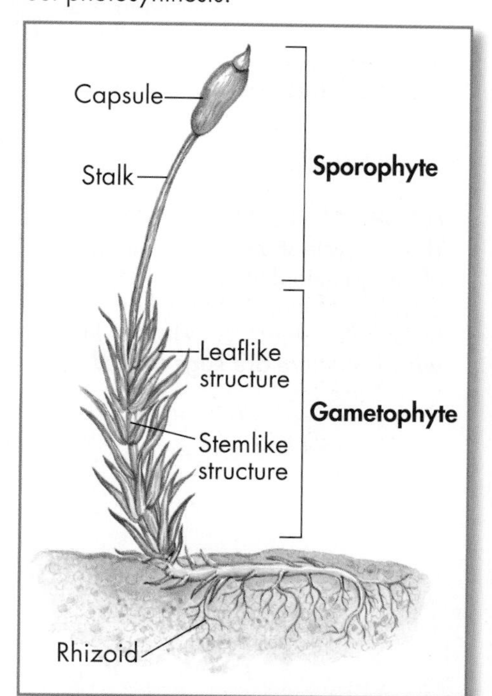

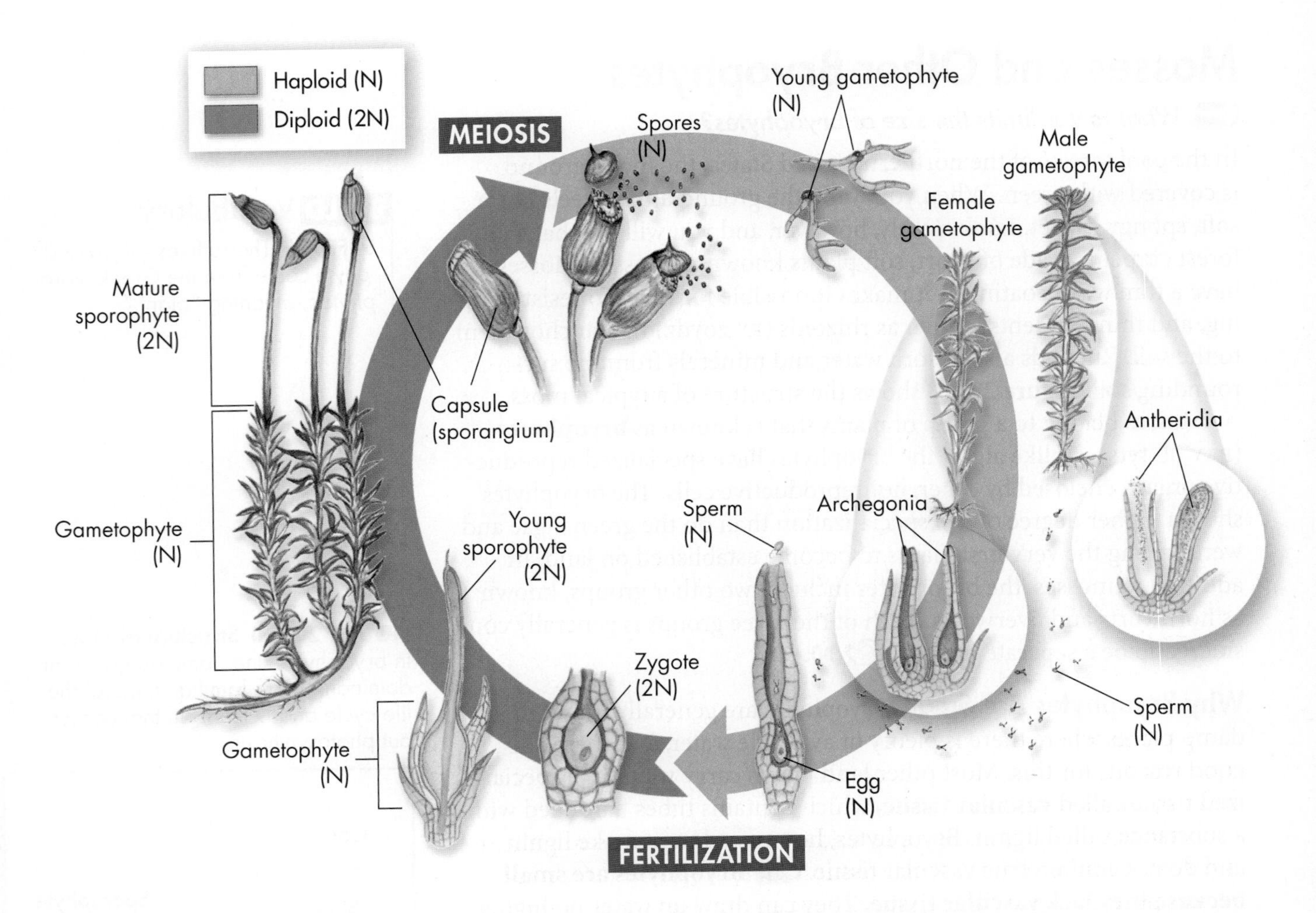

FIGURE 22–11 Moss Life Cycle This life cycle shows the dominance of the gametophyte stage that is typical of mosses and other bryophytes. **Interpret Visuals** ***In which structure are eggs found?***

▶ ***Gametophyte*** When a moss spore lands in a moist place, it sprouts and grows into a tangled mass of green filaments. **Figure 22–11** shows the life cycle of a typical moss. As this young gametophyte grows, it forms rhizoids that grow into the ground and shoots that grow into the air. These shoots grow into the familiar green moss plants.

Gametes are formed in reproductive structures at the tips of the gametophytes. Some bryophyte species produce both sperm and eggs on the same plant, whereas other species produce sperm and eggs on separate plants. Eggs are produced in **archegonia** (ahr kuh GOH nee uh; singular: archegonium). Sperm are produced in **antheridia** (an thur ID ee uh; singular: antheridium). Sperm and egg cells fuse to produce a diploid zygote.

▶ ***Sporophyte*** The zygote marks the beginning of the sporophyte stage of the life cycle. It develops into a multicellular embryo, growing within the body of the gametophyte and depending on it for water and nutrients. Eventually, the sporophyte grows out of the gametophyte, and develops a long stalk ending in a capsule that looks a bit like a salt shaker. The spore capsule is called a **sporangium** (spoh RAN jee um; plural: sporangia). Inside the capsule, haploid spores are produced by meiosis. When the capsule ripens, it opens, and haploid spores are scattered to the wind to start the cycle again.

In Your Notebook *Classify each of these moss structures as gametophyte or sporophyte: sporangium, spore, archegonium, zygote.*

Vascular Plants

How is vascular tissue important?

For millions of years, early plants grew no taller than a meter high because they lacked vascular tissue. Then, about 420 million years ago, something remarkable happened. The small, mosslike plants on land were suddenly joined by taller plants, some of which were as large as small trees. What happened? Fossil evidence shows that these new plants were the first to have a transport system with true vascular tissue. Vascular tissue carries water and nutrients much more efficiently than does any tissue found in bryophytes. With the evolution of vascular tissue, plants were able to grow high above the ground.

Evolution of a Transport System Vascular plants are also known as **tracheophytes** (TRAY kee uh fyts), after a specialized type of water-conducting cell they contain. These cells, called **tracheids** (TRAY kee idz), are hollow tubelike cells with thick cell walls strengthened by lignin, as shown in **Figure 22–12.** Tracheids were one of the great evolutionary innovations of the plant kingdom.

Tracheids are found in **xylem** (ZY lum), a tissue that carries water upward from the roots to every part of a plant. Tracheids are connected end to end like a series of tin cans. Openings between tracheids known as pits allow water to move through a plant more efficiently than by diffusion alone.

Vascular plants also have a second transport tissue called phloem. **Phloem** (FLOH um) transports solutions of nutrients and carbohydrates produced by photosynthesis. Like xylem, the main cells of phloem are long and specialized to move fluids throughout the plant body. **Vascular tissues--xylem and phloem--make it possible for vascular plants to move fluids through their bodies against the force of gravity.**

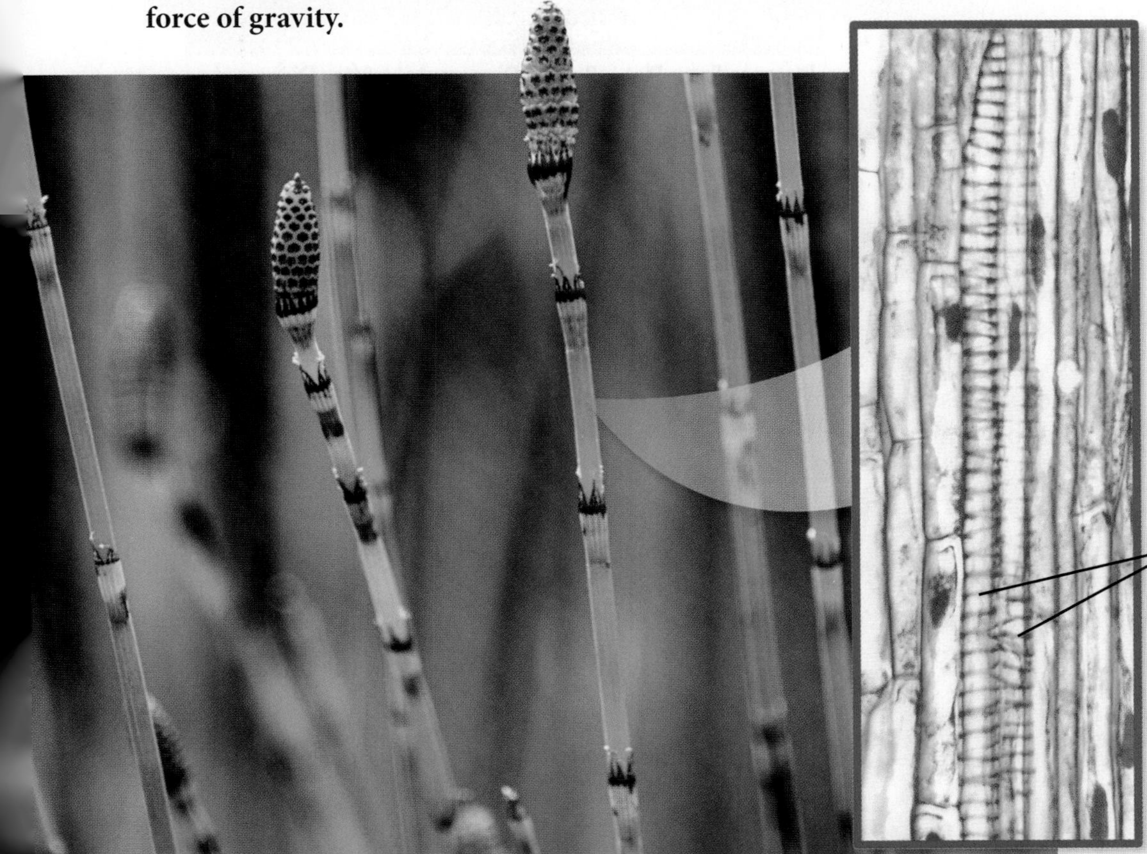

ZOOMING IN

VASCULAR TISSUE

FIGURE 22–12 Horsetails are among the most primitive plant species to have specialized vascular tissue. The micrograph at right shows a much-magnified view of the tracheids. You can see the rings of reinforcing lignin surrounding the tracheids.

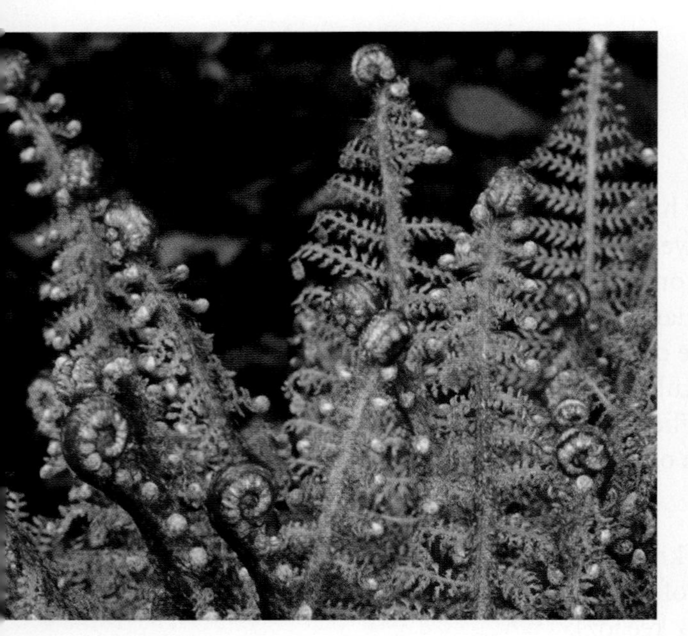

FIGURE 22–13 Structure of a Fern Ferns are easily recognized because of their delicate leaves, called fronds. Before a frond unfurls, it is called a fiddlehead.

Seedless Vascular Plants Although the tracheophytes include all seed-bearing plants, vascular tissue is also found in many groups of plants that do not produce seeds. Among the seedless vascular plants alive today are three phyla commonly known as club mosses, horsetails, and ferns. DOL•23

The most numerous seedless plants are the ferns. More than 11,000 species of ferns are living today. Ferns have true vascular tissues, strong roots, creeping or underground stems called rhizomes (RY zohmz), and large leaves called fronds, shown in **Figure 22–13.** Ferns can thrive in areas with little light. They are most abundant in wet, or at least seasonally wet, habitats.

Life Cycle The large plants we recognize as ferns are actually diploid sporophytes. The fern life cycle is shown in **Figure 22–14.** Ferns and other vascular plants have a life cycle in which the diploid sporophyte is the dominant stage.

In the fern life cycle, spores grow into thin, heart-shaped haploid gametophytes. Although it is tiny, the gametophyte grows independently of the sporophyte. As in bryophytes, sperm and eggs are produced on these gametophytes in antheridia and archegonia, respectively.

Fertilization requires at least a thin film of water, so that the sperm can swim to the eggs. The diploid zygote produced by fertilization immediately begins to develop into a new sporophyte plant. As the sporophyte matures, haploid spores develop on the undersides of the fronds in sporangia, and the cycle begins again.

Analyzing Data

Keeping Ferns in Check

Dennstaedtia punctilobula is a fern that grows on the forest floor and often crowds out tree seedlings, blocking efforts to regrow trees after logging or other work in a forest. To understand the fern better, scientists measured the number of viable fern spores per square centimeter of soil at various distances from a plot of existing fern plants. They counted spores in the soil in July, as the ferns were just beginning to grow; and in November, after they had released their spores.

1. Graph Place the data from the table on a line graph showing the number of spores per square centimeter versus their distance from the plot. Use different colors for the before and after dispersal data points.

Number of Spores in Soil

Distance From Plot of Ferns (meters)	Before Dispersal (July)	After Dispersal (November)
0	14	54
2	16	18
4	5	9
10	10	17
50	2	7

2. Calculate What percentage of the spores after dispersal are found within 4 meters of the parent plants? MATH

3. Interpret Graphs Are spore numbers higher before dispersal or after dispersal? Explain.

4. Draw Conclusions Would cutting down nearby clusters of ferns prevent ferns from invading patches of the forest that have just been cut for timber? Explain your reasoning on the basis of the data.

FIGURE 22–14 Fern Life Cycle In the life cycle of a fern, the dominant and recognizable stage is the diploid sporophyte.
Interpret Visuals ***Are the spores haploid or diploid?***

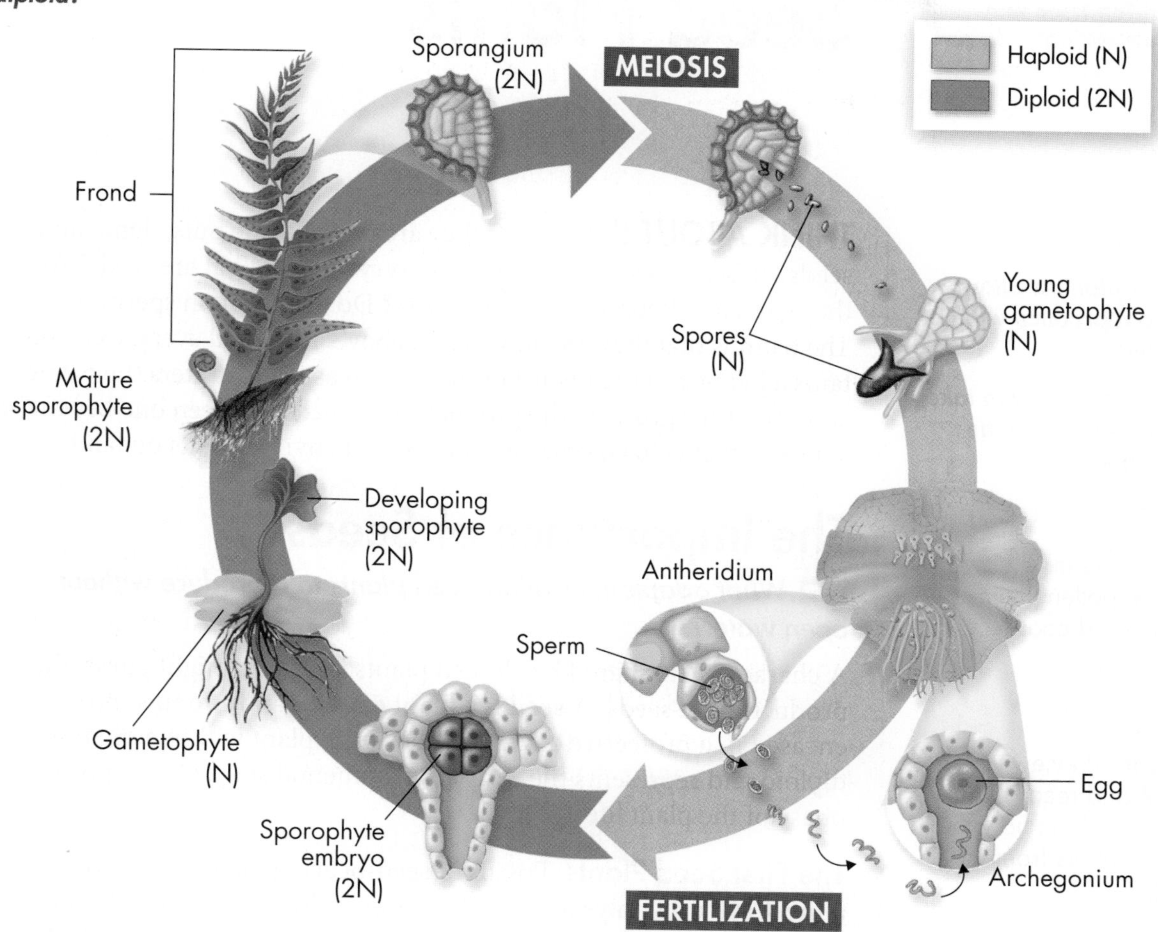

22.2 Assessment

Review Key Concepts

1. a. **Review** In what kind of environments are green algae found?
   b. **Compare and Contrast** How are green algae similar to and different from other plants?
2. a. **Review** Why are bryophytes small?
   b. **Apply Concepts** How is water essential to the life cycle of a bryophyte?
3. a. **Review** What function do vascular tissues allow?
   b. **Infer** The size of plants increased dramatically with the evolution of vascular tissue. How might these two events be related?

BUILD VOCABULARY

4. Find all the word roots for the terms in this lesson that end in *-phyte*. (Recall that the suffix *-phyte* means "plant.") Now, look up the meaning of each word root paired with *-phyte*. Write down the root-suffix translation of each of these words. Which translations help you remember what each word means? Explain.

22.3 Seed Plants

Key Questions

What adaptations allow seed plants to reproduce without standing water?

How does fertilization take place in gymnosperms in the absence of water?

Vocabulary

seed • gymnosperm • angiosperm • pollen grain • pollination • seed coat • ovule • pollen tube

Taking Notes

Preview Visuals Preview **Figure 22–17** and record your first impressions about how the pine life cycle differs from the fern life cycle.

THINK ABOUT IT Whether they are acorns, pine nuts, dandelion seeds, or beans, seeds can be found everywhere. What are seeds? Are they gametes? Reproductive structures? Do they contain sperm or eggs? The truth is that they are none of the above. Each and every seed contains a living plant ready to sprout as soon as it encounters the proper conditions for growth. The production of seeds has been one key to the ability of plants to colonize even the driest environments on land.

The Importance of Seeds

What adaptations allow seed plants to reproduce without open water?

A characteristic shared by all seed plants is, as you might guess, the production of seeds. A **seed** is a plant embryo and a food supply, encased in a protective covering. The living plant within a seed is diploid and represents the early developmental stage of the sporophyte phase of the plant life cycle.

The First Seed Plants Fossils of seed-bearing plants exist from almost 360 million years ago. These fossils document several evolutionary stages in the development of the seed. Similarities in DNA sequences from modern plants provide evidence that today's seed plants are all descended from common ancestors.

The fossil record indicates that ancestors of seed plants evolved new adaptations that enabled them to survive in many environments on dry land. Unlike mosses and ferns, the gametes of seed plants do not need standing water for fertilization. **Adaptations that allow seed plants to reproduce without standing water include a reproductive process that takes place in cones or flowers, the transfer of sperm by pollination, and the protection of embryos in seeds.**

Cones and Flowers In seed plants, the male gametophytes and the female gametophytes grow and mature directly within the sporophyte. The gametophytes usually develop in reproductive structures known as cones or flowers. In fact, seed plants are divided into two groups on the basis of which of these structures they have. Nearly all **gymnosperms** (JIM noh spurmz) bear their seeds directly on the scales of cones. In contrast, flowering plants, or **angiosperms** (AN jee oh spurmz), bear their seeds in flowers inside a layer of tissue that protects the seed. **Figure 22–15** compares the reproductive structures of gymnosperms and angiosperms.

BUILD Vocabulary

WORD ORIGINS The prefix *gymno-* comes from the Greek word *gymnos*, meaning "naked." The prefix *angio-* comes from the Greek word *angeion*, meaning "vessel." The suffix *-sperm* means "seed."

VISUAL SUMMARY

REPRODUCTION IN SEED PLANTS

FIGURE 22–15 The two major groups of seed plants can be distinguished by their reproductive structures. Interpret Visuals ***How is the location of developing seeds different in the two groups?***

Pollen In seed plants, the entire male gametophyte is contained in a tiny structure called a **pollen grain.** Sperm produced by this gametophyte do not swim through water to fertilize the eggs. Instead, pollen grains are carried to the female reproductive structure by wind or animals such as insects. The transfer of pollen from the male reproductive structure to the female reproductive structure is called **pollination.**

Seeds After fertilization, the zygote contained within a seed grows into a tiny plant—the sporophyte embryo. The embryo often stops growing while it is still small and contained within the seed. The embryo can remain in this condition for weeks, months, or even years. A tough **seed coat** surrounds and protects the embryo and keeps the contents of the seed from drying out. Seeds can survive long periods of bitter cold, extreme heat, or drought. The embryo begins to grow when conditions are once again right; it does this by using nutrients from the stored food supply until it can carry out photosynthesis on its own.

MYSTERY CLUE

Samples of material taken from Iceman's digestive tract contained pollen from plant species that grow at different elevations. How could this evidence be used to reconstruct Iceman's movements on his last day alive?

In Your Notebook *Make a Venn diagram that records the shared and distinct characteristics of gymnosperms and angiosperms.*

The Life Cycle of a Gymnosperm

How does fertilization take place in gymnosperms in the absence of water?

The word *gymnosperm* actually means "naked seed." The name reflects the fact that gymnosperms produce seeds that are exposed on the scales within cones. Gymnosperms alive today include relatively rare plants such as cycads and ginkgoes and the much more abundant plants known as conifers, which include pines and firs. DOL•24–DOL•25

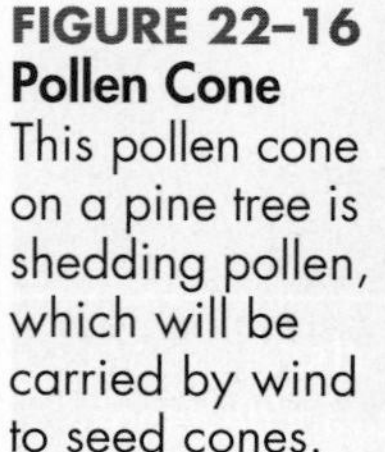

FIGURE 22–16
Pollen Cone
This pollen cone on a pine tree is shedding pollen, which will be carried by wind to seed cones.

Pollen Cones and Seed Cones Reproduction in conifers takes place in cones, which are produced by the mature sporophyte plant. Conifers produce two types of cones: pollen cones and seed cones. Pollen cones, also called male cones, produce the pollen grains. As tiny as it is, a pollen grain makes up the entire male gametophyte stage of the gymnosperm life cycle. One of the haploid nuclei in the pollen grain will divide later to produce two sperm nuclei.

The more familiar seed cones, or female cones, produce female gametophytes. Seed cones are generally much larger than pollen cones. Near the base of each scale of the seed cones are two **ovules** (AHV yoolz), the structures in which the female gametophytes develop. Within the ovules, meiosis produces haploid cells that grow and divide to produce female gametophytes. These gametophytes may contain hundreds or thousands of cells. When mature, each gametophyte contains a few large egg cells, each ready for fertilization by sperm nuclei.

Pollination and Fertilization The conifer life cycle typically takes two years to complete. The life cycle of a pine is shown in **Figure 22–17.** The cycle begins in the spring as male cones release enormous numbers of pollen grains that are carried away by the wind. Some of these pollen grains reach female cones. There, pollen grains are caught in a sticky secretion on the scales of the female cone and pulled inside toward the ovule. **In gymnosperms, the direct transfer of pollen to the female cone allows fertilization to take place without the need for gametes to swim through standing water.**

Development Inside Seeds If a pollen grain lands near an ovule, the grain splits open and begins to grow a structure called a **pollen tube,** which contains two haploid sperm nuclei. Once the pollen tube reaches the newly developed female gametophyte, one sperm nucleus disintegrates; the other fertilizes the egg contained within the female gametophyte. Fertilization produces a diploid zygote, which grows into an embryo—the new sporophyte plant. The embryo is then encased to form a seed. The seed is ready to be scattered by the wind and grow into a new plant.

MYSTERY CLUE

An unfinished bow made from conifer wood was found with Iceman. No other weapons for hunting or self-defense were found with him. How might he have died?

In Your Notebook *Make a flowchart that records the events leading up to fertilization in a gymnosperm.*

FIGURE 22–17 Pine Life Cycle
In the life cycle of pine trees and other gymnosperms, the mature sporophyte trees produce male and female cones containing the gametophytes.
Sequence ***Which sporophyte stage develops immediately after fertilization?***

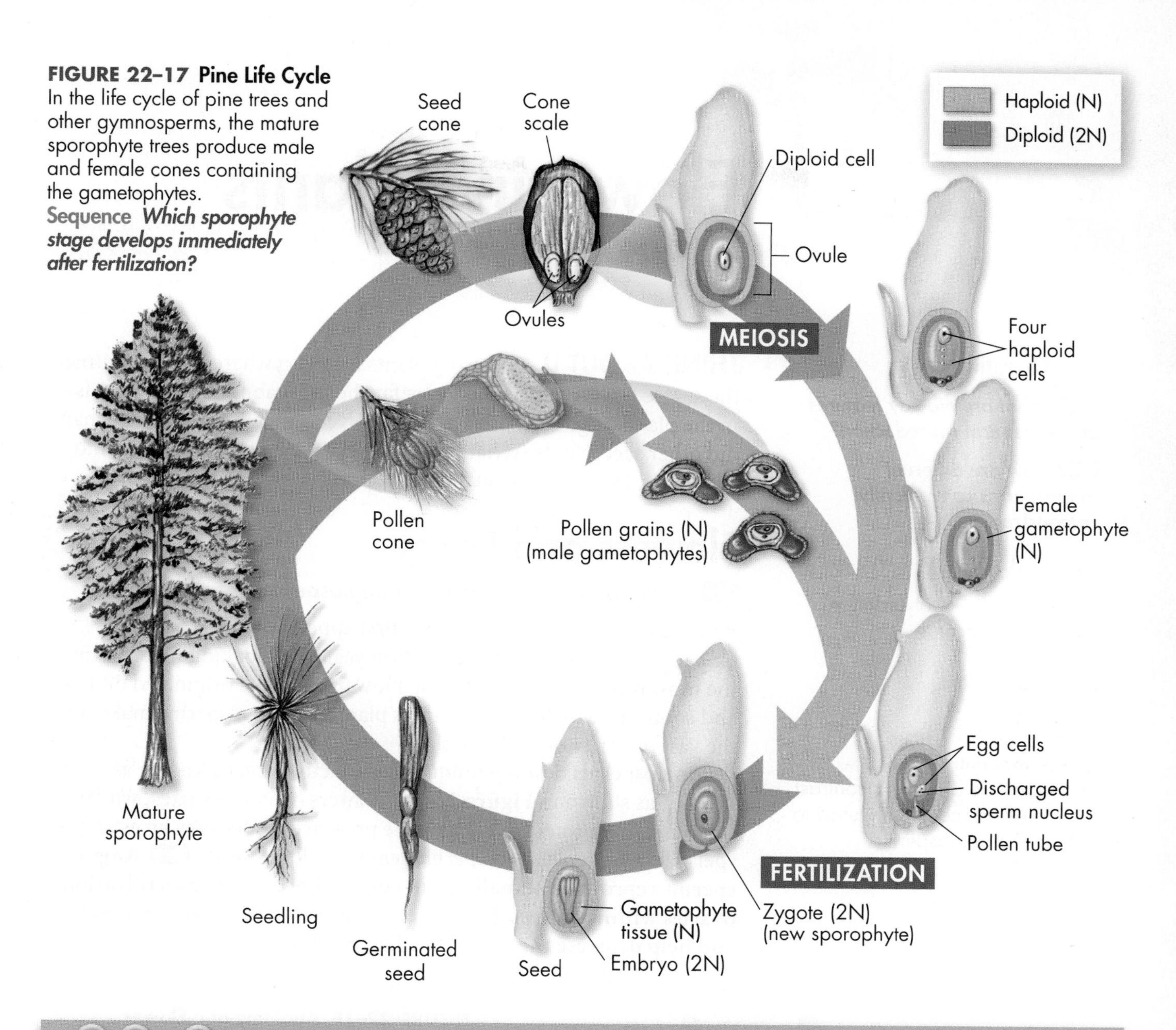

22.3 Assessment

Review Key Concepts

1. **a. Review** List three adaptations of seed plants that allow them to reproduce without open water.
   b. Apply Concepts Pollination is a process that occurs only in seed plants. What process in seedless plants is analogous to pollination?
2. **a. Review** Describe how fertilization takes place in a gymnosperm.
   b. Classify Make a table with two columns—labeled haploid and diploid—and assign each of the following structures from the pine life cycle to the appropriate column: pollen tube, seed cone, embryo, ovule, seedling.

WRITE ABOUT SCIENCE

Creative Writing

3. Design and write an advertisement promoting seeds. Your target audience, or customers, are seedless plants. Include information on how seeds will make their lives easier.

22.4 Flowering Plants

Key Questions

What are the key features of angiosperm reproduction?

How are different angiosperms conveniently categorized?

Vocabulary

ovary • fruit • cotyledon • monocot • dicot • woody plant • herbaceous plant

Taking Notes

Compare/Contrast Table As you read, use a table to contrast three methods commonly used to categorize angiosperms.

THINK ABOUT IT Flowering plants are everywhere. They dominate the surface of the earth and are by far the most abundant organisms in the plant kingdom. And yet they evolved much more recently than did other seed plants. What features of these plants enabled them to take Earth by storm? What are the secrets of their success?

Flowers and Fruits

What are the key features of angiosperm reproduction?

Flowering plants, or angiosperms, first appeared during the Cretaceous Period, about 135 million years ago, making their origin the most recent of all plant phyla. Flowering plants originated on land and soon came to dominate Earth's plant life. Angiosperms make up the vast majority of plant species.

Angiosperms develop unique reproductive organs known as flowers, as shown in **Figure 22–18.** Flowers contain **ovaries,** which surround and protect the seeds. The presence of an ovary gives angiosperms their name: *Angiosperm* means "enclosed seed." **Angiosperms reproduce sexually by means of flowers. After fertilization, ovaries within flowers develop into fruits that surround, protect, and help disperse the seeds.** In Chapter 24, you will explore angiosperm reproduction in more detail.

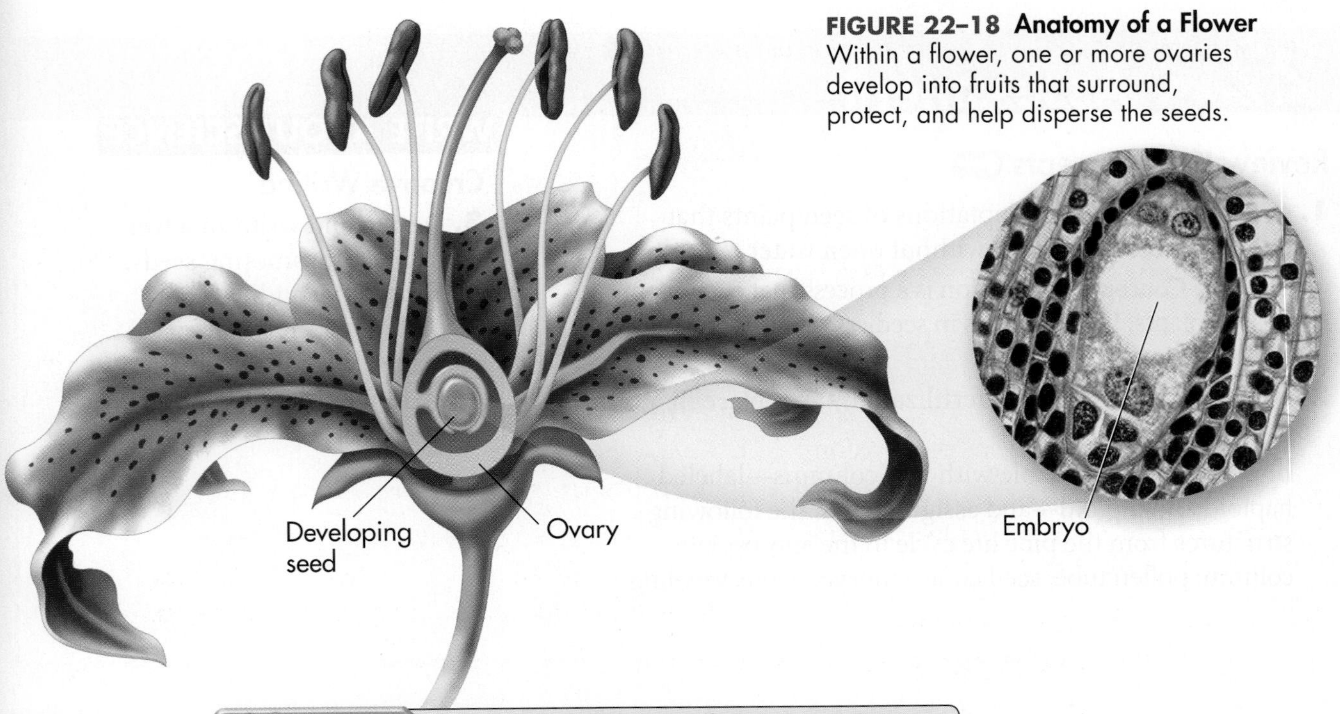

FIGURE 22–18 Anatomy of a Flower Within a flower, one or more ovaries develop into fruits that surround, protect, and help disperse the seeds.

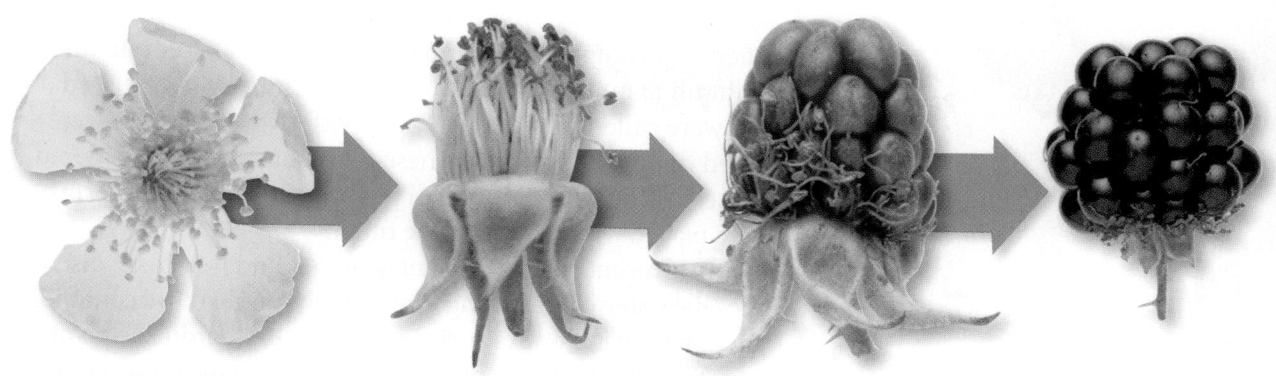

FIGURE 22–19 From Flower to Fruit Following pollination and fertilization, a blackberry flower's multiple ovaries develop into a cluster of many individual fruits.

Advantages of Flowers In general, flowers are an evolutionary advantage to plants because they attract animals such as bees, moths, or hummingbirds. These animals—drawn by the color, scent, or even the shape of the flower—carry pollen with them as they leave. Because these animals go directly from flower to flower, they can carry pollen to the next flower they visit. This means of pollination is much more efficient than the wind pollination of most gymnosperms.

Advantages of Fruits After pollination, the ovary develops into a fruit. In **Figure 22–19,** you can see the progression of development of a blackberry flower into a fruit. The angiosperm **fruit** is a structure containing one or more matured ovaries. The wall of the fruit helps disperse the seeds inside it, carrying them away from the parent plant.

Consider what happens when an animal eats a fleshy fruit, such as a berry. Seeds from the fruit generally enter the animal's digestive system. By the time these seeds leave the digestive system—ready to sprout—the animal may have traveled many kilometers. By using fruit, flowering plants increase the ranges they inhabit, spreading seeds over hundreds of square kilometers. The fruit—a unique feature of angiosperms—is yet another reason for their success.

In Your Notebook *Summarize the function of flowers and fruits in the reproduction of angiosperms.*

Quick Lab
GUIDED INQUIRY

What Forms Do Fruits Take?

1. Use a hand lens to examine a variety of fruits. Write down or draw your observations.
2. Place each fruit in a petri dish and use a scalpel to dissect it. **CAUTION:** *Use care with sharp instruments.*
3. Locate the seeds within each fruit.

Analyze and Conclude

1. **Compare and Contrast** How do the fruits vary in structure?
2. **Infer** For each fruit, infer how its structure affects its function—that is, how might it aid seed dispersal? Explain your answers.

FIGURE 22–20 *Archaefructus*

BUILD Vocabulary

PREFIXES The prefix *mono-* means "one," while the prefix *di-* means "two." A **monocot** embryo has one cotyledon, while a **dicot** has two cotyledons.

Angiosperm Classification The great diversity of angiosperms has made them especially difficult to classify in the scientific sense. For many years, flowering plants were classified according to the number of seed leaves, or **cotyledons** (kaht uh LEED uns), in their embryos. Those with one seed leaf were called **monocots.** Those with two seed leaves were called **dicots.** At one time, these two groups were considered classes within the angiosperm phylum, and all angiosperms were placed in one class or the other.

More recent studies of plant genomes and new fossil discoveries have shown that things are actually a little more complicated than that. For example, in 2002, an extraordinary plant fossil was discovered in northeastern China. Given the name *Archaefructus,* which means "ancient fruit," this organism is the oldest known plant with reproductive organs like those found in modern flowers. It is more ancient than modern-day monocots and dicots and can't be classified as either.

Other recent evidence suggests that *Amborella,* a plant found only on the Pacific island of New Caledonia, belongs to still another ancient lineage of plants. Information gained from the *Amborella* discovery led scientists to place other plants, such as the water lilies, near the base of angiosperm evolution.

Figure 22–21 summarizes one modern view of angiosperm classification. Scientific classification now places the monocots into a single group but places the dicots in a variety of distinct and different categories. This means, of course, that the term *dicot* is no longer used for classification. However, it can still be used to describe many of the characteristics of plant structure, and that is how it is used in this book. DOL•30–DOL•33

FIGURE 22–21 Angiosperm Clades
Five of the major clades of angiosperms are represented here. Scientists are still working out the relationships among these groups.

Amborella Clade Only one species still exists in this oldest branch of angiosperms. Its floral parts have a spiral arrangement.

Water Lily Clade The water lilies are another very old group. Early water lily flowers may have been no more than 1 cm across, in contrast to the large and showy water lilies of today.

Magnoliids This clade contains a wide range of floral diversity, from species with rather small, plain flowers to the dinner-plate sized *Magnolia* flower shown here.

Monocots This clade contains about 20 percent of all angiosperms. Monocots include several important crop species, such as rice, corn, and wheat, as well as orchids, lilies, and irises.

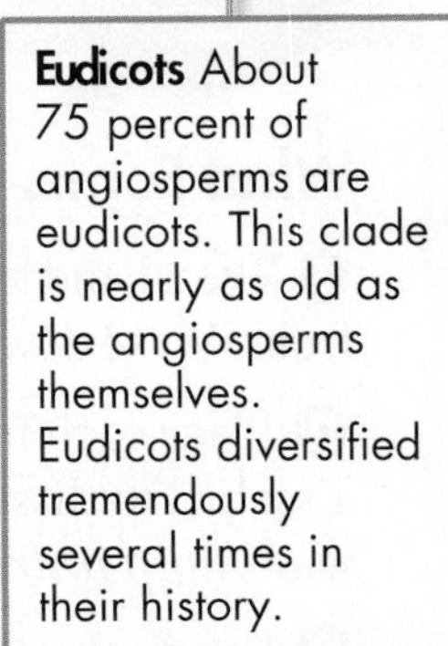

Eudicots About 75 percent of angiosperms are eudicots. This clade is nearly as old as the angiosperms themselves. Eudicots diversified tremendously several times in their history.

Angiosperm Diversity

How are different angiosperms conveniently categorized?

While scientific classification best reflects the evolutionary relationships among flowering plants, many people who work with plants—for example, farmers, gardeners, and foresters—tend to categorize angiosperms using more convenient methods. **Angiosperms are often grouped according to the number of their seed leaves, the strength and composition of their stems, and the number of growing seasons they live.** Naturally, these categories can overlap. An iris, for example, is a nonwoody plant, it has a single seed leaf, and it may live for many years.

MYSTERY CLUE

Fragments of cultivated wheat were found in Iceman's digestive tract. How did scientists confirm it was a monocot? How do you think they knew it was a cultivated form instead of wild?

Monocots and Dicots As **Figure 22–22** shows, angiosperms may be termed monocots or dicots based on the number of seed leaves, or cotyledons, they produce. Although we no longer classify both as scientific groups, the term is still useful. Monocots and dicots differ in characteristics such as the distribution of vascular tissue in stems, roots, and leaves, and the number of petals per flower. Monocots include plants such as corn, wheat, lilies, orchids, and palms. Monocot grasses—especially wheat, corn, and rice—have the important distinction of being the first plants to be cultivated in mass quantities for food. Dicots include roses, clover, tomatoes, oaks, and daisies.

Characteristics of Monocots and Dicots

	Seeds	Leaves	Flowers	Stems	Roots
Monocots	Single cotyledon	Parallel veins	Floral parts often in multiples of 3	Vascular bundles scattered throughout stem	Fibrous roots
Dicots	Two cotyledons	Branched veins	Floral parts often in multiples of 4 or 5	Vascular bundles arranged in a ring	Taproot

FIGURE 22–22 Comparing Monocots and Dicots This table compares the characteristics of monocots and dicots. **Interpret Tables** ***How do the flowers of monocots and dicots typically differ?***

Woody and Herbaceous Plants The flowering plants can also be subdivided into groups according to the characteristics of their stems. One of the most important and noticeable stem characteristics is woodiness. **Woody plants** are made primarily of cells with thick cell walls that support the plant body. Woody plants include trees, shrubs, and vines. Shrubs are typically smaller than trees, and vines have stems that are long and flexible.

Plant stems that are smooth and nonwoody are characteristic of **herbaceous plants** (hur BAY shus). Herbaceous plants do not produce wood as they grow. Examples of herbaceous plants include dandelions, zinnias, petunias, and sunflowers.

Comparing Plants by Life Span

Category	Life Span				Characteristics	Examples
	Year 1	Year 2	Year 3	Year 4		
Annuals					• Grow from seed to maturity, flower, produce seeds, and die in just one growing season	Marigolds, petunias, pansies, zinnias, tomatoes, wheat, cucumbers
Biennials					• Year 1: Sprout and grow very short stems and sometimes leaves • Year 2: Grow new stems and leaves, flower, produce seeds, then die	Parsley, celery, evening primroses, foxgloves
Perennials					• Most have woody stems. • Some have herbaceous stems that die each winter and are replaced in the spring.	Peonies, many grasses, palm trees, maple trees, honeysuckle, asparagus

FIGURE 22–23 Comparing Plants by Life Span Categories of plant life spans include annuals, biennials, and perennials. **Interpret Tables** ***Which flowering plant completes its life cycle in one year?***

Annuals, Biennials, and Perennials If you've ever planted a garden, you know that many flowering plants grow, flower, and die in a single year. Other types of plants continue to grow from year to year. The life span of plants is determined by a combination of genetic and environmental factors. Many long-lived plants continue growing despite yearly environmental fluctuations. However, harsh environmental conditions can shorten the life span of other plants. The characteristics of the three categories of plant life spans—annual, biennial, and perennial—are summarized in **Figure 22–23.**

22.4 Assessment

Review Key Concepts

1. a. **Review** What reproductive structures are unique to angiosperms? Briefly describe the function of each.
   b. **Form a Hypothesis** Which are more likely to be dispersed by animals—the seeds of an angiosperm or seeds of a gymnosperm? Explain your reasoning.
2. a. **Review** What are three common ways to categorize angiosperms?
   b. **Explain** How do these methods of categorization differ from scientific classification methods?
   c. **Form an Opinion** Is it useful or misleading to categorize angiosperms in ways that do not reflect evolutionary relationships? Defend your opinion.

VISUAL THINKING

3. Prepare a display comparing two specific plants—one monocot and one dicot. On this display, show photographs or drawings of the plants and write a brief summary of the basic differences between these two types of angiosperms.

Careers & BIOLOGY

In addition to providing oxygen, plants are a source of food, fiber, and beauty. If you'd like to work with plants, you might want to consider one of the careers below.

FARMER

Farmers grow and harvest food. In the United States, nearly 20 percent of all land is used to grow crops such as corn, soybeans, wheat, and barley. Crop farmers must prepare their land for farming and make decisions about fertilizer use, crop rotation, and pest resistance.

PLANT PATHOLOGIST

Just as animals need doctors, so do plants. Plant pathologists are specialists in plant health. Plant diseases can have a huge economic impact. For example, from 1990 to 2000, one fungal disease destroyed $2.6 billion of wheat in the United States. Using microbiology, soil science, cell biology, genetics, and biochemistry, plant pathologists diagnose and treat plant diseases.

BOTANICAL ILLUSTRATOR

Botanical illustrators provide visuals that help people understand and appreciate biology. Working in a museum, outdoors, in a botanical garden, or at home, botanical illustrators create images of both plants and organisms related to plants. Illustrations may be used in various locations, such as guidebooks, textbooks, or museum displays.

CAREER CLOSE-UP:

Marya C. Roddis, Botanical Illustrator and Educator

Marya Roddis uses her talents as a botanical illustrator to inspire enthusiasm for the natural world. The granddaughter of an ethnobotanist—that is, someone who studies how people use plants—Ms. Roddis was taught to see the value and importance of all living things. Through workshops and after-school activities with local children, Ms. Roddis encourages her students to observe their environment and communicate what they learn. Whether through these educational efforts or her professional guidebook illustrations, Ms. Roddis's goal is to use art as a tool to help students who do not learn well in traditional settings.

"Illustration is a natural part of biology study. The detailed work required to create drawings can lead to success in all areas of scientific work."

WRITING Choose a topic that you have studied so far in biology (such as ecology, cells, or evolution) and explain how illustrations helped you understand the material.

NGSS Smart Guide Introduction to Plants

Disciplinary Core Idea LS4.C Adaptation: How does the environment influence populations of organisms over multiple generations? The five main groups of plants are green algae, bryophytes, seedless vascular plants, gymnosperms, and angiosperms. Over time, plants accumulated adaptations that allowed for success on dry land.

22.1 What Is a Plant?

- The lives of plants center on the need for sunlight, gas exchange, water, and minerals.
- Over time, the demands of life on land favored the evolution of plants more resistant to the drying rays of the sun, more capable of conserving water, and more capable of reproducing without water.
- The life cycle of land plants has two alternating phases, a diploid (2N) phase and a haploid (N) phase. The shift between haploid and diploid is known as the alternation of generations.

alternation of generations 637 • sporophyte 637 • gametophyte 637

Biology.com

Untamed Science Video The Untamed Science biologists interview plant experts to learn about healing chemicals manufactured by plants.

22.2 Seedless Plants 

- Green algae are mostly aquatic. They are found in fresh and salt water, and in some moist areas on land.
- Bryophytes are small because they lack vascular tissue.
- Vascular tissues—xylem and phloem—make it possible for vascular plants to move fluids through their bodies against the force of gravity.

bryophyte 641 • vascular tissue 641 • archegonium 642 • antheridium 642 • sporangium 642 • tracheophyte 643 • tracheid 643 • xylem 643 • phloem 643

Biology.com

Data Analysis Investigate bracken (a type of fern) to see if there is a connection between bracken and stomach cancer.

InterActive Art Review and compare life cycles of vascular and nonvascular plants.

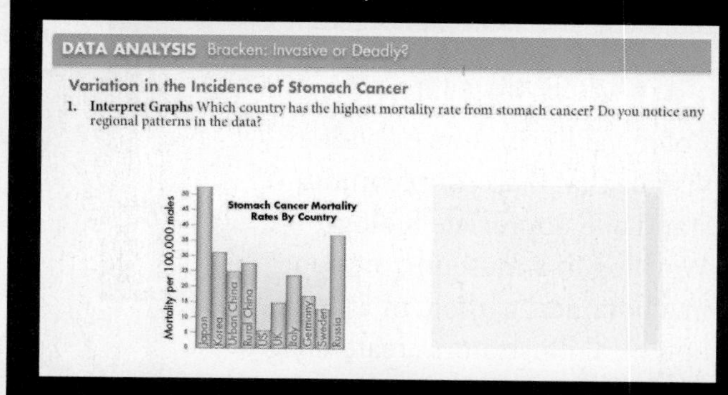

CHECKING YOUR *Scientific Literacy*

Refer to the lesson content and digital assets below as you prepare for your chapter assessment. Then, evaluate your understanding of plants by answering these questions.

1. Scientific and Engineering Practice **Obtaining, Evaluating, and Communicating Information** Perform an Internet search to learn about the plant *Silene stenophylla*. Read about how scientists have grown *Silene stenophylla* plants from 30,000-year-old frozen fruits buried by ancient squirrels. Write a short paper about how scientists grew these plants and how they determined the age of the fruit they found.

2. Crosscutting Concept **Cause and Effect** What effect did the evolution of a vascular system have on the ability of plants to survive in various environments?

3. Craft and Structure Choose a term from the text you read to learn about *Silene stenophylla* for which you do not know the definition. Try to define the term based on the context of what you read. Use a dictionary to check your definition.

4. STEM You are a scientist studying plants. You want to ensure that scientists in the future will be able to study the plants that are in existence today. Describe how you would preserve the seeds. Would you store seeds of gymnosperms differently than seeds of angiosperms?

22.3 Seed Plants

- Adaptations that allow seed plants to reproduce without standing water include a reproductive process that takes place in cones or flowers, the transfer of sperm by pollination, and the protection of embryos in seeds.
- In gymnosperms, the direct transfer of pollen to the female cone allows fertilization to take place without the need for gametes to swim through water.

seed 646 • gymnosperm 646 • angiosperm 646 • pollen grain 647 • pollination 647 • seed coat 647 • ovule 648 • pollen tube 648

Biology.com

Tutor Tube Is this plant a male or a female? Compare animals and plants to understand plant reproduction.

Art in Motion Follow the process of pollination and fertilization in a pine tree to see how a plant embryo is formed.

TUTOR TUBE Sexual Reproduction in Plants—An Animal Perspective

22.4 Flowering Plants

- Angiosperms reproduce sexually by means of flowers. After fertilization, ovaries within flowers develop into fruits that surround, protect, and help disperse the seeds.
- Angiosperms are often grouped according to the number of their seed leaves, the strength and composition of their stems, and the number of growing seasons they live.

ovary 650 • fruit 651 • cotyledon 652 • monocot 652 • dicot 652 • woody plant 653 • herbaceous plant 653

Biology.com

Art Review See how well you can distinguish monocots and dicots.

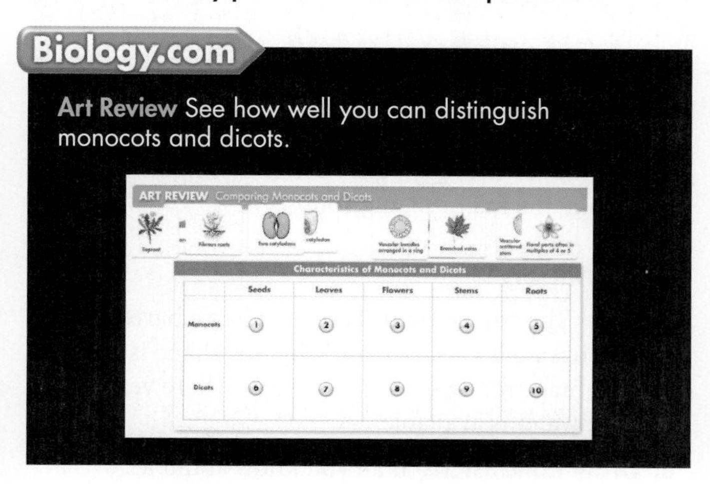

Real-World Lab

Exploring Plant Diversity This chapter lab is available in your lab manual and at Biology.com

22 Assessment

22.1 What Is a Plant?

Understand Key Concepts

1. Which of the following is NOT a characteristic of plants?
   a. eukaryotic cells
   b. cell walls containing chitin
   c. multicellular structure
   d. chlorophyll

2. The first land plants likely evolved from
   a. protists.
   b. green algae.
   c. mosses.
   d. red algae.

3. Two gases that plants must exchange are
   a. oxygen and nitrogen.
   b. carbon dioxide and nitrogen.
   c. oxygen and carbon dioxide.
   d. carbon dioxide and carbon monoxide.

4. Recent changes in the classification of the plant kingdom are based on
   a. studies comparing DNA sequences.
   b. comparison of physical structures.
   c. differences and similarities in life cycles.
   d. whether or not a plant uses seeds to reproduce.

5. **Craft and Structure** Give the term for the process shown below. Then describe what this term means.

6. List the features biologists use to distinguish among the major plant groups.

Think Critically

7. **Infer** The barrel cactus is shaped like a barrel and has no leaves. Explain why this shape is an advantage for a plant that survives where very little water is available.

8. **Draw Conclusions** If all you know about a particular plant is that it lives virtually all of its life as a multicellular haploid organism, what can you conclude about the kind of plant it is?

22.2 Seedless Plants

Understand Key Concepts

9. Under unfavorable conditions, the green alga *Chlamydomonas* reproduces by forming a
   a. haploid zygote.
   b. multicellular sporophyte.
   c. diploid zygote.
   d. multicellular gametophyte.

10. The dominant stage of a moss is the
    a. sporophyte.
    b. protonema.
    c. archegonium.
    d. gametophyte.

11. Water is carried upward from the roots to every part of a vascular plant by
    a. cell walls.
    b. phloem.
    c. cuticle.
    d. xylem.

12. The leaves of ferns are called
    a. sori.
    b. fronds.
    c. rhizomes.
    d. spores.

13. Give two examples of colonial green algae and briefly describe their structure.

14. Describe two ways that a lack of lignin limits the height of bryophytes.

15. In the life cycle of a moss, what environmental conditions are necessary for fertilization?

16. **Craft and Structure** What is a sporangium?

17. What are tracheids? What is their function in a vascular plant?

18. How was the ability to produce lignin significant to the evolution of plants?

19. Compare the structure and function of rhizomes, rhizoids, and roots.

20. Describe a fern gametophyte.

Think Critically

21. **Compare and Contrast** Moss plants are small, but ferns can grow as tall as small trees. Explain why this is so.

22. **Apply Concepts** What environmental conditions are needed for a garden of bryophytes to grow in a New Mexico desert?

22.3 Seed Plants

Understand Key Concepts

23. All of the following are characteristics of gymnosperms EXCEPT
a. vascular tissue. **c.** cones.
b. seeds. **d.** flowers.

24. The male reproductive structures of seed plants are called
a. sperm. **c.** pollen grains.
b. ovules. **d.** sporophytes.

25. The structures on a pine tree that contain the gametophytes are
a. flowers. **c.** sori.
b. sporangia. **d.** cones.

26. The green structure in the diagram is called
a. an embryo.
b. a seed coat.
c. a spore.
d. stored food.

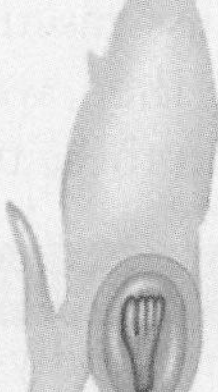

27. What is the function of a seed coat?

28. How do gymnosperm seeds differ from angiosperm seeds?

29. What reproductive adaptations allow conifers to live in dry habitats?

Think Critically

30. **Craft and Structure** Describe the primary similarities and differences between gymnosperms and angiosperms.

31. **Key Ideas and Details** During the age of the dinosaurs, the vast majority of land plants were ferns and mosses. Today, the vast majority are seed plants. Cite evidence from the text and provide an explanation for this change based on the basic requirements of plants.

32. **Apply Concepts** Explain the structure and function of a seed cone and pollen cone. Explain their respective roles in reproduction.

STONE AGE STORYTELLERS

Iceman's story is continually evolving as additional evidence is analyzed. But so far, the plants found with him have revealed a lot. The abundance of chlorophyll in the maple leaves and the species of pollen in his digestive tract both point to a death in late spring. Distinct layers of pollen in his digestive tract suggest that Iceman changed his elevation quite dramatically several times on his final day.

The unfinished bow suggests that Iceman's journey into the mountains was unplanned. Perhaps he was fleeing an enemy? In fact, several years after the bow was found a CT scan of Iceman revealed a stone arrowhead lodged beneath his left shoulder blade and the large gash that the arrowhead left in a major artery.

And what of his society? The primitive wheat in his digestive tract and other intact grains found on his clothes suggest that Iceman's society practiced an early form of agriculture.

1. **Apply Concepts** A clump of moss was also found with Iceman's possessions. Some scientists hypothesize that Iceman used the moss like we use tissues or paper towels today. What property of moss would allow this function? Why is this adaptation necessary for mosses? *Hint*: What kind of tissue do they lack that other land plants have?

2. **Communicate** Write a letter to a friend summarizing the information provided by plant evidence found with Iceman.

3. **Connect to the Big idea** Pollen and seeds are the most reliable plant-related evidence at archeological sites and at modern-day crime scenes because they are long-lasting. Relate this quality to their structure and function in living plants.

22.4 Flowering Plants

Understand Key Concepts

33. In angiosperms, the mature seed is surrounded by a structure called a

a. cone. **c.** fruit.
b. flower. **d.** cotyledon.

34. A plant that has a two-year life cycle is a

a. dicot. **c.** biennial.
b. monocot. **d.** perennial.

35. How do fruits aid in seed dispersal?

36. How does the pattern of veins in a monocot leaf differ from that in a dicot leaf? Draw an example of each.

37. Give one example each of a woody plant and an herbaceous plant.

38. How does the life span of an annual differ from that of a perennial?

Think Critically

39. Compare and Contrast Describe the methods of reproduction and development in the five major plant groups. Include information about the size of the mature plants.

40. Compare and Contrast Compare the size and function of gametophytes in bryophytes, ferns, and seed plants.

Connecting Concepts

Use Science Graphics

41. Classify Study the flower below. Is this plant a monocot or dicot? Explain your answer.

Write About Science

42. Text Types and Purposes Choose a particular group of seedless vascular plants and one group of seed plants. Then, write an essay that compares reproduction in these two groups.

43. Assess the 

At first glance, an oak tree and a zebra hardly seem similar in any way. Describe the characteristics that they do share in common. What are the characteristics of the oak tree that distinguish it from the other kingdoms of living things?

Integration of Knowledge and Ideas

A homeowner has noticed moss growing in the backyard lawn and wants to understand the conditions that cause moss to grow where grass had grown before. Half the homeowner's property is in shade all year, and the other half gets direct sun. The table shown summarizes his observations.

Growth of Moss in Sun and Shade

	Year					
	2003	2004	2005	2006	2007	2008
Area of Moss in Sun (m^2)	0	0	1	2	1	1
Area of Moss in Shade (m^2)	0	2	5	7	6	9

44. Interpret Tables It is clear from the table that moss grows best in areas shaded from the sun. If one of the years in the table had below average rainfall, what would be your guess as to which year that was, based on the data given?

45. Form a Hypothesis What hypothesis about the difference between the shaded and the sunny areas would explain the observed growth of moss?

Standardized Test Prep

Multiple Choice

1. Which of the following is a basic requirement of plants?
   A sunlight
   B carbon dioxide
   C water
   D all of the above

2. What stage in the alternation of generations is represented by fern fronds?
   A sporophyte
   B female gametophyte
   C male gametophyte
   D zygote

3. Which of the following is NOT a characteristic of dicots?
   A branched veins
   B taproot
   C parallel veins
   D two cotyledons

4. Which of the following is a structure associated with gymnosperms?
   A flower
   B cone
   C fruit
   D enclosed seed

5. The mature plant ovary is also referred to as the
   A gymnosperm.
   B pollen grain.
   C fruit.
   D cotyledon.

6. Plants cultivated for food are mostly
   A gymnosperms.
   B woody plants.
   C angiosperms.
   D bryophytes.

Questions 7–9

A group of students placed a sprig of a conifer in a beaker of water. They measured the amount of oxygen given off during a set period of time to determine the rate of photosynthesis. They changed the temperature of the water in the beaker using an ice bucket and a hot plate. Their data are summarized in the graph below.

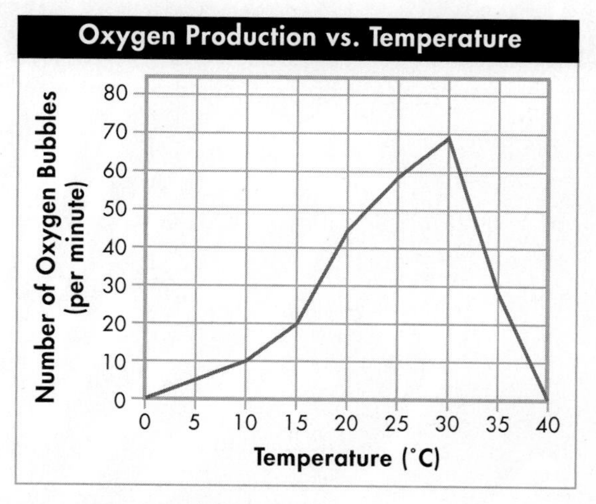

7. What is the independent variable?
   A light intensity
   B temperature
   C oxygen bubbles
   D photosynthesis rate

8. Which variable(s) should the students have held constant?
   A plant type
   B temperature
   C light intensity
   D plant type and light intensity

9. What can you conclude based on the data?
   A The higher the temperature, the more oxygen bubbles are released.
   B There is an optimum temperature for photosynthesis in this species of conifer.
   C All plants are most efficient at 30°C.
   D The lower the temperature, the more oxygen bubbles are released.

Open-Ended Response

10. Explain why seeds were an important adaptation for the success of plants on Earth.

If You Have Trouble With . . .

Question	1	2	3	4	5	6	7	8	9	10
See Lesson	22.1	22.2	22.4	22.3	22.4	22.4	22.3	22.3	22.3	22.3

23 Plant Structure and Function

Big idea Structure and Function

Q: How are cells, tissues, and organs organized into systems that carry out the basic functions of a seed plant?

The leaves of this sundew plant are adapted to capture and digest live prey.

INSIDE:

- 23.1 Specialized Tissues in Plants
- 23.2 Roots
- 23.3 Stems
- 23.4 Leaves
- 23.5 Transport in Plants

BUILDING Scientific Literacy

Deepen your understanding of plant structure and function by engaging in key practices that allow you to make connections across concepts.

NGSS You will develop and use models to demonstrate how the structures of plant organs are related to their functions.

STEM You know that tree rings can tell us the age of a tree, but did you also know that they can provide us with information about Earth's past? Analyze and interpret data related to tree rings.

Common Core You will model with mathematics and write informative texts to describe plant structure and function.

CHAPTER MYSTERY

THE HOLLOW TREE

As you hike through a Central American rain forest on a steamy afternoon on the last day of your tropical vacation, you see many unusual plants and animals. A monkey calls from a distant tree, and a dense fog covers the landscape. Then you stumble on a root and look up. A massive tree stands before you. Its trunk seems to be made up of many intertwined woody branches. Edging closer, you nervously slip your head through one of the larger gaps and look straight up. Inside, you find that the tree is completely hollow.

This tree, a species of fig, is indeed unusual. What happened to the interior of the tree? And how did the tree grow to such a great height if it has no center? As you read this chapter, look for clues that explain the structure of this strange plant. Then, solve the mystery.

Biology.com

Finding the solution to The Hollow Tree mystery is only the beginning. Take a video field trip with the ecogeeks of Untamed Science to see where the mystery leads.

Go online to access additional resources including:
• eText • Flash Cards • Lesson Overviews • Chapter Mystery

23.1 Specialized Tissues in Plants

Key Questions

What are the three principal organs of seed plants?

What are the primary functions of the main tissue systems of seed plants?

How do meristems differ from other plant tissues?

Vocabulary

epidermis • lignin • vessel element • sieve tube element • companion cell • parenchyma • collenchyma • sclerenchyma • meristem • apical meristem

Taking Notes

Concept Map As you read, make a concept map to organize the information in this lesson.

MYSTERY CLUE

The tangled fig "branches" are not actually stems. What are they?

THINK ABOUT IT Have you ever wondered if plants were really alive? Compared to animals, plants don't seem to do much. If you look deep inside a living plant, this first impression of inactivity disappears. Instead, you will find a busy and complex organism. Plants move materials, grow, repair themselves, and constantly respond to the environment. They may act at a pace that seems slow to us, but their cells and tissues work together in remarkably effective ways.

Seed Plant Structure

What are the three principal organs of seed plants?

The cells of a seed plant are organized into different tissues, organs, and systems. **The three principal organs of seed plants are roots, stems, and leaves.** The organs are linked together by systems that run the length of the plant. These systems produce, store, and transport nutrients, and provide physical support and protection.

Roots Roots anchor plants in the ground, holding soil in place and preventing erosion. Root systems often work with soil bacteria and fungi in mutualistic relationships that help the roots absorb water and dissolved nutrients. Roots transport these materials to the rest of the plant, store food, and hold plants upright against forces such as wind and rain.

Stems Plant stems provide a support system for the plant body, a transport system that carries nutrients, and a defensive system that protects the plant against predators and disease. Stems also produce leaves and reproductive organs such as flowers. Whatever the size of a stem, its support system must be strong enough to hold up leaves and branches. The stem's transport system contains tissues that lift water from the roots up to the leaves and carry the products of photosynthesis from the leaves back down to the roots.

Leaves Leaves are the plant's main photosynthetic organs. The broad, flat surfaces of many leaves increase the amount of sunlight plants absorb. Leaves also expose a great deal of tissue to the dryness of the air and, therefore, have adaptations that protect against water loss. Adjustable pores in leaves help conserve water while letting oxygen and carbon dioxide enter and exit the leaf.

In Your Notebook *Relate the three main plant organs back to the basic needs of plants described in Lesson 22.1.*

Quick Lab
GUIDED INQUIRY

What Parts of Plants Do We Eat?

❶ Examine an onion, a potato, and an artichoke. Record your observations as notes and labeled sketches. **CAUTION:** *Do not eat the vegetables.*

❷ Use your observations to classify each vegetable as a root, stem, leaf, or other plant part.

Analyze and Conclude

1. Classify How did you classify the onion? Explain what characteristics you used to make this decision.

2. Infer How did you classify the potato? How is its structure related to its function?

3. Infer How did you classify the artichoke? What does its inner structure tell you about its function?

Plant Tissue Systems

What are the primary functions of the main tissue systems of seed plants?

Within the roots, stems, and leaves of plants are specialized tissue systems, shown in **Figure 23–1.** Plants have three main tissue systems: dermal, vascular, and ground. Dermal tissue covers a plant almost like skin covers you. Vascular tissue forms a system of pipelike cells that help support the plant and serve as its "bloodstream," transporting water and nutrients. Ground tissue produces and stores food. Next, you will see how the cells in these systems compare to one another.

Dermal Tissue Dermal tissue in young plants consists of a single layer of cells called the **epidermis** (ep uh DUR mis). The outer surfaces of epidermal cells are often covered with a thick waxy layer called the cuticle, which protects against water loss. Some epidermal cells have tiny projections known as trichomes (TRY kohmz). Trichomes help protect the leaf and may give the leaf a fuzzy appearance. **Dermal tissue is the protective outer covering of a plant.**

In older plants, dermal tissue may be many cell layers deep and may be covered with bark. In roots, dermal tissue includes root hair cells that help absorb water.

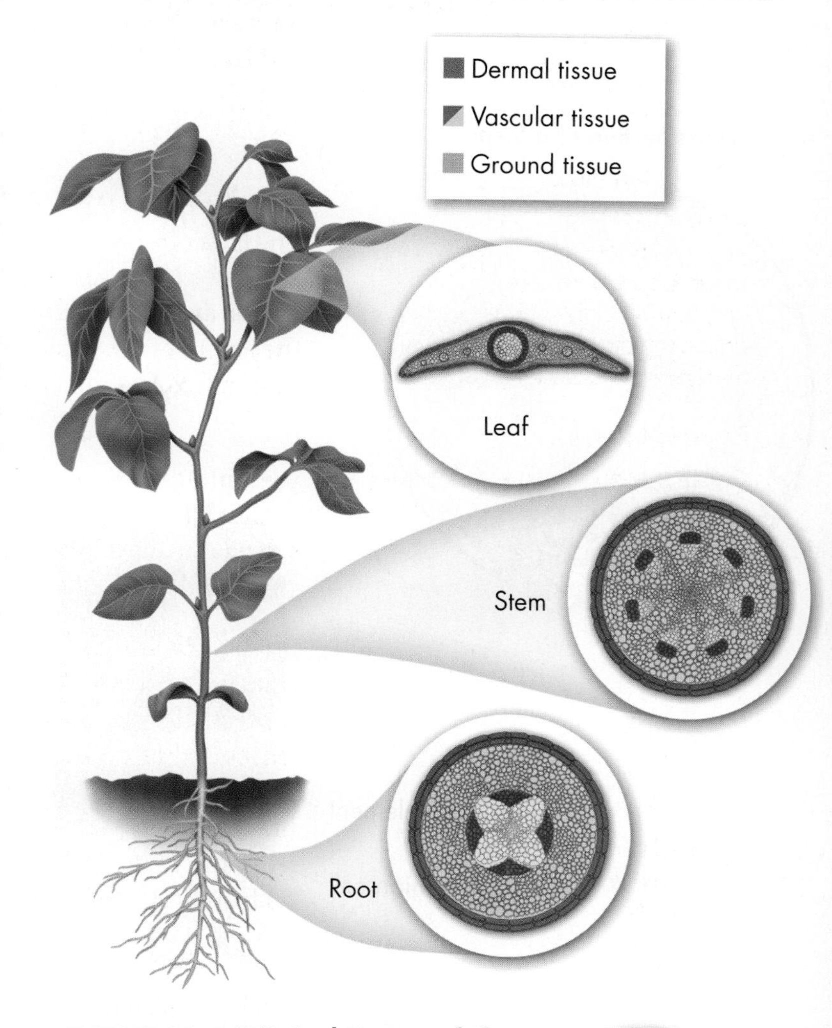

FIGURE 23–1 Principal Organs of Plants These cross sections of the principal organs of seed plants show that all three organs contain dermal tissue, vascular tissue, and ground tissue. **Interpret Visuals** ***Which tissue type is found in the center of a root?***

Vascular Tissue The two kinds of vascular tissue are xylem, a water-conducting tissue, and phloem, a tissue that carries dissolved food. As you can see in **Figure 23–2,** both xylem and phloem consist of long, slender cells that connect almost like sections of pipe. **Vascular tissue supports the plant body and transports water and nutrients throughout the plant.**

▶ ***Xylem: Tracheids*** All seed plants have xylem cells called tracheids. Recall from Chapter 22 that tracheids are long and narrow, with tough cell walls that help to support the plant. As they mature, tracheids die, leaving only their cell walls. These cell walls contain **lignin,** a complex molecule that resists water and gives wood much of its strength. Openings in the walls connect neighboring cells and allow water to flow from cell to cell. Thinner regions of the wall, known as pits, allow water to diffuse from tracheids into surrounding ground tissue. These adaptations allow tracheids to carry water throughout the plant and distribute it to tissues where it is needed.

▶ ***Xylem: Vessel Elements*** In addition to tracheids, angiosperms possess a second form of xylem tissue known as a **vessel element.** Vessel elements are wider than tracheids and are arranged end to end on top of one another like a stack of tin cans. After they mature and die, cell walls at both ends are left with slitlike openings through which water can move freely. In some vessel elements, the end walls disappear altogether, producing a continuous tube.

▶ ***Phloem: Sieve Tube Elements*** Unlike xylem cells, phloem cells are alive at maturity. The main phloem cells are **sieve tube elements,** which are arranged end to end, forming sieve tubes. The end walls of sieve tube elements have many small holes through which nutrients move from cell to cell in a watery stream. As sieve tube elements mature, they lose their nuclei and most other organelles. The remaining organelles hug the inside of the cell wall and are kept alive by companion cells.

▶ ***Phloem: Companion Cells*** The cells that surround sieve tube elements are called **companion cells.** Companion cells keep their nuclei and other organelles through their lifetime. Companion cells support the phloem cells and aid in the movement of substances in and out of the phloem.

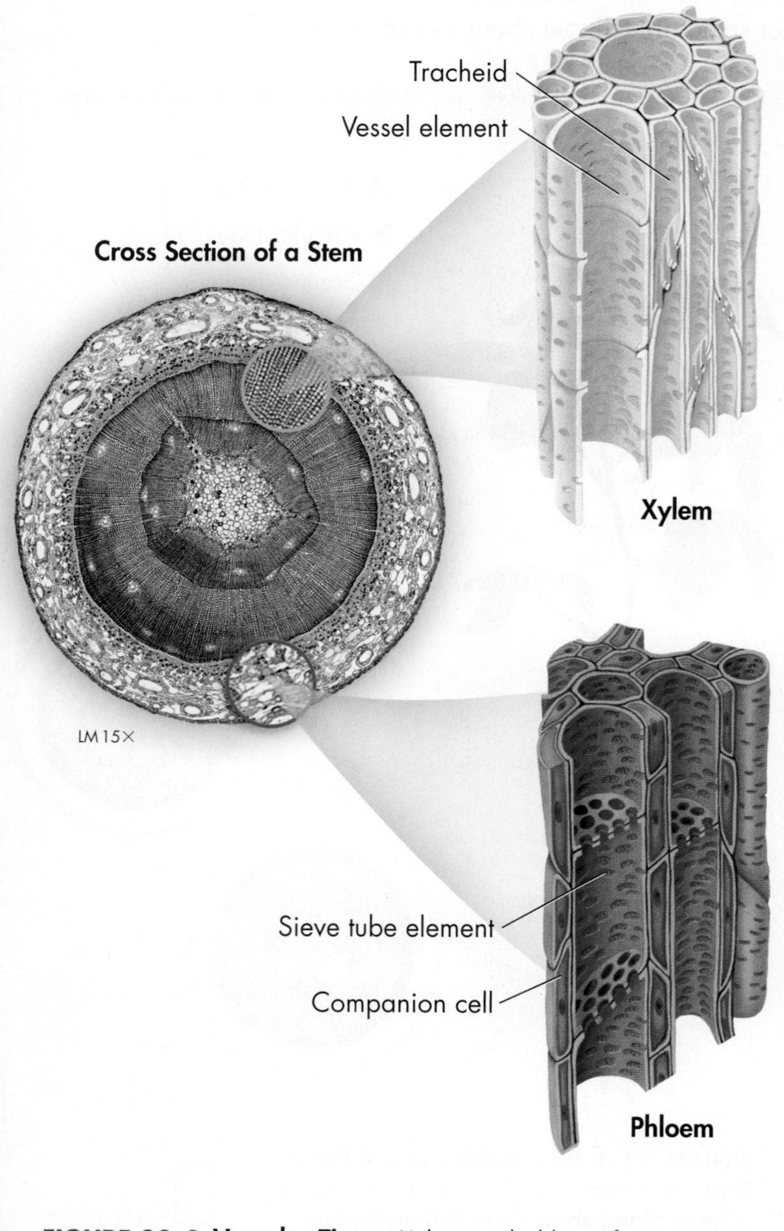

FIGURE 23–2 Vascular Tissue Xylem and phloem form the vascular transport system that moves water and nutrients throughout a plant. **Compare and Contrast** ***How are tracheids and sieve tube elements similar? How are they different?***

Ground Tissue Plant tissue called ground tissue is neither dermal nor vascular. **Ground tissue produces and stores sugars, and contributes to physical support of the plant.** Ground tissue is an important part of food at the dinner table, too. The edible portions of plants like potatoes, squash, and asparagus are mostly ground tissue. Most ground tissue consists of **parenchyma** (puh RENG kih muh). Parenchyma cells have a thin cell wall and a large central vacuole surrounded by a thin layer of cytoplasm. In leaves, these cells contain many chloroplasts and are the site of most of a plant's photosynthesis.

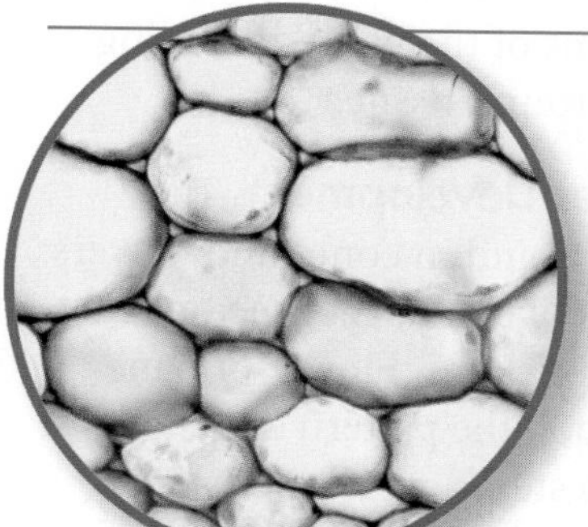

Parenchyma
Thin cell walls

Collenchyma
Thicker cell walls

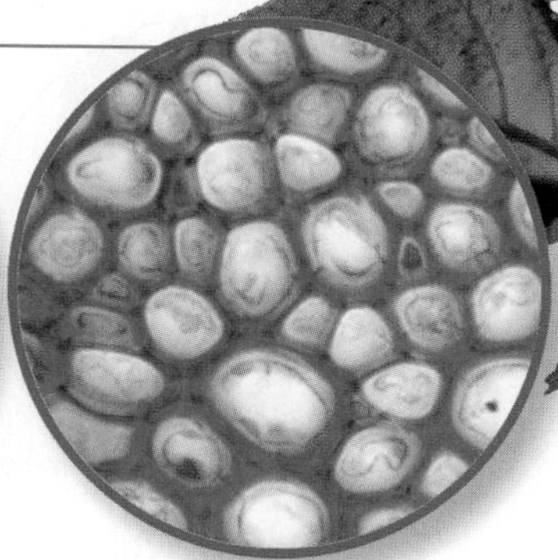

Sclerenchyma
Thickest cell walls

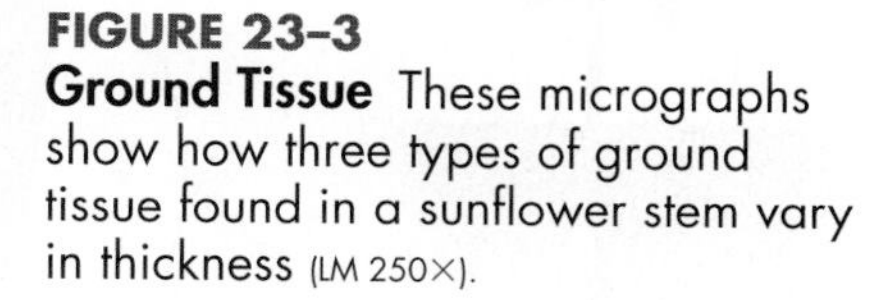

FIGURE 23–3
Ground Tissue These micrographs show how three types of ground tissue found in a sunflower stem vary in thickness (LM 250×).

Ground tissue may also contain two types of cells with thicker cell walls. **Collenchyma** (kuh LENG kih muh) cells have strong, flexible cell walls that help support plant organs. Chains of such cells make up the familiar "strings" of a stalk of celery. **Sclerenchyma** (sklih RENG kih muh) cells have extremely thick, rigid cell walls that make ground tissue such as seed coats tough and strong. Sclerenchyma fibers are used to make rope from hemp, and when you last used a nutcracker to open a walnut, you broke through some really tough sclerenchyma!

In Your Notebook *Make a three-column chart in which to summarize information about the three main tissue systems of plants.*

Plant Growth and Meristems

How do meristems differ from other plant tissues?

When most animals reach adulthood, they stop growing. Not so with most plants. Even the oldest trees produce new leaves and new reproductive organs every year, almost as if they remained "forever young." How do they do it? The secrets of plant growth are found in **meristems,** tissues that, in a sense, really do stay young. **Meristems are regions of unspecialized cells in which mitosis produces new cells that are ready for differentiation.** Meristems are found in places where plants grow rapidly, such as the tips of stems and roots. The undifferentiated cells they produce are very much like the stem cells of animals.

BUILD Vocabulary

RELATED WORD FORMS *Apex* and *apical* are related word forms. *Apex* is a noun meaning the narrowed or pointed end, or tip, and *apical* is an adjective describing something related to or located at the apex.

Apical Meristems Because the tip of a stem or root is known as its apex, meristems in these rapidly growing regions are called **apical meristems.** Unspecialized cells produced in apical meristems divide rapidly as stems and roots increase in length. **Figure 23–4** shows examples of stem and root apical meristems.

At first, the new cells that are pushed out of meristems look very much alike: They are unspecialized and have thin cell walls. Gradually, they develop into mature cells with specialized structures and functions. This process is called differentiation. As the cells differentiate, they produce each of the tissue systems of the plant, including dermal, vascular, and ground tissue.

Meristems and Flower Development The highly specialized cells found in cones and flowers (which are the reproductive organs of seed plants), are also produced in meristems. Flower or cone development begins when the pattern of gene expression changes in a stem's apical meristem. These changes transform the apical meristem of a flowering plant into a floral meristem. Floral meristems produce the tissues of flowers, which include the plant's reproductive organs as well as the colorful petals that surround them.

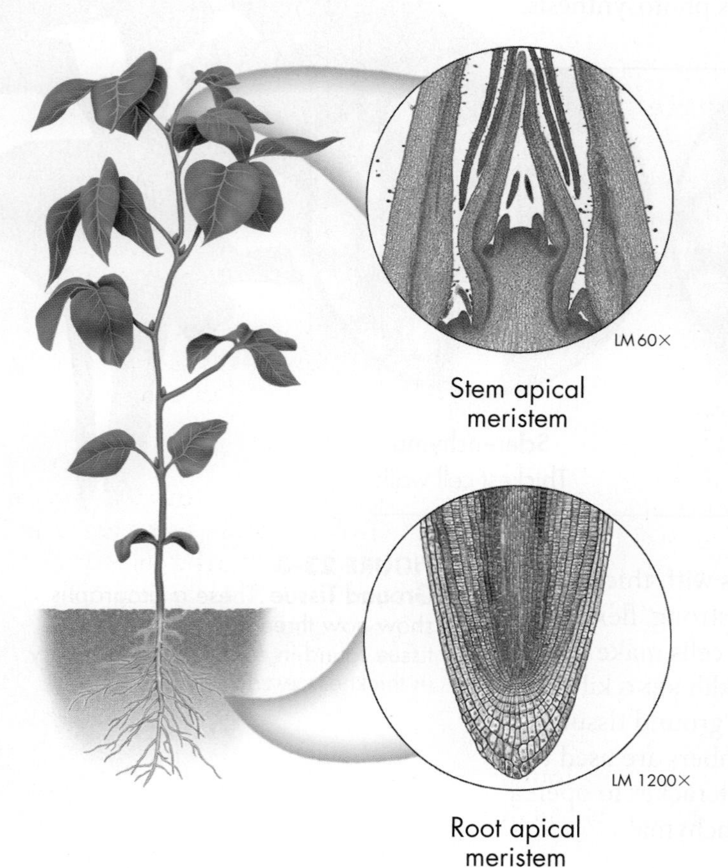

FIGURE 23–4 Apical Meristems Apical meristems are found in the growing tips of stems and roots. Within these meristems, unspecialized cells are produced by mitosis.

23.1 Assessment

Review Key Concepts

1. **a. Review** What are the three main organs of seed plants?
   b. Interpret Diagrams Review **Figure 23–1.** How are the three main organs of seed plants similar in structure?
2. **a. Review** What are the three main tissue systems of plants?
   b. Compare and Contrast How do the main functions of a plant's tissue systems differ?
3. **a. Review** What is the function of meristems?
   b. Form a Hypothesis How might the presence of meristems explain the ability of plants to regenerate from cuttings?

Apply the Big idea

Structure and Function

4. You probably have some knowledge of the human circulatory system. Based on this knowledge, write a paragraph comparing and contrasting the structure and function of the vascular system of a plant to the human circulatory system. *Hint:* Show how the systems are alike and different.

23.2 Roots

THINK ABOUT IT Can you guess how large a typical plant's root system is? Get ready for a surprise if you think that roots are small and insignificant. In a 1937 study of a single rye plant, botanist Howard Dittmer showed that the length of all the branches in the rye plant's root system was an astonishing 623 kilometers (387 miles). The surface area of these roots was more than 600 square meters—130 times greater than the combined areas of its stems and leaves!

Key Questions

What are the main tissues in a mature root?

What are the different functions of roots?

Vocabulary

root hair • cortex • endodermis • vascular cylinder • root cap • Casparian strip

Taking Notes

Outline Before you read, use the headings of the lesson to make an outline about plant roots. As you read, fill in phrases after each heading that provide key information.

Root Structure and Growth

What are the main tissues in a mature root?

As soon as a seed begins to sprout, it puts out its first root to draw water and nutrients from the soil. Other roots soon branch out from this first root, adding length and surface area to the root system. Rapid cell growth pushes the tips of the growing roots into the soil. The new roots provide raw materials for the developing stems and leaves before they emerge from the soil.

Types of Root Systems The two main types of root systems are taproot systems and fibrous root systems, shown in **Figure 23–5.** Taproot systems are found mainly in dicots. Fibrous root systems are found mainly in monocots. Recall from Chapter 22 that monocots and dicots are two categories of flowering plants.

► ***Taproot System*** In some plants, the primary root grows long and thick and gives rise to smaller branch roots. The large primary root is called a taproot. Taproots of oak and hickory trees grow so long that they can reach water several meters down. Carrots, dandelions, and beets have short, thick taproots that store sugars and starches.

► ***Fibrous Root System*** In other plants, such as grasses, the system begins with one primary root. But it is soon replaced by many equally sized branch roots that grow separately from the base of the stem. These fibrous roots branch to such an extent that no single root grows larger than the rest. The extensive fibrous root systems produced by many plants help prevent topsoil from being washed away by heavy rain.

FIGURE 23–5 A Comparison of Two Root Systems Dandelions have a taproot system (left), while grasses have a fibrous root system (right).

ZOOMING IN

ANATOMY OF A ROOT

FIGURE 23–6 A root consists of a central vascular cylinder surrounded by ground tissue and the epidermis.

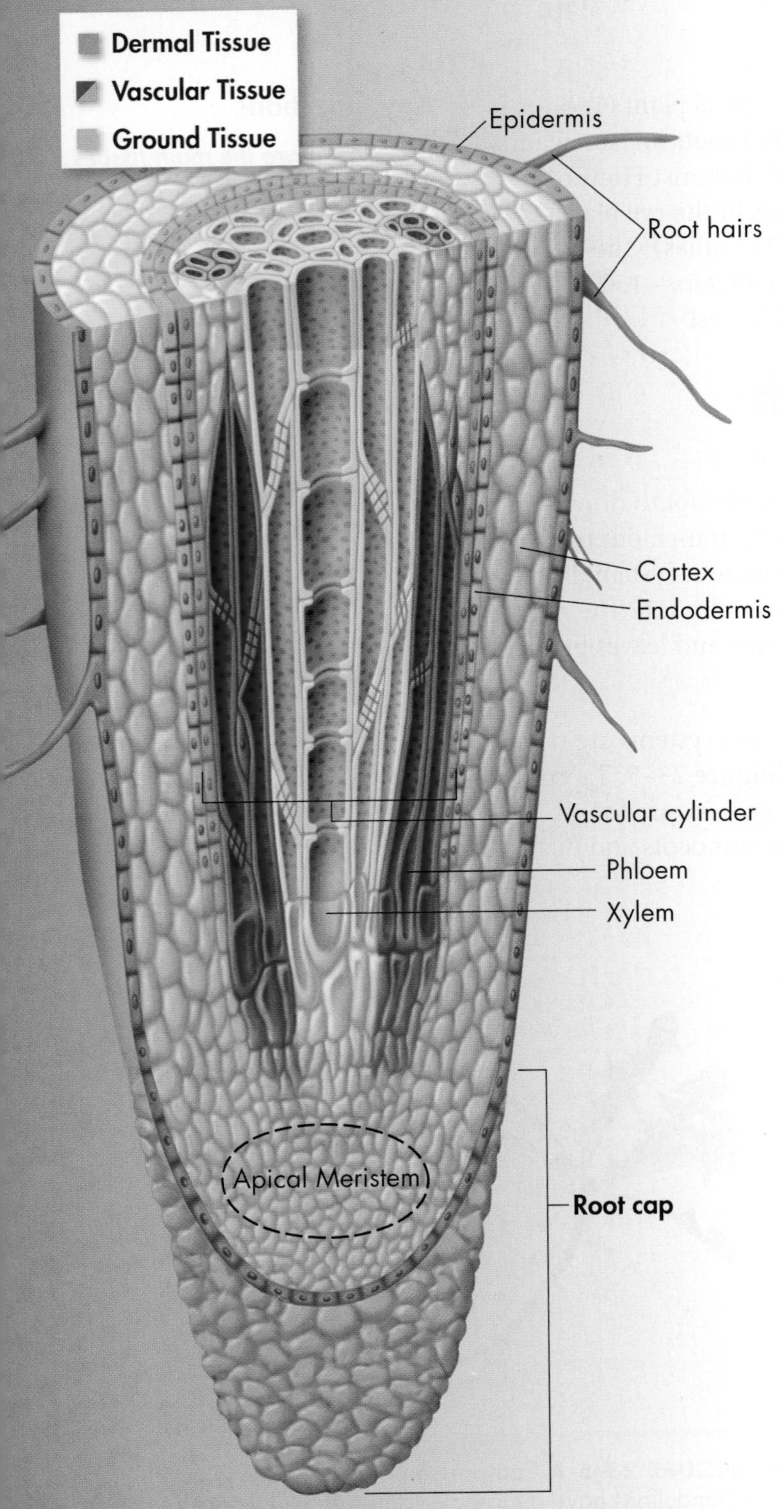

Anatomy of a Root Roots contain cells from the three tissue systems—dermal, vascular, and ground tissue, as shown in **Figure 23–6.** **A mature root has an outside layer, called the epidermis, and also contains vascular tissue and a large area of ground tissue.** The root system plays a key role in water and mineral transport. The cells and tissues of a root are specialized to carry out these functions.

▶ ***Dermal Tissue: Epidermis*** The root's epidermis performs the dual functions of protection and absorption. Its surface is covered with thin cellular projections called **root hairs.** These hairs penetrate the spaces between soil particles and produce a large surface area that allows water and minerals to enter.

▶ ***Ground Tissue*** Just inside the epidermis is a region of ground tissue called the **cortex.** Water and minerals move through the cortex from the epidermis toward the center of the root. The cortex also stores the products of photosynthesis, such as starch.

A layer of ground tissue known as the **endodermis** completely encloses the vascular cylinder. The endodermis, as you will see, plays an essential role in the movement of water and minerals into the center of the root.

▶ ***Vascular Tissue*** At the center of the root, the xylem and phloem together make up a region called the **vascular cylinder.** Dicot roots like the one shown at left have a central column of xylem cells.

▶ ***Apical Meristem*** Roots grow in length when apical meristems produce new cells near the root tips. The root tip is covered by a tough **root cap** that protects the fragile meristem as the root tip forces its way through the soil. As the root grows, the root cap secretes a slippery substance that eases the progress of the root through the soil. Cells at the very tip of the root cap are constantly being scraped away, and new root cap cells are continually added by the meristem.

In Your Notebook *Relate the role of the root cap to that of your outer skin layer, which loses dead cells at a rate of billions of cells per day.*

Root Functions

What are the different functions of roots?

How does a root go about the job of absorbing water and minerals from the soil? Although it might seem to, water does not just "soak" into the root from soil. It takes energy on the part of the plant to absorb water. **Roots support a plant, anchor it in the ground, store food, and absorb water and dissolved nutrients from the soil.**

Uptake of Plant Nutrients An understanding of soil helps explain how plant roots function. Soil is a complex mixture of sand, silt, clay, air, and bits of decaying animal and plant tissue. Soil in different places contains varying amounts of these ingredients. Sandy soil, for example, is made of large particles that retain few nutrients, whereas the finely textured silt and clay soils of the Midwest and southeastern United States are high in nutrients. The ingredients define the soil and determine, to a large extent, the kinds of plants that can grow in it.

To grow, flower, and produce seeds, plants require a variety of inorganic nutrients in addition to carbon dioxide and water. The nutrients needed in largest amounts are nitrogen, phosphorus, potassium, magnesium, sulfur, and calcium. The functions of these essential nutrients within a plant are described in **Figure 23–7.**

In addition to large amounts of these nutrients, small amounts of other nutrients, called trace elements, are just as important. These trace elements include iron, zinc, molybdenum, boron, copper, manganese, and chlorine. As important as they are, excessive amounts of any of these nutrients in soil can also be poisonous to plants.

Essential Plant Nutrients

Nutrient (Chemical Symbol)	Some Roles in Plant	Result of Deficiency
Nitrogen (N)	• Proper leaf growth and color • Synthesis of amino acids, proteins, nucleic acids, and chlorophyll	• Stunted plant growth • Pale yellow leaves ▶
Phosphorus (P)	• Synthesis of DNA • Development of roots, stems, flowers, and seeds	• Poor flowering • Stunted growth
Potassium (K)	• Synthesis of proteins and carbohydrates • Development of roots, stems, and flowers • Resistance to cold and disease	• Weak stems • Stunted roots • Edges of leaves turn brown ▶
Magnesium (Mg)	• Synthesis of chlorophyll	• Thin stems • Mottled, pale leaves
Calcium (Ca)	• Cell growth and division • Cell wall structure • Cellular transport • Enzyme action	• Stunted growth • Curled leaves ▶

FIGURE 23–7 Important Plant Nutrients
Soil contains several nutrients that are essential for plant growth. **Interpret Tables** ***If you notice that a plant is becoming paler and more yellow, what nutrient might it be lacking?***

Active Transport of Dissolved Nutrients The cell membranes of root hairs and other cells in the root epidermis contain active transport proteins. As you know, active transport is a process that uses the energy of ATP to move ions and other materials across membranes. Active transport brings the mineral ions of dissolved nutrients from the soil into the plant. The high concentration of mineral ions in the plant cells causes water molecules to move into the plant by osmosis.

Water Movement by Osmosis You may recall that osmosis is the movement of water across a membrane toward an area where the concentration of dissolved material is higher. By using active transport to accumulate mineral ions from the soil, cells of the root epidermis create conditions under which osmosis causes water to "follow" those ions and flow into the root. Note that the root does not actually pump water. But by pumping mineral ions into its own cells, the end result is almost the same—the water moves from the epidermis through the cortex into the vascular cylinder, as shown in **Figure 23–8.**

BUILD Vocabulary

ACADEMIC WORDS The term **accumulate** means to increase gradually in quantity or number.

Movement Into the Vascular Cylinder Next, the water and dissolved minerals pass the inner boundary of the cortex and move toward the vascular cylinder. The cylinder itself is enclosed by a layer of cortex cells known as the endodermis. The cells of the endodermis are each shaped a bit like a brick. Where these cells meet, their cell walls form a special waterproof zone called a **Casparian strip.** Most of the time, water can diffuse through cell walls, but not here. The strip is almost like a layer of waterproof cement between the bricks in a wall. Imagine many of these bricks placed edge to edge to build a cylinder, with this waterproof cement surrounding each of the bricks. The only way that water and dissolved nutrients could enter that cylinder would be through the bricks themselves.

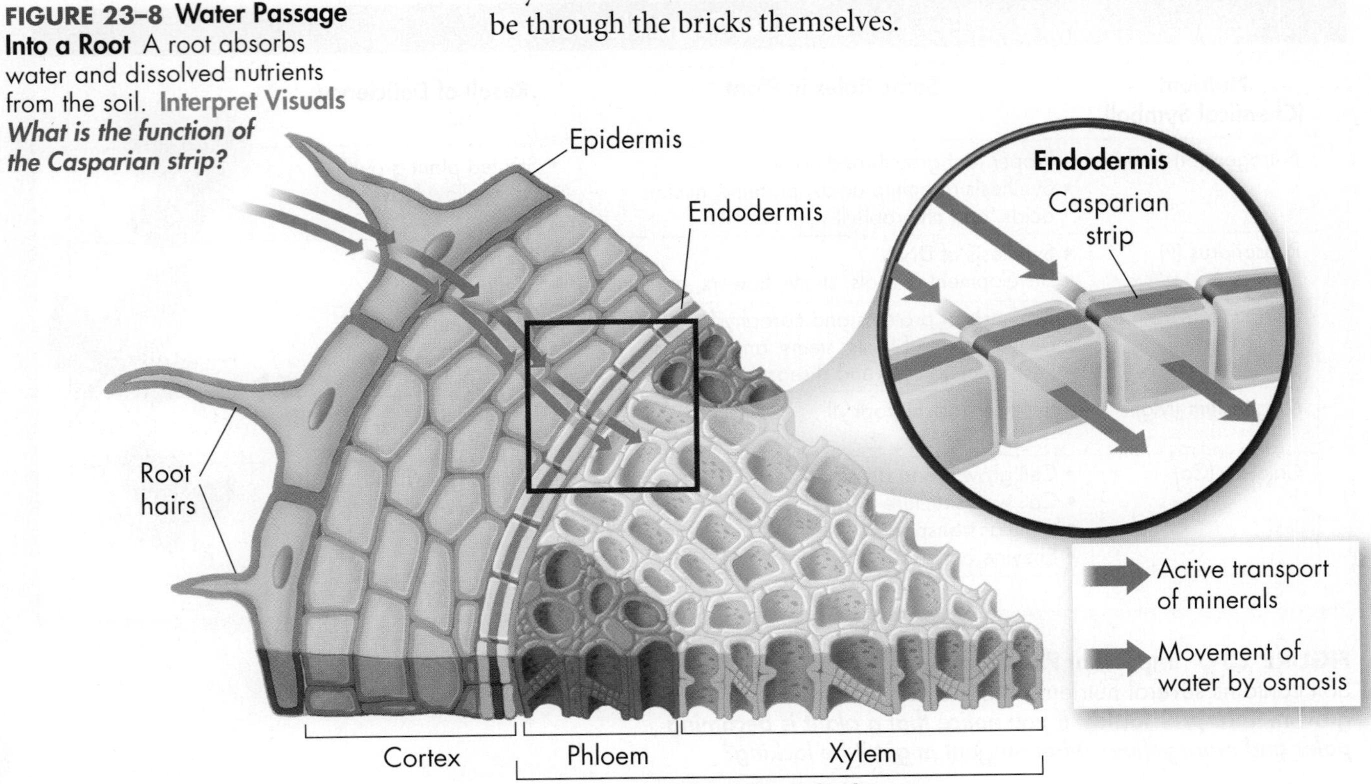

FIGURE 23–8 Water Passage Into a Root A root absorbs water and dissolved nutrients from the soil. **Interpret Visuals** ***What is the function of the Casparian strip?***

The waxy Casparian strip forces water and minerals to move through the cell membranes of endodermis cells rather than in between the cells. This enables the endodermis to filter and control the water and dissolved nutrients that enter the vascular cylinder. More importantly, the Casparian strip ensures that valuable nutrients will not leak back out. As a result, there is a one-way passage of water and nutrients into the vascular cylinder.

Root Pressure Why do plants "need" a system that ensures the one-way movement of water and minerals? That system is how the plant generates enough pressure to move water out of the soil and up into the body of the plant. As minerals are pumped into the vascular cylinder, more and more water follows by osmosis, producing a strong pressure. If the pressure were not contained, roots would expand as they filled with water.

Instead, contained within the Casparian strip, the water has just one place to go—up. Root pressure, produced within the cylinder by active transport, forces water through the vascular cylinder and into the xylem. As more water moves from the cortex into the vascular cylinder, more water in the xylem is forced upward through the root into the stem. In **Figure 23–9,** you can see a demonstration of root pressure in a carrot root.

Root pressure is the starting point for the movement of water through the vascular system of the entire plant. But it is just the beginning. Once you have learned about stems and leaves, you will see how water and other materials are transported within an entire plant.

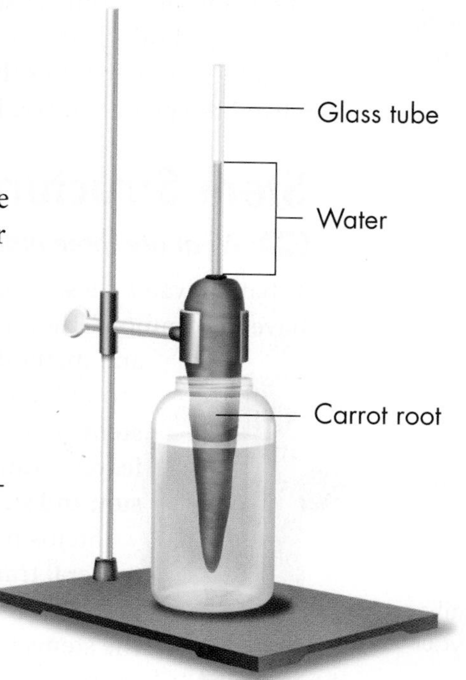

FIGURE 23–9 Root Pressure Demonstration In this setup, a glass tube takes the place of the carrot plant's stem and leaves. As the root absorbs water, root pressure forces water upward into the tube.

23.2 Assessment

Review Key Concepts

1. a. Review How are tissues distributed in a plant root?

b. Compare and Contrast How is the structure of cells in a root's transport system different from the structure of cells making up the epidermis?

2. a. Review Describe the main functions of roots.

b. Explain How is osmosis involved in the absorption of water and nutrients?

c. Apply Concepts Why is it important that the root endodermis permits only a one-way passage of materials?

VISUAL THINKING

3. Draw a diagram to show how roots absorb water and nutrients. Label the diagram and write brief descriptions of the processes shown.

23.3 Stems

Key Questions

What are three main functions of stems?

How do primary growth and secondary growth occur in stems?

Vocabulary

node • bud • vascular bundle • pith • primary growth • secondary growth • vascular cambium • cork cambium • heartwood • sapwood • bark

Taking Notes

Preview Visuals Before you read, preview the art in **Figure 23–14.** Define any familiar terms in your own words, and list any unfamiliar ones. Revise and add to your definitions as you read.

THINK ABOUT IT While visiting the salad bar for lunch, you notice an intriguing range of offerings. After making your basic salad, you decide to add some sliced water chestnuts and bamboo shoots on top. Then you serve yourself some asparagus and potato salad on the side. These good things are all from plants, of course, but can you think of something else that ties them together? They all come from the same part of the plant. Do you have any idea which part?

Stem Structure and Function

What are three main functions of stems?

What do water chestnuts, bamboo shoots, asparagus, and potatoes all have in common? They are all types of stems. Stems vary in size, shape, and method of development. Some grow entirely underground; others reach high into the air. **Aboveground stems have several important functions: Stems produce leaves, branches, and flowers; stems hold leaves up to the sun; and stems transport substances throughout the plant.**

Stems make up an essential part of the water and mineral transport systems of the plant. Xylem and phloem form continuous tubes from the roots through the stems to the leaves. These vascular tissues link all parts of the plant, allowing water, nutrients, and other compounds to be carried throughout the plant. In many plants, stems also function in storage and aid in the process of photosynthesis.

FIGURE 23–10 Cactus Stems
Desert cacti have thick green stems that carry out photosynthesis and are adapted to store water.

Anatomy of a Stem Stems contain the plant's three tissue systems: dermal, vascular, and ground tissue. Stems are surrounded by a layer of epidermal cells that have thick cell walls and a waxy protective coating. Growing stems contain distinct **nodes,** where leaves are attached, as shown in **Figure 23–11.** Small buds are found where leaves attach to the nodes. **Buds** contain apical meristems that can produce new stems and leaves. In larger plants, stems develop woody tissue that helps support leaves and flowers.

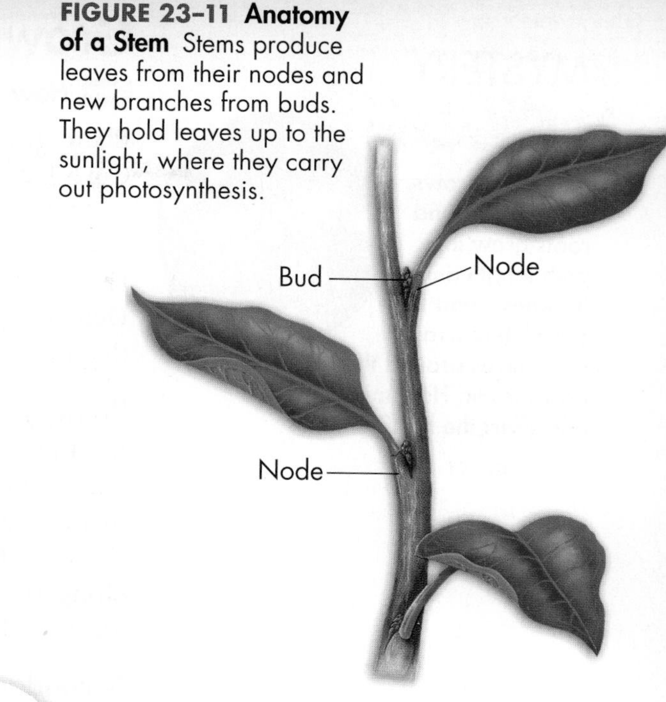

FIGURE 23–11 Anatomy of a Stem Stems produce leaves from their nodes and new branches from buds. They hold leaves up to the sunlight, where they carry out photosynthesis.

Vascular Bundle Patterns The arrangement of tissues in a stem differs among seed plants. In monocots, clusters of xylem and phloem tissue, called **vascular bundles,** are scattered throughout the stem. In most dicots and gymnosperms, vascular bundles are arranged in a cylinder, or ring. For a comparison of monocot and dicot stems, look at **Figure 23–12.**

FIGURE 23–12 Comparing Monocots and Dicots These cross sections through a monocot and dicot stem show their similarities and differences. **Observe** ***How does the arrangement of the vascular bundles differ?***

Cross Section

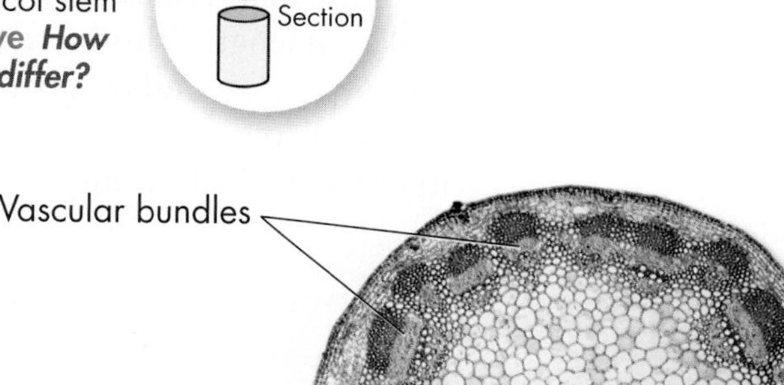
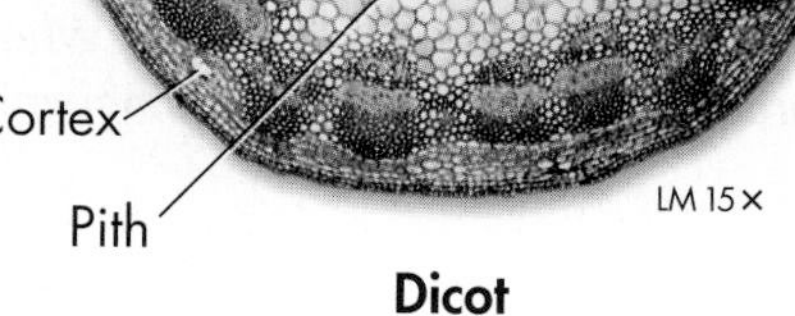

▶ ***Monocot Stems*** The cross section of a young monocot stem shows all three tissue systems clearly. The stem has a distinct epidermis, which encloses ground tissue and a series of vascular bundles. In monocots, vascular bundles are scattered throughout the ground tissue. The ground tissue is fairly uniform, consisting mainly of parenchyma cells.

▶ ***Dicot Stems*** Young dicot stems have vascular bundles, too, but they are generally arranged in an organized, ringlike pattern. The parenchyma cells inside the ring of vascular tissue are known as **pith,** while those outside form the cortex of the stem. These relatively simple tissue patterns become more complex as the plant grows larger and the stem increases in diameter.

In Your Notebook *Create a Venn diagram in which to record similarities and differences in the stem structure of monocots and dicots.*

As the fig grows, its aboveground roots grow in both length and thickness, and they completely wrap themselves around the host's trunk. How might this affect the host?

Growth of Stems

How do primary growth and secondary growth occur in stems?

Plants grow in ways that are very different from how animals grow. Cows have four legs, ants have six, and spiders have eight, but roses and tomatoes don't have a set number of leaves or branches. Unlike animals, the growth of most plants isn't precisely determined. However, plant growth is still carefully controlled and regulated. Depending upon the species, plant growth follows general patterns that produce the characteristic size and shape of the adult plant.

Primary Growth The growth of new cells produced by the apical meristems of roots and stems adds length to the plant. This pattern of growth, occurring at the ends of a plant, is called **primary growth.** The increase in length in a plant due to primary growth from year to year is shown in **Figure 23–13.** **Primary growth of stems is the result of elongation of cells produced in the apical meristem. It takes place in all seed plants.**

Secondary Growth As a plant grows larger, the older stems and roots have more mass to support and more fluid to move through their vascular tissues. As a result, they must increase in thickness as well as in length. This increase in the thickness of stems and roots is known as **secondary growth.** Secondary growth is very common among dicots and nonflowering seed plants such as pines, but it is rare in monocots. This limits the girth of most monocots.

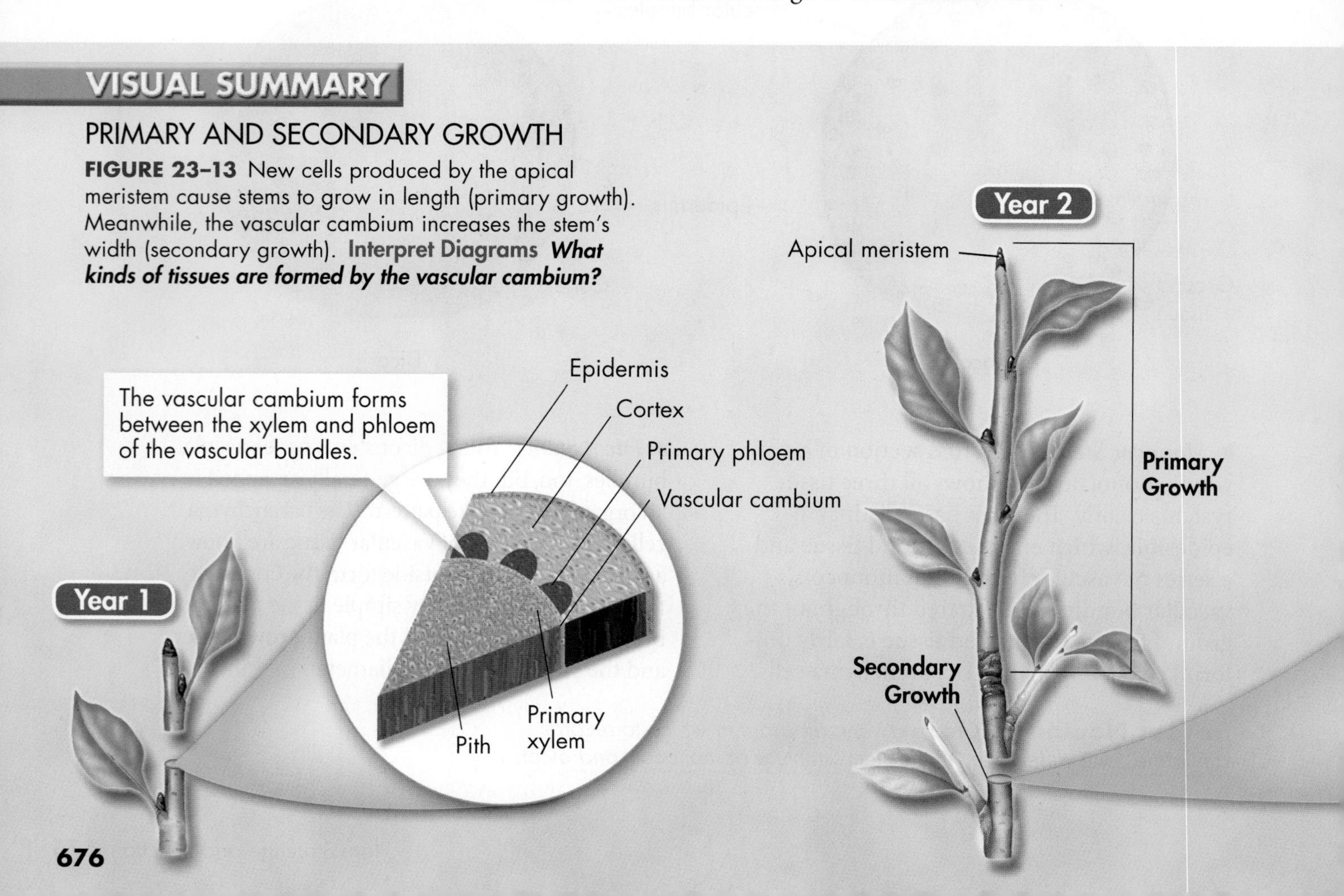

FIGURE 23–13 New cells produced by the apical meristem cause stems to grow in length (primary growth). Meanwhile, the vascular cambium increases the stem's width (secondary growth). **Interpret Diagrams** ***What kinds of tissues are formed by the vascular cambium?***

Unlike monocots, most dicots have meristems within their stems and roots that can produce true secondary growth. This enables many dicots to grow to great heights because the increase in width supports the extra weight. In addition to showing primary growth, **Figure 23–13** illustrates the pattern of secondary growth in a dicot stem.

In conifers and dicots, secondary growth takes place in meristems called the vascular cambium and cork cambium. The **vascular cambium** produces vascular tissues and increases the thickness of stems over time. The **cork cambium** produces the outer covering of stems. Similar types of cambium tissue enable roots to grow. The addition of new tissue in these cambium layers increases the thickness of stems and roots.

Growth From the Vascular Cambium In a young dicot stem, bundles of xylem and phloem are arranged in a ring. Once secondary growth begins, the vascular cambium appears as a thin, cylindrical layer of cells between clusters of vascular tissue. This new meristem forms between the xylem and phloem of each vascular bundle. Divisions in the vascular cambium give rise to new layers of xylem and phloem. As a result, the stem becomes wider. Each year, the cambium continues to produce new layers of vascular tissue, causing the stem to become thicker and thicker.

In Your Notebook *List in sequence all the tissues found in a mature woody stem. Start from the center and move outward.*

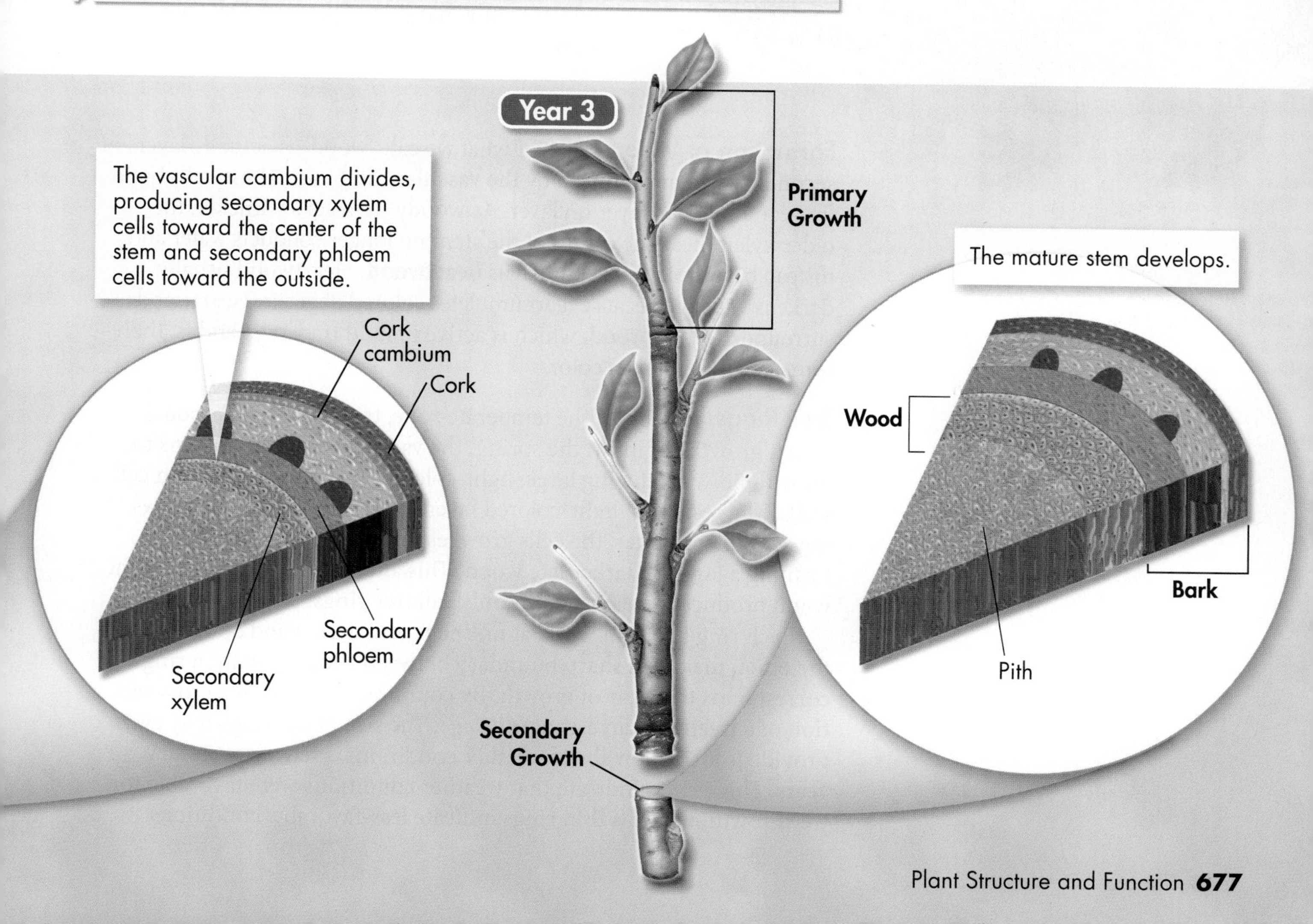

Reading a Tree's History

The analysis of tree rings can help determine information about a tree and the environment in which it grew. A tree's age can be measured by counting its growth rings—each ring is produced by a year of growth. The specific environmental conditions for each year of growth can be inferred by examining the relative width and color of each ring. Use the photograph at left to answer the questions.

1. Calculate Approximately how old was this tree when it was cut down? MATH

2. Infer Areas A and B were both produced by four years of growth, yet they are different widths. What climatic conditions might account for this difference?

3. Interpret Visuals The area at C is blackened from a fire that apparently affected only one side of the tree. Describe how the tree grew after this fire.

Formation of Wood Most of what we call "wood" is actually layers of secondary xylem produced by the vascular cambium. These cells build up year after year, layer on layer. As woody stems grow thicker, the older xylem near the center of the stem no longer conducts water and instead becomes what is known as **heartwood.** Heartwood usually darkens with age because it accumulates colored deposits. Heartwood is surrounded by **sapwood,** which is active in fluid transport and is, therefore, usually lighter in color.

Tree Rings In most of the temperate zone, tree growth is seasonal. When growth begins in the spring, the vascular cambium begins to grow rapidly, producing large, light-colored xylem cells with thin cell walls. The result is a light-colored layer of early wood. As the growing season continues, the cells grow less and have thicker cell walls, forming a layer of darker late wood. This alternation of dark and light wood produces what we commonly call tree rings.

Each ring has light wood at one edge and dark wood at the other, making a sharp boundary between rings. Usually, a ring corresponds to a year of growth. By counting the rings in a cross section of a tree, you can estimate its age. The size of the rings may even provide information about weather conditions, such as wet or dry years. Thick rings indicate that weather conditions were favorable for tree growth, whereas thin rings indicate less-favorable conditions.

FIGURE 23–14 Formation of Wood and Bark This diagram shows the layers of wood and bark in a mature tree that has undergone several years of secondary growth. **Classify** ***Which two tissues are meristems?***

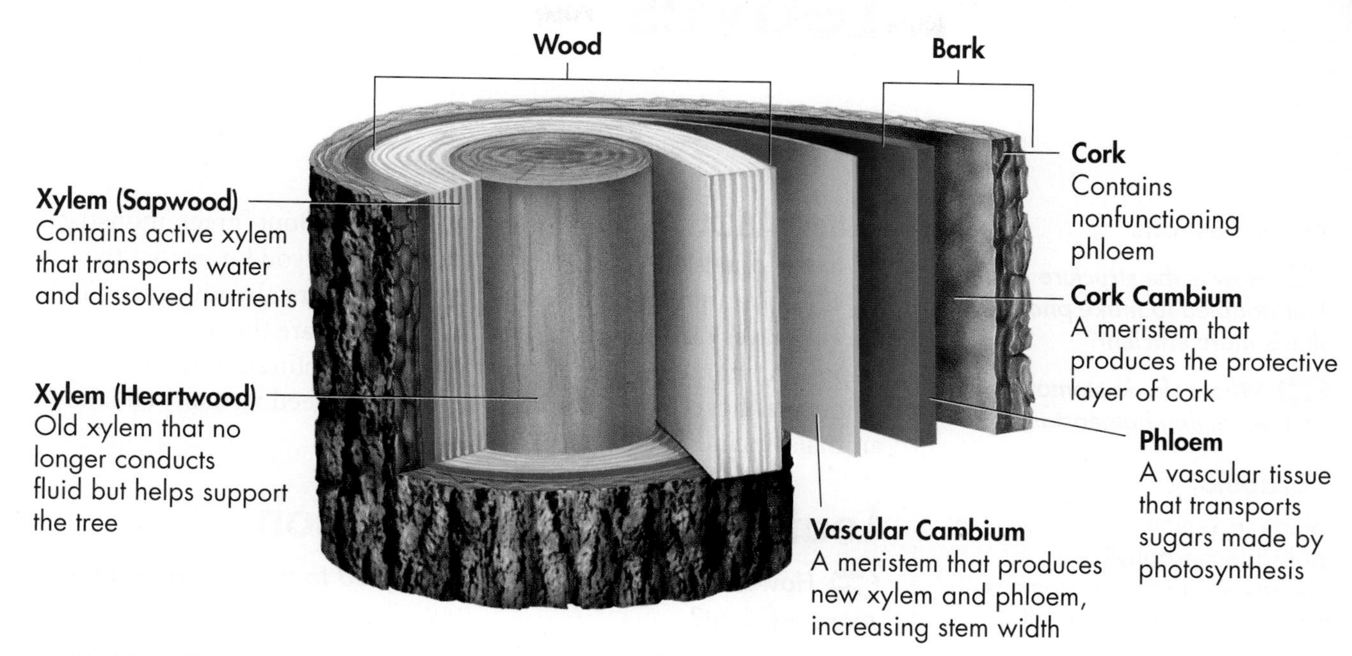

Formation of Bark In a mature stem, all of the tissues found outside the vascular cambium make up the **bark,** as shown in **Figure 23–14.** These tissues include phloem, the cork cambium, and cork. As a tree expands in width, the phloem layer must grow as well. This expansion may cause the oldest tissues to split and fragment as the expanding stem stretches them. The cork cambium surrounds the cortex and produces a thick, protective layer of waterproof cork that prevents the loss of water from the stem. As the stem increases in size, outer layers of dead bark often crack and flake off the tree.

23.3 Assessment

Review Key Concepts

1. a. **Review** What are three important functions of stems?
   b. **Explain** How does the arrangement of vascular bundles in monocot stems differ from that of dicot stems?
   c. **Apply Concepts** How do the functions of a stem relate to the functions of the roots and leaves of a plant?
2. a. **Review** Define primary and secondary growth.
   b. **Explain** Which meristem is involved in primary growth? Which are involved in secondary growth? Explain their roles.
   c. **Predict** Describe what would happen over time to a tree sapling that could grow only taller, not wider.

WRITE ABOUT SCIENCE

Creative Writing

3. Pretend that you are small enough to enter a dicot plant through its root system. Describe what you would see as you traveled into a plant and through one of its stems. Include illustrations to enhance your description. *Hint:* Review the illustrations in this chapter for ideas.

23.4 Leaves

Key Questions

How is the structure of a leaf adapted to make photosynthesis more efficient?

What role do stomata play in maintaining homeostasis?

Vocabulary

blade • petiole • mesophyll • palisade mesophyll • spongy mesophyll • stoma • transpiration • guard cell

Taking Notes

Preview Visuals Before you read the lesson, look at **Figure 23–15.** Locate the three main tissue systems and infer which tissue system makes up the leaf veins.

THINK ABOUT IT We hear a lot these days about "green industry," such as biofuels and material recycling, but did you know that the most important manufacturing sites on Earth are already green? They are the leaves of plants. In a sense, plant leaves are the world's most important manufacturers. Using the energy captured in their leaves, plants make the sugars, starches, and oils that feed virtually all land animals, including us.

Leaf Structure and Function

How is the structure of a leaf adapted to make photosynthesis more efficient?

Recall from Chapter 8 that photosynthesis uses carbon dioxide and water to produce sugars and oxygen. Leaves, therefore, must have a way of obtaining carbon dioxide and water as well as distributing end products. **The structure of a leaf is optimized to absorb light and carry out photosynthesis.**

Anatomy of a Leaf To collect sunlight, most leaves have a thin, flattened part called a **blade.** The flat shape of a leaf blade maximizes the amount of light it can absorb. The blade is attached to the stem by a thin stalk called a **petiole** (PET ee ohl). Like roots and stems, leaves have an outer covering of dermal tissue and inner regions of ground and vascular tissues, as shown in **Figure 23–15.**

▶ ***Dermal Tissue*** Leaves are covered on their top and bottom surfaces by epidermis. Leaf epidermis is made of a layer of tough, irregularly shaped cells with thick outer walls that resist tearing. The epidermis of nearly all leaves is also covered by a waxy cuticle. The cuticle is a waterproof barrier that protects tissues and limits the loss of water through evaporation.

▶ ***Vascular Tissue*** The vascular tissues of leaves are connected directly to the vascular tissues of stems, making them part of the plant's fluid transport system. Xylem and phloem tissues are bundled in leaf veins that run from the stem throughout the leaf.

▶ ***Ground Tissue*** The area between leaf veins is filled with a specialized ground tissue known as **mesophyll** (MES uh fil), where photosynthesis occurs. The sugars produced in mesophyll move to leaf veins, where they enter phloem sieve tubes for transport to the rest of the plant.

MYSTERY CLUE

The mature fig's stems and leaves block sunlight from the host. How might this affect photosynthesis in the host?

ZOOMING IN

ANATOMY OF A LEAF

FIGURE 23–15 Leaves absorb light and carry out most of the photosynthesis in a plant. **Compare and Contrast** ***Compare the structure of the two types of mesophyll cells in a leaf.***

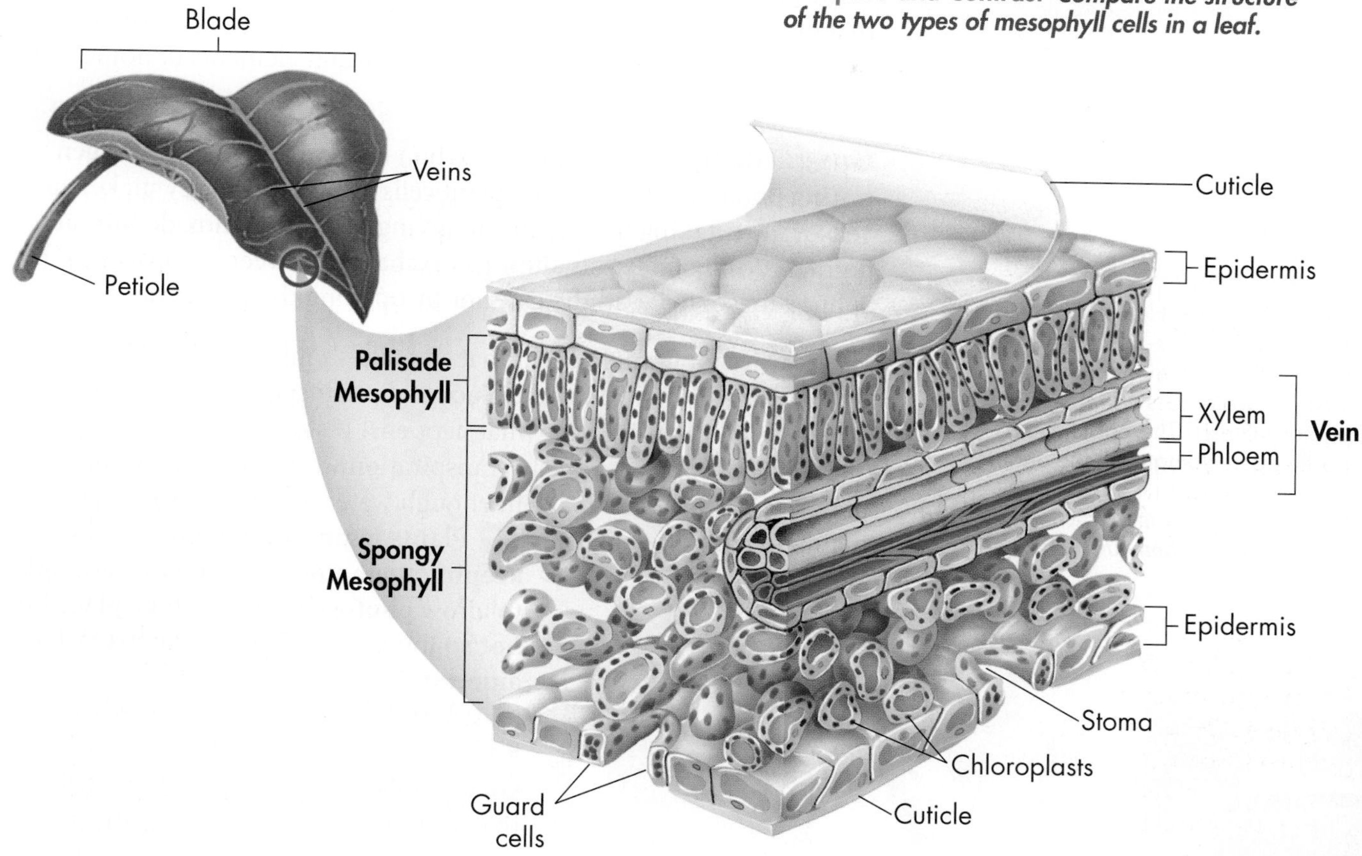

Photosynthesis The mesophyll tissue in most leaves is highly specialized for photosynthesis. Beneath the upper epidermis is a layer of cells called the **palisade mesophyll,** containing closely packed cells that absorb light that enters the leaf. Beneath the palisade layer is a loose tissue called the **spongy mesophyll,** which has many air spaces between its cells. These air spaces connect with the exterior through **stomata** (singular: stoma). Stomata are small openings in the epidermis that allow carbon dioxide, water, and oxygen to diffuse into and out of the leaf.

Transpiration The walls of mesophyll cells are kept moist so that gases can enter and leave the cells easily. The trade-off to this feature is that water evaporates from these surfaces and is lost to the atmosphere. **Transpiration** is the loss of water through leaves. This lost water may be replaced by water drawn into the leaf through xylem vessels in the vascular tissue. Transpiration helps to cool leaves on hot days, but it may also threaten the leaf's survival if water is scarce.

In Your Notebook *Make a two-column table in which you list structures found in a leaf cross section and describe their functions.*

BUILD Vocabulary

WORD ORIGINS **Mesophyll** comes from two Greek words: *meso,* meaning "middle," and *phyllon,* meaning "leaf." **Stomata** comes from the Greek word meaning "mouths."

Gas Exchange and Homeostasis

What role do stomata play in maintaining homeostasis?

You might not think of plants as "breathing" the same way that animals do, but plants need to exchange gases with the atmosphere, too. Plants, in fact, can even be suffocated by lack of oxygen, something that often happens during extensive flooding. A plant's control of gas exchange is actually one of the most important elements of homeostasis for these remarkable organisms.

Gas Exchange Leaves take in carbon dioxide and give off oxygen during photosynthesis. When plant cells use the food they make, the cells respire, taking in oxygen and giving off carbon dioxide (just as animals do). Plant leaves allow gas exchange between air spaces in the spongy mesophyll and the exterior by opening their stomata.

Homeostasis It might seem that stomata should be open all the time, allowing gas exchange to take place and photosynthesis to occur at top speed. However, this is not what happens! If stomata were kept open all the time, water loss due to transpiration would be so great that few plants would be able to take in enough water to survive. So, plants maintain a kind of balance. **Plants maintain homeostasis by keeping their stomata open just enough to allow photosynthesis to take place but not so much that they lose an excessive amount of water.**

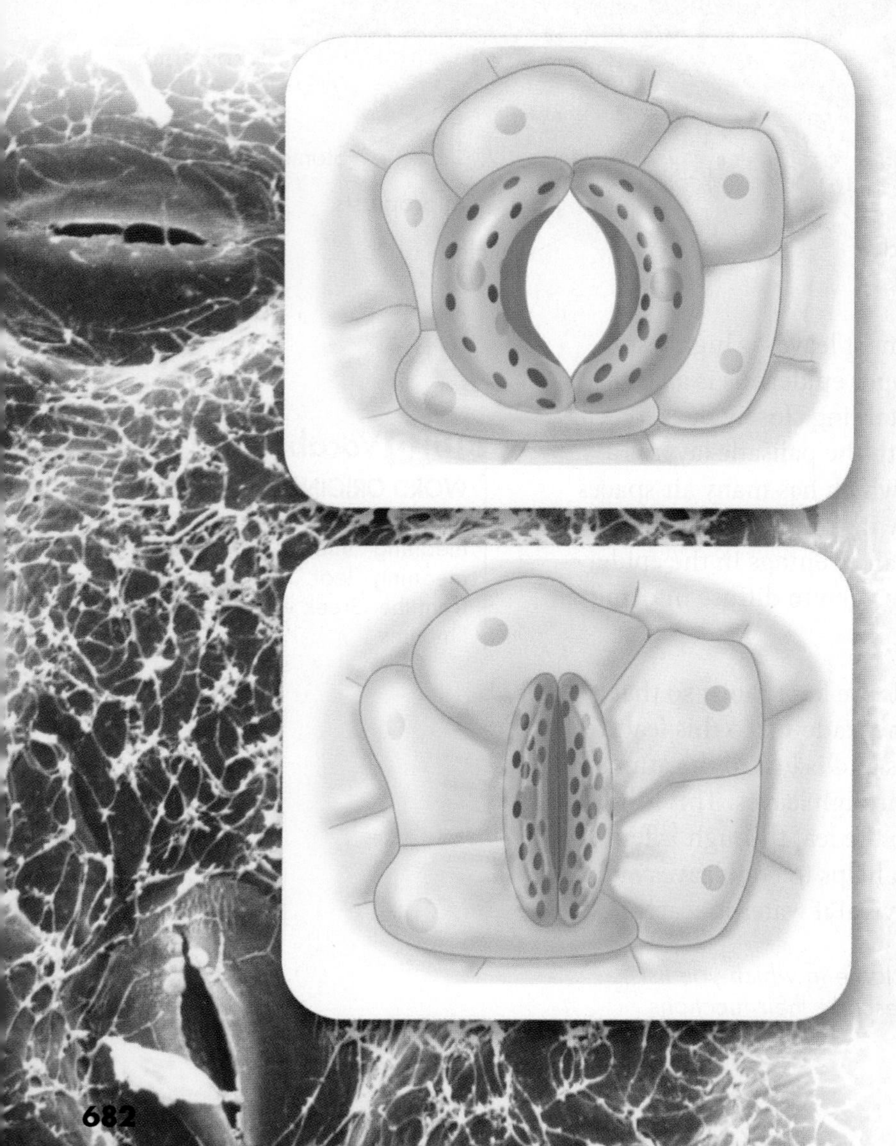

FIGURE 23–16 How Guard Cells Function Plants regulate the opening and closing of their stomata to balance water loss with rates of photosynthesis. The photo shows two partly open stomata on the underside of a camellia leaf (SEM 1500×).
Observe ***How is the structure of guard cells related to their function?***

Guard cells in the epidermis of each leaf are the key to this balancing act. **Guard cells** are highly specialized cells that surround the stomata and control their opening and closing. Guard cells regulate the movement of gases, especially water vapor and carbon dioxide, into and out of leaf tissues.

The stomata open and close in response to changes in water pressure within the guard cells, as shown in **Figure 23–16.** When water is abundant, it flows into the leaf, raising water pressure in the guard cells, which then open the stomata. The thin outer walls of the cells are forced into a curved shape, which pulls the thick inner walls of the guard cells away from one another, opening the stoma. Carbon dioxide can then enter through the stoma, and water is lost by transpiration.

When water is scarce, the opposite occurs. Water pressure within the guard cells decreases, the inner walls pull together, and the stoma closes. This reduces further water loss by limiting transpiration.

Quick Lab
GUIDED INQUIRY

Examining Stomata

1. Obtain different kinds of leaves from your teacher.
2. Spread a thick coating of clear nail polish on the underside of each leaf.
3. Wait about 10 minutes for the polish to dry completely.
4. Attach a strip of clear tape to the polish and gently peel off the tape, lifting the dried polish.
5. Tape the polish to a clean microscope slide and examine under a 400× lens.
6. For each leaf, move the microscope stage so you can count stomata from three distinct fields of view.

Analyze and Conclude

1. **Calculate** What is the average number of stomata per square cm for each leaf? MATH
2. **Graph** Make a graph that compares these averages.
3. **Form a Hypothesis** What could account for differences in stoma density among plants? Write a hypothesis.

In general, stomata are open during the daytime, when photosynthesis is active, and closed at night, when open stomata would only lead to water loss. However, stomata may be closed even in bright sunlight under hot, dry conditions in which water conservation is a matter of life and death. Guard cells respond to conditions in the environment, such as wind and temperature, helping to maintain homeostasis within a leaf.

Transpiration and Wilting Osmotic pressure keeps a plant's leaves and stems rigid, or stiff. High transpiration rates can lead to wilting. Wilting results from the loss of water—and therefore pressure—in a plant's cells. Without this internal pressure to support them, the plant's cell walls bend inward, and the plant's leaves and stems wilt. When a leaf wilts, its stomata close. As a result, transpiration slows down significantly. Thus, wilting helps a plant to conserve water.

FIGURE 23–17 Wilting A plant may wilt when water is scarce.

In Your Notebook *Make a list of molecules that are exchanged through the stomata. Which ones primarily enter the leaf? Which ones primarily exit the leaf?*

Adaptations of Leaves

FIGURE 23–18 The plants shown here grow in different biomes. The leaves of these plants have adaptations to the dry or low-nutrient conditions in which they live.

◀ **Pitcher Plant** The leaf of a pitcher plant is modified to attract and then digest insects and other small prey. Such plants typically live in nutrient-poor soils and rely on animal prey as their source of nitrogen.

▼ **Living Stone** The two leaves of a living stone are adapted for hot, dry conditions. They are rounded, which minimizes the exposure of their surface to the air. They also have very few stomata.

Spruce The narrow leaves of a spruce tree contain a waxy epidermis as well as stomata that are sunken below the surface of the leaf. These adaptations reduce water loss from the leaves. ▶

Cactus Cactus leaves are actually nonphotosynthetic thorns that protect against herbivores. Most of the plant's photosynthesis is carried out in its stems. ▼

23.4 Assessment

Review Key Concepts

1. a. **Review** Describe how the structure of a leaf is adapted to make photosynthesis more efficient.
   b. **Explain** What is the role of the palisade mesophyll in a leaf?
   c. **Form a Hypothesis** The leaves of desert plants often have two or more layers of palisade mesophyll, rather than the single layer that is characteristic of most leaves. How might this modified structure be advantageous to a desert plant?
2. a. **Review** How do stomata help plants maintain homeostasis?
   b. **Predict** Are stomata more likely to be open or closed on a hot day? Explain your answer.

BUILD VOCABULARY

3. The terms *spongy* and *palisade* are adjectives that describe two specific kinds of mesophyll. Look up these words in a dictionary to discover other contexts in which they are used. Then, explain why they are appropriate words for the types of mesophyll they describe.

23.5 Transport in Plants

THINK ABOUT IT Look at a tall tree. Maybe there's one outside your school that's 15 meters high or even taller. Think about how much work it would be to haul water up to the top of that tree. Now think of a giant redwood, a hundred meters high. How does water get to the top?

Water Transport

What are the major forces that transport water in a plant?

Recall that active transport and root pressure cause water to move from soil into plant roots. The pressure created by water entering the tissues of a root can push water upward in a plant stem. However, this pressure does not exert nearly enough force to lift water up into trees. Other forces are much more important.

Transpiration The major force in water transport is provided by the evaporation of water from leaves during transpiration. As water evaporates through open stomata, the cell walls within the leaf begin to dry out. Cell walls contain cellulose, the same material used in paper. As you know, dry paper towels strongly attract water. Similarly, the dry cell walls draw water from cells deeper inside the leaf. The pull extends into vascular tissue so that water is pulled up through xylem.

How important is transpirational pull? On a hot day, even a small tree may lose as much as 100 liters of water to transpiration. The hotter and drier the air, and the windier the day, the greater the amount of water lost. As a result of this water loss, the plant draws up even more water from the roots. **Figure 23–19** shows an analogy for transpirational pull.

Key Questions

What are the major forces that transport water in a plant?

What drives the movement of fluid through phloem tissue in a plant?

Vocabulary

adhesion • capillary action • pressure-flow hypothesis

Taking Notes

Compare/Contrast Table As you read, create a table in which to compare and contrast the functions of xylem and phloem.

VISUAL ANALOGY

TRANSPIRATIONAL PULL

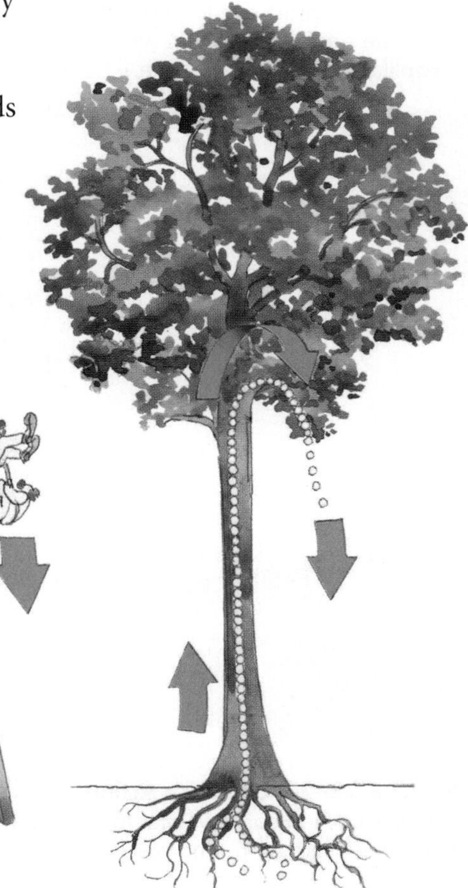

FIGURE 23–19 Imagine a chain of circus clowns who are tied together and climbing a tall ladder. When the first clown reaches the top, he falls off, pulling the clowns behind him up and over the top. Similarly, the chain of water molecules in a plant extends from the leaves down to the roots. As molecules exit leaves through transpiration, they pull up the molecules behind them.

Quick Lab
GUIDED INQUIRY

What Is the Role of Leaves in Transpiration? 

1. Use a scalpel to cut 1 cm off the bottoms of three celery stalks. **CAUTION:** *Use the scalpel with care.*

2. Remove the leaves from one stalk. Use a cotton swab to apply petroleum jelly to both sides of all the leaves on another stalk. Place all three stalks into a plastic container holding about 200 mL of water and several drops of food coloring.

3. Place the plastic container in a sunny location. Observe the celery at the end of the class and the next day. Record your observations each day.

Analyze and Conclude

1. **Observe** In which stalk did the colored water rise the most? The least?

2. **Infer** What effect did the petroleum jelly have on transpiration? What part of the leaf did the petroleum jelly affect?

3. **Draw Conclusions** How are leaves involved in transpiration?

How Cell Walls Pull Water Upward To pull water upward, plants take advantage of some of water's most interesting physical properties. Water molecules are attracted to one another by a force called cohesion. Recall from Chapter 2 that cohesion is the attraction of molecules of the same substance to each other. Water cohesion is especially strong because of the tendency of water molecules to form hydrogen bonds with each other. Water molecules can also form hydrogen bonds with other substances. This results from a force called **adhesion,** which is attraction between unlike molecules.

If you were to place empty glass tubes of various diameters into a dish of water, you would see both cohesion and adhesion at work. The tendency of water to rise in a thin tube is called **capillary action.** Water is attracted to the walls of the tube, and water molecules are attracted to one another. The thinner the tube, the higher the water will rise inside it, as shown in **Figure 23–20.**

BUILD Vocabulary

WORD ORIGINS The word *capillary* comes from the Latin word for "hair." Hairs are long and thin, like the narrow spaces in which **capillary action** takes place.

Putting It All Together What does capillary action have to do with water movement through xylem? Recall that xylem tissue is composed of tracheids and vessel elements that form many hollow, connected tubes. These tubes are lined with cellulose cell walls, to which water adheres very strongly. So, when transpiration removes some water from the exposed walls, strong adhesion forces pull in water from the wet interior of the leaf. That pull is so powerful that it extends even down to the tips of roots and, through them, to the water in the soil. **The combination of transpiration and capillary action are the major forces that move water through the xylem tissues of a plant.**

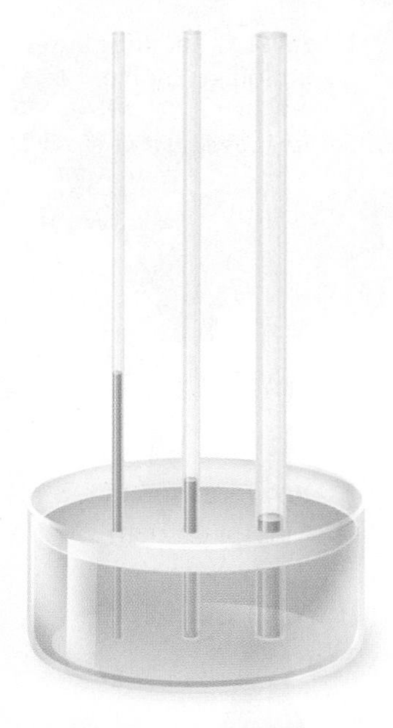

FIGURE 23–20 Capillary Action Capillary action causes water to move much higher in a narrow tube than in a wide tube.

In Your Notebook *Distinguish between the terms* cohesion *and* adhesion *by writing two sentences that use the terms.*

Nutrient Transport

What drives the movement of fluid through phloem tissue in a plant?

How do sugars move in the phloem? The leading explanation of phloem transport is known as the **pressure-flow hypothesis,** shown in **Figure 23–21.** As you know, unlike the cells that form xylem, the sieve tube cells in phloem remain alive. ❶ Active transport moves sugars into the sieve tube from surrounding tissues. ❷ Water then follows by osmosis, creating pressure in the tube at the source of the sugars. ❸ If another region of the plant has a need for sugars, they are actively pumped out of the tube and into the surrounding tissues. Osmosis then causes water to leave the tube, reducing pressure in the tube at such places. The result is a pressure-driven flow of nutrient-rich fluid from the sources of sugars (source cells) to the places in the plants where sugars are used or stored (sink cells). **Changes in nutrient concentration drive the movement of fluid through phloem tissue in directions that meet the nutritional needs of the plant.**

The pressure-flow system gives plants enormous flexibility in responding to changing seasons. During the growing season, sugars from the leaves are directed into ripening fruits or into roots for storage. As the growing season ends, the plant drops its fruits and stores nutrients in the roots. As spring approaches, chemical signals stimulate phloem cells in the roots to pump sugars back into phloem sap. Then the pressure-flow system raises these sugars into stems and leaves to support rapid growth.

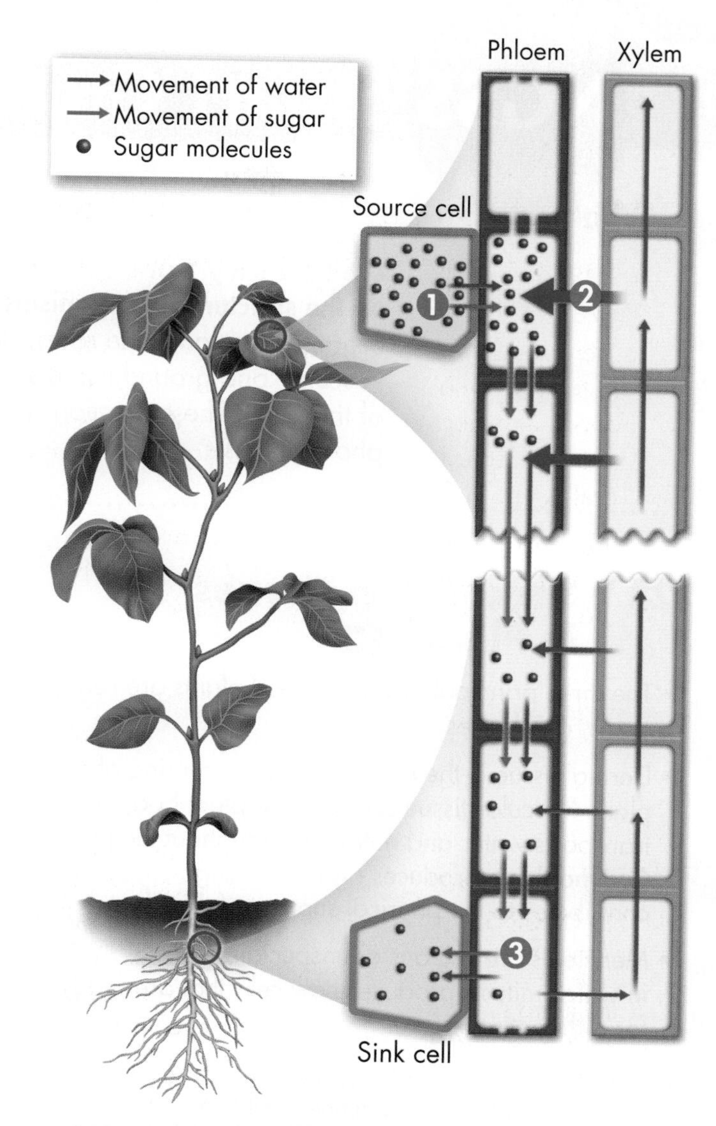

FIGURE 23–21 Pressure-Flow Hypothesis The diagram shows the movement of sugars as explained by the pressure-flow hypothesis. **Relate Cause and Effect** ***How does the movement of sugars affect the movement of water?***

23.5 Assessment

Review Key Concepts

1. a. **Review** What two forces are responsible for 90 percent of the upward flow of water through a plant?

   b. **Predict** If a plant's stomata close on a hot, dry day, how could this affect the plant's rate of photosynthesis?

2. a. **Review** What is the hypothesis that explains the movement of fluid through phloem in a plant?

   b. **Compare and Contrast** Contrast the roles of active and passive transport in the movement of phloem.

Apply the Big idea

Homeostasis

3. Explain how movement of sugars in the phloem contributes to homeostasis in a plant.

NGSS Smart Guide Plant Structure and Function

Disciplinary Core Idea LS1.A Structure and Function: How do the structures of organisms enable life's functions? The main organs of a plant—the roots, stems, and leaves—contain dermal, vascular, and ground tissue systems that carry out the basic functions of the plant. These functions include protection, transport, and photosynthesis.

23.1 Specialized Tissues in Plants

- The three principal organs of seed plants are roots, stems, and leaves.
- Dermal tissue is the protective outer covering of a plant. Vascular tissue supports the plant body and transports water and nutrients throughout the plant. Ground tissue produces and stores sugars and contributes to the physical support of the plant.
- Meristems are regions of unspecialized cells in which mitosis produces new cells that are ready for differentiation.

epidermis 665 • lignin 666 • vessel element 666 • sieve tube element 666 • companion cell 666 • parenchyma 667 • collenchyma 667 • sclerenchyma 667 • meristem 667 • apical meristem 668

Biology.com

Untamed Science Video The Untamed Science crew takes you to several exotic locations to see unique plant structures and adaptations.

23.2 Roots

- A mature root has an outside layer, called the epidermis, and also contains vascular tissue and a large area of ground tissue.
- Roots support a plant, anchor it in the ground, store food, and absorb water and dissolved nutrients from the soil.

root hair 670 • cortex 670 • endodermis 670 • vascular cylinder 670 • root cap 670 • Casparian strip 672

Biology.com

Art in Motion See how plant roots absorb nutrients and water molecules.

23.3 Stems

- Aboveground stems have several important functions: Stems produce leaves, branches, and flowers; stems hold leaves up to the sun; and stems transport substances throughout the plant.
- Primary growth of stems is the result of elongation of cells produced in the apical meristem. It takes place in all seed plants.
- In conifers and dicots, secondary growth takes place in meristems called the vascular cambium and cork cambium.

node 675 • bud 675 • vascular bundle 675 • pith 675 • primary growth 676 • secondary growth 676 • vascular cambium 677 • cork cambium 677 • heartwood 678 • sapwood 678 • bark 679

Biology.com

Tutor Tube Tune into Tutor Tube to see how new tissue growth makes plants taller.

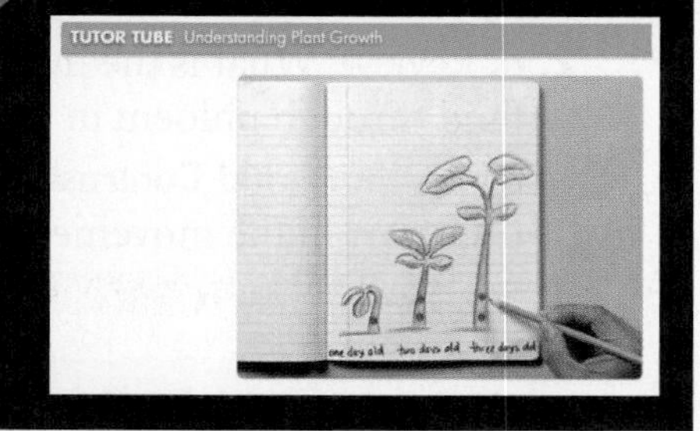

CHECKING YOUR *Scientific Literacy*

Refer to the lesson content and digital assets below as you prepare for your chapter assessment. Then, evaluate your understanding of plant structure and function by answering these questions.

1. **Scientific and Engineering Practice Developing and Using Models** Draw or construct a model of a seed plant. Include the main organs of the plant and show how materials such as water, minerals, oxygen, and carbon dioxide enter and move through the plant.

2. **Crosscutting Concept Structure and Function** Describe how the structures and functions of the organs in your model would change if the plant grew in a wet and cloudy environment.

3. **Model with Mathematics** In some trees, the shapes of the leaves can vary within the tree. Draw a model of a leaf that would most likely be found on the top of an oak tree. Draw a model of a leaf that would be found on the interior of the canopy of the same oak tree. Calculate the surface areas of your leaves. How do they compare?

4. **STEM** You are in charge of irrigation at a botanical garden. Your job is to keep the plants in the garden healthy and looking their best. Outline a plan of how you would keep the plants properly watered. Include in your plan such factors as the frequency and length of irrigation sessions, time of day you would water the plants, and dependence on weather conditions.

23.4 Leaves

- The structure of a leaf is optimized to absorb light and carry out photosynthesis.
- Plants maintain homeostasis by keeping their stomata open just enough to allow photosynthesis to take place but not so much that they lose an excessive amount of water.

blade 680 • petiole 680 • mesophyll 680 • palisade mesophyll 681 • spongy mesophyll 681 • stoma 681 • transpiration 681 • guard cell 682

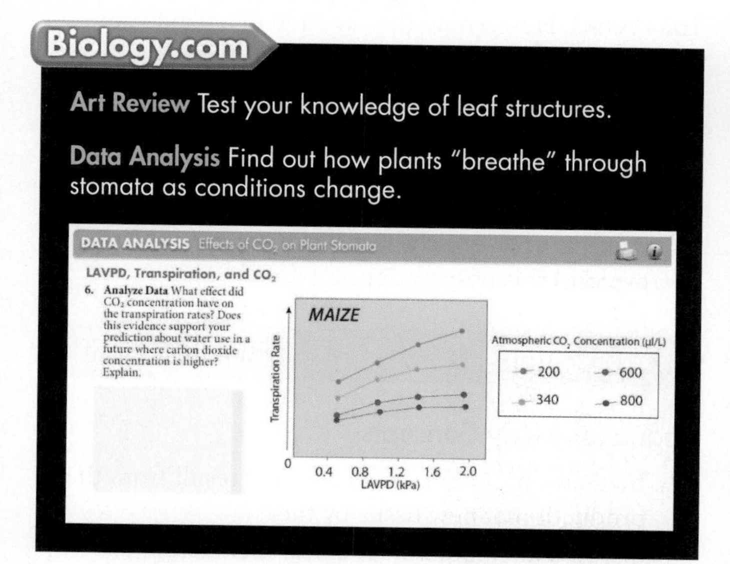

23.5 Transport in Plants

- The combination of transpiration and capillary action are the major forces that move water through the xylem tissues of a plant.
- Changes in nutrient concentration drive the movement of fluid through phloem tissue in directions that meet the nutritional needs of the plant.

adhesion 686 • capillary action 686 • pressure-flow hypothesis 687

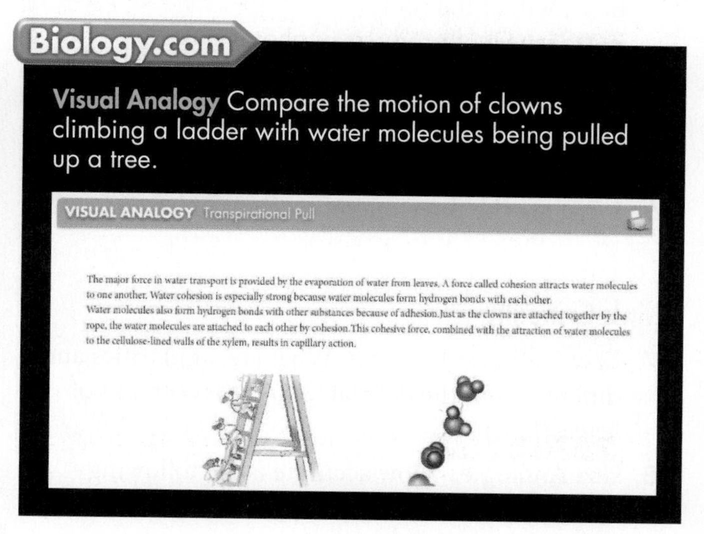

Design Your Own Lab

Identifying Growth Zones in Roots This chapter lab is available in your lab manual and at Biology.com

23 Assessment

23.1 Specialized Tissues in Plants

Understand Key Concepts

1. The plant organ that supports the plant body and carries nutrients between different parts of the plant is the
   a. root. c. leaf.
   b. stem. d. flower.

2. Which type of plant tissue would be found ONLY in the circled areas of the plant shown?
   a. meristem tissue
   b. vascular tissue
   c. dermal tissue
   d. ground tissue

3. Tracheids and vessel elements make up
   a. phloem. c. xylem.
   b. trichomes. d. meristem.

4. Phloem functions primarily in
   a. transport of water.
   b. growth of the root.
   c. transport of products of photosynthesis.
   d. increasing stem width.

5. What is the principal difference between mature xylem and mature phloem cells?

6. What are the three main functions of leaves?

Think Critically

7. **Compare and Contrast** What are similarities and differences in the dermal tissues of roots and of leaves?

8. **Use Analogies** How accurate is the following analogy for describing plant growth?

   Apical meristems cause plants to grow the way a high-rise office building grows during construction.

23.2 Roots

Understand Key Concepts

9. Which of the following are found in roots?
   a. vascular tissue only
   b. ground tissue only
   c. dermal and vascular tissue only
   d. dermal, vascular, and ground tissue

10. As a growing root pushes through the soil, the delicate apical meristem is protected by
    a. a root cap. c. bark.
    b. xylem. d. root hairs.

11. Which of the following is a trace element absorbed by roots?
    a. nitrogen c. zinc
    b. phosphorus d. potassium

12. The waterproof strip that is found in the cell walls of the endodermis is the
    a. vascular cambium. c. Casparian strip.
    b. vascular cylinder. d. cortex.

13. How are root hairs important to plants?

14. What are the primary kinds of material that plants obtain from the soil through roots?

15. What causes root pressure, and what is the function of root pressure for the plant?

Think Critically

16. **Predict** How would the function of a plant root be affected if it lacked a Casparian strip?

17. **Form a Hypothesis** While transplanting a houseplant to a larger pot, you notice that the roots had been very crowded in the old pot. Over the next few weeks, the plant's growth and overall appearance improve greatly. Develop a hypothesis that explains this observation.

23.3 Stems

Understand Key Concepts

18. Increases in the thickness of stems result from the production of new tissue by the
    a. vascular cambium. c. apical meristem.
    b. mesophyll. d. ground tissue.

19. Most water transport in stems takes place in
 a. heartwood. c. phloem.
 b. pith. d. sapwood.

20. From what type of tissue does bark develop?

21. Craft and Structure What is the main difference between monocot and dicot stems?

Think Critically

22. **Infer** If your classmate gave you a slide showing a cross section of a dicot, how would you know whether it was from a root or a stem?

23. **Design an Experiment** Describe an experiment to collect data to find the relationship between a plant's life span and ability to undergo secondary growth.

24. **Apply Concepts** In the art of bonsai, gardeners keep trees small by cutting the roots and tips of the branches. The trunk of the tree, however, continues to increase in width. How do you explain the ever-increasing width of the trunk?

23.4 Leaves

Understand Key Concepts

25. Most of the photosynthetic activity of a leaf takes place in the
 a. vascular bundles.
 b. waxy cuticle and epidermis.
 c. palisade and spongy mesophyll.
 d. guard cells and stomata.

26. Stomata open and close in response to water pressure within
 a. root cells. c. guard cells.
 b. cell walls. d. xylem.

27. What is the function of the epidermis and cuticle layers in a leaf?

28. What needs of the plant are met by controlling the opening and closing of stomata?

Think Critically

29. **Compare and Contrast** Compare the ways in which a cactus and a conifer are adapted to their respective biomes.

30. **Relate Cause and Effect** A plant's stomata are open early on a summer day when the air is cool and moist. By afternoon, when the air is hot and dry, the stomata are closed. Explain this observation.

solve the CHAPTER MYSTERY

THE HOLLOW TREE

The life of a strangler fig starts with a sticky seed deposited on a high tree branch by an animal such as a bird, bat, or monkey. At first, growth is slow because the roots have access only to the few dissolved nutrients found in the rainwater and leaf litter that collect in the crevices of the host's branches. But, after the first roots grow down the host's trunk and enter the ground, the fig's growth rate increases rapidly. The fig sends down many more roots. These roots become tangled and grafted together, crushing the host's bark and constricting the circulation of nutrients inside the phloem.

In addition, the fig's stems and leaves eventually grow taller than the host, shading it from the sun. This makes photosynthesis by the host less efficient. Below ground, the fig's roots compete with the host's roots for limited nutrients in the soil.

This triple punch—strangulation, competition for light, and competition for nutrients—usually kills the host. Left behind is an impressive "hollow" fig tree.

1. **Use Analogies** One scientist has described the strangler fig as a "vegetable octopus." Explain how this analogy relates to the habits of the strangler fig.

2. **Predict** Plants that sprout and grow on top of other plants are called epiphytes. In what biome do you think epiphytes are most common? Explain your prediction.

3. **Connect to the Big idea** How is the structure of the strangler fig different from that of the "typical plant" you studied in this chapter? Compare the advantages and disadvantages of a rain-forest plant that sprouts in the soil versus one that sprouts high off the ground.

23.5 Transport in Plants

Understand Key Concepts

31. The rise of water in a tall plant depends on capillary action and
a. osmosis.
b. evaporation.
c. nutrient transport.
d. transpirational pull.

32. The pressure-flow hypothesis explains
a. water movement in xylem.
b. water and nutrient movement in phloem.
c. water and nutrient movement in xylem.
d. water movement in phloem.

33. Attraction between water molecules and other substances is
a. adhesion.
b. capillary action.
c. transpiration.
d. cohesion.

34. What is the main function of phloem?

35. **Craft and Structure** What are source cells and sink cells?

Think Critically

36. **Relate Cause and Effect** Would transpirational pull be stronger on a hot, humid day or on a hot, dry day? Explain.

37. **Apply Concepts** Why are maple trees tapped for their sugar in the early spring rather than in the summer or autumn?

Connecting Concepts

Use Science Graphics

Use the graph to answer questions 38–40.

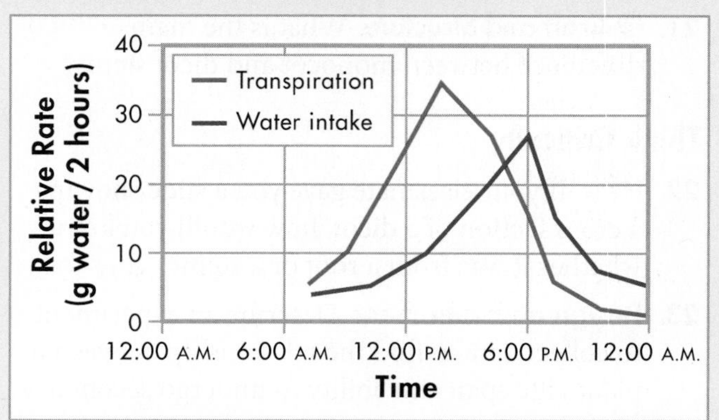

38. **Analyze Data** When is the greatest amount of water lost through transpiration?

39. **Analyze Data** About how many grams of water are lost every two hours when the transpiration curve is at its highest peak?

40. **Draw Conclusions** What is the relationship between transpiration and water intake?

Write About Science

41. **Text Types and Purposes** Explain how the cells in the vascular tissue of a root are specialized for transport of water and minerals.

42. **Assess the Big idea** Describe how several different tissue types in a leaf work together to support a functioning plant organ.

Analyzing Data

Integration of Knowledge and Ideas

A scientist selects a single three-year-old birch tree to study growth patterns. She inserts a nail into the trunk 1.0 m above ground level. At intervals over the next 15 years, she measures the total height of the tree and the circumference of the tree at the point where the nail is sticking out. She records the results, which are shown here.

Tree Age (yr)	Height of Tree (m)	Height of Nail (m)	Tree Circumference (cm)
3	2.1	1.0	9.0
8	4.5	1.0	15.0
13	8.0	1.0	26.0
18	9.0	1.0	29.5

43. **Interpret Tables** In which time interval was tree growth the greatest?
a. 0 to 3 years
b. 3 to 8 years
c. 8 to 13 years
d. 13 to 18 years

44. **Calculate** What was the average growth in the height of the tree from sprouting to age 18?
a. 0.2 meters per year
b. 0.5 meters per year
c. 1.0 meters per year
d. 2.0 meters per year

Standardized Test Prep

Multiple Choice

1. Which of the following cell types is NOT found in a plant's vascular tissue?
   A tracheid
   B vessel element
   C guard cell
   D companion cell

2. Where in a plant does mitosis produce new cells?
   A meristems
   B chloroplasts
   C mesophyll
   D heartwood

3. Which tissues make up tree bark?
   A phloem
   B cork
   C cork cambium
   D all of the above

4. Which is NOT a factor in the movement of water through a plant's vascular tissues?
   A transpiration
   B capillary action
   C osmotic pressure
   D meristems

5. All of the following conduct fluids in a plant EXCEPT
   A heartwood.
   B sapwood.
   C phloem.
   D xylem.

6. Where does most of the photosynthesis occur in a plant?
   A stomata
   B guard cells
   C vascular cambium
   D mesophyll tissue

7. Which of the following structures prevents the backflow of water into the root cortex?
   A palisade mesophyll
   B root cap
   C cambium
   D Casparian strip

8. Which of the following plants has a fibrous root system?
   A dandelion
   B beet
   C radish
   D grass

Questions 9–10

A student compared the average number of stomata on the top side and the underside of the leaves of different plants. Her data are summarized in the table below.

Average Number of Stomata (per square mm)		
Plant	Top Surfaces of Leaves	Bottom Surfaces of Leaves
Pumpkin	29	275
Tomato	12	122
Bean	40	288

9. What generalization can be made based on the data?
   A All plants have more stomata on the top side of their leaves than on the bottom side.
   B Plants have fewer stomata on the top side of their leaves than on the bottom side.
   C Some plants have more stomata on the top side of their leaves than on the bottom side.
   D The number of stomata is the same from plant to plant.

10. Pumpkins, tomatoes, and beans all grow in direct sunlight. Assuming the plants receive plenty of water, stomata on the lower surface of their leaves
    A are always closed.
    B are usually clogged with dust.
    C are unlikely to close at night.
    D stay open during daylight hours.

Open-Ended Response

11. Contrast the functions of xylem and phloem.

If You Have Trouble With . . .											
Question	1	2	3	4	5	6	7	8	9	10	11
See Lesson	23.1	23.1	23.3	23.5	23.3	23.4	23.2	23.2	23.4	23.4	23.5

24 Plant Reproduction and Response

Growth, Development, and Reproduction

Q: How do changes in the environment affect the reproduction, development, and growth of plants?

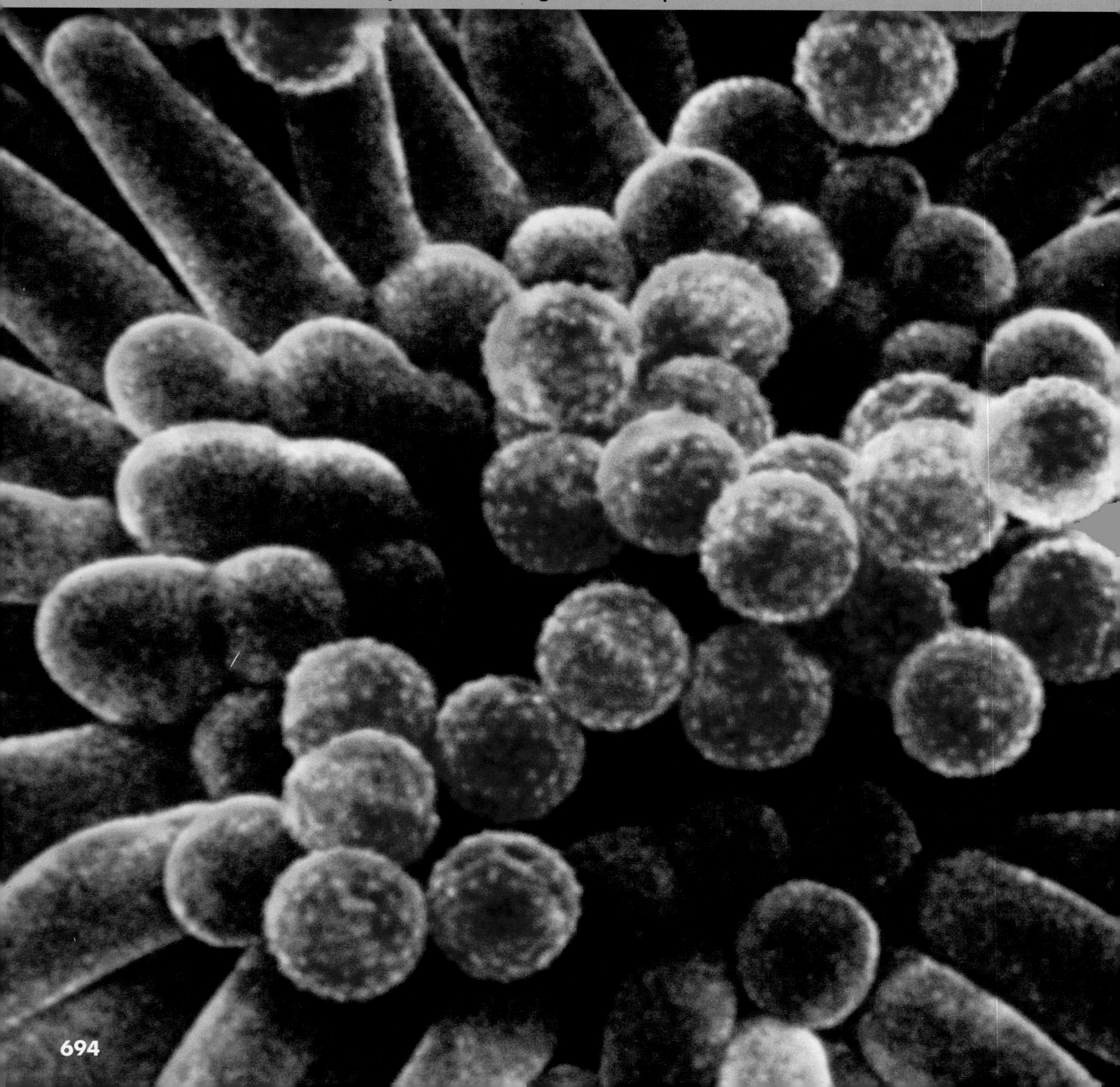

INSIDE:

- 24.1 Reproduction in Flowering Plants
- 24.2 Fruits and Seeds
- 24.3 Plant Hormones
- 24.4 Plants and Humans

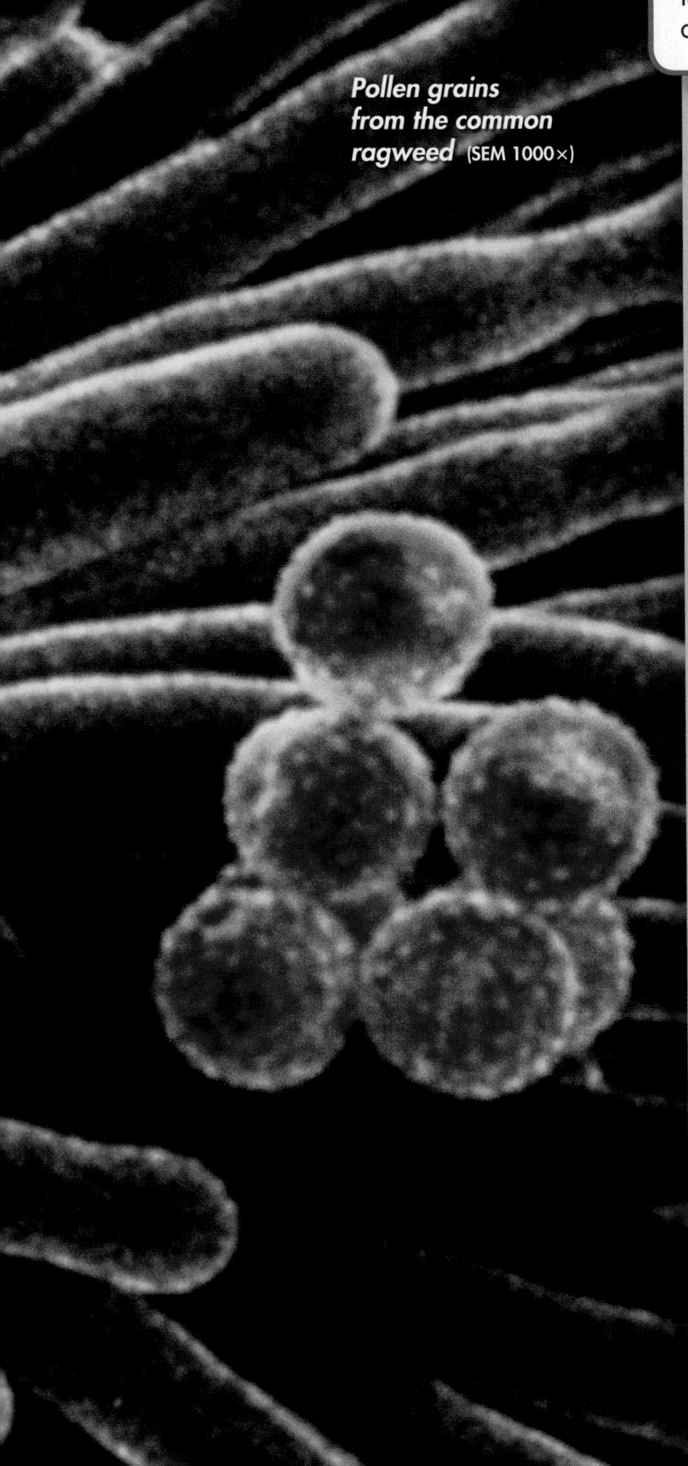

Pollen grains from the common ragweed (SEM 1000×)

BUILDING *Scientific Literacy*

Deepen your understanding of plant reproduction and response by engaging in key practices that allow you to make connections across concepts.

NGSS You will construct explanations to demonstrate how plants grow and respond to their environments.

STEM How long will a ripe banana stay edible? Learn how it is possible to ship fruits around the world without spoiling.

Common Core You will use quantitative reasoning and cite textual evidence to support your analysis of plant reproduction and response.

CHAPTER MYSTERY

THE GREEN LEMONS

For years, a California warehouse had stored freshly picked green lemons before they were shipped to market. The warehouse managers knew that the lemons would be a ripe yellow and ready to ship to market about five days after they arrived. Or so they thought. One year, for safety reasons, they decided to replace the warehouse's kerosene heaters with modern electric ones. Then, to their astonishment, when they began to pack their first shipment of five-day-old lemons, they had to call a halt. The fruit they expected to ship were still a bright, and very unripe, green. What had happened? As you read the chapter, look for clues that provide information about the case of the green lemons. Solve the mystery.

Biology.com

Finding the solution to The Green Lemons mystery is only the beginning. Take a video field trip with the ecogeeks of Untamed Science to see where the mystery leads.

Go online to access additional resources including:
• eText • Flash Cards • Lesson Overviews • Chapter Mystery

24.1 Reproduction in Flowering Plants

Key Questions

What are flowers?

How does fertilization in angiosperms differ from fertilization in other plants?

What is vegetative reproduction?

Vocabulary

stamen • anther • carpel • stigma • pistil • embryo sac • double fertilization • endosperm • vegetative reproduction • grafting

Taking Notes

Two-Column Table Construct a two-column table with the headings, *Male Gametophyte* and *Female Gametophyte*. As you read, take notes on the characteristics of each type of gametophyte.

THINK ABOUT IT What makes a flower beautiful? The symmetry of its petals, its rich colors, and, sometimes, its fragrance. But, at the heart of it, what's behind all this beauty? The answer is, simply, angiosperm sexual reproduction. To a plant, the whole point of a flower is to bring gametes together for reproduction and to protect the resulting zygote and embryo.

The Structure of Flowers

What are flowers?

You may think of flowers as decorative objects that brighten our world, and so they are. However, the presence of so many flowers in the world is visible evidence of something else—the stunning evolutionary success of the angiosperms, or flowering plants. The structure of a typical angiosperm flower is shown in **Figure 24–1.** **Flowers are reproductive organs that are composed of four different kinds of specialized leaves: sepals, petals, stamens, and carpels.**

Sepals and Petals The outermost circle of floral parts contains the sepals (SEE pulz). In many plants, the sepals are green and closely resemble ordinary leaves. Sepals enclose the bud before it opens, and they protect the flower while it is developing. Petals, which are often brightly colored, are found just inside the sepals. The colors, number, and shapes of such petals attract insects and other pollinators to the flower.

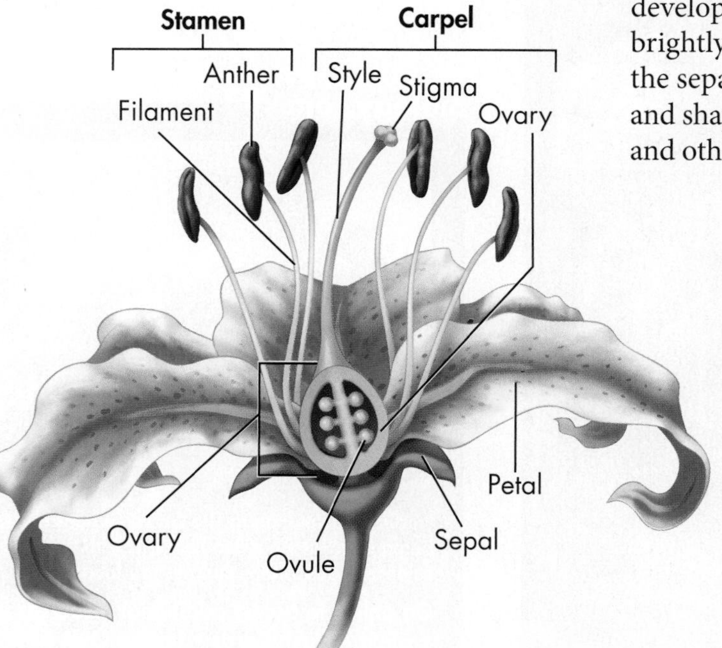

FIGURE 24–1 The Parts of a Flower This diagram shows the parts of a typical flower. The flowers of some angiosperm species, however, do not have all the parts shown here.

Stamens Within the ring of petals are the structures that produce male and female gametophytes. The **stamens** are the male parts of the flower. Each stamen consists of a stalk called a filament with an anther at its tip. **Anthers** are the structures in which pollen grains—the male gametophytes—are produced. In most angiosperm species, the flowers have several stamens. If you rub your hand on the anthers of a flower, a yellow-orange dust may stick to your skin. This dust is made up of thousands of individual pollen grains.

Carpels The innermost floral parts are the carpels. **Carpels** produce and shelter the female gametophytes and, later, seeds. Each carpel has a broad base forming an ovary, which contains one or more ovules where female gametophytes are produced. The diameter of the carpel narrows into a stalk called the style. At the top of the style is a sticky or feathery portion known as the **stigma,** which is specialized to capture pollen. Botanists sometimes call a single carpel or several fused carpels a **pistil.**

In Your Notebook *Make a two-column table labeled* Male *and* Female. *Then list and define the structures that make up a flower in the appropriate column.*

Variety in Flowers Flowers vary greatly in shape, color, and size, as shown in **Figure 24–2.** A typical flowering plant produces both male and female gametophytes. In some species, however, male and female gametophytes are produced on different plants. In some species, many flowers grow close together to form a composite structure that looks like a single flower, as seen in the Queen Anne's lace at right.

FIGURE 24–2 Variety Among Flowers Flowers vary greatly in structure. **Form a Hypothesis** ***How might it be an advantage for a plant to have many flowers clustered in a single structure?***

◀ **Iris** The drooping petal-like structures are in fact modified sepals. The fuzzy yellow stripe running down the center guides bees and other pollinators to the male and female parts at the interior of the flower.

Queen Anne's Lace Some flowerlike structures are actually clusters of many individual flowers. ▶

Passion Flower Some flowers have stamens and pistils you can easily count. In this dramatic flower, five stamens lie beneath three pistils. ▼

Wild Rose This flower has many stamens surrounding a tight cluster of carpels at the center. ▶

Quick Lab
GUIDED INQUIRY

What Is the Structure of a Flower?

1. Examine a flower carefully. Make a detailed drawing of the flower and label as many parts as you can. Note whether the anthers are above or below the stigma.
2. Remove an anther and place it on a slide. While holding the anther with forceps, use the scalpel to cut one or more thin slices across the anther. **CAUTION:** *Be careful with sharp tools. Place the slide on a flat surface before you start cutting.*
3. Lay the slices flat on the microscope slide and add a drop of water and a coverslip. Observe the slices with the microscope at low power. Make a labeled drawing of your observations.
4. Repeat steps 2 and 3 with the ovary.

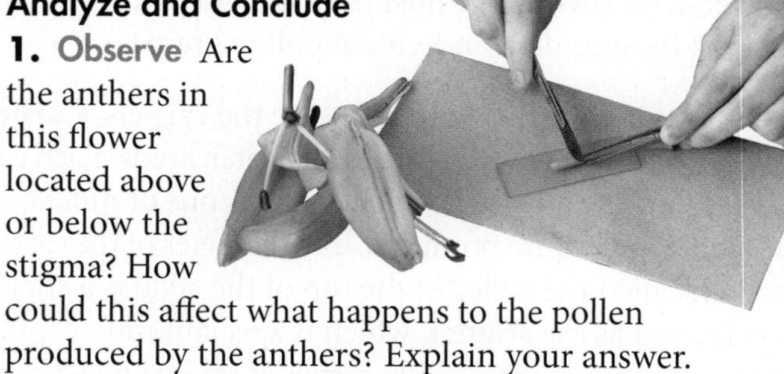

Analyze and Conclude

1. **Observe** Are the anthers in this flower located above or below the stigma? How could this affect what happens to the pollen produced by the anthers? Explain your answer.
2. **Apply Concepts** What structures did you identify in the anther? What is the function of these structures?
3. **Apply Concepts** What structures did you identify in the ovary? What is the function of these structures?
4. **Draw Conclusions** Which parts of the flower will become the seeds? Which parts will become the fruit?

The Angiosperm Life Cycle

How does fertilization in angiosperms differ from fertilization in other plants?

Like other plants, angiosperms have a life cycle that shows an alternation of generations between a diploid sporophyte phase and a haploid gametophyte stage. Recall that in vascular plants, including ferns and gymnosperms, the sporophyte plant is much larger than the gametophyte. This trend continues in angiosperms, where male and female gametophytes live within the tissues of the sporophyte.

Development of Male Gametophytes The male gametophytes—the pollen grains—develop inside anthers. This process is shown in the top half of **Figure 24–3.** First, meiosis produces four haploid spore cells. Each spore undergoes one mitotic division to produce the two haploid nuclei of a single pollen grain. The two nuclei are surrounded by a thick wall that protects the male gametophyte from dryness and damage when it is released. The pollen grains stop growing until they are released from the anther and land on a stigma.

In Your Notebook *Make a flowchart that records the stages of development of an angiosperm's male gametophyte.*

Development of Female Gametophytes While the male gametophytes are forming, female gametophytes develop inside each carpel of a flower. The ovules—the future seeds—are enveloped in a protective ovary—the future fruit.

How do the female gametophytes form? As shown in the bottom half of **Figure 24–3,** a single diploid cell goes through meiosis to produce four haploid cells, three of which disintegrate. The remaining cell undergoes mitosis, producing eight nuclei. These eight nuclei and the surrounding membrane are called the **embryo sac.** The embryo sac, contained within the ovule, makes up the female gametophyte of a flowering plant.

Next, cell walls form around six of the eight nuclei. One of the eight nuclei, near the base of the gametophyte, is the nucleus of the egg—the female gamete. If fertilization takes place, this egg cell will fuse with the male gamete to become the zygote that grows into a new sporophyte plant.

ZOOMING IN

THE DEVELOPMENT OF GAMETOPHYTES

FIGURE 24–3 The diagrams show the development of the male gametophyte inside an anther and the development of the female gametophyte inside a single ovule. **Interpret Visuals** ***In each case—male and female—which cellular process produces the first haploid cell?***

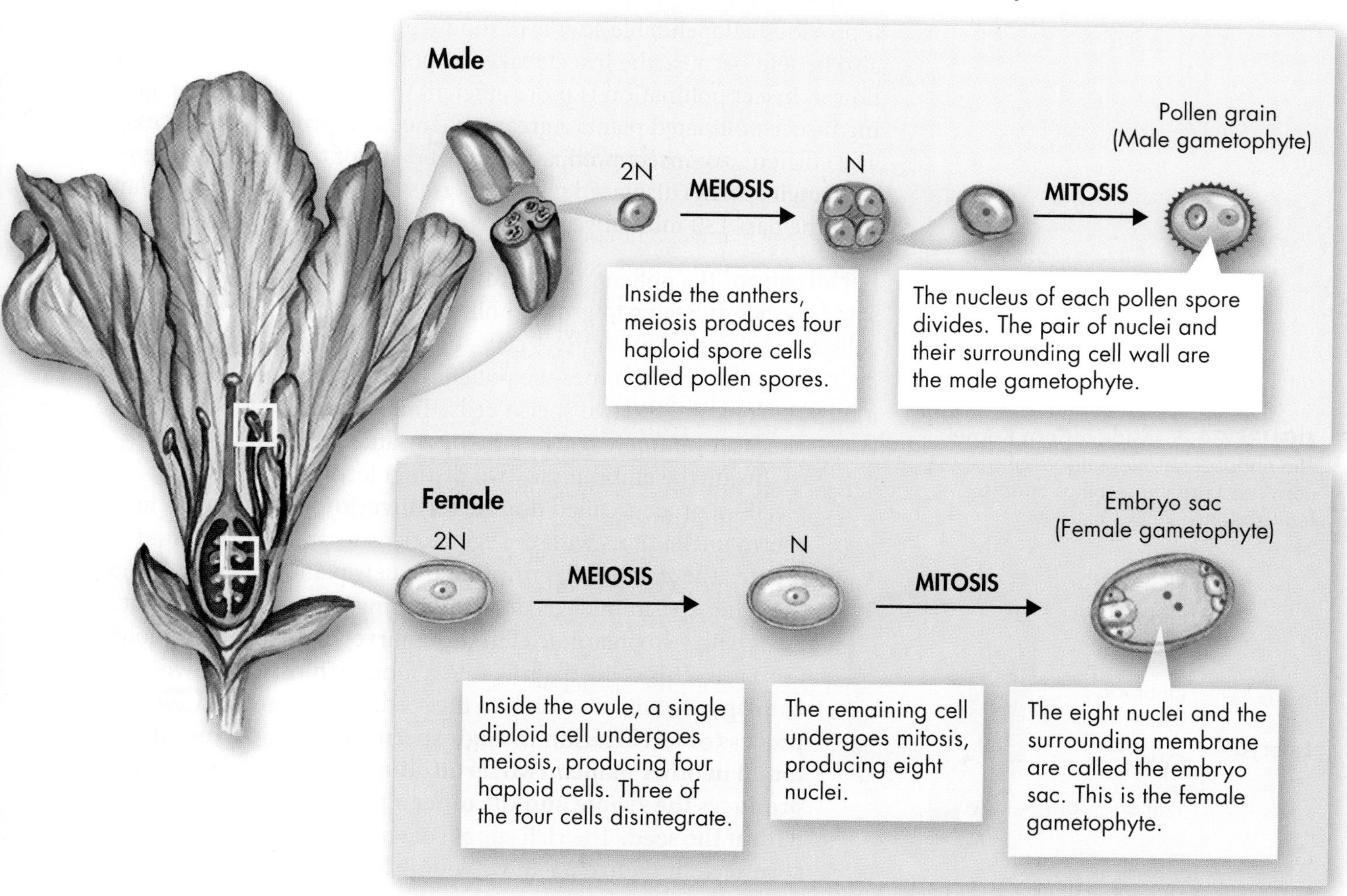

FIGURE 24–4 Pollination The appearance of a flower often indicates how it is pollinated. The flowers of an oak tree (left) are typical of wind-pollinated flowers in that they are small and not very showy but produce vast amounts of pollen. In contrast, many animal-pollinated flowers are large and brightly colored (right).

BUILD Vocabulary

RELATED WORD FORMS Several word forms are derived from the word *pollen*. *Pollination* is the transfer of pollen from one flower to another. A *pollinator* is an animal that moves pollen.

Pollination Pollination is the transfer of pollen to the female portions of the flower. Some angiosperms are wind pollinated, but most are pollinated by animals. These animals, mainly insects, birds, and bats, carry pollen from one flower to another. Because wind pollination is less efficient than animal pollination, wind-pollinated plants, such as the oak tree in **Figure 24–4,** rely on favorable weather and sheer numbers of pollen grains to get pollen from one plant to another. Animal-pollinated plants have a variety of adaptations, such as bright colors and sweet nectar, to attract and reward animals. Animals have evolved body shapes that enable them to reach nectar deep within certain flowers. For example, hummingbirds have long, thin beaks that can probe deep into flowers to reach the nectar supply.

Insect pollination is beneficial to insects and other animals because it provides a dependable source of food—pollen and nectar. Plants also benefit because the insects take the pollen directly from flower to flower. Insect pollination is more efficient than wind pollination, giving insect-pollinated plants a greater chance of reproductive success. The efficiency of insect pollination may be one of the main reasons why angiosperms displaced gymnosperms as the dominant land plants over the past 130 million years.

Fertilization If a pollen grain lands on the stigma of a flower of the same species, it begins to grow a pollen tube. Of the pollen grain's two cells, one cell—the "generative" cell—divides and forms two sperm cells. The other cell becomes the pollen tube. The pollen tube contains a tube nucleus and the two sperm cells. The pollen tube grows into the style, where it eventually reaches the ovary and enters an ovule.

Inside the embryo sac, two distinct fertilizations take place—a process called **double fertilization.** First, one of the sperm nuclei fuses with the egg nucleus to produce a diploid zygote. The zygote will grow into the new plant embryo. Second, the other sperm nucleus does something truly remarkable—it fuses with two polar nuclei in the embryo sac to form a triploid (3N) cell. This cell will grow into a food-rich tissue known as **endosperm,** which nourishes the seedling as it grows. **The process of fertilization in angiosperms is distinct from that found in other plants. Two fertilization events take place—one produces the zygote and the other a tissue, called endosperm, within the seed.** The rich supply of endosperm, as shown in the corn seed in **Figure 24–5,** will nourish the embryo as it grows.

FIGURE 24–5 Inside a Corn Kernel The endosperm and embryo of a corn seed are the result of double fertilization.

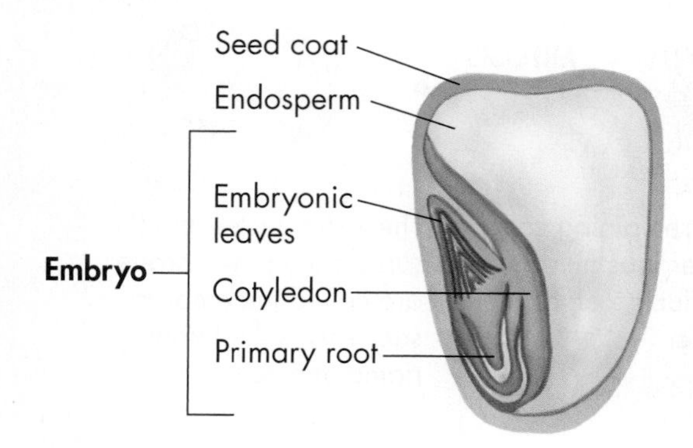

VISUAL SUMMARY

ANGIOSPERM LIFE CYCLE

FIGURE 24–6 In the life cycle of a typical angiosperm, the developing seeds of a flower are protected and nourished inside the ovary. **Relate Cause and Effect** ***Which two structures are the result of fertilization?***

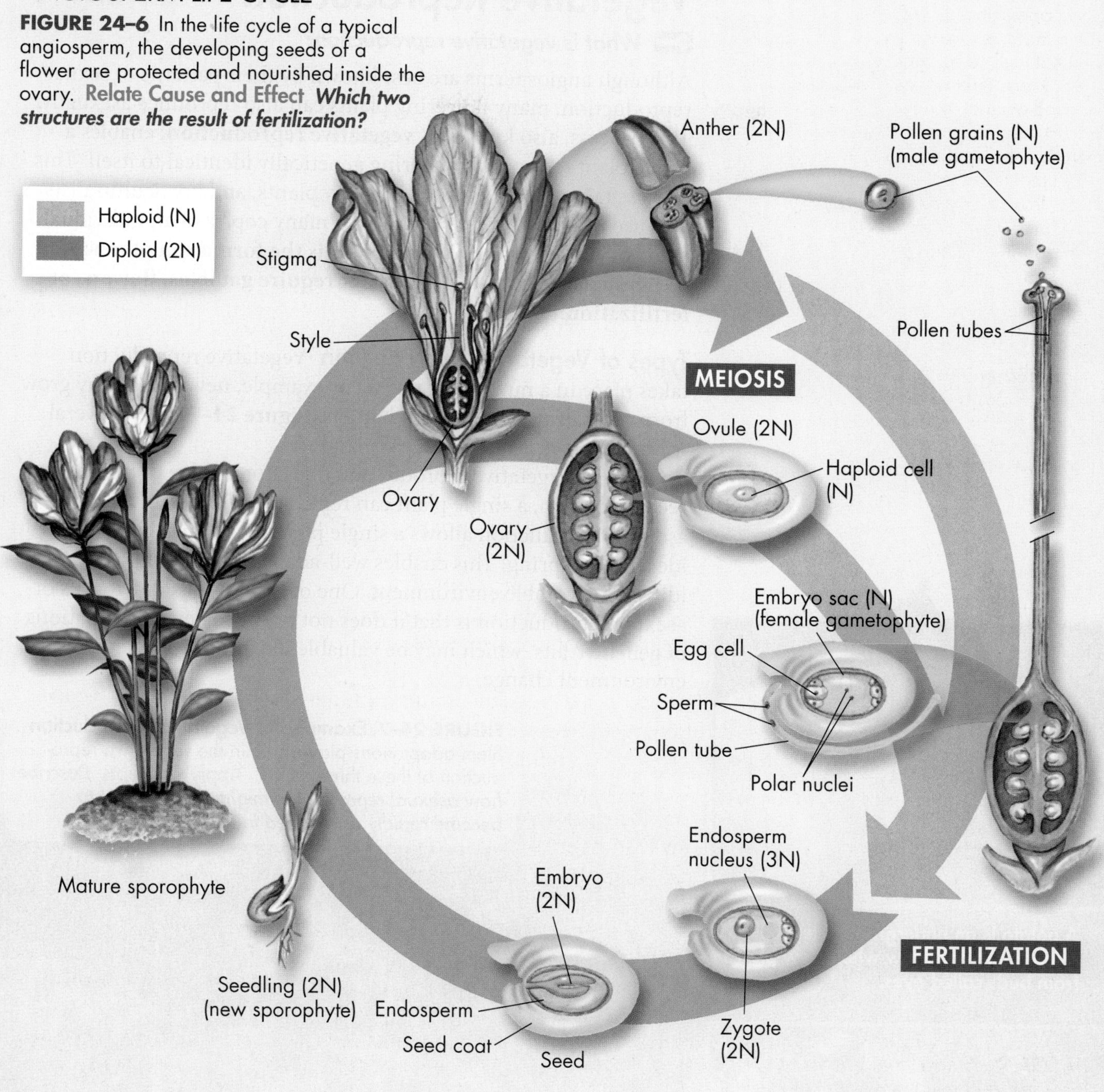

Double fertilization may be another reason why the angiosperms have been so successful. By using endosperm to store food, the flowering plant spends very little energy on producing seeds from ovules until double fertilization has actually taken place. The energy saved can be used to make many more seeds. **Figure 24–6** summarizes the life cycle of a typical angiosperm.

In Your Notebook *Make an outline detailing the key features of angiosperm reproduction.*

Vegetative Reproduction

What is vegetative reproduction?

Although angiosperms are best known by their patterns of sexual reproduction, many flowering plants can also reproduce asexually. This process, also known as **vegetative reproduction,** enables a single plant to produce offspring genetically identical to itself. This process takes place naturally in many plants, and horticulturalists also use it as a technique to produce many copies of an individual plant. **Vegetative reproduction is the formation of new individuals by mitosis. It does not require gametes, flowers, or fertilization.**

Types of Vegetative Reproduction Vegetative reproduction takes place in a number of ways. For example, new plants may grow from roots, leaves, stems, or plantlets. **Figure 24–7** shows several ways plant species reproduce vegetatively.

Because vegetative reproduction does not involve pollination or seed formation, a single plant can reproduce quickly. In addition, asexual reproduction allows a single plant to produce genetically identical offspring. This enables well-adapted individuals to rapidly fill a favorable environment. One of the obvious drawbacks of asexual reproduction is that it does not produce new combinations of genetic traits, which may be valuable if conditions in the physical environment change.

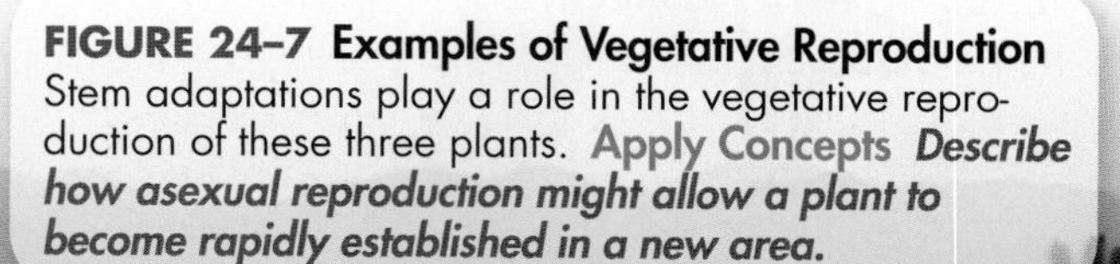

FIGURE 24–7 Examples of Vegetative Reproduction Stem adaptations play a role in the vegetative reproduction of these three plants. **Apply Concepts** ***Describe how asexual reproduction might allow a plant to become rapidly established in a new area.***

▲ A potato is an underground stem called a tuber that can grow whole new plants from buds called *eyes*.

▲ Strawberry plants send out long, trailing stems called stolons. Nodes that rest on the ground produce roots and upright stems and leaves.

Cholla and many other cactus species can reproduce by dropping sections of their stems. The small individuals growing at the base of the larger adults are, in fact, clones.

FIGURE 24–8 Grafting When just starting to bud, a branch from a lemon tree is grafted onto the branch of an established orange tree. Months later, the mature branch bears lemon fruit. Grafting leads to a single plant bearing more than one species of fruit.

Plant Propagation Horticulturists often take advantage of vegetative reproduction. To propagate plants with desirable characteristics, horticulturists use cuttings or grafting to make many identical copies of a plant or to produce offspring from seedless plants.

One of the simplest ways to reproduce plants vegetatively is by cuttings. A grower cuts from the plant a length of stem that includes a number of buds containing meristem tissue. That stem is then partially buried in soil or in a special mixture of nutrients that encourages root formation.

Grafting is a method of propagation used to reproduce seedless plants and varieties of woody plants that cannot be propagated from cuttings. To graft, a piece of stem or a lateral bud is cut from the parent plant and attached to another plant, as shown in **Figure 24–8.** Grafting works only when the two plants are closely related, such as when a bud from a lemon tree is grafted onto an orange tree. Grafting usually works best when plants are dormant, which allows the wounds created by the cut to heal before new growth starts.

MYSTERY CLUE

Whether the lemons grew on a grafted branch or not did not affect the ripening schedule. Rather, a certain variable that changed after the lemons were harvested contributed to the failure of the lemons to ripen. What might it have been?

24.1 Assessment

Review Key Concepts

1. a. **Review** Name and describe four kinds of specialized leaves that make up a flower.

   b. **Classify** Which of the structures of a flower are the male sexual organs? Which are the female organs?

2. a. **Review** Describe the features of fertilization that are characteristic of angiosperms.

   b. **Explain** How is fertilization in angiosperms different from fertilization in other types of plants?

   c. **Apply Concepts** Relate the characteristics of angiosperm reproduction to angiosperm success.

3. a. **Review** Define vegetative reproduction.

   b. **Compare and Contrast** Compare the advantages and disadvantages of sexual reproduction versus asexual reproduction in flowering plants.

VISUAL THINKING

4. Review the life cycle of the green alga *Chlamydomonas* in Lesson 22.2. Make a compare/contrast table comparing alternation of generations in flowering plants and in *Chlamydomonas.* Include which stage (haploid or diploid) of each organism's life cycle is dominant and when meiosis occurs.

24.2 Fruits and Seeds

Key Questions

How do fruits form?

How are seeds dispersed?

What factors influence the dormancy and germination of seeds?

Vocabulary

dormancy
germination

Taking Notes

Flowchart Make a flowchart that shows the process of germination and the factors that influence it. Indicate the differences between monocots and dicots.

THINK ABOUT IT What are fruits, and what purpose do they serve for the plants that produce them? Would it surprise you to learn that if you ate a meal of corn on the cob and baked beans, from the point of view of a biologist, you were actually eating fruit? And have you ever wondered why plants go to the "trouble" of surrounding their seeds with tasty, fleshy fruits like those produced by apples, oranges, and grapes? Here's a hint: You, and all the animals that enjoy those fruits, are being used. Plants may be smarter than you think.

Seed and Fruit Development

How do fruits form?

The development of the seed, which protects and nourishes the plant embryo, contributed greatly to the success of plants on land. The *angiosperm* seed, encased inside a fruit, was an even better adaptation. As you will see, by helping a seed get into the best possible location to start its new life, fruits were immediately favored by natural selection.

Once fertilization of an angiosperm is complete, nutrients flow into the flower tissue and support the development of the growing embryo within the seed. **As angiosperm seeds mature, ovary walls thicken to form a fruit that encloses the developing seeds.** A fruit is simply a matured angiosperm ovary, usually containing seeds. An exception is found in commercially grown fruits that are selectively bred to be seedless, such as some varieties of grapes. Examples of fruits are shown in **Figure 24–9.**

The term *fruit* applies to the sweet things we usually think of as fruits, such as apples, grapes, and strawberries. However, foods such as peas, corn, beans, rice, cucumbers, and tomatoes, which we commonly call vegetables, are also fruits.

The ovary wall surrounding a simple fruit may be fleshy, as it is in grapes and tomatoes, or tough and dry, like the shell that surrounds peanuts. (The peanuts themselves are the seeds.)

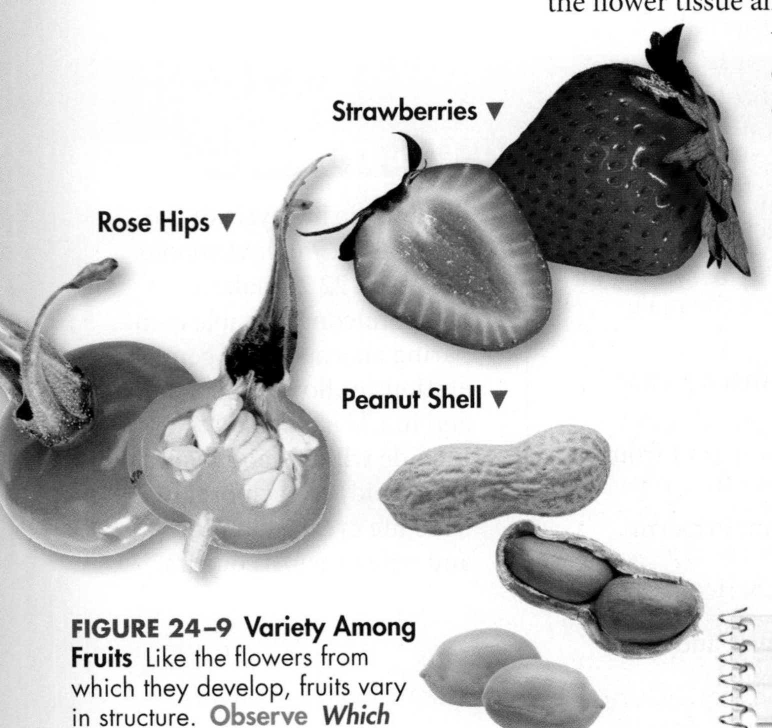

FIGURE 24–9 Variety Among Fruits Like the flowers from which they develop, fruits vary in structure. **Observe** *Which example is a dry fruit?*

In Your Notebook *Make a list of the first ten "vegetables" that come to mind and place a check mark next to ones you think are fruits. Explain why.*

Seed Dispersal

How are seeds dispersed?

What are fleshy fruits for, and why have they been favored by natural selection? They are not there to nourish the seedling—the endosperm does that. So why should these plants have seeds that are wrapped in an additional layer of nutrient-packed tissue? It seems pointless, but in evolutionary terms, it makes all the sense in the world.

Think of the blackberries that grow wild in the forests of North America. Each seed is enclosed in a sweet, juicy fruit, making it a tasty treat for all kinds of animals. What good is such sweetness if all it does is get the seed eaten? Well, believe it or not, that's exactly the point.

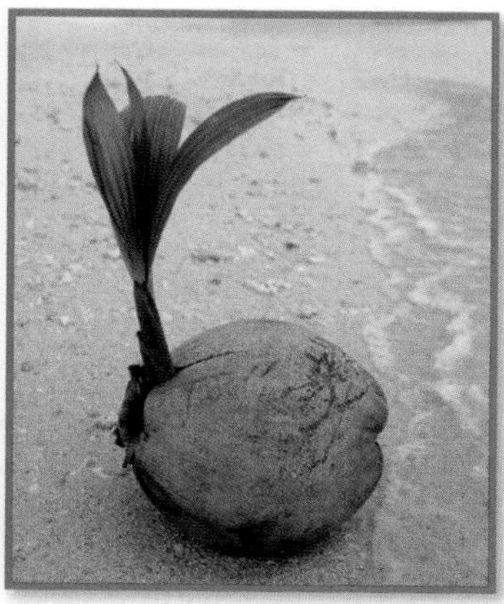

FIGURE 24–10 Mechanisms of Seed Dispersal A Bohemian waxwing feasts on mountain ash berries (left) and will later disperse the seeds in its feces. Parachute-like dandelion fruits catch the wind, carrying the tiny attached seeds far and wide (middle). The buoyant coconut fruit can disperse its seed over great distances of water (right).

Dispersal by Animals The seeds of many plants, especially those encased in sweet, fleshy fruits, are often eaten by animals. The seeds are covered with tough coatings, allowing them to pass through an animal's digestive system unharmed. The seeds then sprout in the feces eliminated from the animal. These fruits provide nutrition for the animal and also help the plant disperse its seeds—often to areas where there is less competition with the parent plants. **Seeds contained in fleshy, nutritious fruits are usually dispersed by animals.** Three mechanisms of seed dispersal are shown in **Figure 24–10.**

Animals also disperse many dry fruits, but not necessarily by eating them. Dry fruits sometimes have burs or hooks that catch in an animal's fur, enabling them to be carried many miles from the parent plant.

Dispersal by Wind and Water Animals are not the only means by which plants can scatter their seeds. Seeds are also adapted for dispersal by wind and water. **Seeds dispersed by wind or water are typically contained in lightweight fruits that allow them to be carried in the air or in buoyant fruits that allow them to float on the surface of the water.** A dandelion seed, for example, is attached to a dry fruit that has a parachute-like structure. This adaptation allows the seed to glide considerable distances away from the parent plant. Some seeds, like the coconut, are dispersed by water. Coconut fruits are buoyant enough to float in seawater for many weeks, enabling the seeds to reach and colonize even remote islands.

Temperature and Seed Germination

Arisaema dracontium—"green dragon"—is a plant that grows from the southern United States to Canada. The graph shows germination rates of *Arisaema* seeds gathered from two locations and stored at two different temperatures.

1. Interpret Graphs What effect does chilling have on germination of seeds from Ontario? How does it affect the seeds from Louisiana?

2. Form a Hypothesis Describe how the different rates of seed germination might be explained in terms of adaptation to the local climate.

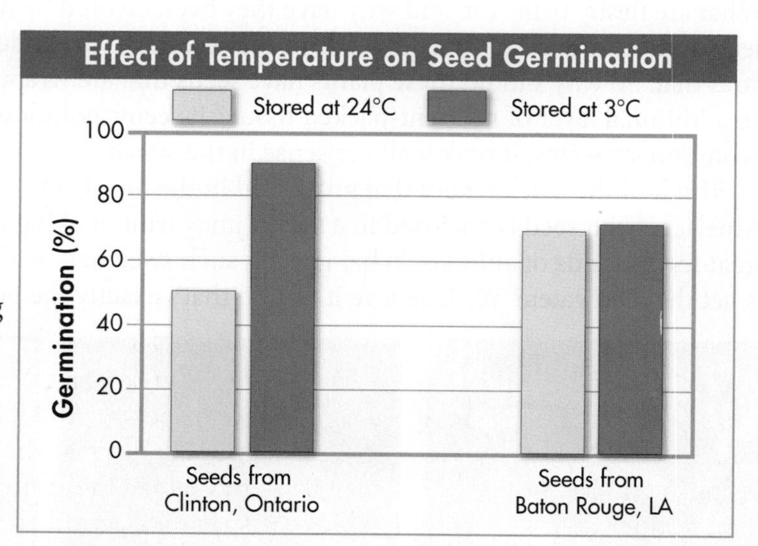

BUILD Vocabulary

WORD ORIGINS The word **dormancy** comes from the Latin word *dormire,* meaning "to sleep."

Seed Dormancy and Germination

What factors influence the dormancy and germination of seeds?

Some seeds sprout so rapidly that they are practically instant plants. Bean seeds are a good example. With proper amounts of water and warmth, a mature bean seed rapidly sprouts and develops into a green plant. But many seeds will not grow when they first mature. Instead, these seeds enter a period of **dormancy,** during which the embryo is alive but not growing. The length of dormancy varies in different species. **Germination** is the resumption of growth of the plant embryo. **Environmental factors such as temperature and moisture can cause a seed to end dormancy and germinate.**

MYSTERY CLUE

What variables in the lemons' environment could have changed with the switch to electric heating? Start a list of variables that affect seed development and fruit ripening. Add to your list as you continue reading the chapter.

How Seeds Germinate Before germinating, seeds absorb water. The absorbed water causes food-storing tissues to swell, cracking open the seed coat. Through the cracked seed coat, the young root emerges and begins to grow. The shoot—the part of the plant that will grow above ground—emerges next.

The Role of Cotyledons Cotyledons are a flowering plant's first leaves. Their job is to store nutrients and then transfer them to the growing embryo as the seed germinates. **Figure 24–11** compares germination in a monocot and a dicot. Monocots have a single cotyledon, which usually remains underground while it passes nutrients to the young plant. The growing monocot shoot emerges from the soil protected by a sheath. In dicots, which have two cotyledons, there is no sheath to protect the tip of the young plant. Instead, the upper end of the shoot bends to form a hook that forces its way through the soil. This protects the delicate tip of the plant, which straightens as it emerges into the sunlight. In some species, the cotyledons appear above ground as the plant emerges, while in others, such as the garden pea, the cotyledons remain underground.

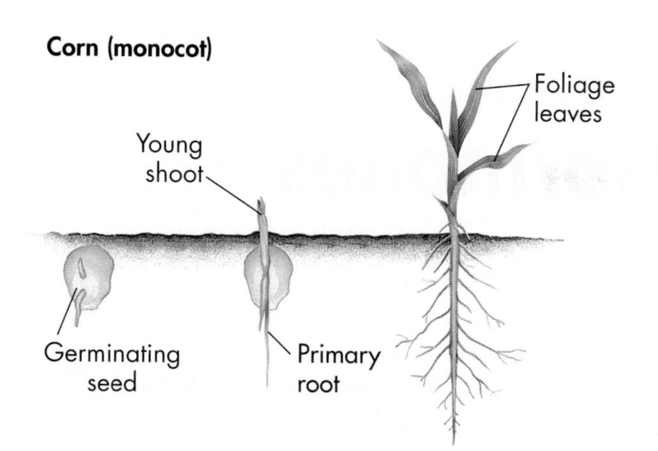

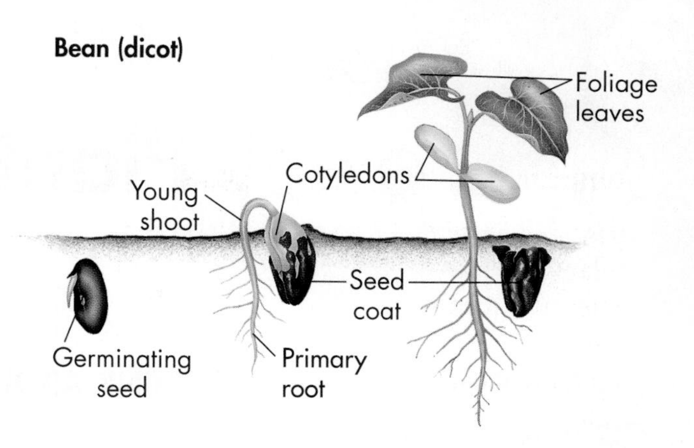

Advantages of Dormancy Seed dormancy can be adaptive in several ways. For one, it can allow for long-distance dispersal. And it also allows seeds to germinate under ideal growth conditions. The seeds of most temperate plants, for example, germinate in the spring, when conditions are best for growth. For some species, a period of cold temperatures during which the seeds are dormant is required before growth can begin. Seeds can easily survive winter cold, but many young green plants cannot. The period of cold that is required is long enough that seeds will not germinate until the dangerous winter season has passed.

Sometimes, only extreme environmental conditions can end seed dormancy. Some pine trees, for example, produce seeds in cones that remain sealed until the high temperatures generated by forest fires cause the cones to open. The high temperature both activates and releases the seeds, allowing the plants to reclaim the forest quickly after a fire.

FIGURE 24–11 Germination: A Comparison The monocot corn seedling (left) grows directly upward, protected by a sheath of tissue that surrounds the developing leaves. In contrast, the garden bean (right) forms a hook in its stem that gently pulls the new plant tissues through the soil. **Predict** ***What might happen to a germinating seedling that lacked such adaptations?***

24.2 Assessment

Review Key Concepts

1. **a. Review** Describe how fruits form.
   b. Infer Is a pumpkin a fruit? Describe any evidence you use in making your inference.
2. **a. Review** Describe two methods of seed dispersal.
   b. Pose Questions A new angiosperm species is discovered. What questions would you ask before predicting how its seeds are dispersed? Explain the rationale for asking each question.
3. **a. Review** Summarize the environmental factors that affect seed germination.
   b. Explain Why is it adaptive for some seeds to remain dormant before they germinate?
   c. Apply Concepts The seeds of a bishop pine germinate only after exposure to the extreme heat of a forest fire. Evaluate the significance of this structural adaptation.

WRITE ABOUT SCIENCE

Creative Writing

4. Imagine that you are writing a children's book about seeds and that you are working on the chapter on dispersal. Write from one to three paragraphs on seed dispersal by wind. *Hint:* Try to include details that you would have found appealing when you were about eight years old.

24.3 Plant Hormones

Key Questions

🔑 *What roles do plant hormones play?*

🔑 *What are some examples of environmental stimuli to which plants respond?*

🔑 *How do plants respond to seasonal changes?*

Vocabulary

hormone • target cell • receptor • auxin • apical dominance • cytokinin • gibberellin • abscisic acid • ethylene • tropism • phototropism • gravitropism • thigmotropism • photoperiodism

Taking Notes

Concept Map As you read, build a concept map summarizing the effects of different hormones on plant growth.

THINK ABOUT IT Plants, like all organisms, are collections of cells. Plants grow in response to factors such as light, moisture, temperature, and gravity. But how do roots "know" to grow down, and how do stems "know" to grow up? How do the tissues of a plant determine the right time of year to produce flowers? In short, how do the collections of cells in a plant manage to act together as a single organism? Is something carrying messages from cell to cell?

Hormones

🔑 ***What roles do plant hormones play?***

Hormones are chemical signals produced by living organisms that affect the growth, activity, and development of cells and tissues. In plants, hormones may act on the same cells in which they are made, or they may travel to different cells and tissues. This is in contrast to *animal* hormones, which typically act at a location some distance away from the cells that produce them.

🔑 **Plant hormones serve as signals that control development of cells, tissues, and organs. They also coordinate responses to the environment.** The two functions fit together well, because plants respond to the environment mainly by changing their development.

The steps in one mechanism of hormone action in plants is shown in **Figure 24–12.** In this case, the hormone moves through the plant from the place where it is produced to the place where it triggers its response.

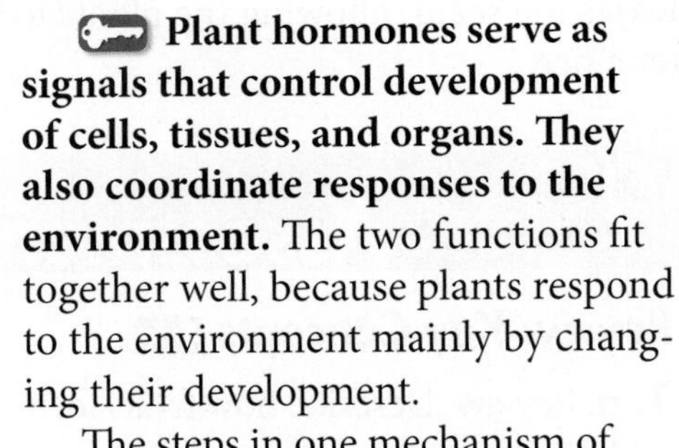

FIGURE 24–12 Hormones and Flower Development In some species, hormone-producing cells in a mature flower release hormones that travel into flower buds and inhibit development. Once the mature flower is done blooming, production of the inhibiting hormone will decline, and the flower bud can then begin its bloom.

How Hormones Act Cells in an organism affected by a particular hormone are called **target cells.** To respond to a hormone, a cell must contain hormone **receptors**—usually proteins—to which hormone molecules bind. The response that results will depend on what kinds of receptors are present in the target cell. One kind of receptor might alter metabolism; a second might speed growth; a third might inhibit cell division. Thus, depending on the receptors present, a given hormone may affect roots differently from stems or flowers—and the effects may change as the developing organs add or remove receptors. Cells that do not contain receptors are generally unaffected by hormones.

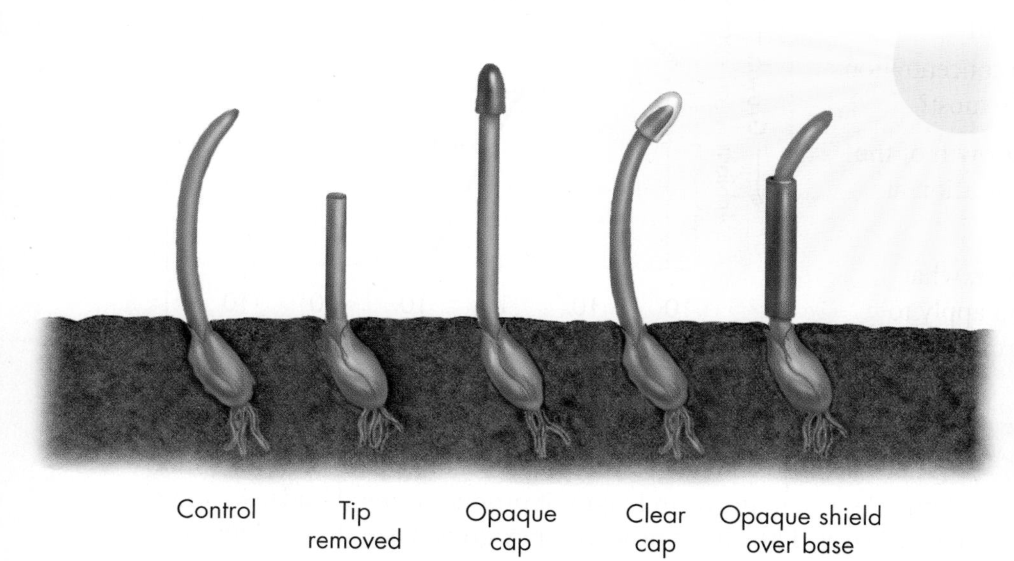

FIGURE 24–13
How Plants Detect Light
The Darwins conducted controlled experiments to determine which region of the plant senses light. When they removed the seedling tip or placed an opaque cap over the tip, they observed no bending toward light. But when they placed a clear cap on the tip or an opaque shield around the base, they observed bending similar to that seen in the control. **Control Variables** ***What variable did the Darwins control for by comparing the results of seedlings treated with a clear cap versus no cap?***

Auxins The first step in the discovery of plant hormones came over a century ago, and was made by a scientist already familiar to you. In 1880, Charles Darwin and his son Francis published the results of a series of experiments exploring the mechanism behind a grass seedling's tendency to bend toward light as it grows.

The results of their experiments, shown in **Figure 24–13,** suggested that the tip of the seedling somehow senses light. The Darwins hypothesized that the tip produces a substance that regulates cell growth. More than forty years later, the regulatory substances produced by the tips of growing plants were identified and named *auxins.* **Auxins** stimulate cell elongation and the growth of new roots, among other roles that they play. They are produced in the shoot apical meristem and transported to the rest of the plant.

▶ ***Auxins and Cell Elongation*** One of the effects of auxins is to stimulate cell elongation, as shown in **Figure 24–14.** In the Darwins' experiment, when light hits one side of the shoot, auxins collect in the shaded part of the shoot. This change in concentration stimulates cells on the dark side to lengthen. As a result, the shoot bends away from the shaded side and toward the light.

FIGURE 24–14 Auxins and Cell Elongation Cells elongate more on the shaded side of the shoot, where there is a higher concentration of auxins.

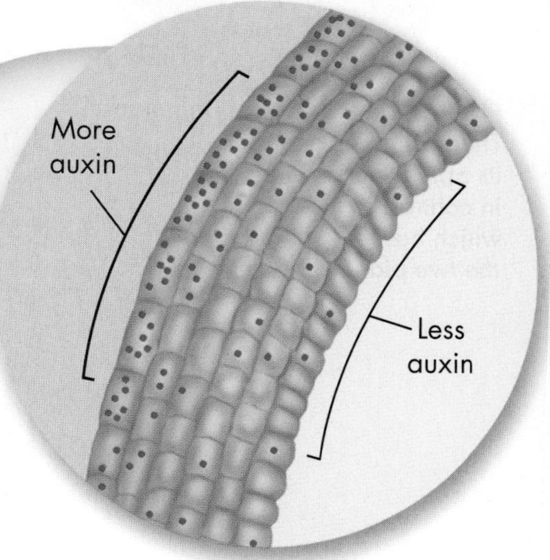

In Your Notebook *Review **Figure 24–13.** Describe how the results led the Darwins to conclude that the tip of the seedling senses light.*

Analyzing Data

Auxins and Plant Growth

This graph shows the results of experiments in which carrot cells were grown in the presence of varying concentrations of auxins. The blue line shows the effects on root growth. The red line shows the effects on stem growth.

1. **Interpret Graphs** At what auxin concentration are the stems stimulated to grow the most?

2. **Interpret Graphs** How is the growth of the roots affected by the auxin concentration at which stems grow the most?

3. **Infer** If you were a carrot farmer, what concentration of auxin should you apply to your fields to produce the largest carrot roots?

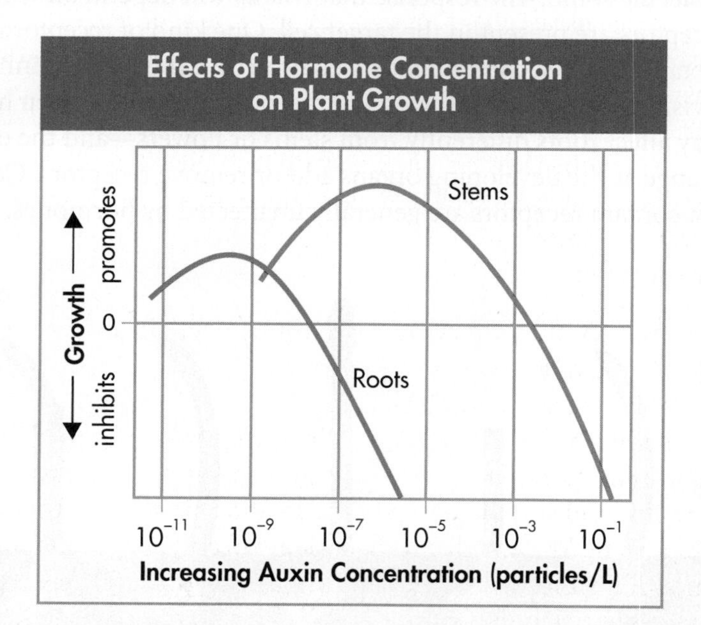

FIGURE 24–15 Apical Dominance The basil plant on the right has had its apical meristem pinched off, in contrast to the plant on the left, which hasn't. **Observe** ***How are the two plants different?***

▶ ***Auxins and Branching*** Auxins also regulate cell division in meristems. As a stem grows in length, it produces lateral buds. As you may have observed, the buds near the apex grow more slowly than those near the base of a plant. The reason for this delay is that growth at the lateral buds is inhibited by auxins. Because auxins move out from the apical meristem, the closer a bud is to the stem's tip, the more it is inhibited. This phenomenon is called **apical dominance.** If you snip off the tip of a plant, these lateral buds begin to grow more quickly. The plant becomes bushier. This is because the apical meristem—the source of the growth-inhibiting auxins—has been eliminated.

Cytokinins **Cytokinins** are plant hormones that are produced in growing roots and in developing fruits and seeds. Cytokinins stimulate cell division, they interact with auxins to help to balance root and shoot growth, and stimulate regeneration of tissues damaged by injury. Cytokinins also delay the aging of leaves and play important roles in the early stages of plant growth.

Cytokinins often produce effects opposite to those of auxins. For example, root tips make cytokinins and send them to shoots; shoot tips make auxins and send them to roots. This exchange of signals can restore lost organs and keep root and shoot growth in balance. Auxins stimulate the initiation of new roots, and they inhibit the initiation and growth of new shoot tips. Cytokinins do just the opposite. So if a tree is cut down, the stump will often make new shoots because auxins have been removed and cytokinins accumulate near the cut.

In Your Notebook *Make a 2×2 table labeled* Shoot *and* Root *across the top and* Auxins *and* Cytokinins *down the side. Then, fill in the effects of these hormones.*

Gibberellins For years, farmers in Japan knew of a disease that weakened rice plants by causing them to grow unusually tall. The plants would flop over and fail to produce a high yield of rice grain. Farmers called the disease the "foolish seedling" disease. In 1926, Japanese biologist Eiichi Kurosawa discovered that a fungus, *Gibberella fujikuroi*, caused this extraordinary growth. His experiments showed that the fungus produced a growth-promoting substance.

In fact, the chemical produced by the fungus mimicked hormones produced naturally by plants. These hormones, called **gibberellins,** stimulate growth and may cause dramatic increases in size, particularly in stems and fruits.

MYSTERY CLUE

Kerosene lamps emit carbon dioxide and ethylene as they burn, while electric heaters do not. Could this fact help explain the delayed ripening of the lemons?

Abscisic Acid Gibberellins also interact with another hormone, abscisic acid, to control seed dormancy. **Abscisic acid** inhibits cell division, thereby halting growth.

Recall that seed dormancy allows the embryo to rest until conditions are good for growth. When seed development is complete, abscisic acid stops the seed's growth and shifts the embryo into a dormant state. The embryo rests until environmental events shift the balance of hormones. Such events may include a strong spring rain that washes abscisic acid away. (Gibberellins do not wash away as easily.) Without the opposing effect of abscisic acid, the gibberellins can signal germination.

Abscisic acid and gibberellins have opposite effects, much like the auxins and cytokinins. **The opposing effects of plant hormones contribute to the balance necessary for homeostasis.**

Ethylene One of the most interesting plant hormones, ethylene, is actually a gas. Fruit tissues release small amounts of the hormone **ethylene,** stimulating fruits to ripen. Ethylene also plays a role in causing plants to seal off and drop organs that are no longer needed. For example, petals drop after flowers have been pollinated, leaves drop in autumn, and fruits drop after they ripen. In each case, ethylene signals cells at the base of the structure to seal off from the rest of the plant by depositing waterproof materials in their walls.

A Summary of Plant Hormones

Hormone	Some of the Effects	Where Found
Auxins	Promote cell elongation and apical dominance; stimulate growth of new roots	Produced in shoot apical meristem and transported elsewhere
Cytokinins	Stimulate cell division; affect root growth and differentiation; may work in opposition to auxins	Growing roots
Gibberellins	Stimulate growth; influence various developmental processes; promote germination	Meristems of shoot, root, and seed embryo
Abscisic acid	Inhibits cell division; promotes seed dormancy	Terminal buds; seeds
Ethylene	Stimulates fruits to ripen; causes plants to seal off and drop unnecessary organs, such as leaves in autumn	Fruit tissues; aging leaves and flowers

FIGURE 24–16
A Summary of Plant Hormones
This table lists some of the effects of the major plant hormones and where the hormones can be found in the plant body.
Interpret Tables ***Name two pairs of hormones that work in opposition to each other.***

Tropisms and Rapid Movements

What are some examples of environmental stimuli to which plants respond?

Like all living things, plants need the power of movement to cope with the environment. Many plant movements are slow, but some are so fast that even animals cannot keep up with them.

BUILD Vocabulary
WORD ORIGINS The word **tropism** comes from a Greek word that means "turning."

Tropisms Plant sensors that detect environmental stimuli signal elongating organs to reorient their growth. These growth responses are called **tropisms.** **Plants respond to environmental stimuli such as light, gravity, and touch.**

FIGURE 24–17 Three Tropisms

▶ ***Light*** The tendency of a plant to grow toward a light source is called **phototropism.** This response can be so quick that young seedlings reorient themselves in a matter of hours. Recall that changes in auxin concentration are responsible for phototropism. Experiments have shown that auxins migrate toward shaded tissue, possibly due to changes in membrane permeability in response to light.

▶ ***Gravity*** Auxins also affect **gravitropism,** the response of a plant to gravity. For reasons still not understood, auxins migrate to the lower sides of horizontal roots and stems. In horizontal stems, the migration causes the stem to bend upright. In horizontal roots, however, the migration causes roots to bend downward.

▶ ***Touch*** Some plants even respond to touch, a process called **thigmotropism.** Vines and climbing plants exhibit thigmotropism when they encounter an object and wrap around it. Other plants, such as grape vines, have extra growths called tendrils that emerge near the base of the leaf and wrap tightly around any object they encounter.

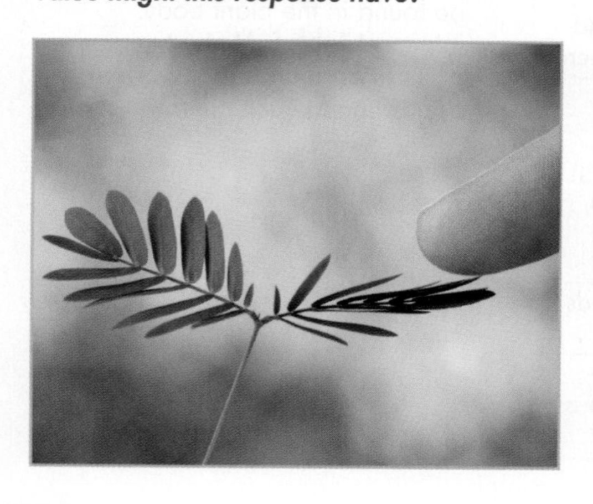

FIGURE 24–18 Rapid Movement The mimosa plant responds to touch by folding in its leaves quickly. This response is produced by decreased osmotic pressure in cells near the base of each leaflet. **Infer** ***What adaptive value might this response have?***

Rapid Movements Some plant responses are so rapid that it would be a mistake to call them tropisms. **Figure 24–18** shows what happens if you touch a leaf of *Mimosa pudica*, appropriately called the "sensitive plant." Within only two or three seconds, its two leaflets fold together completely. The carnivorous Venus' flytrap also demonstrates a rapid response. When an insect lands on a flytrap's leaf, it triggers sensory cells on the inside of the leaf, sending electrical signals from cell to cell. A combination of changes in osmotic pressure and cell wall expansion causes the leaf to snap shut, trapping the insect inside.

Response to Seasons

How do plants respond to seasonal changes?

"To every thing there is a season." Nowhere is this more evident than in the regular cycles of plant growth. Year after year, some plants flower in the spring, others in summer, and still others in the fall. Plants such as chrysanthemums and poinsettias flower when days are short and are therefore called short-day plants. Plants such as spinach and irises flower when days are long and are therefore known as long-day plants.

Photoperiod and Flowering How do all these plants manage to time their flowering so precisely? In the early 1920s, scientists discovered that tobacco plants flower according to their photoperiod, the number of hours of light and darkness they receive. Additional research showed that many other plants also respond to changing photoperiods, a response called **photoperiodism.** This type of response is summarized in **Figure 24–19.** **Photoperiodism is a major factor in the timing of seasonal activities such as flowering and growth.**

It was later discovered that a plant pigment called phytochrome (FYT oh krohm) is responsible for plant responses to photoperiod. Phytochrome absorbs red light and activates a number of signaling pathways within plant cells. By mechanisms that are still not understood completely, plants respond to regular changes in these pathways. These changes determine the patterns of a variety of plant responses.

FIGURE 24–19 Effects of Photoperiod Changes in the photoperiod can affect the seasonal timing of flowering. **Form an Opinion** ***Are "short-day plant" and "long-day plant" the best names for categorizing these plants, or would it be better to name plants after their responses to night length? Explain your reasoning.***

Effect of Photoperiod on Flowering

	Long Day	Short Day	Interrupted Night
	Midnight / Noon	Midnight / Noon	Midnight / Noon
Short-Day Plant		Short-day plants flower only when exposed to an extended period of darkness.	
Long-Day Plant	Long-day plants flower when exposed to a short period of darkness.		Long-day plants also flower if a brief period of light interrupts the darkness–this essentially divides one long night into two short nights.

Winter Dormancy Phytochrome also regulates the changes in activity that prepare many plants for dormancy as winter approaches. Recall that dormancy is the period during which an organism's growth and activity decrease or stop. **As cold weather approaches, deciduous plants turn off photosynthetic pathways, transport materials from leaves to roots, and seal off leaves from the rest of the plant.**

▶ ***Leaf Loss*** In temperate regions, many flowering plants lose their leaves during the colder months. At summer's end, the phytochrome in leaves absorbs less light as days shorten and nights become longer. Auxin production drops, but the production of ethylene increases. The change in the relative amounts of these two hormones starts a series of events that gradually shut down the leaf.

As chlorophyll breaks down, other pigments that have been present all along—including yellow and orange carotenoids—become visible for the first time. The brilliant reds come from freshly made anthocyanin pigments.

▶ ***Changes to Meristems*** Hormones also produce important changes in apical meristems. Instead of continuing to produce leaves, meristems produce thick, waxy scales that form a protective layer around new leaf buds. Enclosed in its coat of scales, a terminal bud can survive the coldest winter days. At the onset of winter, xylem and phloem tissues pump themselves full of ions and organic compounds. The resulting solution acts like antifreeze in a car, preventing the tree's sap from freezing. This is one of several mechanisms plants use to survive the bitter cold.

FIGURE 24–20 Adaptations for Winter In autumn, leaves shut down photosynthesis and fall from deciduous trees. Meanwhile, meristems at the tips of the branches produce thick, waxy scales that cover and protect new stem and leaf buds through the harsh winter.

24.3 Assessment

Review Key Concepts

1. a. **Review** Describe how plant hormones contribute to homeostasis.
   b. **Infer** Why should a person who trims trees for a living know about the effect of apical dominance on the shape of trees? Explain.
2. a. **Review** Give three examples of plant responses to external stimuli.
   b. **Apply Concepts** Using a houseplant, a marker, and a sunny windowsill, describe how you might measure the plant's response to light.
3. a. **Review** Summarize plant responses to seasonal changes.
   b. **Explain** Which type of plant—short-day or long-day—is likely to bloom in the summer? Explain your answer.
   c. **Design an Experiment** How could a garden-store owner determine what light conditions are needed for a particular flowering plant to bloom? Design a controlled experiment to find out.

Apply the Big idea

Evolution

4. Review what you learned about evolution by natural selection in Chapter 16. Then, using what you know about natural selection, describe how plant adaptations for dormancy may have developed over time.

24.4 Plants and Humans

THINK ABOUT IT A stroll through the produce section of a grocery store will convince you that plants are important. Even a medium-sized food store will contain products made from hundreds of different plant species. But which ones are the most important? Are there certain plants that we simply couldn't live without?

Agriculture

Which crops are the major food supply for humans?

The importance of agriculture—the systematic cultivation of plants—should be obvious, even to those of us who live in urban areas and seldom visit a farm. Modern farming is the foundation on which human society is built. North America has some of the richest, most productive cropland in the world. As a result, farmers in the United States and Canada produce so much food that they are able to feed millions of people around the world as well as their own citizens.

Worldwide Patterns Many scholars now trace the beginnings of human civilization to the cultivation of crop plants. Evidence suggests that agriculture developed separately in many parts of the world about 10,000 to 12,000 years ago. Once people discovered how to grow plants for food, the planting and harvesting of crops tended to keep them in one place for much of the year, leading directly to the establishment of social institutions. Even today, agriculture is the principal occupation of more human beings than any other activity.

Thousands of different plants—nearly all of which are angiosperms—are raised for food in various parts of the world. Yet, despite this diversity, much of human society depends upon just a few of these plants. **Worldwide, most people depend on a few crop plants, such as rice, wheat, soybeans, and corn, for the bulk of their food supply.** The same crops are also used to feed livestock.

Key Questions

Which crops are the major food supply for humans?

What are some examples of benefits besides food that humans derive from plants?

Vocabulary

green revolution

Taking Notes

Preview Visuals Preview **Figure 24–24.** Identify what plants provided the raw materials for the products shown in the photos. Then list any other products you can think of that come from plants.

FIGURE 24–21 Plants and Agriculture Rice is a staple crop in China and many nations of Southeast Asia.

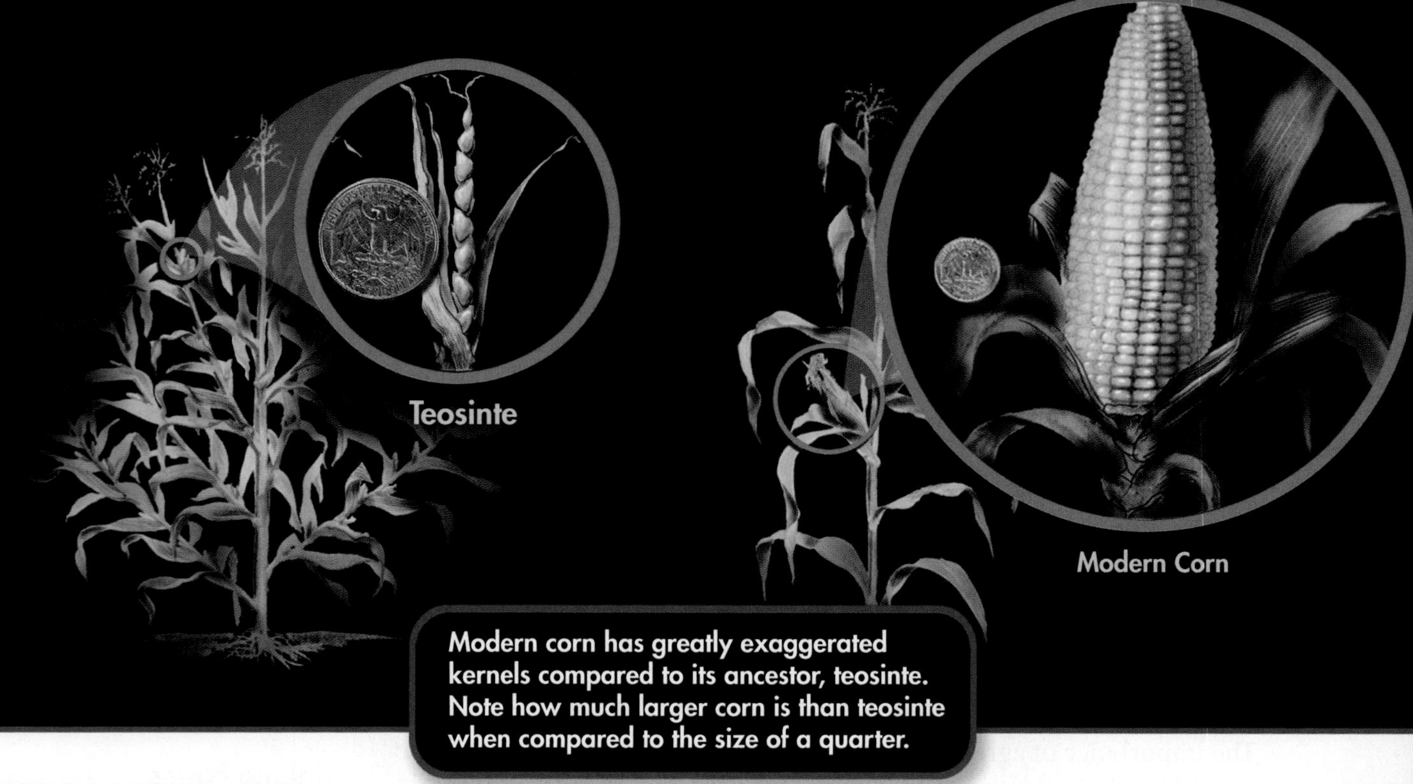

FIGURE 24–22 From Wild Grass to Staple Crop The selective breeding of a wild grass called teosinte about 8000 years ago led to the development of maize and modern corn.

You may not have thought of it this way, but the food we eat from most crop plants is taken from their seeds. For nutrition, most of humanity worldwide depends on the endosperm of only a few carefully cultivated species of grass. The pattern in the United States follows this trend. Roughly 80 percent of all U.S. cropland is used to grow just four crops: wheat, corn, soybeans, and hay. Of these crops, three—wheat, corn, and hay—are derived from grasses.

New Plants The discovery and introduction of new crop plants has frequently changed human history. Before they were discovered in the Americas, many important crops—including corn, peanuts, and potatoes—were unknown in Europe. The introduction of these plants changed European agriculture rapidly. We think of boiled potatoes, for example, as a traditional staple of German and Irish cooking, but 400 years ago, potatoes were new items in the diets of Europeans.

The efficiency of agriculture has been improved through the selective breeding of crop plants and improvements in farming techniques. Recall from Chapter 15 that selective breeding is a method for improving a species by allowing only organisms with certain traits to produce the next generation. The corn grown by Native Americans, for example, was developed more than 8000 years ago from teosinte, a wild grass found in Mexico. Further selective breeding has produced modern-day corn. The changes caused by selective breeding can be very dramatic, as shown in **Figure 24–22.**

In more recent times, other familiar crops have been the product of selective breeding. Sugar beets, the source of most refined sugar from the United States, were produced from the ordinary garden beet using selective breeding. Plants as different as cabbage, broccoli, and Brussels sprouts have been developed from a single species of wild mustard.

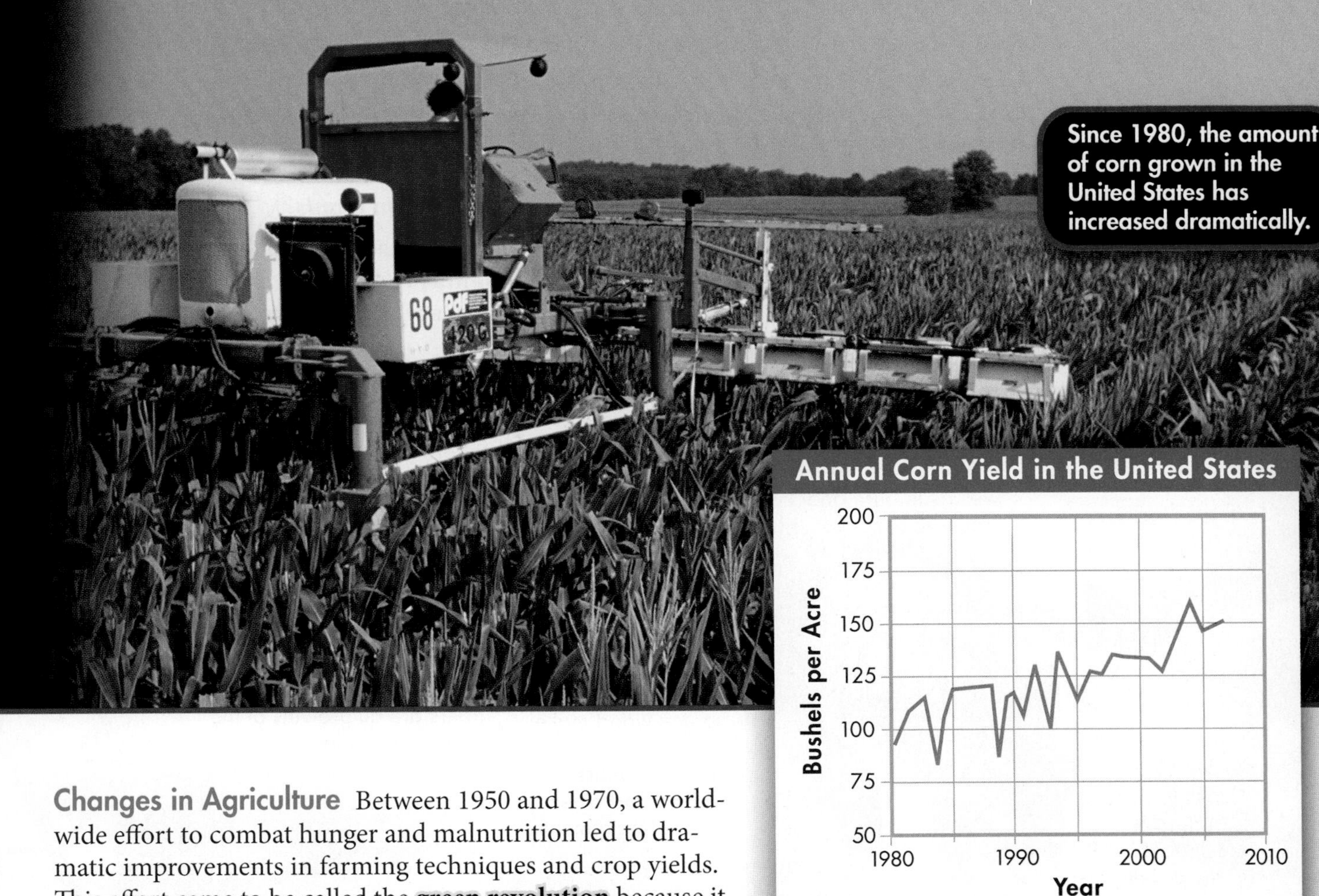

Changes in Agriculture Between 1950 and 1970, a worldwide effort to combat hunger and malnutrition led to dramatic improvements in farming techniques and crop yields. This effort came to be called the **green revolution** because it greatly increased the world's food supply. Green revolution technologies enabled many countries to end chronic food shortages and, in some cases, become exporters of surplus food.

At the heart of the green revolution was the use of high-yield varieties of seed and fertilizer. For thousands of years, farmers have added essential nutrients in the form of natural fertilizers such as animal manure. While some farmers today still use these traditional methods, many farmers use artificial fertilizers.

Fertilizers are labeled with three numbers that reflect the percentage by weight of three elements: nitrogen (N), phosphorus (P), and potassium (K). A bag of garden fertilizer labeled "20-10-5" is 20 percent nitrogen, 10 percent phosphorus, and 5 percent potassium by weight.

Fertilizers and pesticides must be used with great care. Overfertilizing can kill crop plants by putting too high a concentration of salts into the soil. The intensive use of fertilizers can also affect the groundwater. When large amounts of nitrogen- and phosphate-containing fertilizer are used near wetlands and streams, runoff from the fields may contaminate the water. Pesticides can also pose a health risk. Chemical pesticides are poisons, and they have the potential to harm wildlife and leave dangerous chemical residues in food.

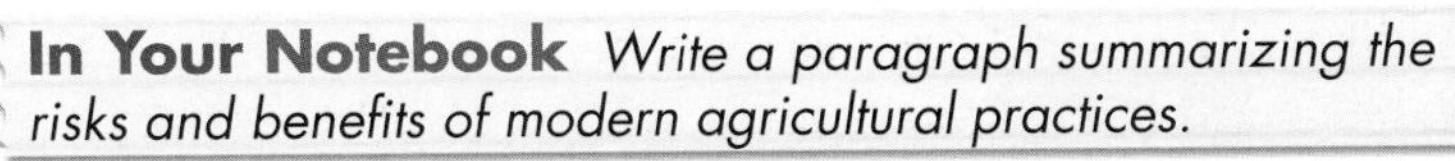

In Your Notebook *Write a paragraph summarizing the risks and benefits of modern agricultural practices.*

FIGURE 24–23 Reading a Fertilizer Label Three numbers typically appear on a fertilizer label. **Apply Concepts** ***Describe what the numbers on this label mean.***

Fiber, Wood, and Medicine

What are some examples of other benefits besides food that humans derive from plants?

Some of the most important uses of plants have nothing to do with food. **Plants produce the raw materials for our homes and clothes, and some of our most powerful and effective medicines.** Some examples of plant products are shown in **Figure 24–24.** If you're reading this page out of the printed book in your classroom, you are turning paper pages made from the conifer forests of North America, possibly sitting on a chair made from oak tree xylem, and probably wearing at least one piece of clothing made from the fibers of the cotton plant.

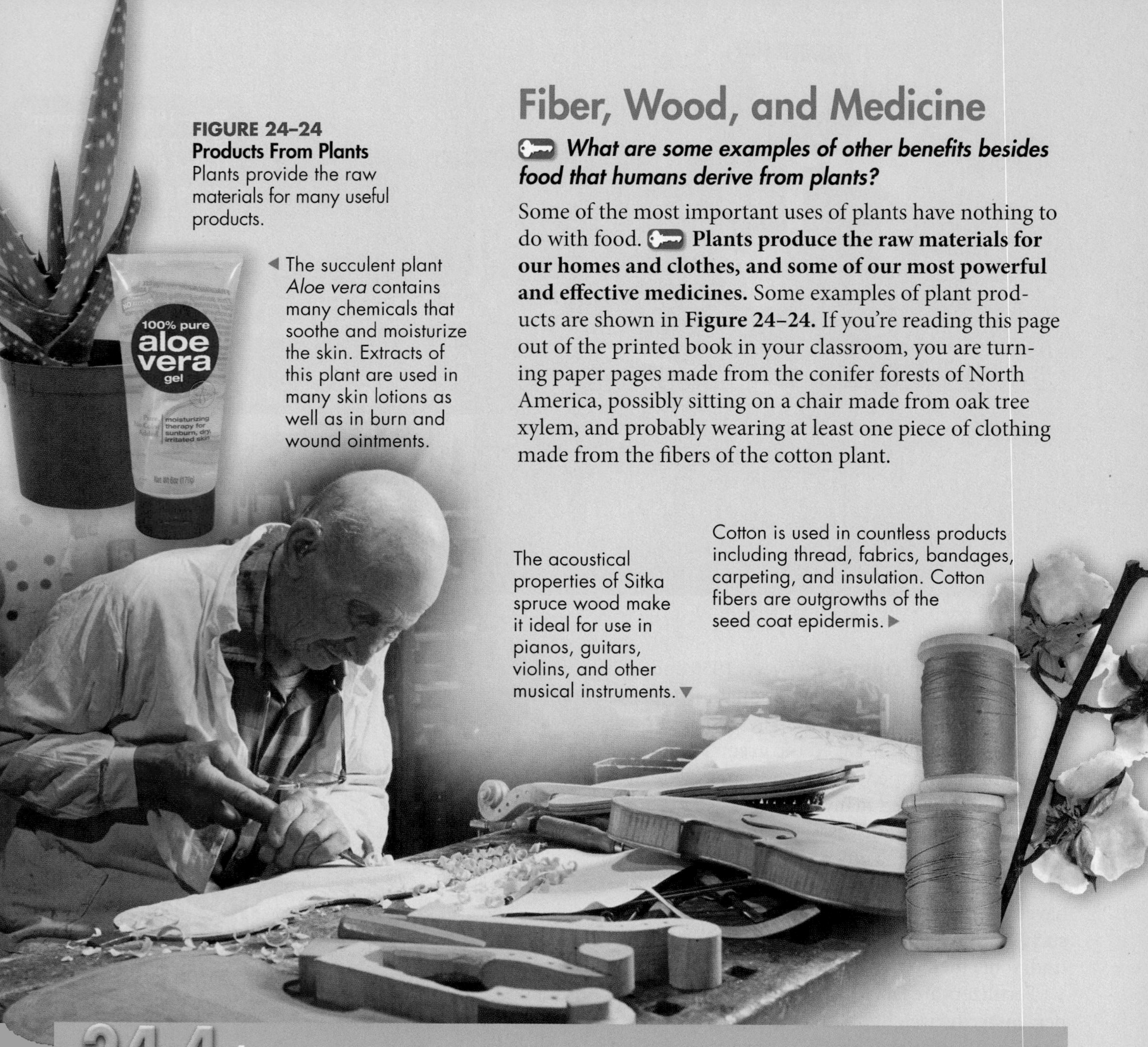

FIGURE 24–24
Products From Plants Plants provide the raw materials for many useful products.

◀ The succulent plant *Aloe vera* contains many chemicals that soothe and moisturize the skin. Extracts of this plant are used in many skin lotions as well as in burn and wound ointments.

The acoustical properties of Sitka spruce wood make it ideal for use in pianos, guitars, violins, and other musical instruments. ▼

Cotton is used in countless products including thread, fabrics, bandages, carpeting, and insulation. Cotton fibers are outgrowths of the seed coat epidermis. ▶

24.4 Assessment

Review Key Concepts

1. a. Review Name four crops that make up the base of the world's food supply.

b. Relate Cause and Effect Describe how selective breeding was used to develop corn from an ancestral grass that looked very different.

2. a. Review Besides food, what other important products are developed from plants?

b. Infer What effect could plant species extinction have on therapeutic drug development?

ANALYZING DATA

Use the line graph from **Figure 24–22** to answer the questions below.

3. Calculate By about how much did the amount of corn produced per acre of farmland increase between 1985 and 2005? MATH

4. Interpret Graphs How would you describe the overall trend in the data?

5. Predict What factors do you think might influence corn production in the next decade?

Biology & HISTORY

WHST.9-10.6 Production and Distribution of Writing, WHST.9-10.7 Research to Build and Present Knowledge. Also RST.9-10.10

The Evolution of Agriculture **More than 10,000 years ago, humans began a gradual transition from hunter-gatherer societies to civilizations that were reliant on crops—many of which are still grown today.**

7500 7000 6500 6000 5500 5000 4500 4000

8000 B.C.

Inhabitants of the Middle East begin to farm wheat. The change from gathering a crop in the wild to farming it eventually contributes to the rise of one of the earliest Middle Eastern civilizations.

5500 B.C.

▲ Barley is grown in the Nile Valley of Egypt. About 2000 years later, farming settlements are united throughout the Nile Valley, and Egyptian culture flourishes.

4500 B.C.

Rice farming becomes well established in southern China, southeast Asia, and northern India. Rice farming spreads widely from these regions, and rice later becomes a major Chinese export.

3500 B.C.

The potato is farmed in the Andes Mountains of South America. Early Andean farmers eventually produce hundreds of different varieties of potatoes by growing them on irrigated terraces built on mountain slopes. ▼

5000 B.C.

People in central Mexico grow a form of corn called maize. Early corncobs are only about an inch long and have a few dozen kernels. The ancestor of corn was a wild grass called teosinte.

7000 B.C.

▲ Chilies and avocados become important additions to the diets of Mesoamerican people. Chilies are used for flavoring foods, and avocados provide vitamins and oils.

WRITING **The domestication of all major crops had a huge impact on the growth of civilizations. Choose one of the crops discussed above and research how that crop contributed to the rise of civilization and culture in the region discussed. Using technology, produce, publish, and update a poster display that summarizes your findings in words, pictures, and other graphics.**

NGSS Smart Guide Plant Reproduction and Response

Disciplinary Core Idea LS1.B Growth and Development of Organisms: How do organisms grow and develop? Plants reproduce, develop, and grow in response to cues from the environment such as light, temperature, and moisture. Hormones manufactured by the plant regulate the plant's response to change.

24.1 Reproduction in Flowering Plants

- Flowers are reproductive organs that are composed of four different kinds of specialized leaves: sepals, petals, stamens, and carpels.
- The process of fertilization in angiosperms is distinct from that found in other plants. Two fertilization events take place—one produces the zygote and the other a tissue, called endosperm, within the seed.
- Vegetative reproduction is the formation of new individuals by mitosis. It does not require gametes, flowers, or fertilization.

stamen 697 • anther 697 • carpel 697 • stigma 697 • pistil 697 • embryo sac 699 • double fertilization 700 • endosperm 700 • vegetative reproduction 702 • grafting 703

Biology.com

Untamed Science Video Using time-lapse videography, the Untamed Science crew reveals how plants move in response to various stimuli.

Art Review Review the structures of a flower.

Art in Motion Watch how meiosis and mitosis produce ova and pollen in an angiosperm.

24.2 Fruits and Seeds 

- As angiosperm seeds mature, ovary walls thicken to form a fruit that encloses the developing seeds.
- Seeds contained in fleshy, nutritious fruits are usually dispersed by animals.
- Seeds dispersed by wind or water are typically contained in lightweight fruits that allow them to be carried in the air or in buoyant fruits that allow them to float on the surface of the water.
- Environmental factors such as temperature and moisture can cause a seed to end dormancy and germinate.

dormancy 706 • germination 706

Biology.com

Taking Notes Make a flowchart that shows the process of germination and the factors that influence it.

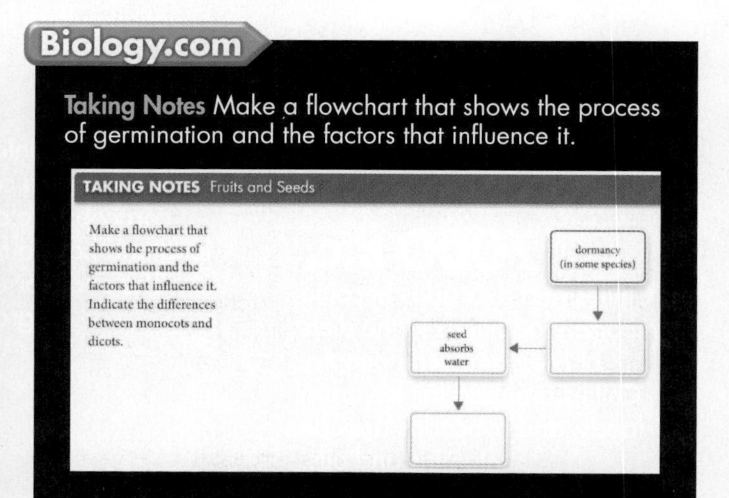

CHECKING YOUR *Scientific Literacy*

Refer to the lesson content and digital assets below as you prepare for your chapter assessment. Then, evaluate your understanding of plant reproduction and response by answering these questions.

1. **Scientific and Engineering Practice Constructing Explanations** Why do seed packets for some types of plants, such as tomatoes, advise gardeners to plant the seeds indoors six to eight weeks prior to the last frost? Would these instructions apply if you live in an area where there is a short winter season? Explain.

2. **Crosscutting Concept Cause and Effect** A news report states that the population of bees in the area has decreased considerably. What effect will this have on a tomato plant crop, if any? Explain. (*Hint:* Tomato plants are self-fertilizing.)

3. **Key Ideas and Details** Cite specific textual evidence to explain the different ways plants respond to their environments.

4. **STEM** You work at a garden center and want to "coax" the chrysanthemum plants to flower early so that customers will be more likely to purchase them. Describe a method to achieve this and explain why your method will work.

24.3 Plant Hormones

- Plant hormones serve as signals that control development of cells, tissues, and organs. They also coordinate responses to the environment.
- The opposing effects of plant hormones contribute to the balance necessary for homeostasis.
- Plants respond to environmental stimuli such as light, gravity, and touch.
- Photoperiodism is a major factor in the timing of seasonal activities such as flowering and growth.
- As cold weather approaches, deciduous plants turn off photosynthetic pathways, transport materials from leaves to roots, and seal off leaves.

hormone 708 • target cell 709 • receptor 709 • auxin 709 • apical dominance 710 • cytokinin 710 • gibberellin 711 • abscisic acid 711 • ethylene 711 • tropism 712 • phototropism 712 • gravitropism 712 • thigmotropism 712 • photoperiodism 713

Biology.com

InterActive Art Change the length of day and see how it affects both short- and long-day plants.

24.4 Plants and Humans

- Worldwide, most people depend on a few crop plants for the bulk of their food supply.
- Plants produce the raw materials for our homes and clothes, and some of our most powerful and effective medicines.

green revolution 717

Biology.com

Data Analysis Analyze potential fuel energy that comes from different plants and different plant parts.

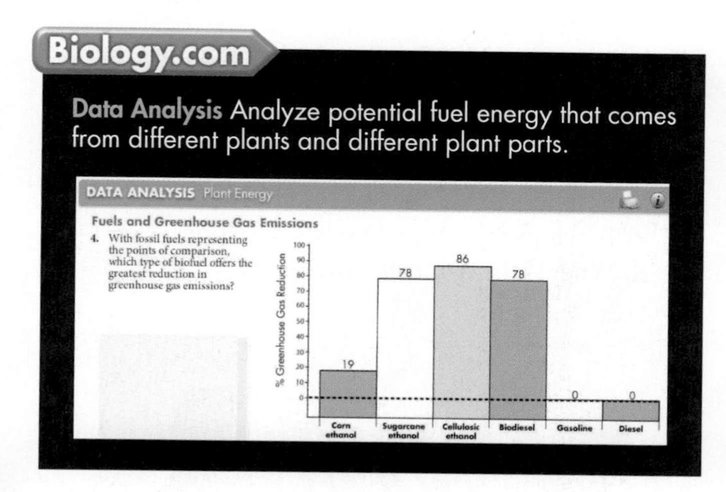

Real-World Lab

Plant Hormones and Leaves This chapter lab is available in your lab manual and at **Biology.com**

24 Assessment

24.1 Reproduction in Flowering Plants

Understand Key Concepts

1. In angiosperms, the structures that produce the male gametophyte are called the
   a. anthers.
   b. sepals.
   c. pollen tubes.
   d. stigmas.

2. Pollination occurs when pollen lands on
   a. the style.
   b. the stigma.
   c. the filament.
   d. the anther.

3. The process in which a single plant produces many offspring genetically identical to itself is
   a. sexual reproduction.
   b. agriculture.
   c. dormancy.
   d. vegetative reproduction.

4. What is a carpel? Where is it located in a typical flower?

5. Describe at least two ways in which pollen is transferred from one plant to another.

6. What are the products of double fertilization? Describe them.

Think Critically

7. **Interpret Visuals** The diagram below shows the parts of a typical flower.
   a. Inside which structure is pollen produced?
   b. What structure is represented by A? What is its function?
   c. In which structure do seeds develop?
   d. What is the name of structure G?

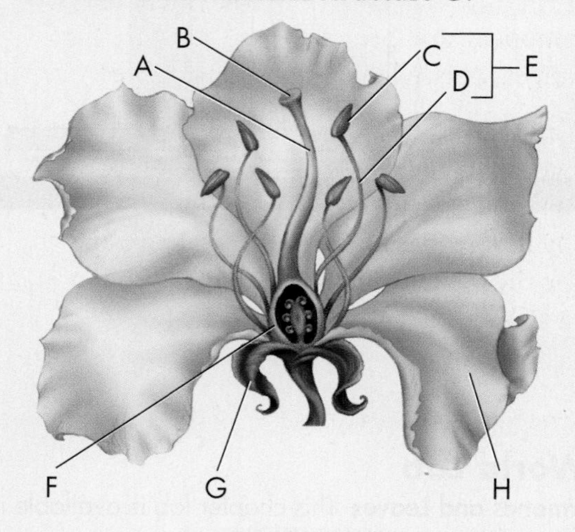

24.2 Fruits and Seeds

Understand Key Concepts

8. The thickened ovary wall of a plant may join with other parts of the flower to become the
   a. fruit.
   b. seed.
   c. endosperm.
   d. cotyledon.

9. The period during which the embryo is alive but not growing is called
   a. fertilization.
   b. vegetative growth.
   c. dormancy.
   d. germination.

10. Give examples of seed dispersal by animal, wind, and water.

11. What is the function of dormancy?

Think Critically

12. **Key Ideas and Details** Some plants form flowers that produce stamens but no carpels. Could fruit form on one of these flowers? Cite textual evidence to support your answer.

13. **Infer** The seeds of lupines, an arctic plant, can remain dormant for thousands of years. Why might this trait be important to a plant in an arctic environment?

14. **Design an Experiment** A friend suggests that seeds do not need cotyledons to grow. You argue that cotyledons are important to seeds. Design a controlled experiment that shows the effect on seed growth of removing cotyledons.

24.3 Plant Hormones

Understand Key Concepts

15. Chemical signals in plants affecting the growth, activity, and development of cells and tissues are called
    a. hormones.
    b. enzymes.
    c. auxins.
    d. phytochromes.

16. Substances that stimulate cell division and cause dormant seeds to sprout are
    a. gibberellins.
    b. auxins.
    c. cytokinins.
    d. phytochromes.

17. Photoperiod is a measurement of
   a. water level.
   b. day length.
   c. gravity.
   d. nutrients.

18. Craft and Structure Explain how auxins act in opposition to cytokinins.

19. Craft and Structure What is a tropism? Give one example of a tropism that affects plant stems and another example of a tropism that affects roots.

20. Describe two different ways in which a plant may respond to changes in photoperiod.

21. Describe what happens to deciduous plants during winter dormancy.

Think Critically

22. **Form a Hypothesis** Describe a particular plant thigmotropism and hypothesize how it benefits the plant.

23. **Infer** Spinach is a long-day plant that grows best with a night length of 10 hours or less. Why is spinach not usually grown near the equator?

24.4 Plants and Humans

Understand Key Concepts

24. The first indications of human agriculture occurred about
   a. 1000 years ago.
   b. 10,000 years ago.
   c. 100,000 years ago.
   d. 1,000,000 years ago.

25. The majority of human plant food comes from plants that are
   a. gymnosperms.
   b. perennials.
   c. angiosperms.
   d. conifers.

26. Give an example of a plant you have eaten in the last 24 hours that you think is a product of selective breeding. Explain why you think so.

Think Critically

27. **Infer** The bulk of human plant foods comes from seeds, which constitute only a small part of the plant body. Explain how this is possible.

28. **Compare and Contrast** Compare and contrast the benefits and the dangers of using pesticides and fertilizers to grow food crops.

29. **Form a Hypothesis** Form a hypothesis to explain why plants are a good source of medicines.

solve the CHAPTER MYSTERY

THE GREEN LEMONS

To solve the mystery of why the lemons remained unripe, growers remembered a story from the nineteenth century. In those days, gas streetlights were commonly used in large cities. A few years after such streetlights were installed, city dwellers noticed that trees growing near the streetlamps had developed short, thick stems, and dropped their leaves much earlier than they should have. It was as if hormone levels in the trees had been affected. And, in fact, the levels had changed. One of the components of the gas used in the lights was ethylene.

Recall from Lesson 24.3 that one effect of ethylene is to stimulate the ripening of fruit. Whether the ethylene is manufactured by the plant or externally, as in the case of a kerosene heater, doesn't matter. Because ethylene is a gas, it can diffuse through the air, cell walls, and membranes of a plant and its fruit. When that source of ethylene was taken away from the picked lemons (when the electric heaters replaced the kerosene ones)—the ripening stimulus was removed, and the lemons stayed green.

1. **Relate Cause and Effect** Tomatoes put in a paper bag with apples ripen much more quickly than those placed in the open air. What would this suggest about the effects of ripe apples on unripe tomatoes?

2. **Propose a Solution** How could farmers, shippers, and produce marketers use the effects of ethylene to their advantage?

3. **Connect to the Big idea** Recent studies have shown that gaseous hormones are involved in a plant's systemwide response to an attack by herbivores such as caterpillars. What might be the benefits of a gaseous hormone in such a situation?

Connecting Concepts

Use Science Graphics

Recall that growth responses of plants to external stimuli are called *tropisms.* A tropism is positive if the affected plant part grows toward the stimulus. The response is negative if the plant part grows away from the stimulus. The experiment shown below was intended to test the effect of gravitropism on plant growth. The conclusion drawn from the experiment was that the plant stems grow upward due to negative gravitropism. Use the diagram to answer questions 30–33.

30. Interpret Visuals Describe the three experimental setups and the result of each.

31. Form a Hypothesis What was the probable hypothesis for this experiment?

32. Interpret Visuals From the experimental setups shown, was the hypothesis successfully tested? Explain.

33. Evaluate and Revise Indicate what kinds of changes you would make to improve this experimental design.

Write About Science

34. © Integration of Knowledge and Ideas Sometimes, people grow houseplants on a windowsill. Books on houseplants often advise giving the plants a one-quarter turn every other week. Write a paragraph explaining why turning the plant is a good idea. (*Hint:* Be sure to include an explanation of tropisms in your answer.)

35. Assess the Big idea Trace the text's explanation of why flowers are the key to the evolutionary success of the angiosperms.

Analyzing Data

Integration of Knowledge and Ideas

In a laboratory experiment, fruits from five different kinds of trees were dropped from a height of 4 meters, and the time it took them to reach the ground was measured. Assume that for every second a fruit falls, it is carried 1.5 meters away from the parent tree.

Fruit Type Versus Dispersal Time

Type of Tree	Average Time (s) for Seed to Fall 4 m
Norway maple	5.2
Silver maple	4.9
White ash	3.1
Shagbark hickory	0.9
Red oak	0.9

36. Draw Conclusions Based on the data and the illustrations of the fruit structures, which of the following is the most reasonable conclusion?

a. Winged seeds carry more nutrition for the growing embryo than do seeds without wings.
b. Wind is not very effective in carrying seeds away from the parent plant.
c. Acorns are more likely to germinate if they fall close to the parent plant.
d. Red oak and hickory depend on factors other than wind to achieve dispersal.

37. Analyze Data Given the same wind, which fruit type is most likely to be carried farthest from the parent tree?

a. Red oak
b. Silver maple
c. Norway maple
d. White ash

Standardized Test Prep

Multiple Choice

1. Where in a flower are pollen grains produced?
A sepals
B carpels
C anthers
D ovary

2. Which part of the flower develops into a fruit?
A pollen tube
B sepals
C stigma
D ovary

3. Which flower structure includes all the others?
A style
B carpel
C stigma
D ovary

4. The trumpet honeysuckle has long, red, narrow tubular flowers. What is the most likely means of pollination?
A wind
B water
C bee
D hummingbird

5. All of the following are fruits EXCEPT
A tomato.
B corn.
C potato.
D cucumber.

6. Seeds that are contained in large, fleshy fruits are usually dispersed by
A animals.
B water.
C wind.
D rotting.

7. Which of the following causes fruit to ripen?
A auxin
B cytokinin
C ethylene
D gibberellin

8. Which is an example of thigmotropism?
A change in leaf color
B climbing vines
C blooming
D photoperiod

Questions 9–10

The results of an experiment are summarized in the art below.

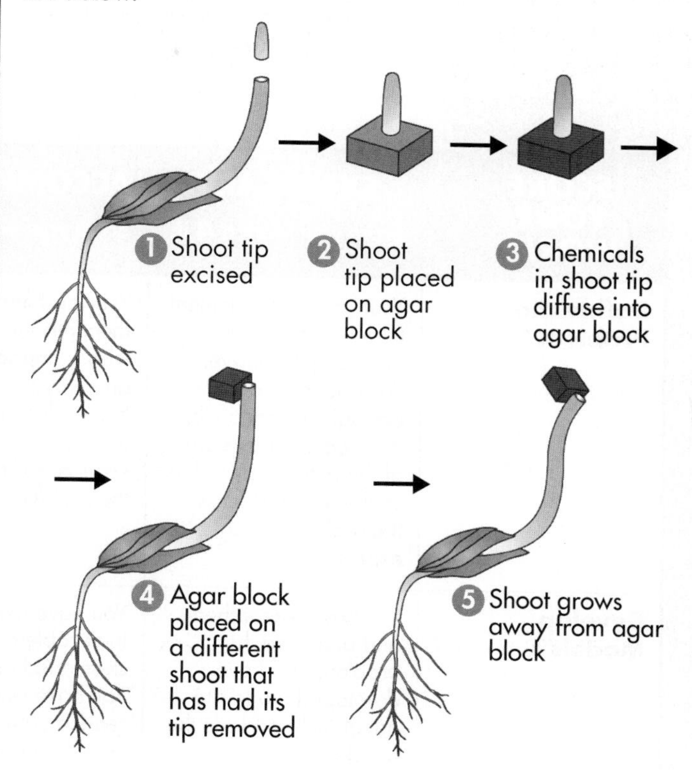

9. Which of the following can be concluded from the results of this experiment alone?
A Hormones are produced in the growing tips of plant roots.
B Plants grow toward the sun due to compounds produced in their stems.
C Agar blocks contain a variety of plant compounds.
D Compounds produced in shoot tips can cause stems to bend.

10. Applying your knowledge of specific plant hormones, explain the results.

Open-Ended Response

11. Describe why seed dormancy is a valuable adaptation that has helped explain the evolutionary success of seed plants.

If You Have Trouble With . . .

Question	1	2	3	4	5	6	7	8	9	10	11
See Lesson	24.1	24.2	24.1	24.1	24.2	24.2	24.3	24.3	24.3	24.3	24.2

NGSS Problem-based Learning

UNIT PROJECT 6

Your Self-Assessment

The rubric below will help you evaluate the design of your green roof system.

SCORE YOUR WORK!	EXEMPLARY Score your work a 4 if:	ACCOMPLISHED Score your work a 3 if:	DEVELOPING Score your work a 2 if:	BEGINNING Score your work a 1 if:
Define the Problem ▶	Your problem statement is well researched and comprehensive. You understand the ecological, economic, and social dimensions of the problem. You feel as if you can discuss the problem as an expert.	Your problem statement identifies the problem, who is impacted by it, and why it is important. You understand the ecological and economic dimensions of the problem.	You have written a problem statement that identifies the problem, but you are still gathering background information.	You have identified what you think is a problem but are unsure how to explain or define it.
Develop Models ▶	You have researched and analyzed the problem and have developed models to adequately address the problem.	You have researched the problem and developed models. You have revised and refined your models based on feedback.	You have developed models, but you are unsure how to compare, evaluate, and improve upon them.	You have developed models but are unsure how they relate to the problem you are trying to solve.
	TIP Even the "best" model will have some trade-offs. If you can't single out one optimal model, present a range of models that are each optimized for a particular scenario.			
Design a Solution ▶	You have specified effective design criteria and constraints and developed a solution that has been refined based on discussion and feedback.	You have specified design criteria and constraints and developed a solution, but you have not obtained feedback.	You have identified some design criteria but no constraints. Your solution has not been refined to optimize the achievement of your design criteria.	You understand the difference between criteria and constraints but are unsure how they apply to your solution.
	TIP Well-defined criteria and constraints will make it easier for you to evaluate possible solutions.			
Communicate Information ▶	You have presented your finished proposal to an audience beyond your class or school. Your design plan can be used as a case study to help promote green roofs.	Your proposal has been discussed, revised, and edited based on feedback. The final draft makes a convincing argument for your design solution.	You have organized your documents and findings into a draft proposal, but you have not obtained feedback on it.	You have notes and materials documenting the design process, but you are unsure how to present them.
	TIP Persuasive ideas can be communicated with the use of tables, graphs, drawings, or models.			

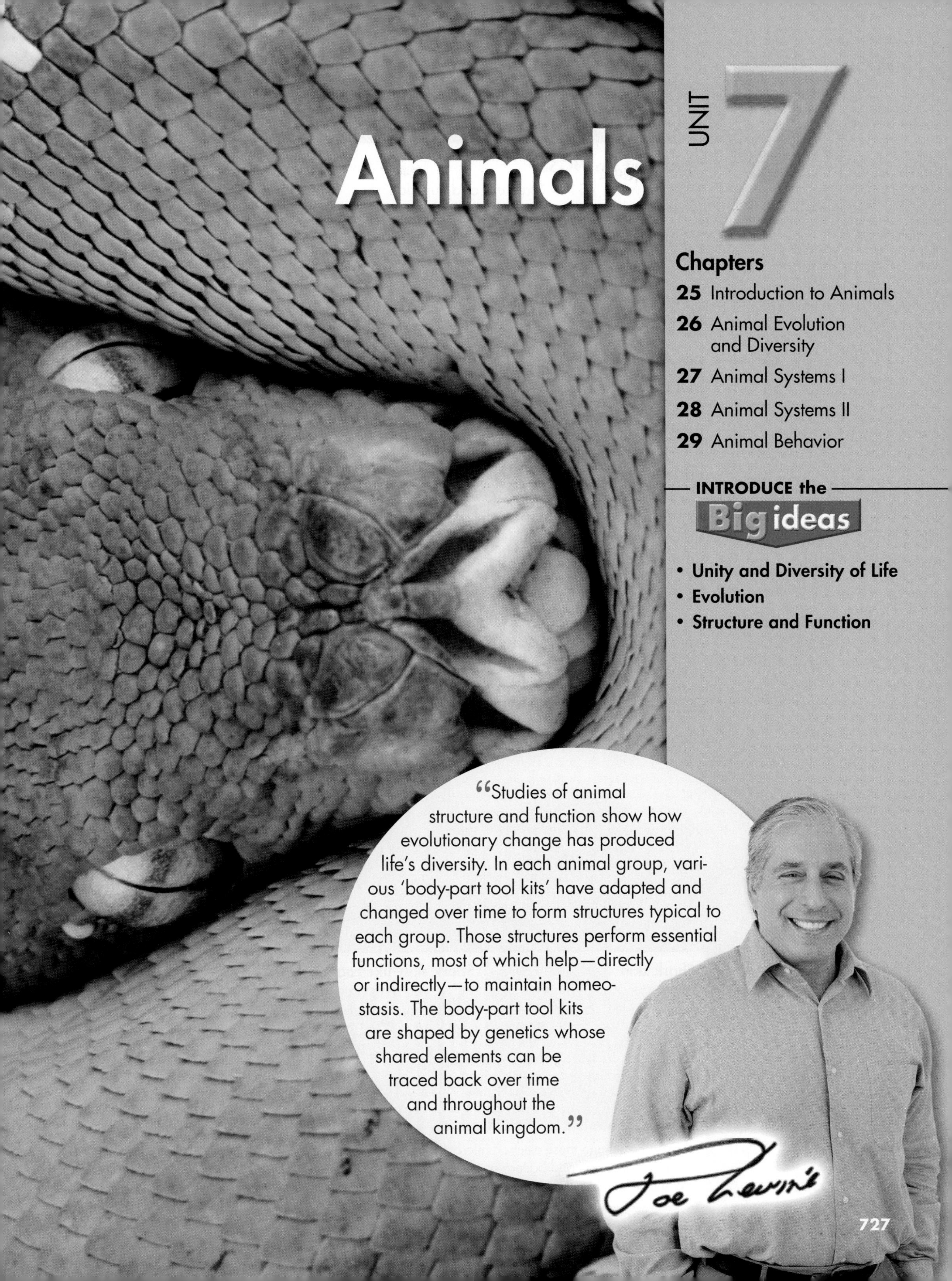

Animals

UNIT 7

Chapters

25 Introduction to Animals

26 Animal Evolution and Diversity

27 Animal Systems I

28 Animal Systems II

29 Animal Behavior

INTRODUCE the **Big ideas**

- **Unity and Diversity of Life**
- **Evolution**
- **Structure and Function**

"Studies of animal structure and function show how evolutionary change has produced life's diversity. In each animal group, various 'body-part tool kits' have adapted and changed over time to form structures typical to each group. Those structures perform essential functions, most of which help—directly or indirectly—to maintain homeostasis. The body-part tool kits are shaped by genetics whose shared elements can be traced back over time and throughout the animal kingdom."

Joe Levine

NGSS Problem-based Learning

UNIT 7 PROJECT

Spotlight on NGSS
- **Core Idea LS1.A** Structure and Function
- **Practice** Developing and Using Models
- **Crosscutting Concept** Structure and Function

Biomimicry

Natural selection has solved many technological and engineering problems, including waterproofing, energy capture, food production, transportation, and packaging. Thus, looking to nature can provide inspiration for innovative designs and solutions to our problems. Biomimicry is the study and imitation of nature in order to design and engineer human-made objects.

For example, consider the inspiration sharks have provided for new technologies. The skin of a shark might look smooth from far away, but up close you can see that it is made up of small scales, called dermal denticles. The grooves on these dermal denticles prevent marine organisms from adhering to the shark. Scientists have recently been inspired by shark skin and have developed surfaces and coatings with features that mimic the dermal denticles. Such coatings can be used on ship hulls to prevent barnacles from attaching to them, making them lighter and more energy efficient. Additionally, these modified surfaces and coatings are also being used in the medical industry to prevent bacteria from adhering to a variety of surfaces.

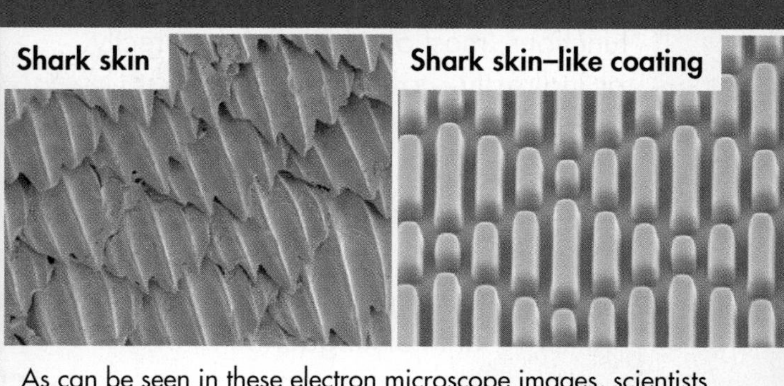

As can be seen in these electron microscope images, scientists mimicked the diamond pattern on shark skin (left) when developing the shark skin–like coating (right).

Your Task: Be Inspired by Nature!

STEM

Take the role of an engineer and design an object that is inspired by a unique trait of an animal. Think about items you use in your daily life, such as electronic devices, books, shoes, and clothing. How can one of these items be improved or redesigned using the principles of biomimicry? Begin by researching different animals and their traits. Identify a specific trait of an animal that would be beneficial in improving an existing item. Develop a model to illustrate your idea. Use the rubric at the end of the unit to evaluate your work for this task.

Define a Problem Select a unique animal trait that you would like to research and evaluate. Begin by asking questions about the structure and function of this trait. What problems does this trait solve for the animal? Write a statement that describes the animal that you chose, its unique trait, and the technological problems that this trait may solve.

Develop Models Apply this trait to an item you use by reimagining the item. Develop ideas for how this trait could solve a problem or limitation of this item. Your idea could be an improvement or alternative to a current solution. Develop your idea with a written description and drawing, or construct a three-dimensional model to illustrate the structure and function of your item. Write your reasoning behind your ideas.

Plan and Carry Out an Investigation What else can you learn about the structure of the animal's trait and its use? Write a plan to identify the information you need to predict how the structure of the animal trait you chose will improve the function of the item you have chosen to modify. Identify the type of data you need to collect for you to be able to improve upon your model. Then, carry out your investigation using Internet resources, books or articles about your animal, or by observing your animal in nature or at a local zoo.

Analyze and Interpret Data Organize your research by creating a table, chart, graph, or another tool. Interpret the data by identifying how your research relates to your model. Do you need any more data to refine your model? Adjust your model based on your investigation and analysis. You can adjust the model itself or write a description of what changes need to be made.

Communicate Information Present your new product. Your presentation can be in the form of a report, video, or website that showcases your design and explains the reasoning behind your design decisions. Ask for feedback about your product from your teacher or classmates.

Biology.com

Go online for resources on biomimicry and engineering that will help you develop your model.

25 Introduction to Animals

Unity and Diversity of Life

Q: What characteristics and traits define animals?

INSIDE:

- 25.1 What Is an Animal?
- 25.2 Animal Body Plans and Evolution

Though they look very different, the hundreds of animal species that make up and live near a coral reef share characteristics common to all animals.

BUILDING *Scientific Literacy*

Deepen your understanding of animals by engaging in key practices that allow you to make connections across concepts.

NGSS You will develop and use models to illustrate the characteristics and traits of animals.

STEM Have you ever wondered what it takes to be a marine biologist? Watch a video and find out!

Common Core You will use quantitative reasoning and draw information from texts to support your analysis of animal characteristics and traits.

CHAPTER MYSTERY

SLIME DAY AT THE BEACH

It was a warm October day in Massachusetts when phone calls started streaming in to beach offices, aquariums, and even 9-1-1 lines. Beaches near Boston were coated with a thick, glistening layer of jellylike ooze. Beachgoers were mystified and worried. Some thought there had been an oil spill, but police and fire personnel verified that it was not oil.

More slimy masses kept washing up onto the seashores. People noticed that some of the gooey blobs appeared to be pulsating with life. When they looked closely, investigators saw that the slime was made up of small, individual critters—each transparent and the size of a fingernail. But what were they? As you read this chapter, look for clues to help you determine what the slime was.

Biology.com

Finding the solution to the Slime Day at the Beach mystery is only the beginning. Take a video field trip with the ecogeeks of Untamed Science to see where the mystery leads.

Go online to access additional resources including:
• eText • Flash Cards • Lesson Overviews • Chapter Mystery

25.1 What Is an Animal?

Key Questions

What characteristics do all animals share?

What characteristics distinguish invertebrates and chordates?

What essential functions must animals perform to survive?

Vocabulary

invertebrate
chordate
notochord
pharyngeal pouch
vertebrate
feedback inhibition

Taking Notes

Outline As you read, make an outline about the features of animals.

For more on the diversity of animals, go to the Visual Guide.
DOL•30–DOL•64

THINK ABOUT IT An osprey circles a salt marsh searching for prey. Suddenly, it dives, extending razor-sharp talons. With a triumphant whistle, it carries a struggling fish back to its young. On the bottom of the bay, worms burrow beneath rocks carpeted with orange sponges. In the air above, mosquitoes swarm, searching for a blood meal. All these different inhabitants of the Atlantic coast are animals.

Characteristics of Animals

What characteristics do all animals share?

All members of the animal kingdom share certain characteristics. Animals are all heterotrophs; they obtain nutrients and energy by eating other organisms. Animals are also multicellular; their bodies are composed of many cells. The cells that make up animal bodies are eukaryotic, containing a nucleus and membrane-bound organelles. Unlike the cells of algae, fungi, and plants, animal cells lack cell walls. **Animals, which are members of the kingdom Animalia, are multicellular, heterotrophic, eukaryotic organisms whose cells lack cell walls.**

Types of Animals

What characteristics distinguish invertebrates and chordates?

Animal diversity is so vast and differences among animals so great that we need to divide these organisms into groups to even begin talking about them. Animals are often classified into two broad categories: invertebrates and chordates.

Invertebrates More than 95 percent of animal species are informally called **invertebrates.** **Invertebrates include all animals that lack a backbone, or vertebral column.** Because this category lumps together organisms that *lack* a characteristic, rather than those that *share* a characteristic, "invertebrates" do not form a clade or any other kind of true category in the system of biological classification. Invertebrates include at least 33 phyla, which are the largest taxonomic groups of animals. Invertebrates include sea stars, worms, jellyfishes, and in sects. They range in size from dust mites to colossal squid more than 14 meters long.
DOL•31–DOL•45.

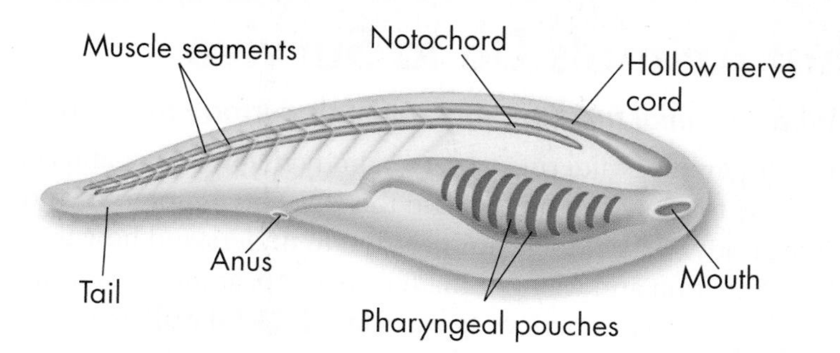

FIGURE 25–1 Characteristics of Chordates All chordates have a dorsal, hollow, nerve cord; a notochord; pharyngeal pouches; and a tail that extends beyond the anus. Some chordates possess all these traits as adults; others possess them only as embryos.

Chordates Fewer than 5 percent of animal species are **chordates,** members of the clade commonly known as Phylum Chordata. **All chordates exhibit four characteristics during at least one stage of life: a dorsal, hollow nerve cord; a notochord; a tail that extends beyond the anus; and pharyngeal** (fuh RIN jee ul) **pouches.** As you see in **Figure 25–1,** the hollow nerve cord runs along the dorsal (back) part of the body. Nerves branch from this cord at intervals. The **notochord** is a long supporting rod that runs through the body just below the nerve cord. Most chordates have a notochord only when they are embryos. At some point in their lives, all chordates have a tail that extends beyond the anus. **Pharyngeal pouches** are paired structures in the throat region, which is also called the pharynx. In some chordates, such as fishes, slits develop that connect pharyngeal pouches to the outside of the body. Pharyngeal pouches may develop into gills used for gas exchange.

Phylum Chordata includes some odd aquatic animals known as nonvertebrate chordates, which lack vertebrae. Most chordates, however, develop a backbone, or vertebral column, constructed of bones called vertebrae (singular: vertebra). Chordates with backbones are called **vertebrates.** Vertebrates include fishes, amphibians, reptiles, birds, and mammals. **DOL•46–DOL•64**

MYSTERY CLUE

Scientists verified that the organisms were young animals that had a stiff rod running along the tail. What does this suggest about the slimy critters?

FIGURE 25–2 Invertebrates and Chordates Both of these animals have fuzzy bodies with wings and both can fly, but the similarities end there. Butterflies are insects, which are invertebrates, and bats are mammals, which are chordates. **Classify** ***Bats have backbones. In which of the two major groups of chordates would you classify bats?***

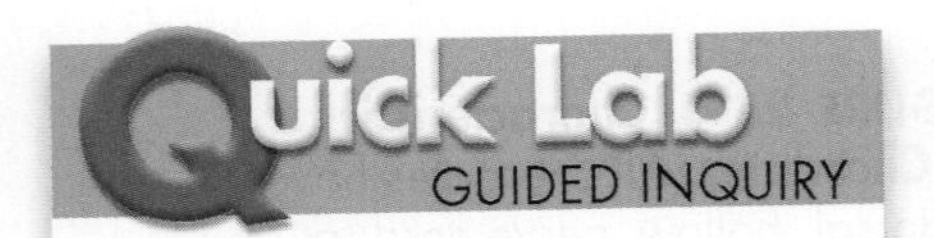

GUIDED INQUIRY

How Hydra Feed

1. Your teacher will provide you with hydra and *Daphnia*, small aquatic invertebrates. Using a dropper pipette, gently place one hydra onto a well slide.
2. Let the hydra adjust to its surroundings for 5 to 10 minutes.
3. Using your dropper, add one *Daphnia* to the slide.
4. Observe the hydra under a microscope.

Analyze and Conclude

1. Observe What happens when the *Daphnia* is added to the same slide as the hydra?

2. Draw Conclusions How do the hydra's tentacles help it to maintain homeostasis?

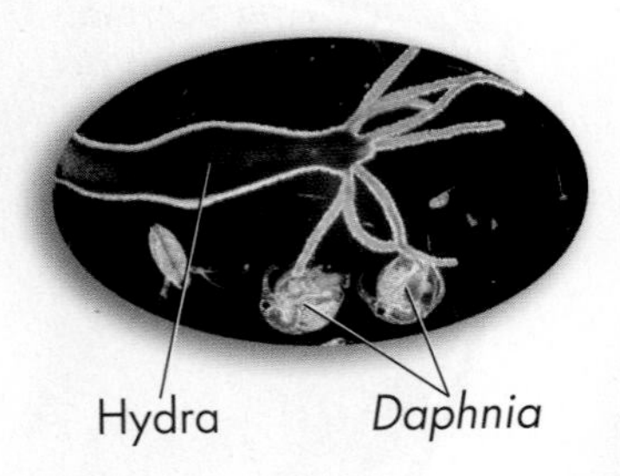

3. Pose Questions Formulate two questions about how the hydra survives in its environment.

What Animals Do to Survive

What essential functions must animals perform to survive?

Animals display a bewildering variety of body shapes, sizes, and colors. The best way to study and understand this diversity is not to memorize all the body parts of these animals, but to understand how the structures function and why. No matter their appearance, all animals must perform similar functions to stay alive. **Like all organisms, animals must maintain homeostasis by gathering and responding to information, obtaining and distributing oxygen and nutrients, and collecting and eliminating carbon dioxide and other wastes. They also reproduce.** The body systems that perform these functions are closely linked to one another. Over time, members of different animal phyla have evolved very different body structures that perform these essential functions. You will study these structures in more detail in Chapters 27 and 28.

Maintaining Homeostasis Recall that all organisms must keep their internal environment relatively stable, a process known as maintaining homeostasis. In animals, maintaining homeostasis is the most important function of all body systems. For example, most reptiles, birds, and mammals cannot excrete excess salt very well. Those that hunt or feed in salt water, such as the marine iguana in **Figure 25–3,** have adaptations that allow them to remove salt from their bodies.

Often, homeostasis is maintained by feedback inhibition. **Feedback inhibition,** or negative feedback, is a system in which the product or result of a process limits the process itself. If your house gets too cold, for example, the thermostat turns on the heat. As heat warms the house, the thermostat turns the heater off. Your body's thermostat works the same way. If you get too cold, you shiver, using muscle activity to generate heat. If you get too hot, you sweat, which helps you lose heat.

In this unit, you will learn about body systems in various animal groups. You will see how different groups have evolved different ways of ensuring their body systems stay in balance.

FIGURE 25–3 Homeostasis Marine iguanas are reptiles that feed in salt water. Reptile excretory systems are not adapted to process salt water. So these reptiles maintain homeostasis by sneezing a combination of salt and nasal mucus you might call "snalt." Snalt sometimes coats their bumpy heads and spiny necks, as you can see in this photo.

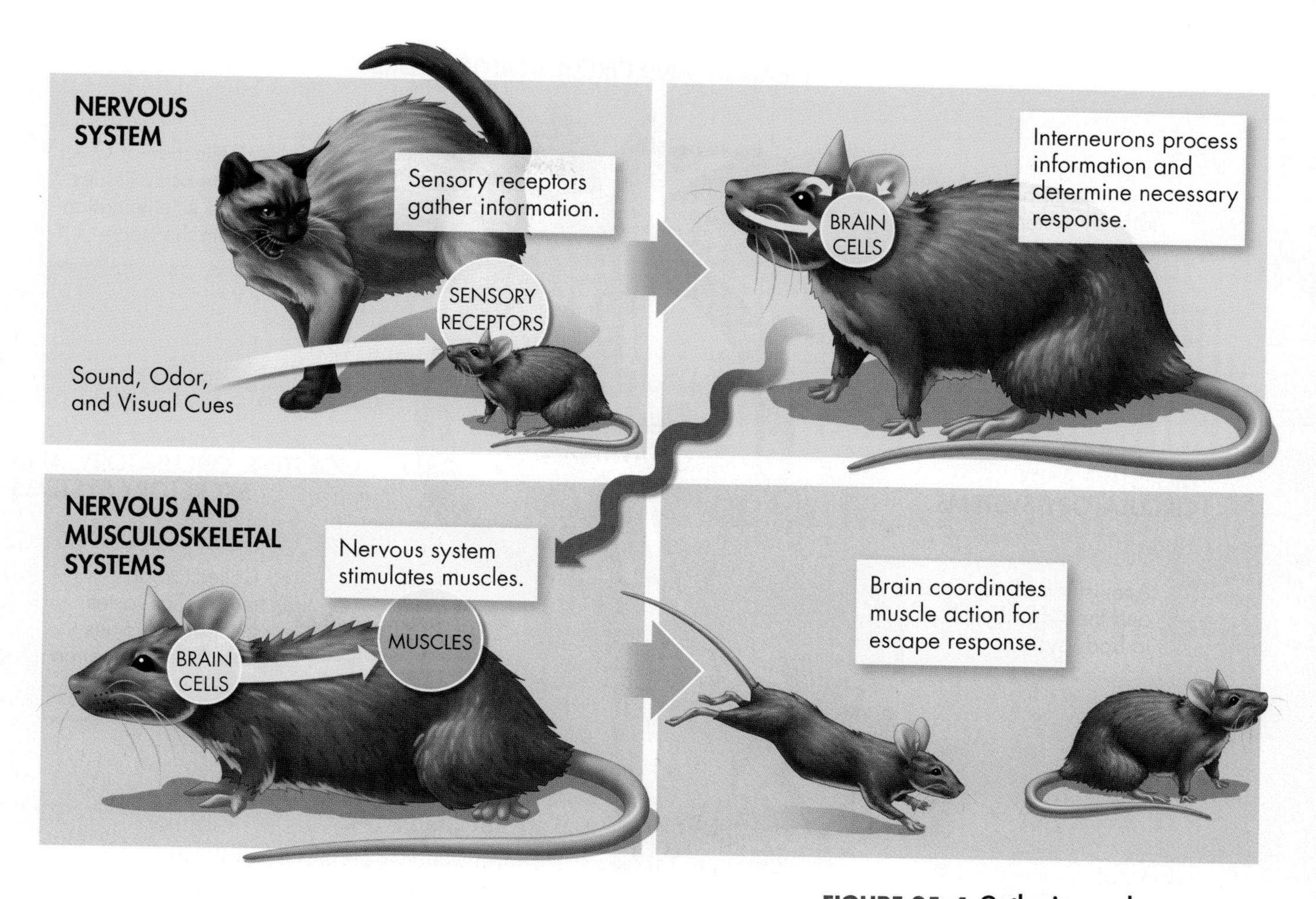

FIGURE 25–4 Gathering and Responding to Information The nervous and muscular systems work together to produce a response. **Predict** ***Would an animal with a malfunctioning nervous system be likely to produce an appropriate muscular response to a predator? Explain.***

Gathering and Responding to Information Complex animals, such as mammals, use several linked body systems to respond to events in their environment, as shown in **Figure 25–4.** The nervous system gathers information using cells called receptors that respond to sound, light, chemicals, and other stimuli. Other nerve cells collect and process that information and determine how to respond. Some invertebrates have only a loose network of nerve cells, with no real center. Other invertebrates and most chordates have large numbers of nerve cells concentrated into a brain.

Animals often respond to the information processed in their nervous system by moving around. Muscle tissue generates force by becoming shorter when stimulated by the nervous system. Muscles work together with some kind of supporting structure called a skeleton to make up the musculoskeletal system. Skeletons vary widely from phylum to phylum. Some invertebrates, such as earthworms, have skeletons that are flexible and function through the use of fluid pressure. Insects and some other invertebrates have external skeletons. The bones of vertebrates form an internal skeleton. For example, the hard shell of a lobster is an external skeleton, while your bones are part of your internal skeleton.

In Your Notebook *Construct a flowchart showing the events in **Figure 25–4** in chronological order.*

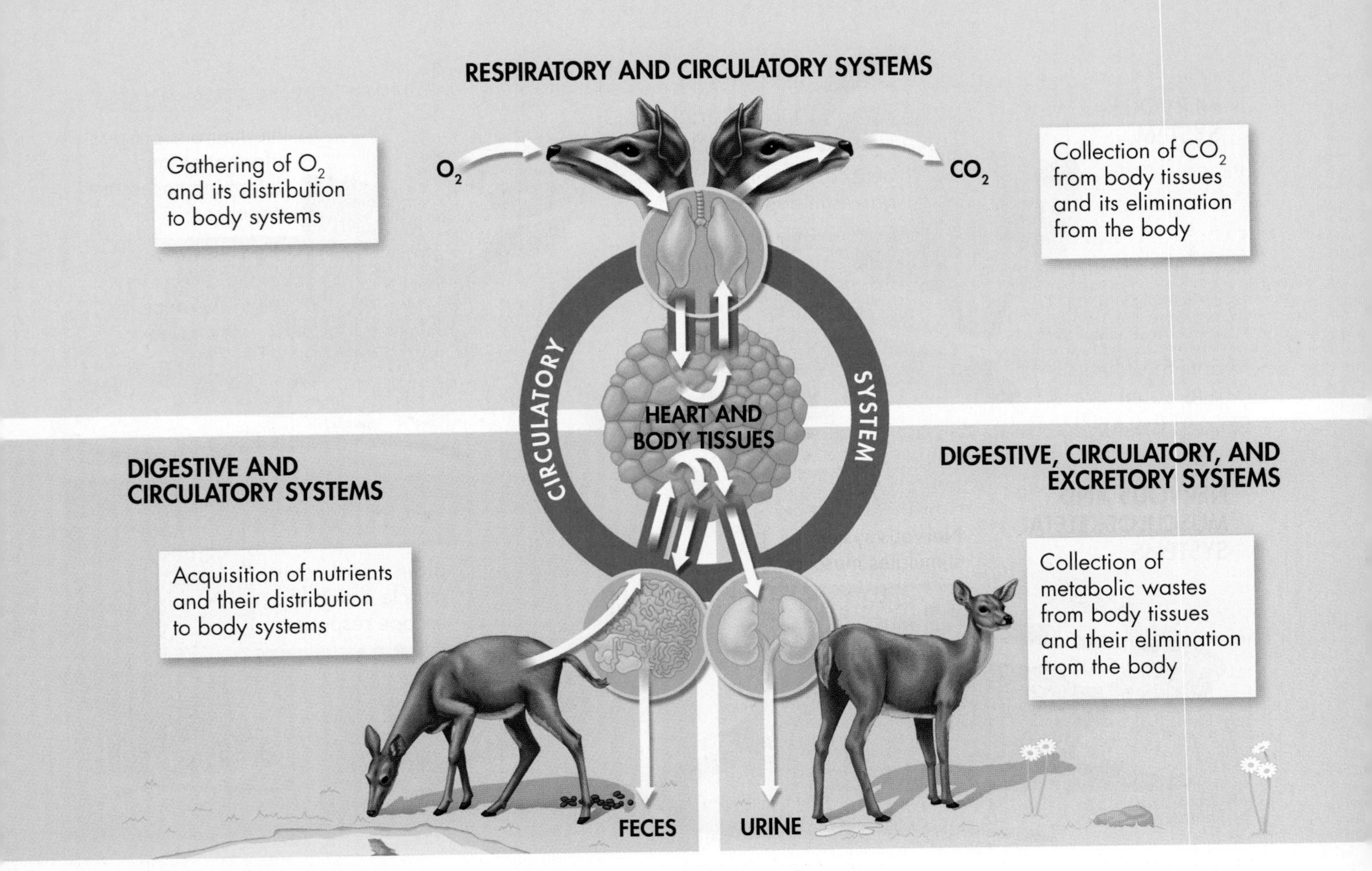

VISUAL SUMMARY

MOVING MATERIALS IN, AROUND, AND OUT OF THE BODY

FIGURE 25–5 The structures of an animal's respiratory, digestive, and excretory systems must work together with those of its circulatory system.

Obtaining and Distributing Oxygen and Nutrients All animals must breathe to obtain oxygen. Small animals that live in water or in wet places can "breathe" by allowing oxygen to diffuse across their skin. Larger animals use a respiratory system based on one of many different kinds of gills, lungs, or air passages. In addition, all animals must eat to obtain nutrients. Most animals have a digestive system that acquires food and breaks it down into forms cells can use.

After acquiring oxygen and nutrients, animals must transport them to cells throughout their bodies. For many animals, this task of transporting oxygen and nutrients requires some kind of circulatory system. Therefore, the structures and functions of respiratory and digestive systems must work together with circulatory systems, as shown in **Figure 25–5.** Among vertebrates, including humans, the circulatory system is especially important in supplying oxygen and nutrients. In humans, for example, brain tissue begins to die within moments if its blood supply is interrupted by a stroke.

Collecting and Eliminating CO_2 and Other Wastes Animals' metabolic processes generate carbon dioxide and other waste products. Some of those waste products contain nitrogen, often in the form of ammonia. Both carbon dioxide and ammonia are toxic in high concentrations. So these wastes must be excreted, or eliminated from the body.

Many animals eliminate carbon dioxide by simply using their respiratory systems. However, most complex animals have a specialized organ system—the excretory system—for eliminating other wastes, such as ammonia. The excretory system concentrates or processes these wastes and either expels them immediately or stores them before eliminating them.

Before waste products can be discharged from the body, they must first be collected from cells throughout body tissues and then delivered to the respiratory or excretory system. Some sort of circulatory system is often necessary to perform these functions. So the collection and elimination of wastes requires close interactions between the structures and functions of three body systems, as shown in **Figure 25–5** on the previous page.

Reproducing Most animals reproduce sexually by producing haploid gametes. Sexual reproduction helps create and maintain genetic diversity, which increases a species' ability to evolve and adapt as the environment changes. Many invertebrates and a few vertebrates can also reproduce asexually. Asexual reproduction usually produces offspring that are genetically identical to the parent. It allows animals to increase their numbers rapidly but does not generate genetic diversity.

FIGURE 25–6 Reproduction Like many vertebrates, this pygmy marsupial frog is caring for her young while they develop. Unlike most animals, she is carrying her eggs on her back!

25.1 Assessment

Review Key Concepts

1. a. **Review** Which characteristics do all animals share?
   b. **Classify** A classmate is looking at a unicellular organism under a microscope. She asks you if it is an animal. What would you say, and why?
2. a. **Review** What is the defining characteristic of invertebrates? What are four characteristics of chordates?
   b. **Explain** Why would you be unlikely to find a notochord in an adult chordate?
   c. **Compare and Contrast** How do vertebrates differ from other chordates?
3. a. **Review** Describe the essential functions performed by all animals.
   b. **Explain** Why must waste products produced by metabolic processes be eliminated from an animal's body?
   c. **Sequence** Which body system delivers waste products to the respiratory and excretory systems?

VISUAL THINKING

4. Make a two-column chart that lists the ways that animals gather and respond to information. In the first column, list each function. In the second column, include a drawing, photograph, or clipping of a structure that performs that function.

RST.9-10.10 Level of Text Complexity

Careers & BIOLOGY

Are you interested in a career with animals? If so, you might be interested in one of the careers below.

ZOO CURATOR

When you think of a zoo worker, you likely picture a keeper feeding animals, right? Zookeepers are not the only people working in zoos, however! Zoo curators are responsible for overseeing a specific part of a zoo's work. There are many different kinds of curators, including research curators, animal curators, and conservation curators. Each contributes to the zoo's mission of wildlife protection and preservation.

BEEKEEPER

More than one quarter of the American diet comes from food plants that are pollinated by bees. Beekeepers maintain beehives and are therefore a vital part of the agriculture business. Bees are rented to farmers for pollination of crops such as almonds, apples, peaches, soybeans, and many types of berries. Beekeepers may also use their hives to produce beeswax and honey.

INVERTEBRATE BIOLOGIST

More than 95 percent of animals lack a backbone. From corals to spiders, earthworms to sea stars, the variety is amazing! Biologists may study invertebrate behavior, evolution, ecology, or anatomy. With so many species to choose from, the research is as varied as the animals themselves.

CAREER CLOSE-UP:

Dr. Scottie Yvette Henderson, Invertebrate Biologist

The strange and diverse creatures of the ocean inspire Dr. Scottie Henderson, an instructor of biology at the University of Puget Sound in Tacoma, Washington. Her current research focuses on tiny, potentially parasitic crabs that infest a clam called *Nuttallia obscurata*. Dr. Henderson and her colleagues are looking at the interactions of the clam and crab to better understand the nature of their symbiotic relationship. Some evidence points to parasitism, but the relationship may be commensal. Nothing, however, is as important to Dr. Henderson as getting her students interested in and excited about science.

"Stop and take a look at the world around you. Biology is exciting! There are many unanswered questions . . . and many questions waiting to be asked."

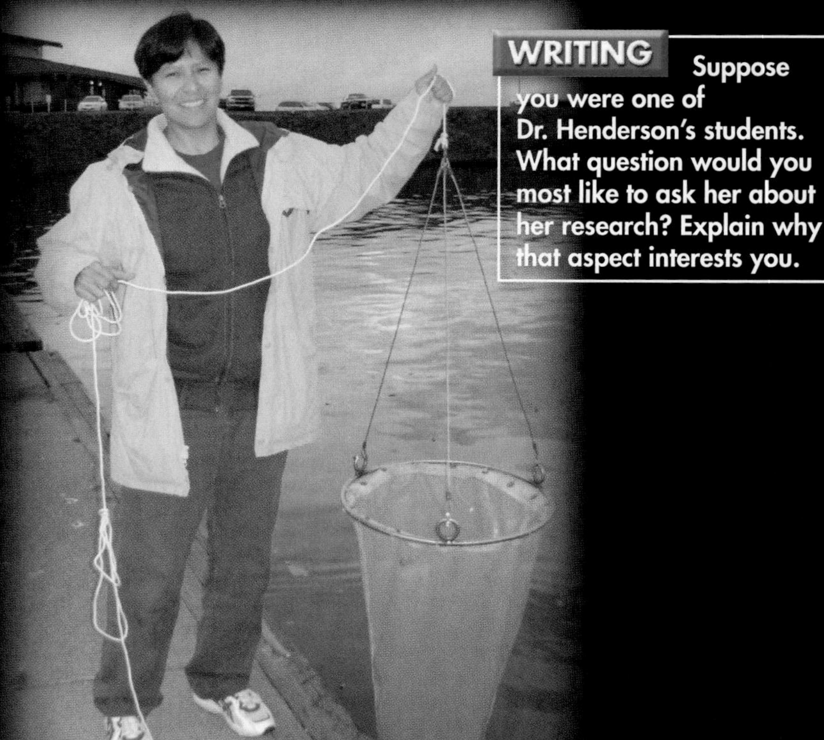

WRITING Suppose you were one of Dr. Henderson's students. What question would you most like to ask her about her research? Explain why that aspect interests you.

25.2 Animal Body Plans and Evolution

Key Questions

- ***What are some features of animal body plans?***
- ***How are animal phyla defined?***

Vocabulary

radial symmetry • bilateral symmetry • endoderm • mesoderm • ectoderm • coelom • pseudocoelom • zygote • blastula • protostome • deuterostome • cephalization

Taking Notes

Concept Map Draw a concept map showing the different features of animal body plans and the different types of each feature.

THINK ABOUT IT Animals alive today have typically been produced by two processes: the development of a multicellular individual from a single fertilized egg cell, and the evolution of a modern species from its ancestors over many millions of years. The history of the evolutionary changes to animal body structures has been known for years. Today, exciting research is revealing how changes in the genes that control embryological development are connected to the evolution of body structures. This research field, often referred to as "evo-devo," is one of the hottest areas in biology today.

Features of Body Plans

What are some features of animal body plans?

Our survey of the animal kingdom focuses on how animal body structures and systems perform life's essential functions. Each animal phylum has a unique organization of particular body structures that is often referred to as a body plan. **Features of animal body plans include levels of organization, body symmetry, differentiation of germ layers, formation of body cavities, patterns of embryological development, segmentation, cephalization, and limb formation.**

Levels of Organization As the first cells of most animals develop, they differentiate into specialized cells that are organized into tissues. Recall that a tissue is a group of cells that perform a similar function. Animals typically have several types of tissues, including epithelial, muscle, connective, and nervous tissues. Epithelial tissues cover body surfaces, inside and out. The epithelial cells that line lung surfaces, for example, have thin, flat structures through which gases can diffuse easily.

Tissues combine during growth and development to form organs. Organs work together to make up organ systems that carry out complex functions. Your digestive system, for example, includes tissues and organs such as your lips, mouth, stomach, intestines, and anus.

FIGURE 25–7 Body Symmetry Animals with radial symmetry have body parts that extend from a central point. Animals with bilateral symmetry have distinct anterior and posterior ends and right and left sides.

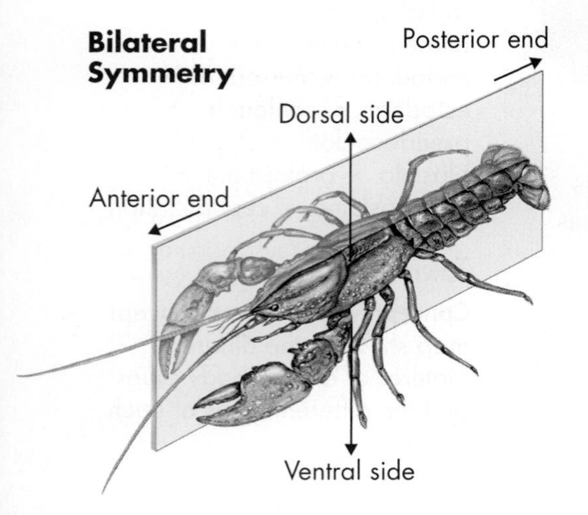

Body Symmetry The bodies of most animals exhibit some type of symmetry. Some animals, such as the sea anemone in **Figure 25–7,** have body parts that extend outward from the center, like the spokes of a bicycle wheel. These animals exhibit **radial symmetry,** in which any number of imaginary planes drawn through the center of the body could divide it into equal halves. The most successful animal groups exhibit **bilateral symmetry,** in which a single imaginary plane divides the body into left and right sides that are mirror images of one another. Animals with bilateral symmetry have a definite front, or anterior, end and a back, or posterior, end. Bilaterally symmetrical animals also have an upper, or dorsal, side and a lower, or ventral, side. When you ride a horse, you are riding on its dorsal side.

Differentiation of Germ Layers During embryological development, the cells of most animal embryos differentiate into three layers called germ layers. Cells of the **endoderm,** or innermost germ layer, develop into the linings of the digestive tract and much of the respiratory system. Cells of the **mesoderm,** or middle layer, give rise to muscles and much of the circulatory, reproductive, and excretory organ systems. The **ectoderm,** or outermost layer, produces sense organs, nerves, and the outer layer of the skin.

Formation of a Body Cavity Most animals have some kind of body cavity—a fluid-filled space between the digestive tract and body wall. A body cavity provides a space in which internal organs can be suspended, and room for those organs to grow. For example, your stomach and other digestive organs are suspended in your body cavity. Most complex animal phyla have a true **coelom** (SEE lum), a body cavity that develops within the mesoderm and is completely lined with tissue derived from mesoderm. Some invertebrates have only a primitive jellylike layer between the ectoderm and endoderm. Other invertebrates lack a body cavity altogether, and are called acoelomates. Still other invertebrate groups have a **pseudocoelom,** which is only partially lined with mesoderm. **Figure 25–8** summarizes the tissue structures of animals with and without coeloms.

FIGURE 25–8 Body Cavities Acoelomates lack a coelom between their body wall and digestive cavity. Pseudocoelomates have body cavities that are partially lined with tissues from the mesoderm.

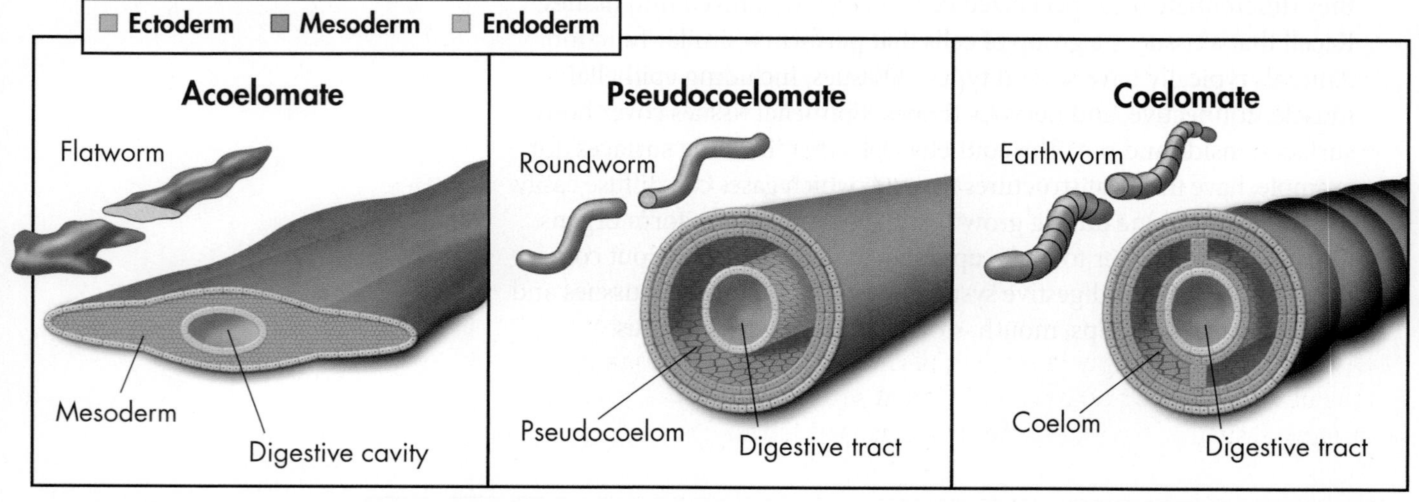

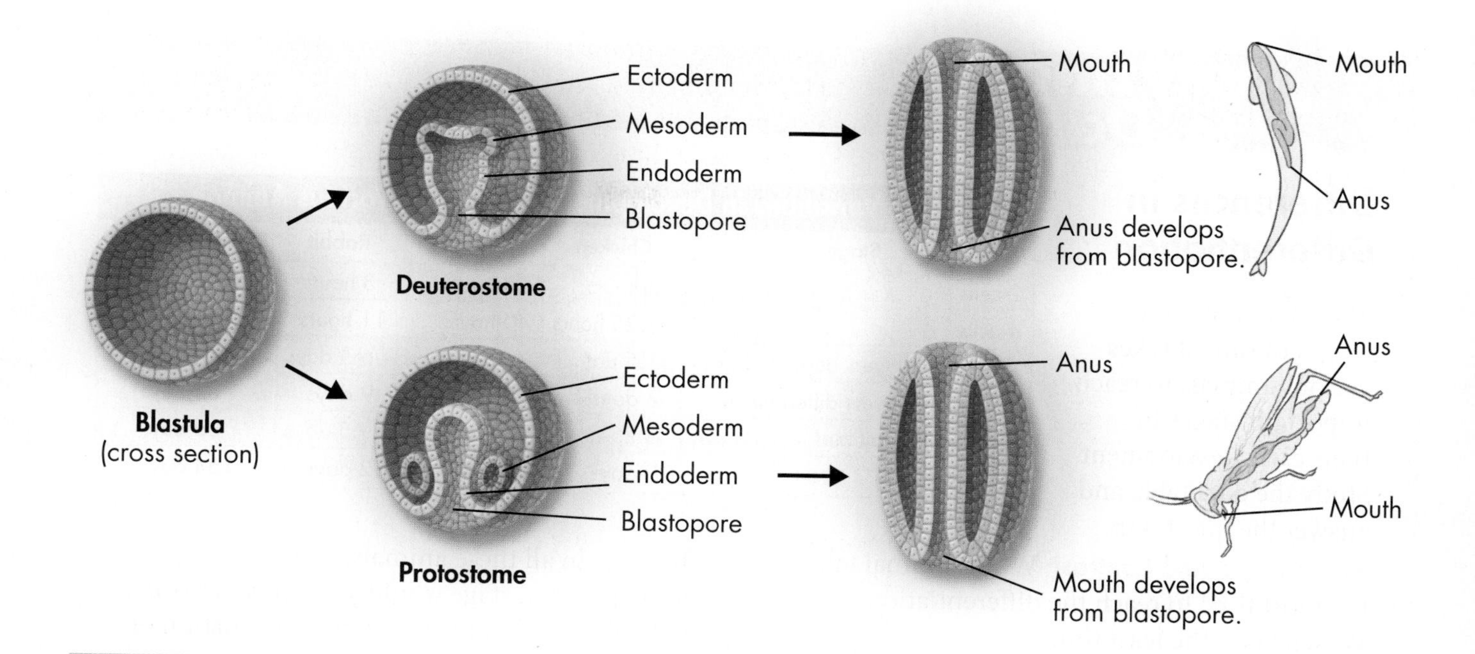

FIGURE 25–9 Blastula and Blastopore Formation During the early development of an animal embryo, a hollow ball of cells called a blastula forms. An opening called a blastopore forms in this ball. In deuterostomes, such as fishes, the blastopore forms an anus. In protostomes, such as grasshoppers, the blastopore develops into the mouth.

Patterns of Embryological Development Every animal that reproduces sexually begins life as a **zygote,** or fertilized egg. As the zygote begins to develop, it forms a **blastula** (BLAS tyoo luh), a hollow ball of cells like an inflated balloon. As the blastula develops, it folds in on itself, as if you were holding the balloon and pushing your thumbs toward the center. This folding changes a ball of cells into an elongated structure with a tube that runs from one end to the other. This tube becomes the digestive tract, as shown in **Figure 25–9.**

At first, this digestive tract has only a single opening to the outside, called a blastopore. An efficient digestive tract, however, needs two openings: a mouth through which food enters and an anus through which wastes leave. In phyla that are **protostomes** (PROH tuh stohms), the blastopore becomes the mouth. In protostomes, including most invertebrates, the anus develops at the opposite end of the tube. In phyla that are **deuterostomes** (DOO tur uh stohms), the blastopore becomes the anus, and the mouth is formed from the second opening that develops. Chordates and echinoderms are deuterostomes. This similarity in development is one of several characteristics that indicate that echinoderms are closely related to chordates.

Segmentation: Repeating Parts As many bilaterally symmetrical animals develop, their bodies become divided into numerous repeated parts, or segments. These animals are said to exhibit segmentation. Segmented animals, such as worms, insects, and vertebrates, typically have at least some internal and external body parts that repeat on each side of the body. Bilateral symmetry and segmentation are found together in many of the most successful animal groups.

Segmentation has been important in animal evolution because of the way genes control the production and growth of body segments. If an organism has segmentation, simple mutations can cause changes in the number of body segments. Different segments can also become specialized, such as having a head or specialized limbs.

Differences in Differentiation

The table shows the length of time it takes various animals to reach important stages in their early development. Study the data table and answer the questions.

Time Variations in Developmental Stages of Various Animals

Stage	Chicken	Hamster	Rabbit	Rhesus Monkey
2 cells	3 hours	16 hours	8 hours	24 hours
4 cells	3.25 hours	40 hours	11 hours	36 hours
Three germ layers begin to form	1.5 days	6.5–7 days	6.5 days	19 days
Three germ layers differentiate	3 days	8 days	9 days	25 days
Formation of tail bud	3.25 days	8.5 days	9.5 days	26 days
Birth/Hatching	22 days	16 days	32 days	164 days

1. Compare and Contrast Which animal takes the most time to reach the differentiation stage? Which takes the least time?

2. Calculate How much longer does it take a rhesus monkey zygote to reach the 4-cell stage than it does a chicken zygote? MATH

3. Infer In all these animals, which developmental stage would you expect to occur first—formation of the coelom or formation of the blastula?

BUILD Vocabulary

SUFFIXES The word **cephalization** has two suffixes: *-ize*, meaning "to become," and *-ation*, meaning "the process of." When these suffixes are added to the root word *cephal-*, meaning "head," the new word means "the process of becoming a head."

Cephalization: Getting a Head Animals with bilateral symmetry typically exhibit **cephalization** (sef uh lih ZAY shun), the concentration of sense organs and nerve cells at their anterior end. This anterior end is often different enough from the rest of the body that it is called a head. The most successful animal groups, including arthropods and vertebrates, exhibit pronounced cephalization.

Close examination of insect and vertebrate embryos shows that their heads are formed by the fusion and specialization of several body segments during development. As those segments fuse, their internal and external parts combine in ways that concentrate sense organs, such as eyes, in the head. Nerve cells that process information and "decide" what the animal should do also become concentrated in the head. Not surprisingly, animals with heads usually move in a "head-first" direction. This is so that the concentration of sense organs and nerve cells comes in contact with new parts of the environment first.

Limb Formation: Legs, Flippers, and Wings Segmented, bilaterally symmetrical animals typically have external appendages on both sides of the body. These appendages vary from simple groups of bristles in some worms, to jointed legs in spiders, wings in dragonflies, and a wide range of limbs, including bird wings, dolphin flippers, and monkey arms. These very different kinds of appendages have evolved several times, and have been lost several times, in various animal groups.

In Your Notebook *Explain in your own words why animals with heads tend to move in a "head-first" direction.*

VISUAL SUMMARY

BODY PLANS

FIGURE 25–10 The body plans of modern invertebrates and chordates suggest evolution from a common ancestor.

Ectoderm
Mesoderm
Endoderm

	 Sponges	 Cnidarians 	 Arthropods 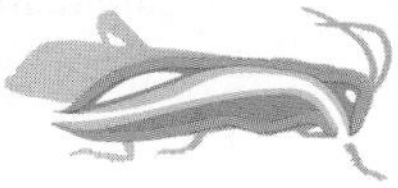	 Roundworms 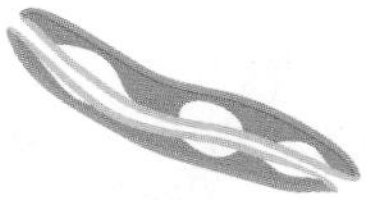	 Flatworms 
Levels of Organization	Specialized cells	Specialized cells, tissues	Specialized cells, tissues, organs	Specialized cells, tissues, organs	Specialized cells, tissues, organs
Body Symmetry	Absent	Radial	Bilateral	Bilateral	Bilateral
Germ Layers	Absent	Two	Three	Three	Three
Body Cavity	–	Acoelom	True coelom	Pseudocoelom	Acoelom
Embryological Development	–	–	Protostome	Protostome	Protostome
Segmentation	Absent	Absent	Present	Absent	Absent
Cephalization	Absent	Absent	Present	Present	Present

Ectoderm
Mesoderm
Endoderm

	 Annelids 	 Mollusks 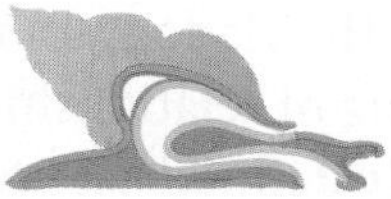	 Echinoderms 	 Chordates 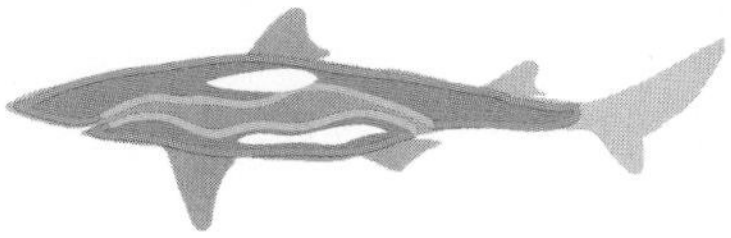
Levels of Organization	Specialized cells, tissues, organs	Specialized cells, tissues, organs	Specialized cells, tissues, organs	Specialized cells, tissues, organs
Body Symmetry	Bilateral	Bilateral	Radial (as adults)	Bilateral
Germ Layers	Three	Three	Three	Three
Body Cavity	True coelom	True coelom	True coelom	True coelom
Embryological Development	Protostome	Protostome	Deuterostome	Deuterostome
Segmentation	Present	Absent	Absent	Present
Cephalization	Present	Present	Absent (as adults)	Present

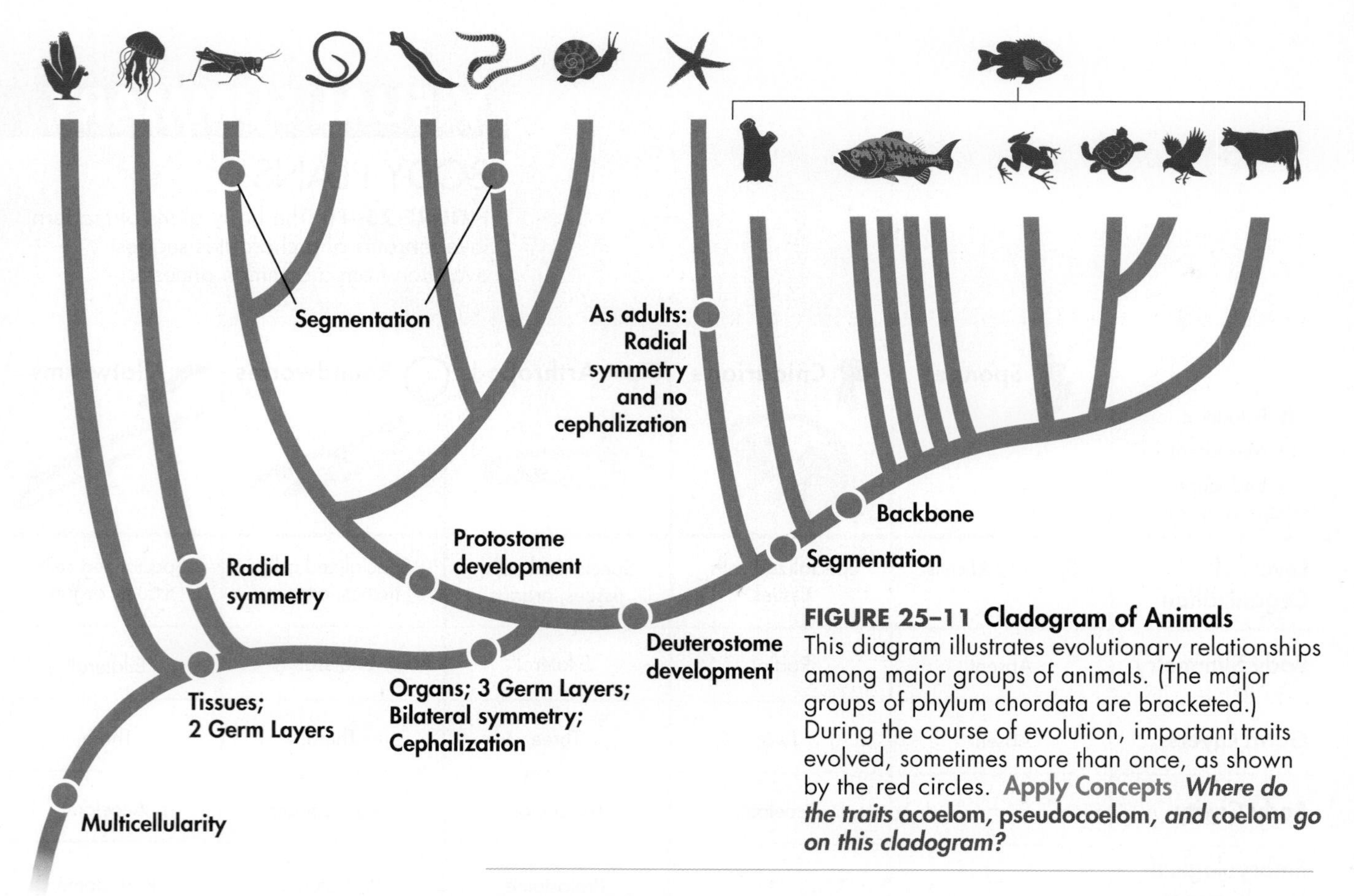

FIGURE 25–11 Cladogram of Animals This diagram illustrates evolutionary relationships among major groups of animals. (The major groups of phylum chordata are bracketed.) During the course of evolution, important traits evolved, sometimes more than once, as shown by the red circles. **Apply Concepts** ***Where do the traits*** **acoelom, pseudocoelom,** ***and*** **coelom** ***go on this cladogram?***

The Cladogram of Animals

How are animal phyla defined?

The features of animal body plans you have just learned about provide information for building the cladogram, or phylogenetic tree, of animals. Recall that the evolutionary history presented in a cladogram represents a set of evolutionary hypotheses based on characteristics of living species, evidence from the fossil record, and comparative genomic studies. The cladogram in **Figure 25–11** presents our current understanding of relationships among animal phyla. **Animal phyla are typically defined according to adult body plans and patterns of embryological development.** For example, the phylum Arthropoda is defined by a body plan that includes bilateral symmetry, segmentation, cephalization, an external skeleton, and jointed legs.

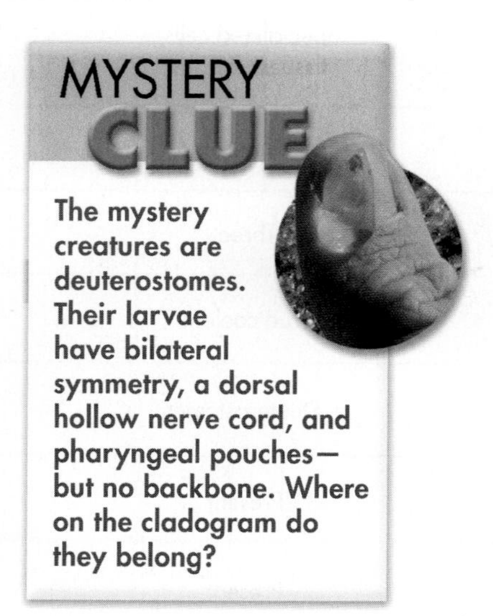

The mystery creatures are deuterostomes. Their larvae have bilateral symmetry, a dorsal hollow nerve cord, and pharyngeal pouches—but no backbone. Where on the cladogram do they belong?

Differences Between Phyla The cladogram of animals indicates the sequence in which important body plan features evolved. Every phylum has a unique combination of ancient traits inherited from its ancestors and new traits found only in that particular phylum. It may be tempting to think of a cladogram as a story about "improvements" from one phylum to the next over time. But that isn't the case. The complicated body systems of vertebrates aren't necessarily better than the "simpler" systems of invertebrates. Any system found in living animals functions well enough to enable those animals to survive and reproduce. For example, most chordate brains are more complex than the brains of flatworms. But flatworm brains obviously work well enough to enable flatworms, as a group, to survive.

Changes Within Phyla: Themes and Variations

Within each phylum, different groups represent different variations on the basic body plan themes that have evolved over time. Land vertebrates, for example, typically have four limbs. Many, such as frogs, walk (or hop) on four limbs that we call "legs." Among birds, the front limbs have evolved into wings. In many primates, the front limbs have evolved into what we call "arms." Both wings and arms evolved through changes in the standard vertebrate forelimb.

FIGURE 25–12 Limb Variations Birds have evolved front limbs specialized as wings, whereas frogs have evolved four "legs."

Evolutionary Experiments In a sense, you can think of each phylum's body plan as an evolutionary "experiment," in which a particular set of body structures performs essential functions. An organism's first appearance represents the beginning of this "experiment." The very first versions of most major animal body plans were established hundreds of millions of years ago, as you'll learn in the next chapter. Ever since that time, each phylum's evolutionary history has shown variations in body plan as species have adapted to changing conditions. If the changes have enabled members of a phylum to survive and reproduce, the phylum still exists. If the body plan hasn't functioned well enough over time, members of the phylum, or particular groups within the phylum, have become extinct.

25.2 Assessment

Review Key Concepts

1. a. **Review** List eight features of animal body plans.
   b. **Infer** How is the embryology of echinoderms similar to that of vertebrates? What might this similarity indicate about their evolutionary relationship?
2. a. **Review** What two features define animal phyla?
   b. **Relate Cause and Effect** What happens to a phylum over time if its body plan doesn't enable its members to survive and reproduce?

WRITE ABOUT SCIENCE

Description

3. Explain the description of a body plan as an evolutionary "experiment." In your explanation, describe the difference between successful and unsuccessful body plans in terms of the different outcomes.

NGSS Smart Guide Introduction to Animals

Disciplinary Core Idea LS1.A Structure and Function: How do the structures of organisms enable life's functions? Animals are multicellular, heterotrophic, eukaryotic organisms whose cells lack cell walls.

25.1 What Is an Animal?

- Animals, members of the kingdom Animalia, are multicellular, heterotrophic, eukaryotic organisms whose cells lack cell walls.
- Invertebrates include all animals that lack a backbone, or vertebral column.
- All chordates exhibit four characteristics during at least one stage of life: a dorsal, hollow nerve cord; a notochord; a tail that extends beyond the anus; and pharyngeal pouches.
- Like all organisms, animals must maintain homeostasis by gathering and responding to information, obtaining and distributing oxygen and nutrients, and collecting and eliminating carbon dioxide and other wastes. They also reproduce.

invertebrate 730 • chordate 731 • notochord 731 • pharyngeal pouch 731 • vertebrate 731 • feedback inhibition 732

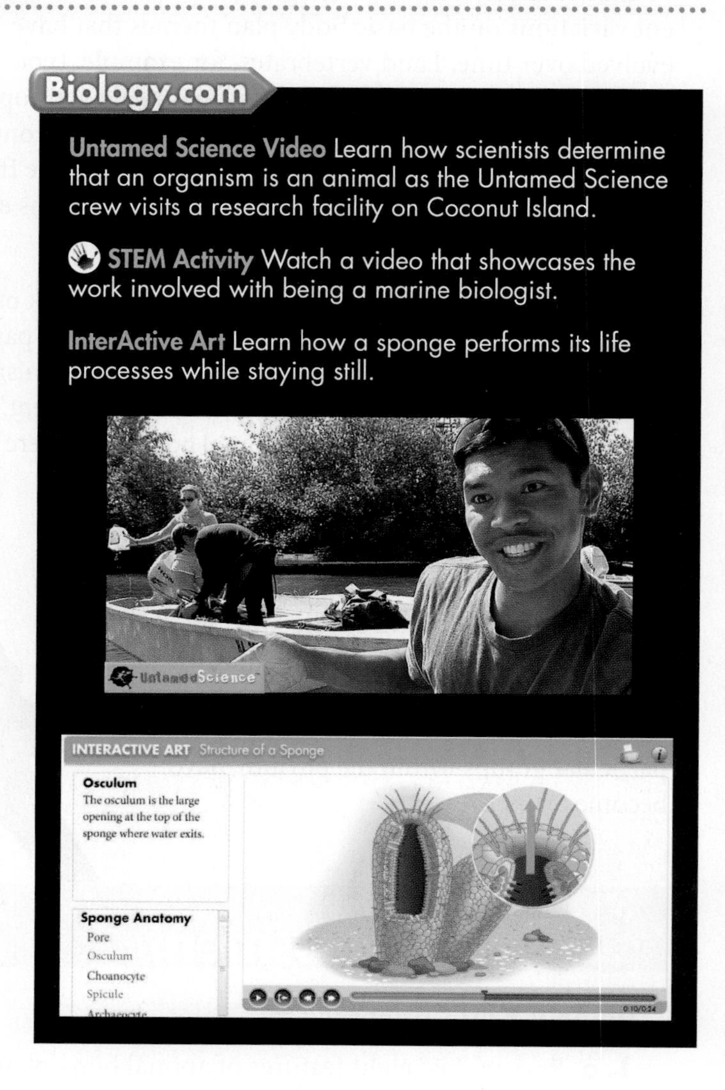

CHECKING YOUR *Scientific Literacy*

Refer to the lesson content and digital assets below as you prepare for your chapter assessment. Then, evaluate your understanding of animals by answering these questions.

1. **Scientific and Engineering Practice Developing and Using Models** Draw an outline of the human body. Use this model to label the different features of the human body plan.
2. **Crosscutting Concept Structure and Function** Explain how the major organ systems of animals work together to aid in maintaining homeostasis.
3. **Research to Build and Present Knowledge** How are animal phyla typically classified? Draw evidence from the text to support your answer.
4. **STEM** Provide an example of a technology that uses feedback inhibition and explain how it works. You may wish to draw a flowchart.

25.2 Animal Body Plans and Evolution

- Features of animal body plans include levels of organization, body symmetry, differentiation of germ layers, formation of body cavities, patterns of embryological development, segmentation, cephalization, and limb formation.
- Animal phyla are typically defined according to adult body plans and patterns of embryological development.

radial symmetry 738 • **bilateral symmetry 738** • **endoderm 738** • **mesoderm 738** • **ectoderm 738** • **coelom 738** • **pseudocoelom 738** • **zygote 739** • **blastula 739** • **protostome 739** • **deuterostome 739** • **cephalization 740**

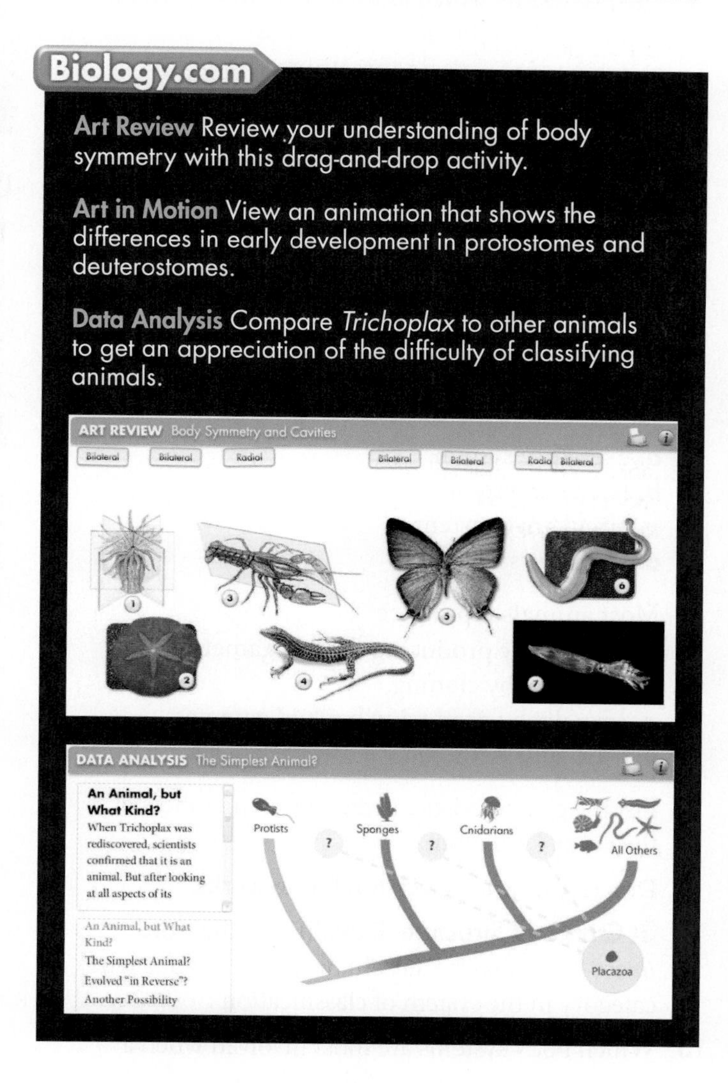

Skills Lab

Comparing Invertebrate Body Plans This chapter lab is available in your lab manual and at Biology.com

25 Assessment

25.1 What Is an Animal?

Understand Key Concepts

1. A multicellular, eukaryotic heterotroph whose cells lack cell walls is a(n)
   a. protist. c. animal.
   b. virus. d. plant.
2. Which of the following is characteristic of all chordates but not found in invertebrates?
   a. a notochord c. a circulatory system
   b. four legs d. an exoskeleton
3. The process by which animals take in oxygen and give off carbon dioxide is known as
   a. responding. c. breathing.
   b. reproducing. d. excreting.
4. Animals that have a backbone, also called a vertebral column, are known as
   a. invertebrates. c. homeostasis.
   b. prokaryotes. d. vertebrates.
5. The job of collecting waste materials from a complex animal's body cells and delivering them to organs that will release them from the body is carried out by the
   a. excretory system.
   b. nervous system.
   c. circulatory system.
   d. digestive system.
6. Most animals reproduce
   a. sexually by producing diploid gametes.
   b. asexually by cloning.
   c. sexually by producing haploid gametes.
   d. asexually by fission.
7. List the characteristics shared by all members of the animal kingdom.
8. Describe how feedback inhibition works.
9. **Craft and Structure** Explain why the word *invertebrate* may be a useful word but is not a true category in the system of classification.
10. Which body systems are most involved when a raccoon discovers that a full trash can is a food source, and it knocks over the can to find the food?

Think Critically

11. **Classify** What characteristic distinguishes vertebrates from nonvertebrate chordates?
12. **Apply Concepts** In what ways do the digestive and respiratory systems depend on the circulatory system to carry out the functions of obtaining nutrients and eliminating wastes?
13. **Compare and Contrast** How does the way animals dispose of carbon dioxide differ from the way they dispose of ammonia?
14. **Relate Cause and Effect** Describe generally how the nervous and musculoskeletal systems of a rabbit react when it sees a predator such as a coyote.

25.2 Animal Body Plans and Evolution

Understand Key Concepts

15. Many animals have body symmetry with left and right sides that are mirror images of each other. This type of symmetry is
    a. radial. c. circular.
    b. bilateral. d. dorsal.
16. The developing embryo shown below is a ___?___, a group that includes ___?___.

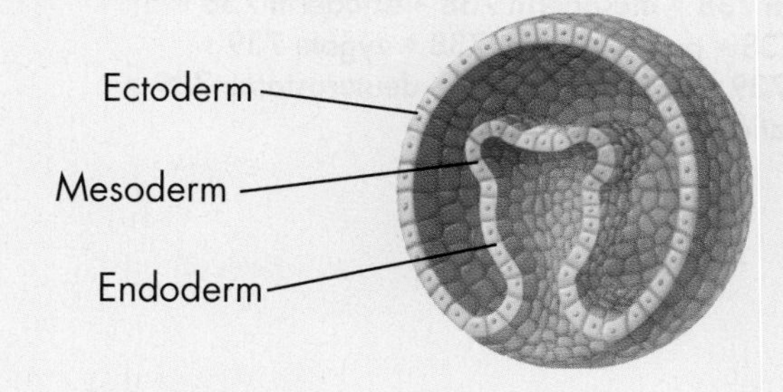

    a. protostome; invertebrates other than echinoderms
    b. protostome; vertebrates
    c. deuterostome; echinoderms and chordates
    d. deuterostome; invertebrates
17. An animal whose mouth is formed from the blastopore is a(n)
    a. deuterostome. c. protostome.
    b. endoderm. d. mesoderm.

18. A concentration of sense organs and nerve cells in the anterior end of the body is known as
 a. fertilization.
 b. cephalization.
 c. symmetry.
 d. multicellularity.

19. Which of the following animals shows radial symmetry?
 a. earthworm
 b. fish
 c. insect
 d. sea anemone

20. Which germ layer produces the nerves and sense organs of animals?
 a. ectoderm
 b. endoderm
 c. mesoderm
 d. periderm

21. Most chordates that live on land have
 a. two limbs.
 b. four limbs.
 c. six limbs.
 d. eight limbs.

22. What is an acoelomate?

23. Describe the major developmental difference that distinguishes protostomes from deuterostomes.

24. What is a blastula?

25. List the three germ layers.

26. Name two body plan characteristics shared by all arthropods and vertebrates.

27. What is one major advantage of cephalization?

Think Critically

28. **Research to Build and Present Knowledge** Why is bilateral symmetry an important development in the evolution of animals? Draw evidence from the text to support your analysis.

29. **Sequence** Rank the following developments in the order of their appearance during evolution: tissues, deuterostome development, multicellularity, protostome development.

30. **Form a Hypothesis** Animals with radial symmetry, such as sea anemones, lack cephalization, while animals with bilateral symmetry have it. State a hypothesis that would explain this observation.

31. **Craft and Structure** Why is it inaccurate to state that the cladogram of animals shows the improvements in body plans that have occurred over time?

solve the CHAPTER MYSTERY

SLIME DAY AT THE BEACH

Although most people had never seen creatures like these before, biologists had no trouble identifying them. They were salps—descendents of the most ancient members of phylum Chordata. Salps belong to a group of chordates called tunicates. As adults, most tunicates live attached to rocks or the seafloor. Salps are unusual among tunicates: The adults are free-swimming. They pump water in through their mouths and out the other end, feeding and propelling themselves through the water at the same time. Salps are usually found in the surface waters of tropical seas, but they can be carried north by the Gulf Stream and are sometimes washed onto beaches by storms.

1. **Compare and Contrast** How are salps different from jellyfish?

2. **Connect to the Big idea** Use the Internet to research salps and other tunicates. Explain why these peculiar-looking animals are classified in the phylum Chordata.

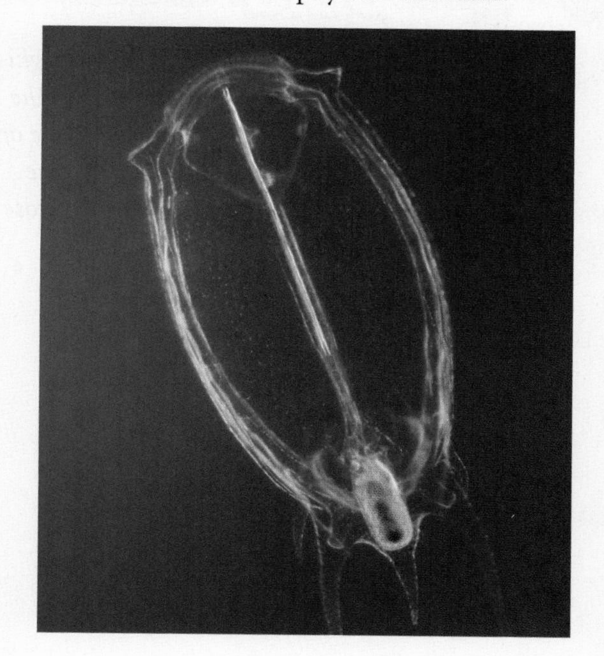

Connecting Concepts

Use Science Graphics

Use the graph to answer questions 32–34.

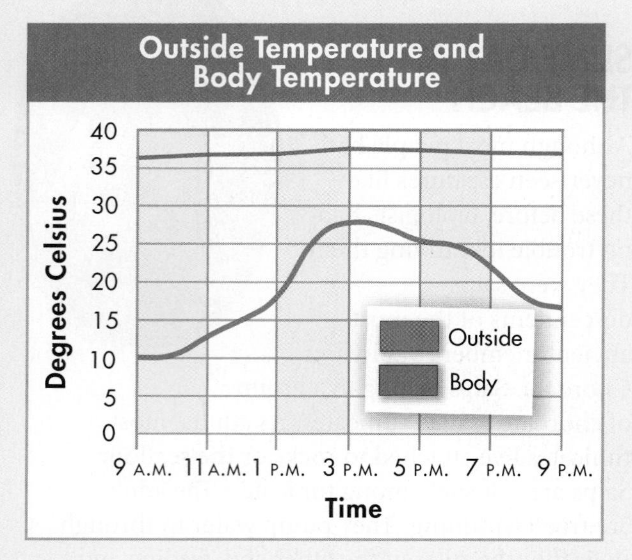

32. **Interpret Graphs** At what time of day is the body temperature closest to that of the outside environment?
33. **Draw Conclusions** What is the relationship between body temperature and the temperature of the environment?
34. **Infer** How do you explain the shape of the graph for body temperature?

Write About Science

35. **Key Ideas and Details** Trace the text's explanation of how animals have achieved such great diversity through cephalization and segmentation.
36. **Craft and Structure** Analyze the author's purpose in providing an explanation of what a cladogram depicts.
37. **Assess the**  If you were presented with a small, living organism, how would you try to determine whether it was an animal?

Integration of Knowledge and Ideas

The human digestive system converts food into glucose, a sugar that the body can use for energy. The following data were collected by taking a sample of blood from a person at various times during the day and measuring the relative volume of glucose in the blood.

Time of Day	Amount of Glucose (mg/100 mL)
9 A.M.	102
10 A.M.	98
11 A.M.	130
12 noon	115
1 P.M.	103
2 P.M.	100
3 P.M.	102

38. **Interpret Tables** During which time interval is it most likely that this person ate a meal?
    - **a.** 9 A.M. to 10 A.M.
    - **b.** 10 A.M. to 11 A.M.
    - **c.** 1 P.M. to 2 P.M.
    - **d.** 2 P.M. to 3 P.M.
39. **Infer** Which value in the table would you expect to be closest to the homeostatic value for the amount of glucose in the blood?
40. **Apply Concepts** Explain how feedback inhibition might be involved in the changing levels of glucose in the blood.

Standardized Test Prep

Multiple Choice

1. Which of the following is a type of tissue that arises in most animals during development?
   A endoderm
   B mesoderm
   C ectoderm
   D all of the above

2. Which of the following is NOT a characteristic of animals?
   A the ability to make their own food
   B the ability to move
   C eukaryotic cells
   D cells that lack cell walls

3. A hollow ball of cells formed after the zygote undergoes division is called a
   A coelom.
   B protostome.
   C deuterostome.
   D blastula.

4. Which trend did NOT occur during invertebrate evolution?
   A specialization of cells
   B development of a notochord
   C bilateral symmetry
   D cephalization

5. What is a function of the excretory system?
   A to supply cells with oxygen and nutrients
   B to rid the body of metabolic wastes
   C to gather information from the environment
   D to break down food

6. Animals often respond to information processed by their nervous system by moving around, using their
   A circulatory system.
   B excretory system.
   C musculoskeletal system.
   D digestive system.

7. The concentration of nerve tissue and organs in one end of the body is called
   A cephalization.
   B segmentation.
   C body symmetry.
   D nerve nets.

Questions 8 and 9

A biology student has two samples of earthworms in soil, as shown below. The student knows that, because the worms' body temperature changes with the environment, the worms in Sample A have a higher body temperature than those in Sample B. The student uses a stereomicroscope to count the number of heartbeats per minute for three worms from each sample.

Sample A: At temperature of worms' soil environment

Sample B: In ice water

8. Look at the student's two samples. What can you conclude?
   A Sample A is the control.
   B Sample B is the control.
   C Either sample can serve as a control.
   D This is not a controlled experiment.

9. The student finds that the worms from Sample A have a faster heart rate than the worms from Sample B. What hypothesis might you form based on this observation?
   A The worms in Sample A are healthier than the worms in Sample B.
   B A decrease in body temperature corresponds to an increase in heart rate.
   C There is no relationship between body temperature and heart rate.
   D A decrease in body temperature corresponds to a decrease in heart rate.

Open-Ended Response

10. What characteristics distinguish invertebrates from nonvertebrate chordates?

If You Have Trouble With . . .

Question	1	2	3	4	5	6	7	8	9	10
See Lesson	25.2	25.1	25.2	25.2	25.1	25.1	25.2	25.1	25.1	25.1

26 Animal Evolution and Diversity

Evolution

Q: How have animals descended from earlier forms through the process of evolution?

The bony skeleton of this gopher snake reveals the snake's evolutionary relationship to other vertebrates.

INSIDE:

- 26.1 Invertebrate Evolution and Diversity
- 26.2 Chordate Evolution and Diversity
- 26.3 Primate Evolution

Deepen your understanding of animal evolution and diversity by engaging in key practices that allow you to make connections across concepts.

NGSS You will construct scientific explanations for the evolution and diversity of animals based on data from the fossil record.

STEM Could wolves have evolved into domesticated dogs? Watch a video to find out. Then, investigate how new technologies provide evidence for evolution.

Common Core You will construct viable arguments and assess the usefulness of your sources in your research on animal evolution and diversity.

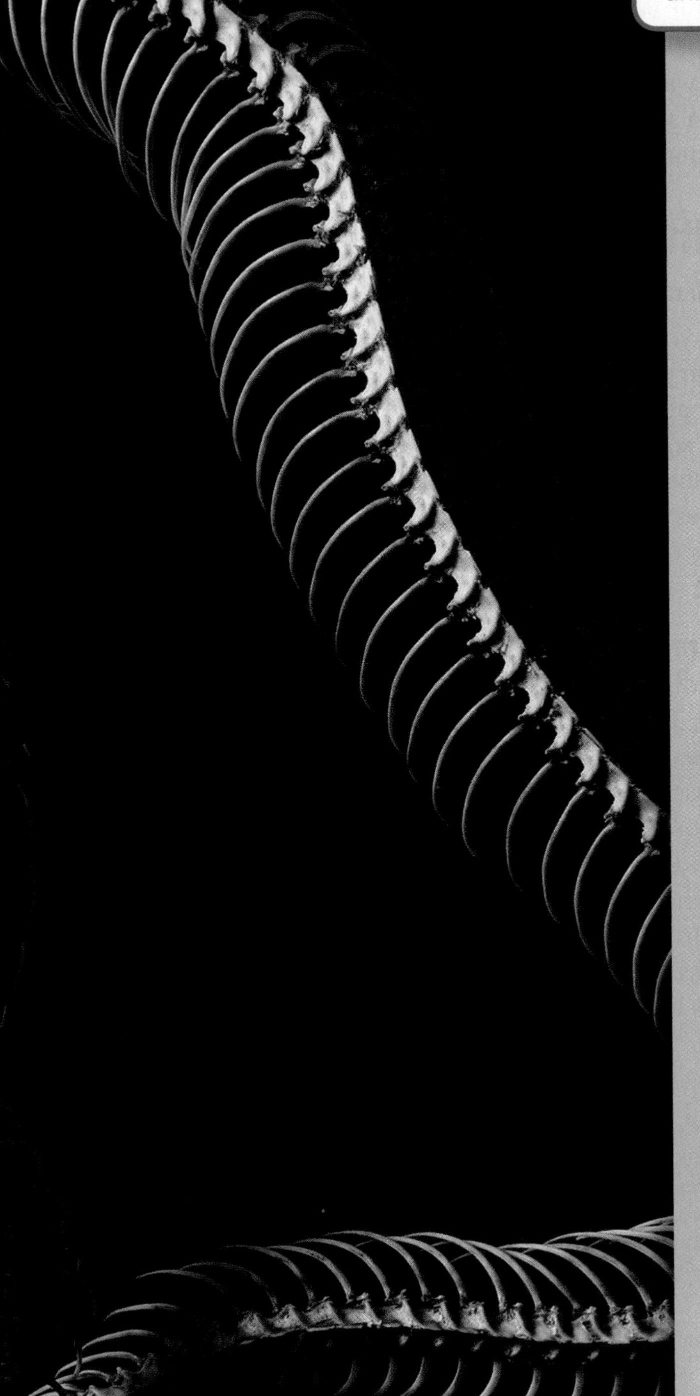

CHAPTER MYSTERY

FOSSIL QUEST

To Josh and Pedro, a fossil hunting trip sounded like a great idea. They would be outside, and the trip would look good on college applications. But where should they go? Their parents were okay with the idea—if they stayed within the United States. Both boys liked dinosaurs and extinct mammals. Josh wanted to search for the very *first* animals, but Pedro wanted to look for the ancestors of birds. Where could they find these fossils . . . and were there any good sites in the United States?

Josh and Pedro needed to figure out in which periods their target animals lived and where they could find rocks of the appropriate ages. As you read this chapter, look for clues to the geologic time periods Josh and Pedro's target animals might have lived in and where the boys might expect to find their fossils. Then, solve the mystery.

Biology.com

Finding the solution to the Fossil Quest mystery is only the beginning. Take a video field trip with the ecogeeks of Untamed Science to see where the mystery leads.

Go online to access additional resources including:
- eText • Flash Cards • Lesson Overviews • Chapter Mystery

26.1 Invertebrate Evolution and Diversity

Key Questions

When did the first animals evolve?

What does the cladogram of invertebrates illustrate?

Vocabulary

appendage
larva
trochophore

Taking Notes

Preview Visuals Before you read, preview the cladogram of invertebrates in **Figure 26–3.** Take note of any questions you have about it and try to answer them as you read.

For more on the diversity of animals, go to the Visual Guide.
DOL•30–DOL•64

THINK ABOUT IT The origins of the first animals are shrouded in mystery. Since Darwin, paleontologists have known, on the basis of fossil evidence, that many modern multicellular phyla first appeared during a geologically brief period called the "Cambrian Explosion," between 530 and 515 million years ago. How did so many kinds of animals evolve so quickly? What simpler forms could they have evolved from? Until recently, few fossils predating the Cambrian Period had been found, so there was no way to answer these questions. Then, over the last few decades, a series of discoveries revolutionized our understanding of early animal evolution.

Origins of the Invertebrates

When did the first animals evolve?

For roughly 3 billion years after the first prokaryotic cells evolved, all prokaryotes and eukaryotes were single-celled. We don't know when the first multicellular animals evolved from single-celled eukaryotes. Several kinds of data support the hypothesis that animals evolved from ancestors they shared with organisms called choanoflagellates (koh AN uh FLAJ uh layts). These are usually single-celled eukaryotes, but they sometimes grow in colonies. They share several characteristics with sponges, the simplest multicellular animals.

Traces of Early Animals Our oldest evidence of multicellular life comes from recently discovered microscopic fossils that are roughly 600 million years old. The first animals were tiny and soft-bodied, so few fossilized bodies exist. Still, recent studies have uncovered incredibly well preserved fossils of eggs and embryos, such as the embryo in **Figure 26–1.** Other fossils from this time period have been identified as parts of sponges and animals similar to jellyfish. Paleontologists have also identified what are called "trace fossils" from this time period. Trace fossils are tracks and burrows made by animals whose body parts weren't fossilized. **Such fossil evidence indicates that the first animals began evolving long before the Cambrian Explosion.**

FIGURE 26–1 Fossil Evidence Fossils such as the 565-million-year-old embryo at left are among the rarest and most valuable treasures that the backbreaking work of hunting for microfossils can yield. (SEM 100×)

The Ediacaran Fauna Some of the most exciting and important discoveries about animal life before the Cambrian Period come from fossils in the Ediacara Hills of Australia. These strange fossils, which date from roughly 565 to about 544 million years ago, have intrigued paleontologists for years. The body plans they show are different from those of anything alive today. They show little evidence of cell, tissue, or organ specialization, and no organization into a front and back end. Some may have had photosynthetic algae living within their bodies. Some were segmented and had bilateral symmetry. Some seem to be related to invertebrates such as jellyfishes and worms. Many of the organisms were flat and lived on the bottom of shallow seas.

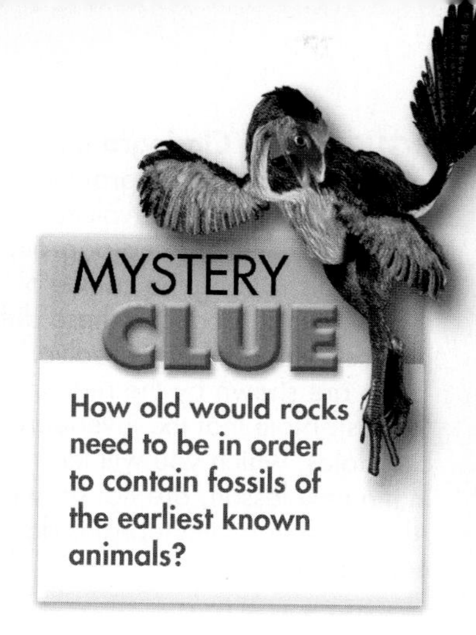

How old would rocks need to be in order to contain fossils of the earliest known animals?

The Cambrian Explosion Fossils from the Cambrian Period, which began about 542 million years ago, paint a fascinating picture of invertebrate life. Two major Cambrian fossil sites are in Chengjiang, China, and in the Burgess Shale of Canada. Cambrian fossils show that over a period of 10–15 million years, animals evolved complex body plans, including specialized cells, tissues, and organs. Many had body symmetry; segmentation; a front and back end; and **appendages,** structures such as legs or antennae protruding from the body. Some Cambrian animals had also evolved shells, skeletons, and other hard body parts. Hard body parts tend to persist longer after an organism dies, so they are more likely to become fossilized.

A number of Cambrian fossils have been identified as ancient members of modern invertebrate phyla, such as the arthropod *Marrella* in **Figure 26–2.** However, some early Cambrian fossils represent extinct groups so peculiar that no one knows what to make of them! Other Cambrian animals appear to be early chordates. By the end of the Cambrian Period, all the basic body plans of modern phyla had been established. Later evolutionary changes, which produced the more familiar body structures of modern animals, involved significant variations on these basic body plans.

Modern Invertebrate Diversity Today, invertebrates are the most abundant animals on Earth. They live in nearly every ecosystem, participate in nearly every food web, and vastly outnumber so-called "higher animals," such as reptiles and mammals.

FIGURE 26–2 Cambrian Animals This Cambrian Period fossil of *Marrella splendens* was found in the Burgess Shale in Canada. The illustration shows what *Marrella* and other Burgess Shale animals may have looked like. **Infer** ***Why do scientists have more detailed data on Cambrian animals than they do on pre-Cambrian animals?***

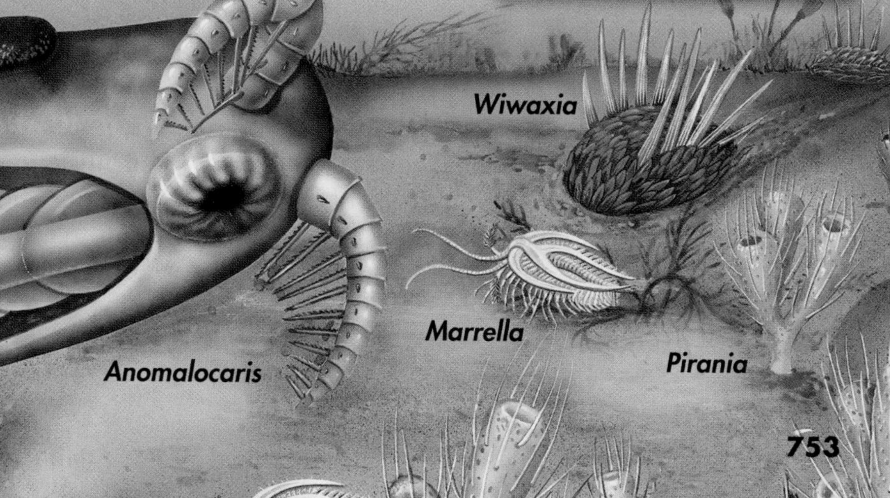

FIGURE 26–3 Cladogram of Invertebrates This diagram shows current hypotheses of evolutionary relationships among major groups of animals. During the course of evolution that produced these different groups, important traits evolved. These are shown by the red circles (nodes). Note that the invertebrate chordates, which you will learn about in the next lesson, are not shown. Also, note that invertebrates do not form a clade.

Cladogram of Invertebrates

What does the cladogram of nonchordate invertebrates illustrate?

Major clades of living invertebrates are shown in **Figure 26–3.** **This cladogram shows current hypotheses about the sequential nature of, and evolutionary relationships among, major living invertebrate groups. This branching, hierarchical classification system, based on similarities and differences shared among groups, also indicates the sequence in which some important features evolved.** These features include body symmetry, cephalization, segmentation, and formation of a coelom. Many of these features evolved in Cambrian animals.

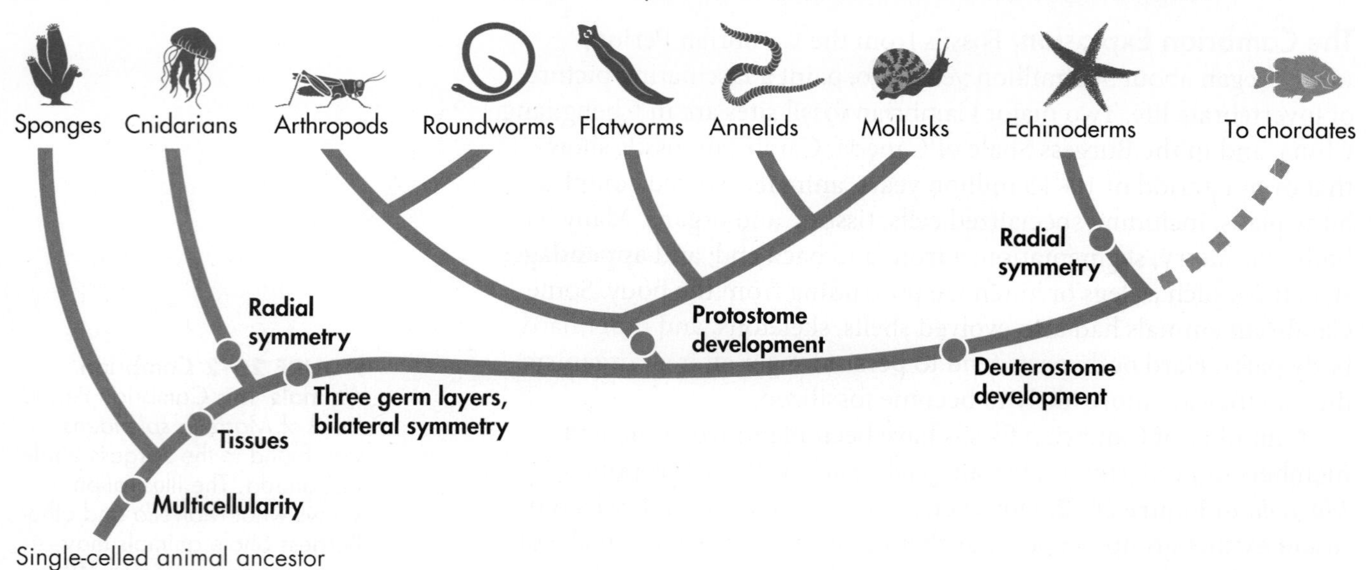

INVERTEBRATE	PHYLUM	DESCRIPTION
Sponges 	Sponges are members of the phylum Porifera (por IHF er uh), which means "pore bearers" in Latin, reflecting the fact that they have tiny openings, or pores, all over their bodies.	Sponges are classified as animals because they are multicellular, heterotrophic, lack cell walls, and contain a few specialized cells. Sponges are the most ancient members of the kingdom Animalia and are among the simplest organisms to be placed in the clade Metazoa, with all other multicellular animals. DOL•31
Cnidarians 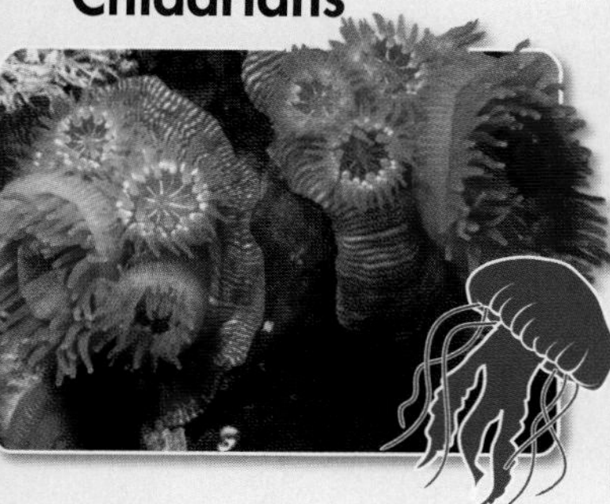	Jellyfishes, sea fans, sea anemones, hydras, and corals are all members of the phylum Cnidaria (ny DAYR ee uh).	Cnidarians are aquatic, soft-bodied, carnivorous, radially symmetrical animals with stinging tentacles arranged in circles around their mouths. Some, such as corals, have skeletons. They are the simplest animals to have body symmetry and specialized tissues. Some cnidarians live as independent individuals. Others live in colonies composed of many individuals. DOL•32–DOL•33

INVERTEBRATE	PHYLUM	DESCRIPTION
Arthropods 	Members of the phylum Arthropoda (ahr THRAHP oh duh) include spiders; centipedes; insects; and crustaceans, such as crabs. *Arthron* means "joint" in Greek, and *podos* means "foot."	Arthropods have bodies divided into segments, a tough external skeleton called an exoskeleton, cephalization, and jointed appendages. Arthropods appeared in the sea about 600 million years ago and have since colonized freshwater habitats, the land, and the air. At least a million species have been identified—more than three times the number of all other animal species combined! DOL•34–DOL•37
Nematodes (Roundworms) 	Members of the phylum Nematoda range in size from microscopic to 1 meter in length.	Nematodes, or roundworms, are unsegmented worms with pseudocoeloms, specialized tissues and organ systems, and digestive tracts with two openings—a mouth and an anus. Some are free-living and inhabit soil or various aquatic habitats. Others are parasites that infect a wide range of plants and animals, including humans. Nematodes were once thought to be closely related to flatworms, annelids, and mollusks but have been found to be more closely related to arthropods. DOL•38
Flatworms 	The phylum Platyhelminthes (plat ih hel MIN theez) contains the flatworms.	Flatworms are soft, unsegmented, flattened worms that have tissues and internal organ systems. They are the simplest animals to have three embryonic germ layers, bilateral symmetry, and cephalization. Most flatworms are no more than a few millimeters thick. Flatworms do not have coeloms. DOL•39
Annelids 	The phylum Annelida (un NEL ih duh) includes earthworms, some exotic-looking marine worms, and parasitic, bloodsucking leeches.	Annelids are worms with segmented bodies and a true coelom lined with tissue derived from mesoderm. The name Annelida is derived from the Latin *annellus*, which means "little ring." The name refers to the ringlike appearance of the body segments of annelids. DOL•40–DOL•41

INVERTEBRATE	PHYLUM	DESCRIPTION
Mollusks 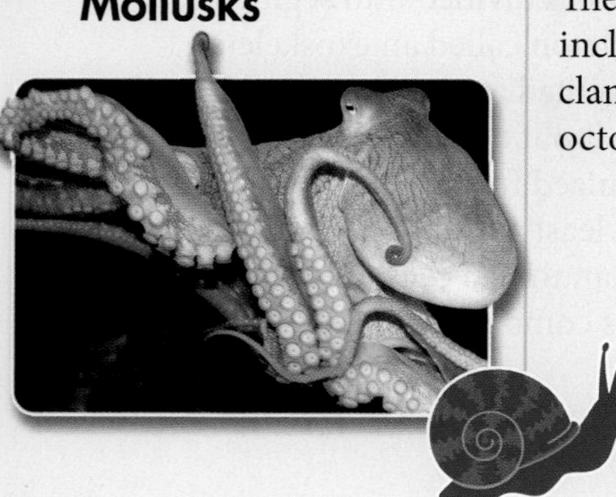	The phylum Mollusca includes snails, slugs, clams, squids, and octopi.	Mollusks are soft-bodied animals that typically have an internal or external shell. Like annelids, mollusks have true coeloms surrounded by mesoderm. They also have complex organ systems. Why are animals as different-looking as snails, clams, and squid in the same phylum? One answer lies in the behavior of their **larvae** (singular: larva), or immature stages. Many mollusks have a free-swimming larval stage called a **trochophore** (TRAHK oh fawr). The trochophore is also characteristic of many annelids, indicating that annelids and mollusks are closely related. DOL•42–DOL•43
Echinoderms 	The phylum Echinodermata (ee KY noh durm aht uh) includes sea stars, sea urchins, and sand dollars, all of which live only in the sea. *Echino-* means "spiny" in Greek, and *dermis* means "skin" in Latin.	Echinoderms have spiny skin and an internal skeleton. They also have a water vascular system—a network of water-filled tubes that include suction-cuplike structures called tube feet, which are used for walking and for gripping prey. Most adult echinoderms exhibit five-part radial symmetry. The skin of an echinoderm is stretched over an internal skeleton of calcium carbonate plates. Although radial symmetry is characteristic of simpler animals such as cnidarians, echinoderms are more closely related to humans and other chordates because they are deuterostomes. DOL•44–DOL•45

26.1 Assessment

Review Key Concepts

1. a. **Review** What was the Cambrian Explosion?
   b. **Explain** When does fossil evidence indicate that the first animals evolved?
   c. **Relate Cause and Effect** What two characteristics of early animals explain the scarcity of animal fossils older than the Cambrian Period?
2. a. **Review** What is a cladogram?
   b. **Explain** What does the cladogram of invertebrates show?
   c. **Sequence** Which body plan feature evolved first—radial symmetry or deuterostome development?

VISUAL THINKING

3. Design a "new" invertebrate. Create an illustration on which you point out its body plan features. Then, show its place on the cladogram of invertebrates, and write a caption explaining how its features helped you decide where it belongs.

26.2 Chordate Evolution and Diversity

THINK ABOUT IT At first glance, fishes, amphibians, reptiles, birds, and mammals appear to be very different. Some have feathers, others have fins. Some fly, others swim or crawl. Yet, all are members of the phylum in which we ourselves are classified—phylum Chordata.

Origins of the Chordates

What are the most ancient chordates?

Chordates are the animals we know best because they are generally large (as animals go), often conspicuous, and strike us as beautiful, impressive, cute, or frightening. Some we keep as pets, others many of us eat as sources of protein. How did all these diverse forms arise?

The Earliest Chordates What were the earliest chordates like? **Embryological studies suggest that the most ancient chordates were related to the ancestors of echinoderms.** The rich Cambrian fossil deposits that record invertebrate history also include some early chordate fossils, such as *Pikaia* (pih KAY uh), which is shown in **Figure 26–4.** When *Pikaia* was first discovered, it was thought to be a worm. Then scientists determined that it had a notochord and paired muscles arranged in a series, like those of simple modern chordates. In 1999, fossil beds from later in the Cambrian Period yielded specimens of *Myllokunmingia* (MY loh kuhn min jee uh), the earliest known vertebrate. These fossils show muscles arranged in a series, traces of fins, sets of feathery gills, a head with paired sense organs, and a skull and skeletal structures likely made of cartilage. **Cartilage** is a strong connective tissue that is softer and more flexible than bone. It supports all or part of a vertebrate's body. In humans, cartilage supports the nose and external ears.

Modern Chordate Diversity Modern chordates are very diverse, consisting of six groups: the nonvertebrate chordates and the five groups of vertebrates—fishes, amphibians, reptiles, birds, and mammals. About 96 percent of all modern chordate species are vertebrates. Among vertebrates, fishes are the largest group by far. Yet, today's chordate species are only a small fraction of the total number of chordates that have existed over time.

DOL•46–DOL•64

Key Questions

What are the most ancient chordates?

What can we learn by studying the cladogram of chordates?

Vocabulary

cartilage
tetrapod

Taking Notes

Venn Diagram Construct a Venn diagram comparing and contrasting nonvertebrate chordates and vertebrates.

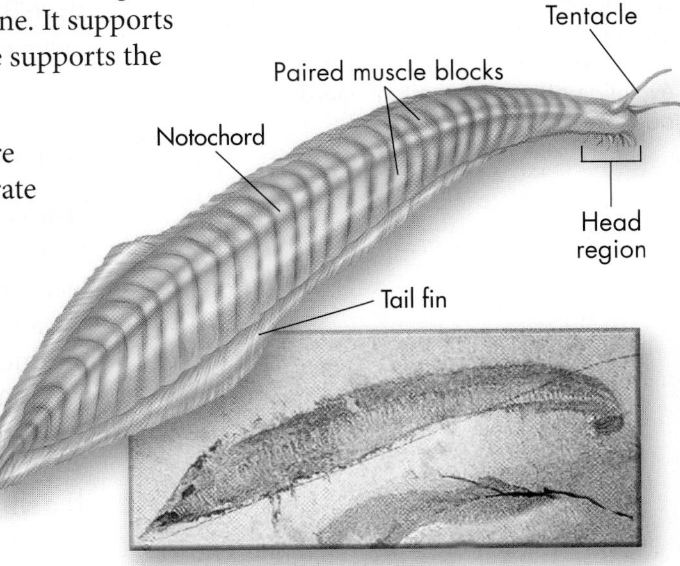

FIGURE 26–4 *Pikaia*, an Early Chordate *Pikaia* is the earliest chordate known from the fossil record. **Classify** ***Which chordate characteristics did Pikaia possess?***

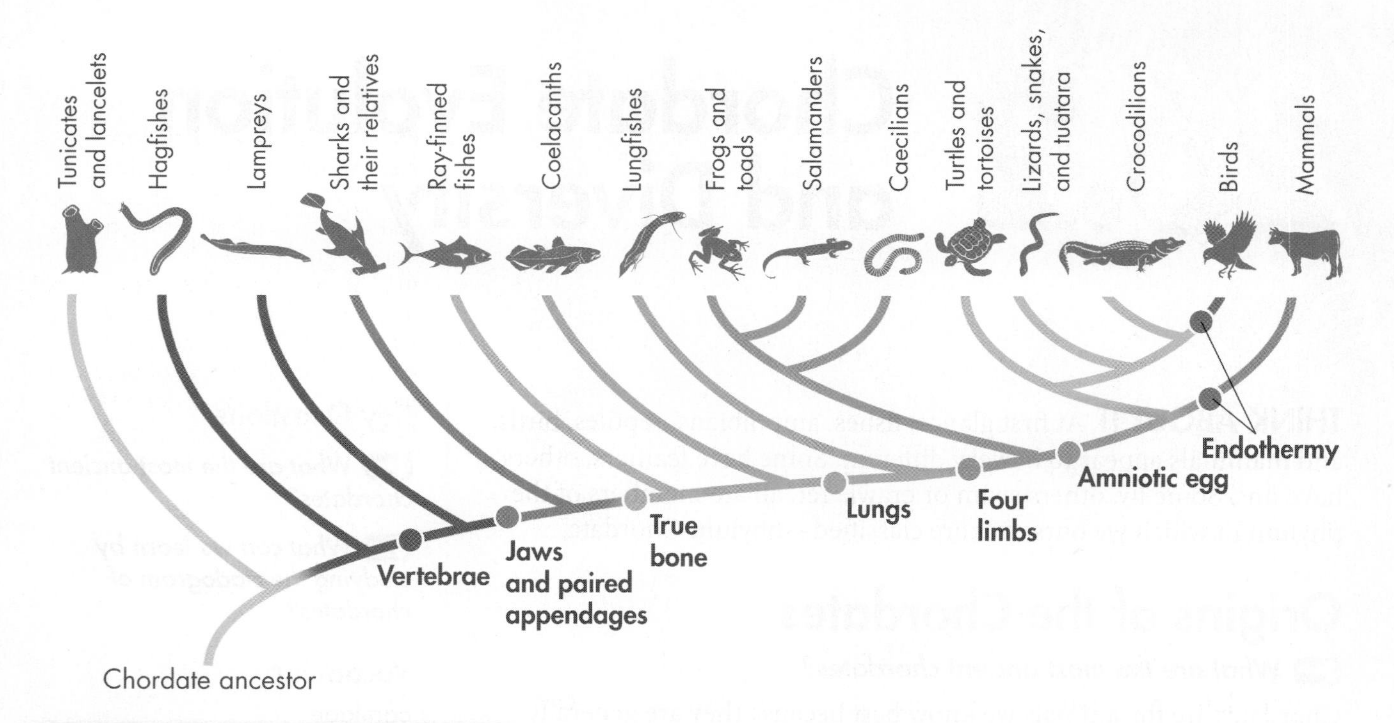

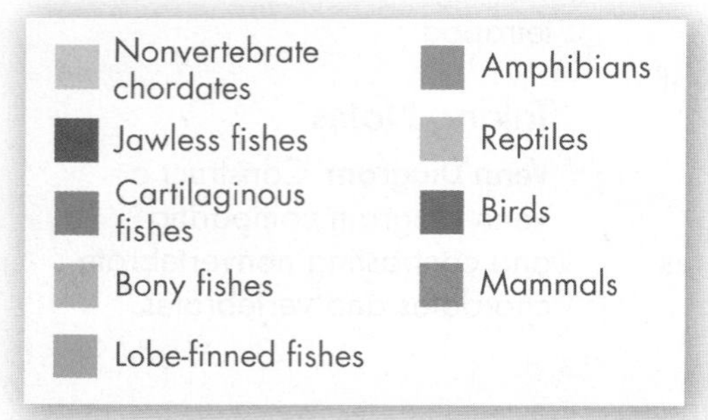

FIGURE 26–5 Cladogram of Chordates The phylum Chordata includes both vertebrates and nonvertebrate chordates. All groups (clades) share a common invertebrate ancestor. This cladogram shows current hypotheses about the evolutionary relationships among living chordate groups. The different colored lines represent the traditional groupings of these animals, as listed in the key. The circles indicate the evolution of some important chordate adaptations.

Cladogram of Chordates

What can we learn by studying the cladogram of chordates?

The hard body structures of many chordates fossilize well, so there is an excellent fossil record of chordate evolutionary history. **The cladogram of chordates presents current hypotheses about relationships among chordate groups. It also shows at which points important vertebrate features, such as jaws and limbs, evolved.** The cladogram of chordates is shown in **Figure 26–5.**

The circles (nodes) in the cladogram represent the appearance of certain adaptive features during chordate evolution. Each time a new adaptation evolved in chordate ancestors, a major adaptive radiation occurred. One notable adaptation, for example, was the development of jaws, which jump-started the adaptive radiation of jawed fishes—now the most diverse chordate group. Other important adaptations include the development of true bone and paired appendages. Refer to the geologic time scale in Chapter 19 as you read about the evolutionary history of chordates.

Nonvertebrate Chordates Two chordate groups lack backbones. These nonvertebrate chordates are tunicates and lancelets. Fossil evidence from the Cambrian Period suggests that the ancestors of living nonvertebrate chordates diverged from the ancestors of vertebrates more than 550 million years ago.

Adult tunicates (subphylum Urochordata) look more like sponges than us. They have neither a notochord nor a tail. But their larval forms have all the key chordate characteristics. The small, fishlike lancelets (subphylum Cephalochordata) live on the sandy ocean bottom. DOL•46–DOL•47

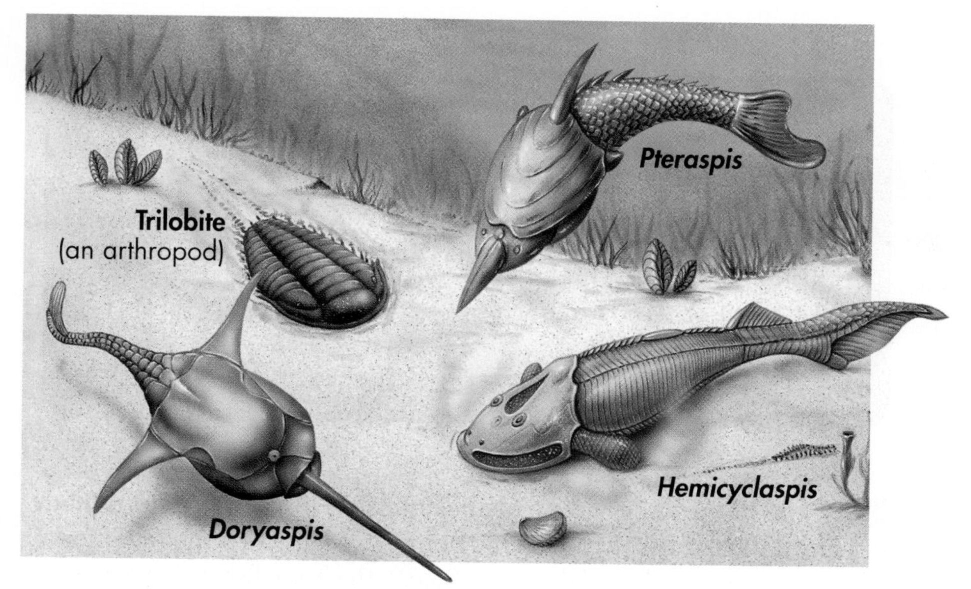

FIGURE 26–6 Ancient Jawless Fishes Ancient jawless fishes, some of which were armored, such as those shown here, lived during the early Devonian Period. Since they lacked jaws, they were limited in their ability to feed and defend themselves. Paired fins, however, gave these fishes control over their movement, and the lineage led to today's hagfishes and lampreys.

Jawless Fishes The earliest fishes appeared in the fossil record during the late Cambrian Period, about 510 million years ago. These odd-looking creatures had no true jaws or teeth, and their skeletons were made of cartilage. During the Ordovician and Silurian periods fishes had a major adaptive radiation. The products of this radiation ruled the seas during the Devonian Period, also called the Age of Fishes. Some armored jawless fishes, such as those in **Figure 26–6,** became extinct at the end of the Devonian, about 360 million years ago. Two other ancient clades of jawless fishes gave rise to the two clades, sometimes called classes, of modern jawless fishes: lampreys and hagfishes. **DOL•48–DOL•51**

Lampreys and hagfishes both lack vertebrae and have notochords as adults. (They do have parts of what could be called a skull, which is one of the reasons they are still classified as vertebrates.) Lampreys are filter feeders as larvae and parasites as adults. Hagfishes have pinkish gray, wormlike bodies, secrete incredible amounts of slime, and tie themselves into knots!

Sharks and Their Relatives Other ancient fishes evolved a revolutionary feeding adaptation: jaws. Jaws hold teeth and muscles, which make it possible to bite and chew plants and other animals.

Early fishes also evolved paired pectoral (anterior) and pelvic (posterior) fins. These fins were attached to limb girdles, which are supporting structures made of cartilage or bone. Paired fins offered more control of body movement, while tail fins and powerful muscles gave greater thrust.

These adaptations launched the adaptive radiation of the class Chondrichthyes (kahn DRIK theez): the sharks, rays, and skates. The Greek word *chondros* means "cartilage," the tissue that makes up the skeletons of these "cartilaginous" fishes. There are hundreds of species of modern sharks, skates, and rays, ranging from predatory carnivores, such as the great white shark in **Figure 26–7,** to shy plankton feeders.

In Your Notebook *How did the evolution of paired fins help early fishes succeed in their environments?*

FIGURE 26–7 Jaws Though about 360 million years separate *Dunkleosteus* (fossil, top) and today's great white shark (bottom), you can easily see the important adaptation they have in common—jaws. But *Dunkleosteus* could have bitten "Jaws" the shark in half! **Pose Questions** ***What question would you ask a researcher about* Dunkleosteus?**

FIGURE 26–8 Bony Fishes Bony fishes, such as the perch this skeleton once belonged to, have skeletons made of true bone.
Compare and Contrast ***List two ways that bony fishes differ from the fishes in Figure 26–6.***

Bony Fishes Another group of ancient fishes evolved skeletons made of hard, calcified tissue called true bone. This launched the radiation of the class Osteichthyes (ahs tee IK theez), the bony fishes. You can see the skeleton of a modern bony fish in **Figure 26–8.** Most modern bony fishes belong to a huge group called ray-finned fishes.

▶ ***Ray-Finned Fishes*** Ray-finned fishes are aquatic vertebrates with skeletons of true bone; most have paired fins, scales, and gills. The name "ray-finned" refers to bony rays connected by a layer of skin to form fins. The fin rays support the skin much as the thin rods in a handheld folding fan support the webbing of the fan. Most fishes you are familiar with, such as eels, goldfish, and catfish, are ray-finned fishes.

▶ ***Lobe-Finned Fishes*** Lobe-finned fishes are a different group of bony fishes that evolved fleshy fins supported by larger, more substantial bones. The few modern fishes that are descendants of ancient lobe-finned fishes include lungfishes and coelacanths (SEE luh kanths). Another group of ancient lobe-finned fishes evolved into the ancestors of four-limbed vertebrates, or **tetrapods.**

VISUAL SUMMARY

FROM FINS TO FEET

FIGURE 26–9 The cladogram shows a few of the animal groups in the evolution of the feet of tetrapods from the fins of ancient bony fish. All of the illustrated animal groups are extinct.

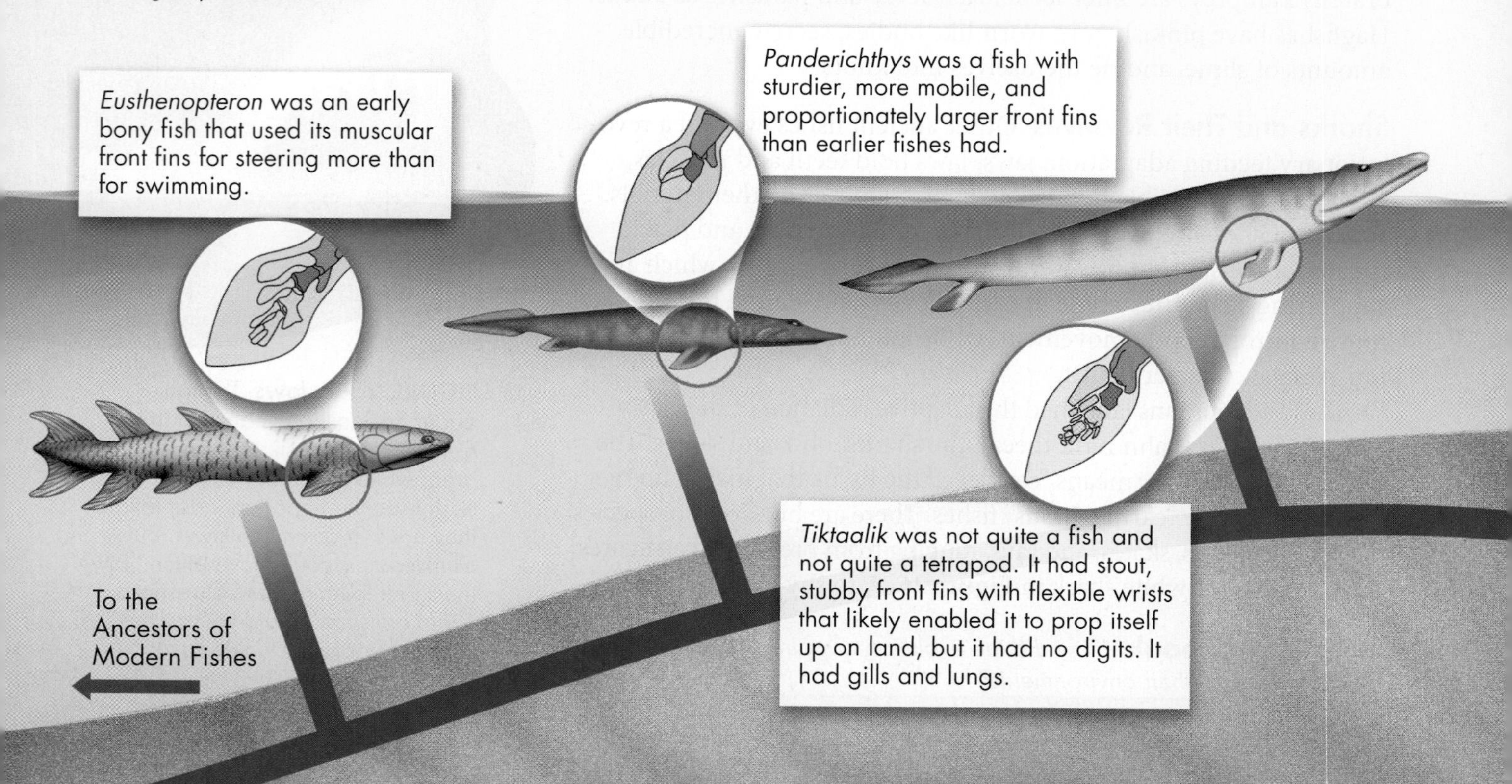

Amphibians The word *amphibian* means "double life," referring to the fact that these animals live in water as larvae but on land as adults. Amphibians are vertebrates that also, with some exceptions, require water for reproduction, breathe with lungs as adults, have moist skin with mucous glands, and lack scales and claws. DOL•52–DOL•53

▶ ***The Unique "Fishapod"*** The general story of early amphibian evolution has been known for years. Several fossils indicate that various lines of lobe-finned fishes evolved sturdier and sturdier appendages, which resembled the limbs of tetrapods. But in recent years, a series of spectacular transitional fossils have been discovered that document in detail the skeletal transformation from lobe-fins to limbs, as shown in **Figure 26–9.** One of the most interesting of these finds is *Tiktaalik*, a fossil of which is shown in **Figure 26–10.** It is an animal with such a mix of fish and tetrapod features that its discoverers informally referred to it as a "fishapod"—part fish, part tetrapod.

▶ ***Terrestrial Adaptations*** Of course, life on land requires more than just legs to crawl around on. Early amphibians also evolved ways to breathe air and protect themselves from drying out, as you will read in Chapters 27 and 28. These adaptations fueled another adaptive radiation. Amphibians became the dominant vertebrates of the warm, swampy Carboniferous Period, about 359 to 300 million years ago. But this success didn't last. Climate changes caused many low, swampy habitats to disappear. Most amphibian groups became extinct by the end of the Permian Period, about 250 million years ago. Only three orders of amphibians survive today—frogs and toads, salamanders, and caecilians (see SIL ee unz).

In Your Notebook *Explain how the climate changes of the Permian Period could have caused the decline of amphibians.*

BUILD Vocabulary

WORD ORIGINS The word **tetrapod** comes from the Greek words "tetra," meaning *four*, and "pod," meaning *foot*.

FIGURE 26–10 *Tiktaalik*, the Fishapod The 375-million-year-old *Tiktaalik* fossil was discovered in Canada in 2004. It is considered a transitional fossil because it shows features of both tetrapods and the fish they evolved from—fins *with* wrist bones; gills *and* lungs. *Tiktaalik* could swim and breathe underwater like a fish OR crawl and breathe out of water like a tetrapod, so its discoverers called it a "fishapod."

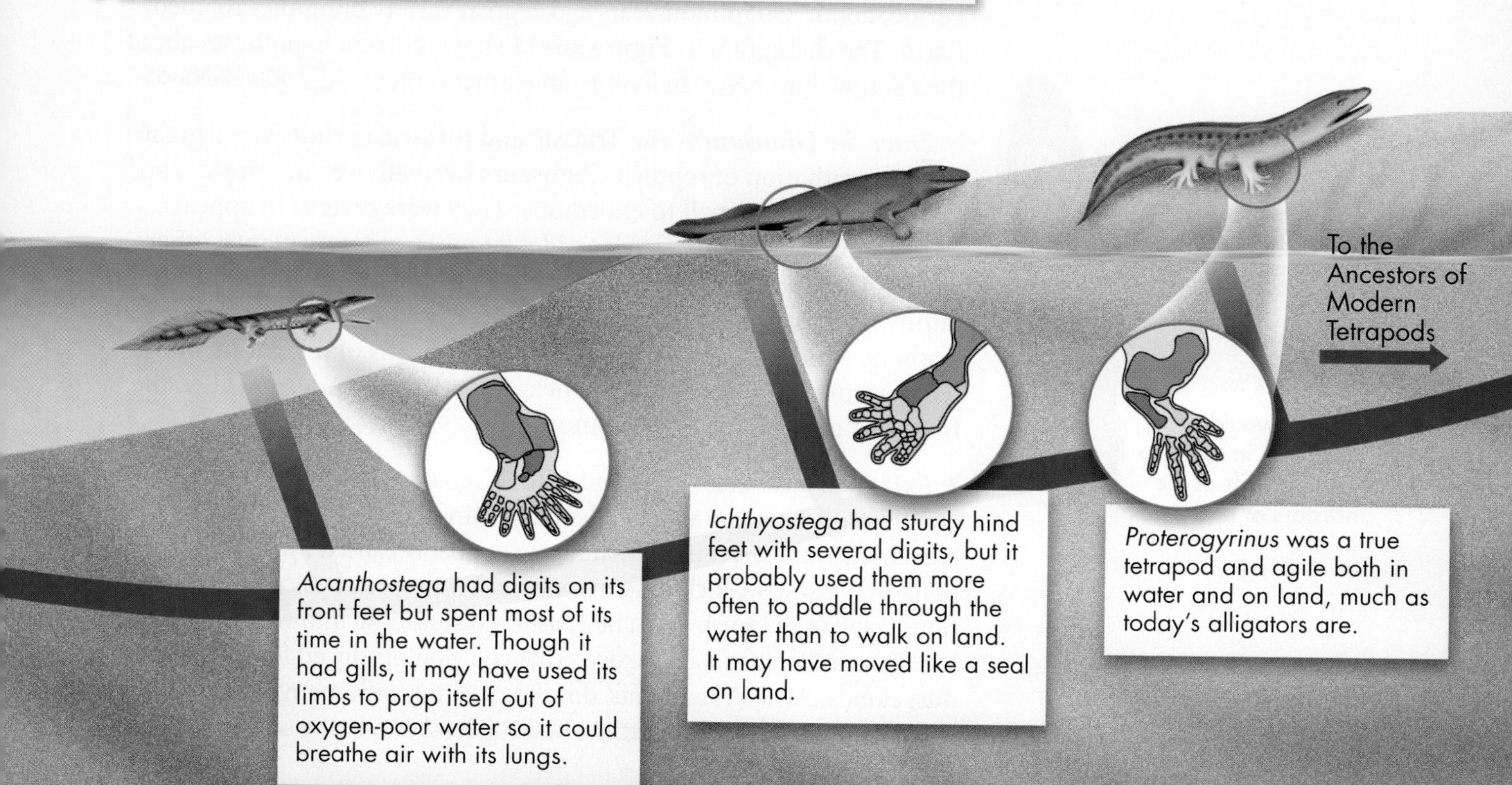

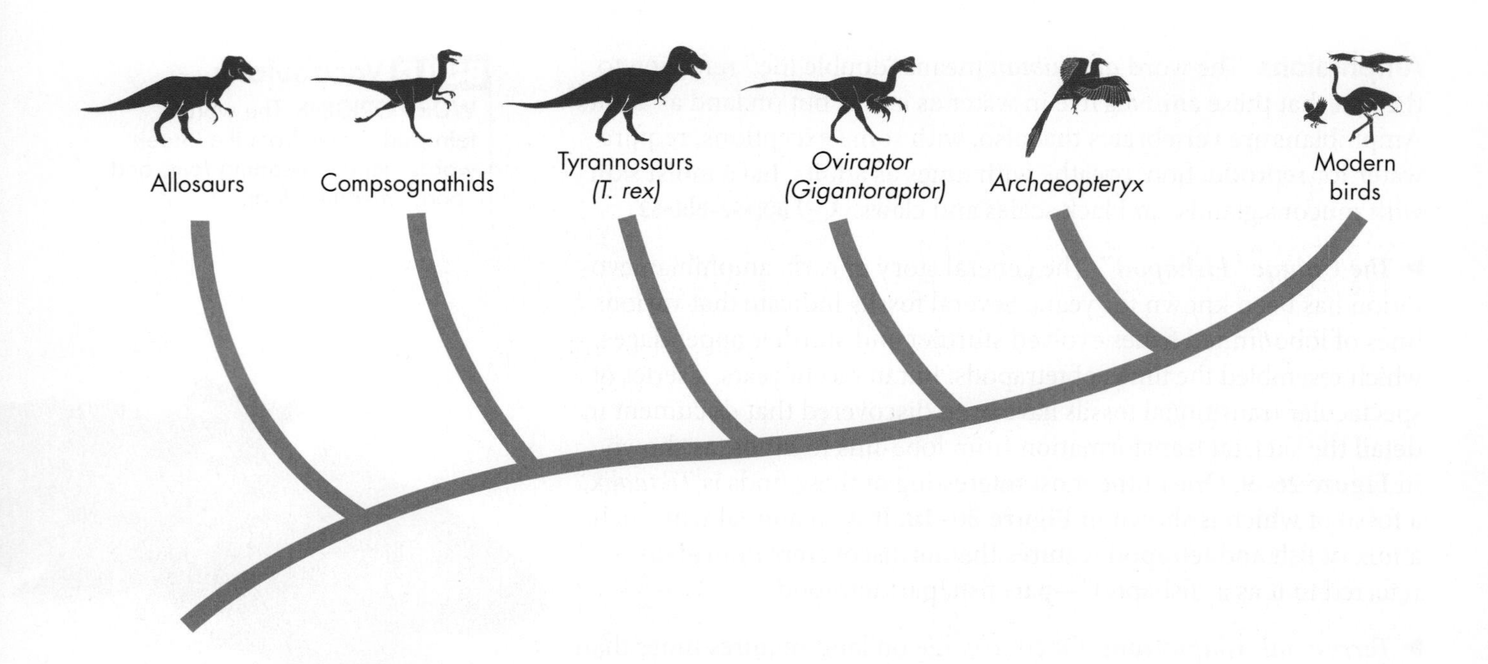

FIGURE 26–11 Evolution of Reptiles and Birds The diagram shows current hypotheses about the evolutionary relationships between living and extinct reptiles. None of the groups shown are direct ancestors of modern reptiles or modern birds.

Reptiles Reptiles, which evolved from ancient amphibians, were the first vertebrates to evolve adaptations to drier conditions. A reptile is a vertebrate with dry, scaly skin, well-developed lungs, strong limbs, and shelled eggs that do not develop in water. Living reptiles are represented by four groups: lizards and snakes, crocodilians, turtles and tortoises, and the tuatara (too uh TAH ruh).

The first known reptile fossil dates back to the Carboniferous Period, 350 million years ago. As the Carboniferous Period ended and the Permian Period began, Earth's climate became cooler and less humid. Many lakes and swamps dried up. Under these drier conditions, the first great adaptive radiation of reptiles began. By the end of the Permian Period, about 250 million years ago, a great variety of reptiles roamed Earth. The cladogram in **Figure 26–11** shows current hypotheses about the relationships between living and extinct reptiles. DOL•54–DOL•55

▶ ***Enter the Dinosaurs*** The Triassic and Jurassic periods saw a great adaptive radiation of reptiles. Dinosaurs lived all over the world, and they ranged from small to enormous. They were diverse in appearance and in habit: Some, such as *Plateosaurus,* ate leafy plants; others, such as *Coelophysis,* were carnivorous. Duckbilled *Maiasaura* lived in family groups and cared for eggs and young. Some dinosaurs even had feathers, which may have first evolved as a means of regulating body temperature. The evolutionary lineage that led to modern birds came from one group of feathered dinosaurs.

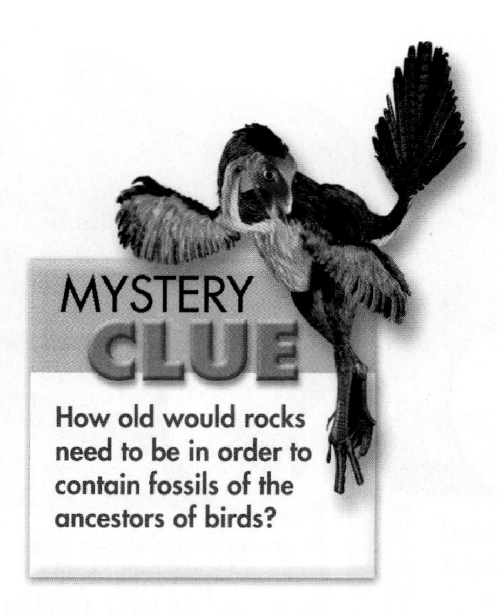

▶ ***Exit the Dinosaurs*** At the end of the Cretaceous Period, about 66 million years ago, a worldwide mass extinction occurred. According to current hypotheses, this extinction was probably caused by a combination of natural disasters, including massive and widespread volcanic eruptions, a fall in sea level, and a huge asteroid smashing into what is now the Yucatán Peninsula in Mexico. That collision produced forest fires and dust clouds. After these events, dinosaurs, along with many other animal and plant groups, became extinct both on land and in the sea.

Birds Today's birds are extremely diverse. Birds are reptiles that regulate their internal body temperature. They have an outer covering of feathers; strong yet lightweight bones; two legs covered with scales that are used for walking or perching; and front limbs modified into wings. DOL•56–DOL•59

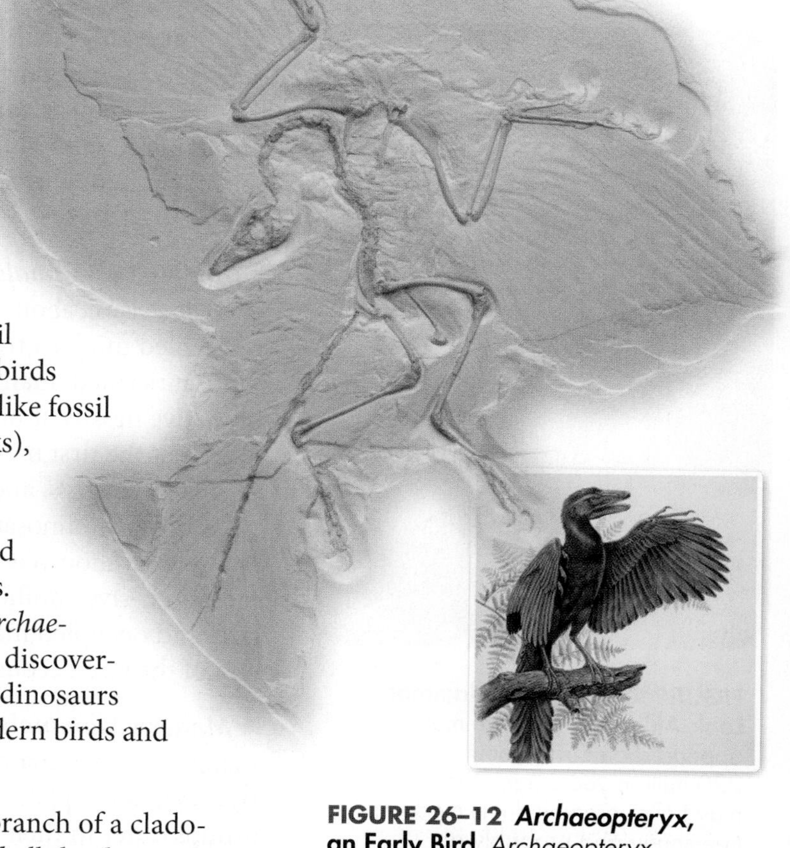

FIGURE 26–12 *Archaeopteryx,* an Early Bird *Archaeopteryx,* shown both in the fossil and artist's conception, was a bird that showed both dinosaur characteristics (teeth, bony tail) and bird characteristics (flight feathers). Because of the weight of its teeth and bony tail and its small breastbone, it might not have been able to fly very well.

▶ ***Bird Roots*** If you've ever wished that dinosaurs were still around, you're in luck! Recent fossil discoveries strongly support the hypothesis that birds evolved from a group of dinosaurs. The first birdlike fossil discovered was *Archaeopteryx* (ahr kee AHP tur iks), from the late Jurassic Period, about 150 million years ago. *Archaeopteryx* looked so much like a small, running dinosaur that it would be classified as a dinosaur except for its highly evolved feathers. You can see a fossil and an artist's conception of *Archaeopteryx* in **Figure 26–12.** A whole series of recent discoveries of well-preserved ancient birds and feathered dinosaurs has done a lot to "connect the dots" between modern birds and their dinosaur ancestors.

▶ ***Bird Classification*** As you recall, a clade is a branch of a cladogram that includes a single common ancestor and all the descendants of that ancestor. If you look back at **Figure 26–11,** you will see that recognizing birds as descendants of dinosaurs should cause a change in their classification. Modern birds by themselves, the traditional class Aves, form a clade within the clade containing dinosaurs. And because the clade containing dinosaurs is part of a larger clade of reptiles, modern birds are also reptiles. The traditional class Reptilia, which includes living reptiles and dinosaurs but *not* birds, however, is not a clade.

Feather Evolution

The information in the table shows the evolution of feathers in some groups of dinosaurs that preceded modern birds.

Group (listed alphabetically)	Feather Status
Allosaurs	None
Archaeopteryx	Flight feathers
Compsognathids	Hairlike feathers
Oviraptors	True feathers
Tyrannosaurs	Branched feathers

1. Organize Data Recalling what you learned about drawing cladograms in Chapter 18, use the information to place these traits correctly on **Figure 26–11.** (Redraw the cladogram in your notebook.)

2. Draw Conclusions Which type of feathers would you expect modern birds to possess?

FIGURE 26–13 Early Mammal Look-Alike The first mammals appeared on Earth about 220 million years ago. They may have resembled this modern tree shrew and probably ate insects.

Mammals Members of the traditional class Mammalia include about 5000 species that range in size from mice to whales. Characteristics unique to mammals include mammary glands in females, which produce milk to nourish young, and hair. Mammals also breathe air, have four-chambered hearts, and regulate their internal body temperature. DOL•60–DOL•64

▶ ***The First Mammals*** True mammals first appeared during the late Triassic Period, about 220 million years ago. They were very small and resembled modern tree shrews, like the one in **Figure 26–13.** While dinosaurs ruled, mammals remained generally small and were probably active mostly at night. New fossils and DNA analyses suggest, however, that the first members of modern mammalian groups, including primates, rodents, and hoofed mammals, evolved during this period. After the great dinosaur extinction at the end of the Cretaceous Period, about 65 million years ago, mammals underwent a long adaptive radiation. Over millions of years, mammals diversified, increased in size, and occupied many niches. The Cenozoic Era, which began at the end of the Cretaceous Period, is usually called the Age of Mammals.

▶ ***Modern Mammals*** By the beginning of the Cenozoic Era, three major groups of mammals had evolved—monotremes (MAHN oh treemz), marsupials (mahr soo pee ulz), and placentals. These three groups differ in their means of reproduction and development.

Only five species of the egg-laying monotremes, including the duckbill platypus, exist today, all in Australia and New Guinea. Marsupials, which include kangaroos, koalas, and wombats, bear live young that usually complete their development in an external pouch. Placental mammals—which include most of the mammals you are familiar with—have embryos that develop further while still inside the mother. After birth, most placental mammals care for their young and nurse them to provide nourishment.

26.2 Assessment

Review Key Concepts

1. a. **Review** Name the group of animals whose ancestors were related to the earliest chordates.
   b. **Compare and Contrast** Why did scientists classify *Pikaia* as a chordate instead of as a worm?
2. a. **Review** What two aspects of evolutionary history does the cladogram of chordates show?
   b. **Explain** How do nonvertebrate chordates differ from other chordates?
   c. **Interpret Visuals** According to **Figure 26–5,** which chordate feature evolved earlier—endothermy or lungs?

Apply the Big idea

Evolution

3. Recall what you learned about plant evolution in Chapter 22. Based on the evolutionary changes shown in the cladograms on pages 636 and 758, identify the first major adaptations that allowed plants and chordates to live on land. In what ways are chordate adaptations to life on land similar to plant adaptations to life on land?

26.3 Primate Evolution

THINK ABOUT IT Carolus Linnaeus placed our species, *Homo sapiens*, in an order he named Primates, which means "first" in Latin. But what are primates "first" in? When primates appeared, there was little to distinguish them from other mammals, aside from an increased ability to use their eyes and front limbs together. As primates evolved, however, several other characteristics became distinctive.

What Is a Primate?

What characteristics do all primates share?

Primates, including lemurs, monkeys, and apes, share several adaptations for a life spent in trees. **In general, a primate is a mammal that has relatively long fingers and toes with nails instead of claws, arms that can rotate around shoulder joints, a strong clavicle, binocular vision, and a well-developed cerebrum.** The lemur in **Figure 26–14** shows many of these characteristics. DOL•64

Fingers, Toes, and Shoulders Primates typically have five flexible fingers and toes on each hand or foot that can curl to grip objects firmly and precisely. This enables many primates to run along tree limbs and swing from branch to branch with ease. In addition, most primates have thumbs and big toes that can move against the other digits. This allows many primates to hold objects firmly in their hands or feet. Primates' arms are well suited for climbing because they can rotate in broad circles around a strong shoulder joint attached to a strong clavicle, or collar bone.

Binocular Vision Many primates have a broad face, so both eyes face forward with overlapping fields of view. This facial structure gives primates excellent binocular vision. **Binocular vision** is the ability to combine visual images from both eyes, providing depth perception and a three-dimensional view of the world. This comes in handy for judging the locations of tree branches, from which many primates swing.

Well-Developed Cerebrum In primates, the "thinking" part of the brain—the cerebrum—is large and intricate. This well-developed cerebrum enables more-complex behaviors than are found in many other mammals. For example, many primate species create elaborate social systems that include extended families, adoption of orphans, and even warfare between rival troops.

Key Questions

What characteristics do all primates share?

What are the major evolutionary groups of primates?

What adaptations enabled later hominine species to walk upright?

***What is the current scientific thinking about the genus* Homo?**

Vocabulary

binocular vision • anthropoid • prehensile tail • hominoid • hominine • bipedal • opposable thumb

Taking Notes

Outline Before you read, outline this lesson. As you read, add details to your outline.

FIGURE 26–14 Primate This lemur displays several primate characteristics—it has flexible fingers and toes, arms that can rotate in broad circles around the shoulder joint, and forward-facing eyes that allow for binocular vision.

Quick Lab
GUIDED INQUIRY

Binocular Vision

❶ Throw a paper ball to your partner, who should try to catch the ball with one hand. Record whether your partner caught the ball.

❷ Now have your partner close one eye. Repeat Step 1.

Analyze and Conclude

1. Use Tables and Graphs Exchange results with other groups. Make a bar graph for the class data comparing the results with both eyes open and one eye shut.

2. Draw Conclusions How is binocular vision useful to primates?

Evolution of Primates

What are the major evolutionary groups of primates?

Humans and other primates evolved from a common ancestor that lived more than 65 million years ago. One recently discovered fossil, *Carpolestes,* which lived 56 million years ago in Wyoming, has been proposed as an example of the first primate. Early in their history, primates split into two groups. **Primates in one of these groups look very little like typical monkeys. This group contains the lemurs and lorises. The other group includes tarsiers and the anthropoids, the group that includes monkeys, great apes, and humans.** Refer to **Figure 26–15** as you read about the evolutionary relationships between these groups.

Lemurs and Lorises With few exceptions, lemurs and lorises are small, nocturnal primates with large eyes adapted to seeing in the dark. Many have long snouts. Living members include the bush babies of Africa, the lemurs of Madagascar, and the lorises of Asia.

Tarsiers and Anthropoids Primates more closely related to humans than to lemurs belong to a different group, members of which have broader faces and widely separated nostrils. This group includes the tarsiers of Asia and the anthropoids. **Anthropoids** (AN thruh poydz), or humanlike primates, include monkeys, great apes, and humans. Anthropoids split into two groups around 45 million years ago, as the continents on which they lived moved apart.

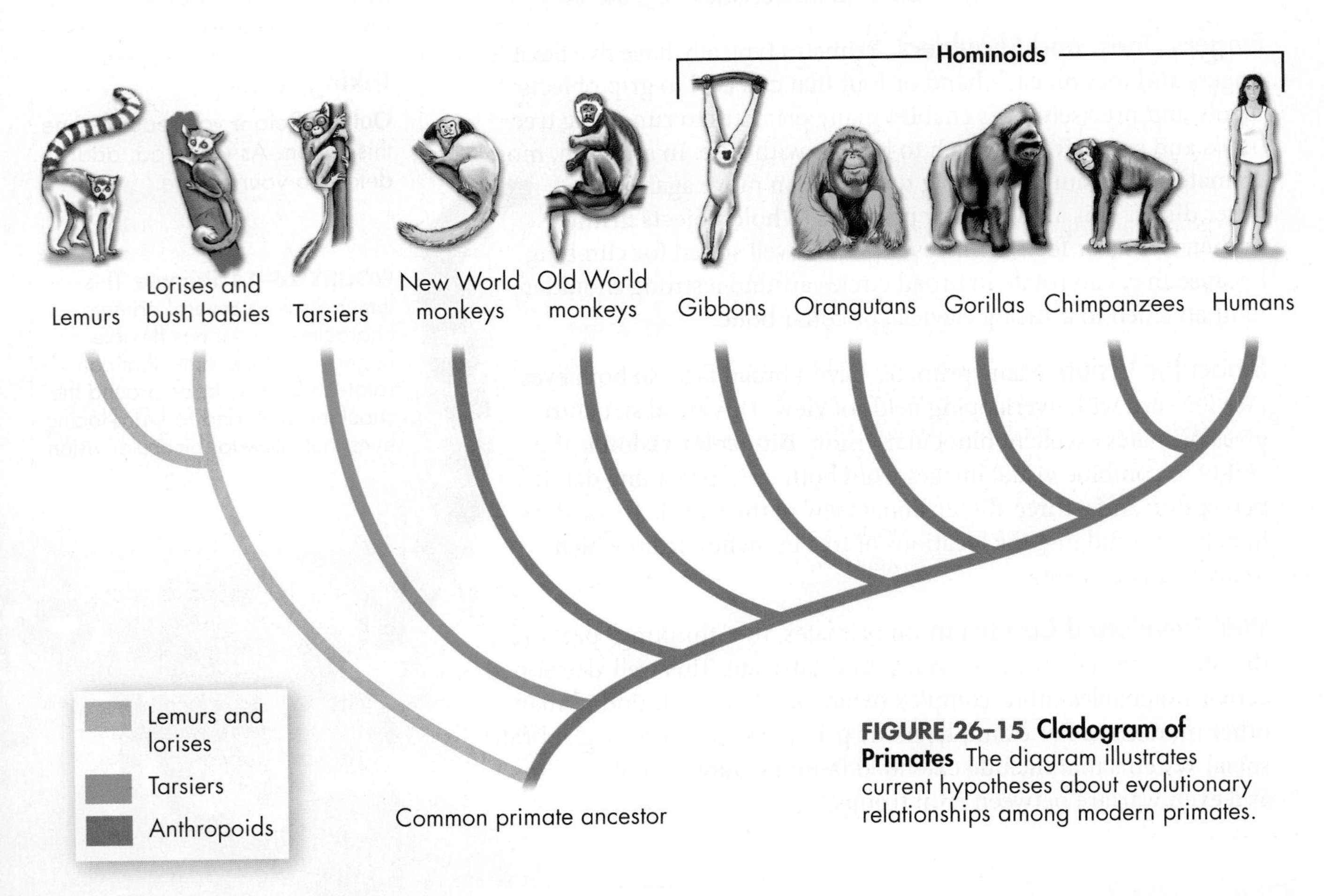

FIGURE 26–15 Cladogram of Primates The diagram illustrates current hypotheses about evolutionary relationships among modern primates.

► ***New World Monkeys*** Members of one anthropoid branch, the New World monkeys, are found in Central and South America. (Europeans used the term *New World* to refer to North and South America.) Members of this group, which includes squirrel monkeys and spider monkeys, live almost entirely in trees. They have long, flexible arms that enable them to swing from branch to branch. New World monkeys also have a long, **prehensile tail** that can coil tightly enough around a branch to serve as a "fifth hand."

► ***Old World Monkeys and Great Apes*** The other anthropoid branch, which evolved in Africa and Asia, includes the Old World monkeys and great apes. Old World monkeys, such as langurs and macaques (muh KAHKS), spend time in trees but lack prehensile tails. Great apes, also called **hominoids,** include gibbons, orangutans, gorillas, chimpanzees, and humans. Recent DNA analyses confirm that, among the great apes, chimpanzees are humans' closest relatives.

Hominine Evolution

What adaptations enabled later hominine species to walk upright?

Between 6 and 7 million years ago, the lineage that led to humans split from the lineage that led to chimpanzees. The hominoids in the lineage that led to humans are called **hominines.** Hominines include modern humans and all other species more closely related to us than to chimpanzees. Hominines evolved the ability to walk upright, grasping thumbs, and large brains. **Figure 26–16** shows some ways in which the skeletons of modern humans differ from those of hominoids such as gorillas. **The skull, neck, spinal column, hip bones, and leg bones of early hominine species changed shape in ways that enabled later species to walk upright.** The evolution of this **bipedal,** or two-footed, locomotion was very important, because it freed both hands to use tools. Meanwhile, the hominine hand evolved an **opposable thumb** that could touch the tips of the fingers, enabling the grasping of objects and the use of tools.

Hominines also evolved much larger brains. The brains of chimpanzees, our closest living relatives, typically range in volume from 280 to 450 cubic centimeters. The brains of *Homo sapiens,* on the other hand, range in size from 1200 to 1600 cubic centimeters! Most of the difference in brain size results from a radically expanded cerebrum.

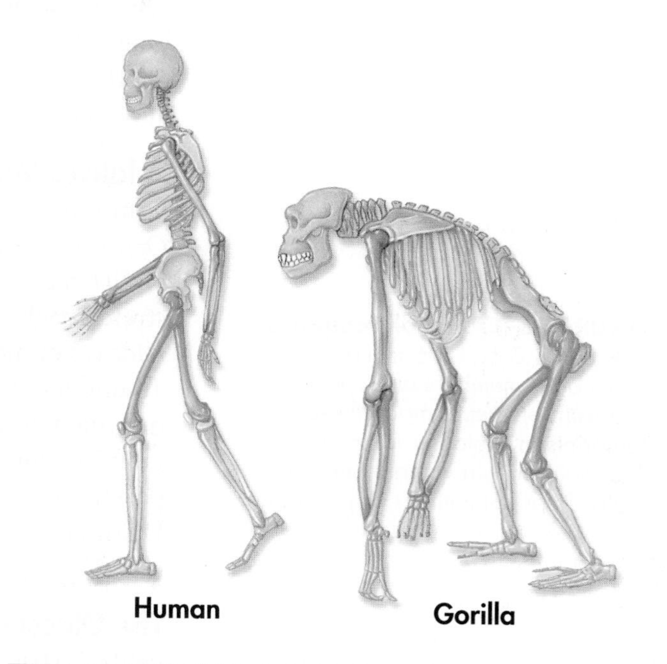

Comparing Human and Gorilla Skeletons

Feature	Human	Gorilla
Skull	Atop S-shaped spine	Atop C-shaped spine
Spinal cord	Exits at bottom of skull	Exits near back of skull
Arms and hands	Arms shorter than legs; hands don't touch ground when walking	Arms longer than legs; hands touch ground when walking
Pelvis	Bowl-shaped	Long and narrow
Thigh bones	Angled inward, directly below body	Angled away from pelvis

FIGURE 26–16 Comparison of Hominoids Modern hominines walk upright on two legs; gorillas use all four limbs. The diagrams show many of the skeletal characteristics that allow hominines to walk upright. **Compare and Contrast** ***According to the chart and illustrations, what are the other skeletal differences between humans and gorillas?***

New Findings and New Questions The study of human ancestors is exciting and constantly changing. Since the 1990s, new discoveries in Africa have doubled the number of known hominine species. Those discoveries also doubled the length of the known hominine fossil record—from 3.5 million years to 7 million years, a time that corresponds closely to the time at which DNA studies suggest that the lineage that led to humans split from the lineage that led to chimpanzees. These new data have enhanced the picture of our species' past. Questions still remain as to how fossil hominines are related to one another—and to humans. In fact, the field is changing so rapidly that all we can present here is a sampling of current hypotheses.

Relatives Versus Ancestors Most paleontologists agree that the hominine fossil record includes seven genera—*Sahelanthropus, Orrorin, Ardipithecus, Australopithecus, Paranthropus, Kenyanthropus,* and *Homo*—and at least 20 species. These diverse hominine fossils stretch back in time roughly 7 million years. All these species are *relatives* of modern humans, but not all of them are human *ancestors*. To understand that distinction, think of your family. Your relatives may include aunts, uncles, cousins, parents, grandparents, and great-grandparents. All of these folks are your relatives, but only your parents, grandparents, and great-grandparents are your ancestors. Distinguishing relatives from ancestors in the hominine family is an ongoing challenge.

FIGURE 26–17 Laetoli Footprints Between 3.8 and 3.6 million years ago, members of a species of *Australopithecus* made these footprints at Laetoli in Tanzania. The footprints show that hominines walked upright millions of years ago.

The Oldest Hominine? In 2002, paleontologists working in north-central Africa discovered a fossil skull roughly 7 million years old. This fossil, called *Sahelanthropus,* is a million years older than any known hominine. *Sahelanthropus* had a brain about the size of that of a modern chimp, but its short, broad face was more like that of a human. Scientists are still debating whether this fossil represents a hominine.

Australopithecus Some early hominine fossil species seem to belong to the lineage that led to modern humans, while others formed separate branches off the main hominine line. One early group of hominines, of the genus *Australopithecus,* lived from about 4 million to about 1.5 million years ago. These hominines were bipedal apes, but their skeletons suggest that they probably spent at least some time in trees. The structure of their teeth suggests a diet rich in fruit.

The best-known of these species is *Australopithecus afarensis,* which lived from roughly 4 million to 2.5 million years ago. The humanlike footprints in **Figure 26–17,** about 3.6 million years old, were probably made by members of this species. *A. afarensis* fossils indicate the species had small brains, so the footprints show that hominines walked bipedally long before large brains evolved. Other fossils of this genus indicate that males were much larger than females. You can see artists' conceptions of young female and adult female *A. afarensis* in **Figure 26–18.**

In Your Notebook *How long ago does DNA evidence suggest that the human lineage split from the chimpanzee lineage?*

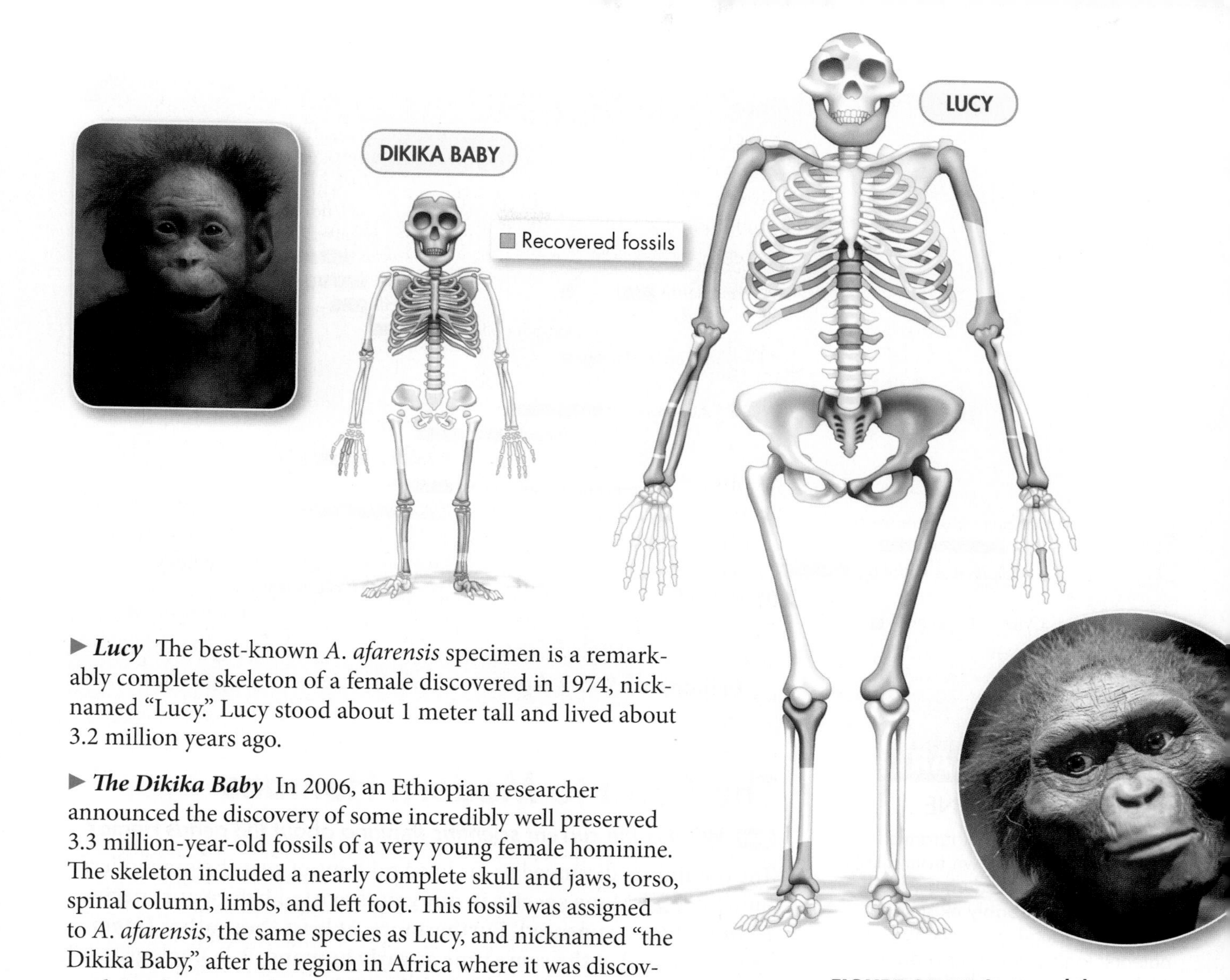

► ***Lucy*** The best-known *A. afarensis* specimen is a remarkably complete skeleton of a female discovered in 1974, nicknamed "Lucy." Lucy stood about 1 meter tall and lived about 3.2 million years ago.

► ***The Dikika Baby*** In 2006, an Ethiopian researcher announced the discovery of some incredibly well preserved 3.3 million-year-old fossils of a very young female hominine. The skeleton included a nearly complete skull and jaws, torso, spinal column, limbs, and left foot. This fossil was assigned to *A. afarensis*, the same species as Lucy, and nicknamed "the Dikika Baby," after the region in Africa where it was discovered. Leg bones confirmed that the Dikika Baby walked bipedally, while her arm and shoulder bones suggest that she would have been a better climber than modern humans. Researchers will be extracting information from these bones for years.

FIGURE 26–18 Lucy and the Dikika Baby "Lucy" and "the Dikika Baby" are nicknames of two very important fossils of the hominine *A. afarensis*. Lucy is a partial skeleton of an adult female. The Dikika Baby is the most-complete fossil yet found of this species. These two fossils were discovered just 6 miles apart in Ethiopia. **Interpret Visuals** ***Given the fossils recovered, which face shape would you expect scientists to be more confident about—the Dikika Baby's or Lucy's?***

Paranthropus Three more-recent species, which grew to the size of well-fed football linebackers, have been placed in their own genus, *Paranthropus*. These *Paranthropus* species had huge, grinding back teeth. Their diets probably included coarse and fibrous plant foods like those eaten by modern gorillas. Paleontologists now place *Paranthropus* on a separate, dead-end branch of our family tree.

Hominine Relationships Researchers once thought that human evolution took place in relatively simple steps in which hominine species, over time, became gradually more humanlike. But it is now clear that a series of hominine adaptive radiations produced a number of species whose relationships are difficult to determine. As a result, what once looked like a simple hominine "family tree" with a single main trunk now looks more like a shrub with multiple trunks.

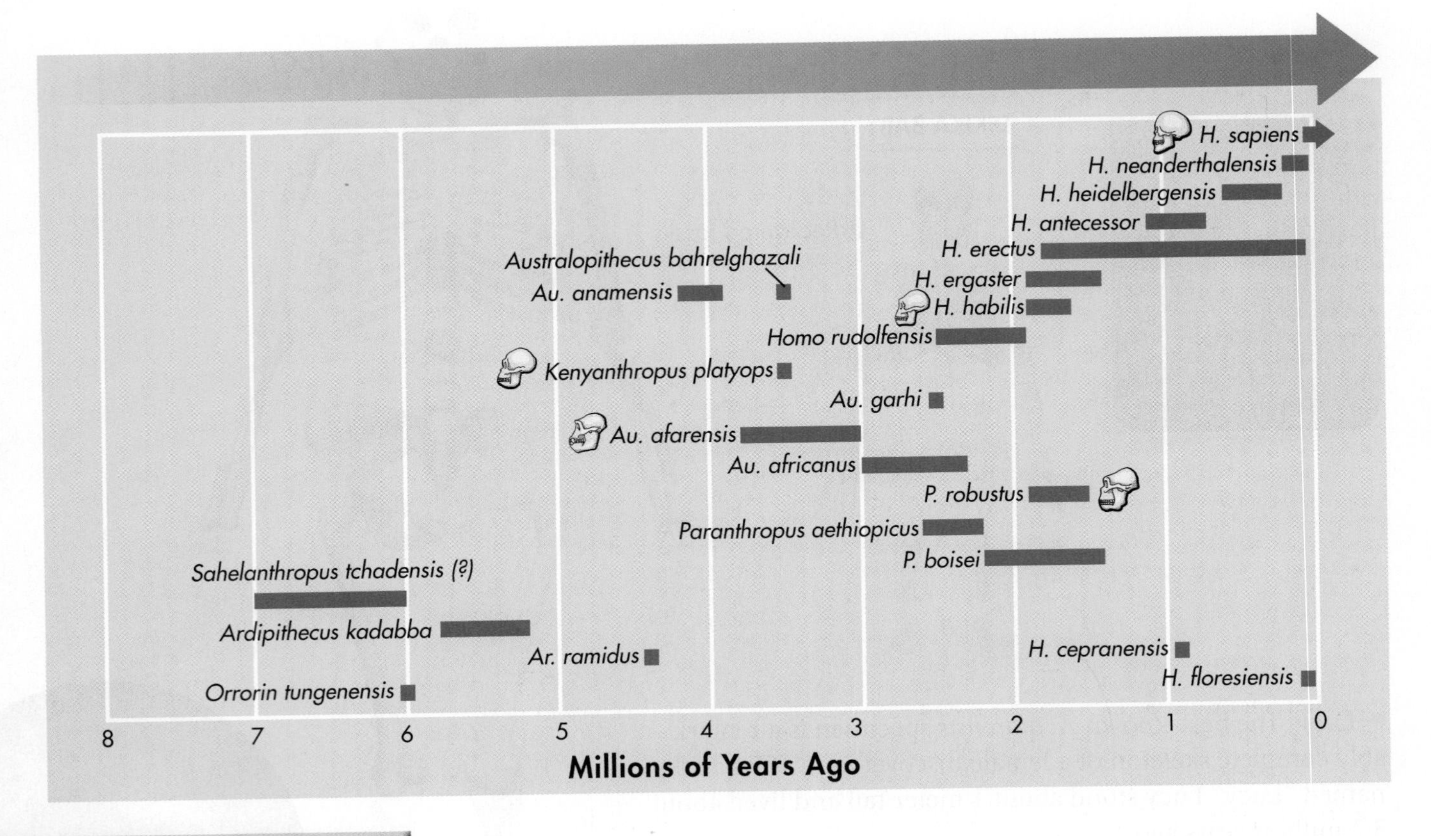

VISUAL SUMMARY

HOMININE TIME LINE

FIGURE 26–19 The diagram shows hominine species known from fossils and the time ranges during which each species probably existed. These time ranges may change as paleontologists gather new data. At this writing, several competing hypotheses present different ideas about how these species are related to one another and to *Homo sapiens*. So far, there is no single, universally accepted hypothesis, so we present these data as a time line, rather than as a cladogram. The fossil record shows that hominine evolution did not proceed along a simple, straight-line transformation of one species into another. Rather, a series of adaptive radiations produced a number of species, several of which display a confusing mix of primitive and modern traits. **Interpret Graphs** ***According to this time line, which species in the genus* Homo *lived at the same time?***

The Road to Modern Humans

***What is the current scientific thinking about the genus* Homo?**

The hominines discussed so far lived millions of years before modern humans. **Many species in our genus existed before our species, *Homo sapiens*, appeared. Furthermore, at least three other *Homo* species existed at the same time as early humans.** Paleontologists still do not completely understand the relationships among species in our own genus.

The Genus *Homo* About 2 million years ago, a new group of hominine species appeared. Several of these fossils resemble modern human bones enough that they have been classified in the genus *Homo*. One set of fossils from this time period was found with tools made of stone and bone, so it was named *Homo habilis* (HAB uh luhs), which means "handy man" in Latin. The earliest fossils that most researchers agree can be definitely assigned to the genus *Homo* have been called *Homo ergaster*. *H. ergaster* was larger than *H. habilis* and had a bigger brain and downward-facing nostrils that resemble those of modern humans. *Homo rudolfensis* appeared before *H. ergaster*, but some researchers choose to classify it in the genus *Australopithecus* instead of *Homo*.

Out of Africa—But When and Who? Researchers agree that our genus originated in Africa and migrated from there to populate the world. But many questions remain. When did hominines first leave Africa? Did more than one species make the trip? Which of those species were human ancestors and which were merely relatives? You can see some of the current hypotheses in **Figure 26–20**.

▶ ***The First to Leave*** Fossil and molecular evidence suggest that some hominines left Africa long before *Homo sapiens* evolved. It also appears that more than one *Homo* species made the trip in waves. Again, researchers differ as to the identity of various fossils, but agree that hominines began migrating out of Africa at least 1.8 million years ago. Hominine remains from that period were found in the Republic of Georgia, which is north of Turkey and far from Africa. Some researchers who have examined those remains argue that they might belong to a smaller-brained *Homo* species, *Homo habilis*.

▶ **Homo erectus *in Asia*** According to some researchers, groups of *Homo erectus* left Africa and traveled all the way across India and through China to Southeast Asia. In fact, some of the oldest known specimens of *H. erectus* were uncovered on the Indonesian island of Java. This suggests that these ancient wanderers spread very rapidly once they left Africa. These *H. erectus* populations continued to survive and evolve across Asia for as long as 1.5 million years.

▶ ***The First* Homo sapiens** Paleontologists have long debated where and when *Homo sapiens* arose. One hypothesis, called the multiregional model, suggests that, in several parts of the world, modern humans evolved independently from widely separated populations of *H. erectus*. Another hypothesis, the "out-of-Africa" model, proposes that modern humans evolved in Africa about 200,000 years ago, migrated out of Africa through the Middle East, and replaced the descendants of earlier hominine species.

Recently, molecular biologists analyzed mitochondrial DNA from living humans around the world to determine when they last shared a common ancestor. The estimated date for that African common ancestor is between 200,000 and 150,000 years ago. More recent DNA data suggest that a small subset of those African ancestors left northeastern Africa between 65,000 and 50,000 years ago to colonize the world. These data strongly support the out-of-Africa model.

BUILD Vocabulary

MULTIPLE MEANINGS The word *sapient* means "wise." It is also used as an adjective referring to *Homo sapiens*.

FIGURE 26–20 Out of Africa Data show that relatives and ancestors of modern humans left Africa in waves. But when—and how far did they travel? By comparing the mitochondrial DNA of living humans and by continuing to study the fossil record, scientists hope to improve our understanding of the complex history of *Homo sapiens*. (Note: Skulls on the map do not indicate that skulls were found at each location.)

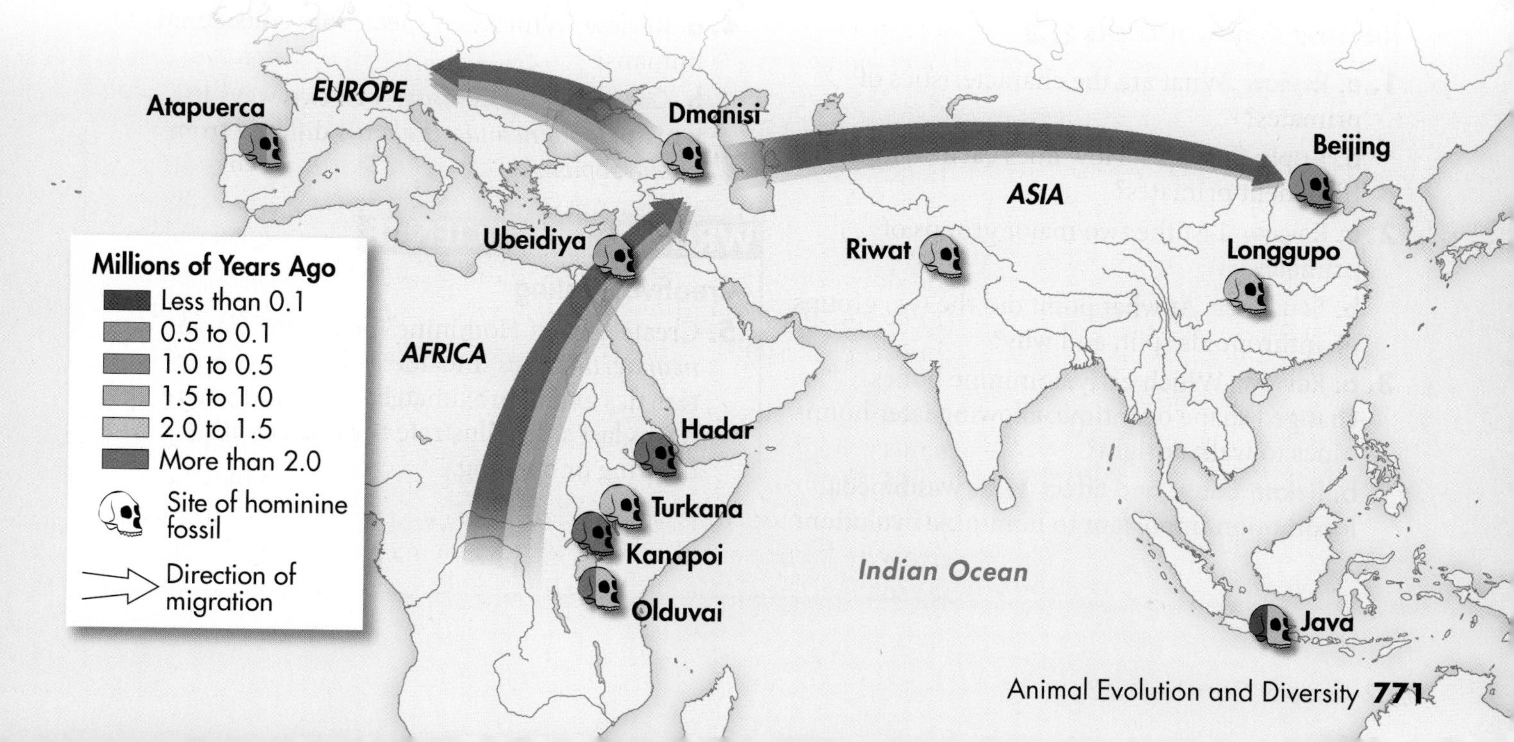

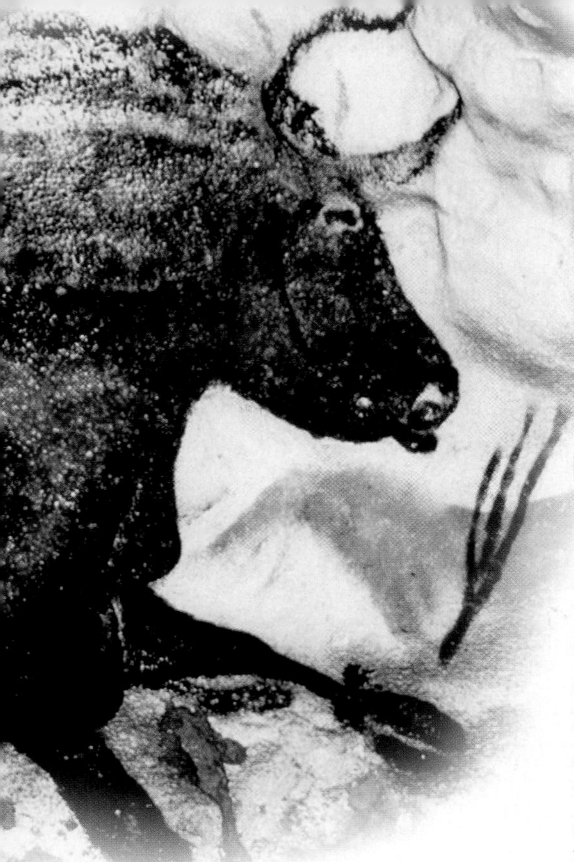

FIGURE 26–21 Cro-Magnon Art This ancient cave painting from France shows the remarkable artistic abilities of Cro-Magnons. **Infer** ***How might these painted images be related to the way in which these early humans lived?***

Modern Humans The story of modern humans over the past 200,000 years involves two main species in the genus *Homo*.

▶ **Homo neanderthalensis** Neanderthals flourished in Europe and western Asia beginning about 200,000 years ago. Evidence suggests that they made stone tools, lived in complex social groups, had controlled use of fire, and were excellent hunters. They buried their dead with simple rituals. Neanderthals survived in parts of Europe until about 28,000 to 24,000 years ago.

▶ ***Modern* Homo sapiens** Anatomically modern *Homo sapiens,* whose skeletons look like those of today's humans, arrived in the Middle East from Africa about 100,000 years ago. By about 50,000 years ago, *H. sapiens* populations were using new technology to make more sophisticated stone blades. They also began to make elaborately worked tools from bones and antlers. They produced spectacular cave paintings and buried their dead with elaborate rituals. In other words, these people, including the group known as Cro-Magnons, began to behave like modern humans.

When *H. sapiens* arrived in the Middle East, they found Neanderthals already living there. Neanderthals and *H. sapiens* lived side by side in the Middle East for about 50,000 years. Groups of modern humans moved into Europe between 40,000 and 32,000 years ago. There, too, *H. sapiens* coexisted alongside Neanderthals for several thousand years. For the last 24,000 years, however, our species has been Earth's only hominine. Why did Neanderthals disappear? Did they interbreed with *H. sapiens*? No one knows for sure. What we do know is that our species, *Homo sapiens,* is the only surviving member of the once large and diverse hominine clade.

26.3 Assessment

Review Key Concepts

1. a. **Review** What are the characteristics of primates?
   b. **Apply Concepts** How does each characteristic benefit primates?
2. a. **Review** List the two major groups of primates.
   b. **Sequence** At what point did the two groups of anthropoids split, and why?
3. a. **Review** Which early hominine bones changed shape over time, allowing later hominines to walk upright?
   b. **Relate Cause and Effect** How was bipedal locomotion important to hominine evolution?
4. a. **Review** Which two species are considered humans?
   b. **Compare and Contrast** List two ways in which *Homo neanderthalensis* differed from *Homo sapiens.*

WRITE ABOUT SCIENCE

Creative Writing

5. Create a "Lost Hominine" poster for *Homo neanderthalensis.* Include its known characteristics and approximately when and where it was last seen. Illustrate the poster with a drawing or clipping.

Biology & HISTORY

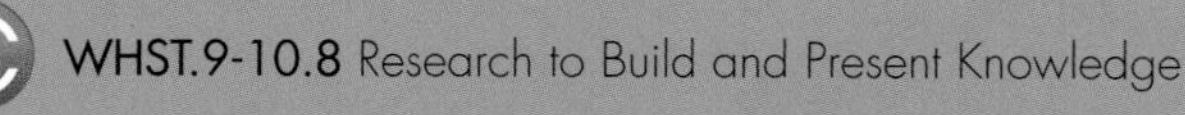

Human-Fossil Seekers **The study of human origins is an exciting search for our past. Piecing together this complicated story requires the skills of many scientists.**

1855 1885 1915 1945 1955 1975 2005 2035

1868
Edouard Lartet
Henry Christy
French geologist Lartet and English banker Christy unearth several ancient human skeletons in a rock shelter called Cro-Magnon in France. These hominine fossils are the first to be classified as *Homo sapiens.*

1886
Marcel de Puydt
Max Lohest
De Puydt and Lohest describe two Neanderthal skeletons found in a cave in Belgium. Their detailed description shows that Neanderthals were an extinct human form, not an abnormal form of modern human.

1924
Raymond Dart
Dart, an Australian anatomist, finds an early hominine fossil—a nearly complete skull of a child—in South Africa. This specimen was placed in a new genus called *Australopithecus.*

1974
Donald Johanson
An American paleontologist and his team find 40 percent of a skeleton of *Australopithecus*, which they call Lucy, in the Afar region of Ethiopia. The skeleton is about 3.2 million years old.

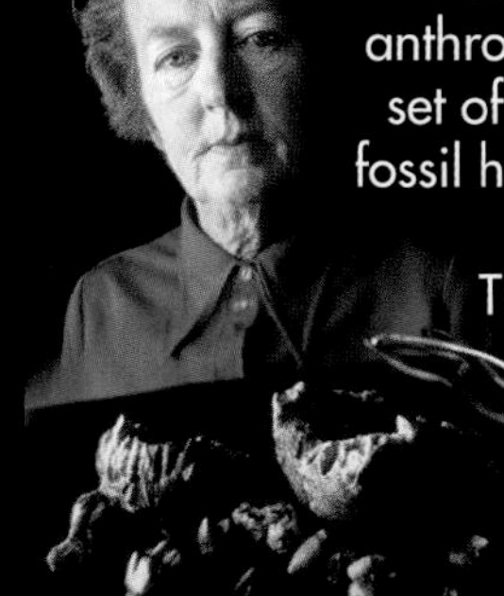

1978
Mary Leakey
Mary Leakey, a British anthropologist, discovers a set of 3.6 million-year-old fossil hominine footprints at Laetoli in Tanzania. The footprints provide evidence that early hominines walked erect on two legs.

2001
Meave Leakey
Nature publishes Meave Leakey's discovery of a 3.5–3.2 million year old skull that may be a human ancestor other than *Australopithecus.*

2002
Ahounta Djimdoumalbaye
Djimdoumalbaye, a college student in Chad, discovers the cranium of what may be the oldest known hominine, *Sahelanthropus tchadensis.*

2006
Zeresenay Alemseged
Zeresenay, an Ethiopian paleoanthropologist, announces his discovery of the fossilized skeleton of a young hominine in the Dikika region of Ethiopia. It is the most complete example of *A. afarensis* ever discovered and is about 3.3 million years old.

WRITING **Use the library or Internet sources suggested by your teacher to research one of these discoveries. Assess the usefulness of each source. Present your research in a poster with images and captions.**

NGSS Smart Guide Animal Evolution and Diversity

Disciplinary Core Idea LS4.A Evidence of Common Ancestry and Diversity: What evidence shows that different species are related? Invertebrates and chordates share common structures that emphasize that all animals descended, with modifications, from common ancestors.

26.1 Invertebrate Evolution and Diversity

- Fossil evidence indicates that the first animals began evolving long before the Cambrian Explosion.
- The cladogram of invertebrates presents current hypotheses about evolutionary relationships among major groups of modern invertebrates. It also indicates the sequence in which some important features evolved.

appendage 753 • larva 756 • trochophore 756

Biology.com

Untamed Science Video Join the Untamed Science crew as they talk with insect experts to better understand why there are more than a million insects.

InterActive Art Build a cladogram of invertebrates.

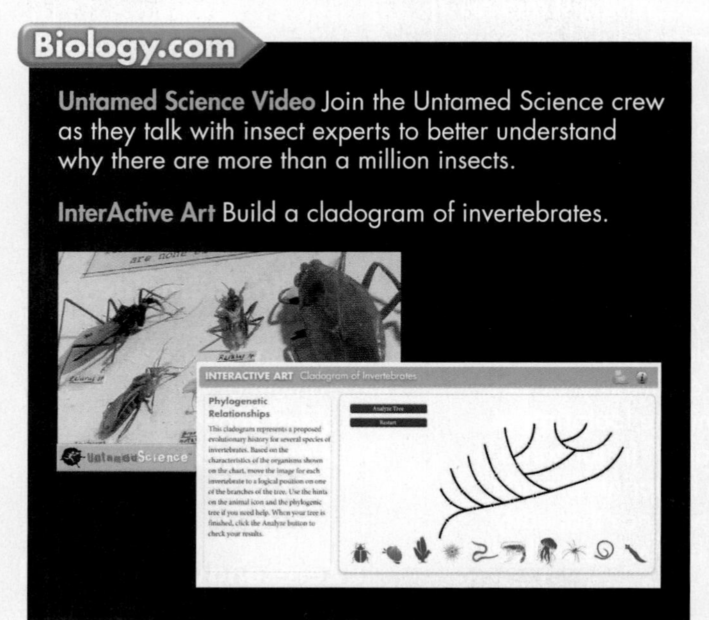

26.2 Chordate Evolution and Diversity

- Embryological studies suggest that the most ancient chordates were related to the ancestors of echinoderms.
- The cladogram of chordates presents current hypotheses about relationships among chordate groups. It also shows at which points important vertebrate features, such as jaws and limbs, evolved.

cartilage 757 • tetrapod 760

Biology.com

STEM Activity Watch a video and investigate some of the technologies scientists use to gather evidence for evolution.

Taking Notes Construct a Venn diagram to compare and contrast nonvertebrate chordates and vertebrates.

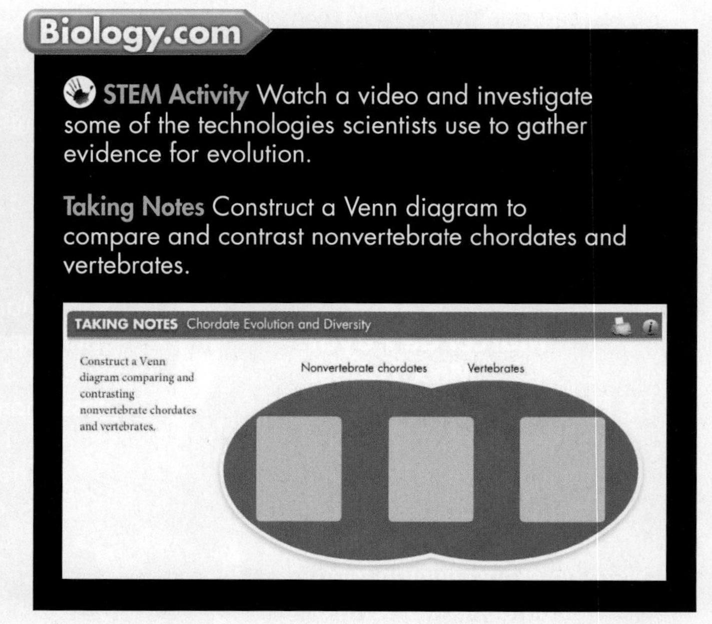

CHECKING YOUR *Scientific Literacy*

Refer to the lesson content and digital assets below as you prepare for your chapter assessment. Then, evaluate your understanding of animal evolution and diversity by answering these questions.

1. Crosscutting Concept **Cause and Effect** Describe an adaptation or multiple adaptations that evolved in chordates and the adaptive radiations that resulted from those adaptations.

2. **Scientific and Engineering Practice Constructing Explanations** Perform research to learn more about the out-of-Africa model of human evolution. Explain how DNA evidence supports this hypothesis on the evolution of *Homo sapiens*.

3. **Research to Build and Present Knowledge** Assess the usefulness of the sources you used in your research on the evolution of humans. Did you gather your information from reputable sources?

4. STEM Investigate and describe the technologies that are used to study animal evolution. These technologies can include methods to study DNA evidence and fossil dating methods, among other technologies.

26.3 Primate Evolution

- A primate is a mammal that has relatively long fingers and toes with nails instead of claws, arms that can rotate around shoulder joints, a strong clavicle, binocular vision, and a well-developed cerebrum.
- Primates in one group look very little like typical monkeys and include lemurs and lorises. The other group includes tarsiers and the anthropoids, the group that includes monkeys and humans.
- The skull, neck, spinal column, hip bones, and leg bones of early hominine species changed shape in ways that enabled later species to walk upright.
- Many species in our genus existed before our species, *Homo sapiens*, appeared. Furthermore, at least three other *Homo* species existed at the same time as early humans.

binocular vision 765 • anthropoid 766 • prehensile tail 767 • hominoid 767 • hominine 767 • bipedal 767 • opposable thumb 767

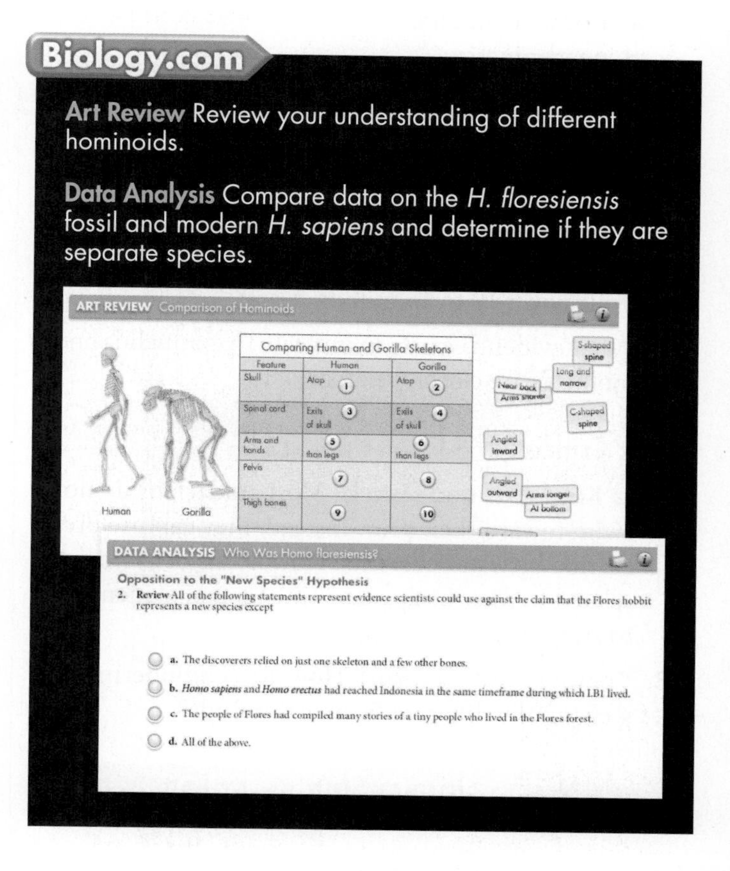

Forensics Lab

Investigating Hominoid Fossils This chapter lab is available in your lab manual and at Biology.com

26 Assessment

26.1 Invertebrate Evolution and Diversity

Understand Key Concepts

1. The ancestors of many modern animal phyla first appeared during the
a. Burgess Period. **c.** Precambrian Era.
b. Cambrian Period. **d.** Ediacaran Period.

2. Animals in the phylum Porifera include
a. chordates. **c.** sponges.
b. sea stars. **d.** sea anemones.

3. Most adult echinoderms show
a. bilateral symmetry.
b. top and bottom symmetry.
c. radial symmetry.
d. no symmetry.

4. Of the following groups, which has the largest number of species by far?
a. arthropods **c.** mollusks
b. annelids **d.** echinoderms

5. What body plan features did Cambrian animals evolve over 10 to 15 million years?

6. What evidence exists to indicate that annelids and mollusks are closely related?

Think Critically

7. Key Ideas and Details Most cnidarians do not swim toward their prey. Instead, they capture prey carried by water currents. How is this behavior related to their body plan? Cite textual evidence to support your answer.

8. Compare and Contrast How are echinoderms structurally different from arthropods?

26.2 Chordate Evolution and Diversity

Understand Key Concepts

9. The evolution of jaws and paired fins was an important development during the rise of
a. tunicates. **c.** fishes.
b. lancelets. **d.** amphibians.

10. Examine the diagrams below. Which of these is a jawed cartilaginous fish?

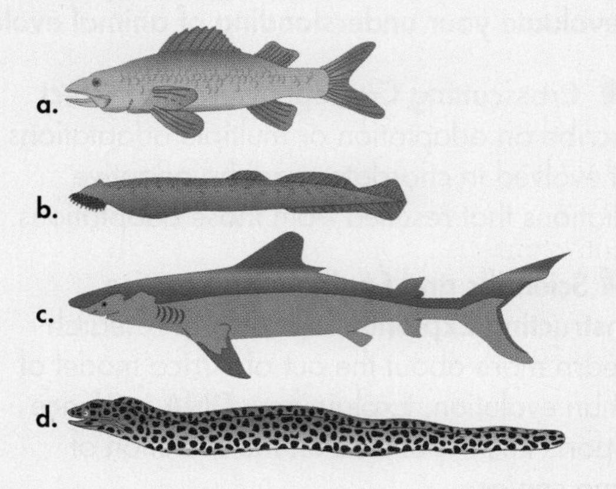

11. Which adaptation is NOT characteristic of reptiles?
a. scaly skin **c.** gills
b. shelled egg **d.** lungs

12. Dinosaurs became extinct at the end of the
a. Cretaceous Period. **c.** Carboniferous Period.
b. Triassic Period. **d.** Permian Period.

13. The single most important characteristic that separates birds from other living animals is the presence of
a. hollow bones. **c.** two legs.
b. feathers. **d.** wings.

14. Which of the following is a placental mammal?
a. duckbill platypus **c.** kangaroo
b. whale **d.** koala

15. Which two major groups of fishes evolved from the early jawed fishes and still survive today?

16. What adaptation enables birds to live in environments that are colder than those in which most reptiles live?

17. Describe how the young of monotremes, marsupials, and placental mammals obtain nourishment.

Think Critically

18. Research to Build and Present Knowledge Which anatomical characteristics of nonvertebrate chordates suggest that, in terms of evolutionary relationships, these animals are more closely related to vertebrates than to other groups of animals? Draw evidence from the text to support your answer.

26.3 Primate Evolution

Understand Key Concepts

19. Anthropoids include monkeys and
    a. lemurs.
    b. lorises.
    c. tarsiers.
    d. humans.

20. Which of the following is a characteristic specific to primates?
    a. body hair
    b. rotation at the shoulder joint
    c. notochord
    d. ability to control body temperature

21. How many hominine species exist today?
    a. one
    b. two
    c. nine
    d. twelve

22. The first hominines appear in the fossil record about
    a. 30,000 years ago.
    b. 100,000 years ago.
    c. 6 to 7 million years ago.
    d. 120 million years ago.

23. What anatomical characteristic allows for the binocular vision that occurs in primates?

24. **Craft and Structure** Describe the adaptations that make some primates successful tree dwellers.

25. List the unique characteristics of hominines. Give an example of a hominine.

Think Critically

26. **Interpret Photos** List three primate characteristics shown by the monkey in the photo.

solve the CHAPTER MYSTERY

FOSSIL QUEST

Josh was working against great odds. "His" fossils would be about 600 million years old. He learned that there are few places where rocks that old haven't been destroyed by geological activity. Most known sites are in China and Australia—none are in the United States. Pedro found better news. Reptiles related to bird ancestors lived around the same time as his favorite dinosaurs, during the Cretaceous Period. There are a number of places where fossils of that age have been found—including the Green River area in Utah. Because both boys like dinosaurs, they joined an Earthwatch teen expedition to the Green River to search for bird ancestors!

1. **Infer** Why is it so much harder to find fossils from the Proterozoic Eon, when the earliest known animals lived, than it is to find fossils from the Cretaceous Period, when the ancestors of birds lived? (*Hint:* See the graph below.)

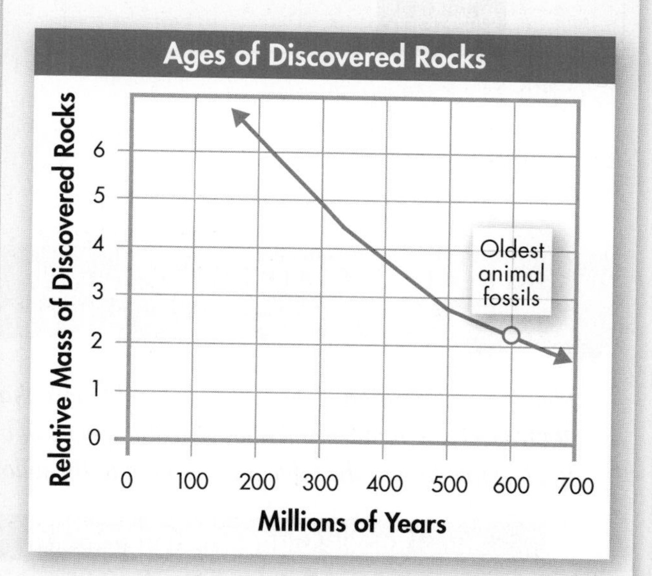

2. **Connect to the Big idea** Starting at the Earthwatch Web site, do an Internet search for fossil hunting expeditions you could join.

Connecting Concepts

Use Science Graphics

The chart below shows the relative numbers of species in four groups of vertebrates over time. The thickness of each band shows the relative number of species in that group. Use the chart to answer questions 27–30.

Era	Period	Number of Species
Cenozoic	Recent	
	Paleogene	
Mesozoic	Cretaceous	
	Jurassic	
	Triassic	
Paleozoic	Permian	
	Carbonif-erous	
	Devonian	
	Silurian	
	Ordovician	

Bony Fishes
Archosaurs
Birds
Mammals
Vertebrates

27. Compare How is this diagram similar to a traditional cladogram? What additional information can be learned from it?

28. Interpret Visuals Which of the groups shown has the greatest number of species today?

29. Infer Archosaurs are a group of reptiles that includes the dinosaurs, pterosaurs, modern crocodiles, and birds. Why do you think that birds are shown separately from the other archosaurs in the diagram?

30. Apply Concepts Describe the trend for each group shown from the beginning of the Mesozoic to today. Which groups were affected by the mass extinction at the end of the Mesozoic? Which groups have experienced adaptive radiations?

Write About Science

31. Key Ideas and Details Write a paragraph in which you summarize, in your own words, what the first animals were like.

32. Description In a paragraph, describe in your own words when the dinosaurs went extinct and what events contributed to the extinction.

33. Explanation Why is it important to estimate the age of a hominine fossil as well as to analyze its structural characteristics?

34. Assess the Big idea Life on Earth began in water. What were some of the major adaptations that animals evolved that allowed them to survive out of water?

Integration of Knowledge and Ideas

Living members of the reptile clade include more than 8000 species of reptiles and about 10,000 species of birds. Review the data in the table and respond to the following.

Reptile Clade Diversity

Group	Estimated Number of Species
Lizards and snakes	8400
Turtles and tortoises	310
Crocodilians	23
Tuataras	2
Birds	10,000

35. Graph Construct a circle graph that presents the data in the table.

36. Evaluate What was your biggest challenge in representing these data in a circle graph?

37. Analyze Data Consider other methods of graphing data. Which type of graph might represent these data in a more helpful way?

38. Graph Graph these data using a method you would consider more helpful to a reader or explain why none would be.

Standardized Test Prep

Multiple Choice

1. Which of the following is NOT a mollusk?
- **A** leech
- **B** squid
- **C** clam
- **D** snail

2. Which of the following invertebrates have segmented bodies?
- **A** flatworms
- **B** roundworms
- **C** cnidarians
- **D** annelids

3. All animals have some form of body symmetry EXCEPT
- **A** sponges.
- **B** jellyfishes.
- **C** worms.
- **D** arthropods.

4. Which of the following groups can be classified as nonvertebrate chordates?
- **A** sponges
- **B** tunicates
- **C** fishes
- **D** all of the above

5. Many scientists think that birds evolved from
- **A** mammal-like reptiles.
- **B** amphibians.
- **C** mammals.
- **D** dinosaurs.

6. Which of the following is NOT a characteristic of reptiles?
- **A** scaly skin
- **B** eggs with shells
- **C** lungs
- **D** mammary glands

7. Which of the following are hominoids?
- **A** all mammals
- **B** all primates
- **C** humans only
- **D** all great apes

8. When did the first true mammals appear?
- **A** Cretaceous Period
- **B** Triassic Period
- **C** Cenozoic Era
- **D** Carboniferous Period

Questions 9–11 Refer to the following cladogram.

Lamprey
Trout
Salamander
Lizard
Wallaby
Human
Placenta
Mammary Glands
Amniotic Egg
Four Limbs
Backbone
Notochord

9. Which characteristic is shared by humans, wallabies, and trout?
- **A** placenta
- **B** notochord
- **C** four limbs
- **D** mammary glands

10. Which animals have the closest evolutionary relationship, as shown by the cladogram?
- **A** humans and wallabies
- **B** humans and lizards
- **C** humans and lampreys
- **D** humans and trout

11. A valid conclusion from this cladogram is that
- **A** salamanders, trout, and lampreys all have a backbone.
- **B** four limbs appeared in vertebrate evolution before the notochord appeared.
- **C** humans and lampreys share a common ancestor.
- **D** mammary glands appeared in vertebrate evolution after the placenta appeared.

Open-Ended Response

12. What is the difference between a hominoid and a hominine?

If You Have Trouble With . . .

Question	1	2	3	4	5	6	7	8	9	10	11	12
See Lesson	26.1	26.1	26.1	26.2	26.2	26.2	26.3	26.2	26.2	26.2	26.2	26.3

27 Animal Systems I

Structure and Function

Q: How do the structures of animals allow them to obtain essential materials and eliminate wastes?

INSIDE:

- 27.1 Feeding and Digestion
- 27.2 Respiration
- 27.3 Circulation
- 27.4 Excretion

Red-billed oxpeckers are carnivores that have a mutualistic relationship with zebras. These birds eat ticks and insects that feed on the zebras, freeing them of these parasites.

BUILDING *Scientific Literacy*

Deepen your understanding of animal systems by engaging in key practices that allow you to make connections across concepts.

NGSS You will develop and use models to demonstrate how the structures of animals allow them to carry out essential life functions.

STEM Learn about the growing industry of bivalve aquaculture. Then, use a simulation to determine the best place to plant clams.

Common Core You will interpret data and write informative texts on animal systems.

CHAPTER MYSTERY

(NEAR) DEATH BY SALT WATER

It started as an adventure. Some college buddies tried their own version of a "survivor" experience. During summer vacation, they were dropped off on an uninhabited tropical island, with minimal supplies. They would be picked up in a few days.

The island was hot and dry, and they discovered that there was no fresh water. They knew that coconuts could provide fluids in the form of coconut "milk." But one group member hated coconuts. He figured he'd get his fluids by drinking salt water. At first, he was fine—although he was thirstier than his friends. Then, he became nauseated and weak. His condition worsened quickly. Soon he was seriously ill—with dizziness, headaches, and an inability to concentrate. His friends began to panic. What was happening? As you read the chapter, look for clues to help you explain the reason for the survivalist's illness. Then, solve the mystery.

Biology.com

Finding the solution to the (Near) Death by Salt Water mystery is only the beginning. Take a video field trip with the ecogeeks of Untamed Science to see where the mystery leads.

Go online to access additional resources including:
- eText • Flash Cards • Lesson Overviews • Chapter Mystery

27.1 Feeding and Digestion

Key Questions

🔑 *How do animals obtain food?*

🔑 *How does digestion occur in animals?*

🔑 *How are mouthparts adapted for different diets?*

Vocabulary

intracellular digestion
extracellular digestion
gastrovascular cavity
digestive tract
rumen

Taking Notes

Outline Before you read, use the headings in this lesson to outline the ways animals obtain and digest food. As you read, add details to your outline.

THINK ABOUT IT From tiny insects that dine on our blood, to bison that feed on prairie grasses, to giant blue whales that feed on plankton, all animals are heterotrophs that obtain nutrients and energy from food. In fact, adaptations for different styles of feeding are a large part of what makes animals so interesting.

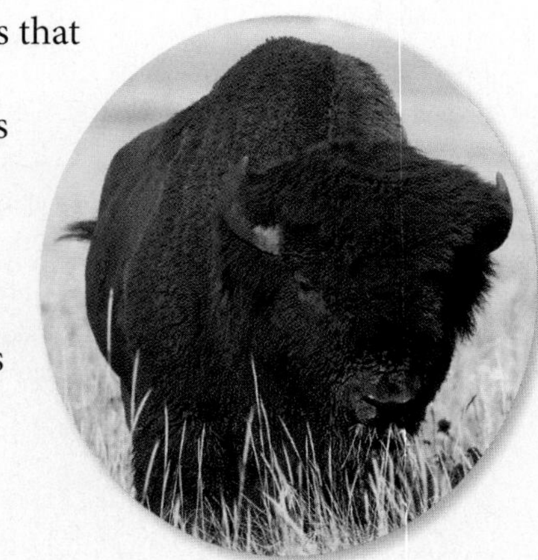

Obtaining Food

🔑 *How do animals obtain food?*

As the old saying goes, you are what you eat. For animals, we can rephrase that as "how you look and act depends on what and how you eat." The converse is also true: What and how you eat depends on how you look and act. To learn why that's true, we'll compare the various ways animals, such as those in **Figure 27–1,** obtain their food.

Filter Feeders Filter feeders strain their food from water. 🔑 **Most filter feeders catch algae and small animals by using modified gills or other structures as nets that filter food items out of water.** Many invertebrate filter feeders are small or colonial organisms, like worms and sponges, that spend their adult lives in a single spot. Many vertebrate filter feeders such as whale sharks and blue whales, on the other hand, are huge, and feed while swimming.

Detritivores Detritus is made up of decaying bits of plant and animal material. 🔑 **Detritivores feed on detritus, often obtaining extra nutrients from the bacteria, algae, and other microorganisms that grow on and around it.** From earthworms on land to a wide range of worms and crustaceans in aquatic habitats, detritivores are essential components of many ecosystems.

Carnivores 🔑 **Carnivores eat other animals.** Mammalian carnivores, such as wolves, use teeth, claws, and speed or stealthy hunting tactics to bring down prey. You probably don't often think about carnivorous invertebrates, but many would be terrifying if they were larger. Some cnidarians paralyze prey with poison-tipped darts, while some spiders immobilize their victims with venomous fangs.

Biology.com Lesson Notes • Visual Analogy • Self-Test • Lesson Assessment

Herbivores **Herbivores eat plants or parts of plants in terrestrial and aquatic habitats.** Some herbivores, such as locusts and cattle, eat leaves, which is not an easy way to make a living! Leaves don't have much nutritional content, are difficult to digest, and can contain poisons or hard particles that wear down teeth. Other herbivores, including birds and many mammals, specialize in eating seeds or fruits, which, in contrast to leaves, are often filled with energy-rich compounds.

BUILD Vocabulary

WORD ORIGINS The word part *-vore* comes from the Latin verb *vorare,* which means "to devour."

Nutritional Symbionts Recall that a symbiosis is the dependency of one species on another. Symbionts are the organisms involved in a symbiosis. **Many animals rely upon symbiosis for their nutritional needs.**

▶ ***Parasitic Symbionts*** Parasites live within or on a host organism, where they feed on tissues or on blood and other body fluids. Some parasites are just nuisances, but many cause serious diseases in humans, livestock, and crop plants. Parasitic flatworms and roundworms afflict millions of people, particularly in the tropics.

▶ ***Mutualistic Symbionts*** In mutualistic relationships, both participants benefit. Reef-building corals depend on symbiotic algae that live within their tissues for most of their energy. Those algae capture solar energy, recycle nutrients, and help corals lay down their calcium carbonate skeletons. The algae, in turn, gain nutrition from the corals' wastes and protection from algae eaters. Also, animals that eat wood or plant leaves rely on microbial symbionts in their guts to digest cellulose.

FIGURE 27–1 Obtaining Food The orca, sea slug, barnacles, and cleaner shrimp obtain their food in different ways.

Carnivore – Orca

Herbivore – Sea Slug

Filter Feeders – Barnacles

Detritivore – Cleaner Shrimp

Analyzing Data

Protein Digestion

A scientist performed an experiment to determine the amount of time needed for a certain carnivorous animal to digest animal protein. He placed pieces of hard-boiled egg white (an animal protein) in a test tube containing hydrochloric acid, water, and the enzyme pepsin, which digests protein. The graph shows the rate at which the egg white was "digested" over a 24-hour period.

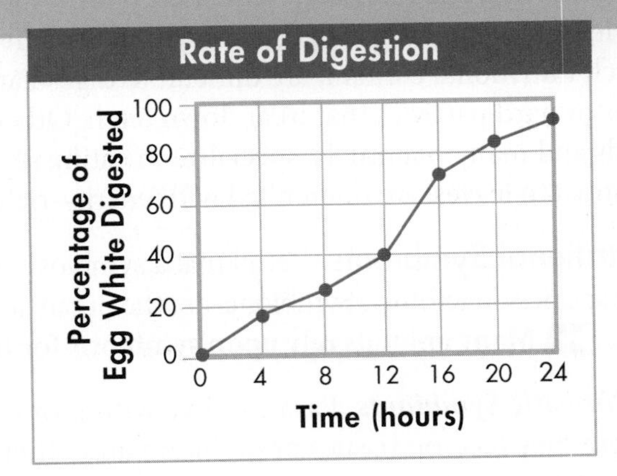

1. Interpret Graphs Describe the trend in the amount of protein digested over time.

2. Analyze Data About how many hours did it take for half of the protein to be digested?

3. Draw Conclusions How would you expect the rate of meat digestion to differ in an animal whose digestive tract had less of the enzyme pepsin?

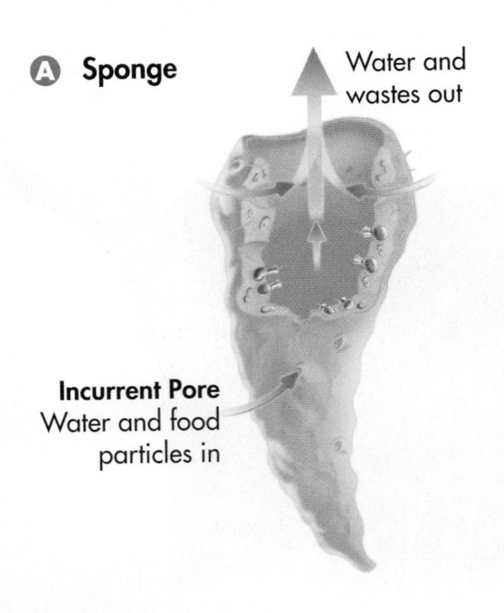

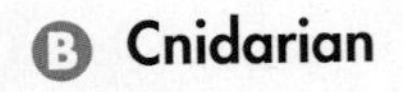

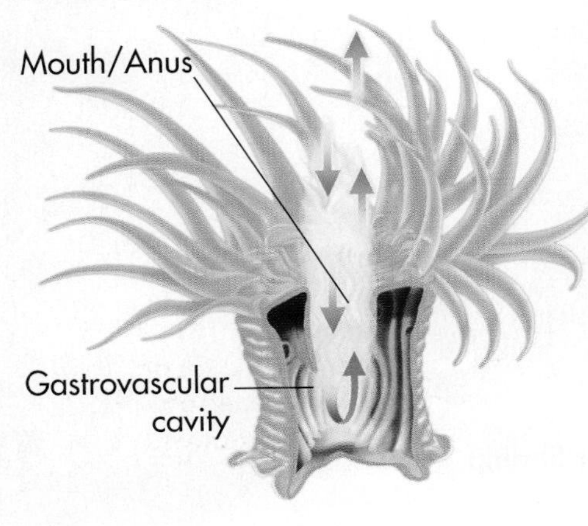

Processing Food

How does digestion occur in animals?

Obtaining food is just the first step. Food must then be broken down, or digested, and absorbed to make energy and nutrients available to body tissues. **Some invertebrates break down food primarily by intracellular digestion, but many animals use extracellular digestion to break down food.** A variety of digestive systems are shown in **Figure 27–2.**

Intracellular Digestion Animals have evolved many ways of digesting and absorbing food. The simplest animals, such as sponges, digest food inside specialized cells that pass nutrients to other cells by diffusion. This digestive process is known as **intracellular digestion.**

Extracellular Digestion Most more-complex animals rely on extracellular digestion. **Extracellular digestion** is the process in which food is broken down outside cells in a digestive system and then absorbed.

▶ ***Gastrovascular Cavities*** Some animals have an interior body space whose tissues carry out digestive and circulatory functions. Some invertebrates, such as cnidarians, have a **gastrovascular cavity** with a single opening through which they both ingest food and expel wastes. Some cells lining the cavity secrete enzymes and absorb digested food. Other cells surround food particles and digest them in vacuoles. Nutrients are then transported to cells throughout the body.

▶ ***Digestive Tracts*** Many invertebrates and all vertebrates, such as birds, digest food in a tube called a **digestive tract,** which has two openings. Food moves in one direction, entering the body through the mouth. Wastes leave through the anus.

One-way digestive tracts often have specialized structures, such as a stomach and intestines, that perform different tasks as food passes through them. You can think of a digestive tract as a kind of "disassembly line" that breaks down food one step at a time. In some animals, the mouth secretes digestive enzymes that start the chemical digestion of food. Then, mechanical digestion may occur as specialized mouthparts or a muscular organ called a gizzard breaks food into small pieces. Then, chemical digestion begins or continues in a stomach that secretes digestive enzymes. Chemical breakdown continues in the intestines, sometimes aided by secretions from other organs such as a liver or pancreas. Intestines also absorb the nutrients released by digestion.

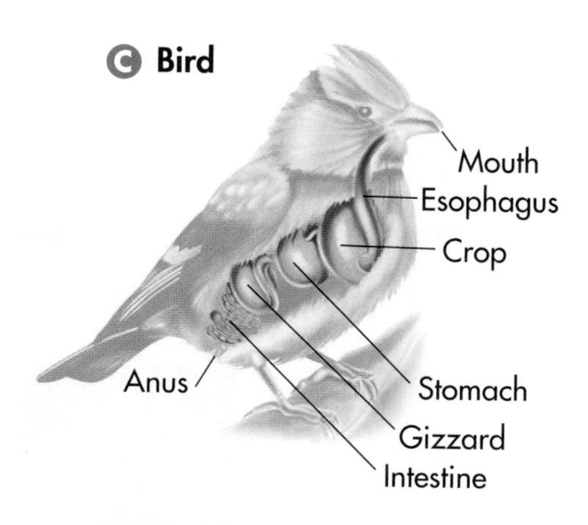

FIGURE 27–2 Digesting Food Animals have different digestive structures with different functions. A The sponge (previous page) has one digestive opening and uses intracellular digestion to process its food. B The cnidarian (previous page) processes its food by extracellular digestion in a gastrovascular cavity. C The bird has a one-way digestive tract with two openings.

▶ ***Solid Waste Disposal*** No matter how efficiently an animal breaks down food and extracts nutrients, some indigestible material will always be left. These solid wastes, or feces, are expelled either through the single digestive opening or through the anus.

Specializations for Different Diets

How are mouthparts adapted for different diets?

The mouthparts and digestive systems of animals have evolved many adaptations to the physical and chemical characteristics of different foods, as shown in **Figure 27–3.** As a window into these specializations, we'll examine adaptations to two food types that are very different physically and chemically: meat and plant leaves.

Specialized Mouthparts Carnivores and leaf-eating herbivores usually have very different mouthparts. These differences are typically related to the different physical characteristics of meat and plant leaves.

VISUAL ANALOGY

SPECIALIZED TEETH

FIGURE 27–3 Mouthparts The specialized jaws and teeth of animals are well adapted to their diets.

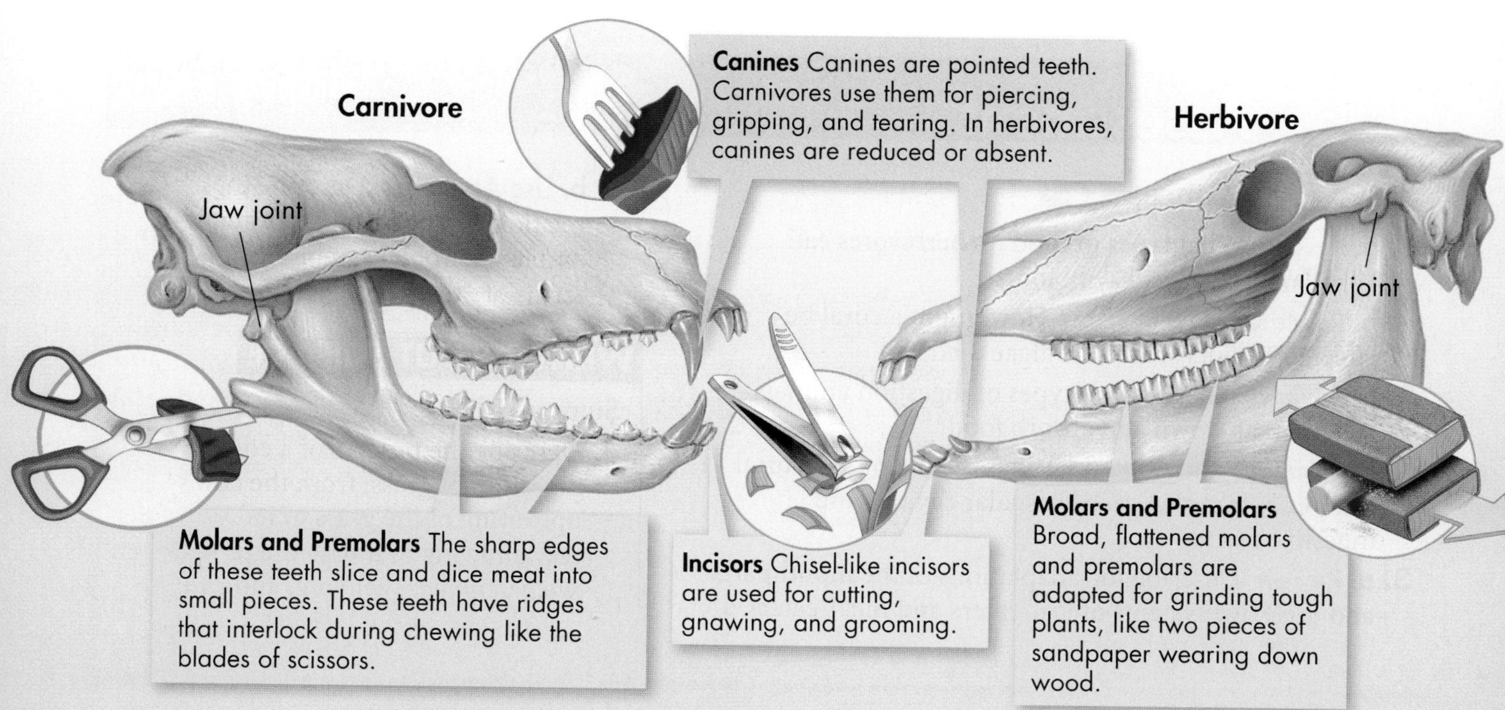

▶ ***Eating Meat*** **Carnivores typically have sharp mouthparts or other structures that can capture food, hold it, and "slice and dice" it into small pieces.** Carnivorous mammals, such as wolves, have sharp teeth that grab, tear, and slice food like knives and scissors would. The jaw bones and muscles of carnivores are adapted for up and down movements that chop meat into small pieces.

▶ ***Eating Plant Leaves*** **Herbivores typically have mouthparts adapted to rasping or grinding.** To digest leaf tissues, herbivores usually need to tear plant cell walls and expose their contents. To do this, many herbivorous invertebrates, from mollusks to insects, have mouthparts that grind and pulverize leaf tissues. Herbivorous mammals, such as the horse in **Figure 27–4,** have front teeth and muscular lips adapted to grabbing and pulling leaves, and flattened molars that grind leaves to a pulp. The jaw bones and muscles of mammalian herbivores are also adapted for side-to-side "grinding" movements.

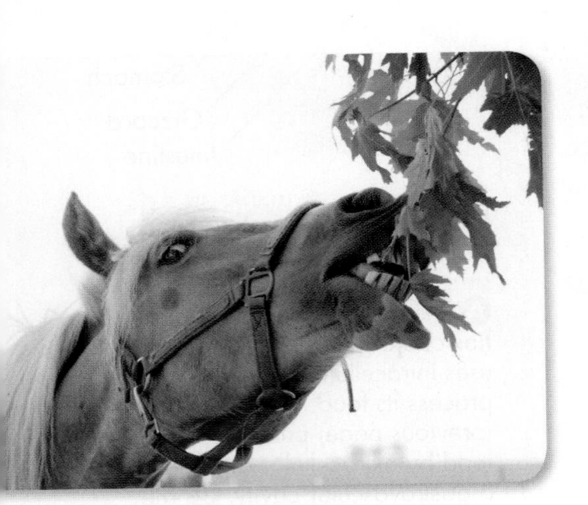

FIGURE 27–4 Eating Plant Leaves The teeth and jaws of herbivores, such as horses, are adapted for pulling, rasping, and grinding plant leaves.

Specialized Digestive Tracts Carnivorous invertebrates and vertebrates typically have short digestive tracts that produce fast-acting, meat-digesting enzymes. These enzymes can digest most cell types found in animal tissues.

No animal produces digestive enzymes that can break down the cellulose in plant tissue, however. Some herbivores have very long intestines or specialized pouches in their digestive tracts that harbor microbial symbionts that digest cellulose. Cattle, for example, have a pouchlike extension of their esophagus called a **rumen** (plural: rumina), in which symbiotic bacteria digest cellulose. Animals with rumina, or ruminants, regurgitate food that has been partially digested in the rumen, chew it again, and reswallow it. This process is called "chewing the cud."

27.1 Assessment

Review Key Concepts

1. a. Review What types of food do herbivores eat? What are nutritional symbionts?

b. Relate Cause and Effect How might a coral be affected if all its symbiotic algae died?

2. a. Review What are two types of digestion animals use to break down and absorb food?

b. Compare and Contrast What is a major structural difference between gastrovascular cavities and digestive tracts?

3. a. Review Describe the adaptations of the mouthparts and digestive systems of leaf-eaters and meat-eaters.

b. Use Analogies Describe the relationship between a ruminant and its microbial symbionts in terms of "teamwork."

WRITE ABOUT SCIENCE

Summary

4. Describe the process of a cow's digestion of grass, from the cow's uprooting of the grass to its reswallowing of it. Use the terms *molar, rumen, symbiont,* and *cud.*

27.2 Respiration

THINK ABOUT IT All animal tissues require oxygen for respiration and produce carbon dioxide as a waste product. For that reason, all animals must obtain oxygen from their environment and release carbon dioxide. In other words, all animals need to "breathe." Humans can drown because our lungs can't extract the oxygen we need from water. Most fishes have the opposite problem; out of water, their gills don't work. How are these different respiratory systems adapted to their different functions?

Key Questions

What characteristics do the respiratory structures of all animals share?

How do aquatic animals breathe?

What respiratory structures enable land animals to breathe?

Vocabulary

gill • lung • alveolus

Taking Notes

Concept Map Draw a concept map showing the characteristics of the lung structures of vertebrates.

Gas Exchange

What characteristics do the respiratory structures of all animals share?

Despite all the amazing things living cells can do, none can actively pump oxygen or carbon dioxide across membranes. Yet, in order to breathe, all animals must exchange oxygen and carbon dioxide with their surroundings. How do they do it? Animals have evolved respiratory structures that promote the movement of these gases in the required directions by passive diffusion.

Gas Diffusion and Membranes As you may recall, substances diffuse from an area of higher concentration to an area of lower concentration. Gases diffuse most efficiently across a thin, moist membrane that is permeable to those gases. The larger the surface area of that membrane, the more diffusion can take place, just as a bumpy paper towel absorbs more liquid than a smooth one does. These physical principles create a set of requirements that respiratory systems must meet, one way or another.

Requirements for Respiration Because of the behavior of gases, all respiratory systems share certain basic characteristics. **Respiratory structures provide a large surface area of moist, selectively permeable membrane. Respiratory structures maintain a difference in the relative concentrations of oxygen and carbon dioxide on either side of the respiratory membrane, promoting diffusion.**

FIGURE 27–5 Requirements for Respiration Respiratory surfaces are moist, so exhaled air contains a lot of moisture. That exhaled moisture condenses into visible "fog" if outside air is cold.

FIGURE 27–6 Respiration With Gills Many aquatic animals, such as fishes, respire with gills, which are thin, selectively permeable membranes. As water passes over the gills, gas exchange is completed within the gill capillaries.

Respiratory Surfaces of Aquatic Animals

How do aquatic animals breathe?

Some aquatic invertebrates, such as cnidarians and some flatworms, are relatively small and have thin-walled bodies whose outer surfaces are always wet. These animals rely on diffusion of oxygen and carbon dioxide through their outer body covering. A few aquatic chordates, including lancelets, some amphibians, and even some sea snakes, rely to varying extents on gas exchange by diffusion across body surfaces.

For large, active animals, however, skin respiration alone is insufficient. **Many aquatic invertebrates and most aquatic chordates other than reptiles and mammals exchange gases through gills.** As shown in **Figure 27–6, gills** are feathery structures that expose a large surface area of thin, selectively permeable membrane to water. Inside gill membranes is a network of tiny, thin-walled blood vessels called capillaries. Many animals, including aquatic mollusks and fishes, actively pump water over their gills as blood flows through inside. This helps maintain differences in oxygen and carbon dioxide concentration that promote diffusion. **Aquatic reptiles and aquatic mammals, such as whales, breathe with lungs and must hold their breath underwater. Lungs** are organs that exchange oxygen and carbon dioxide between blood and air. You will learn more about lungs shortly.

Quick Lab
GUIDED INQUIRY

Breathing in Clams and Crayfishes  

1. *Do not touch the clam or crayfish.* Put a drop of food coloring in the water near a clam's siphons. Observe what happens to the coloring.

2. Put a drop of food coloring in the water near the middle of a crayfish. **CAUTION:** *Keep your fingers away from the crayfish's pincers.* Observe what happens to the coloring.

Analyze and Conclude

1. **Observe** Describe what happened to the coloring in step 1. How does water move through a clam's gills?

2. **Infer** What is the clam's main defense? How is the location of the clam's siphons related to this defense?

3. **Compare and Contrast** What happened in step 2? Compare the flow of water through the gills of clams and crayfishes.

4. **Infer** Unlike many other arthropods, crayfishes have gills. Why do crayfishes need gills?

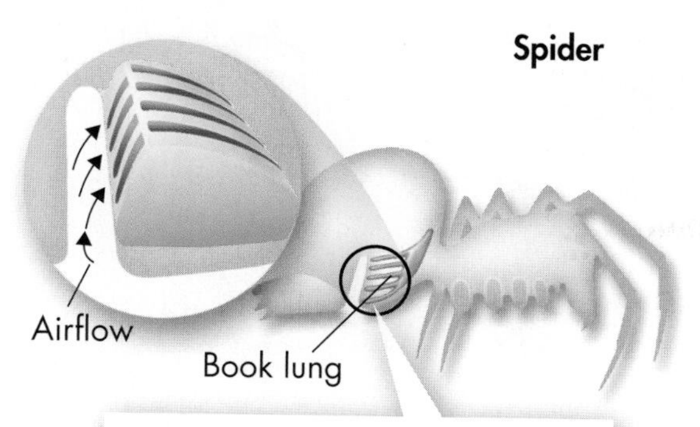

Spiders respire using organs called book lungs, which are made of parallel, sheetlike layers of thin tissues that contain blood vessels.

Insect

Tracheal tubes

Spiracles

In most insects, a system of tracheal tubes extends throughout the body. Air enters and leaves the system through openings in the body surface called spiracles. In some insects, oxygen and carbon dioxide diffuse through the tracheal system, and in and out of body fluids. In other insects, body movements help pump air in and out of the tracheal system.

FIGURE 27–7 Respiratory Structures of Terrestrial Invertebrates Terrestrial invertebrates have a wide variety of respiratory structures, including skin, mantle cavities, book lungs, and tracheal tubes. These structures must stay moist even in the driest of conditions in order to function properly.

Respiratory Surfaces of Terrestrial Animals

What respiratory structures enable land animals to breathe?

Terrestrial animals, as you might have guessed, face a challenge that aquatic animals don't. Terrestrial animals must keep their respiratory membranes moist in dry environments.

Respiratory Surfaces in Land Invertebrates The wide range of body plans among terrestrial invertebrates reveals very different strategies for respiration. **Respiratory structures in terrestrial invertebrates include skin, mantle cavities, book lungs, and tracheal tubes.** Some land invertebrates, such as earthworms, that live in moist environments can respire across their skin, as long as it stays moist. In other invertebrates, such as land snails, respiration is accomplished by the mantle cavity, which is lined with moist tissue and blood vessels. Insects and spiders have more complex respiratory systems, as you can see in **Figure 27–7.**

Lung Structure in Vertebrates Terrestrial vertebrates display a wide range of breathing adaptations. **But all terrestrial vertebrates—reptiles, birds, mammals, and the land stages of most amphibians—breathe with lungs.** Although lung structure in these animals varies, the processes of inhaling and exhaling are similar. Inhaling brings oxygen-rich air through the trachea (TRAY kee uh), or airway, into the lungs. Inside the lungs, oxygen diffuses into the blood through lung capillaries. At the same time, carbon dioxide diffuses out of capillaries into the lungs. Oxygen-poor air is then exhaled.

In Your Notebook *Would you expect dolphins to breathe with gills or lungs? Explain your answer.*

BUILD Vocabulary

MULTIPLE MEANINGS The biological term *respiration* has different, though related, meanings. In animals, it can refer to gas exchange, the intake of oxygen and release of waste gases, or to *cellular respiration*, the cell process that releases energy by breaking down food molecules in the presence of oxygen. Because cellular respiration requires oxygen, the two processes are related.

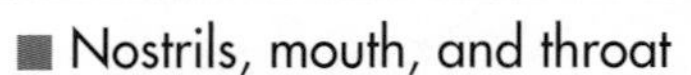

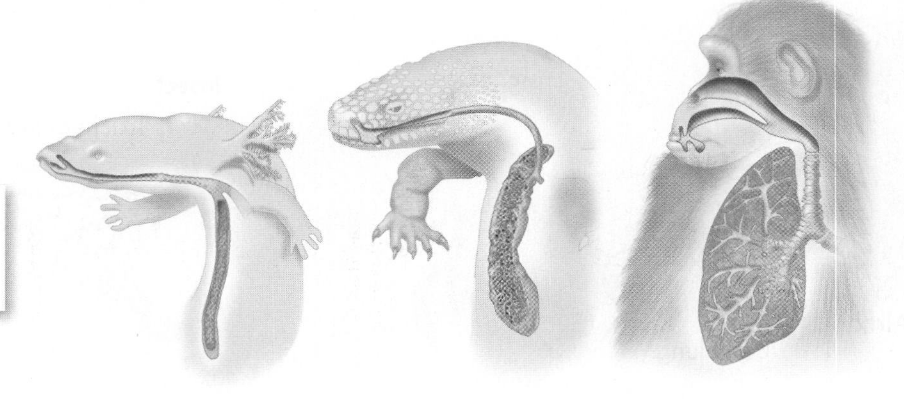

FIGURE 27–8 Lungs Terrestrial vertebrates breathe with lungs. Lungs with a larger surface area can take in more oxygen and release more carbon dioxide. Mammals have the greatest lung surface area among animals. Infer ***Why do mammals require a large surface area with which to process oxygen?***

▶ ***Amphibian, Reptilian, and Mammalian Lungs*** The internal surface area of lungs increases from amphibians to reptiles to mammals, as shown in **Figure 27–8.** A typical amphibian lung is little more than a sac with ridges. Reptilian lungs are often divided into chambers that increase the surface area for gas exchange. Mammalian lungs branch extensively, and their entire volume is filled with bubblelike structures called **alveoli** (al VEE uh ly; singular: alveolus). Alveoli provide an enormous surface area for gas exchange. The structure of mammalian lungs enables mammals to take in the large amounts of oxygen required by their high metabolic rates. However, in the lungs of mammals and most other vertebrates, air moves in and out through the same tracheal passageway. For this reason, some stale, oxygen-poor air is trapped in the lungs. In humans, this stale air is typically equivalent to about one third of the air inhaled in a normal breath.

▶ ***Bird Lungs*** In birds, the lungs are structured so that air flows mostly in only one direction. No stale air gets trapped in the system. A unique system of tubes and air sacs in birds' respiratory systems enables this one-way airflow. Thus, gas exchange surfaces are continuously in contact with fresh air. This highly efficient gas exchange helps birds obtain the oxygen they need to power their flight muscles at high altitudes for long periods of time.

27.2 Assessment

Review Key Concepts

1. a. Review In what ways are the respiratory structures of all animals similar?
   b. Apply Concepts Explain why it is important that respiratory surfaces are moist and selectively permeable.
2. a. Review Which groups of aquatic animals breathe with gills? With lungs?
   b. Relate Cause and Effect Why do some animals actively pump water over their gills?
3. a. Review How do terrestrial invertebrates and terrestrial vertebrates breathe?
   b. Interpret Visuals Contrast the structures of amphibian, reptilian, and mammalian lungs, as shown in **Figure 27–8.**

WRITE ABOUT SCIENCE

Description

4. Describe the events that occur when a mammal respires, including the path of air through its lungs.

27.3 Circulation

THINK ABOUT IT Your mouth takes food into your body, and your digestive tract breaks it down. But how do the energy and nutrients get to your body cells? How does oxygen from your lungs get to your brain and the rest of your body? How do carbon dioxide and wastes generated within your body get eliminated? While some aquatic animals with bodies only a few cells thick rely solely on diffusion to transport materials, most animals rely on a circulatory system.

Key Questions

How do open and closed circulatory systems compare?

How do the patterns of circulation in vertebrates compare?

Vocabulary

heart
open circulatory system
closed circulatory system
atrium
ventricle

Taking Notes

Cycle Diagram As you read, draw a cycle diagram showing a five-step sequence in which blood pumps through a closed, two-loop circulatory system.

Open and Closed Circulatory Systems

How do open and closed circulatory systems compare?

Many animals move blood through their bodies using one or more hearts. A **heart** is a hollow, muscular organ that pumps blood around the body. A heart can be part of either an open or a closed circulatory system.

Open Circulatory Systems Arthropods and most mollusks have **open circulatory systems,** such as the one in **Figure 27–9.** **In an open circulatory system, blood is only partially contained within a system of blood vessels as it travels through the body.** One or more hearts or heartlike organs pump blood through vessels that empty into a system of sinuses, or spongy cavities. There, blood comes into direct contact with body tissues. Blood then collects in another set of sinuses and eventually makes its way back to the heart.

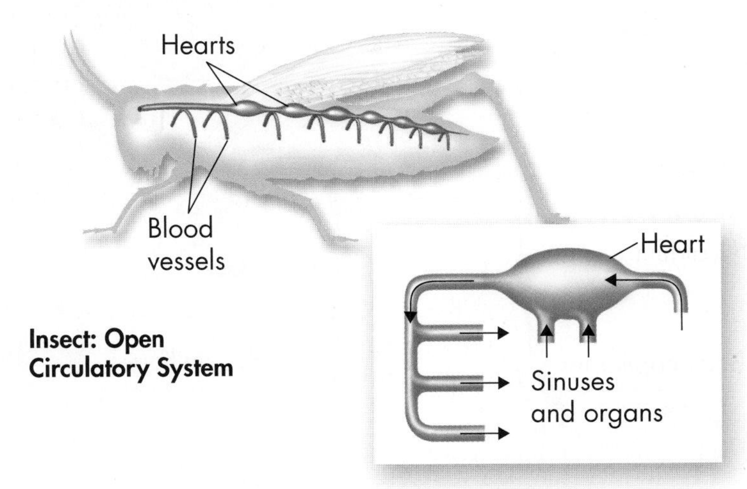

FIGURE 27–9 Open Circulatory System In an open circulatory system, blood is not entirely contained within blood vessels. Grasshoppers, for example, have open circulatory systems in which blood leaves vessels and moves through sinuses before returning to a heart.

FIGURE 27–10 Closed Circulatory System Annelids, such as earthworms, and many more-complex animals have closed circulatory systems. Blood stays within the vessels of a closed circulatory system.

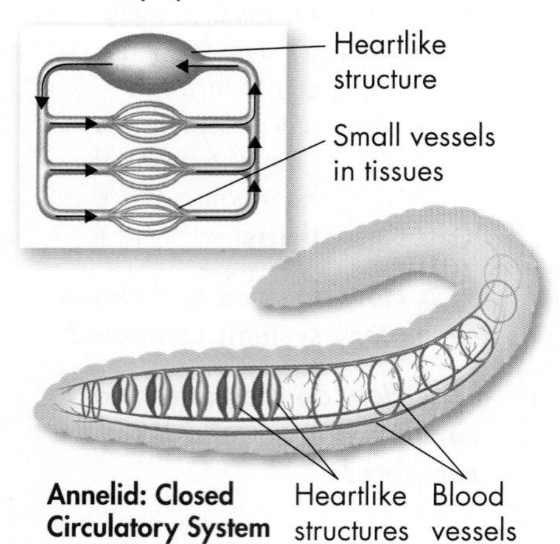

Closed Circulatory Systems Many larger, more active invertebrates, including annelids and some mollusks, and all vertebrates have **closed circulatory systems,** such as the one shown in **Figure 27–10.** **In a closed circulatory system, blood circulates entirely within blood vessels that extend throughout the body.** A heart or heartlike organ forces blood through these vessels. Nutrients and oxygen reach body tissues by diffusing across thin walls of capillaries, the smallest blood vessels. Blood that is completely contained within blood vessels can be pumped under higher pressure, and thus can be circulated more efficiently, than can blood in an open system.

Single- and Double-Loop Circulation

How do the patterns of circulation in vertebrates compare?

As chordates evolved, they developed more-complex organ systems and more-efficient channels for internal transport. You can see two main types of circulatory systems of vertebrates in **Figure 27–11.**

Single-Loop Circulation **Most vertebrates with gills have a single-loop circulatory system with a single pump that forces blood around the body in one direction.** In fishes, for example, the heart consists of two chambers: an atrium and a ventricle. The **atrium** (plural: atria) receives blood from the body. The **ventricle** then pumps blood out of the heart and to the gills. Oxygen-rich blood then travels from the gills to the rest of the body and returns, oxygen-poor, to the atrium.

BUILD Vocabulary

MULTIPLE MEANINGS The word **atrium** has different but parallel meanings in everyday usage and in biology. In everyday usage, it means a large entrance hall. In biology, it means a heart chamber through which blood from the body enters the heart.

Double-Loop Circulation As terrestrial vertebrates evolved into larger and more active forms, their capillary networks became larger. Using a single pump to force blood through the entire system would have been increasingly difficult. This issue was avoided as the lineage of vertebrates that led to reptiles, birds, and mammals evolved. **Most vertebrates that use lungs for respiration have a double-loop, two-pump circulatory system.**

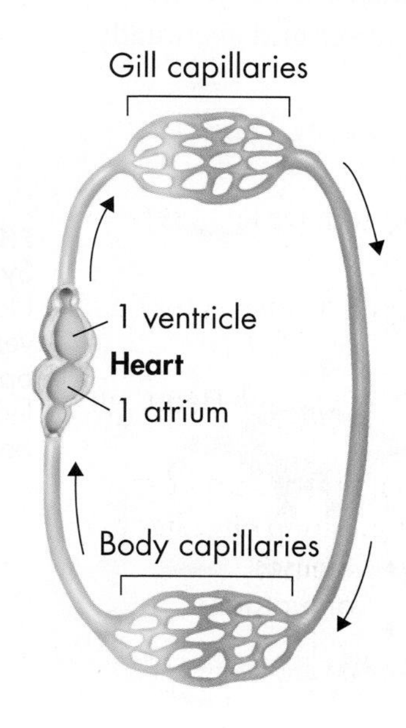

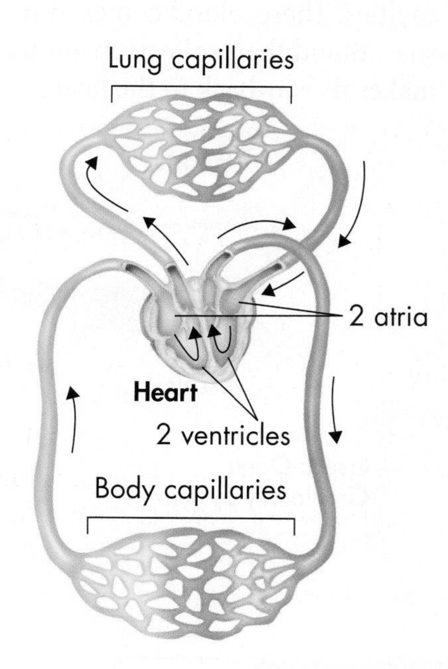

FIGURE 27–11 Single- and Double-Loop Circulation Most vertebrates that use gills for respiration have a single-loop circulatory system that forces blood around the body in one direction (left). Vertebrates that use lungs have a double-loop system (right). (Note that in diagrams of animals' circulatory systems, blood vessels carrying oxygen-rich blood are red, while blood vessels carrying oxygen-poor blood are blue.)

The first loop, powered by one side of the heart, forces oxygen-poor blood from the heart to the lungs. After the blood picks up oxygen (and drops off carbon dioxide) in the lungs, it returns to the heart. Then, the other side of the heart pumps this oxygen-rich blood through the second circulatory loop to the rest of the body. Oxygen-poor blood from the body returns to the heart, and the cycle begins again.

MYSTERY CLUE

Human blood is only about a third as salty as seawater. It needs to circulate through very small capillaries. What might happen if the water content of a person's blood were to drop too low?

Mammalian Heart-Chamber Evolution Four-chambered hearts like those in modern mammals are actually two separate pumps working next to one another. But where did the second pump come from? During chordate evolution, partitions evolved that divided the original two chambers into four. Those partitions transformed one pump into two parallel pumps. The partitions also separated oxygen-rich blood from oxygen-poor blood. We can get an idea of how the partitions evolved by looking at other modern vertebrates.

Amphibian hearts usually have three chambers: two atria and one ventricle. The left atrium receives oxygen-rich blood from the lungs. The right atrium receives oxygen-poor blood from the body. Both atria empty into the ventricle. Some mixing of oxygen-rich and oxygen-poor blood in the ventricle occurs. However, the internal structure of the ventricle directs blood flow so that most oxygen-poor blood goes to the lungs, and most oxygen-rich blood goes to the rest of the body.

Reptilian hearts typically have three chambers. However, most reptiles have a partial partition in their ventricle. Because of this partition, there is even less mixing of oxygen-rich and oxygen-poor blood than there is in amphibian hearts.

FIGURE 27–12 Reptilian Heart Under the armor-like hide of this crocodile lies a heart with two atria and one ventricle.

27.3 Assessment

Review Key Concepts

1. a. Review Describe an open circulatory system. Describe a closed circulatory system.

b. Explain Which groups of animals tend to have each type of circulatory system?

c. Relate Cause and Effect How does having a closed circulatory system benefit a large, active animal?

2. a. Review What are two different patterns of circulation found in vertebrates?

b. Compare and Contrast What is the major structural difference between vertebrates that have single-loop circulatory systems and those that have double-loop systems?

Apply the Big idea

Structure and Function

3. Do you think large, active vertebrates would have been likely to succeed if closed circulatory systems had not evolved? Explain your reasoning.

27.4 Excretion

Key Questions

How do animals manage toxic nitrogenous waste?

How do aquatic animals eliminate wastes?

How do land animals remove wastes while conserving water?

Vocabulary

excretion • kidney • nephridium • Malpighian tubule

Taking Notes

Preview Visuals Note three questions you have about **Figure 27–15**. As you read, try to answer your questions.

THINK ABOUT IT If you think about the first three lessons in this chapter, you'll realize that they are missing something. We've discussed how respiratory systems obtain oxygen and get rid of carbon dioxide. We've also discussed how animals obtain and digest food and get rid of indigestible material. But cellular respiration generates other kinds of wastes that are released into body fluids and that must be eliminated from the body. What are these wastes and how do animals get rid of them?

The Ammonia Problem

How do animals manage toxic nitrogenous waste?

The breakdown of proteins by cells releases a nitrogen-containing, or nitrogenous, waste: ammonia. This creates a problem, because ammonia is poisonous! Even moderate concentrations of ammonia can kill most cells. Animal systems address this difficulty in one of two ways. **Animals either eliminate ammonia from the body quickly or convert it into other nitrogenous compounds that are less toxic.** The elimination of metabolic wastes, such as ammonia, is called **excretion.** Some small animals that live in wet environments rid their bodies of ammonia by allowing it to diffuse out of their body fluids across their skin. Most larger animals, and even some smaller ones that live in dry environments, have excretory systems that process ammonia and eliminate it from the body.

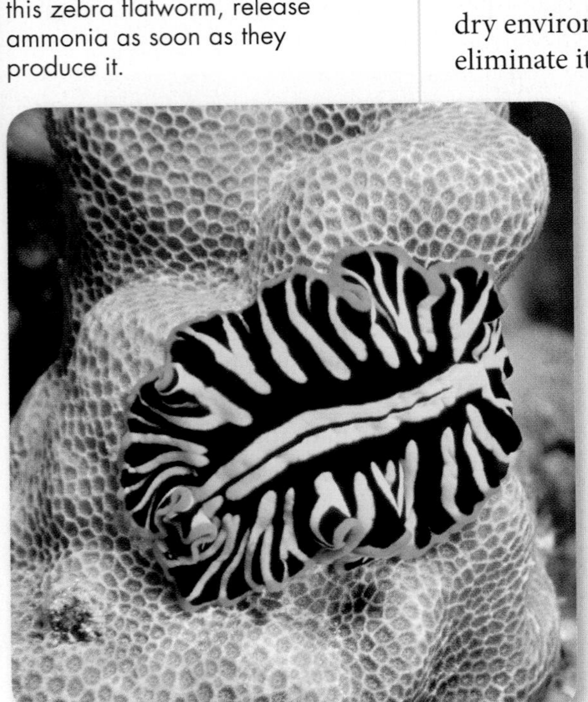

FIGURE 27–13 Ammonia Some aquatic animals, such as this zebra flatworm, release ammonia as soon as they produce it.

Storing Nitrogenous Wastes Animals that cannot dispose of ammonia continuously, as it is produced, have evolved ways to hold, or "store," nitrogenous wastes until they can be eliminated. In most cases, ammonia itself cannot be stored in body fluids, because it is too toxic. Insects, reptiles, and birds typically solve this problem by converting ammonia into a sticky white compound called uric acid, which you can see in **Figure 27–14.** Uric acid is much less toxic than ammonia and is also less soluble in water. Mammals and some amphibians, on the other hand, convert ammonia to a different nitrogenous compound—urea. Like uric acid, urea is less toxic than ammonia, but unlike uric acid, urea is highly soluble in water.

FIGURE 27–14 Other Nitrogenous Compounds Large and/or terrestrial animals either convert ammonia to uric acid and excrete it as sticky white guano, as have these gulls, or they convert ammonia into urea and release it, diluted, as urine.

Maintaining Water Balance Getting rid of any type of nitrogenous waste involves water. For that reason, excretory systems are extremely important in maintaining the proper balance of water in blood and body tissues. In some cases, excretory systems eliminate excess water along with nitrogenous wastes. In other cases, excretory systems must eliminate nitrogenous wastes while conserving water. Many animals use **kidneys** to separate wastes and excess water from blood. This waste and water forms a fluid called urine.

Kidneys perform these functions despite a serious limitation: No living cell can actively pump water across a membrane. Yet kidneys need to separate water from waste products. You may recall that cells can pump ions across their membranes. Kidney cells pump ions from salt to create osmotic gradients. Water then "follows" those ions passively by osmosis. This process works well but leaves kidneys with one weakness: They usually cannot excrete excess salt.

In Your Notebook *Explain how kidneys remove excess water from the blood.*

Humans, like most land-dwelling mammals, have evolved kidneys that are designed to conserve salt, not to get rid of it. How could this have posed a problem for the sick "survivor"?

Excretion in Aquatic Animals

How do aquatic animals eliminate wastes?

Aquatic animals have an advantage in getting rid of nitrogenous wastes because they are surrounded by water. **In general, aquatic animals can allow ammonia to diffuse out of their bodies into surrounding water, which dilutes the ammonia and carries it away.** But aquatic animals still face excretory challenges. Many need to either eliminate water from their bodies or to conserve it, depending on whether they live in fresh or salt water. The excretion issues of aquatic animals are summarized in **Figure 27–15** on the next page.

Fresh Water

The bodies of freshwater animals, such as fishes, contain a higher concentration of salt than the water they live in.

So water moves into their bodies by osmosis, mostly across the gills. Salt diffuses out. If they didn't excrete water, they'd look like water balloons with eyes!

So they excrete water through kidneys that produce lots of watery urine. They don't drink, and they actively pump salt in across their gills.

Salt Water

The bodies of saltwater animals, such as fishes, contain a lower concentration of salt than the water they live in.

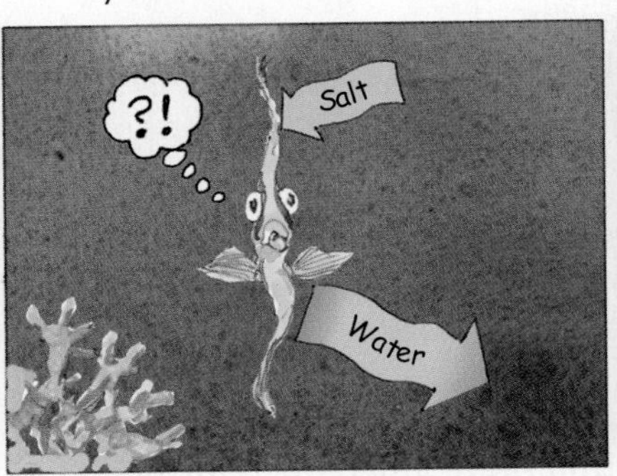

So they lose water through osmosis, and salt diffuses in. If they didn't conserve water and eliminate salt, they'd shrivel up like dead leaves.

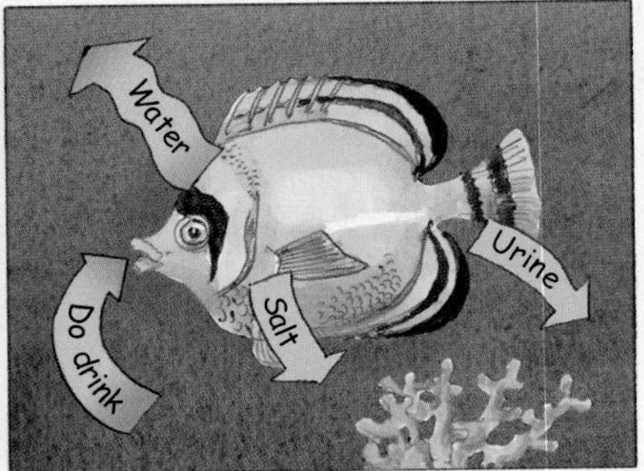

So they conserve water by producing very little concentrated urine. They drink, and they actively pump salt out across their gills.

VISUAL ANALOGY

EXCRETION IN AQUATIC ANIMALS

FIGURE 27–15 All animals must rid their bodies of ammonia while maintaining appropriate water balance. Freshwater and saltwater animals face very different challenges in this respect. **Interpret Visuals** ***What are two ways freshwater fishes avoid looking like "water balloons with eyes"?***

Freshwater Animals Many freshwater invertebrates lose ammonia to their environment by simple diffusion across their skin. Many freshwater fishes and amphibians eliminate ammonia by diffusion across the same gill membranes they use for respiration.

The situation is more complex for some freshwater invertebrates and most freshwater fishes. The concentration of water in their freshwater environments is higher than the concentration of water in their body fluids. So water moves passively into their bodies by osmosis, and salt leaves by diffusion. To help maintain water balance, flatworms have specialized cells called flame cells that remove excess water from body fluids. That water travels through excretory tubules and leaves through pores in the skin. Amphibians and freshwater fishes typically excrete excess water in very dilute urine. Freshwater fishes also pump salt actively inward across their gills.

Saltwater Animals Marine invertebrates and vertebrates typically release ammonia by diffusion across their body surfaces or gill membranes. Many marine invertebrates have body fluids with water concentrations similar to that of the seawater around them. For that reason, these animals have less of a problem with water balance than do freshwater invertebrates. Marine fishes, however, tend to lose water to their surroundings because their bodies are less salty than the water they live in. These animals actively excrete salt across their gills. Their kidneys also produce small quantities of very concentrated urine to conserve water.

Excretion in Terrestrial Animals

How do land animals remove wastes while conserving water?

Land animals also face challenges. In dry environments, they can lose large amounts of water from respiratory membranes that must be kept moist. In addition, they must eliminate nitrogenous wastes in ways that require disposing of water—even though they may not be able to drink water. **Figure 27–16** shows the excretory systems of some terrestrial animals.

Terrestrial Invertebrates **Some terrestrial invertebrates, including annelids and mollusks, produce urine in nephridia.** **Nephridia** (singular: nephridium) are tubelike excretory structures that filter body fluid. Typically, body fluid enters the nephridia through openings called nephrostomes and becomes more concentrated as it moves along the tubes. Urine leaves the body through excretory pores. **Other terrestrial invertebrates, such as insects and arachnids, convert ammonia into uric acid.** Nitrogenous wastes, such as uric acid, are absorbed from body fluids by structures called **Malpighian tubules,** which concentrate the wastes and add them to digestive wastes traveling through the gut. As water is absorbed from these wastes, they form crystals that form a thick paste, which leaves the body through the anus. This paste contains little water, so this process minimizes water loss.

Terrestrial Vertebrates In terrestrial vertebrates, excretion is carried out mostly by the kidneys. **Mammals and land amphibians convert ammonia into urea, which is excreted in urine. In most reptiles and birds, ammonia is converted into uric acid.** Reptiles and birds pass uric acid through ducts into a cavity that also receives digestive wastes from the gut. The walls of this cavity absorb most of the water from the wastes, causing the uric acid to separate out as white crystals. The result is a thick, milky-white paste that you would recognize as "bird droppings."

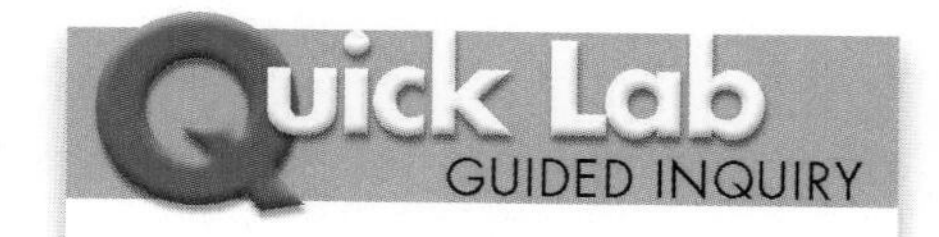

Water and Nitrogen Excretion

1. Label one test tube Urea and the other Uric Acid. Place 2 grams of urea in the one labeled Urea. Place 2 grams of uric acid in the one labeled Uric Acid.
2. Add 15 mL of water to each test tube. Stopper and shake the test tubes for 3 minutes.
3. Observe each test tube. Record your observations.

Analyze and Conclude

1. Observe Which substance—urea or uric acid—is less soluble in water? Explain.

2. Infer Reptiles excrete nitrogenous wastes in the form of uric acid. How does this adaptation help reptiles survive on land?

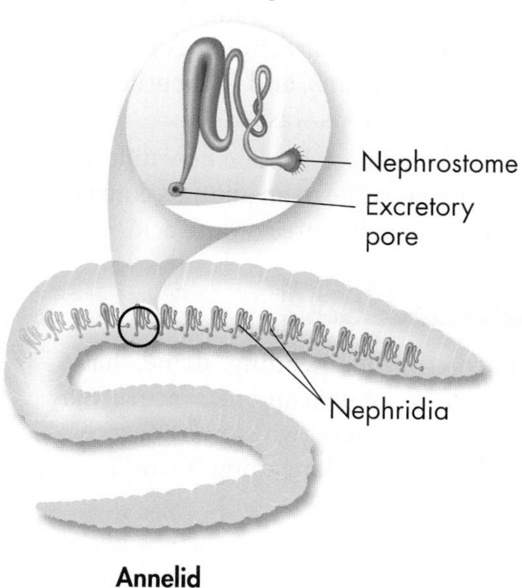

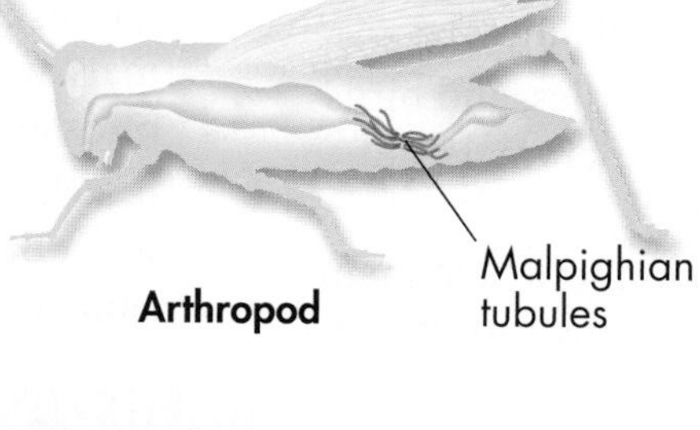

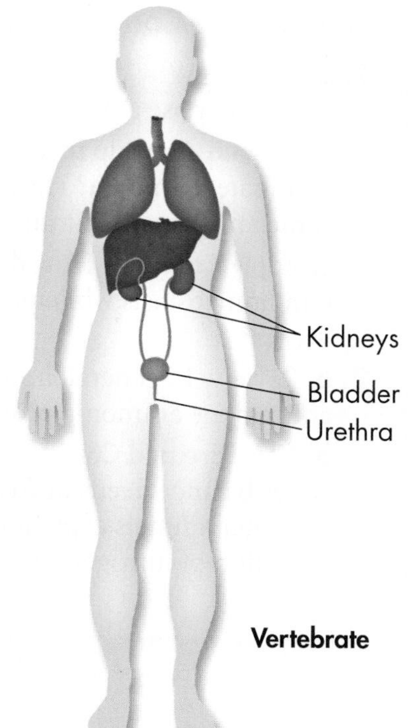

FIGURE 27–16 Excretion in Terrestrial Animals Some terrestrial invertebrates, such as annelids, rid their bodies of ammonia by releasing urine created in their nephridia (left). Some insects and arachnids have Malpighian tubules, which absorb uric acid from body fluids and combine it with digestive wastes (above). In vertebrates, such as humans, excretion is carried out mostly by the kidneys (right).

Adaptations to Extreme Environments The kidneys of most terrestrial vertebrates are remarkable organs, but the way they operate results in some limitations. Most vertebrate kidneys, for example, cannot excrete concentrated salt. That's why most vertebrates cannot survive by drinking seawater. All that extra salt would overwhelm the kidneys, and the animal would die of dehydration. Some marine reptiles and birds, such as the petrel in **Figure 27–17,** have evolved specialized glands in their heads that excrete very concentrated salt solutions. Another remarkable excretory adaptation is found in the kangaroo rats of the American southwest. The kidneys of these desert rodents produce urine that is 25 times more concentrated than their blood! In addition, their intestines are so good at absorbing water that their feces are almost dry.

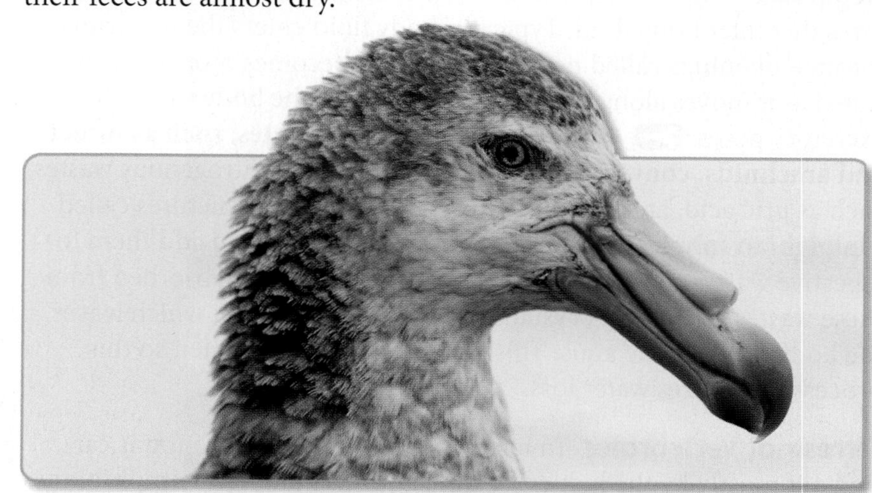

FIGURE 27–17 Excretion Adaptations Some terrestrial animals that spend a large amount of time in salt water, such as this petrel, have special adaptations to excrete excess salt. Specialized salt glands produce a concentrated salt solution, which can sometimes been seen dripping out of their elongated nostrils.

27.4 Assessment

Review Key Concepts

1. a. **Review** Why does the metabolic waste ammonia pose a problem for all animals?
   b. **Explain** How do insects, reptiles, and birds eliminate ammonia? How do mammals and some amphibians eliminate it?
   c. **Apply Concepts** How do kidneys help maintain homeostasis while processing nitrogenous wastes?
2. a. **Review** In general, how do aquatic animals address the ammonia problem?
   b. **Compare and Contrast** How do the differing water balance needs of freshwater animals and saltwater animals explain the difference in their excretion of nitrogenous wastes?
3. a. **Review** In what form do (a) annelids and mollusks, (b) insects and arachnids, (c) mammals and land amphibians, and (d) reptiles and birds excrete nitrogenous wastes?
   b. **Relate Cause and Effect** Explain how differing water balance needs relate to an animal's conversion of ammonia to either urea or uric acid.

BUILD VOCABULARY

4. The Greek word *ouron,* meaning "urine," has led to the root *uro-*, of *urea* and *uric* (acid). Why is it appropriate that these two words are each formed from a root word meaning "urine"?

RST.9-10.2 Key Ideas and Details, RST.9-10.10 Level of Text Complexity

Technology & BIOLOGY

Bioartificial Kidneys

Hundreds of thousands of people suffer from kidney failure. Kidneys eliminate wastes from blood while retaining vital compounds, so if your kidneys fail, you can die, poisoned by your own nitrogenous wastes. Because kidney function is so important, researchers are always working on better replacements for it.

Today's techniques save lives but are far from ideal. Techniques such as artificial kidney dialysis pass a patient's blood through a system of extremely fine tubes that retain blood cells and vital proteins while filtering out wastes, which are discarded. But important compounds are also filtered out, and they must be given back to the patient intravenously.

A technique invented by Dr. H. David Humes at the University of Michigan could eliminate this drawback. It combines the technique above with the latest in biotechnology to produce what is called a renal tubule assist device, or RAD. Tiny tubes (blue in the figure to the right) are lined with a matrix (shown in yellow) on which living cells can grow. The matrix-covered tubes are then "seeded" with cells from kidneys that were donated for transplant but could not be used. When properly cared for, these kidney cells grow one layer thick (shown in pink) to cover the matrix.

Membrane tube

Matrix

Living kidney cells

During treatment, the fluid that would normally be discarded is passed through the inside of these tubes, while the patient's blood passes by on the other side. The living kidney cells then act on the waste fluid, returning valuable compounds like glucose to the blood, and adding the important compounds made in healthy kidneys. Researchers hope that this technique can someday improve the lives of the many people whose kidneys fail.

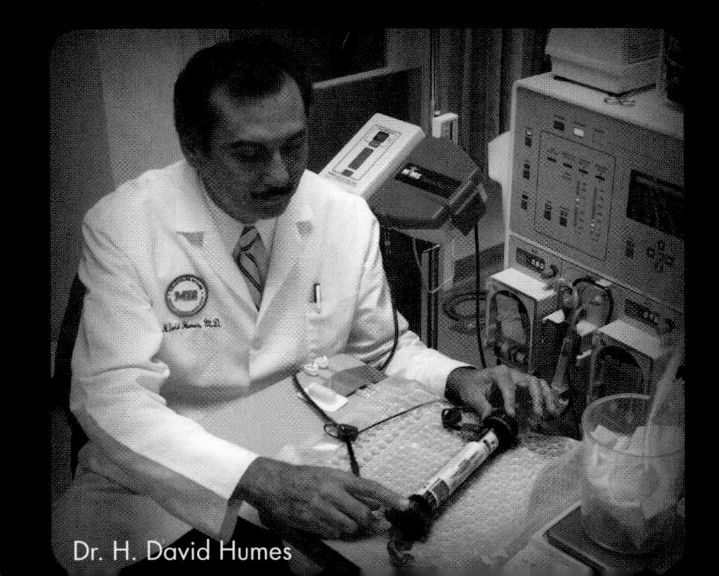
Dr. H. David Humes

WRITING **If possible, interview someone who has undergone kidney dialysis, asking about five questions you have about the experience. Transcribe your interview, and include a short introduction to the topic. Alternately, read about the experience of kidney dialysis, and write a one-page summary.**

NGSS Smart Guide Animal Systems I

Disciplinary Core Idea LS1.A Structure and Function: How do the structures of organisms enable life's functions? The circulatory system transports nutrients from the digestive system and oxygen from the respiratory system to body cells. It then transports cellular waste to the excretory system and carbon dioxide to the respiratory system.

27.1 Feeding and Digestion

- Most filter feeders catch algae and small animals by using modified gills or other structures as nets that filter food items out of water. Detritivores feed on detritus. Carnivores eat other animals. Herbivores eat plants or parts of plants. Nutritional symbionts rely upon symbiosis.
- Some invertebrates break down food primarily through intracellular digestion, but many animals use extracellular digestion.
- Carnivores typically have sharp mouthparts or other structures that capture food, hold it, and cut it into small pieces. Herbivores typically have mouthparts adapted to rasping or grinding.

intracellular digestion 784 • extracellular digestion 784 • gastrovascular cavity 784 • digestive tract 784 • rumen 786

Biology.com

Untamed Science Video Trek carefully with the Untamed Science crew as they get up close and personal with bears to learn about their adaptations.

STEM Activity Learn about the practice of clam farming and use a simulation to plant your own juvenile clams.

Visual Analogy Compare the structure and function of the types of teeth with common objects.

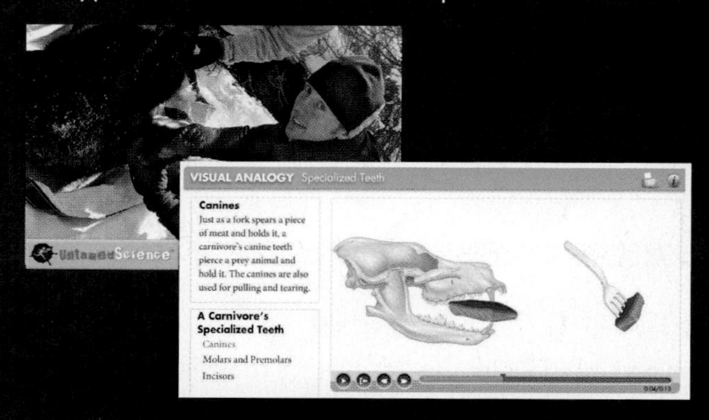

27.2 Respiration

- Respiratory structures provide a large surface area of moist, selectively permeable membrane and maintain a difference in the concentrations of oxygen and carbon dioxide on either side of the respiratory membrane, promoting diffusion.
- Many aquatic invertebrates and most aquatic chordates other than reptiles and mammals exchange gases through gills. Aquatic reptiles and aquatic mammals, such as whales, breathe with lungs and must hold their breath underwater.
- Respiratory structures in terrestrial invertebrates include skin, mantle cavities, book lungs, and tracheal tubes. Terrestrial vertebrates breathe with lungs.

gill 788 • lung 788 • alveolus 790

Biology.com

Art Review Review your knowledge of the different types of respiratory systems with this activity.

Data Analysis Investigate the relationship between body size, tracheal structure, and the amount of atmospheric oxygen to understand why insects were larger in the Paleozoic era than they are today.

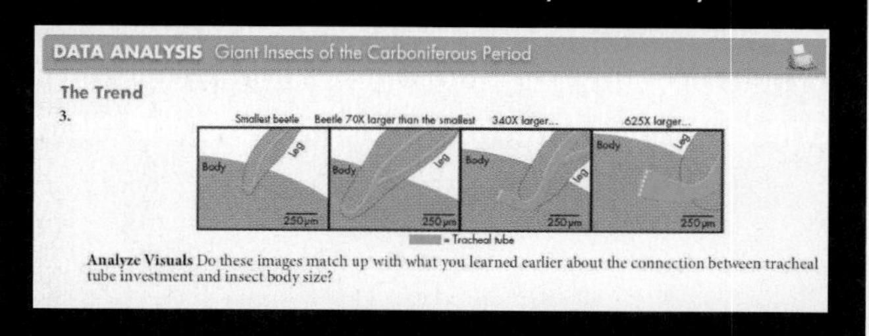

CHECKING YOUR *Scientific Literacy*

Refer to the lesson content and digital assets below as you prepare for your chapter assessment. Then, evaluate your understanding of animal systems by answering these questions.

1. Scientific and Engineering Practice **Developing and Using Models** Choose an animal structure from this chapter that would enable you to carry out a desired function that you otherwise would not be able to do. Draw a diagram to serve as a model of how your body would accommodate such a structure.

2. Crosscutting Concept **Structure and Function** How would the addition of this structure affect the other structures and functions in your body?

3. Text Types and Purposes Write an informative text in which you explain how your digestive, respiratory, circulatory, and excretory systems allow you to maintain homeostasis in your body.

4. STEM You are an engineer assigned with the task of developing an underwater breathing device that does not require an external oxygen tank. List the features that you must incorporate into this device. Also, explain the constraints you need to consider in your design.

27.3 Circulation

- In an open circulatory system, blood is only partially contained within blood vessels. In a closed circulatory system, blood circulates entirely within blood vessels.
- Most vertebrates with gills have a single-loop circulatory system with a single pump that forces blood around the body in one direction. Most vertebrates that use lungs for respiration have a double-loop, two-pump circulatory system.

heart 791 • open circulatory system 791 • closed circulatory system 792 • atrium 792 • ventricle 792

Biology.com

InterActive Art See how single- and double-loop circulation systems compare.

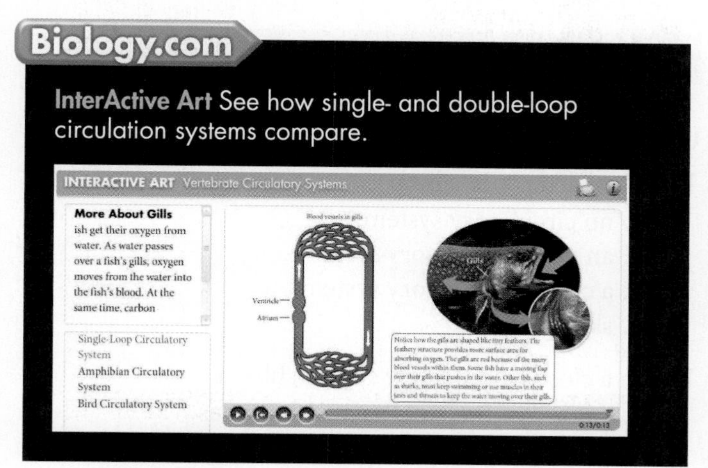

27.4 Excretion

- Animals either eliminate ammonia from the body quickly or convert it into other nitrogenous compounds that are less toxic.
- Aquatic animals allow ammonia to diffuse out of their bodies into surrounding water.
- Some terrestrial invertebrates, including annelids and mollusks, produce urine in nephridia. Other terrestrial invertebrates, such as insects and arachnids, convert ammonia into uric acid. Mammals and land amphibians convert ammonia into urea. Most reptiles and birds convert ammonia into uric acid.

excretion 794 • kidney 795 • nephridium 797 • Malpighian tubule 797

Biology.com

Taking Notes Ask questions you have about Figure 27-15 and try to answer your questions as you read.

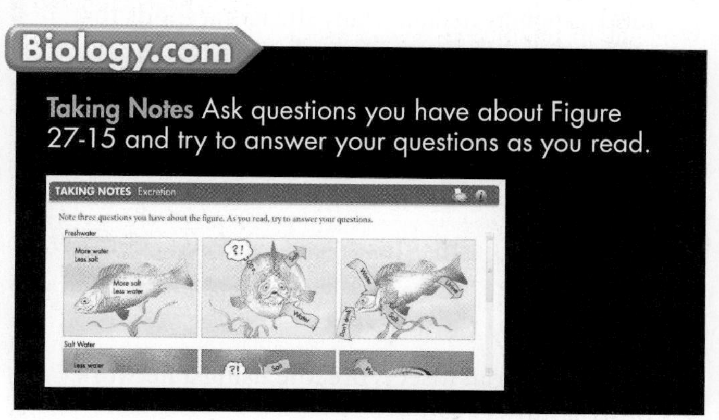

Skills Lab

Anatomy of a Squid This chapter lab is available in your lab manual and at Biology.com

27 Assessment

27.1 Feeding and Digestion

Understand Key Concepts

1. An animal that relies primarily on intracellular digestion is the
   a. sponge. c. dragonfly.
   b. clam. d. earthworm.

2. Animals that obtain food by ingesting decaying bits of plant and animal material are called
   a. herbivores. c. detritivores.
   b. carnivores. d. filter feeders.

3. Algae that live in the bodies of reef-building corals are
   a. parasitic symbionts.
   b. mutualistic symbionts.
   c. occupants that have no effect on the coral animals.
   d. consumed as food by the coral animals.

4. Compare the processes of intracellular and extracellular digestion.

5. Describe the differences between the canine and molar teeth of herbivorous and carnivorous mammals.

6. How do vertebrate filter feeders obtain food?

Think Critically

7. **Classify** You are observing an animal that has a digestive tract. Does this animal practice intracellular digestion or extracellular digestion? Explain your answer.

8. **Pose Questions** Hummingbirds eat high-energy foods, such as nectar. Many ducks eat foods that contain less energy, such as plant leaves. What are some research questions you could investigate to discover more about the diet of a bird species and its energy needs?

27.2 Respiration

Understand Key Concepts

9. Most terrestrial insects breathe using a network of structures called
   a. gills. c. book gills.
   b. tracheal tubes. d. book lungs.

10. In order for the exchange of oxygen and carbon dioxide to take place, an animal's respiratory surfaces must be kept
    a. cold. c. hot.
    b. dry. d. moist.

11. Most fishes exchange gases by pumping water from their mouths
    a. over their gills.
    b. through the lungs.
    c. over their atria.
    d. through their esophagus.

12. Describe two types of respiratory structures found in terrestrial invertebrates.

13. What respiratory structures do all terrestrial vertebrates possess?

14. **Craft and Structure** With what respiratory structures do aquatic reptiles and aquatic mammals breathe? What inconvenience does this cause when they are underwater?

Think Critically

15. **Predict** During heavy rains, earthworms often emerge from their burrows. What might happen to an earthworm if it did not return to its burrow when the ground dried out?

16. **Infer** Land snails have a respiratory structure called a mantle cavity, which is covered with mucus. What might the purpose of the mucus be?

27.3 Circulation

Understand Key Concepts

17. Most arthropods have
    a. no circulatory system.
    b. an open circulatory system.
    c. a closed circulatory system.
    d. skin gills.

18. In a closed circulatory system, blood
    a. comes in direct contact with tissues.
    b. remains within blood vessels.
    c. empties into sinuses.
    d. does not transport oxygen.

19. Most chordates that have gills for respiration have a(n)
   a. double-loop circulatory system.
   b. accessory lung.
   c. single-loop circulatory system.
   d. four-chambered heart.

20. In the gills of aquatic animals, how do the respiratory and circulatory systems interact?

21. Describe the circulatory system of a mammal as open or closed, and state the number of loops and the number of heart chambers.

22. Compare single-loop circulation and double-loop circulation.

Think Critically

23. **Integration of Knowledge and Ideas** The diagrams below represent two kinds of circulatory systems.

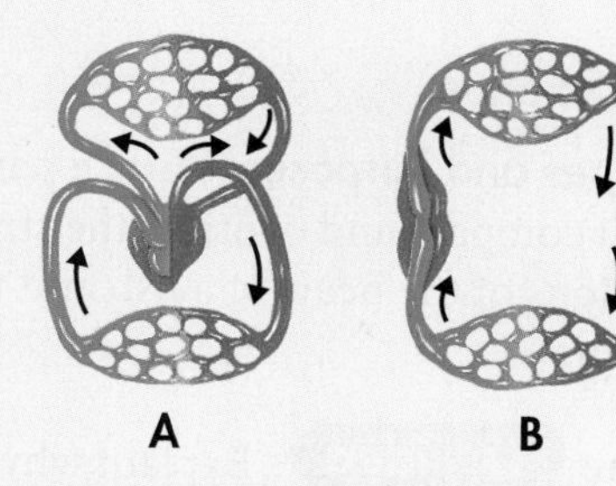

   a. Which diagram illustrates a heart with blood containing carbon dioxide but little oxygen?
   b. Which diagram shows a circulatory system with a four-chambered heart?

24. **Apply Concepts** How do a fish's respiratory and circulatory systems work together to maintain homeostasis in the body as a whole?

27.4 Excretion

Understand Key Concepts

25. The composition of and levels of body fluids in mammals are controlled by the
   a. lungs. **c.** intestine.
   b. kidneys. **d.** heart.

26. The elimination of metabolic wastes from the body is called
   a. excretion. **c.** respiration.
   b. circulation. **d.** digestion.

27. Why do most animals convert ammonia into urea or uric acid?

solve the CHAPTER MYSTERY

(NEAR) DEATH BY SALT WATER

Luckily, the pick-up the group arranged arrived earlier than planned. They rushed the sick man to a hospital, where he was diagnosed with severe dehydration and given water and intravenous fluids. If he had gone much longer without treatment, doctors told his friends, he would have died. What had happened? Why didn't his friends suffer the same problems?

As sailors have known for centuries, humans can't drink seawater for any length of time. But why *can't* we drink seawater?

Because seawater is saltier than human blood and body fluids, drinking it loads the body with excess salt. Human kidneys cannot produce urine with salt concentrations high enough to get rid of that salt efficiently. So the kidneys are forced to excrete more water in urine than the amount of salt water consumed. This lowers body water content to the point that blood literally becomes thicker and harder to push through fine capillary networks. Cells and tissues begin to dehydrate, and fatal kidney failure can result.

1. **Compare and Contrast** While the group member who drank seawater became seriously ill, the other group members experienced some water stress as well. What was going on in their circulatory and excretory systems, and why was it not as serious?

2. **Propose a Solution** If you were marooned on an island that had no fresh water, what would be your plan for getting some?

3. **Connect to the Big idea** Although humans can't drink salt water, and can't exist without fresh water, many marine birds and reptiles can do either or both. Using the Web, research the different strategies other animals use to regulate salt content and water balance.

28. What is the difference in kidney function of freshwater fishes and saltwater fishes?

Think Critically

29. **Infer** The excretory systems of terrestrial invertebrates, such as earthworms, convert ammonia to less toxic substances. Explain why this change is unnecessary in small aquatic invertebrates, such as planarians.

30. **Apply Concepts** Of all the nitrogenous wastes eliminated by animals, uric acid requires the least water to excrete. Why is the production of uric acid an advantage to animals that live on land?

Connecting Concepts

Use Science Graphics

A student conducts an experiment to measure the effect of caffeine on the heart rate of a small pond-water crustacean called Daphnia. The heart of this animal is visible through its transparent shell. With the help of a dissecting microscope, the student counts the heartbeats per minute before and after adding increasing amounts of coffee to the water surrounding the animal. Each data point in the graph at the top right represents the mean of five trials. Use the graph to answer questions 31 and 32.

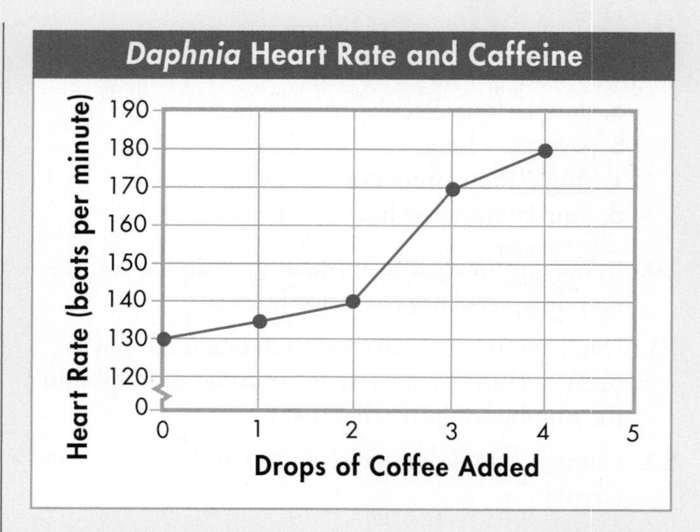

31. **Interpret Graphs** Describe the effect of caffeine on the heart rate of *Daphnia*.

32. **Predict** What would be your prediction of the effect of five or more drops of coffee on the heart rate of *Daphnia*?

Write About Science

33. **Text Types and Purposes** Write a paragraph in which you compare and contrast the structures and functions of the heart of a fish and the heart of a mammal.

34. **Assess the** 

Explain why a digestive tract is a more efficient structure for taking in and processing the food eaten by a large animal than a gastrovascular cavity would be.

Analyzing Data

Integration of Knowledge and Ideas

A researcher conducted an experiment to see how air temperature affects the speed at which a snake can hunt for food. The experimenter placed the snake a fixed distance away from a piece of food and recorded the air temperature. Then, she recorded the time it took for the snake to reach the food. She repeated the experiment four times. Each time, the experimenter changed the air temperature. The data are shown to the right.

35. **Interpret Tables** At what temperature did the snake reach the food the fastest?

36. **Analyze Data** How did the time to reach the food change as the temperature increased?

37. **Draw Conclusions** What conclusion about snake hunting and temperature can you draw from the data?

The Effect of Temperature on Snake Hunting Speed

Temperature (°C)	Time (seconds)
4	51
10	50
15	43
21	37
27	35

Standardized Test Prep

Multiple Choice

1. Animals that live on an animal and feed on its body tissues are called
   A parasites.
   B carnivores.
   C herbivores.
   D detritivores.

2. Examining the teeth of an animal can give information about whether it
   A practices intracellular or extracellular digestion.
   B is a filter feeder or a detritivore.
   C is a nutritional symbiont.
   D is a herbivore or a carnivore.

3. Movement of oxygen and carbon dioxide across a respiratory surface requires
   A that the respiratory surface be moist.
   B active transport by the cells of the respiratory surface.
   C alveoli.
   D an equal concentration of both gases on both sides of the membrane.

4. In an open circulatory system, blood
   A is confined to blood vessels at all times.
   B circulates around body tissues.
   C exchanges gases with lung alveoli.
   D is not required for exchanging gases with body cells.

5. In chordates with four-chambered hearts, there is
   A only one loop in the circulatory system.
   B mixing of oxygen-rich and oxygen-poor blood.
   C partial partition of the ventricle.
   D no mixing of oxygen-rich and oxygen-poor blood.

6. Most reptiles excrete wastes in the form of
   A urea.
   B ammonia.
   C uric acid.
   D toxins.

7. What is a function of the excretory system?
   A to supply cells with oxygen and nutrients
   B to rid the body of metabolic wastes
   C to exchange oxygen and carbon dioxide with the environment
   D to break down food

Questions 8–9

A biology student is investigating the relationship between cricket chirping and air temperature. She catches a cricket and places it in a jar. She leaves the jar outside, and each day she counts the number of chirps during a 15-second period. At the same time, she records the outside temperature near the cricket. Her data for a 5-day period are shown below.

Temperature and Cricket Chirping

Day	Number of Chirps in 15 Seconds	Outside Temperature (ºC)
Monday	31	23
Tuesday	20	16
Wednesday	12	11
Thursday	29	21
Friday	25	19

8. At which of the following temperatures would a cricket be most likely to chirp 9 times in 15 seconds?
   A 10°C
   B 18°C
   C 0°C
   D 25°C

9. What can the student conclude from this experiment?
   A Crickets cannot chirp more than 31 times in 15 seconds.
   B The number of times a cricket chirps decreases when the temperature decreases.
   C The number of times a cricket chirps increases when the temperature decreases.
   D There is no relationship between the number of times a cricket chirps and temperature.

Open-Ended Response

10. Which types of vertebrates have double-loop circulation and which types have single-loop circulation?

If You Have Trouble With . . .

Question	1	2	3	4	5	6	7	8	9	10
See Lesson	27.1	27.1	27.2	27.3	27.3	27.4	27.4	27.3	27.3	27.3

28 Animal Systems II

Big idea

Structure and Function

Q: How do the body systems of animals allow them to collect information about their environments and respond appropriately?

The dense down feathers and shared body heat of these huddled young penguins help them stay warm.

INSIDE:

- 28.1 Response
- 28.2 Movement and Support
- 28.3 Reproduction
- 28.4 Homeostasis

BUILDING *Scientific Literacy*

Deepen your understanding of animal systems by engaging in key practices that allow you to make connections across concepts.

NGSS You will analyze and interpret data about animal systems.

STEM Math and engineering often go hand in hand. Watch a video to see how math and engineering relate to the movement of snakes.

Common Core You will interpret data and write arguments focused on animal systems.

CHAPTER MYSTERY

SHE'S JUST LIKE HER MOTHER!

It was December 2001, and the employees of the shark exhibits at the Henry Doorly Zoo in Omaha, Nebraska, had just discovered that one of their bonnethead sharks had given birth to a female baby bonnethead. They were shocked. For three years, there had been only three bonnethead sharks in the tank in which the baby shark was born. They were all female.

Some female sharks, including bonnetheads, which are related to hammerheads, are known to store sperm for later fertilization. Did this explain how the shark got pregnant? As you read this chapter, look for clues that help explain how the baby bonnethead's mother got pregnant. Also, think about how sharks typically reproduce and the effect of that process on the offspring's genetic material. Then, solve the mystery.

Biology.com

Finding the solution to the She's Just Like Her Mother! mystery is only the beginning. Take a video field trip with the ecogeeks of Untamed Science to see where the mystery leads.

Go online to access additional resources including:
- eText • Flash Cards • Lesson Overviews • Chapter Mystery

28.1 Response

Key Questions

🔑 ***How do animals respond to events around them?***

🔑 ***What are the trends in nervous system evolution?***

🔑 ***What are some types of sensory systems in animals?***

Vocabulary

neuron • stimulus • sensory neuron • interneuron • response • motor neuron • ganglion • cerebrum • cerebellum

Taking Notes

Preview Visuals Before you read, preview the diagram of neural circuits in **Figure 28–1.** Take note of any questions you have about it and try to answer them as you read.

THINK ABOUT IT Imagine that you are at a favorite place—a beach or the basketball court. Think about how the sun and wind feel on your face or how good it feels to make the perfect layup. Now, think about the way you experience that place. You gather information about your surroundings through senses such as vision and hearing. Your nervous system collects that information. Your brain decides how to respond to it. The same is true for all animals—though the structures that perform these functions vary from phylum to phylum.

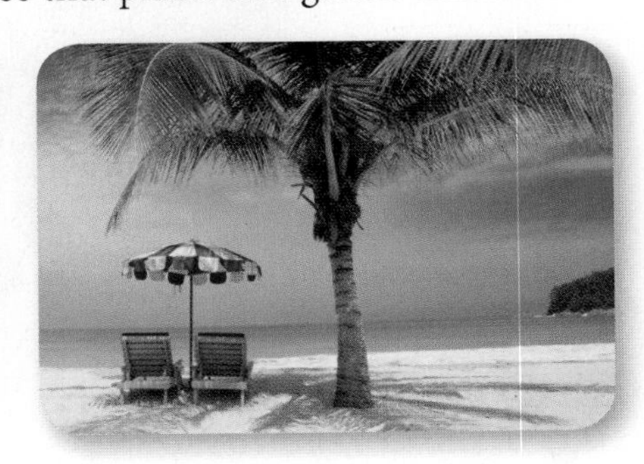

How Animals Respond

🔑 ***How do animals respond to events around them?***

Animals must often respond to events or environmental conditions within seconds, or even tiny fractions of a second. Sometimes they need to catch food. Other times, they need to escape predators. Most animals have evolved specialized nervous systems that enable them to respond to events around them. Nervous systems are composed of specialized nerve cells, or **neurons.** The structure of neurons enables them to receive and pass on information. Working together, neurons acquire information from their surroundings, interpret that information, and then "decide" what to do about it.

Detecting Stimuli Information in the environment that causes an organism to react is called a **stimulus** (plural: stimuli). Chemicals in air or water can stimulate the nervous system. Light or heat can also serve as a stimulus. The sound of your phone ringing on a Friday night is a stimulus to which you might respond by running to answer it!

Animals' ability to detect stimuli depends on specialized cells called **sensory neurons.** Each type of sensory neuron responds to a particular stimulus such as light, heat, or chemicals. Humans share many types of sensory cells with other animals. For that reason, many animals react to stimuli that humans notice, including light, taste, odor, temperature, sound, water, gravity, and pressure. But many animals have types of sensory cells that humans lack. That's one reason why some animals respond to stimuli that humans cannot detect, such as very weak electric currents or Earth's magnetic field.

Processing Information When sensory neurons detect a stimulus, they pass information about it to other nerve cells. Those neurons, which typically pass information to still other neurons, are called **interneurons,** as shown in **Figure 28–1.** Interneurons process information and determine how an animal responds to stimuli.

Does a particular odor mean food . . . or danger? Is the immediate environment too hot, too cold, or just right? The number of interneurons an animal has, and the ways those interneurons process information, determine how flexible and complex an animal's behavior can be.

Some invertebrates, such as cnidarians and worms, have very few interneurons. These animals are capable of only simple responses to stimuli. They may swim toward light or toward a chemical stimulus that signals food. Vertebrates have more highly developed nervous systems with larger numbers of interneurons. The brain is made up of many of these interneurons. That's why the behaviors of vertebrates can be more complex than those of most invertebrates.

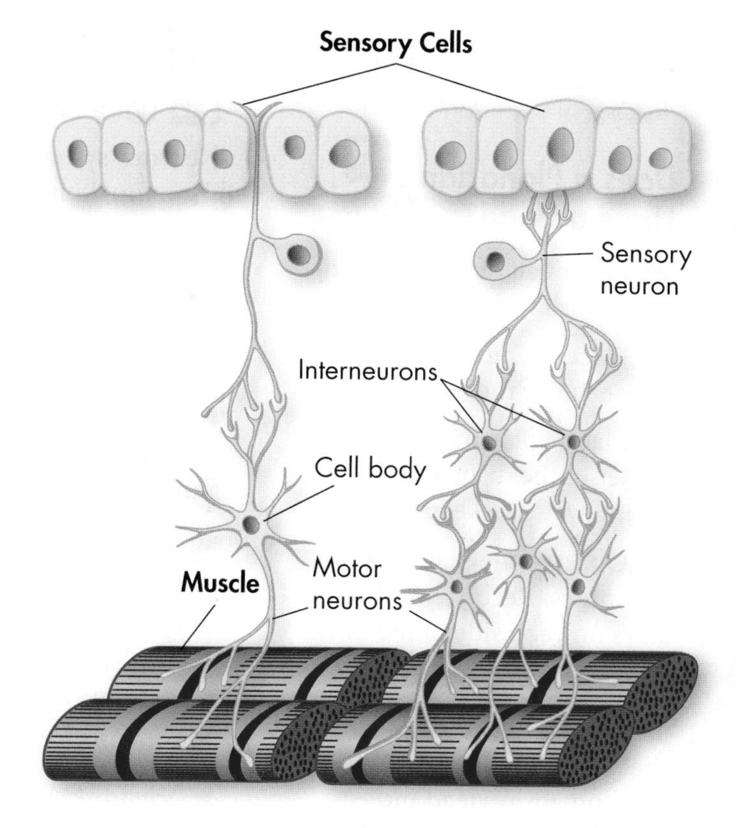

FIGURE 28–1 Neural Circuits In some neural circuits, sensory neurons connect to motor neurons in ways that enable fast but simple responses (left). In others, specialized sensory cells connect to sensory neurons, which connect to interneurons, which connect to motor neurons (right). The more complex a circuit is, the more complex an animal's responses to stimuli can be.

Responding A specific reaction to a stimulus is called a **response.** For example, waking up when you hear the alarm is a response. **When an animal responds to a stimulus, body systems—including the nervous system and the muscular system—work together to generate a response.** Responses to many stimuli are directed by the nervous system. However, those responses are usually carried out by cells or tissues that are not nerve cells. A lion's decision to lunge at prey, as in **Figure 28–2,** is carried out by muscle cells that produce movement. In that case, nerve cells called **motor neurons** carry "directions" from interneurons to muscles. Other responses to environmental conditions may be carried out by other body systems, such as respiratory or circulatory systems.

FIGURE 28–2 Response Mammals, like the lioness and zebra shown here, have complex sensory organs and many interneurons. These animals can therefore process and respond to information in complex ways. Lions stalk and pursue their prey, and zebra try to evade their attackers.

Quick Lab
GUIDED INQUIRY

Does a Planarian Have a Head?

1. Cover half of the outside of a petri dish with black paper.
2. Place a white sheet of paper under the other half.
3. Place a planarian in the center of the dish, and add spring water to keep it moist.
4. Observe the planarian for 2 minutes. Record how long it stays on each side of the dish.

Analyze and Conclude

1. Form a Hypothesis When the planarian moved, did one end always lead the way? Form a hypothesis that explains your observation.

Trends in Nervous System Evolution

What are the trends in nervous system evolution?

Nervous systems vary greatly in organization and complexity across the animal kingdom. **Animal nervous systems exhibit different degrees of cephalization and specialization.**

Invertebrates Invertebrate nervous systems range from simple collections of nerve cells to complex organizations that include many interneurons. You can see some examples in **Figure 28–3.**

▶ ***Nerve Nets, Nerve Cords, and Ganglia*** Cnidarians, such as jellyfishes, have simple nervous systems called nerve nets. As the name implies, nerve nets consist of neurons connected into a netlike arrangement with few specializations. In other radially symmetric invertebrates, echinoderms such as sea stars, for example, some interneurons are grouped together into nerves, or nerve cords, that form a ring around the animals' mouths and stretch out along their arms. In still other invertebrates, a number of interneurons are grouped together into small structures called **ganglia** (singular: ganglion), in which interneurons connect with one another.

▶ ***"Heads"*** As you learned in Chapter 25, bilaterally symmetric animals often exhibit cephalization, the concentration of sensory neurons and interneurons in a "head." Certain flatworms and roundworms show some cephalization. Some cephalopod mollusks and many arthropods show higher degrees of cephalization. In these animals, interneurons form ganglia in several places. Typically, the largest ganglia are located in the head region and are called cerebral ganglia.

▶ ***Brains*** In some species, cerebral ganglia are further organized into a structure called a brain. The brains of some cephalopods, such as octopi, enable complex behavior, including several kinds of learning.

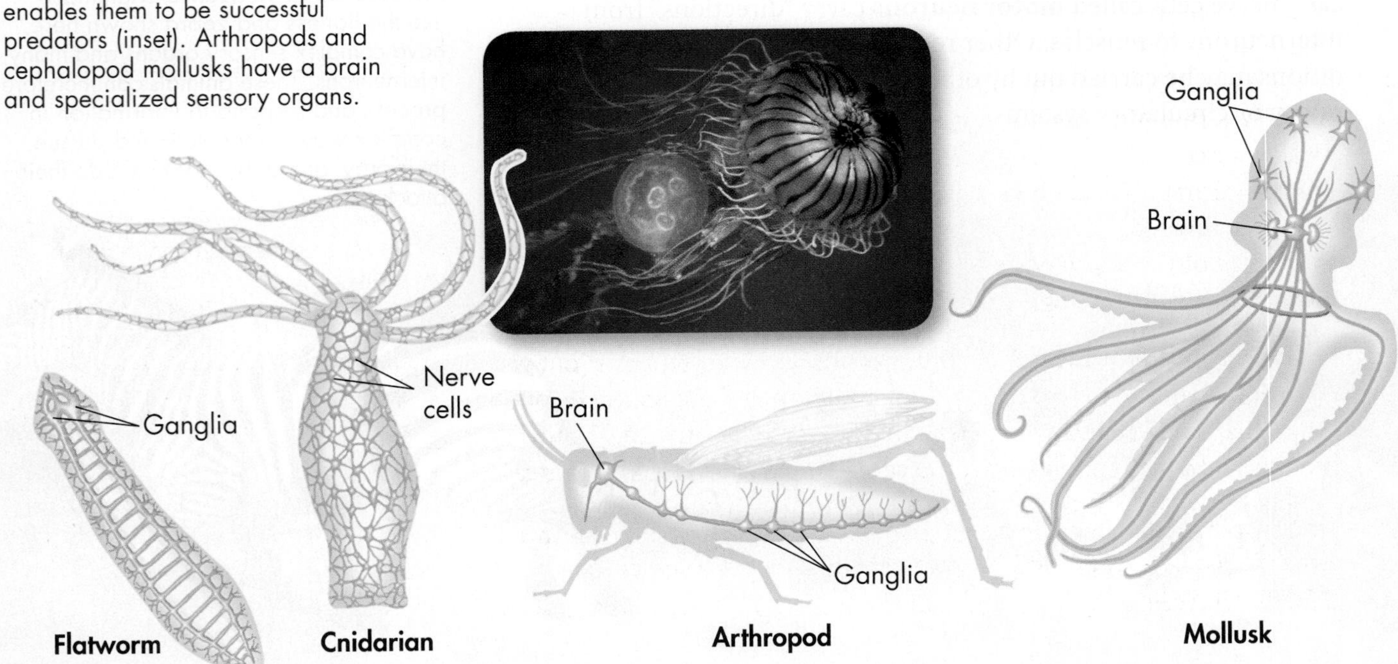

FIGURE 28–3 Invertebrate Nervous Systems Invertebrate nervous systems have different degrees of cephalization and specialization. Flatworms have centralized nervous systems with small ganglia in their heads. Cnidarians have a nerve net which, despite its simplicity, enables them to be successful predators (inset). Arthropods and cephalopod mollusks have a brain and specialized sensory organs.

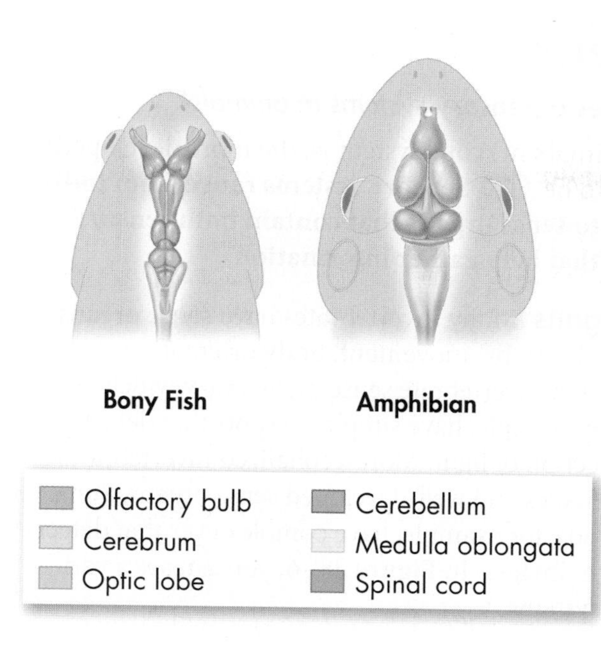
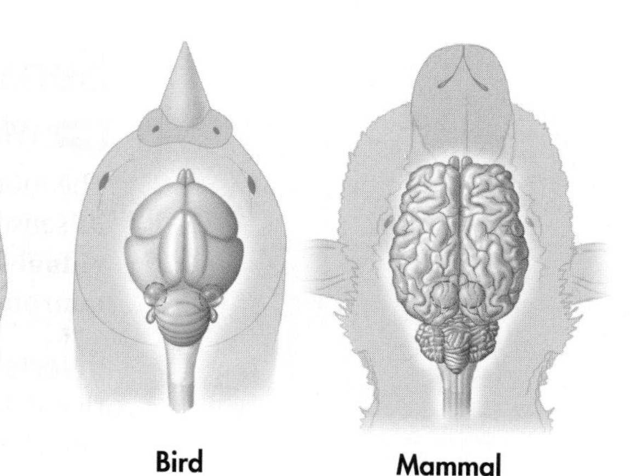

FIGURE 28–4 Vertebrate Brains The cerebrum and cerebellum increase in size from fishes to mammals. In fishes, amphibians, and reptiles, the cerebrum, or "thinking" region, is relatively small. In birds and mammals, and especially in primates, the cerebrum is much larger and may contain folds that increase its surface area. The cerebellum is also most highly developed in birds and mammals.

Chordates Nonvertebrate chordates, which have no vertebrate-type "head" as adults, still have a cerebral ganglion. Vertebrates, on the other hand, show a high degree of cephalization and have highly developed nervous systems. Vertebrate brains are formed from many interneurons within the skull. These interneurons are connected with each other and with sensory neurons and motor neurons in the head and elsewhere in the body. The human brain contains more than 100 billion nerve cells, each of which sends signals to as many as 1000 other nerve cells and receives signals from up to 10,000 more.

▶ ***Parts of the Vertebrate Brain*** Regions of the vertebrate brain include the cerebrum, cerebellum, medulla oblongata, optic lobes, and olfactory bulbs. The **cerebrum** is the "thinking" region of the brain. It receives and interprets sensory information and determines a response. The cerebrum is also involved in learning, memory, and conscious thought. The **cerebellum** coordinates movement and controls balance, while the medulla oblongata controls the functioning of many internal organs. Optic lobes are involved in vision, and olfactory bulbs are involved in the sense of smell. Vertebrate brains are connected to the rest of the body by a thick collection of nerves called a spinal cord, which runs through a tube in the vertebral column.

▶ ***Vertebrate Brain Evolution*** Brain evolution in vertebrates follows a general trend of increasing size and complexity from fishes, through amphibians and reptiles, to birds and mammals. **Figure 28–4** shows how the size and complexity of the cerebrum and cerebellum increase.

In Your Notebook *Construct a figure of speech that explains, in terms of another object, how the folds of the mammalian cerebellum increase its surface area.*

FIGURE 28–5 Not Such a Bird Brain The brains of some chickadee species are so sophisticated that the part responsible for remembering locations gets bigger when the bird stores food in the fall. When winter comes, the tiny bird is better able to find its hundreds of storage places. (In spring, its brain returns to normal size.) **Infer** ***Which of the six main parts of the chickadee brain would you expect to grow in the fall?***

Sensory Systems

What are some types of sensory systems in animals?

The more complex an animal's nervous system is, the more developed its sensory systems tend to be. **Sensory systems range from individual sensory neurons to sense organs that contain both sensory neurons and other cells that help gather information.**

Invertebrate Sense Organs Many invertebrates have sense organs that detect light, sound, vibrations, movement, body orientation, and chemicals in air or water. Invertebrate sense organs vary widely in complexity. Flatworms, for example, have simple eyespots that detect only the presence and direction of light. More-cephalized invertebrates have specialized sensory tissues and well-developed sense organs. Some cephalopods and arthropods, for example, have complex eyes that detect motion and color and form images. In **Figure 28–6,** you can see a variety of invertebrate visual systems.

FIGURE 28–6 Invertebrate Eyes Invertebrate sense organs, such as the eyes shown in the photos, vary greatly in structure and complexity.

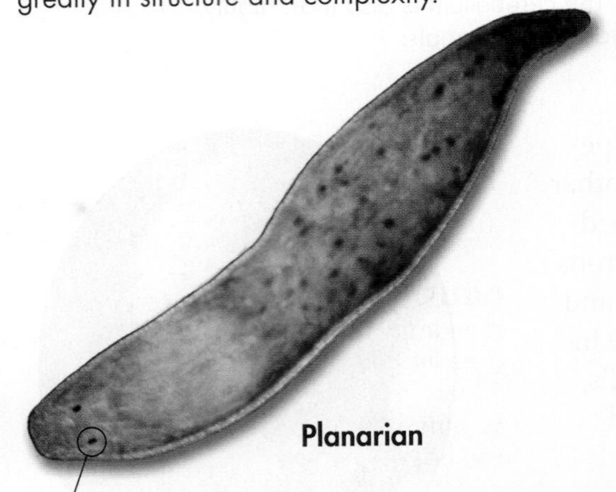

Eyespot: Some animals have eyespots, which are groups of cells that can detect changes in the amount of light (LM 50×).

Simple Eye: The 40–60 simple eyes of a scallop do not form images. They do, however, detect movement well enough to enable the scallop to escape its predators.

Compound Eye: The compound eyes of arthropods are made up of many lenses that detect minute changes in movement and color but produce less-detailed images than human eyes do.

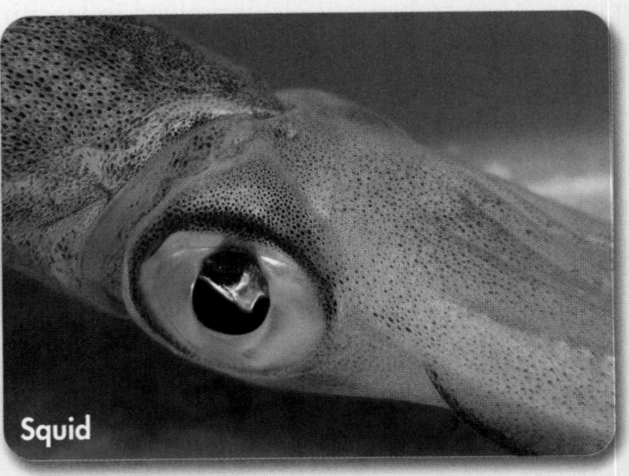

Complex Eye: Octopi and squid have eyes as complex as fishes and humans, though their structures differ.

Chordate Sense Organs Nonvertebrate chordates have few specialized sense organs. In tunicates, sensory cells in and on the siphons and other internal surfaces help control the amount of water passing through the pharynx. Lancelets have a cerebral ganglion with a pair of eyespots that detect light.

In contrast, most vertebrates have highly evolved sense organs. Many vertebrates have very sensitive organs of taste, smell, and hearing. Some sharks, for example, can sense 1 drop of blood in 100 liters of water! And although all mammalian ears have the same basic parts, they differ in their ability to detect sound, as you can see in **Figure 28–7.** In fact, bats and dolphins can even find objects in their environment using echoes of their own high-frequency sounds. A great many species of fishes, amphibians, reptiles, birds, and mammals have color vision that is as good as, or better than, that of humans.

Some species, including certain fishes and the duckbill platypus, can detect weak electric currents in water. Some animals, such as sharks, use this "electric sense" to navigate by detecting electric currents in seawater that are caused by Earth's magnetic field. Other "electric fishes" can create their own electric currents. These fishes use electric pulses to communicate with one another, in much the same way that other animals communicate using sound. Many species that can detect electric currents use the ability to track down prey in dark, murky water. Some birds can detect Earth's magnetic field directly, and they use that ability to navigate during long-distance migrations.

Animal	Hearing Range (Hz)
Tree frog	50–4000
Canary	250–8000
Dog	67–45,000
Bat	2000–110,000
Human	30–23,000
Elephant	16–12,000
Bottlenose dolphin	75–150,000

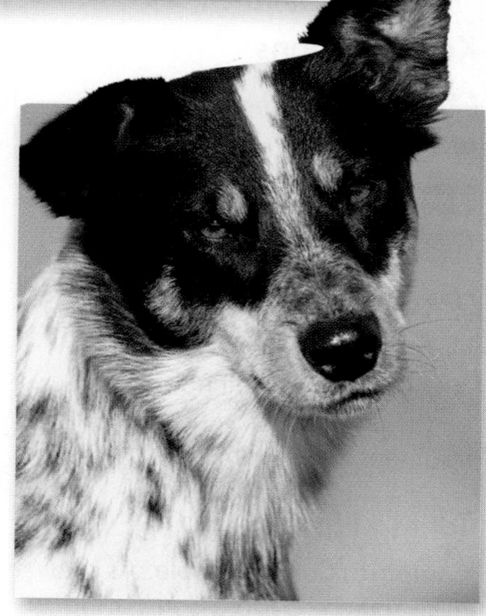

FIGURE 28–7 Vertebrate Hearing Human senses are not necessarily superior to those of other animals. Interpret Tables ***Would you expect to be able to hear the highest pitch a dog can hear? Explain.***

28.1 Assessment

Review Key Concepts

1. a. Review List three body systems that work together to create a response to a stimulus.

b. Explain What is the role of a motor neuron?

c. Sequence What is the correct sequence of the roles played by the following in the response to a stimulus: interneuron, motor neuron, sensory neuron, muscle.

2. a. Review What are two general ways in which nervous systems differ among animal groups?

b. Compare and Contrast Describe the degree of cephalization shown by cnidarians, flatworms, octopi, and vertebrates.

3. a. Review Give an example of an animal with a very simple sensory system and an example of one with a complex sensory system.

b. Infer What is the general relationship between the complexity of an animal's nervous system and that of its sensory system?

WRITE ABOUT SCIENCE

Explanation

4. The compound eyes of insects detect movement better than they distinguish details. How might the ability to detect movement be more important to an insect than the ability to see fine details? (*Hint:* Consider the size of an insect in relation to that of its predators.)

28.2 Movement and Support

Key Questions

What are the three types of skeletons?

How do muscles enable movement?

Vocabulary

hydrostatic skeleton • exoskeleton • molting • endoskeleton • joint • ligament • tendon

Taking Notes

Compare/Contrast Table As you read, create a table comparing and contrasting the three types of skeletons.

THINK ABOUT IT As a dragonfly hovers over a stream, a trout leaps out of the water to catch it. An earthworm wriggles through leaf litter nearby. A falcon streaks overhead, hunting a mouse scampering across a field. All these invertebrates and vertebrates face similar challenges as they move through air or water, or over land. In order to move, animals use different structures that work in similar ways.

Types of Skeletons

What are the three types of skeletons?

To move efficiently, all animals must do two things. First, they must generate physical force. Then, they must somehow apply that force against air, water, or land in order to push or pull themselves around.

Skeletal Support An animal's ability to move efficiently is greatly enhanced by rigid body parts. Legs push against the ground. Bird wings push against air, and fins or flippers apply force against water. Each of these body parts is supported by some sort of skeleton. **Animals have three main kinds of skeletal systems: hydrostatic skeletons, exoskeletons, and endoskeletons.**

▶ ***Hydrostatic Skeletons*** Some invertebrates, such as cnidarians and annelids, have hydrostatic skeletons. The **hydrostatic skeleton** of a cnidarian such as a hydra, for example, consists of fluids held in a gastrovascular cavity that can alter the animal's body shape drastically by working with contractile cells in its body wall. When a hydra closes its mouth and the cells encircling its body wall constrict, the animal elongates and its tentacles extend, as shown in the left photo of **Figure 28–8.** Because water is not compressible, constricting the cavity elongates the animal, somewhat like a water balloon that has been squeezed. A hydra often sits in this position for hours, waiting for prey to swim by. If it is disturbed, its mouth opens, allowing water to flow out, and longitudinal cells in its body wall contract, shortening the body, as in the right photo of **Figure 28–8.**

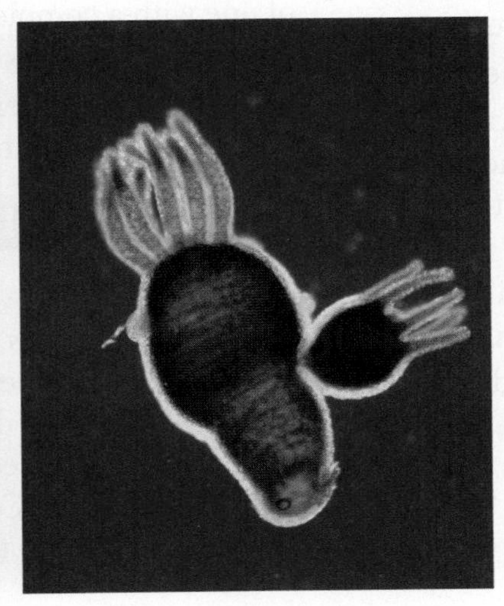

FIGURE 28–8 Hydrostatic Skeleton Some invertebrates, such as this hydra, have hydrostatic skeletons. When a hydra closes its mouth, water trapped in its body causes it to elongate (left). When it opens its mouth again, water is released, and it becomes shorter (right).

▶ ***Exoskeletons*** Many arthropods have exoskeletons, as do most mollusks, such as snails and clams. The **exoskeleton,** or external skeleton, of an arthropod is a hard body covering made of a complex carbohydrate called chitin. Most mollusks have exoskeletons, or shells, made of calcium carbonate.

Jointed exoskeletons enable various arthropods to swim, fly, burrow, walk, crawl, and leap. They can also provide watertight coverings that enable some arthropods to live in Earth's driest places. An exoskeleton can also provide physical protection from predators—as you know if you have ever tried to crack a crab or lobster shell or seen a mollusk withdraw into its shell. Mollusks with two-part shells are called bivalves. Bivalves such as clams can also close their shells to avoid drying out.

But exoskeletons have disadvantages. An external skeleton poses a problem when the animal it belongs to needs to grow. To increase in size, arthropods break out of their exoskeleton and grow a new one, in a process called **molting,** shown in **Figure 28–9.** Exoskeletons are also relatively heavy. The larger arthropods get, the heavier their skeletons become in proportion to their body weight. This is one reason that some science-fiction monsters could never exist. The legs of an elephant-size spider would collapse under the spider's weight!

FIGURE 28–9 Exoskeleton Arthropods such as this cicada periodically "grow out" of their exoskeletons and have to break out of them in order to grow new ones. **Infer** ***How might molting be a dangerous inconvenience?***

▶ ***Endoskeletons*** Echinoderms and vertebrates have endoskeletons. An **endoskeleton** is a structural support system within the body. Sea stars and other echinoderms have an endoskeleton made of calcified plates, as you can see in **Figure 28–10.** These skeletal plates support and protect echinoderms, and also give them a bumpy texture.

Vertebrates have an endoskeleton made of cartilage or a combination of cartilage and bone. Sharks and some other fishes have skeletons made almost entirely of cartilage. In other vertebrates, most of the skeleton is bone. Four-limbed vertebrates also have structures called limb girdles that support limbs and allow the animal to move around.

FIGURE 28–10 Endoskeleton Not every endoskeleton looks like yours! Some invertebrates, including echinoderms such as this sea star, have rigid internal body supports. These supports are not, however, made of bone.

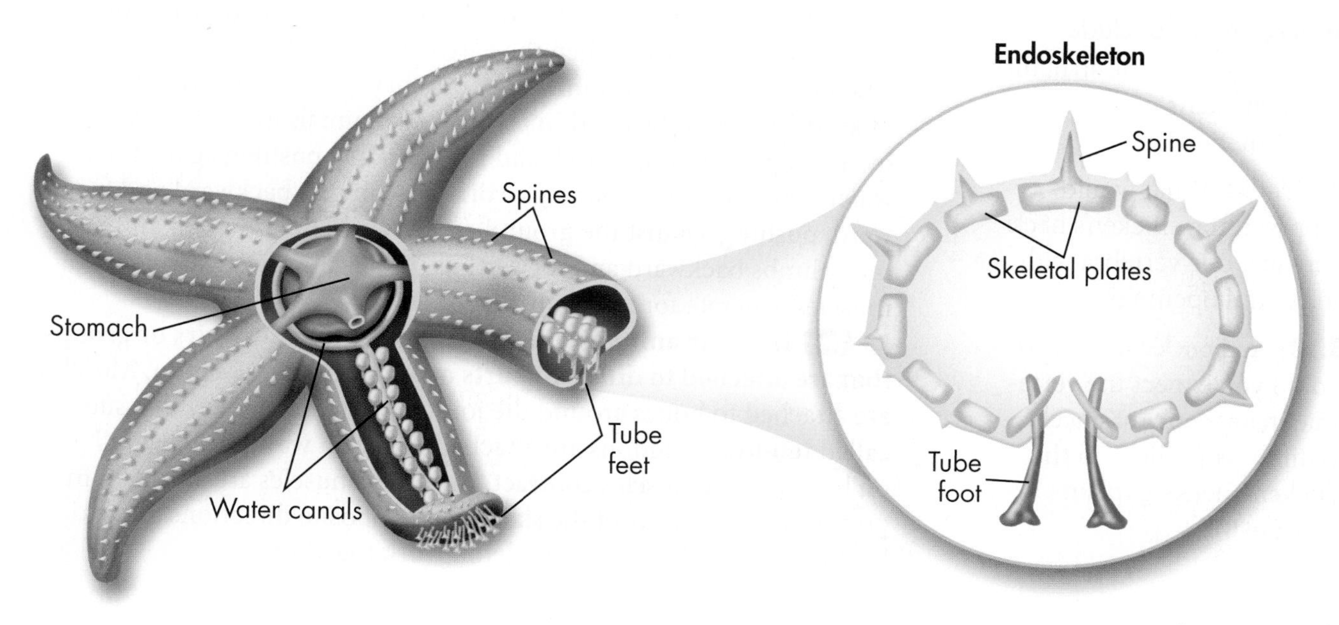

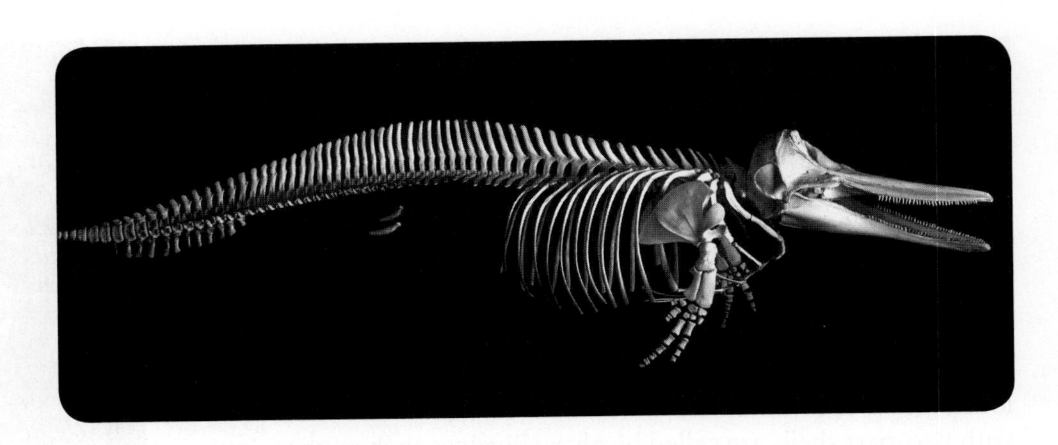

FIGURE 28–11 Vertebrate Skeleton A typical vertebrate skeleton, such as that of this dolphin, is made up mostly of bone.

Evolution has produced a wide range of variations of the vertebrate endoskeleton that enables these animals to swim, fly, burrow, walk, crawl, or leap, but they all provide strong, lightweight support. Of course, because an internal skeleton does not surround the body, it cannot protect an animal the way that an exoskeleton can. On the other hand, an internal skeleton can grow as an animal grows, so the animal does not need to molt. Because endoskeletons are lightweight in proportion to the bodies they support, even land-dwelling vertebrates can grow very large.

Joints If an animal's rigid skeleton were made of one piece, or if its parts were rigidly attached to each other, the animal couldn't move. Arthropods and vertebrates can move because their skeletons are divided into many parts that are connected by **joints.** Joints are places where parts of a skeleton are held together in ways that enable them to move with respect to one another. In vertebrates, bones are connected at joints by strong connective tissues called **ligaments.** Most joints are formed by a combination of ligaments, cartilage, and lubricating joint fluid that enables bones to move without painful friction.

What Are Some Adaptations of Vertebrae?

❶ Obtain a chicken neck from your teacher. Bend the neck back and forth and from side to side.

❷ Insert a dissecting probe into the opening at the top of the neck. What do you observe? **CAUTION:** *Use care with sharp instruments.*

Analyze and Conclude

1. **Infer** How is the structure of the chicken's neck related to its function?
2. **Predict** What would happen if the chicken's neck were just one vertebra with no central opening?
3. **Draw Conclusions** How would you expect the vertebrae in an elephant's neck to differ from those in the chicken's neck? Explain your answer.

Muscles and Movement

How do muscles enable movement?

Muscles are specialized tissues that produce physical force by contracting, or getting shorter, when they are stimulated. Muscles can relax when they aren't being stimulated, but they cannot actively get longer. That presents a problem. Think about how animals move. Fishes swim by moving their bodies back and forth, and by pushing against the water with their fins. Your legs work by swinging backward and forward, pushing against the ground as you walk. But how can animals move limbs backward and forward or push against water or land if muscles generate force in only one direction?

In many animals, muscles work together in pairs or groups that are attached to different parts of a supporting skeleton. Muscles are attached to bones around the joints by tough connective tissue called **tendons.** Tendons are attached in such a way that they pull on bones when muscles contract. Typically, muscles are arranged in groups that pull parts of the skeleton in opposite directions. Here's how muscles and parts of a skeleton work together.

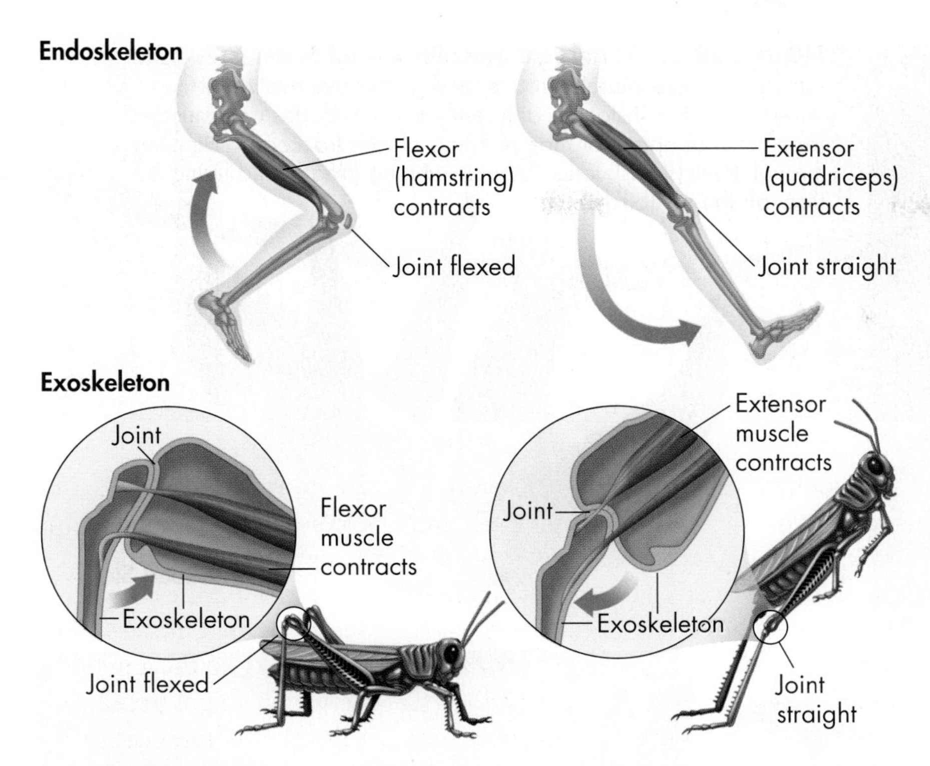

FIGURE 28–12 Muscles and Joints The diagrams show how muscles work with a vertebrate endoskeleton and an arthropod exoskeleton to bend and straighten joints. **Compare and Contrast** ***To what structure are arthropod muscles attached? To what structure are vertebrate muscles attached?***

Movement Arthropod muscles are attached to the inside of the exoskeleton. Vertebrate muscles are attached around the outside of bones. In both cases, different pairs or groups of muscles pull across joints in different directions. As you can see in **Figure 28–12,** when one muscle group contracts, it bends, or flexes, the joint. When the first group relaxes and the second group contracts, the joint straightens.

Vertebrate Muscular and Skeletal Systems An amazing variety of complex combinations of bones, muscle groups, and joints have evolved in vertebrates. In many fishes and snakes, muscles are arranged in blocks on opposite sides of the backbone. These muscle blocks contract in waves that travel down the body, bending it first to one side and then to the other. As these waves of movement travel down the body, they generate thrust. The limbs of many modern amphibians and reptiles stick out sideways from the body, as though the animals were doing push-ups. If you watch these animals move, you will see that many use sideways movements of their backbone to move their limbs forward and backward.

Most mammals stand with their legs straight under them, whether they walk on two legs or four. Mammalian limbs have evolved in ways that enable many different kinds of movement, as you can see in **Figure 28–13** on the next page. The shapes and relative positions of bones and muscles, and the shapes of joints, are linked very closely to the functions they perform. Limbs that are specialized for high-speed running or long-distance jumping have very differently shaped bones, muscles, and joints than limbs adapted for flying, swimming, or manipulating objects. In fact, paleontologists can reconstruct the habits of extinct animals by studying the joints of fossil bones and the places where tendons and ligaments once attached.

FIGURE 28–13 Vertebrate Musculoskeletal Systems A great variety of bones, muscle groups, and joints have evolved in vertebrates. For instance, differently shaped bones and muscles form limbs adapted for manipulating objects (raccoons), climbing through trees (sloths), long-distance jumping (frogs), and flying through the air (birds).

◀ Raccoon

◀ Three-toed Sloth

▲ Tree Frog

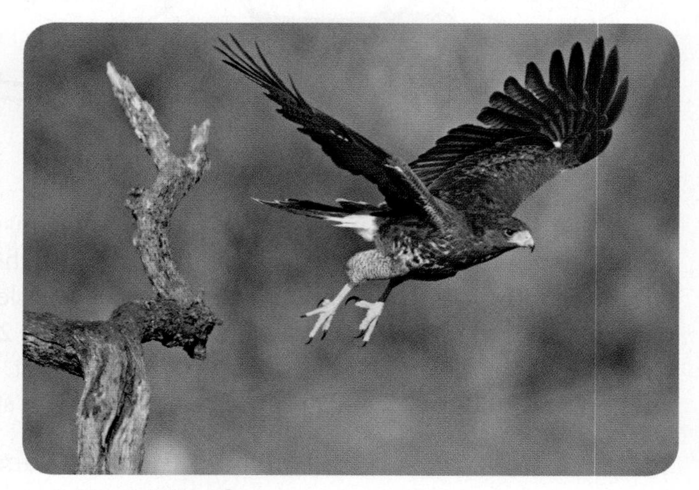

▲ Harris's Hawk

28.2 Assessment

Review Key Concepts

1. a. Review What body structures generate force? With what other body structure do these structures work to enable movement?

   b. Infer Why are the largest land animals vertebrates?

2. a. Review What characteristics are common to the skeletons of all vertebrates?

   b. Form a Hypothesis Suppose that you were to find a vertebrate fossil that showed a joint structure with muscle and tendon relationships similar to that of a squirrel. For which kinds of movement would you predict the animal had been best adapted?

Apply the Big idea

Structure and Function

3. Create a model of a vertebrate or invertebrate joint. Make sure the muscles are attached to the same skeletal structures they would be attached to in a real animal and that the muscles and skeletal structure allow the joint to bend and flex.

28.3 Reproduction

THINK ABOUT IT Sexual reproduction can be dangerous. Just ask a male praying mantis—who may be devoured by his mate. Or a male peacock, whose success in courting a female depends on his growing and lugging around a huge tail that makes it harder for him to escape predators. Or a male emperor penguin, who incubates an egg for months on antarctic ice in temperatures far below zero. Or a female deer, who carries around the ever-increasing weight of her developing young for seven months, while she runs from predators such as coyotes and seeks food for herself and the young she carries. Yet, most animal species engage in sexual reproduction during at least part of their life cycles. Why?

Key Questions

How do asexual and sexual reproduction in animals compare?

How do internal and external fertilization differ?

Where do embryos develop?

How are terrestrial vertebrates adapted to reproduction on land?

Vocabulary

oviparous • ovoviviparous • viviparous • placenta • metamorphosis • nymph • pupa • amniotic egg • mammary gland

Taking Notes

Outline Before you read, use the headings and key concepts in this lesson to make an outline about animal reproduction. As you read, add details to your outline.

Asexual and Sexual Reproduction

How do asexual and sexual reproduction in animals compare?

Many invertebrates and a few chordates can reproduce asexually.

Asexual Reproduction Animals reproduce asexually in many ways. Some cnidarians divide in two. Some animals reproduce through budding, which produces new individuals as outgrowths of the body wall. Females of some species, such as the whiptail lizard in **Figure 28–14,** can reproduce asexually by producing eggs that develop without being fertilized. This process is called parthenogenesis (pahr thuh noh JEN uh sis). Parthenogenesis produces offspring that carry DNA inherited only from their mothers. This means of reproduction occurs in some crustaceans and insects but very rarely in vertebrates.

Asexual reproduction requires only one parent, so individuals in favorable environmental conditions can reproduce rapidly. But since offspring produced asexually carry only a single parent's DNA, they have less genetic diversity than do offspring produced sexually. Lack of genetic diversity can be a disadvantage to a population if its environment changes.

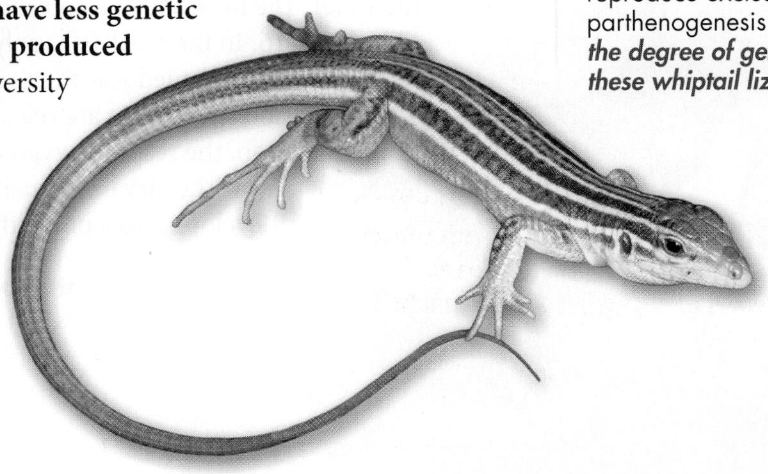

FIGURE 28–14 Parthenogenesis Some whiptail lizard species reproduce exclusively by parthenogenesis. **Infer** ***Describe the degree of genetic diversity in these whiptail lizard species.***

MYSTERY CLUE

When investigators analyzed the baby shark's DNA, they found that it was homozygous for all the traits they examined, including two rare traits. Why was that unusual?

Sexual Reproduction Recall from Chapter 11 that sexual reproduction involves meiosis, the process that produces haploid reproductive cells, or gametes. Gametes carry half the number of chromosomes found in body cells. Typically, male animals produce small gametes, called sperm, which swim. Females produce larger gametes called eggs, which do not swim. When haploid gametes join during fertilization, they produce a zygote that contains the diploid number of chromosomes.

Sexual reproduction maintains genetic diversity in a population by creating individuals with new combinations of genes. Because genetic diversity is the raw material on which natural selection operates, sexually reproducing populations are better able to evolve and adapt to changing environmental conditions. On the other hand, sexual reproduction requires two individuals of different sexes. So, the density of a population must be high enough to allow mates to find each other.

In most animal species that reproduce sexually, each individual is either male or female. Among annelids, mollusks, and fishes, however, some species are hermaphrodites (hur MAF roh dyts), which means that some individuals can be both male and female or can convert from one sex to the other. In some species, individuals can produce eggs and sperm at the same time. Usually, these animals don't fertilize their own eggs, but exchange sperm with another individual. Some species, such as the clownfish in **Figure 28–15,** may change from one sex to the other as they mature.

FIGURE 28–15 Hermaphrodites In this species of clownfish, *Amphiprion percula,* all individuals are born male and change to female as they grow. In some other hermaphroditic species, individuals are born female and change to male as they grow, or are both sexes at the same time.

Reproductive Cycles A number of invertebrates have life cycles that alternate between sexual and asexual reproduction. Parasitic worms and cnidarians alternate between forms that reproduce sexually and forms that reproduce asexually.

Parasitic worms such as blood flukes mature in the body of an infected person, reproduce sexually, and release embryos that pass out of the body in feces. If the embryos reach fresh water, they develop into larvae and infect snails, in which they reproduce asexually. Then the larvae are released, ready to infect another person.

Many cnidarians alternate between two body forms: polyps that grow singly or in colonies and medusas that swim freely in the water. The life cycle of a common jellyfish, *Aurelia,* is shown in **Figure 28–16.** In these jellyfish, polyps produce medusas asexually by budding. The medusas then reproduce sexually by producing eggs and sperm that are released into the water. After fertilization, the resulting zygote grows into a free-swimming larva. The larva eventually attaches to a hard surface and develops into a polyp that may continue the cycle.

In Your Notebook *Explain why a genetically diverse species can adapt more easily to disease and change.*

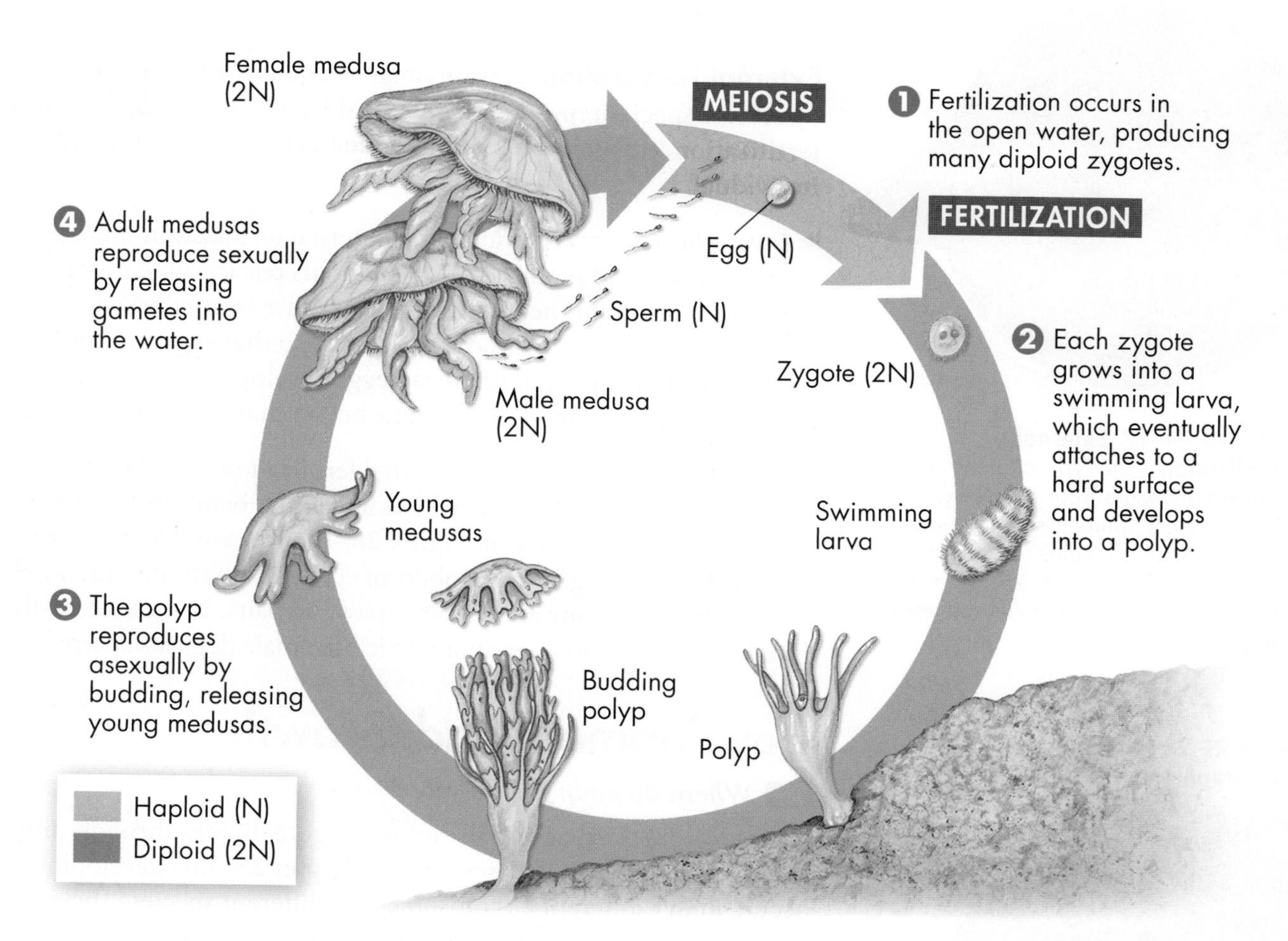

FIGURE 28–16 Alternating Reproductive Cycles The reproductive cycle of *Aurelia,* a jellyfish, alternates between asexual and sexual reproduction. A zygote is produced sexually by medusas and grows into a larva. The larva develops into a polyp that buds, reproducing asexually. The polyp releases a medusa.

Internal and External Fertilization

How do internal and external fertilization differ?

In sexual reproduction, eggs and sperm meet either inside or outside the body of the egg-producing individual. These alternatives are called internal and external fertilization, respectively.

Internal Fertilization Many aquatic animals and nearly all terrestrial animals reproduce by internal fertilization. **During internal fertilization, eggs are fertilized inside the body of the egg-producing individual.**

▶ ***Invertebrates*** Invertebrates that reproduce by internal fertilization range in complexity from sponges to arachnids. The eggs of sponges and some other aquatic animals are fertilized by sperm released by others of their species and taken in from the surrounding water. In many arthropod species, males deposit sperm inside the female's body during mating.

▶ ***Chordates*** Some fishes and amphibians, and all reptiles, birds, and mammals, reproduce by internal fertilization. In some amphibian species, males deposit "sperm packets" into the surrounding environment; females then pick up these packets and take them inside their bodies. In many other chordate species, males have an external sexual organ that deposits sperm inside the female during mating.

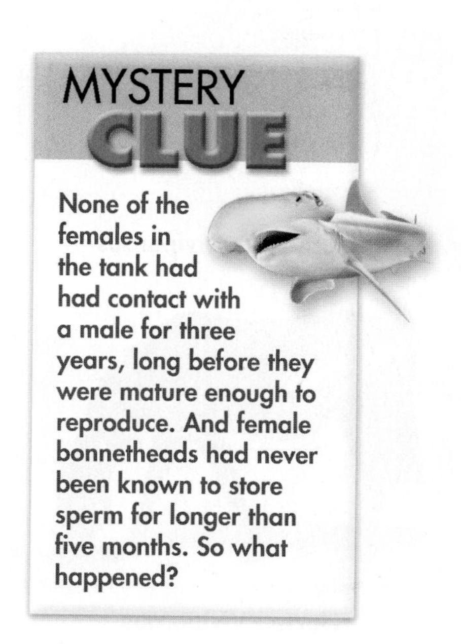

MYSTERY CLUE

None of the females in the tank had had contact with a male for three years, long before they were mature enough to reproduce. And female bonnetheads had never been known to store sperm for longer than five months. So what happened?

FIGURE 28–17 External Fertilization One type of external fertilization results from spawning. When aquatic animals spawn, females release eggs and males release sperm at the same time. **Infer** ***What is the cloudy substance behind this spawning male wrasse?***

External Fertilization A wide range of aquatic invertebrate and vertebrate species reproduce by external fertilization. **In external fertilization, eggs are fertilized outside the body of the egg-producing individual.**

▶ ***Invertebrates*** Invertebrates with external fertilization include corals, worms, and mollusks. These animals release large numbers of eggs and sperm into the water. Gamete release is usually synchronized with tides, phases of the moon, or seasons so that eggs and sperm are present at the same time. Fertilized eggs develop into free-swimming larvae that typically develop for a time before changing into adult form.

▶ ***Chordates*** Chordates with external fertilization include most non-vertebrate chordates and many fishes and amphibians. In some fish species, such as the wrasse in **Figure 28–17,** males and females spawn in a school, releasing large numbers of eggs and sperm into the water. Other fishes and many amphibians spawn in pairs. In these cases, the female usually releases eggs onto which the male deposits sperm.

FIGURE 28–18 Embryo Development

Robin - Oviparous

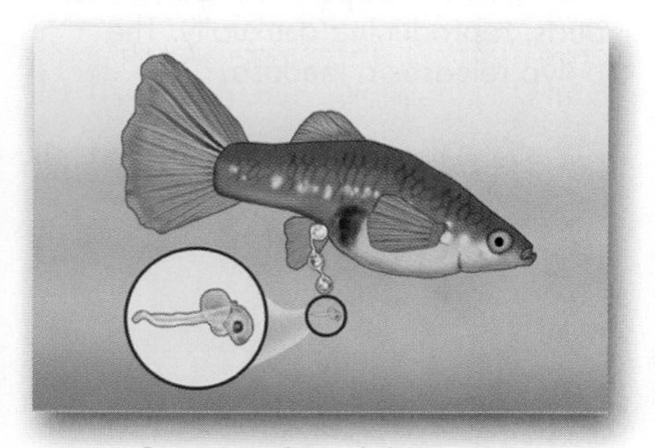

Guppy - Ovoviviparous

Horse - Viviparous

Development and Growth

Where do embryos develop?

After eggs are fertilized, the resulting zygote divides through mitosis and differentiates as described in Chapter 25. This development occurs under different circumstances in different species. The care and protection given to developing embryos also varies widely.

Where Embryos Develop Embryos develop either inside or outside the body of a parent in various ways. **Animals may be oviparous, ovoviviparous, or viviparous.**

▶ ***Oviparous Species*** **Oviparous** (oh VIP uh rus) species are those in which embryos develop in eggs outside the parents' bodies. Most invertebrates, many fishes and amphibians, most reptiles, all birds, and a few odd mammals are oviparous.

▶ ***Ovoviviparous Species*** In **ovoviviparous** (oh voh vy VIP uh rus) species, embryos develop within the mother's body, but they depend entirely on the yolk sac of their eggs. The young do not receive any additional nutrients from the mother. They either hatch within the mother's body or are released immediately before hatching. Young swim freely shortly after hatching. Guppies and other fishes in their family, along with some shark species, are ovoviviparous.

▶ ***Viviparous Species*** **Viviparous** (vy VIP uh rus) species are those in which embryos obtain nutrients from the mother's body during development. Viviparity occurs in most mammals and in some insects, sharks, bony fishes, amphibians, and reptiles. In viviparous insects, and in some sharks and amphibians, young are nourished by secretions produced in the mother's reproductive tract. In placental mammals, young are nourished by a **placenta**—a specialized organ that enables exchange of respiratory gases, nutrients, and wastes between the mother and her developing young.

How Young Develop Most newborn mammals and newly hatched birds and reptiles look a lot like miniature adults. Infant body proportions are different from those of adults, and newborns have more or less hair, fur, or feathers than adults have. But it is pretty clear that a newly hatched snake is not going to grow up to be something totally different, such as an eagle!

For many other groups of animals, however, it's not as clear. As most invertebrates, nonvertebrate chordates, fishes, and amphibians develop, they undergo metamorphosis. **Metamorphosis** is a developmental process that leads to dramatic changes in shape and form.

► ***Aquatic Invertebrates*** Many aquatic invertebrates have a larval stage, which looks nothing like an adult. These larvae often swim or drift in open water before undergoing metamorphosis and assuming their adult form. Members of some phyla, such as cnidarians, have a single larval stage. Other groups, such as crustaceans, may pass through several larval stages before they look like miniature adults.

► ***Terrestrial Invertebrates*** Insects may undergo one of two types of metamorphosis. Some insects, such as grasshoppers, undergo gradual or incomplete metamorphosis, as shown in **Figure 28–19.** Immature forms, or **nymphs** (nimfs), resemble adults, but they lack functional sexual organs and some adult structures such as wings. As they molt several times and grow, nymphs gradually acquire adult structures.

Other insects, such as butterflies, undergo complete metamorphosis. Larvae of these animals look nothing like their parents, and they feed in different ways. Larvae molt and grow, but they change little in appearance. Then they undergo a final molt and change into a **pupa** (PYOO puh; plural: pupae), the stage in which an insect larva develops into an adult. During the pupal stage, the entire body is remodeled inside and out! The adult that emerges looks like a completely different animal. Don't let your familiarity with caterpillars and butterflies dull your wonder at this change. If land vertebrates underwent this kind of metamorphosis, a larva that looks like a snake could, in fact, grow up into an eagle!

FIGURE 28–19 Insect Metamorphosis Insects usually undergo metamorphosis during their growth and development. The chinch bug (left) undergoes incomplete metamorphosis, in which the nymphs look similar to the adults. The ladybug (right) undergoes complete metamorphosis. The developing larva and the pupa look completely different from the adult.

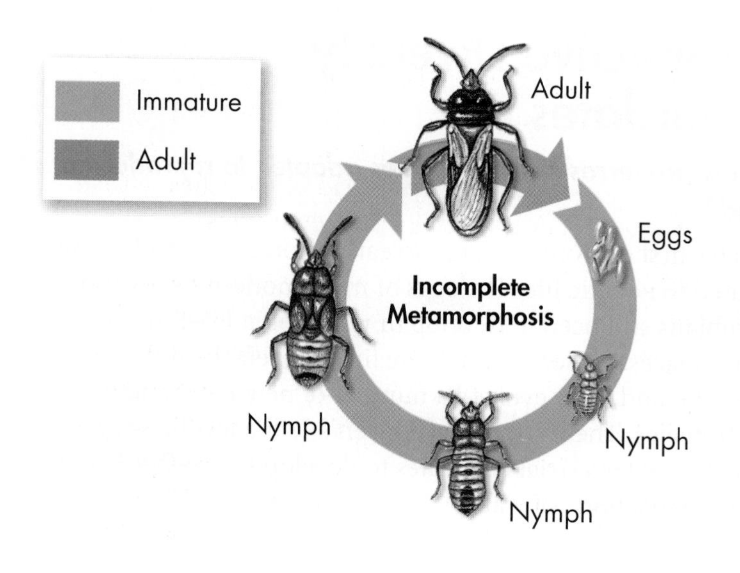

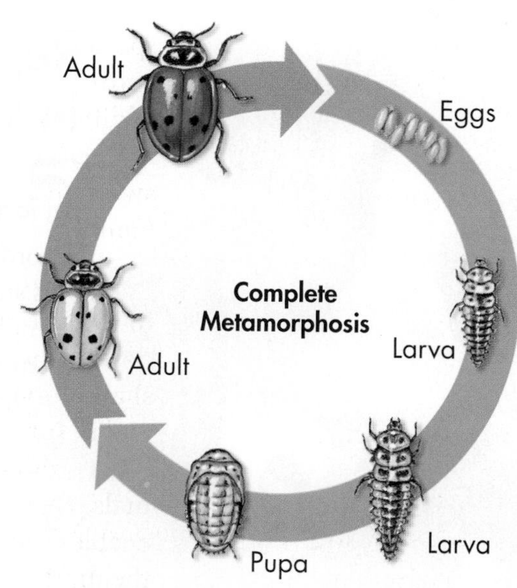

FIGURE 28–20 Amphibian Metamorphosis Amphibians typically begin their lives in the water and metamorphose into adults that live on land. Frog tadpoles, such as the one in the photo, start out with flippers, gills, and a tail and mature into adults that have legs, lungs, and no tail.

Control of metamorphosis in arthropods is accomplished by hormones. Recall that hormones are chemicals produced in one organ of an organism that affect that organism's other tissues and organs. In insects that undergo complete metamorphosis, high levels of a juvenile hormone keep an insect in its larval form. As the insect matures, its production of juvenile hormone decreases. Eventually, the concentration of juvenile hormone drops below a certain threshold. The next time the insect molts, it becomes a pupa. When no juvenile hormone is produced, the insect undergoes a pupa-to-adult molt.

▶ ***Amphibians*** Amphibians typically undergo metamorphosis that is controlled by hormones. This metamorphosis changes amphibians from aquatic young into terrestrial adults. Tadpoles, such as the one in **Figure 28–20,** are one type of amphibian larvae.

In Your Notebook *What chemicals control metamorphosis in arthropods and amphibians?*

Care of Offspring Animals' care of their offspring varies from no care at all to years of nurturing. Most aquatic invertebrates and many fishes and amphibians release large numbers of eggs that they completely ignore. This reproductive strategy succeeds in circumstances favoring populations that disperse and grow rapidly.

But other animals care for their offspring. Some amphibians incubate young in their mouth, on their back, or even in their stomach! Birds and mammals generally care for their young. Maternal care is an important mammalian characteristic, and the bond between mother and young is often very close, as the pandas in **Figure 28–21** demonstrate. Males of many species also help care for young. Parental care helps young survive in crowded, competitive environments. Typically, species that provide intensive or long-term parental care give birth to fewer young than do species that offer no parental care.

FIGURE 28–21 Care of Offspring Long-term, intensive care of offspring is a characteristic of mammals, such as the mother panda in the photo. A wild panda cub will stay with its mother for up to 18 months while she protects it and teaches it how to be a panda.

Reproductive Diversity in Chordates

How are terrestrial vertebrates adapted to reproduction on land?

Chordates first evolved in water, so early chordate reproduction was suited to aquatic life. The eggs of most modern fishes and amphibians still need to develop in water, or at least in very moist places. As some vertebrate lineages left the water to live on land, they evolved a number of new reproductive strategies. These strategies now enable the fertilized eggs of many terrestrial chordates to develop somewhere other than in a body of water.

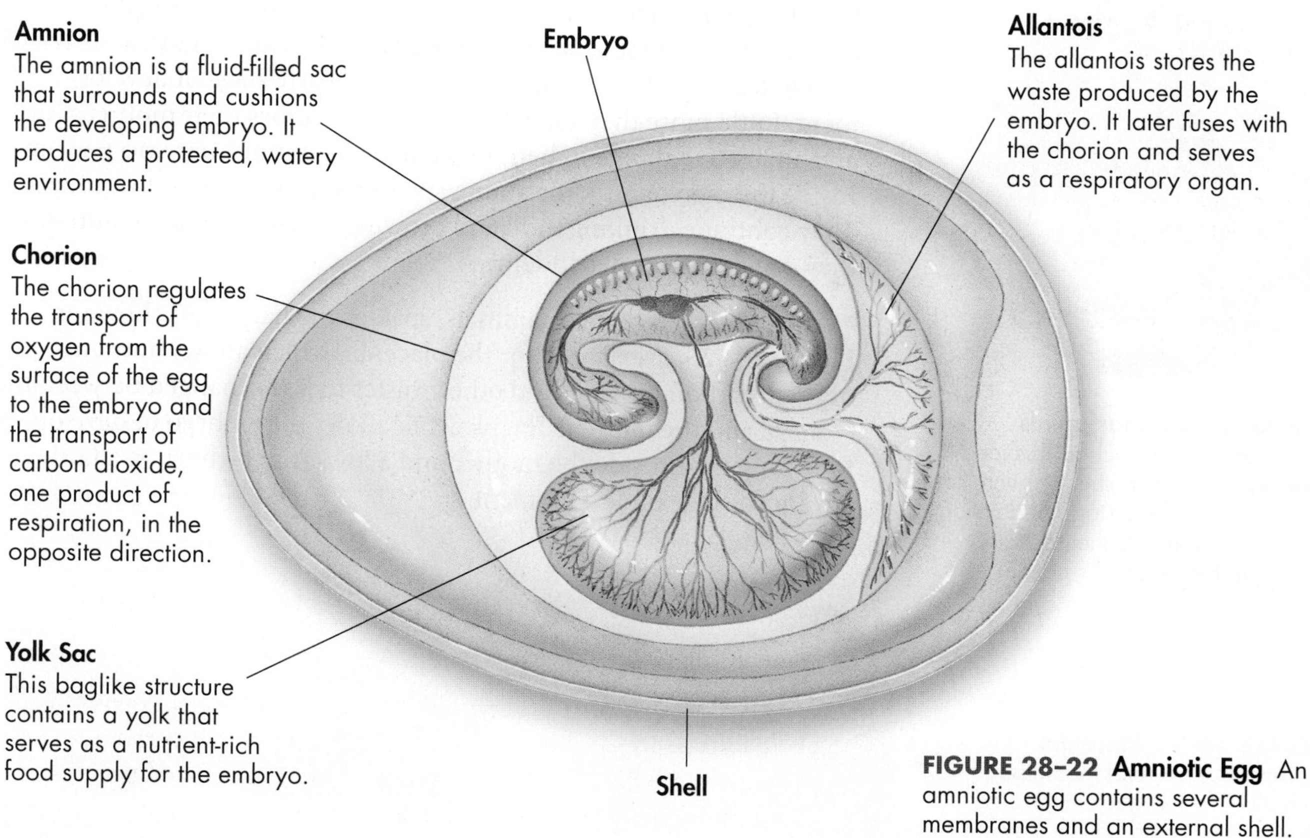

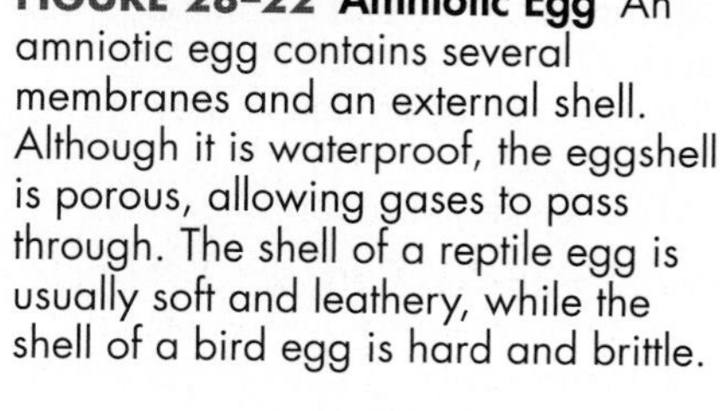

FIGURE 28–22 Amniotic Egg An amniotic egg contains several membranes and an external shell. Although it is waterproof, the eggshell is porous, allowing gases to pass through. The shell of a reptile egg is usually soft and leathery, while the shell of a bird egg is hard and brittle.

The Amniotic Egg **Reptiles, birds, and a few mammals have evolved amniotic eggs in which an embryo can develop outside its mother's body, and out of water, without drying out.** The **amniotic** (am nee AH tik) **egg** is named after the amnion, one of four membranes that surround the developing embryo. The amnion, yolk sac, chorion, and allantois membranes of the amniotic egg, along with its shell, provide a protected environment in which an embryo can develop out of water. You can learn about the functions of the membranes in **Figure 28–22.** The amniotic egg is one of the most important vertebrate adaptations to life on land.

Mammalian Reproductive Strategies Mammals have evolved various adaptations for reproducing and caring for their young. **The three groups of mammals—monotremes, marsupials, and placentals—differ greatly in their means of reproduction and development, but all nourish their young with mother's milk.**

▶ ***Monotremes*** Reproduction in monotremes, such as the echidna in **Figure 28–23,** combines reptilian and mammalian traits. Like a reptile, a female monotreme lays soft-shelled, amniotic eggs that are incubated outside her body. The eggs hatch in about ten days. But like other mammals, young monotremes are nourished by milk produced by the mother's **mammary glands.** Female monotremes secrete milk, not through well-developed nipples like other mammals, but through pores on the surface of the abdomen.

FIGURE 28–23 Monotremes There are only five species of monotremes, four of which are spiny echidnas similar to the one above. Monotremes lay eggs but, like all mammals, feed their young with milk produced by the mother.

FIGURE 28–24 **Marsupials** Marsupial young, such as this joey peeking out of its mother's pouch, are born at a very early stage of development. They complete their development nursing in their mother's pouch.

▶ ***Marsupials*** Marsupials, such as the kangaroos in **Figure 28–24,** bear live young that usually complete their development in an external pouch. Marsupial young are born at a very early stage of development. Little more than embryos, they crawl across their mother's fur and attach to a nipple in her pouch, or marsupium (mar SOO pee um). Inside the marsupium, the young spend months attached to a nipple. They continue drinking milk and growing inside the marsupium until they can survive independently.

▶ ***Placentals*** Placental mammals, such as the harp seals shown in **Figure 28–25,** are named for the placenta, which allows nutrients, oxygen, carbon dioxide, and other wastes to be exchanged between the embryo and the mother. The placenta allows the embryo to develop for a long time inside the mother and allows it to be born at a fairly advanced stage of development.

FIGURE 28–25 **Placental Mammals** Placental mammals, such as harp seals, are nourished through a placenta before they are born and by their mother's milk after they are born.

28.3 Assessment

Review Key Concepts

1. a. **Review** Compare asexual reproduction and sexual reproduction in terms of the genetic diversity resulting from each.
   b. **Infer** Why might sexual reproduction, as opposed to asexual reproduction, produce a population better able to survive disease or environmental changes?
2. a. **Review** Define the two types of fertilization.
   b. **Predict** Why would you expect species that employ external fertilization to reproduce in the water?
3. a. **Review** Define the three ways in which embryos develop.
   b. **Compare and Contrast** What is the difference between a nymph and a pupa?
4. a. **Review** What structure enables reptiles and birds to reproduce outside of water?
   b. **Interpret Visuals** In your own words, describe the functions of two of the membranes shown in **Figure 28–22.**

WRITE ABOUT SCIENCE

Creative Writing

5. Write an advertisement for an amniotic egg. Draw and label the parts of the egg, including each of the membranes and the shell. Describe the purpose of each structure and how it's ideally suited for its function.

28.4 Homeostasis

THINK ABOUT IT A herd of wildebeests plods across Africa's Serengeti Plain. The land is parched, so they are on the move toward greener pastures. They move mechanically, their steps using as little energy as possible. With no food in their guts, their bodies mobilize energy stored in fat deposits for distribution to body tissues. Between drinking holes, their bodies conserve water by producing as little urine as possible. All their body systems work together in a joint effort to survive this difficult passage.

Key Questions

Why is the interdependence of body systems essential?

How do animals control their body temperature?

Vocabulary

endocrine gland • ectotherm • endotherm

Taking Notes

Venn Diagram Draw a Venn diagram comparing and contrasting the temperature control strategies of ectotherms and endotherms.

Interrelationship of Body Systems

Why is the interdependence of body systems essential?

Homeostasis, or controlled internal conditions, is essential to an organism's survival. Wildebeest brain cells, like those of humans, must be kept at a stable temperature and supplied with a steady stream of glucose for energy—even when the animal is under stress. The brain cells must be bathed in fluid with a constant concentration of water and be cleansed of metabolic waste products. These conditions must not dramatically change during droughts, floods, famines, heat, or cold. Failure of homeostasis, even for a few minutes, would lead to permanent brain injury or death.

You've learned about digestive, respiratory, circulatory, excretory, nervous, muscular, and skeletal systems separately. Yet all of these systems are interconnected. **All body systems work together to maintain homeostasis.** In most animals, respiratory and digestive systems would be useless without circulatory systems to distribute oxygen and nutrients. Similarly, the excretory system needs a circulatory system to collect carbon dioxide and nitrogenous wastes from body tissues and deliver them to the lungs and excretory organs. Muscles wouldn't work without a nervous system to direct them and a skeletal system to support them.

In addition to the organ systems that you have already learned about, you will now learn about other body systems, those that fight disease, produce and release chemical controls, and manage body temperature—all to help ensure homeostasis.

FIGURE 28–26 Interrelationship of Body Systems All body systems must work together to keep stressed animals, such as these migrating wildebeests, alive.

Fighting Disease The controlled environment within an animal's body is a comfortable place for hostile invaders as well as for its own cells. Most environments contain disease-causing microorganisms, or pathogens, that may take advantage of steady supplies of oxygen and nutrients intended for body tissues. If pathogens enter the body and grow, they may disrupt homeostasis in ways that cause disease.

Most animals have an immune system that can distinguish between "self" and "other." Once the immune system discovers "others" in the body, it attacks the invaders and works to restore homeostasis. Your body experiences this process regularly, any time you catch a cold or fight off other kinds of infections. During the process, you may develop a fever and feel other effects of the battle going on within your body.

BUILD Vocabulary

WORD ORIGINS Not all hormones are produced by endocrine glands. Erythropoietin is released by the kidneys. It prompts the body to make more red blood cells, which carry oxygen through the body. And when you look at the Greek words *erythropoietin* is made from—*erythros,* meaning "red," and *poiesis,* meaning "a making"—the hormone's function isn't much of a surprise.

Chemical Controls Vertebrates, such as the migrating wildebeest, along with arthropods and many other invertebrates, regulate many body processes using a system of chemical controls. **Endocrine glands** are part of that system. Endocrine glands regulate body activities by releasing hormones into the blood. Hormones are carried by blood or body fluids to organs. Some hormones, as you have learned, control growth, development, and metamorphosis in insects.

Mammals, like other vertebrates, have endocrine glands that are part of an endocrine system. Some hormones control the way the body stores energy or mobilizes it—as in the case of the wildebeests. Other hormones regulate the amount of water in the body and the amount of calcium in bones.

Analyzing Data

Comparing Ectotherms and Endotherms

The graph shows the internal body temperatures maintained by several ectotherms and endotherms at different environmental temperatures.

1. Interpret Graphs Which animal has the highest body temperature when the environmental temperature is between 0°C and 10°C? Which has the lowest body temperature under those conditions?

2. Infer Which animals represented in the graph are ectotherms? Which are endotherms? Explain your answers.

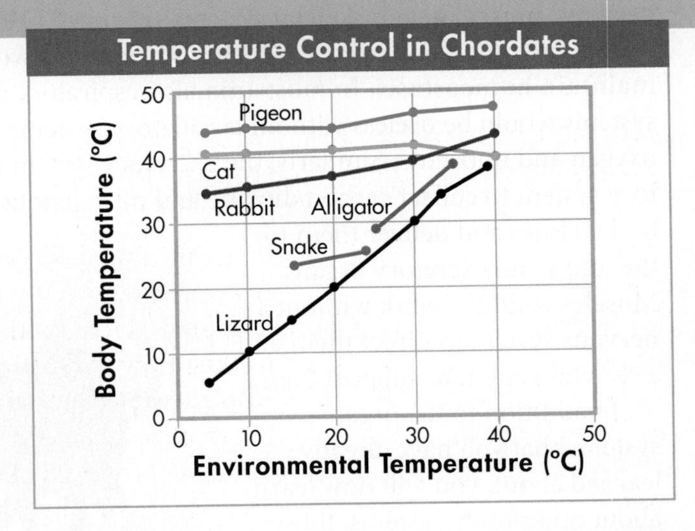

3. Predict If these animals lived in your area, would you expect all of them to be equally active year-round? If not, why not?

Body Temperature Control

How do animals control their body temperature?

Control of body temperature is important for maintaining homeostasis, particularly in areas where temperature varies widely with time of day and with season. Why is temperature control so important? Because many body functions are influenced by temperature. For example, muscles cannot operate if they are too cold or too hot. Cold muscles contract slowly, making an animal slow to react. If muscles get too hot, on the other hand, they may tire easily.

Body temperature control requires three components: a source of heat, a way to conserve heat when necessary, and a method of eliminating excess heat when necessary. An animal may be described as an ectotherm or endotherm based on the structures and behaviors that enable it to control its body temperature.

FIGURE 28–27 Ectotherm This shovel-snouted lizard, an ectotherm, lives in the Namib Desert in Africa, one of the hottest places on Earth. It is regulating its body temperature by stilting—raising its body off the hot sand by performing a sort of push-up. **Infer** ***Do you think stilting is more likely to raise or lower body temperature? Explain.***

Ectotherms On cool, sunny mornings, lizards bask in the sun. This doesn't mean that they are lazy! A lizard is an **ectotherm**—an animal whose regulation of body temperature depends mostly on its relationship to sources of heat outside its body. **Most reptiles, invertebrates, fishes, and amphibians are ectotherms that regulate body temperature primarily by absorbing heat from, or losing heat to, their environment.**

Ectotherms have relatively low metabolic rates when resting, so their bodies don't generate much heat. When active, their muscles generate heat, just as your muscles do. However, most ectotherms lack effective body insulation, so their body heat is easily lost to the environment. That's why ectotherms warm up by basking in the sun. They also have to regulate their body temperature in hot conditions. The lizard in **Figure 28–27** is "stilting" to cool off. Ectotherms also often use underground burrows, where there are fewer temperature extremes. On hot, sunny days, they might seek shelter in a burrow that is cooler than the land surface. On chilly nights, those same burrows are warmer than the surface, enabling the animal to conserve some body heat.

In Your Notebook *Explain in your own words why the word* coldblooded *is an incorrect way to describe an ectotherm.*

Endotherms An **endotherm** is an animal whose body temperature is regulated, at least in part, using heat generated by its body. **Endotherms, such as birds and mammals, have high metabolic rates that generate heat, even when they are resting.** Birds conserve body heat primarily with insulating feathers, such as fluffy down. Mammals use combinations of body fat and hair for insulation. Some birds and most mammals can get rid of excess heat by panting, as the dingo in **Figure 28–28** is doing. Humans sweat to help reduce their body temperature. As sweat evaporates, it removes heat from the skin and the blood in capillaries just under the surface of the skin. Thus, as warm blood flows through the cooled capillaries, it loses heat.

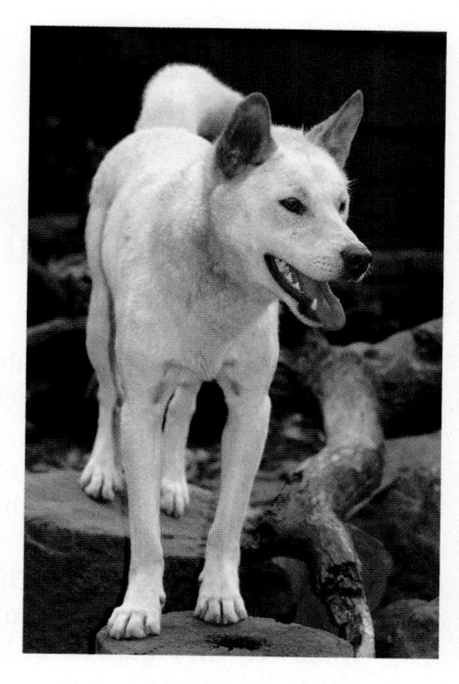

FIGURE 28–28 Endotherm Many endotherms, such as this dingo, pant when they are very warm. Panting allows air to evaporate some of the moisture in the blood-vessel rich mouth and respiratory tract, cooling the blood.

FIGURE 28–29 Endotherm Insulation Like some of their dinosaur ancestors, modern birds such as this northern cardinal use feathers to stay warm. When a bird gets cold, its dense, fluffy undercoat of down feathers stands up and creates spaces next to the bird's skin in which body heat is trapped.

Comparing Ectotherms and Endotherms Ectothermy and endothermy each have advantages and disadvantages in different situations. Endotherms move around easily during cool nights or in cold weather because they generate and conserve body heat. That's how musk oxen live in the tundra and killer whales swim through polar seas. But the high metabolic rate that generates this heat requires a lot of fuel. The amount of food needed to keep a single cow alive would be enough to feed ten cow-sized lizards!

Ectothermic animals need much less food than similarly sized endotherms. In environments where temperatures stay warm and fairly constant, ectothermy is a more energy-efficient strategy. But large ectotherms run into trouble if it gets very cold at night or stays cold for long periods. It takes a long time for a large animal to warm up in the sun after a cold night. That's one reason why most large lizards and amphibians live in tropical or subtropical areas.

Evolution of Temperature Control There is little doubt that the first land vertebrates were ectotherms. But questions remain about when and how often endothermy evolved. Although modern reptiles are ectotherms, a great deal of evidence suggests that at least some dinosaurs were endotherms. Many feathered dinosaur fossils have been discovered recently, suggesting that these animals, like modern birds such as that in **Figure 28–29,** used feathers for insulation. Current evidence suggests that endothermy has evolved at least twice among vertebrates. It evolved once along the lineage of ancient reptiles that led to birds, and once along the lineage of ancient reptiles that led to mammals.

28.4 Assessment

Review Key Concepts

1. a. Review How do the immune system and endocrine glands help to maintain homeostasis?

b. Explain Give an example of how multiple body systems function together to maintain homeostasis.

c. Apply Concepts Describe how the circulatory and endocrine systems of the migrating wildebeests in **Figure 28-26** help them maintain homeostasis.

2. a. Review Define *ectotherm.* Define *endotherm.*

b. Explain Why must an endotherm eat more food than an ectotherm of the same size?

c. Form a Hypothesis How might birds and mammals have evolved different means of insulating their bodies?

VISUAL THINKING

3. Construct a table that compares ectothermy and endothermy. Include the ways body temperature is controlled, relative rates of metabolism, relative amounts of food eaten, advantages, disadvantages, and examples of animals with each method of temperature regulation.

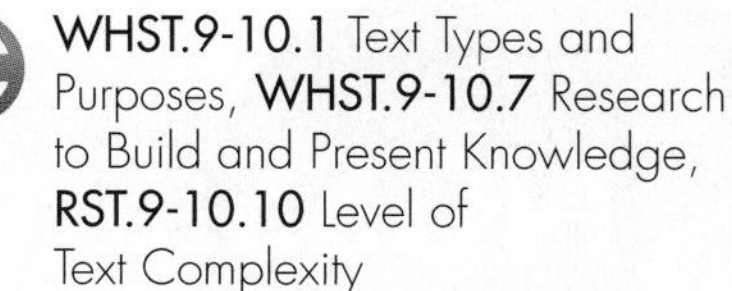

Head for the Hills?

The Miami soccer coach was upset. The Denver team *was* tops in the league. But his players were well-trained—and several had *still* collapsed from fatigue in the second quarter. He knew that "the mile-high city" was aptly named: its air was less dense than his players were accustomed to. But he'd flown his team in three days early. Why didn't that help? He decided to do some research.

He learned that the lower air density in Denver means that every breath has 15 percent less oxygen than it has at sea level. This means that there is less oxygen for the lungs and blood to deliver to muscles. Less oxygen decreases the performance of muscles that work for long periods. The body can adapt to altitude, but it takes about a week—so his strategy of arriving three days early fell short.

High-altitude adaptation includes an increase in the lung's ability to get oxygen into blood, as well as an improvement in the ability of muscle cells to use oxygen. The body's production of active red blood cells also increases, stimulated by low oxygen availability. So people who live or train at high altitudes have an advantage over "flatlanders." This information helps explain why runners from places like Nairobi, Kenya (altitude 5450 feet), compete so well in endurance events.

High-Altitude Training Should Not Be Restricted

Several training regimes legally use the effects of altitude to maximize performance. Some coaches have their players live or train at high altitude. Other players sleep in special tents whose air contains less oxygen. So even teams that live at low altitudes can mimic the effects of high altitude in their training. These techniques cause a natural increase in the body's production of the hormone erythropoietin (EPO), which stimulates the production of red blood cells that carry oxygen.

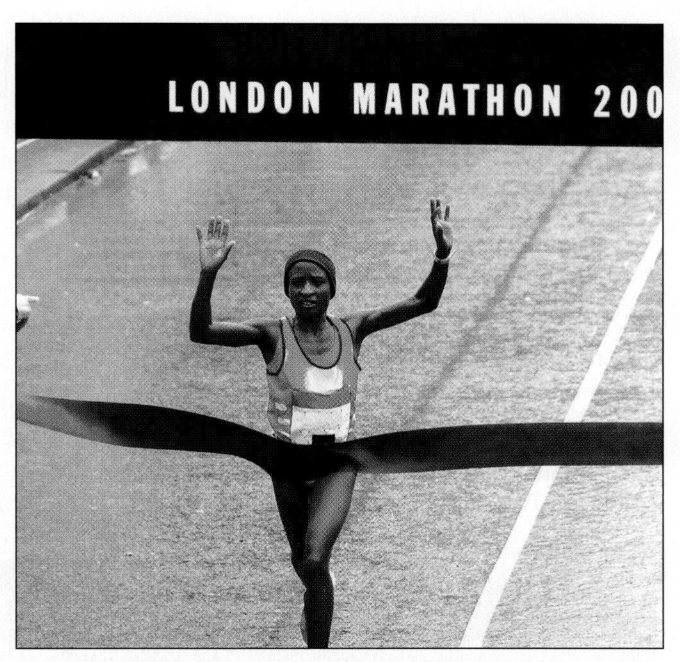

Elite marathoners, such as this Kenyan runner, often live and/or train in high-altitude areas.

High-Altitude Training Should Be Restricted

The injection of extra EPO during training—a biotech method of what is called "blood doping"—is illegal. High-altitude training regimes are just a "natural" way to accomplish exactly what blood doping does. It is unfair and possibly unsafe.

Research and Decide

1. Analyze the Viewpoints Using the Internet, research high-altitude training and "blood doping." Compare and contrast the effects of high-altitude training regimes with the effects of blood doping.

2. Form an Opinion Are current regulations fair to athletes from low-altitude states and countries? Write an argument to support your claim.

NGSS Smart Guide Animal Systems II

Disciplinary Core Idea LS1.A Structure and Function: How do the structures of organisms enable life's functions? Nervous systems collect and process information from the environment and coordinate the responses of muscular, endocrine, immune, and reproductive systems, so that animals can maintain homeostasis and reproduce.

28.1 Response

- When an animal responds to a stimulus, body systems work together to generate a response.
- Animal nervous systems exhibit different degrees of cephalization and specialization.
- Sensory systems range from individual sensory neurons to sense organs.

neuron 808 • stimulus 808 • sensory neuron 808 • interneuron 809 • response 809 • motor neuron 809 • ganglion 810 • cerebrum 811 • cerebellum 811

Biology.com

Untamed Science Video Join the Untamed Science crew as they interview experts to learn more about how the sex of offspring is determined in some animals.

Art Review Review your understanding of vertebrate brains.

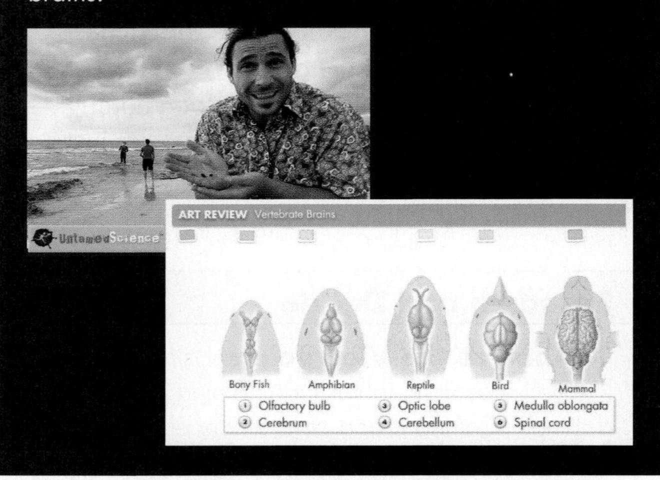

28.2 Movement and Support

- Animals have three main kinds of skeletal systems: hydrostatic skeletons, exoskeletons, and endoskeletons.
- In many animals, muscles work together in pairs or groups that are attached to different parts of a supporting skeleton.

hydrostatic skeleton 814 • exoskeleton 815 • molting 815 • endoskeleton 815 • joint 816 • ligament 816 • tendon 816

Biology.com

STEM Activity Watch a video about snakes to see how mathematics and engineering relate to animal movement.

InterActive Art Look at the structure and function of the water vascular system in a sea star.

Art in Motion Watch an animation that shows the motion of joints in both exoskeletons and endoskeletons.

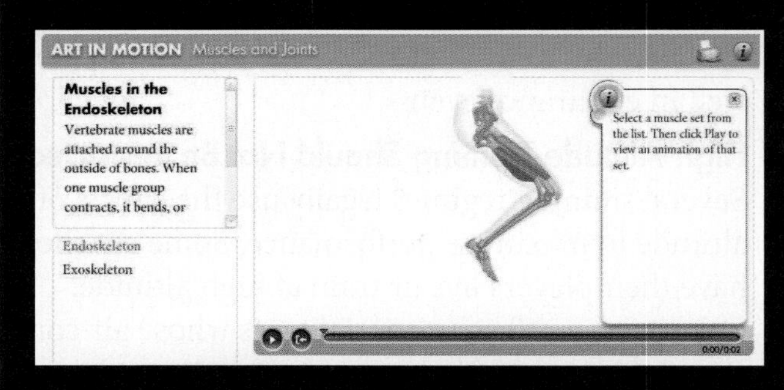

CHECKING YOUR *Scientific Literacy*

Refer to the lesson content and digital assets below as you prepare for your chapter assessment. Then, evaluate your understanding of animal systems by answering these questions.

1. Scientific and Engineering Practice **Analyzing and Interpreting Data** Rank the vertebrate brains shown in Figure 28-4 based on the following criteria: overall size, relative size of the cerebrum, relative size of the cerebellum, and relative sizes of the olfactory bulb and optic lobe. Is there a correlation between the sizes of the different regions of these vertebrate brains and the functions essential to survival of each type of animal? Explain why or why not.

2. Crosscutting Concept **Structure and Function** Compare the structures of exoskeletons and endoskeletons. How do the structures of these skeletons affect the way animals with these types of skeletons move?

3. Text Types and Purposes Should *in vitro* fertilization (IVF) be used to breed mammals in captivity? Write an argument to explain why or why not. Support your claims with relevant, accurate data using credible sources.

4. STEM Describe a technology that detects and responds to a particular type of sensory stimulus, such as light or sound.

28.3 Reproduction

- Asexual reproduction requires only one parent, so individuals may reproduce rapidly. But offspring produced asexually have less genetic diversity than do offspring produced sexually. Sexual reproduction maintains genetic diversity in a population by creating individuals with new combinations of genes.
- During internal fertilization, eggs are fertilized inside the body of the egg-producing individual. During external fertilization, eggs are fertilized outside the body.
- Animals may be oviparous, ovoviviparous, or viviparous.
- Reptiles, birds, and a few mammals have evolved amniotic eggs in which an embryo can develop without drying out. Mammals differ greatly in their means of reproduction and development, but all nourish their young with mother's milk.

oviparous 822 • **ovoviviparous 822** • **viviparous 822** • **placenta 822** • **metamorphosis 823** • **nymph 823** • **pupa 823** • **amniotic egg 825** • **mammary gland 825**

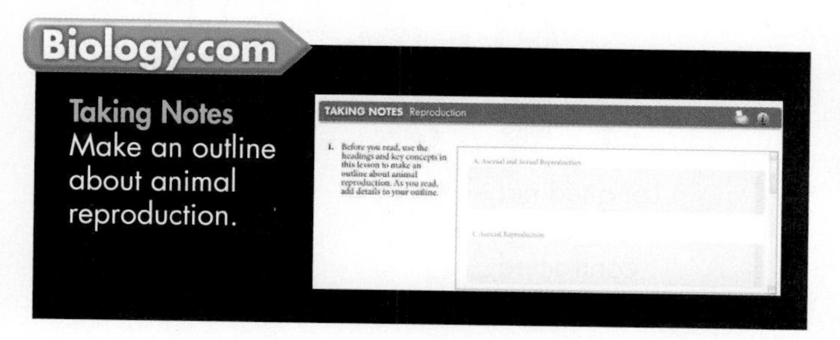

28.4 Homeostasis

- All body systems work together to maintain homeostasis.
- Most reptiles, invertebrates, fishes, and amphibians are ectotherms that regulate body temperature primarily by picking up heat from, or losing heat to, their environment. Endotherms, such as birds and mammals, have high metabolic rates that generate heat, even when they are resting.

endocrine gland 828 • **ectotherm 829** • **endotherm 829**

Biology.com

Data Analysis Investigate some of the ways mammals survive in cold temperatures.

DATA ANALYSIS Winter Survival

Metabolic Activity

4. **Analyze Data** Describe what happens to the metabolic rate as the lower critical temperature is reached and the environment gets even colder.

Physical | Physiological | LCT | NST | ST | Increasing Metabolic Rate | Normal metabolic rate | Decreasing Outside Temperature

Real-World Lab

Comparing Bird and Mammal Bones This chapter lab is available in your lab manual and at Biology.com

28 Assessment

28.1 Response

Understand Key Concepts

1. Information received from the environment that causes an organism to respond is called a
   a. response. c. reaction.
   b. stimulus. d. trigger.

2. The simplest nervous systems are called
   a. cephalopods. c. nerve nets.
   b. motor neurons. d. sensory neurons.

3. In vertebrates, the part of the brain that coordinates body movements is the
   a. olfactory lobe. c. cerebrum.
   b. optic lobe. d. cerebellum.

4. The arrows in this diagram are pointing to which structures?

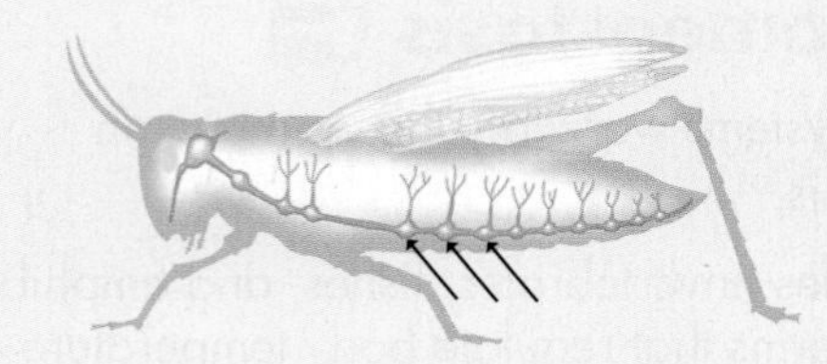

   a. ganglia c. nerve nets
   b. brains d. motor neurons

5. What two major trends in the evolution of the nervous system do invertebrates exhibit?

6. In general, how do the brains of mammals compare with the brains of other vertebrates? What is the significance of that difference?

7. What kinds of environmental stimuli are some animals capable of sensing that humans cannot sense?

Think Critically

8. **Craft and Structure** List the three major types of neurons and compare their roles.

9. **Key Ideas and Details** Suppose a pet dog is having difficulty coordinating its movements. Why might a veterinarian X-ray the dog's brain? Cite textual evidence to support your answer.

28.2 Movement and Support

Understand Key Concepts

10. Which of the following animals uses a hydrostatic skeleton to move?
    a. arthropod c. fish
    b. sponge d. annelid

11. An arthropod's exoskeleton performs all of the following functions EXCEPT
    a. production of gametes.
    b. protection of internal organs.
    c. support of the animal's body.
    d. prevention of loss of body water.

12. Vertebrates have endoskeletons made of
    a. chitin. c. calcium carbonate.
    b. cartilage and/or bone. d. bone only.

13. Muscles generate force
    a. only when they lengthen.
    b. only when they shorten.
    c. when they lengthen or shorten.
    d. all the time.

14. How do a fish's muscles function when the fish swims?

Think Critically

15. **Compare and Contrast** List two advantages and two disadvantages of exoskeletons and endoskeletons.

16. **Apply Concepts** The diagrams below show a type of skeletal system found in invertebrates. What is the name for this type of skeleton? Describe how it functions.

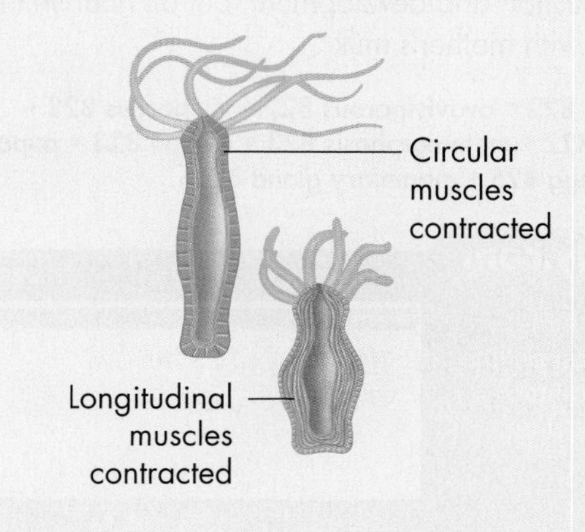

28.3 Reproduction

Understand Key Concepts

17. Individual animals that produce both sperm and eggs are called

a. gametes.
b. hermaphrodites.
c. fragments.
d. buds.

18. A species that lays eggs that develop outside of the mother's body is

a. oviparous.
b. viviparous.
c. ovoviviparous.
d. nonviparous.

19. Which structure in female mammals produces milk to nourish young?

a. kidney
b. pupa
c. mammary gland
d. placenta

20. Which of the following are NOT placental mammals?

a. seals
b. marsupials
c. carnivores
d. primates

21. **Craft and Structure** Describe the life cycle of a typical cnidarian. Be sure to include the alternation of the polyp form with the medusa form.

22. Compare and contrast internal and external fertilization.

23. What survival advantage does the placenta confer on mammals?

Think Critically

24. **Compare and Contrast** Describe the differences between a newborn placental mammal and a newborn marsupial.

25. **Infer** Many mammals care for their young for extended periods of time. This parental behavior does not help the parent survive. Why, then, might extended parental care have been naturally selected for in these species?

28.4 Homeostasis

Understand Key Concepts

26. Stable internal conditions are called

a. homeostasis.
b. ectothermy.
c. endothermy.
d. reactivity.

solve the CHAPTER MYSTERY

SHE'S JUST LIKE HER MOTHER!

In May 2007, the researchers published their conclusions. The baby bonnethead had been produced by a process called automictic parthenogenesis, in which the mother's unfertilized egg divides and the resulting elements fuse to make a zygote with two sets of identical chromosomes. (See below.) That is why the baby shark was homozygous for every trait; she had two identical sets of alleles.

This was the first time researchers had seen parthenogenesis in a cartilaginous fish. It has now been observed in insects and, though rarely, in every vertebrate lineage except mammals.

Normal Fertilization

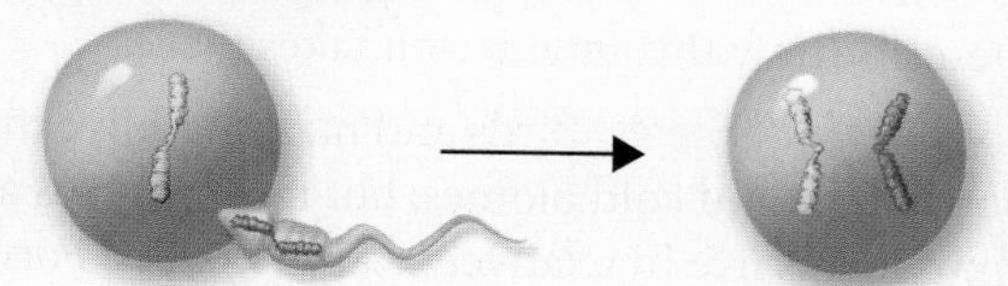

Egg is fertilized with sperm.

Baby shark has one set of chromosomes from each parent.

Automictic Parthenogenesis

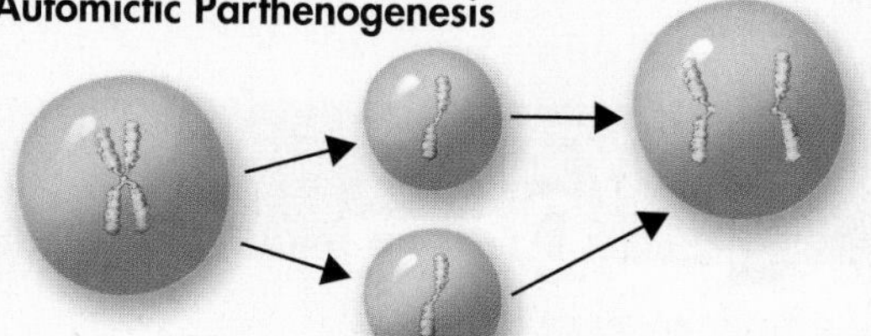

Egg doubles and divides its genetic material.

New cells combine. Baby shark has two sets of identical chromosomes, both from the mother.

1. Connect to the **Big idea** Explain why the baby shark was *not* a clone, or exact genetic copy, of its mother (assuming the mother was conceived by sexual reproduction).

27. The main source of heat for an ectotherm is
  a. its high rate of metabolism.
  b. the environment.
  c. its own body.
  d. its food.

28. Endotherms
  a. control body temperature through behavior.
  b. control body temperature from within.
  c. obtain heat from outside their bodies.
  d. have relatively low rates of metabolism.

29. How do endocrine glands help regulate body activities?

30. What is the function of the immune system?

31. Explain the advantages and disadvantages of ectothermy and endothermy.

32. What does current evidence suggest about the evolution of endothermy?

Think Critically

33. **Apply Concepts** What two body systems interact to deliver hormones to the organs they affect? Describe how this interaction takes place.

34. **Form a Hypothesis** Birds and mammals live in both warm and cold biomes, but most reptiles and amphibians live in relatively warm biomes. Form a hypothesis that would explain this difference.

Connecting Concepts

Use Science Graphics

Use the diagram to answer questions 35 and 36.

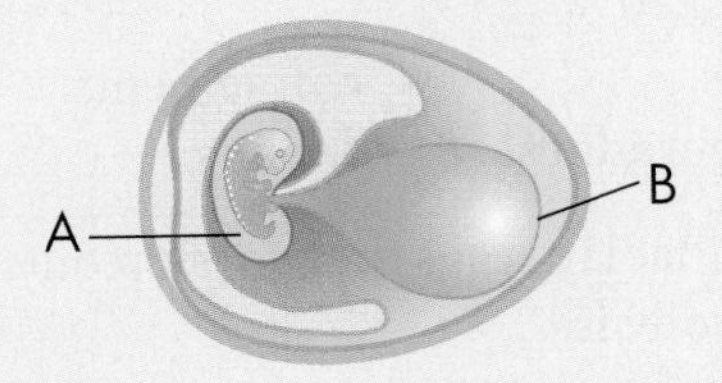

35. **Interpret Visuals** What is the function of the membrane labeled A?

36. **Infer** What is membrane B? What is the function of the structure it surrounds?

Write About Science

37. **Production and Distribution of Writing** Write an essay in which you compare the brain of a fish to the brain of a mammal. In your essay, identify the main parts of the brain and compare and contrast their structures and functions in the two animals. (*Hint:* Before you write, construct a Venn diagram that compares the two brains.)

38. **Assess the**  Give an example of how the nervous system of an ectotherm could initiate behavior that would help the animal regulate its internal temperature.

Integration of Knowledge and Ideas

The following table shows the mass of the heart as a percentage of total body mass for humans and three types of birds.

Heart	Percentage of Total Body Mass
Human	0.42
Vulture	0.57
Sparrow	1.68
Hummingbird	2.37

39. **Interpret Tables** Which animal has the largest heart in proportion to its body mass?

40. **Draw Conclusions** Based on the data, what general conclusion can you draw about the relative heart sizes of humans and birds?

41. **Form a Hypothesis** Form a hypothesis that could explain your conclusion. (*Hint:* Consider the different energy needs of humans and birds.)

Standardized Test Prep

Multiple Choice

1. What part of a vertebrate's brain is the "thinking" region?
A olfactory bulb
B cerebellum
C cerebrum
D medulla oblongata

2. Neurons that receive and send information from and to other neurons are called
A ganglia.
B motor neurons.
C sensory neurons.
D interneurons.

3. The skeleton of a shark is composed primarily of
A bone.
B vertebrae.
C cartilage.
D tendons.

4. Joints between bones of the human skeleton are held together mostly by
A tendons.
B muscles.
C ligaments.
D skin.

5. In oviparous species, embryos
A develop internally.
B obtain nutrients directly from the mother's body.
C obtain nutrients from the external environment.
D develop outside of the body.

6. Most animals reproduce sexually by producing
A buds.
B clones.
C haploid gametes.
D diploid gametes.

7. Maintaining homeostasis in multicellular organisms requires
A a properly functioning heart.
B a nervous system.
C hormones.
D all body systems working together.

8. Which of the following are endothermic?
A fish and amphibians
B mammals and birds
C reptiles and mammals
D all vertebrates

Questions 9–10

Study the illustration of a reptile brain and answer questions 9 and 10.

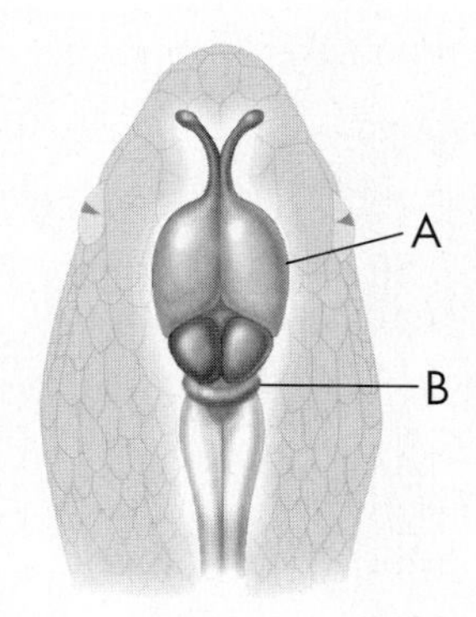

9. What is the name of the structure labeled A?
A cerebrum
B optic lobe
C cerebellum
D olfactory bulb

10. What are some functions of the structure labeled B?
A vision
B control of internal organ functions
C connection of the brain to the rest of the body
D coordination of movement and control of balance

Open-Ended Response

11. Why can't muscles function individually?

If You Have Trouble With . . .

Question	1	2	3	4	5	6	7	8	9	10	11
See Lesson	28.1	28.1	28.2	28.2	28.3	28.3	28.4	28.4	28.1	28.1	28.2

29 Animal Behavior

Evolution

Q: How do animals interact with one another and their environments?

Timber wolves fighting

INSIDE:

- 29.1 Elements of Behavior
- 29.2 Animals in Their Environments

BUILDING *Scientific Literacy*

Deepen your understanding of animal behavior by engaging in key practices that allow you to make connections across concepts.

NGSS You will analyze data of animal behaviors so that you can learn to identify patterns in the learned and social behaviors of animals.

STEM How do wolves behave in a pack? You will find out when you observe wolves in Yellowstone Park.

Common Core You will gather relevant information from authoritative sources in order to write about animal behaviors.

ELEPHANT CALLER ID?

On a hot, dusty afternoon in Africa's Etosha National Park, a group of elephants approaches a watering hole. They begin to drink and splash around. Suddenly, they freeze. One places her trunk flat on the ground, the tip pointing toward her feet. Soon, the elephants clump together defensively, nudging young calves into the center of the group. Some place weight on the front of their feet—as close to standing on tiptoe as elephants get. Most keep their huge ears flattened against their heads. Soon, they shuffle nervously away from the watering hole, staying in defensive formation. Human observers nearby can't see or hear anything that might have alarmed the group. What are these elephants doing—and why? As you read this chapter, look for clues that explain the behavior of this elephant group. Then, solve the mystery.

Biology.com

Finding the solution to the Elephant Caller ID mystery is only the beginning. Take a video field trip with the ecogeeks of Untamed Science to see where the mystery leads.

Go online to access the following including:
- eText • Flash Cards • Lesson Overviews • Chapter Mystery

29.1 Elements of Behavior

Key Questions

What is the significance of behavior in the evolution of animal species?

What is an innate behavior?

What are the major types of learning?

How do many complex behaviors arise?

Vocabulary

behavior • innate behavior • learning • habituation • classical conditioning • operant conditioning • insight learning • imprinting

Taking Notes

Concept Map As you read, create a concept map to organize the information in this section.

MYSTERY CLUE

Researchers have noticed that elephants living elsewhere exhibit behaviors similar to those of the elephants at the watering hole. What does this suggest about the importance of these behaviors?

THINK ABOUT IT At a table by a Caribbean seaside restaurant, a young tourist eats a hamburger, unaware that he's being watched—by an iguana. The lizard crawls closer, looking dangerous. The boy spots him and, with a cry of fear, jumps up onto his chair. But the iguana isn't interested in human toes. This lizard, a shy tree-dwelling species, is a vegetarian. He makes a beeline for the French fries that the boy, in his panic, knocked to the ground. What's so interesting about this scene? These iguanas normally don't approach humans. And they're not known for being as clever as, say, dogs or cats. But this particular male iguana has learned that getting close to humans can mean easy access to food.

Behavior and Evolution

What is the significance of behavior in the evolution of animal species?

The activity of that hungry iguana is just one example of **behavior,** which is usually defined as the way an organism reacts to stimuli in its environment. Some behaviors are as simple as a dog turning its head and pricking up its ears in response to a noise. Other behaviors can be quite complex. For example, some animals wash some of their food before eating it. Usually, behaviors are performed when an animal detects and responds to some sort of stimulus in its environment. The way an animal responds to a stimulus, however, often depends on its internal condition. If our friend the iguana hadn't been hungry, for example, it would probably have kept its distance from the boy and his French fries!

Many behaviors are essential to survival. To survive and reproduce, animals must be able to find and catch food, select habitats, avoid predators, and find mates. Behaviors that make these activities possible are therefore as vital to survival and reproduction as any physical characteristic, such as teeth or claws.

You've learned how physical traits are shaped by instructions coded in an organism's genome. The nervous system, which makes behaviors possible, is clearly influenced by genes. So, it shouldn't surprise you to learn that some behaviors are also influenced by genes and can therefore be inherited. That's why certain behaviors can evolve under the influence of natural selection, just as physical traits do. For example, the genes that code for the behavior of the moth in **Figure 29–1** help the moth escape predators. **If a behavior that is influenced by genes increases an individual's fitness, that behavior will tend to spread through a population. Over many generations, various kinds of adaptive behaviors can play central roles in the survival of populations and species.**

In Your Notebook *In your own words, explain how animal behavior can evolve through natural selection.*

FIGURE 29–1 Anti-Predatory Display Moths of the genus *Automeris* normally rest with their front wings over their hind wings (left). If disturbed, the moth will move its front wings to expose a striking circular pattern on its hind wings (right). This behavior may scare off predators that mistake the moth's hind-wing pattern for the eyes of a predatory owl such as the great horned owl. **Infer** ***Given that the moth displays this wing pattern and doesn't fly away when confronted by a predator, predict two characteristics of the moth's typical predator.***

Innate Behavior

What is an innate behavior?

Why do newly hatched birds beg for food within moments after hatching? How does a spider know to spin its web? These animals are exhibiting **innate behaviors,** also called instincts. **Innate behaviors appear in fully functional form the first time they are performed, even though the animal has had no previous experience with the stimuli to which it responds.** The suckling of a newborn mammal is a classic example of a simple innate behavior. Other innate behaviors, such as the weaving of a spider web or the building of hanging nests by weaver birds, can be quite complex. All innate behaviors depend on patterns of nervous system activity that develop through complex interactions between genes and the environment. Biologists do not yet fully understand just how these interactions occur, but the usefulness of innate behaviors is obvious. They enable animals to perform certain tasks essential to survival without the need for experience.

Learned Behavior

What are the major types of learning?

If all behaviors were innate, animals would have a tough time adapting to unpredictable changes in their environments. (And if all behavior were innate, you wouldn't be reading this book!) Many complex animals live in unpredictable environments, where their fitness can depend on behaviors that can be altered as a result of experience. Acquiring changes in behavior during one's lifetime is called **learning.**

FIGURE 29–2 Learned Behavior
This chimpanzee is using a stick as a tool to "fish" for termites in a termite nest.

Many animals have the ability to learn. Organisms with simple nervous systems, such as sea stars, shrimp, and most other invertebrates, may learn only rarely. Among a few invertebrates, and many chordates, learning is common and occurs under a wide range of circumstances. In animals that care for their young, for example, offspring can learn behaviors from their parents or other caretakers. The chimpanzee in **Figure 29–2** is exhibiting a complex learned behavior—using a tool to gather food. Scientists have identified several different ways of learning. **The four major types of learning are habituation, classical conditioning, operant conditioning, and insight learning.**

Habituation The simplest type of learning is habituation. **Habituation** is a process by which an animal decreases or stops its response to a repetitive stimulus that neither rewards nor harms the animal. Often, learning to ignore a stimulus that offers neither a reward nor a threat can enable an individual to spend its time and energy more efficiently. Consider the common shore ragworm, an invertebrate shown to be capable of simple learning. This animal lives in a sandy tube that it leaves to feed. If a shadow passes overhead, the worm will instantly retreat to the safety of its burrow. Yet, if repeated shadows pass within a short time span, this response quickly subsides. When the worm has learned that the shadow is neither food nor threat, it will stop responding. At this point the worm has been habituated to the stimulus. In **Figure 29–3,** you can see birds becoming habituated to the stimulus of passing cars.

FIGURE 29–3 Habituation
Birds on the side of a road take flight when a car approaches (left). After many cars have passed and not harmed them, these birds (right) have become habituated to cars, and they no longer take flight when one approaches.

Classical Conditioning One evening you sit down to eat a kind of food you've never tried before. But shortly after you start eating, you get sick from a stomach virus. You feel better the next day, but as your parents present you with leftovers of the same food, you feel sick again. From that time on, whenever you smell that particular food, you become nauseated. This is an example of classical conditioning. **Classical conditioning** is a form of learning in which a certain stimulus comes to produce a particular response, usually through association with a positive or negative experience. In this case, the stimulus is the smell of that particular food, and the response is nausea. The food didn't make you sick, but you've been conditioned to associate the smell of that food with illness.

Classical conditioning was first described around 1900 by Russian physiologist Ivan Pavlov, who was studying dogs' responses to food. Pavlov first noted that dogs salivate as an innate response to food. Then Pavlov discovered that if he always rang a bell when he offered food, a dog would salivate whenever it heard a bell, even if no food was present. Pavlov's experiment produced salivation (a response) in reaction to the bell (a stimulus) associated with food.

Operant Conditioning Conditioning is often used to train animals. **Operant conditioning** occurs when an animal learns to behave in a certain way through repeated practice, to receive a reward or avoid punishment. Operant conditioning was first described in the 1940s by the American psychologist B. F. Skinner. Skinner invented a testing procedure that used a box called a "Skinner box." A Skinner box contains a colored button or lever that delivers a food reward when pressed. After an animal is rewarded several times, it learns that it gets food whenever it presses the button or lever. At this point, the animal has learned by operant conditioning how to obtain food. In **Figure 29–4,** you can see how a dog can be trained to ring a bell to be let out of the house.

Operant conditioning is sometimes described as a form of trial-and-error learning. Trial-and-error learning begins with a random behavior that is rewarded in an event called a trial. Most trials result in errors, but occasionally a trial will lead to a reward or punishment.

Insight Learning The most complicated form of learning is insight learning, or reasoning. **Insight learning** occurs when an animal applies something it has already learned to a new situation, without a period of trial and error. For instance, if you are given a new math problem on an exam, you may apply principles you have already learned in the class to solve the problem. Insight learning is common among humans and some other primates. In one experiment, a hungry chimpanzee used insight learning to figure out how to reach a bunch of bananas hanging overhead: It stacked some boxes on top of one another and climbed to the top of the stack. In contrast, if a dog accidentally wraps its leash around a tree, the dog is usually unable to figure out how to free itself.

FIGURE 29–4 Operant Conditioning A dog randomly brushes its tail against a bell hanging on a doorknob (top). The owner responds by opening the door to let the dog outside (middle). After the "ring the bell; open the door" sequence has occurred several times, the dog has learned to ring the bell when it wants to go out (bottom).

FIGURE 29–5 Imprinting in the Wild This wild baby sandhill crane has imprinted on its mother and will follow her in flight.

Complex Behaviors

How do many complex behaviors arise?

Though behaviors may be learned, they often involve significant innate components. **Many complex behaviors combine innate behavior with learning.** Young white-crowned sparrows, for example, have an innate ability to recognize their own species' song and to distinguish it from the songs of other species. To sing their complete species-specific song, however, young birds must hear it sung by adults.

Another example of behavior with both innate and learned components is called imprinting. Some animals, such as birds, recognize and follow the first moving object that they see during a critical time in their early lives. This process is called **imprinting.** How does imprinting involve both innate and learned behavior? The young birds have an innate urge to follow the first moving object they see. But they are not born knowing what that object will look like, so they must learn from experience what to follow. Usually, birds such as cranes imprint on their mother, as shown in **Figure 29–5.**

Quick Lab
GUIDED INQUIRY

What Kind of Learning Is Practice?

1. Draw straight lines on a piece of paper to divide it into several sections of different sizes and shapes. Then, cut the paper into sections along those lines.
2. Shuffle the pieces, and then time another student as he or she tries to reassemble the pieces. Record how long it takes the student to do this task.
3. Repeat step 2 three times. Construct a graph showing how the time needed to assemble the puzzle changed with repeated practice.

Analyze and Conclude

1. Analyze Data Explain the shape of your graph. How did the time needed to reassemble the pieces change with repeated trials?

2. Draw Conclusions What kind of learning was displayed in this activity? Was it habituation, classical conditioning, operant conditioning, or some other kind of learning? Explain your answer.

Once imprinting has occurred, the behavior becomes fixed. Sometimes, the fixed object of imprinting shows up later in life. When baby geese mature, for example, they search for mates who resemble the individual on whom they imprinted as goslings. In nature, this is almost always their mother, and therefore a member of their own species. When humans get involved, odd things are possible. One researcher arranged for baby geese to imprint on him. It was amusing to watch these birds following him around. Amusing, that is, until the birds matured—and began to court the researcher! Sometimes experiments have even caused birds to imprint on objects such as the hand puppet worn by the human in **Figure 29–6.**

Imprinting doesn't have to involve vision. Animals can imprint on sounds, odors, or any other sensory cues. Newly hatched salmon, for example, imprint on the odor of the stream in which they hatch. Young salmon then head out to sea. Years later, when they mature, the salmon remember the odor of their home stream and return there to spawn.

FIGURE 29–6 Imprinting in Captivity Recently hatched cranes raised in captivity imprinted on a red-headed bird puppet worn on the hand of a researcher. Later, that puppet is used to introduce these birds to the wild by guiding them along a migration route that they would normally learn by following their parents.

29.1 Assessment

Review Key Concepts

1. a. Review What is behavior?

b. Apply Concepts How does natural selection affect animal behavior?

2. a. Review How does a newborn animal know exactly "what to do" the moment it is born?

b. Predict What would happen if a newborn kitten did not have the suckling instinct?

3. a. Review What are the four types of learning?

b. Apply Concepts Give an example of how humans learn through classical conditioning.

4. a. Review Which aspect of imprinting is innate? Which aspect is learned?

b. Infer How might isolating a newborn animal from members of its own species affect it?

Apply the Big idea

Evolution

5. Use what you learned in this lesson to explain why behavioral responses are important to the survival of a species.

WHST.9-10.1 Text Types and Purposes, WHST.9-10.8 Research to Build and Present Knowledge. Also RST.9-10.10, WHST.9-10.10

Biology & Society

Should Marine Mammals Be Kept in Captivity?

Many types of marine mammals, including dolphins, killer whales, and seals, are kept in captive display for educational, entertainment, and research purposes. Yet, there is strong debate about whether public display of such animals is ethical. Should we prohibit the capture of marine mammals for public display?

The Viewpoints

Captivity Should Be Prohibited Some people feel that capturing and training marine mammals purely for entertainment purposes is not justified. They believe that because marine mammals are naturally social, with strong family bonds, they are not suited to capture or confinement. These people are concerned that the process of capture disrupts social groups.

Those opposed to the captivity of marine mammals also argue that confinement places the animals in an unnatural situation—one that is monotonous, limited, and unhealthful. In the wild, whales and dolphins travel long distances and dive much deeper than is possible in a shallow display tank. There is also a concern that human interaction with captive marine mammals increases the risk of transmitting diseases to the animals.

Captivity Should Be Allowed Other people believe that we have an obligation to convey knowledge of the natural world to the public by displaying animals and educating ourselves about them. Information obtained by observing captive animals and interacting with them may be helpful in managing their populations in the wild. Many people argue that the adverse effects of captivity are outweighed by the benefits of conservation, an enhanced human appreciation for animals, and the advancement of scientific knowledge. There is also evidence that human interactions with captive dolphins may help people with disabilities, such as autism.

Research and Decide

1. Analyze the Viewpoints To make an informed decision, learn more about this issue by consulting library or Internet resources. Then, list the options for education, entertainment, and research involving marine mammals. What are the benefits? What are the costs?

2. Form an Opinion Should marine mammals be kept in captivity? Are there some instances when captivity is a good solution and other instances when it is not? Write an argument to support your claim.

29.2 Animals in Their Environments

THINK ABOUT IT As twilight falls on a coral reef, its inhabitants act like New York commuters during evening rush hour. Daytime workers, whose "jobs" involve feeding near the reef, head for home. Some form lines and create traffic jams as they jockey for position in "apartment buildings"—cracks, crevices, and caves in the reef where they will rest until dawn. For a time, twilight predators menace any straggling daytime fishes disoriented by the gloom. Then the night shift emerges. These creatures of darkness may have huge, staring eyes—or none at all. They take over the coral metropolis at night. At dawn, the cycle reverses. Critters of the night disappear, and the workday begins.

Key Questions

How do environmental changes affect animal behavior?

How do social behaviors increase an animal's evolutionary fitness?

How do animals communicate with others in their environments?

Vocabulary

circadian rhythm • migration • courtship • territory • aggression • society • kin selection • communication • language

Taking Notes

Outline Before you read, use the headings in this lesson to make an outline about animals in their environments. As you read, add details to your outline.

Behavioral Cycles

How do environmental changes affect animal behavior?

The daily changeover on coral reefs is an example of regular cycles in nature. **Many animals respond to periodic changes in the environment with daily or seasonal cycles of behavior.** Behavioral cycles that occur daily, like those on coral reefs, are called **circadian** (sur KAY dee un) **rhythms.** You sleep at night and attend school during the day in another example of a circadian rhythm.

Other cycles are seasonal. In temperate and polar regions, for example, many species are active during spring, summer, and fall, but enter into a sleeplike state, or dormancy, during winter. In mammals, dormancy is called hibernation. Dormancy allows an animal to survive periods when food and other resources may not be available.

Another type of seasonal behavior is **migration,** the seasonal movement from one environment to another. Many species of animals migrate—often over huge distances. **Figure 29–7** shows the long migration of green sea turtles. Migration allows animals to take advantage of favorable environmental conditions. For example, many songbirds live in tropical regions where temperatures are moderate and food remains available during northern winters. When these birds fly north in the spring, they take advantage of seasonally abundant food and find space to nest and raise their young.

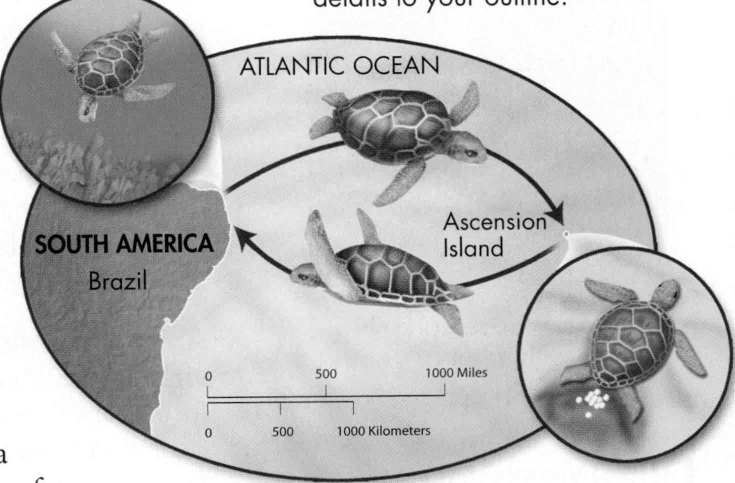

FIGURE 29–7 Seasonal Behavior of Green Sea Turtles Each year, green sea turtles migrate back and forth between their feeding grounds on Brazil's coast and their nesting grounds on Ascension Island.

FIGURE 29–8 Types of Social Behavior Social behaviors such as courtship and territorial marking help define social groups.

▲ Gannets perform a courtship ritual.

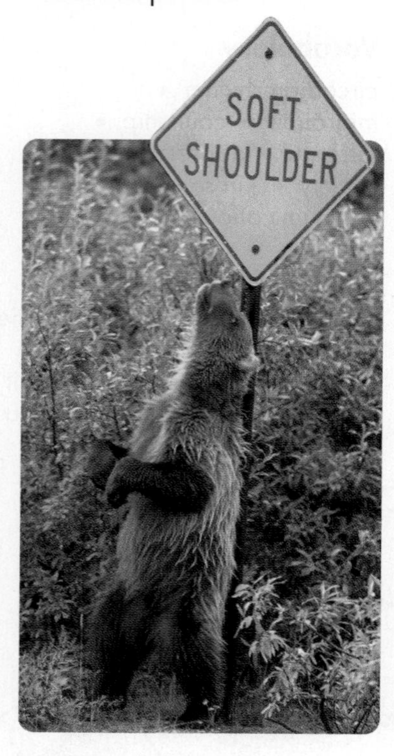

▲ A grizzly bear marks its territory with its fur and scent.

BUILD Vocabulary

WORD ORIGINS The adjective *social* is related to the Latin words *socialis*, meaning "social," and *societas*, meaning "companionship." Animals that are social spend most of their time in a group.

Social Behavior

How do social behaviors increase an animal's evolutionary fitness?

Whenever birds sing, bighorn sheep butt heads, or chimpanzees groom each other, they are engaging in social behavior. Social behaviors include courtship, territoriality, aggression, and the forming of societies. **Choosing mates, defending or claiming territories or resources, and forming social groups can increase evolutionary fitness.**

Courtship For sexually reproducing animal species, evolutionary survival requires individuals to locate and mate with another member of its species at least once. **Courtship** is behavior during which members of one sex (usually males) advertise their willingness to mate, and members of the opposite sex (usually females) choose which mate they will accept. Typically, males send out signals—sounds, visual displays, or chemicals—that attract females. The musical trill of a tree frog, for example, is a breeding call.

In some species, courtship involves an elaborate series of behaviors called rituals. A ritual is a series of behaviors performed the same way by all members of a population for the purpose of communicating. Most rituals consist of specific signals and responses that continue until mating occurs. For example, gannets bond by engaging in "beak-pointing"—intertwining their necks while pointing their beaks to the sky, a behavior you can see in **Figure 29–8.**

Territoriality and Aggression Many animals behave in ways that prevent other individuals from using limited resources. Often, these animals occupy a specific area, or **territory,** that they defend against competitors. Territories usually contain resources, such as food, water, nesting sites, shelter, and potential mates, which are necessary for survival and reproduction. If a rival enters a territory, the "owner" of the territory attacks in an effort to drive the rival away. Grizzly bears will often mark their territories with their fur and scent by scratching their backs on rough surfaces, such as trees or signposts.

While competing for resources, animals may also show **aggression,** threatening behaviors that one animal uses to exert dominance over another. Fights between male sea lions over territory and "harems" of females often leave both rivals bloodied.

Animal Societies An animal **society** is a group of animals of the same species that interact closely and often cooperate. Mammals form many types of societies, which offer a range of advantages. Zebras and other grazers, for example, are safer from predators when they are part of a group than when they are alone. Societies can also improve animals' ability to hunt, to protect their territory, to guard their young, or to fight with rivals. In wild African dog packs, for instance, adult females take turns guarding all the pups in the pack, while the other adults hunt for prey. Macaque, baboon, and other primate societies hunt together, travel in search of new territory, and interact with neighboring societies.

Members of a society are often related to one another. Elephant herds, for example, consist of mothers, aunts, and their offspring. (Males are kicked out when they reach puberty.) The theory of **kin selection** holds that helping relatives can improve an individual's evolutionary fitness because related individuals share a large proportion of their genes. Helping a relative survive therefore increases the chance that the genes an individual shares with that relative will be passed along to offspring.

The most extreme examples of relatedness, and the most complex animal societies (other than human societies), are found among social insects such as ants, bees, and wasps. In social insect colonies, all individuals cooperate to perform extraordinary feats, such as building complex nests. In an ant colony, such as the one in **Figure 29–9**, all workers in the colony are females who are very closely related—which means that they share a large proportion of each others' genes. Worker ants are also sterile. For this reason it is advantageous for them to cooperate to help their "mother" (the queen) reproduce and raise their "sisters" (other workers). Male ants function only to fertilize the queen.

MYSTERY CLUE

When threatened by predators, adult elephants in a herd form a defensive circle around the youngest members. What may have triggered this behavior in the elephants at the watering hole?

FIGURE 29–9 An Ant Society In a leaf-cutter ant society, only a single queen reproduces. Different groups of ants within the colony perform other tasks.

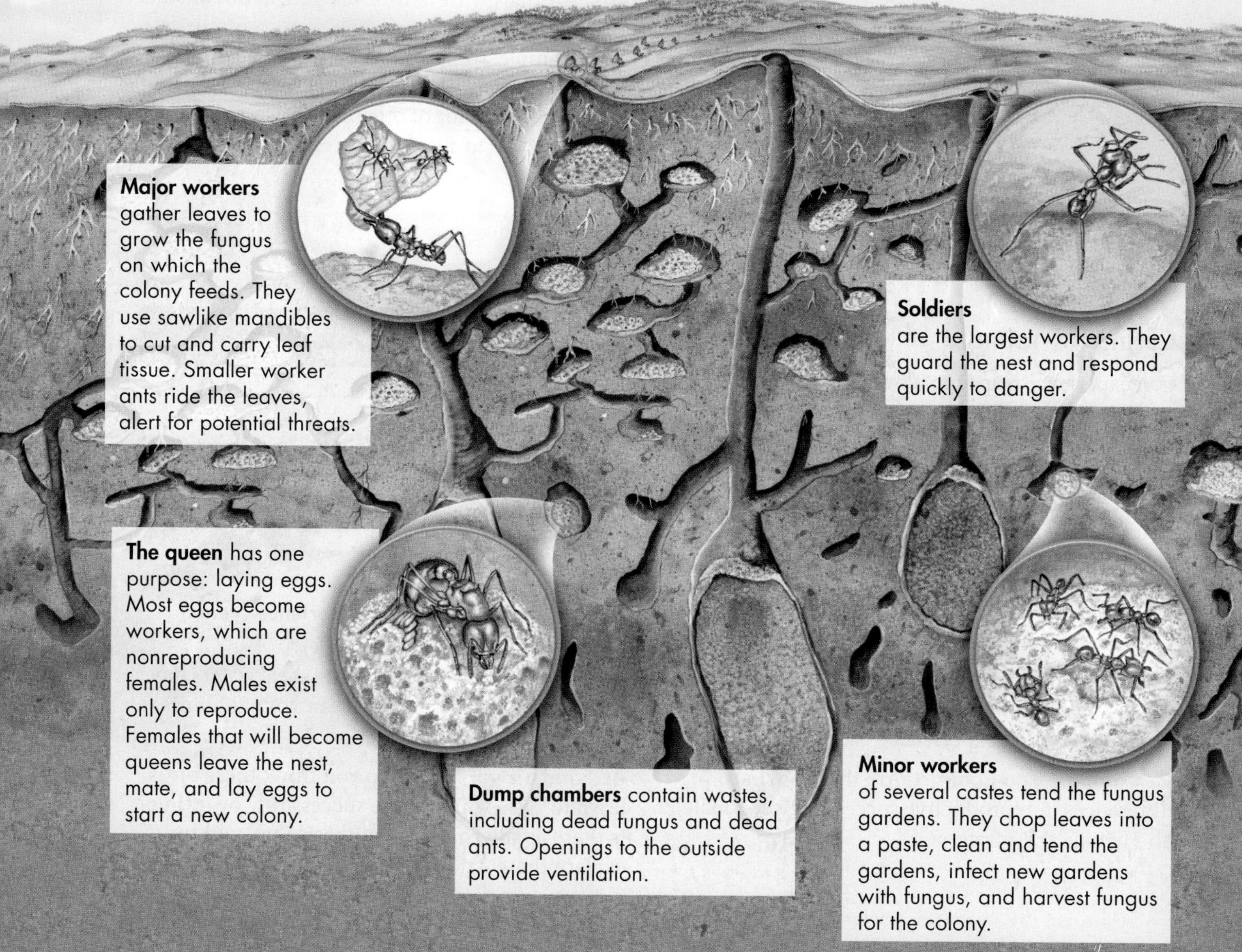

FIGURE 29–10 Types of Communication Different animals rely on different methods of communication to get their messages across. Fireflies, for example, flash a light generated within their bodies to attract mates.

Communication

How do animals communicate with others in their environments?

Because social behavior involves more than one individual, it requires **communication**—the passing of information from one organism to another. **Animals may use a variety of signals to communicate with one another. Some animals are also capable of language.** The specific techniques that animals use depend on the types of stimuli their senses can detect.

Visual Signals Many animals have eyes that sense shapes and colors at least as well as humans do, and they often use visual signals. For example, squids, which have large eyes, change their color to broadcast a variety of signals. In many animal species, males and females have different color patterns, and males use color displays to advertise their readiness to mate. Some animals, such as fireflies, even send signals using light generated within their bodies, as you can see in **Figure 29–10.**

Chemical Signals Animals with well-developed senses of smell, including insects, fishes, and many mammals, can communicate with chemicals. For example, some animals, including lampreys, bees, and ants, release pheromones (FEHR uh mohnz), chemical messengers that affect the behavior of other individuals of the same species, to mark a territory or to signal their readiness to mate.

Analyzing Data

Caring for Young

Can experience help animals other than humans learn to care for their young better? The data at the right are from field studies of a seabird, the short-tailed shearwater. Each pair produces only one egg a year. If that egg breaks or if the chick dies, the egg is not replaced. The graph shows the percentage of eggs that develop into free-flying young, in relation to the breeding experience of the parents. This ratio is called reproductive success.

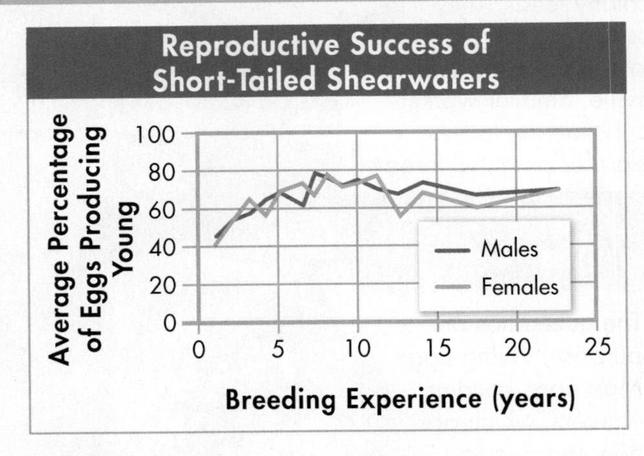

1. **Interpret Tables** What is the approximate success rate of a female shearwater with five years of breeding experience?

2. **Compare and Contrast** Are there obvious differences in reproductive success between male and female shearwaters?

3. **Draw Conclusions** Do older shearwaters have better reproductive success than younger birds have? Explain your answer.

4. **Form a Hypothesis** Do you think these birds learn to raise young more successfully over time? Is there an alternative hypothesis that could explain these data?

Sound Signals Most animal species that have vocal abilities and a good sense of hearing communicate using sound. Some have evolved elaborate systems of communication. Dolphins communicate in the ocean using sound signals. Bottlenose dolphins each have their own unique "signature" whistle that, amazingly, functions to inform others of who is sending the communication. Elephants also make distinctive sounds, both with their vocal apparatus and with their feet, that can identify them. Elephants, and some other animals, can send messages that the recipient feels rather than hears.

FIGURE 29–11 Language Dolphins seem to have a language of their own.

Language The most complicated form of communication is language. **Language** is a system of communication that combines sounds, symbols, and gestures according to rules about sequence and meaning, such as grammar and syntax. Many animals, including elephants, primates, and dolphins such as those in **Figure 29–11,** have complex communication systems. Some even seem to have "words"—calls with specific meanings such as "lions on the prowl." Many species, including honeybees, convey complex information using various kinds of signals. However, untrained animals don't seem to use the rules of grammar and syntax we use to define human language.

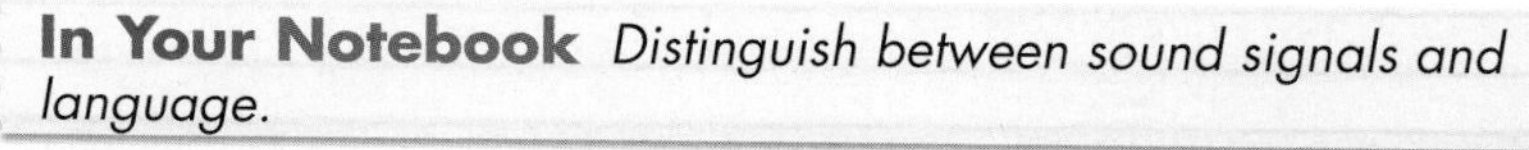

In Your Notebook *Distinguish between sound signals and language.*

29.2 Assessment

Review Key Concepts

1. a. **Review** Name two ways in which animal behavior is related to environmental cycles.

   b. **Apply Concepts** Travelers who travel across time zones often experience what is called jet lag, which includes fatigue and disrupted sleep patterns. How can you explain this condition?

2. a. **Review** List three types of social behavior.

   b. **Relate Cause and Effect** How does membership in a society increase the evolutionary fitness of individuals in the society?

3. a. **Review** What are the main ways in which animals communicate with one another?

   b. **Draw Conclusions** Suppose you discover a new type of animal that is very different in appearance from other animals you have seen. How could observing the sense organs of this animal help you to understand if and how it communicates?

Apply the Big idea

Evolution

4. Explain two ways that an animal's social behavior can influence its evolutionary fitness.

NGSS Smart Guide Animal Behavior

Disciplinary Core Idea LS1.D Information Processing: How do organisms detect, process, and use information about the environment? Animal behavior may be innate, learned, or a combination of both. An animal's behavior is affected by other organisms and changes in its environment.

29.1 Elements of Behavior

- If a behavior that is influenced by genes increases an individual's fitness, that behavior will tend to spread through a population. Over many generations, various kinds of adaptive behaviors can play central roles in the survival of populations and species.
- Innate behaviors appear in fully functional form the first time they are performed, even though the animal has had no previous experience with the stimuli to which it responds. Innate behaviors are also called instincts.
- The four major types of learning are habituation, classical conditioning, operant conditioning, and insight learning.
- Many complex behaviors combine innate behavior with learning.

behavior 840 • innate behavior 841 • learning 842 • habituation 842 • classical conditioning 843 • operant conditioning 843 • insight learning 843 • imprinting 844

Biology.com

Untamed Science Video See how the latest technologies help the explorers from Untamed Science track the movements of different animals.

Tutor Tube Instinct? Sorting out everyday and scientific behavior terms.

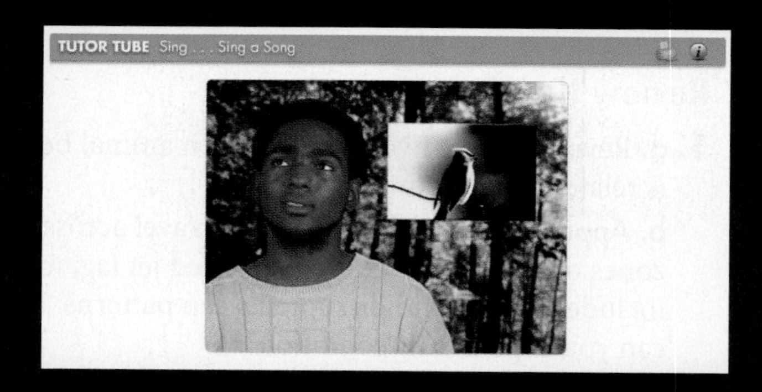

CHECKING YOUR *Scientific Literacy*

Refer to the lesson content and digital assets below as you prepare for your chapter assessment. Then, evaluate your understanding of animal behavior by answering these questions.

1. **Scientific and Engineering Practice Developing and Using Models** Construct models to illustrate the processes of classical conditioning and insight learning. Then, use your models, as well as the models shown in the text, to explain to a friend the different types of learned behaviors.

2. **Crosscutting Concept Patterns** Explain how patterns in the environment, such as daily or seasonal changes, affect an animal's behavior.

3. **Craft and Structure** This chapter explains many different types of animal behaviors. How does the structure of the text in the lessons and chapter help you understand the different types of animal behaviors?

4. **STEM** Researchers have been studying animal behaviors for centuries. New technology sometimes allows researchers to discover more about a process. What types of studies of animal behaviors do you think would benefit most from the use of new technology? Explain your answer.

29.2 Animals in Their Environments

- Many animals respond to periodic changes in the environment with daily or seasonal cycles of behavior.
- Choosing mates, defending or claiming territories or resources, and forming social groups can increase evolutionary fitness.
- Animals may use a variety of signals to communicate with one another. Some animals are also capable of language.

circadian rhythm 847 • migration 847 • courtship 848 • territory 848 • aggression 848 • society 848 • kin selection 849 • communication 850 • language 851

Biology.com

Art in Motion Watch social behavior in a population.

Art Review Identify animal communication strategies.

Data Analysis See how researchers improve understanding of animal migration behavior by improving data collection technology.

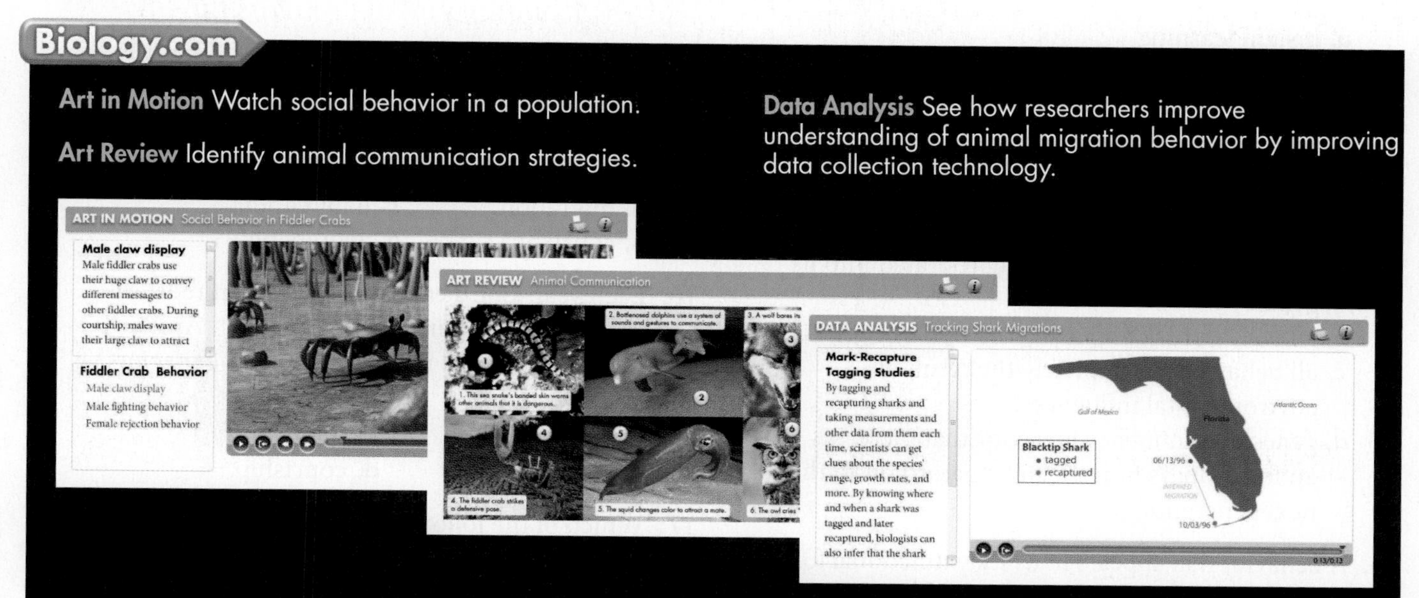

Design Your Own Lab

Termite Tracks This chapter lab is available in your lab manual and at Biology.com

29 Assessment

29.1 Elements of Behavior

Understand Key Concepts

1. The way an organism reacts to stimuli in its environment is called
   a. behavior.
   b. learning.
   c. conditioning.
   d. imprinting.

2. A decrease in response to a stimulus that neither rewards nor harms an animal is called
   a. instinct.
   b. operant conditioning.
   c. habituation.
   d. classical conditioning.

3. Insight learning is common among
   a. dogs.
   b. primates.
   c. birds and insects.
   d. birds only.

4. Many complex behaviors combine innate behavior with learning. Which of these behaviors is shown below?

   a. insight learning
   b. imprinting
   c. classical conditioning
   d. operant conditioning

5. Animal behaviors can evolve through natural selection because
   a. what an animal learns is incorporated into its genes.
   b. all behavior is completely the result of genes.
   c. all behavior is completely the result of environmental influences.
   d. genes that influence behavior that increases an individual's fitness can be passed on to the next generation.

6. A behavior that appears in its fully functional form the first time an animal performs it is
   a. learned.
   b. habituated.
   c. imprinted.
   d. innate.

7. Describe an example of a stimulus and a corresponding response in animal behavior.

8. What is the brain's role in an animal's response to a stimulus?

9. How can habituation contribute to an animal's survival?

10. Describe Pavlov's experiment. What is this type of learning called?

11. What is operant conditioning?

Think Critically

12. **Infer** A baby smiles when her mother comes near. Often, the baby is picked up and cuddled as a result of smiling. Explain what type of learning the baby is showing.

13. **Apply Concepts** Explain how a racehorse's ability to win races is a combination of inherited and learned behaviors.

14. **Production and Distribution of Writing** Choose a kind of animal with which you are familiar. Think of three questions about that animal's behavior that you might ask. Then describe the observations you would need to make in order to answer the questions.

29.2 Animals in Their Environments

Understand Key Concepts

15. A threatening behavior with which an animal exerts dominance over another is
   a. migration.
   b. courtship.
   c. habituation.
   d. aggression.

16. Each year, a bird called the American redstart travels from its winter home in South America to its nesting area in New York. This behavior is called
   a. migration.
   b. competition.
   c. imprinting.
   d. courtship.

17. Which of the following is NOT a type of social behavior?
   a. operant conditioning
   b. territoriality
   c. hunting in a pack
   d. courtship

18. A system of communication that uses meaningful sounds, symbols, or gestures according to specific rules is called
   a. behavior.
   b. language.
   c. competition.
   d. a signature.

19. Because a highway has been constructed through a forest, many of the animals that once lived there have had to move to a different wooded area. Is their move an example of migration? Explain your answer.

20. Identify two ways in which social behavior can benefit an animal.

21. **Craft and Structure** Analyze and explain how aggression and territorial behavior are related.

22. What are pheromones? Give an example of how they are used.

23. What is kin selection?

Think Critically

24. **Integration of Knowledge and Ideas** When temperatures are low and food is scarce, some mammals enter into a state of dormancy. Dormancy is an energy-saving adaptation in which metabolism decreases and, therefore, body temperature declines. The graph below tracks a ground squirrel's body temperature over the course of a year.
   a. Describe the pattern that you observe.
   b. What can you infer about the squirrel's behavior at different times of the year?

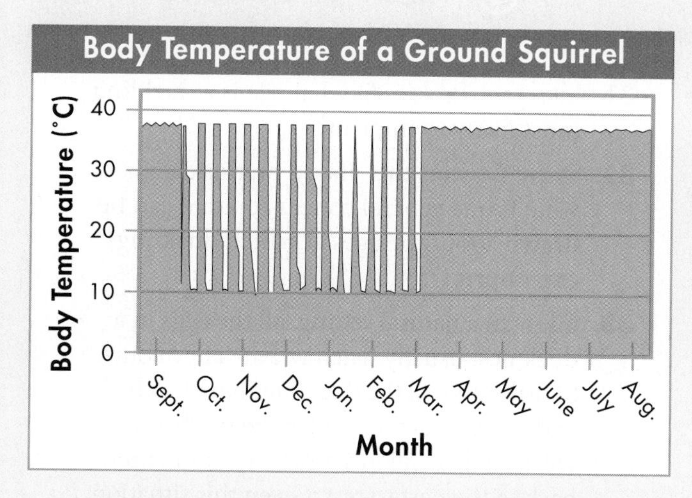

25. **Form a Hypothesis** Although the members of many animal species derive benefits from living in social groups, members of other species live alone. What might be some of the advantages of solitary living?

solve the CHAPTER MYSTERY

ELEPHANT CALLER ID?

Elephants can communicate with loud calls. But they can also communicate with vibrations that can travel for several kilometers. They do so by drumming on the ground with their feet and by making low rumbling calls that contain very low frequency sounds called "infrasound." The air vibrations can travel up to 16 kilometers while the in-ground vibrations can travel up to 32 kilometers.

Elephants detect these vibrations with specialized pads of fat in their feet and receptor cells in their feet and trunks. By standing motionless and pressing their feet and trunks to the ground, the elephants are better able to sense the vibrations. Particular patterns of vibrations can even identify individuals—like caller ID for elephants!

The vibrations can contain greetings, locations of food, and warnings of danger. The elephants at the watering hole had detected a message from another herd of elephants far away: "Warning! Lions!"

1. **Compare and Contrast** How do vibrations in the ground compare with airborne sounds as a means of long-distance communication?
2. **Form a Hypothesis** When researchers play back sounds and vibrations from elephant groups living far from Etosha, the elephants in Etosha don't always react. Why might this be the case?
3. **Craft and Structure** Analyze the author's purpose in choosing elephant behavior as the topic of the Chapter Mystery.
4. **Connect to the Big idea** How do elephants' responses to low-frequency vibrations in their environment affect their survival?

Connecting Concepts

Use Science Graphics

Mice can learn to run through a maze to find a food reward. As they have more practice runs in the maze, they take fewer wrong turns and reach the food more quickly. Twelve mice are put in a maze once a day for 10 days. The mean of their times to reach the food is calculated and plotted as the red line below. The mice are then kept out of the maze for a month. The blue line shows the results of those later trials. Use the graph to answer questions 26–28.

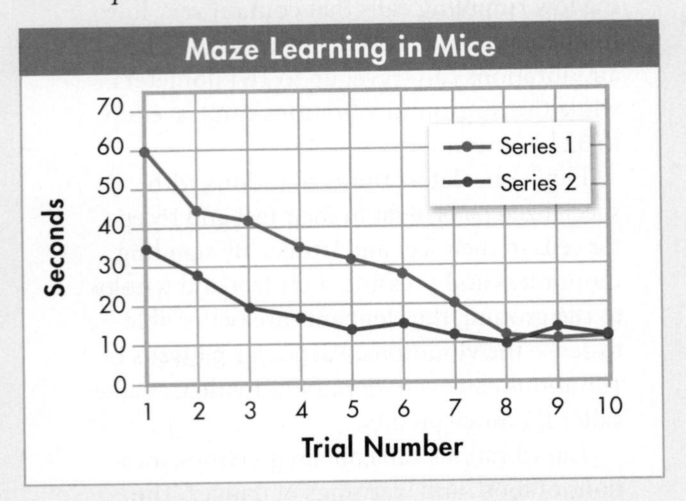

26. **Integration of Knowledge and Ideas** Explain what is happening after trial 8 on both sets of trials.

27. **Pose Questions** After the first set of trials, what kinds of questions might the experimenters have asked that resulted in their performing the second set of trials?

28. **Integration of Knowledge and Ideas** Explain the difference in the shapes of the graphs of the two trials.

Write About Science

29. **Production and Distribution of Writing** Write a paragraph describing something you have learned by insight learning. Explain how you used past knowledge and experience in learning it. (*Hint:* In your paragraph, explain what insight learning is.)

30. **Assess the**  Choose a social behavior in an animal. Describe how the behavior represents a response to the environment that aids in the survival of the species.

Analyzing Data

Integration of Knowledge and Ideas

A researcher found that the likelihood that a duckling would imprint depended on its age. After hatching, ducklings were kept isolated for different lengths of time and then placed with an adult duck. The imprinting data appear in this table.

Age (in hours)	Percent Successful Imprints
2	0
6	0
12	15
15	50
18	20
24	0
28	0

31. **Interpret Tables** At what age is a duckling most likely to imprint?

32. **Draw Conclusions** Based on this data, what is one general conclusion that can be drawn about the ages at which ducklings can imprint?

33. **Infer** In a natural setting, all the eggs in a duck nest usually hatch within a few hours. Shortly after all the ducklings have hatched, and periodically thereafter, the mother duck will lead them all to a food source and then back to the nest to rest. Given this situation, which ducklings are most likely to imprint on their mother?

Standardized Test Prep

Multiple Choice

1. A rat that learns to press a button to get food is exhibiting
   A insight learning. C classical conditioning.
   B operant conditioning. D habituation.

2. A dog that always salivates at the ringing of a bell is exhibiting
   A insight learning. C classical conditioning.
   B operant conditioning. D habituation.

3. A chimpanzee that stacks boxes in order to reach a banana hanging from the ceiling is showing
   A insight learning. C classical conditioning.
   B operant conditioning. D habituation.

4. A bird that stops responding to a repeated warning call when the call is not followed by an attack is showing
   A insight learning. C classical conditioning.
   B operant conditioning. D habituation.

5. Which kind of behavior does NOT involve learning?
   A habituation C imprinting
   B trial and error D instinct

6. A male three-spined stickleback fish will attack male red-bellied sticklebacks and models of fishes that have a red underside. It will not attack males or models lacking a red underside. What can you conclude from the three-spined stickleback's behavior?
   A The stimulus for an attack is a red underside.
   B The stimulus for an attack is aggression.
   C The stimulus for an attack is the presence of a fish with red fins.
   D The stimulus for an attack is the presence of a fish model.

7. Which of the following is NOT an innate behavior?
   A a dog looking for its food dish
   B a baby mammal sucking milk
   C a worm moving away from bright light
   D a spider spinning a web

Questions 8–9

A researcher observed sedge warblers during breeding season. She charted the number of different songs a male bird sang compared to the time it took him to pair with a mate. The graph shows her data.

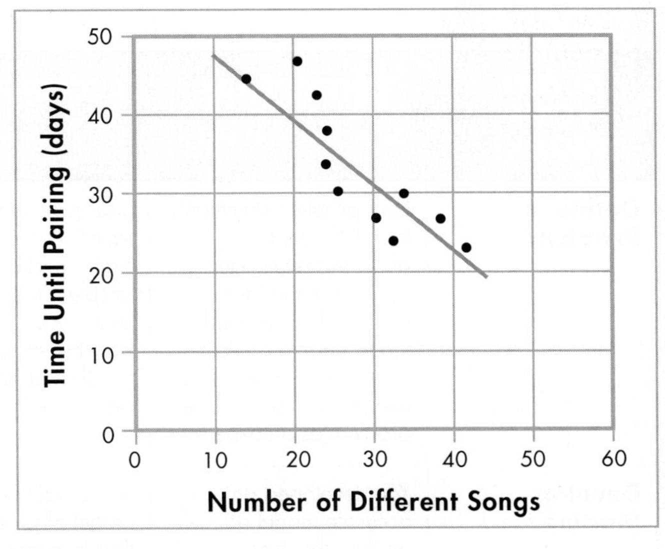

8. The researcher was trying to find out whether there is a correlation between
   A the number of a male bird's songs and the number of offspring.
   B the number of a male bird's songs and his attractiveness to females.
   C a male's age and when he mates.
   D a female's age and when she mates.

9. What can you conclude based on the graph?
   A Males prefer females that do not sing.
   B Females prefer males that do not sing.
   C Males prefer females with a larger number of songs.
   D Females prefer males with a larger number of songs.

Open-Ended Response

10. How does defending a specific territory benefit an animal?

If You Have Trouble With . . .										
Question	1	2	3	4	5	6	7	8	9	10
See Lesson	29.1	29.1	29.1	29.1	29.1	29.1	29.1	29.2	29.2	29.2

NGSS Problem-based Learning

UNIT 7 PROJECT

Your Self-Assessment

The rubric below will help you evaluate the model of your item inspired by nature.

SCORE YOUR WORK!	EXEMPLARY Score your work a 4 if:	ACCOMPLISHED Score your work a 3 if:	DEVELOPING Score your work a 2 if:	BEGINNING Score your work a 1 if:
Define a Problem ▶	Your problem statement is well thought out and comprehensive. You understand the animal, the trait, and the technical problem the trait may solve. You feel you can discuss the problem as an expert.	Your problem statement identifies the animal, the trait, and the technical problem the trait may solve. You understand how the structure and function of animal traits are related.	You have written a problem statement that identifies the animal and the trait, but you are unsure of the technical problem the trait may solve.	You are aware that animals have unique traits but are unable to identify the function that a specific trait serves for an animal.
Develop Models ▶	You developed an accurate model of an item that has been redesigned to incorporate an animal trait. Your model is supported by concrete scientific reasoning.	You developed a model of an item that incorporates an animal trait. You have thoroughly explained the reasoning behind your ideas.	You developed a model, but you are unable to explain the reasoning behind your ideas.	You developed a model, but you are unsure how to incorporate an animal trait into it.
	TIP Be sure your work represents your full idea.			
Plan and Carry Out an Investigation ▶	You formulated a well-defined plan that identifies the information you need to improve your model. You used a variety of reliable sources to gather your data.	You formulated a plan that identifies the information you need to improve your model. You gathered the data you needed from at least one reliable source.	You formulated a plan but had difficulty finding reliable sources from which to gather the data you needed.	You are unsure of the information you need to improve your model. You did not gather data from reliable sources.
Analyze and Interpret Data ▶	You organized your research and collected additional data to solidify your analysis. You applied your data to improve your model.	You organized your research and applied your data to improve your model.	You organized your research but were unsure how to use your data to improve your model.	You are unsure how to organize your data and improve your model.
	TIP Organize your research to identify patterns and relationships in your data.			
Communicate Information ▶	You presented your finished model or prototype to an audience beyond your class or school. Your ideas have the potential to be developed into a commercial product.	Your presentation showcased your model and the reasoning behind your design decisions. You accepted feedback from your peers and revised your model accordingly.	You have shown your model to your teacher and classmates but have not obtained feedback on it.	You are unsure how to present your model to your class.

The Human Body

UNIT 8

Chapters

30 Digestive and Excretory Systems

31 Nervous System

32 Skeletal, Muscular, and Integumentary Systems

33 Circulatory and Respiratory Systems

34 Endocrine and Reproductive Systems

35 Immune System and Disease

INTRODUCE the **Big ideas**

- **Homeostasis**
- **Structure and Function**

"Have you ever thought about the teamwork involved in tying your shoe? Your eyes locate the laces. Muscles, bones, and nerves coordinate an intricate series of maneuvers to pull them tight and tie the knot. In the background, lungs and bloodstream work constantly to bring oxygen and chemical fuel to those muscles and nerves. The body is an incredible machine, but what is most extraordinary is the way in which its systems and organs work together."

NGSS Problem-based Learning

UNIT 8 PROJECT

Spotlight on NGSS
- **Core Idea LS1.A** Structure and Function
- **Practice** Developing and Using Models
- **Crosscutting Concept** Systems and System Models

Body Mechanics

Read the following information about medical technologies. As you learn about human body systems, research how these technologies work and think about how they might be improved or adapted.

The human body is a complex machine made up of many interconnected parts that work together. But what happens if one of these parts is missing or becomes damaged over time? How can you keep the "machine" running?

The mechanics charged with solving this problem include physicians, biologists, chemists, physicists, engineers, inventors, and designers. They collaborate to design solutions that prolong or improve the lives of people with disabilities and chronic illnesses.

Various technologies are available to replace a *part* of the human body or a *function* that a diseased body is no longer able to do on its own. Some examples are shown here.

For people who live with a disability such as limb loss, a prosthetic limb can greatly improve quality of life.

Your Task: Refine the Design

Evaluate a technology modeled on human body systems, and propose a way to improve the design. A useful technology is one that solves a problem. Your task is to research a medical problem and the technology available to solve it. The solution may be mechanical, or it may be a biochemical treatment for a functional problem, such as a hormone deficiency. After studying the nature of the problem and the range of possible solutions, you will suggest an optimized solution. It may be a refinement of an existing solution, or it may be an alternative solution that you design. Use the rubric at the end of the unit to evaluate your work for this task.

Define the Problem Discuss with your teacher the technology you wish to focus on. What is the problem or need that the technology was designed to solve? List the strengths and weaknesses of the technology. Make sure to consider the people who use it. What else might they add to your list? Explain why the current solution to the problem is incomplete.

Identify the requirements that must be met in order to solve the problem or need. These requirements are called design criteria. Then list the factors that might limit the success of your design. These factors are called design constraints.

Design a Solution Generate ideas for solving the problem. How would you design an improvement or alternative to the solution(s) currently in use? Evaluate your ideas for a solution based on the design criteria and constraints that you have specified. Select one or more of these ideas to develop further.

Develop Models Use sketches, physical models, or simulations to help you visualize and test your design. What does the model tell you about the strengths and limitations of your design? Ask your classmates for feedback on your proposed solution, and revise the design as needed. What feedback might you expect from the people in need of your solution? What flaws might they identify based on your model?

Obtain, Evaluate, and Communicate Information Create a Web page or presentation that summarizes your work. With your teacher's help, identify an expert on the problem you are solving, and ask that person to provide feedback on your design. Use the feedback to further improve your design.

Medical implants are devices that are placed surgically inside the body. These include artificial joints (such as knees or hips), artificial heart valves, pacemakers, cochlear implants, and dental implants.

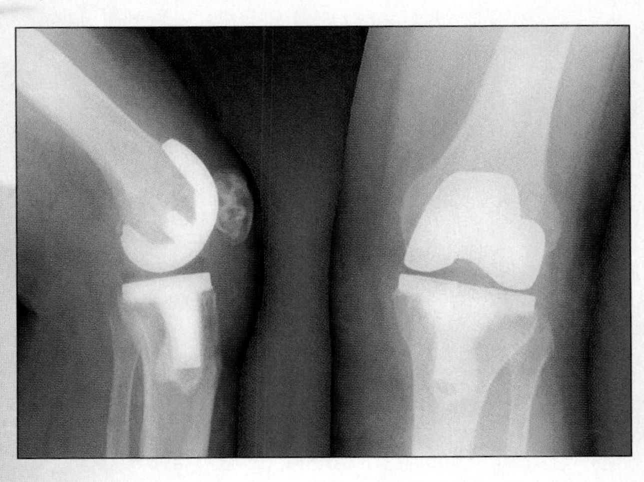

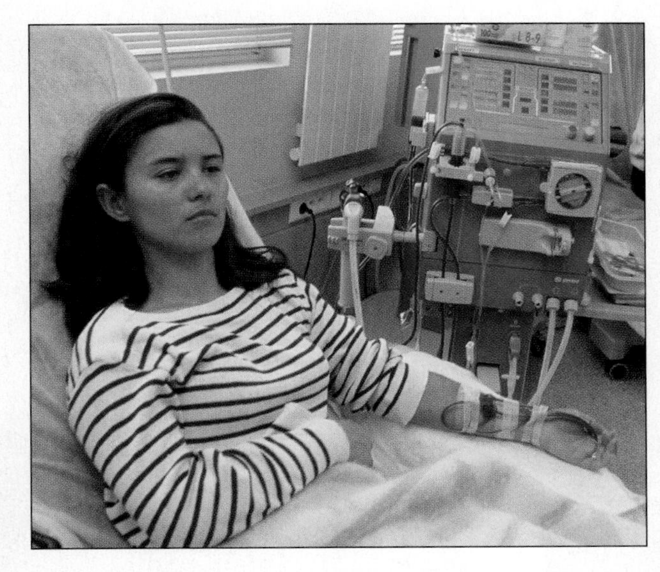

A dialysis machine is a device that cleans and filters the blood. It is used to treat people suffering from kidney failure.

Biology.com

Find resources on medical technologies research to help inform your solution design.

30 Digestive and Excretory Systems

Big idea: Homeostasis

Q: How are the materials that enter and leave your body related to the processes that maintain homeostasis?

Food sellers display their goods at a floating food market on Dal Lake in northern India.

INSIDE:

- 30.1 Organization of the Human Body
- 30.2 Food and Nutrition
- 30.3 The Digestive System
- 30.4 The Excretory System

BUILDING *Scientific Literacy*

Deepen your understanding of the processes of digestion and excretion by engaging in key practices that allow you to make connections across concepts.

NGSS You will use models to explain how the structures and functions of the digestive and excretory systems contribute to homeostasis.

STEM Analyze data that connect what digestive enzymes do with the chemical environments in which they function.

Common Core You will use textual evidence and information from multiple sources to support your analysis of claims related to diet and health.

CHAPTER MYSTERY

THE TELLTALE SAMPLE

On the first day of summer football practice, all players were given a physical. Each student was handed a plastic cup and directed to the restroom. "Please provide me with a sample," the physician requested. The athletes had no idea how much could be learned about their health and behavior from a small urine sample.

Right after handing over their samples, Philip and Seth were sent home and told to drink plenty of water before the next practice. The next day, Andrew was told to see his physician because he could have diabetes. Several days later, another student was dropped from the team for violating the school's antidrug policy. How was all this information obtained? As you read this chapter, look for clues to discover what can be learned about the body by examining what leaves it. Then, solve the mystery.

Biology.com

Finding the solution to The Telltale Sample mystery is only the beginning. Take a video field trip with the ecogeeks of Untamed Science to see where the mystery leads.

Go online to access additional resources including:
• eText • Flash Cards • Lesson Overviews • Chapter Mystery

30.1 Organization of the Human Body

Key Questions

How is the human body organized?

What is homeostasis?

Vocabulary

epithelial tissue
connective tissue
nervous tissue
muscle tissue
homeostasis
feedback inhibition

Taking Notes

Preview Visuals Examine **Figure 30–2.** For each system, describe how you think it interacts with at least one other system.

THINK ABOUT IT The batter slaps a ground ball to the shortstop, who fields it cleanly and throws the ball toward your position—first base. In a single motion, you extend your glove hand, catch the ball, and extend your foot to touch the edge of the base. An easy out, a routine play. But think about how many systems of your body are involved in making this type of "routine" play. How do they all work together?

Organization of the Body

How is the human body organized?

Every cell in the human body is both an independent unit and an interdependent part of a larger community—the entire organism. To complete a winning play, a player at first base has to use her eyes to watch the ball and use her brain to figure out how to position her body. With the support of her bones, muscles move her body to first base. Meanwhile, the player's lungs absorb oxygen, which her blood carries to cells for use during cellular respiration. Her brain monitors the location of the ball and sends signals that guide her glove hand to make the catch.

How can so many individual cells and parts work together so efficiently? One way to answer this question is to study the organization of the human body. **The levels of organization in the body include cells, tissues, organs, and organ systems.** At each level of organization, these parts of the body work together to carry out the major body functions.

Cells A cell is the basic unit of structure and function in living things. As you learned in Chapter 7, individual cells in multicellular organisms tend to be specialized. Specialized cells, such as bone cells, blood cells, and muscle cells, are uniquely suited to perform a particular function.

Tissues A group of cells that perform a single function is called a tissue. There are four basic types of tissue in the human body—epithelial, connective, nervous, and muscle. **Figure 30–1** shows examples of each type of tissue.

	Epithelial Tissue	Connective Tissue	Nervous Tissue	Muscle Tissue
FUNCTIONS	Protection, absorption, and excretion of materials	Binding of epithelial tissue to structures, support, and transport of substances	Receiving and transmitting nerve impulses	Voluntary and involuntary movements
LOCATIONS	Skin, lining of digestive system, certain glands	Under skin, surrounding organs, blood, bones	Brain, spinal cord, and nerves	Skeletal muscles, muscles surrounding digestive tract and blood vessels, the heart

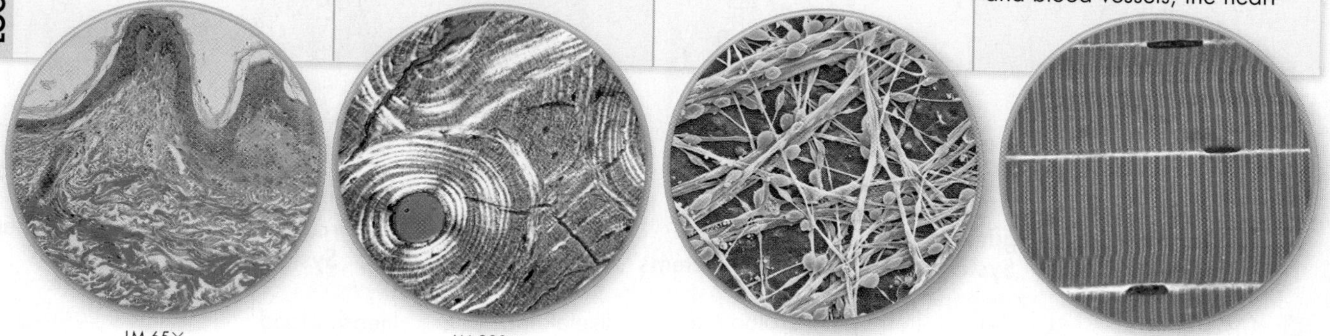

LM 65×

LM 280×

SEM 295×

LM 275×

FIGURE 30–1 Types of Tissues The four major types of tissues in the human body are epithelial tissue, connective tissue, nervous tissue, and muscle tissue. **Predict** ***Which organ may not contain all four types of tissue?***

▶ ***Epithelial Tissue*** The tissue that lines the interior and exterior body surfaces is called **epithelial tissue.** Your skin and the lining of your stomach are both examples of epithelial tissue.

▶ ***Connective Tissue*** A type of tissue that provides support for the body and connects its parts is **connective tissue.** This type of tissue includes fat cells, bone cells, and even blood cells. Many connective tissue cells produce collagen, a long, tough fiber-like protein that is the most common protein in the body. Collagen gives tissues strength and resiliency, helping them to keep their shape even under pressure.

▶ ***Nervous Tissue*** Nerve impulses are transmitted throughout the body by **nervous tissue.** Neurons, the cells that carry these impulses, and glial cells, which surround and protect neurons, are both examples of nervous tissue.

▶ ***Muscle Tissue*** Movements of the body are possible because of **muscle tissue.** Some muscles are responsible for the movements you control, such as the muscles that move your arms and legs. Some muscles are responsible for movements you cannot control, such as the tiny muscles that control the size of the pupil in the eye.

Organs A group of different types of tissues that work together to perform a single function or several related functions is called an organ. The eye is an organ made up of epithelial tissue, nervous tissue, muscle tissue, and connective tissue. As different as these tissues are, they all work together for a single function—sight.

Organ Systems An organ system is a group of organs that perform closely related functions. For example, the brain and spinal cord are organs of the nervous system. The organ systems interact to maintain homeostasis in the body as a whole. The organ systems, along with their structures and main functions, are shown on the next page.

VISUAL SUMMARY

HUMAN BODY SYSTEMS

FIGURE 30–2 Although each of the organ systems shown here has a different set of functions, they all work together, as a whole, to maintain homeostasis.

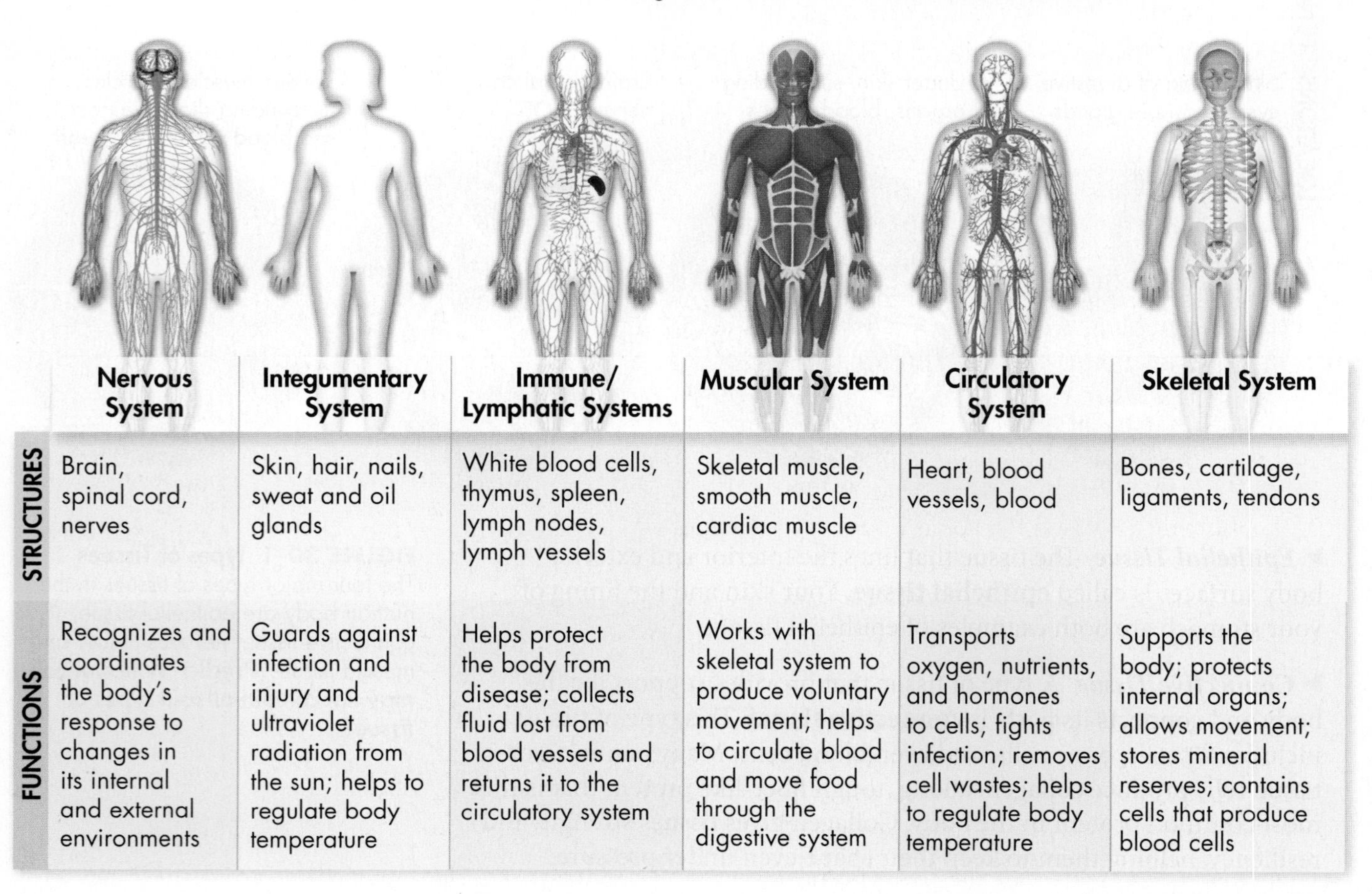

	Nervous System	Integumentary System	Immune/ Lymphatic Systems	Muscular System	Circulatory System	Skeletal System
STRUCTURES	Brain, spinal cord, nerves	Skin, hair, nails, sweat and oil glands	White blood cells, thymus, spleen, lymph nodes, lymph vessels	Skeletal muscle, smooth muscle, cardiac muscle	Heart, blood vessels, blood	Bones, cartilage, ligaments, tendons
FUNCTIONS	Recognizes and coordinates the body's response to changes in its internal and external environments	Guards against infection and injury and ultraviolet radiation from the sun; helps to regulate body temperature	Helps protect the body from disease; collects fluid lost from blood vessels and returns it to the circulatory system	Works with skeletal system to produce voluntary movement; helps to circulate blood and move food through the digestive system	Transports oxygen, nutrients, and hormones to cells; fights infection; removes cell wastes; helps to regulate body temperature	Supports the body; protects internal organs; allows movement; stores mineral reserves; contains cells that produce blood cells

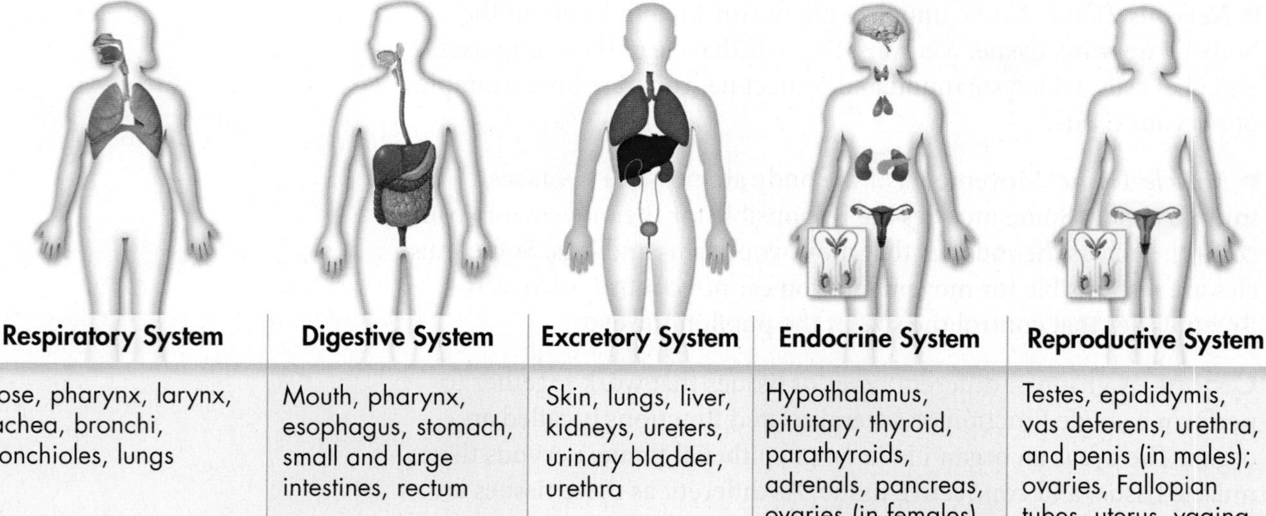

	Respiratory System	Digestive System	Excretory System	Endocrine System	Reproductive System
STRUCTURES	Nose, pharynx, larynx, trachea, bronchi, bronchioles, lungs	Mouth, pharynx, esophagus, stomach, small and large intestines, rectum	Skin, lungs, liver, kidneys, ureters, urinary bladder, urethra	Hypothalamus, pituitary, thyroid, parathyroids, adrenals, pancreas, ovaries (in females), testes (in males)	Testes, epididymis, vas deferens, urethra, and penis (in males); ovaries, Fallopian tubes, uterus, vagina (in females)
FUNCTIONS	Brings in oxygen needed for cellular respiration and removes excess carbon dioxide from the body	Breaks down food; absorbs nutrients; eliminates wastes	Eliminates waste products from the body	Controls growth, development, and metabolism; maintains homeostasis	Produces gametes; in females, nurtures and protects developing embryo

Homeostasis

What is homeostasis?

Some things are easy to observe. When you run or swim or even write the answer to a test question, you can see your body at work. But behind the scenes, your body's systems are working constantly to do something that is difficult to see and that few people appreciate—maintaining a controlled, stable internal environment. This stable environment is called **homeostasis,** which means "similar standing." **Homeostasis describes the relatively constant internal physical and chemical conditions that organisms maintain despite changes in internal and external environments.** Homeostasis may not be obvious, but for a living organism, it's literally a matter of life or death.

Feedback Inhibition If you've ever watched someone driving a car down a relatively straight road, you may have noticed how the person constantly moves the wheel left or right, adjusting direction to keep the vehicle in the middle of the lane. In a certain sense, that's how the systems of the body work, too, keeping internal conditions within a certain range, and never allowing them to go too far to one side or the other.

▶ ***A Nonliving Example*** One way to understand homeostasis is to look at a nonliving system that automatically keeps conditions within a certain range like a home heating system. In most homes, heat is supplied by a furnace that burns oil or natural gas. When the temperature within the house drops below a set point, a thermostat sensor switches the furnace on. Heat produced by the furnace warms the house. When the temperature rises above the set point, the thermostat switches the furnace off, keeping the temperature within a narrow range.

A system like this is said to be controlled by feedback inhibition. **Feedback inhibition,** or negative feedback, is the process in which a stimulus produces a response that opposes the original stimulus. **Figure 30–3** summarizes the feedback inhibition process in a home heating system. When the furnace is switched on, it produces a product (heat) that changes the environment of the house (by raising the air temperature). This environmental change then "feeds back" to "inhibit" the operation of the furnace. In other words, heat from the furnace eventually raises the temperature high enough to trigger a feedback signal that switches the furnace off. Systems controlled by feedback inhibition are generally very stable.

In Your Notebook *Describe another example of a nonliving system that requires constant adjustment.*

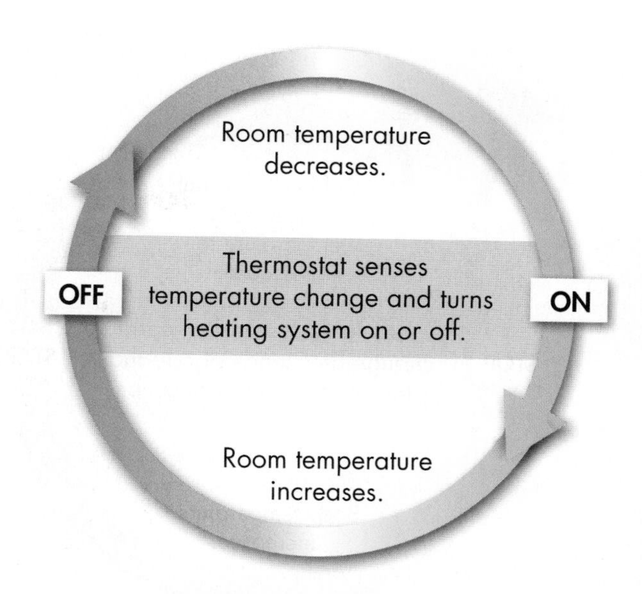

FIGURE 30–3 Feedback Inhibition A home heating system uses a feedback loop to maintain a stable, comfortable environment within a house.
Interpret Diagrams ***What is the stimulus in this feedback loop?***

BUILD Vocabulary

ACADEMIC WORDS The noun **inhibition** means "the act of blocking the action of." Therefore, feedback inhibition refers to a response that blocks further actions of a stimulus.

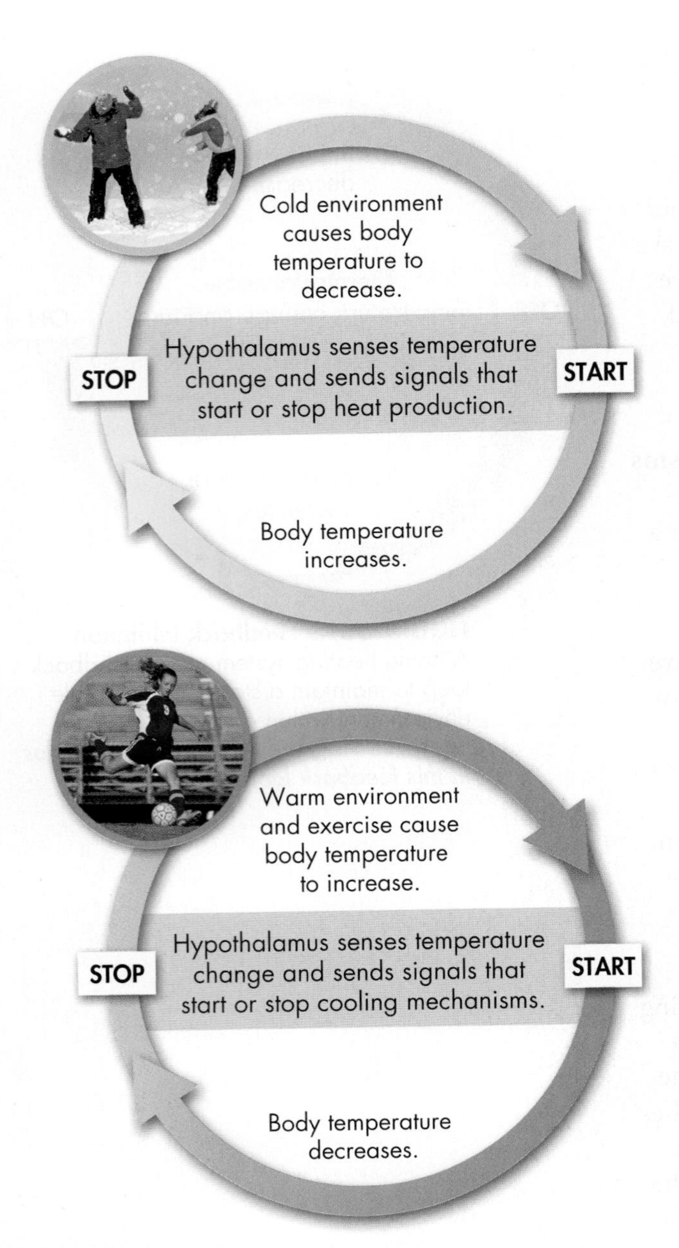

FIGURE 30–4 Body Temperature Control In the human body, temperature is controlled through various feedback inhibition mechanisms. **Infer** ***Why do you think moving around on a cold day helps to keep you warm?***

▶ ***A Living Example*** Could biological systems achieve homeostasis through feedback inhibition? Absolutely. All that is needed is a system that regulates some aspect of the cellular environment and that can respond to feedback from its own activities by switching on or off as needed. Such mechanisms are very common, not only in the human body, but in all forms of life.

One example is the maintenance of body temperature. The body regulates temperature by a mechanism that is remarkably similar to that of a home heating system. You can follow body temperature regulation in **Figure 30–4.** A part of the brain called the hypothalamus contains nerve cells that monitor both the temperature of the skin at the surface of the body and the temperature of organs in the body's core.

If the nerve cells sense that the core temperature has dropped much below 37°C, the hypothalamus produces chemicals that signal cells throughout the body to speed up their activities. Heat produced by this increase in activity, especially cellular respiration, causes a gradual rise in body temperature, which is detected by nerve cells in the hypothalamus.

Have you ever been so cold that you began to shiver? If your body temperature drops well below its normal range, the hypothalamus releases chemicals that signal muscles just below the surface of the skin to contract involuntarily—to "shiver." These muscle contractions release heat, which helps the body temperature to rise toward the normal range.

If body temperature rises too far above 37°C, the hypothalamus slows down cellular activities to reduce heat production. This is one of the reasons you may feel tired and sluggish on a hot day. The body also responds to high temperatures by producing sweat, which helps to cool the body surface by evaporation.

Quick Lab
OPEN-ENDED INQUIRY

Maintaining Temperature 

You will receive a thermometer and three beakers of water at the following temperatures: 25°C, 35°C, and 40°C. Develop a method to keep the temperature of the 35°C water within one degree for a period of fifteen minutes. You may use the contents of the other two beakers.

Analyze and Conclude

1. Compare and Contrast Compare this experiment to what happens in your own body during temperature regulation.

2. Interpret Visuals Make a feedback loop similar to the ones in **Figure 30–4** that shows how feedback inhibition was involved in this activity.

The Liver and Homeostasis The liver is technically part of the digestive system because it produces bile, which aids in the digestion of fats. However, it is also fair to say that the liver is one of the body's most important organs for homeostasis.

For example, when proteins are broken down for energy, ammonia, a toxic byproduct, is produced. The liver quickly converts ammonia to urea, which is much less toxic. The kidneys, as you will read a bit later, then remove urea from the blood. The liver also converts many dangerous substances, including some drugs, into compounds that can be removed from the body safely.

One of the liver's most important roles involves regulating the level of a substance we take almost for granted as something completely harmless—the simple sugar, glucose. Glucose is obtained from the foods we eat, and cells take glucose from the blood to serve as a source of energy for their everyday activities. Naturally, right after a meal, as the body absorbs food molecules, the level of glucose in the blood begins to rise. That's where the liver comes in. By taking glucose out of the blood, it keeps the level of glucose from rising too much. As the body uses glucose for energy, the liver releases stored glucose to keep the level of the sugar from dropping too low.

The liver's role in keeping blood glucose levels within a certain range is critical. Too little glucose, and the cells of the nervous system will slow down to the point that you may lose consciousness and pass out. On the other hand, too much glucose gradually damages cells in the eyes, kidneys, heart, and even the immune system. Abnormally high levels of glucose are associated with a disease called diabetes. In diabetes, changes occur in either the pancreas or body cells that affect the cells' ability to absorb glucose. Diabetes, one of the fastest-growing health problems in the developed world, is the unfortunate result of failure of homeostasis with respect to blood glucose levels.

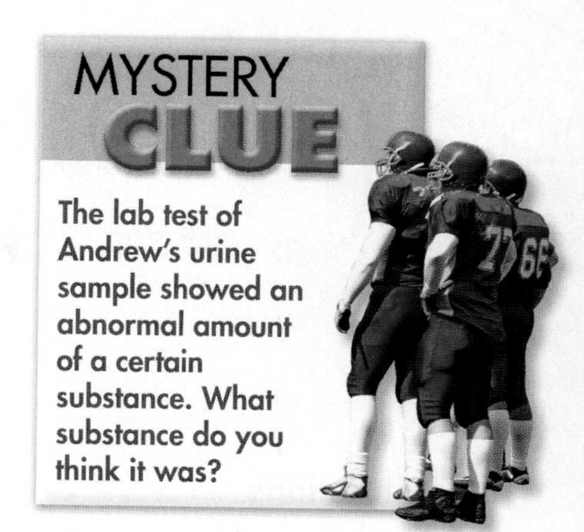

The lab test of Andrew's urine sample showed an abnormal amount of a certain substance. What substance do you think it was?

30.1 Assessment

Review Key Concepts

1. a. **Review** What are the four types of tissues?
   b. **Explain** Describe the function of three organ systems depicted in **Figure 30–2.**
   c. **Classify** Compare the characteristics of two types of tissues. Identify parts of the body that contain these types of tissues.
2. a. **Review** What is homeostasis?
   b. **Explain** What are two roles of the liver in maintaining homeostasis?
   c. **Apply Concepts** Do you think that feelings of hunger and fullness are an example of feedback inhibition? Explain.

VISUAL THINKING

3. Draw a Venn diagram to relate the four basic levels of organization in the human body. Provide at least three examples for each level of organization. *Hint:* Your Venn diagram should have a nesting structure. One set of examples could be skin cells, epithelial tissue, skin, and the integumentary system.

30.2 Food and Nutrition

Key Questions

Why do we need to eat?

What nutrients does your body need?

What is meant by the term "balanced diet"?

Vocabulary

Calorie
carbohydrate
fat
protein
vitamin
mineral

Taking Notes

Outline Before you read, make an outline of the major headings in the lesson. As you read, fill in main ideas and supporting details for each heading.

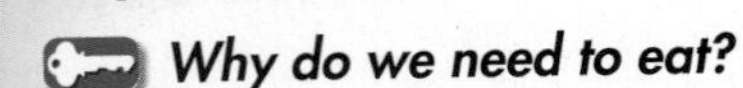

THINK ABOUT IT When you feel hungry, how would you describe the feeling? Do you feel full of energy and ready to go? Or do you feel weak and just a little bit lazy? Why? What do these sensations tell us about the purpose of food in the body?

Food and Energy

Why do we need to eat?

Have you ever wondered why you need food? The most obvious answer is energy. You need energy to climb stairs, lift books, run, and even to think. Just as a car needs gasoline, your body needs fuel, and food is that fuel. **Molecules in food contain chemical energy that cells use to produce ATP. Food also supplies raw materials your body needs to build and repair tissues.**

Energy The energy available in food can be measured in a laboratory in a surprisingly simple way—by burning it! When food is burned, most energy in the food is converted to heat, which is measured in terms of calories. A calorie is the amount of heat needed to raise the temperature of 1 gram of water by 1 degree Celsius. The "Calories" you've heard about in food are actually dietary Calories, written with a capital C. One dietary **Calorie** is equal to 1000 calories, or 1 kilocalorie (kcal). As you may recall, the energy stored in food molecules is released during cellular respiration and used to produce the ATP molecules that power cellular activities.

Raw Materials Chemical pathways, including cellular respiration, can extract energy from almost any type of food. So why does it matter which foods you eat? The reason is that food also supplies the raw materials used to build and repair body tissues. Some of these raw materials are needed to make enzymes, the lipids in cell membranes, and even DNA. In fact, food contains at least 45 substances that the body needs but cannot manufacture. A healthy diet ensures that your body receives all of these required substances.

In Your Notebook *Prepare a table to fill in with information about the nutrients. For each nutrient, include foods in which it is found and describe its role in the body.*

Nutrients

What nutrients does your body need?

Nutrients are substances in food that supply the energy and raw materials your body uses for growth, repair, and maintenance. **The nutrients that the body needs include water, carbohydrates, fats, proteins, vitamins, and minerals.**

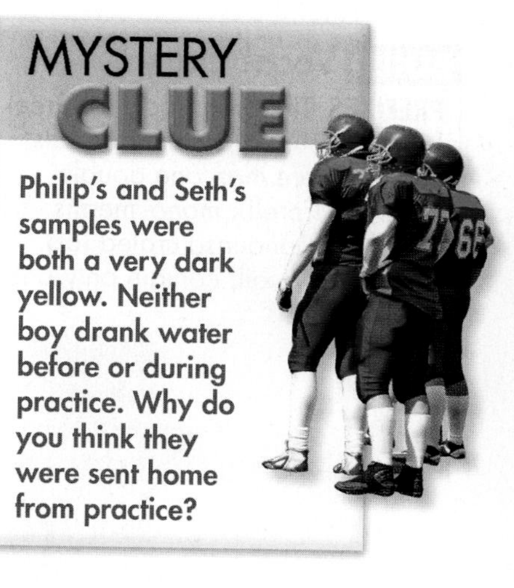

Water The most important nutrient is water. Every cell in the human body needs water because many of the body's processes, including chemical reactions, take place in water. Water makes up the bulk of blood, extracellular fluid, and other bodily fluids. On hot days or when you take part in strenuous exercise, sweat glands remove water from your tissues and release it as sweat on the surface of your body. Water is also lost from the body in urine and with every breath you exhale.

Humans need to drink at least 1 liter of fluid each day. If enough water is not taken in to replace what is lost, dehydration can result. Dehydration leads to problems with many body systems, and under extreme conditions it can be fatal.

Carbohydrates Simple and complex **carbohydrates** are a major source of energy for the body. **Figure 30–5** shows some of the foods that contain carbohydrates. The sugars found in fruits, honey, and sugar cane are simple carbohydrates, or monosaccharides and disaccharides. The starches found in grains, potatoes, and vegetables are complex carbohydrates, or polysaccharides. Starches are broken down by the digestive system into simple sugars. These molecules are absorbed into the blood and carried to cells throughout the body. Excess blood sugar is converted into glycogen, which is stored in the liver and in skeletal muscles. Excess sugar may also be converted to and stored as body fat.

Whole-grain breads, bran, and many fruits and vegetables contain the complex carbohydrate cellulose, often called fiber. Although the human digestive system cannot break down cellulose, you need fiber in your diet. The bulk supplied by fiber helps muscles move food and wastes through your digestive system. Fiber may also have other benefits, such as reducing the risk of heart disease and Type II diabetes.

FIGURE 30–5 Carbohydrates Pastas and cereals are also foods that are rich in carbohydrates. Simple carbohydrates do not have to be digested or broken down. Complex carbohydrates, such as those found in whole grains, must be broken down before they can be used by the body. **Infer** ***Which type of carbohydrate—simple or complex—provides the body with quick energy?***

BUILD Vocabulary

PREFIXES The prefix *poly-* is Greek for "many." Polyunsaturated fats contain more than one double bond. The prefix *mono-* means "single." Monounsaturated fats, such as olive oil, contain only one double bond.

Fats Because our society places great emphasis on a trim appearance, the word "fat" has a bad reputation. But fats, or lipids, are an important part of a healthy diet. **Fats** help the body absorb fat-soluble vitamins and are a part of cell membranes, nerve cells, and certain hormones. Deposits of fat protect and insulate body organs and are a source of stored energy.

Fats usually form when a glycerol molecule combines with fatty acids. Some of these acids, called essential fatty acids, cannot be made in the body and are needed to perform many of fat's functions. Based on the structure of their fatty acid chains, fats are classified as saturated or unsaturated. When there are only single bonds between the carbon atoms in the fatty acids, each carbon atom has the maximum number of hydrogen atoms and the fat is said to be saturated. Most saturated fats, such as butter, are solids at room temperature.

Unsaturated fats have one or more double bonds between carbon atoms, which reduces the number of hydrogen atoms in their fatty acids. Unsaturated fats are usually liquids at room temperature. Because many vegetable oils contain more than one double bond, they are called polyunsaturated.

Food manufacturers often modify unsaturated fats in vegetable oils by adding hydrogen to them. These processed fats are called trans fats. Trans fats are solid at room temperature and have a longer shelf life than unsaturated fats. However, recent studies suggest that trans fats may be associated with serious health concerns, including heart disease.

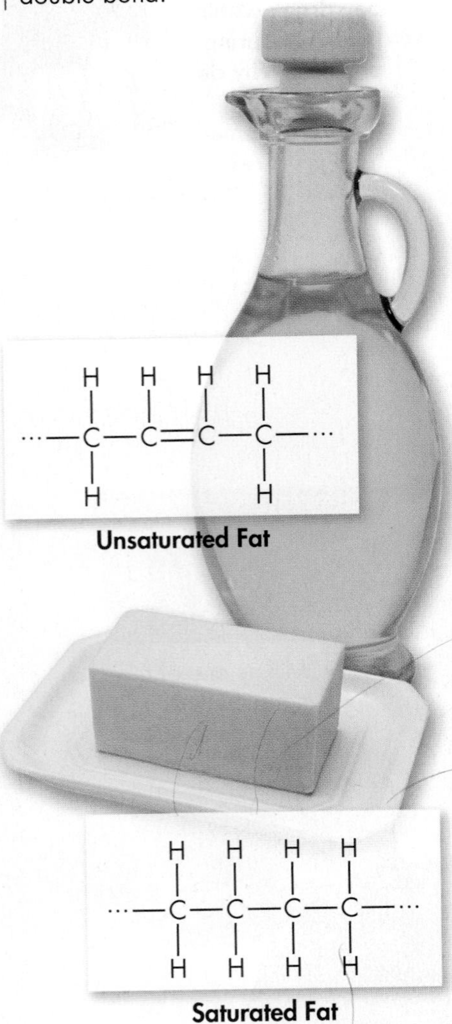

FIGURE 30–6 Fats At room temperature, most saturated fats are solid and most unsaturated fats are liquid. Saturated fats have been tied to many health problems. Consuming limited amounts of unsaturated fats, such as those found in avocados and olive oil, may have some benefits.

Proteins Proteins have a wide variety of roles in the body. **Proteins** supply raw materials for growth and repair of structures such as skin and muscle. Many enzymes that control cellular chemistry by increasing the rates of chemical reactions are made of proteins. Proteins also have regulatory and transport functions. For example, the hormone insulin is a protein that regulates the level of sugar in the blood. Hemoglobin, a protein found in red blood cells, helps transport oxygen. Proteins can also be used as energy sources when other nutrients, such as carbohydrates and fats, are in short supply.

Proteins are polymers of amino acids. The body is able to synthesize only 12 of the 20 amino acids used to make proteins. The other eight are called essential amino acids. Essential amino acids must be obtained from the foods that you eat. Foods shown below, such as meat, fish, eggs, and milk, generally contain all eight essential amino acids. Foods derived from plants, such as grains and beans, do not. People who don't eat animal products must eat a combination of plant foods, such as beans and rice, to obtain all of the essential amino acids.

Vitamins Organic molecules that the body needs in very small amounts are called **vitamins.** Most vitamins are needed by the body to help perform chemical reactions. If we think of proteins, fats, and carbohydrates as the building blocks of the body, then vitamins are the tools that help to put them together. As shown in **Figure 30–7,** most vitamins must be obtained from food. However, the bacteria that live in the large intestine are able to synthesize vitamins K and B_{12}.

There are two types of vitamins: fat-soluble and water-soluble. The fat-soluble vitamins A, D, E, and K can be stored in the fatty tissues of the body. The body can build up small deposits of these vitamins for future use. The water-soluble vitamins, which include vitamin C and the B vitamins, dissolve in water and cannot be stored in the body. Therefore, they should be included in the foods you eat each day.

A diet lacking certain vitamins can have serious health consequences. Eating a variety of foods will supply the daily vitamin needs of most people. Large doses of vitamin supplements do not benefit the body. Excessive amounts of the fat-soluble vitamins A, D, and K can be toxic.

FIGURE 30–7 Vitamins This table lists the food sources and functions of 14 essential vitamins. The fat-soluble vitamins are listed in the first four rows. **Interpret Tables** ***What is the function of vitamin K?***

Vitamins

Vitamin	Sources	Function
A (retinol)	Yellow, orange, and dark-green vegetables; fortified dairy products	Important for growth of skin cells; important for night vision
D (calciferol)	Fish oils, eggs; made by skin when exposed to sunlight; added to dairy products	Promotes bone growth; increases calcium and phosphorus absorption
E (tocopherol)	Green leafy vegetables, seeds, vegetable oils	Antioxidant; prevents cellular damage
K 	Green leafy vegetables; made by bacteria that live in human intestine	Needed for normal blood clotting
B_1 (thiamine)	Whole grains, pork, legumes, milk	Metabolism of carbohydrates
B_2 (riboflavin)	Dairy products, meats, vegetables, whole grains	Growth; energy metabolism
Niacin	Liver, milk, whole grains, nuts, meats, legumes	Important in energy metabolism
B_6 (pyridoxine)	Whole grains, meats, vegetables	Important for amino acid metabolism
Pantothenic acid	Meats, dairy products, whole grains	Needed for energy metabolism
Folic acid 	Legumes, nuts, green leafy vegetables, oranges, broccoli, peas, fortified grains	Involved in nucleic acid metabolism; prevents neural-tube defects
B_{12} (cyanocobalamin)	Meats, eggs, dairy products, enriched cereals	Involved in nucleic acid metabolism; maturation of red blood cells
C (ascorbic acid) 	Citrus fruits, tomatoes, red or green peppers, broccoli, cabbage, strawberries	Maintains cartilage and bone; antioxidant; improves iron absorption; important for healthy gums and wound healing
Biotin	Legumes, vegetables, meat	Coenzyme in synthesis of fat; glycogen formation; amino acid metabolism
Choline	Egg yolk, liver, grains, legumes	Part of phospholipids and neurotransmitters

Important Minerals		
Mineral	**Sources**	**Function**
Calcium	Dairy products, salmon, kale, tofu, collard greens, legumes	Bone and tooth formation; blood clotting; nerve and muscle function
Phosphorus	Dairy products, meats, poultry, grains	Bone and tooth formation; acid-base balance
Iron	Meats, eggs, legumes, whole grains, green leafy vegetables, dried fruit	Component of hemoglobin and of electron carriers used in energy metabolism
Chlorine	Table salt, processed foods	Acid-base balance; formation of gastric juice
Sodium	Table salt, processed foods	Acid-base balance; water balance; nerve and muscle function
Potassium	Meats, dairy products, fruits and vegetables, grains	Acid-base balance; water balance; nerve and muscle function
Magnesium	Whole grains, green leafy vegetables	Activation of enzymes in protein synthesis
Fluorine	Fluoridated drinking water, tea, seafood	Maintenance of bone and tooth structure
Iodine	Seafood, dairy products, iodized salt	Component of thyroid hormones
Zinc	Meats, seafood, grains	Component of certain digestive enzymes

FIGURE 30–8 Minerals A healthful diet should include small amounts of certain minerals to maintain a healthy body. **Infer** ***Why do you think some cities and towns add fluoride to their water supplies?***

Minerals Inorganic nutrients that the body needs, usually in small amounts, are called **minerals. Figure 30–8** lists some of the minerals needed by the body. Calcium, for example, is required to produce the calcium phosphate that makes up bones and teeth. Iron is needed to make hemoglobin, the oxygen-carrying protein in red blood cells. A constant supply of minerals in the diet is needed to replace those lost in sweat, urine, and digestive wastes.

Nutrition and a Balanced Diet

What is meant by the term "balanced diet"?

The science of nutrition—the study of food and its effects on the body—tries to determine how food helps the body meet all of its various needs. Because of the work of nutritionists, many tools have been developed to help people plan healthful diets. **A balanced diet provides nutrients in adequate amounts and enough energy for a person to maintain a healthful weight.**

Balancing Your Diet Food labels can be used to choose healthful foods. Food labels provide general information about nutrition as well as specific information about the product. They can be used to determine if you are consuming enough of some of the important vitamins and minerals.

In Your Notebook *List seven types of information you can learn about a food from its food label.*

Note on the food label shown in **Figure 30–9** that fat contains about 9 Calories per gram, while carbohydrate and protein contain 4 Calories per gram. Why the difference? The carbon atoms in fats generally have more C–H (carbon to hydrogen) bonds than the carbon atoms in carbohydrates or proteins. Oxidizing these C–H bonds releases a great deal of energy. Because of this, oxidizing a gram of fat releases more energy than does oxidizing a gram of protein or carbohydrate, giving fats a greater energy value in Calories per gram.

When using food labels, it is important to remember that Percent Daily Values are based on a 2000-Calorie diet. However, nutrient needs are affected by age, gender, and lifestyle. The daily energy needs of an average-sized teenager who exercises regularly are about 2200 Calories for females and about 2800 Calories for males. People who are more active than average have greater energy needs. When a person stops growing or becomes less active, energy needs decrease.

Maintaining a Healthful Weight Inactive lifestyles and high-Calorie diet are contributing factors to the growing rate of obesity in the United States during the last several decades. Exercising about 30 minutes a day and eating a balanced diet can help maintain a healthful weight. Regular physical activity helps to maintain a healthful weight by burning excess Calories. Other benefits of physical activity include strengthening of the heart, bones, and muscles.

The American Heart Association recommends a diet with a maximum of 30 percent of Calories from fat, of which only 7 percent should be from saturated fats and 1 percent from trans fats. Controlling fat intake is important for several reasons. Foods that contain a high amount of any type of fat are high in Calories. A diet high in saturated fats and trans fats increases the risk for developing heart disease, Type II diabetes, or both.

Nutrition Facts

Serving Size 1 cup (30g)
Servings Per Container About 10

Amount Per Serving
Calories 110 Calories from Fat 17

	% Daily Value*
Total Fat 2g	**3%**
Saturated Fat 0g	**0%**
Trans Fat 0.5g	
Cholesterol 0mg	**0%**
Sodium 280mg	**12%**
Total Carbohydrate 22g	**7%**
Dietary Fiber 3g	**12%**
Sugars 1g	
Protein 3g	

Vitamin A 10%	• Vitamin C 20%
Calcium 4%	• Iron 45%

* Percent Daily Values are based on a 2,000 Calorie diet. Your Daily Values may be higher or lower depending on your calorie needs:

	Calories	2,000	2,500
Total Fat	Less than	65g	80g
Sat. Fat	Less than	20g	25g
Cholesterol	Less than	300mg	300mg
Sodium	Less than	2,400mg	2,400mg
Total Carbohydrate		300g	375g
Fiber		25g	30g

Calories per gram:
Fat 9 • Carbohydrate 4 • Protein 4

Ingredients: Whole grain oats, sugar, salt, milled corn, oat fiber, dried whey, hon

FIGURE 30–9 Food Label Reading food labels can help you track how many Calories you consume in a day and if you are meeting your requirements for important nutrients.

30.2 Assessment

Review Key Concepts

1. a. **Review** What are the two reasons humans need to eat?
   b. **Infer** Foods that contain many Calories but few raw materials are said to contain empty Calories. What do you think the phrase *empty Calories* means?
2. a. **Review** List six nutrients that the body needs.
   b. **Compare and Contrast** How are saturated and unsaturated fats similar? How are they different?
3. a. **Review** How can food labels be used to plan a balanced diet?
   b. **Calculate** One serving of a particular food contains 16 g of carbohydrates, 2 g of protein, and 10 g of fats. Approximately how many Calories does it contain? MATH

ANALYZING DATA

Examine **Figure 30–9** and answer the questions.

4. a. **Calculate** If you ate 2 cups of this product, how many grams of fat would you eat? How many total Calories would you eat? MATH
   b. **Evaluate** This product's packaging advertises that it contains 0 g of trans fat. Does that mean the product contains no trans fat? Explain.

WHST.9-10.1 Text Types and Purposes, WHST.9-10.7 Research to Build and Present Knowledge. Also RST.9-10.10, WHST.9-10.10

Biology & Society

Who Should Solve America's Obesity Problem?

As old subway cars are replaced and new sport stadiums are built, a trend is obvious. Seats are much larger than they used to be. For example, in the old arena of the Indiana Pacers, seats were 18 inches wide. In the new arena, the *smallest* seats are 21 inches wide. Advertisers tout that the seats are more comfortable. But the reality is, larger seats are needed because Americans have become fatter.

From the late 1970s to the early 2000s, the percentage of adults in the United States who are obese increased from 15 percent to 32.9 percent. During the same time period, the percentage of adolescents (ages 12 to 19) who are overweight more than tripled, from 5 percent to 17.4 percent. The trend shows no sign of changing.

The causes for what has been called the "obesity epidemic" seem apparent—a lifestyle of high-Calorie diets and lack of exercise. But the solutions are not so obvious. Many state and local governments have tried to gain control of the epidemic by removing high-Calorie foods from schools. Some people support these efforts, but others believe the government is encroaching too closely on personal lives. Should the government play a role in fighting obesity by controlling the foods served in school?

Many schools throughout the country have replaced vending machines that offered soda and other sugary drinks with those that offer only water, milk, or 100 percent juice.

Viewpoints

The Government Must Play a Role Obesity increases the risk of high blood pressure, Type II diabetes, stroke, arthritis, and some cancers. An increase in the rates of these diseases will strain the healthcare system and affect the economy by reducing the number of healthy adults in the workforce.

Overweight children are likely to become obese adults. Schools should play an active role in limiting students' exposure to high-Calorie foods that are not nutritious.

The Government Should Not Play a Role Food choices are a personal decision. Keeping unhealthful foods out of school will not prepare students for making healthful decisions in the real world. Parents and educators should teach children how to make healthful choices, rather than simply controlling their options.

Research and Decide

1. Evaluate Research changes that have been proposed or made recently in your school to address the obesity epidemic. Have some foods been removed from the cafeteria? Have the offerings in vending machines changed? Explain.

2. Form an Opinion Do you think that recent changes in your school menu, if any, are positive changes? Should more changes, or less, be made? Write an argument to support your claim.

30.3 The Digestive System

THINK ABOUT IT When you're hungry, your whole body needs food. But the only system in the body that food actually enters is the digestive system. So, how does food get to the rest of the body after the process of digestion?

Functions of the Digestive System

What are the functions of the digestive system?

The need for food presents every animal with at least two challenges. The first is how to obtain it. Once an animal has caught or gathered its food, its body faces a new challenge—how to convert the food into useful molecules. In humans and many other animals, this is the job of the digestive system. **The digestive system converts food into small molecules that can be used by the cells of the body. Food is processed by the digestive system in four phases—ingestion, digestion, absorption, and elimination.**

Ingestion Naturally, the first step in digestion is getting food into the system. Ingestion, as the process is called, is the process of putting food into your mouth—the opening to the digestive tract.

Digestion As food passes through the digestive system, it is broken down in two ways—by mechanical and chemical digestion. **Mechanical digestion** is the physical breakdown of large pieces of food into smaller pieces. These smaller pieces can be swallowed and accessed by digestive enzymes. During **chemical digestion,** enzymes break down food into the small molecules the body can use.

Absorption Once food has been broken into small molecules, it can be absorbed by cells in the small intestine. From the small intestine, the molecules enter the circulatory system, which transports them throughout the body.

Elimination The digestive system cannot digest and absorb all the substances in food that enter the body. Some materials, such as cellulose, travel through the large intestine and are eliminated from the body as feces.

Key Questions

What are the functions of the digestive system?

What occurs during digestion?

How are nutrients absorbed and wastes eliminated?

Vocabulary

mechanical digestion • chemical digestion • amylase • esophagus • peristalsis • stomach • pepsin • chyme • small intestine • villus • large intestine

Taking Notes

Flowchart Make a flowchart that shows the route food takes through the digestive system.

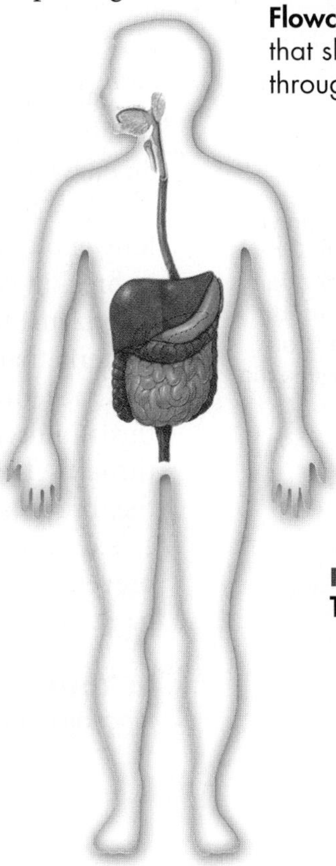

FIGURE 30–10
The Digestive System

The Process of Digestion

What occurs during digestion?

The human digestive system, like those of other chordates, is built around an alimentary canal—a one-way tube that passes through the body. **During digestion, food travels through the mouth, esophagus, stomach, and small intestine. Mechanical digestion and chemical digestion are the two processes by which food is reduced to molecules that can be absorbed.** Both mechanical digestion and chemical digestion start in the mouth.

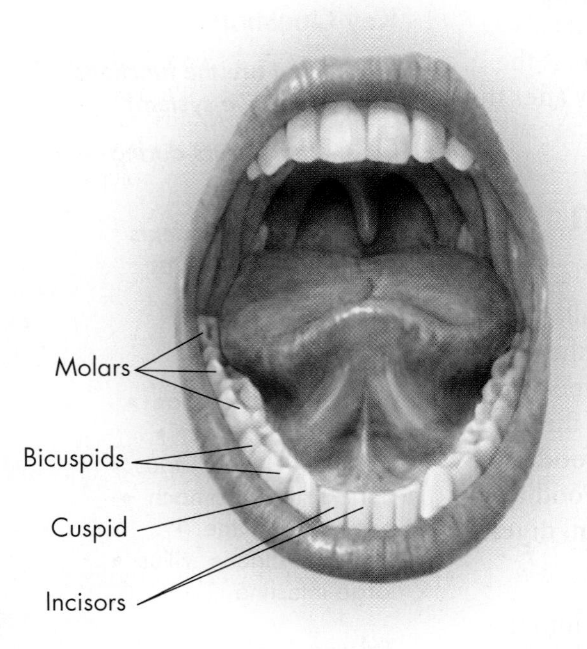

FIGURE 30–11 The Mouth Digestion begins in the mouth, where the tongue, teeth, and saliva form food into a moist lump that can be swallowed. **Infer** ***How do human teeth reflect an omnivorous diet?***

The Mouth As you take a forkful of food into your mouth, the work of the digestive system begins. Teeth and saliva start to work on your food first. Chewing begins the process of mechanical digestion. Chemical digestion begins as digestive enzymes in saliva start the breakdown of complex carbohydrates into smaller molecules.

▶ ***Teeth*** The teeth, shown in **Figure 30–11,** are anchored in the bones of the jaw. The surfaces of the teeth are protected by a coating of mineralized enamel. The teeth do much of the mechanical work of digestion. The incisors, cuspids, and bicuspids cut into and tear at food. The molars grind and crush food into a fine paste that can be swallowed. Meanwhile, your tongue moves food around so that it comes in contact with your teeth.

▶ ***Saliva*** As the teeth cut and grind the food, the salivary glands secrete saliva, which helps to moisten the food and make it easier to chew. The release of saliva is under the control of the nervous system and can be triggered by the scent of food—especially when you are hungry!

Saliva not only eases the passage of food through the digestive system but also begins the process of chemical digestion. Saliva contains an enzyme called **amylase** that begins to break the chemical bonds in starches, forming sugars. If you chew on a starchy food like a cracker long enough, it will begin to taste sweet—the result of amylase's work in breaking down starches into sugars. Saliva also contains lysozyme, an enzyme that fights infection by digesting the cell walls of many bacteria that may enter the mouth with food.

Once food is chewed, the combined actions of the tongue and throat muscles push the clump of food, called a bolus, down the throat. When you swallow, the bolus first enters the area at the back of the throat called the pharynx. As this occurs, a flap of connective tissue called the epiglottis closes over the opening to the trachea. This action prevents food from moving into the air passageways to the lungs as it passes through the pharynx and into the esophagus.

BUILD Vocabulary

WORD ORIGINS The prefix *amyl-* refers to starch and has both Greek (*amylon*) and Latin (*amylum*) origins. The suffix *-ase* is commonly used to indicate that a substance is an enzyme. **Amylase** is an enzyme that acts on starch.

In Your Notebook *Explain in your own words two protective functions of the mouth and throat.*

The Esophagus From the throat, the bolus passes through a tube called the **esophagus** into the stomach. You might think that gravity draws food down through the esophagus, but this is not correct. In fact, you can swallow quite well in zero gravity, as astronauts do, or even while standing on your head. The reason is that contractions of smooth muscles, known as **peristalsis** (pehr uh STAL sis), provide the force that moves food through the esophagus toward the stomach. Peristalsis in the esophagus is shown in **Figure 30–12.**

After food passes into the stomach, a thick ring of muscle called the cardiac sphincter closes the esophagus. This prevents the contents of the stomach from flowing back. Overeating or drinking excess caffeine can cause a backflow of stomach acid into the esophagus. The result is a burning sensation in the center of the chest known as heartburn. Despite its name, heartburn has nothing to do with the heart. Nonetheless, persistent heartburn can cause serious damage to the esophagus and is a reason to visit a doctor.

FIGURE 30–12 Peristalsis Muscles in the walls of the esophagus contract in waves. Each wave pushes the chewed clump of food, or bolus, in front of it. Eventually, the bolus is pushed into the stomach.

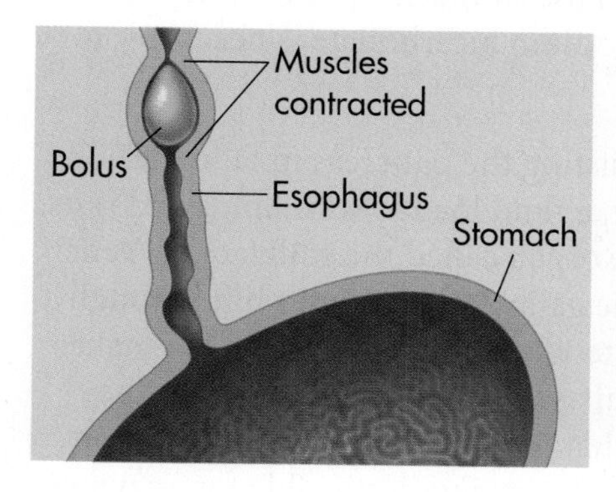

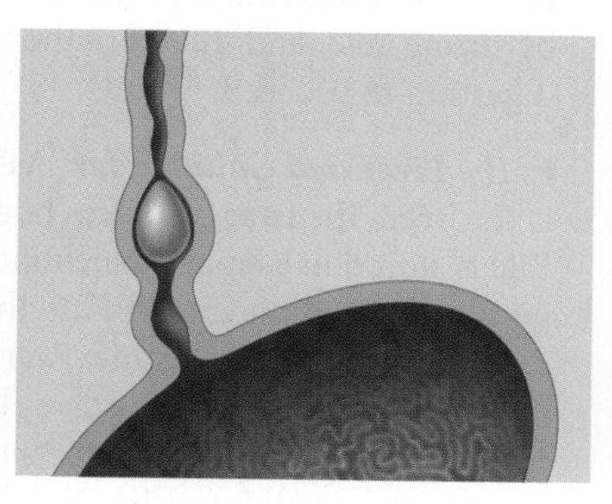

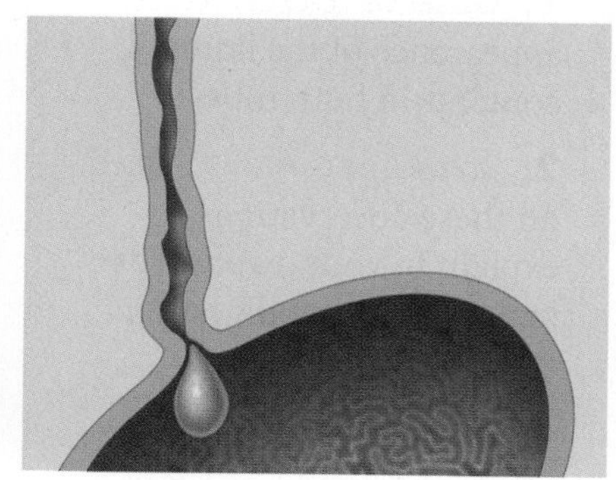

Chemical Digestion in the Stomach The **stomach** is a large muscular sac that continues the chemical and mechanical digestion of food. The lining of the stomach contains millions of microscopic gastric glands that release many substances into the stomach. Some of these glands produce hydrochloric acid. Other glands release an enzyme called pepsin that is activated in and functions best in acidic conditions. **Pepsin** breaks proteins into smaller polypeptide fragments.

Another stomach gland produces mucus, a fluid that lubricates and protects the stomach wall. If this protective layer fails, acids may erode the stomach lining and cause a sore called a peptic ulcer. For years, physicians thought that the primary cause of ulcers was too much stomach acid. They prescribed drugs that reduced symptoms but did not cure ulcers. Scientists have since discovered that most peptic ulcers are the result of infection with the bacterium *Helicobacter pylori.* Most peptic ulcers can now be cured with antibiotics that kill the bacteria.

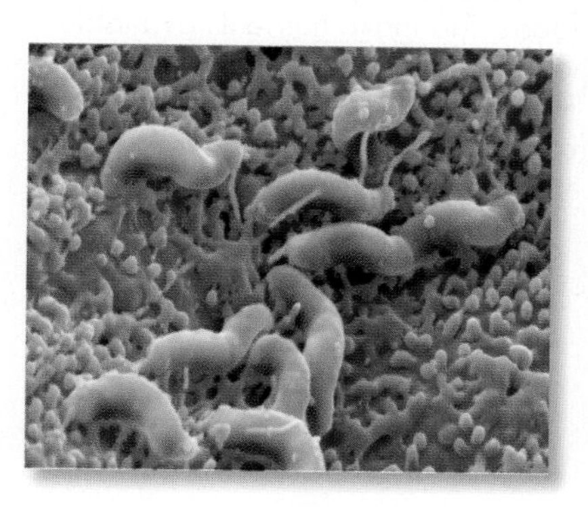

FIGURE 30–13 *Helicobacter pylori* After many years of blaming lifestyle factors for ulcers, researchers discovered that these bacteria are the cause. *H. pylori* burrow into the stomach wall and cause inflammation (SEM 6800×).

Mechanical Digestion in the Stomach Alternating contractions of the stomach's three smooth muscle layers thoroughly churn and mix the swallowed food. The churning causes further breakdown of the chunks of swallowed food and allows enzymes greater access to the food. Gradually, a mixture with an oatmeal-like consistency called **chyme** (KYM) is produced. After an hour or two, the pyloric valve, which is located between the stomach and small intestine, opens, and chyme begins to spurt into the small intestine.

Modeling Bile Action

1. Add 10 mL of water and 2 drops of olive oil into two test tubes.
2. Add 3 mL of a 5 percent liquid soap solution to one test tube.
3. Stir the contents of both tubes. Record your observations.

Analyze and Conclude

1. Observe Describe the appearance of the liquid contents in both tubes.

2. Draw Conclusions Based on these observations, explain in your own words how bile aids fat digestion.

Digestion in the Small Intestine As chyme is pushed through the pyloric valve, it enters the duodenum (doo oh DEE num). The duodenum is the first part of the **small intestine,** and it is where almost all of the digestive enzymes enter the intestine. Most of the chemical digestion and absorption of the food you eat occurs in the small intestine. As chyme enters the duodenum from the stomach, it mixes with enzymes and digestive fluids from the pancreas, the liver, and even the lining of the duodenum itself. The pancreas and liver are shown in **Figure 30–15.**

▶ ***Pancreas*** Just behind the stomach is the pancreas, a gland that serves three important functions. One function is to produce hormones that regulate blood sugar levels. Within the digestive system, the pancreas has two other roles. It produces enzymes that break down carbohydrates, proteins, lipids, and nucleic acids. The pancreas also produces sodium bicarbonate, a base that quickly neutralizes stomach acid as chyme enters the duodenum. The enzymes produced by the pancreas, unlike those produced in the stomach, would be destroyed by strong acid, and therefore the sodium bicarbonate is necessary for digestion to proceed.

▶ ***The Liver and Gallbladder*** Assisting the pancreas in fat digestion is the liver. The liver produces bile, a fluid loaded with lipids and salts. Bile is stored in a small, pouchlike organ called the gallbladder. When fat is present in the duodenum, the gallbladder releases bile through a duct into the small intestine. Fats tend to glob together, which makes fat digestion by enzymes such as lipase difficult. Bile breaks up the globs of fat into smaller droplets that disperse in the watery environment of the small intestine. This action makes it possible for enzymes to reach the smaller fat droplets and break them down.

In Your Notebook *Summarize the two roles of the pancreas in fat digestion.*

FIGURE 30–14 Effects of Digestive Enzymes Digestive enzymes hasten the breakdown of foods and make nutrients available to the body. **Interpret Tables** ***Where in the body does the digestion of carbohydrates begin?***

Effects of Digestive Enzymes

Active Site	Enzyme	Effect on Food
Mouth	Salivary amylase	Breaks down starches into disaccharides
Stomach	Pepsin	Breaks down proteins into large peptides
Small intestine (released from pancreas)	Pancreatic amylase	Continues the breakdown of starch
	Trypsin	Continues the breakdown of protein
	Lipase	Breaks down fat
Small intestine	Maltase, sucrase, lactase	Breaks down remaining disaccharides into monosaccharides
	Peptidase	Breaks down dipeptides into amino acids

THE DIGESTIVE SYSTEM

FIGURE 30–15 Food travels through many organs as it is broken down into nutrients your body can use. The time needed for each organ to perform its role varies based on the type of food consumed.

VISUAL SUMMARY

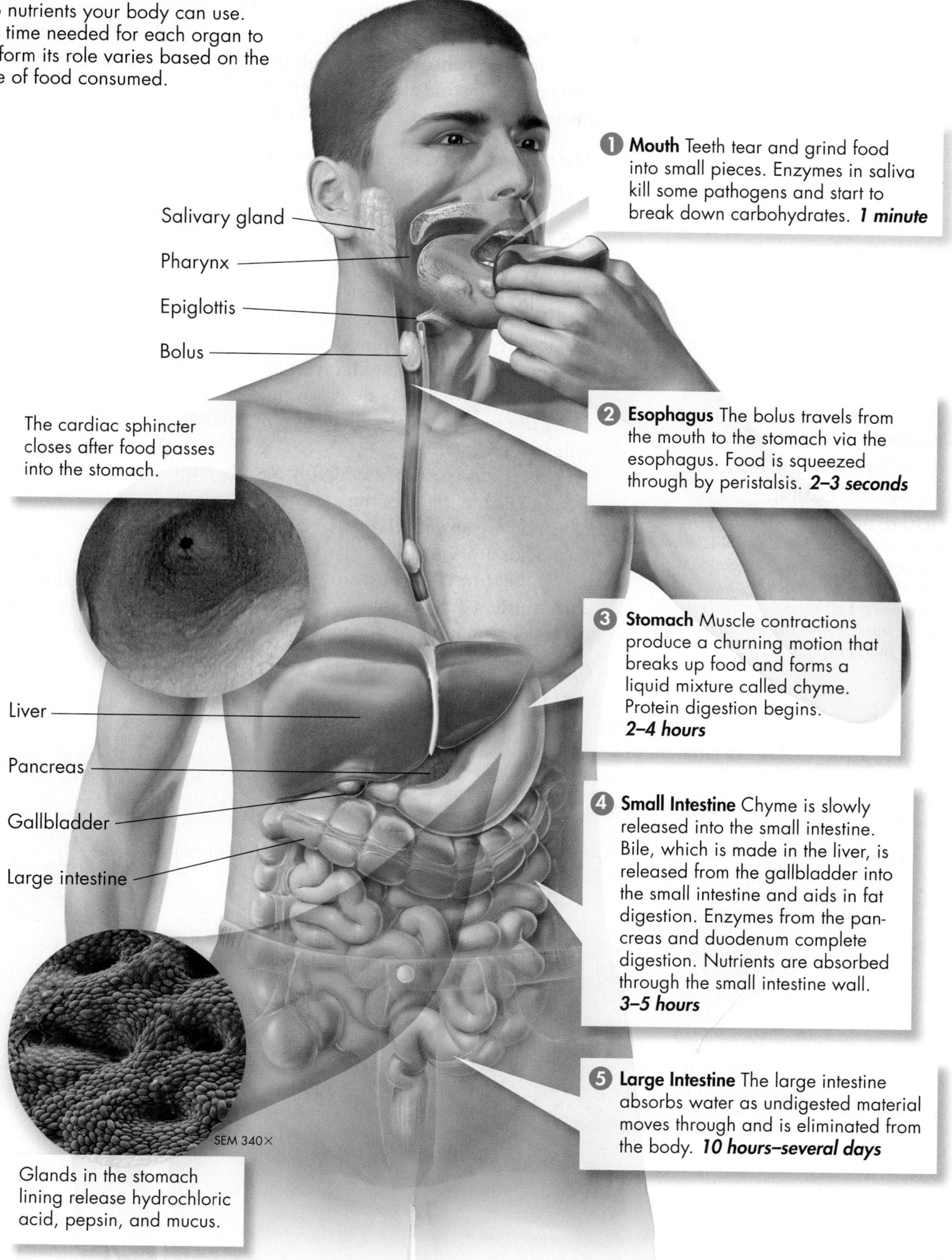

Salivary gland

Pharynx

Epiglottis

Bolus

The cardiac sphincter closes after food passes into the stomach.

Liver

Pancreas

Gallbladder

Large intestine

SEM 340×

Glands in the stomach lining release hydrochloric acid, pepsin, and mucus.

1 **Mouth** Teeth tear and grind food into small pieces. Enzymes in saliva kill some pathogens and start to break down carbohydrates. ***1 minute***

2 **Esophagus** The bolus travels from the mouth to the stomach via the esophagus. Food is squeezed through by peristalsis. ***2–3 seconds***

3 **Stomach** Muscle contractions produce a churning motion that breaks up food and forms a liquid mixture called chyme. Protein digestion begins. ***2–4 hours***

4 **Small Intestine** Chyme is slowly released into the small intestine. Bile, which is made in the liver, is released from the gallbladder into the small intestine and aids in fat digestion. Enzymes from the pancreas and duodenum complete digestion. Nutrients are absorbed through the small intestine wall. ***3–5 hours***

5 **Large Intestine** The large intestine absorbs water as undigested material moves through and is eliminated from the body. ***10 hours–several days***

Absorption and Elimination

How are nutrients absorbed and wastes eliminated?

Once the small intestine has completed the digestive process, nutrients must be absorbed from the alimentary canal. **Most nutrients from food are absorbed through the walls of the small intestine. The large intestine absorbs water and several vitamins and prepares waste for elimination from the body.**

Absorption From the Small Intestine After leaving the duodenum, chyme moves along the rest of the small intestine. By this time, most of the chemical digestion has been completed. The chyme is now a rich mixture of small- and medium-sized nutrient molecules that are ready to be absorbed.

The small intestine is specially adapted for absorption of nutrients. Its folded surface and fingerlike projections provide an enormous surface area for absorption of nutrient molecules. The fingerlike projections, called **villi** (singular: villus), are covered with tiny projections known as microvilli. As slow, wavelike contractions move the chyme along the surface, microvilli absorb nutrients. **Figure 30–16** illustrates villi and microvilli.

Nutrient molecules are rapidly absorbed into the cells lining the small intestine. Most of the products of carbohydrate and protein digestion are absorbed into the capillaries in the villi. Most fats and fatty acids are absorbed by lymph vessels.

By the time chyme is ready to leave the small intestine, it is basically nutrient-free. Complex organic molecules have been digested and absorbed, leaving only water, cellulose, and other undigestible substances behind.

ZOOMING IN

ABSORPTION IN THE SMALL INTESTINE

FIGURE 30–16 The lining of the small intestine consists of folds that are covered with tiny projections called villi. Within each villus there is a network of blood capillaries and lymph vessels that absorb and carry away nutrients.

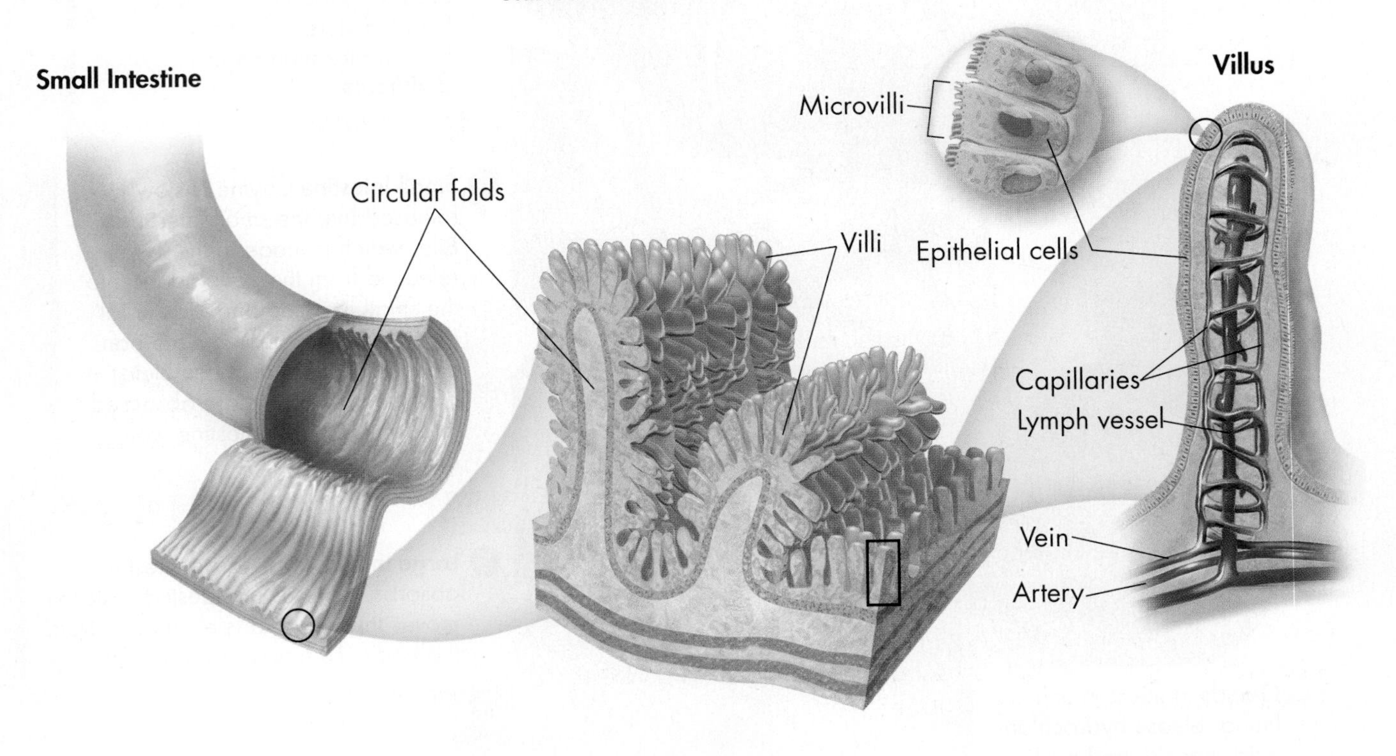

As material leaves the small intestine and enters the large intestine, it passes by a small saclike organ called the appendix. In some mammals, the appendix processes cellulose and other materials. The only time humans notice their appendix is when it becomes clogged and inflamed, causing appendicitis. The remedy for appendicitis is to remove the infected organ by surgery—as quickly as possible—before it can rupture or break open.

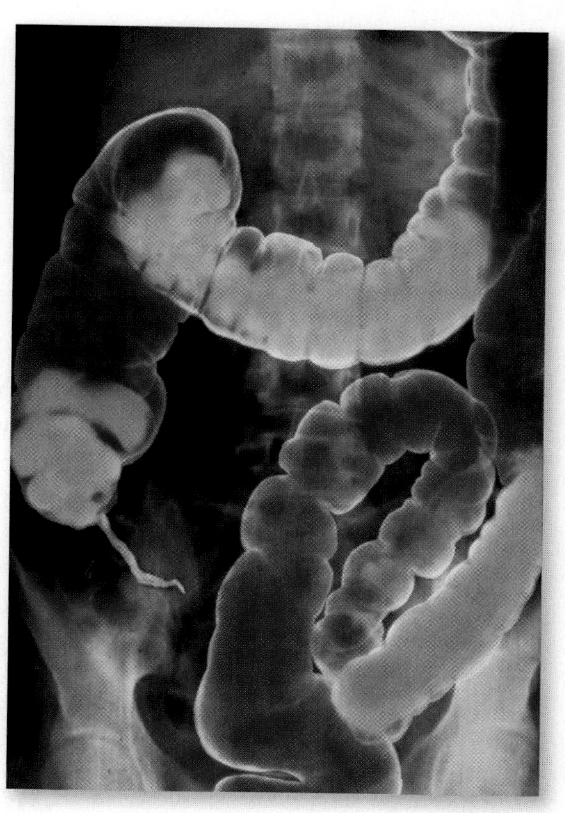

FIGURE 30–17 The Large Intestine This X-ray shows the large intestine and its contents.

Absorption From the Large Intestine When chyme leaves the small intestine, it enters the **large intestine,** or colon. The large intestine is actually much shorter than the small intestine. The large intestine gets its name due to its diameter, which is much greater than the small intestine's diameter. The primary function of the large intestine is to remove water from the undigested material that is left. Water is absorbed quickly across the wall of the large intestine, leaving behind the undigested materials. Rich colonies of bacteria present in the large intestine produce compounds that the body is able to absorb and use, including vitamin K. When large doses of antibiotics are given to fight an infection, they can destroy these bacteria, and vitamin K deficiency can occur.

Elimination The concentrated waste material—the feces—that remains after most of the water has been removed passes into the rectum and is eliminated from the body through the anus. When something happens that interferes with the removal of water by the large intestine, you usually become aware of it right away. If not enough water is absorbed, a condition known as diarrhea occurs. If too much water is absorbed from the undigested materials, a condition known as constipation occurs.

30.3 Assessment

Review Key Concepts

1. a. **Review** Explain the function of the digestive system.
   b. **Compare and Contrast** What is the difference between mechanical digestion and chemical digestion?
2. a. **Review** List the structures that food travels through during digestion and give the function of each.
   b. **Relate Cause and Effect** Some people have a disorder in which their stomach muscles cannot contract and churn food. What effect do you think this has on the length of time food stays in the stomach?
3. a. **Review** Explain how nutrients are absorbed.
   b. **Apply Concepts** What impact do the folds and villi of the small intestine have on absorption?

Apply the Big idea

Matter and Energy

4. How would the rate of digestion be affected if the various organs and glands did not release enzymes? *Hint:* You may wish to refer to Chapter 2 for a review of enzyme action.

30.4 The Excretory System

Key Questions

What is the principal role of the structures of the excretory system?

How do the kidneys clean the blood?

How do the kidneys help maintain homeostasis?

Vocabulary

excretion
ureter
urinary bladder
urethra
nephron
filtration
glomerulus
Bowman's capsule
reabsorption
loop of Henle

Taking Notes

Preview Visuals Examine **Figure 30–19.** What does this Figure reveal about the important functions of the kidneys?

THINK ABOUT IT It's a hot day, and you've been getting thirsty for hours. Finally, you get the chance to go inside, and you gulp down more than a liter of water. The water tastes great, but as you drink, you begin to wonder. Where's all that water going? Will it just dilute your blood, or is something in your body making sure that everything stays in balance?

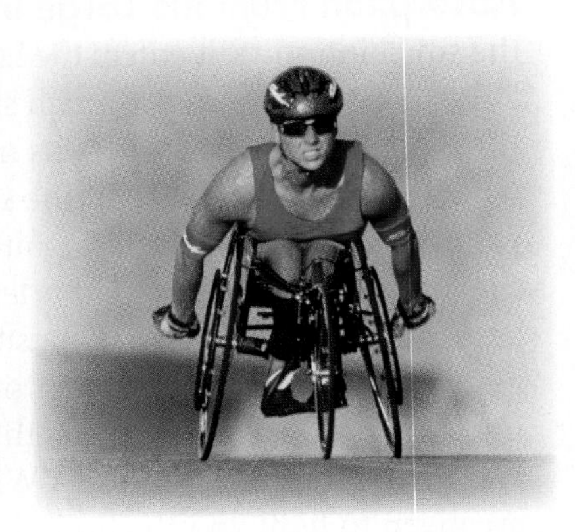

Structures of the Excretory System

What is the principal role of the structures of the excretory system?

The chemistry of the human body is a marvelous thing. An intricate system of checks and balances controls everything from your blood pressure to your body temperature. Nutrients are absorbed, stored, and carefully released when they are needed. However, every living system, including the human body, produces chemical waste products, some of which are so toxic that they will cause death if they are not eliminated.

For example, as a normal consequence of being alive, every cell in the body produces waste compounds, including excess salts and carbon dioxide. Ammonia, one of the most toxic of these waste compounds, is produced when the amino acids from proteins are used for energy. Ammonia is converted to a less toxic compound called urea, but it, too, must be eliminated from the body. The process by which these metabolic wastes are eliminated to maintain homeostasis is called **excretion.** Excretion is one part of the many processes that maintain homeostasis.

The excretory system, which includes the skin, lungs, liver, and kidneys, excretes metabolic wastes from the body. The ureters, urinary bladder, and urethra are also involved in excretion. **Figure 30–18** shows the major organs of excretion.

In Your Notebook *Make a two-column table that lists the organs of excretion in the first column and their function in the second column.*

The Skin The skin excretes excess water, salts, and a small amount of urea in sweat. By releasing sweat in very small amounts, this process eliminates wastes even when you may not think you're sweating.

The Lungs The blood transports carbon dioxide, a waste product of cellular respiration, from the body cells to the lungs. When you exhale, your lungs excrete carbon dioxide and small amounts of water vapor.

The Liver The liver plays many important roles in excretion. As we have seen, one of its principal activities is the conversion of potentially dangerous nitrogen wastes, a product of protein breakdown, into less toxic urea. Urea, which is highly soluble, is then transported through the blood to the kidneys for elimination from the body.

The Kidneys The major organs of excretion are the kidneys, a pair of fist-sized organs located on either side of the spinal column near the lower back. Through a complex filtering process, the kidneys remove excess water, urea, and metabolic wastes from the blood. The kidneys produce and excrete a waste product known as urine. **Ureters** transport urine from the kidneys to the **urinary bladder,** where the urine is stored until it is released through the **urethra.**

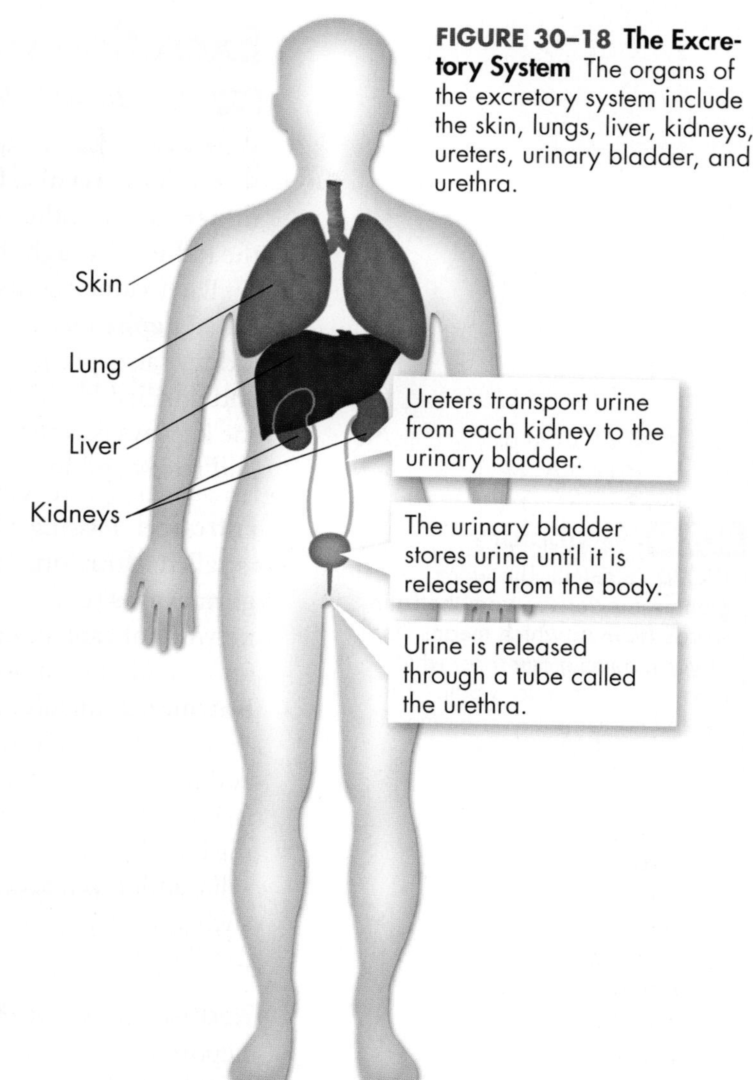

FIGURE 30–18 The Excretory System The organs of the excretory system include the skin, lungs, liver, kidneys, ureters, urinary bladder, and urethra.

Analyzing Data

The Composition of Urine

The kidneys are selective filters. As blood passes through them, urea, other impurities, and excess salts are removed from the blood. But important substances such as water, protein, and glucose remain in circulation. The collected waste products are excreted in urine. The concentrations of certain substances in the blood compared to their concentration in urine reveal the important work of the kidneys.

Concentrations of Selected Substances in Blood and Urine

Substance	Average Concentration in Blood (g/mL)	Average Concentration in Urine (g/mL)
Calcium	0.01	0.02
Glucose	0.10	0.00
Potassium	0.02	0.20
Sodium	0.32	0.60
Urea	0.03	2.00

1. Interpret Data Which substances listed have the highest and lowest concentrations in the blood? Which substances have the highest and lowest concentrations in the urine?

2. Calculate Approximately how many times more concentrated is urea in urine than in the blood? MATH

3. Infer Recall that urea is a byproduct of amino acid breakdown. How might the urea concentration vary in the blood and urine as the result of high protein diets? Explain.

Excretion and the Kidneys

How do the kidneys clean the blood?

What does a kidney do? **As waste-laden blood enters the kidney through the renal artery, the kidney removes urea, excess water and minerals, and other waste products.** The clean, filtered blood leaves the kidney through the renal vein and returns to circulation.

Each kidney contains nearly a million individual processing units called **nephrons.** These nephrons are where most of the work of the kidney takes place—impurities are filtered out, wastes are collected, and purified blood is returned to circulation. Blood purification in the kidneys is complex and involves two distinct processes: filtration and reabsorption.

BUILD Vocabulary

WORD ORIGINS The word **glomerulus** derives from the Latin words *glomus,* which means "ball of yarn," and *glomerare,* which means "to form into a ball." The twisted capillaries of a glomerulus resemble a ball of yarn.

Filtration Passing a liquid or gas through a filter to remove wastes is called **filtration.** The filtration of blood mainly takes place in the **glomerulus** (gloh MUR yoo lus). A glomerulus is a small but dense network of capillaries (very small blood vessels) encased in the upper end of each nephron by a hollow, cup-shaped structure called **Bowman's capsule.** A glomerulus is shown in **Figure 30–19.**

Because the blood is under pressure and the walls of the capillaries and Bowman's capsule are permeable, much of the fluid from the capillaries flows into Bowman's capsule. The material that is filtered from the blood is called the filtrate. The filtrate contains water, urea, glucose, salts, amino acids, and some vitamins. Large substances in the blood, such as proteins and blood cells, are too large to pass through the capillary walls.

Reabsorption Nearly 180 liters of filtrate pass from the blood into nephron tubules every day. That's the equivalent of 90 2-liter bottles of soft drink. Thank goodness, not all of those 180 liters are excreted. In fact, nearly all of the material that moves into Bowman's capsule makes its way back into the blood. The process by which water and dissolved substances are taken back into the blood is called **reabsorption.**

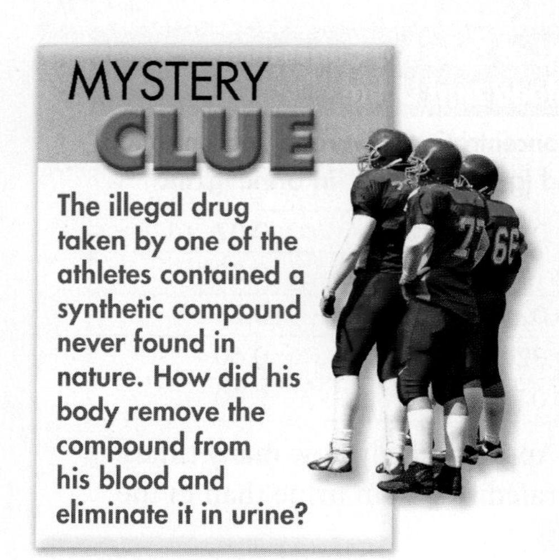

A number of materials, including salts, vitamins, amino acids, fats, and glucose, are removed from the filtrate by active transport and reabsorbed by the capillaries. Because water follows these materials by osmosis, almost 99 percent of the water that enters Bowman's capsule is actually reabsorbed into the blood. In effect, the kidney first throws away nearly everything and then takes back only what the body needs. This is how the kidney is able to remove drugs and toxic compounds from the blood—even chemicals the body has never seen before.

A section of the nephron tubule called the **loop of Henle** is responsible for conserving water and minimizing the volume of the filtrate. The waste material—now called urine—that remains in the tubule is emptied into a collecting duct.

Urine Excretion From the collecting ducts, urine flows to the ureter of each kidney. The ureters carry urine to the urinary bladder for storage until the urine leaves the body through the urethra.

ZOOMING IN

STRUCTURE AND FUNCTION OF THE KIDNEYS

FIGURE 30–19 Kidneys are made up of nephrons. Blood enters the nephron, where impurities are filtered out and emptied into the collecting duct. Purified blood leaves a nephron through a vein. **Interpret Visuals** ***List in order the structures that blood flows through in a kidney.***

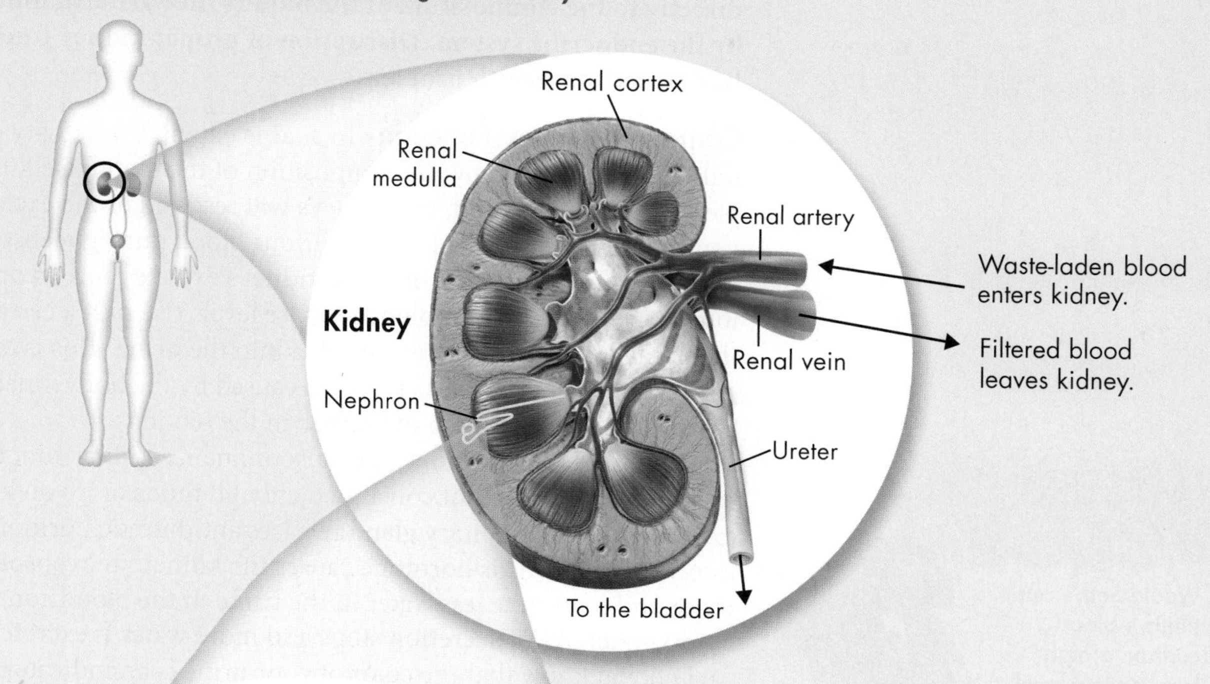

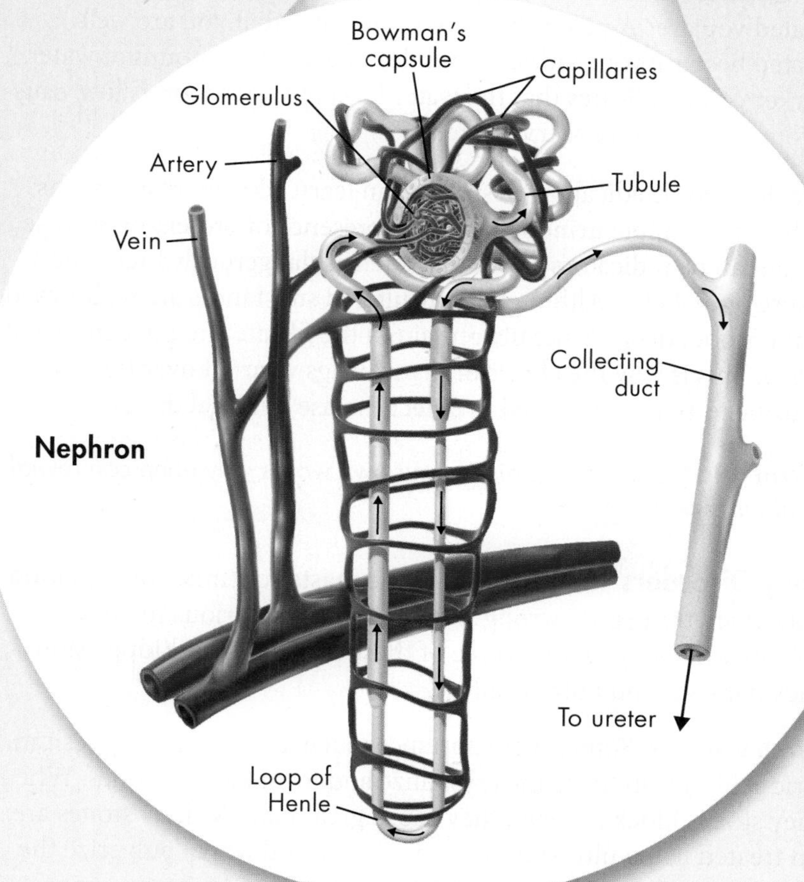

❶ **Filtration** Blood enters a nephron through a capillary. From the glomerulus, filtrate flows into a tubule. Blood cells and large substances remain in the capillary.

❷ **Reabsorption** As the filtrate moves through the tubule, water and many other substances that are important to the body are reabsorbed through capillary walls into the blood.

❸ **Urine Excretion** Once water and other important substances are reclaimed by the blood, the filtrate is called urine. Collecting ducts gather urine and transport it to a ureter.

The Kidneys and Homeostasis

How do the kidneys help maintain homeostasis?

The kidneys play an important role in maintaining homeostasis. Besides removing wastes, the kidneys also maintain blood pH and regulate the water content of the blood. **The kidneys respond directly to the composition of the blood. They are also influenced by the endocrine system. Disruption of proper kidney function can lead to serious health problems.**

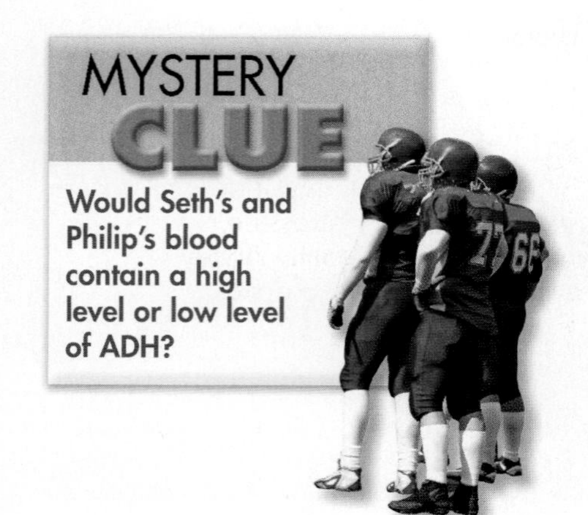

Control of Kidney Function To a large extent, the activity of the kidneys is controlled by the composition of the blood itself. For example, if you eat salty food, the kidneys will respond to the excess salt in your blood by returning less salt to your blood during reabsorption. If the blood is too acidic, then the kidneys excrete more hydrogen ions in the urine. If your blood glucose levels rise past a certain point, the kidneys will even excrete glucose into the urine. This is one of the danger signals of diabetes, a disease caused by the body's inability to control the concentration of glucose in the blood.

Glands release hormones that also influence kidney function. For example, if you have not consumed enough fluids or if you have sweat excessively, your pituitary gland releases antidiuretic hormone (ADH) into your blood. This hormone causes the kidneys to reabsorb more water and to excrete less water in the urine. If the blood contains excess water, ADH secretion stops and more water is excreted.

Did you know that the color of your urine is an indicator of how hydrated you are? A pale yellow color indicates that you are well hydrated because your kidneys are releasing a good amount of water. A darker color indicates that the water level in your blood is low, causing your kidneys to conserve water.

Urine Testing Medical professionals can learn a lot about a person's health from a simple urine sample. The presence of protein or glucose in urine can be indicators of diseases such as dangerously high blood pressure or diabetes. Although many filtered substances are reabsorbed into the blood, drugs generally remain in the filtrate and are eliminated in urine. This is why the effects of many drugs wear off over time and why urine tests are often used to detect the use of illegal drugs.

In Your Notebook *Explain in your own words why urine can reveal a lot about a person's health.*

Kidney Disorders The kidneys are the master chemists of the blood supply. If anything goes wrong with the kidneys, serious medical problems will likely follow. Three of these problems are kidney stones, kidney damage, and kidney failure.

▶ ***Kidney Stones*** Sometimes substances such as calcium, magnesium, or uric acid salts in the urine crystallize and form kidney stones. When kidney stones block a ureter, they cause great pain. Kidney stones are often treated using ultrasound waves. The sound waves pulverize the stones into smaller fragments, which are eliminated with the urine.

▶ ***Kidney Damage*** Many diseases, injuries, and exposure to hazardous substances can lead to impaired kidney function. But most cases of kidney damage in the United States are related to high blood pressure and diabetes. Excessive blood pressure damages the delicate filtering mechanism, and high blood sugar levels cause the kidneys to filter more blood than normal. Over time, the tubules weaken, and the kidneys may fail to keep up with the demands placed upon them.

▶ ***Kidney Failure*** When kidneys can no longer cleanse the blood and maintain a state of homeostasis in the body, a person is said to be in kidney failure. A patient with kidney failure must receive dialysis or undergo a kidney transplant as shown in **Figure 30–20.**

During dialysis, a machine performs the role of the kidneys. The patient's blood is pumped through the machine, cleansed, and pumped back into the body. Although the procedure is painless, it is very time-consuming. Most patients receive dialysis treatments three times a week for about four hours each time. To prevent the buildup of fluid and harmful materials between treatments, patients must restrict their fluid intake and eat foods low in potassium, phosphorus, and salt.

In transplantation, a patient receives a kidney and ureter from a compatible donor. Fortunately for the donor, a person can survive with just one healthy kidney.

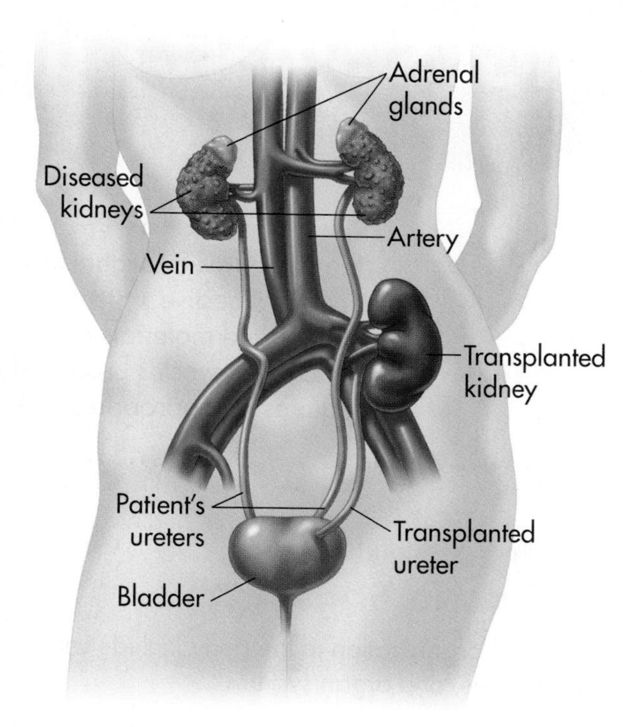

FIGURE 30–20 Kidney Transplantation Unless the patient's diseased kidneys are causing infection or high blood pressure, they are left in place when a healthy kidney and ureter are transplanted from a donor.

30.4 Assessment

Review Key Concepts

1. a. Review List the organs that are involved in excretion.

b. Classify Why is excretion important for homeostasis?

2. a. Review What substances do the kidneys remove from blood?

b. Sequence Explain what happens during filtration, reabsorption, and urine excretion.

3. a. Review Describe how the kidneys help maintain water balance.

b. Apply Concepts Why do you think protein and glucose in the urine are signs of kidney damage?

BUILD VOCABULARY

4. Two words that are often used interchangeably are *excretion* and *secretion.* They have two distinct meanings, however. An excretion is usually a waste product of metabolism that is expelled from an organism. A secretion is a useful substance that is released inside or outside an organism. Name one example each of an excretion and a secretion from this lesson.

NGSS Smart Guide Digestive and Excretory Systems

Disciplinary Core Idea LS1.A Structure and Function: How do the structures of organisms enable life's functions? The structures of your digestive system break down and absorb foods that provide energy and materials to cells. The structures of your excretory system remove the wastes produced during cellular functions. Homeostasis requires an appropriate balance of these inputs and outputs.

30.1 Organization of the Human Body

- The levels of organization in the body include cells, tissues, organs, and organ systems.
- Homeostasis describes the relatively constant internal physical and chemical conditions that organisms maintain despite changes in internal and external environments.

epithelial tissue 863 • connective tissue 863 • nervous tissue 863 • muscle tissue 863 • homeostasis 865 • feedback inhibition 865

Biology.com

Art Review Review your understanding of human body systems.

Tutor Tube Tune into Tutor Tube for another perspective on how homeostasis works in the human body.

ART REVIEW Human Body Systems

Nervous System | Integumentary System | Immune/Lymphatic System | Muscular System

30.2 Food and Nutrition 

- Molecules in food contain chemical energy that cells use to produce ATP. Food also supplies raw materials your body needs to build and repair tissues.
- The nutrients that the body needs include water, carbohydrates, fats, proteins, vitamins, and minerals.
- A balanced diet provides nutrients in adequate amounts and enough energy for a person to maintain a healthful weight.

Calorie 868 • carbohydrate 869 • fat 870 • protein 870 • vitamin 871 • mineral 872

Biology.com

Data Analysis Use nutritional data to plan a lunch that meets your own personal requirements.

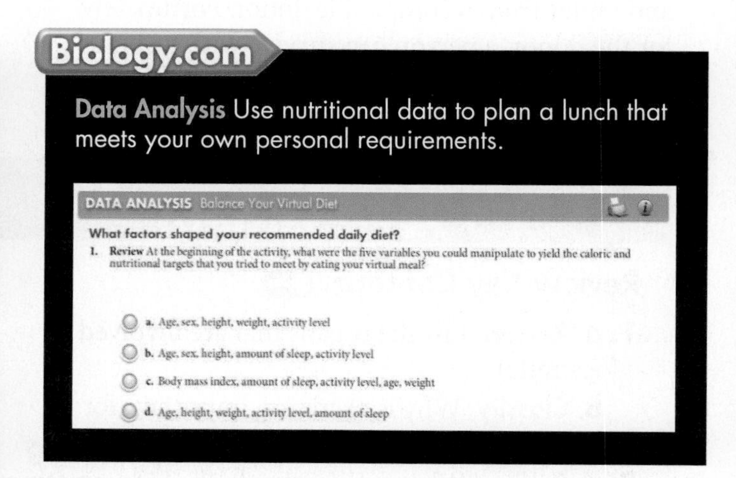

CHECKING YOUR *Scientific Literacy*

Refer to the lesson content and digital assets below as you prepare for your chapter assessment. Then, evaluate your understanding of the digestive and excretory systems by answering these questions.

1. Scientific and Engineering Practice **Developing and Using Models** How useful is a filter in a swimming pool or home aquarium as a model for a nephron? What are the strengths and limitations of this model?

2. Crosscutting Concept **Cause and Effect** Using the feedback mechanisms that regulate body temperature, describe an example of a cause-and-effect relationship that contributes to homeostasis.

3. Text Types and Purposes Use reliable Internet and other appropriate sources to research claims about "fad" diets. Use data and evidence to write a focused argument that develops these claims and evaluates their validity.

4. STEM You are a biomedical engineer assigned to evaluate the efficiency of a prototype for a new kind of artificial kidney. Make a list of the criteria you think are most important to evaluate. Describe the reasons for your choices.

30.3 The Digestive System

- The digestive system converts food into small molecules that can be used by the cells of the body. Food is processed by the digestive system in four phases—ingestion, digestion, absorption, and elimination.
- During digestion, food travels through the mouth, esophagus, stomach, and small intestine. Mechanical digestion and chemical digestion are the two processes by which food is reduced to molecules that can be absorbed.
- Most nutrients from food are absorbed through the walls of the small intestine. The large intestine absorbs water and several vitamins and prepares waste for elimination from the body.

mechanical digestion 875 • chemical digestion 875 • amylase 876 • esophagus 877 • peristalsis 877 • stomach 877 • pepsin 877 • chyme 877 • small intestine 878 • villus 880 • large intestine 881

Biology.com

Untamed Science Video Hold your nose as you join the Untamed Science crew to learn what scientists can discover about animals by investigating their scat.

Art in Motion Watch peristalsis in action in the esophagus.

30.4 The Excretory System

- The excretory system, which includes the skin, lungs, liver, and kidneys, excretes metabolic wastes from the body.
- As waste-laden blood enters the kidney through the renal artery, the kidney removes urea, excess water and minerals, and other waste products.
- The kidneys respond directly to the composition of the blood. They are also influenced by the endocrine system. Disruption of proper kidney function can lead to serious health problems.

excretion 882 • ureter 883 • urinary bladder 883 • urethra 883 • nephron 884 • filtration 884 • glomerulus 884 • Bowman's capsule 884 • reabsorption 884 • loop of Henle 884

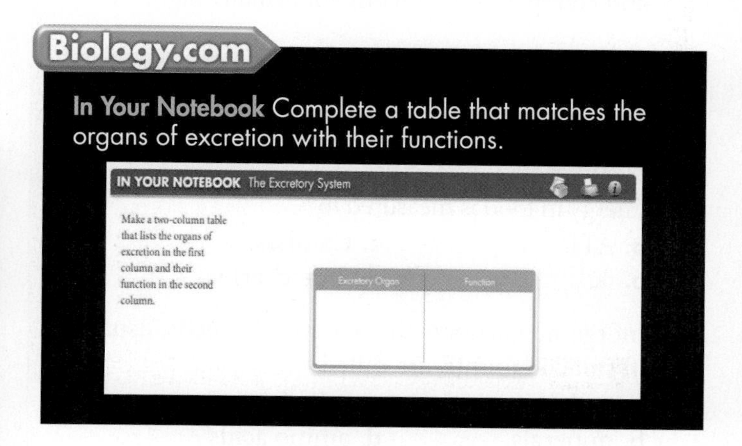

Biology.com

In Your Notebook Complete a table that matches the organs of excretion with their functions.

Real-World Lab

Digestion of Dairy Products This chapter lab is available in your lab manual and at Biology.com

30 Assessment

30.1 Organization of the Human Body

Understand Key Concepts

1. The type of tissue that covers the body, lines internal surfaces, and forms glands is
   a. muscle tissue. c. epithelial tissue.
   b. connective tissue. d. nervous tissue.

2. The process of maintaining a relatively constant internal environment despite changes in the external environment is called
   a. regulation. c. synapse.
   b. homeostasis. d. stimulation.

3. What do all types of tissue have in common?
   a. They are all made of connective tissue.
   b. They are all made of cells.
   c. They are all found in every organ.
   d. They are all made of organs.

4. Why is it important for an organism to maintain homeostasis?

5. Name the four types of tissues and describe one characteristic of each.

Think Critically

6. **Craft and Structure** Would you classify blood as a cell, a tissue, or an organ? Explain.

7. **Key Ideas and Details** Infections may lead to an immune response that results in a high fever. Citing evidence from the text about the action of enzymes, predict what may happen if a person's body temperature remains abnormally high.

30.2 Food and Nutrition

Understand Key Concepts

8. Energy in food is measured in
   a. ATP. c. Calories.
   b. fats. d. disaccharides.

9. Inorganic nutrients that your body needs, usually in small amounts, are called
   a. vitamins. c. proteins.
   b. minerals. d. amino acids.

10. Which nutrients provide the body with energy?

11. In what three ways are proteins important to the body?

Think Critically

12. **Text Types and Purposes** Many food manufacturers have replaced trans fats with other fats that may not have the same level of heart disease risk. Some nutritionists fear that people will think foods such as French fries, doughnuts, and cookies are healthful if they are not made with trans fats. Write a short essay explaining why these foods are still not healthful choices.

13. **Calculate** If a person consumed 2000 Calories while following the typical diet, how many more of those Calories would be from saturated fat than if they were following the recommended diet? MATH

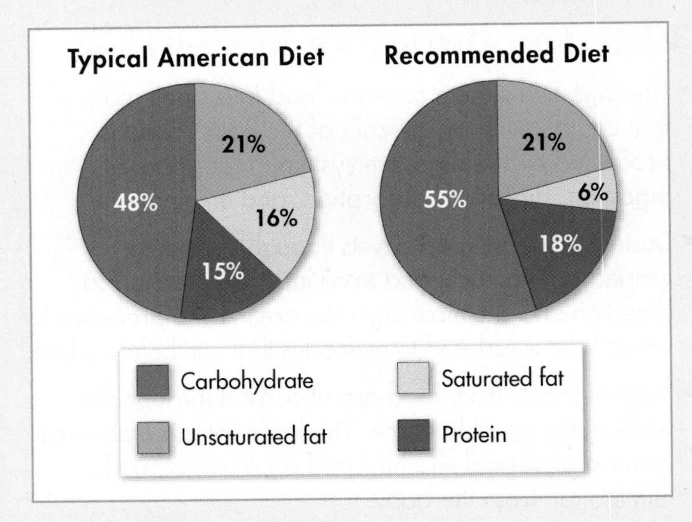

   a. 320 c. 120
   b. 200 d. 100

30.3 The Digestive System

Understand Key Concepts

14. Where does mechanical digestion begin?
    a. the esophagus
    b. the large intestine
    c. the mouth
    d. the small intestine

15. An enzyme in saliva that can break the chemical bonds in starch is
    a. pepsin. c. amylase.
    b. bile. d. chyme.

16. Explain why swallowed food does not normally enter the airway leading to the lungs.

17. What is the importance of enzymes during digestion?

18. Describe the functions of the pancreas.

19. **Key Ideas and Details** Summarize how the structure of the villi is adapted to their function?

Think Critically

20. **Infer** Individuals who have had part, or even all, of their stomachs removed can survive if fed predigested food. Could these individuals also survive without a small intestine? Explain.

21. **Predict** Suppose that your doctor prescribed an antibiotic that killed all the bacteria in your body. What effect would this have on your digestive system?

30.4 The Excretory System

Understand Key Concepts

22. Which of the following is the basic functional unit in a kidney?
    a. nephron.
    b. glomerulus.
    c. Bowman's capsule.
    d. loop of Henle.

23. Urine is excreted from the body through the
    a. ureter.
    b. urinary bladder.
    c. urethra.
    d. renal vein.

24. What is the role of the skin in excretion?

25. What materials are filtered from blood in the kidney? What materials do not leave the bloodstream?

26. How is the water-regulating activity of the kidney controlled?

Think Critically

27. **Apply Concepts** Explain why kidney failure can be a fatal condition.

28. **Infer** When there is too much fluid in the blood, the heart must pump harder. Diuretics are substances that stimulate the kidneys to remove more fluid from the body. Why do you think diuretics are used to treat high blood pressure?

THE TELLTALE SAMPLE

For centuries, people have used urine for clues to health and disease. The Greeks, for example, knew that diabetics had excessive sugar in their urine and called the disease *diabetes mellitus*. Mellitus is the Greek word for honey.

- **Physical Examination** During this step, the color and clarity are examined. The shade of yellow indicates the amount of water being released by the kidneys. Urine of a color other than yellow could indicate the presence of blood. Or, it could simply indicate someone has eaten a lot of beets. Urine should be clear, rather than cloudy.
- **Microscopic Examination** The presence of mucus, white blood cells, or microorganisms in urine indicates a probable infection. Cloudy urine may also be caused by crystals, which could indicate kidney stones or a metabolic problem.
- **Chemical Examination** Hundreds of chemical tests can be performed on urine. Chemical dipsticks are used that change color in the presence of other chemicals. These tests can reveal a lot about kidney function, liver function, and overall homeostasis in the body.

1. **Infer** How does urine reveal so much about the health of the human body?
2. **Production and Distribution of Writing** Most drug urine tests performed for schools do not test for alcohol or tobacco. Write a clear and coherent argument for why you agree or disagree with this policy.
3. **Connect to the Big idea** Ketones are a product of the breakdown of fat for energy. Ketones in the urine can be an indication of diabetes. Why do you think this is?

Connecting Concepts

Use Science Graphics

Pancreatic secretions contain sodium bicarbonate and enzymes. The graph shows the secretions of the pancreas in response to three different substances in chyme. Use the graph to answer questions 29 and 30.

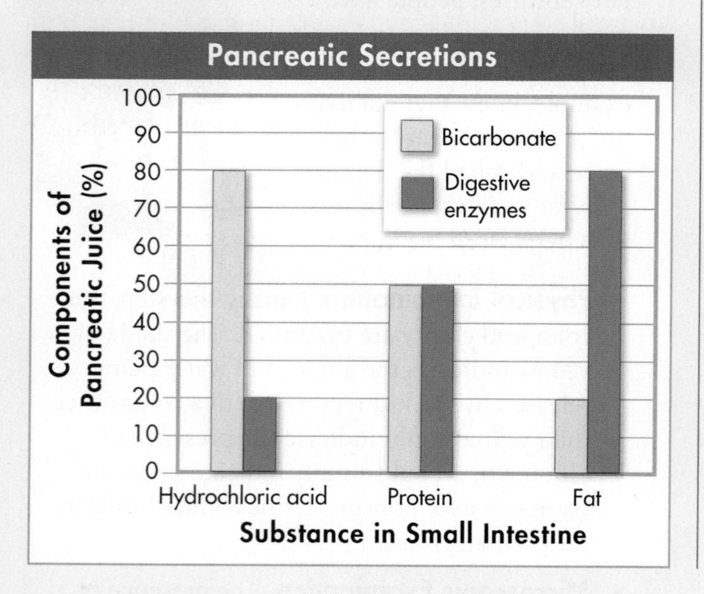

29. **Integration of Knowledge and Ideas** Each pair of bars represents the response of the pancreas to a different variable. What are the three variables?

30. **Integration of Knowledge and Ideas** Compare the composition of pancreatic secretions in the presence of hydrochloric acid and fat.

Write About Science

31. **Production and Distribution of Writing** A children's television workshop wants to explain the process of digestion to young viewers. Work with a team to create a computer presentation that describes the travels of a hamburger and bun through the digestive system. As you work, refine your presentation to include graphics and links to reliable, informative web sites.

32. **Assess the Big idea** Using **Figure 30–2,** choose five body systems that are involved in maintaining homeostasis in your body as you answer these assessment questions. Explain how these five body systems work together.

Analyzing Data

 Integration of Knowledge and Ideas

MyPlate classifies food into five categories: fruits, vegetables, grains, protein, and dairy, plus oils. Personalized eating plans can be found at choosemyplate.gov. The plan shown here contains daily recommendations for Ryan, a 15-year-old male who weighs 140 pounds, is 5 feet 7 inches tall, and is physically active about 30 to 60 minutes a day.

33. **Predict** If Ryan were to join the swim team and have practice for two hours after school every day, what would happen to the number of Calories he could consume? Explain.

34. **Infer** For Ryan to meet his grain requirements, which group of foods would be his BEST choice in a single day?

a. sweetened cereal, pasta, white bread
b. whole-grain bagel, a doughnut, and pasta
c. whole-grain cereal, potato chips, and whole-grain bread
d. oatmeal, whole-grain bread, and a sweet potato

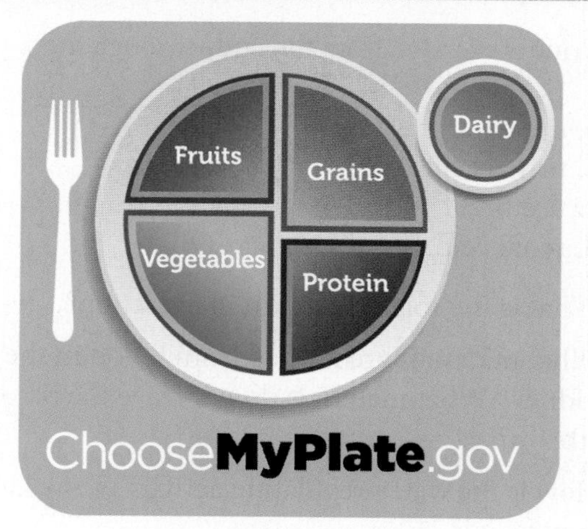

Total Calories: 2800
Exercise: 60 minutes most days

FRUITS	GRAINS	DAIRY
2.5 cups	10 ounces (at least 5 ounces whole grain)	3 cups
VEGETABLES	**PROTEIN**	**OILS**
3.5 cups	7 ounces	8 teaspoons

Standardized Test Prep

Multiple Choice

1. Which of the following is NOT a kind of tissue in the human body?
   A epithelial
   B connective
   C interstitial
   D nervous

2. Each of the following aids in the process of digestion EXCEPT the
   A teeth.
   B saliva.
   C stomach.
   D kidney.

3. In the human body, hydrochloric acid is responsible for the low pH of the contents of the
   A kidney.
   B gallbladder.
   C stomach.
   D liver.

4. Which is NOT a function of the kidneys?
   A removal of waste products from the blood
   B maintenance of blood pH
   C regulation of water content of the blood
   D excretion of carbon dioxide

5. The main function of the digestive system is to
   A break down large molecules into smaller molecules.
   B excrete oxygen and carbon dioxide.
   C synthesize minerals and vitamins needed for a healthy body.
   D remove waste products from the blood.

6. In the kidneys, both useful substances and wastes are removed from the blood by
   A reabsorption.
   B excretion.
   C dialysis.
   D filtration.

7. Which of the following is NOT a role of fats in the body?
   A Deposits of fat act as insulation.
   B They are components of cell membranes.
   C They help with absorption of fat-soluble vitamins.
   D They provide the body with essential amino acids.

Questions 8–9

A student is studying the effect of temperature on the action of an enzyme in stomach fluid. The enzyme digests protein. An investigation was set up using five identical test tubes. Each tube contained 40 mL of stomach fluid and 20 mm of glass tubing filled with gelatin. Each tube was subjected to a different temperature. After 48 hours, the amount of gelatin digested in each tube was measured in millimeters. The results for the five test tubes are shown in the table.

Effect of Temperature on Enzyme Action

Test Tube	Temperature (°C)	Amount of Digestion After 48 Hours
1	2	0.0 mm
2	10	3.0 mm
3	22	4.5 mm
4	37	8.0 mm
5	100	0.0 mm

8. Which is the manipulated (independent) variable in this investigation?
   A gastric fluid
   B length of glass tubing
   C temperature
   D time

9. Another test tube was set up that was identical to the other test tubes and placed at a temperature of 15°C for 48 hours. What amount of digestion would you expect to occur in this test tube?
   A less than 3.0 mm
   B between 3.0 mm and 4.5 mm
   C between 4.5 mm and 8.0 mm
   D more than 8.0 mm

Open-Ended Question

10. Fad diets that boast of rapid weight loss often become popular. Many of these diets involve eating only a limited variety of foods. Explain why these diets are an unhealthful way to lose weight.

If You Have Trouble With . . .

Question	1	2	3	4	5	6	7	8	9	10
See Lesson	30.1	30.3	30.3	30.4	30.3	30.4	30.2	30.3	30.3	30.2

31 Nervous System

Structure and Function

Q: How does the structure of the nervous system allow it to regulate functions in every part of the body?

The sights, sounds, and smells at a ball game provide a fan's nervous system with a lot of stimulation.

INSIDE:

- 31.1 The Neuron
- 31.2 The Central Nervous System
- 31.3 The Peripheral Nervous System
- 31.4 The Senses

Deepen your understanding of the structure and functions of the nervous system by engaging in key practices that allow you to make connections across concepts.

NGSS You will construct and use models to explain relationships among the regions of the brain and its functions.

STEM Can artificial retinas help people who are visually impaired? You will investigate how advances in technology can change the lives of individuals with physical disabilities.

Common Core You will translate technical information expressed in words in the text into visual form to describe the structure of the brain and its functions in the nervous system.

POISONING ON THE HIGH SEAS

From the middle to late 1700s, Captain James Cook commanded several voyages of discovery to the South Pacific for Great Britain. The discovery of new lands brought him many riches; the discoveries of new animals, however, were not always pleasant.

September 7, 1774, was a remarkable day on the HMS *Resolution*. The ship's butcher died from a fall, there was a solar eclipse, and a clerk traded some cloth for a freshly caught fish.

Although Cook ate only a few bites of the fish, within a few hours, the captain felt "an extraordinary weakness" in his limbs, lost all sense of touch, and could not sense the weight of objects. It took eleven days for the men who ate the fish to recover. A pig and dog who ate some of the fish's organs were dead by morning. As you read through this chapter, look for clues as to how eating this fish could produce such deadly effects.

Biology.com

Finding the solution to this mystery is only the beginning. Take a video field trip with the ecogeeks of Untamed Science to see where the mystery leads.

Go online to access additional resources including:
- eText • Flash Cards • Lesson Overviews • Chapter Mystery

31.1 The Neuron

Key Questions

What are the functions of the nervous system?

What is the function of neurons?

How does a nerve impulse begin?

Vocabulary

peripheral nervous system • central nervous system • cell body • dendrite • axon • myelin sheath • resting potential • action potential • threshold • synapse • neurotransmitter

Taking Notes

Outline Before you read, use the green and blue headings to make an outline. As you read, fill in the subtopics and smaller topics. Then, add phrases or a sentence after each to provide key information.

THINK ABOUT IT All of us are aware of the world outside our bodies. How do we know about that world? How do you really know what's happening outside? When you reached for this book and opened it to this page, how did you make these things happen? Even more mysteriously, how did the words on this page that you are reading right now get into your mind? The answers to all these questions are to be found in the nervous system.

Functions of the Nervous System

What are the functions of the nervous system?

The nervous system is our window on the world. **The nervous system collects information about the body's internal and external environment, processes that information, and responds to it.** These functions are accomplished by the peripheral nervous system and the central nervous system. The **peripheral nervous system,** which consists of nerves and supporting cells, collects information about the body's external and internal environment. The **central nervous system,** which consists of the brain and spinal cord, processes that information and creates a response that is delivered to the appropriate part of the body through the peripheral nervous system.

Think about what happens when you search through your backpack for a pencil. Information is sent to your central nervous system about the objects you are touching. Your brain processes the information and determines that the first object you touch is too square to be a pencil. Then your brain sends messages via your peripheral nervous system to the muscles in your hand, commanding them to keep searching.

Imagine the billions of messages that are sent throughout your body at any given moment. The messages may tell you to laugh at a funny joke, or they may tell your brain that it's cold outside. These messages enable the different organs of the body to act together and also to react to conditions in the world around us. How does this communication occur?

Peripheral Nervous System
Gathers information and sends it to the central nervous system

Input

Central Nervous System
Processes the information and forms a response

Output

Peripheral Nervous System
Carries the response of the central nervous system to glands and muscles

FIGURE 31–1 Information Flow in the Nervous System

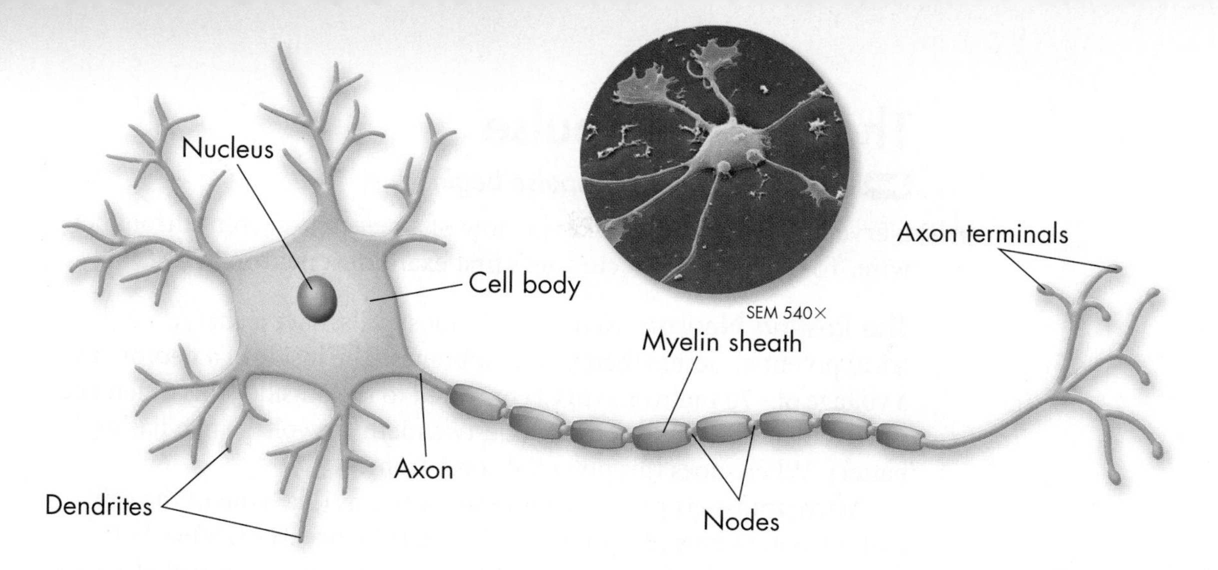

FIGURE 31–2 The Neuron The nervous system controls and coordinates functions throughout the body. The basic unit of the nervous system is the neuron.

Neurons

What is the function of neurons?

The messages carried by the nervous system are electrical signals called impulses. **Nervous system impulses are transmitted by cells called neurons.**

Types of Neurons Neurons can be classified into three types according to the direction in which an impulse travels. Sensory neurons carry impulses from the sense organs, such as the eyes and ears, to the spinal cord and brain. Motor neurons carry impulses from the brain and the spinal cord to muscles and glands. Interneurons do the high-level work. They process information from sensory neurons and then send commands to other interneurons or motor neurons.

Structure of Neurons Although neurons come in many shapes and sizes, they all have certain features in common. As shown in **Figure 31–2,** the largest part of a typical neuron is its **cell body,** which contains the nucleus and much of the cytoplasm.

Spreading out from the cell body are short, branched extensions called dendrites. **Dendrites** receive impulses from other neurons and carry impulses to the cell body. The long fiber that carries impulses away from the cell body is the **axon.** An axon ends in a series of small swellings called axon terminals. Neurons may have dozens of dendrites, but usually they have only one axon. In most animals, axons and dendrites of different neurons are clustered into bundles of fibers called nerves. Some nerves contain fibers from only a few neurons, but others contain hundreds or even thousands of neurons.

In some neurons, the axon is surrounded by an insulating membrane known as the **myelin** (MY uh lin) **sheath.** The myelin sheath that surrounds a single, long axon has many gaps, called nodes, where the axon membrane is exposed. As an impulse moves along the axon, it jumps from one node to the next. This arrangement causes an impulse to travel faster than it would through an axon without a myelin sheath.

BUILD Vocabulary

MULTIPLE MEANINGS The word *terminal* can be a noun or adjective. As a noun, it may refer to a place where information is entered into a computer or a station where people or goods are moved from one place to another. As an adjective, it may describe something that is beyond rescue or placed at the end of a structure.

In Your Notebook *Make a two-column table that lists the structures of a neuron in one column and their functions in the next column.*

The Nerve Impulse

How does a nerve impulse begin?

Nerve impulses are a bit like the flow of an electric current through a wire. To see how this occurs, let's first examine a neuron at rest.

The Resting Neuron Neurons, like most cells, have a charge, or electrical potential, across their cell membranes. The inside of a neuron has a voltage of −70 millivolts (mV) compared to the outside. This difference, or **resting potential,** is roughly one-twentieth the voltage in a flashlight battery. Where does this potential come from?

Active transport proteins pump sodium ions (Na^+) out of the cell and potassium ions (K^+) into it as shown in **Figure 31–3.** Since both ions are positively charged, this alone doesn't produce a potential across the membrane. However, ungated potassium channel proteins make it easier for K^+ ions than Na^+ ions to diffuse back across the membrane. Because there is a higher concentration of K^+ ions inside the cell as a result of active transport, there is a net movement of positively charged K^+ ions out of the cell. As a result, the inside becomes negatively charged compared to the outside, producing the resting potential.

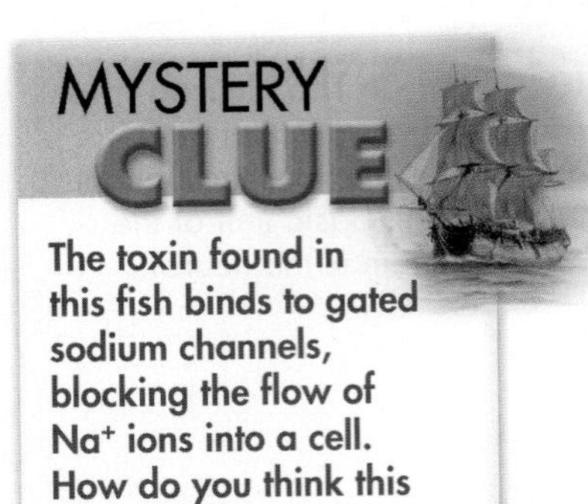

The toxin found in this fish binds to gated sodium channels, blocking the flow of Na^+ ions into a cell. How do you think this might affect muscle movement?

The Moving Impulse A neuron remains in its resting state until it receives a stimulus large enough to start a nerve impulse. **An impulse begins when a neuron is stimulated by another neuron or by the environment.** Once it begins, the impulse travels quickly down the axon away from the cell body toward the axon terminals. In myelinated axons, the impulse moves even more rapidly as its skips from one node to the next.

What actually happens during an impulse? As **Figure 31–5** shows, the impulse itself is a sudden reversal of the resting potential. The neuron cell membrane contains thousands of "gated" ion channels. At the leading edge of an impulse, gated sodium channels open, allowing positively charged Na^+ ions to flow into the cell. The inside of the membrane temporarily becomes more positive than the outside, reversing the resting potential. This reversal of charges, from more negatively charged to more positively charged, is called a nerve impulse, or an **action potential.**

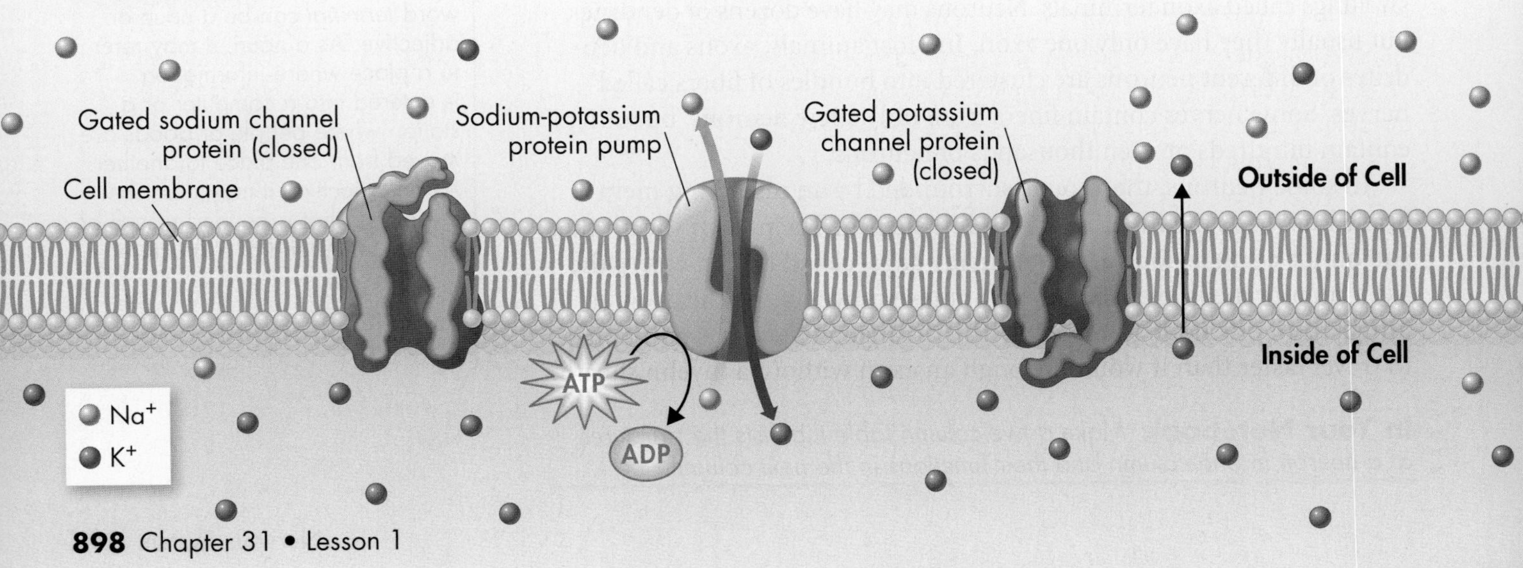

FIGURE 31–3 The Resting Neuron The sodium-potassium pump in the neuron cell membrane uses ATP to pump Na^+ ions out of the cell and to pump K^+ ions in. A small amount of K^+ ions diffuse out of the cell (through ungated channels), but gated channels block Na^+ ions from flowing into the resting neuron. **Apply Concepts** ***Is the action of the sodium-potassium pump an example of diffusion or active transport? Explain.***

VISUAL ANALOGY

A CHAIN REACTION

FIGURE 31–4 With a strong enough push, the fall of one domino leads to the fall of the next. An action potential moves along a neuron in a similar manner. **Use Analogies** ***Compare and contrast how an action potential traveling along an axon is like the fall of a row of dominoes.***

Once the impulse passes, sodium gates close and gated potassium channels open, allowing K^+ ions to flow out. This restores the resting potential so that the neuron is once again negatively charged on the inside. All the while, the sodium-potassium pump keeps working, ensuring that the axon will be ready for more action potentials.

A nerve impulse is self-propagating; that is, the flow of ions at the point of the impulse causes sodium channels just ahead of it to open. This allows the impulse to move rapidly along the axon. You could compare the flow of an impulse to the fall of a row of dominoes. As each domino falls, it causes the next domino to fall.

In Your Notebook *In your own words, summarize what happens across a neuron's membrane when it is at rest and during an action potential.*

Threshold Not all stimuli are capable of starting an impulse. The minimum level of a stimulus that is required to cause an impulse in a neuron is called its **threshold.** Any stimulus that is weaker than the threshold will not produce an impulse. A nerve impulse is an all-or-none response. Either the stimulus produces an impulse, or it does not produce an impulse.

The threshold principle can also be illustrated by using a row of dominoes. If you were to gently press the first domino in a row, it might not move at all. A slightly harder push might make the domino teeter back and forth but not fall. A push strong enough to cause the first domino to fall into the second, and start the whole row falling, is like a threshold stimulus.

If all action potentials have the same strength, how do we sense if a stimulus, like touch or pain, is strong or weak? The brain determines this from the frequency of action potentials. A weak stimulus might produce three or four action potentials per second, while a strong one might result in as many as 100 per second. If you accidentally hit your finger with a hammer, those action potentials fire like mad!

FIGURE 31–5 The Moving Impulse Once an impulse begins, it will continue down an axon until it reaches the end. In an axon with a myelin sheath, the impulse jumps from node to node.

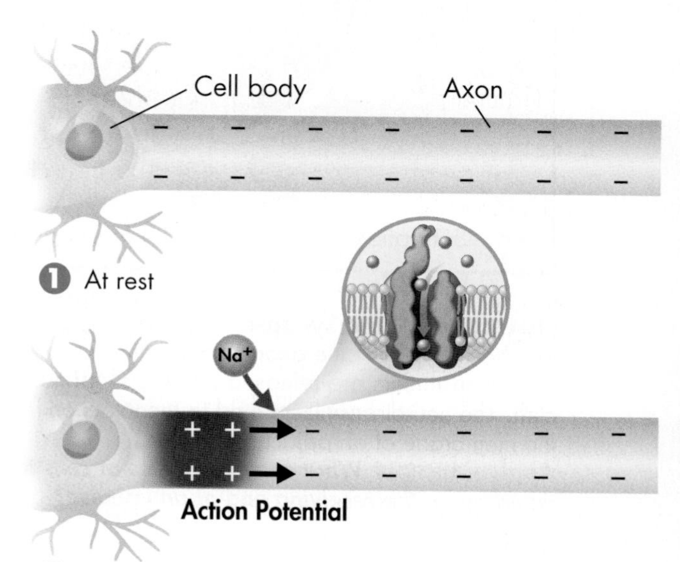

2 At the leading edge of the impulse, gated sodium channels open. Na^+ ions flow into the cell, reversing the potential between the cell membrane and its surroundings. This rapidly moving reversal of charge is called an action potential.

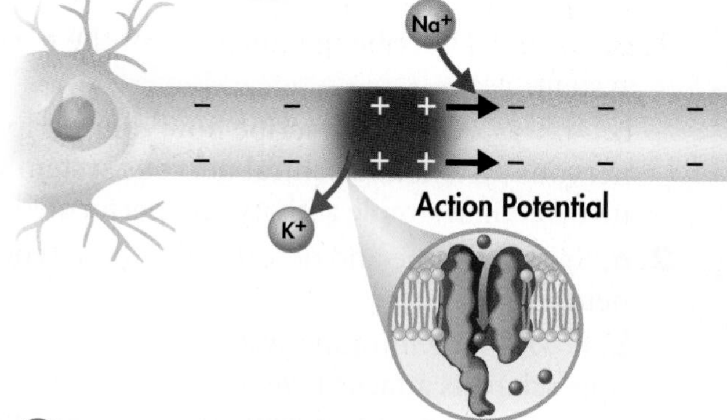

3 As the action potential passes, gated potassium channels open, allowing K^+ ions to flow out and restoring the resting potential inside the axon.

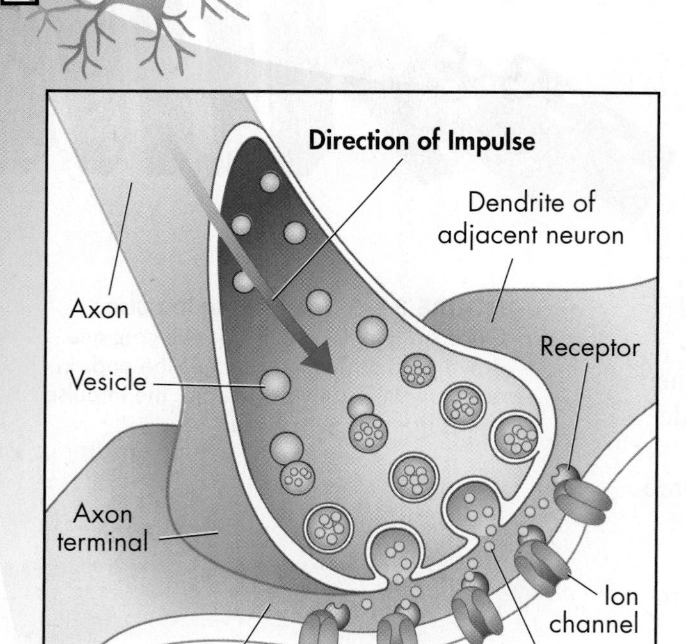

FIGURE 31–6 The Synapse When an impulse reaches the end of the axon of one neuron, neurotransmitters are released into the synaptic cleft. The neurotransmitters bind to receptors on the membrane of an adjacent cell.
Apply Concepts ***What are three types of cells that could be on the receiving end of an impulse?***

The Synapse At the end of the neuron, the impulse reaches an axon terminal, which may pass the impulse along to another cell. A motor neuron, for example, may pass impulses to a muscle cell, causing the muscle cell to contract. The point at which a neuron transfers an impulse to another cell is called a **synapse** (SIN aps). As shown in **Figure 31–6**, a space, called the synaptic cleft, separates the axon terminal from the adjacent cell.

The axon terminal at a synapse contains tiny vesicles filled with neurotransmitters. **Neurotransmitters** are chemicals that transmit an impulse across a synapse to another cell. When an impulse arrives at the synapse, neurotransmitters are released from the axon, diffuse across the synaptic cleft, and bind to receptors on the membrane of the receiving cell. This binding opens ion channels in the membrane of the receiving cell. If the stimulation exceeds the cell's threshold, a new impulse begins.

Once they have done their work, the neurotransmitters are released from the receptors on the cell surface. They are then broken down by enzymes in the synaptic cleft or taken up and recycled by the axon terminal.

31.1 Assessment

Review Key Concepts

1. a. Review Describe the functions of the nervous system.

b. Apply Concepts Describe how your peripheral nervous system and central nervous system were involved in a simple activity you performed today.

2. a. Review Name and describe the three types of neurons.

b. Predict The immune system of people with multiple sclerosis attacks myelin sheaths in the central nervous system. The myelin breaks down and scar tissue may result. How do you think this would affect the transmission of signals from the central nervous system?

3. a. Review What happens when a neuron is stimulated by another neuron?

b. Infer How can the level of pain you feel vary if a stimulus causes an all-or-none response?

VISUAL THINKING

4. Create a flowchart to show the events that occur as a nerve impulse travels from one neuron to the next. Include as much detail as you can. Use your flowchart to explain the process to a classmate.

31.2 The Central Nervous System

THINK ABOUT IT Who's in charge? The nervous system contains billions of neurons, each of them capable of carrying impulses and sending messages. What keeps them from sending impulses everywhere and acting like an unruly mob? Is there a source of order in this complex system, a central place where information is processed, decisions are made, and order is enforced?

The Brain and Spinal Cord

Where does processing of information occur in the nervous system?

The control point of the central nervous system is the brain. **Each of the major areas of the brain—the cerebrum, cerebellum, and brain stem—are responsible for processing and relaying information.** Like the central processing unit of a computer, information processing is the brain's principal task. **Figure 31–8** on the next page provides details about the major areas of the brain.

While most organs in the body function to maintain homeostasis, the brain itself is constantly changed by its interactions with the environment. Sensory experience changes many of the patterns of neuron connections in the brain, and stem cells in the brain produce new neurons throughout life. Many of these new cells originate in regions associated with learning and memory. Far from staying the same, the highly flexible brain reacts to and changes constantly with the world around it.

Most of the neurons that enter and leave the brain do so in a large cluster of neurons and other cells known as the spinal cord. **The spinal cord is the main communication link between the brain and the rest of the body.** The spinal cord is a bit like a major telephone line, carrying thousands of signals at once between the central and peripheral nervous systems. Thirty-one pairs of spinal nerves branch out from the spinal cord, connecting the brain to different parts of the body. Certain kinds of information, including many reflexes, are processed directly in the spinal cord. A **reflex** is a quick, automatic response to a stimulus. The way in which you pull your hand back quickly when pricked by a pin is an example of a reflex.

In Your Notebook *Make a three-column table that lists the major structures of the brain described in **Figure 31–8**, their functions, and how they interact with at least one other brain structure.*

Key Questions

Where does processing of information occur in the nervous system?

How do drugs change the brain and lead to addiction?

Vocabulary

reflex • cerebrum • cerebral cortex • thalamus • hypothalamus • cerebellum • brain stem • dopamine

Taking Notes

Concept Map As you read, construct a concept map that shows how the structures of the central nervous system are related to each other.

FIGURE 31–7 The Central Nervous System The central nervous system consists of the brain and spinal cord.

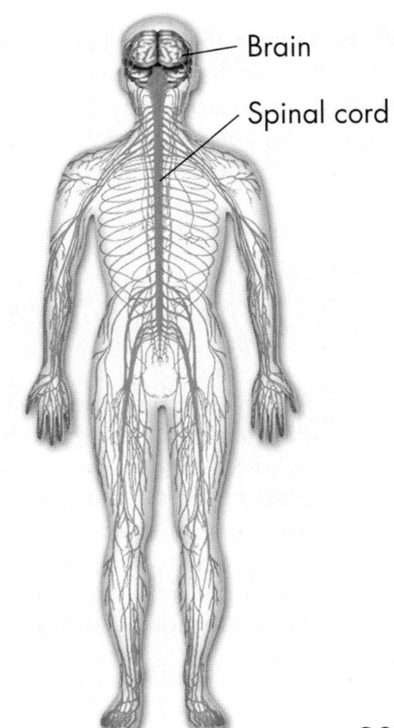

VISUAL SUMMARY

THE BRAIN

FIGURE 31–8 The brain contains billions of neurons and other supporting tissue that process, relay, and form responses to an incomprehensible amount of information every moment. **Infer** ***Which structure of the brain most likely filters information traveling from the spinal cord to the brain?***

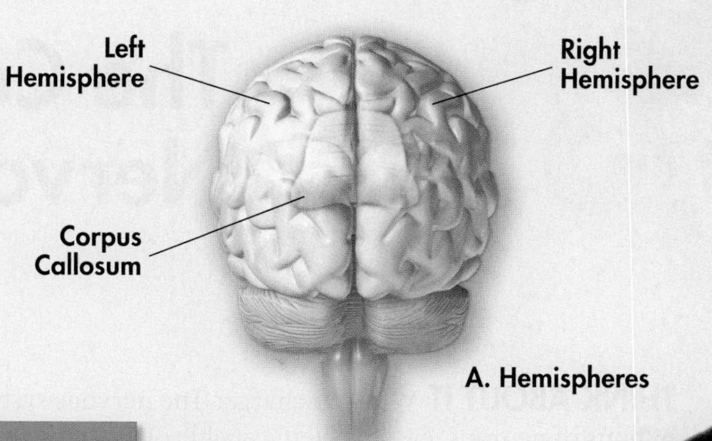

Cerebrum

The largest region of the human brain is the cerebrum. The **cerebrum** is responsible for the voluntary, or conscious, activities of the body. It is also the site of intelligence, learning, and judgment.

Hemispheres As shown in **Figure 31–8A** (a back view of the brain), a deep groove divides the cerebrum into left and right hemispheres. The hemispheres are connected by a band of tissue called the corpus callosum. Remarkably, each hemisphere deals mainly with the opposite side of the body. Sensations from the left side of the body go to the right hemisphere, and those from the right side go to the left hemisphere. Commands to move muscles are delivered in the same way.

As shown in **Figure 31–8B,** each hemisphere is divided into regions called lobes. The four lobes are named for the skull bones that cover them. Each of these lobes are associated with different functions.

Cerebral Cortex The cerebrum consists of two layers. The outer layer of the cerebrum is called the **cerebral cortex** and consists of densely packed nerve cell bodies known as gray matter. The cerebral cortex processes information from the sense organs and controls body movements. It is also where thoughts, plans, and learning abilities are processed. Folds and grooves on the outer surface of the cerebral cortex greatly increase its surface area.

White Matter The inner layer of the cerebrum is known as white matter. Its whitish color comes from bundles of axons with myelin sheaths. These axons may connect different areas of the cerebral cortex, or they may connect the cerebrum to other areas of the brain such as the brain stem.

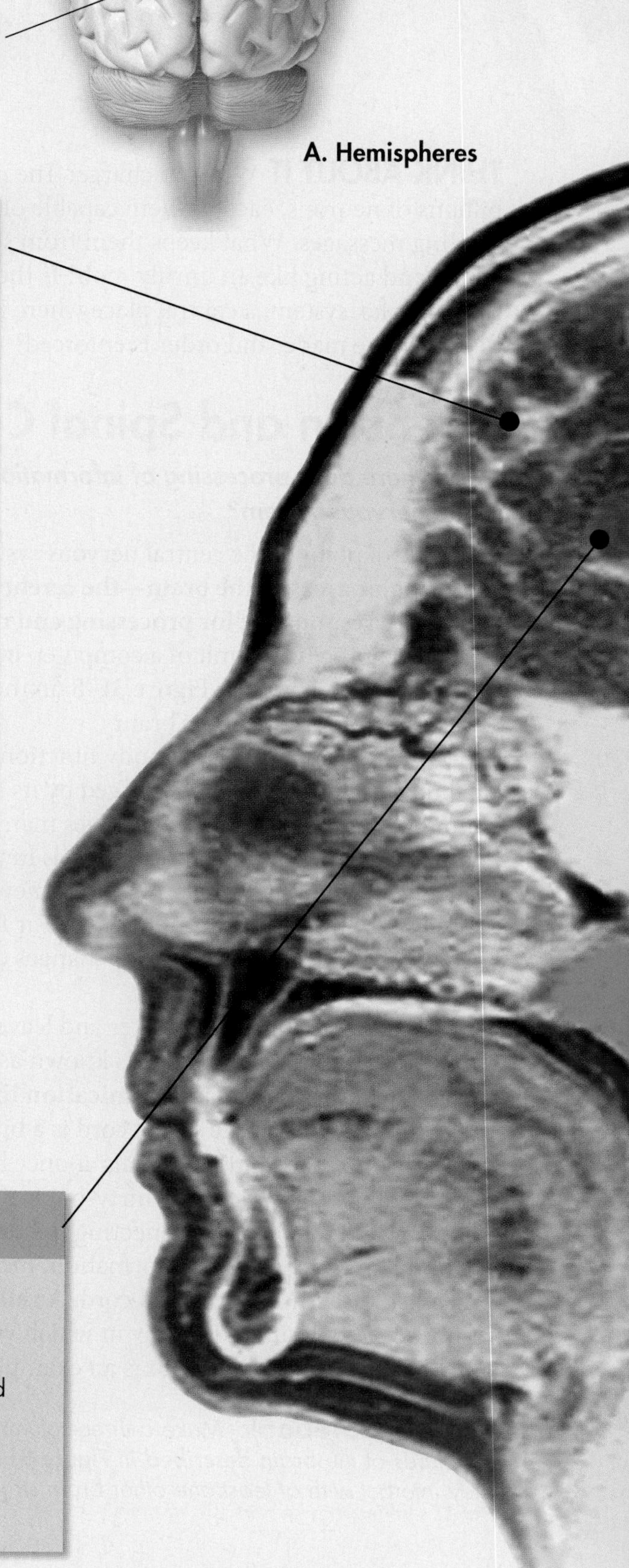

Limbic System

A number of important functions have been linked to the many structures that make up the limbic system including emotion, behavior, and memory. For example, a region deep within the brain called the amygdala (uh MIG duh luh) has been associated with emotional learning, including fear and anxiety, as well as the formation of long-term memories. The limbic system is also associated with the brain's pleasure center, a region that produces feelings of satisfaction and well-being.

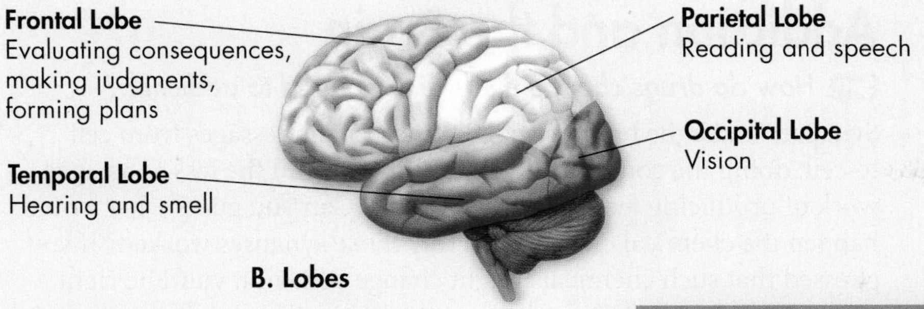

Frontal Lobe
Evaluating consequences, making judgments, forming plans

Parietal Lobe
Reading and speech

Occipital Lobe
Vision

Temporal Lobe
Hearing and smell

B. Lobes

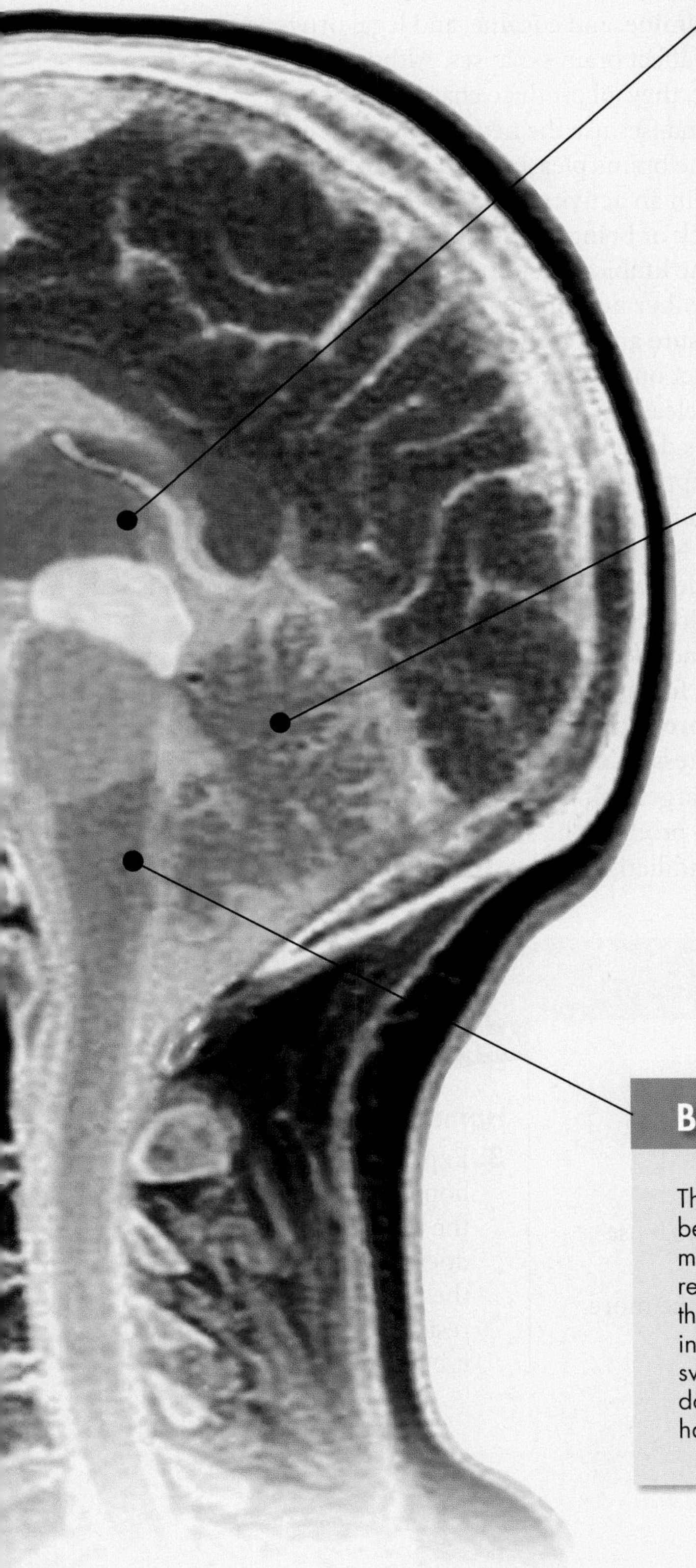

Thalamus and Hypothalamus

The thalamus and hypothalamus are found between the brain stem and the cerebrum. The **thalamus** receives messages from sensory receptors throughout the body and then relays the information to the proper region of the cerebrum for further processing. Just below the thalamus is the hypothalamus. The **hypothalamus** is the control center for recognition and analysis of hunger, thirst, fatigue, anger, and body temperature. The hypothalamus also helps to coordinate the nervous and endocrine systems.

Cerebellum

The second largest region of the brain is the **cerebellum.** Information about muscle and joint position, as well as other sensory inputs, are sent to the cerebellum. Although the commands to move muscles come from the cerebral cortex, sensory information allows the cerebellum to coordinate and balance the actions of these muscles. This enables the body to move gracefully and efficiently.

When you begin any new activity involving muscle coordination, such as hitting a golf ball or threading a needle, it is the cerebellum that actually learns the movements and coordinates the actions of scores of individual muscles when the movement is repeated.

Brain Stem

The **brain stem** connects the brain and spinal cord. Located just below the cerebellum, the brain stem includes three regions—the midbrain, the pons, and the medulla oblongata. Each of these regions regulates the flow of information between the brain and the rest of the body. Some of the body's most important functions—including regulation of blood pressure, heart rate, breathing, and swallowing—are controlled by the brain stem. The brain stem does the work of keeping the body functioning even when you have lost consciousness due to sleep or injury.

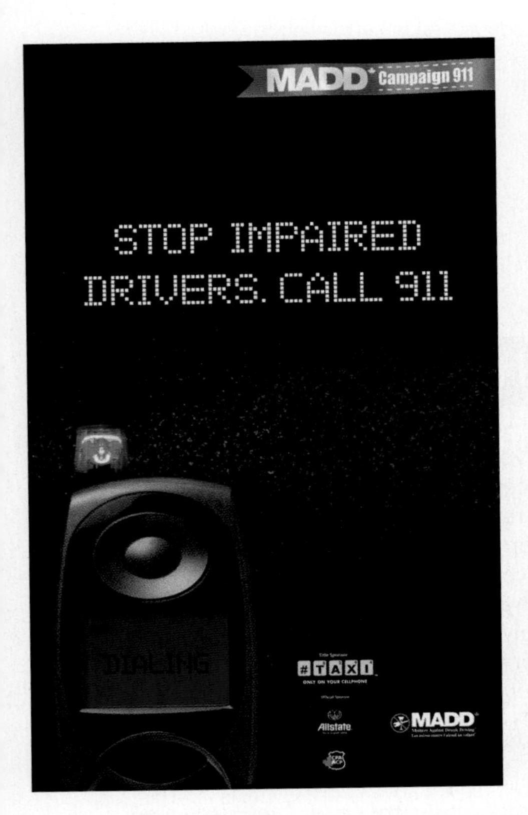

FIGURE 31–9 Drugs and Society The damage to the brain is only the start of the damage that drugs cause. For example, alcohol abuse costs the United States about $185 billion a year in health care costs, treatment services, property damage, and lost productivity.

Addiction and the Brain

How do drugs change the brain and lead to addiction?

Synapses make the brain work by transferring messages from cell to cell, doing the conscious work of thinking and the less conscious work of producing feelings and emotions. Can you guess what would happen if a chemical changed the way those synapses worked? If you guessed that such chemicals might change behavior, you'd be right.

Nearly every addictive substance, including illegal drugs such as heroin, methamphetamine, and cocaine, and legal drugs, such as tobacco and alcohol, affect brain synapses. Although the chemistry of each drug is different, they all produce changes in one particular group of synapses. These synapses use the neurotransmitter **dopamine** and are associated with the brain's pleasure and reward centers.

When we engage in an activity that brings us pleasure, whether it's eating a tasty snack or being praised by a friend, neurons in the hypothalamus and the limbic system release dopamine. Dopamine molecules stimulate other neurons across these synapses, producing the sensation of pleasure and a feeling of wellbeing.

Addictive drugs act on dopamine synapses in a number of ways. Methamphetamine releases a flood of dopamine, producing an instant "high." Cocaine keeps dopamine in the synaptic region longer, intensifying pleasure and suppressing pain. Drugs made from opium poppies, like heroin, stimulate receptors elsewhere in the brain that lead to dopamine release. Nicotine, the addictive substance in tobacco, and alcohol, the most widely abused drug in the United States, also cause increased release of dopamine.

The brain reacts to excessive dopamine levels by reducing the number of receptors for the neurotransmitter. As a result, normal activities no longer produce the sensations of pleasure they once did. Addicts feel depressed and sick without their drugs. Because there are fewer receptors, larger amounts of tobacco, alcohol, and illegal drugs are required to produce the same high. The result is a deeper and deeper spiral of addiction that is difficult to break.

31.2 Assessment

Review Key Concepts

1. a. Review What are the three major regions of the brain?
   b. Describe Explain the role of the spinal cord.
   c. Infer How do reflexes protect the body from injury?
2. a. Review Describe three ways that drugs affect synapses that use the neurotransmitter dopamine.
   b. Apply Concepts Why do many drug users begin to take more and more of the drug they abuse?

Apply the Big idea

Homeostasis

3. Explain the brain's role in homeostasis in regard to the body as a whole. How does homeostasis within the brain differ from the rest of the body? How must it be similar?

RST.9-10.7 Integration of Knowledge and Ideas.
Also RST.9-10.10

Technology & BIOLOGY

Studying the Brain and Addiction

Studies at the National Institute of Drug Abuse (NIDA) have demonstrated why drugs that stimulate dopamine produce a pattern of addiction that is difficult to break. The brain is a flexible organ that responds to its environment and continually adjusts its internal chemistry. When it senses increased levels of dopamine, it adjusts by cutting down on the number of receptors for the neurotransmitter.

NIDA researchers used a powerful imaging technique known as positron emission tomography (PET) to visualize the density of dopamine receptors in brains affected by drug addition, and the results, shown here, are striking. Brains of individuals abusing alcohol and illegal drugs show dramatically lower concentrations of dopamine receptors than the brains of individuals not abusing the drugs.

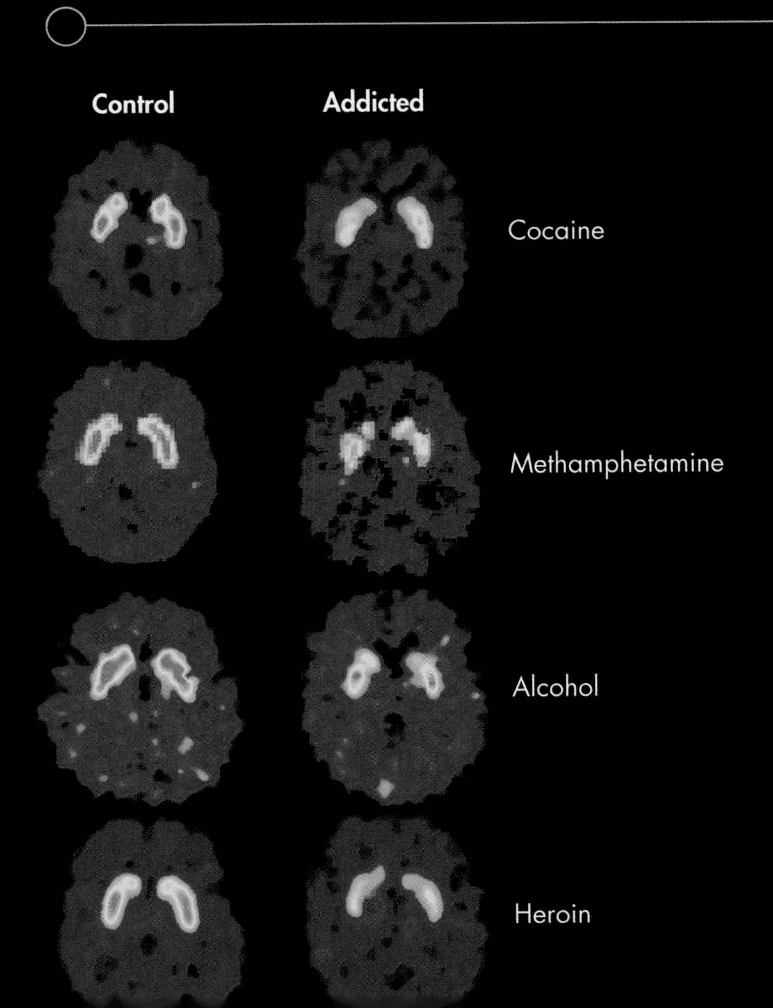

Positron emission tomography (PET) allows researchers to visualize labeled molecules deep inside the body. PET is routinely used to pinpoint regions of cellular activity. To locate dopamine receptors, a molecule that binds to the receptor is labeled with a radioactive isotope of carbon. Within a few minutes, the isotope emits a subatomic particle called a positron. The location of the particle is revealed by gamma rays released when it collides with other particles. By locating thousands of positron emissions, computers can put together detailed images showing the location of the labeled molecules.

◀ In these images, areas of highest dopamine receptor density appear red. Areas of lowest dopamine receptor density appear green.

WRITING **Using the information in this feature, create a poster to discourage peers from using addictive drugs.**

31.3 The Peripheral Nervous System

Key Questions

How does the central nervous system receive sensory information?

How do muscles and glands receive commands from the central nervous system?

Vocabulary

somatic nervous system
reflex arc
autonomic nervous system

Taking Notes

Flowchart As you read, make a flowchart that shows the flow of information between the divisions of the peripheral nervous system and the central nervous system.

THINK ABOUT IT It's all about input and output. No computer is worth much unless it can accept input from the world around it. And, no matter how quickly it calculates, no result is of any meaning unless there's a way to output it. The central nervous system faces the same issues. Can you guess what it uses for input and output devices?

The Sensory Division

How does the central nervous system receive sensory information?

The peripheral nervous system consists of all the nerves and associated cells that are not part of the brain or spinal cord. Cranial nerves go through openings in the skull and stimulate regions of the head and neck. Spinal nerves stimulate the rest of the body. The cell bodies of cranial and spinal nerves are arranged in clusters called ganglia.

The peripheral nervous system, our link with the outside world, consists of two major divisions—the sensory division and the motor division. **The sensory division of the peripheral nervous system transmits impulses from sense organs to the central nervous system.** The motor division transmits impulses from the central nervous system to the muscles and glands.

Sensory receptors are cells that transmit information about changes in the environment—both internal and external. These changes are called stimuli. Sensory receptors can be categorized by the type of stimuli to which they respond. **Figure 31–10** shows the functions and locations of several types of sensory receptors. When stimulated, sensory receptors transmit impulses to sensory neurons. Sensory neurons then transmit impulses to the central nervous system.

FIGURE 31–10 Sensory Receptors Sensory receptors react to a specific stimulus such as light or sound by sending impulses to sensory neurons. **Apply Concepts** *List three types of sensory receptors that are activated when you walk into a busy flower shop.*

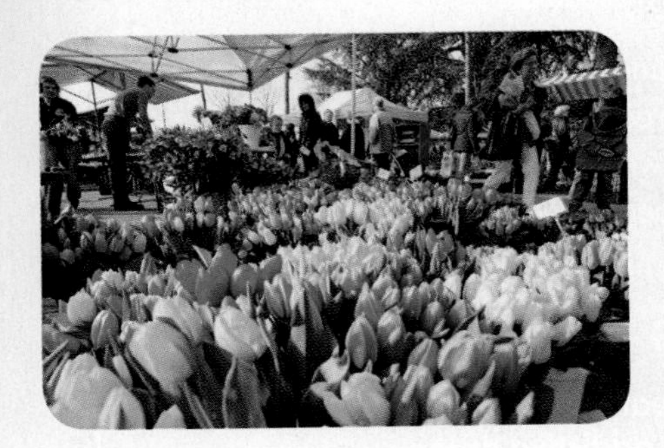

Sensory Receptors

Type	Responds to	Some Locations
Chemoreceptor	Chemicals	Mouth, nose, blood vessels
Photoreceptor	Light	Eyes
Mechanoreceptor	Touch, pressure, vibrations, and stretch	Skin, hair follicles, ears, ligaments, tendons
Thermoreceptor	Temperature changes	Skin, hypothalamus
Pain receptor	Tissue injury	Throughout the body

The Motor Division

How do muscles and glands receive commands from the central nervous system?

The nervous system plays a key role in maintaining homeostasis by coordinating the activities of other systems and organs. Once it has gathered and processed sensory information, the nervous system sends commands to the rest of the body. **The motor division of the peripheral nervous system transmits impulses from the central nervous system to muscles or glands.** These messages are relayed through one of two divisions, the somatic nervous system or the autonomic nervous system.

Somatic Nervous System

The **somatic nervous system** regulates body activities that are under conscious control, such as the movement of skeletal muscles. Most of the time you have control over skeletal muscle movement, but when your body is in danger the central nervous system may take over.

▶***Voluntary Control*** Every time you lift your finger or wiggle your toes, you are using motor neurons of the somatic nervous system. Impulses originating in the brain are carried through the spinal cord where they synapse with the dendrites of motor neurons. The axons from these motor neurons extend from the spinal cord carrying impulses directly to muscles, causing the contractions that produce voluntary movements.

▶***Reflex Arcs*** Although the somatic nervous system is generally considered to be under conscious control, some actions of the system occur automatically. If you accidentally step on a tack with your bare foot, your leg may recoil before you are even aware of the pain.

This rapid response (a reflex) is caused by impulses that travel a pathway known as a **reflex arc,** as shown in **Figure 31–11.** ❶ In this example, sensory receptors react to the sensation of the tack and send an impulse to sensory neurons. ❷ Sensory neurons relay the information to the spinal cord. ❸ An interneuron in the spinal cord processes the information and forms a response. ❹ A motor neuron carries impulses to its effector, a muscle that it stimulates. ❺ The muscle contracts and your leg moves. Meanwhile, impulses carrying information about the injury are sent to your brain. By the time your brain interprets the pain, however, your leg and foot have already moved. The spinal cord does not control all reflexes. Many reflexes that involve structures in your head, such as blinking or sneezing, are controlled by the brain.

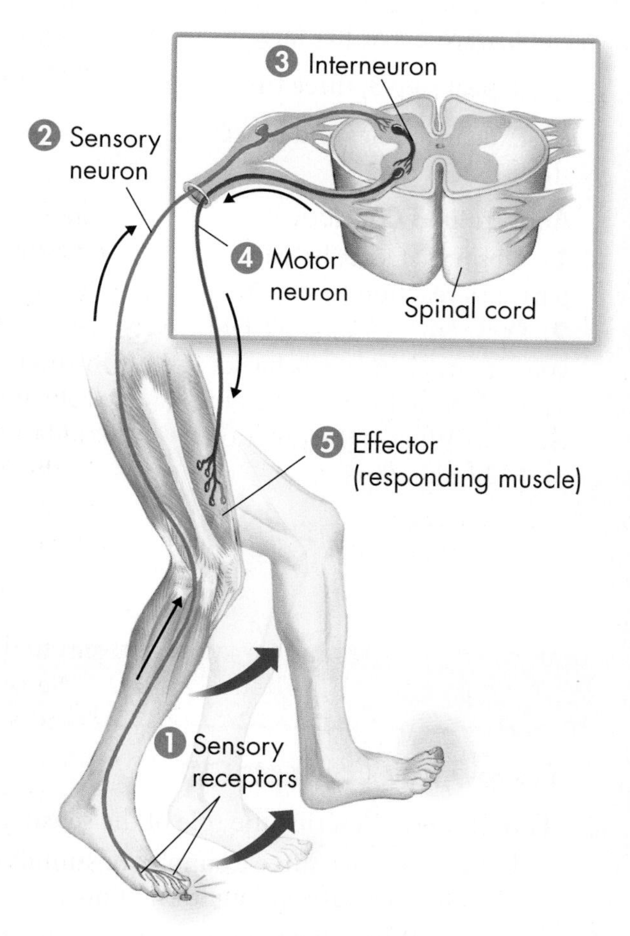

FIGURE 31–11 Reflex Arc When you step on a tack, sensory receptors stimulate a sensory neuron, which relays the signal to an interneuron within the spinal cord. The signal is then sent to a motor neuron, which in turn stimulates a muscle that lifts your leg.

In Your Notebook *In your own words, describe how a reflex arc works. Include the role of the three types of neurons in your description.*

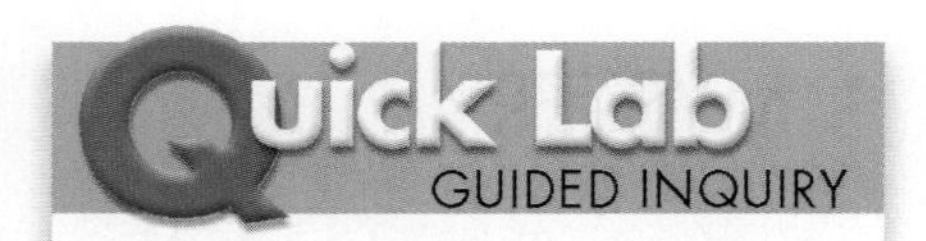

How Do You Respond To an External Stimulus?

❶ Have your partner put on safety goggles.

❷ Crumple up a sheet of scrap paper into a ball.

❸ Watch your partner's eyes carefully as you toss the paper ball toward his or her face.

❹ Repeat step 3, three times.

❺ Exchange roles and repeat steps 1, 3, and 4.

Analyze and Conclude

1. Observe Describe your partner's reaction to step 3.

2. Compare and Contrast Did you see any change in behavior as you repeated step 3? Explain.

3. Infer What is the function of the blink reflex?

Autonomic Nervous System The **autonomic nervous system** regulates activities that are involuntary, or not under conscious control. For instance, when you start to run, the autonomic nervous system speeds up your heart rate and blood flow to the skeletal muscles, stimulates the sweat glands, and slows down the contractions of smooth muscles in the digestive system. You may not be aware of any of these activities, but all of them enable you to run faster and farther.

The autonomic nervous system consists of two equally important parts, the sympathetic nervous system and the parasympathetic nervous system. Why two systems? In general, the sympathetic and parasympathetic systems have opposite effects on each organ they influence. In the same way that a driver must be able to turn the steering wheel both left and right to keep a car on the road, the two systems produce a level of fine control that coordinates organs throughout the body.

For example, heart rate is increased by the sympathetic nervous system but decreased by the parasympathetic nervous system. In general, the sympathetic system prepares the body for intense activity. Its stimulation causes an increase in blood pressure, the release of energy-rich sugar into the blood, and shutting down of activities not related to the body's preparation to "fight or flee" in response to stress. In contrast, the parasympathetic system causes what might be called the "rest and digest" response. It lowers heart rate and blood pressure, activates digestion, and activates pathways that store food molecules in the tissues of the body.

31.3 Assessment

Review Key Concepts

1. a. Review Describe the role of the sensory division.

b. Explain Give three examples of stimuli that your sensory receptors are responding to right now.

c. Infer Which type of sensory receptors most likely responds to a change in blood pressure that causes more force to be exerted on your blood vessels? Explain.

2. a. Review Describe the function of the two parts of the motor division of the peripheral nervous system.

b. Explain Is a reflex part of the central nervous system, the peripheral nervous system, or both?

c. Apply Concepts Describe a situation in which you would expect your sympathetic nervous system to be more active than your parasympathetic nervous system.

Apply the Big idea

Structure and Function

3. Which part of the peripheral nervous system is involved in both innate behaviors and learned behaviors? Explain. (*Hint:* See Lesson 29.1.)

31.4 The Senses

THINK ABOUT IT We live in a world of sensations. Think about how many of your experiences today can only be described in terms of what you felt, tasted, smelled, heard, and saw. Our senses are our link to experiencing the outside world, and we often take them for granted. Think for a moment of the color red. How would you describe the sensation of seeing red, as opposed to blue or green, to someone who was blind? Or, how would you describe the taste of an apple to someone who had never tasted one before? The inputs we get from our senses are almost impossible to describe, and yet we use them every moment of the day.

Key Questions

How does the body sense touch, temperature, and pain?

How are the senses of smell and taste similar?

How do the ears and brain process sounds and maintain balance?

How do the eyes and brain produce vision?

Vocabulary

taste bud • cochlea • semicircular canals • cornea • iris • pupil • lens • retina • rods • cones

Taking Notes

Preview Visuals Before reading, preview **Figure 31–14.** Write down at least two questions you have about the information in the figure.

Touch and Related Senses

How does the body sense touch, temperature, and pain?

Because nearly all regions of the skin are sensitive to touch, your skin can be considered your largest sense organ. **Different sensory receptors in the body respond to touch, temperature, and pain.** All of these receptors are found in your skin, but some are also found in other areas.

Touch Human skin contains at least seven types of sensory receptors, including several that respond to different levels of pressure. Stimulation of these receptors creates the sensation of touch. Not all parts of the body are equally sensitive to touch. The skin on your fingers, as you might expect, has a much higher density of touch receptors than the skin on your back.

Temperature Thermoreceptors are sensory cells that respond to heat and cold. They are found throughout the skin, and also in the hypothalamus, part of the brain that senses blood temperature. Recently, researchers studying the cell membrane proteins that sense heat made an interesting discovery. The chemical substances that make jalapeño peppers taste "hot" actually bind to these very same proteins.

Pain Pain receptors are found throughout the body. Some, especially those in the skin, respond to physical injuries like cutting or tearing. Many tissues also have pain receptors that respond to chemicals released during infection or inflammation. The brain, interestingly, does not have pain receptors. For this reason, patients are often kept conscious during brain surgery, enabling them to tell surgeons what sensations are produced when parts of the brain are stimulated.

MYSTERY CLUE

Based on Cook's symptoms, which of his senses was greatly affected by the toxin? Explain.

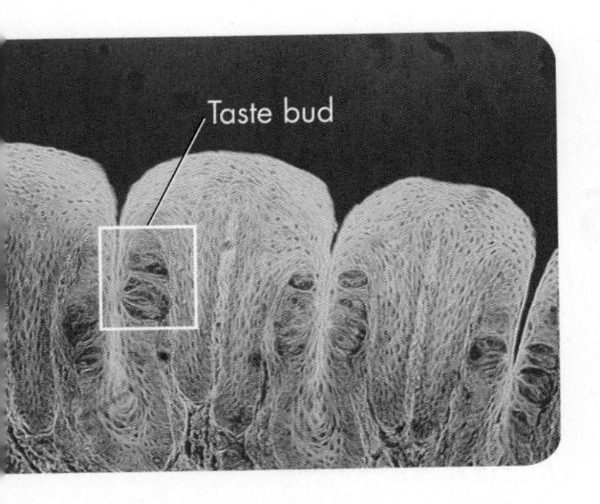

FIGURE 31–12 Taste Buds The surface of the tongue contains many tiny projections. Taste buds line the tops of some of these and line the sides of other projections (LM 80×).

Smell and Taste

How are the senses of smell and taste similar?

You may never have thought of it this way, but your senses of taste and smell actually involve the ability to detect chemicals. Chemical-sensing cells known as chemoreceptors in the nose and mouth are responsible for both of these senses. **Sensations of smell and taste are both the result of impulses sent to the brain by chemoreceptors.**

Your sense of smell is capable of producing thousands of different sensations. In fact, much of what we commonly call the "taste" of food and drink is actually smell. To prove this to yourself, eat a few bites of food while holding your nose. You'll discover that much of the taste of food disappears until you release your nose and breathe freely.

The sense organs that detect taste are the **taste buds.** Most of the taste buds are on the tongue, but a few are found at other locations in the mouth. The surface of the tongue is shown in **Figure 31–12.** Sensory cells in taste buds respond to salty, bitter, sweet, and sour foods. Recently, a fifth kind of taste sensation was identified, now called "umami," from the Japanese word for savory. Umami receptors are strongly stimulated by monosodium glutamate (MSG), a substance often added to Asian foods to enhance their flavor. They are also stimulated by meat and cheese, which typically contain the amino acid glutamate.

In Your Notebook *Explain the relationship between smell and taste.*

Analyzing Data

Sound Intensity

Sound intensity, or loudness, is measured in units called decibels (dB). The threshold of hearing for the human ear is 0 dB. For every 10 dB increase, the sound intensity increases ten times. Sound levels for several sound sources are shown in the bar graph.

Loud noises can permanently damage vibration-sensing cells in the cochlea. Exposure to sounds above 80 dB for several hours at a time can damage hearing. Exposure to sounds about 120 dB for even a few seconds can damage hearing.

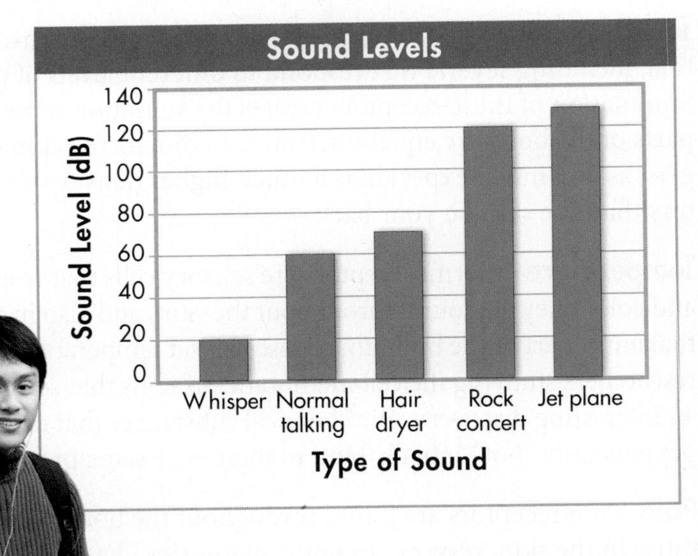

1. Calculate How much more intense is normal talking than a whisper? Explain. MATH

2. Infer Why do you think that hearing damage caused by repeated exposure to loud noises, such as portable music devices set at a high volume, might not reveal itself for many years?

Hearing and Balance

How do the ears and brain process sounds and maintain balance?

The human ear has two sensory functions, one of which, of course, is hearing. The other function is detecting positional changes associated with movement. **Mechanoreceptors found in parts of the ear transmit impulses to the brain. The brain translates the impulses into sound and information about balance.**

Hearing Sound is nothing more than vibrations moving through the air around us. The ears are the sensory organs that can distinguish both the pitch and loudness of those vibrations. The structure of the ear is shown in **Figure 31–13.**

Vibrations enter the ear through the auditory canal and cause the tympanum (TIM puh num), or eardrum, to vibrate. Three tiny bones, commonly called the hammer, anvil, and stirrup, transmit these vibrations to a membrane called the oval window. Vibrations there create pressure waves in the fluid-filled **cochlea** (KAHK lee uh) of the inner ear. The cochlea is lined with tiny hair cells that are pushed back and forth by these pressure waves. In response, the hair cells send nerve impulses to the brain, which processes them as sounds.

Balance Your ears contain structures that help your central nervous system maintain your balance, or equilibrium. Within the inner ear just above the cochlea are three tiny canals. They are called semicircular canals because each forms a half circle. The **semicircular canals** and the two tiny sacs located behind them monitor the position of your body, especially your head, in relation to gravity.

The semicircular canals and the sacs are filled with fluid and lined with hair cells. As the head changes position, the fluid in the canals also changes position. This causes the hair on the hair cells to bend. This action, in turn, sends impulses to the brain that enable it to determine body motion and position.

FIGURE 31–13 The Ear The diagram shows the structures in the ear that transmit sound. The SEM shows hair cells in the inner ear. The motion of these sensitive hair cells produces nerve impulses that travel to the brain through the cochlear nerve. **Predict** ***How would frequent exposure to loud noises that damage hair cells affect a person's threshold for detecting sound?***

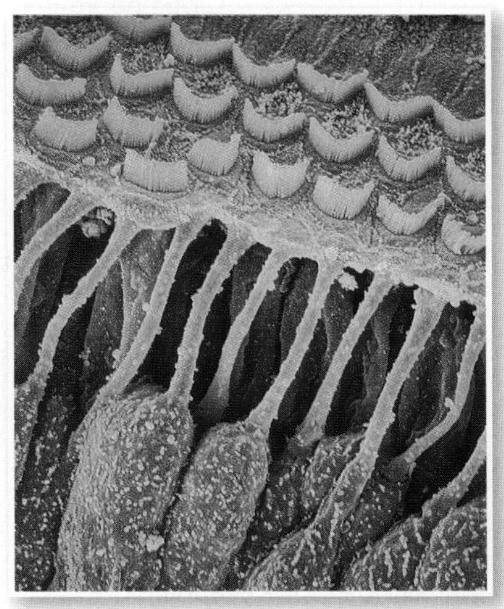

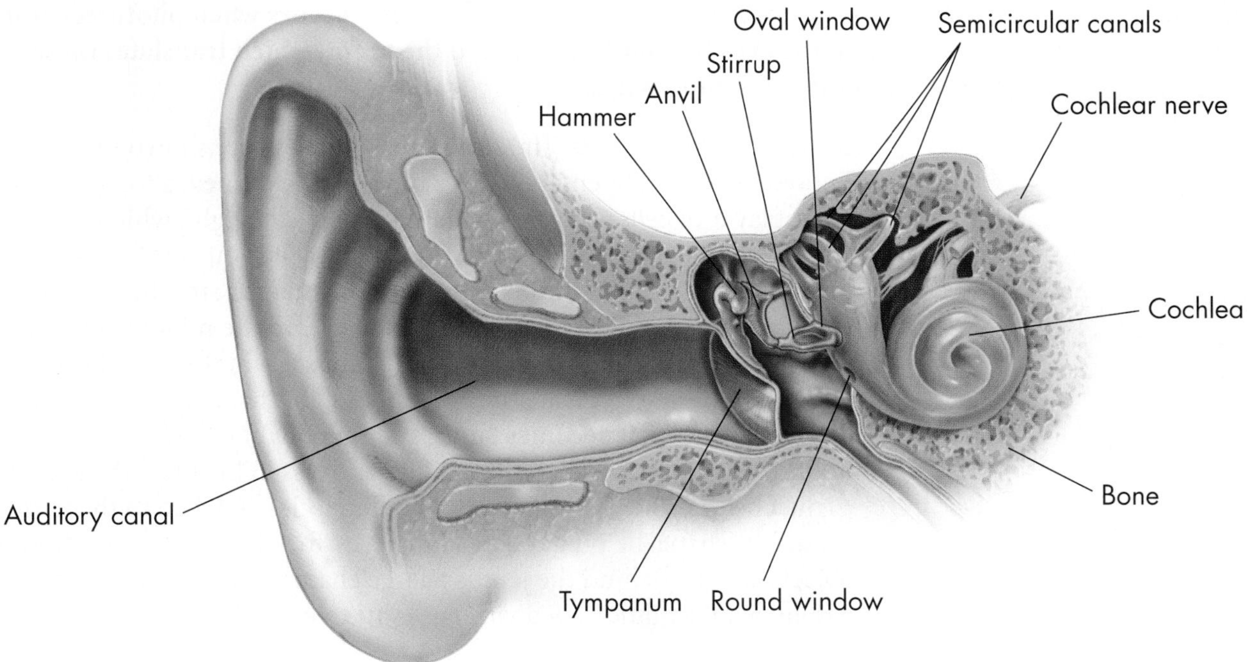

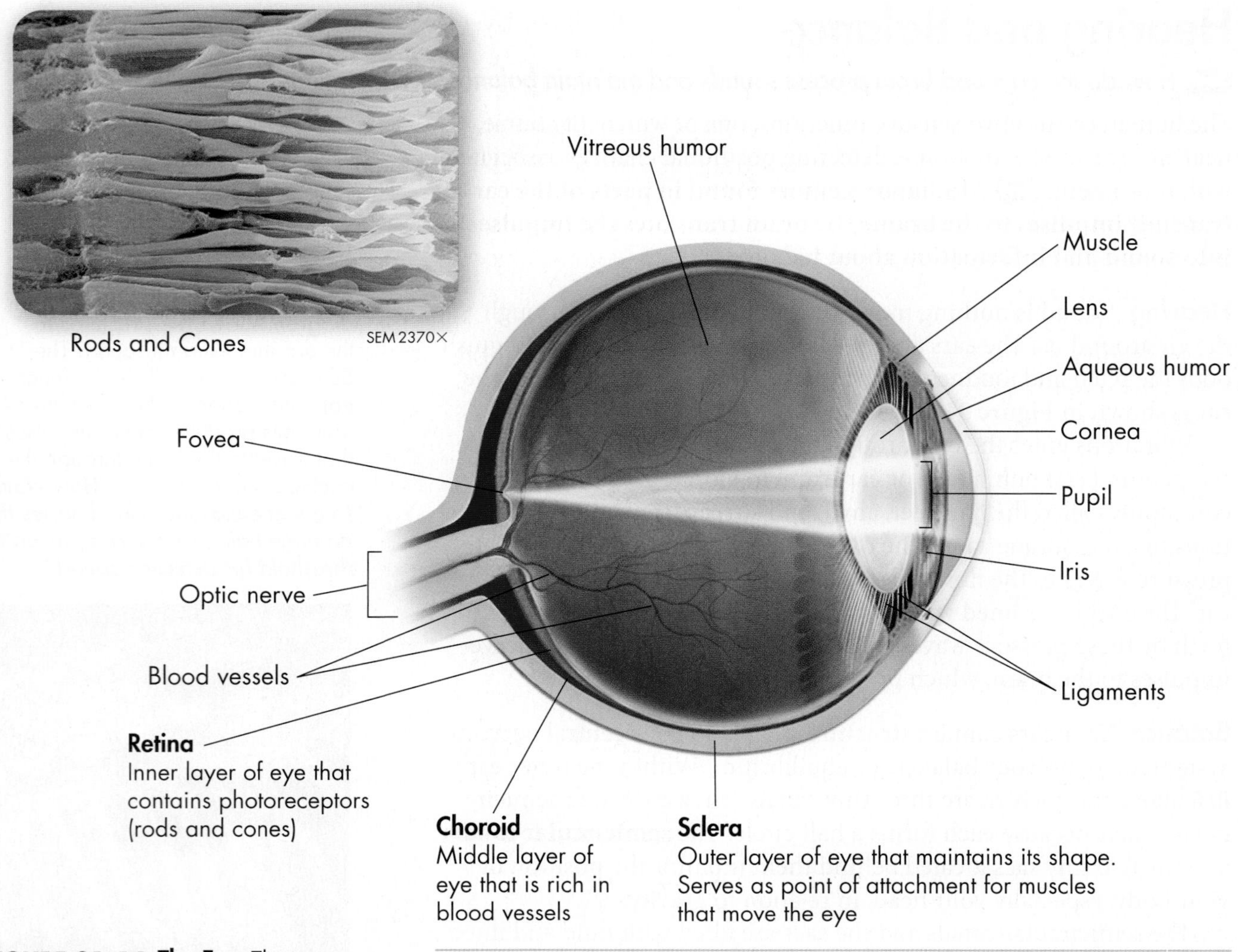

FIGURE 31–14 The Eye The eye is a complicated sense organ. The sclera, choroid, and retina are three layers of tissues that form the inner wall of the eyeball. Interpret Graphics ***What is the function of the sclera?***

Vision

How do the eyes and brain produce vision?

The world around us is bathed in light, and the sense organs we use to detect that light are the eyes. **Vision occurs when photoreceptors in the eyes transmit impulses to the brain, which translates these impulses into images.**

Structures of the Eye The structures of the eye are shown in **Figure 31–14.** Light enters the eye through the **cornea,** a tough transparent layer of cells. The cornea helps to focus the light, which then passes through a chamber filled with a fluid called aqueous (AY kwee us) humor. At the back of the chamber is a disk-shaped structure called the **iris.** The iris is the colored part of the eye. In the middle of the iris is a small opening called the **pupil.** Tiny muscles in the iris adjust the size of the pupil to regulate the amount of light that enters the eye. In dim light, the pupil becomes larger and more light enters the eye. In bright light, the pupil becomes smaller and less light enters the eye.

Just behind the iris is the **lens.** Small muscles attached to the lens change its shape, helping to adjust the eyes' focus to see near or distant objects clearly. Behind the lens is a large chamber filled with a transparent, jellylike fluid called vitreous (VIH tree us) humor.

How You See The lens focuses light onto the **retina,** the inner layer of the eye. Photoreceptors are arranged in a layer in the retina. The photoreceptors convert light energy into nerve impulses that are carried to the brain through the optic nerve. There are two types of photoreceptors: rods and cones. **Rods** are extremely sensitive to light, but they do not distinguish different colors. They only allow us to see black and white. **Cones** are less sensitive than rods, but they do respond to different colors, producing color vision. Cones are concentrated in the fovea, the site of sharpest vision.

The impulses assembled by this complicated layer of interconnected cells leave each eye by way of the optic nerve, which carry the impulses to the appropriate regions of the brain. There are no photoreceptors where the optic nerve passes through the back of the eye, producing a blind spot in part of each image sent to the brain. During the processing of the nerve impulses, the brain fills in the holes of the blind spot with information.

If the eye merely took photographs, the images would be no more detailed than the blurry images taken by an inexpensive camera and would be incomplete. The images we actually see of the world, however, are much more detailed, and the reason is the sophisticated way in which the brain processes and interprets visual information.

BUILD Vocabulary

ACADEMIC WORDS Distinguish may mean "to recognize as different," "to perceive clearly with a sense," or "to show a difference."

In Your Notebook *Make a flowchart that shows the sequence of how light and nerve impulses travel from the outside environment to the brain.*

31.4 Assessment

Review Key Concepts

1. a. **Review** What three types of sensations do receptors in the skin respond to?
   b. **Predict** Do you think that the soles of your feet or the back of your neck has the greater concentration of sensory receptors? Explain.
2. a. **Review** What are the five basic tastes detected by taste buds?
   b. **Apply Concepts** Why can't you taste food when you have a bad cold?
3. a. **Review** Which structures in the ear gather information about the position of your body?
   b. **Apply Concepts** If you spin around for a time, the fluid in your semicircular canals also moves. When you stop suddenly, why do you think you feel like you are still moving?
4. a. **Review** Identify the relationship between the cornea, pupil, lens, retina, and optic nerve and the photoreceptors of the eye.
   b. **Infer** Some people suffer from night blindness. Which type of photoreceptor is likely not functioning correctly? Explain.

WRITE ABOUT SCIENCE

Creative Writing

5. Imagine that you have lost your sense of taste for one day. Write a 3- to 4-paragraph essay describing how the absence of this sense would affect your day.

NGSS Smart Guide Nervous System

Disciplinary Core Idea LS1.D Information Processing: How do organisms detect, process, and use information about the environment? A complicated, but highly organized, network of cells and supporting tissues makes up our nervous system. The nervous system allows us to gather information about our world, process it, and produce responses.

31.1 The Neuron

- The nervous system collects information about the body's internal and external environment, processes that information, and responds to it.
- Nervous system impulses are transmitted by cells called neurons.
- An impulse begins when a neuron is stimulated by another neuron or by the environment.

peripheral nervous system 896 • central nervous system 896 • cell body 897 • dendrite 897 • axon 897 • myelin sheath 897 • resting potential 898 • action potential 898 • threshold 899 • synapse 900 • neurotransmitter 900

Biology.com

Interactive Art Watch a nerve impulse move down a neuron.

Visual Analogy Compare an action potential moving along a neuron to a row of falling dominoes.

31.2 The Central Nervous System

- Each of the major areas of the brain—the cerebrum, cerebellum, and brain stem—is responsible for processing and relaying information.
- The spinal cord is the main communication link between the brain and the rest of the body.
- The brain reacts to excessive dopamine levels by reducing the number of receptors for the neurotransmitter. As a result, normal activities no longer produce the sensations of pleasure they once did.

reflex 901 • cerebrum 902 • cerebral cortex 902 • thalamus 903 • hypothalamus 903 • cerebellum 903 • brain stem 903 • dopamine 904

Biology.com

Taking Notes Construct a concept map showing the relationships among structures of the central nervous system.

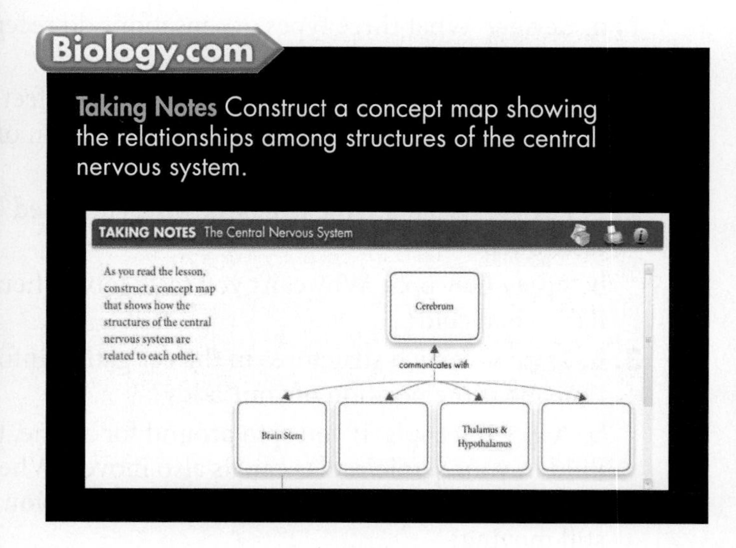

CHECKING YOUR *Scientific Literacy*

Refer to the lesson content and digital assets below as you prepare for your chapter assessment. Then, evaluate your understanding of the nervous system by answering these questions.

1. Scientific and Engineering Practice **Developing and Using Models** The human brain is often compared to a computer. Explain how a computer does and does not model the functions of the brain.

2. Crosscutting Concept **Structure and Function** How do the structures that make up a reflex arc each contribute to the nervous system's ability to detect, process, and use information from the environment?

3. Integration of Knowledge and Ideas Use *Visual Summary: The Brain* in Figure 31-8 to create a table that summarizes the functions of the different regions of the brain.

4. STEM You are a cosmetics scientist asked to hold focus tests for a company that wants to know what shampoo aromas most appeal to customers. Design a plan for the smell tests. Include a description of the test samples, the variables to be controlled, and the data to be collected.

31.3 The Peripheral Nervous System

- The sensory division of the peripheral nervous system transmits impulses from sense organs to the central nervous system.
- The motor division of the peripheral nervous system transmits impulses from the central nervous system to muscles or glands. The motor division is divided into the somatic nervous system and the autonomic nervous system.

somatic nervous system 907 • reflex arc 907 • autonomic nervous system 908

Biology.com

Untamed Science Video Hold on tightly as the Untamed Science crew takes you on a quick tour of how animal toxins affect the body.

Art in Motion View a short animation that shows how a reflex arc works.

ART IN MOTION Reflex Arc

Reflex Arc: A Rapid Response

When you step on something sharp, sensory receptors stimulate a sensory neuron, which relays the signal to an interneuron within the spinal cord.

The signal is then sent to a motor neuron, which in turn stimulates a muscle that lifts your leg.

31.4 The Senses

- Different sensory receptors in the body respond to touch, temperature, and pain.
- Sensations of smell and taste are both the result of impulses sent to the brain by chemoreceptors.
- Mechanoreceptors found in parts of the ear transmit impulses to the brain. The brain translates the impulses into sound and information about balance.
- Vision occurs when photoreceptors in the eyes transmit impulses to the brain, which translates these impulses into images.

taste bud 910 • cochlea 911 • semicircular canals 911 • cornea 912 • iris 912 • pupil 912 • lens 912 • retina 913 • rods 913 • cones 913

Biology.com

Art Review Review your understanding of the structures in the eyes and ears with this drag-and-drop activity.

Data Analysis Investigate the relationship between the color of food and perception of flavor.

STEM Activity View a video about the use of artificial retinas to help visually impaired individuals. Discuss in class the impact of technological advances on populations with physical disabilities.

Real-World Lab

Testing Sensory Receptors for Touch This chapter lab is available in your lab manual and at Biology.com

31 Assessment

31.1 The Neuron

Understand Key Concepts

1. The basic units of structure and function in the nervous system are
   a. neurons.
   b. axons.
   c. dendrites.
   d. neurotransmitters.

2. In the diagram below, the letter A is pointing to the
   a. myelin sheath.
   b. axon.
   c. dendrite.
   d. cell body.

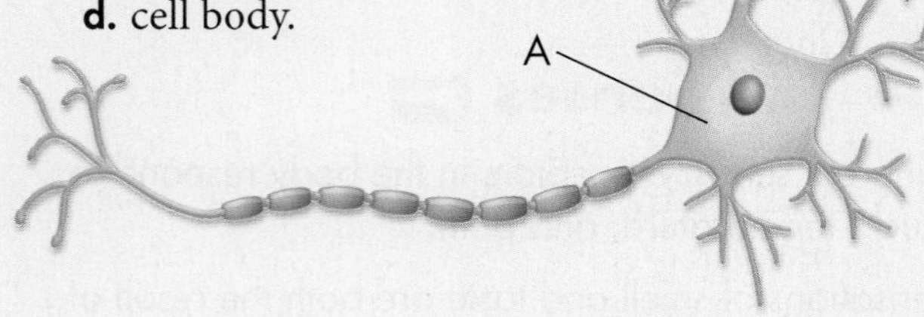

3. The place where a neuron transfers an impulse to another cell is the
   a. synapse.
   b. dendrite.
   c. myelin sheath.
   d. receptor.

4. Name the three types of neurons and describe their function in the nervous system.

5. Describe the movement of sodium and potassium ions during the resting potential.

6. **Craft and Structure** Why can an action potential be described as an all-or-none event?

Think Critically

7. **Key Ideas and Details** Suppose a portion of an axon is cut so that it is no longer connected to its cell body. What effect would that have on the transmission of impulses? Cite textual evidence to support your analysis.

8. **Production and Distribution of Writing** Develop a paragraph to clearly explain how a neuron and an electrical extension cord are similar and how they are different.

31.2 The Central Nervous System

Understand Key Concepts

9. The central nervous system consists of the
   a. sense organs.
   b. reflexes.
   c. brain and spinal cord.
   d. sensory and motor neurons.

10. Voluntary, or conscious, activities of the body are controlled primarily by the
    a. medulla oblongata.
    b. cerebrum.
    c. cerebellum.
    d. brain stem.

11. Methamphetamines are a type of drug that affects the brain by
    a. causing the release of excess dopamine.
    b. blocking the production of dopamine.
    c. increasing the number of dopamine receptors.
    d. increasing the number of synapses in the brain.

12. Describe the structure and function of the cerebrum.

13. Describe the relationship between the brain stem and the spinal cord.

14. How does nicotine influence dopamine receptors in the brain?

Think Critically

15. **Infer** A stroke occurs when blood flow to part of the brain stops due to a clot or broken blood vessel. What conclusion might you make about the location of damage in a person who has difficulty speaking and cannot move many of his muscles on the right side of his body?

16. **Compare and Contrast** In what ways are the effects of methamphetamine and cocaine on the brain similar? How are they different?

31.3 The Peripheral Nervous System

Understand Key Concepts

17. The sympathetic nervous system and the parasympathetic nervous system are specific divisions of the
    a. peripheral nervous system.
    b. central nervous system.
    c. somatic nervous system.
    d. autonomic nervous system.

18. Reflexes are behaviors that
    a. involve only sensory neurons.
    b. are controlled by the autonomic nervous system.
    c. are under conscious control.
    d. occur involuntarily without conscious control.

19. Describe the advantage of a reflex response in the survival of an organism.

20. List the divisions of the autonomic nervous system and give the function of each.

Think Critically

21. **Text Types and Purposes** Design an experiment to determine how time of day may affect reaction time. Formulate a hypothesis and write your procedure. Have your teacher check your experimental plan before you begin.

22. **Apply Concepts** A routine examination by a doctor usually includes a knee-jerk reflex test. What is the purpose of this test? What could the absence of a response indicate?

31.4 The Senses

Understand Key Concepts

23. The semicircular canals and the two tiny sacs located behind them help maintain
    a. night vision.
    b. body position and balance.
    c. respiratory rate.
    d. temperature.

24. The senses of taste and smell involve sensory receptors called
    a. photoreceptors.
    b. chemoreceptors.
    c. thermoreceptors.
    d. mechanoreceptors.

25. The fluid-filled structure in the ear that sends information to the brain about sound is the
    a. tympanum.
    b. oval window.
    c. stirrup.
    d. cochlea.

26. **Key Ideas and Details** Trace the path of light through the eye.

27. What are the functions of rods and cones?

28. Trace the path of sound through the ear.

29. What are the five basic tastes?

solve the CHAPTER MYSTERY

POISONING ON THE HIGH SEAS

Because of a sketch done by the ship's naturalist, Georg Forster, it is suspected that Cook and his men most likely ate *Tetraodon lagocephalus sceleratus*, also known as the Silver-stripe blaasop. Bacteria that live in the fish's liver, gonads, intestines, and skin produce a poison called tetrodotoxin. The poison can remain active even after the fish is cooked at high temperatures. Tetrodotoxin binds to and blocks voltage-gated sodium channels, especially in the peripheral nervous system.

Unlike the men on Cook's ship, today, some Japanese chefs are specially trained to prepare fish—known as pufferfish—that contain this toxin. The dish, known as fugu, is highly prized by diners in exclusive restaurants. The prepared fish have a unique taste and produce a tingling sensation in the mouth and throat when eaten. Improper preparation of fugu can lead to serious consequences for the diner, including death.

Obviously, tetrodotoxin doesn't poison the fish that produce it. Studies of the fish's genome have revealed a mutation in the gene that codes for the structure of sodium channel proteins. The mutation changes the surface shape of the channel, and prevents the toxin from binding to it.

1. **Infer** Based on the location of bacteria in *Tetraodon*, what methods are likely involved in preparing fugu?

2. **Apply Concepts** Describe in your own words why the fish are not affected by their own toxin.

3. **Connect to the Big idea** Some researchers have explored the possibility of using tetrodotoxin to treat severe pain. Assess the text for evidence that suggests why the toxin might be useful in this way. What safety concerns would you have to consider when designing a study to test this possibility?

Think Critically

30. **Infer** What is the advantage of having a greater concentration of touch receptors in the fingers, toes, and face?

31. **Integration of Knowledge and Ideas** The graph below compares age to the nearest distance in centimeters that many people can see an object clearly. Describe the general trend of the graph. At what age does the slope of the graph begin to change rapidly? What do you think might explain this change?

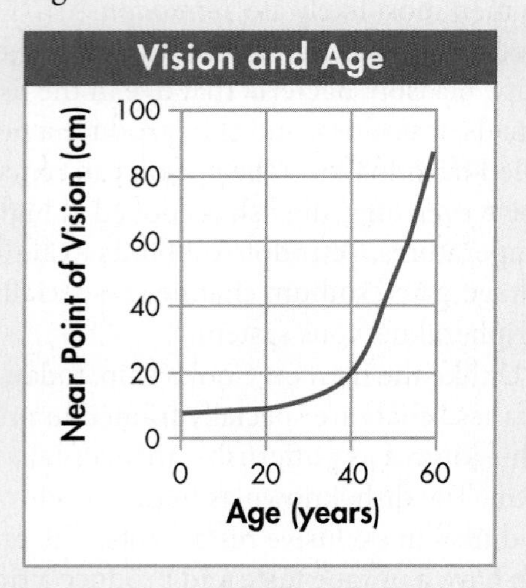

Connecting Concepts

Use Science Graphics

Use the illustration to answer questions 32 and 33.

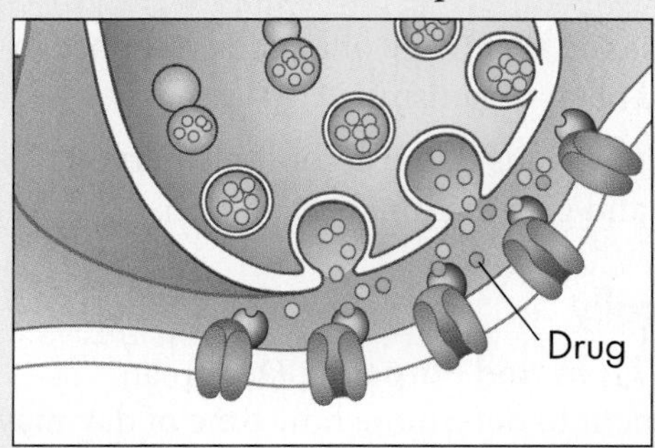

32. **Interpret Visuals** Do you think this illustration shows a drug interfering with enzymes that break down a neurotransmitter at a synapse, or a drug mimicking a neurotransmitter? Explain.

33. **Integration of Knowledge and Ideas** Draw an illustration of a synapse showing the effect of a drug that increases the rate of neurotransmitter secretion.

Write About Science

34. **Range of Writing** Write a paragraph explaining how addiction to drugs, alcohol, or tobacco all have a similar basis in the way they alter the function of the human nervous system.

35. **Assess the Big idea** Describe how the shape and structure of the neuron is related to its function in the nervous system.

Analyzing Data

Integration of Knowledge and Ideas

The graph shows the relationship between neuron diameter and impulse conduction speed in myelinated axons in a mammal. Use the graph to answer questions 36 and 37.

36. **Interpret Graphs** What conclusion can be made from the graph concerning the relationship between the speed of conduction of an action potential and the diameter of an axon?

37. **Calculate** In the reflex arc that involves touching a hot object and pulling away your hand, the impulses must travel a total distance of about 1.5 m. How long does it take for the reflex to occur if the neurons are 5 µm in diameter? MATH

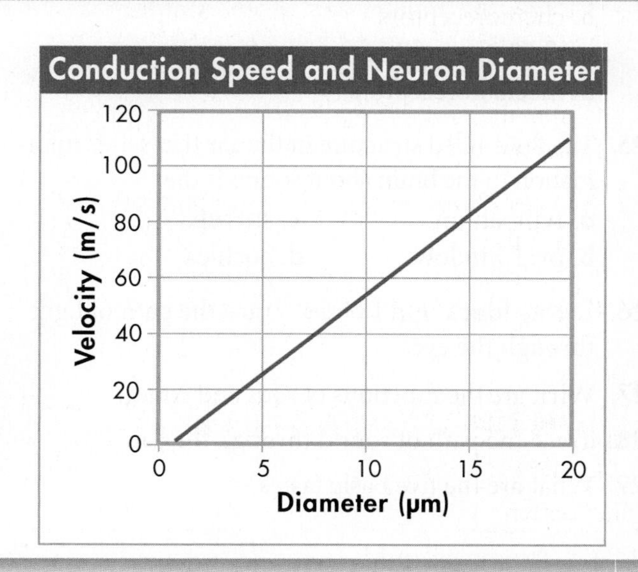

Standardized Test Prep

Multiple Choice

1. The largest and most prominent part of the human brain is the
A cerebrum.
B cerebellum.
C thalamus.
D brain stem.

2. The point of connection between two neurons is called a
A threshold.
B synapse.
C neurotransmitter.
D dendrite.

3. The part of a neuron that carries impulses away from the cell body is called a(n)
A axon.
B dendrite.
C vesicle.
D synapse.

4. The minimum stimulus level that will cause a neuron to produce an action potential is called the
A resting potential.
B impulse.
C threshold.
D synapse.

5. The part of the brain responsible for collecting sensory input from the body and relaying it to appropriate brain centers is the
A limbic system.
B thalamus.
C cerebellum.
D cerebrum.

6. The major function of the spinal cord is
A emotional learning and memory storage.
B control of voluntary muscle movements.
C fine control of detailed muscle movement.
D a principal communication path between the brain and the rest of the body.

7. Involuntary activities carried out throughout the body are the primary responsibility of the
A somatic nervous system.
B autonomic nervous system.
C spinal cord.
D limbic system.

8. The part of the eye that contains photoreceptor cells is the
A cornea.
B iris.
C retina.
D optic nerve.

Questions 9–10

Blood alcohol concentration (BAC) is a measure of the amount of alcohol in the blood per 100 mL of blood. In some states, if a driver has a BAC of 0.08 percent, he or she is considered legally drunk. The table below lists an average BAC as alcohol consumption increases. Use the information in the table to answer the questions.

Blood Alcohol Concentration (Percent)

Drinks in One Hour	Body Mass					
	45 kg	54 kg	63 kg	72 kg	81 kg	90 kg
1	0.04	0.03	0.03	0.02	0.02	0.02
2	0.07	0.06	0.05	0.05	0.04	0.04
3	0.11	0.09	0.08	0.07	0.06	0.06
4	0.14	0.12	0.10	0.09	0.08	0.07
5	0.18	0.15	0.13	0.11	0.10	0.09
6	0.21	0.18	0.15	0.14	0.12	0.11
7	0.25	0.21	0.18	0.16	0.14	0.13
8	0.29	0.24	0.21	0.18	0.16	0.14

9. How many drinks in one hour would cause a 63 kg person to have a BAC of 0.08 percent?
A 1
B 3
C 5
D 7

10. If a 54 kg person had 3 drinks in one hour, what would his or her BAC percentage be?
A 0.06
B 0.08
C 0.09
D 0.11

Open-Ended Response

11. How do the parasympathetic and sympathetic nervous system work together in the body?

If You Have Trouble With . . .

Question	1	2	3	4	5	6	7	8	9	10	11
See Lesson	31.2	31.1	31.1	31.1	31.2	31.2	31.3	31.4	31.2	31.2	31.3

Skeletal, Muscular, and Integumentary Systems

Structure and Function

Q: What systems form the structure of the human body?

The skeletal and muscular systems of this gymnast interact closely as she performs these graceful movements.

INSIDE:

- 32.1 The Skeletal System
- 32.2 The Muscular System
- 32.3 Skin—The Integumentary System

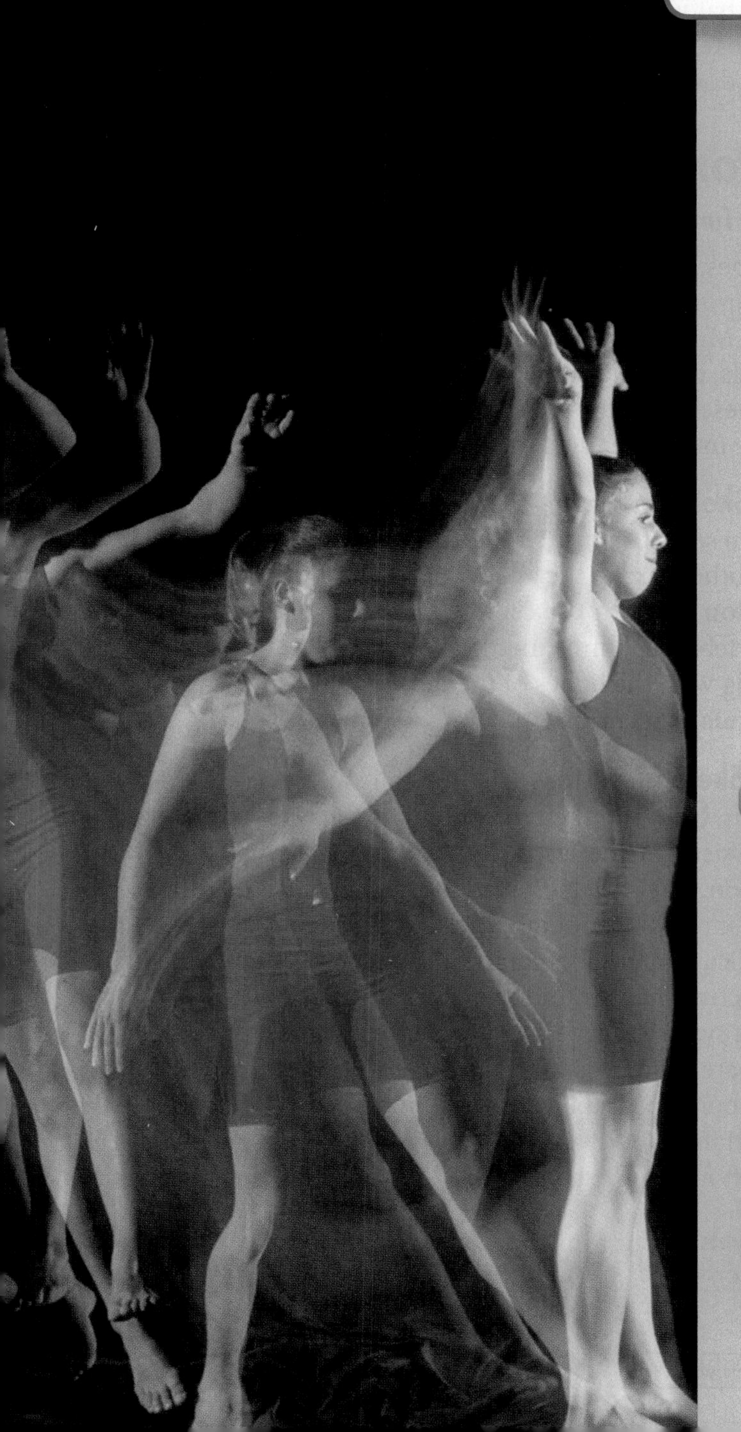

Deepen your understanding of the structures and functions of the skeletal, muscular, and integumentary systems by engaging in key practices that allow you to make connections across concepts.

NGSS You will use models to explain your understanding of interactions between bones and skeletal muscles.

STEM Learn about the kind of work exercise scientists do and how their research translates into practical and commercial uses.

Common Core Based on what you read in this chapter, you will provide an accurate summary of the complex processes involving bone and muscle that produce movement.

THE DEMISE OF A DISEASE

In the early twentieth century, many poorly nourished children living in northern cities of the United States and Europe had very soft, weak bones, a condition called "rickets." These children often had bowed legs, malformed wrists, and many other skeletal problems. Meanwhile, poorly nourished children living in southern cities rarely developed rickets.

During this time, rickets was a health problem that seriously affected many children living in cold climates. No one knew the cause or how to cure the disease. Some people claimed that regular doses of cod liver oil cured rickets, but many considered this folklore.

Scientists were eager to find answers. What was the connection between rickets and the northern climate? And could cod liver oil be a cure? In the chapter, look for clues that helped scientists develop ideas about the cause of rickets. Then, solve the mystery.

Biology.com

Finding the solution to The Demise of a Disease is just the beginning. Take a video field trip with the ecogeeks of Untamed Science to continue exploring your world.

Go online to access additional resources including:
• eText • Flash Cards • Lesson Overviews • Chapter Mystery

32.1 The Skeletal System

Key Questions

What are the functions of the skeletal system?

What is the structure of a typical human bone?

What is the role of joints?

Vocabulary

axial skeleton
appendicular skeleton
Haversian canal
bone marrow
cartilage
ossification
osteoblast
osteocyte
osteoclast
joint
ligament

Taking Notes

Outline Before you read, make an outline with the green and blue headings in the lesson. As you read, fill in main ideas and supporting details for each heading.

THINK ABOUT IT An animal's skeleton is so durable that its bones are often recognizable thousands of years after the animal's death. Bones are so tough and strong, in fact, that it's easy to think of them as though they were nothing more than rigid, lifeless supports for the rest of the body. If that were true, what would happen if one of those supports broke? Broken bones, as you know, can heal. How does that happen? And what does that tell you about the nature of our skeleton?

The Skeleton

What are the functions of the skeletal system?

To retain their shapes, all organisms need some type of structural support. Unicellular organisms have a cytoskeleton that provides structural support. Multicellular animals have cytoskeletons within their individual cells, but a skeleton is needed to provide support for the whole body. These skeletons include the external exoskeletons of arthropods and the internal endoskeletons of vertebrates.

Structure of the Skeleton There are 206 bones in the adult human skeleton. As you can see in **Figure 32–1,** some of these bones are in the axial skeleton and others are in the appendicular skeleton.

The **axial skeleton** supports the central axis of the body. It consists of the skull, the vertebral column, and the rib cage. The bones of the arms and legs, along with the bones of the pelvis and shoulder area, form the **appendicular skeleton.**

Functions of the Skeletal System The skeletal system has many important functions. **The skeleton supports the body, protects internal organs, assists movement, stores minerals, and is a site of blood cell formation.** The skeletal system supports and shapes the body much like an internal wooden frame supports a house. Bones also protect the delicate internal organs of the body. For example, the skull forms a protective shell around the brain.

Bones provide a system of levers on which muscles act to produce movement. Levers are rigid rods that can be moved about a fixed point. In addition, bones contain reserves of minerals, mainly calcium salts that are important to body processes. Finally, new blood cells are produced in the soft marrow tissue that fills cavities in some bones.

In Your Notebook *Use a two-column table to list the roles of the skeletal system and an example of each role.*

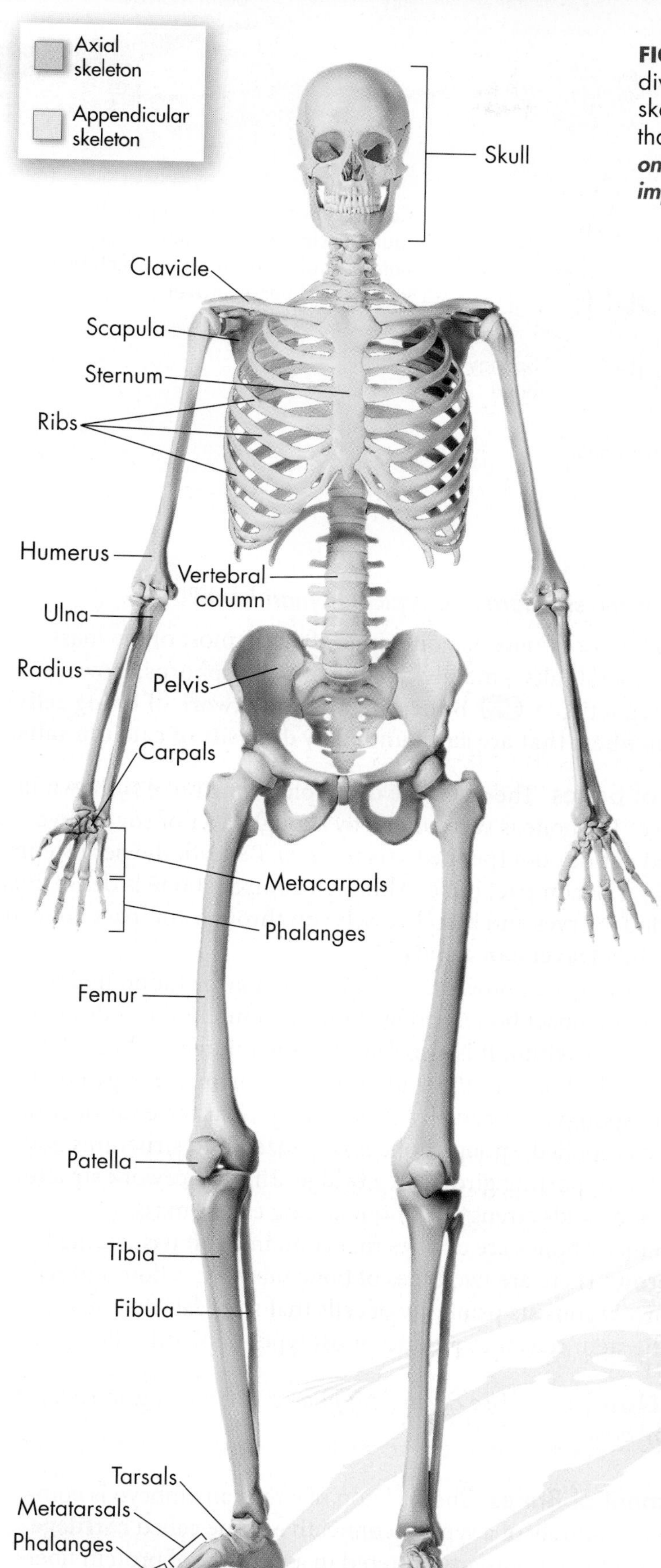

FIGURE 32–1 The Skeleton The human skeleton is divided into the axial skeleton and the appendicular skeleton. The skeleton consists of living tissue that has many roles in the body. **Classify** ***Name one bone structure, other than the rib cage, that is important for protecting internal organs.***

FUNCTIONS OF THE SKELETON

Support The bones of the skeleton support and give shape to the human body.

Protection Bones protect the delicate internal organs of the body. For example, the ribs form a basketlike cage around the heart and lungs.

Movement Bones provide a system of levers on which muscles act to produce movement.

Mineral Storage Bones contain reserves of minerals, including calcium, that are important to many body processes. When blood calcium levels are low, some reserves are released from bones.

Blood Cell Formation Many types of blood cells are produced in soft tissue that fills the internal cavities of some bones.

VISUAL ANALOGY

The human body would collapse without its bony skeleton, just as a house could not stand without its wooden frame.

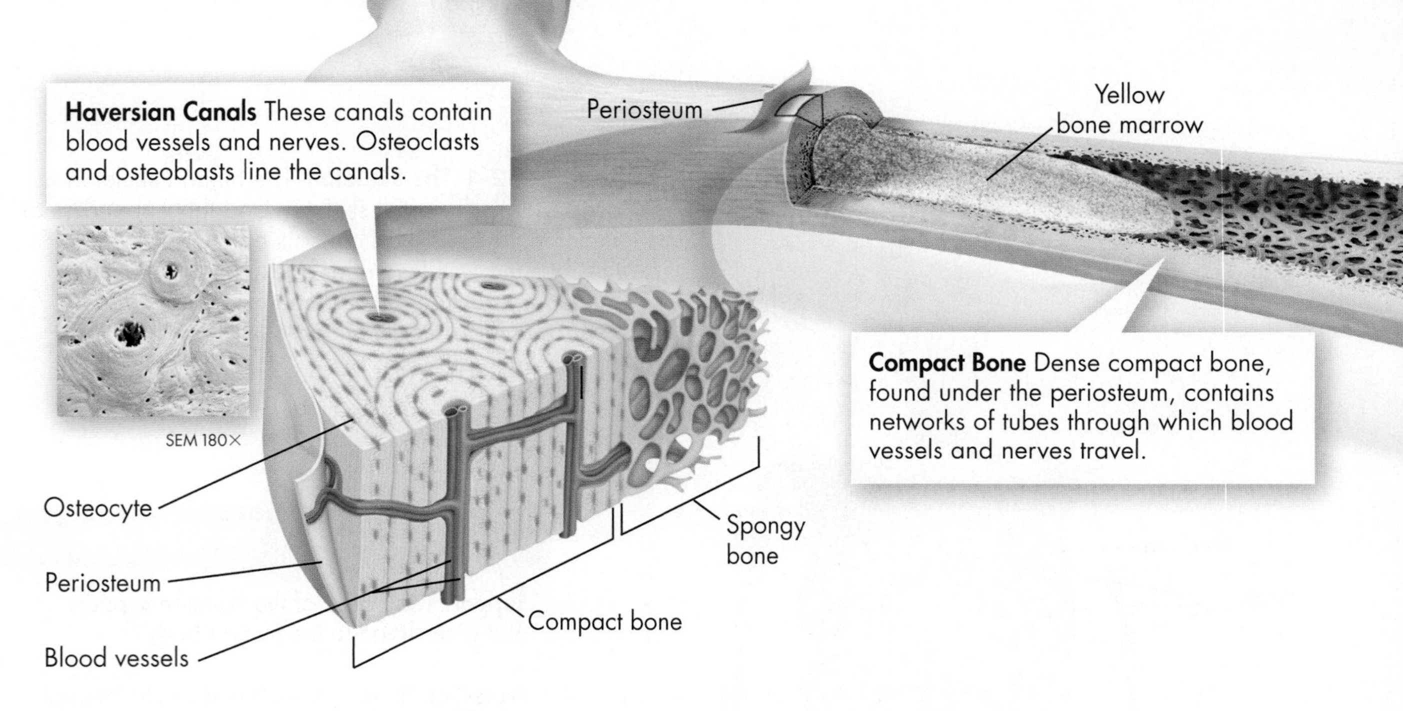

Quick Lab
GUIDED INQUIRY

Observe Calcium Loss

1. Describe the appearance and feel of two chicken bones. Then, place the bones in separate jars.
2. Pour vinegar into one jar until the bone is covered. Pour water into the other jar until the bone is covered. Cover both jars.
3. Check the jars each day for three days. Record your observations.
4. On the third day, remove the bones. Describe their appearance and feel.

Analyze and Conclude

1. Compare and Contrast How do the two bones differ?

2. Infer Vinegar reacts with and removes calcium from bone. Based on your observations, what characteristic of bone can be associated with calcium?

Bones

What is the structure of a typical human bone?

It is easy to think of bones as nonliving. After all, most of the mass of bone is mineral salts—mainly calcium and phosphorus. However, bones are living tissue. **Bones are a solid network of living cells and protein fibers that are surrounded by deposits of calcium salts.**

Structure of Bones The structure of a typical long bone is shown in **Figure 32–2.** The bone is surrounded by a tough layer of connective tissue called periosteum (pehr ee AHS tee um). Beneath the periosteum is a thick layer of compact bone. Although compact bone is dense, it is far from solid. Nerves and blood vessels run through compact bone in channels called **Haversian canals.**

A less dense tissue known as spongy bone may be found under the outer layer of compact bone. Spongy bone is found in the ends of long bones such as the femur. It is also found in the middle of short, flat bones such as the bones in the skull. Despite its name, spongy bone is not soft and spongy; it is actually quite strong. Near the ends of bones where force is applied, spongy bone is organized into structures that resemble the supporting girders in a bridge. This latticework structure in spongy bone adds strength without adding excess mass.

Within many bones are cavities that contain a soft tissue called **bone marrow.** There are two types of bone marrow: yellow and red. Yellow marrow consists primarily of cells that store fat. Red marrow contains the stem cells that produce most types of blood cells.

In Your Notebook *Use a Venn diagram to compare compact bone and spongy bone.*

Development of Bones The skeleton of a human embryo is composed almost entirely of a type of connective tissue called **cartilage.** Cartilage-producing cells are scattered in a network of protein fibers including both tough collagen and flexible elastin.

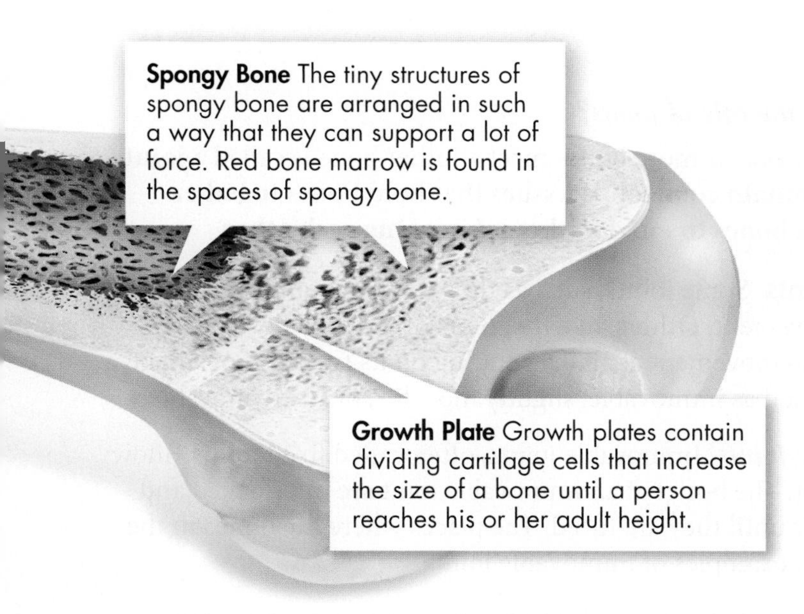

ZOOMING IN

STRUCTURE OF A BONE

FIGURE 32–2 A typical long bone such as the femur contains spongy bone and compact bone. Within compact bone are Haversian canals, which contain blood vessels and nerves. **Infer** ***What could be a result if a child breaks a bone and damages the growth plate?***

Unlike bone, cartilage does not contain blood vessels. Its cells rely on the diffusion of nutrients from the tiny blood vessels in surrounding tissues. Because cartilage is dense and fibrous, it can support weight despite its extreme flexibility.

Cartilage is gradually replaced by bone during the process of bone formation called **ossification** (ahs uh fih KAY shun). Ossification begins up to seven months before birth. Bone tissue forms as cells called **osteoblasts** secrete mineral deposits that replace the cartilage in developing bones. As bone tissue completes its development, most osteoblasts mature into osteocytes. **Osteocytes** help to maintain the minerals in bone tissue and continue to strengthen the growing bone.

Many long bones, including those of the arms and legs, have growth plates at either end. The growth of cartilage at these plates causes the bones to lengthen. Gradually, this cartilage is replaced by bone tissue, and the bones become larger and stronger. During late adolescence or early adulthood, growth plates become completely ossified, and the person "stops growing." Cartilage remains in those parts of the body that are flexible, such as the tip of the nose and the external part of ears. As you will read later, cartilage also cushions the areas where bones meet, such as in the knee.

Bone Remodeling and Repair In many ways, a bone is never finished growing. Bones are remodeled throughout life by small numbers of osteoblasts, which continue to build bone tissue, and **osteoclasts**—cells that break down bone minerals. Both functions are important because they enable bones to remodel and strengthen in response to exercise and stress. Without the continuous breakdown of old bone tissue and buildup of new bone tissue, bones would become brittle and weak. Both types of cells work together to repair broken and damaged bones.

Some older adults, especially women, develop a disorder called osteoporosis. In osteoporosis, osteoclasts break down bone much faster than osteoblasts rebuild it. Osteoporosis leads to weak bones due to excessive decrease in bone density. Research suggests that consuming plenty of calcium and performing weight-bearing exercise such as walking could help to prevent this serious problem.

Joints

What is the role of joints?

A place where one or more bones meet another bone is called a **joint.** **Joints contain connective tissues that hold bones together. Joints permit bones to move without damaging each other.**

Types of Joints Some joints, such as those of the shoulders, allow extensive movement. Others, like the joints of the fully developed skull, allow no movement at all. Depending on its type of movement, a joint is classified as immovable, slightly movable, or freely movable.

▶ ***Immovable Joints*** Immovable joints, often called fixed joints, allow no movement. The bones at an immovable joint are interlocked and grow together until they are fused. The places where the bones in the skull meet are examples of immovable joints.

▶ ***Slightly Movable Joints*** Slightly movable joints permit a small amount of movement. Unlike the bones of immovable joints, the bones of slightly movable joints are separated from each other. The joints between the two bones of the lower leg and the joints between vertebrae are examples of slightly movable joints.

▶ ***Freely Movable Joints*** Freely movable joints permit movement in two or more directions. Freely movable joints are grouped according to the shapes of the surfaces of the adjacent bones. Several types of freely movable joints are shown in **Figure 32–3.**

BUILD Vocabulary

ACADEMIC WORDS The adjective **adjacent** means "lying near" or "next to." Joints can form only at adjacent bones.

FIGURE 32–3 Freely Movable Joints Freely movable joints make actions possible. Many freely movable joints are involved in the movements this gymnast needs for her routine.

Ball-and-Socket Found in the shoulders and hips, these joints allow for movement in many directions. They are the most freely movable joints.

Hinge These joints permit back-and-forth motion, like the opening and closing of a door. They are found in the elbows, knees, and ankles.

Saddle These joints allow one bone to slide in two directions. Saddle joints allow a thumb to move across a palm.

Pivot These joints allow one bone to rotate or turn around another. Pivot joints allow you to turn your arm at your elbow and shake your head to say no.

Structure of Joints In freely movable joints, cartilage covers the surfaces where two bones come together. This protects the bones from damage as they move against each other. The joints are also surrounded by a fibrous joint capsule that helps hold the bones together while still allowing for movement.

The joint capsule consists of two layers. The outer layer forms strips of tough connective tissue called ligaments. **Ligaments,** which hold bones together in a joint, are attached to the membranes that surround bones. The inner layer of the joint capsule, called the synovial (sih NOH vee uhl) cavity, contains cells that produce a substance called synovial fluid. Synovial fluid enables the surfaces of the bones connected at the joint to slide over each other smoothly.

In some freely movable joints, such as the knee shown in **Figure 32–4,** there are small sacs of synovial fluid called bursae (BUR see; singular: bursa). Bursae reduce the friction between the bones of a joint and any tissues they come in contact with. Bursae also act as tiny shock absorbers.

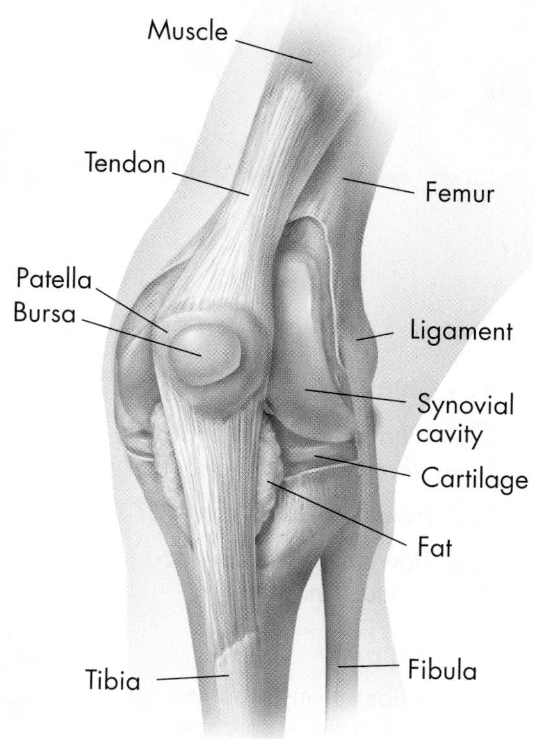

FIGURE 32–4 The Knee The knee joint is protected by cartilage and bursae. Ligaments hold together the four bones that make up the knee joint—the femur, patella, tibia, and fibula. **Infer** ***How do cartilage and bursae help reduce friction?***

Joint Injuries A common injury among young athletes is damage to the anterior cruciate ligament (ACL). This ligament is found in the center of the knee between the femur and tibia. It prevents the tibia from shifting too far forward during movement. ACL damage can be caused by the rapid pivoting, leaping, and forceful contacts that occur when playing sports like basketball and soccer. If the ACL is damaged, the knee becomes unstable and prone to other injuries.

Excessive strain on a joint may produce inflammation, a response in which excess fluid causes swelling, pain, heat, and redness. Inflammation of a bursa is called bursitis.

Wear and tear over the years often leads to osteoarthritis. This disorder develops as the cartilage of often used joints in the fingers, knees, hips, and spine begins to break down. The affected joints become painful and stiff as unprotected bones start to rub together.

32.1 Assessment

Review Key Concepts

1. a. **Review** List the different functions of the skeletal system.
   b. **Predict** If blood calcium levels in a person's body were consistently low due to poor diet, what could the effect be on the person's bones?
2. a. **Review** Describe the structure of a typical bone.
   b. **Infer** Why do you think the amount of cartilage decreases and the amount of bone increases as a baby grows?
3. a. **Review** What is a joint?
   b. **Use Analogies** Which type of freely movable joint would you compare to a doorknob? Explain.

WRITE ABOUT SCIENCE

Creative Writing

4. Use library or Internet resources to learn more about osteoporosis. Then, develop an advertising campaign for the dairy industry based on the relationship between calcium and healthy bone development and maintenance.

32.2 The Muscular System

Key Questions

What are the principal types of muscle tissue?

How do muscles contract?

How do muscle contractions produce movement?

Vocabulary

muscle fiber • myofibril • myosin • actin • sarcomere • neuromuscular junction • acetylcholine • tendon

Taking Notes

Concept Map As you read, make a concept map that shows the relationship among the terms in this section.

THINK ABOUT IT How much of your body do you think is muscle? Ten percent? Maybe fifteen percent, if you're really in shape? As surprising as it might seem, about one third of the mass of an average person's body is muscle, and that's true even if you're not a well-conditioned varsity athlete. What's all that muscle doing? Some of the answers might surprise you.

Muscle Tissue

What are the principal types of muscle tissue?

Despite the fantasies of Hollywood horror films, a skeleton cannot move by itself. That's the job of the muscular system. Naturally, this system includes the large muscles in your arms and legs. However, it also includes thousands of tiny muscles throughout the body that help to regulate blood pressure and move food through the digestive system. In fact, muscles power every movement of the body—from a leap in the air to the hint of a smile.

Muscle tissue is found everywhere in the body—not just right beneath the skin but also deep within the body. Not only is muscle tissue found where you might least expect it, but also there is more than one kind of muscle tissue. **There are three different types of muscle tissue: skeletal, smooth, and cardiac.** Each type of muscle, shown in **Figure 32–6,** is specialized for specific functions in the body. Skeletal muscle is often found, as its name implies, attached to bones, and it is usually under voluntary control. Smooth muscle is found throughout the body and is usually not under voluntary control. Cardiac muscle makes up most of the mass of the heart, and, like smooth muscle, it is not under voluntary control.

In Your Notebook *Make a two-column chart to describe the three types of muscle tissue. Label the first column Type and the second column Function.*

FIGURE 32–5 Muscles in Action This pole-vaulter's skeletal muscles are clearly defined as she propels herself forward.

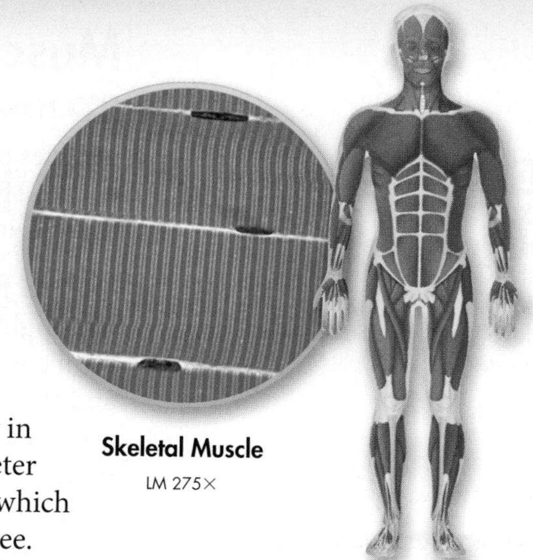

Skeletal Muscles Skeletal muscles are usually attached to bones. They are responsible for such voluntary movements as typing on a keyboard, dancing, or winking an eye. When viewed under a microscope at high magnification, skeletal muscle appears to have alternating light and dark bands called "striations." For this reason, skeletal muscle is said to be striated. Most skeletal muscle movements are consciously controlled by the central nervous system (the brain and spinal cord).

Skeletal muscle cells are large, have many nuclei, and vary in length. The shortest skeletal muscle, which is about 1 millimeter long, is found in the middle ear. The longest skeletal muscle, which may be as long as 30 centimeters, runs from the hip to the knee. Because skeletal muscle cells are long and slender, they are often called **muscle fibers.**

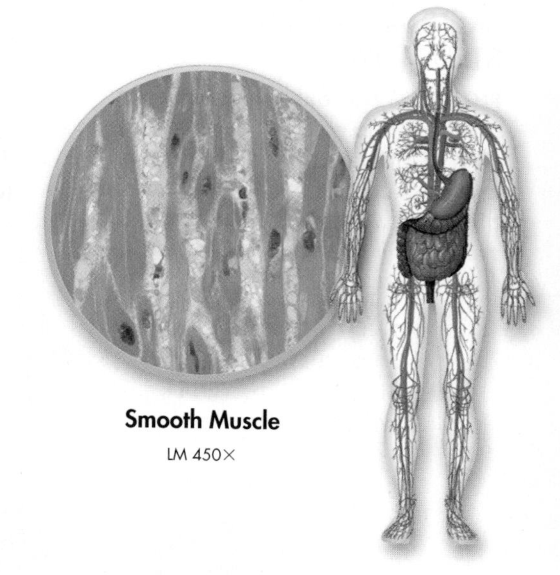

Smooth Muscles Smooth muscle cells are so named because they don't have striations and, therefore, look "smooth" under the microscope. These cells are spindle-shaped and usually have a single nucleus. Smooth muscle movements are usually involuntary. They are found throughout the body and form part of the walls of hollow structures such as the stomach, blood vessels, and intestines. Smooth muscles move food through your digestive tract, control the way blood flows through your circulatory system, and even decrease the size of the pupils of your eyes in bright light. Powerful smooth muscle contractions are also responsible for pushing a baby out of its mother's uterus during childbirth. Most smooth muscle cells can function without direct stimulation by the nervous system. The cells in smooth muscle tissue are connected to one another by gap junctions that allow electrical impulses to travel directly from one muscle cell to a neighboring muscle cell.

Cardiac Muscle Cardiac muscle is found in just one place in the body—the heart. It shares features with both skeletal muscle and smooth muscle. Cardiac muscle is striated like skeletal muscle, although its cells are smaller and usually have just one or two nuclei. Cardiac muscle is similar to smooth muscle because it is not under the direct control of the central nervous system. Like smooth muscle cells, cardiac muscle cells can contract on their own and are connected to their neighbors by gap junctions. You will learn more about cardiac muscle and its role in the function of the heart in Chapter 33.

Cardiac Muscle
LM 370×

FIGURE 32–6 Muscle Tissue The three types of muscle tissue look different under a microscope, but all muscle tissue has the ability to produce movement. **Compare and Contrast** ***What is the key difference between control of skeletal muscle contraction and smooth muscle contraction?***

Muscle Contraction

How do muscles contract?

Muscles produce movements by shortening, or contracting, from end to end. How do cells generate such force? The answer can be found in the way in which two kinds of muscle protein filaments interact.

BUILD Vocabulary

PREFIXES The prefix *myo-* in **myofibril** means "muscle." The word comes from the Greek word for mouse. Artists and sculptors once thought well-developed muscles looked like mice under the skin.

Muscle Fiber Structure Skeletal muscle cells, or fibers, are filled with tightly packed filament bundles called **myofibrils.** Each myofibril contains thick filaments of a protein called **myosin** (MY uh sin) and thin filaments of a protein called **actin.** These filaments are arranged in an overlapping pattern that produces the stripes or striations so visible through a microscope. The thin actin filaments are bound together in areas called Z lines. Two Z lines and the filaments between them make up a unit called a **sarcomere.** **Figure 32–7** shows the structure of a muscle fiber.

The Sliding-Filament Model Myosin and actin filaments are actually tiny force-producing engines. **During a muscle contraction, myosin filaments form cross-bridges with actin filaments. The cross-bridges then change shape, pulling the actin filaments toward the center of the sarcomere.** As shown in **Figure 32–8,** this action decreases the distance between the Z lines, and the fiber shortens. Then the cross-bridge detaches from actin and repeats the cycle by binding to another site on the actin filament. As thick and thin filaments slide past each other, the fiber shortens. For this reason, the process is called the sliding-filament model of muscle contraction.

When hundreds of thousands of myosin cross-bridges repeat these actions, the muscle fiber shortens with considerable force. Contractions like this enable you to run, lift weights, or even turn a page in a book. Because one molecule of ATP supplies just enough energy for one interaction between a myosin cross-bridge and an actin filament, a muscle cell needs plenty of ATP.

VISUAL SUMMARY

SKELETAL MUSCLE STRUCTURE

FIGURE 32–7 Skeletal muscles are made up of bundles of muscle fibers composed of myofibrils. Each myofibril contains actin and myosin filaments. **Interpret Visuals** ***What type of unit are actin and myosin filaments arranged in?***

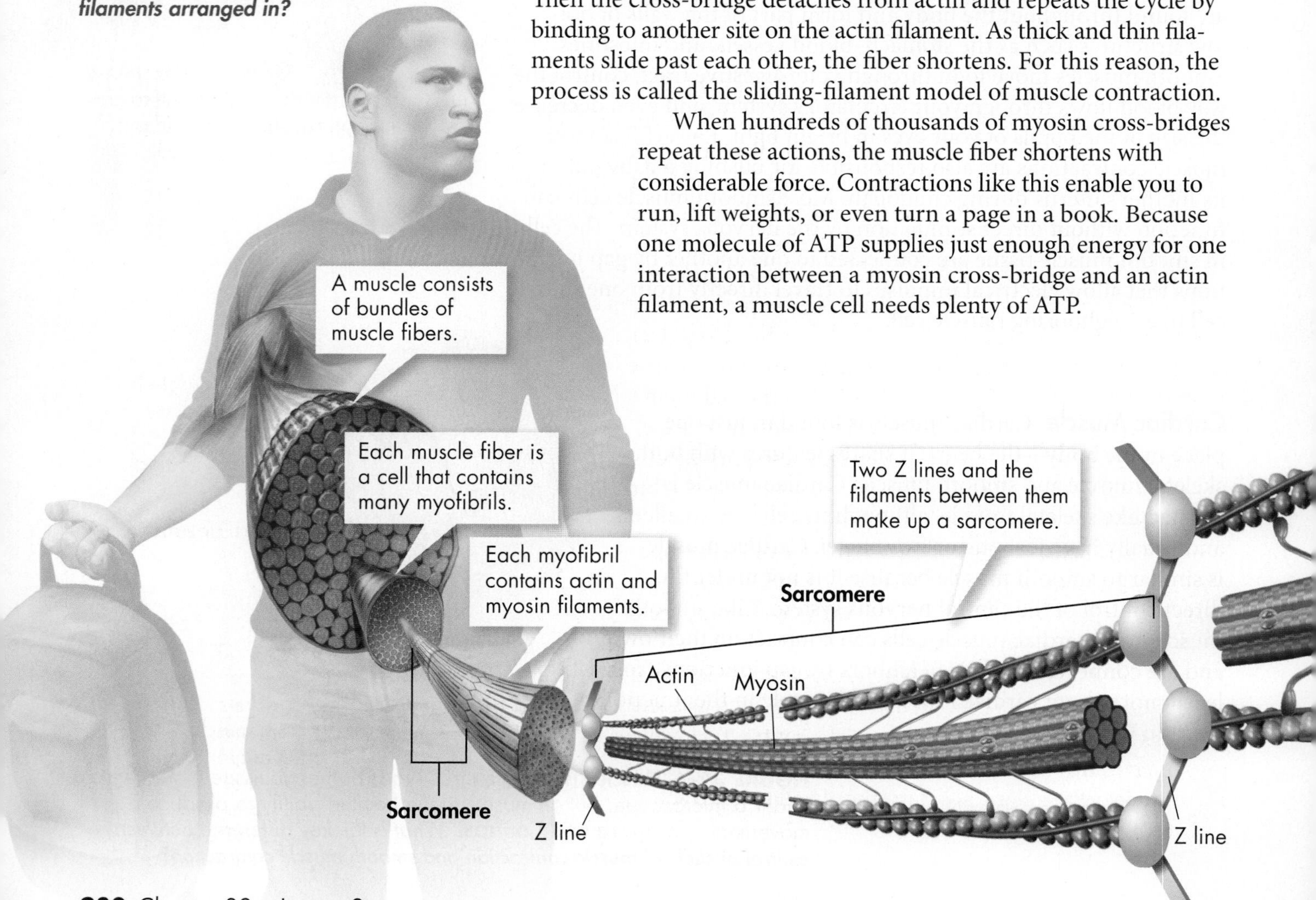

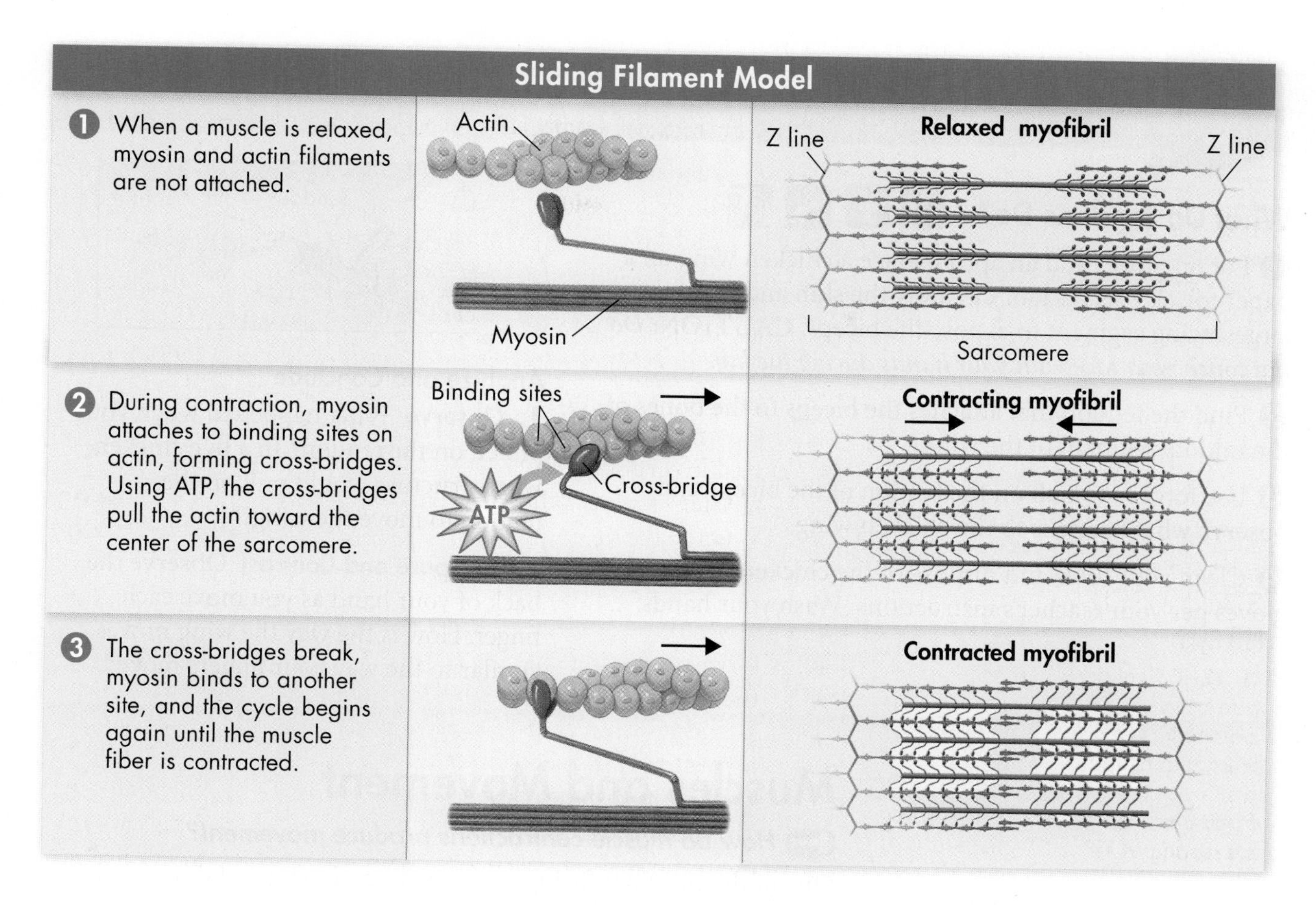

FIGURE 32–8 Sliding-Filament Model During muscle contraction, interaction between myosin filaments and actin filaments causes a muscle fiber to contract.

Control of Muscle Contraction Skeletal muscles are useful only if they contract in a controlled fashion. Remember that motor neurons connect the central nervous system to skeletal muscle cells. Impulses from these motor neurons control the contraction of muscle fibers.

A motor neuron and a skeletal muscle cell meet at a type of synapse known as a **neuromuscular** (noo roh MUS kyoo lur) **junction.** When a motor neuron is stimulated, its axon terminals release a neurotransmitter called **acetylcholine** (as ih til KOH leen). Acetylcholine (ACh) molecules diffuse across the synapse, producing an impulse (action potential) in the cell membrane of the muscle fiber. The impulse causes the release of calcium ions (Ca^{2+}) within the fiber. These ions affect regulatory proteins that allow myosin cross-bridges to bind to actin.

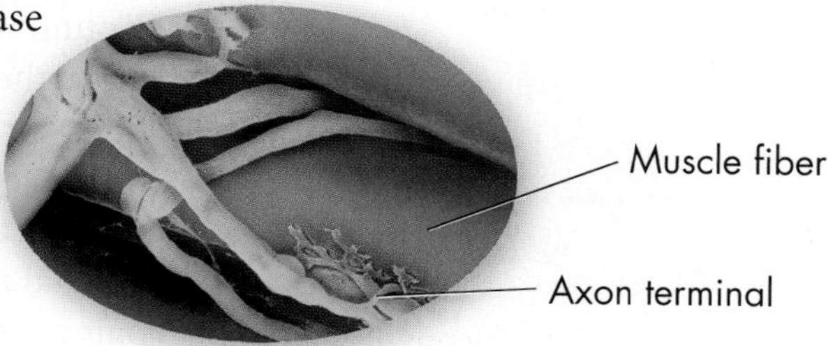

FIGURE 32–9 Neuromuscular Junction

A muscle cell contracts until the release of ACh stops and an enzyme produced at the axon terminal destroys any remaining ACh. Then, the muscle cell pumps Ca^{2+} back into storage, the cross-bridges stop forming, and the contraction ends.

What is the difference between a strong contraction and a weak contraction? When you lift something light, such as a sheet of paper, your brain stimulates only a few cells to contract. However, as you exert maximum effort, such as when lifting your book bag, almost all the muscle cells in your arm are stimulated to contract.

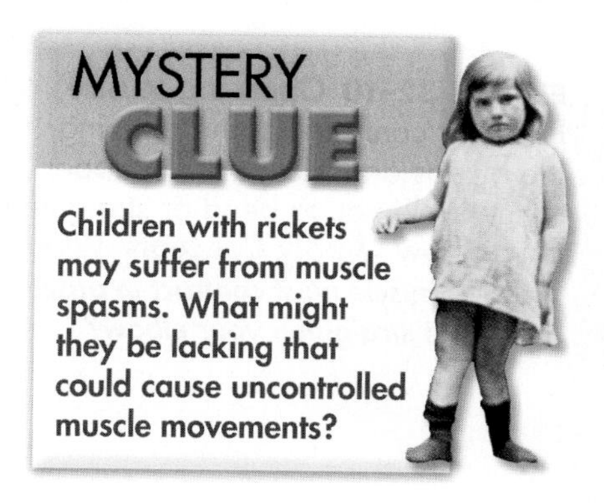

Quick Lab
GUIDED INQUIRY

What Do Tendons Do?

❶ Put on gloves and an apron. Place a chicken wing on a paper towel. Peel back or cut away the skin and fat of the largest wing segment to expose the biceps. **CAUTION:** *Do not touch your face with your hands during the lab.*

❷ Find the tendon that attaches the biceps to the bones of the middle segment of the wing.

❸ Use forceps to pull on the tendon of the biceps and observe what happens to the chicken wing.

❹ Clean your tools and dispose of the chicken wing and gloves per your teacher's instructions. Wash your hands.

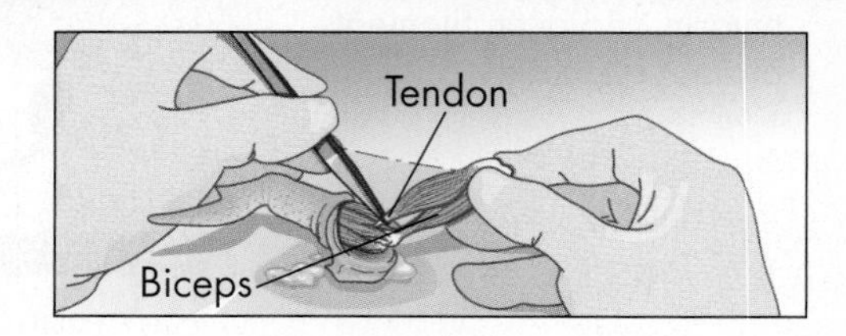

Analyze and Conclude

1. Observe What happened when you pulled on the tendon? In a live chicken, what structure would pull on the tendon to move the wing?

2. Compare and Contrast Observe the back of your hand as you move each finger. How is the way the wing moves similar to the way your fingers move?

Muscles and Movement

How do muscle contractions produce movement?

One of the most confusing concepts to understand about muscles is that they can produce force only by contracting in one direction. Yet, you know from experience that you can use your muscles to push as well as to pull. How is this possible?

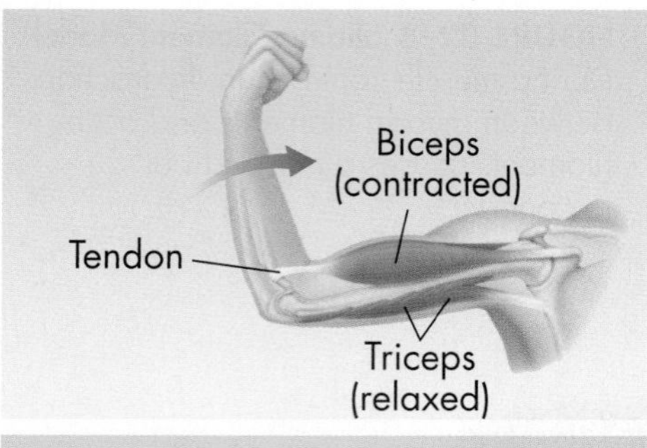

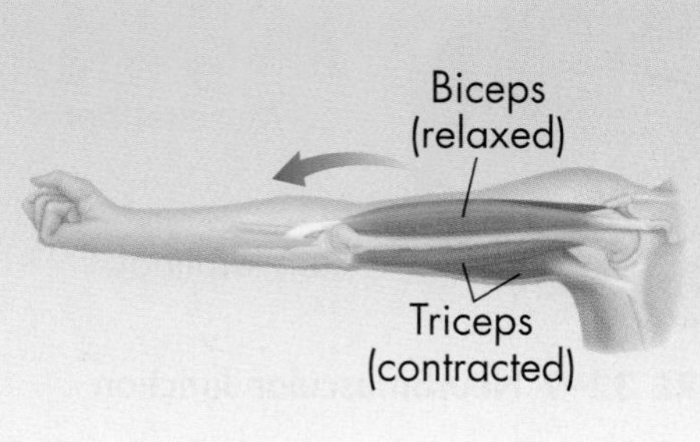

FIGURE 32–10 Opposing Muscle Pairs By contracting and relaxing, the biceps and triceps in the upper arm enable you to bend or straighten your elbow. **Apply Concepts** ***Which skeletal muscle must contract in order for you to straighten your elbow?***

How Muscles and Bones Interact Skeletal muscles are joined to bones by tough connective tissues called **tendons.** Tendons are attached in such a way that they pull on the bones and make them work like levers. The joint functions as a fulcrum—the fixed point around which the lever moves. The muscles provide the force to move the lever. Usually, several muscles that pull in different directions surround each joint. **Skeletal muscles generate force and produce movement by pulling on body parts as they contract.**

We can use our muscles to push as well as to pull because most skeletal muscles work in opposing pairs. When one muscle in the pair contracts, the other muscle in the pair relaxes. The muscles of the upper arm shown in **Figure 32–10** are a good example of this dual action. When the biceps muscle contracts, it bends, or flexes, the elbow joint. When the triceps muscle contracts, it opens, or extends, the elbow joint. A controlled movement requires the involvement of both muscles. To hold a tennis racket or a violin requires a balance of forces between the biceps and the triceps.

This is why the training of athletes and musicians is so difficult. The brain must learn how to work opposing muscle groups in just the right ways to make the involved joints move precisely.

In Your Notebook *Explain in your own words the role of opposing pairs in muscle contraction.*

Types of Muscle Fibers There are two principal types of skeletal muscle fibers—red and white. The types of muscle fibers vary in their specific functions. Red muscle, or slow-twitch muscle, contains many mitochondria. The dark color of red muscle comes from small blood vessels that deliver a rich supply of blood and from an oxygen-storing protein called myoglobin. The abundant mitochondria and generous supply of oxygen allow these fibers to derive their energy through aerobic respiration and work for long periods of time. Red muscle is useful for endurance activities like long-distance running.

White muscle, or fast-twitch muscle, contracts more rapidly and generates more force than does red muscle, but its cells contain few mitochondria and tire quickly. White fibers are useful for activities that require great strength or quick bursts of speed, like sprinting.

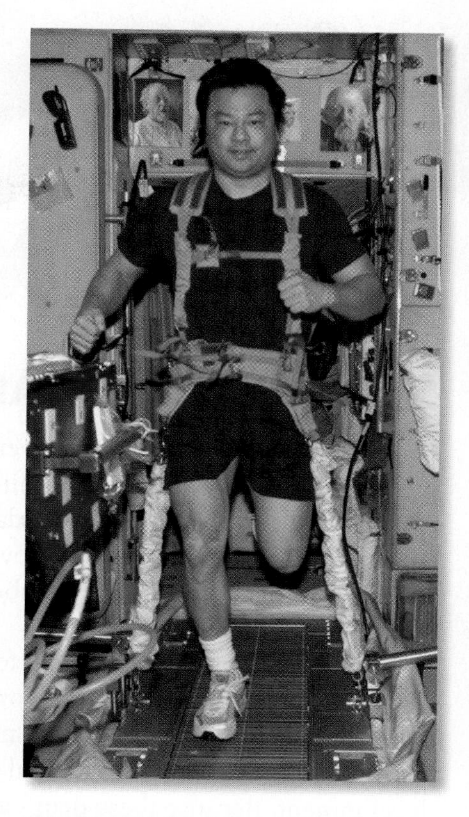

FIGURE 32–11 Preventing Muscle Loss Without gravity, many muscles go unused. An astronaut in space may lose up to 5 percent of muscle mass a week. Exercise helps to maintain muscles—and bones, too.

Exercise and Health Regular exercise is important to maintain muscular strength and flexibility. Muscles that are exercised regularly stay firm and increase in size and strength due to added filaments. Muscles that are not used become weak and can visibly decrease in size. Regular exercise helps to maintain resting muscle tone—a state of partial contraction. Muscle tone is responsible for keeping the back and legs straight and the head upright, even when you are relaxed.

Aerobic exercises—such as running and swimming—place strong demands on the heart and lungs, helping these systems to become more efficient. This, in turn, increases physical endurance—the ability to perform an activity without fatigue. Regular exercise also strengthens your bones, making them thicker and stronger. Strong bones and muscles are less likely to become injured.

Resistance exercises, such as weight lifting, increase muscle size and strength. Over time, weight-training exercises will help to maintain coordination and flexibility.

32.2 Assessment

Review Key Concepts

1. a. **Review** List the three types of muscle tissue.
   b. **Compare and Contrast** Compare and contrast the structure and function of the three types of muscle tissue.
2. a. **Review** What structures make up a skeletal muscle?
   b. **Explain** Describe how a muscle contracts.
   c. **Predict** A type of poisonous gas destroys the enzyme that breaks down acetylcholine. What effect do you think this gas has on the body?
3. a. **Review** Explain the role of tendons in movement.
   b. **Apply Concepts** In training for an Olympic weight-lifting event, which muscle fibers would be the most important to develop?

VISUAL THINKING

4. Create your own model to show how actin filaments slide over myosin filaments during a muscle contraction. Include as much detail in your model as possible.

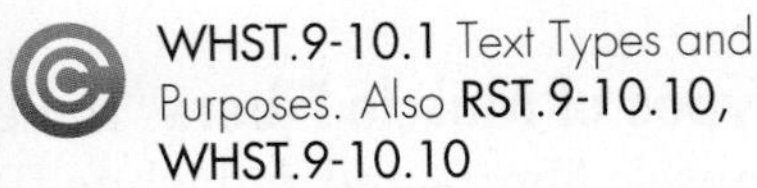

WHST.9-10.1 Text Types and Purposes. Also RST.9-10.10, WHST.9-10.10

Biology & Society

Should Student Athletes Be Tested for Steroids?

Until 1976, East Germany had never won an Olympic gold medal in women's swimming. That year, they won 13 out of the 14 gold medals awarded for swimming. Eventually, it was discovered that the young athletes had been given anabolic steroids without their knowledge.

Anabolic steroids are synthetic forms of the hormone testosterone. They were originally developed to help treat men who could not produce enough of the hormone for normal growth and development. Because these drugs also make it easier for athletes to add muscle mass and recover from workouts, they are sometimes used illegally to improve performance.

In the early part of this century, steroid use emerged as a big controversy in professional baseball. Many people now think that steroid use by professional baseball players was not taken seriously. Legislators and parents argue that this lax attitude has led many young athletes to think that steroid use is acceptable.

Some student athletes use steroids hoping to improve their chance of playing either in college or professionally. However, steroids are not only illegal, they are dangerous. Decades after unknowingly being given steroids, many of the 1976 East German swimmers are suffering from the long-term effects of steroid use, such as tumors, liver disease, heart problems, infertility, and depression. Other, more short-term effects of steroid use include breast development in males, acne, and increased chance of ligament and tendon injury.

Due to the rising rate of steroid use, some states have enacted policies for testing student athletes. But the policies are often controversial.

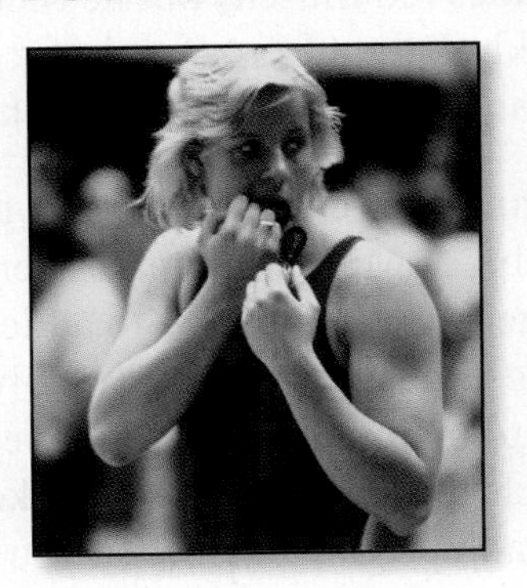

Kornelia Ender was a member of the East German swim team when some swimmers were given steroids without their knowledge. In 1976, she won four gold medals.

Viewpoints

For Testing Student athletes who use steroids risk both their short- and long-term health. Although educating students about the risks of steroids is important, many students will ignore the risks and take their chances. Schools should help to protect these athletes. Also, athletes who do not use steroids should not have to compete against those who do.

Against Testing Steroid testing is more expensive than testing for other drugs, and many schools don't have the funds. Also, there are many ways to "fool" steroid tests, so the tests could be just a waste of money. Although the Supreme Court has ruled that drug testing of students is constitutional, some people still feel that testing violates their privacy rights.

Research and Decide

1. Evaluate Identify additional viewpoints for and against testing high school athletes for steroids. Develop claims and counterclaims fairly for each viewpoint, noting their strengths and limitations.

2. Communicate State your viewpoint on testing student athletes for steroid use. Write a coherent argument to support your claim.

32.3 Skin—The Integumentary System

THINK ABOUT IT What's the largest organ in your body? No, it is not your ears or stomach, or even your lungs or heart. By far the largest human organ is the skin. If that sounds a little strange, it's probably because you're used to taking your skin for granted—it's just the outside of your body, right? Well, the skin has a lot of roles that go beyond just covering your body.

Integumentary System Functions

What are the principal functions of the integumentary system?

The integumentary system includes the skin, hair, and nails. The skin—the major organ of the system—has many different functions, but its most important function is protection. **The integumentary system serves as a barrier against infection and injury, helps to regulate body temperature, removes wastes from the body, gathers information, and produces vitamin D.**

Protection The skin forms a barrier that blocks out pathogens and debris and prevents the body from drying out. The skin also provides protection from the sun's ultraviolet radiation. Nails, which protect the tips of fingers and toes, are also produced by the skin.

Body Temperature Regulation The skin helps to regulate body temperature by releasing excess heat generated by working cells, while keeping in enough heat to maintain normal body temperature. Hair also helps to prevent heat loss from the head.

Excretion Small amounts of sweat are constantly released from your sweat glands. Sweat contains waste products such as urea and salts that need to be excreted from the body.

Information Gathering The skin contains several types of sensory receptors. It serves as the gateway through which sensations such as pressure, heat, cold, and pain are transmitted from the outside environment to the nervous system.

Vitamin D Production One of the skin's most important functions is the production of vitamin D, which is needed for absorption of calcium and phosphorus from the small intestine. Sunlight is needed for one of the chemical reactions that produce vitamin D in skin cells.

Key Concepts

What are the principal functions of the integumentary system?

What are the structures of the integumentary system?

What are some problems that affect the skin?

Vocabulary

epidermis • keratin • melanocyte • melanin • dermis • sebaceous gland • hair follicle

Taking Notes

Preview Visuals Before you read, preview **Figure 32–12.** Make a two-column table. In the first column, list all of the structures labeled in the figure. As you read, fill in the function of each structure in the second column.

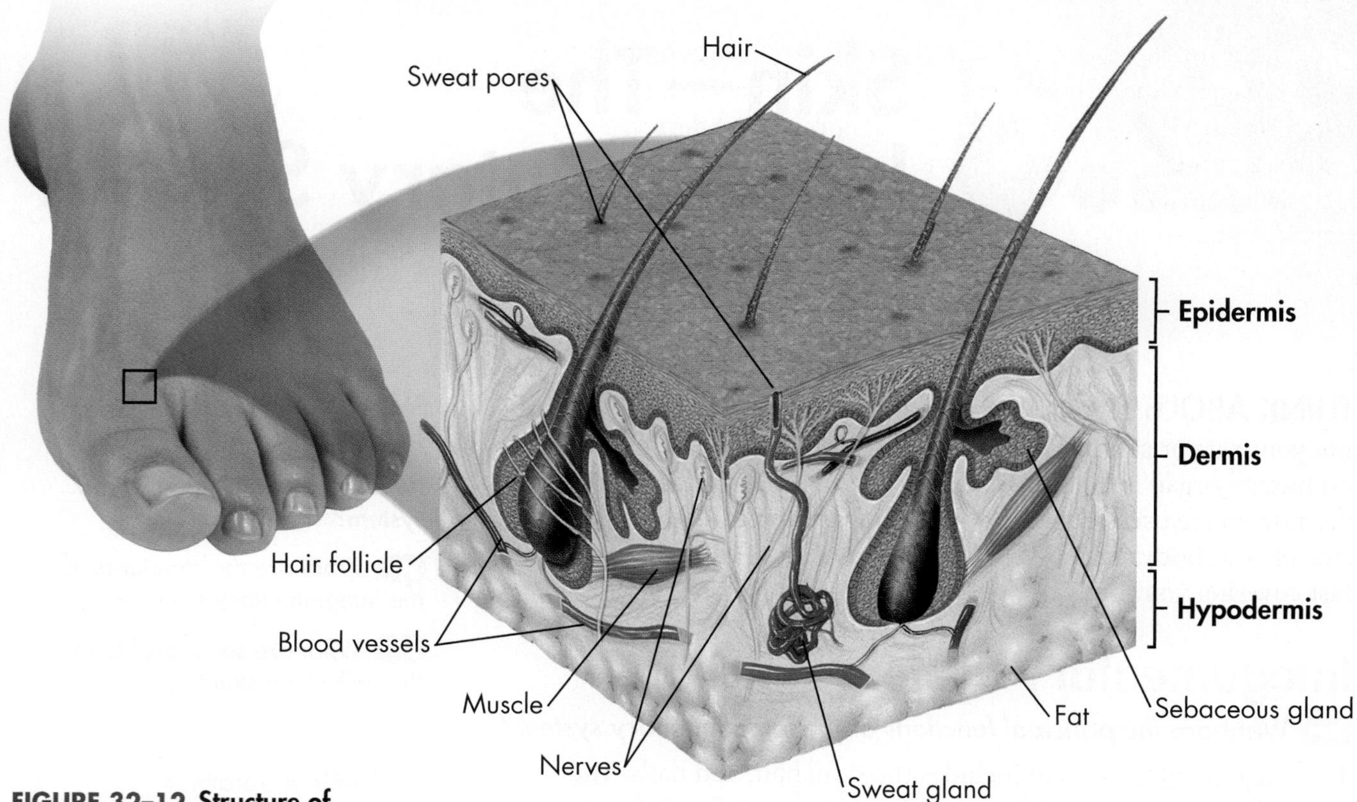

FIGURE 32–12 Structure of the Skin The skin has an outer layer called the epidermis and an inner layer called the dermis. **Infer** ***Why do you think a slight scratch on the surface of the skin does not bleed?***

BUILD Vocabulary

WORD ORIGINS The prefix *epi-* in **epidermis** comes from the Greek word meaning "on" or "upon." *Dermis* derives from the Greek *derma*, meaning "skin."

Integumentary System Structures

What are the structures of the integumentary system?

Many structures are required to fulfill all the functions you just read about. **Skin and its related structures—the hair, nails, and several types of glands—make up the integumentary system.** The skin is made up of two main layers—the epidermis and the dermis. Beneath the dermis is a layer of fat (the hypodermis) and loose connective tissue that helps insulate the body. **Figure 32–12** shows many of the structures that make up the skin.

Epidermis The outer layer of the skin is the **epidermis.** The epidermis has two layers. The outer layer of the epidermis—the layer that you can see—is made up of dead cells. The inner layer of the epidermis is made up of living cells, including stem cells. These cells divide rapidly, producing new skin cells that push older cells to the surface of the skin. As the older cells move upward, they flatten, and their organelles disintegrate. They also begin making **keratin,** a tough, fibrous protein.

Eventually, the older cells die and form a tough, flexible, waterproof covering on the surface of the skin. This outer layer of dead cells is shed or washed away at a surprising rate. Once every four to six weeks, a new layer of dead cells replaces an old layer.

The epidermis also contains **melanocytes** (MEL uh noh cytes), which are cells that produce a dark brown pigment called **melanin.** Melanin helps protect the skin by absorbing ultraviolet rays from the sun. Skin color is directly related to the production of melanin. The melanocytes of people with darker skin produce more melanin than the melanocytes of people with lighter skin produce.

Dermis The **dermis** lies beneath the epidermis and contains the protein collagen, blood vessels, nerve endings, glands, sensory receptors, smooth muscles, and hair follicles. Structures in the dermis interact with other body systems to maintain homeostasis by helping to regulate body temperature. When the body needs to conserve heat on a cold day, the blood vessels in the dermis narrow. This brings blood closer to the body's core and prevents heat from escaping through the skin. On hot days, the blood vessels widen, bringing heat from the body's core to the skin.

Sweat glands in the dermis also aid temperature regulation. Excess heat is released when sweat glands produce perspiration, or sweat. When sweat evaporates, it takes heat away from your body.

The skin also contains **sebaceous** (suh BAY shus) **glands,** which secrete an oily substance called sebum that is released at the surface of the skin. Sebum helps to keep the keratin-rich epidermis flexible and waterproof. Because it is acidic, it can kill bacteria on the surface of the skin.

In Your Notebook *Explain whether the epidermis, the dermis, or both layers are involved in protection and temperature regulation.*

Hair The basic component of human hair and nails is keratin. In other animals, keratin forms a variety of structures, including bull horns, reptile scales, bird feathers, and porcupine quills.

Hair covers almost every exposed surface of the human body and has some important functions. Hair on the head protects the scalp from ultraviolet light from the sun and provides insulation from the cold. Hairs in the nostrils, external ear canals, and around the eyes (in the form of eyelashes) prevent dirt and other particles from entering the body.

Hair is produced by cells at the base of structures called hair follicles. **Hair follicles** are tubelike pockets of epidermal cells that extend into the dermis. New research has shown that hair follicles contain stem cells that help to renew the skin and heal wounds. The hairs shown in **Figure 32–13** are actually large columns of cells that have filled with keratin and then died. Rapid cell growth at the base of the hair follicle causes the hair to grow longer. Hair follicles are in close contact with sebaceous glands. The oily secretions of these glands help hairs stay soft and flexible.

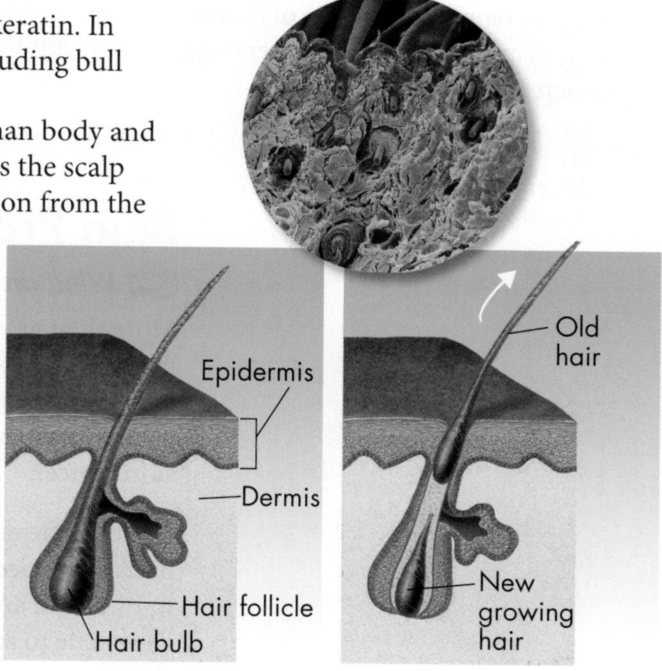

FIGURE 32–13 Hair As a new hair grows, it pushes the old hair out of the follicle. The micrograph shows individual hairs in their follicles.

Nails Nails grow from an area of rapidly dividing cells known as the nail root. The nail roots are located near the tips of the fingers and toes. During cell division, the cells of the nail root fill with keratin and produce a tough, platelike nail that covers and protects the tips of the fingers and toes. Nails grow at an average rate of 3 millimeters per month, with fingernails growing about three times faster than toenails.

The Rising Rate of Melanoma

Over the past several decades, the incidence of some deadly cancers, such as lung cancer, has decreased among people aged 20–54. Some people attribute this to decreasing smoking rates. During the same time period, the incidence of melanoma increased for the same age group. The incidence of both lung cancer and melanoma increases with age. But melanoma is one of the most common cancers in young adults.

What are some possible reasons for this increase? Despite public health efforts, many people still consider tanned skin a sign of health. Also, many people do not use enough sunscreen for it to be effective.

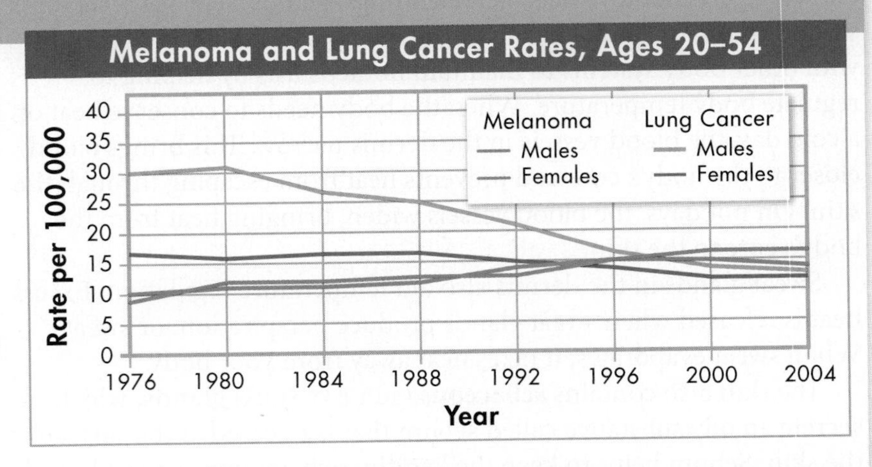

1. **Interpret Graphs** Describe the trend shown in this graph for the incidence of lung cancer and melanoma from 1976 to 2004.

2. **Infer** In what year does the rate of melanoma surpass the rate of lung cancer in men? In women?

3. **Predict** The data are only for a specific age group. If you were to look at similar data for the whole population, how do you think the graph would differ? Explain.

Skin Problems

What are some problems that affect the skin?

More than any other organ, the skin is constantly bombarded by internal and external factors that affect its health. **The skin's constant interaction with the environment can lead to problems of varying degrees of severity. Such problems include acne, hives, and skin cancer.**

Acne Acne develops when sebum and dead skin cells form plugs in hair follicles. Bacteria are often trapped in the plug, which leads to infection and inflammation. Up to 85 percent of people experience acne to some degree during adolescence and young adulthood. One hypothesis about acne suggests that high hormone levels during puberty lead to increased sebum production. There are many treatments for acne that can be purchased over the counter. But if the acne is severe and scarring is likely, a dermatologist—a doctor who specializes in skin care—should be consulted.

Hives Allergic reactions to food or medicine often display themselves as red welts commonly called hives. When the body experiences an allergic reaction, a chemical called histamine may be released. Histamine causes small blood vessels to widen. Fluid can ooze from the vessels into surrounding tissues, which causes the swelling that leads to hives.

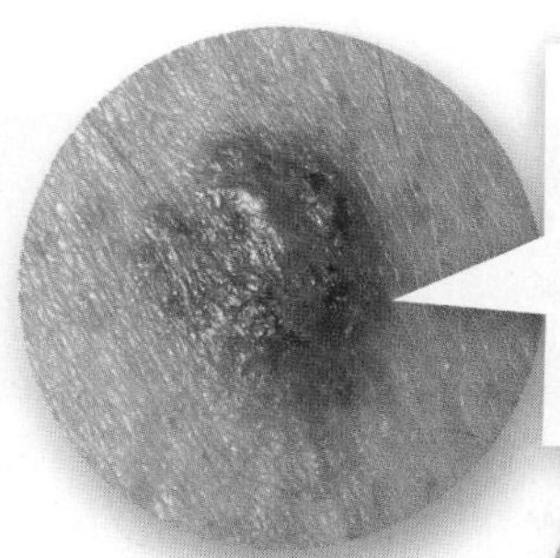

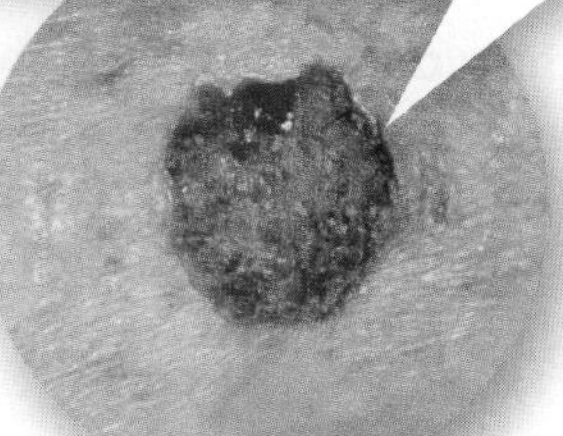

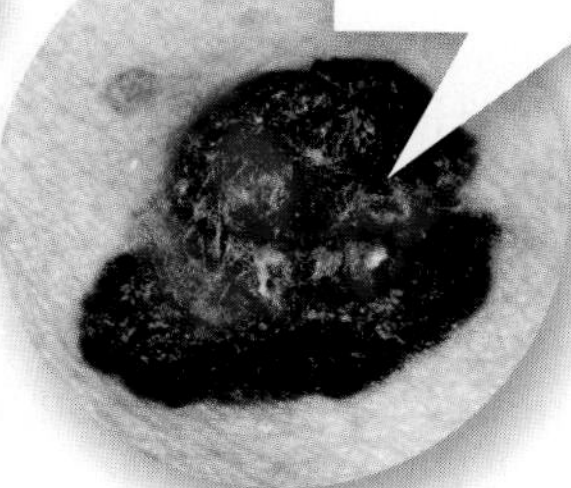

FIGURE 32–14 Skin Cancer Early detection is important in treating skin cancer. Signs of skin cancer may include a sore that does not heal or a sudden change in a mole's appearance. You should also see a doctor if you notice a new mole that is larger than 6 mm, has irregular borders, or is an odd color.

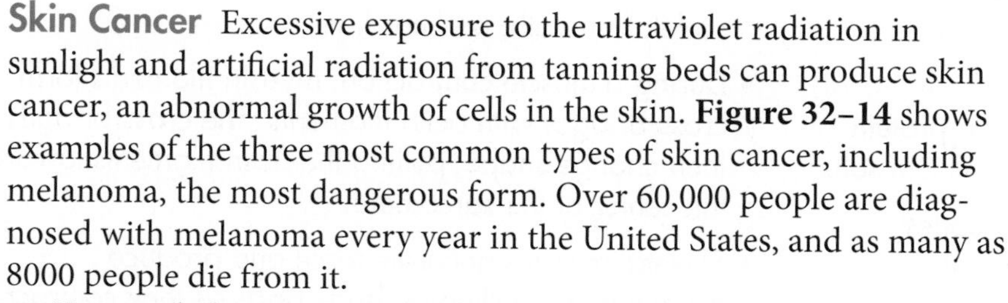

Skin Cancer Excessive exposure to the ultraviolet radiation in sunlight and artificial radiation from tanning beds can produce skin cancer, an abnormal growth of cells in the skin. **Figure 32–14** shows examples of the three most common types of skin cancer, including melanoma, the most dangerous form. Over 60,000 people are diagnosed with melanoma every year in the United States, and as many as 8000 people die from it.

You can help protect yourself from this dangerous disease by avoiding tanning salons and wearing a hat, sunglasses, and protective clothing whenever you plan to spend time outside. In addition, you should always use a sunscreen that protects against both UV-A rays and UV-B rays and that has a sun protection factor (SPF) of at least 15.

In Your Notebook *Summarize the steps you can take to protect your skin from sun damage.*

32.3 Assessment

Review Key Concepts

1. a. **Review** List the functions of the integumentary system.
   b. **Classify** What organs and tissues make up the integumentary system?
2. a. **Review** What structures are found in the epidermis? What structures are found in the dermis?
   b. **Apply Concepts** Explain two ways that the skin can help remove excess heat from the body.
3. a. **Review** What are some ways to reduce your risk of developing skin cancer?
   b. **Sequence** Explain the events that lead to acne.

Apply the Big idea

Structure and Function

4. Compare and contrast the structure and function of the dermal tissue in plants discussed in Chapter 23 with the structures in human skin. *Hint:* You may wish to organize your ideas in a Venn diagram.

NGSS Smart Guide Skeletal, Muscular, and Integumentary Systems

Disciplinary Core Idea LS1.A Structure and Function: How do the structures of organisms enable life's functions? The skeletal, muscular, and integumentary systems all form the structure of the human body. These systems have many functions that include support, movement, protection, and contributing to homeostasis.

32.1 The Skeletal System

- The skeleton supports the body, protects internal organs, assists movement, stores minerals, and is a site of blood cell formation.
- Bones are a solid network of living cells and protein fibers that are surrounded by deposits of calcium salts.
- Joints contain connective tissues that hold bones together. Joints permit bones to move without damaging each other.

axial skeleton 922 • appendicular skeleton 922 • Haversian canal 924 • bone marrow 924 • cartilage 924 • ossification 925 • osteoblast 925 • osteocyte 925 • osteoclast 925 • joint 926 • ligament 927

Biology.com

Untamed Science Video Hold on to your seats as the Untamed Science crew whisks you to NASA to learn about the effect space travel has on an astronaut's bones.

Visual Analogy Explore how the skeleton is like the framework of a house.

Interactive Art Watch how the various joints in the body move.

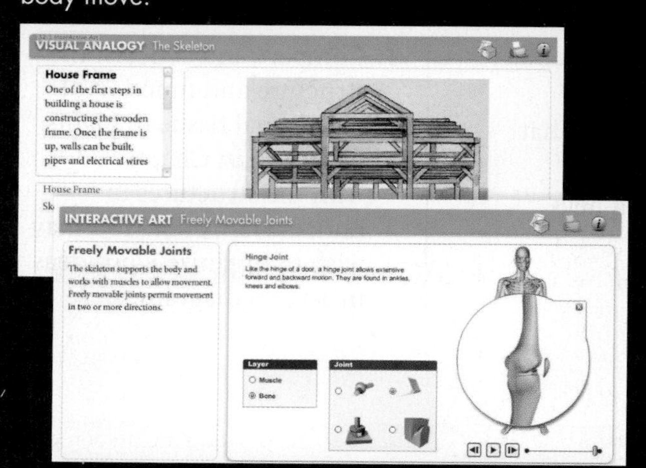

32.2 The Muscular System 

- There are three different types of muscle tissue: skeletal, smooth, and cardiac.
- During a muscle contraction, myosin filaments form cross-bridges with actin filaments. The cross-bridges then change shape, pulling the actin filaments toward the center of the sarcomere.
- Skeletal muscles generate force and produce movement by pulling on body parts as they contract.

muscle fiber 929 • myofibril 930 • myosin 930 • actin 930 • sarcomere 930 • neuromuscular junction 931 • acetylcholine 931 • tendon 932

Biology.com

Art in Motion Watch the process of muscle contraction.

Tutor Tube Watch an analogy to help you learn about the sliding-filament model of muscle contraction.

STEM Activity View a video that showcases the work of exercise scientists and find out how scientific research can be applied to commercial use.

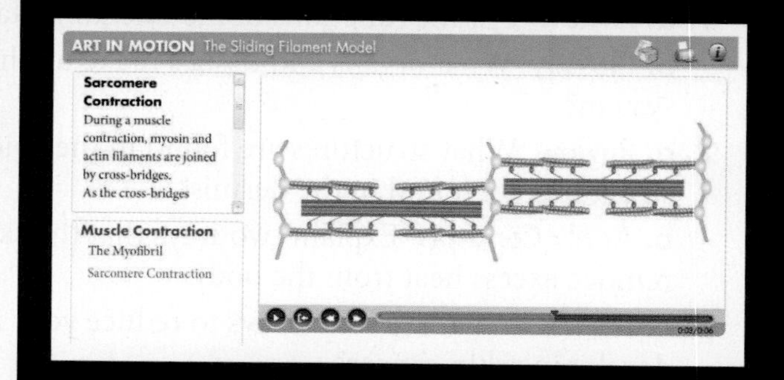

CHECKING YOUR *Scientific Literacy*

Refer to the lesson content and digital assets below as you prepare for your chapter assessment. Then, evaluate your understanding of the skeletal, muscular, and integumentary systems by answering these questions.

1. **Scientific and Engineering Practice Developing and Using Models** A marionette is a type of puppet with moveable joints. Its moving parts can be controlled from above, using strings or wires. Do you think a marionette is a good model for musculo-skeletal movement? Why or why not?

2. Crosscutting Concept **Structure and Function** List five functions of the integumentary system. For each function, write the name of one or more structures shown in Figure 32-12 that help carry out that function.

3. Key Ideas and Details Suppose you are holding your arm straight out in front of you. Summarize the interactions of nerves, muscles, and bones that enable you to bend your arm to touch your nose.

4. STEM You are an engineer tasked with designing crutches for use by children 8 to 10 years old who sustain ankle injuries. Describe the key characteristics of your prototype and the constraints you would need to consider in developing your design.

32.3 Skin—The Integumentary System

- The integumentary system serves as a barrier against infection and injury, helps regulate body temperature, removes wastes from the body, gathers information, and produces vitamin D.
- Skin and its related structures—the hair, nails, and several types of glands—make up the integumentary system.
- The skin's constant interaction with the environment can lead to problems of varying degrees of severity. Such problems include acne, hives, and skin cancer.

epidermis 936 • keratin 936 • melanocyte 936 • melanin 936 • dermis 937 • sebaceous gland 937 • hair follicle 937

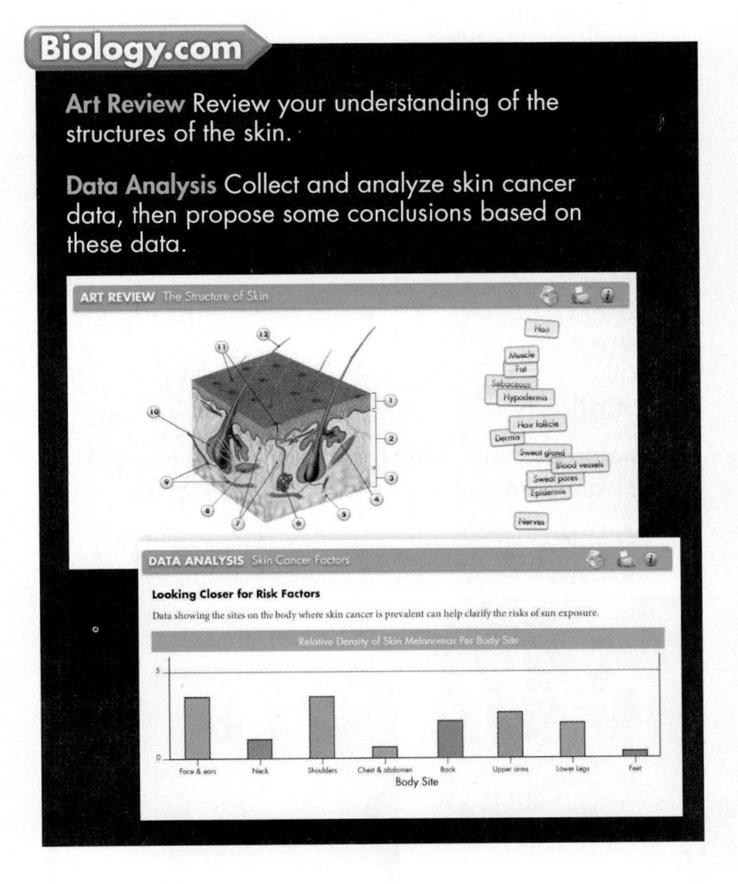

Skills Lab

Comparing Limbs This chapter lab is available in your lab manual and at Biology.com

32 Assessment

32.1 The Skeletal System

Understand Key Concepts

1. The network of tubes that runs through compact bone is called the
   a. periosteum.
   b. joint.
   c. Haversian canals.
   d. marrow.

2. What occurs during ossification?
   a. Bones lose minerals and mass.
   b. Cartilage is replaced by bone.
   c. Vitamin D is synthesized.
   d. Bones fracture more easily.

3. Small sacs of synovial fluid that help reduce friction between the bones of a joint are called
   a. bursae.
   b. ligaments.
   c. tendons.
   d. cartilage.

4. What types of tissues are found in the skeletal system?

5. What is the advantage of spongy bone tissue in the ends of long bones?

6. Draw a diagram of a long bone and label the structures.

7. Which type of freely movable joint allows for the most range of motion?

Think Critically

8. **Interpret Visuals** Which bone sample shows signs of osteoporosis, choice *a* or choice *b*? Explain.

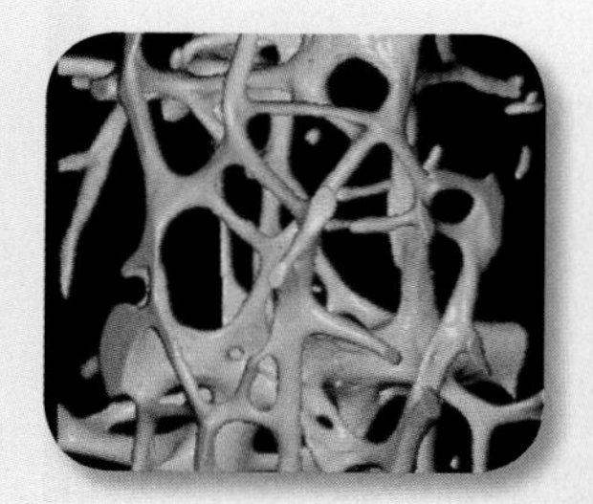

a.

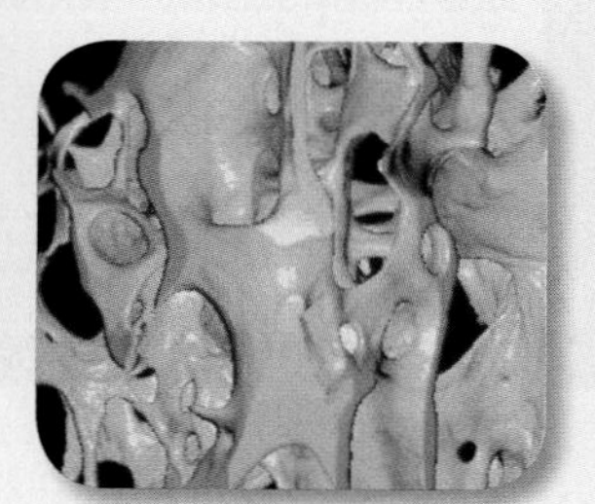

b.

9. **Infer** Disks of rubbery cartilage are found between the individual bones in the spinal column. What function do you think these disks serve?

10. **Craft and Structure** Blood vessels bring oxygen and nutrients to all parts of the body. Ligaments contain fewer blood vessels than some other tissues do. Explain how this difference might relate to the rate of healing in injured ligaments?

11. **Use Models** Suppose you want to build a robotic arm that works the way the human elbow works. Describe or sketch three facts about the elbow that you could use in your planning.

32.2 The Muscular System

Understand Key Concepts

12. In which part of the body would you find striated muscle tissue with relatively small cells that have one or two nuclei?
    a. thigh
    b. stomach
    c. blood vessels
    d. heart

13. Two proteins that are involved in the contraction of muscle are
    a. sarcomere and myofibril.
    b. actin and myosin.
    c. periosteum and cartilage.
    d. ATP and acetylcholine.

14. The point of contact between a motor neuron and a skeletal muscle cell is called a
    a. cross-bridge site.
    b. gap junction.
    c. sarcomere.
    d. neuromuscular junction.

15. Describe the primary function of each of the three types of muscle tissue.

16. **Key Ideas and Details** Use the sliding-filament model to trace how skeletal muscles work.

17. Describe how the release of acetylcholine from a motor neuron affects a muscle cell.

18. **Craft and Structure** Explain the meaning of the statement: "Most skeletal muscles work in opposing pairs."

19. **Craft and Structure** How does the difference between the structure of fast-twitch and slow-twitch muscle fibers relate to their function?

Think Critically

20. **Relate Cause and Effect** Certain bacteria produce a toxin that prevents the release of acetylcholine from the motor neurons. Explain why this can result in a fatal loss of muscle movement.

21. **Research to Build and Present Knowledge** Although exercising can increase your strength and endurance, over-exercising can have adverse effects on the body. Use multiple library and Internet resources to find out about these adverse effects. Summarize your findings in a brief report. Include a list of the sources you use.

32.3 Skin—The Integumentary System

Understand Key Concepts

22. The outer layer of skin is called the
    a. dermis.
    b. keratin.
    c. epidermis.
    d. melanin.

23. Which structure releases a secretion that contributes to the formation of acne?

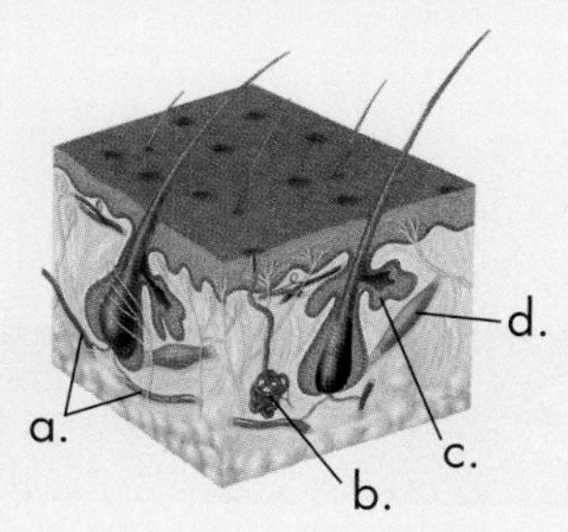

24. Where are new skin cells produced to replace old cells that have been shed?
    a. in the outer layer of the epidermis
    b. in the inner layer of the epidermis
    c. in the dermis
    d. in the sebaceous glands

25. Describe three ways the integumentary system performs the function of protection.

26. Compare the structures of the inner and outer layers of the skin.

27. Describe two ways the skin helps to maintain homeostasis.

28. How do fingernails and toenails grow?

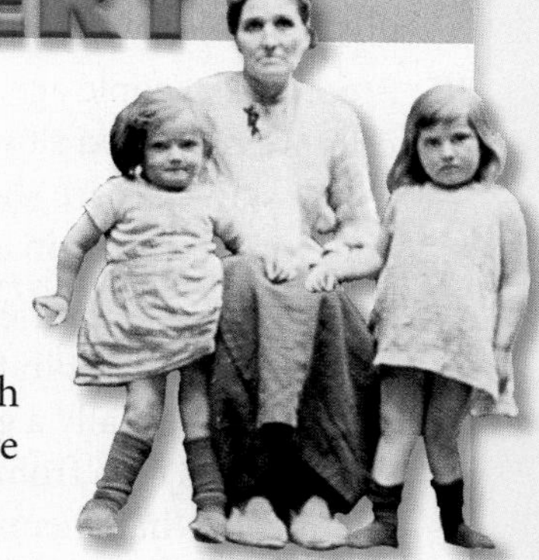

THE DEMISE OF A DISEASE

The search for the cause and a cure for rickets revealed two findings. Both cod liver oil and exposure to ultraviolet light could prevent and cure rickets.

The first finding indicated that cod liver oil contains a nutrient involved in bone health. Starting in the 1930s, many parents in the United States—including the parents of one of this textbook's authors—gave their children a daily dose of bitter cod liver oil.

The second finding indicated that exposure to the sun influences bone health. This explained why children in colder climates were more susceptible to rickets. They had little sun exposure during cold, dark winter months.

But scientists still wondered, what was the connection between cod liver oil and ultraviolet light? How could both treatments result in the same positive outcome?

Through the work of many scientists, we now know that vitamin D is the responsible nutrient in cod liver oil. And, when exposed to ultraviolet light, the skin makes compounds that can be converted to vitamin D. We've also learned that vitamin D helps the body absorb calcium and phosphorus from the digestive system.

Children today are spared cod liver oil doses because vitamin D is added to milk. Rickets is now a rare disease in the United States.

1. **Explain** Why were children in southern cities less likely to develop rickets?

2. **Compare and Contrast** Describe the structure of the bones of a healthy child in comparison to the bones of a child who developed rickets.

3. **Connect to the Big idea** Explain how vitamin D is related to the structure and function of the three systems you learned about in this chapter.

Think Critically

29. Infer A skin callus is a thickening of the epidermis caused by repeated rubbing. Why do people often get calluses on their feet?

30. Predict As people age, the rate at which new skin cells are produced slows down, but the rate at which skin cells are shed does not. What effect do you think this has on a person's skin?

31. Relate Cause and Effect People with albinism have little pigment in their skin, hair, or eyes. Albinism is usually a genetically recessive disease that inhibits cells from producing a particular chemical. What chemical do you think is lacking in people with albinism? Explain.

Connecting Concepts

Use Science Graphics

32. Infer Because cartilage does not appear on X-ray film, it is seen as a clear area between the shaft and the ends of bones. Examine the X-rays. Which hand belongs to the youngest person? Explain.

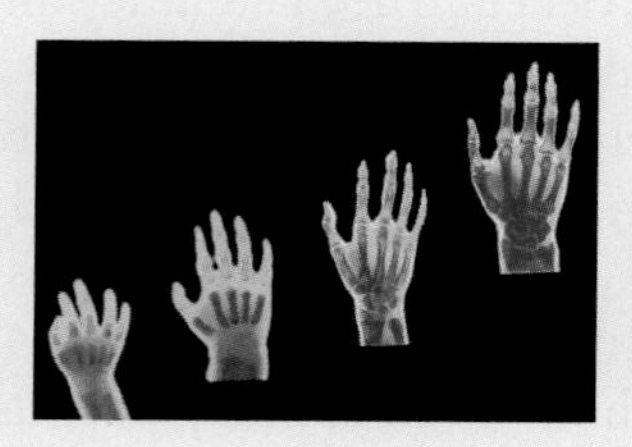

Write About Science

33. Key Ideas and Details Severe burns of large areas of the skin can be a life-threatening injury. Cite evidence from the chapter to explain in a paragraph the greatest risks these burn patients face.

34. Text Types and Purposes Support and movement are the basic functions of the skeletal, muscular, and integumentary systems. In an essay, compare these three body systems with similar structures of a building. Make connections and distinctions, using specific details. For example, which body system has the same function as the girders of a building? How are they similar? How are they different? (*Hint:* To get started, you may want to list shared characteristics.)

35. Assess the Big idea Recall what you learned about the bones of fishes, amphibians, reptiles, birds, and mammals. Compare examples of specific skeletal parts, such as backbones or forelimbs. Relate the bones to the way each animal moves.

Analyzing Data

Integration of Knowledge and Ideas

The UV (ultraviolet) index is a rating system developed by the Environmental Protection Agency. The goal of the index is to inform the public about the level of UV radiation to expect on a given day. Higher numbers indicate higher levels of UV radiation. The table shows the average UV indexes for various cities in the United States.

Average UV Index Values, Selected Locations

Location	Average UV Index Value	
	Winter	Summer
Anchorage, Alaska	< 1	3–4
Atlanta, Georgia	2	8
Honolulu, Hawaii	6	11–12
Miami, Florida	4	10–11
New York, New York	1–2	6–7
Phoenix, Arizona	3	10
Portland, Oregon	1	5–6
St. Louis, Missouri	1–2	7–8

36. Interpret Data A person visiting which of the following cities would need to be the most careful about his or her level of sun exposure?
- **a.** Honolulu in winter
- **b.** Anchorage in summer
- **c.** Atlanta in winter
- **d.** St. Louis in summer

37. Form a Hypothesis Develop a hypothesis that could explain the relationship between geography and average UV index values.

Standardized Test Prep

Multiple Choice

1. What determines differences in skin color among individuals?
A number of melanocytes
B amount of melanin produced by each melanocyte
C amount of keratin in the skin
D amount of sebum produced

2. Smooth muscle is found in the
A walls of blood vessels.
B heart.
C neuromuscular junctions.
D joints.

3. All of the following are important roles of the skeletal system EXCEPT
A protection of internal organs.
B facilitation of movement.
C storage of mineral reserves.
D regulation of body temperature.

4. Which of the following supplies the energy required for muscle contractions?
A myosin
B ATP
C acetylcholine
D actin

5. The tough layer of connective tissue surrounding each bone is called
A tendon.
B ligament.
C periosteum.
D cartilage.

6. Joints that allow one bone to rotate around another are
A gliding joints.
B ball-and-socket joints.
C hinge joints.
D pivot joints.

7. Which of the following is NOT found in skin tissue?
A keratin
B collagen
C marrow
D melanin

8. What is a function of sebum?
A It moistens the hair and skin.
B It gives skin its color.
C It insulates the body.
D It makes nails and hair rigid.

Questions 9–10

As people age, the mineral content of their bone decreases. People who fail to build enough bone in adolescence and young adulthood or who lose bone at a faster than normal rate are at risk for developing osteoporosis. The graph below shows typical bone mass of men and women through most of the life span.

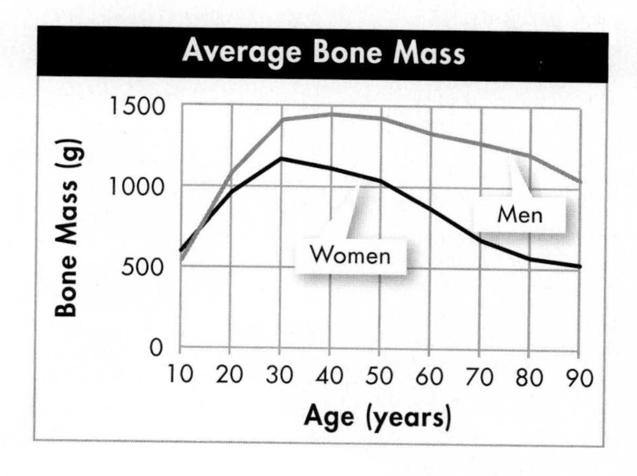

9. Between which ages do both men and women gain bone mass at the highest rate?
A 10–20 years
B 20–30 years
C 30–40 years
D 50–60 years

10. A valid conclusion that can be drawn from this graph is that, on average,
A women lose more bone mass as they age than men do.
B men lose more bone mass as they age than women do.
C women and men lose the same bone mass as they age.
D men continue to gain bone mass as they age.

Open-Ended Response

11. Doctors recommend that women eat calcium-rich foods and get plenty of exercise during adolescence and early adulthood. How could this help prevent osteoporosis later in life?

If You Have Trouble With . . .

Question	1	2	3	4	5	6	7	8	9	10	11
See Lesson	32.3	32.2	32.1	32.2	32.1	32.1	32.3	32.3	32.1	32.1	32.1

33 Circulatory and Respiratory Systems

Structure and Function

Q: How do the structures of the circulatory and respiratory systems allow for their close functional relationship?

Usually, we are not conscious of breathing, but we can control it during activities, such as swimming.

Deepen your understanding of the structures and functions of the circulatory and respiratory systems by engaging in key practices that allow you to make connections across concepts.

NGSS You will plan an investigation among class members that examines the connection between exercise and heart rate.

STEM What factors are involved with replacing a human heart? Investigate the issues that surround the engineering design of artificial hearts and other devices that assist circulation.

Common Core You will translate technical information expressed in words in the text into a chart that diagrams the structure and function of blood.

INSIDE:

- 33.1 The Circulatory System
- 33.2 Blood and the Lymphatic System
- 33.3 The Respiratory System

IN THE BLOOD

At the age of 60, John underwent surgery to reroute blood around blocked vessels in his heart. Since then, he has limited his fat intake and stuck to an exercise program. Still, today he is meeting with his doctor to talk about a new medication that would break up the fatty deposits re-forming in his heart's vessels.

Down the hall, 6-year-old Lila is seeing her doctor today, too. Her vessels are also clogged with fatty deposits, which means she is dangerously close to a heart attack, even at her young age. Both of these patients suffer from a genetic disease that affects a substance transported in blood. What is that disease? And why did it affect them at such different ages? As you read this chapter, look for clues to the identity of this genetic disease and the research that explains it. Then, solve the mystery.

Biology.com

Finding the solution to the In the Blood mystery is just the beginning. Take a video field trip with the ecogeeks of Untamed Science to continue exploring your world.

Go online to access additional resources including:
• eText • Flash Cards • Lesson Overviews • Chapter Mystery

33.1 The Circulatory System

Key Questions

What are the functions of the circulatory system?

How does the heart pump blood through the body?

What are three types of blood vessels?

Vocabulary

myocardium • atrium • ventricle • valve • pulmonary circulation • systemic circulation • pacemaker • artery • capillary • vein

Taking Notes

Preview Visuals Before you read, look at **Figure 33–3.** Make a list of questions about the illustration. As you read, write down the answers.

THINK ABOUT IT "I was about 47 when I collapsed one day at work. There are 22 minutes out of my life that I don't remember. I had gone into cardiac arrest." These are the words of a man who survived a heart attack. Fortunately he received prompt treatment and had successful heart surgery. He continued to live a fairly normal life. He even ran the Boston Marathon! But more than one-third of the 1.2 million Americans who suffer a heart attack each year die. This grim evidence shows that the heart and the circulatory system it powers are vital to life. Why is that so?

Functions of the Circulatory System

What are the functions of the circulatory system?

Some animals have so few cells that all of their cells are in direct contact with the environment. Diffusion and active transport across cell membranes supply the cells with oxygen and nutrients and remove waste products. The human body, however, contains millions of cells that are not in direct contact with the external environment. Because of this, humans need a circulatory system. **The circulatory system transports oxygen, nutrients, and other substances throughout the body, and removes wastes from tissues.**

People who live in large cities face a set of problems like those of the body's cells. City dwellers need food and goods that are produced elsewhere, and they need to get rid of their garbage and other wastes. People need to move around within the city. How are these needs met? By the city's transportation system—a network of streets, highways, and subway or train lines that deliver goods to the city and remove wastes from it. The human body's major transportation system is a closed circulatory system made up of a heart, blood vessels, and blood.

VISUAL ANALOGY

A CITY'S TRANSPORTATION SYSTEM

FIGURE 33–1 The human circulatory system is like the highways and streets of a large city. **Use Analogies** ***Compare the needs of a person living in a large city with the needs of a cell in the body.***

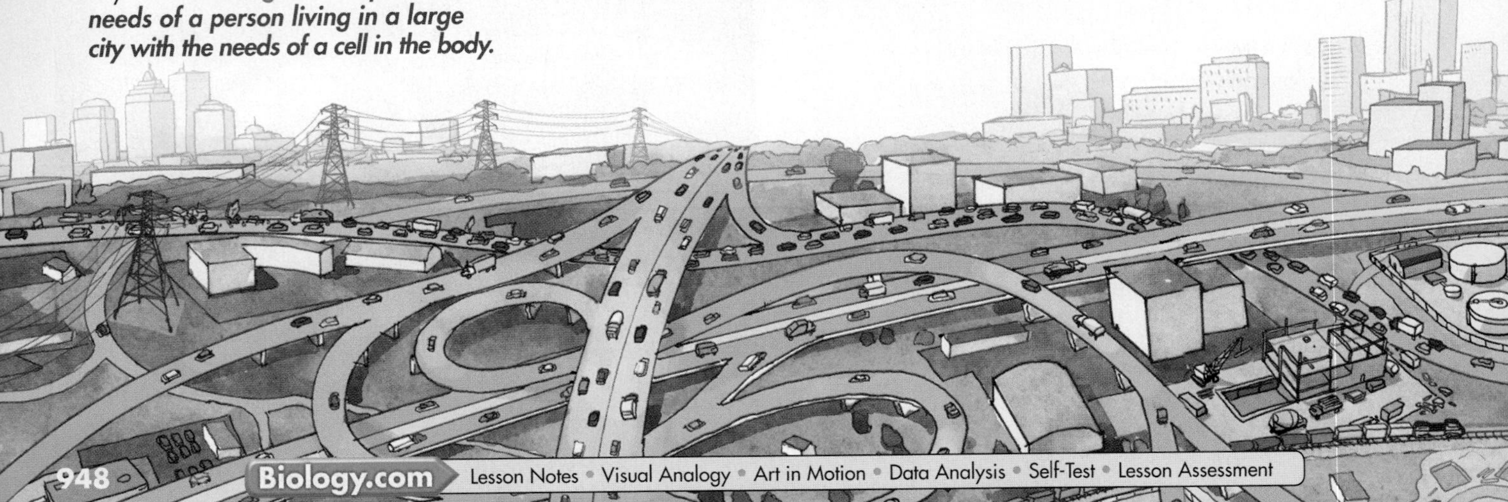

The Heart

How does the heart pump blood through the body?

Much of the time, you're probably not even aware of your heart at work. But when you exercise, you can feel your heart beating near the center of your chest.

Heart Structure Your heart, which is a hollow organ about the size of a clenched fist, is composed almost entirely of muscle. The muscles begin contracting before you are born and stop only when you die. In the walls of the heart, two thin layers of epithelial and connective tissue form a sandwich around a muscle layer called the **myocardium.** **Powerful contractions of the myocardium pump blood through the circulatory system.** An adult's heart contracts on average 72 times a minute, pumping about 70 milliliters of blood with each contraction.

As **Figure 33–2** shows, the heart is divided into four chambers. A wall called the septum separates the right side of the heart from the left side. The septum prevents oxygen-poor and oxygen-rich blood from mixing. On each side of the septum are an upper and lower chamber. Each upper chamber, or **atrium** (plural: atria), receives blood from the body. Each lower chamber, or **ventricle,** pumps blood out of the heart.

In Your Notebook *An Olympic pool contains about 2,000,000 liters of water. In one year, could an average heart pump enough blood to fill an Olympic pool? Explain your answer.*

BUILD Vocabulary

WORD ORIGINS The word *cardiac,* the prefix *cardio-,* and the suffix *-cardium* are all based on the Greek word *kardia,* which means "heart."

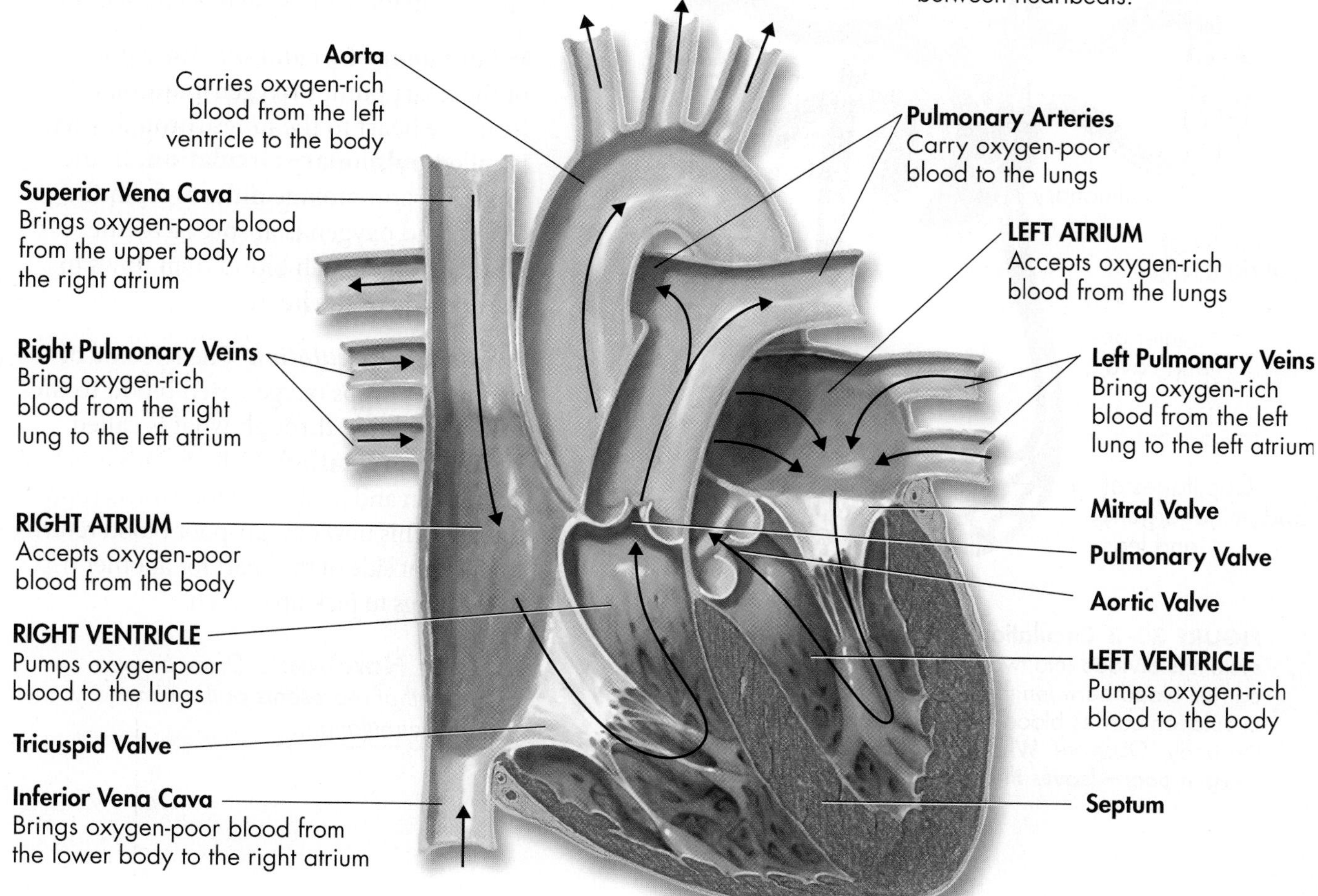

FIGURE 33–2 The Heart The human heart has four chambers: the right atrium, the right ventricle, the left atrium, and the left ventricle. Valves located between the atria and ventricles and between the ventricles and vessels leaving the heart prevent blood from flowing backward between heartbeats.

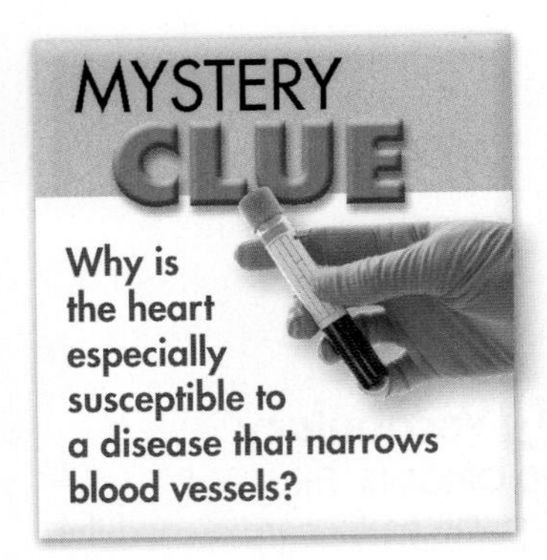

Blood Flow Through the Heart Blood from the body enters the heart through the right atrium; blood from the lungs, through the left atrium. When the atria contract, blood flows into the ventricles. Flaps of connective tissue called **valves** are located between the atria and the ventricles. When blood moves from the atria into the ventricles, those valves open. When the ventricles contract, the valves close, preventing blood from flowing back into the atria. Valves are also located at the exits of each ventricle. This system of valves keeps blood moving through the heart in one direction, like traffic on a one-way street.

The Heart's Blood Supply Heart muscle needs a constant supply of oxygen and nutrients. Surprisingly, the heart gets very little oxygen and nutrients from the blood it pumps through its chambers. Instead, a pair of blood vessels called *coronary arteries,* which branch from the aorta and run through heart tissue, supply blood to the heart muscle. Coronary arteries and the vessels that branch from them are relatively narrow, considering the needs of the heart. If they are blocked, heart muscle cells run out of oxygen and could begin to die. This is what happens during a heart attack, which we discuss in Lesson 33.2.

Circulation Although it is one organ, the heart functions as two pumps. One pump pushes blood to the lungs, while the other pump pushes blood to the rest of the body, as shown in **Figure 33–3.** The two pathways of blood through the body are called pulmonary circulation and systemic circulation.

▶ ***Pulmonary Circulation*** The right side of the heart pumps oxygen-poor blood from the heart to the lungs through what is called **pulmonary circulation.** In the lungs, carbon dioxide diffuses from the blood, and oxygen is absorbed by the blood. Oxygen-rich blood then flows to the left side of the heart.

▶ ***Systemic Circulation*** The left side of the heart pumps oxygen-rich blood to the rest of the body through what is called **systemic circulation.** Cells absorb much of the oxygen and load the blood with carbon dioxide. This now oxygen-poor blood returns to the right side of the heart for another trip to the lungs to pick up oxygen.

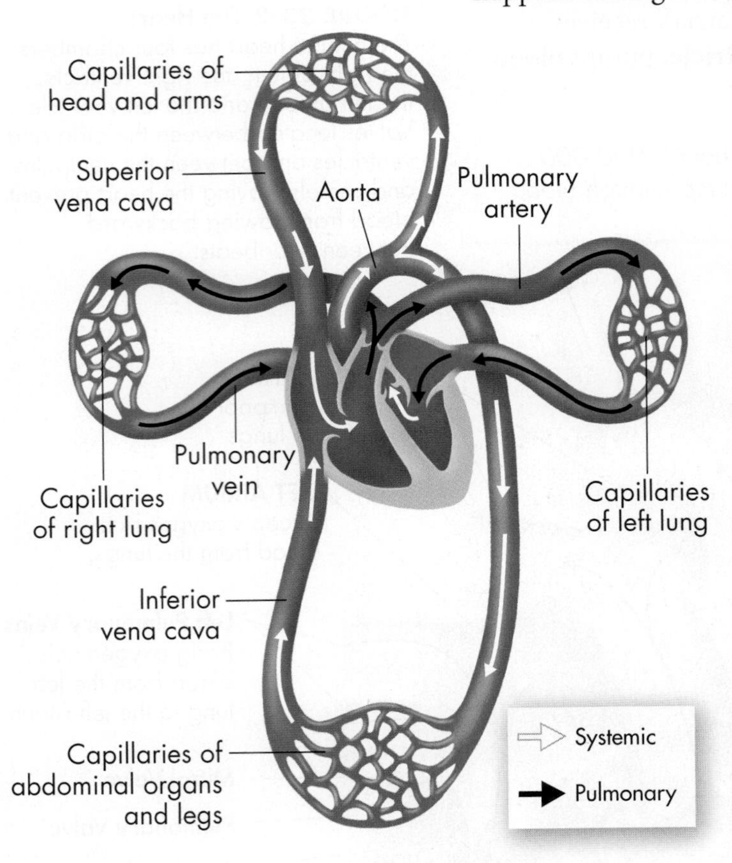

FIGURE 33–3 Circulation Pathways The circulatory system is divided into two pathways. Pulmonary circulation carries blood between the heart and the lungs. Systemic circulation carries blood between the heart and the rest of the body. **Observe** ***What kind of blood—oxygen-rich or oxygen-poor—leaves the lungs and returns to the heart?***

In Your Notebook *Draw a cycle diagram that represents both pulmonary and systemic circulation.*

Heartbeat To be an efficient pump, the heart must beat in an orderly and coordinated way. Two networks of muscle fibers coordinate the heart's pumping action—one in the atria and one in the ventricles. When a single muscle fiber in either network is stimulated, the entire network contracts.

❶ **Atria Contract** Each contraction begins in a small group of cardiac muscle fibers—the sinoatrial node (SA node)—located in the right atrium. The SA node "sets the pace" for the heart, so it is also called the **pacemaker.** When the SA node fires, an electrical impulse spreads through the entire network of muscle fibers in the atria and the atria contract.

❷ **Ventricles Contract** The impulse from the SA node is then picked up by another group of muscle fibers called the atrioventricular node (AV node). Here the impulse is delayed for a fraction of a second while the atria contract and pump blood into the ventricles. Then the AV node produces impulses that spread through the ventricles and cause the ventricles to contract, pumping blood out of the heart. This two-step pattern of contraction—first the atria and then the ventricles—makes the heart an efficient pump.

Control of Heart Rate Your heart rate varies depending on your body's need to take in oxygen and release carbon dioxide. During vigorous exercise, for example, your heart rate could increase to about 200 beats per minute. Heartbeat is not directly controlled by the nervous system, but the autonomic nervous system does influence the activity of the SA node. Neurotransmitters released by the sympathetic nervous system increase heart rate. Those released by the parasympathetic nervous system decrease heart rate.

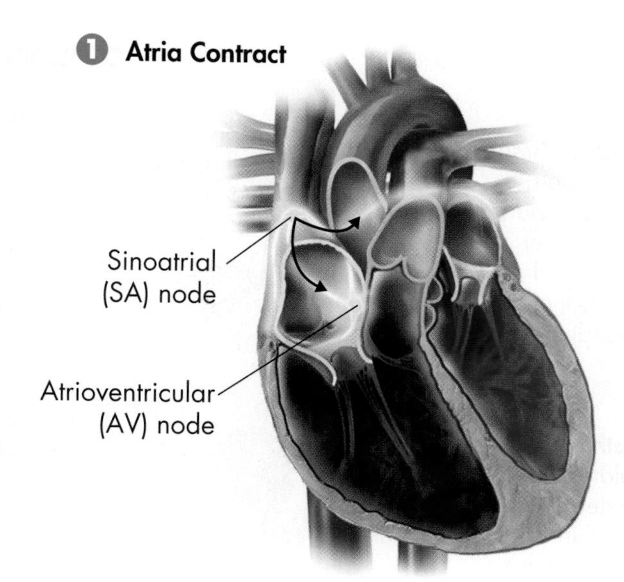

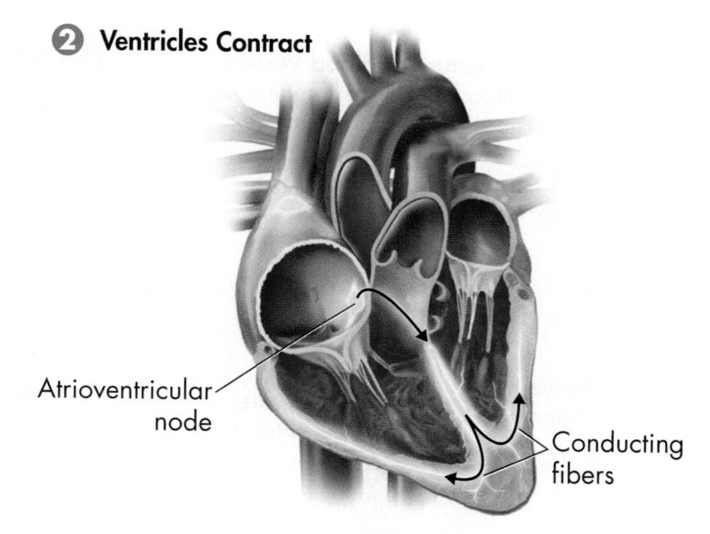

FIGURE 33–4 Heartbeat The SA node generates an impulse that spreads through the atria, causing the muscle fibers to contract and pump blood to the ventricles. The AV node picks up the signal and, after a slight delay, sends an impulse through the ventricles, causing them to contract.

Quick Lab
GUIDED INQUIRY

What Factors Affect Heart Rate?

❶ While sitting, measure your heart rate. Find the pulse in one of your wrists using the first two fingers of your other hand.

❷ Count the number of beats for 15 seconds, and multiply by 4. This gives you the number of beats per minute.

Analyze and Conclude

1. **Predict** What do you think would happen if you stood up? Would your heart rate decrease, increase, or stay the same?

2. **Evaluate** Test your prediction by standing up and measuring your heart rate again. Explain your results.

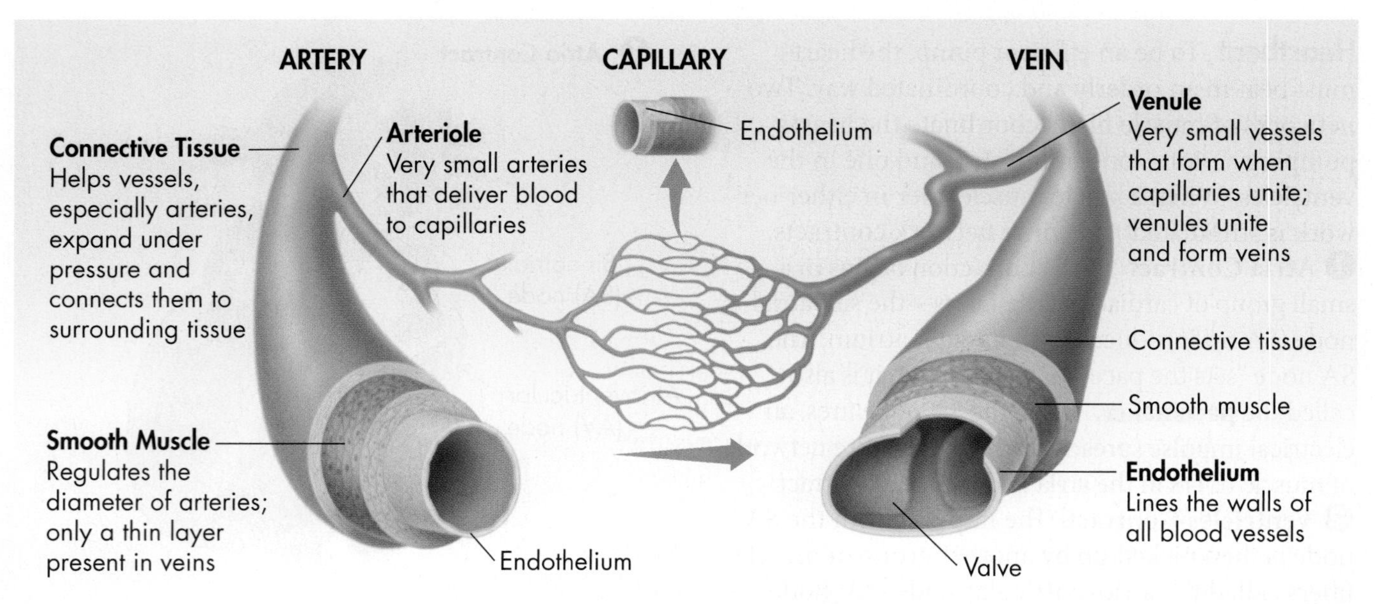

FIGURE 33–5 Structure of Blood Vessels The structure of blood vessel walls contributes to the vessels' functions.

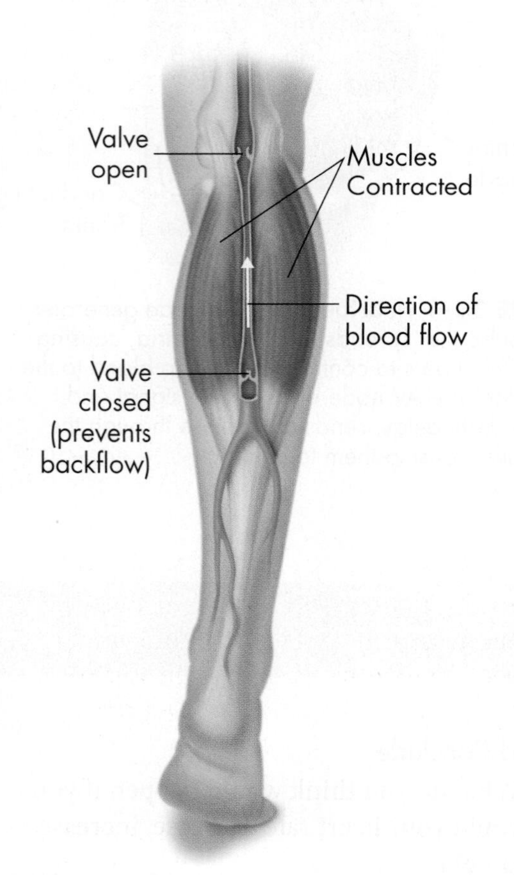

FIGURE 33–6 Blood Flow in Veins The contraction of skeletal muscles helps move blood in veins toward the heart.
Draw Conclusions ***What role do valves play in large veins?***

Blood Vessels

What are three types of blood vessels?

Oxygen-rich blood leaving the left ventricle passes into the aorta. The aorta is the first of a series of vessels that carries blood through the systemic circulation and back to the heart. **As blood flows through the circulatory system, it moves through three types of blood vessels—arteries, capillaries, and veins.**

Arteries **Arteries** are large vessels that carry blood from the heart to the tissues of the body. Arteries are the superhighways of the circulatory system. Except for the pulmonary arteries, all arteries carry oxygen-rich blood. Arteries have thick elastic walls that help them withstand the powerful pressure produced when the heart contracts and pumps blood through them. **Figure 33–5** describes the three layers of tissue found in artery walls—connective tissue, smooth muscle, and endothelium.

Capillaries The smallest blood vessels are the **capillaries.** Capillaries are the side streets and alleys of the circulatory system. Most capillaries are so narrow that blood cells pass through them in single file. Their extremely thin walls allow oxygen and nutrients to diffuse from blood into tissues, and carbon dioxide and other waste products to move from tissues into blood.

Veins After blood passes through the capillaries, it returns to the heart through **veins.** Blood often must flow against gravity through the large veins in your arms and legs. Many veins are located near and between skeletal muscles, as shown in **Figure 33–6.** When you move, the contracting skeletal muscles squeeze the veins, pushing blood toward the heart. Many veins contain valves. The valve that is farthest from the heart closes to ensure blood continues to flow in one direction.

Blood Pressure Like any pump, the heart produces pressure. When it contracts, it produces a wave of fluid pressure in the arteries, known as blood pressure. Although blood pressure falls when the heart relaxes between beats, the system still remains under pressure due to the elasticity of the arterial walls. It's a good thing, too. Without that pressure, blood would stop flowing through the body.

Healthcare workers measure blood pressure with a device called a sphygmomanometer (sfig moh muh NAHM uh tur), an inflatable cuff with a pump and a meter. The cuff is wrapped around the upper arm and inflated until blood flow through the artery that runs down the arm is blocked. As the pressure is released, the healthcare worker listens for a pulse with a stethoscope and records a number from the meter. This number represents the systolic pressure—the force in the arteries when the ventricles contract. When the pulse sound disappears, a second number is recorded. This number represents the diastolic pressure—the force in the arteries when the ventricles relax. A typical blood pressure reading for a healthy teen or adult is below 120/80.

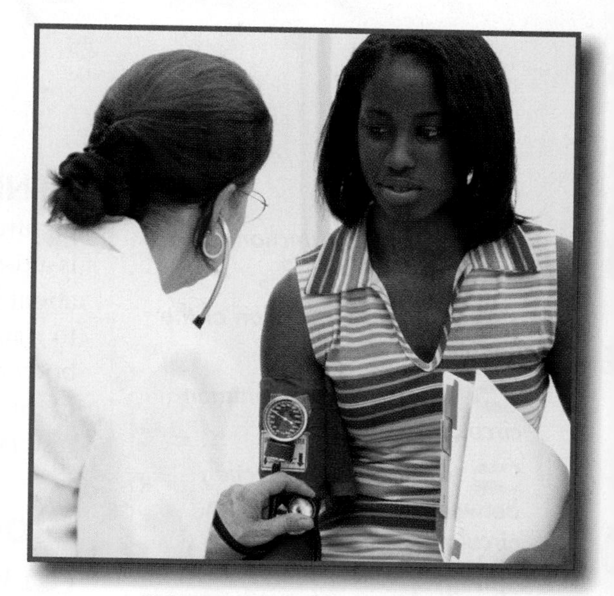

FIGURE 33–7 Measuring Blood Pressure It's important to have your blood pressure measured because blood pressure that is too high or too low can have serious effects on most body systems.

The body regulates blood pressure in a number of ways. Sensory receptors in blood vessels detect blood pressure and send impulses to the brain stem. When blood pressure is high, the autonomic nervous system releases neurotransmitters that relax the smooth muscles in blood vessel walls. When blood pressure is low, neurotransmitters are released that cause the smooth muscles in vessel walls to contract.

The kidneys also regulate blood pressure by affecting the volume of blood. Triggered by hormones produced by the heart and other organs, the kidneys remove more water from the blood and eliminate it in urine when blood pressure is high or conserve more water when blood pressure is low.

33.1 Assessment

Review Key Concepts

1. a. **Review** List the structures of the circulatory system and explain their roles.
   b. **Apply Concepts** Why do humans need a circulatory system?
2. a. **Review** Describe the two paths of blood circulation through the body.
   b. **Relate Cause and Effect** How would damage to the sinoatrial node affect the heart's function?
3. a. **Review** Describe the functions of three types of blood vessels in the circulatory system.
   b. **Infer** If you were standing, would you expect the blood pressure to be higher in your arm or in your leg? Explain your answer. (*Hint:* Think about which area of the body is closer to the source of pressure.)

VISUAL THINKING

4. Trace **Figure 33–2.** Label the four chambers of the heart. Add arrows and labels to indicate how blood flows through the heart.

33.2 Blood and the Lymphatic System

Key Questions

What is the function of each component in blood?

What is the function of the lymphatic system?

What are three common circulatory diseases?

What is the connection between cholesterol and circulatory disease?

Vocabulary

plasma • red blood cell • hemoglobin • white blood cell • platelet • lymph • atherosclerosis

Taking Notes

Outline Before you read, make an outline of the major headings in the lesson. As you read, fill in main ideas and supporting details for each heading.

THINK ABOUT IT When you think about body tissues, you probably picture something with a definite shape, like muscle or skin. But blood is a tissue too—it just happens to be in liquid form! The more you think about blood, the more remarkable its many functions are. In addition to transporting oxygen and fighting disease, it carries substances your body makes and sources of energy such as sugars and fats. In fact, one of the best ways to judge a person's health is—you guessed it—a blood test. How does this unusual tissue perform so many essential functions?

Blood

What is the function of each component in blood?

You might think that the most important function of blood is to serve as the body's transportation system. But the various components of blood also help regulate body temperature, fight infections, and produce clots that help minimize the loss of body fluids from wounds.

Plasma The human body contains 4 to 6 liters of blood. About 55 percent of total blood volume is a straw-colored fluid called **plasma.** **Plasma is about 90 percent water and 10 percent dissolved gases, salts, nutrients, enzymes, hormones, waste products, plasma proteins, cholesterol, and other important compounds.**

The water in plasma helps to control body temperature. Plasma proteins consist of three types—albumin, globulins, and fibrinogen. Albumin and globulins transport substances such as fatty acids, hormones, and vitamins. Albumin also plays an important role in regulating osmotic pressure and blood volume. Some globulins fight viral and bacterial infections. Fibrinogen is necessary for blood to clot.

Red Blood Cells The most numerous cells in blood are **red blood cells,** or erythrocytes (eh RITH roh syts). **The main function of red blood cells is to transport oxygen.** They get their crimson color from the iron in **hemoglobin,** a protein that binds oxygen in the lungs and releases it in capillary networks throughout the body. Then red blood cells transport some carbon dioxide to the lungs.

Red blood cells are disks that are thinner in their center than along their edges. They are produced by cells in red bone marrow. As red blood cells mature and fill with hemoglobin, their nuclei and other organelles are forced out. Red blood cells circulate for an average of 120 days before they are destroyed in the liver and spleen.

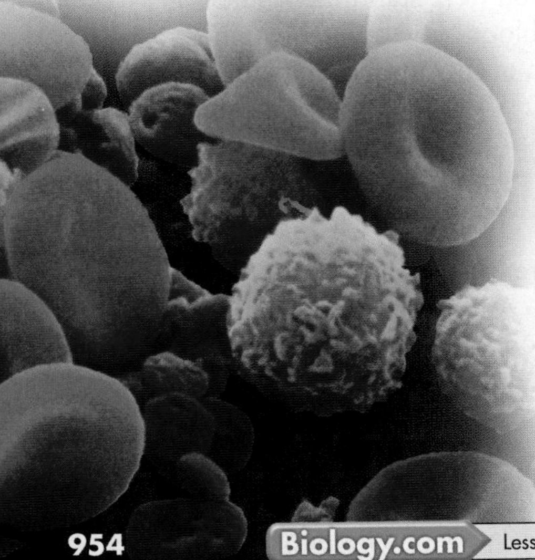

FIGURE 33–8 Blood Cells The micrograph shows red blood cells (red disks), white blood cells (gold orbs), and platelets (pink fragments) (SEM 1866×).

White Blood Cells **White blood cells,** or leukocytes (LOO koh syts), are the "army" of the circulatory system. **White blood cells guard against infection, fight parasites, and attack bacteria.** The body can increase the number of active white blood cells dramatically during a "battle" with foreign invaders. In fact, a sudden increase in white blood cells is a sign that the body is fighting a serious infection. White blood cells are not confined to blood vessels. Many white blood cells can slip through capillary walls to attack foreign organisms.

Different types of white blood cells perform different protective functions. For example, macrophages engulf pathogens. Lymphocytes are involved in the immune response. B lymphocytes produce antibodies that fight infection and provide immunity. T lymphocytes help fight tumors and viruses. You will learn more about lymphocytes and other white blood cells in Chapter 35.

In a healthy person, white blood cells are outnumbered by red blood cells by almost 1000 to 1. Like red blood cells, white blood cells are produced from stem cells in bone marrow. Unlike red blood cells, however, white blood cells keep their nuclei and can live for years.

Platelets Blood loss can be life-threatening. Fortunately, a minor cut or scrape may bleed for a bit, but then the bleeding stops. Why? Because blood clots. **Blood clotting is made possible by plasma proteins and cell fragments called platelets.** The cytoplasm of certain bone marrow cells divides into thousands of small fragments. The fragments, each enclosed in a cell membrane, break off and enter the blood as **platelets.**

When platelets come in contact with the edges of a broken blood vessel, their surface becomes sticky, and they cluster around the wound. These platelets release proteins called clotting factors that start a series of reactions. **Figure 33–9** summarizes one part of the clotting process.

In Your Notebook *Make a flowchart that describes the blood-clotting process.*

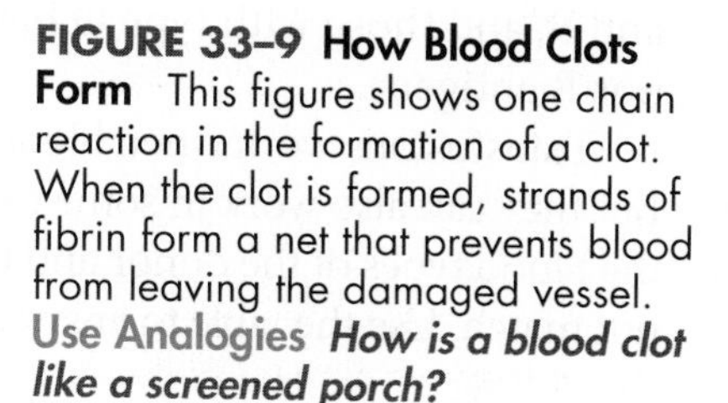

FIGURE 33–9 How Blood Clots Form This figure shows one chain reaction in the formation of a clot. When the clot is formed, strands of fibrin form a net that prevents blood from leaving the damaged vessel. **Use Analogies** ***How is a blood clot like a screened porch?***

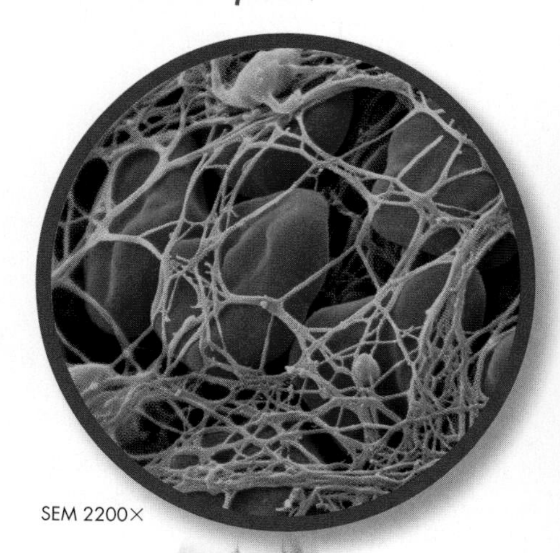

SEM 2200×

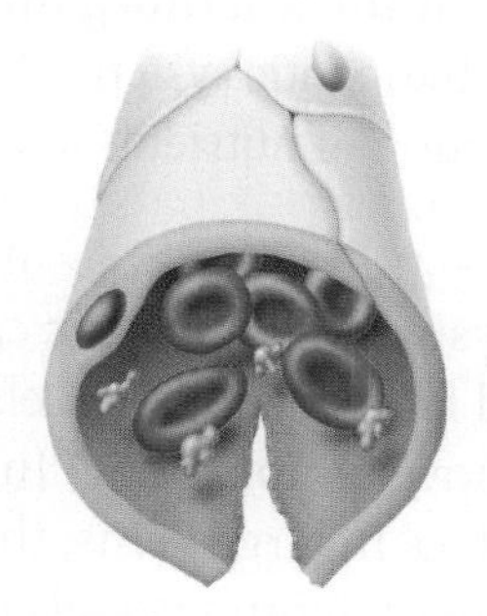

1 Capillary Wall Breaks
A blood vessel is injured by a cut or scrape.

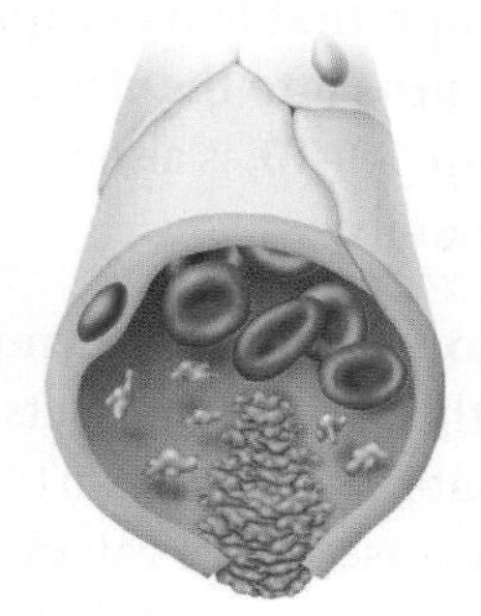

2 Platelets Take Action
Platelets clump at the site and release the clotting factor thromboplastin, which triggers a series of reactions. Thromboplastin converts the protein prothrombin into the enzyme thrombin.

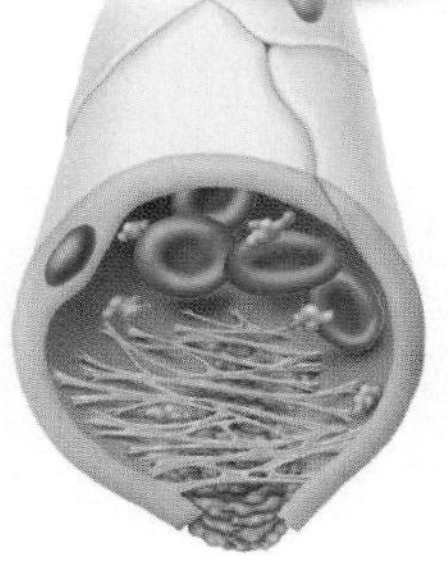

3 Clot Forms
Thrombin converts the soluble plasma protein fibrinogen into insoluble, sticky fibrin filaments, which form the clot. The clot seals the damaged area and prevents further loss of blood.

Blood Transfusions

The first successful transfusion of human blood was carried out in 1818. But many later recipients had severe reactions to transfused blood, and a number died. Today we know why. We inherit one of four blood types—A, B, AB, or O—which are determined by antigens, or the lack of antigens, on our blood cells. Antigens are substances that trigger an immune response. People with blood type A have A antigens on their cells, those with type B have B antigens, those with AB blood have both A and B, and those with type O have neither A nor B antigens.

Transfusions work when blood types match. But they can also work in some cases even when the blood types of the donor and the recipient do not match. Use the table to answer the questions that follow.

Blood Transfusions

Blood Type of Donor	Blood Type of Recipient			
	A	B	AB	O
A	✓	x	✓	x
B	x	✓	✓	x
AB	x	x	✓	x
O	✓	✓	✓	✓

x = Unsuccessful transfusion ✓ = Successful transfusion

1. Draw Conclusions Which blood type is sometimes referred to as the "universal donor"? Which is known as the "universal recipient"?

2. Infer In a transfusion involving blood types A and O, does it matter which blood type is the recipient's and which is the donor's?

3. Apply Concepts Write a brief explanation of the results in the chart using information about phenotypes and genotypes in blood group genes. (*Hint*: Review Lesson 14.1 if needed.)

The Lymphatic System

What is the function of the lymphatic system?

As blood passes through capillaries, some blood cells and components of plasma move through capillary walls and into the fluid between cells, carrying nutrients, dissolved oxygen, and salts. Each day about 3 liters of fluid, and the small particles it contains, leaves the blood. Most of this fluid, known as **lymph,** is reabsorbed into capillaries, but not all of it. The rest goes into the lymphatic system. **The lymphatic system is a network of vessels, nodes, and organs that collects the lymph that leaves capillaries, "screens" it for microorganisms, and returns it to the circulatory system.** The lymphatic system, shown in **Figure 33–10,** is also involved in the absorption of nutrients and in immunity.

Role in Circulation Lymph collects in a system of lymphatic capillaries that slowly conducts it into larger and larger lymph vessels. The lymphatic system doesn't have a pump to move lymph along. Instead, lymph vessels have valves, similar to the valves in large veins, that prevent lymph from flowing backward. Pressure on lymph vessels from surrounding skeletal muscles helps move lymph through the system into larger and larger ducts. These ducts return lymph to the blood through openings in the subclavian veins just below the shoulders. When injury or disease blocks lymphatic vessels, lymph can accumulate in tissues, causing swelling called edema.

Role in Nutrient Absorption The lymphatic system also plays an important role in the absorption of nutrients. A system of lymph vessels runs alongside the intestines. The vessels pick up fats and fat-soluble vitamins from the digestive tract and transport these nutrients into the bloodstream.

Role in Immunity Hundreds of small bean-shaped enlargements—called lymph nodes—are scattered along lymph vessels throughout the body. Lymph nodes act as filters, trapping microorganisms, stray cancer cells, and debris as lymph flows through them. Fleets of white blood cells inside lymph nodes engulf or otherwise destroy this cellular "trash." When large numbers of microorganisms are trapped in lymph nodes, the nodes become enlarged. The "swollen glands" that are symptoms of certain kinds of infections are actually swollen lymph nodes.

The thymus and spleen also play important roles in the immune functions of the lymphatic system. The thymus is located beneath the sternum. T lymphocytes mature in the thymus before they can function in the immune system. The functions of the spleen are similar to those of lymph nodes. However, instead of lymph, blood flows through the spleen, where it is cleansed of microorganisms and other debris. The spleen also removes old or damaged blood cells and stores platelets.

In Your Notebook *Compare and contrast the functions of the circulatory system and the lymphatic system.*

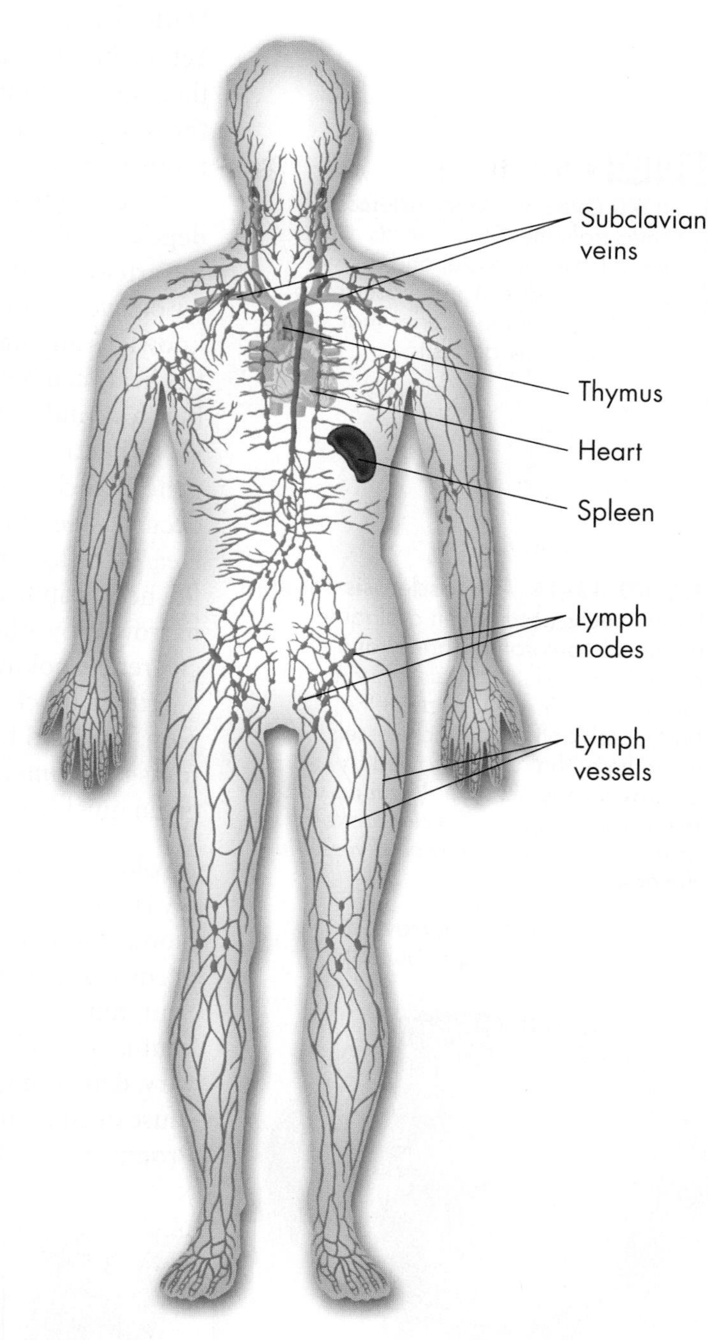

FIGURE 33–10 The Lymphatic System The lymphatic system is a network of vessels, nodes, and organs that recycles fluids from tissues and plays a role in nutrient absorption and immunity. **Infer** ***Why do you think your doctor feels your neck for swollen lymph nodes when you are sick?***

Circulatory System Diseases

What are three common circulatory diseases?

Diseases of the circulatory system can progress for many years before they are discovered. Often the first sign of circulatory problems is an event that affects the heart or brain. Why? Tissues in these vital organs begin to die within moments if their oxygen supply is interrupted. **Three common and serious diseases of the circulatory system are heart disease, stroke, and high blood pressure.** Damage to heart muscle from a heart attack or to the brain from a stroke can be fatal. Individuals with high blood pressure are at higher risk for both heart disease and stroke. Heart disease is the leading cause of death in the United States.

Heart Disease Heart muscle requires a constant supply of oxygen. Yet the heart is supplied with blood by just two coronary arteries and their smaller branches. There are many types of heart disease, but the most common occur when blood flow through these vessels is obstructed.

BUILD Vocabulary

WORD ORIGINS Atherosclerosis comes from the Greek words *athero* (gruel or paste) and *sclerosis* (hardness). Atherosclerosis is hardening of the arteries that results from fatty deposits.

One example is **atherosclerosis,** a condition in which fatty deposits called plaques build up in artery walls and eventually cause the arteries to stiffen. Over time, plaques often bulge into the center of a vessel and restrict blood flow to heart muscle. Chest pain, known as angina, can be a sign of restricted blood flow. Eventually, the heart can be weakened or damaged by oxygen deprivation, leading to a condition called heart failure.

If the cap on a plaque ruptures, a blood clot may form that completely blocks an artery, as shown in **Figure 33–11.** A heart attack occurs as heart muscle cells become damaged and possibly die. Heart attacks can also damage the SA or AV nodes, which can affect the heart's ability to beat in a coordinated way. Arteries severely narrowed by atherosclerosis, the use of drugs such as cocaine, and cigarette smoking can also lead to a heart attack.

Heart attack symptoms include nausea; shortness of breath; chest pain; and pain in the neck, jaw, or left arm. People with these symptoms need *immediate* medical attention. Medication needs to be given quickly to increase blood flow and save heart muscle.

Stroke The sudden death of brain cells when their blood supply is interrupted is called a stroke. Some strokes are caused by a blood clot that blocks a blood vessel in the brain. A stroke can also occur if a weak blood vessel breaks and causes bleeding in the brain. Symptoms of stroke include severe headache, numbness, dizziness, confusion, and trouble seeing or speaking. The results of a stroke vary, depending on which part of the brain it affects. Some strokes cause death. Other strokes may cause paralysis or loss of speech. Prompt medical treatment may lessen the severity of a stroke.

FIGURE 33–11 Atherosclerosis Most heart attacks occur when a plaque ruptures in a coronary artery and a clot forms. Clots can also form in large vessels in other parts of the body, break off, and block vessels in the heart that are narrowed by atherosclerosis. **Predict** ***What do you think would happen if a clot broke off from an artery and blocked a vessel in the brain?***

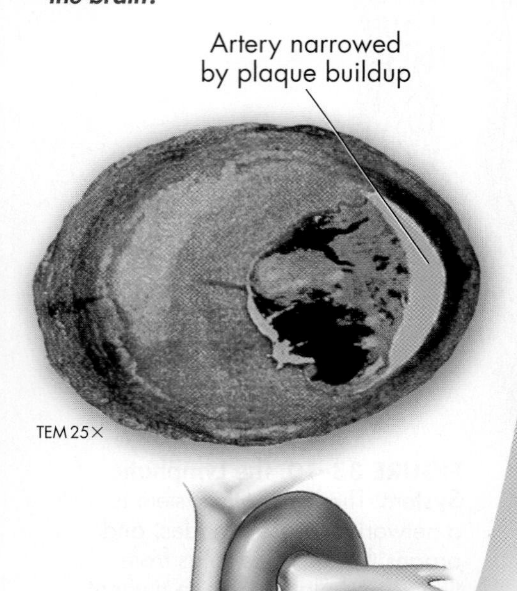

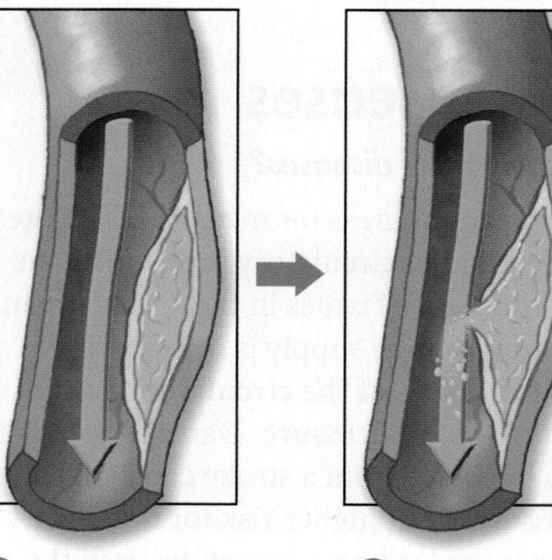

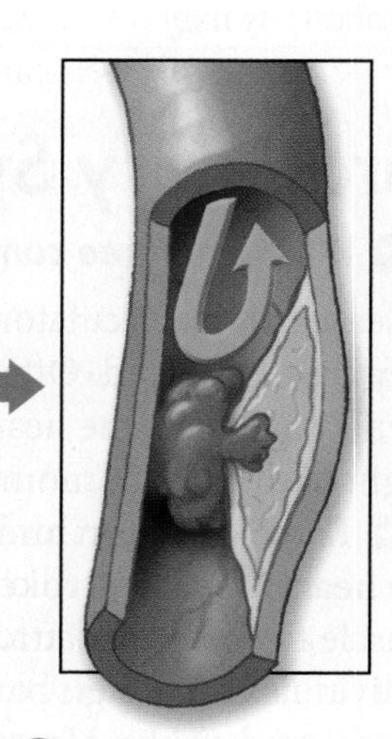

High Blood Pressure High blood pressure, or hypertension, is usually defined as a reading of 140/90 or higher. Because hypertension often has no symptoms, people may have it for years and not know. Meanwhile, heart damage occurs as the heart struggles to push blood through vessels. Hypertension also causes small tears in blood vessels, which sets the stage for atherosclerosis. Likewise, the stiffened arteries that result from atherosclerosis can contribute to high blood pressure. Diet, exercise, and prescription drugs can help control hypertension. Uncontrolled hypertension can lead to heart attack, stroke, and kidney damage.

Risk Factors for Heart Disease and Stroke

Controllable Risk Factors	Uncontrollable Risk Factors
Diet	Age
Exercise	Family history
Weight	Gender (men have more heart attacks)
Not smoking	
High blood cholesterol	
High blood pressure	
Diabetes	

FIGURE 33–12 Risk Factors for Heart Disease and Stroke Some risk factors for heart disease can be controlled because they are related to behavior. For example, people can control their diets and their exercise levels and many can take medication to control diabetes. But other risk factors, such as age and family history, cannot be controlled.

Understanding Circulatory Disease

What is the connection between cholesterol and circulatory disease?

Diseases of the circulatory system do not have a single cause. **Figure 33–12** lists several factors that increase the risk of heart and stroke. Although many risk factors can be controlled, this can be difficult. In some cases, medications may not be accessible or may not be effective. For example, blood cholesterol levels can be difficult to control. But researchers have learned a lot about blood cholesterol levels, their connection to atherosclerosis, and how the condition can be managed.

What Is Cholesterol? Cholesterol is a lipid that is part of animal cell membranes. It is also used in the synthesis of some hormones, bile, and vitamin D. Cholesterol is transported in the blood primarily by two types of lipoproteins—low-density lipoprotein (LDL) and high-density lipoprotein (HDL). LDL is the cholesterol carrier that is most likely to cause trouble in the circulatory system because it becomes part of plaque. HDL, often called good cholesterol, generally transports excess cholesterol from tissues and arteries to the liver for removal from the body.

Measures of a person's blood cholesterol actually are measures of lipoproteins. Normal total blood cholesterol levels range from 100 to 200 milligrams per deciliter (mg/dL). A person's LDL level should be less than 100 mg/dL. A man's HDL level should be greater than 40 mg/dL; a woman's HDL level should be greater than 50 mg/dL.

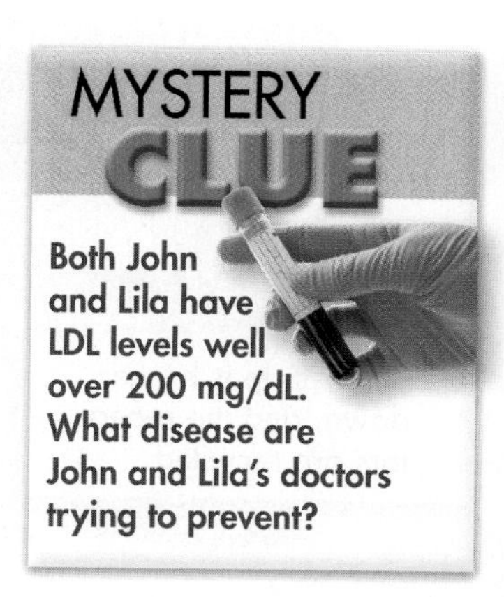

Sources of Cholesterol The liver manufactures cholesterol, which is then transported through the blood to tissues. Humans also consume cholesterol in meat, eggs, dairy products, and fried foods, especially if those foods are high in saturated or trans fats.

Cholesterol and Atherosclerosis Years ago researchers compared cholesterol levels and heart attack rates in different groups of people. In certain villages in Japan and Yugoslavia, the average cholesterol level was 160. In those populations, the heart attack rate was very low—fewer than five attacks for every 1000 men over a ten-year period. In parts of Finland, researchers found mean cholesterol levels of 265. In that population, the heart attack rate was 14 times higher! **Research indicates that high cholesterol levels, along with other risk factors, lead to atherosclerosis and higher risk of heart attack.**

What controls the level of cholesterol in blood? Is there any medical treatment that can lower cholesterol and reduce the risk of atherosclerosis? These questions led researchers Michael Brown and Joseph Goldstein to studies that earned them a Nobel Prize in 1985.

Identifying the LDL Receptor Brown and Goldstein discovered LDL receptors on the cell membrane of liver cells, as shown in **Figure 33–13.** LDL binds to these receptors and then is taken into the cells. Once inside, cholesterol is broken down and then stored or used for making bile or more cholesterol. When blood cholesterol levels are high, liver cells take cholesterol from the blood and do not make it. When blood cholesterol levels are low, the liver produces it.

In Your Notebook *Make a feedback loop to demonstrate the relationship between blood cholesterol levels and healthy liver cells.*

FIGURE 33–13 LDL Receptors When blood LDL levels are high, liver cells with normal LDL receptors take up LDL and use it or store it. However, the liver cells of some people have defective LDL receptors. Those cells cannot remove cholesterol from the blood and do not stop producing cholesterol.

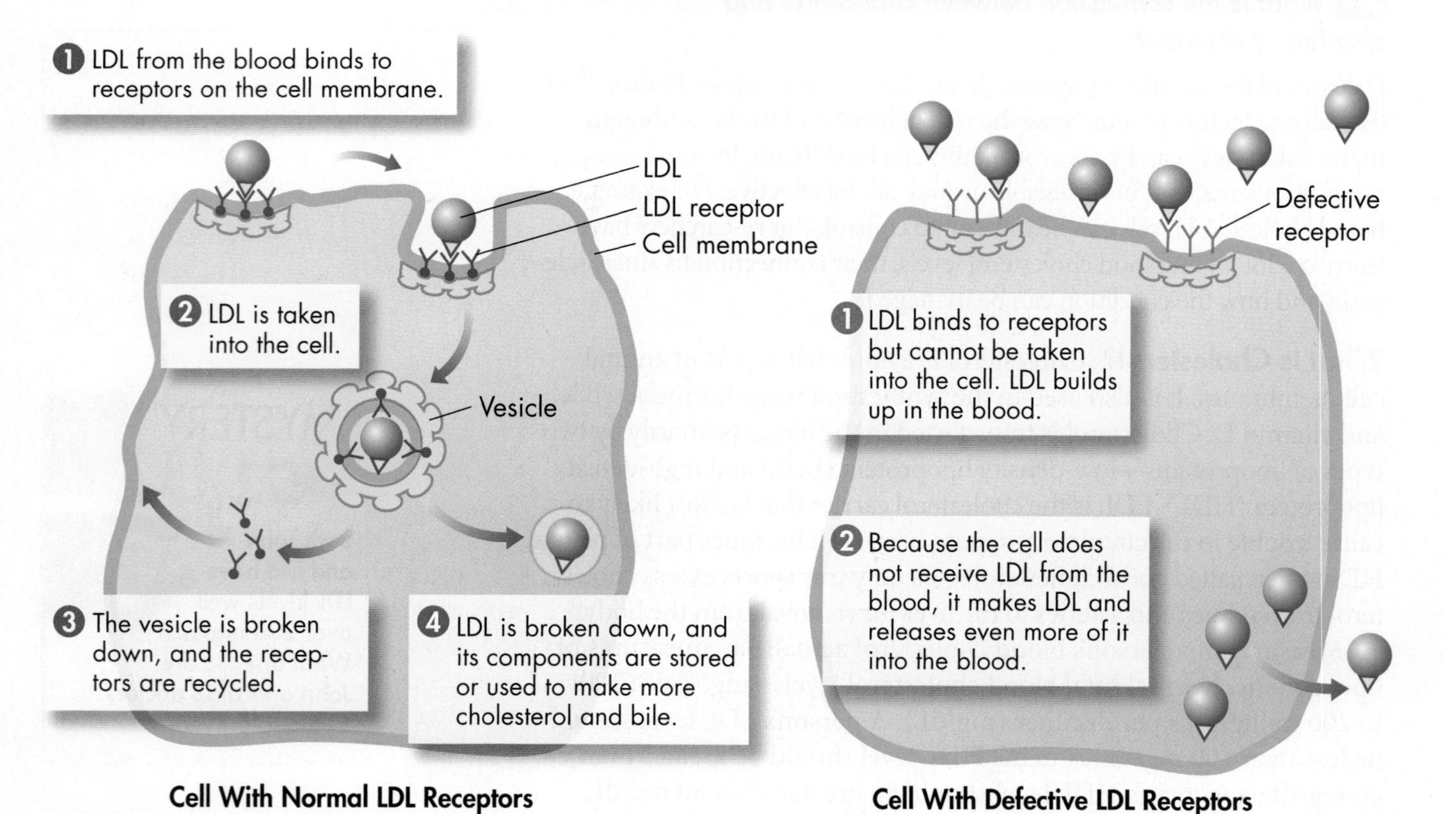

Brown and Goldstein also found that some people carry genes that produce defective LDL receptors. This causes two problems. First, without working LDL receptors, liver cells can't remove cholesterol from blood. Second, these liver cells don't get the signal to stop producing cholesterol. People with defective LDL receptors have very high cholesterol levels, even if they don't eat much cholesterol or fat.

From Genetic Disease to the Public Does understanding this genetic defect help us understand high cholesterol in the general public? Brown and Goldstein learned that people who eat high-fat diets store excess cholesterol in their liver cells. Those cells then stop making LDL receptors and removing cholesterol from blood. The excess cholesterol is then deposited in arteries. So a diet that is high in cholesterol can cause symptoms similar to those of a genetic disease!

Brown and Goldstein's work led to the development of drugs that can help people with high cholesterol. For example, statins block the synthesis of cholesterol in liver cells. This stimulates the liver to produce more LDL receptors, which then remove excess cholesterol from the blood.

Keeping Your Circulatory System Healthy It is much easier to prevent heart disease than to cure it. Prevention starts when you're young, with healthy habits that include a balanced diet, regular exercise, and not smoking. A healthy diet may protect your arteries from atherosclerosis. Exercise strengthens your heart and helps your circulatory system work efficiently. Never starting to smoke will protect your circulatory system from the many dangerous chemicals in tobacco smoke.

MYSTERY CLUE

What genetic defect do both mystery patients carry? Can you think of a genetic reason why Lila's symptoms are so much worse than John's?

33.2 Assessment

Review Key Concepts

1. a. **Review** List the main function of plasma, red blood cells, white blood cells, and platelets.
   b. **Infer** Hemophilia is a genetic disorder that results from a defective protein in the clotting pathway. What do you think happens to a person with hemophilia who has a minor cut?
2. a. **Review** Describe the role of the lymphatic system.
   b. **Compare and Contrast** How are the functions of veins and lymphatic vessels similar? How are they different?
3. a. **Review** What are the risk factors for the three common diseases of the circulatory system?
   b. **Form a Hypothesis** Why do you think atherosclerosis may lead to hypertension?
4. a. **Review** What are two types of cholesterol carriers found in the blood?
   b. **Compare and Contrast** Explain how high blood cholesterol develops in someone with a genetic disorder versus someone who eats a high-fat diet.

WRITE ABOUT SCIENCE

Creative Writing

5. **Research to Build and Present Knowledge** Use library or Internet resources to research the connection between diet and circulatory disease. Write a short commentary for a television news program that explains the connection and demonstrates your understanding of the subject.

RST.9-10.10 Level of Text Complexity, WHST.9-10.10 Range of Writing

Technology & BIOLOGY

Testing for Heart Disease

Ever-improving imaging techniques make it possible for doctors to diagnose heart disease and disorders quickly and without the risk of invasive procedures. None of these tests involves inserting instruments into the body, but they reveal the inner workings of the heart with remarkable accuracy.

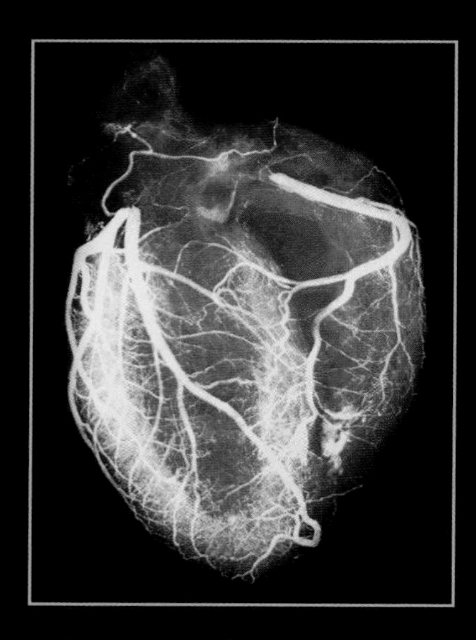

Computed Tomography Angiography

A patient is injected with an iodine-based dye. Then the CT scanner rotates over the patient and takes multiple X-rays of the heart, which a computer uses to form three-dimensional images. The test can show if parts of blood vessels are blocked or damaged. The results can be used to determine what further tests are needed or as a guide for planning surgery.

WRITING **In a paragraph, explain which technique would most likely be used to check for advanced atherosclerosis in a coronary artery.**

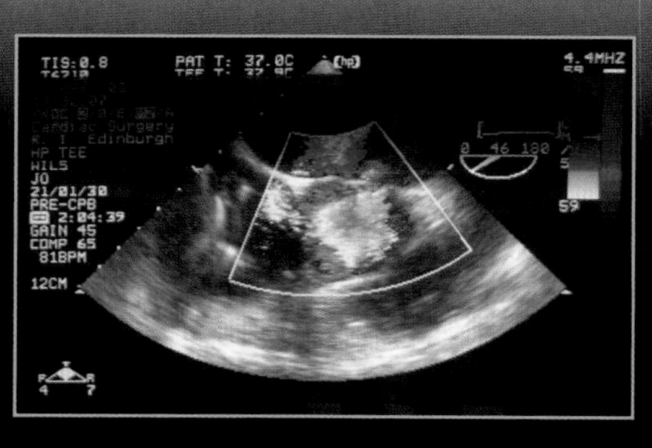

Echocardiography

High-frequency sound waves, transmitted through the chest, are fed into a computer, which analyzes the "echoes" to produce moving images of the heart. This is an especially safe test because it doesn't involve radiation or dyes. The test allows doctors to see the heart in action. It can reveal an enlarged heart, reduced pumping action, and structural problems.

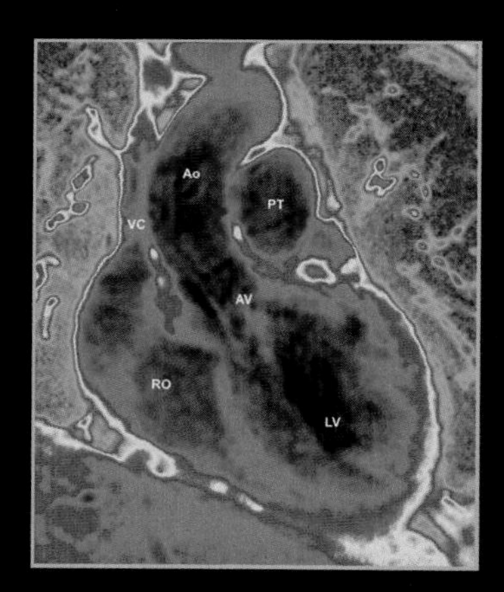

Magnetic Resonance Imaging (MRI)

MRI uses powerful magnets to produce images that are particularly good for examining muscle and other soft tissue. Professionals analyzing MRI images can see the difference between healthy tissue and unhealthy tissue. MRI does not involve radiation or iodine-based dyes. It can be used to assess heart muscle damage caused by a heart attack, birth defects, or abnormal growths.

33.3 The Respiratory System

THINK ABOUT IT When medics examine an unconscious accident victim, one of the first things they do is check whether the person is breathing. This is one way to determine whether there is still a life to save. Why do we make such a close connection between breathing and life? For that matter, why do we need to breathe? All cells in our body, especially brain cells, require a constant supply of oxygen for cellular respiration. Without oxygen, many cells begin to die within minutes. The respiratory system works together with the circulatory system to provide our cells with oxygen. Any interruption in that vital function can be fatal.

Key Questions

What is the function of the respiratory system?

How are oxygen and carbon dioxide exchanged and transported throughout the body?

What mechanisms are involved in breathing?

How does smoking affect the respiratory system?

Vocabulary

pharynx
trachea
larynx
bronchus
alveolus
diaphragm

Taking Notes

Flowchart Make a flowchart that shows the path of air through the respiratory system.

Structures of the Respiratory System

What is the function of the respiratory system?

For organisms, rather than single cells, *respiration* means the process of gas exchange between a body and the environment. **The human respiratory system picks up oxygen from the air we inhale and releases carbon dioxide into the air we exhale.** With each breath, air enters the body through the air passageways and fills the lungs, where gas exchange takes place. The circulatory system links this exchange of gases in the lungs with our body tissues. The respiratory system consists of the nose, pharynx, larynx, trachea, bronchi, and lungs.

Nose The respiratory passageways transport air into some of the most delicate tissues in the body. To keep lung tissue healthy, air entering the respiratory system must be filtered, moistened, and warmed. Hairs lining the entrance to the nasal cavity start the filtering process by trapping large particles. Incoming air is warmed in the inner nasal cavity and sinuses. These areas produce mucus that moistens the air and catches even more dust particles. If you've ever blown your nose after spending time in a dusty environment, you've seen evidence of the way nasal hairs and mucus protect the lungs.

In Your Notebook *In your own words, compare and contrast cellular respiration and respiration at the organism level.*

FIGURE 33–14 Cilia Cilia in the trachea sweep mucus and debris away from the lungs. **Infer** ***What would likely happen to a person's respiratory system if the cilia were damaged by pollutants?***

Pharynx, Larynx, and Trachea Air moves through the nose to a cavity at the back of the mouth called the **pharynx,** or throat. The pharynx serves as a passageway for both air and food. Air moves from the pharynx into the **trachea,** or windpipe. When you swallow food or liquid, a flap of tissue called the epiglottis covers the entrance to the trachea, ensuring that the food or liquid goes into the esophagus.

Between the pharynx and the trachea is the larynx. The **larynx** contains two highly elastic folds of tissue known as the vocal cords. When muscles pull the vocal cords together, the air moving between them causes the cords to vibrate and produce sounds. Your ability to speak, shout, and sing comes from these tissues.

Mucus produced in the trachea continues to trap inhaled particles. Cilia lining the trachea sweep both mucus and trapped particles away from the lungs toward the pharynx. From there, the mucus and particles can be swallowed or spit out. This process helps keep the lungs clean and open for the important work of gas exchange.

Lungs From the trachea, air moves into two large tubes in the chest cavity called **bronchi** (singular: bronchus). Each bronchus leads to one lung. Within each lung, the large bronchus divides into smaller bronchi, which lead to even smaller passageways called bronchioles. Bronchi and bronchioles are surrounded by smooth muscles controlled by the autonomic nervous system. As the muscles contract and relax, they regulate the size of air passageways.

The bronchioles continue to divide until they reach a series of dead ends—millions of tiny air sacs called **alveoli** (singular: alveolus). Air moving through these tubes can be compared to a motorist who takes an exit off an eight-lane highway onto a four-lane highway, makes a turn onto a two-lane road, and then proceeds onto a narrow country lane—which dead-ends. Alveoli are grouped in clusters, like bunches of grapes. A delicate network of capillaries surrounds each alveolus.

BUILD Vocabulary

MULTIPLE MEANINGS Alveolus is also the term for a honeycomb cell in a beehive or a tooth socket in the jaw.

Quick Lab
GUIDED INQUIRY

What's in the Air? 

1. Trace the outline of a microscope slide on graph paper. Repeat four times.
2. Cut out the outlines and tape them to the bottom of five slides.
3. Pick indoor and outdoor spots to place your slides. On the back of each slide, write your initials, the date, and where you will put the slide.
4. Cover the front of each slide with a thin coat of petroleum jelly.
5. Leave the slides in the locations you chose for at least 24 hours.
6. Collect the slides, place them under a microscope, and count the number of particles in ten of the squares on each slide. Record your results.

Analyze and Conclude

1. Observe On which slide did you count the most particles? The fewest?

2. Draw Conclusions Were you surprised by the results? Why or why not?

3. Apply Concepts What structures in your body prevent most of these particles from entering your lungs?

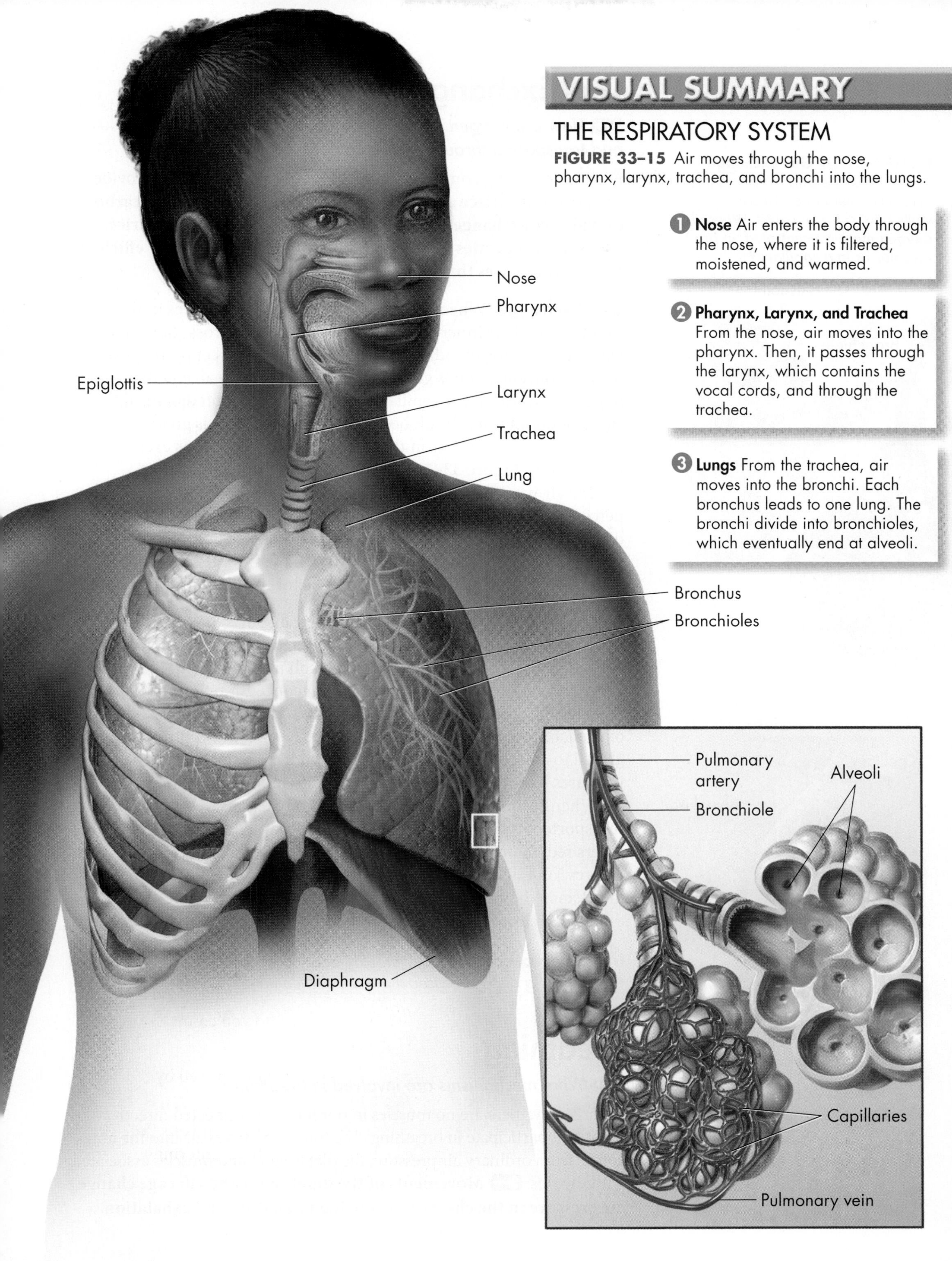

VISUAL SUMMARY

THE RESPIRATORY SYSTEM

FIGURE 33–15 Air moves through the nose, pharynx, larynx, trachea, and bronchi into the lungs.

1. **Nose** Air enters the body through the nose, where it is filtered, moistened, and warmed.

2. **Pharynx, Larynx, and Trachea** From the nose, air moves into the pharynx. Then, it passes through the larynx, which contains the vocal cords, and through the trachea.

3. **Lungs** From the trachea, air moves into the bronchi. Each bronchus leads to one lung. The bronchi divide into bronchioles, which eventually end at alveoli.

FIGURE 33–16 **Gas Exchange** Carbon dioxide and oxygen diffuse across capillary and alveolus walls. **Draw Conclusions** ***Where is oxygen more concentrated, in an alveolus or in a capillary?***

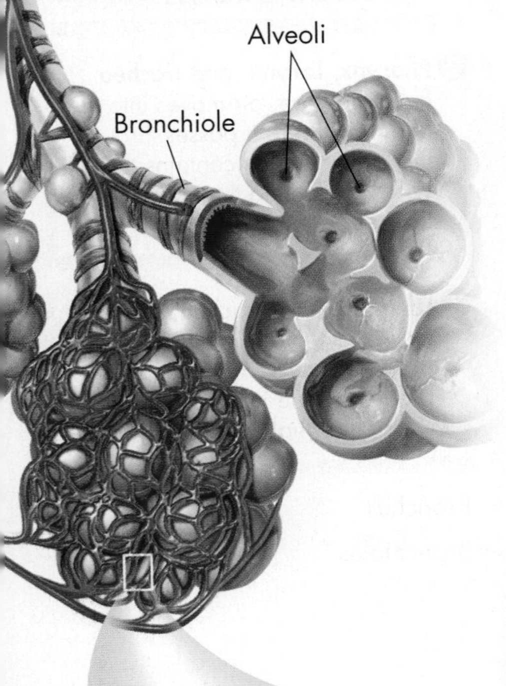

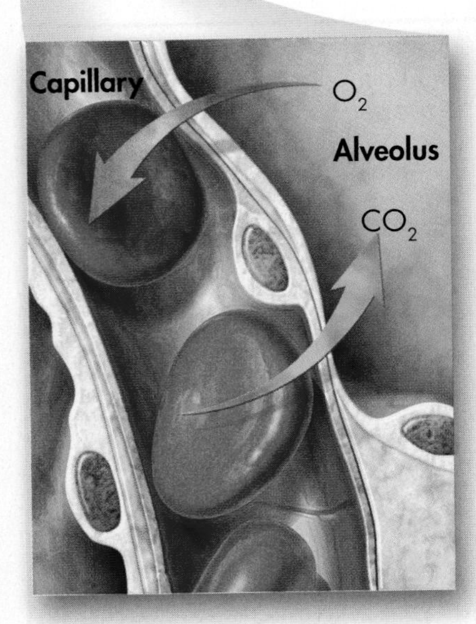

Gas Exchange and Transport

How are oxygen and carbon dioxide exchanged and transported throughout the body?

Each healthy lung contains about 150 million alveoli, which provide an enormous surface area for gas exchange. **Oxygen and carbon dioxide are exchanged across the walls of alveoli and capillaries. Chemical properties of blood and red blood cells allow for efficient transport of gases throughout the body.**

Gas Exchange When air enters alveoli, oxygen dissolves in the moisture on their inner surface and then diffuses across thin capillary walls into the blood. Oxygen diffuses in this direction because the oxygen concentration is greater in the air within the alveoli than it is in the blood within the capillaries. Meanwhile, carbon dioxide diffuses from blood into the alveoli because its concentration is greater in the blood than it is in the air in the alveoli. The process of gas exchange is illustrated in **Figure 33–16.**

The air you inhale usually contains 21 percent oxygen and 0.04 percent carbon dioxide. Exhaled air usually contains less than 15 percent oxygen and 4 percent carbon dioxide. This means your lungs remove about a fourth of the oxygen in the air you inhale and increase the carbon dioxide content of that air by a factor of 100.

Transport Hemoglobin binds with and transports oxygen that diffuses from alveoli to capillaries. It also increases the efficiency of gas exchange. Diffusion of oxygen from alveoli into capillaries is a passive process. That process stops when oxygen concentration in the blood and alveoli is the same. But hemoglobin actively binds to dissolved oxygen, removing it from plasma and enabling diffusion from the alveoli to continue. Hemoglobin binds with so much oxygen that it increases blood's oxygen-carrying capacity more than 60 times.

When carbon dioxide diffuses from body tissues to capillaries, it is transported in the blood in three different ways. Most carbon dioxide enters red blood cells and combines with water, forming carbonic acid. The rest of it dissolves in plasma or binds to hemoglobin and proteins in plasma. These processes are reversed in the lungs, where carbon dioxide is released into alveoli and exhaled.

In Your Notebook *What would happen to the surface area for gas exchange if a disease caused the walls between alveoli to break down?*

Breathing

What mechanisms are involved in breathing?

Surprisingly, there are no muscles in our lungs or connected directly to them that participate in breathing. The force that drives air into the lungs comes from ordinary air pressure, the diaphragm, and muscles associated with the ribs. **Movements of the diaphragm and rib cage change air pressure in the chest cavity during inhalation and exhalation.**

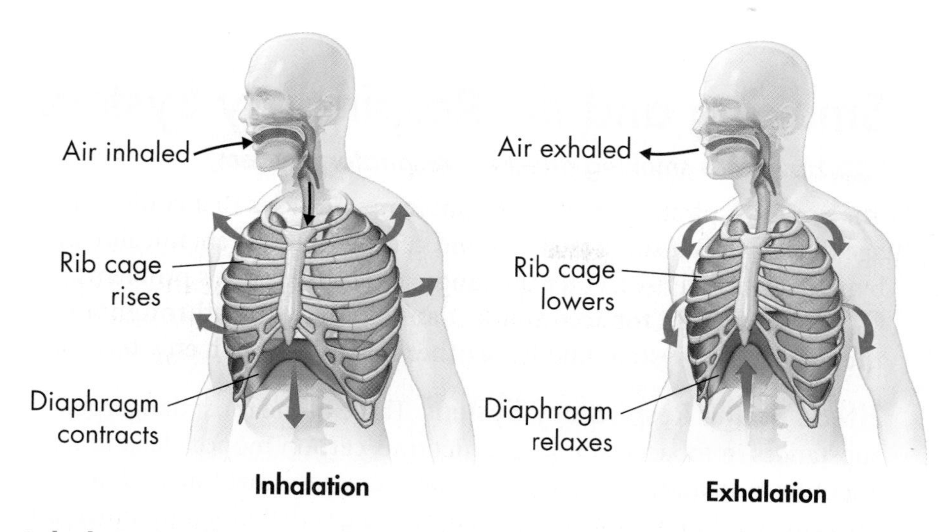

FIGURE 33–17 Breathing During inhalation, the rib cage rises and the diaphragm contracts, increasing the size of the chest cavity. During exhalation, the rib cage lowers and the diaphragm relaxes, decreasing the size of the chest cavity. Humans have some conscious control over breathing—when they swim or play an instrument, for example.

Inhalation The lungs are sealed in two sacs, called pleural membranes, inside the chest cavity. At the bottom of the chest cavity is a large dome-shaped muscle known as the **diaphragm.**

As **Figure 33–17** shows, when you inhale, the diaphragm contracts and flattens. Muscles between the ribs also contract, raising the rib cage. These actions increase the volume of the chest cavity. Because the chest cavity is tightly sealed, this creates a partial vacuum inside the cavity. Atmospheric pressure does the rest, filling the lungs as air rushes into the breathing passages.

Exhalation During ordinary breathing, exhalation is usually passive. Both the rib cage and the diaphragm relax. This relaxation decreases the volume of the chest cavity and makes air pressure in the chest cavity greater than atmospheric pressure. Air rushes back out of the lungs. To blow out a candle, speak, sing, or yell, however, you need more force than passive exhalation provides. The extra force is provided by muscles between the ribs and abdominal muscles, which contract vigorously as the diaphragm relaxes.

The system works only because the chest cavity is sealed. If a wound punctures the chest—even if it does not affect the lungs directly—air may leak into the chest cavity and make breathing impossible. This is one reason chest wounds are always serious.

Breathing and Homeostasis You can control your breathing almost any time you want, to blow up a balloon or to play a trumpet. But this doesn't mean that breathing is purely voluntary. Your nervous system has final control of your breathing muscles whether you are conscious or not. This is why people who drown have water in their lungs. When they lose consciousness, they "breathe" water into their lungs.

Breathing is initiated by the breathing center in the part of the brain stem called the medulla oblongata. Sensory neurons in or near the medulla and in some large blood vessels gather information about carbon dioxide levels in the body and send the information to the breathing center. When stimulated, the breathing center sends nerve impulses that cause the diaphragm and chest muscles to contract, bringing air into the lungs. The higher the blood carbon dioxide level, the stronger the impulses. If the blood carbon dioxide level reaches a critical point, the impulses become so powerful that you cannot keep from breathing.

Smoking and the Respiratory System

How does smoking affect the respiratory system?

The upper respiratory tract filters out many particles that could damage the lungs. But some particles and certain kinds of chemicals can bypass those defenses, enter the lungs, and cause serious problems. **Chemicals in tobacco smoke damage structures throughout the respiratory system and have other negative health effects, too.**

Effects on the Respiratory System Three of the most dangerous substances in tobacco smoke are nicotine, carbon monoxide, and tar. Nicotine is an addictive stimulant that increases heart rate and blood pressure. Carbon monoxide is a poisonous gas that blocks hemoglobin from binding with oxygen, thus interfering with oxygen transport in blood. Tar contains at least 60 compounds known to cause cancer.

Tobacco smoke also paralyzes cilia in the trachea. With the cilia out of action, inhaled particles stick to the walls of the respiratory tract or enter the lungs, and smoke-laden mucus is trapped along the airways. Irritation from accumulated particles and mucus triggers a cough—called a smoker's cough—to clear the airways. Smoking also causes the lining of the respiratory tract to swell, which reduces airflow to the alveoli.

FIGURE 33–18 Effect of Smoking on Lungs Chemicals in cigarette smoke damage cilia in the lungs. Over time, particles build up and lead to respiratory diseases such as chronic bronchitis, emphysema, and lung cancer. The damage that smoking can cause to lungs is visible in the bottom photograph.

Healthy Lung

Smoker's Lung

Diseases Caused by Smoking Damage to the respiratory system from smoking can become permanent and lead to diseases such as chronic bronchitis, emphysema, and lung cancer. Only 30 percent of male smokers live to age 80, but 55 percent of male nonsmokers live to that age. Clearly, smoking reduces life expectancy. The effect of smoking on the lungs can be seen in **Figure 33–18.**

▶ ***Chronic Bronchitis*** In chronic bronchitis, the bronchi become inflamed and clogged with mucus. Smoking even a moderate number of cigarettes on a regular basis can produce chronic bronchitis. Affected people often find simple activities, like climbing stairs, difficult. Treatments can control symptoms, but there is no cure.

▶ ***Emphysema*** Long-term smoking can lead to emphysema (em fuh SEE muh). Emphysema is the loss of elasticity and eventual breakdown of lung tissue. This condition makes breathing difficult. People with emphysema cannot get enough oxygen to the body tissues or rid the body of excess carbon dioxide. There is no cure for emphysema, but it can be treated with medication.

▶ ***Lung Cancer*** Lung cancer is particularly deadly because, by the time it is detected, it usually has spread to other areas of the body. Few people diagnosed with lung cancer live more than five years. About 87 percent of lung cancer deaths are due to smoking.

What Secondhand Smoke Does
Exposes people to cancer-causing chemicals such as formaldehyde, arsenic, and ammonia
Aggravates asthma
Increases incidence of ear infections
Causes sticky platelets and damaged blood vessels
Causes up to 70,000 deaths from heart disease each year

FIGURE 33–19 Secondhand Smoke Effects Smokers not only put their own health at risk, but also the health of their family and friends exposed to their smoke.

Other Effects of Smoking Smoking also has very negative effects on the circulatory system. For example, it raises blood pressure by constricting blood vessels, which forces the heart to work harder to deliver enough oxygen.

Nonsmokers exposed to high levels of secondhand smoke are also at greater risk for respiratory and circulatory system disease. Inhaling the smoke of others is particularly dangerous for young children because their lungs are still developing. Studies now indicate that children of smokers are twice as likely as children of nonsmokers to develop asthma or other respiratory problems. Pregnant women who smoke place their babies at risk for many complications, some of which can lead to lifelong problems.

Whatever the age of a smoker, and no matter how long that person has smoked, his or her health can be improved by quitting. Nicotine is a powerful drug with strong addictive qualities that make it very difficult to quit smoking. Considering the medical dangers and the powerful addiction, the best solution is not to start smoking.

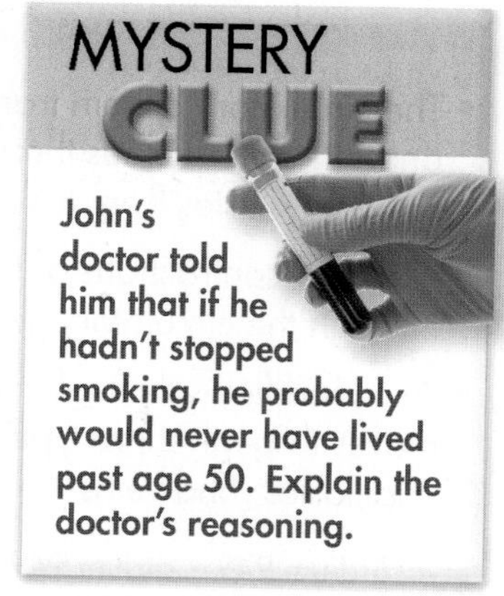

John's doctor told him that if he hadn't stopped smoking, he probably would never have lived past age 50. Explain the doctor's reasoning.

33.3 Assessment

Review Key Concepts

1. a. Review Explain the function of the respiratory system.

b. Use Analogies Explain how a molecule of oxygen flowing through the respiratory system is like a commuter driving home from work.

2. a. Review Describe the process of gas exchange in the lungs.

b. Relate Cause and Effect Carbon monoxide, a poisonous gas, binds to hemoglobin more easily than oxygen does. Based on this information, why do you think that carbon monoxide alarms in homes have saved many lives?

3. a. Review Explain the process of breathing.

b. Infer The brain's breathing center responds to the level of carbon dioxide in the blood, not the level of oxygen. What consequences could this have for people at high altitudes, where oxygen levels are low?

4. a. Review Describe the effects of smoking on the respiratory system.

b. Apply Concepts People with emphysema cannot exhale as much carbon dioxide as people with healthy lungs can. Why do you think this leaves them short of breath?

Apply the Big idea

Structure and Function

5. Compare and contrast human respiration with what you learned about respiration in birds and fish in Chapter 27.

NGSS Smart Guide Circulatory and Respiratory Systems

Disciplinary Core Idea LS1.A Structure and Function: How do the structures of organisms enable life's functions? The functions of the circulatory and respiratory systems are closely connected. Without the circulatory system, oxygen could not be transported from the lungs to the brain and the rest of the body. Without the respiratory system, the powerful cardiac muscles would not receive the oxygen they need to drive the circulatory system.

33.1 The Circulatory System

- The circulatory system transports oxygen, nutrients, and other substances throughout the body, and removes wastes from tissues.
- Powerful contractions of the myocardium pump blood through the circulatory system.
- As blood flows through the circulatory system, it moves through three types of blood vessels—arteries, capillaries, and veins.

myocardium 949 • atrium 949 • ventricle 949 • valve 950 • pulmonary circulation 950 • systemic circulation 950 • pacemaker 951 • artery 952 • capillary 952 • vein 952

Biology.com

Untamed Science Video Bundle up as the Untamed Science crew journeys to cold climates to show us how some animals handle extreme environments.

Visual Analogy Compare the structure and function of the circulatory system to a system of highways and secondary roads.

Art in Motion View a short animation that shows the beating of the heart as well as the transmission of impulses from the SA and AV nodes.

Data Analysis Use electrocardiography to diagnose various heart conditions.

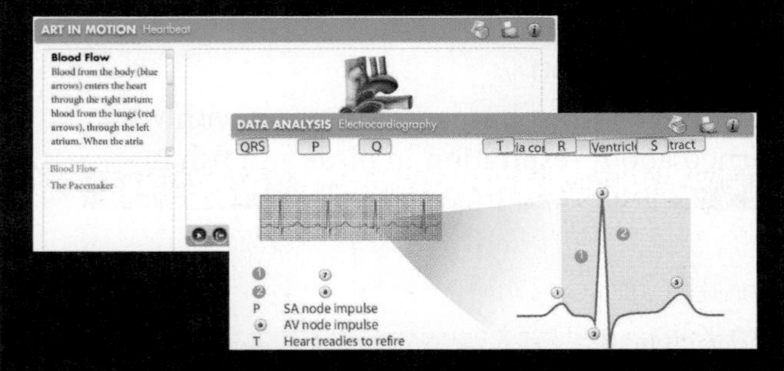

33.2 Blood and the Lymphatic System

- Plasma is about 90 percent water and 10 percent dissolved gases, salts, nutrients, enzymes, hormones, waste products, plasma proteins, cholesterol, and other important compounds.
- The main function of red blood cells is to transport oxygen.
- White blood cells guard against infection, fight parasites, and attack bacteria.
- Blood clotting is made possible by plasma proteins and cell fragments called platelets.
- The lymphatic system is a network of vessels, nodes, and organs that collects the lymph that leaves capillaries, "screens" it for microorganisms, and returns it to the circulatory system.
- Three serious diseases of the circulatory system are heart disease, stroke, and high blood pressure.
- Research indicates that high cholesterol levels, along with other risk factors, lead to atherosclerosis and higher risk of heart attack.

plasma 954 • red blood cell 954 • hemoglobin 954 • white blood cell 955 • platelet 955 • lymph 956 • atherosclerosis 958

Biology.com

In Your Notebook Use a flowchart to review the blood-clotting process.

CHECKING YOUR *Scientific Literacy*

Refer to the lesson content and digital assets below as you prepare for your chapter assessment. Then, evaluate your understanding of the circulatory and respiratory systems by answering these questions.

1. **Scientific and Engineering Practice Planning and Carrying Out Investigations** Plan an investigation to compare heart rates among the members of your class, at rest and during light and moderate exercise. Describe the procedure, the data you would collect, and how you would organize your analysis.

2. Crosscutting Concept **Cause and Effect** What causes a runner's breathing rate to remain rapid for a period of time after he or she has stopped running?

3. Integration of Knowledge and Ideas Use information in the text to construct a concept map that identifies the components of blood and the details about each of their functions.

4. STEM Some scientists think the increasing rate of asthma in the United States is related, in part, to air pollution. List a set of questions you would ask about the problem. Pick one of your questions and describe the kind of quantitative information you would gather to help you better understand the problem.

33.3 The Respiratory System

- The human respiratory system picks up oxygen from the air we inhale and releases carbon dioxide into the air we exhale.
- Oxygen and carbon dioxide are exchanged across the walls of alveoli and capillaries. Chemical properties of blood and red blood cells allow for efficient transport of gases throughout the body.
- Movements of the diaphragm and rib cage change air pressure in the chest cavity during inhalation and exhalation.
- Chemicals in tobacco smoke damage structures throughout the respiratory system and have other negative health effects, too.

pharynx 964 • trachea 964 • larynx 964 • bronchus 964 • alveolus 964 • diaphragm 967

Biology.com

Art Review Review your understanding of the different parts of the respiratory system.

Interactive Art Watch an animation that shows the process of breathing and how the vocal cords produce sound.

ART REVIEW The Respiratory System

INTERACTIVE ART Breathing

Speaking

Two vocal cords stretch across the larynx. When you speak, muscles make the vocal cords contract, narrowing the slit-like opening between them. Air from the lungs rushes through this opening, causing the vocal cords to vibrate. This vibration creates the voice. Use the slider to adjust the vocal cords.

Design Your Own Lab

Tidal Volume and Lung Capacity This chapter lab is available in your lab manual and at Biology.com

33 Assessment

33.1 The Circulatory System

Understand Key Concepts

1. The circulatory system includes the
   a. lungs, heart, and brain.
   b. lungs, blood vessels, and heart.
   c. heart, blood, and blood vessels.
   d. heart, arteries, and veins.

2. The upper chambers of the heart are the
   a. ventricles.
   b. septa.
   c. myocardia.
   d. atria.

3. Blood leaving the heart for the body passes through a large blood vessel called the
   a. aorta.
   b. vena cava.
   c. pulmonary vein.
   d. pulmonary artery.

4. Compare pulmonary circulation and systemic circulation.

5. **Key Ideas and Details** Trace the flow of blood through the heart starting with the right atrium.

6. What is the function of valves in the heart? In what other structures of the circulatory system are valves found?

7. Describe the function of the pacemaker.

8. Describe how the heart beats.

9. Compare the size and structure of arteries, capillaries, and veins.

10. Distinguish between systolic pressure and diastolic pressure.

Think Critically

11. **Text Types and Purposes** Write a scientific procedure for an experiment that determines the amount of time needed for a person's heart rate to return to an at-rest rate after exercise.

12. **Draw Conclusions** Some large veins have one-way valves, which keep blood flowing in one direction. Why don't arteries need similar valves?

33.2 Blood and the Lymphatic System

Understand Key Concepts

13. Cells that protect the body by engulfing foreign cells or producing antibodies are
    a. red blood cells.
    b. cilia.
    c. platelets.
    d. white blood cells.

14. Nutrients and wastes are exchanged with body cells through the walls of
    a. veins.
    b. capillaries.
    c. arteries.
    d. atria.

15. The protein found in red blood cells that transports oxygen is called
    a. hemoglobin.
    b. fibrinogen.
    c. prothrombin.
    d. thrombin.

16. The process shown below is made possible by plasma proteins and cell fragments called
    a. fibrins.
    b. thrombins.
    c. platelets.
    d. lymphocytes.

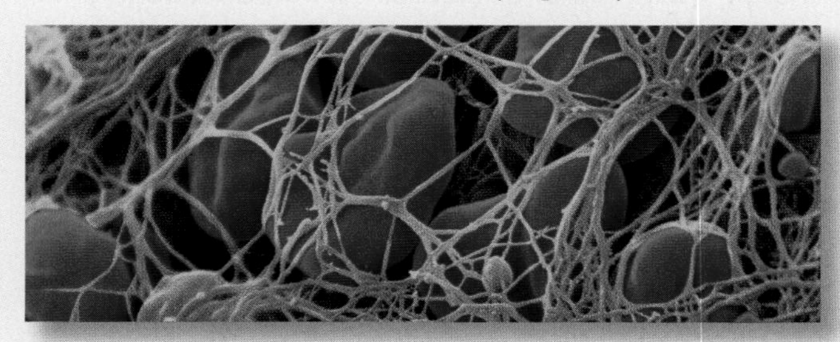

17. Describe the functions of each major component in blood.

18. What are the primary functions of the lymphatic system?

19. **Craft and Structure** Why is LDL known as "bad" cholesterol? Why is HDL known as "good" cholesterol?

Think Critically

20. **Apply Concepts** Why would a person with a low red blood cell count feel tired?

21. **Infer** Aspirin reduces the clot-forming ability of the blood. Why would a doctor prescribe aspirin for someone who has had a stroke?

22. **Predict** Explain how the removal of someone's lymph nodes can affect his or her ability to fight disease.

33.3 The Respiratory System

Understand Key Concepts

23. The tiny hollow air sacs in the lungs where gas exchange takes place are the

a. alveoli. **c.** capillaries.
b. lymph nodes. **d.** bronchioles.

24. Two highly elastic folds of tissue known as the vocal cords are found in the

a. larynx. **c.** trachea.
b. pharynx. **d.** bronchi.

25. The large flat muscle that moves up and down and alters the volume of the chest cavity is the

a. trachea. **c.** diaphragm.
b. epiglottis. **d.** larynx.

26. What part of the brain controls involuntary breathing?

27. What are three dangerous substances in tobacco smoke? Describe how each affects the body.

28. How does emphysema affect the respiratory system?

Think Critically

29. Infer Tobacco smoke can kill white blood cells in the respiratory tract, the cells that help keep the respiratory system clean by consuming debris. How do you think this contributes to the development of smoker's cough?

30. Craft and Structure The table shows the relative blood flow through some organs in the human body—that is, the percentage of blood that flows through a given organ. Through which organ(s) does all of the blood flow? Relate the effect of exercise on blood flow to skeletal muscles.

Blood Flow Through Human Organs

Organ	Percentage of Total Flow
Brain	14%
Heart	5%
Kidneys	22%
Liver	13%
Lungs	100%
Skeletal muscles	18%
Skeletal muscles during exercise	75%

solve the CHAPTER MYSTERY

IN THE BLOOD

Both John and Lila have a genetic disease called familial hypercholesterolemia, which is caused by a gene defect on chromosome 19. John is heterozygous for the disorder. Although his liver cells make a mixture of normal and defective LDL receptors, his blood cholesterol levels were so high that he had serious atherosclerosis by age 35. Most people with this disease have had a heart attack by age 60.

Lila is homozygous for the defective allele—a very rare condition. Her liver cells do not produce any functional LDL receptors. Her atherosclerosis became apparent when she was only 4 years old. Fatty deposits can be seen in the corneas of her eyes and beneath the skin near her elbows and knees.

Research on this genetic defect helped uncover the role of liver cell LDL receptors in regulating blood cholesterol. Researchers then applied that information to cases of high cholesterol among the general public. The result was the development of several new classes of drugs that are helping some people live longer.

1. Predict If Lila were to have a child, what is the likelihood that she would pass on the allele for familial hypercholesterolemia to her child? If John had a son, what is the likelihood that the son inherited the allele?

2. Infer Most heterozygous patients can keep their LDL levels under control with medication that prevents their liver from making cholesterol. But these medications generally do not lower the LDL levels of homozygous patients. Why do you think that is so?

3. Connect to the Big idea If an individual knows that hypercholesterolemia runs in his or her family, what steps can he or she take to live a long and healthy life?

31. **Use Models** Construct a simple stethoscope out of rubber tubing and a metal funnel. Listen for the sounds of air rushing into and out of your lungs and record a description. How does the sound change when you cough?

Connecting Concepts

Use Science Graphics

The following graph is based on pulse rates taken each minute for two students doing the same exercises. The exercises begin at minute 1 and end at minute 8. Use the graph to answer questions 32–34.

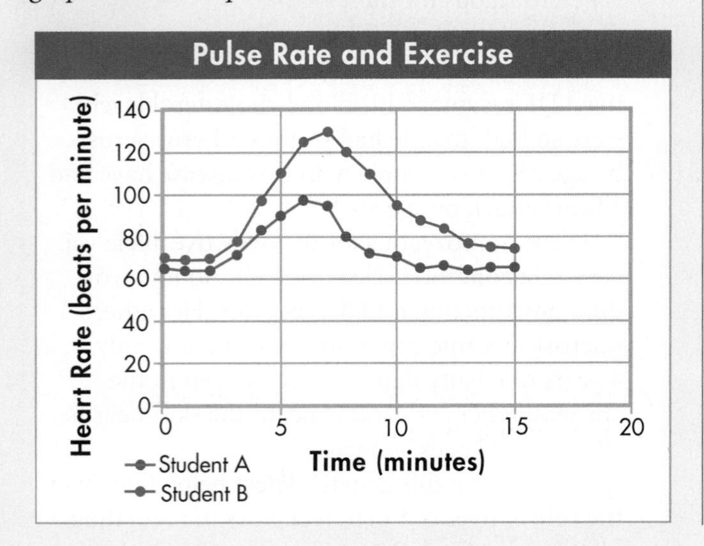

32. **Interpret Graphs** At about which minute did each student reach his or her highest heart rate?

33. **Key Ideas and Details** Which of the two students is most likely in better physical condition? Cite evidence from the graph that supports your answer.

34. **Predict** What other changes in the circulatory and respiratory systems would you expect to take place in the time interval shown?

Write About Science

35. **Production and Distribution of Writing** Make a list of the things you do that affect your circulatory and respiratory systems. Then place a check mark next to those things that are helpful and a circle next to those that are harmful. Use your list to develop, organize, and write a brief essay explaining how you could improve helpful habits and change or break harmful ones.

36. **Assess the Big idea** Describe the relationship between the human circulatory system and the respiratory system. How does the proper functioning of those systems affect other body systems?

Integration of Knowledge and Ideas

High blood pressure is a major risk factor for heart disease in the United States. By age 44, about 25 percent of Americans have high blood pressure, and many of them do not know it. Use the graph to answer questions 37 and 38.

37. **Interpret Graphs** In what age group do women start to have a higher incidence of high blood pressure than men?

38. **Calculate** Between which age groups do you find the largest percentage increase in cases of high blood pressure? MATH
    - a. women between 20–34 and 35–44 years of age
    - b. men between 20–34 and 35–44 years of age
    - c. women between 55–64 and 65–74 years of age
    - d. men between 45–54 and 55–64 years of age

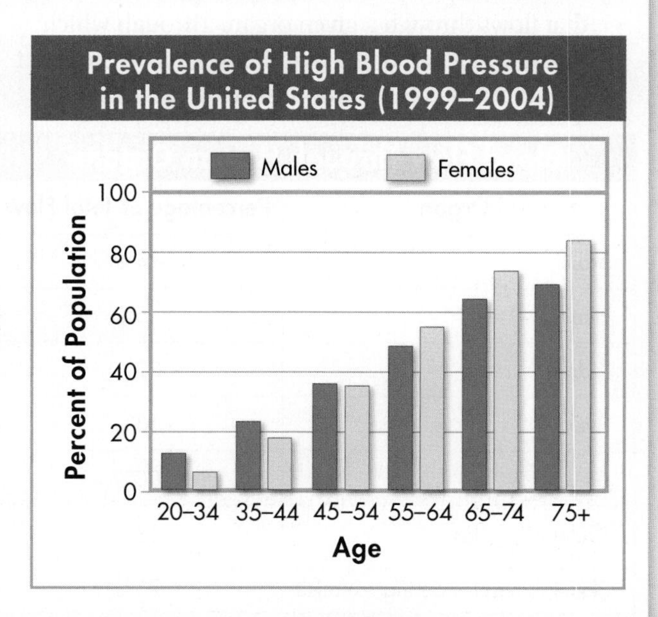

Standardized Test Prep

Multiple Choice

1. In the human heart, oxygen-rich blood would be found in the
A right atrium and the right ventricle.
B right atrium and the left atrium.
C left atrium and the left ventricle.
D right ventricle and the left ventricle.

2. Which statement BEST describes an interaction between the circulatory system and the respiratory system that helps maintain homeostasis?
A Blood plasma transports salts, nutrients, and proteins through the body to keep it healthy.
B The diaphragm and rib cage work together to move air into and out of the lungs.
C Lymph nodes filter out bacteria that could cause disease.
D Blood cells pick up and carry oxygen from the lungs to the body's cells.

3. A heartbeat begins with an impulse from the
A nervous system.
B sinoatrial node.
C atrioventricular node.
D aorta.

4. All of the following are components of human blood EXCEPT
A plasma. **C** phagocytes.
B mucus. **D** platelets.

5. Nicotine in tobacco
A is not addictive.
B lowers blood pressure.
C blocks the transport of oxygen.
D increases heart rate.

6. Antibodies are produced by
A red blood cells.
B platelets.
C B lymphocytes.
D hormones.

Questions 7–10

Use the diagram below to answer the questions that follow.

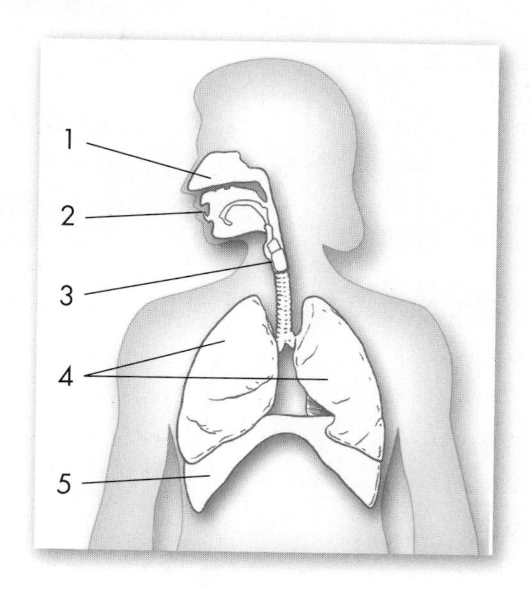

7. Which structure's primary function is to warm and moisten inhaled air?
A 1 **C** 4
B 3 **D** 5

8. Which structure contains the vocal cords?
A 1 **C** 3
B 2 **D** 4

9. Damage to which structure can lead to emphysema?
A 2 **C** 4
B 3 **D** 5

10. Which structure contains alveoli?
A 2 **C** 4
B 3 **D** 5

Open-Ended Response

11. Explain why the risk factors for heart disease and strokes are similar.

If You Have Trouble With . . .

Question	1	2	3	4	5	6	7	8	9	10	11
See Lesson	33.1	33.3	33.1	33.2	33.3	33.2	33.3	33.3	33.3	33.3	33.2

34 Endocrine and Reproductive Systems

Big idea **Homeostasis**

Q: How does the body use chemical signals to maintain homeostasis?

INSIDE:

- 34.1 The Endocrine System
- 34.2 Glands of the Endocrine System
- 34.3 The Reproductive System
- 34.4 Fertilization and Development

This worker's endocrine system is partly responsible for the sweaty palms and racing heart he likely experienced his first day on the job.

BUILDING *Scientific Literacy*

Deepen your understanding of the structures and functions of the endocrine and reproductive systems by engaging in key practices that allow you to make connections across concepts.

NGSS You will communicate your understanding of hormonal interactions by reading and explaining key ideas in the text.

STEM Explore the training, technology, and opportunities that relate to working in the field of biomedical engineering.

Common Core You will analyze the authors' purpose in choosing health factors related to athletic training as the topic for this Chapter Mystery.

CHAPTER MYSTERY

OUT OF STRIDE

Lisa trained hard during spring track and over the summer. But as the new school year approached, she wasn't satisfied. For her cross-country team to win the state championship, she felt that she needed to be faster. A teammate suggested she lose a few pounds. Lisa had already lost weight over the summer, but she decided to lose some more.

In addition to her strenuous workouts, Lisa stopped snacking before practice and avoided high-calorie foods. She did lose weight. But she was always tired. She also noticed that she had not had a menstrual period in four months. The week before the championship meet, she collapsed in pain at practice. She had suffered a stress fracture to her lower leg. Her season was over.

Lisa's doctor told her that all of her symptoms were related. As you read this chapter, look for clues to explain why excessive exercise and dieting had these effects on Lisa. Then, solve the mystery.

Biology.com

Finding the solution to the Out of Stride mystery is just the beginning. Take a video field trip with the ecogeeks of Untamed Science to continue exploring your world.

Go online to access additional resources including:
• eText • Flash Cards • Lesson Overviews • Chapter Mystery

34.1 The Endocrine System

Key Questions

What are the components of the endocrine system?

How do hormones affect cells?

Vocabulary

hormone
target cell
exocrine gland
endocrine gland
prostaglandin

Taking Notes

Compare/Contrast Table As you read, make a table that compares and contrasts the two different types of hormones.

THINK ABOUT IT If you had to get a message to just one or two friends, what would you do? One solution would be to make a telephone call that would carry your message directly to those friends over telephone wires. But what if you wanted to send a message to thousands of people? You could broadcast your message on the radio so that everyone tuned to a particular station could hear it. Just like you, cells send messages, too. They can make a direct call or send out a broadcast.

Hormones and Glands

What are the components of the endocrine system?

Your nervous system works much like a telephone. Many impulses move swiftly over a system of wire-like neurons that carry messages directly from one cell to another. But another system, the endocrine system, is more like a radio, "broadcasting" chemical messages. These chemical messengers, called **hormones,** are released in one part of the body, travel through the blood, and affect cells in other parts of the body. **The endocrine system is made up of glands that release hormones into the blood. Hormones deliver messages throughout the body.** In the same way that a radio broadcast can reach thousands or even millions of people in a large city, hormones can affect almost every cell in the body.

Hormones Hormones act by binding to specific chemical receptors on cell membranes or within cells. Cells that have receptors for a particular hormone are called **target cells.** If a cell does not have receptors for a particular hormone, the hormone has no effect on it.

In general, the body's responses to hormones are slower and longer lasting than its responses to nerve impulses. It may take several minutes, several hours, or even several days for a hormone to have its full effect on its target cells. A nerve impulse, on the other hand, may take only a fraction of a second to reach and affect its target cells.

Many endocrine functions depend on the effects of two opposing hormones. For example, the hormone insulin prompts the liver to convert blood glucose to glycogen and store it. The hormone glucagon prompts the liver to convert glycogen to glucose and release it in the blood. The opposing effects of insulin and glucagon maintain homeostasis by keeping blood glucose levels within a narrow range.

Glands A gland is an organ that produces and releases a substance, or secretion. **Exocrine glands** release their secretions through tube-like structures (called ducts) either out of the body or directly into the digestive system. Exocrine glands include those that release sweat, tears, and digestive enzymes. **Endocrine glands** usually release their secretions (hormones) directly into the blood, which transports the secretions throughout the body. **Figure 34–1** shows the location of the major endocrine glands. Although not usually considered as endocrine glands, other body structures such as bones, fat tissue, the heart, and the small intestine also produce and release hormones.

In Your Notebook *Make a three-column table. Label the columns Gland, Hormone(s), and Function. Fill in the table as you read.*

MYSTERY CLUE

Fat tissue may send signals to the hypothalamus when fat reserves are low. Lisa's body fat percentage dropped from 17 percent to 9 percent. Could this have affected such signals?

FIGURE 34–1 Major Endocrine Glands Endocrine glands produce hormones that affect many parts of the body. **Interpret Graphics** ***What is the function of the pituitary gland?***

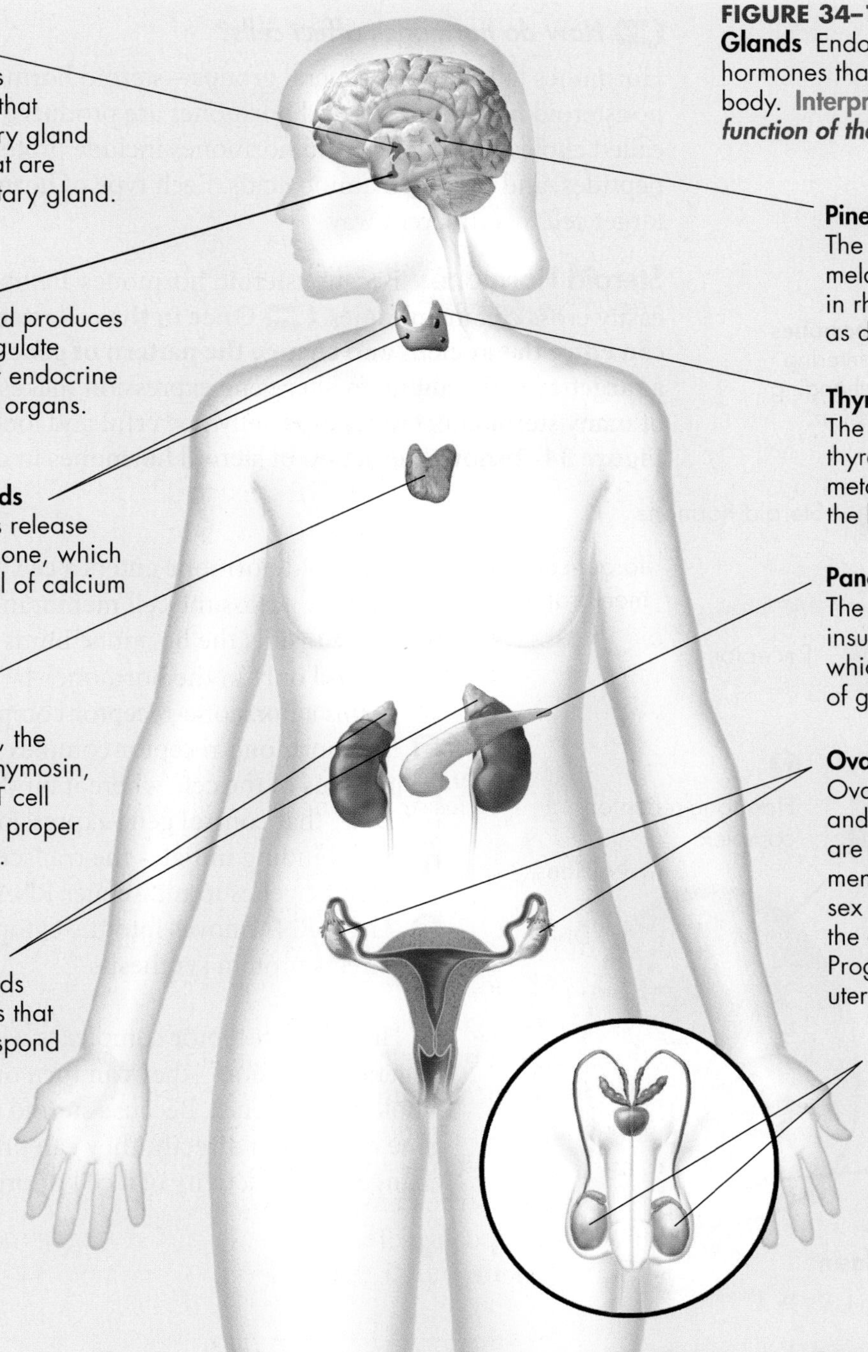

BUILD Vocabulary

WORD ORIGINS Prostaglandins get their name from a gland in the male reproductive system, the prostate, in which they were first discovered.

Prostaglandins The glands of the endocrine system were once thought to be the only organs that produced hormones. However, nearly all cells have been shown to produce small amounts of hormonelike substances called **prostaglandins** (prahs tuh GLAN dinz). Prostaglandins are modified fatty acids that are produced by a wide range of cells. They generally affect only nearby cells and tissues, and thus are sometimes known as "local hormones."

Some prostaglandins cause smooth muscles, such as those in the uterus, bronchioles, and blood vessels, to contract. One group of prostaglandins causes the sensation of pain during most headaches. Aspirin helps to stop the pain of a headache because it inhibits the synthesis of these prostaglandins.

Hormone Action

How do hormones affect cells?

Hormones fall into two general groups—steroid hormones and nonsteroid hormones. Steroid hormones are produced from a lipid called cholesterol. Nonsteroid hormones include proteins, small peptides, and modified amino acids. Each type of hormone acts on a target cell in a different way.

Steroid Hormones Because steroid hormones are lipids, they can easily cross cell membranes. **Once in the cell, steroid hormones can enter the nucleus and change the pattern of gene expression in a target cell.** The ability to alter gene expression makes the effects of many steroid hormones especially powerful and long lasting. **Figure 34–2** shows the action of steroid hormones in cells.

FIGURE 34–2 Steroid Hormones Steroid hormones act by entering the nucleus of a cell and changing the pattern of gene expression.

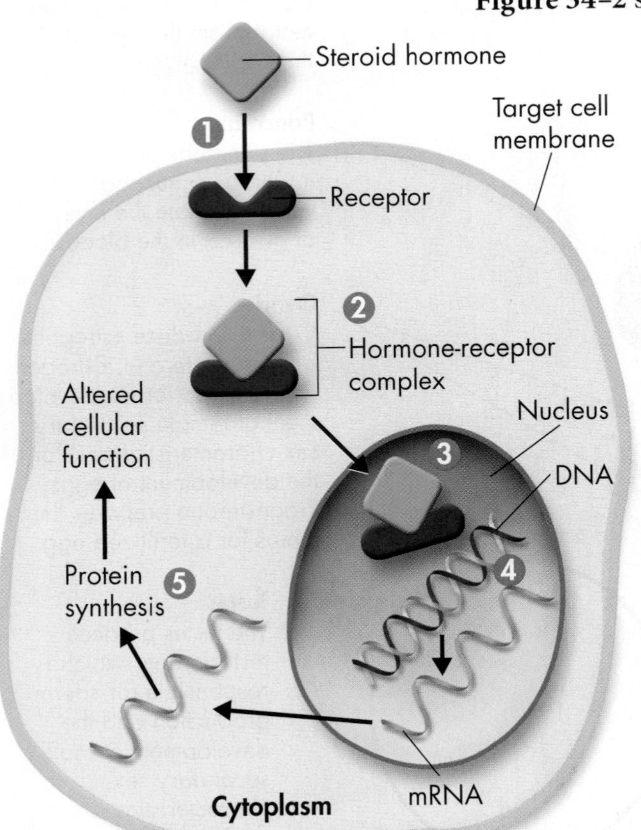

1. A steroid hormone enters a cell by passing directly across the cell membrane.
2. Once inside, the hormone binds to a receptor (found only in the hormone's target cells) and forms a hormone-receptor complex.
3. The hormone-receptor complex enters the nucleus of the cell, where it binds to regions of DNA that control gene expression.
4. This binding initiates the transcription of specific genes to messenger RNA (mRNA).
5. The mRNA moves into the cytoplasm and directs protein synthesis.

Hormone-receptor complexes work as regulators of gene expression—they can turn on or turn off whole sets of genes. Because steroid hormones affect gene expression directly, they can produce dramatic changes in the activity of a cell or organism.

Nonsteroid Hormones Nonsteroid hormones generally cannot pass through the cell membrane of their target cells. **Nonsteroid hormones bind to receptors on cell membranes and cause the release of secondary messengers that affect cell activities. Figure 34–3** shows the action of nonsteroid hormones in cells.

1. A nonsteroid hormone binds to receptors on the cell membrane.
2. The binding of the hormone activates enzymes on the inner surface of the cell membrane.
3. These enzymes release secondary messengers such as calcium ions, nucleotides, and even fatty acids to relay the hormone's message within the cell. One common secondary messenger is cAMP (cyclic AMP), which is produced from ATP.
4. These secondary messengers can activate or inhibit a wide range of cell activities.

Steroid and nonsteroid hormones can have powerful effects on their target cells. It is therefore especially important to understand the ways in which the endocrine system regulates their production and release into the blood.

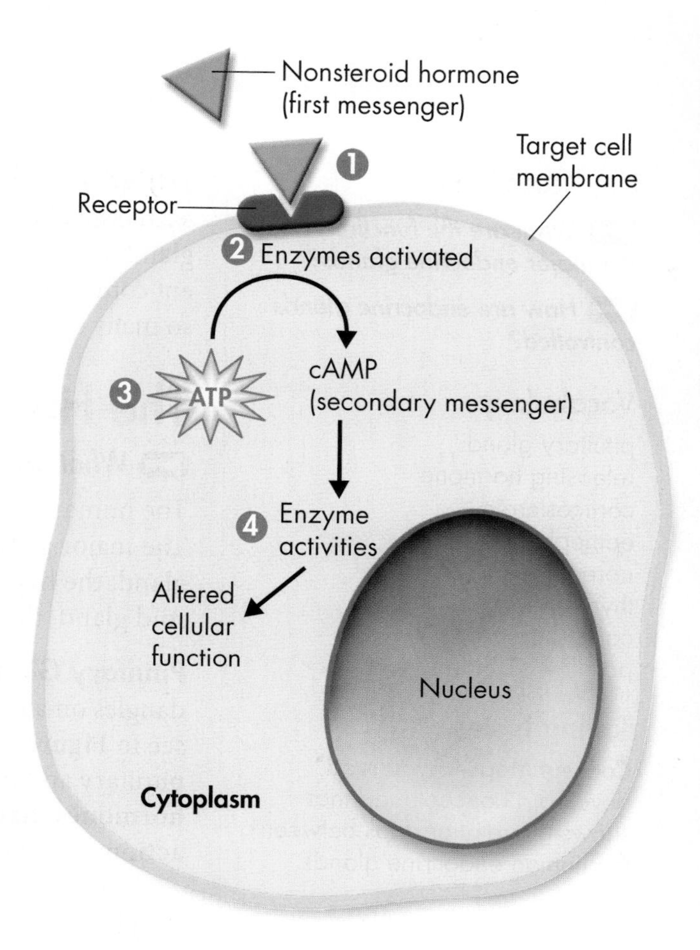

FIGURE 34–3 Nonsteroid Hormones
Nonsteroid hormones bind to receptors on a target cell membrane and cause the release of secondary messengers that affect cell activities.

34.1 Assessment

Review Key Concepts

1. **a. Review** What are the two components of the endocrine system?
   b. Explain Explain the difference between endocrine and exocrine glands.
   c. Compare and Contrast How are hormones and prostaglandins similar? How are they different?
2. **a. Review** Explain how steroid hormones act on a cell.
   b. Explain Explain how nonsteroid hormones act on a cell.
   c. Apply Concepts Use what you learned in Chapter 7 about how materials cross cell membranes to explain the actions of steroid hormones and nonsteroid hormones.

Apply the Big idea

Homeostasis

3. What are the advantages of having both a nervous system and an endocrine system?

34.2 Glands of the Endocrine System

Key Questions

🔑 *What are the functions of the major endocrine glands?*

🔑 *How are endocrine glands controlled?*

Vocabulary

pituitary gland
releasing hormone
corticosteroid
epinephrine
norepinephrine
thyroxine
calcitonin
parathyroid hormone

Taking Notes

Concept Map As you read, develop a concept map that shows the relationships between the human endocrine glands.

FIGURE 34–4 Pituitary Gland The pituitary gland is located below the hypothalamus in the brain. Some of the hormones released by the pituitary control other glands, while others affect other types of tissues.

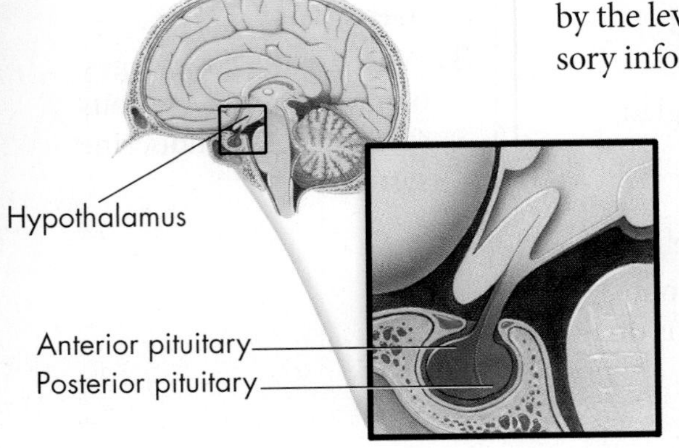

THINK ABOUT IT Organs in most body systems are connected to each other, but that's not the case with the endocrine system. Endocrine glands are scattered throughout the body, many of them with no apparent connection to each other. How does the body control and regulate so many separate organs so that they act together as a single system?

The Human Endocrine Glands

🔑 ***What are the functions of the major endocrine glands?***

The human endocrine system regulates a wide variety of activities. The major glands of the endocrine system include the pituitary gland, the hypothalamus, the adrenal glands, the pancreas, the thyroid gland, the parathyroid glands, and the reproductive glands.

Pituitary Gland The **pituitary gland** is a bean-size structure that dangles on a slender stalk of tissue at the base of the brain. As you can see in **Figure 34–4,** the gland is divided into two parts: the anterior pituitary and the posterior pituitary. 🔑 **The pituitary gland secretes hormones that directly regulate many body functions or control the actions of other endocrine glands.**

Proper function of the pituitary gland is essential. For example, if the gland produces too much growth hormone (GH) during childhood, the body grows too quickly, resulting in a condition called gigantism. Too little GH during childhood causes pituitary dwarfism, which can be treated with GH produced by genetically engineered bacteria.

Hypothalamus The hypothalamus, which is attached to the posterior pituitary, is the link between the central nervous system and the endocrine system. 🔑 **The hypothalamus controls the secretions of the pituitary gland.** The activities of the hypothalamus are influenced by the levels of hormones and other substances in the blood and by sensory information collected by other parts of the central nervous system.

The hypothalamus contains the cell bodies of neurosecretory cells whose axons extend into the posterior pituitary. Antidiuretic hormone, which stimulates the kidney to absorb water, and oxytocin, which stimulates contractions during childbirth, are made in the cell bodies of the hypothalamus and stored in the axons entering the posterior pituitary. When the cell bodies are stimulated, axons in the posterior pituitary release these hormones into the blood.

Anterior Pituitary Gland Hormones

Hormone	Action
Follicle-stimulating hormone (FSH)	Stimulates production of mature eggs in ovaries and sperm in testes
Luteinizing hormone (LH)	Stimulates ovaries and testes; prepares uterus for implantation of fertilized egg
Thyroid-stimulating hormone (TSH)	Stimulates the synthesis and release of thyroxine from the thyroid gland
Adreno-corticotropic hormone (ACTH)	Stimulates release of some hormones from the adrenal cortex
Growth hormone (GH)	Stimulates protein synthesis and growth in cells
Prolactin	Stimulates milk production in nursing mothers
Melanocyte-stimulating hormone (MSH)	Stimulates melanocytes in the skin to increase the production of the pigment melanin

FIGURE 34–5 Anterior Pituitary Hormones The hypothalamus secretes releasing hormones that signal the anterior pituitary to release its hormones.
Classify ***Which of these hormones stimulate other endocrine glands?***

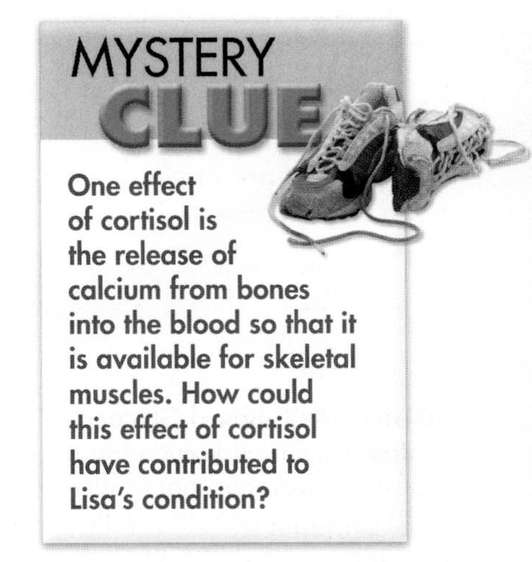

One effect of cortisol is the release of calcium from bones into the blood so that it is available for skeletal muscles. How could this effect of cortisol have contributed to Lisa's condition?

In contrast, the hypothalamus has indirect control of the anterior pituitary. The hypothalamus produces **releasing hormones,** which are secreted into blood vessels leading to the anterior pituitary. The hypothalamus produces a specific releasing hormone that controls the secretion of each anterior pituitary hormone. Hormones released by the anterior pituitary gland are listed in **Figure 34–5.**

Adrenal Glands The adrenal glands are pyramid-shaped structures that sit on top of the kidneys. **The adrenal glands release hormones that help the body prepare for—and deal with—stress.** As shown in **Figure 34–6,** the outer part of the gland is called the adrenal cortex and the inner part is the adrenal medulla.

About 80 percent of an adrenal gland is its adrenal cortex. The adrenal cortex produces more than two dozen steroid hormones called **corticosteroids** (kawr tih koh STEER oydz). One of these hormones, aldosterone (al DAHS tuh rohn), regulates blood volume and pressure. Its release is stimulated by dehydration, excessive bleeding, or Na^+ deficiency. Another hormone, called cortisol, helps control the rate of metabolism of carbohydrates, fats, and proteins. Cortisol is released during physical stress such as intense exercise.

Hormones released from the adrenal medulla produce the heart-pounding, anxious feeling you get when excited or frightened—commonly known as the "fight or flight" response. When you are under this sort of stress, impulses from the sympathetic nervous system stimulate cells in the adrenal medulla to release large amounts of **epinephrine** (commonly referred to as adrenaline) and **norepinephrine.** These hormones increase heart rate and blood pressure. They also cause air passageways to widen, allowing for an increase in oxygen intake, and stimulate the release of extra glucose. If your heart rate speeds up and your hands sweat when you take a test, it's your adrenal medulla at work!

FIGURE 34–6 Adrenal Glands The adrenal glands release hormones that help the body handle stressful situations. The adrenal cortex and adrenal medulla contain different types of tissues and release different hormones.

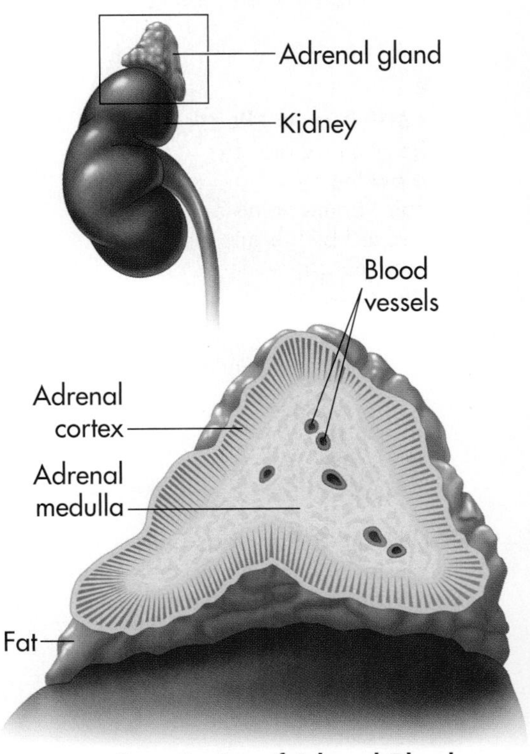

Cross Section of Adrenal Gland

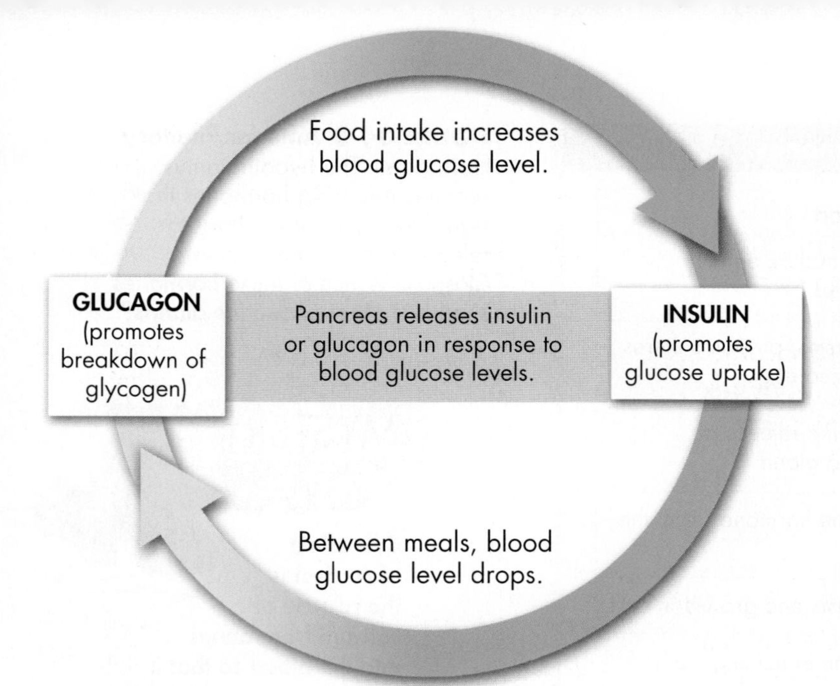

FIGURE 34–7 Blood Glucose Control Insulin and glucagon are opposing hormones that ensure blood glucose levels stay within a normal range. **Infer** ***Explain why this feedback loop does not apply to a person with untreated diabetes.***

Pancreas The pancreas is both an exocrine and an endocrine gland. As an exocrine gland, it releases digestive enzymes that help break down food. However, other cells in the pancreas release hormones into the blood.

The hormone-producing portion of the pancreas consists of clusters of cells. These clusters, which resemble islands, are called the "islets of Langerhans," after their discoverer, German anatomist Paul Langerhans. Each islet contains beta cells, which secrete the hormone insulin, and alpha cells, which secrete the hormone glucagon. **Insulin and glucagon, produced by the pancreas, help to keep the blood glucose level stable.**

▶ ***Blood Glucose Regulation*** When blood glucose levels rise after a person eats, the pancreas releases insulin. Insulin stimulates cells to take glucose out of the blood, which prevents the levels of blood glucose from rising too rapidly and ensures that glucose is stored for future use. Insulin's major target cells are in the liver, skeletal muscles, and fat tissue. The liver and skeletal muscles store glucose as glycogen. In fat tissue, glucose is converted to lipids.

Within one or two hours after a person has eaten, when the level of blood glucose drops, glucagon is released from the pancreas. Glucagon stimulates the liver and skeletal muscle cells to break down glycogen and release glucose into the blood. Glucagon also causes fat cells to break down fats so that they can be converted to glucose. These actions help raise the blood glucose level back to normal. **Figure 34–7** summarizes the insulin and glucagon feedback loop.

▶ ***Diabetes Mellitus*** When the body fails to produce or properly respond to insulin, a condition known as diabetes mellitus occurs. The very high blood glucose levels that result from diabetes can damage almost every system and cell in the body.

There are two types of diabetes mellitus. Type I diabetes is an autoimmune disorder that usually develops in people before the age of 15. The immune system kills beta cells, resulting in little or no secretion of insulin. People with Type I diabetes must follow a strict diet and receive daily doses of insulin to keep their blood glucose level under control.

The second type of diabetes, Type II, most commonly develops in people after the age of 40. People with Type II diabetes produce low to normal amounts of insulin. However, their cells do not properly respond to the hormone because the interaction of insulin receptors and insulin is inefficient. In its early stages, Type II diabetes can often be controlled through diet and exercise. Unfortunately, the incidence of Type II diabetes is rising rapidly in the United States and other countries as a result of increasing obesity, especially among young people.

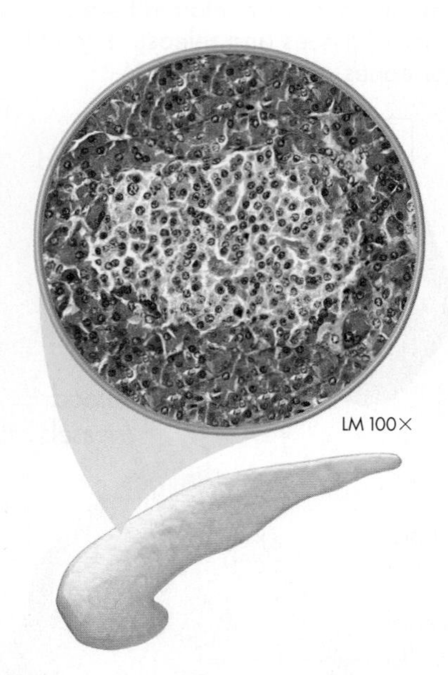

FIGURE 34–8 Pancreas Cells The cluster of light-colored cells is an islet of Langerhans, which contains alpha and beta cells. In Type I diabetes, a person's immune system kills beta cells, which produce insulin.

Thyroid and Parathyroid Glands The thyroid gland is located at the base of the neck and wraps around the upper part of the trachea. **The thyroid gland has a major role in regulating the body's metabolism.** Recall that metabolism is the sum of all the chemical reactions that occur in the body. The thyroid gland produces the hormone **thyroxine,** which increases the metabolic rate of cells throughout the body. Under the influence of thyroxine, cells become more active, use more energy, and produce more heat.

Iodine is needed to produce thyroxine. In parts of the world where diets lack iodine, severe health problems may result. Low levels of thyroxine in iodine-deficient infants produce a condition called cretinism (KREE tuh niz um), in which neither the skeletal system nor the nervous system develops properly. Iodine deficiency usually can be prevented by the addition of small amounts of iodine to table salt or other food items.

Thyroid problems are a fairly common disorder. If the thyroid produces too much thyroxine, a condition called hyperthyroidism occurs. Hyperthyroidism results in nervousness, elevated body temperature, increased blood pressure, and weight loss. Too little thyroxine causes a condition called hypothyroidism. Lower body temperature, lack of energy, and weight gain are signs of this condition. A goiter, as shown in **Figure 34–9,** can be a sign of hypothyroidism.

The thyroid also produces calcitonin, a hormone that reduces blood calcium levels. **Calcitonin** signals the kidneys to reabsorb less calcium from filtrate, inhibits calcium's absorption in the small intestine, and promotes calcium's absorption into bones. Its opposing hormone is parathyroid hormone, which is released by the four parathyroid glands located on the back surface of the thyroid. **Parathyroid hormone** (PTH) increases the calcium levels in the blood by promoting the release of calcium from bone, the reabsorption of calcium in the kidneys, and the uptake of calcium from the digestive system. The actions of PTH promote proper nerve and muscle function and proper bone structure.

In Your Notebook *Summarize how blood-calcium levels are regulated.*

Reproductive Glands The gonads—ovaries and testes—are the body's reproductive glands. **The gonads serve two important functions: the production of gametes and the secretion of sex hormones.** In females, ovaries produce eggs and secrete a group of hormones called estrogens. In males, the testes produce sperm and secrete the hormone testosterone. You'll learn more about the gonads and their hormones in the next lesson.

FIGURE 34–9 Thyroid Gland A goiter is an enlargement of the thyroid gland. A goiter may be the result of iodine deficiency. Without iodine, the thyroid cannot finish producing thyroxine, but its precursor continues to build up in the gland.

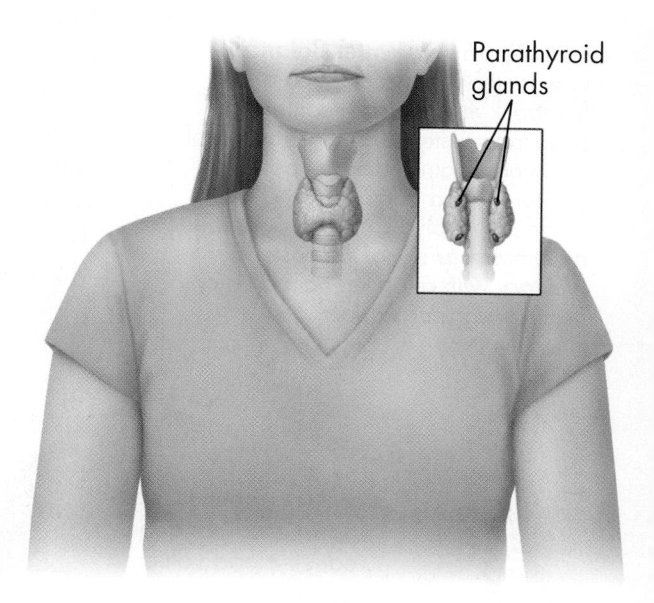

Normal Thyroid

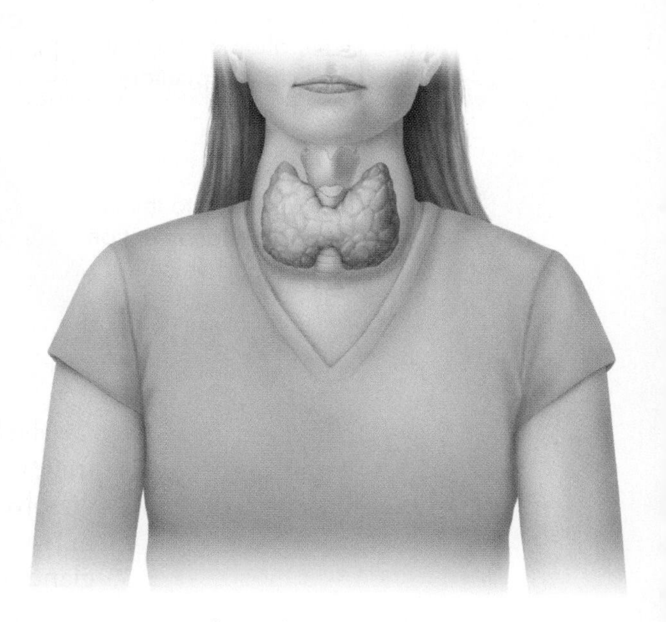

Enlarged Thyroid (Goiter)

Control of the Endocrine System

How are endocrine glands controlled?

Even though the endocrine system is one of the master regulators of the body, it, too, must be controlled. **Like most systems of the body, the endocrine system is regulated by feedback mechanisms that function to maintain homeostasis.**

Recall that feedback inhibition occurs when an increase in any substance "feeds back" to inhibit the process that produced the substance in the first place. Home heating and cooling systems, controlled by thermostats, are examples of mechanical feedback loops. The actions of glands and hormones of the endocrine system are biological examples of the same type of process.

FIGURE 34–10 Water Balance One method by which internal feedback mechanisms regulate the endocrine system is the interaction of the hypothalamus and the posterior pituitary gland in maintaining water balance. Apply Concepts ***Does the hypothalamus signal the posterior pituitary with releasing hormones or nervous signals? Explain.***

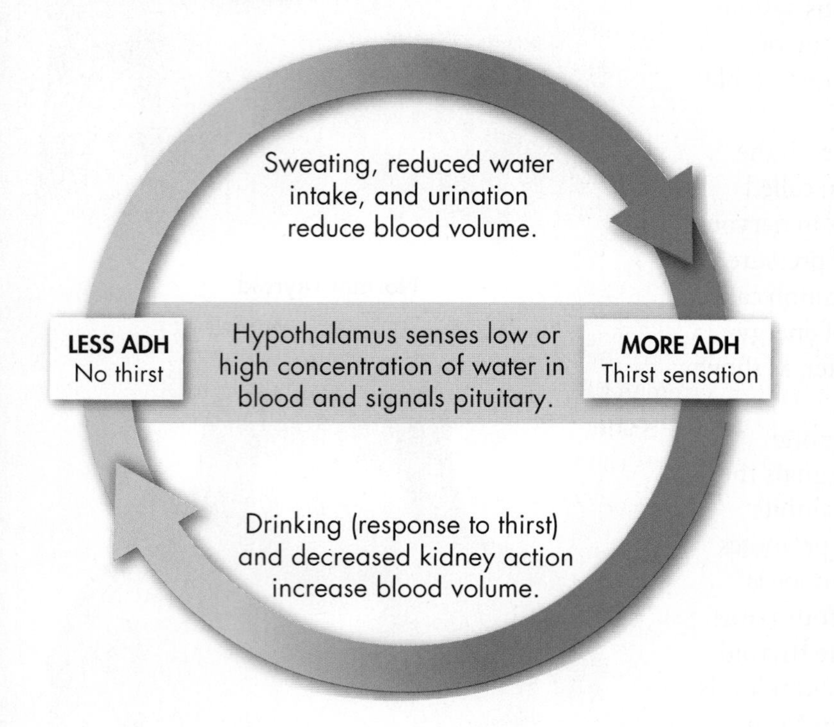

Maintaining Water Balance Homeostatic mechanisms regulate the levels of a wide variety of materials dissolved in the blood and in extracellular fluids. These materials include hydrogen ions; minerals such as sodium, potassium, and calcium; and soluble proteins such as serum albumin, which is found in blood plasma. Most of the time, homeostatic systems operate so smoothly that we are scarcely aware of their existence. However, that is not the case with one of the most important homeostatic processes, the one that regulates the amount of water in the body. **Figure 34–10** illustrates the water balance mechanism.

When you exercise strenuously, you lose water as you sweat. If this water loss continued, your body would soon become dehydrated. Generally, that doesn't happen, because your body's homeostatic mechanisms swing into action.

The hypothalamus contains cells that are sensitive to the concentration of water in the blood. As you lose water, the concentration of dissolved materials in the blood rises. The hypothalamus responds in two ways. First, the hypothalamus signals the posterior pituitary gland to release a hormone called antidiuretic hormone (ADH). ADH molecules are carried by the blood to the kidneys, where the removal of water from the blood is quickly slowed down. Later, you experience a sensation of thirst—a signal that you should drink to restore lost water.

When you finally get around to taking that drink, you might take in a liter of fluid. Most of that water is quickly absorbed into the blood. This volume of water could dilute the blood so much that the equilibrium between the blood and the body cells would be disturbed. Large amounts of water would diffuse across blood vessel walls into body tissues. Body cells would swell with the excess water.

Needless to say, this doesn't happen, because the homeostatic mechanism controlled by the hypothalamus intervenes again. When the water content of the blood rises, the pituitary releases less ADH. In response to lower ADH levels, the kidneys remove water from the blood, restoring the blood to its proper concentration. This homeostatic system sets both upper and lower limits for blood water content. A water deficit stimulates the release of ADH, causing the kidneys to conserve water; an oversupply of water causes the kidneys to eliminate the excess water in urine.

Controlling Metabolism As another example of how internal feedback mechanisms regulate the activity of the endocrine system, let's look at the thyroid gland and its principal hormone, thyroxine. Recall that thyroxine increases the metabolic activity of cells. Does the thyroid gland determine how much thyroxine to release on its own? No, the activity of the thyroid gland is instead controlled by the hypothalamus and the anterior pituitary gland. When the hypothalamus senses that the thyroxine level in the blood is low, it secretes thyrotropin-releasing hormone (TRH), a hormone that stimulates the anterior pituitary to secrete thyroid-stimulating hormone (TSH). TSH stimulates the release of thyroxine by the thyroid gland. High levels of thyroxine in the blood inhibit the secretion of TRH and TSH, which stops the release of additional thyroxine. This feedback loop keeps the level of thyroxine in the blood relatively constant.

The hypothalamus is also sensitive to temperature. When the core body temperature begins to drop, even if the level of thyroxine is normal, the hypothalamus produces extra TRH. The release of TRH stimulates the release of TSH, which stimulates the release of additional thyroxine. Thyroxine increases oxygen consumption and cellular metabolism. The increase in metabolic activity that results helps the body maintain its core temperature even when the outside temperature drops.

BUILD Vocabulary

PREFIXES The prefixes *anti-* and *ante-* can be easily confused. *Anti-*, as in *antidiuretic*, means "against" or "opposite." *Ante-*, as in *anterior*, means "before."

34.2 Assessment

Review Key Concepts

1. **a. Review** Describe the role of each major endocrine gland.

   b. Explain How is the hypothalamus an important part of both the nervous system and the endocrine system?

   c. Compare and Contrast Compare and contrast the two types of diabetes.

2. **a. Review** Explain how the endocrine system helps maintain homeostasis.

   b. Explain On a hot day, you play soccer for an hour and lose a lot of water in sweat. List the steps that your body takes to regain homeostasis.

   c. Predict Suppose the secretion of a certain hormone causes an increase in the concentration of substance *X* in the blood. A low concentration of *X* causes the hormone to be released. What is the effect on the rate of hormone secretion if an abnormal condition causes the level of *X* in the blood to remain very low?

WRITE ABOUT SCIENCE

Creative Writing

3. Create a brochure that describes both types of diabetes. You may wish to include information on risk factors, treatment, and preventive measures that can be taken. Use images from magazines or the Internet to illustrate your brochure.

34.3 The Reproductive System

Key Questions

What effects do estrogens and testosterone have on females and males?

What are the main functions of the male reproductive system?

What are the main functions of the female reproductive system?

What are some of the most commonly reported sexually transmitted diseases?

Vocabulary

puberty • testis • scrotum • seminiferous tubule • epididymis • vas deferens • semen • ovary • menstrual cycle • ovulation • corpus luteum • menstruation • sexually transmitted disease

Taking Notes

Outline Before you read, use the green and blue headings in this lesson to make an outline. As you read, fill in subtopics and phrases to describe the subtopics.

THINK ABOUT IT Among all the systems of the body, the reproductive system is unique. If any other system in the body failed to function, the result would be death. However, an individual can lead a healthful life without reproducing. But is there any other system that is more important for our existence as a species? Without the reproductive system, we could not produce the next generation, and our species would come to an end. So, in a certain sense, this may be the most important system in the body.

Sexual Development

What effects do estrogens and testosterone have on females and males?

At first, male and female human embryos are nearly identical in appearance. Then, during the seventh week of development, the reproductive systems of male and female embryos begin to develop along different lines. The male pattern of development is triggered by the production of testosterone in the gonads of the embryo. In female embryos, testosterone is absent and the female reproductive system develops under the influence of estrogens produced in the embryo's gonads.

Estrogens and testosterone, which have powerful effects on the body, are steroid hormones primarily produced in the gonads. In addition to shaping the sexual development of the embryo, these hormones act on cells and tissues to produce many of the physical characteristics associated with males and females. **In females, the effects of the sex hormones include breast development and a widening of the hips. In males, they result in the growth of facial hair, increased muscular development, and deepening of the voice.**

In childhood, the gonads and the adrenal cortex produce low levels of sex hormones that influence development. However, neither the testes nor the ovaries can produce active reproductive cells until puberty. **Puberty** is a period of rapid growth and sexual maturation during which the reproductive system becomes fully functional. The age at which puberty begins varies considerably among individuals. It usually occurs between the ages of 9 and 15, and, on average, begins about one year earlier in females than in males. Puberty actually begins in the brain, when the hypothalamus signals the pituitary to produce two hormones that affect the gonads—follicle-stimulating hormone (FSH) and luteinizing hormone (LH).

In Your Notebook *Summarize the effects of estrogens on females and testosterone on males.*

The Male Reproductive System

What are the main functions of the male reproductive system?

The release of LH stimulates cells in the testes to produce increased amounts of testosterone. Testosterone causes the male physical changes associated with puberty and, together with FSH, stimulates the development of sperm. **When puberty is complete, the reproductive system is fully functional, meaning that the male can produce and release active sperm.**

Figure 34–11 shows the structures of the male reproductive system. Just before birth (or sometimes just after), the primary male reproductive organs, the **testes** (singular: testis), descend from the abdomen into an external sac called the **scrotum.** The testes remain in the scrotum, outside the body cavity, where the temperature is a few degrees lower than the normal temperature of the body (37°C). The lower temperature is important for proper sperm development.

Sperm Development Within each testis are clusters of hundreds of tiny tubules called **seminiferous** (sem uh NIF ur us) **tubules** where sperm develop. A cross section of one tubule is shown in **Figure 34–11.** Specialized diploid cells within the tubules undergo meiosis and form the haploid nuclei of mature sperm. Recall that a haploid cell contains only a single set of chromosomes.

After they are produced in the seminiferous tubules, sperm are moved into the **epididymis** (ep uh DID ih mis), in which they mature and are stored. From the epididymis, some sperm are moved into a tube called the **vas deferens.** The vas deferens extends upward from the scrotum into the abdominal cavity. Eventually, the vas deferens merges with the urethra, the tube that leads to the outside of the body through the penis.

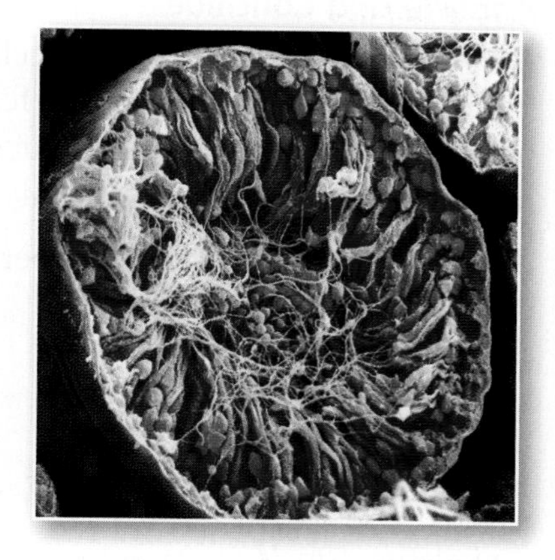

FIGURE 34–11 Male Reproductive System The main structures of the male reproductive system produce and deliver sperm. The micrograph shows a cross section of one tiny seminiferous tubule containing developing sperm (SEM 150×).

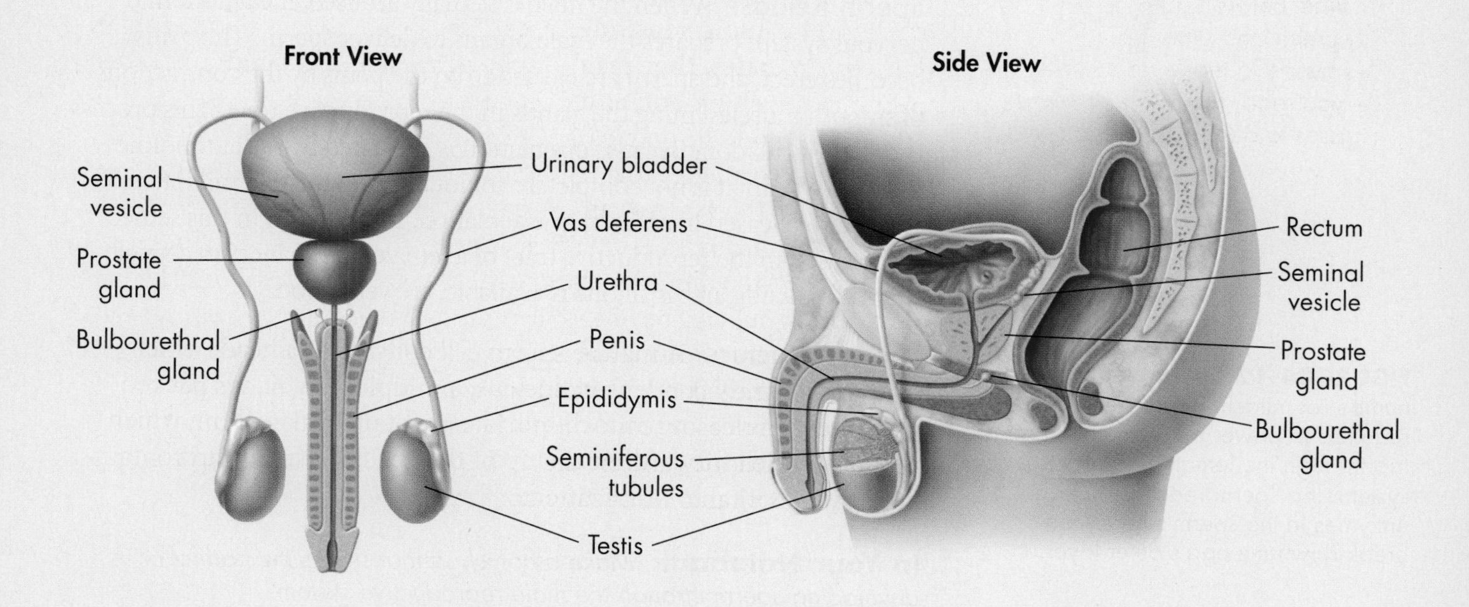

Quick Lab
GUIDED INQUIRY

Tracing Human Gamete Formation

1. Recall that cells in the testes and ovaries undergo meiosis as they form gametes—sperm and eggs.

2. For each letter, indicate how many chromosomes are in the cells at that stage and whether the cells are diploid (2N) or haploid (N). Answers *a.* and *e.* have been provided for you.

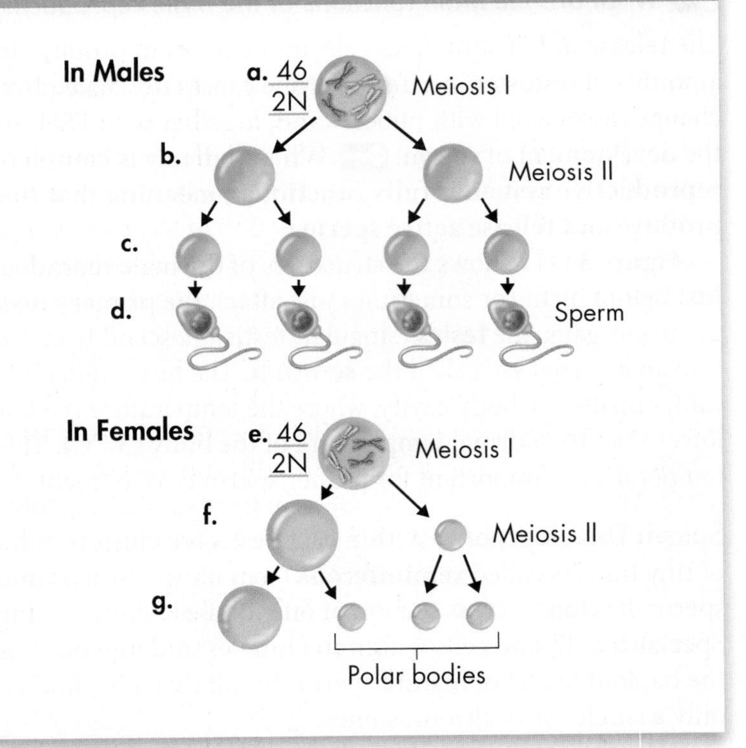

Analyze and Conclude

1. Interpret Visuals For every cell that undergoes meiosis in a male or female, what is the ratio of sperm produced in males to eggs produced in females?

2. Infer What percentage of sperm cells will contain a Y chromosome?

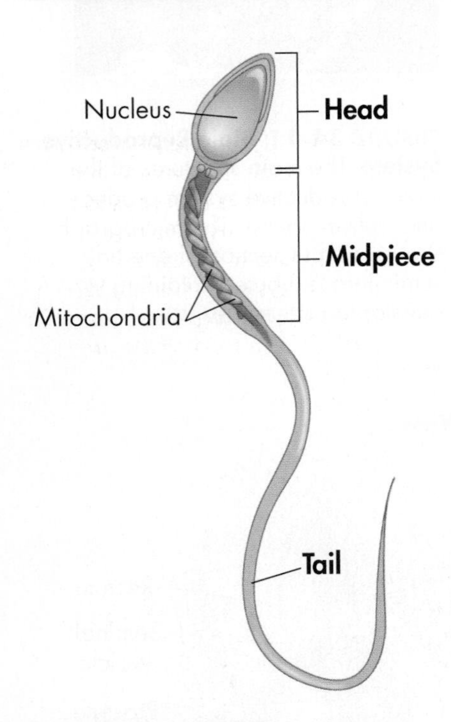

FIGURE 34–12 Sperm A large number of mitochondria are required to power a sperm cell's trip through the female reproductive system. If a sperm reaches an egg, enzymes in the sperm's head can break down the egg's outer layer.

Glands lining the reproductive tract—including the seminal vesicles, the prostate, and the bulbourethral (bul buh yoo REE thrul) glands—produce a nutrient-rich fluid called seminal fluid. The seminal fluid nourishes the sperm and protects them from the acidity of the female reproductive tract. The combination of sperm and seminal fluid is known as **semen.** The number of sperm present in even a few drops of semen is astonishing. Between 50 million and 130 million sperm are present in 1 milliliter of semen. That's about 2.5 million sperm per drop!

Sperm Release When the male is sexually aroused, the autonomic nervous system prepares the male organs to deliver sperm. The penis becomes erect, and sperm are ejected from the penis by the contractions of smooth muscles lining the glands in the reproductive tract. This process is called ejaculation. Because ejaculation is regulated by the autonomic nervous system, it is not completely voluntary. About 2 to 6 milliliters of semen are released in an average ejaculation. If the sperm in this semen are released in the reproductive tract of a female, the chances of a single sperm fertilizing an egg, if one is available, are very good.

Sperm Structure A mature sperm cell consists of a head, which contains a highly condensed nucleus; a midpiece, which is packed with energy-releasing mitochondria; and a tail, or flagellum, which propels the cell forward. At the tip of the head is a small cap containing enzymes vital to fertilization.

In Your Notebook *Make a flowchart that shows the path of developing sperm through the male reproductive system.*

The Female Reproductive System

What are the main functions of the female reproductive system?

The primary reproductive organs of the female are the **ovaries.** As in males, puberty in females starts when the hypothalamus signals the pituitary gland to release FSH and LH. FSH stimulates cells within the ovaries to produce increased amounts of estrogens and to start producing egg cells. **The main function of the female reproductive system is to produce egg cells, or ova (singular: ovum). In addition, the system prepares the female's body to nourish a developing embryo.**

Female Reproductive Structures At puberty, each ovary contains as many as 400,000 primary follicles, which are clusters of cells surrounding a single egg. The function of a follicle is to help an egg mature for release into the reproductive tract, where it may be fertilized by a sperm. Despite the huge number of primary follicles, a female's ovaries release only about 400 mature eggs in her lifetime.

In addition to the ovaries, other structures in the female reproductive system include the Fallopian tubes, uterus, cervix, and the vagina. **Figure 34–13** shows the location of these structures.

The Menstrual Cycle One ovary usually produces and releases one mature ovum every 28 days or so. The process of egg formation and release occurs as part of the **menstrual cycle,** a regular sequence of events involving the ovaries, the lining of the uterus, and the endocrine system. The menstrual cycle is regulated by hormones made by the hypothalamus, pituitary, and ovaries; it is controlled by internal feedback mechanisms.

During the menstrual cycle, an egg develops within a follicle and is released from an ovary. In addition, the uterus is prepared to receive a fertilized egg. If an egg is not fertilized, it is discharged, along with the lining of the uterus. If an egg is fertilized, embryonic development begins and the menstrual cycle ceases. The menstrual cycle includes the follicular phase, ovulation, the luteal phase, and menstruation.

FIGURE 34–13 Female Reproductive System The main function of the female reproductive system is to produce ova. The ovaries are the main organs of the female reproductive system. **Predict** ***Which structure is most likely lined with cilia that push an egg toward the uterus? Explain.***

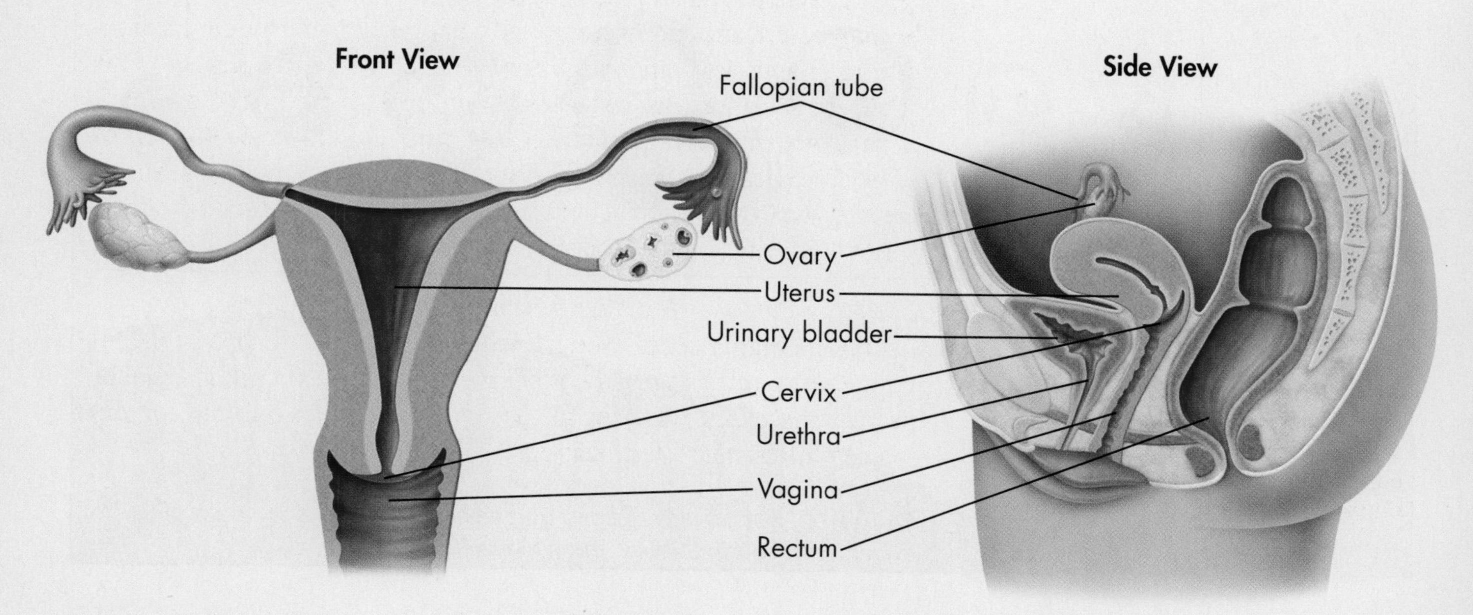

▶ ***Follicular Phase*** As shown in **Figure 34–14,** on day 1 of a menstrual cycle, blood estrogen levels are low. The hypothalamus reacts to low estrogen levels by producing a releasing hormone that stimulates the anterior pituitary to secrete FSH and LH. These two hormones travel to the ovaries, where they cause a follicle to mature. Usually, just a single follicle develops, but sometimes two or even three mature during the same cycle.

As the follicle develops, the cells surrounding the egg enlarge and begin to produce increased amounts of estrogens. This causes the estrogen level in the blood to rise dramatically. High blood estrogen levels cause the hypothalamus to produce less releasing hormone, and the pituitary releases less LH and FSH. Estrogens also cause the lining of the uterus to thicken in preparation for receiving a fertilized egg. The development of an egg during this phase takes about 12 days.

VISUAL SUMMARY

THE MENSTRUAL CYCLE

FIGURE 34–14 The menstrual cycle includes several phases. Notice the changes in hormone levels in the blood, the development of the follicle, and the changes in the uterine lining during the menstrual cycle. **Interpret Diagrams** ***During which phase of the menstrual cycle are estrogen levels the highest?***

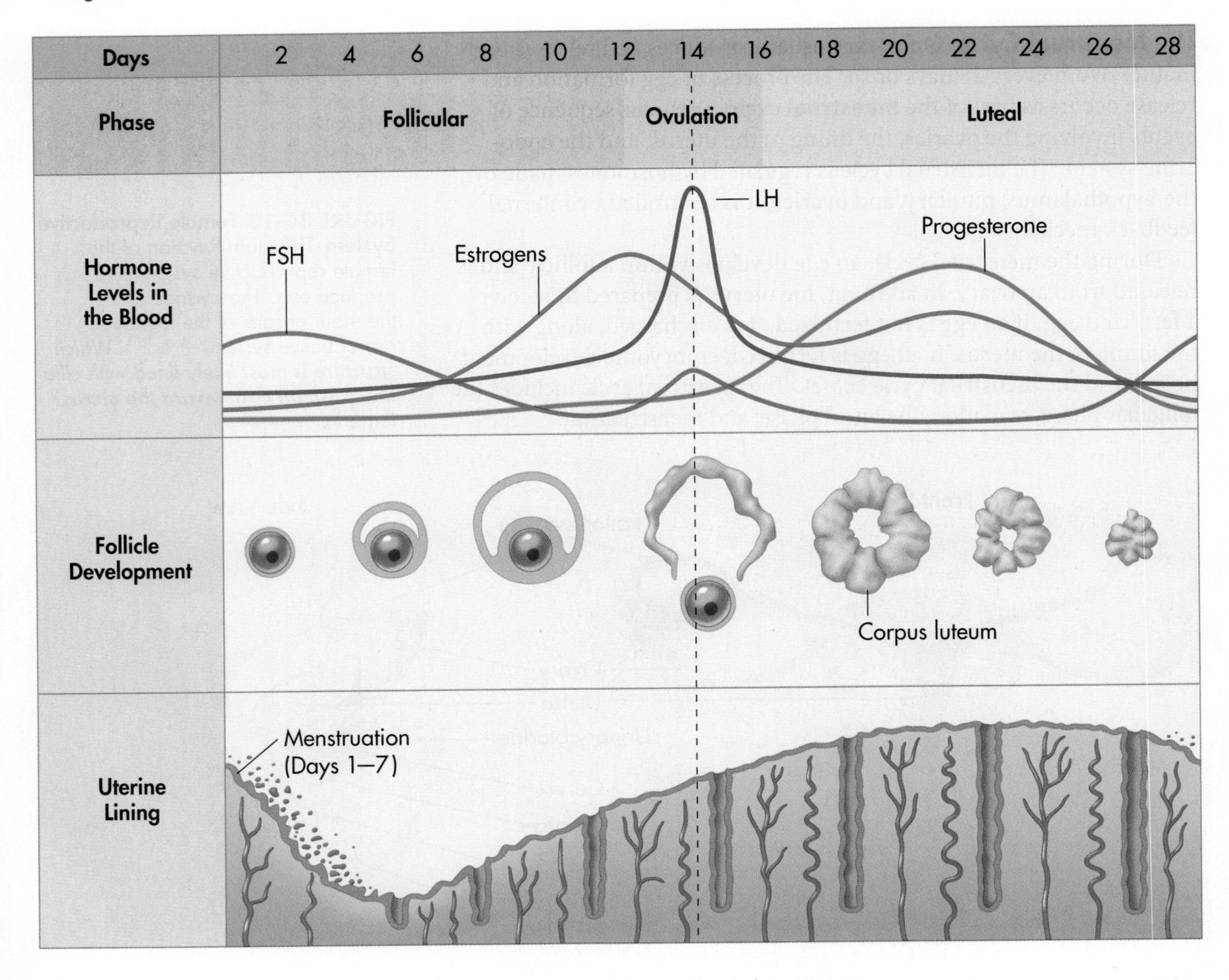

► ***Ovulation*** As the follicle grows, it releases more and more estrogens. When concentrations of these hormones reach a certain level, the hypothalamus reacts by triggering a burst of LH and FSH from the anterior pituitary. The sudden increase in these hormones (especially LH) causes the follicle to rupture. The result is **ovulation,** the release of an egg from the ovary into one of the Fallopian tubes. When released, the egg is stalled in metaphase of meiosis II and will remain that way unless it is fertilized. As the newly released egg is drawn into the Fallopian tube, microscopic cilia push the cell through the fluid-filled tube, toward the uterus.

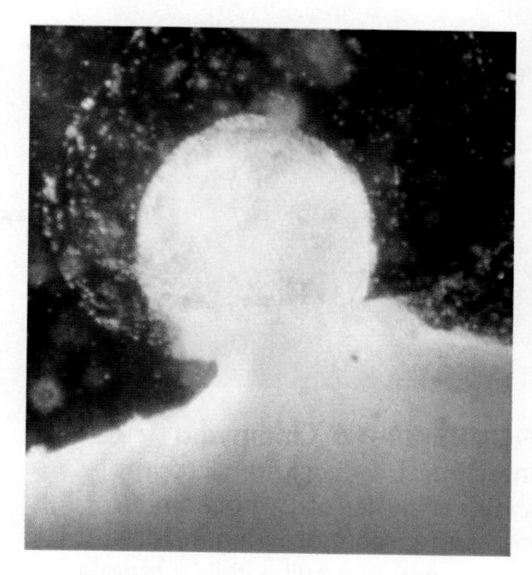

FIGURE 34–15 Ovulation (LM 160×)

► ***Luteal Phase*** The luteal phase begins immediately after ovulation. As the egg moves through the Fallopian tube, the cells of the ruptured follicle change. The follicle turns yellow and is now known as the **corpus luteum** (KAWR pus LOOT ee um), which means "yellow body" in Latin. The corpus luteum continues to release estrogens but also begins to release another steroid hormone called progesterone. Progesterone also stimulates the growth and development of the blood supply and surrounding tissue in the already-thickened uterine lining. The rise in these hormones once again inhibits the production of FSH and LH. Thus, additional follicles do not develop during this cycle.

Unless fertilization occurs and an embryo starts to develop, the fall of LH levels leads to the degeneration of the corpus luteum. Estrogen levels fall, the hypothalamus signals the release of FSH and LH from the anterior pituitary, and the follicular phase begins again.

MYSTERY CLUE

Low fat reserves in women have been associated with low FSH and LH levels. Tests showed that Lisa's blood had very low levels of these hormones. How might this have affected her menstrual cycle?

► ***Menstruation*** At the start of the new follicular phase, low estrogen levels also cause the lining of the uterus to detach from the uterine wall. This tissue, along with blood and the unfertilized egg, are discharged through the vagina. This phase of the cycle is called **menstruation.** Menstruation lasts about three to seven days on average. A new cycle begins with the first day of menstruation.

The menstrual cycle continues, on average, until a female is in her late forties to early fifties. At this time, the production of estrogens declines, and ovulation and menstruation stop. The permanent stopping of the menstrual cycle is called menopause.

BUILD Vocabulary

WORD ORIGINS The word **menstruation** comes from the Latin word *mensis,* meaning "month."

Pregnancy Of course, the menstrual cycle also ceases if a woman becomes pregnant. During the first two days of the luteal phase, immediately following ovulation, the chances that an egg will be fertilized are the greatest. This is usually from 14 to 18 days after the completion of the last menstrual cycle. If a sperm fertilizes an egg, the fertilized egg completes meiosis and immediately undergoes mitosis. After several divisions, a ball of cells will form and implant itself in the lining of the uterus. Within a few days of implantation, the uterus and the growing embryo will release hormones that keep the corpus luteum functioning for several weeks. This allows the lining of the uterus to nourish and protect the developing embryo and prevents the menstrual cycle from starting again.

In Your Notebook *Draw a cycle diagram to represent the phases and days of the menstrual cycle.*

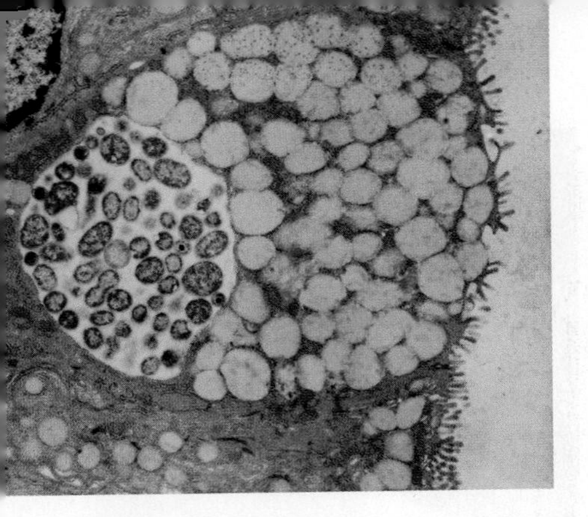

FIGURE 34–16 Chlamydia Infection This electron micrograph shows a cluster of *C. trachomatis* bacteria (green) growing inside a mucus-secreting cell within a female reproductive tract. The bacteria will eventually overwhelm the cell and cause it to burst, allowing the infection to spread (TEM 2400×).

Sexually Transmitted Diseases

What are some of the most commonly reported sexually transmitted diseases?

Diseases that spread by sexual contact, or **sexually transmitted diseases** (STDs), are a serious health problem in the United States. A 2008 study by the Centers for Disease Control and Prevention showed that one in four girls and young women aged 14 to 19 were infected with an STD.

Unfortunately, public health information about STDs has not kept pace with the rate of infection. For example, one might think that the name of the most commonly reported infectious disease in the United States would be a household word, but it isn't. **Chlamydia is not only the most common bacterial STD, it is the most commonly reported bacterial disease in the United States.** Chlamydia, which damages the reproductive tract and can lead to infertility, is caused by a bacterium that is spread by sexual contact. Other bacterial STDs include gonorrhea and syphilis.

Viruses can also cause STDs. **Viral STDs include hepatitis B, genital herpes, genital warts, and AIDS.** Unlike the bacterial STDs, viral infections cannot be treated with antibiotics.

Some viral STDs, such as AIDS, can be fatal. Tens of thousands of people in the United States die from AIDS each year. In addition, the virus that causes genital warts—human papillomavirus (HPV)—is a major cause of cervical cancer in women. Recently, a vaccine has been developed that can prevent some HPV infections. To be effective, the vaccine must be administered before a woman is infected with HPV.

STDs can be avoided. Any sexual contact carries with it the chance of infection. The safest course to follow is to abstain from sexual contact before marriage, and for both partners in a committed relationship to remain faithful. The next safest course is to use a latex condom, but even condoms do not provide 100 percent protection.

34.3 Assessment

Review Key Concepts

1. a. Review Explain what happens during puberty.

b. Compare and Contrast Compare and contrast the sexual development of male embryos to that of female embryos.

2. a. Review Describe the function of the male reproductive system.

b. Sequence Explain how sperm develop.

3. a. Review Describe the functions of the female reproductive system.

b. Interpret Visuals What happens during each stage of the menstrual cycle? *Hint:* Refer to **Figure 34–14.**

4. a. Review Name two STDs caused by bacteria and two caused by viruses.

b. Evaluate Why do you think that young people are especially at risk for STDs?

Apply the Big idea

Cellular Basis of Life

5. Sperm cells contain numerous mitochondria. Use what you learned about mitochondria in Chapter 7 to explain how mitochondria might influence sperm activity.

34.4 Fertilization and Development

THINK ABOUT IT Of all the wonders of the living world, is there anything more remarkable than the formation of a new human being from a single cell? In a sense, we know how this happens. The embryo goes through round after round of cell division, producing the trillions of cells in a newborn baby. Simple enough, it seems. But how do these cells arrange themselves so beautifully into the tissues and organs of the body, and how does an individual cell "know" to become an embryonic skin, heart, or blood cell? These are some of the most important questions in all of biology, and we are only beginning to learn the answers.

Key Questions

What takes place during fertilization and the early stages of human development?

What important events occur during the later stages of human development?

Vocabulary

zygote • blastocyst • implantation • gastrulation • neurulation • placenta • fetus

Taking Notes

Flowchart As you read, draw a flowchart that shows the steps from fertilized egg to newborn baby.

Fertilization and Early Development

What takes place during fertilization and the early stages of human development?

The story of human development begins with the gametes—sperm produced in the testes and egg cells produced in the ovaries. Sperm and egg must meet, so that the two gametes can fuse to form a single cell. With this single cell, the process of development begins. **The fusion of a sperm and egg cell is called fertilization.**

Fertilization During sexual intercourse, sperm are released when semen is ejaculated through the penis into the vagina. Semen is generally released just below the cervix, the opening that connects the vagina to the uterus. Sperm swim actively through the uterus into the Fallopian tubes. Hundreds of millions of sperm are released during an ejaculation. If an egg is present in one of the Fallopian tubes, its chances of being fertilized are good.

The egg is surrounded by a protective layer that contains binding sites to which sperm can attach. The sperm head then releases powerful enzymes that break down the protective layer of the egg. The haploid (N) sperm nucleus enters the haploid egg, and chromosomes from sperm and egg are brought together. Once the two haploid nuclei fuse, a single diploid (2N) nucleus is formed, containing a single set of chromosomes from each parent cell. The fertilized egg is called a **zygote.** At this point the developing human can also be called an embryo.

FIGURE 34–17 Sperm Meet Egg Many sperm usually reach an egg, but only one sperm can successfully break through the egg's protective barrier (SEM 650×).

FIGURE 34–18 Ernest Everett Just One of the great pioneers of cell biology, E.E. Just investigated the process of fertilization. He discovered that changes in an egg's cell membrane prevent more than one sperm from fertilizing an egg.

What prevents more than one sperm from fertilizing an egg? Early in the twentieth century, cell biologist Ernest Everett Just found the answer. The egg cell contains a series of granules just beneath its outer surface. When a sperm enters the egg, the egg reacts by releasing the contents of these granules outside the cell. The material in the granules coats the surface of the egg, forming a barrier that prevents other sperm from attaching to, and entering, the egg.

Multiple Embryos If two eggs are released during the same menstrual cycle and each is fertilized, fraternal twins may result. Fraternal twins are not identical in appearance and may even be different sexes, because each has been formed by the fusion of a different sperm and different egg cell.

Sometimes a single zygote splits apart and produces two genetically identical embryos. These two embryos are called identical twins. Because they result from the same fertilized egg, identical twins are always the same sex.

Implantation While still in the Fallopian tube, the zygote begins to undergo mitosis, as shown in **Figure 34–19.** As the embryo grows, a cavity forms in the center, until the embryo becomes a hollow ball of cells known as a **blastocyst.** About six or seven days after fertilization, the blastocyst attaches to the wall of the uterus and begins to grow into the tissues of the mother. This process is known as **implantation.**

At this point, cells in the blastocyst begin to specialize. This specialization process, called differentiation, results in the development of the various types of tissues in the body. A cluster of cells, known as the inner cell mass, develops within the inner cavity of the blastocyst. The body of the embryo will develop from these cells, while the other cells of the blastocyst will differentiate into some of the tissues that support and protect the embryo.

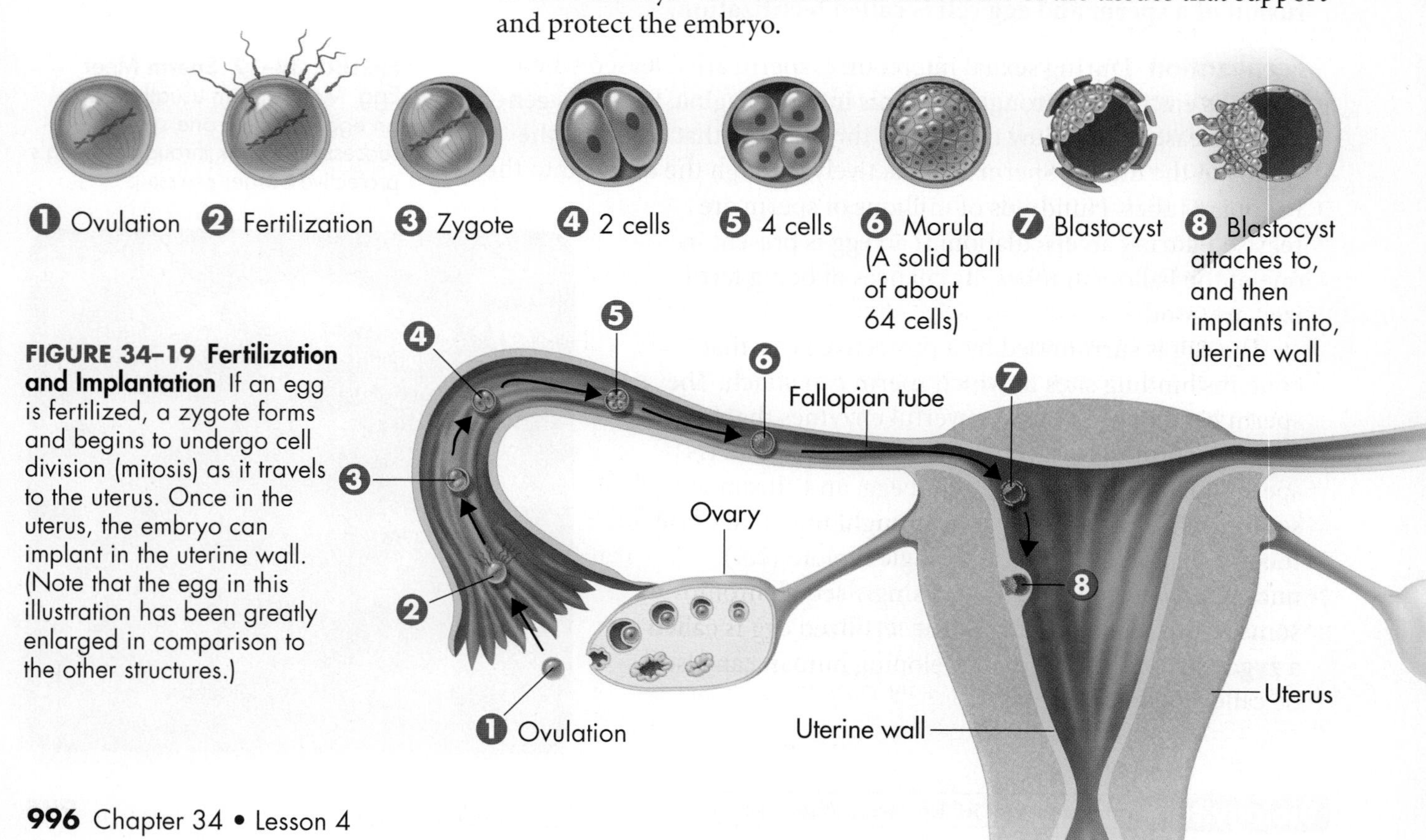

FIGURE 34–19 Fertilization and Implantation If an egg is fertilized, a zygote forms and begins to undergo cell division (mitosis) as it travels to the uterus. Once in the uterus, the embryo can implant in the uterine wall. (Note that the egg in this illustration has been greatly enlarged in comparison to the other structures.)

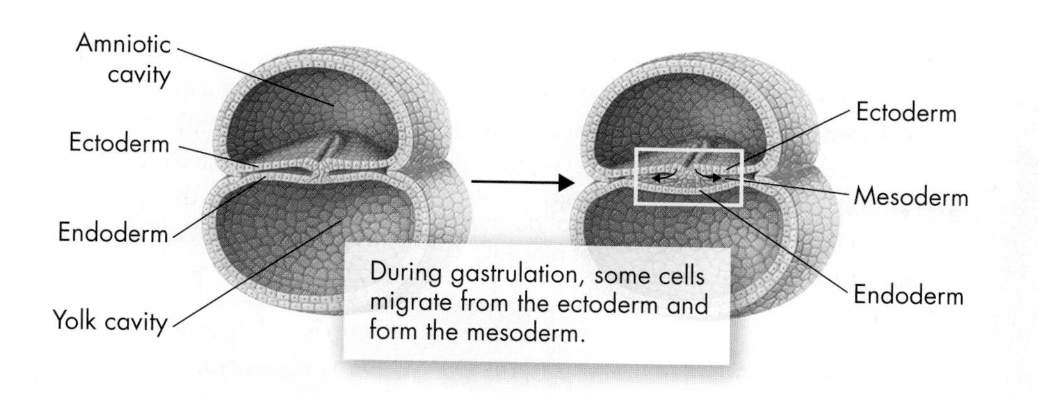

FIGURE 34–20 Gastrulation This stage results in the formation of three cell layers—the ectoderm, mesoderm, and endoderm. The three layers form all of the organs and tissues of the embryo.

Gastrulation As development continues, the embryo begins a series of dramatic changes that will produce the key structures and tissue layers of the body. **Key events in early development include gastrulation, which produces the three cell layers of the embryo, and neurulation, which leads to the formation of the nervous system.** The result of **gastrulation** (gas troo LAY shun) is the formation of three cell layers called the ectoderm, mesoderm, and endoderm. The ectoderm and endoderm form first. The mesoderm is produced by a process of cell migration shown in **Figure 34–20.**

The ectoderm will develop into the skin and the nervous system. Mesoderm cells differentiate and form many of the body's internal structures, including bones, muscle, blood cells, and gonads. Endoderm forms the linings of organs in the digestive system, such as the stomach and intestines, as well as in the respiratory and excretory systems.

Neurulation Gastrulation is followed by another important step in development, neurulation (NUR uh lay shun). **Neurulation,** shown in **Figure 34–21,** is the first step in the development of the nervous system. Shortly after gastrulation is complete, a block of mesodermal tissue begins to differentiate into the notochord. Recall that all chordates possess a notochord at some stage of development. As the notochord develops, the ectoderm near the notochord thickens and forms the neural plate. The raised edges of the neural plate form neural folds and the neural crest. The neural folds gradually move together and form the neural tube, from which the spinal cord and brain will develop. Cells of the neural crest migrate to other locations and become types of nerve cells, skin pigment cells, and other structures such as the lower jaw.

If the neural tube does not close completely, a serious birth defect known as spina bifida can result. Studies show that folic acid (vitamin B_9) can prevent most cases of spina bifida. Because neurulation usually occurs before a woman knows she's pregnant, folic acid is an important nutrient in any woman's diet.

In Your Notebook *Explain in your own words what occurs during neurulation.*

FIGURE 34–21 Neurulation During neurulation, the ectoderm undergoes changes that lead to the formation of a neural tube that develops into the brain and spinal cord. Neural crest cells develop into many types of nerves.

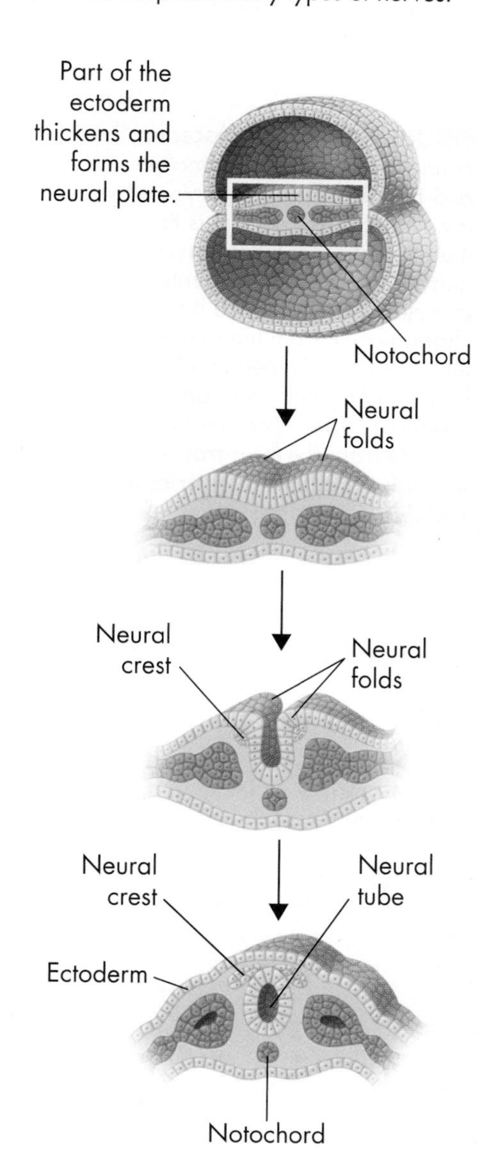

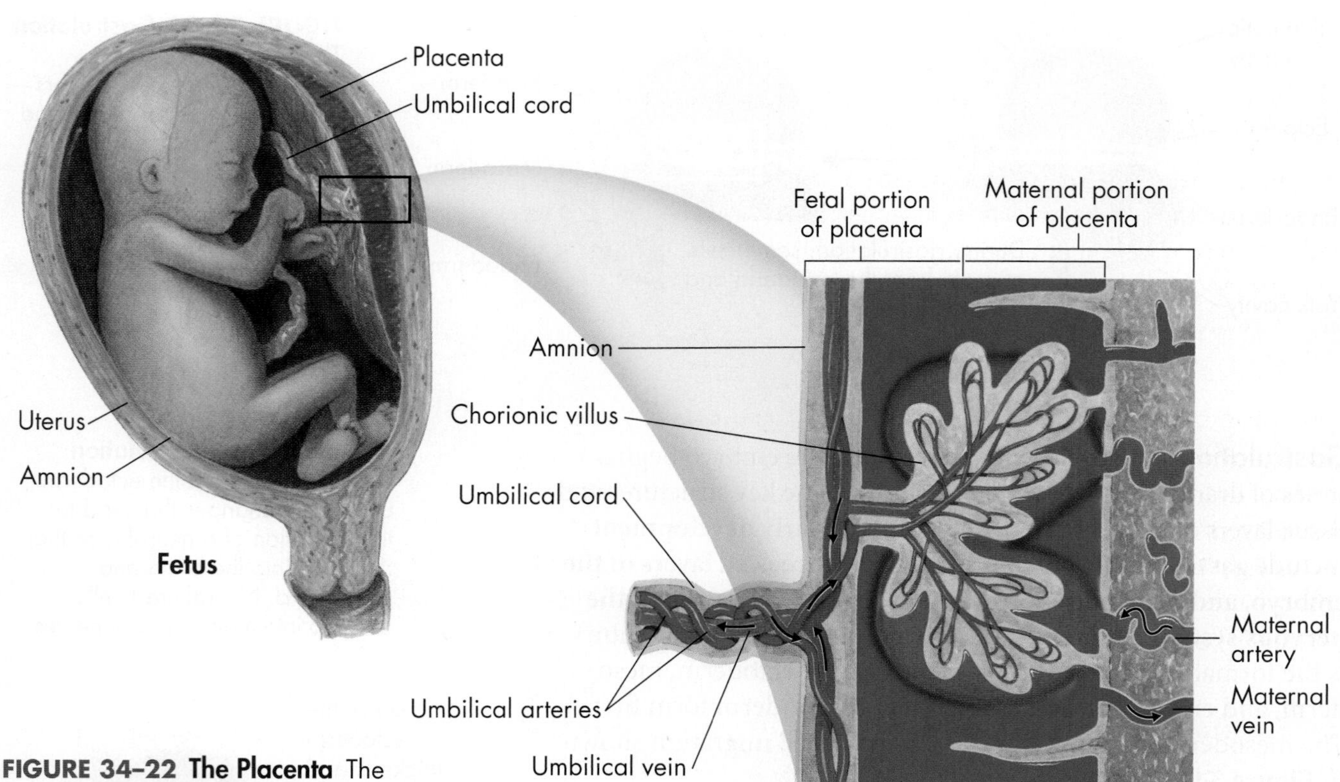

FIGURE 34–22 The Placenta The connection between the mother and the developing embryo or fetus is called the placenta. It is through the placenta that the embryo gets its oxygen and nutrients and excretes wastes. Notice how the chorionic villi from the fetus extend into the mother's uterine lining (indicated by the overlapping brackets). **Infer** ***Does carbon dioxide from the fetus travel through the umbilical arteries or umbilical vein?***

The Placenta As the embryo develops, specialized membranes form to protect and nourish the embryo. The embryo is surrounded by the amnion, a sac filled with amniotic fluid that cushions and protects the developing embryo. Another sac, known as the chorion, forms just outside the amnion. The chorion makes direct contact with the tissues of the uterus. Near the end of the third week of development, small, fingerlike projections called chorionic villi form on the outer surface of the chorion and extend into the uterine lining.

The chorionic villi and uterine lining form a vital organ called the **placenta.** The placenta is the connection between the mother and embryo that acts as the embryo's organ of respiration, nourishment, and excretion. Across this thin barrier, oxygen and nutrients diffuse from the mother's blood to the embryo's blood; carbon dioxide and metabolic wastes diffuse from the embryo's blood to the mother's blood.

The blood of the mother and that of the embryo flow past each other, but they do not mix. The exchange of gases and other substances occurs in the chorionic villi. **Figure 34–22** shows a portion of the placenta. The umbilical cord, which contains two arteries and one vein, connects the embryo to the placenta.

After eight weeks of development, the embryo is called a **fetus.** By the end of three months of development, most of the major organs and tissues of the fetus are fully formed. The fetus may begin to move and show signs of reflexes. The fetus is about 8 centimeters long and has a mass of about 28 grams.

In Your Notebook *Explain in your own words the role of the placenta in human development.*

Later Development

What important events occur during the later stages of human development?

Although most of the tissues and organs of the embryo have been formed after three months of development, many of them are not yet ready to go to work on their own. On average, another six months of development takes place before all of these systems are fully prepared for life outside the uterus.

Months 4–6 **During the fourth, fifth, and sixth months after fertilization, the tissues of the fetus become more complex and specialized, and begin to function.** The fetal heart becomes large enough so that it can be heard with a stethoscope. Bone continues to replace the cartilage that forms the early skeleton. A layer of soft hair grows over the skin of the fetus. As the fetus increases in size, the mother's abdomen swells to accommodate it. The mother begins to feel the fetus moving.

Months 7–9 **During the last three months before birth, the organ systems of the fetus mature, and the fetus grows in size and mass.** The fetus doubles in mass, and the lungs and other organs undergo a series of changes that prepare them for life outside the uterus. The fetus is now able to regulate its body temperature. In addition, the central nervous system and lungs complete their development. **Figure 34–23** shows an embryo and a fetus at different stages of development.

On average, it takes nine months for a fetus to develop fully. Babies born before eight months of development are called premature babies and often have severe breathing problems as a result of incomplete lung development.

FIGURE 34–23 Human Development At 7 weeks, most of the organs of an embryo have begun to form. The heart—the large, dark, rounded structure—is beating. By 14 weeks, the hands, feet, and legs have reached their birth proportions. The eyes, ears, and nose are well developed. At 20 weeks, muscle development has increased, and eyebrows and nails have grown in. When a fetus is full term, it is capable of living on its own.

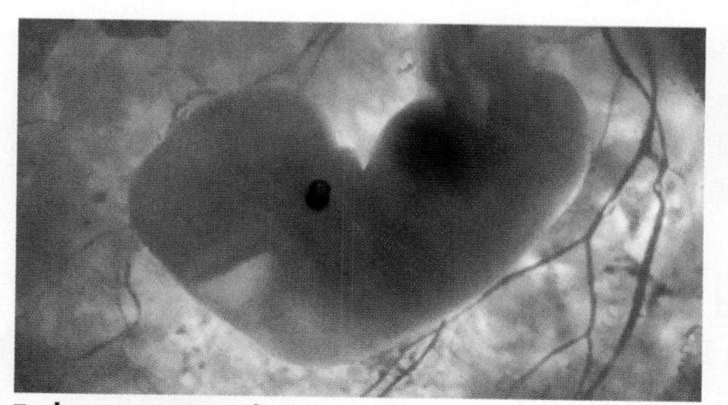

Embryo at 7 Weeks

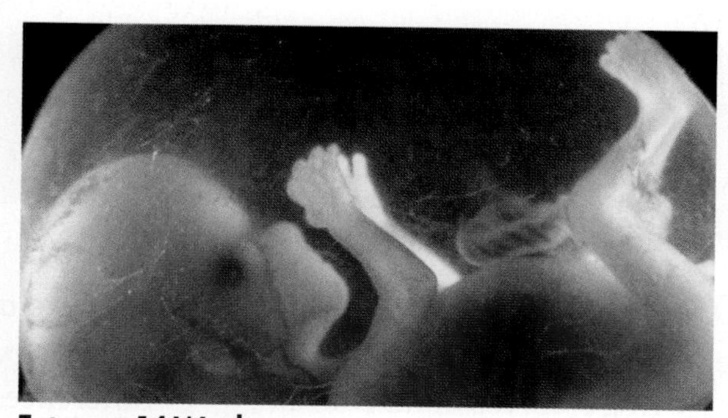

Fetus at 14 Weeks

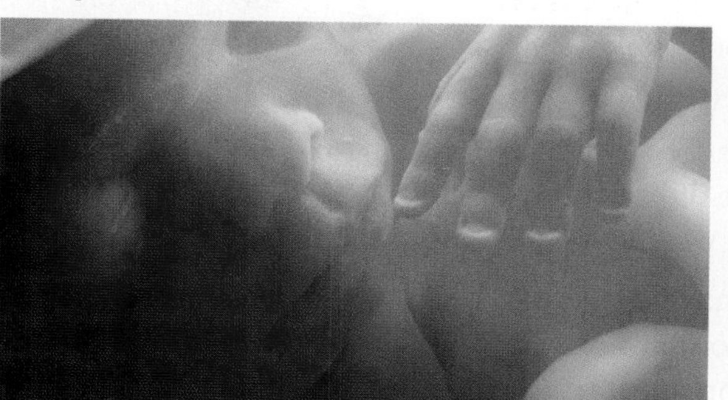

Fetus at 20 Weeks

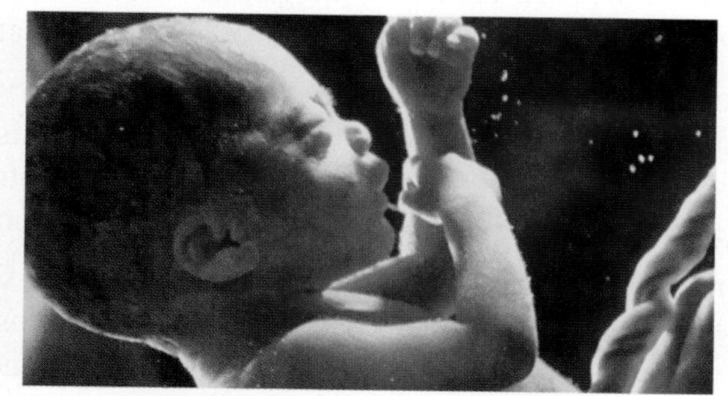

Fetus at Full Term

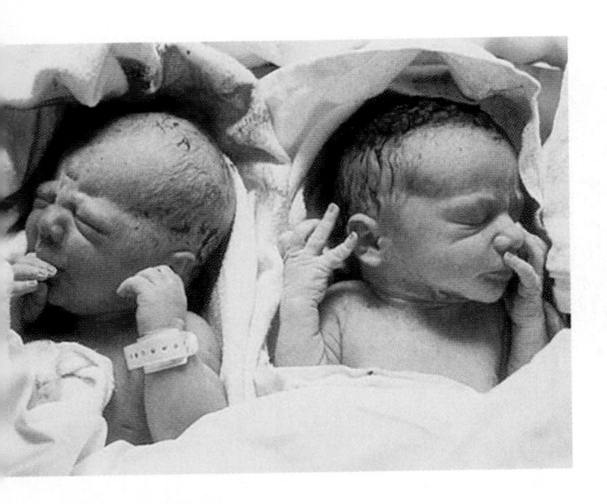

FIGURE 34–24 Newborns Twins, ten minutes after birth, adjusting to life outside the uterus.

Childbirth About nine months after fertilization, the fetus is ready for birth. A complex set of factors triggers the process; one of these factors is the release of the hormone oxytocin from the mother's posterior pituitary gland. Oxytocin affects a group of large involuntary muscles in the uterine wall. As these muscles are stimulated, they begin a series of rhythmic contractions collectively known as labor. As labor progresses, the contractions become more frequent and more powerful. The opening of the cervix expands until it is large enough for the head of the baby to pass through. At some point, the amniotic sac breaks, and the fluid it contains rushes out of the vagina. Contractions of the uterus force the baby, usually head first, out through the vagina.

As the baby meets the outside world, he or she may begin to cough or cry, a process that rids the lungs of fluid. Breathing starts almost immediately, and the blood supply to the placenta begins to dry up. The umbilical cord is clamped and cut, leaving a small piece attached to the baby. This piece will soon dry and fall off, leaving a scar known as the navel—or, its more familiar term, the belly button. In a final series of uterine contractions, the placenta itself and the now-empty amniotic sac are expelled from the uterus as the afterbirth.

The baby now begins an independent existence. Most newborns are remarkably hardy. Their systems quickly make the switch to life outside the uterus, supplying their own oxygen, excreting wastes on their own, and maintaining their own body temperatures.

The interaction of the mother's reproductive and endocrine systems does not end at childbirth. Within a few hours after birth, the pituitary hormone prolactin stimulates the production of milk in the breast tissues of the mother. The nutrients present in that milk contain everything the baby needs for growth and development during the first few months of life.

Quick Lab
GUIDED INQUIRY

Embryonic Development

1. Use a dropper pipette to transfer several early-stage frog embryos in water to a depression slide. **CAUTION:** *Handle glass slides with care.*

2. Look at the embryos under the dissecting microscope at low power. Sketch what you see.

3. Look at the prepared slides of the early embryonic stages of a frog. Make sketches of what you see.

Analyze and Conclude

1. **Observe** Describe any differences you saw among the cells. At what stage is cell differentiation visible?

2. **Observe** Were you able to see a distinct body plan? At what stage did the body plan become visible?

3. **Draw Conclusions** Describe any organs you saw. At what stage did specific organs form?

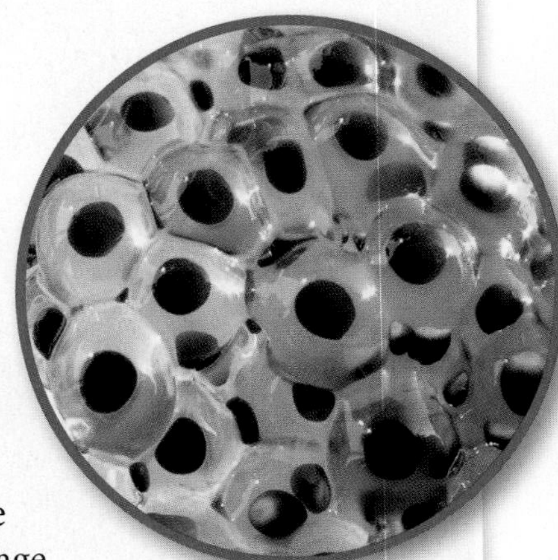

Frog Embryos

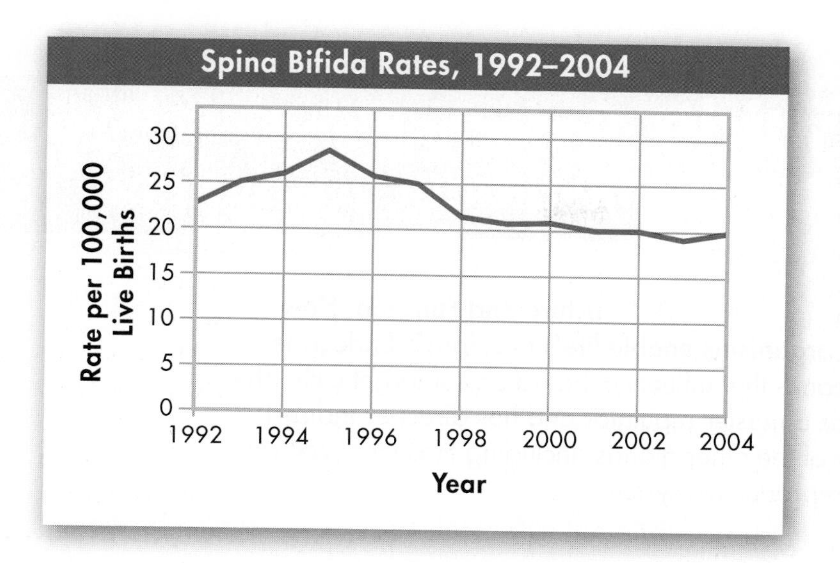

FIGURE 34–25 Preventing Spina Bifida In 1993, the U.S. Public Health Service recommended that women consume 4 mg of folic acid per day. Between 1996 and 1998, manufacturers of enriched grain products began to add folic acid to their products. Interpret Graphs ***Is there any indication that increase in folic acid intake had an effect on the rate of spina bifida cases?***

Infant and Maternal Health Although the placenta acts as a barrier to many harmful or disease-causing agents, some do pass through this barrier and affect the health of the embryo. The virus that causes AIDS can infect the developing fetus, and the virus responsible for rubella (German measles) can cause birth defects. Alcohol can permanently injure the nervous system, and drugs such as heroin and cocaine can cause drug addiction in newborn babies. Smoking during pregnancy can double the risk of low weight at birth, leading to other severe health problems. There is no substitute for professional medical care during pregnancy nor for responsible behavior on the part of the pregnant woman to protect the life within her.

From 1970 to 2000, the infant mortality rate in the United States decreased by about 65 percent. Many factors, including more women seeking early prenatal care and advances in medical technology, contributed to this decrease. **Figure 34–25** shows how one recent public health initiative affected the incidence of a serious birth defect—spina bifida.

34.4 Assessment

Review Key Concepts

1. a. Review Describe the process of fertilization.
b. Explain What is the role of the placenta?
c. Relate Cause and Effect How do the outcomes of gastrulation and neurulation contribute to human development?

2. a. Review List developments that occur in the fetus during months 4–6 and months 7–9.
b. Explain What is oxytocin, and what is its role in childbirth?
c. Apply Concepts Why do you think doctors recommend that women avoid many medications during pregnancy?

ANALYZING DATA

Review **Figure 34–25** and answer the questions.

3. a. Interpret Graphs In what years did the incidence of spina bifida show the greatest decline?
b. Relate Cause and Effect What reasons could explain this decline in the incidence of spina bifida?

NGSS Smart Guide Endocrine and Reproductive Systems

Disciplinary Core Idea LS1.A Structure and Function: How do the structures of organisms enable life's functions? Endocrine glands release hormones that influence the actions of target cells. The hypothalamus acts as a master regulator and has direct or indirect influence over many of the other glands, including those that control the functions of the reproductive system.

34.1 The Endocrine System

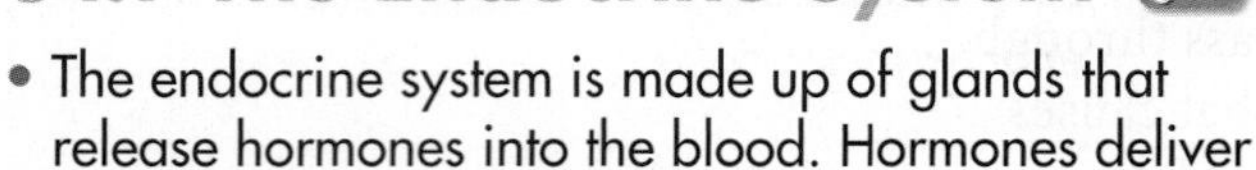

- The endocrine system is made up of glands that release hormones into the blood. Hormones deliver messages throughout the body.
- Steroid hormones can easily cross cell membranes. Once inside the nucleus, they change the pattern of gene expression in target cells.
- Nonsteroid hormones bind to receptors on cell membranes and cause the release of secondary messengers that affect cell activities.

hormone 978 • target cell 978 • exocrine gland 979 • endocrine gland 979 • prostaglandin 980

Biology.com

Untamed Science Video The Untamed Science crew helps us better understand the role epinephrine plays in regulating our response to fear and danger.

Art Review Review your understanding of the major endocrine glands in the body.

Art in Motion Watch how steroid and nonsteroid hormones act differently on cells.

ART REVIEW Major Endocrine Glands

ART IN MOTION The Endocrine System

Steroid Hormones
This binding initiates the transcription of specific genes to messenger RNA (mRNA). The mRNA moves into the cytoplasm and directs protein

Steroid Hormones
Nonsteroid Hormones

Steroid hormone
Receptor

34.2 Glands of the Endocrine System

- The pituitary gland secretes hormones that directly regulate many body functions or control the actions of other endocrine glands.
- The hypothalamus controls the secretions of the pituitary gland.
- The adrenal glands release hormones that help the body prepare for—and deal with—stress.
- Insulin and glucagon help keep the blood glucose level stable.
- The thyroid gland has a major role in regulating the body's metabolism.
- The two functions of gonads are the production of gametes and the secretion of sex hormones.
- Like most systems of the body, the endocrine system is regulated by feedback mechanisms that function to maintain homeostasis.

pituitary gland 982 • releasing hormone 983 • corticosteroid 983 • epinephrine 983 • norepinephrine 983 • thyroxine 985 • calcitonin 985 • parathyroid hormone 985

Biology.com

Data Analysis Analyze data about the risk factors and effects of diabetes.

DATA ANALYSIS Obesity and Diabetes

A Growing Problem

In the past three decades, the incidence of diabetes in the United States has risen sharply. In 1980, an estimated 5.6 million people had diabetes. By 2006, the number had more than tripled to 16.8 million. Researchers estimate that by 2030, 30.3 million Americans will have diabetes.

Scientists want to understand why people develop diabetes. Researchers have found that obesity is one risk factor for developing type II diabetes, the type that often develops during adulthood.

But, what types of data analysis led to this conclusion?

CHECKING YOUR *Scientific Literacy*

Refer to the lesson content and digital assets below as you prepare for your chapter assessment. Then, evaluate your understanding of the endocrine and reproductive systems by answering these questions.

1. Scientific and Engineering Practice **Obtaining, Evaluating, and Communicating Information** Review the text discussion about hormonal interactions that regulate the menstrual cycle. Describe the feedback mechanisms that control the steps of the cycle and identify which mechanisms are examples of feedback inhibition.

2. Crosscutting Concept **Stability and Change** Using examples you learned about in this chapter, explain how hormones both maintain stability in the body and cause changes in response to environmental conditions.

3. Craft and Structure Why do you think an explanation of the female athlete triad was chosen by the textbook authors as a topic for the Chapter Mystery?

4. STEM Individuals with diabetes must test their blood glucose levels regularly. Research the glucose detection technologies that are available for home use, and describe how they compare in effectiveness, cost, and ease of use.

34.3 The Reproductive System

- In females, the effects of the sex hormones include breast development and a widening of the hips. In males, they result in the growth of facial hair, increased muscular development, and a deepening of the voice.
- The main functions of the male reproductive system are to produce and deliver sperm.
- The main function of the female reproductive system is to produce egg cells. The system also prepares the female's body to nourish an embryo.
- Chlamydia is the most common bacterial STD in the United States. Viral STDs include hepatitis B, genital herpes, genital warts, and AIDS.

puberty 988 • testis 989 • scrotum 989 • seminiferous tubule 989 • epididymis 989 • vas deferens 989 • semen 990 • ovary 991 • menstrual cycle 991 • ovulation 993 • corpus luteum 993 • menstruation 993 • sexually transmitted disease 994

Biology.com

In Your Notebook Review your understanding of the phases of the menstrual cycle.

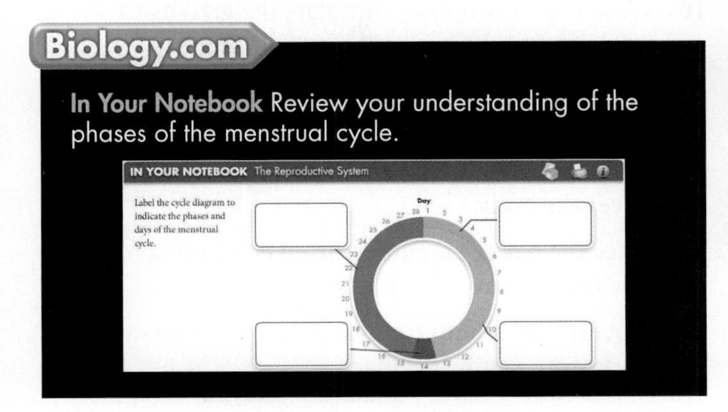

34.4 Fertilization and Development

- The fusion of a sperm and egg cell is called fertilization.
- Gastrulation produces the three cell layers of the embryo. Neurulation leads to the formation of the nervous system.
- During the fourth, fifth, and sixth months after fertilization, the tissues of the fetus become more complex and specialized.
- During the last three months before birth, the organ systems of the fetus mature, and the fetus grows in size and mass.

zygote 995 • blastocyst 996 • implantation 996 • gastrulation 997 • neurulation 997 • placenta 998 • fetus 998

Biology.com

Taking Notes Construct a flowchart that shows the developmental steps that lead from fertilization to birth.

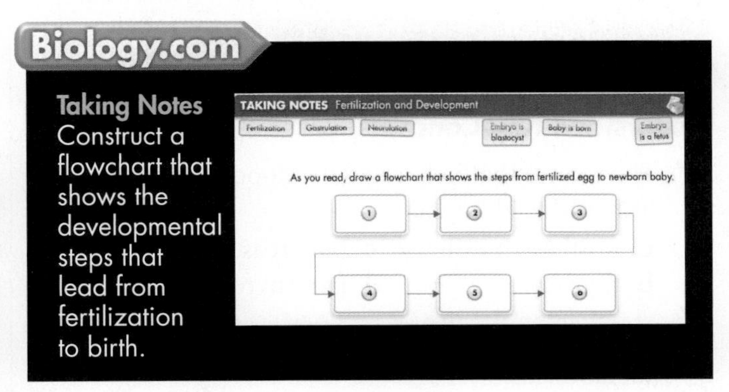

Forensics Lab

Diagnosing Endocrine Disorders This chapter lab is available in your lab manual and at Biology.com

34 Assessment

34.1 The Endocrine System

Understand Key Concepts

1. Which choice is a chemical messenger that can directly influence gene expression?
   a. nonsteroid hormone   c. ATP
   b. steroid hormone   d. cAMP
2. A modified fatty acid that is released by a cell and affects local cells and tissues is likely a(n)
   a. nonsteroid hormone.
   b. steroid hormone.
   c. prostaglandin.
   d. exocrine secretion.
3. **Key Ideas and Details** What is the relationship between a hormone and a target cell? Cite a specific example from the text to explain your answer.
4. Many body functions are influenced by the action of two hormones with opposing effects. Why are such pairs of hormones useful?

Think Critically

5. **Use Analogies** In many areas during rush hour, radio stations broadcast traffic reports. How are traffic reports similar to hormones? How do the reports act as a feedback control mechanism to control the flow of traffic?
6. **Infer** After a hormone is secreted by a gland, the circulatory system transports it all through the body. Why doesn't every cell respond to the hormone?

34.2 Glands of the Endocrine System

Understand Key Concepts

7. Hormones that help regulate blood calcium levels are produced by the
   a. posterior pituitary.   c. pancreas.
   b. thymus.   d. parathyroid gland.
8. Which hormone influences a person's rate of metabolism?
   a. PTH   c. thyroxine
   b. aldosterone   d. calcitonin
9. How does a feedback mechanism regulate the activity of the endocrine system?
10. How does the secretion of epinephrine prepare the body to handle emergencies?
11. What happens if blood glucose levels are not properly regulated?

Think Critically

12. **Apply Concepts** The heartbeat of a swimmer was found to increase significantly both before and during a swim meet. Explain why this could happen.
13. **Form a Hypothesis** Iodine is required to complete the production of thyroxine. Why do you think the thyroid gland enlarges in response to iodine deficiency?

34.3 The Reproductive System

Understand Key Concepts

14. The diagram shows the female reproductive system. Which structure is indicated by the *X*?

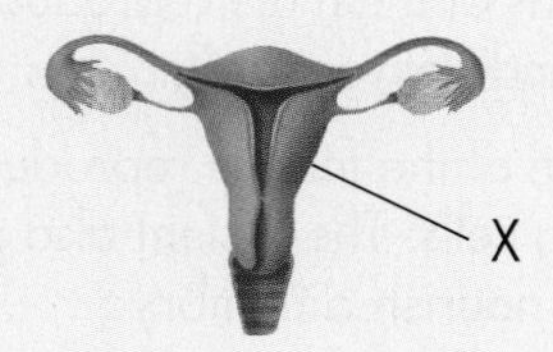

   a. uterus   c. ovary
   b. Fallopian tube   d. cervix
15. Which male reproductive structure releases sperm into the urethra?
   a. epididymis   c. prostate gland
   b. vas deferens   d. testis
16. Which two hormones stimulate the gonads to produce their hormones?
17. List the secondary sex characteristics that appear in males and females at puberty.
18. Trace the path of a sperm from a testis until it leaves the body.
19. Trace the path of an unfertilized egg from a follicle until it leaves the body.
20. Provide one example of how the menstrual cycle works by negative feedback.

Think Critically

21. Apply Concepts Describe how each of the following represents an adaptation that helps to ensure successful fertilization: seminal fluid; production and release of millions of sperm; cilia lining the Fallopian tubes; flagellum of a sperm.

22. Craft and Structure Predict the effects that insufficient amounts of FSH and LH would have on the menstrual cycle.

34.4 Fertilization and Development

Understand Key Concepts

23. Another name for a fertilized egg is a
- **a.** gastrula.
- **b.** placenta.
- **c.** zygote.
- **d.** blastocyst.

24. Fertilization usually occurs in the
- **a.** uterus.
- **b.** vagina.
- **c.** Fallopian tube.
- **d.** ovary.

25. After the eighth week of development, the human embryo is known as a(n)
- **a.** zygote.
- **b.** infant.
- **c.** fetus.
- **d.** morula.

26. Key Ideas and Details Trace the development of a zygote from fertilization through implantation.

27. Explain the importance of the three layers that form during gastrulation.

28. What is the function of the placenta?

29. Describe what happens during childbirth.

Think Critically

30. Apply Concepts The placenta develops from tissues produced by both the embryo and the uterus. How does the structure of the placenta prevent the mother's blood from mixing with the blood of the developing embryo?

31. Infer Occasionally, a zygote does not move into the uterus but attaches to the wall of a Fallopian tube. Why might this be a very dangerous situation for the mother?

32. Draw Conclusions Explain why the suppression of the menstrual cycle is important to the success of a full-term pregnancy.

solve the CHAPTER MYSTERY

OUT OF STRIDE

Although a healthful diet and exercise contribute to maintaining a healthy body, a balance between the two is important. Lisa lost this balance, and the reactions from her endocrine system led to a disorder known as the female athlete triad. The triad consists of three factors:

- **Disordered Eating** During her quest to become a faster runner, Lisa did not provide her body with enough nutrients and energy to support all of its functions.
- **Amenorrhea** Lack of menstrual cycles for three or more months is called amenorrhea. Lisa's hypothalamus responded to low energy levels by not signaling the pituitary to release FSH and LH. As a result, her menstrual cycle ceased and estrogen levels dropped.
- **Weakened Bones** Lisa's bones lost more calcium than normal because of high cortisol and low estrogen levels. This calcium loss, along with the low-calcium levels from her poor diet, led to weakened bones, which are at risk for stress fractures.

The problems associated with the female athlete triad are related to inadequate nutrition. Lisa used more energy and nutrients than she took in. The reaction of her endocrine system was normal, but it had negative effects on her health.

1. Relate Cause and Effect Explain why the menstrual cycle cannot continue without FSH and LH.

2. Sequence Make a flowchart to describe the factors of the female athlete triad.

3. Infer Why do you think that women who have gone through menopause are at risk for osteoporosis—a weakening of the bones due to calcium loss?

4. Connect to the Big idea Explain the three factors that led to Lisa's weakened bones. In a paragraph, propose ways Lisa can prevent this from happening again.

Connecting Concepts

Use Science Graphics

The graph shows the levels of glucose in the blood of two people during a five-hour period immediately following the ingestion of a typical meal. Use the graph to answer questions 33 and 34.

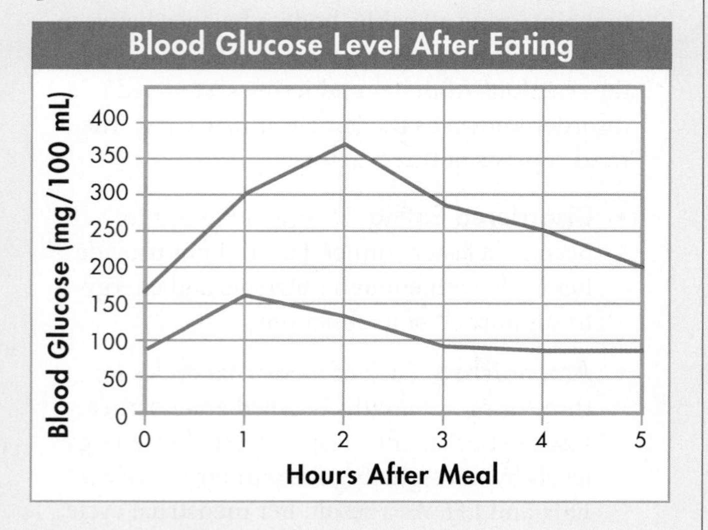

33. Interpret Graphs How long does it take the blood glucose level of the person represented by the blue line to return to a homeostatic value?

34. Integration of Knowledge and Ideas Which line represents a person who may have diabetes? Which line represents a person who does not have diabetes? Explain your answers.

Write About Science

35. Research to Build and Present Knowledge Anabolic steroids are synthetic versions of testosterone. Although anabolic steroids have important medical uses, they can damage the body if abused. Gather relevant information from multiple authoritative print and digital sources to find out more about anabolic steroids. Then write an article for your school newspaper describing steroids' harmful effects. Use your own words to integrate the information into your article. Cite your sources.

36. Assess the Big idea Choose one of the endocrine glands and, using language from the text, describe how that gland is involved in a feedback mechanism that maintains homeostasis.

Integration of Knowledge and Ideas

These graphs show the number of multiple births since 1980. The first graph shows births of twins. The second graph shows multiple births consisting of three or more babies. Use the graphs to answer questions 37 and 38.

37. Interpret Graphs Which birth rate showed the greatest percentage increase from 1980 to 2005?

a. twins
b. triplets or more
c. They increased by the same percentage.
d. It's impossible to tell from the data.

38. Calculate Approximately how many times greater was the number of twin births compared with the number of births of triplets or more in the year 1995? (*Hint:* Note that the numbers on the *y*-axes are different scales.)

a. two times
b. ten times
c. fifteen times
d. twenty times

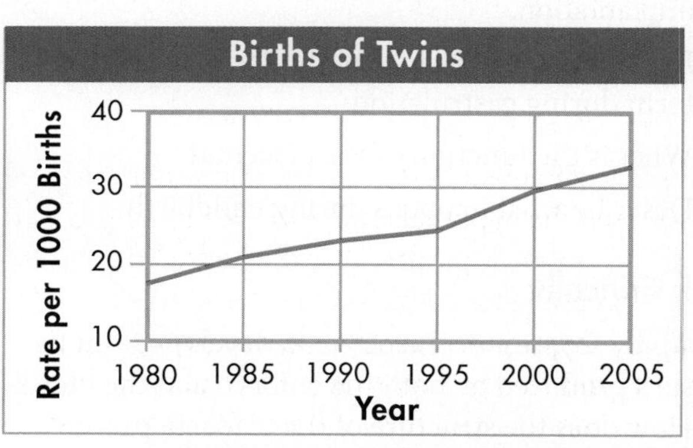

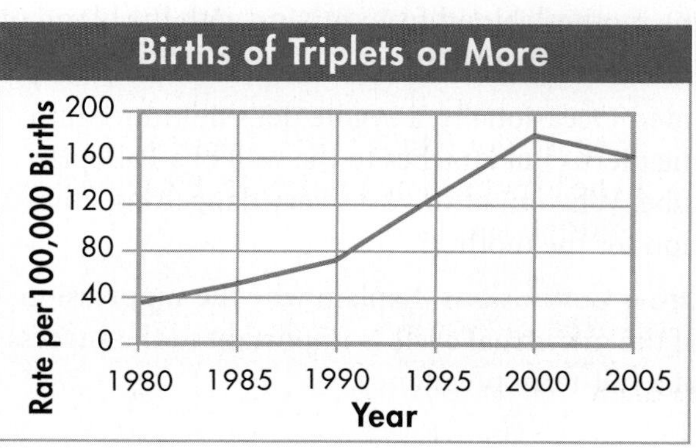

Standardized Test Prep

Multiple Choice

1. Which sequence correctly describes the route sperm take through the human male reproductive system?
   A vas deferens, urethra, epididymis
   B epididymis, vas deferens, urethra
   C vas deferens, epididymis, urethra
   D urethra, epididymis, vas deferens

2. Each of these terms refers to a stage in the human menstrual cycle EXCEPT
   A ovulation.
   B luteal phase.
   C corpus phase.
   D follicular phase.

3. Which of the following is NOT an endocrine gland?
   A pituitary gland
   B parathyroid gland
   C sweat gland
   D adrenal gland

4. During which stage of embryonic development does the neural tube form?
   A implantation
   B gastrulation
   C neurulation
   D fertilization

5. Which of the following is where an egg matures prior to release into the reproductive tract?
   A follicle
   B blastocyst
   C ovary
   D ovum

6. The structure(s) in the male reproductive system that stores mature sperm until they are released by the male reproductive system is (are) the
   A vas deferens.
   B penis.
   C seminiferous tubules.
   D epididymis.

7. Which statement best describes the relationship between the hypothalamus and the pituitary gland?
   A The anterior pituitary gland makes hormones that are released by the hypothalamus.
   B The hypothalamus produces releasing hormones that promote the release of particular hormones from the anterior pituitary.
   C The hypothalamus produces releasing hormones that promote the release of particular hormones from the posterior pituitary.
   D The posterior pituitary sends nervous signals to the hypothalamus to prompt the release of hormones.

Questions 8–11

The diagram below shows the female endocrine system. Use the diagram to answer the questions.

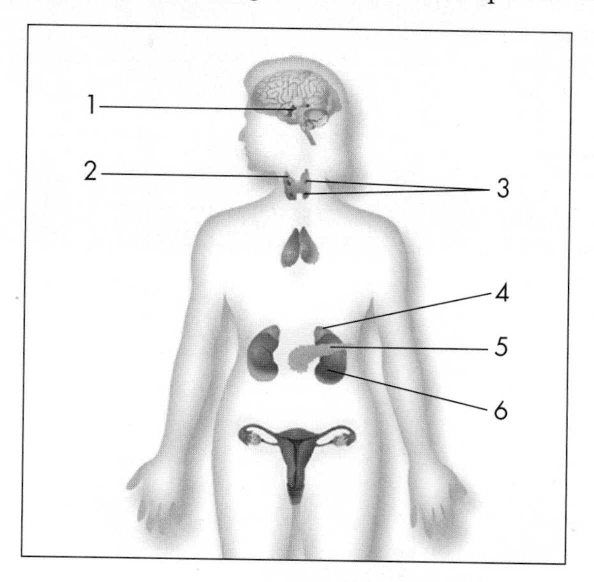

8. Which gland helps the body prepare for and deal with stress?
   A 1
   B 2
   C 4
   D 6

9. Which gland is both an endocrine and an exocrine gland?
   A 2
   B 3
   C 4
   D 5

10. Which gland secretes growth hormone?
    A 1
    B 2
    C 3
    D 5

11. Which gland secretes thyroxine?
    A 1
    B 2
    C 3
    D 4

Open-Ended Response

12. In a paragraph, describe the difference between the origin of fraternal and identical twins.

If You Have Trouble With . . .

Question	1	2	3	4	5	6	7	8	9	10	11	12
See Lesson	34.3	34.3	34.1	34.4	34.3	34.3	34.2	34.2	34.2	34.2	34.2	34.4

35 Immune System and Disease

Big idea **Homeostasis**

Q: How does the body fight against invading organisms that may disrupt homeostasis?

German soldiers on their way to a possible bird flu outbreak

INSIDE:

- 35.1 Infectious Disease
- 35.2 Defenses Against Infection
- 35.3 Fighting Infectious Disease
- 35.4 Immune System Disorders

BUILDING *Scientific Literacy*

Deepen your understanding of the structures and functions of the immune system by engaging in key practices that allow you to make connections across concepts.

NGSS You will formulate a set of questions that could be used to design an inquiry about the emergence and spread of an infectious disease.

STEM Investigate the technologies that help keep healthcare providers protected against the spread of blood-borne pathogens. Then, design a technology of your own.

Common Core You will write a clear and coherent plan for an informational campaign to boost public health practices among the students in your school.

CHAPTER MYSTERY

THE SEARCH FOR A CAUSE

In 1975, researcher Allen Steere faced a medical mystery. Thirty-nine children and several adults living in one small area of Connecticut were suffering from joint pain and inflammation. At first glance, the children's symptoms looked like a rare form of childhood arthritis. And the adults' symptoms seemed to indicate age-related arthritis. But Steere thought it unlikely that there would be so many cases of childhood and age-related arthritis in a small population, in such a short period of time.

Steere looked for another explanation. The patients all lived in small towns and rural areas. Their symptoms all started at more or less the same time of year. Could these patients be suffering from a previously unreported infectious disease?

Biology.com

Finding the solution to this medical mystery is only the beginning. Take a video field trip with the ecogeeks of Untamed Science to continue exploring your world.

Go online to access additional resources including:
• eText • Flash Cards • Lesson Overviews • Chapter Mystery

35.1 Infectious Disease

Key Questions

What causes infectious disease?

How are infectious diseases spread?

Vocabulary

infectious disease
germ theory of disease
Koch's postulates
zoonosis
vector

Taking Notes

Two-Column Table Use a two-column table to list the ways diseases are spread and describe each way.

THINK ABOUT IT For thousands of years, people believed that diseases were caused by curses, evil spirits, or vapors rising from foul marshes or dead plants and animals. In fact, malaria was named after the Italian words *mal aria*, meaning "bad air." This isn't all that surprising, because, until microscopes were invented, most causes of disease were invisible to the human eye!

Causes of Infectious Disease

What causes infectious disease?

During the mid-nineteenth century, French chemist Louis Pasteur and German bacteriologist Robert Koch established a scientific explanation for infectious disease. Pasteur's and Koch's observations and experiments led them to conclude that **infectious diseases** occur when microorganisms cause physiological changes that disrupt normal body functions. Microorganisms were commonly called "germs," so this conclusion was called the **germ theory of disease.** That's unfortunate now, because the word *germ* has no scientific meaning.

Agents of Disease If *germ* isn't a scientific term, how should we describe the causes of infectious disease? **Infectious diseases can be caused by viruses, bacteria, fungi, "protists", and parasites.** Except for parasites, most of these disease-causing microorganisms are called pathogens. **Figure 35–1** provides more information and examples of pathogens and parasites.

FIGURE 35–1 Examples of Agents of Disease Infectious diseases are caused by pathogens and parasites—organisms that invade a body and disrupt its normal functions.

Viruses
Characteristics: nonliving, replicate by inserting their genetic material into a host cell and taking over many of the host cell's functions
Diseases Caused: common cold, influenza, chickenpox, warts
▼ ***Influenza Virus*, Strain taken from a Beijing 1993 epidemic** (TEM 120,000×)

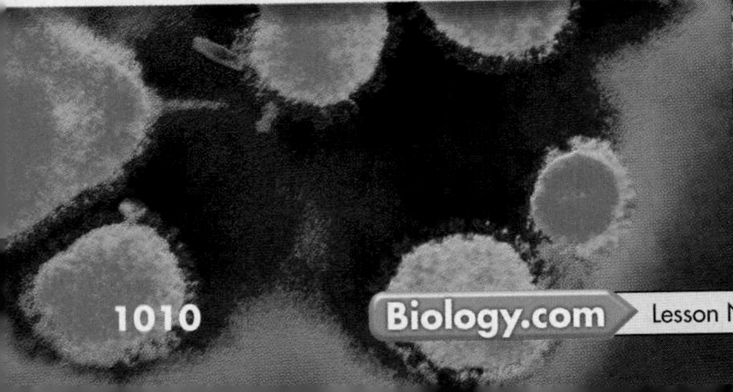

Bacteria
Characteristics: break down the tissues of an infected organism for food, or release toxins that interfere with normal activity in the host
Diseases Caused: streptococcus infections, diphtheria, botulism, anthrax
▼ ***Mycobacterium* causes tuberculosis** (SEM 10,600×)

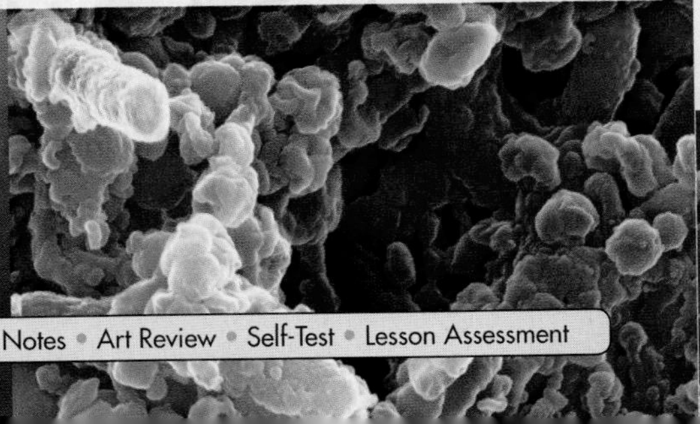

Fungi
Characteristics: cause infections on the surface of the skin, mouth, throat, fingernails, and toenails; dangerous infections may spread from lungs to other organs
Diseases Caused: ringworm, thrush
▼ ***Trichophyton interdigitale* causes athlete's foot** (SEM 2800×)

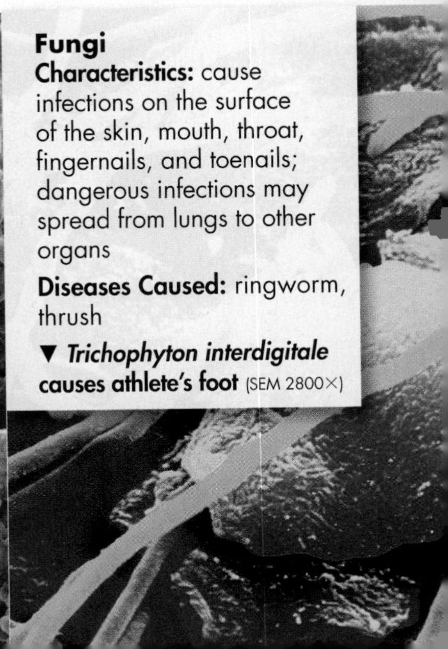

Koch's Postulates Koch's studies with bacteria led him to develop rules for identifying the microorganism that causes a specific disease. These rules are known as **Koch's postulates.**

1. The pathogen must always be found in the body of a sick organism and should not be found in a healthy one.

2. The pathogen must be isolated and grown in the laboratory in pure culture.

3. When the cultured pathogens are introduced into a healthy host, they should cause the same disease that infected the original host.

4. The injected pathogen must be isolated from the second host. It should be identical to the original pathogen.

Koch's ideas played such a vital role in the development of modern medicine that he was awarded a Nobel Prize in 1905. Today, we know that there can be exceptions to these rules, but they remain important guidelines for identifying the causes of new and emerging diseases.

Symbionts vs. Pathogens Parts of the human body provide excellent habitats for microorganisms. Fortunately, most microorganisms that take advantage of our hospitality are symbionts that are either harmless or actually beneficial. Yeast and bacteria grow in the mouth and throat without causing trouble. Bacteria in the large intestine help with digestion and produce vitamins. In fact, if all your cells disappeared, the outlines of your body and digestive tract would still be recognizable—as a ghostly outline of microorganisms!

What's the difference between harmless microorganisms and pathogens that cause disease? The "good guys" obtain nutrients, grow, and reproduce without disturbing normal body functions. The "bad guys" cause problems in various ways. Some viruses and bacteria directly destroy the cells of their host. Other bacteria and single-celled parasites release poisons that kill the host's cells or interfere with their normal functions. Parasitic worms may block blood flow through blood vessels or organs, take up the host's nutrients, or disrupt other body functions.

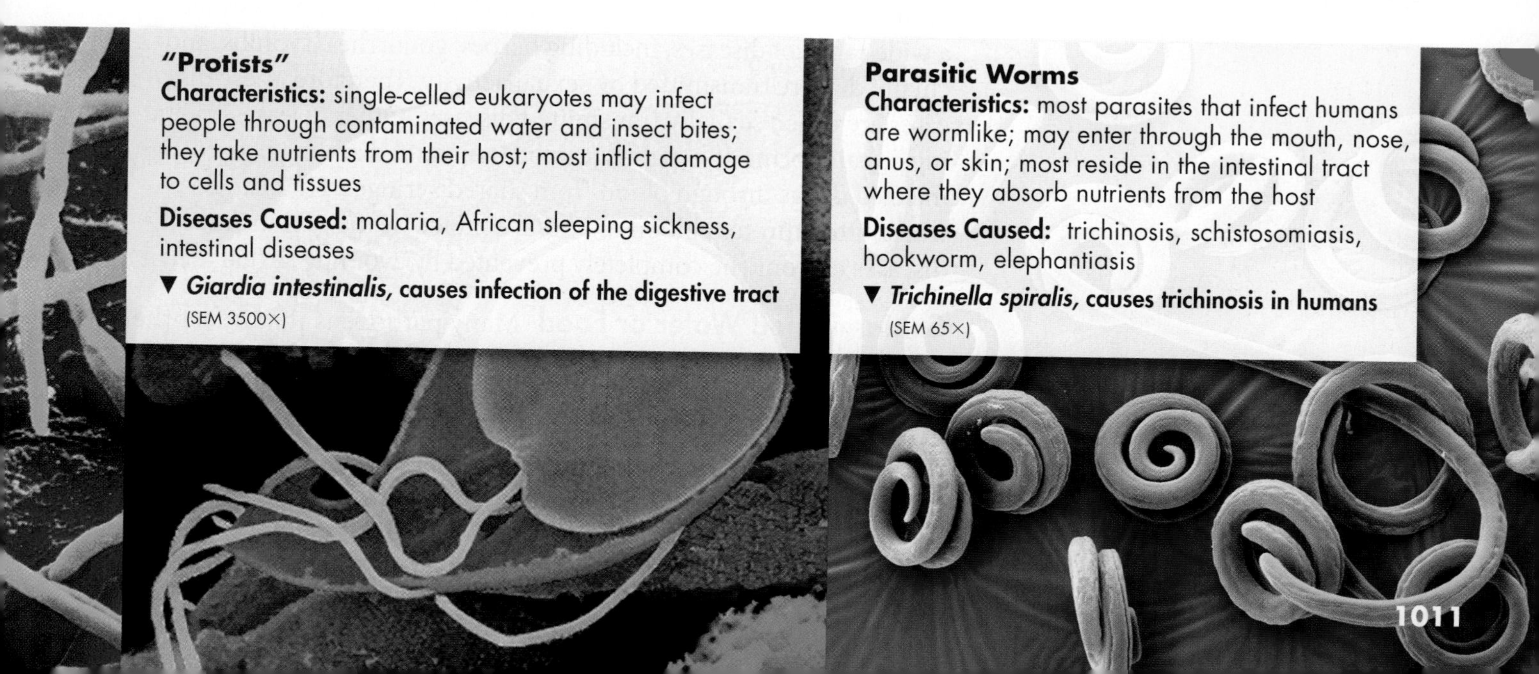

"Protists"

Characteristics: single-celled eukaryotes may infect people through contaminated water and insect bites; they take nutrients from their host; most inflict damage to cells and tissues

Diseases Caused: malaria, African sleeping sickness, intestinal diseases

▼ ***Giardia intestinalis*, causes infection of the digestive tract** (SEM 3500×)

Parasitic Worms

Characteristics: most parasites that infect humans are wormlike; may enter through the mouth, nose, anus, or skin; most reside in the intestinal tract where they absorb nutrients from the host

Diseases Caused: trichinosis, schistosomiasis, hookworm, elephantiasis

▼ ***Trichinella spiralis*, causes trichinosis in humans** (SEM 65×)

How Diseases Spread

How are infectious diseases spread?

Infectious diseases can be spread in a number of ways. **Some diseases are spread through coughing, sneezing, physical contact, or exchange of body fluids. Some diseases are spread through contaminated water or food. Still other diseases are spread to humans from infected animals.**

Pathogens are often spread by symptoms of disease, such as sneezing, coughing, or diarrhea. In many cases, these symptoms are changes in host behavior that help pathogens spread and infect new hosts! After all, if a virus infects only one host, that virus will die when the host's immune system kills it or when the host dies. For that reason, natural selection favors pathogens with adaptations that help them spread from host to host.

FIGURE 35–2 Sneezing Some infectious diseases are spread from person to person by sneezing. Thousands of pathogen particles can be released in a sneeze. **Infer** ***Why is it more beneficial to sneeze into a tissue rather than covering your mouth with your hand?***

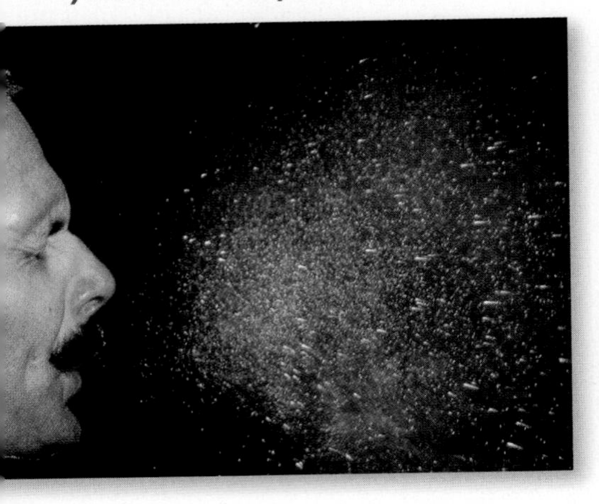

Coughing, Sneezing, and Physical Contact Many bacteria and viruses that infect the nose, throat, or respiratory tract are spread by indirect contact. Coughing and sneezing releases thousands of tiny droplets that can be inhaled by other people. Those droplets also settle on objects such as doorknobs. If you touch those objects and then touch your mouth or nose, you can transfer the pathogens to a new home! Thus, the ability of a flu virus or a tuberculosis bacterium to cause a host to sneeze or cough is an adaptation that increases transmission of the pathogen from one host to another.

Other pathogens, including drug-resistant staphylococci that cause skin infections, can be transferred by almost any kind of body-to-body contact. They can also be transferred by contact with towels or certain kinds of sports equipment.

Minimizing transmission of these diseases is surprisingly simple. The most important means of infection control is thorough and frequent hand washing. If you have a cold or flu, cover your mouth with a tissue when you cough or sneeze, and wash your hands regularly.

BUILD Vocabulary

PREFIXES The prefix *trans-*, used in words such as *transmission, transferred,* and *transportation,* comes from the Latin *trans-* which means "across" or "beyond."

Exchange of Body Fluids Some pathogens require specific kinds of direct contact to be transferred from host to host. For example, a wide range of diseases, including herpes, gonorrhea, syphilis, and chlamydia, are transmitted by sexual activity. Therefore, these diseases are called sexually transmitted diseases. Other diseases, including certain forms of hepatitis, can be transmitted among users of injected drugs through blood from shared syringes. HIV can be transmitted through blood or sexual contact. Sexually transmitted diseases can only be completely prevented by avoiding sexual activity.

Contaminated Water or Food Many pathogens that infect the digestive tract are spread through water contaminated with feces from infected people or other animals. Symptoms of these diseases often include serious diarrhea. This is another adaptation that helps pathogens spread from one host to another, especially in places with poor sanitation.

Contaminated water may be consumed, or it may carry pathogens onto fruits or vegetables. If those foods are eaten without being washed thoroughly, infection can result. In recent years, several disease outbreaks have been traced to transmission through packaged salad greens.

Bacteria of several kinds are commonly present in seafood and uncooked meat, especially ground meat. If meats and seafood are not stored and cooked properly, illness can result.

MYSTERY CLUE

Many of the sick children remembered receiving strange insect bites that summer, which developed into rashes. What clue did this give Steere?

Zoonoses: The Animal Connection Many diseases that have made headlines in recent years thrive in both human and other animal hosts. Any disease that can be transmitted from animals to humans is called a **zoonosis** (plural: zoonoses). Mad cow disease, severe acute respiratory syndrome (SARS), West Nile virus, Lyme disease, Ebola, and bird flu are all zoonoses. Transmission can occur in various ways. Sometimes an animal carries, or transfers, zoonotic diseases from an animal host to a human host. These carriers, called **vectors,** transport the pathogen but usually do not get sick themselves. In other cases, infection may occur when a person is bitten by an infected animal, consumes the meat of an infected animal, or comes in close contact with an infected animal's wastes or secretions.

FIGURE 35–3 Vectors Vectors are animals that harbor a pathogen. The pathogen may spread to a human through the bite of the vector, or when a person eats the vector.

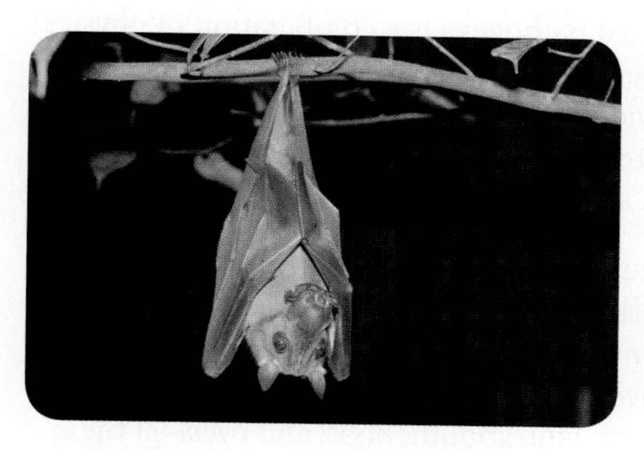

▲ **Fruit bat that may carry the Ebola virus**

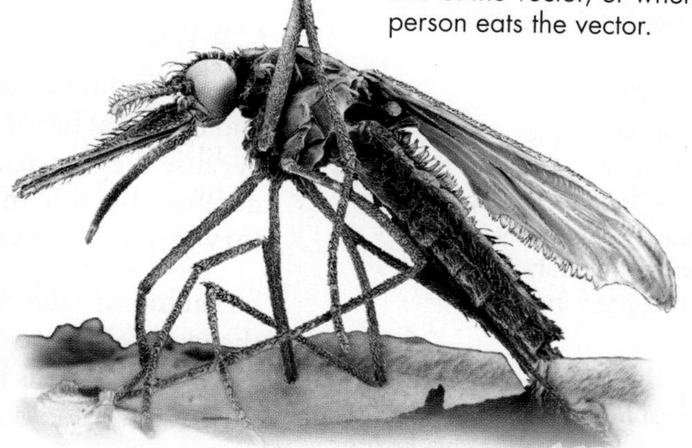

▲ **Mosquito that transfers West Nile virus from birds to humans**

35.1 Assessment

Review Key Concepts

1. a. Review List the types of organisms that can cause disease.

b. Explain What are ways that pathogens can cause disease in their hosts?

c. Infer If a researcher introduced a suspected pathogen into many healthy hosts, but none of them became sick, what could this indicate?

2. a. Review What are the ways in which infectious diseases are spread?

b. Explain How do vectors contribute to the spread of disease?

c. Apply Concepts Why do you think it's a beneficial adaptation for a pathogen to make its host very sick without killing the host? (*Hint:* Think about how viruses replicate.)

WRITE ABOUT SCIENCE

Description

3. Animals infected with the virus that causes rabies often salivate excessively and are apt to bite other animals even when unprovoked. In a paragraph, explain how these symptoms lead to the spread of the virus.

35.2 Defenses Against Infection

Key Questions

What are the body's nonspecific defenses against pathogens?

What is the function of the immune system's specific defenses?

What are the body's specific defenses against pathogens?

Vocabulary

inflammatory response • histamine • interferon • fever • immune response • antigen • antibody • humoral immunity • cell-mediated immunity

Taking Notes

Concept Map Use the green and blue headings in this lesson to make a concept map. Add details to your map as you read.

FIGURE 35–4 Nonspecific Defenses On the walls of the trachea, pollen grains (yellow) are trapped by mucus and carried by hair-like cilia away from the lungs (SEM 500×).

THINK ABOUT IT With pathogens all around us, it might seem amazing that most of us aren't sick most of the time. Why are we usually free from infections, and why do we usually recover from pathogens that do infect us? One reason is that our bodies have an incredibly powerful and adaptable series of defenses that protect us against a wide range of pathogens.

Nonspecific Defenses

What are the body's nonspecific defenses against pathogens?

The body's first defense against pathogens is a combination of physical and chemical barriers. These barriers are called nonspecific defenses because they act against a wide range of pathogens. **Nonspecific defenses include the skin, tears and other secretions, the inflammatory response, interferons, and fever.**

First Line of Defense The most widespread nonspecific defense is the physical barrier we call skin. Very few pathogens can penetrate the layers of dead cells that form the skin's surface.

But your skin doesn't cover your entire body. Pathogens could easily enter your body through your mouth, nose, and eyes—if these tissues weren't protected by other nonspecific defenses. For example, saliva, mucus, and tears contain lysozyme, an enzyme that breaks down bacterial cell walls. Mucus in your nose and throat traps pathogens. Then, cilia push the mucous-trapped pathogens away from your lungs. Stomach secretions destroy many pathogens that are swallowed.

Second Line of Defense If pathogens make it into the body, through a cut in the skin, for example, the body's second line of defense swings into action. These mechanisms include the inflammatory response, the actions of interferons, and fever.

▶ ***Inflammatory Response*** The **inflammatory response** gets its name because it causes infected areas to become red and painful, or inflamed. As shown in **Figure 35–5,** the response begins when pathogens stimulate cells called mast cells to release chemicals known as histamines.

Histamines increase the flow of blood and fluids to the affected area. Fluid leaking from expanded blood vessels causes the area to swell. White blood cells move from blood vessels into infected tissues. Many of these white blood cells are phagocytes, which engulf and destroy bacteria. All this activity around a wound may cause a local rise in temperature. That's why a wounded area sometimes feels warm.

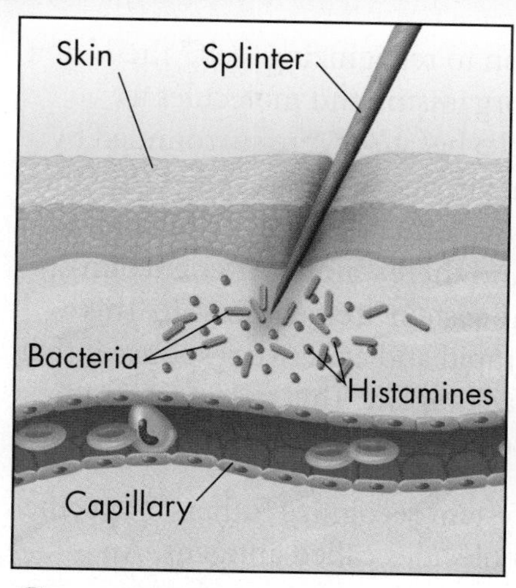

1 In response to the wound and invading pathogens, mast cells release histamines, which stimulate increased blood flow to the area.

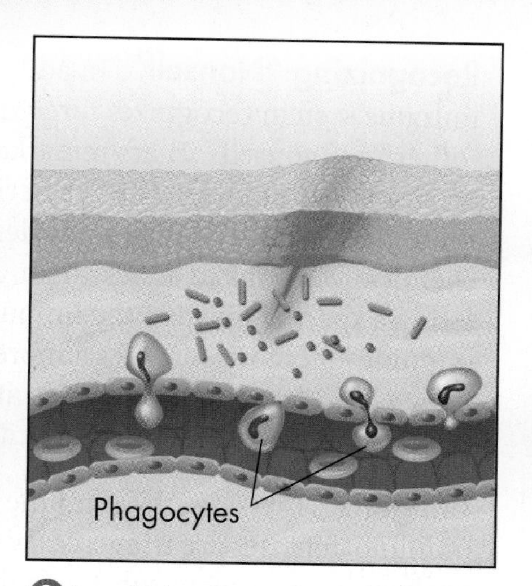

2 Local blood vessels dilate. Fluid leaves the capillaries and causes swelling. Phagocytes move into the tissue.

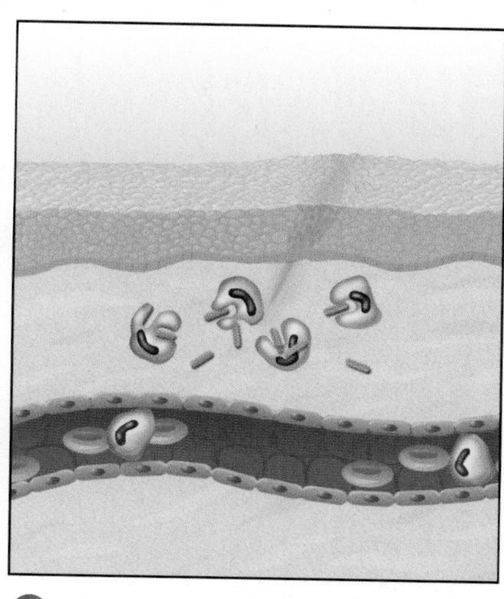

3 Phagocytes engulf and destroy the bacteria and damaged cells.

FIGURE 35–5 Inflammatory Response The inflammatory response is a nonspecific defense reaction to tissue damage caused by injury or infection. When pathogens enter the body, phagocytes move into the area and engulf the pathogens. Infer ***What part of the inflammatory response leads to redness around a wounded area?***

▶ ***Interferons*** When viruses infect body cells, certain host cells produce proteins that inhibit synthesis of viral proteins. Scientists named these proteins **interferons** because they "interfere" with viral growth. By slowing down the production of new viruses, interferons "buy time" for specific immune defenses to respond and fight the infection.

▶ ***Fever*** The immune system also releases chemicals that increase body temperature, producing a **fever.** Increased body temperature may slow down or stop the growth of some pathogens. Higher body temperature also speeds up several parts of the immune response.

In Your Notebook *Develop an analogy that compares the body's nonspecific defenses to a large building's security system.*

Specific Defenses: The Immune System

What is the function of the immune system's specific defenses?

The main function of the immune system's specific defenses is easy to describe but complex to explain. **The immune system's specific defenses distinguish between "self" and "other," and they inactivate or kill any foreign substance or cell that enters the body.** Unlike the nonspecific defenses, which respond to the general threat of infection, specific defenses respond to a particular pathogen.

Recognizing "Self" A healthy immune system recognizes all cells and proteins that belong in the body, and treats these cells and proteins as "self." It recognizes chemical markers that act like a secret password that says, "I belong here. Don't attack me!" Because genes program the passwords, no two individuals—except identical twins—ever use the same one. This ability to recognize "self" is essential, because the immune system controls powerful cellular and chemical weapons that could cause problems if turned against the body's own cells.

Recognizing "Nonself" In addition to recognizing "self," the immune system recognizes foreign organisms and molecules as "other," or "nonself." That's remarkable, because we're surrounded by an almost infinite variety of bacteria, viruses, and parasites. Once the immune system recognizes invaders as "others," it uses cellular and chemical weapons to attack them. And there's more. After encountering a specific invader, the immune system "remembers" it. This immune "memory" enables a more rapid and effective response if that same pathogen, or a similar one, attacks again. This specific recognition, response, and memory are called the **immune response.**

Antigens How does the immune system recognize "others"? Specific immune defenses are triggered by molecules called antigens. An **antigen** is any foreign substance that can stimulate an immune response. Typically, antigens are located on the outer surfaces of bacteria, viruses, or parasites. The immune system responds to antigens by increasing the number of cells that either attack the invaders directly or that produce proteins called antibodies.

The main role of **antibodies** is to tag antigens for destruction by immune cells. Antibodies may be attached to particular immune cells or may be free-floating in plasma. The body makes up to 10 billion different antibodies. The shape of each type of antibody allows it to bind to one specific antigen.

FIGURE 35–6 B Lymphocyte

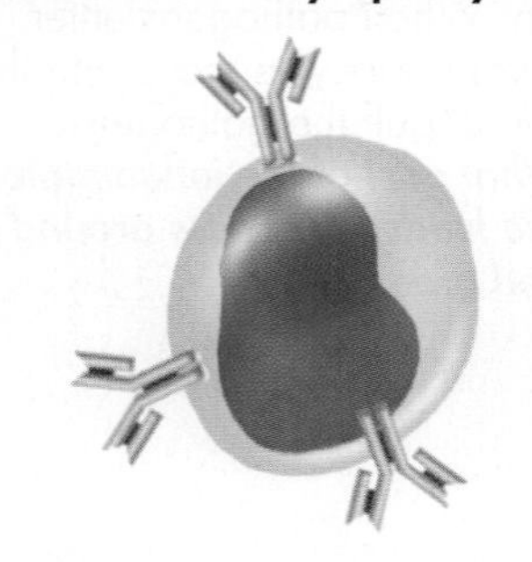

FIGURE 35–7 T Lymphocyte

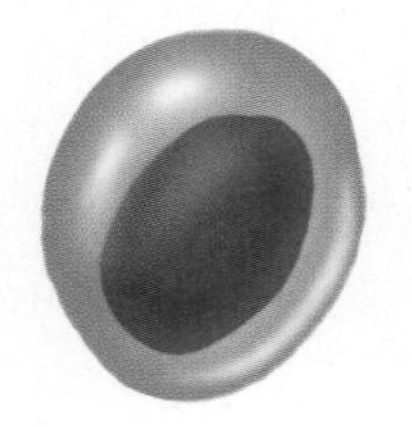

Lymphocytes The immune system guards the entire body, which means its cells must travel throughout the body. The main working cells of the immune response are B lymphocytes (B cells) and T lymphocytes (T cells). B cells are produced in, and mature in, red bone marrow. T cells are produced in the bone marrow but mature in the thymus—an endocrine gland. Each B cell and T cell is capable of recognizing *one* specific antigen. A person's genes determine the particular B and T cells that are produced. When mature, both types of cells travel to lymph nodes and the spleen, where they will encounter antigens.

Although both types of cells recognize antigens, they go about it differently. B cells, with their embedded antibodies, discover antigens in body fluids. T cells must be presented with an antigen by infected body cells or immune cells that have encountered antigens.

The Immune System in Action

What are the body's specific defenses against pathogens?

B and T cells continually search the body for antigens or signs of antigens. **The specific immune response has two main styles of action: humoral immunity and cell-mediated immunity.**

Humoral Immunity The part of the immune response called **humoral immunity** depends on the action of antibodies that circulate in the blood and lymph. This response is activated when antibodies embedded on a few existing B cells bind to antigens on the surface of an invading pathogen.

BUILD Vocabulary

WORD ORIGINS The word *humor* comes from the Latin word for moisture. Body fluids such as blood, lymph, and hormones are sometimes referred to as humors. **Humoral immunity** refers to the immune response that happens in body fluids.

How does this binding occur? As shown in **Figure 35–8**, an antibody is shaped like the letter Y and has two identical antigen-binding sites. The shapes of the binding sites enable an antibody to recognize a specific antigen with a complementary shape.

When an antigen binds to an antibody carried by a B cell, T cells stimulate the B cell to grow and divide rapidly. That growth and division produces many B cells of two types: plasma cells and memory B cells.

▶ ***Plasma Cells*** Plasma cells produce and release antibodies that are carried through the bloodstream. These antibodies recognize and bind to free-floating antigens or to antigens on the surfaces of pathogens. When antibodies bind to antigens, they act like signal flags to other parts of the immune system. Several types of cells and proteins respond to that signal by attacking and destroying invaders. Some types of antibodies can disable invaders until they are destroyed.

A healthy adult can produce about 10 billion different types of antibodies, each of which can bind to a different type of antigen! This antibody diversity enables the immune system to respond to virtually any kind of "other" that enters the body.

In Your Notebook *It is a common misconception that the immune system cannot combat pathogens it has not encountered before. In a paragraph, explain why that statement is not true.*

▶ ***Memory B Cells*** Plasma cells die after an infection is gone. But some B cells that recognize a particular antigen remain alive. These cells, called memory B cells, react quickly if the same pathogen enters the body again. Memory B cells rapidly produce new plasma cells to battle the returning pathogen. This secondary response occurs much faster than the first response to a pathogen. Immune memory helps provide long-term immunity to certain diseases and is the reason that vaccinations work. **Figure 35–11** summarizes the first and second response of humoral immunity.

FIGURE 35–8 Antibody Structure

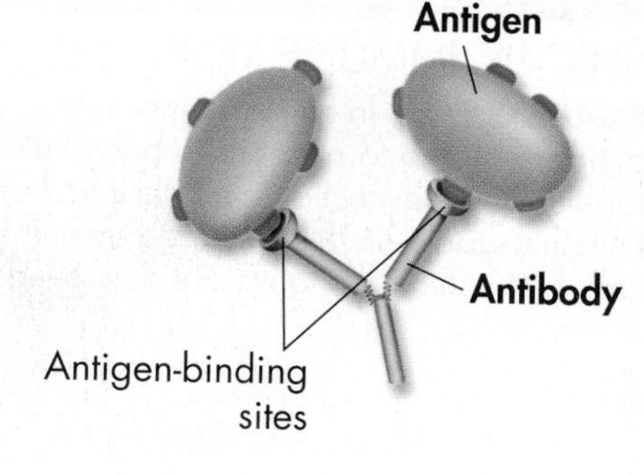

FIGURE 35–9 Plasma Cells

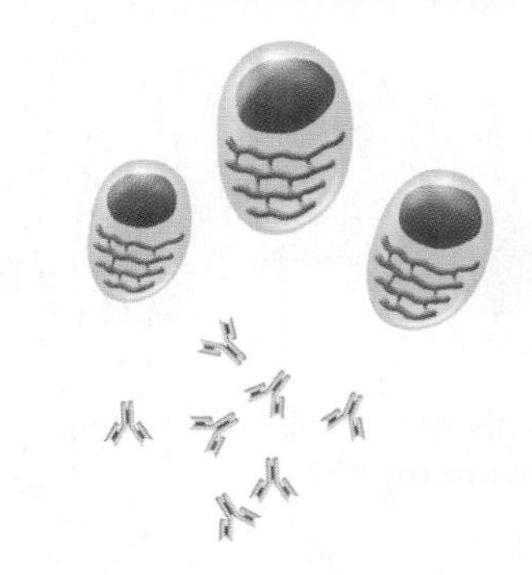

FIGURE 35–10 Memory B Cells

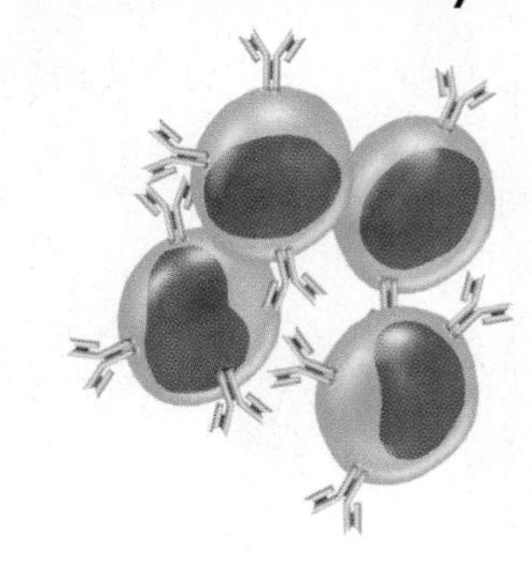

Analyzing Data

Immune System "Memory"

Antibody concentration in a person's blood reveals the difference between the first and second immune response. Day 1 indicates the first exposure to Antigen A. Day 28 marks a second exposure to Antigen A and the first exposure to Antigen B.

1. Interpret Graphs After first exposure to an antigen, about how long does it take for antibodies to reach a detectable level?

2. Infer What could explain the significant increase in antibodies to A seen after Day 30?

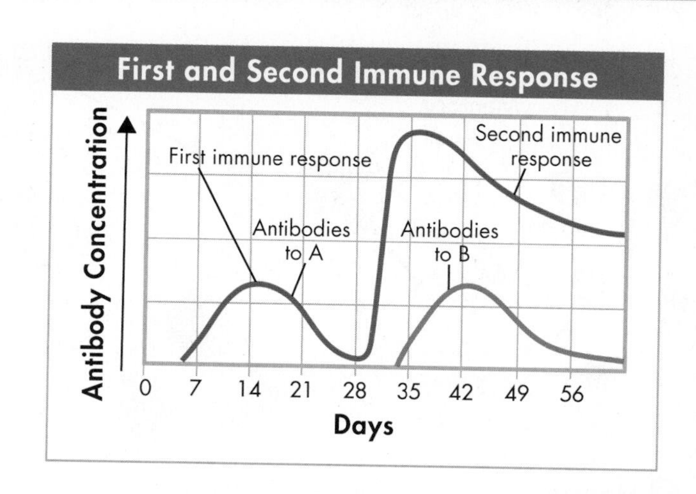

VISUAL SUMMARY

SPECIFIC IMMUNE RESPONSE

FIGURE 35–11 In humoral immunity, antibodies bind to antigens in body fluids and tag them for destruction by other parts of the immune system. In cell-mediated immunity, body cells that contain antigens are destroyed.

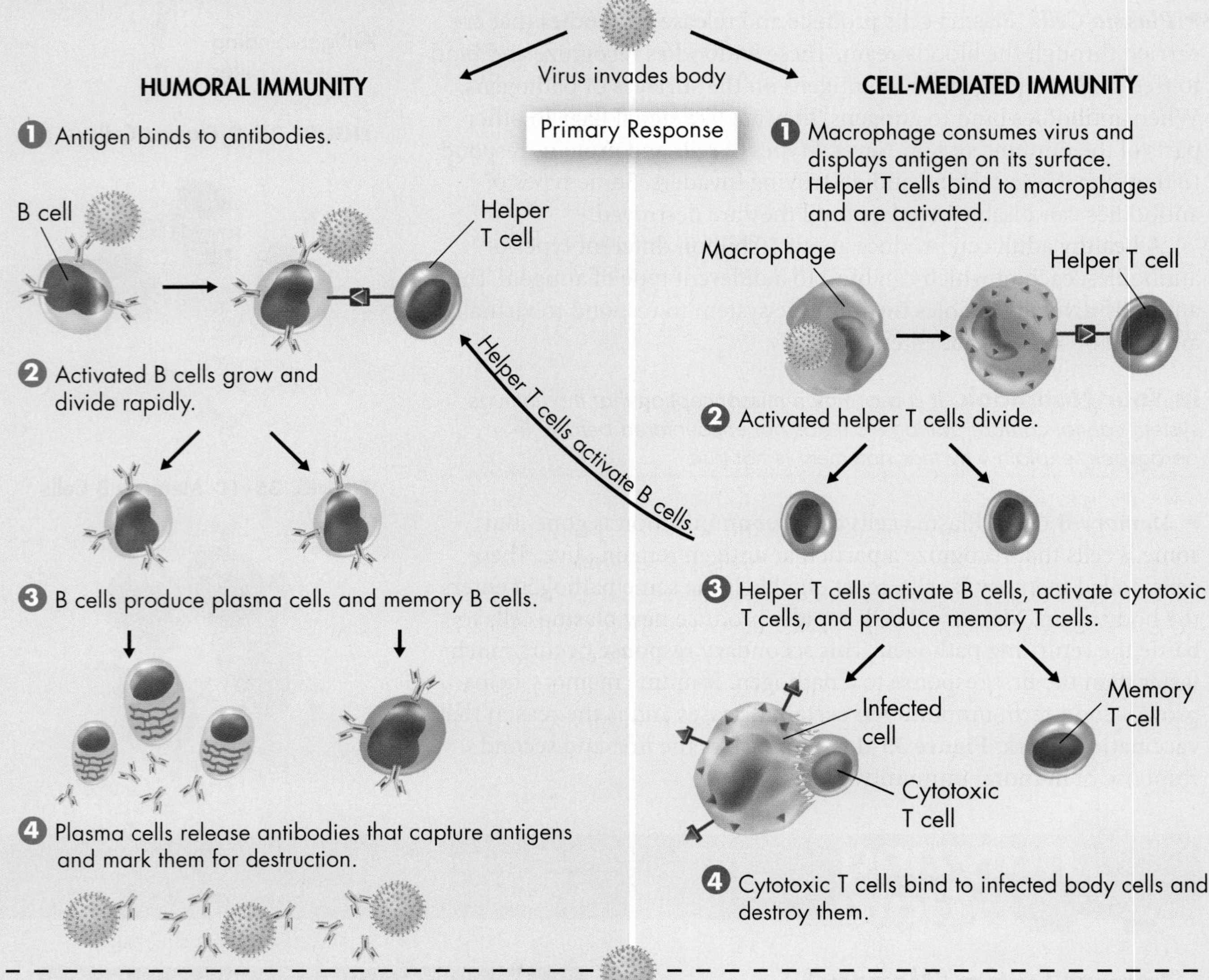

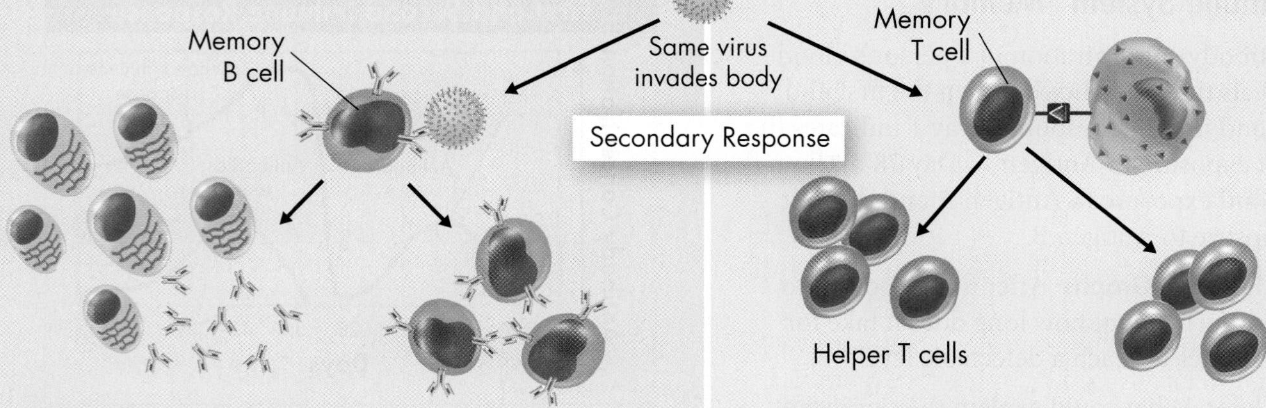

Cell-Mediated Immunity Another part of the immune response, which depends on the action of macrophages and several types of T cells, is called **cell-mediated immunity.** This part of the immune system defends the body against some viruses, fungi, and single-celled pathogens that do their dirty work inside body cells. T cells also protect the body from its own cells if they become cancerous.

When a cell is infected by a pathogen or when a macrophage consumes a pathogen, the cell displays a portion of the antigen on the outer surface of its membrane. This membrane attachment is a signal to circulating T cells called helper T cells. Activated helper T cells divide into more helper T cells, which go on to activate B cells, activate cytotoxic T cells, and produce memory T cells.

Cytotoxic T cells hunt down body cells infected with a particular antigen and kill the cells. They kill infected cells by puncturing their membranes or initiating apoptosis (programmed cell death). Memory helper T cells enable the immune system to respond quickly if the same pathogen enters the body again.

Another type of T cell, called suppressor T cells, helps to keep the immune system in check. They inhibit the immune response once an infection is under control. They may also be involved in preventing autoimmune diseases.

Although cytotoxic T cells are helpful in the immune system, they make the acceptance of organ transplants difficult. When an organ is transplanted from one person to another, the normal response of the recipient's immune system would be to recognize it as nonself. T cells and proteins would damage and destroy the transplanted organ. This process is known as rejection. To prevent organ rejection, doctors search for a donor whose cell markers are nearly identical to the cell markers of the recipient. Still, organ recipients must take drugs—usually for the rest of their lives—to suppress the cell-mediated immune response.

FIGURE 35–12 Cytotoxic T cell

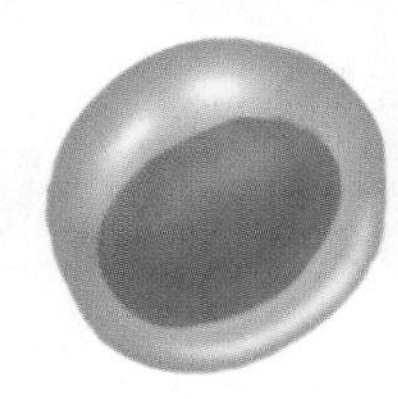

FIGURE 35–13 Memory T cell

35.2 Assessment

Review Key Concepts

1. a. **Review** List the body's nonspecific defenses against pathogens.
   b. **Sequence** Describe the steps of the inflammatory response.
2. a. **Review** How does the immune system identify a pathogen?
   b. **Compare and Contrast** How are the roles of B and T cells different? How are their roles similar?
3. a. **Review** What are the two main styles of action of the specific immune response?
   b. **Apply Concepts** Why would a disease that destroys helper T cells also compromise the humoral response?

VISUAL THINKING

4. These two T cells are attached to a cancer cell. What type of immune response are these cells a part of?

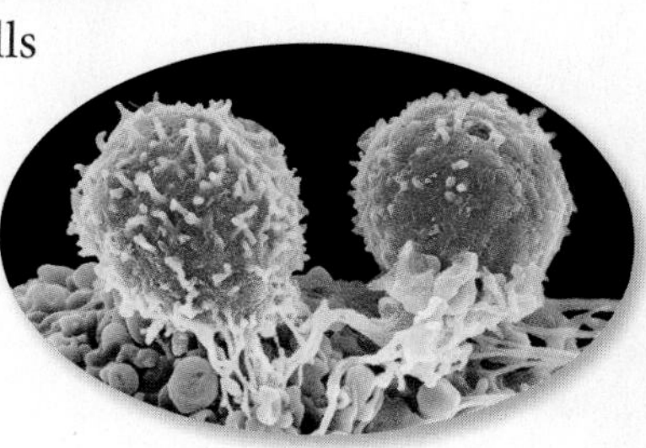

35.3 Fighting Infectious Disease

Key Questions

How do vaccines and externally produced antibodies fight disease?

How do public health measures and medications fight disease?

Why have patterns of infectious diseases changed?

Vocabulary

vaccination
active immunity
passive immunity

Taking Notes

Venn Diagram Make a Venn diagram that compares and contrasts active and passive immunity.

THINK ABOUT IT More than 200 years ago, English physician Edward Jenner noted that milkmaids who contracted a mild disease called cowpox didn't develop smallpox. At the time, smallpox was a widespread disease that killed many people. Jenner wondered, could people be protected from smallpox by deliberately infecting them with cowpox?

Acquired Immunity

How do vaccines and externally produced antibodies fight disease?

Jenner performed a bold experiment. He put fluid from a cowpox patient's sore into a small cut he made on the arm of a young boy named James Phipps. As expected, James developed mild cowpox. Two months later, Jenner injected James with fluid from a smallpox infection. Fortunately for James (and Jenner!), the boy didn't develop smallpox. His cowpox infection had protected him from smallpox infection. Ever since that time, the injection of a weakened form of a pathogen, or of a similar but less dangerous pathogen, to produce immunity has been known as a **vaccination.** The term comes from the Latin word *vacca*, meaning "cow," as a reminder of Jenner's work.

Active Immunity Today, we understand how vaccination works. **Vaccination stimulates the immune system with an antigen. The immune system produces memory B cells and memory T cells that quicken and strengthen the body's response to repeated infection.** This kind of immunity, called **active immunity,** may develop as a result of natural exposure to an antigen (fighting an infection) or from deliberate exposure to the antigen (through a vaccine).

Passive Immunity Disease can be prevented in another way. **Antibodies produced against a pathogen by other individuals or animals can be used to produce temporary immunity.** If externally produced antibodies are introduced into a person's blood, the result is **passive immunity.** Passive immunity lasts only a short time because the immune system eventually destroys the foreign antibodies.

Passive immunity can also occur naturally or by deliberate exposure. Natural passive immunity occurs when antibodies are passed from a pregnant woman to the fetus (across the placenta), or to an infant through breast milk. For some diseases, antibodies from humans or animals can be injected into an individual. For example, people who have been bitten by rabid animals are injected with antibodies for the rabies virus.

FIGURE 35–14 Jenner Vaccinating James Phipps

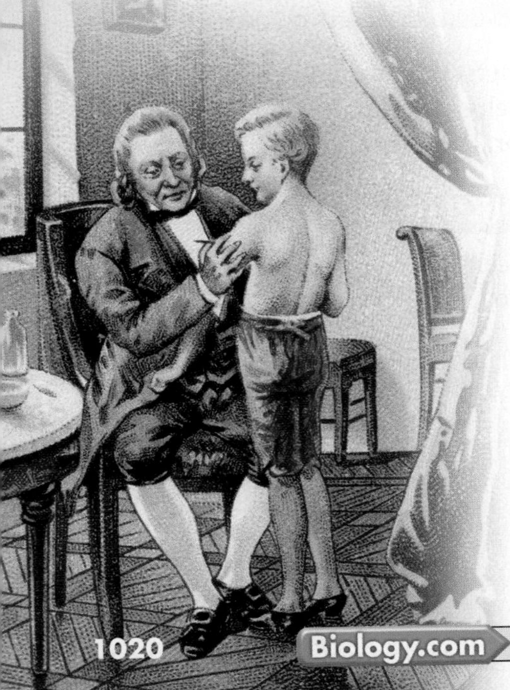

Quick Lab
GUIDED INQUIRY

How Do Diseases Spread? 

❶ Your teacher has placed a fluorescent material in the classroom to simulate a virus. Keep track of the people and objects you touch. Then, use a UV lamp to check for the "virus" on your hands, objects, and people you have touched since entering the classroom. **CAUTION:** *Do not look directly at the UV light.*

❷ Exchange results with your classmates to determine how the "virus" spread through the classroom. Wash your hands with soap and warm water.

Analyze and Conclude

1. **Infer** What can you infer about how the "virus" spread through the classroom?

2. **Apply Concepts** How does thorough hand washing help prevent the spread of diseases?

Public Health and Medications

How do public health measures and medications fight disease?

In 1900, more than 30 percent of deaths in the United States were caused by infectious disease. In 2005, less than 5 percent of deaths were caused by infectious disease. Two factors that contributed to this change are public health measures and the development of medications.

Public Health Measures When humans live in large groups, behavior, cleanliness of food and water supplies, and sanitation all influence the spread of disease. The field of public health offers services and advice that help provide healthy conditions. **Public health measures help prevent disease by monitoring and regulating food and water supplies, promoting vaccination, and promoting behaviors that avoid infection.** Promoting childhood vaccinations and providing clean drinking water are two important public health activities that have greatly reduced the spread of many diseases that once killed many people.

Medications Prevention of infectious disease is not always possible. Medications, such as antibiotics and antiviral drugs, are other weapons that can fight pathogens. **Antibiotics can kill bacteria, and some antiviral medications can slow down viral activity.**

The term *antibiotic* refers to a compound that kills bacteria without harming its host. In 1928, Alexander Fleming was the first scientist to discover an antibiotic. Fleming noticed that a mold, *Penicillium notatum*, seemed to produce something that inhibited bacterial growth. Research determined that this "something" was a compound Fleming named penicillin. Researchers learned to mass-produce penicillin just in time for it to save thousands of World War II soldiers. Since then, dozens of antibiotics have saved countless numbers of lives.

Antibiotics have no effect on viruses. However, antiviral drugs have been developed to fight certain viral infections. These drugs generally inhibit the ability of viruses to invade cells or to multiply once inside cells.

In Your Notebook *How does your school promote public health?*

FIGURE 35–15 Broad Street Pump In 1854, through investigation that included interviewing residents and mapping, Dr. John Snow learned that the source of a London cholera outbreak was a water pump like this replica. This is a major event in the history of public health.

If Steere's patients were helped by antibiotics, what clue would this have given him about the disease's pathogen?

FIGURE 35–16 Causes of Emerging Disease Illegally imported animals can lead to the spread of emerging disease. **A.** In 2003, dormice and other rodents from Africa spread monkeypox to prairie dogs in the United States, which then infected humans. **B.** The spread of SARS also has been associated with the wild animal trade.

A. Dormouse

B. Students wearing masks to protect them from diseases such as SARS

New and Re-Emerging Diseases

Why have patterns of infectious diseases changed?

By 1980, many people thought that medicine had conquered infectious disease. Vaccination and other public health measures had wiped out polio in the United States and had eliminated smallpox globally. Antibiotics seemed to have bacterial diseases under control. Some exotic diseases remained in the tropics, but researchers were confident that epidemics would soon be history. Unfortunately, they were wrong.

In recent decades, a host of new diseases have appeared, including AIDS, SARS, hantavirus, monkeypox, West Nile virus, Ebola, and avian influenza ("bird flu"). Other diseases that people thought were under control are re-emerging as a threat and spreading to new areas. What's going on?

Changing Interactions With Animals **Two major reasons for the emergence of new diseases are the ongoing merging of human and animal habitats and the increase in the exotic animal trade.** As people clear new areas of land and as environments change, people come in contact with different animals and different pathogens. Exotic animal trade, for pets and food, has also given pathogens new opportunities to jump from animals to humans. Both monkeypox and SARS are thought to have started this way. Pathogens are also evolving in ways that enable them to infect different hosts.

Misuse of Medications **Misuse of medications has led to the re-emergence of diseases that many people thought were under control.** For example, many strains of the pathogens that cause tuberculosis and malaria are evolving resistance to a wide variety of antibiotics and other medications. In addition, diseases such as measles are making a comeback because some people fail to follow vaccination recommendations.

35.3 Assessment

Review Key Concepts

1. a. **Review** Explain how vaccinations and externally produced antibodies help the immune system fight disease.
   b. **Compare and Contrast** Describe the difference between active and passive immunity.
2. a. **Review** What are the goals of public health measures?
   b. **Relate Cause and Effect** Why is it important to discern if a sickness is caused by a bacterium or a virus?
3. a. **Review** Describe two major contributing factors involved in the spread of new and re-emerging diseases.
   b. **Infer** How do you think the ease of global travel has affected the spread of emerging diseases? Explain.

Apply the Big idea

Science as a Way of Knowing

4. **Research to Build and Present Knowledge** Getting vaccinated is much safer than getting the disease that the vaccine prevents. However, vaccines are capable of causing side effects. As a class, arrange a debate that addresses the benefits and risks of vaccinations. Use multiple, reliable sources to synthesize information to support your arguments.

Biology & HISTORY

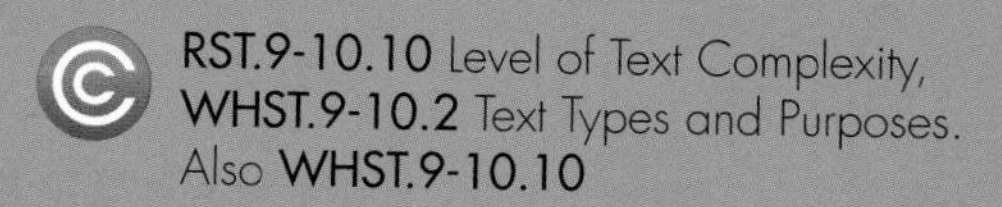

RST.9-10.10 Level of Text Complexity, WHST.9-10.2 Text Types and Purposes. Also WHST.9-10.10

Emerging Diseases Due to factors such as changing interactions with animals and misuse of medications, the problem of infectious disease is far from solved.

1967
Surgeon General William H. Stewart announces, "It is time to close the book on infectious diseases."

1975
Lyme disease is first documented in the United States.

1976
First outbreak of Ebola occurs in the Democratic Republic of the Congo.

1981
First reports surface of illness later identified as AIDS in Los Angeles.

1983
HIV is identified as the cause of AIDS.

1986
Researchers discover bovine spongiform encephalopathy (BSE), commonly called mad cow disease, in cattle in Britain.

1996
The British government admits that humans can contract BSE from eating infected beef.

2002
First SARS outbreak occurs in China's Guangdong province.

2003
The United States reports its first case of mad cow disease in Washington State.

CDC reports cases of monkeypox in people who handled infected prairie dogs.

Avian Influenza A strain H5N1 spreads through domestic poultry in Asia.

2005
CDC reports that 7.8 percent of tuberculosis cases in the U.S. are resistant to the first-line drug used to treat it.

2007
Fourteen countries have reported a total of 351 confirmed human cases of avian influenza (H5N1) and 219 deaths.

WRITING In a short essay, discuss why Surgeon General Stewart would have been confident in his 1967 announcement. Then discuss two factors that have contributed to the comeback of infectious disease.

35.4 Immune System Disorders

Key Questions

How can misguided immune responses cause problems?

What causes AIDS and how is it spread?

Vocabulary

allergy
asthma

Taking Notes

Outline Before you read, make an outline of the major headings in the lesson. As you read, fill in main ideas and supporting details for each heading.

THINK ABOUT IT A healthy immune system accurately distinguishes "self" from "other" and responds appropriately to dangerous invaders in the body. Sometimes, however, the immune system's weaponry is misdirected at the body's own cells. Other times, the immune system itself is disabled by disease. What happens in these cases?

When the Immune System Overreacts

How can misguided immune responses cause problems?

The immune systems of some people overreact to harmless antigens, such as pollen, dust mites, mold, pet dander, and possibly their own cells. **A strong immune response to harmless antigens can produce allergies, asthma, and autoimmune disease.**

Allergies Antigens that cause allergic reactions are called allergens. When allergens enter the body of people affected by **allergies,** they trigger an inflammatory response by causing mast cells to release histamines. If this response occurs in the respiratory system, it increases mucus production and causes sneezing, watery eyes, a runny nose, and other irritations. Drugs called antihistamines help relieve allergy symptoms by counteracting the effects of histamines.

FIGURE 35–17 Allergens Pet dander, dead skin shed from cats and dogs, is a common allergen (SEM 40×).

Asthma Allergic reactions in the respiratory system can create a dangerous condition called asthma. **Asthma** is a chronic disease in which air passages narrow, causing wheezing, coughing, and difficulty breathing. Both hereditary and environmental factors influence asthma symptoms. Asthma attacks can be triggered by respiratory infections, exercise, emotional stress, and certain medications. Other triggers include cold or dry air, pollen, dust, tobacco smoke, pollution, molds, and pet dander.

Asthma is serious and can be life-threatening. If treatment is not started early enough or if medications are not taken properly, severe asthma can lead to permanent damage or destruction of lung tissue. There is no cure, but people with asthma can sometimes control the condition. If the attacks are caused by an allergen, tests can identify which allergens cause the problem. Inhaled medications can relax smooth muscles around the airways and relieve asthma symptoms.

In Your Notebook *Sometimes allergies are described as "overreactions of the immune system." Explain what that phrase means.*

Food Allergies

About four percent of Americans have food allergies. Eight foods account for 90 percent of all food allergies—milk, eggs, peanuts, tree nuts, wheat, soy, fish, and shellfish. Approximately 30,000 emergency-room visits and 150–200 deaths each year can be attributed to food allergies. Most of the deaths are due to peanut allergies. The graph shows the percentage of children who had allergies from 1998–2006.

1. Analyze Data Discuss the general trend of food allergies for both age groups.

Food Allergies Among Children in the United States

Age	1998–2000	2001–2003	2004–2006
0–4	3.8%	4.2%	4.6%
5–17	3.3%	3.4%	3.9%

2. Calculate Which age group shows the greatest change from 1998–2006? What is the percent change in both age groups? MATH

3. Infer Propose a reason why more children age 4 and under have allergies than children age 5–17.

Autoimmune Diseases Sometimes a disease occurs in which the immune system fails to properly recognize "self," and attacks cells or compounds in the body as though they were pathogens. **When the immune system attacks the body's own cells, it produces an autoimmune disease.** Examples of autoimmune diseases are Type I diabetes, rheumatoid arthritis, and lupus.

In Type I diabetes, antibodies attack insulin-producing cells in the pancreas. In rheumatoid arthritis, antibodies attack tissues around joints. Lupus is an autoimmune disease in which antibodies attack organs and tissues causing areas of chronic inflammation throughout the body.

Some autoimmune diseases can be treated with medications that alleviate specific symptoms. For example, people with Type I diabetes can take insulin. Other autoimmune diseases are treated with medications that suppress the immune response. However, these medications also decrease the normal immune response and must be monitored.

BUILD Vocabulary

ACADEMIC WORDS Alleviate is a verb that means "to lessen" or "to relieve." It comes from the Latin *ad-* (to) and *-levis* (light in weight).

HIV and AIDS

What causes AIDS and how is it spread?

During the late 1970s, physicians began reporting serious infections produced by microorganisms that didn't normally cause disease. Previously healthy people began to suffer from *Pneumocystis carinii* pneumonia, Kaposi sarcoma (a rare form of skin cancer), and fungal infections of the mouth and throat. Because these diseases are normally prevented by a healthy immune response, doctors concluded that these patients must have weakened immune systems. Diseases that attack a person with a weakened immune system are called opportunistic diseases. Researchers concluded that these illnesses were symptoms of a new disorder they called acquired immunodeficiency syndrome (AIDS). Research eventually revealed that this "syndrome" was an infectious disease caused by a pathogen new to science.

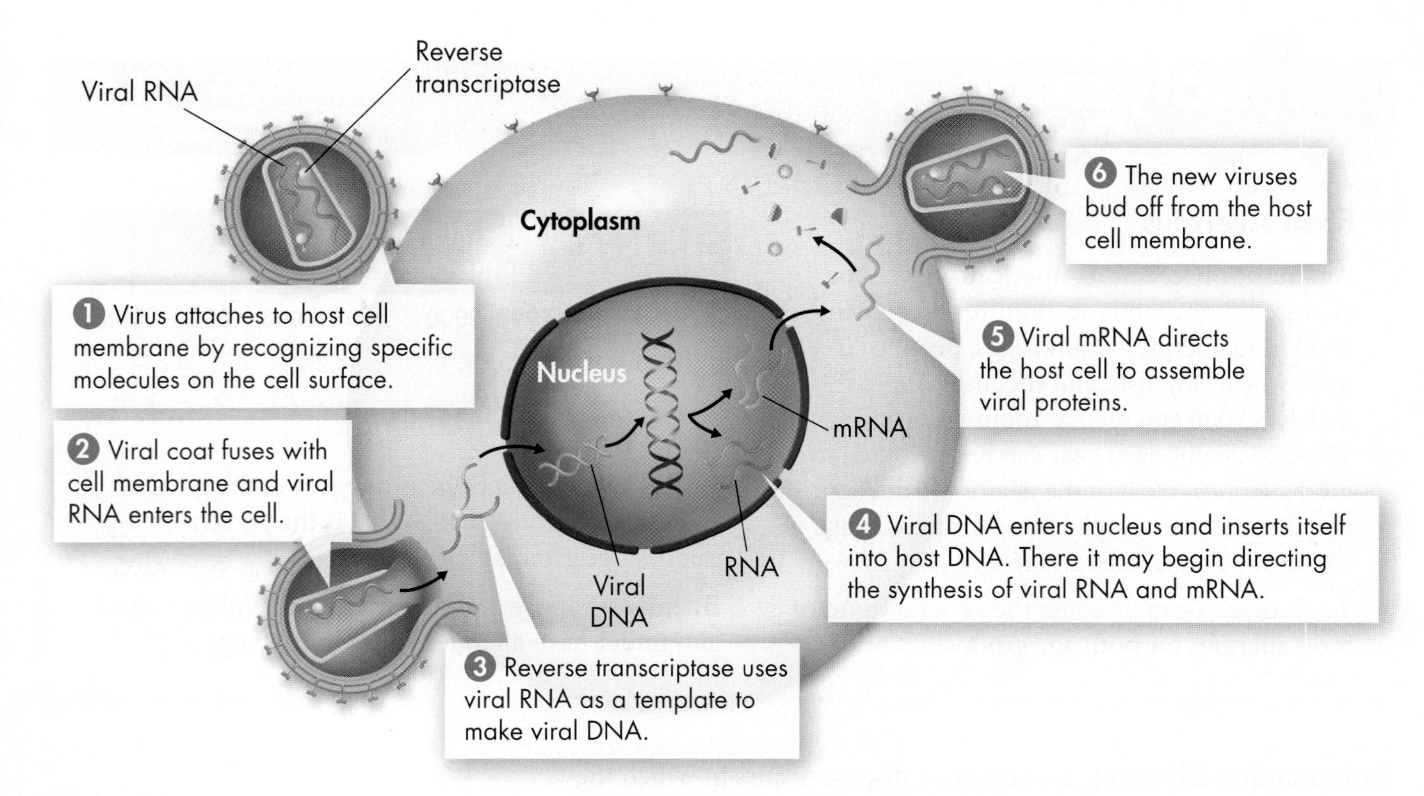

FIGURE 35–18 HIV Infection HIV travels through the blood, where it binds to receptors on helper T cells. Inside the cell, the viral DNA directs the cell to produce many new viruses. These new viruses are quickly released back into the blood, where they infect more cells. **Apply Concepts** ***In what steps are changes to HIV's genetic information most likely to occur?***

HIV **In 1983, researchers identified the cause of AIDS—a virus they called human immunodeficiency virus (HIV).** HIV is deadly for two reasons. First, HIV can hide from the defenses of the immune system. Second, HIV attacks key cells within the immune system, leaving the body with inadequate protection against other pathogens.

HIV is a retrovirus that carries its genetic information in RNA, rather than DNA. When HIV attacks a cell, it binds to receptor molecules on the cell membrane and inserts its contents into the cell. **Figure 35–18** explains how HIV replicates inside a host cell.

Target: T Cells Among HIV's main targets are helper T cells—the command centers of the specific immune response. Over time, HIV destroys more and more T cells, crippling the ability of the immune system to fight HIV and other pathogens. The progression of HIV infection can be monitored by counting helper T cells. The fewer helper T cells, the more advanced the disease, and the more susceptible the body becomes to other diseases. When an HIV-infected person's T cell count reaches about one sixth the normal level, he or she is diagnosed with AIDS.

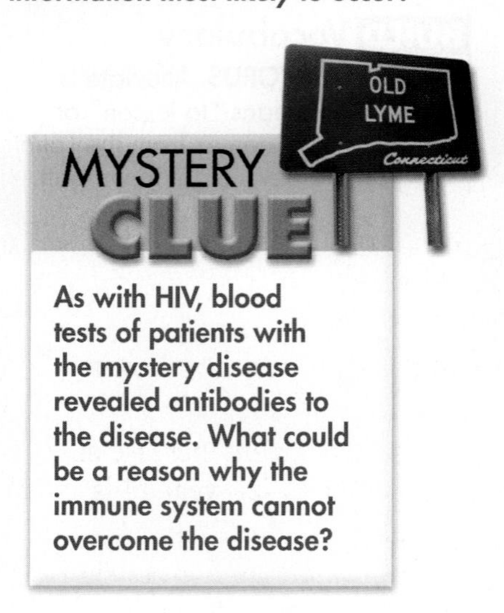

As with HIV, blood tests of patients with the mystery disease revealed antibodies to the disease. What could be a reason why the immune system cannot overcome the disease?

HIV Transmission Although HIV is deadly, it is not easily transmitted. It is not transmitted through coughing, sneezing, sharing clothes, or other forms of casual contact. HIV can only be transmitted through contact with infected blood, semen, vaginal secretions, or breast milk. The four main ways that HIV is transmitted are sexual intercourse with an infected person; sharing needles with an infected person; contact with infected blood or blood products; or from an infected mother to her child during pregnancy, birth, or breast-feeding.

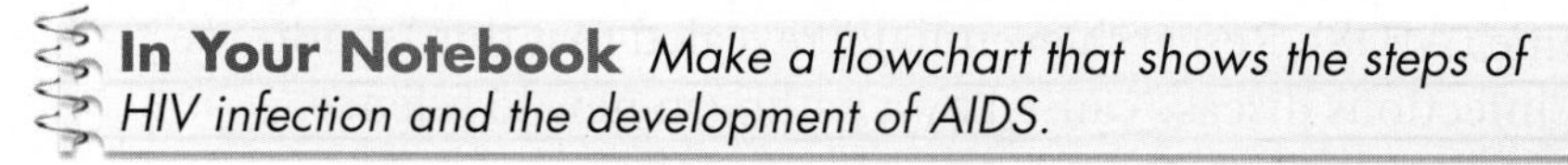
In Your Notebook *Make a flowchart that shows the steps of HIV infection and the development of AIDS.*

Preventing HIV Infection You can choose behaviors that reduce your risk of becoming infected with HIV. **The only no-risk behavior with respect to HIV transmission is abstinence from sexual activity and intravenous drug use.** Within a committed relationship, such as marriage, sexual fidelity between two uninfected partners presents the least risk of becoming infected with HIV. People who share needles to inject themselves with drugs are at a high risk for contracting HIV. For this reason, people who have sex with drug abusers are also at high risk. Before 1985, HIV was transmitted to some patients through transfusions of infected blood or blood products. But, such cases have been virtually eliminated by screening the blood supply for HIV antibodies and by discouraging potentially infected individuals from donating blood.

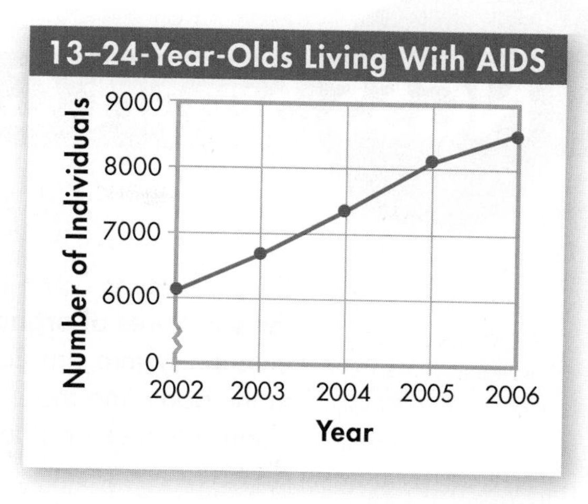

FIGURE 35–19 Adolescents and Young Adults Living With AIDS in the United States

Can AIDS Be Cured? At present, there is no cure for AIDS. A steady stream of new drugs makes it possible to survive HIV infection for years. Unfortunately, HIV mutates and evolves rapidly. For this reason, the virus has evolved into many strains that are resistant to most drugs used against them. No one has developed a vaccine that offers protection for any length of time.

At present, the only way to control the virus is to use a combination of expensive drugs that fight the virus in several ways. Current drugs interfere with the enzymes HIV uses to insert its RNA into a host cell, to convert RNA to DNA, and to integrate its DNA into the host's DNA. Because of these drugs, more people infected with HIV in the United States are living with HIV rather than dying from it. In many parts of Africa and Asia, however, these expensive drugs are not available.

Unfortunately, the knowledge that HIV can be treated (though not cured) has given some people the misconception that HIV infection is not serious. That idea is dead wrong.

35.4 Assessment

Review Key Concepts

1. **a. Review** What happens during an allergy attack? What happens in an autoimmune disease?

   b. Apply Concepts In treating asthma, the first thing many physicians do is ask patients to list times and places they have experienced attacks. Why do you think physicians do this?

2. **a. Review** What is the virus that causes AIDS? Describe how it is spread.

   b. Infer Why is it difficult for a person with HIV to fight off other infections?

ANALYZING DATA

Review **Figure 35–19** and answer the following questions.

3. **a. Interpret Data** What percent increase in AIDS cases occurred in 13–24-year-olds from 2002 to 2006? MATH

   b. Draw Conclusions What are two conclusions that you could draw regarding the increasing number of adolescents and young adults living with AIDS?

NGSS Smart Guide Immune System and Disease

Disciplinary Core Idea LS1.A Structure and Function: How do the structures of organisms enable life's functions? The cells of the immune system can distinguish between cells and proteins that belong in the body and those that do not. Immune cells and accompanying chemicals seek and destroy antigens and pathogens that can cause disease.

35.1 Infectious Disease

- Infectious diseases can be caused by viruses, bacteria, fungi, "protists," and parasites.
- Some diseases are spread through coughing, sneezing, physical contact, or exchange of body fluids. Some diseases are spread through contaminated water or food. Still other diseases are spread to humans from infected animals.

infectious disease 1010 • germ theory of disease 1010 • Koch's postulates 1011 • zoonosis 1013 • vector 1013

Biology.com

Untamed Science Video Be careful what you touch as you follow the Untamed Science crew on a journey through human allergies.

Virtual Lab Use the Microscopy and Systematics lab benches to learn more about interactions between parasites and their hosts.

Taking Notes Complete a table that summarizes the ways in which diseases spread.

UntamedScience

TAKING NOTES Infectious Disease

Use a two-column table to list the ways diseases are spread and describe each way.

How Disease is Spread	Description

35.2 Defenses Against Infection

- Nonspecific defenses include the skin, tears and other secretions, the inflammatory response, interferons, and fever.
- The immune system's specific defenses distinguish between "self" and "other," and they inactivate or kill any foreign substance or cell that enters the body.
- The specific immune response has two main styles of action: humoral immunity and cell-mediated immunity.

inflammatory response 1014 • histamine 1014 • interferon 1015 • fever 1015 • immune response 1016 • antigen 1016 • antibody 1016 • humoral immunity 1016 • cell-mediated immunity 1019

Biology.com

Taking Notes Construct a concept map that describes the body's defenses against infection.

TAKING NOTES Defenses Against Infection

Use the green and blue headings in this lesson to make a concept map. Add details to your map as you read.

Defenses Against Infection

include

CHECKING YOUR *Scientific Literacy*

Refer to the lesson content and digital assets below as you prepare for your chapter assessment. Then, evaluate your understanding of the immune system and disease by answering these questions.

1. Scientific and Engineering Practice **Asking Questions and Defining Problems** If you were part of a team investigating a new or re-emerging infectious disease, what questions would you ask to guide the investigation and search for a way to stop the spread of the disease?

2. Crosscutting Concept **Cause and Effect** What effect does the body's ability to distinguish between "self" and "nonself" have on the responses of the immune system?

3. Production and Distribution of Writing Design a poster campaign for your school that will promote ways for students to reduce the spread of colds and flu.

4. STEM Many people use gel, liquid, or foam hand sanitizers to kill bacteria and prevent the spread of some diseases. Construct arguments for and against the widespread use of this technology.

35.3 Fighting Infectious Disease

- Vaccination stimulates the immune system with an antigen. The immune system produces memory B cells and memory T cells that quicken and strengthen the body's response to repeated infection.
- Antibodies produced against a pathogen by other individuals or animals can be used to produce temporary immunity.
- Public health measures help prevent disease by monitoring and regulating food and water supplies, promoting vaccination, and promoting behaviors that avoid infection.
- Antibiotics can kill bacteria, and some antiviral medications can slow down viral activity.
- Two major reasons for the emergence of new diseases are the ongoing merging of human and animal habitats and the increase in the exotic animal trade.
- Misuse of medications has led to the re-emergence of diseases that many people thought were under control.

vaccination 1020 • **active immunity 1020** • **passive immunity 1020**

Biology.com

Data Analysis Analyze the pros and cons of a clean, "germ-free" lifestyle.

35.4 Immune System Disorders

- A strong immune response to harmless antigens can produce allergies, asthma, and autoimmune disease.
- When the immune system attacks the body's own cells, it produces an autoimmune disease.
- In 1983, researchers identified the cause of AIDS—a virus they called human immunodeficiency virus (HIV).
- The only no-risk behavior with respect to HIV transmission is abstinence from sexual activity and intravenous drug use.

allergy 1024 • **asthma 1024**

Biology.com

Art in Motion View an animation of HIV infecting a cell.

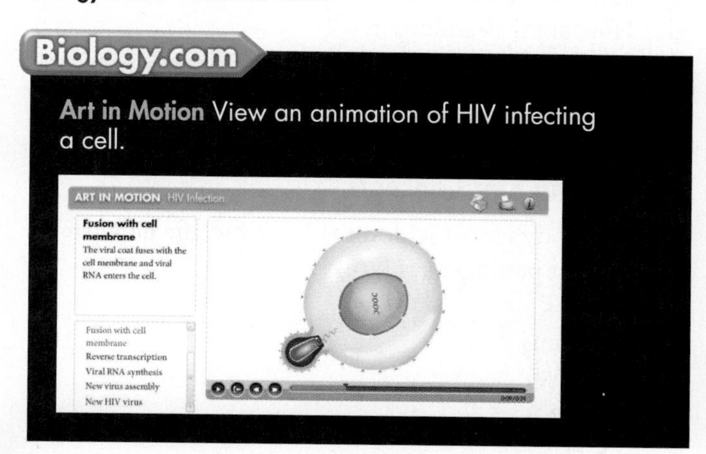

Forensics Lab

Detecting Lyme Disease This chapter lab is available in your lab manual and at **Biology.com**

35 Assessment

35.1 Infectious Diseases

Understand Key Concepts

1. Any change, other than an injury, that disrupts the normal functions of a person's body systems is a
   a. disease.
   b. pathogen.
   c. toxin.
   d. vector.
2. Disease-causing agents such as viruses, bacteria, and fungi are known as
   a. antibodies.
   b. antigens.
   c. pathogens.
   d. toxins.
3. What is the germ theory of disease?
4. What do researchers use Koch's postulates to determine?
5. List five types of agents that can produce infectious disease. Give an example of a disease that each specific pathogen may cause.
6. What is a zoonosis?
7. What are some ways by which the spread of disease can be prevented?

Think Critically

8. **Infer** Why is the fourth step of Koch's postulates necessary to prove that a disease is caused by a specific pathogen?
9. **Craft and Structure** In what way are symbiotic organisms that live on or in the human body similar to pathogens that may take up residence? How are they different?

35.2 Defenses Against Infection

Understand Key Concepts

10. The body's most widespread nonspecific defense against pathogens is (are)
    a. tears.
    b. mucus.
    c. saliva.
    d. skin.
11. A nonspecific defense reaction to tissue damage caused by injury or infection is known as
    a. the inflammatory response.
    b. active immunity.
    c. cell-mediated immunity.
    d. passive immunity.
12. What are antibodies? Describe their form and function.
13. Describe the roles of helper T cells and cytotoxic T cells.
14. Distinguish between humoral immunity and cell-mediated immunity.

Think Critically

15. **Infer** Many people become alarmed if they have a slight fever. Why might a slight fever that lasts no more than a few days be beneficial?
16. **Craft and Structure** How does the secondary response to an antigen differ from the primary response to an antigen?

35.3 Fighting Infectious Disease

Understand Key Concepts

17. Injecting antibodies from an animal to help prevent a disease from occurring in a human is called
    a. active immunity.
    b. passive immunity.
    c. antibiotic therapy.
    d. vaccination.
18. What is a common goal of researchers who develop antibiotics and antiviral drugs?
    a. to kill bacteria
    b. to prevent infections
    c. to stop pathogens without harming host cells
    d. to kill viruses
19. Who discovered the first antibiotic and how did he discover it?
20. **Key Ideas and Details** Summarize two ways that public health has influenced the prevention of infectious disease.

21. Describe how passive immunity to a disease is obtained and why it lasts for only a short period of time.

22. What are two major contributing factors to emerging diseases?

Think Critically

23. **Text Types and Purposes** Edward Jenner developed his smallpox vaccine in 1796. Jenner tested his theory that infection with cowpox could prevent smallpox on a young boy. Do you think Jenner was justified in using the child as a test subject? Could this experiment be conducted today? Write a well-constructed argument to support your answer.

24. **Infer** It is not always easy to determine if a patient has a bacterial infection or a viral infection. How could this contribute to the misuse of medications?

35.4 Immune System Disorders

Understand Key Concepts

25. A strong response by a person's immune system to a harmless antigen in the environment is called
    - **a.** cell-mediated immunity.
    - **b.** an allergy.
    - **c.** inflammatory response.
    - **d.** an autoimmune disease.

26. The main target cells of HIV are
    - **a.** insulin-producing cells in the pancreas.
    - **b.** T lymphocytes.
    - **c.** B lymphocytes.
    - **d.** cells in the liver.

27. **Craft and Structure** Explain why allergies are not classified as autoimmune diseases.

28. Describe the specific action of HIV that makes an infected person unable to fight off other infections.

Think Critically

29. **Research to Build and Present Knowledge** Draw evidence from the text to analyze why a second bee sting is more dangerous than the first for a person who is allergic to bee stings.

30. **Infer** Reverse transcriptase is not a very accurate enzyme. How could this contribute to the rapid evolution of drug resistance in HIV?

solve the CHAPTER MYSTERY

LM 10×

THE SEARCH FOR A CAUSE

The disease of unknown cause was named Lyme disease, after the town of Lyme, Connecticut, where many of the patients lived. Steere's investigation was helped by a researcher who isolated a bacterium called *Borrelia burgdorferi* from deer ticks. The ticks had been captured in the area where patients lived. Steere found the same bacterium in the patients. Could this bacterium be the cause of Lyme disease?

For ethical reasons, Steere could not infect healthy people with the bacterium, but he did infect healthy laboratory mice. The mice developed arthritis and other symptoms that were similar to the Lyme disease patients' symptoms. Steere recovered bacteria from sick mice and injected them into healthy mice, which then also developed the disease.

Now researchers know that a bite from a deer tick carrying *B. burgdorferi* may transmit the bacterium. *B. burgdorferi* can "swim" through tissues around tick bites, causing the spreading rash that some patients reported. The bacterium then seems to infect many types of cells, including macrophages, nerve cells, and muscle cells. Some *B. burgdorferi* proteins resemble proteins in the myelin sheaths around some nerve cells. This may cause an autoimmune response that leads to arthritis and other problems that persist after the infection is gone.

1. **Explain** What set of rules did Steere use to determine if *B. burgdorferi* was the pathogen responsible for Lyme disease?

2. **Infer** Lyme disease patients who are quickly treated with antibiotics usually do much better than those who are treated later. Why do you think this is the case?

3. **Connect to the Big idea** Deer and deer ticks thrive in wooded areas that grow back after the areas have been cleared and at the edges of woodlands. How might suburban development contribute to an increase in Lyme disease?

Connecting Concepts

Use Science Graphics

John Snow made a map similar to the one below to help him determine the source of the cholera outbreak in London. The dots represent the locations of people who died of cholera. The Xs represent pumps. Use the map to answer questions 31 and 32.

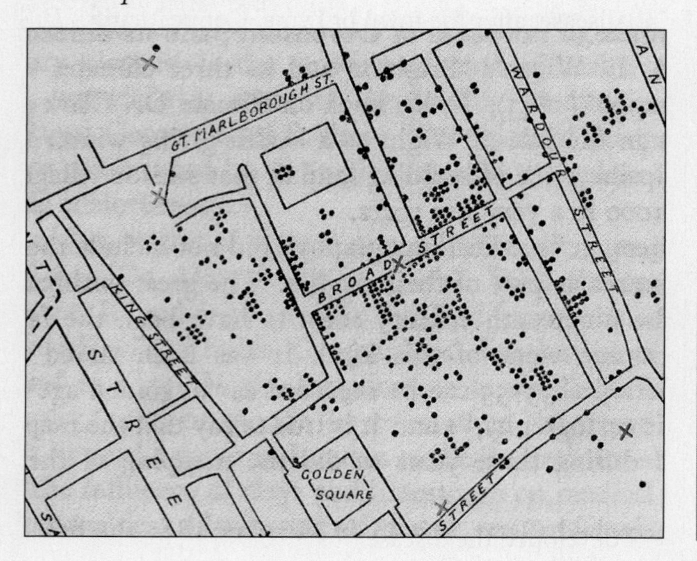

31. **Key Ideas and Details** Which pump do you think Snow determined was most likely the source of the cholera outbreak? Cite evidence from the map to explain your answer.

32. **Apply Concepts** Do you think a map such as this one could be used to discover the source of a food poisoning outbreak? Explain.

Write About Science

33. **Production and Distribution of Writing** The ability of bacteria to resist antibiotics has become an increasing public health problem. This problem is due to the overuse and misuse of antibiotics. Suppose that one of your friends always takes antibiotics when he or she is sick. Write a letter to your friend explaining the problem of antibiotic resistance.

34. **Assess the Big idea** Explain how the germ theory of disease led people to develop very simple methods that could prevent the transmission of many diseases.

Integration of Knowledge and Ideas

The graph shows the number of cases of the viral disease, polio, in the world from 1980 until 2004. It also shows the percentage of the world's population that was vaccinated for the disease.

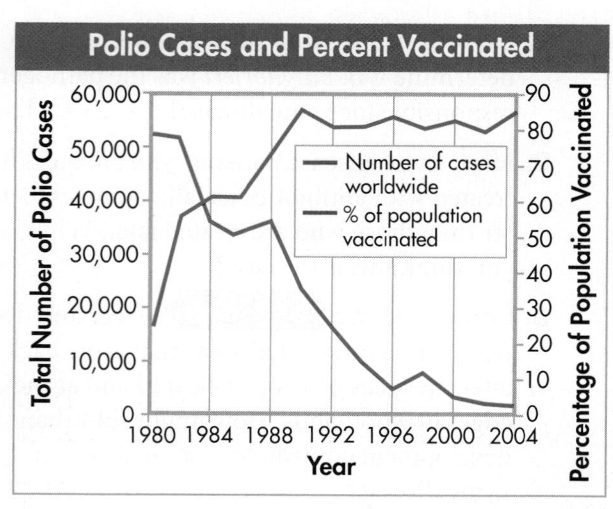

35. **Interpret Graphs** During which of the following time periods was there the greatest drop in the number of polio cases around the world?

a. between 1980 and 1982
b. between 1984 and 1988
c. between 1988 and 1996
d. between 1996 and 2004

36. **Draw Conclusions** Which of the following is the most reasonable conclusion to draw from the data shown in the graph?

a. As the number of people vaccinated increases, the number of polio cases stays constant.
b. As the number of people vaccinated increases, the number of polio cases has increased.
c. As the number of people vaccinated increases, the number of polio cases has decreased.
d. Polio has been eliminated as a disease, and vaccination is no longer necessary.

Standardized Test Prep

Multiple Choice

1. All of the following prevent pathogens from entering the human body EXCEPT
A red blood cells. **C** mucus.
B tears. **D** skin.

2. Which of the following is NOT part of the inflammatory response?
A White blood cells rush to infected tissues.
B Blood vessels near the wound shrink.
C Phagocytes engulf and destroy pathogens.
D The wound becomes red.

3. What is the role of a vector in the spread of disease?
A A vector is an inanimate object, such as a doorknob, where pathogens may collect.
B A vector must infect a host for its life cycle to continue.
C Vectors usually do not suffer from the infection, they just spread it from host to host.
D A vector is a pathogen.

4. Which type of lymphocyte produces antibodies that are released into the bloodstream?
A cytotoxic T cells **C** phagocytes
B helper T cells **D** plasma cells

5. Which of the following is NOT a white blood cell?
A interferon **C** cytotoxic T cell
B macrophage **D** lymphocyte

6. Which is an example of naturally occurring passive immunity?
A vaccination
B exposure to a disease
C an infant consuming antibodies in breast milk
D antibodies are injected from another person

7. How do medications help a person with asthma?
A Antihistamines counteract the effects of histamines.
B They suppress the immune system.
C They increase mucus production in the lungs.
D They relax smooth muscles around airways.

Questions 8–9

A researcher measured the concentrations of HIV and T cells in 120 HIV-infected patients over a period of 10 years. Her data are summarized in the graph.

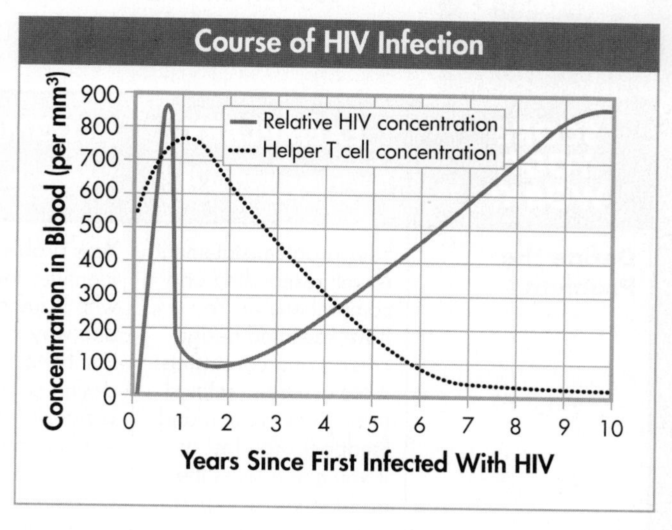

8. What happened to the HIV concentration over years 2 through 9?
A It stayed about the same, then suddenly increased.
B It stayed about the same, then suddenly decreased.
C It steadily increased.
D It steadily decreased.

9. What is probably responsible for the change in HIV concentration during the first year?
A immune response
B inflammatory response
C passive immunity
D HIV stopped replicating

Open-Ended Response

10. Explain why some symptoms of disease such as coughing and sneezing are advantageous to the pathogen that causes the disease.

If You Have Trouble With . . .

Question	1	2	3	4	5	6	7	8	9	10
See Lesson	35.2	35.2	35.1	35.2	35.2	35.3	35.4	35.4	35.4	35.1

NGSS Problem-based Learning

UNIT 8 PROJECT

Your Self-Assessment

The rubric below will help you evaluate your work as you design your solution and communicate your findings.

SCORE YOUR WORK!	EXEMPLARY Score your work a 4 if:	ACCOMPLISHED Score your work a 3 if:	DEVELOPING Score your work a 2 if:	BEGINNING Score your work a 1 if:
Define the Problem ▶	Your problem statement is well researched and comprehensive. You have specified design criteria and constraints that have been refined through discussion and feedback. You feel as if you can discuss the problem as an expert.	Your problem statement identifies the problem, who is impacted by it, and why it is important. You have specified design criteria and constraints that your solution must meet.	You have written a statement that identifies the problem, and you have identified some basic design criteria for a solution.	You have identified what you think is a problem but are unsure how to explain or define it.
	TIP Ask, *"What* is the problem? *Why* is it important? *Who* is affected by it?"			
Design a Solution ▶	You have generated ideas for solutions, evaluated them, and refined them. You understand that different solutions may be optimized for different design criteria.	You have generated ideas for solutions and evaluated them based on design criteria and constraints.	You have generated ideas for possible solutions, but you are not sure how to compare or evaluate them.	You have brainstormed an idea for a possible solution.
Develop Models ▶	You have refined your design through iterative rounds of sketching or modeling. You have identified and discussed the strengths and limitations of your design.	You have sketched or modeled your design, analyzed it, and identified strengths and limitations of the design.	You have sketched or modeled a design for a possible solution.	You have identified a possible solution that can be realized through modeling and/or testing.
	TIP The model development process is iterative, which means you won't get it right on the first try. Keep asking questions and evaluating your ideas, and try to steadily improve them.			
Obtain, Evaluate, and Communicate Information ▶	You have presented and discussed your final design solution with an audience beyond your class and/or school.	You have presented your design proposal and have received feedback from students and teachers.	You have organized your documents and findings but have not shared or presented them.	You have compiled notes and materials documenting the design process, but you are unsure how to present them.

ECOSYSTEMS: INTERACTIONS, ENERGY, AND DYNAMICS

Science and Engineering Practices
Designing Solutions

Self-Sustaining Habitats

Have you ever seen a movie where people live in settlements on the moon or on other planets? It's actually not such a far-fetched idea. Scientists have been investigating the possibilities of extra-terrestrial living for many years. But how can it happen without the biotic and abiotic factors found on Earth?

A self-sustaining habitat is one in which all life-sustaining elements—oxygen, food, and water—and waste disposal are provided in a closed system. Researchers in the United States and around the world have been studying, planning, designing, and testing such habitats. One of the first self-sustaining habitats on land, Biosphere 2, was developed in the 1980s and 1990s by a United States research team with the help of international leaders in the areas of ecology, biology, engineering, and other sciences. Built in the Arizona desert, Biosphere 2 attempted to duplicate Earth's atmosphere and ecosystems. It was designed to provide its inhabitants with their food, water, air, and waste disposal for a two-year duration. In 1991 and again in 1994, humans were sealed inside the glass building to determine if the habitat was indeed self-sustaining.

While living in Biosphere 2, the inhabitants studied the biotic and abiotic factors of a number of Earth's ecosystems. The 3000 species living amongst Biosphere's seven biomes provided researchers with data about Earth that was difficult to discover in any other way. Today, the University of Arizona owns the structure and uses it primarily for research.

Designing Solutions

Research self-sustaining habitats in space and under water. Evaluate the designs for how well they balance and control essential biotic and abiotic factors and for how food supply, waste disposal, and recycling issues are handled in the habitats. Then, using what you learned, work in a group to design a space or under water habitat. Include the elements of food supply, carbon dioxide reduction, oxygen production, water purification, power generation, and waste management. Include a key to distinguish between biotic and abiotic factors.

FROM MOLECULES TO ORGANISMS: STRUCTURE AND FUNCTION

Science and Engineering Practices
Asking Questions

Hydroponics

The United Nations estimates that the world's total population will reach 8.5 billion by 2030. That number represents a lot of mouths to feed! An important question is how can enough food be produced to feed 8.5 billion people every day. One answer may be hydroponics.

Hydroponics is a technology that allows plants to grow in nutrient-rich water instead of nutrient-rich soil. Although the initial cost of growing plants hydroponically is greater than that of growing plants in soil, hydroponic systems require no weeding and no pesticides, and they actually use less water than growing plants in soil. In addition, hydroponic systems make it easier to control growing conditions such as wind, watering, and nutrient concentrations, and there is no need to worry about factors like soil texture. Hydroponic crops have high yields because roots do not have to grow extensively to obtain and absorb nutrients; and, with smaller root systems, more plants can be grown in less space than is needed to grow plants in soil.

Photosynthesis is an essential process for all plants. For plants to photosynthesize they need carbon dioxide, water, nutrients, and light. As long as these essentials can be made available, food crops can be grown in urban areas as well as rural areas.

Ask Questions

Suppose you are setting up a hydroponic system to grow tomato plants in a large basement. You might start with a general question such as *What do tomato plants need to photosynthesize?* Then you would need to refine your question—for example, by breaking it down into more specific questions that you can research or test: *Can tomato plants grow with artificial light? How many hours of light do they need every day?* Write specific questions like these to help you determine the ideal conditions for growing tomatoes and the best way to provide those conditions. If you can't find an answer to a question, plan an investigation that would provide some answers. Assume that there is enough money to buy the necessary supplies. Create a chart that shows the sequence of the questions you asked and the answers you found (or your suggestion for finding the answer). Present your findings to the class.

HEREDITY: INHERITANCE AND VARIATION OF TRAITS

Science and Engineering Practices
Mathematical and Computational Thinking

Exponential Growth

The ability to manipulate DNA has resulted in powerful tools to combat disease, improve crops, solve crimes, and research living organisms. At the heart of many biotechnology applications, from DNA sequencing to the genetic manipulation of plants, animals, and microorganisms, is the polymerase chain reaction (PCR). This reaction is extremely useful because DNA is often available in very small amounts, and PCR allows researchers to make millions of copies of specific DNA fragments. It is effective because the number of copies of the original DNA fragment increases *exponentially*.

The term "exponential" is often used when a variable increases rapidly, but a fast increase isn't necessarily exponential, and an exponential increase isn't necessarily fast (in fact, it can start out very slowly!). When there is true exponential growth, the variable increases more and more quickly as its value increases. This kind of growth is often seen when an invasive species begins to reproduce at an alarming rate or when a new virus spreads uncontrollably.

In PCR, the original DNA fragments are put through a series of procedures that doubles their number. The cycle is repeated over and over until enough copies have been produced. Under ideal conditions, every available DNA fragment would be copied perfectly in every cycle, and every copy would be preserved so that it can be copied in the next cycle. In reality, there are losses due to several factors, including mutations and imperfect copying. However, it is instructive to understand what happens under ideal circumstances.

Use Mathematical and Computational Thinking

If ideal conditions for PCR are assumed, the final number of available DNA fragments (N). can be easily calculated. The only pieces of information required are the initial number of fragments (A) and the number of replication cycles (n). These values can be plugged into the following exponential equation:

$$N = A(2^n).$$

Suppose that you start out with just 5 DNA fragments.

- How many fragments will you have after 2 cycles? After 5 cycles? After 10 cycles?
- How many cycles do you need to reach a goal of one million fragments?
- Construct a graph of N as a function of n. Provide appropriate labels.

BIOLOGICAL EVOLUTION: UNITY AND DIVERSITY

Science and Engineering Practices
Engaging in Argument From Evidence

Robotics and Evolution

According to Charles Darwin's theory of evolution, the fittest individuals are more likely to survive and reproduce, and species change over time as a result. It is difficult to observe this process directly in living organisms because the changes are very slow; therefore, some researchers are using technology tools to model Darwin's theory, test its predictions, and gain insight into evolutionary patterns without having to observe living organisms over many generations. These tools include robots.

Robots can be designed to interact with their environment on their own, without human assistance. Such robots are autonomous—they can move freely and manipulate their environment, simulating the behavior of living organisms. To build autonomous robots, scientists use principles of biology and behavioral science, as well as advanced technologies such as genetic algorithms and neural networks. Robots can also be programmed to exchange snippets of their computer code with other robots, just as living organisms may exchange bits of genetic information during reproduction. The slightly altered code can be used in different ways to build new "generations" of robots, whose functionality will therefore be slightly different from that of their "parents."

A robot that shares such important similarities with living organisms can be used to model biological evolution. By tracking the activity, changes, successes and failures of their robotic "creatures," scientists hope to better understand the forces that drive the evolution of living organisms.

Engage in Argument From Evidence

Investigate the work of Professor John Long and his "TadRos" (or another recent attempt to use robots to simulate evolution). Choose one conclusion of the research and write a review of this conclusion. Your review should include the evidence presented by the authors to support their conclusion, as well as any evidence you can find in related fields like genetics or paleontology. You should also include any evidence you find that seems to be *against* the conclusion.

A Visual Guide to
The Diversity of Life

▲ *The Chambered Nautilus, found today in the Pacific Ocean, is one of the few living representatives of a group that once flourished in ancient seas 265 million years before the dinosaurs evolved. This Visual Guide will give you a glimpse of life's great variety and evolutionary history.*

A Visual Guide to The Diversity of Life

CONTENTS

How to Use This Guide DOL•3
The Tree of Life DOL•4
Bacteria DOL•6
- Proteobacteria DOL•7
- Spirochaetes DOL•7
- Actinobacteria DOL•7
- Cyanobacteria DOL•7

Archaea DOL•8
- Crenarchaeotes DOL•9
- Euryarchaeotes DOL•9
- Korarchaeotes DOL•9
- Nanoarchaeotes DOL•9

Protists DOL•10
- Excavates DOL•11
- Chromalveolates DOL•12
- Cercozoa, Foraminiferans, and Radiolarians DOL•14
- Rhodophytes DOL•15
- Amoebozoa DOL•15
- Choanozoa DOL•15

Fungi DOL•16
- Basidiomycetes DOL•17
- Ascomycetes DOL•18
- Zygomycetes DOL•19
- Chytrids DOL•19

Plants DOL•20
- Green Algae DOL•21
- Bryophytes DOL•22
- Seedless Vascular Plants DOL•23
- Gymnosperms DOL•24
- Angiosperms DOL•26

Animals DOL•30
- Porifera (Sponges) DOL•31
- Cnidarians DOL•32
- Arthropods DOL•34
- Nematodes (Roundworms) DOL•38
- Platyhelminthes (Flatworms) DOL•39
- Annelids (Segmented Worms) DOL•40
- Mollusks DOL•42
- Echinoderms DOL•44
- Nonvertebrate Chordates DOL•46
- Fishes DOL•48
- Amphibians DOL•52
- Reptiles DOL•54
- Birds DOL•56
- Mammals DOL•60

HOW TO USE THIS GUIDE

Use this visual reference tool to explore the classification and characteristics of organisms, including their habitats, ecology, behavior, and other important facts. This guide reflects the latest understandings about phylogenetic relationships within the three domains of life. Divided into six color-coded sections, the Visual Guide begins with a brief survey through the Bacteria and Archaea domains. It next discusses the major groups of protists, fungi, and plants. The final section provides information on nine animal phyla.

1. See how the group of organisms relates to others on the tree of life.

2. Learn about the general characteristics that all members of the group share.

3. Discover the members of the group and learn about their traits.

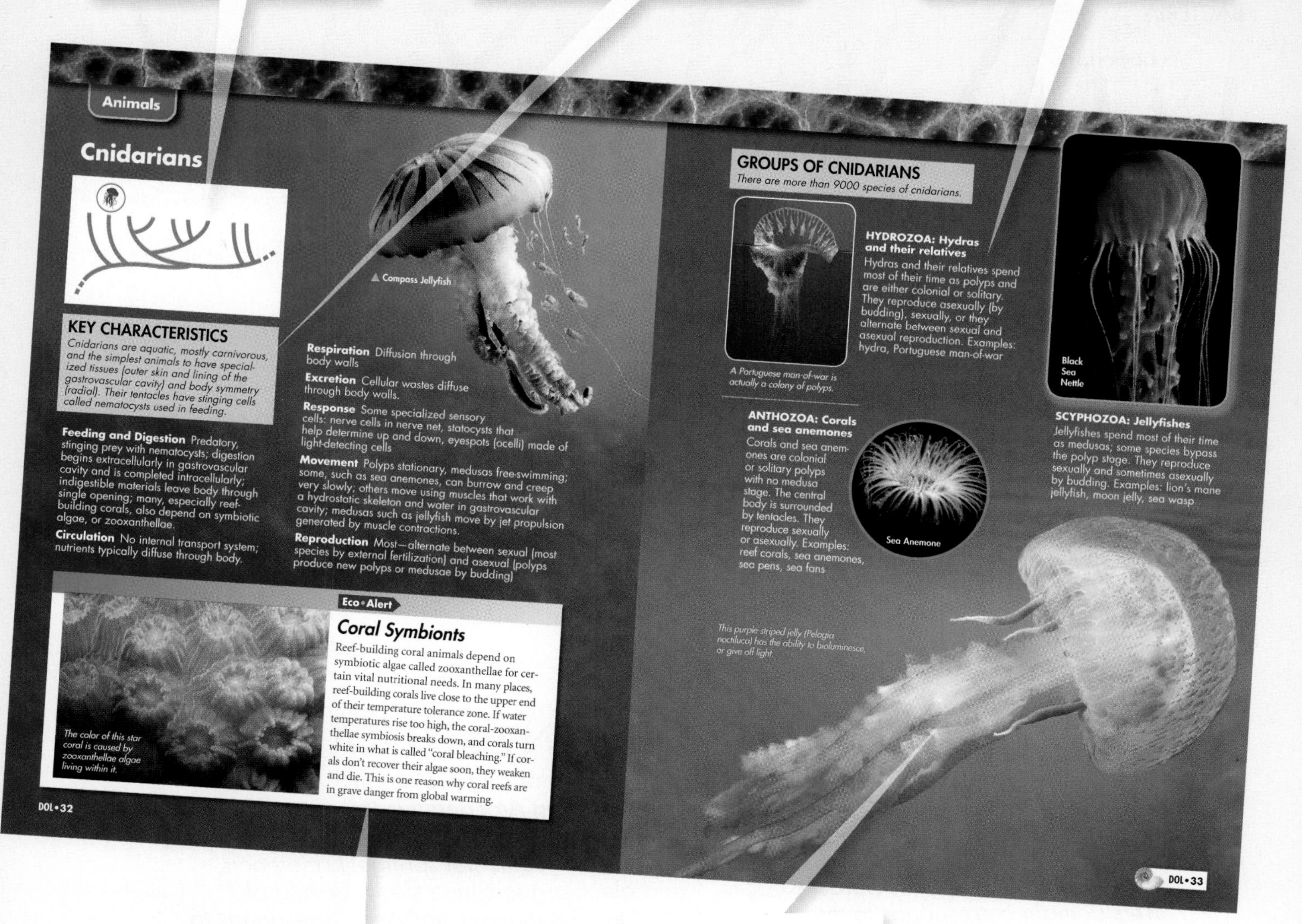

Animals

Cnidarians

KEY CHARACTERISTICS

Cnidarians are aquatic, mostly carnivorous, and the simplest animals to have specialized tissues (outer skin and lining of the gastrovascular cavity) and body symmetry (radial). Their tentacles have stinging cells called nematocysts used in feeding.

Feeding and Digestion Predatory, stinging prey with nematocysts; digestion begins extracellularly in gastrovascular cavity and is completed intracellularly; indigestible materials leave body through single opening; many, especially reef-building corals, also depend on symbiotic algae, or zooxanthellae.

Circulation No internal transport system; nutrients typically diffuse through body.

Respiration Diffusion through body walls

Excretion Cellular wastes diffuse through body walls.

Response Some specialized sensory cells: nerve cells in nerve net, statocysts that help determine up and down, eyespots (ocelli) made of light-detecting cells

Movement Polyps stationary, medusas free-swimming; some, such as sea anemones, can burrow and creep very slowly; others move using muscles that work with a hydrostatic skeleton and water in gastrovascular cavity; medusas such as jellyfish move by jet propulsion generated by muscle contractions.

Reproduction Most—alternate between sexual (most species by external fertilization) and asexual (polyps produce new polyps or medusae by budding)

▲ Compass Jellyfish

Eco • Alert

Coral Symbionts

Reef-building coral animals depend on symbiotic algae called zooxanthellae for certain vital nutritional needs. In many places, reef-building corals live close to the upper end of their temperature tolerance zone. If water temperatures rise too high, the coral-zooxanthellae symbiosis breaks down, and corals turn white in what is called "coral bleaching." If corals don't recover their algae soon, they weaken and die. This is one reason why coral reefs are in grave danger from global warming.

The color of this star coral is caused by zooxanthellae algae living within it.

DOL • 32

GROUPS OF CNIDARIANS

There are more than 9000 species of cnidarians.

HYDROZOA: Hydras and their relatives
Hydras and their relatives spend most of their time as polyps and are either colonial or solitary. They reproduce asexually (by budding), sexually, or they alternate between sexual and asexual reproduction. Examples: hydra, Portuguese man-of-war

A Portuguese man-of-war is actually a colony of polyps.

ANTHOZOA: Corals and sea anemones
Corals and sea anemones are colonial or solitary polyps with no medusa stage. The central body is surrounded by tentacles. They reproduce sexually or asexually. Examples: reef corals, sea anemones, sea pens, sea fans

Sea Anemone

Black Sea Nettle

SCYPHOZOA: Jellyfishes
Jellyfishes spend most of their time as medusas; some species bypass the polyp stage. They reproduce sexually and sometimes asexually by budding. Examples: lion's mane jellyfish, moon jelly, sea wasp

This purple-striped jelly (Pelagia noctiluca) has the ability to bioluminesce, or give off light.

DOL • 33

4. Investigate current news and interesting facts about the group.

5. See photographs of representative animals within each group.

THE TREE OF LIFE

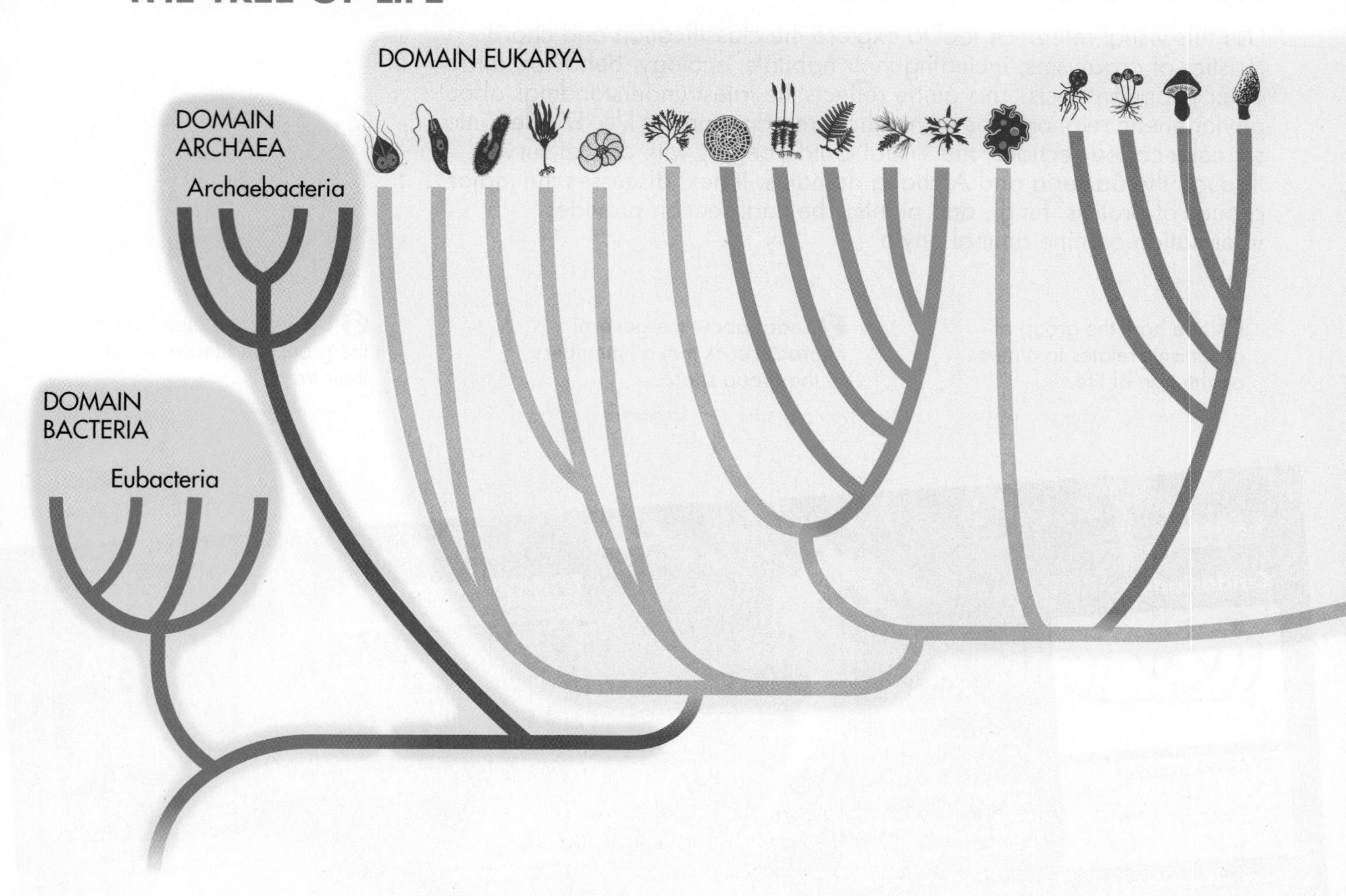

Before you begin your tour through the kingdoms of life, review this big picture from Chapter 18. The pages that follow will give you a glimpse of the incredible diversity found within each of the "branches" shown here.

DOMAIN BACTERIA

Members of the domain Bacteria are unicellular and prokaryotic. The bacteria are ecologically diverse, ranging from free-living soil organisms to deadly parasites. This domain corresponds to the kingdom Eubacteria.

DOMAIN ARCHAEA

Also unicellular and prokaryotic, members of the domain Archaea live in some of the most extreme environments you can imagine, including volcanic hot springs, brine pools, and black organic mud totally devoid of oxygen. The domain Archaea corresponds to the kingdom Archaebacteria.

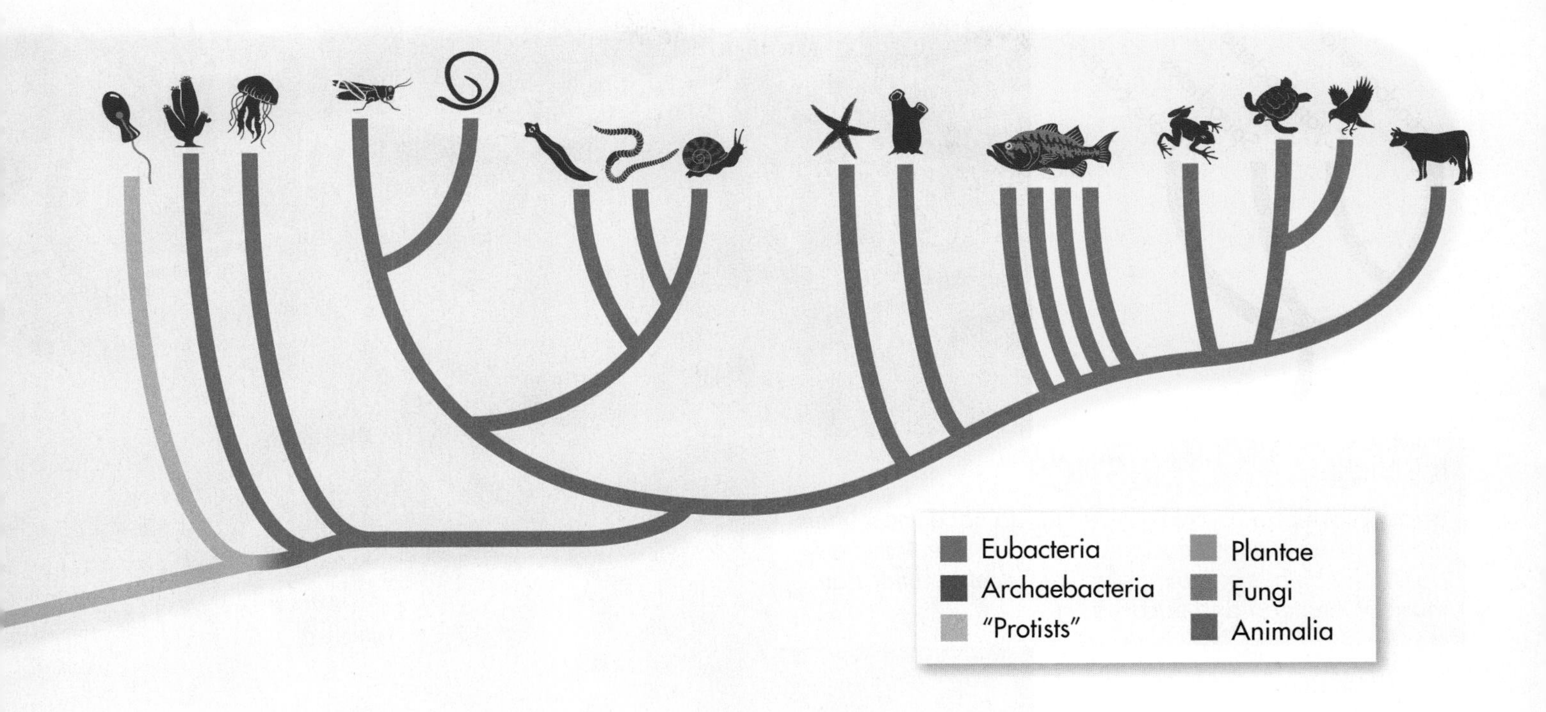

DOMAIN EUKARYA

The domain Eukarya consists of all organisms that have cells with nuclei. It is organized into the four remaining kingdoms of the six-kingdom system: Protista, Fungi, Plantae, and Animalia.

THE "PROTISTS"

Notice that the branches for the kingdom Protista are not together in one area, as is the case with the other kingdoms. In fact, recent molecular studies and cladistic analyses have shown that "eukaryotes formerly known as Protista" do not form a single clade. Current cladistic analysis divides these organisms into at least six clades. They cannot, therefore, be properly placed into a single taxon.

FUNGI

Members of the kingdom Fungi are heterotrophs. Most feed on dead or decaying organic matter. The most recognizable fungi, including mushrooms, are multicellular. Some fungi, such as yeasts, are unicellular.

PLANTS

Members of the kingdom Plantae are autotrophs that carry out photosynthesis. Plants have cell walls that contain cellulose. Plants are nonmotile—they cannot move from place to place.

ANIMALS

Members of the kingdom Animalia are multicellular and heterotrophic. Animal cells do not have cell walls. Most animals can move about, at least for some part of their life cycle.

Bacteria

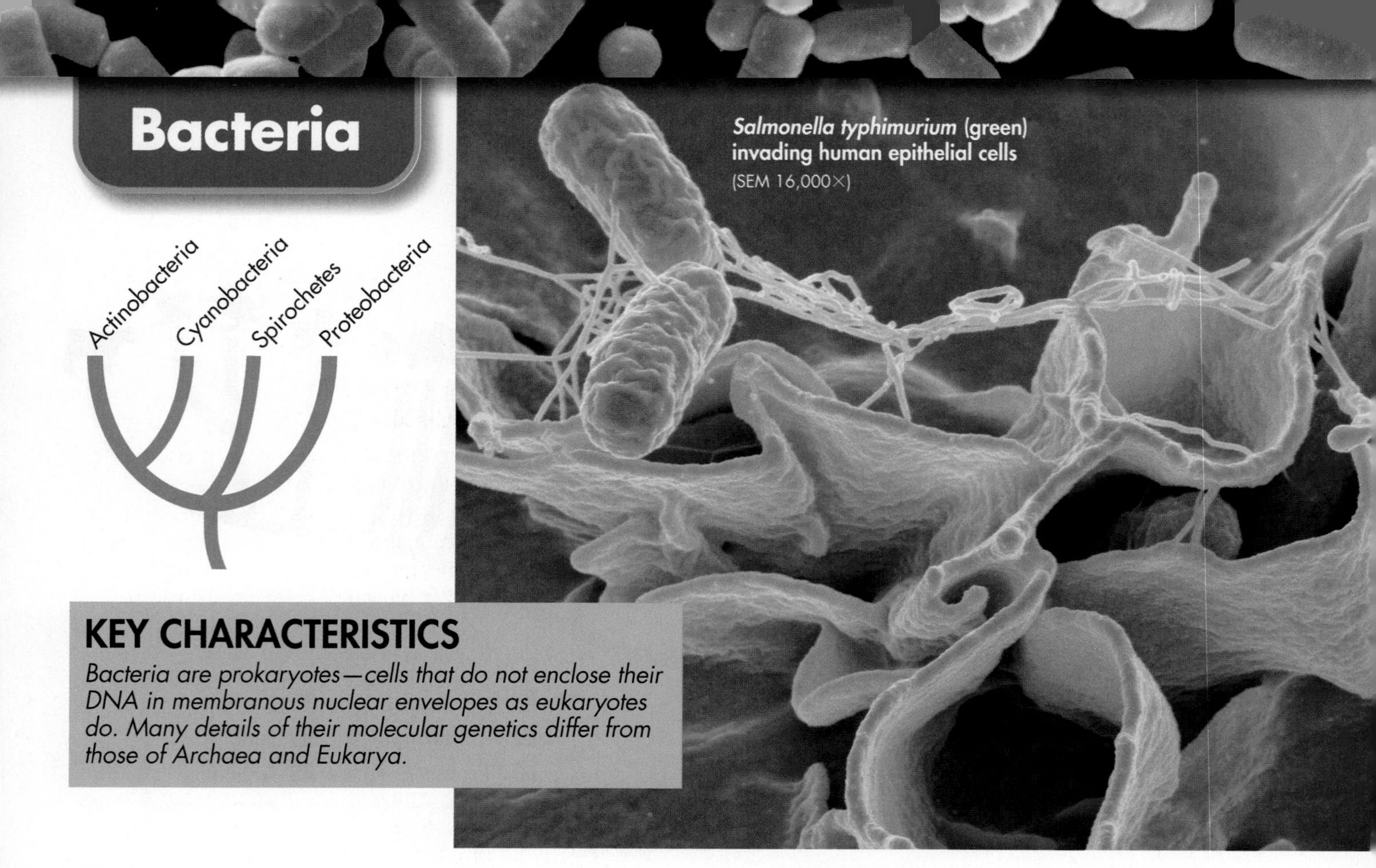

Salmonella typhimurium (green) invading human epithelial cells (SEM 16,000×)

KEY CHARACTERISTICS

Bacteria are prokaryotes—cells that do not enclose their DNA in membranous nuclear envelopes as eukaryotes do. Many details of their molecular genetics differ from those of Archaea and Eukarya.

Cell Structure Variety of cell shapes, including spherical, rodlike, and spiral; most have cell walls containing peptidoglycan. Few if any have internal organelles. Some have external flagella for cell movement.

Genetic Organization All essential genes are in one large DNA double helix that has its ends joined to form a closed loop. Smaller loops of DNA (plasmids) may carry nonessential genes. Simultaneous transcription and translation; introns generally not present; histone proteins absent

Reproduction By binary fission; no true sexual reproduction; some achieve recombination by conjugation.

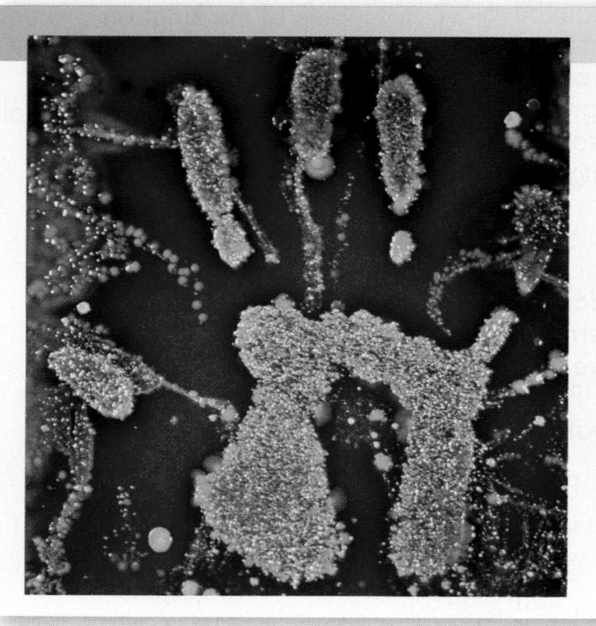

Did You Know?

A World of Bacteria

Putting Bacteria in Proper Perspective

"Planet of the Bacteria" was the title of an essay by the late Stephen Jay Gould. He pointed out that the dominant life forms on planet Earth aren't humans, or animals, or plants. They are bacteria. They were here first, and they inhabit more places on the planet than any other form of life. In fact, bacteria make up roughly 10 percent of our own dry body weight! In terms of biomass and importance to the planet, bacteria truly do rule this planet. They, not we, are number one.

◀ *The bacterial colonies shown here are growing in the print of a human hand on agar gel.*

GROUPS OF BACTERIA

There is no generally agreed phylogeny for the bacteria. Included here are some of the major groups within the domain.

PROTEOBACTERIA

This large and diverse clade of bacteria includes *Escherichia* (*E. coli*), *Salmonella*, *Helicobacter*, and the nitrogen-fixing soil bacterium *Rhizobium*.

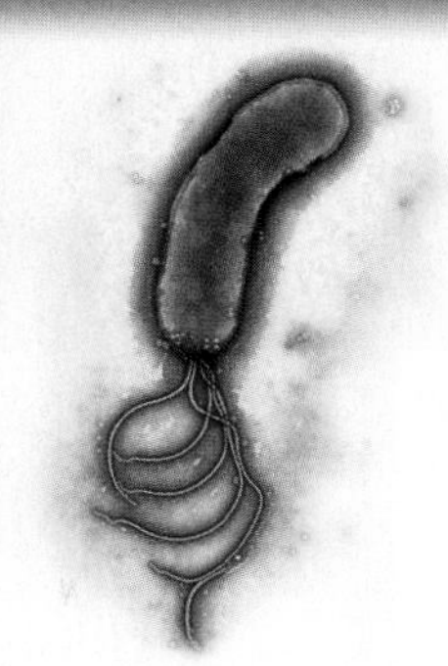

◀ Helicobacter pylori *is rod-shaped and has several flagella used for movement. This bacterium infects the stomach lining and causes ulcers in some people.* (TEM 7100×)

The spiral-shaped bacterium that causes syphilis is Treponema pallidum. (SEM 10,000×) ▼

SPIROCHAETES

The spirochaetes (SPY roh keets) are named for their distinctive spiral shape. They move in a corkscrew-like fashion, twisting along as they are propelled by flagella on both ends of the cell. Most are free-living, but a few cause serious diseases, including syphilis, Lyme disease, and leptospirosis.

ACTINOBACTERIA

A large number of soil bacteria belong to this group. Some form long filaments. Members include the *Streptomyces* and *Actinomyces*, which are natural producers of many antibiotics, including streptomycin. A related group is the *Firmicutes*. The *Firmicutes* include *Bacillus anthracis* (anthrax), *Clostridia* (tetanus and botulism), and *Bacillus thuringensis*, which produces a powerful insecticide used for genetic engineering in plants.

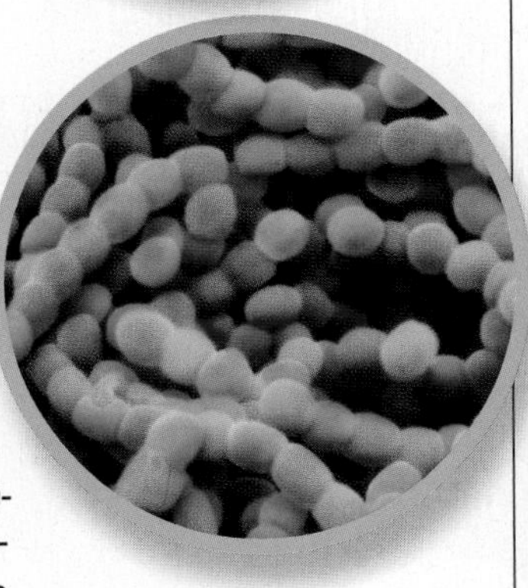

▲ *Chains of spores of soil bacteria, genus* Streptomyces (SEM 3400×)

CYANOBACTERIA

The cyanobacteria are photosynthetic prokaryotes that were once called "blue-green algae." They are among the oldest organisms on Earth, having been identified in rocks dating to more than 3 billion years ago. They are found in salt water and fresh water, in the soil, and even on the surfaces of damp rocks. They are the only organisms on Earth that are able to fix carbon and nitrogen under aerobic conditions, and this enables them to play critical roles in the global ecosystem, where they serve as key sources of carbon and nitrogen.

▼ *Many cyanobacteria form long filaments of attached cells, like those shown here (genus* Lyngbya, SEM 540×).

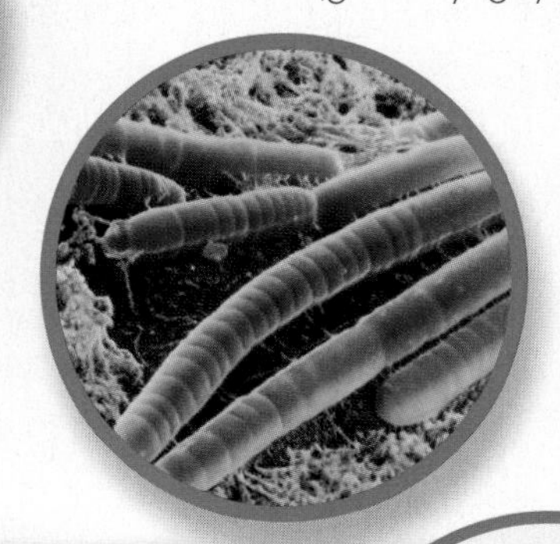

• A Closer Look

The Gram Stain

A Microbiologist's Quick Diagnostic

The Gram stain, developed by the nineteenth-century Danish physician Hans Christian Gram, allows microbiologists to categorize bacteria quickly into one of two groups based on their cell wall composition. Gram-positive bacteria lack a membrane outside the cell wall and take up the stain easily. Gram-negative bacteria, on the other hand, have an outer membrane of lipids and carbohydrates that prevents them from absorbing the gram stain. Many gram-negative bacteria are found among the proteobacteria. On the other hand, actinobacteria are mostly gram-positive.

Gram-positive bacteria appear purple after staining, while gram-negative bacteria appear pink. (LM 1000×) ▶

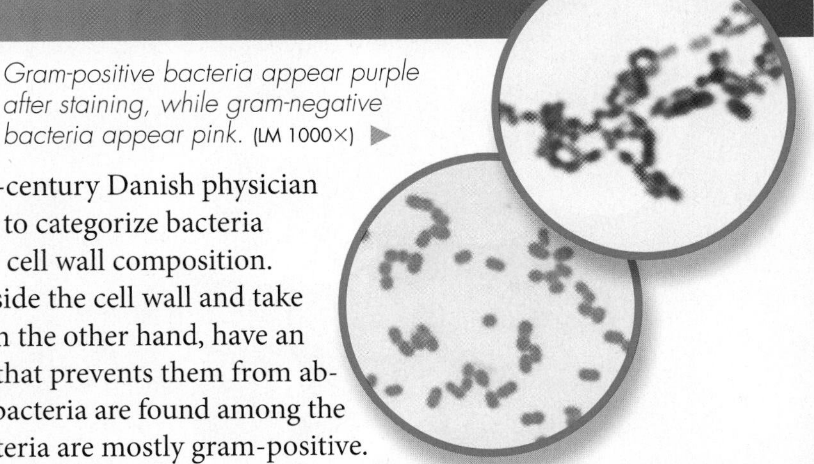

Archaea

KEY CHARACTERISTICS

Archaea are prokaryotes that differ from bacteria in so many details of structure and metabolism that they are viewed as a different domain than bacteria. Genetically, they have more in common with eukaryotes than with bacteria. Their cell walls do not contain peptidoglycan.

Cell Structure Cells similar to those of bacteria in appearance; many have flagella that are different in structure and biochemical composition from bacterial flagella. Cell membrane lipids also different from those of bacteria; few internal organelles

Genetic Organization As in bacteria, all essential genes are in one large DNA double helix that has its ends joined to form a closed loop. Proteins responsible for transcription and translation are similar to those of eukaryotes. Also like eukaryotes, most species contain introns, and all species contain DNA-binding histone proteins.

Reproduction By binary fission; no true sexual reproduction, but some achieve recombination by conjugation.

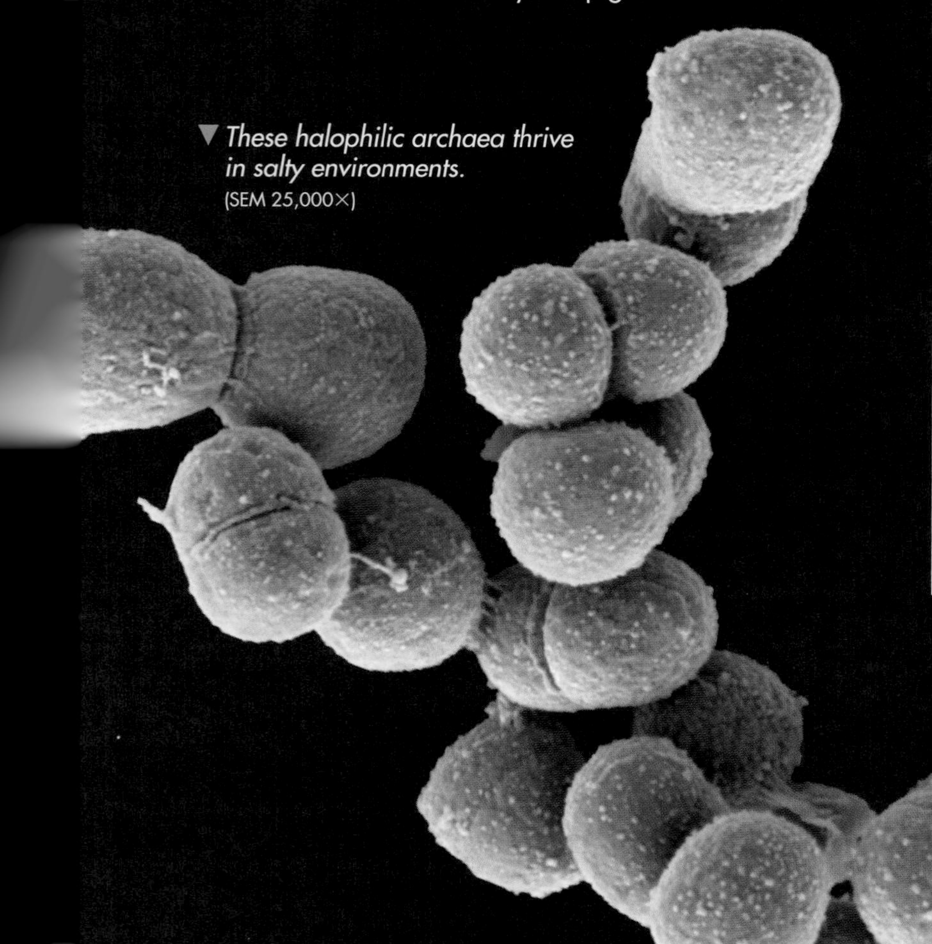

▼ *These halophilic archaea thrive in salty environments.* (SEM 25,000×)

▲ *The volcano Solfatara, near Naples, Italy, is home to many archaea in the genus* Sulfolobus.

• Did You Know?

Hot Enough for You?

The Original Extremists

Way before extreme sports and extreme reality TV shows came the archaea—the original and ultimate extremists. When archaea were first discovered, biologists called them *extremophiles,* a term that literally means "lovers of the extreme." For many archaea, the name still fits. In fact, they have proven especially difficult to grow in the lab, since they require such extreme temperatures and dangerous chemical conditions to thrive. One species will grow only in sulfuric acid! Archaea found in deep-sea ocean vents thrive in temperatures exceeding 100° Celsius, while others enjoy life in the frigid waters of the Arctic.

GROUPS OF ARCHAEA

To date, four major clades of archaea have been identified. Biologists continue to debate how these clades are related to one another.

CRENARCHAEOTES

The crenarchaeotes (kren AHR kee ohts) include organisms that live in the hottest and most acidic environments known. Most of the known species have been isolated from thermal vents and hot springs—the prefix *cren-* means "spring." Some species grow using organic compounds as energy sources, but others fix carbon from carbon dioxide, using hydrogen or sulfur to provide chemical energy.

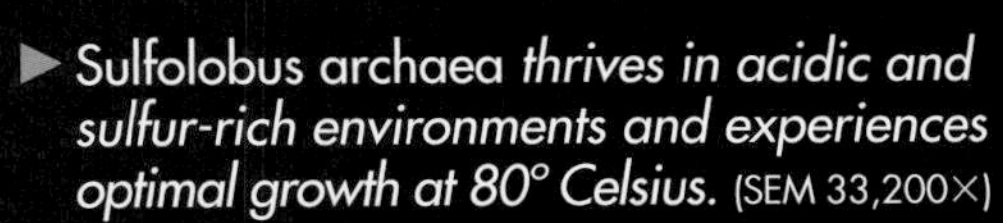

▶ Sulfolobus archaea *thrives in acidic and sulfur-rich environments and experiences optimal growth at 80° Celsius.* (SEM 33,200×)

KORARCHAEOTES

Scientists recently discovered the korarchaeote (kawr AHR kee oht) lineage in Obsidian Pool, Yellowstone National Park, and have since discovered more species in Iceland. Their DNA sequences place them apart from other archaea. The korarchaeotes may in fact be one of the least-evolved lineages of modern life that has been detected in nature so far.

▲ *Korarchaeotes from Obsidian Pool are shown in a lab culture with other microbes from their community.* (SEM 6000×)

NANOARCHAEOTES

Only a single species of this group has been discovered, in 2002, attached to a much larger crenarchaeote! Nanoarchaeotes (na noh AHR kee ohts) grow in hot vents near the coastal regions of the ocean and show definite molecular differences from other archaea. More researc is needed to characterize thi group, but what is known is that they have the smallest known genome of any organism.

▼ *The newly discovered* Nanoa chaeum equitans *(smaller cell is shown attached to its host, genus* Ignicoccus *(larger cells* (LM 2000×)

▼ Colony of *Methanosarcina* archaea (SEM 40,000×)

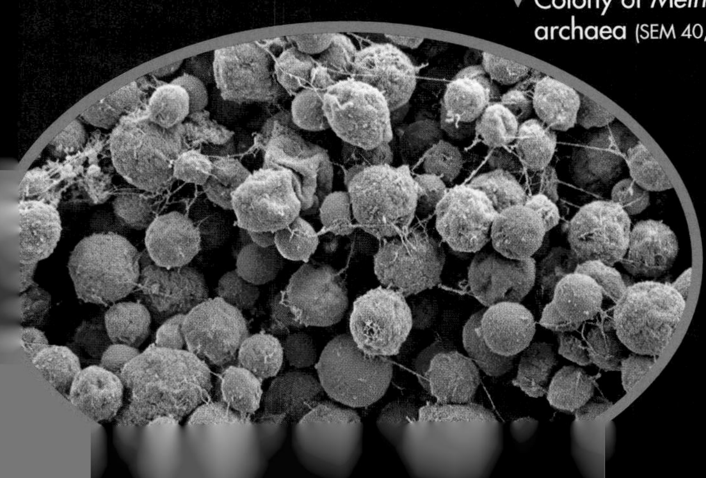

EURYARCHAEOTES

The euryarchaeotes (yoor ee AHR kee ohts) are a very diverse group of archaea, living in a broac range of habitats. The prefix *eury-* comes from a Greek word meaning "broad." The methanoger are a major group of euryarchaeotes that play essential roles in the environment. They help to break down organic compounds in oxygen-poor environments, releasing methane gas in the process. Another group, the *Halobacteria*, are foun in salt ponds, where the concentration of sodiur chloride approaches saturation.

Protists

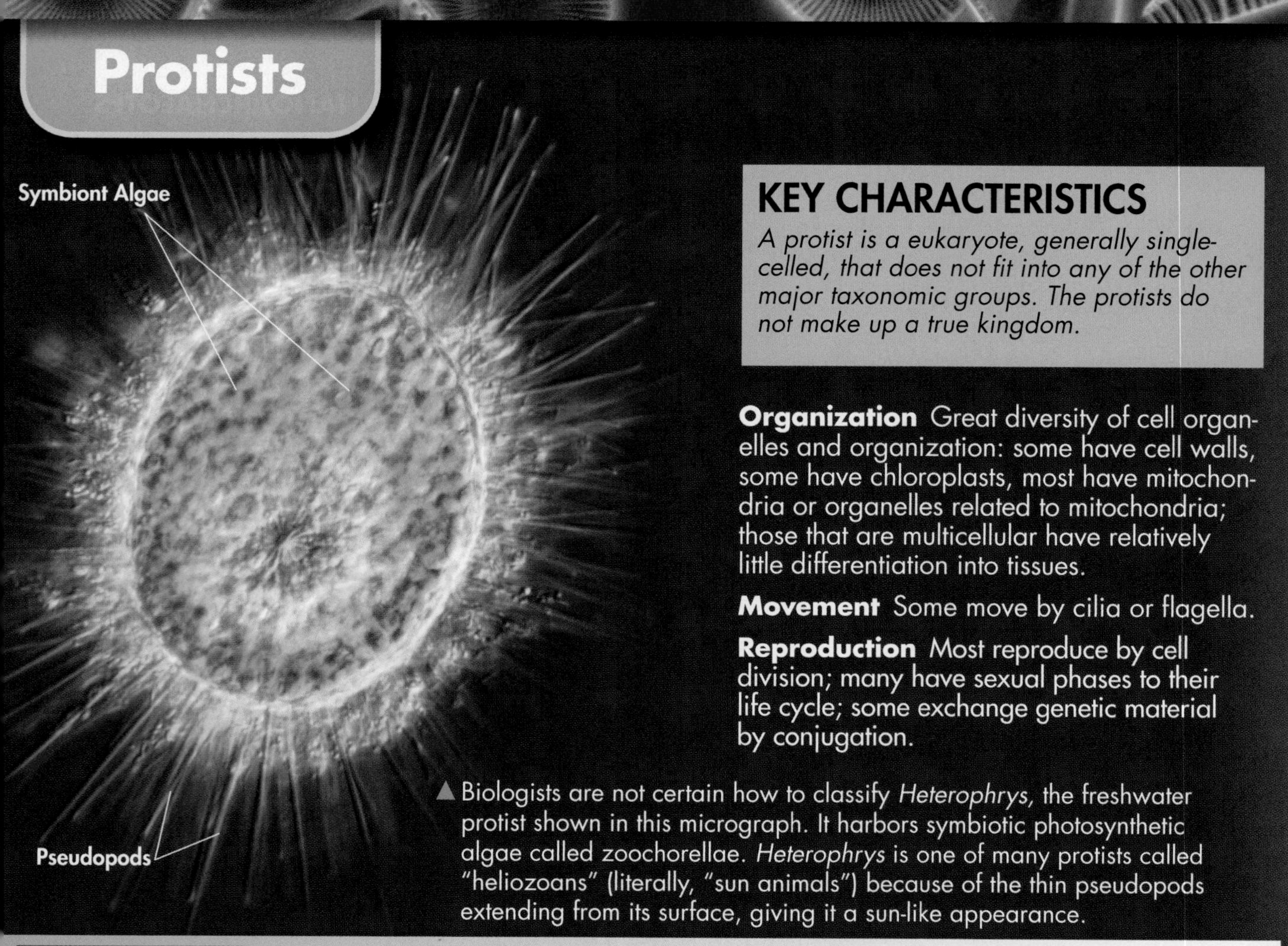

KEY CHARACTERISTICS

A protist is a eukaryote, generally single-celled, that does not fit into any of the other major taxonomic groups. The protists do not make up a true kingdom.

Organization Great diversity of cell organelles and organization: some have cell walls, some have chloroplasts, most have mitochondria or organelles related to mitochondria; those that are multicellular have relatively little differentiation into tissues.

Movement Some move by cilia or flagella.

Reproduction Most reproduce by cell division; many have sexual phases to their life cycle; some exchange genetic material by conjugation.

▲ Biologists are not certain how to classify *Heterophrys*, the freshwater protist shown in this micrograph. It harbors symbiotic photosynthetic algae called zoochorellae. *Heterophrys* is one of many protists called "heliozoans" (literally, "sun animals") because of the thin pseudopods extending from its surface, giving it a sun-like appearance.

• **Did You Know?**

The Kingdom That Isn't

The Challenges of Classifying Protists

Biologists traditionally classified protists by splitting them into funguslike, plantlike, and animal-like groups. This seemed to work for a while, but when they studied protists more carefully with new research tools, including genome-level molecular analysis, this traditional system simply fell apart.

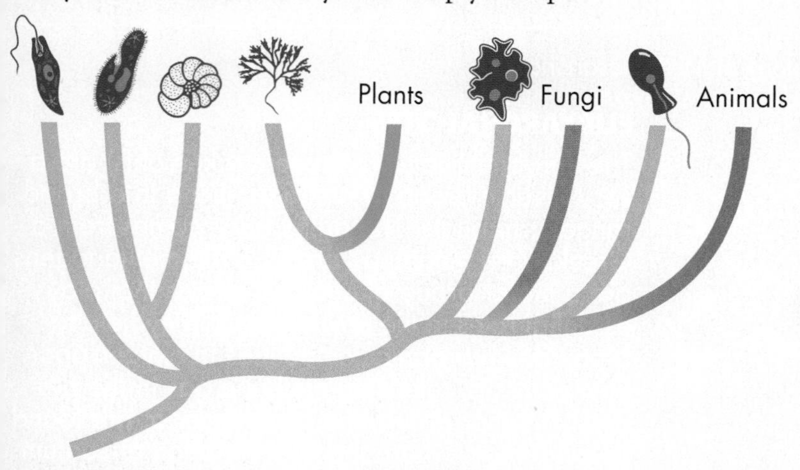

Biologists now think that protists shouldn't be classified as a kingdom at all. In fact, when scientists look for the deepest and most fundamental divisions among eukaryotes, they find that all of those divisions are within the protists themselves, not between protists and other eukaryotes. Starting over, biologists could simply use those divisions to define newer, more accurate "kingdoms," but that might cause new problems. For one thing, it would lump two of the traditional kingdoms (animals and fungi) together, and it would leave a handful of kingdoms that contain only unicellular organisms. There is no perfect solution to this problem. Here, "protists" are considered a kingdom for the sake of convenience, but keep in mind that their differences are really too great for any single kingdom to contain.

Excavates

KEY CHARACTERISTICS

Excavates (EKS kuh vayts) have a characteristic feeding groove, usually supported by microtubules. Most have flagella. A few lack mitochondria and are unable to carry out oxidative phosphorylation, although they do possess remnants of the organelle.

GROUPS OF EXCAVATES

The excavates include a wide diversity of protists, from free-living photosynthesizers to some of humankind's most notorious pathogens.

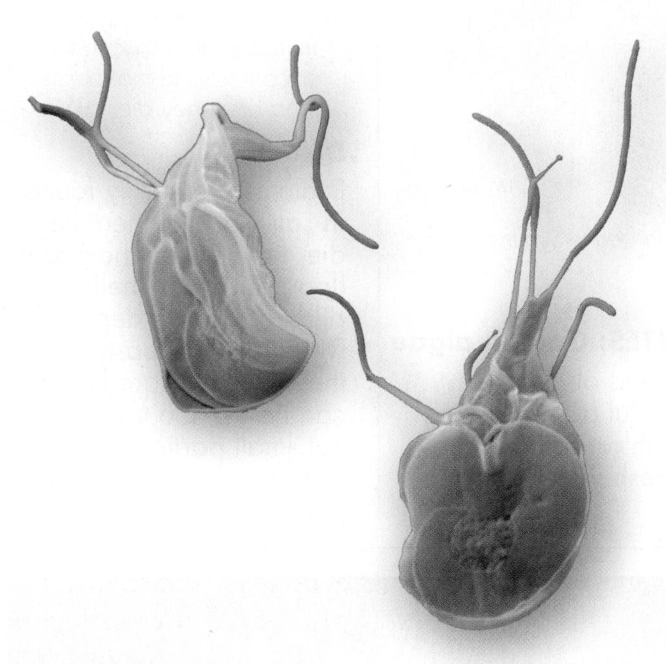

▲ *The diplomonad* Giardia *is a dangerous intestinal parasite that frequently contaminates freshwater streams.* Giardia *infections are common in wildlife and pet dogs and cats.* (SEM 1800×)

DIPLOMONADS

These organisms get their name from the fact that they possess two distinct and different nuclei (from Greek, diplo = double). The double nuclei probably derived from an ancient symbiotic event in which one species was engulfed by another. Cells contain multiple flagella, usually arranged around the body of the cell. Most species of diplomonads are parasitic.

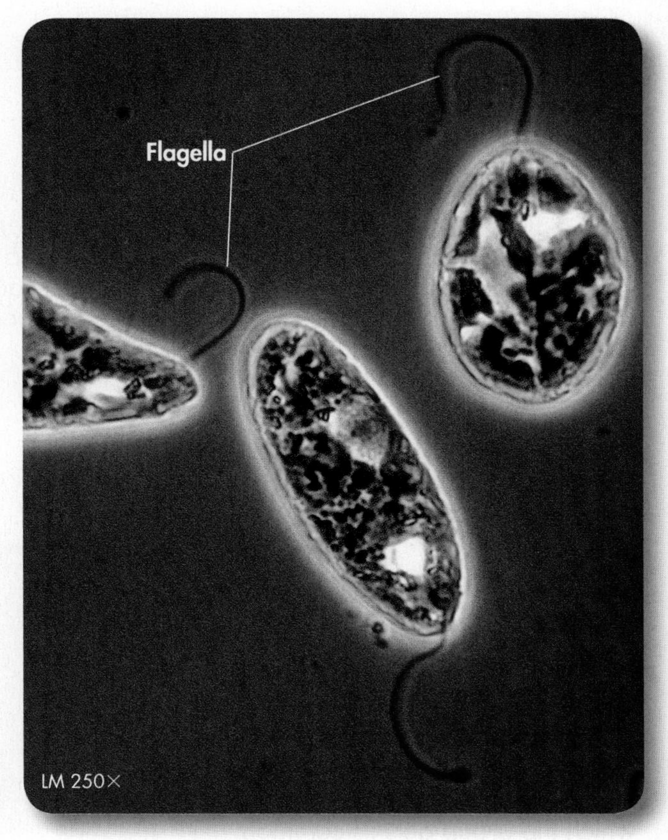

▲ *Photosynthetic* Euglena gracilis *is commonly found in lakes and ponds.*

DISCICRISTATES

Discicristates (disk ee KRIS tayts) are named for the disc-shaped cristae present in their mitochondria. Some species are photosynthetic and free-living, such as *Euglena*, while others are dangerous parasites.

▼ *The ribbonlike cells of* Trypanosoma brucei *cause African sleeping sickness. The parasitic protist is transmitted by tsetse flies to humans, where it infects the blood, lymph, and spinal fluid. Severe nervous system damage and death are the usual result.* (SEM 6700×)

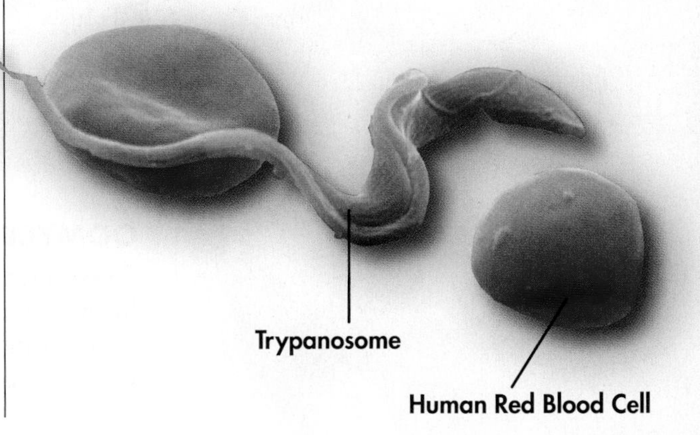

SEM 280×

Chromalveolates

KEY CHARACTERISTICS

Chromalveolates (krohm AL vee uh layts) get their name from alveoli, flattened vesicles that line the cell membrane. The prefix chromo-*, meaning "pigment," reflects evidence that members of this clade share a common ancestor that had accessory pigments used in photosynthesis.*

GROUPS OF CHROMALVEOLATES

The chromalveolates are one of the largest and most diverse groups of eukaryotes.

PHAEOPHYTES: Brown algae

Phaeophytes (FAY uh fyts) are mostly found in salt water. They are some of the most abundant and visible of the algae. Most species contain fucoxanthin, a greenish-brown pigment from which the group gets its common name. The multicellular brown alga known as giant kelp can grow as large as 60 meters in length.

▼ *Brown algae in genus* Fucus *are commonly found in tidepools and on rocky shorelines of the United States.*

LM 200×

▲ *This species, in genus* Synura*, is a colonial alga.*

CHRYSOPHYTES: Golden algae

Chrysophytes (KRIS oh fyts) are known for colorful accessory pigments in their chloroplasts. Most are found in fresh water and are photosynthetic.

SEM 1000×

▲ *Diatoms often produce intricate shells made from silicon dioxide that persist long after they die.*

DIATOMS

Diatoms are mostly found in salt water. When they die, they sink to the ocean floor, and their shells pile up in large deposits. Diatomaceous earth, as these deposits are known, can be used to screen out small particles, and is often used in swimming pool filters.

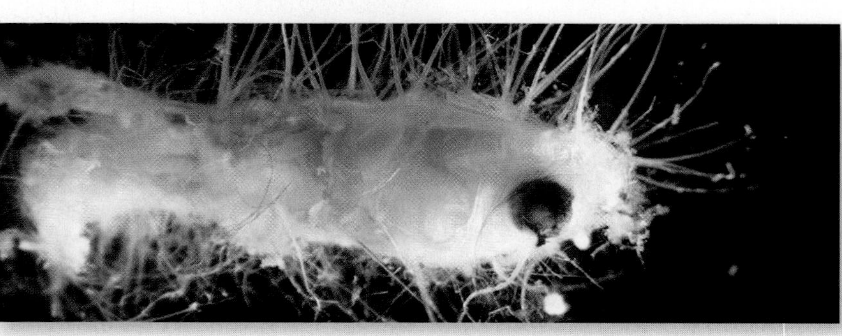

▲ *Water molds growing on a dead goldfish*

OOMYCETES: Water molds

These nonphotosynthetic organisms are often confused with fungi. Oomycetes (oh uh MY seed eez) typically produce fuzzy mats of material on dead or decaying animals and plants. Oomycetes are also responsible for a number of serious plant diseases, including potato blight, sudden oak death, and ink disease, which infects the American chestnut tree.

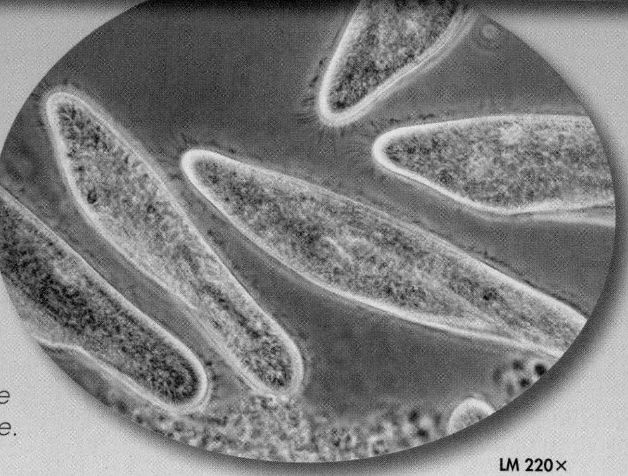

Paramecium multimicronucleatum is the largest paramecium, with cells that are visible to the naked eye.

LM 220×

CILIATES

These common organisms may contain hundreds or even thousands of short cilia extending from the surface of the cell. The cilia propel the ciliate through the water, and may sweep food particles into a gullet. Ciliates are large compared to other protists, with some cells exceeding 1 mm in length.

DINOFLAGELLATES

Dinoflagellates are photosynthetic protists found in both fresh and salt water. Their name comes from their two distinct flagella, usually oriented at right angles to each other. Roughly half of dinoflagellate species are photosynthetic; the other half live as heterotrophs. Many dinoflagellate species are luminescent, and when agitated by sudden movement in the water, give off light.

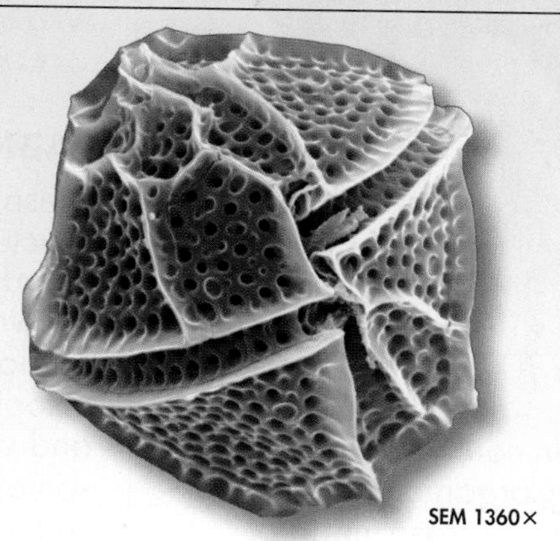

SEM 1360×

▲ *The two flagella of dinoflagellates originate in grooves within thick plates of cellulose that resemble a cross shape, as shown here (genus* Protoperidinium*).*

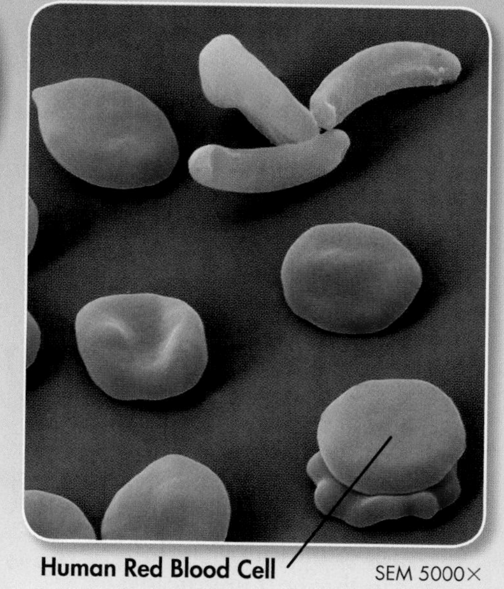

SEM 5000×

▲ *Apicomplexans in genus* Plasmodium *are mosquito-borne parasites. Shown in green are the remnants of a red blood cell that burst when plasmodia reproduced inside.*

APICOMPLEXANS

The apicomplexans (AYP ih kum plek sunz) are named for a unique organelle near one end of the cell known as the apical complex. This structure contains vesicles with enzymes that allow apicomplexans to enter other cells and take up residence as parasites.

Eco • Alert

Toxic Blooms

Dangerous Dinoflagellates

Great blooms of the dinoflagellates *Gonyaulax* and *Karenia* have occurred in recent years on the East Coast of the United States, although scientists are not sure of the reason. These blooms are known as "red tides." *Gonyaulax* and *Karenia* produce a toxin that can become amplified in the food chain when filter-feeding shellfish such as oysters contentrate it in their tissues. Eating shellfish from water affected by red tide can cause serious illness, paralysis, and even death.

▲ *A red tide containing toxic dinoflagellates*

Cercozoa, Foraminiferans, and Radiolarians

There is no single morphological characteristic that unites this diverse trio, but many have extensions of cytoplasm called pseudopods and many produce protective shells. The grouping together of Cercozoa, Foraminifera, and Radiolaria is based almost entirely on molecular analyses and not on morphology.

FORAMINIFERANS

Foraminifera (fawr uh min IF uh ra) produce intricate and beautiful shells that differ from species to species. Slender pseudopods that emerge through tiny holes in the shell enable them to capture food, including bacteria. As many as 4000 species exist.

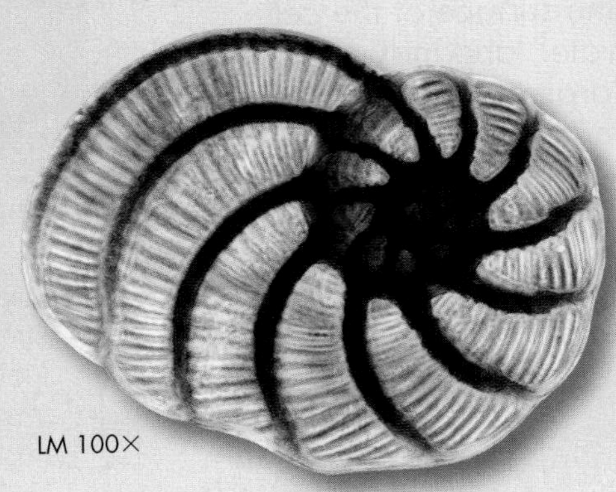

▼ *Peneroplis pertusus* has a spiral-shaped shell.

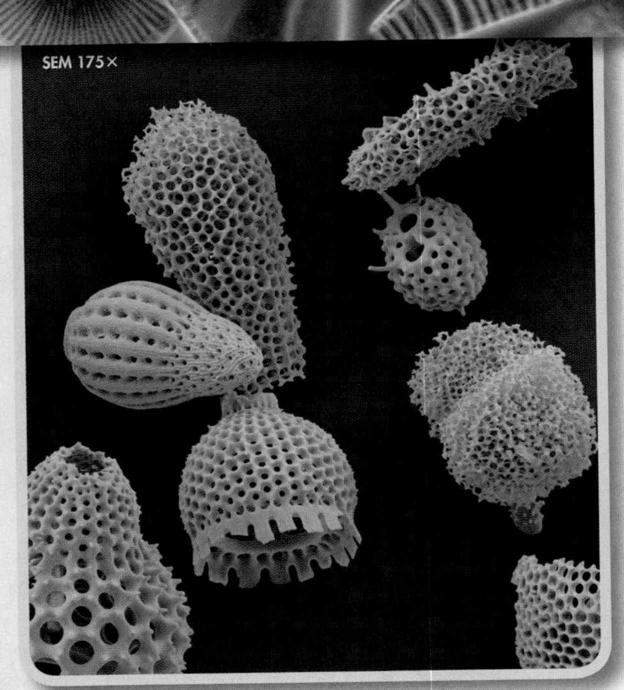

▲ Radiolarian shells are composed of silica or strontium sulfate.

RADIOLARIANS

These organisms have an intricate structure in which the nucleus is found in an inner region of the cell known as the endoplasm. The outer portion of the cell, known as the ectoplasm, contains lipid droplets and vacuoles. These organisms sometimes form symbiotic relationships with photosynthetic algae, from which they obtain food.

CERCOZOA

Members of this clade are common in soil, where they feed on bacteria as well as decaying organic matter. Many have flagella, and some produce scales made of silica that protect their surfaces.

• A Look Back in Time

Foraminiferan Fossils

Ancient Climates Revealed

Abundant fossils of foraminiferans have been found in sediments dating to the Cambrian period (560 million years ago). For decades, oil companies have taken advantage of these ancient fossils to locate the sediments most likely to contain oil, but now there is another use for them—measuring the sea temperature of ancient Earth. Foraminiferans take dissolved oxygen from seawater to make the calcium carbonate ($CaCO_3$) in their shells, and when they do so, they take up two isotopes of oxygen, ^{16}O and ^{18}O. Because water made from ^{16}O is less dense, more of it evaporates into the atmosphere when the seas are warm—increasing the amount of ^{18}O in the remaining seawater, and in the fossil shells. The ratio between ^{16}O and ^{18}O in these fossils allows scientists to study the history of seawater temperature, as shown in the graph above.

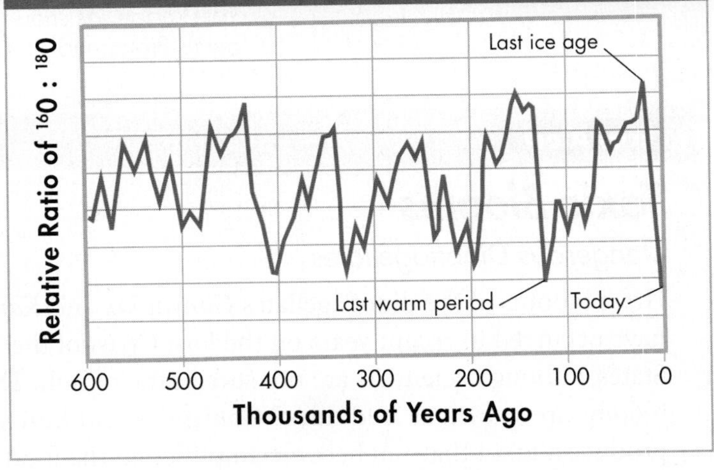

Rhodophytes

Also known as the red algae, these organisms get their name (from Greek, *rhodo* = red and *phyte* = plant) from reddish accessory pigments called phycobilins (fy koh BIL inz). These highly efficient pigments enable red algae to grow anywhere from the ocean's surface to depths as great as 268 meters. Most species are multicellular. Rhodophytes are the sister group to kingdom Plantae.

▼ Some things that we call seaweeds, such as this rhodophyte, are actually protists

Amoebozoa

Members of the Amoebozoa (uh MEE boh zoh ah) are amoebalike organisms that move by means of cytoplasmic streaming, also called amoeboid movement, using pseudopods.

▼ *This solitary amoeba,* Penardia mutabilis, *has very slender pseudopods.*

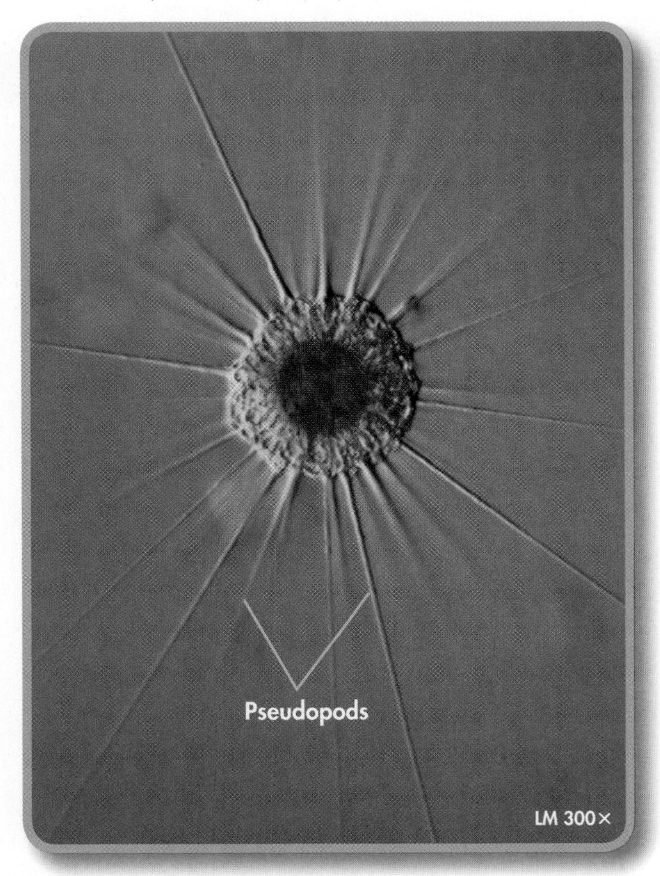

▼ *Slime molds live as single microscopic amoebas in the soil, but aggregate into a colony when conditions are right, forming a multicellular fruiting body. In this image, some of the fruiting bodies have burst, releasing spores.*

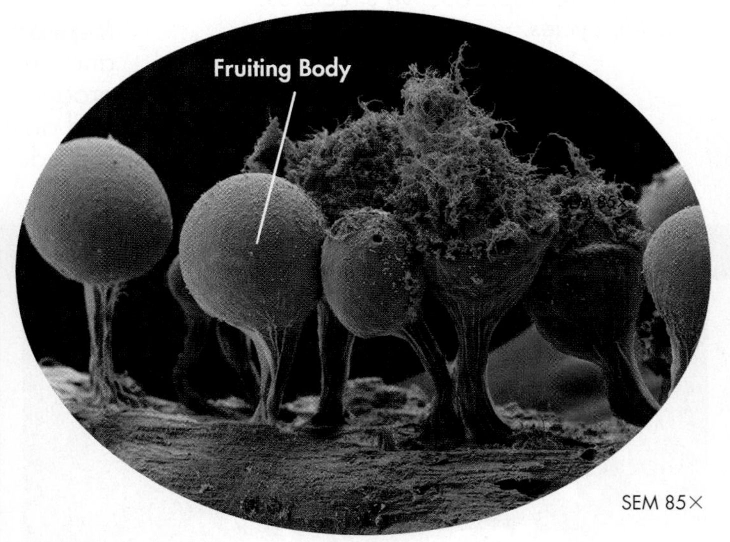

SEM 85×

Choanozoa

Members of the clade Choanozoa (koh AN uh zoh uh) can be solitary or colonial and are found in aquatic environments around the world. This clade is the sister group to kingdom Animalia.

Choanoflagellates are a major group in the clade Choanozoa. They get their name from a collar of cytoplasm that surrounds their single flagellum (from Greek, *choano* = collar). Many species trap food within the collar and ingest it.

Fungi

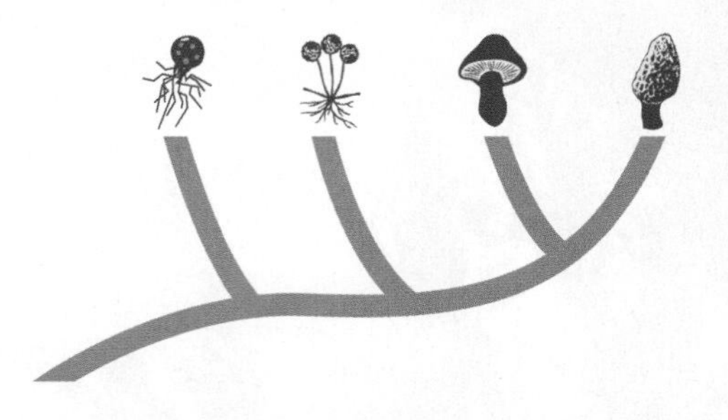

▲ *Stinkhorn fungus (genus* Dictyophora*)*

KEY CHARACTERISTICS

Fungi are heterotrophic eukaryotes with cell walls that contain chitin. Fungi were once thought to be plants that had lost their chloroplasts. It is now clear, however, that they are much more closely related to animals than to plants. More than 100,000 species of fungi are known. Distinctions among the phyla are made on the basis of DNA comparisons, cell structure, reproductive structures, and life cycles.

Organization Some are unicellular yeasts, but most have a multicellular body called a mycelium that consists of one or more slender, branching cells called hyphae.

Feeding and Digestion Obtain food by extracellular digestion and absorption

Reproduction Most have sexual phases to their life cycle and are haploid at most points during the cycle. Most produce tough, asexual spores, which are easily dispersed and able to endure harsh environmental conditions. Asexual reproduction by budding and splitting is also common.

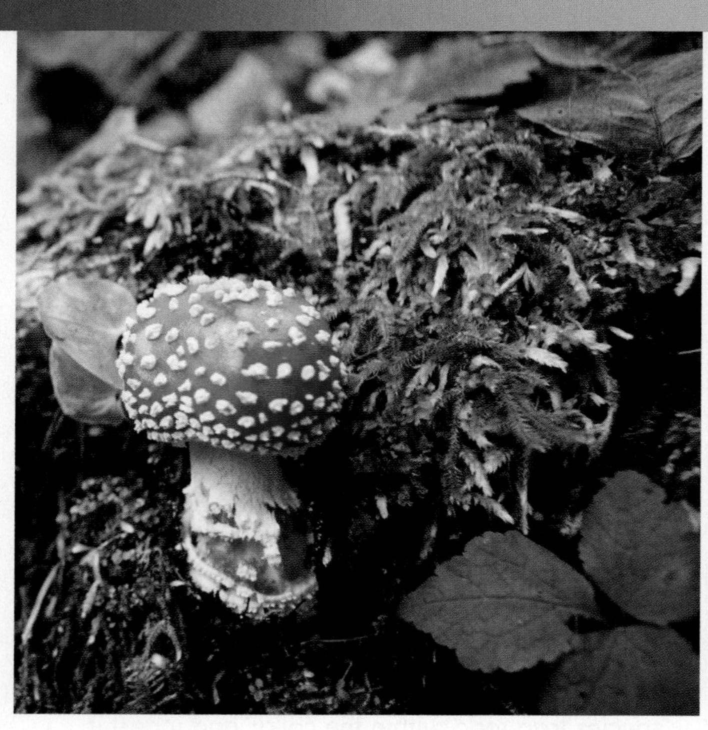

▲ *Fly agaric (*Amanita muscaria*) is poisonous to humans.*

• A Closer Look

Consumers Beware!

Edible and Inedible Mushrooms

Many types of fungi have long been considered delicacies, and several different species of mushrooms are cultivated for food. You may have already tasted sliced mushrooms on pizza, feasted on delicious sautéed portobello mushrooms, or eaten shiitake mushrooms. When properly cooked and prepared, domestic mushrooms are tasty and nutritious.

Wild mushrooms are a different story: Although some are edible, many are poisonous. Because many species of poisonous mushrooms look almost identical to edible mushrooms, you should never pick or eat any mushrooms found in the wild. Instead, mushroom gathering should be left to experts who can positively identify each mushroom they collect. The result of eating a poisonous mushroom can be severe illness, or even death.

Basidiomycetes

The basidiomycetes, or club fungi, are named for the basidium (buh SID ee um; plural: basidia). The basidium is a reproductive cell that resembles a club.

Life Cycle Basidiomycetes undergo what is probably the most elaborate life cycle of all the fungi, shown below.

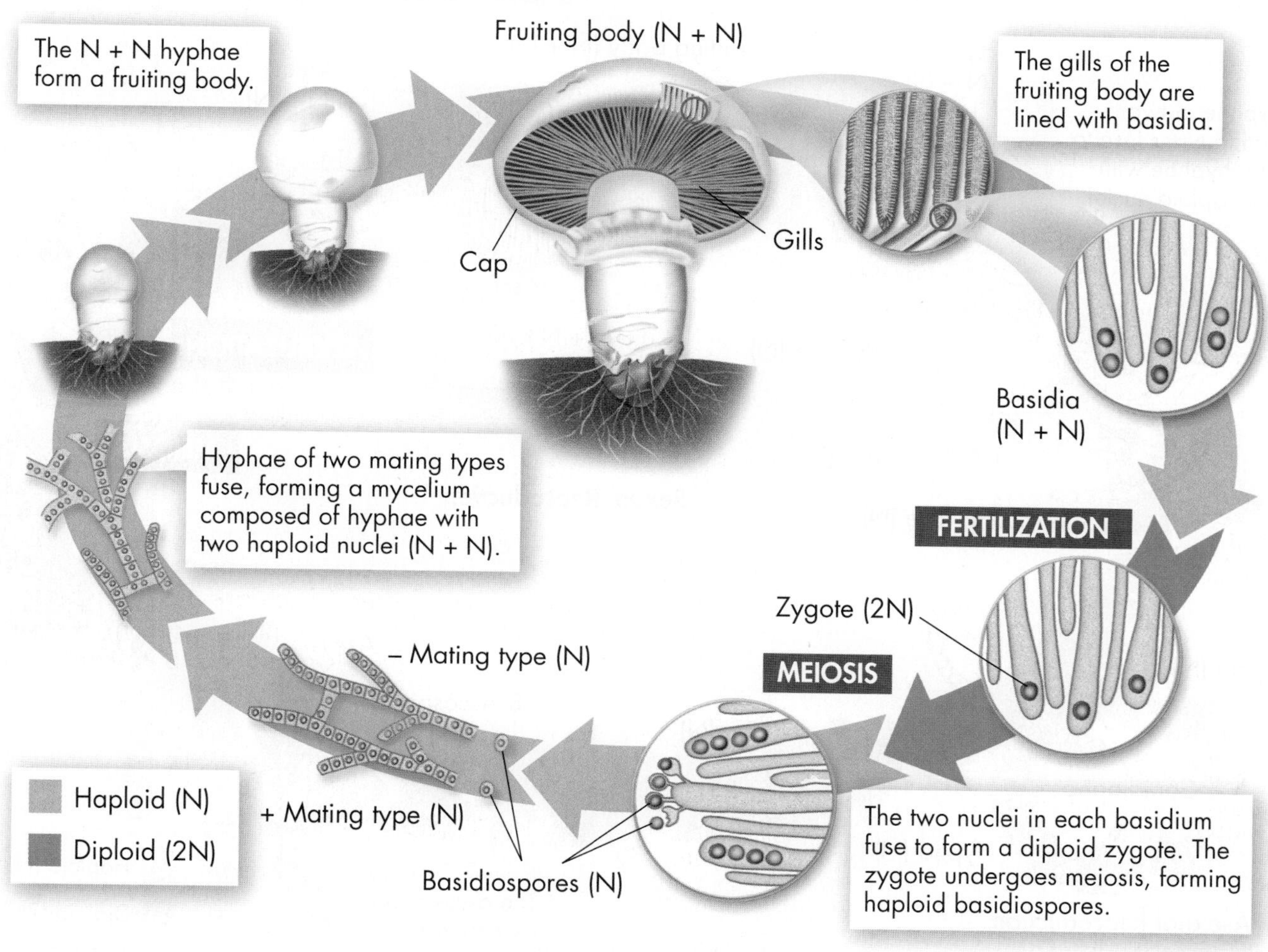

Diversity More than 26,000 species of basidiomycetes have been described, roughly a quarter of all known fungal species. Examples include the stinkhorn and fly agaric mushrooms shown on the previous page, and the shelf fungus and puffball at right.

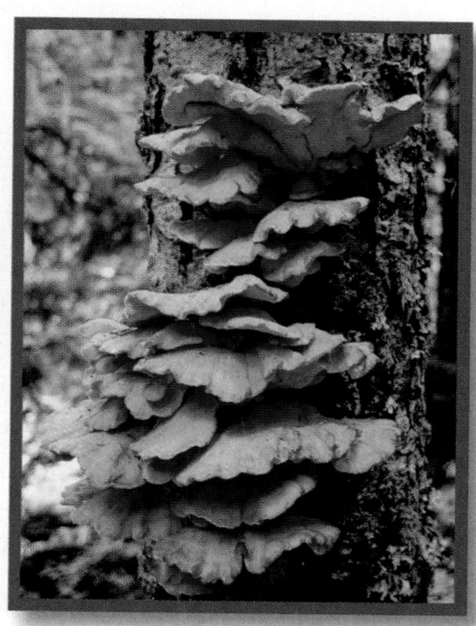

► *Shelf fungi (Polypore family) often grow on the sides of dead or dying trees.*

▼ *A puffball releases its spores.*

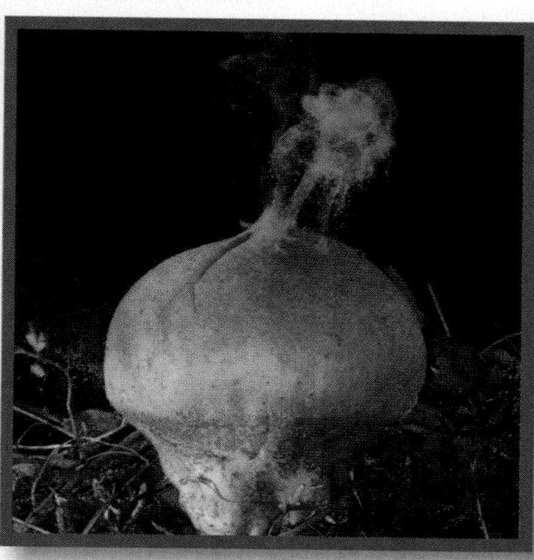

Ascomycetes

The ascomycetes, or sac fungi, are named for the ascus (AS kus), a saclike reproductive structure that contains spores.

Life Cycle The ascomycete life cycle includes an asexual phase, in which haploid spores are released from structures called conidiophores, and a sexual phase.

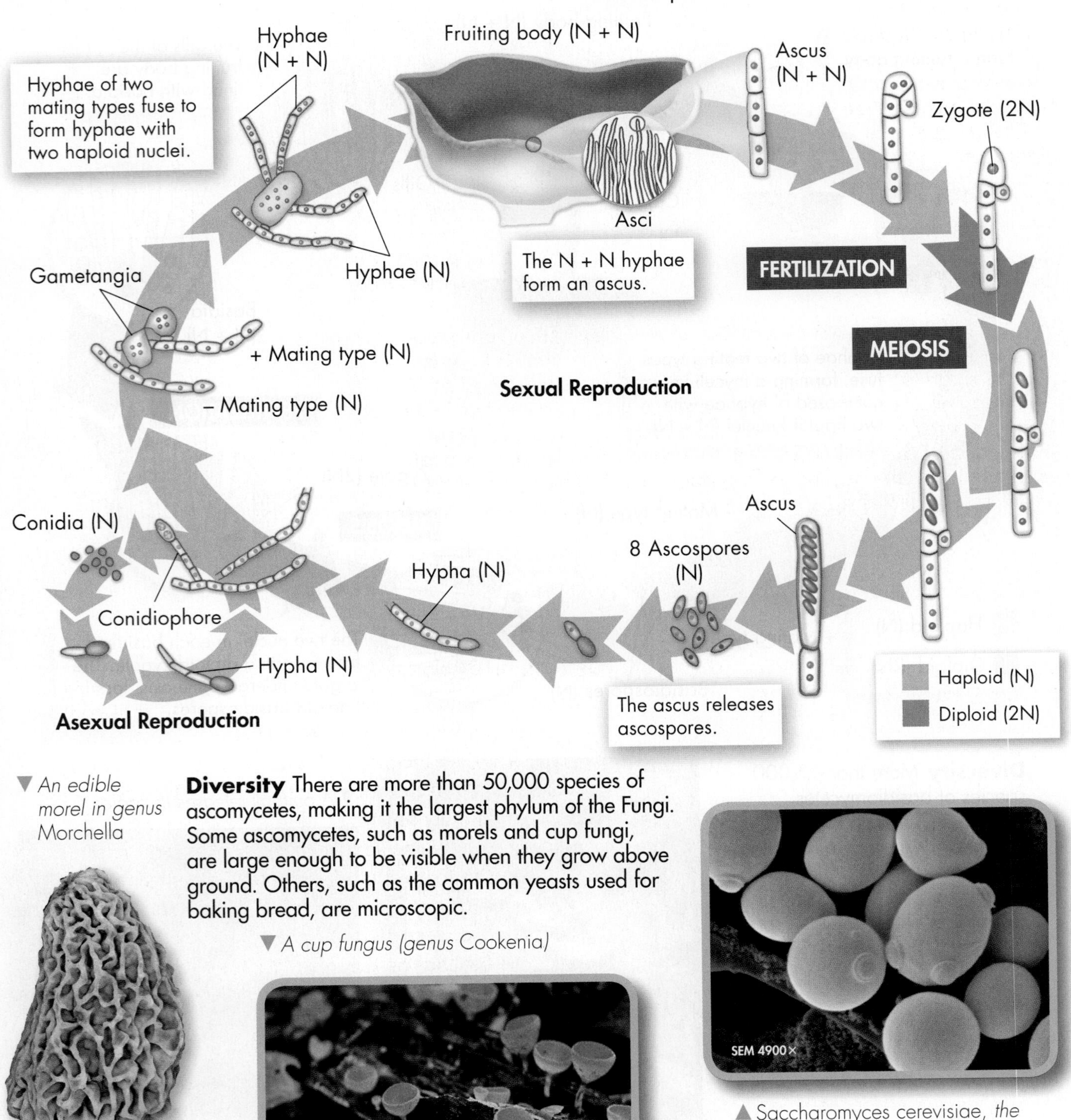

▼ *An edible morel in genus* Morchella

Diversity There are more than 50,000 species of ascomycetes, making it the largest phylum of the Fungi. Some ascomycetes, such as morels and cup fungi, are large enough to be visible when they grow above ground. Others, such as the common yeasts used for baking bread, are microscopic.

▼ *A cup fungus (genus* Cookenia*)*

SEM 4900×

▲ Saccharomyces cerevisiae, *the yeast used to raise bread dough, is a unicellular ascomycete that reproduces asexually by budding.*

Zygomycetes

The hyphae of zygomycetes generally lack cross walls between cells. Zygomycetes get their name from the sexual phase of their reproductive cycle, which involves a structure called a zygosporangium that forms between the hyphae of two different mating types. One group within the zygomycetes, the Glomales, form symbiotic mycorrhizae (my koh RY zee) with plant roots.

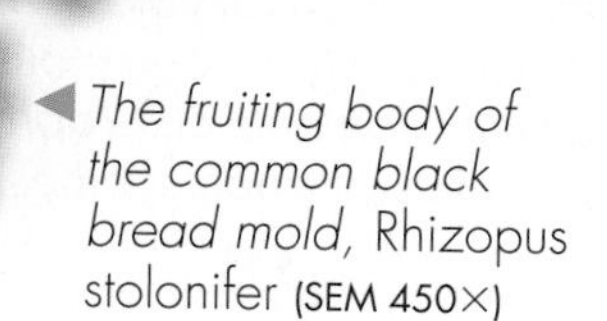

◀ *The fruiting body of the common black bread mold,* Rhizopus stolonifer (SEM 450×)

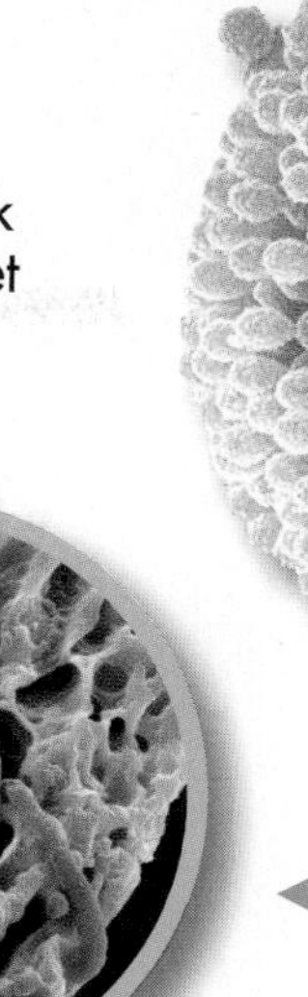

◀ *This micrograph shows mycorrhizal fungi in symbiosis with soybean roots. The soybean plant provides nutrient sugars to the fungus, while the fungus provides water and essential minerals to the plant.* (SEM 200×)

Chytrids

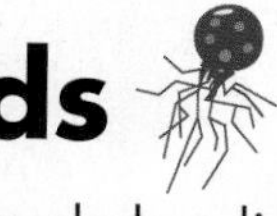

Members of this phylum live in water or moist soil. Their reproductive cells have flagella, making them the only fungi known to have a motile stage to their life cycle. Chytrids are especially good at digesting cellulose, the material of plant cell walls—some live in the digestive systems of cows and deer, helping them to digest plant matter. Others are pathogens—certain chytrids have recently been associated with the decline of frog populations around the world. About 1000 species are known, many of them recently discovered.

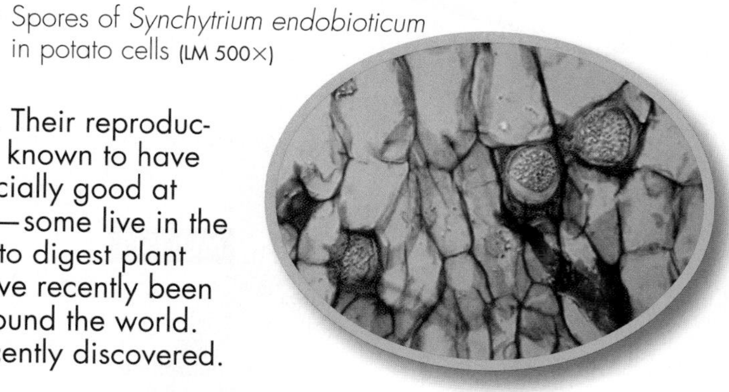

▶ Spores of *Synchytrium endobioticum* in potato cells (LM 500×)

Eco • Alert

Look to the Lichens

Lichens as Bio-Indicators

Lichens are mutualistic associations between a fungus, usually an ascomycete, and a photosynthetic organism, usually an alga. They are incredibly durable, and have even been reported to survive in the vacuum of space. However, they are also incredibly sensitive indicators of the state of the atmosphere. In particular, when sulfur dioxide is released into the atmosphere, it often reacts with water to form acids (including sulfuric acid) that pollute rainfall. Lichens can be severely damaged by acidic rainfall, although the degree of damage depends on the substrate upon which they grow. Lichens disappear first from the bark of pine and fir trees, which are themselves somewhat acidic. Lichens on elms, which have alkaline bark, are the last to go. By carefully monitoring the health of lichen populations of various trees, scientists can use these remarkable organisms as low-tech monitors for the health of the environment.

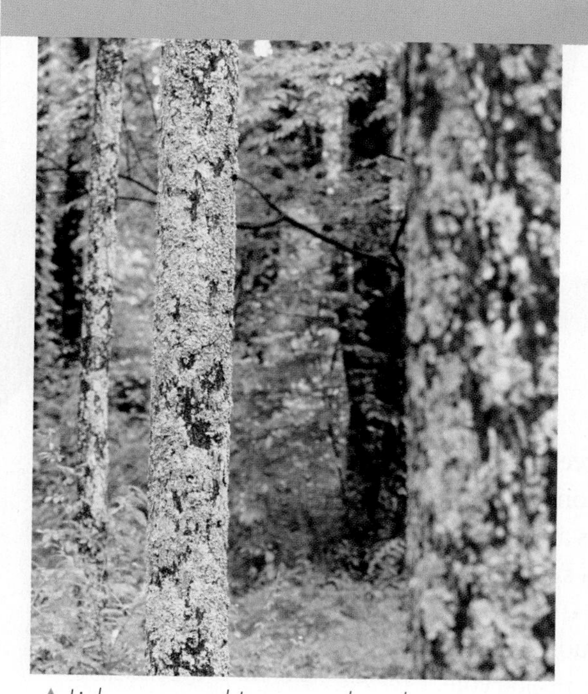

▲ *Lichen-covered Japanese beech*

Plants

KEY CHARACTERISTICS

Plants are eukaryotes with cell walls composed of cellulose. Plants carry out photosynthesis using the green pigments chlorophyll a and b, and they store the products of photosynthesis as starch.

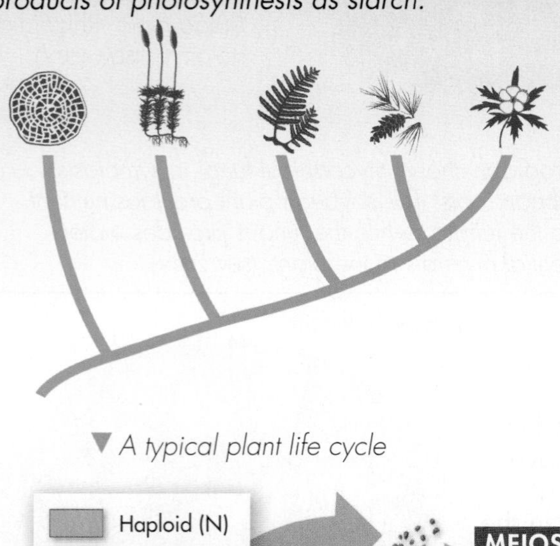

▶ *A banana plant in bloom*

▼ *A typical plant life cycle*

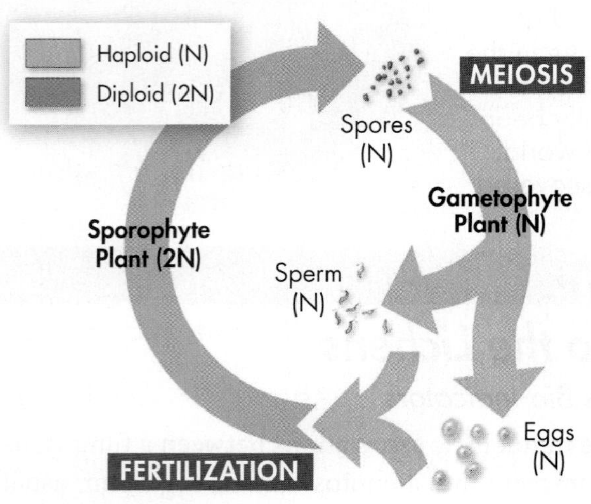

• A Closer Look

Prokaryotes Within

The Origins of Chloroplasts

Chloroplasts, which contain their own DNA, are found in all green plants, but where did they come from? In 1905, the Russian botanist Konstantin Mereschkowsky, noticing the similarities between chloroplasts and cyanobacteria, proposed that these organelles originated from a symbiotic relationship formed with the ancestors of today's plants.

This hypothesis still holds up very well today. New DNA studies suggest that all chloroplasts are descended from a single photosynthetic prokaryote, closely related to today's cyanobacteria.

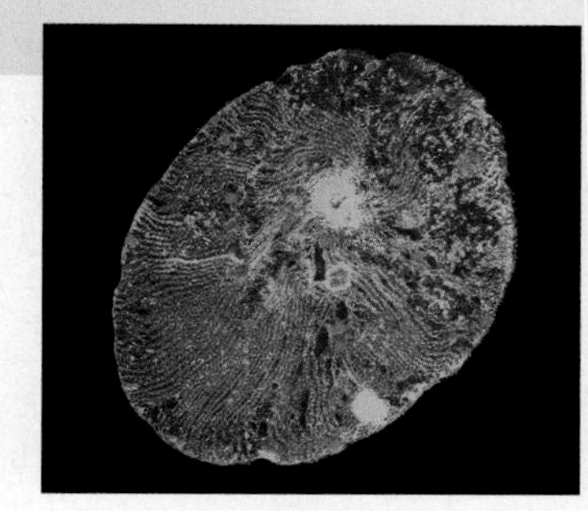

The photosynthetic membranes (shown in green) visible in this thin section of a cyanobacterium resemble the thylakoid membranes of plant cell chloroplasts. (TEM 14,000×)

Green Algae

KEY CHARACTERISTICS

The green algae are plants that do not make embryos. All other plants form embryos as part of their life cycle. The green algae include both unicellular and multicellular species, and they are primarily aquatic.

Organization Single cells, colonies, and a few truly multicellular species

Movement Many swim using whiplike flagella.

Water Transport Water diffuses in from the environment.

Reproduction Asexual and sexual, with gametes and spores; some species show alternation of generations.

▲ *Clumps of* Spirogyra, *a filamentous green alga, are commonly called water silk or mermaid's tresses.*

GROUPS OF GREEN ALGAE

The three most diverse groups of green algae are profiled below.

CHLOROPHYTES: Classic green algae

These algae usually live as single cells, like *Chlamydomonas*, or in colonies, like *Volvox*. They are found in both fresh and salt water, and some species are even known to live in arctic snowbanks.

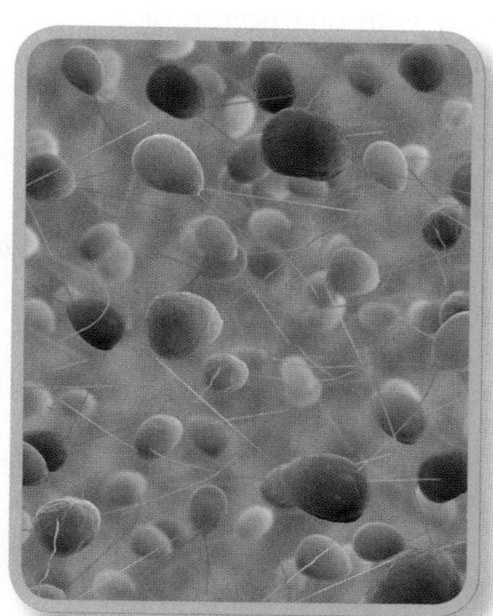
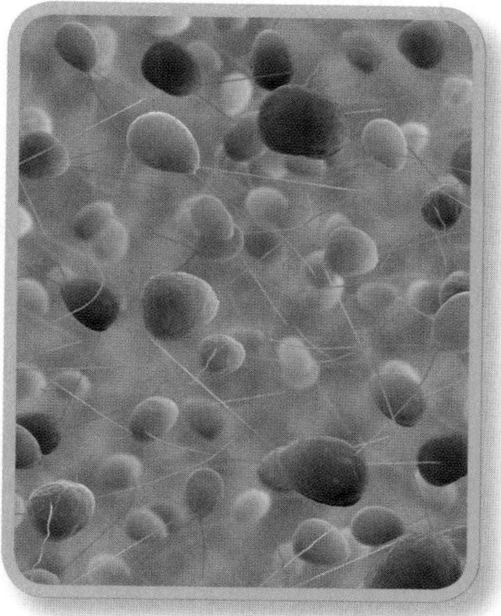

► Chlamydomonas *is a unicellular green alga. Each cell has two flagella, which are used in movement.* (SEM 3000×)

ULVOPHYTES: Sea lettuces

The ulvophytes are large organisms composed of hundreds or thousands of cells. Most form large, flattened green sheets and are often simply called seaweed. They show both haploid and diploid phases in their life cycle, but in many species, such as the common sea lettuce, *Ulva*, it is difficult to tell the two phases apart.

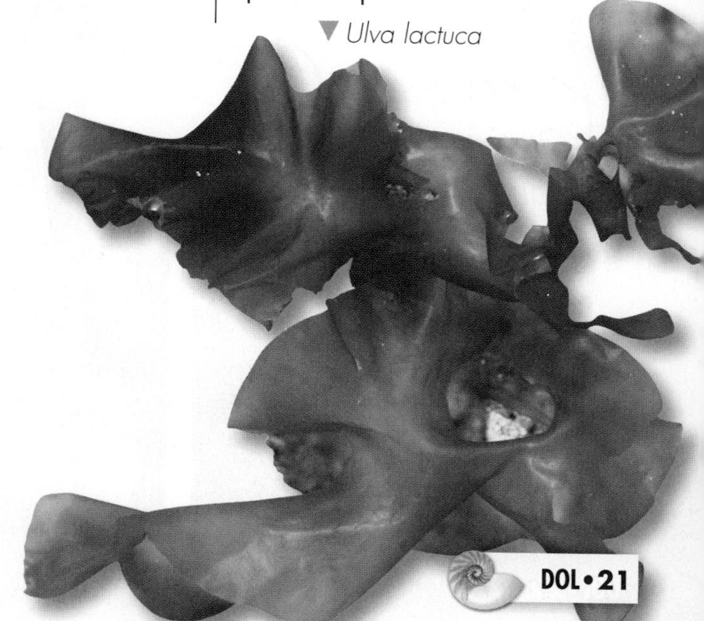

▼ *Ulva lactuca*

CHAROPHYTES: Stoneworts

Among the green algae, the charophytes (KAHR uh fyts) are the closest relatives of more complicated plants. They are mostly freshwater species. Their branching filaments may be anchored to the substrate by thin rhizoids.

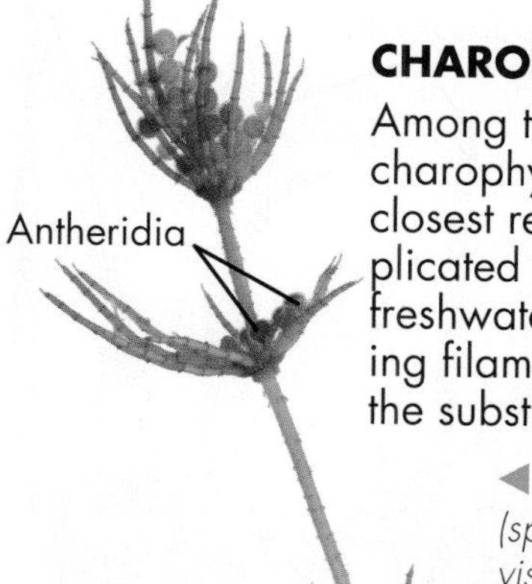

◄ Chara *with antheridia (sperm-producing structures) visible*

Bryophytes

KEY CHARACTERISTICS

Bryophytes (BRY oh fyts), found mostly on land, are multicellular plants that lack true vascular tissue. This lack of vascular tissue limits their height to just a few centimeters and restricts them to moist soils.

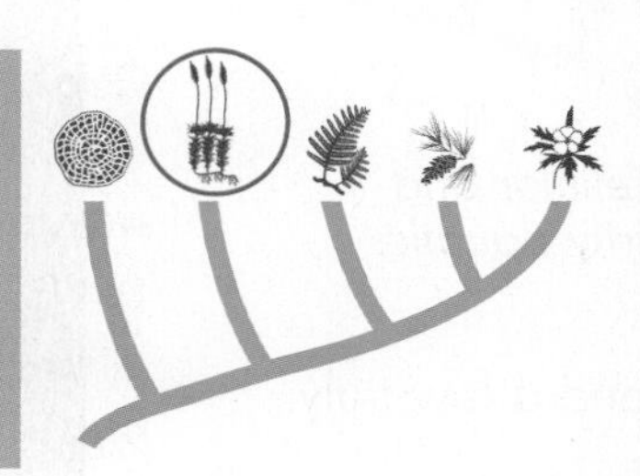

▲ *Mosses thrive in shady, damp locations, such as along the banks of this Oregon creek.*

Organization Complex and specialized tissues, including protective external layers and rhizoids

Movement Adults stationary; male gametes swim to egg cells using flagella.

Water Transport Diffusion from cell to cell; in some mosses, water flows through specialized tissue.

Reproduction All reproduce sexually with alternation of generations, producing gametes and spores. Most reproduce asexually, too. The gametophyte stage is dominant, with the sporophyte stage dependent on the gametophyte.

GROUPS OF BRYOPHYTES

Although they are listed together here, the three major groups of bryophytes are now considered to have evolved independently from each other.

MOSSES: Classic bryophytes

Mosses are found on damp, well-shaded soil, and occasionally along the sides of tree trunks.

Sporophyte

LIVERWORTS

Liverworts are flat, almost leaflike plants that grow on the damp forest floor. Sporophytes are small and grow on the underside of female gametophytes.

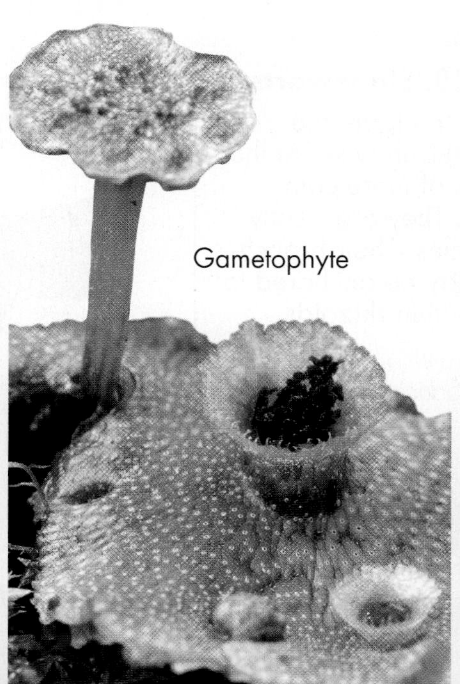

HORNWORTS

Hornworts get their name from their sporophytes, tiny green structures resembling horns. Like other bryophytes, hornworts are found mostly in damp, well-shaded areas. Only about 100 species are known.

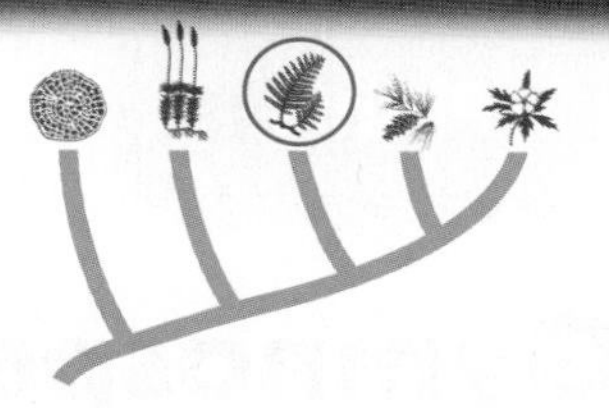

Seedless Vascular Plants

KEY CHARACTERISTICS

This informal grouping lumps together all the plants that have true vascular tissue but lack seeds. Vascular tissue is a key adaptation to life on land. By carrying water and food throughout plant structures, vascular tissue permitted the evolution of roots and tree-size plants, and it allowed plants to spread into dry areas of land.

Organization Complex and specialized tissues, including true roots, stems, and leaves

Movement Adults stationary; male gametes swim to egg cells using flagella.

Water Transport Through vascular tissue

Reproduction Alternation of generations, producing spores, eggs, and swimming sperm; the sporophyte stage is dominant, but the sporophyte is not dependent on the gametophyte as it is in bryophytes.

GROUPS OF SEEDLESS VASCULAR PLANTS

Besides the flowering plants, these organisms make up the most diverse collection of land plants, with more than 10,000 known species.

FERNS

Ferns are common and abundant. Because they need standing water to reproduce, ferns are generally found in areas that are damp at least part of the year. The sporophyte phase of the life cycle is dominant. Spores are produced in prominent clusters known as sori (SOH ry) on the undersides of leaves.

▼ *Polypodium vulgare*

CLUB MOSSES

Not really mosses, these vascular plants are also called lycopods (LY koh pahdz). These plants were especially abundant during the Carboniferous Period 360 to 290 million years ago, when they grew as large as trees. Today, their remains make up a large part of coal deposits mined for fuel.

▼ *The small club moss known as* Lycopodium *can be found growing on the forest floor throughout the temperate regions of North America. They look like tiny pine trees at first glance, but they are, in fact, small, seedless plants.*

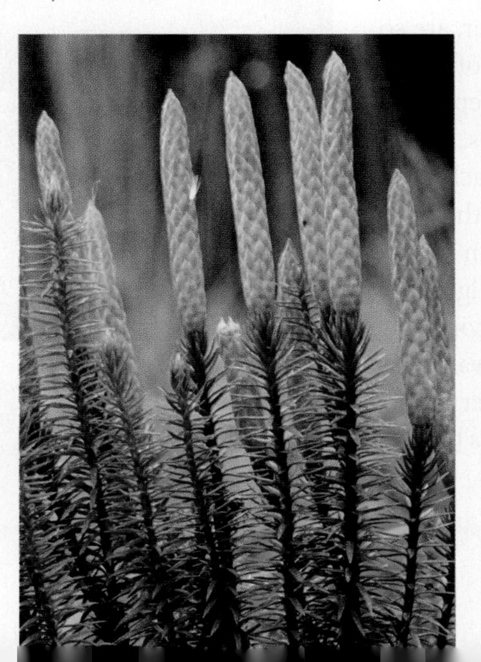

HORSETAILS

Only a single living genus of horsetails is known, *Equisetum* (ek wi SEET um). They get their name due to their resemblance to horses' tails. Today, only 25 species are known, confined to wet areas of soil. But horsetails were once much more diverse, larger in size, and abundant. Abrasive silica, found in many horsetails, was used in colonial times as a scouring powder to help clean pots and pans.

▼ *Equisetum*

Gymnosperms

KEY CHARACTERISTICS

Gymnosperms are seed-bearing vascular plants whose seeds are exposed to the environment, rather than being enclosed in a fruit. The seeds are usually located on the scales of cones.

Organization True roots, stems, and leaves

Movement Adults stationary; within pollen grains, male gametophytes drift in air or are carried by animals to female structures, where they release sperm that move to eggs.

Water Transport Through vascular tissue

Reproduction Sexual; alternation of generations; the sporophyte stage is dominant. Female gametophytes live within the parent sporophyte. Because pollen grains carry sperm to eggs, open water is not needed for fertilization.

Some bristlecone pines are thousands of years old, such as this one growing in Nevada.

Rising From the Ashes

Fire's Role in Seed Germination

We generally think of forest fires as being natural disasters, and that's typically true. Some gymnosperm species, however, are so well adapted to the arid conditions of the American West that they actually depend upon such fires to spread their seeds.

The best-known example is the Jack pine, *Pinus banksiana*. Its seed cones are thick and heat resistant. When engulfed in a fire, its seeds escape damage. The fire's high heat helps to open the outer coat of the cone, enabling the seeds to pop out afterward. As a result, Jack pines are among the very first plants to repopulate a forest that has been damaged by fire.

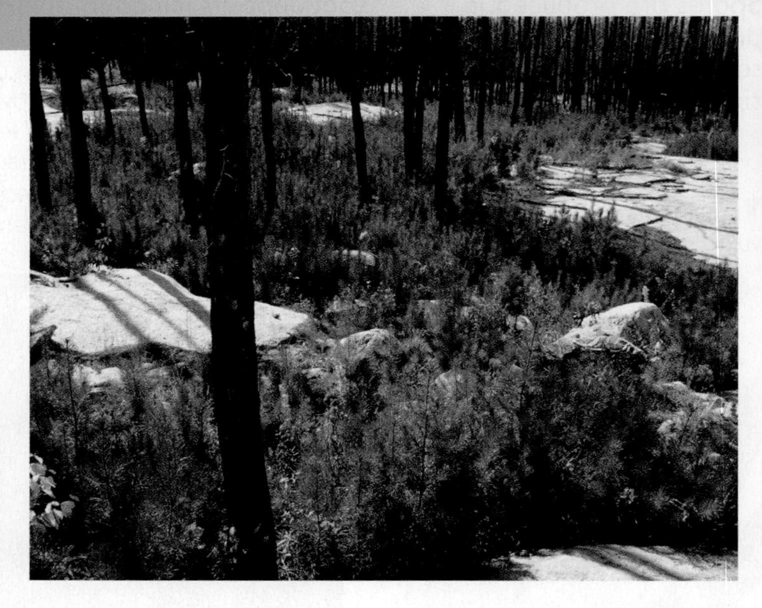

The high heat of a forest fire opens the cones of the Jack pines, releasing their seeds. In this photograph, Jack pine seedlings are growing among the charred remnants of mature trees that burned in a forest fire.

GROUPS OF GYMNOSPERMS

There are four groups of gymnosperms, representing about 800 species in total.

CONIFERS

Conifers are by far the most diverse group of living gymnosperms, represented by nearly 700 species worldwide. They include the common pine, spruce, fir, and redwood trees that make up a large share of the forests in the temperate regions of the world. Conifers have enormous economic importance. Their wood is used for residential building, to manufacture paper, and as a source of heat. Compounds from their resins are used for a variety of industrial purposes.

▲ *Most conifers retain their leaves year-round.*

CYCADS

Cycads (SY kads) are beautiful palmlike plants that have large cones. Cycads first appeared in the fossil record during the Triassic Period, 225 million years ago. Huge forests of cycads thrived when dinosaurs roamed Earth. Today, only nine genera of cycads exist. Cycads can be found growing naturally in tropical and subtropical places such as Mexico, the West Indies, Florida, and parts of Asia, Africa, and Australia.

► *A Sago Palm,* Cycas revoluta

▲ *Ginkgoes are often planted in urban settings, where their toughness and resistance to air pollution make them popular shade trees.*

GINKGOES

Ginkgoes (GING kohs) were common when dinosaurs were alive, but today the group contains only one species, *Ginkgo biloba*. The living *Ginkgo* species looks similar to its fossil ancestors—in fact, *G. biloba* may be one of the oldest seed plant species alive today.

GNETOPHYTES

About 70 present-day species of gnetophytes (NET oh fyts) are known, placed in just three genera. The reproductive scales of these plants are clustered in cones.

► Welwitschia mirabilis, *an inhabitant of the Namibian desert in southwestern Africa, is one of the most remarkable gnetophytes. Its huge leathery leaves grow continuously and spread across the ground.*

Angiosperms

KEY CHARACTERISTICS

Angiosperms are plants that bear seeds in a closed ovary. The ovary is part of a reproductive organ known as a flower. Seeds are formed in a double fertilization event, which forms a diploid embryo and a triploid endosperm tissue. As seeds mature, ovaries develop into fruits that help to disperse the seeds.

Organization True roots, stems, and leaves

Movement Adults stationary; within pollen grains, male gametophytes drift in air or are carried by animals to female structures, where they release sperm that move to eggs.

Water Transport Through vascular tissue

Reproduction Sexual, with alternation of generations; also asexual. The sporophyte stage is dominant. Female gametophytes live within the parent sporophyte. Pollen carries sperm to eggs, so open water is not needed for fertilization.

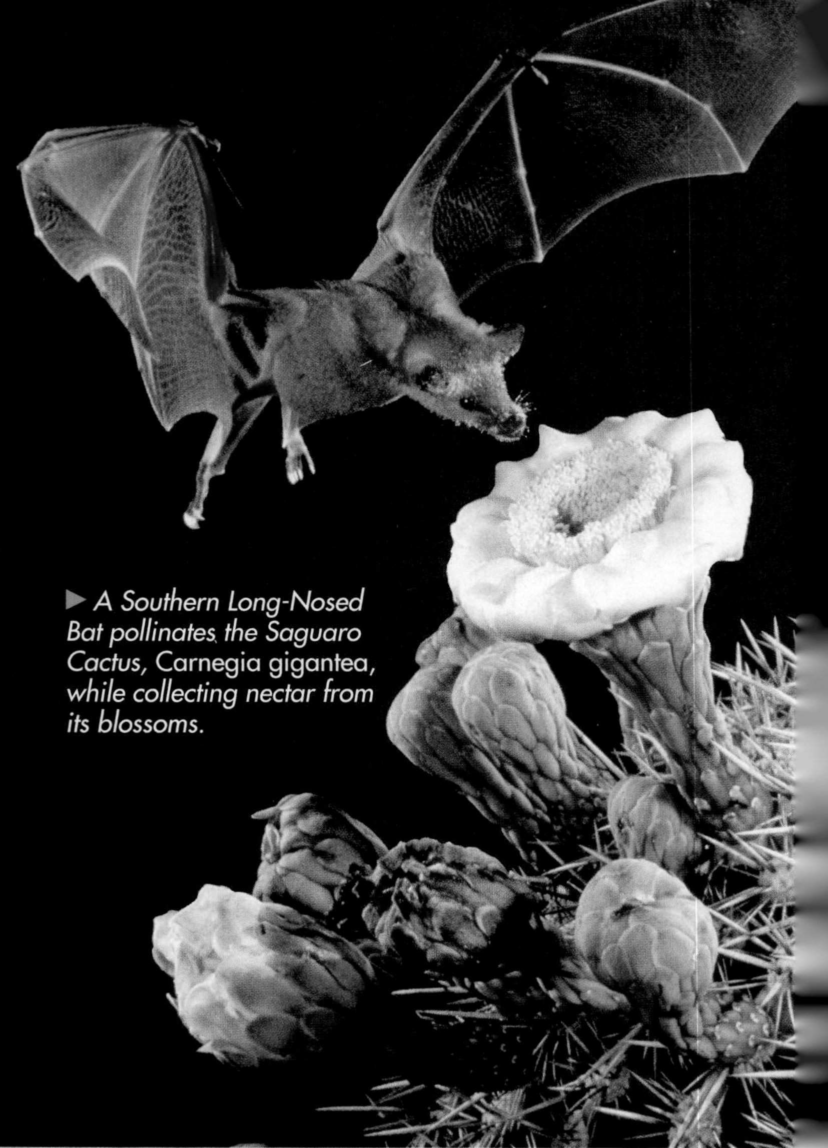

▶ *A Southern Long-Nosed Bat pollinates the Saguaro Cactus,* Carnegia gigantea, *while collecting nectar from its blossoms.*

• A Closer Look

Whatever Happened to Monocots and Dicots?

Traditionally, flowering plants have been divided into just two groups, monocots and dicots, based on the number of seed leaves in their embryos. Today, however, molecular studies have shown that the dicots aren't really one group. Some of the most primitive flowering plants (like *Amborella*) are dicots, and so are some of the most advanced flowering plants, while the monocots fall right in between. So, while monocots are indeed a single group, the term *dicots* is now just an informal, though still useful, grouping.

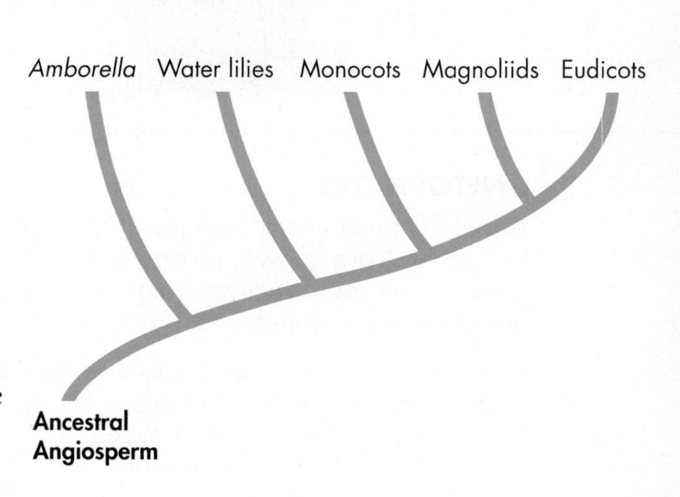

GROUPS OF ANGIOSPERMS

The great majority of plant species—over 260,000—are angiosperms.

▲ *Water lilies are aquatic plants that produce flowers and leaves, which float on the surface of the water.*

NYMPHAEACEAE: Water lilies

About 50 species of water lilies are known, and they are of special interest to plant taxonomists. Their DNA and flower structure suggest that they are, along with *Amborella*, one of the earliest groups to have split off from the main line of flowering plant evolution. Examples of water lilies are found throughout the world.

MAGNOLIIDS:
Magnolia trees and others

The most famous genus of these plants is *Magnolia*, which includes nearly 200 species. Laurels and tulip poplars are also magnoliids (mag NOH lee ids). Because of their flower structure, magnoliids were once thought to be nearly as primitive as water lilies. Genetic studies now suggest that they split off from the rest of the angiosperm line after monocots and, therefore, do not represent the earliest flowering plants.

▼ *The Tulip Poplar is a long, straight tree often used as wood for telephone poles. Its flowers are greenish and shaped like tulips.*

► *Magnolia trees produce conspicuous flowers, which contain multiple stamens and multiple pistils.*

AMBORELLA

Amborella does not represent a group of plants but instead just a single species found only on the island of New Caledonia in the South Pacific Ocean. DNA studies show that *Amborella* is equally separated from all other flowering plants living today, suggesting that it is descended from plants that split off from the main line of flowering plant evolution as long ago as 100 million years.

▲ *The flowers of* Amborella trichopoda *are simpler than those of most other plants, and the species has a number of features that place it at the very base of flowering plant evolution.*

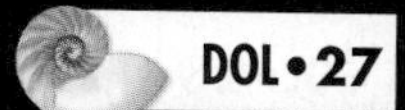

GROUPS OF ANGIOSPERMS CONTINUED...

MONOCOTS

The monocots include an estimated 65,000 species, roughly 20 percent of all flowering plants. They get their name from the single seed leaf found in monocot embryos, and they include some of the plants that are most important to human cultures. Monocots grown as crops account for a majority of the food produced by agriculture. These crops include wheat, rice, barley, corn, and sugar cane. Common grasses are monocots, as are onions, bananas, orchids, coconut palms, tulips, and irises.

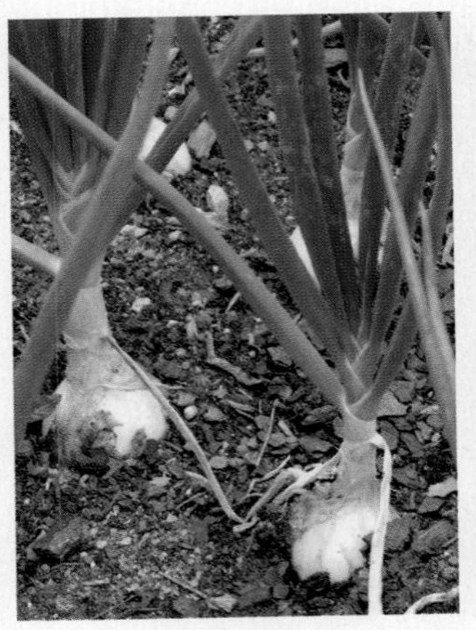

▲ Onions are just one of many examples of monocot crop species.

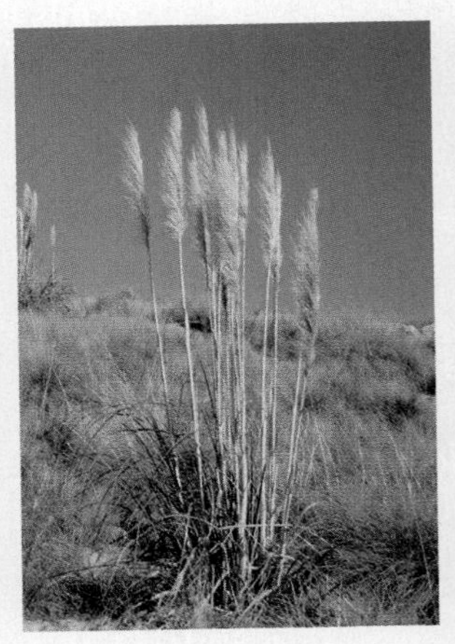

▲ This African hillside is dotted with clumps of Wild Pampas Grass.

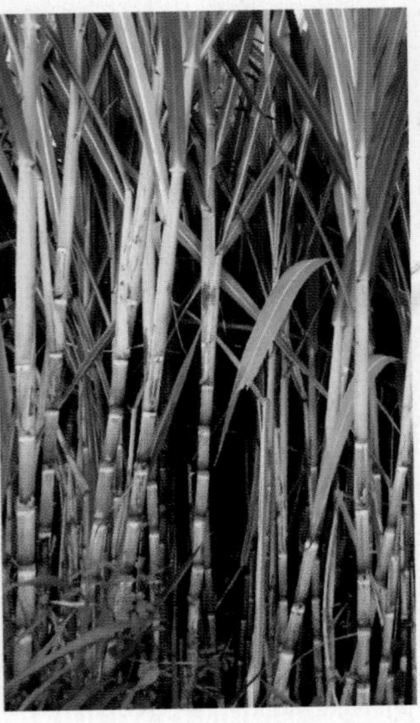

▲ Many orchid species are grown by enthusiasts for their rare beauty. Notice the aerial roots on this specimen, which grows as an epiphyte in its natural environment.

◀ After harvest, sugar cane regrows without being planted again for several cycles.

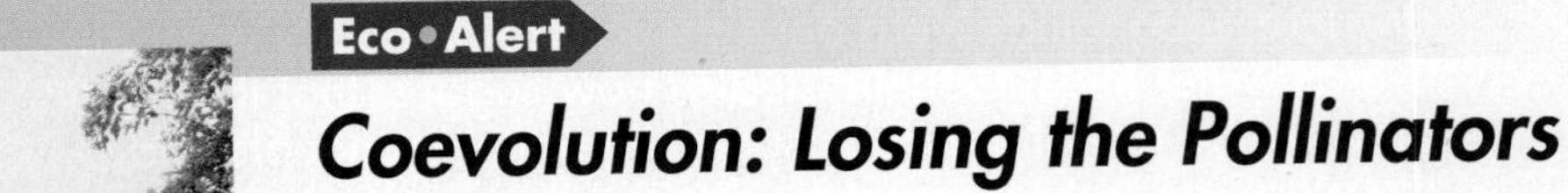

Eco • Alert

Coevolution: Losing the Pollinators

The successes of flowering plants are clearly due to coevolution with their insect pollinators. Common honey bees are among the most important of these, because they gather nectar from the flowers of hundreds of plant species and spread pollen from plant to plant as they go.

Unfortunately, beekeepers around the world, including the United States, are facing a serious crisis. "Colony collapse disorder," as beekeepers describe it, causes bees to fly away from the hive and either never return, or return only to weaken and die. The disease threatens to affect scores of important crops, which depend upon bees to produce fruit and seeds. Suspicion has centered on a fungus or a virus that might spread from colony to colony, but at this point there is no definitive cause or cure.

EUDICOTS: "TRUE DICOTS"

Eudicots (YOO dy kahts) account for about 75 percent of all angiosperm species. The name means "true dicots," and these plants are the ones usually given as examples of dicot stem, leaf, and flower structure. Eudicots have distinctive pollen grains with three grooves on their surfaces, and DNA studies strongly support their classification in a single group. They include a number of important subgroups, five of which are described here.

Ranunculales

The ranunculales subgroup (ruh NUNH kyu lay les) includes, and is named after, buttercups (genus *Ranunculus*). Also included in this subgroup are a number of well-known flowers such as columbines, poppies, barberries, and moonseed.

▼ Rocky Mountain Columbine

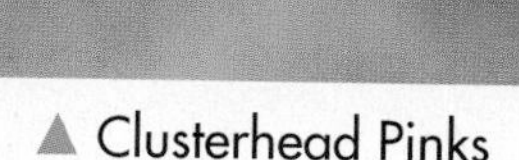

▲ Clusterhead Pinks

Caryophyllales

Cacti are probably the most well-known plants in the caryophyllales subgroup (KAR ee oh fy lay les). Pinks and carnations, spinach, rhubarb, and insect-eating plants, such as sundews and pitcher plants, are also members.

Saxifragales

Plants in the saxifragales (SAK suh frij ay les) subgroup include peonies, witch hazel, gooseberries, and coral bells.

Rosids

The rosids include, as you might expect, the roses. However, this subgroup also includes many popular fruits, such as oranges, raspberries, strawberries, and apples. Some of the best-known trees, including poplars, willows, and maples, are also members.

▲ Orange

◄ Peony

Asterids

The nearly 80,000 asterid species include sunflowers, azaleas, snapdragons, blueberries, tomatoes, and potatoes.

▼ *The flower heads in a field of sunflowers all track the sun as it moves across the sky; thus, they all face the same direction.*

Animals

KEY CHARACTERISTICS

Animals are multicellular, heterotrophic, eukaryotic organisms whose cells lack cell walls.

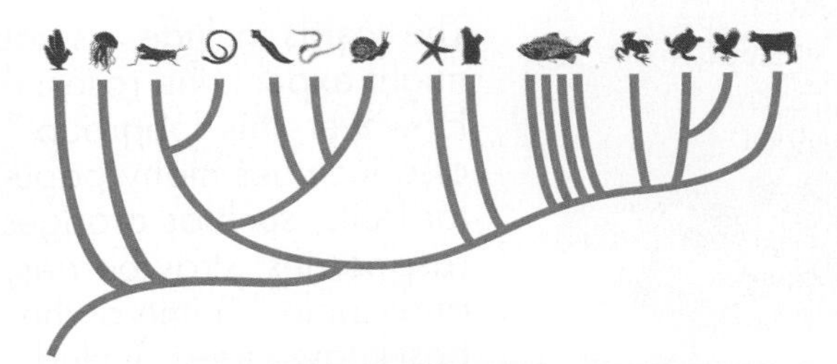

• A Closer Look

A Common Ancestor

Recent molecular studies and cladistic analyses recognize the clade Choanozoa to be the true sister group to all Metazoa—multicellular animals. Choanozoa is one group of organisms formerly called "protists" and is named for choanoflagellates (art and photo right), single-celled, colonial organisms that look like certain cells of sponges and flatworms. Evidence suggests that the choanoflagellates alive today are the best living examples of what the last common ancestor of metazoans looked like.

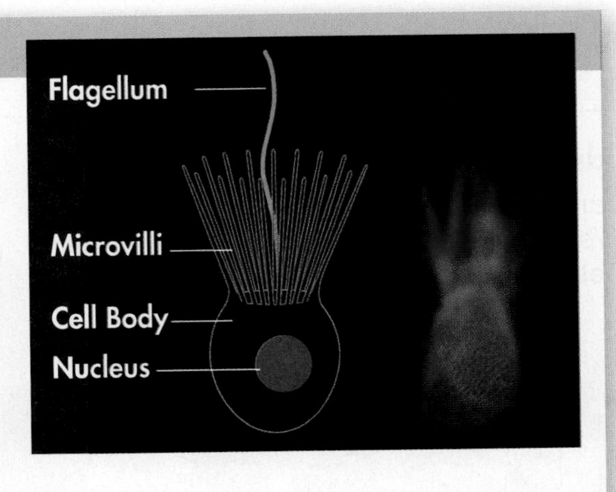

Porifera (Sponges)

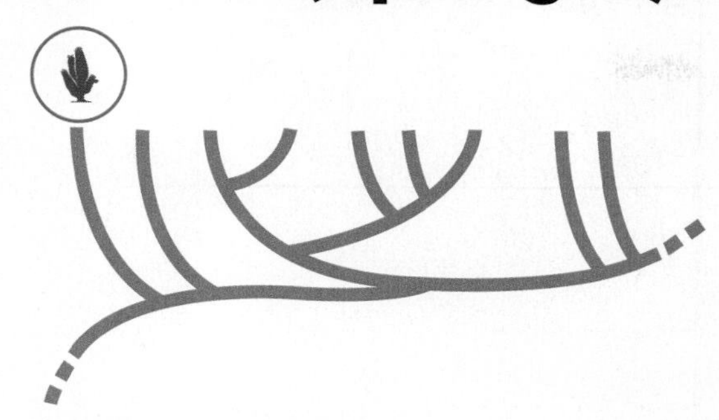

KEY CHARACTERISTICS

Sponges are the simplest animals. They are classified as animals because they are multicellular, heterotrophic, lack cell walls, and have some specialized cells. They are aquatic, lack true tissues and organs, and have internal skeletons of spongin and/or spicules of calcium carbonate or silica. Sponges have no body symmetry.

Feeding and Digestion Filter feeders; intracellular digestion.

Circulation Via flow of water through body

Respiration Oxygen diffuses from water into cells as water flows through body.

Excretion Wastes diffuse from cells into water as water flows through body.

Response No nervous system; little capacity to respond to environmental changes.

Movement Juveniles drift or swim freely; adults are stationary.

Reproduction Most—sexual with internal fertilization; water flowing out of sponge disperses sperm, which fertilizes eggs inside sponge(s); may reproduce asexually by budding or producing gemmules.

GROUPS OF SPONGES

There are more than 5000 species of sponges; most are marine. Three major groups are described below.

DEMOSPONGIAE: Typical sponges

More than 90 percent of all living sponge species are in this group, including the few freshwater species. They have skeletons made of spongin, a flexible protein. Some species have silica spicules. Examples: Yellow Sponge, bath sponges, Carnivorous Mediterranean Sponge, tube sponges

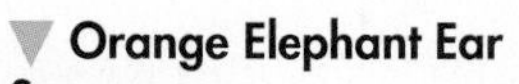

Orange Elephant Ear Sponge

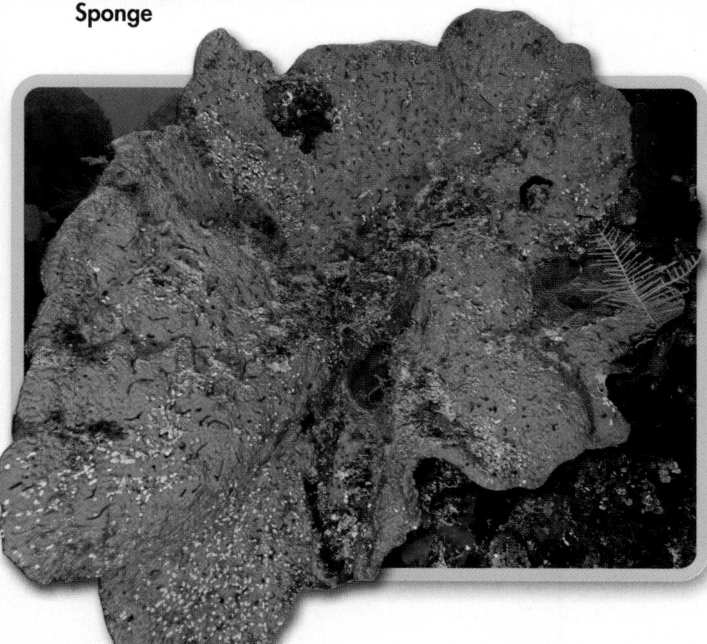

HEXACTINELLIDA: Glass sponges

Glass sponges live in the deep ocean and are especially abundant in the Antarctic. They are called "glass" sponges because their skeletons are made of glasslike silica spicules. Examples: Venus's Flower Basket, Cloud Sponge

Glass Sponge

CALCAREA: Calcareous sponges

Calcareous sponges live in shallow, tropical marine waters and are the only sponges with calcium carbonate spicules. Example: *Clathrina*

Yellow Tubular Sponge

Cnidarians

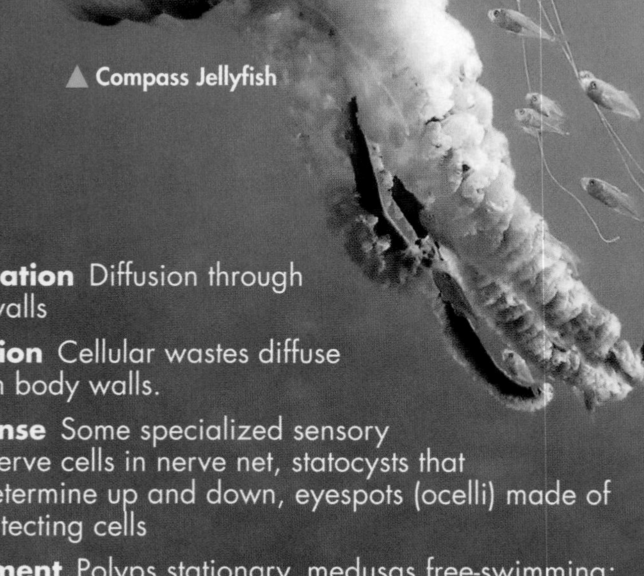

▲ Compass Jellyfish

KEY CHARACTERISTICS

Cnidarians are aquatic, mostly carnivorous, and the simplest animals to have specialized tissues (outer skin and lining of the gastrovascular cavity) and body symmetry (radial). Their tentacles have stinging cells called nematocysts used in feeding.

Feeding and Digestion Predatory, stinging prey with nematocysts; digestion begins extracellularly in gastrovascular cavity and is completed intracellularly; indigestible materials leave body through single opening; many, especially reef-building corals, also depend on symbiotic algae, or zooxanthellae.

Circulation No internal transport system; nutrients typically diffuse through body.

Respiration Diffusion through body walls

Excretion Cellular wastes diffuse through body walls.

Response Some specialized sensory cells: nerve cells in nerve net, statocysts that help determine up and down, eyespots (ocelli) made of light-detecting cells

Movement Polyps stationary, medusas free-swimming; some, such as sea anemones, can burrow and creep very slowly; others move using muscles that work with a hydrostatic skeleton and water in gastrovascular cavity; medusas such as jellyfish move by jet propulsion generated by muscle contractions.

Reproduction Most—alternate between sexual (most species by external fertilization) and asexual (polyps produce new polyps or medusae by budding)

Eco • Alert

The color of this star coral is caused by zooxanthellae algae living within it.

Coral Symbionts

Reef-building coral animals depend on symbiotic algae called zooxanthellae for certain vital nutritional needs. In many places, reef-building corals live close to the upper end of their temperature tolerance zone. If water temperatures rise too high, the coral-zooxanthellae symbiosis breaks down, and corals turn white in what is called "coral bleaching." If corals don't recover their algae soon, they weaken and die. This is one reason why coral reefs are in grave danger from global warming.

GROUPS OF CNIDARIANS

There are more than 9000 species of cnidarians.

A Portuguese man-of-war is actually a colony of polyps.

HYDROZOA: Hydras and their relatives

Hydras and their relatives spend most of their time as polyps and are either colonial or solitary. They reproduce asexually (by budding), sexually, or they alternate between sexual and asexual reproduction. Examples: hydra, Portuguese man-of-war

Black Sea Nettle

SCYPHOZOA: Jellyfishes

Jellyfishes spend most of their time as medusas; some species bypass the polyp stage. They reproduce sexually and sometimes asexually by budding. Examples: lion's mane jellyfish, moon jelly, sea wasp

ANTHOZOA: Corals and sea anemones

Corals and sea anemones are colonial or solitary polyps with no medusa stage. The central body is surrounded by tentacles. They reproduce sexually or asexually. Examples: reef corals, sea anemones, sea pens, sea fans

Sea Anemone

This purple-striped jelly (Pelagia noctiluca) has the ability to bioluminesce, or give off light.

Arthropods

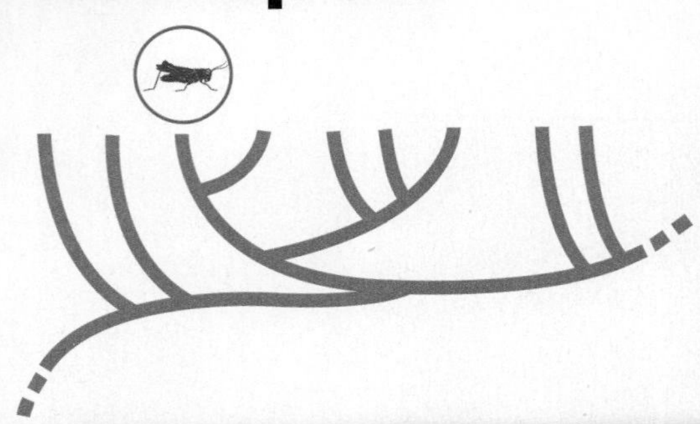

KEY CHARACTERISTICS

Arthropods are the most diverse of all multicellular organisms. They have segmented bodies and jointed appendages. They are supported by tough exoskeletons made of chitin, which they periodically shed as they grow. Arthropods are coelomate protostomes with bilateral symmetry.

Eco • Alert

Beetle Damage

You probably know that some insects can seriously damage crop plants. But insects affect plants in natural habitats, too. One example is the mountain pine beetle, which is dramatically extending its range. Global warming appears to be enabling the beetle to survive farther north, and at higher altitudes, than it used to. The new beetle infestation is causing extensive damage to northern and high-altitude forests in North America. The death of millions of acres of trees has resulted in the release of large amounts of carbon dioxide, a greenhouse gas, into the atmosphere. You can see the sort of damage the beetles cause in the photo at right.

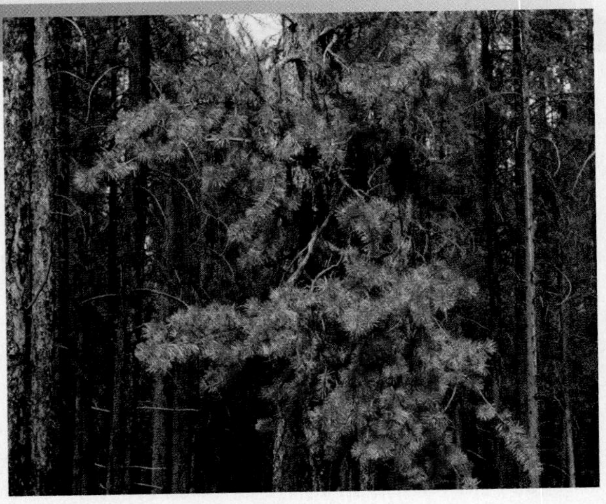

▲ *Mountain pine beetle damage to pine trees in White River National Forest, Colorado*

Feeding and Digestion Extremely diverse: herbivores, carnivores, detritivores, parasites, bloodsuckers, scavengers, filter feeders; digestive system with two openings; many feeding specializations in different groups

Circulation Open circulatory system with heart and arteries

Respiration Terrestrial—tracheal tubes or book lungs; aquatic—gills or book gills (horseshoe crabs)

Excretion Terrestrial—Malpighian tubules; aquatic—diffusion into water

Response Well-developed nervous system with brain; sophisticated sense organs

Movement Muscles attached internally to jointed exoskeletons

Reproduction Usually sexual, although some species may reproduce asexually under certain circumstances; many undergo metamorphosis during development

Most animals, including this land crab, are arthropods.

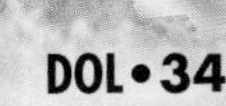

GROUPS OF ARTHROPODS

Phylum Arthropoda contains more known species than any other phylum. Scientists have identified more than 1,000,000 arthropod species, and some scientists expect there are millions yet to be identified. Arthropods are classified based on the number and structure of body segments and appendages.

CRUSTACEA: Crustaceans

There are crustacean species in almost every habitat, but most are aquatic, and most of these are marine. They have two or three body sections, two pairs of antennae, and chewing mouthparts called mandibles. Many have a carapace, or "shell," that covers part or all of the body. Examples: crabs, lobsters, crayfish, pill bugs, water fleas, barnacles

▲ **Lobster**

CHELICERATA: Chelicerates

Living chelicerates include horseshoe crabs and arachnids. (Their extinct relatives include trilobites and giant "sea-scorpions.") Most living chelicerates are terrestrial. The body is composed of two parts—the cephalothorax and abdomen. The first pair of appendages are specialized feeding structures called chelicerae. Chelicerates have no antennae.

Horseshoe crabs are actually more closely related to spiders than to crabs!

Merostomata: Horseshoe crabs

The class Merostomata once included many species, but only four species of horseshoe crab survive today. All are marine. They have five pairs of walking legs and a long, spinelike tail.

Arachnida: Arachnids

The vast majority of arachnids are terrestrial. They have four pairs of walking legs and no tail. Examples: spiders, ticks, mites, scorpions, daddy longlegs

▲ **Mexican Fireleg Tarantula**

Animals

UNIRAMIA: Uniramians

Most uniramians are terrestrial, although some are aquatic for all or part of their lives. They have one pair of antennae, mandibles, and unbranched appendages. Uniramians include at least three fourths of all known animal species!

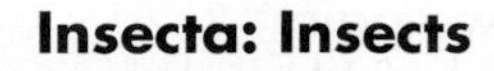

Uniramians include centipedes, millipedes, and insects—more than three fourths of all known animal species, including this green snaketail dragonfly.

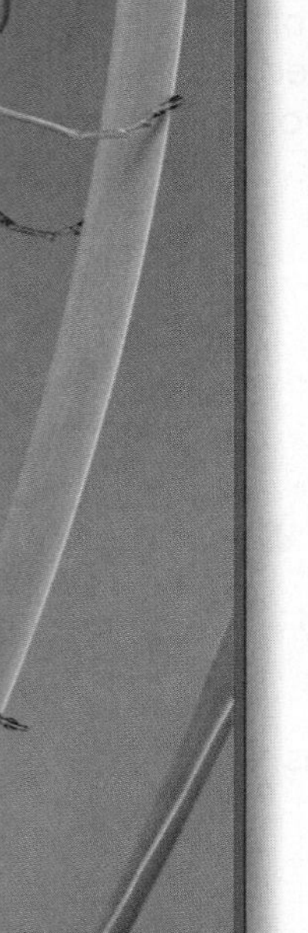

Praying Mantis

Insecta: Insects

There are more than 1,000,000 insect species in more than 25 orders. An insect body is divided into three parts—head, thorax, and abdomen. Insects have three pairs of legs and usually one or two pairs of wings attached to the thorax. Some insects undergo complete metamorphosis. Examples: termites, ants, beetles, dragonflies, flies, moths, grasshoppers

The death's head hawk moth is named for the skull-like shape on the adult's head (above). Like many insects, this moth undergoes complete metamorphosis, during which the larva (below), or caterpillar, turns into a pupa, and, eventually, an adult moth.

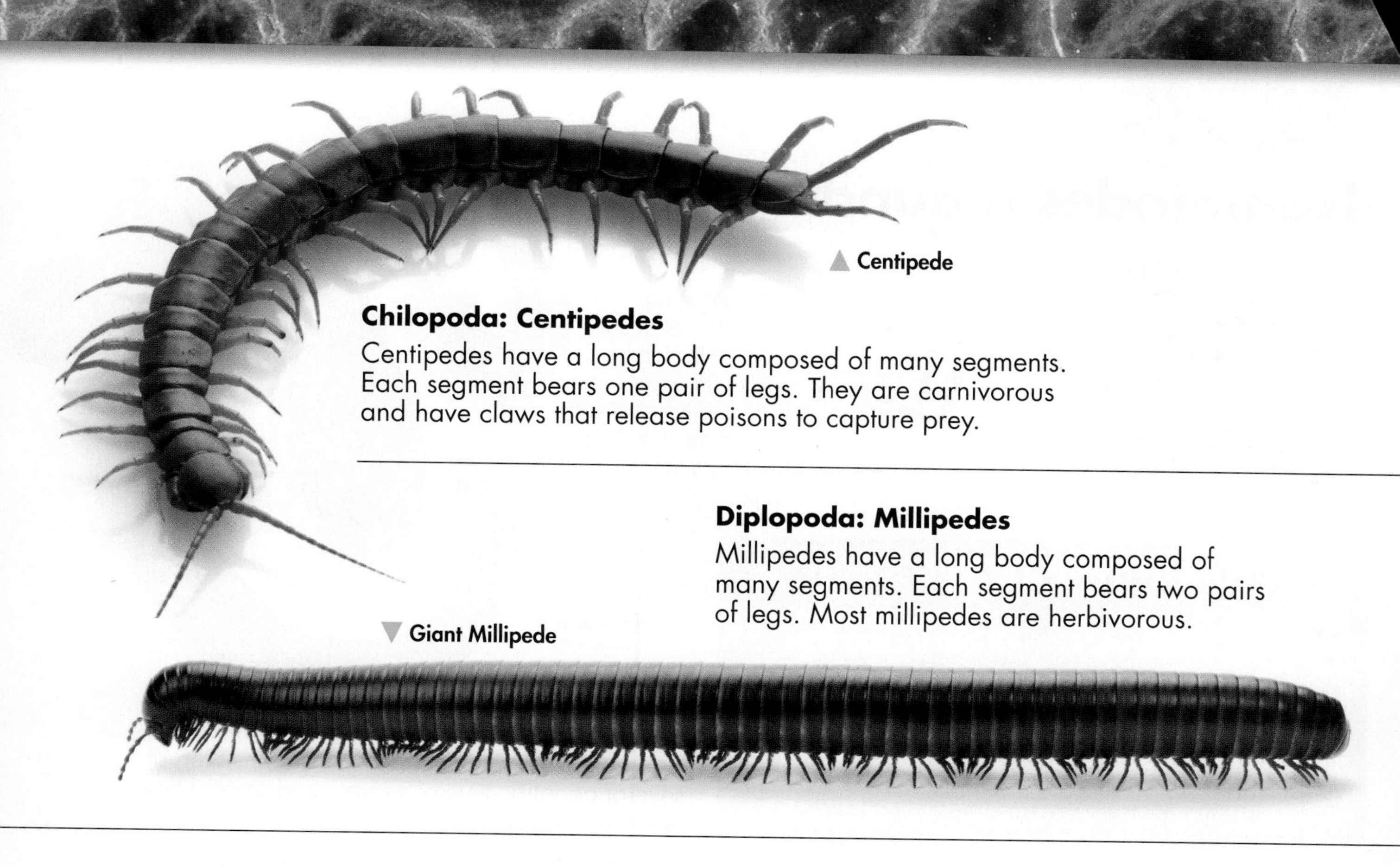

Chilopoda: Centipedes

Centipedes have a long body composed of many segments. Each segment bears one pair of legs. They are carnivorous and have claws that release poisons to capture prey.

Diplopoda: Millipedes

Millipedes have a long body composed of many segments. Each segment bears two pairs of legs. Most millipedes are herbivorous.

Insecta (continued)

Many "bugs" benefit humans. For example, ladybugs (which are not all "ladies") eat garden pests and bees pollinate plants. Insects such as praying mantises, katydids, flies, moths, beetles, and ants also have important roles in ecosystems.

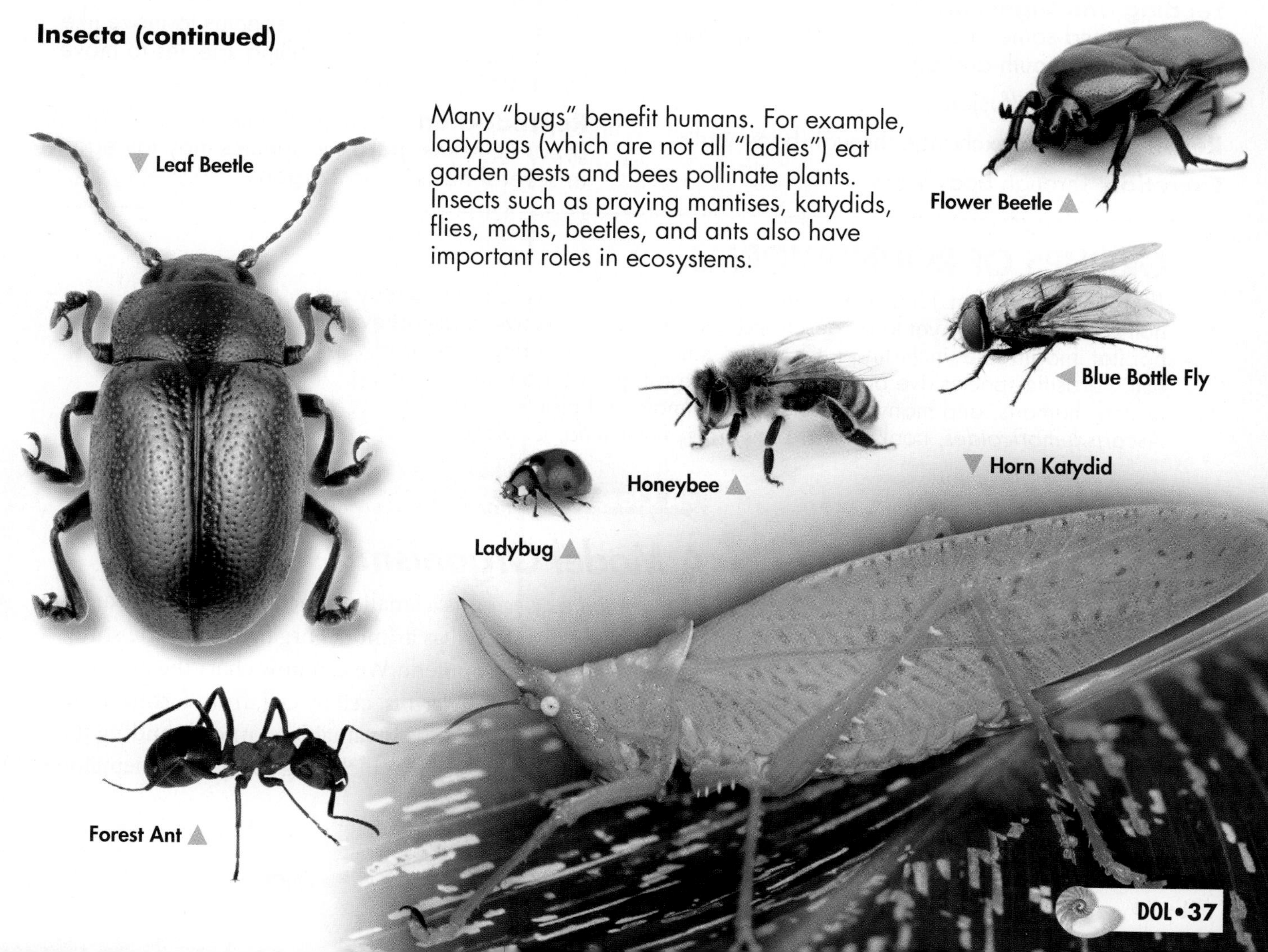

Nematodes (Roundworms)

Pinworms can infest the intestinal tract of humans. Although anyone can become infected with pinworms, infection is most common in children aged 5 to 10.

Pinworm (colorized SEM)

KEY CHARACTERISTICS

Nematodes, or roundworms, are unsegmented worms with a tough outer cuticle, which they shed as they grow. This "molting" is one reason that nematodes are now considered more closely related to arthropods than to other wormlike animals. Nematodes are the simplest animals to have a "one-way" digestive system through which food passes from mouth to anus. They are protostomes and have a pseudocoelom.

Feeding and Digestion Some predators, some parasites, and some decomposers; one-way digestive tract with mouth and anus

Circulation By diffusion

Respiration Gas exchange through body walls

Excretion Through body walls

Response Simple nervous system consisting of several ganglia, several nerves, and several types of sense organs

Movement Muscles work with hydrostatic skeleton, enabling aquatic species to move like water snakes and soil-dwelling species to move by thrashing around.

Reproduction Sexual with internal fertilization; separate sexes; parasitic species may lay eggs in several hosts or host organs.

GROUPS OF ROUNDWORMS

There are more than 15,000 known species of roundworms, and there may be half a million species yet to be described. Free-living species live in almost every habitat imaginable, including fresh water, salt water, hot springs, ice, and soil. Parasitic species live on or inside a wide range of organisms, including insects, humans, and many domesticated animals and plants. Examples: *Ascaris lumbricoides*, hookworms, pinworms, *Trichinella*, *C. elegans*

A Closer Look

A Model Organism?

Caenorhabditis elegans is a small soil nematode. Fifty years ago, this species was selected as a "model organism" for the study of genetics and development. We can now chart the growth and development of *C. elegans*, cell by cell, from fertilization to adult. This information is invaluable in understanding the development of other species—including many other nematodes that cause serious disease.

C. elegans (LM 64×)

Platyhelminthes (Flatworms)

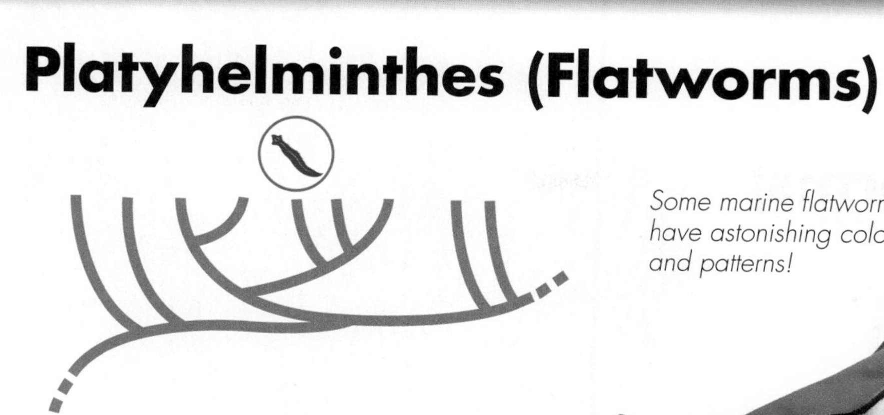

Some marine flatworms have astonishing colors and patterns!

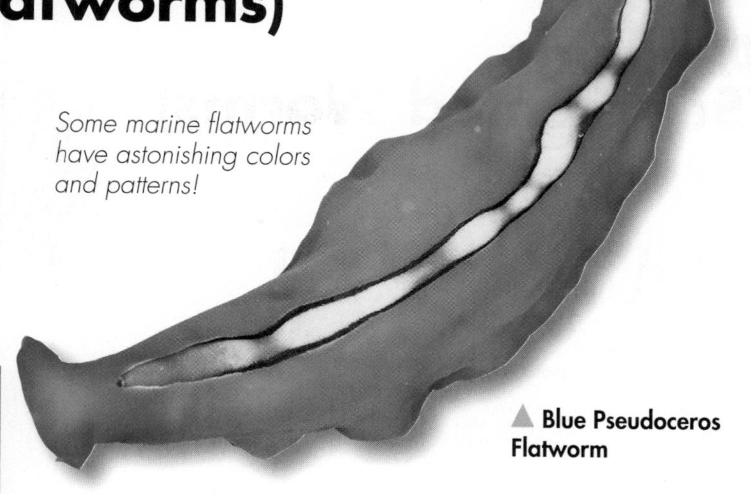

▲ Blue Pseudoceros Flatworm

KEY CHARACTERISTICS

Flatworms are soft worms with tissues and internal organ systems. They are the simplest animals to have three embryonic germ layers, bilateral symmetry, and cephalization. They are acoelomates.

Feeding and Digestion Free-living—predators or scavengers that suck food in through a pharynx and digest it in a system that has one opening. Parasitic—feed on blood, tissue fluids, or cell pieces of the host, using simpler digestive systems than free-living species have. Tapeworms, which absorb nutrients from food that the host has already digested, have no digestive system.

Circulation By diffusion

Respiration Gas exchange by diffusion

Excretion Some—flame cells remove excess water and may remove metabolic wastes such as ammonia and urea. Many flame cells are connected to tubules that release substances through pores in the skin.

Response Free-living—several ganglia connected by nerve cords that run through the body, along with eyespots and other specialized sensory cells; parasitic—simpler nervous system than free-living forms have

Movement Free-living—using cilia and muscle cells.

Reproduction Free-living—most are hermaphrodites that reproduce sexually with internal fertilization; parasitic—commonly reproduce asexually by fission but also often reproduce sexually

GROUPS OF FLATWORMS

Flatworms are an amazingly diverse group of worms that include more than 20,000 species. They have historically been placed into three classes, but these taxa now appear not to be true clades, and will probably change.

TREMATODA: Flukes

Most flukes are parasites that infect internal organs of their hosts, but some infect external parts such as skin or gills. The life cycle typically involves more than one host or organ. Examples: *Schistosoma*, liver fluke

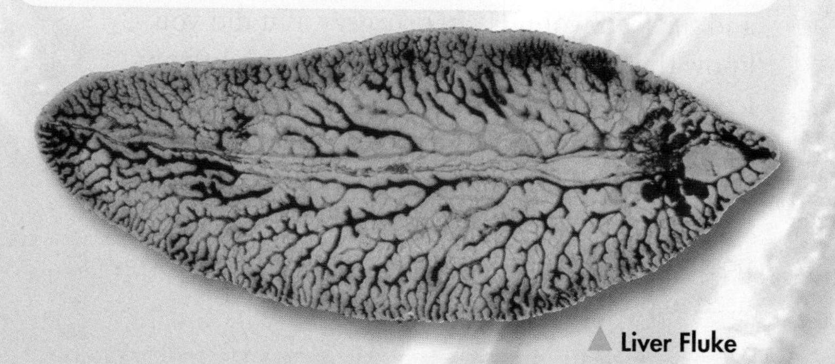

▲ Liver Fluke

TURBELLARIA: Turbellarians

Turbellarians are free-living aquatic and terrestrial predators and scavengers. Many are colorful marine species. Examples: planarians, polyclad flatworm

CESTODA: Tapeworms

Tapeworms are very long intestinal parasites that lack a digestive system and absorb nutrients directly through their body walls. The tapeworm body is composed of many repeated sections (proglottids) that contain both male and female reproductive organs.

Annelids (Segmented Worms)

Peacock worms, whose feather-shaped gills look somewhat like peacock feathers, are marine annelids, or polychaetes.

KEY CHARACTERISTICS

Annelids are coelomate protostome worms whose bodies are composed of segments separated by internal partitions. The annelid digestive system has two openings.

Feeding and Digestion Filter feeders, carnivores, or parasites; many obtain food using a muscular pharynx, often equipped with "teeth"; widely varied digestive systems—some, such as earthworms, have complex digestive tracts.

Circulation Closed circulatory system with dorsal and ventral blood vessels; dorsal vessel pumps blood like a heart.

Respiration Aquatic—gills; terrestrial—skin

Excretion Digestive waste exits through anus; nitrogenous wastes eliminated by nephridia

Response Nervous system includes a rudimentary brain and several nerve cords; sense organs best-developed in free-living saltwater species

Movement Hydrostatic skeleton based on sealed body segments surrounded by longitudinal and circular muscles; many annelids have appendages that enable movement.

Reproduction Most—sexual, some through external fertilization with separate sexes, but others are simultaneous hermaphrodites that exchange sperm; most have a trochophore larval stage

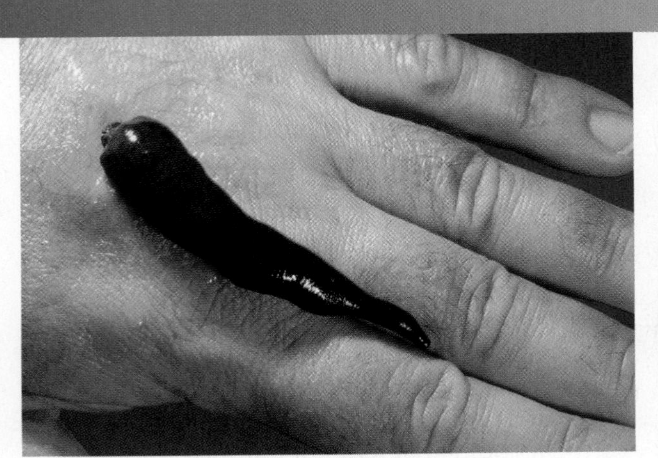

▲ *Leech* (Hirudo medicinalis) *drawing blood from a hand*

• Did You Know?

Not-So-Modern Medicine

You may have heard that medieval healers used leeches to remove "excess" blood from patients and to clean wounds after surgery. But did you know that leeches—or at least compounds from leech saliva—have a place in modern medicine? Leech saliva contains the protein hirudin, which prevents blood from clotting. Some surgeons use leeches to relieve pressure caused by blood that pools in tissues after plastic surgery. Hirudin is also used to prevent unwanted blood clots.

Feather-Duster Worms

GROUPS OF ANNELIDS

There are more than 15,000 species of annelids.

HIRUDINEA: Leeches

Most leeches live in fresh water. They lack appendages. Leeches may be carnivores or blood-sucking external parasites. Example: medicinal leech *(Hirudo medicinalis)*

Earthworm

OLIGOCHAETA: Oligochaetes

Oligochaetes live in soil or fresh water. They lack appendages. Some use setae for movement but have fewer than polychaetes.
Examples: *Tubifex*, earthworms

POLYCHAETA: Polychaetes

Polychaetes live in salt water; many move with paddle-like appendages called parapodia tipped with bristle-like setae. Examples: sandworms, bloodworms, fanworms, feather-duster worms

The white, bristle-like structures on the sides of this bearded fireworm are setae.

Mollusks

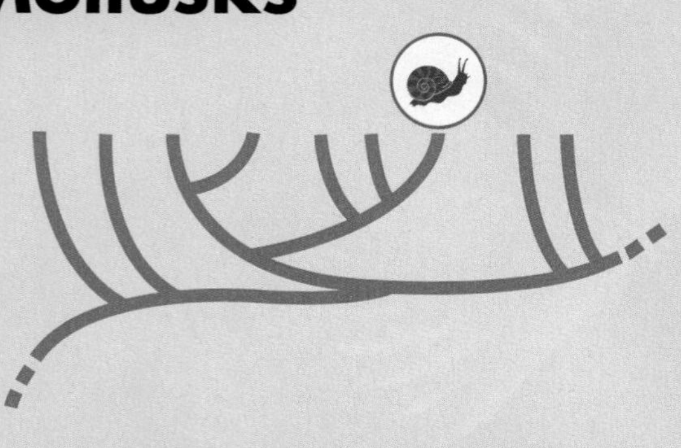

KEY CHARACTERISTICS

Mollusks have soft bodies that typically include a muscular foot. Body forms vary greatly. Many mollusks possess a hard shell secreted by the mantle, but in some, the only hard structure is internal. Mollusks are coelomate protostomes with bilateral symmetry.

Feeding and Digestion Digestive system with two openings; diverse feeding styles—mollusks can be herbivores, carnivores, filter feeders, detritivores, or parasites

Circulation Snails and clams—open circulatory system; octopi and squid—closed circulatory system

Respiration Aquatic mollusks—gills inside the mantle cavity; land mollusks—a saclike mantle cavity whose large, moist surface area is lined with blood vessels.

Excretion Body cells release ammonia into the blood, which nephridia remove and release outside the body.

Response Complexity of nervous system varies greatly; extremely simple in clams, but complex in some octopi.

Movement Varies greatly, by group. Some never move as adults, while others are very fast swimmers.

Reproduction Sexual; many aquatic species have a free-swimming trochophore larval stage.

Colossal Squid

• Did You Know?

The Colossal Squid

The colossal squid, the largest of all mollusks, has the largest eyes of any known animal. One 8-meter-long, 450-kilogram specimen of the species *Mesonychoteuthis hamiltoni* had eyes 28 centimeters across—larger than most dinner plates! The lens of this huge eye was the size of an orange.

GROUPS OF MOLLUSKS

Mollusks are traditionally divided into several classes based on characteristics of the foot and the shell; specialists estimate that there are between 50,000 and 200,000 species of mollusks alive today.

Giant Clam

Garden Snail

Chambered Nautilus

BIVALVIA: Bivalves

Bivalves are aquatic. They have a two-part hinged shell and a wedge-shaped foot. They are mostly stationary as adults. Some burrow in mud or sand; others attach to rocks. Most are filter feeders that use gill siphons to take in water that carries food. Clams have open circulatory systems. Bivalves have the simplest nervous systems among mollusks. Examples: clams, oysters, scallops, mussels

GASTROPODA: Gastropods

There are both terrestrial and aquatic gastropods. Most have a single spiral, chambered shell. Gastropods use a broad, muscular foot to move and have a distinct head region. Snails and slugs feed with a structure called a radula that usually works like sandpaper. Some species are predators whose harpoon-shaped radula carries deadly venom. They have open circulatory systems. Many gastropod species are cross-fertilizing hermaphrodites. Examples: snails, slugs, nudibranchs, sea hares

CEPHALOPODA: Cephalopods

Cephalopods live in salt water. The cephalopod has a highly developed brain and sense organs. The head is attached to a single foot, which is divided into tentacles. They have closed circulatory systems. Octopi use beaklike jaws for feeding; a few are venomous. Cephalopods have the most complex nervous systems among mollusks; octopi have complex behavior and have shown the ability to learn in laboratory settings. Examples: octopi, squids, nautilus, cuttlefish

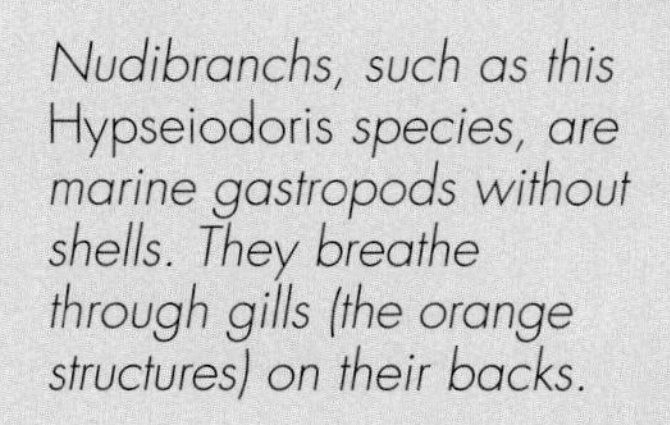

Nudibranchs, such as this Hypseiodoris *species, are marine gastropods without shells. They breathe through gills (the orange structures) on their backs.*

Echinoderms

KEY CHARACTERISTICS

Echinoderms are marine animals that have spiny skin surrounding an endoskeleton. Their unique water vascular system includes tube feet with suction-cuplike ends used in moving and feeding. The water vascular system also plays a role in respiration, circulation, and excretion. Echinoderms are coelomate deuterostomes. Adults exhibit 5-part radial symmetry.

Feeding and Digestion Method varies by group—echinoderms can be filter feeders, detritivores, herbivores, or carnivores.

Circulation Via fluid in the coelom, a rudimentary system of vessels, and the water vascular system

Respiration Gas exchange is carried out by surfaces of tube feet, and, in many species, by skin gills.

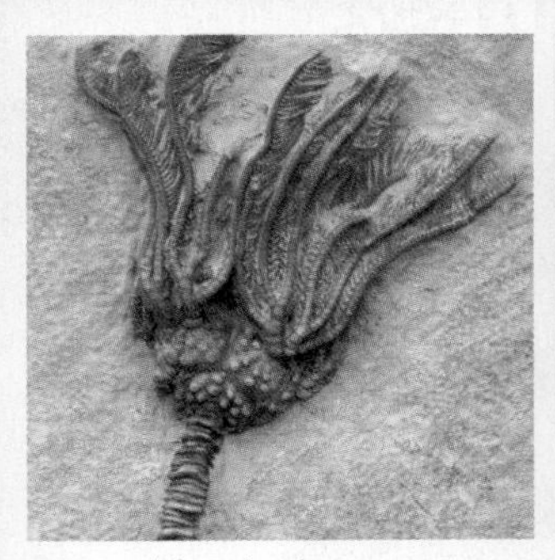

Crinoid fossil, about 400 million years old

Living modern crinoid (feather star)

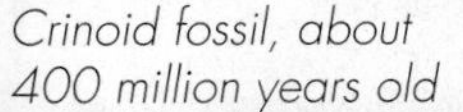

• A Look Back in Time

Crinoids Then and Now

Echinoderms have a long fossil record that dates all the way back to the Cambrian Period. Although these animals have been evolving for millions of years, some fossil crinoids look a great deal like living crinoids.

Excretion Digestive wastes released through anus; nitrogenous cellular wastes excreted as ammonia through tube feet and skin gills.

Response Minimal nervous system; nerve ring is connected to body sections by radial nerves; most have scattered sensory cells that detect light, gravity, and chemicals secreted by prey.

Movement In most, tube feet work with endoskeleton to enable locomotion.

Reproduction Sexual, with external fertilization; larvae have bilateral symmetry, unlike adults.

You can't miss the 5-part radial symmetry of this red mesh sea star moving across a coral reef.

GROUPS OF ECHINODERMS

There are more than 7000 species of echinoderms.

CRINOIDEA: Crinoids

Crinoids are filter feeders; some use tube feet along feathery arms to capture plankton. The mouth and anus are on the upper surface of the body disk. Some are stationary as adults while others can "walk" using short "arms" on the lower body surface. Examples: sea lily, feather star

▶ Feeding Crinoid

◀ Sea Star

ASTEROIDEA: Sea stars

Sea stars are bottom dwellers whose star-shaped bodies have flexible joints. They are carnivorous—the stomach pushes through the mouth onto the body tissues of prey and pours out digestive enzymes. The stomach then retracts with the partially digested prey; digestion is completed inside the body. Examples: crown-of-thorns sea star, sunstar

▼ Basket Star

OPHIUROIDEA: Ophiuroids

Ophiuroids have small body disks, long, armored arms, and flexible joints. Most are filter feeders or detritivores. Examples: brittle star, basket star

ECHINOIDEA: Echinoids

Echinoids lack arms. Their endoskeletons are rigid and boxlike and covered with movable spines. Most echinoids are herbivores or detritivores that use five-part jawlike structures to scrape algae from rocks. Examples: sea urchin, sand dollar, sea biscuit

▼ *Sea urchins grazing on kelp*

▼ Sea cucumber

HOLOTHUROIDEA: Sea cucumbers

Sea cucumbers have a cylindrical, rubbery body with a reduced endoskeleton and no arms. They typically lie on their side and move along the ocean floor by the combined action of tube feet and body-wall muscles. These filter feeders or detritivores use a set of retractable feeding tentacles on one end to take in sand and detritus, from which they glean food.

Nonvertebrate Chordates

Tunicates are chordates named for the colorful tunic-like covering the adults have. As larvae, tunicates have all the characteristics of chordates, as well as bilateral symmetry, but as adults, they look very, very different.

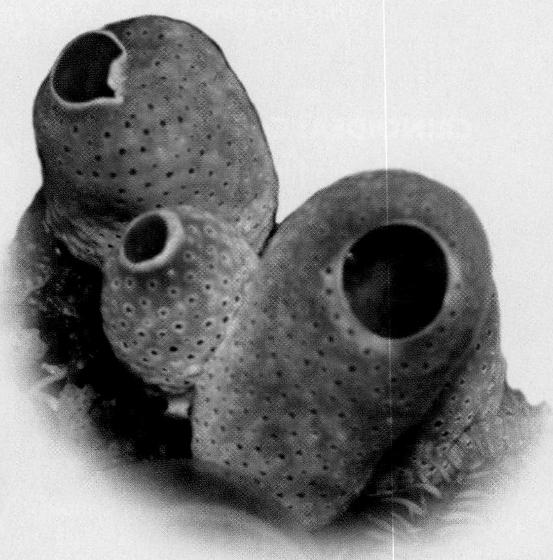

KEY CHARACTERISTICS

The nonvertebrate chordates are the only chordates that lack a backbone. Like other chordates, they have a nerve cord, notochord, pharyngeal pouches, and a tail at some point during development. They are coelomate deuterostomes. The two subphyla, tunicates and lancelets, differ significantly.

Feeding and Digestion Filter feeders; tunicates—in most, water carrying food particles enters through an incurrent siphon; food is strained out in the pharynx and passed to the digestive system; lancelets—mucus in the pharynx catches food particles carried in by water, which are then carried into digestive tract

Circulation Closed; tunicates—heart pumps blood by "wringing out," and flow periodically reverses direction; lancelets—no heart, but blood vessels pump blood through body in one direction

Respiration Tunicates—gas exchange occurs in the gills and across other body surfaces; lancelets—through pharynx and body surfaces

Excretion Tunicates—most through excurrent siphon; lancelets—flame cells in nephridia release water and nitrogenous wastes into the atrium and out through an opening called an atriopore

Response Cerebral ganglion, few specialized sensory organs; tunicates—sensory cells in and on the siphons and other internal surfaces help control the amount of water passing through the pharynx; lancelets—a pair of eyespots detect light

Movement Tunicates—free-swimming larvae, but most are stationary as adults; lancelets—no appendages: they move by contracting muscles paired on either side of the body

Reproduction Tunicates—most sexual and hermaphroditic with external fertilization, but some reproduce by budding; most have free-swimming tadpole-like larvae that metamorphose into adults; lancelets—sexual with external fertilization

Eco • Alert

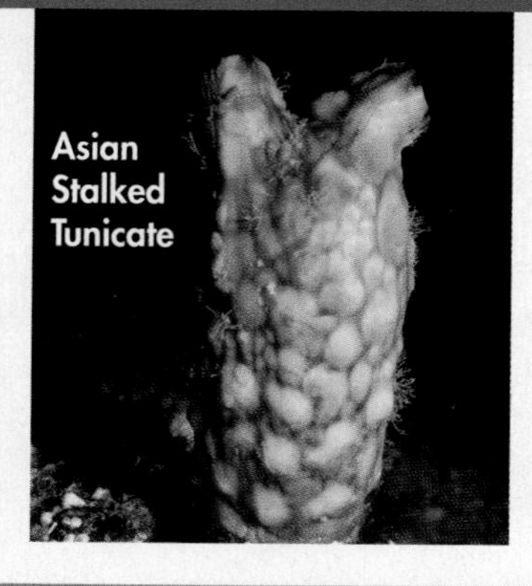

Asian Stalked Tunicate

Out-of-Control Tunicates

You've never heard of them, but Asian stalked tunicates are disrupting marine ecosystems in Washington State; Prince Edward Island, Canada; and elsewhere. Tunicate larvae are carried in the ballast water of freight ships and discharged wherever the ships make port. There, away from their usual predators, the tunicates grow out of control, smothering shellfish beds and covering boats, docks, and underwater equipment. Researchers are still trying to figure out how to control them.

GROUPS OF NONVERTEBRATE CHORDATES

There are two major groups of nonvertebrate chordates: tunicates and lancelets (sometimes called amphioxus).

Two lancelets, Branchiostoma lanceolatum, *poking out of sand.*

CEPHALOCHORDATA: Lancelets

Lancelets are fishlike animals that have bilateral symmetry and live in salt water. They are filter feeders and have no internal skeleton. Example: *Branchiostoma*

UROCHORDATA: Tunicates

Tunicates are filter feeders that live in salt water. Most adults have a tough outer covering ("tunic") and no body symmetry; most display chordate features and bilateral symmetry only during larval stages. Many adults are stationary; some are free-swimming. Examples: sea squirts, sea peaches, salps

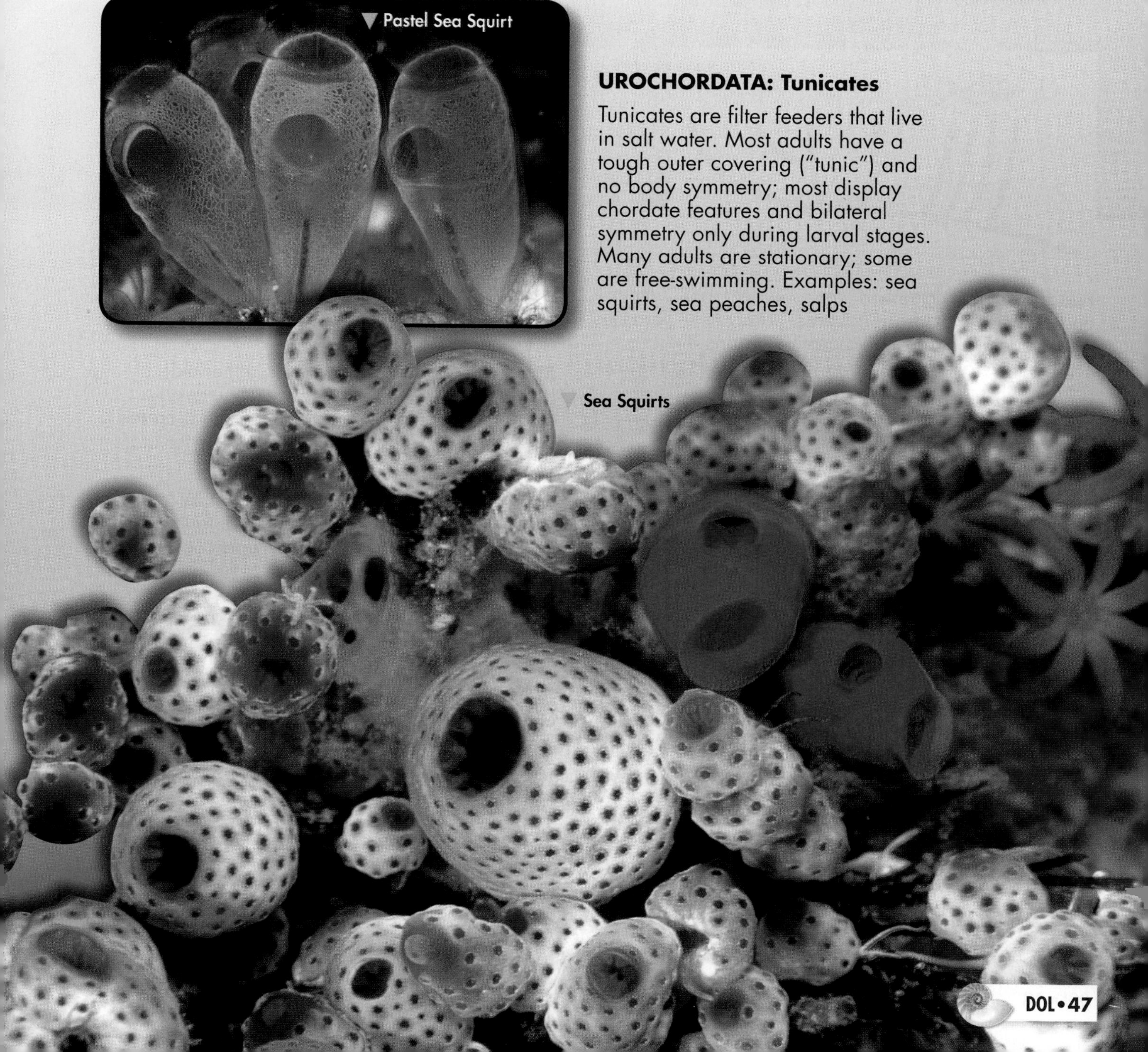
▼ Pastel Sea Squirt

▼ Sea Squirts

Fishes

KEY CHARACTERISTICS

The word fish is used informally to describe aquatic vertebrates that look similar even though they belong to several different clades, because all are adapted to life in water. Most vertebrates we call fishes have paired fins, scales, and gills.

Feeding and Digestion Varies widely, both within and between groups: herbivores, carnivores, parasites, filter feeders, detritivores; digestive organs often include specialized teeth and jaws, crop, esophagus, stomach, liver, pancreas

Circulation Closed, single-loop circulatory system; two-chambered heart

Respiration Gills; some have specialized lungs or other adaptations that enable them to obtain oxygen from air.

Excretion Diffusion across gill membranes; kidneys

Response Brain with many parts; highly developed sense organs, including lateral line system

Movement Paired muscles on either side of backbone; many have highly maneuverable fins; the largest groups have two sets of paired fins; some have a gas-filled swim bladder that regulates buoyancy.

Reproduction Methods vary within and between groups: external or internal fertilization; oviparous, ovoviviparous, or viviparous

• A Look Back in Time

Live Birth in Devonian Seas

You might think that live birth is a recent addition to chordate diversity. Guess again. Recent fossil finds of fishes from the Devonian Period show that at least one group of fishes was already bearing live young 380 million years ago. Two incredibly well preserved fossils, including that of the fish *Materpiscis*, show the remains of young with umbilical cords still attached to their mother's bodies. This is the earliest fossil evidence of viviparity in vertebrates.

▲ *Artist's conception of* Materpiscis *giving birth*

GROUPS OF FISHES

Fishes are the largest group of vertebrates, including more than 30,000 species. Evolutionary classification of these animals is still a work in progress; many traditional groups are not clades. "Fishes" actually represent several ancient clades, one of which includes tetrapods, or four-limbed vertebrates. Fishes, as we treat them here, include two groups of jawless fishes (hagfishes and lampreys), cartilaginous fishes, and bony fishes.

Sweetlips are, despite their funny faces, easily recognizable as fish.

"JAWLESS FISHES"

Hagfishes and lampreys make up separate clades, but their bodies share common features that distinguish them from other fishes. They have no jaws, lack vertebrae, and their skeletons are made of fiber and cartilage.

PETROMYZONTIDA: Lampreys

Lampreys are mostly filter feeders as larvae and parasites as adults. The head of an adult lamprey is taken up almost completely by a circular, tooth-bearing, sucking disk with a round mouth. Adult lampreys typically attach themselves to fishes. They hold on to their hosts using the teeth in their sucking disk and then scrape away at the skin with a rasping tongue. Lampreys then suck up their host's tissues and body fluids. Because lampreys feed mostly on blood, they are called "vampires of the sea."

Lamprey mouth

▲ Lamprey

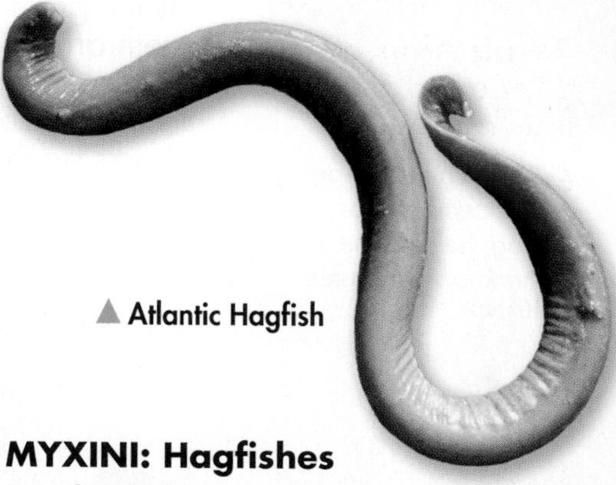

▲ Atlantic Hagfish

MYXINI: Hagfishes

Hagfishes have pinkish gray wormlike bodies and four or six short tentacles around their mouths. They retain notochords as adults. Hagfishes lack image-forming eyes, but have light-detecting sensors scattered around their bodies. They feed on dead and dying animals using a rasping tongue that scrapes away layers of flesh.

Tiger Shark

CHONDRICHTHYES: Cartilaginous Fishes

Members of this clade are considered "cartilaginous" because they lack true bone; their skeletons are built entirely of cartilage. Most cartilaginous fishes also have tough scales, which make their skin as rough as sandpaper.

Holocephalans: Chimaeras

Chimaeras have smooth skin that lacks scales. Most have just a few platelike, grinding teeth and a venomous spine located in front of the dorsal fin. Examples: ghostfish, ratfish, rabbitfish

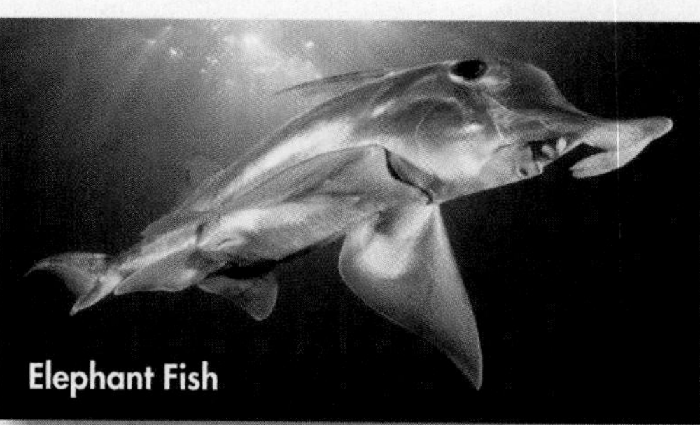

Elephant Fish

Elasmobranchii: Sharks, skates, and rays

Sharks, skates, and rays are very diverse, but all have skin covered with toothlike scales known as dermal denticles. Elasmobranchii make up the vast majority of living cartilaginous fish species.

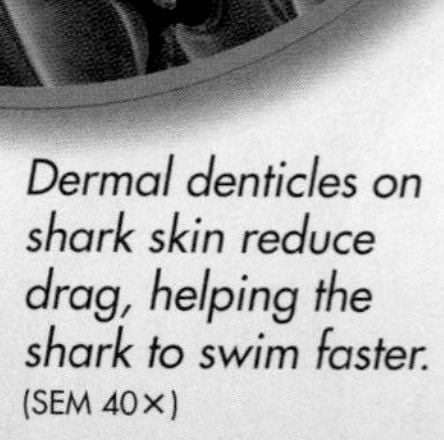

Dermal denticles on shark skin reduce drag, helping the shark to swim faster. (SEM 40×)

Galeomorphi: Sharks

Most of the 350 or so shark species have large, curved asymmetrical tails, torpedo-shaped bodies, and pointed snouts with a mouth underneath. Predatory sharks, such as the great white, have many teeth arranged in rows. As teeth in the front rows are worn out or lost, new teeth replace them. Some sharks go through 20,000 teeth in their lifetime! Other sharks are filter feeders, and some species have flat teeth for crushing mollusk and crustacean shells. Examples: great white shark, whale shark, hammerhead shark

Hammerhead Shark

Squalomorphi: Skates and rays

Skates and rays have diverse feeding habits. Some feed on bottom-dwelling invertebrates by using their mouths as powerful vacuums. Others filter-feed on plankton. When not feeding or swimming, many skates and rays cover themselves with a thin layer of sand and rest on the ocean floor. Example: stingray

Blue-Spotted Stingray

OSTEICHTHYES: BONY FISHES

The skeletons of these vertebrates are made of true bone. This clade includes the ancestors and living members of all "higher" vertebrate groups—including tetrapods.

Rainbow Trout

Actinopterygii: Ray-finned fishes

Almost all living bony fishes, such as these rainbow trout, belong to a huge group called ray-finned fishes. The name *ray-finned* refers to the slender bony rays that are connected to one another by a layer of skin to form fins.

Coelacanth

Sarcopterygii: Lobe-finned fishes

Seven living species of bony fishes, including lungfishes and coelacanths, are classified as lobe-finned fishes. Lungfishes live in fresh water; coelacanths live in salt water. The fleshy fins of lobe-finned fishes are supported by strong bones rather than rays. Some of these bones are homologous to the limb bones of land vertebrates. Examples: lungfish, coelacanths

This clade includes the ancestors of tetrapods, which means, that all living tetrapods (including us!) are Sarcopterygians. As a result, the bony-fish clade includes almost half of all chordate species.

Amphibians

Marsupial Frog

KEY CHARACTERISTICS

The word amphibian *means "double life," an apt name for these vertebrates, most of which live in water as larvae and on land as adults. Most adult amphibians breathe with lungs, lack scales and claws, and have moist skin that contains mucous glands.*

Feeding and Digestion Tadpoles—usually filter feeders or herbivores with long, coiled intestines to digest plant material; adults—carnivores with shorter intestines for processing meat

Circulation Double-loop system with three-chambered heart

Respiration Larvae breathe through skin and gills; most adult species have lungs, though a few use gills; lungless salamanders breathe through their mouth-cavity lining and skin.

Excretion Kidneys produce urine.

Response Well-developed nervous and sensory systems; organs include protective nictitating membrane over moveable eyes, tympanic membranes, lateral line system

Movement Larvae have tails; adults have limbs (except caecilians); some have specialized toes for climbing.

Reproduction Most lay eggs without shells that are fertilized externally; most undergo metamorphosis from aquatic tadpole larvae that breathe with gills to land-dwelling adults, which usually have lungs and limbs.

Eco • Alert

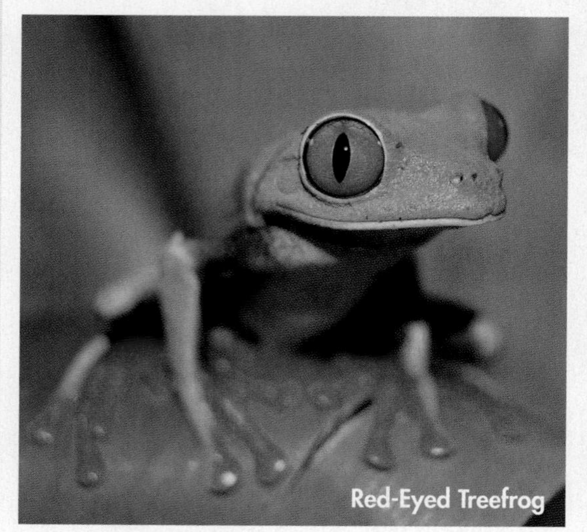
Red-Eyed Treefrog

The Frogs Are Disappearing!

For several decades, scientists have noticed that amphibian populations worldwide have been decreasing, and a number of species have become extinct. Scientists have not yet pinpointed a single cause for this problem. It is, however, becoming clear that amphibians are susceptible to a variety of environmental threats, including habitat loss, ozone depletion, acid rain, water pollution, fungal infections, and introduced aquatic predators.

To better understand this decline, biologists worldwide have been focusing their efforts and sharing data about amphibian populations. One amphibian-monitoring program covers all of North America.

GROUPS OF AMPHIBIANS

The three orders of amphibians include more than 6000 species, roughly 5000 of which are frogs and toads.

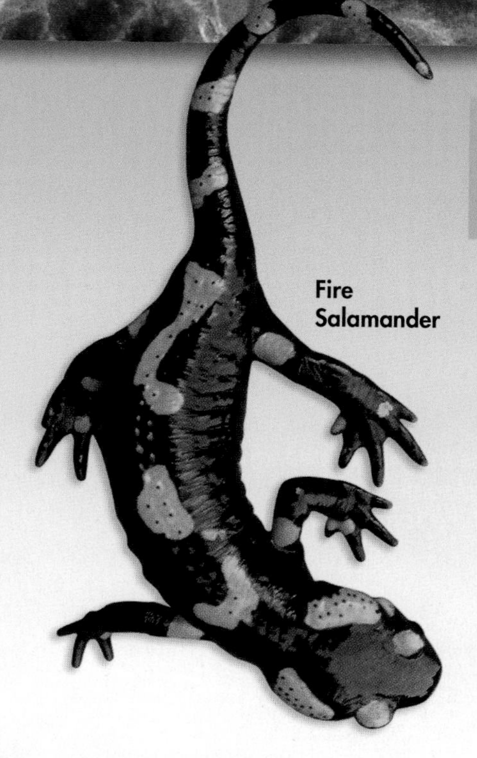
Fire Salamander

URODELA: Salamanders and newts

Salamanders and newts have long bodies and tails. Most also have four legs. All are carnivores. Adults usually live in moist woods, where they tunnel under rocks and rotting logs. Some salamanders, such as the mud puppy, keep their gills as adults and live in water all their lives. Examples: barred tiger salamander, red eft

American Toad

ANURA: Frogs and toads

Adult frogs and toads are amphibians without tails that can jump. Frogs tend to have long legs and make long jumps, whereas toads have shorter legs that limit them to shorter hops. Frogs are generally more dependent on bodies of fresh water than toads, which may live in moist woods or even deserts. Examples: treefrogs, leopard frog, American toad, spadefoot toad

APODA: Caecilians

The least-known and most unusual amphibians are the legless caecilians. They have tentacles, and many have fishlike scales embedded in their skin—which shows that not all amphibians fit the general definition. Caecilians live in water or burrow in moist soil or sediment, feeding on small invertebrates such as termites. Examples: ringed caecilian, yellow-striped caecilian

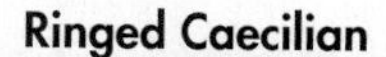

Ringed Caecilian

▶ *Because amphibian eggs must develop in water, most amphibians live in moist climates. Some, such as this alpine newt, live on cool, rainy mountain slopes.*

Reptiles

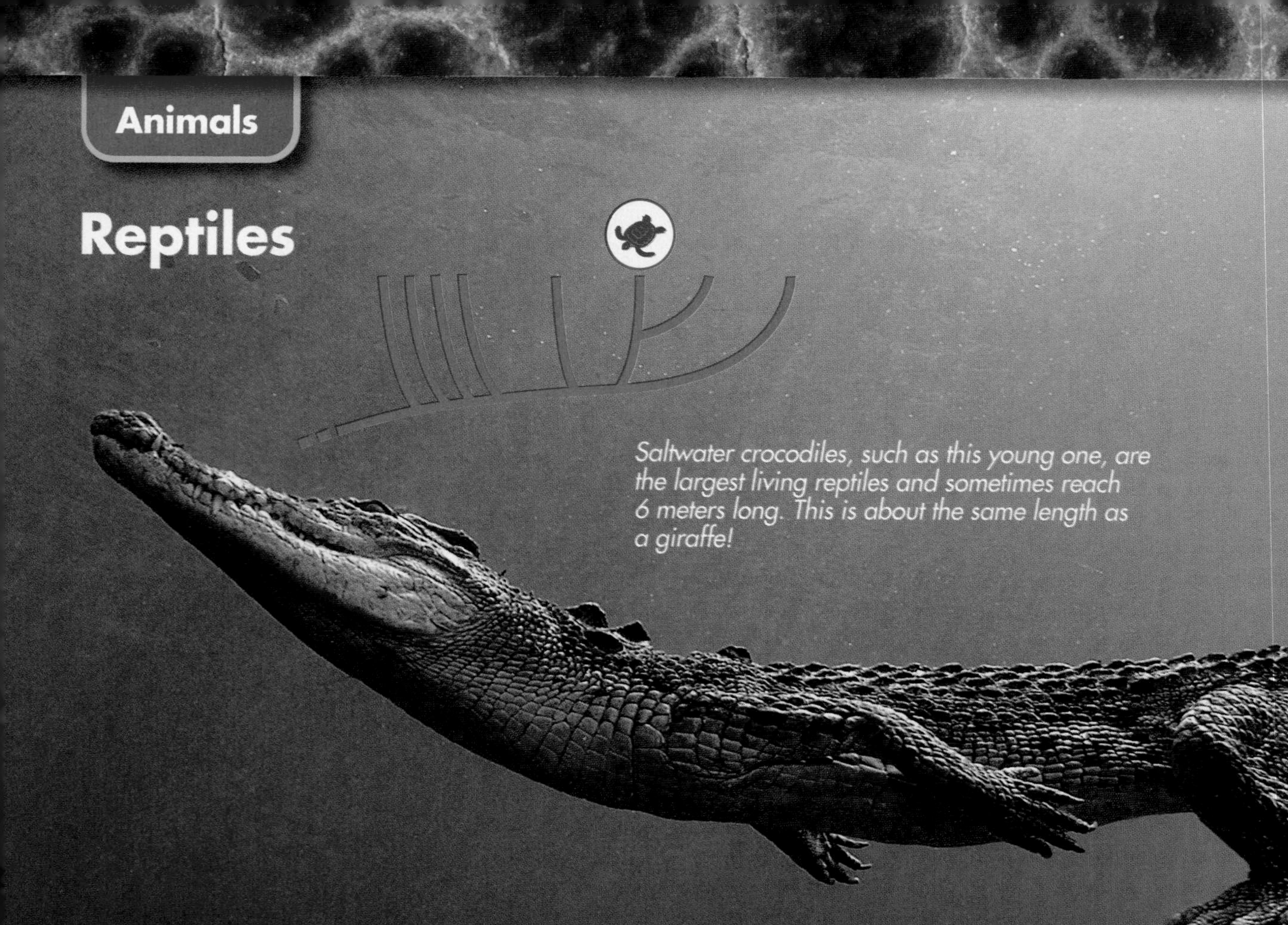

Saltwater crocodiles, such as this young one, are the largest living reptiles and sometimes reach 6 meters long. This is about the same length as a giraffe!

KEY CHARACTERISTICS OF REPTILES

Living reptiles, traditionally classified in the class Reptilia, are ectothermic vertebrates with dry, scaly skin; lungs; and amniotic eggs. Modern evolutionary classification now recognizes a larger clade Reptilia that includes living reptiles, extinct dinosaurs, and birds—the living descendants of one dinosaur group.

Feeding and Digestion Feeding methods vary by group; digestive systems—herbivores have long digestive systems to break down plant materials; carnivores may swallow prey whole

Circulation Two loops; heart with two atria and one or two ventricles

Respiration Spongy lungs provide large surface area for gas exchange; lungs operated by muscles and moveable ribs

Excretion Kidneys; urine contains ammonia or uric acid

Response Brain; well-developed senses including, in some species, infrared detectors that can spot warm-bodied prey in the dark

Movement Strong limbs (except snakes)

Reproduction Internal fertilization via cloaca; amniotic egg with leathery shell

Eco • Alert

Calling Doctor 'Gator!

You might think of alligators mostly as killing machines, but their blood may soon provide medicines that can save lives. An alligator's immune system works quite differently from our own. Proteins in their white blood cells can kill multidrug resistant bacteria, disease-causing yeasts, and even HIV. Remarkably, these proteins work against pathogens to which the animals have never been exposed. Researchers are currently sequencing the genes for these proteins and hope to develop them into human medicines in the near future.

GROUPS OF REPTILES

There are nearly 9000 species of reptiles (not including birds).

SPHENODONTA: Tuataras

The tuatara, found only on a few small islands off the coast of New Zealand, is the only living member of this group. Tuataras resemble lizards in some ways, but they lack external ears and retain primitive scales.

Tuatara

SQUAMATA: Lizards, snakes, and relatives

There are more than 8000 species of lizards and snakes. Most lizards have legs, clawed toes, and external ears. Some lizards have evolved highly specialized structures, such as glands in the lower jaw that produce venom. Snakes are legless; they have lost both pairs of legs through evolution. Examples: iguanas, milk snake, coral snake

Leopard Gecko

Leopard Tortoise

TESTUDINE: Turtles and tortoises

Turtles and tortoises have a shell built into their skeleton. Most can pull their heads and legs into the shell for protection. Instead of teeth, these reptiles have hornlike ridges covering their jaws equipped with sharp beaklike tips. Strong limbs can lift their body off the ground when walking or, in the case of sea turtles, can drag their body across a sandy shore to lay eggs. Examples: snapping turtles, green sea turtles, Galápagos tortoise

ARCHOSAURS: Crocodilians; pterosaurs and dinosaurs (extinct); and birds

This clade includes some of the most spectacular animals that have ever lived. The extinct dinosaurs and pterosaurs (flying reptiles), whose adaptive radiations produced some of the largest animals ever to walk Earth or fly above it, are the closest relatives of birds. Living crocodilians are short-legged and have long and typically broad snouts. They are fierce carnivorous predators, but the females are attentive mothers. Crocodilians live only in regions where the climate remains warm year-round. We discuss birds separately. Examples: extinct types: *Tyrannosaurus*, *Pteranodon*; living types: alligators, crocodiles, caimans, and birds (see following pages)

Spectacled Caiman

Animals

Birds

Today, only birds have feathers. These delicate, intricately interlocking and beautiful structures keep birds warm and cool and enable most to fly.

Common Kingfisher

KEY CHARACTERISTICS OF BIRDS

Birds, once placed in a class of their own, are now recognized as endothermic reptiles with feathers and hard-shelled, amniotic eggs that are descended from dinosaurs. Birds have two scaly legs and front limbs modified into wings, which enable most species to fly.

Feeding and Digestion No teeth; bills adapted to widely varied foods, including insects, seeds, fruits, nectar, fish, meat; organs of the digestive system include crop, gizzard, cloaca

Circulation Two loops with four-chambered heart; separation of oxygen-rich and oxygen-poor blood

Respiration Constant, one-way flow of air through lungs and air sacs increases the efficiency of gas exchange and supports high metabolic rate

Excretion Kidneys remove nitrogenous wastes from blood, converting them to uric acid, which is excreted through cloaca

Response Brain with large optic lobes and enlarged cerebellum; highly evolved sense organs including, in some species, eyes that can see ultraviolet light

Movement Skeleton made up of lightweight, hollow bones with internal struts for strength; powerful muscles; most fly

Reproduction Internal fertilization via cloaca; amniotic egg with hard, brittle shell; depending on species, newly hatched young may be precocial—downy-feathered chicks able to move around and feed themselves, or altricial—bare-skinned and totally dependent on their parents

• A Look Back in Time

Birds of a Feather

Fossils recently discovered in lake beds in China have greatly expanded our understanding of bird evolution. One exciting discovery was that of a four-winged dinosaur named *Microraptor gui* from about 125 million years ago. *Microraptor gui*, which was related to *Tyrannosaurus rex*, had feathers on both its wings *and* its legs, so some researchers hypothesize that it flew like a biplane! This and other fossils show that several lineages of dinosaurs and ancient birds evolved various kinds of feathers over millions of years.

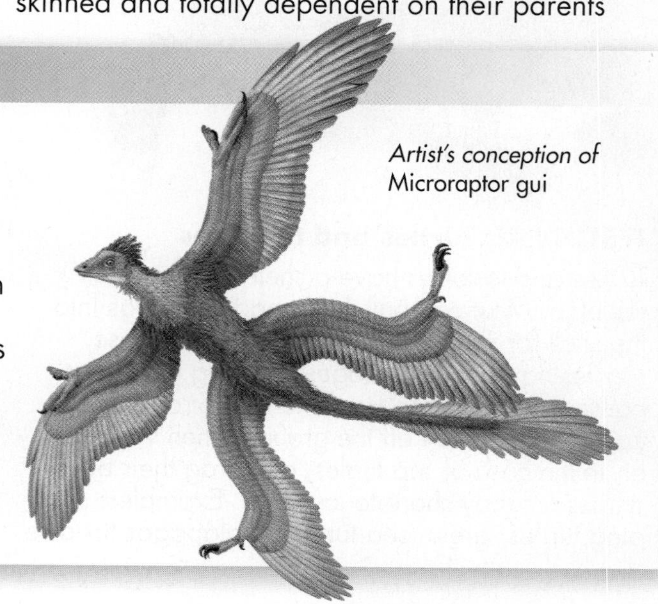

Artist's conception of Microraptor gui

GROUPS OF BIRDS

Evolutionary classification of living birds is still a work in progress, as different techniques and analyses produce different results. There are about 10,000 species. The groups described below illustrate some of the diversity of birds.

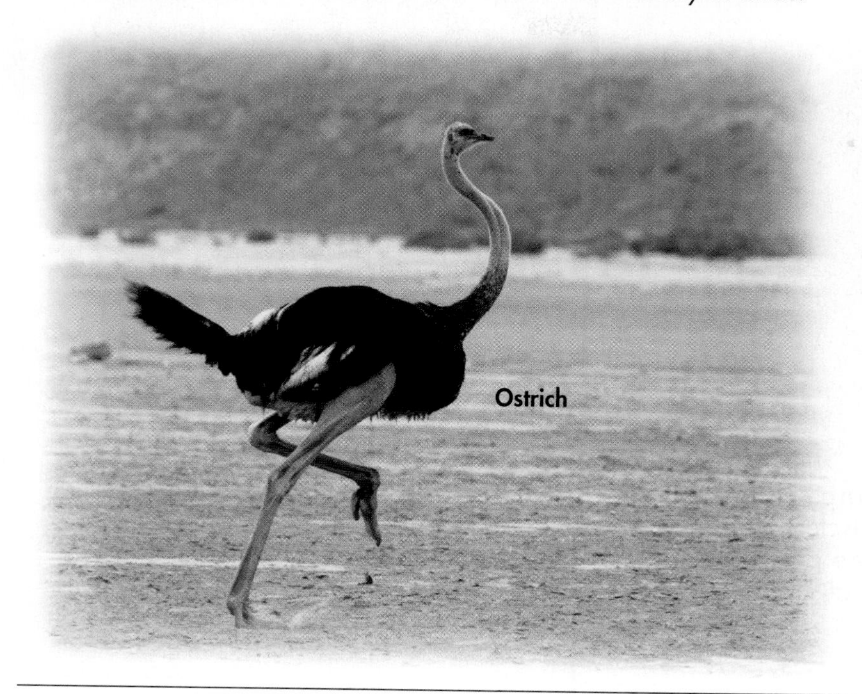
Ostrich

PALEOGNATHAE: Ostriches, emus, kiwis, and relatives

This group represents an early branch of the bird family tree that is separate from all other living birds. This clade includes the largest birds alive today. Ostriches can be 2.5 meters tall and weigh 130 kilograms! Kiwis, however, are only about the size of chickens. Roughly a dozen living species are scattered throughout the Southern Hemisphere. All are flightless, but the larger species can run very fast. They generally eat a variety of plant material, insects, and other small invertebrates. Examples: ostrich, emus, brown kiwi, greater rhea, dwarf cassowary

SPHENISCIDAE: Penguins

These flightless birds of the Southern Hemisphere are adapted to extreme cold and hunting in water. Though they cannot fly, they use their wings as flippers when they swim. Penguins have more feathers per square centimeter than any other bird; this density allows them to repel water and conserve heat effectively. Some species form large colonies. Examples: emperor penguin, chinstrap penguin, king penguin

King Penguins

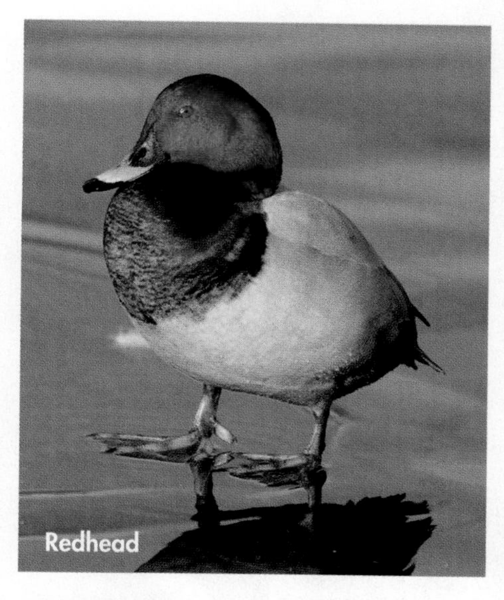
Redhead

ANATIDAE: Ducks, geese, and swans

These birds spend much of their time feeding in bodies of water. Webbed feet enable them to paddle efficiently across the surface of the water. Most fly well, however, and many species migrate thousands of kilometers between breeding and resting locations. Examples: redhead, Ross's goose, trumpeter swan

Ferruginous Hawk

FALCONIDAE AND ACCIPITRIDAE:
Falcons, eagles, and hawks
These fierce predators, often called raptors, typically have powerful hooked bills, large wingspans, and sharp talons. Raptors have powerful flight muscles and keen eyesight, enabling them to see prey at a distance. Examples: Eurasian kestrel, golden eagle, Galápagos hawk

Great-Spotted Woodpecker

PICIDAE AND RAMPHASTIDAE:
Woodpeckers and toucans
Woodpeckers are tree-dwelling birds with two toes in front and two in back. (Most birds have three in front and one in back; the two-and-two arrangement makes moving up and down tree trunks easier.) Woodpeckers are typically carnivores that eat insects and their larvae. Toucans usually use their huge, often colorful bills to eat fruit. Examples: black woodpecker, keel-billed toucan

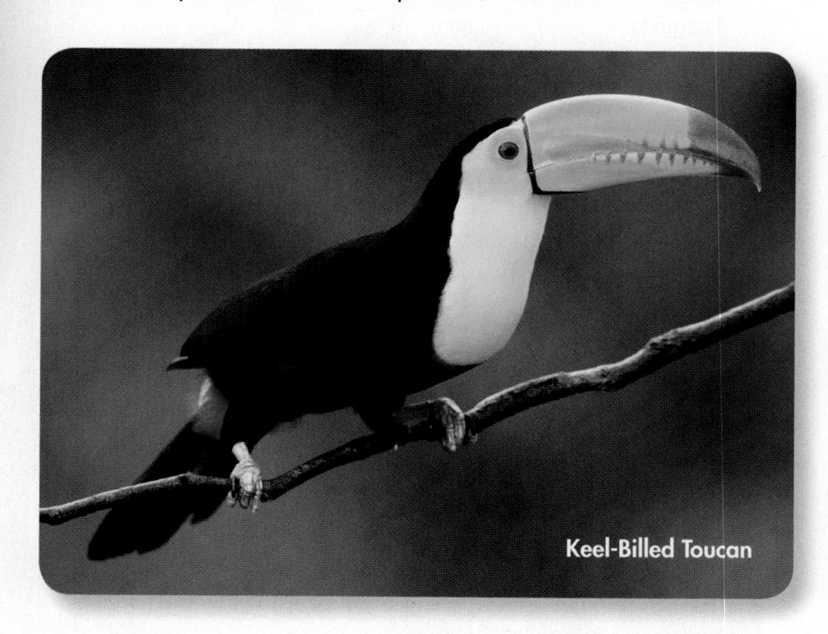

Keel-Billed Toucan

PASSERIFORMES: Passerines

Also called perching birds, this is by far the largest and most diverse group of birds, with about 5000 species. Most are songbirds. Examples: flycatchers, mockingbirds, cardinals, crows, chickadees, and finches.

Hooded Warbler

Scarlet Tanager

Blue Grosbeak

Lark Sparrow

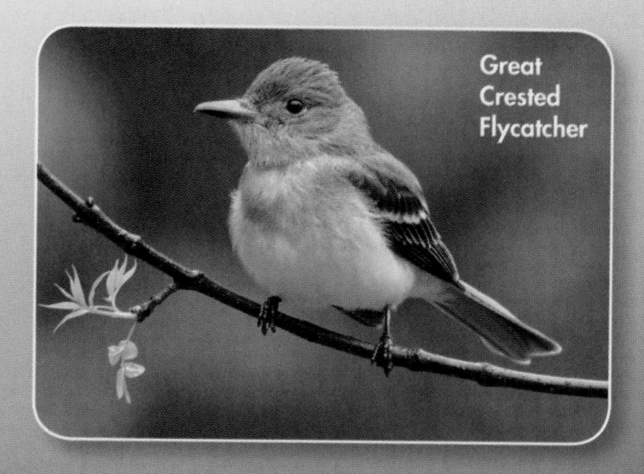

Great Crested Flycatcher

Mammals

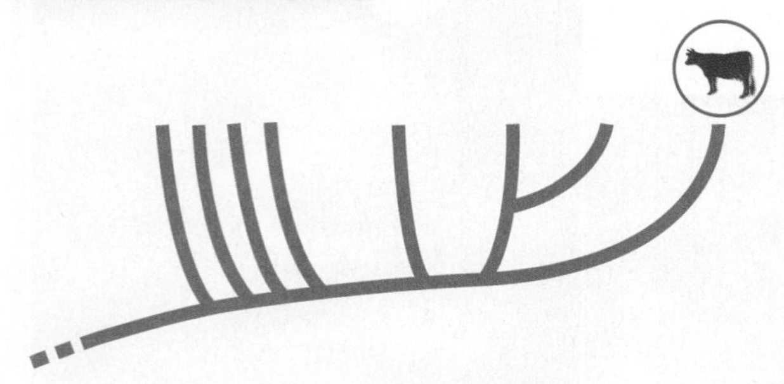

KEY CHARACTERISTICS

Mammals are endothermic vertebrates with hair and mammary glands that produce milk to nourish their young.

Feeding and Digestion Diet varies with group; foods range from seeds, fruits, and leaves to insects, fish, meat, and even blood; teeth, jaws, and digestive organs are adapted to diet

Circulation Two loops; four-chambered heart; separation of oxygen-rich and oxygen-poor blood

Respiration Lungs controlled by two sets of muscles.

Excretion Highly evolved kidneys filter urea from blood and produce urine.

Response Most highly evolved brain of all animals; keen senses

Movement Flexible backbone; variations in limb bones and muscles enable wide range of movement across groups: from burrowing and crawling to walking, running, hopping, and flying

Reproduction Internal fertilization; developmental process varies with group (monotreme, marsupial, placental)

• Did You Know?

Platypus: Mix-and-Match Genome

The duckbill platypus has such an odd mix of reptile and mammal features that some scientists thought the first specimens were hoaxes produced by sticking parts of different animals together! Recent genome studies have revealed an equally odd mix of reptilian and mammalian genes. Genes for reptile-like vision, the production of egg yolk, and the production of venom link the platypus to reptiles. Genes for the production of milk link it to other mammals. The evidence provides confirmation that this monotreme represents a truly ancient lineage, one from the time close to that at which mammals branched off from reptiles.

GROUPS OF MAMMALS

The three living groups of mammals are the monotremes, the marsupials, and the placentals. There are about 5000 species of mammals, usually divided into about 26 orders, most of which are placentals. There is only one order of monotremes.

This young moose is enjoying a moment of independence from its mother. Mammals provide intensive parental care to their young.

Short-Beaked Echidna

MONOTREMATA: Monotremes

Monotremes—egg-laying mammals—share two important characteristics with reptiles. First, the digestive, reproductive, and urinary systems of monotremes all open into a cloaca similar to that of reptiles. Second, monotreme development is similar to that of reptiles. Like a reptile, a female monotreme lays soft-shelled eggs incubated outside her body. The eggs hatch in about ten days. Unlike reptiles, however, young monotremes are nourished by mother's milk, which they lick from pores on the surface of her abdomen. Only five monotreme species exist today, all in Australia and New Guinea. Examples: duckbill platypus, echidnas

---

MARSUPIALIA: Marsupials

Marsupials bear live young at an extremely early stage of development. A fertilized egg develops into an embryo inside the mother's reproductive tract. The embryo is then "born" in what would be an embryonic stage for most other mammals. It crawls across its mother's fur and attaches to a nipple that, in most species, is located in a pouch called the marsupium. The embryo spends several months attached to the nipple. It continues to nurse until it can survive on its own. Examples: kangaroos, wallabies, wombats, opossums

Wombat

PLACENTALIA: Placental Mammals

Placental mammals are the mammals with which you are most familiar. This group gets its name from a structure called the placenta, which is formed when the embryo's tissues join with tissues within the mother's body. Nutrients, gases, and wastes are exchanged between embryo and mother through the placenta. Development may take as little as a few weeks (mice), to as long as two years (elephants). After birth, most placental mammals care for their young and provide them with nourishment by nursing. Examples: mice, cats, dogs, seals, whales, elephants, humans

Chiroptera: Bats

These are the only mammals capable of true flight. There are more than 900 species of bats! They eat mostly insects or fruit and nectar, although a few species feed on the blood of other vertebrates. Examples: fruit bats, little brown myotis, vampire bat

Epauletted Bat, roosting

Lioness attacking Greater Kudu

Carnivora: Carnivores

Many members of this group, such as tigers and hyenas, chase or stalk prey by running or pouncing, then kill with sharp teeth and claws. Dogs, bears, and other members of this group may eat plants as well as meat. Examples: dogs, cats, skunks, seals, bears

Sirenia: Sirenians

Sirenians are herbivores that live in rivers, bays, and warm coastal waters scattered throughout the world. These large, slow-moving mammals lead fully aquatic lives. Examples: manatees, dugongs

Manatee mother and nursing calf

Four-Toed Hedgehog mother and baby

Insectivora: Insectivores

These insect eaters have long, narrow snouts and sharp claws that are well suited for digging. Examples: shrews, moles, hedgehogs

Perissodactyla: Hoofed, odd-toed mammals

This group is made up of hoofed animals with an odd number of toes on each foot. Like artiodactyls, this group contains mostly large, grazing animals. Examples: horses, zebras, rhinoceroses

Tapir hoof

Central American Tapir

Artiodactyla: Hoofed, even-toed mammals

These large, grazing, hoofed mammals have an even number of toes on each foot. Examples: cattle, sheep, pigs, hippopotami

Giraffe Hoof

Angolan Giraffes

Marmot

Rodentia: Rodents

Rodents have a single pair of long, curved incisor teeth in both their upper and lower jaws, used for gnawing wood and other tough plant material. Examples: rats, squirrels, porcupines

Cetacea: Cetaceans

Like sirenians, cetaceans—the group that includes whales and dolphins—are adapted to underwater life, yet must come to the surface to breathe. Most cetaceans live and breed in the ocean. Examples: whales, dolphins

Atlantic Spotted Dolphin

European Hare

Lagomorpha: Rabbit, hares, and pikas

Lagomorphs are entirely herbivorous. They differ from rodents by having two pairs of incisors in the upper jaw. Most lagomorphs have hind legs that are adapted for leaping.

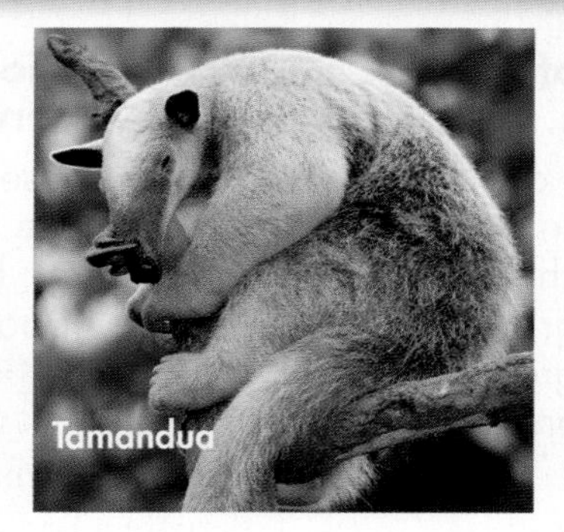
Tamandua

Xenarthra: Edentates

The word *edentate* means "toothless," which refers to the fact that some members of this group (sloths and anteaters) have simple teeth without enamel or no teeth at all. Armadillos, however, have more teeth than most other mammals! Examples: sloths, anteaters, armadillos

Proboscidea: Elephants

These are the mammals with trunks. Some time ago, this group went through an extensive adaptive radiation that produced many species, including mastodons and mammoths, which are now extinct. Only two species, the Asian Elephant and the African elephant, survive today.

Asian Elephant and calf

Primates: Lemurs, monkeys, apes, humans, and relatives

Members of this group are closely related to ancient insectivores but have a highly developed cerebrum and complex behaviors.

Sifaka

Tarsier

Langur

Baboon and baby

Orangutan

Gorilla

Chimpanzee

Appendix A Science Skills

Data Tables and Graphs

How can you make sense of the data from a science experiment? The first step is to organize the data. You can organize data in data tables and graphs to help you interpret them.

Data Tables

You have gathered your materials and set up your experiment. But before you start, you need to plan a way to record what happens during the experiment. By creating a data table, you can record your observations and measurements in an orderly way.

Suppose, for example, that a scientist conducted an experiment to find out how many kilocalories people of different body masses burned while performing various activities for 30 minutes. The data table below shows the results.

Notice in this data table that the independent variable (body mass) is the heading of the first column. The dependent variable (for Experiment 1, the number of kilocalories burned while bicycling for 30 minutes) is the heading of the next column. Additional columns were added for related experiments.

Calories Burned in 30 Minutes

Body Mass	Experiment 1: Bicycling	Experiment 2: Playing Basketball	Experiment 3: Watching Television
30 kg	60 Calories	120 Calories	21 Calories
40 kg	77 Calories	164 Calories	27 Calories
50 kg	95 Calories	206 Calories	33 Calories
60 kg	114 Calories	248 Calories	38 Calories

Bar Graphs

A bar graph is useful for comparing data from two or more distinct categories. In this example, pancreatic secretions in the small intestine are shown. To create a bar graph, follow these steps.

1. On graph paper, draw a horizontal, or x-axis, and a vertical, or y-axis.
2. Write the names of the categories (the independent variable) along one axis, usually the horizontal axis. You may put the categories on the vertical axis if that graph shape better fits on your page. Label the axis.
3. Label the other axis with the name of the dependent variable and the unit of measurement. Then, create a scale along that axis by marking off equally spaced numbers that cover the range of the data values.
4. For each category, draw a solid bar at the appropriate value. Then, fill in the space from the bar to the axis representing the independent variable. Make all the bars the same width.
5. Add a title that describes the graph.

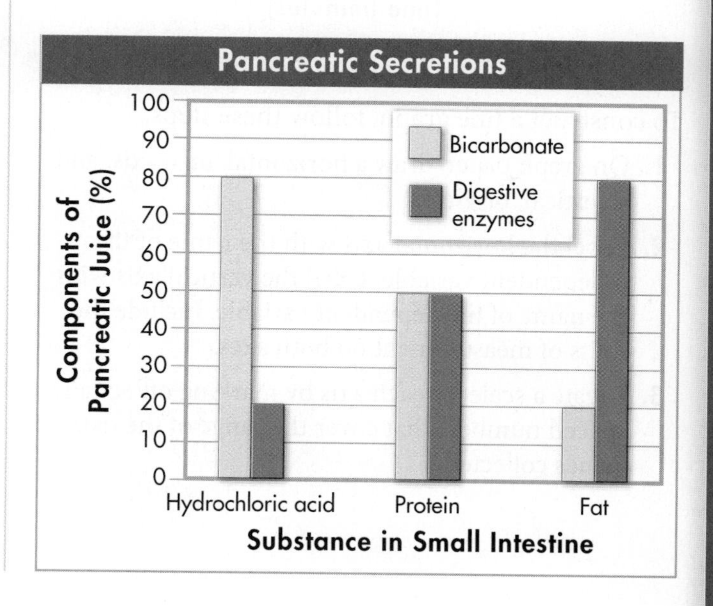

Line Graphs

A line graph is used to display data that show how the dependent variable changes in response to manipulations of the independent variable. You can use a line graph when your independent variable is continuous, that is, when there are other points between the ones that you tested. For example, the graph below shows how the growth of a bacterial population is related to time. The graph shows that the number of bacteria approximately doubles every 20 minutes. Line graphs are powerful tools because they also allow you to estimate values for conditions that you did not test in the experiment.

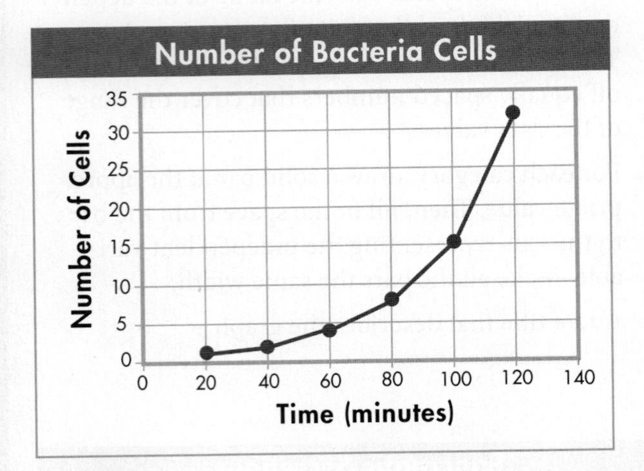

To construct a line graph, follow these steps.

1. On graph paper, draw a horizontal, or x-axis, and a vertical, or y-axis.
2. Label the horizontal axis with the name of the independent variable. Label the vertical axis with the name of the dependent variable. Include the units of measurement on both axes.
3. Create a scale on each axis by marking off equally spaced numbers that cover the range of the data values collected.
4. Plot a point on the graph for each data value. To do this, follow an imaginary vertical line extending up from the horizontal axis for an independent variable value. Then, follow an imaginary horizontal line extending across from the vertical axis at the value of the associated dependent variable. Plot a point where the two lines intersect. Repeat until all your data values are plotted.
5. Connect the plotted points with a solid line. Not all graphs are linear, so you may discover that it is more appropriate to draw a curve to connect the points.

The data in the graph at the left fit neatly on a smooth curve. But if you were to connect each data point on the graph below, you would have a mess that yielded little useful information. In some cases, it may be most useful to draw a line that shows the general trend of the plotted points. This type of line is often called a line of best fit. Such a line runs as closely as possible to all the points and allows you to make generalizations or predictions based on the data. Some points will fall above or below a line of best fit.

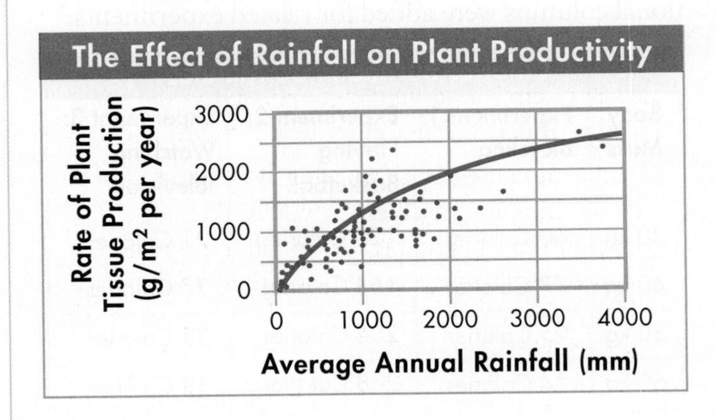

Circle Graphs

Circle graphs, or pie charts, display data as parts of a whole. Like bar graphs, circle graphs can be used to display data that fall into separate categories. Unlike bar graphs, however, circle graphs can only be used when you have data for all the categories that make up a given group. The circle, or "pie," represents 100 percent of a group, while the sectors, or slices, represent the percentages of each category that make up that group. The example below compares the different blood groups found in the U.S. population.

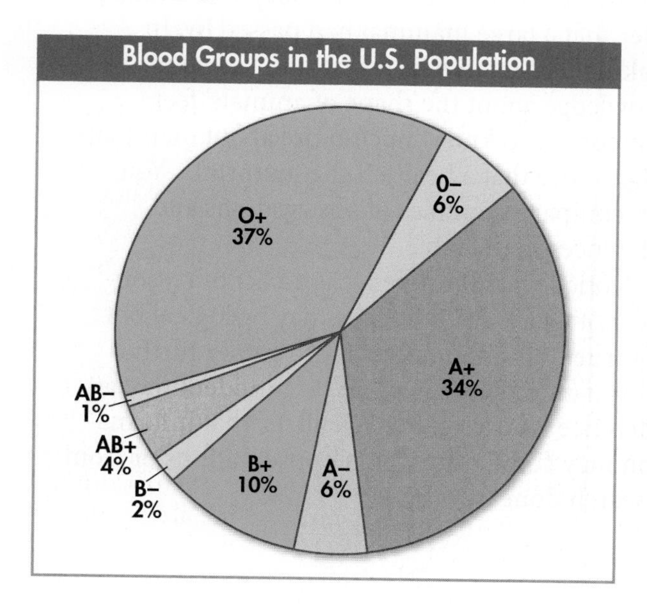

To construct a circle graph, follow these steps.

1. Draw a circle and mark the center. Then, draw a radius line from the center to the circle's edge.
2. Determine the size of a sector of the graph by calculating the number of degrees that correspond to a percentage you wish to represent. For example, in the graph shown, B^+ makes up 10 percent of all blood groups; 360 degrees $\times$ 0.10 = 36 degrees.
3. With a protractor fixed at the center of the circle, measure the angle—in this example, 36 degrees—from the existing radius, and draw a second radius at that point. Label the sector with its category and the percentage of the whole it represents. Repeat for each of the other categories, measuring each sector from the previous radius so the sectors don't overlap.
4. For easier reading, shade each sector differently.
5. Add a title that describes the graph.

Reading Diagrams

In scientific figures showing a cut-away of a structure, the diagram or photograph is showing the structure from a particular angle. Look for clues throughout this book that will help you interpret the view being shown.

Cross Sections

A cross section shows a horizontal cut through the middle of a structure. This icon will help you locate cross sections.

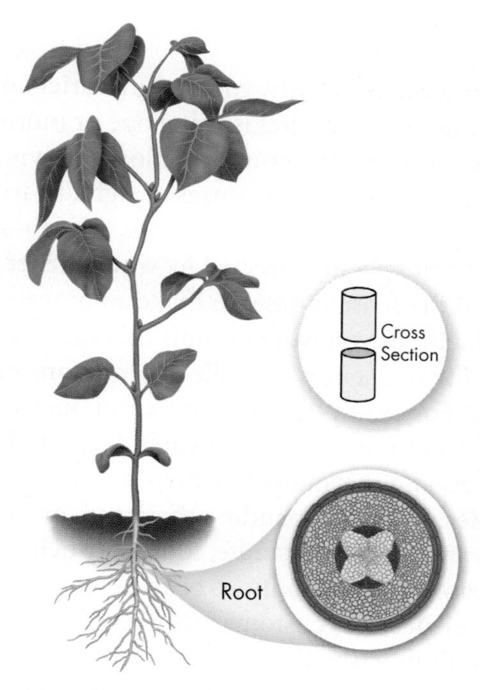

Longitudinal Sections

A longitudinal section shows a vertical cut through the middle of a structure. This icon will help you locate longitudinal sections.

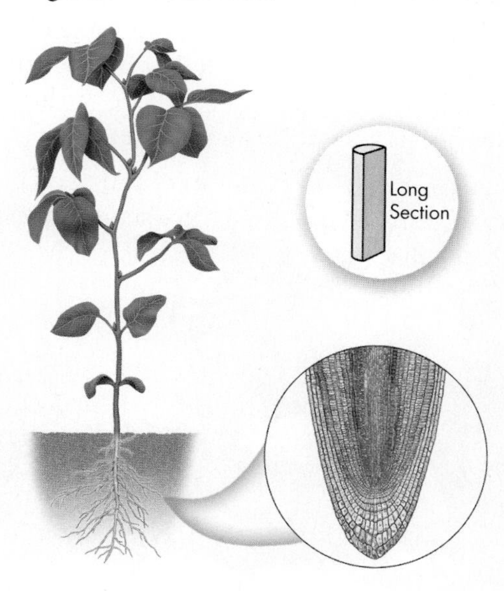

Basic Process Skills

During a biology course, you often carry out short lab activities as well as lengthier experiments. Here are some skills that you will use.

Observing

In every science activity, you make a variety of observations. Observing is using one or more of the five senses to gather information. Many observations involve the senses of sight, hearing, touch, and smell. On rare occasions in a lab—but only when explicitly directed by your teacher—you may use the sense of taste to make an observation.

Sometimes you will use tools that increase the power of your senses or make observations more precise. For example, hand lenses and microscopes enable you to see things in greater detail. Rulers, balances, and thermometers help you measure key variables. Besides expanding the senses or making observations more accurate, tools may help eliminate personal opinions or preferences.

In science, it is customary to record your observations at the time they are made, usually by writing or drawing in a notebook. You may also make records by using computers, digital cameras, video cameras, and other tools. As a rule, scientists keep complete accounts of their observations, often using tables to organize their observations.

Inferring

In science, as in daily life, observations are usually followed by inferences. Inferring is interpreting an observation or statement based on prior knowledge.

For example, suppose you're on a mountain hike and you see footprints like the ones illustrated below. Based on their size and shape, you might infer that a large mammal had passed by. In making that inference, you would use your knowledge about the shape of animals' feet. Someone who knew much more about mammals might infer that a bear left the footprints. You can compare examples of observations and inferences in the table.

Notice that an inference is an act of reasoning, not a fact. An inference may be logical but not true. It is often necessary to gather further information before you can be confident that an inference is correct. For scientists, that information may come from further observations or from research done by others.

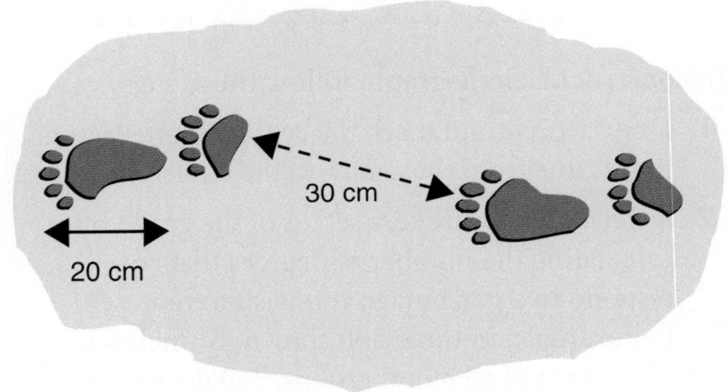

Comparing Observations and Inferences

Sample Observations	Sample Inferences
The footprints in the soil each have five toes.	An animal made the footprints.
The larger footprints are about 20 cm long.	A bear made the footprints.
The space between each pair of footprints is about 30 cm.	The animal was walking, not running.

As you study biology, you may make different types of inferences. For example, you may generalize about all cases based on information about some cases: *All the plant roots I've observed grow downward, so I infer that all roots grow downward.* You may determine that one factor or event was caused by another factor or event: *The bacteria died after I applied bleach, so I infer that bleach kills bacteria.* Predictions may be another type of inference.

Predicting

People often make predictions, but their statements about the future could be either guesses or inferences. In science, a prediction is an inference about a future event based on evidence, experience, or knowledge. For example, you can say, *On the first day of next month, it will be sunny.* If your statement is based on evidence of weather patterns in the area, then the prediction is scientific. If the statement was made without considering any evidence, it's just a guess.

Predictions play a major role in science because they provide a way to test ideas. If scientists understand an event or the properties of a particular object, they should be able to make accurate predictions about that event or object. Some predictions can be tested simply by making observations. At other times, carefully designed experiments are needed.

Classifying

If you have ever heard people debate whether a tomato is a fruit or a vegetable, you've heard an argument about classification. Classifying is the process of grouping items that are alike according to some organizing idea or system. Classifying occurs in every branch of science, but it is especially important in biology because living things are so numerous and diverse.

You may have the chance to practice classifying in different ways. Sometimes you will place objects into groups using an established system. At other times, you may create a system of your own by examining a variety of objects and identifying their properties.

Classification can have different purposes. Sometimes it's done just to keep things organized, to make lab supplies easy to find, for example. More often, though, classification helps scientists understand living things better and discover relationships among them. For example, one way biologists determine how groups of vertebrates are related is to compare their bones. Biologists classify certain animal parts as bone or muscle and then investigate how they work together.

Using Models

Some cities refuse to approve any new buildings that could cast shadows on a popular park. As architects plan buildings in such locations, they use models that can show where a proposed building's shadow will fall at any time of day in any season of the year. A model is a mental or physical representation of an object, process, or event. In science, models are usually made to help people understand natural objects and processes.

Models can be varied. Mental models, such as mathematical equations, can represent some kinds of ideas or processes. For example, the equation for the surface area of a sphere can model the surface of Earth, enabling scientists to determine its size. Physical models can be made of a huge variety of materials; they can be two dimensional (flat) or three dimensional (having depth). In biology, a drawing of a molecule or a cell is a typical two-dimensional model. Common three-dimensional models include a representation of a DNA molecule and a plastic skeleton of an animal.

Physical models can also be made "to scale," which means they are in proportion to the actual object. Something very large, such as an area of land being studied, can be shown at 1/100 of its actual size. A tiny organism can be shown at 100 times its size.

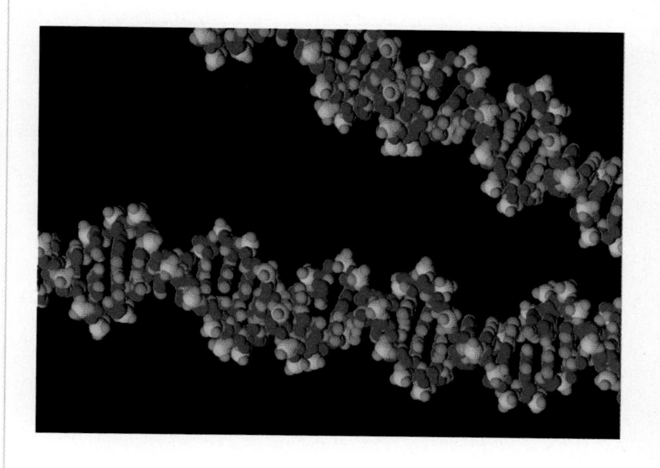

Organizing Information

When you study or want to communicate facts and ideas, you may find it helpful to organize information visually. Here are some common graphic organizers you can use. Notice that each type of organizer is useful for specific types of information.

Flowcharts

A flowchart can help you represent the order in which a set of events has occurred or should occur. Flowcharts are useful for outlining the steps in a procedure or stages in a process with a definite beginning and end.

To make a flowchart, list the steps in the process you want to represent and count the steps. Then, create the appropriate number of boxes, starting at the top of a page or on the left. Write a brief description of the first event in the first box, and then fill in the other steps, box by box. Link each box to the next event in the process with an arrow.

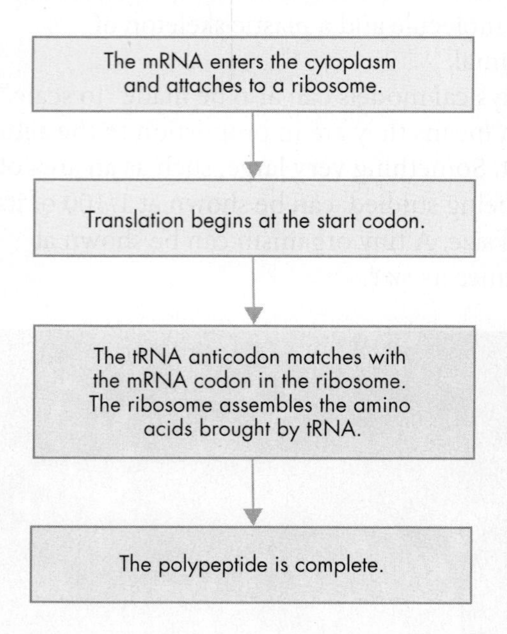

Concept Maps

Concept maps can help you organize a topic that has many subtopics. A concept map begins with a main idea and shows how it can be broken down into specific topics. It makes the ideas easier to understand by presenting their relationships visually.

You construct a concept map by placing the concept words (usually nouns) in ovals and connecting the ovals with linking words. The most general concept usually is placed at the top of the map or in the center. The content of the other ovals becomes more specific as you move away from the main concept. The linking words, which describe the relationship between the linked concepts, are written on a line between two ovals. If you follow any string of concepts and linking words down through a map, they should sound almost like a sentence.

Some concept maps may also include linking words that connect a concept in one branch to another branch. Such connections, called cross-linkages, show more complex interrelationships.

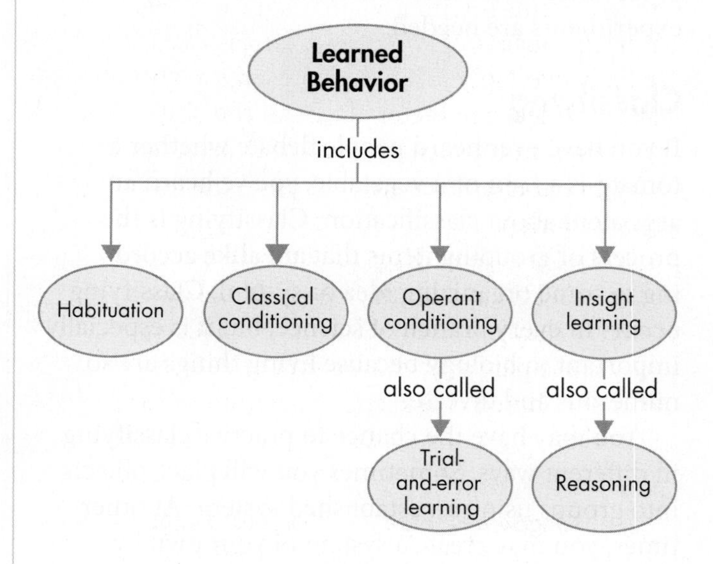

Compare/Contrast Tables

Compare/contrast tables are useful for showing the similarities and differences between two or more objects or processes. The table provides an organized framework for making comparisons based on specific characteristics.

To create a compare/contrast table, list the items to be compared across the top of the table. List the characteristics that will form the basis of your comparison in the column on the left. Complete the table by filling in information for each item.

Comparing Fermentation and Cellular Respiration

Characteristic	Fermentation	Cellular Respiration
Starting reactants	Glucose	Glucose, oxygen
Pathways involved	Glycolysis, several others	Glycolysis, Krebs cycle, electron transport
End products	CO_2 and alcohol *or* CO_2 and lactic acid	CO_2, H_2O
Number of ATP molecules produced	2	36

Venn Diagrams

Another way to show similarities and differences between items is with a Venn diagram. A Venn diagram consists of two or more ovals that partially overlap. Each oval represents a particular object or idea. Characteristics that the objects share are written in the area of overlap. Differences or unique characteristics are written in the areas that do not overlap.

To create a Venn diagram, draw two overlapping ovals. Label them with the names of the objects or the ideas they represent. Write the unique characteristics in the part of each oval that does not overlap. Write the shared characteristics within the area of overlap.

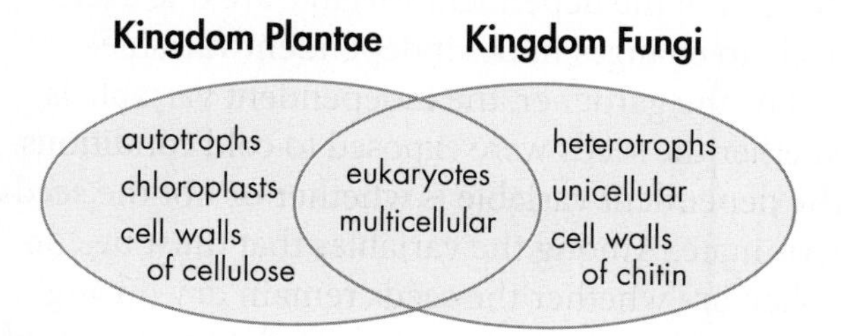

Cycle Diagrams

A cycle diagram shows a sequence of events that is continuous, or cyclical. A continuous sequence does not have a beginning or an end; instead, each event in the process leads to another event. The diagram shows the order of the events.

To create a cycle diagram, list the events in the process and count them. Draw one box for each event, placing the boxes around an imaginary circle. Write one of the events in an oval, and then draw an arrow to the next oval, moving clockwise. Continue to fill in the boxes and link them with arrows until the descriptions form a continuous circle.

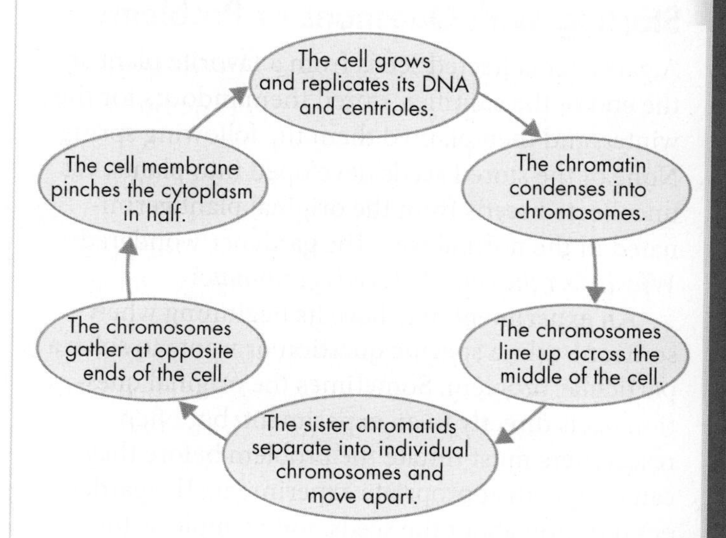

Conducting an Experiment

A science experiment is a procedure designed to test a prediction. Some types of experiments are fairly simple to design. Others may require ingenious problem solving.

Starting With Questions or Problems

A gardener collected seeds from a favorite plant at the end of the summer, stored them indoors for the winter, and then planted them the following spring. None of the stored seeds developed into plants, yet uncollected seeds from the original plant germinated in the normal way. The gardener wondered: *Why didn't the collected seeds germinate?*

An experiment may have its beginning when someone asks a specific question or wants to solve a particular problem. Sometimes the original question leads directly to an experiment, but often researchers must restate the problem before they can design an appropriate experiment. The gardener's question about the seeds, for example, is too broad to be tested by an experiment, because there are so many possible answers. To narrow the topic, the gardener might think about related questions: *Were the seeds I collected different from the uncollected seeds? Did I try to germinate them in poor soil or with insufficient light or water? Did storing the seeds indoors ruin them in some way?*

Developing a Hypothesis

In science, a question about an object or event is answered by developing a possible explanation called a hypothesis. The hypothesis may be developed after long thought and research, or it may come to a scientist "in a flash." How a hypothesis is formed doesn't matter; it can be useful as long as it leads to predictions that can be tested.

The gardener decided to focus on the fact that the nongerminating seeds were stored in the warm conditions of a heated house. That premise led the person to propose this hypothesis: *Seeds require a period of low temperatures in order to germinate.*

The next step is to make a prediction based on the hypothesis, for example: *If seeds are stored indoors in cold conditions, they will germinate in the same way as seeds left outdoors during the winter.* Notice that the prediction suggests the basic idea for an experiment.

Designing an Experiment

A carefully designed experiment can test a prediction in a reliable way, ruling out other possible explanations. As scientists plan their experimental procedures, they pay particular attention to the factors that must be controlled.

The gardener decided to study three groups of seeds: (1) some that would be left outdoors throughout the winter, (2) some that would be brought indoors and kept at room temperature, and (3) some that would be brought indoors and kept cold.

Controlling Variables

As researchers design an experiment, they identify the variables, factors that can change. Some common variables include mass, volume, time, temperature, light, and the presence or absence of specific materials. An experiment involves three categories of variables. The factor that scientists purposely change is called the independent variable. An independent variable is also known as a manipulated variable. The factor that may change because of the independent variable and that scientists want to observe is called the dependent variable. A dependent variable is also known as a responding variable. Factors that scientists purposely keep the same are called controlled variables. Controlling variables enables researchers to conclude that the changes in the dependent variable are due exclusively to changes in the independent variable.

For the gardener, the independent variable is whether the seeds were exposed to cold conditions. The dependent variable is whether or not the seeds germinate. Among the variables that must be controlled are whether the seeds remain dry during storage, when the seeds are planted, the amount of water the seeds receive, and the type of soil used.

Interpreting Data

The observations and measurements that are made in an experiment are called data. Scientists usually record data in an orderly way. When an experiment is finished, the researcher analyzes the data for trends or patterns, often by doing calculations or making graphs, to determine whether the results support the hypothesis.

For example, after planting the seeds in the spring, the gardener counted the seeds that germinated and found these results: None of the seeds kept at room temperature germinated, 80 percent of the seeds kept in the freezer germinated, and 85 percent of the seeds left outdoors during the winter germinated. The trend was clear: The gardener's prediction appeared to be correct.

To be sure that the results of an experiment are correct, scientists review their data critically, looking for possible sources of error. Here, *error* refers to differences between the observed results and the true values. Experimental error can result from human mistakes or problems with equipment. It can also occur when the small group of objects studied does not accurately represent the whole group. For example, if some of the gardener's seeds had been exposed to a herbicide, the data might not reflect the true seed germination pattern.

Drawing Conclusions

If researchers are confident that their data are reliable, they make a final statement summarizing their results. That statement, called the conclusion, indicates whether the data support or refute the hypothesis. The gardener's conclusion was this: *Some seeds must undergo a period of freezing in order to germinate.* A conclusion is considered valid if it is a logical interpretation of reliable data.

Following Up an Experiment

When an experiment has been completed, one or more events often follow. Researchers may repeat the experiment to verify the results. They may publish the experiment so that others can evaluate and replicate their procedures. They may compare their conclusion with the discoveries made by other scientists. And they may raise new questions that lead to new experiments. For example, *Are the spores of fungi affected by temperature as these seeds were?*

Researching other discoveries about seeds would show that some other types of plants in temperate zones require periods of freezing before they germinate. Biologists infer that this pattern makes it less likely the seeds will germinate before winter, thus increasing the chances that the young plants will survive.

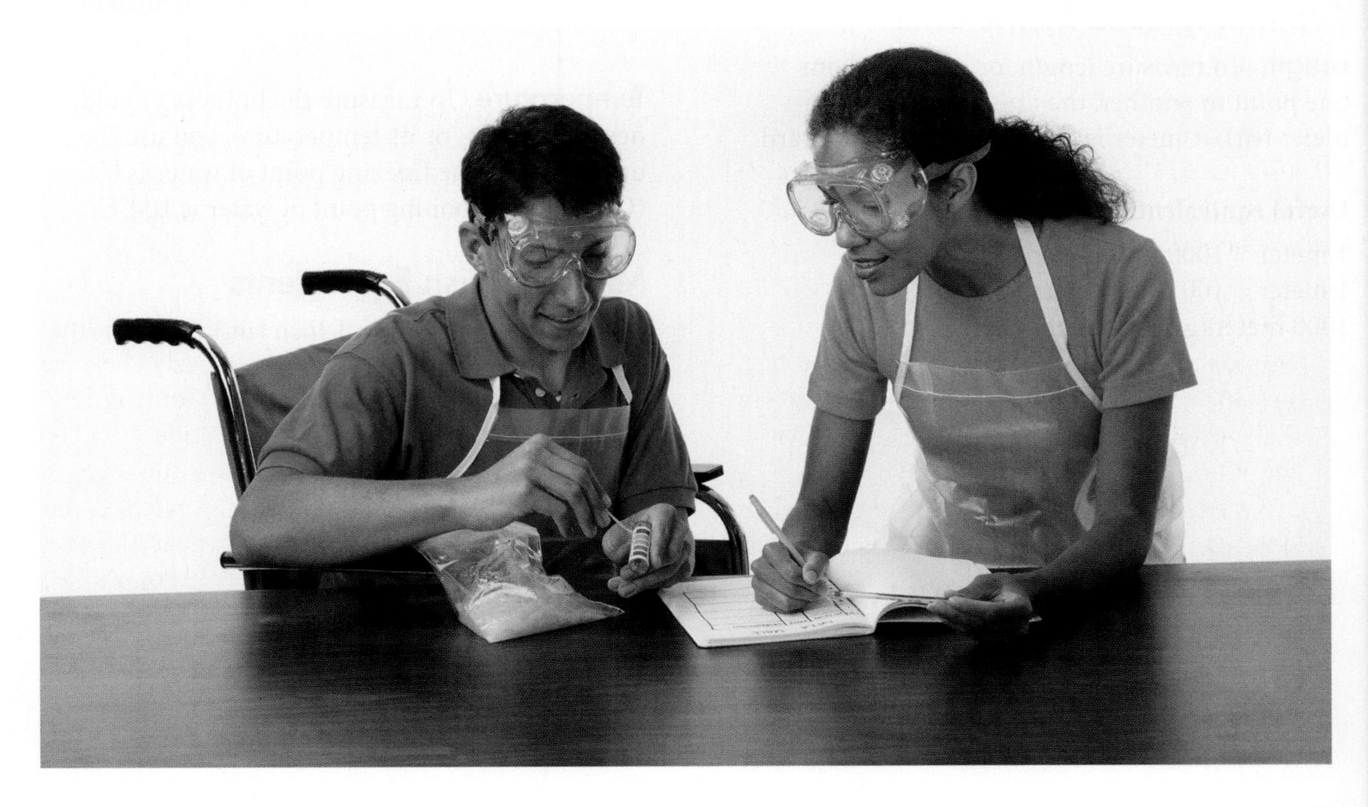

The Metric System

The standard system of measurement used by scientists throughout the world is known as the International System of Units, abbreviated as SI (Système International d'Unités, in French). It is based on units of 10. Each unit is 10 times larger or 10 times smaller than the next unit. The table lists the prefixes used to name the most common SI units.

Common SI Prefixes		
Prefix	**Symbol**	**Meaning**
kilo-	k	1000
hecto-	h	100
deka-	da	10
deci-	d	0.1 (one tenth)
centi-	c	0.01 (one hundredth)
milli-	m	0.001 (one thousandth)

Commonly Used Metric Units

Length To measure length, or distance from one point to another, the unit of measure is a meter (m). A meter is slightly longer than a yard.

Useful equivalents:

1 meter = 1000 millimeters (mm)
1 meter = 100 centimeters (cm)
1000 meters = 1 kilometer (km)

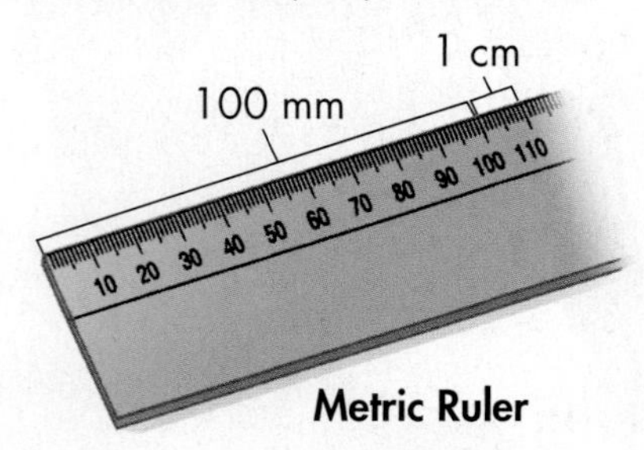

Metric Ruler

Volume To measure the volume of a liquid, or the amount of space an object takes up, the unit of measure is a liter (L). A liter is slightly more than a quart.

Useful equivalents:

1 liter = 1000 milliliters (mL)

Mass To measure the mass, or the amount of matter in an object, the unit of measure is the gram (g). A paper clip has a mass equal to about one gram.

Useful equivalents:

1000 grams = 1 kilogram (kg)

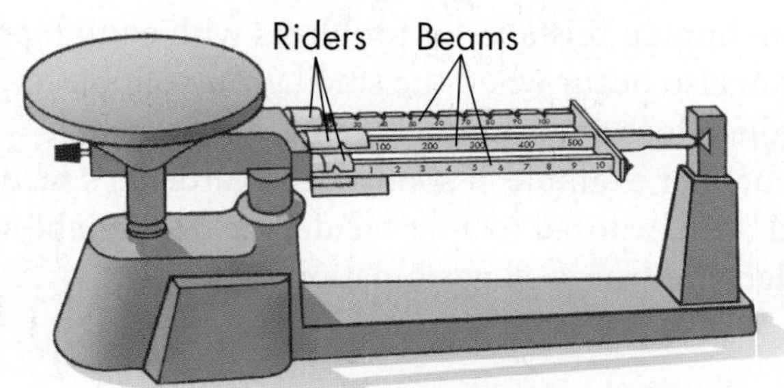

Triple-Beam Balance

Temperature To measure the hotness or coldness of an item, or its temperature, you use the unit degrees. The freezing point of water is 0°C (Celsius). The boiling point of water is 100°C.

Metric-English Equivalents

2.54 centimeters (cm) = 1 inch (in.)
1 meter (m) = 39.37 inches (in.)
1 kilometer (km) = 0.62 miles (mi)
1 liter (L) = 1.06 quarts (qt)
236 milliliters (mL) = 1 cup (c)
1 kilogram (kg) = 2.2 pounds (lb)
28.3 grams (g) = 1 ounce (oz)
°C = 5/9 × (°F – 32)

Safety Symbols

These symbols appear in laboratory activities to alert you to possible dangers and to remind you to work carefully.

Safety Goggles Always wear safety goggles to protect your eyes during any activity involving chemicals, flames or heating, or the possibility of flying objects, particles, or substances.

Lab Apron Wear a laboratory apron to protect your skin and clothing from injury.

Plastic Gloves Wear disposable plastic gloves to protect yourself from contact with chemicals or organisms that could be harmful. Keep your hands away from your face, and dispose of the gloves according to your teacher's instructions at the end of the activity.

Breakage Handle breakable materials such as thermometers and glassware with care. Do not touch broken glass.

Heat-Resistant Gloves Use an oven mitt or other hand protection when handling hot materials. Hot plates, hot water, and glassware can cause burns. Never touch hot objects with your bare hands.

Heating Use a clamp or tongs to hold hot objects. Do not touch hot objects with your bare hands.

Sharp Object Scissors, scalpels, pins, and knives are sharp. They can cut or puncture your skin. Always direct sharp edges and points away from yourself and others. Use sharp instruments only as directed.

Electric Shock Avoid the possibility of electric shock. Never use electrical equipment around water or when the equipment or your hands are wet. Be sure cords are untangled and cannot trip anyone. Disconnect equipment when it is not in use.

Corrosive Chemical This symbol indicates the presence of an acid or other corrosive chemical. Avoid getting the chemical on your skin or clothing, or in your eyes. Do not inhale the vapors. Wash your hands when you are finished with the activity.

Poison Do not let any poisonous chemical get on your skin, and do not inhale its vapor. Wash your hands when you are finished with the activity.

Flames Tie back loose hair and clothing, and put on safety goggles before working with fire. Follow instructions from your teacher about lighting and extinguishing flames.

No Flames Flammable materials may be present. Make sure there are no flames, sparks, or exposed sources of heat present.

Fumes Poisonous or unpleasant vapors may be produced. Work in a ventilated area or, if available, in a fume hood. Avoid inhaling a vapor directly. Test an odor only when directed to do so by your teacher, using a wafting motion to direct the vapor toward your nose.

Physical Safety This activity involves physical movement. Use caution to avoid injuring yourself or others. Follow instructions from your teacher. Alert your teacher if there is any reason that you should not participate in the activity.

Animal Safety Treat live animals with care to avoid injuring the animals or yourself. Working with animal parts or preserved animals may also require caution. Wash your hands when you are finished with the activity.

Plant Safety Handle plants only as your teacher directs. If you are allergic to any plants used in an activity, tell your teacher before the activity begins. Avoid touching poisonous plants and plants with thorns.

Disposal Chemicals and other materials used in the activity must be disposed of safely. Follow the instructions from your teacher.

Hand Washing Wash your hands thoroughly when finished with the activity. Use soap and warm water. Lather both sides of your hands and between your fingers. Rinse well.

General Safety Awareness You may see this symbol when none of the symbols described earlier applies. In this case, follow the specific instructions provided. You may also see this symbol when you are asked to design your own experiment. Do not start your experiment until your teacher has approved your plan.

Science Safety Rules

Working in the laboratory can be an exciting experience, but it can also be dangerous if proper safety rules are not followed at all times. To prepare yourself for a safe year in the laboratory, read the following safety rules. Make sure that you understand each rule. Ask your teacher to explain any rules you don't understand.

Dress Code

1. Many materials in the laboratory can cause eye injury. To protect yourself from possible injury, wear safety goggles whenever you are working with chemicals, burners, or any substance that might get into your eyes. Avoid wearing contact lenses in the laboratory. Tell your teacher if you need to wear contact lenses to see clearly, and ask if there are any safety precautions you should observe.
2. Wear a laboratory apron or coat whenever you are working with chemicals or heated substances.
3. Tie back long hair to keep it away from any chemicals, burners, candles, or other laboratory equipment.
4. Before working in the laboratory, remove or tie back any article of clothing or jewelry that can hang down and touch chemicals and flames.

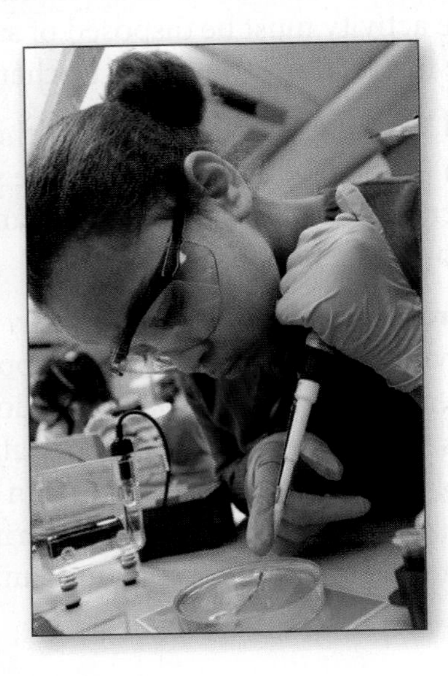

General Safety Rules and First Aid

5. Read all directions for an experiment several times. Follow the directions exactly as they are written. If you are in doubt about any part of the experiment, ask your teacher for assistance.
6. Never perform investigations your teacher has not authorized. Do not use any equipment unless your teacher is in the lab.
7. Never handle equipment unless you have specific permission.
8. Take care not to spill any material in the laboratory. If spills occur, ask your teacher immediately about the proper cleanup procedure. Never pour chemicals or other substances into the sink or trash container.
9. Never eat or drink in, or bring food into, the laboratory.
10. Immediately report all accidents, no matter how minor, to your teacher.
11. Learn what to do in case of specific accidents, such as getting acid in your eyes or on your skin. (Rinse acids off your body with lots of water.)
12. Be aware of the location of the first-aid kit. Your teacher should administer any required first aid due to injury. Your teacher may send you to the school nurse or call a physician.
13. Know where and how to report an accident or fire. Find out the location of the fire extinguisher, fire alarm, and phone. Report any fires to your teacher at once.

Heating and Fire Safety

14. Never use a heat source such as a candle or burner without wearing safety goggles.
15. Never heat a chemical you are not instructed to heat. A chemical that is harmless when cool can be dangerous when heated.
16. Maintain a clean work area and keep all materials away from flames. Be sure that there are no open containers of flammable liquids in the laboratory when flames are being used.
17. Never reach across a flame.

18. Make sure you know how to light a Bunsen burner. (Your teacher will demonstrate the proper procedure for lighting a burner.) If the flame leaps out of a burner toward you, turn the gas off immediately. Do not touch the burner. It may be hot. Never leave a lighted burner unattended!
19. When you are heating a test tube or bottle, point the opening away from yourself and others. Chemicals can splash or boil out of a heated test tube.
20. Never heat a closed container. The expanding hot air, vapors, or other gases inside may blow the container apart, causing it to injure you or others.
21. Never pick up a container that has been heated without first holding the back of your hand near it. If you can feel the heat on the back of your hand, the container may be too hot to handle. Use a clamp or tongs when handling hot containers or wear heat-resistant gloves if appropriate.

Using Chemicals Safely

22. Never mix chemicals for "the fun of it." You might produce a dangerous, possibly explosive substance.
23. Many chemicals are poisonous. Never touch, taste, or smell a chemical that you do not know for certain is harmless. If you are instructed to smell fumes in an experiment, gently wave your hand over the opening of the container and direct the fumes toward your nose. Do not inhale the fumes directly from the container.
24. Use only those chemicals needed in the investigation. Keep all container lids closed when a chemical is not being used. Notify your teacher whenever chemicals are spilled.
25. Dispose of all chemicals as instructed by your teacher. To avoid contamination, never return chemicals to their original containers.
26. Be extra careful when working with acids or bases. Pour such chemicals from one container to another over the sink, not over your work area.
27. When diluting an acid, pour the acid into water. Never pour water into the acid.
28. If any acids or bases get on your skin or clothing, rinse them with water. Immediately notify your teacher of any acid or base spill.

Using Glassware Safely

29. Never heat glassware that is not thoroughly dry. Use a wire screen to protect glassware from any flame.
30. Keep in mind that hot glassware will not appear hot. Never pick up glassware without first checking to see if it is hot.
31. Never use broken or chipped glassware. If glassware breaks, notify your teacher and dispose of the glassware in the proper trash container.
32. Never eat or drink from laboratory glassware. Thoroughly clean glassware before putting it away.

Using Sharp Instruments

33. Handle scalpels or razor blades with extreme care. Never cut material toward you; cut away from you.
34. Notify your teacher immediately if you cut yourself when in the laboratory.

Working With Live Organisms

35. No experiments that will cause pain, discomfort, or harm to animals should be done in the classroom or at home.
36. Your teacher will instruct you how to handle each species that is brought into the classroom. Animals should be handled only if necessary. Special handling is required if an animal is excited or frightened, pregnant, feeding, or with its young.
37. Clean your hands thoroughly after handling any organisms or materials, including animals or cages containing animals.

End-of-Experiment Rules

38. When an experiment is completed, clean up your work area and return all equipment to its proper place.
39. Wash your hands with soap and warm water before and after every experiment.
40. Turn off all burners before leaving the laboratory. Check that the gas line leading to the burner is off as well.

Use of the Microscope

The microscope used in most biology classes, the compound microscope, contains a combination of lenses. The eyepiece lens is located in the top portion of the microscope. This lens usually has a magnification of 10×. Other lenses, called objective lenses, are at the bottom of the body tube on the revolving nosepiece. By rotating the nosepiece, you can select the objective through which you will view your specimen.

The shortest objective is a low-power magnifier, usually 10×. The longer ones are of high power, usually up to 40× or 43×. The magnification is marked on the objective. To determine the total magnification, multiply the magnifying power of the eyepiece by the magnifying power of the objective. For example, with a 10× eyepiece and a 40× objective, the total magnification is 10 × 40 = 400×.

Learning the name, function, and location of each of the microscope's parts is necessary for proper use. Use the following procedures when working with the microscope.

1. Carry the microscope by placing one hand beneath the base and grasping the arm of the microscope with the other hand.
2. Gently place the microscope on the lab table with the arm facing you. The microscope's base should be resting evenly on the table, approximately 10 cm from the table's edge.
3. Raise the body tube by turning the coarse adjustment knob until the objective lens is about 2 cm above the opening of the stage.
4. Rotate the nosepiece so that the low-power objective (10×) is directly in line with the body tube. A click indicates that the lens is in line with the opening of the stage.
5. Look through the eyepiece and switch on the lamp or adjust the mirror so that a circle of light can be seen. This is the field of view. Moving the lever of the diaphragm permits a greater or smaller amount of light to come through the opening of the stage.
6. Place a prepared slide on the stage so that the specimen is over the center of the opening. Use the stage clips to hold the slide in place.
7. Look at the microscope from the side. Carefully turn the coarse adjustment knob to lower the body tube until the low-power objective almost touches the slide or until the body tube can no longer be moved. Do not allow the objective to touch the slide.
8. Look through the eyepiece and observe the specimen. If the field of view is out of focus, use the coarse adjustment knob to raise the body tube while looking through the eyepiece. **CAUTION:** *To prevent damage to the slide and the objective, do not lower the body tube using the coarse adjustment while looking through the the eyepiece.* Focus the image as best you can with the coarse adjustment knob. Then, use the fine adjustment knob to focus the image more sharply. Keep both eyes open when viewing a specimen. This helps prevent eyestrain.

1. **Eyepiece:** Contains a magnifying lens.
2. **Arm:** Supports the body tube.
3. **Low-power objective:** Provides a magnification of 10x.
4. **Stage:** Supports the slide being observed.
5. **Opening of the stage:** Permits light to pass up to the eyepiece.
6. **Fine adjustment knob:** Moves the body tube slightly to adjust the image.
7. **Coarse adjustment knob:** Moves the body tube to focus the image.
8. **Base:** Supports the microscope.
9. **Illuminator:** Produces light or reflects light up toward the eyepiece.
10. **Diaphragm:** Regulates the amount of light passing up toward the eyepiece.
11. **Stageclips:** Hold the slide in place.
12. **High-power objective:** Provides a magnification of 40x.
13. **Nosepiece:** Holds the objectives and can be rotated to change the magnification.
14. **Body tube:** Maintains the proper distance between the eyepiece and the objectives.

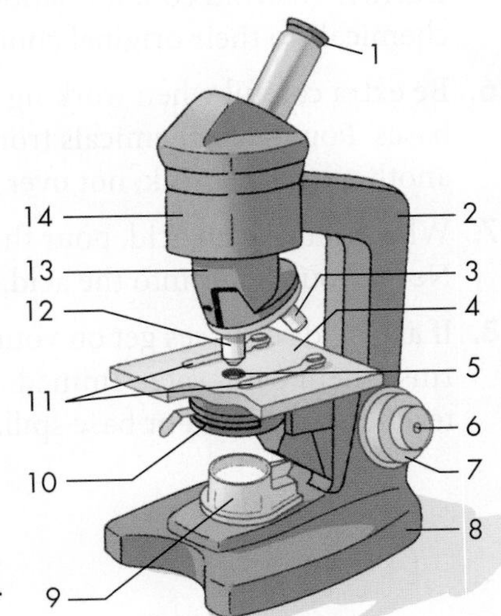

9. Adjust the lever of the diaphragm to allow the right amount of light to enter.
10. To change the magnification, rotate the nosepiece until the desired objective is in line with the body tube and clicks into place.
11. Look through the eyepiece and use the fine adjustment knob to bring the image into focus.
12. After every use, remove the slide. Return the low-power objective into place in line with the body tube. Clean the stage of the microscope and the lenses with lens paper. Do not use other types of paper to clean the lenses; they may scratch the lenses.

Preparing a Wet-Mount Slide

1. Obtain a clean microscope slide and a coverslip. A coverslip is very thin, permitting the objective lens to be lowered very close to the specimen.
2. Place the specimen in the middle of the microscope slide. The specimen must be thin enough for light to pass through it.
3. Using a dropper pipette, place a drop of water on the specimen.

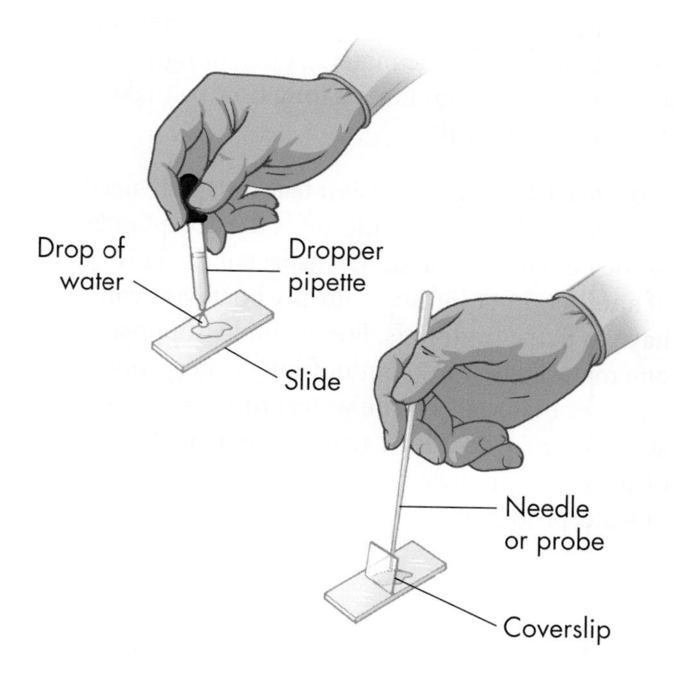

4. Lower one edge of the coverslip so that it touches the side of the drop of water at about a 45° angle. The water will spread evenly along the edge of the coverslip. Using a dissecting needle or probe, slowly lower the coverslip over the specimen and water as shown in the drawing. Try not to trap any air bubbles under the coverslip. If air bubbles are present, gently tap the surface of the coverslip over the air bubble with a pencil eraser.
5. Remove any excess water around the edge of the coverslip with a paper towel. If the specimen begins to dry out, add a drop of water at the edge of the coverslip.

Staining Techniques

1. Obtain a clean microscope slide and coverslip.
2. Place the specimen in the middle of the microscope slide.
3. Using a dropper pipette, place a drop of water on the specimen. Place the coverslip so that its edge touches the drop of water at a 45° angle. After the water spreads along the edge of the coverslip, use a dissecting needle or probe to lower the coverslip over the specimen.
4. Add a drop of stain at the edge of the coverslip. Using forceps, touch a small piece of lens paper or paper towel to the opposite edge of the coverslip, as shown in the drawing. The paper causes the stain to be drawn under the coverslip and to stain the cells in the specimen.

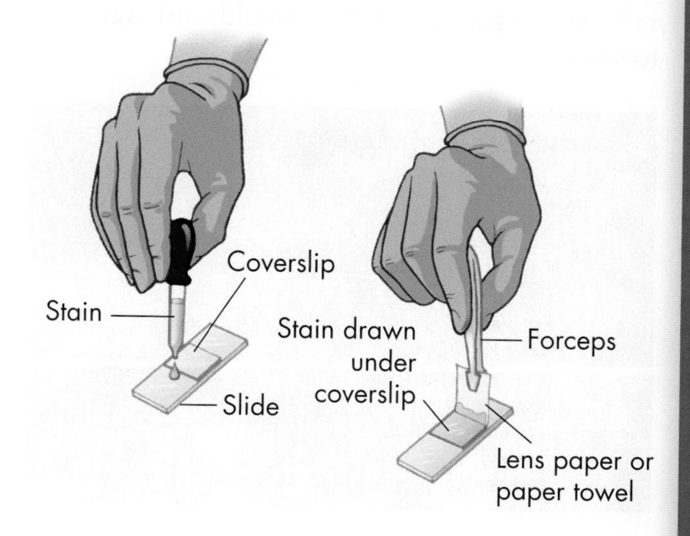

Appendix C Technology & Design

Engineers are people who use scientific and technological knowledge to solve practical problems. To design new products, engineers usually follow the process described here, even though they may not follow these steps in the exact order.

Identify a Need

Before engineers begin designing a new product, they must first identify the need they are trying to meet. For example, suppose you are a member of a design team in a company that makes toys. Your team has identified a need: a toy boat that is inexpensive and easy to assemble.

Research the Problem

Engineers often begin by gathering information that will help them with their new design. This research may include finding articles in books, in magazines, or on the Internet. It may also include talking to other engineers who have solved similar problems. Engineers also often perform experiments related to the product they want to design.

For your toy boat, you could look at toys that are similar to the one you want to design. You might do research on the Internet. You could also test some materials to see whether they would work well in a toy boat.

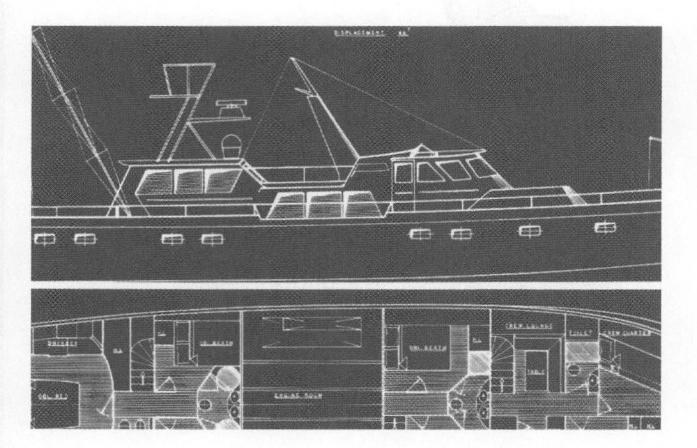

Design a Solution

Research gives engineers information that helps them design a product. When engineers design new products, they usually work in teams.

Generating Ideas Often, design teams hold brainstorming meetings in which any team member can contribute ideas. Brainstorming is a creative process in which one team member's suggestions can spark ideas in other group members. Brainstorming can lead to new approaches to solving a design problem.

Evaluating Constraints During brainstorming, a design team will often come up with several possible designs. The team must then evaluate each one.

As part of their evaluation, engineers consider constraints. Constraints are factors that limit or restrict a product design. Physical characteristics, such as the properties of materials used to make your toy boat, are constraints. Cost and time are also constraints. If the materials in a design cost a lot, or if the design takes a long time to make, it may be impractical.

Making Trade-offs Design teams usually need to make trade-offs. A trade-off is the acceptance of the benefits of one design aspect at the cost of another. In designing your toy boat, you will have to make trade-offs. For example, suppose one material is sturdy but not fully waterproof. Another material is more waterproof, but breakable. You may decide to give up the benefit of sturdiness in order to obtain the benefit of waterproofing.

Build and Evaluate a Prototype

Once the team has chosen a design plan, the engineers build a prototype of the product. A prototype is a working model used to test a design. Engineers evaluate the prototype to see whether it works well, is easy to operate, is safe to use, and holds up to repeated use.

Think of your toy boat. What would the prototype be like? Of what materials would it be made? How would you test it?

Troubleshoot and Redesign

Few prototypes work perfectly, which is why they need to be tested. Once a design team has tested a prototype, the members analyze the results and identify any problems. The team then tries to troubleshoot, or fix the weaknesses in the design. For example, if your toy boat leaks or wobbles, the boat should be redesigned to eliminate those problems.

Communicate the Solution

A team needs to communicate the final design to the people who will manufacture the product. To do this, teams may use sketches, detailed drawings, computer simulations, and written descriptions.

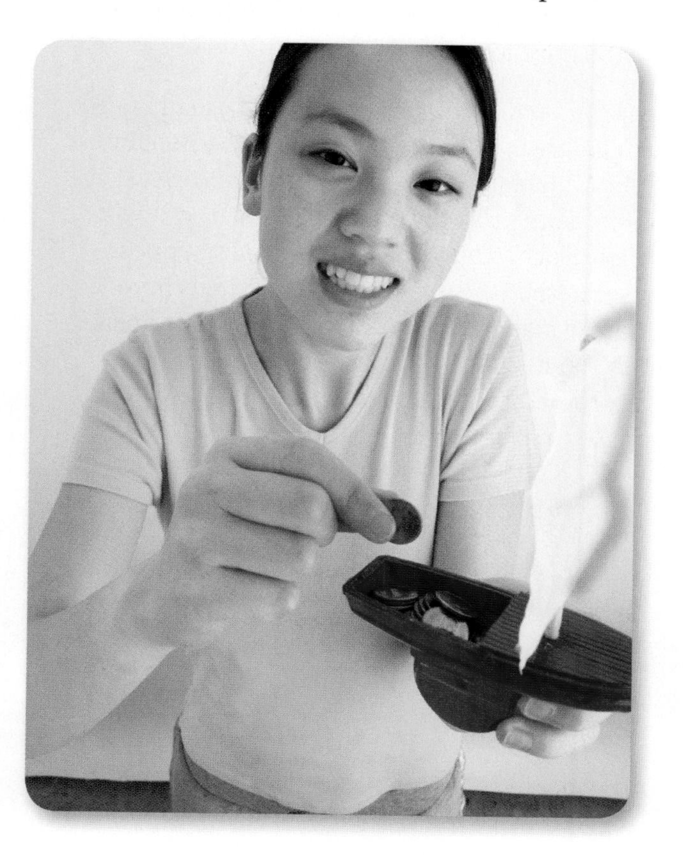

Activity

You can use the technology design process to design and build a toy boat.

Research and Investigate

1. Visit the library or go online to research toy boats.
2. Investigate how a toy boat can be powered, including wind, rubber bands, or baking soda and vinegar.
3. Brainstorm materials, shapes, and steering methods for your boat.

Design and Build

4. Based on your research, design a toy boat that
   - is made of readily available materials
   - is no larger than 15 cm long and 10 cm wide
   - includes a power system, a rudder, and a cargo area
   - travels 2 meters in a straight line while carrying a load of 20 pennies
5. Sketch your design and write a step-by-step plan for building your boat. After your teacher approves your plan, build your boat.

Evaluate and Redesign

6. Test your boat, evaluate the results, and identify any technological design problems in your boat.
7. Based on your evaluation, redesign your toy boat so it performs better.
8. As a class, compare the test results for each boat. Choose the model that best meets the needs of the toy company.

Appendix D Math Skills

Scientists use math to organize, analyze, and present data. This appendix will help you review some basic math skills.

Formulas and Equations

Formulas and equations are used in many areas of science. Both formulas and equations show the relationships between quantities. Any numerical sentence that contains at least one variable and at least one mathematical operator is called an equation. A formula is a type of equation that states the relationship between unknown quantities represented by variables.

For example, Speed = Distance ÷ Time is a formula, because no matter what values are inserted, speed is always equal to distance divided by time. The relationship between the variables does not change.

Example

Follow these steps to convert a temperature measurement of 50°F to Celsius.

1. Determine the formula that shows the relationship between these quantities.
   $°F = (9/5 \times °C) + 32°F$
2. Insert values you know into the formula.
   $50°F = (9/5 \times °C) + 32°F$
3. Solve the resulting equation.
   $50°F - 32°F = (9/5 \times °C)$
   $18°F = 9/5 \times °C$
   $18°F \times 5/9 = 10°C$

Applying Formulas and Equations

There are many applications of formulas in science. The example described below uses a formula to calculate density.

Example

Follow these steps to calculate the density of an object that has a mass of 45 g and a volume of 30 cm^3.

1. Determine the formula that shows the relationship between these quantities.
   Density = Mass/Volume
2. Insert values you know into the formula.
   Density = 45 g/30 cm^3
3. Solve the resulting equation.
   Density = 1.5 g/cm^3

Mean, Median, and Mode

The mean is the average, or the sum of the data divided by the number of data items. The middle number in a set of ordered data is called the median. The mode is the number that appears most often in a set of data.

Example

A scientist counted the number of distinct songs sung by seven different male birds and collected the data shown below.

Male Bird Songs

Bird	A	B	C	D	E	F	G
Number of Songs	36	29	40	35	28	36	27

To determine the mean number of songs, find the sum of the songs sung by all the male birds and divide by the number of male birds.

$$\text{Mean} = 231/7 = 33 \text{ songs}$$

To find the median number of songs, arrange the data items in numerical order and identify the number in the middle.

27 28 29 35 36 36 40

The number in the middle is 35, so the median number of songs is 35.

The mode is the value that appears most frequently. In the data, 36 appears twice, while every other item appears only once. Therefore, 36 is the mode.

Estimation

An estimate is a reasonable approximation of a numerical value. Estimates are made based on careful assumptions and known information.

Scientists use estimates in biology for two primary reasons: when an exact count or calculation cannot be made or is impractical to make, and to make reasonable approximations of answers that will be calculated or measured later.

One method for estimation used in biology is sampling. In sampling, the number of organisms in a small area (a sample) is multiplied to estimate the number of organisms in a larger area.

Example

Follow these steps to use sampling to estimate the total number of birds in the photo.

1. Count the birds in the highlighted area of the photo. In the highlighted area of the photo, there are 36 birds.
2. Determine the portion of the entire photo represented by the highlighted area. In this case, the highlighted area is 1/6 of the total area.
3. Calculate your estimate by multiplying the number of birds in the sample area by 6 (because the entire photo is 6 times as large as the sample area). A reasonable estimate of the total number of birds is 36×6, or 216 birds.

HINT: Estimates and calculated answers are rarely exactly the same. However, a large difference between an estimated answer and a calculated answer indicates there may be a problem with the estimate or calculation.

Using Measurements in Calculations

Density is an example of a value that is calculated using two measurements. Density represents the amount of mass in a particular volume of a substance. The units used for density are grams per milliliter (g/mL) or grams per cubic centimeter (g/cm^3). Density is calculated by dividing an object's mass by its volume.

Example

Follow these steps to calculate the density of an object.

1. Measure and record the mass of an object in grams.
2. Measure and record the volume of an object in mL or cm^3.
3. Use the following formula to calculate density:

$$\text{Density} = \text{Mass/Volume}$$

Effects of Measurement Errors

Density is calculated using two measured values. An error in the measurement of either mass or volume will result in the calculation of an incorrect density.

Example

A student measured the mass of an object as 2.5 g and its volume as 2.0 cm^3. The actual mass of the object is 3.5 g; the actual volume is 2.0 cm^3. What is the effect of the measurement error on the calculation of density?

Follow these steps to determine the effect of a measurement error on calculation.

1. Determine the density using the student's measurements.
   Density = Mass/Volume
   Density = 2.5 g/2.0 cm^3
   Density = 1.25 g/cm^3
2. Determine the density using the actual values.
   Density = Mass/Volume
   Density = 3.5 g/2.0 cm^3
   Density = 1.75 g/cm^3
3. Compare the calculated and the actual values.

In this case, a measurement of mass that was less than the actual value resulted in a calculated value for the density that was less than the actual density.

Accuracy

The accuracy of a measurement is its closeness to the actual value. Measurements that are accurate are close to the actual value.

Both clocks on this page show a time of 3:00. Suppose, though, that these clocks had not been changed to reflect daylight savings time. The time shown on the clocks would be inaccurate. On the other hand, if the actual time is 3:00, these clocks would be accurate.

Precision

Precision describes the exactness of a measurement. The clocks shown on this page differ in precision. The analog clock measures time to the nearest minute. The digital clock measures time to the nearest second. Time is measured more precisely by the digital clock than by the analog clock.

Comparing Accuracy and Precision

There is a difference between accuracy and precision. Measurements can be accurate (close to the actual value) but not precise. Measurements can also be precise but not accurate. When making scientific measurements, both accuracy and precision are important. Accurate and precise measurements result from the careful use of high-quality measuring tools.

Significant Figures

Significant figures are all of the digits that are known in a measurement, plus one additional digit, which is an estimate. In the figure below, the length of a turtle's shell is being measured using a centimeter ruler. The ruler has unnumbered divisions representing millimeters. In this case, two numbers can be determined exactly: the number of centimeters and the number of millimeters. One additional digit can be estimated. So, the measurement of this turtle's shell can be recorded with three significant figures as 8.80 centimeters.

Rules for Significant Digits
Follow these rules to determine the number of significant figures in a number.

All nonzero numbers are significant.

Example: 3217 has four significant digits.

Zeros are significant if
- They are between nonzero digits. Example: 509
- They follow a decimal point and a nonzero digit. Example: 7.00

Zeros are not significant if
- They follow nonzero digits in a number without a decimal. Example: 7000
- They precede nonzero digits in a number with a decimal. Example: 0.0098

Calculating With Significant Figures

When measurements are added or subtracted, the precision of the result is determined by the precision of the least-precise measurement. The result may need to be rounded so the number of digits after the decimal is the same as the least-precise measurement.

Example
Follow these steps to determine the correct number of significant figures when adding 4.51 g, 3.27 g, and 6.0 g.

1. Determine which measurement is reported with the least degree of precision. In this case, the least-precise measurement, 6.0 g, has one digit after the decimal point.
2. The result must be rounded so that it also has one digit after the decimal point. After rounding, the result of this calculation is 13.8 g.

When measurements are multiplied or divided, the answer must have the same number of significant figures as the measurement with the fewest number of significant figures.

Example
Follow these steps to determine the correct number of significant figures when multiplying 120 m by 6.32 m.

1. Determine the number of significant figures in each of the measurements. In this case, the measurement 120 m has two significant figures; the measurement 6.32 m has three significant figures.
2. The result must be rounded to have only two significant figures. After rounding, the result of this calculation is 760 m^2.

Scientific Notation

In science, measurements are often very large or very small. Using scientific notation makes these large and small numbers easier to work with.

Using scientific notation requires an understanding of exponents and bases. When a number is expressed as a base and an exponent, the base is the number that is used as a factor. The exponent tells how many times the base is multiplied by itself. For example, the number 25 can be expressed as a base and an exponent in the following way:

$$25 = 5 \times 5 = 5^2$$

In the example above, 5 is the base and 2 is the exponent. In scientific notation, the base is always the number 10. The exponent tells how many times the number 10 is multiplied by itself.

A number written in scientific notation is expressed as the product of two factors, a number between 1 and 10 and the number 10 with an exponent. For example, the number 51,000 can be expressed in scientific notation. To find the first factor, move the decimal to obtain a number between 1 and 10. In this case, the number is 5.1. The exponent can be determined by counting the number of places the decimal point was moved. The decimal point was moved four places to the left. So, 51,000 expressed in scientific notation is 5.1×10^4.

Numbers that are less than one can also be expressed in scientific notation. In the case of numbers less than one, the decimal point must be moved to the right to obtain a number between 1 and 10. For example, in the number 0.000098, the decimal point must move five places to the right to obtain the number 9.8. When the decimal point is moved to the right, the exponent is negative. So, 0.000098 expressed in scientific notation is 9.8×10^{-5}.

Calculating With Scientific Notation

Numbers expressed in scientific notation can be used in calculations. When adding or subtracting numbers expressed in scientific notation, the first factors must be rewritten so the exponents are the same.

Example

Follow these steps to add $(4.30 \times 10^4) + (2.1 \times 10^3)$.

1. Move the decimal point in one of the expressions so the exponents are the same.
   $(43.0 \times 10^3) + (2.1 \times 10^3)$
2. Add the first factors, keeping the value of the exponents the same.
   $(43.0 \times 10^3) + (2.1 \times 10^3) = 45.1 \times 10^3$
3. Move the decimal point so the first factor is expressed as the product of a number between and 1 and 10 and an exponent with base 10.
   $45.1 \times 10^3 = 4.51 \times 10^4$

When numbers expressed in scientific notation are multiplied, the exponents are added. When numbers expressed in scientific notation are divided, the exponents are subtracted.

Example

Use the following steps to determine the area of a rectangular field that has a length of 1.5×10^3 meters and a width of 3.2×10^2 meters.

1. Write down the expressions to be multiplied.
   $(1.5 \times 10^3 \text{ m})(3.2 \times 10^2 \text{ m})$
2. Multiply the first factors, add the exponents, and multiply any units.
   $= (1.5 \times 3.2)(10^{3+2}) \text{ m} \times \text{m}$
   $= 4.8 \times 10^5 \text{ m}^2$

Dimensional Analysis

Scientific problems and calculations often involve unit conversions, or changes from one unit to another. Dimensional analysis is a method of unit conversion.

Suppose you were counting a pile of pennies. If there were 197 pennies in the pile, how many dollars would the pennies be worth? To determine the answer, you need to know the conversion factor between pennies and dollars. A conversion factor simply shows how two units are related. In this case, the conversion factor is 100 pennies = 1 dollar. Determining that 197 pennies is equal to $1.97 is an example of a unit conversion.

In dimensional analysis, the conversion factor is usually expressed as a fraction. Remember that the two values in any conversion factor are equal to one another. So, the two values form a fraction with the value of 1. Look at the example below to see how dimensional analysis can be applied to an everyday problem.

Example

A student walked 1.5 kilometers as part of a school fitness program. How many meters did the student walk?

1. 1.5 km = _?_ m
2. 1 km = 1000 m
3. 1000 m/1 km
4. 1.5 km × 1000 m/1 km = 1500 m (cross out "km" in two places); 1.5 km = 1500 m

1. 1.5 km = ? m
2. 1 km = 1,000 m
3. $\frac{1{,}000 \text{ m}}{1 \text{ km}}$
4. $1.5 \not{\text{km}} \times \frac{1{,}000 \text{ m}}{1 \not{\text{km}}} = 1{,}500 \text{ m}$

1.5 km = 1,500 m

Applying Dimensional Analysis

There are many applications of dimensional analysis in science. The example below demonstrates the use of dimensional analysis to convert units.

Example

The average teenage girl needs about 2200 kilocalories of energy from food each day. How many calories is this equivalent to?

Use the following steps to convert kilocalories to calories.

1. Determine the conversion factor that relates the two units.
   1 kilocalorie = 1000 calories
2. Write the conversion factor in the form of a fraction.
   1000 calories/1 kilocalorie
3. Multiply the measurement by the conversion factor.
   2200 kilocalories × 1000 calories/1 kilocalorie = 2,200,000 calories

Appendix E Periodic Table

Periodic Table of the Elements

Representative Elements: Alkali metals; Alkaline earth metals; Other metals; Metalloids; Nonmetals; Noble gases

Transition Elements: Transition metals; Inner transition metals

C Solid; Br Liquid; He Gas; Tc Not found in nature

Key: 13 — Atomic number; 2 8 3 — Electrons in each energy level; Al — Element symbol; Aluminum — Element name; 26.982 — Atomic mass†

†*The atomic masses in parentheses are the mass numbers of the longest-lived isotope of elements for which a standard atomic mass cannot be defined.*

Each cell: atomic number, symbol, name, atomic mass (electrons in each energy level).

Period	1 1A	2 2A	3 3B	4 4B	5 5B	6 6B	7 7B	8	9 8B	10	11 1B	12 2B	13 3A	14 4A	15 5A	16 6A	17 7A	18 8A
1	1 H Hydrogen 1.0079 (1)																	2 He Helium 4.0026 (2)
2	3 Li Lithium 6.941 (2 1)	4 Be Beryllium 9.0122 (2 2)											5 B Boron 10.81 (2 3)	6 C Carbon 12.011 (2 4)	7 N Nitrogen 14.007 (2 5)	8 O Oxygen 15.999 (2 6)	9 F Fluorine 18.998 (2 7)	10 Ne Neon 20.179 (2 8)
3	11 Na Sodium 22.990 (2 8 1)	12 Mg Magnesium 24.305 (2 8 2)											13 Al Aluminum 26.982 (2 8 3)	14 Si Silicon 28.086 (2 8 4)	15 P Phosphorus 30.974 (2 8 5)	16 S Sulfur 32.06 (2 8 6)	17 Cl Chlorine 35.453 (2 8 7)	18 Ar Argon 39.948 (2 8 8)
4	19 K Potassium 39.098 (2 8 8 1)	20 Ca Calcium 40.08 (2 8 8 2)	21 Sc Scandium 44.956 (2 8 9 2)	22 Ti Titanium 47.90 (2 8 10 2)	23 V Vanadium 50.941 (2 8 11 2)	24 Cr Chromium 51.996 (2 8 13 1)	25 Mn Manganese 54.938 (2 8 13 2)	26 Fe Iron 55.847 (2 8 14 2)	27 Co Cobalt 58.933 (2 8 15 2)	28 Ni Nickel 58.71 (2 8 16 2)	29 Cu Copper 63.546 (2 8 18 1)	30 Zn Zinc 65.38 (2 8 18 2)	31 Ga Gallium 69.72 (2 8 18 3)	32 Ge Germanium 72.59 (2 8 18 4)	33 As Arsenic 74.922 (2 8 18 5)	34 Se Selenium 78.96 (2 8 18 6)	35 Br Bromine 79.904 (2 8 18 7)	36 Kr Krypton 83.80 (2 8 18 8)
5	37 Rb Rubidium 85.468 (2 8 18 8 1)	38 Sr Strontium 87.62 (2 8 18 8 2)	39 Y Yttrium 88.906 (2 8 18 9 2)	40 Zr Zirconium 91.22 (2 8 18 10 2)	41 Nb Niobium 92.906 (2 8 18 12 1)	42 Mo Molybdenum 95.94 (2 8 18 13 1)	43 Tc Technetium (98) (2 8 18 14 1)	44 Ru Ruthenium 101.07 (2 8 18 15 1)	45 Rh Rhodium 102.91 (2 8 18 16 1)	46 Pd Palladium 106.4 (2 8 18 18)	47 Ag Silver 107.87 (2 8 18 18 1)	48 Cd Cadmium 112.41 (2 8 18 18 2)	49 In Indium 114.82 (2 8 18 18 3)	50 Sn Tin 118.69 (2 8 18 18 4)	51 Sb Antimony 121.75 (2 8 18 18 5)	52 Te Tellurium 127.60 (2 8 18 18 6)	53 I Iodine 126.90 (2 8 18 18 7)	54 Xe Xenon 131.30 (2 8 18 18 8)
6	55 Cs Cesium 132.91 (2 8 18 18 8 1)	56 Ba Barium 137.33 (2 8 18 18 8 2)	71 Lu Lutetium 174.97 (2 8 18 32 9 2)	72 Hf Hafnium 178.49 (2 8 18 32 10 2)	73 Ta Tantalum 180.95 (2 8 18 32 11 2)	74 W Tungsten 183.85 (2 8 18 32 12 2)	75 Re Rhenium 186.21 (2 8 18 32 13 2)	76 Os Osmium 190.2 (2 8 18 32 14 2)	77 Ir Iridium 192.22 (2 8 18 32 15 2)	78 Pt Platinum 195.09 (2 8 18 32 17 1)	79 Au Gold 196.97 (2 8 18 32 18 1)	80 Hg Mercury 200.59 (2 8 18 32 18 2)	81 Tl Thallium 204.37 (2 8 18 32 18 3)	82 Pb Lead 207.2 (2 8 18 32 18 4)	83 Bi Bismuth 208.98 (2 8 18 32 18 5)	84 Po Polonium (209) (2 8 18 32 18 6)	85 At Astatine (210) (2 8 18 32 18 7)	86 Rn Radon (222) (2 8 18 32 18 8)
7	87 Fr Francium (223) (2 8 18 32 18 8 1)	88 Ra Radium (226) (2 8 18 32 18 8 2)	103 Lr Lawrencium (262) (2 8 18 32 32 9 2)	104 Rf Rutherfordium (261) (2 8 18 32 32 10 2)	105 Db Dubnium (262) (2 8 18 32 32 11 2)	106 Sg Seaborgium (263) (2 8 18 32 32 12 2)	107 Bh Bohrium (264) (2 8 18 32 32 13 2)	108 Hs Hassium (265) (2 8 18 32 32 14 2)	109 Mt Meitnerium (268) (2 8 18 32 32 15 2)	110 Ds Darmstadtium (269) (2 8 18 32 32 17 1)	111 Rg Roentgenium (272) (2 8 18 32 32 18 1)	112 Cn Copernicium (277) (2 8 18 32 32 18 2)	113 Nh Nihonium (286) (2 8 18 32 32 18 3)	114 Fl Flerovium (289) (2 8 18 32 32 18 4)	115 Mc Moscovium (289) (2 8 18 32 32 18 5)	116 Lv Livermorium (293) (2 8 18 32 32 18 6)	117 Ts Tennessine (294) (2 8 18 32 32 18 7)	118 Og Oganesson (294) (2 8 18 32 32 18 8)

Elements 104–118 are the transactinide elements.

Lanthanide Series

57 La Lanthanum 138.91 (2 8 18 18 9 2)	58 Ce Cerium 140.12 (2 8 18 20 8 2)	59 Pr Praseodymium 140.91 (2 8 18 21 8 2)	60 Nd Neodymium 144.24 (2 8 18 22 8 2)	61 Pm Promethium (145) (2 8 18 23 8 2)	62 Sm Samarium 150.4 (2 8 18 24 8 2)	63 Eu Europium 151.96 (2 8 18 25 8 2)	64 Gd Gadolinium 157.25 (2 8 18 25 9 2)	65 Tb Terbium 158.93 (2 8 18 27 8 2)	66 Dy Dysprosium 162.50 (2 8 18 28 8 2)	67 Ho Holmium 164.93 (2 8 18 29 8 2)	68 Er Erbium 167.26 (2 8 18 30 8 2)	69 Tm Thulium 168.93 (2 8 18 31 8 2)	70 Yb Ytterbium 173.04 (2 8 18 32 8 2)

Actinide Series

89 Ac Actinium (227) (2 8 18 32 18 9 2)	90 Th Thorium 232.04 (2 8 18 32 18 10 2)	91 Pa Protactinium 231.04 (2 8 18 32 20 9 2)	92 U Uranium 238.03 (2 8 18 32 21 9 2)	93 Np Neptunium (237) (2 8 18 32 22 9 2)	94 Pu Plutonium (244) (2 8 18 32 24 8 2)	95 Am Americium (243) (2 8 18 32 25 8 2)	96 Cm Curium (247) (2 8 18 32 25 9 2)	97 Bk Berkelium (247) (2 8 18 32 27 8 2)	98 Cf Californium (251) (2 8 18 32 28 8 2)	99 Es Einsteinium (252) (2 8 18 32 29 8 2)	100 Fm Fermium (257) (2 8 18 32 30 8 2)	101 Md Mendelevium (258) (2 8 18 32 31 8 2)	102 No Nobelium (259) (2 8 18 32 32 8 2)

Glossary

A

abiotic factor: physical, or nonliving, factor that shapes an ecosystem (66)
factor abiótico: factor físico, o inanimado, que da forma a un ecosistema

abscisic acid: plant hormone that inhibits cell division and, therefore, growth (711)
ácido abscísico: hormona vegetal que inhibe la división celular y, por ende, el crecimiento

acetylcholine: neurotransmitter that produces an impulse in a muscle cell (931)
acetilcolina: neurotransmisor que produce un impulso en una célula muscular

acid: compound that forms hydrogen ions (H^+) in solution; a solution with a pH of less than 7 (44)
ácido: compuesto que en una solución produce iones hidrógeno (H^+); una solución con un pH inferior a 7

acid rain: rain containing nitric and sulfuric acids (164)
lluvia ácida: lluvia que contiene ácido nítrico y ácido sulfúrico

actin: thin filament of protein found in muscles (930)
actina: microfilamento de proteína que se halla en los músculos

action potential: reversal of charges across the cell membrane of a neuron; also called a nerve impulse (898)
potencial de acción: inversión de las cargas a través de la membrana de una neurona; también llamado impulso nervioso

activation energy: energy that is needed to get a reaction started (51)
energía de activación: energía necesaria para que comience una reacción

active immunity: immunity that develops as a result of natural or deliberate exposure to an antigen (1020)
inmunidad activa: inmunidad que se desarrolla a consecuencia de la exposición natural o deliberada a un antígeno

adaptation: heritable characteristic that increases an organism's ability to survive and reproduce in an environment (461)
adaptación: característica heredable que aumenta la capacidad de un organismo de sobrevivir y reproducirse en un medio ambiente

adaptive radiation: process by which a single species or a small group of species evolves into several different forms that live in different ways (550)
radiación adaptativa: proceso mediante el cual una sola especie o un grupo pequeño de especies evoluciona y da lugar a diferentes seres que viven de diversas maneras

adenosine triphosphate (ATP): compound used by cells to store and release energy (226)
trifosfato de adenosina (ATP): compuesto utilizado por las células para almacenar y liberar energía

adhesion: force of attraction between different kinds of molecules (41, 686)
adhesión: fuerza de atracción entre diferentes tipos de moléculas

aerobic: process that requires oxygen (252)
aeróbico: proceso que requiere oxígeno

age structure: number of males and females of each age in a population (131)
estructura etaria: número de machos y de hembras de cada edad en una población

aggression: threatening behavior that one animal uses to exert dominance over another animal (848)
agresión: comportamiento amenazador que emplea un animal para ejercer control sobre otro animal

algal bloom: increase in the amount of algae and other producers that results from a large input of a limiting nutrient (611)
florecimiento de algas: aumento de la cantidad de algas y otros productores debido a una gran entrada de un nutriente limitante

allele: one of a number of different forms of a gene (310)
alelo: cada una de las diversas formas de un gen

allele frequency: number of times that an allele occurs in a gene pool compared with the number of alleles in that pool for the same gene (483)
frecuencia alélica: número de veces que aparece un alelo en un caudal genético, comparado con la cantidad de alelos en ese caudal para el mismo gen

allergy: overreaction of the immune system to an antigen (1024)
alergia: reacción exagerada del sistema inmune ante un antígeno

alternation of generations: life cycle that has two alternating phases—a haploid (N) phase and diploid (2N) phase (608, 637)
alternancia de generaciones: ciclo vital con dos fases que se alternan, una fase haploide (N) y una fase diploide (2N)

alveolus (pl. alveoli): one of many tiny air sacs at the end of a bronchiole in the lungs that provides surface area for gas exchange to occur (790, 964)
alvéolos: pequeños sacos, ubicados en las terminaciones de los bronquiolos pulmonares, que proporcionan una superficie en la que tiene lugar el intercambio gaseoso

amino acid: compound with an amino group on one end and a carboxyl group on the other end (48)
aminoácido: compuesto que contiene un grupo amino en un extremo y un grupo carboxilo en el otro extremo

Glossary *(continued)*

amniotic egg: egg composed of shell and membranes that creates a protected environment in which the embryo can develop out of water (825)
huevo amniota: huevo formado por una cáscara y membranas que crea un ambiente protegido en el cual el embrión puede desarrollarse en un medio seco

amylase: enzyme in saliva that breaks the chemical bonds in starches (876)
amilasa: enzima de la saliva que fragmenta los enlaces químicos de los almidones

anaerobic: process that does not require oxygen (252)
anaeróbico: proceso que no requiere oxígeno

analogous structures: body parts that share a common function, but not structure (469)
estructuras análogas: partes del cuerpo que tienen la misma función, mas no la misma estructura

anaphase: phase of mitosis in which the chromosomes separate and move to opposite ends of the cell (283)
anafase: fase de la mitosis en la cual los cromosomas se separan y se desplazan hacia los extremos opuestos de la célula

angiosperm: group of seed plants that bear their seeds within a layer of tissue that protects the seed; also called flowering plant (646)
angiospermas: grupo de plantas con semillas, que están protegidas con una capa de tejido. Se conocen también como plantas que florecen.

anther: flower structure in which pollen grains are produced (697)
antera: estructura de la flor en la cual se generan los granos de polen

antheridium (pl. antheridia): male reproductive structure in some plants that produces sperm (642)
anteridio: en algunas plantas, estructura reproductora masculina que produce esperma (anterozoides)

anthropoid: primate group made up of monkeys, apes, and humans (766)
antropoide: grupo de primates constituido por monos, simios y humanos

antibiotic: group of drugs used to block the growth and reproduction of bacterial pathogens (588)
antibiótico: grupo de drogas utilizadas para bloquear el desarrollo y la reproducción de organismos patógenos bacterianos

antibody: protein that either attacks antigens directly or produces antigen-binding proteins (1016)
anticuerpo: proteína que ataca directamente a los antígenos o produce proteínas que se unen a los antígenos

anticodon: group of three bases on a tRNA molecule that are complementary to the three bases of a codon of mRNA (369)
anticodón: grupo de tres bases en una molécula de ARN de transferencia que son complementarias a las tres bases de un codón de ARN mensajero

antigen: any substance that triggers an immune response (1016)
antígeno: cualquier sustancia que provoca una respuesta inmune

aphotic zone: dark layer of the oceans below the photic zone where sunlight does not penetrate (117)
zona afótica: sección oscura de los océanos donde no penetra la luz solar, situada debajo de la zona fótica

apical dominance: phenomenon in which the closer a bud is to the stem's tip, the more its growth is inhibited (710)
dominancia apical: fenómeno por el cual cuanto más cerca de la punta del tallo está un brote, más se inhibe su crecimiento

apical meristem: group of unspecialized cells that divide to produce increased length of stems and roots (668)
meristemo apical: grupo de células no especializadas que se dividen para producir un aumento en la longitud de tallos y raíces

apoptosis: process of programmed cell death (288)
apoptosis: proceso de muerte celular programada

appendage: structure, such as a leg or antenna that extends from the body wall (753)
apéndice: estructura, como una pierna o una antena, que se proyecta desde la superficie corporal

appendicular skeleton: the bones of the arms and legs along with the bones of the pelvis and shoulder area (922)
esqueleto apendicular: los huesos de los brazos y de las piernas junto con los huesos de la pelvis y del área de los hombros

aquaculture: raising of aquatic organisms for human consumption (176)
acuicultura: cría de organismos acuáticos para el consumo humano

aquaporin: water channel protein in a cell (210)
acuaporina: proteína que canaliza el agua en una célula

Archaea: domain consisting of unicellular prokaryotes that have cell walls that do not contain peptidoglycan; corresponds to the kingdom Archaebacteria (526)
Arqueas: dominio formado por procariotas unicelulares cuyas paredes celulares no contienen peptidoglicano; corresponden al reino de las Arqueabacterias

archegonium (pl. archegonia): structure in plants that produces egg cells (642)
arquegonio: estructura de las plantas que produce óvulos

artery: large blood vessel that carries blood away from the heart to the tissues of the body (952)
  arteria: vaso sanguíneo grande que transporta la sangre desde el corazón a los tejidos del cuerpo

artificial selection: selective breeding of plants and animals to promote the occurrence of desirable traits in offspring (458)
  selección artificial: cría selectiva de plantas y animales para fomentar la ocurrencia de rasgos deseados en la progenie

asexual reproduction: process of reproduction involving a single parent that results in offspring that are genetically identical to the parent (19, 277)
  reproducción asexual: proceso de reproducción que involucra a un único progenitor y da por resultado descendencia genéticamente idéntica a ese progenitor

asthma: chronic respiratory disease in which air passages narrow, causing wheezing, coughing, and difficulty breathing (1024)
  asma: enfermedad respiratoria crónica en la cual las vías respiratorias se estrechan, provocando jadeos, tos y dificultad para respirar

atherosclerosis: condition in which fatty deposits called plaque build up inside artery walls and eventually cause the arteries to stiffen (958)
  arteriosclerosis o ateroesclerosis: enfermedad en la cual se acumulan depósitos de grasa llamados placas en el interior de las paredes arteriales que, con el tiempo, causan un endurecimiento de las arterias

atom: the basic unit of matter (34)
  átomo: unidad básica de la materia

ATP synthase: cluster of proteins that span the cell membrane and allow hydrogen ions (H^+) to pass through it (237)
  ATP sintasa: complejo de proteínas unidas a la membrana celular que permiten el paso de los iones de hidrógeno (H^+) a través de ella

atrium (pl. atria): upper chamber of the heart that receives blood from the rest of the body (792, 949)
  aurícula: cavidad superior del corazón que recibe sangre del resto del cuerpo

autonomic nervous system: part of the peripheral nervous system that regulates activities that are involuntary, or not under conscious control; made up of the sympathetic and parasympathetic subdivisions (908)
  sistema nervioso autónomo: parte del sistema nervioso periférico que regula las actividades involuntarias, o que son independientes de la conciencia; está compuesto por las subdivisiones simpática y parasimpática

autosome: chromosome that is not a sex chromosome; also called autosomal chromosome (393)
  autosoma: cromosoma que no es un cromosoma sexual; también llamado cromosoma autosómico

autotroph: organism that is able to capture energy from sunlight or chemicals and use it to produce its own food from inorganic compounds; also called a producer (69, 228)
  autótrofo: organismo capaz de atrapar la energía de la luz solar o de las sustancias químicas y utilizarla para producir su propio alimento a partir de compuestos inorgánicos; también llamado productor

auxin: regulatory substance produced in the tip of a growing plant that stimulates cell elongation and the growth of new roots (709)
  auxina: sustancia reguladora producida en la punta de una planta en crecimiento que estimula el alargamiento celular y el crecimiento de raíces nuevas

axial skeleton: skeleton that supports the central axis of the body; consists of the skull, vertebral column, and the rib cage (922)
  esqueleto axial: esqueleto que sostiene al eje central del cuerpo; consiste en el cráneo, la columna vertebral y la caja torácica

axon: long fiber that carries impulses away from the cell body of a neuron (897)
  axón: fibra larga que lleva los impulsos desde el cuerpo celular de una neurona

B

bacillus (pl. bacilli): rod-shaped prokaryote (582)
  bacilo: procariota con forma de bastón

background extinction: extinction caused by slow and steady process of natural selection (548)
  extinción de fondo: extinción causada por un proceso lento y continuo de selección natural

Bacteria: domain of unicellular prokaryotes that have cell walls containing peptidoglycan; corresponds to the kingdom eubacteria (525)
  Bacteria: pertenece al dominio de los unicelulares procariota cuyas paredes celulares contienen peptidoglicano; corresponde al reino de las Eubacterias

bacteriophage: kind of virus that infects bacteria (340, 575)
  bacteriófago: clase de virus que infecta a las bacterias

bark: tissues that are found outside the vascular cambium, including the phloem, cork cambium, and cork (679)
  corteza: tejidos que se hallan fuera del cámbium vascular, incluidos el floema, el cámbium suberoso y el corcho

base: compound that produces hydroxide ions (OH^-) in solution; solution with a pH of more than 7 (44)
  base: compuesto que en una solución produce iones hidróxido (OH^-); una solución con un pH superior a 7

Glossary *(continued)*

base pairing: principle that bonds in DNA can form only between adenine and thymine and between guanine and cytosine (348)
 apareamiento de bases: principio que establece que los enlaces en el ADN sólo pueden formarse entre adenina y timina y entre guanina y citocina

behavior: manner in which an organism reacts to changes in its internal condition or external environment (840)
 comportamiento: manera en que un organismo reacciona a los cambios que ocurren en su condición interna o en el medio ambiente externo

behavioral isolation: form of reproductive isolation in which two populations develop differences in courtship rituals or other behaviors that prevent them from breeding (495)
 aislamiento conductual: forma de aislamiento reproductivo en la cual dos poblaciones desarrollan diferencias en sus rituales de cortejo o en otros comportamientos que evitan que se apareen

benthos: organisms that live attached to or near the bottom of lakes, streams, or oceans (117)
 bentos: organismos que viven adheridos al fondo, o cerca del fondo, de lagos, arroyos u océanos

bias: particular preference or point of view that is personal, rather than scientific (14)
 parcialidad: preferencia especial o punto de vista que es personal en lugar de ser científico

bilateral symmetry: body plan in which a single imaginary line can divide the body into left and right sides that are mirror images of each other (738)
 simetría bilateral: diseño corporal en el cual una línea imaginaria divide al cuerpo en dos lados, izquierdo y derecho, que son imágenes reflejas una del otra

binary fission: type of asexual reproduction in which an organism replicates its DNA and divides in half, producing two identical daughter cells (583)
 fisión binaria: tipo de reproducción asexual en la cual un organismo replica su ADN, se divide por la mitad y produce dos células hijas idénticas

binocular vision: ability to merge visual images from both eyes, providing depth perception and a three-dimensional view of the world (765)
 visión binocular: capacidad de fusionar las imágenes visuales provenientes de ambos ojos, lo cual proporciona una percepción profunda y una visión tridimensional del mundo

binomial nomenclature: classification system in which each species is assigned a two-part scientific name (512)
 nomenclatura binaria: sistema de clasificación en el cual a cada especie se le asigna un nombre científico que consta de dos partes

biodiversity: total of the variety of organisms in the biosphere; also called biological diversity (166)
 biodiversidad: totalidad de los distintos organismos que se hallan en la biósfera; también denominada diversidad biológica

biogeochemical cycle: process in which elements, chemical compounds, and other forms of matter are passed from one organism to another and from one part of the biosphere to another (79)
 ciclo biogeoquímico: proceso en el cual los elementos, los compuestos químicos y otras formas de materia pasan de un organismo a otro y de una parte de la biósfera a otra

biogeography: study of past and present distribution of organisms (465)
 biogeografía: estudio de la distribución pasada y presente de los organismos

bioinformatics: application of mathematics and computer science to store, retrieve, and analyze biological data (407)
 bioinformática: aplicación de las matemáticas y de la informática para almacenar, recuperar y analizar información biológica

biological magnification: increasing concentration of a harmful substance in organisms at higher trophic levels in a food chain or food web (161)
 bioacumulación: concentración creciente de sustancias perjudiciales en los organismos de los niveles tróficos más elevados de una cadena o red alimentaria

biology: scientific study of life (17)
 biología: estudio científico de la vida

biomass: total amount of living tissue within a given trophic level (78)
 biomasa: cantidad total de tejido vivo dentro de un nivel trófico dado

biome: a group of ecosystems that share similar climates and typical organisms (65)
 bioma: un grupo de ecosistemas que comparten climas similares y organismos típicos

biosphere: part of Earth in which life exists including land, water, and air or atmosphere (21, 64)
 biósfera: parte de la Tierra en la cual existe vida, y que incluye el suelo, el agua y el aire o atmósfera

biotechnology: process of manipulating organisms, cells, or molecules, to produce specific products (419)
 biotecnología: proceso de manipular organismos, células o moléculas con el fin de obtener productos específicos

biotic factor: any living part of the environment with which an organism might interact (66)
 factor biótico: cualquier parte viva del medio ambiente con la cual un organismo podría interaccionar

bipedal: term used to refer to two-foot locomotion (767)
 bípedo: término utilizado para referirse a la locomoción sobre dos pies

blade: thin, flattened part of a plant leaf (680)
lámina foliar o limbo: parte delgada y plana de la hoja de una planta

blastocyst: stage of early development in mammals that consists of a hollow ball of cells (294, 996)
blastocisto: etapa temprana del desarrollo de los mamíferos que consiste en una bola hueca formada por una capa de células

blastula: hollow ball of cells that develops when a zygote undergoes a series of cell divisions (739)
blástula: esfera hueca de células que se desarrolla cuando un cigoto atraviesa una serie de divisiones celulares

bone marrow: soft tissue found in bone cavities (924)
médula ósea: tejido blando que se halla en las cavidades de los huesos

bottleneck effect: a change in allele frequency following a dramatic reduction in the size of a population (490)
efecto cuello de botella: un cambio en la frecuencia alélica que resulta cuando el tamaño de una población reduce drásticamente

Bowman's capsule: cuplike structure that encases the glomerulus; collects filtrate from the blood (884)
cápsula de Bowman: estructura en forma de taza que encierra al glomérulo; recoge los filtrados provenientes de la sangre

brain stem: structure that connects the brain and spinal cord; includes the medulla oblongata and the pons (903)
tronco cerebral: estructura que conecta al cerebro con la médula espinal; incluye el bulbo raquídeo y el puente de Varolio

bronchus (pl. bronchi): one of two large tubes in the chest cavity that leads from the trachea to the lungs (964)
bronquio: cada uno de los dos conductos largos ubicados en la cavidad torácica que parten desde la tráquea y llegan a los pulmones

bryophyte: group of plants that have specialized reproductive organs but lack vascular tissue; includes mosses and their relatives (641)
briofitas: grupo de plantas que tienen órganos reproductores especializados pero carecen de tejido vascular; incluyen a los musgos y sus congéneres

bud: plant structure containing apical meristem tissue that can produce new stems and leaves (675)
yema o gema: estructura de las plantas que contiene tejido del meristemo apical y puede producir nuevos tallos y hojas

buffer: compound that prevents sharp, sudden changes in pH (44)
solución amortiguadora: compuesto que evita cambios bruscos y repentinos en el pH

C

calcitonin: hormone produced by the thyroid that reduces blood calcium levels (985)
calcitonina: hormona producida por la tiroides que reduce los niveles de calcio en la sangre

Calorie: measure of heat energy in food; equivalent to 1000 calories (868)
Caloría: medida de la energía térmica de los alimentos, equivalente a 1000 calorías

calorie: amount of energy needed to raise the temperature of 1 gram of water by 1 degree Celsius (250)
caloría: cantidad de energía necesaria para elevar la temperatura de 1 gramo de agua en 1 grado Celsius

Calvin cycle: light-independent reactions of photosynthesis in which energy from ATP and NADPH is used to build high-energy compounds such as sugar (238)
ciclo de Calvin: reacciones de la fotosíntesis independientes de la luz en las cuales se utiliza la energía del ATP y del NADPH para elaborar compuestos con alto contenido energético, como el azúcar

cancer: disorder in which some of the body's cells lose the ability to control growth (289)
cáncer: enfermedad en la cual algunas de las células del cuerpo pierden la capacidad de controlar su crecimiento

canopy: dense covering formed by the leafy tops of tall rain forest trees (112)
dosel forestal: cubierta densa formada por las copas de los árboles altos del bosque tropical

capillary: smallest blood vessel; brings nutrients and oxygen to the tissues and absorbs carbon dioxide and waste products (952)
capilar: más pequeño de los vaso sanguíneo más pequeño; lleva nutrientes y oxígeno a los tejidos y absorbe dióxido de carbono y productos de desecho

capillary action: tendency of water to rise in a thin tube (686)
capilaridad: tendencia del agua a ascender en un tubo delgado

capsid: protein coat surrounding a virus (575)
cápsida: cobertura de proteínas que rodea a un virus

carbohydrate: compound made up of carbon, hydrogen, and oxygen atoms; type of nutrient that is the major source of energy for the body (46, 869)
hidrato de carbono: compuesto formado por átomos de carbono, hidrógeno y oxígeno; tipo de nutriente que es la fuente principal de energía para el cuerpo

carnivore: organism that obtains energy by eating animals (71)
carnívoro: organismo que obtiene energía al comer otros animales

Glossary *(continued)*

carpel: innermost part of a flower that produces and shelters the female gametophytes (697)
carpelo: parte interna de una flor que produce y alberga los gametofitos femeninos

carrying capacity: largest number of individuals of a particular species that a particular environment can support (135)
capacidad de carga: mayor cantidad de individuos de una especie en particular que un medio ambiente específico puede mantener

cartilage: type of connective tissue that supports the body and is softer and more flexible than bone (757, 924)
cartílago: tipo de tejido conectivo que sostiene al cuerpo y es más blando y flexible que el hueso

Casparian strip: waterproof strip that surrounds plant endodermal cells and is involved in the one-way passage of materials into the vascular cylinder in plant roots (672)
banda de Caspary: banda impermeable que rodea a las células endodérmicas de las plantas y participa en el transporte unidireccional de las sustancias hacia el interior del cilindro vascular de las raíces de las plantas

catalyst: substance that speeds up the rate of a chemical reaction (52)
catalizador: sustancia que acelera la velocidad de una reacción química

cell: basic unit of all forms of life (191)
célula: unidad básica de todas las formas de vida

cell body: largest part of a typical neuron; contains the nucleus and much of the cytoplasm (897)
cuerpo celular: parte más grande de una neurona típica; que contiene el núcleo y gran parte del citoplasma

cell cycle: series of events in which a cell grows, prepares for division, and divides to form two daughter cells (280)
ciclo celular: serie de sucesos en los cuales una célula crece, se prepara para dividirse y se divide para formar dos células hijas

cell division: process by which a cell divides into two new daughter cells (276)
división celular: proceso por el cual una célula se divide en dos células hijas nuevas

cell membrane: thin, flexible barrier that surrounds all cells; regulates what enters and leaves the cell (193)
membrana celular: barrera flexible y delgada que rodea a todas las células; regula lo que entra y sale de la célula

cell theory: fundamental concept of biology that states that all living things are composed of cells; that cells are the basic units of structure and function in living things; and that new cells are produced from existing cells (191)
teoría celular: concepto fundamental de la Biología que establece que todos los seres vivos están compuestos por células; que las células son las unidades básicas estructurales y funcionales de los seres vivos; y que las células nuevas se producen a partir de células existentes

cell wall: strong, supporting layer around the cell membrane in some cells (203)
pared celular: capa resistente que sirve de sostén y está situada alrededor de la membrana celular de algunas células

cell-mediated immunity: immune response that defends the body against viruses, fungi, and abnormal cancer cells inside living cells (1019)
inmunidad celular: respuesta inmune que desde las células defiende al cuerpo contra virus, hongos y células anormales cancerígenas

cellular respiration: process that releases energy by breaking down glucose and other food molecules in the presence of oxygen (281)
respiración celular: proceso que libera energía al descomponer la glucosa y otras moléculas de los alimentos en presencia de oxígeno

central nervous system: includes the brain and spinal cord; processes information and creates a response that it delivers to the body (896)
sistema nervioso central: incluye el cerebro y la médula espinal; procesa información y genera una respuesta que es enviada al cuerpo

centriole: structure in an animal cell that helps to organize cell division (199, 282)
centríolo: estructura de una célula animal que contribuye a organizar la división celular

centromere: region of a chromosome where the two sister chromatids attach (282)
centrómero: región de un cromosoma donde se unen las dos cromátidas hermanas

cephalization: concentration of sense organs and nerve cells at the anterior end of an animal (740)
cefalización: concentración de órganos sensoriales y células nerviosas en el extremo anterior de un animal

cerebellum: part of the brain that coordinates movement and controls balance (811, 903)
cerebelo: parte del encéfalo que coordina el movimiento y controla el equilibrio

cerebral cortex: outer layer of the cerebrum of a mammal's brain; center of thinking and other complex behaviors (902)
corteza cerebral: capa externa del cerebro de un mamífero; centro del raciocinio y otros comportamientos complejos

cerebrum: part of the brain responsible for voluntary activities of the body; "thinking" region of the brain (811, 902)
cerebro: parte del encéfalo responsable de las actividades voluntarias del cuerpo; región "pensante" del encéfalo

chemical digestion: process by which enzymes break down food into small molecules that the body can use (875)
digestión química: proceso por el cual las enzimas descomponen los alimentos en moléculas pequeñas que el cuerpo puede utilizar

chemical reaction: process that changes, or transforms, one set of chemicals into another set of chemicals (50)
reacción química: proceso que cambia, o transforma, un grupo de sustancias químicas en otro grupo de sustancias químicas

chemosynthesis: process in which chemical energy is used to produce carbohydrates (70)
quimiosíntesis: proceso en el cual la energía química se utiliza para producir hidratos de carbono

chitin: complex carbohydrate that makes up the cell walls of fungi; also found in the external skeletons of arthropods (618)
quitina: hidrato de carbono complejo que forma las paredes celulares de los hongos; también se halla en los esqueletos externos de los artrópodos

chlorophyll: principal pigment of plants and other photosynthetic organisms (230)
clorofila: pigmento fundamental de las plantas y de otros organismos fotosintéticos

chloroplast: organelle found in cells of plants and some other organisms that captures the energy from sunlight and converts it into chemical energy (202)
cloroplasto: orgánulo de las células de las plantas y de otros organismos que captura la energía de la luz solar y la convierte en energía química

chordate: animal that has, for at least one stage of its life, a dorsal, hollow nerve cord, a notochord, a tail that extends beyond the anus, and pharyngeal pouches (731)
cordado: animal que, al menos durante una etapa de su vida, tiene un cordón nervioso hueco y dorsal, un notocordio, una cola que se prolonga más allá del ano y bolsas faríngeas

chromatid: one of two identical "sister" parts of a duplicated chromosome (282)
cromátida: una de las dos partes "hermanas" idénticas de un cromosoma duplicado

chromatin: substance found in eukaryotic chromosomes that consists of DNA tightly coiled around histones (280)
cromatina: sustancia que se halla en los cromosomas eucarióticos y que consiste en ADN enrollado apretadamente alrededor de las histonas

chromosome: threadlike structure of DNA and protein that contains genetic information; in eukaryotes, chromosomes are found in the nucleus; in prokaryotes, they are found in the cytoplasm (279)
cromosoma: estructura larga de ADN y proteína, con forma de hilo, que posee información genética; en los eucariotas, los cromosomas están dentro del núcleo; en los procariotas, los cromosomas están en el citoplasma

chyme: mixture of enzymes and partially-digested food (877)
quimo: mezcla de enzimas y alimentos parcialmente digeridos

cilium (pl. cilia): short hairlike projection that produces movement (607)
cilio: pequeña prolongación parecida a un pelo que produce movimiento

circadian rhythm: behavioral cycles that occur daily (847)
ritmo circadiano: ciclos conductuales que ocurren diariamente

clade: evolutionary branch of a cladogram that includes a single ancestor and all its descendants (516)
clado: rama evolutiva de un cladograma que incluye a un único ancestro y a todos sus descendientes

cladogram: diagram depicting patterns of shared characteristics among species (517)
cladograma: diagrama que representa patrones de características compartidas entre especies

class: in classification, a group of closely related orders (514)
clase: en la clasificación, un grupo de varios órdenes relacionados estrechamente

classical conditioning: type of learning that occurs when an animal makes a mental connection between a stimulus and some kind of reward or punishment (843)
condicionamiento clásico: tipo de aprendizaje que ocurre cuando un animal realiza una conexión mental entre un estímulo y algún tipo de recompensa o castigo

climate: average year-to-year conditions of temperature and precipitation in an area over a long period of time (96)
clima: promedio anual de las condiciones de temperatura y precipitación en un área durante un largo período de tiempo

clone: member of a population of genetically identical cells produced from a single cell (427)
clon: miembro de una población de células genéticamente idénticas producidas a partir de una célula única

closed circulatory system: type of circulatory system in which blood circulates entirely within blood vessels that extend throughout the body (792)
sistema circulatorio cerrado: tipo de sistema circulatorio en el cual la sangre circula completamente dentro de los vasos sanguíneos que se extienden por todo el cuerpo

coccus (pl. cocci): spherical prokaryote (582)
coco: procariota de forma esférica

cochlea: fluid-filled part of inner ear; contains nerve cells that detect sound (911)
cóclea: parte del oído interno llena de fluidos; contiene las células nerviosas que detectan el sonido

codominance: situation in which the phenotypes produced by both alleles are completely expressed (319)
codominancia: situación en la cual los fenotipos producidos por ambos alelos están expresados completamente

Glossary *(continued)*

codon: group of three nucleotide bases in mRNA that specify a particular amino acid to be incorporated into a protein (366)
codón: grupo de tres bases de nucleótidos en el RNA mensajero que especifican la incorporación de un aminoácido en particular en una proteína

coelom: body cavity lined with mesoderm (738)
celoma: cavidad corporal revestida de mesodermo

coevolution: process by which two species evolve in response to changes in each other over time (551)
coevolución: proceso por el cual dos especies evolucionan en respuesta a cambios mutuos en el transcurso del tiempo

cohesion: attraction between molecules of the same substance (41)
cohesión: atracción entre moléculas de la misma sustancia

collenchyma: in plants, type of ground tissue that has strong, flexible cell walls; helps support larger plants (667)
colénquima: en las plantas, tipo de tejido fundamental que tiene paredes celulares fuertes y flexibles; contribuye a sostener las plantas más grandes

commensalism: symbiotic relationship in which one organism benefits and the other is neither helped nor harmed (104)
comensalismo: relación simbiótica en la cual un organismo se beneficia y el otro ni se beneficia ni sufre daño

communication: passing of information from one organism to another (850)
comunicación: traspaso de información desde un organismo a otro

community: assemblage of different populations that live together in a defined area (64)
comunidad: conjunto de varias poblaciones que viven juntas en un área definida

companion cell: in plants, phloem cell that surrounds sieve tube elements (666)
célula anexa: en las plantas, célula del floema que rodea a los vasos cribosos

competitive exclusion principle: principle that states that no two species can occupy the same niche in the same habitat at the same time (101)
principio de exclusión competitiva: principio que afirma que dos especies no pueden ocupar el mismo nicho en el mismo hábitat al mismo tiempo

compound: substance formed by the chemical combination of two or more elements in definite proportions (36)
compuesto: sustancia formada por la combinación química de dos o más elementos en proporciones definidas

cone: in the eye, photoreceptor that responds to light of different colors, producing color vision (913)
cono: en el ojo, receptor de luz que responde a la luz de diferentes colores, produciendo la visión a color

coniferous: term used to refer to trees that produce seed-bearing cones and have thin leaves shaped like needles (114)
coníferas: término utilizado para referirse a los árboles que producen conos portadores de semillas y que tienen hojas delgadas con forma de aguja

conjugation: process in which paramecia and some prokaryotes exchange genetic information (583, 608)
conjugación: proceso mediante el cual los paramecios y algunos procariotas intercambian información genética

connective tissue: type of tissue that provides support for the body and connects its parts (863)
tejido conectivo: tipo de tejido que proporciona sostén al cuerpo y conecta sus partes

consumer: organism that relies on other organisms for its energy and food supply; also called a heterotroph (71)
consumidor: organismo que depende de otros organismos para obtener su energía y su provisión de alimentos; también llamado heterótrofo

control group: group in an experiment that is exposed to the same conditions as the experimental group except for one independent variable (7)
grupo de control: en un experimento, grupo que está expuesto a las mismas condiciones que el grupo experimental, excepto por una variable independiente

controlled experiment: experiment in which only one variable is changed (7)
experimento controlado: experimento en el cual sólo se cambia una variable

convergent evolution: process by which unrelated organisms independently evolve similarities when adapting to similar environments (551)
evolución convergente: proceso mediante el cual organismos no relacionados evolucionan independientemente hacia caracteres similares cuando se adaptan a ambientes parecidos

cork cambium: meristematic tissue that produces the outer covering of stems during secondary growth of a plant (677)
cámbium suberoso: tejido del meristemo que produce la cubierta exterior de los tallos durante el crecimiento secundario de una planta

cornea: tough transparent layer of the eye through which light enters (912)
córnea: membrana dura y transparente del ojo a través de la cual entra la luz

corpus luteum: name given to a follicle after ovulation because of its yellow color (993)
cuerpo lúteo: nombre dado a un folículo después de la ovulación debido a su color amarillo

cortex: in plants, region of ground tissue just inside the root through which water and minerals move (670)
corteza radicular: en las plantas, región de tejido fundamental situada en el interior de la raíz a través de la cual pasan el agua y los minerales

corticosteroid: steroid hormone produced by the adrenal cortex (983)
corticosteroide o corticoide: hormona esteroídica producida por la corteza de las glándulas adrenales

cotyledon: first leaf or first pair of leaves produced by the embryo of a seed plant (652)
cotiledón: primera hoja o primer par de hojas producidas por el embrión de una planta fanerógama

courtship: type of behavior in which an animal sends out stimuli in order to attract a member of the opposite sex (848)
cortejo: tipo de comportamiento en el cual un animal emite estímulos para atraer a un miembro del sexo opuesto

covalent bond: type of bond between atoms in which the electrons are shared (37)
enlace covalente: tipo de enlace entre átomos en el cual se comparten los electrones

crossing-over: process in which homologous chromosomes exchange portions of their chromatids during meiosis (324)
entrecruzamiento: proceso por el cual los cromosomas homólogos intercambian partes de sus cromátidas durante la meiosis

cyclin: one of a family of proteins that regulates the cell cycle in eukaryotic cells (286)
ciclina: un componente de la familia de proteínas que regulan el ciclo celular de las células eucariotas

cytokinesis: division of the cytoplasm to form two separate daughter cells (282)
citocinesis: división del citoplasma para formar dos células hijas separadas

cytokinin: plant hormone produced in growing roots and in developing fruits and seeds (710)
citoquinina: hormona vegetal que se genera en las raíces en crecimiento y en los frutos y semillas en desarrollo

cytoplasm: in eukaryotic cells, all cellular contents outside the nucleus; in prokaryotic cells, all of the cells' contents (196)
citoplasma: en una célula eucariota, todo el contenido celular fuera del núcleo; en las células procariotas, todo el contenido de las células

cytoskeleton: network of protein filaments in a eukaryotic cell that gives the cell its shape and internal organization and is involved in movement (199)
citoesqueleto: en una célula eucariota, red de filamentos proteínicos que otorga a la célula su forma y su organización interna y participa en el movimiento

D

data: evidence; information gathered from observations (8)
datos: evidencia; información reunida a partir de observaciones

deciduous: term used to refer to a type of tree that sheds its leaves during a particular season each year (112)
caduco: término utilizado para referirse a un tipo de árbol que pierde sus hojas cada año durante una estación en particular

decomposer: organism that breaks down and obtains energy from dead organic matter (71)
descomponedor: organismo que descompone y obtiene energía de la materia orgánica muerta

deforestation: destruction of forests (159)
deforestación: destrucción de los bosques

demographic transition: change in a population from high birth and death rates to low birth and death rates (144)
transición demográfica: en una población, cambio de índices de nacimiento y mortalidad altos a índices de nacimiento y mortalidad bajos

demography: scientific study of human populations (143)
demografía: estudio científico de las poblaciones humanas

dendrite: extension of the cell body of a neuron that carries impulses from the environment or from other neurons toward the cell body (897)
dendrita: prolongación del cuerpo celular de una neurona que transporta impulsos desde el medio ambiente o desde otras neuronas hacia el cuerpo celular

denitrification: process by which bacteria convert nitrates into nitrogen gas (84)
desnitrificación: proceso por el cual las bacterias del suelo convierten los nitratos en gas nitrógeno

density-dependent limiting factor: limiting factor that depends on population density (138)
factor limitante dependiente de la densidad: factor limitante que depende de la densidad de la población

density-independent limiting factor: limiting factor that affects all populations in similar ways, regardless of the population density (140)
factor limitante independiente de la densidad: factor limitante que afecta a todas las poblaciones de manera similar, sin importar la densidad de la población

deoxyribonucleic acid (DNA): genetic material that organisms inherit from their parents (18)
ácido desoxirribonucleico (ADN): material genético que los organismos heredan de sus padres

dependent variable: variable that is observed and that changes in response to the independent variable; also called the responding variable (7)
variable dependiente: variable que está siendo observada y cambia en respuesta a la variable independiente; también llamada variable de respuesta

derived character: trait that appears in recent parts of a lineage, but not in its older members (518)
carácter derivado: rasgo que aparece en los descendientes recientes de un linaje, pero no en sus miembros más viejos

dermis: layer of skin found beneath the epidermis (937)
dermis: capa de la piel situada debajo de la epidermis

desertification: lower land productivity caused by overfarming, overgrazing, seasonal drought, and climate change (159)
desertificación: disminución de la productividad de la tierra debido al cultivo y al pastoreo excesivo, a la sequía estacional y al cambio climático

detritivore: organism that feeds on plant and animal remains and other dead matter (71)
detritívoro: organismo que se alimenta de restos animales y vegetales y demás materia orgánica muerta

deuterostome: group of animals in which the blastopore becomes an anus, and the mouth is formed from the second opening that develops (739)
deuteróstomos: grupo de animales en los cuales el blastoporo se convierte en ano y la boca se forma a partir del desarrollo de una segunda abertura

diaphragm: large flat muscle at the bottom of the chest cavity that helps with breathing (967)
diafragma: músculo plano y grande ubicado en la parte inferior de la cavidad torácica que participa en la respiración

dicot: angiosperm with two seed leaves in its ovary (652)
dicotiledónea: angiosperma con dos cotiledones (hojas embrionarias) en su ovario

differentiation: process in which cells become specialized in structure and function (293, 381)
diferenciación: proceso en el cual las células se especializan en estructura y función

diffusion: process by which particles tend to move from an area where they are more concentrated to an area where they are less concentrated (208)
difusión: proceso por el cual las partículas tienden a desplazarse desde un área donde están más concentradas hacia un área donde están menos concentradas

digestive tract: tube that begins at the mouth and ends at the anus (784)
tracto digestivo: tubo que comienza en la boca y termina en el ano

diploid: term used to refer to a cell that contains two sets of homologous chromosomes (323)
diploide: término utilizado para referirse a una célula que contiene dos series de cromosomas homólogos

directional selection: form of natural selection in which individuals at one end of a distribution curve have higher fitness than individuals in the middle or at the other end of the curve (489)
selección direccional: forma de selección natural en la cual los individuos que se hallan en un extremo de la curva de distribución poseen una mayor capacidad de adaptación que los individuos que se hallan en el centro o en el otro extremo de la curva

disruptive selection: natural selection in which individuals at the upper and lower ends of the curve have higher fitness than individuals near the middle of the curve (489)
selección disruptiva: forma de selección natural en la cual los individuos que se hallan en los extremos superior e inferior de la curva poseen una mayor capacidad de adaptación que los individuos que se hallan cerca del centro de la curva

DNA fingerprinting: tool used by biologists that analyzes an individual's unique collection of DNA restriction fragments; used to determine whether two samples of genetic material are from the same person (433)
prueba de ADN: herramienta utilizada por los biólogos mediante la cual se analiza el conjunto de los fragmentos de restricción de ADN exclusivo de cada individuo; utilizada para determinar si dos muestras de material genético pertenecen a la misma persona; también llamada huella genética o análisis de ADN

DNA microarray: glass slide or silicon chip that carries thousands of different kinds of single-stranded DNA fragments arranged in a grid. A DNA microarray is used to detect and measure the expression of thousands of genes at one time (432)
chip de ADN: superficie de vidrio o chip de silicona que contiene miles de diferentes tipos de fragmentos de ADN de una sola cadena dispuestos en una cuadrícula. Un chip de ADN se utiliza para detectar y medir la expresión de miles de genes a la vez

DNA polymerase: principal enzyme involved in DNA replication (351)
ADN polimerasa: enzima fundamental involucrada en la replicación del ADN

domain: larger, more inclusive taxonomic category than a kingdom (525)
dominio: categoría taxonómica más amplia e inclusiva que un reino

dopamine: neurotransmitter that is associated with the brain's pleasure and reward centers (904)
dopamina: neurotransmisor que está asociado con los centros de placer y de recompensa del cerebro

dormancy: period of time during which a plant embryo is alive but not growing (706)
latencia: período de tiempo durante el cual un embrión vegetal está vivo pero no crece

double fertilization: process of fertilization in angiosperms in which the first event produces the zygote, and the second, the endosperm within the seed (700)
doble fertilización: proceso de fecundación de las angiospermas en el cual se produce, en el primer suceso el cigoto y en el segundo, el endospermo dentro de la semilla

E

ecological footprint: total amount of functioning ecosystem needed both to provide the resources a human population uses and to absorb the wastes that population generates (173)
huella ecológica: cantidad total de ecosistema en funcionamiento necesaria para proporcionar los recursos que utiliza una población humana y para absorber los residuos que genera esa población

ecological hot spot: small geographic area where significant numbers of habitats and species are in immediate danger of extinction (171)
zona de conflicto ecológico: área geográfica pequeña donde cantidades importantes de hábitats y especies se hallan en peligro de extinción inmediato

ecological pyramid: illustration of the relative amounts of energy or matter contained within each trophic level in a given food chain or food web (77)
pirámide ecológica: ilustración de las cantidades relativas de energía o materia contenidas dentro de cada nivel trófico en una cadena o red alimenticia dada

ecological succession: series of gradual changes that occur in a community following a disturbance (106)
sucesión ecológica: serie de cambios graduales que ocurren en una comunidad después de una alteración

ecology: scientific study of interactions among organisms and between organisms and their environment (65)
ecología: estudio científico de las interacciones entre organismos y entre los organismos y su medio ambiente

ecosystem: all the organisms that live in a place, together with their nonliving environment (65)
ecosistema: todos los organismos que viven en un lugar, junto con su medio ambiente inanimado

ecosystem diversity: variety of habitats, communities, and ecological processes in the biosphere (166)
diversidad de ecosistemas: variedad de hábitats, comunidades y procesos ecológicos que existen en la biósfera

ectoderm: outermost germ layer; produces sense organs, nerves, and outer layer of skin (738)
ectodermo: capa embrionaria más externa; desarrolla órganos sensoriales, nervios y la capa exterior de la piel

ectotherm: animal whose body temperature is determined by the temperature of its environment (829)
animal de sangre fría: animal cuya temperatura corporal está determinada por la temperatura de su medio ambiente

electron: negatively charged particle; located in the space surrounding the nucleus (34)
electrón: partícula con carga negativa; ubicada en el espacio que rodea al núcleo

electron transport chain: series of electron carrier proteins that shuttle high-energy electrons during ATP-generating reactions (236)
cadena de transporte de electrones: serie de proteínas transportadoras que llevan electrones de alta energía, durante las reacciones generadoras de ATP

element: pure substance that consists entirely of one type of atom (35)
elemento: sustancia pura que consiste íntegramente en un tipo de átomo

embryo: developing stage of a multicellular organism (292)
embrión: una de las etapas de desarrollo de un organismo multicelular

embryo sac: female gametophyte within the ovule of a flowering plant (699)
saco embrionario: gametofito femenino dentro del óvulo de una planta que produce flores

emerging disease: disease that appears in the population for the first time, or an old disease that suddenly becomes harder to control (590)
enfermedad emergente: enfermedad que aparece en una población por primera vez o una enfermedad antigua que de pronto se vuelve más difícil de controlar

emigration: movement of individuals out of an area (132)
emigración: desplazamiento de individuos fuera de un área

endocrine gland: gland that releases its secretions (hormones) directly into the blood, which transports the secretions to other areas of the body (828, 979)
glándula endocrina: glándula que vierte sus secreciones (hormonas) directamente en la sangre, para ser transportadas a otras áreas del cuerpo

endoderm: innermost germ layer; develops into the linings of the digestive tract and much of the respiratory system (738)
endodermo: capa embrionaria más interna, a partir de la cual se desarrollan los revestimientos del tracto digestivo y gran parte del sistema respiratorio

endodermis: in plants, layer of ground tissue that completely encloses the vascular cylinder (670)
endodermis: en las plantas, un capa de tejido fundamental que envuelve completamente al cilindro vascular

endoplasmic reticulum: internal membrane system found in eukaryotic cells; place where lipid components of the cell membrane are assembled (200)
retículo endoplasmático: sistema de membranas internas de las células eucariotas; lugar donde se reúnen los componentes lipídicos de la membrana celular

endoskeleton: internal skeleton; structural support system within the body of an animal (815)
endoesqueleto: esqueleto interno; sistema estructural de sostén dentro del cuerpo de un animal

Glossary *(continued)*

endosperm: food-rich tissue that nourishes a seedling as it grows (700)
endospermo: tejido nutritivo que alimenta a una plántula a medida que crece

endospore: structure produced by prokaryotes in unfavorable conditions; a thick internal wall that encloses the DNA and a portion of the cytoplasm (583)
endospora: estructura producida por los procariotas en condiciones desfavorables; una gruesa pared interna que encierra al ADN y a una parte del citoplasma

endosymbiotic theory: theory that proposes that eukaryotic cells formed from a symbiotic relationship among several different prokaryotic cells (556)
teoría endosimbiótica: teoría que propone que las células eucariotas se formaron a partir de una relación simbiótica entre varias células procariotas distintas

endotherm: animal whose body temperature is regulated, at least in part, using heat generated within its body (829)
endotermo: animal cuya temperatura corporal se regula, al menos en parte, utilizando el calor generado dentro de su cuerpo

enzyme: protein catalyst that speeds up the rate of specific biological reactions (52)
enzima: proteína catalizadora que acelera la velocidad de reacciones biológicas específicas

epidermis: in plants, single layer of cells that makes up dermal tissue (665); in humans, the outer layer of the skin (936)
epidermis: en las plantas, única capa de células que forma el tejido dérmico; en los seres humanos, la capa exterior de la piel

epididymis: organ in the male reproductive system in which sperm mature and are stored (989)
epidídimo: órgano del sistema reproductor masculino en el cual el esperma madura y se almacena

epinephrine: hormone released by the adrenal glands that increases heart rate and blood pressure and prepares the body for intense physical activity; also called adrenaline (983)
epinefrina: hormona liberada por las glándulas adrenales que aumenta la frecuencia cardíaca y la presión sanguínea y prepara al cuerpo para una actividad física intensa; también llamada adrenalina

epithelial tissue: type of tissue that lines the interior and exterior body surfaces (863)
tejido epitelial: tipo de tejido que reviste el interior y el exterior de las superficies del cuerpo

era: major division of geologic time; usually divided into two or more periods (543)
era: división principal del tiempo geológico; usualmente dividida en dos o más períodos

esophagus: tube connecting the mouth to the stomach (877)
esófago: tubo que conecta la boca con el estómago

estuary: kind of wetland formed where a river meets the ocean (119)
estuario: tipo de humedal que se forma donde un río se une al océano

ethylene: plant hormone that stimulates fruits to ripen (711)
etileno: hormona vegetal que estimula la maduración de los frutos

Eukarya: domain consisting of all organisms that have a nucleus; includes protists, plants, fungi, and animals (526)
Eukarya (eucariontes): dominio compuesto por todos los organismos que tienen un núcleo; incluye a los protistas, las plantas, los hongos y los animales

eukaryote: organism whose cells contain a nucleus (193)
eucariota: organismo cuyas células contienen un núcleo

evolution: change over time; the process by which modern organisms have descended from ancient organisms (450)
evolución: cambio en el transcurso del tiempo; el proceso por el cual los organismos actuales se derivaron de los organismos antiguos

excretion: process by which metabolic wastes are eliminated from the body (794, 882)
excreción: proceso por el cual se eliminan del cuerpo los residuos metabólicos

exocrine gland: gland that releases its secretions, through tubelike structures called ducts, directly into an organ or out of the body (979)
glándula exocrina: glándula que vierte sus secreciones directamente a un órgano o al exterior del cuerpo a través de estructuras tubulares denominadas conductos

exon: expressed sequence of DNA; codes for a protein (365)
exón: secuencia expresada de ADN; codifica una porción específica de una proteína

exoskeleton: external skeleton; tough external covering that protects and supports the body of many invertebrates (815)
exoesqueleto: esqueleto externo; cubierta externa dura que protege y sostiene el cuerpo de muchos invertebrados

exponential growth: growth pattern in which the individuals in a population reproduce at a constant rate (132)
crecimiento exponencial: patrón de crecimiento en el cual los individuos de una población se reproducen a una tasa constante

extinct: term used to refer to a species that has died out and has no living members (538)

extinto: término utilizado para referirse a una especie que ha desaparecido y de la que ninguno de sus miembros está vivo

extracellular digestion: type of digestion in which food is broken down outside the cells in a digestive system and then absorbed (784)

digestión extracelular: tipo de digestión en la cual el alimento es degradado fuera de las células dentro de un sistema digestivo y luego se absorbe

F

facilitated diffusion: process of diffusion in which molecules pass across the membrane through cell membrane channels (209)

difusión facilitada: proceso de difusión en el cual las moléculas atraviesan la membrana a través de los canales de la membrana celular

family: in classification, group of similar genera (513)

familia: en la clasificación, grupo de géneros similares

fat: lipid; made up of fatty acids and glycerol; type of nutrient that protects body organs, insulates the body, and stores energy (870)

grasa: lípido; compuesto de ácidos grasos y glicerina; tipo de nutriente que protege a los órganos del cuerpo, actúa como aislante térmico y almacena energía

feedback inhibition: process in which a stimulus produces a response that opposes the original stimulus; also called negative feedback (732, 865)

inhibición de la retroalimentación: proceso en el cual un estímulo produce una respuesta que se opone al estímulo original; también llamada retroalimentación negativa

fermentation: process by which cells release energy in the absence of oxygen (262)

fermentación: proceso por el cual las células liberan energía en ausencia de oxígeno

fertilization: process in sexual reproduction in which male and female reproductive cells join to form a new cell (309)

fecundación: proceso de la reproducción sexual en el cual las células reproductoras masculinas y femeninas se unen para formar una célula nueva

fetus: a human embryo after eight weeks of development (998)

feto: un embrión humano después de ocho semanas de desarrollo

fever: increased body temperature that occurs in response to infection (1015)

fiebre: temperatura corporal elevada que se produce como respuesta a una infección

filtration: process of passing a liquid or gas through a filter to remove wastes (884)

filtración: proceso de hacer pasar un líquido o un gas a través de un filtro para quitar los residuos

fitness: how well an organism can survive and reproduce in its environment (461)

aptitud: capacidad de un organismo para sobrevivir y reproducirse en su medio ambiente

flagellum (pl. flagella): structure used by protists for movement; produces movement in a wavelike motion (607)

flagelo: estructura utilizada por los protistas para desplazarse; produce un desplazamiento con un movimiento semejante al de una onda

food chain: series of steps in an ecosystem in which organisms transfer energy by eating and being eaten (73)

cadena alimenticia: serie de pasos en un ecosistema, en que los organismos transfieren energía al alimentarse y al servir de alimento

food vacuole: small cavity in the cytoplasm of a protist that temporarily stores food (612)

vacuola alimenticia: pequeña cavidad situada en el citoplasma de los protistas que almacena alimentos por algún tiempo

food web: network of complex interactions formed by the feeding relationships among the various organisms in an ecosystem (74)

red alimenticia: red de interacciones complejas constituida por las relaciones alimenticias entre los varios organismos de un ecosistema

forensics: scientific study of crime scene evidence (433)

ciencias forenses: estudio científico de las pruebas en la escena del crimen

fossil: preserved remains or traces of ancient organisms (452)

fósil: restos conservados o vestigios de organismos antiguos

founder effect: change in allele frequencies as a result of the migration of a small subgroup of a population (490)

efecto fundador: cambio en las frecuencias alélicas como consecuencia de la migración de un subgrupo pequeño de una población

frameshift mutation: mutation that shifts the "reading frame" of the genetic message by inserting or deleting a nucleotide (373)

mutación de corrimiento de estructura: mutación que cambia el "marco de lectura" del mensaje genético insertando o eliminando un nucleótido

fruit: structure in angiosperms that contains one or more matured ovaries (651)

fruto: estructura de las Angiospermas que contiene uno o más ovarios maduros

fruiting body: reproductive structure of a fungus that grows from the mycelium (619)

cuerpo fructífero: estructura reproductora de los hongos que se desarrolla a partir del micelio

G

gamete: sex cell (312)
gameto: célula sexual

gametophyte: gamete-producing plant; multicellular haploid phase of a plant life cycle (637)
gametofito: planta que produce gametos; fase haploide multicelular del ciclo vital de una planta

ganglion (pl. ganglia): group of interneurons (810)
ganglio nervioso: grupo de interneuronas

gastrovascular cavity: digestive chamber with a single opening (784)
cavidad gastrovascular: cámara digestiva con una sola apertura

gastrulation: process of cell migration that results in the formation of the three cell layers—the ectoderm, the mesoderm, and the endoderm (997)
gastrulación: proceso de migración celular que da por resultado la formación de las tres capas celulares—el ectodermo, el mesodermo y el endodermo

gel electrophoresis: procedure used to separate and analyze DNA fragments by placing a mixture of DNA fragments at one end of a porous gel and applying an electrical voltage to the gel (404)
electroforesis en gel: procedimiento utilizado para separar y analizar fragmentos de ADN colocando una mezcla de fragmentos de ADN en un extremo de un gel poroso y aplicando al gel un voltaje eléctrico

gene: sequence of DNA that codes for a protein and thus determines a trait; factor that is passed from parent to offspring (310)
gen: secuencia de ADN que contiene el código de una proteína y por lo tanto determina un rasgo; factor que se transmite de un progenitor a su descendencia

gene expression: process by which a gene produces its product and the product carries out its function (370)
expresión génica: proceso por el cual un gen produce su producto y el producto lleva a cabo su función

gene pool: all the genes, including all the different alleles for each gene, that are present in a population at any one time (483)
caudal de genes: todos los genes, incluidos todos los alelos diferentes para cada gen, que están presentes en una población en un momento dado

gene therapy: process of changing a gene to treat a medical disease or disorder. An absent or faulty gene is replaced by a normal working gene. (431)
terapia genética o génica: proceso en el cual se cambia un gen para tratar una enfermedad o una afección médica. Se reemplaza un gen ausente o defectuoso con un gen de funcionamiento normal.

genetic code: collection of codons of mRNA, each of which directs the incorporation of a particular amino acid into a protein during protein synthesis (366)
código genético: conjunto de codones del ARN mensajero, cada uno de los cuales dirige la incorporación de un aminoácido en particular a una proteína durante la síntesis proteica

genetic diversity: sum total of all the different forms of genetic information carried by a particular species, or by all organisms on Earth (166)
diversidad genética: suma de todas las distintas formas de información genética portadas por una especie en particular, o por todos los organismos de la Tierra

genetic drift: random change in allele frequency caused by a series of chance occurrences that cause an allele to become more or less common in a population (490)
tendencia genética: alteración al azar de la frecuencia alélica causada por una serie de acontecimientos aleatorios que hacen que un alelo se vuelva más o menos común en una población

genetic equilibrium: situation in which allele frequencies in a population remain the same (491)
equilibrio genético: situación en la cual las frecuencias alélicas de una población se mantienen iguales

genetic marker: alleles that produce detectable phenotypic differences useful in genetic analysis (425)
marcador genético: alelos que producen diferencias fenotípicas detectables, útiles en el análisis genético

genetics: scientific study of heredity (308)
genética: estudio científico de la herencia

genome: entire set of genetic information that an organism carries in its DNA (392)
genoma: todo el conjunto de información genética que un organismo transporta en su ADN

genomics: study of whole genomes, including genes and their functions (407)
genómica: estudio integral de los genomas, incluyendo los genes y sus funciones

genotype: genetic makeup of an organism (315)
genotipo: composición genética de un organismo

genus: group of closely related species; the first part of the scientific name in binomial nomenclature (512)
género: grupo de especies relacionadas estrechamente; la primera parte del nombre científico en la nomenclatura binaria

geographic isolation: form of reproductive isolation in which two populations are separated by geographic barriers such as rivers, mountains, or bodies of water, leading to the formation of two separate subspecies (495)

aislamiento geográfico: forma de aislamiento reproductivo en el cual dos poblaciones están separadas por barreras geográficas como ríos, montañas o masas de agua, dando lugar a la formación de dos subespecies distintas

geologic time scale: timeline used to represent Earth's history (542)

escala de tiempo geológico: línea cronológica utilizada para representar la historia de la Tierra

germ theory of disease: idea that infectious diseases are caused by microorganisms (1010)

teoría microbiana de la enfermedad: idea de que las enfermedades infecciosas son causadas por microorganismos

germination: resumption of growth of the plant embryo following dormancy (706)

germinación: reanudación del crecimiento del embrión de la planta después de la latencia

gibberellin: plant hormone that stimulates growth and may cause dramatic increases in size (711)

giberelina: hormona de las plantas que estimula el crecimiento y puede causar aumentos significativos de tamaño

gill: feathery structure specialized for the exchange of gases with water (788)

branquia: estructura tegumentaria especializada en el intercambio de los gases con el agua

global warming: increase in the average temperatures on Earth (177)

calentamiento global: aumento del promedio de temperatura en la Tierra

glomerulus: small network of capillaries encased in the upper end of the nephron; where filtration of the blood takes place (884)

glomérulo: pequeña red de capilares encerrados en el extremo superior del nefrón; donde tiene lugar la filtración de la sangre

glycolysis: first set of reactions in cellular respiration in which a molecule of glucose is broken into two molecules of pyruvic acid (254)

glicólisis: primer conjunto de reacciones en la respiración celular, en las cuales una molécula de glucosa se descompone en dos moléculas de ácido pirúvico

Golgi apparatus: organelle in cells that modifies, sorts, and packages proteins and other materials from the endoplasmic reticulum for storage in the cell or release outside the cell (201)

aparato de Golgi: orgánulo de las células que modifica, clasifica y agrupa las proteínas y otras sustancias provenientes del retículo endoplasmático para almacenarlas en la célula o enviarlas fuera de la célula

gradualism: the evolution of a species by gradual accumulation of small genetic changes over long periods of time (549)

gradualismo: evolución de una especie por la acumulación gradual de pequeños cambios genéticos ocurridos en el transcurso de largos períodos de tiempo

grafting: method of propagation used to reproduce seedless plants and varieties of woody plants that cannot be propagated from cuttings (703)

injerto: método de propagación utilizado para reproducir plantas sin semillas y algunas variedades de plantas leñosas que no pueden propagarse a partir de esquejes

gravitropism: response of a plant to the force of gravity (712)

geotropismo: respuesta de una planta a la fuerza de la gravedad

green revolution: development of highly productive crop strains and use of modern agriculture techniques to increase yields of food crops (717)

revolución verde: el desarrollo de variedades de cultivos altamente productivos y el uso de técnicas agrícolas modernas para aumentar el rendimiento de los cultivos

greenhouse effect: process in which certain gases (carbon dioxide, methane, and water vapor) trap sunlight energy in Earth's atmosphere as heat (97)

efecto invernadero: proceso mediante el cual ciertos gases (dióxido de carbono, metano y vapor de agua) atrapan la energía de la luz solar en la atmósfera terrestre en forma de calor

growth factor: one of a group of external regulatory proteins that stimulate the growth and division of cells (287)

factor de crecimiento: una de las proteínas del grupo de proteínas reguladoras externas que estimulan el crecimiento y la división de las células

guard cell: specialized cell in the epidermis of plants that controls the opening and closing of stomata (682)

célula de guarda (o célula oclusiva): célula especializada de la epidermis vegetal que controla la apertura y el cierre de los estomas

gullet: indentation in one side of a ciliate that allows food to enter the cell (612)

citofaringe: hendidura a un costado de un ciliado que permite que los alimentos entren a la célula

gymnosperm: group of seed plants that bear their seeds directly on the scales of cones (646)

Gimnospermas: grupo de plantas fanerógamas que tienen sus semillas directamente sobre las escamas de los conos

H

habitat: area where an organism lives, including the biotic and abiotic factors that affect it (99)

hábitat: área donde vive un organismo, incluidos los factores bióticos y abióticos que lo afectan

Glossary *(continued)*

habitat fragmentation: splitting of ecosystems into pieces (168)
fragmentación del hábitat: la ruptura, o separación en partes, de los ecosistemas

habituation: type of learning in which an animal decreases or stops its response to a repetitive stimulus that neither rewards nor harms the animal (842)
habituación: tipo de aprendizaje en el cual un animal disminuye o cancela su respuesta ante un estímulo repetido que no recompensa ni castiga al animal

hair follicle: tubelike pockets of epidermal cells that extend into the dermis; cells at the base of hair follicles produce hair (937)
folículo piloso: sacos tubulares de las células epidérmicas que se prolongan hacia el interior de la dermis; las células situadas en la base de los folículos pilosos, producen pelo

half life: length of time required for half of the radioactive atoms in a sample to decay (540)
vida media: período de tiempo requerido para que se desintegre la mitad de los átomos radiactivos de una muestra

haploid: term used to refer to a cell that contains only a single set of genes (323)
haploide: tipo de célula que posee un solo juego de cromosomas

Hardy-Weinberg principle: principle that states that allele frequencies in a population remain constant unless one or more factors cause those frequencies to change (491)
principio de Hardy-Weinberg: el principio que afirma que las frecuencias alélicas de una población permanecen constantes a menos que uno o más factores ocasionen que esas frecuencias cambien

Haversian canal: one of a network of tubes running through compact bone that contains blood vessels and nerves (924)
conducto de Havers: uno de los tubos de una red que recorre longitudinalmente el hueso compacto y contiene vasos sanguíneos y nervios

heart: hollow muscular organ that pumps blood throughout the body (791)
corazón: órgano muscular hueco que bombea la sangre a todo el cuerpo

heartwood: in a woody stem, the older xylem near the center of the stem that no longer conducts water (678)
duramen: en un tallo leñoso, el xilema más viejo situado cerca del centro del tallo que ya no conduce agua

hemoglobin: iron-containing protein in red blood cells that binds oxygen and transports it to the body (954)
hemoglobina: proteína de los glóbulos rojos que contiene hierro, fija el oxígeno y lo transporta al organismo

herbaceous plant: type of plant that has smooth and nonwoody stems; includes dandelions, zinnias, petunias, and sunflowers (653)
planta herbácea: tipo de planta que tiene tallos blandos y no leñosos; incluye dientes de león, cinias, petunias y girasoles

herbivore: organism that obtains energy by eating only plants (71)
herbívoro: organismo que obtiene energía alimentándose solo de plantas

herbivory: interaction in which one animal (the herbivore) feeds on producers (such as plants) (102)
herbivorismo: interacción en la cual un animal (el herbívoro) se alimenta de productores (como las plantas)

heterotroph: organism that obtains food by consuming other living things; also called a consumer (71, 228)
heterótrofo: organismo que obtiene su alimento consumiendo otros seres vivos; también llamado consumidor

heterozygous: having two different alleles for a particular gene (314)
heterocigota: que tiene dos alelos diferentes para un gen dado

histamine: chemical released by mast cells that increases the flow of blood and fluids to the infected area during an inflammatory response (1014)
histamina: sustancia química liberada por los mastocitos que aumenta el flujo de la sangre y los fluidos hacia el área infectada durante una respuesta inflamatoria

homeobox gene: The homeobox is a DNA sequence of approximately 130 base pairs, found in many homeotic genes that regulate development. Genes containing this sequence are known as homeobox genes, and they code for transcription factors, proteins that bind to DNA, and they also regulate the expression of other genes. (382)
gen homeobox: el homeobox es una secuencia de ADN de aproximadamente 130 pares de bases, presente en muchos genes homeóticos que regulan el desarrollo. Los genes que contienen esta secuencia se denominan genes homeobox y codifican los factores de transcripción, las proteínas que se adhieren al ADN y regulan la expresión de otros genes

homeostasis: relatively constant internal physical and chemical conditions that organisms maintain (19, 214, 865)
homeostasis: las condiciones internas, químicas y físicas, que los organismos mantienen relativamente constantes

homeotic gene: a class of regulatory genes that determine the identity of body parts and regions in an animal embryo. Mutations in these genes can transform one body part into another (382)

gen homeótico: tipo de genes reguladores que determinan la identidad de las partes y regiones del cuerpo en un embrión animal. Las mutaciones de estos genes pueden transformar una parte del cuerpo en otra

hominine: hominoid lineage that led to humans (767)

homínino: linaje hominoide que dio lugar a los seres humanos

hominoid: group of anthropoids that includes gibbons, orangutans, gorillas, chimpanzees, and humans (767)

homínido: grupo de antropoides que incluye a los gibones, orangutanes, gorilas, chimpacés y seres humanos

homologous: term used to refer to chromosomes in which one set comes from the male parent and one set comes from the female parent (323)

homólogos: término utilizado para referirse a los cromosomas en los que un juego proviene del progenitor masculino y un juego proviene del progenitor femenino

homologous structures: structures that are similar in different species of common ancestry (468)

estructuras homólogas: estructuras que son similares en distintas especies que tienen un ancestro común

homozygous: having two identical alleles for a particular gene (314)

homocigota: que tiene dos alelos idénticos para un gen dado

hormone: chemical produced in one part of an organism that affects another part of the same organism (708, 978)

hormona: sustancia química producida en una parte de un organismo que afecta a otra parte del mismo organismo

Hox gene: a group of homeotic genes clustered together that determine the head to tail identity of body parts in animals. All hox genes contain the homeobox DNA sequence. (382)

gen Hox: grupo de genes homeóticos agrupados en un conjunto que determinan la identidad posicional de las partes del cuerpo de los animales. Todos los genes Hox contienen la secuencia de ADN homeobox

humoral immunity: immunity against antigens in body fluids, such as blood and lymph (1016)

inmunidad humoral: inmunidad contra los antígenos presentes en los fluidos corporales, como la sangre y la linfa

humus: material formed from decaying leaves and other organic matter (114)

humus: material formado a partir de hojas en descomposición y otros materiales orgánicos

hybrid: offspring of crosses between parents with different traits (309)

híbrido: descendencia del cruce entre progenitores que tienen rasgos diferentes

hybridization: breeding technique that involves crossing dissimilar individuals to bring together the best traits of both organisms (419)

hibridación: técnica de cría que consiste en cruzar individuos diferentes para reunir los mejores rasgos de ambos organismos

hydrogen bond: weak attraction between a hydrogen atom and another atom (41)

enlace de hidrógeno: atracción débil entre un átomo de hidrógeno y otro átomo

hydrostatic skeleton: skeleton made of fluid-filled body segments that work with muscles to allow the animal to move (814)

esqueleto hidrostático: esqueleto constituido por segmentos corporales llenos de fluido que trabajan con los músculos para permitir el movimiento del animal

hypertonic: when comparing two solutions, the solution with the greater concentration of solutes (210)

hipertónica: al comparar dos soluciones, la solución que tiene la mayor concentración de solutos

hypha (pl. hyphae): one of many long, slender filaments that makes up the body of a fungus (619)

hifa: uno de muchos filamentos largos y delgados que componen el cuerpo de un hongo

hypothalamus: structure of the brain that acts as a control center for recognition and analysis of hunger, thirst, fatigue, anger, and body temperature (903)

hipotálamo: estructura del cerebro que funciona como un centro de control para el reconocimiento y el análisis del hambre, la sed, la fatiga, el enojo y la temperatura corporal

hypothesis: possible explanation for a set of observations or possible answer to a scientific question (7)

hipótesis: explicación posible para un conjunto de observaciones o respuesta posible a una pregunta científica

hypotonic: when comparing two solutions, the solution with the lesser concentration of solutes (210)

hipotónica: al comparar dos soluciones, la solución que tiene la menor concentración de solutos

I

immigration: movement of individuals into an area occupied by an existing population (132)

inmigración: desplazamiento de individuos a un área ocupada por una población ya existente

immune response: the body's specific recognition, response, and memory to a pathogen attack (1016)

respuesta inmune: reconocimiento, respuesta y memoria específicos que tiene el cuerpo respecto al ataque de un organismo patógeno

implantation: process in which the blastocyst attaches to the wall of the uterus (996)

implantación: proceso en el cual la blástula se adhiere a la pared del útero

imprinting: type of behavior based on early experience; once imprinting has occurred, the behavior cannot be changed (844)
impronta: tipo de comportamiento basado en las primeras experiencias; una vez que ocurre la impronta, el comportamiento no puede cambiarse

inbreeding: continued breeding of individuals with similar characteristics to maintain the derived characteristics of a kind of organism (419)
endogamia: la cría continua de individuos con características semejantes para mantener las características derivadas de un tipo de organismo

incomplete dominance: situation in which one allele is not completely dominant over another allele (319)
dominancia incompleta: situación en la cual un alelo no es completamente dominante sobre otro alelo

independent assortment: one of Mendel's principles that states that genes for different traits can segregate independently during the formation of gametes (317)
distribución independiente: uno de los principios de Mendel que establece que los genes para rasgos diferentes pueden segregarse independientemente durante la formación de los gametos

independent variable: factor in a controlled experiment that is deliberately changed; also called manipulated variable (7)
variable independiente: en un experimento controlado, el factor que se modifica a propósito; también llamada variable manipulada

index fossil: distinctive fossil that is used to compare the relative ages of fossils (540)
fósil guía: fósil distintivo usado para comparar las edades relativas de los fósiles

infectious disease: disease caused by a microorganism that disrupts normal body functions (1010)
enfermedad infecciosa: enfermedad causada por un microorganismo que altera las funciones normales del cuerpo

inference: a logical interpretation based on prior knowledge and experience (7)
inferencia: interpretación lógica basada en la experiencia y en conocimientos previos

inflammatory response: nonspecific defense reaction to tissue damage caused by injury or infection (1014)
respuesta inflamatoria: reacción defensiva no específica al daño causado a los tejidos por una herida o una infección

innate behavior: type of behavior in which the behavior appears in fully functional form the first time it is performed even though the animal has had no previous experience with the stimuli to which it responds; also called instinct (841)
comportamiento innato: tipo de comportamiento en el cual la conducta aparece en forma completamente funcional la primera vez que se lleva a cabo, aunque el animal no tenga ninguna experiencia previa con los estímulos a los que responde; también llamado instinto

insight learning: type of behavior in which an animal applies something it has already learned to a new situation, without a period of trial and error; also called reasoning (843)
aprendizaje por discernimiento: tipo de comportamiento en el cual un animal aplica algo que ya ha aprendido a una situación nueva, sin un período de ensayo y error; también llamado razonamiento

interferon: one of a group of proteins that help cells resist viral infection (1015)
interferón: un tipo de proteína que ayuda a las células a combatir las infecciones virales

interneuron: type of neuron that processes information and may relay information to motor neurons (809)
interneurona: tipo de neurona que procesa información y la puede transmitir para estimular las neuronas

interphase: period of the cell cycle between cell divisions (281)
interfase: período del ciclo celular entre las divisiones celulares

intracellular digestion: type of digestion in which food is digested inside specialized cells that pass nutrients to other cells by diffusion (784)
digestión intracelular: tipo de digestión en la cual los alimentos se digieren dentro de células especializadas que pasan los nutrientes a otras células mediante difusión

intron: sequence of DNA that is not involved in coding for a protein (365)
intrón: secuencia de ADN que no participa en la codificación de una proteína

invertebrate: animal that lacks a backbone, or vertebral column (730)
invertebrado: animal que carece de columna vertebral

ion: atom that has a positive or negative charge (37)
ion: átomo que tiene una carga positiva o negativa

ionic bond: chemical bond formed when one or more electrons are transferred from one atom to another (37)
enlace iónico: enlace químico que se forma cuando uno o más electrones se transfieren de un átomo a otro

iris: colored part of the eye (912)
iris: parte coloreada del ojo

isotonic: when the concentration of two solutions is the same (210)
isotónica: cuando la concentración de dos soluciones es la misma

isotope: one of several forms of a single element, which contains the same number of protons but different numbers of neutrons (35)
isótopo: cada una de las diferentes formas de un único elemento, que contiene la misma cantidad de protones pero cantidades distintas de neutrones

J

joint: place where one bone attaches to another bone (816, 926)
articulación: sitio donde un hueso se une a otro hueso

K

karyotype: micrograph of the complete diploid set of chromosomes grouped together in pairs, arranged in order of decreasing size (392)
cariotipo: micrografía de la totalidad del conjunto diploide de cromosomas agrupados en pares, ordenados por tamaño decreciente

keratin: tough fibrous protein found in skin (936)
queratina: proteína fibrosa y resistente que se halla en la piel

keystone species: single species that is not usually abundant in a community yet exerts strong control on the structure of a community (103)
especie clave: especie que habitualmente no es abundante en una comunidad y sin embargo ejerce un fuerte control sobre la estructura de esa comunidad

kidney: an organ of excretion that separates wastes and excess water from the blood (795)
riñón: órgano excretor que separa los residuos y el exceso de agua de la sangre

kin selection: theory that states that helping relatives can improve an individual's evolutionary fitness because related individuals share a large proportion of their genes (849)
selección de parentesco: teoría que enuncia que ayudar a los congéneres puede mejorar la aptitud evolutiva de un individuo porque los individuos emparentados comparten una gran parte de sus genes

kingdom: largest and most inclusive group in Linnaean classification (514)
reino: grupo más grande e inclusivo del sistema de clasificación inventado por Linneo

Koch's postulates: set of guidelines developed by Koch that helps identify the microorganism that causes a specific disease (1011)
postulados de Koch: conjunto de pautas desarrollado por Koch que ayuda a identificar al microorganismo que causa una enfermedad específica

Krebs cycle: second stage of cellular respiration in which pyruvic acid is broken down into carbon dioxide in a series of energy-extracting reactions (256)
ciclo de Krebs: segunda fase de la respiración celular en la cual el ácido pirúvico se descompone en dióxido de carbono en una serie de reacciones que liberan energía

L

language: system of communication that combines sounds, symbols, and gestures according to a set of rules about sequence and meaning, such as grammar and syntax (851)
lenguaje: sistema de comunicación que combina sonidos, símbolos y gestos según un conjunto de reglas sobre la secuencia y el significado, como la gramática y la sintaxis

large intestine: organ in the digestive system that removes water from the undigested material that passes through it; also called colon (881)
intestino grueso: órgano del sistema digestivo que extrae el agua del material no digerido que pasa por él; también llamado colon

larva (pl. larvae): immature stage of some organisms (756)
larva: etapa inmadura de un organismo

larynx: structure in the throat that contains the vocal cords (964)
laringe: órgano situado en la garganta que contiene las cuerdas vocales

learning: changes in behavior as a result of experience (842)
aprendizaje: cambios en el comportamiento a consecuencia de la experiencia

lens: structure in the eye that focuses light rays on the retina (912)
cristalino: estructura del ojo que enfoca los rayos luminosos en la retina

lichen: symbiotic association between a fungus and a photosynthetic organism (623)
liquen: asociación simbiótica entre un hongo y un organismo fotosintético

ligament: tough connective tissue that holds bones together in a joint (816, 927)
ligamento: tejido conectivo resistente que mantiene unidos a los huesos en una articulación

light-dependent reactions: set of reactions in photosynthesis that use energy from light to produce ATP and NADPH (233)
reacciones dependientes de la luz: en la fotosíntesis, conjunto de reacciones que emplean la energía proveniente de la luz para producir ATP y NADPH

Glossary *(continued)*

light-independent reactions: set of reactions in photosynthesis that do not require light; energy from ATP and NADPH is used to build high-energy compounds such as sugar; also called the Calvin cycle (233)
 reacciones independientes de la luz: en la fotosíntesis, conjunto de reacciones que no necesitan luz; la energía proveniente del ATP y del NADPH se emplea para construir compuestos con gran contenido energético, como el azúcar; también llamado ciclo de Calvin

lignin: substance in vascular plants that makes cell walls rigid (666)
 lignina: sustancia de las plantas vasculares que hace rígidas a las paredes celulares

limiting factor: factor that causes population growth to decrease (137)
 factor limitante: un factor que hace disminuir el crecimiento de la población

limiting nutrient: single essential nutrient that limits productivity in an ecosystem (85)
 nutriente limitante: un solo nutriente esencial que limita la productividad de un ecosistema

lipid: macromolecule made mostly from carbon and hydrogen atoms; includes fats, oils, and waxes (47)
 lípido: macromolécula compuesta principalmente por átomos de carbono e hidrógeno; incluye las grasas, los aceites y las ceras

lipid bilayer: flexible double-layered sheet that makes up the cell membrane and forms a barrier between the cell and its surroundings (204)
 bicapa lipídica: lámina flexible de dos capas que constituye la membrana celular y forma una barrera entre la célula y su entorno

logistic growth: growth pattern in which a population's growth slows and then stops following a period of exponential growth (135)
 crecimiento logístico: patrón de crecimiento en el cual el desarrollo de una población se reduce y luego se detiene después de un período de crecimiento exponencial

loop of Henle: section of the nephron tubule that is responsible for conserving water and minimizing the volume of the filtrate (884)
 asa de Henle: una sección del túbulo de nefrón responsable de conservar el agua y minimizar el volumen del material filtrado

lung: respiratory organ; place where gases are exchanged between the blood and inhaled air (788)
 pulmón: órgano respiratorio; lugar donde se intercambian los gases entre la sangre y el aire inhalado

lymph: fluid that is filtered out of the blood (956)
 linfa: fluido procedente de la sangre

lysogenic infection: type of infection in which a virus embeds its DNA into the DNA of the host cell and is replicated along with the host cell's DNA (577)
 infección lisogénica: tipo de infección en la cual un virus inserta su ADN en el ADN de la célula huésped y se replica junto con el ADN de dicha célula huésped

lysosome: cell organelle that breaks down lipids, carbohydrates, and proteins into small molecules that can used by the rest of the cell (198)
 lisosoma: orgánulo celular que descompone los lípidos, los hidratos de carbono y las proteínas en moléculas pequeñas que pueden ser utilizadas por el resto de la célula

lytic infection: type of infection in which a virus enters a cell, makes copies of itself, and causes the cell to burst (576)
 infección lítica: tipo de infección en la cual un virus penetra una célula, hace copias de sí mismo y provoca la ruptura o muerte celular

M

macroevolutionary patterns: changes in anatomy, phylogeny, ecology, and behavior that take place in clades larger than a single species (546)
 patrones de macroevolución: cambios que ocurren en la anatomía, filogenia, ecología y comportamiento de clados que abarcan a más de una especie

Malpighian tubule: structure in most terrestrial arthropods that concentrates the uric acid and adds it to digestive wastes (797)
 túbulo de Malpighi: estructura de la mayoría de los artrópodos terrestres que concentra el ácido úrico y lo incorpora a los residuos digestivos

mammary gland: gland in female mammals that produces milk to nourish the young (825)
 glándula mamaria: glándula de las hembras de los mamíferos que produce leche para alimentar a las crías

mass extinction: event during which many species become extinct during a relatively short period of time (548)
 extinción masiva: suceso durante el cual se extinguen muchas especies durante un período de tiempo relativamente corto

matrix: innermost compartment of the mitochondrion (256)
 matriz: compartimento más interno de la mitocondria

mechanical digestion: physical breakdown of large pieces of food into smaller pieces (875)
 digestión mecánica: descomposición física de grandes pedazos de comida en pedazos más pequeños

meiosis: process in which the number of chromosomes per cell is cut in half through the separation of homologous chromosomes in a diploid cell (324)
meiosis: proceso por el cual el número de cromosomas por célula se reduce a la mitad mediante la separación de los cromosomas homólogos de una célula diploide

melanin: dark brown pigment in the skin that helps protect the skin by absorbing ultraviolet rays (936)
melanina: pigmento marrón oscuro de la piel que contribuye a protegerla al absorber los rayos ultravioletas

melanocyte: cell in the skin that produces a dark brown pigment called melanin (936)
melanocito: célula de la piel que produce un pigmento marrón oscuro llamado melanina

menstrual cycle: regular sequence of events in which an egg develops and is released from the body (991)
ciclo menstrual: secuencia regular de sucesos en la cual un huevo se desarrolla y se elimina del cuerpo

menstruation: discharge of blood and the unfertilized egg from the body (993)
menstruación: descarga de sangre y del huevo no fertilizado del cuerpo

meristem: regions of unspecialized cells responsible for continuing growth throughout a plant's lifetime (667)
meristemos: regiones de células no especializadas responsables del crecimiento continuo de una planta durante su vida

mesoderm: middle germ layer; develops into muscles, and much of the circulatory, reproductive, and excretory systems (738)
mesodermo: capa embrionaria media; se desarrolla para dar lugar a los músculos y gran parte de los sistemas circulatorio, reproductor y excretor

mesophyll: specialized ground tissue found in leaves; performs most of a plant's photosynthesis (680)
mesófilo: tejido fundamental especializado que se halla en las hojas; realiza la mayor parte de la fotosíntesis de una planta

messenger RNA (mRNA): type of RNA that carries copies of instructions for the assembly of amino acids into proteins from DNA to the rest of the cell (363)
ARN mensajero: tipo de ARN que transporta copias de las instrucciones para el ensamblaje de los aminoácidos en proteínas, desde el ADN al resto de la célula

metabolism: the combination of chemical reactions through which an organism builds up or breaks down materials (19)
metabolismo: la combinación de reacciones químicas a través de las cuales un organismo acumula o desintegra materiales

metamorphosis: process of changes in shape and form of a larva into an adult (823)
metamorfosis: proceso de cambios en la estructura y forma de una larva hasta que se convierte en adulto

metaphase: phase of mitosis in which the chromosomes line up across the center of the cell (282)
metafase: fase de la mitosis en la cual los cromosomas se alinean a través del centro de la célula

microclimate: environmental conditions within a small area that differs significantly from the climate of the surrounding area (96)
microclima: condiciones medioambientales de un área pequeña que difieren significativamente del clima del área circundante

migration: seasonal behavior resulting in the movement from one environment to another (847)
migración: comportamiento estacional que da por resultado el desplazamiento desde un medio ambiente a otro

mineral: inorganic nutrient the body needs, usually in small amounts (872)
mineral: nutriente inorgánico que el cuerpo necesita, usualmente en pequeñas cantidades

mitochondrion: cell organelle that converts the chemical energy stored in food into compounds that are more convenient for the cell to use (202)
mitocondria: orgánulo celular que convierte la energía química almacenada en los alimentos en compuestos más apropiados para que la célula los use

mitosis: part of eukaryotic cell division during which the cell nucleus divides (282)
mitosis: fase de la división de las células eucariotas durante la cual se divide el núcleo celular

mixture: material composed of two or more elements or compounds that are physically mixed together but not chemically combined (42)
mezcla: material compuesto por dos o más elementos o compuestos que están mezclados físicamente pero no están combinados químicamente

molecular clock: method used by researchers that uses mutation rates in DNA to estimate the length of time that two species have been evolving independently (498)
reloj molecular: método de investigación que emplea las tasas de mutación del ADN para estimar el lapso de tiempo en que dos especies han evolucionado independientemente

molecule: smallest unit of most compounds that displays all the properties of that compound (37)
molécula: la unidad más pequeña de la mayoría de los compuestos que exhibe todas las propiedades de ese compuesto

molting: process of shedding an exoskeleton and growing a new one (815)
muda: proceso de desprendimiento de un exoesqueleto y el crecimiento de uno nuevo

monocot: angiosperm with one seed leaf in its ovary (652)
monocotiledónea: angiosperma con un cotiledón (hoja embrionaria) en su ovario

monoculture: farming strategy of planting a single, highly productive crop year after year (155)
 monocultivo: estrategia agrícola que consiste en plantar año tras año un único cultivo altamente productivo

monomer: small chemical unit that makes up a polymer (46)
 monómero: pequeña unidad química que forma un polímero

monophyletic group: group that consists of a single ancestral species and all its descendants and excludes any organisms that are not descended from that common ancestor (516)
 grupo monofilético: grupo que consiste en una especie con un único ancestro y todos sus descendientes y excluye a todos los organismos que no descienden de ese ancestro común

monosaccharide: simple sugar molecule (46)
 monosacárido: molécula de azúcar simple

motor neuron: type of nerve cell that carries directions from interneurons to either muscle cells or glands (809)
 neurona motora: tipo de célula nerviosa que lleva las instrucciones provenientes de las interneuronas a las células musculares o las glándulas

multiple alleles: a gene that has more than two alleles (320)
 alelos múltiples: un gen que tiene más de dos alelos

multipotent: cell with limited potential to develop into many types of differentiated cells (295)
 multipotentes: células con potencial limitado para generar muchos tipos de células diferenciadas

muscle fiber: long slender skeletal muscle cells (929)
 fibra muscular: células largas y delgadas de los músculos esqueléticos

muscle tissue: type of tissue that makes movements of the body possible (863)
 tejido muscular: tipo de tejido que hace posibles los movimientos del cuerpo

mutagen: chemical or physical agents in the environment that interact with DNA and may cause a mutation (375)
 mutágeno: agentes físicos o químicos del medioambiente que interaccionan con el ADN y pueden causar una mutación

mutation: change in the genetic material of a cell (372)
 mutación: cambio en el material genético de una célula

mutualism: symbiotic relationship in which both species benefit from the relationship (103)
 mutualismo: relación simbiótica en la cual ambas especies se benefician

mycelium (pl. mycelia): densely branched network of the hyphae of a fungus (619)
 micelio: la red de filamentos muy ramificados de las hifas de un hongo

mycorrhiza (pl. mycorrhizae): symbiotic association of plant roots and fungi (624)
 micorriza: asociación simbiótica entre las raíces de las plantas y los hongos

myelin sheath: insulating membrane surrounding the axon in some neurons (897)
 vaina de mielina: membrana aislante que rodea al axón de algunas neuronas

myocardium: thick middle muscle layer of the heart (949)
 miocardio: capa media, gruesa y musculosa del corazón

myofibril: tightly packed filament bundles found within skeletal muscle fibers (930)
 miofibrilla: manojos de filamentos muy apretados que se hallan dentro de las fibras de los músculos esqueléticos

myosin: thick filament of protein found in skeletal muscle cells (930)
 miosina: filamento grueso de proteína que se halla en las células de los músculos esqueléticos

N

NAD^+ (nicotinamide adenine dinucleotide): electron carrier involved in glycolysis (255)
 NAD^+ (dinucleótido de nicotinamida adenina): transportador de electrones que participa en la glucólisis

$NADP^+$ (nicotinamide adenine dinucleotide phosphate): carrier molecule that transfers high-energy electrons from chlorophyll to other molecules (232)
 $NADP^+$ (fosfato de dinucleótido de nicotinamida adenina): molécula transportadora de electrones que transfiere electrones de alta energía desde la clorofila a otras moléculas

natural selection: process by which organisms that are most suited to their environment survive and reproduce most successfully; also called survival of the fittest (463)
 selección natural: proceso por el cual los organismos más adaptados a su medioambiente sobreviven y se reproducen más exitosamente; también llamada supervivencia del más apto

nephridium (pl. nephridia): excretory structure of an annelid that filters body fluid (797)
 nefridio: estructura excretora de los anélidos que filtra el fluido corporal

nephron: blood-filtering structure in the kidneys in which impurities are filtered out, wastes are collected, and purified blood is returned to the circulation (884)
 nefrón: estructura filtradora de la sangre en los riñones, en la cual se filtran las impurezas, se recogen los desechos y la sangre purificada se devuelve a la circulación

nervous tissue: type of tissue that transmits nerve impulses throughout the body (863)
tejido nervioso: tipo de tejido que transmite los impulsos nerviosos por el cuerpo

neuromuscular junction: the point of contact between a motor neuron and a skeletal muscle cell (931)
unión neuromuscular: el punto de contacto entre una neurona motora y una célula de un músculo esquelético

neuron: nerve cell; specialized for carrying messages throughout the nervous system (808)
neurona: célula nerviosa; especializada en conducir mensajes a través del sistema nervioso

neurotransmitter: chemical used by a neuron to transmit an impulse across a synapse to another cell (900)
neurotransmisor: sustancia química utilizada por una neurona para transmitir un impulso a otra célula a través de una sinapsis

neurulation: the first step in the development of the nervous system (997)
neurulación: primer paso en el desarrollo del sistema nervioso

niche: full range of physical and biological conditions in which an organism lives and the way in which the organism uses those conditions (100)
nicho: toda la variedad de condiciones biológicas y físicas en las que vive un organismo y la manera en la que dicho organismo utiliza esas condiciones

nitrogen fixation: process of converting nitrogen gas into nitrogen compounds that plants can absorb and use (84)
fijación de nitrógeno: el proceso por el cual el gas nitrógeno se convierte en los compuestos nitrogenados que las plantas pueden absorber y utilizar

node: part on a growing stem where a leaf is attached (675)
nudo: parte de un tallo en crecimiento donde está adherida una hoja

nondisjunction: error in meiosis in which the homologous chromosomes fail to separate properly (401)
no disyunción: error que ocurre durante la meiosis, en el que cromosomas homólogos no logran separarse adecuadamente

nonrenewable resource: resource that cannot be replenished by a natural process within a reasonable amount of time (157)
recurso no renovable: recurso que no se puede reponer mediante un proceso natural dentro de un período de tiempo razonable

norepinephrine: hormone released by the adrenal glands that increases heart rate and blood pressure and prepares the body for intense physical activity (983)
norepinefrina o noradrenalina: hormona liberada por las glándulas adrenales que aumenta la frecuencia cardíaca y la presión sanguínea y prepara al cuerpo para realizar actividad física intensa

notochord: long supporting rod that runs through a chordate's body just below the nerve cord (731)
notocordio: extenso bastón de apoyo que se extiende a lo largo del cuerpo de los cordados, justo por debajo del cordón nervioso

nucleic acid: macromolecules containing hydrogen, oxygen, nitrogen, carbon, and phosphorus (48)
ácido nucleico: macromoléculas que contienen hidrógeno, oxígeno, nitrógeno, carbono y fósforo

nucleotide: subunit of which nucleic acids are composed; made up of a 5-carbon sugar, a phosphate group, and a nitrogenous base (48)
nucleótido: subunidad que constituye los ácidos nucleicos; compuesta de un azúcar de 5 carbonos, un grupo fosfato y una base nitrogenada

nucleus: the center of an atom, which contains the protons and neutrons (34); in cells, structure that contains the cell's genetic material in the form of DNA (193)
núcleo: el centro de un átomo, contiene los protones y los neutrones; en las células, la estructura que contiene el material genético de la célula en forma de ADN

nutrient: chemical substance that an organism needs to sustain life (82)
nutriente: sustancia química que un organismo necesita para continuar con vida

nymph: immature form of an animal that resembles the adult form but lacks functional sexual organs (823)
ninfa: forma inmadura de un animal que se parece a la forma adulta, pero carece de órganos sexuales funcionales

observation: process of noticing and describing events or processes in a careful, orderly way (6)
observación: el método de percibir y describir sucesos o procesos de manera atenta y ordenada

omnivore: organism that obtains energy by eating both plants and animals (71)
omnívoro: organismo que obtiene energía alimentándose de plantas y animales

open circulatory system: type of circulatory system in which blood is only partially contained within a system of blood vessels as it travels through the body (791)
sistema circulatorio abierto: tipo de sistema circulatorio en el cual la sangre, cuando fluye por el cuerpo, está solo parcialmente contenida dentro de un sistema de vasos sanguíneos

operant conditioning: type of learning in which an animal learns to behave in a certain way through repeated practice, to receive a reward or avoid punishment (843)
acondicionamiento operante: tipo de aprendizaje en el cual un animal aprende a comportarse de cierta manera mediante una práctica repetida, para recibir una recompensa o evitar un castigo

operator: short DNA region, adjacent to the promoter of a prokaryotic operon, that binds repressor proteins responsible for controlling the rate of transcription of the operon (378)
operador: pequeña región de ADN, adyacente al promotor del operón de una procariota, que une las proteínas represoras responsables de controlar la tasa de transcripción del operón

operon: in prokaryotes, a group of adjacent genes that share a common operator and promoter and are transcribed into a single mRNA (377)
operón: en las procariotas, grupo de genes adyacentes que comparten un operador y un promotor en común y que son transcritas a un solo ARN mensajero

opposable thumb: thumb that enables grasping objects and using tools (767)
pulgar oponible o prensible: un pulgar que permite aferrar objetos y utilizar herramientas

order: in classification, a group of closely related families (513)
orden: en la clasificación, un grupo de familias relacionadas estrechamente

organ: group of tissues that work together to perform closely related functions (216)
órgano: grupo de tejidos que trabajan juntos para realizar funciones estrechamente relacionadas

organ system: group of organs that work together to perform a specific function (216)
sistema de órganos: grupo de órganos que trabajan juntos para realizar una función específica

organelle: specialized structure that performs important cellular functions within a cell (196)
orgánulo: estructura especializada que realiza funciones celulares importantes dentro de una célula

osmosis: diffusion of water through a selectively permeable membrane (210)
ósmosis: la difusión de agua a través de una membrana de permeabilidad selectiva

osmotic pressure: pressure that must be applied to prevent osmotic movement across a selectively permeable membrane (211)
presión osmótica: la presión que debe aplicarse para evitar el movimiento osmótico a través de una membrana de permeabilidad selectiva

ossification: process of bone formation during which cartilage is replaced by bone (925)
osificación: el proceso de formación de hueso durante el cual el cartílago es reemplazado por hueso

osteoblast: bone cell that secretes mineral deposits that replace the cartilage in developing bones (925)
osteoblasto: célula ósea que secreta depósitos minerales que reemplazan al cartílago de los huesos en desarrollo

osteoclast: bone cell that breaks down bone minerals (925)
osteoclasto: célula ósea que degrada los minerales óseos

osteocyte: bone cell that helps maintain the minerals in bone tissue and continue to strengthen the growing bone (925)
osteocito: célula ósea que ayuda a conservar los minerales en el tejido óseo y continúa fortaleciendo al hueso en crecimiento

ovary: in plants, the structure that surrounds and protects seeds (650); in animals, the primary female reproductive organ; produces eggs (991)
ovario: en las plantas, la estructura que rodea a las semillas y las protege; órgano reproductor femenino fundamental en los animales; produce huevos

oviparous: species in which embryos develop in eggs outside a parent's body (822)
ovíparo: especie animal en la cual los embriones se desarrollan en huevos fuera del cuerpo del progenitor

ovoviparous: species in which the embryos develop within the mother's body but depend entirely on the yolk sac of their eggs (822)
ovovíparo: especie animal en la cual los embriones se desarrollan dentro del cuerpo de la madre, pero dependen completamente del saco vitelino de sus huevos

ovulation: the release of a mature egg from the ovary into one of the Fallopian tubes (993)
ovulación: liberación de un huevo maduro desde el ovario a una de las trompas de Falopio

ovule: structure in seed cones in which the female gametophytes develop (648)
óvulo: estructura de las semillas coníferas donde se desarrollan los gametos femeninos

ozone layer: atmospheric layer in which ozone gas is relatively concentrated; protects life on Earth from harmful ultraviolet rays in sunlight (175)
capa de ozono: capa atmosférica en la cual el gas ozono se encuentra relativamente concentrado; protege a los seres vivos de la Tierra de los perjudiciales rayos ultravioletas de la luz solar

P

pacemaker: small group of cardiac muscle fibers that maintains the heart's pumping rhythm by setting the rate at which the heart contracts; the sinoatrial (SA) node (951)
marcapasos: grupo pequeño de fibras musculares cardíacas que mantiene el ritmo de bombeo del corazón estableciendo la frecuencia a la que se contrae el corazón; el nodo sinusal

paleontologist: scientist who studies fossils (539)
paleontólogo: científico que estudia los fósiles

palisade mesophyll: layer of cells under the upper epidermis of a leaf (681)
mesófilo en empalizada: capa de células situada bajo la epidermis superior de una hoja

parasitism: symbiotic relationship in which one organism lives on or inside another organism and harms it (104)
parasitismo: relación simbiótica en la cual un organismo vive sobre otro organismo o en su interior y lo perjudica

parathyroid hormone (PTH): hormone produced by parathyroid gland that increases calcium levels in the blood (985)
hormona de la paratiroides: hormona producida por la glándula paratiroides que aumenta los niveles de calcio en la sangre

parenchyma: main type of ground tissue in plants that contains cells with thin cell walls and large central vacuoles (667)
parénquima: tipo principal de tejido fundamental de las plantas que contiene células con paredes celulares delgadas y vacuolas centrales grandes

passive immunity: temporary immunity that develops as a result of natural or deliberate exposure to an antibody (1020)
inmunidad pasiva: inmunidad transitoria que se desarrolla a consecuencia de una exposición natural o deliberada a un anticuerpo

pathogen: disease-causing agent (586)
patógeno: agente que causa una enfermedad

pedigree: chart that shows the presence or absence of a trait according to the relationships within a family across several generations (396)
árbol genealógico: diagrama que muestra la presencia o ausencia de un rasgo de acuerdo con las relaciones intrafamiliares a través de varias generaciones

pepsin: enzyme that breaks down proteins into smaller polypeptide fragments (877)
pepsina: enzima que descompone las proteínas en fragmentos de polipéptidos más pequeños

period: division of geologic time into which eras are subdivided (543)
período: división del tiempo geológico en la que se subdividen las eras

peripheral nervous system: network of nerves and supporting cells that carries signals into and out of the central nervous system (896)
sistema nervioso periférico: red de nervios y células de apoyo que transporta señales hacia y desde el sistema nervioso central

peristalsis: contractions of smooth muscles that provide the force that moves food through the esophagus toward the stomach (877)
peristalsis: contracciones de los músculos lisos que proporcionan la fuerza que hace avanzar los alimentos a través del esófago hacia el estómago

permafrost: layer of permanently frozen subsoil found in the tundra (115)
permacongelamiento: capa de subsuelo congelado en forma permanente que se halla en la tundra

petiole: thin stalk that connects the blade of a leaf to a stem (680)
pecíolo: pedúnculo delgado que une la lámina de una hoja con un tallo

pH scale: scale with values from 0 to 14, used to measure the concentration of H^+ ions in a solution; a pH of 0 to 7 is acidic, a pH of 7 is neutral, and a pH of 7 to 14 is basic (43)
escala del pH: escala con valores de 0 a 14, utilizada para medir la concentración de iones H^+ en una solución; un pH de 0 a 7 es ácido, un pH de 7 es neutro y un pH de 7 a 14 es básico

pharyngeal pouch: one of a pair of structures in the throat region of a chordate (731)
bolsa faríngea: cada una de las dos estructuras situadas en la región de la garganta de los cordados

pharynx: tube at the back of the mouth that serves as a passageway for both air and food; also called the throat (964)
faringe: tubo situado a continuación de la boca que sirve de conducto para que pasen el aire y los alimentos; también llamada garganta

phenotype: physical characteristics of an organism (315)
fenotipo: características físicas de un organismo

phloem: vascular tissue that transports solutions of nutrients and carbohydrates produced by photosynthesis through the plant (643)
floema: tejido vascular que transporta por toda la planta las soluciones de nutrientes e hidratos de carbono producidos en la fotosíntesis

photic zone: sunlight region near the surface of water (117)
zona fótica: región cerca de la superficie del mar en la que penetra la luz solar

photoperiodism: a plant response to the relative lengths of light and darkness (713)
fotoperiodismo: la respuesta de una planta a los tiempos relativos de luz y oscuridad

photosynthesis: process used by plants and other autotrophs to capture light energy and use it to power chemical reactions that convert carbon dioxide and water into oxygen and energy-rich carbohydrates such as sugars and starches (70, 228)
fotosíntesis: proceso empleado por las plantas y otros organismos autótrofos para atrapar la energía luminosa y utilizarla para impulsar reacciones químicas que convierten el dióxido de carbono y el agua en oxígeno e hidratos de carbono de gran contenido energético, como azúcares y almidones

photosystem: cluster of chlorophyll and proteins found in thylakoids (235)
fotosistema: conjunto de clorofila y proteínas que se hallan en los tilacoides

phototropism: tendency of a plant to grow toward a light source (712)
fototropismo: la tendencia de una planta a crecer hacia una fuente de luz

phylogeny: the evolutionary history of a lineage (516)
filogenia: historia evolutiva del linaje

phylum (pl. phyla): in classification, a group of closely related classes (514)
filo: en la clasificación, un grupo de clases estrechamente relacionadas

phytoplankton: photosynthetic algae found near the surface of the ocean (73)
fitoplancton: algas fotosintéticas que se hallan cerca de la superficie del océano

pigment: light-absorbing molecule used by plants to gather the sun's energy (230)
pigmento: moléculas que absorben la luz, empleadas por las plantas para recolectar la energía solar

pioneer species: first species to populate an area during succession (107)
especies pioneras: las primeras especies en poblar un área durante la sucesión ecológica

pistil: single carpel or several fused carpels; contains the ovary, style, and stigma (697)
pistilo: un único carpelo o varios carpelos unidos; contiene el ovario, el estilo y el estigma

pith: parenchyma cells inside the ring of vascular tissue in dicot stems (675)
médula: en los tallos de las dicotiledóneas, las células parenquimatosas ubicadas en el interior del anillo de tejido vascular

pituitary gland: small gland found near the base of the skull that secretes hormones that directly regulate many body functions and controls the actions of several other endocrine glands (982)
glándula pituitaria: pequeña glándula situada cerca de la base del cráneo que secreta hormonas que regulan directamente muchas funciones corporales y controla las acciones de varias otras glándulas endocrinas

placenta: specialized organ in placental mammals through which respiratory gases, nutrients, and wastes are exchanged between the mother and her developing young (822, 998)
placenta: órgano especializado de los mamíferos placentarios a través del cual se intercambian los gases respiratorios, los nutrientes y los residuos entre la madre y su cría en desarrollo

plankton: microscopic organisms that live in aquatic environments; includes both phytoplankton and zooplankton (119)
plancton: organismos microscópicos que viven en medios ambientes acuáticos; incluye el fitoplancton y el zooplancton

plasma: straw-colored liquid portion of the blood (954)
plasma: parte líquida de la sangre de color amarillento

plasmid: small, circular piece of DNA located in the cytoplasm of many bacteria (424)
plásmido: pequeña porción circular de ADN ubicada en el citoplasma de muchas bacterias

plasmodium: amoeboid feeding stage in the life cycle of a plasmodial slime mold (613)
plasmodio: etapa de alimentación ameboide del ciclo vital de los mohos mucilaginosos

plate tectonics: geologic processes, such as continental drift, volcanoes, and earthquakes, resulting from plate movement (544)
tectónica de placas: procesos geológicos, como la deriva continental, los volcanes y los terremotos, que son consecuencia de los movimientos de las placas

platelet: cell fragment released by bone marrow that helps in blood clotting (955)
plaqueta: fragmento celular liberado por la médula espinal que interviene en la coagulación de la sangre

pluripotent: cells that are capable of developing into most, but not all, of the body's cell types (294)
pluripotentes: células capaces de convertirse en la mayoría de células del cuerpo, pero no en todas

point mutation: gene mutation in which a single base pair in DNA has been changed (373)
mutación puntual: mutación genética en la cual se ha modificado un único par de bases en el ADN

pollen grain: structure that contains the entire male gametophyte in seed plants (647)
grano de polen: la estructura que contiene a todo el gametofito masculino en las plantas fanerógamas

pollen tube: structure in a plant that contains two haploid sperm nuclei (648)
tubo polínico: en una planta, estructura que contiene dos núcleos espermáticos haploides

pollination: transfer of pollen from the male reproductive structure to the female reproductive structure (647)
polinización: transferencia de polen desde la estructura reproductora masculina hacia la estructura reproductora femenina

pollutant: harmful material that can enter the biosphere through the land, air, or water (160)
contaminante: material nocivo que puede ingresar en la biósfera a través de la tierra, el aire o el agua

polygenic trait: trait controlled by two or more genes (320, 486)
rasgo poligénico: rasgo controlado por dos o más genes

polymer: molecules composed of many monomers; makes up macromolecules (46)
polímero: molécula compuesta por muchos monómeros; forma macromoléculas

polymerase chain reaction (PCR): the technique used by biologists to make many copies of a particular gene (423)

reacción en cadena de la polímerasa (PCR): técnica usada por los biólogos para hacer muchas copias de un gen específico

polypeptide: long chain of amino acids that makes proteins (366)

polipéptido: cadena larga de aminoácidos que constituye las proteínas

polyploidy: condition in which an organism has extra sets of chromosomes (376)

poliploidía: condición en la cual un organismo tiene grupos adicionales de cromosomas

population: group of individuals of the same species that live in the same area (64)

población: grupo de individuos de la misma especie que viven en la misma área

population density: number of individuals per unit area (131)

densidad de población: número de individuos que viven por unidad de superficie

predation: interaction in which one organism (the predator) captures and feeds on another organism (the prey) (102)

depredación: interacción en la cual un organismo (el predador) captura y come a otro organismo (la presa)

prehensile tail: long tail that can coil tightly enough around a branch to serve as a "fifth hand" (767)

cola prensil: cola larga que puede enrollarse apretadamente alrededor de una rama, y que puede usarse como una "quinta mano"

pressure-flow hypothesis: hypothesis that explains the method by which phloem sap is transported through the plant from a sugar "source" to a sugar "sink" (687)

teoría de flujo por presión: teoría que explica el método por el cual la savia del floema recorre la planta desde una "fuente" de azúcar hacia un "vertedero" de azúcar

primary growth: pattern of growth that takes place at the tips and shoots of a plant (676)

crecimiento primario: patrón de crecimiento que tiene lugar en las puntas y en los brotes de una planta

primary producer: first producer of energy-rich compounds that are later used by other organisms (69)

productor primario: los primeros productores de compuestos ricos en energía que luego son utilizados por otros organismos

primary succession: succession that occurs in an area in which no trace of a previous community is present (106)

sucesión primaria: sucesión que ocurre en un área en la cual no hay rastros de la presencia de una comunidad anterior

principle of dominance: Mendel's second conclusion, which states that some alleles are dominant and others are recessive (310)

principio de dominancia: segunda conclusión de Mendel, que establece que algunos alelos son dominantes y otros son recesivos

prion: protein particles that cause disease (592)

prión: partículas de proteína que causan enfermedades

probability: likelihood that a particular event will occur (313)

probabilidad: la posibilidad de que ocurra un suceso dado

product: elements or compounds produced by a chemical reaction (50)

producto: elemento o compuesto producido por una reacción química

prokaryote: unicellular organism that lacks a nucleus (193, 580)

procariota: organismo unicelular que carece de núcleo

promoter: specific region of a gene where RNA polymerase can bind and begin transcription (365)

promotor: región específica de un gen en donde la ARN polimerasa puede unirse e iniciar la transcripción

prophage: bacteriophage DNA that is embedded in the bacterial host's DNA (577)

profago: ADN del bacteriófago que está alojado en el interior del ADN del huésped bacteriano

prophase: first and longest phase of mitosis in which the genetic material inside the nucleus condenses and the chromosomes become visible (282)

profase: primera y más prolongada fase de la mitosis, en la cual el material genético dentro del interior del núcleo se condensa y los cromosomas se hacen visibles

prostaglandin: modified fatty acids that are produced by a wide range of cells; generally affect only nearby cells and tissues (980)

prostaglandina: ácidos grasos modificados que son producidos por una amplia gama de células; generalmente afectan solo a las células y tejidos cercanos

protein: macromolecule that contains carbon, hydrogen, oxygen, and nitrogen; needed by the body for growth and repair (48, 870)

proteína: macromolécula que contiene carbono, hidrógeno, oxígeno y nitrógeno; necesaria para el crecimiento y reparación del cuerpo

protostome: an animal whose mouth is formed from the blastopore (739)

protóstomo: animal cuya boca se desarrolla a partir del blastoporo

pseudocoelom: body cavity that is only partially lined with mesoderm (738)

pseudoceloma o falso celoma: cavidad corporal que está revestida sólo parcialmente con mesodermo

Glossary *(continued)*

pseudopod: temporary cytoplasmic projection used by some protists for movement (606)
seudópodo: prolongación citoplasmática transitoria utilizada por algunos protistas para moverse

puberty: period of rapid growth and sexual maturation during which the reproductive system becomes fully functional (988)
pubertad: período de crecimiento rápido y de maduración sexual durante el cual el sistema reproductor se vuelve completamente funcional

pulmonary circulation: path of circulation between the heart and lungs (950)
circulación pulmonar: recorrido de la circulación entre el corazón y los pulmones

punctuated equilibrium: pattern of evolution in which long stable periods are interrupted by brief periods of more rapid change (549)
equilibrio interrumpido: patrón de evolución en el cual los largos períodos de estabilidad se ven interrumpidos por breves períodos de cambio más rápido

Punnett square: diagram that can be used to predict the genotype and phenotype combinations of a genetic cross (315)
cuadro de Punnett: un diagrama que puede utilizarse para predecir las combinaciones de genotipos y fenotipos en un cruce genético

pupa: stage in complete metamorphosis in which the larva develops into an adult (823)
pupa: etapa de la metamorfosis completa en la cual la larva se convierte en un adulto

pupil: small opening in the iris that admits light into the eye (912)
pupila: pequeña abertura en el iris que deja pasar la luz al ojo

Q, R

radial symmetry: body plan in which any number of imaginary planes drawn through the center of the body could divide it into equal halves (738)
simetría radial: diseño corporal en el cual cualquier número de ejes imaginarios dibujados a través del centro del cuerpo lo dividirá en mitades iguales

radiometric dating: method for determining the age of a sample from the amount of a radioactive isotope to the nonradioactive isotope of the same element in a sample (540)
datación radiométrica: método para determinar la edad de una muestra a partir de la cantidad de isótopo radioactivo en relación a la de isótopo no radiactivo del mismo elemento en dicha muestra

reabsorption: process by which water and dissolved substances are taken back into the blood (884)
reabsorción: proceso por el cual el agua y las sustancias disueltas regresan a la sangre

reactant: elements or compounds that enter into a chemical reaction (50)
reactante: elemento o compuesto que participa en una reacción química

receptor: on or in a cell, a specific protein to whose shape fits that of a specific molecular messenger, such as a hormone (217, 709)
receptor: proteína específica que puede encontrarse en la membrana celular o dentro de la célula, cuya forma se corresponde con la de un mensajero molecular específico, por ejemplo una hormona

recombinant DNA: DNA produced by combining DNA from different sources (424)
ADN recombinante: ADN producido por la combinación de ADN de orígenes diferentes

red blood cell: blood cell containing hemoglobin that carries oxygen (954)
glóbulo rojo: célula sanguínea que contiene hemoglobina y transporta oxígeno

reflex: quick, automatic response to a stimulus (901)
reflejo: respuesta rápida y automática a un estímulo

reflex arc: the sensory receptor, sensory neuron, motor neuron, and effector that are involved in a quick response to a stimulus (907)
arco reflejo: el receptor sensorial, la neurona sensorial, la neurona motora y el efector que participan en una respuesta rápida a un estímulo

relative dating: method of determining the age of a fossil by comparing its placement with that of fossils in other rock layers (540)
datación relativa: método para determinar la edad de un fósil comparando su ubicación con la de los fósiles hallados en otras capas de roca

releasing hormone: hormone produced by the hypothalamus that makes the anterior pituitary secrete hormones (983)
hormona liberadora: hormona producida por el hipotálamo que hace que la glándula pituitaria anterior secrete hormonas (983)

renewable resource: resource that can be produced or replaced by healthy ecosystem functions (157)
recurso renovable: recurso que se puede producir o reemplazar mediante el funcionamiento saludable del ecosistema

replication: process of copying DNA prior to cell division (350)
replicación: proceso de copia de ADN previo a la división celular

reproductive isolation: separation of a species or population so that they no longer interbreed and evolve into two separate species (494)
aislamiento reproductor: separación de una especie o de una población de tal manera que ya no pueden aparearse y evolucionan hasta formar dos especies separadas

resource: any necessity of life, such as water, nutrients, light, food, or space (100)
recurso: todo lo necesario para la vida, como agua, nutrientes, luz, alimento o espacio

response: specific reaction to a stimulus (809)
respuesta: reacción específica a un estímulo

resting potential: electrical charge across the cell membrane of a resting neuron (898)
potencial de reposo: carga eléctrica que pasa a través de la membrana celular de una neurona en reposo

restriction enzyme: enzyme that cuts DNA at a sequence of nucleotides (403)
enzima restrictiva: enzima que corta el ADN en una secuencia de nucleótidos

retina: innermost layer of the eye; contains photoreceptors (913)
retina: membrana más interna del ojo; contiene receptores susceptibles a la luz

retrovirus: RNA virus that contains RNA as its genetic information (578)
retrovirus: ARN viral cuya información genética está contenida en el ARN

ribonucleic acid (RNA): single-stranded nucleic acid that contains the sugar ribose (362)
ácido ribonucleico (ARN): hebra única de ácido nucleico que contiene el azúcar ribose

ribosomal RNA (rRNA): type of RNA that combines with proteins to form ribosomes (363)
ARN ribosomal: tipo de ARN que se combina con proteínas para formar los ribosomas

ribosome: cell organelle consisting of RNA and protein found throughout the cytoplasm in a cell; the site of protein synthesis (200)
ribosoma: orgánulo celular formado por ARN y proteína que se halla en el citoplasma de una célula; lugar donde se sintetizan las proteínas

RNA interference (RNAi): introduction of double-stranded RNA into a cell to inhibit gene expression (380)
ARN de interferencia: introducción de un ARN de doble hebra en una célula para inhibir la expresión de genes específicos

RNA polymerase: enzyme that links together the growing chain of RNA nucleotides during transcription using a DNA strand as a template (364)
ARN polimerasa: enzima que enlaza los nucleótidos de la cadena de ARN en crecimiento durante la transcripción, usando una secuencia de ADN como patrón o molde

rod: photoreceptor in the eyes that is sensitive to light but can't distinguish color (913)
bastoncillo: receptor ubicado en los ojos que es susceptible a la luz, pero que no puede distinguir el color

root cap: tough covering of the root tip that protects the meristem (670)
cofia: cubierta dura de la punta de las raíces que protege al meristemo

root hair: small hairs on a root that produce a large surface area through which water and minerals can enter (670)
pelo radicular: pelos pequeños sobre una raíz que producen una superficie extensa a través de la cual pueden penetrar el agua y los minerales

rumen: stomach chamber in cows and related animals in which symbiotic bacteria digest cellulose (786)
panza: cavidad del estómago de las vacas y otros rumiantes en la cual las bacterias simbióticas digieren la celulosa

S

sapwood: in a woody stem, the layer of secondary phloem that surrounds the heartwood; usually active in fluid transport (678)
albura: en un tallo leñoso, la capa de floema secundario que rodea al duramen; participa usualmente en el transporte de fluidos

sarcomere: unit of muscle contraction; composed of two z-lines and the filaments between them (930)
sarcómero: unidad de contracción muscular; compuesto por dos líneas "z" y los filamentos que hay entre ellas

scavenger: animal that consumes the carcasses of other animals (71)
carroñero: animal que consume los cadáveres de otros animales

science: organized way of gathering and analyzing evidence about the natural world (5)
ciencia: manera organizada de reunir y analizar la información sobre el mundo natural

sclerenchyma: type of ground tissue with extremely thick, rigid cell walls that make ground tissue tough and strong (667)
esclerénquima: tipo de tejido fundamental con células extremadamente rígidas y gruesas que lo hacen fuerte y resistente

scrotum: external sac that houses the testes (989)
escroto: bolsa externa que contiene a los testículos

sebaceous gland: gland in the skin that secretes sebum (oily secretion) (937)
glándula sebácea: glándula de la piel que secreta sebo (secreción oleosa)

secondary growth: type of growth in dicots in which the stems increase in thickness (676)
crecimiento secundario: tipo de crecimiento de las dicotiledóneas en el cual los tallos aumentan su grosor

secondary succession: type of succession that occurs in an area that was only partially destroyed by disturbances (107)
sucesión secundaria: tipo de sucesión que ocurre en un área destruida sólo parcialmente por alteraciones

seed: plant embryo and a food supply encased in a protective covering (646)
semilla: embrión vegetal y fuente de alimento encerrada en una cubierta protectora

seed coat: tough covering that surrounds and protects the plant embryo and keeps the contents of the seed from drying out (647)
envoltura de la semilla: cubierta dura que rodea y protege al embrión de la planta y evita que el contenido de la semilla se seque

segregation: separation of alleles during gamete formation (312)
segregación: separación de los alelos durante la formación de gametos

selective breeding: method of breeding that allows only those organisms with desired characteristics to produce the next generation (418)
reproducción selectiva o selección artificial: método de reproducción que sólo permite la producción de una nueva generación a aquellos organismos con características deseadas

selectively permeable: property of biological membranes that allows some substances to pass across it while others cannot; also called semipermeable membrane (205)
permeabilidad selectiva: propiedad de las membranas biológicas que permite que algunas sustancias pasen a través de ellas mientras que otras no pueden hacerlo; también llamada membrana semipermeable

semen: the combination of sperm and seminal fluid (990)
semen: combinación de esperma y de fluido seminal

semicircular canal: one of three structures in the inner ear that monitor the position of the body in relation to gravity (911)
canal semicircular: una de las tres estructuras ubicadas en el oído interno que controlan la posición del cuerpo en relación con la fuerza de la gravedad

seminiferous tubule: one of hundreds of tubules in each testis in which sperm develop (989)
túbulo seminífero: uno de los cientos de túbulos situados en cada testículo, en los cuales se produce el esperma

sensory neuron: type of nerve cell that receives information from sensory receptors and conveys signals to central nervous system (808)
neurona sensorial: tipo de célula nerviosa que recibe información de los receptores sensoriales y transmite señales al sistema nervioso central

sex chromosome: one of two chromosomes that determines an individual's sex (393)
cromosoma sexual: uno de los pares de cromosomas que determina el sexo de un individuo

sex-linked gene: gene located on a sex chromosome (395)
gen ligado al sexo: gen situado en un cromosoma sexual

sexual reproduction: type of reproduction in which cells from two parents unite to form the first cell of a new organism (19, 277)
reproducción sexual: tipo de reproducción en la cual las células de dos progenitores se unen para formar la primera célula de un nuevo organismo

sexual selection: when individuals select mates based on heritable traits (492)
selección sexual: cuando un individuo elige a su pareja sexual atraído por sus rasgos heredables

sexually transmitted disease (STD): disease that is spread from person to person by sexual contact (994)
enfermedad de transmisión sexual (ETS): enfermedad que se transmite de una persona a otra por contacto sexual

sieve tube element: continuous tube through the plant phloem cells, which are arranged end to end (666)
tubo crivoso: tubo continuo que atraviesa las células del floema vegetal, que están puestas una junto a otra

single-gene trait: trait controlled by one gene that has two alleles (485)
rasgo de un único gen (monogénico): rasgo controlado por un gen que tiene dos alelos

small intestine: digestive organ in which most chemical digestion and absorption of food takes place (878)
intestino delgado: órgano digestivo en el cual tiene lugar la mayor parte de la digestión química y la absorción de los alimentos

smog: gray-brown haze formed by a mixture of chemicals (163)
esmog: neblina marrón grisácea formada por una mezcla de compuestos químicos

society: group of closely related animals of the same species that work together for the benefit of the group (848)
sociedad: grupo de animales de la misma especie, estrechamente relacionados, que trabajan juntos para el beneficio del grupo

solute: substance that is dissolved in a solution (42)
soluto: sustancia que está disuelta en una solución

solution: type of mixture in which all the components are evenly distributed (42)
solución: tipo de mezcla en la cual todos los compuestos están distribuidos de forma homogénea

solvent: dissolving substance in a solution (42)
disolvente: sustancia que disuelve una solución

somatic nervous system: part of the peripheral nervous system that carries signals to and from skeletal muscles (907)
sistema nervioso somático: parte del sistema nervioso periférico que conduce señales hacia y desde los músculos esqueléticos

speciation: formation of a new species (494)
especiación: formación de una nueva especie

species: a group of similar organisms that can breed and produce fertile offspring (64, 494)
especie: un grupo de organismos similares que pueden reproducirse y producir una descendencia fértil

species diversity: number of different species that make up a particular area (166)
diversidad de especies: número de especies diferentes que forman un área determinada

spirillum (pl. spirilla): spiral or corkscrew-shaped prokaryote (582)
espirilo: procariota con forma helicoidal o espiral

spongy mesophyll: layer of loose tissue found beneath the palisade mesophyll in a leaf (681)
mesófilo esponjoso: capa de tejido suelto situado debajo del mesófilo en empalizada de una hoja

sporangium (pl. sporangia): spore capsule in which haploid spores are produced by meiosis (609, 642)
esporangio: cápsula en la cual se producen las esporas haploides mediante meiosis

spore: in prokaryotes, protists, and fungi, any of a variety of thick-walled life cycle stages capable of surviving unfavorable conditions (607)
espora: en los procariotas, los protistas y los hongos, cada una de las células que, en un momento de su ciclo de vida, produce una membrana gruesa y resistente capaz de sobrevivir en condiciones desfavorables

sporophyte: spore-producing plant; the multicellular diploid phase of a plant life cycle (637)
esporofito: planta productora de esporas; la fase diploide multicelular del ciclo vital de una planta

stabilizing selection: form of natural selection in which individuals near the center of a distribution curve have higher fitness than individuals at either end of the curve (489)
selección estabilizadora: forma de selección natural en la cual los individuos situados cerca del centro de una curva de distribución tienen mayor aptitud que los individuos que se hallan en cualquiera de los extremos de la curva

stamen: male part of a flower; contains the anther and filament (697)
estambre: parte masculina de una flor; contiene la antera y el filamento

stem cell: unspecialized cell that can give rise to one or more types of specialized cells (295)
célula troncal: célula no especializada que puede originar uno o más tipos de células especializadas

stigma: sticky part at the top of style; specialized to capture pollen (697)
estigma: parte pegajosa situada en la parte superior del estilo; especializado en atrapar el polen

stimulus (pl. stimuli): signal to which an organism responds (18, 808)
estímulo: señal a la cual responde un organismo

stoma (pl. stomata): small opening in the epidermis of a plant that allows carbon dioxide, water, and oxygen to diffuse into and out of the leaf (681)
estoma: pequeña abertura en la epidermis de una planta que permite que el dióxido de carbono, el agua y el oxígeno entren y salgan de la hoja

stomach: large muscular sac that continues the mechanical and chemical digestion of food (877)
estómago: gran bolsa muscular que continúa la digestión mecánica y química de los alimentos

stroma: fluid portion of the chloroplast; outside of the thylakoids (231)
estroma: parte fluida del cloroplasto; en el exterior de los tilacoides

substrate: reactant of an enzyme-catalyzed reaction (52)
sustrato: reactante de una reacción catalizada por enzimas

suspension: mixture of water and nondissolved material (42)
suspensión: mezcla de agua y material no disuelto

sustainable development: strategy for using natural resources without depleting them and for providing human needs without causing long-term environmental harm (157)
desarrollo sostenible: estrategia para utilizar los recursos naturales sin agotarlos y para satisfacer las necesidades humanas sin causar daños ambientales a largo plazo

symbiosis: relationship in which two species live close together (103)
simbiosis: relación en la cual dos especies viven en estrecha asociación

synapse: point at which a neuron can transfer an impulse to another cell (900)
sinapsis: punto en el cual una neurona puede transferir un impulso a otra célula

systematics: study of the diversity of life and the evolutionary relationships between organisms (512)
sistemática: estudio de la diversidad de la vida y de las relaciones evolutivas entre los organismos

systemic circulation: path of circulation between the heart and the rest of the body (950)
circulación sistémica: recorrido de la circulación entre el corazón y el resto del cuerpo

T

taiga: biome with long cold winters and a few months of warm weather; dominated by coniferous evergreens; also called boreal forest (114)
  taiga: bioma con inviernos largos y fríos y pocos meses de tiempo cálido; dominado por coníferas de hojas perennes; también llamada bosque boreal

target cell: cell that has a receptor for a particular hormone (709, 978)
  célula diana o célula blanco: célula que posee un receptor para una hormona determinada

taste bud: sense organs that detect taste (910)
  papila gustativa: órgano sensorial que percibe los sabores

taxon (pl. taxa): group or level of organization into which organisms are classified (512)
  taxón: grupo o nivel de organización en que se clasifican los organismos

telomere: repetitive DNA at the end of a eukaryotic chromosome (352)
  telómero: ADN repetitivo situado en el extremo de un cromosoma eucariota

telophase: phase of mitosis in which the distinct individual chromosomes begin to spread out into a tangle of chromatin (283)
  telofase: fase de la mitosis en la cual los distintos cromosomas individuales comienzan a separarse y a formar hebras de cromatina

temporal isolation: form of reproductive isolation in which two or more species reproduces at different times (495)
  aislamiento temporal: forma de aislamiento reproductivo en la cual dos o más especies se reproducen en épocas diferentes

tendon: tough connective tissue that connects skeletal muscles to bones (816, 932)
  tendón: tejido conectivo resistente que une los músculos esqueléticos a los huesos

territory: a specific area occupied and protected by an animal or group of animals (848)
  territorio: área específica ocupada y protegida por un animal o un grupo de animales

testis (pl. testes): primary male reproductive organ; produces sperm (989)
  testículo: órgano reproductor masculino fundamental; produce esperma

tetrad: structure containing four chromatids that forms during meiosis (324)
  tétrada: estructura con cuatro cromátidas que se forma durante la meiosis

tetrapod: vertebrate with four limbs (760)
  tetrápode: vertebrado con quatro membros

thalamus: brain structure that receives messages from the sense organs and relays the information to the proper region of the cerebrum for further processing (903)
  tálamo: estructura cerebral que recibe mensajes de los órganos sensoriales y transmite la información a la región adecuada del cerebro para su procesamiento ulterior

theory: well-tested explanation that unifies a broad range of observations and hypotheses, and enables scientists to make accurate predications about new situations (13)
  teoría: explicación basada en pruebas que unifica una amplia gama de observaciones e hipótesis; permite que los científicos hagan predicciones exactas ante situaciones nuevas

thigmotropism: response of a plant to touch (712)
  tigmotropismo: respuesta de una planta al tacto

threshold: minimum level of a stimulus that is required to cause an impulse (899)
  umbral: nivel mínimo que debe tener un estímulo para causar un impulso

thylakoid: saclike photosynthetic membranes found in chloroplasts (231)
  tilacoide: membranas fotosintéticas con forma de bolsa situadas en los cloroplastos

thyroxine: hormone produced by the thyroid gland, which increases the metabolic rate of cells throughout the body (985)
  tiroxina: hormona producida por la glándula tiroides que aumenta el metabolismo de las células de todo el cuerpo

tissue: group of similar cells that perform a particular function (216)
  tejido: grupo de células similares que realizan una función en particular

tolerance: ability of an organism to survive and reproduce under circumstances that differ from their optimal conditions (99)
  tolerancia: capacidad de un organismo de sobrevivir y reproducirse en circunstancias que difieren de sus condiciones óptimas

totipotent: cells that are able to develop into any type of cell found in the body (including the cells that make up the extraembryonic membranes and placenta) (294)
  totipotentes: células capaces de convertirse en cualquier tipo de célula del cuerpo (incluidas las células que forman las membranas situadas fuera del embrión y la placenta)

trachea: tube that connects the larynx to the bronchi; also called the windpipe (964)
  tráquea: tubo que conecta a la laringe con los bronquios

tracheid: hollow plant cell in xylem with thick cell walls strengthened by lignin (643)
  traqueida: célula vegetal ahuecada del xilema con paredes celulares gruesas, fortalecida por la lignina

tracheophyte: vascular plant (643)
traqueófita: planta vascular

trait: specific characteristic of an individual (309)
rasgo: característica específica de un individuo

transcription: synthesis of an RNA molecule from a DNA template (364)
transcripción: síntesis de una molécula de ARN a partir de una secuencia de ADN

transfer RNA (tRNA): type of RNA that carries each amino acid to a ribosome during protein synthesis (363)
ARN de transferencia: tipo de ARN que transporta a cada aminoácido hasta un ribosoma durante la síntesis de proteínas

transformation: process in which one strain of bacteria is changed by a gene or genes from another strain of bacteria (339)
transformación: proceso en el cual una cepa de bacterias es transformada por uno o más genes provenientes de otra cepa de bacterias

transgenic: term used to refer to an organism that contains genes from other organisms (426)
transgénico: término utilizado para referirse a un organismo que contiene genes provenientes de otros organismos

translation: process by which the sequence of bases of an mRNA is converted into the sequence of amino acids of a protein (368)
traducción (genética): proceso por el cual la secuencia de bases de un ARN mensajero se convierte en la secuencia de aminoácidos de una proteína

transpiration: loss of water from a plant through its leaves (681)
transpiración: pérdida del agua de una planta a través de sus hojas

trochophore: free-swimming larval stage of an aquatic mollusk (756)
trocófora: estado larvario de un molusco acuático durante el cual puede nadar libremente

trophic level: each step in a food chain or food web (77)
nivel trófico: cada paso en una cadena o red alimenticia

tropism: movement of a plant toward or away from stimuli (712)
tropismo: movimiento de una planta hacia los estímulos o en dirección opuesta a ellos

tumor: mass of rapidly dividing cells that can damage surrounding tissue (289)
tumor: masa de células que se dividen rápidamente y pueden dañar al tejido circundante

U

understory: layer in a rain forest found underneath the canopy formed by shorter trees and vines (112)
sotobosque: en un bosque tropical, la capa de vegetación que se halla bajo el dosel forestal, formada por árboles más bajos y enredaderas

ureter: tube that carries urine from a kidney to the urinary bladder (883)
uréter: conducto que transporta la orina del riñón a la vejiga urinaria

urethra: tube through which urine leaves the body (883)
uretra: conducto por donde la orina sale del cuerpo

urinary bladder: saclike organ in which urine is stored before being excreted (883)
vejiga urinaria: órgano en forma de bolsa en el cual se almacena la orina antes de ser excretada

V

vaccination: injection of a weakened, or a similar but less dangerous, pathogen to produce immunity (1020)
vacunación: inyección de un patógeno debilitado o similar al original, pero menos peligroso, para producir inmunidad

vaccine: preparation of weakened or killed pathogens used to produce immunity to a disease (588)
vacuna: preparación hecha con organismos patógenos debilitados o muertos que se utiliza para producir inmunidad a una enfermedad

vacuole: cell organelle that stores materials such as water, salts, proteins, and carbohydrates (198)
vacuola: orgánulo celular que almacena sustancias como agua, sales, proteínas e hidratos de carbono

valve: flap of connective tissue located between an atrium and a ventricle, or in a vein, that prevents backflow of blood (950)
válvula: pliegue de tejido conectivo ubicado entre una aurícula y un ventrículo, o en una vena, que impide el retroceso de la sangre

van der Waals force: slight attraction that develops between oppositely charged regions of nearby molecules (38)
fuerzas de van der Waals: atracción leve que se desarrolla entre las regiones con cargas opuestas de moléculas cercanas

vas deferens: tube that carries sperm from the epididymis to the urethra (989)
conducto deferente: tubo que transporta el esperma desde el epidídimo a la uretra

vascular bundle: clusters of xylem and phloem tissue in stems (675)
hacecillo vascular: manojo de tejidos del xilema y del floema en los tallos

vascular cambium: meristem that produces vascular tissues and increases the thickness of stems (677)
cámbium vascular: meristemo que produce tejidos vasculares y aumenta el grosor de los tallos

Glossary *(continued)*

vascular cylinder: central region of a root that includes the vascular tissues—xylem and phloem (670)
  cilindro vascular: región central de una raíz que incluye a los tejidos vasculares xilema y floema

vascular tissue: specialized tissue in plants that carries water and nutrients (641)
  tejido vascular: tejido especializado de las plantas que transporta agua y nutrientes

vector: animal that transports a pathogen to a human (1013)
  vector: animal que transmite un patógeno a un ser humano

vegetative reproduction: method of asexual reproduction in plants, which enables a single plant to produce offspring that are genetically identical to itself (702)
  reproducción vegetativa: método de reproducción asexual de las plantas que permite que una única planta produzca descendencia genéticamente idéntica a sí misma

vein: blood vessel that carries blood from the body back to the heart (952)
  vena: vaso sanguíneo que transporta la sangre del cuerpo de regreso al corazón

ventricle: lower chamber of the heart that pumps blood out of heart to the rest of the body (792, 949)
  ventrículo: cavidad inferior del corazón que bombea la sangre fuera del corazón hacia el resto del cuerpo

vertebrate: animal that has a backbone (731)
  vertebrado: animal que posee una columna vertebral

vessel element: type of xylem cell that forms part of a continuous tube through which water can move (666)
  elemento vascular (o vaso): tipo de célula del xilema que forma parte de un tubo continuo a través del cual el agua puede desplazarse

vestigial structure: structure that is inherited from ancestors but has lost much or all of its original function (469)
  estructura vestigial: estructura heredada de los ancestros que ha perdido su función original en gran parte o por completo

villus (pl. villi): fingerlike projection in the small intestine that aids in the absorption of nutrient molecules (880)
  vellosidad: proyección en forma de dedo en el intestino delgado que contribuye a la absorción de las moléculas nutrientes

virus: particle made of proteins, nucleic acids, and sometimes lipids that can replicate only by infecting living cells (574)
  virus: partícula compuesta por proteínas, ácidos nucleicos y, a veces, lípidos, que puede replicarse sólo infectando células vivas

vitamin: organic molecule that helps regulate body processes (871)
  vitamina: molécula orgánica que ayuda a regular los procesos corporales

viviparous: animals that bear live young that are nourished directly by the mother's body as they develop (822)
  vivíparo: animal que da a luz crías vivas que se nutren directamente dentro del cuerpo de la madre mientras se desarrollan

W

weather: day-to-day conditions of the atmosphere, including temperature, precipitation, and other factors (96)
  tiempo: condiciones diarias de la atmósfera, entre las que se incluyen la temperatura, la precipitación y otros factores

wetland: ecosystem in which water either covers the soil or is present at or near the surface for at least part of the year (119)
  humedal: ecosistema en el cual el agua cubre el suelo o está presente en la superficie durante al menos una parte del año

white blood cell: type of blood cell that guards against infection, fights parasites, and attacks bacteria (955)
  glóbulo blanco: tipo de célula sanguínea que protege de las infecciones, combate a los parásitos y ataca a las bacterias

woody plant: type of plant made primarily of cells with thick cell walls that support the plant body; includes trees, shrubs, and vines (653)
  planta leñosa: tipo de planta constituida fundamentalmente por células con paredes celulares gruesas que sostienen el cuerpo de la planta; en este tipo se incluyen los árboles, arbustos y vides

X, Y, Z

xylem: vascular tissue that carries water upward from the roots to every part of a plant (643)
  xilema: tejido vascular que transporta el agua hacia arriba, desde las raíces a cada parte de una planta

zoonosis (pl. zoonoses): disease transmitted from animal to human (1013)
  zoonosis: enfermedad transmitida por un animal a un ser humano

zooplankton: small free-floating animals that form part of plankton (76)
  zooplancton: pequeños animales que flotan libremente y forman parte del plancton

zygote: fertilized egg (325, 739, 995)
  cigoto: huevo fertilizado

Index

A

Abiotic factors, **66**–67, 100, 111
Abscisic acid, **711**
Absorption, 875
Acetylcholine, **931**
Acetyl-CoA, 256
Acid rain, **164**
Acids, **44**
amino, **48**–49, 84, 366, 870
fatty, 47–48
nucleic, **48,** 344, 403, 436
Acne, 938
Acoelomates, 738
Acquired characteristics, 456
Acquired immune deficiency syndrome (AIDS), 578, 588, 994, 1001, 1023, 1025–1027
Acquired immunity, 1020
Actin, 199, 606, **930**
Action potential, **898**–899
Activation energy, **51**–52
Active immunity, **1020**
Active site, 53
Active transport, **212**–213, 227
in plants, 635, 641, 666, 672–673, 685–687
Adaptations, **461**
to biomes, 112–115
in chordates, 758–759, 761–762, 825
evolutionary, 487
excretory, 798
to high altitude, 831
of leaves, 684
of mouthparts, 785–786
and natural selection, 461–464
in pathogens, 1012
seasonal, 714
of seed plants, 646
Adaptive radiation, **550**–551
Addiction, 904–905
Adenine, 344–345, 348, 366
Adenosine diphosphate (ADP), 227, 235
Adenosine triphosphate (ATP), 48, **226**–227
ATP synthase, **237**, 258
and cellular respiration, 252, 254–260
and exercise, 264–265
and fermentation, 262–263
and muscle contraction, 930
and photosynthesis, 235–239
Adhesion, **41, 686**
Adrenal glands, 979, **983**
Adreno-corticotropic hormone (ACTH), 983
Adult stem cells, 295, 297
Aerobic respiration, **252**
African sleeping sickness, 615
Age of Fishes, 759
Age structure of populations, **131,** 144
Aggression, **848**
Agriculture, 155, 715–717, 719
AIDS, 578, 588, 994, 1001, 1023, 1025–1027
Air pollution, 163–165
Albumin, 954
Alcohol, 904
Alcoholic fermentation, 263
Aldosterone, 983
Alemseged, Zeresenay, 773
Algae, 610
brown, 527, 602, 604, 611
in food chains and webs, 73–74
green, 528, 610, 623, 634, 636–640
and photosynthesis, 70
phytoplankton, 73, 117
red, 528, 611
symbiotic, 783
unicellular, 214
Algal bloom, **611**
Alkaline solution, 44
Allantois, 825
Alleles, **310,** 482. *See also* Genetics
allele frequencies, **483,** 490–492
dominant and recessive, 310–312, 318, 394
and gene linkage, 328–329
multiple and codominant, **320,** 394
and phenotypes, 488
segregation of, **312,** 314, 318
and traits, 397, 485–486
Allergies, **1024**
Alternation of generations, **608, 637**
Alveoli, 790, **964**–966
Amebic dysentery, 615
Amenorrhea, 1005
Amino acids, **48**–49, 870
and genetic code, 366
in nitrogen cycle, 84
Ammonia, 585, 794–795, 867, 882
Amnion, 825, 998
Amniota, 518–519
Amniotic egg, **825**
Amoebas, 213, 606, 608, 612
Amoeboid movement, 606
Amphibians, 761, 790, 793, 796
brains of, 811
fertilization in, 821–822
metamorphosis in, 824
Amygdala, 902
Amylase, **876**
Anaerobic respiration, **252,** 262
Angina, 958
Angiosperms, **646,** 647, 650–654
classification of, 652–653
double fertilization in, **700**–701
fruit development in, 704
life cycle of, 698–701
structure of, 696–697
types of, 653–654
vessel elements in, **666**
Animal behavior, 840–851
and climate change, 178
communication, 850–851
complex, 844–845
cycles of, 847
and evolution, 840–841, 848–849
innate, **841,** 844–845
learned, 842–843
social, 848–849
Animalia, 514, 523–524, 528, 730
Animals, 730–743. *See also* Chordates; Invertebrates; Vertebrates
animal society, **848**–849
asexual reproduction in, 277, 735, **819**
body plans of, 737–741
cell differentiation in, 293
cells of, 203, 206–207, 211, 215
characteristics of, 730
cladogram of, 742–743
cloned, 427, 429
cytokinesis, 284
development of young, 823–824
differentiation stages in, 740
embryo development in, 738–739, 822
and emerging diseases, 1022
genetically modified, 429–430
homeostasis in, 732, 827–830
hormones of, 708
and language, 851
parental care in, 824
and pollination, 700
response to stimuli, 733, 808–809
and seed dispersal, 651, **705**
sexual reproduction in, 735, **820**–822
transgenic, 426, 429–430
Animal systems
circulation, 734, **791**–793, 864
digestive, 784–786

Index *continued*

Animal systems (cont'd)
- endocrine, 828
- excretion, 734–735, **794**–798
- feeding and digestion, 734, 782–786
- and homeostasis, 732, 827–830
- muscular, 733, **816**–818
- nervous, 733, 808–811, 864, 951
- reproductive, 277, 735, 819–826
- respiration, 734–735, 787–790, 864
- sensory, **812**–813
- skeletal, 733, **814**–816

Annelids, 755–756, 792
Anterior cruciate ligament (ACL), 927
Antheridia, **642**
Anthers, **697**
Anthrax, 583, 589
Anthropoids, **766,** 766–767
Antibiotics, **588,** 591, 1021
Antibodies, **1016**–1017
Anticodons, **369**
Antidiuretic hormone (ADH), 886, 982, 986–987
Antigens, 394, **1016**
Antihistamines, 1024
Antiparallel strands, 347
Anus, 739
Aorta, 952
Aphotic zone, **117,** 121
Apical dominance, **710**
Apical meristems, **668,** 670, 714
Apoptosis, **288,** 1019
Appendages, **755**
Appendicular skeleton, **922**
Appendix, 881
Aquaculture, **176**
Aquaporins, **210**
Aquatic animals
- excretion in, 795–796
- larval stage of, 823
- respiratory systems of, 788

Aquatic ecosystems, 117–121
- biotic and abiotic factors in, 66–67
- changes in, 63
- energy production in, 70
- estuaries, **119**
- food chains and webs in, 73–76
- freshwater, 118–119
- marine, 120–121
- nutrient limitation in, 86
- underwater conditions in, 117–118

Aqueous humor, 912
Archaea, 524, **526,** 580–581
Archaebacteria, 524, 526
Archaeopteryx, 763
Archean Eon, 542
Archegonia, **642**
Arteries, **952**
Arthropods, 755, 815, 817, 821
Artificial selection, 457–**458,** 461
Asexual reproduction, **19, 277**–278
- in animals, 735, **819**
- in fungi, 621
- parasitic worms, 820
- in plants, 277, 640, 702–703
- in prokaryotes, 281
- in protists, 608–609

Aspirin, 980
Association of Zoos and Aquariums (AZA), 170
Asthma, 163, **1024**
Atherosclerosis, **958**–960
Athlete's foot, 623, 1010
Atmospheric resources, 163–165
Atomic number, 35
Atoms, **34**–38
ATP. *See* Adenosine triphosphate (ATP)
ATP synthase, **237,** 258
Atrioventricular (AV) node, 951, 958
Atrium, **792**–793, **949**–951
Aurelia, 820–821
Australopithecus afarensis, 768–769, 773
Autoimmune diseases, 1025
Autonomic nervous system, **908**
Autosomes, **393**
Autotrophic protists, 610–611
Autotrophs, **69,** 117, **228,** 250
Auxins, **709**–712
Avery, Oswald, 340, 349
Avian influenza, 591, 1013, 1023
Axial skeleton, **922**
Axon, **897**
Axon terminal, 897, 900

B

Bacilli, **582**
Backbones, 731
Background extinction, **548**
Bacteria, 214
- bacterial meningitis, 587
- and cell organelles, 557
- chemosynthetic, 70
- classification of, 524–526
- diseases caused by, 586–588, 1010–1011
- early images of, 190
- gene expression in, 377
- growth of, 133
- lateral gene transfer in, 485
- and mutations, 375, 420
- photosynthetic, 70, 545, 555
- and recombinant DNA, 424–425
- viral infections of, 340–341, 575–577

Bacteria, domain, 524–**525,** 580–581
Bacterial transformation, **339**–340
Bacteriophages, **340**–341, **575**–577
Balance, 911
Balanced diet, 872–873
Bar coding, DNA, 529
Bark, **679**
Barr body, 396
Basal cell carcinoma, 939
Base pairing, **348,** 350
Bases, **44**
Beagle, 450–451
Beak size, 473
Beans, 706
Bears, 512, 848
Beekeeper, 736
Behavior, **767, 841,** 844–845. *See also* Animal behavior
Behavioral isolation, **495,** 497
Beijerinck, Martinus, 574
Benign tumors, 289
Benthic zone, 117, 121
Benthos, **117**
Beta-carotene, 430
Beta cells, 984
Beta-globin, 398
Bias, **14**
Biennial plants, 654
Bilateral symmetry, **738**–739
Bile, 878
Binary fission, 281, **583**
Binocular vision, **765**
Binomial nomenclature, **511**
Biodiversity, **166**–171. *See also* Diversity
- conservation, 170–171
- patterns of, 451–453, 465
- threats to, 168–170
- types of, 166–167

Biogeochemical cycles, **79,** 163
Biogeography, **465**
Bioinformatics, **407,** 422
Biological magnification, **161**
Biology, 17–25, 80
- defined, **17**
- fields of, 22–23
- measurement in, 24
- molecular, **23,** 370
- and safety, 25
- themes of, 20–21

Biomass, **78**
Biomes, **65,** 110–116

Biosphere, **21, 64**. *See also* Ecology; Ecosystems
Biotechnology, **23, 419,** 436–439. *See also* Genetic engineering
Biotic factors, **66**–67, 100
Bipedal locomotion, **767**
Bird flu, 591, 1013, 1023
Birds
behavioral isolation in, 495
brains of, 811
digestion in, 785
evolution of, 520, 547, 762–763
excretion in, 797
imprinting in, 844–845
learned behavior in, 842
lungs of, 790
and migration, 813, **847**
and natural selection, 472–473, 489, 496–497
and resource sharing, 101
temperature control in, 829–830
Birthrate, 132, 142–143
Bisphenol-A (BPA), 16
Bivalves, 815
Blade of leaf, **680**
Blastocyst, **294, 996**
Blastopore, 739
Blastula, 272–273, **739**
Blood, 42, 954–956. *See also* Circulatory system, human
carbon dioxide removal from, 50, 52
cell formation, 922–923
clotting, 955
flukes, 820
groups, 394
and kidneys, 884, 886–887
pH of, 44
pressure, 887, 953, 959
red blood cells, **954**–955
transfusions, 956
types, 320, 956
vessels, 952–953
white blood cells, **955**
B lymphocytes, 955, 1016–1017
Body cavity, 738
Body plans of animals, 737–741
Body temperature, 828–830, 866, 935, 937, 987
Bolus, 876
Bonds, chemical, 36–38
carbon, 45
covalent, **37,** 344
hydrogen, **41,** 348
ionic, **37**
Bone marrow, **924**
Bones, 922, **924**–925
Bony fishes, 760
Boreal forests, 114
Botanical illustrator, 655
Bottleneck effect, **490**
Botulism, 586
Bovine spongiform encephalopathy (BSE), 573, 1023
Bowman's capsule, **884**
Brain, human, 901–905
Brains, 810–811
Brain stem, **903**
Bread mold, 620–621
Breathing, 966–967
Breeding
artificial selection, 457–**458**
inbreeding, **419**
selective, **418**–420, 716
Bronchi, **964**
Bronchioles, 964
Bronchitis, 968
Brown algae, 527, 602, 604, 611
Bryophytes, **641**–642
Bt toxin, 428
Buds of plant, **675**
Buffers, **44**
Bulbourethral gland, 990
Bulk transport, 213
Burbank, Luther, 419
Burgess Shale, 753
Bursae, 927
Bursitis, 927

C

C. elegans, 293–294, 381
Caecilians, 761
Calcitonin, **985**
Calcium, 82, 925, 985
Calcium ions (Ca^{2+}), 931
Calcium phosphate, 872
Calorie, **250, 868**
Calvin, Melvin, 229, 238
Calvin cycle, **238**–239
Cambium, **677**–679
Cambrian Explosion, 753, 758–759
Cambrian Period, 542, 543, 560
Camouflage, 113–114
CAM plants, 241
Cancer, **289**–290
lung, 938, 968
skin, 337, 357, 938–939, 1025
viral, 588
Canines, 785
Canopy, **112**
Capillaries, 788, 792, **952,** 964–966
Capillary action, 41, **686**
Capsid, **575**
Carbohydrates, **46**–47, 250–251, **869**
Carbon, 35, 45
Carbon credits, 171
Carbon cycle, 82–83
Carbon dating, 541
Carbon dioxide
atmospheric, 164, 169, 178
in blood, 52
and breathing, 966–967
and cellular respiration, 253
and climate, 545
and nutrient cycles, 82–83
and photosynthesis, 239, 241, 253
removal from bloodstream, 50, 52
Carbonic anhydrase, 52
Carboniferous Period, 543, 561, 761–762
Carbon monoxide, 968
Cardiac muscle, **928**–929
Carnivores, **71, 782,** 785–786
Carpels, **697**
Carrying capacity
of biosphere, 155
of species, **135**
Cartilage, **757, 924**–925
Casparian strip, **672**–673
Catalysts, **52**
Cell, 20, **191, 862**
active transport, **212**–213, 227
animal, 203, 206–207, 211, 215
artificial, 435
B and T, 955, 957, 1016–1019
beta, 984
Casparian strip, **672**–673
cell membranes, **193,** 203–204, 209–213
cell plate, 284
cell stains, 191
cell theory, **191**
companion, **666**
daughter, 276, 280, 325, 327–328
diploid and haploid, **323**–328
discovery of, 190–191
elongation of, 709
flame, 796
and food molecules, 250
glial, 863
in ground tissue of plants, 667
guard, **682**–683
and homeostasis, **214**–217
human, 392–393, 862
mast, 1014
in meristems, **667**–668
and microscopes, 190–192
migration, 997
multipotent, **295**
mutations of, 375
organelles, **196,** 198–202
passive transport, **209**–211

Index *continued*

Cell (cont'd)
- plant, 203, 206–207, 211, 215
- plasma, 1017
- pluripotent, **294**
- red blood, **954**–955
- RNA synthesis in, 364–365
- size of, 193, 274–276
- in skin, 936
- specialization of, 215, 380
- stem, 294–297
- structure of, 196–207
- target, **709, 978**
- totipotent, **294**
- tumors, **289**
- and viruses, 579
- walls of, **203**
- white blood, **955**

Cell body, neuron, **897**

Cell cycle, **280**–282
- and apoptosis, **288**
- eukaryotic, 281
- growth factors, **287**
- phases of, 281–282
- prokaryotic, 281
- regulating, 286–290

Cell differentiation, 215, 292–297, 380–381
- defined, **293**
- and environment, 383
- and Hox genes, 382
- in meristems, 668
- in plants, 292
- stem cells, 294–297

Cell division, 191, 274–290. *See also* Cell cycle; Meiosis
- and cancer, 290
- and cell size, 274–276
- and chromosomes, **279**–280, 282–283
- controls on, 286–288
- cytokinesis, **282,** 284
- defined, **276**
- and interphase, **281**
- mitosis, **282**–285, 328
- and reproduction, 277–278

Cell-mediated immunity, 1018–**1019**

Cellular junctions, 217

Cellular respiration, 250–260
- aerobic and anaerobic, **252,** 260
- in animals, 787–790
- defined, **251**
- efficiency of, 256
- and electron transport, 258
- and exercise, 265
- glycolysis, 252, **254**–255
- Krebs cycle, 252, **256**–260
- overview of, 251–252
- and photosynthesis, 253
- stages of, 251

Cellulase, 614

Cellulose, 47, 869

Cenozoic Era, 563, 764

Centipedes, 755

Central nervous system, **896,** 901–904

Centrioles, **199, 282**

Centromere, **282**

Cephalization, **740,** 810–811

Cephalopods, 810

Cerebellum, **811, 903**

Cerebral cortex, **902**

Cerebral ganglia, 810–811

Cerebrum, 765, **811, 902**

Cervix, 995

Chargaff, Erwin, 344, 348–349

Chase, Martha, 340–341, 349

Chemical digestion, 785, **875**–878

Chemical reactions, **50**–51, 80

Chemiosmosis, 237, 258

Chemistry of life, 34–53
- atoms, **34**
- bonding, 36–38
- carbon compounds, 45–49
- chemical reactions, **50**–51, 80
- compounds, **36**–38
- elements and isotopes, **35**
- enzymes, **52**–53

Chemoreceptors, 910

Chemosynthesis, **70**

Chemosynthetic organisms, 117, 121

Chemotherapy, 290

Chesapeake Bay, 119

Chicken pox, 588

Childbirth, 1000

Chitin, **618,** 815

Chlamydia, **994**

Chlamydomonas, 640

Chlorine, 35, 37

Chlorofluorocarbons (CFCs), 175

Chlorophyll, 202, 230–232

Chloroplasts, **202,** 231, 557

Choanoflagellates, 752

Cholesterol, 319, 959–961, 980

Chordata, 513, 731

Chordates, **730,** 788. *See also* Invertebrates; Vertebrates
- adaptations in, 758–759, 761–762, 825
- cladogram of, 758–763
- embryological development in, 739
- evolution of, 757–764, 792
- fertilization in, 821–822
- nonvertebrate, 731, 758, 811, 813, 822
- reproductive diversity in, 824–826
- sense organs, 813

Chorion, 825, 998

Christy, Henry, 773

Chromatid, **282**

Chromatin, 197, **280,** 352

Chromosomes, 197, **279**–280
- artificial, 424
- autosomal, **393**
- and cell division, 282–283
- chromosomal mutations, 372, 374–375
- disorders of, 401
- eukaryotic, 280, 343, 352–353
- and gene linkage, 328–329
- homologous, **323**–325, 327
- human, 392–397
- karyotypes, **392**–393
- and polyploid plants, 376, 420
- polyploidy, 376
- prokaryotic, 279
- sex, **393**
- telomeres, **352**
- X and Y chromosomes, 393–396, 401, 434

Chyme, **877,** 880

Cilia, 199, **607,** 964

Ciliates, 607

Circadian rhythms, **847**

Circulatory system, human, 734, **948**–953
- and cholesterol, 959–961
- diseases of, 957–959
- heart, **791,** 793, 949–951
- and respiratory system, 963

Circulatory systems, animal, 734, **791**–793, 864

Citric acid cycle, 256

Clades, **516**–520, 546–547, 763

Cladograms, **517**–520
- of animals, 742–743
- bird lineage, 547
- of chordates, 758–763
- of invertebrates, 754–756
- of primates, 766–767

Class, **514**

Classical conditioning, **843**

Classification, 510–528
- binomial nomenclature, **512**
- clades, **516**–520, 546–547, 763
- class, **514**
- and DNA, 521–522
- domains, **525**–528

Classification (cont'd)
evolutionary, 516–520
family, **513**
of fossils, 546
of fungi, 621
genus, **512**
kingdoms, **514,** 523–525
Linnaean system, **513**–515
order, **513**
phylum, **514**
of plants, 634, 637, 641, 652–653
of prokaryotes, 580–581
of protists, 602–604
systematics, **512**–513, 516–517
three-domain system, 524–528
Climate, **96**–98, 110
and biomes, **65,** 110–116
change, 170, **177**–179
diagram, 111
evolution of, 544
Climax communities, 108–109
Clone, **427,** 429
Closed circulatory systems, **792**
Clotting, blood, 955
Cnidarians, 754, 784, 810, 814, 820
Cocaine, 904
Cocci, **582**
Cochlea, **911**
Codominance, **319,** 394
Codons, **366**–367
Coelom, **738**
Coenzyme A, 256
Coevolution, **551**–552
Cohesion, **41,** 686
Collagen, 863
Collenchyma, **667**
Collins, Francis, 349, 402
Colon, 881
Colorblindness, 395
Commensalism, **104**
Common ancestors, **464,** 468–469, 516–521
Common cold, 578, 588
Communities, **64,** 100–102, 108–109
Companion cells, **666**
Competition, ecological
and communities, 100–102
and natural selection, 473, 497
and population density, 138
Competitive exclusion principle, **101**
Complementary strands, 350
Complex carbohydrates, 47
Compounds, chemical, **36**–38
Computed Tomography Angiography, 962
Conditioning. *See also* Animal behavior
classical, **843**
operant, **843**
Cones, 646, 648, 668
Cones (of eye), **913**
Coniferous forests, 114
Conifers, **114,** 648, 677
Conjugation, **583,** 591, **608**
Connective tissue, **863**
Conservation, 162, 170–171, 176. *See also* Resources, natural
Constipation, 881
Consumers, **71**–72
Continental drift, 544–545
Continental shelf, 121
Contour plowing, 160
Contractile vacuole, 198
Control group, **7**
Controlled experiment, **7,** 9
Convergent evolution, **551**
Cook, James, 895, 917
Coral reefs, 121, 611, 728–729, 847
Cork, 190
Cork cambium, **677,** 679
Cornea, **912**
Corn smut, 622
Coronary arteries, 950, 958
Corpus callosum, 902
Corpus luteum, **993**
Cortex, **670**
Corticosteroids, **983**
Cortisol, 983
Cotyledons, **652**–653, 706
Coughing, 1012
Courtship behavior, **848**
Covalent bonds, **37,** 344
Cowpox, 1020
C4 plants, 241
Crassulacean Acid Metabolism (CAM), 241
Creatine supplements, 261
Creativity, 10
Cretaceous Period, 548, 562, 762, 764
Cretinism, 985
Creutzfeld-Jacob Disease, 597
Crick, Francis, 346–350, 362
Crocodilians, 762
Cro-Magnons, 772
Crop plants, 428–429, 437–438, 622, 715–717
Crop rotation, 160
Crossing-over, **324,** 329, 484, 499
Cross-pollination, 309
Crustaceans, 501, 755
Crutzen, Paul J., 175
Cryptosporidium, 607, 615
Currents, ocean, 98, 118
Cyanobacteria, 70, 584, 610, 623
Cycles
carbon, 82–83
cell, **280**–282, 286–290
matter, 79–86
nitrogen, 82–83
nutrients, 82–86
phosphorous, 85
water, 81
Cyclic AMP (cAMP), 981
Cyclins, **286,** 288
Cystic fibrosis (CF), 399–400, 431
Cytochrome *c*, 471
Cytokinesis, **282,** 284, 324
Cytokinins, **710**
Cytoplasm, **196,** 580
Cytosine, 344–345, 348, 366
Cytoskeleton, **199,** 922
Cytotoxic T cells, 1019

D

Dart, Raymond, 773
Darwin, Charles, 13, 137, 143, 450–473, 482, 549, 552, 709
Darwin, Francis, 709
Data, **8**
Dating techniques, 466, 540–541
Daughter cells, 276, 280, 325, 327–328
Death rate, 132, 142–143
Decibels (dB), 910
Deciduous plants, **112,** 714
Decomposers, **71,** 74, 584
Decomposition, 622
Deforestation, **159**
Dehydration, 189, 869
Democritus, 34
Demographic transition, **144**
Demography, **143**. *See also* Population growth
Dendrites, **897**
Denitrification, **84**
Density-dependent limiting factors, **137**–141
Density-independent limiting factors, **137,** 140–141
Deoxyribonucleic acid. *See* DNA
Deoxyribose, 344, 362
Dependent variable, **7**
De Puydt, Marcel, 773
Derived characters, **518**–519, 521
Dermal tissue, **665,** 670, 680
Dermis, **937**
Descent with modification, 464, 516
Desertification, **159**

Index *continued*

Deserts, 113
Detritivores, **71,** 74, **782**
Detritus, 782
Deuterostomes, **739**
Devonian Period, 561, 759
Diabetes mellitus, 867, 886, 984, 1025
Dialysis, 887
Diaphragm, 966–**967**
Diarrhea, 881
Diastolic pressure, 953
Diatoms, 603, 604
Dicer enzyme, 380–381
Dichloro diphenyl trichloroethane (DDT), 161, 169
Dichotomous key, **511**
Dicots, **652**–653, 669, 675, 677, 706–707
Diet, balanced, 872–873
Differentiation, **381**–382. *See also* Cell differentiation
Diffusion, **208**–211, 787
Digestion
  animal, 734, 784–786
  chemical, 785, **875**–878
  mechanical, 785, **875**–877
Digestive system, human, 737, 864, **875**–881
  absorption and elimination, 880–881
  digestive enzymes, 876–878
  digestive process, **875**–878
Digestive tract, 739, **784**–785
Dihybrid cross, 317
Dikika Baby, 769, 773
Dinoflagellates, 611
Dinosaurs, 520, 548, 550, 762–763, 830
Diphtheria, 586
Diploid cells, **323**–328
Diploid (2N) phase, **637**
Directional selection, **489**
Disaccharides, 46, 869
Diseases
  of animals, 622
  autoimmune, 1025
  bacterial, 586–588, 1010–1011
  circulatory system, 957–961
  emerging, **590**–592, 1022–1023
  genetic, 947, 961, 973
  infectious, **23, 1010**–1023
  intestinal, 615
  of plants, 622
  and public health, 1021
  sexually transmitted (STDs), **994,** 1012
  and transgenic organisms, 430
  viral, 588–589, 994, 1010–1011
Disruptive selection, **489**
Distribution of populations, 131
Dittmer, Howard, 669
Diversity, 21. *See also* Biodiversity
  of chordates, 731, 757–764
  ecosystem, **166**
  and extinction, 548
  genetic, **166**–168, 820
  of invertebrates, 730, 753–756
  species, **166,** 168
Djimdoumalbaye, Ahounta, 773
DNA, 18, 23, 48, 340–353. *See also* Human heredity
  and cell size, 274, 276
  chromosomes, **279**–280
  and classification, 521–522, 529
  components of, **344**–345
  crossing-over, 484, 499
  DNA fingerprinting, **433**–434, 443
  DNA microarray, **432**
  DNA polymerase, **351,** 404, 423
  DNA replication, 281, **350**–353, 374–375, 424–425
  double-helix model of, **347**–348
  in eukaryotic cells, 193, 352–353
  evolution of, 554–555
  extraction, 403
  functions of, 342–343
  and genetic disorders, 398
  as genetic material, 340–341
  and hominine evolution, 771
  manipulating, 403–405
  mtDNA, 434
  mutation rates in, 498–499
  and privacy, 437
  in prokaryotic cells, 193, 352–353
  recombinant, 421–425, 430
  and RNA, 362–365
  sequencing, 404–409
Domains, **525**–528
Dominance
  codominance, **319**
  incomplete, **319**
  principle of, **310,** 318
Dominant alleles, 310–312, 318, 394
Dopamine, **904**–905
Dormancy, plant, **706**–707, 714, 847
Double fertilization, **700**–701
Double Helix, The, 346
Double-helix model, **347**–348
Double-loop circulatory systems, **792**–793
Down syndrome, 401
Drip irrigation, 162
Drosophila melanogaster, 318, 323, 328, 361, 382, 387, 490, 501
Drug addiction, 904–905
Dry forests, 112
Dunkleosteus, 759
Duodenum, 878
Dwarfism, 430, 982
Dynamic interaction. *See* Interdependence

E

E. coli, 377, 383, 421, 572–573, 581
Ears, 911
Earth
  age of, 454–455, 467
  early history of, 553–555
  evolution of, 544–545
  geologic time scale, **542**–543
Earthworms, 755, 792
Ebola, 1013, 1023
Echinoderms, 739, 756–757, 810, 815
Echocardiography, 962
Ecological pyramids, **77–78**
Ecology, **23,** 64–**65**
  case studies, 175–179
  defined, **65**
  disturbance of, 76, 80, 106–109
  ecological footprint, **173**–174
  ecological hot spot, **171**
  ecological succession, **106**–109
  global, **22**
  methods of studying, 68
  and sustainability, 157, 174
Ecosystems, **65.** *See also* Aquatic ecosystems
  biomes, **65,** 110–116
  and climate, **96**–98
  competition in, 100–103
  diversity of, **166**
  goods and services, 156–157
  limiting factors in, 137
  niches in, 99–**100**
  preserving, 170–171
  recycling within, 79–86
  succession in, **106**–109
  symbioses in, 103–104
Ectoderm, **738,** 997
Ectotherm, **829**–830
Edema, 956
Ediacaran fauna, 753
Egg, 325, 825. *See also* Reproductive system, human
Ejaculation, 990
Elbow, 932
Electric currents, 813

Electron, **34,** 36–38
Electron carriers, **232,** 236
Electron microscopes, 192
Electron transport chain
and cellular respiration, 258
and photosynthesis, **236,** 252
Elements, **35**
Elimination, 875
Embryological development, 469, 500–501, 737, 739
Embryonic stem cells, 272–273, 295–297
Embryos, **292**
animal, 738–739, 822
fossils of, 752
and gene regulation, 380–382
plant, 647
Embryos, human
cartilage in, 924–925
development of, 988, 995–1001
implantation of, **996**
multiple, 996
Embryo sac, **699**
Emerging diseases, **590**–592, 1022–1023
Emigration, **132**
Emphysema, 968
Endangered species, 169–170
Endocrine glands, **828, 979,** 984
Endocrine system, 828, 864, **978**–987
Endocytosis, 212–213
Endoderm, **738,** 997
Endodermis, **670,** 672–673
Endoplastic reticulum, **200**–201, 248–249
Endoskeleton, **815**–817, 922
Endosperm, **700**–701
Endospore, **583**
Endosymbiotic theory, 202, **556**–557
Endotherm, **829**–830
Energy, 20
activation, **51**–52
and ATP, 226–227
from autotrophs, 71
and carbohydrates, 46
and cellular respiration, 260
in chemical reactions, 51
from chemosynthesis, 70
and chlorophyll, 230–231
consumers of, 71
ecological pyramids of, 77
and exercise, 264–265, 873
and food, 250, 868
in food chains and webs, 73–76
heat, 41
and oxygen, 252
from photosynthesis, 70
producers of, 69–70
in prokaryotes, 582
solar, 97
Environment. *See also* Adaptations; Ecology; Ecosystems
and animal behavior, 847
biotic and abiotic, 66–67
disturbance of, 76, 80, 106–109
and evolution, 544–545, 547–548
and gene expression, 321, 383
and human activity, 154–156, 173–174, 183
and mutations, 375
and survival, 278, 473, 491
Environmental Protection Agency (EPA), 173
Enzymes, **52–53**
digestive, 785–786, 876–878
DNA polymerase, **351**
restriction, **403**–405, 421
Epidermis, **665,** 670, **936**
Epididymis, **989**
Epigenetics, 409
Epiglottis, 876, 964
Epinephrine, **983**
Epiphytic plants, 112
Epithelial tissue, 737, **863**
Era, geologic, **543**
Erythrocytes, 954
Escherichia coli, 377, 383, 421, 572–573, 581
Esophagus, **877**
Essential amino acids, 870
Essential fatty acids, 870
Estivation, 112
Estrogens, **988,** 992
Estuary, **119**
Ethics, 14, 296, 438–439
Ethylene, **711**
Eubacteria, 524–525
Euglena, 603, 604
Eukarya, 524, **526**
Eukaryotes, **193**–194, 521
cell cycle of, 281
cell structure of, 196–207
DNA replication in, 352–353
electron transport chain in, 258
eukaryotic chromosomes, 280, 343, 352–353
gene regulation in, 379–381
and genetic variation, 557
origin of, 556–557
and protists, 605
single-celled, 523, 602, 752
transcription in, 364
unicellular, 214, 526–527
viral infections of, 577
Everglades, 119
"Evo-devo," 500, 737
Evolution, 13, 19, 21, **450**. *See also* Natural selection
adaptive radiation, **550**–551
and animal behavior, 840–841, 848–849
of animals, 742–743
and artificial selection, 457–**458**
of bacteria and viruses, 591
of birds, 520, 547, 762–763
of chordates, 757–764, 792
and classification, 516–522
coevolution, **551**–552
common ancestors, 464, 468–469, 516–521, 546
convergent, **551**
descent with modification, 464, 516
of DNA and RNA, 554–555
of Earth, 544–545
of endothermy, 830
and environment, 544–544, 547–548
of eukaryotic cells, 556–557
and fossil record, 466–467
and gene pools, 482–483
and genetic drift, **490**
and genetic equilibrium, 491–492
hominine, **767**–772
and Hox genes, 500–501
of infectious diseases, 23
of insects, 487
of invertebrates, 752–756
macroevolution, 546–547
of mammals, 468
of mitochondria, 557
molecular, 498–501
of multicellular organisms, 558, 640, 752
of nervous systems, 810–811
of organic molecules, 554
of plants, 636–640, 646, 650
of populations, 483–492
of primates, **765**–772
of prokaryotic cells, 556–557
of protists, 605
rate of, 549
and sexual reproduction, 558
and speciation, **494**–497, 517, 546–547
of vertebrate brains, 811
Excretion, **882**
Excretory systems, 734–735, **794**–798, 864, **882**–887
Exercise
and energy, 264–265, 873
and health, 933

Index *continued*

Exhalation, 967
Exocrine glands, **979,** 984
Exocytosis, 212–213
Exons, **365**
Exoskeleton, **815,** 817, 922
Experiment, controlled, **7,** 9
Experimentation, ecological, 68
Exponential growth, **132**–133
Extinction, 168–170, **538,** 546–548, 762
Extracellular digestion, **784**
Eyes
  in animals, 850
  in fruit flies, 361, 387
  human, 912–913
  invertebrate, 812
  vertebrate, 813

F

Facilitated diffusion, **209**–211
FAD (flavine adenine dinucleotide), 256
$FADH_2$, 256
Fallopian tubes, 991, 993
Family, **513**
Farmer, 655
Fast-twitch muscle, 933
Fats, 250–251, **870,** 873
Fat-soluble vitamins, 871
Fatty acids, 47–48
Feathers, 762
Feces, 881
Feedback inhibition, **732, 865,** 986
Female cone, 648
Female reproductive system, **991**–993. *See also* Reproductive system, human
Fermentation, 252, **262**–263, 265
  alcoholic, 263
  lactic acid, 263, 265, 269
Ferns, 638, 644–645
Fertilization, 325, **995**–996. *See also* Reproduction
  double, **700**–701
  external, **822**
  internal, **821**
  of pea plants, **309**
Fertilizers, 84, 86, 717
Fetal development, 999
Fetus, **998**
Fever, **1015**
Fiber, 869
Fibrinogen, 954
Fibrous root system, 669
Fight or flight response, 983
Filter feeders, **782**
Filtration of blood, **884**
Fins, 759
Fish, 500, 757, 759–761
  brains of, 811
  circulatory systems of, 792
  excretion in, 795–796
  fertilization in, 821–822
  ghost, 33
  jawless, 759
  respiratory systems of, 788
Fishapod, 761
Fitness, **461**. *See also* Natural selection
Flagella, 199, 581, **607**
Flagellates, 607
Flame cells, 796
Flatworms, 755, 783, 794, 796, 810, 812
Flavine adenine dinucleotide (FAD), 256
Fleming, Alexander, 1021
Flowering plants, 650–654, 696–703
Flowers, 646, 650–651
  and hormones, 708
  and meristems, 668
  and photoperiod, **713**
  and pollinators, 552
  structure of, **696**–697
Fluid mosaic model, 205
Fluorescence microscopy, 191, 291
Folic acid, 997
Follicle stimulating hormone (FSH), 983, 988–989, 991–993
Follicular phase, 992
Food, **868**
  allergies, 1025
  chains, **73**–76
  and digestive system, 875–881
  and energy, 250, 868
  and fats, 250–251, 873
  genetically modified, 437–438
  labels, 872–873
  nutrients in, 869–673
  webs, **74**–76
Food vacuole, **612**
Forensics, **433**
Forensic scientist, 322
Forests, 109, 112, 114, 159
Forster, Georg, 917
Fossil fuels, 82, 178
Fossil preparator, 559
Fossils, 453, 538–545
  adaptive radiations in, 550
  of birds, 763
  Cambrian, 753, 758–759
  chordate, 757–759, 761–762
  classification of, 546
  and continental drift, 545
  dating, 541
  Ediacaran fauna, 753
  of eggs and embryos, 752
  and evolutionary theory, 466–467
  hominine, 768–769, 773
  index, **540**
  microfossils, 555
  of plants, 636, 638, 646, 652
  trace, 752
  types of, 538–539
Founder effect, **490,** 496
Four-chambered heart, 793
Frameshift mutations, **373**
Franklin, Rosalind, 346–349
Fraternal twins, 996
Freshwater ecosystems, 118–119
Freshwater resources, 160–162
Frogs, 383, 600, 735, 761, 818, 848
Fronds, 644
Frontal lobe, 903
Fructose, 46
Fruit, **651, 704**
Fruit flies, 318, 323, 328, 361, 382, 387, 490
Fruiting body, **619**
Fungi, 523–524, 527, **618**–625, 1010
  classification of, 621
  lichens, 107, **623**
  mycorrhizae, **624**–625
  parasitic, 622–623
  reproduction in, 621

G

Galactose, 46, 378
Galapágos finches, 453, 471–473, 496–497
Galapágos islands, 452
Gallbladder, **878**
Gametes, **312,** 325
  and gene sets, 323
  human, 985, 990, 995
  of plants, 637
Gametophytes, **637**–638, 646, 697–699
Ganglia, **810,** 906
Gas exchange, 635, 682, 787, 966
Gastrovascular cavity, **784**
Gastrulation, **997**
Gel electrophoresis, **404**–405, 422
Gene flow, 492

Genes, 18, **310**
alleles of, **310,** 482
and behavior, 841
and cancers, 289
and chromosomes, 323
as derived characters, 521
and DNA, 340–341
and DNA microarrays, 432
gene duplication, 499–500
gene expression, **370**–371, 377–383, 980
gene families, 500
gene pools, **483,** 490
gene regulation, 377–383
gene therapy, **431**
homeobox, **382**
homeotic, **382**
homologous, 471
Hox, **382,** 471, 500–501
identifying, 406
lateral gene transfer, 485
linkage, 328–329
mapping, 328–329
MC1R, 394
mutations of, 372–376, 484
and phenotypes, 485–486
and proteins, 370
sex-linked, **395**
Ubx, 501
Genetic engineering, 418–439
in agriculture and industry, 428–429
ethics of, 438–439
in health and medicine, 430–432
personal identification, 433–434
and privacy, 402, 436–437
recombinant DNA, 421–425
and safety, 437–438
selective breeding, **418**–420, 716
transgenic organisms, 426–427
Genetic Information Non-discrimination Act, 402, 409, 437
Genetics, **308**–329. *See also* Alleles; Meiosis
codominance, **319,** 394
dominant and recessive alleles, 310–312, 318
genetic code, **366**–367, 370, 470
genetic diseases, 947, 961, 973
genetic disorders, 395, 398–401
genetic diversity, **166**–168, 820
genetic drift, **490**
genetic equilibrium, **491**–492
genetic marker, **425**
genetic recombination, 484
genetic testing, 402, 431
genetic variation, 419–420, 482–486, 558
incomplete dominance, **319**
multiple alleles, **320**
principle of dominance, **310,** 318
principle of independent assortment, **317**–318, 328–329
segregation of alleles, **312,** 314, 318
traits, **309, 320**
Genital herpes, 994
Genital warts, 994
Genome, human, **392**–393, 403–409, 436
Genomics, **23, 407,** 435
Genotypes, **315,** 321, 397–398, 407, 482
Genus, **511,** 516
Geographic isolation, **495,** 496
Geological processes, 80
Geologic time scale, **542**–543
German measles, 1001
Germination, **706**–707
Germ layers, 738
Germ theory of disease, 586, **1010**
Gibberellins, **711**
Gibson, Daniel G., 435
Gigantism, 982
Gills, 788
Gizzard, 785
Glands, **978**–979, 982–987
adrenal, 979, **983**
endocrine and exocrine, **979,** 984
hypothalamus, 866, **903,** 979, **982,** 986–987
pancreas, **878,** 979, **984**
pineal, **979**
pituitary, 886, 979, **982**–983
reproductive, 985
thymus, 957, **979,** 1016
thyroid, 979, **985,** 987
Glial cells, 863
Global ecology, **22**
Global warming, **177**
Globins, 500
Globulins, 954
Glomerulus, **884**
Glucagon, 978, **984**
Glucose, 46–47, 209, 251, 378, 867
Glycogen, 47
Glycolysis, 252, **254**–255, 262
Goiter, 985
Goldstein, Joseph, 960–961
Golgi apparatus, **201**
Gonads, **985**
Gonorrhea, 994
Gradualism, **549**
Grafting, **703**
Grana, 231
Grant, Peter and Rosemary, 472–473, 496
Grassland, 112–113
Gravitropism, **712**
Gray matter, 902
Great Plains, 158–159
Green algae, 528, 610, 623, 634, 636–640
Green fluorescent protein (GFP), 422
Greenhouse effect, **97,** 163–164, 178, 545
Griffith, Frederick, 338–340, 349, 423
Ground tissue, 665, **666,** 670, 680
Growth hormone (GH), 982–983
Growth plates, 925
Guanine, 344–345, 348, 366
Guard cells, **682**–683
Guerin, Camille, 589
Gullet, **612**
Gymnosperms, **646**–649

H

Habitat, **99**
Habitat fragmentation, **168**
Habituation, **842**
Hair, 937
color, 394, 396–397
follicles, **937**
Half-life, **540**–541
Haploid cells, **323**–328
Haploid (N) phase, **637**
Haplotypes, 407
Hardy-Weinberg principle, **491**
Haversian canals, **924**
Hearing, 911
Heart, **791,** 793, 949–951
Heartburn, 877
Heart disease, 167, 957–958, 962
Heartwood, **678**
Heat
capacity, 41
greenhouse effect, 97
transport, 98
Helicobacter pylori, 877
Helix, 346
Helmont, Jan van, 225, 229, 245
Helper T cells, 1019, 1026
Hemoglobin, 49, 57, 375, 398, 870, 872, **954,** 966
Hepatitis B, 588–589, 994
Herbaceous plants, **653**
Herbicides, 428, 438
Herbivores, **71, 783,** 785–786
Herbivory, **102,** 138–139
Heredity, 20, 308, 318. *See also* Genetics; Human heredity
Hermaphrodites, 820
Heroin, 904

Herpes, 994
Hershey, Alfred, 340–341, 349
Heterotrophic protists, 612–613
Heterotrophs, **71, 228,** 250, 527
Heterozygous organisms, **314**
Hibernation, 847
High-altitude adaptation, 831
High-density lipoprotein (HDL), 959
Histamines, 938, **1014**
Histones, 280, 352
HIV, 578, 589, 1023, 1026–1027
Homeobox genes, **382**
Homeostasis, **19**–20
  in animals, 732, 827–830
  and blood glucose levels, 978, 984
  and breathing, 967
  and cells, **214**–217
  and endocrine system, 828, 986–987
  in humans, **865**–867, 886–887
  and kidneys, 886–887
  and pH, 44
  of plants, 682–683, 711
  and skin, 935, 937
Homeotic genes, **382**
Hominines, **767**–772
Hominoids, **767**
Homo erectus, 771
Homo ergaster, 770
Homo habilis, 770–771
Homologous chromosomes, **323**–325, 327
Homologous proteins, 471
Homologous structures, **468**–469
Homo neanderthalensis, 772, 773
Homo sapiens, 767, 770, 772, 773
Homozygous organisms, **314**
Hooke, Robert, 190
Hormone receptors, **709,** 980
Hormones, **978**
  animal, 708
  and homeostasis, 828
  local, 980
  and metamorphosis, 824
  parathyroid, **985**
  plant, **708**–714
  prostaglandins, **980**
  releasing, **983**
  sex, **988**
  steroid and nonsteroid, 379, **980–981**
Hornworts, 641
Hoxc8, 470
Hox genes, **382,** 471, 500–501
Human activity, 154–156, 173–174, 183. *See also* Resources, natural
Human body
  circulatory system, 734, 791, 793, **948**–953, 957–961, 963
  digestive system, 737, 864, **875**–881
  embryonic development, 988, 995–999
  endocrine system, 828, 864, **978**–987
  excretory system, **882**–887
  fetal development and childbirth, 999–1001
  and homeostasis, **865**–867, 886–887
  integumentary system, 864, **935**–939
  levels of organization of, 862–864
  lymphatic system, 864, **956**–957
  muscular system, 864, **928**–933
  nervous system, 896–908
  and nutrients, 868–873, 880
  organs in, 863
  reproductive system, 864, **988**–993
  respiratory system, **963**–969
  sense organs, 909–913
  skeletal system, 864, **922**–927
  specialized cells in, 862
  tissues in, 862–863
  and water, 869, 986
Human Genome Project, **406**–409, 430
Human growth hormone (HGH), 3, 29, 430
Human heredity, 392–409
  genetic disorders, 395, 398–401
  and genome, **392**–393, 403–409
  karyotypes, **392**–393
  pedigrees, **396**–397
  transmission of traits, 394–396
Human immunodeficiency virus (HIV), 578, 589, 1023, 1026–1027
Human papillomavirus (HPV), 588–589, 994
Human populations. *See also* Populations
  growth patterns of, 143–145
  history of, 142–143
  impact of, 80, 84, 109, 139, 141
Humoral immunity, **1016**–1018
Huntington's disease, 399
Hutton, James, 454–455, 459, 467
Hybridization, **419**
Hybrids, **309**
Hydras, 754, 814
Hydrochloric acid, 44
Hydrogen, 35, 51
Hydrogen bonds, **41,** 348
Hydrogen ions, 43–44
  and cellular respiration, 258
  and photosynthesis, 236–237
Hydrophilic molecules, 204
Hydrophobic molecules, 204
Hydrostatic skeleton, **814**
Hypercholesterolemia, 973
Hypertension, 887, 959
Hyperthyroidism, 985
Hypertonic solutions, **210**–211
Hyphae, **619**
Hypodermis, 936
Hyponatremia, 221
Hypothalamus, 866, **903,** 979, **982,** 986–987
Hypothesis, **7,** 9
Hypothyroidism, 985
Hypotonic solutions, **210**–211

I

Iceman, 633, 659
Ichthyosaur, 536–537
Identical twins, 996
Immigration, **132**
Immune response, **1016**–1018
Immune system, 828, 864
  acquired immunity, 1020
  cell-mediated immunity, 1018–**1019**
  disorders of, 1024–1027
  humoral immunity, **1016**–1018
  and infectious diseases, 1015–1019
Immunity, 588
Implantation of embryo, **996**
Imprinting, **844**–845
Impulses, 897–900
Inbreeding, **419**
Incisors, 785
Incomplete dominance, **319**
Independent assortment, principle of, **317**–318, 328–329
Independent variable, **7**
Index fossils, **540**
Induced pluripotent stem cells (iPS cells), 297
Infant mortality, 1001
Infectious diseases, **23, 1010**–1023. *See also* Diseases,
  causes of, 1010–1011
  and immune system, 1015–1019
  nonspecific defenses against, 1014–1015
  spread of, 1012–1013, 1021–1023
Inference, **7**
Inflammation, 927
Inflammatory response, **1014**
Influenza, 575, 588–589, 591, 1010
Ingenhousz, Jan, 229

Ingestion, 875
Inhalation, 967
Inheritance, 308, 310, 318, 456, 461. *See also* Genetics; Human heredity
Innate behavior, **841,** 844–845
Inorganic chemistry, 45
Insecticides, 161, 428, 438
Insects, 755
colonies of, 849
and Hox genes, 501
metamorphosis in, 823–824
mutations in, 376
and plant evolution, 552
and pollination, 700
respiratory system of, 789
Insight learning, **843**
Insulin, 870, 978, **984**
Integrated pest management (IPM), 162
Integumentary system, 864, **935**–939
Interdependence, 21
of biosphere, 64–65
in food chains and webs, 73–76
Interferons, **1015**
Intergovernmental Panel on Climate Change (IPCC), 177–178
International HapMap Project, 407
International System of Units (SI), 24
Interneurons, **809,** 897
Interphase, **281,** 324
Intertidal zone, 120
Intestinal diseases, 615
Intestine
large, **881**
small, **878,** 880–881
Intracellular digestion, **784**
Introduced species, 169
Introns, **365**
Invasive species, 136
Invertebrate biologist, 736
Invertebrates, **730**. *See also* Chordates; Vertebrates
body cavities of, 738
cladogram of, 754–756
development of young, 823–824
evolution of, 752–753
fertilization in, 821–822
nervous systems of, 810
respiratory systems of, 788–789
response to stimuli in, 809
sense organs of, 812
skeletons of, 814
Iodine, 985
Ionic bonds, **37**
Ions, **37,** 43–44
Iris, 653, 697
Irish Potato Famine, 601, 629
Iris (of eye), **912**
Islets of Langerhans, 984
Isolation, reproductive, **494**–495
Isotonic solutions, **210**–211
Isotopes, **35**
Ivanovski, Dmitri, 574
Iwata, So, 229

J

Jawless fishes, 759
Jaws, 758–759
Jellyfish, 422, 754, 820–821
Jenner, Edward, 589, 1020
Johanson, Donald, 773
Joints, **816**–817, **926**–927
Jurassic Period, 562, 762
Just, Ernest Everett, 996

K

Karposi sarcoma, 1025
Karyotypes, **392**–393
Kenyanthropus, 768
Keratin, **936**
Keystone species, **103,** 167
Kidneys, **795,** 867, 883–887, 986–987
artificial, 799
and blood pressure, 884, 886–887, 953
and salt, 798
Kilocalorie, 250
Kingdoms, **514,** 523–525
Kin selection, **849**
Klinefelter's syndrome, 401
Koch, Robert, 1010–1011
Koch's postulates, **1011**
Krebs, Hans, 256
Krebs cycle, 252, **256**–260

L

Laboratory technician, 195
Lac operon, 377–378, 383
Lac repressor, 378
Lactic acid fermentation, 263, 265, 269
Lactobacillus, 585
Lactose, 378, 383
Laetoli footprints, 768, 773
Lamarck, Jean-Baptiste, 456, 459, 460
Lampreys, 759, 850
Lancelets, 758, 813
Land plants, 637
Langerhans, Paul, 984
Language, **851**
Large intestine, **881**
Lartet, Edouard, 773
Larvae, **756,** 823
Larynx, **964**
Lateral gene transfer, 485
Leaf loss, 714
Leakey, Maeve, 773
Leakey, Mary, 773
Learning, **842**–845. *See also* Animal behavior
Leaves of plants, **664,** 680–684
Leeches, 104, 755
Lens, **912**
Leucine, 367
Leukocytes, 955
Levels of organization
animals, 737
ecological, 65
human body, 862–864
multicellular organisms, 216–217
proteins, 49
Lewis, Edward B., 382
Lichens, 107, **623**
Life cycle of plants, 637
angiosperms, 698–701
ferns, 644–645
green algae, 640
gymnosperms, 648–649
mosses, 641–642
Life span of plants, 654
Ligaments, **816, 927**
Light. *See also* Sunlight
absorption, 230
and photosynthesis, 240
plant response to, 709, 712–713
Light-dependent reactions, **233,** 235–237
Light-independent reactions, **233,** 238–239
Light microscopes, 191–192
Lignin, 641, 643, **666**
Limb formation, 468–469, 740, 743
Limb girdles, 815
Limbic system, 902
Limiting nutrient, **85**
Linnaean classification system, 512–514
Linnaeus, Carolus, 510–513, 515, 516
Lipid bilayer, **204**–205
Lipids, **47,** 204, 870
Lipoproteins, 959
Liver, 201, 867, **878,** 883, 960–961
Liverworts, 641
Living things, 17–19
Lobes, 902
Local hormones, 980
Logistic growth, 134–**135**
Lohest, Max, 773
Long-day plants, 713
Loop of Henle, **884**
Loudness, 910

Index *continued*

Low-density lipoprotein (LDL), 959–961
"Lucy," 769, 773
Lung cancer, 938, 968
Lungs, 788–790, 883, 950, 964, 966–969
Lupus, 1025
Luteal phase, 993
Luteinizing hormone (LH), 983, 988–989, 991–993
Lyell, Charles, 454–455, 459
Lyme disease, 587, 1013, 1023, 1031
Lymph, **956**
Lymphatic system, 864, **956**–957
Lymph nodes, 957
Lymphocytes, 955, 1016
Lyon, Mary, 396
Lysogenic infections, **577**
Lysosomes, **198**
Lysozyme, 429, 876, 1014
Lytic infections, **576**–577

M

Macleod, Colin, 349
Macroevolution, **546**–547
Macromolecules, 46–49, 250, 344
Macrophages, 955
Mad cow disease, 573, 597, 1023
Magnetic Resonance Imaging (MRI), 962
Malaria, 400, 616, 1010
Male cone, 648
Male reproductive system, **988–990.** *See also* Reproductive system, human
Malignant tumors, 289
Malpighian tubules, **797**
Malthus, Thomas, 142, 457, 459, 460
Mammalia, 518–519, 520
Mammals, **764**
  adaptive radiation of, 550
  brains of, 811
  ears of, 813
  evolution of, 468
  marine, 846
  muscular systems of, 817
  reproductive strategies of, 825–826
  respiratory system of, 790
Mammary glands, **825**
Mangrove swamps, 12, 119
Manipulated variable, 7
Marcus, Rudolph, 229
Margulis, Lynn, 557
Marine biologist, 105
Marine ecosystems, 120–121
Marsupials, 764, 826
Marsupium, 826
Mass extinction, **548**
Mass number, 35
Mast cells, 1014
Matrix, **256**
Matter, cycles of, 20, 79–86
Mayer, Julius Robert, 229
McCarty, Maclyn, 349
MC1R gene, 394
Measurement, scientific, 24
Mechanical digestion, 785, **875**–877
Mechanoreceptors, 911
Medications, 1021
Medicine, 430–432
Medulla oblongata, 811, 903, 967
Meiosis, **324**–329, 820. *See also* Sexual reproduction
  in angiosperms, 698–699
  crossing-over, **324,** 329, 484, 499
  meiosis I, 324–325
  meiosis II, 325
  and mitosis, 326–328
  nondisjunction in, 401
Melanin, **936**
Melanocytes, **936**
Melanocyte-stimulating hormone (MSH), 983
Melanoma, 938–939
Mello, Craig, 381
Membranes
  cell, **193,** 203–204, 209–213
  and diffusion, 787
  thylakoid, 236–237
Memory B cells, 1017–1019
Mendel, Gregor, 308–318, 329, 349, 370, 482
Meningitis, 587
Menopause, 993
Menstrual cycle, **991**–993
Menstruation, **993**
Mercury, 161
Meristems, **667**–668, 670, 677
Mesoderm, **738,** 997
Mesophyll, **680**–681
Mesozoic Era, 544, 562
Messenger RNA (mRNA), **363,** 368–370
Metabolism, **19,** 985
Metamorphosis, 383, **823**–824
Methamphetamine, 904
Methicillin-resistant *Staphylococcus aureus* (MRSA), 591
Methionine, 369
Metric system, **24**
Microclimates, **96**
Microfilaments, 199
Microfossils, 555
Micrographs, 192
MicroRNA (miRNA), 380–381
Microscopes, 190–192
Microscopist, 195
Microspheres, 554
Microtubules, 199
Microvilli, 880
Midbrain, 903
Migration, 813, **847,** 997
Miller, Stanley, 554
Minerals, **872**
  in bones, 922–923
  and plants, 635
Mitochondria, **202,** 216, 248–249, 252, 521
  evolution of, 557
  matrix of, **256**
  in muscle, 933
Mitochondrial DNA (mtDNA), 434
Mitosis, **282**–285. *See also* Asexual reproduction
  in angiosperms, 698–699
  DNA replication, 353
  and meiosis, 326–328
  in protists, 608
Mitotic spindle, 199
Mixture, **42**
Modeling, ecological, 68
Model systems, 308
Molars, 785
Molds, 603, 604, 608–609, 613, 620–621
Molecular biology, **23,** 370
Molecular clocks, **498**–499
Molecular evolution, 498–501
Molecular transport, 212
Molecules, **37**
  electron carrier, **232**
  evolution of, 554
  homologous, 470–471
  hydrophilic/hydrophobic, 204
  macromolecules, 46–49, 250, 344
  polar, 40–41
  water, 37, 40–41
Molina, Mario, 175
Mollusks, 756, 810, 815
Molting, **815**
Monera, 523–524
Monkeys, 766–767
Monocots, **652**–653, 669, 675–676, 706–707

Monoculture, **155**
Monohybrid cross, 317
Monomers, **46**
Monophyletic group, **516**
Monosaccharides, **46,** 869
Monotremes, 764, 825
Montreal Protocol, 175
Morgan, Thomas Hunt, 318, 323, 328–329, 349
Mosquito, 616–617, 812
Mosses, 638, 641–642
Motor neurons, **809,** 897, 900, 931
Mountain ranges, 116
Mount Saint Helens, 106, 109
Mouth, 739, 876
Mouthparts, 785–786
Mucus, 877, 964, 1014
Mullis, Kary, 423
Multicellular organisms, 527–528
  cells of, 215–217
  evolution of, 558, 640, 752
Multiple alleles, **320**
Multipotent cells, **295**
Muscle contraction, 930–932
Muscle fibers, **929**
Muscle tissue, **863,** 928–929
Muscular systems
  animals, 733, **816**–818
  humans, 864, **928**–933
Museum guide, 559
Mushrooms, 618–619, 622
Mutagens, **375**
Mutant, 420
Mutations, **372**–376, 484, 491
  bacterial, 420
  chromosomal, 372, 374–375
  effects of, 374–376
  gene, 372–376, 484
  and genetic variation, 419–420
  and molecular clocks, 498–499
  neutral, 484, 498
  in prokaryotes, 583
Mutualism, **103,** 614, 623–625
Mutualistic symbionts, 783
Mycelium, **619**
Mycobacterium tuberculosis, 583, 586, 1010
Mycorrhizae, **624**–625
Myelin sheath, **897**
Myocardium, **949**
Myofibrils, **930**
Myoglobin, 933
Myosin, **930**

N

NAD^+, **255**–256, 262–263
NADH, 255–256, 262–263
$NADP^+$, **232**–233, 235–237
NADPH, 232, 235–239
Nails, human, 937
National Institute of Drug Abuse (NIDA), 905
Natural resources. *See* Resources, natural
Natural selection, 460–464. *See also* Evolution
  and adaptations, 461–464
  and beak size, 472–473, 496–497
  and competition, 473, 497
  defined, 462
  directional, **489**
  disruptive, **489**
  and extinction, 548
  and genetic diversity, 820
  and phenotypes, 482–483, 488–489
  on polygenic traits, 488–489
  on single-gene traits, 488
Navigation, 813
Neanderthals, 772, 773
Negative feedback, 732, 865
Nematodes, 755
Neogene Period, 563
Nephridia, **797**
Nephrons, **884**
Nephrostomes, 797
Nerve impulses, 897–900
Nerve nets, 810
Nerves, 897
Nervous system, human, 896–908
  autonomic, **908**
  central, **896,** 901–904
  peripheral, **896,** 906–908
  sensory receptors, 906
  somatic, **907**
  sympathetic/parasympathetic, 908
Nervous systems, animal, 733, 808–811, 864, 951
Nervous tissue, **863**
Neuromuscular junction, **931**
Neurons, **808**–809, 863, 897–898, 931
Neurotransmitters, **900**
Neurulation, **997**
Neutral mutations, 484, 498
Neutrons, **34**–35
Niches, 99–**100**
Nicotinamide adenine dinucleotide (NAD^+), **255**
Nicotinamide adenine dinucleotide phosphate ($NADP^+$), **232**–233, 235–237
Nicotine, 904, 968
Nitrogen cycle, 84
Nitrogen fixation, **84,** 585
Nitrogenous bases, 344–345
Nitrogenous waste, 794–795
Nodes of plant stem, **675**
Nondisjunction, **401**
Nonpoint source pollution, 160
Nonrenewable resources, **157**
Nonsteroid hormones, **980**
Nonvertebrate chordates, 731, 758, 811, 813, 822
Norepinephrine, **983**
Northwestern coniferous forests, 114
Nose, 963
Notochord, **731**
Nucleic acids, **48,** 344, 403, 436
Nucleolus, 197
Nucleosomes, 280, 352
Nucleotides, **48,** 344–345, 357
Nucleus, **34, 193**–194, **197**
Nutrient cycles, 82–86
  carbon, 82–83
  limitations of, 85–86
  nitrogen, 82–83
  phosphorous, 82–83
Nutrients, **82, 869**
  absorption of, 957
  and human body, 868–873, 880
  limiting, **85**
  and plant growth, 160, 671, 687
Nymphs, **823**

Obesity, 874
Observation
  ecological, 68
  scientific, **6**
Occipital lobe, 903
Occupation of organism, 100
Ocean currents, 98, 118
Oceans, 120–121. *See also* Aquatic ecosystems
Oken, Lorenz, 191
Omnivores, **71**
1000 Genomes Project, 409
On the Origin of Species, 460
Oomycetes, 608
Open circulatory systems, **791**
Operant conditioning, **843**
Operator (O), **378**
Operon, **377**–378
Opposable thumb, **767**
Optic lobes, 811
Optic nerve, 913
Order, **513**
Ordovician Period, 560, 759

Index *continued*

Organelles, **196,** 198–202
Organic chemistry, 45. *See also* Chemistry of life
Organic molecules, 554
Organisms, interactions of, **65**
Organ rejection, 1019
Organs
cellular, **216**
human, 863
Organ systems
cellular, **216**
human, 863
Osmosis, **210**–211
and kidneys, 795
in plants, 672
Osmotic pressure, **211**
Ossification, **925**
Osteoarthritis, **925**
Osteoblasts, **925**
Osteoclasts, **925**
Osteocytes, **925**
Osteoporosis, **925**
Ova, 202, **991**–993, 995–996
Ovaries
of flowers, **650,** 697
human, 979, **991**
Overfishing, 176
Oviparous species, **822**
Ovoviviparous species, **822**
Ovulation, **993**
Ovules, **648,** 697
Oxygen, 35
accumulation of, 555
and electron transport, 258
in glycolysis, 255
molecules of, 37
and photosynthesis, 236, 253
and respiration, 249, 251–253, 966–967
Oxytocin, 982, 1000
Ozone, 163
Ozone layer, **175**

P

Pacemaker, **951**
Pain receptors, 909
Paleontologist, 538, **539,** 559
Paleozoic Era, 560–561
Palisade mesophyll, **681**
Pancreas, **878,** 979, **984**
Paramecium, 198, 608, 612
Parasitic symbionts, 783, 1011
Parasitism, **104,** 140, 579
in fungi, 622–623
in protists, 615–616
in worms, 820, 1011
Parasympathetic nervous system, 908
Parathyroid gland, 979, **985**
Parathyroid hormone, **985**
Parenchyma, **667**
Parental care, 824
Parietal lobe, 903
Park ranger, 105
Parthenogenesis, 819, 835
Particulates, 164
Passion flower, 697
Passive immunity, **1020**
Passive transport, **209**–211
Pasteur, Louis, 586, 589, 1010
Patents, 436, 438
Pathogens, **586,** 828, 1010–1011. *See also* Infectious diseases
Pathologist, 195
Pavlov, Ivan, 843
Pax6 gene, 387
Pedigrees, **396**–397
Peer review, 12
Penicillin, 591, 1021
Pepsin, 53, **877**
Peptidoglycan, 525, 581
Perennial plants, 654
Period, geologic, **543**
Periodic Table, 35
Periosteum, 924
Peripheral nervous system, **896,** 906–908
Peristalsis, **877**
Permafrost, **115**
Permian Period, 561, 762
Perspiration, 883, 935, 937
Pesticides, 161, 487, 717
Petals, **696**
Petiole, **680**
Phagocytosis, 213, 1014–1015
Phanerozoic Eon, 542–543, 560–563
Pharyngeal pouches, **731**
Pharynx, 731, 876, **964**
Phenotypes, **315**
alleles and, 488
and genes, 485–486
and genotypes, 321, 398
and natural selection, 482–483, 488–489
Phenylalanine, 369, 399
Pheromones, 850
Phipps, James, 1020
Phloem, **643,** 666
Phosphorus cycle, 85
Photic zone, **117,** 121
Photoperiod, **713**
Photoreceptors, 912–913
Photosynthesis, **70,** 224–241
and adenosine triphosphate (ATP), **226**–227, 235–239
and autotrophs, **228**
C4 and CAM plants, 113, 241
and carbon dioxide, 239, 241, 253
and cellular respiration, 253
and chlorophyll, **230**–232
and chloroplasts, 202, 231
defined, **228**
factors affecting, 240–241
and light-dependent reactions, 233, 235–237
and light-independent reactions, 233, 238–239
and mesophyll, 681
overview of, 232–233
and plants, 70, 113, 230–231
and prokaryotes, 584
rate of, 240
Photosynthetic bacteria, 70, 555
Photosynthetic protists, 610–611
Photosystems, **235**–236
Phototropism, **712**
pH, **43**–44, 53
Phylogenetic systematics, 516–517
Phylogeny, **516**
Phylum, **514**
Phytochrome, 713–714
Phytoplankton, **73,** 117, 611
Pigments, **230**–231
Pikaia, 757
Pili, 581
Pineal gland, **979**
Pinocytosis, 213
Pioneer species, **107**
Pistil, **697**
Pith in plant stem, **675**
Pituitary gland, 886, 979, **982**–983
Placenta, **822,** 826, **998**
Placentals, 764, 826
Planarian, 812
Plankton, **118**
Plantae, 523–524, 528, 634, 637
Plant breeder, 322
Plant pathologist, 655
Plant reproduction, 696–718
flowering plants, 696–703
seed plants, 646–647, 650, 704–707
vegetative, **702**–703
Plants, 634–654. *See also* Angiosperms
asexual reproduction in, 277, 640, 702–703
branching in, 710

Plants (cont'd)
and cell differentiation, 292
cells of, 203, 206–207, 211, 215
characteristics of, 634–635
classification of, 634, 637, 641, 652–653
crop, 428–429, 437–438, 715–717
cytokinesis, 284
diseases of, 622
epiphytic, 112
flowering, 650–654, 696–703
fossils of, 636, 638, 646, 652
fruit of, **651, 704**
genetically modified, 428, 430
growth of, 676–679, 687
gymnosperms, **646**–648
and herbivory, **102,** 138–139, 552
history and evolution of, 636–640, 646, 650
homeostasis in, 682–683, 711
hormones in, **708**–714
and humans, 715–718
leaves of, **664,** 680–684
life cycle of, 637, 640–642, 644–645, 648–649
life span of, 654
light detection in, 709
and mycorrhizae, 624–625
and nutrients, 160, 671, 687
and photosynthesis, 70, 113, 230–231
polyploidy in, **376,** 420
and polysaccharides, 47
roots of, **664,** 669–673
and seasonal change, 713–714
seed, 638, 646–649, 664–668, 704–707
seedless, 639–645
stems of, **664,** 674–679
temporal isolation in, 495
tissues in, 641, 643–644, 665–668, 670, 680
transgenic, 426, 428, 430
transport in, 635, 641, 666, 672–673, 685–687
tropisms, **712**
vacuoles in, 198
and water, 635, 641, 672–673, 685–687
Plaques, 958
Plasma, **954**
Plasma cells, 1017
Plasma membrane, 193
Plasmids, **424**–425
Plasmodium, **613**
Plasmodium, 607, 616–617, 1011
Platelets, **955**
Plate tectonics, **544**–545
Pleural membranes, 967
Pluripotent cells, **294**
Point mutations, **373**
Point source pollution, 160
Polar molecules, 40–41
Polar zones, 97
Poliovirus, 588–589
Pollen cone, 648
Pollen grain, 215, 325, **647,** 698
Pollen tube, **648,** 700
Pollination, 309, 552, **647**–648, 651, 700
Pollutants, **160**–165, 169
Polychlorinated biphenyls (PCBs), 161
Polygenic traits, **320, 486,** 488–489
Polymeraze chain reaction (PCR), **423**
Polymerization, 46
Polymers, **46**
Polypeptides, 48, **366**
Polyploidy, **376,** 420
Polysaccharides, 47, 869
Polyunsaturated fats, 47, 870
Pons, 903
Population geneticist, 322
Population growth, 130–145
exponential, **132**–133
factors affecting, 132
human, 142–145
limiting factors, **137**–141
logistic, 134–**135**
rate of, 131
Populations, **64**
age structure of, **131,** 144
alleles in, 483
density of, **131,** 138–140
describing, 130–131
distribution of, 131
evolution of, 483–492
frequency of phenotypes in, 485–486, 488–489
and gene pools, 483, 490
and genetic diversity, 820
and genetic drift, **490**
and genetic equilibrium, **491**–492
geographic range of, 131
overcrowding in, 140
Porifera, 754
Positron emission tomography (PET), 905
Prasher, Douglas, 422
Precambrian Time, 543
Precipitation, 112
Predation, **102,** 138–139
Pregnancy, 993
Prehensile tail, **767**
Premolars, 785
Pressure-flow hypothesis, **687**
Priestly, Jacob, 229
Primary growth, **676**
Primary producers, **69**
Primary succession, **106**–107
Primates, **765**–772
Principle of dominance, **310,** 318
Principle of independent assortment, **317**–318, 328–329
Principles of Geology, 454–455
Prions, **592,** 597
Probability, **313**–314
Producers, 584
Products, **50**
Progesterone, 993
Prokaryotes, **193**–194, 197, 523, **580**–585. *See also* Bacteria
archaea, 524, **526,** 580–581
asexual reproduction in, 281
cell cycle of, 281
classification of, 580–581
DNA replication in, 352–353
electron transport chain in, 258
evolution of, 556–557
in food production, 263
gene regulation in, 377–378
microfossils of, 555
mutations in, 583
photosynthetic, 584
prokaryotic chromosomes, 279
reproduction in, 583
structure and function of, 582–585
transcription in, 364
unicellular, 214, 525–526
Prolactin, 983, 1000
Promoters, **365**
Prophage, **577**
Prostaglandins, **980**
Prostate gland, 990
Proteinoid microspheres, 554
Proteins, **48–49, 870**
cell production of, 200–201
collagen, 863
cyclins, **286,** 288
food value of, 250–251
and genes, 370
histones, 280
homologous, 471
levels of organization, 49
protein carriers, 209
protein pumps, 212
protein synthesis, 363, 366–370
regulatory, 287
and ribosomes, **200**
in skeletal muscles, 930
Proterozoic Eon, 542, 543
Protista, 523–524, 526–527, 602

Index *continued*

Protists, 526–527, **602**–617, 1011
  autotrophic, 610–611
  classification of, 602–604
  evolution of, 605
  heterotrophic, 612–613
  movement of, 606–607
  reproduction in, 608–609
  symbiotic, 614–616
Protons, **34**–35
Protostomes, **739**
Protozoan, 214
Prusiner, Stanley, 592
Pseudocoelom, **738**
Pseudopods, **606**
Puberty, **988**
Public health, 1021
Pulmonary circulation, **950**
Punctuated equilibrium, **549**
Punnett squares, **315**–316
Pupa, **823**
Pupil, **912**
Pyloric valve, 877
Pyramids of biomass, 78
Pyramids of energy, 77
Pyramids of numbers, 78
Pyruvic acid, 252, 254, 256

Q

Qualitative data, 8
Quantitative data, 8
Quaternary Period, 542, 563

R

Rabies, 589
Radial symmetry, **738,** 756
Radioactive isotopes, **35**
Radioactivity, 466
Radiometric dating, **540**–541
Ragweed, 694–695
Rain forests, 109, 112
Random mating, 492
Reabsorption, **884**
Reactants, **50**
Receptors, **217**
Recessive alleles, 310–312, 318, 394
Recombinant DNA, 421–**424,** 430
Red blood cells, **954**–955
Red muscle, 933
Red tide, 611
Reflex, **901**
Reflex arc, **907**
Regenerative medicine, 296
Regional climates, 110
Regulatory proteins, 287
Rejection, organ, 1019
Relative dating, **540**
Releasing hormones, **983**
Renewable resources, **157**
Replication, DNA, **350**–353
Reproduction, 19–20. *See also* Asexual reproduction; Sexual reproduction
  in animals, 277, 735, 819–826
  and cell division, 277–278
  in flowering plants, 696–703
  in fungi, 621
  in mammals, 825–826
  in prokaryotes, 583
  in protists, 608–609
  reproductive success, 850
  in seed plants, 646–647, 650, 704–707
  vegetative, **702**–703
Reproductive isolation, **494**–495, 497
Reproductive system, human, 864
  embryonic development, 988, 995–1001
  female, **991**–993
  glands of, 985
  male, **988**–990
Reptiles, 468, 762, 790, 793, 811
Resistance training, 933
Resources, natural, **100**–101, 158–165. *See also* Biodiversity
  atmospheric, 163–165
  freshwater, 160–162
  renewable/nonrenewable, **157**
  soil, 158–160
Respiration, 249, 251–253, 963, 966–967. *See also* Cellular respiration
Respiratory system, human, **963**–969
Respiratory systems, animal, 734–735, 789–790, 864
Responding variable, 7
Response, **809**
Resting potential, **898**–899
Restriction enzymes, **403**–405, 421
Retina, **913**
Retroviruses, **578**
Rh blood group, 394
Rheumatoid arthritis, 1025
Rhizoids, 641
Rhizomes, 644
Ribonucleic acid (RNA). *See* RNA
Ribose, 362
Ribosomal RNA (rRNA), **363,** 370
Ribosomes, **200,** 363, 368–370
Rickets, 921, 943
RNA, 48, **362**–371
  and DNA, 362–365
  evolution of, 554–555
  miRNA, 380–381
  mRNA, **363,** 368–370
  RNA interference, **380**–381
  RNA polymerase, **364**–365
  transcription, **364,** 368, 377–379
  translation, **368**–370
  tRNA, **363,** 368–370
  viruses, 578
Rods, **913**
Roots of plants, **664,** 669–673
  anatomy of, 670
  functions of, 671–673
  length of, 669
  root cap, **670**
  root hairs, **670**
  types of, 669
Roundworms, 755, 783, 810
Rowland, F. Sherwood, 175
Rubella, 1001
Rumen, **786**

S

Sabin, Albert, 589
Safety studies, 16
Saliva, 876
Salk, Jonas, 589
Salt, 36–37, 42, 798
Salt marshes, 6–8, 11–12, 119
Sapwood, **678**
Sarcomere, **930**
SARS, 590, 1013, 1022–1023
Saturated fats, 47, 870
Scavengers, **71**
Schleiden, Matthias, 191
Schopenhauer, Arthur, 6
Schwann, Theodor, 191
Science, 4–15
  attitudes of, 10
  defined, **5**
  goals of, 5
  as knowing, 21
  measurement in, 24
  methodology of, 6–9
  peer review, 12
  scientific theories, **13**
  and society, 14–15
Sclerenchyma, **667**
Scrotum, **989**
Sebaceous glands, **937**
Sebum, 937–938
Secondary growth, **676**–677
Secondary messengers, 981
Secondary succession, **106**–107

Sedimentary rock, 539–540
Seed, **646**
Seed coat, **647**
Seed cone, 648
Seed dispersal, 651, **705**
Seedless plants, 639–645
Seed plants, 638, 646–649
meristems, **667**–668
reproduction in, 646–647, 650, 704–707
structure of, 664–665
tissue systems of, 665–668
Segmentation, 739
Segregation of alleles, **312,** 314, 318
Selection. *See* Natural selection
Selective breeding, **418**–420, 716
Selectively permeable membranes, **205**
Self-pollination, 309
Semen, **990**
Semicircular canals, **911**
Seminal fluid, 990
Seminal vesicles, 990
Seminiferous tubules, **989**
Sensation, 935
Sense organs, human, 909–913
Sensory neurons, **808**–809, 897
Sensory receptors, 906
Sensory systems, animal, **812**–813
Sepals, **696**
Septum, 949
Severe acute respiratory syndrome (SARS), 590, 1013, 1022–1023
Sewage, 162
Seward, William H., 1023
Sex chromosomes, **393**
Sex hormones, **988**
Sex-linked gene, **395**
Sexual development, 988
Sexually transmitted diseases (STDs), **994,** 1012
Sexual reproduction, **19, 277**–278. *See also* Plant reproduction
and allele frequency, 492
in animals, 735, **820**–822
and evolution, 558
female reproductive system, **991**–993
fertilization, **995**–996
in fungi, 621
genetic recombination in, 484
male reproductive system, **988**–990
in protists, 608–609
Short-day plants, 713
Shrubland, 112–113
Sickle cell disease, 375, 391, 398, 400, 413
Sieve tube elements, **666,** 687
Silencing complex, 380–381
Silurian Period, 560, 759
Simard, Suzanne, 625
Single-celled eukaryotes, 523, 602, 752. *See also* Protists
Single-gene traits, **485,** 488
Single-loop circulatory systems, **792**
Single nucleotide polymorphisms (SNPs), 407
Sinoatrial (SA) node, 951, 958
Sinuses, 791
Sister chromatid, 282
Skeletal muscle, **928**–929
Skeletal systems, 733, **814**–816, 864, **922**–927
Skeleton, 754
Skepticism, 10
Skin, 883, **936**. *See also* Integumentary system
cancer, 337, 357, 938–939, 1025
color, 394
as sense organ, 909
Skinner, B. F., 843
Skinner box, 843
Sliding-filament model, 930–931
Slime molds, 603, 604, 613
Slow-twitch muscle, 933
Small intestine, **878,** 880–881
Smallpox, 589, 1020
Smell, 910
Smog, **163**
Smoking, 968–969
Smooth muscle, **928**–929
Sneezing, 1012
Sodium, 35, 37
Sodium bicarbonate, 878
Sodium chloride, 36–37, 42
Sodium-potassium pump, 899–900
Soil, 671
erosion, **159**–160
resources, 158–160
Soil ecosystems
nitrogen cycle in, 84
nutrient limitation in, 86
Solar energy. *See also* Sunlight
and climate, 97
and photosynthesis, 70
Solute, **42**
Solution, **42–43**
Solvent, **42**
Somatic nervous system, **907**
Sound intensity, 910
Southern, Edward, 422
Southern blotting, 422
Speciation, **494**–497, 517, 546–547
Galapágos finches, 496–497
and reproductive isolation, 494–495
Species, **64, 494,** 509, 533
and acquired characteristics, 456
carrying capacity of, **135**
diversity, **166,** 168
endangered, 169–170
and genus, 516
introduced, 169
invasive, 136
keystone, **103,** 167
naming, 510–511
and niches, 100–101
oviparous/ovoviviparous, **822**
pioneer, **107**
tolerance of, **99**
Species survival plans (SSPs), 170
Spectrum, visible, 230
Sperm, 325, 989–990, 995–996
Sphygmomanometer, 953
Spina bifida, 997, 1001
Spinal cord, 811, 901
Spirilla, **582**
Spleen, 957
Sponges, 752, 754, 784, 821
Spongy bone, 924
Spongy mesophyll, **681**
Sporangium, **609, 642**
Spores, **607**
Sporophytes, **637**–638, 698
Squamous cell carcinoma, 939
Stabilizing selection, **489**
Staining, cell, 191
Stamens, **697**
Stanley, Wendell, 574
Starches, 46, 869
Start codon, 367
Statins, 961
Steere, Allen, 1009, 1031
Stem cells, **295**–297
Stems of plants, **664,** 674–679
Steroid hormones, 379, **980**
Steroids, 47, 934
Stigma, **697**
Stimulus, **18,** 733, **808**–809, 906
Stomach, **877**
Stomata, **681**–683
Stop codon, 367
Strep throat, 587
Stroke, 958
Stroma, **231**
Struggle for existence, 460, 462–463
Sturtevant, Alfred, 329
Style, 697
Subatomic particles, 34
Substrates, **52**–53
Succession, ecological, **106**–109
Sucrose, 46

Index *continued*

Sugars
carbohydrates, 46, 869
in human body, 867
and photosynthesis, 239
in plants, 687
Sunlight, 230. *See also* Light
and aquatic ecosystems, 117
and plants, 635
and skin, 337, 357, 935–936, 939
Superbugs, 591
Suppressor T cells, 1019
Surface tension, 41
Survival of the fittest, 461–463
Survival strategies, 278
Suspension, **42**
Sustainable development, 156–**157**, 160, 174, 176
Sutton, Walter, 349
Sweating, 828, 883, 935, 937
Symbionts, 783, 1011
Symbiosis, **103**–104, 783
Symbiotic protists, 614–616
Symmetry, body, 738
Sympathetic nervous system, 908
Synapse, **900**
Synovial cavity, 927
Synthetic genome, 435
Syphilis, 994
Systematics, **512,** 516–517
Systemic circulation, **950**
Systolic pressure, 953

T

Taiga, **114**
Taproot system, 669
Tar, 968
Target cells, **709, 978**
Taste buds, **910**
TATA box, 379
Taxon, **512,** 516
Technology, 11
Teeth, 876
Telomerase, 352
Telomeres, **352**
Temperate forests, 114
Temperate zones, 97
Temperature
body, 828–830, 866, 935, 937, 987
and butterfly wing color, 321
and enzymes, 53
and extinction, 170
global warming, **177**
and photosynthesis, 240
and seed germination, 706–707
sensory response to, 909
and water depth, 118
Temporal isolation, **495**
Temporal lobe, 903
Tendons, **816, 932**
Terracing, 160
Territorial behavior, **848**
Tertiary Period, 563
Testes, 979, **989**
Testosterone, **988**–989
Tetanus, 587
Tetrad, **324**
Tetrapoda, 518–519
Tetrapods, **760**
Thalamus, **903**
Theory, **13**
Thermoreceptors, 909
Thigmotropism, **712**
Thomas, Lewis, 64
Threshold, **899**
Throat, 964
Thylakoid membranes, 236–237
Thylakoids, **231,** 233, 235
Thymine, 344–345, 348, 362
Thymus, 957, **979,** 1016
Thyroid gland, 979, **985,** 987
Thyroid-stimulating hormone (TSH), 983, 987
Thyrotropin-releasing hormone (TRH), 987
Thyroxine, **985,** 987
Tiktaalik, 761
Tissues
animal, 737
cell, **216**
Tissues, human, 862–863
connective, **863**
epithelial, **863**
muscle, **863**
nervous, **863**
Tissues, plant. *See also* Vascular tissue of plants
dermal, **665,** 670, 680
ground, 665, **666,** 670, 680
meristems, **667**–668
T lymphocytes, 955, 957, 1016–1019
Tobacco, 713, 968–969
Tobacco mosaic disease, 574–575
Tolerance of species, **99**
Topsoil, 158
Totipotent cells, **294**
Touch, 909
Toxins, 586
Trace elements, 671
Trace fossils, 752
Trachea, 789, **964**
Trachea epithelium, 215
Tracheids, **643,** 666
Tracheophytes, **643–644**
Traits, **309**
and alleles, 397, 485–486
and environment, 321
polygenic, **320, 486,** 488–489
single-gene, **485,** 488
transmission of, 394–396
Transcription, **364,** 368, 377–378
Transcription factors, 379
Trans fats, 870
Transfer RNA (tRNA), **363,** 368–370
Transformation, bacterial, **339**–340
Transgenic organisms, **426**–430
Translation, **368**–370
Transpiration, **681**–683, 685–686
Transport
active, **212**–213, 227
passive, **209**–211
in plants, 635, 641, 666, 672–673, 685–687
Tree of life, **23**
Trees, 653
classifying, 515
growth of, 678–679
and mycorrhizae, 624–625
rings of, 678
Trial-and-error learning, 843
Triassic Period, 562, 762, 764
Trichomes, 665
Trilobites, 540
Trisomy, 401
Trochophore, **756**
Trophic levels, 77–78
Tropical rain forests, 109, 112
Tropical zone, 97
Tropisms, **712**
Tryptophan, 367
Tuatara, 762
Tuber, 702
Tuberculosis, 583, 586–587, 589, 1010, 1023
Tubulins, 199
Tumors, **289**
Tundra, 115
Tunicates, 747, 758, 813
Turkish Angola, 416–417
Turner's syndrome, 401
Twins, 996
Two-factor cross, 317
Tympanum, 911
Type 1 diabetes, 1025
Typhoid, 400

U

Ubx gene, 501
Ulcers, 877
Ultraviolet (UV) light, 337, 344, 357
Umami receptors, 910
Umbilical cord, 998
Understory, **112**
Unicellular organisms, 214, 525–527
Unsaturated fats, 47, 870
Uracil, 362, 366
Urea, 794, 797, 882–883
Ureters, **883**
Urethra, **883,** 989
Urey, Harold, 554
Uric acid, 794, 797
Urinary bladder, **883**
Urine, 795, 797, 883–884, 886

V

Vaccination, 1017, **1020**
Vaccine, **588**–589, 593
Vacuoles, **198**
Valence electrons, 36
Valves, **950**
Van der Waals forces, **38**
Van Leeuwenhoek, Anton, 190
Variation, 419–420, 457–458, 460, 462–463, 482–486. *See also* Natural selection
Vascular tissue of plants, **641,** 643–644, 665–666, 680
  vascular bundles, **675**
  vascular cambium, **677**–679
  vascular cylinder, **670,** 672–673
Vas deferens, **989**
Vectors, **1013**
Vegetables, 704
Vegetative reproduction, **702**–703
Veins, human, **952**
Veins, leaf, 680
Venter, Craig, 349
Ventricle, **792**–793, **949**–951
Venus' flytrap, 712
Vertebrae, 731
Vertebrates, **731,** 757, 789–790. *See also* Chordates; Invertebrates
  brains of, 811
  cephalization in, 740
  circulatory systems of, 734, 792–793
  homologous structures in, 469
  limb formation in, 743
  muscular systems of, 816–818
  nervous systems of, 809
  sense organs, 813
  skeletons of, 733, 815–818
Vesicles, 198
Vessel elements, **666**
Vestigial structures, **469**
Villi, **880**
Viral sexually transmitted diseases (STDs), **994**
Virchow, Rudolf, 191
Viruses, **574**–579
  bacterial, 340–341
  and cells, 579
  discovery of, 574
  diseases caused by, 588–589, 994, 1010–1011
  and medication, 1021
  structure and composition of, 575
  viral infections, 576–578
Visible spectrum, 230
Vision, 912–913
Vitamin D production, 935
Vitamin K deficiency, 881
Vitamins, **871**
Vitreous humor, 912
Viviparous species, **822**
Vocal cords, 964

W

Wallace, Alfred Russel, 459, 460
Water
  adhesion, **686**
  atomic composition of, 36
  capillary action, **686**
  cohesion, 686
  cycle, 81
  dehydration, 189, 869
  and human body, 869, 986
  intoxication, 221
  molecules of, 37, 40–41
  osmosis, **210**–211, 672, 795
  and photosynthesis, 240
  and plants, 635, 641, 672–673, 685–687
  properties of, 40–43
  quality and pollution, 156, 160–162
  and seed germination, 706
  solutions and suspensions, **42**
  and spread of disease, 1012–1013
Water mold, 608–609, 613
Watershed, 162
Water-soluble vitamins, 871
Water transport in plants, 666, 685–686
Watson, James, 349–350, 362
Weather, **96**
West Nile virus, 588, 1013
Wetlands, **119,** 156
Wheat rust, 622
White blood cells, **955**
White matter, 902
White muscle, 933
Wildlife photographer, 105
Wilmut, Ian, 427
Wilting, 683
Wind
  and heat transport, 98
  pollination, 700
Wood, 678–679
Woodland, 113
Woody plants, **653**
Worms
  earthworms, 755, 792
  flatworms, 755, 783, 794, 796, 810, 812
  marine, 755
  parasitic, 820, 1011
  roundworms, 755, 783, 810
  tapeworms, 104

X

X chromosomes, 393–396, 401
X-ray diffraction, 346
Xylem, **643,** 666

Y

Y chromosomes, 393–395, 434
Yamanaka, Shinya, 297
Yeasts, 192, 214, 263, 278, 424, 585, 619, 623
Yolk sac, 825

Z

Zoo curator, 736
Zoonosis, **1013**
Zooplankton, **76,** 117
Zygote, **325,** 637, **739, 995**–996

Credits

Staff Credits Jennifer Angel, Amy C. Austin, Laura Baselice, Neil Benjamin, Diane Braff, Lisa Brown, Ken Chang, Daniel Clem, Kristen Ellis, Julia Gecha, Ellen Granter, Anne Jones, Kathilynn Kalina, Stephanie Keep, Toby Klang, Alyse Kondrat, Beth Kun, Kimberly Fekany Lee, Ranida Touranont McKneally, Anne McLaughlin, Rich McMahon, Maria Milczarek, Laura Morgenthau, Michelle Reyes, Rashid Ross, Laurel Smith, Lisa Smith-Ruvalcaba, Ted Smykal, Emily Soltanoff, Cheryl Steinecker, Berkley Wilson

Additional Credits Lisa Furtado Clark, Amy Hamel, Courtenay Kelley, Hilary L. Maybaum

Front Cover, Spine Ralph A. Clevenger/Corbis; **Back Cover** hotshotsworldwide/Fotolia

(For credits for custom pages, refer to copyright page.) **iii** (T, B) Stew Milne; **x** (TL) ©Ralph A. Clevenger/Corbis, (C) PhotoLibrary Group, Inc.; **xii** (TC) Photo Researchers, Inc.; **xiii** (BL, TC) Colin Keates/Courtesy of the Natural History Museum, London/©DK Images; **xiv** (TC, TR) Peter Chapwick/©DK Images, (TL) Southhampton General Hospital/Science Photo Library/Photo Researchers, Inc.; **xv** (TR) iStockphoto; **xvi** (B) Alamy Images, (C) Dennis MacDonald/PhotoEdit, Inc.; **xvii** (TL) David M. Phillips/Photo Researchers, Inc.; **xxii** (BR) Stew Milne. **1a** (B) slowfish/Shutterstock, (CL) akarapong/Shutterstock; **1b** (CR) Mareck CECH/Shutterstock; **1** (L) Keren Su/Corbis, (BR) Stew Milne; **2** (B) ©ClassicStock/Alamy Images; **3** (TR) Jupiter Images; **4** (T) ©ClassicStock/Alamy Images, (B) ©Suzanne Long/Alamy Images; **5** (TR) ©Helio & Van Ingen/NHPA/Photoshot; **6** (T) Michael Melford/National Geographic/Getty Images; **8** (CL) Jupiter Images, (L) Michael Melford/National Geographic/Getty Images; **10** (TL) ©ClassicStock/Alamy Images; **11** (CR) Jupiter Images, (B) Ron Chapple Stock/Photolibrary Group, Inc.; **12** (B) Dani Yeske/Peter Arnold/PhotoLibrary Group, Inc.; **14** (BL) Pascal Goetgheluck/Photo Researchers, Inc.; **15** Romeo Gocad/AFP/Getty Images; **16** (TR, TL, TCR, TCL) SuperStock; **17** (TL) ©ClassicStock/Alamy Images, (BR) Masa Ushioda/Image Quest Marine; **18** (Bkgrd) ©Martin Rugner/Pixtal/AGE Fotostock, (Inset, BR) Cathy Keifer/Shutterstock, (BL) Nigel Cattlin/Alamy Images; **19** (Inset, Upper MR, Inset, BR) Biophoto Associates/Photo Researchers, Inc., (BR) Custom Medical Stock Photo, (CR) Ilyashenko Oleksiy/Shutterstock, (TR) John Marshall/Corbis; **21** (TR) Jupiter Images, (B) PhotoLibrary Group, Inc.; **22** (C) ©Worldwide Picture Library/Alamy Images; **23** (TR) Jim Richardson/Corbis, (Upper MR) Joseph Nettis/Photo Researchers, Inc., (BCR) Michael Nichols/National Geographic/Getty Images, (Lower M) Photo by Scott Bauer/Courtesy of USDA, Agricultural Research Service, (BR) Tek Image/Photo Researchers, Inc.; **24** (B) ©Chinafotopres-US/Sipa/NewsCom; **25** (TR) Brand X/Jupiter Images; **28** (TL) Anthony Bannister/Gallo/Corbis; **29** (TR) Jupiter Images; **32** (C) Hans Strand/Corbis; **33** (TR) ©Julian Gutt/AFP/Getty Images/NewsCom; **34** (TL) Hans Strand/Corbis; **35** (TR) ©MarcelClemens/Shutterstock; **37** (BR) ©Julian Gutt/AFP/Getty Images/NewsCom; **38** (C) ©Natural Visions/Alamy Images, (CR) Andrew Syred/Photo Researchers, Inc., (TL) Martin Harvey/Corbis; **39** (BR) ©fivespots/Shutterstock, (BL) Courtesy of Harvard-MIT Division of Health Sciences and Technology; **40** (TL) Hans Strand/Corbis; **41** (BC) Richard Megna/Fundamental Photographs, NYC; **42** (TL) ©Julian Gutt/AFP/Getty Images/NewsCom; **43** (CR) fanelie rosier/iStockphoto, (TCR) Milos Luzanin/iStockphoto, (TR) Nancy Louie/iStockphoto, (BCR) Noam Armonn/iStockphoto; **45** (TL) Hans Strand/Corbis; **47** (BC) ©Valentyn Volkov/Shutterstock; **50** (TL) Hans Strand/Corbis; **53** (TR) ©Julian Gutt/AFP/Getty Images/NewsCom, (CL) Thomas A. Steitz/Howard Hughes Medical Institute Yale University; **57** (TR) ©Julian Gutt/AFP/Getty Images/NewsCom; **60** (TL) Keren Su/Corbis; **61a** (CR) Kerstin Layer/mauritius images GmbH/Alamy Images, (CL) Shutterstock; **61b** (C) Andy Z./Shutterstock, (CL) Pi-Lens/Shutterstock; **61** (L) ©konmesa/Shutterstock, (BR) Stew Milne; **62** (B) Stan Osolinski/Oxford Scientific/Jupiter Images; **63** (TR) Jeff Rotman Photography; **64** (BR) ©nik wheeler/Alamy Images, (BC) Anup Shah/WILDLIFE/PhotoLibrary Group, Inc., (TL) Stan Osolinski/Oxford Scientific/Jupiter Images, (BL) Tim Graham/Getty Images; **65** (BL) Dmitri Kessel/Time Life Pictures/Getty Images, (BR) NASA Goddard Space Flight Center/Reto Stöckli (land surface, shallow water, clouds). Enhancements: Robert Simmon (ocean color, compositing, 3D globes, animation)/NASA, (BC) Planetary Visions Ltd./Photo Researchers, Inc.; **67** (TR) Jeff Rotman Photography; **68** (TL) ©Benjamin Alblach Galan/Shutterstock; **69** (BR) ©Cathy Keifer/Shutterstock, (TL) Stan Osolinski/Oxford Scientific/Jupiter Images; **70** (R) ©Science Source/Photo Researchers, Inc., (BL) Vincenzo Lombardo/Photodisc/Getty Images; **71** (TCR) ©Anna Yu/Alamy Images, (BC) ©MJ Prototype/Shutterstock, (CL) Carol Farneti Foster/Oxford Scientific/PhotoLibrary Group, Inc., (TCL) Kevin Schafer/Peter Arnold/PhotoLibrary Group, Inc., (BR) Project Amazonas, Inc., (CR) Roy Toft/National Geographic Creative/Getty Images, (Bkgd) Will & Deni McIntyre/Stone/Getty Images; **72** (CL) Jeff Rotman Photography, (T) Woodt Stock/Alamy Images; **73** (TL) Stan Osolinski/Oxford Scientific/Jupiter Images; **76** (TL) Jeff Rotman Photography; **79** (TL) Stan Osolinski/Oxford Scientific/Jupiter Images; **80** (CL) Adrian Dorst/Peter Arnold/PhotoLibrary Group, Inc., (BC) Corbis/Photolibrary Group, Inc., (TCL) Francesco Ruggeri/Getty Images, (TC) Image Source/Photolibrary Group, Inc., (BL) Jim Wark/Peter Arnold/PhotoLibrary Group, Inc., (TL) Joel Sartore/Getty Images, (CC) Kalish Dimaggio/Flirt Collection/PhotoLibrary Group, Inc., (Upper MR) NASA; **82** (TL) ©Olga_i/Shutterstock; **84** (BL) Jeff Rotman Photography; **87** (BL) Goddard Space Flight Center, Scientific Visualization Studio/NASA, (CR) Goddard Space Flight Center, The Sea/WIFS Project and GeoEye, Scientific Visualization Studio/NASA; **91** (TR) Jeff Rotman Photography; **94** (B) ©Bryan Busovicki/Shutterstock; **95** (TR) Nature Picture Library; **96** (TL) ©Bryan Busovicki/Shutterstock, (TR) NOAA/ZUMA/Corbis; **99** (TL) ©Bryan Busovicki/Shutterstock; **100** (BL) ©Karel Gallas/Shutterstock; **102** (TCL) ©Pi-Lens/Shutterstock; **103** (BR) ©Hal Beral/Corbis, (TR) Nature Picture Library; **104** (TL) Mark Taylor/Nature Picture Library, (C) Todd Pusser/Nature Picture Library; **105** (B) Stan Tekiela; **106** (TL) ©Bryan Busovicki/Shutterstock; **108** (BL, BCL) Orlando Carrasquillo, El Yunque National Forest, Ecosystem Management Team; **109** (TR) ©Gary Braasch/Corbis, (CC) ©Macduff Everton/Corbis, (CR) Anthony P. Bolante/Reuters/Corbis, (BR) Hallmark Institute/PhotoLibrary Group, Inc.; **110** (TL) ©Bryan Busovicki/Shutterstock; **112** (T) Juan Carlos Muñoz/AGE Fotostock, (BL) Michele Burgess/SuperStock, (CL) Staffan Widstrand/Nature Picture Library; **113** (T) ©Danita Delimont/Alamy Images, (C) ©JLImages/Alamy, (BL) Debra Behr/Alamy Images; **114** (TL) ©Natural Selection Jerry Whaley/Design Pics/Corbis, (C) ©Radius Images/Alamy, (BL) AGE Fotostock/SuperStock; **115** ©ImageState/Alamy Images; **116** (B) ©All Canada Photos/Alamy Images, (TL) Nature Picture Library; **117** (TL) ©Bryan Busovicki/Shutterstock, (BL) Corbis/SuperStock; **118** (BL) ©cappi thompson/Shutterstock, (BC) David Noton/Nature Picture Library, (CL) Nature Picture Library, (BR) Philippe Clement/Nature Picture Library; **119** (BCL) ©irishman/Shutterstock, (BL) ©Mark Hodnett/Fotolia, (BR) AGE Fotostock/SuperStock, (BCR) Cindy Ruggieri; **121** Dante Fenolio/Photo Researchers, Inc.; **124** (BL) Anthony Bannister/Gallo Images/Corbis; **125** (TR) Nature Picture Library; **128** (B) ©FLPA/Alamy Images; **129** (TR) Geoff Dann/©DK Images; **130** (TL) ©FLPA/Alamy Images, (B) Stephen Ausmus/Courtesy of USDA, Agricultural Research Service; **131** (CR) ©Juniors Bildarchiv/Alamy Images, (TR) SuperStock; **132** (BL) Geoff Dann/©DK Images; **133** (BR) ©Ewan Chesser/Shutterstock, (CR) SciMAT/Photo Researchers, Inc.; **134** (B) Digital Vision/Thinkstock; **136** (TR, TL, TCR, TCL) SuperStock, (CR) Wolfgang Pölzer/Alamy Images; **137** (TL) ©FLPA/Alamy Images; **138** (TL) John Conrad/Corbis; **139** (T) D. Robert & Lorri Franz/Corbis, WILDLIFE/Peter Arnold Inc./PhotoLibrary Group, Inc.; **140** (BL) Geoff Dann/©DK Images, (TL) Les Stocker/PhotoLibrary Group, Inc.; **141** (TR) Xinhua Photo/Jiang Yi/©Associated Press; **142** (TL) ©FLPA/Alamy Images; **143** (CR, BR) FOTOGRAF/Still Pictures/PhotoLibrary Group, Inc.; **145** (T) image100/Photolibrary Group, Inc.; **147** (CR) Xinhua Photo/Jiang Yi/©Associated Press; **149** (TR) Geoff Dann/©DK Images; **152** (B) NSSDC/NASA; **153** (TR) eStock Photo; **154** (BR, BC) Douglas Peebles/eStock Photo, (BL) eStock Photo, (TL) NSSDC/NASA; **155** (B) ©Elena Elisseeva/Shutterstock; **156** (B) Gary Sullivan/The Wetlands Initiative archives, (BCL) The Wetlands Initiative archives; **157** (CR) Bobby Model/Getty Images, (TR) Maurizio Borgese/Hemis/Corbis; **158** (BL) ©Associated Press, (TL) NSSDC/NASA; **159** (BR) eStock Photo; **160** (TL) Jim Richardson/Corbis; **162** (CL) Richard T. Nowitz/Corbis, (B) Sy Djibril/Panapress/MAXPPP/NewsCom; **163** (TR) Jason Lee/Reuters Media; **164** (BL) ©Petr Vopena/Shutterstock, (BCL) Adam Hart-Davis/Photo Researchers, Inc.; **166** (TL) NSSDC/NASA, (BL) Penn State University/©Associated Press; **167** (BR) Marcos G. Meider/SuperStock, (BC) Noel Hendrickson/Getty Images, (BL) Paolo Aguilar/epa/Corbis; **168** (B) Creatas/Photolibrary Group, Inc., (TL) eStock Photo; **169** (BR) eStock Photo, (T) MARIANA BAZO/Reuters/Landov LLC; **170** (BL) ©ZUMA Press/NewsCom; **171** (BR) Barbara Walton/epa/Corbis; **173** (TL) NSSDC/NASA; **175** (CL) Images by Greg Shirah, NASA Goddard Space Flight Center Scientific Visualization Studio/NASA, (BCL) Jerry Mason/Photo Researchers, Inc.; **176** (BL) Brian J. Skerry/National Geographic Image Collection, Jeffrey L. Rotman/Corbis; **177** (B) Armin Rose/Shutterstock; **178** (TR) ©Laura Romin & Larry Dalton/Alamy Images; **179** (C) Charmagne Leung/California Academy of Sciences, Special Collections, (TR) Mark Fairhurst/UPPA/Photoshot/NewsCom, (TCR) Realimage/Alamy Images; **181** (B) ©Elena Elisseeva/Shutterstock; **183** (TR) eStock Photo; **186** (TL) ©konmesa/Shutterstock; **187a** (B) Brian Chase/Shutterstock; **187** (L) Clouds Hill Imaging Ltd./Corbis, (BR) Stew Milne; **188** (B) Dr. Robert Berdan; **189** (TR) Simple Stock Shots/AGE Fotostock; **190** (TL) Dr. Robert Berdan, (BL) Science & Society Picture Library; **191** (BR) Biophoto Associates/Photo Researchers, Inc., (TR) Grafissimo/iStockphoto; **192** (TCL) ©Stevens Frederic/Sipa/NewsCom, (BC) Dr. Gopal Murti/Photo Researchers, Inc., (BL) Michael Abbey/Photo Researchers, Inc., (BR) SciMAT/Photo Researchers, Inc.; **193** (CR) Simple Stock Shots/AGE Fotostock; **195** (BR) Dr. Tanasa Osborne; **196** (TL) Dr. Robert Berdan; **198** (BL) Biophoto Associates/Photo Researchers, Inc., (BR) Eric Grave/Science Photo Library/Photo Researchers, Inc.; **199** (T) Dr. Torsten Wittmann/Photo Researchers, Inc., (BR) Omikron/Photo Researchers, Inc.; **202** (BC) Photo Researchers, Inc.; **204** Science Photo Library/Custom Medical Stock Photo; **208** (TL) Dr. Robert Berdan, (BL) Simple Stock Shots/AGE Fotostock, (TCR) Vince Streano/Stone/Getty Images; **213** (CR) NIBSC/Photo Researchers, Inc.; **214** (TL) Dr. Robert Berdan, (BL) Volker Steger/Christian Bardele/SPL/Photo Researchers, Inc.; **215** (BR, BL) Ed Reschke/Peter Arnold/PhotoLibrary Group, Inc.; **217** Don W. Fawcett/Photo Researchers, Inc.; **221** (TR) Alan Bailey/Rubberball/Getty Images/Photolibrary Group, Inc.; **224** (B) Perennou Nuridsany/Photo Researchers, Inc.; **225** (TR) Bridgeman Art Library; **226** (TL) Perennou Nuridsany/Photo Researchers, Inc.; **228** (TL) Bridgeman Art Library, (CR) Michael Lohmann/Glow Images; **229** Biophoto Associates/Photo Researchers, Inc., Imperial College London, James Barber, The Royal Society, (TL) The Granger Collection, New York/©The Granger Collection, NY, (BL) The Print Collector/Alamy Images; **230** (TL) Perennou Nuridsany/Photo Researchers, Inc.; **231** (B) Andy Small/GAP Photos/Getty Images; **232** (BR) Bridgeman Art Library; **235** (BR) Doable/A. Collection/amana images/Getty Images, (TL) Perennou Nuridsany/Photo Researchers, Inc.; **236** (TL) Imagewerks/Getty Images; **239** (TR) Bridgeman Art Library; **241** (T) Dorn1530/Shutterstock; **243** (CR) Imagewerks/Getty Images; **244** (BC) John T. Fowler/Alamy Images; **245** (TR) Bridgeman Art Library; **248** (B) Professor Pietro M. Motta/Photo Researchers, Inc.; **249** (TR) Kike Calvo/V&W/Image Quest Marine; **250** (TL) Professor Pietro M. Motta/Photo Researchers, Inc.; **251** (BR) Martin Jacobs/Stock Food Creative/Getty Images; **252** (BL) Kike Calvo/V&W/Image Quest Marine; **254** (TL) Professor Pietro M. Motta/Photo Researchers, Inc.; **256** (TL) Kike Calvo/V&W/Image Quest Marine; **261** (CR) Photos/Jupiter Images, (TR, TL, T) SuperStock; **262** (TL) Professor Pietro M. Motta/Photo Researchers, Inc.; **264** (BL) Brand X Pictures/Photolibrary Group, Inc.; **265** (CR) Fotosonline/Alamy, (TR) Kike Calvo/V&W/Image Quest Marine; **267** (BR) Brand X Pictures/Photolibrary Group, Inc.; **269** (TR) Michael Nolan/Peter Arnold/PhotoLibrary Group, Inc.; **272** (B) Michael Abbey/Photo Researchers, Inc.; **273** (TR) Joe McDonald/Corbis; **274** (R) Art Wolfe/Photo Researchers, Inc., (TL) Michael Abbey/Photo Researchers, Inc.; **277** (CL) CNRI/Photo Researchers, Inc., (R) Ed Reschke/PhotoLibrary Group, Inc., (C) Oxford Scientific Films/PhotoLibrary Group, Inc.; **278** (BR, BC) Getty Images, (TL) Joe McDonald/Corbis; **279** (TL) Michael Abbey/Photo Researchers, Inc.; **283** (BR) Ed Reschke/Peter Arnold, Inc./PhotoLibrary Group, Inc.; **284** (C) Dr. Gopal Murti/Photo Researchers, Inc., (TL) Joe McDonald/Corbis, (CR) Wood/Custom Medical Stock Photo; **285** (TR, TL, TC, BR, BL, BC) Ed Reschke/Peter Arnold/PhotoLibrary Group, Inc.; **286** (TL) Michael Abbey/Photo Researchers, Inc.; **287** (TR) Joe McDonald/Corbis, (BL) Design Pics Inc./Alamy, (BR) Scott Camazine/Photo Researchers, Inc.; **288** (Bkgd) National Institutes of Health, (Inset) Paul Martin/Wellcome Images; **291** (TL) Dwight Smith/Shutterstock, (BR, BL) Max Planck Institute for Molecular Genetics, (TR) Volker Steger/Peter Arnold, Inc./Alamy Images; **292** (Inset, BR) ©John Durham/Photo Researchers, Inc., (BR) DK Images, (BCL) Jim Haseloff/Wellcome Images, (TL) Michael Abbey/Photo Researchers, Inc., (Inset, BL) Professor Ray F. Evert/University of Wisconsin; **294** (BL) Joe McDonald/Corbis; **297** (TR) Rick Wood/Rapport Press/NewsCom; **301** (TR) Joe McDonald/Corbis; **304** (TL) Clouds Hill Imaging Ltd./Corbis; **305a** (BL) Ruta Saulyte-Laurinaviciene/Shutterstock; **305b** (BL) AFP/NewsCom, (BR) Charles Bennett/Associated Press; **305** (L) BSIP LAURENT/Photo Researchers, Inc., (BR) Stew Milne; **306** (B) Biosphoto/Labat J.-M. & Rouquette F./Peter Arnold Inc/PhotoLibrary Group, Inc.; **307** (TR) blickwinkel/Alamy Images; **308** (BL) Bettman/Corbis, (TL) Biosphoto/Labat J.-M. & Rouquette F./Peter Arnold Inc/PhotoLibrary Group, Inc.; **310** (TL) blickwinkel/Alamy Images; **313** (TL) Biosphoto/Labat J.-M. & Rouquette F./Peter Arnold Inc/PhotoLibrary Group, Inc., (B) Brand X Pictures/Jupiter Images; **318** (CL) Maximilian Weinzierl/Alamy Images; **319** (TL) Biosphoto/Labat J.-M. & Rouquette F./Peter Arnold Inc/PhotoLibrary Group, Inc., (TCR, BL) Christopher Burrows/Alamy Images; **320** (BL) blickwinkel/Alamy Images; **321** (TCR) Alvin E. Staffan/Photo Researchers, Inc., (TR) Robert Shantz/Alamy Images; **322** (TR) Photo courtesy Sofia Cleland; **323** (TL) Biosphoto/Labat J.-M. & Rouquette F./Peter Arnold Inc/PhotoLibrary Group, Inc.; **329** (TR) blickwinkel/Alamy Images; **333** (TR) blickwinkel/Alamy Images, (TL) Ryerson Clark/iStockphoto; **336** (B) Charles C. Benton/Kite Aerial Photography; **337** (BR) aliciahh/iStockphoto; **338** (TL) Charles C. Benton/Kite Aerial Photography; **340** (BL) ©Eye of Science/Photo Researchers, Inc.; **344** (BR) aliciahh/iStockphoto, (TL) Charles C. Benton/Kite Aerial Photography; **346** (TC) Erwin Chargaff; courtesy of University Archives/Columbia University in the City of New York, (TL) National Portrait Gallery, London/National Portrait Gallery, London, (TR) Oregon State University Libraries Special Collections; **347** (T) ©A. Barrington Brown/Photo Researchers, Inc., (BR) aliciahh/iStockphoto, (TR) Kenneth Eward/BioGrafx/Photo Researchers, Inc., (TCR) Wellcome Images; **349** (TC) arlindo71/iStockphoto, (CL) Biophoto Associates/Photo Researchers, Inc., (CL) GlobalP/iStockphoto, (BR) Reprinted by permission from Macmillan Publishers Ltd: Nature, February 15, 2001, Volume 409, Number 6822. Copyright ©2001/Nature Magazine; **350** (TL) Charles C. Benton/Kite Aerial Photography; **351** (TCR) aliciahh/iStockphoto, (BR) Dr. Gopal Murti/Photo Researchers, Inc.; **352** (CL) Dr. Hesed Padilla-Nash/National Cancer Institute; **355** (BL) Oregon State University Libraries Special Collections; **357** (TCR) aliciahh/iStockphoto; **360** (B) Jan Daly/Shutterstock; **361** (TR) EYE OF SCIENCE/SPL/Photo Researchers, Inc.; **362** (TL) Jan Daly/Shutterstock; **366** (TL) Jan Daly/Shutterstock; **369** (CR) MRC Lab of Molecular Biology/Wellcome Images; **370** (CL) EYE OF SCIENCE/SPL/Photo Researchers, Inc.; **372** (BC) Bob Gibbons/Photo Researchers, Inc., (TL) Jan Daly/Shutterstock, (BR) Tony Camacho/Photo Researchers, Inc.; **375** (BR) Eye of Science/Photo Researchers, Inc.; **376** (TL) travismanley/iStockphoto; **377** (BR) G. Murti/Photo Researchers, Inc., (TL) Jan Daly/Shutterstock; **379** (BR) EYE OF SCIENCE/SPL/Photo Researchers, Inc.; **381** Steve Gschmeissner/Photo Researchers, Inc.; **382** (BL) EYE OF SCIENCE/SPL/Photo Researchers, Inc.; **383** (CR) ©Randy M. Ury/Corbis, (TR) Robert Clay/Alamy Images, (TCR) WILDLIFE GmbH/Alamy Images; **387** (TR) EYE OF SCIENCE/SPL/Photo Researchers, Inc.; **390** (B) Digital Vision/Photolibrary Group, Inc.; **391** (TR) Sebastian Kaulitzki/Fotolia; **392** (BL) CNRI/Photo Researchers, Inc., (TL) Digital Vision/Photolibrary Group, Inc.; **393** (TL) Image Source/Getty Images; **394** (TL) CMCD Visual Symbols Library; **395** (TR) Sebastian Kaulitzki/Fotolia; **396** (TL) Dave King/©DK Images; **398** (TL) Digital Vision/Photolibrary Group, Inc.; **399** (BR) Sebastian Kaulitzki/Fotolia; **402** (CR) Phanie/Photo Researchers, Inc., (TR, TL, TCR, TCL) SuperStock; **403** (C) Barry Rosenthal/Getty Images, (TL) Digital Vision/Photolibrary Group, Inc.; **407** (BL) arlindo71/iStockphoto, (BC) Kenneth Eward/BioGrafx/Photo Researchers, Inc., (BCL) Photo Researchers, Inc., (BR) PR NEWSWIRE/©Associated Press, (TR) Sebastian Kaulitzki/Fotolia; **408** (TCL) Reprinted by permission from Macmillan Publishers Ltd: Nature, February 15, 2001, Volume 409, Number 6822. Copyright ©2001/Nature Magazine; **410** (TR) ©Associated Press; **411** (CR) Phanie/Photo Researchers, Inc.; **412** Phanie/Photo Researchers, Inc.; **413** (TR) Sebastian Kaulitzki/Fotolia; **416** (B) Yonhap News/YNA/NewsCom; **417** (TR) David Parker/Photo Researchers, Inc.; **418** (B) Corbis Royalty Free/Jupiter Images, (TCR) Thinkstock/Jupiter Images, (TL) Yonhap News/

YNA/NewsCom; **419** (BR) Gilbert S Grant/Photo Researchers/Getty Images; **420** (TL) Jose F. Poblete/Corbis; **421** (BR) David Parker/Photo Researchers, Inc., (TC) Photo Researchers, Inc., (TL) Yonhap News/YNA/NewsCom; **422** (TL) Wernher Krutein/Photovault; **424** (BL) Torunn Berge/Photo Researchers, Inc.; **427** (CR) PA/Files/©AP Images; **428** (B) AGStock/Alamy Images, (T) Yonhap News/YNA/NewsCom; **429** (TCR) Katrina Brown/iStockphoto, (CR) University of California, Davis; **430** MIKE ALQUINTO/EPA/NewsCom; **431** (BR) Courtesy of Mickie Gelsinger; **432** (CL) Dr. Malcolm Campbell and the GCAT Program of Davidson College.; **433** Index Stock/SuperStock; **434** (TL) David Parker/Photo Researchers, Inc.; **435** (BL, BCR) J. Craig Venter Institute; **436** (TL) Yonhap News/YNA/NewsCom; **437** (TR) Charles Smith/Corbis, (BR) David Parker/Photo Researchers, Inc.; **439** (T) GloFish, LLC; **443** (TR) David Parker/Photo Researchers, Inc.; **446** (TL) BSIP LAURENT/Photo Researchers, Inc.; **447a** (BL) Henrik Lehnerer/Alamy; **447b** (B) Dean Pennala; **447** (L) Frans Lanting Studio/Alamy Images, (BR) Stew Milne; **448** (B) Chip Clark/Smithsonian National Museum of Natural History; **449** (TR) Chris Johns/National Geographic Image Collection; **450** (B) Chip Clark/Smithsonian National Museum of Natural History, (TCR) Science Source/Photo Researchers, Inc.; **451** (BL)/©DK Images; **452** (TCR) CarolineTolsma/Shutterstock, (BL) Chris Johns/National Geographic Image Collection, (TR) Pete Oxford Danita Delimont Photography/NewsCom; **454** (B) Andrew Bayda/Fotolia, (T) Chip Clark/Smithsonian National Museum of Natural History; **455** (CR) Sergio Piumatti, (BCR) The Natural History Museum, London; **456** (CL) Shutterstock; **457** (TR) Snark/Art Resource, NY; **458** (TL, TC, CL) Wolsenholme, D. (fl.1862)/Down House, Kent, UK/Bridgeman Art Library; **459** (CR, BR) ©The Granger Collection, NY, (CL, C) Heritage Image Partnership/Imagestate Media; **460** (TL) Chip Clark/Smithsonian National Museum of Natural History; **461** (TL) Barry Mansell/Nature Picture Library, (CR) Hal Beral/V&W/The Image Works, Inc., (BR) Markus Varesvuo/Nature Picture Library, (BL) Steve Kaufman/Peter Arnold/PhotoLibrary Group, Inc.; **463** (TR) Chris Johns/National Geographic Image Collection, (BR) Maria Mosolova/Photo Researchers, Inc., (BL) Photodisc/SuperStock; **464** (TL) Cambridge University Library, (BR) JacoBecker/Shutterstock; **465** (TL) Chip Clark/Smithsonian National Museum of Natural History, (BR) Chris Johns/National Geographic Image Collection; **467** (TR) J. G. M. Thewissen/neomed.edu/The Thewissen Lab; **469** (TC) 4nature/Peter Arnold Inc./PhotoLibrary Group, Inc., (TR) Eye of Science/Photo Researchers, Inc., (L) Wayne Lynch/AGE Fotostock; **471** (TCR) Thermal Biology Institute, (TR) Used by permission of Globe Pequot Press. From Seen and Unseen, by Kathy B. Sheehan, David J. Patterson, Brett Leigh Dicks, and Joan M Henson, ©2005. Photo by David J. Patterson; **477** (TR) Chris Johns/National Geographic Image Collection; **480** (B) Imagebroker RF/Photolibrary Group, Inc.; **482** (TL) Imagebroker RF/Photolibrary Group, Inc., (BL) Jupiterimages/Jupiter Images; **484** (T) Brand X/SuperStock; **485** (B) Premaphotos/Alamy Images; **486** (TL) Rare Book Division or Print Collection. Miriam and Ira D. Wallach Division of Arts, Prints, and Photographs. Astor Lennox and Tilden Foundations./New York Public Library Picture Collection; **487** (TL) Imagebroker RF/Photolibrary Group, Inc., (C) Paul Grebliunas/Getty Images; **488** (BL) Charlie Ott/Photo Researchers, Inc.; **491** (BR) Liz Leyden/iStockphoto; **492** (TL) Ashley Cooper/Corbis; **493** (C) CDC, (TR, TL, TCR, TCL) SuperStock; **494** (TL) Imagebroker RF/Photolibrary Group, Inc.; **495** (TR) Corbis/SuperStock, (TC) Rick & Nora Bowers/Alamy Images, (Inset R) Thomas & Pat Leeson/Photo Researchers, Inc.; **498** (TL) Imagebroker RF/Photolibrary Group, Inc.; **503** (CR) Rick & Nora Bowers/Alamy Images; **505** (BCR) James Cavallini/Custom Medical Stock Photo; **508** (B) Chip Clark/Smithsonian National Museum of Natural History; **509** (TR) Kevin Schafer/AGE Fotostock, (BR) Mark Hamblin/AGE Fotostock; **510** (TL) Chip Clark/Smithsonian National Museum of Natural History, (BL) Corbis/SuperStock; **511** (TR, TCR, CC, BR, BL, BCR) Matthew Ward/©DK Images, (CR) Peter Chadwick/©DK Images; **512** (BR) Hans Strand/AGE Fotostock, (TR) Mark Hamblin/AGE Fotostock, (BL) Paul Debois/AGE Fotostock; **513** (CR) Roslin, Alexander (1718–93)/Chateau de Versailles, France, Lauros/Giraudon/Bridgeman Art Library, (TL) S-dmit/iStockphoto; **514** (TCR) Kare Telnes/Image Quest Marine, (CL) W. Perry Conway/Corbis; **515** (CR) AGE Fotostock/SuperStock, (C) Richard Waters/iStockphoto; **516** (TL) Chip Clark/Smithsonian National Museum of Natural History; **518** (BR) blickwinkel/Alamy Images, (C) Martin Harvey/Alamy Images, (L) Tom Smith/Shutterstock; **521** (TR) AfriPics/Alamy Images, (CR) B.K. Wheeler/VIREO, (BR) Chris Zwaenepoel/iStockphoto; **522** (TCL) Corbis/SuperStock, (CR) Frank Leung/iStockphoto, (TCR) Josep Pena Llorens/iStockphoto, (TL) Kevin Schafer/AGE Fotostock, (CL) Raven Regan/Robertstock; **523** (TL) Chip Clark/Smithsonian National Museum of Natural History; **525** (BR) Science Source/Photo Researchers, Inc.; **526** (B) Dr. Stan Erlandsen; Dr. Dennis Feely/CDC, (CL) Eye of Science/Photo Researchers, Inc.; **527** (BR) Du?an Zidar/Fotolia; **528** (TL) Shutterstock; **529** (BR) Biodiversity Institute of Ontario, (BC) Derek Dammann/iStockphoto, (BL) Tim Zurowski/Glow Images; **533** (TCR) Kevin Schafer/AGE Fotostock, (TR) Mark Hamblin/AGE Fotostock; **536** (B) The Natural History Museum, London; **537** (TR) Marvin Dembinsky Photo Associates/Alamy Images; **538** (BR) Howard Grey/Getty Images, (BL) Lester V. Bergman/Corbis, (TL) The Natural History Museum, London; **539** (BL) ©Jason Edwards/Getty Images; **543** (CR) Marvin Dembinsky Photo Associates/Alamy Images; **545** (BR) Danita Delimont/Alamy Images; **546** (B) Francois Gohier/Photo Researchers, Inc., (TL) The Natural History Museum, London; **548** (CL) Marvin Dembinsky Photo Associates/Alamy Images; **552** (TL) malko/Fotolia; **553** (BL) AGE Fotostock/SuperStock, (TL) The Natural History Museum, London; **555** (BCR) Science Photo Library/Photo Researchers, Inc.; **559** (BR) Kristi Curry Rogers; **560** (CL) Chase Studio/Photo Researchers, Inc., (TR, TC, BR, BL) Colin Keates/©DK Images, (CR) DK Images/©DK Images; **561** (BCR) Colin Keates/Natural History Museum, London/©DK Images, (TCR) Interfoto Pressebildagentur/Alamy Images, (TR) Kathie Atkinson/Oxford Scientific Films/PhotoLibrary Group, Inc., (BL) Kevin Schafer/Corbis, (BR) Laurie O'Keefe/Photo Researchers, Inc., (TL) Luis Rey/©DK Images, (CL) Publiphoto/Photo Researchers, Inc.; **562** (CL) blickwinkel/Alamy Images, (TR) BrunoStock/AGE Fotostock, (TL) Colin Keates/©DK Images, (CR) DeAgostini Picture Library/Getty Images, (CC, BC) John Cancalosi/Peter Arnold/PhotoLibrary Group, Inc., (BR) John Downs/©DK Images, (BL) Oxford Scientific/PhotoLibrary Group, Inc.; **563** (TR) Colin Keates/©DK Images, (BL) Colin Keates/Courtesy of the Natural History Museum, London/©DK Images, (CR) Corbis/Photolibrary Group, Inc., (BR) Hemis/Alamy Images, (CL) Robin Carter/©DK Images; **567** (TR) Marvin Dembinsky Photo Associates/Alamy Images; **570** (TL) Frans Lanting Studio/Alamy Images; **571b** (B) Diane Cook and Len Jenshel/Getty Images; **571** (BR) Stew Milne, (L) Studio Eye/Corbis; **572** (B) Ted Horowitz/Corbis; **573** (TR) Peter Cade/Getty Images; **574** (TR) Norm Thomas/Photo Researchers, Inc., (BL) Peter Cade/Getty Images, (TL) Ted Horowitz/Corbis; **575** (CR) Dr. O. Bradfute/FOTOGRAF/Peter Arnold/PhotoLibrary Group, Inc., (BR) James Cavallini/Photo Researchers, Inc., (TR) M. Wurtz/Science Photo Library/Photo Researchers, Inc.; **579** (CR) Peter Cade/Getty Images; **580** (TCR) diego cervo/iStockphoto, (TR) Steve Gschmeissner/Photo Researchers, Inc., (TL) Ted Horowitz/Corbis; **581** (BR) Brand X/Jupiter Images, (TR) CNRI/Science Photo Library/Photo Researchers, Inc.; **582** (B) Alfred Pasieka/Photo Researchers, Inc., (TL) Eye of Science/Photo Researchers, Inc., (TC) S. Lowry/Univ Ulster/Stone/Getty Images, (TR) Scott Camazine/Photo Researchers, Inc.; **583** (TR) A.B. Dowsett/Science Photo Library/Photo Researchers, Inc., (C) Dr. Linda M. Stannard, University of Cape Town/Photo Researchers, Inc., (BC) Tony Freeman/PhotoEdit, Inc.; **584** (B) ©Dustie/Shutterstock, (Inset, R, Inset, BR) David M. Phillips/Photo Researchers, Inc., (Inset, BL) Robert Knauft/Biology Pics/Photo Researchers, Inc., (CL) Steve Gschmeissner/Science Photo Library/Photo Researchers, Inc.; **585** David M Dennis/PhotoLibrary Group, Inc.; **586** (BL) Peter Cade/Getty Images, (TL) Ted Horowitz/Corbis; **587** (BR) JGI Getty Images; (TR) Juergen Berger/Science Photo Library/Photo Researchers, Inc.; **588** (BR) Scott Camazine/Alamy Images; **589** (T) Bettmann/Corbis, (BCR) Saturn Stills/Science Photo Library/Photo Researchers, Inc., (Inset, CL) SCIENCE PHOTO LIBRARY/Photolibrary Group, Inc.; **590** (CL) Reuters/Corbis; **591** (CR) Southhampton General Hospital/Science Photo Library/Photo Researchers, Inc.; **592** (CL) Peter Cade/Getty Images; **593** (CR) Michael Newman/PhotoEdit, Inc., (TR, TL, TCR, TCL) SuperStock; **595** (BL) David M. Phillips/Photo Researchers, Inc.; **596** (CR) Alfred Pasieka/Science Photo Library/Photo Researchers, Inc., (CL) Dr. Kari Lounatmaa/Photo Researchers, Inc., (TR) Eye of Science/Science Photo Library/Photo Researchers, Inc., (TL) W. Wurtz/Science Photo Library/Photo Researchers, Inc.; **597** (TR) Peter Cade/Getty Images; **598** (TR) Stanley Maloy/San Diego State University; **600** (B) Buiten-Beeld/Alamy Images; **601** (TR) Nigel Cattlin/Photo Researchers, Inc.; **602** (TL) Buiten-Beeld/Alamy Images, (BL) steve bly/Alamy Images, (BR) Steve Gschmeissner/Photo Researchers, Inc.; **603** (BR) Eye of Science/Photo Researchers, Inc., (BC) M. I. Walker/Photo Researchers, Inc., (BL) Steve Gschmeissner/Photo Researchers, Inc.; **604** Arterra Picture Library/Alamy Images, (TR) Eye of Science/Photo Researchers, Inc., (TL) M. I. Walker/Photo Researchers, Inc., (TCR) Steve Gschmeissner/Photo Researchers, Inc.; **605** (TR) Andrew H. Knoll/Andrew Knoll/Harvard University; **606** (TL) Buiten-Beeld/Alamy Images, (CR, C, BL) Michael Abbey/Photo Researchers, Inc.; **607** Nigel Cattlin/Photo Researchers, Inc.; **610** (TCR) Bob Gibbons/Photo Researchers, Inc., (TL) Buiten-Beeld/Alamy Images; **611** (BR) Bill Bachman/Photo Researchers, Inc., (TCR) Carlos Villoch/Image Quest Marine, (Upper BR) Howard Hall/PhotoLibrary Group, Inc., (T) jocrebbin/Shutterstock; **612** (TCL) Eric Grave/Photo Researchers, Inc.; **613** (BR) Dr. Fred M. Rhoades, (T) Nigel Cattlin/Photo Researchers, Inc.; **614** (B) David R. Frazier/Photolibrary, Inc/Photo Researchers, Inc., (BCR) Eric V. Grave/Photo Researchers, Inc., (BCL) George D. Lepp/Corbis; **615** (CR) Andrew Paul Leonard/Photo Researchers, Inc., (TR) Eye of Science/Photo Researchers, Inc., (CC) Eye of Science/Photo Researchers, Inc., (TL) John McConnico/©Associated Press; **617** (BR) Susan Meiselas/©Magnum Photos; **618** Alamy, (TL) Buiten-Beeld/Alamy Images, (BL) Mantonature/iStockphoto; **619** Nigel Cattlin/Photo Researchers, Inc.; **621** (T) Nigel Cattlin/Photo Researchers, Inc.; **622** (BR) Emanuele Biggi/PhotoLibrary Group, Inc., (BCL) Inga Spence/Photo Researchers, Inc., (TL) Michael P. Gadomski/Photo Researchers, Inc.; **623** (BR) Neil Hardwick/Alamy Images; **625** (TL, T) Bill Heath, (Inset, R) Dr. Michael Castellano/USDA Forest Service; **627** Michael Abbey/Photo Researchers, Inc.; **629** (CR) Andrew Syred/SPL/Photo Researchers, Inc., (T) Nigel Cattlin/Photo Researchers, Inc.; **630** (TR) Compliments of Mycorrhizal Applications, Inc.; **632** (B) Craig Tuttle/Corbis; **633** (TR) Corbis; **634** (BR) ©Digital Vision/Alamy, (TL) Craig Tuttle/Corbis; **636** (TL) Custom Medical Stock Photo; **637** (TR) Corbis; **639** (BR, BC) Chase Studio/Photo Researchers, Inc., (TL) Craig Tuttle/Corbis; **640** (BL) Perennou Nuridsany/Photo Researchers, Inc., (BC) Roland Birke/Peter Arnold Inc./PhotoLibrary Group, Inc.; **641** Steve Heap/Shutterstock; **643** (BC)/Dr. David T. Webb, (BL) Howard Rice/Garden Picture Library/PhotoLibrary Group, Inc.; **644** (TL) J S Sira/Garden Picture Library/PhotoLibrary Group, Inc.; **646** (TL) Craig Tuttle/Corbis; **647** (CC) ©IngridHS/Fotolia, (TCL) Blanche Branton/Shutterstock, (BR) Corbis, (TCR) Custom Medical Stock Photo, (TL) M.Gabriel/Peter Arnold Inc./PhotoLibrary Group, Inc., (TR) Manfred Kage/Peter Arnold Inc./PhotoLibrary Group, Inc., (BCR) Matthew Ward/©DK Images, (CR) Peter Chadwick/©DK Images, (TC) Walter H. Hodge/Peter Arnold Inc./PhotoLibrary Group, Inc.; **648** (BL) Corbis, (TL) Perennou Nuridsany/Photo Researchers, Inc.; **650** (TL) Craig Tuttle/Corbis, Ed Reschke/PhotoLibrary Group, Inc.; **651** (BR) Getty Images, (TR) Peter Chadwick/©DK Images, (TCR) Peter Chadwick/©DK Images, (TCL) Peter Chapwick/©DK Images, (TL) Peter Chadwick/©DK Images; **652** (TL) ©David Dilcher, (BCR) Julien Grondin/iStockphoto, (BCL) Pavel Bortel/iStockphoto, (BL) Sangtae Kim, University of Florida. Photo courtesy the Botanical Society of America./Botanical Society of America, (BR) Skip Odonnell/iStockphoto, (BC) The Garden Picture Library/Alamy Images; **653** (TR) Corbis; **655** (BR) Marya C. Roddis; **657** (BL) Perennou Nuridsany/Photo Researchers, Inc.; **659** (TR) Photo Archives of the South Tyrol Museum of Archaeology; **660** Casey Chinn Photography/Shutterstock; **662** (B) Perennou Nuridsany/Science Photo Library/Photo Researchers, Inc.; **663** (TR) Ted Mead/PhotoLibrary Group, Inc.; **664** (TL) Perennou Nuridsany/Science Photo Library/Photo Researchers, Inc., (BL) Ted Mead/PhotoLibrary Group, Inc.; **666** (CL) Ray F. Evert/University of Wisconsin; **667** (TCR, TCL, TC) Biophoto Associates/Photo Researchers, Inc., (TR) Judy Lawrence/Alamy, (BR) Ted Mead/PhotoLibrary Group, Inc.; **668** (T, B) Ray F. Evert/University of Wisconsin; **669** (BR) Derek Croucher/Alamy Images, (TL) Perennou Nuridsany/Science Photo Library/Photo Researchers, Inc., (BC) Tom Biegalski/Shutterstock; **671** (TR) Ted Mead/PhotoLibrary Group, Inc.; **674** (B) Martine Mouchy/PhotoLibrary Group, Inc., (TL) Perennou Nuridsany/Science Photo Library/Photo Researchers, Inc.; **675** Ed Reschke/Peter Arnold/PhotoLibrary Group, Inc.; **676** (TL) Ted Mead/PhotoLibrary Group, Inc.; **678** (TL) Manfred Kage/Peter Arnold/PhotoLibrary Group, Inc.; **680** (TL) Perennou Nuridsany/Science Photo Library/Photo Researchers, Inc., (BL) Ted Mead/PhotoLibrary Group, Inc.; **682** Biophoto Associates/Photo Researchers, Inc.; **683** (BR) ©Nigel Cattlin/Photo Researchers, Inc.; **684** (TL) David Kirkland/PhotoLibrary Group, Inc., (BL) Garden World Images/AGE Fotostock, (CR) Muriel Hazan/BIOS/Peter Arnold/PhotoLibrary Group, Inc., (TR) Peter Miller/Getty Images; **685** (TL) Perennou Nuridsany/Science Photo Library/Photo Researchers, Inc.; **691** Ted Mead/PhotoLibrary Group, Inc.; **694** Biophoto Associates/Photo Researchers, Inc.; **695** (TR) Martin Perez/iStockphoto; **696** (T) Biophoto Associates/Photo Researchers, Inc.; **697** (CR) ©iofoto/Shutterstock, (BR) Brian Durell/SuperStock, (BC) James Urbach/SuperStock, (BL) Mark Bolton/Getty Images; **700** (TR) ©yulia eniseyskaya/Fotolia, (TL) Wermter, C./Peter Arnold Inc./PhotoLibrary Group, Inc.; **702** (CL) Ed Reschke/Peter Arnold/PhotoLibrary Group, Inc., (BL) Michael P. Gadomski/Photo Researchers, Inc., (B) Richard Cummins/SuperStock; **703** (CR) Martin Perez/iStockphoto, (TL) National Trust Photographic Library/Stephen Robson/The Image Works, Inc., (TR) Tristan Lafranchis/Peter Arnold/PhotoLibrary Group, Inc.; **704** (T) Biophoto Associates/Photo Researchers, Inc., (BCL) Corbis/SuperStock, (BC) Sergio33/Shutterstock, (BL) Westend61/Alamy; **705** (CL) ©S. R. Miller/Fotolia, (CR) Peter Lilja/AGE Fotostock, (C) Werner H. Muller/Peter Arnold Inc./PhotoLibrary Group, Inc.; **706** (BL) Martin Perez/iStockphoto; **708** Biophoto Associates/Photo Researchers, Inc.; **710** (CL, CC) Walter Chandoha Photography; **711** (TR) Martin Perez/iStockphoto; **712** Martin Shields/Photo Researchers, Inc.; **714** (TCL) ©Hallgerd/Fotolia, (TL) Mark Bolton/Corbis; **715** (T) Biophoto Associates/Photo Researchers, Inc., (B) Keren Su/Corbis; **716** (TL) Nicolle Rager Fuller/National Science Foundation; **717** (TR) Bill Barksdale/Corbis, (TL) Nicolle Rager Fuller/National Science Foundation; **718** (BCR) Brand X/SuperStock, (CL) Christophe Boisvieux/Corbis, (TL) Ingram Publishing/SuperStock, (CR) Mats Tooming/iStockphoto; **719** (BCL) Brand X/SuperStock, (BR) EITAN ABRAMOVICH/Getty Images, (CL) Purestock/SuperStock, (TC) Visual Arts Library (London)/Alamy Images; **723** (TR) Martin Perez/iStockphoto; **726** (TL) Studio Eye/Corbis; **727a** (BL, BC) Sharklet Technologies, (Bkgd) Richard Carey/Fotolia; **727** (L) Martin Harvey/Corbis, (BR) Stew Milne; **728** (B) Jeff Hunter/The Image Bank/Getty Images; **729** (TR) Kelly McGann Hebert; **730** (TL) Jeff Hunter/The Image Bank/Getty Images; **731** (BL) ©Ocean/Corbis, (BR) BIOS/Peter Arnold/PhotoLibrary Group, Inc., (TR) Kelly McGann Hebert; **732** (BC) ©Stubblefield Photography/Shutterstock, (CL) Tierbild Okapia/Photo Researchers, Inc.; **735** (CR) ©Michael Fogden/Photoshot; **736** Courtesy, Dr. Scottie Yvette Henderson; **737** (TL) Jeff Hunter/The Image Bank/Getty Images, (C) Michael Durham/Nature Picture Library; **742** (BR) Kelly McGann Hebert; **743** (CR) BIO Klein & Hubert/Peter Arnold/PhotoLibrary Group, Inc., (TR) Tom Brakefield/Stockbyte/Thinkstock; **745** (BL) ©Stubblefield Photography/Shutterstock; **747** (TR) Kelly McGann Hebert, (BR) Peter Parks/Image Quest Marine; **750** (B) Editions Xavier Barral; **751** (TR) National Geographic Image Collection; **752** (TL) Editions Xavier Barral, (BL) Shuhai Xiao/Virginia Tech; **753** (BCR) Chip Clark/Smithsonian National Museum of Natural History, (TR) National Geographic Image Collection; **754** (BL) ©Kerry L. Werry/Shutterstock, (BCL) ©WaterFrame/Alamy Images; **755** (BL) ©John Anderson/Fotolia, (BCL) ©Premaphotos/Alamy Images, (TL) iStockphoto, (TCL) Kim Taylor/©DK Images; **756** (TL) ©Clarence Alford/Fotolia, (CL) Roger Steene/Image Quest Marine; **757** (TL) Editions Xavier Barral, (BR) Simon Conway Morris/Simon Conway Morris; **759** (BR) ©Photoshot Holdings Ltd/Alamy Images, Chip Clark/Smithsonian National Museum of Natural History; **760** (T) Enzo & Paolo Ragazzini/Corbis; **761** (TR) Anne Ryan/Polaris Images; **762** (BL) National Geographic Image Collection; **763** (CR) DeAgostini/Getty Images, (T) Natural History Museum/DK Images; **764** (TL) ©Bruce Coleman Inc./Alamy Images; **765** (TL) Editions Xavier Barral, (BR) Profimedia/SuperStock; **768** (BL) John Reader/Photo Researchers, Inc.; **769** (CR) Photo by Craig Hartley, (TL) Sarah Leen and Rebecca Hale/National Geographic Image Collection, Zeresenay Alemseged; **772** (TL) SuperStock; **773** (TC, CL, BC) John Reader/Photo Researchers, Inc.; **775** (BL) Chip Clark/Smithsonian National Museum of Natural History; **777** (BL) Jany Sauvanet/Photo Researchers, Inc., (TR) National Geographic Image Collection; **780** (B) ©Robert Harding Picture Library Ltd/Alamy Images; **781** (TR) iStockphoto; **782** (TL) ©Robert Harding Picture Library Ltd/Alamy Images, (TCR) Radius Images/Photolibrary Group, Inc.; **783** (CL, BR) ©WaterFrame/Alamy Images, (BL) Nancy Sefton/Photo Researchers, Inc., (CR) PHONE PHONE - Auteurs Danna Patricia/Peter Arnold Inc./PhotoLibrary Group, Inc.; **786** (TCL) Kim Karpeles/AGE Fotostock; **787** (TL) ©Robert Harding Picture Library Ltd/Alamy Images, (BR) Paul Nicklen/National Geographic/Getty Images; **791** (TL) ©Robert Harding Picture Library Ltd/Alamy Images; **793** (TR) iStockphoto, (CR) Jerry Young/©DK Images; **794** (TL) ©Robert Harding Picture Library Ltd/Alamy Images, (BL) Masa Ushioda/Image Quest Marine; **795** (T) ©Oliver Smart/Alamy Images, (CR) iStockphoto; **798** (CR) ©Westend61 GmbH/Alamy; **799** (BR) Courtesy of Daniel S. Cutler; **803** (TR) iStockphoto; **806** (B) ©Papilio/Alamy Images; **807** (TR) Image Quest Marine; **808** (TL) ©Papilio/Alamy Images, (CR) Dmitry Ersler/iStockphoto; **809** (B) ©Steve Bloom Images/Alamy Images; **810** (BC) UNEP/The Image Works, Inc.; **811** ©nialat/Fotolia; **812** (BC) ©Thierry Berrod/Photo Researchers, Inc., (BR) John F White/SuperStock, (BL) Larry West/Photo Researchers, Inc., (CL) Oxford Scientific/PhotoLibrary Group, Inc., (CR) Roger Steene/Image Quest Marine; **813** (CR) Dale C. Spartas/Corbis; **814** (BL, BC) ©Biophoto Associates/Photo Researchers, Inc., (TL) ©Papilio/

Credits *continued*

Alamy Images; **815** (TR) ©Ishbukar Yalilfatar/Shutterstock; **816** (T) Patrick Gries/Editions Xavier Barral; **818** (TL) ©Angelika Mothrath/Fotolia, (CL) Frank Greenaway/©DK Images, (CR) Greg Lasley Nature Photography, (TR) Jerry Young/©DK Images; **819** (TL) ©Papilio/Alamy Images, (BC) ©Rick & Nora Bowers/ Alamy Images; **820** (BL) Carlos Villoch 2004/Image Quest Marine, (TL) Image Quest Marine; **821** (BR) Image Quest Marine; **822** (TR) Alamy Images; **824** (BL) ©Keren Su/China Span/Alamy Images, (TL) ©Photoshot Holdings Ltd/Alamy Images; **825** (BR) ©nimade/Fotolia; **826** (TL) ©idiz/Shutterstock, (C) Tom Brakefield/Corbis; **827** (T) ©Papilio/Alamy Images, (BR) Corbis; **829** (TR) ©Bruce Coleman Inc./Alamy Images, (BR) ©r-o-x-o-r/Fotolia; **830** (TL) ©Steve Byland/Fotolia; **831** (CR) David Bebber/Reuters/Corbis, (TR, TL, TCR, TCL) SuperStock; **835** (TR) Image Quest Marine; **838** (B) ©Juniors Bildarchiv/Alamy Images; **839** (TR) Ariadne Van Zandbergen/ africanpictures; **840** (TL) ©Juniors Bildarchiv/Alamy Images, (BL) Ariadne Van Zandbergen/africanpictures, (TR) Polka Dot Images/SuperStock; **841** (C) Jeff Lepore/Photo Researchers, Inc., (CL) Photo Researchers, Inc., (CR) David Davis/Shutterstock; **842** (TL) ©Steve Bloom Images/Alamy Images; **844** (TC) ©NaturePL/SuperStock, (CL) Markus Botzek/Bridge/Corbis, (T) Slawomir Jastrzebski/iStockphoto; **845** (TR) Jason Hahn Photography, (TC) Operation Migration, Inc., (T) Slawomir Jastrzebski/iStockphoto; **846** (BR) ©vlad_g/Fotolia, (TR, TL, T) SuperStock; **847** (TL) ©Juniors Bildarchiv/Alamy Images; **848** (BL) Corbis/SuperStock, (TL) M Delpho/PhotoLibrary Group, Inc.; **849** (TR) Ariadne Van Zandbergen/africanpictures; **850** (TL) ©Anita P. Peppers/Fotolia; **851** (TR) ©petrock/Fotolia; **853** (CR) ©Anita P. Peppers/Fotolia, (BL) ©Steve Bloom Images/ Alamy Images; **855** (TR) Ariadne Van Zandbergen/africanpictures; **858** (TL) Martin Harvey/Corbis; **859a** (BR) Michael Svoboda/ Getty Images, (BL) Miriam Maslo/Photo Researchers, Inc.; **859b** (BR) BSIP SA/Alamy Images; **859** (L) ©PNC/Blend Images/ Corbis, (BR) Stew Milne; **860** (B) Reuters/Corbis; **861** (TR) Cristian Ciobanu/Fotolia; **862** (TCR) DennisMacDonald/ PhotoEdit, Inc., (TL) Reuters/Corbis; **863** (R) Ed Reschke/ PhotoLibrary Group, Inc., (TL) Image Source/Photolibrary Group, Inc., (TCL) Manfred Kage/PhotoLibrary Group, Inc., (CR) Thomas Deerinck, NCMIR/Photo Researchers, Inc.; **866** (CL) CLS Design/Shutterstock, (TL) Greg Ceo/Taxi/Getty Images; **867** (TR) Cristian Ciobanu/Fotolia; **868** (TCR) Jupiter Images/Pixland/Alamy, (TL) Reuters/Corbis; **869** (BR) Corel, (TR) Cristian Ciobanu/Fotolia, (BL) Gord Horne/iStockphoto, (CR) Photodisc/Getty Images; **870** (TL) Elena Schweitze/ iStockphoto, (B) Emilio Ereza/Alamy, (CL) Image Source Pink/ Jupiter Images; **871** (TCL) FoodPix/Jupiter Images, (CL) Maximilian Stock Ltd/photocuisine/Corbis, (BL) Potapova Valeriya/iStockphoto; **872** (TC) Diane Diederich/iStockphoto, (BC) James Steidl/iStockphoto, (CC) Leander/Alamy Images; **874** (CR) ©Associated Press, (TR, TL, TCR, TCL) SuperStock; **875** (TL) Reuters/Corbis; **877** (BR) Eye of Science/Photo Researchers, Inc.; **879** (CL) Dr. David M. Martin/Photo Researchers, Inc., (BL) Meckes/Otawa/SPL/Photo Researchers, Inc.; **881** (TR) CNRI/Photo Researchers, Inc.; **882** (TL) Reuters/ Corbis, (TR) Richard Hamilton Smith/Corbis; **884** (BL) Cristian Ciobanu/Fotolia; **886** (CL) Cristian Ciobanu/Fotolia; **889** (BL) Jupiter Images/Pixland/Alamy; **891** (TR) Cristian Ciobanu/ Fotolia; **894** (B) Jonathan Daniel/Getty Images; **895** (TR) As for Resolution - From the series 'Captain James Cook, Son of the Land, Master of the Sea' by Robin Brooks/Black Dog Studios/ Bridgeman Art Library; **896** (TL) Jonathan Daniel/Getty Images; **897** (TC) Juergen Berger/Photo Researchers, Inc.; **898** (CL) As for Resolution - From the series 'Captain James Cook, Son of the Land, Master of the Sea' by Robin Brooks/Black Dog Studios/ Bridgeman Art Library; **901** (TL) Jonathan Daniel/Getty Images; **902** (B) BSIP/Photo Researchers, Inc.; **904** (TL) MADD Canada, used with permission.; **905** (BL) NIH/NIDA/National Institutes of Health; **906** (BL) Alessandro Della Bella/Keystone/Corbis, (TL) Jonathan Daniel/Getty Images; **907** (TR) As for Resolution - From the series 'Captain James Cook, Son of the Land, Master of the Sea' by Robin Brooks/Black Dog Studios/Bridgeman Art Library; **909** (BR) As for Resolution - From the series 'Captain James Cook, Son of the Land, Master of the Sea' by Robin Brooks/Black Dog Studios/Bridgeman Art Library, (TL) Jonathan Daniel/Getty Images; **910** (TL) Dr. Cecil H. Fox/Photo Researchers, Inc., (BC) Red Chopsticks/Getty Images; **911** (CR) Susumu Nishinag/Photo Researchers, Inc.; **912** (TL) Ralph C. Eagle Jr./Photo Researchers, Inc.; **915** (TCR) Alessandro Della Bella/Keystone/Corbis; **917** (TR) As for Resolution - From the series 'Captain James Cook, Son of the Land, Master of the Sea' by Robin Brooks/Black Dog Studios/Bridgeman Art Library; **920** (B) Joe McNally/Getty Images; **921** (TR) Radcliffe Institute, Harvard University/Schlesinger Library; **922** (TL) Joe McNally/ Getty Images; **924** (TL) Andrew Syred/Photo Researchers, Inc.; **925** (CR) Radcliffe Institute, Harvard University/Schlesinger Library; **926** (B) Photo AltoRF/TIPS North America; **928** (BL) Jim McIsaac/Getty Images, (TL) Joe McNally/Getty Images; **929** (BR) Ed Reschke/Peter Arnold/PhotoLibrary Group, Inc., (T) Ed Reschke/PhotoLibrary Group, Inc., (CR) Oxford Scientific Films/ PhotoLibrary Group, Inc.; **931** (BCR) Don W. Fawcett/Photo Researchers, Inc., (BR) Radcliffe Institute, Harvard University/ Schlesinger Library; **933** (TR) NASA; **934** (CR) Allsport UK/ Getty Images, (TR, TL, TCR, TCL) SuperStock; **935** (TL) Joe McNally/Getty Images, (BR) Radcliffe Institute, Harvard University/Schlesinger Library; **936** (TL) Hemera Technologies/ Thinkstock; **937** (CR) Steve Gschmeissner/Science Photo Library/Photo Researchers, Inc.; **939** (TL, TC) Dr. P. Marazzi/ Science Photo Library/Photo Researchers, Inc., (R) JAMES STEVENSON/SCIENCE PHOTO LIB/Custom Medical Stock Photo; **941** (TR) NASA, (BL) Photo AltoRF/TIPS North America; **942** (B) European Synchrotron Radiation Facility/ CREATIS/Photo Researchers, Inc.; **943** (TR) Radcliffe Institute, Harvard University/Schlesinger Library; **944** (CL) Salisbury District Hospital/Science Photo Library/Photo Researchers, Inc.; **946** (B) ©Steve Hix/Somos Images/Corbis; **947** (TR) iStockphoto; **948** (TL) ©Steve Hix/Somos Images/Corbis; **950** (TL) iStockphoto; **953** Creatas/Jupiter Images; **954** (TL) ©Steve Hix/Somos Images/Corbis, (BL) National Cancer Institute/ Science Photo Library/Photo Researchers, Inc.; **955** David M. Phillips/Photo Researchers, Inc.; **958** GJLP/Photo Researchers, Inc.; **959** (BR) iStockphoto, (CR) Michelaubryphoto/Fotolio; **961** (TR) iStockphoto; **962** (BR) BSIP/Photo Researchers, Inc., (TR, CL) Science Photo Library/Photo Researchers, Inc.; **963** (TL) ©Steve Hix/Somos Images/Corbis, (TC) Bob Peterson/Upper Cut Images/Getty Images; **964** (TL) ©Susumu Nishinaga/Photo Researchers, Inc.; **967** (BR) David Madison/Photodisc/Alamy; **968** Arthur Glauberman/Photo Researchers, Inc.; **969** (CR) iStockphoto, (TR) Simon Marcus/PhotoLibrary Group, Inc.; **971** (TR) National Cancer Institute/Science Photo Library/Photo Researchers, Inc.; **972** David M. Phillips/Photo Researchers, Inc.; **973** (TR) iStockphoto; **976** Michael J. Doolittle/The Image Works, Inc.; **977** (TR) iStockphoto; **978** Alamy, (L) Michael J. Doolittle/The Image Works, Inc.; **979** (TR) iStockphoto; **982** Michael J. Doolittle/The Image Works, Inc.; **983** (TR) iStockphoto; **984** Astrid & Hanns-Frieder Michler/Photo Researchers, Inc.; **988** Michael J. Doolittle/The Image Works, Inc.; **989** (CR) David H. Phillips/Photo Researchers, Inc.; **993** (TR) Claude Edelman/Photo Researchers, Inc., (CR) iStockphoto; **994** (TL) David M. Phillips/Photo Researchers, Inc.; **995** (T) Michael J. Doolittle/The Image Works, Inc., (BR) Yorgos Nikas/ Stone/Getty Images; **996** (TL) Scurlock Studio Records, Archives Center, National Museum of American History, Behring Center/ National Museum of American History/Smithsonian Institution; **999** (CR) Claude Edelmann/Photo Researchers, Inc., (BL) Neil Bromhall/Photo Researchers, Inc., (CL, BR) Petit Format/Photo Researchers, Inc.; **1000** Peter Baxter/Shutterstock, (TL) Simon Puschmann/Bilderberg/Aurora; **1005** (TR) iStockphoto; **1008** (B) Thomas Haentzschel/©Associated Press; **1009** (TR) Christopher S. Wheeler; **1010** (BR) ©Biophoto Associates/Photo Researchers, Inc., (BL) NIBSC/SPL/Photo Researchers, Inc., SPL/ Photo Researchers, Inc., (TL) Thomas Haentzschel/©Associated Press; **1011** (BR) Eye of Science/Photo Researchers, Inc., (BL) SPL/Photo Researchers, Inc.; **1012** Wood/Custom Medical Stock Photo; **1013** (CR) ©Biophoto Associates/Photo Researchers, Inc., (TR) Christopher S. Wheeler, (CL) HUGH MAYNARD/Nature Picture Library; **1014** Steve Gschmeissner/Photo Researchers, Inc., (TL) Thomas Haentzschel/©Associated Press; **1019** (BR) Photo Researchers, Inc.; **1020** (BL) Mary Evans Picture Library/ Photo Researchers, Inc., (TL) Thomas Haentzschel/©Associated Press; **1021** (BR) Christopher S. Wheeler, (CR) The John Snow Archive and Research Companion/MATRIX/Michigan State University; **1022** (CL) ©Goh Chai Hin/AFP/NewsCom, (TL) Frank Greenaway/DK/Getty Images; **1023** (CL) CAMR/A.Barry Dowswtt/Photo Researchers, Inc., (TR) Kevin P. Casey /Corbis, (BC) Travel Ink/Gallo Images/Getty Images; **1024** (BL) Eric Isselée/iStockphoto, (TL) Thomas Haentzschel/©Associated Press, (BCL) Volker Steger/Photo Researchers, Inc.; **1026** (BCR) Christopher S. Wheeler; **1029** (CL) ©Biophoto Associates/Photo Researchers, Inc.; **1031** (T) Science Source/Photo Researchers, Inc.; **1032** (CL) Royal Geographical Society, London; **1034** (TL) ©PNC/Blend Images/Corbis; **DOL•1** (B) 2003 James D. Watt/ Image Quest Marine, (T) Harald Theissen/Photolibrary Group, Inc.; **DOL•6** (BL) Science Pictures Ltd./Photo Researchers, Inc., (TR) Science Source/Photo Researchers, Inc., (Header Bar) Scimat/Photo Researchers, Inc.; **DOL•7** (TR) BIOMEDICAL IMAGING UNIT, SOUTHAMPTON/Photo Researchers, Inc., (BR) CDC, (CR, CC) Microfield Scientific Ltd/Photo Researchers, Inc., (TC) Science Source/Photo Researchers, Inc.; **DOL•8** (Header Bar) Dr. Gerhard Wanner, (B) Eye of Science/ Photo Researchers, Inc., (TR) Stephen & Donna O'Meara/Photo Researchers, Inc.; **DOL•9** (Inset CC) Dr. Gerhard Wanner, (CR) Dr. Karl Stetter, University of Regensberg, (TC, B) Eye of Science/Photo Researchers, Inc.; **DOL•10** (TL) M. I. Walker/ Photo Researchers, Inc., (Header Bar) Scenics & Science/Alamy Images; **DOL•11** (CL, BR) Eye of Science/Photo Researchers, Inc., (TR) Roland Birke/Peter Arnold Images/PhotoLibrary Group, Inc.; **DOL•12** (BL) ©Daniel Clem, (C) M. I. Walker/ Photo Researchers, Inc., (BR) Noble Proctor/Photo Researchers, Inc., (TR) Steve Gschmeissner/Photo Researchers, Inc., (TCR) Susumu Nishinaga/Photo Researchers, Inc.; **DOL•13** (C) Biophoto Associates/Photo Researchers, Inc., (TR) Eye of Science/Photo Researchers, Inc., (BR) Karen Gowlett-Holmes/ PhotoLibrary Group, Inc., (TC) Michael Abbey/Photo Researchers, Inc.; **DOL•14** (L) Eric V. Grav/Photo Researchers, Inc., (R) Eye of Science/Photo Researchers, Inc.; **DOL•15** (T) Alain Le Toquin/PhotoLibrary Group, Inc., (CR) Eye of Science/ Photo Researchers, Inc., (BL) M. I. Walker/Photo Researchers, Inc.; **DOL•16** (T) ©Pi-Lens/Shutterstock, (TR) ©vilainecrevette/ Shutterstock, (Header Bar) Biophoto Associates/Photo Researchers, Inc., (BL) Thomas Northcut/Thinkstock; **DOL•17** (BR) Bartomeu Borrell/PhotoLibrary Group, Inc., (BC) Fletcher & Baylis/Photo Researchers, Inc.; **DOL•18** (BC) ©MJ Prototype/ Shutterstock, (T) ©Pi-Lens/Shutterstock, (BR) Custom Medical Stock Photo, (BL) Scott Camazine/Photo Researchers, Inc.; **DOL•19** (TR, CR) Biophoto Associates/Photo Researchers, Inc., (TL) Eye of Science/Photo Researchers, Inc., (BL) Shunsuke Yamamoto Photography/Thinkstock; **DOL•20** (BR) A. Barry Dowsett/Photo Researchers, Inc., (T) Creatas/Photolibrary Group, Inc., (TR) Naturfoto Honal/Corbis; **DOL•21** (C) ©Andrew Syred/Photo Researchers, Inc., (TR) Dr. Jeremy Burgess/Photo Researchers, Inc., (BR) LAURIE CAMPBELL/ NHPA/Photoshot, (BL) Lee W. Wilcox; **DOL•22** (BC) ©blickwinkel/Alamy Images, (TR) Craig Tuttle/Corbis, (BR) Lee W. Wilcox, (BL) Peter Chadwick/©DK Images; **DOL•23** (BC) ©imagebroker/Alamy, (BCL) DR GARY BROWN/GardenWorld Images, (BL) L THOMAS/GardenWorld Images, (BR) Robert HENNO/Alamy Images; **DOL•24** (BR) ©All Canada Photos/ Alamy Images, (BCR) Bettmann/Corbis, (TR) Rob Blakers/ PhotoLibrary Group, Inc.; **DOL•25** (CR) ©Saruri/Fotolia, (CL) Alan and Linda Detrick/Photo Researchers, Inc., (BR) Gerald Hoberman/PhotoLibrary Group, Inc., (TR) Peter Chadwick/©DK Images; **DOL•26** (TR) Dr. Merlin D. Tuttle/ Photo Researchers, Inc.; **DOL•27** (BR) Florida Images/Alamy Images, (BL) HUBAUT Damien/AGE Fotostock, (CR) Sangtae Kim/Botanical Society of America, (TL) xyno6//iStockphoto; **DOL•28** (CL) ©Anne Kitzman/Shutterstock, (BR) ©Dani Vincek/Shutterstock, (CR) ©margouillat photo/Shutterstock, (BL) KirsanovV/iStockphoto, (CC) Mafuta/iStockphoto, (TR) Trevor Sims/GardenWorld Images; **DOL•29** (CC) ©Laura Gangi Pond/Shutterstock, (CC) Casey Chinn Photography/ Shutterstock, (B) Creatas/SuperStock, (TR) Martin Siepmann/ AGE Fotostock, (CL) Michael Valdez/iStockphoto, (BCR) Stephen Hayward/©DK Images; **DOL•30** (TC) ©wyssu/Fotolia, (T) Corbis/SuperStock, (BR) Figure:Stephen Fairclough;Image, Monika Abedin/King Lab; **DOL•31** (BR) Daniel L. Geiger/ SNAP/Alamy Images, (C) Jez Tryner/Image Quest Marine, (BL) Masa Ushioda/Image Quest Marine; **DOL•32** (TR) ©vilainecrevette/Shutterstock, (B) Andrew J. Martinez/Photo Researchers, Inc.; **DOL•33** (TR) ©Kippy Spilker/Shutterstock, (B) ©vilainecrevette/Shutterstock, (TL) Peter Parks/Image Quest Marine, (C) Wernher Krutein/Photovault; **DOL•34** (B) age fotostock/SuperStock, (TCR) Ed Andrieski/©Associated Press; **DOL•35** (BR) ©Eric Isselee/Shutterstock, (BL) Colin Keates/©DK Images, (T) Justine Gecewicz/iStockphoto; **DOL•36** (BL) ©Cathy Keifer/Shutterstock, (BR) ©Florian Andronache/ Shutterstock, (T) ©Hintau Aliaksei/Shutterstock, (CR) ©jewo55/ Fotolia; **DOL•37** (BCL) ©Alex Staroseltsev/Shutterstock, (CL) ©alslutsky/Shutterstock, (BC) ©Dani Vincek/Shutterstock, (CR) ©Eric Isselee/Shutterstock, (TCR) ©fivespots/Shutterstock, (T) ©James Steidl/Shutterstock, (BR) ©Luis Louro/Alamy Images, (BL) ©Potapov Alexander/Shutterstock, (BCR) ©vladimirdavydov/Fotolia; **DOL•38** (TR) ©Steve Gschmeissner/ Photo Researchers, Inc., (BL, Bkgrd) Photo Researchers, Inc.; **DOL•39** (TR) 2007 Jez Tryner/Image Quest Marine, (BL) Volker Steger/Photo Researchers, Inc.; **DOL•40** (TR) Kåre Telnes/Image Quest Marine, (BL) Leslie Newman & Andrew Flowers/Photo Researchers, Inc.; **DOL•41** (BL,) ©John Anderson/Fotolia, (BR) ©Kuttelvaserova/Shutterstock, (T) Carlos Villoch, 2004/Image Quest Marine; **DOL•42** (Bkgrd) ©SuperStock/Alamy Images, (TR) 2002 Peter Herring/Image Quest Marine; **DOL•43** (TL) ©Ingvars Birznieks/Shutterstock, (TR) ©SuperStock/Alamy Images, (B) Roger Steene/Image Quest Marine, (TC) superclic/ Alamy Images; **DOL•44** (B) Jeff Rotman/Nature Picture Library, (TR) Jez Tryner/Image Quest Marine, (TC) Kaj R. Svensson/ Photo Researchers, Inc.; **DOL•45** (TL) Jez Tryner/Image Quest Marine, (CR) Joe Dovala/PhotoLibrary Group, Inc., (TR) Michelle Pacitto/Shutterstock, (BL) Roger Steene/Image Quest Marine, (CL) WaterFrame/Alamy Images; **DOL•46** (BL) age fotostock/SuperStock, (TR) John A. Anderson/Shutterstock; **DOL•47** (TR) Heather Angel/Natural Visions, (TCL) Roger Steene/Image Quest Marine, (B) T.Burnett-K.Palmer/V&W/ Image Quest Marine; **DOL•48** (BL) Brian Choo, (T) James D. Watt/Image Quest Marine, (BL) Nature Magazine; **DOL•49** (BR) blickwinkel/Alamy Images, (BL) Peter Batson/Image Quest Marine, (BC) REUTERS/Handout/Reuters Media; **DOL•50** (CL) EYE OF SCIENCE/SCIENCE PHOTO LIBRARY/Photo Researchers, Inc., (BR) Filip Nowicki/Fotolia, (TL) James D. Watt/Image Quest Marine, (TR) Kelvin Aitken/V&W/Image Quest Marine, Mark Doherty/Shutterstock; **DOL•51** (T) Robert Nystrom/Shutterstock, (BL) Tom McHugh/Photo Researchers, Inc.; **DOL•52** (TR) Dr. Morley Read/Shutterstock, (BL) HJFOTOS/Fotolia; **DOL•53** (C) Alessandro Mancini/Alamy Images, (TL) Fotografik/Fotolia, (BR) Juniors Bildarchiv/Glow Images, (TR) Kevin Snair/SuperStock; **DOL•54** (T) WaterFrame/ Alamy Images; **DOL•55** (TR) DK Images, (C) erllre/Fotolia, (BCL) fivespots/Shutterstock, Itinerant Lens/Shutterstock, (BR) Johan_R/Shutterstock; **DOL•56** (T) BERNARD CASTELEIN/ Nature Picture Library, (BR) Journal Nature/Portia Sloan/Getty Images; **DOL•57** (TL) age fotostock/SuperStock, (BCR) imagebroker/Alamy Images, (BL) kwest/Shutterstock; **DOL•58** (BR) edurivero/iStockphoto, (BL) FomaA/Fotolia, (T) Stephen Mcsweeny/Shutterstock; **DOL•59** (CC) Cal Vornberger/Alamy, (TL) Glenn Bartley/Glow Images, (CR) Martha Marks/ Shutterstock, (BL) Michael G. Mill/Shutterstock, (TR) Stubblefield Photography/Shutterstock; **DOL•60** (BL) Dave Watts/Nature Picture Library, (T) Julie Lubick/Shutterstock; **DOL•61** (CL) Steven David Miller/Nature Picture Library, (BR) Timothy Craig Lubck/Shutterstock; **DOL•62** (BR) Eric Isselée/ Shutterstock, (TCR) Gallo Images/Corbis, (CL) Ingo Arndt/ Nature Picture Library, (BL) Liquid Productions, LLC/ Shutterstock; **DOL•63** (TL) Jim Clare/Nature Picture Library, (TCL) Jupiter Images, (TCR) Mogens Trolle/Shutterstock, (B) Reinhard Dirscherl/AGE Fotostock, (TR) Sean Tilden/Alamy Images, (CC) SYLVIE BIGONI/Fotolia; **DOL•64** (TCL) Animal/ Shutterstock, (CL) BlueOrange Studio/Shutterstock, (BR) Creatas/Photolibrary Group, Inc., (TCR) Elio Della Ferrera/ Nature Picture Library, (TC) Karel Gallas/Shutterstock, (BL) Kjersti Joergensen/Shutterstock, (TR) Lockwood & Dattatri/ Nature Picture Library, (TL) Matthijs Wetterauw/Shutterstock, (BC) Prisma/SuperStock, (CR) Stéphane Bidouze/Shutterstock; **A-04** (BL) ©Suzanne Long/Alamy Images; **A-05** (BR) Kenneth Eward/BioGrafx/Photo Researchers, Inc.; **A-12** (BL) Peter Casolino/Alamy Images; **A-16** (BL) Tanton Yachts; **A-19** (BR) Kevin Fleming/Corbis; **A-20** (BCR) Tom Pantages.